Comets II

Comets II

**M. C. Festou
H. U. Keller
H. A. Weaver**

Editors

With 88 collaborating authors

Foreword by Richard P. Binzel

THE UNIVERSITY OF ARIZONA PRESS
Tucson

in collaboration with

LUNAR AND PLANETARY INSTITUTE
Houston

About the front cover:

A fragmenting comet. Many cometary nuclei, especially Sun-grazers, break up into several fragments. This view shows such a comet from the vicinity of one of the larger fragments, furiously jetting gas and dust from its freshly exposed surface. Another of the large fragments, which by now has moved very far away from its sibling, is seen to the left with its own coma and tail. The planet Venus provides a nice backdrop to this dynamic cometary phenomenon. *Painting by William K. Hartmann, Planetary Science Institute, Tucson, Arizona.*

About the back cover:

The main objective of cometary science is to determine the nature, origin, and evolution of comets. Most Earth-based images of comets cannot show the source of cometary phenomena, namely the nucleus, but they do place comets among the most beautiful objects in the sky. Spacecraft images of comets are needed to reveal the true nature of the nucleus. *Top:* This Hubble Space Telescope image of Comet 1995 O1 (Hale-Bopp) was taken on September 26, 1995, when the comet was about 6.5 AU from both the Sun and the Earth. The comet had just undergone a spectacular outburst and was releasing dust in a sprinkler-like, spiral-shaped pattern. C/Hale-Bopp became one of the brightest comets of the past century with gas and dust production rates about 20 times those of the famous 1P/Halley. Its unseen nucleus has a diameter estimated at 50 to 70 kilometers. *Image courtesy of NASA and the European Space Agency. Bottom:* The Stardust spacecraft penetrated deep into the coma of Comet 81P/Wild 2 and imaged the nucleus, which has a diameter of approximately 5 kilometers, at a resolution of about 20 meters. This image was taken at a range of 255 kilometers on January 2, 2004, and reveals a nucleus that is remarkably rough and complex with surface features that include pinnacles and large crater-like depressions. *Image courtesy of the Stardust team, NASA, and the Jet Propulsion Laboratory.*

The University of Arizona Press
in collaboration with the Lunar and Planetary Institute
© 2004 The Arizona Board of Regents
All rights reserved
∞ This book is printed on acid-free, archival-quality paper.
Manufactured in the United States of America

10 09 08 07 06 05   6 5 4 3 2 1

Library of Congress Cataloging-in-Publication Data

Comets II / editors, Michel C. Festou, H. Uwe Keller, and
Harold A. Weaver.
        p. cm. — (Space science series)
Includes bibliographical references and index.
ISBN 0-8165-2450-5 (cloth : alk. paper)
1. Comets. I. Title: Comets 2. II. Title: Comets two. III. Festou, M.
IV. Keller, H. U. V. Weaver, Harold A., 1953–
QB721.C42 2005
523.6—dc22
2004024020

Dedicated to Dr. Fred L. Whipple
(1906–2004)

Dr. Fred Lawrence Whipple, the world's foremost authority on comets, died on August 30, 2004, in Cambridge, Massachusetts. He was 97.

Whipple received his undergraduate degree in mathematics from the University of California at Los Angeles before declaring that math was too boring. Astronomy, for him, held a universe of possibilities. In 1931 he earned a doctorate in astronomy from the University of California at Berkeley.

Whipple immediately took a position at the Harvard College Observatory, eventually becoming the Phillips Professor of Astronomy there. In 1955 he took over the directorship of the Smithsonian Astrophysical Observatory (SAO), which he helped move to Cambridge from Washington, DC. When he retired from the directorship in 1973, the SAO was merged with the Harvard College Observatory to create the Harvard-Smithsonian Center for Astrophysics. In large part due to his vigor and talent, the center is today considered one of the world's top organizations for astronomical research. He was the first to hire a young Carl Sagan for the faculty at Harvard, before Sagan joined the Cornell faculty. Joseph Veverka, the current chair of Cornell's astronomy department, was Whipple's last graduate student at Harvard.

One of Whipple's early research activities involved countering the idea that meteors had hyperbolic orbits. Since he was put in charge of Harvard's observing, he took the opportunity to examine the Harvard patrol plates and discovered six comets.

At mid-century the "sandbank model" idea that comets were loose aggregates of material was prevalent. Whipple changed all that with his suggestion that a comet had to have a nucleus that is a "dirty snowball" of water ice, ammonia, methane, and carbon dioxide with embedded meteoritic dust. This prediction was verified in 1986 when the Giotto mission's Halley Multicolour Camera Team, of which Whipple was a member, obtained the first images showing a nucleus with jets emerging from it. Whipple's recognition that the outgassing of the rotating nucleus would create a rocketlike thrust that could explain the nongravitational anomalies observed in cometary orbital motion, as well as explain the persistence of cometary spectra, was a conceptual leap that is considered one of the most important contributions to solar system studies in the twentieth century.

An early believer in space exploration, he invented the "Whipple shield," a device that shields spacecraft from meteoroids and other interplanetary projectiles. He was credited with creating the first space telescope, the Orbiting Astronomical Observatory. As SAO director he helped to create the Multiple Mirror Telescope on Mt. Hopkins, Arizona, a site now named the Fred L. Whipple Observatory. At the age of 92, he was selected to serve on NASA's Comet Nucleus Tour (CONTOUR), becoming the oldest researcher to accept an active role on a space mission. Had CONTOUR been successful, Whipple would have lived to realize his dream of seeing his favorite comet — Comet Encke — up close during CONTOUR's scheduled flyby in November 2003.

"I'm an engineer at heart," Whipple later said. "I've been able to judge what instruments will work and what can be built. That's been the secret to my success."

Photo courtesy of the Harvard-Smithsonian Center for Astrophysics.

Contents

PART I: A GLOBAL VIEW ON COMETARY SCIENCE

PART II: FROM THE INTERSTELLAR MEDIUM TO THE SOLAR NEBULA

PART III: NATURE AND EVOLUTION OF COMETARY ORBITS

PART VI: DUST AND PLASMA

PART VII: INTERRELATIONS

Collaborating Authors

Scientific Organizing Committee

The editors thank the following scientists for their assistance in the planning stages of this book:

D. Bockelée-Morvan
A. Cochran
M. Combi
M. Duncan
M. Fulle
W. Huebner
W. M. Irvine

J. I. Lunine
K. Meech
T. Mukai
D. Prialnik
H. Rauer
M. Sykes

Acknowledgment of Reviewers

The editors gratefully acknowledge the following individuals, as well as several anonymous reviewers, for their time and effort in reviewing chapters for this volume:

Dominique Bockelée-Morvan
Alan Boss
Donald Brownlee
Humberto Campins
Maria Teresa Capria
Steven B. Charnley
Anita L. Cochran
Michael Combi
Thomas Cravens
Gabriele Cremonese
Jacques Crovisier
Dale P. Cruikshank
Jeff Cuzzi
Daniel D. Durda
Anne Dutrey
Thérèse Encrenaz
Julio A. Fernández
Yanga Fernandez
Alan Fitzsimmons
G. Forti

Perry A. Gerakines
Paul F. Goldsmith
Joe Hahn
Lee Hartmann
Wing Ip
William Irvine
Peter Jenniskens
David Jewitt
Klaus Jockers
Laurent Jorda
William Julian
Vasili Kharchenko
Nikolai Kiselev
Roger Knacke
Norbert Kömle
Philippe Lamy
A.-Chantal Levasseur-Regourd
Harold F. Levison
Aigen Li
Javier Licandro

Renu Malhotra
Vince Mannings
Brian G. Marsden
Neil McBride
William B. McKinnon
Karen Meech
Alessandro Morbidelli
Thomas Quinn
David Schleicher
Rita Schulz
Don Shemansky
Karoly Szego
Gonzalo Tancredi
Jeremy Tatum
Imre Toth
Malcolm Walmsley
Paul Weissman
Diane H. Wooden
Susan Wyckoff

Foreword

Comet science stands at the threshold of a new era. Long gone are the ages when comets terrified and mystified human civilizations. Comets now captivate substantial scientific attention and garner widespread public interest. Our hard-earned knowledge and understanding of what comets are, where they originate, and how they evolve has reached an amazing level of sophistication, as evidenced within these pages. Yet fundamentally, we have so much still to learn about comets. The goal of *Comets II* is to serve as the foundation upon which the next era of comet science can be built.

Within each Space Science Series volume authors are challenged to convey concisely what we know, how we know it, and where we go from here. Our intended audience is new students entering with unfettered new ideas, as well as current planetary science practitioners who have the experience to recognize new connections across broad disciplines. Progress can be made by bringing clarity to existing paradigms, by forging new links among related or seemingly disparate areas, or most exciting, by overturning components of our current understanding through new observational or theoretical discoveries. On the horizon we can foresee these advances in comprehending comets also bringing us key insights about the formation of our own world, our own solar system, the interstellar medium, stellar disks, extrasolar planets, and more. *Comets II* strives to enable such advancement while also serving as a benchmark for measuring our progress.

Rocket science is difficult. Spacecraft exploration of comets is even more so. Ongoing and undaunted *in situ* exploration of comets holds tremendous promise for ushering in the new era we seek. *Stardust* is bringing us our first dust samples from a specific comet (81P/Wild 2), *Deep Impact* looks to make the first test of the structure of a cometary nucleus (9P/Tempel 1), and *Rosetta* promises to make the first extended *in situ* analysis of both the nucleus and coma environment of a comet (67P/Churyumov-Gerasimenko). Further bold steps are easy to envision but challenging to accomplish, such as a returned sample from the surface and a cryogenically preserved returned sample from the deep interior. Perseverance, always a necessary element in science, will continue to be one of the most important ingredients for success in cometary exploration.

Unshrouding the well-guarded mysteries of comets and the challenge of their exploration requires extensive international collaboration. Cometary science enjoys perhaps the broadest international participation of any planetary science field, well exemplified by the international distribution of the editors and contributors to *Comets II*. All deserve congratulations and thanks for the toil, dedication, brilliance, cooperation, and resolve necessary to bring this project to conclusion. Especially deserving are Renée Dotson and co-workers of the Lunar and Planetary Institute, who brought this volume to physical reality. Through their effort they have wrought this gateway to the present and future of comet science. We welcome all who enter.

Richard P. Binzel
Space Science Series General Editor
Lexington, Massachusetts
July 2004

Preface

During the last third of the twentieth century, comets have progressively risen in prominence. No longer are they simply "minor bodies" from the outer solar system. As our knowledge of their physical nature, chemical composition, likely formation sites, and dynamical history have improved, comets have emerged as a "Rosetta stone," providing key information on the origin and evolution of the solar system. Inside the snow line, the building blocks in the early solar system were very similar to the asteroids, and very much like the comets outside this limit. While the terrestrial planets are the end products of planetary formation in the inner solar system, we had, until very recently, only indirect clues that icy bodies acccumulated to form the cores of the giant planets. Now, with the discovery of a population of large objects beyond the orbit of Neptune, presumably composed of ices and rocks, the evidence is even stronger that comet-like objects played a key role in the formation of the gaseous planets. Some of the comets that formed in the vicinity of the giant planets were ejected from that region very early during the solar system's evolution, and a small fraction of them are returning now as "new comets," probably changed very little during their long residence in the Oort cloud. On the other hand, some fragments of transneptunian objects are currently being transported into the inner solar system by planetary perturbations, where they are observed as Centaurs and short-period comets. Spacecraft missions, advanced Earth-based observation techniques, and state-of-the-art numerical models are used to reveal the details of the important messages from the past that comets carry. Yet the history of the solar system predates the formation of the planets and the Sun and begins with the formation of the atoms, molecules, and solid particles that were incorporated into the nebula from which our planetary system emerged. The evolutionary path between the interstellar medium materials formed more than 4.6 billion years ago and the comets that we study today is very complex. This book brings together the latest information to present our best current understanding of the properties of the presolar matter; where, how, and when it was processed; and finally how comets formed and evolved, both dynamically and physico-chemically. Our approach when designing the content of *Comets II* was to produce a book composed of truly complementary chapters and written by a team of more than 100 international authorities on comets. All aspects of cometary science are presented and discussed with the clear intention that the newcomer to the field, student or young researcher, will find all the basic ideas that a textbook would provide. The complexity of the problems examined is also fully presented and discussed, and the experienced researcher will find a comprehensive presentation of the status of cometary science at the beginning of the twenty-first century. An extensive glossary is included that both defines many of the terms used in cometary science and signals where conceptual difficulties might exist. The glossary was prepared by the editors, many *Comets II* authors, and Dr. Jeremy Tatum. We are deeply indebted to the many authors and referees for their efforts in helping to create *Comets II*. We also thank the members of the *Comets II* scientific organizing committee for their assistance in shaping the structure and content of the book. The Editors would also like to recognize the critical contribution of the *Comets II* technical editor Renée Dotson, whose invaluable assistance and pleasant demeanor made life as easy as possible for the editors, authors, and referees.

Michel C. Festou, H. Uwe Keller, and Harold A. Weaver

Part I:
A Global View on Cometary Science

A Brief Conceptual History of Cometary Science

M. C. Festou
Observatoire Midi-Pyrénées

H. U. Keller
Max-Planck Institute für Aeronomie

H. A. Weaver
Applied Physics Laboratory of The Johns Hopkins University

The history of cometary astronomy can be naturally divided into five major periods, with each transition marked by an important new insight. Before 1600, comets were usually viewed as heavenly omens, or possibly meteorological phenomena in the terrestrial atmosphere, and were not yet clearly established as astronomical bodies. Then followed two centuries of mostly positional measurements triggered by the stunning discovery of the universal law of gravitational attraction. Two highlights from this period, which lasted until the early nineteenth century, were the successful prediction of the March 1759 return of 1P/Halley's comet and the discovery of the nongravitational motion of Comet 2P/Encke. The era of cometary physics began with the passage of 1P/Halley in 1835, when spatial structures in a comet were described in detail for the first time. The year 1950 marked the emergence of the modern picture of comets as an ensemble of solar system objects composed of primordial ice and dust, generally on long-period orbits and shaped by their interactions with the solar radiation field and the solar wind. Finally, the space missions to Comet 21P/Giacobini-Zinner in 1985, and especially to 1P/Halley in 1986, provided the first *in situ* measurements and the first images of a cometary nucleus. While these *in situ* observations significantly improved our understanding of cometary phenomena, they also posed many new questions for which we are still seeking answers. In this introductory chapter, we only briefly discuss the pre-modern observations of cometary phenomena, which are already well described in the monograph by *Yeomans* (1991), and focus instead on the advances in cometary science during the past 65 years or so, especially on the developments since the publication of the *Comets* book in 1982.

1. PRE-MODERN ERA: BEFORE 1950

The word "comet" comes from the Greek "kometes," which literally means "long-haired," but the earliest extant records of cometary observations date from around 1000 BC in China (*Ho*, 1962) and probably from about the same time in Chaldea. Ideas about the true nature of comets are available from the time of the rise of Hellenistic natural philosophy around 550 BC, when the Pythagoreans considered comets to be wandering planets that were infrequently seen, mostly near the horizon in the morning or evening skies. In his *Meteorology* (ca. ~330 BC), Aristotle relegated comets to the lowest, sublunary sphere in his system of spherical shells and described them as "dry and warm exhalations" in the atmosphere. There is no mention of comets in Ptolemy's *Almagest*, presumably because he did not consider them to be of celestial origin, but he described them in astrological terms in his *Tetrabiblos*. The Aristotelian ideas about planets and comets were upheld for an entire millennium during which there was little scientific advancement in the field of astronomy. The first doubts about the Aristotelian view seem to have been expressed by Aquinas and by Bacon in his *Opus Tertium* from 1267, although both men, like most of their contemporaries, strongly believed that comets were evil omens.

Toscanelli observed 1P/Halley in 1456 and several other comets between 1433 and 1472 with improved accuracy, inaugurating the renaissance of European observational astronomy after the long post-Aristotle scientific freeze. *Brahe*'s (1578) exceptionally accurate observations initiated a new era for observational astronomy, as he beautifully demonstrated that the horizontal parallax of the bright Comet C/1577 V1 was certainly smaller than 15 arcminutes, corresponding to a distance in excess of 230 Earth radii, consequently farther away than the Moon. This result raised the question of how comets move, and the suggestion that they orbit in highly elongated ellipses was made in 1610 by the amateur astronomer Lower (*Rigaud*, 1833). At about the same time, Hooke and Borelli suggested that cometary orbits could be parabolic. Dörffel was the first to state specifically that the two bright comets seen in 1680 and 1681 (C/1680 V1) were one and the same, before and after its perihelion passage, and that it moved along a parabola with the Sun as its focus, thus providing an explanation for the fact that many comets were seen in pairs, one in the morning and the other in the evening. Newton, in his *Principia*

(*Newton,* 1687), developed the brilliant tool that could link all these observations together. He applied his new theory of gravitation to show that the comet of 1680 (C/1680 V1) moved in an elliptical, albeit almost parabolic, orbit and that it passed only about 0.00154 AU above the surface of the Sun. *Halley* (1705) computed the orbits of a dozen well-observed comets and demonstrated the periodic nature of the bright comet of 1682 (1P/1682 Q1). "Halley's comet" was telescopically recovered by Palitzsch in December 1758, which conclusively proved the validity of Newton's law of gravity out to the distance of the aphelion at 36 AU, more than three times the distance of Saturn, the outermost planet known at that time.

Cometary astronomy in the eighteenth century witnessed the gradual improvement in techniques for orbital computations (e.g., *Brandt,* 1985; *Yeomans,* 1991), and by the beginning of the nineteenth century this had become a rather straightforward task, albeit a somewhat arduous one when planetary perturbations had to be taken into account. Some basic features of the orbital distribution of comets were established, e.g., the extremely broad range of orbital periods. While some comets were found to have orbits virtually indistinguishable from parabolas, others were clearly confined to the inner solar system inside Jupiter's orbit. As time passed, a concentration of comets moving in similar orbits with fairly low inclinations and with aphelia close to Jupiter's orbit became more and more obvious, giving rise to the concept of the Jupiter family of comets. Two ideas were proposed to explain the existence of this family: Either there was a continual ejection of comets from Jupiter (*Lagrange,* 1814), or there was a mechanism of dynamical evolution, called "capture," whereby the comets would become concentrated into such orbits (*Laplace,* 1816). It was soon recognized that comets in general, and members of the Jupiter family in particular, suffer by far their largest orbital perturbations due to the action by Jupiter. The restricted Sun-Jupiter-comet three-body problem consequently offered an interesting approximation for the study of their dynamical behavior.

In 1835, 1P/Halley became the first comet for which spatially detailed structures were extensively observed. In particular, Herschel, Bessel, and Struve described the presence of jets, cones, and streamers [cf. reproductions appearing in *Donn et al.* (1986)]. This led *Bessel* (1836b), following *Olbers* (1812), to postulate the ejection of solid particles in the direction of the Sun, and that these particles were somehow forced back into a "tail" by an unknown repulsive force acting in the anti-sunward direction. The connection between the Sun and comet tails had been suspected for a long time but never expressed so clearly before. Comets were now identified as physical entities, and not solely masses circulating around the Sun (described later as "visible nothings" by Struve because of their inability to leave any sign of their existence — e.g., against the solar disk or in front of stars — except for their response to the Sun's gravitational pull). Arago directed his polarimeter toward the tail of Comet Tralles (C/1819 N1) and found the light

to be polarized, hence it was "reflected" sunlight. He could not relate this observation to the presence of solid particles or gases because their respective effects on solar light were then unknown. The observation by Bond in 1862 of Comet Donati (C/1858 L1) revealed that the plane of polarization of the light was the Sun-Earth-comet plane, thus definitively demonstrating the solar origin of the scattered light. Around the same time, observations of Comet Tebbutt of 1861 (C/1861 J1, the great comet) by Secchi showed that the center of coma light was not polarized while the outer coma was. Other observations indicated different results, which we interpret today as due to the varying gas-to-dust ratio in the coma. As publications during this period demonstrate, these observations were ahead of their time and could not be properly interpreted. In particular, it is striking to observe that even though in 1940 the polarization mechanisms were all understood ["reflection" or "scattering" by particles, a surface, or gases (see *Öhman,* 1939, 1941)], no mention of "solid particles" as constituents of cometary comae was made. In his review of 1942, *Bobrovnikoff* (1942) mentions "meteoric dust" as a possible participant in producing comet spectra, since the nucleus must be made of meteoritic material, but he also said that "the continuous spectrum is still a mystery."

The link between comets and meteors was made by *Schiaparelli* (1866, 1867), who observed that the Perseid and Leonid meteor streams coincided with the orbits of Comets 109P/Swift-Tuttle (1862 III) and 55P/Tempel-Tuttle (1866 I) respectively. This was the proof that comets were indeed losing solid particles. Bredikhin (quoted by *Jaegermann,* 1903) further developed the comet tail theory based on an *ad hoc* repulsive force from the Sun that varied with the square of the heliocentric distance. This became known as the Bessel-Bredikhin mechanical model and was widely used. *Finson and Probstein* (1968a,b) published a gas dynamic model describing the gas-dust interaction and the solar light-dust interaction that is still in widespread use today. *Eddington* (1910) introduced the fountain model of particle ejection, in which parabolas represent the outer envelopes of particle trajectories emitted from the sunlit hemisphere of the nucleus. The repulsive force acting on the particles was identified by *Arrhenius* (1900) as the pressure exerted by sunlight. The corresponding theory was further developed by *Schwarzschild* (1901) and extended to molecules by *Debye* (1909). (At this point, it is worth mentioning that neither observations nor ejection models suggested that comets were losing material from the nightside.)

After Halley, *Encke* (1820) was the second to successfully predict the return of a comet (in 1822). Comet 2P/Encke has the shortest period of all known comets, 3.3 years, which provides similar Sun-Earth-comet configurations every 10 years. The comet was subsequently found to arrive systematically at perihelion about 0.1 days earlier than predicted, even when taking planetary perturbations into account. Inspired by his observations of an asymmetric distribution of luminous matter in the head of Comet Halley in 1835, *Bessel* (1836a) interpreted this as a Sun-oriented asymmetric outflow and suggested that a nongravitational effect might arise due to the rocket-type impulse imparted by such an outflow, possibly explaining perihelion shifts such as those observed for Comet 2P/Encke. It would take over a century for this superb idea to be fully accepted by the scientific community because the existence of a solid body at the center of the coma was far from being unanimously accepted. In fact, the theory that comets were a swarm of solid particles was the most favored by scientists at that time.

The first spectroscopic observations of the gas component of comets were made by *Donati* (1864) and *Huggins* (1868), who visually compared the spectra of Comets Tempel (C/1864 N1) and Tempel-Tuttle (55P/1865 Y1), respectively, with flame spectra. They found that the bands seen in the comet and in the flame were similar. Huggins also visually noted the presence of a broad continuum, which he identified with reflected sunlight. Sunlight was known since the work of Fraunhofer to be characterized by the presence of numerous absorption lines, in particular the strong H and K lines in the UV region and the Na D lines in the yellow-orange region. Huggins was not a professional astronomer. Nor was Usherwood, who in 1858 took the first photograph of a comet, C/1858 L1 Donati. The bands recorded by Huggins, known as the "carbon" or "Swan" bands, were found in all subsequent observations of comets. The Swan bands so strongly dominated cometary spectra that carbon was immediately believed to be an important constituent of comets. While Draper was taking the second-ever image of a comet, C/Tebbutt (1881 K1, the great comet), Huggins was recording the first photographic spectrum of a comet when observing the same object through a slit spectrograph designed to record stellar spectra. Photography and spectroscopy soon became the standard way of studying comets. In 1882, the Na D doublet was identified in the (bright) comets of that year that passed close to the Sun. A few other emissions were seen, but those would not be identified for a few more decades. In particular, spectra of the tail were taken at the beginning of the twentieth century in Comets Daniel (C/1907 L2) and Morehouse (C/1908 R1). The N_2^+ emission was found near 3914 Å in the tail spectrum of the latter comet, but it was likely an atmospheric feature as, in the light of modern observations, a careful subtraction of the telluric lines is required to allow any firm identification of these faint emissions. In addition, such identification work is extremely difficult when working with photographic prism objective spectra.

Baldet (1926) published a comprehensive catalog of the spectra of 40 comets obtained since 1864, together with a complete bibliography of all comets observed until that time by spectroscopy. This work, and that of Nicholas Bobrovnikoff on the 1910 apparition of Comet 1P/Halley, which appeared five years later (*Bobrovnikoff*, 1931), are the first two comprehensive papers of cometary physics published in the twentieth century. Because only very bright comets were studied at that time, all observed comets appeared to have very similar spectra, although differences were qualitatively noted, in particular the relative strength of the con-

tinuous and discrete (dubbed "emission lines") components of the spectrum. These differences were not understood, since neither the emission mechanism of the light nor coma abundances were known. *Schwarzschild and Kron* (1911) studied the intensity distribution in P/Halley's straight tail during the 1910 passage and suggested that the emission could be explained by the effect of absorption of solar light, followed by its reemission, i.e., fluorescence. This naturally led to the key result obtained by Polydore Swings (*Swings*, 1941), who solved the long-standing problem of why the violet CN bands (3875 Å) in cometary spectra did not resemble CN laboratory spectra. Because of the presence of absorption lines in the solar spectrum, the intensity at the exciting wavelength depends on the Doppler shift caused by the comet's motion relative to the Sun, which thus determines the strength of the emission lines in the comet's spectrum; this phenomenon is known today as the "Swings effect." Many papers written during this time describe the spectral properties of specific features in comets, such as the "central point" or "jets," but are of little value today because of the poor spatial resolution of the observations.

2. MODERN ERA: AFTER 1950

A major evolution in cometary science took place in 1950–1951 with the formulation of three very important ideas within a short timespan. The icy-conglomerate ("dirty snowball") model of the cometary nucleus was proposed by *Whipple* (1950). Then, from dynamical studies of the distribution of semimajor axes of comets, came the identification by *Oort* (1950) of a distant population of comets now known as the Oort cloud. Finally, *Biermann* (1951) gave the correct explanation for the motions of features in cometary plasma tails caused by their interactions with a flow of charged particles emanating from the Sun's surface (i.e., the solar wind). None of these ideas resulted directly from new observational evidence, and important parts of them had been proposed earlier by other scientists, but this was the first time that the known facts were effectively combined, leading to a comprehensive description. A new picture of comets, and the existence of a family of celestial bodies, were suddenly revealed at the same time.

2.1. The Icy-Conglomerate Model of the Nucleus

In a series of enlightening papers published between 1932 and 1939, Wurm suggested that since the radicals and ions observed in cometary comae were not chemically stable, these species must be created by photochemistry of more stable molecules residing inside the nucleus [see, e.g., the reviews by *Wurm* (1943) and *Swings* (1943) and references therein]. It is Wurm who first proposed the concept of "parent" compounds/molecules in the nucleus, while "daughter" species would be created in the coma by photochemistry. Spectroscopic studies revealed the nature of the daughter species but not the parents. In the 1940s, Swings developed his ideas based on Wurm's reasoning, and the key role of these two scientists appears to have been overlooked in the later literature. The presence of CO or CO_2, C_2N_2, CH_4, N_2, and NH_3 in the comet was invoked on the basis that CO^+, CN, CH, CO_2^+, N_2^+, and NH emissions were found in cometary spectra. H_2O was also considered as a potential parent molecule, following the discovery of the OH 3090 Å UV emission in 1941. Additional molecules (e.g., C_2) were required since all known coma species could not be explained from the above molecules. In 1948, Swings came very close to proposing an icy model for the nucleus by suggesting that the above-mentioned molecules could exist in the solid state in the nucleus (*Swings*, 1943, 1948). (We note here that quantitative arguments were largely missing from the discussion because of the lack of appropriate observational material and laboratory measurements on the parent molecule properties. The situation significantly improved in that regard about 30 years later.)

A key question was how molecules were stored in the cometary material. Since the middle of the nineteenth century, a great deal of research had concentrated on understanding the nature of the central source of gas and dust in comets, and the alternate theory to Lyttleton's sandbank model (*Lyttleton*, 1948) was that an invisible solid nucleus was the source of all observed cometary material. *Swings* (1942) suggested that molecules similar to those found in meteorites could possibly be stored in the nucleus by occlusion. Independently (probably) of Swings, *Levin* (1943) developed his desorption theory from the surface of meteoritic material to demonstrate that his sandbank model for the nucleus had a solid basis. An average desorption heat of 6000 cal/mole was deduced from the interpretation of the brightness law followed by comets as the heliocentric distance changes, which was in agreement with laboratory data for the above-mentioned cometary molecules (in the 2,000–10,000 cal/mole range). However, the amount of material that could be desorbed from a swarm of particles with an expected cometary mass could not explain the persistence of comae over several months during single passages, and consequently the survival of comets such as 1P/Halley or 2P/Encke for centuries and millennia. Seeing-limited observations of comets passing near Earth showed a central, unresolved light source of dimensional upper limits in the 10–100-km range (cf. review by *Richter*, 1963). Upper limits to cometary masses had also been estimated from the absence of evidence for mutual gravitational attraction of the components of Comet P/Biela in 1846, or of any influence on the Earth's orbit for very close passages, like those of Comet P/Lexell (D/1770 L1). Masses in the 10^{12}–10^{17} kg range were surmised (*Whipple*, 1961).

In an attempt to synthesize all known facts about the cometary nucleus, and with particular attention to the long-standing problem of explaining the nongravitational perihelion passage delays, *Whipple* (1950, 1951) laid the foundations for the modern model of a solid nucleus. Building on the ideas of *Laplace* (1813) and *Bessel* (1836a) (cf. *Levin*, 1985), *Whipple* (1950, 1951) described the nucleus as a mixture of ices from which the gases in the coma are produced by sublimation in increasing quantities as the comet approaches the Sun and the surface temperature of the nucleus rises. Meteoritic dust is also released from the nucleus when these ices evaporate, hence the famous expression "icy conglomerate." The relative proportions of the various ices were discussed only qualitatively; not only was the nature of all the parent species needed to explain the unknown coma composition, but Whipple's main concern was to explain the non-Keplerian motion of Comet 2P/Encke. Nevertheless, Whipple's model was hugely influential because of its ability to successfully explain within a single conceptual framework many observed cometary phenomena, such as (1) the large gas production rates [200 kg/s of C_2 for 1P/Halley in 1910 derived by *Wurm* (1943)], for which the desorption model was totally inadequate; (2) the observed jet-like structures in the coma and the erratic activity, impossible to produce if the source of gas and dust was a cloud of particles; (3) the observed nongravitational forces by means of the momentum transfer by the outflow of gas from the nucleus, the sign of the net effect on the orbital motion being dependent on the orientation of the nucleu spin vector; (4) the fact that most comets that pass extremely close to the Sun, e.g., the Kreutz Sun-grazing comet group, apparently survive such approaches; and (5) the fact that comets are the sources of meteor streams. Items (2)–(4) gave particularly strong arguments for a solid nucleus rather than a sandbank structure, a model that had other difficulties in addition to failing to explain the above-mentioned points.

Although Whipple's model quickly won general acceptance, there were some shortcomings. The main one was described by Whipple himself, namely, the large differences among the latent heats of vaporization of the various ices he thought might be present in the nucleus. He also remarked that the low vapor pressure of water was a serious problem when explaining the observed presence of the OH emission far from the Sun. As a result, the highly volatile material should be removed from the surface layer of the nucleus long before perihelion, in contradiction with the observation of radicals and ions such as CH and CH^+ near the Sun. This objection was tentatively removed when *Delsemme and Swings* (1952) noticed that almost all parent molecules (except NH_3) required to explain the observed radicals and ions in comets could coexist in the nucleus in the form of solid clathrate hydrates of H_2O, where the volatile "guest molecule" occupies a cage in the H_2O crystal lattice. From the stoichiometry of hexagonal ice, the mean value of the occupation number is somewhat smaller than 6. In this way, the highly volatile material does not disappear too rapidly, and is also freed together with less-volatile molecules, which explains why the spectrum remains more or less similar throughout a comet's apparition. Delsemme and Swings' ideas also implied that comets contain mostly water molecules, something that was not proven at that time, nor even discussed in any detail.

2.2.   The Oort Cloud

As a consequence of the nineteenth century work on cometary orbits and the discovery of nongravitational forces, dynamical studies of individual comets were carried out during the first decades of the twentieth century, with particular attention paid to the influence of planetary perturbations. The earlier results on the non-existence of interplanetary ether were confirmed and statistical considerations about the distribution and dynamical origin of comets naturally followed, including the question of whether or not some comets have "original" hyperbolic orbits (reciprocal semimajor axis $1/a_{orig} < 0$), which would mean that they were not members of the solar system. The work by *Strömgren* (1914, 1947) and colleagues demonstrated that there were no original hyperbolic orbits among the observed comets; all apparently hyperbolic orbits were actually perturbed into those states by planetary effects, mainly the influence of Jupiter. Comets were not coming from interstellar space. *Sinding* (1937) produced a list of values of $1/a_{orig}$ for 21 long-period comets, which, together with the work by *van Woerkom* (1948), formed the basis for Oort's famous paper (*Oort,* 1950) on the existence of a cometary population residing in the outer reaches of the solar system. The idea of a cloud of distant hypothetical comets, stable against stellar perturbations, necessary to explain the fact that many comets had $a_{orig} > 10,000$ AU, had been expressed earlier by *Öpik* (1932). *Oort* (1950) deduced the existence of such a cloud by studying the actual distribution of the semimajor axes of 19 observed comets. He discussed the important excess of long-period comets with $1/a_{orig} < 10^{-4}$ AU^{-1}, i.e., with aphelia beyond 20,000 AU, and concluded that while comets can remain in stable orbits out to distances of about 200,000 AU, some of them could from time to time be diverted inward by the perturbations of passing stars. These stellar perturbations should have randomized the orbital inclinations of the comet orbits over the age of the solar system. Oort estimated that the cloud should contain about 2×10^{11} comets to explain the discovery rate of new comets each year. With a mean cometary mass of 10^{13} kg, the total mass of the Oort cloud would be $\sim 2 \times 10^{24}$ kg, or ~ 0.3 M$_\oplus$.

Based on *van Woerkom*'s (1948) theory of the orbital diffusion caused by planetary perturbations, Oort found that the number of comets with $1/a_{orig} \leq 10^{-4}$ AU^{-1} is much larger than what would be expected from the population of long-period elliptical orbits, which suggested that many of the comets become unobservable after their first passage through the inner solar system. This observational fact still does not have a universally accepted explanation. In a subsequent study, *Oort and Schmidt* (1951) distinguished between "new comets" (those making their first visit near the Sun's neighborhood and the planets) and "old comets" (those returning on much less elongated elliptic orbits). The former appeared to be dustier and to brighten more slowly than the latter. These kinds of analyses were revisited later to include the role of interstellar clouds and the galactic tide in delivering new comets to the planetary region. However, the basic concept of the Oort cloud as an outer halo of the solar system has been substantiated by later studies, based on continuously improved cometary orbits by Marsden and co-workers [*Marsden and Sekanina* (1973), *Marsden et al.*

(1978), and the catalogs of cometary orbits regularly published by the Central Bureau for Astronomical Telegrams].

Our ideas regarding the long-term dynamics of the Oort cloud have evolved considerably over the years. While passages of individual stars dominated the discussion in early investigations, the tidal effects of the galaxy as a whole have become recognized in the last two decades as the prime mechanism providing new comets to the inner solar system from the outer part of the cloud. The dramatic effects that might follow upon close encounters with massive perturbers, such as giant molecular clouds (*Biermann and Lüst,* 1977), also received a great deal of attention.

Although Oort explored the possibility of having an inner cloud extending inside of 20,000 AU, his preferred model was a thick shell of comets near the outskirts of the solar system. The idea of an inner Oort cloud was proposed later by *Hills* (1981) for two reasons: (1) to explain the replenishment of the Oort cloud inevitably depopulated by the above-mentioned perturbations, and (2) to provide another source region for Oort cloud comets besides scattering by the giant planets, which is a very inefficient process. These ideas are currently being completely revisited to incorporate the existence and role of the recently discovered transneptunian belt and the various subpopulations of transneptunian objects.

In most theories, Oort cloud comets formed in the Jupiter-Neptune region. The mass of the solar nebula and the time of formation of the planets are key ingredients of Monte Carlo simulations that investigate the transfer of comets from the region of the outer planets to the Oort cloud. The number of comets currently residing in the Oort cloud required to explain the discovery rate of new comets is on the order of $1–2 \times 10^{12}$ (*Heisler,* 1990; *Weissman,* 1996). Recent simulations indicate that comets that were formed in the Saturn-Uranus region currently make up the bulk of the surviving Oort cloud comets, whereas few comets formed near Jupiter are left [see *Dones et al.* (2004) for a current perspective on the complex formation and evolution of the Oort cloud].

2.3. Ion Tails and the Solar Wind

Because the tails of comets can be so impressive, both to the layman and to the professional astronomer, they have been the subject of many investigations. Astronomers of all eras have been struck by the fact that the tail appearances may vary dramatically from one comet to another. In the early twentieth century, it was deduced from the study of the motion of kinks and knots in the tail that the antisolar force acting on straight tails was enormous, up to 1000 times the solar gravity. As early as 1859, *Carrington* (1859) suspected a physical connection between a major solar flare and enhanced magnetic activity on Earth some hours later. Ideas about the possible existence of a stream of particles from the Sun, perhaps electrically charged, emerged toward the end of the nineteenth century, in particular to explain the excitation of molecules and ions observed in cometary comae. It was also found that cometary ion tails (formerly called Type I tails) develop closer to the Sun than dust tails (formerly called Type II tails; the comet tail type is directly related to the strength of the repulsive force that obviously varies smoothly, hence the difficulty of any categorization scheme). However, it was only 50 years later that *Hoffmeister* (1943) provided the crucial observations of a gas tail aberration of 6°, i.e., the angle between the observed tail and the antisolar direction. This was correctly interpreted by *Biermann* (1951) in terms of an interaction between the cometary ions in the tail and the solar wind (hereafter SW), a continuous stream of electrically charged particles from the Sun with velocities of several hundred kilometers per second. His derived plasma densities were unrealistically high, since electrons were thought to accelerate the cometary ions. *Alfvén* (1957) removed this discrepancy by introducing the role of a "frozen" interplanetary magnetic field, which is carried along with the SW particles. Until space probes could study the SW *in situ*, cometary ion tails were the only well-distributed SW probes in interplanetary space and they largely remain so for regions outside the ecliptic plane. The existence of sector boundaries in the SW explains the separation of ion tails from the head of comets; this is one of many phenomena that indicate that the SW properties may change on very short timescales, as described in numerous papers published beginning in 1968 by Brandt and collaborators.

3. SPACE MISSIONS TO COMETS

Following the increased interest in comets that arose in the late 1970s, due in large part to the predicted return of 1P/Halley, another giant leap in our understanding of comet phenomena occurred in March 1986 when six spacecraft (henceforth "S/C") made *in situ* observations of this comet. There are undoubtedly "pre-Halley" and "post-Halley" eras (much as historians describe the transition from the Dark Ages to the Renaissance period); for the first time, the nucleus of a comet was seen. However, the first cometary encounter took place six months earlier — on September 11, 1985 — when the International Cometary Explorer (ICE) passed through the tail of Comet 21P/Giacobini-Zinner, ~8000 km from the nucleus. The main results were the confirmation of the plasma tail model, indications about the ion composition, and the detection of a neutral current sheet at the center of the tail. ICE continued on to register the effects of 1P/Halley on the interplanetary medium from a distance of 28×10^6 km sunward.

Five spacecraft encountered 1P/Halley in early 1986: *Vega 1* (March 6, closest approach distance of 8890 km), *Suisei* (March 8, 150,000 km), *Vega 2* (March 9, 8030 km), *Sakigake* (March 11, 7×10^6 km), and *Giotto* (March 14, 600 km). Concurrently, an unparalleled, long-term Earth-based observational effort was coordinated by the International Halley Watch (IHW) (*Newburn and Rahe*, 1990). The IHW archive, with more than 25 GB of data, was released in December 1992 (*International Halley Watch*, 1992), and the associated summary volume (*Sekanina and Fry*, 1991)

contains detailed information about the data obtained within the various IHW networks. The observations were made by both professionals and amateurs in all wavebands from the ultraviolet (UV) at 120 nm to the radio at 18 cm. It was particularly fruitful to combine space- and Earth-based observations for calibration and long-term monitoring purposes. The earlier cometary models could be tested and refined with the aid of the *in situ* measurements, leading to many new insights.

The nucleus was observed at close range for the first time; it was found to be larger (equivalent radius 5.5 km) and darker (albedo ~4%) than expected. In the *Giotto* images, surface features (craters, ridges, mountains, etc.) and source regions were observed (*Keller et al.*, 1988). There were no signs of nightside outgassing. The coma was found to be highly structured on all scales (jets, shells, ion streamers, etc.) and the gaseous component was analyzed *in situ* by mass spectroscopy. Signals at atomic masses of 1 and from 12 to ~55 amu were detected. H_2O was confirmed to represent 85% by weight of the gas phase (see further discussion below), and the likely presence of large organic polymeric molecules was indicated. The dust was analyzed by size and composition and there was an unexpectedly high fraction of very small grains, down to the sensitivity limit of ~10^{-19} kg. Particles rich in metals and silicates were found as expected, but particles rich in H, O, C, and N ("CHONs") were seen for the first time and were thought to be related to the smallest grains mentioned above (*Kissel et al.*, 1986). The integrated mass loss experienced by the nucleus at this passage, on the order of 4×10^{11} kg (but very uncertain) was ~0.5% of the total mass of the nucleus, estimated at $1-3 \times 10^{14}$ kg. Nucleus images taken *in situ* and ground-based observations were not sufficient to unambiguously determine the complex (excited) rotational state of the nucleus. A cavity devoid of magnetic field was detected within 5000 km of the nucleus. The various predicted plasma effects were confirmed, including the existence of a bow shock, and the adjacent interplanetary medium was found to be kinematically and magnetically extremely turbulent. Some of the species invoked to explain the mass spectra were produced by gas phase reactions in the coma, as anticipated a decade earlier by *Oppenheimer* (1975).

The fast flybys of Comets 19P/Borrelly in September 2001 (NASA's *Deep Space 1* mission) and 81P/Wild 2 in January 2004 (NASA's *Stardust* mission) have produced two new images of cometary nuclei. In some respects, these two nuclei are very similar to that of 1P/Halley, i.e., all are very dark objects with complex surface structures. There appear to be some significant differences among the three nuclei (e.g., the more spherical shape and possibly "younger" surface of 81P/Wild 2), but the different spatial resolutions of the three investigations may account for some of the apparent diversity. 81P/Wild 2 is covered by what will probably soon be called "erosion craters," but we must await the full publication of the results, and probably *in situ* investigations of other nuclei in the future, to understand how these craters are activated and how long they survive. Despite the advances enabled by the spacecraft encounters, our knowledge of the composition of the non-icy component will have to await new *in situ* measurements or, better yet, the return to Earth of a coma dust sample, very likely in 2007 from the *Stardust* experiment.

4. THE INVENTORY OF COMETARY VOLATILES AND COMPARATIVE COMETOLOGY

Although number density estimates for cometary comae had been derived since the time of Wurm's investigations in the 1930s, the figures obtained were rather uncertain and their reliability limited by the lack of quantitative information on the excitation mechanisms for the observed emissions. Thus, it is not too surprising that, continuing the earlier investigations by Swings and McKellar, most spectroscopic studies between 1950 and 1970 were devoted to a never-ending attempt to discover and identify new emission lines and bands, as well as unraveling the structure of the ro-vibrational bands of the comet radicals and ions. In that regard, special reference must here be made to the numerous and important contributions from the "Liège school," reviews of which are given by *Swings* (1956) and *Arpigny* (1965). During this epoch, rather complete and fairly accurate models of the fluorescence of the CN, CH, OH, and C_2 radicals were built. The advent of high-resolution spectroscopy in the late 1950s allowed the identification of many unknown lines, most of which were due to C_2 and NH_2. Despite these efforts, it is worth noting that thousands of lines in the optical spectra of comets remain unidentified even today; the most likely candidate molecules responsible for these emissions are S_2, CO^+, CO_2^+, and C_3 in the near-UV, C_2 and NH_2 in the optical, and NH_2 and H_2O^+ in the optical infrared (IR). Many new lines have been recently discovered in the near-IR and radio regions (the submillimeter region is also becoming increasingly accessible), and these domains eagerly await a new generation of cometary spectroscopists.

5. WATER AS THE MAIN CONSTITUENT OF COMETS

In 1958, high-resolution spectroscopy allowed the separation of the terrestrial oxygen lines from the cometary ones and also led to the definitive confirmation of the presence of the isotopic lines of ^{13}C, long suspected to be present in comets. The detection of the [O I] red lines in Comet Mrkos (1957 V) (*Swings and Greenstein*, 1958) created a completely new problem: It was demonstrated by *Wurm* (1963) that if fluorescence was responsible for this emission, then very large amounts of O must be present in the coma, dwarfing the amount of C (e.g., C_2). Thus, Wurm proposed a different mechanism, photodestruction of an O-bearing species, to produce atomic oxygen in an excited state. The idea that some coma species may be produced directly into an excited state can be traced back to *McKellar* (1943), but this

suggestion was not explored in detail until 1964 (*Biermann and Trefftz*, 1964). Their work led to the prediction that photodissociation of parent molecules is the main production mechanism of the excited O atoms. The resulting production rates of the parent molecules were estimated to log $Q(s^{-1}) \sim 30$–31 in case of a bright comet, much larger than those of the parents of CO^+, CN, or C_2. However, it is worth noting that in the mid 1960s, the evaluation by *Arpigny* (1965) that the total content of the coma in OH radicals was similar to that of CN or C_2 was not used to discuss the possible abundance of water in comets. This figure was actually considered as being on the low side by *Biermann* (1967, 1968), who concluded that cometary nuclei must consist of molecules of H bound to C, O, and N atoms with the consequence that a huge cloud of H atoms must surround comets. What is striking in the literature of those times is that OH and O are treated as two independent species. The most often-quoted parent molecules were — as in 1950 — H_2O, CO, CO_2, CH_4, and NH_3, and unsuccessful attempts were made in the mid 1970s to find the two latter molecules in spectra of Comet Kohoutek (C/1973 E1). In this context, the discovery in 1970 by the Orbiting Astronomical Observatory (OAO-2) and the Orbiting Geophysical Observatory (OGO-5) of huge Ly-α haloes of neutral H (>1.5×10^7 km) around Comets Tago-Sato-Kosaka (C/1969 T1) and Bennett (C/1969 Y1) (see *Code and Savage*, 1972) did not come as a complete surprise. However, linking these H atoms to water only came later, when the relative abundances of H and OH were investigated in the same comets (see below).

An unknown ion was observed in Comet C/1973 E1(Kohoutek) by *Herbig* (1973) and *Benvenuti and Wurm* (1974). *Herzberg and Lew* (1974) had just obtained the first laboratory spectra of the H_2O^+ ion and tentatively identified this ion as the source of the new cometary emission. The same emission was later found in cometary spectra recorded as early as 1942 [e.g., data given in *Swings et al.* (1943)]. Although the water ions are profusely injected in comet tails, their presence there is not conspicuous, which is a clear indication that the ion is rapidly lost, unlike the other tail ions, especially CO^+. The main loss mechanism is a charge exchange reaction with water molecules leading to the formation of the H_3O^+ ion (*Aikin*, 1974), which itself is likely destroyed in electron recombination reactions. H_3O^+ was indeed found to be one of the main ions in the comae of 21P/Giacobini-Zinner and 1P/Halley during the later *in situ* investigations. Although the presence of the H_2O^+ ion provided strong evidence for the presence of H_2O in the nucleus, observations of this ion have not generally been used to derive H_2O production rates in comets because of the many difficulties in producing accurate models for the ion distributions in cometary comae.

Although the OH emission band at 3090 Å was first identified in Comet C/1941 I (Cunningham) by *Swings* (1941), the first reliable OH production rate measurements only date from the early 1970s (*Code et al.*, 1972; *Blamont and Festou*, 1974; *Keller and Lillie*, 1974). The analysis of the Ly-α isophotes of Comet C/1969 Y1 (Bennett) by *Ber-*

taux et al. (1973) showed that the velocity of the H atoms was about 8 km s^{-1}. Following an investigation of the photolysis of water molecules by sunlight, these authors suggested the possibility that the majority of the observed H atoms were the result of the dissociation of OH radicals. Keller and co-workers reached similar conclusions in a series of independent papers: *Keller* (1971) discussed the possibility that the observed H atoms in Comet C/1969 Y1 (Bennett) might arise from the direct dissociation of water, ideas that he further developed later (*Keller*, 1973a,b). However, these investigations, as well as that of *Bertaux et al.* (1973), were limited by the fact that the parameters governing the water photolysis were not well known at that time. *Blamont and Festou* (1974) measured both the unknown scale length of OH and the production rate of that radical in Comet C/1973 E1 (Kohoutek). *Keller and Lillie* (1974) also measured the scale length of OH (in Comet Bennett) and found a value in complete agreement with that found for Comet Kohoutek. An important clue that H$_2$O was the main source of both the H atoms and the OH radicals came when the velocity of the H atoms was measured directly from Copernicus observations (*Drake et al.*, 1976) and, indirectly, from the analysis of the velocity of H atoms from Ly-α observations (cf. review by *Keller*, 1976), and was found to be fully consistent with the water photolysis scheme. For the first time, based on an experimental study of the photolysis of water molecules and simultaneous measurements of the H and O production rates, it was demonstrated that water was the likely parent of most of the H atoms and the OH radicals.

Subsequent systematic observations of OH, H, and O emissions in more comets using the International Ultraviolet Explorer (IUE) (*Weaver et al.*, 1981a,b) further strengthened the case for H$_2$O as the dominant cometary volatile. Beginning with Comet C/1979 Y1 (Bradfield), a long series of high-quality observations of the UV spectra of comets was obtained with the IUE in programs led by A'Hearn, Feldman, and Festou, from which a self-consistent set of OH production rates was derived (e.g., *Festou and Feldman*, 1987). About 50 comets were observed during the period from 1978 through 1995, and OH production rates were systematically derived for all of them. A comprehensive theory of OH fluorescence, which has been used to interpret the UV observations, was developed by *Schleicher and A'Hearn* (1982, 1988).

Following the discovery of the 18-cm maser emission of OH (*Biraud et al.*, 1974; *Turner*, 1974), radio OH emission has been monitored by the Nançay observatory in over 50 comets by Crovisier and collaborators (*Crovisier et al.*, 2002), with observations continuing to the present day with improved sensitivity. These OH monitoring programs established without a doubt the ubiquity of water as the dominant volatile constituent in comets; no comet was found to be deficient in water. In addition, many parameters of the OH radical are derived from 18-cm observations, and information on the kinematics in the coma as well as the determination of the OH production rates are obtained on a regular basis. The methodology for determining OH velocity profiles was worked out by *Bockelée-Morvan and Gérard*

(1984). The detailed mechanism by which comets emit OH photons at radio wavelengths was investigated by *Despois et al.* (1981).

The radio observations of OH have become the main source of water production rates since 1996. Including both the radio and UV OH observations, water production rates have been derived for about 100 comets. Comets were often followed during a significant fraction of their orbits. The radio and UV determinations of the water production rates do not always agree, as has been discussed by *Schloerb* (1988, 1989), but these large databases are still extremely useful.

H$_2$O itself was not definitively detected until its strong IR ro-vibrational emissions were measured by *Mumma et al.* (1986) in the coma of 1P/Halley during observations from the Kuiper Airborne Observatory, and later from the *Vega* flyby spacecraft (*Combes et al.*, 1986). The water molecule was also directly detected in 1P/Halley using the neutral mass spectrometer on the *Giotto* spacecraft (*Krankowsky et al.*, 1986). Non-resonance fluorescence emissions of water at IR wavelengths can now be used rather routinely to monitor water production rates in comets (cf. *Dello Russo et al.*, 2000), but the number of comets observed in this way is still rather small, at least compared to the number whose OH emission has been monitored at radio wavelengths.

There are multiple production pathways for H atoms. Their excitation by the solar Ly-α line; their interaction with the SW, molecules, and ions; and their kinematics are hard to model. Water production rates can nevertheless be derived from the observation of H lines, as demonstrated in the insightful investigation of the H I UV emission in comets of *Richter et al.* (2000), which summarizes the most recent work on this topic. One often overlooked conclusion that can be drawn from the many observations of the H comae of comets is that, even with the modeling errors and calibration uncertainties, the production of non-water H-bearing species probably can be no larger than ~20–30% of the H$_2$O production rate for most comets observed within 1 AU of the Sun, which leaves CO, CO$_2$, and a few hydrocarbon molecules, as we shall see below, as the main candidates to supply, after H$_2$O, the bulk of the remaining volatile portion of the cometary nucleus.

6. OTHER COMETARY PARENT MOLECULES

In 1970, the known optical emissions were from daughter or granddaughter species (e.g., C$_2$, C$_3$, CN, CH, O, NH, NH$_2$) with uncertain and not very abundant progenitors in the nucleus. Only the O atom was thought to be possibly as abundant as the newly discovered dominant H and OH. For compositional research to develop, technological advances that broadened the wavelength range of the observations were required. The UV window was the first to be explored, followed a few years later by the IR region, and slightly later by the radio region.

Feldman and his collaborators recorded high-quality and high-sensitivity UV spectra of comets during sounding rocket observations of C/1973 E1 (Kohoutek) (*Feldman et al.*,

1974) and C/1975 V1-A (West) (*Feldman and Brune,* 1976). The latter observations provided the first detection of CO in a comet and demonstrated that this molecule was one of the most abundant in comets, although we now know that the amount of CO varies greatly from comet to comet. The CO UV emission has been observed in nearly every bright comet since then, first by the IUE and subsequently by the Hubble Space Telescope (HST) and the Far Ultraviolet Spectroscopic Explorer (FUSE). The UV observations also provide access to two other potential parent molecules: the short-lived S_2, which was discovered during IUE observations of C/1983 H1 (IRAS-Araki-Alcock) (*A'Hearn et al.,* 1982), and CO_2, which can be indirectly probed via emission in the forbidden CO Cameron band emission (*Weaver et al.,* 1994).

The 1970s witnessed the development of systematic, quantitative observations of optical cometary emissions by means of photoelectric narrow-band filter photometry (by A'Hearn, Schleicher, Millis, and their collaborators) and CCD spectroscopy (by several groups led by Cochran, Newburn, and Fink). A review of the early observations and the observing techniques is given by *A'Hearn* (1983). In the early 1980s spectrophotometry developed rapidly when fast detectors became available. This method provides both a good separation of band or line emissions and spatial information on the distribution of coma species. In parallel, numerous theoretical studies, aimed at calculating the fluorescence efficiencies of the coma radicals and ions, resulted in the establishment of reliable conversions of observed surface brightnesses into column densities of the different species. The last step in the data analysis process is then the derivation of gas production rates. A systematic survey of the principal optical emissions from 85 comets produced the first evidence for the existence of compositional families among the comets (*A'Hearn et al.,* 1995).

A real breakthrough in the detection of parent species occurred in the mid 1980s with the development of new instrumentation and techniques at IR and radio wavelengths, which could be used to detect ro-vibrational and pure rotational emissions from molecules. Thus, parent species can now be observed directly, leading to more direct information than that obtained from their destruction products. The apparition of two exceptionally bright comets in the mid 1990s, C/1996 B2 (Hyakutake) and C/1995 O1 (Hale-Bopp), together with improvements in instrumentation and the use of the Infrared Space Observatory (ISO), permitted the discovery of more than two dozen parent molecules (*Brooke et al.,* 1996; *Mumma et al.,* 1996; *Crovisier et al.,* 1997; *Bockelée-Morvan et al.,* 2000), among which is the fairly abundant CO_2 molecule, first detected in 1P/Halley both spectroscopically at IR wavelengths (*Combes et al.,* 1986) and with a neutral mass spectrometer (*Krankowsky et al.,* 1986) and suspected to be present in comets for decades because of the well-known CO_2^+ bands. Subsequent observations of other comets are leading to interesting compositional intercomparisons (*Mumma et al.,* 2003; *Biver et al.,* 2002), although caution regarding the interpretation of compositional trends is advised because of the small number statistics for most species. *Bockelée-Morvan et al.* (2004)

gives a detailed account of the era that began after the passage of Comet 1P/Halley when groundbased IR spectrometers and radio telescopes were systematically used to investigate cometary composition.

7. COMETARY ORBITS

From the perspective of cometary dynamics, the modern era is defined by the advent of efficient and powerful computers. For the first time, numerical simulations of the orbital evolution of comets over the age of the solar system, including the gravitational influences of close encounters with Jupiter and other planets, stars, and interstellar clouds, have been performed. Computers also revolutionized the work on orbit determination and allowed the linkage of past and recent apparitions of observed comets, as well as the preparation of ephemerides for upcoming apparitions, even for long-lost comets. Whereas Oort had been working on a small sample of comets to build his theory, *Marsden et al.* (1978) improved the earlier statistics by using 200 well-determined long-period orbits. They found a concentration of inverse semimajor axes corresponding to an average aphelion distance ≤60,000 AU for q > 2 AU, only about half as remote as Oort's original distance. Besides the apparently abnormal fading of new comets (required because Oort's peak is too high), a major problem remained: the apparent overabundance of Jupiter-family comets. *Everhart* (1972) found a possible route of direct transfer from the Oort cloud via jovian perturbations at repeated encounters with the planet, but the efficiency of this transfer was too low to account for the observed number of Jupiter-family comets. An alternative scenario came from orbital integrations of the observed comets by *Kazimirchak-Polonskaya* (1972): The comets might not be captured by Jupiter alone, but rather by a stepwise process involving all the giant planets. The modern solution is that a disk-like source of comets in the outer reaches of the solar system is required to explain the properties of the Jupiter family of comets (now called "ecliptic comets," adopting Levison's 1996 taxonomy of comet orbits). Kazimirchak-Polonskaya's process naturally explains the existence of the Centaur's family.

A major step forward during this period dealt with the modeling of nongravitational effects in cometary motions. *Marsden* (1969) introduced a nongravitational force into the Newtonian equations of motion with simple expressions for the radial and transverse components in the orbital plane. These involved a function of the heliocentric distance r expressing a standard "force law," multiplied by a coefficient whose value was determined along with the osculating orbital elements by minimizing the residuals of the fit to positional observations. The radial coefficient was called A_1 and the transverse A_2. It was recognized that the model might not be physically realistic and that more meaningful parameters might be derived from a more general formalism, but attempts in this direction were unsuccessful (*Marsden,* 1970). The final update of the model was made in 1973 (*Marsden et al.,* 1973), stimulated by calculations of the H_2O sublimation rate as a function of r (*Delsemme and Miller,* 1971).

This was taken as the model force law, expressed as an algebraic function g(r) whose parameters were chosen to fit Delsemme and Miller's results. Eventually, more realistic models were constructed for the jet force resulting from asymmetric H_2O outgassing, including the heat flow in the surface layers of the nucleus (*Rickman and Froeschlé*, 1983). As a result it was found that the true force law might be very different from the g(r) formula. These efforts led *Rickman* (1989) and *Sagdeev et al.* (1988) to the first evaluation of the density of a comet that, not too surprisingly, was quite low and implied a high porosity for the nucleus. The new image of a comet nucleus after the *in situ* exploration of 1P/Halley, and the intensive efforts of Crifo and collaborators to model the near-nucleus environment, has opened the door to improved models, as described in *Yeomans et al.* (2004). These efforts will never be completely successful until we finally understand what the expression "activity of comets" really means.

8.   THE TRANSNEPTUNIAN BELT AS A COMET RESERVOIR

Around 1950, the Kant-Laplace nebular hypothesis for the origin of the solar system was reconsidered in the light of the chemical compositions of the planets and their variation with heliocentric distance. *Edgeworth* (1949) and *Kuiper* (1949, 1951) argued that it is unlikely for the solar nebula to have ended abruptly at the position of Neptune's orbit, and thus a large population of planet precursors with a generally icy composition had to exist outside the region of the giant planets. *Kuiper* (1951) claimed that such bodies could be identified with Whipple's cometary nuclei and suggested that Pluto's gravitational action (its mass was then thought to be in the 0.11 M_s range) might have scattered the objects into Neptune's zone of influence, whereupon ejection into the Oort cloud would ensue. Both Kuiper and Edgeworth suggested that the original comet belt just beyond Neptune might still be intact. *Fernández* (1980) was the first to predict the existence of such a belt in a quantitative way, and he demonstrated that this belt is probably the principal source of the ecliptic comets. This latter idea was later expanded by Duncan and collaborators; *Duncan et al.* (2004) provides the current perspective on this subject.

The discovery of the transneptunian object 1992 QB1 by Jewitt and Luu in 1992 provided dramatic observational evidence that the basic hypothesis of Kuiper and Edgeworth might be correct. The number of observed "Kuiper belt objects" (KBOs) is now approaching 1000, and the total population likely exceeds 10^5 members with sizes larger than 100 km. The structure of the belt, its dynamical subclasses, and its evolution are currently the subject of intense research activity by the community.

We also note that dynamical investigations produce new results that seem to contradict the "conventional wisdom" every decade or so, with some evidence that the pace of change is accelerating as computers continue to get faster

and new analysis techniques are employed. Thus, we should not be surprised to find in five years that the ideas presented in *Comets II* on the origin and evolution of the Kuiper belt and Oort cloud have changed significantly.

9.   FINAL REMARKS

During the last few decades, theories based on the original ideas of Kant and Laplace have obtained the status of the "standard" theory for the formation of the solar system. Numerical calculations and a wealth of new information on the structure and composition of planetary bodies, as well as data gathered from astrophysical studies of circumstellar disks and extra-solar-system planets, seem to leave little room for alternate theories.

Data from comets have strengthened our understanding of the formation and evolution of the solar nebula and have helped us to see the intimate connections between our solar system, the interstellar medium, and extra-solar-system planets. However, these connections are not always clear and easy to recognize and many mysteries still remain. In particular, our knowledge of the composition and physical structure of cometary nuclei remain rather primitive in many respects, and we can expect surprising discoveries with ever-increasingly sophisticated observations. The imminent return of a sample of cometary matter from 81P/Wild 2 is especially anticipated, as those results will almost certainly guide our future exploration of comets. On the other hand, the *Stardust* results will likely leave many questions about the icy composition of cometary nuclei, which should be better addressed by the *Rosetta* mission, due to rendezvous with Comet 61P/Churyumov-Gerasimenko in 2015.

Nevertheless, we offer *Comets II* as the best compendium of cometary knowledge for the next decade. With the help of our colleagues in the cometary community, we have attempted to put together a volume that presents a logical and comprehensive treatment of cometary science. The structure of the book was laid out with the clear objective of covering most areas of cometary science through a set of contiguous, non-overlapping chapters. In a sense, it is a book written by a team of about 100 collaborative authors. We have included at the beginning of this volume a series of chapters that describe what may happen to interstellar materials that are used to make up a planetary system. Obviously, the path between interstellar molecules/grains and comets is quite long and complicated. It is this long journey and what happens along the way that is described in this book. Enjoy!

REFERENCES

A'Hearn M. F. (1983) Photometry of comets. In *Solar System Photometry Handbook* (R. M. Genet, ed.), pp. 3-1 to 3-33. Willman Bell Inc., Richmond.

A'Hearn M. F., Dwek E., Feldman P. D., Millis R. L., Schleicher D. G., Thompson D. T., and Tokunaga A. T. (1982) The grains and gas in comet Bowell 1980b. In *Cometary Exploration, Vol. II* (T. I. Gombosi, ed.), pp. 159–166. Central Research

Institute for Physics, Hungarian Academy of Sciences.

A'Hearn M. F., Millis R. L., Schleicher D. G., Osip D. J., and Birch P. V. (1995) The ensemble properties of comets: Results from narrowband photometry of 85 comets, 1976–1992. *Icarus, 118*, 223–270.

Aikin A. C. (1974) Cometary coma ions. *Astrophys., J., 193*, 263–264.

Alfvén H. (1957) On the theory of comet tails. *Tellus, 9*, 92–96.

Arpigny C. (1965) Spectra of comets and their interpretation. *Annu. Rev. Astron. Astrophys., 3*, 351–376.

Arrhenius S. A. (1900) Über die Ursache der Nordlicher. *Phys. Zeitschr., 2*, 81–86.

Baldet F. (1926) Recherches sur la constitution des comètes et sur les spectres du carbone. Thèse, Faculté des Sciences, Paris. Annu. Obs. Astron. Phys. Meudon Vol. 7. 109 pp.

Benvenuti P. and Wurm K. (1974) Spectroscopic observations of Comet Kohoutek (1973f). *Astron. Astrophys., 31*, 121–122.

Bertaux J. L., Blamont J. E., and Festou M. (1973) Interpretation of hydrogen Lyman-alpha observations of Comets Bennett and Encke. *Astron. Astrophys., 25*, 415–430.

Bessel F. W. (1836a) Bemerkungen über mögliche Unzulänglichkeit der die Anziehungen allein berücksichtigenden Theorie der Kometen. Von Herrn Geheimen-Rath u Ritter Bessel. *Astron. Nachr., 13*, 345–350.

Bessel F. W. (1836b) Beobachtungen uber die physische Beschaffenheit des Halley's schen Kometen und dadurch veranlasste Bemerkungen. *Astron. Nachr., 13*, 185–232.

Biermann L. (1951) Kometenschweife und solare Korpuskularstrahlung. *Z. Astrophys., 29*, 274–286.

Biermann L. (1967) On the cometary plasmas. In *Proceedings of ESRIN Study Group Plasma in Space and in the Laboratory,* Frascati, May–June 1966, pp. 181–184. ESRO SP-20.

Biermann L. (1968) *On the Emission of Atomic Hydrogen in Comets.* Jila Report No. 93, University of Colorado, Boulder. 12 pp.

Biermann L. and Lüst R. (1977) Über die Richtungsverteilung der Geschwindigkeiten in der "Oort"schen Wolke (der das Sonnensystem umgebenden Kometenkerne). *Sitz. Ber. Bayer. Ak. Wiss. Mat-Nat. Klas., 11*, 159–166.

Biermann L. and Trefftz E. (1964) Über die Mechanismen der Ionisation und der Anregung in Kometenatmosphären. Mit 9 Textabbildungen. *Z. Astrophys., 59*, 1–28.

Biraud F., Bourgois G., Crovisier J., Fillit R., Gérard E., and Kazès I. (1974) OH observations of comet Kohoutek (1973f) at 18 cm wavelength. *Astron. Astrophys., 34*, 163–166.

Biver N., Bockelée-Morvan D., Crovisier J., Colom P., Henry F., Moreno R., Paubert G., Despois D., Lis D. C. (2002) Chemical composition diversity among 24 comets observed at radio wavelengths. *Earth Moon Planets, 90(1)*, 323–333.

Blamont J. E. and Festou M. (1974) Observations of the comet Kohoutek (1973f) in the resonance light ($A^2\Sigma^+ – X^2\Pi_i$) of the OH. *Icarus, 23*, 538–544.

Bobrovnikoff N. T. (1931) Halley's comet in 1910. *Publ. Lick Obs., 17(II)*, 309–318.

Bobrovnikoff N. T. (1942) Physical theory of comets in the light of spectroscopic data. *Rev. Mod. Phys., 14*, 164–178.

Bockelée-Morvan D. and Gérard E. (1984) Radio observations of the hydroxyl radical in comets with high spectral resolution. kinematics and asymmetries of the OH coma in C/Meier (1978 XXI), C/Bradfield (1979 X), and C/Austin (1982g). *Astron. Astrophys., 131*, 111–122.

Bockelée-Morvan D., Lis D. C., Wink J. E., Despois D., Crovisier J., Bachiller R., Benford D. J., Biver N., Colom P., Davies J. K., Gérard E., Germain B., Houde M., Mehringer D., Moreno R., Paubert G., Phillips T. G., and Rauer H. (2000) New molecules found in comet C/1995 O1 (Hale-Bopp). Investigating the link between cometary and interstellar material. *Astron. Astrophys., 353*, 1101–1114.

Bockelée-Morvan D., Crovisier J., Mumma M. J., and Weaver H. A. (2004) The composition of cometary volatiles. In *Comets II* (M. C. Festou et al., eds.), this volume. Univ. of Arizona, Tucson.

Brahe T. (1578) Manuscript (in German). Codex Vind. 10689. (Translated in 1986 by J. Brager and N. Henningsen, E. Eilertsen Publ., Copenhagen.)

Brandt L. (1985) Zur Geschichte des Kometenproblems von Newton bis Olbers. *Sterne und Weltraum, 7*, 372–376.

Brooke T. Y., Tokunaga A. T., Weaver H. A., Crovisier J., Bockelée-Morvan D., and Crisp D. (1996) Detection of acetylene in the infrared spectrum of Comet Hyakutake. *Nature, 383*, 606–608.

Carrington R. C. (1859) Description of a singular appearance seen in the Sun on September 1, 1859. *Mon. Not. R. Astron. Soc., 20*, 13–15.

Code A. D. and Savage B. D. (1972) Orbiting Astronomical Observatory: Review of scientific results. *Science, 177*, 213–221.

Code A. D., Houck T. E., and Lillie C. F. (1972) Ultraviolet observations of comets. In *The Scientific Results from OAO-2* (A. D. Code, ed.), pp. 109–114. NASA SP-310, Washington, DC.

Combes M., Moroz V. I., Crifo J. F., Lamarre J. M., Charra J., Sanko N. F., Soufflot A., Bibring J. P., Cazes S., Coron N., Crovisier J., Emerich C., Encrenaz T., Gispert R., Grigoriev A. V., Guiyot G., Krasnopolsky V. A., Nikolsky Yu. V., and Rocard F. (1986) Infrared sounding of comet Halley from Vega 1. *Nature, 321*, 266–268.

Crovisier J., Leech K., Bockelée-Morvan D., Brooke T. Y., Hanner M. S., Altieri B., Keller H. U., and Lellouch E. (1997) The spectrum of Comet Hale-Bopp (C/1995 O1) observed with the Infrared Space Observatory at 2.9 AU from the Sun. *Science, 275*, 1904–1907.

Crovisier J., Colom P., Gérard E., Bockelée-Morvan D., and Borgeois G. (2002) Observations at Nançay of the OH 18-cm lines in comets. The data base. Observations made from 1982 to 1999. *Astron. Astrophys., 393*, 1053–1064.

Debye P. (1909) Der Lichdruck auf Kugeln von beliebigem Material. *Ann. Phys. Leipzig (4), 30*, 57.

Dello Russo N., Mumma M. J., DiSanti M. A., Magee-Sauer K., Novak R., and Rettig T. W. (2000) Water production and release in Comet C/1995 O1 Hale-Bopp. *Icarus, 143*, 324–337.

Delsemme A. H. and Miller D. C. (1971) Physico-chemical phenomena in comets — III. The continuum of comet Burnham (1960 II). *Planet. Space Sci., 19*, 1229–1257.

Delsemme A. H. and Swings P. (1952) Hydrates de gaz dans les noyaux cométaires et les grains interstellaires. *Ann. Astrophys., 15*, 1–6.

Despois D., Gérard E., Crovisier J., and Kazès I. (1981) The OH radical in comets — Observation and analysis of the hyperfine microwave transitions at 1667 MHz and 1665 MHz. *Astron. Astrophys., 99*, 320–340.

Donati G. B. (1864) Schreiben des Herrn Prof. Donati an den Herausgeber. *Astron. Nachr., 62*, 375.

Dones L., Weissman P. R., Levison H. F., and Duncan M. J. (2004)

Oort cloud formation and dynamics. In *Comets II* (M. C. Festou et al., eds.), this volume. Univ. of Arizona, Tucson.

Donn B., Rahe J., and Brandt J. C. (1986) *Atlas of Comet Halley 1910.* NASA SP-488, Washington, DC. 600 pp.

Drake J. F., Jenkins E. B., Bertaux J. L., Festou M., and Keller H. U. (1976) Lyman-alpha observations of Comet Kohoutek 1973 XII with Copernicus. *Astrophys. J., 209,* 302–311.

Duncan M., Levison H., and Dones L. (2004) Dynamical evolution of ecliptic comets. In *Comets II* (M. C. Festou et al., eds.), this volume. Univ. of Arizona, Tucson.

Eddington A. S. (1910) The envelopes of Comet Morehouse (1908c). *Mon. Not. R. Astron. Soc., 70,* 442–458.

Edgeworth K. E. (1949) The origin and evolution of the solar system. *Mon. Not. R. Astron. Soc., 109,* 600–609.

Encke J. F. (1820) Ueber die Bahn des Pons'schen Kometen (86 Olb.) nebst Berechnung seines Laufs bei seiner nächsten Wiederkehr in Jahre 1822. *Berliner Astr. Jahrbuch für 1823,* pp. 211–223.

Everhart E. (1972) The origin of short-period comets. *Astrophys. Lett., 10,* 131–135.

Feldman P. D. and Brune W. H. (1976) Carbon production in comet West 1975n. *Astrophys. J. Lett., 209,* L45–L48.

Feldman P. D., Takacs P. Z., Fastie W. G., and Donn B. (1974) Rocket U.V. spectrophotometry of Comet Kouhoutek. *Science, 185,* 705–707.

Fernández J. (1980) On the existence of a comet belt beyond Neptune. *Mon. Not. R. Astron. Soc., 192,* 481–491.

Festou M. C. and Feldman P. D. (1987) Comets. In *Exploring the Universe with the IUE Satellite* (Y. Kondo and W. Wamsteker, eds.), pp. 101–118. Reidel, Dordrecht.

Finson M. L. and Probstein R. F. (1968a) A theory of dust comets. I. Model and equations. *Astrophys. J., 154,* 327–352.

Finson M. L. and Probstein R. F. (1968b) A theory of dust comets. II. Results for comet Arend-Roland. *Astrophys. J., 154,* 353–380.

Halley E. (1705) *Astronomiae Cometicae Synopsis,* Oxford. (Translated as *A Synopsis of the Astronomy of Comets.* 1705, London.) 24 pp.

Heisler J. (1990) Monte Carlo simulations of the Oort comet cloud. *Icarus, 88,* 104–121.

Herbig G. H. (1973) *Comet Kohoutek (1973 f).* IAU Circular No. 2596.

Herzberg G. and Lew H. (1974) Tentative identification of the H₂O⁺ ion in Comet Kohoutek. *Astron. Astrophys., 31,* 123–124.

Hills J. G. (1981) Comet showers and the steady-state infall of comets from the Oort cloud. *Astrophys. J., 86,* 1730–1740.

Ho P.-Y. (1962) Ancient and mediaeval observations of comets and novae in Chinese sources. *Vist. Astr., 5,* 127–225.

Hoffmeister C. (1943) Physikalische Untersuchungen an Kometen. I. Die Beziehungen des primären Schweifstrahls zum Radiusvektor. *Z. Astrophys., 22,* 265–285.

Huggins W. (1868) Further observations on the spectra of some of the stars and nebulae, with an attempt to determine therefrom whether these bodies are moving towards or from the Earth, also observations on the spectra of the Sun and of Comet II, 1868. *Proc. R. Soc., 15,* 529–564.

International Halley Watch (1992) *Comet Halley Archive.* CD-ROM Vol. USA_NASA_IHY_HAL_0001_TO_0023. NASA Goddard Space Flight Center, Greenbelt, Maryland.

Jaegermann R. (1903) *Prof. Th. Bredikhin's Mechanische Untersuchungen Über Cometenformen.* In *Systematischer Darstellung von R. Jaegermann.* St. Petersburg, Russia. 500 pp.

Kazimirchak-Polonskaya E. I. (1972) The major planets as powerful transformers of cometary orbits. In *The Motion, Evolution of Orbits, and Origin of Comets* (G. A. Chebotarev et al., eds.), pp. 373–397. Reidel, Dordrecht.

Keller H. U. (1971) Wasserstoff als Dissoziationsprodukt in Kometen. *Mitt. Astron. Gesell., 30,* 143–148.

Keller H. U. (1973a) Lyman-alpha radiation in the hydrogen atmospheres of comets: A model with multiple scattering. *Astron. Astrophys., 23,* 269–280.

Keller H. U. (1973b) Hydrogen production rates of Comet Beunett (1969i) in the first half of April 1970. *Astron. Astrophys., 27,* 51–57.

Keller H. U. (1976) The interpretations of ultraviolet observations of comets. *Space. Sci. Rev., 18,* 641–684.

Keller H. U. and Lillie C. F. (1974) The scale length of OH and the production rates of H and OH in Comet Bennett (1970 II). *Astron. Astrophys., 34,* 187–196.

Keller H. U., Kramm R., and Thomas N. (1988) Surface features on the nucleus of Comet Halley. *Nature, 331,* 227–231.

Kissel J., Sagdeev R. Z., Bertaux J. L., Angarov V. N., Audouze J., Blamont J. E., Büchler K., Evlanov E. N., Fechtig H., Fomenkova M. N., von Hoerner H., Inogamov N. A., Khromov V. N., Knabe W., Krueger F. R., Langevin Y., Leonas V. B., Levasseur-Regourd A. C., Managadze G. G., Podkolzin S. N., Shapiro V. D., Tabaldyev S. R., and Zubkov B. V. (1986) Composition of comet Halley dust particles from VEGA observations. *Nature, 321,* 280–282.

Krankowsky D., Lämmerzahl P., Herrwerth I., Woweries J., Eberhardt P., Dolder U., Herrmann U., Schulte W., Berthelier J. J., Illiano J. M., Hodge R. R., and Hoffman J. H. (1986) In situ gas and ion measurements at Comet Halley. *Nature, 321,* 326–329.

Kuiper G.P. (1949) The law of planetary and satellite distances. *Astrophys. J., 109,* 308–313.

Kuiper G. P. (1951) On the origin of the solar system. In *Astrophysics* (J. A. Hynek, ed.), pp. 357–424. McGraw Hill, New York.

Lagrange J. L. (1814) Sur les comètes, Connais. *Temps, 216,* and *Additions à la Connaissance des Temps,* p. 213.

Laplace P. S. (1813) *Exposition du Système du Monde,* 4th edition, Paris.

Laplace P. S. (1816) Sur l'origine des comètes. *Additions à la Connaissance des Temps,* pp. 211–218.

Levin B. J. (1943) Gas evolution from the nucleus of a comet as related to the variations in its absolute brightness. *Comp. Rend. (Dokl.) Acad. Sci. URSS, 38,* 72–74.

Levin B. J. (1985) Erratum — Laplace Bessel and the icy model of cometary nuclei. *Astron. Q., 5(18),* 113.

Lyttleton R. A. (1948) On the origin of comets. *The Observatory, 72,* 33–35.

Marsden B. G. (1969) Comets and nongravitational forces. II. *Astron. J., 74,* 720–734.

Marsden B. G. (1970) Comets and nongravitational forces. III. *Astron. J., 75,* 75–84.

Marsden B. G., and Sekanina Z. (1973) On the distribution of "original" orbits of comets of large perihelion distance. *Astron. J., 78,* 1118–1124.

Marsden B. G., Sekanina Z., and Yeomans D. K. (1973) Comets and nongravitational forces. V. *Astron. J., 78,* 211–225.

Marsden B. G., Sekanina Z., and Everhart E. (1978) New osculating orbits for 110 comets and analysis of original orbits for 200 comets. *Astron. J., 83,* 64–71.

McKellar A.(1943) Rotational distribution of CH molecules in the nucleus of Comet Cunningham (1940c). *Astrophys. J., 98,* 1–5.

Mumma M. J., Weaver H. A., Larson H. P., Davis D. S., and Williams M. (1986) Detection of water vapor in Halley's comet. *Science, 232,* 1523–1528.

Mumma M. J., Disanti M. A., dello Russo N., Fomenkova M., Magee-Sauer K., Kaminski C. D., and Xie D. X. (1996) Detection of abundant ethane and methane, along with carbon monoxide and water, in Comet C/1996 B2 Hyakutake: Evidence for interstellar origin. *Science, 272,* 1310–1314.

Mumma M. J., Disanti M. A., dello Russo N., Magee-Sauer K., Gibb E., and Novak R. (2003) Remote infrared observations of parent volatiles in comets: A window on the early solar system. *Adv. Space Res., 31(12),* 2563–2575.

Newburn R. and Rahe J. (1990) The International Halley Watch. In *Comet Halley: Investigations, Results, Interpretations, Vol. 1 — Organization, Plasma, Gas* (J. W. Mason, ed.), pp. 23–32. Ellis Horwood Library of Space Science and Space Technology, Series in Astronomy, Chichester, England.

Newton Sir I. (1687) *Philosophiae Naturalis Principia Mathematica.* Jussu Societatis Regiae ac typis Josephi Streatti, London. 510 pp.

Öhman Y. (1939) On some observations made with a modified Pickering polarigraph. *Mon. Not. R. Astron. Soc., 99,* 624–633.

Öhman Y. (1941) *Measurements of Polarization in the Spectra of Comet Cunningham (1940c) and Comet Paraskevopoulos (1941c).* Ann. Obs. Stockholm, Vol. 13, No. 11. 15 pp.

Olbers H. W. M. (1812) Ueber den Schweif des grossen Cometen von 1811 (On the tail of the great Comet of 1811). *Mon. Corr. z. Beförd. d. Erd u. Himmels-Kunde, 25,* 3.

Oort J. H. (1950) The structure of the cloud of comets surrounding the solar system and a hypothesis concerning its origin. *Bull. Astr. Inst. Netherl., 11,* 91–110.

Oort J. H. and Schmidt M. (1951) Differences between new and old comets. *Bull. Astron. Inst. Netherl., 11,* 259–270.

Oppenheimer M. (1975) Gas phase chemistry in comets. *Astrophys. J., 196,* 251–259.

Öpik E. J. (1932) Notes on stellar perturbations of nearly parabolic orbits. *Proc. Am. Acad. Astron. Sci., 67,* 169–182.

Rickman H. (1989) The nucleus of Comet Halley — Surface structure, mean density, gas and dust production. *Adv. Space Res., 9(3),* 59–71.

Rickman H. and Froeschlé C. (1983) Model calculations of nongravitational effects on Comet P/Halley. In *Cometary Exploration, Vol. III* (T. I. Gombosi, ed.), pp. 109–114. Central Research Institute for Physics, Hungarian Academy of Sciences.

Richter N. B. (1963) *The Nature of Comets.* Methuen, London. 221 pp.

Richter K., Combi M. R., Keller H. U., and Meier R. R. (2000) Multiple scattering of hydrogen Lyα radiation in the coma of Comet Hyakutake (C/1996 B2). *Astrophys. J., 531,* 599–611.

Rigaud S. P. (1833) Supplement to Dr. Bradley's miscellaneous work, Oxford. Reprinted in 1972 in *Sources of Science Series, No. 97,* Johnson Reprint Corp., New York.

Sagdeev R. Z., Elyasberg P. E., and Moroz V. I. (1988) Is the nucleus of Comet Halley a low density body? *Nature, 331,* 240–242.

Schiaparelli G. V. (1866) Intorno al corso ed all' origine probabile delle stelle meteoriche. *Bull. Meteorol. Roma, V* (1866), 97–106,

113–121, 124–125, 129–131; *VI* (1867), 9–10.

Schiaparelli G. V. (1867) Sur la relation qui existe entre les comètes et les étoiles filantes. Par M. G. V. Schiaparelli. *Astron. Nachr., 68,* 331–332.

Schleicher D. G. and A'Hearn M. F. (1982) OH fluorescence in comets — Fluorescence efficiency of the ultraviolet bands. *Astrophys. J., 258,* 864–877.

Schleicher D. G. and A'Hearn M. F. (1988) The fluorescence of cometary OH. *Astrophys. J., 331,* 1058–1077.

Schloerb F. P. (1988) Collisional quenching of cometary emission in the 18 centimeter OH transitions. *Astron. J., 332,* 524–530.

Schloerb F. P. (1989) UV and radio observations of OH in comets. In *Asteroids, Comets, Meteors III* (C.-I. Lagerkvist et al., eds.,) pp. 431–434. Uppsala Observatory, Sweden.

Schwarzschild K. (1901) Der Druck der Lichtes auf kleine Kugeln und die Arrhenius'sche Theorie der Kometen schweife. *Sitz. Ber. Akad. München, 31,* 293.

Schwarzschild K. and Kron E. (1911) On the distribution of brightness in the tail of Halley's comet. *Astrophys. J., 34,* 342–352.

Sekanina Z. and Fry L. (1991) *The Comet Halley Archive Summary Volume.* Jet Propulsion Laboratory, Pasadena. 332 pp.

Sinding E. (1937) Zur Bestimmung der ursprüngen Gestalt parabelnaher Kometenbahnen. *Astron. Nachr., 261,* 458–460.

Strömgren E. (1914) Über der Ursprung der Kometen. *Publ. Obs. Copenhagen, 19,* 189–250.

Strömgren E. (1947) The short-period comets and the hypothesis of their capture by the major planets. *Danske Vidensk Selsk. Mat.-fys. Medd., 24(5).* 10 pp.

Swings P. (1941) Complex structure of cometary bands tentatively ascribed to the contour of the solar spectrum. *Lick Obs. Bull., XIX,* 131–136.

Swings P. (1942) Molecular bands in cometary spectra. Identifications. *Rev. Mod. Phys., 14,* 190–194.

Swings P. (1943) Cometary spectra (council report on the progress at astronomy). *Mon. Not. R. Astron. Soc., 103,* 86–87.

Swings P. (1948) Le spectre de la Comète d'Encke 1947. *Ann. Astrophys., 11,* 124–136.

Swings P. (1956) The spectra of the comets. *Vist. Astron., 2,* 958–981.

Swings P. and Greenstein J. E. (1958) Présence des raies interdites de l'oxygène dans les spectres cométaires. *C. R. Acad. Sci. Paris, 246(4),* 511–513.

Swings P., McKellar A., and Minkowski R. (1943) Cometary emission spectra in the visual region. *Astrophys. J., 98,* 142–152.

Turner B. E. (1974) Detection of OH at 18-centimeter wavelength in Comet Kohoutek (1973f). *Astrophys. J. Lett., 189,* L137–L139.

van Woerkom A. J. J. (1948) On the origin of comets. *Bull. Astron. Inst. Netherl., 10,* 445–472.

Weaver H. A., Feldman P. D., Festou M. C., and A'Hearn M. F. (1981a) Water production models for Comet Bradfield 1979 X. *Astrophys. J., 251,* 809–819.

Weaver H. A., Feldman P. D., Festou M. C., A'Hearn M. F., and Keller H. U. (1981b) IUE observations of faint comets. *Icarus, 47,* 449–463.

Weaver H. A., Feldman P. D., McPhate J. B., A'Hearn M. F., Arpigny C., and Smith T. E. (1994) Detection of CO Cameron band emission in Comet P/Hartley 2 (1991 XV) with the Hubble Space Telescope. *Astrophys. J., 422,* 374–380.

Weissman P. R. (1996) The Oort Cloud. In *Completing the Inven-*

tory of the Solar System (T. W. Rettig and J. M. Hahn, eds.), pp. 265–288. ASP Conference Series 107, Astronomical Society of the Pacific, San Francisco.

Whipple F. L. (1950) A comet model. I. The acceleration of Comet Encke. *Astrophys. J., 111,* 375–394.

Whipple F. L. (1951) A comet model. II. Physical relations for comets and meteors. *Astrophys. J., 113,* 464–474.

Whipple F. L. (1961) Comets: Problems of the cometary nucleus. *Astron. J., 66,* 375–380.

Wilkening L. L., ed. (1982) *Comets.* Univ. of Arizona, Tucson. 766 pp.

Wurm K. (1943) Die Natur der Kometen. *Mitt. Hamburg. Sternw., 8(51),* 57–92.

Wurm K. (1963) The physics of comets. In *The Moon, Meteorites and Comets* (B. M. Middlehurst and G. P. Kuiper, eds.), pp. 573–617. Univ. of Chicago, Chicago.

Yeomans D. K. (1991) *Comets: A Chronological History of Observation, Science, Myth, and Folklore.* Wiley, New York. 485 pp.

Yeomans D. K., Chodas P. W., Sitarski G., Szutowicz S., and Królikowska M. (2004) Cometary orbit determination and nongravitational forces. In *Comets II* (M. C. Festou et al., eds.), this volume. Univ. of Arizona, Tucson.

Cometary Science: The Present and Future

Michael F. A'Hearn
University of Maryland

1. INTRODUCTION

The subsequent chapters in this book provide a comprehensive study of our present knowledge of comets, from the interstellar medium, through formation of the solar system and the present day, to the death of comets. This chapter will make no attempt to summarize these chapters. Rather, based on the material in the following chapters, this chapter will ask about the high-level state of our knowledge in major areas. Is our knowledge mostly speculation based on fragmentary data? Is our knowledge mature in terms of data but immature in interpretation? Is the entire area mature with a full understanding of the implications for the larger fields of science? A natural outgrowth of this approach is to ask where we might be a decade into the future. What would we like to know? What are we likely to know? Where might the big surprises lie?

No references (except one noncometary reference) are given in this chapter because the topics are all covered in more detail elsewhere in this book and those chapters contain a far more appropriate set of primary references than could be provided here. Some of the areas, notably dynamics, are also surveyed at a higher level later in this book. Other scientists will likely disagree with some of the conclusions and speculations presented here, but the main purpose of the chapter is not to be definitive but to stimulate the reader to think about future directions. As such, it is more important to be provocative than to be definitive.

2. BASIC PHYSICAL PROPERTIES

One of the fundamental anomalies of cometary studies is how far we have come without knowing basic physical parameters like the sizes of the nuclei. The true shapes, sizes, and albedos of nuclei come only from *in situ* imaging and this is necessarily limited to a small number of comets. Because of the small size and infrequent close approaches to Earth, the radar studies that have contributed so much to asteroid sizes and shapes have been comparatively ineffective in elucidating the same parameters for comets. Fortunately, we are now beginning to get reasonable estimates of the size distribution, albeit mostly from optical data, which require an assumed albedo to yield a size. Many of the data, furthermore, consist of only single observations rather than complete rotational lightcurves, which adds scatter to the distribution. Because of the limited size of the dataset available now, the slope of the distribution is not accurately known but we are beginning to get reasonable estimates. In

fact, binning the data in uniform size ranges rather than using each individual object implies a steeper slope (–2.5) for the cumulative size distribution than is generally accepted (–1.5 to –2.0), suggesting that our sample size is limiting our precision. What we do not know yet is whether the flattening of the cumulative distribution at sizes below 2 km is due mostly to selection effects or mostly to a real dearth of small comets. Physical arguments suggest that most small "comets" (much less than a kilometer) must be either dormant or extinct and thus not recognizable as comets, and this means that we should expect a turnover at some point not too far from where it is observed. Provided sufficient telescope time can be made available, this is an area in which we can expect a final answer within the decade, although statistics may still not be good enough to determine the differences in size distribution for certain dynamical classes even though differences are expected, e.g., between comets from the Oort cloud and comets from the Kuiper belt.

Separating the size from the albedo is still a major challenge with fewer than two dozen comets having both parameters well known. The availability of the Spitzer Space Telescope (SST, née SIRTF) can lead to a large sample of comets for which the size and albedo are independently determined. The only question is whether enough observing time will be made available in the 5- to 7-year expected lifetime of the SST. Rotational lightcurves, if done in sufficient detail and from different aspects, can provide the convex hull of the body and the rotational state. However, the necessity to carry out these observations either at very high spatial resolution or when the comets are far from the Sun and thus possibly inactive makes it likely that, unlike the case for asteroids, it will not be practical to get complete rotational lightcurves for more comets than the ones of special interest, such as spacecraft targets.

The amazing aspect of the basic properties of comets, and surprisingly little realized outside the community of cometary specialists, is that we still do not have a single, measured mass for a cometary nucleus, and thus not a single, measured density. Clever use of the nongravitational acceleration of comets has led to a series of estimates that appear to be converging on densities around 0.5 g cm^{-3}, but this is still an extremely model-dependent result and thus uncertain by as much as half an order of magnitude. We will not have a directly measured mass for a typical cometary nucleus until there is a rendezvous mission, although one could in principle measure a mass for an unusually large comet, such as Chiron or even Hale-Bopp, with a slow flyby provided drag by the coma can be separated out. The first rendezvous mission currently underway is *Rosetta*, which will not measure the mass of P/Churyumov-Gerasimenko until 2014. There is some hope that one or another of the smaller missions being proposed might be selected and arrive at its target earlier.

Determining interior structure, which would be invaluable for our understanding of formation and evolution, will be limited to what can be learned from the *Deep Impact* mission until there is a rendezvous mission and/or a soft lander, again something that is not currently scheduled to occur for another decade. *Deep Impact* will excavate a large crater with an artificial meteorite impact in order to study the outermost tens of meters. Suggestions of chemical heterogeneity come from remote sensing. We know from D/Shoemaker-Levy 9 (SL9) that, at least for scales comparable to a kilometer, the strength is $<10^3$ dyn/cm^2 and rubble-pile models with similarly low strength at smaller scales successfully describe various phenomena, but we do not have any model-independent, direct constraints on the strength other than SL9.

Although our knowledge of basic physical properties is sparse now, we expect to have a mature understanding of the properties not related to the mass well within the next decade. Our knowledge of mass and density will still be immature, although we should have good numbers for one or two bodies.

3. CHEMICAL COMPOSITION

Although the separation between volatiles and refractories is not rigid, it is convenient to think of the composition in these terms, the volatiles being those species seen in the gas phase and the refractories being those seen in the solid phase, for which mineralogy and crystal structure are also important. Clearly some species can be considered in both ways and, in the future, we can expect that many more species will be studied both as solids and as gases. Our knowledge of composition is limited almost entirely to the coma, the spectra of nuclei being almost featureless, a notable exception being two very weak features seen in the *Deep Space 1* spectra of P/Borrelly. Other than that, the surfaces of comets are known only to be very dark, presumably from a combination of particle shadowing due to porosity and the inclusion of very dark, carbonaceous material as one of the abundant components at the surface. We are thus faced with the problem of deciding the extent to which the composition in the coma is representative of the composition in the nucleus.

3.1. Volatiles

Our knowledge of the composition of the gaseous coma is quite extensive, coming from remote sensing at wavelengths from the X-ray to the radio and from *in situ* measurements with a mass spectrometer primarily at P/Halley and to a minor extent in the tail of P/Giacobini-Zinner. In the data available to date, the *in situ* measurements provide coverage of all species up to a given mass, but with insufficient mass resolution to uniquely separate, e.g., N_2 from CO. On the other hand, remote sensing at the highest spectral resolution can easily separate all species but suffers from incompleteness of coverage, depending on dipole moments of the molecules, on lifetimes, and on excitation conditions. There are roughly 80 species firmly identified in comets, but this is almost certainly a very incomplete list. All but one of these (S_2) are also seen in the interstellar medium, but the converse is not true and the abundance ratios, in general, are not similar. The *in situ* measurements with mass spectrometers show a continuum of all masses at higher masses and these have generally not been deconvolved to individual species. Although the majority of new species recently have been found at infrared and millimeter wavelengths, new species continue to be found at shorter wavelengths as well.

Unfortunately, many of the identified species have had abundances measured only in one or a few comets, so that one has no sense of whether or not there is wide variation from comet to comet that might be correlated with origin or evolution. At optical wavelengths there does appear to be a correlation of abundances of C_2 and C_3 (relative to H_2O) with place of origin but these are the only species, other than CN and possibly NH and NH_2, for which there are data on a sufficiently large number of comets to study such correlations reliably. Even where we have a correlation, the mechanism for producing the correlation is not understood beyond speculation. While the field of gaseous abundances is mature in many ways, there are still many discoveries and measurements to be made, particularly in expanding the infrared and millimeter-wave measurements to a large ensemble of comets but also in identifying new species since the list of unidentified lines seen in comets is extremely long.

Interpretation of the chemistry of the coma is further limited by the fact that most of the species observed, including virtually all the easily observed species at optical and ultraviolet wavelengths, are clearly fragments of larger molecules that existed in the nucleus. Extensive chemical models of the coma, including many hundreds of reactions, have been constructed by several authors. To the extent that processes other than photodissociation and photoionization matter, these models are sensitive to the physical conditions in the coma, primarily the density and kinetic temperature as a function of distance from the surface. Furthermore, processes such as photodissociation are sources of heating in the coma and even the shape of the nucleus may play a major role in the spatial profile of density and temperature. The feedback between photochemistry and physical conditions has been calculated only for water, but other species could also affect the physical conditions. Furthermore, there is some likelihood that reactions involving excited states (electronically excited molecules and/or molecules with excess kinetic energy) may be important in producing some species and these reactions have generally not been included in the calculations with large chemical networks. The net result is that in only a very few cases have the chemical pathways been reliably traced from observed species to parent molecules. Several species thought on chemical grounds to be parent molecules directly from the nucleus have even been shown to have spatial profiles, implying that they are produced from other species, probably by thermal or photodesorption from grains, at some distance from the nucleus. Furthermore, as was made especially clear with the advent of C/Hale-Bopp, the relative abundances of species vary dramatically with heliocentric distance even in a single comet. Again, a few cases of variation can be explained in terms of processes in the coma, but most are not explained at all and in virtually no case is there consensus that we can correct for the variation with heliocentric distance adequately to say something definitive about nuclear abundances.

In the area of evolution of nuclear ices, the theoreticians have far outstripped the observers and there are extensive models of the depletion and migration of nuclear ices due to successive perihelion passages. These models have been used to explain the asymmetries in visual lightcurves, but there are insufficient observational data on the predominant ices to properly test any of the models. The predictions of the models cover a wide range so detailed measurements of ice with depth in a cometary nucleus would easily discriminate among the models.

In the next decade we can anticipate numerous discoveries of previously unknown species. A few will come with traditional telescopes and instruments when there is a suitably bright comet. More will come from new facilities like the Atacama Large Millimeter Array (ALMA) and SST. Observations with the spectrometer on the *Deep Impact* flyby spacecraft may provide new species seen only very close to the nucleus, but the limited sensitivity is not expected to show new molecules with typical spatial profiles.

Our knowledge of volatile abundances in the coma is reasonably mature, but, despite considerable important work, our interpretation of these abundances in terms of the nuclear abundances is primitive.

3.2. Refractories

The refractory species are much less well known. Remote sensing has brought us primarily the silicate feature, including identification of crystalline olivine, and specifically Mg-rich crystalline olivine in addition to amorphous olivine and pyroxene. *In situ* measurements of grains at P/Halley brought us CHON particles, but the specific chemical composition of the particles can not readily be inferred from those measurements. The presence of CHON particles filled a major gap in our understanding of the overall abundances, since combining these particles with the volatiles leads to more or less solar abundances for all but the lightest elements and the most volatile species, such as N_2 and the noble gases. Remote sensing has also brought us the CH-stretch feature in the near-infrared. Much of this is from formaldehyde and methanol, but there may be a more refractory component as well. We probably have a large number of refractory cometary particles in our collection of micrometeorites, but the evidence that they are cometary is mostly circumstantial, including, e.g., stratospheric particles collected during the Leonid meteor storm. At least in part because of difficulty in associating micrometeorites with specific comets, the micrometeorites have not yet led to any significant constraints on comets. Rather, the particles have been associated with comets at least in part because they resemble what we think ought to come from comets.

On the other hand, in January 2006 the *Stardust* mission will return a large number of refractory particles that are unambiguously from P/Wild 2, as well as others that are almost certainly interstellar. This should enable us to determine which of the micrometeorites are truly from comets and it should give us the first true measurements of the actual distribution of particle composition, size, and mineralogy. There may be some selection effects in which particles can be lifted from the surface to be collected by the *Stardust* spacecraft, e.g., due to chemical differences correlated with size and/or with stickiness, but these selection effects are small compared to the advance that will be achieved from analyzing these particles in the laboratory. Particles on the surface of a nucleus will be analyzed *in situ* by *Rosetta* a decade hence, while contextual information is simultaneously gathered and this will provide even greater advances. In particular, we can hope to understand what fraction of the solid grains were brought directly from the interstellar medium and perhaps the conditions under which other grains condensed in the protoplanetary disk.

Our knowledge of the refractory composition is far more primitive than our knowledge of the volatiles, but already we are unable to explain the variation from comet to comet of crystalline olivine vs. disordered silicates.

4. EVOLUTIONARY EFFECTS

4.1. Dynamical

Our understanding of the orbital evolution of comets seems clear at some high level — formation from Jupiter outward, followed by ejection to the Oort cloud or capture into the giant planets for comets formed inside Neptune or by successive gravitational captures leading ultimately to Jupiter's family of comets for comets formed beyond Neptune, possibly involving some time in the scattered disk population. The details, however, are not well understood. For example, the relative proportions of Oort cloud comets formed at different distances from the Sun, while calculable with current models of planetary evolution, are sensitive to the models for the formation of the solar system as a whole and these are not well constrained. The injection of comets from the Oort cloud to the inner solar system is also understood in general but not in quantitative detail.

Our simulations of the orbits for comets with small (<3 AU) perihelia are quite good and we have detailed, but model-dependent, simulations of the nongravitational forces acting on comets that appear quite reliable. The results of the nongravitational forces are also reasonably well understood. The models of the nongravitational forces for comets seem somewhat *ad hoc* in the cases of comets for which the nongravitational parameters change significantly from one apparition to the next. While this can probably be traced to changes in the angular momentum vector induced by the torques of the jets themselves, the simulations do not appear to provide unique solutions to the problem.

Closely related to the precession-induced changes in nongravitational accelerations is the effect of torques on total angular momentum. Rotational periods are generally longer than for comparably sized asteroids and this may be due to the influence of outgassing torques. However, there is only one well-determined case of excited state rotation, namely P/Halley, although torques from outgassing jets should be quite capable of producing excited-state rotation in many cases. Is the lack of other comets in excited-state rotation due to the fact that the phenomenon is rare or due to the limited nature of the data on rotational state for most comets? This author thinks that it is mostly the latter — the data needed to show excited state rotation are very extensive unless the data include *in situ* images that show the rotational orientation accurately over many rotations.

4.2. Physical

The physical evolution of comets is not well understood at all, even though everyone agrees that some comets, such as P/Encke, are very evolved. There is surprisingly little systematic, observable difference between comets that appear to have had very different evolutionary histories. The chemistry is similar, there is a wide range of gas/dust ratios for various evolutionary states, and there are insufficient data on the nuclei of any but Jupiter-family comets to make any sensible comparison.

Statistical arguments show that short-period comets must somehow become inactive on a timescale comparable to or less than their dynamical lifetime. Similar statistical arguments suggest that a large fraction of dynamically new comets from the Oort cloud must break up or disperse by other means on their first approach to the Sun. Direct photometric evidence shows that the surviving dynamically new comets from the Oort cloud behave differently on approach to the Sun than do any other comets, including these very same comets on their first departure from the inner solar system. This last effect is generally understood to be due to the loss of the outer layer of the comet, which had been so irradiated by cosmic rays over 4.5 G.y. that it was chemically unstable, by explosive release at some large heliocentric distance as the comet first enters the planetary region. Is this photometric difference related to the inferred breakup of dynamically new comets? All other aspects of evolution are even less well understood.

There are many models for the evolution of the outer layers of cometary nuclei, but they consider different processes and lead to a wide variety of possible scenarios for the evolution, depending not only on the orbital properties and the initial mix of ices and dust but also on the actual processes assumed to dominate in the models. As noted above, the theoretical modelers have far outstripped the ob-

servational data in the evolution of the nucleus. The cohesiveness of the refractory material, the porosity at various stages of the evolution, the effective thermal conductivity, the choice of the initial abundances of the ices, and even the tortuosity of the pores are all critical parameters in understanding the evolution. Relatively little work has been done on the evolution of the refractory components, other than calculations of the parameters that affect the amount of dust lifted off the surface. There should be, for example, differences between the surface refractory solids and those in the interior due to different densities and the consequent difference in the effect of drag forces. Similarly, there should be differences in composition due to different types of cohesiveness or ability to stick together in larger aggregates. However, we have no data on such selection effects and will not have it until we have measurements on the surface of a nucleus.

Perhaps the most dramatic form of physical evolution is the breakup of comets. Breakups range from releasing one or more discrete fragments, which usually disappear on timescales of an orbital period or even much less, through repeated release of small fragments (C/Hyakutake 1996 B2), to complete dispersal of the entire comet (C/LINEAR 1999 S4). Statistically there appears to be a few percent chance of any given comet breaking up on its passage through the inner solar system. The statistics of observed breakups are not yet good enough to know whether there are differences among dynamical groups of comets or to verify the suggestion above that very many dynamically new comets might break up on their first approach to the planetary region. Except in a few cases where the disruption can be reliably explained by tidal forces, e.g., D/Shoemaker-Levy 9, the mechanism for breakups is not at all understood, several alternative scenarios having been proposed.

The fundamental dichotomy in our work with comets is that we claim to use them to study the conditions in the early solar system but we do not know how to separate out the evolutionary effects to reveal the primordial conditions.

5. ORIGIN AND THE EARLY SOLAR SYSTEM

The key questions in this area are (1) whether interstellar ices survived the accretion shock and were incorporated directly into comets, (2) whether any chemical reactions (either in the gas phase or on grain surfaces) were important in that part of the accretion disk in which comets formed, and therefore, (3) whether or not the details of the abundances of ices in comets are good constraints on the conditions in the early solar system. Laboratory experiments have shown, for example, that if condensation is the only important process, the relative abundances of common ices (H_2O, CO_2, CO) are very sensitive to the temperature. But this assumes that the material is initially in the gas phase, a condition that is not satisfied if any interstellar ices survive until they are incorporated in comets. This condition, of course, may vary with distance from the Sun. In particular, comets formed in the vicinity of Jupiter are much more likely to have formed from locally condensed ices, particularly if they form in the outer parts of the protojovian disk.

Models of the early solar system predict varying amounts of radial mixing of material. This probably implies mixing at a macroscopic scale of cometesimals that accreted at different locations in the protoplanetary disk. Studying the heterogeneity of cometary nuclei at scales of tens to hundreds of meters can thus provide key information on the degree of radial mixing in the early solar system. The data from remote sensing are only suggestive of heterogeneity and insufficient thus far to provide useful constraints.

6. COMETS AND TERRESTRIAL PLANETS

The role of comets at the terrestrial planets is still unclear, both in the early solar system and today. Did comets deliver most of Earth's water and organics? The difference in D/H ratios between comets and terrestrial ocean water (SMOW) strongly argues at first glance against comets being the primary source. There are ways around this since comets that formed near Jupiter and were scattered into the inner solar system at a very early stage might well have had different D/H ratios than do today's Oort cloud comets (the only ones for which D/H has been measured). Other authors have argued that the water could have been delivered by asteroids containing hydrated minerals. This appears to be a question that is wide open today, although a cometary origin for both water and organics appears more likely to this author.

The role of comets today is also unclear. It has been argued that the Chicxulub crater was formed by a comet, although others have argued that it was formed by a carbonaceous chondrite. This specific example aside, there is no doubt that comets impact Earth and the other terrestrial planets today. Recent estimates suggest that they are a relatively small contributor to the flux of impactors, and it appears to this author that future estimates will continue to show them as a rather small fraction of the impact flux, except insofar as the near-Earth-object (NEO) population includes a large subpopulation of dormant or extinct comets.

7. BREAKTHROUGH GOALS

In the next one or two decades, we can anticipate numerous breakthroughs in cometary science. It is instructive, however, to divide these breakthroughs into two different types: the predictable, paradigm-settling measurements and the serendipitous, paradigm-changing measurements. In the former category we can put the definitive questions that we ask when proposing a major investigation, while in the latter we can only speculate.

Among the predictable breakthroughs, we can expect *Stardust* to enable us to relate cometary dust to the dust captured in the stratosphere and to the zodiacal dust. We can realistically expect that it will even tell us a lot about the formation of comets and the preservation of interstellar refractory grains. We can expect SST to pin down the albedos of many cometary nuclei and also of many TNOs and thus to dramatically improve our knowledge of the size distribution of both types of objects. There may be surprises in the values of the albedos, just as many people were surprised when Comets P/Arend-Rigaux, P/Neujmin 1, and P/Halley all turned out to be very dark, but this kind of surprise can be anticipated and used to argue for observing time.

The conceptually utterly simple experiment of *Deep Impact* will dramatically narrow the uncertainty in our understanding of the structure of cometary nuclei. This is a case in which we can scope the range of plausible outcomes and eliminate all but one or two immediately after the experiment, thus constraining the properties of the nucleus. The *Rosetta* mission, unless beaten by a shorter-lifetime mission selected in the near future, will provide us with the first direct measurement of the mass (and low-order moments) of a cometary nucleus and thus of its density. *Rosetta* will also provide breakthrough tomographic measurements of a nucleus to understand for the first time the large-scale heterogeneity. It will also provide unprecedented information on "how a comet works." We can anticipate that there will likely be one or more small cometary missions selected for flight in the next half decade so that there would be major results well within the next decade and a half.

Proposals have already been solicited for a mission to return a cold sample from the surface layers of a cometary nucleus. Such a mission would provide even more tremendous advances than those from *Stardust*. Being able to measure the details of icy grains, the intimacy of mixing between ices and refractories, and the relative abundances of different volatiles will be a tremendous advance. Such a mission will be a great complement to *Rosetta*. The next logical step after return of a surface sample is to return a cold sample from deep (tens of meters) inside a nucleus that preserves the chemical and crystalline form of the ices sampled, thus providing details on what will be hinted at by *Deep Impact*. Such a mission would provide crucial information on the scale at which different ices are mixed in whatever layers can be probed, from the evolved layers near the surface where we can understand the transport of volatiles as the comet evolves to the more nearly primordial layers below the thermal wave where we might understand the condensation process. All these space missions will be invaluable in helping us interpret the remote sensing data from far more comets than we can ever visit with spacecraft.

We can also expect major advances from a large-aperture, dedicated, survey telescope, whether it be an expanded version of the Pan Stars array of telescopes being built by the University of Hawai'i or the Large Synoptic Survey Telescope (LSST) being designed by the U.S. National Optical Astronomy Observatory and several partners. Depending on how the survey is implemented, particularly depending on how strongly the search strategy emphasizes NEOs, we could expect major advances in our understanding of the size distribution and the orbital distribution of comets and in the size distribution of transneptunian objects (TNOs). If some survey goes faint enough, we might even be able to provide a good estimate of what fraction of Jupiter-family comets are primordial bodies from the Kuiper belt as opposed to being fragments of larger bodies in the Kuiper belt.

We will certainly have many discoveries of newly identified chemical species from ALMA, because of both its superb sensitivity and its superb spatial resolution. Similar discoveries should come from the tremendously increased sensitivity and reduced beam dilution of the Large Millimeter Telescope being built in Mexico by the University of Massachusetts and the Mexican Instituto Nacional de Astrofísica, Optica, y Electrónica.

Even dramatic increases in computational power and/or algorithm design could lead to breakthroughs. One such advance might be the ability to integrate a very large number of orbits in order to assess with proper statistics, based on the uncertainty in the observed orbit, the probability that any given Jupiter-family comet had, at an earlier time in its orbital evolution, been for some time in an orbit with smaller perihelion distance. Such a calculation has been done, for example, for Chiron but with only a small set of possible, current orbital elements, and the recent discovery of "keyholes" in the uncertainty space of the orbits of NEOs suggests that our understanding of cometary orbital evolution will require far more computation than has been done for any comet thus far. In another computational arena, one can certainly expect major advances in our ability to combine many physical processes into a single simulation in order to better understand the origin of the comets and the outer planets. Calculations that incorporate, simultaneously and with all feedback loops, the complete gravitational field, the network of gas phase, and surficial chemistry including kinetic inhibition, radiative processes, and magnetohydrodynamics are tremendously difficult. Carrying out such calculations would be a major step forward in understanding the formation of comets and the solar system.

While some cometary scientists might think it heresy, it seems likely that we might learn more about comets from certain missions to other bodies than from many types of missions to comets. In particular, detailed, *in situ* studies of or return of samples from a jovian Trojan might tell us directly about the primordial comets that formed near Jupiter's orbit. A similar mission to a classical Kuiper belt object might tell us about the primordial state (except for collisions) of cometary material now in Jupiter-family comets. This is not to say that they would be more revealing than any mission to a comet, since missions that explore new parameter space at a comet will be extremely valuable.

8. SERENDIPITY

Turning to the truly serendipitous, paradigm-altering discoveries, *Harwit* (1984) has written eloquently about the nature of dramatic new discoveries in astronomy and shown very well that nearly all the dramatic, new discoveries have come from making measurements in new domains. These are entirely different from the breakthroughs described immediately above in that there is no way to predict the area in which these surprises might occur. In astronomy these new domains have been either previously unobserved spectral ranges or order-of-magnitude improvements in sensitivity or in spectral or spatial resolution. Astronomy as a whole has the entire universe to study so there is more scope for serendipitous discovery than in the relatively narrow domain of cometary science, but the principle is still the same and still important.

Cometary science has gained from serendipitous discoveries in the past. The use of new wavelength domains led, at the shortest wavelengths, to the discovery of high fluxes of X-rays from comets and thus to new emission mechanisms and, at the longest wavelengths, to the discovery of the ultraviolet-pumped maser of OH at 21 cm in comets. The application of radar led to the discovery of clouds of large particles (more than a centimeter) in surprising dynamical situations in more than one comet. None of these would be considered paradigm-altering, i.e., none of these changed our picture of the role of comets in the solar system, but they were surprising, serendipitous results. There are more new domains of measurement to be applied to comets than there are for most of astronomy and this compensates in part for the fact that we are considering a much smaller piece of the universe. In addition to the new domains of measurement available to traditional astronomy, the new domains of measurement for cometary science certainly include all the new types of experiments and measurements, both microscopic and macroscopic, that can be carried out *in situ*, as well as new domains of computational space.

Thus we can expect serendipitous discoveries, whether new populations of comets, new physical processes, or a new picture of how comets work, as long as we continue to apply dramatically new techniques to studying comets. There is no way to predict the areas in which area the most exciting such discoveries will occur.

REFERENCES

Harwit M. (1984) *Cosmic Discovery: The Search, Scope, and Heritage of Astronomy.* Massachusetts Institute of Technology, Cambridge. 334 pp.

Part II:

From the Interstellar Medium to the Solar Nebula

The Cycle of Matter in Our Galaxy:
From Clouds to Comets

William M. Irvine
University of Massachusetts

Jonathan I. Lunine
University of Arizona

The processing of grains from original interstellar material to that contained in comets occurred under a complex range of conditions extending from the diverse environments within molecular clouds to the protoplanetary disk itself. Grain surface chemistry at very low temperatures gives way to grain growth, heating and partial sublimation, recondensation, and then agglomeration into cometary-sized bodies and larger. While the overall direction of the evolution can be sketched, little in the way of direct observations is available once collapse occurs, so that it remains difficult to describe specifically the temperature-pressure-composition histories of grains that will eventually find their way into comets.

1. INTRODUCTION

All cometary matter was once, of course, interstellar. But unresolved is whether comets still contain demonstrable and significant signatures of the interstellar material from which they formed, either in terms of preserved or partially altered molecular constituents or in the isotopic ratios found in cometary molecules. This fundamental issue has been discussed, e.g., by *Ehrenfreund et al.* (2002, 2004), *Charnley et al.* (2002), *Irvine and Bergin* (2000), *Irvine et al.* (2000), *Bockelée-Morvan et al.* (2000), *Crovisier and Encrenaz* (2000), and *Fegley* (1999). One can imagine two extreme scenarios in this regard: first, that comets are conglomerates of essentially unprocessed interstellar grains, as proposed by *Greenberg* (1982, 1998); second, that the formation of the solar nebula was a sufficiently energetic process, even in its outer portions, that presolar molecules were completely destroyed and that the infalling material was homogenized (*Lewis,* 1972). In the latter case the chemical composition of dense interstellar clouds is not directly relevant to the chemistry of comets; in the former situation, exactly the opposite is true. As is usually the case in science, the truth probably lies somewhere between these two extreme views.

To understand the extent to which cometary material has been reprocessed from its initial interstellar state requires that we understand the coupled physical and chemical processes at work in the various environments leading from the interstellar medium to the modern Kuiper belt and Oort cloud — the main dynamical reservoirs of comets today. The chapters in this section of the book consider each of these environments in turn, from molecular clouds (*Wooden et al.,* 2004), to the formation and evolution of planet-forming circumstellar disks (*Boss,* 2004), the observational properties of disks and their interactions with the central star (*Dutrey et al.,* 2004), models of the formation and evolution of solids from the grains to cometesimals (*Weidenschilling,* 2004), and the coupled physics and chemistry within the disk (*Lunine and Gautier,* 2004). This progression represents a temporal sequence, but it also represents a dramatic reduction in the spatial scales of the evolution — and consequently a steep increase in the difficulty of the constraining observations. We know a lot about the broad nature of chemistry and physical processes in the vast sweep of the interstellar clouds; we know much less in a direct sense about the details of the chemical and physical processes affecting grains on spatial scales of astronomical units in planet-forming disks. But the chemical and isotopic evidence from small bodies in our solar system, including comets, provides a detailed local perspective provided we can correctly interpret it. In briefly recapitulating this cycle of matter from molecular clouds to comets, we will emphasize some of the major outstanding gaps in our current knowledge.

2. THE INTERSTELLAR MEDIUM, INCLUDING MOLECULAR CLOUDS

Cometary dust begins with the nucleosynthetic production of heavy elements in stars, both during their main sequence and the sometimes-explosive terminal phases of evolution. Novae, supernovae, and the winds of asymptotic giant branch (AGB) stars (the state arrived at by stars less than 10 $M_\odot$ when all the H and He is exhausted in the stellar core) enrich the interstellar medium in new elements. Because the interstellar medium is mixed, and matter is cycled through multiple generations of stars, the elements and the stable isotopes are a mixture of the products of stars of a variety of masses and stellar generations (the latter determined by the heavy element abundance at the time of the star's formation).

Grains may form in a variety of environments, but the circumstellar envelopes of AGB stars are particularly important for forming both silicate and carbonaceous grains as the C/O ratio varies over time in the star's envelope. Once formed, grains do not simply persist until their incorporation in comets; rather, they are cycled many times from gaseous to condensed phases. Grains formed in the diffuse interstellar medium may be modified or evaporated by ultraviolet photolysis powered by the intense radiation from newly formed stars. Shocks generated by supernova explosions — the end product of massive star formation — may destroy grains at large distances from the original site of the explosion, with grains reforming both in the wakes of the shocks and in the much denser environments of molecular clouds.

A detailed characterization of the chemical nature and physical structure of grains has not been achieved. The nature of the circumstellar and interstellar silicates, including their composition and crystal structure, is complex and is relevant to the origin of cometary silicates (see *Wooden et al.,* 2004). For the more volatile elements, it seems unavoidable that the grains are the primary reservoir for C and a major reservoir for O, and very likely that there is a continuous size range from large carbonaceous molecules (PAHs; see below) to particulate grains (cf. *Whittet,* 2003; *van Dishoeck and Blake,* 1998). Given the difficulty of characterizing terrestrial kerogen, the lack of consensus on the nature of the organic component of interstellar grains is perhaps not surprising.

Mixing processes in the galaxy must be fairly efficient, for the vast majority of material in interplanetary dust particles (IDPs) — perhaps the best samples of the non-ice phases of comets available — is isotopically of so-called "solar" composition. Although isotopic anomalies are dramatic and informative of the details of solar system formation (*Cameron,* 2002), the material exhibiting these typically represent only 1% of the material present (see *Wooden et al.,* 2004).

Chemistry in the interstellar medium is complex because of the enormous range of environments contained therein. In the diffuse interstellar medium (H density 1–100 cm^{-3}, temperature ~100 K), gas phase chemistry is dominated by the abundant ultraviolet radiation pumped out by massive O and B stars and largely unimpeded over stellar distances. The size of the molecules formed is generally limited because of the instability of long chains and other complex structures against UV dissociation. There is a relatively high abundance of atomic (ionized or neutral) C, N, S, H, etc. However, the molecular phase in the diffuse clouds is dominated by polycyclic aromatic hydrocarbons (PAHs), which are relatively stable against UV dissociation and may be a very important, if not dominant, agent in determining the thermal balance of the interstellar medium. Astoundingly, they may contain up to 10% of the C in the galaxy (although some estimates are an order of magnitude lower), and as a group represent the third most abundant molecule after molecular H and CO (*Wooden et al.,* 2004). They are the most plausible candidate species for producing the diffuse interstellar bands, whose origin has long been a mystery (e.g., *Herbig,* 1995). However, no single PAH has definitively been identified in the interstellar medium, and the precise mix of species and the (presumably environment-dependent) ratio of neutral to ionized molecules remain unknown. Polycyclic aromatic hydrocarbons crop up in a multitude of environments, from barbeque grills to martian meteorites to the interstellar medium; they should be present as well in cometary matter. Because of their ubiquity and ease of formation, they are rather ambiguous signposts to the details of C chemistry, and in the diffuse interstellar medium are only the most abundant of a menagerie of C forms such as chains, fullerenes, and (rarely) diamonds.

The cold, dense (n > 200 cm^{-3}) clouds of the interstellar medium are sites of star formation, most likely of low mass stars, and are characterized by extremely low temperatures around 10–20 K. External ultraviolet radiation is excluded, but penetrating cosmic rays provide a population of ions and electrons that provide what little heating there is. Under such conditions, very tiny (0.1-μm radius) grains and hence extensive surface chemistry dominate, and it is here that some of the very-low-temperature features of cometary matter, such as the para-ortho ratio in water, may be locked in (*Mumma,* 1997). The dominant C form in the gas is CO, and observations of nearby clouds such as TMC-1 and the perhaps more typical L134N indicate a rich suite of hydrocarbons, nitriles, and acids — foreshadowing a complexity of C-based cometary chemistry that has yet to be fully explored (*Wooden et al.,* 2004). Unlike the major reservoir for C in the grains, the bulk of the N in these clouds is probably in the gas phase as N_2, although direct observational confirmation is limited.

Water is present in the dense clouds, but is frozen out on grains where it is the dominant ice component. This water ice may be the principal O reservoir in these regions, although quantitative assessment is difficult (cf. *Bergin et al.,* 2000). Water ice forms predominately on more refractory grains through H-atom additions to O atoms, rather than direct condensation in what is still an extremely tenuous medium. Mechanisms for returning molecules to the gas from cold (10 K) grains are highly uncertain, although a variety of means have been proposed. Thermal and cosmic-ray-powered sublimation and chemical cycling of water likely occur in portions of the cloud. Thermal processing of the water aided by exothermic chemical reactions could sublimate the ice, which then would reattach to grains by adsorption. In any case the ice would be expected to be amorphous. What other types of chemistry might occur, and the isotopic fractionation achievable (for example, in H, C, N), depend on the details of very-low-temperature chemical processes on grains that are extremely challenging to replicate under normal laboratory conditions. *Wooden et al.* (2004) claim that clathrate formation has been inferred to occur in dense clouds, although the crystalline form — what is normally called a clathrate — is likely kinetically inhibited at 20 K in favor of direct adsorption. The abundance of ammonia (NH_3) in these ice mantles is also somewhat uncertain.

Low-mass and high-mass star formation lead to warmer clouds and "hot cores" respectively. The hot cores provide a wealth of information on condensed species in colder environments because such species are sublimated at the high hot-core temperatures, while still protected against the free-radical chemistry abundant in high-ultraviolet environments. In the Orion hot core, for example, H_2O, H_3N, H_2S, CH_3OH, and HCN are all more abundant than in cold clouds. Together with CO and HCHO, this mix of species is "familiar" to us from the cometary point of view — the hot-core sublimated mix contains the species seen in cometary comae, if not quite in the same relative abundances. Regardless of whatever other processing occurs in the protoplanetary disk, the existence of some familiar species emphasizes the link between cometary and interstellar (molecular cloud) grains (cf. *Irvine and Bergin*, 2000; *Tielens*, 2001). This link is even more evident in the large isotopic fractionation measured for H in cometary H_2O and HCN, which reflects that seen in molecular species in interstellar clouds, although not always to the same extent (see *Ehrenfreund et al.*, 2004; *Irvine et al.*, 2000). While the environments of low-mass star formation typically do not exhibit the same relative abundances, this is surely because temperatures are not high enough to sublimate the material except in the observationally inaccessible, optically thick, inner portions of the collapsing clumps. Some recent high-spatial-resolution observations do reveal hot-core-type molecules in the hot gas close to the low-mass protostar IRAS 16293-2422 (*Cazaux et al.*, 2003).

3. FROM COLLAPSING CORES TO DISKS

The loss of magnetic support as cloud clumps become denser and the neutral molecules "slip through" the much-smaller ion population is thought to initiate the collapse toward formation of a low-mass protostar and, depending on the clump's angular momentum, a protoplanetary disk [outflows from massive stars may also trigger clump collapse; see *Boss* (2004) for a comparison of the two processes]. Once initiated, the collapse leads to elevated temperatures in the center eventually sufficient to trigger thermonuclear fusion, and the protostar feeds on the gaseous and solid material falling inward over a period on the order of 1 million years. If a disk forms (*Boss*, 2004), then the interaction of the disk with the surrounding collapsing cloud is complex. Infalling grains, subjected to increasing density, grow in size well past that of the tiny interstellar material. However, heating of the grains by direct exposure to accretion shocks and drag (*Lunine et al.*, 1991; *Chick and Cassen*, 1997) occurs, and can modify the grain composition or destroy it entirely by sublimation.

The survival of interstellar grains during the infall is assured only for those that fall into the outermost parts of the disk, where the shock is weak and infall velocities low. Grain material that is sublimated may readsorb or condense on the grain residues in the relatively cool and high-density gas of the disk midplane, but the resulting distribution of various molecular species in the recondensation may differ from that of the original grain. Inward, progressively more refractory species are sublimated during infall; in the inner regions of the disk it is likely that all but the most refractory material is returned to the gas phase.

Given that interstellar (as opposed to circumstellar) silicates appear to be amorphous, the nature of cometary silicates, which include both crystalline and amorphous material, presents a challenge. The prevailing view is that the amorphous component probably represents surviving interstellar material, presumably similar to the GEMS (glass imbedded with metal and sulfides) found in IDPs, while the crystalline silicates are probably produced or at least annealed in the solar nebula. The issue is discussed in detail in *Ehrenfreund et al.* (2004).

As the protostar grows, ignites fusion reactions, and develops a strong bipolar outflow roughly perpendicular to the disk, other chemistry begins to affect the grains. Ultraviolet irradiation again becomes an issue for gas and grains that end up at large distances from the disk center, where they may be wafted upward periodically to the surface of the flared disk and hence be exposed to the protostellar emission by virtue of their altitude above the midplane. The bipolar flow encounters the surrounding molecular cloud in the form of a wind-cloud shock (*Wooden et al.*, 2004), where additional heating and energetic chemistry may occur.

There are, however, major uncertainties concerning the physical nature of the disk. As *Boss* (2004) points out, the detailed thermal, density, and turbulent structure cannot yet be predicted, and there is a serious lack of observational data (*Dutrey et al.*, 2004), reflecting the need for higher-angular-resolution studies that will only become possible with future facilities such as the Atacama Large Millimeter Array (ALMA) telescope and the James Webb Space Telescope (JWST). There is, for example, no agreement on the mechanisms for transfer of mass and angular momentum in the disk, a problem related to uncertainties in the character of the "viscosity" that is assumed to exist and is modeled through the parameter α (hence, the so-called α-disk models). Moreover, the process of giant planet formation is still controversial. All this is relevant to comets, since it describes the environment in which volatile species condense/sublimate, material is transferred radially, etc.

Furthermore, an even greater uncertainty exists as to whether low-mass stars like the Sun might in fact form in clusters where massive stars form, such as the Orion molecular cloud, which would be suffused with ultraviolet radiation from surrounding young protostars, shocks, and infusion of material from supernovae (*Boss*, 2004), leading to a much more active chemistry than in the more traditional picture of relatively isolated star formation. Even if they formed in a less-energetic environment that might characterize more isolated low-mass star formation, *Dutrey et al.* (2004) point out that many T Tauri stars are members of binary or multiple systems, complicating the traditional approach to the physics of the disks. Although the physics of some of these processes has been modeled, little attention has been paid to the connection to the resulting grain chemistry and isotopic reequilibration at various distances from the protostar. Furthermore, direct observation of the material in the collapsing core is exceedingly difficult as spatial scales shrink from parsecs to tens or hundreds of astronomical units and densities rise by many orders of magnitude relative to the average (10^3–10^5 cm^{-3}) in the molecular cloud itself.

4. EVOLUTION OF GRAINS WITHIN THE PROTOPLANETARY DISK

Assuming that the protoplanetary disk is not destroyed on very short timescales by the external molecular cloud environment, gas phase processes drive chemistry in the disk over timescales on the order of 1 million years or more, based on astronomical observation of gas in disks (*Dutrey et al.*, 2004). Gravitational and magnetic torques presumably launch waves in the gas and dust, compressing and heating material and perhaps in some disks forming giant gaseous planets that may or may not be analogs of Jupiter and Saturn. These torques could be strong enough to melt solids in certain parts of the disk, one explanation offered for the chondrules seen in meteorites; strong gas dynamical heating of material falling into the inner portion of the disk could also lead to such melting. Turbulent viscosity within the disk heats the midplane even in the absence of stellar radiation blocked by the optically thick conditions. Excursions in the accretion rate may lead to cooling and then sudden heating events akin to the observed "FU Orionis" phenomenon. The star itself, in blowing a bipolar flow vertically, also clears out the innermost part of the disk, subjecting adjacent material to extreme heating.

The strong radial temperature gradient imposed on the protoplanetary disk by turbulent dissipation ensures a radial gradation in the condensation and stabilization in grains of species of differing volatility. Hence the most refractory silicates (e.g., corundum) condense closest to the protostar, followed by the less-refractory Mg silicates and Fe, then a complex series of S compounds, followed by water ice at several astronomical units or beyond. The condensation front of water ice — sometimes called the snowline — is crucial because it represents a steep outward increase in the surface density of solids in the disk, with potential implication for the formation timescale of giant planets if these were seeded by initial formation of solid cores. But from the point of view of cometary grains and their composition, the condensation of water ice adds complexity to the history. We are forced to consider two sources of water ice and

trapped volatiles for comets: water ice that survived infall into the disk and water ice formed in the disk by condensation. The mechanism of volatile trapping and the total volatile content are likely to be very different for the two sources, but both may be relevant since the source region for some of the Oort cloud comets could be the Jupiter-Saturn feeding zone, which was not far from the snowline.

The *Galileo* probe measurements of supersolar abundances in the jovian atmosphere do not rule out a primarily solar nebula condensed ("native") vs. interstellar (*Owen et al.*, 1999) origin for the water ice that carried these volatiles into growing Jupiter, because we do not yet know the O (hence H_2O) abundance of the bulk planet (*Gautier et al.*, 2001). It is highly likely that these two kinds of water ice both existed in the protoplanetary disk, because it is hard to avoid sublimation and recondensation of some of the water, but equally difficult to argue that all the infalling grains were heated sufficiently to completely sublimate the water ice phases. Cometary tests of just how much native vs. interstellar water ice is present must rely on the isotopic composition of the water ice and trapped volatiles, and perhaps the ortho-para ratio of the H in the water. Unfortunately, definitive predictions of these properties tied to the disk evolution are difficult to make.

Within protoplanetary disks, the chemistry and properties of grains will be altered in the immediate environment of formation of giant planets. Depending upon the details of such planet formation, the balance of oxidized vs. reduced C and N species will be altered in the gas very close to the giant planets (*Mousis et al.*, 2002; *Irvine et al.*, 2000). More reduced species may find their way into solids forming near the planets, these solids then are transported outward through the main disk, or grains may be destroyed entirely by catastrophically rapid giant planet formation (*Mayer et al.*, 2002). In our own solar system, the latter seems unlikely, because abundant solid material remained to populate the Kuiper belt and the Oort cloud. Nonetheless, the gaps that giant planets open during their formation reduce the gas density and perturb the orbits of larger solid bodies, while inward and outward of the gap higher densities might have perturbed the volatile content of the icy planetesimals.

The growth of comet-sized planetesimals, their dynamical histories as they were ejected from the region of giant planet formation and portions of the Kuiper belt, and the resulting compositional distinctions between comets residing in different dynamical reservoirs were rather complex (*Weidenschilling*, 2004). This evolution occurred as the disk transitioned from gas-rich to gas-poor, a process that was complex and may differ in terms of timescale from one protoplanetary disk to another. The growth from centimeter- to kilometer-sized bodies is particularly poorly understood, as it is unclear theoretically in which circumstances collisions among the objects will lead to a net gain in the mass of the larger partner (*Weidenschilling*, 2004). The vertical as well as the radial structure of the disk may be important in particle growth, but no three-dimensional simulations have yet been carried out.

During the gas-rich phase, interactions of comet-sized small bodies with each other and with the giant planets was strongly affected by the presence of the gas, both from a dynamical point of view and with respect to the incorporation of additional volatiles in the water ice of the small bodies (*Lunine and Gautier*, 2004). As the gas dissipated, additional volatile incorporation ceased and the dynamical interactions of the cometary bodies and the planets simplified.

5. COMETS IN THE REMNANT DISK

The disk remaining after much of the gas is removed — a process that might take 10^7 yr or more, although the massive gas phase is likely to be a factor of 10 shorter — consists of solids ranging from dust to giant planets. Planet formation is not complete, however, because the formation of the terrestrial planets in our system evidently took several tens of millions of years, based on isotopic evidence (*Yin et al.*, 2002; *Kleine et al.*, 2002) and dynamical simulations (*Chambers and Wetherill*, 1998). The process during this time was the pumping up of the eccentricities of an initially quiescent disk, inward of Jupiter, dominated in mass by bodies from Mercury- to Mars-sized. As the orbits of these perturbed "planets" crossed, collisions and net growth occurred. Meanwhile the giant planets were perturbing the orbits of more distant icy bodies — the massive host of what we would today call comets — leading both to outward ejection and movement inward on eccentric orbits (*Morbidelli et al.*, 2000).

Many of these comets struck the growing terrestrial planets over millions of years, supplying organic compounds and water, but comets were not the dominant supplier of water to Earth if the isotopic evidence is taken at face value. The D-to-H ratio of measured long-period comets is twice that of the Earth's oceans (*Meier et al.*, 1998). Furthermore, the dynamical simulations themselves suggest that large bodies in an asteroid belt much more heavily populated than today were the principal contributors to the ocean of the Earth (*Morbidelli et al.*, 2000), although the same conclusion need not hold for the waters of Mars. Nonetheless, the accretion of cometary material by the Earth was a direct result of the process by which the giant planets gravitationally cleared much of the outer solar system of small icy bodies, relegating these to the Oort cloud.

At the same time, modest migration of the giant planets — particularly Neptune — sculpted the orbits of comets remaining beyond 30 AU into what are now the classical and resonant disks of the Kuiper belt, and also ejected comets inward into the so-called scattered Kuiper belt disk (*Luu and Jewitt*, 2002). These gravitational interactions with the giant planets — and for the Oort cloud comets, with the surrounding environment of young stars formed in the same epoch as the Sun — represent another set of perturbations on the thermal and radiation environment of the comets. Beyond

this time, comets existed much as they exist today — in different dynamical reservoirs, some perturbed inward toward the Sun to sublimate, possibly break up, and generate wonder in the skies of Earth for humans who would arise 4.55 billion years after the solar system's formation.

6. EXPULSION OF COMETARY MATTER AT THE END OF THE SUN'S MAIN-SEQUENCE EVOLUTION

The Sun is approximately halfway through its main-sequence life, and at the end of this time will expand to become a red giant, with an outer envelope at the orbit of the Earth. The initiation and then termination of He fusion in the red giant's core will trigger a second collapse and expansion, this time to a red supergiant with an outer envelope at the orbit of Mars. A subsequent series of "flashes" and thermal pulses will eject half the mass of the Sun to space, leaving behind a white dwarf and stripping the atmosphere of Saturn's moon Titan (*Lorenz et al.*, 1997). The luminosity excursions during this time will sublimate the ices from Kuiper belt comets, and the loss of half the Sun will free much of the Oort cloud from the gravitational bonds of the solar system — an echo of the postformation ejection of cometary bodies from the region of the giant planets. Cometary material, in the gaseous and solid phases, will travel freely through the interstellar medium, and the cycling of matter that 10 billion years before became the Sun and solar system will have been closed.

7. FUTURE OBSERVATIONS

It is neither unkind nor disrespectful to refer to the above as a kind of fairy tale grounded in a spotty tapestry of observations glued together by dynamical and chemical modeling. Seminal spacecraft observations, such as the ESA *Giotto* compositional studies, have been supplemented by increasingly powerful groundbased studies. Nonetheless, it is exceedingly difficult to penetrate collapsing cores or to tease apart the details of planet-forming disks on scales of astronomical units. Remnant disks are easier, but spatial resolution is still a problem. Future cometary measurements from, most spectacularly, *Rosetta*, are discussed elsewhere in this book, but upcoming ground- and space-based facilities for examining astronomical structures involved in planet formation are equally exciting (*Dutrey et al.*, 2004). ALMA will see details in planet-forming disks on scales of a few astronomical units, and will be able to sensitively probe the gas composition and grain properties. The Space Infrared Telescope Facility (SIRTF) will track the colder dust in remnant and planet-forming disks with sufficient sensitivity to perhaps indicate the effect of massive planets on the dust populations, while the JWST will apply more powerful angular resolution and midinfrared sensitivity to probe the compositional gradations in the grains of remnant disks. These and other facilities will severely test our ideas about the process of planet formation, and how interstellar grains become — through many different evolutionary pathways — the stuff of cometary nuclei.

Acknowledgments.   J.L. and W.I. acknowledge the support of the NASA Origins of Solar Systems, Astrobiology, and Planetary Astronomy Programs in the preparation of this chapter. We are grateful for the helpful comments from two anonymous reviewers.

REFERENCES

Bergin E. A., Melnick G. J., Stauffer J. R., Ashby M. L. N., Chin G., Erickson N. R., Goldsmith P. F., Harwit M., Howe J. E., Kleiner S. C., Koch D. G., Neufeld D. A., Patten B. M., Plume R., Scheider R., Snell R. L., Tolls V., Wang Z., Winnewisser G., and Zhang Y. F. (2000) Implications of Submillimeter Wave Astronomy Satellite observations for interstellar chemistry and star formation. *Astrophys. J. Lett., 539,* L129–L132.

Bockelée-Morvan D., Lis D. C., Wink J. E., Despois D., Crovisier J., Bachiller R., Benford D. J., Biver N., Colom P., Davies J. K., Gérard E., Germain B., Houde M., Mehringer D., Moreno R., Paubert G., Phillips T. G., and Rauer H. (2000) New molecules found in Comet C/1995 O1 (Hale-Bopp): Investigating the link between cometary and interstellar material. *Astron. Astrophys., 353,* 1101–1114.

Boss A. P. (2004) From molecular clouds to circumstellar disks. In *Comets II* (M. C. Festou et al., eds.), this volume. Univ. of Arizona, Tucson.

Cameron A. G. W. (2002) Birth of a solar system. *Nature, 418,* 924–925.

Cazaux S., Tielens A. G. G. M., Ceccarelli C., Castets A., Wakelam V., Caux E., Parise B., and Teyssier D. (2003) The hot core around the low-mass protostar IRAS 16293-2422: Scoundrels rule! *Astrophys. J. Lett., 593,* L51–L55.

Chambers J. E. and Wetherill G. W. (1998) Making the terrestrial planets: N-body integrations of planetary embryos in three dimensions. *Icarus, 136,* 304–327.

Charnley S. B., Rodgers S. D., Kuan Y.-J., and Huang H.-C. (2002) Biomolecules in the interstellar medium and in comets. *Adv. Space Res., 30,* 1419–1431.

Chick K. M. and Cassen P. (1997) Thermal processing of interstellar dust grains in the primitive solar environment. *Astrophys. J., 477,* 398–409.

Crovisier J. and Encrenaz T. (2000) *Comet Science.* Cambridge Univ., Cambridge. 173 pp.

Dutrey A., Lecavelier des Etangs A., and Augereau J.-C. (2004) The observation of circumstellar disks: Dust and gas components. In *Comets II* (M. C. Festou et al., eds.), this volume. Univ. of Arizona, Tucson.

Ehrenfreund P., Irvine W., Becker L., Blank J., Colangeli L., Derenne S., Despois D., Dutrey A., Fraaije H., Lazcano A., Owen T., and Robert F. (2002) Astrophysical and astrochemical insights into the origin of life. *Rept. Prog. Phys., 65,* 1427–1487.

Ehrenfreund P., Charnley S. B., and Wooden D. H. (2004) From interstellar material to cometary particles and molecules. In *Comets II* (M. C. Festou et al., eds.), this volume. Univ. of Arizona, Tucson.

Fegley B. Jr. (1999) Chemical and physical processing of presolar materials in the solar nebula and the implications for preservation of presolar materials in comets. *Space Science Rev., 90,* 239–252.

Gautier D., Hersant F., Mousis O., and Lunine J. I. (2001) Enrichments in volatiles in Jupiter: A new interpretation of the Galileo measurements. *Astrophys. J. Lett., 550,* L227–L230 (erratum *559,* L183).

Greenberg J. M. (1982) What are comets made of? A model based on interstellar dust. In *Comets* (L. L. Wilkening, ed.), pp. 131–163. Univ. of Arizona, Tucson.

Greenberg J. M. (1998) Making a comet nucleus. *Astron. Astrophys., 330,* 375–380.

Herbig G. H. (1995) The diffuse interstellar bands. *Annu. Rev. Astron. Astrophys., 33,* 19–73.

Irvine W. M. and Bergin E. A. (2000) Molecules in comets: An ISM-solar system connection? In *Astrochemistry: From Molecular Clouds to Planetary Systems* (Y. C. Minh and E. van Dishoeck, eds.), pp. 447–460. IAU Symposium 197, Astronomical Society of the Pacific, San Francisco.

Irvine W. M., Schloerb F. P., Crovisier J., Fegley B. Jr., and Mumma M. J. (2000) Comets: A link between interstellar and nebular chemistry. In *Protostars and Planets IV* (V. Mannings et al., eds.), pp. 1159–1200. Univ. of Arizona, Tucson.

Kleine T., Münker C., Mezger K., and Palme H. (2002) Rapid accretion and early core formation on asteroids and the terrestrial planets from Hf–W chronometry. *Nature, 418,* 952–955.

Lewis J. S. (1972) Low temperature condensation from the solar nebula. *Icarus, 16,* 241–252.

Lorenz R. D., Lunine J. I., and McKay C. P. (1997) Titan under a red giant sun: A new kind of "habitable" moon. *Geophys. Res. Lett., 24,* 2905–2908.

Lunine J. I. and Gautier D. (2004) Coupled physical and chemical evolution of volatiles in the protoplanetary disk: A tale of three elements. In *Comets II* (M. C. Festou et al., eds.), this volume. Univ. of Arizona, Tucson.

Lunine J. I., Engel S., Rizk B., and Horanyi M. (1991) Sublimation and reformation of icy grains in the primitive solar nebula. *Icarus, 94,* 333–343.

Luu J. X. and Jewitt D. C. (2002) Kuiper belt objects: Relics from the accretion disk of the Sun. *Annu. Rev. Astron. Astrophys., 40,* 63–101.

Mayer L., Quinn T., Wadsley J., and Stadel J. (2002) Formation of giant planets by fragmentation of protoplanetary disks. *Science, 298,* 1756–1759.

Meier R., Owen T. C., Matthews H. E., Jewitt D. C., Bockelée-Morvan D., Biver N., Crovisier J., and Gautier D. (1998) A determination of the HDO/H$_2$O ratio in comet C/1995 O1 (Hale-Bopp). *Science, 279,* 842–844.

Morbidelli A., Chambers J., Lunine J. I., Petit J. M., Robert F., Valsecchi G. B., and Cyr K. E. (2000) Source regions and timescales for the delivery of water on Earth. *Meteoritics & Planet. Sci., 35,* 1309–1320.

Mousis O., Gautier D., and Bockelée-Morvan D. (2002) An evolutionary turbulent model of Saturn's subnebula: Implications for the origin of the atmosphere of Titan. *Icarus, 156,* 162–175.

Mumma M. J. (1997) Organic volatiles in comets: Their relation to interstellar ice and solar nebula material. In *From Stardust to Planetesimals* (Y. J. Pendleton and A. G. G. M. Tielens, eds.), pp. 369–396. ASP Conference Series 122, Astronomical Society of the Pacific, San Francisco.

Owen T., Mahaffy P., Niemann H. B., Atreya S. K., Donahue T. M., Bar-Nun A., and de Pater I. (1999) A new constraint on the formation of giant planets. *Nature, 402,* 269–270.

Tielens A. G. G. M. (2001) The composition of circumstellar and interstellar dust. In *Tetons 4: Galactic Structure, Stars, and the Interstellar Medium* (C. E. Woodward et al., eds.), pp. 92–122.

ASP Conference Series 231, Astronomical Society of the Pacific, San Francisco.

van Dishoeck E. F. and Blake G. A. (1998) Chemical evolution of star-forming regions. *Annu. Rev. Astron. Astrophys., 36,* 317–368.

Weidenschilling S. (2004) From icy grains to comets. In *Comets II* (M. C. Festou et al., eds.), this volume. Univ. of Arizona, Tucson.

Whittet D. H. B. (2003) *Dust in the Galactic Environment, 2nd edition.* Institute of Physics, Bristol. 390 pp.

Wooden D. H., Charnley S. B., and Ehrenfreund P. (2004) Composition and evolution of interstellar clouds. In *Comets II* (M. C. Festou et al., eds.), this volume. Univ. of Arizona, Tucson.

Yin Q., Jacobsen S. B., Yamashita K., Blichert-Toft J., Télouk P., and Albaréde F. (2002) A short timescale for terrestrial planet formation from Hf–W chronometry of meteorites. *Nature, 418,* 949–952.

Composition and Evolution of Interstellar Clouds

D. H. Wooden
NASA Ames Research Center

S. B. Charnley
NASA Ames Research Center

P. Ehrenfreund
Leiden Observatory

In this chapter we describe how elements have been and are still being formed in the galaxy and how they are transformed into the reservoir of materials present in protostellar environments. We discuss the global cycle of matter from stars, where nucleosynthesis produces heavy elements that are ejected through explosions and winds into the interstellar medium (ISM), through the formation and evolution of interstellar cloud material. In diffuse clouds, low-energy cosmic rays impact silicate grains, amorphizing crystals, and UV photons easily penetrate, sponsoring a simple photochemistry. In dense cold molecular clouds, cosmic rays penetrate, driving a chemistry where neutral-neutral reactions and ion-molecule reactions increase the complexity of molecules in icy grain mantles. In the coldest, densest prestellar cores within molecular clouds, all available heavy elements are depleted onto grains. Dense cores collapse to form protostars and the protostars heat the surrounding infalling matter and release molecules previously frozen in ices into the gas phase, sponsoring a rich gas-phase chemistry. Some material from the cold regions and from hot or warm cores within molecular clouds probably survives to be incorporated into the protoplanetary disks as interstellar matter. For diffuse clouds, for molecular clouds, and for dense hot cores and dense warm cores, the physiochemical processes that occur within the gas and solid state materials are discussed in detail.

1. GALACTIC INTERSTELLAR MEDIUM

1.1. Overview: Cycle of Matter from Stars through the Interstellar Medium to the Solar System

As with our Sun, new stars form in the dense cores of quiescent cold molecular clouds (*Boss,* 2004) from interstellar materials. To a minor extent, pre-main-sequence stars replenish the interstellar medium (ISM) with material through their bipolar outflows and jets. The ISM primarily becomes enriched with nucleosynthesized "metals", i.e., elements heavier than H and He, through supernovae (SNe) explosions of high-mass stars (SNe type II) and of primary stars in a low-mass binary systems (SNe type I). Other sources of enrichment include the massive winds of low-mass asymptotic giant branch (AGB) stars and novae (e.g., *Jones,* 2001; *Chiappini et al.,* 2003; *Wheeler et al.,* 1989). Supernovae explosions (*McKee and Ostriker,* 1977), the UV photons from massive O and B stars (*Wolfire et al.,* 2003), and, to a lesser extent, AGB stellar winds inject energy into the ISM. This energy is deposited in shocks and generates turbulence that acts on many different length scales to bring together, compress, and even shear apart enhancements in the interstellar gas density. Turbulent energy is degraded efficiently into thermal energy when turbulence is concentrated in small

volumes such as in shocks and in small intermittent regions of velocity shear where viscous dissipation occurs (*Vázquez-Semandeni et al.,* 2000). In the ISM, gas is processed rapidly through a wide range of temperatures, densities, and ionization stages, as given in Table 1, under the influence of turbulent and thermal processes, pressure gradients, and magnetic and gravitational forces. Interstellar clouds comprise definable structures in the ISM, but only represent two of five ISM components (section 1.2). Stars enrich the ISM (section 2) with the gas and dust that eventually contributes to the formation of new star systems after cooling and passing through interstellar cloud phases. Processes that contribute to increasing the complexity of solid-state and molecular materials are introduced by *Irvine and Lunine* (2004) and discussed here in detail; these processes primarily occur in interstellar clouds, i.e., in diffuse clouds (section 3) and molecular clouds (section 4), at low temperatures (≤ 100 K). In interstellar clouds, molecules and solid-state materials are more protected from the destruction mechanisms — UV irradiation, cosmic rays, fast electrons — prevalent in the highly energetic intercloud environment of the ISM of the galaxy. In the dense, hot high-mass and warm low-mass protostellar cores (section 5) that only comprise tiny fractions of the mass of molecular clouds, a rich gas-phase chemistry occurs that increases the complexity of the materials infalling onto

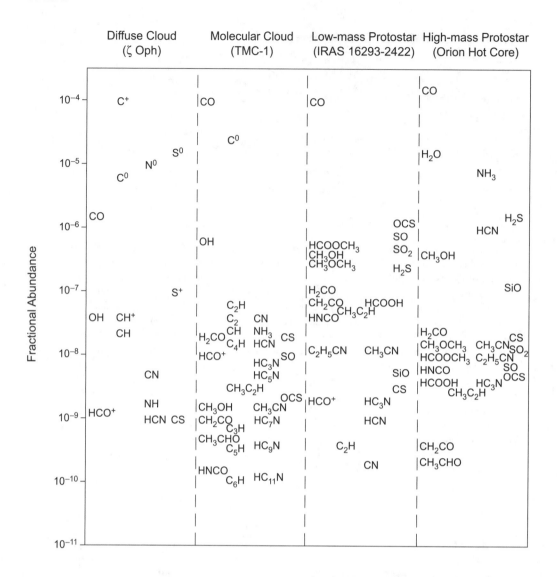

Fig. 1. Representative fractional abundances of ions, atoms, and molecules in different regions of the interstellar medium. The fractional abundances for ζ Oph that sample a line of sight through diffuse clouds are relative to the total abundance of H. Fractional abundances listed for the other objects are relative to H_2. The species included are for comparative purposes and are not a complete inventory of these sources.

the protostellar disks. The increase in the complexity of molecules from diffuse clouds, through molecular clouds to star-forming cores, is demonstrated from left to right in Fig. 1. A fraction of these ISM materials survive to be incorporated into comets (*Ehrenfreund et al.,* 2004).

1.2. Components of the Interstellar Medium

Of the total mass and volume of the galaxy, the ISM constitutes only ~10–15% of the mass but most of the volume. The mass of the ISM is concentrated within a thin disk that extends to ~25–30 kpc and has a vertical scale height of ~400–600 pc [cf. *Ferrière* (2001) for an extensive review on the galactic ISM]. About half the mass of the ISM is in discrete interstellar clouds that occupy ~1–2% of the interstellar volume. Interstellar clouds are concentrated along the

spiral arms but also occur in interarm regions and in the galactic halo above and below the galactic plane.

Physical conditions in the ISM lead to the description of ISM as the coexistence and interaction of five components (included in Table 1): (1) the hot ionized medium (HIM), also called the coronal gas; (2) the warm ionized medium (WIM) containing ionized atomic hydrogen atoms (H^+), referred to in other works as the diffuse ionized gas (DIG); (3) the warm neutral medium (WNM), containing neutral atomic hydrogen atoms (H^0 or H); (4) the atomic cold neutral medium (CNM), hereafter referred to as diffuse clouds, dominated by H but containing some molecular hydrogen (H_2); and (5) the molecular CNM, dominated by H_2, hereafter referred to as molecular clouds or dark clouds.

The components of the ISM (*Spitzer,* 1985) characterized by significantly different temperatures — hot, warm,

TABLE 1.   Interstellar medium components and their physical properties.

ISM Component	Common Designations	T (K)	Density* (cm^{-3})	State of Hydrogen	x_e[†]	Typical Diagnostics
Hot ionized medium (HIM)	Coronal gas	10^6	0.003	H^+	1	X-ray emission, UV absorption
Warm ionized medium (WIM)	Diffuse ionized gas (DIG)	10^4	>10	H^+	1	Optical, UV, IR, $H\alpha$, H II regions
Warm neutral medium (WNM)	Intercloud H I	8×10^3–10^4	0.1	H^0	0.1	21-cm emission
Atomic cold neutral medium (CNM)	Diffuse clouds	100	10–100	$H^0 + H_2$	~10^{-3}	21-cm absorption, 3.4-μm absorption, UV absorption
Molecular cold neutral medium (CNM)	Molecular clouds Dense clouds Dark clouds	0–50 1	10^3–10^5	H_2	10^{-7}–10^{-8}	Extinction, far-IR, radio: CO, CS, NH_3, H_2CO, HCO^+
Molecular hot cores	Protostellar cores	100–300	>10^6	H_2	≤10^{-8}	Rovibrational emission CH_3CN, CH_3OH, NH_3, HCN, SO_2

*Density of H_2 in molecular clouds and cores (n_{H_2}), otherwise density of H (n_H).
[†]Ionization fraction (x_e).

and cold — result from the balance between energy injected (primarily) by supernova shock waves and radiative losses (*Spitzer,* 1990; *Ferrière,* 2001; *Wolfire et al.,* 2003). The gas atoms and ions are excited by collisions with electrons and other atoms and ions and by absorption of UV photons. The gas is cooled by electronic transitions of highly ionized heavy elements in the hot component, by electronic transitions of singly ionized and neutral atoms in the warm ionized and warm neutral components, and by vibrational and rotational modes of molecules in the cold atomic and cold molecular components of the ISM.

Ionized hydrogen (H II) gas constitutes the WIM, where gas temperatures of 10^4 K and higher pervade, and $Ly\alpha$ and $H\alpha$ photons arise from the H^+ recombination cascade. Neutral atomic hydrogen (H I) gas and neutral (e.g., C, N, S) and singly ionized atoms (e.g., C^+, S^+) constitute the WNM, i.e., the low-density intercloud medium ($n_H \approx 0.2$–0.5 cm^{-3}, 6000 K ≤ Tg ≤ 10,000 K, A_V ≤ 0.5 mag). H I gas exists in diffuse clouds of rarefied atomic CNM ($n_H \approx 20$–50 cm^{-3}, 50 K ≤ Tg ≤ 100 K, $0.5 \le A_V \le 1$ mag). The dominant heating process of both the WNM and the diffuse CNM is by collisions with photoelectrons ejected by polycyclic aromatic hydrocarbon (PAH) macromolecules (*Bakes and Tielens,* 1994, 1998). The WNM is cooled by far-IR forbidden line emission of the [C II] 158-μm and [O I] 63-μm lines, by recombination of electrons onto grains, and by $Ly\alpha$ emission from recombining H^+ atoms. The CNM is primarily cooled through the [C II] 158-μm line emission (*Wolfire et al.,* 2003). The atomic CNM and molecular CNM are distinct structures in the ISM, constituting the diffuse clouds and molecular clouds respectively, where gas-grain and gas-gas

interactions increase the complexity of molecular materials that are of interest to understanding the formation of cometary materials. Therefore, we will concentrate on the processes in the interstellar clouds in the following discussion, as well as in other sections of this chapter.

Diffuse clouds have sheet-like or filamentary structures (*Heiles,* 1967), and are often seen as cocoons around cold, giant molecular clouds and smaller molecular (dark) clouds (dense molecular CNM with n_H > 100 cm^{-3}, 10 K ≤ Tg ≤ 20 K, A_V ≥ 5 mag). Heating of molecular clouds is primarily by cosmic rays, and at molecular cloud surfaces by grain photoelectric heating and collisions with excited H_2 molecules. Cooling is primarily by CO molecular line emission. Gas temperatures in molecular clouds also are affected by collisions with dust grains that either cool or warm the gas, depending on the density (*Burke and Hollenbach,* 1983) and on the freezing out of molecules (coolants) from the gas phase (*Goldsmith,* 2001).

Translucent clouds are intermediaries ($1 < A_V < 5$ mag) between the cold atomic diffuse clouds and cold molecular clouds, and are most easily distinguished in the galactic halo at high galactic latitudes (*Hartmann et al.,* 1998) and by H_2, CH, CN, CO (*Rachford et al.,* 2002), and H_2CO molecules (*Magnani and Onello,* 1995). We do not discuss these translucent clouds in detail; see *Turner* (2000) for a discussion of their properties and *Ingalls et al.* (2002) for recent modeling of heating and cooling processes. Figure 1 summarizes the fractional abundances of gas-phase ions, atoms, and molecules in four different components of the ISM. Note the increased complexity of molecules present in hot and warm dense molecular cloud cores.

1.3. Interstellar Medium Structures

A large fraction of the volume of the Milky Way is filled with the HIM, i.e., a tenuous, ionized coronal gas (*McKee and Ostriker,* 1977). Supernovae explosions generate the coronal gas (*Spitzer,* 1990) when their ejecta collide with and shock the surrounding medium. Expanding supernova ejecta sweep up ambient ISM material, compressing it into rapidly expanding shells that cool quickly due to their high densities, possibly becoming molecular gas after ~10^6 yr (*McCray and Kafatos,* 1987). Upon collision with a massive interstellar cloud the supernova remnant slows and breaks up into fragments, mixing with the interstellar clouds. Ions, primarily from H (87%), He (10%), ~1% metals (C, N, O), and ~2% electrons, are accelerated in supernova shocks and are the major source of relativistic particles in the ISM commonly referred to as cosmic rays. The cosmic-ray spectrum (1–1000 MeV) is the main energy source for ionizing the molecular gas (*Cravens and Dalgarno,* 1978; *Cesarsky and Völk,* 1978). The low-energy cosmic-ray spectrum is not measured, because the solar wind with its magnetic field deflects these particles, but is instead inferred from the cosmic-ray ionization rate deduced from diffuse cloud chemical models (section 3.6). There is a high probability that 1-MeV cosmic rays will be absorbed close to their place of origin [within ~2 pc (*Spitzer and Jenkins,* 1975)]. In fact, the cosmic rays (E ≤ 100 MeV) that ionize the WNM and CNM do not travel far from their supernova sources (*Kulsrad and Cesarsky,* 1971). EUV (13.6 eV ≤ hν ≤ 100 eV) and X-ray radiation dominate the ionization of the intercloud WNM. The gas in the WNM dominates the EUV and X-ray opacity. FUV (6 eV ≤ hν < 13.6 eV) radiation dominates the heating of the WNM and the photoelectric heating of the CNM diffuse clouds and molecular cloud boundaries. The dust in the CNM clouds dominates the FUV opacity.

Massive stars often form in clusters, and their explosions supply enough of a continuum of energy to form supershells or superbubbles. The HIM that fills ~20% of the galactic ISM local to the Sun is mostly seen as superbubbles (*Ferrière,* 1998). The expansion of superbubbles may extend beyond the local scale height of the galaxy and form chimneys, expelling HIM material into the galactic halo. As the expelled material gains height, the gas cools and forms high- and intermediate-velocity diffuse clouds in the halo (e.g., *Richter et al.,* 2003). These diffuse clouds then fall back down onto the galactic disk and contribute even more energy to turbulent excitation of the ISM than supernovae; this convective cycle is called the galactic fountain (*Shapiro and Field,* 1976). Through Hα emission line studies, the WIM is observed to persist not only in the galactic plane in patches, filaments, and loops, but also in the galactic halo. The escape of UV photons from massive stars through superbubbles and chimneys accounts for the energy required to maintain the WIM in the halo (*Dove et al.,* 2000). From the temperatures and densities determined for the ionized and neutral components, the thermal pressures of the HIM and the WIM exceed the WNM by factors of ~3–15 and ~2, respectively. The turbulence and thermal forces in the HIM and WIM, i.e., the shocks and pressures from ionized gas, can sweep up and compress WNM or shear apart CNM diffuse clouds. Hence, primarily supernovae, and to a lesser extent stellar winds, are responsible for the turbulent nature of the ISM, the multiple components of the ISM, and the structures of the interstellar clouds.

Diffuse clouds are thought to form when streams of WNM H I collide (e.g., *Ballesteros-Paredes et al.,* 1999b). Widespread warm H I gas (WNM at 6,000–10,000 K) exists in the space between cold H I diffuse clouds (CNM at 50–100 K), as observed by H I 21-cm emission and absorption lines, respectively (cf. *Ferrière,* 2001). The density ratio between the WNM and diffuse clouds is approximately the inverse of the temperature ratio, supporting the view that the atomic WNM and CNM are in rough thermal pressure equilibrium (*Boulares and Cox,* 1990; *Wolfire et al.,* 2003). However, recent arguments have been made that diffuse clouds are not pressure confined by the WNM but rather result from turbulent density fluctuations in colliding gas streams (*Ballesteros-Paredes et al.,* 1999b). Such nonthermal pressures due to turbulent motions, magnetic fields, and cosmic rays are required to produce the observed vertical scale height of the WNM component (*Wolfire et al.,* 2003) and maintain sufficient pressures such that the neutral component of the ISM occurs in the galaxy as two phases — WNM and CNM. The balance between heating and cooling processes in the neutral medium, either through turbulent dynamics (*Ballesteros-Paredes et al.,* 1999b) or in equilibrium conditions (*Wolfire et al.,* 2003), generates a relationship between pressure and neutral hydrogen density that is double-valued for a range of pressures: The neutral two-phase ISM consists of either warm and rarefied (WNM) or cold and dense (CNM) material (*Goldsmith et al.,* 1969). The two phases are in thermal equilibrium but not hydrostatic equilibrium. The kinematics of atomic gas and molecular gas are similar (*Grabelsky et al.,* 1987), and the morphological distinction between WNM and CNM is often not clear: Warm H I gas is seen as an intercloud medium and diffuse clouds are seen to have H I halos. Massive stars form in molecular clouds that have diffuse cloud cocoons that are embedded within the intercloud WNM, so much of the mass of the CNM and the WNM is in photodissociation (PDR) regions illuminated by UV photons from massive stars (*Hollenbach and Tielens,* 1999).

Most of the mass of H I is in the CNM as diffuse clouds even though most of the volume of the ISM containing H I is in the rarefied intercloud WNM (*Dickey and Lockman,* 1990). Furthermore, most of the mass of CNM clouds is in molecular clouds. Specifically, most of the mass of gas in the ISM is in giant molecular clouds (GMCs) that each span tens to hundreds of parsecs and contain 10^5–10^6 $M_\odot$ of gas. GMCs with masses greater than ~10^4 $M_\odot$ are gravitationally bound [e.g., for the outer galaxy (*Heyer et al.,* 2001)]. By number, most molecular clouds are less massive than 10^4 $M_\odot$ and have internal turbulent motions in excess of their gravitational binding energies (cf. *Heyer et al.,* 2001). The formation and existence of these gravitationally unbound molecular clouds is thought to be a result of a highly energetic turbulent ISM in which diffuse cloud material is compressed through supernova shocks, colliding H I streams, spiral density waves, and gravity. All molecular clouds including GMCs fill only 1–2% of the volume of the ISM.

1.4. Interstellar Cloud Lifetimes

The lifetimes of diffuse clouds and molecular clouds is a rapidly evolving subject. A couple of decades ago, concepts of molecular cloud formation were motivated to explain the formation of GMCs because that is where most of the molecular gas resides. The formation of molecular clouds by ballistic aggregation of smaller clouds on timescales of 4×10^7 yr was considered (*Field and Hutchins, 1968*). This view was superseded by the concept that large-scale magnetic Parker instabilities formed molecular clouds and produced lifetimes of $\sim 2 \times 10^7$ yr (*Blitz and Shu, 1980*). Based on the supposition that the entire GMC could form stars, the long lifetimes of GMCs were consistent with the observed slow rate at which molecular gas turns into stars. Mechanisms were evoked to delay the onset of star formation in GMCs, including ambipolar diffusion of magnetic flux (*Shu et al., 1987*) and turbulent cloud support (e.g., *Nakano, 1998*). We now know that only a small fraction of a GMC, or of any molecular cloud, has sufficient density to be protostellar cores (*Boss, 2004; Elmegreen, 2000*). Therefore the observed low star formation rate is a result of the small volume fraction of molecular clouds that are dense enough to be protostellar cores (*Elmegreen, 2000*).

Modern models of diffuse cloud formation involve gravitational instabilities and turbulence (e.g., *Vázquez-Semandeni et al., 2000*). The structure of the ISM is very fragmentary: Even on the largest scales, GMCs are seen as fragments inside H I clouds (*Grabelsky et al., 1987*). Structures inside GMCs are created by turbulence and are of similar fractal dimension to larger structures (*Falgarone et al., 1991*). It takes about a volume of WNM 1 kpc in diameter to make a CNM GMC. If this material is brought together within several tens of parsecs, the density of the gas is higher than average (10^3 cm^{-3}), but most of it is WNM. Gravity might pull this material together in a dynamical time (G$\rho^{-1/2}$) of $\sim 10^7$ yr (*Elmegreen, 2000*). However, turbulent processes stimulated by gravity, supernovae shocks, collisions with supernova remnant shells, spiral density waves (e.g., *Elmegreen, 1979*), or collisions with other clouds can speed up this process. In this view, molecular clouds are an intermediate-scale manifestation of the turbulent cascade of energy from its injection into the HIM and WIM to small dissipative scales (*Vázquez-Semandeni, 2004*).

The lifetimes of diffuse clouds are not known. There is less mass in diffuse clouds than in molecular clouds, so under equilibrium conditions, the lifetimes of diffuse clouds would be shorter. In the local cloud containing the solar neighborhood, the fraction of ISM in molecular clouds is less than half. The self-gravitating or star-forming fraction of molecular clouds is significantly less than half. In the turbulent scenario, the diffuse clouds that are not dense enough or massive enough to be self-gravitating would have

to live longer than the star-forming molecular clouds. That is, the rarefied CNM that takes the form of diffuse clouds could stay diffuse for a long time between its successive incorporations into dense CNM as GMCs or smaller molecular clouds. Each individual diffuse cloud is probably buffeted by turbulence and changed from diffuse cloud to molecular clouds in only a short time. Molecular clouds become diffuse clouds when dispersed by star formation or turbulent forces. Thus, much of the volume of the H I gas is retained in diffuse clouds.

The rapid formation of molecular clouds within diffuse clouds is thought to occur on the order of a few million years (*Ballesteros-Paredes et al.,* 1999a,b; *Hartmann et al.,* 2001). Most numerical simulations of diffuse cloud formation, however, are not followed to high enough densities (100 cm^{-3}) to confirm that a large fraction of the H I gas is turned to H$_2$ on timescales of $\sim 10^6$ yr (*Ballesteros-Paredes et al.,* 1999b; *Ostriker et al.,* 2001). The formation of H$_2$ (section 3) occurs when a sufficient column of H I shields the H$_2$ against UV dissociation [A_V(min) $\approx$ 0.5–1 (*van Dishoeck and Blake,* 1998)] and when there is a sufficient concentration of dust on whose surfaces H$_2$ forms (cf. *Richter et al.,* 2003). Molecular hydrogen formation can occur in diffuse clouds in the galactic halo with volume densities of only $n_H \approx 30$ cm^{-3} because the UV field is not as intense as in the galactic plane (*Richter et al.,* 2003). In the galactic plane, H$_2$ can form under equilibrium conditions in $\sim 10^6$ yr when both the densities are sufficient ($n_H \approx 100$ cm^{-3}) and the surrounding H I intercloud medium has sufficient H I column depth. These conditions occur for diffuse clouds that encompass $\sim 10^4$ M$_\odot$ of gas and are gravitationally bound (*Hartmann et al.,* 2001). Many gravitationally unbound smaller molecular clouds exist, however, at densities of $n_H \approx 100$ cm^{-3}; these clouds contain H$_2$ and CO but are without massive H I cocoons. The conundrum is that timescales for H$_2$ formation appear to be faster than the timescales constrained by equilibrium processes (*Cazaux and Tielens,* 2002) at the densities observed for molecular clouds. *Pavlovski et al.* (2002) find that H$_2$ formation occurs in shock-induced transient higher-density enhancements more rapidly than expected from the average density of the gas. Three-dimensional computations of decaying supersonic turbulence in molecular gas show the complete destruction and the rapid reformation of H$_2$ in filaments and clumps within a diffuse cloud structure. These regions of turbulent compression then relax and the H$_2$ becomes relatively evenly distributed throughout the cloud. Therefore, molecular clouds may form out of diffuse cloud material several times faster in a turbulent ISM (*Pavlovski et al.,* 2002), hence making the rapid formation and existence of gravitationally unbound molecular clouds of $\sim 10^3$ M$_\odot$, such as the Chameleon I–III and ρ Ophiuchi molecular clouds, theoretically realizable.

1.5. Rapid Star Formation

The rapid formation of molecular clouds, of stars within molecular clouds, and of molecular cloud dispersal, is a controversial issue. For detailed discussions on the subject, see recent excellent reviews by *Boss* (2004), *Mac Low and Klessen* (2003), and *Larson* (2003). After a molecular cloud forms, star formation appears to proceed quickly in about a cloud crossing time, and the crossing time depends on the turbulent velocities in the cloud (*Elmegreen*, 2000; *Hartmann et al.*, 2001; *Vázquez-Semandeni*, 2004). The duration of star formation corresponds to the size of the region: Spreads of less than ~4×10^6 yr in age are deduced for star clusters, the Orion region has a spread of ages of 10^7 yr, and the Gould's Belt a spread of 3×10^7 yr (*Elmegreen*, 2000). Careful application of pre-main-sequence theoretical stellar evolution tracks to stellar associations in Taurus shows that there is a 10^7-yr spread in the ages of the association members, but the majority of stars are $\leq 4 \times 10^6$ yr and the fewer older stars tend to lie on the cloud boundaries (*Palla and Stahler*, 2002). Many stellar associations do not show evidence for intermittent stellar formation activity in their derived stellar ages but do show over the past ~10^7 yr an acceleration in the rate of stellar births during the most recent ~4×10^6 yr (*Palla and Stahler*, 2000). This is as expected because the rate of star formation accelerates as the cloud contracts (B. G. Elmegreen, personal communication, 2003). Thus the rapid-star-formation scenario is invoked to explain the absence of molecular gas from clusters more than ~4×10^6 yr in age, and the — still controversial (*Palla and Stahler*, 2002) — lack of an age dispersion in T Tauri stars of more than ~4×10^6 yr (*Hartmann et al.*, 2001). Given the concept that small regions collapse to form stars and star clusters in ~$1–3 \times 10^6$ yr during the lifetime of the larger complex, then rapid molecular cloud formation (~10^6 yr) is followed by rapid dissolution of molecular gas back to diffuse cloud material by the energy injected by young, newly formed stars or by turbulent motions. Shorter lifetimes of molecular clouds (~$1–4 \times 10^6$ yr) also help to resolve the many problems for interstellar chemistry posed by longer lifetimes (~10^7 yr) (section 4.3).

Molecular clouds are very inhomogeneous in density, containing density-enhanced "clumps" (approximately a few parsecs, 10^3 H_2 cm^{-3}), and within these clumps contain smaller, particularly dense, self-gravitating "cores" (~0.1 pc, $\geq 10^5$ H_2 cm^{-3}) (*Larson*, 2003). Clumps are thought to be the precursors of star clusters and small groups of stars, although most clumps are not as a whole self-gravitating (*Gammie et al.*, 2003). In the view of rapid cloud formation, protostellar core masses result from a rapid sampling of existing cloud structures (clumps and cores) (*Elmegreen*, 2000). Evidence for this lies in the fact that the distribution of masses of newly formed stars, i.e., the initial mass function, is similar to the distribution of masses of protostellar cores in molecular clouds (*Motte et al.*, 1998; *Testi and Sargent*, 1998; *Luhman and Rieke*, 1999). Protostellar cores formed by turbulent processes are supercritical, i.e., their gravitational energies available for collapse are greater than their internal thermal and magnetic energies (*Mac Low and Klessen*, 2003). This concept is supported by the fact that the existence of a subcritical core has yet to be convincingly demonstrated (*Nakano*, 1998).

Most of the molecular mass is in GMCs; GMCs contain a wide range of core masses (*Nakano et al.*, 1995) and are the birth sites of stellar clusters containing high-, intermediate-, and low-mass stars. By number, most of the low-mass stars form in regions of high-mass star formation. Low-mass stars also form in lower mass molecular clouds in the galactic plane, such as the ~10^4-$M_\odot$ Taurus molecular cloud complex that spans ~30 pc. The formation of our proto-Sun and the solar system is often thought to have formed in a Taurus-like cloud because T Tauri stars are proto-Sun analogs, although the high levels of radioactive isotopes in primitive bodies suggests the solar system formed in a region of high-mass star formation (*Boss*, 2004; *Vanhala and Boss*, 2000). Complex gas phase chemistry occurs in protostellar cores because the reservoir of infalling material is rapidly made more complex: Gases trapped in the solid phase in icy grain mantles are warmed by the protostar and released into the gas phase. High-mass protostars are more luminous than low-mass protostars and therefore heat the surrounding infalling core material to a greater extent, producing hot cores, while low-mass protostars produce warm cores. In so far that this infalling core material has not yet passed through the protostellar disk, the rich gas-phase chemistry that occurs in hot and warm cores is considered part of the reservoir of interstellar cloud materials that contributes to the protoplanetary disk and the formation of cometary materials.

1.6. Cycling of the Elements through Diffuse Clouds and Molecular Clouds

Stars that form in quiescent dense cold molecular clouds heat and energize the ISM by ejecting matter. The ISM is continually enriched in newly nucleosynthesized heavy elements primarily through supernovae and AGB star winds (section 2.1). Much of the matter that is shed into the ISM occurs through the circumstellar envelopes (CSE) of late-type stars where the physical conditions are conducive to the condensation of dust grains (section 2.2–2.5). Dust grains are also thought to condense in supernovae and novae (e.g., *Jones*, 2000). Most of the heavy elements are therefore injected into the ISM as dust grains.

The dust grains in interstellar clouds, however, are not necessarily those produced by AGB stars and supernovae. Interstellar medium processes contribute to the rapid destruction of dust grains, including shocks produced by supernovae and turbulence, the same turbulence that can act to form interstellar clouds (section 1.4). Grain destruction processes are most efficient in the warm neutral component (WNM) of the ISM where gas densities are moderate ($n_H \approx$ 0.25 cm^{-3}) and temperatures are high (8×10^3–10^4 K) (*Jones et al.*, 1996; *McKee et al.*, 1987). In shocks, collisions occur between grains and gas atoms and ions, and between the dust grains themselves. Gas-grain collisions result in erosion and sputtering of grains. Grain-grain collisions result in fragmentation and, for the larger-velocity collisions, vaporization. From the passage through supernovae shocks, grain lifetimes are predicted to be shortened to 6×10^8 and 4×10^8 yr for carbonaceous and silicate grains respectively (*Jones et al.*, 1996), about 50 times too short compared to a nominal grain lifetime of 2×10^{10} yr required to maintain

>90% of the silicates ejected by stars as particles in the ISM. Fragmentation is about 10 times faster than grain destruction by sputtering and vaporization (*Jones et al.,* 1994, 1996; *Jones,* 2000), rapidly removing grains larger than ~0.05 μm from the ISM grain size distribution. However, grains as large as 1 μm are observed in the ISM. Thus, most ISM dust grains are not stardust but grains that formed in the ISM (section 3.2) (*Jones et al.,* 1994, 1996, 1997; *Tielens,* 1998; *Jones,* 2000, 2001; *Draine,* 2003). The timescales for the cycling of the elements through the diffuse clouds and molecular clouds can therefore be quantified by studying the evolution of the ISM dust-grain population.

In the diffuse clouds, the compositions of the dust grains are studied by deducing from observed absorption lines the fraction of each element that appears depleted from the gas phase (see *Sembach and Savage,* 1996). The timescales for the cycling of the elements through diffuse clouds appears to be rapid because of the uniformity of depletion patterns in diffuse clouds of similar physical conditions (*Jones,* 2000). For example, no galactic vertical gradients in the elemental depletion patterns are seen, so the galactic fountain (section 1.2) effectively mixes supernovae-enriched material in the disk with the halo on timescales of 10^7–10^8 yr (*Jenkins and Wallerstein,* 1996). Also, the observed extreme depletions of supernovae elements (Ca, Ti, Fe) only can occur if >99.9% of all interstellar gas is cycled at least once through the dust-forming circumstellar envelope of a cool star (*Jenkins,* 1987).

The cycle of matter between the different components of the ISM is rapid. The resident times spent in different components are set by various processes such as supernovae-driven and turbulence-driven shock dissipation in the HIM, WIM, and WNM, and turbulent compression, velocity shearing, and H_2 formation in the CNM. Supernovae shocks that sweep up ambient interstellar matter may form molecular clouds in times as short as 10^6 yr (*McCray and Kafatos,* 1987). Since most of the molecular mass is in giant molecular clouds that form massive stars, the timescales for massive star formation and dissipation (~4×10^6 yr; section 1.5) drive the cycling times for the WNM and CNM components of the ISM. If molecular clouds form in times as short as ~1–3×10^6 yr (section 1.4), then diffuse cloud material probably passes rapidly into and out of the colder, denser molecular cloud structures. Theoretical models now predict that diffuse cloud material that was previously molecular cloud material will have an enriched gas-phase chemistry compared to diffuse cloud material that formed from WNM (*Bettens and Herbst,* 1996; *Price et al.,* 2003). The chemistry of diffuse clouds may help to trace their history.

2. FORMATION OF THE ELEMENTS AND DUST GRAINS IN STARS

2.1. Nucleosynthesis

Beyond the large amounts of H, some He, and small amounts of Li created during the Big Bang, other elements that constitute cometary bodies, the planets, and our Sun were produced via nucleosynthesis in the interiors of stars or in the explosive nucleosynthesis of novae and supernovae. Nuclear burning sequences are subjects of detailed study (*Clayton,* 1983; *Wheeler et al.,* 1989) and differ considerably for low- to intermediate-mass stars (*Renzini and Voli,* 1981) and high-mass stars (*Woosley and Weaver,* 1995; *Thielemann et al.,* 1996).

In brief, during a star's life on the main sequence, H fuses into He primarily through three temperature-dependent processes: two proton-proton reactions and the CNO bi-cycle. In the CNO cycle, He is a primary product and N is a secondary product formed at the expense of C and O already present in the stellar interior. After time on the main sequence, H is exhausted in the core and a He core is formed. This He core will begin nuclear burning into C through the triple-α-particle reaction when the core has sufficiently contracted and heated to temperatures of ~10^8 K. During the time that H is burning only in a shell around the preignited He core, the stellar envelope expands and the star transits from the main sequence to the red giant branch of the luminosity-temperature or Hertzprung-Russell (HR) diagram (*Willson,* 2000; *Iben,* 1974). The He core ignites and is later exhausted; the star expands again due to He shell burning and outer H shell burning as an asymptotic giant branch (AGB) star. During this AGB phase of low- to intermediate-mass stars when the expanded envelopes have low surface gravity, stars shed material through their winds and eject significant amounts of ^{4}He, ^{12}C, ^{13}C, and ^{14}N into the ISM. An AGB star is observed to undergo substantial mass loss in the form of a stellar wind. In cooler outer regions of the stellar atmosphere, elements form molecules, and these gas phase molecules condense into solid particles at temperatures lower than about 2000 K. The ratio of C to O (C/O) in the gas determines the mineralogy of the dust that forms (*Tsuji,* 1973), as will be discussed below (sections 2.2–2.4). Stellar radiation pressure on these dust grains drives them outward and gas-grain collisions drag the gas along with the dust. This leads to the existence of a massive circumstellar envelope (CSE) and a slow AGB stellar wind that enriches the surrounding ISM with molecules and dust.

The evolution of the stellar structure is critical to the yields of C and N (*Busso et al.,* 1999; *Chiappini et al.,* 2003). Carbon and N are primary products during the third dredge-up stage on the AGB if nuclear burning ("hot bottom burning") at the base of the convective envelope is sufficient. During thermal pulses in ^{13}C-rich pockets, slow neutron capture through the "s-process" leads to elements such as ^{14}N, ^{22}Ne, ^{25}Mg, and rare-earth elements including Sr, Y, Zr, Ba, La, Ce, Nd, Pr, Sm, Eu, and in the presence of Fe seed nuclei slow neutron capture uniquely creates very heavy nuclei such as ^{134}Ba, ^{152}Ga, and ^{164}Er. For a description of the dependence of nucleosynthesis in AGB stars on stellar structure and metallicity, see the review by *Busso et al.* (1999).

Iron is produced when long-lived white dwarf low-mass stars in binary star systems eventually explode as type Ia supernovae. Oxygen is produced when short-lived massive stars explode as type II supernovae (SNe II). In a SNe II all the burning shells (the star's "onion skin structure") are expelled back into the ISM. The massive star's elemental abundances from outside to inside echo the burning cycles: H, He, C, O, Ne, Mg, S, and Si. Silicon burns to Fe, and as Fe is the most stable of all the elements, it therefore cannot be fused into another form. Once an Fe core forms, the star no longer produces enough photons to support itself against gravity, and the core collapses. The collapsing core reaches such high densities and temperatures that nuclear statistical equilibrium occurs: All elements break down into nucleons and build back up to Fe (*Woosley*, 1986). At even higher densities and temperatures, all matter turns into neutrons, and the implosion is halted at the neutron star core. Then the star that is collapsing onto the neutron core "bounces" and the massive star explodes. If the neutron core is larger than the Chandrasekhar mass limit of $1.2\,M_\odot$, a black hole forms. In the inner Si- and O-rich layers of the star that are expanding at 2500–5000 km s^{-1}, explosive nucleosynthesis occurs, where the capture of neutrons by nuclei occurs in the rapid "r-process" to build up the Fe-peak elements and the interesting radioactive nuclear chronometers ^{232}Th, ^{235}U, ^{238}U, and ^{244}Pu. Explosive O burning leads to Si, S, Ar, and Ca. Explosive Si burning also leads to Si, S, Ar, Ca, as well as stable Fe and Ni, and radioactive Co that decays to radioactive Ni that then decays to stable Fe (*Woosley*, 1988). Massive stars contribute most of the O, and significant amounts of He and C to the galaxy (*Chiappini et al.*, 2003), although there are significantly fewer massive stars than low-mass stars.

The reservoir of elements out of which the solar system formed is the result of the mixing of nucleosynthetic products from many generations of stars of different masses. Galactic chemical evolution (GCE) is modeled by folding together the history of star formation in the Milky Way; the initial mass function that describes the relative number of stars of a given initial stellar mass; the increase of stellar metallicity with time; the spatial distribution of stars in the galaxy; and the nucleosynthetic yields as functions of stellar mass, structure, and metallicity (*Matteucci*, 2001; *Chiappini et al.*, 2003). In the Milky Way 4.5×10^9 yr ago, GCE had created the reservoir of elements of "solar composition." Even though GCE reveals that most of the O and Si originates from SNe II, the isotopically anomalous dust grains from SNe II are very few in number in meteorites (*Yin et al.*, 2002; *Nittler et al.*, 1996). In fact, in interplanetary dust particles (IDPs), O-isotopic anomalies are now being found that are attributable to AGB stars (*Messenger*, 2000; *Messenger et al.*, 2002, 2003). This isotopically anomalous O found in silicate grains within IDPs has a high (1%) abundance by mass. Within the measurement errors, the other 99% of the IDPs are composed of approximately solar composition materials. Therefore, even the more primitive dust grains in our solar system indicate that most of the heavy elements locked in dust grains lost the isotopic signatures of their birth sites prior to entering the solar nebula. It has been hypothesized that dust grains are evaporated and recondensed in supernova shocks in the galaxy (*Jones et al.*, 1996). Furthermore, dust grains probably reform in molecular clouds (*Dominik and Tielens*, 1997). The evidence in the

most primitive interplanetary dust grains, of probable cometary origin, is that the material that came into the solar nebula primarily was of approximate solar composition.

2.2. Asymptotic Giant Branch Star Circumstellar Envelopes

The circumstellar envelope (CSE) of an AGB star provides the ideal environment for complex silicates and carbonaceous material to grow and polymerize (see *Draine*, 2003, for a review). The chemistry of the dust that condenses depends on the C/O ratio. The molecule CO is very tightly bound and consumes all available C or O, whichever is less abundant. If there is C left over after the formation of CO, then a C-based dust chemistry occurs. Conversely, if there is leftover O, O-rich silicate dust forms. Early in the AGB phase, when C/O ratios are less than unity, silicate dust forms, as is widely observed in the dust shells around these stars (*Molster et al.*, 2002a,b,c). The chemistry of the O-rich dust follows condensation pathways for silicates, aluminum oxides, and alumino-silicates. Later in the AGB phase, sufficient newly synthesized C is dredged up from the stellar interior to increase the atmospheric C/O ratio above unity. In this case, carbonaceous dust is formed. The chemistry of C-dust formation is believed to be similar to that which produces soot in terrestrial combustion, and important chemical intermediates are the PAH molecules (*Frenklach and Feigelson*, 1989; *Tielens and Charnley*, 1997). Distinctive IR emission bands from hydrocarbon grains are detected in C-rich AGB stars, but the aromatic infrared bands (AIBs) from PAHs are not detected prob-

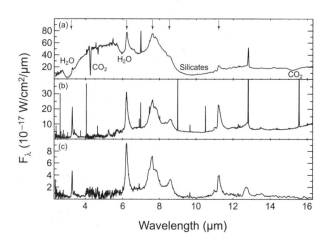

Fig. 2. PAH emission (arrows) in different lines of sight through the ISM. **(a)** Mon R2 IRS2 depicts a line of sight toward a massive protostar, exhibiting dust grains with ice mantles (H_2O, CO_2), and a UV-illuminated region where the gas is photodissociated and PAH emission is evident. **(b)** Orion Bar, which is a photodissociation region; the chemistry is dominated by UV and the PAH emission dominates the IR spectra. **(c)** PAH emission is also seen in the environs of the reflection nebula NGC 2023. Figure courtesy of J. Keane and E. Peeters.

ably because AGB stars are too cool to produce the UV photons that excite the PAHs. In the global cycle of matter in the galaxy, PAHs are small enough to survive supernovae shocks (*Papoular,* 2003) and UV exposure in the WIM component of the ISM, thereby enabling them to be the carriers of the AIB features (see Fig. 2; section 3.4) in protostellar regions (e.g., Mon R2 IRS2), photodissociation regions around massive stars (e.g., the Orion Bar), and reflection nebulae (e.g., NGC 2023). Thus, stellar mass loss during the AGB phase is an important source of silicate and C-rich dust for the ISM (*Willson,* 2000).

2.3. Oxygen-rich Asymptotic Giant Branch Silicate and Oxide Dust

Oxygen-rich dust grains in the ISM are recognized to be primarily Fe-bearing amorphous, i.e., disordered, silicate dust grains (*Li and Greenberg,* 1997). Oxygen-rich AGB stars are recognized through their IR spectra to be the main producers of these amorphous silicates. Silicates largely fall into two mineral families: olivine $(Mg,Fe)_2SiO_4$ and pyroxene $(Mg,Fe)SiO_3$ (cf. *Hanner and Bradley,* 2004). The Mg-pure end members of the olivine group and pyroxene group are forsterite and enstatite respectively. The "ISM" 9.7-µm silicate absorption band is attributed to Fe-bearing amorphous olivine (*Li and Greenberg,* 1997) or to Fe-bearing olivines and pyroxenes in a ratio of ~5 : 1 (*Kemper et al.,* 2004). In the past five years, analysis of IR spectra of O-rich AGB stars obtained with the short wavelength spectrometer (SWS) onboard the Infrared Space Observatory (ISO) reveals other amorphous forms and crystalline forms of silicates and oxides. In particular, O-rich AGB stars also produce Mg-rich crystalline olivine and Mg-rich crystalline pyroxene grains in varying abundance ratios (*Molster et al.,* 2002a,b,c). The high Mg-content of the crystalline silicates is confirmed by the wavelengths of the resonant peaks (*Fabian et al.,* 2001a). In O-rich AGB stars characterized by high mass loss rates of $~10^{-4}$ $M_\odot$ yr^{-1}, i.e., the OH/IR stars, at most a few percent of the total mass of dust ejected can be crystalline silicates, while the preponderance is amorphous silicates (*Kemper et al.,* 2001). In O-rich AGB stars with low mass-loss rates of $~10^{-6}$ $M_\odot$ yr^{-1} and CSEs that are less optically thick in their IR continuum than the OH/IR stars, i.e., the Mira variable stars, the ISO SWS detection limits constrain the crystalline silicates to be ~40% or less of the total mass ejected (*Kemper et al.,* 2001). More stringent limits will be determined for Miras by Spitzer Space Telescope observations. The crystalline silicates are not seen, however, in the diffuse ISM: Constraints of <5% and <0.5% are determined from the ISM 9.7-µm feature and the 9.7-µm feature toward the galactic center by *Li and Draine* (2002) and *Kemper et al.* (2004) respectively. Silicate crystals are efficiently transformed to amorphous silicate structures in the ISM, probably through cosmic-ray impacts during their residence in diffuse clouds (section 3.3).

Oxygen-rich AGB stars with more rarefied CSEs than Miras show emission features of either amorphous silicates, or oxides, or both (*Cami,* 2002). The spectrally broad amorphous silicate bands (with central wavelengths at ~9.7 µm and spanning ~8–12.5 µm) are the most common bands seen in O-rich AGB stars, with the 13-µm feature being second. Postulated carriers for the 13-µm feature include corundum (e.g., *Onaka et al.,* 1989), silica (*Speck et al.,* 2000), and spinel (*Posch et al.,* 1999; *Fabian et al.,* 2001b). Spinel $(MgAl_2O_4)$ can account for the rather common 13-µm feature, as well as the 16.8-µm and 32-µm features in the ISO SWS spectra (*Cami,* 2002). Alumina (Al_2O_3), in the case of G Her, matches the profile of 11-µm feature. Magnesiowüstite $(Mg_{0.1}Fe_{0.9}O)$ is identified as the carrier of the 19.5-µm feature (*Posch et al.,* 2002; *Cami,* 2002).

The silicates and the Al oxides follow two distinct condensation pathways because Si is much more abundant than Al. These thermodynamic condensation sequences were developed for the solar nebula (*Grossman,* 1972). In an AGB outflow that originates at the stellar photosphere and expands and cools, the first grains to condense are those with the highest condensation temperature: corundum (Al_2O_3) at 1760 K. After corundum condenses, aluminosilicates condense, including Ca-bearing gehlenite $(Ca_2Al_2SiO_7)$. Gehlenite then reacts with the Mg in the gas at about 1550 K to form spinel $(MgAl_2O_4)$ (*Speck et al.,* 2000). Isotopically anomalous presolar spinel grains from AGB stars are found in meteorites (*Nittler et al.,* 1997).

In the same parcel of gas, the silicates condense: Magnesium-rich olivine condenses first as forsterite at about 1440 K. By reactions with SiO molecules in the gas phase, cooling of the forsterite leads to the formation of enstatite. At high temperatures the incorporation of Fe into crystalline silicate minerals is inhibited thermodynamically (*Grossman,* 1972). When the gas temperatures drop below the crystallization temperature of ~1000 K, conditions prevail where Fe can be incorporated into the minerals and amorphous forms of silicates can grow. This is the mechanism suggested for the formation of Fe-bearing amorphous silicates in AGB stars (*Tielens et al.,* 1997). Iron is present in the gas phase and condenses as pure Fe grains at temperatures of about 900 K (*Gail and Sedlmayr,* 1999), significantly cooler than the Mg-rich silicates. Pure Fe grains, even though spectrally featureless, have been invoked to explain the hot near-IR continuum in AGB stars (*Kemper et al.,* 2002).

Not all minerals appear to follow a thermodynamic sequence. Magnesiowüstite (MgFeO) is not expected to form, yet it is seen in spectra of O-rich AGB stars (*Cami,* 2002). Similarly, the preponderant amorphous silicates in the CSE of AGB stars are not a consequence of a thermodynamic condensation sequence. A quantitative explanation of their formation will probably require the application of a kinetic condensation theory that includes surface exchange processes (*Gail and Sedlmayr,* 1999).

2.4. Asymptotic Giant Branch Carbon-rich Dust

The extended convective envelopes of AGB stars evolve from O-rich to C-rich during the third dredge-up phase when C is nucleosynthesized at the base of the CSE and brought to the stellar surface. In the CSE of C-rich AGB stars, a complex carbon chemistry occurs that is analogous to carbon soot formation in a candle flame or in industrial smoke stacks. An active acetylene (C_2H_2) chemistry appears to be the starting point for the development of hexagonal aromatic rings of C atoms, the structure of which mimics the cross-section of a honeycomb. These aromatic rings probably react further to form large aromatic networks such as soot. The kinetic theory for PAH and soot formation is a subject of current astrophysics research (*Cherchneff et al.,* 1991; *Cadwell et al.,* 1994; *Hudgins et al.,* 2001). These soot particles also contain aliphatic bonds, i.e., nonaromatic C–C, C=C, C≡C, and C–H bonds.

At intermediate radii in the AGB CSE a transient photochemistry occurs (*Glassgold,* 1996), but by the time the CSE edge is reached, all small volatile molecules that formed in this region, as well as at the photosphere (e.g., HC_3N, SO_2, H_2CO, CO, H_2, C_2H_2, SiO), are completely destroyed and hence play no part in interstellar cloud chemistry. The solid particles from AGB CSE, upon entering the warm ionized and warm neutral components of the ISM, can be structurally altered by large fluxes of UV photons (*Mennella et al.,* 2001) and fast H atoms (*Mennella et al.,* 1999) present there (*Chiar et al.,* 1998; *Pendleton and Allamandola,* 2002).

After many thermal pulses the AGB star has shed much of its extended envelope and transitions to the protoplanetary nebula phase. The white dwarf central star begins to be revealed, its strong UV radiation field dissipating and revealing a carbon chemistry that is dominated by aliphatic carbon materials (*Chiar et al.,* 1998). The aliphatic bonds emit through the 3.4-μm feature (e.g., *Chiar et al.,* 1998; *Goto et al.,* 2003). The 6.9-μm band is also observed, which is characteristic of aliphatic-aromatic hybrid compounds containing methylene (–CH_2) substructures (*Hrivnak et al.,* 2000). As the CSE is becoming rarefied and excited by the wind and the UV photons of the hot white dwarf stellar core, UV photons destroy the aliphatic bonds (*Menella et al.,* 2001). Some of this aliphatic material formed in AGB CSE may survive the intense galactic UV field and interstellar shocks to be incorporated into the diffuse clouds, as evidenced by the strong similarities between the shapes and substructures in the 3.4-μm features in the protoplanetary nebula CRL 618 and in the line of sight to the galactic center (*Chiar et al.,* 1998, 2000). In the mere few thousand years during which a planetary nebulae emerges, however, the hydrocarbon dust becomes dehydrogenated and the aliphatic bonds are transformed to aromatic bonds. The greatest evolution of the PAH features occurs in the rapid post-AGB protoplanetary nebula phase (cf. *Mennella et al.,* 2003). Ultraviolet light from AGB stars is insufficient to excite the PAHs, and so the PAHs are not directly observable at their birth sites in the AGB CSE.

In addition to soot (hydrocarbon grains containing aromatic and aliphatic bonds), silicon carbide (SiC) forms in the C-rich AGB CSE. Silicon carbide grains have two primary crystallographic forms: α-SiC and β-SiC. Only β-SiC is spectroscopically detected in C-rich AGB stars (*Speck et al.,* 1999). Silicon carbide grains are only a minor component of the ISM, constituting less than 5% of the submicrometer dust

that contributes to the interstellar extinction curve (*Whittet et al.,* 1990). All solid grains larger than a few micrometers lack spectroscopic signatures, so large SiC grains may be present in the ISM but are not spectroscopically detected. Presolar SiC grains constitute all the β-SiC structure (*Daulton et al.,* 2002), have isotopic anomalies indicative of AGB CSEs, and are typically larger than 1 μm in size (*Amari et al.,* 2001). Perhaps only the largest SiC grains survive their journey through the ISM to be incorporated into meteorites.

3. PHYSIOCHEMICAL PROCESSES IN DIFFUSE CLOUDS

3.1. Overview

In this section we summarize the composition of and the important chemical processes operating in diffuse clouds of gas and dust. (Note that the lines of sight to background stars frequently sample many diffuse clouds, so diffuse cloud matter often is called the diffuse interstellar medium, or DISM.) Cosmic dust forms in the atmospheres of evolved stars (sections 2.2–2.4). Subsequent complex evolution of dust and gas is driven by interactions with gas and by processes such as heating, UV radiation, shocks, and cosmic-ray (energetic ion) bombardment. However, during this journey only the most refractory compounds survive, while most of the simple species are destroyed. Supernova remnants are a source of galactic cosmic rays, and these play an important chemical role throughout all phases of the ISM, including diffuse clouds. Diffuse interstellar clouds are subject to dynamical phenomena such as supernovae-driven shocks and cloud-cloud collisions. As the interstellar plasma is highly compressible, sound waves and magnetohydrodynamic waves readily steepen into shock waves; the energy available from these shocks opens up many pathways in a high-temperature chemistry (*Flower and Pineau des Forêts,* 1998). The dissipation of interstellar turbulence has also been considered as potentially important fo driving chemical reactions that cannot take place at the low average kinetic temperature of the diffuse medium (*Falgarone et al.,* 1995; *Joulain et al.,* 1998).

Diffuse clouds have temperatures of around 70–100 K that are the result of the thermal balance between heating and radiative cooling. Diffuse cloud gas is heated through photoelectron emission (section 1.2) from dust grains, PAHs (section 2.1), and C atoms. Cooling is predominately through radiation from fine-structure transitions of C^+ and O; these are excited by collisions with H atoms and with the electrons produced from the photoionization of C atoms.

A diffuse cloud is characterized by a density of ~1–100 H atoms cm^{-3}, a gas temperature of ~100 K, and an active photochemistry that forms, destroys, and shapes interstellar matter. As they have optical depths less than unity, diffuse clouds are almost unshielded from the UV radiation produced by massive O and B stars. These UV photons can easily penetrate almost the entire extent of a diffuse cloud and lead to a gas phase composition that is dominated by photochemical reactions, and this, in turn, only permits the formation of fairly simple molecules. Spectroscopy from the UV to radio wavelengths reveals a diversity of solid-state species and gas molecules that exist in the diffuse clouds. Several diatomic molecules such as CO, CH, CN, CH^+, and H_2 have been identified in diffuse clouds (*Lucas and Liszt,* 2000). Simple polyatomic species such as hydrogen cyanide (HCN), formylium (HCO^+), H_2CO, and cyclopropenylidene (C_3H_2) are observed in diffuse clouds and in translucent clouds (section 1.2) (*Turner,* 2000), which are interstellar clouds of a slightly higher extinction and density than diffuse clouds. In contrast, dense molecular clouds whose interiors are shielded from the ISM UV radiation field are the sites where more complex molecules form (see Fig. 1; section 4).

3.2. Depletion of Heavy Elements into Dust Grains in Diffuse Clouds

Dust grains form in the cool CSE of AGB stars, in novae, and in supernovae ejecta (e.g., *Draine,* 2003; *Jones,* 2001) (sections 2.2–2.4). Fragmentation and sputtering of dust grains in low- and high-velocity shocks in the HIM and WIM release heavy elements into the gas phase and selectively enhance the abundances of very small grains (e.g., *Zagury et al.,* 1998). In diffuse clouds, however, the heavy elements (except N and S) primarily reside in dust grains as deduced from the observed depletion of the heavy-element abundances from the gas phase relative to cosmic abundances (for a review, see *Savage and Sembach,* 1996). Thus dust destruction in the hot and warm components of the ISM is followed by reformation and aggregation in diffuse clouds (*Jones,* 2001), and the grains quickly become chemically and isotopically homogenized (*Jones,* 2000). Interstellar medium grain size distribution ranges from 0.001- to 0.05-μm grains (PAHs, HACs, amorphous carbons) up to 1-μm grains (amorphous silicate). To account for the larger grains in the ISM grain size distribution, grain growth and aggregation is inferred to occur in diffuse clouds, molecular clouds, and dense cores. Grains aggregated in diffuse clouds are likely to be highly porous and fractal in structure because coagulation processes do not make highly compact grains (*Weidenschilling and Ruzmaikina,* 1994). Evidence for coagulation into fluffy aggregates in a translucent ($A_V \approx 4$) filamentary cloud in the Taurus region is shown through analysis of multiwavelength emission data (*Stepnik et al.,* 2001). Grains may coagulate into larger grains at the boundaries of diffuse and molecular clouds (*Miville-Deschênes et al.,* 2002) and within dense molecular clouds (*Bianchi et al.,* 2003).

Dust grain compositions are measured through the depletion studies of diffuse clouds, and grain mineralogical types are then deduced from the relative ratios of the depleted elements. Depletion measurements indicate four groups of dust-forming elements in the following order, where each group represents about a factor of 10 drop in abundance (*Jones,* 2000): (1) C and O; (2) Mg, Si, and Fe; (3) Na, Al, Ca, and Ni; and (4) K, Ti, Cr, Mn, and Co. Oxygen is not measured along the lines of sight of the halo clouds, but it is assumed that O resides in the dust. Of the rare earth

elements, over 70% are contained in the dust. The patterns of elemental depletions for different diffuse cloud lines of sight also show that the more depleted the element, the higher the condensation temperature of the dust species that the element constitutes. There are two chemically distinct grain populations. Some investigators refer to these dust populations as "core" and "mantle" (*Savage and Sembach,* 1996), or "more-refractory" and "less-refractory" dust components (*Jones,* 2000). Only the more-refractory dust grains survive in the harsher environments of high-velocity clouds in the galactic halo. These more-refractory grains are primarily composed of O, Mg, Fe, and Si with smaller amounts of Ni, Cr, and Mn. The deduced mineralogic composition of the grains depends on the relative abundances of the elements, which in turn depends on the abundances taken as solar or cosmic. Currently, there are three different references for cosmic abundance: (1) meteorites (*Anders and Grevesse,* 1989; *Grevesse and Noels,* 1993), (2) B stars in the solar neighborhood (*Snow and Witt,* 1996, and references therein), and (3) that deduced for the ISM (*Snow and Witt,* 1996). Depending on the three different cosmic abundance references, *Jones* (2000) deduces the more-refractory dust component in the diffuse medium to be either (1) an Fe-rich olivine-type silicate, (2) a mixed silicate and oxide, or (3) an Fe oxide. *Savage and Sembach* (1996) deduce the more-refractory dust component to be a combination of olivine-type silicate ($Mg_y,Fe_{1-y})_2SiO_4$), oxides (MgO, Fe_2O_3, Fe_3O_4), and pure Fe grains. Both more- and less-refractory grains appear to exist in the diffuse clouds in the galactic disk. By subtracting the abundances of the more-refractory component from the total, the less-refractory grain component is deduced to consist of Mg-rich olivine-type grains (*Jones,* 2000; *Savage and Sembach,* 1996).

Moreover, not all Si appears to be bound in silicates and oxides (*Jones,* 2000). This can be interpreted to mean that Mg and Fe exists in forms independent of Si (*Sofia et al.,* 1994; *Savage and Sembach,* 1996). On the other hand, a continuous evolution of the chemistry of the grains may occur by the preferential erosion of Si with respect to Mg, and the further erosion of both Si and Mg with respect to Fe (*Jones,* 2000). The erosion of Mg with respect to Fe is expected for sputtering (*Jones et al.,* 1994, 1996; *Jones,* 2000). The enhanced erosion of Si with respect to Mg may occur with low-energy (~4 KeV) cosmic-ray bombardment (*Carrez et al.,* 2002) in diffuse clouds (section 3.3).

For a given level of depletion in different diffuse clouds, the elemental variations are small. Perhaps the relative elemental abundances are established early in the life of the dust grain and are not significantly altered. The maintenance of grain composition over the lifetime of a grain is surprising because theoretical models predict that shocks of only ~100 km s^{-1} are required to release enough material from grains to explain the abundances derived for the halo clouds (*Sembach and Savage,* 1996), and this in turn suggests significant reprocessing. Dust grain recondensation following the passage of supernova shocks (*Jones et al.,* 1994, 1996) and formation of grains in molecular clouds (*Dominik and Tielens,* 1997) are suggested mechanisms to account for the long lifetime of interstellar dust grains. If significant fractions of ISM dust are destroyed and reaccreted, it is difficult to see how pure silicates, uncontaminated by C, could form (*Jones,* 2000). In diffuse clouds, grains may form in non-thermal-equilibrium conditions because temperatures and pressures are low compared to AGB CSEs. In higher-density regions, accretion onto preexisting grains is faster than in lower-density regions, and grain mantles can form under conditions far from thermodynamic equilibrium. Grain formation may occur through kinetic processes (e.g., *Gail and Sedlmayr,* 1999). Low-temperature grain-formation processes also have been suggested (*Sembach and Savage,* 1996), but the details of these processes are as yet unknown.

Only the most refractory materials, i.e., minerals that have high condensation and vaporization temperatures, appear to have survived intact through the hot and warm components of the ISM to the CNM. These presolar grains are identified by their anomalous isotopic ratios and are found in meteorites and in IDPs (*Hanner and Bradley,* 2004; *Sykes et al.,* 2004). Dust grains that have isotopically anomalous presolar signatures of their birth sites include diamonds (*Lewis et al.,* 1987) and TiC grains (*Nittler et al.,* 1996) that condensed in the ejecta of SNe II, SiC grains that primarily formed in AGB CSE (*Lewis et al.,* 1994), graphite grains that formed in the CSEs of massive stars (*Hoppe et al.,* 1992), oxide grains that formed in the O-rich CSE of red giant stars and AGB stars (*Nittler et al.,* 1997), and silicates from AGB stars (*Messenger,* 2000; *Messenger et al.,* 2003). Presolar silicate grains currently are not found in meteorites, and are <1% by mass of IDPs. Out of more than 1000 subgrains measured in 9 IDPs, 6 subgrains have extreme isotopic $^{17}O/^{16}O$ and $^{18}O/^{16}O$ ratios: Three are of unknown origin, two are amorphous silicates [glass with embedded metal and sulfides, or GEMS (*Bradley,* 1994; *Hanner and Bradley,* 2004)], and one is a Mg-rich crystalline olivine (forsterite) (*Messenger et al.,* 2003). The dust grains mentioned above that are known to be presolar by their isotopic signatures do not represent the typical grains in the ISM that are composed primarily of silicates, metal oxides (from depletion studies), amorphous C, and aromatic hydrocarbons. At this time, the only link that exists between presolar grains and the silicates deduced from diffuse clouds depletion studies are the GEMS (*Bradley,* 1994). Recall that of the nine IDPs measured by *Messenger et al.* (2003), 99% of the mass falls in an error ellipse centered on solar composition, albeit the measurement errors on these submicrometer samples are still large, and this supports the suggestion by *Jones* (2000) that grain materials are rapidly homogenized through destruction-formation processes in the ISM.

3.3. Cosmic-Ray (Ion) Bombardment of Silicate Dust in Diffuse Clouds

Oxygen-rich AGB stars produce a fraction of their silicate dust in crystalline forms (~4–40%) (sections 2.2–2.3). In the ISM the fraction of crystalline silicates is very low, however [<5%; *Li and Draine* (2002); <0.5%, *Kemper et al.*

(2004)], and is insufficient to account for the fraction of crystalline silicates deduced for comets (*Hanner and Bradley,* 2004; *Ehrenfreund et al.,* 2004). Therefore silicate crystals are efficiently amorphized in the ISM. Laboratory experiments demonstrate the bombardment of crystalline silicates by energetic ions amorphizes the silicates, which simulates cosmic-ray impacts in ISM shocks (*Demyk et al.,* 2001).

When cosmic rays impact grains, significant processing occurs. Most cosmic rays are H^+ and He^+ ions. Recent laboratory measurements show that the following changes occur to Fe-bearing olivine crystals when bombarded by 4-KeV He^+ ions (*Carrez et al.,* 2002): (1) The crystalline structure changes to a disordered, amorphous structure; (2) the porosity is increased; (3) Fe is reduced from its original stoichiometric inclusion in the mineral lattice to embedded nanophase Fe; and (4) Si is ejected, changing the chemistry from olivine to pyroxene. When higher-energy He^+ ions of 50 KeV are used, similar changes occur to the structure of the mineral as occurred with the 4-KeV He^+ ions, but the chemistry [i.e., (4) in the above list] is unaltered (*Jäger et al.,* 2003). Through depletion studies of diffuse clouds we know that less Si (more Mg and Fe) in the grains exists than is consistent with silicate mineral stoichiometry. Cosmic-ray bombardment of silicate grains by ~4-KeV He^+ ions is a feasible explanation for the observed preferential ejection of Si from the grains (section 3.2). Cosmic-ray bombardment of Fe-bearing silicate minerals also serves to reduce the Fe from its incorporation in the mineral lattice to nanophase Fe grains embedded within a Mg-rich silicate (*Carrez et al.,* 2002). This reduction mechanism explains the presence of nanophase Fe in GEMS in IDPs (section 3.2) (*Bradley,* 1994; *Brownlee et al.,* 2000). Thus cosmic-ray bombardment of silicates in diffuse clouds may be a viable process for explaining the lack of crystalline silicates in the ISM, the observed properties of the more- and less-refractory grains in diffuse clouds (section 3.2), and the amorphous silicate component of cometary IDPs (*Hanner and Bradley,* 2004; *Ehrenfreund et al.,* 2004; *Wooden,* 2002).

The cosmic-ray energies that are utilized in the laboratory to change the properties of silicates only penetrate the diffuse clouds. It is the higher-energy cosmic rays (1–100 MeV, section 1.3) that penetrate the dense molecular clouds and ionize the gas, contributing to the coupling between the gas and the magnetic field and to ion-molecule chemistry (*Goldsmith and Langer,* 1978). Thus it is the lower energy (1–100 keV) cosmic rays that penetrate diffuse clouds that can amorphize silicates, and therefore change the composition of the silicates in a manner necessary to explain the depletion studies of the diffuse clouds (*Jones,* 2000).

3.4. Carbonaceous Dust and Macromolecules in the Diffuse Clouds

By observing the "extinction" of starlight toward stars we can trace the nature of interstellar dust particles. The so-called interstellar extinction curve samples the spectroscopic absorption and emission features of interstellar dust from the UV to IR wavelengths. Two of the most relevant signatures of interstellar dust are observed at 217.5 nm [called the "ultraviolet extinction bump" (e.g., *Fitzpatrick and Massa*, 1990)] and 3.4 μm (*Pendleton et al.*, 1994). The 217.5-nm feature is well modeled by UV-irradiated nano-sized amorphous carbon (AC) or hydrogenated amorphous carbon (HAC) (*Mennella et al.*, 1998) or by a bimodal distribution of hydrocarbon particles with a range of hydrogenation and a dehydrogenated macromolecule coronene (a compact PAH $C_{24}H_{12}$) (*Duley and Seahra*, 1999). Other materials such as carbon black of different composition, fullerenes, and PAHs do not show a good match with the UV extinction bump (*Cataldo*, 2002). Due to constraints on the abundance of elemental C in the ISM (*Sofia et al.*, 1997; *Gnacinski*, 2000), C atoms in sp^2 bonds and C atoms in C–H bonds in the same hydrocarbon grains may be responsible for the 217.5-nm and 3.4-μm features respectively. The 3.4-μm stretching band and the 6.85- and 7.25-μm bending modes [e.g., toward the galactic center (*Chiar et al.*, 2000)] are attributed to the C–H bonds in hydrocarbon grains that occur in –CH_2 (methylene) and –CH_3 (methyl) aliphatic groups. Carriers of the interstellar 3.4-μm absorption feature exist in material ejected from some C-rich evolved stars, i.e., the protoplanetary nebula CRL 618 (*Chiar et al.*, 1998). The intense galactic UV field (*Mennella et al.*, 2001) and cosmic rays (*Mennella et al.*, 2003), however, rapidly destroy aliphatic bonds. The aliphatic bonds in hydrocarbon grains are probably formed primarily in diffuse clouds (section 3.5).

Carbonaceous dust in the diffuse medium appears to be highly aromatic in nature, consisting of aromatic hydrocarbon moieties bonded by weak van der Waals forces and aliphatic hydrocarbon bridges (cf. *Pendleton and Allamandola*, 2002). Among the most abundant C-based species in diffuse clouds are PAHs (see *Puget and Léger*, 1989). Polycyclic aromatic hydrocarbons contain 1–10% of the total C in the Milky Way, and are the next most abundant molecules after H_2 and CO. They play a vital role in the heating and cooling of the WNM, diffuse clouds, and surfaces of molecular clouds (*Salama et al.*, 1996; *Wolfire et al.*, 1995, 2003). Ultraviolet photons more energetic than 13.6 eV ionize H and produce energetic electrons that, in turn, heat and ionize the gas in photodissociation regions. Ultraviolet photons less energetic than 13.6 eV would pass practically unattenuated through space if it were not for the PAHs. When PAHs absorb UV photons, both energetic electrons are ejected that collisionally heat the gas (*Bakes and Tielens*, 1994), and energy retained by the molecules is transferred to vibration modes that then radiate through strong thermal-IR emission bands. These aromatic infrared bands (AIBs) result from C–H stretching and bending modes and C–C ring-stretching modes (*Tielens et al.*, 1999) and have characteristic wavelengths (see Fig. 2): 3.3 μm, 11.3 μm, and 12.5 μm for the C–H bonds on the periphery of the PAH, and 6.2 μm, 7.7 μm, and 8.6 μm for the PAH skeletal C–C bonds. The relative band strengths depend on the size of the PAH macromolecule and on its negatively charged, neutral, or positively charged ionization state (*Bakes et al.*, 2001a,b).

The AIBs are ubiquitous in the diffuse ISM in locations where UV photons can excite them. Polycyclic aromatic hydrocarbon emission arises from photodissociation regions (PDRs) where UV light ionizes and excites the gas and dust (e.g., the Great Nebula in Orion) and in reflection nebulae (e.g., NGC 2023), and where massive stars are illuminating the edges of molecular clouds, as shown in Fig. 2. The AIB bands also are found in external galaxies. Only PAHs in the gas phase provide the necessary properties for internal energy conversion after absorbing a photon, in order to emit at this wavelength range. Though little is known about the exact PAH species present in those environments, their abundance and size distribution can be estimated (*Boulanger et al.*, 1998). Laboratory experiments and theoretical calculations of PAHs have revealed important details about their charge state and structural properties (*Allamandola et al.*, 1999; *Van Kerckhoven et al.*, 2000; *Bakes et al.*, 2001a,b).

A long-standing spectroscopic mystery in the diffuse medium is the diffuse interstellar bands (DIBs). More than 300 DIBs, narrow and broad bands, and many of them of weak intensity, can be observed toward hot stars throughout the diffuse interstellar medium. Substructures detected in some of the narrow, strong DIBs strongly suggest a gas-phase origin of the carrier molecules. Consequently, good candidates are abundant and stable C-bearing macromolecules that reside ubiquitously in the diffuse cloud gas (*Ehrenfreund and Charnley*, 2000). Polycyclic aromatic hydrocarbons are therefore among the most promising carrier candidates (*Salama et al.*, 1996). The same unidentified absorption bands are also observed in extragalactic targets (*Ehrenfreund et al.*, 2002).

The harsh conditions in the diffuse component of the ISM, e.g., a UV radiation field of 10^8 photons cm^{-2} s^{-1}, determines the chemistry in such a way that large stable species may either stay intact, change their charge state, become dehydrogenated, or get partially destroyed. Small unstable species are rapidly destroyed. Small amounts of cosmic C probably are incorporated into species such as carbon chains, diamonds, and fullerenes (*Ehrenfreund and Charnley*, 2000). The presence of such small species in the diffuse clouds, including short carbon chains, thus indicates an efficient formation mechanism. Grain fragmentation by shocks and subsequent release of subunits into the gas phase can also provide a source of relatively large and stable gas-phase species (e.g., *Papoular*, 2003). Among those, fullerenes have not been unambiguously identified. Recently it has been shown that UV radiation and γ-radiation cause the oligomerization or polymerization of fullerenes in the solid state. Continuous ion irradiation causes the complete degradation of the fullerene molecules into carbonaceous matter that resembles diamond-like carbon (*Cataldo et al.*, 2002). This raises doubts as to whether fullerenes can ever be observed as an interstellar dust component. In contrast, PAHs have large UV cross-sections and can survive in harsh UV environments. Due to their high photostability, PAHs are among the only free-flying gas-phase molecules that can survive passage through the harsh UV environment of the

WIM and WNM components of the ISM to the diffuse clouds.

3.5. Ultraviolet and Cosmic-Ray Processing of Hydrocarbon Dust in the Diffuse Clouds

The 3.4-μm absorption bands are seen along many lines of sight through the diffuse medium (*Pendleton et al.*, 1994; *Rawlings et al.*, 2003). The 3.4-μm feature was well measured in emission in the protoplanetary nebulae CRL 618 (*Chiar et al.*, 1998) and shown to match the 3.4-μm absorption feature toward the galactic center, suggesting that at least some of the hydrocarbons in diffuse clouds originate as stardust (*Chiar et al.*, 1998, 2000). Some AGB hydrocarbon dust grains are likely to survive ISM shocks and the transition to diffuse clouds (*Jones et al.*, 1996; *Papoular*, 2003), where, if they maintain or regain their degree of hydrogenation, they will contribute to the 3.4-μm aliphatic feature. Grain destruction is so efficient in the ISM (section 1.6), however, that most AGB carbonaceous grains probably readily lose the memory of their birth sites and a new equilibrium for the formation of aliphatic hydrocarbon bonds is set up in diffuse clouds (*Mennella et al.*, 2002).

A long history of laboratory studies exists on the formation of organic residues from the UV photolysis of ices deposited at low temperatures. These organic residues have aliphatic bonds and so the carriers of the 3.4-μm feature in the diffuse clouds were thought to be the products of energetic processing of icy grain mantles within dense molecular clouds (e.g., *Greenberg et al.*, 1995). Several lines of evidence show that the carriers of the 3.4-μm feature are more consistent with plasma-processed pure hydrocarbons than with energetically processed organic residues: (1) The 3.4-μm feature toward the galactic center is unpolarized (*Adamson et al.*, 1999), while the 9.7-μm silicate feature is polarized, so the 3.4-μm feature is not a mantle on silicate cores; (2) the 3.4-μm feature is not observed in dense clouds, requiring that at least 55% of the C–H bonds seen in diffuse clouds be absent from dense molecular clouds (*Muñoz Caro et al.*, 2001), the presumed sites of their formation; and (3) by comparison with laboratory analogs, carriers of the 3.4-μm feature have little O and N, which is uncharacteristic of organic residues.

The carrier of 3.4-μm aliphatic bonds in diffuse clouds plausibly is a consequence of the equilibrium between destruction of C–H bonds via UV photons and cosmic rays and the rapid formation of C–H bonds by the collisions with abundant atomic H atoms. The C–H bond destruction rates for diffuse clouds environments are deduced from experiments of UV irradiation and 30-KeV He$^+$ ion bombardment of hydrocarbon grains. Competition between the formation and destruction results in saturation of C–H bonds in ~10^4 yr, i.e., short times compared to the lifetime of material in diffuse clouds (*Mennella et al.*, 2002, 2003).

Assessment of formation and destruction rates of C–H bonds in different CNM environments provides an explanation for the observed presence of the 3.4-μm feature in the diffuse medium and its absence from the dense medium (*Mennella et al.*, 2003). Between the two destruction mechanisms, UV photons dominate over cosmic rays in diffuse clouds while cosmic rays dominate over UV photons in dense molecular clouds. In molecular clouds, the same cosmic rays (1–100 MeV) that ionize the gas also break C–H bonds. C–H bonds are destroyed and not readily formed: C–H bond formation is stifled by the absence of H atoms in the gas (hydrogen is in H$_2$ molecules) and by ice mantles on grains that inhibit C–H bond formation. The C–H bond destruction rates in molecular clouds are estimated from the cosmic-ray ionization rates deduced from molecular cloud chemical models (section 4).

3.6. Gas-Phase Chemical Reactions in the Diffuse Clouds

Here the key reaction processes occurring in interstellar chemistry are summarized in the context of diffuse clouds (e.g., *van Dishoeck and Black*, 1988). Many of the basic chemical processes that operate in diffuse clouds also operate, to a greater or lesser extent, in other regions of the ISM and in protostellar environments. Additional processes are described below where appropriate. Diffuse cloud chemistry predominantly forms simple molecules as a direct consequence of the high UV flux, as shown in Fig. 1. In diffuse clouds most of the nitrogen is present as N^0 and N$^+$; in dense molecular clouds nitrogen is mostly in the form of N$_2$. Recent observations indicate the presence of several polyatomic species in diffuse clouds, the origin of which is uncertain at present (*Lucas and Liszt*, 2000).

Molecular hydrogen plays a key role in the gas-phase chemistry but cannot be produced from it since the low densities and temperatures of the diffuse clouds mean that three-body reaction pathways to H$_2$ are excluded. Such reactions can form H$_2$ in dense, high-energy environments such as the early universe or in the winds of novae and supernovae (*Rawlings et al.*, 1993). Instead, in diffuse clouds interstellar H$_2$ is produced when H atoms collide, stick, migrate, and react on the surfaces of dust grains (*Hollenbach and Salpeter*, 1971). This catalytic process releases about 4.6 eV of energy, so the H$_2$ formed is ejected into the gas and also contributes to heating the gas. Experimental studies of H$_2$ formation on surfaces analogous to those believed to be present in the diffuse clouds indicate that H$^+$ recombination appears to be slower (*Pirronello et al.*, 1997, 1999) than predicted by theory (*Hollenbach and Salpeter*, 1971). This may raise difficulties for producing H$_2$ in diffuse clouds. Nevertheless, the H/H$_2$ ratio that is obtained in diffuse clouds is a balance between H$_2$ formation on dust and UV photodestruction; the latter is controlled nonlinearly by H$_2$ self-shielding. Although NH formation by grain-surface chemistry has been suggested (*Crawford and Williams*, 1997), the role of surface catalysis for other interstellar molecules is highly uncertain at present.

The forms in which the major heavy elements are present are based largely on their ionization potentials and the energy spectrum of the interstellar UV field. Atoms with ionization potentials less than 13.6 eV are readily photoionized, i.e.,

$$C + \nu \rightarrow C^+ + e$$

Thus, whereas C, S, and various refractory metals (e.g., Fe, Mg, Na) are almost completely ionized, there are insufficient photons energetic enough to ionize O and N. Cosmic-ray particles can ionize H and H_2, and He atoms, to H^+, H_2^+, and He^+. The H_2^+ ions can subsequently react rapidly with H_2 molecules to produce H_3^+ and this ion readily transfers a proton to other atomic and molecular species with greater proton affinities than H_2. The electrons produced in these ionizations are lost through radiative recombination reactions such as

$$C^+ + e \rightarrow C + \nu$$

or in dissociative recombination reactions such as

$$H_3^+ + e \rightarrow H_2 + H, \text{ or } 3H$$

The charge transfer process

$$H^+ + O \rightarrow O^+ + H$$

followed by

$$O^+ + H_2 \rightarrow OH^+ + H$$

initiates a sequence of exchange reactions with H_2 that ends with the dissociative recombination of H_3O^+ producing OH and H_2O. Ion-molecule reactions also partially contribute to CO production in diffuse clouds through

$$C^+ + OH \rightarrow CO^+ + H$$

$$CO^+ + H_2 \rightarrow HCO^+ + H$$

$$HCO^+ + e \rightarrow CO + H$$

The precise value of the appropriate rate coefficient for H_3^+ recombination is at present controversial (*McCall et al.,* 2002). Exothermic ion-molecule processes typically proceed at the Langevin (collisional) rate. It should be noted, however, that a large population of PAH molecules can qualitatively alter the chemistry of interstellar clouds. In diffuse clouds, where a substantial abundance of PAHs is required to heat the gas through the photoelectric effect (e.g., *Bakes and Tielens,* 1994, 1998), positive atomic ions are destroyed more efficiently by PAHs than by electrons (*Lepp et al.,* 1988; *Liszt,* 2003). If a substantial PAH population exists in molecular clouds, then atomic and molecular cations are lost primarily through mutual neutralization with PAH anions (*Lepp and Dalgarno,* 1988).

Radiative association can initiate a limited hydrocarbon chemistry starting from

$$C^+ + H_2 \rightarrow CH_2^+ + \nu$$

and leading to CH and C_2. Neutral-neutral exchange reactions can also be important for producing simple molecules

$$C + OH \rightarrow CO + H$$

$$N + CH \rightarrow CN + H$$

The *in situ* production of complex molecules within diffuse clouds is inhibited by the efficient photodestruction of diatomic and triatomic molecules, e.g.,

$$OH + \nu \rightarrow O + H$$

Of course, other chemical reactions have to occur, and we now discuss these in the context of the so-called "CH+ problem" (*Williams,* 1992; *Gredel,* 1999). Chemical models of the diffuse interstellar medium can reproduce the observed column densities of many species (*van Dishoeck and Black,* 1988). However, to date no model has been able to reproduce the observed abundance of CH+ while obeying all the other observational constraints. The obvious formation reaction is highly endoergic by 4640 K

$$C^+ + H_2 \rightarrow CH^+ + H \qquad (1)$$

and will not proceed at normal diffuse clouds temperatures. When diffuse interstellar gas is heated and compressed by shock waves, many chemical reactions that are endothermic, or possess activation energy barriers, can occur in the post-shock gas (e.g., *Mitchell and Deveau,* 1983). For example, the reaction above (equation (1)) and

$$O + H_2 \rightarrow OH + H$$

$$C + H_2 \rightarrow CH + H$$

However, hydrodynamic shocks have not been shown to be able to resolve the CH+ issue without violating the constraints on, e.g., the observed OH abundance and the rotational populations of H_2 (e.g., *Williams,* 1992).

As interstellar clouds are magnetized and only partially ionized, slow shock waves can exhibit more complex structures than simple shock discontinuities (e.g., *Roberge and Draine,* 1990). The fact that some magnetohydrodynamic waves can travel faster than the sound speed in these plasmas means that there is a substantial difference in the velocities of the ions and neutrals. This effect gives rise to C-shocks in which all variables are continuous across the shock structure (see *Draine,* 1980). C-shocks also have been pro-

TABLE 2.   Ice composition* toward interstellar sources[†].

Molecule	W33A High-Mass Protostar	NGC7538 IRS9 High-Mass Protostar	Elias 29 Low-Mass Protostar	Elias 16 Field Star
H_2O	100	100	100	100
CO	9	16	5.6	25
CO_2	14	20	22	15
CH_4	2	2	<1.6	—
CH_3OH	22	5	<4	<3.4
H_2CO	1.7–7[‡]	5	—	—
OCS	0.3	0.05	<0.08	—
NH_3	15	13	<9.2	<6
C_2H_6	—	<0.4	—	—
HCOOH	0.4–2[‡]	3	—	—
OCN^-	3	1	<0.24	<0.4
HCN	<3	—	—	—

*Abundances by number relative to water ice.

[†]Adapted from *Ehrenfreund and Charnley* (2000).

[‡]Abundances with large uncertainties are shown as a range of values.

posed to produce CH+ (*Pineau des Forêts et al.,* 1986; *Draine and Katz,* 1986) but with mixed success when compared to observation (*Gredel et al.,* 1993; *Flower and Pineau des Forêts,* 1998; *Gredel et al.,* 2002).

Finally, while dissipation of interstellar turbulence occurs in shocks, it also occurs intermittently in small regions of high-velocity shear (*Falgarone et al.,* 1995), and the associated heating has been suggested as important for driving the reaction shown in equation (1) (*Joulain et al.,* 1998). In both cases, the high heating efficiency occurs because the energy available from dissipation of turbulence is degraded to thermal energy within a small volume. At present, the CH+ problem remains unsolved, but it is clear that solving the origin of CH+ will significantly impact the study of the physics and chemistry of diffuse clouds.

4.   PHYSIOCHEMICAL PROCESSES IN COLD DENSE MOLECULAR CLOUDS

4.1.   Overview

In this section we describe the gaseous and solid-state composition of dark, dense molecular clouds, the regions where high- and low-mass stars form. Considering an evolutionary chemical sequence originating from AGB envelopes and the diffuse ISM, molecular clouds are the point where interstellar chemistry starts to have direct relevance for cometary composition (e.g., *Langer et al.,* 2000). Specifically, molecular ice mantles (see Table 2) can form on siliceous and carbonaceous dust grains, and processing of these ices opens up many more chemical pathways (see Fig. 3).

Molecular clouds possess higher densities of gas and dust than diffuse clouds, their presumed precursors. Cosmic rays penetrate into the deepest cloud interiors and pro-

duce H_3^+, He+, and electrons that heat the gas. This results in a rich gas phase chemistry (*van Dishoeck et al.,* 1993). Dust grains shield the inner regions of the cloud from external UV radiation and most of the photons capable of dissociating H_2 (and other molecules) are absorbed at cloud surfaces. Hence, grain catalysis can convert almost all the available H to H_2 (section 3.6). As most of the gas phase C is present as CO, these clouds cool principally by molecular rotational emission from CO molecules excited by collisions with H_2 and He. The gas and dust temperatures are tightly restricted to lie in a range around 10 K (*Goldsmith and Langer,* 1978). At lower densities, the gas and dust are not thermally well coupled and depletion of coolant species may increase the gas temperature (*Goldsmith,* 2001). Although external UV photons do not significantly penetrate these clouds, there probably exists a weak UV flux throughout molecular cloud interiors. This flux derives from excitation of H_2 by energetic electrons produced in primary cosmic-ray impacts. The

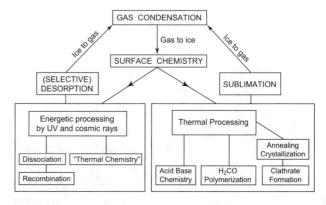

Fig. 3.   Physiochemical processes experienced by ice and gas in dense clouds (*Ehrenfreund and Charnley,* 2000).

subsequent UV emission spectrum can photodissociate and photoionize interstellar molecules (*Prasad and Tarafdar,* 1983), and can significantly influence the abundances of some species (e.g., atomic C) (*Gredel et al.,* 1987).

4.2. Gas-Phase Chemistry

Most of the chemical reactions described for diffuse clouds (section 3.6) also play a role in molecular cloud chemistry. Cosmic-ray ionizations drive a chemistry where ion-molecule and neutral-neutral reactions convert an appreciable fraction of the heavy elements to molecular forms. Thus, one finds that most C exists as CO, although a significant fraction of atomic C is present (C/CO ≈ 0.1). The dominant form of O is not well determined in these sources but is understood to be atomic, from chemical models and the observed lack of O_2 (*Goldsmith et al.,* 2000, 2002). Based on observed N_2H^+ abundances, almost all the available nitrogen is in N_2. Sulfur constitutes a puzzle, as its elemental depletion with respect to the diffuse ISM, and its major form, are unknown (e.g., *Charnley et al.,* 2001a; *Scappini et al.,* 2003). Many S-bearing molecules are detected and this suggests that atomic S is probably the major repository of this element. Molecules containing refractory metals (e.g., Fe, Mg, Na) are not present and metals are understood to be completely depleted on/in dust grains.

Proton transfer reactions produce many molecular ions and observations of these (HCO^+, DCO^+) can be used to estimate the electron density or the abundances of undetectable molecules (e.g., N_2H^+ to trace N_2). It appears that many neutral-neutral reactions have significant rate coefficients at low temperatures (*Chastaing et al.,* 2001). Reactions involving various hydrocarbons (e.g., C_2H_2 acetylene) with C atoms and with the CN radical are particularly important in producing many of the long carbon-chain compounds (*Cherchneff and Glassgold,* 1993), e.g., cyanoacetylene

$$CN + C_2H_2 \rightarrow HC_3N + H$$

These reactions can generate many of the higher cyanopolyenes [i.e., HC_5N, HC_7N (*Dickens et al.,* 2001), HC_9N, and $HC_{11}N$], as well as various hydrocarbon chains (e.g., C_4H, C_6H, C_8H), that are detected in molecular clouds, such as TMC-1 in Taurus (*Pratap et al.,* 1997; *Markwick et al.,* 2000) (Fig. 1). Several important species are known in star-forming clouds but are not shown in Fig. 1 (e.g., atomic O and CO_2). Some species, undoubtedly present in molecular clouds, are difficult or impossible to detect directly in sources like TMC-1 (e.g., S, O, N_2, CO_2 and C_2H_2). It is worth noting that significant compositional differences are found among molecular clouds. For example, L134N does not contain large abundances of carbon-chain molecules (*Pagani et al.,* 2003; *Dickens et al.,* 2000) and may be more typical than TMC-1 (*Dickens et al.,* 2001).

The extremely cold temperatures in these dense clouds mean that molecular zero-point energies can be an important factor in the gas-phase kinetics. Molecular clouds are there-

fore regions where molecules exhibiting substantial fractionation of isotopes of H, C, and N can form. Deuterium can become enhanced in H_3^+ since the reverse of the reaction

$$H_3^+ + HD \rightarrow H_2D^+ + H_2$$

is very slow at 10 K. Similar reactions also occur for some hydrocarbon ions such as CH_3^+ and CH_5^+. The H_2D^+ ions formed can initiate the distribution of D atoms throughout the molecular chemistry (e.g., *Millar et al.,* 1989, 2000). Isotopic exchange involving $^{13}C^+$ and ^{12}CO leads to enhancements of ^{13}C in CO (*Langer et al.,* 1984) and a general decrease of $^{13}C/^{12}C$ in other molecules. Chemical fractionation of ^{15}N also occurs in ion-neutral reactions (*Terzieva and Herbst,* 2000) but the highest $^{15}N/^{14}N$ ratios theoretically possible in interstellar molecules may require special conditions apart from very low temperatures [such as CO depleted but N_2 not depleted in the gas phase (*Charnley and Rodgers,* 2002)]. The extreme sensitivity of these fractionation reactions to the gas temperature is further strong evidence of the importance of cold cloud chemistry for understanding the origin of the enhanced isotopic fractionation found in primitive solar system material such as comets and meteorites.

4.3. Gas-Grain Interactions

Siliceous and carbonaceous micrometer- and submicrometer-sized dust particles that are produced in the outflows of late-type AGB stars provide a catalytic surface for a variety of reactions to occur when they are dispersed throughout molecular clouds. Atoms and molecules strike and stick to the surfaces of these cold dust grains and this process leads to the formation of an amorphous, mixed, molecular ice mantle covering the siliceous/carbonaceous grains. Table 2 shows the observed composition of interstellar ices. The forming ice mantles consist of molecules directly accreted from the gas (e.g., CO) and molecules formed through chemical reactions at about 10 K. These ices are susceptible to various kinds of energetic processing: cosmic-ray impacts, heating near protostars, and perhaps UV photolysis.

At typical molecular cloud densities, particles are accreted from the gas at the rate of about one per day. The sticking coefficients for most of these species at 10 K are calculated to be close to unity (e.g., *Leitch-Devlin and Williams,* 1985). On the surface, heavy particles such as CO are relatively immobile. Atoms can diffuse and react with the immobile species, as well as among themselves. Hydrogen atoms can rapidly explore the surface by quantum-mechanical tunnelling. Experiments show that, once a monolayer has formed, quantum diffusion by H atoms is very rapid (*Manico et al.,* 2001). An H atom can therefore scan the entire surface to find any available co-reactants. Heavy atoms (e.g., C and O) can diffuse by thermal hopping (e.g., *Tielens and Allamandola,* 1987). Slowly, the accreted atoms react chemically, are converted to various molecules, and as a result, ice mantles are formed.

Unabated accretion, i.e., in the absence of an efficient means of returning molecules to the gas, would completely remove the heavy element component from the molecular gas on an accretion timescale of $3 \times 10^9 \, n_H^{-1}$ yr, i.e., about 10^5 yr at a density of $n_H \approx 10^4 \, cm^{-3}$ (e.g., *Brown and Charnley*, 1990). Hence, if molecular clouds are dynamically much older than this, there must be some desorption from the grain surfaces to prevent this "accretion catastrophe." However, the precise nature of the putative desorption mechanism (or mechanisms) is not yet unambiguously identified. Several candidate mechanisms have been proposed, including evaporation following grain heating by cosmic rays, mantle explosions, and ejection upon molecule formation (see *Willacy and Millar*, 1998). As molecular clouds are also regions of star formation, dynamical events also have been proposed for removing and reforming the molecular ices within the accretion timescale; these include sputtering in low-velocity shock waves (*Bergin et al.*, 1998; *Charnley et al.*, 2001a) and grain-grain collisions induced by clump collisions or wave motions (*Markwick et al.*, 2000; *Dickens et al.*, 2001).

The "accretion catastrophe" is most evident in a dynamical scenario where molecular clouds are long-lived ($\sim 10^7$ yr), and where low-mass star formation is controlled quasistatically by ambipolar diffusion, the relative drift between ions and neutrals in partially ionized plasma. Ambipolar diffusion leads to interstellar cloud material losing magnetic support against gravity on characteristic timescales of $\sim 5 \times 10^6$ yr that are determined by the electron density. The electron density is set by the ionization rate that is, in turn, determined from the chemistry of molecular cloud material (e.g., *Shu et al.*, 1987, 1993). An enormous effort has been made over the past two decades to reconcile observed molecular abundances and the results of chemical models with such long cloud lifetimes, i.e., to reconcile the apparent "chemical youth" of objects many millions of years old. However, when viewed in the context of the more recent dynamical scenario (section 1.4) where molecular cloud formation and star formation are controlled by supernovae-driven supersonic turbulence (e.g., *Mac Low and Klessen*, 2003; *Larson*, 2003), and the shorter cloud lifetimes ($\sim 1-3 \times 10^6$ yr) inferred from observations (*Elmegreen*, 2000; *Hartmann et al.*, 2001), the accretion catastrophe problem may be resolvable. In this picture, most interstellar cloud materials are in a low-density, primarily atomic phase as diffuse clouds and hence can have accretion timescales longer than the molecular cloud lifetimes. The long lifetimes of diffuse cloud material will not greatly affect the molecular composition of molecular clouds.

Dense prestellar cores evolve roughly on free-fall timescales and have shorter lifetimes than the molecular clouds in which they lie. The fact that substantial selective depletions (e.g., CO relative to N_2) are observed in these regions (*Bergin et al.*, 2002) indicates that a viable surface desorption mechanism is still needed. However, this mechanism eventually should be overwhelmed in the most dense, coldest regions. It is very likely that the central regions of prestellar cores pass through a phase where all the available

heavy atoms have been incorporated into molecular ices. Short cloud lifetimes (section 1.4), and the inference that molecular clouds are indeed chemically young, could also provide a natural explanation for other astrochemical problems, such as the high C^0/O_2 ratio that is observed to be widespread in galactic clouds (*Goldsmith et al.*, 2000). However, the existence of small-scale (~ 0.01 pc) molecular differentiation in dense clouds still presents a major challenge for chemical models (*Dickens et al.*, 2001; *Takakuwa et al.*, 2000).

Observations of dense cores within molecular clouds show strong evidence for substantial depletions (*Bergin et al.*, 2002); this suggests whatever desorption mechanism operates at lower densities is overwhelmed in dense cold cores.

4.4. Observations of Astronomical Ices

The protostars that form in molecular clouds are natural IR background sources, which can be utilized to perform solid-state spectroscopy on the column of gas along the line-of-sight. Groundbased observations performed since the 1970s demonstrated the presence of abundant water ice as well as CO ice (e.g., *Whittet*, 1993) (see Fig. 2a). The absorption features of water ice and silicates observed at 3 μm and 10 μm, respectively, dominate the spectrum toward stars in dense molecular clouds. Those bands are broad and often saturated and are the reason why the inventory of ices will never be complete. Many species, in particular NH_3 and CH_3OH as well as species with a C=O group, are either blended or masked by those dominant spectral bands. This strongly biases abundance determinations.

Other species remained undetected until sophisticated IR satellites provided the full range of spectral data from 1 to 200 μm. The ISO provided us with very exciting spectra that allowed us to compile an inventory of interstellar ice species and measure their abundances in various interstellar environments [for reviews on ISO data, see *Ehrenfreund and Schutte* (2000) and *Gibb et al.* (2000)]. ISO identified two distinct ice layer compositions: hydrogen-rich ices ("polar" H_2O-dominated with traces of CO, CO_2, CH_4, NH_3, CH_3OH, HCOOH, and H_2CO) and the more volatile "apolar" CO-dominated ices (with small admixtures of O_2 and N_2). The build up of apolar ice layers occurs at successively lower temperatures than polar ice layers on grain mantles. Abundances and inventories of ice species toward low- and high-mass protostars can be found in Table 2.

The solid CO band at 4.67 μm can be well studied by groundbased observations. Laboratory studies reveal that the CO band profile is strongly influenced by neighboring species in the ice and acts therefore as a good tracer of the overall grain-mantle composition (*Ehrenfreund et al.*, 1996). Recent high-resolution measurements of CO using 8-m class telescopes reveal a multicomponent structure. *Boogert et al.* (2002a) observed the M-band spectrum of the class I protostar L1489 IRS in the Taurus molecular cloud. The CO band profile showed a third component apart from CO in "polar" ices (CO mixed with H_2O) and CO in "apolar" ices (see above). The high-spectral-resolution observations show that the apolar component has two distinct components, likely due to pure CO and CO mixed with CO_2, O_2, and/or N_2. *Pontoppidan et al.* (2003a) conclude from recent ground-based observations of more than 40 targets that the CO band profile can be fitted with a three-component model that includes pure CO, CO embedded in water ice, and a third component that is attributed to the longitudinal component of the vibrational transition in pure crystalline CO ice (appearing when the background source is linearly polarized). This three-component model applied in varying ratios provides an excellent fit to all bands observed, and indicates a rather simple universal chemistry in star-forming regions with 60–90 % of the CO in a nearly pure form (*Pontoppidan et al.*, 2003a). The freezeout of CO in a circumstellar disk around the edge-on class I object CRBR 2422.8-3423 has been recently observed with the ISAAC instrument on the Very Large Telescope (VLT) by *Thi et al.* (2002) with the highest abundance observed so far.

In high-spectral-resolution spectra, *Boogert et al.* (2002a) detected toward the massive protostar NGC 7538 IRS 9 a narrow absorption feature at 4.779 µm (2092.3 cm^{-1}) attributed to the vibrational stretching mode of the ^{13}CO isotope in pure CO icy grain mantles. This is the first detection of ^{13}CO in icy grain mantles in the ISM. A ratio of $^{12}CO/^{13}CO = 71 \pm 15$ (3σ) was deduced, in good agreement with gas-phase CO studies ($^{12}CO/^{13}CO \approx 77$) and the solid $^{12}CO_2/^{13}CO_2$ ratio of 80 ± 11 found in the same line of sight (*Boogert et al.*, 2002b). The ratio is confirmed by the ratio observed in the low-mass star IRS 51, namely a $^{12}CO_2/^{13}CO_2$ ratio of 68 ± 10 (*Pontoppidan et al.*, 2003a).

The abundance of NH_3 in interstellar grain mantles has been a hotly debated subject for a long time (e.g., *Gibb et al.*, 2001). The reported abundances for NH_3 relative to H_2O range from 5–15%. The most recent VLT data confirm abundances on the lower end of this scale. *Dartois et al.* (2002) report ~7% for the NH_3/H_2O ratio toward the evolved massive protostars GL 989 and GL 2136, derived from a band at 3.47 µm attributed to ammonia hydrate. *Taban et al.* (2003) observe the 2.21-µm band of solid NH_3 and provide only an upper limit toward W33A; their limit is ~5% for the NH_3/H_2O ratio.

The lines of sight toward star-forming regions consist of regions with strongly varying conditions. Temperatures are low in molecular cloud clumps but can be very high in hot core regions close to the star, where ices are completely sublimated. The spectra that sample the different regions in the line of sight toward star forming regions need to be deconvolved using additional information about the gas phase composition and geometry of the region. For example, Fig. 2a shows spectra of material in the line of sight to a massive protostar that intercepts both cold cloud ices and UV-illuminated PAHs. In the past few years, ground-based telescopes of the 8–10-m class have allowed us to observe some ice species with unprecedented spectral resolution. The spectral resolution also enables us to study lines of sight toward low-mass protostars, which usually were too faint to be observed with the ISO satellite. VLT and KECK II, equipped with the ISAAC and NIRSPEC spectrographs, respectively, have delivered exciting new data on ice species

such as CO (and ^{13}CO), NH_3, and CH_3OH (*Thi et al.*, 2002; *Boogert et al.*, 2002b; *Dartois et al.*, 2002; *Pontoppidan et al.*, 2003a,b; *Taban et al.*, 2003). The CH_3OH abundance relative to water ice measured in star-forming regions lies between 0% and 30%. Two high-mass protostars have been identified with very large CH_3OH abundances, namely up to 25% relative to water ice (*Dartois et al.*, 1999). New VLT data of the 3.52-µm band (C-H stretching mode of CH_3OH) also show abundances up to 20% toward low-mass stars in the Serpens cloud (*Pontopiddan et al.*, 2003b).

4.5. Solid-State Chemical Reactions on Dust

Astrochemical theories suggest that only a few classes of reactions appear to be necessary to form most of the molecules observed in ices (*Herbst*, 2000). Several exothermic H atom additions to C, O, N, and S produce methane (CH_4), water, ammonia (NH_3), and hydrogen sulphide (H_2S). Atom additions to closed-shell molecules such as CO possess substantial activation-energy barriers. However, due to their small mass, H and D atoms could saturate these molecules by tunnelling through this barrier (*Tielens and Hagen*, 1982; *Tielens*, 1983). Hydrogenation of CO has been suggested to be the source of the large abundances of solid methanol (CH_3OH) seen in many lines of sight toward protostars (e.g., *Tielens and Charnley*, 1997; *Charnley et al.*, 1997; *Caselli et al.*, 2002), and the enormous D/H ratios observed in both formaldehyde (H_2CO) and methanol in protostellar cores where CO ices have been sublimated to the gas phase (*Loinard et al.*, 2000; *Parise et al.*, 2002). Additionally, CO_2 could form by O atom addition to CO at 10 K (*Tielens and Hagen*, 1982). Based on these mechanisms many large organic molecules may form on grains (e.g., *Charnley*, 1997b).

Thermal processing close to the protostar leads to molecular diffusion, structural changes within the ice matrix, and subsequently to sublimation. Ice segregation, and possibly even clathrate formation, has been observed in dense clouds (*Ehrenfreund et al.*, 1998). Simple thermal processing of ice mixtures can itself produce new molecules. In the laboratory, it has been shown that the heating of ice mixtures containing formaldehyde [H_2CO, also called methanal) and ammonia (NH_3) results in polymerization of the formaldehyde into polyoxymethylene (POM, (-CH_2-O-)$_n$] (*Schutte et al.*, 1993). Extensive and detailed laboratory studies have been undertaken to study the chemistry of interstellar ice analogs (*Cottin et al.*, 1999). For many years these studies were restricted to either UV photolysis (e.g., *Allamandola et al.*, 1997) or proton irradiation (e.g., *Moore and Hudson*, 1998) of bulk ices; in both cases the effects of warming the ices are usually considered ("thermal chemistry," Fig. 3). The strongly attenuated UV flux in such dense environments strongly limits photolysis processes. In contrast, cosmic rays can penetrate dense molecular clouds and effect the structure and composition of the ices, including sputtering of icy grain mantles (*Cottin et al.*, 2001).

Atom addition reactions on grain surfaces are necessary to initially form water ice mantles, as gas phase ion-molecule reactions cannot produce large quantities of water. However, it is only recently that atom reactions on analog surfaces have come to be studied in detail (e.g., *Hiraoka et al.,* 1998; *Pironello et al.,* 1999; *Watanabe and Kouchi,* 2002). From the existing laboratory data it is possible to make comparisons of the relative efficiency of each chemical process (photolysis, radiolysis, and atom additions) in producing the major observed mantle molecules from CO. Both radiolysis (i.e., proton irradiation) and photolysis easily produce CO_2. Recent experiments show that, although oxidation of CO by O atoms has a small activation energy barrier, this reaction can form CO_2 (*Roser et al.,* 2001). However, in these experiments CO_2 is only produced efficiently when the reacting CO and O atoms are covered by a layer of H_2O molecules and warmed. This appears to increase the migration rates in the ice and suggests that the longer migration times available on grain surfaces in dense clouds, times much longer than can reasonably be studied in the laboratory, may make this reaction the source of the observed CO_2. Also, cold H additions in ices containing acetylene (C_2H_2) have been proposed as the origin of the ethane (C_2H_6) found in comets (*Mumma et al.,* 1996) and experiments support this idea (*Hiraoka et al.,* 2000). As yet, ethane is not detected in the ISM but these experiments suggest it has an interstellar origin (see *Ehrenfreund et al.,* 2004).

Ultraviolet photolysis of ice mantles produces radicals that may migrate and react (see Fig. 3). However, experiments show that UV processing of H_2O/CH_4 mixtures is very inefficient at producing methanol (CH_3OH) (*d'Hendecourt et al.,* 1986). In fact, experiments show that methanol is readily decomposed to formaldehyde under photolysis (*Allamandola et al.,* 1988). By contrast, radiolysis of dirty ice mixtures can produce methanol in astronomically interesting amounts (*Hudson and Moore,* 1999). Initial reports that cold H atom additions to CO could produce methanol at low temperatures were positive (*Hiraoka et al.,* 1994, 1998) but were not supported by more recent experiments by the same group (*Hiraoka et al.,* 2002). However, recent laboratory experiments (*Watanabe and Kouchi,* 2002; *Watanabe et al.,* 2003) strongly argue for CH_3OH production via this mechanism.

Experiments involving H[+] irradiation (radiolysis) are designed to simulate cosmic-ray bombardment (see Fig. 3). Radiolysis proceeds by ionization of H_2O to H_2O^+, from which a proton then transfers to water to form protonated water, H_3O^+. Subsequent dissociative electron recombination of protonated water produces a population of energetic, reactive H atoms and hydroxyl (OH) molecules in the ice (*Moore and Hudson,* 1998). Experiments show that radiolysis of H_2O/CO ice mixtures can produce high abundances of H_2CO and CH_3OH (*Hudson and Moore,* 1999), although formic acid (HCOOH, also called methanoic acid) is overproduced relative to the $HCOOH/CH_3OH$ ratio observed in interstellar ices. Radiolysis of ice mixtures containing CO and C_2H_2 also produces ethane, as well as several other putative mantle molecules such as CH_3CHO and C_2H_5OH (*Hudson and Moore,* 1997; *Moore and Hudson,* 1998).

One interstellar molecule that certainly originates in energetically processed ices is the carrier of the so-called "XCN" 4.62-μm absorption feature, observed in the diffuse clouds (*Pendleton et al.,* 1999), toward protostars (e.g., *Tegler et al.,* 1995), and, most recently, in the nearby dusty starburst AGN galaxy NGC 4945 (*Spoon et al.,* 2003). Laboratory experiments indicate that "XCN" is produced both by photolysis (*Lacy et al.,* 1984; *Bernstein et al.,* 2000) and by radiolysis (*Hudson and Moore,* 2000; *Palumbo et al.,* 2000). At present, the best candidate for the identity of this carrier appears to be the OCN^- ion, formed in a solid-state acid-base reaction between NH_3 and isocyanic acid (HNCO) (see *Novozamsky et al.,* 2001, and references therein).

4.6. Theoretical Modeling of Solid-State Reactions

There have been many attempts to model low-temperature grain-surface chemical reactions (e.g., *Allen and Robinson,* 1977; *Tielens and Charnley,* 1997; *Herbst,* 2000). Heterogeneous catalytic chemistry on grain surfaces occurs primarily through the Langmuir-Hinshelwood mechanism. Here reactive species arrive from the gas, stick to the surface, and then migrate until they encounter another species with which they can react (section 4.3); the product molecule may either remain on the surface or be desorbed. In deterministic theoretical models it is relatively straightforward to quantitatively treat the accretion and desorption of surface species (e.g., *Charnley,* 1997c). However, to correctly model surface diffusion and reaction, a fully stochastic treatment involving solution of the associated master equation for the surface populations is necessary (*Charnley,* 1998). A number of recent papers have modeled surface chemistry in this way, although the adopted methods of solution have differed considerably (*Charnley,* 2001; *Biham et al.,* 2001; *Green et al.,* 2001; *Stantcheva et al.,* 2002; *Caselli et al.,* 2002). However, in marked contrast to the amount of effort expended in trying to model surface reactions, there has been almost no theoretical modeling of the kinetics associated with the photolysis or radiolysis of the bulk ice mantle (however, see *Ruffle and Herbst,* 2001; *Woon,* 2002).

5. PHYSIOCHEMICAL PROCESSES IN DENSE STAR-FORMING CORES

5.1. Overview

Understanding the physics and chemistry of low-mass star formation is most relevant for solar system studies. Modeling well-studied sources at various evolutionary stages enables us to understand how much processing interstellar material experienced as it became incorporated into the nascent protosolar nebula (*van Dishoeck and Blake,* 1998; *Ehrenfreund and Charnley,* 2000).

Star formation occurs rapidly in molecular clouds, probably within a million years (section 1.4). The mass scale and the occurrence of isolated stars, binaries, stellar clusters, and associations are determined by the local and global competition between self-gravity and the turbulent velocity field in interstellar clouds. Massive stars may form from a single isolated core (*Stahler et al.,* 2000), or may grow in stellar clusters through coalescence of two or more low-mass and intermediate-mass stars (*Clark et al.,* 2000; *Bonnell and Bate,* 2002). Formation of single and binary low-mass stars can proceed from collapse and fragmentation of less-massive regions of lower density (*Mac Low and Klessen,* 2003).

The chemistry of its surroundings is dramatically influenced by the protostar. A central aspect in understanding the chemical evolution prior to and following star formation appears to be the prestellar accretion of molecules onto grains and their subsequent removal, which occurs in a hot (or warm) dense core of the molecular cloud surrounding the accreting protostar. Due to the higher intrinsic line fluxes, and stronger IR emission, most previous studies of protostellar gas and solid-phase compositions have focused on high-mass star-forming hot cores, such as Orion A and Sagittarius B2 (*Johansson et al.,* 1984; *Cummins et al.,* 1986; *Blake et al.,* 1987; *Turner,* 1991; *Smith et al.,* 1989). However, it is now becoming clear that low-mass protostars also pass through a warm core phase (e.g., *Schoier et al.,* 2002; *Cazaux et al.,* 2003). In this section we present an overview of the chemistry, and associated molecular morphology that develops around an accreting protostar.

5.2. Hot Cores Around Massive Protostars

Hot molecular cores are most commonly identified with the earliest phases of massive star formation (*Churchwell,* 2002). Hot cores contain a young protostar embedded within its dense cocoon of gas and dust and the most sophisticated chemical models have been developed to account for the observations of these massive hot cores (*Doty et al.,* 2002; *Rodgers and Charnley,* 2003). The elevated temperatures and densities in these small regions (<0.1 pc, see Table 1) produce a chemical composition markedly distinct from other regions of the ISM as shown in Fig. 1. Note that in Fig. 1 the abundances listed for IRAS 16293 are for the warm inner component as determined by *Schoier et al.* (2002) except for CO, HCO+, CN, HCN, C_2H, C_3H_2, and CS. The C_2H_5CN entry and the remarkably high abundances of $HCOOCH_3$, CH_3OCH_3, and HCOOH are taken from *Cazaux et al.* (2003). Many large organic molecules are also detected in hot molecular cores but have been omitted from Fig. 1; these include ethanol, glycolaldehyde, ethylene glycol, acetone, acetic acid, and glycine (*Ohishi et al.,* 1995; *Hollis et al.,* 2000, 2002; *Snyder et al.,* 2002; *Remijan et al.,* 2003; *Kuan et al.,* 2003). It is now understood that hot core chemistry is mainly the result of the volatile contents of the ice mantles being deposited into the gas via desorption. Evidence for this are the high abundances of saturated molecules (water, ammonia, methane) and enhanced D/H ratios, both of which could only have been set

in much colder conditions (section 4). Similar chemical processes also can be expected to play a key role in low-mass star-formation environments, and so here a brief overview of chemistry in massive star formation is given.

The basic chemistry of massive star formation can be best described by distinguishing between processes that occur in a cold prestellar phase and those that occur in a hot phase after a protostar has formed (*Brown et al.,* 1988). The cold phase includes the quasistatic chemical evolution in molecular clouds (section 4) and eventually an isothermal gravitational collapse to a small, cold (10 K), dense core. Rapid protostellar heating of the core to hot core temperatures (100–300 K) induces thermal evaporation (sublimation) of icy grain mantles and hot core chemistry.

During the isothermal collapse, gas-grain collision times become so short that most of the heavy atoms and molecules in the center of the cloud core are in the solid state (section 4). There is recent observational evidence for the disappearance of CO and N_2 molecules in the central region of dense cores (e.g., *Bergin et al.,* 2002; *Bacmann et al.,* 2002; *Caselli et al.,* 2003). Subsequent protostellar heating returns the volatile ices to the gas phase. Radioastronomy permits the ice composition to be studied in more detail when in the gas phase than is possible by direct IR absorption studies of the solid phase using field stars behind molecular clouds, or toward embedded protostars (e.g., *Charnley et al.,* 2001b). For example, the recently discovered gas-phase organics, including vinyl alcohol (*Turner and Apponi,* 2001), glycolaldehyde (*Hollis et al.,* 2000), and ethylene glycol (*Hollis et al.,* 2002), are inferred to be present in the solid-phase on grains only in trace amounts.

In the collapse phase, the warming and subsequent release of the grain material can be used to constrain the nature of grain-surface reactions. Identifying the origin of specific molecules (and classes of molecules) can provide important information on the composition of the prestellar core and on the nature of grain-surface chemistry. There are three ways in which the molecules observed in hot cores can originate: First, molecules produced in the cold prestellar phase (e.g., CO) that accreted on grains and formed ices can simply be returned to a hot environment. Second, other atoms and radicals can stick to grains and take part in grain-surface reactions to form mixed molecular ices (section 4). However, not all the complex molecules observed in hot cores are products of grain-surface chemistry. It transpires that a highly transient chemistry can also occur in the gas during the hot core phase, and therefore the third formation pathway for hot core molecules is *in situ* molecule formation (*Charnley et al.,* 1992; *Caselli et al.,* 1993). For example, alkyl cation transfer reactions involving surface-formed alcohols can produce many larger organic molecules, such as ethers (e.g., *Charnley et al.,* 1995; *Charnley,* 1997b). Hot core chemistry is more accurately described than that of warm cores because the initial conditions (i.e., the grain mantle composition) are constrained better, and because the evolution lasts less than ~10^5 yr or so.

Between individual cores, a strong chemical differentiation is observed between O-bearing and N-bearing molecules. This chemical differentiation is modeled and attributed to differences in temperature and evolutionary state of the core: Hotter cores tend to have higher abundances of N-bearing molecules (*Rodgers and Charnley,* 2001). Models of the hot-phase chemistry have been constructed for the chemistries of second-row elements (P, Si, S), which are particularly sensitive to the temperature in the hot gas (*Charnley and Millar,* 1994; *Mackay,* 1995; *Charnley,* 1997a). Observations of S-bearing molecules may be particularly useful as molecular clocks for star-formation timescales (*Hatchell et al.,* 1998; *Buckle and Fuller,* 2003; *Wakelam et al.,* 2003). Ice mantles also can be sputtered in shock waves, and postshock chemistry could play a role in either initiating, or contributing to, hot core chemical evolution (*Charnley and Kaufman,* 2000; *Viti et al.,* 2001). ISO SWS observations of solid and gaseous CO_2 toward many protostars indicate that shock processing may be common (*Boonman et al.,* 2003). Finally, hot cores surrounding massive protostars eventually evolve into ultracompact HII regions (*Churchwell,* 2002) and photodissociation regions where the strong UV field and X-rays rapidly destroy molecules (*Tielens and Hollenbach,* 1985; *Maloney et al.,* 1996).

5.3. Warm Cores Around Low-Mass Protostars

At the heart of a low-mass star-forming core lie the protostar and its surrounding disk, from which comets, asteroids, and planets eventually form. Observations show that many of the chemical characteristics seen in hot cores — small-scale differentiation, shock tracers, high D/H ratios, and the presence of putative mantle organic molecules — are also present toward low-mass sources (Fig. 1) (*van Dishoeck et al.,* 1993; *McMullin et al.,* 1994; *Langer et al.,* 2000; *Loinard et al.,* 2001; *Schoier et al.,* 2002; *Cazaux et al.,* 2003). Many of the chemical processes described above (section 5.2) should come into play in determining the overall chemical morphology and evolutionary state of these regions. Hence, elucidating the major physiochemical processes that occur around massive protostars may allow us to determine the likely chemical structure of the low-mass protostellar core in which the Sun formed. Figure 4 shows schematically the main physical regions in the environment of a low-mass accreting protostar (see *Mumma et al.,* 1993; *Lunine et al.,* 2000). Several regions exist where the local physical conditions will drive a specific chemical evolution. Interstellar materials from the ~0.1-pc dense core, and in particular from the innermost core heated by the protostar, may become incorporated into the protoplanetary disk after passing through the accretion shock at the boundary between the disk and the infalling core. However, the innermost core regions including accretion shock, which are most important for understanding the processing/connection of interstellar and cometary materials, have not yet been observed directly in great detail; this will become possible

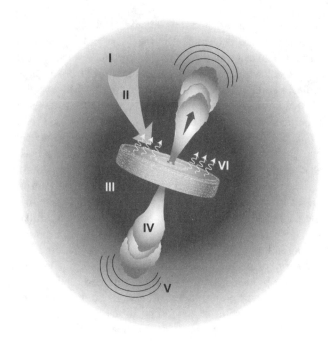

Fig. 4. Chemistry around a low-mass protostar. In the environment of a low-mass protostar, physical processes determine the chemistry of distinct regions in the protostellar core. Distinct physical regions are characterized by different molecular tracers and include Region I — cold cloud, HCO^+ and N_2H^+; Region II — infall, N_2H^+; Region III — sphere of thermal influence, $(CH_3)_2O$; Region IV — bipolar outflow (CO); Region V — wind-cloud bow shock, H_2S, SO, SO_2, SiO, CH_3OH, D_2CO; and Region VI — accretion shock at the disk surface, CS. Details of the chemical processes active in each region are discussed in the text (section 5). Artwork by J. Woebcke.

with the Atacama Large Millimetre Array (ALMA) (*Wootten,* 2001). Below we discuss the physiochemical properties of each of the six regions identified in Fig. 4 around a low-mass protostar.

Region I: Cold cloud. The chemistry in this region is essentially that which is outlined for cold, dense molecular clouds (section 4). Molecular tracers of the cold cloud include HCO^+ and N_2H^+.

Region II: Infall. The initial density structure and the precise details of the subsequent collapse are the subject of current debate (*Larson,* 2003; *Mac Low and Klessen,* 2003). It is clear, however, that when gravity dominates over support by thermal, turbulent, and magnetic pressures, a dynamic core collapse will occur and cold material from the infalling envelope will become incorporated into the protostar and its accretion disk. These infall motions can be detected spectroscopically from molecular line observations (*Evans,* 1999; *Myers et al.,* 2000). Chemical models utilizing various collapse scenarios (*Shu,* 1977; *Larson,* 1969; *Penston,* 1969) have been constructed (*Rawlings et al.,* 1992; *Aikawa et al.,* 2001, 2003; *Rodgers and Charnley,* 2003). A common molecular tracer of infall is N_2H^+.

Region III: The "sphere of thermal influence." As material approaches the protostellar disk, the accretion luminosity of the disk heats the infalling dust and gas. The dust and gas temperatures at any radii are controlled by the mass accretion physics (*Adams and Shu*, 1985; *Ceccarelli et al.*, 1996) and this should therefore be reflected in the chemical evolution of the core. The chemistry is strongly radially dependent in the protostellar cocoon since the molecular desorption rate depends exponentially on the local dust temperature, as well as on the abundances and binding energies of molecules present in the grain mantles (see *Ehrenfreund et al.*, 1998; *Rodgers and Charnley*, 2003). As interstellar gas and dust falls down toward the protostar, initially only the most volatile molecules are efficiently desorbed (CO, N_2, O_2, CH_4) and the chemistry is similar to that of molecular clouds (*Brown and Charnley*, 1991). Eventually, those remaining molecules with binding energies less than water and ammonia (e.g., CH_3OH, H_2CO, C_2H_2) are desorbed. In the inner regions, all the volatile mantle molecules have been removed and water, methanol, and ammonia are present in the gas at high abundances (see *van der Tak et al.*, 2000). In these innermost regions the chemical evolution should most strongly resemble that of massive hot cores (section 5.2). A molecular tracer of the sphere of thermal influence is $(CH_3)_2O$.

Region IV: Bipolar outflow. Accretion of infalling interstellar material onto the protostellar disk appears to be intimately connected with the presence of atomic jets and bipolar molecular (i.e., CO) outflows (*Konigl and Pudritz*, 2000). Low-velocity bipolar flows (less than about 25 km s^{-1}) are believed to be molecular gas entrained from the ambient cloud by the higher velocity jets (typically 150–400 km s^{-1}). The jets themselves are probably driven through magnetohydrodynamic processes in the accretion disk (e.g., *Konigl and Pudritz*, 2000; *Shu et al.*, 2000). Although hydrogen in the jets is primarily atomic, some molecule formation can proceed within jets (*Glassgold et al.*, 1991) and this, for example, could be the origin of some of the high-velocity CO in the outflows. Alternatively, as ambient cloud material is entrained, atomic material from the jet can turbulently mix with surrounding molecular gas and an active chemistry can occur along the jet edge (*Taylor and Raga*, 1995).

Region V: Wind-cloud bow shock. Bipolar outflows drive strong shock waves into the surrounding natal cloud material. The associated high temperatures and compression permit many endothermic chemical reactions to occur. Nonthermal sputtering of icy grain mantles by neutral atoms of He and H and H_2 removes icy grain mantles and erodes refractory grain cores (*Draine et al.*, 1983; *Flower and Pineau des Forêts*, 1994). Silicon-bearing molecules are observed to be highly depleted in cold molecular clouds (*Herbst et al.*, 1989) so common chemical signposts of shock activity in star-forming regions are the greatly enhanced abundances of gaseous SiO, produced by the sputtering of silicate grains (*Schilke et al.*, 1997; *Garay et al.*, 2002). The shocked outflow lobes of low-mass protostars,

such as L1157, are observed to exhibit a particularly rich chemical composition — a mixture of sputtered solid material and the products of high-temperature reactions in the postshock gas (*Bachiller et al.*, 2001). Molecular tracers of the wind-cloud bow shock include H_2S, SO, SO_2, CH_3OH, D_2CO, and, as mentioned above, SiO.

Region VI: Disk accretion shock. Gravitational collapse of a rotating core will generally lead to the formation of an accretion disk (*Cassen and Moosman*, 1981; *Terebey et al.*, 1984; *Boss*, 2004). Infalling material is decelerated in an accretion shock at the disk surface where interstellar chemistry effectively ends. Shock chemistry and other processes associated with disk accretion begin to dominate the chemistry of the material first entering the nebula (*Lunine et al.*, 1991; *Chick and Cassen*, 1997). *Neufeld and Hollenbach* (1994) have modeled the disk accretion shock as a dissociative "J-shock" and determined the regions of the disk where various interstellar materials (refractory metals, refractory and volatile organics, and ice) would be vaporized. The smallest accretion shock speeds and preshock densities favor the survival of the most volatile materials; this occurs in the outermost region of the disk. Apart from a few CS observations (*Blake et al.*, 1992; *Walker et al.*, 1994; *Velusamy et al.*, 2002), thus far there have been no detailed observational studies of this critical region, and so the chemical composition of the (interstellar) material in this phase is relatively uncharacterized.

6. SUMMARY

The ISM is turbulent and interstellar cloud structures are very filamentary and fragmentary. Stars that form in relatively quiescent cold dense molecular cloud cores then inject energy into the ISM at the ends of their lives via explosions or winds. Supernovae-driven shocks accelerate cosmic rays and generate turbulence. Gas is rapidly processed from very high temperatures (10^6 K) and rarefied densities (≤ 1 cm^{-3}) to low temperatures ($\leq 10^2$ K) and moderately high densities ($n_H > 10$–10^2 cm^{-3}) as turbulence concentrates matter into shocks and small intermittent regions of high-velocity shear where viscous dissipation and heating and cooling processes occur. Turbulent compression of H I in the WNM, driven by supernova shocks, colliding H I streams, spiral density waves, or gravity, leads to the rapid formation of cold H I diffuse clouds. Turbulence may also speed up the formation of H_2 on grain surfaces in intermittent shock-induced density enhancements ($n_H \gg 10^3$ cm^{-3}), enabling the formation of molecular clouds on short timescales (~10^6 yr). Most of the galaxy's molecular gas is in giant molecular clouds that fill only 1–2% of the volume of the ISM. Rapidly, stars form out of molecular cloud cores that encompass only a small fraction of the volume of a molecular cloud. Massive stars only live a few million years, so molecular cloud material is dispersed in ~4×10^6 yr from the energy injected by the winds and jets of young stars or by turbulence. Diffuse cloud material is converted to molecular clouds and back to diffuse clouds on relatively short timescales. Physiochemical processes in diffuse clouds and molecular clouds increase the complexity of solid-state and molecular materials, a fraction of which survives the protostellar collapse to be incorporated into comets.

Stars enrich the ISM with nucleosynthesized elements: SNe II contribute O, SNe Ia contribute Fe, and AGB stars contribute C and N. Supernovae also provide the energy for the turbulence in the ISM. Most of the heavy elements are deposited into the ISM as dust grains. In particular, O-rich AGB circumstellar envelopes (CSE) produce silicates, oxides, and alumino-silicates, and C-rich AGB CSEs produce amorphous carbon, hydrocarbons with aromatic moieties (i.e., PAHs) and aliphatic bonds, and silicon carbide. However, dust grains are rapidly destroyed in the ISM by supernovae shocks and recondense in postshock gas or form in diffuse clouds by processes that as yet are unknown. Furthermore, most aliphatic bonds in hydrocarbon dust from AGB stars are destroyed by interstellar UV photons or by UV photons from hot white dwarf stellar cores when AGB stars become planetary nebulae over the course of a few thousand years. Aliphatic bonds probably are reformed in diffuse clouds when H atoms impinge on carbonaceous dust grains. Cosmic-ray bombardment of silicate grains in diffuse clouds selectively removes primarily Si and secondarily Mg from the grains, and efficiently converts silicate crystals to amorphous silicates. Only a minor fraction of the most refractory stardust grains survive their passage through the ISM to be incorporated as presolar grains in meteorites and cometary IDPs.

Galactic UV photons also destroy all small volatile molecules formed in the AGB CSE and thus simple molecules must be reformed in diffuse clouds. The only molecules to survive the passage through the WIM and WNM to diffuse clouds are PAHs. Photoelectric heating of the diffuse clouds and molecular cloud boundaries occurs when UV photons are absorbed by PAHs.

Diffuse clouds are unshielded from UV and photochemistry only permits the formation of fairly simple molecules. Many major heavy elements that are not depleted into grains are singly ionized: C^+, S^+, Fe^+, Mg^+, and Na^+; only O and N are neutral. This means that many neutral-neutral chemical reactions that occur in dense clouds are not viable in diffuse clouds, as exemplified by the "CH^+ problem." Neutral H_2 reacts with C^+ to form CH and C_2. Cosmic rays ionize H and H_2 that can then subsequently react rapidly to produce other ions and species such as OH. Ion-molecule reactions also contribute to the formation of CO in diffuse clouds.

Cosmic rays penetrate dense molecular clouds where UV photons do not, ionizing and heating the gas, promoting ion-molecule chemistry, and affecting the composition of icy mantles on dust grains. Ion-molecule and neutral-neutral reactions convert an appreciable fraction of the available heavy elements into molecules. This process begins with the freezing out of atoms from the gas onto silicate and carbonaceous grain surfaces. Ices are observed along the lines of sight through molecular clouds to protostars: CO-rich ice mantles with O_2 and N_2 and H_2O-rich ice mantles with traces of CO, CO_2, methane (CH_4), ammonia (NH_3), methanol (CH_3OH), formic acid or methanoic acid (HCOOH), and formaldehyde or methanal (H_2CO). A few classes of reactions are needed to form most of the molecules observed in ices. Reactions beginning with simple hydrocarbons such as acetylene (C_2H_2) form long C-chain molecules. Furthermore, many large organic molecules can form on grain surfaces. Low kinetic temperatures favor the fractionation of isotopes of H, C, and N. In order to maintain molecules in the gas phase, desorption from grain surfaces must occur, possibly due to grain heating by cosmic rays, mantle explosions, sputtering in low-velocity shocks, or grain-grain collisions.

Ice mantles consisting of molecules accreted from the gas are made more complex by energetic processing by cosmic rays, heating in protostellar cores, and perhaps UV photolysis. Although UV photons cannot penetrate deep into molecular clouds, UV photons are created in molecular clouds by H_2 excited by energetic electrons produced by primary cosmic-ray interactions. In particular, the carrier of the "XCN" band is produced by photolysis and/or radiolysis (proton irradiation). The complex molecules formed on cold grain surfaces are revealed in hot and warm protostellar cores when these molecules are returned to the gas phase via desorption.

In a collapsing molecular cloud core the protostar dramatically influences the chemistry of its surroundings. Hot cores and warm cores surround high-mass and low-mass protostars respectively. Molecules are evaporated from icy grain mantles, react on grain surfaces to form mixed molecular ices, and participate in a highly transient chemistry in which molecules form through gas-gas reactions. A strong chemical differentiation occurs between O-bearing and N-bearing molecules with temperature and time: Hotter cores tend to have high abundances of N-bearing molecules. The chemical processes occurring in hot cores are discussed, and many of the same processes apply to warm cores. The chemistries that delineate six different regions around a low-mass protostar are described and shown in Fig. 4. The molecular cloud core collapses (N_2H^+) and heats the surrounding material as traced by $(CH_3)_2O$. Bipolar CO outflows and jets shock the ambient cloud core material as traced by H_2S, SO, SO_2, CH_3OH, SiO, and D_2CO. Interstellar material falling onto the protoplanetary disk passes through an accretion shock traced by CS emission. The physical conditions of the accretion shock region in the outer disk will become better constrained by future high-spatial-resolution, high-sensitivity observations using ALMA.

Interstellar chemistry, i.e., physiochemical processes in diffuse clouds and molecular clouds, sets the chemical boundary conditions on the composition of the material that was initially available for incorporation into the protosolar nebula, and thereafter into cometary matter. Understanding the degree of modification of these pristine interstellar molecules and solids as they passed through the accretion shock in the outer disk, to subsequently take part in nebular chemistry, will permit detailed comparison of the composition of comets with ISM material, and hence an appreciation of the chemical diversity among comets (*Ehrenfreund et al.,* 2004).

Acknowledgments. The authors collectively thank P. Goldsmith, W. Irvine, and an anonymous reviewer for their time and comments on this chapter. Discussions with B. Elmegreen (D.H.W.) and L. Allamandola (S.B.C.) are gratefully appreciated. Studies at NASA Ames (D.H.W.) of dust in comets and pre-main-sequence stars are supported by NASA's Planetary Astronomy and Origins of Solar Systems Programs. Studies of theoretical astrochemistry at NASA Ames (S.B.C.) is supported by NASA's Planetary Atmospheres and Origins of Solar Systems Programs through funds allocated by NASA Ames under Interchange No. NCC2-1412 and by the NASA Astrobiology Institute. Laboratory astrochemistry and astrobiology (P.E.) is supported by NWO (VI), SRON, and ESA. We are grateful to S. Rodgers for preparation of Fig. 1. We thank E. Peeters and J. Keane for the ISO reduced data and preparation of Fig. 2.

REFERENCES

Adams F. C. and Shu F. H. (1985) Infrared emission from protostars. *Astrophys. J., 296,* 655–669.

Adamson A. J., Whittet D. C. B., Chrysostomou A., Hough J. H., Aitken D. K., Wright G. S., and Roche P. F. (1999) Spectropolarimetric constraints on the nature of the 3.4 micron absorber in the interstellar medium. *Astrophys. J., 512,* 224–229.

Aikawa Y., Ohashi N., Inutsuka S., Herbst E., and Takakuwa S. (2001) Molecular evolution in collapsing prestellar cores. *Astrophys. J., 552,* 639–653.

Aikawa Y., Ohashi N., and Herbst E. (2003) Molecular evolution in collapsing prestellar cores. II. The effect of grain-surface reactions. *Astrophys. J., 593,* 906–924.

Allamandola L. J., Sandford S. A., and Valero G. J. (1988) Photochemical and thermal evolution of interstellar/precometary ice analogs. *Icarus, 76,* 225–252.

Allamandola L. J., Bernstein M. P., and Sandford S. A. (1997) Photochemical evolution of interstellar/precometary organic material. In *Astronomical and Biochemical Origins and the Search for Life in the Universe* (C. B. Cosmovici et al., eds.), pp. 23–47. Editrice Compositori, Bologna.

Allamandola L. J., Hudgins D. M., and Sandford S. A. (1999) Modeling the unidentified infrared emission with combinations of polycyclic aromatic hydrocarbons. *Astrophys. J. Lett., 511,* L115–L119.

Allen M. and Robinson G. W. (1977) The molecular composition of dense interstellar clouds. *Astrophys. J., 212,* 396–415.

Amari S., Nittler L. R., Zinner E., Lodders K., and Lewis R. S. (2001) Presolar SiC grains of type A and B: Their isotopic compositions and stellar origins. *Astrophys. J., 559,* 463–498.

Anders E. and Grevesse N. (1989) Abundances of elements: Meteoritic and solar. *Geochim. Cosmochim. Acta, 53,* 197–214.

Bachiller R., Perez Gutierrez M., Kumar M. S. N., and Tafalla M. (2001) Chemically active outflow L 1157. *Astron. Astrophys., 372,* 899–912.

Bacmann A., Lefloch B., Ceccarelli C., Castets A., Steinacker J., and Loinard L. (2002) The degree of CO depletion in pre-stellar cores. *Astron. Astrophys., 389,* L6–L10.

Bakes E. L. O. and Tielens A. G. G. M. (1994) The photoelectric heating mechanism for very small graphitic grains and polycyclic aromatic hydrocarbons. *Astrophys. J., 427,* 822–838.

Bakes E. L. O. and Tielens A. G. G. M. (1998) The effects of polycyclic aromatic hydrocarbons on the chemistry of photodissociation regions. *Astrophys. J., 499,* 258–266.

Bakes E. L. O., Tielens A. G. G. M., and Bauschlicher C. W. Jr. (2001a) Theoretical modeling of infrared emission from neutral and charged polycyclic aromatic hydrocarbons. I. *Astrophys. J., 556,* 501–514.

Bakes E. L. O., Tielens A. G. G. M., and Bauschlicher C. W. Jr. (2001b) Theoretical modeling of infrared emission from neutral and charged polycyclic aromatic hydrocarbons. II. *Astrophys. J., 560,* 261–271.

Ballesteros-Paredes J., Hartmann L., and Vázquez-Semandeni E. (1999a) Turbulent flow-driven molecular cloud formation: A solution to the post-T Tauri problem? *Astrophys. J., 527,* 285–297.

Ballesteros-Paredes J., Vázquez-Semandeni E., and Scalo J. (1999b) Clouds as turbulent density fluctuations: Implications for pressure confinement and spectral line data interpretation. *Astrophys. J., 515,* 286–303.

Bergin E. A., Neufeld D. A., and Melnick G. J. (1998) The postshock chemical lifetimes of outflow tracers and a possible new mechanism to produce water ice mantles. *Astrophys. J., 499,* 777–792.

Bergin E. A., Alves J., Huard T., and Lada C. J. (2002) N_2H^+ and $C^{18}O$ depletion in a cold dark cloud. *Astrophys. J. Lett., 570,* L101–L104.

Bernstein M. P., Sandford S. A., and Allamandola L. J. (2000) H, C, N, and O isotopic substitution studies of the 2165 wavenumber (4.62 micron) "XCN" feature produced by ultraviolet photolysis of mixed molecular ices. *Astrophys. J., 542,* 894–897.

Bettens R. P. A. and Herbst E. (1996) The abundance of very large hydrocarbons and carbon clusters in the diffuse interstellar medium. *Astrophys. J., 468,* 686–693.

Bianchi S., Gonçalves J., Albrecht M., Caselli P., Chini R., Galli D., and Walmsley M. (2003) Dust emissivity in the submm/mm: SCUBA and SIMBA observations of Barnard 68. *Astron. Astrophys., 399,* L43–L46.

Biham O., Furman I., Pirronello V., and Vidali G. (2001) Master equation for hydrogen recombination on grain surfaces. *Astrophys. J., 553,* 595–603.

Blake G. A., Sutton E. C., Masson C. R., and Phillips T. G. (1987) Molecular abundances in OMC-1 — The chemical composition of interstellar molecular clouds and the influence of massive star formation. *Astrophys. J., 315,* 621–645.

Blake G. A., van Dishoeck E. F., and Sargent A. I. (1992) Chemistry in circumstellar disks — CS toward HL Tauri. *Astrophys. J. Lett., 391,* L99–L103.

Blitz L. and Shu F. H. (1980) The origin and lifetime of giant molecular cloud complexes. *Astrophys. J., 238,* 148–157.

Bonnell I. A. and Bate M. R. (2002) Accretion in stellar clusters and the collisional formation of massive stars. *Mon. Not. R. Astron. Soc., 336,* 659–669.

Boss A. P. (2004) From molecular clouds to circumstellar disks. In *Comets II* (M. C. Festou et al., eds.), this volume. Univ. of Arizona, Tucson.

Boogert A. C. A., Hogerheijde M. R., Ceccarelli C., Tielens A. G. G. M., van Dishoeck E. F., Blake G. A., Latter W. B., and Motte F. (2002a) The environment and nature of the class I protostar Elias 29: Molecular gas observations and the location of ices. *Astrophys. J., 570,* 708–723.

Boogert A. C. A., Blake G. A., and Tielens A. G. G. M. (2002b) High-resolution 4.7 micron Keck/NIRSPEC spectra of protostars. II. Detection of the ^{13}CO isotope in icy grain mantles. *Astrophys. J., 577,* 271–280.

Boonman A. M. S., van Dishoeck E. F., Lahuis F., and Doty S. D. (2003) Gas-phase CO_2 toward massive protostars. *Astron. Astrophys., 399,* 1063–1072.

Boulanger F., Boissel P., Cesarsky D., and Ryter C. (1998) The shape of the unidentified infra-red bands: Analytical fit to ISOCAM spectra. *Astron. Astrophys., 339,* 194–200.

Boulares A. and Cox D. P. (1990) Galactic hydrostatic equilibrium with magnetic tension and cosmic-ray diffusion. *Astrophys. J., 365,* 544–558.

Bradley J. P. (1994) Chemically anomalous, preaccretionally irradiated grains in interplanetary dust from comets. *Science, 265,* 925–929.

Brown P. D. and Charnley S. B. (1990) Chemical models of interstellar gas-grain processes. I — Modeling and the effect of accretion on gas abundances and mantle composition in dense clouds. *Mon. Not. R. Astron. Soc., 244,* 432–443.

Brown P. D. and Charnley S. B. (1991) Chemical models of interstellar gas-grain processes. II — The effect of grain-catalysed methane on gas phase evolution. *Mon. Not. R. Astron. Soc., 249,* 69–75.

Brown P. D., Charnley S. B., and Millar T. J. (1988) A model of the chemistry in hot molecular cores. *Mon. Not. R. Astron. Soc., 231,* 409–417.

Brownlee D. E., Joswiak D. J., Bradley J. P., Gezo J. C., and Hill H. G. M. (2000) Spatially resolved acid dissolution of IDPs. The state of carbon and the abundance of diamonds in the dust (abstract). In *Lunar and Planetary Science XXXI,* Abstract #1921. Lunar and Planetary Institute, Houston (CD-ROM).

Buckle J. V. and Fuller G. A. (2003) Sulphur-bearing species as chemical clocks for low mass protostars? *Astron. Astrophys., 399,* 567–581.

Burke B. F. and Hollenbach D. J. (1983) The gas-grain interaction in the interstellar medium — Thermal accommodation and trapping. *Astrophys. J., 265,* 223–234.

Busso M., Gallino R., and Wasserburg G. J. (1999) Nucleosynthesis in asymptotic giant branch stars: Relevance for galactic enrichment and solar system formation. *Annu. Rev. Astron. Astrophys., 37,* 239–309.

Cadwell B. J., Wang H., Feigelson E. D., and Frenklach M. (1994) Induced nucleation of carbon dust in red giant stars. *Astrophys. J., 429,* 285–299.

Cami J. (2002) Molecular gas and dust around evolved stars. Ph.D. thesis, Univ. of Amsterdam. 294 pp.

Carrez P., Demyk K., Cordier P., Gengembre L., Crimblot J., d'Hendecourt L., Jones A. P., and Leroux H. (2002) Low-energy He ion irradiation-induced amorphization and chemical changes in olivine: Insights for silicate dust evolution in the interstellar medium. *Meteoritics & Planet. Sci., 37,* 1599–1614.

Caselli P., Hasegawa T. I., and Herbst E. (1993) Chemical differentiation between star-forming regions — The Orion hot core and compact ridge. *Astrophys. J., 408,* 548–558.

Caselli P., Stantcheva T., Shalabiea O., Shematovich V. I., and Herbst E. (2002) Deuterium fractionation on interstellar grains studied with modified rate equations and a Monte Carlo approach. *Planet. Space Sci., 50,* 1257–1266.

Caselli P., van der Tak F. F. S., Ceccarelli C., and Bacmann A. (2003) Abundant H_2D^+ in the pre-stellar core L1544. *Astron.*

Astrophys., 403, L37–L41.

Cassen P. and Moosman A. (1981) On the formation of protostellar disks. *Icarus, 48,* 353–376.

Cataldo F. (2002) An investigation on the optical properties of carbon black, fullerite, and other carbonaceous materials in relation to the spectrum of interstellar extinction of light. *Fullerenes, Nanotubes, Carbon Nanostructures, 10/2,* 155–170.

Cataldo F., Baratta G. A., and Strazzulla G. (2002) He$^+$ ion bombardment of C_{60} fullerene: An FT-IR and Raman Study. *Fullerenes, Nanotubes Carbon Nanostructures, 10/3,* 197–206.

Cazaux S. and Tielens A. G. G. M. (2002) Molecular hydrogen formation in the interstellar medium. *Astrophys. J. Lett., 575,* L29–L32.

Cazaux S., Tielens A. G. G. M., Ceccarelli C., Castets A., Wakelam V. Caux E., Parise B., and Teyssier D. (2003) The hot core around the low-mass protostar IRAS 16293-2422: Scoundrels rule! *Astrophys. J. Lett., 593,* L51–L55.

Ceccarelli C., Hollenbach D. J., and Tielens A. G. G. M. (1996) Far-infrared line emission from collapsing protostellar envelopes. *Astrophys. J., 471,* 400–426.

Cesarsky C. J. and Völk H. J. (1978) Cosmic ray penetration into molecular clouds. *Astron. Astrophys., 70,* 367–377.

Charnley S. B. (1997a) Sulfuretted molecules in hot cores. *Astrophys. J., 481,* 396–405.

Charnley S. B. (1997b) On the nature of interstellar organic chemistry. In *Astronomical and Biochemical Origins and the Search for Life in the Universe* (C. B. Cosmovici et al., eds.), pp. 89–96. Editrice Compositori, Bologna.

Charnley S. B. (1997c) Chemical models of interstellar gas-grain processes. III — Molecular depletion in NGC 2024. *Mon. Not. R. Astron. Soc., 291,* 455–460.

Charnley S. B. (1998) Stochastic astrochemical kinetics. *Astrophys. J. Lett., 509,* L121–L124.

Charnley S. B. (2001) Stochastic theory of molecule formation on dust. *Astrophys. J. Lett., 562,* L99–L102.

Charnley S. B. and Kaufman M. J. (2000) Carbon dioxide in star-forming regions. *Astrophys. J. Lett., 529,* L111–L114.

Charnley S. B. and Millar T. J. (1994) The chemistry of phosphorus in hot molecular cores. *Mon. Not. R. Astron. Soc., 270,* 570–574.

Charnley S. B. and Rodgers S. D. (2002) The end of interstellar chemistry as the origin of nitrogen in comets and meteorites. *Astrophys. J. Lett., 569,* L133–L137.

Charnley S. B., Tielens A. G. G. M., and Millar T. J. (1992) On the molecular complexity of the hot cores in Orion A — Grain surface chemistry as 'The last refuge of the scoundrel'. *Astrophys. J. Lett., 399,* L71–L74.

Charnley S. B., Kress M. E., Tielens A. G. G. M., and Millar T. J. (1995) Interstellar alcohols. *Astrophys. J., 448,* 232–239.

Charnley S. B., Tielens A. G. G. M., and Rodgers S. D. (1997) Deuterated methanol in the Orion compact ridge. *Astrophys. J. Lett., 482,* L203–L206.

Charnley S. B., Rodgers, S. D., and Ehrenfreund P. (2001a) Gas-grain chemical models of star-forming molecular clouds as constrained by ISO and SWAS observations. *Astron. Astrophys., 378,* 1024–1036.

Charnley S. B., Ehrenfreund P., and Kuan Y- J. (2001b) Spectroscopic diagnostics of organic chemistry in the protostellar environment. *Spectrochim. Acta, Part A: Molecular and Biomolecular Spectroscopy, 57,* 685–704.

Chastaing D., Le Picard S. D., Sims I. R., and Smith I. W. M. (2001) Rate coefficients for the reactions of C(^3P_J) atoms with C_2H_2, C_2H_4, $CH_3C = CH$ and $H_2C = C = CH_2$ at temperatures down to 15 K. *Astron. Astrophys., 365,* 241–247.

Cherchneff I. and Glassgold A. E. (1993) The formation of carbon chain molecules in IRC+10216. *Astrophys. J. Lett., 419,* L41–L44.

Cherchneff I., Barker J. R., and Tielens A. G. G. M. (1991) Polycyclic aromatic hydrocarbons optical properties and contribution to the acceleration of stellar outflows. *Astrophys. J., 377,* 541–552.

Chiappini C., Romano D., and Matteucci F. (2003) Oxygen, carbon, and nitrogen evolution in galaxies. *Mon. Not. R. Astron. Soc., 339,* 68–81.

Chiar J. E., Pendleton Y. J., Geballe T. R., and Tielens A. G. G. M. (1998) Near-infrared spectroscopy of the proto-planetary nebula CRL 618 and the origin of the hydrocarbon dust component in the interstellar medium. *Astrophys. J., 507,* 281–286.

Chiar J. E., Tielens A. G. G. M., Whittet D. C. B., Schutte W. A., Boogert A. C. A., Lutz D., van Dishoeck E. F., and Bernstein M. P. (2000) The composition and distribution of dust along the line of sight toward the galactic center. *Astrophys. J., 537,* 749–762.

Chick K. M. and Cassen P. (1997) Thermal processing of interstellar dust grains in the primitive solar environment. *Astrophys. J., 477,* 398–409.

Churchwell E. (2002) Ultra-compact HII regions and massive star formation. *Annu. Rev. Astron. Astrophys., 40,* 27–62.

Clarke C. J., Bonnell I. A., and Hillenbrand L. A. (2000) The formation of stellar clusters. In *Protostars and Planets IV* (V. Mannings et al., eds.), pp. 151–177. Univ. of Arizona, Tucson.

Clayton D. D. (1983) *Principles of Stellar Evolution and Nucleosynthesis.* Univ. of Chicago, Chicago. 612 pp.

Cottin H., Gazeau M. C., and Raulin F. (1999) Cometary organic chemistry: A review from observations, numerical and experimental simulations. *Planet. Space Sci., 47,* 1141–1162.

Cottin H., Szopa C., and Moore M. H. (2001) Production of hexamethylenetetramine in photolyzed and irradiated interstellar cometary ice analogs. *Astrophys. J. Lett., 561,* L139–L142.

Cravens T. E. and Dalgarno A. (1978) Ionization, dissociation, and heating efficiencies of cosmic rays in a gas of molecular hydrogen. *Astrophys. J., 219,* 750–752.

Crawford I. A. and Williams D. A. (1997) Detection of interstellar NH towards Zeta Ophiuchi by means of ultra-high-resolution spectroscopy. *Mon. Not. R. Astron. Soc., 291,* L53–L56.

Cummins S. E., Linke R. A., and Thaddeus P. (1986) A survey of the millimeter-wave spectrum of Sagittarius B2. *Astrophys. J. Suppl. Ser., 60,* 819–878.

Dartois E., Schutte W. A., Geballe T. R., Demyk K., Ehrenfreund P., and d'Hendecourt L. (1999) Methanol: The second most abundant ice species towards the high-mass protostars RAFGL7009S and W 33A. *Astron. Astrophys., 342,* L32–35.

Dartois E., d'Hendecourt L., Thi W., Pontoppidan K. M., and van Dishoeck E. F. (2002) Combined VLT ISAAC/ISO SWS spectroscopy of two protostellar sources: The importance of minor solid state features. *Astron. Astrophys., 394,* 1057–1068

Daulton T. L., Bernatowicz T. J., Lewis R. S., Messenger S., Stadermann F. J., and Amari S. (2002) Polytype distribution in presolar SiC: Microstructural characterization by transmission electron microscopy (abstract). In *Lunar and Planetary Science XXXIII,* Abstract #1127. Lunar and Planetary Institute, Houston (CD-ROM).

Demyk K., Carrez Ph., Leroux H., Cordier P., Jones A. P., Borg J., Quirico E., Raynal P. I., and d'Hendecourt L. (2001) Structural and chemical alteration of crystalline olivine under low energy He+ irradiation. *Astron. Astrophys., 362,* L38–L41.

d'Hendecourt L. B., Allamandola L. J., Grim R. J. A., and Greenberg J. M. (1986) Time-dependent chemistry in dense molecular clouds. II — Ultraviolet photoprocessing and infrared spectroscopy of grain mantles. *Astron. Astrophys., 158,* 119–134.

Dickens J., Irvine W. M., Snell R. L., Bergin E. A., Schloerb F. P., Pratap P., and Miralles M. P. (2000) A study of the physics and chemistry of L134N. *Astrophys. J., 542,* 870–889.

Dickens J., Langer W. D., and Velusamy T. (2001) Small-scale abundance variations in TMC-1: Dynamics and hydrocarbon chemistry. *Astrophys. J., 558,* 693–701.

Dickey J. M. and Lockman F. J. (1990) HI in the galaxy. *Annu. Rev. Astron. Astrophys., 28,* 215–261.

Dominik C. and Tielens A. G. G. M. (1997) The physics of dust coagulation and the structure of dust aggregates in space. *Astrophys. J., 480,* 647–673.

Doty S. D., van Dishoeck E. F., van der Tak F. F. S., and Boonman A. M. S. (2002) Chemistry as a probe of the structures and evolution of massive star-forming regions. *Astron. Astrophys., 389,* 446–463.

Dove J. B., Shull J. M., and Ferrara A. (2000) The escape of ionizing photons from OB associations in disk galaxies: Radiation transfer through superbubbles. *Astrophys. J., 531,* 846–860.

Draine B. T. (1980) Interstellar shock waves with magnetic precursors. *Astrophys. J., 241,* 1021–1038.

Draine B. T. (2003) Interstellar dust grains. *Annu. Rev. Astron. Astrophys., 41,* 241–289.

Draine B. T. and Katz N. (1986) Magnetohydrodynamic shocks in diffuse clouds. II — Production of CH+, OH, CH, and other species. *Astrophys. J., 310,* 392–407.

Draine B. T., Roberge W. G., and Dalgarno A. (1983) Magnetohydrodynamic shock waves in molecular clouds. *Astrophys. J., 264,* 485–507.

Duley W. W. and Seahra S. S. (1999) 2175 Å and 3.4 micron absorption bands and carbon depletion in the diffuse interstellar medium. *Astrophys. J. Lett., 522,* L129–L132.

Ehrenfreund P. and Charnley S. B. (2000) Organic molecules in the interstellar medium, comets, and meteorites: A voyage from dark clouds to the early earth. *Annu. Rev. Astron. Astrophys., 38,* 427–483.

Ehrenfreund P. and Schutte W. A. (2000) Infrared observations of interstellar ices. In *Astrochemistry: From Molecular Clouds to Planetary Systems* (Y. C. Minh and E. F. van Dishoeck, eds.), pp. 135–146. IAU Symposium 197, Astronomical Society of the Pacific, San Francisco.

Ehrenfreund P., Boogert A. C. A., Gerakines P. A., Jansen D. J., Schutte W. A., Tielens A. G. G. M., and van Dishoeck E. F. (1996) A laboratory database of solid CO and CO_2 for ISO. *Astron. Astrophys., 315,* L314–L344.

Ehrenfreund P., Dartois E., Demyk K., and d'Hendecourt L. (1998) Ice segregation toward massive protostars. *Astron. Astrophys., 339,* L17–20.

Ehrenfreund P., Cami J., Jiménez-Vicente J., Foing B. H., Kaper L., van der Meer A., Cox N., d'Hendecourt L., Maier J. P., Salama F., Sarre P. J., Snow T. P., and Sonnentrucker P. (2002) Detection of diffuse interstellar bands in the Magellanic clouds. *Astrophys. J. Lett., 576,* L117–L120.

Ehrenfreund P., Charnley S. B., and Wooden D. H. (2004) From interstellar material to comet particles and molecules. In *Comets II* (M. C. Festou et al., eds.), this volume. Univ. of Arizona, Tucson.

Elmegreen B. G. (1979) Gravitational collapse in dust lanes and the appearance of spiral structure in galaxies. *Astrophys. J., 231,* 372–383.

Elmegreen B. G. (2000) Star formation in a crossing time. *Astrophys. J., 530*, 277–281.

Evans N. (1999) Physical conditions in regions of star formation. *Annu. Rev. Astron. Astrophys., 37*, 311–362.

Fabian D., Henning Th., Jäger C., Mutschke H., Dorschner J., and Wehrhan O. (2001a) Steps towards interstellar silicate mineralogy. VI. Dependence of crystalline olivine IR spectra on iron content and particle shape. *Astron. Astrophys., 378*, 228–238.

Fabian D., Posch T., Mutschke H., Kerschbaum F., and Dorschner J. (2001b) Infrared optical properties of spinels. A study of the carrier of the 13, 17, and 32 μm emission features observed in ISO-SWS spectra of oxygen–rich AGB stars. *Astron. Astrophys., 373*, 1125–1138.

Falgarone E., Phillips T. G., and Walker C. K. (1991) The edges of molecular clouds: Fractal boundaries and density structure. *Astrophys. J., 378*, 186–201.

Falgarone E., Pineau des Forêts G., and Roueff E. (1995) Chemical signatures of the intermittency of turbulence in low density interstellar clouds. *Astron. Astrophys., 300*, 870–880.

Ferrière K. (1998) The hot gas filling factor in our galaxy. *Astrophys. J., 503*, 700–716.

Ferrière K. M. (2001) The interstellar environment of our galaxy. *Rev. Mod. Phys., 73*, 1031–1066.

Field G. B. and Hutchins J. (1968) A statistical model of interstellar clouds. II. Effect of varying clouds cross-sections and velocities. *Astrophys. J., 153*, 737–742.

Fitzpatrick E. L. and Massa D. (1990) An analysis of the shapes of ultraviolet extinction curves. III — An atlas of ultraviolet extinction curves. *Astrophys. J. Suppl. Ser., 72*, 163–189.

Flower D. R. and Pineau des Forêts G. (1994) Grain/mantle erosion in magnetohydrodynamic shocks. *Mon. Not. R. Astron. Soc., 268*, 724–732.

Flower D. R. and Pineau des Forêts G. (1998) C-type shocks in the interstellar medium: Profiles of CH+ and CH absorption lines. *Mon. Not. R. Astron. Soc., 297*, 1182–1188.

Frenklach M. and Feigelson E. D. (1989) Formation of polycyclic aromatic hydrocarbons in circumstellar envelopes. *Astrophys. J., 341*, 372–384.

Gail H.-P. and Sedlmayr E. (1999) Mineral formation in stellar winds. I. Condensation sequence of silicates and iron grains in stationary oxygen rich outflows. *Astron. Astrophys., 347*, 594–616.

Gammie C. F., Lin Y.-T., Stone J. M., and Ostriker E. C. (2003) Analysis of clumps in molecular cloud models: Mass spectrum, shapes, alignment, and rotation. *Astrophys. J., 592*, 203–216.

Garay G., Mardones D., Rodriguez L. F., Caselli P., and Bourke T. L. (2002) Methanol and silicon monoxide observations toward bipolar outflows associated with class 0 objects. *Astrophys. J., 567*, 980–998.

Gibb E., Whittet D. C. B., Schutte W. A., Boogert A. C. A., Chiar J., Ehrenfreund P., Gerakines P. A., Keane J. V., Tielens A. G. G. M., van Dishoeck E. F., and Kerkhof O. (2000) An inventory of interstellar ices toward the embedded protostar W33A. *Astrophys. J., 536*, 347–356.

Gibb E. L., Whittet D. C. B., and Chiar J. E. (2001) Searching for ammonia in grain mantles toward massive young stellar objects. *Astrophys. J., 558*, 702–716.

Glassgold A. E. (1996) Circumstellar photochemistry. *Annu. Rev. Astron. Astrophys., 34*, 241–278.

Glassgold A. E., Mamon G. A., and Huggins P. J. (1991) The formation of molecules in protostellar winds. *Astrophys. J., 373*, 254–265.

Gnacinski P. (2000) Interstellar carbon abundance. *Acta Astron., 50*, 133–149.

Goldsmith D. W. (2001) Molecular depletion and thermal balance in dark cloud cores. *Astrophys. J., 557*, 736–746.

Goldsmith D. W., Habing J. J., and Field G. B. (1969) Thermal properties of interstellar gas heated by cosmic rays. *Astrophys. J., 158*, 173–184.

Goldsmith P. F. and Langer W. D. (1978) Molecular cooling and thermal balance of dense interstellar clouds. *Astrophys. J., 222*, 881–895.

Goldsmith P. F., Melnick G. J., Bergin E. A., Howe J. E., Snell R. L., Neufeld D. A., Harwit M., Ashby M. L. N., Patten B. M., Kleiner S. C., Plume R., Stauffer J. R., Tolls V., Wang Z., Zhang Y. F., Erickson N. R., Koch D. G.; Schieder R., Winnewisser G., and Chin G. (2000) O_2 in interstellar molecular clouds. *Astrophys. J. Lett., 539*, L123–L127.

Goldsmith P. F., Li D., Bergin E. A., Melnick G. J., Tolls V., Howe J. E., Snell R. L., and Neufeld D. A. (2002) Tentative detection of molecular oxygen in the rho Ophiuchi cloud. *Astrophys. J., 576*, 814–831.

Goto M., Gaessler W., Hayano Y., Iye M., Kamata Y., Iye M, Kamata Y., Kanzawa T., Kobayashi N., Minowa Y., Saint-Jacques D. J., Takami H., Takato N., and Terada H. (2003) Spatially resolved 3 micron spectroscopy of IRAS 22272+5435: Formation and evolution of aliphatic hydrocarbon dust in protoplanetary nebulae. *Astrophys. J., 589*, 419–429.

Grabelsky D. A., Cohen R. S., Bronfman L., Thaddeus P., and May J. (1987) Molecular clouds in the Carina Arm: Large-scale properties of molecular gas and comparison with HI. *Astrophys. J., 315*, 122–141.

Gredel R. (1999) Interstellar C_2 absorption lines towards CH+ forming regions. *Astron. Astrophys., 351*, 657–668.

Gredel R., Lepp, S., and Dalgarno A. (1987) The C/CO ratio in dense interstellar clouds. *Astrophys. J. Lett., 323*, L137–L139.

Gredel R., van Dishoeck E. F., and Black J. H. (1993) The abundance of CH+ in translucent molecular clouds — Further tests of shock models. *Astron. Astrophys., 269*, 477–495.

Gredel R., Pineau des Forêts G., and Federman S. R. (2002) Interstellar CN toward CH+-forming regions. *Astron. Astrophys., 389*, 993–1014.

Green N. J. B., Toniazzo T., Pilling M. J., Ruffle D. P., Bell N., and Hartquist T. (2001) A stochastic approach to grain surface chemical kinetics. *Astron. Astrophys., 375*, 1111–1119.

Greenberg J. M., Li A., Mendoza-Gomez C. X., Schutte W. A., Gerakines P. A., and de Groot M. (1995) Approaching the interstellar grain organic refractory component. *Astrophys. J. Lett., 455*, L177–L180.

Grevesse N. and Noels A. (1993) Photospheric abundances. In *Origin and Evolution of the Elements* (N. Prantzos et al., eds.), pp. 15–25. Cambridge Univ., New York.

Grossman L. (1972) Condensation in the primitive solar nebula. *Geochim. Cosmochim. Acta, 39*, 47–64.

Hanner M. S. and Bradley J. P. (2004) Composition and mineralogy of cometary dust. In *Comets II* (M. C. Festou et al., eds.), this volume. Univ. of Arizona, Tucson.

Hartmann D., Magnani L., and Thaddeus P. (1998) A survey of high-latitude molecular gas in the northern hemisphere. *Astrophys. J., 492*, 205–212.

Hartmann L., Ballesteros-Paredes J., and Bergin E. A. (2001) Rapid formation of molecular clouds and stars in the solar neighborhood. *Astrophys. J., 562*, 852–868.

Hatchell J., Thompson M. A., Millar T. J., and MacDonald G. H. (1998) Sulphur chemistry and evolution in hot cores. *Astron.*

Astrophys., 338, 713–722.

Heiles C. (1967) Observations of the spatial structure of interstellar hydrogen. I. High-resolution observations of a small region. *Astrophys. J. Suppl. Ser., 15,* 97–130.

Herbst E. (2000) Models of gas-grain chemistry in star-forming regions. In *Astrochemistry: From Molecular Clouds to Planetary Systems* (Y. C. Minh and E. F. van Dishoeck, eds.), pp. 147–159. IAU Symposium 197, Astronomical Society of the Pacific, San Francisco.

Herbst E., Millar T. J., Wlodek S., and Bohme D. K. (1989) The chemistry of silicon in dense interstellar clouds. *Astron. Astrophys., 222,* 205–210.

Heyer M. H., Carpenter J. M., and Snell R. L. (2001) The equilibrium state of molecular regions in the outer galaxy. *Astrophys. J., 551,* 852–866.

Hiraoka K., Ohashi N., Kihara Y., Yamamoto K., Sato T., and Yamashita A. (1994) Formation of formaldehyde and methanol from the reactions of H atoms with solid CO at 10–20K. *Chem. Phys. Lett., 229,* 408–414.

Hiraoka K., Miyagoshi T., Takayama T., Yamamoto K., and Kihara Y. (1998) Gas-grain processes for the formation of CH_4 and H_2O: Reactions of H atoms with C, O, and CO in the solid phase at 12 K. *Astrophys. J., 498,* 710–715.

Hiraoka K., Takayama T., Euchi A., Handa H., and Sato T. (2000) Study of the reactions of H and D atoms with solid C_2H_2, C_2H_4, and C_2H_6 at cryogenic temperatures. *Astrophys. J., 532,* 1029–1037.

Hiraoka K., Sato T., Sato S., Sogoshi, N., Yokoyama T., Takashima H., and Kitagawa S. (2002) Formation of formaldehyde by the tunneling reaction of H with solid CO at 10 K revisited. *Astrophys. J., 577,* 265–270.

Hollenbach D. J. and Salpeter E. E. (1971) Surface recombination of hydrogen molecules. *Astrophys. J., 163,* 155–164.

Hollenbach D. J. and Tielens A. G. G. M. (1999) Photodissociation regions in the interstellar medium of galaxies. *Rev. Mod. Phys., 71,* 173–230.

Hollis J. M., Lovas F. J., and Jewell P. R. (2000) Interstellar glycolaldehyde: The first sugar. *Astrophys. J. Lett., 540,* L107–L110.

Hollis J. M., Lovas F. J., Jewell P. R., and Coudert L. H. (2002) Interstellar antifreeze: Ethylene glycol. *Astrophys. J. Lett., 571,* L59–L62.

Hoppe P., Amari S., Zinner E., and Lewis R. S. (1992) Large oxygen isotopic anomalies in graphite grains from the Murchison meteorite. *Meteoritics, 27,* 235.

Hrivnak B. J., Volk K., and Kwok S. (2000) 2–45 Micron infrared spectroscopy of carbon-rich proto-planetary nebulae. *Astrophys. J., 535,* 275–292.

Hudgins D. M., Bauschlicher C. W. Jr., and Allamandola L. J. (2001) Closed-shell polycyclic aromatic hydrocarbon cations: A new category of interstellar polycyclic aromatic hydrocarbons. *Spectrochim. Acta, Part A: Molecular and Biomolecular Spectroscopy, 57,* 907–930.

Hudson R. L. and Moore M. H. (1997) Hydrocarbon radiation chemistry in ices of cometary relevance. *Icarus, 126,* 233–235.

Hudson R. L. and Moore M. H. (1999) Laboratory studies of the formation of methanol and other organic molecules by water + carbon monoxide radiolysis: Relevance to comets, icy satellites, and interstellar ices. *Icarus, 140,* 451–461.

Hudson R. L. and Moore M. H. (2000) New experiments and interpretations concerning the "XCN" band in interstellar ice analogues. *Astron. Astrophys., 357,* 787–792.

Iben I. Jr. (1974) Post main sequence evolution of single stars.

Annu. Rev. Astron. Astrophys., 12, 215–256.

Ingalls J. G., Reach W. T., and Bania T. M. (2002) Photoelectric heating and [C II] cooling of high galactic translucent clouds. *Astrophys. J., 579,* 289–303.

Irvine W. and Lunine J. (2004) The cycle of matter in the galaxy. In *Comets II* (M. C. Festou et al., eds.), this volume. Univ. of Arizona, Tucson.

Jäger C., Dorschner J., Mutschke H., Posch Th., and Henning Th. (2003) Steps towards interstellar silicate mineralogy. VII. Spectral properties and crystallization behavior of magnesium silicates produced by the sol-gel method. *Astron. Astrophys., 408,* 193–204.

Jenkins E. B. (1987) Elemental abundances in the interstellar atomic material. In *Interstellar Processes* (D. J. Hollenbach and H. A. Thronson, eds.), pp. 533–559. Reidel, Dordrecht.

Jenkins E. B. and Wallerstein G. (1996) Hubble Space Telescope observations of interstellar in three high-latitude clouds. *Astrophys. J., 440,* 227–240.

Johansson L. E. B., Andersson C., Ellder J., Friberg P., Hjalmarson A., Hoglund B., Irvine W. M., Olofsson H., and Rydbeck G. (1984) Spectral scan of Orion A and IRC+10216 from 72 to 91 GHz. *Astron. Astrophys., 130,* 227–256.

Jones A. P. (2000) Depletion patterns and dust evolution in the interstellar medium. *J. Geophys. Res., 105,* 10257–10268.

Jones A. P. (2001) Interstellar and circumstellar grain formation and survival. *Philos. Trans. R. Soc. London, A359,* 1961–1972.

Jones A. P., Tielens A. G. G. M., Hollenbach D. J., and McKee C. F. (1994) Grain shattering in shocks in the interstellar medium. *Astrophys. J., 433,* 797–810.

Jones A. P., Tielens A. G. G. M., and Hollenbach D. J. (1996) Grain shattering in shocks: The interstellar grain size distribution. *Astrophys. J., 469,* 740–764.

Joulain K., Falgarone E., Pineau Des Forêts G., and Flower D. (1998) Non-equilibrium chemistry in the dissipative structures of interstellar turbulence. *Astron. Astrophys., 340,* 241–256.

Kemper F., Waters L. B. F. M., de Koter A., and Tielens A. G. G. M. (2001) Crystallinity versus mass-loss rates in asymptotic giant branch stars. *Astron. Astrophys., 369,* 132–141.

Kemper F., de Koter A., Waters L B. F. M., Bouwman J., and Tielens A. G. G. M. (2002) Dust and the spectral energy distribution of the OH/IR star OH 127.8+0.0: Evidence for circumstellar iron. *Astron. Astrophys., 384,* 585–593.

Kemper F., Vriend W. J., and Tielens A. G. G. M. (2004) On the degree of crystallinity of silicates in the diffuse interstellar medium. *Astrophys. J.,* in press.

Konigl A. and Pudritz R. E. (2000) Disk winds and the accretion-outflow connection. In *Protostars and Planets IV* (V. Mannings et al., eds.), pp. 759–788. Univ. of Arizona, Tucson.

Kuan Y.-J., Charnley S. B., Huang H.-C., Tseng W.-L., and Kisiel Z. (2003) Interstellar glycine. *Astrophys. J., 593,* 848–867.

Kulsrud R. M. and Cesarsky C. J. (1971) The effectiveness of instabilities for the confinement of high energy cosmic rays in the galactic disk. *Astrophys J. Lett., 8,* L189.

Lacy J. H., Baas F., Allamandola L. J., van de Bul, C. E. P., Persson S. E., McGregor P. J., Lonsdale C. J., and Geballe T. R. (1984) 4.6 micron absorption features due to solid phase CO and cyano group molecules toward compact infrared sources. *Astrophys. J., 276,* 533–543.

Langer W. D., Graedel T. E., Frerking M. A., and Armentrout P. B. (1984) Carbon and oxygen isotope fractionation in dense interstellar clouds. *Astrophys. J., 277,* 581–604.

Langer W. D., van Dishoeck E. F., Bergin E. A., Blake G. A., Tielens A. G. G. M., Velusamy T., and Whittet D. C. B. (2000)

Chemical evolution of protostellar matter. In *Protostars and Planets IV* (V. Mannings et al., eds.), pp. 29–57. Univ. of Arizona, Tucson.

Larson R. B. (1969) Numerical calculations of the dynamics of collapsing proto-star. *Mon. Not. R. Astron. Soc., 145,* 271–295.

Larson R. B. (2003) The physics of star formation. *Rept. Prog. Phys., 66,* 1651–1701.

Leitch-Devlin M. A. and Williams D. A. (1985) Sticking coefficient for atoms and molecules at the surfaces of interstellar dust grains. *Mon. Not. R. Astron. Soc., 213,* 295–306.

Lepp S. and Dalgarno A. (1988) Polycyclic aromatic hydrocarbons in interstellar chemistry. *Astrophys. J., 324,* 553–556.

Lepp S., Dalgarno A., van Dishoeck E. F., and Black J. H. (1988) Large molecules in diffuse interstellar clouds. *Astrophys. J., 329,* 418–424.

Lewis R. S., Ming T., Wacker J. F., Anders E., and Steel E. (1987) Interstellar diamonds in meteorites. *Nature, 326,* 160–162.

Lewis R. S., Amari S., and Anders E. (1994) Interstellar grains in meteorites. II. SiC and its noble gases. *Geochim. Cosmochim. Acta, 58,* 417–494.

Li A. and Draine B. T. (2002) Are silicon nanoparticles an interstellar dust component? *Astrophys. J., 564,* 803–812.

Li A. and Greenberg J. M. (1997) A unified model of interstellar dust. *Astron. Astrophys., 323,* 566–584.

Liszt H. S. (2003) Gas-phase recombination, grain neutralization and cosmic-ray ionization in diffuse gas. *Astron. Astrophys., 398,* 621–630.

Loinard L., Castets A., Ceccarelli C., Tielens A. G. G. M., Faur A., Caux E., and Duvert G. (2000) The enormous abundance of D_2CO in IRAS 16293–2422. *Astron. Astrophys., 359,* 1169–1174.

Loinard L., Castets A., Ceccarelli C., Caux E., and Tielens A. G. G. M. (2001) Doubly deuterated molecular species in protostellar environments. *Astrophys. J. Lett., 552,* L163–L166.

Lucas R. and Liszt H. S. (2000) Comparative chemistry of diffuse clouds. I. C_2H and C_3H_2. *Astron. Astrophys., 358,* 1069–1076.

Luhman K. L. and Rieke G. H. (1999) Low-mass star formation and the initial mass function in the ρ Ophiuchi cloud core. *Astrophys. J., 525,* 440–465.

Lunine J. I., Engel S., Rizk B., and Horanyi M. (1991) Sublimation and reformation of icy grains in the primitive solar nebula. *Icarus, 94,* 333–344.

Lunine J. I., Owen T. C., and Brown R. H. (2000) The outer solar system: Chemical constraints at low temperatures on planet formation. In *Protostars and Planets IV* (V. Mannings et al., eds.), pp. 1055–1080. Univ. of Arizona, Tucson.

Mac Low M.-M. and Klessen R. S. (2004) Control of star formation by supersonic turbulence. *Rev. Mod. Phys., 76,* 125–194.

Mackay D. D. S. (1995) The chemistry of silicon in hot molecular cores. *Mon. Not. R. Astron. Soc., 274,* 694–700.

Maloney P. R., Hollenbach D. J., and Tielens A. G. G. M. (1996) X-ray-irradiated molecular gas. I. Physical processes and general results. *Astrophys. J., 466,* 561–584.

Magnani L. and Onello J. S. (1995) A new method for determining the CO to H conversion factor for translucent clouds. *Astrophys. J., 443,* 169–180.

Manico G., Ragun G., Pirronello V., Roser J. E., and Vidali G. (2001) Laboratory measurements of molecular hydrogen formation on amorphous water ice. *Astrophys. J. Lett., 548,* L253–L256.

Markwick A. J., Millar T. J., and Charnley S. B. (2000) On the abundance gradients of organic molecules along the TMC-1

ridge. *Astrophys. J., 535,* 256–265.

Matteucci F. (2001) The formation of chemical elements and their abundances in the solar system. *Earth Moon Planets, 85,* 245–252.

McCall B. J., Hinkle K. H., Geballe T. R., Moriarty-Schieven G. H., Evans N. J. II, Kawaguchi K., Takano S., Smith V. V., and Oka T. (2002) Observations of H_3^+ in the diffuse interstellar medium. *Astrophys. J., 567,* 391–406.

McCray R. and Kafatos M. (1987) Supershells and propagating star formation. *Astrophys. J., 317,* 190–196.

McKee C. F. and Ostriker J. P. (1977) A theory of the interstellar medium — Three components regulated by supernova explosions in an inhomogeneous substrate. *Astrophys. J., 218,* 148–169.

McKee C. F., Hollenbach D. J., Seab C. G., and Tielens A. G. G. M. (1987) The structure of the time-dependent interstellar shocks and grain destruction in the interstellar medium. *Astrophys. J., 318,* 674–701.

McMullin J. P., Mundy L. G., Wilking B. A., Hezel T., and Blake G. A. (1994) Structure and chemistry in the northwestern condensation of the Serpens molecular cloud core. *Astrophys. J., 424,* 222–236.

Mennella V., Colangeli L., Bussoletti E., Palumbo P., and Rotundi A. (1998) A new approach to the puzzle of the ultraviolet interstellar extinction bump. *Astrophys. J. Lett., 507,* L177–L180.

Mennella V., Brucato J. R., Colangeli L., and Palumbo P. (1999) Activation of the 3.4 micron band in carbon grains by exposure to atomic hydrogen. *Astrophys. J. Lett., 524,* L71–L74.

Mennella V., Muñoz Caro G. M., Ruiterkamp R., Schutte W. A., Greenberg J. M., Brucato J. R., and Colangeli L. (2001) UV photodestruction of CH bonds and the evolution of the 3.4 μm feature carrier. II. The case of hydrogenated carbon grains. *Astron. Astrophys., 367,* 355–361.

Mennella V., Brucato J. R., Colangeli L., and Palumbo P. (2002) C–H bond formation in carbon grains by exposure to atomic hydrogen: The evolution of the carrier of the interstellar 3.4 micron band. *Astrophys. J., 569,* 531–540.

Mennella V., Baratta G. A., Esposito A., Ferini G., and Pendleton Y. J. (2003) The effects of ion irradiation on the evolution of the carrier of the 3.4 micron interstellar absorption band. *Astrophys. J., 587,* 727–738.

Messenger S. (2000) Identification of molecular-cloud material in interplanetary dust particles. *Nature, 404,* 968–971.

Messenger S., Keller L. P., and Walker R. M. (2002) Discovery of abundant interstellar silicates in cluster IDPs (abstract). In *Lunar and Planetary Science XXXIII,* Abstract #1887. Lunar and Planetary Institute, Houston (CD-ROM).

Messenger S., Keller L. P., Stadermann F. J., Walker R. M., and Zinner E. (2003) Samples of stars beyond the solar system: Silicate grains in interplanetary dust. *Science, 300,* 105–108.

Millar T. J., Bennett A., and Herbst E. (1989) Deuterium fractionation in dense interstellar clouds. *Astrophys. J., 340,* 906–920.

Millar T. J, Roberts H., Markwick A. J., and Charnley S. B. (2000) The role of H_2D^+ in the deuteration of interstellar molecules. *Philos. Trans. R. Soc. London, A358,* 2535–2547.

Mitchell G. F. and Deveau T. J. (1983) Effects of a shock on the molecular composition of a diffuse interstellar cloud. *Astrophys. J., 266,* 646–661.

Miville-Deschênes M.-A., Boulanger F., Joncas G., and Falgarone E. (2002) ISOCAM observations of the Ursa Major cirrus: Evidence for large abundance variations of small dust grains. *Astron. Astrophys., 381,* 209–218.

Molster F. J., Waters L. B. F. M., Tielens A. G. G. M., and Barlow M. J. (2002a) Crystalline silicate dust around evolved stars. I. The sample stars. *Astron. Astrophys., 382,* 184–221.

Molster F. J., Waters L. B. F. M., and Tielens A. G. G. M. (2002b) Crystalline silicate dust around evolved stars. II. The crystalline silicate complexes. *Astron. Astrophys., 382,* 222–240.

Molster F. J., Waters L. B. F. M., Tielens A. G. G. M., Koike C., and Chihara H. (2002c) Crystalline silicate dust around evolved stars. III. A correlations study of crystalline silicate features. *Astron. Astrophys., 382,* 241–255.

Moore M. H. and Hudson R. L. (1998) Infrared study of ion-irradiated water-ice mixtures with hydrocarbons relevant to comets. *Icarus, 135,* 518–527.

Motte F., Andre P., and Neri R. (1998) The initial conditions of star formation in the rho Ophiuchi main cloud: Wide-field millimeter continuum mapping. *Astron. Astrophys., 336,* 150–172.

Mumma M. J., Weissman P. R., and Stern S. A. (1993) Comets and the origin of the solar system. In *Protostars and Planets III* (E. H. Levy and J. I. Lunine, eds.), pp. 1177–1252. Univ. of Arizona, Tucson.

Mumma M. J., Disanti M. A., dello Russo N., Fomenkova M., Magee-Sauer K., Kaminski C. D., and Xie D. X. (1996) Detection of abundant ethane and methane, along with carbon monoxide and water, in comet C/1996 B2 Hyakutake: Evidence for interstellar origin. *Science, 272,* 1310–1314.

Muñoz Caro G. M., Ruiterkamp R., Schutte W. A., Greenberg J. M., and Mennella V. (2001) UV photodestruction of CH bonds and the evolution of the 3.4 μm feature carrier. I. The case of aliphatic and aromatic molecular species. *Astron. Astrophys., 367,* 347–354.

Myers P. C., Evans N. J. II, and Ohashi N. (2000) Observations of infall in star-forming regions. In *Protostars and Planets IV* (V. Mannings et al., eds.), pp. 217–245. Univ. of Arizona, Tucson.

Nakano T. (1998) Star formation in magnetic clouds. *Astrophys. J., 494,* 587–604.

Nakano T., Hasegawa T., and Norman C. (1995) The mass of a star formed in a cloud core: Theory and its application to the Orion A cloud. *Astrophys. J., 450,* 183–195.

Neufeld D. and Hollenbach D. J. (1994) Dense molecular shocks and accretion onto protostellar disks. *Astrophys. J., 428,* 170–185.

Nittler L. R., Amari S., Zinner E., Woosley S. E., and Lewis R. S. (1996) Extinct [44]Ti in presolar graphite and SiC: Proof of a supernova origin. *Astrophys. J. Lett., 462,* L31–L34.

Nittler L. R., Alexander C. M. O'D., Gao X., Walker R. M., and Zinner E. (1997) Stellar sapphires: The properties and origins of presolar Al_2O_3 in meteorites. *Astrophys. J., 483,* 475–495.

Novozamsky J. H., Schutte W. A., and Keane J. V. (2001) Further evidence for the assignment of the XCN band in astrophysical ice. *Astron. Astrophys., 379,* 588–591.

Ohishi M., Ishikawa S.-I., Yamamoto S., Saito S., and Amano T. (1995) The detection and mapping observations of C_2H_5OH in Orion Kleinmann-Low. *Astrophys. J. Lett., 446,* L43–L46.

Onaka T., de Jong T., and Willems F. J. (1989) A study of M Mira variables based on IRAS LRS observations. I. — Dust formation in the circumstellar shell. *Astron. Astrophys., 218,* 169–179.

Ostriker E. C., Stone J. M., and Gammie C. F. (2001) Density, velocity, and magnetic field structure in turbulent molecular cloud models. *Astrophys. J., 546,* 980–1005.

Pagani L., Lagache G., Bacmann A., Motte F., Cambrésy L., Fich M., Teyssier D., Miville-Deschênes M.-A., Pardo J.-R., Apponi A. J., and Stepnik B. (2003) L183 (L134N) revisited. I. The very cold core and the ridge. *Astron. Astrophys., 406,* L59–L62.

Palla F. and Stahler S. W. (2000) Accelerating star formation in clusters and associations. *Astrophys. J., 540,* 255–270.

Palla F. and Stahler S. W. (2002) Star formation in space and time: Taurus-Auriga. *Astrophys. J., 581,* 1194–1203.

Palumbo M. E., Pendleton Y. J., and Strazzulla G. (2000) Hydrogen isotopic substitution studies of the 2165 wavenumber (4.62 micron) "XCN" feature produced by ion bombardment. *Astrophys. J., 542,* 890–893.

Papoular R. (2003) Collisions between carbonaceous grains in the interstellar medium. *Astron. Astrophys.,* in press.

Parise B., Ceccarelli C., Tielens A. G. G. M., Herbst E., Lefloch B., Caux E., Castets A., Mukhopadhyay I., Pagani L., and Loinard L. (2002) Detection of doubly-deuterated methanol in the solar-type protostar IRAS 16293–2422. *Astron. Astrophys., 393,* L49–L53.

Pavlovski G., Smith M. D., Mac Low M.-M., and Rosen A. (2002) Hydrodynamical simulations of the decay of high-speed molecular turbulence — I. Dense molecular regions. *Mon. Not. R. Astron. Soc., 337,* 477–487.

Pendleton Y. J. and Allamandola L. J. (2002) The organic refractory material in the diffuse interstellar medium: Mid-infrared spectroscopic constraints. *Astrophys. J. Suppl. Ser., 138,* 75–98.

Pendleton Y. J., Sandford S. A., Allamandola L. J., Tielens A. G. G. M., and Sellgren K. (1994) Near-infrared absorption spectroscopy of interstellar hydrocarbon grains. *Astrophys. J., 437,* 683–696.

Pendleton Y. J., Tielens A. G. G. M., Tokunaga A. T., and Bernstein M. P. (1999) The interstellar 4.62 micron band. *Astrophys. J., 513,* 294–304.

Penston M. V. (1969) Dynamics of self-gravitating gaseous spheres — III. Analytical results in the free-fall of isothermal cases. *Mon. Not. R. Astron. Soc., 144,* 425–448.

Pineau des Forêts G., Flower D. R., Hartquist T. W., and Dalgarno A. (1986) Theoretical studies of interstellar molecular shocks. III. The formation of CH^+ in diffuse clouds. *Mon. Not. R. Astron. Soc., 220,* 801–824.

Pironello V., Biham O., Liu C., Shen L., and Vidali G. (1997) Efficiency of molecular hydrogen formation on silicates. *Astrophys. J. Lett., 483,* L131–L134.

Pironello V., Liu C., Roser J. E., and Vidali G. (1999) Measurements of molecular hydrogen formation on carbonaceous grains. *Astron. Astrophys., 344,* 681–686.

Pontoppidan K. M., Fraser H. J., Dartois E., Thi W. F., van Dishoeck E. F., Boogert A. C., d'Hendecourt L., Tielens A. G. G. M., and Bisschop S. E. (2003a) A 3–5 micron VLT spectroscopic survey of embedded young low mass stars I: Structure of the CO ice. *Astron. Astrophys., 408,* 981.

Pontoppidan K. M., Dartois E., van Dishoeck E. F., Thi W. F., and d'Hendecourt L. (2003b) Detection of abundant solid methanol toward young low mass stars. *Astron. Astrophys., 404,* L17–L21.

Posch T., Kerschbaum F., Mutschke H, Fabian D., Dorschner J., and Hron J. (1999) On the origin of the 13 μm feature. A study of ISO-SWS spectra of oxygen-rich AGB stars. *Astron. Astrophys., 352,* 609–618.

Posch Th., Kerschbaum F., Mutschke H., Dorschner J., and Jäger C. (2002) On the origin of the 19.5 μm feature. Identifying circumstellar Mg-Fe-oxides. *Astron. Astrophys., 393,* L7–L10.

Prasad S. S. and Tarafdar S. P. (1983) UV radiation field inside dense clouds — Its possible existence and chemical implications. *Astrophys. J., 267*, 603–609.

Pratap P., Dickens J. E., Snell R. L., Miralles M. P., Bergin E. A., Irvine W. M., and Schloerb F. P. (1997) A study of the physics and chemistry of TMC-1. *Astrophys. J., 486*, 862–885.

Price R. J., Viti S., and Williams D. A. (2003) On the origin of diffuse clouds. *Mon. Not. R. Astron. Soc., 343*, 1257–1262.

Puget J. L. and Léger A. (1989) A new component of the interstellar matter — Small grains and large aromatic molecules. *Annu. Rev. Astron. Astrophys., 27*, 161–198.

Rachford B. L., Snow T. P., Tumlinson J., Shull J. M., Blair W. P., Ferlet R., Friedman S. D., Gry C., Jenkins E. B., Morton D. C., Savage B. D., Sonnentrucker P., Vidal-Madjar A., Welty D. E., and York D. G. (2002) A far ultraviolet spectroscopic explorer survey of interstellar molecular hydrogen in translucent clouds. *Astrophys. J., 577*, 221–244.

Rawlings J. M. C., Hartquist T. W., Menten K. M., and Williams D. A. (1992) Direct diagnosis of infall in collapsing protostars. I. The theoretical identification of molecular species with broad velocity distributions. *Mon. Not. R. Astron. Soc., 255*, 471–485.

Rawlings J. M. C., Drew J. E., and Barlow M. J. (1993) Excited hydrogen and the formation of molecular hydrogen via associative ionization. I. Physical processes and outflows from young stellar objects. *Mon. Not. R. Astron. Soc., 265*, 968–982.

Rawlings M. G., Adamson A. J., and Whittet D. C. B. (2003) Infrared and visual interstellar absorption features towards heavily reddened field stars. *Mon. Not. R. Astron. Soc., 341*, 1121–1140.

Remijan A., Snyder L. E., Friedel D. N., Liu S.-Y., and Shah R. Y. (2003) Survey of acetic acid toward hot molecular cores. *Astrophys. J., 590*, 314–332.

Renzini A. and Voli M. (1981) Advanced evolutionary stages of intermediate-mass stars. I. Evolution of surface compositions. *Astron. Astrophys., 94*, 175–193

Richter P., Walker B. P., Savage B. D., and Sembach K. R. (2003) A far ultraviolet spectroscopic explorer survey of molecular hydrogen in intermediate-velocity clouds in the Milky Way halo. *Astrophys. J., 586*, 230–248.

Roberge W. G. and Draine B. T. (1990) A new class of solutions for interstellar magnetohydrodynamic shock waves. *Astrophys. J., 350*, 700–721.

Rodgers S. D. and Charnley S. B. (2001) Chemical differentiation in regions of massive star formation. *Astrophys. J., 546*, 324–329.

Rodgers S. D. and Charnley S. B. (2003) Chemical evolution in protostellar envelopes — Cocoon chemistry. *Astrophys. J., 585*, 355–371.

Roser J. E., Vidali G., Manico G., and Pirronello V. (2001) Formation of carbon dioxide by surface reactions on ices in the interstellar medium. *Astrophys. J. Lett., 555*, L61–L64.

Ruffle D. P. and Herbst E. (2001) New models of interstellar gas-grain chemistry. III. Solid CO_2. *Mon. Not. R. Astron. Soc., 322*, 770–778.

Salama F., Bakes E. L. O., Allamandola L. J., and Tielens A. G. G. M. (1996) Assessment of the polycyclic aromatic hydrocarbons — diffuse interstellar band proposal. *Astrophys. J., 458*, 621–636.

Savage K. R. and Sembach B. D. (1996) Interstellar abundances from absorption-line observations with the Hubble Space Telescope. *Annu. Rev. Astron. Astrophys., 34*, 279–329.

Scappini F., Cecchi-Pestellini C., Smith H., Klemperer W., and Dalgarno A. (2003) Hydrated sulphuric acid in dense molecular clouds. *Mon. Not. R. Astron. Soc., 41*, 657–661.

Schilke P., Walmsley C. M., Pineau des Forêts G., and Flower D. R. (1997) SiO production in interstellar shocks. *Astron. Astrophys., 321*, 293–304.

Schoier F. L., Jorgensen J. K., van Dishoeck E. F., and Blake G. A. (2002) Does IRAS 16293–2422 have a hot core? Chemical inventory and abundance changes in its protostellar environment. *Astron. Astrophys., 390*, 1001–1021.

Schutte W. A., Allamandola L. J., and Sandford S. A. (1993) An experimental study of the organic molecules produced in cometary and interstellar ice analogs by thermal formaldehyde reactions. *Icarus, 104*, 118–137.

Sembach B. D. and Savage K. R. (1996) The gas and dust abundances of diffuse halo clouds in the Milky Way. *Astrophys. J., 457*, 211–227.

Shapiro P. R. and Field G. B. (1976) Consequences of a new hot component of the interstellar medium. *Astrophys. J., 205*, 762–765.

Shu F. H. (1977) Self-similar collapse of isothermal spheres and star formation. *Astrophys. J., 214*, 488–497.

Shu F. H., Adams F. C., and Lizano S. (1987) Star formation in molecular clouds — Observation and theory. *Annu. Rev. Astron. Astrophys., 25*, 23–81.

Shu F. H., Najita J., Galli D., Ostriker E., and Lizano S. (1993) The collapse of clouds and the formation and evolution of stars and disks. In *Protostars and Planets* (E. H. Levy and J. I. Lunine, eds.), pp. 3–45. Univ. of Arizona, Tucson.

Shu F. H., Najita J. R., Shang H., and Li Z.-Y. (2000) X-winds theory and observations. In *Protostars and Planets IV* (V. Mannings et al., eds.), pp. 789–814. Univ. of Arizona, Tucson.

Smith R. G., Sellgren K., and Tokunaga A. T. (1989) Absorption features in the 3 micron spectra of protostars. *Astrophys. J., 344*, 413–426.

Snow T. P. and Witt A. N. (1996) Interstellar depletions updated; where all the atoms went. *Astrophys. J. Lett., 468*, L65–L68.

Snyder L. E., Lovas F. J., Mehringer D. M., Miao Y., Kuan Y.-J., Hollis J. M., and Jewell P. R. (2002) Confirmation of interstellar acetone. *Astrophys. J., 578*, 245–255.

Sofia U. J., Cardelli J. A., and Savage B. D. (1994) The abundant elements in interstellar dust. *Astrophys. J., 430*, 650–666.

Sofia U. J., Cardelli J. A., Guerin K. P., and Meyer D. M. (1997) Carbon in the diffuse interstellar medium. *Astrophys. J. Lett., 482*, L105–L108.

Speck A. K., Hofmeister A. M., and Barlow M. J. (1999) The silicon carbide problem: Astronomical and meteoritic evidence. *Astrophys. J. Lett., 513*, L87–L90.

Speck A. K., Barlow M. J., Sylvester R. J., and Hofmeister A. M. (2000) Dust features in the 10-μm infrared spectra of oxygen-rich evolved stars. *Astron. Astrophys. Suppl. Ser., 146*, 437–464.

Spitzer L. Jr. (1985) Average density along interstellar lines of sight. *Astrophys. J. Lett., 290*, L21–L24.

Spitzer L. Jr. (1990) Theories of the hot interstellar gas. *Annu. Rev. Astron. Astrophys., 28*, 71–101.

Spitzer L. Jr. and Jenkins E. B. (1975) Ultraviolet studies of the interstellar gas. *Annu. Rev. Astron. Astrophys., 13*, 133–164.

Spoon H., Moorwood A., Pontoppidan K. M., Cami J., Kregel M., Lutz D., and Tielens A. G. G. M. (2003) Detection of strongly processes ice in the central starburst of NGC 4945. *Astron. Astrophys., 402*, 499–507.

Stahler S. W., Palla F., and Ho P. T. P. (2000) The formation of massive stars. In *Protostars and Planets IV* (V. Mannings et al., eds.), pp. 327–354. Univ. of Arizona, Tucson.

Stantcheva T., Shematovich V. I., and Herbst E. (2002) On the master equation approach to diffusive grain-surface chemistry: The H, O, CO system. *Astron. Astrophys., 391*, 1069–1080.

Stepnik B., Abergel A., Bernard J.-P., Boulanger F., Canbrésy L., Giard M., Jones A., Lagache G., Lamarre J.-M., Mény C., Pajot F., Le Peintre F., Ristorcelli I., Serra G., and Torre J.-P. (2001) Evolution of the dust properties in a translucent cloud. In *The Promise of the Herschel Space Observatory* (G. L. Pilbratt et al., eds.), pp. 269–272. ESA SP-460, Noordwijk, The Netherlands.

Sykes M. V., Grün E., Reach W. T., and Jenniskens P. (2004) The interplanetary dust complex and comets. In *Comets II* (M. C. Festou et al., eds.), this volume. Univ. of Arizona, Tucson.

Takakuwa S., Mikami H., Saito M., and Hirano N. (2000) A comparison of the spatial distribution of $H^{13}CO^+$, CH_3OH, and $C^{34}S$ emission and its implication in Heiles Cloud 2. *Astrophys. J., 542,* 367–379.

Taban I. M., Schutte W. A., Pontoppidan K. M., and van Dishoeck E. F. (2003) Stringent upper limits to the solid NH_3 abundance towards W33A from near-IR spectroscopy with the Very Large Telescope. *Astron. Astrophys., 399,* 169–175.

Taylor S. D. and Raga A. C. (1995) Molecular mixing layers in stellar outflows. *Astron. Astrophys., 296,* 823–832.

Tegler S. C., Weintraub D. A., Rettig T. W., Pendleton Y. J., Whittet D. C. B., and Kulesa C. A. (1995) Evidence for chemical processing of precometary icy grains in circumstellar environments of pre-main-sequence stars. *Astrophys. J., 439,* 279–287.

Terebey S., Shu F. H., and Cassen P. (1984) The collapse of the cores of slowly rotating isothermal clouds. *Astrophys. J., 286,* 529–551.

Terzieva R. and Herbst E. (2000) The possibility of nitrogen isotopic fractionation in interstellar clouds. *Mon. Not. R. Astron. Soc., 317,* 563–568.

Testi L. and Sargent A. I. (1998) Star formation in clusters: A survey of compact millimeter-wave sources in the Serpens core. *Astrophys. J. Lett., 508,* L91–L94.

Thi W. F., Pontoppidan K. M., van Dishoeck E. F., Dartois E., and d'Hendecourt L. (2002) Detection of abundant solid CO in the disk around CRBR 2422.8–3423. *Astron. Astrophys., 394,* L27–L30.

Thielemann F.-K., Nomoto K., and Hashimoto M. (1996) Core-collapse super-novae and their ejecta. *Astrophys. J., 460,* 408–436.

Tielens A. G. G. M. (1983) Surface chemistry of deuterated molecules. *Astron. Astrophys., 119,* 177–184.

Tielens A. G. G. M. (1998) Interstellar depletions and the life cycle of interstellar dust. *Astrophys. J., 499,* 267–272.

Tielens A. G. G. M. and Allamandola L. J. (1987) Composition, structure, and chemistry of interstellar dust. In *Interstellar Processes* (D. J. Hollenbach and H. A. Thronson, eds.), pp. 397–469. Reidel, Dordrecht.

Tielens A. G. G. M. and Charnley S. B. (1997) Circumstellar and interstellar synthesis of organic molecules. *Origins Life Evol. Biosph., 27,* 23–51.

Tielens A. G. G. M. and Hagen W. (1982) Model calculations of the molecular composition of interstellar grain mantles. *Astron. Astrophys., 114,* 245–260.

Tielens A. G. G. M. and Hollenbach D. J. (1985) Photodissociation regions. I. — Basic model. II. — A model for the Orion photodissociation region. *Astrophys. J., 291,* 722–754.

Tielens A. G. G. M., Waters L. B. F. M., Molster F. J., and Justtanont K. (1997) Circumstellar silicate mineralogy. *Astrophys. Space Sci., 255,* 415–426.

Tielens A. G. G. M., Hony S., van Kerckhoven C., and Peeters E. (1999) Interstellar and circumstellar PAHs. In *The Universe As Seen by ISO* (P. Cox et al., eds.), pp. 579–588. ESA SP-427, Noordwijk, The Netherlands.

Turner B. E. (1991) A molecular line survey of Sagittarius B2 and Orion-KL from 70 to 115 GHz. *Astrophys. J. Suppl. Ser., 76,* 617–686.

Turner B. E. (2000) A common gas-phase chemistry for diffuse, translucent, and dense clouds. *Astrophys. J., 542,* 837–860.

Turner B. E. and Apponi A. (2001) Microwave detection of interstellar vinyl alcohol, CH_2CHOH. *Astrophys. J. Lett., 561,* L207–L210.

Tsuji T. (1973) Molecular abundances in stellar atmospheres. *Astron. Astrophys., 23,* 411–431.

van der Tak F. F. S., van Dishoeck E. F., and Caselli P. (2000) Abundance profiles of CH_3OH and H_2CO toward massive young stars as tests of gas-grain chemical models. *Astron. Astrophys., 361,* 327–339.

van Dishoeck E. F. and Black J. H. (1988) Diffuse cloud chemistry. In *Rate Coefficients in Astrochemistry* (T. J. Millar and D. A. Williams, eds.), pp. 209–237. Kluwer, Dordrecht.

van Dishoeck E. F. and Blake G. A. (1998) Chemical evolution of star-forming regions. *Annu. Rev. Astron. Astrophys., 36,* 317–368.

van Dishoeck E. F., Blake G. A., Draine B. T., and Lunine J. I. (1993) The chemical evolution of protostellar and protoplanetary matter. In *Protostars and Planets III* (E. H. Levy and J. I. Lunine, eds.), pp. 163–241. Univ. of Arizona, Tucson.

Vanhala H. A. T. and Boss A. P. (2000) Injection of radioactivities into the presolar cloud: Convergence testing. *Astrophys. J., 538,* 911–921.

Van Kerckhoven C., Hony S., Peeters E., Tielens A. G. G. M., Allamandola L. J., Hudgins D. M., Cox P., Roelfsema P. R., Voors R. H. M., Waelkens C., Waters L. B. F. M., and Wesselius P. R. (2000) The C-C-C bending modes of PAHs: A new emission plateau from 15 to 20 μm. *Astron. Astrophys., 357,* 1013–1019.

Vázquez-Semandeni E. (2004) The turbulent star formation model. Outline and tests. In *IAU Symposium 221: Star Formation at High Angular Resolution* (M. Burton et al., eds.), in press. Astronomical Society of the Pacific, San Francisco.

Vázquez-Semandeni E., Ostriker E. C., Passot T., Gammie C. F., and Stone J. M. (2000) Compressible MHD turbulence: Implications for molecular cloud and star formation. In *Protostars and Planets IV* (V. Mannings et al., eds.), pp. 3–28. Univ. of Arizona, Tucson.

Velusamy T., Langer W. D., and Goldsmith P. F. (2002) Tracing the infall and the accretion shock in the protostellar disk: L1157. *Astrophys. J. Lett., 565,* L43–L46.

Viti S., Caselli P., Hartquist T. W., and Williams D. A. (2001) Chemical signatures of shocks in hot cores. *Astron. Astrophys., 370,* 1017–1025.

Wakelam V., Castets A., Ceccarelli C., Lefloch B., Caux E., and Pagani L. (2003) Sulphur-bearing species in the star forming region L1689N. *Astron. Astrophys.,* in press.

Walker C. K., Maloney P. R., and Serabyn E. (1994) Vibrationally excited CS: A new probe of conditions in young protostellar systems. *Astrophys. J. Lett., 437,* L127–L130.

Watanabe N. and Kouchi A. (2002) Efficient formation of formaldehyde and methanol by the addition of hydrogen atoms to CO in H_2O-CO ice at 10 K. *Astrophys. J. Lett., 571,* L173–L176.

Watanabe N., Shiraki T., and Kouchi A. (2003) The dependence of H_2CO and CH_3OH formation on the temperature and thickness of H_2O-CO ice during the successive hydrogenation of

CO. *Astrophys. J. Lett., 588,* L121–L124.

Weidenschilling S. J. and Ruzmaikina T. V. (1994) Coagulation of grains in static and collapsing protostellar clouds. *Astrophys. J., 430,* 713–726.

Wheeler C. J., Sneden C., and Truran J. W. Jr. (1989) Abundance ratios as a function of metallicity. *Annu. Rev. Astron. Astrophys., 27,* 279–349.

Willacy K. and Millar T. J. (1998) Desorption processes and the deuterium fractionation in molecular clouds. *Mon. Not. R. Astron. Soc., 298,* 562–568.

Williams D. A. (1992) The chemistry of interstellar CH⁺ — The contribution of Bates and Spitzer (1951). *Planet. Space Sci., 40,* 1683–1693.

Whittet D. C. B. (1993) Observations of molecular ices. In *Dust and Chemistry in Astronomy* (T. J. Millar and D. A. Williams, eds.), pp. 9–35. IOP Publications Ltd., Bristol.

Whittet D. C. B., Duley W. W., and Martin P. G. (1990) On the abundance of silicon carbide in the interstellar medium. *Mon. Not. R. Astron. Soc., 244,* 427–431.

Wheeler J. C., Sneden C., and Truran J. W. Jr. (1989) Abundance ratios as a function of metallicity. *Annu. Rev. Astron. Astrophys., 27,* 279–349.

Willson L. A. (2000) Mass loss from cool stars: Impact on the evolution of stars and stellar populations. *Annu. Rev. Astron. Astrophys., 38,* 573–611.

Wolfire M. G., Hollenbach D., McKee C. F., Tielens A. G. G. M., and Bakes E. L. O. (1995) The neutral atomic phases of the interstellar medium. *Astrophys. J., 443,* 152–168.

Wolfire M. G., McKee C. F., Hollenbach D., and Tielens A. G. G. M. (2003) Neutral atomic phases of the interstellar medium in the galaxy. *Astrophys. J., 587,* 278–311.

Wooden D. H. (2002) Comet grains: Their IR emission and their relation to ISM grains. *Earth Moon Planets, 89,* 247–287.

Woon D. E. (2002) Pathways to glycine and other amino acids in ultraviolet-irradiated astrophysical ices determined via quantum chemical modeling. *Astrophys. J. Lett., 571,* L177–180.

Woosley S. E. (1986) Nucleosynthesis and stellar evolution. In *Saas-Fee Advanced Course 16: Nucleosynthesis and Chemical Evolution,* given at the Swiss Society for Astronomy and Astrophysics (SSAA) (B. Hauck et al., eds.), pp. 1–195. Geneva Observatory, Sauverny, Switzerland.

Woosley S. E. (1988) SN1987A — After the peak. *Astrophys. J., 330,* 218–253.

Woosley S. E. and Weaver T. A. (1995) The evolution and explosion of massive stars. II. Explosive hydrodynamics and nucleosynthesis. *Astrophys. J. Suppl. Ser., 101,* 181–235.

Wootten H. A., ed. (2001) *Science with the Atacama Large Millimeter Array.* Astronomical Society of the Pacific, San Francisco. 365 pp.

Yin Q., Jacobsen S. B., and Yamashita K. (2002) Diverse supernova sources of pre-solar material inferred from molybdenum isotopes in meteorites. *Nature, 416,* 881–883.

Zagury F., Jones A., and Boulanger F. (1998) Dust composition in the low density medium around Spica. *Lecture Notes Phys., 506,* 385–388.

From Molecular Clouds to Circumstellar Disks

Alan P. Boss
Carnegie Institution of Washington

We now have examples of nearly all the phases of evolution of a dense molecular cloud core into a young star, and at least a provisional theoretical understanding of the star formation process. However, our understanding of the processes by which the material leftover from the star formation process is converted into planets and comets is much less secure, both observationally and theoretically. Major questions remain unanswered, such as the processes by which proto-planetary disks transport mass and angular momentum, and the formation mechanism of the solar system's gas and ice giant planets. Indeed, the latter question raises the issue of whether the solar system formed in a relatively benign region of low-mass star formation, similar to Taurus, or in a more violent region, similar to Orion. Evidently much remains to be learned before we can claim to understand the origin of the solar system.

1. INTRODUCTION

By their very nature as primitive bodies that have never experienced prolonged metamorphism or strong heating, comets are believed to preserve much of the record of the formational processes that led to the origin of our solar system. Presolar grains contained in comets undoubtedly carry much of the history of galactic nucleosynthesis in their isotopic abundances, while fragile molecular species in comets speak of the cold, dark clouds that form new generations of stars, and of processes in the circumstellar disk from which the solar system originated about 4.6 G.y. ago. Much of the motivation for observational and cosmochemical studies of meteorities and comets stems from the desire to use the results to constrain or otherwise illuminate the physical and chemical conditions in the solar nebula, the Sun's circumstellar disk, in the hope of learning more about the processes that led to the formation of our planetary system.

In addition to studies of comets, there are important lessons to be learned about the planet formation process from astrophysical observations of young stellar objects and their accompanying protoplanetary disks (*Dutrey et al.,* 2004), from the discovery of other planetary systems, and from theoretical models. A key resource for learning more about these subjects is the compendium volume *Protostars and Planets IV* (*Mannings et al.,* 2000). The present chapter is adapted and modified from a recent review article about the solar nebula, focusing on issues of interest to cosmochemists (*Boss,* 2004). Here we address topics of more interest for comets.

1.1. Implications from the Discovery of Extrasolar Planetary Systems

Prior to 1995, the solar system was the only known example of a planetary system in orbit around a Sun-like star, and the question of the uniqueness of the solar system was more of a philosophical than a scientific matter. Somewhat improbable formation mechanisms could be plausibly argued to have operated in the solar nebula, so long as only one such planetary system was known to exist. The discovery of the first extrasolar planets (*Mayor and Queloz,* 1995; *Marcy and Butler,* 1996) has forever changed that viewpoint. While the planetary census of the solar neighborhood is still underway, in the next few years we will have a good idea about the prevalence of planetary systems with long-period Jupiter-mass planets, systems that might be close analogs to our own. We already know that in some cases, the planetary formation process produces planets with masses considerably larger than that of Jupiter, moving on highly eccentric orbits in the very region of the planetary system where we find the terrestrial planets and asteroids in the solar system.

This shocking fact is a sobering lesson that the outcome of the planet formation and evolution process need not resemble our system. Understanding the reasons for such wide variations in outcome remains a central issue in planet formation theory (*Weidenschilling,* 2004). Until we can claim to have a general understanding of planet formation processes, our understanding of the origin of any single system, including the solar system, must be considered provisional. Given that the Oort cloud comets are thought to have formed in the same general region of the nebula as the giant planets, the specific question of the formation mechanism of the solar system's giant planets is of critical interest for cometary science.

1.2. Theoretical Progress in Understanding Basic Physics of the Solar Nebula

Theoreticians specializing in planet formation have been energized by the discovery of extrasolar planetary systems, often to the detriment of further understanding of the origin of the solar system. Nevertheless, the rejuvenation of the field engendered by these discoveries will have a major effect there as well, as new theories designed for extrasolar systems are applied to our own. For example, the disk instability mechanism for gas-giant-planet formation (*Boss, 1997*) may be equally applicable to extrasolar protoplanetary disks or to the solar nebula. The importance of understanding the basic physics of the solar nebula has led to innovative new approaches to old problems, such as the X-wind mechanism for processing of solids (*Shu et al., 1996*), and the rediscovery of the magneto-rotational instability as a mechanism for disk evolution (*Balbus and Hawley, 1991*). Coupled with ever-increasing computational power, which permits brute-force numerical solutions of otherwise intractable problems, theoretical models of the varied processes in the solar nebula have made considerable progress, although the need for guidance from observations and laboratory studies cannot be overestimated.

2. MOLECULAR CLOUD COLLAPSE

The protosun and solar nebula were formed by the self-gravitational collapse of a dense molecular cloud core, much as we see new stars being formed today in regions of active star formation. The formation of the solar nebula was largely an initial value problem: i.e., given detailed knowledge of the particular dense molecular cloud core that was the presolar cloud, one could in principle calculate the flow of gas and dust subject to the known laws of physics and thus predict the basic outcome. Specific details of the outcome cannot be predicted, however, as there appears to be an inevitable amount of stochastic, chaotic evolution involved, e.g., in the orbital motions of any ensemble of gravitationally interacting particles. Nevertheless, we expect that at least the gross features of the solar nebula should be predictable from theoretical models of cloud collapse, constrained by astronomical observations of young stellar objects. For this reason, the physical structure of likely precollapse clouds is of interest with regard to inferring the formation mechanism of the protosun and the structure of the accompanying solar nebula.

2.1. Observations of Precollapse Clouds

Astronomical observations at long wavelengths (e.g., millimeter) are able to probe deep within interstellar clouds of gas and dust that are opaque at short wavelengths (e.g., visible wavelengths). These clouds are composed primarily of H_2 gas, He, and molecules such as CO, hence the term molecular clouds. About 1% by mass of these clouds is in the form of submicrometer-sized dust grains, with about another 1% composed of gaseous molecules and atoms of elements heavier than He. Regions of active star formation are located within molecular clouds and complexes ranging in mass from a few solar masses to >1,000,000 $M_\odot$. This association of young stars with molecular clouds is the most obvious manifestation of the fact that stars form from these clouds. Many of the densest regions of these clouds were found to contain embedded infrared objects, i.e., newly formed stars whose light is scattered, absorbed, and reemit-

ted at infrared wavelengths in the process of exiting the placental cloud core. Such cores have already succeeded in forming stars. Initial conditions for the collapse of the presolar cloud can be more profitably ascertained from observations of dense cloud cores that do not appear to contain embedded infrared objects; i.e., precollapse cloud cores.

Precollapse cloud cores are composed of cold molecular gas with temperatures in the range of about 7 K to 15 K, and with gas densities of about 1000 to 100,000 molecules per cubic centimeter. Some clouds may be denser yet, but this is hard to determine because of the limited density ranges for which suitable molecular tracers are abundant (typically isotopes of CO and H_3N). Masses of these clouds range from roughly 1 $M_\odot$ to thousands of solar masses, with the distribution of clump masses fitting a power law such that most of the clumps are of low mass, as is also true of stars in general. In fact, one recent estimate of the mass distribution of precollapse clouds in Taurus is so similar to the initial mass function of stars, that it appears that the stellar mass distribution may be determined primarily by processes occurring prior to the formation of precollapse clouds (*Onishi et al., 2002*). The cloud properties described below are used to constrain the initial conditions for hydrodynamical models of the collapse of cloud cores.

Large radio telescopes have enabled high-spatial-resolution mapping of precollapse clouds and the determination of their interior density structure. While such clouds undoubtedly vary in all three space dimensions, the observations are typically averaged in angle to yield an equivalent, spherically symmetric density profile. These radial density profiles have shown that precollapse clouds typically have flat density profiles near their centers, as is to be expected for a cloud that has not yet collapsed to form a star (*Bacmann et al., 2000*), surrounded by an envelope with a steeply declining profile that could be fit with a power law. The density profile thus resembles that of a Gaussian distribution, or more precisely, the profile of the Bonnor-Ebert sphere, which is the equilibrium configuration for an isothermal gas cloud (*Alves et al., 2001*).

While precollapse clouds often have a complicated appearance, attempts have been made to approximate their shapes with simple geometries. Triaxial spheroids seem to be required in general, although most lower-mass clouds appear to be more nearly oblate than prolate (*Jones et al., 2001*). On the larger scale, prolate shapes seem to give a better fit than oblate spheroids. Another study found that the observations could be fit with a distribution of prolate spheroids with axis ratios of 0.54 (*Curry, 2002*), and argued that the prolate shapes derived from the filamentary nature of the parent clouds. We shall see that the precollapse cloud's shape is an important factor for the outcome of the protostellar collapse phase.

Precollapse clouds have significant interior velocity fields that appear to be a mixture of turbulence derived from fast stellar winds and outflows, and magnetohydrodynamic waves associated with the ambient magnetic field. In addition, there may be evidence for a systematic shift in velocities across one axis of the cloud, which can be interpreted as solid-body rotation around that axis. When estimated in this manner, typical rotation rates are found to be below the level needed for cloud support by centrifugal force, yet large enough to result in considerable rotational flattening once cloud collapse begins. Ratios of rotational to gravitational energy in dense cloud cores range from 0.002 to 1, with a typical value being 0.02 (*Goodman et al.,* 1993). The presence of a net angular momentum for the cloud is essential for the eventual formation of a centrifugally supported circumstellar disk.

2.2. Onset of Collapse Phase

Dense cloud cores are supported against their own self-gravity by a combination of turbulent motions, magnetic fields, thermal (gas) pressure, and centrifugal force, in roughly decreasing order of importance. Turbulent motions inevitably dissipate over timescales that are comparable to a cloud's freefall time (the time over which an idealized, pressureless sphere of gas of initially uniform density would collapse to form a star), once the source of the turbulence is removed. For a dense cloud core, freefall times are on the order of 0.1 m.y. However, dense clouds do not collapse on this timescale, because once turbulence decays, magnetic fields provide support against self-gravity.

Magnetic field strengths in dense clouds are measured by Zeeman splitting of molecular lines, and found to be large enough (about 10–1000 µG) to be capable of supporting dense clouds, provided that both static magnetic fields and magnetohydrodynamic waves are present (*Crutcher,* 1999). Field strengths are found to depend on the density to roughly the 1/2 power, as is predicted to be the case if ambipolar diffusion controls the cloud's dynamics (*Mouschovias,* 1991). Ambipolar diffusion is the process of slippage of the primarily neutral gas molecules past the ions, to which the magnetic field lines are effectively attached. This process occurs over timescales of a few million years or more for dense cloud cores, and inevitably leads to the loss of sufficient magnetic field support such that the slow inward contraction of the cloud turns into a rapid, dynamic collapse phase, when the magnetic field is no longer in control. This is generally believed to be the process through which stars in regions of low-mass star formation begin their life, the "standard model" of star formation (*Shu et al.,* 1987).

Recently this standard model has been challenged by evidence for short cloud lifetimes and a highly dynamic star formation process driven by large-scale outflows (*Hartmann et al.,* 2001). In regions of high-mass star formation, where the great majority of stars are believed to form (Fig. 1), quiescent star formation of the type envisioned in the standard model occurs only until the phase when high-mass stars begin to form and evolve. The process of high-mass star formation is not understood as well as that of low-mass stars, but observations make it clear that events such as the supernova explosions that terminate the life of massive stars can result in the triggering of star formation in neighbor-

ing molecular clouds that are swept up and compressed by the expanding supernova shock front (*Preibisch and Zinnecker,* 1999). Even strong protostellar outflows are capable of triggering the collapse of neighboring dense cloud cores (*Foster and Boss,* 1996). A supernova shock-triggered origin for the presolar cloud has been advanced as a likely source of the short-lived radioisotopes (e.g., ^{26}Al) that existed in the early solar nebula (*Cameron and Truran,* 1977). Detailed models of shock-triggered collapse have shown that injection of shock-front material containing the ^{26}Al into the collapsing protostellar cloud can occur, provided that the shock speed is on the order of 25 km/s (*Boss,* 1995), as is appropriate for a moderately distant supernova or for the wind from an evolved red giant star.

2.3. Outcome of Collapse Phase

Once a cloud begins to collapse as a result of ambipolar diffusion or triggering by a shock wave, supersonic inward motions develop and soon result in the formation of an optically thick first core, with a size on the order of 10 AU. This central core is supported primarily by the thermal pressure of the H_2 gas, while the remainder of the cloud continues to fall onto the core. For a 1-$M_\odot$ cloud, this core has a mass of about 0.01 $M_\odot$. Once the central temperature reaches about 2000 K, thermal energy goes into dissociating the H molecules, lowering the thermal pressure and leading to a second collapse phase, during which the first core disappears and a second, final core is formed at the center, with a radius a few times that of the Sun. This core then accretes mass from the infalling cloud over a timescale of about 1 m.y. (*Larson,* 1969). In the presence of rotation or magnetic fields, however, the cloud becomes flattened into a pancake, and may then fragment into two or more protostars. At this point, we cannot reliably predict what sort of dense cloud core will form in precisely what sort of star or stellar system, much less what sorts of planetary systems will accompany them, but certain general trends are evident.

The standard model pertains only to formation of single stars, whereas most stars are known to be members of binary or multiple star systems. There is growing observational evidence that multiple star formation may be the rule, rather than the exception (*Reipurth,* 2000). If so, then it may be that single stars like the Sun are formed in multiple protostar systems, only to be ejected soon thereafter as a result of the decay of the multiple system into an orbitally stable configuration (*Bate et al.,* 2002). In that case, the solar nebula would have been subject to strong tidal forces during the close encounters with other protostars prior to its ejection from the multiple system. This hypothesis has not been investigated in detail (but see *Kobrick and Kaula,* 1979). Detailed models of the collapse of magnetic cloud cores, starting from initial conditions defined by observations of molecular clouds, show that while initially prolate cores tend to fragment into binary protostars, initially oblate clouds form multiple protostar systems that are highly unstable and likely to eject single protostars and their disks

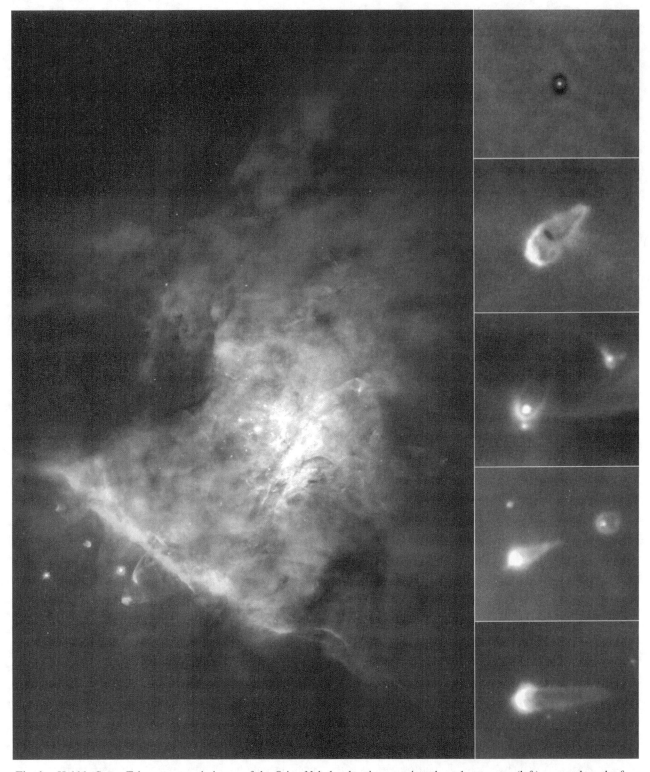

Fig. 1. Hubble Space Telescope mosaic image of the Orion Nebula, showing a region about 1 pc across (left) centered on the four massive stars of the Orion Trapezium. Also shown (right) are individual circumstellar disks orbiting around young stars in Orion, with a typical size of about 1000 AU. Many of these disks are being photoevaporated by UV emission from the Trapezium stars. [Photo credits: C. R. O'Dell and S. K. Wong (Rice University) and NASA. Hubble image STScI-PRC95-45a.]

(*Boss,* 2002a). Surprisingly, magnetic fields were found to enhance the tendency for a collapsing cloud to fragment by helping to prevent the formation of a single central mass concentration of the type assumed to form in the standard model of star formation.

In the case of nonmagnetic clouds, where thermal pressure and rotation dominate, single protostars can result from the collapse of dense cloud cores that are rotating slowly enough to avoid the formation of a large-scale protostellar disk that could then fragment into a binary system (*Boss and Myhill,* 1995). Alternatively, the collapse of an initially strongly centrally condensed (power-law), nonmagnetic cloud leads to the formation of a single central body (*Yorke and Bodenheimer,* 1999). However, considering that most cloud cores are believed to be supported to a significant extent by magnetic fields, the applicability of these results is uncertain. In the case of shock-triggered collapse, calculations have shown that weakly magnetic clouds seem to form single protostars when triggering occurs after the core has already contracted toward high central densities (*Vanhala and Cameron,* 1998).

In the case of the nonmagnetic collapse of a spherical cloud (*Yorke and Bodenheimer,* 1999), the protostar that forms is orbited by a protostellar disk with a similar mass. When angular momentum is transported outward by assumed gravitational torques, and therefore mass is transported inward onto the protostar, the amount of mass remaining in the disk is still so large that most of this matter must eventually be accreted by the protostar through other processes. Hence the disk at this phase must still be considered a protostellar disk, not a relatively late phase, protoplanetary disk where any objects that form have some hope of survival in orbit. Thus even in the relatively simple case of nonmagnetic clouds, it is not yet possible to compute the expected detailed structure of a protoplanetary disk, starting from the initial conditions of a dense cloud core. Calculations starting from less-idealized initial conditions, such as a segment of an infinite sheet (Fig. 2), suffer from the same limitations (*Boss and Hartmann,* 2001).

Because of the complications of multiple protostar formation, magnetic field support, possible shock-wave triggering, and angular momentum transport in the disk during the cloud infall phase, among others, a definitive theoretical model for the collapse of the presolar cloud has not yet emerged.

2.4. Observations of Star-forming Regions

Observations of star-forming regions have advanced our understanding of the star formation process considerably in the last few decades. We now can study examples of nearly all phases of the evolution of a dense molecular cloud core into a nearly fully formed star (i.e., the ~1-$M_\odot$ T Tauri stars). As a result, the theory of star formation is relatively mature, with future progress expected to center on defining the role played by binary and multiple stars and on refining observations of known phases of evolution.

Protostellar evolution can be conveniently subdivided into six phases that form a sequence in time. The usual starting point is the precollapse cloud, which collapses to form the first protostellar core, which is then defined to be a Class I object. The first core collapses to form the final, second core, or Class 0 object, which has a core mass less than that of the infalling envelope. Class I, II, and III objects (*Lada and Shu,* 1990) are defined in terms of their spectral energy distributions at midinfrared wavelengths, where the emission is diagnostic of the amount of cold, circumstellar dust. Class I objects are optically invisible,

infrared protostars with so much dust emission that the circumstellar gas mass is on the order of 0.1 $M_\odot$ or more. Class II objects have less dust emission, and a gas mass of about 0.01 $M_\odot$. Class II objects are usually optically visible, T Tauri stars, where most of the circumstellar gas resides in a disk rather than in the surrounding envelope. Class III objects are weak-line T Tauri stars, with only trace amounts of circumstellar gas and dust. While these classes imply a progression in time from objects with more to less gas emission, the time for this to occur for any given object is highly variable: Some Class III objects appear to be only 0.1 m.y. old, while some Class II objects have ages of several million years, based on theoretical models of the evolution of stellar luminosities and surface temperatures. Evidence for dust disks has been found around even older stars, such as Beta Pictoris, with an age of about 10 m.y., although its disk mass is much smaller than that of even Class III objects. Stars with such "debris disks" are often classified as Class III objects.

Multiple examples of all these phases of protostellar evolution have been found, with the exception of the short-lived Class I objects, which have not yet been detected. It is noteworthy that observations of protostars and young stars find a higher frequency of binary and multiple systems than is the case for mature stars, implying the orbital decay of many of these young systems (*Reipurth,* 2000; *Smith et al.,* 2000).

2.5. Interactions of Young Stars with Their Disks

A remarkable aspect of young stellar objects is the presence of strong molecular outflows for essentially all young stellar objects, even the Class 0 objects. This means that at the same time that matter is still accreting onto the protostar, it is also losing mass through a vigorous wind directed in a bipolar manner in both directions along the presumed rotation axis of the protostar/disk system (Fig. 3). In fact, the energy needed to drive this wind appears to be derived from mass accretion by the protostar, as observed wind momenta are correlated with protostellar luminosities and with the amount of mass in the infalling envelope (*Andrè,* 1997).

There are two competing mechanisms for driving bipolar outflows, both of which depend on magnetic fields to sling ionized gas outward and to remove angular momentum from the star/disk system (see *Shu et al.,* 2000; *Königl and Pudritz,* 2000). One mechanism is the X-wind model, where coronal winds from the central star and from the inner edge of the accretion disk join together to form the magnetized X-wind, launched from an orbital radius of a few stellar radii. The other mechanism is a disk wind, launched from the surface of the disk over a much larger range of distances, from less than 1 AU to as far away as 100 AU or so. In both mechanisms, centrifugal support of the disk gas makes it easier to launch this material outward, and bipolar flows develop in the directions perpendicular to the disk, because the disk forces the outflow into these preferred directions. Because it derives from radii deeper within the star's gravi-

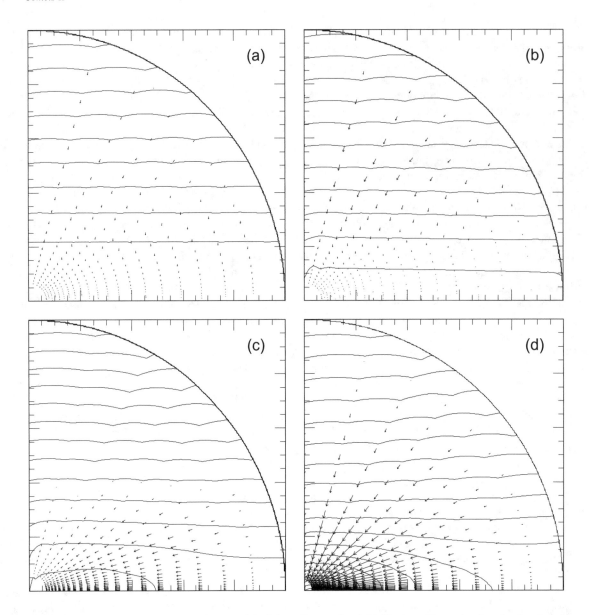

Fig. 2. Density contours and velocity vectors for a model of the collapse of an initially sheet-like molecular cloud to form a star and circumstellar disk (*Boss and Hartmann,* 2001). Four times are shown: **(a)** $t = 0.0\ t_{ff}$, $\rho_{max} = 2.1 \times 10^{-18}$ g cm^{-3}, $v_{max} = 6.6 \times 10^{-3}$ cm s^{-1}; **(b)** $t = 1.4\ t_{ff}$, $\rho_{max} = 3.2 \times 10^{-17}$ g cm^{-3}, $v_{max} = 5.8 \times 10^{3}$ cm s^{-1}; **(c)** $t = 5.7\ t_{ff}$, $\rho_{max} = 1.3 \times 10^{-14}$ g cm^{-3}, $v_{max} = 1.7 \times 10^{3}$ cm s^{-1}; **(d)** $t = 8.6\ t_{ff}$, $\rho_{max} = 2.0 \times 10^{-13}$ g cm^{-3}, $v_{max} = 2.1 \times 10^{4}$ cm s^{-1}. Density contours correspond to changes by a factor of 2. Region shown has a radius of 5700 AU ~ 0.03 pc. Only one quadrant of the two-dimensional calculation is shown, which has symmetry above and below the midplane (bottom border) and around the cloud's rotation and symmetry axis (left hand border). The initially flat sheet contracts and then collapses vertically and radially to form a central protostar with a mass of about 0.1 M$_\odot$ (unseen, at lower lefthand corner of plot), orbited by a flattened circumstellar disk and an infalling envelope with modest infall velocities on the largest scale.

tational potential well, an X-wind is energetically favored over a disk wind. However, observations show that during FU Orionis-type outbursts in young stars, mass is added onto the central star so rapidly that the X-wind region is probably crushed out of existence, implying that the strong outflows that still occur during these outbursts must be caused by an extended disk wind (*Hartmann and Kenyon,* 1996).

All T Tauri stars are believed to experience FU Orionis outbursts, so disk winds may be the primary driver of bipolar outflows, at least in the early FU Orionis phase of evolution. There is a strong correlation between the amount of mass available for accretion onto the disks and the amount of momentum in the outflow (*Bontemps et al.,* 1996), suggesting that disk mass accretion is directly related to outflow energetics. It is unclear at present what effect an X-wind or a disk wind would have on the planet formation process, beyond being responsible for the loss of energy (and angular momentum in the latter case), as the winds are thought to be launched either very close to the protostar in the former case, or from the disk's surface in the latter case.

The simple picture of the solar nebula being removed by a spherically symmetric T Tauri wind has long since

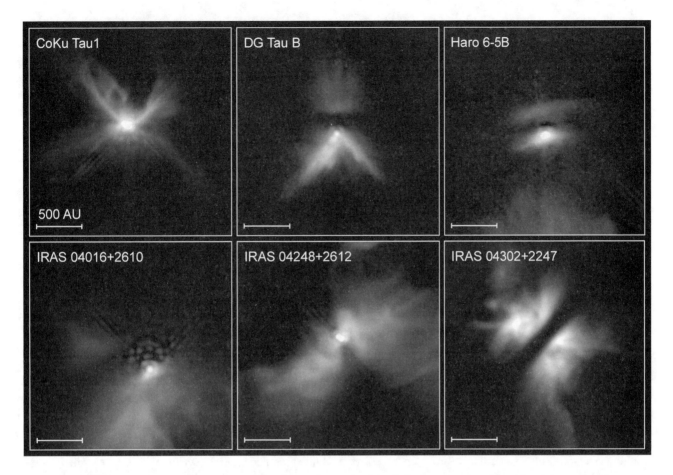

Fig. 3. Hubble Space Telescope images of circumstellar disks in the Taurus star-forming region, taken with the NICMOS camera. The images reveal the disks to be roughly hourglass-shaped configurations with openings threaded by molecular outflows (unseen), extending about 500 AU across. The central stars are mostly hidden from view by the nearly edge-on disks (dark stripes), but their light is reflected off the top and bottom surfaces of the disks and by the infalling envelopes of gas and dust. [Photo credits: D. Padgett (IPAC/Caltech), W. Brandner (IPAC), K. Staplefeldt (JPL), and NASA. Hubble image STScI-PRC1999-05a.]

been supplanted by the realization that young stars have directed, bipolar outflows that do not sweep over most of the disk. However, mature stars like the Sun do have approximately isotropic winds, so there must be some transition phase where the bipolar star/disk wind evolves into a more spherically symmetric stellar wind. Presumably this enhanced stellar wind would eventually scour any remaining gas and dust from the system. In addition, Poynting-Robertson drag and radiation pressure are able to remove the smaller dust grains around older stars, such as Beta Pictoris, the former by orbital decay inward onto the star, and the latter by being driven.

3. EVOLUTION OF CIRCUMSTELLAR DISKS

On theoretical grounds, even an initially highly centrally condensed (i.e., power-law-density profile) cloud core is likely to collapse to form a protostar surrounded by a fairly massive protostellar disk and envelope. Currently available observations of disks around young stars (e.g., *Dutrey et al.,* 2004) imply that at early ages, disk masses are not always a significant fraction of the protostar's mass. How-

ever, these observations are unable to probe the innermost regions (i.e., within 50 AU or so), because of limited spatial resolution, so the true amount of disk mass at early phases remains uncertain. Nevertheless, the expectation is that protostellar disks must somehow transport most of their mass inward to be accreted by the protostar, eventually evolving into protoplanetary disks, where planetary bodies should be able to form and survive their subsequent interactions with the disk. This process occurs even as collapse of the presolar cloud onto the growing disk continues, adding significant amounts of mass and angular momentum. The transition point from a protostellar disk to a protoplanetary disk is not clear, and the physical mechanisms responsible for disk evolution in either of these two phases remain uncertain, although progress seems to have been made in ruling out several proposed mechanisms.

3.1. Angular Momentum Transport Mechanisms

The basic theory of the evolution of an accretion disk can be derived by assuming that there is some physical mechanism operating that results in an effective viscosity of the gas. Because the intrinsic molecular viscosity of H_2

gas is far too small to have an appreciable effect on disk evolution in a reasonable amount of time, theorists have sought other sources for an effective viscosity, such as turbulence. In a fully turbulent flow, the effective viscosity can be equal to the molecular viscosity multiplied by a large factor: the ratio of the Reynolds number of the disk (about 10 billion) to the critical Reynolds number for the onset of turbulence (about 1000), or a factor of about 10 million. Under very general conditions, it can be shown (*Lynden-Bell and Pringle,* 1974) that a viscous disk will evolve in such a manner as to transport most of its mass inward, thereby becoming more tightly gravitationally bound, and minimizing the total energy of the system. In order to conserve angular momentum, this means that angular momentum must be transported outward along with a small fraction of the mass, so that the accretion disk expands outside some radius. The loss of significant angular momentum by centrifugally launched winds somewhat relieves this need for the accretion disk to expand; this additional angular momentum sink was not recognized when the theory was first developed (note, however, that in the case of an X-wind, relatively little angular momentum can be lost by the X-wind). While the basic physics of a viscous accretion disk is fairly well developed, the physical mechanism(s) responsible for disk evolution remain contentious.

Given the high Reynolds number of a protoplanetary disk, one might expect that a turbulent cascade of energy would occur and result in an effective turbulent viscosity that might be sufficient to drive disk evolution. However, because of the strong differential rotation in a Keplerian disk, a high Reynolds number is not a sufficient condition for fully developed turbulence. Instead, the Rayleigh criterion, which applies to rotating fluids but is not strictly applicable to the solar nebula, suggests that Keplerian disks are stable with respect to turbulence.

While differential rotation may inhibit convective motions in the radial direction in a disk, motions parallel to the rotation axis are relatively unaffected by rotation. In a disk where heat is being generated near the midplane, and where dust grains are the dominant source of opacity, the disk is likely to be unstable to convective motions in the vertical direction, which carry the heat away from the disk's midplane and deposit it close to the disk's surface, where it can be radiated away. Convective instability was conjectured to lead to sufficiently robust turbulence for the resulting turbulent viscosity to be large enough to drive disk evolution (*Lin and Papaloizou,* 1980), a seemingly attractive, self-consistent scenario that has motivated much of the work on viscous evolution of the solar nebula. However, three-dimensional hydrodynamical models of vertically convectively unstable disks have shown that the convective cells that result are sheared by differential rotation to such an extent that the net transport of angular momentum is very small, and may even be in the wrong direction (see *Stone et al.,* 2000). As a result, convectively driven disk evolution does not seem to be a major driver. In addition, heating of the surface of the disk by radiation from the central protostar will also act to suppress vertical convection.

It has also been suggested that finite-amplitude (nonlinear) disturbances to Keplerian flow could result in a self-sustaining shear instability that would produce significant turbulence (*Dubrulle,* 1993). However, when three-dimensional hydrodynamical models were again used to investigate this possibility, it was found that the initially assumed turbulent motions decayed rather than grew (*Stone et al.,* 2000). Evidently purely hydrodynamical turbulence can neither grow spontaneously nor be self-sustained upon being excited by an external perturbation.

In spite of these discouraging results for hydrodynamical turbulence, another possibility remains and is under investigation (*Klahr and Bodenheimer,* 2003), that of a global baroclinic instability. In this mechanism, turbulence results in essence from steep temperature gradients in the radial direction, which then battle centrifugal effects head-on. Three-dimensional hydrodynamical models imply that this mechanism can drive inward mass transport and outward angular momentum transport, as desired. However, models by Klahr with a different numerical code have reached different conclusions, so the situation regarding this mechanism is unclear.

Rossby waves occur in planetary atmospheres as a result of shearing motions and can produce large-scale vortices such as the Great Red Spot on Jupiter. Rossby waves have been proposed to occur in the solar nebula as a result of Keplerian rotation coupled with a source of vortices. While prograde rotation (cyclonic) vortices are quickly dissipated by the background Keplerian flow, retrograde (anticyclonic) vortices are able to survive for longer periods of time (*Godon and Livio,* 1999). Rossby waves have been advanced as a significant source of angular momentum transport in the disk (*Li et al.,* 2001). Rossby vortices could serve as sites for concentrating dust particles, but the difficulty in forming the vortices in the first place, coupled with their eventual decay, makes this otherwise attractive idea somewhat dubious (*Godon and Livio,* 2000). In addition, the restriction of these numerical studies to thin, two-dimensional disk models, where refraction of the waves away from the midplane is not possible, suggests that in a fully three-dimensional calculation, Rossby waves may be less vigorous than in the thin disk calculations (*Stone et al.,* 2000).

While a purely hydrodynamical source for turbulence has not yet been demonstrated, the situation is much different when magnetohydrodynamical (MHD) effects are considered in a shearing, Keplerian disk. In this case, the Rayleigh criterion for stability can be shown to be irrelevant: Provided only that the angular velocity of the disk decreases with radius, even an infinitesimal magnetic field will grow at the expense of the shear motions, a fact that had been noted by S. Chandrasekhar in a 1960 paper but was largely ignored until it was rediscovered some 30 years later.

Balbus and Hawley (1991) pointed out that in the presence of rotational shear, even a small magnetic field will grow on a very short timescale. The basic reason is that magnetic field lines can act like rubber bands, linking two parcels of ionized gas. The parcel that is closer to the protosun will orbit faster than the other, increasing its distance from the other parcel. This leads to stretching of the magnetic field lines linking the parcel, and so to a retarding force on the forward motion of the inner parcel. This force transfers angular momentum from the inner parcel to the outer parcel, which means that the inner parcel must fall farther inward toward the protosun, increasing its angular velocity, and therefore leading to even more stretching of the field lines and increased magnetic forces. Because of this positive feedback, extremely rapid growth of an infinitesimal seed field occurs. Consequently, the magnetic field soon grows so large and tangled that its subsequent turbulent evolution must be computed with a fully nonlinear, multidimensional MHD code.

Three-dimensional MHD models of a small region in the solar nebula (*Hawley et al.,* 1995) have shown that, as expected, a tiny seed magnetic field soon grows and results in a turbulent disk where the turbulence is maintained by the magnetic instability. In addition, the magnetic turbulence results in a net outward flow of angular momentum, as desired. The magnetic field grows to a certain value and then oscillates about that mean value, depending on the assumed initial field geometry, which is large enough to result in relatively vigorous angular momentum transport. While promising, these studies of the magneto-rotational instability (MRI) are presently restricted to small regions of the nebula, and the global response of the disk to this instability remains to be determined.

Magneto-rotational instability is a powerful phenomenon, but is limited to affecting nebula regions where there is sufficient ionization for the magnetic field, which is coupled only to the ions, to have an effect on the neutral atoms and molecules. The MRI studies described above all assume ideal MHD, i.e., a fully ionized plasma, where the the magnetic field is frozen into the fluid. At the midplane of the solar nebula, however, the fractional ionization is expected to be quite low in the planetary region. Both ambipolar diffusion and resistivity (ohmic dissipation) are effective at limiting magnetic field strengths and suppressing MRI-driven turbulence, but a fractional ionization of only about 1 ion per 1000 billion atoms is sufficient for MRI to proceed in spite of ambipolar diffusion and ohmic dissipation. Close to the protosun, disk temperatures are certainly high enough for thermal ionization to create an ionization fraction greater than this, and thus to maintain full-blown MRI turbulence. Given that a temperature of at least 1400 K is necessary, MRI instability may be limited to the innermost 0.2 AU or so in quiescent phases, or as far out as about 1 AU during rapid mass accretion phases (*Boss,* 1998; *Stone et al.,* 2000).

At greater distances, disk temperatures are too low for thermal ionization to be effective. Cosmic rays were thought to be able to ionize the outer regions of the nebula, but the fact that bipolar outflows are likely to be magnetically driven means that cosmic rays may have a difficult time reaching the disk midplane (*Dolginov and Stepinski,* 1994). However, the coronae of young stars are known to be prolific emitters of hard X-rays, which can penetrate the bipolar outflow and reach the disk surface at distances of about

1 AU or so, where they are attenuated (*Glassgold et al.,* 1997). As a result, the solar nebula is likely to be a layered accretion disk (*Gammie,* 1996), where MRI turbulence results in inward mass transport within thin, lightly ionized surface layers, while the layers below the surface do not participate in MRI-driven transport. Thus the bulk of the disk, from just below the surface to the midplane, is expected to be a magnetically dead zone. Layered accretion is thought to be capable of driving mass inflow at a rate of about 1 $M_{\odot}$ in 100 m.y., sufficient to account for observed mass accretion rates in quiescent T Tauri stars.

The remaining possibility for large-scale mass transport in the solar nebula is gravitational torques. The likelihood that much of the solar nebula was a magnetically dead zone where MRI transport was ineffective leads to the suggestion that there might be regions where inward MRI mass transport would cease, leading to a local pileup of mass, which might then cause at least a local gravitational instability of the disk (*Gammie,* 1996). In addition, there is observational and theoretical evidence that protostellar disks tend to start their lives with sufficient mass to be gravitationally unstable in their cooler regions, leading to the formation of nonaxisymmetric structure and hence the action of gravitational torques, and that these torques may be the dominant transport mechanism in early phases of evolution (*Yorke and Bodenheimer,* 1999).

In order for gravitational torques to be effective, a protostellar disk or the solar nebula must be significantly nonaxisymmetric, e.g., threaded by clumps of gas, or by spiral arms, much like a spiral galaxy. In that case, trailing spiral structures, which inevitably form as a result of Keplerian shear, will result in the desired outward transport of angular momentum. This is because in a Keplerian disk, an initial bar-shaped density perturbation will be sheared into a trailing spiral arm configuration. The inner end of the bar rotates faster than the outer end and therefore moves ahead of the outer end. Because of the gravitational attraction between the inner end and the outer end, the inner end will have a component of this gravitational force in the backward direction, while the outer end will feel an equal and opposite force in the forward direction. The inner end will thus lose orbital angular momentum, while the outer end gains this angular momentum. As a result, the inner end falls closer to the protosun, while the outer end moves farther away, with a net outward transport of angular momentum.

Models of the growth of nonaxisymmetry during the collapse and formation of protostellar disks show that large-scale bars and spirals can form with the potential to transfer most of the disk angular momentum outward on timescales as short as 1000 yr to 0.1 m.y. (*Boss,* 1989), sufficiently fast to allow protostellar disks to transport most of their mass inward onto the protostar and thereby evolve into protoplanetary disks.

Early numerical models of the evolution of a gravitationally unstable disk (e.g., *Cassen et al.,* 1981) suggested that a disk would have to be comparable in mass to the central protostar in order to be unstable, i.e., gravitational instability could occur in protostellar, but not in protoplanetary disks. Analytical work on the growth of spiral density waves implied that for a 1-$M_\odot$ star, gravitational instability could occur in a disk with a mass as low as 0.19 $M_\odot$ (*Shu et al.,* 1990). Recent three-dimensional hydrodynamical models have shown that vigorous gravitational instability can occur in a disk with a mass of 0.1 $M_\odot$ or even less, in orbit around a 1-$M_\odot$ star (*Boss,* 2000), because of the expected low midplane temperatures (about 30 K) in the outer disk implied by cometary compositions (*Kawakita et al.,* 2001) and by observations of disks (*D'Alessio et al.,* 2001). Similar models with a complete thermodynamical treatment (*Boss,* 2002b), including convective transport and radiative transfer, show that a marginally gravitationally unstable solar nebula develops a robust pattern of clumps and spiral arms, persisting for many disk rotation periods, and resulting in episodic mass accretion rates onto the central protosun that vary between accreting 1 $M_\odot$ in 10 m.y. to as short as 1000 yr. The latter rates appear to be high enough to account for FU Orionis outbursts.

Because angular momentum transport by a strongly gravitationally unstable disk is so rapid, it is unlikely that protostellar or protoplanetary disks are ever strongly gravitationally unstable, because they can probably evolve away from such a strongly unstable state faster than they can be driven into it by, e.g., accretion of more mass from an infalling envelope or radiative cooling. As a result, it is much more likely that a disk will approach gravitational instability from a marginally unstable state (*Cassen et al.,* 1981). Accordingly, recent models of gravitationally unstable disks have focused almost exclusively on marginally gravitationally unstable disks (e.g., *Boss,* 2000), where primarily the outer disk, beyond about 5 AU, participates in the instability. Inside about 5 AU, disk temperatures appear to be too high for an instability to grow there, although these inner regions may still be subject to shock fronts driven by clumps and spiral arms in the gravitationally unstable, outer region. One-armed spiral density waves can propagate right down to the stellar surface. Gravitational forces are intrinsically global in nature, and their effect on different regions of the nebula can be expected to be highly variable in both space and time. On the other hand, turbulent viscosity is a local process that is usually assumed to operate more or less equally efficiently throughout a disk. As a result, it is unclear if gravitational effects can be faithfully modeled as a single, effective viscosity capable of driving disk evolution in the manner envisioned by *Lynden-Bell and Pringle* (1974). Nevertheless, efforts have been made to try to quantify the expected strength of gravitational torques in this manner (*Lin and Pringle,* 1987). Three-dimensional models of marginally gravitationally unstable disks imply that such an effective viscosity is indeed large and comparable to that in MRI models (*Laughlin and Bodenheimer,* 1994).

3.2. Evolution of the Solar Nebula

Given an effective source of viscosity, in principle the time evolution of the solar nebula can be calculated in great detail, at least in the context of the viscous accretion disk model. The strength of an effective viscosity is usually quantified by the α parameter. α is often defined in various ways, but typically α is defined to the constant that when multiplied by the sound speed and the vertical scale height of the disk (two convenient measures of a typical velocity and length scale), yields the effective viscosity of the disk (*Lynden-Bell and Pringle,* 1974). Three-dimensional MHD models of the MRI imply a typical MRI α of about 0.005 to 0.5 (*Stone et al.,* 2000). Similarly, three-dimensional models of marginally gravitationally unstable disks imply an α of about 0.03 (*Laughlin and Bodenheimer,* 1994). Steady mass accretion at the low rates found in quiescent T Tauri stars requires an α of about 0.01 (*Calvet et al.,* 2000), in rough agreement with these estimates. Once planets have formed and become massive enough to open gaps in their surrounding disks, their orbital evolution becomes tied to that of the gas. As the gaseous disk is transported inward by viscous accretion, these planets must also migrate inward. The perils of orbital migration for planetary formation and evolution are addressed in *Ward and Hahn* (2000) and *Lin et al.* (2000). Here we limit ourselves to considering the evolution of dust and gas prior to the formation of planetary-sized bodies.

The generation of viscous accretion disk models was an active area of research during the period when convective instability was believed to be an effective source of viscosity. *Ruden and Pollack* (1991) constructed models where convective instability was assumed to control the evolution, so that in regions where the disk became optically thin and thus convectively stable, the effective viscosity vanished. Starting with an α of about 0.01, they found that disks evolved for about 1 m.y. before becoming optically thin, often leaving behind a disk with a mass of about 0.1 $M_\odot$. Midplane temperatures at 1 AU dropped precipitously from about 1500 K initially to about 20 K when convection ceased and the disk was optically thin at that radius. Similarly dramatic temperature drops occur throughout the disk in these models, and the outer regions of the models eventually became gravitationally unstable as a result.

Given that convective instability is no longer considered to be a possible driver of disk evolution, the *Ruden and Pollack* (1991) models are interesting, but not likely to be applicable to the solar nebula. Unfortunately, little effort has gone into generating detailed viscous accretion models in the interim: The theoretical focus seems to have been more on the question of determining which mechanisms are contenders for disk evolution than on the question of the resulting disk evolution. In particular, the realization that the MRI mechanism is likely to have operated only in the magnetically active surface layers of the disk, and not in the magnetically dead bulk of the disk, presents a formidable technical challenge for viscous accretion disk models, which have usually been based on the assumption that the nebula can be represented by a thin, axisymmetric disk (e.g., *Ruden and Pollack,* 1991), greatly simplifying the numerical solution. The need for consideration of the vertical as well as the radial structure of the disk, and possibly the azimuthal (nonaxisymmetric) structure as well, points toward the requirement of a three-dimensional magnetohydrodynamical calculation of the entire disk. Such a calculation has not been performed, and even the MRI calculations performed to date on small regions of a disk can only be carried forward in time for a small fraction of the expected lifetime of the disk.

Some progress has been made in two-dimensional hydrodynamical models of a thick disk evolving under the action of a globally defined α viscosity, representing the effects of torques in a marginally gravitationally unstable disk (*Yorke and Bodenheimer,* 1999), but in these models the evolution eventually slows down and leaves behind a fairly massive protostellar disk after 10 m.y., with a radius on the order of 100 AU.

One aspect of particular interest about viscous accretion disk models is the evolution of solid particles, both in terms of their thermal processing and in terms of their transport in the nebula. Interstellar dust grains are small enough (submicrometer-sized) to remain well-coupled to the gas, so they will move along with the gas. During this phase, the gas and dust may undergo trajectories that are outward at first, as the disk accretes matter from the infalling envelope and expands by outward angular momentum transport, followed by inward motion once accretion stops and the disk continues to accrete onto the protostar (*Cassen,* 1996). Once collisional coagulation gets underway and grain growth begins, solid particles begin to move with respect to the gas, suffering gas drag and additional radial migration as a result (*Weidenschilling,* 1988, 2004).

The bulk compositions of the bodies in the inner solar system show a marked depletion of volatile elements compared to the solar composition. *Cassen* (2001) has shown that these volatile depletions can be explained as a result of the condensation of hot gases and coagulation of the resulting refractory dust grains into solids that are decoupled from the gas through the rapid growth of kilometer-sized planetesimals. The volatile elements remain in gaseous form at these temperatures, and so avoid being incorporated into the planetesimals that will eventually form the terrestrial planets and asteroids. In order for this process to work, significant regions of the nebula must have been hot enough at the midplane to keep volatiles in the gaseous form, a situation that would characterize the nebula when mass accretion rates were on the order of 1 $M_\odot$ in less than 10 m.y. The volatile gases would then be removed from the terrestrial planet region by viscous accretion onto the protosun. The postulated rapid growth from dust grains to kilometer-sized bodies required by this scenario appears to be possible (*Woolum and Cassen,* 1999).

3.3. Marginally Gravitationally Unstable Disks

Given the apparent limitation of MRI-driven accretion to the surfaces of protoplanetary disks, it would appear that gravitational torques may have to be responsible for the evolution in the bulk of the disk. In addition, there are strong theoretical reasons why gravitational torques may be effective, including the difficulty in forming the gas giant planets by the conventional means of core accretion. The standard model for Jupiter formation by core accretion envisions a nebula that has a surface density high enough for a solid core to form within about 8 m.y. through runaway accretion (*Pollack et al.,* 1996). However, such a nebula is likely to be marginally gravitationally unstable, a situation that could result in the rapid formation of gas giant planets in a few thousand years by the formation of self-gravitating clumps of gas and dust (*Boss,* 1997, 2000, 2002b).

In their pioneering study of a marginally unstable disk, *Laughlin and Bodenheimer* (1994) found strong spiral arm formation, but no clumps, presumably in large part as a result of the limited spatial resolution that was computationally possible at the time (up to 25,000 particles). Recent work has shown (*Boss,* 2000) that when a million or more grid points are included, three-dimensional hydrodynamic models of marginally gravitationally unstable disks demonstrate the persistant formation of self-gravitating clumps, although even these models do not appear to have sufficient spatial resolution to follow the high-density clumps indefinitely in time. Regardless of whether or not such disk instability models can lead to gas-giant-planet formation, the likelihood that the solar nebula was at least episodically marginally gravitationally unstable has important implications for cosmochemistry.

Perhaps the most well-known, unsolved problem in cosmochemistry is the question of the mechanism whereby dust grain aggregates were thermally processed to form chondrules and some rounded refractory inclusions. Chondrule compositions and textures require rapid heating and somewhat slower cooling for their explanation; a globally hot nebula is inconsistent with these requirements (*Cassen,* 2001). A wide variety of mechanisms has been proposed and generally discarded, but recent work seems to have largely solved the problem (*Desch and Connolly,* 2002). In a marginally gravitationally unstable nebula, clumps and spiral arms at about 8 AU will drive one-armed spiral arms into the inner nebula that at times result in shock fronts orientated roughly perpendicular to the orbits of bodies in the asteroidal region. Because of the tendency toward corotation in self-gravitating structures, this will lead to solids encountering a shock front at speeds on the order of 10 km/s, sufficiently high to result in postshock temperatures of about 3000 K. Detailed one-dimensional models of heating and cooling processes in such a shock front have shown that shock speeds around 7 km/s are optimal for matching chondrule cooling rates and therefore textures (*Desch and Connolly,* 2002).

If disk evolution near the midplane is largely controlled by gravitational torques rather than by a turbulent process such as MRI or convection, then mixing processes might be profoundly different as a result. Gravitational torques could potentially result in matter flowing through the disk without being rapidly homogenized through mixing by turbulence. As a result, spatially heterogeneous regions of the disk might persist for some amount of time, if they were formed in the first place by processes such as the triggered injection of shock-wave material (*Vanhala and Boss,* 2000) or the spraying and size-sorting of solids processed by an X-wind onto the surface of the nebula (*Shu et al.,* 1996). However, because convective motions appear to play an important part in cooling the disk midplane in recent models of disk instability (*Boss,* 2002b), it is unclear if gravitational torques could act in isolation without interference from convective motions or other sources of turbulence. At any rate, spatially heterogeneous regions might only last for a short fraction of the nebular lifetime, requiring rapid coagulation and growth of kilometer-sized bodies if evidence of this phase is to be preserved.

If an X-wind is responsible for driving bipolar outflows, then there are possibly important implications for the thermal processing of solids and the production of short-lived radioisotopes (*Shu et al.,* 1996, 2001). The basic idea is that some of the solids that spiral inward and approach the boundary layer between the solar nebula and the protosun will be lifted upward by the same magnetically driven wind that powers bipolar outflows. While close to the protosun, these solids will be subject to heating by the solar radiation field and to spallation by particles from solar flares. Following this processing, the solids will be lofted onto size-sorted trajectories that return them to the surface of the solar nebula at several AU or beyond. Note, however, that if a disk wind operates along with or instead of an X-wind, then any solids lofted from the inner region may be unable to return directly to the asteroid region around 2.5 AU, as the disk wind being launched at those same distances may prevent their infall onto the disk. Lofting of solids to distances greater than 2.5 AU, e.g., to where the Oort cloud comets presumably formed, might also be prohibited by a disk wind operating closer to the protosun.

The X-wind model has also been advanced as a means for thermal processing of chondrules (*Shu et al.,* 1996). While it has not yet been possible to calculate the detailed thermal history of chondrule precursors to see if the required impulsive heating and slower cooling rates can be matched, the fact that this mechanism implies size-sorting of the particles as they are lofted upward and return to the nebula on ballistic trajectories means that small dust particles and chondrule-sized particles that were thermally processed together in the X-wind region will eventually be separated upon their return to the nebula (*Desch and Connolly,* 2002). As a result, it is difficult to explain the fact that chondrules and the fine-grained matrix in which they reside are chemically complementary: Their combined, bulk composition is roughly solar, although individually they are not. This seems to indicate that thermal processing in a more closed system, such as occurs with shock-wave processing within the nebula, may be a better means to explain chondrule thermal processing.

3.4. New Scenario for Solar System Formation

While formation as a single star in an isolated, dense cloud core is usually imagined for the presolar cloud, in reality there are very few examples of isolated star formation. Most stars form in regions of high-mass star formation, similar to Orion (Fig. 1), with a smaller fraction forming in smaller clusters of low-mass stars, like Taurus (Fig. 3). The radiation environment differs considerably between these two extremes, with Taurus being relatively benign, and with Orion being flooded with ultraviolet (UV) radiation once massive stars begin to form (*Hollenbach et al.,* 2000). Even in Taurus, though, individual young stars emit UV and X-ray radiations at levels considerably greater than mature stars (*Feigelson and Montmerle,* 1999). Ultraviolet radiation from the protosun has been suggested as a means of removing the residual gas from the outermost solar nebula (i.e., beyond about 10 AU) through photoevaporation of H atoms (*Shu et al.,* 1993), a process estimated to require about 10 m.y.

Ultraviolet radiation from a nearby massive star has been invoked as a means to photoevaporate more rapidly the gas in the outer solar nebula (beyond about 10 AU) and then to form the ice giant planets by photoevaporating the gaseous envelopes of the outermost gas giant protoplanets (*Boss et al.,* 2002), as may be happening in the protoplanetary disks seen in Hubble images of the Orion nebula cluster (Fig. 1). The Orion proplyds imply a considerably different evolution scenario in the outer disk than is the case for isolated disks in regions like Taurus (Fig. 3), where the disks appear to be classic cases of symmetric, circumstellar disks with perpendicular outflows, similar to that envisioned in simple theoretical models (e.g., Fig. 2).

The full implications for cometary formation and dynamics in the Orion scenario remains to be elucidated, but clearly the Orion environment would lead to much shorter disk lifetimes than in regions like Taurus, yet such disks may still be able to yield planetary systems similar to our own. Considering the timescale for growth to kilometer-sized bodies (*Weidenschilling,* 2004), this should not adversely impact the formation of comets. Shortening the outer disk lifetime may lead to more rapid depletion of the inner disk that it otherwise feeds, however, and so may help to prevent subsequent loss of the inner planets by inward orbital migration driven by disk interactions. The process of ejection of the solar nebula from a region of high-mass star formation (prior to the eventual supernova explosion that accompanies high-mass star formation) might even help explain the high eccentricities and inclinations found for Kuiper belt comets (*Ida et al.,* 2000).

Comets subjected to the intense UV radiation of an Orion-like environment will undergo photochemical processing, leading to the production of a thick layer of organic compounds, containing, e.g., amino acids (*Bernstein et al.,* 2002), which will act as an effective sunblock against subsequent radiation processing. Thus the Orion scenario could offer a means of getting a head start on the prebiotic chemistry that is necessary for the origin of life. If correct, this hypothesis could be used to argue that planetary systems similar to our own, even to the extent of supporting life, might well be common in the galaxy.

Acknowledgments. This work was partially supported by the NASA Planetary Geology and Geophysics Program under grant

NAG 5-10201, and by the National Science Foundation under grant AST-99-83530.

REFERENCES

Alves J. F., Lada C. J., and Lada E. A. (2001) Internal structure of a cold dark molecular cloud inferred from the extinction of background starlight. *Nature, 409,* 159–161.

Andrè P. (1997) The evolution of flows and protostars. In *Herbig-Haro Flows and the Birth of Low Mass Stars* (B. Reipurth and C. Bertout, eds.), pp. 483–494. Kluwer, Dordrecht.

Bacmann A., Andrè P., Puget J.-L., Abergel A., Bontemps S., and Ward-Thompson D. (2000) An ISOCAM absorption survey of pre-stellar cloud cores. *Astron. Astrophys., 361,* 555–580.

Balbus S. A. and Hawley J. F. (1991) A powerful local shear instability in weakly magnetized disks. I. Linear analysis. *Astrophys. J., 376,* 214–222.

Bate M. R., Bonnell I. A., and Bromm V. (2002) The formation mechanism of brown dwarfs. *Mon. Not. R. Astron. Soc., 332,* L65–L68.

Bernstein M. P., Dworkin J. P., Sandford S. A., Cooper G. W., and Allamandola L. J. (2002) Racemic amino acids from the ultraviolet photolysis of interstellar ice analogues. *Nature, 416,* 401–403.

Bontemps S., Andrè P., Terebey S., and Cabrit S. (1996) Evolution of outflow activity around low-mass embedded young stellar objects. *Astron. Astrophys., 311,* 858–872.

Boss A. P. (1989) Evolution of the solar nebula I. Nonaxisymmetric structure during nebula formation. *Astrophys. J., 345,* 554–571.

Boss A. P. (1995) Collapse and fragmentation of molecular cloud cores. II. Collapse induced by stellar shock waves. *Astrophys. J., 439,* 224–236.

Boss A. P. (1997) Giant planet formation by gravitational instability. *Science, 276,* 1836–1839.

Boss A. P. (1998) Temperatures in protoplanetary disks. *Annu. Rev. Earth Planet. Sci., 26,* 53–80.

Boss A. P. (2000) Possible rapid gas giant planet formation in the solar nebula and other protoplanetary disks. *Astrophys. J. Lett., 536,* L101–L104.

Boss A. P. (2002a) Collapse and fragmentation of molecular cloud cores. VII. Magnetic fields and multiple protostar formation. *Astrophys. J., 568,* 743–753.

Boss A. P. (2002b) Evolution of the solar nebula. V. Disk instabilities with varied thermodynamics. *Astrophys. J., 576,* 462–472.

Boss A. P. (2004) The solar nebula. In *Treatise on Geochemistry: Volume 1. Meteorites, Planets, and Comets* (A. Davis, ed.), in press. Elsevier, Oxford.

Boss A. P. and Hartmann L. (2001) Protostellar collapse in a rotating, self-gravitating sheet. *Astrophys. J., 562,* 842–851.

Boss A. P. and Myhill E. A. (1995) Collapse and fragmentation of molecular cloud cores. III. Initial differential rotation. *Astrophys. J., 451,* 218–224.

Boss A. P., Wetherill G. W., and Haghighipour N. (2002) Rapid formation of ice giant planets. *Icarus, 156,* 291–295.

Calvet N., Hartmann L., and Strom S. E. (2000) Evolution of disk accretion. In *Protostars and Planets IV* (V. Mannings et al., eds.), pp. 377–399. Univ. of Arizona, Tucson.

Cameron A. G. W. and Truran J. W. (1977) The supernova trigger for formation of the solar system. *Icarus, 30,* 447–461.

Cassen P. M., Smith B. F., Miller R., and Reynolds R. T. (1981) Numerical experiments on the stability of preplanetary disks.

Cassen P. (1996) Models for the fractionation of moderately volatile elements in the solar nebula. *Meteoritics & Planet. Sci., 31,* 793–806.

Cassen P. (2001) Nebula thermal evolution and the properties of primitive planetary materials. *Meteoritics & Planet. Sci., 36,* 671–700.

Crutcher R. M. (1999) Magnetic fields in molecular clouds: Observations confront theory. *Astrophys. J., 520,* 706–713.

Curry C. L. (2002) Shapes of molecular cloud cores and the filamentary mode of star formation. *Astrophys. J., 576,* 849–859.

D'Alessio P., Calvet N., and Hartmann L. (2001) Accretion disks around young objects. III. Grain growth. *Astrophys. J., 553,* 321–334.

Desch S. J., and Connolly H. C. (2002) A model of the thermal processing of particles in solar nebula shocks: Application to the cooling rates of chondrules. *Meteoritics & Planet. Sci., 37,* 183–207.

Dolginov A. Z. and Stepinski T. F. (1994) Are cosmic rays effective for ionization of protoplanetary disks? *Astrophys. J., 427,* 377–383.

Dubrulle B. (1993) Differential rotation as a source of angular momentum transfer in the solar nebula. *Icarus, 106,* 59–76.

Dutrey A., Lecavelier des Etangs A., and Augereau J.-C. (2004) The observation of circumstellar disks: Dust and gas components. In *Comets II* (M. C. Festou et al., eds.), this volume. Univ. of Arizona, Tucson.

Feigelson E. D. and Montmerle T. (1999) High energy processes in young stellar objects. *Annu. Rev. Astron. Astrophys., 37,* 363–408.

Foster P. N. and Boss A. P. (1996) Triggering star formation with stellar ejecta. *Astrophys. J., 468,* 784–796.

Gammie C. F. (1996) Layered accretion in T Tauri disks. *Astrophys. J., 457,* 355–362.

Glassgold A. E., Najita J., and Igea J. (1997) X-ray ionization of protoplanetary disks. *Astrophys. J., 480,* 344–350.

Godon P. and Livio M. (1999) On the nonlinear hydrodynamic stability of thin Keplerian disks. *Astrophys. J., 521,* 319–327.

Godon P. and Livio M. (2000) The formation and role of vortices in protoplanetary disks. *Astrophys. J., 537,* 396–404.

Goodman A. A., Benson P. J., Fuller G. A., and Myers P. C. (1993) Dense cores in dark clouds. VIII. Velocity gradients. *Astrophys. J., 406,* 528–547.

Hartmann L. and Kenyon S. J. (1996) The FU Orionis phenomenon. *Annu. Rev. Astron. Astrophys., 34,* 207–240.

Hartmann L., Ballesteros-Paredes J., and Bergin E. A. (2001) Rapid formation of molecular clouds and stars in the solar neighborhood. *Astrophys. J., 562,* 852–868.

Hawley J. F., Gammie C. F., and Balbus S. A. (1995) Local three-dimensional magnetohydrodynamic simulations of accretion disks. *Astrophys. J., 440,* 742–763.

Hollenbach D. J., Yorke H. W., and Johnstone D. (2000). Disk dispersal around young stars. In *Protostars and Planets IV* (V. Mannings et al., eds.), pp. 401–428. Univ. of Arizona, Tucson.

Ida S., Larwood J., and Burkert A. (2000) Evidence for early stellar encounters in the orbital distribution of Edgeworth-Kuiper belt objects. *Astrophys. J., 528,* 351–356.

Jones C. E., Basu S., and Dubinski J. (2001) Intrinsic shapes of molecular cloud cores. *Astrophys. J., 551,* 387–393.

Kawakita H., Watanabe J., Ando H., Aoki W., Fuse T., Honda S., Izumiura H., Kajino T., Kambe E., Kawanomoto S., Sato B., Takada-Hidai M., and Takeda Y. (2001) The spin temperature of NH_3 in Comet C/1999S4 (LINEAR). *Science, 294,* 1089–1091.

Klahr H. H. and Bodenheimer P. (2003) Turbulence in accretion disks: Vorticity generation and and angular momentum transport in disks via the global baroclinic instability. *Astrophys. J., 582,* 869–892.

Kobrick M. and Kaula W. M. (1979) A tidal theory for the origin of the solar nebula. *Moon and Planets, 20,* 61–101.

Königl A. and Pudritz R. E. (2000) Disk winds and the accretion-outflow connection. In *Protostars and Planets IV* (V. Mannings et al., eds.), pp. 759–787. Univ. of Arizona, Tucson.

Lada C. J. and Shu F. H. (1990) The formation of sunlike stars. *Science, 248,* 564–572.

Larson R. B. (1969) Numerical calculations of the dynamics of a collapsing proto-star. *Mon. Not. R. Astron. Soc., 145,* 271–295.

Laughlin G. and Bodenheimer P. (1994) Nonaxisymmetric evolution in protostellar disks. *Astrophys. J., 436,* 335–354.

Li H., Colgate S. A., Wendroff B., and Liska R. (2001) Rossby wave instability of thin accretion disks. III. Nonlinear simulations. *Astrophys J., 551,* 874–896.

Lin D. N. C. and Papaloizou J. (1980) On the structure and evolution of the primordial solar nebula. *Mon. Not. R. Astron. Soc., 191,* 37–48.

Lin D. N. C. and Pringle J. E. (1987) A viscosity prescription for a self-gravitating accretion disk. *Mon. Not. R. Astron. Soc., 225,* 607–613.

Lin D. N. C., Papaloizou J. C. B., Terquem C., Bryden G., and Ida S. (2000) Orbital evolution and planet-star tidal interaction. In *Protostars and Planets IV* (V. Mannings et al., eds.), pp. 1111–1134. Univ. of Arizona, Tucson.

Lynden-Bell D. and Pringle J. E. (1974) The evolution of viscous disks and the origin of the nebular variables. *Mon. Not. R. Astron. Soc., 168,* 603–637.

Mannings V., Boss A. P., and Russell S. S., eds. (2000) *Protostars and Planets IV.* Univ. of Arizona, Tucson. 1422 pp.

Marcy G. W. and Butler R. P. (1996) A planetary companion to 70 Virginis. *Astrophys J. Lett., 464,* L147–L151.

Mayor M. and Queloz D. (1995) A Jupiter-mass companion to a solar-type star. *Nature, 378,* 355–359.

Mouschovias T. Ch. (1991) Magnetic braking, ambipolar diffusion, cloud cores, and star formation: Natural length scales and protostellar masses. *Astrophys. J., 373,* 169–186.

Onishi T., Mizuno A., Kawamura A., Tachihara K., and Fukui Y. (2002) A complete search for dense cloud cores in Taurus. *Astrophys. J., 575,* 950–973.

Pollack J. B., Hubickyj O., Bodenheimer P., Lissauer J. J., Podolak M., and Greenzweig Y. (1996) Formation of the giant planets by concurrent accretion of solids and gas. *Icarus, 124,* 62–85.

Preibisch T. and Zinnecker H. (1999) The history of low-mass star formation in the Upper Scorpius OB association. *Astron. J., 117,* 2381–2397.

Reipurth B. (2000) Disintegrating multiple systems in early stellar evolution. *Astron. J., 120,* 3177–3191.

Ruden S. P. and Pollack J. B. (1991) The dynamical evolution of the protosolar nebula. *Astrophys. J., 375,* 740–760.

Shu F. H., Adams F. C., and Lizano S. (1987) Star formation in molecular clouds: Observation and theory. *Annu. Rev. Astron. Astrophys., 25,* 23–72.

Shu F. H., Tremaine S., Adams F. C., and Ruden S. P. (1990) Sling amplification and eccentric gravitational instabilities in gaseous disks. *Astrophys. J., 358,* 495–514.

Shu F. H., Johnstone D., and Hollenbach D. (1993) Photoevaporation of the solar nebula and the formation of the giant planets. *Icarus, 106,* 92–101.

Shu F. H., Shang H., and Lee T. (1996) Toward an astrophysical theory of chondrites. *Science, 271,* 1545–1552.

Shu F. H., Najita J. R., Shang H., and Li Z.-Y. (2000) X-winds: Theory and observations. In *Protostars and Planets IV* (V. Mannings et al., eds.), pp. 789–813. Univ. of Arizona, Tucson.

Shu F. H., Shang H., Gounelle M., Glassgold A. E., and Lee T. (2001) The origin of chondrules and refractory inclusions in chondritic meteorites. *Astrophys. J., 548,* 1029–1050.

Smith K. W., Bonnell I. A., Emerson J. P., and Jenness T. (2000) NGC 1333/IRAS 4: A multiple star formation laboratory. *Mon. Not. R. Astron. Soc., 319,* 991–1000.

Stone J. M., Gammie C. F., Balbus S. A., and Hawley J. F. (2000) Transport processes in protostellar disks. In *Protostars and Planets IV* (V. Mannings et al., eds.), pp. 589–611. Univ. of Arizona, Tucson.

Vanhala H. A. T. and Cameron A. G. W. (1998) Numerical simulations of triggered star formation. I. Collapse of dense molecular cloud cores. *Astrophys. J., 508,* 291–307.

Vanhala H. A. T. and Boss A. P. (2000) Injection of radioactivities into the presolar cloud: Convergence testing. *Astrophys. J., 538,* 911–921.

Ward W. R. and Hahn J. M. (2000) Disk-planet interactions and the formation of planetary systems. In *Protostars and Planets IV* (V. Mannings et al., eds.), pp. 1135–1155. Univ. of Arizona, Tucson.

Weidenschilling S. J. (1988) Formation processes and time scales for meteorite parent bodies. In *Meteorites and the Early Solar System* (J. F. Kerridge and M. S. Matthews, eds.), pp. 348–371. Univ. of Arizona, Tucson.

Weidenschilling S. J. (2004) From icy grains to comets. In *Comets II* (M. C. Festou et al., eds.), this volume. Univ. of Arizona, Tucson.

Woolum D. S. and Cassen P. (1999) Astronomical constraints on nebula temperatures: Implications for planetesimal formation. *Meteoritics & Planet. Sci., 34,* 897–907.

Yorke H. W. and Bodenheimer P. (1999) The formation of protostellar disks. III. The influence of gravitationally induced angular momentum transport on disk structure and appearance. *Astrophys. J., 525,* 330–342.

Observation of Circumstellar Disks: Dust and Gas Components

Anne Dutrey
Observatoire de Bordeaux

Alain Lecavelier des Etangs
Institut d'Astrophysique de Paris

Jean-Charles Augereau
Service D'Astrophysique du CEA-Saclay and Institut d'Astrophysique de Paris

Since the 1990s, protoplanetary disks and planetary disks have been intensively observed from optical to the millimeter wavelengths, and numerous models have been developed to investigate their gas and dust properties and dynamics. These studies remain empirical and rely on poor statistics, with only a few well-known objects. However, the late phases of the stellar formation are among the most critical for the formation of planetary systems. Therefore, we believe it is timely to tentatively summarize the observed properties of circumstellar disks around young stars from the protoplanetary to the planetary phases. Our main goal is to present the physical properties that are considered to be observationally robust and to show their main physical differences associated with an evolutionary scheme. We also describe areas that are still poorly understood, such as how protoplanetary disks disappear, leading to planetary disks and eventually planets.

1. INTRODUCTION

Before the 1980s, the existence of protoplanetary disks of gas and dust around stars similar to the young Sun (4.5 b.y. ago) was inferred from the theory of stellar formation (e.g., *Shakura and Suynaev,* 1973), the knowledge of our own planetary system, and dedicated models of the protosolar nebula. The discovery of the first bipolar outflow in L1551 in 1980 drastically changed the view of the stellar formation. In the meantime, optical polarimetric observations by *Elsasser and Staude* (1978) revealed the existence of elongated and flattened circumstellar dust material around some pre-main-sequence (PMS) stars such as the low-mass T Tauri stars. The T Tauri are understood to be similar to the Sun when it was about 10^6 yr old. A few years later, observations from the InfraRed Astronomical Satellite (IRAS) found significant infrared (IR) excesses around many T Tauri stars, showing the existence of cold circumstellar dust (*Rucinski,* 1985). More surprisingly, IRAS also showed the existence of weak IR excess around main-sequence stars such as Vega, ε Eridani, or β Pictoris (*Aumann et al.,* 1985). These exciting discoveries motivated several groups to model the spectral energy distribution (SED) of T Tauri stars (e.g., *Adams et al.,* 1987) and Vega-like stars (*Harper et al.,* 1984).

On one hand, for PMS stars, the emerging scenario was the confirmation of the existence of circumstellar disks orbiting the T Tauri stars, the gas and dust being residual from the molecular cloud that formed the central star (*Shu et al.,* 1987). Since such disks contain enough gas (H_2) to allow, in theory, formation of giant planets, they are often called protoplanetary disks. During this phase, the dust emission is optically thick in the near-IR (NIR) and the central young star still accretes from its disk. Such disks are also naturally called accretion disks.

On the other hand, images of β Pictoris by *Smith and Terrile* (1984) demonstrated that Vega-type stars or old PMS stars can also be surrounded by optically thin dusty disks. These disks were called debris disks or later planetary disks, because planetesimals should be present and indirect evidence of planets was found in some of them (β Pictoris).

In this chapter, we review the current observational knowledge of circumstellar disks from the domain of the ultraviolet (UV) to the millimeter (mm). We discuss in sections 2 and 3 the properties of protoplanetary disks found around young low-mass (T Tauri) and intermediate-mass (Herbig AeBe) stars. In section 4, we summarize the properties of transition disks that still have some gas component but also have almost optically thin dust emission in the NIR, objects that are thought to be in the phase of dissipating their primary gas and dust. We present the properties of optically thin dust disks orbiting old PMS, zero-age-main-sequence (ZAMS) or Vega-type stars, such as the β Pictoris debris disk in section 5. We conclude by reviewing future instruments and their usefulness for studying such objects.

2. PROTOPLANETARY DISKS: T TAURI STARS

Following the standard classification (see *Boss,* 2004), T Tauri stars typically present the SEDs of Class II objects.

Evidence for disk features around these young stars comes principally from the following observational considerations:

1. A flat and geometrically thin distribution accounts for the SED (produced by the dust emission) from the optical to the millimeter (including the IR excess) because the extinction toward most T Tauri stars is very low.

2. In the 1990s, adaptive optics (AO) systems on ground-based telescopes and the Hubble Space Telescope (HST) began to image these disks. Dust grains at the disk surface scatter the stellar light, revealing the disk geometry of circumstellar material, as in the case of HH30 (see Plate 2).

3. By mapping the CO J = 1–0 and J = 2–1 line emission from the gas, large millimeter arrays such as OVRO or the IRAM interferometers clearly demonstrate that the circumstellar material has a flattened structure and is in Keplerian rotation.

Most T Tauri stars form in binary or multiple systems (*Mathieu et al.*, 2000), and many observational results show that binarity strongly affects the dust and gas distribution as a result of tidal truncations. The material can be in a circumbinary ring as in the GG Tau disk (*Dutrey et al.*, 1994) or confined in small, truncated, circumstellar disks. However, for simplicity, we will focus here on properties of disks encountered around stars known as single.

2.1. Mass Accretion Rates

T Tauri stars with IR excess usually present optical emission lines (e.g., *Edwards et al.*, 1994). Studies of these lines reveal that the stars are still accreting/ejecting material from their disk even if the main ejection/accretion phase is over. When they present strong H_α emission lines (equivalent linewidth $W_{H_\alpha} \geq 10$ Å), they are called classical line T Tauri stars (CTTs).

Observations show that the mass accretion rate (and the mass ejection rate) decreases from the protostars to the Class III phases by several orders of magnitude (*Hartmann et al.*, 1998; *Boss*, 2004). Despite the uncertainties resulting from the various observational methods and tracers used for measuring both rates, there is a clear correlation between mass loss and mass accretion. Typical values for the mass accretion rate of a few 10^{-5} $M_\odot$/yr are found for Class 0 objects, while T Tauri stars have lower values around 10^{-8} $M_\odot$/yr. Some protostars such as the FU Orionis objects even exhibit episodic outbursts with accretion rates as high as a few 10^{-4} $M_\odot$/yr (*Hartmann and Kenyon*, 1996). Assuming a mass accretion rate of ~10^{-8} $M_\odot$/yr, a T Tauri star of 0.5 $M_\odot$ would accrete only 0.01 $M_\odot$ in 1 m.y. Hence, most of the accretion must occur in the protostellar phase.

2.2. Modeling the Spectral Energy Distribution

Global properties of disks can be inferred from the SED, but because of the lack of angular resolution, the results are strongly model dependent and there is usually no unique solution. Moreover, many stars are binaries and SEDs are not always individually resolved, leading to possible misinterpretations.

Small dust particles (of radius a) are efficient absorbers of wavelengths radiation with $\lambda \leq a$. In equilibrium between heating and cooling, at longer wavelengths they reemit a continuous spectrum that closely resembles a thermal spectrum. At short wavelengths, the scattering of the stellar light by dust grains can dominate the spectrum, with the limit between the scattering and the thermal regimes being around ~3–5 μm. Very close to the star (~0.1–5 AU), the temperature is sufficiently high (~500–1000 K), and the NIR/optical continuum can be dominated by the thermal emission of very hot grains.

Spectral energy distributions can be reproduced by models of disks (e.g., *Pringle*, 1981; *Hartman*, 1998) that assume that (1) the disk is reprocessing the stellar light (passive disk) or (2) the disk is heated by viscous dissipation (active disk). In both models, since there is no vertical flow, the motions are circular and remain Keplerian [$v(r) = \sqrt{GM_*/r}$, where M_* is the stellar mass]. As a consequence the disk, in hydrostatic equilibrium, is geometrically thin with $H \ll r$ where H is the disk scale height. For viscous disks, the viscosity ν is usually expressed by the so-called α parameter linked to ν by $\nu = \alpha c_s H$ where c_s is the sound speed. The accretion remains subsonic with $v/r \sim \alpha c_s H/r \ll c_s$ and $\alpha \sim 0.01$.

Chiang and Goldreich (1997) have also developed a model of the passive disk in which the optically thin upper layer of the disk is superheated above the blackbody equilibrium temperature by the stellar light impinging on the disk, which produces a kind of disk atmosphere. Both viscous heating and superheated layers seem to be necessary to properly take the observations into account (*D'Alessio et al.*, 1998, 1999).

Many observers interpret the SEDs assuming the disk is geometrically thin with simple power law dependencies vs. radius for the surface density [$\Sigma(r) = \Sigma_o(r/r_o)^{-p}$] and the temperature [$T_k(r) = T_o(r/r_o)^{-q}$]. In this case, there is no assumption about the origin of the heating mechanism, and the results are compared to more sophisticated models similar to those described above.

Dust disks are usually optically thick up, to $\lambda \sim 100$–200 μm, allowing us to trace the disk temperature. In active disks as in passive geometrically thin disks, the radial dependence of the temperature follows $T_k(r) \propto r^{-0.75}$. *Beckwith et al.* (1990) have found that typical temperature laws encountered in T Tauri disks are more likely given by $\propto r^{-0.65-0.5}$. However, *Kenyon and Hartmann* (1987) have shown that for a flaring disk, the temperature profile should be as shallow as q = 0.5, closer to the observed values.

2.3. Dust Content

Longward of ~100 μm wavelength, the dust emission becomes optically thin. Observations at these wavelengths are adequately explained by a dust absorption coefficient following $\kappa_\nu = \kappa_o[\nu/10^{12}(Hz)]^\beta$ with $\beta \simeq 0.5$–1 and $\kappa_o = 0.1$ cm² g⁻¹ of gas + dust (with a gas to dust ratio of 100) (*Beckwith et al.*, 1990). The spectral index is significantly lower than in molecular clouds, where $\beta \simeq 2$. Compared to values found in molecular clouds, both β and κ_o show that a significant fraction of the grains have evolved and started to aggregate (*Henning and Stognienko,* 1996; *Beckwith et al.,* 2000), grains may even have fractal structures (*Wright,* 1987). Dust encountered in protoplanetary disks seems to be a mixture of silicate and amorphous carbon covered by icy mantles (*Pollack et al.,* 1994). The exact composition is poorly known; recent VLT observations of broadband absorption features in the NIR begin to put quantitative constraints on solid species located on grain mantles such as CO or H_2O (*Dartois et al.,* 2002; *Thi et al.,* 2002). These results also confirm that many molecules may have condensed from gas phases of grains (see also sections 2.4 and 3) in the cold (~20–15 K) outer portion of the disk. Very close to the star, the dust mantles may be significantly different since the disk is hotter (~1000 K at 0.1 AU).

Optical/NIR interferometry is a powerful tool to trace the innermost disk. *Monnier and Millan-Gabet* (2002) have observed several disks around T Tauri and Herbig Ae/Be stars. They found that the observed inner disk sizes (r_{in} ~ 0.1 AU) of T Tauri stars are consistent with the presence of an optically thin cavity for a NIR emission arising from silicate grains of sizes a ≥ 0.5–1 μm that are heated close to their temperature of sublimation.

At NIR and optical wavelengths, small dust particles also scatter the stellar light that impinges upon the disk surface, producing reflection nebulas imaged by optical telescopes. Plate 2 shows in false color the HH30 disk seen edge-on, which appears as a dark lane. The star ejects a jet perpendicular to the disk plane and is highly obscured by the material along the line-of-sight (visual extinction up to A_v ≥ 30 mag). Only the disk surface or atmosphere (*Chiang and Goldreich,* 1997) is seen. Due to the high opacity, these data cannot allow us to estimate the dust mass distribution without making *a priori* (or external) assumptions about the vertical distribution. However, grains with a typical size of around a ~ 0.05–1 μm are responsible for scattering (*Close et al.,* 1998; *Mac Cabe et al.,* 2002). Since forward scattering is easier to produce than the backward scattering (e.g., GG Tau) (*Roddier et al.,* 1996), the disk inclination usually provides a simple explanation for the observed brightness asymmetry.

Estimating the gas and dust mass of these disks is done by several methods (see also section 2.4) but is quite uncertain. Analyzing the SEDs in the optically thin part of the spectrum leads to T Tauri disk masses (gas + dust) ranging from 0.1 to 0.001 $M_\odot$ (*Beckwith et al.,* 1990). These determinations suffer from many uncertainties such as the value of κ_o and the gas-to-dust ratio, which is usually assumed to have an interstellar value of 100. Moreover, the inner part of the disk is still optically thick (up to radii of ~10–30 AU at 3 mm). Resolved images of the thermal dust emission obtained with millimeter arrays allow a separation of possible opacity and spectral index effects. This procedure also estimates the surface density radial profile $\Sigma(r) = \Sigma_0(r/r_0)^{-p}$. Typical values of p are around 1–1.5 (*Dutrey et al.,* 1996). Such low values imply that the reservoir of the mass is in the outer disk traced by submillimeter images. Assuming a single surface density distribution from 0.1 to 500 AU, a disk

with p = 1 has only 10% of its mass located within r = 10 AU, while with p = 1.9, the same disk has 50% of its mass within the same radius.

In summary, a significant fraction of the grains in protoplanetary disks appear to be more evolved than in molecular clouds, and coagulation processes have already started. However, the vertical dust distribution is not yet constrained. Moreover, observations at a given frequency are mainly sensitive to grains of size a ≤ a few λ (absorption or diffusion cross-sections cannot significantly exceed the geometrical cross-section, even for more complex grain features and aggregates) (*Pollack et al.,* 1994; *Krügel and Siebermorgen,* 1994). Only multiwavelength analysis of resolved images, from the optical to the centimeter domains, should allow us to reach conclusions about the dust sedimentation. Estimating the mass of the disks remains difficult. However, continuum millimeter images suggest that the reservoir of mass is located at a large distance, r ≥ 30–50 AU. Using a gas-to-dust ratio of 100, analyses of the dust emission indicate that the total (dust + gas) disk masses are in the range 0.001–0.1 $M_\odot$.

2.4. Gas Content

In protoplanetary disks, molecular abundances are defined with respect to H_2 since this is the main (gas) component, and the gas-to-dust ratio, which is not yet measured, is assumed to be 100, as in the molecular clouds. Several groups (*Thi et al.,* 2001b; *Richter et al.,* 2002) have recently started direct investigation of H_2 (thanks to its quadrupolar transitions in the near- and mid-IR region), but the most-used tracer of the gas phase remains CO.

Carbon monoxide is the most abundant molecule after molecular hydrogen. Its first rotation lines are observable with current millimeter interferometers, allowing astronomers to trace the properties of outer gas disks. The current sensitivity of millimeter arrays is limited and does not allow the observation of CO lines for r ≤ 30–50 AU. Since the density is very high ($n(H_2)$ ≥ 10^6 cm^{-3}), the J = 1–0 and J = 2–1 CO lines are thermalized by collision with H_2 in the whole disk. Hence, a simple model of Keplerian disk assuming LTE conditions is sufficient to derive the CO disk properties (*Dutrey et al.,* 1994).

CO maps reveal that disks are in Keplerian rotation (*Koerner et al.,* 1993) and that many disks in Taurus-Auriga clouds are large, with typical radii R_{out} ≃ 300–800 AU (see *Simon et al.,* 2000, their Fig. 1). Comparing resolved CO maps to disk models by performing a χ^2 minimization of the disk parameters (*Guilloteau and Dutrey,* 1998, see their Fig. 1) provides useful information about the density and temperature distributions. The temperature radial profiles deduced from ^{12}CO images are consistent with stellar heating in flared disks and the turbulence appears to be small, less than 0.1 km s^{-1} (*Dutrey et al.,* 2004). Since the ^{12}CO and ^{13}CO J = 1–0 and J = 2–1 have different opacities, they sample different disk layers. A global analysis of these lines permits the derivation of the vertical temperature gradient. *Dartois et al.* (2003) have shown that in the

DM Tau disk, the "CO disk surface," traced by ^{12}CO, is located around ~3 H above the disk midplane; the ^{13}CO J = 2–1 samples material at about 1 H, while J = 1–0 is representative of the disk midplane. They also deduce a vertical kinetic gradient that is in agreement with disk models (e.g., *D'Alessio et al., 1999*); the midplane is cooler (~13 K) than the CO disk surface (~30 K at 100 AU). This appears in the region of the disk where the dust is still optically thick to the stellar radiation while it is already optically thin to its own emission, around r ~ 50–200 AU in the DM Tau case. Beyond r ≥ 200 AU, where the dust becomes optically thin to both processes, the temperature profile appears vertically isothermal.

A significant fraction of the DM Tau disk has a temperature below the CO freezeout point (17 K), but enough CO in the gas phase remains to allow the J = 2–1 line of the main isotope to be optically thick. The chemical behavior of molecules and coupling between gas and dust are poorly known. There have been only a few attempts to survey many molecules in protoplanetary disks (*Dutrey et al., 1997*; *Kastner et al., 1997*; *van Zadelhoff et al., 2001*). Today, in addition to ^{13}CO and $C^{18}O$, only the more abundant species after the carbon monoxide are detectable, such as HCO^+, CS, HCN, CN, HNC, H_2CO, C_2H, and DCO^+. By studying the excitation conditions of the various transitions observed in the DM Tau disk, *Dutrey et al.* (1997) deduced molecular abundances indicating large depletion factors, ranging from 5 for CO to 100 for H_2CO and HCN, with respect to the abundances in the TMC1 cloud. They also directly measured the H_2 density; the total disk mass they estimated is a factor of 7 smaller than the total mass measured from the thermal dust emission. Both results suffer from uncertainties; only a more detailed analysis will allow the conclusion that the gas-to-dust ratio is lower than 100, even if such behavior is expected.

The chemistry of the gas phase in the inner disk is poorly constrained because of limited sensitivity (*Najita et al., 2000*). Models of nebulae irradiated by stellar radiation, including X-ray emission, suggest a complex chemistry (*Glassgold et al., 1997*), even at relatively large radii (r ≥ 50 AU) (*Najita et al., 2001*).

Plate 1 summarizes the observable properties of a protoplanetary disk encountered around a T Tauri star of 0.5 $M_\odot$ and located at a distance of 150 pc.

2. 5. Illustration Through HH30

The HH30 observations, shown in Plate 2, give one of the most complete pictures of the material surrounding a PMS star. Optical and millimeter observations clearly trace the same physical object. In Plate 2, the HH30 dust disk observed in the optical by the HST (*Burrows et al., 1996*) is close to edge-on and appears as a dark lane; only the disk surface or atmosphere is bright. The jet emission is also seen, perpendicular to the disk plane. J. Pety (personal communication, 2003) has superimposed in contours to this image the blueshifted and redshifted integrated emission (with respect to the systemic velocity) of the ^{13}CO J = 2–1 line

in the disk. The CO disk extends as far as the dust disk and the velocity gradient is along the major disk axis, as expected for rotation. ^{12}CO J = 2–1 emission is also observed within the jet (extreme velocity). In the HH30 case the low angular resolution of the data does not allow one to separate between the ^{12}CO J = 2–1 emission associated with the outflow, the cloud, and the disk. This is a common problem that cannot be fully solved by current interferometric observations. Selecting sources that are located in a region devoid of CO emission of the molecular cloud minimizes the confusion.

Since the disk is seen edge-on, the vertical distribution can be estimated from the optical observations of the dust (*Burrows et al., 1996*). The results are somewhat model dependent, but one can conclude that the disk is pressure supported (dominated by the central star) with the best fit given by $H(r) \propto r^{1.45}$. The authors estimate the surface density law to be $\Sigma(r) \propto 1/r$ and the disk mass ~6 × 10^{-3} $M_\odot$. CO observations reveal that the disk is in Keplerian rotation around a central star of mass 0.5 $M_\odot$ and has an outer radius of R_{out} = 440 AU. Interestingly, this is in agreement with the best fit of the optical data, which gives R_{out} ~ 400 AU.

2.6. Proplyds

The disks we described so far are found in low-mass star-forming regions such as the Taurus-Auriga clouds. Physical properties of disks surrounding low-mass stars born inside clusters forming massive stars may be significantly different because they are exposed to the strong ambient UV field generated by nearby OB stars and can be photoevaporated by it (*Johnstone, 1998*). This is the case for the proplyds, which are protoplanetary disks seen in silhouette against the strong H II region associated to the Trapezium cluster in Orion A (e.g., *McCaughrean and O'Dell, 1996*).

3. DISKS AROUND INTERMEDIATE-MASS STARS: HERBIG Ae/Be STARS

With masses in the range 2–8 $M_\odot$, Herbig Ae/Be (HAeBe) stars, massive counterparts of T Tauri stars, are the progenitors of A and B main-sequence stars. Since they are more luminous and massive than the T Tauri stars, the surrounding material is submitted to stronger UV and optical stellar flux.

Several Herbig Ae stars are isolated but located in nearby star-forming regions (Taurus, R Oph); their observed properties can be directly compared with those of T Tauri stars. This is not the case for Herbig Be stars because most of them are located at larger and uncertain distances (D ≥ 500–800 pc). Therefore, in this section we will mainly discuss isolated Herbig Ae stars.

Like T Tauri stars, SEDs of HAeBe stars exhibit strong IR excesses. Optical and NIR observations (*Grady et al., 1999*) reveal that many of these objects are surrounded by large reflection nebulae (e.g., more than 1000 AU for the A0 star AB Auriga), revealing envelopes or halos (*Leinert et al., 2001*). However, there is now clear evidence that

Herbig Ae stars are also surrounded by disks. In particular, resolved CO maps from millimeter arrays reveal that the circumstellar material is also in Keplerian rotation [e.g., MWC480, an A4 star (*Manning et al.*, 1997) and HD 34282, an A0 star (*Pietu et al.*, 2003)]. Millimeter continuum surveys also suggest that the total surrounding mass may have a tendency to increase with the stellar mass (see *Natta et al.*, 2000, their Fig. 1). So far, one of the more massive Keplerian disks ($\sim$0.11 $M_\odot$) has been found around an A0 Herbig Ae star: HD 34282 (*Pietu et al.*, 2003).

Since the medium is hotter than for T Tauri stars, one would expect different behavior, in particular, a rich chemistry. The limited sensitivity at present of molecular surveys at millimeter wavelengths does not allow one to distinguish significant differences, and outer disks ($r \geq 50$ AU) of Herbig Ae stars appear similar to "cold" outer disks found around T Tauri stars.

However, most of the differences should appear in the warm material, closer to the star. Optical/NIR interferometric observations by *Monnier and Millan-Gabet* (2002) revealed that the observed inner radius of disks is usually larger for HAeBe stars than for T Tauri stars. This is understood in term of truncation by dust sublimation close to the star. They also found that grain sizes are similar for T Tauri and HAeBe stars. *Dullemond et al.* (2001) have shown that direct irradiation of Herbig Ae disks at their inner radius can explain the bump observed at IR wavelengths in the SEDs of Herbig Ae stars. In a few cases, as for HD 100546, direct detection of H_2 with the Far Ultraviolet Spectroscopic Explorer (FUSE) (*Lecavelier des Etangs et al.*, 2003) reveals the existence of warm ($T \simeq 500$ K) molecular gas close to the star ($r \sim 0.5$–1 AU).

The fact that HAeBe stars are hotter has also favored the use of the Infrared Space Observatory (ISO) to characterize the geometry and the dust composition of the disk close to the star. *Bouwman et al.* (2000) have performed a detailed spectroscopic study from 2 to 200 μm of the circumstellar material surrounding AB Auriga (A0) and HD 163296 (A1). Their analysis of the SEDs, assuming an optically thin dust model, has revealed the existence of both hot ($T \sim 1000$ K) and cold ($T \sim 100$ K, most of the mass) dust components, while the NIR emission at 2 μm can be explained by the presence of metallic iron grains. As in T Tauri disks, substantial grain growth has occured, with grain size up to $\sim$0.1–1 mm. It is also important to note that comparisons of the ISO spectrum of the Herbig Ae star HD 100546 with those of Comet Hale-Bopp have revealed many similarities (*Waelkens et al.*, 1999).

3.1. MWC 480: Similarities and Differences with a T Tauri Disk

MWC 480 is located at $D = 140$ pc, in the Auriga cloud. CO observations by *Manning et al.* (1997) have revealed that the surrounding disk is in Keplerian rotation around an A4 star (*Simon et al.*, 2000). The disk is large ($R_{out} \simeq 600$ AU) and inclined by about 35° along the line of sight. The stellar mass is around $\sim$2 $M_\odot$. The temperature deduced

from the optically thick ^{12}CO J = 2–1 line (which probes about three scale heights above the disk midplane) is $T \sim 60$ K at $R = 100$ AU; this is significantly larger than the temperature of $\sim$30 K found for T Tauri disks using the same tracer. Interestingly, the disk is not detected in the NIR in scattered light. *Augereau et al.* (2001a) have deduced that either the dust emission at 1.6 μm is optically too thin to be detected, or there is a blob of optically thick material close to the star that hides the outer disk from the stellar radiation. Knowledge of the disk scale height is required to choose between these possibilities. The existence of such blobs is also favored by SED models of several Herbig Ae stars (*Meeus et al.*, 2001, their Fig. 8).

Near-infrared and CO/millimeter detections are not necessarily linked: The T Tauri DM Tau has the best known disk at millimeter wavelengths, but the disk was only recently detected in the NIR by performing deep integration with the HST (*Grady et al.*, 2003). Keeping this in mind, the MWC 480 disk appears very similar to a T Tauri disk, at least for the outer part ($r \geq 50$ AU), although it is hotter and perhaps somewhat more massive.

3.2. UX Orionis Phenomenon

Some HAeBe stars, such as UX Orionis, have a very complex spectroscopic, photometric, and polarimetric variability that has been, in some cases, monitored for years. The variability has usually a short periodicity on the order of $\sim$1 yr and can be as deep as two magnitudes in the V band.

It is very tempting to link this phenomenon to planetary formation; the variability could be caused by clumps of material (such as clouds of protocomets) located in the innermost disk and orbiting the star. *Natta and Whitney* (2000) have developed a model in which a screen of dust sporadically obscures the star; this happens when the disk is tilted by about 45°–68° along the line of sight. One clearly needs more sensitive multiwavelength data at high angular resolution on a large sample of Herbig Ae stars to distinguish among the various models.

4. FROM PROTOPLANETARY TO PLANETARY DISKS

On one side, one finds massive gaseous protoplanetary disks around T Tauri and Herbig Ae stars, and on the other, one finds dusty planetary disks around young main-sequence stars. A natural question is then whether there are disks in an intermediate state. If so, what are their observational characteristics? Limited by the sensitivity of current telescopes, we know of only a few examples of objects that can be considered to be "transition" disks.

4.1. Surprising Case of BP Tau

BP Tau is often considered as the prototype of CTTs. It has a high accretion rate of $\sim 3 \times 10^{-8}$ $M_\odot$/yr from its circumstellar disk, which produces strong excess emission in the ultraviolet, visible, and NIR (*Gullbring et al.*, 1998). It is also very young [6×10^5 yr (*Gullbring et al.*, 1998)]. Despite these strong CTT characteristics, its millimeter properties are very different from those of other T Tauri stars surrounded by CO disks.

Recent CO J = 2–1 and continuum at 1.3-mm images from the Institut de Radio Astronomie Millimétrique (IRAM) array have revealed a weak and small CO and dust disk (*Simon et al.*, 2000). With a radius of about ≃120 AU, the disk is small and in Keplerian rotation around a (1.3 ± 0.2) (D/140 pc) $M_\odot$ mass star. A deeper analysis of these CO J = 2–1 data (*Dutrey et al.*, 2003) also shows that the J = 2–1 transition is marginally optically thin, contrary to what is observed in other T Tauri disks. The disk mass, estimated from the millimeter continuum emission by assuming a gas-to-dust ratio of 100, is very small (1.2×10^{-3} $M_\odot$) (for comparison, a factor of 10 below the minimum initial mass of the solar nebula). By reference to the mass deduced from the continuum, the CO depletion factor can be estimated; this leads to a factor as high as ~150 with respect to H_2. Even taking into account possible uncertainties such as a lower gas-to-dust ratio or a higher value for the dust absorption coefficient, the CO depletion remains high compared to other CO disks. Finally, the kinetic temperature derived from the CO data is also relatively high, about ~50 K at 100 AU.

Both the relatively high temperature and the low disk mass suggest that a significant fraction of the disk might be superheated (above the blackbody temperature), similar to a disk atmosphere (e.g., *Chiang and Goldreich*, 1997) (see also section 2). With reasonable assumptions for the dust grain properties and surface density, one can then estimate the fraction of small grains (a ≃ 0.1 μm) still present in the disk to reach $\tau_V = 1$ in the visible at the disk midplane. Since it corresponds to a total mass of small grains of about 10% of the total mass of dust derived from the millimeter continuum data (1.2×10^{-5} $M_\odot$), this is not incompatible with the current data, but should be confirmed by optical and NIR observations.

It is also interesting that the CO content of BP Tau is too high to result from evaporation of protocomets [falling evaporating body (FEB) model; see also section 5]. Considering the total number of CO molecules in the disk and the CO evaporation rate of an active comet such as Hale-Bopp, a few times 10^{11} large comets similar to Hale-Bopp would be simultaneously required to explain the amount of CO gas present in the BP Tau disk. This is well above the number of FEBs falling on β Pictoris per year (a few hundred).

Taken together, the unusual millimeter properties suggest that BP Tau may be a transient object in the phase of clearing out its outer disk.

4.2. Ambiguous Case of HD 141569

HD 141569 is a B9 star located at ~100 pc. The position of the star close to the ZAMS in the HR diagram and the presence of an IR circumstellar excess lead many authors to classify it as an HAeBe star. This was reinforced by the presence of circumstellar gas later detected by *Zuckerman et al.* (1995) and by the identification of emission features

tentatively attributed to PAH (*Sylvester and Skinner*, 1996, and references therein) that are frequently observed in SEDs of HAeBe (e.g., *Meeus et al.*, 2001). But the lack of excess in the NIR, the lack of photometric variability, the faint intrinsic measured polarimetry (*Yudin*, 2000), and most importantly, the low disk-to-star luminosity ratio (8.4×10^{-3}) correspond to the description of a Vega-like star. Both HAeBe and Vega-like classes show a large spread of ages. With an age of 5 m.y. (*Weinberger et al.*, 2000), HD 141569 falls at the common edge of the two categories.

HD 141569 is among the few stars that show a spatially resolved optically thin dust disk in the NIR. Contrary to the β Pictoris disk, the inclination of the HD 141569 disk on the line of sight offers the opportunity to investigate both the radial and azimuthal profiles of the dust surface density in great detail. Using coronagraphic techniques, the HST identified a complex dust structure seen in scattered light of about 10 times our Kuiper belt size (~500 AU) (*Augereau et al.*, 1999). Mid-infrared thermal emission observations only partly compensate for the lack of constraints on the innermost regions (<1" ≡ 100 AU) masked by the HST coronagraph (*Fisher et al.*, 2000). Both data help to sketch out the overall dust distribution, as summarized in Fig. 4 of *Marsh et al.* (2002). The dust appears depleted inside ~150 AU compared to the outer regions. The outer disk has a complex shape dominated by two nonaxisymmetric and not accurately concentric wide annulii at 200 and 325 AU (*Mouillet et al.*, 2001) respectively. Interestingly, the furthest ring is made of grains smaller than the blowout size limit, which theoretically points out the presence of cold gas in the outer disk (*Boccaletti et al.*, 2003). Surprising, an arc that is radially thin but azimuthally extended over ~90° is located at about 250 AU, precisely between the two major ring-like structures. This information about the disk morphology is very valuable because it indicates the impact of internal (planets?) and/or external gravitational perturbations (stellar companions?) (*Augereau and Papaloizou*, 2004).

The detection of a substantial amount of cold gas associated with an optically thin dust disk is also unusual (*Zuckerman et al.*, 1995). Recent millimeter interferometric observations of HD 141569 reveal the gaseous counterpart of the extended disk resolved in scattered light (see Plate 3). The CO gas in rotation shows a velocity gradient consistent with the major axis of the optical disk. Interestingly, hot gas (CO) is also detected by high-resolution mid-IR spectroscopy, which reveals the gaseous content of the inner disk (*Brittain and Rettig*, 2002) at a few tens of AU from the stars.

The HD 141569 disk possesses NIR properties close to those of the β Pictoris disk and mid-IR and millimeter properties close to those of an Herbig Ae disk, so it seems reasonable to consider it as a transition disk.

4.3. Puzzle of Weak-Line T Tauri Stars

Weak-line T Tauri stars (WTTs) have a SED that presents a weak IR excess; unlike CTTs, they do not exhibit strong optical emission lines (with $W_{H_\alpha} \leq 10$ Å). As such, they are usually considered to be the evolved counterpart of the

CTTs stars and are classified as Class III objects surrounded by an optically thin NIR disk (a few ~10^7 yr).

However, several studies show that a significant fraction of the WTTs have ages on the same order as those of CTTs (*Stahler and Walter,* 1993; *Grosso et al.,* 2000). Among them, one interesting example is the case of V836 Tau: This star presents the optical properties of WTT stars; its millimeter characteristics are very similar to those of BP Tau since it is also surrounded by a compact CO disk (*Duvert et al.,* 2000). Its observed properties are also very similar to those of the BP Tau disk. More recently, *Bary et al.* (2002) have reported H_2 detection around DOAr 21, a WTT located in ρ Oph that is even more puzzling.

We have only a few examples of transition disks, and each of them exhibits very different properties that are dependent upon the observational approach (optical vs. millimeter/submillimeter observations). This clearly demonstrates that only multiwavelength studies can allow the retrieval of the physical properties of these objects. The scenario by which massive disks dissipate and perhaps form planets is poorly constrained today.

5. PLANETARY DISKS

Since the lifetime of massive protoplanetary disks is observed to be less than a few ~10^7 years, we should not *a priori* expect disk structure beyond that age. As circumstellar disks evolve, their mass decreases. When the disks dissipate, the material becomes less bright, less dense, and apparently more difficult to detect. Infrared excess detection of material around nearby main-sequence stars (*Aumann et al.,* 1985) leads to the conclusion that the lifetime of the thin disks is longer than that of massive disks. Since the time spent at these late stages is longer, this provides the opportunity to detect evolved disks in the solar neighborhood (less than ~100 parsecs) around stars older than few 10^7 years. These disks are less dense than protoplanetary disks, but their proximity allows us to observe them in great detail. The disks seen around main-sequence stars are now believed to be the visible part of more massive systems in which most of the mass is preserved in the form of planetesimals and even planets. There, planetary formation is either at the end or already finished (*Lagrange et al.,* 2000).

The duration of the "planetary disk" phenomenon is so long (see below) that by nature they are obviously not the remaining material of the protoplanetary disks. It is now clear that these planetary disks have been replenished with material from a preexisting reservoir. As we will see below, the basic process needed to sustain these disks is based on the release of dust and/or gas by colliding asteroids and/or by evaporating planetesimals. These disks are thus also described as debris disks or second-generation disks.

5.1. Dust in Planetary Disks

First detected by IRAS because of their infrared excess, the dust component of planetary disks is the easiest part to detect. The spectral energy distribution of main-sequence

stars with circumstellar material shows infrared excess above ~10 μm, from which it is possible to have some indication about the dust size, spatial distribution, and total mass, or simply information about the fraction of stars harboring such planetary disks (*Backman and Paresce,* 1993). More recent ISO surveys of nearby stars show that about 20% of the stars have infrared excess attributed to circumstellar material (*Dominik,* 1999) with a typical lifetime of about 400 m.y. (*Habing et al.,* 1999, 2001).

For an extremely small fraction of these disks, it is possible to image the dust. These images are produced by the scattered light at visible wavelengths, or by images of the infrared thermal emission of the warm part of the disk at 10 or 20 μm. The first historical image of such a disk was the image of the disk of β Pictoris obtained with coronographic observations of the scattered light (*Smith and Terrile,* 1984). Imaging is difficult; indeed, β Pictoris remained the only disk imaged until the late 1990s (see, e.g., *Lecavelier des Etangs et al.,* 1993; *Kalas and Jewitt,* 1995; *Mouillet et al.,* 1997a; *Heap et al.,* 2000), when new instruments allowed observers to image a few other disks by the detection of the scattered light (*Schneider et al.,* 1999) (Fig. 1) or by the detection of the thermal emission in the infrared [around HR 4796 (*Jawardhana et al.,* 1998; *Koerner et al.,* 1998)] or in the submillimeter [images of Vega, Fomalhaut, and β Pictoris have been obtained by *Holland et al.* (1998), and images of ε Eridani by *Greaves et al.* (1998)].

Observations of the dust provide important constraints on the spatial structure of the disks. For example, the morphology of the β Pictoris disk and the inferred spatial distribution have been analyzed in great detail (*Artymowicz et al.,* 1989; *Kalas and Jewitt,* 1995). Images revealed unexpected properties, such as the presence of a break at ~120 AU in the radial distribution of the dust, the warp of the disk plane, and various asymmetries. All asymmetric features are often attributed to gravitational perturbation of massive bodies such as Jupiter-mass planets (*Lecavelier des Etangs et al.,* 1996; *Lecavelier des Etangs,* 1997b; *Augereau et al.,* 2001b).

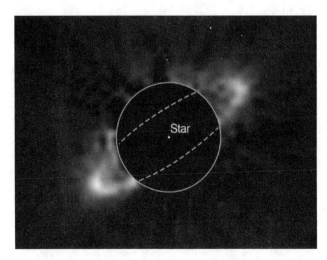

Fig. 1.   Image of dust scattered light in the HR 4796 ring observed with the Hubble Space Telescope (*Schneider et al.,* 1999).

The most intriguing characteristic of the inner portion of the β Pictoris disk is the so-called "warp," which consists of a change of the inclination of the midplane of the disk inside about 80 AU. This warp is explained by the presence of a planet on an inclined orbit with the same inclination as the tilt of the inner disk (*Mouillet et al.*, 1997b). The measured warp distance allows one to constrain $M_p \times D_p^2$, where M_p and D_p are the mass and the distance of the perturbing planet. In the case of β Pictoris we have

$$M_p D_p^2 \approx 2 \times 10^{-3} \, M_* (10 \text{ AU})^2 (t/10^7 \text{ yr})^{-1}$$

If the age of the system is t ~ 2×10^7 yr (*Barrado y Navas-cués et al.*, 1999), then a Jupiter-mass planet at 10 AU and inclined by 5° from the disk plane can easily explain the observed warp.

Similarly, in the case of HR 4796, sharp truncation of the ring structure is observed (Fig. 1) (*Schneider et al.*, 1999). The presence of a ring is expected to shed light on the physical phenomena that occurs at places where planets are typically supposed to form. However, there is still no general agreement on the interpretation of this truncation, which could be produced by the gravitational perturbations of a planet (e.g., *Wyatt et al.*, 1999) or by drag of the dust by the gas component of the disk (*Klahr and Lin*, 2001; *Takeushi and Artymowicz*, 2001).

A key point concerning disks around main-sequence stars is that the lifetime of the dust is shorter than the age of these systems (see *Artymowicz*, 1997, p. 206, Fig. 8). In the β Pictoris disk, dust particles are destroyed by collisions between grains, which produce submicrometer debris quickly eliminated by the radiation pressure. In the less-dense disks, such as the ε Eridani ring (*Greaves et al.*, 1998), the Pointing-Robertson drag is the dominant process, which also eliminates the dust on a timescale shorter than the disk age. As the lifetime of the dust is very short, one must consider that the observed dust is continuously resupplied (*Backman and Paresce*, 1993). It is generally thought that the debris disks are substantial disks of colliding planetesimals [see *Backman et al.* (1995) for an analogy with a collision in the solar system Kuiper belt]. These disks can thus be considered as the signature of complete planetary systems that, like our own solar system, contain interplanetary dust, asteroid-like kilometer-sized bodies, and probably comets and planets. The visible part of these disks is only the fraction of material having the largest cross section, showing the presence of invisible but more massive objects.

5.2. Gas Disks Around Main-Sequence Stars

Although apparently more difficult to interpret, the gas component of the planetary disks provides the opportunity for the most detailed modeling of circumstellar processes. In particular, the β Pictoris spectroscopic variability is now well explained in many details by the evaporation of cometary objects close to the star (see section 5.3).

In contrary to emission from more massive disks, emission lines from the gaseous planetary disks are, in most cases, below the detection limit of current instruments. Molecular transitions at millimeter wavelengths are too faint to be detected. They give only upper limits on the gas content (*Dent et al.*, 1995; *Liseau*, 1999). Detections of infrared emission of H_2 at 17 and 28 μm around main-sequence stars have been reported by ISO (*Thi et al.*, 2001a,b). However, these detections have been challenged by groundbased observations at 17 μm, which show no detection with three times better sensitivity (*Richter et al.*, 2002). FUSE observations also showed that if the ISO detection of H_2 emission around β Pictoris is real, then this H_2 is not distributed widely throughout the disk (*Lecavelier des Etangs et al.*, 2001). The HST detection of Fe II emission lines in the disk of β Pictoris is marginal and yet to be confirmed (*Lecavelier des Etangs et al.*, 2000). Finally, the only strong detection of emission from the gaseous component in a planetary disk has been performed by *Olofsson et al.* (2001). With high-resolution spectroscopy of the β Pictoris disk, they clearly detected the resonantly scattered Na emission through the Na I doublet line at 5990 and 5996 Å. The gas can be traced from less than 30 AU to at least 140 AU from the central star. Unfortunately, this observation remains unique. This definitely presents a new field of observation with an original technique. Although emission spectroscopy of tenuous gas disks is difficult, absorption spectroscopy is much more sensitive. In planetary disks seen nearly edge-on, the central star can be used as a continuum source, and the detection of absorption lines offers the opportunity to scrutinize the gaseous content in details.

A few months after the discovery of the dust disk around β Pictoris, its gaseous counterpart was discovered through the Ca II absorption lines (see Fig. 2) (*Hobbs et al.*, 1985; *Vidal-Madjar et al.*, 1986). Because one absorbing component is seen identically in all observations and at the same radial velocity as the star (20 km s^{-1}), it is called the "stable" gas component. This stable gas is composed of small amounts of neutral Na and Fe, as well as large amounts of singly ionized species such as Ca^+, Fe^+ Mg^+, Mn^+, and Al^+. The overall composition is close to solar (*Lagrange et al.*, 1998). Ultraviolet spectroscopy leads to the detection of two very peculiar elements: C I and the CO molecule, which both have short lifetimes. On the other hand, the OH molecule was not detected with a relatively tight upper limit (*Vidal-Madjar et al.*, 1994). The numerous electronic transitions of CO in the ultraviolet give constrains on the column density, temperature (~20 K), and a very unusual isotopic ratio $^{12}CO/^{13}CO = 15 \pm 2$ (*Jolly et al.*, 1998; *Roberge et al.*, 2000).

Circumstellar gas signatures similar to those of β Pictoris ones are also seen around some other main-sequence stars. It should be noted, however, that in the rare cases where a Ca II (or, e.g., Fe II) line at the star radial velocity has been detected toward other stars like HR 10, these stars have been identified because they also show either spectral variability, or the presence of over-ionized species, or redshifted optically thick absorption lines. This lack of detection of only the circumstellar absorption at the systemic velocity may be caused by possible confusion with the interstellar medium. The first main-sequence star discovered to have a

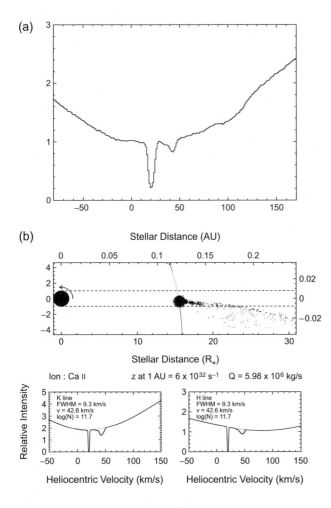

Fig. 2. **(a)** Ca II spectrum of β Pictoris (*Lecavelier des Etangs,* 2000). The stable component is visible at 20 km s⁻¹ heliocentric velocity. This spectrum shows two variable redshifted components: one sharp absorption at low velocity and one broad absorption line at high velocity (>100 km s⁻¹). **(b)** Simulation of Ca II absorption lines produced by a FEB at large distance. The line is redshifted and sharp, with a profile very similar to the profile of the redshifted variable lines observed in the β Pictoris spectrum.

et al., 1997b). Links between these gas and dust features have yet to be understood.

5.3. Beta Pictoris Disk: A Cometary Disk

The β Pictoris disk has certainly been the most fruitful for surprises and discoveries. In addition to the gaseous stable component, variable absorption features have been detected and surveyed since 1984 (Fig. 2). Features that are slowly variable are confined most of the time to one or two components redshifted by 10–30 km s⁻¹ relative to the star. Although such structures seem to be changing in both velocity and strength (by about ±10 km s⁻¹ in velocity and large factors in strength), they nevertheless remain very comparable during a few consecutive hours, often from one day to the next and even sometime over weeks.

Some other components present strong variability, in particular in the Mg II and Al III lines. These components are also observed in the Ca II and Fe II line as weak and broad absorptions spread over a few tens of kilometers per second. The changes are observed on a very short timescale, hours or even less. These features are mostly strongly redshifted, with shifts that could reach 300–400 km s⁻¹. These highly varying features were detected only in ionized species, including highly (over-)ionized ones like Al⁺⁺ and C⁺⁺⁺, completely unexpected in such a relatively "cool" stellar environment. The rapid changes make these features difficult to track.

These spectral variations are interpreted with a scenario of star-grazing comets. The presence of strong redshifted ionized gas is difficult to understand, since the very high radiation pressure should expel it very quickly. The variable absorption lines are almost always redshifted, corresponding to infalling gas seen in absorption against the stellar continuum (in an edge-on disk). The gas must then be injected with very high inward radial velocity. There is only one simple way to produce this situation, namely the evaporation of grains moving toward the star. Since the radiation pressure acts on grains, they must be injected with high velocity, through the evaporation from more massive bodies for which the gravitation is much larger that the radiation force. This model has been developed in great detail (*Beust et al.,* 1990, 1991a,b), and can be referred to as the "evaporation of star-grazing comets." Indeed, the strong variability in the circumstellar lines of the ionized elements like Fe⁺, Al⁺⁺, C⁺⁺⁺, and Mg⁺ is now attributed to the evaporation of kilometer-sized, "cometary-like" bodies falling toward the star; this is the "falling evaporating bodies (FEB)" scenario. The over-ionized variable species Al⁺⁺ and C⁺⁺⁺ cannot be produced by photoionization, but *Beust and Tagger* (1993) showed that they can be formed by collisional ionization in the coma surrounding these FEBs.

Among the different phenomena that can be explained by this model, one can stress the observation of abnormal line ratio in doublet features. For example, although the Mg II doublet has an intrinsic oscillator strength ratio of 2, the measured ratio is exactly 1, even for unsaturated lines (*Vidal-Madjar et al.,* 1994). This proves that the absorbing

very similar spectroscopic behavior to β Pictoris is HR 10 (*Lagrange et al.,* 1990). This star shows variable redshifted or blueshifted absorption lines (*Lagrange et al.,* 1990; *Welsh et al.,* 1998), and a central component seems relatively stable. Very highly excited levels of Fe⁺ have been detected, proving that the gas is not interstellar but circumstellar. Redshifted optically thick Mg II lines have also been detected and interpreted as small clouds of excited gas falling toward the star (*Lecavelier des Etangs,* 1998).

51 Oph is an interesting case because circumstellar dust is present simultaneously with the gas: 51 Oph presents a complex system with dusty infrared excess due to cold dust, silicate emission features, absorption lines by overionized species (*Grady and Silvis,* 1993), Fe⁺ at excited levels, an abnormal Mg II line-intensity ratio, and finally a possible detection of C I in the circumstellar gas (*Lecavelier des Etangs*

gas cloud is optically thick (ratio = 1) but does not cover the total stellar disk (the lines are not saturated). This behavior is directly explained by the FEB model, which produces clumpiness of the absorbing clouds (*Beust et al.,* 1989). Similar ratios have also been detected in redshifted absorptions toward other "β Pictoris-like stars."

With several hundreds of FEBs per year, the frequency is several orders of magnitudes higher than that of Sun-grazing comets in the solar system. Planetary perturbations are thought to be the process responsible. Direct scattering by close encounters with a massive planet does not seem to be efficient unless the planet eccentricity is very high (*Beust et al.,* 1991b). *Beust and Morbidelli* (1996) proposed a generic model based on a mean-motion resonance with a single massive planet on a moderately eccentric orbit (e ≈ 0.05). Indeed, a test particle trapped in the 4 : 1 resonance with such a planet becomes star-grazing after ≈10,000 planetary revolutions. This model explains not only the preferred infall direction, but also the radial velocity-distance relation observed in the FEBs.

Many other stars show redshifted absorption lines (HR 2174, 2 And, etc.). All these detections of redshifted absorption lines raise the question of the explanation for the quasi-absence of blueshifted events. In the β Pictoris case, it is believed that the orbits of the star-grazing comets always present approximately the same angle to the observer (because the gas is seen in absorption against the stellar continuum in an edge-on disk). However, when observed on several stars, some blueshifted orientation should be expected. Given the very impressive fit between the β Pictoris observations and the FEB model, it is likely that the β Pictoris FEB phenomenon is somehow particular, and that other stars present either real infall on the star or their evaporating bodies may be generally destroyed before they reach the periastron (*Grinin et al.,* 1996). In the latter case, the process needed to place these bodies on very eccentric orbits in less than one orbital timescale remains to be discovered.

Another class of cometary-like object might also be present around β Pictoris. Collision of planetesimals are believed to continuously resupply most of the dusty disks around main-sequence stars. In the case of β Pictoris, a significant part of the disk can be also produced by the evaporation of kilometer-sized bodies located at several tens of AU from the central star (*Lecavelier des Etangs et al.,* 1996). Indeed, in the β Pictoris disk, CO evaporates below 120 AU from the star. If bodies enter that region, they start to evaporate and eject dust particles. These particles are subsequently spread outward in the whole disk by the radiation pressure. The distribution of their eccentric orbits gives a dust surface density similar to that observed around β Pictoris. This alternative scenario for the production of dust in the β Pictoris disk easily explains any asymmetry even at large distances, because a planet in the inner disk can influence the distribution of nearby parent bodies, producing the outward spread of dust (*Lecavelier des Etangs,* 1998).

The observed CO/dust ratio is another argument in favor of this scenario. An important characteristic of the β Pictoris disk is the presence of cold CO and C (*Vidal-Madjar et al.,* 1994; *Roberge et al.,* 2000). CO and C I are destroyed by ultraviolet interstellar photons (extreme UV flux from the star is negligible). Like the dust, they have lifetimes shorter than the age of the star ($t_{CO} \sim t_{CI} \sim 200$ yr). A mechanism must replenish CO with a mass rate of $\dot{M}_{CO} \sim 10^{11}$ kg s^{-1}. The corresponding dust/CO supplying rate is $\dot{M}_{dust}/\dot{M}_{CO} \approx 1$. This is very similar to the dust/CO ratio in the material supplied by evaporation in the solar system. This similarity provides an indication that the β Pictoris dust disk could be supplied by evaporating bodies orbiting at several tens of AUs from the star, much in the same way as Chiron evaporates at dozen of AUs from the Sun.

5.4. Toward a Global Picture

Disks around main-sequence stars are probably related to the presence of young planetary systems in a phase of strong activity. They show that the planetary systems are still active and evolve after their formation.

There are still many unknowns. This new field of astronomy is still a collection of different objects that do not correspond to an evolutionary scheme, and a global picture is still to be created. It is clear that the many different names used for these disks, which we elected to call the "planetary disks," show that there is not an unique understanding of the phenomenon. Some authors refer to the "Vega phenomenon," often to explain the infrared excess. The term "β Pictoris phenomenon," used because of the number of different phenomena observed around that star, is even more confusing. "Kuiper disk," "cometary disk," or alternatively "debris disk" refer to evidence concerning the different origins of these disks. It is not yet clear if the many pieces shown here correspond to the same puzzle. New observations and theoretical works will be needed to resolve these issues in the next decade.

6. FUTURE OBSERVATIONS

The examples given in the previous sections clearly demonstrate that the frontier between the different classes of objects is not well constrained, partly because the statistics are still too poor to provide a quantitative understanding of some of the observed properties. Since very few disks have been resolved so far, deriving a timescale and a detailed evolutionary scheme from protoplanetary to planetary disks remains speculative. Moreover, when it comes to the protoplanetary phase, our understanding is crudely limited to the cold outer disks (r ≥ 50 AU).

Throughout this chapter, the examples used have illustrated that only multiwavelength studies of disk properties will allow astronomers to properly incorporate in their models all the physical processes involved in the observed phenomena.

6.1. Challenge for Protoplanetary/Transition Disks

In protoplanetary disks, most of the material lies in the relatively cold outer part of the disks. Hence resolved submillimeter observations, obtained with large millimeter arrays such as ALMA, will provide the best tool to investigate this reservoir of mass (r > 20–50 AU). In particular, ALMA will observe large samples, providing statistics on disk properties and frequency. Of course, to study the hotter inner disk, where planets form (0.1 ≥ r ≥ 10 AU), optical, NIR, and mid-IR interferometry techniques are required. As soon as they will be able to produce images (even with a few baseline numbers), instruments such as AMBER and MIDI on the Very Large Telescope Interferometer (VLTI), or the Optical Hawaiian Array for Nanoradian Astronomy (OHANA), will add to our understanding of the dust properties and composition. Images (or a reasonable UV coverage) are necessary in order to choose between all the existing models of dust disks; in particular, they should allow us to disentangle the geometry, temperature, and opacity effects. Figure 3 summarizes which part of the disk can be investigated, depending on the instrument used. A combination similar to "ALMA and VLTI" would efficiently sample the global disk properties. A necessary step to understand how planets form is to view the gaps that are created by protoplanets. For this purpose, images are definitely required, either at submillimeter or NIR wavelengths. In its large configuration, ALMA will have baselines up to 14 km, providing an angular resolution of ~0.03" (or 4 AU at the Taurus distance) at λ = 1.3 mm. Hydrodynamics coupled to radiative transfer simulations of the dust emission at 350 GHz by *Wolf et al.* (2002) show that ALMA will be able to resolve a gap created by a proto-Jupiter, located at 5 AU from a star at D = 150 pc. Concerning the gas content, in spectral lines near λ = 1.3 mm, ALMA will be about 30 times more sensitive than the IRAM array and quantitative chemical studies could begin. Multitransition analyses would even allow observers to measure abundance gradients in the

disk. Protoplanetary disks are indeed H_2 disks, and direct investigation of the H_2 distribution and mass remains the most direct way to study how protoplanetary disks dissipate. This domain will strongly benefit from satellites such as the Space Infrared Telescope Facility (SIRTF)/Spitzer Space Telescope (SST) and the James Webb Space Telescope (JWST). Finally, the current knowledge of protoplanetary disks is biased by the sensitivity limitations, and we can image only the brighter disks. Disk clearing is poorly constrained. The ALMA sensitivity will allow many other objects similar to BP Tau, and even optically thin dust disks in the NIR, to be imaged in the submillimeter.

6.2. Challenge for Planetary Disks

By extrapolation from ISO results on the occurrence of debris disks around main-sequence stars, and according to the Hipparcos catalog, one can predict $\sim 10^2$ and $\sim 10^3$ planetary dust disks around (hypothetical) stars younger than about 0.5 G.y. within 20 pc and 50 pc radii of the Sun respectively. These stars are close enough that they are not limited by angular resolution considerations, but their disks are simply too tenuous to be detected with current instruments. This points out a crucial need for an enhancement in sensitivity combined with high-angular-resolution techniques. Precise disk shapes, fine structures, and asymmetries as revealed by imaging may indeed be signposts of undetected gravitational companions, such as planets perturbing the underlying disk of kilometer-sized bodies that release the observed short-lived dust.

The observational techniques discussed below are summarized in Plate 4. In the NIR, high-contrast imaging with single-aperture telescopes is required to detect faint dust disks very close to a bright star. For instance, new generations of adaptive optics systems and new concepts of coronagraphic masks are currently under study, with the prime goal being detection of faint objects from the ground, ideally down to planets (e.g., VLT/Planet Finder). At these wavelengths (but also in the mid-IR), the innermost regions of planetary disks will nevertheless remain unreachable without the help of interferometry (e.g., Keck I and VLTI). Resolving the material within the very first AU around young main-sequence stars and ultimately producing images by NIR and mid-IR interferometry are attractive challenges in the near future. In the mid-IR, single-aperture telescopes suffer an unavoidable decrease in spatial resolution. The predominant gain at these wavelengths will mostly come from future spacebased telescopes, particularly the 6-m JWST and SST, with increases of two or three orders of magnitudes in detection thresholds. While current high-resolution imagers in the mid-IR are limited to disks around nearby stars younger than a few tens of millions of years, JWST's Mid-Infrared Instrument (MIRI) should resolve debris disks around 1-G.y.-old A-type stars at 50 pc. In the millimeter, ALMA will permit the detection of an unresolved (but optically thin) clump of dust of 10^{-2} $M_\oplus$ orbiting a star up to 100 pc away in only 1 h of observing time (see also *ALMA*, 2004).

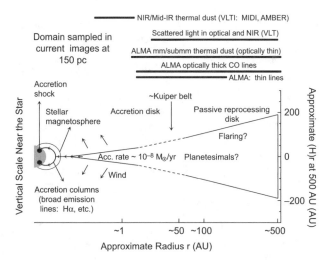

Fig. 3. This montage summarizes the observable properties of a protoplanetary disk encountered around a T Tauri star located at 150 pc. This illustrates which region of the disk is sampled depending on the telescope in use.

More and more observational efforts will certainly be focused on the gas in order to provide a better understanding of its timescales and dissipation processes. Since the gas content of young planetary disks is yet badly constrained, anticipating future results might be very speculative. However, the ALMA sensitivity would allow to detect, in 1 h of observing time, a CO column density of a few 10^{12} cm^{-2} in a beam of 2.5" (planetary disks are close to us, and hence angularly extended) assuming a linewidth of 3 km/s. This is well below the CO column density of 10^{15} cm^{-2} detected by *Vidal-Madjar et al.* (1994) in β Pictoris. Depending on the gas geometry, the distance and the evolutionary status of the disk, ALMA would allow astronomers to place some important constraints on various gas model distributions in planetary disks.

Acknowledgments. We thank A. Vidal-Madjar for fruitful comments. Figure 3 has been reproduced with the kind permission of H. Beust. A.D. would like to thank S. Guilloteau for a long term and fruitful collaboration. M. Simon is also acknowledged for a careful reading of the manuscript. Many thanks to J. Pety, who provided material for HH30 (Plate 2 and recent millimeter results). J.C.A. thanks the CNES for financial support.

REFERENCES

Adams F. C., Lada C. J., and Shu F. H. (1987) Spectral evolution of young stellar objects. *Astrophys. J., 312,* 788.

ALMA (2004) *Science with ALMA.* Available on line at http://www.eso.org/projects/alma/science/.

Artymowicz P. (1997) β Pictoris: An early solar system? *Annu. Rev. Earth Planet. Sci., 25,* 175.

Artymowicz P., Burrows C., and Paresce F. (1989) The structure of the β Pictoris circumstellar disk from combined IRAS and coronagraphic observations. *Astrophys. J., 337,* 494–513.

Augereau J. C. and Papaloizou J. (2004) Structuring the HD 141569 A circumstellar dust disk. Impact of eccentric bound stellar companions. *Astron. Astrophys., 414,* 1153.

Augereau J. C., Lagrange A. M., Mouillet D., and Ménard F. (1999) HST/NICMOS2 observations of the HD 141569 A circumstellar disk. *Astron. Astrophys., 350,* L51.

Augereau J. C., Lagrange A. M., Mouillet D., and Ménard F. (2001a) HST/NICMOS2 coronagraphic observations of the circumstellar environment of three old PMS stars: HD 100546, SAO 206462 and MWC 480. *Astron. Astrophys., 365,* 78.

Augereau J. C., Nelson R. P., Lagrange A. M., Papaloizou J. C. B., and Mouillet D. (2001b) Dynamical modeling of large scale asymmetries in the β Pictoris dust disk. *Astron. Astrophys., 370,* 447–455.

Aumann H. H. (1985) IRAS observations of matter around nearby stars. *Publ. Astron. Soc. Pacific, 97,* 885–891.

Backman D. E. and Paresce F. (1993) Main sequence stars with circumstellar solid material: The Vega phenomenon. In *Protostars and Planets III* (E. H. Levy and J. I. Lunine, eds.), pp. 1253–1304. Univ. of Arizona, Tucson.

Backman D. E., Dasgupta A., and Stencel R. E. (1995) Model of a Kuiper belt small grain population and resulting far-infrared emission. *Astrophys. J. Lett., 450,* L35.

Barrado y Navascués D., Stauffer J. R., Song I., and Caillault J.-P. (1999) The age of β Pictoris. *Astrophys. J. Lett., 520,* L123–L126.

Bary J. S., Weintraub D. A., and Kastner J. H. (2002) Detection of molecular hydrogen orbiting a "naked" T Tauri star. *Astrophys. J. Lett., 576,* L73.

Beckwith S. V. W., Sargent A. I., Chini R. S., and Guesten R. (1990) A survey for circumstellar disks around young stellar objects. *Astron. J., 99,* 924.

Beckwith S. V. W., Henning T., and Nakagawa Y. (2000) Dust properties and assembly of large particles in protoplanetary disks. In *Protostars and Planets IV* (V. Manning et al., eds.), p. 533. Univ. of Arizona, Tucson.

Beust H. and Morbidelli A. (1996) Mean-motion resonances as a source for infalling comets toward β Pictoris. *Icarus, 120,* 358–370.

Beust H. and Tagger M. (1993) A hydrodynamical model for infalling evaporating bodies in the β Pictoris circumstellar disk. *Icarus, 106,* 42.

Beust H., Lagrange-Henri A. M., Vidal-Madjar A., and Ferlet R. (1989) The β Pictoris circumstellar disk. IX — Theoretical results on the infall velocities of Ca II, Al III, and Mg II. *Astron. Astrophys., 223,* 304–312.

Beust H., Vidal-Madjar A., Ferlet R., and Lagrange-Henri A. M. (1990) The β Pictoris circumstellar disk. X — Numerical simulations of infalling evaporating bodies. *Astron. Astrophys., 236,* 202–216.

Beust H., Vidal-Madjar A., Ferlet R., and Lagrange-Henri A. M. (1991a) The β Pictoris circumstellar disk. XI — New Ca II absorption features reproduced numerically. *Astron. Astrophys., 241,* 488–492.

Beust H., Vidal-Madjar A., and Ferlet R. (1991b) The β Pictoris protoplanetary system. XII — Planetary perturbations in the disk and star-grazing bodies. *Astron. Astrophys., 247,* 505–515.

Boccaletti A., Augereau J. C., Marchis F., and Hahn J. (2003) Ground-based near-infrared imaging of the HD 141569 circumstellar disk. *Astrophys. J., 585,* 494.

Boss A. P. (2004) From molecular clouds to circumstellar disks. In *Comets II* (M. C. Festou et al., eds.), this volume. Univ. of Arizona, Tucson.

Bouwman J., de Koter A., van den Ancker M. E., and Waters L. B. F. M. (2000) The composition of the circumstellar dust around the Herbig Ae stars AB Aur and HD 163296. *Astron. Astrophys., 360,* 213.

Brittain S. D. and Rettig T. W. (2002) CO and H$_3^+$ in the protoplanetary disk around the star HD141569. *Nature, 418,* 57.

Burrows C. J. and 15 colleagues (1996) Hubble Space Telescope observations of the disk and jet of HH30. *Astrophys. J., 473,* 437.

Chiang E. I. and Goldreich P. (1997) Spectral energy distributions of T Tauri stars with passive circumstellar disks. *Astrophys. J., 490,* 368.

Close L., Dutrey A., Roddier F., Guilloteau S., Roddier C., Northcott M., Ménard F., Duvert G., Graves J. E., amd Potter D. (1998) Adaptive optics imaging of the circumbinary disk around the T Tauri binary UY Aurigae: Estimates of the binary mass and circumbinary dust grain size distribution. *Astrophys. J., 499,* 883.

D'Alessio P., Canto J., Calvet N., and Lizano S. (1998) Accretion disks around young objects. I. The detailed vertical structure. *Astrophys. J., 500,* 411.

D'Alessio P., Calvet N., Hartmann L., Lizano S., and Cantó J. (1999) Accretion disks around young objects. II. Tests of well-mixed models with ISM dust. *Astrophys. J., 527,* 893.

Dartois E., d'Hendecourt L., Thi W., Pontoppidan K. M., and van Dishoeck E. F. (2002) Combined VLT ISAAC/ISO SWS spectroscopy of two protostellar sources. The importance of minor solid state features. *Astron. Astrophys., 394,* 1057.

Dartois E., Dutrey A., and Guilloteau S. (2003) Structure of the DM Tau outer disk: Probing the vertical kinetic temperature gradient. *Astron. Astrophys., 399,* 773.

Dent W. R. F., Greaves J. S., Mannings V., Coulson I. M., and Walther D. M. (1995) A search for molecular gas components in prototypal Vega-excess systems. *Mon. Not. R. Astron. Soc., 277,* L25–L29.

Dominik C. (1999) ISO Observations of Vega-like stars. In *Solid Interstellar Matter: The ISO Revolution* (L. d'Hendecourt et al., eds.), p. 277. EDP Sciences/Springer-Verlag, Berlin.

Dullemond C. P., Dominik C., and Natta A. (2001) Passive irradiated circumstellar disks with an inner hole. *Astrophys. J., 560,* 957.

Dutrey A., Guilloteau S., and Simon M. (1994) Images of the GG Tauri rotating ring. *Astron. Astrophys., 286,* 149.

Dutrey A., Guilloteau S., Duvert G., Prato L., Simon M., Schuster K., and Ménard F. (1996) Dust and gas distribution around T Tauri stars in Taurus-Auriga. I. Interferometric 2.7 mm continuum and ^{13}CO J = 1–0 observations. *Astron. Astrophys., 309,* 493.

Dutrey A., Guilloteau S., and Guélin M. (1997) Chemistry of protosolar-like nebulae: The molecular content of the DM Tau and GG Tau disks. *Astron. Astrophys., 317,* L55.

Dutrey A., Guilloteau S., and Simon M. (2003) The BP Tau disk: A missing link between Class II and III objects? *Astron. Astrophys., 402,* 1003.

Duvert G., Guilloteau S., Ménard F., Simon M., and Dutrey A. (2000) A search for extended disks around weak-lined T Tauri stars. *Astron. Astrophys., 355,* 165.

Edwards S., Hartigan P., Ghandour L., and Andrulis C. (1994) Spectroscopic evidence for magnetospheric accretion in classical T Tauri stars. *Astron. J., 108,* 1056.

Elsasser H. and Staude H. J. (1978) On the polarization of young stellar objects. *Astron. Astrophys., 70,* L3.

Fisher R. S., Telesco C. M., Piña R. K., Knacke R. F., and Wyatt M. C. (2000) Detection of extended thermal infrared emission around the Vega-like source HD 141569. *Astrophys. J. Lett., 532,* L141.

Glassgold A. E., Najita J., and Igea J. (1997) X-ray ionization of protoplanetary disks: Erratum. *Astrophys. J., 485,* 920.

Grady C. A. and Silvis J. M. S. (1993) The circumstellar gas surrounding 51 Ophiuchi — A candidate proto-planetary system similar to β Pictoris. *Astrophys. J. Lett., 402,* L61–L64.

Grady C. A., Woodgate B., Bruhweiler F. C., Boggess A., Plait P., Lindler D. J., Clampin M., and Kalas P. (1999) Hubble Space Telescope imaging spectrograph coronagraphic imaging of the Herbig AE star AB Aurigae. *Astrophys. J. Lett., 523,* L151.

Grady C. A., Woodgate B., Stapelfeldt K., Padgett D., Stecklum B., Henning T., Grinin V., Quirrenbach A., and Clampin M. (2003) HST/STIS coronagraphic imaging of the disk of DM Tauri. *Bull. Am. Astron. Soc., 34,* 1137.

Greaves J. S., Holland W. S., Moriarty-Schieven G., Jenness T., Dent W. R. F., Zuckerman B., McCarthy C., Webb R. A., Butner H. M., Gear W. K., and Walker H. J. (1998) A dust ring around ε Eridani: Analog to the young solar system. *Astrophys. J. Lett., 506,* L133–L137.

Grinin V., Natta A., and Tambovtseva L. (1996) Evaporation of star-grazing bodies in the vicinity of UX Ori-type stars. *Astron. Astrophys., 313,* 857–865.

Grosso N., Montmerle T., Bontemps S., André P., and Feigelson E. D. (2000) X-rays and regions of star formation: A combined ROSAT-HRI/near-to-mid IR study of the ρ Oph dark cloud.

Astron. Astrophys., 359, 113.

Guilloteau S. and Dutrey A. (1998) Physical parameters of the Keplerian protoplanetary disk of DM Tauri. *Astron. Astrophys., 339,* 467.

Gullbring E., Hartmann L., Briceno C., Calvet N. (1998) Disk accretion rates for T Tauri stars. *Astrophys. J., 492,* 323.

Habing H. J., Dominik C., Jourdain de Muizon M., Kessler M. F., Laureijs R. J., Leech K., Metcalfe L., Salama A., Siebenmorgen R., and Trams N. (1999) Disappearance of stellar debris disks around main-sequence stars after 400 million years. *Nature, 401,* 456–458.

Habing H. J., Dominik C., Jourdain de Muizon M., Laureijs R. J., Kessler M. F., Leech K., Metcalfe L., Salama A., Siebenmorgen R., Trams N., and Bouchet P. (2001) Incidence and survival of remnant disks around main-sequence stars. *Astron. Astrophys., 365,* 545–561.

Harper D. A., Loewenstein R. F., and Davidson J. A. (1984) On the nature of the material surrounding Vega. *Astrophys. J., 285,* 808.

Hartmann L. (1998) *Accretion Processes in Star Formation.* Cambridge Astrophysics Series Vol. 32, Cambridge Univ., Cambridge.

Hartmann L. and Kenyon S. J. (1996) The FU Orionis phenomenon. *Annu. Rev. Astron. Astrophys., 34,* 207.

Hartmann L., Calvet N., Gullbring E., and D'Alessio P. (1998) Accretion and the evolution of T Tauri disks. *Astrophys. J., 495,* 385.

Heap S. R., Lindler D. J., Lanz T. M., Cornett R. H., Hubeny I., Maran S. P., and Woodgate B. (2000) Space Telescope imaging spectrograph coronagraphic observations of β Pictoris. *Astrophys. J., 539,* 435–444.

Henning T. and Stognienko R. (1996) Dust opacities for protoplanetary accretion disks: Influence of dust aggregates. *Astron. Astrophys., 311,* 291.

Hobbs L. M., Vidal-Madjar A., Ferlet R., Albert C. E., and Gry C. (1985) The gaseous component of the disk around β Pictoris. *Astrophys. J. Lett., 293,* L29–L33.

Holland W. S., Greaves J. S., Zuckerman B., Webb R. A., McCarthy C., Coulson I. M., Walther D. M., Dent W. R. F., Gear W. K., and Robson I. (1998) Submillimetre images of dusty debris around nearby stars. *Nature, 392,* 788–790.

Jayawardhana R., Fisher S., Hartmann L., Telesco C., Pina R., and Fazio G. (1998) A dust disk surrounding the young A star HR 4796A. *Astrophys. J. Lett., 503,* L79.

Johnstone D., Hollenbach D., and Bally J. (1998) Photoevaporation of disks and clumps by nearby massive stars: Application to disk destruction in the Orion Nebula. *Astrophys. J., 499,* 758.

Jolly A., McPhate J. B., Lecavelier A., Lagrange A. M., Lemaire J. L., Feldman P. D., Vidal Madjar A., Ferlet R., Malmasson D., and Rostas F. (1998) HST–GHRS observations of CO and C I fill in the β-Pictoris circumstellar disk. *Astron. Astrophys., 329,* 1028–1034.

Kalas P. and Jewitt D. (1995) Asymmetries in the β Pictoris dust disk. *Astron. J., 110,* 794.

Kastner J., Zuckerman B., Weintraub D. A., and Forveille T. (1997) X-ray and molecular emission from the nearest region of recent star formation. *Science, 277,* 67.

Kenyon S. J. and Hartmann L. (1987) Spectral energy distributions of T Tauri stars — Disk flaring and limits on accretion. *Astrophys. J., 323,* 714.

Klahr H. H. and Lin D. N. C. (2001) Dust distribution in gas disks: A model for the ring around HR 4796A. *Astrophys. J., 554,* 1095–1109.

Koerner D. W., Sargent A. I., and Beckwith S. W. W. (1993) A rotating gaseous disk around the T Tauri star GM Aurigae. *Icarus, 106,* 2.

Koerner D. W., Ressler M. E., Werner M. W., and Backman D. E. (1998) Mid-infrared imaging of a circumstellar disk around HR 4796: Mapping the debris of planetary formation. *Astrophys. J. Lett., 503,* L83.

Krügel E. and Siebenmorgen R. (1994) Dust in protostellar cores and stellar disks. *Astron. Astrophys., 288,* 929.

Lagrange A.-M., Beust H., Mouillet D., Deleuil M., Feldman P. D., Ferlet R., Hobbs L., Lecavelier des Etangs A., Lissauer J. J., McGrath M. A., McPhate J. B., Spyromilio J., Tobin W., and Vidal-Madjar A. (1998) The β Pictoris circumstellar disk. XXIV. Clues to the origin of the stable gas. *Astron. Astrophys., 330,* 1091–1108.

Lagrange A.-M., Backman D. E., and Artymowicz P. (2000) Planetary material around main-sequence stars. In *Protostars and Planets IV* (V. Manning et al., eds.), p. 639. Univ. of Arizona, Tucson.

Lagrange-Henri A. M., Beust H., Ferlet R., Vidal-Madjar A., and Hobbs L. M. (1990) HR 10 — A new β Pictoris-like star? *Astron. Astrophys., 227,* L13–L16.

Lecavelier des Etangs A. (1998) Planetary migration and sources of dust in the β Pictoris disk. *Astron. Astrophys., 337,* 501–511.

Lecavelier des Etangs A. (2000) Models of β Pictoris and disks around main-sequence stars (invited review). In *Disk, Planetesimals, and Planets* (F. Garzón et al., eds.), p. 308. ASP Conference Series 219, Astronomical Society of the Pacific, San Francisco.

Lecavelier des Etangs A., Perrin G., Ferlet R., Vidal Madjar A., Colas F., Buil C., Sevre F., Arlot J. E., Beust H., Lagrange-Henri A. M., Lecacheux J., Deleuil M., and Gry C. (1993) Observation of the central part of the β-Pictoris disk with an anti-blooming CCD. *Astron. Astrophys., 274,* 877.

Lecavelier des Etangs A., Vidal-Madjar A., and Ferlet R. (1996) Dust distribution in disks supplied by small bodies: Is the β Pictoris disk a gigantic multi-cometary tail? *Astron. Astrophys., 307,* 542–550.

Lecavelier des Etangs A., Vidal-Madjar A., Backman D. E., Deleuil M., Lagrange A.-M., Lissauer J. J., Ferlet R., Beust H., and Mouillet D. (1997a) Discovery of C I around 51 Ophiuchi. *Astron. Astrophys., 321,* L39–L42.

Lecavelier des Etangs A., Deleuil M., Vidal-Madjar A., Lagrange-Henri A.-M., Backman D., Lissauer J. J., Ferlet R., Beust H., and Mouillet D. (1997b) HST-GHRS observations of candidate β Pictoris-like circumstellar gaseous disks. *Astron. Astrophys., 325,* 228–236.

Lecavelier des Etangs A., Hobbs L. M., Vidal-Madjar A., Beust H., Feldman P. D., Ferlet R., Lagrange A.-M., Moos W., and McGrath M. (2000) Possible emission lines from the gaseous β Pictoris disk. *Astron. Astrophys., 356,* 691–694.

Lecavelier des Etangs A., Vidal-Madjar A., Roberge A., Feldman P. D., Deleuil M., André M., Blair W. P., Bouret J.-C., Désert J.-M., Ferlet R., Friedman S., Hébrard G., Lemoine M., and Moos H. W. (2001) Deficiency of molecular hydrogen in the disk of β Pictoris. *Nature, 412,* 706–708.

Lecavelier des Etangs A., Deleuil M., Vidal-Madjar A., Roberge A., Le Petit F., Hébrard G., Ferlet R., Feldman P.D., Désert J.-M., and Bouret J.-C. (2003) FUSE observations of H₂ around the Herbig AeBe stars HD 100546 and HD 163296. *Astron. Astrophys., 407,* 935.

Leinert Ch., Haas M., Abraham P., and Richichi A. (2001) Halos around Herbig Ae/Be stars — More common than for the less

massive T Tauri stars. *Astron. Astrophys., 375,* 927.

Liseau R. (1999) Molecular line observations of southern main-sequence stars with dust disks: α PS A, β Pic, ε ERI and HR 4796 A. Does the low gas content of the β PIC and ε ERI disks hint of planets? *Astron. Astrophys., 348,* 133–138.

Mannings V. and Sargent A. I. (1997) A high-resolution study of gas and dust around young intermediate-mass stars: Evidence for circumstellar disks in Herbig AE systems. *Astrophys. J., 490,* 792.

Marsh K. A., Silverstone M. D., Becklin E. E., Koerner D. W., Werner M. W., Weinberger A. J., and Ressler M. E. (2002) Mid-infrared images of the debris disk around HD 141569. *Astrophys. J., 573,* 425.

Mathieu R. D., Ghez A. M., Jensen E. L. N., and Simon M. (2000) Young binary stars and associated disks. In *Protostars and Planets IV* (V. Mannings et al., eds.), p. 703. Univ. of Arizona, Tucson.

McCabe C., Duchêne G., and Ghez A. M. (2002) NICMOS images of the GG Tauri circumbinary disk. *Astrophys. J., 575,* 974.

McCaughrean M. J. and O'Dell C. R. (1996) Direct imaging of circumstellar disks in the Orion Nebula. *Astron. J., 111,* 1977.

Meeus G., Waters L. B. F. M., Bouwman J., van den Ancker M. E., Waelkens C., and Malfait K. (2001) ISO spectroscopy of circumstellar dust in 14 Herbig Ae/Be systems: Towards an understanding of dust processing. *Astron. Astrophys., 365,* 476.

Monnier J. and Millan-Gabet R. (2002) On the interferometric sizes of young stellar objects. *Astrophys. J., 579,* 694.

Mouillet D., Lagrange A.-M., Beuzit J.-L., and Renaud N. (1997a) A stellar coronograph for the COME-ON-PLUS adaptive optics system. II. First astronomical results. *Astron. Astrophys., 324,* 1083–1090.

Mouillet D., Larwood J. D., Papaloizou J. C. B., and Lagrange A. M. (1997b) A planet on an inclined orbit as an explanation of the warp in the β Pictoris disc. *Mon. Not. R. Astron. Soc., 292,* 896.

Mouillet D., Lagrange A. M., Augereau J. C., and Ménard F. (2001) Asymmetries in the HD 141569 circumstellar disk. *Astron. Astrophys., 372,* L61.

Najita J. R., Edwards S., Basri G., and Carr J. (2000) Spectroscopy of inner protoplanetary disks and the star-disk interface. In *Protostars and Planets IV* (V. Mannings et al., eds.), p. 457. Univ. of Arizona, Tucson.

Najita J. R., Bergin E. A., and Ullom J. N. (2001) X-ray desorption of molecules from grains in protoplanetary disks. *Astrophys. J., 561,* 880.

Natta A. and Whitney B. A. (2000) Models of scattered light in UXORs. *Astron. Astrophys., 364,* 633.

Natta A., Grinin V., and Mannings V. (2000) Properties and evolution of disks around pre-main-sequence stars of intermediate mass. In *Protostars and Planets IV* (V. Mannings et al., eds.), p. 559. Univ. of Arizona, Tucson.

Olofsson G., Liseau R., and Brandeker A. (2001) Widespread atomic gas emission reveals the rotation of the β Pictoris disk. *Astrophys. J. Lett., 563,* L77–L80.

Piétu V., Dutrey A., and Kahane C. (2003) A Keplerian disk around the Herbig Ae star HD 34282. *Astron. Astrophys., 398,* 565.

Pollack J., Hollenbach D., Beckwith S., Simonelli D. P., Roush T., and Fong W. (1994) Composition and radiative properties of grains in molecular clouds and accretion disks. *Astrophys. J., 421,* 615.

Pringle J. E. (1981) Accretion discs in astrophysics. *Annu. Rev.*

Astron. Astrophys., 19, 137.

Richter M. J., Jaffe D. T., Blake G. A., and Lacy J. H. (2002) Looking for pure rotational H_2 emission from protoplanetary disks. *Astrophys. J. Lett., 572,* L161–L164.

Roberge A., Feldman P. D., Lagrange A. M., Vidal-Madjar A., Ferlet R., Jolly A., Lemaire J. L., and Rostas F. (2000) High-resolution Hubble Space Telescope STIS spectra of CI and CO in the β Pictoris circumstellar disk. *Astrophys. J., 538,* 904–910.

Roddier C., Roddier F., Northcott M. J., Graves J. E., and Jim K. (1996) Adaptive optics imaging of GG Tauri: Optical detection of the circumbinary ring. *Astrophys. J., 463,* 326.

Rucinski S. M. (1985) IRAS observations of T Tauri and post-T Tauri stars. *Astron. J., 90,* 2321.

Schneider G., Smith B. A., Becklin E. E., Koerner D. W., Meier R., Hines D. C., Lowrance P. J., Terrile R. J., Thompson R. I., and Rieke M. (1999) NICMOS imaging of the HR 4796A circumstellar disk. *Astrophys. J. Lett., 513,* L127–L130.

Shakura N. I. and Sunyaev R. A. (1973) Black holes in binary systems. Observational appearance. *Astron. Astrophys., 24,* 337.

Shu F. H., Adams F. C., and Lizano S. (1987) Star formation in molecular clouds — Observation and theory. *Annu. Rev. Astron. Astrophys., 25,* 23.

Simon M., Dutrey A., and Guilloteau S. (2000) Dynamical masses of T Tauri stars and calibration of pre-main-sequence evolution. *Astrophys. J., 545,* 1034.

Smith B. A. and Terrile R. J. (1984) A circumstellar disk around β Pictoris. *Science, 226,* 1421–1424.

Stahler S. W. and Walter F. M. (1993) Pre-main-sequence evolution and the birth population. In *Protostars and Planets III* (E. Levy and J. I. Lunine, eds.), pp. 405–428. Univ. of Arizona, Tucson.

Stapelfeldt K. R., Burrows C. J., Koerner D., Krist J., Watson A. M., Trauger J. T., and the WFPC2 IDT (1995) WFPC2 imaging of GM Aurigae: A circumstellar disk seen in scattered light. *Bull. Am. Astron. Soc., 27,* 1446.

Sylvester R. J. and Skinner C. J. (1996) Optical, infrared and milli-metre-wave properties of Vega-like systems — II. Radiative transfer modelling. *Mon. Not. R. Astron. Soc., 283,* 457.

Takeuchi T. and Artymowicz P. (2001) Dust migration and mor-phology in optically thin circumstellar gas disks. *Astrophys. J., 557,* 990–1006.

Thi W. F., Blake G. A., van Dishoeck E. F., van Zadelhoff G. J., Horn J. M. M., Becklin E. E., Mannings V., Sargent A. I., van den Ancker M. E., and Natta A. (2001a) Substantial reservoirs of molecular hydrogen in the debris disks around young stars. *Nature, 409,* 60–63.

Thi W. F., van Dishoeck E. F., Blake G. A., van Zadelhoff G. J., Horn J., Becklin E. E., Mannings V., Sargent A. I., van den Ancker M. E., Natta A., and Kessler J. (2001b) H_2 and CO emission from disks around T Tauri and Herbig Ae pre-main sequence stars and from debris disks around young stars: Warm and cold circumstellar gas. *Astrophys. J., 561,* 1074–1094.

Thi W. F., Pontoppidan K. M., van Dishoeck E. F., Dartois E., and d'Hendecourt L. (2002) Detection of abundant solid CO in the disk around CRBR 2422.8-3423. *Astron. Astrophys., 394,* L27.

van Zadelhoff G.-J., van Dishoeck E. F., Thi W.-F., and Blake G. A. (2001) Submillimeter lines from circumstellar disks around pre-main sequence stars. *Astron. Astrophys., 377,* 566.

Vidal-Madjar A., Ferlet R., Hobbs L. M., Gry C., and Albert C. E. (1986) The circumstellar gas cloud around β Pictoris. II. *Astron. Astrophys., 167,* 325–332.

Vidal-Madjar A., Lagrange-Henri A.-M., Feldman P. D., Beust H., Lissauer J. J., Deleuil M., Ferlet R., Gry C., Hobbs L. M., McGrath M. A., McPhate J. B., and Moos H. W. (1994) HST-GHRS observations of β Pictoris: Additional evidence for in-falling comets. *Astron. Astrophys., 290,* 245–258.

Waelkens C., Malfait K., and Waters L. B. F. M. (1999) Comet Hale-Bopp, circumstellar dust, and the interstellar medium. *Earth Moon Planets, 79,* 265.

Weinberger A. J., Rich R. M., Becklin E. E., Zuckerman B., and Matthews K. (2000) Stellar companions and the age of HD 141569 and its circumstellar disk. *Astrophys. J., 544,* 937.

Welsh B. Y., Craig N., Crawford I. A., and Price R. J. (1998) β Pic-like circumstellar disk gas surrounding HR 10 and HD 85905. *Astron. Astrophys., 338,* 674–682.

Wolf S., Gueth F., Henning T., and Kley W. (2002) Detecting planets in protoplanetary disks: A prospective study. *Astrophys. J. Lett., 566,* L97.

Wright E. (1987) Long-wavelength absorption by fractal dust grains. *Astrophys. J., 320,* 818.

Wyatt M. C., Dermott S. F., Telesco C. M., Fisher R. S., Grogan K., Holmes E. K., and Piña R. K. (1999) How observations of circumstellar disk asymmetries can reveal hidden planets: Peri-center glow and its application to the HR 4796 disk. *Astrophys. J., 527,* 918–944.

Yudin R. V. (2000) Analysis of correlations between polarimetric and photometric characteristics of young stars. A new approach to the problem after eleven years' study. *Astron. Astrophys. Suppl. Ser., 144,* 285.

Zuckerman B., Forveille T., and Kastner J. H. (1995) Inhibition of giant planet formation by rapid gas depletion around young stars. *Nature, 373,* 494.

From Icy Grains to Comets

S. J. Weidenschilling
Planetary Science Institute

Comets formed from icy grains in the outer region of the solar nebula. Their coagulation into macroscopic bodies was driven by differential motions induced by nebular gas drag. The hierarchical growth by collisions produced "rubble pile" structures with sizes up to ~100 km on timescales on the order of 1 m.y. Two-dimensional models of this growth, including orbital decay due to drag, show radial mixing that lessens the tendency seen in one-dimensional models for components of a single characteristic size. Radial migration causes redistribution of condensed matter in the outer nebula, and produces a sharp outer edge to the Kuiper belt.

1. INTRODUCTION

Cometary nuclei are planetesimals that formed in the outer reaches of the solar nebula. Presumably, they were produced by the same process that formed planetesimals in the region of the terrestrial planets and the asteroid belt, but incorporated volatiles (notably water ice) that were in solid form in the cold outer nebula. While comets may not be pristine, they are probably the least-altered objects surviving from the origin of the solar system. While much can be learned about their formation from their chemistry, they may also provide a unique record of the physical processes involved in their accretion. The material now present in any comet originally existed in the solar nebula as microscopic grains, probably a mixture of surviving interstellar grains and nebular condensates. Somehow, these submicrometer-sized particles were assembled into bodies of sizes at least tens to hundreds of kilometers. It is clear that comets are not uniform aggregates of grains, but have structure on larger scales. They display complex behavior that varies both temporally and spatially (outbursts and jetting), and implies inhomogeneities on scales of tens to hundreds of meters (*Mumma et al.,* 1993; *Weissman et al.,* 2004). On the other hand, imaging of the nuclei of Comets Halley and Borrelly at comparable resolution did not reveal obvious larger structural units; although both bodies were irregular in shape, they did not appear to be lumpy on kilometer scales. Comets are structurally weak, as demonstrated by shedding of fragments, occasional splitting, tidal disruption of Shoemaker-Levy 9 during its encounter with Jupiter (*Asphaug and Benz,* 1996), and the spontaneous disruption of Comet LINEAR (*Weaver et al.,* 2001). The observed properties of nuclei are consistent with "rubble pile" structures with components ~100 m in size that are very weakly bonded, or perhaps held together only by gravity. These properties are the expected result of formation by accretion in the solar nebula.

2. PARTICLE MOTIONS IN THE SOLAR NEBULA

The motions of solid particles in the solar nebula are dominated by drag forces due to gas; this is true even in the outermost region, where the density is low, and solids are relatively more abundant due to condensation of volatiles at low temperatures. The radial pressure gradient partially supports the gas against the Sun's gravity, causing it to rotate at slightly less than the local Kepler velocity (*Whipple,* 1972). The fractional deviation from Keplerian rotation is approximately the ratio of the thermal energy of the gas to its orbital kinetic energy. One can show that ΔV, the difference between the gas velocity and Kepler velocity V_k, is proportional to the temperature T and the square root of the heliocentric distance R (*Weidenschilling,* 1977). A typical magnitude for ΔV is ~50 m s^{-1} for plausible nebular models. As T decreases with R, ΔV does not vary strongly; for a plausible temperature gradient of $T \propto R^{-1/2}$, ΔV is independent of R. Thus, the deviation from Keplerian motion is larger in proportion to the orbital velocity at larger heliocentric distances. Typically, $\Delta V/V_k$ is a few times 10^{-3} in the region of the terrestrial planets, but can exceed 10^{-2} beyond Neptune's distance.

Solid particles are not supported by pressure forces. As a consequence, no particle can be at rest with respect to the gas, but always has some components of radial and transverse velocity. Their magnitudes depend on the particle size (more precisely, area/mass ratio) and drag law (*Adachi et al.,* 1976; *Weidenschilling,* 1977). A small particle moves with the angular velocity of the gas (negligible transverse component), but drifts radially inward at a rate that increases with size. A large body pursues a Kepler orbit, experiencing a transverse "headwind" that causes its orbit to decay; the rate of decay decreases with size. The peak radial velocity, equal to ΔV, occurs at the transition between these regimes. The size at which this peak velocity is reached

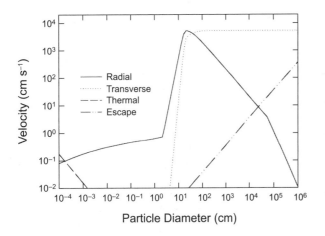

Fig. 1. Particle velocities as functions of size in a model solar nebula. The nebular parameters at 31.5 AU are: gas density 1.5×10^{-13} g cm^{-3}, T = 57 K, ΔV = 52 m s^{-1}. Particles are assumed to have fractal structures to diameter 1 cm, and density 1 g cm^{-3} at d > 20 cm. Shown are radial and transverse components of systematic motions relative to the gas, thermal velocity, and gravitational escape velocity.

does not vary strongly with heliocentric distance, and is typically on the order of 1 m (Fig. 1).

The size dependence of the drag-induced velocity means that any bodies that are not identical move relative to each other and any ensemble of bodies having a range of sizes will experience collisions. For small bodies (d ≤ 1 m), only radial drift is significant, and their relative velocities will be the difference between their radial velocities (far from the central plane of the nebula, their settling velocities may also be significant). Large bodies have high transverse velocity relative to the gas, but the same magnitude, ΔV, for all such bodies, so the relative velocity between any pair of large bodies is essentially the difference between their radial velocities. The relative velocity between a small and large body is essentially the latter's transverse velocity.

At the low temperatures of a few tens of degrees or less in the outer nebula, thermal motions dominate for micrometer-sized grains, but are negligible for larger bodies. Mutual gravitational perturbations are significant for bodies larger than about 1 km. Between those limits, gas drag dominates. If turbulence is present, it can also induce relative motions. However, strong turbulence is unlikely to persist at large heliocentric distances. In the outer nebula, the surface density may be low enough to allow ionization by cosmic rays, driving magnetorotational instability (*Sano et al.,* 2000), but the low gas density also favors dissipation by ambipolar diffusion. The intensity of turbulence generated by this mechanism is unclear. In any case, the largest eddies would have turnover timescales, imposed by the nebula's rotation, comparable to the Kepler period. Bodies larger than about 1 m cannot respond to turbulent fluctuations on such timescales, while motions of smaller bodies are correlated, leading to relative velocities smaller than

the turbulent velocity (*Weidenschilling and Cuzzi,* 1993). Unless the turbulence is very strong, the dominant source of relative velocities is differential motion due to gas drag.

3. COLLISIONAL COAGULATION

If collisions between particles are to result in growth, there must be some mechanism that allows them to stick together. Perfect sticking is not required (and is unlikely under most conditions for real materials), but at least some collisions must yield net growth for bodies of all sizes. This is not a problem for the initial stage of coagulation, involving micrometer-sized grains. For small particles, the relative velocities (thermal or drift) are low enough to allow sticking by van der Waals surface forces. Theoretical models (*Dominik and Tielens,* 1997) and laboratory experiments (*Blum and Wurm,* 2000) confirm that coagulation under such conditions produces fluffy, fractal-like aggregates with densities that decrease with increasing size.

Under zero-gravity conditions in the solar nebula, this process may produce gossamer structures of macroscopic (approximately centimeter-sized) dimensions. However, this type of growth cannot continue indefinitely; as the aggregates become larger their drift velocities increase, and collisions become energetic enough to allow some compaction by rearranging bonds between grains. As the aggregates become denser, their drift speeds increase further.

Bodies of comparable size will collide rarely, and only at low velocities. As shown in the simulations of *Weidenschilling* (1997), most collisions occur between bodies of quite different sizes. This circumstance favors growth, because the smaller "projectile" does not deliver enough kinetic energy to disrupt the larger "target," unless the latter has very low impact strength. However, although a low probability of disruption is necessary for growth, it is not sufficient. The increase in drift velocities with size leads to a problem: As can be seen from Figs. 1 and 2, meter-sized bodies will have velocities relative to centimeter-sized particles as large as ΔV, i.e., tens of meters per second [if the particles settle into a dense layer in the central plane of the nebula, collective motion of particles and gas reduces the effective value of ΔV (cf. *Nakagawa et al.,* 1986)]. If such impacts do not result in net mass gain, then growth might stall before meter-sized bodies could form. If collisions (or some other process) can produce bodies larger than this critical threshold, then further growth to kilometer size, at which gravity can contribute to sticking, is assured. The centimeter-to-kilometer range is the critical stage of collisional accretion.

It is not clear whether impacts at such speeds would result in net gain or loss of mass. This question cannot be answered definitively, as we do not know the mechanical properties and impact strength of cometary material. *Sirono and Greenberg* (2000) estimated the tensile and compressive strengths of porous aggregates of icy grains, and concluded that compaction would dissipate a large fraction of energy during collisions, and would produce merged bodies

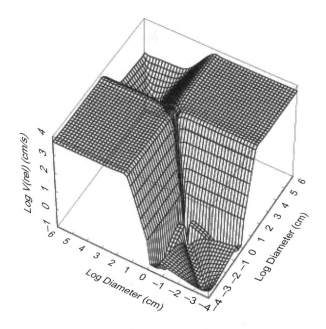

Fig. 2. Contour plot of relative velocities between particles for the same parameters assumed in Fig. 1. Particles of sizes ≤1 cm have very low relative velocities, but bodies larger than ~1 m have velocities ~50 m s^{-1}. The "valley" in the central region shows that bodies of nearly equal size have low relative velocities due to drag. Gravitational stirring dominates at sizes larger than about 1 km.

with substantial cohesion. *Bridges et al.* (1996) conducted low-speed impact experiments involving frost-covered bodies. They concluded that the observed sticking was due to interpenetration of irregular surfaces, and suggested that frost (of water ice and/or other volatiles) aided aggregation in the solar nebula. While this could not be the only mechanism (planetesimals evidently formed in the inner nebula, at temperatures too high for frosts), the mechanical properties of ices, especially their low elastic moduli, may have aided accretion in the outer solar system. *Weidenschilling* (1997) emphasized that most collisions driven by gas drag would involve a small body impacting a much larger one. If the bodies are porous aggregates, then the smaller one might become embedded in the larger (it is not necessary to assume that the projectile survives intact). Weidenschilling assumed that an impact added the projectile's mass to the target, but removed an amount of material proportional to the impact energy; this is equivalent to the assumption that there is a critical impact velocity with net gain (loss) of mass below (above) that threshold. An accreting body would experience impacts by smaller particles having a range of sizes and relative velocities; as mentioned above, bodies of comparable size would collide more gently. Whether a given body would grow or erode would depend not only on the impact strength of the bodies, but on the size distribution of the ensemble, which would itself be a function of their strength. The impact strength assumed by Weidenschilling resulted in net growth, but the range of allowable parameters remains to be determined.

Wurm et al. (2001) suggested that accretion of meter-sized bodies would be aided by aerodynamic forces acting on small grains. Grains would impact on the leading side, brought by the "headwind" due to the body's motion through the gas. Grains or small aggregates striking at velocities ~ΔV would rebound (or displace grains from the target) at speeds lower by perhaps an order of magnitude. The gas flow would reverse their motion and return them to the body's surface, reimpacting at speeds low enough for sticking by surface forces between the grains. This mechanism could aid accretion at R ~ 1 AU, where plausible values for the gas density yield trajectories of only a few centimeters for bouncing grains. However, it would not be effective in the outer nebula, where impact speeds are comparable, but the gas density is lower by orders of magnitude; the "turnaround distance" for grains would be correspondingly larger, and their probability of reimpact on a meter-sized target would be minuscule. Thus, particle coagulation in the region of comet formation appears to require effective sticking in collisions.

4. GRAVITATIONAL INSTABILITY

With the uncertainties attendant upon the messy and poorly constrained mechanism of collisional sticking, the possibility of forming planetesimals by gravitational instability remains enticing. In the "classical" instability scenario (*Safronov*, 1969; *Goldreich and Ward*, 1973), small particles settle to form a layer in the midplane of the solar nebula. When this layer reaches a critical density it becomes unstable to density perturbations, which produce condensations that collapse under their own gravity into solid bodies. However, bodies formed by this mechanism in the outer nebula would be much larger than typical comets, and would not have any structure on macroscopic scales. In any case, gravitational instability does not eliminate the need for particle coagulation. As pointed out by *Weidenschilling* (1980), before the particle layer can reach the critical density, it drags the entrained gas at a velocity higher than that of the pressure-supported nebula. The resulting shear flow is unstable, and becomes turbulent, preventing further settling. A layer of small (centimeter-sized or smaller) particles cannot attain the critical density.

If the bodies can accrete to sizes large enough to decouple from the shear-induced turbulence (≥1 m), then they can settle enough for the layer to attain the critical density. This condition is necessary for instability, but is not sufficient. The velocity dispersion must also be small enough to allow density perturbations to grow. When the particle velocities are controlled by gas drag, they are not isotropic. As bodies of this size are too large to be stirred by turbulence, but too small for gravitational perturbations to be effective, their vertical velocity dispersion is very small; however, they may have a significant dispersion in radial velocity. If all particles were identical, then they would have the same radial velocity due to drag, but any real ensemble of particles produced by coagulation should have a range of sizes and corresponding velocities. The resulting dispersion inhibits gravitational instability (*Weidenschilling, 1995*). *Ward* (2000) showed that if the particle layer was bimodal with two different velocities, density perturbations in the populations would be decoupled. For bodies larger than about a meter, the radial velocity decreases with size (Fig. 1); the dispersion will diminish as the bodies grow, and the mean radial velocity decreases. The conditions for gravitational instability (density and velocity dispersion) can be attained, but only after collisional coagulation has produced bodies with mean size ≥10 m. As this is larger than the size of maximum radial velocity, there is no obstacle to further growth by collisions; gravitational instability becomes unnecessary by the time it becomes feasible. The instability may still occur at that point, but its outcome will be different from the classical model. Bodies tens of meters in size are poorly damped by either drag or collisions, so the layer maintains a significant velocity dispersion, which allows only density perturbations of long wavelength to grow (*Weidenschilling, 1995*). The layer may break up into self-gravitating clusters of particles, but these condensations are too large (i.e., have too much angular momentum due to the rotation of the nebula) to collapse directly into solid objects. Such condensations would probably be transient features, which would be torn apart by differential rotation. It is suggestive that for plausible values of the nebula's surface density, the characteristic wavelength of instability corresponds to condensation masses equal to those of compact bodies with diameters ~100 km, about the size of typical Kuiper belt objects. However, this may be coincidental; it will be shown below that bodies of this size could grow by collisions within the lifetime of the solar nebula.

5. MODELING THE GROWTH OF COMETESIMALS

5.1. One-Dimensional Models

Weidenschilling (1997) applied a numerical model of particle coagulation and settling in the solar nebula to cometary formation; details of the model are described therein, and will be summarized only briefly here. That model is one-dimensional in the vertical direction, treating the particle population in a series of levels at a single heliocentric distance. In each level, the particle size distribution is modeled by the mass in logarithmic diameter bins from 10^{-4} cm to >100 km. The evolution of the size distribution is computed in each of a series of levels, with collision rates due to thermal motion, differential settling, and radial and transverse motions due to gas drag. Particle densities are assumed to vary with size, to simulate fractal structure of small aggregates and compaction of macroscopic bodies. Collisional outcomes depend on particle sizes and impact velocities, consistent with an assumed impact strength. At each timestep, the number of collisions between particles of various sizes and the consequent changes in the size distribution are computed within each level. Bodies are transferred between levels, downward toward the midplane at rates proportional to their settling velocities, and by diffusion upward and downward along concentration gradients where turbulence is present. Formation of a dense midplane layer results in shear-induced turbulence, and the structure of the layer is determined by a balance between settling and turbulent diffusion, with gravitational stirring included for bodies large enough to decouple from the gas. The model parameters were chosen for a low-mass solar nebula at 30 AU, with surface density of solids and gas of 0.4 and 29 g cm^{-2}, respectively. From the assumed initial state, with all solids present as micrometer-sized grains with a uniform solids/gas mass ratio, the ensemble of particles evolved on a timescale of few times 10^5 yr, or a few thousand orbital periods. This evolution could be divided into rather distinct stages. During the first few times 10^4 yr, thermal coagulation produced low-density, fractal-like aggregates, with sizes ~10^{-2} cm, at all levels. As the aggregates in the upper levels settled toward the midplane, larger ones grew by sweeping up smaller ones; this led to rapid growth and "rainout" of approximately centimeter-sized aggregates, which formed a layer with density greater than that of the gas. The shear between this layer and the surrounding pressure-supported gas produced turbulence, which prevented settling of the small aggregates within the layer. However, collisional growth continued, and a thinner sublayer of larger bodies developed. This sublayer attained a density much higher than the classical threshold for gravitational instability (solids/gas ratio ~10^3), but the velocity dispersion was large enough to prevent instability until the mean size was tens of meters. With the assumption that instabilities in a poorly damped system would not collapse, the collisional evolution was continued until bodies tens of kilometers in size accreted, after a model time of 2.5×10^5 yr. By that point, gravitational stirring became effective and the velocity dispersion increased, with a significant vertical component. This caused the particle layer to become thicker, and its density decreased; presumably still larger bodies would continue to grow by gravitational accretion.

The size distribution of the accreting bodies (Fig. 3) developed a distinct peak in the size range of tens to hundreds of meters. The cause of this peak was the dependence of radial velocity on size (Fig. 1) and the fact that collisions were due primarily to differences in radial velocity. The bodies with the largest velocities were quickly depleted by impacting larger bodies, opening a "valley" in the size distribution at meter sizes. Bodies in the size range from tens of meters to about a kilometer still had velocities larger than their escape velocities, so there was no gravitational enhancement of their collision rates. Because their radial velocities decreased with size, the larger bodies grew more slowly, allowing smaller ones to catch up, and keeping the peak narrow. Eventually, the largest bodies began to have significant gravitational enhancement of their collision rates, and the peak became broader. During this stage of growth the mean collision velocities decreased with increasing size, reaching a minimum value at the transition from drag-con-

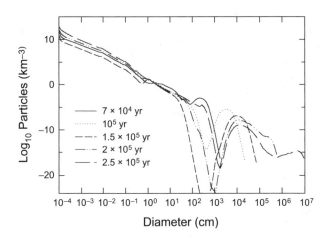

Fig. 3. Computed size distribution in the nebular midplane at various times, using the one-dimensional model of *Weidenschilling* (1997). After 10^5 yr, a "valley" forms in the range 1–10 m; bodies of this size are near the peak in radial velocity, and are rapidly depleted by collisions with larger bodies. Mass piles up in a peak at sizes ~100 m, where the radial velocity and collision rate are decreasing with size.

trolled growth to gravitational accretion. This transition occurred when kilometer-sized bodies were accreting bodies ~100 m in size. Larger bodies experienced more energetic collisions as impact velocities became dominated by their escape velocities. *Weidenschilling* (1997) speculated that cometary nuclei originated as "rubble piles" that were rather well-compacted at meter scale, but with more void space on larger scales, with a tendency to preserve structure on the scale of ~100 m.

The results of the one-dimensional coagulation model are consistent with observable physical properties of comets. However, that model has one significant shortcoming. The peak radial velocity of ~50 m s^{-1} corresponds to 1 AU/century; this is approached only by bodies in a narrow range of sizes (roughly 0.1–10 m), but even the lower radial velocities of larger and smaller bodies could still result in substantial migration on the computed growth timescale. The one-dimensional model conserves mass locally; it includes the effects of radial velocities on the collision rate, but does not account for movement of mass into or out of the region of calculation at a given heliocentric distance. This movement has two effects on the physical modeling: At any location the total mass may increase or decrease relative to the starting value, and the size distribution may be altered by different rates of migration by bodies of various sizes. In addition, there may be compositional effects as material that condensed at a range of heliocentric distances and temperatures is mixed during accretion.

Stepinski and co-workers (*Stepinski and Valageas*, 1996, 1997; *Kornet et al.*, 2001) recognized that radial migration due to gas drag could be significant. They produced a model for particle coagulation and migration in a circumstellar

accretion disk. The properties of the particle layer are averaged through its thickness, so their model is effectively one-dimensional in the radial direction, and complementary to that of *Weidenschilling* (1997). They do not compute a size distribution, but only a mean particle size as a function of time at a given radius. This approach neglects differential drift motions among bodies of different sizes, and allows only collisions due to turbulence, so their model breaks down for disks with low turbulence. Despite these limitations, they demonstrated that orbital decay due to gas drag during particle growth could substantially alter the radial distribution of solids relative to the gas during the ~10^6–10^7-yr lifetime of the solar nebula.

5.2. Two-Dimensional Models

In order to overcome the limitations of one-dimensional approaches, Weidenschilling (in preparation, 2003) has developed a fully two-dimensional model of planetesimal formation. Its operation is similar to the one-dimensional model described above, but the vertical structure of the particle layer and its size distribution are computed in a series of zones over a range of heliocentric distance. Bodies are transferred between zones at rates corresponding to their radial velocities. The model also includes collective radial motion of the particle layer due to turbulent shear stress acting in the boundary layer (*Goldreich and Ward*, 1973) and radial diffusion by turbulence, but for macroscopic bodies these are generally unimportant compared to orbital decay due to the drag of the "headwind." Application of this two-dimensional model reveals some significant differences from the one-dimensional case.

Figure 4 shows the size distribution in the range resulting from a two-dimensional simulation from 30 to 90 AU, at model time of 5×10^5 yr. The innermost zone at this time corresponds to the final stage reached by the one-dimensional simulation in Fig. 3. While the size distributions are similar, in the two-dimensional case the "valley" around ~1 m is much shallower, and the peak at ~100 m is broader and more subdued. The apparent reason for this difference is the dependence of the growth time on heliocentric distance. It can be seen that large bodies form more slowly at larger values of R, due to the lower surface density. Recall that in the one-dimensional model, once bodies with sizes of tens of meters formed, meter-sized bodies were depleted by colliding with the larger ones due to their high radial velocities; this is the cause of the gap in the size distribution.

In the two-dimensional model, meter-sized bodies that form at larger distances migrate inward, continually replenishing the population at that size and filling the gap. The addition of these bodies also keeps the median size smaller. The model for gravitational stirring scales the out-of-plane random velocity to the escape velocity of the median-sized body in the midplane; the smaller median size in the two-dimensional case results in a more flattened particle layer than in the one-dimensional model.

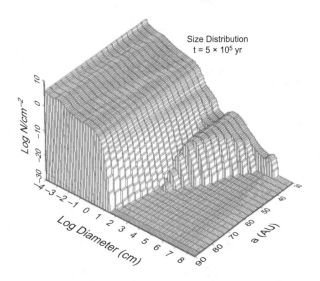

Fig. 4. Results of a two-dimensional simulation performed in 20 zones from 30 to 90 AU, with an R^{-1} variation in surface density. The assumed surface density in the innermost zone is about half that in the one-dimensional simulation, approximately doubling the evolution timescale. After 5×10^5 yr, the innermost zone contains bodies larger than 100 km, while no bodies larger than 1 km have formed beyond ~40 AU. In the inner zones, the "valley" at ~1–10 m is less distinct due to inward migration of bodies of this size that formed at larger distances.

The gaseous component of the solar nebula is assumed to remain constant, but radial migration results in significant redistribution of the solid matter (Fig. 5). The inner zones are initially depleted by orbital decay of the first-formed meter-sized bodies, After ~2×10^5 yr the surface density in the inner region rises again due to the formation

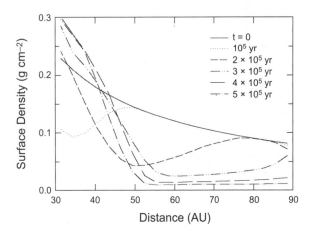

Fig. 5. Evolution of the surface density of solids for the case shown in Fig. 4. At $t = 0$, when all condensed matter is in the form of dust, the surface density varies as R^{-1}. Inside ~45 AU, the growth of meter-sized bodies causes initial depletion by orbital decay, but this loss is replenished by matter coming in from greater distances that is captured by larger bodies. The region beyond ~45 AU suffers net depletion; the surface density and mean particle size decline steeply from 30 to 50 AU.

of kilometer-sized or larger bodies that resist orbital decay. These grow by catching the smaller bodies migrating inward from larger distances. The gradient of the surface density of solids becomes much steeper, and develops a rather sharp "edge" in the range 40–50 AU, at about half the radius of the nebula. The reason for this behavior is due to the fact that the peak radial velocity induced by gas drag, and the size at which this peak occurs, do not vary significantly with heliocentric distance. Most of the migration occurs during growth from about 0.1 to 10 m diameter. This growth takes longer at the lower densities found at greater distances, so a body that begins to accrete at a larger distance moves inward by a greater amount. Empirically, the growth time varies approximately as $R^{3/2}$, so the fractional change of distance increases with R, causing the surface density gradient to steepen. Because the particle layer is highly flattened, small bodies migrating inward are efficiently captured by larger bodies that have stopped their orbital decay. Mass tends to pile up at the outermost distance where kilometer-sized bodies grow, which further steepens the gradient of surface density. A similar "pileup" occurs in the radial one-dimensional model of *Stepinski and Valageas* (1997), so this effect is not sensitive to details of the model. This transition is also accompanied by a steep decline in the mean size of planetesimals. It can be seen in Fig. 4 that no bodies larger than ~1 km have accreted beyond 50 AU. Continuation of the simulation to later times would not produce any large bodies at this distance, as the surface density of solid material is too low for further growth.

This simulation assumed a laminar nebula, i.e., the only turbulence is that produced locally near the central plane by shear between the particle layer and the surrounding gas. If there is an additional source of turbulence, e.g., magneto-rotational instability, then its main effect would be to increase the thickness of the particle layer. The lower density would cause accretion to be slower, allowing more migration during growth. Thus, the initial extent of the nebula would have to be even greater to yield an edge at this distance.

6. FORMATION TIMESCALES

Motions induced by drag lead to relatively rapid formation of sizable bodies, compared with purely gravitational accretion in the absence of gas. *Kenyon* (2002) modeled planetesimal accretion in the Kuiper belt. His simulations included gravitational stirring and fragmentation; most did not include gas drag, or considered only damping of eccentricities and inclinations without orbital decay. The assumed surface density of the planetesimal swarm was comparable to the cases discussed above, but the initial size distribution was quite different: a power law with most of the mass in bodies of radius ~100 m. Kenyon's model required ~10 m.y. to accrete 100-km bodies while drag-driven accretion produces such in <1 m.y. in regions that are not depleted by radial migration. Part of this difference can be ascribed to the starting conditions. Gravitational accretion would be faster for smaller initial sizes, which produce less stirring and lower velocities, with earlier transition to runaway growth. However, gravitational stirring is isotropic, producing out-of-plane velocities that thicken the layer and reduce the collision rate as the mean size increases. Drag-induced motions are more favorable to accretion because they are parallel to the nebular midplane. Turbulent stirring is ineffective for meter-sized and larger bodies, so the particle layer can attain high density, while the radial and transverse velocities are high enough to allow frequent collisions. In contrast, the inner zones in the simulation experience a net increase in surface density of solids over the initial value, allowing more rapid growth there.

The rapid formation of 100-km-sized bodies may have consequences for their thermal evolution. If short-lived radionuclides such as ^{26}Al were present in the outer part of the solar nebula, accretion times ~1 m.y. imply that Kuiper belt objects of this size probably experienced enough heating to undergo melting and differentiation in their interiors (*Prialnik et al., 2004*). In that case, they might be rather compact bodies with low porosity and moderately high density and strength. However, modeling of the collisional evolution of the Kuiper belt since its formation (*Farinella et al., 2000*) shows that fragmentation of such large bodies is rare, and implies that most objects with sizes of a few kilometer are derived from smaller parent bodies a few tens of kilometers in size. The typical cometary nucleus formed as a rubble pile, and has remained as such throughout the history of the solar system.

7. CONCLUSIONS

Most of the results of the one-dimensional particle coagulation/settling model remain valid: Cometary nuclei formed by collisional coagulation driven by differential motions due to gas drag in the solar nebula, with "rubble pile" structures resulting from hierarchical accretion. The timescale for growth of bodies as large as typical Kuiper belt objects (~100 m) is ~1 m.y., well within the estimated lifetime of the solar nebula.

The one-dimensional results also suggested that comets should be composed of components with a characteristic size ~100 m. The two-dimensional model alters that picture: Radial migration mixes bodies of different sizes that formed at different heliocentric distances; there should be less of a deficit of meter-sized components, and a size distribution of subunits within a given nucleus with more resemblance to a power law. The variety of component sizes would result in a smaller fraction of void space within a nucleus than that produced by a single size. There may still be some preference for 100-m subunits, as this size is subjected to the lowest impact velocity during accretion. The smaller (meter-sized) components are accreted at higher impact velocities than larger ones, and may be more compacted at such scales.

There is compelling observational evidence that the Kuiper belt has an abrupt outer edge in the range ~45–50 AU (*Trujillo and Brown, 2001*). This feature cannot be accounted for by gravitational accretion models that do not include radial migration, as growth timescales and sizes vary too gradually with heliocentric distance to produce such a sharp transition (*Kenyon and Luu, 1998, 1999*). While there are other possible explanations for truncation of the Kuiper belt, such as a stellar encounter during formation of the solar system (*Kobayashi and Ida, 2001*), drag-induced migration during planetesimal formation provides a plausible — perhaps unavoidable — mechanism for producing a sharp edge. The increasing difficulty of accretion with distance also puts in doubt the existence of a massive primordial belt of material in the region ~50–100 AU, as suggested by *Stern* (1996). The solar nebula probably had to extend to such a distance simply to account for the present extent of the Kuiper belt, but most of the condensed matter originally present there was removed by gas drag before it could form bodies of significant size. Bodies in the region of the current Kuiper belt (40–50 AU) probably contain a substantial fraction of material that originally condensed (or survived from the presolar cloud) at much larger distances, possibly beyond 100 AU. In the denser inner region of the nebula, accretion times were shorter and the migration distance was less. Comets that formed in the region of the giant planets probably comprise material derived from a smaller range of heliocentric distances.

REFERENCES

Adachi I., Hayashi C., and Nakazawa K. (1976) The gas drag effect on the elliptical motion of a solid body in the primordial solar nebula. *Prog. Theor. Phys., 56,* 1756–1771.

Asphaug E. and Benz W. (1996) Size, density, and structure of Comet Shoemaker-Levy 9 inferred from the physics of tidal breakup. *Icarus, 121,* 225–248.

Blum J. and Wurm G. (2000) Experiments on sticking, restructuring, and fragmentation of preplanetary dust aggregates. *Icarus, 143,* 138–146.

Bridges F., Supulver K., Lin D. N. C., Knight R., and Zafra M. (1996) Energy loss and sticking mechanisms in particle aggregation in planetesimal formation. *Icarus, 123,* 422–435.

Dominik C. and Tielens A. G. G. M. (1997) The physics of dust coagulation and the structure of dust aggregates in space. *Astrophys. J., 480,* 647–673.

Farinella P., Davis D. R., and Stern S. A. (2000) Formation and collisional evolution of the Edgeworth-Kuiper belt. In *Protostars and Planets III* (E. Levy and J. Lunine, eds.), pp. 1255–1082. Univ. of Arizona, Tucson.

Goldreich P. and Ward W. R. (1973) The formation of planetesimals. *Astrophys. J., 183,* 1051–1061.

Kenyon S. (2002) Planet formation in the outer solar system. *Publ. Astron. Soc. Pac., 114,* 265–283.

Kenyon S. and Luu J. (1998) Accretion in the early Kuiper belt. 1. Coagulation and velocity evolutions. *Astron. J., 115,* 2136–2160.

Kenyon S. and Luu J. (1999) Accretion in the early Kuiper belt. II. Fragmentation. *Astron. J., 188,* 1101–1119.

Kobayashi H. and Ida S. (2001) The effects of a stellar encounter on a planetesimal disk. *Icarus, 153,* 416–429.

Kornet K., Stepinski T., and Rozyczka M. (2001) Diversity of planetary systems from evolution of solids in protoplanetary disks. *Astron. Astrophys., 378,* 180–191.

Mumma M., Weissman P., and Stern S. A. (1993) Comets and the origin of the solar system: Reading the Rosetta stone. In *Proto-stars and Planets III* (E. Levy and J. Lunine, eds.), pp. 1177–1252. Univ. of Arizona, Tucson.

Nakagawa Y., Sekiya M., and Hayashi C. (1986) Settling and growth of dust particles in a laminar phase of a low mass solar nebula. *Icarus, 67,* 375–390.

Prialnik D., Benkhoff J., and Podolak M. (2004) Modeling comet nuclei. In *Comets II* (M. C. Festou et al., eds.), this volume. Univ. of Arizona, Tucson.

Safronov V. S. (1969) *Evolyutsiya Doplanetnogo Oblake i Obrazovanie Zemli i Planet* (*Evolution of the Protoplanetary Cloud and Formation of the Earth and Planets*). Nauka, Moscow. 244 pp. (Translated in NASA TTF–677.)

Sano T., Miyama S., Umebayashi T., and Nakano T. (2000) Magnetorotational instability in protoplanetary disks. II. Ionization state and unstable regions. *Astrophys. J., 543,* 486–501.

Sirono S. and Greenberg J. M. (2000) Do cometesimal collisions lead to bound rubble piles or to aggregates held together by gravity? *Icarus, 145,* 230–238.

Stepinski T. and Valageas P. (1996) Global evolution of solid matter in turbulent protoplanetary disks. I. Aerodynamics of solid particles. *Astron. Astrophys., 309,* 301–312.

Stepinski T. and Valageas P. (1997) Global evolution of solid matter in turbulent protoplanetary disks. II. Development of icy planetesimals. *Astron. Astrophys., 319,* 1007–1019.

Stern A. (1996) On the collisional environment, accretion time scales, and architecture of the massive, primordial Kuiper belt. *Astron. J., 112,* 1203–1211.

Trujillo C. and Brown M. (2001) The radial distribution of the Kuiper belt. *Astrophys. J. Lett., 554,* L95–L98.

Ward W. R. (2000) On planetesimal formation: The role of collective particle behavior. In *Origin of the Earth and Moon* (R. M. Canup and K. Righter, eds.), pp. 75–84. Univ. of Arizona, Tucson.

Weaver H. A. and 20 colleagues (2001) HST and VLT investigations of the fragments of comet C/1999 S4 (LINEAR). *Science, 292,* 1329–1333.

Weidenschilling S. J. (1977) Aerodynamics of solid bodies in the solar nebula. *Mon. Not. R. Astron. Soc., 180,* 57–70.

Weidenschilling S. J. (1980) Dust to planetesimals: Settling and coagulation in the solar nebula. *Icarus, 44,* 172–189.

Weidenschilling S. J. (1995) Can gravitational instability form planetesimals? *Icarus, 116,* 433–435.

Weidenschilling S. J. (1997) The origin of comets in the solar nebula: A unified model. *Icarus, 127,* 290–306.

Weidenschilling S. J. and Cuzzi J. N. (1993) Formation of planetesimals in the solar nebula. In *Protostars and Planets III* (E. Levy and J. Lunine, eds.), pp. 1031–1060. Univ. of Arizona, Tucson.

Weissman P., Asphaug E., and Lowry S. (2004) Structure and density of cometary nuclei. In *Comets II* (M. C. Festou et al., eds.), this volume. Univ. of Arizona, Tucson.

Whipple F. L. (1972) On certain aerodynamic processes for asteroids and comets. In *From Plasma to Planet* (A. Elvius, ed.), pp. 211–232. Wiley, New York.

Wurm G., Blum J., and Colwell J. (2001) A new mechanism relevant to the formation of planetesimals in the solar nebula. *Icarus, 151,* 318–321.

Coupled Physical and Chemical Evolution of Volatiles in the Protoplanetary Disk: A Tale of Three Elements

Jonathan I. Lunine

University of Arizona

Daniel Gautier

Observatoire de Paris-Meudon

Volatiles contained within comets have been subjected to a range of physical and chemical properties within the protoplanetary disk out of which the solar system formed. Here we focus on three elements — O, N, and S — that occur in molecular forms of widely varying volatility. These molecular forms play distinct roles in the disk, ranging from fundamentally determining opacity and oxidation state (water), to being a passive tracer of the O chemistry (hydrogen sulfide) and the amount of unprocessed molecular cloud material incorporated in grains (ammonia). The interactions among the various molecular species are complex, reflecting numerous physical processes in the disk, only some of which are well enough understood to model in detail with confidence.

1. THE IMPORTANCE OF VOLATILES AS TRACERS OF DISK PROPERTIES

Some of the details of planet formation are hidden in the abundances of the elements and isotopes, now present in the planets themselves, in the small bodies of the solar system, and in the meteorites. Constraints imposed by these abundances are more powerful if they are based on multiple elements, or more than one isotopic ratio. Models that simulate the history of volatiles that may condense or adsorb in different ways as a function of position or time in the disk are difficult to construct because they must consider processes occurring on a broad range of spatial and temporal scales. Yet, a full understanding of the processes by which our solar system formed will not be gained until these processes are quantified to a sufficient extent, and this has yet to be achieved.

Oxygen, S, and N are elements that are abundant (Table 1) and of particular importance. Water drives the oxidation state of the disk where it occurs in the gas phase, and is a primary planet-building material in the outer part of the disk. The icy nature of comets, many of the moons of the giant planets, and of the Kuiper belt attest to H_2O ice as a principal planet-building material. Water is dominant as a solid, and thus it affects the energy balance of the protoplanetary disk, rather than being a passive tracer. Sulfur's complex chemistry is coupled to the H_2O abundance in the inner disk; the diversity of volatilities of S compounds themselves mean that some will be found in the rocky (refractory) primitive material of the chondrites, while others (notably H_2S) will be trapped in H_2O ice. H_2S was also found in a number of comets originating from the Oort cloud. This variety means that S has left clues to the assembly of the building blocks of planets and comets throughout the solar system, and is thus of especial value. Nitrogen is a key element because it is well measured in comets and in Jupiter and may have existed in planetesimals as highly volatile molecular nitrogen, N_2, or as ammonia, NH_3 (or related compounds), or both. Molecular nitrogen is extremely volatile; NH_3 is only modestly more volatile than H_2O ice. Which of the two forms was more abundant in outer solar system solids can be constrained by jovian and cometary data, and in turn this provides constraints on the conditions under which the icy planetesimals that seeded the giant planets formed.

Galileo probe and remote sensing data from flybys, orbiters, and Earth-based observatories provide a detailed set of elemental abundances for Jupiter unrivalled by that for any of the other giant planets (*Atreya et al.,* 2003), and illustrate the motivation for trying to understand the coupling of dynamics and chemistry for multiple volatile species. Noble gas (Ar, Kr, Xe) and three major element (S, N, C) abundances in the sensible atmosphere are remarkably uniform, being roughly 2–4 times the solar value; He is solar, while Ne and O are depleted relative to solar (*Niemann et al.,* 1998). The elemental O depletion is in fact a depletion of H_2O, because H_2O is the primary carrier of O in Jupiter. The depletion is almost certainly associated with the meteorology of the atmosphere through which the probe flew (*Atreya et al.,* 2003). Observations of jovian moist convection by the *Galileo* solid-state imaging (SSI) camera as well as spectra obtained by the *Galileo* Near-Infrared Mapping Spectrometer (NIMS) support a deep H_2O abundance in excess of solar (*Gierasch et al.,* 2000). [The Ne depletion is more problematic, since the abundance is much less than one would expect for a simple solar composition gas, and

TABLE 1.   Solar elemental abundances (*Cox,* 2000).

Element	Abundance*
Oxygen	8.5×10^{-4}
Carbon	3.6×10^{-4}
Nitrogen	1.1×10^{-4}
Magnesium	3.8×10^{-5}
Silicon	3.5×10^{-5}
Iron	3.2×10^{-5}
Sulfur	1.8×10^{-5}

*Abundances are normalized to that of atomic H.

condensation is excluded for this volatile noble gas. It is possible that the Ne is dissolved in an immiscible phase of He deep in Jupiter's interior (*Roulston and Stevenson,* 1995).]

Two models to explain the pattern of major elements and noble gases in Jupiter have been offered. Both are based on the accepted inference that Jupiter's interior is heavily enriched in "heavy" (non-H or non-He) elements relative to solar values (*Fortney and Hubbard,* 2003). The source of the heavy elements cannot be the gas in the disk itself, which supplies the heavy elements in solar proportion to H and He. It must come, instead, from a condensed phase of dust and planetesimals that accreted onto Jupiter as the giant planet grew (*Pollack et al.,* 1996). The two models draw on different sources of solids to achieve the heavy element abundance pattern seen in the Jupiter atmosphere. Because of the high abundance of O and hence H_2O in the protoplanetary disk (which is assumed to have solar elemental composition overall), H_2O ice is likely to have been the principle carrier of the volatiles that supplied the heavy elements. In one model Jupiter accreted solid material formed in a cold molecular cloud environment in which volatiles were indiscriminately adsorbed onto amorphous H_2O ice at temperatures below 30 K; this material was delivered by infall to the protoplanetary disk and accreted by Jupiter (*Owen et al.,* 1999). We call this the "amorphous" model. The second ("clathrate") model invokes formation of planetesimals local to the formation zone of Jupiter in a cooling disk, so that volatiles are progressively trapped in clathrate hydrate from 150 K down to 38 K (*Gautier et al.,* 2001). (In this chapter we use the terms "protoplanetary disk" or just "disk" rather than "nebula" or "solar nebula." These last two terms refer to the epoch in the formation of our own solar system when the protoplanetary disk was largely gaseous, and it is this part of the evolution we focus on in the present chapter. For clarity we prefer the more astronomical term "disk" here.)

Because the trapping conditions and hence mechanism of volatile trapping in the H_2O ice differs in the two models, the ratio of H_2O ice to trapped volatiles differs in the two cases. The amorphous model predicts that, in order to produce the observed enrichments of 2–4 times solar in S, N, C, and the heavier noble gases, enough H_2O must have entered Jupiter to enrich the elemental O abundance by a factor of 4 relative to solar (*Atreya et al.,* 2003). The clathrate model, because it is much less efficient in trapping volatiles, predicts a much larger accretion of H_2O to explain the pattern of heavy elements, leading to an elemental O abundance 9 times solar (*Gautier et al.,* 2001). Determination of the deep H_2O abundance will have to await a deep probe or microwave sounding mission.

The two models also differ in their assumptions about the elements N and S. The amorphous model explains the jovian abundance of H_2S, hence the element S, in the context of the overall enrichment pattern, starting with a solar abundance of S in the gas phase from which the ices were formed. In order to fit the observations, the clathrate model requires that S in the gas phase in the vicinity of Jupiter, where the clathrate formed, was about half the solar value. The principal S compound in the gas phase in either model is assumed to have been hydrogen sulfide (H_2S). As discussed below, the depletion around 5 AU required by the clathration model could be explained by a combination of radial redistribution of H_2O, chemical reactions in the hot inner disk, and outward transport by turbulent mixing. In the amorphous model for Jupiter, most of the N is molecular (N_2); in the clathrate model it is both in N_2 and in NH_3.

Thus the three elements O, N, and S are coupled in terms of the molecular forms required for consistency with jovian composition under these two diverse views. But they are necessarily also coupled to the details of the physical processes by which they came to be trapped in the solids that seeded Jupiter. In the remainder of this short chapter we focus on protoplanetary disk processes that have affected the partitioning of O, S, and N among compounds and phases, the interactions (chemical and physical) among these compounds and phases, and the resulting record left behind in comets. The examination is necessarily sketchy, intended to give a flavor for the state of knowledge rather than provide a complete model for the co-evolution of volatiles and disk. In section 2 the infall of primitive solid matter to the disk, and the resulting sublimation of volatiles, is examined. Transport and chemistry within the disk itself is the subject of section 3, while the overall implications for cometary and planetary formation are covered in section 4. Section 5 provides a brief list of upcoming missions and groundbased facilities that can provide additional data to test some of the ideas engendered by existing data and modeling.

2. INFALL AND MODIFICATION OF VOLATILES

The protoplanetary disk out of which the planets formed was almost certainly the product of the collapse of a larger region of material within a molecular cloud complex (*Dutrey et al.,* 1997). To what extent this occurred in a low- vs. high-mass star-forming region has recently become controversial again as a result of the timescale problem for the formation of the giant planets — particularly Uranus and Neptune (*Boss,* 2003). However, in either case molecular cloud material was accelerated to large velocities as it fell into the nascent protoplanetary disk, and as a result solids were heated and perhaps sublimated before reaching the protoplanetary disk. Indeed, even before this the material was likely warmed by radiation from the proto-Sun, residing as it was far above the center of the disk and hence outside of the shielding effects of the optically thick disk itself.

The extent to which heating and sublimation of molecular cloud grains occurred during infall is a matter of both theoretical and observation debate. From the theoretical point of view, imagine a grain coming into the protoplanetary disk from a region high above the midplane. Analyses of protoplanetary disk models indicate that the presence of shocks will generally create a dependence of the gas density on altitude above the disk midplane that decouples all but the smallest grains from the gas, leading to heating and sublimation of the grains. Turbulence in the gas associated with the collapse process enhances the collision rate between grains and, if collisions are not predominantly destructive, growth to 1-μm solid particles or 100-μm fluffy aggregates (from an assumed starting size of 0.1 μm) is possible (*Weidenschilling and Ruzmaikina,* 1994). Existence of millimeter-sized grains is suggested from the spectral energy distribution of some young stars (*Beckwith et al.,* 2000), although populations of small grains are certainly present as well (*Li and Lunine,* 2003). It is thus likely that some amount of grain-gas decoupling, and consequent gas-dynamical (i.e., roughly, frictional) heating of the grains occurs in the disk (*Ziglina and Ruzmaikina,* 1999). Direct radiative heating of the grains associated with certain kinds of shock boundaries is also possible.

Thus grains were heated during their passage into the protoplanetary disk, but the extent to which this happened depends on several factors, including the final distance of the grains from the proto-Sun but most importantly the accretion rate (*Neufeld and Hollenbach,* 1994). Because the accretion rate controls the disk luminosity as well as the strength of the shock, the problem is a coupled one that must be solved numerically (*Chick and Cassen,* 1997). Two endmember cases have been considered in the literature. For a disk in a relatively low-luminosity state (total luminosity of disk plus protostar equal to 1–2 times that of the Sun), silicates, metals, and refractory organics survive the heating within the 1-AU terrestrial planet region; volatile organics survive to within a few AU, and H_2O ice would be fully vaporized within 5 AU. Alternatively, higher accretion rates onto the disk are possible (*Chick and Cassen,* 1997), for which the luminosity may be 1 to 2 orders of magnitude higher, and then H_2O ice grains could be vaporized to a distance of 30 AU.

Of concern here is the fate of the volatiles contained within the H_2O ice. The original grains likely were composed of amorphous ice in which volatile molecular species were adsorbed. Thus the sublimation rate during heating should be largely controlled by the H_2O ice latent heat as well as the grain size, which determines the thermal emissivity. Since the grain size and extent of fluffiness are uncertain, large variations in the extent to which grains destined for the outer solar system (beyond 5 AU) sublimate are obtained in numerical models, but the general pattern is nearly complete sublimation of H_2O and more volatile species in the Jupiter-Saturn region and very little at orbits corresponding to the Kuiper belt (*Lunine et al.,* 1991).

Sublimated material reincorporated in grains according to the local temperature-pressure environment. At 5 AU temperatures declined relatively slowly during the time of giant planet formation, and hence much of the sublimated H_2O ice recondensed not as amorphous ice, but rather in crystalline form (*Kouchi et al.,* 1994). Volatiles released from amorphous ice by sublimation were progressively trapped in the crystalline ice as temperatures dropped, likely in the form of clathrate hydrate. (*Gautier et al.,* 2001). At 30 AU, persistently low temperatures ensured rapid recondensation of amorphous H_2O ice with simultaneous trapping of volatiles in the newly formed grains. Thus, the physical effects on grains of radiative and gas-dynamical heating associated with infall suggest an important dichotomy between H_2O ice and trapped volatiles at 5 AU, vs. that at 30 AU, with intermediate properties in the intervening realm of the outer planets.

To what extent is this dichotomy reflected in the properties of comets, whose sources range over the entire outer solar system realm from 5 AU to 30 AU and beyond? Because the compositional and isotopic information for short-period comets, the origin of which appears to lie in the Kuiper belt, is rather poor compared to that for long-period comets, which formed well inward of 30 AU, comparison is difficult. The abundance of elemental N is modestly depleted in comets, and that of N_2 and related volatile N species strongly depleted (see *Bockelée-Morvan et al.,* 2004). One possible explanation for this depletion is that temperatures in the region where the grains of long-period comets formed was relatively high, cooling slowly, and the consequent incorporation of volatiles involved a fractionation process akin to, or in fact corresponding to, the formation of clathrate hydrates (*Iro et al.,* 2003). In such a process, even though temperatures become low enough to allow the trapping of N in H_2O ice, the competition for void spaces among species favors carbon monoxide over N_2, so that, if H_2O ice is not abundant enough to clathrate both species, the latter is preferentially excluded from the icy grains. Much of the N in comets is then the result of condensation of NH_3 into the H_2O ice as a stochiometric hydrate, with the N_2 being secondary or absent.

This picture has a number of implications. First, the progressive trapping of volatile species in the crystalline clathrate hydrate requires that the icy component of molecular cloud grains, with their indiscriminant mix of volatile and nonvolatile species, largely or fully sublimated in the 5–10-AU region. Comets formed of grains from this region — Oort cloud comets — should exhibit a compositional pattern in which the relative abundances of the most volatile species are altered from molecular cloud values. Short-period comets, on the other hand, which had their origin in the Kuiper belt at 30 AU and beyond, where sublimation was absent or limited, ought to have a grain composition that reflects a cold molecular cloud origin.

Second, the model of *Iro et al.* (2003) weakly suggests a variation in the H_2O ice abundance at 5 AU and beyond relative to that obtained assuming the solar composition of elemental O, because the CO abundance seems highly variable from comet to comet. Iro et al. explain the CO variation as a consequence of the altered number of adsorption sites available for CO caused by fluctuating amounts of H_2O ice. In fact, the clathrate trapping model to explain the abundance of volatiles in Jupiter requires that the amount of H_2O ice at 5 AU be double the value implied by a solar abundance of elemental O, while the amount of S be half its solar elemental value (*Gautier et al.*, 2001), and the N values in several well-measured comets also require an elevated H_2O abundance (Fig. 1). Thus, the N abundance in long-period comets is tied, albeit indirectly, to the history of H_2O and S-bearing species in the 5 AU region. A mechanism for enriching H_2O ice in the 5 AU region, namely through cold trapping of H_2O vapor mixed outward from the hot inner disk as suggested originally by *Morfill and Volk* (1984), is presented in the next section.

Third, if the histories of cometary grains embedded within long- and short-period comets are as different as this picture implies, then we expect distinct isotopic ratios be-tween the two classes of comets as well. In particular, D/H in three long-period comets is twice that of terrestrial ocean H_2O (*Meier et al.*, 1998b; *Bockelée-Morvan et al.*, 1998), which imposes stringent constraints on the source of the Earth's oceans in dynamical models (*Morbidelli et al.*, 2000). However, these models assume a constant D/H ratio in H_2O in comets from the long-period comets of the giant planet realm to the short-period Kuiper belt comets at 30 AU. The lack of grain heating and sublimation at 30 AU suggests this assumption may not be realistic; perhaps, e.g., D/H in H_2O in short-period comets is even larger than in the long-period bodies.

3. EFFECTS OF TRANSPORT WITHIN THE DISK ON VOLATILE DISTRIBUTION

The complex evolution of volatile species does not end with the sublimation and reformation of solid grains. Transport processes in the disk bring grains and gas-phase species into warmer realms close to the disk center, and then outward again. Those grains that cross the phase boundary of a major component of their composition will experience dramatic changes, e.g., the sublimation of H_2O ice. Gas phase species that cross boundaries corresponding to changes in thermodynamically (or kinetically) preferred molecular species will have their compositions altered in complex ways — this is especially the case with S. In the present section we describe the coupling between the physics of transport in the disk and the chemistry/thermodynamics of phase transitions and reactions for H_2O and S. In the preceding section we established that the N abundance and its molecular form in grains is tied to the H_2O abundance directly, and to S indirectly, so at the end of the section we more closely examine possible similar effects for the molecular carriers of elemental N. The question of transport processes in disks, specifically their nature and origin in dissipative processes, is a complex one for which the reader is referred to other reviews (*Stone et al.*, 2000).

3.1. Water

The condensation of H_2O is a fundamentally important process in protoplanetary disks, affecting the visible and infrared opacity, the abundance of solids, and the accretion rate of planets embedded in the disk. But the gaseous and solid abundances of H_2O as a function of the distance from the disk center are altered as well by condensation when radial transport is considered (*Morfill and Volk*, 1984). In particular, imagine that some mixing process acts to carry H_2O vapor along a radius at the disk midplane. The process could involve macroscopic motions associated with large-scale turbulence (turbulent diffusion) or instabilities of some sort. (From a practical point of view, microscopic diffusion in the disk is too slow to generate the effects considered here.) Absent condensation, adsorption, or chemical creation/destruction, the transport mechanism will not alter

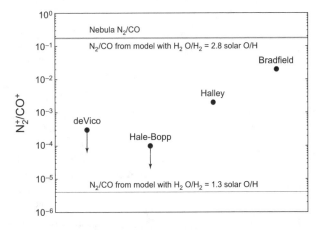

Fig. 1. The N_2^+ to CO^+ ratios [nearly equal to the N_2/CO values] measured in four long-period comets. Bracketing these values are the predicted N_2/CO ratios from the *Iro et al.* (2003) model with (bottom line) a roughly solar H_2O-to-H_2 ratio and (top line) a H_2O-to-H ratio 2.8 times solar. See the text for discussion of a mechanism for enriching the H_2O abundance in the outer protoplanetary disk at the expense of the inner disk. The gas-phase N_2/CO value in the protoplanetary disk — unaltered by grain trapping — is just above but nearly coincident with the 2.8 times solar line. Thus roughly doubling the H_2O abundance in the region of long-period comet formation could explain the N_2 to CO ratio in comets. Figure from *Iro et al.* (2003); for a discussion of the relationship of the observed ions to the total neutral populations of N_2 and CO, see *Bockelée-Morvan et al.* (2004).

the radial distribution of the vapor abundance because any transient spatial gradients created by the motions are quickly eliminated by the same mechanisms. In the presence of chemical/thermodynamic processes, the outcome is quite different.

We consider condensation as a simple sink, or loss process, for H_2O occurring at a discrete radial distance from the Sun at a given time in the evolution of the protoplanetary disk. Condensation of a particular species occurs when the partial pressure of that species exceeds the saturation vapor pressure; since the latter is steeply dependent on temperature, this criterion simply translates into a distance from the disk center at which the species is saturated. For example, the H_2O condensation front (sometimes called the snowline) might occur at 5 AU at a given time in the disk, then move inward to 3 or 4 AU as the disk ages and cools. There is no special requirement on the properties of the disk to possess such a condensation front, other than that the temperature must decline with radial distance and time. Water vapor inward of the condensation front will be transported radially by whatever process, and when it passes through the snowline it will condense out in the form of microscopic grains. (Supercooling of the vapor is a distinct possibility because of the low temperatures involved, ~150–160 K at typical gas pressures in the disk. However, this does not change the argument; it simply means the effective snowline is pushed outward somewhat relative to the formal thermodynamic line that is defined as above.)

In principle the condensation changes nothing, because the microscopic grains of ice are embedded in the gas and move with it, just as does the H_2O vapor itself. But the grains grow, and decouple from the disk gas when they reach radii exceeding on the order of 0.1 cm (*Cyr et al.*, 1998). When this happens, although grains will spiral inward as a result of gas drag, the effective inward radial velocity is smaller than the turbulent velocity. The inward radial drift velocity declines as the grain size increases. Hence, the grains effectively are "left behind" in the 5-AU region, while the H_2O vapor continues to obey the standard diffusive mechanics associated with the gas phase. The snowline becomes a cold trap, and over time the H_2O abundance inward of that line declines (*Stevenson and Lunine*, 1988). The precise history of the decline depends upon detailed assumptions about grain growth rates, mechanisms of vapor transport, and position of the snowline, among others (*Cyr et al.*, 1998). Disruptive collisions presumably play an important role, considering that dusty circumstellar disks are observed at ages of several millions of years (*Beckwith et al.*, 2000). A typical model result for a cylindrically symmetric disk, with transport processes in the disk parameterized as turbulent diffusion, is shown in Fig. 2. Define the "starting value" as the abundance of H_2O in the absence of condensation processes, a value determined by the elemental O abundance and partitioning of O among H_2O, CO and other compounds (*Cyr et al.*, 1999). Water vapor abundances inward of the snowline can drop to well

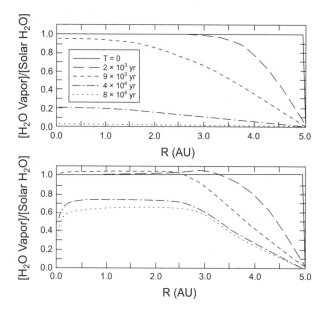

Fig. 2. Water vapor abundance is plotted as a function of distance in a cylindrically-symmetric and turbulent protoplanetary disk, where the snowline is initially at 5 AU, for various times in the evolution of the disk. Top panel assumes that ice grains formed at the snowline do not drift inward. Bottom panel allows inward drift of the grains as they grow and decouple from the gas; the snowline evolves inward as well. "Solar H_2O" refers to the starting abundance of H_2O, uniform through the disk, and based on the solar O abundance. From *Cyr et al.* (1998).

under half of the starting value as disk transport processes deliver H_2O to the snowline. This H_2O is accumulated beyond the snowline, leading to an excess of H_2O ice there. Water abundances double the starting value may be accumulated in that region.

The implications of the redistribution of H_2O are interesting. The peculiar geochemistry of the enstatite meteorites might be a signature of the depletion of H_2O inward of the disk snowline (*Hutson and Ruzicka*, 2000). The abundance of organics increases dramatically and the mix of organic species changes in the inner disk with decreasing O (H_2O) abundance (*Cyr et al.*, 1999). The long timescales required for the formation of Jupiter in the core accretion model (*Lunine et al.*, 2003) might be reduced dramatically if the abundance of solids were enhanced over solar (*Lissauer*, 1987), as would happen were H_2O ice to accumulate beyond the snowline (*Stevenson and Lunine*, 1988). An enrichment in H_2O ice of a factor of 2 beyond the snowline is required if the crystalline clathrate model of the enrichment of Jupiter is to be consistent with the interior abundances of the major elements in the solar system's largest planet, as noted above. Likewise, the depletion of N in comets can be explained by the same model if H_2O ice were enriched where the long-period comets formed. Whether the enrichment of ice at the snowline is carried far enough beyond 5 AU to make this plausible has not been quantitatively evaluated.

Alternatively, one might argue that the protoplanetary disk was quiescent enough that H_2O ice delivered from the molecular cloud did not sublimate in the 5 AU region. While this appears to be a difficult proposition based on the infall models described in the last section, it cannot be ruled out, and in particular "contamination" of the 5–10-AU region from cold solid matter delivered relatively pristine to more remote disk regions, then transported inward, is another intriguing possibility that bears further study.

To assess whether indeed the evolution of H_2O in the disk included establishment of a cold trap, with enrichment beyond and depletion inward of that point, we must examine other volatiles to assess whether the signature of this process might be written upon their abundances.

3.2. Sulfur

Sulfur occurred in a variety of chemical forms of widely varying volatility in the protoplanetary disk. The most volatile abundant form was likely H_2S, and it is this molecule that would have represented the bulk of the S trapped in H_2O ice. The S abundance in nine long-period comets in the form of H_2S seems to vary from 0.4% to 1.5% relative to H_2O ice, and this is generally less than what is predicted from the trapping of a solar complement of H_2S in clathrate hydrate in the disk (*Iro et al., 2003*). Likewise, the S abundance in Jupiter cannot be reproduced by the model unless H_2S is depleted in the feeding zone (the region from which the planet acquires material) of Jupiter relative to solar abundance (*Gautier et al., 2001*), but here the constraint is more severe: The depletion must be relative to the background H_2. Thus simply augmenting the amount of H_2O ice by — for example — cold-trapping is not sufficient.

[Note that the mass M of gas within a ring of width ΔR, centered at a heliocentric distance R, is defined by $dM/dR = 2\pi R q \Sigma$, where Σ is the surface density of H and q the mixing ratio of the considered species. Since the radial dependence of surface density is weaker than r^{-2} (*Beckwith et al., 2000*), the mass of a molecular species with constant mixing ratio as a function of radial distance increases as one moves outward in the disk. Hence most of its mass is located in the outer disk [see, for example, Fig. 14 of *Hersant et al.* (2001)]. Therefore, the mass of S that was trapped in solid form in the feeding zone of Jupiter, and that subsequently enriched the planet in elemental S, was small compared to the total mass of S in the disk.]

A depletion of H_2S in the feeding zone implies a loss process for H_2S somewhere else in the disk — most plausibly in the inner region. Indeed, some elemental S is trapped in the warm inner disk in refractory forms such as FeS (troilite), SiS, MgS, CaS, and other related compounds. Such S-metal combinations occur fairly rapidly at disk temperature above about 500 K, and one might imagine some of the S being excluded from regions farther out in the disk by this process if turbulent diffusion causes significant ra-

dial mixing. The data in Table 1 suggests that this process is too efficient; there is enough Fe to remove all the S and prevent the formation of any but a very small trace of H_2S. However, many of the elements that would otherwise react with S (Si, Mg) are locked up by reaction with O. Hence H_2S is the primary carrier of S and has essentially the full solar abundance of the elemental S itself.

The situation is different when the depletion of H_2O inward of the snowline is considered, and hence the elemental O abundance there is reduced. In an extremely-O-poor case (10% solar), Si, Mg, and other elements are available to bind with S as SiS and MgS, and thus drive down the abundance of H_2S to very low levels (*Pasek et al., 2003*). The molecule HS also coexists with the SiS and MgS, but is still only 10% or less of the total S abundance. [Interestingly, SiS is seen in the mass-loss envelopes of carbon stars; that is, stars with low O abundance relative to C (*Bieging, 2001*).] Given the much lower volatility of compounds like SiS and MgS compared to HS and H_2S, only the latter two are likely to be seen in the gas phase of the cold outer disk where the giant planets formed. At intermediate O abundances, perhaps more likely to obtain in the disk, the chemistry becomes even more complex, with CaS and MgS "fingers" over narrow ranges of temperature and hence radial distance (*Pasek et al., 2003*).

What this complex chemistry implies for the H_2S abundance in the outer part of the disk requires folding in calculations of the kinetics of the S reactions (*Fegley, 2000*), including the "poisoning" of the iron grains via the buildup of layers of sulfides (*Lauretta et al., 1996*), and a realistic transport model in a turbulent disk. This has not yet been done. Since the chemical calculations suggest that the time-dependence of the H_2O — hence O — abundance in the inner disk might act to destroy H_2S there, the S abundance in the region of Jupiter- and comet-formation could have been depressed as well. More generally, the histories of O-bearing and S-bearing volatile species in the disk are strongly coupled.

3.3. Nitrogen

Little has been done to quantify the coupling to the O abundance of the chemistry of N-bearing species, in particular, NH_3 and N_2. Like carbon monoxide and methane, there is a relationship between the abundances of the oxidized and reduced forms dependent on temperature and H pressure. At very low temperature (200–300 K) the reduced (NH_3) form of N is preferred, but the conversion timescale between the oxidized and reduced phases is so long at such temperatures that, in the disk, the oxidized form (N_2) predominated (*Prinn, 1993*). That NH_3 is seen in comets (*Wyckoff et al., 1991*) suggests a source from the nascent molecular cloud, where NH_3 is known to exist (*van Dishoeck and Blake, 1998*); other sources associated with giant planet formation seem less promising (*Mousis et al., 2002*). The presence of a mix of NH_3 and N_2 derived all or in part from molecular cloud values is not in conflict with the grain sublimation processes considered above; while both are liberated into the gas phase by the sublimation process, ambient temperatures in the disk are too low for chemical exchange of N between N_2 and NH_3 to occur. Hence the total molecular ratio (that in the gas and the grains) is largely preserved, with the exception of a small amount of NH_3 transported to the very innermost, hot part of the disk where reaction kinetics are fast enough to convert it to N_2.

Because NH_3 readily forms so-called stochiometric hydrates with H_2O, there is no condensation front of solid NH_3 in the same sense as for H_2O. Ammonia hydrates form in H_2O ice grains at temperatures above 70 K for typical disk temperature-pressure profiles (*Iro et al.,* 2003). Kinetic inhibition of this process is unlikely because the NH_3 abundance is significantly less than that of H_2O ice (i.e., a few percent), so that relatively little fresh ice must be exposed to the gas to induce hydrate formation. The infall of NH_3 in molecular cloud grains over a broad range of disk semimajor axes, and the stability of the molecule against transformation to N_2, makes it unlikely that a well-defined gradient in the NH_3 abundance was present in icy grains that remained cooler than 70 K.

4. IMPLICATIONS FOR WATER ICE, SULFUR, AND NITROGEN COMPOUNDS IN COMETS

The processes discussed above are a subset of those that acted in the solar system's protoplanetary disk on species of varying volatility. They illustrate that the processes of condensation, sublimation, adsorption, and chemical reactions worked in a coupled fashion with dynamical mechanisms of mixing and heating in the disk. Grains experienced varying extents of heating and destruction during disk infall, and then generation of H_2O ice at the snowline engendered a complex history of fluctuating H_2O vapor levels in the inner disk. These in turn affected the chemistry of the disk in a wide variety of ways — the abundance of organics, the relative abundance of metal sulfides to H_2S, and the abundance of H_2O ice — hence giant planet formation timescales. Further, the abundance of H_2O ice beyond 5 AU determined the abundance of N_2 trapped in grains destined for comets and the giant planets, through the availability of void sites in the recondensed crystalline ice.

Isotopic evidence for these effects in comets is scanty. The D/H value in H_2O in long-period comets is elevated over that in carbonaceous chondrites (*Robert,* 2001), but less extreme than D/H in other molecules observed in molecular clouds (*Bockelée-Morvan et al.,* 1998). The enrichment has been used to argue that the H_2O in comets has not been subjected to warm temperatures in the inner disk, where re-equilibration with H and hence lowering of D/H might occur. However, grain sublimation and vapor cycling through the disk might have modestly reduced the D/H value relative to some starting value in the molecular cloud.

The ortho-to-para ratio in H in cometary H_2O is somewhat more constraining, challenging any model that heats cometary H_2O in the disk to temperatures significantly above 30–50 K. At face value, the comets for which this has been measured — Halley, Hale-Bopp, Hartley-2, and Wilson — ought to be composed of ice that has not seen the inner protoplanetary disk. However, the kinetics of the exchange are very poorly known, and there is also a possibility that the ortho/para value is reset later in the comae of comets by ion chemistry (*Irvine et al.,* 2000).

The N abundance in comets seems to be well explained by a model in which N_2 is not indiscriminately trapped in amorphous ice, but instead must compete for void sites with other volatiles. This too implies a thermal cycling of the H_2O in the grains, up to a temperature that might be inconsistent with the ortho-to-para ratio if the latter reflects the disk environment of the H_2O ice grains.

The S abundance in icy grains is also coupled to the H_2O abundance — more accurately, the oxidation state — in the inner disk, because the latter determines the amount of H_2S in the gas phase vs. the abundance of other S-bearing compounds. Since H_2S appears to be the dominant S-bearing species in comets but is below the solar S/O ratio (*Irvine et al.,* 2000), it may have been depleted by chemical processing in the inner disk. Radial mixing brings H_2S supplied from the molecular cloud inward, destroys some of it, then mixes the H_2S depleted gas outward. How much H_2S depletion occurs — i.e., how much of the original molecular cloud complement is altered in this way — must await more detailed models in which not only are the chemistry and physics coupled, but the time history of the gas oxidation state is folded in as well. All we can say now is that a depletion of S at 5 AU, implied by *one* interpretation of the *Galileo* data, seems to be a plausible outcome of S chemistry in the inner disk (*Pasek et al.,* 2003).

5. FUTURE KEY DATASETS

In this chapter we have sketched a coupled history of the volatile carriers of three elements: O, N, and S. There is, of course, much more about cometary chemistry than just these three elements, and in particular the story of the organics is a rich one that is covered elsewhere in this book. But the origin and evolution of the compounds bearing these elements provides a sufficiently complex and coupled picture that a list of measurements required to test some of the ideas presented here (and those not presented as well!) is daunting. Such measurements will require spacecraft sampling of comets, telescopic observations from the ground, and Earth- or solar-orbiting telescopes covering wavelengths not accessible from the ground.

Direct sampling of comets is required to measure the abundances of relatively refractory species containing N and S, as well as determine isotopic ratios that are impractical to measure spectroscopically. Deuterium abundances in H-bearing species other than H_2O ice, with the exception of a few like HCN (*Meier et al.,* 1998a), remain poorly known, and would help constrain the thermal history of the H_2O ice. Likewise, measurement of the O isotopes — much more difficult spectroscopically — would be of value in this regard. We also wish to know how the volatiles are trapped in comets, although the thermal cycling of comets as they pass close to the Sun might have destroyed this record within a few thermal skin depths of the cometary surface. Incorporation in clathrate hydrates, adsorption in amorphous ice, or the existence of the ices of the volatile species themselves (such as NH_3) constrain the temperature history of the ice, provided these phases are indeed from the time of cometary grain formation. Sampling of noble gases would provide a record of the trapped abundances of chemically inert elements for comparison with the jovian record. Measurement of refractory S-bearing species in the silicate grains will illuminate the high-temperature S chemistry and help sketch out the source of the silicates as well. These goals and others require the sampling of cometary silicates by *Stardust*, and the daringly complex *in situ* studies of *Rosetta*.

Groundbased studies will continue to provide basic elemental, molecular, and isotopic information on long-period comets as they pass close to the Earth. With the largest groundbased systems (the European Southern Observatory's Very Large Telescope) and even larger ones anticipated for the next decade (the proposed 30-m Giant Segmented Mirror Telescope) it may be possible to make similar studies of short-period comets. Is D/H in H_2O the same in the short-period comets as in the long-period comets? Do the compositional differences between short- and long-period comets lay out a story consistent with the presumed colder formation environment of the Kuiper belt vs. the giant planet region? Can one discern the effects of substantial grain sublimation and reformation in the long-period comets against the background of a more primitive grain mix in the short-period comets? Many of the clues are accessible from the ground, but the infrared signatures of composition and isotopic ratios may require the combined large aperture and high-infrared sensitivity of the cryogenic James Webb (Next-Generation) Space Telescope, to be deployed in solar orbit after 2011. This system will also be of great value in mapping the composition of the distant, and cold, bodies resident in the Kuiper belt, providing a systematic look at the composition of what are presumed to be the larger counterparts of short-period comets.

Beyond the infrared, the groundbased Atacoma Large Millimeter Array will provide sensitive isotopic data on comets and Kuiper belt objects, as well as compositional information on planet-forming and debris disks in the solar neighborhood. Ultraviolet studies of cometary comae will further elaborate the chemistry that both illuminates cometary composition and alters isotopic and chemical systematics relative to the nucleus. The timescale for acquiring the ground- and spacebased datasets described here stretch well into the next decade and perhaps beyond. But these ambitious plans seem sufficiently well formulated to lend plausibility to the notion that a history of volatiles and grains, from molecular clouds to comets and planets, will be a legacy of the first two decades of twenty-first century space science.

Acknowledgments. Some of the work described herein was supported by the NASA Origins of Solar Systems Program.

REFERENCES

Atreya S. K., Mahaffy P. R., Niemann H. B., Wong M. H., and Owen T. C. (2003) Composition and origin of the atmosphere of Jupiter — an update, and implications for the extrasolar giant planets. *Planet. Space Sci., 51,* 105–112.

Beckwith S. V. W., Henning T., and Nakagawa Y. (2000) Dust properties and assembly of large particles in protoplanetary disks. In *Protostars and Planets IV* (V. Mannings et al., eds.), pp. 533–558. Univ. of Arizona, Tucson.

Bieging J. (2001) High dynamic range images of silicon sulfide toward IRC+10216. In *Science with the Atacama Large Millimeter Array (ALMA),* pp. 365–368. Astronomical Society of the Pacific, San Francisco.

Bockelée-Morvan D., Gautier D., Lis D. C., Young K., Keene J., Phillips T. G., Owen T., Crovisier J., Goldsmith P. F., Bergin E. A., Despois D., and Wooten A. (1998) Deuterated water in comet C/1996 B2 (Hyakutake) and its implications for the origin of comets. *Icarus, 133,* 147–162.

Bockelée-Morvan D., Crovisier J., Mumma M. J., and Weaver H. A. (2004) The composition of cometary volatiles. In *Comets II* (M. C. Festou et al., eds.), this volume. Univ. of Arizona, Tucson.

Boss A. P. (2003) Formation of gas and ice giant planets. *Earth Planet. Sci. Lett., 202,* 513–523.

Chick K. M. and Cassen P. (1997) Thermal processing of interstellar dust grains in the primitive solar environment. *Astrophys. J., 477,* 398–409.

Cox A. N. (2000) *Allen's Astrophysical Quantities.* American Institute of Physics, New York. 719 pp.

Cyr K. E., Lunine J. I., and Sharp C. (1999) Effects of the redistribution of water in the solar nebula on nebular chemistry. *J. Geophys. Res., 104,* 19003–19014.

Cyr K. E., Sears W. D., and Lunine J. I. (1998) Distribution and evolution of water ice in the solar nebula: Implications for solar system body formation. *Icarus, 135,* 537–548.

Dutrey R. M., Guilloteau S., and Guelin M. (1997) Chemistry of protosolar-like nebulae: The molecular content of the DM Tau and GG Tau disks. *Astron. Astrophys., 317,* L55–L58.

Fegley B. (2000) Kinetics of gas-grain reactions in the solar nebula. *Space Sci. Rev., 92,* 200.

Fortney J. J. and Hubbard W. B. (2003) Phase separation in giant planets: Inhomogeneous evolution of Saturn. *Icarus, 164,* 228–243.

Gautier D., Hersant F., Mousis O., and Lunine J. I. (2001) Enrichments in volatiles in Jupiter: A new interpretation of the Galileo measurements. *Astrophys. J. Lett., 550,* L227–L230 (erratum *559,* L183).

Gierasch P. J., Ingersoll A. P., Banfield D., Ewald S. P., Helfenstein P., Simon-Miller A., Vasaveda A., Breneman H. H., Senske D. A., and the Galileo Imaging Team (2000) Observation of moist convection in Jupiter's atmosphere. *Nature, 403,* 628–630.

Hersant F., Gautier D., and Huré F. (2001) A two-dimensional model for the primordial nebula constrained by d/h measurements in the solar system: Implications for the formation of giant planets. *Astrophys. J., 554,* 391–407.

Hutson M. and Ruzicka A. (2000) A multi-step model for the origin of E3 (enstatite) chondrites. *Meteoritics & Planet. Sci., 35,* 601–608.

Iro N., Gautier D., Hersant F., Bockelée-Morvan D., and Lunine J. I. (2003) An interpretation of the nitrogen deficiency in comets. *Icarus, 161,* 511–532.

Irvine W. M., Crovisier J., Fegley B., and Mumma M. J. (2000) Comets: A link between interstellar and nebular chemistry. In *Protostars and Planets IV* (V. Mannings et al., eds.), pp. 1159–1200. Univ. of Arizona, Tucson.

Kouchi A., Yamamoto Y., Kozasa T., Kuroda T., and Greenberg J. M. (1994) Conditions for condensation and preservation of amorphous ice and crystallinity of astrophysical ices. *Astron. Astrophys., 290,* 1009–1018.

Lauretta D., Kremser D. T., and Fegley B. (1996) The rate of iron sulfide formation in the solar nebula. *Icarus, 122,* 288–315.

Li A. and Lunine J. I. (2003) Modeling the infrared emission from the HR 4796A disk. *Astrophys. J., 590,* 368–378.

Lissauer J. J. (1987) Timescales for planetary accretion and the structure of the protoplanetary disk. *Icarus, 69,* 249–265.

Lunine J. I., Coradini A., Gautier D., Owen T., and Wuchterl G. (2003) The origin of Jupiter. In *Jupiter* (F. Bagenal, ed.), in press. Cambridge Univ., London.

Lunine J. I., Engel S., Rizk B., and Horanyi M. (1991) Sublimation and reformation of icy grains in the primitive solar nebula. *Icarus, 94,* 333–343.

Meier R., Owen T. C., Jewitt D. C., Matthews H. E., Senay M., Biver N., Bockelée-Morvan D., Crovisier J., and Gautier D. (1998a) Deuterium in comet C/1995 O1 (Hale-Bopp): Detection of DCN. *Science, 279,* 1707–1710.

Meier R., Owen T. C., Matthews H. E., Jewitt D. C., Bockelée-Morvan D., Biver N., Crovisier J., and Gautier D. (1998b) A determination of the HDO/H$_2$O ratio in comet C/1995 O1 (Hale-Bopp). *Science, 279,* 842–844.

Morbidelli A., Chambers J., Lunine J. I., Petit J. M., Robert F., Valsecchi G. B., and Cyr K. E. (2000) Source regions and timescales for the delivery of water on Earth. *Meteoritics & Planet. Sci., 35,* 1309–1320.

Morfill G. E. and Volk H. J. (1984) Transport of dust and vapor and chemical fractionation in the early protosolar cloud. *Astrophys. J., 287,* 371–395.

Mousis O., Gautier D., and Bockelée-Morvan D. (2002) An evolutionary turbulent model of Saturn's subnebula: Implications for the origin of the atmosphere of Titan. *Icarus, 156,* 162–175.

Neufeld D. A. and Hollenbach D. J. (1994) Dense molecular shocks and accretion onto protostellar disks. *Astrophys. J., 428,* 170–185.

Niemann H. B., Atreya S. K., Carignan G., Donahue T. M., Haberman J. A., Harpold D. M., Hartle R. E., Hunten D. M., Kasprzak W. T., Mahaffy P. R., Owen T. C., and Way S. (1998) The composition of the Jovian atmosphere as determined by the Galileo probe mass spectrometer. *J. Geophys. Res., 103,* 22831–22846.

Owen T., Mahaffy P., Niemann H. B., Atreya S. K., Donahue T. M., Bar-Nun A., and de Pater I. (1999) A new constraint on the formation of giant planets. *Nature, 402,* 269–270.

Pasek M., Milsom D., Ciesla F., Sharp C., Lauretta D., and Lunine J. I. (2003) Sulfur chemistry in the solar nebula. *Icarus,* in press.

Pollack J. B., Hubickyi O., Bodenheimer P., Lissauer J. J., Podolak M., and Greenzweig Y. (1996) Formation of the giant planets by concurrent accretion of solids and gas. *Icarus, 124,* 62–85.

Prinn R. G. (1993) Chemistry and evolution of gaseous circumstellar disks. In *Protostars and Planets III* (E. H. Levy and J. I. Lunine, eds.), pp. 1005–1028. Univ. of Arizona, Tucson.

Robert F. (2001) The origin of water on Earth. *Science, 293,* 1056–1058.

Roulston M. S. and Stevenson D. J. (1995) Prediction of neon depletion in Jupiter's atmosphere. *Eos Trans. AGU, 76,* 343.

Stevenson D. J. and Lunine J. I. (1988) Rapid formation of Jupiter by diffusive redistribution of water vapor in the solar nebula. *Icarus, 75,* 146–155.

Stone J. M., Gammie C. F., Balbus S. A., and Hawley S. F. (2000) Transport processes in protostellar disks. In *Protostars and Planets IV* (V. Mannings et al., eds.), pp. 589–611. Univ. of Arizona, Tucson.

van Dishoeck E. F. and Blake G. A. (1998) Chemical evolution of star-forming regions. *Annu. Rev. Astron. Astrophys., 36,* 317–368.

Weidenschilling S. J. and Ruzmaikina T. V. (1994) Coagulation of grains in static and collapsing protostellar clouds. *Astrophys. J., 430,* 713–726.

Wyckoff S., Tegler S. C., and Engel L. (1991) Ammonia abundances in four comets. *Astrophys. J., 368,* 279–286.

Ziglina I. N. and Ruzmaikina T. V. (1999) The influx of interstellar matter onto a protoplanetary disk. *Astron. Vestn., 25,* 53–60.

From Interstellar Material to Cometary Particles and Molecules

P. Ehrenfreund
Leiden Observatory

S. B. Charnley and D. H. Wooden
NASA Ames Research Center

The birthplace of stars, planets, and small bodies are molecular clouds consisting mainly of H and He gas as well as tiny amounts of solid particles. Dust and gas from such an interstellar cloud collapsed to form a central condensation that became the Sun, as well as a surrounding disk — the solar nebula. In this protoplanetary disk innumerable submicrometer particles — icy and refractory in nature — agglomerated to larger planetesimals and subsequently into planets. Comets were formed from remnant inner disk material that was not incorporated into planets. Comets assembled beyond the orbit of Jupiter and may therefore provide a record of some pristine material from the parent interstellar cloud. Our knowledge of the composition of comets is predominantly based on evaporation of volatile species and thermal emission from siliceous and carbonaceous dust when bright comets approach the Sun. The investigation of outgassing curves from bright comets has provided a general link with abundances of ices and gas phase molecules detected in dense interstellar clouds. Theoretical models indicate that bulk material in cometary nuclei is stratified in density, porosity, and composition and contains coexisting ice phases, possibly clathrates and trapped gases. It is therefore apparent that the outgassing of species from cometary nuclei will not essentially mimic that expected from pure sublimation of ices of interstellar composition. The current lack of astronomical data related to solar-type star-forming regions, and the statistically small sample of comets studied to date, are the major obstacles to be overcome before a coherent link between interstellar and solar system material can be established. This chapter attempts to compile the current knowledge on the connection between interstellar and cometary material, based on observations of interstellar dust and gas, observations of cometary volatiles, simulation experiments, and the analysis of extraterrestrial matter.

1. INTRODUCTION

In the 1950s comets were described by *Whipple* (1950) as "dirty snowballs" made up of water ice and small rock particles. Cometary observations and related investigations between then and now strongly improved our knowledge and view on these precious and most pristine solar system objects. For a historical perspective on comets the readers are referred to *Festou et al.* (1993a,b, 2004). The apparition of unusually bright comets in the last 2–3 decades and the observations using advanced instrumentation allowed us to get important insights into the coma chemistry (*Bockelée-Morvan et al.*, 2004; *Hanner and Bradley*, 2004) and their relation to the parent interstellar cloud (*Irvine and Lunine*, 2004).

The origin of comets and the content of pristine interstellar material incorporated in them are far from being understood. The assumption that cometary nuclei are aggregates of pristine interstellar ices and dust (*Greenberg*, 1982; *Li and Greenberg*, 1997) is clearly a gross simplification. Processing of infalling interstellar medium (ISM) material within the solar nebula must have occurred in one way or another. A pivotal epoch of processing is when interstellar material is first accreted. However, no detailed studies of the chemistry at the accretion shock exist (cf. *Neufeld and Hollenbach*, 1994). With this caveat, we can consider two limiting cases of processing. First, interstellar material entering far out in the nebula, where the accretion shock is weak (100 AU), should be largely unmodified when first incorporated. In this case, chemical reactions will proceed as this material is transported radially inward (*Aikawa et al.*, 1999; *Gail*, 2001). Second, accretion closer to the protosun will lead to complete dissociation and destruction of interstellar molecules. In this case, the molecules will subsequently form solely from nebular chemistry (e.g., *Fegley and Prinn*, 1989) and these can be radially mixed outward. Today results indicate that presolar material has been chemically and physically processed according to the distance from the protostar (*Chick and Cassen*, 1997; *Fegley*, 1999). Comets are therefore a mixture of interstellar and processed material and their initial composition will differ according to their place of formation.

The "long-period" comets probably formed across the giant planet formation region (5–40 AU) with the majority of them originating from the Uranus-Neptune region. For a detailed discussion of cometary reservoirs and cometary evolution the readers are referred to *Jewitt* (2004), *Dones et al.* (2004), and *Meech and Svoreň* (2004). Given the gradient in physical conditions expected across this region of the nebula, chemical diversity in this comet population is to be expected, as has been inferred for short-period comets (*A'Hearn et al.,* 1995). Processing and dynamical exchange of icy planetesimals in the comet-forming region could have contributed to chemical heterogeneity, or may have "homogenized" cometary nuclei before their ejection to the Oort cloud (*Weissman,* 1999), where they could have experienced further material processing and evolution (*Stern,* 2003). Remote infrared observations of parent volatiles in six Oort cloud comets (including Hyakutake and Hale-Bopp) showed that they have similar volatile organic compositions (*Mumma et al.,* 2003), appear to be more pristine, and display a reasonable similarity to interstellar cloud material.

The noble gas differentiation, deuterium enrichment, and low ortho-para ratio measured in water are all consistent with formation of precometary ices at ~30 K, similar to the conditions in the Uranus-Neptune region (*Bockelée-Morvan et al.,* 2004). Comet-forming material in the Jupiter-Saturn region (5–10 AU) experienced higher temperatures and may also have been exposed to a much higher degree of radiation processing before its assembly into comets. Extrapolating from the case of C/1999 S4 (LINEAR), there is some evidence that such comets should exhibit depletions and different abundance ratios compared to those formed at 30 AU and beyond (*Mumma et al.,* 2001a).

Comets are made of silicates (~25%), organic refractory material (~25%), small carbonaceous molecules (few percent), and ~50% water ice with small admixtures of other ice species (*Greenberg,* 1998). Molecular ices and the gases released upon sublimation, silicate dust, and solid-state carbonaceous materials are the major components of dusty cometary comae that can be studied by astronomical observations and through laboratory simulations (*Rodgers et al.,* 2004; *Hanner and Bradley,* 2004; *Colangeli et al.,* 2004). Studying these compounds in comets, in the ISM, and in meteoritic materials allows us to reveal which processes occurred in the ISM and which occurred during the formation of the solar nebula. For example, the appearance of Mg-rich crystalline silicates in some comets and the scarcity of crystalline silicates in the ISM [<5%, *Li and Greenberg* (1997), <0.5%, *Kemper et al.* (2004)] indicates cometary crystalline silicates are grains that condensed at high temperatures [~1400 K, *Grossman* (1972)] or were annealed from amorphous silicates at somewhat lower temperatures [>900 K, *Fabian et al.* (2000)] in the solar nebula. The origin of cometary solid-state carbonaceous materials is less clearly defined than the origin of the siliceous materials. The bulk of solid-state cosmic C in the interstellar medium is in an unknown form. There is strong evidence that amorphous C and similar macromolecular material takes up most of the

C in the interstellar medium (*Ehrenfreund and Charnley,* 2000; *Mennella et al.,* 1998). The same trend is observed in meteorites, where macromolecular material takes up more than 80% of the C (*Sephton et al.,* 2000; *Gardinier et al.,* 2000). The link between macromolecular C in the solar system and the interstellar macromolecular C is yet to be understood, but it is tempting to assume that such a material is also present in comets.

Cometary nuclei are a highly porous agglomeration of grains of ice and dust and they appear stratified in density, porosity, ice phases, and strength. For a detailed discussion on cometary structure and properties, the readers are referred to *Prialnik et al.* (2004) and *Weissman et al.* (2004). The nuclear ice component probably consists of different coexisting ice phases, including amorphous ice, crystalline ice, and clathrates, with gases trapped in bulk and clathrate hydrates (see Fig. 12 in *Prialnik et al.,* 2004). The structure of the nucleus and the internal processes lead to certain sublimation characteristics and outbursts of ice, dust, and gas when the nucleus approaches the Sun (*Meech and Svoreň,* 2004; *Prialnik et al.,* 2004; *Colangeli et al.,* 2004). Monitoring the sublimation pattern of cometary material provides a powerful tool to obtain more information on their composition. However, from observations it is not possible to achieve a direct correspondence between the heliocentric distance and the volatility of species as defined by their sublimation temperatures, e.g., as pure ices (*Capria,* 2002). The size distribution of grains in cometary bulk material extends over several orders of magnitude, including the pores in between.

The effect of pore size distribution seems to be important for the thermal conductivity. *Shoshany et al.* (2002) show in their models that the thermal conductivity is lowered by several orders of magnitude at high porosities. The physics of the processes responsible for driving sublimation and outbursts from the interior of cometary nuclei is described by *Prialnik et al.* (2004). Heat waves from the surface or internal radioactive heating provide the energy within the cometary nucleus for water ice crystallization, which releases latent heat. The trapped gaseous species, which are released during the crystallization process, move through the pores and carry along small, detached dust particles. Another source of gas in the interior can be sublimation of volatile ices from the pore walls. Once gas is released from the ice in the interior of the nucleus, its pressure will cause it to flow to the surface. Free gases present in the cometary interior are expected to affect the thermal and mechanical structure of the nucleus. The internal pressure may surpass the tensile strength of the fragile grainy configuration. This results in cracking of the porous matrix and subsequent outbursts of gas and dust (*Prialnik,* 2002). Internal pressure buildup by gas, thermal stress, and rotation may cause disruption of the fragile material. On the contrary, sintering (increase of contact area between grains — the Hertz factor — due to heating or compaction) and pore blocking due to larger grains as well as recondensation of volatiles may reconsolidate material. The complexity of internal processes may produce an individual pattern for each comet.

Large-scale groundbased simulation experiments (e.g., KOSI Kometen-Simulation) have been performed to study the evolution of cometary nuclei (*Kochan et al.*, 1998; *Colangeli et al.*, 2004). Sublimation experiments with ice-mineral mixtures showed the metamorphosis of ice, the reduction of volatiles in surface layers, and the formation of a porous low-density (0.1 g/cm^3) dust mantle (*Grün et al.*, 1993). The formation of a dust mantle on the surface and a system of ice layers below the mantle from the different admixed materials have been detected after the insolation of the artificial comet. Those experiments allowed for studies of the mechanisms for heat transfer between the comet surface and its interior, compositional structural and isotopic changes that occur near the comet's surface, and the mechanisms of the ejection of dust and ice grains from the surface (*Kochan et al.*, 1998; *Colangeli et al.*, 2004). In the following sections we will emphasize the discussion on the main components that provide insights into the interstellar–solar system connection and emphasize the processes that could contribute (or not) to modifying interstellar material as it becomes incorporated into comets.

2. TRACING INTERSTELLAR CLOUD MATERIAL IN COMETS

In cometary science, a central issue is elucidating and quantifying which aspects of their chemical composition and heterogeneity can be attributed to being either pristine or partially processed interstellar material, or material formed purely from nebular processes (*Irvine et al.*, 2000; *Irvine and Lunine*, 2004). Astronomical observations now allow us to follow in detail the chemical evolution of pristine interstellar material in a system analogous to that from which the protosolar nebula formed (*Mannings et al.*, 2000; *van Dishoeck and Blake*, 1998; *Ehrenfreund and Charnley*, 2000). The meteoritic record, interplanetary dust particles (IDPs), and asteroidal observations all indicate that some interstellar material underwent a very significant degree of processing in the protosolar nebula (e.g., *Ehrenfreund et al.*, 2002). Observations of a cometary volatile inventory that is largely consistent with the interstellar inventory, and dust grains indicative of nebular processing, are evidence for contributions from at least two sources. However, in many cases, it is not a simple matter to discern which source is responsible for a particular chemical or physical characteristic; a plausible theoretical argument can usually be made for each. Here we discuss several distinctive features of cometary composition. This is done in the context of the processes that may have acted to modify interstellar material incorporated in the protosolar nebula (see section 5.3 of *Wooden et al.*, 2004). We attempt to evaluate the likelihood of each feature being symptomatic of either interstellar origin or complete nebular reprocessing, or some intermediate between these extremes.

For comets we can only compare their volatile composition directly with that of interstellar ices. However, it appears that, in the dense core from which the protosun

formed, chemical conditions eventually evolved to the point where almost all the heavy volatile material had condensed as ice onto dust grains (*Brown et al.*, 1988; *Bergin et al.*, 2002). As cold interstellar gas and ice-mantled dust grains collapsed onto the protosolar nebula, they were heated by radiation and gas-grain drag (e.g., *Lunine*, 1989). Nonrotating collapse calculations indicate that infall timescales can become shorter than most chemical timescales. This results in material from the cool envelope collapsing onto the central protostar without significant chemical alteration (*Rodgers and Charnley*, 2003). However, because of rotation of the dense cloud core, infalling parcels of gas and dust far from the axis of rotation would have had sufficient angular momentum to move along ballistic trajectories and become incorporated in a disk, rather than fall directly into the protosun (*Cassen and Moosman*, 1981; *Terebey et al.*, 1984).

Infalling matter passes through the accretion shock where significant processing can occur. Far from the protosun, the lower nebular (preshock) densities and slower shock speeds meant less processing of interstellar material. In this case, processing may simply involve removal and postshock recondensation of the ices (e.g., *Lunine*, 1989). Approaching the protostar, higher postshock temperatures and enhanced UV fields led to increasingly hostile conditions for the survival of interstellar ices, volatile and refractory organics, and refractory dust grains (*Neufeld and Hollenbach*, 1994; *Chick and Cassen*, 1997).

Compositional and isotopic evidence from analysis of meteorites and IDPs suggests, however, that some volatile interstellar molecules may have entered the nebula relatively unscathed (*Messenger*, 2000; *Irvine et al.*, 2000). The phase of large-scale accretion of molecular cloud material lasted around a few hundred thousand years; most of this was consumed by the protosun (*Cameron*, 1995). Comets began to be assembled in the 5–100-AU region of the nebula toward the end of the stage of nebular disk dissipation (lasting about 50,000 years), when viscous effects dominated nebular evolution. During this stage, the interstellar mass accretion rate was probably about 10–100 times less than that occurring in the initial phase of accretion. Comet formation ended after a further 1–2 m.y., when solar accumulation almost finished and when giant planet formation (at 5–40 AU) was almost complete.

During disk dissipation, there was large-scale inward transport of most of the gas and dust and outward transport of most of the angular momentum (e.g., *Ruden and Pollack*, 1991). Turbulent motions, whether convective or magnetohydrodynamic in origin, produced an outward diffusion of material from the inner nebula (*Morfill and Volk*, 1984). This led to radial mixing of the products of two chemistries (*Irvine et al.*, 2000; *Markwick and Charnley*, 2004). The cold outer protosolar nebula, where accretion favors retention of ISM integrity, was in fact a chemically active region (e.g., *Aikawa and Herbst*, 1999a; *Aikawa et al.*, 1999). Cosmic rays (beyond about 10 AU) and other sources of ionization such as X-rays (*Glassgold et al.*, 1997), UV photons (*Willacy and Langer*, 2000), and radioactive decay [e.g.,

[26]Al and [60]Fe; *Finocchi and Gail* (1997)] can drive a nonequilibrium chemistry involving ion-molecule and neutral-neutral reactions. In the hot inner nebula (within about 1 AU), material can be completely destroyed and lose its interstellar integrity (*Fegley and Prinn*, 1989). The composition of this region is governed by a gas-grain chemistry in thermodynamic equilibrium. There is little direct chemical knowledge about the 5–40-AU regions of disks where comets are formed. Infrared observations of CO can probe the hot innermost regions of disks but these regions are not at present easily accessible to radioastronomical observations (*Dutrey et al.*, 2004).

Radial mixing offers a means of transporting crystalline silicates, be they condensates or annealed grains, outward from the inner nebula into the 5–40-AU region where they were incorporated into comets (*Gail*, 2001; *Bockelée-Morvan et al.*, 2002). However, nebular shocks at around 5–10 AU are another possible candidate (*Harker and Desch*, 2002) and, if correct, would place constraints on the efficiency of radial mixing and thermal convection in the nebula. Shocks in icy regions of the nebula have also been proposed as the origin of chondritic fine-grained phyllosilicates (*Ciesla et al.*, 2003). The detailed effects of such shocks on the volatile organic inventory are probably extensive, and need to be explored to coherently assess nebular processing of cometary materials. Inward transport of the products of "interstellar chemistry," or of similar chemistry in the cold disk, could account for the similarities between the volatile inventory of comets and these products. The key question is therefore how much radial mixing actually occurred (see *Lunine*, 1997; *Fegley*, 1999, 2000; *Lunine and Gautier*, 2004). For a detailed discussion on the physical and chemical processes of disks the readers are referred to *Dutrey et al.* (2004), *Boss* (2004), and *Lunine and Gauthier* (2004).

2.1. Is There Interstellar Cloud Material in Comets?

The gases observed in cometary comae originate from the nuclear ices and so offer insight into the nucleus composition. Coma molecules can either have a "native" source, and so have been sublimed directly from the nucleus itself, or they may appear throughout the coma from an "extended" source, probably due to the decomposition of large organic particles or molecules. For a detailed discussion on cometary volatiles and coma chemistry, the readers are referred to *Bockelée-Morvan et al.* (2004) and *Rodgers et al.* (2004). A comparison with the inventory of interstellar ices, as well as with the gases found in dark molecular clouds and in regions of massive and low-mass star formation, suggests a direct link. Apparent evidence for the retention of an interstellar origin are the facts that the major ice components, and many of the trace molecular species, are also found in cometary comae. However, as briefly discussed here, there are differences in the relative abundances of some cometary species when compared to their interstellar values. Thus, there is some circumstantial evidence for processing of the

precursor interstellar volatiles; the location, epoch, and source of this processing is largely to be determined.

Silicate and C-based micrometer-sized dust particles that are produced in the outflows of late-type stars provide a catalytic surface for a variety of reactions to occur when they are dispersed throughout the molecular cloud (*Ehrenfreund et al.*, 2003). In dense clouds, ices are stable and efficiently formed on the surface of such dust particles. The formation of interstellar ices has been discussed in *Wooden et al.* (2004). An extended inventory of interstellar ice species has only been established for bright high-mass star-forming regions (e.g., *Gibb et al.*, 2000). Whether those abundances are relevant for a comparison with cometary composition in low-mass systems is strongly questioned. There are only incomplete datasets available for solar-type stars. Most of them are characterized by a low flux and were therefore difficult to observe by satellites such as the Infrared Space Observatory (ISO). Groundbased observations allow us only to observe small wavelength regions in telluric windows (*Boogert et al.*, 2002). Table 1 lists solid state abundances measured in high- and low-mass (solar-type) star-forming regions in comparison with cometary volatiles.

As discussed by *Wooden et al.* (2004), the spectrum of interstellar clouds is very rich and shows some very strong bands that mask the signature of a number of other spe-

TABLE 1.   Interstellar ice abundances measured toward high- and low-mass (solar-type) star-forming regions (*Gibb et al.*, 2000, *Nummelin et al.*, 2001, *Pontoppidan et al.*, 2003, *Taban et al.*, 2003) are compared to measured abundances of cometary volatiles in comet Halley, Hyakutake, Hale-Bopp, Lee, LINEAR and Ikeya-Zhang (see *Bockelée-Morvan et al.*, 2004, Table 1).

	High-mass Stars	Solar-type Stars	Comet Average
H_2O	100	100	100
CO	9–20	5–50	1.8–30
CO_2	12–20	12–37	3–6
CH_3OH	0–22	0–25	1.8–2.5
CH_4	1–2	<1	0.14–1.5
H_2CO	1.5–7	—	0.4–4
OCS	0–0.3	<0.08	0.1–0.4
NH_3	0–5	—	0.5–1.5
HCOOH	0.4–3	—	0.09
C_2H_6	<0.4	—	0.11–0.67
HCN	<3	—	0.1–0.3
C_2H_2	—	—	0.1–0.5

The strong diversity of abundances among the main species CO, CO_2, and CH_3OH, even within high- and low-mass star-forming regions, hampers the search for an interstellar/cometary link. The lack of data for the low flux solar-type stars adds to those uncertainties. In contrast to cometary observations, there is no evidence for the presence of C_2H_6, HCN, and C_2H_2 nor S-bearing species (apart from tiny abundances of OCS) in interstellar ices. H_2CO and HCOOH are only tentatively measured in interstellar ices and need a more firm abundance determination.

cies. After H_2O and CO (which may have multiple sources in comets), species such as CO_2, CH_3OH, OCS, NH_3, and CH_4 seem to be among the only molecules that could provide constraints on an interstellar/cometary link. Most of the other weak features may escape detection or their abundance remains poorly determined in the ISM. This applies to C_2H_2, C_2H_6, and HCN, of which none have been observed in the interstellar solid phase. The only molecules that show a similar abundance in comets and some low-mass star-forming regions [such as "Elias 29" (*Boogert et al.*, 2000)] are CH_4 and CH_3OH.

The ice phases within interstellar grain mantles will — when incorporated into comets — contribute to the outgassing properties, as do the structures of cometary nuclei that are stratified in temperature, porosity, density, and composition (*Prialnik et al.*, 2004). Interstellar icy grains are characterized by different ice phases, amorphous, crystalline, segregated boundary layers and possibly clathrates (*Ehrenfreund et al.*, 1999).

In particular, some CO, CO_2, and CH_3OH seem to be present in pure form [e.g., "apolar" CO layers in cold clouds or segregated boundary phases of CO_2 and CH_3OH in warmer cloud regions (*Ehrenfreund et al.*, 1999)], which would allow sublimation at a much lower temperature than when trapped in an H_2O ice matrix. Sublimation of such species at large heliocentric distances could provide evidence for such ice layers. No attempt has ever been made to correlate the outgassing pattern of cometary volatiles with the different ice phases present in interstellar grain mantles.

Carbon dioxide and CH_3OH are the most important ice species (after H_2O and CO) that can be used as a tracer for the interstellar/cometary link. ISO identified CO_2 ice as one of the major components in interstellar ice mantles with an average abundance of ~15–20% relative to water ice. Recently, CO_2 ice abundances of up to 37% (relative to water ice) have been measured toward low-mass protostars (*Nummelin et al.*, 2001). Whereas CO_2 ice is ubiquitous in the interstellar medium (every target measured has CO_2 abundances above 10% relative to water ice), the measured abundance of CH_3OH is highly variable and in fact is undetected toward many sources. Among high-mass and low-mass protostars, the CH_3OH abundance ranges from small upper limits to 25% relative to water ice (*Dartois et al.*, 1999; *Pontoppidan et al.*, 2003). The cometary abundance of methanol is appreciable but generally much smaller (~2%) when compared to interstellar ices (5–25%). Similarly, CO_2 appears to be much less abundant in comets than in molecular clouds (*Feldman et al.*, 2004). Both these observations suggest either partial degradation of the interstellar molecules or complete production of them in the solar nebula. If energetic processing of molecular ices is an efficient means of forming CO_2 (*Wooden et al.*, 2004), and given the more energetic environment of the protosolar nebula, it is surprising that CO_2 is not at least as abundant as in molecular clouds. This may rule out energetic processing of ices in the outer nebula as the source of these compounds. *Gail* (2002) has shown that radial mixing of oxidized C dust from

the inner nebula could not explain the cometary CH_3OH abundances. In this region an additional possible source of methanol is Fischer-Tropsch-type (FTT) synthesis (*Fegley and Prinn*, 1989). Perhaps the simplest explanation is that these molecules are interstellar but have had their original populations partially depleted in the nebula. For example, CO_2 (and H_2CO) are very susceptible to destruction by H atoms in warm gas (*Charnley and Kaufman*, 2000). This may point to CO_2 molecules being partially destroyed at the accretion shock or in nebular shock waves. The situation is less clear for CH_3OH since the direct detection of methanol in a protostellar disk [at <100 AU (*Goldsmith et al.*, 1999)] indicates a large fractional abundance of 3×10^{-7}, apparently at the disk surface, but a lower value deeper in the disk of 2×10^{-8}. This lower value may simply arise from depletion onto dust, in which case the abundance of methanol in any precometary ices could in fact be similar to that found in comets.

The current interstellar ammonia abundances, which are much lower than originally estimated (*Taban et al.*, 2003), are more compatible with cometary measurements (see Table 1). Both interstellar and nebular chemistries will produce N_2 efficiently (e.g., Owen et al., 2001). Thus perhaps one of the most perplexing problems in making a definite connection between the major interstellar volatiles and those in comets is widespread depletion/lack of N_2 as measured by N_2^+ observations of the latter (*Cochran et al.*, 2000; *Cochran*, 2002). This may simply be a volatility issue.

Molecular N may have been more readily evaporated relative to CO due to modest warming of precometary ices (e.g., *Irvine et al.*, 2000). A further possibility is that CO was selectively trapped during formation of clathrate hydrates, whereas N_2 was not (*Iro et al.*, 2003). Alternatively, this deficiency could have an origin in the prestellar phase (*Charnley and Rodgers*, 2002), in which case (and if production and/or mixing of N_2 in the nebula is inefficient) all comets may to some degree show this deficiency relative to the ISM value.

The problem in comparing the abundances of S-bearing compounds is that the cometary parents CS_2 and H_2S are not detected in interstellar ices; although gaseous H_2S is observed to be abundant in star-forming regions. CS_2 and S_2 are unknown in the ISM. The most abundant of the interstellar S-bearing compounds have all been detected in comets. However, some of them are photoproducts of other compounds (e.g., CS_2 and CS, SO and SO_2), or may have a distributed (polymeric?) source (e.g., CS, OCS). These considerations make it very unlikely that cometary sulfuretted species represent pristine interstellar material.

The simple hydrocarbons CH_4, C_2H_2, and C_2H_6 also present ambiguity. Methane is the only molecule among these three that is present in interstellar ices (~1% relative to water ice); acetylene has a cometary abundance similar to the interstellar gas (e.g., *Lahuis and van Dishoeck*, 2000). It is therefore tempting to identify them as being of interstellar origin where ethane forms by reduction of acetylene on cold (10 K) grains. Alternatively, these hydrocarbons could have an origin in nebular chemistry, probably involving some form of gas-grain chemistry. For example, *Gail* (2002) showed that mixing of oxidized C dust could explain the abundances of CH_4 and C_2H_2 in comets, but not the presence of C_2H_6. If this particular nebular chemistry was the origin of these hydrocarbons, then it requires outward radial mixing of CH_4 and C_2H_2 and inward transport of C_2H_6 from the outer nebula, where it perhaps formed on grains. Alternatively, FTT synthesis would require that they all be transported outward (*Fegley and Prinn,* 1989). Unfortunately, it is difficult to detect C_2H_6 in the ISM and therefore to definitively decide on the origin of these compounds in cometary matter.

Apart from the most abundant species, some organic molecules found in molecular clouds have also been identified in the coma of Hale-Bopp (*Bockelée-Morvan et al.,* 2000, 2004). Simulations of interstellar ice analogs show increasing complexity on the molecular scale when energetic and thermal processing is applied (*Allamandola et al.,* 1997; *Moore and Hudson,* 1998; *Cottin et al.,* 2001a; *Colangeli et al.,* 2004). Those complex organics that have been observed in laboratory spectra of processed ice mixtures may or may not be present at small abundance in interstellar grain mantles. However, with current astronomical instrumentation they cannot be observed in the interstellar solid phase. Their signatures will in most cases be too weak relative to the continuum or they will be masked by the presence of other strong bands. Thus, there will never be definite proof for the presence of specific, large organic species in interstellar ice mantles; some indications may come from gas-phase measurements in hot cores that sample the material evaporating from icy particles.

For molecules whose formation involves gas-grain chemistry, we are generally comparing the abundances measured toward regions of massive star formation, so-called hot molecular cores. Such regions may also exist around low-mass cores (*Schoier et al.,* 2002) but these are generally less-well characterized.

The difficulties inherent in connecting observed cometary volatiles with interstellar molecular cloud material can be illustrated for the case of the HNC molecule. Cometary HNC was originally discovered in the coma of Hyakutake (*Irvine et al.,* 1996); the high HNC/HCN ratio provided strong evidence that this HNC was preserved interstellar material. Subsequent calculations showed that the HNC/HCN ratios in Hyakutake and Hale-Bopp could apparently be produced by chemical reactions in the coma (*Rodgers and Charnley,* 1998; *Irvine et al.,* 1998). Further measurements of HNC/HCN ratios in other comets, when confronted with these models, demonstrated that gas-phase chemistry cannot in general be the origin of HNC. The most likely source of HNC in the coma is the decomposition of a large organic, perhaps polymeric, compound (*Rodgers and Charnley,* 2001a; *Irvine et al.,* 2003).

The C chain radicals, C_2 and C_3, are widespread in comets and their abundances indicate a marked variation with the Tisserand parameter (*A'Hearn et al.,* 1995). Identification of their chemical parent(s) is an important problem whose resolution should shed light on chemical differentiation and the place of origin of particular comets. *Helbert et al.* (2002) have shown that the abundances of these molecules could be derived from C_2H_2, CH_3CCH, and C_3H_8 in a coma chemistry driven by electron-impact dissociations. However, methylacetylene and propane have not yet been detected in comets. In this picture, the same uncertainties in distinguishing interstellar and nebular contributions to the hydrocarbons persist. Synthesis of long, unsaturated, C-chain molecules appears to be one signature of interstellar organic chemistry. There have been tentative detections of C_4H and its possible parent C_4H_2 in Halley and Ikeya-Zhang respectively (*Geiss et al.,* 1999; *Magee-Sauer et al.,* 2002). The detection of a large suite of long C-chain molecules in comets, similar to the large ones found in molecular clouds, would be strong evidence for an interstellar origin of these organics.

Of the many other small interstellar organics also found in comets, HNCO, NH_2CHO, CH_3CHO, and HCOOH are expected, along with H_2CO and CH_3OH, to be products of interstellar grain-surface chemistry (*Charnley,* 1997). Although there are differences between the interstellar and cometary abundances, the fact that these molecules are associated with ices does suggest that they are products of interstellar chemistry. As these surface reactions involve low-temperature H-atom additions, they cannot proceed efficiently on surfaces warmer than about 15 K since the H-atom residence time then becomes shorter than the timescales to migrate and react, as shown experimentally (*Watanabe et al.,* 2003). This does not represent a problem in 10-K molecular clouds, but greatly constrains the region of the nebula where such a chemistry could occur. Any differences between the interstellar and cometary abundances may then be due to selective processing in the nebula (see the discussion on methanol above).

The other simple cometary organics, CH_3CN, HC_3N and $HCOOCH_3$, cannot be formed in the coma, and current theories of their formation require gas-phase reactions (*Rodgers and Charnley,* 2001b). The cometary abundances of CH_3CN and HC_3N are consistent with them being either the products of interstellar chemistry or similar processes in the outer nebula. However, interstellar $HCOOCH_3$ is believed to form around protostars in an ion-molecule chemistry driven by evaporated ice mantles (*Blake et al.,* 1987). The methyl formate in comets may then have originated in a sequence involving the desorption of ices, an intervening period of ion-molecule chemistry, followed by recondensation. These conditions may have occurred upon the first accretion of interstellar material, or perhaps during alternating inward and outward radial mixing of interstellar material in the nebula.

In summary, volatile cometary material shows a general qualitative link with the interstellar ice phase and the inventory of identified interstellar molecules. It is unlikely that these similarities reflect the wholesale incorporation of unaltered interstellar ices into comets (cf. *Greenberg,* 1982). As infalling interstellar material is decelerated at the accretion shock, it experiences chemical alteration to varying degrees, depending upon the epoch and position of entry into the nebula. These changes range from those associated with simple evaporation-recondensation of ices in the outer disk regions, to complete molecular dissociation in the inner ones. We may also expect that the (interstellar) chemical clock can be further reset during the assembly of cometesimals in the nebula. It is now realized that the chemistry occurring in the outer regions of protoplanetary disks closely resembles that of dense interstellar regions. Hence, the pristinity of interstellar matter accreted under even the most benign conditions can become further adulterated. Elucidating in detail how the pristine inventory of available interstellar volatiles can be corrupted will require careful modeling of nebular chemistry, subject to the constraints that will be provided by future observations of disk and comet composition.

2.2. Isotopic Fractionation: An Interstellar Signature in Comets?

In cold interstellar clouds, both gas-phase and grain-surface chemistries can lead to enhanced isotopic fractionation in molecules (*Wooden et al.*, 2004). In comets, several molecular isotopic ratios have been measured and these can, in principle, provide important cosmogenic information (Table 3 in *Bockelée-Morvan et al.*, 2004). It must, however, be stressed that conclusions drawn from such observations may be biased due to the limited data available.

Similar $^{18}O/^{16}O$ ratios, both close to the terrestrial value, were measured in Halley and Ikeya-Zhang (*Balsiger et al.*, 1995; *Lecacheux et al.*, 2003). The $^{13}C/^{12}C$ ratio has been measured in C_2, CN, and HCN for several comets and this is also apparently terrestrial (*Wyckoff et al.*, 2000). From this data, one may draw the conclusion that either the natal cloud was of solar composition, or these ratios were set in the protosolar nebula. However, the absolute interstellar fractionation expected in C and O isotopes is generally much less than in other isotopes (*Langer et al.*, 1984). Calculations demonstrate that C-isotopic fractionation by ion-molecule reactions selectively enhances ^{13}C in CO, whereas derived cometary $^{13}C/^{12}C$ ratios are not based on isotopes of CO.

Thus, it cannot be ruled out that other cometary molecules, perhaps derived from CO on dust, may possess higher $^{13}C/^{12}C$ ratios. The ^{13}C enhancements found in IDPs and carbonaceous chondrites may have an origin in these species (e.g., *Cronin and Chang*, 1993).

Enhanced D fractionation is observed in many interstellar molecules. The measured D/H ratios range from about 1×10^{-4} for water to 0.5 for formaldehyde (*Ceccarelli*, 2002). Recently, there has been growing evidence for "superdeuteration" in some molecules (H_2CO, NH_3), exhibited in high D/H ratios and the presence of multideuterated species: D_2CO, ND_3, CHD_2OH (*Loinard et al.*, 2000; *van der Tak et al.*, 2002; *Parise et al.*, 2002). Interestingly, this "superdeuteration" is observed in either low-mass prestellar cores, where CO depletion onto dust may be responsible (*van der Tak et al.*, 2002), or in cores where low-mass star formation has already occurred and the products of grain-surface

deuteration are in the gas phase, e.g., in IRAS 16293-2422 (*Loinard et al.*, 2000).

The fractionation ratio HDO/H_2O has only been measured in three comets (Halley, Hyakutake, and Hale-Bopp), where it was found to be around 3×10^{-4}. In Hale-Bopp, the DCN/HCN ratio was determined to be about 0.002. Coma chemistry models demonstrate that these ratios are truly those of the material residing in the nucleus (*Rodgers and Charnley*, 2002). Gas-phase water D/H ratios measured in massive star-forming regions are generally low, around 0.0003, and comparable with the cometary values. However, recent determinations of HDO/H_2O in Orion suggest it could be higher: ~0.004–0.011 (*Pardo et al.*, 2001). Searches for HDO ice in low-mass protostellar cores yield only an upper limit on HDO/H_2O of about 0.02 (*Parise et al.*, 2002). Surveys of low-mass protostellar cores consistently yield DCN/HCN ratios about a factor of 30 larger than the cometary ratio (*Roberts et al.*, 2002), the latter of which is comparable to the value found in massive cores like Orion (0.003).

Thus, it is difficult to elucidate the physical conditions under which cometary ices formed (and hence the probable location) by making analogies with interstellar sources, at least for these two molecules. The D/H ratios measured in massive cores appear to resemble cometary values best, but low-mass cores are more physically relevant to protoplanetary nebulae and thence, presumably, to comet formation. It is possible that high molecular D/H ratios existed in the protosolar natal cloud core and these were diluted by chemical reactions during accretion and within the nebular disk. Ideally, one would wish to compare D fractionation between the envelope and disk of a forming protostar. Thus far, the only detection of a deuterated molecule in a disk, DCO^+ in TW Hya, yields a D/H ratio of 0.035 (*van Dishoeck et al.*, 2003). This is comparable with that found in low-mass cores but larger than found in massive cores. However, one must be careful in drawing conclusions based on a molecular ion. This observation only provides information on the potential of disk chemistry to deuterate molecules (e.g., *Aikawa and Herbst*, 1999b) and gives no direct connection to the neutral molecules observed in cometary nuclei.

Thus, albeit based on just three measurements, comets appear to be less deuterated than the material from which the protosolar nebula probably formed. This conclusion comes with the important caveat that the interstellar molecules with the most distinctive D fractionation have not yet been measured in comets (i.e., isotopomers of formaldehyde, ammonia, and methanol).

There is some evidence that this may be the case since it has been suggested that there may be differentiation in the D/H ratios between nuclear HCN and the DCN and HCN released into the coma from outgassing grains (*Blake et al.*, 1999).

Furthermore, the carbonaceous component of some IDPs, probably of cometary origin, exhibit D/H ratios close to the interstellar range and higher than that found for HCN [D/H of 0.008 in "Dragonfly" (*Messenger*, 2000)]. Assuming that similarly lower cometary D/H ratios will be found in formaldehyde, ammonia, and methanol, the D/H ratios in comets could result from "erosion" of the pristine D/H ratios at the accretion shock or in nebular shocks. Ion-molecule chemistry in the outer nebula could also act to lower the original interstellar fractionation (*Aikawa and Herbst,* 1999b). Alternatively, it has been proposed that lowering of the water fractionation may proceed by neutral exchange processes in the nebula (*Drouart et al.,* 1999), however, the value of the solid HDO/H_2O ratio assumed in this model is at present controversial (*Texeira et al.,* 1999; *Dartois et al.,* 2003). Important goals for future observations should be to determine the D/H ratios of selected molecules, such as water and HCN, in disks, and also to measure D/H in cometary molecules whose interstellar counterparts are known to have large values (e.g., formaldehyde).

Radio observations of HCN in Hale-Bopp indicate a ^{14}N/^{15}N ratio of around 300 (e.g., *Jewitt et al.,* 1997; *Crovisier and Bockelée-Morvan,* 1999); as with C and O, this is consistent with a terrestrial value (270). Recent measurements of C^{15}N and C^{14}N in Comets Hale-Bopp and C/2000 WM1 (LINEAR) near perihelion by *Arpigny et al.* (2003) indicate that C^{14}N/C^{15}N is about 130, and significantly different from that in HCN. For both molecules (HCN and CN) the ^{12}C/^{13}C ratios were found to be terrestrial. Some CN is certainly coming from photodissociation of HCN but the observations of *Arpigny et al.* (2003) indicate that there must also be another, probably polymeric, parent for CN that is much more highly fractionated in ^{15}N. Such low ^{14}N/^{15}N ratios are also detected in IDPs (*Messenger,* 2000) and can be explained by interstellar chemistry theories (*Charnley and Rodgers,* 2002). This emphasizes the crucial point for understanding the origin of cometary isotopic fractionation — one must attempt to measure isotopic ratios of several different molecules.

2.3. Silicates

Dust grains are entrained in the flow of escaping gases from the nucleus. Dust grains in the coma scatter and absorb sunlight, reemitting the absorbed energy in the thermal infrared. At 1 AU, the highly refractory silicate grains are warmed to radiative equilibrium temperatures well below their melting or annealing temperatures. Therefore, the properties of the silicate grain component of comae dust probe the mineralogy and crystallinity of the silicate dust in the nucleus. Cosmic-ray bombardment of surfaces of long-period comets while in the Oort cloud has the potential to alter the properties of the silicates on the nuclear surface. Collisional evolution of short-period comets in the Kuiper belt and UV irradiation during their many perihelion passages may also alter the grain properties. Comparing the dust properties in long- and short-period comets compares the effect of minimal vs. significant parent-body processing.

As discussed in *Hanner and Bradley* (2004), IR spectra of comets and laboratory studies of cometary interplanetary dust particles show that silicates in comets are dominated by two minerals — olivine and pyroxene or $(Mg_y,Fe_{(1-y)})_2SiO_4$ and $(Mg_x,Fe_{(1-x)})SiO_3$ — and are in two forms: amorphous and crystalline (*Colangeli et al.,* 2004).

Thermal emission models of Comet 19P/Borrelly, one of the few short-period comets with a silicate feature, fit either a grain population dominated by discrete amorphous pyroxene and amorphous C grains (*Hanner et al.,* 1996) or core-mantle aggregate particles consisting of amorphous olivine cores with amorphous C mantles (*Li and Greenberg,* 1998a). Amorphous pyroxene and olivine are found in long period comets in varying proportions (*Hanner et al.,* 1994; *Hanner and Bradley,* 2004). In the core-mantle aggregate model (*Hage and Greenberg,* 1990), long-period comets are fitted with organic refractory mantles (*Li and Greenberg,* 1998b), while short-period Comet Borrelly is better fitted with amorphous C mantles (*Li and Greenberg,* 1998a). These comparisons suggest that parent-body processing and UV irradiation do not change the silicate mineralogy while it may carbonize the organic grain component (*Li and Greenberg,* 1998a). In highly active long-period comets we can compare the grain properties of particles dredged up from deeper layers of the nucleus and entrained in jets with those particles released into the coma from the nuclear surface. In Comet Hale-Bopp, grain mineralogy as revealed through the shape of the 10-μm spectral resonances was uniform to within the measurement uncertainties at different positions in the coma at a given epoch (*Hayward et al.,* 2000), except for a drop in the crystalline pyroxene resonance in the sunward direction (*Harker et al.,* 2002). The grain temperatures and the optical and near-IR polarization were significantly enhanced in Hale-Bopp's jets (*Hadamcik and Levasseur-Regourd,* 2003), indicating a difference in grain morphology or size (*Levasseur-Regourd et al.,* 2002). Analysis of Hale-Bopp suggests that differences between surface and jet particles can be attributed to grain morphology but not silicate mineralogy. Studying the silicate mineralogy and crystallinity therefore probes interstellar and nebular processes affecting silicate dust grains prior to their incorporation into comets (*Hanner et al.,* 1996). The evidence, although not unanimous, forms a consensus that cometary amorphous silicates are of probable interstellar origin while crystalline silicates are of probable nebular origin (*Hanner and Bradley,* 2004; *Wooden,* 2002).

Amorphous silicates in comets probably are relic grains from the interstellar medium (cf. *Hanner and Bradley,* 2004). Formation of amorphous silicates requires very rapid cooling that is probable for an asymptotic giant branch (AGB) stellar outflow (*Wooden et al.,* 2004; *Tielens et al.,* 1997) but formation is unlikely under conditions in the solar nebula (*Yoneda and Grossman,* 1995). Amorphous or glassy forms of minerals are relatively rare in meteoritic materials compared to crystalline forms. Iron-bearing amorphous olivine is fitted to the interstellar 10-μm absorption feature and interstellar extinction curve (*Li and Greenberg,* 1997). Iron-bearing amorphous olivine and pyroxene in a 5:1 ratio are fitted to the absorption feature in the line-of-sight to the galactic center (*Kemper et al.,* 2004). By the relative absence of amorphous silicates from highly processed solar nebula materials (chondrules) and their ubiquity in the ISM, amorphous silicates in comets are probably interstellar.

Further evidence of the interstellar source for cometary amorphous silicates arises from the laboratory studies of chondritic porous interplanetary dust particles (CP IDPs), which are of probable cometary origin (*Bradley*, 1988; *Hanner and Bradley*, 2004). Chondritic porous IDPs are aggregates of crystalline and amorphous silicates, the so-called "GEMS" (*Hanner and Bradley*, 2004), in a matrix of carbonaceous materials (see section 2.4). Irradiation of GEMS by high-energy particles during their residence time in the ISM has been invoked to explain radiation tracks, compositional gradients with GEMS' radii, and the presence of nanophase Fe (*Bradley*, 1994).

Laboratory experiments demonstrate that ion bombardment of 4-KeV He$^+$ ions at fluxes typical of the ISM shocks can amorphize crystalline olivine, increase porosity, reduce the Fe from its stoichiometric inclusion in the mineral to embedded nanophase particles, and damage the chemistry by reducing the O/Si and Mg/Si ratios, effectively increasing pyroxene at the expense of olivine (*Carrez et al.*, 2002). Ion bombardment by 50-KeV He$^+$ ions, characteristic of supernovae shocks, does amorphize the crystalline silicates but leaves the chemistry unaltered (*Jäger et al.*, 2003). Thus, ion bombardment can create the highly radiation damaged, Mg-rich GEMS with nanophase Fe inclusions (*Brownlee et al.*, 2000) in CP IDPs whose spectra are very similar to the amorphous silicates observed in cometary emission spectra (*Hanner and Bradley*, 2004; *Bradley et al.*, 1999a,b). Therefore, morphology and mineralogy indicates that the cometary amorphous silicates have an interstellar origin.

Halley's Fe/Mg ratio is 0.52 and the solar value is 0.84 (*Jessberger et al.*, 1988), so Halley's Fe content is close to solar. However, only 30% of the Fe is in silicates while 70% is in FeS and Fe grains. Halley's Fe grains are highly reduced, as there is <1% FeO. This is also characteristic of the iron within GEMS in CP IDPs where it exists as nanophase Fe or FeS and where the nanophase Fe has been attributed to ion bombardment in the interstellar medium.

The crystalline silicates in comets are of probable nebular origin. The existence of Mg-rich crystalline silicates in comets is revealed by the sharp resonances in cometary 10-μm spectra (*Hanner and Bradley*, 2004; *Harker et al.*, 2002; *Wooden*, 2002). Comet Hale-Bopp's observed 10-μm silicate feature is best-fitted by Mg-rich pyroxene "Cluster" (*Messenger*, 2000) CP IDPs (*Wooden et al.*, 2000). In fact, all crystalline silicates observed by the ISO short wavelength spectrometer (SWS) in comets, stellar outflows, and protoplanetary disks are Mg-rich (*Bouwman et al.*, 2001; *Molster et al.*, 2002a). Figure 1 presents from the ISO-SWS database the best comparison between a pre-main-sequence Herbig Ae star with a protoplanetary disk, an O-rich AGB star, and Comet Hale-Bopp. The clear contrast of strong crystalline peaks is seen in many but not all Herbig Ae stars (*Meeus et al.*, 2001). Crystalline silicates are preferably detected in binary AGB stars that possess disks (*Molster et al.*,

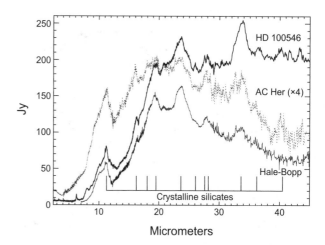

Fig. 1. Infrared Space Observatory (ISO) short wavelength spectrometer (SWS) spectra of the pre-main-sequence Herbig Ae star HD 100546 (*Malfait et al.*, 1998), the O-rich post-AGB binary star AC Her (*Molster et al.*, 2002 a,b,c), and Comet Hale-Bopp (*Crovisier et al.*, 1997), showing the strong similarities in their spectral features. The Mg-rich crystalline silicate resonances are identified. Note the far-IR spectra are dominated by isolated crystalline silicate features of Mg-rich crystalline olivine with a few features of Mg-rich crystalline pyroxene.

2002a), where crystallization is attributed to a low-temperature mechanism that occurs over longer times in the spherical outflows of single AGB stars (*Molster et al.*, 1999).

In the solar system, only silicate crystals in CP IDPs and in Antarctic micrometeorites have such high Mg contents (*Bradley et al.*, 1999c) as deduced for cometary crystals. Comet Halley's silicates are also very Mg-rich compared to chondrules and meteoritic materials. The range of Fe contents in silicate crystals in chondrules is due to the rapid melting of grain aggregates at high temperatures followed by rapid cooling (*Delany*, 1995; *Kracher et al.*, 1984). Since cometary crystals have very little Fe, these crystals have not suffered chondrule-like postformation heating events in the nebula and represent primitive solar nebula materials (*Wooden et al.*, 2000).

The appearance of Mg-rich crystalline silicates in some comets indicates that cometary crystalline silicates are grains that condensed at high temperatures [~1400 K (*Grossman*, 1972; *Hanner and Bradley*, 2004)] or were annealed from amorphous silicates at somewhat lower temperatures [>900 K (*Hallenbeck et al.*, 1998, 2000; *Fabian et al.*, 2000; *Hanner and Bradley*, 2004)] in the solar nebula. If annealing processes occurred in the inner nebula (*Mousis et al.*, 2002; *Bockelée-Morvan et al.*, 2002), then such temperatures occurred early (<300,000 yr) and radial diffusion rapidly transported a uniform concentration of crystals to beyond the snow line where comets formed (5–40 AU) (*Bockelée-Morvan et al.*, 2002). Annealing also may have occurred in shocks in the nebula in the 5–10-AU region, producing local enhancements in the crystalline concentration (*Harker and Desch,* 2002). Annealing temperatures are not reached in the accretion shock between outer disk surface and the prenatal cloud (*Chick and Cassen,* 1997; *Neufeld and Hollenbach,* 1994). Future observations may reveal differences in the crystalline-to-amorphous ratios between long- and short-period comets that will help to constrain solar nebula models. The detection of cometary crystals in long- and short-period comets, however, may not only depend on their concentration but also on grain properties such as grain size and porosity (*Hanner and Bradley,* 2004).

Challenges to the interpretation that cometary amorphous silicates are interstellar and cometary crystals are primitive solar nebula grains comes from recent discoveries using the nanoSIMS (secondary ion mass spectrometry). NanoSIMS is an ion microprobe, allowing elemental and isotopic analysis of small features of solid samples. Six out of more than 1000 subgrains in nine CP IDPs have anomalous $^{17}O/^{16}O$ and $^{18}O/^{16}O$ ratios and clearly carry presolar isotopic signatures (*Messenger et al.,* 2003). Three of the six presolar grains are identified with mineral phases and have AGB isotopic ratios: one forsterite crystal and two GEMS. About 1% by mass of the CP IDPs have distinct presolar signatures. The detection of a presolar forsterite crystal (*Messenger,* 2000) contradicts the concept that Mg-rich cometary crystals are solar nebula grains. The mass fraction, however, is within the range estimated for the ISM [<0.5% (*Kemper et al.,* 2004)]. The measured mass fraction of presolar GEMS, however, is significantly less than the cometary high mass fraction of amorphous silicates of probable interstellar origin [*Hanner and Bradley* (2004); Hale-Bopp, *Harker et al.* (2002)]. At this time, the nanoSIMS instrument detects only the subgrains with large isotopic anomalies. Higher-precision measurements on CP IDPs are required to improve our understanding of the range of isotopic ratios that are considered to be interstellar and of the processes in the ISM that may alter the signatures of the origins of dust grains.

The presence of silicate crystals in comets implies mixing of high-temperature and low-temperature materials in the comet-forming zone. The presence of silicate crystals in comets implies that processed nebular materials were incorporated with volatile-rich icy material. This implies a significant degree of radial transport and/or mixing in the nebula. In CP IDPs of probable cometary origin, the close proximity of oxidized, reduced, and metallic mineral phases, i.e., mineral phases in contact with but far from equilibrium with each other, indicates that these micrometer-sized aggregates of submicrometer units suffered minimal alteration after the grains accreted their constituent parts. "The fact that (post-accretion) alteration of aggregate IDPs hardly reached thermodynamic equilibrium at sub-micrometer scales supports the view that energy for alteration was either scarce or unavailable for sufficiently long periods of time, or both" (*Rietmeijer,* 1998). Porosity distinguishes primitive but not necessarily presolar origin. The high porosity of cometary grains, including CP IDPs (*Rietmeijer,* 1998) and the high porosity deduced from thermal emission models of Hale-

Bopp (*Harker et al.,* 2002; *Li and Greenberg,* 1997), is similar to what is plausible for both interstellar material (*Iati et al.,* 2001; *Vaidya and Gupta,* 1999) and grain aggregates in the solar nebula (*Dominik and Tielens,* 1997). Thus, grains of probable cometary origin are aggregates of silicate and carbonaceous materials that have seen significantly different environments in the solar nebula prior to both their aggregation into a single porous particle and their incorporation into comets.

2.4. Carbonaceous Matter

Our knowledge on carbonaceous material of comets is rather limited. *In situ* measurements of a few nanograms of Halley's coma show C-rich grains that are apparently components of various types, including pure C particles, polycyclic aromatic hydrocarbons (PAHs), branched aliphatic hydrocarbons, C-O or C-N polymers, and more complex compounds containing all four C, H, O, and N atoms (*Fomenkova,* 1997). CHON organic compounds were first described as being similar to kerogens (*Jessberger et al.,* 1988). Heteropolymers or complex organic molecules also are proposed (*Kissel et al.,* 1997). Some of the species remain very speculative due to the limited resolution of the mass spectrometers that flew through Halley's coma.

The *in situ* measurements of Halley reveal that 25% by mass are siliceous-free CHON grains, 25% are carbonaceous-free silicate grains, and 50% are mixed carbonaceous and silicate grains. Mixed grains of greater mass exist in the innermost parts of the coma, while carbonaceous-free silicate grains are dominant further out in the coma (*Fomenkova,* 1999). This suggests that the organic material is the "glue" that holds the silicates together and that the organic material desorbs in the inner coma (*Boehnhardt et al.,* 1990; *Fomenkova,* 1997), freeing the isolated silicate mineral phases. This scenario is also proposed for Comet Hale-Bopp near perihelion (≤1.5 AU) when its amorphous grains showed an increase in porosity, a steeper size distribution (more smaller grains), and slightly higher relative mass fraction of crystalline silicates (*Harker et al.,* 2002). Furthermore, within this same heliocentric range (≤1.5 AU), Comet Hale-Bopp has a strong distributed source of CO, i.e., the CO spatial distribution is more radially extended than the distribution of the dust and other gas components; only 50% of the CO originated from the nucleus (*DiSanti et al.,* 2001).

The other 50% of the CO is deduced to arise from the desorption of an unknown organic component of the dust. A possible candidate for the distributed CO is polyoxymethylene (POM), which is a polymer of formaldehyde (*Boehnhardt et al.,* 1990). The desorption of POM first into formaldehyde followed by photodegradation into CO would, however, create far more formaldehyde (*Cottin et al.,* 2001b) than detected in Hale-Bopp's coma (*Bockelée-Morvan et al.,* 2000). Following the heliocentric dependence of both dust properties and distributed CO sources is currently one of the best observational strategies for studying the organic component of cometary dust.

Infrared observations of dusty comae ubiquitously detect a strong near-IR spectrally featureless thermal emission that is well-fitted by amorphous C. The spectroscopic detection of solid state CHON particles if they are made of kerogen-like material is difficult in the near- and mid-IR because kerogen is about 30 times less absorptive than Fe-bearing amorphous silicates, and even less absorptive than amorphous C; spectroscopic resonances are weak compared to the other grain species. Unidentified gas phase lines are seen in comets at very high spectral resolution in the near-IR (*Magee-Sauer et al.,* 2002; *Mumma et al.,* 2001b) and may be relevant to the mystery of the form of cometary organic gas-phase molecules. The detection of phenantrene ($C_{14}H_{10}$), a gas-phase PAH macromolecule, in Halley is suggested based on UV spectroscopic data (*Moreels et al.,* 1994). The emission band at 3.28 μm in a few comets suggests the presence of aromatics (*Bockelée-Morvan et al.,* 1995; *Colangeli et al.,* 2004); the unambiguous detection of the 3.28-μm feature in moderate resolution IR spectra requires the deconvolution of emission from gas species in the same spectral range. No PAHs have been detected in the ISO IR spectra of C/1995 O1 Hale-Bopp (*Crovisier et al.,* 1997, *Crovisier,* 1999), which may be due to the large heliocentric distance at the time of observation. Polycyclic aromatic hydrocarbons are very strong absorbers of UV radiation and emitters of IR photons, and as such, represent the observable "tip of the iceberg" of cometary organics.

Chondritic porous IDPs are cometary grains that contain both presolar and solar isotopic materials. [For in-depth discussions on the connection between properties of CP IDPs and comets, see *Hanner and Bradley* (2004), *Sykes et al.* (2004), *Wooden* (2002), and *Messenger and Walker* (1998).] Chondritic porous IDPs have high entrance velocities relative to asteroidal particle trajectories (into Earth's stratosphere where they are collected by high-flying aircraft), high porosity and fragility (indicative of minimal processing in the solar nebula), high Mg contents (in comparison with meteoritic materials), contain D-rich organic material (*Keller et al.,* 2000, 2002) and isotopic anomalies in N and C in the C phase (*Messenger,* 2000), and contain highly radiation damaged amorphous silicate spherules (GEMS) in a C-rich matrix (*Bradley et al.,* 1999a). In fact, most of the CP IDPs have remarkably high C abundances, typically several times higher than those of CI chondrites. Carbon is so abundant that it can be directly observed in ultramicrotome sections where it is often seen as regions of pure amorphous C covering areas as large as 1 μm across (*Brownlee et al.,* 2002). Infrared spectroscopy of the ~3.4-μm CH-stretching region in CP IDPs shows the presence of aliphatic hydrocarbons (*Brownlee et al.,* 2000; *Flynn et al.,* 2002). Recent laboratory data show that the abundant organic materials, specifically the aliphatic and aromatic materials, in the CP IDP "Dragonfly" are responsible for the extremely high D/H relative to terrestrial material and indicate a presolar origin (*Keller et al.,* 2002).

In this same "Dragonfly" particle, the CO carbonyl stretch is observed (*Flynn et al.,* 2002), indicating the possible presence of functional groups associated with aldehydes, ketones, and organic acids.

Amorphous C in comets and CP IDPs is likely of interstellar origin. *In situ* measurements of Comet Halley reveal both discrete and mixed mineral assemblages. Of the total grains, ~8–10% by mass is elemental C that is of AGB origin based on its $^{13}C/^{12}C$ ratio (*Fomenkova,* 1999). Amorphous C is invoked to fit the near-IR emission in comets (e.g., *Hanner et al.,* 1994; *Harker et al.,* 2002), although highly absorbing organic refractory mantles on silicate cores can also produce this observed emission in some comets (*Greenberg and Hage,* 1990; *Li and Greenberg,* 1998b). Of Halley's particles, the elemental C component best represents the amorphous C that is abundant in CP IDPs and that is invoked to produce the near-IR thermal emission from comets [20% by mass of the submicrometer grains in Hale-Bopp (*Harker et al.,* 2002)]. Models of the near-IR reflectance spectra of Centaurs and Kuiper belt objects utilize, by number, an amorphous C abundance of 1–20% (D. Cruikshank, personal communication, 2003). Elemental C is, however, ≪1% by mass in carbonaceous chondrites and 3–5% of aqueously altered primitive meteorites (*Brearley and Jones,* 1998). The depletion of amorphous C in inner solar system bodies relative to outer solar system bodies may be a result of the oxygenation of C into CO and CO_2 in the chondrule-formation process (*Ash et al.,* 1998; *Wooden,* 2002; *Cuzzi et al.,* 2003).

Comparing the soluble fraction of carbonaceous meteorites with cometary volatiles indicates that CI chondrites can be strongly related to comets. From the analyses of amino acids in different meteorites it was recently concluded that the formation of an extensive number of amino acids made through processes such as Strecker-Cyanohydrin synthesis and Michael addition, as observed in the Murchison meteorite, requires the presence of a number of aldehydes and ketones, as well as ammonia, water, HC_3N, and HCN (*Ehrenfreund et al.,* 2001). All the small molecules required to make amino acids are in the current inventory of cometary volatiles. However, no ketones and only formaldehyde and acetaldehyde (0.02%) are detected in comets (*Bockelée-Morvan et al.,* 2004). The low number of amino acids and peculiar abundance ratio in CI chondrites (such as Orgueil) is more compatible with cometary chemistry than with chemistry on asteroidal bodies (*Ehrenfreund et al.,* 2001).

In the ISM, the distribution of C is still an unsolved question. In dense molecular cloud material a reasonable fraction of C is incorporated into CO gas (20%) and a small percentage (~5%) is present in C-bearing ice species (discussed in section 2.1). Diffuse interstellar clouds are exposed to UV radiation and show very low levels of CO gas. In such environments about 15% of the cosmic C is attributed to PAHs. Polycyclic aromatic hydrocarbons seem to be prevalent in the diffuse interstellar medium and on the edges of molecular clouds but may or may not be present within molecular clouds. This leaves a large fraction (>50%) of the cosmic C unaccounted for in the interstellar medium.

Laboratory simulations in combination with interstellar observations argue that this missing C is incorporated into solid-state macromolecular C (cf. Fig. 17 of *Pendleton and Allamandola,* 2002) such as amorphous and hydrogenated amorphous C (see *Colangeli et al.,* 2004; *Ehrenfreund and Charnley,* 2000; *Mennella et al.,* 1998).

Though many different forms of C have been discussed, hydrogenated (and dehydrogenated) amorphous C provide currently the best fit to observations of the UV bump at 220 nm in the interstellar extinction curve and simultaneously the best quantitative solution for current dust models (*Mennella et al.,* 1998). Note that graphite has been popularly invoked in the past to explain the UV 220-nm bump (*Hoyle and Wickramasinghe,* 1999; *Li and Draine,* 2001); graphite does not comprise a significant mass fraction of CP IDPs (D. Brownlee, personal communication, 2003).

Carbonaceous chondrites are known to contain a substantial amount of C, up to 3% by weight, and exhibit a range of thermal and aqueous alteration believed to have occurred on their parent bodies. The major part of this C, namely up to 90%, corresponds to a macromolecular organic fraction (*Hayes,* 1967). Solid-state ^{13}C nuclear magnetic resonance (NMR) investigation of the macromolecular material reveals a high level of branching of the aliphatic chains and shows that the aromatic units are highly substituted, especially in the Murchison meteorite (*Gardinier et al.,* 2000). Given that macromolecular C is inferred to constitute more than half the C in the interstellar medium, constitutes ~ 80% of the C in meteorites, and is present in comets, there probably exists a lineage between these reservoirs. Many small (organic) molecules observed in cometary comae originate wholly or partially from the decomposition of much larger molecules/particles. There is at present, however, no direct or indirect evidence that large polymers similar to those possibly present in comets (e.g., POM, PACM, HCN-polymers) reside in the interstellar medium. Observations of interstellar organic compounds, laboratory simulations, and the analysis of extraterrestrial samples offer insights into the molecular forms of the cometary C. We await future *in situ* measurements and the return of cometary samples by the *Stardust* mission to give a more conclusive answer.

3. CONCLUSIONS

The composition of comets provides important clues on the processes that occurred during the formation of our solar system. However, comets certainly evolve chemically and physically as they visit the inner solar system frequently and they are exposed to processing in their storage location. Any proposed similarities between interstellar and cometary material can be tested by astronomical observations, laboratory simulations, and the analysis of extraterrestrial samples, such as IDPs and carbonaceous chondrites. Comparison of interstellar ice abundances to cometary volatiles shows a possible link between them. The physical properties of ice,

the main component of comets, depend on the structural parameters of the material, such as porosity, grain size, material strength, and local density. Organic molecules may act as glue within ice-dust mixtures that also enhances material strength and thermal conductivity. Sublimation of material from cometary nuclei is triggered by a complex system of internal processes (*Prialnik et al.,* 2004) and, even if these molecular ices were present in the same relative proportions as interstellar ices, it is unlikely that the cometary outgassing pattern would accurately reflect this.

The infrared signatures of silicates in comets and in circumstellar regions, as well as analyses of isotopic ratios in IDP silicates, have greatly improved our knowledge of dust in the early solar system and its link to that of comets. The existence of crystalline silicates in comets, as revealed by cometary IR spectra and their apparent scarcity in the ISM, has been invoked as an argument for mixing of "high-temperature" and "low-temperature" materials in the comet-forming zone. In particular, the appearance of Mg-rich crystalline silicates in some comets indicates that these grains originally condensed at high temperatures or were annealed from amorphous silicates at somewhat lower temperatures in the solar nebula. This is proof that processed nebular materials were incorporated into comets.

The overall picture shows that comets are a mixture of interstellar and nebular components (see Fig. 2). Based on the molecular data available in sample comet populations, there is evidence both for chemical heterogeneity, as expected, and also for some degree of chemical homogeneity. Cometary comae are now known to contain many molecules identified in the interstellar medium; the majority of these species emanate from the nuclear materials. These facts are highly suggestive of cometary nuclei containing appreciable fractions of pristine or partially modified interstellar molecules, but are by no means definitive proof of a direct heritage. Compositional similarities do not provide sufficiently accurate constraints; differences may be more illuminating. Cometary nuclei contain at least two molecules not yet identified in molecular clouds: CS_2 and C_2H_6. It would be of interest to definitively rule out the presence of some common, relatively abundant, interstellar molecules. Current upper limits on possible candidates, such as dimethyl ether, are not stringent enough (*Bockelée-Morvan et al.,* 2004).

The molecular isotopic fractionation, measured in D and heavy N, both suggest an origin in chemistry at very low temperatures. Whether this occurred in cold molecular clouds or in the protosolar nebula cannot be decided based on the available data. It could be argued that the D/H ratios in water and hydrogen cyanide are not particularly strong discriminants of formation site. Detection of cometary molecules with D/H ratios above 10%, as well as evidence for multideuteration, would significantly favor a direct interstellar origin (*Ceccarelli,* 2002); however, the current upper limits on D/H in cometary formaldehyde and methanol (Table 3 of *Bockelée-Morvan et al.,* 2004), although meager,

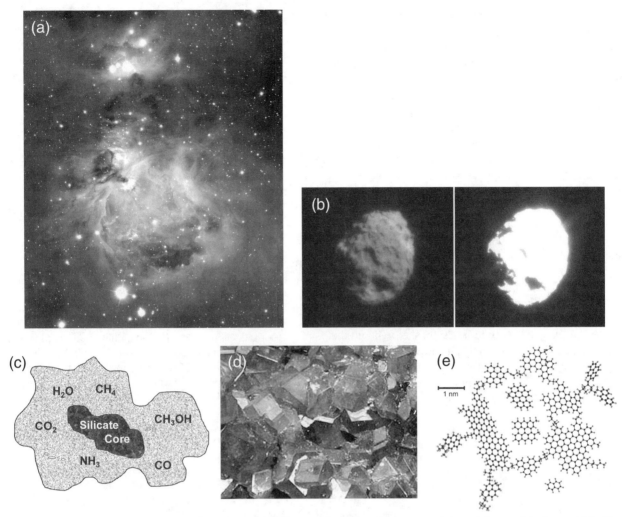

Fig. 2. Comets and their interstellar heritage. **(a)** Interstellar clouds collapse to form stars and protoplanetary disks from which planets, comets and asteroids form. Ices, amorphous silicates, and organics can be observed in dense and diffuse clouds. In protoplanetary disks the infalling material is modified according to the distance from the protostar. Ultraviolet radiation and cosmic rays, shocks, and turbulent mixing alter the original interstellar material before planetary and cometary formation. Image of Orion, courtesy of Robert Gendler. **(b)** Comets are a mixture of interstellar and nebular components; the degree of mixing may be individual for each comet. Their composition, as inferred from observations and the Comet Halley flyby missions, indicate ~50% ice (predominantly water), 25% silicates (amorphous and crystalline), and organic refractory material. Image of Comet Wild 2 Stardust encounter, courtesy of the Stardust Team, Jet Propulsion Laboratory/NASA. **(c)** Interstellar icy grains are characterized by different ice phases — amorphous, crystalline, segregated boundary layers, and possibly clathrates — and besides water, contain partly substantial amounts of CO_2, CO, and CH_3OH. Sublimation of material from cometary nuclei is triggered by a complex system of internal processes and, even if these molecular ices were present in the same relative proportions as interstellar ices, it is unlikely that the cometary outgassing pattern would accurately reflect this. **(d)** Cometary amorphous silicates are of probable interstellar origin while crystalline silicates condensed at high temperatures or were annealed from amorphous silicates at somewhat lower temperatures in the solar nebula. Image courtesy of S. Balm. **(e)** Macromolecular carbon is inferred to constitute more than half of the carbon in the interstellar medium and ~80% of the carbon in meteorites. A lineage between those reservoirs is expected, and such material should also exist in comets. Credit: *Pendleton and Allamandola* (2002).

suggest otherwise. It is therefore of considerable importance to know if low D/H fractionation ratios, of at most a few percent, are common in a statistically larger sample of comets.

Future scientific endeavors aimed at improving our knowledge of the interstellar/solar system connection will include ground- and spacebased telescopes using new and sensitive instrumentation. Such observations should focus on the isotopic ratios in comets, the detection of more cometary organics, and the compilation of detailed chemical inventories of solar-type star-forming regions and their corresponding disks. Theoretical models of the chemical processes associated with the accretion of molecular cloud material and radial mixing in the early solar nebula are vital for determining the starting material from which comets formed. Theoretical models of comet nuclei should aim to reproduce the observed pattern in order to deduce internal properties of comet nuclei that are inaccessible to observations (*Prialnik et al.,* 2004, and Fig. 13 therein). As emphasized by *Colangeli et al.* (2004), laboratory simulations on mixtures of refractory matter and ice will offer a new perspective in the interpretation of cometary observations.

The analysis of extraterrestrial materials (in particular carbonaceous chondrites) remains a crucial method to study refractory matter, including silicates and organics. Isotopic data remain the most important tool to establish a link between regions in interstellar clouds and small solar system bodies. Recent isotopic measurements have allowed the identification of presolar silicates in IDPs; their absence in meteorites indicates that they could not survive parent-body alteration (*Messenger et al.,* 2002). New sensitive techniques, such as the nanoSIMS ion microprobe, can probe isotopic ratios within tiny particles and will contribute significantly to our knowledge of presolar materials. Future observations of more comets will enable a broader consensus on cometary diversity to be established. The ultimate goal of understanding cometary physics and chemistry, and their relation to the parent interstellar cloud, will be attained by future space missions performing *in situ* experiments and possibly bringing a cometary sample back to Earth.

Acknowledgments. Theoretical astrochemistry at NASA Ames Research Center (S.B.C.) is supported by NASA's Exobiology, Planetary Atmospheres, and Origins of Solar Systems Programs through funds allocated by NASA Ames under Interchange No. NCC2-1412. P.E. is supported by VI/NWO, SRON, and ESA. We thank F. Molster for preparation of Fig. 1. We are grateful to L. Colangeli, J. Crovisier, M. Hanner, W. Irvine, and D. Prialnik for comments and discussion.

REFERENCES

A'Hearn M. F., Millis R. L., Schleicher D. G., Osip D. J., and Birch P. V. (1995) The ensemble properties of comets: Results from narrowband photometry of 85 comets, 1976–1992. *Icarus, 118,* 223–270.

Aikawa Y. and Herbst E. (1999a) Molecular evolution in protoplanetary disks. Two-dimensional distributions and column densities of gaseous molecules. *Astron. Astrophys., 351,* 233–246.

Aikawa Y. and Herbst E. (1999b) Deuterium fractionation in protoplanetary disks. *Astrophys. J., 526,* 314–326

Aikawa Y., Umebayashi T., Nakano T., and Miyama S. (1999) Evolution of molecular abundances in protoplanetary disks with accretion flow. *Astrophys. J., 519,* 705–725.

Allamandola L. J., Bernstein M. P., and Sandford S. A. (1997) Photochemical evolution of interstellar/precometary organic material. In *Astronomical and Biochemical Origins and the Search for Life in the Universe* (C. B. Cosmovici et al., eds.), pp. 23–47. Editrice Compositori, Bologna.

Arpigny C., Jehin E., Manfroid J., Hutesmékers D., Zucconi J.-M., Schulz R., and Stüwe J. A. (2003) Anomalous nitrogen isotope ratio in comets. *Science, 301,* 1522–1524.

Ash R. D., Connolly H. C. Jr., Alexander C. M. O'D., Macpherson G. J., and Rumble D. III (1998) Oxygen isotope ratios of natu-

ral and synthetic chondrules: Evidence for insitu reduction by carbon. *Meteoritics & Planet. Sci., 33,* A11.

Balsiger H., Altwegg K., and Geiss J. (1995) D/H and O^{-18}/O^{-16} ratio in the hydronium ion and in neutral water from in situ ion measurements in comet Halley. *J. Geophys. Res., 100,* 5827–5834.

Bergin Edwin A., Alves J., Huard T., and Lada C. J. (2002) N_2H_+ and $C^{18}O$ depletion in a cold dark cloud. *Astrophys. J. Lett., 570,* L101–L104.

Blake G. A., Sutton E. C., Masson C. R., and Phillips T. G. (1987) Molecular abundances in OMC-1 — The chemical composition of interstellar molecular clouds and the influence of massive star formation. *Astrophys. J., 315,* 621–645.

Blake G. A., Qi C., Hogerheijde M. R., Gurwell M. A., and Muhleman D. O. (1999) Sublimation from icy jets as a probe of the interstellar volatile content of comets. *Nature, 398,* 213–216.

Bockelée-Morvan D., Brooke T. Y., and Crovisier J. (1995) On the origin of the 3.2 to 3.6-micron emission features in comets. *Icarus, 116,* 18–39.

Bockelée-Morvan D., Lis D. C., Wink J. E., Despois D., Crovisier J., Bachiller R., Benford D. J., Biver N., Colom P., Davies J. K., Gérard E., Germain B., Houde M., Mehringer D., Moreno R., Paubert G., Phillips T. G., and Rauer H. (2000) New molecules found in comet C/1995 O1 (Hale-Bopp). Investigating the link between cometary and interstellar material. *Astron. Astrophys., 353,* 1101–1114.

Bockelée-Morvan D., Gautier D., Hersant F., Hure J.-M., and Robert F. (2002) Turbulent radial mixing in the solar nebula as the source of crystalline silicates in comets. *Astron. Astrophys., 384,* 1107–1118.

Bockelée-Morvan D., Crovisier J., Mumma M. J., and Weaver H. A. (2004) The composition of cometary volatiles. In *Comets II* (M. C. Festou et al., eds.), this volume. Univ. of Arizona, Tucson.

Boehnhardt H., Fechtig H., and Vanysek V. (1990) The possible role of organic polymers in the structure and fragmentation of dust in the coma of comet P/Halley. *Astron. Astrophys., 231,* 543–547.

Boogert A. C. A., Tielens A. G. G. M., Ceccarelli C., Boonman A. M. S., van Dishoeck E. F., Keane J. V., Whittet D. C. B., and de Graauw Th. (2000) Infrared observations of hot gas and cold ice toward the low mass protostar Elias 29. *Astron. Astrophys., 360,* 683–698.

Boogert A. C. A., Blake G. A., and Tielens A. G. G. M. (2002) High-resolution 4.7 micron Keck/NIRSPEC spectra of protostars. II. Detection of the ^{13}CO isotope in icy grain mantles. *Astrophys. J., 577,* 271–280.

Boss A. (2004) From molecular clouds to circumstellar disks. In *Comets II* (M. C. Festou et al., eds.), this volume. Univ. of Arizona, Tucson.

Bouwman J., Meeus G., de Koter A., Hony S., Dominik C., and Waters L. B. F. M. (2001) Processing of silicate dust grains in Herbig Ae/Be systems. *Astron. Astrophys., 375,* 950–962.

Bradley J. P. (1988) Analysis of chondritic interplanetary dust thinsections. *Geochim. Cosmochim. Acta, 52,* 889–900.

Bradley J. P. (1994) Chemically anomalous, pre-accretionally irradiated grains in interplanetary dust from comets. *Science, 265,* 925–929.

Bradley J. P., Keller L. P., Snow T. P., Hanner M. S., Flynn G. J., Gezo J. C., Clemett S. J., Brownlee D. E., and Bowey J. E. (1999a) An infrared spectral match between GEMS and interstellar grains. *Science, 285,* 1716–1718.

Bradley J. P., Keller L. P., Gezo J., Snow T., Flynn G. J., Brownlee D. E., and Bowey J. (1999b) The 10 and 18 micrometer silicate features of GEMS: Comparison with astronomical silicates (abstract). In *Lunar and Planetary Science XXX,* Abstract #1835. Lunar and Planetary Institute, Houston (CD-ROM).

Bradley J. P., Snow T. P., Brownlee D. E., and Hanner M. S. (1999c) Mg-rich olivine and pyroxene grains in primitive meteoritic materials: Comparison with crystalline silicate data from ISO. In *Solid Interstellar Matter: The ISO Revolution, Les Houches Workshop,* February 2–6, 1998 (L. d'Hendecourt et al., eds.), p. 298. EDP Sciences and Springer-Verlag, Berlin.

Brearley A. J. and Jones R. H. (1998) Chondritic meteorites. In *Planetary Materials: Reviews in Mineralogy, Vol. 36* (J. J. Papike, ed.), pp. 3.1 to 3.370. Mineralogical Society of America, Washington, DC.

Brown P. D., Charnley S. B., and Millar T. J. (1988) A model of the chemistry in hot molecular cores. *Mon. Not. R. Astron. Soc., 231,* 409–417.

Brownlee D. E., Joswiak D. J., Bradley J. P., Gezo J. C., and Hill H. G. M. (2000) Spatially resolved acid dissolution of IDPs: The state of carbon and the abundance of diamonds in the dust (abstract). In *Lunar and Planetary Science XXXI,* Abstract #1921. Lunar and Planetary Institute, Houston (CD-ROM).

Brownlee D. E., Joswiak D. J., Kress M. E., Matrajt G., Taylor S., and Bradley J. (2002) Carbonaceous matter in microgram and nanogram meteoroids. *Meteoritics & Planet. Sci., 37,* A26.

Cameron A. G. W. (1995) The first ten million years in the solar nebula. *Meteoritics, 30,* 133–161.

Capria M. T. (2002) Sublimation mechanisms of comet nuclei. *Earth Moon Planets, 89,* 161–177.

Carrez P., Demyk K., Cordier P., Gengembre L., Grimblot J., d'Hendecourt L., Jones A., and Leroux H. (2002) Low-energy helium ion irradiation-induced amorphization and chemical changes in olivine: Insights for silicate dust evolution in the interstellar medium. *Meteoritics & Planet. Sci., 37,* 1599–1614.

Cassen P. and Moosman A. (1981) On the formation of protostellar disks. *Icarus, 48,* 353–376.

Ceccarelli C. (2002) Millimeter and infrared observations of deuterated molecules. *Planet. Space Sci., 50,* 1267–1273.

Charnley S. B. (1997) On the nature of interstellar organic chemistry. In *Astronomical and Biochemical Origins and the Search for Life in the Universe* (C. B. Cosmovici et al. eds.), pp. 89–96. Editrice Compositori, Bologna.

Charnley S. B. and Kaufman M. J. (2000) Carbon dioxide in starforming regions. *Astrophys. J. Lett., 529,* L111–L114.

Charnley S. B. and Rodgers S. D. (2002) The end of interstellar chemistry as the origin of nitrogen in comets and meteorites. *Astrophys. J. Lett., 569,* L133–L137.

Chick K. and Cassen P. (1997) Thermal processing of interstellar dust grains in the primitive solar environment. *Astrophys. J., 477,* 398–409.

Ciesla F. J., Lauretta D. S., Cohen B. A., and Hood L. L. (2003) A nebular origin for chondritic fine-grained phyllosilicates. *Science, 299,* 549–552.

Cochran A. L. (2002) A Search for N_2^+ in Spectra of Comet C/2002 C1 (Ikeya-Zhang). *Astrophys. J. Lett., 576,* L165–L168.

Cochran A. L., Cochran W. D., and Barker E. S. (2000) N_2^+ and CO^+ in Comets 122P/1995 S1 (deVico) and C/1995 O1 (Hale-Bopp). *Icarus, 146,* 583.

Colangeli L., Brucato J. R., Bar-Nun A., Hudson R. L., and Moore M. H. (2004) Laboratory experiments on cometary materials. In *Comets II* (M. C. Festou et al., eds.), this volume. Univ. of Arizona, Tucson.

Cottin H., Szopa C., and Moore M. H. (2001a) Production of hexamethylenetetramine in photolyzed and irradiated interstellar cometary ice analogs. *Astrophys. J. Lett., 561,* L139–L142.

Cottin H., Gazeau M. C., Benilan Y., and Raulin F. (2001b) Polyoxymethylene as parent molecule for the formaldehyde extended source in Comet Halley. *Astrophys. J., 556,* 417–420.

Cronin J. R. and Chang S. (1993) Organic matter in meteorites: Molecular and isotopic analysis of the Murchison meteorite. In *The Chemistry of Life's Origins* (J. M. Greenberg et al., eds.), pp. 209–258. Kluwer, Dordrecht.

Crovisier J. (1999) Infrared observations of volatile molecules in Comet Hale-Bopp. *Earth Moon Planets, 79,* 125–143.

Crovisier J. and Bockelée-Morvan D. (1999) Remote observations of the composition of cometary volatiles. *Space Sci. Rev., 90,* 19–32.

Crovisier J., Leech K., Bockelée-Morvan D., Brooke T. Y., Hanner M. S., Altieri B., Keller H. U., and Lellouch E. (1997) The spectrum of Comet Hale-Bopp (C/1995 O1) observed with the Infrared Space Observatory at 2.9 AU from the Sun. *Science, 275,* 1915–1918.

Cuzzi J., Davis S. S., and Dobrovolskis A. R. (2003) Radial drift, evaporation, and diffusion: Enhancement and redistribution of silicates, water, and other condensibles in the nebula (abstract). In *Lunar and Planetary Science XXXV,* Abstract #2702. Lunar and Planetary Institute, Houston (CD-ROM).

Dartois E., Schutte W. A., Geballe T. R., Demyk K., Ehrenfreund P., and d'Hendencourt L. (1999) Methanol: The second most abundant ice species towards the high-mass protostars RAFGL7009S and W 33A. *Astron. Astrophys., 342,* L32–L35.

Dartois E., Thi W.-F., Geballe T. R., Deboffle D., d'Hendecourt L., and van Dishoeck E. (2003) Revisiting the solid HDO/H_2O abundances. *Astron. Astrophys., 399,* 1009–1020.

Delany J. S. (1995) Nonthermal initiation of nucleation and chondrule texture development. *Meteoritics, 30,* 501.

DiSanti M. A., Mumma M. J., Russo N. D., and Magee-Sauer K. (2001) Carbon monoxide production and excitation in Comet C/1995 O1 (Hale-Bopp): Isolation of native and distributed CO sources. *Icarus, 153,* 361–390.

Dominik C. and Tielens A. G. G. M. (1997) The physics of dust coagulation and the structure of dust aggregates in space. *Astrophys. J., 480,* 647–673.

Dones L., Weissman P. R., Levison H. F., and Duncan M. J. (2004) Oort cloud formation and dynamics. In *Comets II* (M. C. Festou et al., eds.), this volume. Univ. of Arizona, Tucson.

Drouart A., Dubrulle B., Gautier D., and Robert F. (1999) Structure and transport in the solar nebula from constraints on deuterium enrichment and giant planets formation. *Icarus, 40,* 129–155.

Dutrey A, Lecavelier des Etangs A., and Augereau J.-C. (2004) The observation of circumstellar disks: Dust and gas components. In *Comets II* (M. C. Festou et al., eds.), this volume. Univ. of Arizona, Tucson.

Ehrenfreund P. and Charnley S. B. (2000) Organic molecules in the interstellar medium, comets, and meteorites: A voyage from dark clouds to the early Earth. *Annu. Rev. Astron. Astrophys., 38,* 427–483.

Ehrenfreund P., Kerkhof O., Schutte W. A., Boogert A. C. A., Gerakines P. A., Dartois E., D'Hendecourt L., Tielens A. G. G. M., van Dishoeck E. F., and Whittet D. C. B. (1999) Laboratory studies of thermally processed H_2O-CH_3OH-CO_2

ice mixtures and their astrophysical implications. *Astron. Astrophys., 350,* 240–253.

Ehrenfreund P., Glavin D. P., Botta O., Cooper G., and Bada J. L. (2001) *Proc. Natl. Acad. Sci., 98,* 2138–2141.

Ehrenfreund P. and 12 colleagues (2002) Astrophysical and astrochemical insights into the origin of life. *Rept. Progr. Phys., 65,* 1427–1487.

Ehrenfreund P., Fraser H. J., Blum J., Cartwright J. H. E., García-Ruiz J. M., Hadamcik E., Levasseur-Regourd A. C., Price S., Prodi F., and Sarkissian A. (2003) Physics and chemistry of icy particles in the universe: Answers from microgravity. *Planet. Space Sci., 51,* 473–494.

Fabian D., Jäger C., Henning Th., Dorschner J., and Mutschke H. (2000) Steps towards interstellar silicate mineralogy. V. Thermal evolution of amorphous magnesium silicates and silica. *Astron. Astrophys., 364,* 282–292.

Fegley B. (1999) Chemical and physical processing of presolar materials in the solar nebula and the implications for preservation of presolar materials in comets. *Space Sci. Rev., 90,* 239–252.

Fegley B. (2000) Kinetics of gas-grain reactions in the solar nebula. *Space Sci. Rev., 92,* 177–200.

Fegley B. and Prinn R. G. (1989) Solar nebula chemistry — Implications for volatiles in the solar system. In *The Formation and Evolution of Planetary Systems* (H. A. Weaver and L. Danly, eds.), pp. 171–205. Cambridge Univ., New York.

Feldman P. D., Cochran A. L., and Combi M. R. (2004) Spectroscopic investigations of fragment species in the coma. In *Comets II* (M. C. Festou et al., eds.), this volume. Univ. of Arizona, Tucson.

Festou M. C., Rickman H., and West R. M. (1993a) Comets. I – Concepts and observations. *Astron. Astrophys. Rev., 4,* 363–447.

Festou M. C., Rickman H., and West R. M. (1993b) Comets. 2: Models, evolution, origin and outlook. *Astron. Astrophys. Rev., 5,* 37–163.

Festou M. C., Keller H. U., and Weaver H. A. (2004) A brief conceptual history of cometary science. In *Comets II* (M. C. Festou et al., eds.), this volume. Univ. of Arizona, Tucson.

Finocchi F. and Gail H.-P. (1997) Chemical reactions in protoplanetary accretion disks. III. The role of ionisation processes. *Astron. Astrophys., 327,* 825–844.

Flynn G. J., Keller L. P., Joswiak D., and Brownlee D. E. (2002) Infrared analysis of organic carbon in anhydrous and hydrated interplanetary dust particles: FTIR identification of carbonyl ($C=O$) in IDPs (abstract). In *Lunar and Planetary Science XXXIII,* Abstract #1320. Lunar and Planetary Institute, Houston (CD-ROM).

Fomenkova M. N. (1997) Organic components of cometary dust. In *From Stardust to Planetesimals* (Y. J. Pendelton and A. G. G. M. Tielens, eds.), pp. 415–421. ASP Conference Series 122, Astronomical Society of the Pacific, San Francisco.

Fomenkova M. N. (1999) On the organic refractory component of cometary dust. *Space Science Rev., 90,* 109–114.

Gail H.-P. (2001) Radial mixing in protoplanetary accretion disks. I. Stationary disc models with annealing and carbon combustion. *Astron. Astrophys., 378,* 192–213.

Gail H.-P. (2002) Radial mixing in protoplanetary accretion disks. III. Carbon dust oxidation and abundance of hydrocarbons in comets. *Astron. Astrophys., 390,* 253–265.

Gardinier A., Derenne S., Robert F., Behar F., Largeau C., and Maquet J. (2000) Solid state CP/MAS ^{13}C NMR of the insolu-

ble organic matter of the Orgueil and Murchison meteorites: Quantitative study. *Earth Planet Sci. Lett., 184,* 9–21.

Geiss J., Altwegg K., Balsiger H., and Graf S. (1999) Rare atoms, molecules and radicals in the coma of P/Halley. *Space Sci. Rev., 90,* 253–268.

Gibb E., Whittet D. C. B., Schutte W. A., Chiar J., Ehrenfreund P., Gerakines P. A., Keane J. V., Tielens A. G. G. M., van Dishoeck E. F., and Kerkhof O. (2000) An inventory of interstellar ices toward the embedded protostar W33A. *Astrophys. J., 536,* 347–356.

Glassgold A. E., Najita J. and Igea J. (1997) X-ray ionization of protoplanetary disks. *Astrophys. J., 480,* 344–350.

Goldsmith P. F., Langer W. D., and Velusamy T. (1999) Detection of methanol in a class 0 protostellar disk. *Astrophys. J. Lett., 519,* L173–L176.

Greenberg J. M. (1982) What are comets made of — A model based on interstellar dust. In *Comets* (L. L Wilkening, ed.), pp. 131–162. Univ. of Arizona, Tucson.

Greenberg J. M. (1998) Making a comet nucleus. *Astron. Astrophys., 330,* 375–380.

Greenberg J. M. and Hage J. I. (1990) From interstellar dust to comets — A unification of observational constraints. *Astrophys. J., 361,* 260–274.

Grossman L. (1972) Condensation in the primitive solar nebula. *Geochim. Cosmochim. Acta, 39,* 47–64.

Grün E., Gebhard J., Bar-Nun A., Benkhoff J., Dueren H., Eich G., Hische R., Huebner W. F., Keller H. U., and Klees G. (1993) Development of a dust mantle on the surface of an insolated ice-dust mixture — Results from the KOSI-9 experiment. *J. Geophys. Res., 98,* 15091–15104.

Hadamcik E. and Levasseur-Regourd A. C. (2003) Dust evolution of comet C/1995 O1 (Hale-Bopp) by imaging polarimetric observations. *Astron. Astrophys., 403,* 757–768.

Hage J. I. and Greenberg J. M. (1990) A model for the optical properties of porous grains. *Astrophys. J., 361,* 251–259.

Hallenbeck S. L., Nuth J. A., and Daukantas P. L. (1998) Mid-infrared spectral evolution of amorphous magnesium silicate smokes annealed in vacuum: Comparison to cometary spectra. *Icarus, 131,* 198–209.

Hallenbeck S. L., Nuth J. A. III, and Nelson R. N. (2000) Evolving optical properties of annealing silicate grains: From amorphous condensate to crystalline mineral. *Astrophys. J., 535,* 247–255.

Hanner M. S. and Bradley J. P. (2004) Chemical and mineralogy of cometary dust. In *Comets II* (M. C. Festou et al., eds.), this volume. Univ. of Arizona, Tucson.

Hanner M. S., Lynch D. K., and Russell R. W. (1994) 8–13 micron spectra of comets and the composition of silicate grains. *Astrophys. J., 425,* 274–285.

Hanner M. S., Lynch D. K., Russell R. W., Hackwell J. A., Kellogg R., and Blaney D. (1996) Mid-infrared spectra of Comets P/Borrelly, P/Faye, and P/Schaumasse. *Icarus, 124,* 344–351.

Harker D. E. and Desch S. (2002) Annealing of silicate dust by nebular shocks at 10 AU. *Astrophys. J. Lett., 565,* L109–L112.

Harker D. E., Wooden D. H., Woodward C. E., and Lisse C. M. (2002) Grain properties of Comet C/1995 O1 (Hale-Bopp). *Astrophys. J., 580,* 579–597.

Hayes J. M. (1967) Organic constituents of meteorites: A review. *Geochim. Cosmochim. Acta, 31,* 1395–1440.

Hayward T. L., Hanner M. S., and Sekanina Z. (2000) Thermal infrared imaging and spectroscopy of Comet Hale-Bopp (C/

1995 O1). *Astrophys. J., 538,* 428–455.

Helbert J., Rauer H., Boice D., and Huebner W. (2002) Observations of the C_2 and C_3 radicals and possible implications for the formation region of comets. In *Asteroids, Comets, Meteors 2002, Abstract Book*, p. 175.

Hoyle F. and Wickramasinghe N. C. (1999) A model for interstellar extinction. *Astrophys. Space Sci., 268,* 263–271.

Iati M. A., Cecchi-Pestallini C., Williams D. A., Borghese F., Denti P., Saiji R., and Aiello S. (2001) Porous interstellar grains. *Mon. Not. R. Astron. Soc., 322,* 749–756.

Iro N., Gautier D., Hersant F., Bockelée-Morvan D., and Lunine J. I. (2003) An interpretation of the nitrogen deficiency in comets. *Icarus, 161,* 511–532.

Irvine W. M. and Lunine J. I. (2004) The cycle of matter in our galaxy: From clouds to comets. In *Comets II* (M. C. Festou et al., eds.), this volume. Univ. of Arizona, Tucson.

Irvine W. M., Bockelée-Morvan D., Lis D. C., Matthews H. E., Biver N., Crovisier J., Davies J. K., Dent W. R. F., Gautier D., Godfrey P. D., Keene J., Lowell A. J., Owen T. C., Phillips T. G., Rauer H., Schloerb F. P., Senay M., and Young K. (1996) Spectroscopic evidence for interstellar ices in Comet Hyakutake. *Nature, 383,* 418–420.

Irvine W. M., Bergin E. A., Dickens J. E., Jewitt D., Lovell A. J., Matthews H. E., Schloerb F. P., and Senay M. (1998) Chemical processing in the coma as the source of cometary HNC. *Nature, 393,* 547–550.

Irvine W. M., Crovisier J., Fegley B., and Mumma M. J. (2000) Comets: A link between interstellar and nebular chemistry. In *Protostars and Planets IV* (V. Mannings et al., eds.), pp. 1159–1200. Univ. of Arizona, Tucson.

Irvine W. M., McGonagle D., Bergman P., Lowe T. B., Matthews H. E., Nummelin A., and Owen T. C. (2003) HCN and HCN in Comets C/2000 WM1 (LINEAR) and C/2002 C1 (Ikeya-Zhang). *Orig. Life Evol. Biosph., 33,* 609–619.

Jäger C., Fabian D., Schrempel F., Dorschner J., Henning Th., and Wesch W. (2003) Structural processing of enstatite by ion bombardment. *Astron. Astrophys., 410,* 57–65.

Jessberger E., Cristoforidis A., and Kissel J. (1988) Aspects of the major element composition of Halley's dust. *Nature, 332,* 691–695.

Jewitt D. C. (2004) From cradle to grave: The rise and demise of the comets. In *Comets II* (M. C. Festou et al., eds.), this volume. Univ. of Arizona, Tucson.

Jewitt D. C., Matthews H. E., Owen T., and Meier R. (1997) Measurements of $^{12}C/^{13}C$, $^{14}N/^{15}N$ and $^{32}S/^{34}S$ ratios in Comet Hale-Bopp (C/1995 O1). *Science, 278,* 90–93.

Keller L. P., Messenger S., Flynn G. J., Jacobsen C., and Wirick S. (2000) Chemical and petrographic studies of molecular cloud materials preserved in interplanetary dust. *Meteoritics & Planet. Sci., 35,* A86.

Keller L. P., Messenger S., Flynn G. J., Wirick S., and Jacobsen C. (2002) Analysis of a deuterium hotspot in an interplanetary dust particle: Implications for the carrier of the hydrogen isotopic anomalies in IDPs (abstract). In *Lunar and Planetary Science XXXIII*, Abstract #1869. Lunar and Planetary Institute, Houston (CD-ROM).

Kemper F., Vriend W. J., and Tielens A. G. G. M. (2004) The absence of crystalline silicates in the diffuse interstellar medium. *Astrophys. J.,* in press.

Kissel J., Krueger F. R., and Roessler K. (1997) Organic chemistry in comets from remote and in situ observations. In *Comets and the Origins and Evolution of Life* (P. J. Thomas et al., eds.),

pp. 69–110. Springer-Verlag, Berlin.

Kochan H. W., Huebner W. F., and Sears D. (1998) Simulation experiments with cometary analogous material. *Earth Moon Planets, 80,* 369–411.

Kracher A., Scott E. R. D., and Keil K. (1984) Relict and other anomalous grains in chondrules — Implications for chondrule formation. *Proc. Lunar Planet. Sci. Conf. 14th*, in *J. Geophys. Res., 89,* B559–B566.

Lahuis F. and van Dishoeck E. F. (2000) ISO-SWS spectroscopy of gas-phase C_2H_2 and HCN toward massive young stellar objects. *Astron. Astrophys., 355,* 699–712.

Langer W. D., Graedel T. E., Frerking M. A., and Armentrout P. B. (1984) Carbon and oxygen isotope fractionation in dense interstellar clouds. *Astrophys. J., 277,* 581–604.

Lecacheux A. and 21 colleagues (2003) Observations of water in comets with Odin. *Astron. Astrophys., 402,* L55–L58.

Levasseur-Regourd A. C., Hadmcik E., and Gaulme P. (2002) Physical properties of dust in comets, as compared to those of asteroidal and interplanetary material. In *Proceedings of Asteroids, Comets, Meteors 2002* (B. Warmbein, ed.), pp. 541–544. ESA SP-500, Noordwijk, The Netherlands.

Li A. and Draine B. T. (2001) Infrared emission from interstellar dust. II. The diffuse interstellar medium. *Astrophys. J., 554,* 773–802.

Li A. and Greenberg J. M. (1997) A unified model of interstellar dust. *Astron. Astrophys., 323,* 566–584.

Li A. and Greenberg J. M. (1998a) The dust properties of a short period comet: Comet P/Borrelly. *Astron. Astrophys., 338,* 364–370.

Li A. and Greenberg J. M. (1998b) From interstellar dust to comets: Infrared emission from Comet Hale-Bopp (C/1995 O1). *Astrophys. J. Lett., 498,* L83–L87.

Loinard L., Castets A., Ceccarelli C., Tielens A. G. G. M., Faur A., Caux, E., and Duvert G. (2000) The enormous abundance of D_2CO in IRAS 16293–2422. *Astron. Astrophys., 359,* 1169–1174.

Lunine J. I. (1989) Primitive bodies: Molecular abundances in Comet Halley as probes of cometary formation environments. In *The Formation and Evolution of Planetary Systems* (H. A. Weaver and L. Danly, eds.), pp. 213–242. Cambridge Univ., New York.

Lunine J. I. (1997) Physics and chemistry of the solar nebula. *Orig. Life Evol. Biosph., 27,* 205–224.

Lunine J. I. and Gautier D. (2004) Coupled physical and chemical evolution of volatiles in the protoplanetary disk. In *Comets II* (M. C. Festou et al., eds.), this volume. Univ. of Arizona, Tucson.

Magee-Sauer K., Dello Russo N., DiSanti M. A., Gibb E. L., and Mumma M. J. (2002) CSHELL observations of Comet C/2002 C1 (Ikeya-Zhang) in the 3.0-micron region. *Bull. Am. Astron. Soc., 34,* 868.

Malfait K., Waelkens C., Waters L. B. F. M., Vandenbussche B., Huygen E., and de Graauw M. S. (1998) The spectrum of the young star HD 100546 observed with the Infrared Space Observatory. *Astron. Astrophys., 332,* L25–L28.

Mannings V., Boss A. P., and Russell S. S., eds. (2000) *Protostars and Planets IV.* Univ. of Arizona, Tucson. 1422 pp.

Markwick A. J. and Charnley S. B. (2004) Physics and chemistry of protoplanetary disks: Relation to primitive solar system material. In *Astrobiology: Future Perspectives* (P. Ehrenfreund et al., eds.), in press. Kluwer, Dordrecht.

Meech K. J. and Svoreň J. (2004) Physical and chemical evolution of cometary nuclei. In *Comets II* (M. C. Festou et al., eds.), this volume. Univ. of Arizona, Tucson.

Meeus G., Waters L. B. F. M., Bouwman J., van den Ancker M. E., Waelkens C., and Malfait K. (2001) ISO spectroscopy of circumstellar dust in 14 Herbig Ae/Be systems: Towards an understanding of dust processing. *Astron. Astrophys., 365,* 476–490.

Mennella V., Colangeli L., Bussoletti E., Palumbo P., and Rotundi A. (1998) A new approach to the puzzle of the ultraviolet interstellar extinction bump. *Astrophys. J., 507,* L177–L180.

Messenger S. (2000) Identification of molecular-cloud material in interplanetary dust particles. *Nature, 404,* 968–971.

Messenger S. and Walker R. M. (1998) Possible association of isotopically anomalous cluster IDPs with Comet Schwassmann-Wachman 3 (abstract). In *Lunar and Planetary Science XXIX,* Abstract #1906. Lunar and Planetary Institute, Houston (CD-ROM).

Messenger S., Keller L. P., and Walker R. M. (2002) Discovery of abundant interstellar silicates in cluster IDPs (abstract). In *Lunar and Planetary Science XXXIII,* Abstract #1887. Lunar and Planetary Institute, Houston (CD-ROM).

Messenger S., Keller L. P., Stadermann F. J., Walker R. M., and Zinner E. (2003) Samples of stars beyond the solar system: Silicate grains in interplanetary dust. *Science, 300,* 105–108.

Molster F. J., Yamamura I., Waters L. B. F. M., Tielens A. G. G. M., de Graauw Th., de Jong T., de Koter A., Malfait K., van den Ancker M. E., Van Winckel H., Voors R. H. M., and Waelkens C. (1999) Low-temperature crystallization of silicate dust in circumstellar disks. *Nature, 401,* 563.

Molster F. J., Waters L. B. F. M., Tielens A. G. G. M., and Barlow M. J. (2002a) Crystalline silicate dust around evolved stars. I. The sample stars. *Astron. Astrophys., 382,* 184–221.

Molster F. J., Waters L. B. F. M., and Tielens A. G. G. M. (2002b) Crystalline silicate dust around evolved stars. II. The crystalline silicate complexes. *Astron. Astrophys., 382,* 222–240.

Molster F. J., Waters L. B. F. M., Tielens A. G. G. M., Koike C., and Chihara H. (2002c) Crystalline silicate dust around evolved stars. III. A correlations study of crystalline silicate features. *Astron. Astrophys., 382,* 241–255.

Moore M. H. and Hudson R. L. (1998) Infrared study of ion-irradiated water-ice mixtures with hydrocarbons relevant to comets. *Icarus, 135,* 518–527.

Moreels G., Clairemidi J., Hermine P., Brechignac P., and Rousselot P. (1994) Detection of a polycyclic aromatic molecule in comet P/Halley. *Astron. Astrophys., 282,* 643–656.

Morfill G. E. and Volk H. J. (1984) Transport of dust and vapor and chemical fractionation in the early protosolar cloud. *Astrophys. J., 287,* 371–395.

Mousis O., Gautier D., and Bockelée-Morvan D. (2002) An evolutionary turbulent model of Saturn's subnebula: Implications for the origin of the atmosphere of Titan. *Icarus, 156,* 162–175.

Mumma M. J., Dello Russo N., DiSanti M. A., Magee-Sauer K., Novak R. E., Brittain S., Rettig T., McLean I. S., Reuter D. C., and Xu Li-H. (2001a) The startling organic composition of C/1999 S4 (LINEAR): A comet formed near Jupiter? *Science, 292,* 1334–1339.

Mumma M. J., McLean I. S., DiSanti M. A., Larkin J. E., Dello Russo N., Magee-Sauer K., Becklin E. E., Bida T., Chaffee F., Conrad A. R., Figer D. F., Gilbert A. M., Graham J. R., Levenson N. A., Novak R. E., Reuter D. C., Teplitz H. I., Wilcox M. K., and Xu Li-H. (2001b) A survey of organic volatile species in Comet C/1999 H1 (Lee) using NIRSPEC at the Keck Observatory. *Astrophys. J., 546,* 1183–1193.

Mumma M. J., DiSanti M. A., Dello Russo N., Magee-Sauer K., Gibb E., and Novak R. (2003) Remote infrared observations of parent volatiles in comets: A window on the early solar system. *Adv. Space Res., 31,* 2463–2575.

Neufeld D. and Hollenbach D. J. (1994) Dense molecular shocks and accretion onto protostellar disks. *Astrophys. J., 428,* 170–185.

Nummelin A., Whittet D. C. B., Gibb E. L., Gerakines P. A., and Chiar J. E. (2001) Solid carbon dioxide in regions of low-mass star formation. *Astrophys. J., 558,* 185–193.

Owen T. C., Mahaffy P. R., Niemann H. B., Atreya S., and Wong M. (2001) Protosolar nitrogen. *Astrophys. J. Lett., 553,* L77–L79.

Pardo J. R., Cernicharo J., Herpin F., Kawamura J., Kooi J., and Phillips T. G. (2001) Deuterium enhancement in water towards Orion IRC2 deduced from HDO lines above 800 GHz. *Astrophys. J., 562,* 799–803.

Parise B., Ceccarelli C., Tielens A. G. G. M., Herbst E., Lefloch B., Caux E., Castets A., Mukhopadhyay I., Pagani L., and Loinard L. (2002) Detection of doubly-deuterated methanol in the solar-type protostar IRAS 16293–2422. *Astron. Astrophys., 393,* L49–L53.

Pendleton Y. J. and Allamandola L. J. (2002) The organic refractory material in the diffuse interstellar medium: Mid-infrared spectroscopic constraints. *Astrophys. J. Suppl., 138,* 75–98.

Pontoppidan K. M., Dartois E., van Dishoeck E. F., Thi W.-F., and d'Hendecourt L. (2003) Detection of abundant solid methanol toward young low mass stars. *Astron. Astrophys., 404,* L17–L20.

Prialnik D. (2002) Modeling the comet nucleus interior. *Earth Moon Planets, 89,* 27–52.

Prialnik D., Benkhoff J., and Podolak M. (2004) Modeling the structure and activity of comet nuclei. In *Comets II* (M. C. Festou et al., eds.), this volume. Univ. of Arizona, Tucson.

Rietmeijer F. J. M. (1998) Interplanetary dust particles. In *Planetary Materials: Reviews in Mineralogy, Vol. 36* (J. J. Papike, ed.), pp. 2.1 to 2.95. Mineralogical Society of America, Washington, DC.

Roberts H., Fuller G. A., Millar T. J., Hatchell J., and Buckle J. V. (2002) A survey of [HDCO]/[H$_2$CO] and [DCN]/[HCN] ratios towards low-mass protostellar cores. *Astron. Astrophys., 381,* 1026–1038.

Rodgers S. D. and Charnley S. B. (1998) HNC and HCN in comets. *Astrophys. J. Lett., 501,* L227–L230.

Rodgers S. D. and Charnley S. B. (2001a) On the origin of HNC in Comet Lee. *Mon. Not. R. Astron. Soc., 323,* 84–92.

Rodgers S. D. and Charnley S. B. (2001b) Organic synthesis in the coma of Comet Hale-Bopp? *Mon. Not. R. Astron. Soc., 320,* L61–L64.

Rodgers S. D. and Charnley S. B. (2002) A model of the chemistry in cometary comae: Deuterated molecules. *Mon. Not. R. Astron. Soc., 330,* 660–674.

Rodgers S. D. and Charnley S. B. (2003) Chemical evolution in protostellar envelopes — Cocoon chemistry. *Astrophys. J., 585,* 355–371.

Rodgers S. D., Charnley S. B., Huebner W. F., and Boice D. C. (2004) Physical processes and chemical reactions in cometary comae. In *Comets II* (M. C. Festou et al., eds.), this volume. Univ. of Arizona, Tucson.

Ruden S. P. and Pollack J. B. (1991) The dynamical evolution of the protosolar nebula. *Astrophys. J., 375,* 740–760.

Schoier F. L., Jorgensen J. K., van Dishoeck E. F., and Blake G. A.

(2002) Does IRAS 16293–2422 have a hot core? Chemical inventory and abundance changes in its protostellar environment. *Astron. Astrophys., 390*, 1001–1021.

Sephton M., Pillinger C. T., and Gilmour I. (2000) Aromatic moieties in meteoritic macromolecular m aterials: Analyses by hydrous pyrolysis and ^{13}C of individual compounds. *Geochim. Cosmochim. Acta, 64*, 321–328.

Shoshany Y., Prialnik D., and Podolak M. (2002) Monte Carlo modeling of the thermal conductivity of porous cometary ice. *Icarus, 157*, 219–227.

Stern S. A. (2003) The evolution of comets in the Oort Cloud and Kuiper Belt. *Nature, 424*, 639–642.

Sykes M. V., Grün E., Reach W. T., and Jenniskens P. (2004) The interplanetary dust complex and comets. In *Comets II* (M. C. Festou et al., eds.), this volume. Univ. of Arizona, Tucson.

Taban I. M., Schutte W. A., Pontoppidan K. M., and van Dishoeck E. F. (2003) Stringent upper limits to the solid NH$_3$ abundance towards W 33A from near-IR spectroscopy with the Very Large Telescope. *Astron. Astrophys., 399*, 169–175.

Teixeira T. C., Devlin J. P., Buch V., and Emerson J. P. (1999) Discovery of solid HDO in grain mantles. *Astron. Astrophys., 347*, L19–L22.

Terebey S., Shu F. H., and Cassen P. (1984) The collapse of the cores of slowly rotating isothermal clouds. *Astrophys. J., 286*, 529–551.

Tielens A. G. G. M., Waters L. B. F. M., Molster F. J., and Justtanont K. (1997) Circumstellar silicate mineralogy. *Astrophys. Space Sci., 255*, 415–426.

Vaidya D. B. and Gupta R. (1999) Interstellar extinction by porous grains. *Astron. Astrophys., 348*, 594–599.

van der Tak F. F. S., Schilke P., Muller H. S. P., Lis D. C., Phillips T. G., Gerin M., and Roueff E. (2002) Triply deuterated ammonia in NGC 1333. *Astron. Astrophys., 388*, L53–L56.

van Dishoeck E. F. and Blake G. A. (1998) Chemical evolution of star-forming regions. *Annu. Rev. Astron. Astrophys., 36*, 317–368.

van Dishoeck E. F., Thi W.-F., and van Zadelhoff G.-J. (2003) Detection of DCO$^+$ in a circumstellar disk. *Astron. Astrophys., 400*, L1–L4.

Watanabe N., Shiraki T., and Kouchi A. (2003) The dependence of H$_2$CO and CH$_3$OH formation on the temperature and thickness of H$_2$O-CO ice during the successive hydrogenation of CO. *Astrophys. J. Lett., 588*, L121–L124.

Weissman P. (1999) Diversity of comets: Formation zones and dynamical paths. *Space Science Rev., 90*, 301–311.

Weissman P. R., Asphaug E., and Lowry S. C. (2004) Structure and density of cometary nuclei. In *Comets II* (M. C. Festou et al., eds.), this volume. Univ. of Arizona, Tucson.

Whipple F. L. (1950) A comet model. I. The acceleration of Comet Encke. *Astrophys. J., 111*, 375–394.

Willacy K. and Langer W. D. (2000) The importance of photoprocessing in protoplanetary disks. *Astrophys. J., 544*, 903–920.

Wooden D. H. (2002) Comet grains: Their IR emission and their relation to ISM grains. *Earth Moon Planets, 89*, 247–287.

Wooden D. H., Butner H. M., Harker D. E., and Woodward C. E. (2000) Mg-rich silicate crystals in Comet Hale-Bopp: ISM relics or solar nebula condensates? *Icarus, 143*, 126–137.

Wooden D. H., Charnley S. B., and Ehrenfreund P. (2004) Composition and evolution of interstellar clouds. In *Comets II* (M. C. Festou et al., eds.), this volume. Univ. of Arizona, Tucson.

Wyckoff S., Kleine M., Peterson B. A., Wehinger P. A., and Ziurys L. M. (2000) Carbon isotope abundances in comets. *Astrophys. J., 535*, 991–999.

Yoneda S. and Grossman L. (1995) Condensation of CaO-MgO-Al$_2$O$_3$-SiO$_2$ liquids from cosmic gases. *Geochim. Cosmochim. Acta, 59*, 3413–3444.

Part III:

The Nature and Evolution of Cometary Orbits

Cometary Orbit Determination and Nongravitational Forces

D. K. Yeomans and P. W. Chodas
Jet Propulsion Laboratory/California Institute of Technology

G. Sitarski, S. Szutowicz, and M. Królikowska
Space Research Centre of the Polish Academy of Sciences

The accuracies of the orbits and ephemerides for active comets are most often limited by imperfectly modeled rocket-like accelerations experienced by active comets as a result of the outgassing cometary nucleus near perihelion. The standard nongravitational acceleration model proposed by *Marsden et al.* (1973) has been updated by allowing the nucleus outgassing to act asymmetrically with respect to perihelion, providing for time-dependent effects through the precession of the cometary nucleus and accounting for the outgassing from discrete surface areas on a rotating nucleus. While the most accurate nongravitational models will likely require a detailed *a priori* knowledge of a comet's surface activity and rotation characteristics, it is becoming possible to use only astrometric data to actually solve for some of the parameters that describe the comet's outgassing and rotational characteristics.

1. ASTROMETRY AND THE ORBIT DETERMINATION PROCESS FOR COMETS

The accuracy of the orbit-determination process for comets depends on a number of factors, including the accuracy of the astrometric data, the interval over which these data are available, the extent to which the position of the photometric image represents a comet's true center-of-mass, the incorporation of planetary and asteroidal perturbations, the accuracy of the numerical integration and differential correction processes, and especially the correct modeling of the accelerations due to the comet's outgassing (i.e., nongravitational effects). These latter nongravitational effects, which are due largely to the recoil effect from the vaporization of water ices from the cometary nucleus, will be discussed in later sections.

1.1. Astrometric Data

Optical astrometric positions for celestial objects most often consist of pairs of right ascension and declination values for a given time; these positions are determined using measured offsets from neighboring stars whose positions have been accurately determined. Thus an object's astrometric accuracy depends upon the measurement technique and the accuracy of the reference star catalog employed. For modern astrometric positions, the highest accuracy is achieved when the star catalog has been reduced with respect to Hipparcos reference star positions. One example of this type of catalog is the star catalog (UCAC2) from the United States Naval Observatory (USNO), which includes more than 60,000,000 stars, thus providing the high star density necessary for most astrometric reductions. This catalog goes as deep as magnitude 16 and has an astrometric accuracy better than 0.1 arcsec (http://ad.usno.navy.mil).

Since cometary observations are influenced by random errors, astrometric observations should be selected and weighted before they are used for orbit improvement. Bielicki has elaborated objective criteria for the selection of cometary observations on the assumption that the distribution of observational errors is normal (*Bielicki and Sitarski,* 1991). Bielicki applied the process of selecting and weighting the observations to those observations belonging to one apparition of the comet and thus he could obtain a value of the mean residual for observations of this apparition. Using the values of the mean residuals found for each of several apparitions, one can calculate in advance the mean residual *a priori*, μ_{apr}, which can then be compared with the mean residual *a posteriori* μ_{apo} that results from the model of the comet's motion when linking a number of apparitions. A mathematical model of the comet's motion used for the linkage of several apparitions may be regarded as satisfactory if $\mu_{apo} \approx \mu_{apr}$.

The most powerful astrometric optical positional data for improving an object's orbit are taken when the object is closest to Earth. Unfortunately, active comets have substantial coma, or atmospheres, in the inner solar system and these atmospheres have optical depths that prevent a direct observation of the comet's nucleus. As a result, the observer normally must assume that the object's photometric center (center-of-light) is also the comet's center-of-mass, and this is not often the case. While linking a long data interval for Comet 26P/Grigg-Skjellerup, *Sitarski* (1991) found it necessary to adjust the observations using a radial offset that varied as the inverse cube of the heliocentric distance. *Yeomans* (1994) outlined a procedure whereby this center-of-mass/center-of-light offset at 1 AU from the Sun was included within the orbit-determination process. This offset (S_o) is assumed to vary along the comet-Sun line with an inverse square dependence on heliocentric distance. For Comet Halley, the solution for the offset at 1 AU was about 850 km. In most cases, this offset is not easily determined from the available astrometric data. Of course, rather than trying to account for inaccurate observations of a comet's center-of-mass, it would be preferable if groundbased optical observations actually observed the comet's true position in the first place. In this regard, *Chesley et al.* (2001) compared the groundbased optical astrometric positions for Comet 19P/Borrelly in September 2001 with spacebased observations of the comet's true nucleus taken by the *Deep Space 1*'s optical imaging cameras prior to its close flyby on September 22, 2001. They concluded that the true nucleus position of the comet was more accurately defined by groundbased observations if the brightest pixel were used rather than positions based upon a best-fitting two-dimensional Gaussian fit to the photometric image. That is, the comet's center-of-light should not be assumed to be the comet's center-of-mass. We recommend that observers use the "brightest pixel" technique for reporting cometary astrometric observations.

While the extraordinary power of radar observations to refine an asteroid's orbit has been documented several times (for example, see *Yeomans and Chodas*, 1987; *Yeomans et al.*, 1992), radar observations are available for only five comets (http://ssd.jpl.nasa.gov/radar_data.html). This paucity of radar observations for comets is primarily due to the infrequency with which comets pass close enough to Earth.

1.2. Cometary Equations of Motion and the Orbit Determination Process

The orbit-determination process is a linearized, weighted least-squares estimation algorithm, in which astrometric observations are used to improve an existing orbit (*Lawson and Hanson*, 1974). At each time step of the numerical integration process, the dynamic model should include gravitational perturbations due to the planets and the larger minor planets, relativistic effects, and the accelerations due to the nongravitational effects. The partial derivatives necessary for adjusting the initial conditions should be integrated along with the object's equations of motion. The cometary equations of motion can be written

$$\frac{d^2\mathbf{r}}{dt^2} = -\frac{k^2\mathbf{r}}{r^3} + k^2\sum_j m_j \left[\frac{(\mathbf{r}_j - \mathbf{r})}{|\mathbf{r}_j - \mathbf{r}|^3} - \frac{\mathbf{r}_j}{r_j^3} \right]$$

$$+ \frac{k^2}{c^2 r^3}\left[\frac{4k^2\mathbf{r}}{r} - (\dot{\mathbf{r}} \cdot \dot{\mathbf{r}})\mathbf{r} + 4(\mathbf{r} \cdot \dot{\mathbf{r}})\dot{\mathbf{r}} \right] \quad (1)$$

$$+ A_1 g(r)\mathbf{r}/r + A_2 g(r)\mathbf{t} + A_3 g(r)\mathbf{n}$$

The first term on the righthand side of the equation is the solar acceleration where the Sun's mass has been taken as

unity (*Marsden et al.*, 1973). The second term represents both the direct effects of the perturbing bodies on the comet and the indirect effects of the perturbing bodies upon the Sun. The perturbing bodies are often taken to be the planets and the three most massive minor planets. The second line of the equation represents one form of the relativistic effects (*Anderson et al.*, 1975) that should be included because for many objects with small semimajor axes and large eccentricities, these effects introduce a nonnegligible radial acceleration toward the Sun (*Sitarski*, 1983, 1992c). In addition, these effects are required to maintain consistency with the planetary ephemeris (*Shahid-Saless and Yeomans*, 1994). The accelerations are given in astronomical units/(ephemeris day)^2; k is the Gaussian constant; m_j = the masses of the planets and Ceres, Pallas, and Vesta; $\mathbf{r}$ and r = |$\mathbf{r}$| are the heliocentric position vector and distance of the comet respectively; r_j = |$\mathbf{r}_j$| are the planetary distances from the center of the Sun; and c is the speed of light in AU per day. The third line of this equation gives the standard-model expressions for the outgassing accelerations acting on the comet in the radial (Sun-comet), transverse, and normal directions. These so-called nongravitational effects are discussed in the following sections.

2. HISTORICAL INTRODUCTION TO NONGRAVITATIONAL EFFECTS

Comet 2P/Encke has played a central role in the historical evolution of ideas concerning the rocket-like thrusting of an outgassing cometary nucleus and the attempts to model these so-called nongravitational accelerations. First discovered by Pierre Mechain in 1786, Comet Encke was rediscovered by Caroline Herschel in 1795 and discovered yet again by Jean-Louis Pons in 1805 and in 1818. Johann Encke provided the numerical computations to show that the comets discovered in 1786, 1795, 1805, and 1818 were one and the same object returning to perihelion at 3.3-year intervals. After noting that the comet returned to perihelion a few hours earlier than his predictions, *Encke* (1823) postulated that the comet moved under the influence of a resisting medium that he envisaged as an extension of the Sun's atmosphere or the debris of cometary and planetary atmospheres remaining in space. Encke's resisting medium allowed him to successfully predict the perihelion returns for the comet between 1825 and 1858.

Although the resisting medium theory seemed to be the contemporary consensus opinion, *Bessel* (1836) noted that a comet expelling material in a radial sunward direction would suffer a recoil force, and if the expulsion of material did not take place symmetrically with respect to perihelion, there would be a shortening or lengthening of the comet's period depending on whether the comet expelled more material before or after perihelion (see equation (2)). Although Bessel did not identify the physical mechanism with water vaporization from the nucleus, his basic concept of cometary nongravitational forces would ultimately prove to be correct.

In the second half of the nineteenth century, the motion of Comet Encke did not seem to behave in strict accordance with Encke's resisting medium hypothesis, and several alternate mechanisms were introduced to explain the phenomena (see *Yeomans, 1991; Sekanina, 1991a*). The final blows to the resisting medium came when *Kamienski* (1933) and *Recht* (1940) found uniform decreases in the mean motion of period Comets 14P/Wolf and 6P/d'Arrest respectively. With the discovery of mean motions that decreased with time, as well as some that increased, a successful hypothesis had to explain both phenomena. A resisting medium could only cause the latter phenomena.

The breakthrough work that allowed a proper modeling of the nongravitational effects on comets came with Whipple's introduction of his icy conglomerate model for the cometary nucleus (*Whipple*, 1950, 1951). Part of his motivation for this model was to explain the so-called nongravitational accelerations that were evident in the motion of Comet Encke and many other active periodic comets. That is, even after all the gravitational perturbations of the planets were taken into account, the observations of many active comets could not be well represented without the introduction of additional so-called nongravitational effects into the dynamical model. These effects are brought about by cometary activity when the sublimating ices transfer momentum to the nucleus. The nongravitational effects become most evident as deviations in a comet's perihelion passage when compared with a purely gravitational orbit, and these deviations are typically a fraction of a day per apparition, although for Comet 1P/Halley it is as large as four days.

2.1. Symmetrical Nongravitational Force Model

Whipple noted that for an active, rotating, icy cometary nucleus, a thermal lag between cometary noon and the time of maximum outgassing would introduce a transverse acceleration into a comet's motion. In an attempt to model these effects, *Marsden* (1968, 1969) first introduced a semi-empirical nongravitational acceleration model using what are now termed Style I nongravitational parameters. Style II parameters were added when *Marsden et al.* (1973) introduced what has become the standard, or symmetric, nongravitational acceleration model for cometary motions; a rotating cometary nucleus is assumed to undergo vaporization from water ice that acts symmetrically with respect to perihelion. That is, at the same heliocentric distance before and after perihelion, the cometary nucleus experiences the same nongravitational acceleration. The expressions for these nongravitational accelerations can be written

$$A_1 g(r)\mathbf{r}/r + A_2 g(r)\mathbf{t}$$

where

$$g(r) = \alpha(r/r_0)^{-m}(1 + (r/r_0)^n)^{-k}$$

The scale distance r_0 is the heliocentric distance inside which

the bulk of solar insolation goes to sublimating the comet's ices. For water ice, $r_0 = 2.808$ AU and the normalizing constant $\alpha = 0.111262$. The exponents m, n, and k equal 2.15, 5.093, and 4.6142 respectively. The nongravitational acceleration is represented by a radial term, $A_1 g(r)$, and a transverse term, $A_2 g(r)$, in the equations of motion. The radial unit vector ($\mathbf{r}/r$) is defined outward along the Sun-comet line, while the transverse unit vector ($\mathbf{t}$) is directed normal to $\mathbf{r}/r$, in the orbit plane, and in the general direction of the comet's motion. An acceleration component normal to the orbit plane, $A_3 g(r)$, is also present for most active comets, but its periodic nature often makes it difficult to determine because we are usually solving for an average nongravitational acceleration effect over three or more apparitions. If the comet's nucleus were not rotating, the outgassing in this model would always be toward the Sun and the resulting nongravitational acceleration would act only in the antisolar direction. The rotation of the nucleus, however, coupled with a thermal lag angle (η) between the nucleus subsolar point and the point on the nucleus where there is maximum outgassing, introduces a transverse acceleration component in either the direction of the comet's motion or contrary to it — depending upon the nucleus rotation direction.

Equation (2) represents the time derivative of the comet's orbital semimajor axis (a) as a result of radial and transverse perturbing accelerations (R_p, T_p)

$$\frac{da}{dt} = \frac{2[(e\sin\nu)R_p + (p/r)T_p]}{n\sqrt{1-e^2}} \qquad (2)$$

In this equation, n, e, ν, and r denote, respectively, the orbital mean motion, eccentricity, true anomaly, and the comet's heliocentric distance, while p is the orbital semilatus rectum, $a(1 - e^2)$.

Because of the thermal lag angle, a comet in direct rotation will have a positive transverse nongravitational acceleration component, and from equation (2), it is apparent that the comet's orbital semimajor axis will increase with time (its orbital energy will increase). Likewise, a comet in retrograde rotation will be acted upon by a negative T_p and its semimajor axis will decrease with time. Because the nongravitational acceleration is assumed to act symmetrically with respect to perihelion, the time-averaged effect of the periodic radial acceleration cancels out.

When introducing the standard model, *Marsden et al.* (1973) included possible time dependences in the transverse parameter (A_2). Subsequently, however, the standard nongravitational acceleration model was most often used solving only for the constant radial and transverse parameters (A_1 and A_2) over data intervals short enough that neglected time dependences did not cause systematic trends in the residuals. Solutions for the nongravitational parameters usually require astrometric data from at least three apparitions, and one can empirically determine their change with time by comparing the nongravitational parameters determined from several of these three-apparition solutions.

The standard nongravitational parameters can be expressed as function of time by $A_i(t) = A \cdot C_i(t)$, i = 1, 2, 3, where $A = (A_1^2 + A_2^2 + A_3^2)^{1/2}$, and $C_i(t)$ are direction cosines for the nongravitational force acting on the rotating cometary nucleus. The direction cosines C_i, derived by *Sekanina* (1981) have a form

$$C_1 = \cos\eta + (1 - \cos\eta) \cdot \sin^2 I \cdot \sin^2\lambda$$
$$C_2 = \sin\eta \cdot \cos I + (1 - \cos\eta) \cdot \sin^2 I \cdot \sin\lambda \cdot \cos\lambda$$
$$C_3 = [\sin\eta \cdot \cos\lambda - (1 - \cos\eta) \cdot \cos I \cdot \sin\lambda] \sin\lambda$$

where η is the lag angle, I = the equatorial obliquity, $\lambda = \nu + \phi$, ν is the true anomaly of the comet, ϕ = the cometocentric longitude of the Sun at perihelion; the time dependence of $C_i(t)$ is given by the true anomaly $\nu(t)$. Thus three parameters A_1, A_2, A_3 can be replaced by four parameters A, η, I, ϕ, which should be determined along with the corrections to the six orbital elements in the orbit-determination process. The angles η, I, ϕ describing the direction of the nongravitational force vector in orbital coordinates are presented in Fig. 1.

Usually the radial and transverse components of the nongravitational acceleration parameters (i.e., A_1 and A_2) are determined when investigating the motion of short-period comets. In some cases the parameter A_3 also has a meaningful contribution to the successful orbital solution. When investigating the nongravitational motion of comets using the parameters A, η, I, ϕ it is necessary to first determine values

of A_1, A_2, A_3, even if A_3 is poorly determined, to estimate preliminary values of A, η, I, ϕ. These four parameters can then be improved together with six orbital elements by the least-squares correction process (*Sitarski*, 1990).

Largely because of its success in allowing accurate ephemeris predictions, the standard nongravitational force model has been in use for three decades. More recently, it has become understood that, while this model is often successful in representing the astrometric observational data and allowing the computation of accurate ephemeris predictions, the standard model does not provide a completely accurate representation of the actual processes taking place in the cometary nucleus.

Froeschlé and Rickman (1986) and *Rickman and Froeschlé* (1986) used theoretical calculations to examine the secular evolution of the nongravitational parameters as a function of the heliocentric distance for various kinds of short-period comets and different assumed thermal inertias. In general, their values of these parameters did not correspond to those computed from the standard model. In fact, there was such a wide variation in the respective behavior of the A_1, A_2, and A_3 parameters that no generally applicable model for the nongravitational effects was suggested. They noted that improved models would likely have to include the effects of rotation pole orientation and seasonal heat flows.

For a more comprehensive outline of the earlier work on cometary nongravitational forces, the reader is directed to previously published reviews (*Marsden*, 1968, 1969, 1985; *Marsden et al.*, 1973; *Yeomans*, 1994).

3. MODIFICATIONS TO THE STANDARD NONGRAVITATIONAL ACCELERATION MODEL

3.1. Normal Nongravitational Parameter A_3

A nongravitational acceleration acting normal to the comet's orbit plane will affect the longitude of the ascending node and the orbital inclination, but neither of these perturbations is secular. Since these perturbations are modulated by either sine or cosine functions of the true anomaly, much of the nongravitational perturbations upon the two orbital elements would average to zero even if the normal perturbative forces remain positive or negative throughout the orbit. *Sekanina* (1993c) noted that a meaningful solution for the normal nongravitational parameter (A_3) would be possible only for the special case where the perturbations upon the ascending node and the inclination yield a similar value of A_3. Solutions for the A_3 parameter are often not useful but there are some notable exceptions. Meaningful values of A_3 were obtained for the 1808–1988 apparitions of 26P/Grigg-Skjellerup, the 1906–1991 apparitions of 97P/Metcalf-Brewington, and the 1958–1977 apparitions of 22P/Kopff (*Sitarski*, 1991, 1992a; *Rickman et al.*, 1987a). Over the four returns of periodic Comet 71P/Clark, *Nakano* (1992) found a value for A_3 with a formal uncertainty of only 3% and *Sekanina* (1993c) suggested that for this comet, the

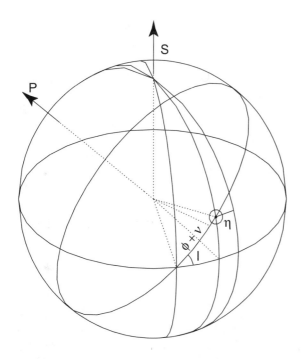

Fig. 1. Orientation of a spherical rotating nucleus. P is the northern orbital pole, S the northern pole of rotation. Angle I is the obliquity of the orbit plane to the comet equator, ϕ the solar longitude at perihelion, ν the true anomaly. The maximum outgassing is shifted behind the subsolar meridian by the lag angle η.

effective nongravitational perturbations on the ascending node and the orbital inclination are about equal. The rotation axis at perihelion is located in the plane defined by the Sun-comet line at perihelion and perpendicular to the comet's orbit plane. The rotation axis is inclined about 45° with respect to the orbit plane. As seen from the comet, the rotation axis at perihelion would be pointing in the general direction of the Sun but about 45° above it. In this configuration, the cometary outgassing produces a noncanceling, nongravitational thrust in a normal direction and hence a valid solution for A_3. In general, the extreme values for A_3 are reached when the single active region is located at a rotation pole and when the obliquity of the orbit plane to the equatorial plane is near 50° or 130°. The principal active vent for Comet 19P/Borrelly is nearly aligned with the rotation pole and *Chesley and Yeomans* (2002) found that an inclusion of the A_3 parameter in their solutions for this comet's orbit significantly improved the prediction ephemeris for the successful flyby of the *Deep Space 1* spacecraft on September 22, 2001.

3.2. Asymmetric Nongravitational Force Models

From equation (2), we note that if the outgassing is asymmetric with respect to perihelion, a purely radial thrust can introduce a secular change in a comet's semimajor axis. This asymmetric nongravitational thrusting was first suggested by *Bessel* (1836), and the modern version of this idea has become known as the asymmetric nongravitational force model. Using the asymmetric light curve of Comet Halley, *Yeomans* (1984) attempted to employ the nucleus rotation parameters introduced by *Sekanina* (1981) to improve the nongravitational force model for Comet Halley. For this latter model, the outgassing was assumed to result from a subsolar active area, which is defined by its cometocentric solar longitude at perihelion (ϕ), the obliquity of the nucleus equator with respect to the orbit plane (I), and by a thermal lag angle (η) measured from the cometary subsolar point to the point of maximum outgassing. Although the optimum lag angle and obliquity turned out to be small, in apparent agreement with subsequent results, the orbital solution did not improve upon the standard nongravitational force model. In an influential work, *Rickman* (1986) pointed out that the asymmetric outgassing observed for Comets 1P/Halley and 22P/Kopff were the likely cause for the nongravitational effects noted for these comets.

Photometric observations of comets suggest that the brightness behaviour of comets near the Sun is sometimes strongly asymmetric with respect to perihelion. *Festou et al.* (1990) established an important statistical correlation between the nongravitational effects and the perihelion asymmetries of the gas production curves. A linear relationship was devised between the nongravitational perturbation of the orbital period, ΔP, and the difference, E, between the integrated gas production rate before and after perihelion. The light curve can, in most cases, be used to indicate whether orbital energy is being added or subtracted as a result of the cometary outgassing. To include this fact in orbital computations, *Sekanina* (1988b) proposed taking into account the asymmetric outgassing of the comet's nucleus with respect to perihelion by replacing the g(r) function with g(r') where r'(t) = r (t − τ). Thus the maximum value of g(r), describing the comet's maximum activity, is shifted by τ with respect to the perihelion time.

Using Sekanina's idea, *Yeomans and Chodas* (1989) analyzed the influence of asymmetric cometary activity with respect to perihelion upon the orbit-determination process for several comets. They examined the nongravitational motion of a number of periodic comets and found that the asymmetric nongravitational acceleration model usually improved an orbital solution when compared to the standard symmetric model. To find the best solution for an individual comet, the authors varied the value of τ in some ranges to obtain the best fit to the astrometric observations. However, τ may be treated as an additional nongravitational parameter that can be determined along with other parameters by the least-squares method.

Sitarski (1994a) elaborated upon the method of determining the values of τ as an additional parameter in the solution and for some comets repeated the cases done by *Yeomans and Chodas* (1989), confirming their conclusions. The asymmetric model of the comet's outgassing can considerably change the values of A_1 and A_2 computed using the standard symmetric model. In long intervals of time the shift τ may change its value and sign. In the case of Comet 6P/d'Arrest, it is about 40 d and is stable in many observed apparitions of the comet, but in the case of Comet 21P/Giacobini-Zinner τ = +14 days during 1959–1973 and τ = −13 d in 1972–1987. Similarly in the case of Comet 22P/Kopff, τ changed its sign within the interval 1906–1990. Assuming that $\tau = \tau_0 + d\tau/dt \cdot (t - t_0)$, where the new parameter $d\tau/dt$ denotes the daily change of τ, Sitarski found it possible to link all the apparitions of Comet Kopff over the interval 1906–1990, determining, along with the orbital elements, five nongravitational parameters: $A_1, A_2, A_3, \tau_0, d\tau/dt$.

Comet 6P/d'Arrest exhibits nearly time-independent nongravitational effects and its visual light curves show an extraordinary asymmetry with respect to perihelion with a peak about 40–50 d after perihelion. Replacement of the standard function g(r) by the observed light curve led to a satisfactory orbital linkage of the positional observations over the intervals 1910–1989 (*Szutowicz and Rickman,* 1993) and 1851–1995 (*Szutowicz,* 1999b).

3.3. Linear Precession Model (Spherically Symmetric Nucleus)

One of the limitations of the standard nongravitational acceleration model is the lack of any long-term time dependence that would allow the nongravitational parameters (A_1, A_2, A_3) to change with time. Traditionally, this limitation has been handled by solving for the nongravitational parameters using limited sets of observations that cover consecutive time intervals. Three apparitions is usually the minimum number of apparitions that allow a meaningful solution for the nongravitational parameters, so consecutive sets of three apparitions are often used for the orbit solutions when these nongravitational parameters are changing with time.

Partly to account for the long-term variations in the nongravitational parameters, *Whipple and Sekanina* (1979) and *Sekanina* (1981, 1984) introduced a model in which sublimating ices would provide a nongravitational force that was not aligned with the nucleus' center-of-mass and hence would (except for locations on the equator and poles) exert a precessional torque on the rotation pole. Coupled with a lag angle between cometary noon and the time of the peak outgassing, this model can introduce a time-varying nongravitational effect in a natural manner.

While the model was first applied to fit the secular decrease of the nongravitational acceleration of Comet 2P/Encke (*Whipple and Sekanina,* 1979), a slightly modified version of this model was developed later and applied to a number of other comets (*Sekanina,* 1984, 1985a,b; *Sekanina and Yeomans,* 1985).

Making the assumption that the spin axis of the rotating nucleus should precess, one can take into account linear terms of the precessional motion of the spin axis by assuming that $\phi(t) = \phi_0 + d\phi/dt \cdot (t - t_0)$ and $I(t) = I_0 + dI/dt \cdot (t - t_0)$, where $d\phi/dt$ and dI/dt are constant. Thus, six nongravitational parameters (A, η, I, ϕ, $d\phi/dt$, dI/dt) must be determined along with corrections to the six orbital elements within the orbit-determination process. On that assumption, *Bielicki and Sitarski* (1991) linked four apparitions of Comet 64P/Swift-Gehrels in the interval 1889–1991. For the standard model with A_1, A_2, A_3 they got the mean residual *a posteriori* $\mu_{apo} = 1".94$ and for the model with linear precession $\mu_{apo} = 1".75$, whereas the mean residual *a priori* $\mu_{apr} = 1".67$. For Comet 26P/Grigg-Skjellerup, *Sitarski* (1991) obtained a satisfactory result linking 16 apparitions of the comet over the 1808–1988 interval using the model of a rotating cometary nucleus with linear precession, and also including a displacement of the photometric center from the center-of-mass. In that case $\mu_{apr} = 1".31$ but $\mu_{apo} = 1".60$.

3.4. Linear Precession Within an Asymmetric Nongravitational Acceleration Model

While using the linear precession model, one may also include a shift τ as an additional nongravitational parameter to account for an asymmetry of the comet's outgassing. This approach made it possible to link seven apparitions of Comet 45P/Honda-Mrkos-Pajdušáková over the interval 1948–1990 and the seven returns to the Sun of Comet 51P/Harrington in 1953–1994 (*Sitarski,* 1995, 1996). For Comet 22P/Kopff, *Sitarski* (1994b) found a successful solution when linking 13 apparitions of the comet over the 1906–1990 interval. However, he had to take into account a complicated function of the shift $\tau(t)$ that, according to the earlier investigations, had changed its sign within the interval considered (*Yeomans and Chodas,* 1989; *Sitarski,* 1994a). Thus,

ten nongravitational parameters (A, η, I_0, dI/dt, ϕ_0, $d\phi/dt$, τ_0, $d\tau/dt$, t_1, and t_2) were determined along with six corrections to the orbital elements. Among those parameters it was possible to determine two critical moments for $\tau(t)$:

$$t_1 = 1936.40 \pm 0.32 \text{ and } t_2 = 1970.92 \pm 0.24$$

when

$$\tau_1 = +27.63 \text{ d for } t \leq t_1 \text{ and } \tau_2 = -39.65 \text{ d for } t \geq t_2$$

but

$$\tau = \tau_0 + d\tau/dt \cdot (t - t_{osc})$$

when

$$t_1 < t < t_2 \text{ and } t_0 = (-2.27 \pm 0.57) \text{ d}$$

for the osculating epoch $t_{osc} = 1951$ December 20.0.

There are examples of successful linkages of many apparitions of short-period comets using nongravitational models of motion that include the rotating and precessing cometary nucleus assuming a constant secular change of I and ϕ (*Sitarski,* 1991, 1994b). However, the linear model for the precession of the spin axis of the nucleus should be considered as only a first approximation and it is often unsuitable for extrapolations over long time intervals, especially when dI/dt is assumed to be constant.

The numerical models of nongravitational motion successfully linking many apparitions of short-period comets are verified if extrapolations of their motions can be used to recover the comets close to the predictions at subsequent apparitions.

3.5. Forced Precession Model (Nonspherical Nucleus)

Various interpretations have been proposed in an effort to understand long-term variations in nongravitational perturbations. One of them is based upon the concept of a forced precession of a nonspherical cometary nucleus, caused by torques associated with the jet force of outgassing. The phenomenon of the spin-axis precession of the cometary rotating nucleus could explain the variations of A_2 with time for Comet 22P/Kopff as found by *Yeomans* (1974), who investigated the long-term nongravitational motion of the comet. *Sekanina* (1984) used values of A_2 obtained by *Yeomans* (1974) to determine a forced precession model for the rotating oblate nucleus of the comet. Thus *Sekanina* (1984) showed a relationship between the physical parameters of the nucleus and its nongravitational behavior.

Assuming that the angles I and ϕ are functions of time as a result of the forced precession due to the asymmetric gas ejection, *Sekanina* (1984) derived formulae for changes of the spin-axis orientation of the cometary nucleus. The following formulae for the time-dependence of I and φ were adopted for use in orbital computations (*Królikowska et al.,* 1998a)

$$I = I_0 + \int_0^t dt \cdot \dot{\phi} \cdot \cos(\alpha + \eta)$$

$$\phi = \phi_0 - \int_0^t dt \cdot \dot{\phi} \cdot \sin(\alpha + \eta)/\sin I$$

$$\dot{\phi} = A \cdot f_p \cdot g(r) \cdot (2 - s) \cdot \sin\psi \cdot \cos\psi \cdot (1 - S_1 \sin^2\psi)^{1/2}(1 - S \sin^2\psi)^{-3/2}$$

where $\dot{\phi}$ is the precession rate of the spin axis, and ψ and α are the cometographic latitude and longitude of the subsolar point respectively. They are given by $\sin\psi = \sin I \cdot \sin\lambda$, and $\tan\alpha = \tan\lambda \cdot \cos I$, respectively. S and S_1 are defined as $S = s \cdot (2 - s)$, $S_1 = S \cdot (2 - S)$, and s denotes the nucleus oblateness (s = $1 - R_b/R_a$, where R_a and R_b are the equatorial and polar radii of the nucleus respectively).

The direction cosines $C_i(t)$ for i = 1, 2, 3 have a more complex form than those derived by *Sekanina* (1981) for the spherically symmetric rotating nucleus since they are modified by terms containing the oblateness s. Variations of I and φ depend on s and on the precession factor f_p, which is connected with the torque factor f_{tor}, introduced by *Sekanina* (1984), by the relation $f_p = s \cdot f_{tor}$. Preliminary estimates of A, η, I_0, ϕ_0 have to be determined from the standard constant parameters A_i, while an initial estimate of the precession factor f_p can be determined by setting $f_p = 0$ and s = 0; the values of the six parameters A, η, I_0, ϕ_0, f_p, and s can then be determined along with the six corrections to the orbital elements in the iterative orbit improvement process. The time shift τ could also be included as an additional parameter taking the more universal function g(r'), r'(t) = r(t − τ) instead of the symmetric function g(r).

Sekanina's forced precession model has been used for investigations into the long-term nongravitational motion of Comets 26P/Grigg-Skjellerup and 45P/Honda-Mrkos-Pajdušáková (*Sitarski,* 1992b, 1995). In both cases values of the oblateness were determined. It was found that s = +0.437 ± 0.014 for Honda-Mrkos-Pajdušáková, and s = −0.373 ± 0.065 for Grigg-Skjellerup. This implies that the nucleus of Honda-Mrkos-Pajdušáková is oblate but for Grigg-Skjellerup the nucleus is a prolate spheroid rotating around its longer axis. Solutions for the forced precession models can be extended to determine the time variations of the angles I and φ. Figure 2 presents plots of I(t) for both comets. The noticeable peaks for Comet Grigg-Skjellerup correspond to rapid changes of I(t) during its perihelion passages. The sudden changes of I(t) for Comet Honda-Mrkos-Pajdušáková are due to close approaches of the comet to Jupiter (e.g., in March 1983 to within 0.111 AU). For Comet Honda-Mrkos-Pajdušáková, *Sitarski* (1995) compared two solutions for the rotating nucleus, with the linear precession and with the forced precession: Whereas the linear preces-

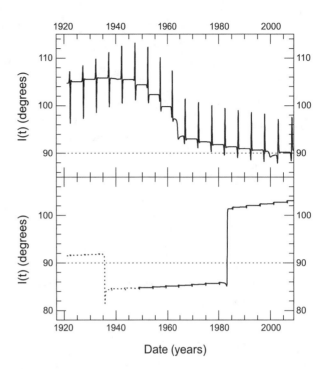

Fig. 2. Temporal variation of angle I due to the spin-axis forced precession of the comet's nucleus for two short-period comets: 26P/Grigg-Skjellerup (top), and 45P/Honda-Mrkos-Pajdušáková (bottom). The Grigg-Skjellerup forced precession model was constructed on the basis of all positional observations taken during 1922–1991 (*Sitarski,* 1992b), and the model for Comet Honda-Mrkos-Pajdušáková was derived from the time interval 1948–1990 (*Sitarski,* 1995). Dashed parts of the curve for Honda-Mrkos-Pajdušáková indicate variation of I before the comet's discovery. Dotted horizontal lines divide models with prograde rotation (I < 90°) from models with retrograde rotation (I > 90°) The same scale on the vertical axes allows one to compare the dramatically different amplitudes of the I variations due to the forced precession alone. Models of these two comets are very different. Forced precession models give a fast precession of the slightly prolate nucleus of Grigg-Skjellerup and a very slow precession for the considerably oblate nucleus of Honda-Mrkos-Pajdušáková (see also Table 1).

sion model could link all the observations with an rms residual of 2".21, the forced precession solution gave an improved rms residual of 1".41.

Królikowska et al. (1998b) applied the forced precession model of the rotating cometary nucleus to examine the nongravitational motion of three comets (30P/Reinmuth 1, 37P/Forbes, and 43P/Wolf-Harrington) over similar intervals of about 70 years. For Comets Reinmuth 1 and Wolf-Harrington they found satisfactory solutions for the pure forced precession model, but for Comet Forbes they had to include an additional parameter, the time shift τ = −9.75 ± 0.76 d. They concluded that the nucleus of Comet Reinmuth 1 is oblate (s = +0.198 ± 0.013) while the nuclei of Comets Forbes and Wolf-Harrington are prolate (s = −0.047 ± 0.005

TABLE 1.   Orbital elements and nucleus physical parameters arising from
the forced precession models for six short-period comets.

	Prolate Spheroid Models			Oblate Spheroid Models		
	37P/Forbes	26P/Grigg-Skjellerup	43P/Wolf-Harrington	46P/Wirtanen	21P/Giacobini-Zinner	45P/Honda-Mrkos-Pajdušáková
Orbital elements						
T	19990504.3758	20021129.7266	19970929.4341	20020826.6370	19981121.3214	20010415.4468
q	1.44602	1.11787	1.58183	1.056769	1.03375	0.53063
e	0.56811	0.63271	0.54398	0.65780	0.70647	0.82476
ω	310°.70	1°.62	187°.13	356°40	172°.54	323°.69
Ω	334°.37	211°.74	254°.76	82°.17	195°.40	91°.36
i	7°.16	22°35	18°.51	11.74	31°.86	3°.89
Forced precession parameters						
A	0.5584	0.01746	+0.3634	0.6780	0.3944	0.3068
η	11°.89	28°.67	6°.36	15°.48	6°.53	12°.16
I_0	121°.65	95°.63	130°.32	145°.15	29°.81	102°.92
ϕ_0	12°.68	338°.20	136°.84	357°.58	260°.85	158°.96
f_p	−0.4623	−2.422	−0.3136	0.1234	0.3460	0.04210
s	−0.0589	−0.0588	−0.0451	0.1019	0.2926	0.4396
τ	−8.33	—	−11.69	−23.48	−49.38	−13.65

Orbital elements are given for the epoch of the last perihelion passage T; angular elements ω, Ω, and i are refer to the equinox J2000.0. Nongravitational parameter A is in units of 10^{-8} AU/day^2, the precession factor f_p is in units of 10^7 d/AU, and the time shift τ is in days.

and s = −0.195 ± 0.032 respectively). The orbital elements and nucleus physical parameters arising from the forced precession model for six short-period comets are presented in Table 1.

3.6.   Erratic Comets

Comet 32P/Comas-Solá belongs to a group of comets dubbed "erratic" by *Marsden and Sekanina* (1971). For these comets, long-term nongravitational effects are irregular and sometimes their values change rapidly. Nongravitational effects in the motion of Comet 21P/Giacobini-Zinner show an irregular behavior in time if we observe values of the parameter A_2 as determined by linking three consecutive apparitions of the comet: In the period 1900–1999, A_2 changed its sign after 1959 (*Yeomans,* 1971). *Sekanina* (1985a) examined the nongravitational motion of Comet Giacobini-Zinner, trying to explain its erratic character by the precessional motion of the spin axis of the comet's nucleus. However, he had to assume an unrealistically large oblateness for the nucleus equal to 0.88. *Sekanina* (1985b) found a similarly unacceptable solution for Comet Comas-Solá, although in this case the irregular behavior of the comet was less dramatic than for Comet Giacobini-Zinner.

Królikowska et al. (2001) investigated the motion of six erratic comets: 16P/Brooks 2, 21P/Giacobini-Zinner, 31P/Schwassmann-Wachmann 2, 32P/Comas Solá, 37P/Forbes, and 43P/Wolf-Harrington. They showed it was possible to link all apparitions of each comet on the basis of a forced precession model with physically reasonable parameters. Hence, one may conclude that the forced precession model of the rotating nonspherical cometary nucleus can explain variations of the nongravitational effects observed in erratic comets (see Fig. 3). However, it was sometimes necessary to introduce several additional parameters — which to some extent simulated the wild behavior of the comet — to obtain a reasonable solution for the numerical model of the comet's motion. For example, to link the observations of Comet Giacobini-Zinner over the 1900–1999 interval, the authors had to include $A^{(1)}$ before 1956 and $A^{(2)}$ after 1956 (instead of one parameter A), τ_1 before 1956, τ_2 between 1956 and 1969, τ_3 between 1969 and 1989, and τ_4 after 1989 (see Fig. 4). Eleven nongravitational parameters [$A^{(1)}$, $A^{(2)}$, η, I_0, ϕ_0, f_p, s, τ_1, τ_2, τ_3, and τ_4] were therefore necessary to link 1589 astrometric observations over a 100-yr interval. It should be noted that the determined oblateness s = +0.2926 ± 0.0055 for the nucleus of Comet Giacobini-Zinner now seems physically reasonable.

3.7.   Case Study of Comet Comas Solá

Królikowska et al. (1998a) studied the nongravitational motion of Comet 32P/Comas Solá during the 1927–1996 interval. This comet had been investigated earlier by *Sekanina* (1985b), who applied the forced precession model and found that this comet precessed more rapidly than any other known comet. His model required a large oblateness of the nucleus, s = 0.57, and according to Sekanina gave "intolerably large perturbations" in the spin-axis obliquity, which changed rapidly by about 90° after 1952. *Królikowska et al.* (1998a) used new observations from the comet's apparitions in 1987 and 1996 and employed the forced precession model fit to 582 observations. They found three almost equivalent models, two with the oblate nucleus and one with

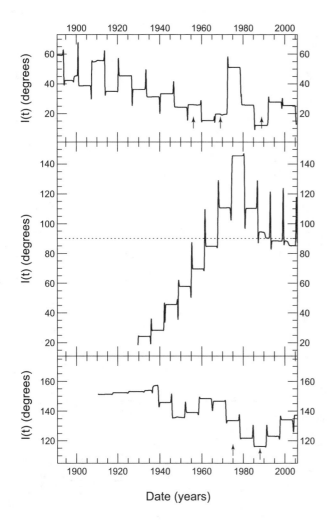

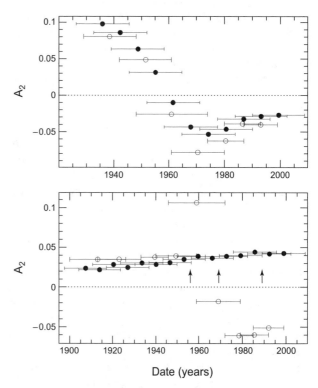

Fig. 3. Temporal variations of the angle I due to the spin-axis forced precession of the comet's nucleus for three erratic comets: 21P/Giacobini-Zinner (top), 37P/Forbes (middle), and 43P/Wolf-Harrington (bottom). The forced precession models were constructed on the basis of all the positional observations taken before the year 2000. There are 9 apparitions of Comets 37P/Forbes and 43P/Wolf-Harrington, and 13 apparitions (almost 100 years) for Comet 21P/Giacobini-Zinner. The same scale on the vertical axes allows one to compare the amplitudes of the I variations. The dotted horizontal line divides the models with prograde rotation (I < 90°) from the models with retrograde rotation (I > 90°). The forced precession models give a slightly prolate shape for the nucleus of Forbes and Wolf-Harrington, and quite an oblate shape for the nucleus of Giacobini-Zinner (see also Table 1). The upright arrows have the same meaning as in Fig. 4.

Fig. 4. Temporal variations in the nongravitational parameter A_2 for two erratic comets: 37P/Forbes (top) and 21P/Giacobini-Zinner (bottom). The open circles represent values of A_2 determined as constants within sets of at least three consecutive apparitions. These circles lie in the middle of the time intervals (shown as thin solid horizontal lines) taken for the calculations. The solid circles are mean values of A_2 (averaged over three consecutive revolutions around the Sun) resulting from the forced precession models. The thick horizontal lines represent the time intervals taken into account for these averaged A_2 values. Upright arrows (Giacobini-Zinner case) indicate the assumed moments of changes for the derived parameters: τ [time shift of the maximum of g(r) with respect to the perihelion time] and A [level of activity; for a detailed description of the Giacobini-Zinner model, see *Królikowska et al.* (2001)]. These postulated moments for the discontinuities of τ (and A) were necessary to obtain a satisfactory forced precession model for Comet Giacobini-Zinner, and they model the actual changes of displacement of maximum activity with respect to perihelion (and the level of activity) for almost 100 years of observations.

the prolate nucleus. The oblate solutions (s = 0.35) were to some extent similar to Sekanina's solution, but I(t) now showed rather moderate variations without rapid jumps. In all cases it was necessary to include two additional values of the time shift, $τ_1$ before 1940, and $τ_2$ after 1940. The best solution appeared to be that for the prolate nucleus where s = –0.105 ± 0.024 and $τ_1$ = –55.05 ± 3.79 d. A solution for $τ_2$ = +8.13 d was found by changing its value to find the best fit to the observations. Three solutions could be compared by their rms mean residuals. In the oblate nucleus

cases it amounted to 2".11, whereas it dropped somewhat to 2".05 for the prolate case: The *a priori* mean residual was 1".41.

3.8. Case Study of Comet Wirtanen

Investigations into the motion of Comet 46P/Wirtanen are especially interesting because this comet is one of a few that have been considered for space flight rendezvous missions. Until early 2003, it was the target body for the European Space Agency's *Rosetta* comet rendezvous mission. The comet, discovered in 1948, had only 67 positional observations through 1991. In that period, the comet experienced two close approaches to Jupiter, to within 0.28 AU in 1972 and to within 0.47 AU in 1984, and both encounters changed the comet's orbit considerably. *Królikowska and Sitarski* (1996) undertook a preliminary investigation of the comet's nongravitational motion, applying the model of the rotating and precessing cometary nucleus with linear precession of the spin axis. They found that the comet's nucleus should be oblate (f_p was positive), but the poor observational material did not allow a determination for the value of s. The 1995–1997 apparition of Comet Wirtanen yielded 247 new positional observations, and *Królikowska and Szutowicz* (1999) again studied the comet's motion based on Sekanina's forced precession model of the rotating cometary nucleus. They were able to determine the value of the oblateness s = +0.1019 ± 0.0342. However, to satisfactorily adjust the solution to all the observations, they had to introduce some additional parameters to the numerical model of the comet's motion: the time shift τ = –23.48 ± 1.25 d, and instead of the single parameter A, two parameters were required, A^I = +0.802 ± 0.009 before 1989.0, and A^{II} = +0.678 ± 0.007 after 1989.0 (in the units of 10^{-8} AU/d^2). This was the best solution (among others considered in their paper) representing the observations with an rms residual equal to 1".59 (the mean residual *a priori* was 1".38). Figure 5 shows the time dependence of I(t) and φ(t) as well as the components of the nongravitational force per unit mass F_i(t), extrapolated to 2015.

3.9. Nongravitational Accelerations Due to Discrete Source Regions

The traditional view of nearly uniform outgassing from any part of the nucleus surface when it is exposed to solar insolation contrasts with the concept of localized active regions. Closeup images of the nucleus of Comet 1P/Halley taken by the spacecraft *Giotto* in March 1986 revealed a few distinct dust jets, emanating from the sunlit side. Similar distinct dust jets were evident in Comet 19P/Borrelly when the *Deep Space 1* spacecraft flew past this comet on September 22, 2001 (*Soderblom et al.*, 2002). The local outgassing restricted to a few "active regions" evolving into craters has recently been incorporated into the physical models of comets (*Colwell et al.*, 1990; *Colwell*, 1997). The distribution of jets around the nucleus and their contribution to the total production rate also has implications for the nongravitational effects in a comet's orbital motion. For a nucleus with discrete outgassing regions (spotty nucleus), the maximum sublimation rate will take place when the subsolar point is closest to an active region, which may not occur at perihelion.

The effects of discrete outgassing on the shape of the gas production curve and on the nongravitational parameters were discussed in detail by *Sekanina* (1991b, 1993a,c). In *Sekanina*'s model (1988a) the absorbed solar energy is spent on sublimation and thermal re-radiation, but not on

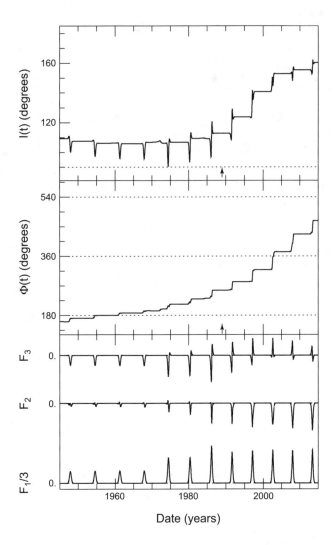

Fig. 5. Forced precession model of Comet 46P/Wirtanen based on an almost 50-yr interval of positional observations. For model details see *Królikowska and Szutowicz* (1999). Temporal variations are presented for the angle I (top), φ (middle), and components F_1, F_2, F_3 of the nongravitational force per unit mass **F** (bottom) due to the spin-axis forced precession of the comet's nucleus. These latter three components are only meant to convey the qualitative changes in their magnitudes and directions near each perihelion passage. This model gives an oblate shape of the nucleus with P_{rot}/R_a = 4.9 ± 1.4 h/km, where R_a denotes the equatorial radius. Assuming 6.0 ± 0.3 h for the rotational period P_{rot} (*Lamy et al.*, 1998), the range of 0.9–1.7 km for the nucleus radius (consistent with the photometric observations) is obtained. The vertical arrow denotes the time (1989) when the modeled levels of outgassing activity and perihelion asymmetry are assumed to change.

heat conduction into the nucleus. Hence there is no need for a thermal lag angle. The sublimation rate from a unit surface area on the nucleus is expressed as a function of heliocentric distance and the Sun's zenith distance. The model, as applied to Comet 2P/Encke (*Sekanina*, 1988a,b, 1991a), allowed him to interpret its observed sunward fan-like coma as an effect of the northern and southern localized vents on the comet's nucleus ($+55°$N, $-75°$S); the comet's spin axis was fixed. *Sekanina* (1991b, 1993a) noted examples of the rotation-averaged sublimation rates of point-like sources at various locations on the nucleus surface calculated for a spherical nucleus with various axial positions in an unperturbed heliocentric orbit. For the spotty model of the nucleus, the meaning of the nongravitational parameters is different than for the standard model. For the spotty model, there is a correlation between the sign of A_2, the asymmetry of the production curve, and the location of the active regions. In the standard model the sense of rotation was directly correlated with the sign of A_2 but this need not be true for the spotty model. Furthermore, a negative value of A_1, unrealistic in the standard model, can correspond to a circumpolar or high-latitude active source and certain combinations of the spin-axis orientation. Erratic discontinuities in the nongravitational perturbations for three comets and their long-term changes in A_2 were interpreted by *Sekanina* (1993b) as the initiation of new active areas or the deactivation of existing ones on the nucleus surface. Thus the proper modeling of the nongravitational effects should contain information on the true characteristics and locations of the comet's active areas.

The rotational-averaged orbital components of the nongravitational acceleration for a nucleus with active regions were adopted for orbital computations and introduced directly into the equations of a comet's motion by *Szutowicz* (2000). The lifetime of each active region was limited by time because of its activation and deactivation. The parameters of the model are three angles η, I, ϕ characterizing the nucleus, the cometocentric latitude of the jth active region β_j, and the constants A_j are proportional to S_j/M, where S_j is the outgassing area of the jth source and M is the nucleus mass. These parameters can be determined along with the osculating orbital elements in the orbit-determination process. The first attempt to introduce the spotty nucleus model into orbital calculations was made to explain a dramatic jump of the nongravitational effects in the motion of Comet 71P/Clark (*Szutowicz*, 1999a). The spotty nucleus model was successfully used to link all observations of Comet 46P/Wirtanen spanning the interval 1947–1997 (*Szutowicz*, 1999b) and of 43P/Wolf-Harrington over the period 1925–1997 (*Szutowicz*, 2000), with the rms residuals equal to 1".57 and 1".39 respectively. For the former comet, the temporal variations of the activity level are responsible for its nongravitational behavior, whereas the orbital solutions for the latter one involved a redistribution of the active areas. From orbital solutions for 43P/Wolf-Harrington, it follows that the northern region ($\sim38°$N) was persistently active and the profile of the comet's activity was modified by the initiation and disappearance of two southern regions ($\sim-4°$S, $-51°$S) in 1965, 1978, and 1991. According to the solutions, the time variation of the modeled sublimation rates is accompanied by parallel changes in the visual light curve. A contradictory result was obtained in the case of Comet 46P/

Wirtanen. The modeled gas production rate peaks before perihelion in accordance with the nongravitational acceleration ($A_2 < 0$), but this is not consistent with the observed light curve that peaks about one week after perihelion (*Rickman and Jorda,* 1998). Clearly there is not a one-to-one correspondence between a comet's visual light curve and its gas production rates and it is the latter that controls the nongravitational effects.

Chesley (2002) used the discrete jet model, averaged over one rotation period, to verify the two main active source regions of 2P/Encke that *Sekanina* (1988a) had suggested (two sources at latitudes $55°$N and $-75°$S). He found that these two source regions, together with Sekanina's pole direction, provided a significant improvement over the standard nongravitational acceleration model in both the goodness of the orbit fit as well as in the orbit's predictive capability. Going further, *Chesley* (2002) found that a pole position around RA = $220°$ and Dec. = $+40°$ provided an even better orbit fit and prediction than those obtained using the pole suggested by Sekanina.

It is a rare opportunity when one can test a cometary nongravitational acceleration model with the help of astrometric imaging data provided by spacecraft. *Chesley* (2002) noted that since the principal active vent for Comet 19P/Borrelly was aligned with the rotation pole, this pole direction could be independently determined using the values of A_1, A_2, A_3, τ; the determined pole direction (RA and Dec. = $208°$, $-4°$) agreed with rotation pole values determined from long-term photometric studies (e.g., *Farnham and Cochran,* 2003) and had reasonable agreement with the pole position determined from the spacecraft images themselves (*Soderblom et al.,* 2002).

4. INFERRING MASSES AND BULK DENSITIES OF THE NUCLEUS USING NONGRAVITATIONAL EFFECTS

It is important to note that there are no direct determinations for the mass or density of any comet and this is likely to remain the situation until a spacecraft rendezvous mission is carried out. Nevertheless, there have been many studies suggesting that comets are rather low-density and porous structures.

Yau et al. (1994) found that the observations of Comet 109P/Swift-Tuttle in 69 B.C. and in A.D. 188, 1737, 1862, and 1992–1993 were consistent with the complete absence of nongravitational effects in this comet's motion and that there have been no obvious changes in this comet's absolute magnitude over two millennia. Because Comet Swift-Tuttle's absolute magnitude has not changed significantly and there is a lack of significant nongravitational effects over the same period, constraints can be placed upon the model for this comet's nucleus. At 1 AU from the Sun, the outgassing activity of Swift-Tuttle is comparable with that of Comet Halley at the same heliocentric distance. Yet Comet Halley experiences an increase in its orbital period of four days per revolution due to nongravitational effects while Swift-Tuttle has no perceptible change in its period. If the mass of Swift-Tuttle were significantly larger than Halley's, however, one would not expect to be able to detect a nongravitational acceleration in its orbital motion. Based upon an analysis of their respective meteor stream characteristics, *Hughes and McBride* (1989) concluded that the mass of Comet Swift-Tuttle is about 10 times larger than Comet Halley.

Rickman (1986) pointed out that radial outgassing forces that act asymmetrically with respect to perihelion were the likely cause of the nongravitational effects upon Comets Halley and Kopff, and he went on to make estimates of their masses and bulk densities. He noted that the water production curves for Halley and other comets show an asymmetry with respect to perihelion and therefore the effect of the radial component, integrated over one orbital period, would be nonzero. The nucleus masses were estimated by comparing nongravitational parameters with the rocket-like forces expected from the gas-production curves. In turn, the gas-production curves were determined from the light curves using an empirical relationship developed by M. Festou. The bulk density for Comet Halley was estimated to be 0.1–0.2 g/cm³ and that for Kopff was lower still. *Rickman et al.* (1987b) continued this type of analysis and estimated masses for 29 short-period comets. The change in the total orbital period per revolution results from the sum of the contributions from the radial and transverse rocket effects. The mass of each comet was determined as a function of its estimated thermal and rotational properties. While bulk densities for individual comets are very uncertain, the bulk densities of these objects as a group were estimated to be less than 0.5 g/cm³, suggesting that the cometary nucleus is a very porous structure. This type of analysis depends upon the assumption that there is a correlation between the light curve and the assumed gas-production curve, that thermal lag angles are present, and that the surface of each object has an unmantled free sublimating area. Using a similar approach for Comet Halley, *Sagdeev et al.* (1988) estimated a bulk density of 0.6 g/cm³ with error bars of +0.9 and −0.4 g/cm³. After a rather complete discussion of the method and the uncertainties involved in this type of analysis, *Peale* (1989) concluded that it is difficult to provide a meaningful constraint for the bulk density of Comet Halley.

The orbital nongravitational perturbations combined with the observed gas production rates were also employed to estimate the masses of the two long-period comets: C/1995 O1 Hale-Bopp and C/1996 B2 Hyakutake (*Szutowicz et al.*, 2002a,b). The observed water production rates as a function of the heliocentric distance were included into the nongravitational model represented by parameters: A (A is proportional to Q_m/M), η, I, ϕ, where Q_m is the observed sublimation rate of water at 1 AU. The derived masses divided by the volumes of both comets give bulk densities as low as 0.1 g/cm³ for Comet Hale-Bopp and 0.2 g/cm³ for Comet Hyakutake. This kind of solution is limited by the thermal properties of the nucleus (i.e., the mean outflow velocity of the molecules), the excess of the production rate

due to the ice halo in the coma, and some stochastic perturbations like splitting or outbursts noted for both comets. The latter may cause stronger nongravitational effects than expected from the water sublimation rate alone.

Farnham and Cochran (2002) estimated the mass of Comet 19P/Borrelly by separately computing the nongravitational force and acceleration acting upon the comet's nucleus. The force was computed from the observed gas production rate and the emission velocity while the nongravitational acceleration was forthcoming from the orbital computations. Dividing the comet's determined mass by the volume estimated from the *Deep Space 1* spacecraft imaging provided a bulk density for the comet's nucleus of 0.49 (−0.20, +0.34) g/cm³.

Davidsson and Gutiérrez (2003) deduced a bulk density for 19P/Borrelly of 0.10–0.30 g/cm³ by modeling the comet's observed water production rates with nucleus surface activity maps based upon sophisticated thermophysical models. By requiring that the model reproduce the nongravitational secular changes in the argument of perihelion and the longitude of the ascending node, they were able to tighten the bulk density range to 0.18–0.30 g/cm³.

5. NONGRAVITATIONAL EFFECTS AND THE SOURCE REGION FOR LONG-PERIOD COMETS

Marsden and collaborators first reported the detections of nongravitational forces in the motion of long-period comets in the late 1960s and early 1970s. In the *Catalogue of Cometary Orbits*, *Marsden and Williams* (2001) gave nongravitational parameters A_1 and A_2 for 23 long-period comets. To determine these parameters, the authors applied the standard nongravitational model. It became evident that nongravitational accelerations affecting the orbital motion of long-period comets (at 1 AU from the Sun) are a few to 10 times larger than the similar accelerations detected for short-period comets (*Marsden et al.*, 1973). In the last decade two famous long-period comets (1995 O1 Hale-Bopp and 1996 B2 Hyakutake) offered unique opportunities for detailed investigation of the nongravitational effects on their orbital motion. The nongravitational effects play an essential role in the dynamical evolution of both comets. In particular, the future orbital evolution, including nongravitational effects, gives a significantly higher probability of these comets being ejected from the solar system than does the pure gravitational orbital motion (*Szutowicz et al.*, 2002a,b; *Królikowska*, 2002).

Nongravitational effects also play a role in the identification of hyperbolic comets. The problem of a negative tail in the distribution of the reciprocals of original semimajor axis ($1/a_{ori}$) has been discussed in detail by many authors. *Marsden et al.* (1973) speculated that neglecting the nongravitational effects tends to produce more hyperbolic original orbits than really is the case. Subsequently, many authors (*Yabushita*, 1991; *Bolatto et al.*, 1995) considered the nongravitational perturbation in a comet's energy per orbital revolution and concluded that these perturbations are too small to explain the negative excess of original binding energy of "hyperbolic" comets. Misleading results can be obtained when the same osculating orbit is used as the initial orbit for backward integrations using nongravitational effects and without these effects. Let us consider that the same osculating orbit is integrated backward twice, first by setting the nongravitational terms equal to zero and then by including the nongravitational accelerations. The differences in the resulting original reciprocals will then be significantly smaller than 10^{-4} AU^{-1}. At first look, this seems to suggest that nongravitational effects provide only modest changes in the true original value for $1/a_{ori}$. However, it is important to realize that an osculating nongravitational orbit determined from a set of observations would not be the same orbit as one determined from these same observations but under the assumption of purely gravitational motion. *Królikowska* (2001) investigated the problem of the original orbits of 33 comets considered as "hyperbolic" comets (in the pure gravitational case). For the 16 cases for which solutions for nongravitational effects could be carried out, she showed that for almost all cases, the original orbits changed from hyperbolic to elliptic. For the two comets for which the original orbits remained hyperbolic (1996 E1, 1996 N1), their original orbits became less hyperbolic as a result of solving for nongravitational effects and the resulting negative original $1/a_{ori}$ values were rather modest. The tendency for negative, original $1/a_{ori}$ values to become positive when nongravitational effects are taken into consideration is due primarily to changes in the orbital eccentricity. Królikowska concluded that the nongravitational effects could significantly affect the Oort peak for comets with perihelion distance smaller than 3 AU.

At the other extreme from the nearly parabolic orbits from the Oort cloud comet lays 2P/Encke with its shortest known cometary orbital period of 3.3 yr. The current orbit of 2P/Encke is completely interior to Jupiter's orbit and is gravitationally decoupled from that planet. In trying to explain how Comet Encke arrived at this stable orbital position, *Steel and Asher* (1996) noted that nongravitational effects on the comet, some four times larger than those that have recently been operative, would be enough to evolve the comet from its current orbit into Jupiter-crossing orbits and hence the same mechanism could have dropped the comet into its current orbit. The nongravitational effects can cause the comet to drift across the jovian and saturnian mean-motion resonances in the asteroid belt and even if these nongravitational effects do not act in the same direction for extended periods of time, they can still strongly modify the orbit of an Encke-like object.

6. SUMMARY

Cometary orbit-determination problems are dominated by the proper modeling of the so-called nongravitational perturbations that are due to the rocket-like thrusting of the outgassing cometary nucleus. Modern astrometric positions, particularly those that are referenced to Hipparcos-based star catalogs and where the brightest pixel is employed as

the true position of the cometary nucleus, are usually accurate to the subarcsecond level. Yet multiple apparition orbital solutions for active short-period comets cannot often provide a root mean square (rms) residual (observed minus computed observational position) that is subarcsecond. It is the improper modeling of the nongravitational effects that is the largest problem by far.

Beginning with Encke's first suggestion of an interplanetary resisting medium to explain the anomalous motion of the comet that bears his name, there have been many different models put forward to explain the accelerations in the motions of active comets that are not due to the gravitational perturbations of neighboring planets or asteroids. Although the notion of an icy conglomerate model for a cometary nucleus (*Whipple,* 1950, 1951) is still in basic agreement with the observations, there have been a number of recent modifications and refinements to this model. Largely as a result of the impressive images of Comet 1P/Halley's nucleus taken by the *Giotto* spacecraft and those taken more recently of 19P/Borrelly by the *Deep Space 1* spacecraft, a "vent" model whereby the outgassing activity takes place from discrete active areas has replaced the picture of an outgassing sunlit hemisphere.

It seems likely that each comet has its own set of peculiar jets located at various places on its surface and operating at different strengths so that a completely accurate model for a particular comet's nongravitational effects would require a detailed knowledge of the comet's surface outgassing features and rotation characteristics. Since this knowledge is available only for those few comets that are visited by spacecraft, orbit practitioners will have to be content with generic models that approximate the true situation. In this regard the recent advances in bringing forth the asymmetric nongravitational acceleration models, the forced nucleus precession models, and especially the discrete active source models are promising.

Many of the existing models solve for physical characteristics of the cometary nucleus (e.g., oblateness, thermal lag angles, positions of sources, and the rotation pole). However, the formal uncertainties computed for these quantities that arise from the orbit-determination process alone must be considered lower limits rather than realistic values. Whenever possible, these quantities should be confirmed using spacecraft observations or a long series of photometric groundbased observations. For example, the oblateness values and rotation pole positions derived from the orbit-determination process should be checked against the aspect ratios and pole positions determined from groundbased photometric studies. As the astrometric datasets improve and lengthen and as the modeling of the cometary nongravitational effects becomes more realistic, there remains the strong possibility that some physical characteristics of comets will soon be accurately determined from the orbit-determination process alone. We are already beginning to see signs that this is the case.

Acknowledgments. A portion of this work was carried out by the Jet Propulsion Laboratory/California Institute of Technology under contract to NASA.

REFERENCES

Anderson J. D., Esposito R. B., Martin W., and Muhlman D. O. (1975) Experimental test of general relativity using time-delay data from Mariner 6 and Mariner 7. *Astrophys. J., 200*, 221–233.

Bessel F. W. (1836) Bemerkungen über mögliche Unzulänglichkeit der die Anziehungen allein berücksichtigenden Theorie der Kometen. *Astron. Nach., 13*, 345–350.

Bielicki M. and Sitarski G. (1991) Nongravitational motion of Comet P/Swift-Gehrels. *Acta Astron., 41*, 309–323.

Bolatto A. D., Fernández J. A., and Carballo G. F. (1995) Asymmetric nongravitational forces in long-period comets. *Planet. Space Sci., 43*, 709–716.

Chesley S. R. (2002) Modeling comet nongravitational accelerations with discrete rotating jets. *Bull. Am. Astron. Soc., 34*, 869–870.

Chesley S. R. and Yeomans D. K. (2002) Alternate approaches to estimating comet nongravitational accelerations (abstract). In *Comets, Asteroids, Meteors 2002*, p. 117. Berlin, Germany.

Chesley S. R., Chodas P. W., Keesey M. S., Bhaskaran S., Owen W. M. Jr., Yeomans D. K., Garrard G. J., Monet A. K. B., and Stone R. C. (2001) Comet 19P/Borrelly orbit determination for the DS1 flyby. *Bull. Am. Astron. Soc., 33*, 1090.

Colwell J. E. (1997) Comet lightcurves: Effects of active regions and topography. *Icarus, 125*, 406–415.

Colwell J. E., Jakosky B. M., Sandor B. J., and Stern S. A. (1990) Evolution of topography on comets. II. Icy craters and trenches. *Icarus, 85*, 205–215.

Davidsson B. J. R. and Gutiérrez P. J. (2003) An estimate of the nucleus density of comet 19P/Borrelly. *Bull. Am. Astron. Soc., 35*, 969.

Encke J. F. (1823) Fortgesetzte nachricht über den Pons'schen kometen. In *Berliner Astronomische jahrbuch fur das jahr 1826*, pp. 124–140. Berlin, Germany

Farnham T. L. and Cochran A. L. (2002) A McDonald observatory study of comet 19P/Borrelly: Placing the Deep Space 1 observations into a broader context. *Icarus, 160*, 398–418.

Festou M., Rickman H., and Kamel I. (1990) Using light-curve asymmetries to predict comet returns. *Nature, 345*, 235–238.

Froeschlé C. and Rickman H. (1986) Model calculations of nongravitational forces on short period comets. I. Low-obliquity case. *Astron. Astrophys., 170*, 145–160.

Hughes D. W. and McBride N. (1989) The mass of meteoroid streams. *Mon. Not. R. Astron. Soc., 240*, 73–79.

Kamienski M. (1933) Über die bewegung des kometen Wolf I in dem zeitraume 1884–1919. *Acta Astron., Ser. A, 3*, 1–56.

Królikowska M. (2001) A study of the original orbits of "hyperbolic" comets. *Astron. Astrophys., 376*, 316–324.

Królikowska M. (2002) Dynamical evolution of five long-period comets with evident non-gravitational effects. In *Proceedings of Asteroids, Comets, Meteors — ACM 2002* (B. Warmbein, ed.), pp. 629-631. ESA SP-500, Noordwijk, Netherlands.

Królikowska M. and Sitarski G. (1996) Evolution of the orbit of comet 46P/Wirtanen during 1947–2013. *Astron. Astrophys., 310*, 992–998.

Królikowska M. and Szutowicz S. (1999) Oblate spheroid model of nucleus of comet 46P/Wirtanen. *Astron. Astrophys., 343*, 997–1000.

Królikowska M., Sitarski G., and Szutowicz S. (1998a) Model of the nongravitational motion for Comet 32P/Comas-Solá. *Astron. Astrophys., 335*, 757–764.

Królikowska M., Sitarski G., and Szutowicz S. (1998b) Forced precession model for three periodic comets: 30P/Reinmuth 1, 37P/Forbes, and 43P/Wolf-Harrington. *Acta Astron., 48*, 91–102.

Królikowska M., Sitarski G., and Szutowicz S. (2001) Forced precession models for six erratic comets. *Astron. Astrophys., 368*, 676–688.

Lamy P. L., Toth I., Jorda L., Weaver H. A., and A'Hearn M. A. (1998) The nucleus and inner coma of comet 46P/Wirtanen. *Astron. Astrophys., 335*, L25.

Lawson C. L. and Hanson R. J. (1974) *Solving Least Squares Problems.* Prentice-Hall, Englewood Cliffs, New Jersey. 340 pp.

Marsden B. G. (1968) Comets and nongravitational forces. *Astron. J., 73*, 367–379.

Marsden B. G. (1969) Comets and nongravitational forces II. *Astron. J., 74*, 720–734.

Marsden B. G. (1985) Nongravitational forces on comets: The first fifteen years. In *Dynamics of Comets: Their Origin and Evolution* (A. Carusi and G. B. Valsecchi, eds.), pp. 343–352. Astrophysics and Space Science Library Vol. 115, Reidel, Dordrecht.

Marsden B. G. and Sekanina Z. (1971) Comets and nongravitational forces. IV. *Astron. J., 76*, 1135–1197.

Marsden B. G. and Williams G. V. (2001) *Catalogue of Cometary Orbits, 13th edition.* Smithsonian Astrophysical Observatory, Cambridge.

Marsden B. G., Sekanina Z., and Yeomans D. K. (1973) Comets and nongravitational forces. V. *Astron. J., 78*, 211–225.

Nakano S. (1992) *Periodic Comet Clark.* Minor Planet Circular 20122.

Peale S. J. (1989) On the density of Halley's comet. *Icarus, 82*, 36–49.

Recht A. W. (1940) An investigation of the motion of periodic comet d'Arrest (1851 II). *Astron. J., 48*, 65–80.

Rickman H. (1986) Masses and densities of Comets Halley and Kopff. In *Proceedings of the Workshop on The Comet Nucleus Sample Return Mission* (O. Melita, ed.), pp. 195–205. ESA SP-249, Noordwijk, The Netherlands.

Rickman H. and Froeschlé C. (1986) Model calculations of nongravitational forces on short period comets. II. High obliquity case. *Astron. Astrophys., 170*, 161–166.

Rickman H., Sitarski G., and Todorovic-Juchniewicz B. (1987a) Nongravitational motion of comet P/Kopff during 1958–1983. *Astron. Astrophys., 188*, 206–211.

Rickman H., Kamel L., Festou M. C., and Froeschlé C. (1987b) Estimates of masses, volumes and densities of short-period comet nuclei. In *Symposium on the Diversity and Similarity of Comets* (E. J. Rolfe and B. Battrick, eds.), pp. 471–481. ESA SP-278, Noordwijk, The Netherlands.

Rickman H. and Jorda L. (1998) Comet 46P/Wirtanen, the target of the Rosetta Mission. *Adv. Space Res., 21*, 1491–1504.

Sagdeev R. Z., Elyasberg P. E., and Moroz V. I. (1988) Is the nucleus of Comet Halley a low density body? *Nature, 331*, 240–242.

Sekanina Z. (1981) Rotation and precession of cometary nuclei. *Annu. Rev. Earth Planet. Sci., 9*, 113–145.

Sekanina Z. (1984) Precession model for the nucleus of periodic Comet Kopff. *Astron. J., 89*, 1573–1583.

Sekanina Z. (1985a) Precession model for the nucleus of periodic comet Giacobini-Zinner. *Astron. J., 90*, 827–845.

Sekanina Z. (1985b) Nucleus precession of periodic Comet Comas Solá. *Astron. J., 90*, 1370–1381.

Sekanina Z. (1988a) Outgassing asymmetry of periodic Comet

Encke. I — Apparitions 1924–1984. *Astron. J., 95,* 911–924.

Sekanina Z. (1988b) Outgassing asymmetry of periodic Comet Encke. II — Apparitions 1868–1918 and a study of the nucleus evolution. *Astron. J., 95,* 1455–1475.

Sekanina Z. (1991a) Encke, the comet. *J. R. Astron. Soc. Canada, 85,* 324–376.

Sekanina Z. (1991b) Cometary activity, discrete outgassing areas, and dust-jet formation. In *Comets in the Post-Halley Era* (R. L. Newburn et al., eds.), pp. 769–823. Astrophysics and Space Science Library Vol. 167, Kluwer, Dordrecht.

Sekanina Z. (1993a) Effects of discrete-source outgassing on motions of periodic comets and discontinuous orbital anomalies. *Astron. J., 105,* 702–735.

Sekanina Z. (1993b) Orbital anomalies of the periodic comets Brorsen, Finlay and Schwassmann-Wachmann 2. *Astron. Astrophys., 271,* 630–644.

Sekanina Z. (1993c) Nongravitational motions of comets: Components of the recoil force normal to orbital plane. *Astron. Astrophys., 277,* 265–282.

Sekanina Z. and Yeomans D. K. (1985) Orbital motion, nucleus precession, and splitting of periodic Comet Brooks 2. *Astron. J., 90,* 2335–2352.

Shahid-Saless B. and Yeomans D. K. (1994) Relativistic effects on the motions of asteroids and comets. *Astron. J., 107,* 1885–1889.

Sitarski G. (1983) Effects of general relativity in the motions of minor planets and comets. *Acta Astron., 33,* 295–304.

Sitarski G. (1990) Determination of angular parameters of a rotating cometary nucleus basing on positional observations of the comet. *Acta Astron., 40,* 405–417.

Sitarski G. (1991) Linkage of all the apparitions of the periodic Comet Grigg-Skjellerup during 1808–1988. *Acta Astron., 41,* 237–253.

Sitarski G. (1992a) Motion of Comet P/Metcalf-Brewington (1906 VI = 1991a). *Acta Astron., 42,* 49–57.

Sitarski G. (1992b) On the rotating nucleus of Comet P/Grigg-Skjellerup. *Acta Astron., 42,* 59–65.

Sitarski G. (1992c) On the relativistic motion of (1566) Icarus. *Astron. J., 104,* 1226–1229.

Sitarski G. (1994a) On the perihelion asymmetry for investigations of the nongravitational motion of comets. *Acta Astron., 44,* 91–98.

Sitarski G. (1994b) The nongravitational motion of Comet P/Kopff during 1906–1991. *Acta Astron., 44,* 417–426.

Sitarski G. (1995) Motion of Comet 45P/Honda-Mrkos-Pajdušáková. *Acta Astron., 45,* 763–770.

Sitarski G. (1996) The nongravitational motion of Comet 51P/Harrington. *Acta Astron., 46,* 29–35.

Soderblom L. A., Becker T. L., Bennett G., Boice A. C., Britt D. T., Brown R. H., Buratti B. J., Isbell C., Giese B., Hare T., Hicks M. D., Howington-Kraus E., Kirk R. L., Lee M., Nelson R. M., Oberst J., Owen T. C., Rayman M. D., Sandel B. R., Stern N., Thomas S. A., Yelle, R. V. (2002) Observations of comet 19P/Borrelly by the miniature integrated camera and spectrometer aboard Deep Space 1. *Science, 296,* 1087–1091.

Steel D. I. and Asher D. J. (1996) On the origin of comet Encke. *Mon. Not. R. Astron. Soc., 281,* 937–944.

Szutowicz S. (1999a) What happened with comet 71P/Clark? In *Evolution and Source Regions of Asteroids and Comets* (J. Svoreň et al., eds.), pp. 259–264. IAU Colloquium 173.

Szutowicz S. (1999b) Analysis of physical properties of a cometary nucleus based on the nongravitational perturbations in a comet's orbital motion. Ph.D. thesis, Nicolaus Copernicus Astronomical Center, Warsaw (in Polish).

Szutowicz S. (2000) Active regions on the surface of Comet 43P/Wolf-Harrington determined from its nongravitational effects. *Astron. Astrophys., 363,* 323–334.

Szutowicz S. and Rickman H. (1993) Investigation of the nongravitational motion of Comet P/d'Arrest. In *Proceedings of the Conference on Astrometry and Celestial Mechanic* (K. Kurzynska et al., eds.), pp. 425–426. Poznan, Poland.

Szutowicz S., Królikowska M., and Sitarski S. (2002a) Modelling of the non-gravitational motion of the Comet C/1996 B2 Hyakutake. In *Proceedings of Asteroids, Comets, Meteors — ACM 2002* (B. Warmbein, ed.), pp. 633–636. ESA SP-500. Noordwijk, Netherlands.

Szutowicz S., Królikowska M., and Sitarski S. (2002b) A study of the non-gravitational effects of the comet C/1995 O1 Hale-Bopp. *Earth Moon Planets, 90,* 119–130.

Whipple F. L. (1950) A comet model. I. The acceleration of comet Encke. *Astron. J., 111,* 375–394.

Whipple F. L. (1951) A comet model II. Physical relations for comets and meteors. *Astrophys. J., 113,* 464–474.

Whipple F. L. and Sekanina Z. (1979) Comet Encke: Precession of the spin axis, nongravitational motion, and sublimation. *Astron. J., 84,* 1894–1909.

Yabushita S. (1991) Maximum nongravitational acceleration due to out-gassing cometary nuclei. *Earth Moon Planets, 52,* 87–92.

Yau K., Yeomans D. K., and Weissman P. R. (1994) The past and future motion of P/Swift-Tuttle. *Mon. Not. R. Astron. Soc., 266,* 305–316.

Yeomans D. K. (1971) Nongravitational forces affecting the motions of periodic comets Giacobini-Zinner and Borrelly. *Astron. J., 76,* 83–86.

Yeomans D. K. (1974) The nongravitational motion of comet Kopff. *Publ. Astron. Soc. Pacific, 86,* 125–127.

Yeomans D. K. (1984) The orbits of Comets Halley and Giacobini-Zinner. In *Cometary Astrometry* (D. K. Yeomans et al., eds.), pp. 167–175. JPL Publication 84-82, Jet Propulsion Laboratory, Pasadena.

Yeomans D. K. (1991) *Comets: A Chronological History of Observation, Science, Myth and Folklore.* Wiley, New York. 485 pp.

Yeomans D. K. (1994) A review of comets and nongravitational forces. In *Asteroids, Comets, Meteors 1993* (A. Milani et al., eds.), pp. 241–254.

Yeomans D. K. and Chodas P. W. (1987) Radar astrometry of near-Earth asteroids. *Astron. J., 94,* 189–200.

Yeomans D. K. and Chodas P. W. (1989) An asymmetric outgassing model for cometary nongravitational accelerations. *Astron. J., 98,* 1083–1093.

Yeomans D. K., Chodas P. W., Keesey M. S., Ostro S. J., Chandler J. F., and Shapiro I. I. (1992) Asteroid and comet orbits using radar data. *Astron. J., 103,* 303–317.

Oort Cloud Formation and Dynamics

Luke Dones
Southwest Research Institute

Paul R. Weissman
Jet Propulsion Laboratory

Harold F. Levison
Southwest Research Institute

Martin J. Duncan
Queen's University

The Oort cloud is the primary source of the "nearly isotropic" comets, which include new and returning long-period comets and Halley-type comets. We focus on the following topics: (1) the orbital distribution of known comets and the cometary "fading" problem; (2) the population and mass of the Oort cloud, including the hypothetical inner Oort cloud; (3) the number of Oort cloud comets that survive from the origin of the solar system to the present time, and the timescale for building the Oort cloud; (4) the relative importance of different regions of the protoplanetary disk in populating the Oort cloud; and (5) current constraints on the structure of the Oort cloud and future prospects for learning more about its structure.

1. INTRODUCTION

"They have observed Ninety-three different Comets, and settled their Periods with great Exactness. If this be true, (and they affirm it with great Confidence) it is much to be wished that their Observations were made publick, whereby the Theory of Comets, which at present is very lame and defective, might be brought to the same Perfection with other Parts of Astronomy."
— Jonathan Swift, *Gulliver's Travels* (1726)

Recorded observations of comets stretch back more than 2000 years (*Kronk, 1999*). For example, *Yau et al.* (1994) showed that a comet noted in Chinese records in the year 69 B.C. was 109P/Swift-Tuttle, which most recently passed perihelion in 1992. However, it is only in the last 400 years that comets have been generally accepted as astronomical, as opposed to atmospheric, phenomena (e.g., *Bailey et al.,* 1990; *Yeomans,* 1991). Even so, learned opinion until the mid-twentieth century was divided on whether comets were interlopers from interstellar space (Kepler, Laplace, and Lyttleton) or members of the solar system (Halley, Kant, and Öpik).

By the mid-nineteenth century, it was well established that most comets have orbits larger than the orbits of the known planets. *Lardner* (1853) stated " . . . we are in possession of the elements of the motions of 207 comets. It appears that 40 move in ellipses, 7 in hyperbolas, and 160 in parabolas." Lardner further divided the comets on elliptical orbits into three categories that roughly correspond to what we would now call Jupiter-family comets (JFCs),

Halley-type comets (HTCs), and returning long-period comets (LPCs) (e.g., *Levison,* 1996). The hyperbolic and parabolic orbits, in turn, represent "new" long-period comets (*Oort,* 1950; *Levison,* 1996). Lardner also noted that, with the exception of JFCs, there were roughly equal numbers of objects that revolved prograde (in the same direction as the planets) and retrograde around the Sun. [Note that the traditional dividing line between JFCs and HTCs is an orbital period P of 20 yr, with JFCs having P < 20 yr (semimajor axes, a, less than 7.4 AU), and HTCs having 20 yr ≤ P < 200 yr (7.4 AU ≤ a ≤ 34.2 AU). *Levison* (1996) introduced the term "nearly isotropic comets" (NICs), which he divided into HTCs, which he took to have semimajor axes a < 40 AU, and the LPCs, which he defined to have a > 40 AU. The upper limit on the semimajor axis of an HTC at 34.2 AU or 40 AU is somewhat arbitrary, but *Chambers* (1997) later showed that there *is* a dynamical basis for this upper limit. Chambers demonstrated that NICs with a < 22.5–39.6 AU, with the upper limit depending upon orbital inclination, can be trapped in mean-motion resonances with Jupiter, while bodies with larger orbits generally cannot.]

Newton (1891) and *van Woerkom* (1948) performed early studies of the effects of gravitational perturbations by Jupiter on cometary orbits. In particular, van Woerkom showed in detail that the observed distribution of cometary orbital energies was inconsistent with an interstellar origin for comets. It then fell to Oort, who had supervised the latter stages of van Woerkom's thesis work (*Blaauw and Schmidt,* 1993), to put the picture together. In his classic 1950 paper, Oort wrote "There is no reasonable escape, I believe, from the conclusion that the comets have always belonged to the solar system. They must then form a huge cloud, extending . . . to distances of at least 150,000 A.U., and possibly still further."

Interestingly, speculations by *Halley* (1705) in his famous *Synopsis of the Astronomy of Comets* can be interpreted as inferring a distant comet cloud. Halley was only able to fit parabolic elements to the 24 comet orbits he derived, but he argued that the orbits would prove to be elliptical, writing, "For so their Number will be determinate and, perhaps, not so very great. Besides, the Space between the Sun and the fix'd Stars is so immense that there is Room enough for a Comet to revolve, tho' the Period of its Revolution be vastly long."

This review will discuss what the observed cometary orbital distribution reveals about the structure of the spherical cloud of comets that now bears Oort's name, and the results of new dynamical simulations of the Oort cloud's formation and subsequent evolution. In section 2 we describe the Oort cloud hypothesis and the evidence for why we believe that there indeed is an Oort cloud. We then review studies of the population and dynamics of the Oort cloud. In section 3 we discuss the hypothetical inner Oort cloud, which has been proposed to possibly contain more comets than the classical Oort cloud, and "comet showers" that might result from a stellar passage through the inner cloud. In section 4 we focus on modern studies of the formation of the Oort cloud, assuming that comets started as planetesimals within the planetary region. In section 5 we discuss constraints on the Oort cloud based upon observations of comets and the impact record of the solar system, and describe future prospects for improving our understanding of the structure of the Oort cloud. Section 6 summarizes our conclusions. We refer the reader to other chapters in this book for discussions of related topics, particularly the overview by Rickman and the chapters by Weidenschilling, Rickman, Yeomans et al., Morbidelli and Brown, Duncan et al., Harmon et al., Boehnhardt, Weissman et al., and Jewitt.

2. POPULATION AND DYNAMICS OF THE OORT CLOUD

2.1. Oort Cloud Hypothesis

We first give an overview of how we believe that LPCs attained orbits at vast distances from the Sun, remained in such orbits for billions of years, and then came close enough to the Sun that they began to sublimate actively.

The early solar system is believed to have consisted of the planets, with their current masses and orbits, and a large number of remnant small solid bodies ("planetesimals") between and slightly beyond the orbits of the planets. We will assume there is no remaining gas in the solar nebula, and will only discuss planetesimals in the region of the giant planets, which we will take to be 4–40 AU, where the planetesimals were likely to contain volatiles such as water ice. Even if the small bodies started on orbits that did not cross the orbits of any of the planets, distant perturbations by the planets, particularly at resonances, would have excited most

of the planetesimals onto planet-crossing orbits in 10 m.y. or less (*Gladman and Duncan*, 1990; *Holman and Wisdom*, 1993; *Levison and Duncan*, 1993; *Grazier et al.*, 1999a,b). The major exception to this rule is in the Kuiper belt, where some orbits remain stable for billions of years (*Holman and Wisdom*, 1993; *Duncan et al.*, 1995; *Kuchner et al.*, 2002). [Here, we define the Kuiper belt to encompass small bodies on low-eccentricity orbits with semimajor axes, a, greater than 35 AU. Long-term orbital integrations indicate that *some* small bodies on near-circular orbits with a ≥ 35 AU are stable for the age of the solar system. For example, *Duncan et al.* (1995) found a stable region between 36 and 39 AU, while *Kuchner et al.* (2002) found that most objects with a ≥ 44 AU are stable for 4 G.y. The stability of objects with a between 39 and 44 AU depends upon their initial eccentricities and inclinations. There are few stable orbits for small bodies with a ≤ 35 AU, except for Trojans of Jupiter and Neptune (*Nesvorný and Dones*, 2002) and main-belt asteroids.] This quasistability explains the existence of the Kuiper belt and the low-inclination ("ecliptic") comets, which include the scattered disk, Centaurs, and JFCs (*Duncan et al.*, 2004).

The LPCs, by contrast, are thought to derive from the planetesimals that did *not* remain on stable orbits, but became planet-crossing. The first stage in placing a comet in the Oort cloud is that planetary perturbations pumped up the orbital energy (i.e., semimajor axis) of a planetesimal, while its perihelion distance q remained nearly constant. If the planets had been the only perturbers, this process would have continued, in general, until the planetesimal's orbit became so large that it became unbound from the solar system, and thereafter wandered interstellar space. However, the very reason that a comet's orbit becomes unbound at large distances — the presence of stars and other matter in the solar neighborhood that exert a gravitational force comparable to that from the Sun — provides a possible stabilizing mechanism. *Öpik* (1932) and *Oort* (1950) pointed out that once the comet's orbit becomes large enough, passing stars affect it. (As we describe below, gas in the solar neighborhood now appears to be a slightly stronger perturber of the Oort cloud than stars.) In fractional terms, stars change cometary perihelion distances much more than they change the overall size of the orbit. (This is a consequence of the long lever arm and slow speed of comets on highly eccentric orbits near aphelion.) If passing stars can lift a comet's perihelion out of the planetary region before the planets can eject it from the solar system, the comet will attain an orbit in the Oort cloud. The characteristic size of the Oort cloud is set by the condition that the timescale for changes in the cometary semimajor axis is comparable to the timescale for changes in perihelion distance due to passing stars. In essence, the comet must be perturbed to a semimajor axis large enough that the orbit is significantly perturbed by passing stars, but not so large that the orbit is too weakly bound to the solar system and the comet escapes. This condition yields a cloud of comets with semimajor axes on the order of 10,000 to 100,000 AU (*Tremaine*, 1993; see also *Heisler and Tremaine*, 1986; and *Duncan et al.*, 1987). The trajectories of the stars are randomly oriented in space, so stellar perturbations eventually cause the comets to attain a nearly isotropic velocity distribution, with a median inclination to the ecliptic of 90° and a median eccentricity of 0.7. Subsequently, passing stars *reduce* the perihelion distances of a small fraction of these comets so that they reenter the planetary region and potentially become observable.

The above description is similar to Oort's vision of the comet cloud. However, less than half the local galactic mass density is provided by stars, the rest being in gas, brown dwarfs, and possibly a small amount of "dark matter" (*Holmberg and Flynn*, 2000). We thus now recognize that the smooth long-term effect of the total amount of nearby galactic matter, i.e., the "galactic tide," perturbs comets somewhat more strongly than do passing stars. The galactic tide causes cometary perihelion distances to cycle outward from the planetary region and back inward again on timescales as long as billions of years (*Heisler and Tremaine*, 1986). In addition, rare, but large, perturbers such as giant molecular clouds (GMCs) may be important for the long-term stability of the Oort cloud.

Dynamically "new" comets typically come from distances of tens of thousands of AU, thereby giving the appearance of an inner edge to the Oort cloud. *Hills* (1981) showed that this apparent inner edge could result from an observational selection effect. The magnitude of the change in perihelion distance per orbit, Δq, of a comet due to either galactic tides or passing stars is a strong function of semimajor axis (a), proportional to $a^{7/2}$. A dynamically new comet with perihelion interior to Jupiter's orbit must have had $q \gtrsim 10$ AU on its previous orbit; otherwise, during the comet's last passage through perihelion, Jupiter and/or Saturn would have likely given it a large energy kick (typically much larger than the comet's orbital binding energy) that would either capture it to a much shorter period orbit or eject it to interstellar space.

If we assume that a comet must come within 3 AU of the Sun to become active and thus observable, Δq must be at least $\sim 10 - 3$ AU $= 7$ AU. It can be shown that, because of the steep dependence of Δq on a, this condition implies that $a \gtrsim 28{,}000$ AU (*Levison et al.*, 2001). Comets with semimajor axes of a few thousand AU could, in principle, be much more numerous than comets from tens of thousands of AU, but they normally would not pass within the orbits of Jupiter and Saturn because of the "Jupiter barrier." Such "inner Oort cloud" comets would only enter the inner solar system following an unusually strong perturbation, such as a close stellar passage. Determining the population of the hypothetical inner Oort cloud is a major goal of modern studies of the formation of the cometary cloud. We now turn to a more detailed discussion of what is known about the orbits of LPCs.

2.2. Observed Orbital Distribution

Figure 1 illustrates the orbital distribution of the 386 single-apparition LPCs whose energies are given in the 2003 *Catalogue of Cometary Orbits* (*Marsden and Will-*

iams, 2003). [The catalog contains 1516 single-apparition comets. Of these, only 386, or about one-quarter of the total, had observations that enable one to solve for the comet's energy. For the other comets, the fits assume a parabolic orbit.] We first introduce some terminology and notation. The symbol a represents the semimajor axis of a comet. The quantity actually determined in orbit solutions is $E \equiv 1/a$, which has units of AU^{-1}. (We will assume these units implicitly in the discussion below.) E, which we will informally refer to as "energy," is a measure of how strongly a comet is held by the Sun. We distinguish three values of E, which we denote E_i, E_o, and E_f. [For "original" and "future" orbits of LPCs (see below), a is computed with respect to the center or mass, or barycenter, of the solar system, while "osculating" orbits of objects in the inner solar system are computed with respect to the Sun.] These de-

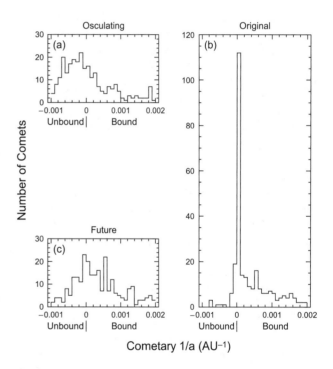

Fig. 1. Distribution of cometary orbital energies, $E \equiv 1/a$, where a is the comet's orbital semimajor axis in AU, from the 2003 *Catalogue of Cometary Orbits*. Only comets with -0.001 AU $< E \leq 0.002$ AU are shown, i.e., only comets whose orbits are apparently weakly unbound ($E < 0$) or weakly bound to the solar system (a $\geq$ 500 AU). The catalog contains 386 "single-apparition" LPCs for which the orbital energy could be determined. Of these, 268, 254, and 251 occupy the bins shown in (**a**), (**b**), and (**c**), respectively. All panels have the same horizontal and vertical scales. (**a**) Osculating value of "energy," E_i. (**b**) Original value of energy, E_o. (**c**) Future value of energy, E_f. The Oort cloud spike is not evident when the histogram is plotted in terms of the orbital energies during or after the comets' passages through the planetary region [(**a**) and (**c**)]. However, when planetary perturbations are "removed" by calculating the comet's orbit before it entered the planetary region, a spike of comets with a > 10,000 AU is evident [(**b**)]. The median semimajor axis of the comets shown in the spike is 27,000 AU. See text for further discussion.

note, respectively, the osculating (i.e., instantaneous) value of the comet's 1/a value when it is passing through the planetary region; the comet's original 1/a before it entered the planetary region (as determined by orbital integration); and the comet's future 1/a after it passes outside the planetary region. A comet with E > 0 is bound to the Sun, i.e., it follows an elliptical orbit. A comet with E < 0 is on a hyperbolic orbit and will escape the solar system on its current orbit; colloquially, such a comet is called "ejected." (Note that a comet's orbital energy per unit mass is $-GM_\odot/2a$, so the sign convention for E is the opposite of that used for orbital energy.) We will also use the symbols q and i to denote a comet's perihelion distance and orbital inclination to the ecliptic, respectively.

Osculating orbits of LPCs passing through the planetary region (Fig. 1a) indicated that many of the orbits were slightly hyperbolic, suggesting that those comets were approaching the solar system from interstellar space. However, when the orbits were integrated backward in time to well before the comets entered the planetary system, yielding the original inverse semimajor axis (denoted as E_0), the distribution changed radically (Fig. 1b). The E_0 distribution is marked by a sharp "spike" of comets at near-zero but bound energies, representing orbits with semimajor axes exceeding 10^4 AU; a low, continuous distribution of more tightly bound orbits; and a few apparently hyperbolic orbits. This is clearly not a random distribution.

Oort recognized that the spike had to be the source of the LPCs, a vast, roughly spherical cloud of comets at distances greater than 10^4 AU from the Sun, but still gravitationally bound to it. [Some researchers have noted that *Öpik* (1932) anticipated Oort's work by studying the effects of stellar perturbations on distant meteoroid and comet orbits, 18 years earlier. Öpik suggested that stellar perturbations would raise the perihelia of comets, resulting in a cloud of objects surrounding the solar system. However, he specifically rejected the idea that comets in the cloud could ever be observed, even indirectly, because he did not recognize that stellar perturbations would also cause some orbits to diffuse back into the planetary region. Öpik concluded that the observed LPCs came from aphelion distances of only 1500–2000 AU. Though *Öpik*'s (1932) paper was a pioneering work on stellar perturbations, it did not identify the cometary cloud as the source of the LPCs or relate the observed orbits to the dynamical theory.] Oort showed that comets in the cloud are so far from the Sun that perturbations from random passing stars can change their orbits and occasionally send some comets back into the planetary system. Oort's accomplishment in defining the source of the LPCs is particularly impressive when one considers that it was based on only 19 well-determined cometary orbits, compared with the 386 high-quality orbits in the 2003 catalog.

In Fig. 1b, about 30% (112) of all 386 comets have $0 \leq E_0 \leq 10^{-4}$. [Another 87 of the comets have $1 \times 10^{-4} < E_0 \leq 2 \times 10^{-3}$, with 132 comets off-scale to the right (i.e., with semimajor axes <500 AU); see Fig. 2.] This region, which corresponds to semimajor axes >10^4 AU, is the spike that

led Oort to postulate the existence of the Oort cloud. Oort suggested that most new comets have aphelion distances of 50,000–150,000 AU, i.e., semimajor axes of 25,000–75,000 AU. More recent determinations give values about half as large for the typical semimajor axes of new comets. The median semimajor axis of the 143 comets in the 2003 *Catalogue of Cometary Orbits* with $E_0 \leq 10^{-4}$ (including those with E_0 slightly less than 0, see below) is 36,000 AU, and is 27,000 AU for the 112 comets with $0 \leq E_0 \leq 10^{-4}$. Even these estimates of the typical value of the semimajor axes may be too large, since these orbit fits do not take into account nongravitational forces.

Thirty-one comets shown in Fig. 1b have $E_0 < 0$. Taken at face value, these comets could be intruders just passing through the solar system. It is more likely that most or all of these comets actually follow elliptical orbits, and that the "hyperbolic" orbits are a consequence of observational errors and/or inexact modeling of nongravitational forces. [If the comets with $E_0 < 0$ were interstellar in origin, they would likely have speeds at "infinity" comparable to the velocity dispersion of disk stars, or tens of kilometers per second (Fig. 1 in *McGlynn and Chapman*, 1989). Such a velocity would imply $E_0 \sim -1$, much larger than the most negative value of E_0 measured for any comet (*Wiegert*, 1996).]

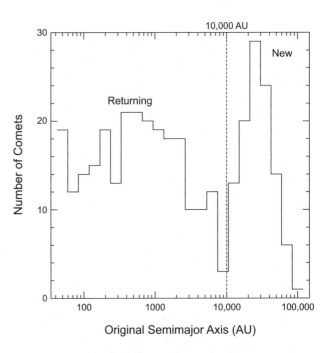

Fig. 2. Distribution of original semimajor axes, a_0. Comets are sometimes classified as "new" or "returning," depending on whether a_0 is greater than or less than 10,000 AU, respectively. However, this classification is crude. To determine whether a particular comet is "new," it must be integrated backward one orbit, under the influence of the Sun and planets, galactic tides, and possibly nongravitational forces and nearby stars (*Dybczyński*, 2001).

Nongravitational forces make orbits (both "hyperbolic" and elliptical) appear more eccentric (i.e., larger) than they actually are (*Marsden et al.*, 1973, 1978; *Królikowska*, 2001). *Marsden et al.* (1973) estimate that if nongravitational forces were correctly accounted for, the average E_o value of a comet would increase by 2×10^{-5}. [This correction is up to 100 times larger for some "hyperbolic" comets (*Królikowska*, 2001).] If this correction applied for all new comets, a comet with a nominal E_o value of 1×10^{-5}, corresponding to a semimajor axis of 100,000 AU, would actually have a semimajor axis of 33,000 AU, and a comet with a nominal a_o of 20,000 AU would have a true semimajor axis of 14,000 AU.

In addition, *Marsden et al.* (1978) showed that orbital fits that neglect nongravitational forces give systematically larger "original" semimajor axes for comets with smaller perihelion distances, for which nongravitational forces are typically more important. They derived an empirical relation $\langle E_o \rangle = (4.63 - 2.37/q) \times 10^{-5}$ for the average original semimajor axis for new comets with a perihelion distance of q measured in AU. In the limit of large q, for which nongravitational forces are less important, this relation gives an average original semimajor axis $a_{ave} = 1/4.63 \times 10^{-5} = 21,600$ AU. Since new comets have e ~ 1, this implies a typical aphelion distance of 43,200 AU and a time-averaged distance $a_{ave} (1 + \frac{1}{2}e^2)$ ~ 32,000 AU. Thus a typical Oort cloud comet resides some 30,000 AU from the Sun, which it circles once every 3 m.y.

Simulations by *Heisler* (1990) predict that during times of low comet flux, the energies of new comets should be peaked near $E_o = 3.5 \times 10^{-5}$, i.e., at a semimajor axis near 29,000 AU. Heisler assumed a local mass density of 0.185 $M_\odot$/pc^3. If the currently accepted value of 0.1 $M_\odot$/pc^3 is assumed instead (see discussion in *Levison et al.*, 2001), the peak semimajor axis should be near 34,000 AU. Thus Heisler's model predicts a semimajor axis that is larger than the inferred location of the peak. This discrepancy could result from errors in orbit determination, contamination of the "new" comet population with dynamically old comets with a > 10^4 AU, or, as Heisler proposed, could indicate that we are presently undergoing a weak comet shower (section 3). Neither the models nor data are yet adequate to determine which explanation is correct.

Figure 1c shows the "future" orbits of the comets; 96 of 386 (25%) are slightly hyperbolic, indicating that they will not return to the planetary region again and will leave the solar system. On their first pass through the planetary system, the distant, random perturbations by Jupiter and the other giant planets eject roughly half the "new" comets to interstellar space, while capturing the other half to smaller, more tightly bound, less-eccentric orbits (*van Woerkom*, 1948); see below. Only about 5% of the new comets are returned to Oort cloud distances of 10^4–10^5 AU (*Weissman*, 1979). On subsequent returns the comets continue to random-walk in orbital energy until they are ejected, are destroyed by one of several poorly understood physical mechanisms (see section 2.3), are captured to a "short-period"

orbit with a revolution period less than 200 yr, or collide with the Sun or a planet. [There are two classes of short-period comets, HTCs and JFCs. The HTCs encompass 41 known objects with a median orbital period of 70.5 yr and a median inclination of 64°. Some or most HTCs may originate in the Oort cloud (*Levison et al.*, 2001). The JFCs include 236 known comets with a median orbital period of 7.5 yr and an median inclination of 11°. By contrast with HTCs, most JFCs probably do not originate in the Oort cloud. The small inclinations of the JFCs argued for the existence of a low-inclination source region, i.e., the Kuiper belt (*Joss*, 1973; *Fernández*, 1980a; *Duncan et al.*, 1988, *Quinn et al.*, 1990; *Fernández and Gallardo*, 1994), but it now appears likely that most JFCs arise from the related structure called the scattered disk (*Duncan et al.*, 2004). The numerical data listed here were derived from the tables of HTCs and JFCs in the Web page of Y. Fernández (http://www.ifa.hawaii.edu/~yan/cometlist.html).]

In Fig. 2 we show the bound comets with a < 10^5 AU on a logarithmic scale. This plot indicates that there are about twice as many comets with "original" semimajor axes (a) ranging from tens to thousands of AU, compared to the number with a > 10^4 AU. Those with a < 10^4 AU are often called "returning" comets; those with a > 10^4 AU are called dynamically "new" comets. The reason for this terminology is as follows. The median value of E_o for the new comets is $1/27,000 = 3.7 \times 10^{-5}$. The magnitude of the typical energy change, $|\Delta E|$, which these comets undergo in one perihelion passage is ~10^{-3}, i.e., more than an order of magnitude larger (*Marsden and Williams*, 2003; cf. *van Woerkom*, 1948; *Everhart*, 1968; *Everhart and Raghavan*, 1970). Since $|\Delta E| \gg E_o$, about half the comets have $E_f = E_o - \Delta E$ ~ $-\Delta E$ and the other half have $E_f = E_o + \Delta E$ ~ $+\Delta E$. The former are ejected from the solar system; the other half are captured onto more tightly bound orbits with a ~ $1/\Delta E$, i.e., semimajor axes of a few thousand AU.

Thus comets with original values of a > 10^4 AU are unlikely to have passed within the orbits of Jupiter and Saturn, that is, within 10 AU of the Sun, in their recent past. (By contrast, the perturbations due to Uranus and Neptune are much smaller. Typical energy perturbations are proportional to M_p/a_p, where M_p is the planet's mass and a_p is its semimajor axis.) The condition that a > 10^4 AU is only a rough criterion for a "new" comet, since the distribution of energy changes is broad and centered on zero. *Dybczyński* (2001) gives a detailed analysis of the past histories of LPCs with well-determined orbits. Some 55% of the observed new comets (statistically consistent with the expected 50%) have $E_f < 0$ and will not return. Only 7% of the returning comets are ejected on their current apparition; since most have recently traversed the planetary region a number of times, they typically have $E_o > |\Delta E|$ (*Quinn et al.*, 1990; *Wiegert and Tremaine*, 1999). The most tightly bound comet in the plot has a = 40.7 AU and an orbital period $40.7^{3/2} = 260$ yr. (Conventionally, LPCs have been taken to be those with orbital periods greater than 200 yr, since until the discovery of Comet 153P = C/2002 C1 Ikeya-Zhang = C/1661

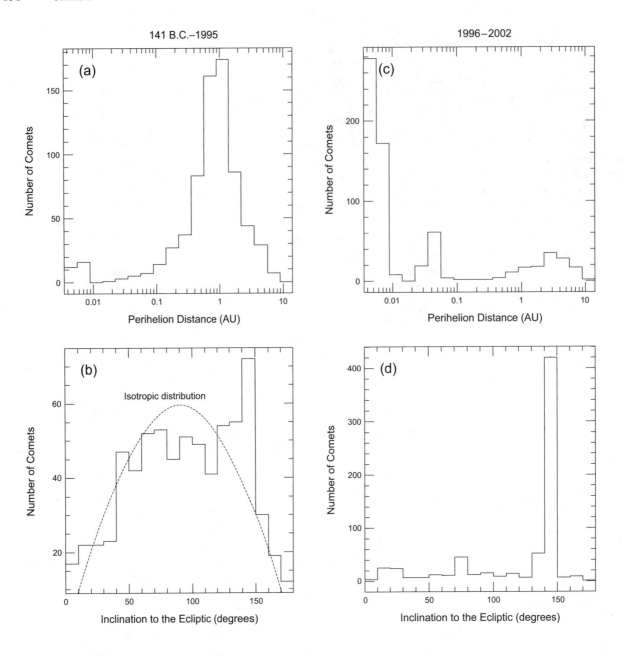

Fig. 3. Distribution of perihelion distances and inclinations to the ecliptic for the 706 historical (i.e., through 1995) single-apparition comets (left panels) and the 680 recent (1996–February 2003) single-apparition comets (right panels) from the 2003 *Catalogue of Cometary Orbits*. **(b)** shows a fit to an isotropic distribution for the non-Sun-grazers. The righthand panels are labeled 1996–2002 because the 2003 *Catalogue* is only complete through 2002, although it does include a few comets discovered early in 2003. The differences between the left and right panels reflect observational selection effects in the discovery of comets. There was a distinct change in the way comets were discovered in the mid-1990s because of (1) the discovery of numerous small Sun-grazing comets by the SOHO spacecraft, which was launched in December 1995, and (2) increased numbers of discoveries of all classes of comets by the automated searches for near-Earth objects begun around the same time.

C1 Hevelius, apparitions of a comet with P > 200 yr had never been definitively linked.)

Figure 3 shows the distribution of perihelion distances, q, and inclination to the ecliptic, i, for the 1386 "single-apparition" LPCs tabulated by *Marsden and Williams* (2003). The lefthand panels plot 706 "historical" comets, starting with C/-146 P1 and ending with C/1995 Y1. The righthand panels show 680 recent comets, starting with C/1996 A2 and ending with C/2003 B1. First consider the

historical comets. The observed perihelion distribution (Fig. 3a) is peaked near q = 1 AU because of two factors with opposite dependences on heliocentric distance. First, historically, comets have only been discovered if they passed well within the orbit of Jupiter (5.2 AU), since water sublimes more readily (and hence cometary activity is more vigorous) when comets are closer to the Sun (*Marsden et al.*, 1973). For comets with q ≤ 3 AU, the total brightness of an "average" comet typically scales as $R^{-4}\Delta^{-2}$, as compared with $R^{-2}\Delta^{-2}$ for a bare nucleus, where R is the comet's distance from the Sun and Δ is its distance from Earth. Thus comets that closely approach the Sun or Earth are brighter and therefore easier to discover (*Everhart*, 1967a). Second, dynamical models suggest that the intrinsic number of comets per unit perihelion distance probably increases with increasing q throughout the entire planetary region (*Weissman*, 1985; *Dones et al.*, 2004).

Figure 3a also shows a smaller peak of comets with perihelion distances $\lesssim$0.01 AU (i.e., less than or approximately twice the radius of the Sun). These are Sun-grazing comets that have likely been driven onto small-q orbits by the secular perturbations of the planets (*Bailey et al.*, 1992). Most of these comets are members of the Kreutz family, which may be the remnants of a comet that broke up near perihelion (*Marsden*, 1967, 1989).

Figure 3b shows the inclination distribution of the historical comets, which roughly resembles an isotropic distribution (dashed curve). *Everhart* (1967b) showed that the departures from an isotropic distribution due to observational selection effects are small, with a ~ 10% preference for discovery of retrograde comets. This indicates that the observable Oort cloud is roughly spherical. The excess of comets with $140° \leq i \leq 150°$ is primarily due to the Kreutz Sun-grazers.

Figure 3c shows the perihelion distribution of comets discovered since 1996. About two-thirds of the comets are Sun-grazers (q < 0.01 AU), with a secondary peak centered near 0.04 AU due to the Meyer, Marsden, and Kracht "near Sun" groups (*Marsden and Meyer*, 2002) [see the Web pages of M. Meyer (http://www.comethunter.de/groups.html), J. Shanklin (http://www.ast.cam.ac.uk/~jds/kreutz.htm), and the U.S. Naval Research Laboratory (http://ares.nrl.navy.mil/sungrazer/)]. Almost all these comets have been discovered by the Solar and Heliospheric Observatory (SOHO) spacecraft (*Biesecker et al.*, 2002). From their apparent failure to survive perihelion passage, the SOHO Sun-grazers must be $\leq$0.1 km in diameter (*Weissman*, 1983; *Iseli et al.*, 2002).

In contrast to the historical discoveries, the distribution of recently discovered comets with q > 0.1 AU peaks not near 1 AU, but rather about 3 AU from the Sun. The outward march of this peak indicates that discovery of LPCs with large perihelia is still severely incomplete. For example, *Hughes* (2001) concludes that LPCs are still being missed beyond 2.5 AU. The advent of electronic detectors and automated near-Earth object surveys has recently led to the discovery of a few LPCs with the largest perihelion distances ever found, including C/1999 J2 Skiff (LONEOS survey, q = 7.11 AU), C/2000 A1 Montani (Spacewatch, q = 9.74 AU), and C/2003 A2 Gleason (Spacewatch, q = 11.43 AU). (See http://www.ifa.hawaii.edu/~yan/cometlist.html for a list of comets with q > 5 AU.) Inferring the true perihelion distribution for active comets at large q would require correcting for observational biases in comet discoveries. Performing this correction is difficult (and has not yet been attempted) because dynamically new comets are often anomalously bright at large heliocentric distances (*Oort and Schmidt*,

1951) on the inbound leg of their orbits, possibly because of sublimation of some type of ice such as CO that is more volatile than water ice.

Finally, Figure 3d shows the inclination distribution of the recently discovered comets. The peaks centered near 20, 75, and 145° are due to the Marsden/Kracht, Meyer, and Kreutz groups of Sun-grazers, respectively.

2.3. Cometary Fading and Disruption

Oort pointed out in his 1950 paper that the number of returning comets in the low continuous distribution (the "returning" comets) decayed at larger values of E. That is, as comets random-walked away from the Oort cloud spike, the height of the low continuous distribution declined more rapidly than could be explained by a purely dynamical model using planetary and stellar perturbations. [In a simple model that considers only the effects of planetary perturbations, and in which comets survive for an infinite length of time, the energy distribution of returning comets should be a constant for energies large compared to the magnitude of a typical energy perturbation (see *Oort*, 1950; *Lyttleton and Hammersley*, 1964) (Fig. 4).] This problem is commonly referred to as "cometary fading," although "fading" is a misnomer as it implies a gradual decline of activity. In fact, it is still not clear what the exact mechanism for fading is. Three physical mechanisms have been proposed to explain the failure to observe as many returning comets as are expected (*Weissman*, 1980a; *Weissman et al.*, 2002). These include (1) random disruption or splitting due to, e.g., thermal stresses, rotational bursting, impacts by other small bodies, or tidal disruption (*Boehnhardt*, 2002, 2004); (2) loss of all volatiles; and (3) formation of a nonvolatile crust or mantle on the nucleus surface (*Whipple*, 1950; *Brin and Mendis*, 1979; *Fanale and Salvail*, 1984). In these three cases, the comet is referred to as, respectively, "disrupted," "extinct," or "dormant." Recently, *Levison et al.* (2002) argued that spontaneous, catastrophic disruption of comets was the dominant physical loss mechanism for returning comets. In any case, the "fading" mechanism must be a physical one; the missing comets cannot be removed by currently known dynamical processes alone (*Weissman*, 1979, 1982; *Wiegert and Tremaine*, 1999).

Oort handled fading by introducing a factor, k, where k is "the probability that a comet is disrupted during a perihelion passage." (Note that Oort specifically called this "disruption" rather than "fading.") Oort adopted a value of k = 0.019, or 0.017 if "short-period" comets with orbital periods <50 yr were omitted. However, Oort found that this value removed comets too rapidly from the system, and thus suggested a slightly lower value of k = 0.014.

In Fig. 4 we show the distribution of original energies, E_o, for the 386 LPCs with the best-determined orbits. If there were no "fading" (section 2.3), the distribution of E_o should be approximately constant for $E_o \gg 10^{-4}$ (dotted line). The actual distribution (solid line) lies far below the expected distribution, implying that many of the comets must have "faded." Models in which surviving comets have

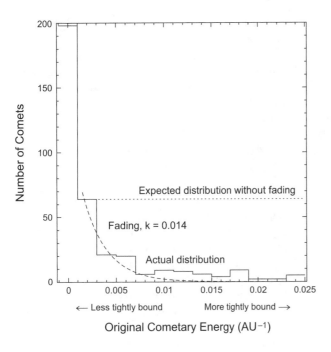

Fig. 4. Distribution of original cometary orbital energy for all 386 comets in the 2003 *Catalogue of Cometary Orbits* with well-determined energies. The histogram represents the observed distribution. The dashed curves give theoretical distributions with and without cometary "fading." *Oort* (1950) developed a simple model in which the number of "old" comets passing perihelion in a given period of time with energy E is proportional to $e^{-\alpha E}$ in the limit of large E. In this expression $\alpha = \sqrt{\frac{4k}{\pi(1-k)}}$; the unit of E is the mean magnitude of the energy perturbation per orbit produced by the planets, which we have taken to be 3.3×10^{-4} AU^{-1}; and k is the probability of disruption per orbit, which we have taken to be either 0 (flat curve) or 0.014 (declining curve). If no fading is assumed, the observed curve is far below the prediction of the model, while when a finite probability of fading is assumed, the model agrees somewhat better with observations. More elaborate fading models (*Weissman*, 1979, 1980a; *Wiegert and Tremaine*, 1999) are in good agreement with the observations, but the results are not necessarily unique. Although many authors have modeled the fading problem, there is still not a definitive physical explanation for fading. See section 2.3 for further discussion.

a constant probability of disruption per perihelion passage (dashed curve) or in which the number decays as a power law in the number of apparitions provide reasonable fits to the actual distribution.

Whipple (1962) treated the problem somewhat differently, modeling the expected cumulative "lifetime" distribution of the LPCs as a power law, $L^{-\kappa}$, where L is the number of returns that the comet makes. Whipple found that $\kappa = 0.7$ with an upper limit on the order of 10^4 returns gave the best fit to the observed orbital data.

Weissman (1979) was the first to use a Monte Carlo simulation to derive the expected cometary orbital energy distribution, including realistic models of the expected loss rate due to a variety of physical destruction mechanisms

(*Weissman*, 1980a). Weissman's simulations contained a parameterization that accounted for cometary perturbations by Jupiter and Saturn, and comets were also perturbed by random passing stars and nongravitational forces. Comets were removed by collisions, random disruption (splitting), and loss of volatiles (sublimation of ices). A fairly good match to the observed E_o distribution in Figs. 1b and 4 was obtained. By tuning such a model to improve the fit, some insight into the possible physical and dynamical loss mechanisms was obtained. Weissman's best fit was with a model in which 10% of dynamically new comets randomly disrupted on their first return and 4% of returning comets disrupted on each subsequent return, with 15% of all comets being immune to disruption.

Wiegert and Tremaine (1999) (see also *Bailey,* 1984) investigated the fading problem by means of direct numerical integrations that included the gravitational effects of the Sun, the four giant planets, and the "disk" component of the galactic tide (see below). They carefully examined the effects of nongravitational forces on comets, as well as the gravitational forces from a hypothetical solar companion or circumsolar disk 100–1000 AU from the Sun. However, like previous authors, Wiegert and Tremaine found that the observed E_o distribution could only be explained if some physical loss process was invoked. They found that they could match the observed E_o distribution if the fraction of comets remaining observable after L passages was proportional to $L^{-0.6 \pm 0.1}$, consistent with the fading law proposed by *Whipple* (1962), or if ~95% of LPCs remain active for only ~6 returns and the remainder last indefinitely.

Historically, most comets have been discovered by amateurs. Determining the true population of comets requires a detailed understanding of observational bias, i.e., the probability that a comet with a specified brightness and orbit will be discovered. Many sources of bias have been identified, but have generally not been modeled in detail (*Everhart*, 1967a,b; *Kresák*, 1982; *Horner and Evans*, 2002; *Jupp et al.*, 2003). In recent years, telescopic surveys that primarily discover asteroids have discovered both active comets and inactive objects on comet-like orbits, which are sometimes called Damocloids. For example, the Near Earth Asteroid Tracking (NEAT) system discovered 1996 PW (*Helin et al.*, 1996), an object of asteroidal appearance that has a = 287 AU, q = 2.5 AU, and i = 30° (*Weissman and Levison*, 1997a). Discoveries by surveys are much better characterized than discoveries by amateurs, particularly for bodies that show little or no cometary activity. Using statistical models of discoveries of inactive (extinct or dormant) comets by surveys (*Jedicke et al.*, 2003), *Levison et al.* (2002) calculated the number of inactive, nearly isotropic comets (NICs) that should be present in the inner solar system. Their study used orbital distribution models from *Wiegert and Tremaine* (1999) and *Levison et al.* (2001) that assumed no disruption of comets. *Levison et al.* (2002) then compared the model results to the 11 candidate dormant NICs (mostly HTCs) that had been discovered as of December 3, 2001. Dynamical models that assume that comets merely stop outgassing predict that surveys should have discovered ~100 times more inactive NICs than are actually seen. Thus, as comets evolve inward from the Oort cloud, 99% of them become unobservable, presumably by breaking into much smaller pieces that rapidly dissipate.

A complication in modeling fading arises because Oort cloud comets on their first perihelion passage are often anomalously bright at large heliocentric distances compared to missing "returning" comets (*Oort and Schmidt*, 1951; *Donn*, 1977; *Whipple*, 1978), and thus their probability of discovery is considerably enhanced. Suggested mechanisms for this effect include a veneer of volatiles accreted from the interstellar medium and lost on the first perihelion passage near the Sun (*Whipple*, 1978), blow-off of a primordial cosmic-ray-processed nucleus crust (*Johnson et al.*, 1987), or the amorphous-to-crystalline water ice phase transformation that occurs at about 5 AU inbound on the first perihelion passage (*Prialnik and Bar-Nun*, 1987). When these Oort cloud comets return, they are generally not observed unless they come within about 3 AU of the Sun, where water ice can begin to sublimate at a sufficient rate to produce an easily visible coma (*Marsden and Sekanina*, 1973). This is illustrated in Fig. 5. The failure to observe many returning LPCs with q > 3 AU is likely to be an observational selection effect, as there is no known physical and/or dynamical mechanism for preferentially removing them. Thus, in comparing the heights of the E_o spike and low distribution, one should only consider comets with q < 3 AU. Considering only comets with q < 3 AU slightly alleviates the fading problem; the ratio of the number of returning to new comets is now 2.5, compared with 1.7 when all 386 high-quality orbits are used. Nonetheless, a returning-to-new ratio of 2.5 is still more than 10 times smaller than predicted by models without fading (*Wiegert and Tremaine*, 1999).

2.4. Population and Mass of the Oort Cloud

To account for the observed flux of dynamically new LPCs, which he assumed to be about 1 per year within 1.5 AU of the Sun, Oort estimated that the population of the cometary cloud was 1.9×10^{11} objects. Oort stated that a "plausible estimate . . . of the average mass of a comet . . . is perhaps about 10^{16} g . . . uncertain by one or two factors of 10." For an assumed density of 0.6 g/cm³, a cometary mass of 10^{16} g corresponds to a diameter of 3.2 km. More recent dynamical models (*Heisler*, 1990; *Weissman*, 1990a) have produced somewhat higher estimates of the number of comets in the Oort cloud, by up to an order of magnitude. These larger numbers come about in part from higher estimates of the flux of LPCs throughout the planetary system, and in part from a recognition of the role of the giant planets in blocking the diffusion of cometary orbits back into the planetary region (*Weissman*, 1985). Comets perturbed inward to perihelia near the orbits of Jupiter and Saturn will likely be ejected from the solar system before they can diffuse to smaller perihelia where they can be observed. Thus,

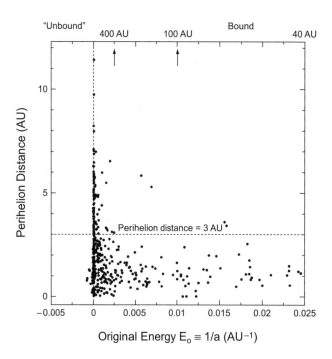

Fig. 5. Scatter diagram in original orbital energy and perihelion distance for the observed LPCs. The vertical band of comets at near-zero E_o is comets making their first perihelion passage from the Oort cloud. Comets diffuse left and right in the diagram as a result of planetary perturbations, primarily by Jupiter (in general, planetary perturbations do not significantly alter either the perihelion distance or the inclination of LPC orbits). Comets perturbed to negative values of E_o escape the solar system. Note the low number of LPCs with perihelion distances q > 3 AU and values of $E_o > 10^{-4}$. This deficit is likely an observational selection effect due to the inability of these comets to generate visible comae through water ice sublimation. Water ice sublimates poorly beyond 3 AU from the Sun. Data from *Marsden and Williams* (2003).

the terrestrial planets region is undersupplied in LPCs as compared with the outer planets region. This effect is known as the "Jupiter barrier." We return to this topic in section 5.

Heisler (1990) performed a sophisticated Monte Carlo simulation of the evolution of the Oort cloud, assuming it had formed with the centrally condensed density profile found by *Duncan et al.* (1987) (hereafter *DQT87*; see section 4). Assuming a new comet flux of 2.1 comets/year with q < 1 AU and "absolute magnitude" $H_{10} < 11$, *Heisler* (1990) inferred that the present-day Oort cloud contains 5×10^{11} comets with a > 20,000 AU and $H_{10} < 11$. *Weissman* (1996) relates H_{10}, which is a measure of a comet's total brightness that is generally dominated by coma, to cometary masses, using 1P/Halley to calibrate the relation (see also *Harmon et al.*, 2004). According to *Weissman* (1996), the diameter and mass of a comet with $H_{10} = 11$ are 2.3 km and 4×10^{15} g respectively. Assuming a broken power-law cometary size distribution from *Everhart* (1967b) (see also *Weissman and Levison*, 1997b), and assuming that a comet's luminosity at a standard distance is proportional to its mass, *Weissman*

(1996) infers that the average mass of a comet is 4×10^{16} g. Using *Heisler*'s (1990) modeled population, this implies a present-day mass of 2×10^{28} g or 3.3 $M_\oplus$ in comets with a > 20,000 AU. *Weissman* (1996) estimated that there are 1×10^{12} comets with a > 20,000 AU and H_{10} < 11, giving a mass for the outer Oort cloud (comets with a > 20,000 AU) of 7 $M_\oplus$. Weissman then assumed, based on *DQT87*, that the inner Oort cloud (a < 20,000 AU) contains about 5 times as much mass as the outer Oort cloud, giving a total present-day Oort cloud mass of 38 $M_\oplus$. However, this estimate is based upon a formation model and not on observations, since (1) comets from the hypothetical inner Oort cloud are not perturbed into the planetary region except during strong comet showers, which only occur some 2% of the time (*Heisler*, 1990), and (2) we are not presently undergoing a strong comet shower (*Weissman*, 1993). We further discuss the population of the inner Oort cloud in sections 3 and 4.

2.5. Oort Cloud Perturbers

Since first proposed in 1950, Oort's vision of a cometary cloud gently stirred by perturbations from distant passing stars has evolved considerably. Additional perturbers have been recognized: GMCs in the galaxy, which were unknown before 1970 (*Biermann*, 1978; *Clube and Napier,* 1982), and the galactic gravitational field itself, in particular the tidal field of the galactic disk (*Byl*, 1983, 1986; *Harrington*, 1985; *Heisler and Tremaine*, 1986). GMC encounters are rare, occurring with a mean interval of perhaps $3–4 \times 10^8$ yr, but can result in major perturbations on the orbits of comets in the Oort cloud. *Hut and Tremaine* (1985) showed that the integrated effect of molecular clouds on the Oort cloud over the history of the solar system is roughly equal to the integrated effects of all stellar passages. Atomic clouds have much smaller effects on the Oort cloud than do stars or molecular clouds (*Hut and Tremaine*, 1985).

The galactic field sets the limits on the outer dimensions of the Oort cloud. The cloud can be roughly described as a prolate spheroid with the long axis oriented toward the galactic center (*Antonov and Latyshev*, 1972; *Smoluchowski and Torbett*, 1984). Maximum semimajor axes are about 1×10^5 AU (i.e., 0.5 pc, or almost 40% the distance to the nearest star) for direct orbits in the galactic plane, decreasing to about 8×10^4 AU for orbits perpendicular to the galactic plane, and increasing to almost 1.2×10^5 AU for retrograde orbits (opposite to galactic rotation).

In addition, stars will occasionally pass directly through the Oort cloud, ejecting some comets and severely perturbing the orbits of others (*Hills*, 1981). A star passage drills a narrow tunnel through the Oort cloud, ejecting all comets within a radius of ~450 AU, for a 1 $M_\odot$ star passing at a speed of 20 km s^{-1} (*Weissman*, 1980b). Over the history of the solar system, Weissman estimated that passing stars have ejected about 10% of the Oort cloud population. The ejected comets will all be positioned close to the path of the perturbing star, as will many of the comets that are thrown into the planetary system in a "cometary shower" (*Weissman*, 1980b; *Dybczyński*, 2002a,b). An extremely close stellar

encounter (interior to the inner edge of the Oort cloud) can, in principle, eject a large fraction of the comets in the entire cloud, because the star pulls the Sun away from the cloud (*Heisler et al.*, 1987). Such drastic encounters have probably not occurred in the past 4 b.y., but may have taken place in the early solar system if the Sun formed in a cluster.

García-Sánchez et al. (1999, 2001; see also *Frogel and Gould*, 1998) used Hipparcos and groundbased data to search for stars that have encountered or will encounter the solar system during a 20-m.y. interval centered on the present. Correcting for incompleteness, *García-Sánchez et al.* (2001) estimate that 11.7 ± 1.3 stellar systems pass within 1 pc (~200,000 AU) of the Sun per million years, so that ~50,000 such encounters should have occurred over the history of the solar system if the Sun had always occupied its current galactic orbit and environment. However, 73% of these encounters are with M dwarfs, which have masses less than 0.4 $M_\odot$. Strong comet showers are generally caused by stars with masses ~1 $M_\odot$. Passages through the Oort cloud by M dwarfs and brown dwarfs typically produce little change in the cometary influx to the planetary region (*Heisler et al.*, 1987).

It is now established that the galactic disk is the major perturber of the Oort cloud at most times (*Harrington*, 1985; *Byl*, 1986; *Heisler and Tremaine*, 1986; *Delsemme*, 1987), though stars and probably GMCs still play an important role in repeatedly randomizing the cometary orbits. Galactic tidal perturbations peak for orbits with their line of apsides at galactic latitudes of ±45° and go to zero at the galactic equator and poles. *Delsemme* (1987) showed that the distribution of galactic latitudes of the aphelion directions of the observed LPCs mimics that dependence. Although a lack of comet discoveries near the galactic equator could be the result of observational selection effects (e.g., confusion with galactic nebulae), the lack of comets near the poles appears to confirm the importance of the galactic tidal field on the Oort cloud.

The galactic tide causes the cometary perihelia to oscillate on timescales on the order of 1 b.y. (*Heisler and Tremaine*, 1986; *DQT87*). In general, the effect of the tide is stronger than that of passing stars because (1) the typical magnitude of galactic tidal perturbations is greater than the perturbation from stars for comets at a particular semimajor axis; and (2) the tide produces a regular stepping inward of cometary perihelia, in contrast to the random-walk nature of stellar perturbations. As a result, tides bring comets into the observable region more efficiently, making it somewhat easier to overcome the dynamical barrier that Jupiter and Saturn present to cometary diffusion into the inner planets region.

Hut and Tremaine (1985) estimated that the dynamical half-life of comets in the Oort cloud due to the effects of passing stars is about 3 G.y. at 25,000 AU and about 1 G.y. at 50,000 AU (see also *Weinberg et al.*, 1987). *Hut and Tremaine* (1985) estimated that the effects of GMCs on the Oort cloud are comparable to those of stars, though there are many uncertainties in how to treat clouds. Thus, due to stellar perturbations, only about 5% of the comets should survive at 50,000 AU for 4.5 G.y., while 5% should survive at 30,000 AU if the effects of clouds are included. Some authors have estimated even shorter lifetimes (e.g., *Bailey,* 1986). This led to suggestions that the observable, "outer" Oort cloud must be replenished, for example, by capture of comets from interstellar space, as suggested by *Clube and Napier* (1984). However, cometary capture is an unlikely process because a three-body gravitational interaction is required to dissipate the excess hyperbolic energy. *Valtonen and Innanen* (1982) and *Valtonen* (1983) showed that the probability of capture is proportional to V_∞^{-7} for $V_\infty \geq 1$ km/s, where V_∞ is the hyperbolic excess velocity. Capture is possible at encounter velocities ≤ 1 km s^{-1}, but is highly unlikely at the Sun's velocity of ~20 km s^{-1} relative to the local standard of rest (*Mignard,* 2000).

More plausibly, the outer Oort cloud could be resupplied from an inner Oort cloud reservoir, i.e., comets in orbits closer to the Sun (*Hills,* 1981; *Bailey,* 1983) that are pumped up by passing stars to replace the lost comets. However, due to uncertainties in cloud parameters and the history of the solar orbit, it may be premature to conclude that the outer Oort cloud has been so strongly depleted during its lifetime that a *massive* inner Oort cloud is *required* to replenish the outer cloud. In particular, existing models of the effect of molecular clouds on the Oort cloud make highly idealized assumptions about the structure of molecular clouds, and are sensitive to assumptions about the history of the Sun's orbit (e.g., the extent of its motion out of the galactic plane). Finally, molecular clouds are part of a "fractal" or "multifractal" continuum of structure in the interstellar medium (*Chappell and Scalo,* 2001). The resulting spatial and temporal correlations in interstellar gas density will result in a much different spectrum of gravitational potential fluctuations experienced by the Oort cloud, compared to an interstellar model that has clouds distributed independently and randomly (J. Scalo, personal communication, 2003). We now turn to a more detailed discussion of the hypothetical inner cloud.

3. INNER OORT CLOUD AND COMET SHOWERS

In Oort's original model, he assumed that the velocity distribution of comets in the Oort cloud is given by an isotropic distribution of the form $f(v) = 3v^2/v_{max}^3$ for $v < v_{max}$ and $f(v) = 0$ for $v > v_{max}$. The velocity v_{max} is a function of distance from the Sun, r, determined by an assumed outer edge of the cloud at distance R_0. Specifically,

$$v_{max} = \sqrt{\frac{2GM_\odot}{R_0}\left(\frac{R_0}{r} - 1\right)}$$

with limiting cases

$$v_{max} \rightarrow \sqrt{2GM_\odot/r}$$

(i.e., the local escape velocity) for $r \ll R_0$ and $v_{max} \rightarrow 0$ for $r \rightarrow R_0$. Assuming that the Oort cloud is in equilibrium (i.e.,

the "pressure" due to the random motions of the comets balances the inward attraction due to solar gravity), this assumed velocity distribution determines the density profile n(r) (comets/AU^3) in the Oort cloud (see, e.g., *Spitzer,* 1987; *Binney and Tremaine,* 1987). Oort's profile is given by $n(r) \propto (R_0/r - 1)^{3/2}$ (*Oort,* 1950; *Bailey,* 1983; *Bailey et al.,* 1990). For $r \ll R_0$, $n(r) \propto r^{-\gamma}$, with $\gamma \approx 1.5$. (The median cometary distance in this model is $0.35 R_0$; at this distance, the effective value of γ is ~1.7.) Density distributions with $\gamma < 3$ have most of the mass in the outer regions of the cloud, so Oort's model predicts that there should be few comets with $r \ll R_0$, i.e., the population of the inner Oort cloud should be small. However, Oort's assumption of an isotropic velocity distribution may not be valid in the inner parts of the cloud. For instance, if the orbits are predominantly radial (i.e., orbital eccentricities ~1), γ should be ~3.5, implying a centrally condensed cloud.

Hills (1981) showed that the apparent inner edge of the Oort cloud at a semimajor axis $a = a_I \approx (1-2) \times 10^4$ AU could be a selection effect due to the rarity of close stellar passages capable of perturbing comets with $a < a_I$. Hills speculated that $\gamma \geq 4$, so that many comets (and perhaps the great majority of comets) might reside in the unseen inner Oort cloud at semimajor axes of a few thousand AU. Besides its possible role as a reservoir that could replenish the outer cloud after it was stripped by a GMC (*Clube and Napier,* 1984), inner Oort cloud comets might be an important source of impactors on the giant planets and their satellites (*Shoemaker and Wolfe,* 1984; *Bailey and Stagg,* 1988; see also *Weissman,* 1986, and section 5). However, the density profile of the Oort cloud is not known *a priori*, but depends in large part upon the formation process.

During rare passages of stars through the inner Oort cloud, comet showers could result (*Hills,* 1981; *Heisler et al.,* 1987; *Fernández,* 1992; *Dybczyński,* 2002a,b). *Heisler* (1990) simulated the LPC flux from the Oort cloud into the planetary region, under the influence of stellar perturbations and a constant galactic tide. She found that the flux is constant within the statistical limits of her dynamical model, except when a major perturbation of the cometary orbits occurs as a result of a penetrating stellar passage. A hypothetical example of the flux vs. time into the terrestrial planets region (q < 2 AU) from *Heisler* (1990) is shown in Fig. 6.

The extreme increases in the cometary flux caused by a penetrating stellar passage through the inner Oort cloud are of particular interest. *Hut et al.* (1987) used a Monte Carlo simulation to show that a 1 M$_\odot$ star passage at 20 km s^{-1} at 3000 AU from the Sun would perturb a shower of ~5 × 10^8 comets into Earth-crossing orbits, raising the expected impact rate by a factor of 300 or more, and lasting 2–3 × 10^6 yr (this model assumed a massive inner Oort cloud with a population five times that of the outer cloud, as predicted by *DQT87*). Comets from the inner Oort cloud make an average of 8.5 returns each (allowing for disruption) during a major cometary shower. The flux is very high, in part, because the shower comets from the inner Oort cloud start from shorter period orbits than outer Oort cloud comets, with typical periods in the inner cloud of 2–5 × 10^5 yr vs.

10,000 < a < 40,000 AU; q < 2 AU

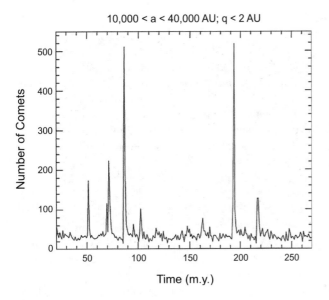

Fig. 6. Number of new LPCs from the Oort cloud entering the terrestrial planets region, q < 2 AU, vs. time, based on a Monte Carlo simulation that included random passing stars and galactic tidal perturbations. The large spikes are comet showers due to random stars penetrating the Oort cloud. From *Heisler* (1990).

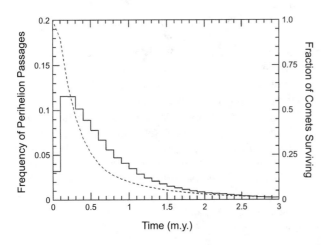

Fig. 7. Dynamical evolution of a shower of comets from the inner Oort cloud due to a close, penetrating stellar passage at 20 km s^{-1} at 3000 AU from the Sun. The solid histogram is the number of comets (arbitrary units) crossing Earth's orbit vs. time; the dashed curve is the fraction of the original shower comets still evolving in the system. On the order of 5×10^8 comets brighter than $H_{10} = 11$ are expected to be thrown into Earth-crossing orbits by the 1 $M_\odot$ star's passage. Roughly 10 of these comets would be expected to strike Earth. From *Hut et al.* (1987).

$3–5 \times 10^6$ yr in the outer cloud. Returning comets tend to be perturbed to even shorter period orbits, $\sim10^3–10^5$ yr. They thus make many returns in a relatively short period of time. The temporal profile and fraction of surviving comets for a major cometary shower as found by *Hut et al.* (1987)

are shown in Fig. 7. The dynamical evolution of cometary showers was also modeled by *Fernández and Ip* (1987). *Farley et al.* (1998) presented the best evidence to date that at least one comet shower has occurred in the past. Specifically, they showed that the flux to Earth of extraterrestrial ^{3}He, a tracer of interplanetary dust, increased for 2.5 m.y., centered near the time of the large Popigai and Chesapeake Bay impacts some 36 m.y. ago and the late Eocene extinction event. However, it is possible that some other mechanism [e.g., an "asteroid shower" following the catastrophic disruption of a main-belt asteroid (*Zappalá et al.*, 1998)] also might have produced the signature detected by *Farley et al.* (1998).

Fortunately, major cometary showers, as a result of deep (q ≲ 3×10^3 AU), penetrating stellar encounters, are rare, occurring perhaps once every 4×10^8 yr. Cometary showers should also occur with a similar frequency due to random encounters with GMCs, but with possibly an order of magnitude less total flux into the planetary region (*Morris and Muller*, 1986). Lesser showers from more distant, but still penetrating stellar passages at heliocentric distances $\sim10^4$ AU occur more frequently, on the order of every 4×10^7 yr (*Dybczyński*, 2002a,b; *Matese and Lissauer*, 2002). If there is a massive inner Oort cloud, random cometary showers may actually dominate the time-averaged LPC flux through the planetary region (*Weissman*, 1990b).

The suggestion that both biological extinction events (*Raup and Sepkoski*, 1984) and impact craters (*Alvarez and Muller*, 1984) on the Earth repeat with a period of approximately 26 m.y. led to several hypotheses that invoked periodic cometary showers as the cause of the extinctions. These hypotheses involved (1) a dwarf companion star to the Sun ("Nemesis") in a distant, eccentric, 26-m.y. period orbit (corresponding to a ~ 90,000 AU) with its perihelion deep in the Oort cloud (*Whitmire and Jackson*, 1984; *Davis et al.*, 1984); (2) a tenth planet circulating in a highly inclined orbit at about 150 AU from the Sun with a precession period of 26 m.y., so that it periodically passed through a transneptunian disk of small bodies (*Whitmire and Matese*, 1985); or (3) the solar system's epicyclic motion above and below the galactic plane. In this last scenario, GMC encounters would occur near the times of galactic plane crossings (*Rampino and Stothers*, 1984), which occur every 26–37 m.y. (*Bahcall and Bahcall*, 1985). The apparent coincidence between galactic plane crossings by the solar system and terrestrial extinction boundaries was originally pointed out by *Innanen et al.* (1978). The Sun's galactic motion was also suggested as the clock mechanism by *Schwartz and James* (1984), although they only speculated about the underlying physical mechanism leading to the extinctions.

A variety of dynamical problems have been identified with each of these hypotheses, and no evidence in support of any of them has been found. As a result, periodic comet shower hypotheses have not gained wide acceptance and are generally discounted today, although *Muller* (2002) recently proposed a modified version of the Nemesis companion-star hypothesis. More detailed discussions of the relevant issues can be found in *Shoemaker and Wolfe* (1986), *Tremaine* (1986), and *Weissman* (1986). Questions have also been raised about the reality of the periodicity in the fossil extinction record. Criticism has been made of the statistical techniques used to claim that the periodicity is significant (*Hoffman*, 1985; *Heisler and Tremaine*, 1989; *Jetsu and Pelt*, 2000), and of the accuracy of the dated tie-points in the geologic record, particularly prior to 140 m.y. ago (*Shoemaker and Wolfe*, 1986).

Variations in the cometary flux into the planetary region as the Sun revolves around its galactic orbit are still the subject of research. For example, the solar system undergoes a near-harmonic motion above and below the galactic plane (*Matese et al.*, 1995; *Nurmi et al.*, 2001). This motion currently carries the planetary system some 50–90 pc out of the plane, comparable to the scale height of the disk (*Bahcall and Bahcall*, 1985). The full period of the oscillation is ~52–74 m.y. *Matese et al.* (1995) showed that this causes the cometary flux to vary sinusoidally by a factor of 2.5–4 over that period, with the maximum flux occurring just after passage through the galactic plane. However, the dynamical model of Matese et al. did not include stellar perturbations. The solar system has passed through the galactic plane in the last few million years, so the current steady-state flux is likely near a local maximum.

4. SIMULATIONS OF THE FORMATION OF THE OORT CLOUD

In his 1950 paper, Oort did not consider the formation of the comet cloud in detail, but speculated

"It seems a reasonable hypothesis to assume that the comets originated together with the minor planets, and that those fragments whose orbits deviated so much from circles between the orbits of Mars and Jupiter that they became subject to large perturbations by the planets, were diffused away by these perturbations, and that, as a consequence of the added effect of the perturbations by stars, part of these fragments gave rise to the formation of the large cloud of comets which we observe today."

Oort proposed the asteroid belt as the source region for the LPCs on the grounds that (1) asteroids and cometary nuclei are fundamentally similar in nature and (2) the asteroid belt was the only stable reservoir of small bodies in the planetary region known at that time. *Kuiper* (1951) was the first to propose that the icy nature of comets required that they be from a more distant part of the solar system, among the orbits of the giant planets. Thus, ever since Oort and Kuiper's work, the roles of the four giant planets in populating the comet cloud have been debated. *Kuiper* (1951) proposed that Pluto, which was then thought to have a mass similar to that of Mars or the Earth, scattered comets that formed between 38 and 50 AU (i.e., in the Kuiper belt!) onto Neptune-crossing orbits, after which Neptune, and to a lesser extent the other giant planets, placed comets in the Oort cloud. [*Stern and Weissman* (2001) have recently argued that the primordial Kuiper belt at heliocentric distances

<35 AU might have been an important source of Oort cloud comets. The *Dones et al.* (2004) simulations bear out this conclusion. However, in these models, perturbations due to Pluto are not important.]

Later work (*Whipple*, 1964; *Safronov*, 1969, 1972) indicated that Jupiter and Saturn tended to eject comets from the solar system, rather than placing them in the Oort cloud. The kinder, gentler perturbations by Neptune and Uranus (if these planets were assumed to be fully formed) thus appeared to be more effective in populating the cloud. However, their role was unclear because the ice giants took a very long time to form in Safronov's orderly accretion scenario. *Fernández* (1978) used a Monte Carlo, Öpik-type code, which assumes that close encounters with planets dominate the orbital evolution of a small body, to calculate the probability that a comet would collide with a planet, be ejected from the solar system, or reach a near-parabolic orbit (i.e., an orbit of a body that might end up in the Oort cloud). He suggested that "Neptune, and perhaps Uranus, could have supplied an important fraction of the total mass of the cometary cloud." *Fernández* (1980b) extended this work by following the subsequent evolution of comets on plausible near-parabolic orbits for bodies that had formed in the Uranus-Neptune region ($5000 \leq a \leq 50{,}000$ AU; 20 AU $\leq q \leq 30$ AU; $i \leq 20°$). He included the effects of passing stars using an impulse approximation and included perturbations by the four giant planets by direct integration for comets that passed within 50 AU of the Sun. Fernández concluded that about 10% of the bodies scattered by Uranus and Neptune would occupy the Oort cloud at present, and that the implied amount of mass scattered by the ice giants was cosmogonically reasonable.

Shoemaker and Wolfe (1984) performed an Öpik-type simulation to follow the ejection of Uranus-Neptune planetesimals to the Oort cloud, including the effects of stellar perturbations for orbits with aphelia >500 AU. They found that ~9% of the original population survived over the history of the solar system, with ~90% of those comets in orbits with semimajor axes between 500 and 20,000 AU; 85% of the latter group had semimajor axes <10,000 AU. Shoemaker and Wolfe also found that the perihelion distribution of the comets was peaked just outside the orbit of Neptune, and estimated a total cloud mass of 100 to 200 $M_\oplus$. Unfortunately, their work was published only in an extended abstract, so the details of their modeling are not known.

The first study using direct numerical integrations to model the formation of the Oort cloud was that of *Duncan et al.* (1987; hereafter *DQT87*). To save computing time, *DQT87* began their simulations with comets on low-inclination, but highly eccentric, orbits in the region of the giant planets (initial semimajor axes, a_0, of 2000 AU and initial perihelion distances, q_0, uniformly distributed between 5 and 35 AU). Gravitational perturbations due to the giant planets and the disk (z) component of the Galactic tide were included (see below). A Monte Carlo scheme from *Heisler et al.* (1987) was used to simulate the effects of stellar encounters. Molecular clouds were not included.

DQT87's main results included the following: (1) The Oort cloud has a sharp inner edge at a heliocentric distance r ~ 3000 AU. (2) For 3000 AU ≤ r ≤ 50,000 AU, the number density of the Oort cloud falls steeply with increasing r, going roughly as $r^{-3.5}$. Thus the Oort cloud is centrally condensed, with roughly 4–5 times as many comets in the inner Oort cloud (a ≤ 20,000 AU) as in the classical outer Oort cloud. (3) The present-day inclination distribution should be approximately isotropic in the outer Oort cloud and most of the inner Oort cloud. The innermost part of the inner Oort cloud, interior to 6000 AU, may still be slightly flattened. (4) Comets with $q_0 \gtrsim 15$ AU are much more likely to reach the Oort cloud and survive for billions of years than are comets with smaller initial perihelia. For example, only 2% of the comets with $q_0 = 5$ AU should occupy the Oort cloud at present, while 24% of the comets with $q_0 = 15$ AU and 41% with $q_0 = 35$ AU should do so. This result appeared to confirm that Neptune and Uranus, which have semimajor axes of 30 and 19 AU, respectively, are primarily responsible for placing comets in the Oort cloud. However, this finding can be questioned, since the highly eccentric starting orbits had the consequence of pinning the perihelion distances of the comets at early stages. This, in turn, allowed Neptune and Uranus to populate the Oort cloud efficiently because they could not lose objects to the control of Jupiter and Saturn.

Dones et al. (2004; hereafter *DLDW*) repeated the study of *DQT87*, starting with "comets" with semimajor axes between 4 and 40 AU and initially small eccentricities and inclinations. These initial conditions are more realistic than the highly eccentric starting orbits assumed by *DQT87*. *DLDW* integrated the orbits of 3000 comets for times up to 4 b.y. under the gravitational influence of the Sun, the four giant planets, the galaxy, and random passing stars. Their model of the galaxy included both the "disk" and "radial" components of the galactic tide. The disk tide is proportional to the local density of matter in the solar neighborhood and exerts a force perpendicular to the galactic plane, while the radial tide exerts a force within the galactic plane. These simulations did not include other perturbers such as molecular clouds, a possible dense early environment if the Sun formed in a cluster (*Gaidos*, 1995; *Fernández*, 1997), or the effects of gas drag (*de la Fuente Marcos and de la Fuente Marcos*, 2002; *Higuchi et al.*, 2002).

DLDW performed two sets of runs with dynamically "cold" and "warm" initial conditions. The results were very similar, so we will focus on the "cold" runs, which included 2000 particles with root-mean-square initial eccentricity, e_0, and inclination to the invariable plane, i_0, equal to 0.02 and 0.01 radians, respectively. *DLDW* assumed that the Sun resided in its present galactic environment during the formation of the Oort cloud.

We will take the results of these calculations at 4 G.y. to refer to the present time. For a comet to be considered a member of the Oort cloud, we require that its perihelion distance exceeded 45 AU at some point in the calculation.

For the "cold" runs, the percentage of objects that were integrated that currently occupy the classical "outer" Oort cloud (20,000 AU ≤ a < 200,000 AU) is only 2.5%, about a factor of 3 smaller than found by *DQT87*. The percentage of objects in the inner Oort cloud (2000 AU ≤ a < 20,000 AU) is 2.7%, almost an order of magnitude smaller than calculated by *DQT87*. This result holds because most comets that begin in the Uranus-Neptune zone evolve inward and are ejected from the solar system by Jupiter or Saturn. A small fraction are placed in the Oort cloud, most often by Saturn. However, all four of the giant planets place comets in the Oort cloud. The Oort cloud is built in two distinct stages in the *DLDW* model. In the first few tens of millions of years, the Oort cloud is built by Jupiter and Saturn, which deliver comets to the outer Oort cloud. After this time, the Oort cloud is built mainly by Neptune and Uranus, with the population peaking about 800 m.y. after the beginning of the simulation (Fig. 8). Objects that enter the Oort cloud during this second phase typically first spend time in the "scattered disk" [45 AU ≤ a < 2000 AU, with perihelion distance <45 AU at all times (*Duncan and Levison,* 1997)] and then end up in the inner Oort cloud.

Plates 5 and 6 show the formation of the Oort cloud in terms of the orbital evolution in semimajor axis as a func-

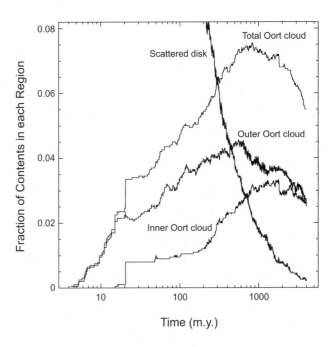

Fig. 8. Fraction of original cometary population placed in the inner and outer Oort clouds and in the scattered disk in the *DLDW* simulation. In these simulations the outer Oort cloud, which is originally populated by comets injected by Jupiter and Saturn, forms more rapidly than the inner Oort cloud, which is primarily populated by comets injected by Uranus and Neptune. The simulation predicts that at present, the populations of the inner and outer Oort clouds are comparable.

tion of perihelion distance and inclination to the invariable plane of the solar system, respectively. We show six "frames" from the *DLDW* integrations at various times in the calculations. (Animations showing these data every 1 m.y. throughout the simulation can be viewed at http://www.boulder.swri.edu/~luke.) Points in these plots are color-coded by their formation location a_0: Jupiter region comets (a_0 between 4 and 8 AU) are magenta triangles; Saturn region comets (8–15 AU) are blue triangles; Uranus region comets (15–24 AU) are green circles; Neptune region comets (24–35 AU) are red circles; Kuiper belt comets (35–40 AU) are black circles.

Plate 5a (0 m.y.) shows that the particles start with very small eccentricities, as represented by the diagonal line of particles that extends from ~4 to 40 AU. After 1 m.y. (Plate 5b), the giant planets, particularly Jupiter and Saturn, have scattered many comets into very eccentric orbits with perihelia still in the region of the giant planets. After 1 m.y., 76% of the test particles remain. Of the 24% lost in the first million years, most were ejected from the solar system by Jupiter or Saturn.

At 10 m.y. (Plate 5c), we see the beginning of the formation of the Oort cloud. Some particles with a ≥ 30,000 AU have had their perihelia raised out of the planetary region by galactic tides and the effects of passing stars. In all, 48% of the particles remain. At 100 m.y. (Plate 5d), the Oort cloud has begun to assume its current form. Twenty-eight percent of the particles remain; 4.7% are in the Oort cloud, with the rest in the planetary region or scattered disk. From 100 m.y. to 1000 m.y. (Plate 5e), particles continue to enter the Oort cloud from the scattered disk. The total number of particles continues to decline — 15% remain — but the population in the Oort cloud peaks around 835 m.y. At 1000 m.y., 7.3% of the comets are in the Oort cloud. Finally, at 4000 m.y. (Plate 5f), the structure of the Oort cloud remains nearly the same as at 1000 m.y., but its population has declined slightly. In total, 11% of the particles that *DLDW* integrated remain. Of these, about half revolve on orbits in the planetary region (i.e., a < 45 AU), primarily in the Kuiper belt, that have changed little. Most of the other survivors reside in the Oort cloud, with nearly equal numbers of comets in the inner and outer clouds.

Plate 6 shows the evolution of the particles' inclinations. Plate 6a (0 m.y.) shows that the particles' inclinations to the invariable plane are initially small. After 1 m.y. (Plate 6b), the planets have scattered the comets into moderately inclined orbits. After 10 m.y. (Plate 6c), the particles with a ≥ 30,000 AU have been perturbed by galactic tides and stars into a nearly isotropic distribution of inclinations. As time continues (Plates 6d–6f), tides affect the inclinations of particles closer to the Sun, so that at 4000 m.y. inclinations are clearly isotropic for a ≥ 7000 AU.

We now return to the issue of how centrally condensed the Oort cloud is. Recall that *DQT87* found a density profile n(r) ∝ $r^{-\gamma}$ with γ ~ 3.5 for 3000 AU < r < 50,000 AU, so that in their model most comets reside in the (normally unobservable) inner Oort cloud. If we fit the entire Oort cloud at 4 G.y. in the *DLDW* model to a single power law, we find γ ~ 3, shallower than the value found by *DQT87*. The shallow slope probably results because all the giant planets inject comets into the Oort cloud, even though most *formed* beyond 20 AU. A value of γ ~ 3 implies that the inner and outer Oort clouds contain comparable numbers of comets at present in this model.

Finally, Fig. 8 shows the time evolution of the populations of the Oort cloud and scattered disk in the simulation. The scattered disk is initially populated by comets scattered by Jupiter and Saturn, and peaks in number at 10 m.y. (off-scale on the plot). The predicted population of the scattered disk in this model at the present time is roughly 10% the population of the Oort cloud.

Likewise, the Oort cloud grows rapidly in the first few tens of millions of years due to comets injected by Jupiter and Saturn, and then undergoes a very prolonged period of growth, primarily due to Uranus and Neptune, with the peak population occurring around 800 m.y. From 150 m.y. to 4000 m.y., the fraction of comets in the Oort cloud ranges between 5% and 7.6%.

Figure 8 also shows the populations of the inner and outer Oort clouds individually. The population of the outer Oort cloud peaks around 600 m.y. The inner Oort cloud peaks around 1.8 G.y. Because of the faster decline of the outer Oort cloud, the ratio of numbers of inner to outer Oort cloud comets increases with time, to 1.1 at present. Nonetheless, this ratio is much smaller than was given by *DQT87*, who found 4–5 times more comets in the inner Oort cloud than in the outer Oort cloud. Only 2.5% of the comets that were initially in the simulation occupy the outer Oort cloud at 4 G.y.

At face value, the low efficiency of Oort cloud formation in the *DLDW* simulation implies a massive primordial protoplanetary disk. Assuming an outer Oort cloud population of 5×10^{11}–1×10^{12} comets (*Heisler*, 1990; *Weissman*, 1996) and an average cometary mass of 4×10^{16} g (section 2.4), the original mass in planetesimals between 4 and 40 AU was ~150–300 $M_\oplus$, some 3–6 times the mass in solids in a "minimum-mass" solar nebula. This amount of mass likely would have produced excessive migration of the giant planets and/or formation of additional giant planets (*Hahn and Malhotra*, 1999; *Thommes et al.*, 2002; *Gomes et al.*, 2004). Since cometary masses are not well determined, it is not yet clear whether the large disk mass inferred by *DLDW* presents a real problem.

The results of the *DLDW* simulations appear inconsistent with observations in another way. The population of the scattered disk that *DLDW* predict, on the order of 10% of the population of the Oort cloud, is much larger than the inferred actual population of the scattered disk (*Trujillo et al.*, 2000). Finally, the *DLDW* model of the Oort cloud appears to be inconsistent with a model of the orbital distribution of the HTCs by *Levison et al.* (2001). Although the class of HTCs includes some objects on retrograde orbits, such as Halley itself, the observed HTCs with perihelia <1.3 AU have a median inclination i_{med} of only 58°. *Levison et al.* (2001), who took i_{med} = 45°, using the data available at that time, showed that the HTCs must originate in a somewhat flattened source region. Since the outer Oort cloud is known to be roughly isotropic, *Levison et al.* (2001) assumed that most HTCs must come from a flattened *inner* Oort cloud. However, because of the "Jupiter barrier" (section 5), the inner Oort cloud must contain many more comets than the outer Oort cloud to provide enough HTCs. By contrast, in the models of *DLDW*, the inner Oort cloud does not contain such a large population, nor is it particularly flattened. This discrepancy suggests some deficiency in one of the models. For example, the inner Oort cloud may not be the source of the HTCs, or the *DLDW* model may not be realistic enough because it neglects processes that were important in the early solar system.

The assumptions of the *DLDW* model are highly idealized. Most importantly, the formation of the Oort cloud needs to be studied in the context of a realistic model for planet formation. That is, the planets were still forming during at least the early stages of the formation of the Oort cloud. Planetary migration in the early solar system (*Fernández and Ip*, 1984) appears to have been important in shaping the Kuiper belt (*Malhotra*, 1995; *Gomes*, 2003; *Levison and Morbidelli*, 2003; *Gomes et al.*, 2004), and the same is likely true for the Oort cloud. Uranus and Neptune may even have formed in the Jupiter-Saturn region (*Thommes et al.*, 1999, 2002), likely changing the fraction of comets that ended up in the Oort cloud (see section 2).

Tremaine (1993), *Gaidos* (1995), *Fernández* (1997), *Eggers et al.* (1997, 1998), *Eggers* (1999), and *Fernández and Brunini* (2000) have discussed star formation in different galactic environments. These authors point out that the Sun may have formed in a denser environment than it now occupies (i.e., in a molecular cloud or star cluster), and found that a more tightly bound Oort cloud would form. For example, *Eggers* (1999) modeled the formation of the Oort cloud, assuming that the Sun spent its first 20 m.y. in a star cluster with an initial density of 1000 or 10 stars/pc³, as compared to the current density of ~0.1 stars/pc³. The resulting cloud is produced primarily by Jupiter and Saturn, and its density peaks at a heliocentric distance of 6000–7000 AU in the 10 stars/pc³ case or at <1000 AU in the 1000 stars/pc³ case. After the cluster dispersed, Uranus and Neptune would have placed comets in the cloud in more or less the same way as they do with the Sun in its current environment.

If the Sun remained in a dense environment for too long, the resulting Oort cloud might not be stable, and the orbits of Uranus and Neptune would have become eccentric and/ or inclined (*Gaidos*, 1995; *Ida et al.*, 2000; *Adams and Laughlin*, 2001; *Levison et al.*, 2004). Drag due to residual gas from the solar nebula may have been important in the formation of the Oort cloud (*de la Fuente Marcos and de la Fuente Marcos*, 2002; *Higuchi et al.*, 2002). Collisions may have been important in determining which regions of the protoplanetary disk could populate the Oort cloud (*Stern and Weissman*, 2001; *Stern*, 2003; *Charnoz and Morbidelli*, 2003).

5. CONSTRAINTS ON THE STRUCTURE OF THE OORT CLOUD

Barely one decade after the discovery of the first Kuiper belt object (besides Pluto), the number of known KBOs is comparable to the number of LPCs that have been discovered in recorded history. Full-sky surveys will likely tilt the balance decisively in favor of the Kuiper belt in the near future (*Jewitt*, 2004). This disparity is, of course, a consequence of the much greater distance to the Oort cloud and the r^{-4} heliocentric brightness dependence for distant bodies seen in reflected light. Thus a 200-km-diameter body with an apparent magnitude of 23 in the Kuiper belt at 40 AU would have a magnitude of 42 in the inner Oort cloud at 3000 AU, and a more typical 2-km comet at 20,000 AU, assuming an albedo of 0.04, would have a magnitude of 60. Thus direct imaging of comets at Oort cloud distances will not be possible in the foreseeable future.

There is no substitute for just counting comets in the Oort cloud. In principle, comets, especially in the inner cloud, could be detected when they occult stars (*Bailey*, 1976). G- or K-type main-sequence stars at a distance of 1 kpc (roughly twice the distance to the stars in Orion's belt) have visual magnitudes of ~15–17. If the brightness of a million such stars (about 10% of the number within 1 kpc) can be monitored, occultations by 30-km comets at a distance of 3000 AU can be detected in principle (e.g., *Axelrod et al.*, 1992; *Brown and Webster*, 1997). The Taiwan-America Occultation Survey (TAOS) project, which will search for occultations by KBOs, will soon come online (*Roques and Moncuquet*, 2000; *Cooray and Farmer*, 2003; *Cooray*, 2003), and detections of Oort cloud comets remain a long-term goal for occultation surveys (C. Alcock, personal communication, 2003).

For the present, our best hope is to try to infer the structure of the Oort cloud from the orbital distribution of known comets. This is a difficult exercise because of the numerous biases affecting discovery (section 2.2), and most importantly, because the "Jupiter barrier" severely limits the number of new comets from the inner Oort cloud that come within about 10 AU of the Sun (section 3).

Bailey (1983) finds that for a ≥ 28,000 AU, the Oort cloud has a density profile n(a) ∝ a$^{-\gamma}$, with γ = 2.4 ± 0.2. This assumes that the probability of discovery per year for a comet with a perihelion distance q well interior to Jupiter's orbit goes inversely as the comet's orbital period, which is plausible. However, Bailey's fits are based on only 37 "new" comets, a subset of those discussed by *Marsden et al.* (1978), with well-determined ("Class I") orbits and q > 2 AU. (The condition on perihelion distance is imposed in order to minimize unmodeled nongravitational effects.) At face value, Bailey's result implies an outer Oort cloud that is more centrally condensed than in Oort's original model (γ ~

1.5) and less centrally condensed than in *DQT87* or *DLDW*, both of whom find $\gamma \sim 3.5$ in the outer cloud. However, since systematic effects due to nongravitational forces, even for comets with q > 2 AU, and unknown observational biases might be important, it will be important to reevaluate Bailey's result by using a more homogeneous dataset.

To better constrain the Oort cloud, we need well-defined surveys that detect a large number of dynamically new LPCs with perihelia beyond Saturn, i.e., with q ≥ 10 AU. (By "comets," we mean bodies in highly eccentric long-period orbits. Such bodies may or may not be active.) The typical energy perturbations produced on comets by Uranus and Neptune are 10–100 times smaller than those produced by Jupiter and Saturn (*Everhart,* 1968; *Fernández,* 1981; *Weissman,* 1985; *Duncan et al.,* 1987), so comets with q ≥ 10 AU suffer no "Uranus barrier" or "Neptune barrier" to produce a bias against comets from the inner Oort cloud.

A key aspect of surveys is having a long enough observational arc to be certain that an object *is* a long-period comet. At present, there is only one known LPC with q > 10 AU, comet C/2003 A2 (Gleason), which was discovered during Spacewatch observations taken in January 2003. At the time of discovery the comet's magnitude was 20 and its heliocentric distance was 11.5 AU, a record. The IAU Circular reporting the discovery noted that the comet's inclination was only 8°, and stated "It seems likely that the object is a Centaur, showing cometary activity as (2060) = 95P/Chiron has shown near perihelion" (*Gleason et al.,* 2003). However, a later fit incorporating prediscovery observations indicated that C/2003 A2 is apparently a bona fide dynamically new comet (*Green,* 2003).

Planned surveys should discover many LPCs with perihelia beyond 10 AU. The Large Synoptic Survey Telescope (LSST) was endorsed as a recommended "major initiative" by the most recent U.S. Decadal Survey in Astronomy and Astrophysics (*National Research Council Astrophysics Survey Committee,* 2001). This 6–8-m optical telescope would survey much of the visible sky weekly down to 24th magnitude, beginning about one decade from now. Its objectives include studies of small bodies in the solar system (*Tyson,* 2002). In the shorter term, Pan-STARRS, a system consisting of four 1.8-m telescopes, is planning to begin operations by 2007. *Jewitt* (2004) estimates that Pan-STARRS will discover at least 400 comets per year (albeit mostly ecliptic comets), including many with large perihelia. It also may provide interesting constraints on the number of interstellar comets passing through the solar system. *Horner and Evans* (2002) note that the GAIA astrometric satellite, which is scheduled to be launched in 2010, is expected to cover about 200 LPCs each year.

The final approach we will discuss for constraining the population of the Oort cloud involves the impact history of the planets and their satellites. At present, ecliptic comets appear to dominate impacts with the giant planets and their inner satellites (*Zahnle et al.,* 1998, 2003), while asteroids dominate on Earth and the other terrestrial planets (*Shoe-*

maker, 1983; *Bailey and Stagg,* 1988; *Shoemaker et al.,* 1990; *Weissman,* 1990b; *McKinnon et al.,* 1997; *Levison et al.,* 2002; *Morbidelli et al.,* 2002; cf. *Rickman et al.,* 2001). *Zahnle et al.* (2003) estimate that ~1% of the impacts on Jupiter are produced by NICs, including both active and dormant cometary impactors. However, this percentage is higher for distant satellites, because the NICs experience less gravitational focusing than do ecliptic comets. For example, *Zahnle et al.* (2003) suggest that NICs produce about 30% of the 10-km craters on Jupiter's prograde irregular satellite Himalia.

The rate of impacts on a planet by LPCs is $\mathcal{R} = \dot{N}(q)\langle p \rangle$, where $\dot{N}(q)$ is the number of comets that pass perihelion within distance q of the Sun per year, and $\langle p \rangle$ is the mean impact probability of the comets with the planet per orbit of the comet. The biggest uncertainty in determining impact rates is in the cumulative perihelion distribution, $\dot{N}(q)$. The perihelion distribution is only well-constrained for 0.5 AU ≤ q ≤ 2.5 AU; over this range the number of comets per AU rises with q. *Zahnle et al.* (2003) assumed $\dot{N}(q) \propto q^2$ throughout the region of the giant planets. However, because they are partly or entirely exterior to the "Jupiter barrier," the saturnian, uranian, neptunian, and Pluto/Charon systems are also subject to impacts by comets that originate in the inner Oort cloud (*Bailey and Stagg,* 1988; *Weissman and Stern,* 1994). *DLDW* find $\dot{N}(q) \propto q^3$ in the giant planets region (also see *Fernández,* 1982; *Weissman,* 1985). As a result of this steeper dependence on q, *DLDW* estimate that LPCs could contribute some 10% of the present-day impacts on Saturn, Uranus, and Neptune, and could dominate the impact rate by comets on the irregular satellites of these planets. [For some irregular satellites, collisions with other such satellites probably dominate the current rates (*Nesvorný et al.,* 2003, 2004).]

Unfortunately, it is not straightforward to place limits on the population of the Oort cloud with the observed impact record. However, the existence of distant irregular satellites of the giant planets, with sizes as small as 1 km, does constrain the population of impactors that have traversed the planetary systems since the irregulars formed. Small satellites are easier to disrupt, and their orbital periods are so long that they cannot reaccrete after a catastrophic disruption event. *Nesvorný et al.* (2004) have used arguments of this sort to rule out some combinations of total mass and size distribution for the residual disk of planetesimals that remained after the giant planets formed.

Finally, if a very strong comet shower takes place due to the passage of a solar-mass star through the inner Oort cloud, the Jupiter barrier is temporarily eliminated, and a large flux of comets will enter the entire solar system, including the region of the terrestrial planets. The number of comets expected to strike Earth during such a shower is proportional to the number of comets in the inner Oort cloud, so the cratering record of Earth can be used to constrain the population of the inner cloud. During the Phanerozoic (the last 543 m.y.), about one or two major showers would be expected, given the known frequency at which stars pass near the Sun (section 3). If the population of the inner cloud were greater than about 100 times the population of the outer cloud, even a single very strong shower would produce more craters than Earth's record allows (*Shoemaker*, 1983; *Grieve and Shoemaker*, 1994; *Hughes*, 2000), and most of the known craters on Earth would have formed during a period lasting only a few million years. This constraint refers to craters tens of kilometers in diameter. There is some evidence that LPC nuclei have a flatter (i.e., more top-heavy) size distribution than do asteroids (*Shoemaker et al.*, 1990; *Levison et al.*, 2002; *Weissman and Lowry*, 2003), so considering only the largest known craters during the last half-billion years on Earth might yield a tighter constraint. As we noted in section 3, the Popigai and Chesapeake Bay craters (~100 km and 85 km in diameter, respectively), do seem to be associated with a comet shower 36 m.y. ago (*Farley et al.*, 1998).

6. SUMMARY

Oort's picture of a near-spherical cloud of comets at distances of tens of thousands of AU is still valid. An Oort cloud of about this size is a natural consequence of the interplay between scattering of planetesimals by the giant planets and tidal torquing by the galaxy and random passing stars. The formation of the Oort cloud is likely to be a protracted process, with the population peaking about 1 b.y. after the planets formed. The observed orbital energy distribution of LPCs requires that comets "fade," perhaps by undergoing spontaneous, catastrophic disruption. The best estimate of the current number of comets in the "outer" Oort cloud (a > 20,000 AU) is 5×10^{11}–1×10^{12} (*Heisler*, 1990; *Weissman*, 1991). Nominally, this estimate refers to comets with diameters and masses greater than 2.3 km and 4×10^{15} g, respectively. However, the relation between cometary brightness and mass is not well understood. Thus the total mass of even the outer Oort cloud is not well-determined.

The sample of new comets that reach the region of the terrestrial planets is biased to objects with a ≥ 20,000–30,000 AU because of the "Jupiter barrier." Thus the population of the inner Oort cloud, at distances of thousands of AU, remains uncertain. Recent simulations suggest that the population of the inner Oort cloud is comparable to that of the outer Oort cloud (*Dones et al.*, 2004), but more realistic simulations are needed. Rare passages of solar-type stars through the inner Oort cloud produce comet showers on all the planets. One such shower appears to have taken place 36 m.y. ago (*Farley et al.*, 1998).

Our present knowledge of the Oort cloud is much like the highly incomplete picture of the Kuiper belt we had one decade ago, after only a few objects had been discovered in the belt. In the next few decades, optical discoveries of comets at distances beyond 10 AU and direct detections by stellar occultations will provide a much better understanding of the inner cloud. Future models of Oort cloud formation will build upon recent advances in our understanding of the Kuiper belt to consider processes such as planetary migration, growth, and collisions (*Morbidelli and Brown*,

2004). Most of the planetesimals that once orbited the Sun were probably ejected from the solar system. If most stars form comet clouds in the same way the Sun did, detection of bona fide interstellar comets is likely in the near future (*McGlynn and Chapman*, 1989; *Sen and Rama*, 1993). Millennia after mankind first wondered what comets were, we are on the verge of glimpsing their home.

Acknowledgments. We thank C. Alcock, W. Bottke, S. Charnoz, A. Cooray, P. Dybczyński, V. Emel'yanenko, T. Lee, A. Morbidelli, D. Nesvorný, P. Nurmi, J. Scalo, A. Stern, and K. Zahnle for discussions, and M. Festou, J. Fernández, and an anonymous reviewer for helpful comments. The abstract service of the NASA Astrophysics Data System was indispensable in the preparation of this chapter. We acknowledge grants from the NASA Planetary Geology and Geophysics program (L.D. and P.R.W.), the NASA Origins of Solar Systems Program (H.F.L.), and NSERC (M.J.D.). This work was performed in part at the Jet Propulsion Laboratory under contract with NASA.

REFERENCES

Adams F. C. and Laughlin G. (2001) Constraints on the birth aggregate of the solar system. *Icarus, 150,* 151–162.

Alvarez W. and Muller R. A. (1984) Evidence from crater ages for periodic impacts on the earth. *Nature, 308,* 718–720.

Antonov V. A. and Latyshev I. N. (1972) Determination of the form of the Oort cometary cloud as the Hill surface in the galactic field. In *The Motion, Evolution of Orbits, and Origin of Comets* (G. A. Chebotarev et al., eds.), pp. 341–345. Reidel, Dordrecht.

Axelrod T. S., Alcock C., Cook K. H., and Park H.-S. (1992) A direct census of the Oort cloud with a robotic telescope. In *Robotic Telescopes in the 1990s* (A. V. Filippenko, ed.), pp. 171–181. ASP Conference Series 34, Astronomical Society of the Pacific, San Francisco.

Bahcall J. N. and Bahcall S. (1985) The Sun's motion perpendicular to the galactic plane. *Nature, 316,* 706–708.

Bailey M. E. (1976) Can "invisible" bodies be observed in the solar system? *Nature, 259,* 290–291.

Bailey M. E. (1983) The structure and evolution of the solar system comet cloud. *Mon. Not. R. Astron. Soc., 204,* 603–633.

Bailey M. E. (1984) The steady-state 1/a distribution and the problem of cometary fading. *Mon. Not. R. Astron. Soc., 211,* 347–368.

Bailey M. E. (1986) The mean energy transfer rate to comets in the Oort cloud and implications for cometary origins. *Mon. Not. R. Astron. Soc., 218,* 1–30.

Bailey M. E. and Stagg C. R. (1988) Cratering constraints on the inner Oort cloud — Steady-state models. *Mon. Not. R. Astron. Soc., 235,* 1–32.

Bailey M. E., Clube S. V. M., and Napier W. M. (1990) *The Origin of Comets*. Pergamon, Oxford. 599 pp.

Bailey M. E., Chambers J. E., and Hahn G. (1992) Origin of sungrazers — A frequent cometary end-state. *Astron. Astrophys., 257,* 315–322.

Biermann L. (1978) Dense interstellar clouds and comets. In *Astronomical Papers Dedicated to Bengt Stromgren* (A. Reiz and T. Anderson, eds.), pp. 327–335. Copenhagen Observatory, Copenhagen.

Biesecker D. A., Lamy P., St. Cyr O. C., Llebaria A., and Howard R. A. (2002) Sungrazing comets discovered with the SOHO/ LASCO coronagraphs 1996–1998. *Icarus, 157,* 323–348.

Binney J. and Tremaine S. (1987) *Galactic Dynamics*. Princeton Univ., Princeton, New Jersey. 747 pp.

Blaauw A. and Schmidt M. (1993) Jan Hendrik Oort (1900–1992). *Publ. Astron. Soc. Pacific, 105,* 681–685.

Boehnhardt H. (2002) Comet splitting — Observations and model scenarios. *Earth Moon Planets, 89,* 91–115.

Boehnhardt H. (2004) Split comets. In *Comets II* (M. C. Festou et al., eds.), this volume. Univ. of Arizona, Tucson.

Brin G. D. and Mendis D. A. (1979) Dust release and mantle development in comets. *Astrophys. J., 229,* 402–408.

Brown M. J. I. and Webster R. L. (1997) Occultations by Kuiper belt objects. *Mon. Not. R. Astron. Soc., 289,* 783–786.

Byl J. (1983) Galactic perturbations on nearly parabolic cometary orbits. *Moon and Planets, 29,* 121–137.

Byl J. (1986) The effect of the galaxy on cometary orbits. *Earth Moon Planets, 36,* 263–273.

Chambers J. E. (1997) Why Halley-types resonate but long-period comets don't: A dynamical distinction between short- and long-period comets. *Icarus, 125,* 32–38.

Chappell D. and Scalo J. (2001) Multifractal scaling, geometrical diversity, and hierarchical structure in the cool interstellar medium. *Astrophys. J., 551,* 712–729.

Charnoz S. and Morbidelli A. (2003) Coupling dynamical and collisional evolution of small bodies: An application to the early ejection of planetesimals from the Jupiter-Saturn region. *Icarus, 166,* 141–156.

Clube S. V. M. and Napier W. M. (1982) Spiral arms, comets and terrestrial catastrophism. *Quart. J. R. Astron. Soc., 23,* 45–66.

Clube S. V. M. and Napier W. M. (1984) Comet capture from molecular clouds: A dynamical constraint on star and planet formation. *Mon. Not. R. Astron. Soc., 208,* 575–588.

Cooray A. (2003) Kuiper Belt object sizes and distances from occultation observations. *Astrophys. J. Lett., 589,* L97–L100.

Cooray A. and Farmer A. J. (2003) Occultation searches for Kuiper Belt objects. *Astrophys. J., 587,* L125–L128.

Davis M., Hut P., and Muller R. A. (1984) Extinction of species by periodic comet showers. *Nature, 308,* 715–717.

de la Fuente Marcos C. and de la Fuente Marcos R. (2002) On the origin of comet C/1999 S4 LINEAR. *Astron. Astrophys., 395,* 697–704.

Delsemme A. H. (1987) Galactic tides affect the Oort cloud: An observational confirmation. *Astron. Astrophys., 187,* 913–918.

Dones L., Levison H. F., Duncan M. J., and Weissman P. R. (2004) Simulations of the formation of the Oort cloud. I. The reference model. *Icarus,* in press.

Donn B.(1977) A comparison of the composition of new and evolved comets. In *Comets, Asteroids, Meteorites: Interrelations, Evolution and Origins* (A. H. Delsemme, ed.), pp. 15–23. IAU Colloquium 39, Univ. of Toledo, Ohio.

Duncan M. J. and Levison H. F. (1997) A scattered comet disk and the origin of Jupiter family comets. *Science, 276,* 1670–1672.

Duncan M., Quinn T., and Tremaine S. (1987) The formation and extent of the solar system comet cloud. *Astron. J., 94,* 1330–1338.

Duncan M., Quinn T., and Tremaine S. (1988) The origin of short-period comets. *Astrophys. J. Lett., 328,* L69–L73.

Duncan M. J., Levison H. F., and Budd S. M. (1995) The dynamical structure of the Kuiper belt. *Astron. J., 110,* 3073–3081 + 3 color plates.

Duncan M. J., Levison H. F., and Dones L. (2004) Dynamical evolution of ecliptic comets. In *Comets II* (M. C. Festou et al., eds.), this volume. Univ. of Arizona, Tucson.

Dybczyński P. A. (2001) Dynamical history of the observed long-period comets. *Astron. Astrophys., 375,* 643–650.

Dybczyński P. A. (2002a) On the asymmetry of the distribution of observable comets induced by a star passage through the Oort cloud. *Astron. Astrophys., 383,* 1049–1053.

Dybczyński P. A. (2002b) Simulating observable comets. I. The effects of a single stellar passage through or near the Oort cometary cloud. *Astron. Astrophys., 396,* 283–292.

Eggers S. (1999) Cometary dynamics during the formation of the solar system. Ph.D. thesis, Max-Planck-Institut für Aeronomie, Katlenburg-Lindau, Germany.

Eggers S., Keller H. U., Kroupa P., and Markiewicz W. J. (1997) Origin and dynamics of comets and star formation. *Planet. Space Sci., 45,* 1099–1104.

Eggers S., Keller H. U., Markiewicz W. J., and Kroupa P. (1998) Cometary dynamics in a star cluster (abstract). In *Astronomische Gesellschaft Meeting Abstracts, 14,* p. 5.

Everhart E. (1967a) Comet discoveries and observational selection. *Astron. J., 72,* 716–726.

Everhart E. (1967b) Intrinsic distributions of cometary perihelia and magnitudes. *Astron. J., 72,* 1002–1011.

Everhart E. (1968) Change in total energy of comets passing through the solar system. *Astron. J., 73,* 1039–1052.

Everhart E. and Raghavan N. (1970) Changes in total energy for 392 long-period comets, 1800–1970. *Astron. J., 75,* 258–272.

Fanale F. P. and Salvail J. R. (1984) An idealized short-period comet model: Surface insolation, H_2O flux, dust flux, and mantle evolution. *Icarus, 60,* 476–511.

Farley K.A., Montanari A., Shoemaker E. M., and Shoemaker C. S. (1998) Geochemical evidence for a comet shower in the late Eocene. *Science, 280,* 1250–1253.

Fernández J. A. (1978) Mass removed by the outer planets in the early solar system. *Icarus, 34,* 173–181.

Fernández J. A. (1980a) On the existence of a comet belt beyond Neptune. *Mon. Not. R. Astron. Soc., 192,* 481–491.

Fernández J. A. (1980b) Evolution of comet orbits under the perturbing influence of the giant planets and nearby stars. *Icarus, 42,* 406–421.

Fernández J. A. (1981) New and evolved comets in the solar system. *Astron. Astrophys., 96,* 26–35.

Fernández J. A. (1982) Comet showers. In *Chaos, Resonance and Collective Dynamical Phenomena in the Solar System* (S. Ferraz-Mello, ed.), pp. 239–254. Kluwer, Dordrecht.

Fernández J. A. (1997) The formation of the Oort cloud and the primitive galactic environment. *Icarus, 129,* 106–119.

Fernández J. A. and Brunini A. (2000) The buildup of a tightly bound comet cloud around an early Sun immersed in a dense galactic environment: Numerical experiments. *Icarus, 145,* 580–590.

Fernández J. A. and Gallardo T. (1994) The transfer of comet from parabolic orbits to short-period orbits: Numerical studies. *Astron. Astrophys., 281,* 911–922.

Fernández J. A. and Ip W.-H (1984) Some dynamical aspects of the accretion of Uranus and Neptune — The exchange of orbital angular momentum with planetesimals. *Icarus, 58,* 109–120.

Fernández J. A. and Ip W.-H (1987) Time dependent injection of Oort cloud comets into Earth-crossing orbits. *Icarus, 71,* 46–56.

Frogel J. A. and Gould A. (1998) No death star — for now. *Astrophys. J. Lett., 499,* L219–L222.

Gaidos E. J. (1995) Paleodynamics: Solar system formation and the early environment of the sun. *Icarus, 114,* 258–268.

García-Sánchez J., Preston R. A., Jones D. L., Weissman P. R., Lestrade J. F., Latham D. W., and Stefanik R. P. (1999) Stellar encounters with the Oort cloud based on Hipparcos data.

Astron. J., *117*, 1042–1055. Erratum in *Astron. J.*, *118*, 600.

García-Sánchez J., Weissman P. R., Preston R. A., Jones D. L., Lestrade J. F., Latham D. W., Stefanik R. P., and Paredes J. M. (2001) Stellar encounters with the solar system. *Astron. Astrophys.*, *379*, 634–659.

Gladman B. and Duncan M. (1990) On the fates of minor bodies in the outer solar system. *Astron. J.*, *100*, 1680–1693.

Gleason A. E., Scotti J. V., Durig D. T., Fry H. H., Zoltowski F. B., Ticha J., Tichy M., Gehrels T., and Ries J. G. (2003) *Comet C/2003 A2*. IAU Circular 8049.

Gomes R. S. (2003) The origin of the Kuiper Belt high-inclination population. *Icarus*, *161*, 404–418.

Gomes R. S., Morbidelli A., and Levison H. F. (2004) Planetary migration in a planetesimal disk: Why did Neptune stop at 30 AU? *Icarus*, in press.

Grazier K. R., Newman W. I., Kaula W. M., and Hyman J. M. (1999a) Dynamical evolution of planetesimals in the outer solar system. I. The Jupiter/Saturn zone. *Icarus*, *140*, 341–352.

Grazier K. R., Newman W. I., Varadi F., Kaula W. M., and Hyman J. M. (1999b) Dynamical evolution of planetesimals in the outer solar system. II. The Saturn/Uranus and Uranus/Neptune zones. *Icarus*, *140*, 353–368.

Green D. W. E. (2003) *Comet C/2003 A2 (Gleason)*. IAU Circular 8067.

Grieve R. A. F. and Shoemaker E. M (1994) The record of past impacts on Earth. In *Hazards Due to Comets and Asteroids* (T. Gehrels, ed.), pp. 417–462. Univ. of Arizona, Tucson.

Hahn J. M. and Malhotra R. (1999) Orbital evolution of planets embedded in a planetesimal disk. *Astron. J.*, *117*, 3041–3053.

Halley E. (1705) *A Synopsis of the Astronomy of Comets*. London. 24 pp.

Harmon J. K., Nolan M. C., Ostro S. J., and Campbell D. B. (2004) Radar studies of comet nuclei and grain comae. In *Comets II* (M. C. Festou et al., eds.), this volume. Univ. of Arizona, Tucson.

Harrington R. S. (1985) Implications of the observed distributions of very long period comet orbits. *Icarus*, *61*, 60–62.

Heisler J. (1990) Monte Carlo simulations of the Oort comet cloud. *Icarus*, *88*, 104–121.

Heisler J. and Tremaine S. (1986) The influence of the galactic tidal field on the Oort comet cloud. *Icarus*, *65*, 13–26.

Heisler J. and Tremaine S. (1989) How dating uncertainties affect the detection of periodicity in extinctions and craters. *Icarus*, *77*, 213–219.

Heisler J., Tremaine S. C., and Alcock C. (1987) The frequency and intensity of comet showers from the Oort cloud. *Icarus*, *70*, 269–288.

Helin E. F. and 19 colleagues (1996) *1996 PW*. Minor Planet Circular 1996-P03.

Higuchi A., Kokubo E., and Mukai T. (2002) Cometary dynamics: Migration due to gas drag and scattering by protoplanets. In *Proceedings of Asteroids, Comets, Meteors-ACM 2002* (B. Warmbein, ed.), pp. 453–456. ESA SP-500, Noordwijk, Netherlands.

Hills J. G. (1981) Comet showers and the steady-state infall of comets from the Oort cloud. *Astron. J.*, *86*, 1730–1740.

Hoffman A. (1985) Patterns of family extinction depend on definition and geologic timescale. *Nature*, *315*, 659–662.

Holman M. J. and Wisdom J. (1993) Dynamical stability in the outer solar system and the delivery of short period comets. *Astron. J.*, *105*, 1987–1999.

Holmberg J. and Flynn C. (2000) The local density of matter mapped by Hipparcos. *Mon. Not. R. Astron. Soc.*, *313*, 209–216.

Horner J. and Evans N. W. (2002) Biases in cometary catalogues and Planet X. *Mon. Not. R. Astron. Soc.*, *335*, 641–654.

Hughes D. W. (2000) A new approach to the calculation of the cratering rate of the Earth over the last 125 ± 20 m.y. *Mon. Not. R. Astron. Soc.*, *317*, 429–437.

Hughes D. W. (2001) The magnitude distribution, perihelion distribution and flux of long-period comets. *Mon. Not. R. Astron. Soc.*, *326*, 515–523.

Hut P. and Tremaine S. (1985) Have interstellar clouds disrupted the Oort comet cloud? *Astron. J.*, *90*, 1548–1557.

Hut P., Alvarez W., Elder W. P., Kauffman E. G., Hansen T., Keller G., Shoemaker E. M., and Weissman P. R. (1987) Comet showers as a cause of mass extinction. *Nature*, *329*, 118–126.

Ida S., Bryden G., Lin D. N. C., and Tanaka H. (2000) Orbital migration of Neptune and orbital distribution of trans-Neptunian objects. *Astrophys. J.*, *534*, 428–445.

Innanen K. A., Patrick A. T., and Duley W. W. (1978) The interaction of the spiral density wave and the Sun's galactic orbit. *Astrophys. Space Sci.*, *57*, 511–515.

Iseli M., Küppers M., Benz W., and Bochsler P. (2002) Sungrazing comets: Properties of nuclei and in situ detectability of cometary ions at 1 AU. *Icarus*, *155*, 350–364.

Jedicke R., Morbidelli A., Spahr T., Petit J., and Bottke W. F. (2003) Earth and space-based NEO survey simulations: Prospects for achieving the Spaceguard goal. *Icarus*, *161*, 17–33.

Jetsu L. and Pelt J. (2000) Spurious periods in the terrestrial impact crater record. *Astron. Astrophys.*, *353*, 409–418.

Jewitt D. (2004). Project Pan-STARRS and the outer solar system. *Earth Moon Planets*, in press.

Johnson R. E., Cooper J. F., Lanzerotti L. J., and Strazzula G. (1987) Radiation formation of a non-volatile comet crust. *Astron. Astrophys.*, *187*, 889–892.

Joss P. C. (1973) On the origin of short-period comets. *Astron. Astrophys.*, *25*, 271–273.

Jupp P. E., Kim P. T., Koo J.-Y., and Wiegert P. (2003) The intrinsic distribution and selection bias of long-period cometary orbits. *J. Am. Stat. Assoc.*, *98*, 515–521.

Kresák L. (1982) Comet discoveries, statistics, and observational selection. In *Comets* (L. L. Wilkening, ed.), pp. 56–82. Univ. of Arizona, Tucson.

Królikowska M.(2001) A study of the original orbits of "hyperbolic" comets. *Astron. Astrophys.*, *376*, 316–324.

Kronk G. W. (1999) *Cometography: A Catalog of Comets. Volume I: Ancient to 1799*. Cambridge Univ., New York. 563 pp.

Kuchner M. J., Brown M. E., and Holman M. (2002) Long-term dynamics and the orbital inclinations of the classical Kuiper belt objects. *Astron. J.*, *124*, 1221–1230.

Kuiper G. P. (1951) On the origin of the solar system. In *Proceedings of a Topical Symposium, Commemorating the 50th Anniversary of the Yerkes Observatory and Half a Century of Progress in Astrophysics* (J. A. Hynek, ed.), pp. 357–424. McGraw-Hill, New York.

Lardner D. (1853) On the classification of comets and the distribution of their orbits in space. *Mon. Not. R. Astron. Soc.*, *13*, 188–192.

Levison H. F. (1996) Comet taxonomy. In *Completing the Inventory of the Solar System* (T. W. Rettig and J. M. Hahn, eds.), pp. 173–191. ASP Conference Series 107, Astronomical Society of the Pacific, San Francisco.

Levison H. F. and Duncan M. J. (1993) The gravitational sculpting of the Kuiper belt. *Astrophys. J. Lett.*, *406*, L35–L38.

Levison H. F. and Morbidelli A. (2003) Forming the Kuiper belt by the outward transport of bodies during Neptune's migration. *Nature, 426*, 419–421.

Levison H. F., Dones L., and Duncan M. J. (2001) The origin of Halley-type comets: Probing the inner Oort cloud. *Astron. J., 121*, 2253–2267.

Levison H. F., Morbidelli A., Dones L., Jedicke R., Wiegert P. A., and Bottke W. F. (2002) The mass disruption of Oort cloud comets. *Science, 296*, 2212–2215.

Levison H. F., Morbidelli A., and Dones L. (2004) Sculpting the Kuiper belt by a stellar encounter: Constraints from the Oort cloud. *Icarus*, in press.

Lyttleton R. A. and Hammersley J. M. (1964) The loss of long-period comets from the solar system. *Mon. Not. R. Astron. Soc., 127*, 257–272.

Malhotra R. (1995) The origin of Pluto's orbit: Implications for the solar system beyond Neptune. *Astron. J., 110*, 420–429.

Marsden B. G. (1967) The sungrazing comet group. *Astron. J., 72*, 1170–1183.

Marsden B. G (1989) The sungrazing comet group. II. *Astron. J., 98*, 2306–2321.

Marsden B. G. and Meyer M. (2002) *Non-Kreutz Near-Sun Comet Groups.* IAU Circular 7832.

Marsden B. G. and Sekanina Z. (1973) On the distribution of "original" orbits of comets of large perihelion distance. *Astron. J., 78*, 1118–1124.

Marsden B. G. and Williams G. V. (2003) *Catalogue of Cometary Orbits*, 15th edition. Smithsonian Astrophysical Observatory, Cambridge. 169 pp.

Marsden B. G., Sekanina Z., and Yeomans D. K. (1973) Comets and nongravitational forces. V. *Astron. J., 78*, 211–225.

Marsden B. G., Sekanina Z., and Everhart E. (1978) New osculating orbits for 110 comets and the analysis of the original orbits of 200 comets. *Astron. J., 83*, 64–71.

Matese J. J. and Lissauer J. J. (2002) Characteristics and frequency of weak stellar impulses of the Oort cloud. *Icarus, 157*, 228–240.

Matese J. J., Whitman P. G., Innanen K. A., and Valtonen M. J. (1995) Periodic modulation of the Oort cloud comet flux by the adiabatically changing galactic tide. *Icarus, 116*, 255–268.

McGlynn T. A. and Chapman R. D. (1989) On the nondetection of extrasolar comets. *Astrophys. J. Lett., 346*, L105–L108.

McKinnon W. B., Zahnle K. J., Ivanov B. A., and Melosh H. J. (1997) Cratering on Venus: Models and observations. In *Venus II* (S. W. Bougher et al., eds.), pp. 969–1014. Univ. of Arizona, Tucson.

Mignard F. (2000) Local galactic kinematics from Hipparcos proper motions. *Astron. Astrophys., 354*, 522–536.

Morbidelli A. and Brown M. E. (2004) The Kuiper belt and the primordial evolution of the solar system. In *Comets II* (M. C. Festou et al., eds.), this volume. Univ. of Arizona, Tucson.

Morbidelli A., Jedicke R., Bottke W. F., Michel P., and Tedesco E. F. (2002) From magnitudes to diameters: The albedo distribution of near Earth objects and the Earth collision hazard. *Icarus, 158*, 329–342.

Morris D. E. and Muller R. A. (1986) Tidal gravitational forces: The infall of "new" comets and comet showers. *Icarus, 65*, 1–12.

Muller R. A. (2002) Measurement of the lunar impact record for the past 3.5 b.y. and implications for the Nemesis theory. In *Catastrophic Events and Mass Extinctions: Impacts and Beyond* (C. Koeberl and K. G. MacLeod, eds.), pp. 659–665. GSA Special Paper 356, Geological Society of America, Boulder, Colorado.

National Research Council Astrophysics Survey Committee (2001) *Astronomy and Astrophysics in the New Millennium.* National Academies Press, Washington, DC. 175 pp.

Nesvorný D. and Dones L. (2002) How long-lived are the hypothetical Trojan populations of Saturn, Uranus, and Neptune? *Icarus, 160*, 271–288.

Nesvorný D., Alvarellos J. L. A., Dones L., and Levison H. F. (2003) Orbital and collisional evolution of the irregular satellites. *Astron. J., 126*, 398–429.

Nesvorný D., Beaugé C., and Dones L. (2004) Collisional origin of families of irregular satellites. *Astron. J., 127*, 1768–1783.

Newton H. A. (1891) Capture of comets by planets. *Astron. J., 11*, 73–75.

Nurmi P., Valtonen M. J., and Zheng J. Q. (2001) Periodic variation of Oort cloud flux and cometary impacts on the Earth and Jupiter. *Mon. Not. R. Astron. Soc., 327*, 1367–1376.

Oort J. H. (1950) The structure of the cloud of comets surrounding the solar system and a hypothesis concerning its origin. *Bull. Astron. Inst. Neth. 11*, 91–110.

Oort J. H. and Schmidt M. (1951) Differences between new and old comets. *Bull. Astron. Inst. Neth. 11*, 259–269.

Öpik E. (1932) Note on stellar perturbations of nearly parabolic orbits. *Proc. Am. Acad. Arts Sci., 67*, 169–183.

Prialnik D. and Bar-Nun A. (1987) On the evolution and activity of cometary nuclei. *Astrophys. J., 313*, 893–905.

Quinn T., Tremaine S., and Duncan M. (1990) Planetary perturbations and the origins of short-period comets. *Astrophys. J., 355*, 667–679.

Rampino M. R. and Stothers R. B. (1984) Terrestrial mass extinctions, cometary impacts, and the sun's motion perpendicular to the galactic plane. *Nature, 308*, 709–712.

Raup D. M. and Sepkoski J. J. (1984) Periodicity of extinctions in the geologic past. *Proc. Natl. Acad. Sci. USA, 81*, 801–805.

Rickman H., Fernández J. A., Tancredi G., and Licandro J. (2001) The cometary contribution to planetary impact rates. In *Collisional Processes in the Solar System* (M. Ya. Marov and H. Rickman, eds.), pp. 131–142. Astrophysics and Space Science Library, Vol. 261, Kluwer, Dordrecht.

Roques F. and Moncuquet M. (2000) A detection method for small Kuiper Belt objects: The search for stellar occultations. *Icarus, 147*, 530–544.

Safronov V. S. (1969) *Evolution of the Protoplanetary Cloud and Formation of the Earth and Planets*, Moscow, Nauka (English translation 1972).

Safronov V. S. (1972) Ejection of bodies from the solar system in the course of the accumulation of the giant planets and the formation of the cometary cloud. In *The Motion, Evolution of Orbits, and Origin of Comets* (A. Chebotarev et al., eds.), pp. 329–334. IAU Symposium 45. Reidel, Dordrecht.

Schwartz R. D. and James P. B. (1984) Periodic mass extinctions and the Sun's oscillation about the galactic plane. *Nature, 308*, 712–713.

Sen A. K. and Rama N. C. (1993) On the missing interstellar comets. *Astron. Astrophys., 275*, 298–300.

Shoemaker E. M. (1983) Asteroid and comet bombardment of the earth. *Annu. Rev. Earth Planet. Sci., 11*, 461–494.

Shoemaker E. M. and Wolfe R. F. (1984) Evolution of the Uranus-Neptune planetesimal swarm (abstract). In *Lunar and Planetary Science XXV*, pp. 780–781. Lunar and Planetary Institute, Houston.

Shoemaker E. M. and Wolfe R. F. (1986) Mass extinctions, crater ages, and comet showers. In *The Galaxy and the Solar System* (R. Smoluchowski et al., eds.), pp. 338–386. Univ. of Arizona, Tucson.

Shoemaker E. M., Wolfe R. F., and Shoemaker C. S. (1990) Asteroid and comet flux in the neighborhood of Earth. In *Global Catastrophes in Earth History: An Interdisciplinary Conference on Impact, Volcanism and Mass Mortality* (V. L. Sharpton and P. D. Ward, eds.), pp. 155–170. GSA Special Paper 247. Geological Society of America, Boulder, Colorado.

Smoluchowski R. and Torbett M. (1984) The boundary of the solar system. *Nature, 311,* 38–39.

Spitzer L. (1987) *Dynamical Evolution of Globular Clusters.* Princeton Univ., Princeton, New Jersey. 191 pp.

Stern S. A. (2003) The evolution of comets in the Oort cloud and Kuiper belt. *Nature, 424,* 639–642.

Stern S. A. and Weissman P. R. (2001) Rapid collisional evolution of comets during the formation of the Oort cloud. *Nature, 409,* 589–591.

Swift J. (1726) *Gulliver's Travels.* Full text available on line at http://www.jaffebros.com/lee/gulliver/index.html.

Thommes E. W., Duncan M. J., and Levison H. F. (1999) The formation of Uranus and Neptune in the Jupiter-Saturn region of the solar system. *Nature, 402,* 635–638.

Thommes E. W., Duncan M. J., and Levison H. F. (2002) The formation of Uranus and Neptune among Jupiter and Saturn. *Astron. J., 123,* 2862–2883.

Tremaine S. (1986) Is there evidence for a solar companion? In *The Galaxy and the Solar System* (R. Smoluchowski et al., eds.), pp. 409–416. Univ. of Arizona, Tucson.

Tremaine S. (1993) The distribution of comets around stars. In *Planets Around Pulsars* (J. A. Phillips et al., eds.), pp. 335–344. ASP Conference Series 36, Astronomical Society of the Pacific, San Francisco.

Trujillo C. A., Jewitt D. C., and Luu J. X. (2000) Population of the scattered Kuiper belt. *Astrophys. J. Lett., 529,* L103–L106.

Tyson J. A. (2002) Large Synoptic Survey Telescope: Overview. In *Survey and Other Telescope Technologies and Discoveries* (J. A. Tyson and S. Wolff, eds.), pp. 10–20. SPIE Proceedings Vol. 4836, SPIE — The International Society for Optical Engineering, Bellingham, Washington.

Valtonen M. J. (1983) On the capture of comets into the inner solar system. *Observatory, 103,* 1–4.

Valtonen M. J. and Innanen K. A. (1982) The capture of interstellar comets. *Astrophys. J., 255,* 307–315.

van Woerkom A. F. F. (1948) On the origin of comets. *Bull. Astron. Inst. Neth., 10,* 445–472.

Weinberg M. D., Shapiro S. L., and Wasserman I. (1987) The dynamical fate of wide binaries in the solar neighborhood. *Astrophys. J., 312,* 367–389.

Weissman P. R. (1979) Physical and dynamical evolution of long-period comets. In *Dynamics of the Solar System* (R. L. Duncombe, ed.), pp. 277–282. Reidel, Dordrecht.

Weissman P. R. (1980a) Physical loss of long-period comets. *Astron. Astrophys., 85,* 191–196.

Weissman P. R. (1990b) Stellar perturbations of the cometary cloud. *Nature, 288,* 242–243.

Weissman P. R. (1982) Dynamical history of the Oort cloud. In *Comets* (L. L. Wilkening, ed.), pp. 637–658. Univ. of Arizona, Tucson.

Weissman P. R. (1983) Cometary impacts with the sun — Physical and dynamical considerations. *Icarus, 55,* 448–454.

Weissman P. R. (1985) Dynamical evolution of the Oort cloud. In *Dynamics of Comets: Their Origin and Evolution* (A. Carusi and G. B. Valsecchi, eds.), pp. 87–96. IAU Colloquium 83, Reidel, Dordrecht.

Weissman P. R. (1986) The Oort cloud and the galaxy: Dynamical

interactions. In *The Galaxy and the Solar System.* (R. Smoluchowski et al., eds.), pp. 204–237. Univ. of Arizona, Tucson.

Weissman P. R. (1990a) The Oort cloud. *Nature, 344,* 825–830.

Weissman P. R. (1990b) The cometary impactor flux at the Earth. In *Global Catastrophes in Earth History* (V. L. Sharpton and P. D. Ward, eds.), pp. 171–180. GSA Special Paper 247, Geological Society of America, Boulder, Colorado.

Weissman P. R. (1991) Dynamical history of the Oort cloud. In *Comets in the Post-Halley Era* (R. L. Newburn Jr. et al., eds.), pp. 463–486. Astrophysics and Space Science Library Vol. 167, Kluwer, Dordrecht.

Weissman P. R. (1993) No, we are not in a cometary shower (abstract). *Bull. Am. Astron. Soc., 25,* 1063.

Weissman P. R. (1996) The Oort cloud. In *Completing the Inventory of the Solar System* (T. W. Rettig and J. M. Hahn, eds.), pp. 265–288. ASP Conference Series 107, Astronomical Society of the Pacific, San Francisco.

Weissman P. R. and Levison H. F. (1997a) Origin and evolution of the unusual object 1996 PW: Asteroids from the Oort cloud? *Astrophys. J. Lett., 488,* L133–L136.

Weissman P. R. and Levison H. F. (1997b) The population of the trans-Neptunian region: The Pluto-Charon environment. In *Pluto and Charon* (S. A. Stern and D. J. Tholen, eds.), pp. 559–604 + 3 color plates. Univ. of Arizona, Tucson.

Weissman P. R. and Lowry S. C. (2003) The size distribution of Jupiter-family cometary nuclei (abstract). In *Lunar and Planetary Science XXXIV,* Abstract #2003. Lunar and Planetary Institute, Houston (CD-ROM).

Weissman P. R. and Stern S. A. (1994) The impactor flux in the Pluto-Charon system. *Icarus, 111,* 378–386.

Weissman P. R., Bottke W. F., and Levison H. F. (2002) Evolution of comets into asteroids. In *Asteroids III* (W. F. Bottke Jr. et al., eds.), pp. 669–686. Univ. of Arizona, Tucson.

Whipple F. L. (1950) A comet model. I. The acceleration of comet Encke. *Astrophys. J., 111,* 375–394.

Whipple F. L. (1962) On the distribution of semimajor axes among comet orbits. *Astron. J., 67,* 1–9.

Whipple F. L. (1964) Evidence for a comet belt beyond Neptune. *Proc. Natl. Acad. Sci. USA, 51,* 711–718.

Whipple F. L. (1978) Cometary brightness variation and nucleus structure. *Moon and Planets, 18,* 343–359.

Whitmire D. P. and Jackson A. A. (1984) Are periodic mass extinctions driven by a distant solar companion? *Nature, 308,* 713–715.

Whitmire D. P. and Matese J. J. (1985) Periodic comet showers and planet X. *Nature, 313,* 36–38.

Wiegert P. (1996) The evolution of long-period comets. Ph.D. thesis, Univ. of Toronto, Toronto.

Wiegert P. and Tremaine S. (1999) The evolution of long-period comets. *Icarus, 137,* 84–121.

Yau K., Yeomans D., and Weissman P. (1994) The past and future motion of Comet P/Swift-Tuttle. *Mon. Not. R. Astron. Soc., 266,* 305–316.

Yeomans D. K. (1991) *Comets. A Chronological History of Observation, Science, Myth, and Folklore.* Wiley, New York. 485 pp.

Zahnle K., Dones L., and Levison H. F. (1998) Cratering rates on the galilean satellites. *Icarus, 136,* 202–222.

Zahnle K., Schenk P., Levison H., and Dones L. (2003) Cratering rates in the outer solar system. *Icarus, 163,* 263–289.

Zappalà V., Cellino A., Gladman B. J., Manley S., and Migliorini F. (1998) Note: Asteroid showers on Earth after family breakup events. *Icarus, 134,* 176–179.

The Kuiper Belt and the Primordial
Evolution of the Solar System

A. Morbidelli

Observatoire de la Côte d'Azur

M. E. Brown

California Institute of Technology

We discuss the structure of the Kuiper belt as it can be inferred from the first decade of observations. In particular, we focus on its most intriguing properties — the mass deficit, the inclination distribution, and the apparent existence of an outer edge and a correlation among inclinations, colors, and sizes — which clearly show that the belt has lost the pristine structure of a dynamically cold protoplanetary disk. Understanding how the Kuiper belt acquired its present structure will provide insight into the formation of the outer planetary system and its early evolution. We critically review the scenarios that have been proposed so far for the primordial sculpting of the belt. None of them can explain in a single model all the observed properties; the real history of the Kuiper belt probably requires a combination of some of the proposed mechanisms.

1. INTRODUCTION

When Edgeworth and Kuiper conjectured the existence of a belt of small bodies beyond Neptune — now known as the Kuiper belt — they certainly were imagining a disk of planetesimals that preserved the pristine conditions of the protoplanetary disk. However, since the first discoveries of transneptunian objects, astronomers have realized that this picture is not correct: The disk has been affected by a number of processes that have altered its original structure. The Kuiper belt may thus provide us with many clues to understand what happened in the outer solar system during the primordial ages. Potentially, the Kuiper belt might teach us more about the formation of the giant planets than the planets themselves. And, as in a domino effect, a better knowledge of giant-planet formation would inevitably boost our understanding of the subsequent formation of the terrestrial planets. Consequently, Kuiper belt research is now considered a top priority in modern planetary science.

A decade after the discovery of 1992 QB$_1$ (*Jewitt and Luu*, 1993), we now know 770 transneptunian objects (semimajor axis a > 30 AU) (all numbers are as of March 3, 2003). Of these, 362 have been observed during at least two oppositions, and 239 during at least three oppositions. Observations at two and three oppositions are necessary for the Minor Planet Center to compute the objects' orbital elements with moderate and good accuracy respectively. Therefore, the transneptunian population is gradually taking shape, and we can start to seriously examine the Kuiper belt structure and learn what it has to teach us. We should not forget, however, that our view of the transneptunian population is still partial and is strongly biased by a number of factors, some of which cannot be easily modeled.

A primary goal of this chapter is to present the orbital structure of the Kuiper belt as it stands based on the current observations. We start in section 2 by presenting the various subclasses that constitute the transneptunian population. Then in section 3 we describe some striking properties of the population, such as its mass deficit, inclination excitation, radial extent and a puzzling correlation between orbital elements and physical properties. In section 4 we finally review the models that have been proposed so far on the primordial sculpting of the Kuiper belt. Some of these models date from the very beginning of Kuiper belt science, when only a handful of objects were known, and have been at least partially invalidated by the new data. Paradoxically, however, as the data increase in number and quality, it becomes increasingly difficult to explain all the properties of the Kuiper belt in the framework of a single scenario. The conclusions are in section 5.

2. TRANSNEPTUNIAN POPULATIONS

The transneptunian population is "traditionally" subdivided in two subpopulations: the scattered disk and the Kuiper belt. The definition of these subpopulations is not uniform, as the Minor Planet Center and various authors often use slightly different criteria. Here we propose and discuss a categorization based on the dynamics of the objects and their relevance to the reconstruction of the primordial evolution of the outer solar system.

In principle, one would like to call the Kuiper belt the population of objects that, even if characterized by chaotic dynamics, do not suffer close encounters with Neptune and thus do not undergo macroscopic migration in semimajor axis. Conversely, the bodies that are transported in semi-

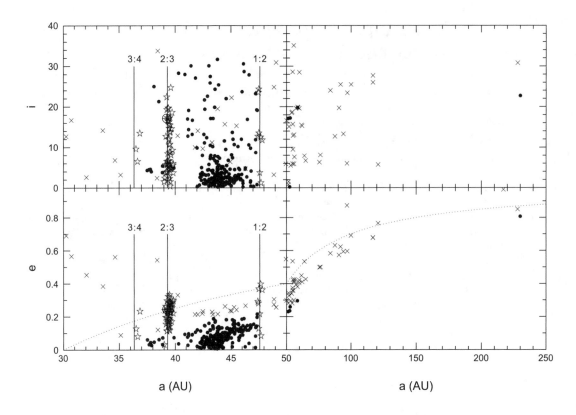

Fig. 1. The orbital distribution of multiopposition transneptunian bodies, as of March 3, 2003. Scattered disk bodies are represented as a cross, classical Kuiper belt bodies as dots, and resonant bodies as stars. In the absence of long-term numerical integrations of the evolution of all the objects and because of the uncertainties in the orbital elements, it is possible that some bodies could have been misclassified. The figure should thus be considered as an indicative representation of the various subgroups that compose the transneptunian population. The dotted curve denotes q = 30 AU. The vertical solid lines mark the locations of the 3:4, 2:3, and 1:2 mean-motion resonances with Neptune. The orbit of Pluto is represented by a crossed circle.

major axis by close and distant encounters with Neptune would constitute the scattered disk. The problem with precisely dividing the transneptunian population into Kuiper belt or scattered disk is related to timescale. On what timescale should we see semimajor axis migration resulting in the classification of an object in the scattered disk? The question is relevant, because it is possible for bodies trapped in resonances to significantly change their perihelion distance and pass from a scattering phase to a nonscattering phase (and vice versa) numerous times over the age of the solar system.

For this reason, we prefer to link the definition of the scattered disk to its formation mechanism. We refer to the scattered disk as the region of orbital space that can be visited by bodies that have encountered Neptune within a Hill's radius at least once during the age of the solar system, assuming no substantial modification of the planetary orbits. We then refer to the Kuiper belt as the complement of the scattered disk in the a > 30 AU region.

To categorize the observed transneptunian bodies into scattered disk and Kuiper belt, we refer to previous work on the dynamics of transneptunian bodies in the framework of the current architecture of the planetary system. For the a < 50 AU region, we use the results by *Duncan et al.* (1995) and *Kuchner et al.* (2002), who numerically mapped the re-

gions of the (a, e, i) space with 32 < a < 50 AU that can lead to a Neptune-encountering orbit within 4 G.y. Because dynamics are reversible, these are also the regions that can be visited by a body after having encountered the planet. Therefore, according to our definition, they constitute the scattered disk. For the a > 50 AU region, we use the results by *Levison and Duncan* (1997) and *Duncan and Levison* (1997), who followed for a time span of another 4 G.y. the evolutions of the particles that encountered Neptune in *Duncan et al.* (1995). Despite the fact that the initial conditions did not cover all possible configurations, we can reasonably assume that these integrations cumulatively show the regions of the orbital space that can be possibly visited by bodies transported to a > 50 AU by Neptune encounters. Again, according to our definition, these regions constitute the scattered disk.

In Fig. 1 we show the (a, e, i) distribution of the transneptunian bodies that have been observed during at least two oppositions. The bodies that belong to the scattered disk according to our criterion are represented as crosses, while Kuiper belt bodies are represented by dots and stars (see explanation of the difference below).

We believe that our definition of scattered disk and Kuiper belt is meaningful for what concerns the major goal of Kuiper belt science, i.e., to reconstruct the primordial evolution of the outer solar system. In fact, all bodies in the solar system must have been formed on orbits with very small eccentricities and inclinations, typical of an accretion disk. In the framework of the current architecture of the solar system, the current orbits of scattered disk bodies might have started with quasicircular orbits in Neptune's zone by pure dynamical evolution. Therefore, they do not provide any relevant clue to uncover the primordial architecture. The opposite is true for the orbits of the Kuiper belt objects with nonnegligible eccentricity and/or inclination. Their existence reveals that some excitation mechanism that is no longer at work occurred in the past (see section 4).

In this respect, the existence of Kuiper belt bodies with a > 50 AU on highly eccentric orbits is particularly important (five objects in Fig. 1, although our classification is uncertain for the reasons explained in the figure caption). Among them, 2000 CR_{105} (a = 230 AU, perihelion distance q = 44.17 AU, and inclination i = 22.7°) is a challenge by itself concerning the explanation of its origin. We call these objects extended scattered disk objects for two reasons: (1) they do not belong to the scattered disk according to our definition but are very close to its boundary and (2) a body of ~300 km like 2000 CR_{105} presumably formed much closer to the Sun, where the accretion timescale was sufficiently short (*Stern*, 1996), implying that it has been subsequently transported in semimajor axis until reaching its current location. This hypothesis suggests that in the past the true scattered disk extended well beyond its present boundary in perihelion distance. Given that the observational biases rapidly become more severe with increasing perihelion distance and semimajor axis, the currently known extended scattered disk objects may be the tip of the iceberg, e.g., the emerging representatives of a conspicuous population, possibly outnumbering the scattered disk population (*Gladman et al.*, 2002).

In addition to the extended scattered disk, we distinguish two other subpopulations of the Kuiper belt. We refer to the Kuiper belt bodies that are located in some major mean-motion resonance with Neptune [essentially the 3:4, 2:3, and 1:2 resonances (star symbols in Fig. 1) but also the 2:5 resonance (see *Chiang et al.*, 2003)] as the resonant population. It is well known that mean-motion resonances offer a protection mechanism against close encounters with the resonant planet (*Cohen and Hubbard*, 1965). For this reason, the resonant population — which, as part of the Kuiper belt, by definition must not encounter Neptune within the age of the solar system — can have perihelion distances much smaller than the other Kuiper belt objects, and even Neptune-crossing orbits (q < 30 AU) as in the case of Pluto. The bodies in the 2:3 resonance are often called Plutinos because of the analogy of their orbit with that of Pluto. We call the collection of Kuiper belt objects with a < 50 AU that are not in any notable resonant configuration the classical belt. Because they are not protected from close encounters with Neptune by any resonance, the stability criterion confines them to the region with small to moderate eccentricity, typically on orbits with q > 35 AU. The adjective "classical" is justified because, among all subpopula-

tions, this is the one whose orbital properties are the most similar to those expected for the Kuiper belt prior to the first discoveries. We note, however, that the classical population is not that "classical." Although moderate, the eccentricities are larger than those that should characterize a protoplanetary disk. Moreover, several bodies have very large inclinations (see section 3.2). Finally, the total mass is only a small fraction of the expected pristine mass in that region (section 3.1). All these elements indicate that the classical belt has also been affected by some primordial excitation and depletion mechanism(s).

3. STRUCTURE OF THE KUIPER BELT

3.1. Missing Mass of the Kuiper Belt

The original argument followed by *Kuiper* (1951) to conjecture the existence of a band of small planetesimals beyond Neptune was related to the mass distribution in the outer solar system. The minimum mass solar nebula inferred from the total planetary mass (plus lost volatiles) smoothly declines from the orbit of Jupiter until the orbit of Neptune (see Fig. 2); why should it abruptly drop beyond the last planet? However, while Kuiper's conjecture on the existence of a transneptunian belt is correct, the total mass in the 30–50-AU range inferred from observations is two orders of magnitude smaller than the one he expected.

Kuiper's argument is not the only indication that the mass of the primordial Kuiper belt had to be significantly larger.

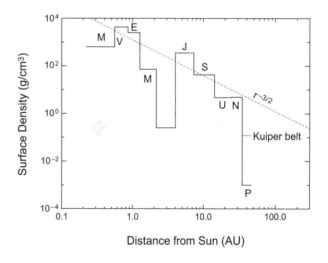

Fig. 2. The mass distribution of the solar nebula inferred from the masses of the planets augmented by the mass needed to bring the observed material to solar composition (data from *Lewis*, 1995). The surface density in the Kuiper belt has been computed assuming a current mass of ~0.1 $M_\oplus$ (*Jewitt et al.*, 1996; *Chiang and Brown*, 1999; *Trujillo et al.*, 2001; *Gladman et al.*, 2001) in the 42–48-AU annulus, and scaling the result by a factor of 70 in order to account for the inferred primordial local ratio between volatiles and solids. The estimate of the total mass in the Kuiper belt overwhelms that of Pluto, but still does not bring the mass to the extrapolation of the ~$r^{-3/2}$ line.

Further evidence for a massive primordial Kuiper belt was uncovered by *Stern* (1995), who found that the objects currently in the Kuiper belt were incapable of having formed in the present environment: Collisions are sufficiently infrequent that 100-km objects cannot be built by pairwise accretion of the current population over the age of the solar system. Moreover, owing to the large eccentricities and inclinations of Kuiper belt objects — and consequently to their high encounter velocities — collisions that do occur tend to be erosive rather than accretional, making bodies smaller rather than larger. Stern suggested that the resolution of this dilemma is that the primordial Kuiper belt was both more massive and dynamically colder, so that more collisions occurred, and they were gentler and therefore generally accretional.

Following this idea, detailed modeling of accretion in a massive primordial Kuiper belt was performed by *Stern* (1996), *Stern and Colwell* (1997a,b), and *Kenyon and Luu* (1998, 1999a,b). While each model includes different aspects of the relevant physics of accretion, fragmentation, and velocity evolution, the basic results are in approximate agreement. First, with ~10 $M_\oplus$ (Earth mass) or more of solid material in an annulus from about 35 to 50 AU on very low eccentricity orbits (e ≤ 0.001), all models naturally produce a few objects on the order of the size of Pluto and approximately the right number of ~100-km objects, on a timescale ranging from several 10^7 to several 10^8 yr. The models suggest that the majority of mass in the disk was in bodies approximately 10 km and smaller. The accretion stopped when the formation of Neptune or other dynamical phenomena (see section 4) began to induce eccentricities and inclinations in the population high enough to move the collisional evolution from the accretional to the erosive regime (*Stern*, 1996). A massive and dynamically cold primordial Kuiper belt is also required by the models that attempt to explain the formation of the observed numerous binary Kuiper belt objects (*Goldreich et al.*, 2002; *Weidenshilling*, 2002).

Therefore, the general formation picture of an initial massive Kuiper belt appears secure. However, a fundamental question remains to be addressed: How did the initial mass disappear? Collisions can grind bodies down to dust particles, which are subsequently transported away from the belt by radiation pressure and/or Poynting Robertson drag, causing a net mass loss. The major works on the collisional erosion of a massive primordial belt have been done by *Stern and Colwell* (1997b) and *Davis and Farinella* (1997, 1998), achieving similar conclusions (for a review, see *Farinella et al.*, 2000). As long as the planetesimal disk was characterized by small eccentricities and inclinations, the collisional activity could only moderately reduce the mass of the belt. However, when the eccentricities and inclinations became comparable to those currently observed, bodies smaller than 50–100 km in diameter could be effectively destroyed. The total amount of mass loss depends on the primordial size distribution. To reduce the total mass from 30 $M_\oplus$ to a fraction of an Earth mass, the primordial size

distribution had to be steep enough that essentially all the mass was carried by these small bodies, while the number of bodies larger than ~100 km had to be basically equal to the present number. Although this outcome seems consistent with the suggestions of the accretional models, there is circumstantial (but nonetheless compelling) evidence suggesting the primordial existence of a much larger number of large bodies (*Stern*, 1991). The creation of Pluto-Charon likely required the impact of two approximately similar-sized bodies that would be the two largest currently known bodies in the Kuiper belt. The probability that the two largest bodies in the belt would collide and create Pluto-Charon is vanishingly small, arguing that many bodies of this size must have been present and subsequently disappeared. Similarly, the existence of Triton and the large obliquity of Neptune are best explained by the existence at one time of many large bodies being scattered through the Neptune system. The elimination of these large bodies (if they existed in the Kuiper belt) could not be due to the collisional activity, but requires a dynamical explanation.

Another constraint against the collisional grinding scenario is provided by the preservation of the binary Kuiper belt objects. The Kuiper belt binaries have large separations, so it can be easily computed that the impact on the satellite of a projectile 100 times less massive with a speed of 1 km/s would give the former an impulse velocity sufficient to escape to an unbound orbit. If the collisional activity was strong enough to cause an effective reduction of the overall mass of the Kuiper belt, these kinds of collisions had to be extremely common, so we would not expect a significant fraction of widely separated binary objects in the current remaining population (*Petit and Mousis*, 2003.)

Understanding the ultimate fate of the 99% of the initial Kuiper belt mass that is no longer in the Kuiper belt is the first step in reconstructing the history of the outer solar system.

3.2. Excitation of the Kuiper Belt

An important clue to the history of the early outer solar system is the dynamical excitation of the Kuiper belt. While eccentricities and inclinations of resonant and scattered objects are expected to have been affected by interactions with Neptune, those of the classical objects should have suffered no such excitation. Nonetheless, the confirmed classical belt objects have an inclination range up to at least 32° and an eccentricity range up to 0.2, significantly higher than expected from a primordial disk, even accounting for mutual gravitational stirring.

The observed distributions of eccentricities and inclinations in the Kuiper belt are highly biased. High-eccentricity objects have closer approaches to the Sun and thus become brighter and more easily detected. High-inclination objects spend little time at the low latitudes at which most surveys take place, while low-inclination objects spend zero time at the high latitudes where some searches have occurred. (Latitude and inclination are defined with respect to the invariable plane, which is a better representation for the plane of the Kuiper belt than is the ecliptic.)

Determination of the eccentricity distribution of the Kuiper belt requires disentanglement of eccentricity and semimajor axis, which is only possible for objects with well-determined orbits for which a well-characterized sample of sufficient size is not yet available. Determination of the inclination distribution, however, is much simpler because the inclination of an object is well determined even after a small number of observations, and the latitude of discovery of each object is a known quantity. Using these facts, *Brown* (2001) developed general methods for debiasing object discoveries to discern the underlying inclination distribution. The simplest method removes the latitude-of-discovery biases by considering only objects discovered within 1° of the invariable plane equator and weights each object by sin(i), where i is the inclination of each object, to account for the proportional fraction of time that objects of different inclination spend at the equator (strictly speaking, one should use only objects found precisely at the equator; expanding to 1° around the equator greatly increases the sample size while biasing the sample slightly against objects with inclinations between 0° and 1°). An important decision to be made in constructing this inclination distribution is the choice of which objects to include in the sample. One option is to use only confirmed classical objects, i.e., those that have been observed at least two oppositions and for which the orbit is reasonably assured of fitting the definition of the classical Kuiper belt as defined above. The possibility exists that these objects are biased in some way against unusual objects that escape recovery at a second opposition because of unexpected orbits, but we expect that this bias is likely to be in the direction of underreporting high-inclination objects. On the other hand, past experience has shown that if we use all confirmed and unconfirmed classical bodies, we pollute the sample with misclassified resonant and scattered objects, which generally have higher inclinations and therefore artificially inflate the inclination distribution of the classical belt. We therefore chose to use only confirmed classical belt bodies, with the caveat that some high-inclination objects might be missing. Figure 3 shows the inclination distribution of the classical Kuiper belt derived from this method. This method has the advantage that it is simple and model independent, but the disadvantage that it makes no use of the information contained in high-latitude surveys where most of the high-inclination objects are discovered. For example, the highest-inclination classical belt body found within 1° of the equator has an inclination of 10.1°, while an object with an inclination of 31.9° has been found at a latitude of 11.2°. The two high-inclination points in Fig. 3 attempt to partially correct this deficiency by using discoveries of objects between 3° and 6° latitude to define the high-inclination end of the inclination distribution, using equation (3) of *Brown* (2001). Observations at these latitudes miss all objects with lower inclinations, but we can linearly scale the high-latitude distribution to match the low-latitude distribution in the region where

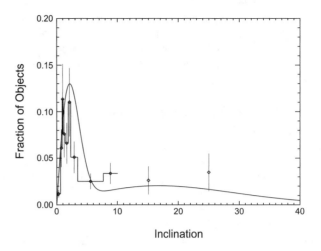

Fig. 3. The inclination distribution of the classical Kuiper belt. The points with error bars show the model-independent estimate constructed from a limited subset of confirmed classical belt bodies, while the smooth line shows the best-fit two-population model.

they are both valid from 6° to 10° and retrieve a correctly relatively calibrated high-inclination distribution.

Brown (2001) developed a more general method to use all objects simultaneously by comparing inclinations of all objects to those found from Monte Carlo observations of simple model inclination distributions at the latitudes of discovery. The simplest reasonable model distribution has a form where f(i)di, the number of objects between inclinations i and i + di, is proportional to $\sin(i) \exp(-i^2/2\sigma^2)$ di where σ is a measure of the excitation of the population. The resonant and scattered objects are both well fit by this functional form with $\sigma = 10° \pm 2°$ and $20° \pm 4°$ respectively. The best single Gaussian fit for the confirmed classical belt objects can be ruled out at a high level of confidence; the observed inclination distribution of the classical Kuiper belt is more complex than can be described by the simplest model. Guided by Fig. 3, we make the assumption that the inclination distribution between about 0° and 3° appears adequately described by a single Gaussian times sine inclination, and search for a functional form to describe the higher-inclination objects. The next simplest functional form is one with a second Gaussian added to the distribution: $f(i)di = \sin(i) [a_1 \exp(-i^2/2\sigma_1^2) + a_2 \exp(-i^2/2\sigma_2^2)]di$. The best fit to the two-Gaussian model, found by modeling the latitudes and inclinations of all confirmed classical belt objects, has parameters $a_1 = 96.4$, $a_2 = 3.6$, $\sigma_1 = 1.8$, and $\sigma_2 = 12$ and is shown in Fig. 3. For this model ~60% of the objects reside in the high-inclination population.

A clear feature of this modeled distribution is the presence of distinct high- and low-inclination populations. While *Brown* (2001) concluded that not enough data existed at the time to determine if the two populations were truly distinct or if the model fit forced an artificial appearance of two populations, the larger amount of data now available, and shown in the model-independent analysis of Fig. 3, confirms that the distinction between the populations is real. The sharp drop around 4° is independent of any model, while the extended distribution to 30° is demanded by the presence of objects with these inclinations.

3.3. Physical Evidence for Two Populations in the Classical Belt

The existence of two distinct classical Kuiper belt populations, which we will call the hot ($i > 4°$) and cold ($i < 4°$) classical populations, could be caused in one of two general ways. Either a subset of an initially dynamically cold population was excited, leading to the creation of the hot classical population, or the populations are truly distinct and formed separately. One manner in which we can attempt to determine which of these scenarios is more likely is to examine the physical properties of the two classical populations. If the objects in the hot and cold populations are physically different, it is less likely that they were initially part of the same population.

The first suggestion of a physical difference between the hot and the cold classical objects came from *Levison and Stern* (2001), who noted that the intrinsically brightest classical belt objects (those with lowest absolute magnitudes) are preferentially found with high inclination. *Trujillo and Brown* (2003) have recently verified this conclusion in a bias-independent manner from a survey for bright objects that covered ~70% of the ecliptic and found many hot classical objects but few cold classical objects.

The second possible physical difference between hot and cold classical Kuiper belt objects is their colors, which relates (in an unknown way) to surface composition. Several possible correlations between orbital parameters and color were suggested by *Tegler and Romanishin* (2000) and further investigated by *Doressoundiram et al.* (2001). The issue was clarified by *Trujillo and Brown* (2002), who quantitatively showed that for the classical belt, inclination, and no other independent orbital parameter, is correlated with color. In essence, the low-inclination classical objects tend to be redder than higher-inclination objects. *Hainaut and Delsanti* (2002) have compiled a list of all published Kuiper belt colors that more than doubles the sample of *Trujillo and Brown* (2002). A plot of color vs. inclination for the classical belt objects in this expanded sample (Fig. 4) confirms the correlation between color and inclination. This expanded sample also conclusively demonstrates that no other independent dynamical correlations occur, although the fact that the low-inclination red classical objects also have low eccentricities, and therefore high perihelia, causes an apparent correlation with perihelion distance as well.

More interestingly, we see that the colors naturally divide into distinct low-inclination and high-inclination populations at precisely the location of the divide between the hot and cold classical objects. These populations differ at a 99.9% confidence level. Interestingly, the cold classical population also differs in color from the Plutinos and the

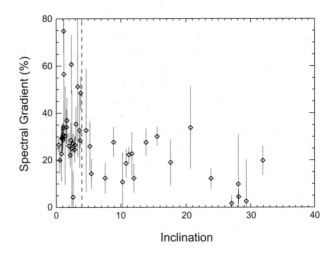

Fig. 4. Color gradient vs. inclination in the classical Kuiper belt. Color gradient is the slope of the spectrum, in % per 100 nm, with 0% being neutral and large numbers being red. The hot and cold classical objects have significantly different distributions of color.

scattered objects at the 99.8% and 99.9% confidence level respectively, while the hot classical population appears identical in color to these other populations. The possibility remains, however, that the colors of the objects, rather than being markers of different populations, are actually caused by the different inclinations. *Stern* (2002), for example, has suggested that the higher average impact velocities of the high-inclination objects will cause large-scale resurfacing by fresh water ice, which could be blue to neutral in color. If this hypothesis were correct, however, we would also expect to see correlations between colors and semimajor axis or eccentricity, which also determine impact velocities. These correlations do not exist. We would also expect to see correlations between color and inclination within the hot and cold populations. Again, these correlations do not exist. Finally, we would expect to see correlations between color and inclination or semimajor axis or eccentricity for all populations, not just the classical belt objects. Once again, no such correlations exist. While collisional resurfacing of bodies may indeed affect colors, there is clearly no causal relationship between average impact velocity and color (*Thébault and Doressoundiram*, 2003). In summary, the significant color and size differences between the hot and cold classical objects implies that these two populations are physically different in addition to being dynamically distinct. A confirmation of the surface composition differences between the hot and cold populations could be made with infrared reflectance spectroscopy, but to date no spectrum of a cold classical Kuiper belt object has been published.

3.4. Radial Extent of the Kuiper Belt

Another important property of interest for understanding the primordial evolution of the Kuiper belt is its radial extent. While initial expectations were that the mass of the

Kuiper belt should smoothly decrease with heliocentric distance — or perhaps even increase in number density by a factor of ~100 back to the level of the extrapolation of the minimum mass solar nebula beyond the region of Neptune's influence (*Stern*, 1996) — the lack of detection of objects beyond about 50 AU soon began to suggest a dropoff in number density (*Dones*, 1997; *Jewitt et al.*, 1998; *Chiang and Brown*, 1999; *Trujillo et al.*, 2001; *Allen et al.*, 2001). It was often argued that this lack of detections was the consequence of a simple observational bias caused by the extreme faintness of objects at greater distances from the Sun (*Gladman et al.*, 1998), but *Allen et al.* (2001, 2002) showed convincingly that for a fixed absolute magnitude, the number of objects with semimajor axis <50 AU was larger than the number >50 AU and thus some density decrease was present.

Determination of the magnitude of the density drop beyond 50 AU was hampered by the small numbers of objects and thus weak statistics in individual surveys. *Trujillo and Brown* (2001) developed a method to use all detected objects to estimate a radial distribution of the Kuiper belt. The method relies on the fact that the heliocentric distance (not semimajor axis) of objects, like the inclination, is well determined in a small number of observations, and that within ~100 AU surveys have no biases against discovering distant objects other than the intrinsic radial distribution and the easily quantifiable brightness decrease with distance. Thus, at a particular distance, a magnitude m_0 will correspond to a particular object size s, but, assuming a power-law differential size distribution, each detection of an object of size s can be converted to an equivalent number n of objects of size s_0 by $n = (s/s_0)^{q-1}$ where q is the differential power-law size index. Thus the observed radial distribution of objects with magnitude m_0, $O(r,m_0)dr$ can be converted to the true radial distribution of objects of size s_0 by

$$R(r,s_0)dr = O(r,m_0)dr \left[\frac{r(r-1)10^{(m-24.55)/5}}{15.60s_0} \right]^{q-1}$$

where albedos of 4% are assumed, but only apply as a scaling factor. Measured values of q for the Kuiper belt have ranged from 3.5 to 4.8 (for a review, see *Trujillo and Brown*, 2001). We will assume the steepest currently proposed value of q = 4.45 (*Gladman et al.*, 2001), which puts the strongest constraints on the existence of distant objects.

Figure 5 shows the total equivalent number of 100-km objects as a function of distance implied by the detection of the known transneptunian objects. One small improvement has been made to the *Trujillo and Brown* (2001) method. The power-law size distribution is only assumed to be valid from 50 to 1000 km in diameter, corresponding to an expected break in the power law at some small diameter (*Kenyon and Luu*, 1999a) and a maximum object's size. The effect of this change is to only use objects between magnitudes 22.7 and 24.8, which makes the analysis only valid from 30 (where a 50-km object would be magnitude 24.8)

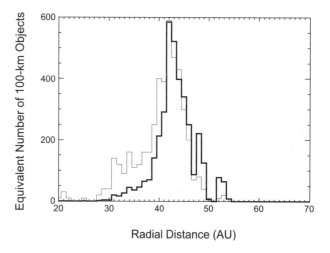

Fig. 5. The radial distribution of the Kuiper belt. The light line shows the observed number of transneptunian objects per AU interval (× 10), while the thick bold line shows the true radial distribution inferred from this observed distribution taking into account biases due to brightness, distance, and size of the object. All discovered transneptunian objects are considered in this analysis, regardless of their dynamical class.

to 80 AU (where a 1000-km object would be magnitude 22.7). Changes in the maximum object size assumed, s_{max}, are equivalent to changing the outer limit of the validity of the analysis by 80 AU $(s_{max}/1000\text{ km})^{1/2}$. Alternatively, one could further restrict the magnitude limits considered to limit the maximum size while maintaining validity to a particular distance. Different choices of minimum and maximum diameters have little effect on the final result unless extreme values for the maximum are chosen.

The analysis clearly shows that the known Kuiper belt is a localized increase in number density. Several implicit assumptions go into the above method, but only extreme changes in these assumptions substantially change the results. For example, a change in the object size distribution beyond 43 AU could mimic a drop in object number density, but only if, by 50 AU, the distribution is so extreme that most of the mass is either in a few (undiscovered) large objects or a large number of (too faint) small objects. A physical reason for such a change is not apparent. Likewise, a lowering of albedo beyond 50 AU could make it appear as if there were a drop in number density, but, again, such a lowering is not physically motivated. A change in the inclination distribution beyond 50 AU could have the effect of hiding objects if they are concentrated in low-inclination orbits close to the invariable plane, but repeating the analysis considering only objects found within 1° of the invariable plane still shows the sharp drop. While changing these assumptions could indeed invalidate the analysis method above, the much simpler conclusion is that the number density of the Kuiper belt peaks strongly at 42 AU and quickly drops off beyond that point.

While the *Trujillo and Brown* (2001) method is good at giving an indication of the radial structure of the Kuiper belt where objects have been found, it is less useful for determining upper limits to the detection of objects where none have been found. A simple extension, however, allows us to easily test hypothetical radial distributions against the known observations by looking at observed radial distributions of all objects found at a particular magnitude m_0 independent of any knowledge of how these objects were found. Assume a true radial distribution of objects $R(r)dr$ and again assume the above power law differential size distribution and maximum size. For magnitudes between m and $m + dm$, we can construct the expected observed radial distribution of all objects found at that magnitude, $o(r,m)drdm$, by

$$o(r,m)drdm = R(r)dr \left[\frac{r(r-1)10^{(m-24.55)/5}}{15.6s_0} \right]^{-q+1} dm$$

where r ranges from that where the object of brightness m has a size of 50 km to that where the object of brightness m has a size of s_{max}. The overall expected observed radial distribution is then simply the sum of $o(r,m)$ over the values of m corresponding to all detected objects. We can then apply a K-S test to determine the probability that the observed radial distribution could have come from the modeled radial distribution. We first apply this test to determine the magnitude of the dropoff beyond 42 AU. Standard assumptions about the initial solar nebula suggest a surface density drop off of $r^{-3/2}$. Figure 6 shows the observed radial distribution of objects compared to that expected if the surface density of objects dropped off as $r^{-3/2}$ beyond

42 AU. This distribution can be ruled out at the many-sigma levels. Assuming that the surface density drops as some power law, we model a range of different distribution $r^{-\alpha}$ and find a best fit of $\alpha = 11 \pm 4$ where the error bars are 3σ. This radial decay function should presumably hold up to ~60 AU, beyond which we expect to encounter a much flatter distribution due to the scattered disk objects.

It has been conjectured that beyond some range of Neptune's influence the number density of Kuiper belt objects could increase back up to the level expected for the minimum mass solar nebula (*Stern,* 1996; see section 3.1). We therefore model a case where the Kuiper belt from 42 to 60 AU falls off as r^{-11} but beyond that the belt reappears at a certain distance δ with a number density found by extrapolating the $r^{-3/2}$ power law from the peak density at 42 AU and multiplying by 100 to compensate for the mass depletion of the classical belt (Fig. 6). Such a model of the radial distribution of the Kuiper belt can be ruled out at the 3σ level for all δ less than 115 AU (around this distance biases due to the slow motions of these objects also become important, so few conclusions can be drawn from the current data about objects beyond this distance). If the model is slightly modified to make the maximum object mass proportional to the surface density at a particular radius, a 100-times resumption of the Kuiper belt can be ruled out inside 94 AU. Similar models can be made where a gap in the Kuiper belt exists at the presently observed location but the belt resumes at some distance with no extra enhancement in number density. These models can be ruled out inside 60 AU at a 99% confidence level.

While all these results are necessarily assumption dependent, several straightforward interpretations are apparent. First, the number density of Kuiper belt objects drops sharply from its peak at around 42 AU. Second, a distant Kuiper belt with a mass approaching that of the minimum mass solar nebula is ruled out inside at least ~100 AU. And finally, a resumption of the Kuiper belt at a density of about 1% expected from a minimum mass solar nebula is ruled out inside ~60 AU.

4. PRIMORDIAL SCULPTING OF THE KUIPER BELT

The previous section makes it clear than the Kuiper belt has lost its accretional disklike primordial structure, sometime during the solar system history. The goal of modelers is to find the scenario, or the combination of compatible scenarios, that can explain how the Kuiper belt acquired the structural properties discussed above. Achieving this goal would probably shed light on the primordial architecture of the planetary system and its evolution.

Several scenarios have been proposed so far. Some of the Kuiper belt properties discussed in section 3 were not yet known when some of these scenarios have been first presented. Therefore in the following — going beyond the original analysis of the authors — we attempt a critical reevaluation of the scenarios, challenging them with all the aspects enumerated in the previous section. We divide the proposed

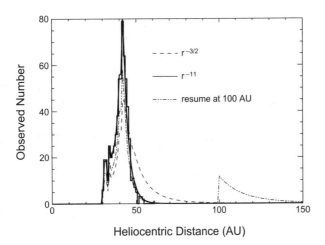

Fig. 6. The observed radial distribution of Kuiper belt objects (solid histogram) compared to observed radial distributions expected for models where the surface density of Kuiper belt objects decreases by $r^{-3/2}$ beyond 42 AU (dashed curve), where the surface density decreases by r^{-11} beyond 42 AU (solid curve), and where the surface density at 100 AU increases by a factor of 100 to the value expected from an extrapolation of the minimum mass solar nebula (dashed-dotted curve).

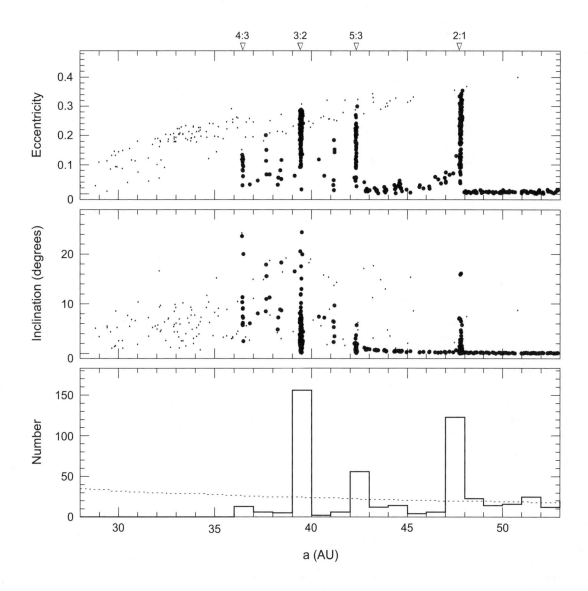

Fig. 7. Final distribution of the Kuiper belt bodies according to the sweeping resonances scenario (courtesy of R. Malhotra). The simulation is done by numerical integrating, over a 200-m.y. timespan, the evolution of 800 test particles on initial quasicircular and coplanar orbits. The planets are forced to migrate (Jupiter: –0.2 AU; Saturn: 0.8 AU; Uranus: 3 AU; Neptune: 7 AU) and reach their current orbits on an exponential timescale of 4 m.y. Large solid dots represent "surviving" particles (i.e., those that have not suffered any planetary close encounters during the integration time); small dots represent the "removed" particles at the time of their close encounter with a planet. In the lowest panel, the solid line is the histogram of semimajor axis of the "surviving" particles; the dotted line is the initial distribution.

scenarios in three groups: (1) those invoking sweeping resonances, which offer a view of gentle evolution of the primordial solar system; (2) those invoking the action of massive scatterers (lost planets or passing stars), which offer an opposite view of violent and chaotic primordial evolution; and (3) those aimed at building the Kuiper belt as the superposition of two populations with distinctive dynamical histories, somehow combining the scenarios in groups (1) and (2).

4.1. Resonance Sweeping Scenarios

Fernández and Ip (1984) showed that, while scattering primordial planetesimals, Neptune should have migrated

outward. *Malhotra* (1993, 1995) realized that, following Neptune's migration, the mean-motion resonances with Neptune also migrated outward, sweeping the primordial Kuiper belt until they reached their present position. From adiabatic theory (*Henrard,* 1982), most of the Kuiper belt objects swept by a mean-motion resonance would have been captured into resonance; they would have subsequently followed the resonance in its migration, while increasing their eccentricity. This model accounts for the existence of the large number of Kuiper belt objects in the 2:3 mean-motion resonance with Neptune (and also in other resonances) and explains their large eccentricities (see Fig. 7). Reproducing the observed range of eccentricities of the resonant bodies requires that Neptune migrated by 7 AU.

Malhotra's (1993, 1995) simulations also showed that the bodies captured in the 2:3 resonance can acquire large inclinations, comparable to that of Pluto and other objects. The mechanisms that excite the inclination during the capture process have been investigated in detail by *Gomes* (2000). The author concluded that, although large inclinations can be achieved, the resulting proportion between the number of high-inclination vs. low-inclination bodies and their distribution in the eccentricity vs. inclination plane do not reproduce the observations very well.

The mechanism of adiabatic capture into resonance requires that Neptune's migration happened very smoothly. If Neptune had encountered a significant number of large bodies (1 $M_\oplus$ or more), its jerky migration would have jeopardized capture into resonances. *Hahn and Malhotra* (1999), who simulated Neptune's migration using a disk of lunar- to martian-mass planetesimals, did not obtain any permanent capture. The precise constraints set by the capture process on the size distribution of the largest disk's planetesimals have never been quantitatively computed, but they are likely to be severe.

In the mean-motion resonance sweeping model the eccentricities and inclinations of the nonresonant bodies are also excited by the passage of many weak resonances, but the excitation that does occur is too small to account for those observed (compare Fig. 7 with Fig. 1). Some other mechanism (like those discussed below) must also have acted to produce the observed overall orbital excitation of the Kuiper belt. The question of whether this other mechanism acted before or after the resonance sweeping and capture process is unresolved. Had it occurred afterward, it would have probably ejected most of the previously captured objects from the resonances (not necessarily a problem if the number of captured bodies was large enough). Had it happened before, then the mean-motion resonances would have had to capture particles from an excited disk. Another long-debated question concerning the sweeping model is the relative proportion between the number of bodies in the 2:3 and 1:2 resonances. The original simulations by Malhotra indicated that the population in the 1:2 resonance should be comparable to — if not greater than — that in the 2:3 resonance. This prediction seemed to be in conflict with the absence of observed bodies in the former resonance at that time. *Ida et al.* (2000a) showed that the proportion between the two populations is very sensitive to Neptune's migration rate and that the small number of 1:2 resonant bodies, suggested by the lack of observations, would just be indicative of a fast migration (10^5–10^6 yr timescale). Since then, five objects have been discovered in or close to the 1:2 resonance (given orbital uncertainties it is not yet possible to guarantee that all of them are really inside the resonance). There is no general consensus on the debiased ratio between the populations in the 2:3 and 1:2 resonances, because the debiasing is necessarily model-dependent and the current data on the population of the 1:2 resonance are sparse. *Trujillo et al.* (2001) estimated a 2:3 to 1:2 ratio close to 1/2, while *Chiang and Jordan* (2002)

obtained a ratio closer to 3. *Chiang and Jordan* (2002) also noted that the positions of the five potential 1:2 resonant objects are unusually located with respect to a reference frame rotating with Neptune, which may also have implications for migration rates and capture mechanisms.

The migration of secular resonances could also have contributed to the excitation of the eccentricities and inclinations of Kuiper belt bodies. Secular resonances occur when the precession rates of the orbits of the bodies are in simple ratio with the precession rates of the orbits of the planets. There are several reasons to think that secular resonances could have been in different locations in the past and migrated to their current location at about 40–42 AU. A gradual mass loss of the belt due to collisional activity, the growth of Neptune's mass, and Neptune's orbital migration would have moved the secular resonance with Neptune's perihelion outward. Levison et al. (personal communication, 1997) found that the Kuiper belt interior to 42 AU would have suffered a strong eccentricity excitation. However, the quantitative simulations show that the orbital distribution of the surviving bodies in the 2:3 resonance would not be similar to the observed one: The eccentricities of most simulated bodies would range between 0.05 and 0.1, while those of the observed Plutinos are between 0.1 and 0.3. Also, in this model there is basically no eccentricity and inclination excitation for the Kuiper belt bodies with a > 42 AU, in contrast with what is observed.

The dissipation of the primordial nebula would also have caused the migration of the secular resonances. *Nagasawa and Ida* (2000) showed that the secular resonances involving the precession rates of the perihelion longitudes would have migrated from beyond 50 AU to their current position during the nebula dispersion. This could have caused eccentricity excitation of the Kuiper belt in the 40–50 AU region. In addition, if the midplane of the nebula was not orthogonal to the total angular momentum vector of the planetary system, a secular resonance involving the precession rates of the node longitudes would also have swept the Kuiper belt, causing inclination excitation. The magnitude of the eccentricity and inclination excitation depends on the timescale of the nebula dissipation. A dissipation timescale of ~10^7 yr is required in order to excite the eccentricities up to 0.2–0.3 and the inclinations up to 20°–30°. But if Neptune was at about 20 AU at the time of the nebula disappearance — as required by the mean-motion resonance sweeping model — the dissipation timescale should have been ~10^8 yr, suspiciously lengthy with respect to what is expected from current theories and observations on the evolution of protoplanetary disks. A major failure of the model is that, because only one nodal secular resonance sweeps the belt, all the Kuiper belt bodies acquire orbits with comparably large inclinations. In other words, the model does not reproduce the observed spread of inclinations, nor their bimodal distribution. No correlation between inclination and size or color can be explained either. The same is not true for the eccentricities, because the belt is swept by several perihelion resonances, which causes a spread in the final

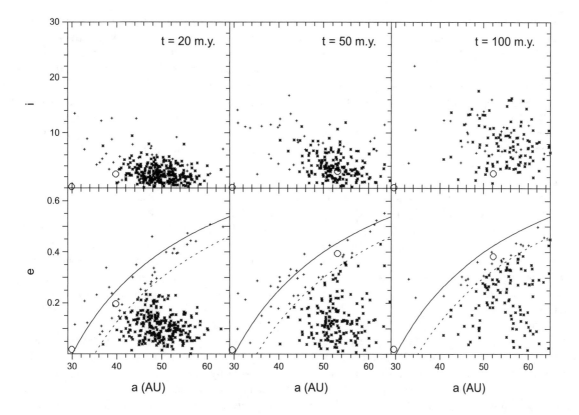

Fig. 8. Snap shots of the Kuiper belt under the scattering action of a 1-$M_\oplus$ planetesimal, itself evolving in the scattered disk. The bold and the dash curves denote q = 30 AU and q = 35 AU respectively. The latter approximately defines the present limit for stability in the Kuiper belt beyond 42 AU, and therefore marks the transition from the classical belt to the scattered disk. The test particles (initially 500, uniformly distributed on circular and coplanar orbits between 35 and 55 AU) are plotted as an asterisk if q > 35 AU, and as a cross otherwise. Neptune and the scattered planetesimal are shown by open circles. From *Petit et al.* (1999).

values. The secular resonance sweeping model cannot explain the existence of significant populations in mean-motion resonances, so that the mean-motion resonance sweeping model would still need to be invoked.

None of the models discussed above can explain the existence of the edge of the belt at ~50 AU.

4.2. Scattering Scenarios

A radically different view has been proposed by *Morbidelli and Valsecchi* (1997), who first proposed that massive Neptune scattered planetesimals (mass on the order of 1 $M_\oplus$), temporarily on Kuiper belt-crossing orbits, could have excited by close encounters the eccentricities and inclinations of the majority of Kuiper belt objects. This idea has been investigated in details by *Petit et al.* (1999), who made direct numerical simulations of the effects of scattered massive planetesimals on test particles representing the initially dynamically cold Kuiper belt. Figure 8 shows snapshots of the status of the Kuiper belt after 20, 50, and 100 m.y. respectively of evolution of an Earth-mass planetesimal in the scattered disk. The test particles (initially 500) were assumed at start on circular and coplanar orbits between 35 and 55 AU. The simulation shows that in the

inner belt (a < 40 AU) less than 1% of the bodies are found on what would become "stable orbits" once the massive planetesimal is dynamically removed. The depletion factor in the 40–47 AU region is 74% after 20 m.y., 91% after 50 m.y., and 96% after 100 m.y. This would completely explain the mass deficit of the current belt. However, beyond 50 AU, ~50% of the original test particles are found on "stable orbits" (q > 35 AU) after 100 m.y., which is inconsistent with the observed "edge." In general terms, this model implies a quite steep positive gradient of the number density of bodies vs. semimajor axis, which is not observed in the real population. In particular, the relative population in the 2:3 resonance would be much smaller than that (4% of the classical belt population) claimed from observations by *Trujillo et al.* (2001). In *Petit et al.* (1999) simulations the median eccentricity and inclination of the survivors after 100 m.y. are 0.19 and 8.6° in the 40–47 AU region and 0.27 and 7.4° beyond 47 AU. If the eccentricity distribution correctly reproduces the observations, the inclination distribution is not bimodal, and completely misses objects with inclination larger than 20°, in contradiction with the observations. No correlation between inclination and colors could be explained within the framework of this model.

A variant of the Petit et al. scenario has been invoked by *Brunini and Melita* (2002) to explain the apparent edge of the Kuiper belt at 50 AU. They showed with numerical simulations that a Mars-sized body residing for 1 G.y. on an orbit with a ~ 60 AU and e ~ 0.15–0.2 could have scattered into Neptune-crossing orbits most of the Kuiper belt bodies originally in the 50–70 AU range, leaving this region strongly depleted and dynamically excited. Such a massive body should have been a former Neptune-scattered planetesimal that decoupled from Neptune due to the dynamical friction exerted by the initially massive Kuiper belt. The orbital distribution inside ~50 AU is not severely affected by the massive planetesimal once on its decoupled orbit at a ~ 60 AU (see also *Melita et al., 2002*). However, a strong dynamical excitation could be obtained during the transfer phase, when the massive planetesimal was transported by Neptune encounters toward a ~ 60 AU, similar to what happens in the *Petit et al.* (1999) simulations. Some of the simulations by *Brunini and Melita* (2002) that include this transfer phase lead to an (a, e) distribution that is perfectly consistent with what is currently observed in the classical belt in terms of mass depletion, eccentricity excitation, and outer edge (see, e.g., their Fig. 10). The corresponding inclination distribution is not explicitly discussed, but it is less excited than in the *Petit et al.* (1999) scenario (M. Melita, personal communication, 2002). Similarly, a correlation between inclination and size or color cannot be reproduced by this mechanism, and a distinctive Plutino population is not formed. Finally, our numerical simulations show that a 1-$M_\oplus$ planet in the Kuiper belt cannot transport bodies up to 200 AU or more by gravitational scattering. Therefore, neither the *Petit et al.* (1999) scenario nor that of *Brunini and Melita* (2002) can explain the origin of the orbit of objects such as 2000 CR_{105}.

A potential problem of the Brunini and Melita scenario is that, once the massive body is decoupled from Neptune, there are no evident dynamical mechanisms that would ensure its later removal from the system. In other words, the massive body should still be present, somewhere in the ~50–70-AU region. A Mars-sized body with 4% albedo at 70 AU would have apparent magnitude brighter than 20, so that, if its inclination is small ($i < 10°$), as expected if the body got trapped in the Kuiper belt by dynamical friction, it is unlikely that it escaped detection in the numerous wide-field ecliptic surveys that have been performed up to now, and in particular in that led by *Trujillo and Brown* (2003).

Another severe problem, for both the *Petit et al.* (1999) and *Brunini and Melita* (2002) scenarios — as well as for any other scenario that attempts to explain the mass depletion of the Kuiper belt by the dynamical ejection of a substantial fraction of Kuiper belt bodies to Neptune-crossing orbit — is that Neptune would have migrated well beyond 30 AU. In *Hahn and Malhotra* (1999) simulations, a 50 $M_\oplus$ disk between 10 and 60 AU drives Neptune to ~30 AU. In this process, Neptune interacts only with the mass in the 10–35-AU disk (about 25 $M_\oplus$), and a massive Kuiper belt remains beyond Neptune. But if the Kuiper belt had been excited to Neptune-crossing orbit, then Neptune would have interacted with the full 50-$M_\oplus$ disk and therefore would have migrated much further. [This fact was not noticed by the simulations of *Petit et al.* (1999) and *Brunini and Melita* (2002), because the former considered a Kuiper belt of massless particles and the latter a Kuiper belt whose total mass was only ~1 $M_\oplus$.] To limit Neptune's migration at 30 AU, the total mass of the disk, including the Kuiper belt, should have been significantly smaller. Our simulations show that even a disk of 15 $M_\oplus$ between 10 and 50 AU, once excited to Neptune-crossing orbit, would drive Neptune too far. Therefore, the scenario of a massive body scattered by Neptune through the Kuiper belt is viable only if the primordial mass of the belt was significantly smaller than usually accepted (accounting only for a few Earth masses).

Motivated by the observation that the eccentricity of the classical belt bodies on average increases with semimajor axis (a fact certainly enhanced by the observational biases, which strongly favor the discovery of bodies with small perihelion distances), *Ida et al.* (2000b) suggested that the structure of the classical belt records the footprint of the close encounter with a passing star. In that paper and in the followup work by *Kobayashi and Ida* (2001), the resulting eccentricities and inclinations were computed as a function of a/D, where a is the original body's semimajor axis and D is the heliocentric distance of the stellar encounter, for various choices of the stellar parameters (inclination, mass, and eccentricity). The eccentricity distribution in the classical belt suggested to the authors a stellar encounter at about ~150 AU. The same parameters, however, do not lead to an inclination excitation comparable to the observed one. The latter would require a stellar passage at ~100 AU or less. From Kobayashi and Ida simulations we argue that a bimodal inclination distribution could be possibly obtained, but a quantitative fit to the debiased distribution discussed in section 3.2 has never been attempted. A stellar encounter at ~100 AU would make most of the classical belt bodies so eccentric to intersect the orbit of Neptune. Therefore, it would explain not only the dynamical excitation of the belt (although a quantitative comparison with the observed distributions has never been done) but also its mass depletion, but would encounter the same problem discussed about concerning Neptune's migration.

Melita et al. (2002) showed that a stellar passage at about 200 AU would be sufficient to explain the edge of the classical belt at 50 AU. An interesting constraint on the time at which such an encounter occurred is set by the existence of the Oort cloud. *Levison et al.* (2003) show that the encounter had to occur much earlier than ~10 m.y. after the formation of Uranus and Neptune, otherwise most of the existing Oort cloud would have been ejected to interstellar space and many of the planetesimals in the scattered disk would have had their perihelion distance lifted beyond Neptune, decoupling from the planet. As a consequence, the extended scattered disk population, with a > 50 AU and 40 < q < 50 AU, would have had a mass comparable or larger than that of the resulting Oort cloud, hardly compatible with the few detections of extended scattered disk objects performed up to now. An encounter with a star during the first million years from planetary formation is a likely event if the Sun formed in a stellar cluster (*Bate et al., 2003*). At such an early time, presumably the Kuiper belt objects were not yet fully formed (*Stern, 1996; Kenyon and Luu, 1998*). In this case, the edge of the belt would be at a heliocentric distance corresponding to a postencounter eccentricity excitation of ~0.05, a threshold value below which collisional damping is efficient and accretion can recover, and beyond which the objects rapidly grind down to dust (*Kenyon and Bromley, 2002*). The edge-forming stellar encounter could not be responsible for the origin of the peculiar orbit of 2000 CR_{105}. In fact, such a close encounter would also produce a relative overabundance of bodies with perihelion distance similar to that of 2000 CR_{105} but with semimajor axis in the 50–200-AU range. These bodies have never been discovered despite the more favorable observational biases. In order that only bodies with a > 200 AU have their perihelion distance lifted, a second stellar passage at about 800 AU is required (*Morbidelli and Levison, 2003*). Interestingly, from the analysis of the Hipparcos data, *Garcia-Sanchez et al.* (2001) concluded that, with the current stellar environment, the closest encounter with a star during the age of the solar system would be at ~900 AU.

4.3. Scenarios for a Two-Component Kuiper Belt

None of the scenarios discussed above successfully reproduce the existence of a cold and a hot population in the classical belt (see section 3.2–3.3) and the correlation between inclination and sizes and colors. The reason is obvious. All these scenarios start with a unique population (the primordial, dynamically cold Kuiper belt). From a unique population, it is very difficult to produce two populations with distinct orbital properties. Even in the case where it might be possible (as in the stellar encounter scenario), the orbital histories of gray bodies cannot differ statistically from those of the red bodies, because the dynamics do not depend on the physical properties. The correlations between colors and inclination can be explained only by postulating that the hot and cold populations of the current Kuiper belt originally formed in distinctive places in the solar system. The scenario suggested by *Levison and Stern* (2001) is that initially the protoplanetary disk in the Uranus-Neptune region and beyond was uniformly dynamically cold, with physical properties that varied with heliocentric distance. Then, a dynamical violent event cleared the inner region of the disk, dynamically scattering the inner disk objects outward. In the scattering process, large inclinations were acquired. Most of these objects have been dynamically eliminated, or persist as members of the scattered disk. However, a few of these objects somehow were deposited in the main Kuiper belt, becoming the hot population of the classical belt currently observed.

Two dynamical scenarios have been proposed so far to explain how planetesimals in the Uranus-Neptune zone

could be permanently trapped in the Kuiper belt. *Thommes et al.* (1999) proposed a radical view of the primordial architecture of our outer solar system in which Uranus and Neptune formed in the Jupiter-Saturn zone. In their simulations, Uranus and Neptune were rapidly scattered outward by Jupiter, where the interaction with the massive disk of planetesimals damped their eccentricities and inclinations by dynamical friction; as a consequence, the planets escaped from the scattering action of Jupiter before ejection on hyperbolic orbit could occur. In about 50% of the cases, the final states resembled the current structure of the outer solar system, with four planets roughly at the correct locations. In this scenario Neptune experienced a high-eccentricity phase lasting for a few million years, during which its aphelion distance was larger than the current one. The planetesimals scattered by Neptune during the dynamical friction process therefore formed a scattered disk that extended well beyond its current perihelion distance boundary. When Neptune's eccentricity decreased down to its present value, the large-q part of the scattered disk became "fossilized," being unable to closely interact with Neptune again. This scenario therefore explains how a population of bodies, originally formed in the inner part of the disk, could be trapped in the classical belt. However, the inclination excitation of this population, although relevant, is smaller than that of the observed hot population. This is probably due to the fact that Neptune's eccentricity is rapidly damped, so that the particles undergo Neptune's scattering action for only a few million years, too short a timescale to acquire large inclinations. For the same reason, the "fossilized" scattered disk does not extend very far in semimajor axis, so that objects like 2000 CR_{105} are not produced in this scenario. Also, the high eccentricity of Neptune would destabilize the bodies in the 2:3 resonance, so that the Plutinos could have been captured only after Neptune's eccentricity damping, during a final quiescent phase of radial migration similar to that in *Malhotra*'s (1993, 1995) scenario. Nevertheless, a Plutino population was never formed in the *Thommes et al.* (1999) simulations, possibly because Neptune's migration was too jerky owing to the encounters with the massive bodies used in the numerical representation of the disk.

Gomes (2003) revisited *Malhotra*'s (1993, 1995) model. Like *Hahn and Malhotra* (1999), he attempted to simulate Neptune's migration, starting from about 15 AU, by the interaction with a massive planetesimal disk extending from beyond Neptune's initial position. But, taking advantage of the improved computer technology, he used 10,000 particles to simulate the disk population, with individual masses roughly equal to twice the mass of Pluto, while Hahn and Malhotra used only 1000 particles with lunar to martian masses. In his simulations, during its migration Neptune scattered the planetesimals and formed a massive scattered disk. Some of the scattered bodies decoupled from the planet by decreasing their eccentricity through the interaction with some secular or mean-motion resonance. If Neptune had not been migrating, as in *Duncan and Levison* (1997) integrations, the decoupled phases would have been

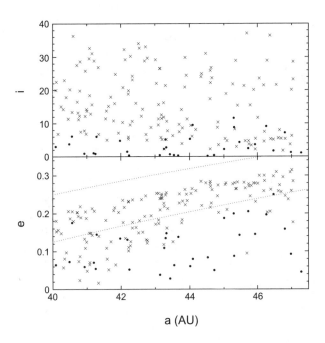

Fig. 9. The orbital distribution in the classical belt according to *Gomes'* (2003) simulations. The dots denote the local population, which is only moderately dynamically excited. The crosses denote the bodies that were originally inside 30 AU. Therefore, the resulting Kuiper belt population is the superposition of a dynamically cold population and of a dynamically hot population, which gives a bimodal inclination distribution comparable to that observed. The dotted curves in the eccentricity vs. semimajor axis plot correspond to q = 30 AU and q = 35 AU. Courtesy of R. Gomes.

transient, because the eccentricity would have eventually increased back to Neptune-crossing values, the dynamics being reversible. But Neptune's migration broke the reversibility, and some of the decoupled bodies managed to escape from the resonances, and remained permanently trapped in the Kuiper belt. As shown in Fig. 9, the current Kuiper belt would therefore be the result of the superposition of these bodies with the local population, originally formed beyond 30 AU and only moderately excited [by the resonance sweeping mechanism, as in *Hahn and Malhotra* (1999)]. Unlike in *Thommes et al.* (1999) simulations, the migration mechanism is sufficiently slow (several 10^7 yr) that the scattered particles have the time to acquire very large inclinations, consistent with the observed hot population. The resulting inclination distribution of the bodies in the classical belt is bimodal, and quantitatively reproduces the debiased inclination distribution computed by *Brown* (2001) from the observations. For the same reason (longer timescale) the extended scattered disk in *Gomes'* (2003) simulations reaches much larger semimajor axes than in *Thommes et al.* (1999) integrations. Although bodies on orbits similar to that of 2000 CR_{105} are not obtained in the nominal simulations, other tests done in *Gomes* (2003) are suggestive that such orbits could be achieved in the framework of the same scenario.

A significant Plutino population is also created in Gomes' simulations. This population is also the result of the superposition of the population coming from Neptune's region with that formed further away and captured by the 2:3 resonance during the sweeping process. Assuming that the bodies' sizes and colors varied in the primordial disk with heliocentric distance, this process would explain why the Plutinos, scattered objects, and hot classical belt objects, which mostly come from regions inside ~30 AU, all appear to have identical color distributions and similar maximum sizes, while only the cold classical population, the only objects actually formed in the transneptunian region, has a different distribution in color and size.

Of all the models discussed in this paper, Gomes' scenario is the one that seems to best account for the observed properties of the classical belt. A few open questions persist, though. The first concerns the mass deficit of the Kuiper belt. In Gomes' simulations about 0.2% of the bodies initially in the Neptune-swept disk remained in the Kuiper belt at the end of Neptune's migration. Assuming that the primordial disk was ~100 $M_\oplus$, this is very compatible with the estimated current mass of the Kuiper belt. But the local population was only moderately excited and not dynamically depleted, so it should have preserved most of its primordial mass. The latter should have been several Earth masses, in order to allow the growth of ~100-km bodies within a reasonable timescale (*Stern, 1996*). How did this local population lose its mass? This problem is also unresolved for the *Thommes et al.* (1999) scenario. The only plausible answer seems to be the collisional erosion scenario, but it has the limitations discussed in section 3.1. Quantitative simulations need to be done. A second problem, also common to the Thommes et al. scenario, is the existence of the Kuiper belt edge at 50 AU. In fact, in neither scenario is significant depletion of the pristine population beyond this threshold obtained. A third problem with *Gomes'* (2003) scenario concerns Neptune's migration. Why did it stop at 30 AU? There is no simple explanation within the model, so Gomes had to artificially impose the end of Neptune's migration by abruptly dropping the mass surface density of the disk at ~30 AU. A possibility is that, by the time that Neptune reached that position, the disk beyond 30 AU had already been severely depleted by collisions. A second possibility is that something (a massive planetary embryo, a stellar encounter, collisional grinding?) opened a gap in the disk at about 30 AU, so that Neptune ran out of material and could not further sustain its migration.

5. CONCLUSIONS AND PERSPECTIVES

Ten years of dedicated surveys have revealed unexpected and intriguing properties of the transneptunian population, such as the existence of a large number of bodies trapped in mean-motion resonances, the overall mass deficit, the large orbital eccentricities and inclinations, and the apparent existence of an outer edge at ~50 AU and a correlation among inclinations, sizes, and colors. Understanding how the Kuiper belt acquired all these properties would probably constrain several aspects of the formation of the outer planetary system and of its primordial evolution.

Up to now, a portfolio of scenarios have been proposed by theoreticians. None of them can account for all the observations alone, and the solution of the Kuiper belt primordial sculpting problem probably requires a sapient combination of the proposed models. The Malhotra-Gomes scenario on the effects of planetary migration does a quite good job at reproducing the observed orbital distribution inside 50 AU. The apparent edge of the belt at 50 AU might be explained by a very early stellar encounter at 150–200 AU. The origin of the peculiar orbit of 2000 CR$_{105}$ could be due to a later stellar encounter at ~800 AU.

The most mysterious feature that remains unexplained in this combination of scenarios is the mass deficit of the cold classical belt. As discussed in this chapter, the mass depletion cannot be explained by the ejection of most of the pristine bodies to Neptune-crossing orbit, because in this case the planet would have migrated well beyond 30 AU. But the collisional grinding scenario also seems problematic, because it requires a peculiar size distribution in the primordial population and relative encounter velocities that are larger than those that characterize the objects of the cold population; moreover, an intense collisional activity would have hardly preserved the widely separated binaries that are frequently observed in the current population.

A possible solution to this problem has been recently proposed by *Levison and Morbidelli* (2003). In their scenario, the primordial edge of the massive protoplanetary disk was somewhere around 30–35 AU and the entire Kuiper belt population — not only the hot component as in *Gomes'* (2003) scenario — formed within this limit and was transported to its current location during Neptune's migration. The transport process of the cold population was different from the one found by Gomes for the hot population. These bodies were trapped in the 1:2 resonance with Neptune and transported outward within the resonance, until they were progressively released due to the nonsmoothness of the planetary migration. In the standard adiabatic migration scenario (*Malhotra*, 1995) there would be a resulting correlation between the eccentricity and the semimajor axis of the released bodies. However, this correlation is broken by a secular resonance embedded in the 1:2 mean-motion resonance. This secular resonance is generated because the precession rate of Neptune's orbit is modified by the torque exerted by the massive protoplanetary disk that drives the migration. Simulations of this process match the observed (a, e) distribution of the cold population fairly well, while the initially small inclinations are only very moderately perturbed. In this scenario, the small mass of the current Kuiper belt population is simply due to the fact that presumably only a small fraction of the massive disk population was initially trapped in the 1:2 resonance and released on stable nonresonant orbits. The preservation of the binary objects is not a problem because these objects were moved out of the massive disk in which they formed by a gentle dynamical process. The final position of Neptune would simply reflect the primitive truncation of the protoplanetary disk. A bigger problem is the explanation of the different physical properties of the cold and hot populations, because they both originated within 35 AU, although in somewhat different parts of the disk. At the time of this writing, this innovative model has not yet been critically debated within the community of experts. But this scenario offers a simple prediction that will be confirmed or denied by future observations: The edge of the cold classical belt is exactly at the location of the 1:2 resonance.

Kuiper belt science is a rapidly evolving field. New observations change our view of the belt every year. Since the discovery of the first transneptunian object 10 years ago, several review papers have been written, and most of them are already obsolete. No doubt this will also be the fate of this chapter, but it can be hoped that the ideas presented here can continue to guide us in the direction of further understanding of what present observations of the Kuiper belt can tell us about the formation and evolution of the outer solar system.

REFERENCES

Allen R. L., Bernstein G. M., and Malhotra R. (2001) The edge of the solar system. *Astroph. J. Lett., 549,* L241–L244.

Allen R. L., Bernstein G. M., and Malhotra R. (2002) Observational limits on a distant cold Kuiper belt. *Astron. J., 124,* 2949–2954.

Bate M. R., Bonnell I. A., and Bromm V. (2003) The formation of a star cluster: Predicting the properties of stars and brown dwarfs. *Mon. Not. R. Astron. Soc., 339,* 577–599.

Brown M. (2001) The inclination distribution of the Kuiper belt. *Astron. J., 121,* 2804–2814.

Brunini A. and Melita M. (2002) The existence of a planet beyond 50 AU and the orbital distribution of the classical Edgeworth-Kuiper-belt objects. *Icarus, 160,* 32–43.

Chiang E. I. and Brown M. E. (1999) Keck pencil-beam survey for faint Kuiper belt objects. *Astron. J., 118,* 1411–1422.

Chiang E. I. and Jordan A. B. (2002) On the Plutinos and Twotinos of the Kuiper belt. *Astron. J., 124,* 3430–3444.

Chiang E. I., Jordan A. B., Millis R. L., Buie M. W., Wasserman L. H., Elliot J. L., Kern S. D., Trilling D. E., Meech K. J., and Wagner R. M. (2003) Resonance occupation in the Kuiper belt: Case examples of the 5:2 and Trojan resonances. *Astron. J., 126,* 430–443.

Cohen C. J. and Hubbard E. C. (1965) The orbit of Pluto. *The Observatory, 85,* 43–44.

Davis D. R. and Farinella P. (1997) Collisional evolution of Edgeworth-Kuiper belt objects. *Icarus, 125,* 50–60.

Davis D. R. and Farinella P. (1998) Collisional erosion of a massive Edgeworth-Kuiper belt: Constraints on the initial population (abstract). In *Lunar and Planetary Science XIX,* pp. 1437–1438. Lunar and Planetary Institute, Houston.

Dones L. (1997) Origin and evolution of the Kuiper belt. In *From Stardust to Planetesim*als (Y. J. Pendleton and A. G. G. M. Tielens, eds.), p. 347. ASP Conference Series 122.

Doressoundiram A., Barucci M. A., Romon J., and Veillet C. (2001) Multicolor photometry of trans-Neptunian objects. *Icarus, 154,* 277–286.

Duncan M. J. and Levison H. F. (1997) Scattered comet disk and the origin of Jupiter family comets. *Science, 276,* 1670–1672.

Duncan M. J., Levison H. F., and Budd S. M. (1995) The long-term stability of orbits in the Kuiper belt. *Astron. J., 110,* 3073–3083.

Farinella P., Davis D. R., and Stern S. A. (2000) Formation and collisional evolution of the Edgeworth-Kuiper belt. In *Protostars and Planets IV* (V. Mannings et al., eds.), pp. 125–133. Univ. of Arizona, Tucson.

Fernández J. A. and Ip W. H. (1984) Some dynamical aspects of the accretion of Uranus and Neptune — The exchange of orbital angular momentum with planetesimals. *Icarus, 58,* 109–120.

Garcia-Sanchez J., Weissman P. R., Preston R. A., Jones D. L., Lestrade J. F., Latham D. W., Stefanik R. P., and Paredes J. M. (2001) Stellar encounters with the solar system. *Astron. Astrophys., 379,* 634–659.

Gladman B., Kavelaars J. J., Nicholson P. D., Loredo T. J., and Burns J. A. (1998) Pencil-beam surveys for faint trans-Neptunian objects. *Astron. J., 116,* 2042–2054.

Gladman B., Kavelaars J. J., Petit J. M., Morbidelli A., Holman M. J., and Loredo Y. (2001) The structure of the Kuiper belt: Size distribution and radial extent. *Astron. J., 122,* 1051–1066.

Gladman B., Holman M., Grav T., Kaavelars J. J., Nicholson P., Aksnes K., and Petit J. M. (2002) Evidence for an extended scattered disk. *Icarus, 157,* 269–279.

Goldreich P., Lithwick Y., and Sari R. (2002) Formation of Kuiper-belt binaries by dynamical friction and three-body encounters. *Nature, 420,* 643–646.

Gomes R. S. (2000) Planetary migration and Plutino orbital inclinations. *Astron. J., 120,* 2695–2707.

Gomes R. S. (2003) The origin of the Kuiper belt high inclination population. *Icarus, 161,* 404–418.

Hahn J. M. and Malhotra R. (1999) Orbital evolution of planets embedded in a planetesimal disk. *Astron. J., 117,* 3041–3053.

Hainaut O. and Delsanti A. (2002) *MBOSS Magnitude and Colors.* Available on line at http://www.sc.eso.org/~ohainaut/MBOSS/.

Henrard J. (1982) Capture into resonance — An extension of the use of adiabatic invariants. *Cel. Mech., 27,* 3–22.

Ida S., Bryden G., Lin D. N., and Tanaka H. (2000a) Orbital migration of Neptune and orbital distribution of trans-Neptunian objects. *Astrophys. J., 534,* 428–445.

Ida S., Larwood J., and Burkert A. (2000b) Evidence for early stellar encounters in the orbital distribution of Edgeworth-Kuiper belt objects. *Astrophys. J., 528,* 351–356.

Jewitt D. C. and Luu J. X. (1993) Discovery of the candidate Kuiper belt object 1992 QB1. *Nature, 362,* 730–732.

Jewitt D., Luu J., and Chen J. (1996) The Mauna-Kea-Cerro-Totlolo (MKCT) Kuiper belt and Centaur survey. *Astron. J., 112,* 1225–1232.

Jewitt D., Luu J., and Trujillo C. (1998) Large Kuiper belt objects: The Mauna Kea 8K CCD survey. *Astron. J., 115,* 2125–2135.

Kenyon S. J. and Luu J. X. (1998) Accretion in the early Kuiper belt: I. Coagulation and velocity evolution. *Astron. J., 115,* 2136–2160.

Kenyon S. J. and Luu J. X. (1999a) Accretion in the early Kuiper belt: II. Fragmentation. *Astron. J., 118,* 1101–1119.

Kenyon S. J. and Luu J. X. (1999b) Accretion in the early outer solar system. *Astrophys. J., 526,* 465–470.

Kenyon S. J. and Bromley B. C. (2002) Collisional cascades in planetesimal disks. I. Stellar flybys. *Astron. J., 123,* 1757–1775.

Kobayashi H. and Ida S. (2001) The effects of a stellar encounter on a planetesimal disk. *Icarus, 153,* 416–429.

Kuchner M. J., Brown M. E., and Holman M. (2002) Long-term dynamics and the orbital inclinations of the classical Kuiper belt objects. *Astron. J., 124,* 1221–1230.

Kuiper G. P. (1951) On the origin of the solar system. In *Astrophysics* (J. A. Hynek, ed.), p. 357. McGraw-Hill, New York.

Levison H. F. and Duncan M. J. (1997) From the Kuiper belt to Jupiter-family comets: The spatial distribution of ecliptic comets. *Icarus, 127,* 13–32.

Levison H. F. and Morbidelli A. (2003) Pushing out the Kuiper belt. *Nature,* in press.

Levison H. F. and Stern S. A. (2001) On the size dependence of the inclination distribution of the main Kuiper belt. *Astron. J., 121,* 1730–1735.

Levison H. F., Morbidelli A., and Dones L. (2003) Forming the outer edge of the Kuiper belt by a stellar encounter: Constraints from the Oort cloud. *Icarus,* in press.

Lewis J. S. (1995) *Physics and Chemistry of the Solar System.* Academic, San Diego. 556 pp.

Malhotra R. (1993) The origin of Pluto's peculiar orbit. *Nature, 365,* 819–821.

Malhotra R. (1995) The origin of Pluto's orbit: Implications for the solar system beyond Neptune. *Astron. J., 110,* 420–432.

Melita M., Larwood J., Collander-Brown S., Fitzsimmons A., Williams I. P., and Brunini A. (2002) The edge of the Edgeworth-Kuiper belt: Stellar encounter, trans-Plutonian planet or outer limit of the primordial solar nebula? In *Asteroids, Comets and Meteors: ACM 2002 — Proceedings of an International Conference* (B. Warmbein, ed.), pp. 305–308. ESA SP-500, Noordwijk, The Netherlands.

Morbidelli A. and Valsecchi G. B. (1997) Neptune scattered planetesimals could have sculpted the primordial Edgeworth-Kuiper belt. *Icarus, 128,* 464–468.

Morbidelli A. and Levison H. F. (2003) Scenarios for the origin of an extended trans-Neptunian scattered disk. *Icarus,* in press.

Nagasawa M. and Ida S. (2000) Sweeping secular resonances in the Kuiper belt caused by depletion of the solar nebula. *Astron. J., 120,* 3311–3322.

Petit J. M. and Mousis O. (2003) KBO binaries: How numerous were they? *Icarus,* in press.

Petit J. M., Morbidelli A., and Valsecchi G. B. (1999) Large scattered planetesimals and the excitation of the small body belts. *Icarus, 141,* 367–387.

Stern S. A. (1991) On the number of planets in the outer solar system — Evidence of a substantial population of 1000 km bodies. *Icarus, 90,* 271–281.

Stern S. A. (1995) Collisional time scales in the Kuiper disk and their implications. *Astron. J., 110,* 856–868.

Stern S. A. (1996) On the collisional environment, accretion time scales, and architecture of the massive, primordial Kuiper belt. *Astron. J., 112,* 1203–1210.

Stern S. A. (2002) Evidence for a collisonal mechanism affecting Kuiper belt object colors. *Astron. J., 124,* 2297–2299.

Stern S. A. and Colwell J. E. (1997a) Accretion in the Edgeworth-Kuiper belt: Forming 100–1000 km radius bodies at 30 AU and beyond. *Astron. J., 114,* 841–849.

Stern S. A. and Colwell J. E. (1997b) Collisional erosion in the primordial Edgeworth-Kuiper belt and the generation of the 30–50 AU Kuiper gap. *Astrophys. J., 490,* 879–885.

Tegler S. C. and Romanishin W. (2000) Extremely red Kuiper-belt objects in near-circular orbits beyond 40 AU. *Nature, 407,* 979–981.

Thébault P. and Doressoundiram A. (2003) A numerical test of the collisional resurfacing scenario. Could collisional activity explain the spatial distribution of color-index within the Kuiper belt? *Icarus, 162,* 27–37.

Thommes E. W., Duncan M. J., and Levison H. F. (1999) The formation of Uranus and Neptune in the Jupiter-Saturn region of the solar system. *Nature, 402,* 635–638.

Trujillo C. A. and Brown M. E. (2001) The radial distribution of the Kuiper belt. *Astrophys. J., 554,* 95–98.

Trujillo C. A. and Brown M. E. (2002) A correlation between inclination and color in the classical Kuiper belt. *Astrophys. J., 566,* 125–128.

Trujillo C. A. and Brown M. E. (2003) The Caltech Wide Area Sky Survey: Beyond (50000) Quaoar. In *Proceedings of the First Decadal Review of the Edgeworth-Kuiper Belt Meeting in Antofagasta, Chile,* in press.

Trujillo C. A., Jewitt D. C., and Luu J. X. (2001) Properties of the trans-Neptunian belt: Statistics from the Canada-France-Hawaii Telescope Survey. *Astron. J., 122,* 457–473.

Weidenschilling S. (2002) On the origin of binary transneptunian objects. *Icarus, 160,* 212–215.

Dynamical Evolution of Ecliptic Comets

Martin Duncan
Queen's University

Harold Levison and Luke Dones
Southwest Research Institute

Ecliptic comets are those with T > 2, where T is the Tisserand parameter with respect to Jupiter. In this chapter, we review the enormous progress that has been made in our understanding of the dynamical evolution of these bodies. We begin by reviewing the evidence that Jupiter-family comets (JFCs; those with 2 < T < 3) form a dynamically distinct class of comets that originate in a flattened disk beyond Neptune. We present a model for the distribution of comets throughout the JFC and Centaur regions that is consistent with current observations, although further observations and numerical simulations in the Centaur region are called for. We then discuss dynamical results (since confirmed by observations) that a significant amount of material that was scattered by Neptune during the early stages of planet formation could persist today in the form of a "scattered disk" of bodies with highly eccentric orbits beyond Neptune. We describe the dynamical mechanisms believed responsible for the longevity of the surviving bodies and argue that if objects in the Kuiper belt and scattered disk have similar size distributions, then the scattered disk is likely to be the primary source of JFCs and Centaurs. Finally, we describe the importance of understanding the ecliptic comet population for the purposes of determining impact rates on the satellites of the giant planets and of age determinations of the satellite surfaces. We present tables of impact rates based on the best currently available analyses. Further refinements of these rates and age determinations await better observations of the Centaur population (including its size distribution), as well as a better understanding of the formation and early dynamical evolution of the outer solar system.

1. INTRODUCTION

The dynamical and physical lifetimes of most observed comets are short compared to the age of the solar system. Thus, comets must be coming from some reservoir or reservoirs that slowly allow comets to leak out to regions where they can be detected. Although these reservoir(s) must be stable enough to retain a significant number of objects for billions of years, there must also be some currently active mechanisms that transport objects into the regions where they are more easily detected.

It is currently believed that there are three main sources of the known comets: the Oort cloud, the Kuiper belt, and the scattered comet disk. The first of these is discussed in *Dones et al.* (2004). Here we discuss the Kuiper belt and scattered disk with regard to their role as the source of ecliptic comets. The structure of these cometary reservoirs are discussed in more detail in *Morbidelli and Brown* (2004).

This chapter is organized as follows. In the next section we briefly review comet taxonomy and terminology. In section 3 we review numerical integrations of the orbits of observed Jupiter-family comets, and in section 4 we summarize results of simulations of the transport of comets from the transneptunian region into the region typically inhabited by JFCs. Section 5 reviews the dynamical properties of the long-lived tail of the distribution of Neptune-scattered comets (the "scattered disk"). In section 6, the impact

rates of ecliptic comets on planets and their moons are discussed, while section 7 contains a brief summary of the state of understanding of the origin of JFCs and the dynamical evolution of ecliptic comets.

2. COMETARY TAXONOMY

Historically, cometary taxonomy was based on orbital period, with comets of orbital period shorter than 200 yr being termed "short-period comets" and those with periods less than 20 yr being further subdivided into the class called "Jupiter-family" comets. However, numerical integrations such as those described below show that under such a scheme a given short-period comet typically shifts in and out of the "Jupiter-family class" many times during its dynamical evolution. *Carusi and Valsecchi* (1987) first suggested that since the Tisserand parameter does not vary substantially during a typical comet's lifetime, a taxonomy based on this parameter might be more appropriate than one based on orbital period.

Recall that the Tisserand parameter with respect to Jupiter, T, is defined as

$$T = a_J/a + 2\sqrt{(1 - e^2)a/a_J}\cos(i) \qquad (1)$$

where a_J is Jupiter's semimajor axis, and a, e, and i refer to an object's semimajor axis, eccentricity, and inclination respectively. It is an approximation to the Jacobi constant, which is an integral of the motion in the circular restricted three-body problem. It is also a measure of the relative velocity between a comet and Jupiter during close encounters, $v_{rel} = v_c\sqrt{3 - T}$, where v_c is Jupiter's velocity about the Sun. Objects with T close to, but smaller than, 3 have very slow, and thus very strong, encounters with Jupiter. Objects with T > 3 cannot cross Jupiter's orbit in the circular restricted case, being confined to orbits either totally interior or totally exterior to Jupiter.

For this chapter, we will adopt a taxonomic scheme based on that of *Levison* (1996), in which the most significant division is based on the Tisserand parameter. In this scheme comets with T > 2 are designated *ecliptic* comets because most members have small inclinations. As we shall see below, these objects most likely originate in the Kuiper belt (*Edgeworth*, 1949; *Kuiper*, 1951; *Fernández*, 1980; *Duncan et al.*, 1988) or the *scattered disk* (*Torbett*, 1989; *Duncan and Levison*, 1997). Comets with T < 2, which are believed to be mainly comets from the Oort cloud (*Oort*, 1950; *Everhart*, 1977) are designated *nearly isotropic* comets, reflecting their inclination distribution (see *Dones et al.*, 2004). Independent of other classifications, a comet is said to be "visible" if its perihelion distance is less than 2.5 AU.

Ecliptic comets can be further subdivided into three groups. Comets with 2 < T < 3 are mainly on Jupiter-crossing orbits and are dynamically dominated by that planet. We call these *Jupiter-family* comets (hereafter called JFCs). Comets with T > 3 (not Jupiter-crossing) are not considered members of the Jupiter family. A comet that has T > 3 and $a > a_J$ (orbit is exterior to Jupiter) is designated as *Chiron-type* or a *Centaur*. A comet that has T > 3 and $a < a_J$ is designated an *Encke-type*. Note that this combination of T and a implies that the orbit of this object is entirely interior to Jupiter, i.e., the aphelion distance is less than a_J. However, it may be too severe to use a strict criteria of T < 3 since there are a few comets with T slightly larger than 3 that dynamically belong to the Jupiter family (i.e., they are not decoupled from Jupiter; they suffer frequent long-lasting encounters with the planet). These are the comets sometimes called the "quasi-Hilda" type (see, e.g., *Kresák*, 1979; *Tancredi et al.*, 1990). This is a very interesting group since these objects experience frequent temporary satellite capture and occasional low-velocity impacts with Jupiter (like Shoemaker-Levy 9) and Jupiter's moons.

3.   DYNAMICS OF OBSERVED JUPITER-FAMILY COMETS

Historically, researchers were interested in the origin and evolution of the group of observed comets known as "short-period" comets, (i.e., those with periods less than 200 yr). The upper two panels of Fig. 1 show the distributions of semimajor axis a and inclination i for these comets vs. Tisserand parameter T. The dashed lines in both plots represent the boundaries of the Jupiter family according to our

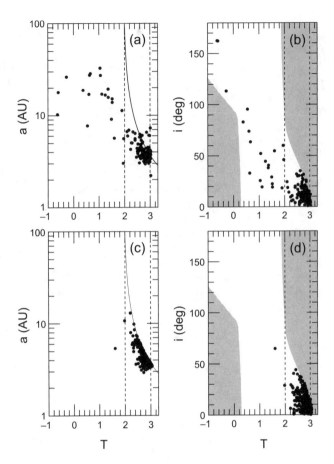

Fig. 1.   (a) The semimajor axes, a, of observed comets with periods less than 200 yr as a function of their Tisserand parameter with respect to Jupiter, T. The dashed vertical lines represent the boundaries of the Jupiter family at T = 2 and T = 3. Objects falling above the solid curve must have perihelia greater than 2.5 AU. (b) The inclination, i, of the same set of comets as in (a) as a function of T. Again, the dashed lines represent the boundaries of the Jupiter family. The shaded areas represent regions that are physically unattainable due to the relationship between i and T assuming $q \leq 2.5$ AU and $Q \geq a_J$. (c) The same as (a) except for simulated comets when they first become "visible," i.e., when their perihelia first drop below 2.5 AU. (d) The same as (b) except for simulated comets when they first become visible. From *Levison and Duncan* (1997).

taxonomy. The strong clumpings of JFCs at values of T just less than 3 and at low inclinations are important clues to their origins, as we shall describe in section 4.

Pioneering integrations of the dynamical evolution of short-period comets spanning roughly 1000 yr were performed by *Carusi et al.* (1985) and *Carusi and Valsecchi* (1987). Further important numerical integrations in the early 1990s include the "COSMO-DICE" project (*Nakamura and Yoshikawa*, 1992a,b), as well as the work of Tancredi and Rickman (*Tancredi and Rickman*, 1992; *Rickman*, 1992; *Tancredi*, 1994), *Lindgren* (1992), and *Emel'yanenko* (1992).

A comprehensive set of long-term integrations spanning up to 10^7 yr of the dynamical evolution of short-period comets was performed by *Levison and Duncan* (1994) (hereafter called *LD94*). In view of the chaotic nature of each individual orbit, *LD94* integrated 4 orbits per comet (each with slightly different initial orbital elements) for the 160 short-period comets known in 1991. The orbits of the Sun, planets (except Mercury and Pluto), and comets were integrated forward and backward in time for 10^7 yr. A comet was followed until it either became unbound from the Sun and reached a distance of 150 AU or became a Sun-grazer with pericenter less than 2 solar radii. Although the chaotic behavior of individual orbits precluded an accurate determination of the long-term fate of any individual comet, it was appropriate to extract statistical information from this sample that should resemble the evolution of the real ensemble of comets.

LD94 found that the Tisserand parameter, T, does not vary substantially for most observed short-period comets: Less than 8% of comets moved in or out of the JFC class during the integration, and most of those that changed tended to remain near the Tisserand dividing line throughout. Thus, the JFCs were found to be a dynamically distinct class within the short-period comets.

The currently observed perihelion distance distribution of comets is strongly peaked toward small values. This is most likely due to a strong observational bias against the discovery of comets with large perihelion distances because they are less active and do not pass close to the Earth. *Quinn et al.* (1990) define a "visible" comet as one with q ≤ 2.5 AU. If a comet has a q greater than this value then, they argue, it is not likely to become bright enough to be discovered. Indeed, only 14% of the known short-period comets have q larger than this value, despite the fact that there is a large region of phase space with q > 2.5 AU available to objects being scattered by Jupiter. For this review, we adopt this definition of visibility.

It was found that in the forward integration, 92% of comets were ejected from the solar system, and that ≈6% were destroyed by becoming Sun-grazers. However, *LD94* did not differentiate between JFCs and Halley-type comets (HTCs) (T < 2, a < 40 AU) when considering Sun-grazers. A subsequent reexamination of those integrations found that half the objects that became Sun-grazers were Halley type, although HTCs represented only ~10% of the comets integrated in *LD94*. A full 30% of the HTCs became Sun-grazers, while only 3% of the JFCs shared the same fate in the integrations. Indeed, only ~1% of the objects that became visible eventually became Sun-grazers.

The median lifetime of all known short-period comets from the current time to ultimate destruction or ejection is approximately 4.5×10^5 yr: The median lifetime of JFCs is 3.25×10^5 yr. The median number of times that JFCs changed from orbits with q < 2.5 AU to q > 2.5 AU was 10. A typical comet spent less than 7% (median value) of its dynamical lifetime with q < 2.5 AU. Since objects with perihelia beyond 2.5 AU are difficult to detect, this implies that there are more than 10 times more undetected JFCs than there are visible JFCs. Of those visible now, half will evolve to states with q > 2.5 AU in roughly 10^3 yr.

The very flattened inclination distribution of JFCs (i.e., their concentration at small inclinations) was found to become more distended as it aged. Since JFCs are a dynamically distinct class, they must have an inclination distribution, when they first become visible, that is even more flattened than that currently observed. Given this evidence that the JFCs originate in a flattened distribution such as the Kuiper belt, we explore the dynamical transport from the transneptunian reservoirs in the next section.

4. FROM THE TRANSNEPTUNIAN REGION TO JUPITER-FAMILY COMETS

We turn now to the study of the origin and dynamical evolution of the class of objects that we referred to in section 2 as "ecliptic comets." As noted in section 3, the observed JFCs (which make up the bulk of the active ecliptic comets) have a very flattened inclination distribution with a median inclination of only 11°. In the last 15 years or so, research attempting to explain this inclination distribution has been extremely active. Indeed, attempting to understand these comets stimulated one of the most important discoveries in planetary science in the last half of the twentieth century — the discovery of the Kuiper belt.

Ecliptic comets were originally thought to originate from nearly isotropic comets that had been captured into short-period orbits by gravitational encounters with the planets (*Newton*, 1891; *Everhart*, 1977; *Bailey*, 1986). *Joss* [(1973), see later work by *Fernández and Gallardo* (1994) and *Levison et al.* (2001)] argued that this process is too inefficient, and *Fernández* (1980) suggested that a belt of distant icy planetesimals beyond Neptune could serve as a more efficient source of most of these comets. *Duncan et al.* (1988) strengthened this argument by performing dynamical simulations that showed that a cometary source beyond Neptune with small inclinations to the ecliptic was far more consistent with the observed orbits of most of these comets than the randomly distributed inclinations of comets in the Oort cloud (see also *Quinn et al.*, 1990). They named this source the Kuiper belt.

The size, extent, and eccentricity of the cometary orbits in this belt were left as open questions in *Duncan et al.* (1988). Traditionally, this work was used to imply that the source of these comets is a primordial population of low eccentricity, moderately low-inclination objects beyond Neptune (in what *we* call the Kuiper belt in this review). An alternative interpretation is that this disk was made up of objects on moderately low-inclination, highly eccentric orbits beyond Neptune (in what *we* call the scattered disk). The details of these interpretations are discussed below. Nevertheless, the first *transneptunian object* (after Pluto and Charon) was discovered in 1992 (*Luu and Jewitt*, 1993);

now more than 800 transneptunian objects are known (for current lists see http://cfa-www.harvard.edu.edu/ps/lists/TNOs.html and http://cfa-www.harvard.edu.edu/iau/lists/Centaurs.html).

To date the only comprehensive simulation of the transport of objects from the transneptunian region to the inner solar system is that of *Levison and Duncan* (1997) (hereafter *LD97*). They assumed that the source region was a dynamically very cold primordial Kuiper belt and that there were enough objects leaking out of this belt, due to the gravitational effects of the planets (e.g., *Levison and Duncan,* 1993; *Holman and Wisdom,* 1993; *Nesvorny and Roig,* 2000; *Kuchner et al.,* 2002) to supply the JFCs. Although this assumption is limiting in some respects (which we discuss below), we believe that there are properties of the dynamics of these objects that are independent of the details of the source reservoir. Thus, we discuss the results of *LD97* in some detail.

LD97 performed numerical orbital integrations of 2200 massless particles as they evolved from Neptune-encountering orbits in the Kuiper belt for times up to a billion years or until they either impacted a massive body or were ejected from the solar system. The initial orbits for these particles were chosen from a previous set of integrations of objects that were initially in low-eccentricity, low-inclination orbits in the Kuiper belt but evolved onto Neptune-crossing orbits on timescales between 1 and 4 G.y. (*Duncan et al.,* 1995). The median inclination of the particles at the time of encountering Neptune was 4°.

LD97 found that as objects evolve inward from the Kuiper belt, they tend to be under the dynamical control of just one planet. That planet will scatter the comets inward and outward in a random walk, typically handing them off to the planet directly interior or exterior to it. Therefore, the comets tend to have eccentricities of about 0.25 between handoffs. However, once they have been scattered into the inner solar system by Jupiter, they can have much larger eccentricities as they evolve outward.

Plate 7 shows the evolution of a typical particle in the perihelion distance (q)–aphelion distance (Q) plane, as it evolves from the Kuiper belt (q > 30 AU) to a visible JFC (the most populous of the visible ecliptic comets). The positions are joined by blue lines until the particle first became "visible" (q < 2.5 AU) and are linked in red thereafter. Initially, the particle spent considerable time with perihelion near the orbit of Neptune (30 AU) and aphelion well beyond the planetary system. However, once its perihelion dropped to Uranus' location, this particle, which was chosen at random from *LD97*'s integrations, clearly shows the handoff behavior described above. It evolved at relatively small eccentricity to visibility (cf. the lines of constant eccentricity e = 0.2 and 0.3 on Plate 7) and it spent considerable time with perihelion or aphelion near the semimajor axis of one of the three outer planets. Its postvisibility phase is reasonably typical of JFCs, with much larger eccentricities than the previsibility comets and perihelion distances near Jupiter or Saturn.

Plate 8 shows the distribution of the ecliptic comets derived from the simulations of *LD97*, assuming that the rate of objects leaving the Kuiper belt has remained approximately constant over the last ~10^8 yr and that there was no huge influx at early epochs. The figure is a contour plot of the fraction of comets per square AU in perihelion-aphelion (q–Q) space. The figure was generated by binning the q–Q values of all the comets at all output points in the integrations, which occurred once every 10^4 yr before a comet became visible and once every 1000 yr after it become visible. The resulting matrix was then normalized so that the total number is 1. Note that we are not plotting, say, the relative number of comets per square AU at a given distance from the Sun, but rather presenting the more abstract density contours on a grid with pericentric distance q on one axis and apocentric distance Q on the other. Also shown in Plate 8 are curves of constant eccentricity and semimajor axis.

There are two well-defined regions in Plate 8. Beyond approximately Q = 7 AU, there is a ridge of high density extending diagonally from the upper right to the center of the plot, near e ≈ 0.25. The peak of this ridge is near e = 0.2 at large Q and tends to increase to e ≈ 0.3 near Q ~ 7 AU. As discussed in *LD97*, the eccentricities of objects that are between the planets and not yet under Jupiter's control are expected to be about 0.25 due to the constraints of the Tisserand parameter. The peak density in this ridge drops by almost 2 orders of magnitude as it moves inward, having a minimum where the semimajor axes of the comets are the same as Jupiter's. This population extends inward of Jupiter's orbit and terminates near the 2:1 mean-motion resonance with Jupiter. Indeed, we find that objects can be forced onto nearly circular orbits with semimajor axes as small as ~4 AU. This inner edge is coincident with the inner edge of Jupiter's "crossing zone" at 3.88 AU. *Gladman and Duncan* (1990) defined Jupiter's "crossing zone" as the region in which objects on initially circular orbits can become Jupiter-crossing. Since this process is time reversible, it is not then surprising that Jupiter can drive comets into nearly circular orbits in this region.

Inside of Q ≈ 7 AU the character of the distribution is quite different. Here there is a ridge of high density extending vertically in the figure at Q ~ 5–6 AU that extends over a wide range of perihelion distances. Objects in this region are the JFCs. This characteristic of a very narrow distribution in Q is seen in the real JFCs and is again a result of the narrow range in T. Taken together, this distribution of comets produces a surface distribution shown as the dashed curve in Fig. 2.

The median dynamical lifetime of the ecliptic comets was found to be 4.5×10^7 yr. (This is the time from the first encounter with Neptune to ejection from the solar system, placement in the Oort cloud, which they took to include comets with semimajor axes >1000 AU, or impact with the Sun or a planet). *LD97* found that about 30% of the objects in the integrations became visible comets (q < 2.5 AU). Of those that became visible, 99.7% were JFCs at the time of first visibility.

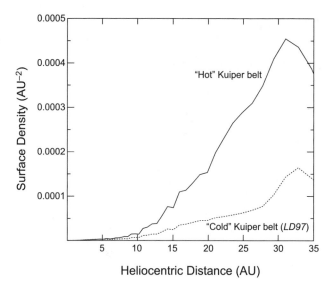

Fig. 2. The predicted surface density of Centaurs and JFCs as a function of heliocentric distance. The dashed curve represents objects initially on low-inclination Kuiper belt orbits in the integrations of *LD97*. The solid curve is from a small number of objects with initial inclinations of roughly 25°.

In order to compare the results of the simulations to the distribution of real comets, Figs. 1c and 1d show the orbital element distribution of all visible comets in the simulations of *LD97* when they first became visible. When making a comparison between the real and simulated JFCs in Fig. 1 it is important to note that there are about five times as many points in the simulated population as in the observed population. Thus, outlying data points in the simulated population are less significant than they may otherwise seem. The distribution of comets in the simulation matches that for the real Jupiter-family remarkably well. Therefore, the simulations show that the Kuiper belt is an excellent potential source for the JFCs. However, the Kuiper belt does not produce very many HTCs (at least initially, see below). This result is consistent with the results of earlier work (*Quinn et al.*, 1990) that the Kuiper belt is the main source of JFCs, whereas another reservoir (possibly the inner Oort cloud) is the most likely source for most of the HTCs (see *Dones et al.*, 2004).

An interesting aspect of the distributions shown in Fig. 1 is the very narrow range in T that JFCs occupy. Although we define a JFC as one with $2 < T < 3$, the median T for the family is 2.8. The narrow range in T is related to small inclinations to the ecliptic seen for these comets: An object on a Jupiter-crossing orbit with $T > 2.8$ and $q < 2.5$ AU must have an inclination less than ~26°.

The inclination distribution of JFCs when they first become visible is more concentrated to small inclinations than the observed population and was found to become more distended as it aged. Indeed, this model predicts an inclina-

tion distribution more extended than the observations unless fading due to physical evolution is included. From this *LD97* estimated that the physical lifetime of JFCs is between 3000 and 25,000 yr. The most likely value is 12,000 yr.

With this estimate of the physical lifetime, *LD97* were able to calibrate the total number of ecliptic comets by using the number of observed JFCs. They thereby estimated that there are roughly a million ecliptic comets with semimajor axes less than 30 AU. This estimate refers to objects that would produce active comets with total absolute magnitudes brighter then 9 if they came within 2.5 AU of the Sun. We discuss the conversion from limits based on absolute magnitudes to those based on physical radii in section 5.

In the simulations of *LD97*, when the ecliptic comets first became visible they were almost entirely members of the Jupiter family. Although a small fraction of these comets switched to visible HTCs, the orbital element distribution of *LD97*'s simulated HTCs is not consistent with the observed distribution. In particular, the semimajor axes of the simulated comets are generally too small. Since *LD97* found that it takes at least 10^5 yr and usually over 10^6 yr to become a visible HTC after the comet first becomes visible, most of these comets have likely become extinct since this is longer then the typical physical lifetime estimated above. Thus, although the Kuiper belt can be the source of at least some of the HTCs, initially low-inclination bodies encountering Neptune are unlikely to provide a significant number of them.

Finally, we consider the Encke-type comets, which are low-inclination comets totally interior to Jupiter's orbit (see section 2). Comet 2P/Encke is the only active member of this population. Although 107P/Wilson-Harrington has a T consistent with this class, it is probably not a comet (*Bottke et al.*, 2002). However, there are several kilometer-sized asteroids known to be in similar orbits (*Asher et al.*, 1993). These small "asteroids" could be extinct comets. Numerical integrations of its orbit show that 2P/Encke will hit the Sun in only 10^5–10^6 yr (*Levison and Duncan*, 1994) due to its close association with secular resonances (*Valsecchi et al.*, 1995).

The *LD97* integrations did not produce any comets similar to 2P/Encke, but included neither the effects of the terrestrial planets nor nongravitational effects. Some of these effects were considered by *Steel and Asher* (1996), *Harris and Bailey* (1998), and *Asher et al.* (2001). *Fernández et al.* (2002) integrated the orbits of a sample of real JFCs with the terrestrial planets and nongravitational forces included. They found that they could produce objects on Encke-like orbits, but only when strong nongravitational forces (as strong as or greater than those that are estimated to be currently acting on Encke) are included.

5. DYNAMICS OF THE SCATTERED DISK

Perhaps the most interesting result of the *LD97* simulations was that about 5% of the particles survived the length of the integration (10^9 yr). All the survivors had semimajor

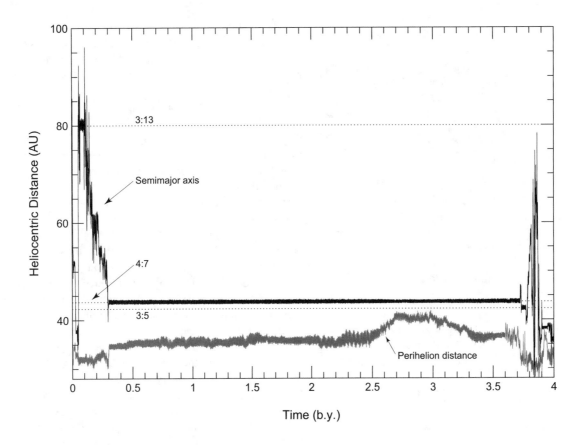

Fig. 3. The temporal behavior of a long-lived member of the scattered disk. The black curve shows the behavior of the comet's semimajor axis. The gray curve shows the perihelion distance. The three dotted curves show the location of the 3:13, 4:7, and 3:5 mean-motion resonances with Neptune. From *Duncan and Levison* (1997).

axes outside the orbit of Neptune. This result implied that there may be a significant population of objects with highly eccentric orbits in an extended disk beyond the orbit of Neptune — a "scattered" comet disk. Although, the idea of the existence of a scattered comet disk dates to *Fernández and Ip* (1983) and *Torbett* (1989), it was the modern work of *LD97* and the followup paper, *Duncan and Levison* (1997) (hereafter *DL97*), that demonstrated that the scattered disk should exist. *DL97* extended the *LD97* integrations to 4×10^9 yr. It was found that 1% of the particles remained in orbits beyond Neptune after 4 b.y. So, if at early times, there was a significant amount of material from the region between Uranus and Neptune or the inner Kuiper belt that evolved onto Neptune-crossing orbits, then there could be a significant amount of this material remaining today. [The first scattered disk object was discovered by *Luu et al.* (1997) shortly after this prediction.]

DL97 found that some of the long-lived objects were scattered to very long-period orbits where encounters with Neptune became infrequent. However, at any given time, the majority of them were interior to 100 AU. Their longevity is due in large part to their being temporarily trapped in or near mean-motion resonances with Neptune. The "stickiness" of the mean-motion resonances, first mentioned

by *Holman and Wisdom* (1993), leads to an overall distribution of semimajor axes for the particles that is peaked near the locations of many of the mean-motion resonances with Neptune. Occasionally, the longevity is enhanced by the presence of the Kozai resonance (*Kozai,* 1962).

In all long-lived cases, particles had their perihelion distances increased so that close encounters with Neptune no longer occurred. Frequently, these increases in perihelion distance were associated with trapping in a mean-motion resonance, although in many cases it has not yet been possible to identify the exact process that was involved. On occasion, the perihelion distance can become large, but 81% of scattered disk objects in the simulations have perihelia between 32 and 36 AU.

Figure 3 shows the dynamical behavior of a typical long-lived particle. This object initially underwent a random walk in semimajor axis due to encounters with Neptune. At about 7×10^7 yr it was temporarily trapped in Neptune's 3:13 mean-motion resonance for about 5×10^7 yr. It then performed a random walk in semimajor axis until about 3×10^8 yr, when it was trapped in the 4:7 mean-motion resonance, where it remained for 3.4×10^9 yr. Notice the increase in the perihelion distance near the time of capture. While trapped in this resonance, the particle's eccentricity became as small as 0.04. After leaving the 4:7, it was trapped temporarily in Neptune's 3:5 mean-motion resonance for ~5 × 10^8 yr and then went through a random walk in semimajor axis for the remainder of the simulation.

DL97 estimate the number of scattered disk objects that their model would require if it was the sole source of the JFCs. They first computed the simulated distribution of comets throughout the solar system at the current epoch (assuming the disk was created 4 × 10^9 yr ago, with the distribution averaged over the last 10^9 yr for better statistical accuracy). They found that the ratio of scattered disk objects to visible JFCs (those with a perihelion distance <2.5 AU) is 1.3 × 10^6. Since there are currently estimated to be 500 visible JFCs (*LD97*; see also *Fernández et al.,* 1999), there are presently ~6 × 10^8 comets in the scattered disk if this model is the sole source of the JFCs. Figure 4 shows the spatial distribution for this model.

The above estimate is not particularly physically meaningful because it refers to comets with absolute magnitudes, H_T, brighter than 9 when they are visible. A more interesting measure is the number of comets greater than some diameter, typically set to 1 km. Unfortunately, this is a non-

trivial conversion, since there is no good correlation between the absolute magnitude of a comet (which is based on a comet's activity) and its size (for discussions, see, e.g., *Levison et al.,* 2001; *Zahnle et al.,* 1998). Examples of possible values for the number of comets in the scattered disk with diameters, D, >1 km range from ~5 × 10^8 [following the calibration of *Bailey et al.* (1994)], through ~2 × 10^9 [following the calibration of *Weissman* (1990)], to ~5 × 10^9 (following *Levison et al.'s* (2001) estimate, based on *Kary and Dones* (1996)].

If the scattered disk is the source of the Centaurs and JFCs, then the estimate of its population has implications for planet formation. Since we employ the calibration of *Weissman* (1990) for this discussion, we note that there are significant uncertainties in these numbers. It was shown above that ~1% of the objects in the scattered disk remain after 4 × 10^9 yr in the simulations of *DL97*, and that ~2 × 10^9 comets with D > 1 km are currently required to supply all the JFCs with the adopted calibration of *Weissman* (1990). Thus, a scattered comet disk requires an initial population of only 2 × 10^{11} comets [or ~1 M$_\oplus$ (*Weissman,* 1990)] on Neptune-encountering orbits. Since planet formation is unlikely to have been 100% efficient, the original disk could have resulted from the scattering of even a small fraction of the tens of Earth masses of cometary material that must have populated the outer solar system in order to have formed Uranus and Neptune. Thus, a disk that supplies the current JFCs and Centaurs appears to have properties consistent with that expected from our current understanding of planet formation.

The first scattered disk object discovered was 1996 TL$_{66}$, found in October 1996 by Jane Luu and colleagues (*Luu et al.,* 1997). Current observations indicate that it has a semimajor axis of 85 AU, a perihelion of 35 AU, and an inclination of 24°. Dozens of other objects are currently known (see http://cfa-www.harvard.edu/iau/lists/Centaurs.html for a complete list.) The total number in 100-km-sized scattered disk objects is estimated to be between 20,000 and 50,000 (*Trujillo et al.,* 2000), comparable to the number of similar-sized Kuiper belt objects interior to 50 AU. The number in comet-sized (1–10-km) bodies remains to be observationally measured. However, using the size distribution of *Weissman and Levison* (1997) and *Trujillo et al.'s* (2000) number, we find that we should expect between 2 × 10^9 and 6 × 10^9 objects with D > 1 km. This number is consistent with our estimates above for a scattered disk source of JFCs.

Indeed, we believe that it can now be argued that the scattered disk is the primary source of the Centaurs and JFCs. If the Kuiper belt were the source, we would expect that the number of objects in the Kuiper belt would be more than an order of magnitude larger than the number of objects in the scattered disk because the average dynamical lifetime of scattered disk objects [2 × 10^8 yr (*Levison et al.,* 2001)] is much shorter than that of a Kuiper belt object (*Duncan et al.,* 1995; *Kuchner et al.,* 2002). However, we have just noted that *Trujillo et al.* (2000) found that the two populations of 100-km objects are comparable. Assuming

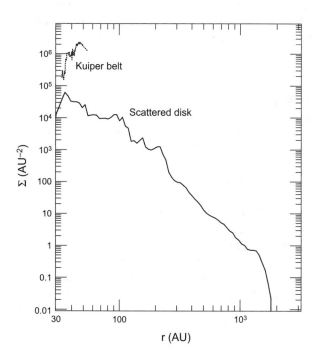

Fig. 4. The surface density of comets beyond Neptune for two different models of the source of JFCs. The dotted curve is the required surface density assuming that a dynamically cold Kuiper belt is the current source (*Levison and Duncan,* 1997). There are 7 × 10^9 comets with H_T < 9 in this distribution between 30 and 50 AU. This curve ends at 50 AU because the models are unconstrained beyond this point and not because it is believed that there are no comets there. The solid curve is the model of *Duncan and Levison* (1997) assuming the scattered disk produced by their integrations is the sole source of the JFCs. There are estimated to be 6 × 10^8 comets with H_T < 9 currently in this distribution (see text). From *Duncan and Levison* (1997).

that objects in the scattered disk and the Kuiper belt have similar size distributions, the fact that most scattered disk objects are on less stable orbits (see *Emel'yanenko et al.,* 2003) and become planet-crossing at a greater rate implies that the scattered disk is the primary source of the JFCs.

This conclusion is also supported by the recent work of *Morbidelli et al.* (2004), who performed a detailed comparison between the two different dynamical models for the scattered disk. They found that the orbital element distribution of observed scattered disk objects is not consistent with a scattered disk that is in dynamical equilibrium with the Kuiper belt. Such a model predicts many more scattered disk objects with semimajor axes between 50 and 60 AU (compared to objects with larger semimajor axes) than are actually observed. However, a model invoking an ancient scattered disk, previously much more massive, is in good agreement with the observed scattered disk. Again, we can conclude that the scattered disk is old and is the most likely source of the Centaurs and JFCs.

Very recently, strong evidence has emerged for the existence of an "extended scattered disk" (*Gladman et al.,* 2002). This disk is comprised of bodies such as 2000 CR$_{105}$ that have very large semimajor axes and perihelia outside 40 AU. These objects cannot have been placed on such orbits by strong gravitational scattering off any of the giant planets in their current orbits. The mass in this population is extremely uncertain due to the difficulty in finding such distant objects in the first place and due to the difficulty of determining their orbits without frequent and well-timed recovery observations. Their cosmogonic implications are significant and are discussed in *Morbidelli and Brown* (2004).

6. PLANETARY IMPACT RATES FROM ECLIPTIC COMETS

An important aspect of the ecliptic comet population is that it is the primary source of impactors on the satellites of the giant planets (*Zahnle et al.,* 1998). Thus, if it is possible to determine the total number and orbital distribution of these objects, it should be possible to estimate the ages of the satellite surfaces. Given the importance of this topic, we dedicate this section to this issue.

As described in section 4, the best set of integrations for this purpose remains those presented in *LD97.* However, before we proceed with a discussion of their results with regard to impact rates, we first need to ask whether the *LD97* model is a good representation of the Centaur (i.e., impacting) population. We do this by comparing the orbital element distribution of *LD97*'s synthetic Centaurs to real objects. Unfortunately, the data suffer from significant observational biases in semimajor axis (a), inclination (i), and perihelion distance (q). Thus, to compare the *LD97* model to observations, we first run the model through a synthetic survey simulator. Here we use one developed by *Morbidelli et al.* (2004) for the transneptunian population. This simulator, in turn, was based on the work of *Brown* (2001).

Our basic procedure is as follows. We generate a set of synthetic Centaurs from the integration of *LD97,* restrict-ing ourselves to objects with 5.2 < a < 30 AU. For each of these we randomly choose an absolute magnitude, the distribution of which is consistent with what is observed in the Kuiper belt [N(<H) ∝ 10$^{\alpha H}$, where H is the absolute magnitude, N is the total number, and α = 0.7 (*Gladman et al.,* 2001; *Trujillo et al.,* 2001). We run these objects through the survey simulator, which calculates the probability that each object would have been discovered by the surveys that have thus far discovered real Centaurs. From this, we can generate an estimate of what the model predicts for the orbital element distribution of the known Centaurs. This result is represented by the dark solid curves in Fig. 5. The dotted curves show the distribution of real, multiopposition, Centaurs. The distributions in a, i, and q are very similar. Thus, we can conclude that the models of *LD97* do indeed match the distribution of the currently observed Centaurs.

However, this result is a necessary but not sufficient condition, because the observations may not sample enough of the Centaurs to be a significant constraint on the overall distribution. For example, *LD97* studied the evolution of a small number of objects leaving the Kuiper belt at inclinations of ~25°. The solid curve in Fig. 2 shows the surface density distribution of this model compared to the standard *LD97* model shown by the dotted curve. Note that there is roughly a factor of three difference between the curves in the outer solar system and a factor of 2 at Saturn. However, they agree at Jupiter.

We have run this "hot" model of the Centaurs through our synthetic survey simulator and the predicted orbital element distribution of discovered objects is shown as the blue curves in Fig. 5. There is reasonably good agreement between this model, the original model, and the observations. The only exception is in the inclinations where objects in this model have slightly (but not significantly) larger inclinations. However, we conclude that the observations cannot yet distinguish between these two reasonable models of the Centaurs. This fact should be considered when interpreting the impacts rates we present below.

Having said this, we return to the issue of impact rates. Included in the analysis presented in *LD97* was a crude estimate of the impact rates on the planets from ecliptic comets. Subsequently, *Levison et al.* (2000) have used the results of *LD97* to present a more precise and detailed analysis of the impact rates and characteristics of the impacting bodies. *Levison et al.* (2000) used three different methods to compute the impact rates on the giant planets (see section 2 of that paper for details). The main results of these calculations are given as the last three columns in their Table 1, which is reproduced here (see Table 1). For this discussion, we note that the authors recommend using the last column (Öpik II) as the standard impact rate: The variation in the entries for each planet should be viewed as a measure of the uncertainty in the rate given in the last column.

The impact rates were calculated assuming that the rate at which comets evolve from their source region to the ecliptic comet population has been constant. This assumption is clearly not correct, although the authors believe that it does

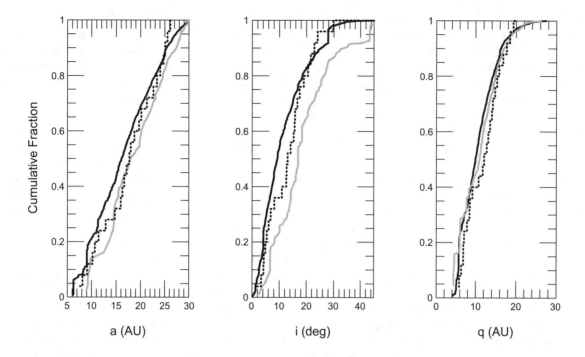

Fig. 5. The orbital element distribution of the Centaurs. The dark solid curves are those derived from the simulations of *LD97*, while the gray curves are derived from a subset that started with objects leaving the Kuiper belt with inclinations of roughly 25°. Both the solid curves were generated using the synthetic survey simulator of *Morbidelli et al.* (2004). The dotted curves show the distribution of observed, multiopposition, Centaurs.

TABLE 1.   Impact rates of ecliptic comets with $H_T < 9$ on the planets according to *Levison et al.* (2000).

Planet	Number of Impacts Stage 1	Number of Impacts Stage 2	Rate Direct (comets/yr)	Rate Öpik I (comets/yr)	Rate Öpik II (comets/yr)
Jupiter	7	114	6.3×10^{-4}	5.0×10^{-4}	6.5×10^{-4}
Saturn	4	17	1.9×10^{-4}	2.0×10^{-4}	2.7×10^{-4}
Uranus	10	0	3.4×10^{-4}	1.2×10^{-4}	1.6×10^{-4}
Neptune	10	0	3.4×10^{-4}	2.6×10^{-4}	3.5×10^{-4}
Mercury	—	—	—	4.8×10^{-9}	6.1×10^{-9}
Venus	—	—	—	4.2×10^{-8}	5.4×10^{-8}
Earth	—	—	—	6.2×10^{-8}	8.0×10^{-8}
Mars	—	—	—	1.4×10^{-8}	1.8×10^{-8}

not have a large effect on the estimates of the current impact rates. However, most likely these impact rates were significantly higher in the distant past. For example, if *DL97*'s model of the scattered disk is correct, the impact rates on the giant planet satellites 2×10^9 yr ago were roughly a factor of 2 larger than current rates. Any attempt to use the impact rates to estimate the ages of satellites should take this into account.

The impact rates given in Table 1 are calibrated to active JFCs with absolute magnitudes, H_T, brighter than 9. As describe above, a more common and standardized measure of impact rates is to present them scaled to the number of objects with diameters greater than 1 km striking a planet

per year. In order to do this, *Levison et al.* (2000) find that the values presented in Table 1 should be multiplied by a factor of ~5, although there are great uncertainties in this number. For example, *Bottke et al.* (2002) use data from the Spacewatch Near Earth Asteroid survey to estimate that this scale factor should be 1.7 ± 1.2 rather than 5. However, this assumes that 100% of all JFCs become inactive rather than disintegrate once their active lifetime is over. If two-thirds of the JFC population self-destructed, the values of the scale factor would agree.

Zahnle et al. (1998, 2001, 2003) computed impact rates on the satellites of Jupiter, Saturn, Uranus, and Neptune, and on Pluto/Charon. Ecliptic comets appear to be the main impactors on all these bodies. In these papers, the (relative) spatial distribution of ecliptic comets was taken from the integrations of *LD97* and *DL97*. However, the authors attempted to perform a much more careful analysis of the uncertainties in the total number of comets in the system.

Galileo observations of the galilean satellites indicate a paucity of small craters relative to an extrapolation of the distribution at larger sizes (see *Schenk et al.,* 2003, for a discussion.) The most reasonable interpretation of this result is that there are relatively few subkilometer comets in the JFC population (at least at Jupiter), which may have important implications for the origins of these objects. Recently, *Zahnle et al.* (2003) argued that this relative paucity must exist all the way out to 30 AU because if not, lifetimes against collisional disruption by ecliptic comets would be $\ll 10^9$ yr for a number of small moons around Saturn, Uranus, and Neptune.

However, if subkilometer comets really are underabundant in the outer solar system, this implies that either (1) there are few such bodies in the source regions — the scattered disk and/or (less likely) the Kuiper belt — or that (2) most small comets are destroyed at heliocentric distances greater than 30 AU. Both these explanations have problems. The former seems unlikely because the transneptunian population is generally believed to have been collisional in the early days of the solar system, and therefore should have produced many small fragments (*Stern,* 1995; *Kenyon,* 2002; *Stern and Weissman,* 2001). The latter is unlikely because there is no obvious way to destroy the comets by thermal or other effects at such great distances from the Sun. However, the size distribution of comet precursors likely will not be known until transneptunian occultation experiments such as TAOS (*Alcock et al.,* 2003) are discovering kilometer-sized bodies in the region.

7. SUMMARY

In the past 15 years there has been enormous progress in our understanding of the dynamical evolution of ecliptic comets, largely due to the ability of researchers to perform orbital integrations of large numbers of comets for timescales on the order of the age of the solar system. The study of the dynamics of comets has helped lead the way to important discoveries about new structures in the solar system, including the Kuiper belt and the scattered disk.

We began this chapter by reviewing the evidence from numerical integrations that observed that JFCs form a dynamically distinct class of comets. Results were discussed that showed that JFCs, upon first becoming visible, have an inclination distribution that is even more flattened than that currently observed. We then reviewed the compelling dynamical evidence that JFCs and Centaurs originate in a flattened disk beyond Neptune.

A model was described for the distribution of comets throughout the JFC and Centaur regions. The integrations that produced the distribution (*LD97*) assumed that the original orbits of the comets had very low inclinations, which

more recent observations suggest is not the case for a large fraction of the transneptunian population. However, we presented evidence in this review that suggested that many of the properties of the distribution (such as the distribution interior to Jupiter) were independent of the detailed structure of the source reservoir. Indeed, we showed that the predicted Centaur orbital distributions in *LD97* are consistent with the distribution of the currently observed, multiopposition Centaurs once observational biases are included. We also noted, however, that the study of a small number of objects leaving the Kuiper belt at inclinations of approximately 25° also produced a Centaur distribution in rough agreement with the observations despite differing in surface density by a factor of 2–3 from the dynamically "colder" model in the outer planetary region. It is clear that further observations, coupled with a better understanding of the formation and early dynamical evolution of the outer solar system, will be required before these uncertainties can be reduced.

We have presented the dynamical results (since confirmed by observation) that a significant amount of material that was scattered by Neptune during the early stages of planet formation could persist today in the form of a "scattered disk" of bodies with highly eccentric orbits beyond Neptune. We described the dynamical mechanisms believed responsible for the longevity of the subset of objects that have survived to the current epoch. The results of the integrations suggest that the dynamical lifetime of scattered disk objects is much less than that of most currently observed Kuiper belt objects. Since the numbers of objects inferred from observations in the two groups are comparable, then if the two groups have similar size distributions, we argue that the scattered disk is likely to be the primary source of JFCs and Centaurs.

Finally, we have described the importance of understanding the ecliptic comet population for the purposes of determining impact rates on the satellites of the giant planets and of age determinations of the satellite surfaces. We have presented tables of impact rates based on the best currently available analyses. Further refinements of these rates and age determinations await better observations of the Centaur population (including its size distribution), as well as a better understanding of the formation and early dynamical evolution of the outer solar system.

REFERENCES

Alcock C. and 17 colleagues (2003) The Taiwanese-American Occultation Survey. In *Proceedings of the First Decadal Review of the Kuiper Belt* (J. Davies and L. Barrera, eds.). Special issue of *Earth Moon Planets, 92,* in press.

Asher D. J., Clube S. V. M., and Steel D. I. (1993) Asteroids in the Taurid complex. *Mon. Not. R. Astron. Soc., 264,* 93.

Asher D. J., Bailey M. E., and Steel D. I. (2001) The role of nongravitational forces in decoupling orbits from Jupiter. In *Collisional Processes in the Solar System* (M. Marov and H. Rickman, eds.), pp. 121–130. Kluwer, Dordrecht.

Bailey M. E. (1986) The near-parabolic flux and the origin of short-period comets. *Nature, 324,* 350–352.

Bailey M., Clube S., Hahn G., Napier W., and Valsecchi G. (1994) Hazards due to giant comets: Climate and short-term catastrophism. In *Hazards Due to Comets and Asteroids* (T. Gehrels et al., eds.), p. 479. Univ. of Arizona, Tucson.

Bottke W., Morbidelli A., Jedicke R., Petit J.-M., Levison H., Michel P., and Metcalfe T. (2002) Debiased orbital and absolute magnitude distribution of the near-earth objects. *Icarus, 156,* 399–433.

Brown M. (2001) The inclination distribution of the Kuiper Belt. *Astron. J., 121,* 2804–2814.

Carusi A. and Valsecchi G. (1987) Dynamical evolution of short-period comets. In *Interplanetary Matter* (Z. Ceplecha and P. Pecina, eds.), pp. 21–28. Astronicky ustav CSAV, Praha.

Carusi A., Kresák, L., Perozzi E., and Valsecchi G. (1985) *Long-Term Evolution of Short-Period Comets.* Adam Hilger, Bristol. 272 pp.

Dones L., Weissman P. R., Levison H. F., and Duncan M. J. (2004) Oort cloud formation and dynamics. In *Comets II* (M. C. Festou et al., eds.), this volume. Univ. of Arizona, Tucson.

Duncan M. and Levison H. (1997) A scattered comet disk and the origin of Jupiter family comets. *Science, 276,* 1670–1672.

Duncan M., Quinn T., and Tremaine S. (1988) The origin of short-period comets. *Astrophys. J. Lett., 328,* L69–L73.

Duncan M., Levison H., and Budd M. (1995) The dynamical structure of the Kuiper Belt. *Astron. J., 110,* 3073.

Edgeworth K. (1949) The origin and evolution of the Solar System. *Mon. Not. R. Astron. Soc., 109,* 600–609.

Emel'yanenko V. (1992) Evolution of orbits of comets having close encounters with Jupiter. I — Analysis of the effect of errors on the initial system of elements. II — Analysis of the effect of Jupiter's oblateness. *Astron. Vestnik, 26,* 24.

Emel'yanenko V., Asher D., and Bailey M. (2003) A new class of trans-neptunian objects in high-eccentricity orbits. *Mon. Not. R. Astron. Soc., 338,* 443–451.

Everhart E. (1977) The evolution of comet orbits as perturbed by Uranus and Neptune. In *Comets, Asteroids, Meteors* (A. H. Delsemme, ed.), pp. 99–104. Univ. of Toledo, Ohio.

Fernández J. A. (1980) On the existence of a comet belt beyond Neptune. *Mon. Not. R. Astron. Soc., 192,* 481–491.

Fernández J. A. and Gallardo T. (1994) The transfer of comets from parabolic orbits to short-period orbits: Numerical studies. *Astron. Astrophys., 281,* 911–922.

Fernández J. A. and Ip W.-H. (1983) On the time evolution of the cometary influx in the region of the terrestrial planets. *Icarus, 54,* 377–387.

Fernández J. A., Tancredi G., Rickman H., and Licandro J. (1999) Are there many inactive Jupiter-family comets among the near-earth asteroid population? *Astron. Astrophys., 352,* 327–340.

Fernández J. A., Gallardo T., and Brunini A. (2002) Are there many inactive Jupiter-family comets among the near-earth asteroid population? *Icarus, 159,* 358–368.

Gladman B. and Duncan M. (1990) On the fates of minor bodies in the outer solar system. *Astron. J., 100,* 1680–1693.

Gladman B., Kavelaars J. J., Petit J.-M., Morbidelli A., Holman M. J., and Loredo T. (2001) Evidence for an extended scattered disk. *Astron. J., 122,* 1051–1066.

Gladman B., Holman M. J., Grav T., Kavelaars J. J., Nicholson P., Aksnes K., and Petit J.-M. (2002) Evidence for an extended scattered disk. *Icarus, 157,* 269–271.

Harris N. W. and Bailey M. (1998) Dynamical evolution of com-

etary asteroids. *Mon. Not. R. Astron. Soc., 297,* 1227–1236.

Holman M. J. and Wisdom J. (1993) Dynamical stability in the outer solar system and the delivery of short period comets. *Astron. J., 105,* 1987–1999.

Joss P. (1973) On the origin of short-period comets. *Astron. Astrophys., 25,* 271.

Kary D. and Dones L. (1996) Capture statistics of short-period comets: Implications for comet D/Shoemaker-Levy 9. *Icarus, 121,* 207–224.

Kenyon S. J. (2002) Planet formation in the outer solar system. *Publ. Astron. Soc. Pacific, 114,* 265–283.

Kozai Y. (1962) Secular perturbations of asteroids with high inclination and eccentricity. *Astron. J., 67,* 591.

Kresák L. (1979) Dynamical interrelations among comets and asteroids. In *Asteroids* (T. Gehrels, ed.), pp 289–309. Univ. of Arizona, Tucson.

Kuchner M. J., Brown M. E., and Holman M. (2002) Long-term dynamics and the orbital inclinations of the classical Kuiper Belt objects. *Astron. J., 124,* 1221–1230.

Kuiper G. P. (1951) O the origin of the solar system. In *Astrophysics: A Topical Symposium* (J. A. Hynek, ed.), p. 357. McGraw-Hill, New York.

Levison H. (1996) Comet taxonomy. In *Completing the Inventory of the Solar System* (T. W. Rettig and J. M. Hahn, eds.), pp. 173–191. Astronomical Society of the Pacific, San Francisco.

Levison H. and Duncan M. (1993) The gravitational sculpting of the Kuiper Belt. *Astrophys. J., 406,* 35–38.

Levison H. and Duncan M. (1994) The long-term dynamical behavior of short-period comets. *Icarus, 108,* 18–36.

Levison H. and Duncan M. (1997) From the Kuiper Belt to Jupiter-family comets: The spatial distribution of ecliptic comets. *Icarus, 127,* 13–32.

Levison H. F., Duncan M. J., Zahnle K., Holman M., and Dones L. (2000) NOTE: Planetary impact rates from ecliptic comets. *Icarus, 143,* 415–420.

Levison H. F., Dones H., and Duncan M. J. (2001) The origin of Halley-type comets: Probing the inner Oort cloud. *Astron. J., 121,* 2253–2267.

Lindgren M. (1992) Dynamical timescales in the Jupiter family. In *Asteroids, Comets, Meteors 1991* (A. W. Harris and E. Bowell, eds.), pp. 371–374. Lunar and Planetary Institute, Houston.

Luu J. and Jewitt D. (1993) Discovery of the candidate Kuiper Belt object 1992 QB1. *Nature, 362,* 730–732.

Luu J., Jewitt D., Trujillo C. A., Hergenrother C. W., Che J., and Offutt W. B. (1997) A new dynamical class in the trans-neptunian solar system. *Nature, 387,* 573.

Morbidelli A. and Brown M. E. (2004) The Kuiper belt and the primordial evolution of the solar system. In *Comets II* (M. C. Festou et al., eds.), this volume. Univ. of Arizona, Tucson.

Morbidelli A., Emel'yanenko V., and Levison H. F. (2004) Origin and orbital distribution of the trans-Neptunian scattered disk. *Mon. Not. R. Astron. Soc.,* in press.

Nakamura T. and Yoshikawa M. (1992a) Long-term orbital evolution of short-period comets found in project "Cosmo-Dice". In *Asteroids, Comets, Meteors 1991* (A. W. Harris and E. Bowell, eds.), p. 433. Lunar and Planetary Institute, Houston.

Nakamura T. and Yoshikawa M. (1992b) Invisible comets on evolutionary track of short-period comets. *Cel. Mech. Dyn. Astron., 54,* 261–266.

Nesvorny D. and Roig F. (2000) Mean motion resonances in the trans-neptunian region. I. The 2:3 resonance with Neptune. *Icarus, 148,* 282–300.

Newton H. A. (1891) Capture of comets by planets. *Astron. J., 11,* 73–75.

Oort J. H. (1950) The structure of the cloud of comets surrounding the Solar System and a hypothesis concerning its origin. *Bull. Astron. Inst. Neth., 11,* 91–110.

Quinn T. R., Tremaine S., and Duncan M. J. (1990) Planetary perturbations and the origins of short-period comets. *Astrophys. J., 355,* 667–679.

Rickman H. (1992) Structure and evolution of the Jupiter family. *Cel. Mech. Dyn. Astron., 54,* 63–69.

Schenk P. M., Chapman C. R., Zahnle K., and Moore J. (2003) Ages and interiors: The cratering record of the galilean satellites. In *Jupiter: The Planet, Satellites, and Magnetosphere* (F. Bagenal et al., eds.), in press. Cambridge Univ., New York.

Steel D. I. and Asher D. J. (1996) On the origin of Comet Encke. *Mon. Not. R. Astron. Soc., 281,* 937–944.

Stern S. A. (1995) Collisional time scales in the Kuiper disk and their implications. *Astron. J., 110,* 856.

Stern S. A. and Weissman P. (2001) Rapid collisional evolution of comets during the formation of the Oort cloud. *Nature, 409,* 589–591.

Tancredi G. (1994) Physical and dynamical evolution of Jupiter family comets: Simulations based on the observed sample. *Planet. Space Sci., 42,* 421–433.

Tancredi G. and Rickman H. (1992) The evolution of Jupiter family comets over 2000 years. In *Chaos, Resonance and Collective Dynamical Phenomena in the Solar System* (S. Ferraz-Mello, ed.), p. 269. IAU Symposium 152, Kluwer, Dordrecht.

Tancredi G., Lindgren M., and Rickman H. (1990) Temporary satellite capture and orbital evolution of Comet P/Helin-Roman-Crockett. *Astron. Astrophys., 239,* 375–380.

Torbett M. (1989) Chaotic motion in a comet disk beyond Neptune — The delivery of short-period comets. *Astron. J., 98,* 1477–1481.

Trujillo C. A., Jewitt D. C., and Luu J. X. (2000) Population of the scattered Kuiper Belt. *Astrophys. J., 529,* 102–106.

Trujillo C. A., Jewitt D. C., and Luu J. X. (2001) Properties of the trans-Neptunian belt: Statistics from the CFHT survey. *Astron. J., 122,* 457–473.

Valsecchi G. B., Morbidelli A., Gonczi R., Farinella P., Froeschlé Ch., and Froeschlé C. (1995) The dynamics of objects in orbits resembling that of P/Encke. *Icarus, 118,* 169–180.

Weissman P. R. (1990) The cometary impactor flux at the earth. In *Global Catastrophes in Earth History* (V. L. Sharpton and P. D. Ward, eds.), pp. 171–180. GSA Special Paper 247.

Weissman P. R. and Levison H. (1997) The population of the trans-neptunian region The Pluto-Charon environment. In *Pluto and Charon* (S. A. Stern and D. J. Tholen, eds.), p. 559. Univ. of Arizona, Tucson.

Zahnle K., Dones L., and Levison H. F. (1998) Cratering rates on the Galilean satellites. *Icarus, 136,* 202–222.

Zahnle K., Schenk P., Sobieszczk S., Dones L., and Levison H. F. (2001) Differential cratering of synchronously rotating satellites by ecliptic comets. *Icarus, 153,* 111–129.

Zahnle K., Schenk P., Levison H., and Dones L. (2003) Cratering rates in the outer solar system. *Icarus, 163,* 263–289.

Current Questions in Cometary Dynamics

Hans Rickman

Uppsala Astronomical Observatory

We summarize the most important findings reported, and unanswered questions identified, in the review chapters on cometary dynamics in this book. Comments are also offered on the significance of these problems — and thus of cometary dynamics — for making progress on several major issues of solar system cosmogony.

1. INTRODUCTION

Cometary dynamics involves the study of transfer processes between widely separated regions of the solar system. Essentially, it aims to understand how cometary orbits change so that comets are brought between outer and inner ranges of heliocentric distance (r). These provide, respectively, cold storage of the cometary volatiles for billions of years, and then the possibility to exhibit cometary activity while consuming those volatiles. One thus talks of distant reservoirs that are still active today, supplying new comets for their first entries into orbits plunging into the water sublimation zone (this has classically been considered as r ≲ 2.5–3 AU, but under some circumstances H_2O may sublimate at significantly larger distances, and modern observational techniques have allowed the discovery of many comets with perihelia around 5–10 AU).

Of course, our interest in cometary dynamics extends well beyond the currently operating mechanisms for bringing comets into "observable" orbits (i.e., orbits with small enough perihelion distance). It also incorporates the past evolution, and even formation, of the distant reservoirs. Thus the scope is ambitious enough to involve an attempt to link the orbital statistics of observed comets to critical features of solar system cosmogony such as the origin of the giant planets and the transneptunian disk. Even though such studies are by necessity limited to statistical properties of the transfer processes, there is also the intriguing challenge of understanding the places of certain particular comets within the resulting scenarios. Examples are the comets for which D/H ratios and a range of molecular abundances have been observed (1P/Halley, 1995 O1 Hale-Bopp, 1996 B2 Hyakutake), and the target comet of ESA's *Rosetta* mission (67P/Churyumov-Gerasimenko).

Most of the progress of research in these areas, reported in the chapters of this book, is of relatively recent origin. In particular, it has important connections to the discovery and characterization of the Edgeworth-Kuiper belt during the last decade. But this book also describes much progress in the classical area of cometary dynamics that seeks to explain the fine details of observed cometary motions and link as many apparitions of periodic comets as possible, while accurately representing all observations. There is thus a trend toward

better astrometric accuracy and improved coverage of the orbits. Moreover, this is getting used along with data on gas production rates and outflow kinematics to construct realistic models of the jet force, from which physical properties of the nuclei like their masses may be deduced.

We realize the important interconnection between cometary physics and dynamics. It is not possible to fully understand the thermophysical or chemical characteristics of a cometary nucleus without reference to its orbital history and formation region, and the details of the orbital motions provide a promising tool to probe structural and evolutionary properties of cometary nuclei — primarily density or porosity and dust coverage. In fact, cometary motions are affected by nongravitational forces that obviously depend on the physical properties of the nuclei, showing that cometary dynamics is influenced by physical properties as well. Moreover, the fate of an individual comet arriving from the Oort cloud may take entirely different courses, depending on the details of the nongravitational effects. Finally, dynamical models for the transfer of comets into "observable" orbits have to be judged against the distribution of observed comets, and this involves a judgement of both observational biases and physical evolutionary effects. Hence cometary dynamics cannot be understood without reference to cometary physics.

The rest of this paper will focus on some of the most interesting aspects of cometary dynamics, attempting to highlight a number of unresolved problems and identify needs for continued research in the future. For references to published literature the reader should primarily consult the full review papers in this book, since in many cases the relevant papers will not be cited here.

2. ORIGIN OF COMETARY POPULATIONS

2.1. Infeed and Capture

It is now considered a well-established fact that the Jupiter family (i.e., short-period comets with Tisserand parameters T in the range 2 < T < 3) predominantly originates in multiplanet captures from the transneptunian population (see *Jewitt*, 2004; *Duncan et al.*, 2004). By contrast, the contribution from the Oort cloud, while not nil, should be rather small. Currently the most interesting question seems to concern the relative contributions of two possible source populations: the "classical Kuiper belt" or the "scattered disk" (*Duncan et al.*, 2004).

The classical Kuiper belt is a population of objects that formed well outside Neptune's orbit and stayed there for the age of the solar system; this population has undergone strong collisional evolution. The present infeed of such objects into Neptune-crossing orbits is due to resonant interactions with the giant planets that increase the eccentricities over a very long timescale, and those interactions should affect only a fraction of the entire population.

On the other hand, the scattered disk consists of objects that started out from encounters with Neptune and subsequently reached a temporary asylum by resonant decrease of eccentricity (and thus increase of perihelion distance). In this case, the present infeed is to be seen as the inevitable return of the refugees, and essentially the whole population will eventually be affected, even though at present only a part of it is likely to be concerned.

Estimates based on the number of discoveries of objects with diameters $D \geq 100$ km indicate that the two populations are about equally numerous. If this holds true down to $D \leq 10$ km, their relative contributions to both the Centaurs and the Jupiter family should be in inverse proportion to the typical timescales of their respective infeed mechanisms. The argument of *Duncan et al.* (2004) is that the infeed timescale is much shorter for the scattered disk, which is hence the dominating source of the Centaurs and the Jupiter family.

This conclusion needs to be further scrutinized, as more observations will yield improved estimates of population sizes for both the Jupiter family, the Centaurs, the Kuiper belt, and the scattered disk. In particular, the extrapolation of numbers of objects from $D \geq 100$ km to $D \leq 10$ km is risky, especially if collisionally evolved and unevolved populations are compared. At present, the estimated number (~100) of $D > 100$ km Centaurs appears on the small side compared with the infeed rates expected from the two transneptunian reservoirs (10^{-5}–10^{-4} yr^{-1} for a population of ~10^4 $D >$ 100 km objects and infeed timescales of 10^8–10^9 yr), given a Centaur dynamical lifetime of ~10^7 yr.

However, on the one hand, better knowledge of the population sizes may in fact remove any trace of a discrepancy, and on the other hand, it also seems necessary to spend further effort on constraining the dynamical timescales. For instance, if large fractions of the Kuiper belt and scattered disk populations are immune to infeed toward Neptune over much longer timescales, it is not even certain that the two sources are rich enough to explain the Centaurs — or, indeed, the Jupiter-family comets.

Whether or not the current conclusion in favor of the scattered disk as the principal source holds up against such efforts, the final answer is bound to have great cosmogonic significance. This is because the scattered disk, presumably, is not collisionally evolved to the same degree as the classical Kuiper belt — otherwise it would be hard to explain

the current population size. Hence, while Jupiter-family comets derived from the Kuiper belt would be collisional fragments (*Farinella and Davis*, 1996), those coming from the scattered disk may still be collisionally unevolved. As a consequence, the findings of the *Rosetta* mission will have to be viewed in a framework that is quite different, depending on which is the actual source population.

Of course, it is of great importance to establish what chemical effects the collisional evolution would have. To what degree is the structure and composition of cometary ice different, comparing a primordial nucleus with one that is a collisional fragment? Studies of differentiation effects in large transneptunian objects (*Choi et al.*, 2002) shed some light on this, but more work is needed.

There is also a third source, however, as already mentioned. The Oort cloud, and especially its inner core (with semimajor axes a in the approximate range $5000 < a < 20{,}000$ AU), is certainly contributing some — as yet unknown — fraction of the Jupiter-family comets as well as the Centaurs. Its contribution to the Halley-type comets (orbital periods $P < 200$ yr and $T < 2$) is undisputed, but the apparent preference for low-inclination orbits of Halley-type comets is a problem, if the classical Oort cloud with its isotropic orbital distribution is considered as the main source. Thus a flattened inner Oort cloud has also been suggested as the principal source of Halley-type comets (*Levison et al.*, 2001), but a full treatment of the infeed and capture from the inner Oort cloud remains to be made.

2.2. Origin of Source Populations

Considerable progress has been made in the modeling of planetesimal scattering from the accretion zone of the giant planets, and hence some insights into the formation of the Oort cloud, Kuiper belt, and scattered disk have also been reached (see *Morbidelli and Brown*, 2004; *Dones et al.*, 2004). Based on the most realistic simulations available so far, it seems that the inner and classical parts of the Oort cloud should be about equal in mass and number of comets, and that the inner Oort cloud should be only slightly "flattened," i.e., show only a slight preference for prograde over retrograde orbits (*Dones et al.*, 2004). However, effects that are likely significant have been neglected, so the picture is not yet complete.

Thus a comprehensive study of Oort cloud formation in combination with giant planet accretion, and with the young solar system placed in a dense environment of surrounding stars, remains to be performed. In parallel, understanding the origin of the Kuiper belt and scattered disk is also presenting a growing challenge. Recent discoveries or suggestions of a mixed Kuiper belt structure involving dynamically hot and cold components with different size and color distributions, with an outer edge at a ~ 50 AU, are prompting investigations of more complex formation scenarios than used by the early models (*Morbidelli and Brown*, 2004).

There may still be a long way to go, and some of the observational data — e.g., on size distributions and frequency of occurrence of binaries — may need to be extended. But the outcome in terms of a more solid picture of how the transneptunian structures formed along with the giant planets is eagerly awaited, to say the least!

3. NONGRAVITATIONAL EFFECTS

Among recent progress in the treatment of nongravitational effects, let us first mention the self-consistent treatment of such effects when calculating both the osculating orbit near perihelion and the original orbit before entry into the planetary system (see *Yeomans et al.,* 2004). It has been found that the original reciprocal semimajor axes $(1/a)_{orig}$ thus derived may differ quite significantly from the corresponding quantities derived assuming purely gravitational motion, and thus the negative reciprocals found for a small set of long-period comets may be mostly explained away by this effect (*Królikowska,* 2001). Important as that may be, it is equally important to realize that the comets with positive $(1/a)_{orig}$ are subject to nongravitational forces as well. Neglecting this, one may significantly affect the width of the Oort peak (*Królikowska,* 2001) and the apparent "inner edge" of the infeed of new comets. Hence, the apparent discrepancy between the inner edge at $a \simeq 20,000$ AU and the theoretically expected one at 28,000 AU (*Levison et al.,* 2001) may not be real, or the problem may be aggravated — further research is necessary.

A related problem, also of importance, is what influence the nongravitational effects may have on the capture of Oort cloud comets into short-period, Halley-type orbits. This concerns comets with small perihelion distances q — typically, $q \lesssim 2.5$ AU. From the fact that the typical, indirect jovian perturbation of !/a for such a comet (*Rickman et al.,* 2001) is much larger than the typical nongravitational perturbation, one would expect the latter to contribute very little. But there is an important difference between the random-walk nature of the jovian perturbations and the systematic progression of the nongravitational effect. Imagine that at least half the comets experience a decrease of the semimajor axis — for instance, due to an excess of outgassing on the preperihelion branch of the orbit. In the absence of a nongravitational effect the comets start their random walk in 1/a dangerously close to the parabolic limit, and it is a well-known fact that this is a severely limiting factor for the capture efficiency (*Everhart,* 1972). Even a slight tendency to walk away from this ejection limit might then have an important consequence by delaying the ejections and thus increasing the chances for a decisive capture event by direct gravitational interaction with Jupiter.

This possibility remains to be investigated. Let us only add that, should it be the case that nongravitational effects may indeed influence the capture of Oort cloud comets significantly, there will also be a dependence on physical properties. For instance, comets with small nuclei will be preferentially affected.

The progress that has been achieved in detailed characterization of the nongravitational effects is likely to continue in the future. However, real breakthroughs may require the advent of new telescopes and instrumentation. When 30–100-m telescopes come on line, imaging of the innermost coma regions should, by resolving the dust jet structures, allow the measurement of much more accurate positions of the nuclei than previously possible. Likewise, this will yield additional information on the main direction of outflow, thus further helping to constrain the nongravitational force models.

When it comes to estimating nuclear masses, a main reason for optimism is the prospect of measuring the nongravitational precession of the perihelia with considerable accuracy (see *Davidsson and Gutiérrez,* 2003). In the framework of the standard model these are expressed by the parameter A_1. This parameter has a significant advantage over A_2 (which similarly measures the delay of perihelion passage) in that the effect, to first order, does not depend on the perihelion asymmetry of the gas production curve but only on the total amount of gas produced from the nucleus. Hence, facing uncertainties over the actual amount of this perihelion asymmetry, it is easier to interpret the A_1 effect than the A_2 effect.

4. CONCLUSIONS

As in the past, cometary dynamics continues to tackle problems of fundamental significance for understanding the formation and evolution of the solar system. We have highlighted several examples of further progress to be expected, as both numerical simulations and observations continue to improve. This will involve sharpening our picture of how the giant planets formed and the transneptunian disk and Oort cloud were shaped. It will also allow a better understanding of how comets formed and evolved into their present orbits, and — hopefully — a more solid framework for interpreting the host of physical and chemical data already obtained and likely forthcoming with *Rosetta* and other space missions.

REFERENCES

Choi Y.-J., Cohen M., Merk R., and Prialnik D. (2002) Long-term evolution of objects in the Kuiper belt zone — Effects of insolation and radiogenic heating. *Icarus, 160,* 300–312.

Davidsson B. J. R. and Gutiérrez P. J. (2003) Estimating the nucleus density of comet 19P/Borrelly. *Icarus,* in press.

Dones L., Weissman P. R., Levison H. F., and Duncan M. J. (2004) Oort cloud formation and dynamics. In *Comets II* (M. C. Festou et al., eds.), this volume. Univ. of Arizona, Tucson.

Duncan M., Levison H., and Dones L. (2004) Dynamical evolution of ecliptic comets. In *Comets II* (M. C. Festou et al., eds.), this volume. Univ. of Arizona, Tucson.

Everhart E. (1972) The origin of short-period comets. *Astrophys. Lett., 10,* 131–135.

Farinella P. and Davis D. R. (1996) Short period comets: Primordial bodies or collisional fragments? *Science, 273,* 938–941.

Jewitt D. C. (2004) From cradle to grave: The rise and demise of the comets. In *Comets II* (M. C. Festou et al., eds.), this volume. Univ. of Arizona, Tucson.

Królikowska M. (2001) A study of the original orbits of "hyperbolic" comets. *Astron. Astrophys., 376,* 316–324.

Levison H. F., Dones L., and Duncan M. J. (2001) The origin of Halley type comets: Probing the inner Oort cloud. *Astron. J., 121,* 2253–2267.

Morbidelli A. and Brown M. E. (2004) The Kuiper belt and the primordial evolution of the solar system. In *Comets II* (M. C. Festou et al., eds.), this volume. Univ. of Arizona, Tucson.

Rickman H., Valsecchi G. B., and Froeschlé Cl. (2001) From the Oort cloud to observable short-period comets. I. The initial stage of cometary capture. *Mon. Not. R. Astron. Soc., 325,* 1303–1311.

Yeomans D. K., Chodas P. W., Szutowicz S., Sitarski G., and Królikowska M. (2004) Cometary orbit determination and nongravitational forces. In *Comets II* (M. C. Festou et al., eds.), this volume. Univ. of Arizona, Tucson.

Part IV:
The Nucleus

In Situ Observations of Cometary Nuclei

H. U. Keller
Max-Planck-Institut für Aeronomie

D. Britt
University of Central Florida

B. J. Buratti
Jet Propulsion Laboratory

N. Thomas
Max-Planck-Institut für Aeronomie
(now at University of Bern)

It is only through close spacecraft encounters that cometary nuclei can be resolved and their properties determined with complete confidence. At the time of writing, only two nuclei (those of Comets 1P/Halley and 19P/Borrelly) have been observed, both by rapid flyby missions. The camera systems onboard these missions have revealed single, solid, dark, lumpy, and elongated nuclei. The infrared systems gave surface temperatures well above the free sublimation temperature of water ice and close to blackbody temperatures. The observed nuclei were much more similar than they were different. In both cases, significant topography was evident, possibly reflecting the objects' sublimation histories. Dust emission was restricted to active regions and jets in the inner comae were prevalent. Active regions may have been slightly brighter than inert areas but the reflectance was still very low. No activity from the nightside was found. In this chapter, the observations are presented and comparisons are made between Comets Halley and Borrelly. A paradigm for the structure of cometary nuclei is also described that implies that the nonvolatile component defines the characteristics of nuclei and that high porosity, large-scale inhomogeneity, and moderate tensile strength are common features.

1. INTRODUCTION

The nuclei of most comets are too small to be resolved by Earth-based telescopes. Even on the rare occasions when a large, possibly resolvable, long-period comet, such as Comet Hale-Bopp (C/1995 O1), enters the inner solar system, the nucleus is obscured from view by the dust coma. Hence, the only means of studying the details of a cometary nucleus is by using interplanetary space probes.

Spacecraft passages to within 10,000 km allow many different techniques to diagnose the properties of the nucleus. The most obvious is high-resolution imaging. However, there are several other remote sensing techniques that could give important information. Only visible and infrared spectroscopy have been used, giving estimates of the surface temperatures of both Comet 1P/Halley and Comet 19P/Borrelly. Indirect measurements through analysis of the volatile (gas) and nonvolatile (dust) components *in situ*, for example, are also vital to our understanding of cometary nuclei since they give information on the composition and structure of the nucleus. Ion mass spectrometers have been particularly useful in this respect for our current understanding.

It is important to emphasize that prior to the spacecraft encounters with Comet 1P/Halley in 1986, the existence of a nucleus was merely inferred from coarse observations. While the idea that a single, small, solid body was at the center of a comet's activity (*Whipple,* 1950) was widely accepted by the scientific community, it was only with the arrival of the Russian *Vega 1* and 2 and European Space Agency's *Giotto* spacecraft at Comet Halley in 1986 that this could be confirmed and other concepts [e.g., the "sandbank" model of *Lyttleton* (1953)] could finally be rejected. It was to be 15 years before another image of a cometary nucleus would be acquired, when NASA's *Deep Space 1* (*DS1*), a technology development mission, successfully imaged the nucleus of Comet 19P/Borrelly in September 2001. Remarkably, the two nuclei observed by these missions were extremely similar.

Comet Halley is the most prominent member of comets on highly inclined (162.24°) and eccentric (0.967) orbits, which are thought to have been members of the Oort cloud. Its period was 76.0 yr and its perihelion distance 0.587 AU. It is one of the most active short-period comets. Therefore, its appearance during the space age triggered the launch of

TABLE 1.   Cometary flybys.

Spacecraft	Closest Approach Distance (km)	Date and Time of Closest Approach	Flyby Velocity (km s⁻¹)	Heliocentric Distance (AU)	Phase Angle of Approach (degrees)	Best Pixel Scale Obtained (m/pixel)	Comment on Imaging Systems and Data
Vega 1	8890	06.03.1986 07:20:06	79.2	0.792	134		Out of focus (FWHM = 10 pixels)
Vega 2	8030	09.03.1986 07:20:00	76.8	0.834	121		Saturated on nucleus
Giotto	596	14.03.1986 00:03:02	68.4	0.89	107.2	38	Little three-dimensional information because of reset 9 s before closest approach
Deep Space 1	2171	22.09.2001	16.5	1.36	88.0	47	Some stereo information, minimum phase angle 52°, no color

the "Halley Armada," comprising five spacecraft that encountered the comet in spring 1986. The Japanese probes, Suisei and Sakegaki (*Hirao and Itoh,* 1987), did not penetrate the comet's inner coma and did not carry experiments for studying the nucleus. The *Vega 1* and *2* spacecraft made encounters on March 6 and 9, 1986, respectively (Table 1). Just after midnight on March 14, 1986, the *Giotto* spacecraft made its closest approach (596 km). The *Vega* and *Giotto* spacecraft all carried sophisticated remote sensing experiments for determination of the properties of the nucleus, and we discuss those results in turn in section 2. In section 3, we discuss the results on the nucleus of Comet 10P/Borrelly obtained from the *DS1* mission.

2.   COMET HALLEY'S NUCLEUS

2.1.   *Vega* Observations

The *Vega* cameras imaged the nucleus throughout their encounters with Comet Halley. However, both imaging systems experienced severe problems. The *Vega 1* television system (TVS) system was out of focus. The point-spread function (PSF) of the instrument was subsequently found to be at least 10 pixels full-width half-maximum (FWHM), caused by a displacement of the detector with respect to the focal plane of about 0.5 mm (*Abergel and Bertaux,* 1995). This effect degraded the effective resolution from around 150 m (at closest approach) to 1.5 km at best and made the images appear extremely fuzzy. While the nucleus was resolved and observed over a range of phase angles, a huge amount of work (e.g., *Merényi et al.,* 1990) had to be invested to correct the images for the degraded PSF and to derive the basic shape of the nucleus. This has proven to be important since the orientation of Comet Halley's nucleus at the time of the *Vega 1* encounter provides a strict constraint on models of the rotational state of the comet (*Belton,* 1990) (see below).

The *Vega 2* TVS also experienced problems. The images are saturated on the nucleus and the data of the coma that were returned are limited in dynamic range to effectively only 5 bits (32 gray levels) digital resolution. While these data are of little interest for direct studies of the nucleus surface, they do provide some information on the near-nucleus jet structures of Comet Halley (Fig. 1). In particular, the data indicate a sunward "fan" of dust emission (*Larson*

Fig. 1.   The *Vega 2* image (#1690) from the *Vega* atlas of Comet 1P/Halley (*Szegö et al.,* 1995) is cleaned and geometrically corrected. It shows the nucleus and its vicinity from a distance of 8030 km (near closest approach) at a phase angle of 28° on March 9, 1986.

et al., 1987) that may have originated from a quasi-linear "crack" in the surface (see also *Szegö et al.,* 1995). The data therefore suggest that activity is restricted to "active regions" (confirmed by *Giotto* images). The lack of structure in the comae of some short-period comets [e.g., Comet 4P/Faye (*Lamy et al.,* 1996)] may indicate a more homogeneously active surface may be appropriate for some nuclei, but *Vega, Giotto,* and *DS1* images clearly show that this is not the case for Comets Halley and Borrelly.

Further support for limited areas of activity comes from the *Vega* IKS experiment, which determined the surface temperature of Comet Halley's nucleus. Temperatures in excess of 350 K were recorded (*Emerich et al.,* 1987a,b). A surface undergoing free sublimation of water ice (at 1 AU heliocentric distance) can only reach an equilibrium temperature of about 220 K even for low albedo values. The IKS measurement indicates that significant parts of Comet Halley's surface were inactive. Care in the interpretation is necessary, however, because micrometer-sized dust particles from the surface rise in temperature rapidly once ejected. If the optical depth, τ, of these particles approaches 1, the effect is to mask the thermal emission from the (possibly) lower temperature surface.

Indeed, the first impressions of the *Vega* TVS data suggested $\tau \approx 1$ (based, to some extent, on the fuzziness of the pictures now attributed to defocusing). This temperature measurement has not been rediscussed subsequently, but support for high surface temperatures came from the *DS1* flyby (*Soderblom et al.,* 2002) where optical depth was not a problem because the activity of Comet Borrelly was more than one order of magnitude lower than that of Comet Halley (at the encounters). These high temperatures confirm that activity is restricted and most of the surface of both comets is inert.

After the *Vega* encounters the size and shape of the nucleus of Comet Halley was not obvious. It was not even clear that there was only a single nucleus. False color images artificially cropped at certain isophote levels [e.g., cover images of *Sagdeev* (1988) and *Szegö et al.* (1995)] are strongly misleading if interpreted as showing the nucleus. It took the observations of the subsequent *Giotto* flyby to provide the basic information for the interpretation of the *Vega* images.

2.2. *Giotto* Observations

2.2.1. *Imaging by the Halley Multicolour Camera.*
More than 2000 images of Comet Halley's coma and nucleus were acquired by the Halley Multicolour Camera (HMC) onboard the *Giotto* spacecraft. A detailed description of the observations and results is given by *Keller et al.* (1995). During approach, the phase angle was 107° and changed only slightly up to the last good image taken from a distance of 2000 km with a resolution of 45 m/pixel. The *Giotto* spacecraft was then hit by large dust particles and lost contact to ground. A representative sample of images is shown in Fig. 2.

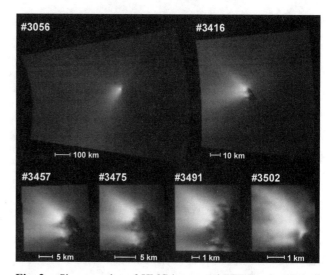

Fig. 2. Six examples of HMC images of P/Halley in original frame sizes. Image #3056 was taken 1814 s (distance to nucleus 124,000 km) and image #3502 was taken 31 s (2200 km) before closest approach.

2.2.2. *Bulk properties of the nucleus.*
In the HMC images only ~25% of the surface area accessible to the camera is illuminated by the Sun, owing to the large phase angle. Fortuitously, the outline of the dark limb is visible against the illuminated dust in the background. This unique circumstance provided a good enough constraint that the fuzzy *Vega* images taken from different solar and rotational phase angles could be interpreted and the bulk properties (volume) of the nucleus could be determined. The maximum length of the nucleus from *Vega* images was 15.3 km. A comparison with the length seen by HMC (14.2 ± 0.3 km) requires the long axis of the nucleus to be 22° above or below the image plane. A major effort of the *Vega* team went into defining the orientation and illuminated outline of the nucleus for both of the flybys (*Merényi et al.,* 1990; *Stooke and Abergel,* 1991). Additional constraints come from the period(s) of the cometary brightness fluctuations derived from Earth-based observations (see also section 2.2.5). Various solutions of the rotation axis and period(s) were suggested [more elaborate interpretations come from *Belton et al.* (1991) and *Szegö et al.* (1995)], but none satisfies all the constraints. The solution by *Belton et al.* (1991) needs the "thick and the thin" ends of the nucleus on the *Vega 1* images interchanged from the orientation derived by the *Vega* team. In addition, the distribution of active areas on the nucleus surface does not satisfy the constraints derived from HMC images. A best fit triaxial ellipsoid with 7.2, 7.22, and 15.3 km for the axes was derived by *Merényi et al.* (1990) with an estimated error of 0.5 km in each figure. Taking into account the deviations from this ellipsoid (with a volume of 420 km³) led to an estimated volume of 365 km³ and an overall surface of 294 km² for Comet Halley's nucleus. Combining this volume with a mass estimate of $1-3 \times 10^{14}$ kg (*Rickman,* 1989) determined from nongravitational forces yields a density of the nucleus of 550 ± 250 kg m⁻³. Other estimates of the

density yield a wider range, not excluding the "intuitive" value of 1000 kg m^{-3} (*Sagdeev et al., 1988; Peale, 1989*).

A surprisingly (at that time) low geometric albedo of $0.04^{+0.02}_{-0.01}$ of the nucleus was derived from *Vega* images (*Sagdeev et al., 1986*) assuming a Moon-like phase function. Similar values are found for other comets from groundbased visible and IR observations (*Keller and Jorda, 2002; Lamy et al., 2004*). Comets are among the darkest objects of the solar system. This albedo of 0.04, measured directly for the first time, has been widely used as a canonical value to determine sizes from photometric observations. The reflectivity of the illuminated surface derived from HMC images for a phase angle of 107° was found to be less than 0.6% (*Keller et al., 1986*). Fitting the observed reflectivity (I/F) across the illuminated surface seen in the HMC images confirmed the Moon-like phase function and yielded a reflectivity at zero phase angle between 0.05 and 0.08, in reasonable agreement with estimates of the peak reflectivity (*Keller et al., 1995; Thomas and Keller, 1989*).

The color of the nucleus was found to be slightly reddish with a gradient of 6 (±3)% per 100 nm between 440 nm and 810 nm (*Thomas and Keller, 1989*), similar to P-type asteroids. The variation of the reflectivity over the visible surface was moderate (*Keller, 1989*), somewhat in contrast to the results of the more detailed observations of the nucleus of Comet Borrelly (see section 3.1). No "icy" patches could be seen, either on HMC or on *Vega* images. Active areas may possibly be slightly brighter than their surroundings, but the increased dust density could confuse the issue. Pure water ice on the surface can be ruled out. However, small contaminations of the ice with carbon suffice to reduce the reflectivity to the observed low value.

2.2.3. Topography and morphology. Two spheres, a larger one on the south end (Fig. 3), connected to each other creating a "waist" would be a higher mode approximation than the ellipsoid. The near 2:1 elongation of the nucleus is rather typical for cometary nuclei (*Keller and Jorda, 2002; Lamy et al., 2004*). Prominent large-scale features are

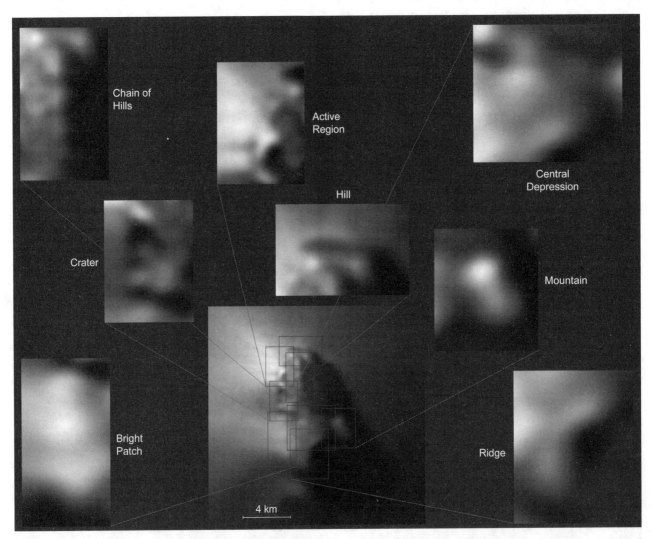

Fig. 3. Features on the surface of the nucleus of Comet 1P/Halley. Sections of the composite image (center bottom) are extracted and expanded by a factor of 3 to show, in detail, notable features on the nucleus mentioned in the text. Nonlinear enhancement has been applied to provide improved contrast. From *Keller et al.* (1988).

the northeastern limb that follows a straight line and terminates in an almost rectangular corner (*duck tail*) that protrudes by $\Delta R/R = 0.3$ above the radius of the best-fit ellipsoid. The terminator on the south (morning) side of the *central depression* paralleled by a bright band (*ridge*) indicates a large-scale feature such as a terrace. The central depression tapers toward the *mountain*. Its illuminated tip lies about 900 m above the best-fit ellipsoid. For more details, see *Keller et al.* (1995).

The roughness of the nucleus is visible down to the resolution limit of the HMC observations (45 m/pixel). The *chain of hills* are an example for the typical scalelength of 0.5–1 km; others are the structures inside the *crater*. It covers a projected area of 12 km² and its depth was estimated to be only 200 m (*Schwarz et al.*, 1986). Most topographic features may be shallow because of the large solar zenith angle (long shadows) during the HMC observations. The strongly irregular shape, the protrusions, the topographic features, the high porosity and low gravity of the nucleus, and the predominance of nonvolatile material all suggest that the surface morphology is characterized by roughness down to small scales (*Kührt et al.*, 1997).

2.2.4. Activity.

2.2.4.1. Overall activity: Activity characterized by dust jets or cones can be directly observed at the illuminated area just below the northern tip of the nucleus around the subsolar point and in direction roughly toward the Sun. In this region the maximum brightness of the images is observed just above the limb. The strongest jet, however, does not originate from this location. At radial distances from the surface lager than the radius of the nucleus, the maximum of the dust column density shifts to an azimuth about 40° south of the projected comet-Sun direction, the direction of the strongest dust jet. Dust emission into this direction is about three times stronger than the subsolar jet and dominates the overall shape of the dust coma (compare the images of Fig. 2). This strong jet originates from the illuminated hemisphere turned away from the observer (*Thomas and Keller*, 1988). A third rather weak jet is directed (in projection) about 90° off the comet-Sun direction toward the north. The overall shape of the dust isophotes can be well modeled by the superposition of these three jets with a cone width of ~40° (FWHM). *Belton et al.* (1991) identified five jets from groundbased, *Vega*, and *Giotto* images. The position of their main jet, however, is not in agreement with the HMC observations.

2.2.4.2. Structures and filaments and topography: The extent of the visible active area covers about 3 km along the bright limb. Here the highest-resolution images taken show structures on the surface and in the dust jet above it. Narrow filamentary structures can be discerned starting at the surface with footprints about 500 m in diameter. Some of these filaments can be followed out to more than 100 km (*Thomas and Keller*, 1987a,b). Overall, more than 15 narrow jets and filaments, some strongly collimated with opening angles of a few degrees, were revealed by image processing (Fig. 4). Some of the filament directions cross each

Fig. 4. The directions of "filaments" seen in the dust emission. The filaments are small inhomogeneities (500 m in diameter at their source). This fine structure in the emission would have been far too faint to be seen by simultaneous groundbased observers. The filaments appear to criss-cross each other.

other, obviously due to the influence of topography (*Thomas et al.*, 1988; *Huebner et al.*, 1988). A few filaments point away from the Sun, emerging behind the dark limb. They probably originate from the small insolated sliver of the nucleus apparently pointing, in projection, in the antisolar direction. No indications of activity on the nightside of the nucleus were found.

The interaction of jet features is also evidenced by the curved dark area in the dust just in front of the lower end of the *bright patch*. A three-dimensional gas dynamics calculation confirms this interaction, mainly caused by the concavity (Fig. 3) at the waist (*Crifo and Rodionov*, 1999). In a series of papers, *Crifo et al.* (2002) and *Rodionov et al.* (2002) have modeled the observations based on the assumption of a uniformly active homogenous nuclear surface rather than on limited active areas within a predominantly inert surface. They assume that the dust production is proportional to the insolation. This leads to a strong concentration of the dust production (and density) toward the Sun-comet line and in the sunward hemisphere. The ratio of the integrated dust of the sunward to that of the antisunward hemisphere would then be about an order of magnitude larger than observed (section 3.2) (see *Keller et al.*, 1995). While Comet Halley was active enough that the hypothesis of a homogenous surface activity could be justified and tested, the jets observed during the flyby of Comet Borrelly (see section 3.3) obviously cannot be explained by a homogenously active surface.

The limited width (FWHM ≈ 40°) of the major jets suggests that they do not originate from flat or convex active areas on an otherwise inert surface. Shallow indentations, however, like the *crater* or larger concave topography (like the *bright patch*) suffice to collimate the dust (*Keller et al.,* 1992). A crude axisymmetric gas-dynamics model that describes the acceleration of dust particles (typically 10 μm) from the surface (*Knollenberg,* 1994) was used to simulate these jets. The narrow fine filaments with opening angles of a few degrees cannot be explained by cavities that would be too deep for the Sun to reach their bottoms. Rather than by an enhancement of activity, the filaments can be formed by reduction of activity in the center of an active area simulating strong (axisymmetric) interactions of shock fronts (*Keller et al.,* 1995).

2.2.5. Nucleus rotation. Shortly after the encounters with Comet Halley, the rotation period of the nucleus was derived by comparing the various images during the three flybys. In a first-order approach, a stable rotation around the axis of maximum inertia (perpendicular to the long axis) was assumed (*Wilhelm,* 1987; *Sagdeev et al.,* 1989). Fits were found for a period slightly above 50 h (2.2 d). Ground-based observations of the coma brightness variations yielded a period of about 7 d, but dynamical features (jets, shells) were in agreement with the 2.2-d periodicity. It is now widely assumed that the spin state of Comet Halley is excited, i.e., that the rotation is not in its energetic minimum and includes nutation (*Sagdeev et al.,* 1989; *Samarasinha and A'Hearn,* 1991; *Belton et al.,* 1991). There is no common understanding of the details (*Keller and Jorda,* 2002). Three flybys and a long series of groundbased observations were not sufficient to pin down the rotational parameters.

3. COMET BORRELLY'S NUCLEUS

The next spacecraft close encounter with a comet occurred on September 22, 2001 when the *DS1* spacecraft flew by the Comet 19P/Borrelly. The *DS1* mission was primarily an engineering test of a solar-electric ion-propulsion system, but part of the mission goals were to test spacecraft instruments and software with close flybys of small bodies. The primary data for this paper comes from imagery taken by the Miniature Integrated Camera and Spectrometer (MICAS) instrument, which included a 1024 × 1024 frame-transfer CCD. The mission plan was to fly by Comet Borrelly with a miss distance of approximately 2000 km on the sunward side at a relative speed of 16.5 km s⁻¹ (*Soderblom et al.,* 2002). Because MICAS was in a fixed orientation on the spacecraft, the whole spacecraft was rotated to keep Comet Borrelly in the field of view.

During the 90 minutes before closest approach, 52 visible wavelength images were taken with MICAS at solar phase angles between 88° and 52°. Shown in Fig. 5 is the highest-resolution image, taken at 3556 km from the comet, with a resolution of 47 m/pixel. Because Comet Borrelly has a long rotation period of 25 ± 0.5 h, *DS1* saw essentially only the illuminated part of one hemisphere of the comet. The full shape and volume of the nucleus could not be revealed. Matching images taken between solar phase angles of about 60° and 52° provides stereo pairs and re-

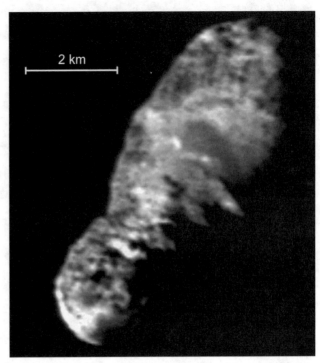

Fig. 5. The highest-resolution image of the Comet 19/P Borrelly. This image was taken at 3556 km from the comet and has a scale of 47 m/pixel.

veals coarse topographic information. Based on this digital terrain model, morphological and photometric information of the surfaces of various terrains can be derived. These more detailed analyses and interpretations have recently been published in a special volume of *Icarus* devoted mainly to the *DS1* flyby of Comet Borrelly. Here we provide a summary of those results.

Comet 19P/Borrelly is a Jupiter-family comet with an orbital period of 6.86 yr, a semimajor axis of 3.61 AU, an inclination of 30.24°, and a perihelion distance of 1.359 AU. The flyby occurred eight days after perihelion while the comet was crossing the ecliptic. *DS1* imagery showed an 8.0 ± 0.1 km × 3.15 ± 0.08 km object shaped like a left footprint with a heel at the bottom of Fig. 5 and the sole toward the top. The rotation axis derived from Earth-based observations corresponds to the short axis exiting near the central mesa (see section 3.2). The pole obliquity and orbital longitude are 102.7° ± 0.5° and 146° ± 1°, corresponding to RA = 214.01° and DEC = –5.07° (*Schleicher et al.,* 2003). The pole is pointed sunward, with a subsolar latitude of ~60° during the encounter (*Soderblom et al.,* 2002). Comet Borrelly has an average disk integrated geometric albedo of 0.029 ± 0.006 (*Buratti et al.,* 2004), even slightly lower than the value measured for Comet Halley (see section 2.2.2). The albedo values of Comet Halley and Comet Borrelly are comparable to those of other dark bodies in the solar system, including the low-albedo regions of Iapetus (*Buratti and Mosher,* 1995), the uranian rings (*Ockert et al.,* 1987), and the lowest-albedo C-type asteroids (*Tedesco et al.,* 1989), including several in comet-like orbits (*Fernández et al.,* 2001).

3.1. Disk-Integrated Photometry

The *DS1* encounter with Comet Borrelly enabled the first photometric modeling of a cometary nucleus. Physical attributes of the surface of the nucleus, including the compaction state of the optically active portion of the regolith and the macroscopic roughness, can be derived by fitting photometric models to the observed brightness as a function of viewing geometry. The nucleus must be observed over a range of solar phase angles to perform this type of analysis.

Figure 6 shows a disk-integrated solar phase curve of Comet Borrelly created from Earth-based observations (*Lamy et al.,* 1998; *Rauer et al.,* 1999) and spacecraft measurements, along with a disk-integrated fit to Hapke's photometric model (*Hapke,* 1981, 1984, 1986). The derived single scattering albedo w = 0.020 and the asymmetry of the phase function g = −0.45 led to an opposition surge amplitude of B_0 = 1.0 and low compaction indicated by the parameter h = 0.0084. The mean slope angle of 20 relates to surface roughness on scales ranging from clumps of particles to mountains.

Comet Borrelly's nucleus has surface physical properties similar to those of C-type asteroids (*Helfenstein and Veverka,* 1989). The single-scattering albedo is lower than measured for any other body. The phase integral derived from the data in Fig. 6 is 0.27 ± 0.01, to yield a Bond albedo of 0.009 ± 0.02, again the lowest of any object in the solar system so far measured. These values, however, depend critically on the few Earth-based measurements.

The extremely low albedo values require high microporosity of the surface that traps the light very efficiently. Appropriate modeling will have to show whether these values can be reached with realistic physical properties.

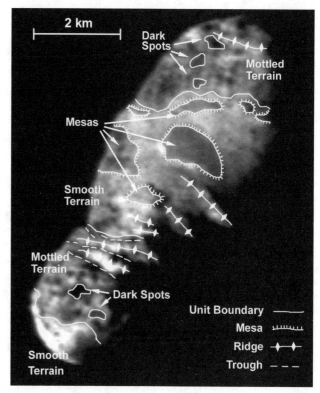

Fig. 7. Unit and feature map of Borrelly based on analysis of the MICAS imagery, including stereo pairs.

3.2. Surface Morphology

Comet Borrelly has a complex surface with a range of morphological features. Interpretations of these features based on analysis of the MICAS imagery, including stereo pairs, are shown in Fig. 7. Four major morphological units can be discerned — dark spots, mottled terrain, mesas, and bright terrain — and two surface features — ridges and fractures. One of the most interesting (but expected) results is the absence of impact craters, commonly associated with small bodies. The upper limit for their diameters is 200 m (*Soderblom et al.,* 2002). While there are a number quasicircular depressions visible, they are most abundant in the mottled terrain and have roughly similar diameters and sometimes regular spacing (cf. the chain of hills on the surface of Comet Halley). Our analysis suggests that they may be sublimation features. Using a simple shape model, a Lommel-Seeliger photometric function, and the phase curve illustrated in Fig. 6, a map of normal reflectances (*Buratti et al.,* 2004) illustrates variegations up to a factor of almost four in albedo (from 0.012 to 0.045) that are correlated with geologic terrains and features. For low-albedo objects such as Comet Borrelly, normal reflectance and geometric albedo are equivalent.

Dark spots: These are the darkest areas on the comet, with a geometric albedo around 0.015. Photometric profiles of the dark spots confirm that they are not shadowed and have photometric properties similar to the mottled terrain (*Nelson et al.,* 2004; *Buratti et al.,* 2004). Dark spots appear to overlie the mottled terrain and hence are the strati-

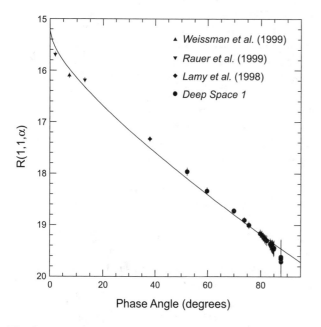

Fig. 6. The disk-integrated solar phase curve of Borrelly created from groundbased, spacebased (HST), and spacecraft (*DS1*) observations.

graphically highest unit on Comet Borrelly. They probably represent the oldest surface lags.

Mottled terrain: The mottled terrain is stratigraphically below the dark spots and consists of areas rough at pixel resolution with depressions, troughs, hills, and ridges. The terrain is dominated by a mixture of quasicircular depressions and low hills. The quasi-circular depressions are about 200–300 m in diameter and are most common on the heel portion of the comet. Low hills tend to be roughly aligned along the long axis of the comet and spaced approximately 300–400 m apart (cf. the chain of hills on Comet Halley; see section 2.2.3). The morphology and albedo variations suggest that the mottled terrain represents older surface lag deposits that have been subjected to extensive sublimation erosion leading to terrain softening and collapse (*Britt et al.,* 2004).

Mesas: Mesas consist of several areas of steep, bright-appearing slopes surrounding darker, flat tops. These features are primarily in the central portion of the comet and appear, along with the smooth terrain, to be associated with some of the active jets. Mesa formation is probably driven, like terrestrial mesas, by erosion (sublimation) on the steep slopes. The mesa slopes are probably one of the most freshly exposed areas on the comet and may be a source of significant gas/dust loss.

Smooth terrain: Photometric analysis suggests that this unit is slightly rougher than average at subpixel scales with geometric albedos of typically 0.032 and in some spots as high as 0.045 (*Buratti et al.,* 2004). The fine pattern of albedo variegations may indicate areas of differential activity and/or surface age as part of the resurfacing processes from dust ejection.

Ridges and fractures: Digital terrain models indicate that the area of the heel is canted about 15° relative to the sole (*Soderblom et al.,* 2002; *Oberst et al.,* 2004). Most of the ridges and fractures are associated with the boundary of this canted area. These ridges of 1–2 km in visible length are oriented normal to the long axis of the comet and could indicate compressional shortening (*Britt et al.,* 2004). If this interpretation is correct, the features require some tensile strength of the nucleus.

3.3. Jets and Active Areas

DS1 observed dust and gas activity including collimated jets and fans (*Soderblom et al.,* 2002). The largest central jet, called the α jet, is a dusty beam a few kilometers wide at the comet (cf. active areas of Comet Halley; see section 2.2.4.1), extending out to at least 100 km and canted 30° from the comet-Sun line (Fig. 8). This feature appears to emanate from the broad central area of the comet, which includes the mesas and the smooth terrain. There are several smaller parallel jets, called the β jets, which are about 200–400 m at the base [cf. filaments of Comet Halley (section 2.2.4.2)], about 4–6 km in length, and canted about 15° from the direction of the α jet. The fan feature is diffuse dust apparently emanating from the smooth terrain unit at the end of Comet Borrelly's heel and oriented roughly along the Sun-comet line. About 35% of the comet's dust produc-

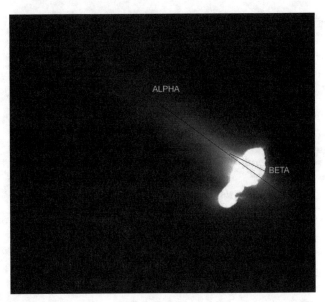

Fig. 8. The dominant dust emission on the sunward side is divided into α and β jets. The α jet is aligned at the core of the main jet. The β jet shown here is one of several roughly parallel smaller collimated jets. The range of this image is about 4825 km.

tion is accounted for by the jets, about 20% is from the fan, and about 15% is in other fans. About 30% of the dust appears above the nightside hemisphere (*Boice et al.,* 2002). This may well be material emitted from the dayside hemisphere, appearing on the nightside due to projection effects. The comet's active area has been estimated at approximately 8% of the total surface (*Boice,* 2002).

Hubble Space Telescope (HST) observations estimated the water production rate to $3.0 \pm 0.6 \times 10^{28}$ s^{-1} at the time of encounter, or about 600 kg s^{-1} (*Weaver et al.,* 2003). Integrated over Comet Borrelly's orbit the water mass loss per apparition would be approximately 2×10^{10} kg. Adding a similar amount of dust suggests an average erosion of the total surface between 0.5 and 1 m per apparition based on a density of the nucleus ≤ 1.000 kg m^{-3}. Water sublimation is strong enough to remove up to 10 m of surface layers at active areas (*Huebner et al.,* 1986). For instance, the mesa slopes could retreat 10–20 m per apparition (*Britt et al.,* 2004). This level of erosion makes active cometary surfaces one of the most dynamic and rapidly changing features in the solar system.

4.   COMPARISONS OF COMETS HALLEY, BORRELLY, AND ASTEROIDS

4.1.   Comets Halley and Borrelly

The *Vega, Giotto,* and *DS1* measurements of cometary nuclei are far more similar than they are different. Both Comets Halley and Borrelly are irregularly shaped, very dark, and active over minor fractions of their surfaces (producing "jets"), and have surface temperatures close to those expected for a blackbody.

The nuclei show surface features inconsistent with a uniformly shrinking ellipsoid (or snowball) and some tensile strength is apparent in protrusions such as the mountain or duck tail of Comet Halley or the canted low end of Comet Borrelly (see section 3.2). It is interesting that observations of structures have been interpreted by the *DS1* team using geological analogs — an approach not adopted by the *Giotto* team, for example. The stereo coverage resulting in a digital terrain model, better solar illumination, and less interference of the smaller dust production provide for a more detailed analysis of the surface features and morphology in the case of Comet Borrelly. Nonetheless, here too the similarities are more striking than the differences. The dark spots and ridges seen on Comet Borrelly (Fig. 7) are very similar to the chain of hills seen at Comet Halley (Fig. 3). The smooth terrain on Comet Borrelly is probably associated with activity and looks rather like the central depression on Comet Halley, which also shows evidence of activity (at least on its sunward extreme). The elevations on Comet Borrelly may be similar to the mountain feature on Comet Halley, while the mottled terrain is reminiscent of the region between the big active area and the chain of hills on Comet Halley.

Active regions appear to have slightly higher albedos than inactive regions, but nowhere does the albedo exceed 0.045 at the resolution of the images (around 50–100 m/pixel). It cannot be completely ruled out that there are small areas within these active regions that have (high) albedos closer to that of pure ice. The inhomogeneity of the surface activity and topography clearly influence the structure of the inner coma, making it difficult to extract information about the nature of the source, the initial acceleration, and particle fragmentation in the flow (*Ho et al.*, 2003). While some preliminary conclusions on the structure of the source region may be possible, neither dataset is really good enough to distinguish between different surface emission models.

The similarities of both cometary nuclei are striking. There is no hint that comets originating from the Oort cloud (Comet Halley) look different from nuclei originating in the Edgeworth-Kuiper belt (Comet Borrelly) [see *Dones et al.* (2004) and *Duncan et al.* (2004) for discussions on these two cometary source regions].

The fortuitous observations of the complete outline of Comet Halley's nucleus, including its unilluminated parts, by HMC and the *Vega 1* and *2* images provide a reasonable shape model and the overall volume of the nucleus. This information is completely missing in the *DS1* data. It took three flybys to model the complex rotation of Comet Halley, even though the quality of the *Vega* images was not good enough to provide sufficient reference points for uncontroversial interpretations. No information on the rotation of Comet Borrelly can be expected from the *DS1* encounter.

4.2. Comet and Asteroid Surfaces

The images of Comets Halley and Borrelly highlight the similarities and differences between comets and other small bodies like asteroids and small moons. Photometric analysis of Comet Borrelly's nucleus suggests that it has a regolith and that its surface looks similar to that of asteroids. However, the processes at work on the two classes of bodies

are very different. Comets Halley and Borrelly are characterized by a lack of impact craters and the existence of complex and sublimation-driven erosional features such as mesas, hills, and mottled terrain. Disk-resolved analysis of Comet Borrelly's roughness and particle phase function suggests that the comet does not get rougher with age, and that regions of the comet are infilled with or mantled by native dust. The surfaces of asteroids are dominated by impact craters. The energy to drive the asteroidal erosion process comes from episodic impact collisions, so this is necessarily a very slow process compared, e.g., to terrestrial erosive processes. Cometary erosion is driven by sublimation of volatiles during the cometary perihelion passage around the Sun. For Jupiter-family comets like Borrelly with frequent perihelion passages, sublimation-driven erosion alters landforms at rates that would be fast even by terrestrial standards. Fundamentally, surfaces of comets are dominated by sublimation while the surfaces of asteroids are dominated by impacts.

4.3. The Nucleus Paradigm

The spacecraft flybys revealed dark, evolved, solid cometary nuclei. The nonvolatile compounds outweigh the ice (*McDonnell et al.*, 1991), quite in contrast to what was considered before the Comet Halley flybys when an upper limit for the dust to gas ratio of 0.3 was used for the engineering model (*Divine*, 1981). Hence, their overall physical properties are better characterized by the nonvolatile component and not by (water) ice (*Keller*, 1987). The extremely low reflectivity argues for a surface of high porosity in accord with the low bulk density of the whole nucleus. The low tensile strength and porosity of the material is also reflected in the frequent fragmentation of the dust particles leaving the nucleus within the gas stream (*Thomas and Keller*, 1990).

The limited areas of activity can hardly be discerned from the generally inert surface, but they do seem to be slightly more reflective. An active region on a Jupiter-family comet with a perihelion distance near 1 AU will lose about 5–10 m depth of surface layer per revolution around the Sun (*Huebner et al.*, 1986). Consequently, the interior (material with volatiles that surfaces in active areas) looks similar to the material of the (inert) surface. The nucleus consists of a porous dust matrix that is in parts enriched with volatiles producing the activity. The inert surface layers are a crust of depleted matrix material that can form large topographic protrusions (the duck tail and mountain on Comet Halley and features such as mesas on Comet Borrelly). The fact that the surface temperature of the inert regions (comprising 80–90% of the surface area) reaches that of a blackbody is then unsurprising. In an active region, any loose mantle of dust (regolith) that might form would be blown away during perihelion passage (*Kührt and Keller*, 1994).

The present nuclei of Comets Halley and Borrelly are the small remnants of frequent splittings and shedding of blocks of material. Estimates of the mass in Halley's associated meteor streams show that its original mass 2000 to 3000 revolutions ago was 5–10 times bigger (*Hughes*, 1985).

The splitting is facilitated by the physical inhomogeneity of the nucleus agglomerated from subnuclei. The observed topography with typical scalelengths from 0.5 to 1 km (chain of hills, mottled terrain, dark spots) to the highly elongated shapes of the nuclei of Comets Halley and Borrelly, with indications of a "waist," reflects their inner structure.

The difficult detection and first characterization of the nucleus of Comet Halley in 1986 provided the fundamental data for our understanding of the nature of comets. It took 15 years to confirm these observations and to extend the conclusions to a second object.

Comet Halley is the most productive short-period comet, but yet only a minor fraction of its surface is active. Typical Jupiter-family comets display activity levels one or two magnitudes less than this, e.g., Comet Borrelly. How is activity at this low level maintained over many revolutions around the Sun? How does activity really work? The physical explanation of this phenomenon is a key to our understanding of comets. The answer will require new missions where the onset and details of activity can be studied in depth.

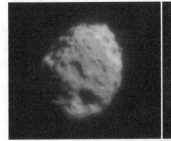

Fig. 9. The nucleus of Comet Wild 2 is shown in this image taken by the *Stardust* nagivation camera during the spacecraft's closest approach to the comet on January 2, 2004. The largest visible dimension is about 5 km. The image was taken within a distance of 500 km of the comet's nucleus. (The image was produced for distribution via the Web, and does not represent the final quality.) From the NASA *Stardust* Web page (Principal Investigator D. Brownlee, University of Washington, Seattle).

5. OUTLOOK

While Earth-based observations are now able to determine the sizes, approximate shapes, and possibly rotational characteristics of cometary nuclei (and hence may provide statistics), the next major step in the study of cometary nuclei will come from future space missions dedicated to their detailed investigation. The first of these will be NASA's Discovery mission, *Stardust*, which will provide information on the nonvolatile composition of the nucleus of Comet 81P/Wild 2. The physical structure of the particles, along with information on their strength, size, and shape, will provide us with some understanding of how structures on the nucleus can form and how the nucleus as a whole came together. The emphasis of this mission is on the sample return. Nevertheless, the navigation camera produced images of the nucleus of unprecedented quality, showing a rough surface with a large number of crater-like features (Fig. 9).

The *Deep Impact* mission will crash a block of copper into Comet 9P/Tempel 1 and study the effects of the impact. This will tell us about the tensile strength and porosity of the nucleus. The impact will also expose pristine material from the interior of the nucleus and we will be able to assess its chemical characteristics with a broad range of analytical instruments onboard the spacecraft. It is a testament to our lack of knowledge of the physical properties of cometary nuclei that many widely different scenarios for the impact are still being considered by the flight team as plausible.

Whatever the mission, the flybys of Comet Halley and of Comet Borrelly clearly show that for the interpretation of surface properties and physical characteristics of the nuclei, high-resolution imaging from different angles and at different phase angles is required. The quality of the in-

formation can be significantly improved with digital terrain models derived from stereoscopic views. The overall size and shape of a cometary nucleus as well as its rotation parameters can only be accessed by multiple flybys or, even better, by a comet rendezvous where the spacecraft orbits the nucleus.

The European Space Agency's *Rosetta* mission is such a rendezvous mission. Originally, the *Rosetta* spacecraft was to be launched in January 2002 to meet with Comet 46P/Wirtanen at 3.5 AU from the Sun. Difficulties with the launcher required a new orientation of the mission. The second launch attempt was successful in March 2004. Its new goal is Comet 67P/Churyumov-Gerasimenko. The *Rosetta* spacecraft will deposit a science package on the surface of the nucleus and then continue to monitor the nucleus right through perihelion. If successful, this ambitious mission would study the nucleus surface down to a resolution of 2 cm/pixel from orbit and provide detailed measurements at even higher resolution from the lander. A strong complement of remote sensing (cameras, spectrometers, tomographic radio experiment) and *in situ* (ion and neutral mass spectrometers, dust analyzer, charge particle analyzer) instruments will observe the nucleus and its activity.

Activity could be monitored, compositional changes followed, surface temperatures tracked, and the internal structure assessed. It is the activity that characterizes comets and leads us to think of them as relics from the formation of the solar system. Hence, the emphasis *Rosetta* places on studying the nuclear activity should be highly rewarded.

In the distant future, the ultimate objective will be a sample return mission that can place strong constraints on the models of cometary origin and formation and, hence, on studies of the evolution of the solar system as a whole.

REFERENCES

Abergel A. and Bertaux J.-L. (1995) Calibration on the ground and in space. In *Images of the Nucleus of Comet Halley, Vol. 2* (R. Reinhard and B. Battrick, eds.), pp. 43–52. ESA SP-1127,

Noordwijk, The Netherlands.

Belton M. J. S. (1990) Rationalization of Halley's periods: Evidence for an inhomogeneous nucleus? *Icarus, 86,* 30–51.

Belton M. J. S., Julian W. H., Anderson A. J., and Mueller B. E. A. (1991) The spin state and homogeneity of Comet Halley's nucleus. *Icarus, 93,* 183–193.

Boice D. C., Britt D. T., Nelson R. M., Sandel B. R., Soderblom L. A., Thomas N., and Yelle R. V. (2002) The near-nucleus environment of 19/P Borrelly during the Deep Space One encounter (abstract). In *Lunar and Planetary Science XXXIII,* Abstract #1810. Lunar and Planetary Institute, Houston (CD-ROM).

Britt D. T, Buratti B. J., Boice D. C., Nelson R. M., Campins H., Thomas N., Soderblom L. A., and Oberst J. (2004) The morphology and surface processes of Comet 19P/Borrelly. *Icarus, 167,* 45–53.

Buratti B. J. and Mosher J. A. (1995) The dark side of Iapetus: Additional evidence for an exogenous origin. *Icarus, 115,* 219–227.

Buratti B. J., Hicks M. D., Oberst J., Soderblom L. A., Britt D. T., and Hillier J. K. (2004) Deep Space 1 photometry of the nucleus of Comet P/19 Borrelly. *Icarus, 167,* 16–29.

Crifo J. F. and Rodionov A. V. (1999) Modelling the circumnuclear coma of comets: Objectives, methods and recent results. *Planet. Space Sci., 47,* 797–826.

Crifo J. F., Rodionov A. V., Szegö K., and Fulle M. (2002) Challenging a paradigm: Do we need active and inactive areas to account for near-nuclear jet activity? *Earth Moon Planets, 90,* 227–238.

Divine N. (1981) Numerical models for Halley dust environments. In *The Comet Halley Dust & Gas Environment* (B. Battrick and E. Swallow, eds.), pp. 25–30. ESA SP-174, Noordwijk, The Netherlands.

Dones L., Weissman P. R., Levison H. F., and Duncan M. J. (2004) Oort cloud formation and dynamics. In *Comets II* (M. C. Festou et al., eds.), this volume. Univ. of Arizona, Tucson.

Duncan M., Levison H., and Dones L. (2004) Dynamical evolution of ecliptic comets. In *Comets II* (M. C. Festou et al., eds.), this volume. Univ. of Arizona, Tucson.

Emerich C. and 12 colleagues (1987a) Temperature of the nucleus of comet Halley. In *Symposium on the Diversity and Similarity of Comets* (E. J. Rolfe and B. Battrick, eds.), pp. 703–706. ESA SP-278, Noordwijk, The Netherlands.

Emerich C. and 11 colleagues (1987b) Temperature and size of the nucleus of comet P/Halley deduced from IKS infrared Vega-1 measurements. *Astron. Astrophys., 187,* 839–842.

Fernández Y. R., Jewitt C. D., and Sheppard S. S. (2001) Low albedos among extinct comet candidates. *Astrophys. J. Lett., 553,* L197–L200.

Hapke B. (1981) Bidirectional reflectance spectroscopy. I. Theory. *J. Geophys. Res., 86,* 3039–3054.

Hapke B. (1984) Bidirectional reflectance spectroscopy. 3. Correction for macroscopic roughness. *Icarus, 59,* 41–59.

Hapke B. (1986) Bidirectional reflectance spectroscopy. 4. The extinction coefficient and the opposition effect. *Icarus, 67,* 264–280.

Helfenstein P. and Veverka J. (1989) Physical characterization of asteroid surfaces from photometric analysis. In *Asteroids II* (R. Binzel et al., eds.), pp. 557–593. Univ. of Arizona, Tucson.

Hirao K. and Itoh T. (1987) The Sakigake/Suisei encounter with comet P/Halley. *Astron. Astrophys., 187,* 39–46.

Ho T.-M., Thomas N., Boice D. C., Köllein C., and Soderblom L. A. (2003) Comparative study of the dust emission of 19P/

Borrelly (Deep Space 1) and 1P/Halley. *Adv. Space Res., 31(12),* 2583–2589.

Huebner W. F., Keller H. U., Wilhelm K., Whipple F. L., Delamere W. A., Reitsema H. J., and Schmidt H. U. (1986) Dust-gas interaction deduced from Halley Multicolour Camera observations. In *20th ESLAB Symposium on the Exploration of Halley's Comet, Vol. 2* (B. Battrick et al., eds.), pp. 363–364. ESA SP-250, Noordwijk, The Netherlands.

Huebner W. F., Boice D. C., Reitsema H. J., Delamere W. A., and Whipple F. L. (1988) A model for intensity profiles of dust jets near the nucleus of Comet Halley. *Icarus, 76,* 78–88.

Hughes D. W. (1985) The size, mass, mass loss and age of Halley's comet. *Mon. Not. R. Astron. Soc., 211,* 103–109.

Keller H. U. (1987) The nucleus of Comet Halley. In *Symposium on the Diversity and Similarity of Comets* (B. Battrick et al., eds.), pp. 447–454. ESA SP-278, Noordwijk, The Netherlands.

Keller H. U. (1989) Comets — Dirty snowballs or icy dirtballs? In *Proceedings of an International Workshop on Physics and Mechanics of Cometary Materials* (J. Hunt and T. D. Guyeme, eds.), pp. 39–45. ESA SP-302, Noordwijk, The Netherlands.

Keller H. U. and Jorda L. (2002) The morphology of cometary nuclei. In *The Century of Space Science* (J. A. M. Bleeker et al., eds.), pp. 1235–1275. Kluwer, Dordrecht.

Keller H. U. and 17 colleagues (1986) First Halley Multicolour Camera imaging results from Giotto. *Nature, 321,* 320–326.

Keller H. U., Kramm R., and Thomas N. (1988) Surface features on the nucleus of comet Halley. *Nature, 331,* 227–231.

Keller H. U., Markiewicz W. J., and Knollenberg J. (1992) Dust emission from cometary craters: Collimation of particle trajectories. *Bull. Am. Astron. Soc., 24(3),* 1018.

Keller H. U., Curdt W., Kramm J. R., and Thomas N. (1995) Images obtained by the Halley Multicolour Camera (HMC) on board the Giotto spacecraft. In *Images of the Nucleus of Comet Halley, Vol. 1* (R. Reinhard et al., eds.), pp. 1–252. ESA SP-1127, Noordwijk, The Netherlands.

Knollenberg J. (1994) Modellrechnungen zur Staubverteilung in der inneren Koma von Kometen unter spezieller Berücksichtigung der HMC-Daten der Giotto-mission. Ph.D. thesis, Georg-August-Universität, Göttingen, Germany.

Kührt E. and Keller H. U. (1994) The formation of cometary surface crusts. *Icarus, 109,* 121–132.

Kührt E., Knollenberg J., and Keller H. U. (1997) Physical risks of landing on a cometary nucleus. *Planet. Space Sci., 45,* 665–680.

Lamy P. L., Toth I., Grün E., Keller H. U., Sekanina Z., and West R. M. (1996) Observations of comet P/Faye 1991 XXI with the Planetary Camera of the Hubble Space Telescope. *Icarus, 119,* 370–384.

Lamy P. L., Toth I., and Weaver H. A. (1998) Hubble Space Telescope observations of the nucleus and inner coma of comet 19P/1904 Y2 (Borrelly). *Astron. Astrophys., 337,* 945–954.

Lamy P. L., Toth I., Fernández Y. R., and Weaver H. A. (2004) The sizes, shapes, albedos, and colors of cometary nuclei. In *Comets II* (M. C. Festou et al., eds.), this volume. Univ. of Arizona, Tucson.

Larson S., Sekanina Z., Levy D., Tapia S., and Senay M. (1987) Comet P/Halley near/nucleus phenomena in 1986. *Astron. Astrophys., 337,* 639–644.

Lyttleton R. A. (1953) *The Comets and Their Origin.* Cambridge Univ., Cambridge. 173 pp.

McDonnell J. A. M., Lamy P. L., and Pankiewicz G. S. (1991) Physical properties of cometary dust. In *Comets in the Post-Halley Era, Vol. 2* (R. L Newburn Jr. et al., eds.), pp. 1043–1073. Kluwer, Dordrecht.

Merényi E., Földy L., Szegö K., Toth I., and Kondor A. (1990) The landscape of comet Halley. *Icarus, 86,* 9–20.

Nelson R. M., Soderblom L. A., and Hapke B. W. (2004) Are the circular, dark features on Comet Borrelly's surface albedo variations or pits? *Icarus, 167,* 37–44.

Oberst J., Giese B., Howington-Kraus E., Kirk R., Soderblom L., Buratti B., Hicks M., Nelson R., and Britt D. (2004) The nucleus of Comet Borrelly: A study of morphology and surface brightness. *Icarus, 167,* 70–79.

Ockert M. E., Cuzzi J. N., Porco C. C., and Johnson T. V. (1987) Uranian ring photometry: Results from Voyager 2. *J. Geophys. Res., 92,* 14969–14978.

Peale S. J. (1989) On the density of Halley's comet. *Icarus, 82,* 36–50.

Rauer H., Hahn G., Harris A., Helbert J., Mottola S., and Oberst J. (1999) Nuclear parameters of comet P/Borrelly. *Bull. Am. Astron. Soc., 31,* 1131.

Rickmann H. (1989) The nucleus of comet Halley: Surface structure, mean density, gas and dust production. *Adv. Space Res., 9(3),* 59–71.

Rodionov A. V., Crifo J.-F., Szegö K., Lagerros J., and Fulle M. (2002) An advanced physical model of cometary activity. *Planet. Space Sci., 50,* 983–1024.

Sagdeev R. Z. (1988) Soviet space science. *Physics Today, 41(5),* 30–38.

Sagdeev R. Z., Avenesov G., Shamis V. A., Ziman Y. L., Krasikov V. A., Tarnopolsky V. A., Kuzmin A. A., Szegö K., Merényi E., and Smith B. A. (1986) TV experiment in Vega mission: Image processing technique and some results. In *20th ESLAB Symposium on the Exploration of Halley's Comet, Vol. 2* (B. Battrick et al., eds.), pp. 295–305. ESA SP-250, Noordwijk, The Netherlands.

Sagdeev R. Z., Elyasberg P. E., and Moroz V. I. (1988) Is the nucleus of comet Halley a low density body? *Nature, 331,* 240–242.

Sagdeev R. Z., Szegö K. Smith B. A., Larson S., Merényi E., Kondor A., and Toth I. (1989) The rotation of comet P/Halley. *Astron. J., 97,* 546–551.

Samarasinha N. H. and A'Hearn M. F. (1991) Observational and dynamical constraints on the rotation of Comet P/Halley. *Icarus, 93,* 194–225.

Schleicher D. G., Woodney L. M., and Millis R. L. (2003) Comet 19P/Borrelly at multiple apparitions: Seasonal variations in gas production and dust morphology. *Icarus, 162,* 415–442.

Schwarz G. and 11 colleagues (1986) Detailed analysis of a surface feature on Comet Halley. In *20th ESLAB Symposium on the Exploration of Halley's Comet, Vol. 2* (B. Battrick et al., eds.), pp. 371–374. ESA SP-250, Noordwijk, The Netherlands.

Soderblom L. A. and 21 colleagues (2002) Observations of Comet 19/P Borrelly by the Miniature Integrated Camera and Spectrometer aboard Deep Space 1. *Science, 296,* 1087–1091.

Stooke P. J. and Abergel A. (1991) Morphology of the nucleus of comet P/Halley. *Astron. Astrophys., 248,* 656–668.

Szegö K. (1995) Discussion of the scientific results derived from the near-nucleus images. In *Images of the Nucleus of Comet Halley, Vol. 2* (R. Reinhard and B. Battrick, eds.), pp. 68–80. ESA SP-1127, Noordwijk, The Netherlands.

Szegö K., Sagdeev F. L., Whipple A., Abergel J. L., Bertaux E., Merényi E., Szalaim S., and Varhalmi L. (1995) Images obtained by the Television System (TVS) on board the Vega spacecraft. In *Images of the Nucleus of Comet Halley, Vol. 2* (R. Reinhard and B. Battrick, eds.), pp. 1–255. ESA SP-1127, Noordwijk, The Netherlands.

Thomas N. and Keller H. U. (1987a) Comet Halley's near-nucleus jet activity. In *Symposium on the Diversity and Similarity of Comets* (E. J. Rolfe and B. Battrick, eds.), pp. 337–342. ESA SP-278, Noordwijk, The Netherlands.

Thomas N. and Keller H. U. (1987b) Fine dust structures in the emission of comet P/Halley observed by the Halley Multicolour Camera on board Giotto. *Astron. Astrophys., 187,* 843–846.

Thomas N. and Keller H. U. (1988) Global distribution of dust in the inner coma of comet P/Halley observed by the Halley Multicolor Camera. In *Dust in the Universe* (M. E. Bailey and D. A. Williams, eds.), pp. 540–541. Cambridge Univ., Cambridge.

Thomas N. and Keller H. U. (1989) The colour of comet P/Halley's nucleus and dust. *Astron. Astrophys., 213,* 487–494.

Thomas N. and Keller H. U. (1990) Interpretation of the inner coma observations of comet P/Halley by the Halley Multicolour Camera. *Ann. Geophys., 8,* 147–166.

Thomas N., Boice D. C., Huebner W. F., and Keller H. U. (1988) Intensity profiles of dust near extended sources on comet Halley. *Nature, 332,* 51–52.

Weaver H. A., Stern S. A., and Parker J. W. (2003) Hubble Space Telescope STIS observations of Comet 19P/Borrelly during the Deep Space 1 encounter. *Astron. J., 126,* 444–451.

Whipple F. L. (1950) A comet model I. The acceleration of Comet Encke. *Astrophys. J., 111,* 375–394.

Wilhelm K. (1987) Rotation and procession of comet Halley. *Nature, 327,* 27–30.

The Sizes, Shapes, Albedos, and Colors of Cometary Nuclei

Philippe L. Lamy
*Laboratoire d'Astronomie Spatiale du
Centre National de la Recherche Scientifique*

Imre Toth
Konkoly Observatory

Yanga R. Fernández
Institute for Astronomy of the University of Hawai'i

Harold A. Weaver
Applied Physics Laboratory of The Johns Hopkins University

We critically review the data on the sizes, shapes, albedos, and colors of cometary nuclei. Reliable sizes have been determined for 65 ecliptic comets (ECs) and 13 nearly isotropic comets (NICs). The effective radii fall in the range 0.2–15 km for the ECs and 1.6–37 km for the NICs. We note that several nuclei recently measured by the Hubble Space Telescope are subkilometer in radius, and that only 5 of the 65 well-measured EC nuclei have effective radii larger than 5 km. We estimate that the cumulative size distribution (CSD) of the ECs obeys a single power law with an exponent $q_S = 1.9 \pm 0.3$ down to a radius of ~1.6 km. Below this value there is an apparent deficiency of nuclei, possibly owing to observational bias and/or mass loss. When augmented by 21 near-Earth objects (NEOs) that are thought to be extinct ECs, the CSD flattens to $q_S = 1.6 \pm 0.2$. The cumulative size distribution of NICs remains ill-defined because of the limited statistical basis compared to ECs. The axial ratios a/b of the measured nuclei of ECs have a median value of ~1.5 and rarely exceed a value of 2, although it must be noted that the observed a/b values are often lower limits because of uncertainties in the aspect angle. The range of rotational periods extends from 5 to 70 h. The lower limit is significantly larger than that of main-belt asteroids and NEOs (~2.2 h, excluding the monolithic fast rotators), and this has implications for the bulk density of cometary nuclei. By combining rotation and shape data when available, we find a lower limit of 0.6 g cm^{-3} for the nucleus bulk density to ensure stability against centrifugal disruption. Cometary nuclei are very dark objects with globally averaged albedos falling within a very restricted range: 0.02–0.06, and possibly even narrower. (B-V), (V-R), and (R-I) color indices indicate that, on average, the color of cometary nuclei is redder than the color of the Sun. There is, however, a large diversity of colors, ranging from slightly blue to very red. While two comets have well-characterized phase functions with a slope of 0.04 mag deg^{-1}, there is evidence for steeper (2P/Encke, 48P/Johnson) and shallower (28P/Neujmin 1) functions, so that the observed range is 0.025–0.06 mag deg^{-1}. The study of the physical properties of cometary nuclei is still in its infancy, with many unresolved issues, but significant progress is expected in the near future from current and new facilities, both ground-based and spaceborne.

1. INTRODUCTION

1.1. Motivation for Studying Cometary Nuclei

There are many reasons why the investigation of cometary nuclei can advance our understanding of the solar system, and this topic is discussed in detail by *Weidenschilling* (2004) and by *Lunine and Gautier* (2004). Briefly, cometary nuclei are the most primitive observable objects remaining from the era of planetary formation. As such, they provide information on the thermophysical conditions of the protoplanetary disk and on the formation mechanism for the icy

planetesimals from which the cores of the outer planets were built. Furthermore, the physical evolution of cometary nuclei over the past 4.6 G.y. must be explained within the context of any unified theory of the solar system, and comparative studies of cometary nuclei and dynamically related bodies [e.g., transneptunian objects and Centaurs (see *Jewitt,* 2004)] should provide insights into the physical and collisional histories of these objects.

Through impacts over the age of the solar system, cometary nuclei have significantly affected the formation and evolution of planetary atmospheres and have provided an important source of volatiles, including water and organic material, to the terrestrial planets. Interest has been building recently in the contribution of cometary nuclei to the Earth impact hazard, which has previously focused mainly on asteroids. Another important motivation for studying cometary nuclei is that their bulk properties may dictate what steps should be taken for hazard mitigation in the event of a predicted collision.

1.2. Origin and Evolution of Cometary Nuclei

Various scenarios have been proposed to explain how cometary nuclei formed from the microscopic grains within the dusty disk of the solar nebula (*Weidenschilling*, 2004). Different formation mechanisms may have been operational at different places within the nebula, and this may have led to diversity in the physical properties of cometary nuclei depending on where they formed. Even if there was a common formation mechanism for all cometary nuclei, diversity could persist because of differences in the physical and chemical conditions at different heliocentric distances (e.g., collisional environment, chemical composition, radiation environment, etc.).

Dynamical arguments support the hypothesis that cometary nuclei originate from at least two different regions of the solar system: The vast majority of the ecliptic comets are thought to be collisional fragments of Kuiper belt objects [the so-called transneptunian objects (see *Duncan et al.*, 2004; *Barucci et al.*, 2004)], while most of the long-period and Halley-type comets probably formed in the vicinity of the giant planets and were subsequently ejected to the Oort cloud where they were stored for most of their lifetimes (*Dones et al.*, 2004).

We follow the classification scheme proposed by *Levison* (1996) and distinguish between ecliptic comets (ECs) and nearly isotropic comets (NICs). This scheme is not profoundly different from the historical tradition, but it has the merit of being based on strict dynamical parameters, namely the Tisserand parameters of comets that are (nearly) constants of motion with respect to Jupiter. ECs have $2 \leq T_J \leq 3$ and are equivalent to the Jupiter-family comets (JFCs), including 2P/Encke (although it is now practically decoupled from Jupiter). ECs in general have orbital periods less than 20 years, hence the quasi-correspondence with the population of short-period comets. NICs have $T_J < 2$ and group together the Halley-type comets (which have a lower limit on their periods, in general, of 20 years) and the long-period comets (old and new). Based on their dynamical histories, the population of NICs is further divided into two subpopulations: (1) dynamically new NICs, which are on their first pass through the inner solar system and typically have semimajor axes, a, greater than $\sim 10^4$ AU, and (2) returning NICs, which have previously passed through the inner solar system and typically have a $\leq 10^4$ AU. The returning NICs are further divided into two subclasses: external returning comets (ERCs) with periods greater than 200 years, and Halley-type comets (HTCs) with orbital periods less than 200 years.

We note that short-period Comets 8P/Tuttle ($P_{orb} = 13.51$ yr, $T_J = 1.623$), 96P/Machholz 1 ($P_{orb} = 5.26$ yr, $T_J = 1.953$), and 126P/IRAS ($P_{orb} = 13.29$ yr, $T_J = 1.987$) are now classified as NICs.

Because of their different origin, the question arises as to whether the two populations (ECs and NICs) have intrinsically different physical properties, or whether they reflect a continuous spectrum of planetesimals in the early solar system, making them more similar than different. The nuclei of ECs suffer significant heating episodes during their frequent passages through the inner solar system, where sublimation processes erode the surface layers, devolatilize the interior, and possibly alter the shape and structure of the nucleus. *Weissman* (1980) showed that ~10% of the NICs split on their first perihelion passage and *Levison et al.* (2002) suggested that 99% of them are disrupted sometime during their dynamical evolution. This may suggest different physical properties for the ECs and the NICs, although convincing, direct evidence of such differences has not yet been found.

In summary, there are a variety of processes associated with the formation and evolution of comets that could affect the physical properties of cometary nuclei. There should be no expectation that comets form a homogeneous group with respect to their physical properties, and it will be interesting to investigate possible correlations of those properties with the comet's place of origin and its subsequent history.

1.3. Historical Perspective

To understand how far we have progressed in the study of cometary nuclei, we summarize briefly some of the important results of the twentieth century. As is commonly done, we define the border between "pre-history" and "recent history" to coincide with the publication of the classic paper on cometary nuclei by *Whipple* (1950).

1.3.1. Pre-history. Before 1950, the paradigm governing the cometary "nucleus" did not involve a central, monolithic body. Rather, the "nucleus" was envisaged as an unbound agglomeration of meteoritic solids. In this sandbank model, described by *Lyttleton* (1953, 1963), all the particles comprising a comet were on independent but very similar orbits, and there was no gravitational binding. This model was consistent with the observations of many cometary phenomena, i.e., the morphological complexity of the inner comae of comets, as well as (qualitatively) the odd behaviors that comae sometimes display. The term "nucleus" itself was used with imprecision, as noted by, e.g., *Bobrovnikoff* (1931) and *Vorontsov-Velyaminov* (1946). Most often, the "nucleus" merely referred to the peak in the surface brightness distribution, which is frequently called the "central condensation."

The basic misconception of this era was a drastic overestimation of the typical size of cometary nuclei, which most researchers thought were tens of kilometers in size. Reports of observers resolving disks of the "nuclei" probably provided the motivation for this misconception. In hindsight, we now recognize that observers were merely seeing the steeply sloped surface brightness distribution of the inner coma. However, a few researchers thought cometary nuclei were monoliths, as small as 1 km in diameter or even less, on the basis of the starlike appearance of Comet 7P/Pons-Winnecke when observed close to Earth in 1927 (*Slipher*, 1927; Baldet, quoted by *Vorontsov-Velyaminov*, 1946). Further impediments to a proper understanding of cometary nuclei were the poorly constrained albedo and phase-darkening behavior. The idea of a nucleus, or dust grains for that matter, with a very low albedo did not become acceptable until after the spacecraft flybys of 1P/Halley in 1986.

1.3.2. Recent history. The "dirty snowball" model proposed by *Whipple* (1950) envisaged the nucleus as "a conglomerate of ices . . . combined in a conglomerate with meteoric materials." Two significant improvements over the sandbank idea were the model's ability to adequately explain both the cometary nongravitational motion and the gas production rate.

With this paradigm established, the future interpretation of data established the relatively (compared to pre-1950) small sizes of nuclei. Photographic data taken by *Roemer* (1965, 1966, 1968) set constraints on the sizes of many nuclei, although at this time the albedo was still thought to be much higher than the currently accepted mean. Furthermore, there was still the problem of unresolved comae around distant comets. Generally, Roemer's photographic observations were not taken at sufficiently large heliocentric distances for the comets to be inactive, and they were significantly contaminated by unresolved coma. *Delsemme and Rud* (1973) tackled the problem of albedo by comparing the nuclear brightness far from the Sun and the gas production rate close to the Sun, and derived albedos that seemed to confirm the high values of conventional wisdom. However, we now know that nuclear sizes based on cometary activity are lower limits, making the derived albedos upper limits, owing to the fact that typically only a small fraction of the nucleus surface is active.

Several significant steps forward were taken in the 1980s. Simultaneous thermal-infrared and optical measurements were made (discussed in section 3.3), establishing that nuclear albedos were low. In 1983 Comet IRAS-Araki-Alcock made an extremely close approach to Earth, and the synthesis of data using modern observational techniques resulted in a fairly complete description of that nucleus [size, albedo, shape, and rotation (*Sekanina*, 1988)]. Finally, the flotilla of spacecraft flying by Comet 1P/Halley confirmed beyond any doubt that a single, solid body lies at the center of a comet.

The past decade has witnessed a major observational effort to study cometary nuclei using medium to large ground-based telescopes and space telescopes outfitted with charge-coupled device (CCD) detectors, and this has resulted in a wealth of new data. Indeed, most of our understanding of cometary nuclei as a population has been derived from observations made during the past decade.

1.4. Observing Cometary Nuclei

The overarching observational goal for studies of cometary nuclei is to understand their ensemble properties, which is accomplished in several ways. The most common observations involve visible-wavelength photometry, from which the color and the product of the cross-section and albedo can be measured. More detailed observations at these wavelengths, such as time series of data at multiple epochs, provide clues on the shape and rotation state of the nucleus. Observations at wavelengths longer than ~5 μm, in the thermal-infrared, provide data on both the size and albedo and constrain the thermal properties of the bulk material comprising the nucleus.

Contrary to popular belief, the optical depth, τ, of most cometary comae is generally small enough to allow direct detection of the nucleus, in principle. Possible exceptions are unusually active comets, such as C/1995 O1 (Hale-Bopp), for which τ may approach unity. For most comets, the real problem lies with the intrinsic faintness of the nucleus relative to the light scattered from dust grains in the coma, i.e., the contrast is usually too small to distinguish the nucleus clearly. Historically, planetary astronomers have attempted to overcome this obstacle by observing comets at large heliocentric distances, when the nucleus was assumed to be inactive and coma-free. On the one hand, the activity level at large heliocentric distances is often so low that most of the observed light can be attributed to reflection from the nucleus. On the other hand, many comets are known to be conspicuously active at large heliocentric distances, preventing such an observational approach. Another approach was to observe only relatively nearby, very low activity comets, whose dust production rates were so small that the nucleus clearly stood out even when the spatial resolution was only hundreds of kilometers, but this works well for only a handful of objects. Spacecraft encounters, of course, are the best way to obtain detailed information on the physical properties of cometary nuclei, and we have learned much from the spectacular encounter images of 1P/Halley and 19P/Borrelly (see *Keller et al.*, 2004). While spacecraft encounters provide "ground truth" that cannot be obtained any other way, this approach is necessarily limited to a small number of objects and cannot be used to determine the properties of cometary nuclei as a population. Fortunately, recent improvements in the resolution capabilities and sensitivities of ground- and spacebased telescopes now allow us to study the physical properties of a large number of cometary nuclei, even in the presence of substantial coma.

1.5. Scope of this Chapter

Within the context of this chapter, "physical properties" refer to the size, shape, albedo, and color of the nucleus. All these properties can be observed directly, and in section 2 we describe in detail the techniques for doing so. In section 3, we summarize the results for each individual comet for which data are available because a comprehensive and critical evaluation of the results cannot be obtained from any other reference. After discussing techniques and results, in section 4 we synthesize all the data to estimate the distribution of sizes, shapes, colors, and rotational periods of cometary nuclei as a population. In section 5 we discuss some outstanding, unresolved issues in the study of cometary nuclei and comment on the direction of future research on the physical properties of cometary nuclei. Some short, concluding remarks comprise section 6.

Some physical properties of cometary nuclei are *not* covered in this chapter. The structure, strength, and bulk density are especially important, but, in general, these can only be estimated indirectly, as discussed by *Weissman et al.* (2004) and by *Boehnhardt* (2004). Although we summarize results on the rotational periods of cometary nuclei, mainly because they are obtained from the same light curve data used to measure shapes, a comprehensive discussion of the rotational properties is given by *Samarasinha et al.* (2004). The physical nature of the ice and dust contained within cometary nuclei (e.g., crystalline vs. amorphous ice, thermal conductivities and heat capacities of the ice and dust, etc.) is very poorly constrained observationally and is discussed from a modeling perspective by *Prialnik et al.* (2004). Finally, the very interesting question of how comets are related to the other minor bodies in the outer solar system is not treated here, but rather is covered separately by *Jewitt* (2004).

2. TECHNIQUES FOR DETECTING AND CHARACTERIZING COMETARY NUCLEI

2.1. General Considerations

Cometary nuclei are certainly among the most difficult objects of the solar system to detect and characterize, usually suffering from the dual problem of being faint and immersed in a coma. The techniques for their study are those first developed for the investigation of asteroids, but with the additional complexity caused by the presence of a coma. The primary technique, visible-wavelength imaging, uses reflected sunlight and takes advantage of high-performance detectors like CCDs. This technique has been most successful for relatively large and/or very low activity nuclei at large heliocentric distances, and for comets observed at close range and with sufficient spatial resolution to separate unambiguously the nuclear and coma signals. The pros and cons of these two cases will be discussed below. A third method using this technique, the *in situ* spacecraft investigation, is discussed by *Keller et al.* (2004) and will not be addressed here.

A second technique relies on the detection of thermal emission from the nucleus. The situation in this case is less favorable than for the reflected light because of the gener-

ally fainter signals, high thermal background with ground-based facilities, and inferior performance of IR detectors. Usually one has no choice but to observe the nuclei at close range, usually exploiting a close encounter with Earth. As with observations of reflected sunlight, a coma will usually be present and must be taken into account. Before the age of large-area infrared array detectors, this was difficult and so, again, very low activity comets were the most popular targets. However, new and improved thermal detectors, such as those on the Space Infrared Telescope Facility (SIRTF), will relax the limitations of the technique.

The sample of objects for which the thermal emission at radio wavelengths may be detected is even more restricted than in the infrared. The nucleus must be exceedingly close (e.g., C/1983 H1 IRAS-Araki-Alcock) or exceedingly large (e.g., C/1995 O1 Hale-Bopp). In addition, radar observations have a Δ^{-4} limitation, where Δ is the geocentric distance, and only rarely do comets pass close enough to the Earth to permit radar measurements of the nucleus (*Harmon et al.*, 2004).

Finally, we discuss rarely performed stellar occultation observations, which have the potential to provide detailed shape information on nuclei and their inner comae.

2.2. Using the Reflected Light

2.2.1. Observations. Detecting the solar light reflected by cometary nuclei remains the most powerful and efficient method to determine their size and to study their properties. However, this technique requires knowledge of the albedo and phase law, as discussed below.

At large heliocentric distances, e.g., $r_h \gtrsim 4$ AU, the activity of most ecliptic comets is very weak, and the coma may become sufficiently faint (or possibly nonexistent) to reveal the "bare" nucleus. Thus, the best strategy for these comets generally is to observe near aphelion. However, there are two main problems: (1) the geometric conditions (large r_h and Δ) usually result in a very faint nuclear signal, and (2) the criterion used to decide the nonexistence of a coma, namely the stellar appearance of the nucleus, is not robust because an unresolved coma can still contribute substantially to the observed signal. The most well-known example is 2P/Encke, which has been anomalously bright at almost every observed aphelion (*Fernández et al.*, 2000, and references therein).

For the NICs, cometary activity can continue well beyond this rough boundary for the ecliptic comets, probably due to the higher abundance of ices more volatile than water, such as CO. Many comets are known to be active beyond 5 AU (e.g., *Szabó et al.*, 2001; *Licandro et al.*, 2000; *Lowry and Fitzsimmons*, 2001) and even beyond 10 AU (*Meech*, 1992), such as 1P/Halley (*West et al.*, 1991) and C/1995 O1 Hale-Bopp. The poor spatial resolution when observing such objects at these distances makes accounting for the coma's contribution highly problematic. Once these long-active comets finally do deactivate, the intrinsic faintness of the nuclear signals generally limits the observations to snapshots in one (R) or two (V, R) bands, often with large uncertainties on the (V-R) color index.

Nevertheless, in a few cases multiple observations have been secured allowing the construction of a (sometimes partial) light curve, which can be used to investigate the shape and rotational state of the nucleus. Despite these limitations, this approach has been pursued by several groups of ground-based observers and has produced valuable data on the physical properties of cometary nuclei. In addition, and quite recently, near-infrared spectra of a few weakly active nuclei have been obtained using large telescopes in an attempt to detect spectral signatures (e.g., water ice and minerals). Currently, only Centaurs (e.g., Chiron, Chariklo) present convincing cases of detection of water ice on their surface.

An entirely different approach has been pioneered by Lamy and co-workers (e.g., *Lamy and Toth,* 1995, *Lamy et al.,* 1998a,b, 1999a, 2001b, 2002) and is based on the very high spatial resolution offered by the Hubble Space Telescope (HST). The basic rationale is that, while the nuclear signal is preserved in the point spread function (PSF) of the telescope, the signal from the coma, an extended source, is diluted as the spatial resolution increases. The contrast between the nucleus and the coma is maximized by observing comets at their minimum geocentric distance. A model for the surface brightness distribution of the nucleus plus coma is constructed and compared to the observed brightness distribution to estimate the signal from the nucleus. The brightness distribution of the comet is modeled as

$$B(\rho) = [k_n \delta(\rho) + coma] \otimes PSF \quad (1)$$

where ρ is the projected distance from the nucleus, $\delta(\rho)$ is the Dirac delta function, $\otimes$ is the convolution operator, and PSF is the point spread function of the telescope. The first term is the contribution of the nucleus, i.e., the PSF scaled by the factor k_n. The coma can be modeled by any function that provides a reasonable representation of the real coma, e.g., the canonical k_c/ρ inverse power law, where k_c is a scaling factor, or a generalized k_c/ρ^a, or a more complex function containing radial and azimuthal variations such as implemented for the asymmetric and structured comae of 19P/Borrelly and Hale-Bopp (C/1995 O1). The scaling factor k_n, the subpixel locations of the nucleus (x_n, y_n), and the parameters of the coma model (e.g., k_c, a) are determined individually on each image by minimizing the residuals between the synthetic and the observed images. The fits are performed either on the azimuthally averaged radial profiles, or on X and Y profiles, or on the full image. The instrumental magnitudes are calculated by integrating the scaled PSFs and are transformed to Johnson-Kron-Cousins magnitudes.

2.2.2. Interpretation of the observations. Once the magnitude, m, of the nucleus has been determined, the standard technique introduced by *Russell* (1916) is used to retrieve its physical properties. Russell's original formula, devised for asteroids observed at large phase angles, has been conveniently reformulated by *Jewitt* (1991) and, in the case of a spherical object, is given by

$$p\Phi(\alpha)r_n^2 = 2.238 \times 10^{22} r_h^2 \Delta^2 10^{0.4(m_\odot - m)} \quad (2)$$

where m, p, α, and $\Phi(\alpha)$ are respectively the apparent magnitude, the geometric albedo, and the phase angle (Sun-comet-observer angle) and phase function $\Phi(\alpha)$ of the nucleus in the same spectral band (e.g., V or R); $m_\odot$ is the magnitude of the Sun (V = –26.75, R = –27.09) in the same spectral band; r_h and Δ are respectively the heliocentric and geocentric distances of the nucleus (both in AU); and r_n is the radius of the nucleus (in meters). Observers often proceed in two steps, introducing first the absolute magnitude, H, of the nucleus (i.e., the magnitude at $r_h = \Delta = 1$ AU, $\alpha = 0°$)

$$H = m - 5\log r_h \Delta - \alpha\beta \quad (3)$$

where the phase function is given by

$$-2.5\log[\Phi(\alpha)] = \alpha\beta \quad (4)$$

and then incorporating the relationship between r_n (in meters) and p

$$r_n = \frac{1.496 \times 10^{11}}{\sqrt{p}} 10^{0.2(m_\odot - H)} \quad (5)$$

A linear phase coefficient $\beta = 0.04$ mag/deg is generally used, with an estimated uncertainty of ±0.02 mag/deg. In fact, a value $\beta = 0.06$ mag/deg has been obtained for 2P/Encke (*Fernández et al.,* 2000) and 48P/Johnson (*Jewitt and Sheppard,* 2003). For observations at small phase angles, the impact of the phase angle effect on the nuclear magnitude is small, but it becomes overwhelming at large phase angles (e.g., a correction of 2 mag to the nuclear magnitude and a factor 2.5 to the radius at $\alpha = 50°$). Finally, once an albedo is assumed (generally $p_V = p_R = 0.04$), or is independently determined, the radius r_n of the nucleus can be calculated. An uncertainty of ±0.017 on the albedo appears realistic, at least for ecliptic comets (see section 4.3 below), and has an impact of ~20% on the value of the radius. In summary, for nuclei observed at small phase angles and whose physical properties are not too unusual ($\beta = 0.04 ± 0.02$ mag/deg and $p = 0.04 ± 0.017$), the measurement of its magnitude offers a robust determination of its radius, at least of one of its cross-sections in the case of single (i.e., "snapshot") observations.

2.3. Using the Thermal Emission

2.3.1. Observations. The asteroid community has been using radiometry for over 30 years (e.g., *Allen,* 1971) to derive robust sizes and albedos. The application of this method to cometary nuclei began in 1984, i.e., *before* the 1P/Halley apparition (*Campins et al.,* 1987), and has been used in earnest since the mid-1990s with the advent of array-detectors sensitive to radiation in the 10–20-μm range.

For datasets of outstanding quality — high signal and multiple wavelengths — it is also possible to constrain various fundamental parameters of the the nucleus, such as thermal inertia and surface roughness (see *Campins and Fernández*, 2003). If multiepoch data are obtained, the thermal phase behavior of the nucleus may be deduced. If time series of IR data are taken simultaneously with visible-wavelength photometry, the existence of large-scale albedo spots on the surface may be discovered. Observations at very long (millimeter or centimeter) wavelengths provide clues on the emissivity of the bulk material in the nucleus (i.e., subsurface).

Unfortunately, the difficulties of observing cometary nuclei usually prevent one from obtaining such a robust dataset. The two main problems are related to the usual observational paradigm: When the nucleus is close to Earth and bright, it is often shrouded in coma, but when it is far from the Sun and less active, it is often too faint. Thus, traditionally the best nuclei to observe are those that are weakly active and/or large or nearby. Work by *Campins et al.* (1987), *Millis et al.* (1988), and *A'Hearn et al.* (1989) are excellent examples of successful observations of just such special comets.

The techniques applied at visible wavelengths to deal with the effects of the coma can also be applied to the thermal IR images. Both cases require excellent spatial resolution, but a complication is that, for the ideal case of diffraction-limited observations where the width of the PSF is proportional to wavelength, the thermal radiation and reflected light sample different scales of the inner coma. Thus, in this case when the dust opacity is constant with wavelength, or decreases with wavelength slower than λ^{-1}, the nucleus-to-coma contrast ratio will generally be larger for the observations at visible wavelengths compared to those made at thermal wavelengths. Since groundbased data in the two wavelength regimes often have similar spatial resolutions owing to the effects of atmospheric seeing, the problem of sampling different spatial scales of the coma usually only applies to spacecraft data. Despite these difficulties, *Jorda et al.* (2000), *Lamy et al.* (2002), and *Groussin et al.* (2003) successfully used the Infrared Space Observatory (ISO) to detect and characterize several nuclei in the 10-μm region, taking advantage of the much better sensitivity to thermal emission resulting from the absence of the warm terrestrial atmosphere.

While typical groundbased thermal measurements are made in the 10-μm atmospheric window (and, less frequently, in the 5- and 20-μm windows), the submillimeter, millimeter, and centimeter windows have been exploited to detect the thermal radio continua of a few very bright nuclei — namely Hale-Bopp (reviewed by *Fernández*, 2003) and IRAS-Araki-Alcock (*Altenhoff et al.*, 1983).

2.3.2. Interpretation and analysis. Once the thermal continuum flux density F_{th} has been measured, it can be interpreted via the equation

$$F_{th}(\lambda) =$$
$$\varepsilon_{th} \iint B_\nu[T(r_h, pq, \eta, \varepsilon_{th}, \theta, \phi), \lambda]d\phi d\cos\theta r_n^2 \frac{\Phi_{th}}{\pi \Delta^2} \qquad (6)$$

where Φ_{th} is the phase function at thermal wavelengths, p is the geometric albedo at reflected wavelengths, B_ν is the Planck function, ε_{th} is the emissivity at thermal wavelengths, η is a factor to account for infrared beaming (see *Spencer et al.*, 1989), and T is the temperature. The temperature itself is a function of r_h, p, η, ε_{th}, the surface cometographic coordinates, θ and ϕ, and the phase integral q, which links the geometric and Bond albedos. *Buratti et al.* (2004) derived $q \simeq 0.3$ for 19P/Borrelly. Traditionally the largest sources of error in this modeling effort came from Φ_{th} and η. Φ_{th} was often parameterized as a function of phase angle, α, such that $-2.5 \log\Phi_{th} \propto \alpha$, but recently the more sophisticated approach of explicitly calculating the surface integral of Planck emission over the Earth-facing hemisphere has become preferable (*Harris*, 1998; *Lamy et al.*, 2002). The beaming parameter η, however, is still largely unconstrained for comets and remains the largest uncertainty; we are only beginning to understand the variety of values possible for near-Earth asteroids comparable in size to the cometary nuclei (e.g., *Delbo et al.*, 2003).

For objects with low albedos, such as cometary nuclei, r_n can be determined to good accuracy from their thermal flux density, provided the observations are secured at low phase angles. This is because the thermal emissivity is close to 1, so the thermal emission does not depend strongly on the assumed value for ε_{th}. This is to be contrasted with the visible case, where the flux is proportional to the geometric albedo, which is very small and can, in principle, vary by a large factor. Fortunately, the range of values measured for the geometric albedo seems to be rather limited (see the previous section), which means that accurate values for the nuclear radius can be derived solely from the visible data as well. In section 2.7, we discuss the measurements of the albedo.

One important caveat to this formulation is that it assumes the nucleus is spherical. Not only does this make r_n an "effective" radius instead of a true radius, but r_n applies only to the Earth-facing cross section at the time the data were taken. Observations over a rotation period are generally needed to constrain the "mean" effective radius. It is, of course, possible to implement the equations to handle a nucleus of ellipsoidal or even arbitrary shape, although frequently the quality of the data does not warrant such an action. Early work by *Brown* (1985) demonstrated how ellipticity of the nucleus can affect the measured fluxes. More recently, *Gutiérrez et al.* (2001) have investigated how arbitrary shapes and variegated surface-ice/surface-dust ratios can affect the thermal behavior of nuclei.

The critical step for this method is to calculate a surface temperature map $T(\theta,\phi)$ of the nucleus for the time at which it was observed. This can be done using a thermal model, the fundamental parameters of which are the rotation period and the thermal inertia (the square root of the product of the conductivity, heat capacity, and bulk density). For most datasets, one of two commonly used thermal models are usually employed. One, for slow-rotators (a.k.a. "standard thermal model"), applies if the rotation is so slow, or the thermal inertia is so low, that every point on the surface is in instantaneous equilibrium with the impinging solar radiation. The other, for rapid-rotators (a.k.a. "isothermal latitude model"), applies if the rotation is so fast, or the thermal inertia is so high, that a surface element does not appreciably cool as it spins away from local noon and out of sunlight. This model also assumes that the rotation axis is perpendicular to the Sun-Earth-object plane. (For an axis that points at the Sun, the two models predict the same temperature map.) Note that the terms "slow-" and "rapid-rotator" are slightly misleading, in that the thermal inertia is usually the physical quantity that determines the thermal behavior. Thus, two cometary nuclei with identical and long rotation periods, but vastly different thermal inertias, may not necessarily both be "slow-rotators."

Furthermore, small bodies in the outer planets region, at ~10 AU or beyond, can behave like rapid-rotators even if their rotational periods are long. This is because thermal radiation scales as T^4 and when T is low enough, those bodies do not cool substantially during nighttime.

In practice, there are few objects in the inner solar system that behave thermally as rapid-rotators, so the slow-rotator model is often employed as the default. Of the cometary nuclei that have been studied, nearly all appear to behave as slow-rotators. The only possible (unconfirmed) exception so far is the very low activity Comet 107P/Wilson-Harrington (*Campins et al.,* 1995). Among the asteroids, one notable rapid-rotator is (3200) Phaethon (*Green et al.,* 1985), which may be a dormant or extinct cometary nucleus. Whether or not the thermal inertias of all highly evolved comets are low remains to be seen. *Campins and Fernández* (2003) give some upper limits to the thermal inertias of a few nuclei, but, for the most part, these limits are roughly an order of magnitude higher than the expected values.

The applicability of the slow- or rapid-rotator model can be quantified by the parameter Θ, introduced by *Spencer et al.* (1989), which is

$$\Theta = \frac{\Gamma\sqrt{\omega}}{\varepsilon\sigma T_{ss}^3} \qquad (7)$$

where Γ is the thermal inertia, ω is the rotational angular frequency, σ is the Stefan-Boltzmann constant, and T_{ss} is the temperature at the subsolar point. Ideal slow-rotators have $\Theta = 0$; rapid-rotators, $\Theta = \infty$. Since Θ depends so steeply on the subsolar temperature, cometary nuclei that mimic slow-rotators near perihelion could conceivably act more like rapid-rotators at aphelion. Due to sensitivity limitations in the mid-IR, at the time of this writing there have been no detections of cometary nuclei at large heliocentric distances, so currently the problem is moot. However, SIRTF

is expected to detect comets out to ~5 AU from the Sun, so the interpretation of radiometry must proceed with caution.

Enhancements to the thermal modeling can be made and are justified when there are measurements of the nucleus's thermal continuum at many wavelengths. At the very minimum, the 10-μm vs. 20-μm color can be used to discriminate between slow-rotators and fast-rotators. A further tack is to recognize that comets have a significant near-surface ice component (unlike the asteroids) that is sublimating away and thus probably affects their thermal behavior. The "mixed model" introduced by Lamy and co-workers (*Lamy et al.,* 2002; *Groussin et al.,* 2003; *Groussin and Lamy,* 2003a) employs a water-ice sublimation term when calculating the surface temperature map. The effect is to provide a generally cooler nucleus than otherwise implied by the standard slow-rotator model. The thermal inertia itself can be roughly constrained with this method. For example, very low values of the thermal inertia, about one-fifth that of the Moon, have been derived for Centaurs Chiron and Chariklo by *Groussin and Lamy* (2003b). Naturally, even more detailed models of nuclear structure and thermal behavior are possible, and these are discussed in *Prialnik et al.* (2004).

2.4. Combining Reflected Light and Thermal Emission

If visible and thermal IR observations are performed simultaneously, then it is possible to solve independently for the radius and the albedo of the nucleus using equations (2) and (6) as system with two unknowns, p and r_n. This method has been implemented for a handful of nuclei (see section 3.3). In practice, and as emphasized in the above section, r_n is determined by the thermal constraint (i.e., equation (6)); consequently the visible constraint (i.e., equation (2)) yields the albedo. An illustration of this practical implementation is given by *Lamy et al.* (2002) for the case of 22P/Kopff.

2.5. Light Curves

The light curve (by which we mean the short-timescale series of photometric measurements, not the orbit-timescale study of activity as a function of r_h) provides information on the shape and rotational period of a cometary nucleus. Only observations at visible (reflected light) and infrared (thermal emission) wavelengths are presently capable of producing such light curves. Very much like the case for asteroids, the periodic temporal variation of the brightness is interpreted in terms of the rotation of an elongated body. Light curves of sufficient length have been obtained for only a few comets (e.g., 2P/Encke), and the interpretation is frequently difficult (e.g., multiple solutions for the rotational period may be found), but the situation has improved with recent datasets that show periods much more clearly (e.g., *Lowry and Weissman,* 2003; *Jewitt and Sheppard,* 2003). *Samarasinha et al.* (2004) discuss these problems in some detail.

One extra complication is the possibility that the nucleus has a nonuniform albedo, which would add a non-shape-related component to the temporal brightness variations. Indeed, spacecraft imaging of 19P/Borrelly revealed some evidence of surface variations (*Soderblom et al.,* 2002), although they are difficult to separate from topography effects because of the modest spatial resolution of the images; see *Nelson et al.* (2004) for a discussion of this problem. The possibility of large-scale albedo features on the surface of the nucleus can be ruled out if visible and thermal light curves are obtained simultaneously. Such light curves will be in-phase for shape-dependent rotational modulation and out-of-phase for albedo-dependent modulation. Generally, however, the subject is often disregarded simply because datasets are rarely of sufficient quality to draw definite conclusions.

The default case is to analyze the temporal variation in terms of the varying apparent cross-section of a rotating, elongated nucleus. All observations available so far are consistent with, and interpreted as, rotation of a prolate spheroid (with semiaxes a and b = c) around one of the short axes. In a few cases, independent constraints on b and c have been obtained. The projected area of a spheroid in simple rotation is given by

$$S = \pi ab^2[(\sin^2\phi/a^2 + \cos^2\phi/b^2)\sin^2\varepsilon + \cos^2\varepsilon/b^2]^{1/2} \quad (8)$$

where ϕ is the rotation angle and ε is the angle between the spin vector of the nucleus and the direction to the Earth. Figure 1 displays the ratio S_{min}/S_{max} (also expressed in magnitude variation, Δm) as a function of a/b and ε. If the orientation of the spin axis is independently constrained, for example by the shape of the coma (*Sekanina,* 1987), the amplitude of the light curve yields the a/b ratio. Together with the absolute magnitude, corresponding to either the minimum or maximum projected areas, one can obtain a solution for the spheroidal shape of the nucleus. Generally, ε is not known, so that only a minimum value of a/b can only be obtained, corresponding to $\varepsilon = 90°$. The situation is even more difficult for "snapshot" observations, as the effective radius, $r_{n,a}$, which represents the instantaneous projected area, will range between $\sqrt{ab}$ and b. For an axial ratio of 2, $r_{n,a} = 0.707\sqrt{ab}$, i.e., within 30% of the maximum value. The problem is, however, less serious than the above simple analysis tends to imply because the temporal aspect very much helps. As illustrated by the light curve of 19P/Borrelly (Fig. 8 of *Lamy et al.,* 1998b), the fraction of time during which the small cross-section is seen is comparatively very short and may even be missed if the time resolution of the observations is not adequate. Consequently, a rotating spheroid displays a cross-section close to its maximum most of the time. As discussed by *Weissman and Lowry* (2003), the integration over all possible (random) orientations and rotational phases shows that the average projected area remains a large fraction $\kappa_n\pi ab$ of its maximum value πab: $\kappa_n = 0.924$ for a/b = 1.5, $\kappa_n = 0.892$ for a/b = 2, and $\kappa_n = 0.866$ for a/b = 3. For the effective radius, $r_{n,a}$, given by the instantaneous projected area, the scaling varies as $\sqrt{\kappa_n}$. For a typical axial ratio a/b = 2, a snapshot observation will, on average, lead to $r_{n,a} = 0.945\sqrt{ab}$, i.e., within 5.5% of the maximum value $\sqrt{ab}$. Even more important for questions such as the size distribution function is the *effective* radius, $r_{n,v}$, that of the sphere having the same volume (or mass) as the spheroid, via $r_{n,v}^3 = ab^2$. The ratio $r_{n,v}/r_{n,a}$ remains close to 1 with value of 0.972 for a/b = 1.5, 0.943 for a/b = 2, and 0.895 for a/b = 3. To summarize, the radius calculated from an observed, apparent projected area will give, on average, an excellent estimate of the effective radius of the equivalent sphere. Note that the averaging with respect to rotational phase is implicitly done when authors average their data values that are too scarce to construct a credible light curve.

A light curve does not strictly give access to a projected area. In the visible, the bidirectional reflectance comes into play but will not be a problem if the scattering properties are homogeneous over the nuclear surface. In the thermal infrared, it is the two-dimensional distribution of temperature over the surface that comes into play, and that is certainly not homogeneous. For example, *Brown*'s (1985) nonspherical thermal model predicts that the amplitude of the light curve will be larger in the infrared than in the visible, an effect apparently observed on 10P/Tempel 2 (*A'Hearn et al.,* 1989). With the question of how to interpret the light curve of cometary nuclei still in its infancy, interpretations beyond the simple spheroidal model discussed above are not warranted. Complex effects, such as shadowing and unilluminated areas, cannot yet be handled properly but have already been noted [e.g., the skewness of the light curve of 9P/Tempel 1 (*Lamy et al.,* 2001b)].

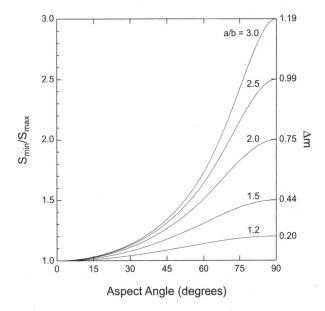

Fig. 1. The minimum to maximum projected area (S_{min}/S_{max}) of rotating prolate ellipsoids with different axial ratios (values are marked near the curves) plotted vs. the aspect angle. The corresponding light curve amplitudes are also indicated (Δm).

2.6. Radar Observations

Studies of cometary nuclei using radar are discussed by *Harmon et al.* (2004). For completeness, we provide here a brief outline of the method. Basically, one sends a burst of microwaves of known power towards a nucleus and measures the power of the returned echo. There have been six such detections of nuclei, although the signal-to-noise is better than 4 in only two cases, C/1983 H1 (IRAS-Araki-Alcock) and C/1996 B2 (Hyakutake). There are as yet no delay-Doppler "images" of a cometary nucleus, as are now being routinely created for close-approaching near-Earth asteroids. The main reason is the scarcity of comets that can overcome the Δ^{-4} dependence for detectability. In terms of the properties of nuclei, the radar data have mostly been used to constrain the radar-albedo and density, via arguments related to the bulk dielectric properties of cometary nuclei and their response to microwaves. The radar albedos are apparently similar to the visible-wavelength albedos. The bulk densities range between 0.5 and 1.5 times that of water, values that are not unexpected.

2.7. Occultations

Occultations are frequently used to constrain the shape and size of asteroids, and can provide a direct test of the validity of other methods, such as radiometry (section 2.3). In principle, the same could be done for comets. An occultation trace may also have wings, owing to nonnegligible optical depth in the inner coma, and this could provide information on dusty gas hydrodynamics in the inner coma and the location of active regions on the surface. In practice, this method is limited by the difficulty of locating a nucleus accurately within a surface brightness distribution that is dominated by coma for most astrometric observations, and in finding suitable stars that are occulted. Even a subarcsecond positional error perpendicular to the proper motion can shift the path of the comet's shadow on Earth by hundreds or thousands of kilometers. Given that the nucleus in question is typically on the order of 1–10 km across, the difficulty of obtaining a successful observation becomes apparent. Moreover, obtaining the ideal dataset with multiple chords through the nucleus requires tight spatial sampling across the predicted path, which can be logistically difficult with limited labor resources and equipment.

Currently the most useful occultation event observed is one by the weakly active and large Centaur Chiron (*Bus et al.,* 1996). One chord through most of the nucleus and one possibly grazing chord were observed, and this constrained the radius to be at least 90 ± 7 km. Chiron's ellipsoid is within 10% of spherical, so this is thought to be a robust lower limit. Groundbased radiometric data (*Campins et al.,* 1994; *Fernández et al.,* 2002a) currently imply a radius of ~80 km, while space (ISO) data give 71 ± 5 km (*Groussin and Lamy,* 2003b), so the agreement is not really satisfactory.

Another occultation event, this one by C/1995 O1 Hale-Bopp, was reported by *Y. Fernández et al.* (1999). Only one chord was measured, and, if real, it is impossible to tell if this chord went through the nucleus, grazed the nucleus, or passed near by. The optical depth of the inner coma may have been significant (i.e., approaching unity) for several tens of kilometers (cf. *Weaver and Lamy,* 1997). With some reasonable assumptions about the dust outflow, the occultation constrained the (assumed-spherical) radius of the nucleus to be no larger than 48 km. With further restrictive assumptions, the upper limit is ~30 km.

There are several other occultations by comets mentioned in the literature, but none of them have probed the nucleus.

2.8. Measuring the Albedo

One must be careful to specify what is meant by the term "albedo" because many different definitions are used. In this paper, we report values for the geometric albedo (p), which is defined as the zero-phase, disk-integrated reflectance relative to that produced by a "perfect" diffusing disk (cf. *Hanner et al.,* 1981). Sometimes the Bond albedo (A) is used instead of the geometric albedo; this is just the fraction of incident light that is scattered in all directions and is related to the geometric albedo by

$$A = pq \qquad (9)$$

where q is the phase integral, which is given by (cf. *Russell,* 1916; *Allen,* 1976)

$$q = 2 \int \Phi(\alpha) \sin(\alpha) d\alpha \qquad (10)$$

where $\Phi(\alpha)$ is the disk-integrated, normalized phase function and α is the phase angle. Note that both albedos are functions of wavelength. In the energy balance equation for the surface of the nucleus, one must calculate the quantity

$$\frac{\int F_{\odot}(\lambda) A(\lambda) d\lambda}{\int F_{\odot}(\lambda) d\lambda} \qquad (11)$$

which *Clark et al.* (1999) defines as the bolometric Bond albedo A_B, and which is wavelength independent. However, for A that varies only slightly with wavelength; e.g., for a gray object, the value at a particular wavelength will suffice and $A \simeq A_B$.

Various complications arise when attempting to derive photometric properties from disk-resolved imagery of the nucleus. So far, we have such data on Comets 1P/Halley and 19P/Borrelly. For an unresolved image of a nucleus, we work with a body that has a subsolar point, and the geometric albedo is simply the true albedo at that point. In that case, we also employ a phase-darkening function to describe the photometric behavior from our (nonzero phase) vantage. For a resolved element of area on the surface of a nucleus, there is likely no subsolar point, and we must account for sunlight impinging on the element with some nonzero zenith angle. Thus, an understanding of the scattering is crucial to disentangle albedo and scattering effects. In a few cases [e.g., asteroid Eros (*Clark et al.,* 2002)], one has a full shape model of the object in question, and then one can use (1) the observed disk-resolved photometry and (2) the known scattering geometry as a function of position on the nucleus to derive fundamental scattering parameters. Most commonly, the formulation presented by *Hapke* (1986) is used.

The most straightforward, assumption-free way to obtain the geometric albedo from resolved imaging of the nucleus is to combine the projected area S known from resolved images with remote photometry of the unresolved nucleus, which gives pS as discussed in section 2.2. Then the ratio unambiguously yields the geometric albedo p of the nucleus. In practice, this usually requires a good understanding of the rotational state, shape, and phase function of the nucleus, since in most cases the resolved imaging and the groundbased data will have been obtained at different epochs. The resolved imaging must be matched to the remote viewing, which may be at a different aspect angle, and certainly one needs to match the rotational state and the projected area. Application of this procedure to the nucleus of 19P/Borrelly will be discussed in section 3.3.1 below.

Returning to disk-integrated (unresolved) data on a cometary nucleus, the radiometric method (section 2.3) is currently the most common way to derive the visible and near-IR geometric albedo. Whereas one usually only needs the IR equation to obtain a good estimate of the nuclear radius (since almost all the incident energy is absorbed and then thermally reradiated), the full method — solving both equations for the two unknowns — is required in order to have a confident, robust albedo measurement. Simultaneity, or an understanding of the rotation state, is also critical.

Another method involves deriving the radar albedo from radar echoes, and assuming that the reflectivity of the nucleus at centimeter wavelengths is similar to that at visible wavelengths. The existing radar data are discussed by *Harmon et al.* (2004). Generally, radar albedos seem to be as dark as their visible counterparts, but the quality of the radar data on cometary nuclei are not yet good enough to make a robust comparison between albedos in the two wavelength regimes.

3. PROPERTIES OF COMETARY NUCLEI

We now present a detailed discussion of the available data on the physical properties of cometary nuclei: size, shape, albedo, color, and rotational period. The bulk of these data comes from six sources, which we briefly describe below. Additional sources will be introduced when discussing individual comets.

3.1. Main Sources of the Data

3.1.1. Scotti (unpublished data, 1995). The largest dataset obtained by a single observer with the same instrument (the 91-cm Spacewatch Schmidt telescope at Kitt Peak equipped with a CCD camera) has unfortunately never been

published, but a short note entitled *Comet Nuclear Magnitudes,* dated January 14, 1995 (hereafter denoted *Sc*) has been widely circulated in the cometary community. Scotti applied a rudimentary technique to subtract the coma assuming a constant surface brightness inward from a thin annulus having a radius of a few times the radius of the seeing disk [a more detailed description and an evaluation of this method are given by *Tancredi et al.* (2000)]. This is obviously an oversimplification, and it always leads to an overestimation of the brightness of the nucleus by an amount that depends entirely on its activity. Scotti produced a table giving the absolute magnitude of 62 cometary nuclei from which he calculated a radius, assuming an albedo of 0.03 (they are reproduced here but scaled to an albedo of 0.04), convincingly demonstrating that the bulk of them are very small bodies with sizes of a few kilometers. These results have further been very useful to estimate the exposure times for spacebased observations.

3.1.2. Lamy and Toth (1995), Lamy et al. (1996, 1998a,b, 1999a,b, 2000, 2001a,b, 2002, 2003), Jorda et al. (2000), Groussin et al. (2003, 2004), Toth et al. (2003). The approach employed by this group (hereafter denoted *La+*), which is to use the high spatial resolution of the HST to photometrically resolve the nucleus, has already been described in section 2.2.1. Except for the few cases of complex comae, such as that of Hale-Bopp (C/1995 O1), the residuals between the observed and modeled images are usually very small, typically a few percent of the signal in the brightest pixel. Figure 2 illustrates the solution obtained from the HST observations of 19P/Borrelly (a spheroid), in comparison with the best *in situ* image obtained by the camera on the *Deep Space 1* spacecraft. Thirty-one nuclei have been detected during the HST observations, all active except for 9P/Tempel 1, which was observed at r_h = 4.48 AU. For 18

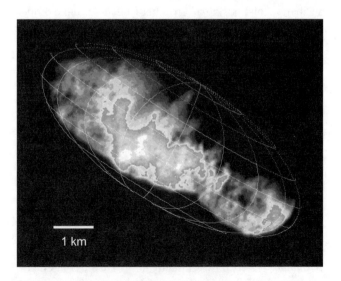

Fig. 2. The prolate spheroid model of the nucleus of Comet 19P/Borrelly derived from the HST observations made in 1994 is verified by the best *in situ* image taken by the *Deep Space 1* spacecraft in 2001 (cf. *Lamy et al.,* 1998b; *Soderblom et al.,* 2002).

comets only snapshot observations were obtained (i.e., one HST orbit per comet), while light curves were measured for 13 nuclei (8 HST orbits per comet, except 6 for 19P/Borrelly and 11 for 67P/Churyumov-Gerasimenko). The derived nuclear magnitudes were converted to standard R-band values, which were then used to derive sizes by adopting an albedo of $p_R = 0.04$ and a phase coefficient of $\beta = 0.04$ mag deg^{-1}.

3.1.3. Licandro et al. (2000). This group (hereafter denoted *Li+*) has used several groundbased telescopes to observe 18 comets at large heliocentric distances to minimize possible coma contributions. No attempt has been made to subtract the contribution from the coma, although seven comets were conspicuously active (six of them at $r_h > 4$ AU). The 11 others were deemed inactive on the basis of their stellar appearance. The nuclear sizes were derived from the V magnitudes assuming $p_V = 0.04$ and $\beta = 0.04$ mag deg^{-1}. For reasons of consistency, we prefer using R magnitudes in the discussion below and in the tables presented later. Accordingly, we converted V magnitudes to R values assuming $(V-R) = 0.5$, the median value for the ecliptic comets (*Toth and Lamy*, 2000), with the following exceptions: (1) 52P/Harrington-Abell, which was observed in the R band; and (2) 49P/Ashbrook-Jackson, 74P/Smirnova-Chernykh, and 96P/Machholz 1, for which $(V-R)$ was independently determined.

3.1.4. Lowry et al. (1999, 2003a,b), Lowry and Fitzsimmons (2001), Lowry and Weissman (2003). This group (hereafter denoted *Lo+*) uses a method similar to that described above, i.e., groundbased observations at large r_h. Out of 73 comets targeted, only 28 were deemed inactive on the basis of their stellar appearance, and the measured R magnitudes were converted to sizes using $p_R = 0.04$ and $\beta = 0.035$ mag deg^{-1}. As most of the comets were observed at small phase angles, the difference between using $\beta = 0.035$ and $\beta = 0.04$ is usually small (e.g., a 2.3% increase in the radius for $\alpha = 10°$), but will be applied for consistency. The remaining 45 comets were either active or were not detected, so that only an upper limit on the size of the nucleus could be obtained.

3.1.5. Meech et al. (2004). This group (hereafter denoted *Me+*) observed 16 JFCs and 1 HTC (109P/Swift-Tuttle) with the Keck telescope using the method described above. They concluded that 11 JFCs, as well as 109P, were inactive based on their stellar appearance. Their V and R magnitudes were converted to sizes using $p_V = p_R = 0.04$ and $\beta = 0.04$ mag deg^{-1}. This leads to two different values for the radius (except for 9P/Tempel 1), and we only compiled those corresponding to the R magnitudes for reasons of consistency. For 9P, we transformed the V to R magnitudes using $(V-R) = 0.5$ and obtained $r_n = 3.04$ km. The remaining five comets were active and only an upper limit to the size could be obtained. In a separate program, this group used the Wide-Field Camera of the HST to search for five NICs at geocentric distances ranging from 20 to 29 AU, but none were detected and only upper limits could be placed on the sizes of their nuclei.

3.1.6. Tancredi et al. (2000). This group (hereafter denoted *Ta+*) has compiled a set of 3990 measurements of "nuclear" magnitudes obtained from a variety of sources, mainly the *Comet Light Curve Catalogue* of *Kamél* (1990, 1992), the Minor Planet Center database (thus including the results of Scotti presented above), the IAU Circulars, the *International Comet Quarterly*, and various scientific articles. The bulk of their analysis consisted of scrutinizing these inhomogeneous data and making sense of them. They rejected all magnitudes determined visually, performed various corrections, and plotted the resulting heliocentric brightnesses. Their "best estimates" of the absolute visual magnitudes $H_N = V(1,1,0)$ generally corresponds to the faintest observed magnitudes and were used to derive sizes using $p_V = 0.04$ and a coefficient $\beta = 0.04$ mag deg^{-1} to correct for phase angle. They have introduced four quality classes (QC) that roughly quantify the uncertainties affecting the nuclear magnitudes, from ±0.3 mag (QC1) to ±1 mag (QC4). The respective numbers of nuclei are 9 for QC1, 18 for QC2, 37 for QC3, and 41 for QC4 for a total of 105 nuclei. It is readily seen that the bulk of the sizes belongs to the lowest quality classes. An updated version of this catalog has been presented by G. Tancredi at the Asteroids, Comets, Meteors 2002 conference and has been kindly made available to us. Those results are included in Table 1, but we have no means of assessing the quality of these improved determinations.

3.2. Sizes, Shapes, and Rotational Properties

In this section, we provide short summaries of the physical properties of individual comets. First we treat the ECs, and then we discuss the NICs. Unless otherwise stated, r_n is used as the generic notation for the radius of a cometary nucleus, while $r_{n,v}$ and $r_{n,a}$ refer to two "effective radii": $r_{n,v}$ refers to the radius of the sphere having the same volume as the observed object, and $r_{n,a}$ refers to the radius of the disk having the same projected area as the observed object. The albedo measurements are discussed separately in the next section.

3.2.1. Ecliptic comets (ECs). 2P/Encke: A robust radiometric measurement of the size is reported by *Fernández et al.* (2000): 2.4 ± 0.3 km. This number is consistent with an earlier radiometrically derived upper limit of 2.9 km by *Campins* (1988). Visible-wavelength estimates of the size have frequently suffered from the spatially unresolved coma that this comet displays at nearly every aphelion. *Fernández et al.* (2000), updating a compilation by *Sekanina* (1976), review the "nuclear" magnitudes that have been published since the 1960s and find that the data having the least coma contamination are from HST in 1997 (published in that same paper), by *Garradd* (1997) in 1997, and by *Jewitt and Meech* (1987) in 1986. *Jewitt and Meech* (1987) state that the maximum radius of the nucleus is $2.8 \leq r_n \leq 6.4$ km, assuming albedos between 0.02 and 0.10, which is consistent with the later radiometric observations. Some photographic photometry was useful in constraining the absolute magnitude of the nucleus, e.g., observations by *Van Biesbroeck* (1962) in

1960 and by Roemer in 1973 (reported by *Marsden*, 1974) and 1974 (reported by *Marsden and Roemer*, 1978). Finally *Fernández et al.* (2000) report an attempt to reconcile all published light curves to derive a shape; the lower limit on one of the axial ratios was found to be 2.6, indicating a very elongated nucleus. The rotational period was constrained in the 1980s by *Jewitt and Meech* (1987) and *Luu and Jewitt* (1990a), using time series of CCD photometry. A period of P = 15.08 h (or possibly $\frac{3}{2}$P = 22.62 h) satisfied all the data. However, more recent data from 2001 and 2002 presented by *Fernández et al.* (2002c) and *Lowry et al.* (2003a,b) indicate that the dominant periodicity may have changed in the intervening years (or was poorly measured in the past). Currently, a period near P = 11.01 or 2P = 22.02 h (close to the above value of 22.62 h) fits the data best. Furthermore, the dominant periodicities from the 1980s are not consistent with the most recent data. The situation hints that the nucleus of 2P/Encke may perhaps be in a complex rotation state (cf. *Belton*, 2000), although further investigations are necessary before a definite conclusion can be drawn.

4P/Faye: Observed with the HST by *La+* in October–November 1991 and in February 2000. We favor the value r_n = 1.8 km obtained in 2000 with the aberration-free HST. Reexamination of the 1991 observations obtained with the aberrated HST indicates that the signal from the nucleus was overestimated.

6P/d'Arrest: A snapshot observation by *Me+* at r_h = 5.4 AU of the inactive nucleus yields r_n = 1.71 km. *Lo+* first determined an upper limit of 2.1 km and later obtained a partial light curve. They derived a mean effective radius of $r_{n,v}$ = 1.6 ± 0.06 km (scaled to β = 0.04 mag deg^{-1}) and a/b > 1.18 ± 0.08. The size determinations from *Me+*, *Lo+*, and *Ta+* (1.5 km) are in good agreement but are considerably smaller than previous estimates of 3.5 km by *Campins and Schleicher* (1995) using IR photometry and of 2.7 km by K. Meech (unpublished data). Determinations of the rotational period have been reported by *Fay and Wisniewski* (1978), 5.17 h; *Lowry and Weissman* (2003), 7.2 ± 0.12 h; and *Gutiérrez et al.* (2003), 6.67 ± 0.03 h. The apparent discrepancies have been thoroughfully analyzed by the latter authors who concluded that, if all these measurements are correct, a change in the period has taken place or the nucleus is in a complex rotational mode. We adopt 7.0 h as a reasonable estimate for the present period of 6P.

7P/Pons-Winnecke: A snapshot observation by *Lo+*, when the comet was apparently inactive at r_h = 5.58 AU, yields r_n = 2.6 ± 0.1 km.

9P/Tempel 1: The two extreme cross-sections observed by *La+* give r_n = 2.8 and 3.3 km. Converting the V magnitude measured by *Me+* to an R magnitude using an average (V-R) = 0.52 yields r_n = 3.07 km, in excellent agreement with the above range and with the upper limit of 3.2 ± 0.1 km determined by *Lo+* without any correction. We very much doubt that the subsequent comatic correction introduced by *Lo+* and the resulting r_n = 2.3 km are correct. As discussed above, comatic corrections of groundbased images remain highly problematic. The brightness obtained by *La+* and *Weissman et al.* (1999) lead to spheroidal solutions

for the nucleus, assuming an aspect angle of ~90°, that are in remarkable agreement: a = 3.8–3.9 km and b = 2.8–2.9 km. *Fernández et al.* (2003) obtained simultaneous visible and near-infrared observations at r_h = 2.55 AU outbound while the comet was still quite active. Two different methods were used to correct for the contribution of the coma (≈15%) in their mid-infrared measurements and their interpretation using the standard thermal model leads to a radius of the maximum cross-section of the nucleus of 3.0 ± 0.2 km. Combining this result with that of *La+* gives p_R = 0.048 ± 0.007, significantly different (but still within the respective uncertainties) from the value p_R = 0.072 ± 0.016 derived by *Fernández et al.* (2003). Their large albedo is probably a consequence of their inability to properly account for the large coma contribution in their visible observations. Combining the above albedo p_R = 0.048 with the measurements of *La+* and *Weissman et al.* (1999) leads to a spheroidal solution with a = 3.5 km and b = 2.6 km and an effective radius $r_{n,v}$ = 2.9 km. From their partial light curve, *La+* extrapolated a rotational period in the range of ~25–33 h. *Fernández et al.* (2003) found that a longer period ~41 h (1.71 d) does not contradict their observations obtained at three different epochs.

10P/Tempel 2: An early, in-depth investigation led *Sekanina* (1987) to constrain the orientation of the rotation axis of the nucleus and its gross physical properties. 10P was extensively observed by *A'Hearn et al.* (1989), who combined optical and infrared photometry, and by *Jewitt and Luu* (1989), who performed CCD photometry from aphelion (thus convincingly detecting a bare nucleus) to perihelion. Their interpretations converge to a spheroidal nucleus with a = 8–8.15 km and b = c = 4–4.3 km with an albedo p_R = 0.024 ± 0.005 and a rotational period of ~9 h. The effective radii are $r_{n,a}$ = 5.7–5.9 km and $r_{n,v}$ = 5.0–5.3 km. The revised values by *Campins et al.* (1995) remain in agreement with these results. Various snapshot observations are also in agreement with these results assuming the above albedo of 0.024 except as noted: *Mueller* (1992), r_n = 5.9 km; *Mueller and Ferrin* (1996), r_n = 5.2 km (p = 0.022); *La+*, r_n = 5.9 km; *Me+*, r_n = 6.4 km. The value of *Ta+* scaled to p_R = 0.024, i.e., r_n = 3.7 km, is inconsistent with the above results.

14P/Wolf: A snapshot observation by *Lo+*, when the comet was apparently inactive at r_h = 3.98 AU, yields r_n = 2.33 ± 0.12 km. The enormous scatter of the data points of *Ta+* makes their estimate of r_n = 1.3 km highly uncertain.

15P/Finlay, 16P/Brooks 2: The large scatter of the data points of *Ta+* makes their estimates highly uncertain.

17P/Holmes: A snapshot HST observation by *La+* yields r_n = 1.71 km.

19P/Borrelly: A complete solution was first proposed by *La+*: a = 4.4 ± 0.3 km, b = 1.8 ± 0.15 km assuming an albedo of 0.04 (see section 3.3.1 for a discussion of this issue). The *in situ* observations of *Deep Space 1* (*Soderblom et al.*, 2002; *Buratti et al.*, 2004) yield a = 4.0 ± 0.05 km and b = 1.6 ± 0.04 km. The above determination gives $r_{n,a}$ = 2.5 km and $r_{n,v}$ = 2.2 km. The snapshot results of *Lo+* and *Weissman et al.* (1999) are consistent with the above solution. The rotational periods $P = 25.0 \pm 0.5$ h found by *La+* and that obtained by *Mueller and Samarasinha* (2001), $P = 26$ h, are in excellent agreement.

21P/Giacobini-Zinner: A snapshot observation by *Mueller* (1992) at $R_h = 3.75$ AU gives $r_n = 2.1$ km (using an albedo of 0.04) and $a/b > 1.5$. The heliocentric light curve of *Ta+* indicates that the comet was still active at that distance and that their estimate $r_n = 1.0$ km derived from observations beyond 4.5 AU is reasonable. A rotational period of 9.5 ± 0.2 h has been reported by *Leibowitz and Brosch* (1986).

22P/Kopff: Combined visible and infrared photometry (*La+*) leads to $r_n = 1.67 \pm 0.18$ km and $p_V = 0.042 \pm 0.006$ ($p_R = 0.047$). The slightly different value of $r_n = 1.52$ km, reported by *Jorda et al.* (2000), resulted from an early, less-elaborate analysis of the same data. The visible light curve built from the eight HST observations spanning ~12 h had a small range of 0.14 ± 0.07 mag and could not constrain the rotational state. The snapshot observation at $r_h = 5.11$ AU by *Lo+*, when scaled to $p_R = 0.047$, gives $r_n = 1.65 \pm 0.1$ km, in remarkable agreement with the above results, as well as with the value estimated by *Ta+* ($r_h = 1.8$ km). A partial light curve recently obtained by *Lo+* at $r_h = 4.49$ AU clearly suggests a rotational period of 12.30 ± 0.8 h and an amplitude range of 0.55 ± 0.07 mag, corresponding to a minimum axial ratio of 1.66 ± 0.11 and a mean effective radius $r_n = 2.76 \pm 0.12$ km (scaled to $p_R = 0.047$ and $\beta = 0.04$ mag deg^{-1}). As discussed by *Lowry and Weissman* (2003), their solution is totally inconsistent with the above results (the pole-on view assumed for the *La+* observations must have resulted in a near-maximum cross section assuming a nucleus in simple rotation) but is consistent with unpublished results by K. Meech obtained at $r_h = 4.73$ AU: $r_n = 2.8$ km and $P_{rot} = 12.91$ h. In an early study of 22P, *Sekanina* (1984) found $P = 9.4 \pm 1.3$ h. At this stage, it is impossible to reconcile the two groups of observations without considering more complex solutions for the rotational state and the shape of the body. For the time being, we keep the self-consistent solution of *La+* for the size and albedo and the values of a/b and P_{rot} from *Lo+*.

24P/Schaumasse: The large scatter in the data used by *Ta+* makes their estimate highly uncertain.

26P/Grigg-Skjellerup: A stellarlike nucleus was observed by *Boehnhardt et al.* (1999) and by *Li+*, and they derived radius values of 1.44 ± 0.05 and 1.57 km respectively. However, a nondetection by *Lowry et al.* (1999) places an upper limit on the radius of 1.2 ± 0.1 km. The graph presented by *Ta+* suggests an inactive nucleus beyond ~2 AU, and their estimated radius of 1.3 km seems reasonable. A spheroid with $a \sim 2.2$ km and $b \sim 1$ km (i.e., $a/b = 2.2$) would be consistent with all the above results. This solution corresponds to $r_{n,v} = 1.3$ km. Radar observations yielded a lower limit $r_n > 0.4$ km (*Kamoun et al.*, 1982, 1999).

28P/Neujmin 1: This low-activity nucleus has been extensively studied and is the second largest EC in the sample. *Campins et al.* (1987) combined visible and infrared photometry when the comet made a relatively close encounter with Earth but was at a large phase angle (~30°). The maximum value of the infrared flux leads to $r_n = 10.6 \pm 0.5$ km and $p_V = 0.026$. Their single minimum value is not consistent with a light curve having $P = 12.75$ h (*Delahodde et al.*, 2001) and, in addition, brings the albedo down to an unrealistic value of $p_V = 0.016$. Visible photometry yields $r_n = 10.0$ km (*Jewitt and Meech* 1988) with $p_R = 0.03 \pm 0.01$ and $r_n = 11.4$ km (*Me+*). The visible albedo $p_V = 0.026$ and the color (V-R) = 0.45 leads to $p_R = 0.04$, so that the above values of r_n need not be scaled. Assuming that (*Me+*) observed the largest projected area πab, and that $a/b = 1.5$ [in fact, a lower limit obtained by *Delahodde et al.* (2001)], we obtain $a = 14.0$ km, $b = 9.3$ km, and $r_{n,v} = 10.7$ km, which is our current best estimate. A detailed analysis of a large set of observations led *Delahodde et al.* (2001) to determine a rotational period of 12.75 ± 0.03 h, in good agreement with previous measurements.

29P/Schwassmann-Wachmann 1: This is the largest EC in our sample, but there is some confusion regarding the classification of this comet. With a Tisserand parameter $T_J = 2.983$, it qualifies as an EC but its orbit also satisfies the strict definition of Centaurs given by *Jewitt and Kalas* (1998): $q \geq 5$ AU and $a \leq 30$ AU (corresponding to the orbits of Jupiter and Neptune respectively). Thus duality arises because the criteria for the two classifications are not consistent, T_J for ECs and q and a for the Centaurs. The current perihelion of 29P is less than 0.3 AU outside Jupiter's aphelion, and it could easily be perturbed into a fully crossing orbit in the near future. On the other hand, its albedo of 0.13 ± 0.04 (*Cruikshank and Brown*, 1983), if correct, is totally atypical of cometary nuclei (see section 3.3) while being common among Centaurs (*Barucci et al.*, 2004). Various estimates of its radius based on visible magnitudes range from 21 to 52 km assuming an albedo of 0.04. *Cruikshank and Brown* (1983) combined thermal measurements at 20 μm and visible photometry to obtain $r_n = 20.0 \pm 2.5$ km and $p_V = 0.13 \pm 0.04$. They correctly noted that the size is controlled by the infrared measurements, while the albedo is controlled by the visible magnitude, which was *estimated*. *Meech et al.* (1993) argued that 29P is probably never totally inactive and attempted to estimate the coma contribution by measuring the total nucleus + coma signal in apertures of different sizes. They determined a minimum axial ratio of 2.6 and, assuming $p_R = 0.04$ and $\beta = 0.04$ mag deg^{-1}, a rotationally averaged radius of $r_n = 15.4 \pm 0.2$ km, a value that we presently select. However, using an albedo of 0.13 reduces this value to 8.6 ± 0.1 km. For the rotational state of 29P, we adopt the simple rotation with a period of 14.0 h (*Meech et al.*, 1993), consistent with the rough estimate of 10 h reported by *Luu and Jewitt* (1993). However, the former authors determined a second period of 32.2 h, implying a complex state of rotation.

31P/Schwassmann-Wachmann 2: Observed as a starlike object at $r_h = 4.58$ AU by *Luu and Jewitt* (1992), who derived a radius of 3.1 km, a minimum axial ratio of 1.6, and a rotational period of 5.58 ± 0.03 h.

33P/Daniel: The value of *Ta+* looks questionable because of the large scatter in the data.

36P/Whipple: A snapshot observation of a starlike nucleus at $r_h = 4.43$ AU by *Lo+* give $r_n = 2.28 \pm 0.21$ km, which is consistent with the value quoted by *Ta+*, $r_n = 2.3$ km.

37P/Forbes: The stellarlike appearance at $r_h = 3.59$ AU led *Li+* to derive $r_n = 1.1$ km, close to the value of 1.0 km estimated by *Ta+*. *La+* obtained a slightly smaller value, $r_n = 0.81$ km. If the above authors observed different extreme cross-sections, the spheroidal solution leads to a = 1.38 km, b = 0.8 km (not unrealistic since a/b = 1.73), and $r_{n,v} = 0.96$ km.

39P/Oterma: The heliocentric light curve reported by *Ta+* does not allow a reliable derivation of the size.

40P/Väisälä 1: Undetected by *Lo+*, thus giving an upper limit $r_n < 3.6 \pm 0.2$ km. The estimate of $r_n = 1.5$ km by *Ta+* is reasonable, although an error bar of ±1 km is warranted given the large scatter in the data.

41P/Tuttle-Giacobini-Krešák: The heliocentric light curve of *Ta+* indicates that this is a very small nucleus; $r_n = 0.7$ km is probably a good estimate, but even this may only be an upper limit.

42P/Neujmin 3: There is too much scatter in the data used by *Ta+* to derive a reliable size of the nucleus. *Krešák et al.* (1984) reported that this comet and 53P/Van Biesbroeck are fragments from a parent comet that split in March 1845.

43P/Wolf-Harrington: Two determinations have been reported by *Lo+* from observations at $r_h = 4.87$ AU, $r_n = 3.3 \pm 0.7$ km, and at $r_h = 4.46$ AU, $r_n = 3.4 \pm 0.2$ km, when the comet had a stellar appearance. At $r_h = 3.04$ AU outbound, the comet was very active, displaying both a coma and a tail, leading *Li+* to impose $r_n \ll 3.1$ km. On the following inbound branch, the comet was reported active at $r_h = 3.9$ AU (*Hainaut et al.,* 1996). The graph of *Ta+* convincingly shows a monotonic decrease of brightness as r_h increases up to 4 AU, the faintest value yielding $r_n = 1.8$ km. A spheroidal solution based on the two above extreme cross-sections leads to a = 6.4 km, b = 1.8 km, and a/b = 3.6, which would be unusually large. Pending further observations, we are inclined to think that the large values of *Lo+* are not correct and that $r_n = 1.8$ km (*Ta+*) is a realistic estimate. Finally, it must be noted that 43P has undergone major orbital changes in the recent past (e.g., q decreased from 2.5 to 1.5 AU in 1936), and this could explain surges of vigorous activity thereafter.

44P/Reinmuth 2: A snapshot observation by *La+* gives $r_n = 1.61$ km. The estimate of 1.5 km by *Ta+* (but with considerable scatter in the data), and the upper limit of 3.1 km from *Lo+*, are consistent with that choice.

45P/Honda-Mrkos-Pajdušáková: *La+* obtained a mean value of $r_n = 0.34 \pm 0.01$ km from observations performed on two consecutive days, making this nucleus one of the smallest ever observed. However, they pointed out that the potentially large systematic error because of the large phase angle ($\alpha \simeq 90°$) during the observations. In fact, if 45P/HMP is as phase darkened as 2P/Encke and 48P/Johnson, then a linear phase coefficient $\beta = 0.06$ mag deg^{-1} should be applied instead of the standard value of 0.04 mag deg^{-1}, and

this leads to $r_n = 0.78$ km. The snapshot observation of *Lo+* gives a much larger value of $r_n = 1.34 \pm 0.55$ km, but the large error bar means that r_n could be as small as 0.79–0.82 km if $\beta = 0.06$ mag deg^{-1} is applied, in agreement with the above revised value. We conclude that $r_n = 0.8$ km is probably the best estimate for the time-being.

46P/Wirtanen: It was marginally detected on CCD frames by *Boehnhardt et al.* (1997) at $r_h = 4.6$ AU, giving an upper limit of the radius of 0.8 km (assuming p = 0.04) and a probable value of 0.69 km. The first unambiguous detection of the nucleus was by *La+*, giving $r_n = 0.62 \pm 0.02$ km (R band), a/b ≥ 1.2, and $P_{rot} = 6.0 \pm 0.3$ h. VLT observations by *Boehnhardt et al.* (2002) gave $r_n = 0.56 \pm 0.04$ km, a/b ≥ 1.4 ± 0.01, and a partial light curve in agreement with the above period. A slightly larger value of $r_n = 0.7$ km is reported by both *Meech et al.* (2000) and *Ta+*, while upper limits are given by *Lo+* and *Me+*. CCD photometry of the already active comet suggested a possible period of 7.6 h (*Meech et al.,* 1997).

47P/Ashbrook-Jackson: A partial light curve was obtained by *La+* (2001), giving a mean radius $r_n = 2.8$ km, a/b ≥ 1.4, and $P_{rot} \geq 44.5$ h. The nucleus appears inactive near aphelion (stellar appearance), so that the determination of *Li+*, $r_n = 3.1$ km, and the estimate of *Ta+*, $r_n = 2.9$ km, are in agreement taking into account the fact that this nucleus is elongated.

48P/Johnson: The nucleus was reported active at $r_h = 3.36$ AU by *Lo+*, thus only giving $r_n \leq 3.5$ km. Measurements by *Li+* at smaller r_h were certainly contaminated by a coma, although they claimed a stellar appearance, and this would explain their large value of $r_n = 3.7$ km. Several months of observations of a starlike nucleus at $r_h \sim 4$ AU allowed *Jewitt and Sheppard* (2003) to secure a fairly complete lightcurve and to derive a spheroidal solution with a = 3.5 and b = 2.6 km (yielding $r_{n,v} = 2.87$ km) and a/b ≥ 1.35 and $P_{rot} = 29.0 \pm 0.04$ h.

49P/Arend-Rigaux: This is a nearly extinct nucleus that has been extensively studied by combined visible and infrared photometry. *Tokunaga and Hanner* (1985) reported a size of $r_n = 4.8 \pm 0.4$ km and a geometric albedo of 0.05 ± 0.01 at 1.25 μm. *Brooke and Knacke* (1986) determined $r_n = 5.1 \pm 1.1$ km and $p_V = 0.02 \pm 0.01$. *Veeder et al.* (1987) found the nucleus to be elongated with equivalent radii of 5.1 and 3.8 km and $p_V = 0.03$. The in-depth investigation by *Millis et al.* (1988) resulted in a more accurate determination of the size and shape: a = 6.5 km and b = 4 km (a/b = 1.63), an albedo of $p_V = 0.028$, and a rotational period of P = 13.47 h. The observational data have been reanalyzed by *Campins et al.* (1995) using new parameters for the thermal model and they give an effective radius of 4.6 ± 0.2 km for a sphere having the maximum projected area πab and a geometric albedo of 0.04 ± 0.01. Keeping a/b = 1.63 from *Millis et al.* (1988), we obtained a = 5.9 km, b = 3.6 km, and $r_{n,v} = 4.24 \pm 0.2$ km. From R-band CCD photometry, *Lo+* reported two determinations of the radius, 3.8 ± 0.1 and 4.0 ± 0.11 km, assuming an albedo of 0.04. With the exception of the value reported by *Ta+* all the above results are consistent, the results of *Millis et al.* (1988) as corrected by *Campins et al.* (1995) providing the most detailed description. Thus we use $r_{n,v} = 4.24$ km and $p_V = 0.04 \pm 0.01$. *Jewitt and Meech* (1985) obtained a light curve from which they obtained two possible rotational periods, 9.58 ± 0.8 and 6.78 ± 0.08 h.

50P/Arend: A snapshot observation by *La+* gave $r_n = 0.95$ km. The revised estimate of *Ta+*, $r_n = 1.0$ km (compared to the original value of 3.0 km), is in good agreement with the above result.

51P/Harrington: *Lo+* give an upper limit of 1.9 km. Although consistent with it, the value of $r_n = 1.4$ km by *Ta+* cannot be considered reliable because of the large scatter in the data. Recent CCD images taken by *Manteca* (2001) show that this comet has split again into two components. A similar splitting was recorded at the 1994 apparition by *Scotti* (1994), who found a double nucleus on Spacewatch images. A detailed analysis of the astrometric data and of the circumstances of the splitting is still in progress (*Sekanina,* 2001).

52P/Harrington-Abell: *Li+* obtained $r_n = 1.4$ km from two images of a starlike nucleus at $r_h = 2.83$ AU. The fainter magnitude reported by *Carlson* (1990) corresponds to $r_n = 1.0$ km, in close agreement with the value selected by *Ta+*, $r_n = 1.1$ km. A spheroidal solution assuming that the above observations correspond to extreme cross-sections yields a = 2, b = 1 km (i.e., a/b = 2), and $r_{n,v} = 1.3$ km.

53P/Van Biesbroeck: *Me+* derived $r_n = 3.33$ km from a snapshot observation at $r_h = 8.31$ AU (i.e., close to aphelion) in agreement with the result $r_n \ll 6.7$ km of *Li+*. The comet is known to be active out to 6 AU, as illustrated by the very erratic heliocentric light curve of *Ta+*. *Krešák et al.* (1984) reported that this comet and 42P/Neujmin 3 are fragments of a parent comet that split in March 1845.

54P/de Vico-Swift: Undetected by *Lo+* at $r_h = 5.39$ AU (aphelion), they obtained an upper limit of 2.1 km.

56P/Slaughter-Burnham: A snapshot observation at $r_h = 7.42$ AU by *Me+* of the inactive nucleus gives $r_n = 1.56$ km. This is in good agreement with the estimate of *Ta+*, $r_n = 1.5$ km.

57P/du Toit-Neujmin-Delporte: *Lo+* determined an upper limit of 1.1 km. The considerable scatter in the heliocentric light curve of *Ta+* makes their estimate of $r_n = 1.6$ km unreliable. The comet has recently split: Two fragments were first discovered (cf. *Marsden,* 2002), followed by 18 more (*Fernández et al., 2002b*). A preliminary analysis of this event has been reported by *Sekanina* (2002a,b).

58P/Jackson-Neujmin: There is too much scatter in the light curve of *Ta+* to estimate a size.

59P/Kearns-Kwee: A snapshot observation by *La+* yields $r_n = 0.79$ km. Such a small nucleus, active out to at least 4.2 AU, would be very difficult to detect from the ground.

60P/Tsuchinshan 2: Certainly a very small nucleus ($r_n < 1$ km) and *Ta+* estimated $r_n = 0.8$ km, but the compiled data in their plot have wide scatter. The nucleus may be as small as ~0.5 km.

61P/Shajn-Schaldach: A partial rotational light curve has been obtained by *La+* giving a mean radius $r_n =$

0.64 km, a/b > 1.3, and $P_{rot} \geq 18$ h. At the time of the HST observations, $r_h = 2.96$ AU, the comet was still very active, the nucleus and the coma contributing equally to the signal in the peak pixel. The snapshot observation of *Lo+* at $r_h = 4.4$ AU, performed under nonphotometric conditions, gives $r_n = 0.92 \pm 0.24$ km. The low end, $r_n = 0.68$ km, is consistent with the result of *La+*. The comet may still have been weakly active at 4.4 AU.

62P/Tsuchinshan 1: Certainly a very small nucleus ($r_n < 1$ km) and (*Ta+*) estimated $r_n = 0.8$ km, but the compiled data in their plot have wide scatter, as in the case of 60P.

63P/Wild 1: A snapshot observation by *La+* gives $r_n = 1.45$ km. The nondetection by *Lowry and Fitzsimmons* (2001), which results in an upper limit $r_n \leq 0.6$ km, is therefore puzzling. Invoking a highly elongated spheroid to reconcile the two observations would be rather artificial.

64P/Swift-Gehrels: A starlike nucleus detected at $r_h = 3.63$ AU by *Li+* gave $r_h = 1.6 \pm 0.1$ km, which is consistent with the upper limit $r_h \leq 1.9$ km obtained by *Lo+* and with the estimate by *Ta+* of $r_n = 1.7$ km.

65P/Gunn: This comet is very active out to aphelion, so that only upper limits were obtained, $r_n \ll 11.7$ km (*Li+*) and $r_n \leq 8.8$ km (*Lo+*). The estimate proposed by *Ta+* is $r_n = 4.8$ km.

67P/Churyumov-Gerasimenko: A rotational light curve has been obtained by *La+* giving a mean radius $r_n = 1.98 \pm 0.02$ km, a/b > 1.3, and $P_{rot} = 12.3 \pm 0.27$ h. This comet was undetected by *Lo+* at aphelion (5.72 AU), thus imposing $r_n \leq 2.9$ km. Observations at 4.87 and 4.97 AU by *Mueller* (1992) give $r_n = 2.8 \pm 0.1$ km (scaled to $p_R = 0.04$) and a/b > 1.7. The heliocentric light curve is well-behaved and shows that the comet is inactive beyond 4.5 AU and the revised estimate $r_n = 2.0$ km by *Ta+* agrees with the result of *La+*.

68P/Klemola: The well-behaved heliocentric light curve produced by *Ta+* suggests that their value $r_n = 2.2$ km is a good estimate.

69P/Taylor: This comet was found to be active at $r_h = 4.03$ AU by *Lo+*, who obtained an upper limit of 3.4 km.

70P/Kojima: A partial rotational light curve was obtained by *La+*, giving a mean radius $r_n = 1.86$ km, a/b > 1.1, and $P_{rot} \geq 22$ h. The large scatter in the data at $r_h > 3.4$ AU makes the estimate of *Ta+*, $r_n = 1.2$ km, rather arbitrary.

71P/Clark: A snapshot observation by *La+* at $r_h = 2.72$ AU gives $r_n = 0.68 \pm 0.07$ km, which is consistent with the nondetection at $r_h = 4.4$ AU by *Lo+* ($r_n \leq 0.9$ km). The observations by *Me+* at aphelion ($r_h = 4.67$ AU) yields $r_n = 1.31 \pm 0.04$ km, similar to the value estimated by *Ta+*. A spheroid with a = 2.13 km and b = 0.75 km could reconcile the two determinations within the error bars, but has a very large axis ratio, a/b ≥ 2.85, and further requires that *La+* and *Lo+* observed the smallest cross-section while *Me+* observed the largest one. This nucleus certainly deserves further observations.

73P/Schwassmann-Wachmann 3: The nucleus was reported as split into two components by *Schuller* (1930), but there was no other independent report. In 1994, the comet was reported active at $r_h = 3.03$ AU, and *Boehnhardt et al.*

(1999) derived $r_n < 1.26$ km. The principal nucleus further split into at least three components in the autumn of 1995. Undetected in 1998 by *Lowry and Fitzsimmons* (2001) at $r_h = 5.03$ AU, they obtained an upper limit for the largest component of $r_n < 0.9$ km. Fragment C was detected by the HST at $r_h = 3.25$ AU and *Toth et al.* (2003) derived $r_n = 0.68 \pm 0.04$ km and a/b > 1.16.

74P/Smirnova-Chernykh: A partial rotational light curve was obtained by *La+* giving a mean radius of $r_n = 2.23 \pm 0.1$ km, a/b > 1.14, and P ~ 20 h. At $r_h = 3.56$ AU, the comet was still very active, the nucleus and the coma contributing equally to the signal in the peak pixel. This explains why *Li+* ($r_h = 4.57$ AU) and *Lo+* ($r_h = 4.61$ AU) obtained only upper limits, $r_n \ll 11.2$ km and $r_n \leq 7.1 \pm 1.1$ km, respectively. The value $r_n = 6$ km estimated by *Ta+* is totally arbitrary.

75P/Kohoutek: A nondetection by *Lo+* gives $r_n \leq 1.5$ km, while the estimate by *Ta+* is $r_n = 1.8$ km.

76P/West-Kohoutek-Ikemura: A partial rotational light curve was obtained by *La+* at $r_h = 3.09$ AU giving a mean radius $r_n = 0.33 \pm 0.03$ km, a/b > 1.47, and $P_{rot} \sim 13$ h. The comet was still active, explaining the much larger value estimated by *Ta+*.

77P/Longmore: There is too much scatter in the data of *Ta+* to obtain a reliable estimate.

78P/Gehrels 2:   A snapshot observation of a starlike nucleus at $r_h = 5.46$ AU *Lo+* yields $r_n = 1.42 \pm 0.12$ km. The heliocentric light curve of *Ta+* displays a lot of scatter at 3.5 AU, suggesting that the comet is still active at that distance.

79P/du Toit-Hartley: A snapshot observation at $r_h = 4.74$ AU by *Lo+* revealed an inactive nucleus whose radius is $r_n = 1.4 \pm 0.3$ km.

81P/Wild 2:   Observed while the nucleus was inactive at $r_h = 4.7$ AU by *Meech and Newburn* (1998), they determined $r_n = 2.0 \pm 0.04$ km and that the nucleus is fairly spherical, or has a relatively long period. Still inactive at $r_h = 4.25$ AU inbound, *Lo+* obtained $r_n = 2.0 \pm 0.3$ km. The comet was, however, found active at $r_h = 4.34$ AU outbound, so that *Li+* put an upper limit $r_n \ll 5.7$ km. Finally, *Fernández* (1999) obtained infrared images at 10.6 μm when the comet was at $r_h = 1.85$ AU, and therefore active. Although the measured flux was probably dominated by coma, that author applied the standard thermal model for asteroids to derive $r_n < 3.0 \pm 0.6$ km. A nearly spherical nucleus with $r_n = 2$ km is the most probable solution. This comet is the target of the *Stardust* mission, which will fly by its nucleus in January 2004.

82P/Gehrels 3:   A partial rotational light curve was obtained by *La+* at $r_h = 3.73$ AU giving a mean radius $r_n = 0.73 \pm 0.02$ km, a/b > 1.6, and $P_{rot} \sim 50$ h. The comet appears to be active all along its orbit, so that *Li+* could only determine $r_n < 3.0$ km.

83P/Russell 1:   A nondetection by *Lo+* at $r_h = 3.01$ AU gives $r_n \leq 0.5$ km.

84P/Giclas: A snapshot observation by *La+* gives $r_n = 0.90 \pm 0.05$ km.

86P/Wild 3:   A partial rotational light curve was obtained by *La+* at $r_h = 2.32$ AU giving a mean radius $r_n = 0.43 \pm$ 0.02 km, a/b > 1.35, and an ill-defined period. A snapshot observation at 4.95 AU by *Me+* gives $r_n = 0.65 \pm 0.03$ km. These two determinations are inconsistent but satisfy the condition $r_n \leq 0.9$ km (*Lo+*). The unpublished value of 3.1 km suggested by Meech is apparently unjustified.

87P/Bus: The light curve obtained by *La+* at $r_h = 2.45$ AU gives a mean radius $r_n = 0.28 \pm 0.01$ km, a/b > 2.20, and $P_{rot} = 25$ h. Two upper limits have been reported, $r_n \leq 0.6$ km at $r_h = 4.32$ AU (undetected) by *Lo+* and $r_n < 3.42$ km at $r_h = 4.77$ AU (close to aphelion) by *Me+* (a coma was present). It is probably impossible to detect this nucleus from the ground.

88P/Howell, 89P/Russell 2, 90P/Gehrels 1, 91P/Russell 3, 94P/Russell 4:   There is too much scatter in the data presented by *Ta+* to obtain reliable size estimates. 89P was found active at $R_h = 3.04$ AU by *Lo+* leading to $r_n < 2.2$ km, and possibly $r_n \leq 1.3 \pm 0.3$ km after removing the comatic contribution.

92P/Sanguin: Two snapshot observations of a starlike nucleus, one at $r_h = 8.57$ AU by *Me+*, the other at $r_h = 4.46$ AU by *Lo+*, give $r_n = 1.19$ km and $r_n = 1.7 \pm 0.63$ km respectively. These values are consistent owing to the large uncertainty in the latter value.

97P/Metcalf-Brewington: A starlike nucleus (with possibly a faint coma) detected at $r_h = 3.67$ AU by *Li+* gives $r_n = 1.5 \pm 0.16$ km, which, strictly speaking, should be considered an upper limit. Observed at 4.76 AU inbound by *Lo+*, it was found inactive resulting in $r_n = 2.18 \pm 0.41$ km. An intermediate size of $r_n = 1.7$ km is compatible with these two determinations, taking into account the error bars.

98P/Takamizawa: Observed at $r_h = 3.78$ AU by *Li+* as a trailed object, their determination of $r_n = 3.7$ km cannot be considered reliable. The heliocentric light curve of *Ta+* suggests that this comet is weakly active; their radius $r_n = 2.4$ km has now been revised to $r_n = 3$ km.

99P/Kowal 1:   There is too much scatter in the data presented by *Ta+* to obtain a reliable estimate.

100P/Hartley 1:   Undetected by *Lo+* at $R_h = 3.94$ AU, they derived an upper limit $r_n < 1.2$ km.

101P/Chernykh: There is two much scatter in the data presented by *Ta+* to obtain a reliable estimate. *Luu and Jewitt* (1991) discovered that this comet has split.

103P/Hartley 2:   The thermal flux of the nucleus was measured at 11.5 μm using ISOCAM on ISO. The preliminary determination $r_n = 0.58$ km (*Jorda et al.,* 2000) has now been revised to $r_n = 0.71 \pm 0.13$ km by *Groussin et al.* (2003), which is consistent with the upper limits of *Li+*, $r_n \ll 5.3$ and of *Lo+* $r_n \leq 5.8$ km.

104P/Kowal 2:   A snapshot observation of a starlike nucleus at $r_h = 3.94$ AU by *Lo+* gives $r_n = 1.0 \pm 0.5$ km.

105P/Singer-Brewster: There is too much scatter in the data presented by *Ta+* to obtain a reliable estimate.

106P/Schuster: A snapshot observation by *La+* gives $r_n = 0.94 \pm 0.05$ km, which is quite close to the value in the revised catalog of *Ta+*, $r_n = 0.8$ km.

107P/Wilson-Harrington: The identification of this object as a comet remains problematic as discussed by *Weissman et al.* (2003) and it was in fact first classified as a near-Earth asteroid. It has a Tisserand parameter slightly in excess of 3 ($T_J = 3.084$) but an orbit typical of ECs (a = 2.643 AU, e = 0.621, i = 2.78°). In their dynamical analysis, *Bottke et al.* (2002) assign it only a 4% probability of being of cometary origin; they find the most probable source to be the outer main belt (65%). Its activity was observed on only one night, on two Palomar photographic plates taken in 1949, and the object is trailed on both images; no activity was detected on plates taken three nights later. Subsequent searches for cometary activity have all been negative (e.g., *Chamberlin et al.*, 1996). 107P was observed simultaneously in the near and thermal infrared by *Campins et al.* (1995). Using first the STM, they obtained $r_n = 1.3 \pm 0.16$ km and $p_J = 0.10 \pm 0.02$, a somewhat surprising value. Thus, they favored the ILM solution, which gives $r_n = 2.0 \pm 0.25$ km and $p_J = 0.05 \pm 0.01$. Since the color is very neutral, this value holds as well for the V and R bands. Visible snapshot observations have been reported by *Lo+* giving $r_n = 1.78 \pm 0.03$ km and *Me+* giving $r_n = 1.96 \pm 0.02$ km. There also exists an unpublished value of $r_n = 2.0$ km by K. Meech. The rotational state has been investigated by *Osip et al.* (1995), who found $P_{rot} = 6.1 \pm 0.05$ h and $a/b \geq 1.2$. If we scale the above *Lo+* and *Me+* values to $p_R = 0.05$, we get $r_n = 1.59 \pm 0.03$ and $r_n = 1.75 \pm 0.02$, respectively. A spheroidal solution with a = 1.9 km and b = 1.6 km with $p_R = 0.05$ is then compatible with all above results, implying an obliquity of 90°, and that *Campins et al.* (1995) and *Me+* observed the largest cross-section while *Lo+* observed the smallest. We then obtain $r_{n,v} = 1.7$ km.

110P/Hartley 3:   A complete rotational light curve has been obtained by *La+* giving a mean radius $r_n = 2.15 \pm 0.05$ km, a/b > 1.3, and $P_{rot} \sim 10$ h.

111P/Helin-Roman-Crockett: This comet was undetected by *Lo+* at $r_h = 4.35$ AU, thus imposing $r_n \leq 1.5$ km. NTT observations made at $r_h = 4.56$ AU by *Delahodde* (2003) give a subkilometer radius of $r_n = 0.6 \pm 0.3$ km.

112P/Urata-Niijima: A snapshot HST observation at $r_h = 2.30$ AU by *La+* gives $r_n = 0.90 \pm 0.05$ km. The estimate by *Ta+* is 0.7 km, but the compiled data in their plot have wide scatter.

113P/Spitaler: This comet was undetected by *Lo+* at $r_h = 4.22$ AU, thus imposing $r_n \leq 2.0$ km. The estimate of *Ta+*, $r_n = 1.1$ km, seems plausible.

114P/Wiseman-Skiff: A snapshot HST observation at $r_H = 1.57$ AU by *La+* gives $r_n = 0.78 \pm 0.04$ km.

115P/Maury: *Me+* observed a starlike nucleus at $r_h = 5.34$ AU and give $r_n = 1.11$ km.

116P/Wild 4:   The heliocentric light curve of *Ta+* suggests that the comet could be inactive at $r_h = 4$ AU. A radius of 3.5 km, recently revised to 3.0 km, is probably a good estimate, pending further observations.

117P/Helin-Roman-Alu: The value of *Ta+*, $r_n = 3.5$ km, comes from measurements at aphelion. It is unclear whether the comet is inactive then.

118P/Shoemaker-Levy 4:   A starlike nucleus detected at $r_h = 4.71$ AU by *Lo+* gives $r_n = 2.4 \pm 0.2$ km similar to the value estimated by *Ta+*, but the compiled data in their plot have wide scatter.

119P/Parker-Hartley: This comet was still very active at $r_h = 3.42$ AU when observed by *Lo+*, who obtained an upper limit $r_n < 7.4 \pm 0.2$ km and then refined that to $r_n < 4.0 \pm 0.6$ km after estimating the comatic contribution. As illustrated by the heliocentric light curve of *Ta+*, the comet is simply too active out to 4 AU to make a sensible estimate of r_n.

120P/Mueller 1:   A starlike nucleus detected at $r_h = 3.08$ AU by *Lo+* gives $r_n = 1.5$ km. The plot of *Ta+* shows considerable scatter of the magnitudes at that distance, indicating that the comet could well be active.

121P/Shoemaker-Holt 2:   A starlike nucleus detected at $r_h = 5.03$ AU by *Lo+* gives $r_n = 1.62 \pm 0.57$ km.

123P/West-Hartley, 124P/Mrkos, 125P/Spacewatch: There is too much scatter in the data presented by *Ta+* to obtain reliable estimates.

128P/Shoemaker-Holt 1:   Two snapshot observations of a starlike nucleus at $r_h = 4.99$ AU by *Lo+* give compatible results, $r_n = 2.48 \pm 0.1$ km and $r_n = 2.12 \pm 0.18$ km, which are both consistent with the upper limit $r_n < 4.0$ km when the comet was observed at $r_h = 3.66$ AU by *Lo+*, when it was still active. We adopt an average value $r_{n,v} = 2.3$ km.

130P/McNaught-Hughes, 131P/Mueller 2, 132P/Helin-Roman-Alu 2, 134P/Kowal-Vavrová, 135P/Shoemaker-Levy 8, 136P/Mueller 3:   There is too much scatter in the data presented by *Ta+* to extract reliable estimates.

137P/Shoemaker-Levy 2:   This comet was observed at $r_h = 4.24$ AU by *Li+*, but various technical problems make their determination $r_n = 4.5$ km questionable. This is confirmed by the upper limit $r_n \leq 3.4$ km found by *Lo+* when they observed the comet at $r_h = 2.29$ AU, while it was still active. The heliocentric light curve of *Ta+* indicates that the magnitude at $r_h = 5$ AU may provide a good estimate, $r_n = 2.9$ km.

138P/Shoemaker-Levy 7:   There is too much scatter in the data presented by *Ta+* to extract a reliable estimate.

143P/Kowal-Mrkos: A starlike nucleus was observed by *Jewitt et al.* (2003) from $r_h = 3.4$ to 4.0 AU and their almost complete lightcurve clearly suggests a rotational period of 17.21 ± 0.1 h and an amplitude range of 0.45 ± 0.05 mag, corresponding to a minimum axial ratio of 1.49 ± 0.05. Assuming an albedo of 0.04 and using the phase coefficient determined by these authors $\beta = 0.043 \pm 0.014$ mag deg^{-1}, the spheroidal solution has semiaxes a = 7.0 and b = 4.7 km, yielding $r_{n,v} = 5.4$ km. Note that the effective radius $r_n = 5.7 \pm 0.6$ km reported by the above authors was derived using a *Bowell et al.* (1989)-type phase curve having G = 0.15, which has an opposition effect of about 0.2 mag above the linear phase law.

147P/Kushida-Muramatsu: The nucleus of this comet is the smallest of all the objects cataloged to date. From observations with the HST at $r_h = 2.83$ AU over a 13-h time interval, *La+* found a very small, $r_n = 0.21 \pm 0.01$ km, highly active nucleus with a/b > 1.53, and a possible rotational period of 9.5 h. It is therefore not surprising that *Lo+* could not detect the comet at $r_h = 4.11$ AU, thus imposing $r_n \leq 2.0$ km.

152P/Helin-Lawrence: This comet is still active at aphelion (r_h = 5.85 AU), and the smallest upper limit is presently $r_n \leq 1.74$ km (*Me+*).

P/1993 W1 (Mueller 5), P/1994 A1 (Kushida), P/1994 J3 (Shoemaker 4), P/1995 A1 (Jedicke), P/1996 A1 (Jedicke), P/1997 C1 (Gehrels), P/1997 G1 (Montani), P/1997 V1 (Larsen): There is too much scatter in the data presented by *Tancredi et al.* (2000) to extract a reliable estimate.

3.2.2. Nearly isotropic comets (NICs). 1P/Halley: The size and shape of its nucleus were determined from *in situ* imaging made by the *Vega 1,2* and *Giotto* spacecrafts in 1986 (*Sagdeev et al.,* 1986a,b; *Keller et al.,* 1986, 1987, 1994; *Keller,* 1990; see also *Keller et al.,* 2004). The nucleus is an elongated, irregularly shaped body approximated by an ellipsoid with semiaxes (a × b × c) of 7.21 ± 0.15 × 3.7 ± 0.1 × 3.7 ± 0.1 km (*Giotto*) and 7.65 ± 0.25 × 3.61 ± 0.25 × 3.61 ± 0.25 km (*Vega 1,2*). It is in nonprincipal axis rotation, and there was a long dispute over whether the nucleus rotates in the "short-axis mode" (SAM) or "long-axis mode" (LAM) (*Sagdeev et al.,* 1989; *Peale and Lissauer,* 1989; *Abergel and Bertaux,* 1990; *Belton,* 1990; *Belton et al.,* 1991; *Samarasinha and A'Hearn,* 1991). In the modes identified as most likely, the long axis conducts a 3.7-d precessional motion around the space-fixed vector of the total rotational angular momentum, while the nucleus also rotates around the long axis with a 7.3-d period.

8P/Tuttle: A single value r_n = 7.8 km was reported by *Li+* when the comet was at r_h = 6.29 AU and appeared inactive.

55P/Tempel-Tuttle: Its effective radius of 1.8 km and a minimum value of 1.5 for the axial ratio were derived from HST WFPC2 and ISO ISOCAM observations (*La+*). Groundbased observers determined similar sizes, e.g., *Hainaut et al.* (1998) and P. Weissman and B. Buratti (personal communication, 2003) in the visible and *Fernández* (1999) in the midinfrared. There is still no data published for the rotational period of the nucleus.

96P/Machholz 1: A starlike nucleus detected at r_h = 4.83 AU by *Li+* gives r_n = 3.5 km. There are unpublished data by K. Meech giving r_h = 2.8 km, a/b > 1.4, and P_{rot} = 6.38 h. A spheroid with a = 4.3 km and b = 2.8 km (a/b ~ 1.5) would be consistent with the two results above, yielding $r_{n,v}$ = 3.2 km.

109P/Swift-Tuttle: A mean effective radius of 11.8 km was determined from groundbased CCD photometry (*O'Ceallaigh et al.,* 1995) at r_h = 5.3 AU outbound in the presence of a weak coma. Later, at r_h ~ 5.8 AU outbound, the nucleus had a stellar appearance and *Boehnhardt et al.* (1996) determined two comparable values of the radius, 12.2 and 12.5 ± 0.3 km at a time interval of 5 d. *Meech et al.* (2004) observed this comet at 14.5 AU and derived an effective radius of 13.7 km. A radius of 15.0 ± 3.0 km was estimated from groundbased IR photometry (*Fomenkova et al.,* 1995). An average radius of 13.0 km appears realistic. The rotational period of the nucleus has been determined by *Sekanina* (1981) from the recurrent pattern of coma jets on 1862 photographs, P_{rot} = 66.5 h, and on 1992 CCD images by

Jorda and Lecacheux (1992), ~69.6 h, *Yoshida et al.* (1993), 69.4 ± 0.24 h, and *Boehnhardt and Birkle* (1994), 67.08 h.

126P/IRAS: The thermal flux of the nucleus was measured at 11.5 µm using ISOCAM. The preliminary determination r_n = 1.43 km (*Jorda et al.,* 2000) has now been refined to r_n = 1.57 ± 0.14 km by *Groussin et al.* (2003).

P/1991 L3 (Levy): A stellar appearance at r_h = 3.1 AU led *Fitzsimmons and Williams* (1994) to consider that they observed a bare nucleus, shortly after it had ceased outgassing. They determined r_n = 5.8 ± 0.1 km, a/b > 1.3, and P_{rot} = 8.34 h.

C/1983 H1 (IRAS-Araki-Alcock): Extensive observations in the visible, infrared, radio, and radar wavelength ranges were performed when it passed near Earth on 11 May 1983. The radar and radio observations of *Altenhoff et al.* (1983), *Goldstein et al.* (1984), *Irvine et al.* (1984), and *Harmon et al.* (1989) converge to a nonspherical nucleus with a radius is in the range 2.5–6.0 km (the a/b ratio could not be determined) and a rotation period is in the range 24–72 h. From a study of the temporal variation of its asymmetric coma, *Watanabe* (1987) and later *Pittichová* (1997) estimated that the period lies in the range 18–170 h. From a synthesis of visible, infrared, and radar observations, *Sekanina* (1988) derived a prolate spheroid nucleus with a = 8, b = c = 3.5 km, and a rotational period of 51.3 h. Infrared observations and a simple thermal model, assuming a constant temperature for the surface of the nucleus, were used to derive that the radius was in the range of 3.6–5.0 km (*Feierberg et al.,* 1984; *Hanner et al.,* 1985; *Brown et al.,* 1985). *Groussin et al.* (2004) reexamined the interpretation of all visible, infrared, and radio observations and using their thermal model, they derived an equivalent radius of r_n = 3.5 ± 0.5 km.

C/1983 J1 (Sugano-Saigusa-Fujikawa): *Hanner et al.* (1987) obtained a value of 0.37 km for the average radius of the nucleus from infrared spectroscopic observations. This result was a clear indication that cometary nuclei, including NICs, could be subkilometer-sized bodies.

C/1995 O1 (Hale-Bopp): *Weaver and Lamy* (1997) and *Fernández* (2003) reviewed all the data pertaining to the size of the nucleus. The former review discusses all wavelengths, while the latter focuses on infrared and radio observations. The dominant visible-wavelength dataset is from HST. The spatial resolution and image quality were sufficient to obtain photometric extractions of the nucleus, from which a radius of ~35 km was derived (*Weaver and Lamy,* 1997). The dominant radiometric datasets are from ISO (*Jorda et al.,* 2000), the Very Large Array (VLA) (*Fernández,* 1999), the Owens Valley Radio Observatory (OVRO) (*Qi,* 1998), and the Institut de Radioastronomie Millimétrique (IRAM) (*Altenhoff et al.,* 1999). Generally, the radio data suggest a smaller nucleus than implied by the infrared data. A compromise solution by *Fernández* (2003) was to argue that (1) the subsurface layer sampled by the radio observations was cooler and/or less emissive than expected, and (2) there was some excess dust not accounted for in the infrared photometry. This would shift the radii from the two wavelength regimes toward each other and leads to a radius of 30 ± 10 km, consistent with the HST results. The rotational period of the nucleus was determined by two different methods: 11.34 ± 0.02 h was derived by *Licandro et al.* (1998) and 11.35 ± 0.04 h by *Jorda et al.* (1999) (see *Samarasinha et al., 2004*). An extensive, groundbased CCD imaging observational campaign (*Farnham et al., 1999*) showed a systematic motion of the rotational pole of the nucleus, and this was interpreted as resulting from precession due to complex rotation. However, *Samarasinha* (2000) showed that there are no effects due to precession in the observed coma morphology.

C/1996 B2 (Hyakutake): Optical, infrared, and radar observations were performed during its close approach to Earth. *Lisse et al.* (1999) estimated a nuclear radius of 2.4 ± 0.5 km from the infrared and optical data. Radar observations revealed a clear detection of the nucleus, but an extremely small radar albedo of 0.011 is required to be consistent with the infrared data (*Harmon et al., 1997, 2004*). If the radar albedo is 0.04, the radius of the nucleus derived from the radar detection drops to only ~1.3 km. Early observations showed a fast rotation period of 6.30 ± 0.03 h, which was later refined by *Schleicher and Osip* (2002) to 6.273 ± 0.007 h. This NIC underwent a partial fragmentation as large fragments (~10–20 m in diameter) were observed traveling away from the nucleus with a velocity of ~10 m s^{-1} (*Lecacheux et al., 1996; Desvoivres et al., 2000*).

C/1983 O1 (Černis), C/1984 K1 (Shoemaker), C/1986 P1 (Wilson), C/1987 H1 (Shoemaker), C/1987 F1 (Torres), C/1988 B1 (Shoemaker), C/1997 T1 (Utsonomiya): Only upper limits are reported for the nuclear radii of these NICs by *Me+*. An upper limit for the radius of C/1997 T1 is also reported by *Fernández* (1999) from infrared measurements.

C/1999 S4 (LINEAR): This comet underwent catastrophic fragmentation in July 2000. Lower limits for the size of the nucleus prior to disruption were derived indirectly from the long-term monitoring of the water production rate: $r_n \geq 0.375$ km by *Mäkinen et al.* (2001), and $r_n \geq 0.44$ km by *Farnham et al.* (2001).

C/2001 OG108 (LONEOS): First classified as an asteroid of the Damocloid group, it developed a small amount of cometary activity as it approached perihelion and was subsequently reclassified as a comet. Simultaneous optical and thermal observations by *Abell et al.* (2003) give an effective radius of 8.9 ± 0.7 km and a visual albedo $p_V = 0.03 \pm 0.005$. Their composite lightcurve indicates a simple rotation with a period of 57.19 ± 0.5 h and a minimum axial ration of 1.5. The spheroidal solution assuming a/b = 1.3 has a = 10.1 km and b = c = 7.9 km.

Essentially all the best data on the sizes and shapes of cometary nuclei are summarized in Tables 1, 2, and 3. The column labeled $r_{n,v}$ displays what we consider to be the most reliable value of the effective radius, as defined in section 3.2. An absence of value means that, in our opinion, a reliable determination does not yet exist.

3.3. Albedo

3.3.1. Ecliptic comets (ECs).
2P/Encke: Using radiometry, *Fernández et al.* (2000) report 0.046 ± 0.023 for the V band.

9P/Tempel 1: Using radiometry, *Fernández et al.* (2003) report 0.072 ± 0.016 for the R band. However, as discussed in section 3.2.1, this large value probably results from coma contaminated visible magnitudes. The most likely value is $p_R = 0.048 \pm 0.007$.

10P/Tempel 2: From radiometry of the nucleus, *A'Hearn et al.* (1989) report an albedo of $0.022^{+0.004}_{-0.006}$ for a wavelength of 4845 Å. *Tokunaga et al.* (1992) report a near-infrared (1.25 to 2.20 μm) albedo of 0.04–0.07, which is consistent with the reddening of the nucleus in this wavelength regime compared to the visible.

19P/Borrelly: *Buratti et al.* (2004) used disk-resolved imaging of the nucleus obtained by the Miniature Integrated Camera and Spectrometer (MICAS) instrument on the *Deep Space 1 (DS1)* mission, and a scattering model based on the *Hapke* (1986) formalism, to calculate a disk-integrated geometric albedo of 0.029 ± 0.006. Table 3 of *Buratti et al.* (2004) indicates that this is the p_V value, but we think that it in fact corresponds to the R band for two reasons. First, the *DS1* images have an effective wavelength of 0.66 μm, and second, the albedo was derived from the absolute R magnitudes, R(1,1,α). Variations are, however, observed on the surface of the nucleus, and the two main types of terrains, smooth and mottled, exhibit mean normal reflectances of 0.03 and 0.022. The above albedo is lower than that assumed by *La+* (0.04) but, as discussed by *Buratti et al.* (2004), the respective uncertainties in the HST and *DS1* measurements make the two results fully consistent. This justifies the superposition of the prolate spheroid model derived from the HST observations and a *DS1* image displayed in Fig. 2.

22P/Kopff: Using radiometry, *Lamy et al.* (2002) report 0.042 ± 0.006 for the V band.

28P/Neujmin 1: As discussed in section 3.2.1, the maximum value of the infrared flux measured by *Campins et al.* (1987) leads to $p_V = 0.026$. This value and the color (V-R) = 0.45 leads to $p_R = 0.04$, in good agreement with the value $p_R = 0.03 \pm 0.01$ determined by *Jewitt and Meech* (1988).

29P/Schwassmann-Wachmann 1: Using radiometry, *Cruikshank and Brown* (1983) report $p_V = 0.13 \pm 0.04$. As emphasized in section 3.2, this value is controlled by the visible magnitude, which was *estimated*.

49P/Arend-Rigaux: The albedo has been constrained by many groups, all using groundbased radiometry. The results of *Millis et al.* (1988), revised by *Campins et al.* (1994), give $p_V = 0.04 \pm 0.01$. In the near-infrared (specifically J band), measurements by *Tokunaga and Hanner* (1985), 0.054 ± 0.010, and by *Brooke and Knacke* (1986), 0.03 ± 0.01, are consistent with the value at visible wavelengths.

107P/Wilson-Harrington: Using radiometry, *Campins*

TABLE 1.   Nuclei of the ecliptic comets (ECs).

Comet	Effective radius (km)								a/b (min)	P_{rot} (h)
	La+	*Lo+*	*Li+*	*Me+*	*Sc*	Others	*Ta+*	$r_{n,v}$		
2P/Encke	—	4.4	—	—	3.2	2.4 3.1 4.5	2.4(1.3)	2.4	2.6	11.
4P/Faye	1.77	—	—	—	2.3	—	2.2(1.7)	1.77	1.25	—
6P/d'Arrest	—	1.6	—	1.71	—	3.5	1.5	1.6	1.2	7.0
7P/Pons-Winnecke	—	2.6	—	—	—	—	1.5	2.6	—	—
9P/Tempel 1	3.13	2.4	—	3.07	3.2	3.32	2.3(1.9)	3.1	1.40	41.0
10P/Tempel 2	4.63	—	—	4.93	4.1	3.1 5.9	2.9	5.3	1.7	9.0
14P/Wolf	—	2.33	—	—	2.0	—	1.3	2.33	—	—
15P/Finlay	—	—	—	—	—	—	0.9	—	—	—
16P/Brooks 2	—	—	—	—	—	—	1.7	—	—	—
17P/Holmes	1.71	—	—	—	—	—	2.0(1.6)	1.71	—	—
19P/Borrelly	2.4	1.9	—	—	—	2.4 2.50	3.0(2.2)	2.2	2.5	25.0
21P/Giacobini-Zinner	—	—	—	—	—	1.9	1.0	1.0	1.5	9.5
22P/Kopff	1.67	1.8	—	<2.9	—	2.46	1.8	1.67	1.7	12.30
24P/Schaumasse	—	—	—	—	1.1	—	0.8	—	—	—
26P/Grigg-Skjellerup	—	≤1.5	1.5	—	1.9	1.44	1.3(1.2)	1.3	1.10	—
28P/Neujmin 1	—	—	—	11.44	—	10.22 10.6	9.1	10.7	1.50	12.75
29P/Schwassmann-Wachmann 1	—	—	—	15.4	—	20.0	—	15.4	2.6	14(32.3)
30P/Reinmuth 1	—	≤3.8	—	—	3.4	—	1.3(1.0)	—	—	—
31P/Schwassmann-Wachmann 2	—	—	—	—	3.2	3.1	3.2	3.1	1.6	5.58
32P/Comas Solá	—	—	—	—	3.6	—	— (2.1)	—	—	—
33P/Daniel	—	—	—	—	1.1	—	0.9	—	—	—
36P/Whipple	—	2.32	—	—	2.8	—	2.3(1.9)	2.32	—	—
37P/Forbes	0.81	—	1.1	—	2.0	—	1.0	0.96	—	—
39P/Oterma	—	—	—	—	—	—	9.1(3.2)	—	—	—
40P/Väisälä 1	—	≤3.6	—	—	1.8	—	1.5	—	—	—
41P/Tuttle-Giacobini-Kresák	—	—	—	—	—	—	0.7	0.70	—	—
42P/Neujmin 3	—	—	—	—	1.0	—	0.6	—	—	—
43P/Wolf-Harrington	—	3.4	≪3.1	—	—	—	1.8	1.8	—	—
44P/Reinmuth 2	1.61	≤3.0	—	—	1.8	—	1.5	1.61	—	—
45P/Honda-Mrkos-Pajdušáková	0.34	1.34	—	—	1.1	—	0.5(0.3)	0.8	1.30	—
46P/Wirtanen	0.62	≤2.6	—	<1.66	—	0.56 0.7	0.7(0.6)	0.60	1.20	6.0
47P/Ashbrook-Jackson	2.8	≤6.1	3.1	—	4.8	—	2.9(2.5)	2.8	1.4	>44
48P/Johnson	—	≤3.5	3.7	—	—	2.87	2.2	2.87	1.35	29.0
49P/Arend-Rigaux	—	4.6	—	—	3.9	4.8 5.1	3.2	4.24	1.63	13.47
50P/Arend	0.95	—	—	—	—	—	3.0(1.0)	0.95	—	—
51P/Harrington	—	≤1.9	—	—	2.1	—	1.4(0.2)	—	—	—
52P/Harrington-Abell	—	—	1.4	—	1.3	1.0	1.1	1.3	—	—
53P/Van Biesbroeck	—	—	≪6.7	3.33	3.9	—	3.8(3.3)	3.33	—	—
54P/de Vico-Swift	—	≤2.1	—	—	—	—	—	—	—	—
56P/Slaughter-Burnham	—	—	—	1.56	—	—	1.5	1.56	—	—
57P/du Toit-Neujmin-Delporte	—	≤1.1	—	—	—	—	1.6	—	—	—
58P/Jackson-Neujmin	—	—	—	—	—	—	0.6	—	—	—
59P/Kearns-Kwee	0.79	—	—	—	2.0	—	1.1	0.79	—	—
60P/Tsuchinshan 2	—	—	—	—	—	—	0.8	—	—	—
61P/Shajn-Schaldach	0.64	0.92	—	—	1.0	—	1.1(1.0)	0.64	1.27	>18
62P/Tsuchinshan 1	—	—	—	—	—	—	0.8	—	—	—
63P/Wild 1	1.45	≤0.6	—	—	—	—	1.5	1.45	—	—
64P/Swift-Gehrels	—	≤1.9	1.6	—	—	—	1.7(2.2)	1.6	—	—
65P/Gunn	—	≤8.8	≪11.7	—	—	—	4.8	—	—	—
67P/Churyumov-Gerasimenko	1.98	≤2.9	—	—	—	2.8	2.5(2.0)	2.0	1.3	12.3
68P/Klemola	—	—	—	—	—	—	2.2	2.2	—	—
69P/Taylor	—	≤3.4	—	—	—	—	2.9	—	—	—
70P/Kojima	1.86	—	—	—	1.3	—	1.2	1.86	1.10	>22
71P/Clark	0.68	≤0.9	—	1.31	—	—	1.3(0.8)	0.68	—	—
72P/Denning-Fujikawa	—	—	—	—	—	—	— (0.8)	—	—	—
73P/Schwassmann-Wachmann 3	0.68*	≤0.9	—	—	— (1.3)	<1.3	1.0	—	1.16*	—
74P/Smirnova-Chernykh	2.23	≤12.7	≪11.2	—	—	—	6.0	2.23	1.14	>20

TABLE 1. (continued).

Comet	Effective radius (km)								a/b (min)	P_{rot} (h)
	La+	Lo+	Li+	Me+	Sc	Others	Ta+	$r_{n,v}$		
75P/Kohoutek	—	≤1.5	—	—	2.0	—	1.8	—	—	—
76P/West-Kohoutek-Ikemura	0.33	—	—	—	1.6	—	1.3	0.33	1.47	>13
77P/Longmore	—	—	—	—	—	—	2.4	—	—	—
78P/Gehrels 2	—	1.42	—	—	—	—	2.1	1.42	—	—
79P/du Toit-Hartley	—	1.4	—	—	1.9	—	1.2	1.4	—	—
81P/Wild 2	—	2.0	≪5.7	—	—	2.0	2.2(2.0)	2.0	—	—
82P/Gehrels 3	0.73	—	<3.0	—	2.1	—	2.0	0.73	1.6	>50
83P/Russell 1	—	≤0.5	—	—	—	—	—	—	—	—
84P/Giclas	0.90	—	—	—	1.3	—	1.4(1.2)	0.90	—	—
86P/Wild 3	0.43	≤0.9	—	0.65	1.3	3.1	0.9	0.43	1.35	>11
87P/Bus	0.28	≤0.6	—	<3.42	2.0	—	1.3	0.28	2.20	>25
88P/Howell	—	—	—	—	1.9	—	1.1(1.0)	—	—	—
89P/Russell 2	—	≤2.2	—	—	1.2	—	1.1	—	—	—
90P/Gehrels 1	—	—	—	—	3.4	—	2.8	—	—	—
91P/Russell 3	—	—	—	—	—	—	1.3	—	—	—
92P/Sanguin	—	1.73	—	1.19	—	—	—	1.19	—	—
94P/Russell 4	—	—	—	—	—	—	1.9	—	—	—
97P/Metcalf-Brewington	—	2.2	1.5	—	—	—	1.3	1.7	—	—
98P/Takamizawa	—	—	3.7	—	2.3	—	2.4(3.0)	—	—	—
99P/Kowal 1	—	—	—	—	4.4	—	4.8	—	—	—
100P/Hartley 1	—	<1.2	—	—	—	—	1.3	—	—	—
101P/Chernykh	—	—	—	—	2.4	—	2.2	—	—	—
103P/Hartley 2	0.8	≤5.8	≪5.3	—	2.4	—	3.8	0.8	—	—
104P/Kowal 2	—	1.0	—	—	—	—	—	1.0	—	—
105P/Singer-Brewster	—	—	—	—	1.0	—	1.0(0.8)	—	—	—
106P/Schuster	0.94	—	—	—	—	—	0.8	0.94	—	—
107P/Wilson-Harrington	—	1.77	—	1.96	—	2.0	—	1.7	1.2	6.10
108P/Ciffreo	—	—	—	—	1.4	—	— (1.1)	—	—	—
110P/Hartley 3	2.15	—	—	—	2.4	—	1.9	2.15	1.30	10
111P/Helin-Roman-Crockett	—	≤1.5	—	—	2.4	0.6	1.5	0.6	—	—
112P/Urata-Niijima	0.90	—	—	—	0.9	—	0.7	0.90	—	—
113P/Spitaler	—	≤2.0	—	—	1.0	—	1.1	1.10	—	—
114P/Wiseman-Skiff	0.78	—	—	—	—	—	— (0.8)	0.78	—	—
115P/Maury	—	—	—	1.11	—	0.8	1.11	—	—	—
116P/Wild 4	—	—	—	—	—	—	3.5(3.0)	—	—	—
117P/Helin-Roman-Alu 1	—	—	—	—	3.9	—	3.5)	—	—	—
118P/Shoemaker-Levy 4	—	2.4	—	—	—	—	1.7	2.4	—	—
119P/Parker-Hartley	—	≤4.0	—	—	—	—	2.5(2.1)	—	—	—
120P/Mueller 1	—	1.5	—	—	1.9	—	0.8	1.5	—	—
121P/Shoemaker-Holt 2	—	1.62	—	—	—	—	2.6	1.62	—	—
123P/West-Hartley	—	—	—	—	—	—	2.2(1.7)	—	—	—
124P/Mrkos	—	—	—	—	—	—	1.6	—	—	—
125P/Spacewatch	—	—	—	—	1.0	—	0.8	0.80	—	—
128P/Shoemaker-Holt 1	—	2.12 2.48	—	—	—	—	2.0	2.3	—	—
129P/Shoemaker-Levy 3	—	—	—	—	—	—	— (2.4)	—	—	—
130P/McNaught-Hughes	—	—	—	—	1.8	—	1.7(1.5)	—	—	—
131P/Mueller 2	—	—	—	—	—	—	0.8	—	—	—
132P/Helin-Roman-Alu 2	—	—	—	—	—	—	0.9	—	—	—
134P/Koval-Vávrová	—	—	—	—	—	—	1.4	—	—	—
135P/Shoemaker-Levy 8	—	—	—	—	1.6	—	1.5(1.3)	—	—	—
136P/Mueller 3	—	—	—	—	—	—	1.9(1.5)	—	—	—
137P/Shoemaker-Levy 2	—	≤3.4	4.5	—	—	—	2.9	2.90	—	—
138P/Shoemaker-Levy 7	—	—	—	—	—	—	0.8(1.0)	—	—	—
139P/Väisälä-Oterma	—	≤4.6	—	—	—	—	2.6	—	—	—
140P/Bowell-Skiff	—	—	—	—	—	—	— (2.3)	—	—	—
141P/Machholz 2	—	—	—	—	—	—	— (1.0)	—	—	—
143P/Kowal-Mrkos	—	—	—	—	—	5.7	— (2.6)	5.4	1.5	17.2

TABLE 1.   (continued).

Comet	Effective radius (km)							$r_{n,v}$	a/b (min)	P_{rot} (h)
	La+	Lo+	Li+	Me+	Sc	Others	Ta+			
144P/Kushida	—	—	—	—	—	—	— (1.2)	—	—	—
147P/Kushida-Muramatsu	0.21	≤2.0	—	—	—	—	2.3(1.9)	0.21	1.53	9.5
148P/Anderson-LINEAR	—	—	—	—	—	—	— (2.1)	—	—	—
152P/Helin-Lawrence	—	≤6.0	—	<1.74	—	—	4.6	—	—	—
154P/Brewington	—	—	—	—	—	—	1.5	—	—	—
P/1993 W1 (Mueller 5)	—	—	—	—	—	—	2.1	—	—	—
P/1994 A1 (Kushida)	—	—	—	—	—	—	1.2	—	—	—
P/1994 J3 (Shoemaker 4)	—	—	—	—	—	—	3.3	—	—	—
P/1995 A1 (Jedicke)	—	—	—	—	—	—	3.0	—	—	—
P/1996 A1 (Jedicke)	—	—	—	—	—	—	5.0	—	—	—
P/1997 C1 (Gehrels)	—	—	—	—	—	—	2.3	—	—	—
P/1997 G1 (Montani)	—	—	—	—	—	—	2.5	—	—	—
P/1997 V1 (Larsen)	—	—	—	—	—	—	3.6	—	—	—
P/1998 S1 (LINEAR-Mueller)	—	—	—	—	—	—	— (4.2)	—	—	—
P/1999 D1 (Hermann)	—	—	—	—	—	—	— (0.7)	—	—	—
P/1999 RO28 (LONEOS)	—	—	—	—	—	—	— (0.1)	—	—	—

*Fragment C.

See text for the references. New radii given by *Ta+* are in brackets.

TABLE 2.   Nuclei of the nearly isotropic comets (NICs).

Comet	Effective radius (km)						$r_{n,v}$	a/b (min)	P_{rot} (h)
	La+	Lo+	Li+	Me+	Sc	Others			
1P/Halley*	—	—	—	—	—	—	5.5	2.0	52.8; 177.6
8P/Tuttle	—	—	7.8	—	—	—	7.8	—	—
55P/Tempel-Tuttle	1.80	—	—	—	—	1.8	1.80	1.50	—
96P/Machholz 1	—	—	3.5	—	—	2.8	3.2	1.4	6.38
109P/Swift-Tuttle	—	—	—	13.7	—	11.8–12.5	13.0	—	69.4
126P/IRAS	1.57	—	—	—	—	—	1.57	—	—
P/1991 L3 (Levy)	—	—	—	—	—	5.8	5.8	1.3	8.34
C/1983 H1 (IRAS-Araki-Alcock)	—	—	—	—	—	3.5–3.7	3.5	—	51.0
C/1983 J1 (Sugano-Saigusa-Fujikawa)	—	—	—	—	—	0.37	0.37	—	—
C/1983 O1 (Černis)	—	—	—	<10.5	—	—	—	—	—
C/1984 K1 (Shoemaker)	—	—	—	<6.4	30.6	—	—	—	—
C/1984 U1 (Shoemaker)	—	—	—	<6.4	29.3	—	—	—	—
C/1986 P1 (Wilson)	—	—	—	—	16.1	<6.0	—	—	—
C/1987 A1 (Levy)	—	—	—	<4.0	2.1	—	—	—	—
C/1987 H1 (Shoemaker)	—	—	—	<4.0	26.7	—	—	—	—
C/1987 F1 (Torres)	—	—	—	<5.9	—	—	—	—	—
C/1988 B1 (Shoemaker)	—	—	—	<6.1	16.1	—	—	—	—
C/1988 C1 (Maury-Phinney)	—	—	—	<6.1	1.1	—	—	—	—
C/1995 O1 (Hale-Bopp)	37	—	—	—	—	30	37	2.6	11.34
C/1996 B2 (Hyakutake)	—	—	—	—	—	2.4	2.4	—	6.27
C/1997 T1 (Utsonomiya)	—	—	—	—	—	<5.8	—	—	—
C/1999 S4 (LINEAR)†	—	—	—	—	—	0.4	0.4	—	—
C/2001 OG108 (LONEOS)	—	—	—	—	—	8.9	8.9	1.3	57.19

* See Table 3. The two periods correspond to the SAM and LAM rotations.
† Nucleus size prior to breakup.

See text for the references.

TABLE 3.   Cometary nuclei with known shape and size.

Comet	a × b × c (km × km × km)	1 : a/b : a/c	Notes*
1P/Halley	7.65 ± 0.25 × 3.61 ± 0.25 × 3.61 ± 0.25	1 : 2.13 : 2.13	[1]
	7.21 ± 0.15 × 3.7 ± 0.1 × 3.7 ± 0.1	1 : 1.95 : 1.95	[2]
10P/Tempel 2	8 × 4 × 4	1 : 2.0, c = b	[3]
	8.2 × 4.9 × 3.5	1 : 1.67 : 2.34	[4]
19P/Borrelly	4.0 ± 0.1 × 1.60 ± 0.02 × 1.60 ± 0.02	1 : 2.5, c = b	[5]
	4.4 ± 0.15 × 1.80 ± 0.08 × 1.80 ± 0.08	1 : 2.4, c = b	[6]

*Notes: [1] *Vega 1, 2*, TVS *in situ* imaging (*Merényi et al.*, 1990); [2] *Giotto* HMC *in situ* imaging (*Keller et al.*, 1994); [3] groundbased CCD photometry (*Jewitt and Luu*, 1989); [4] groundbased observations and modeling (*Sekanina*, 1989); [5] *Deep Space 1* MICAS *in situ* imaging (*Buratti et al.*, 2004); [6] HST WFPC2 high-precision photometry (*Lamy et al.*, 1998b).

et al. (1995) report $p_J = 0.10 \pm 0.02$ using the STM and $p_J = 0.05 \pm 0.01$ using the ILM, the latter value being favored.

3.3.2. Nearly isotropic comets (NICs). 1P/Halley: Resolved imaging of the nucleus led to a value of $0.04^{+0.02}_{-0.01}$, irrespective of the spectral bands "VIS," "RED," or "NIR" of the *Vega 1,2* cameras (*Sagdeev et al.*, 1986a).

55P/Tempel-Tuttle: Using radiometry, *Fernández* (1999) and *Jorda et al.* (2000) both arrived at similar values for the R band: 0.06 ± 0.025 for the former, 0.045 for the latter.

109P/Swift-Tuttle: Fomenkova et al. (1995) used radiometry to estimate a nuclear size, from which the large-heliocentric distance observations by *O'Ceallaigh et al.* (1995) may be used to derive an approximate albedo of about 0.02–0.04 in the R band.

C/1983 H1 (IRAS-Araki-Alcock): Extensive datasets at many wavelengths allowed *Sekanina* (1988) to create a unified model of the properties of the nucleus. The implied albedo in the V band is 0.02 ± 0.01. *Groussin et al.* (2004) have reanalyzed these data and obtained a slightly larger value of 0.03 ± 0.01.

C/1995 O1 (Hale-Bopp): While more data were obtained on this comet than any other, the albedo derivation is problematic owing to the comet's strong coma swamping the nucleus during the whole apparition to date. *Campins and Fernández* (2003) combine the results of *Jorda et al.* (2000) and *Fernández* (1999), who both used radiometry and find a compromise (but very unconstrained) value of 0.04 ± 0.03.

C/2001 OG108 (LONEOS): Using radiometry, *Abell et al.* (2003) report 0.03 ± 0.005 for the V band.

Table 4 summarizes these results on the albedo measurements of cometary nuclei.

3.4.   Colors

Colors by themselves do not provide much information on the physical properties of cometary nuclei, but the distribution of colors compared to other solar system objects, or the correlation of colors with other parameters (e.g., size, orbital parameters, ...), has the potential of offering independent clues on the origin and evolution of these objects

TABLE 4.   Albedos of cometary nuclei.

Comet	Geometric Albedo	λ
Ecliptic Comets		
2P/Encke	0.046 ± 0.023	V
9P/Tempel 1	0.05 ± 0.02	R
10P/Tempel 2	$0.022^{+0.004}_{-0.006}$	4845
10P/Tempel 2	0.04–0.07	JHK
19P/Borrelly	0.03	
22P/Kopff	0.042 ± 0.006	V
28P/Neujmin 1	0.026	V
28P/Neujmin 1	0.03 ± 0.01	R
49P/Arend-Rigaux	0.04 ± 0.01	V
49P/Arend-Rigaux	0.054 ± 0.010	J
49P/Arend-Rigaux	0.03 ± 0.01	J
107P/Wilson-Harrington	0.05 ± 0.01	J
Nearly Isotropic Comets		
1P/Halley	$0.04^{+0.02}_{-0.01}$	V, R, I
55P/Tempel-Tuttle	0.06 ± 0.025	R
55P/Tempel-Tuttle	0.045	R
109P/Swift-Tuttle	0.02–0.04	R
C/1983 H1 IRAS-Araki-Alcock	0.03 ± 0.01	V
C/1995 O1 Hale-Bopp	0.04 ± 0.03	
C/2001 OG108 (LONEOS)	0.03 ± 0.005	V

λ = band or wavelength (in Å) to which albedo applies. References are given in the text.

and their interrelationships. Near-infrared spectroscopic observations of cometary nuclei have been attempted in an effort to detect the spectral signals of water ice and silicates, as successfully performed on several KBOs and Centaurs (e.g., *Jewitt and Luu*, 2001). Finally, we discuss the few thermal spectra obtained so far and their implications for the surface temperature of cometary nuclei.

3.4.1. Broadband colors and reflectivity: The most common color characterization comes from color indices, e.g., (B-V), (V-R), (V-I), etc. As discussed in section 3.1.2, magnitudes of nuclei observed at large heliocentric distances are very faint, and this often leads to large uncertainties in the indices. Continuum spectra of a few nuclei have been obtained, and they can be parameterized using the normalized reflectivity gradient $S' = dS/d\lambda/<S>$, where S is the reflectivity (object flux density divided by the flux density of the Sun at the same wavelength, λ) and $<S>$ is the mean value of the reflectivity in the wavelength range over which $dS/d\lambda$ is computed (*Luu and Jewitt*, 1990b). The gradient, S', is used to express the percentage change in the strength of the continuum per 1000 Å (%/1000 Å). Broadband color indices can also be converted to a normalized reflectivity gradient using the following relation (*Luu and Jewitt*, 1990b)

$$(V\text{-}R)_n = (V\text{-}R)_\odot + 2.5 \log \frac{2 + S'\Delta\lambda}{2 - S'\Delta\lambda} \qquad (12)$$

where $(V\text{-}R)_n$ and $(V\text{-}R)_\odot$ are the color indices of the nucleus and the Sun respectively and $\Delta\lambda$ is the difference between the effective wavelengths of the two filters.

The quantity S' remains of interest as long as it is constant over a sufficiently large spectral interval. This is rarely the case and different values in different spectral intervals must be introduced. Then the S' values become strictly equivalent to the color indices via the above equation.

Table 5 is an updated version of Table 3 in *Jewitt* (2002), summarizing all presently available data on the colors of cometary nuclei. It incorporates recent results from *Meech et al.* (2004), except for the nuclei of 22P/Kopff, 46P/Wirtanen, 87P/Bus, and P/1993 K2 (Helin-Lawrence), which were active at the time of observations, from the compilation of *Hainaut and Delsanti* (2002), and from *Lowry and Weissman* (2003). We comment below on some of the results, starting with the ECs.

2P/Encke: Note the accurate (V-R) from spectrophotometry. A value (V-R) = 0.46 ± 0.02 is consistent with all the data and their respective error bars.

6P/d'Arrest: (V-R) = 0.56 ± 0.02 is consistent with the results of *Jewitt* (2002) and *Meech et al.* (2004), while the value reported by *Lowry and Weissman* (2003), (V-R) = 0.33 ± 0.09, is well outside the above uncertainty. (R-I) = 0.45 ± 0.04 from *Jewitt* (2002) is, however, consistent with the results reported by *Lowry and Weissman* (2003), 0.33 ± 0.12. At the 2σ level, the two values of (B-V) agree, and we adopt (B-V) = $0.85^{+0.2}_{-0.07}$.

10P/Tempel 2: (V-R) = 0.56 ± 0.01 is consistent with the three measurements.

14P/Wolf, 19P/Borrelly: Note the large uncertainties, making these measurements of limited value.

28P/Neujmin 1: There is an excellent agreement on the (V-R) color of this large and inactive nucleus. Taking the average of all measurements leads to (V-R) = 0.47 ± 0.20, making it similar to D-type asteroids (*Campins et al.*, 1987; *Fitzsimmons et al.*, 1994) and most Trojans (*Jewitt and Luu*, 1990).

45P/Honda-Mrkos-Pajdušáková: In addition to the results included in Table 5, *Lamy et al.* (1999a) reported the first (U-B) index ever measured on a comet nucleus, (U-B) = 0.68 ± 0.04.

48P/Johnson: Note the large uncertainty of 0.3 on (V-R).

49P/Arend-Rigaux: The result of (V-R) = 0.47 ± 0.01, obtained by both filter photometry and spectrophotometry on this inactive nucleus, appears to be extremely accurate and reliable. Note the large uncertainty on the (R-I) reported by *Lowry et al.* (2003a), 0.54 ± 0.14, which makes it compatible with the result of *Millis et al.* (1988), (R-I) = 0.43 ± 0.02.

86P/Wild 3: The surprising result of (V-R) = 0.12 ± 0.14 makes this nucleus a very blue object, although the error bar is quite large.

107P/Wilson-Harrington: We favor the spectrophotometric result (V-R) = 0.31 ± 0.03 of *Chamberlin et al.* (1996), which is intermediate between the two available photometric results.

There are only three NIC nuclei for which color information is available: 1P/Halley, 96P/Machholz 1, and C/2001 OG108 (LONEOS).

1P/Halley: From *in situ* imaging by the *Giotto* HMC, *Thomas and Keller* (1989) determined a constant reflectivity gradient S' = 6 ± 3 per 1000 Å in the range 440–810 nm leading to (B-V) = 0.72 ± 0.04, (V-R) = 0.41 ± 0.03, (V-I) = 0.80 ± 0.09, and (R-I) = 0.39 ± 0.06.

96P/Machholz 1: The two available measurements are not consistent at the 1σ level.

C/2001 OG108 (LONEOS): Measurements of (B-V), (V-R), and (R-I) have been reported by *Abell et al.* (2003).

3.4.2. Visible and near-infrared spectra. In principle, spectral analysis is the most effective way to characterize the surface properties of the cometary nuclei. First, it yields high-accuracy color information as presented above, and second, it offers the possibility of detecting solid-state absorption bands, namely those of water ice and minerals. With one exception, and contrary to the case for several Centaurs and KBOs, this expectation has failed to materialize, reinforcing for the time being the value of color information. In addition to the general difficulties of detecting cometary nuclei, spectral observations face the additional problem of very low signals per spectral element. This explains the paucity of groundbased nuclear spectra, mostly restricted to (nearly) inactive nuclei, and the clear superiority of *in situ* spectral observations.

2P/Encke, 10P/Tempel 2, 21P/Giacobini-Zinner, 49P/Arend-Rigaux: Visible spectra obtained at large r_h with a spectral resolution of 10–20 Å are presented by *Luu* (1993). No absorption features were detected, except for a downturn feature in the blue part of the spectrum of 21P, reminiscent of chondritic spectra.

19P/Borrelly: The short-wavelength infrared imaging spectrometer (SWIR) onboard *DS1* secured 45 scans spectra of the nucleus in the 1.3–2.6-μm range (*Soderblom et al.*, 2002). They reveal a strong positive slope toward the red and a single absorption feature at ~2.39 μm, whose origin is unknown (fits of various hydrocarbons were attempted, but none were satisfactory).

28P/Neujmin 1: Observations at large r_h in the spectral range 0.9–2.4 μm by *Campins et al.* (2001) do not show a

TABLE 5.   Colors of cometary nuclei.

Comet	(B-V)	(V-R)	(R-I)	Photometry*	References
Solar colors	0.65	0.35	0.28		
Ecliptic Comets					
2P/Encke	0.78 ± 0.02	0.48 ± 0.02	—	S	LJ90
	—	0.43 ± 0.05	—	F	J02
	—	0.38 ± 0.06	—	F	J02
	—	0.37 ± 0.09	—	F	HD02
6P/d'Arrest	0.78 ± 0.04	0.54 ± 0.04	0.45 ± 0.04	F	J02
	—	0.62 ± 0.08	—	F	M+02
	1.08 ± 0.12	0.33 ± 0.09	0.33 ± 0.12	F	LW03
10P/Tempel 2	—	0.53 ± 0.03	—	F	JM88
	—	0.58 ± 0.03	—	S	JL89
	—	0.56 ± 0.02	—	F	M+02
14P/Wolf	—	0.02 ± 0.22	0.25 ± 0.35	F	Lo+03
19P/Borrelly	—	0.25 ± 0.78	—	F	Lo+03
21P/Giacobini-Zinner	0.80 ± 0.03	0.50 ± 0.02	—	S	L93
22P/Kopff	0.77 ± 0.05	0.50 ± 0.08	0.42 ± 0.03	F	La+02
26P/Grigg-Skjellerup	—	0.42 ± 0.10	—	F	B+99
28P/Neujmin 1	—	0.46 ± 0.04	—	F	Ca+87
	—	0.45 ± 0.05	—	F	JM88
	—	0.50 ± 0.04	—	F	JM88
	—	0.45 ± 0.05	—	F	D+01
	—	0.48 ± 0.06	—	F	M+02
45P/Honda-Mrkos-Pajdušáková	1.12 ± 0.03	0.44 ± 0.03	0.20 ± 0.03	F	La+99
46P/Wirtanen	—	0.45 ± 0.10	—	F	La+98a
48P/Johnson	—	0.50 ± 0.30	—	F	Li+00
49P/Arend-Rigaux	0.77 ± 0.03	0.47 ± 0.01	0.43 ± 0.02	F	M+88
	—	0.47 ± 0.01	—	S	L93
	—	0.40 ± 0.30	—	F	Li+00
	—	0.49 ± 0.11	0.54 ± 0.14	F	Lo+03
53P/Van Biesbroeck	—	0.34 ± 0.08	—	F	M+02
73P/SW3	—	0.48 ± 0.17	—	F	B+99
86P/Wild 3	—	0.12 ± 0.14	—	F	M+02
107P/Wilson-Harrington	—	0.31 ± 0.03	—	S	Ch+96
	—	0.41 ± 0.02	—	F	M+02
	0.61 ± 0.05	0.20 ± 0.04	—	F	LW03
	0.75 ± 0.06	—	—	F	LW03
143P/Kowal-Mrkos	0.84 ± 0.02	0.58 ± 0.02	0.55 ± 0.02	F	J+03
	0.80 ± 0.02	0.58 ± 0.02	0.57 ± 0.02	F	J+03
Nearly Isotropic Comets					
1P/Halley	0.72 ± 0.04	0.41 ± 0.03	0.39 ± 0.06	F/HMC	TK89
96P/Machholz 1	—	0.43 ± 0.03	—	F	M+02
	—	0.30 ± 0.05	—	F	Li+00
C/2001 OG108 (LONEOS)	0.76 ± 0.03	0.46 ± 0.02	0.44 ± 0.03	F	A+03

*F = filter photometry, S = spectrophotometry; HMC = *in situ* measurements by the *Giotto* Halley Multicolour Camera.

References: A+03 (*Abell et al., 2003*); B+99 (*Boehnhardt et al., 1999*); Ca+87 (*Campins et al., 1987*); Ch+96 (*Chamberlin et al., 1996*); D+01 (*Delahodde et al., 2001*); HD02 (*Hainaut and Delsanti, 2002*); JM88 (*Jewitt and Meech, 1988*); JL89 (*Jewitt and Luu, 1989*); J+03 (*Jewitt et al., 2003*); LJ90 (*Luu and Jewitt, 1990a*); L93 (*Luu, 1993*); La+98a (*Lamy et al., 1998a*); La+02 (*Lamy et al., 2002*); Li+00 (*Licandro et al., 2000*); Lo+03 (*Lowry et al., 2003a*); LW03 (*Lowry and Weissman, 2003*); M+88 (*Millis et al., 1988*); M+02 (*Meech et al., 2004*); TK (*Thomas and Keller, 1989*).

water ice signature, a result consistent with earlier observations by *Campins et al.* (1987) and recent observations by *Licandro et al.* (2002).

82P/Gehrels 3:   A featureless red spectrum in the range 0.4–0.98 μm, with a resolution of 30 Å, has been obtained by *De Sanctis et al.* (2000). This spectrum is very similar to those of D-type asteroids.

90P/Gehrels 1:   Observations at large r_h in the spectral range 0.9–2.4 μm by *Delahodde et al.* (2002) show the absence of spectral signatures.

107P/Wilson-Harrington: A featureless spectrum in the range 0.38–0.62 μm with a resolution of 5 Å has been obtained by *Chamberlin et al.* (1996).

124P/Mrkos: Observed by *Licandro et al.* (2003) while inactive at r_h = 1.85 AU, its near-infrared (0.9–2.3 μm), low-resolution spectrum is featureless and slightly redder than the Sun, resembling that of a D-type asteroid.

C/2001 OG108 (LONEOS):   Observed by *Abell et al.* (2003) while inactive, its near-infrared (0.7–2.5 μm) is featureless, slightly redder than the Sun, resembling that of a D-type asteroid.

3.4.3. Thermal spectrum — Surface temperatures.   The surface temperature of the nucleus has been measured for only two comets, 1P/Halley and 19P/Borrelly, thanks to *in situ* observations.

1P/Halley: The infrared radiation of its nucleus was measured at r_h = 0.8 AU by the IKS spectrometer onboard the *Vega 1* spacecraft in two wavelengths bands, 7–10 and 9–14 μm (*Combes et al.,* 1986). The temperature was obtained with two independent and different methods, and the most probable maximum value lies in the range 360–400 K. The hottest region was not at the subsolar point, and the angular thermal lag was about 20° (*Emerich et al.,* 1987). These results suggest that a large fraction of the nucleus surface of 1P/Halley is inactive and not cooled by sublimating ices or evolving gases. The surface may be a lag deposit crust, or perhaps a radiation processed mantle.

19P/Borrelly: The spectra recorded by SWIR onboard *DS1* at r_h = 1.36 AU (*Soderblom et al.,* 2002) permitted a determination of the temperature at the two tips of the elongated nucleus: 300 K and 345 K. These high temperatures are consistent with the absence of water ice bands (cf. section 3.4.2) and, as for 1P/Halley, suggest that a large fraction of the nuclear surface is inactive [only ~10% of its surface is active according to *Lamy et al.* (1998b)].

3.5.   Phase Function

In the above sections, we have highlighted the importance of the phase function $\Phi(\alpha)$ in the determination of the size of cometary nuclei and emphasized that it remains a nonnegligible source of uncertainty. Aside from this technical aspect of the data reduction, the phase function of an atmosphereless body offers a powerful means for investigating the properties of its surface (e.g., roughness and single-particle albedo). Typically, the phase angle data are fit to a parametric model, for instance that of *Hapke* (1993). Of particular interest is the opposition effect, which is neglected when using simple phase laws, such as the one introduced in section 3.2. In addition to the intrinsic difficulties of observing cometary nuclei, the determination of the phase function further requires observations at different phase angles, each one having to be corrected for the effect of the rotation of the nucleus. Ideally, this requires determining the light curve at each phase angle, so that the measurements may be phased to the same rotational position, say the maxima of the light curves (*Delahodde et al.,* 2001).

2P/Encke: A detailed analysis of recent, original measurements and a large collection of historical data led *Fernández et al.* (2000) to derive β = 0.06 mag deg^{-1}. This very steep phase function makes 2P/Encke one of the most phase-darkened objects in the solar system and implies a very rough surface.

9P/Tempel 1:   *Fernández et al.* (2003) estimated a phase coefficient β = 0.07 mag deg^{-1} that is poorly constrained. It is indeed unlikely that such a steep phase function is correct.

19P/Borrelly: Combining the disk-integrated magnitudes calculated from the *DS1* images with the HST (*Lamy et al.,* 1998b) and groundbased measurements (*Rauer et al.,* 1999), *Soderblom et al.* (2002) and *Buratti et al.* (2004) determined $\Phi(\alpha)$ over a large range of phase angle, from 3° to 88°. The phase curve is very similar to that of the dark C-type asteroid 253 Mathilde (*Clark et al.,* 1999). Except for a minor opposition effect restricted to α ≤ 3°, this phase curve is well approximated by a constant linear phase coefficient β = 0.04 mag deg^{-1} over the interval 3°–90°.

28P/Neujmin 1:   *Jewitt and Meech* (1987) determined a phase coefficient β = 0.034 ± 0.012 mag deg^{-1}. *Delahodde et al.* (2001) obtained phase coverage extending from 0.8° to 19° and have been able to correct several data points for the effects of rotation. A linear phase coefficient β = 0.025 ± 0.006 mag deg^{-1} applies down to α ~ 5°. At smaller phase angles, the function steepens and a strong opposition effect appears at α ≤ 1.5°. This effect, comparable to those found on medium albedo p_V ~ 0.15 M-type asteroids and icy satellites, is quite surprising for a cometary nuclei. As surface ice is excluded on such a low-activity nucleus, a high surface porosity could perhaps be invoked, but this possible interpretation has not been investigated.

45P/Honda-Mrkos-Pajdušáková: As discussed in section 3.2, there is a distinct possibility that the HST and groundbased observations can be reconciled by a steep phase function with β = 0.06 mag deg^{-1}, similar to that of 2P/Encke and 48P/Johnson.

48P/Johnson: Observations of a starlike nucleus at phase angles between 6° and 16° led *Jewitt and Sheppard* (2003) to derive β = 0.0592 mag deg^{-1}.

55P/Tempel-Tuttle: The combination of HST and groundbased observations allowed *Lamy et al.* (2004) to obtain the phase function in the interval 3°–55° and to derive a linear phase coefficient β = 0.041 mag deg^{-1}, similar to those of 19P/Borrelly and asteroid Mathilde.

143P/Kowal-Mrkos: Observations of a starlike nucleus at phase angles between 5° and 12.7° led *Jewitt et al.* (2003) to derive a linear phase coefficient β = 0.043 ± 0.0014 mag deg⁻¹.

3.6. Satellites of Cometary Nuclei

The detection of a satellite companion to a cometary nucleus, and the determination of its orbit, would be of unique value as it would provide access to the mass of the primary. If the mass of the nucleus is known, and if the size is independently derived, then the mean bulk density and porosity can be calculated, providing insight into the internal properties of the nucleus.

There are various processes leading to the formation of binary systems among small bodies. In the case of cometary nuclei, a companion could be primordial or could result from the capture of a large fragment ejected by the nucleus; the capture of an external object appears unlikely. To be of value in the sense described above, a satellite must be sufficiently large to allow its detection and should travel on a stable orbit for some time. However, the motion of such a possibly active object around a rotating, nonspherical, and active primary is extremely complex, and is in fact a major concern for the *Rosetta* orbiter. We review below the few cases where a companion may have been directly or indirectly detected.

17P/1892 V1 (Holmes):   In late 1892, this comet underwent a major outburst (leading to its discovery), then faded by 7–8 mag, and flared up again by 6 mag a couple of months later. *Whipple* (1983, 1984) proposed that a satellite could produce this double burst: first a grazing encounter on 4.6 November 1892 and a final impact on 16.3 January 1893. Several details of this scenario explain the observations rather well. *Whipple* (1999) estimated the crushing strength (compressive strength, force/area) from the momentum transfer during the collision of the secondary with the primary nucleus, and it ranges from 4.2 × 10³ to 5.9 × 10⁵ dynes cm⁻², corresponding to mean bulk densities of 0.2 and 1.5 g cm⁻³ respectively. The idea of an hypothetical satellite, however, remains highly speculative.

26P/Grigg-Skjellerup: During the *Giotto* flyby of this comet in 1992, the *in situ* optical probe experiment (OPE) recorded several "spikes." One of them was interpreted by *McBride et al.* (1997) as an object 10–100 m in radius sporting a weak dust coma. However, it is not clear whether this object was in a bound orbit, or slowly traveling away after possibly separating from the nucleus.

C/1995 O1 (Hale-Bopp):   From his analysis of HST WFPC2 images taken in May–October 1996, *Sekanina* (1998a) reported the detection of a companion that could be bound. He estimated a mass ratio of ≈0.1, a semimajor axis of ≈180 km, and a period of 2–3 d for a primary nucleus of radius ≈35 km. Our analysis of the same set of images using a fully anisotropic coma model (*Weaver and Lamy*, 1997; *Toth et al.*, 1999) does not support this detec-

tion. These latter authors conclude that the "satellite" is probably due to the residual signal when fitting an oversimplified elliptical coma model to the real, highly structured coma of Hale-Bopp. Such artifacts have been found in another analysis performed by *Sekanina* (1995), namely that of Comet D/Shoemaker-Levy 9: Clumps of positive residuals were identified as fragments, while clumps of identical but negative residuals were also present.

Adaptive optics observations with the ESO 3.6-m telescope in the near-infrared performed on 6 November 1997 possibly revealed the presence of a satellite: *Marchis et al.* (1999) discussed the pros and cons of this interpretation, but did not reach a clear and firm conclusion. HST images taken on the same day, and on other days, with the Space Telescope Imaging Spectrograph (STIS) do not reveal any obvious companion (*Weaver et al.*, 1999) at the location and magnitude expected if the groundbased detection were real. If there is a satellite, then either it remained within 1 STIS pixel (0.05 arcsec) of the primary for more than three months, or the HST observers were unlucky and observed near the time of an orbital transit event (i.e., when the two objects appear to move across each other).

Possible additional evidence for a companion of Comet Hale-Bopp comes from the analysis of the complex morphology of its coma (jets and halos). *Vasundhara and Chakraborty* (1999) and *Sekanina* (1998b) have noted difficulties in explaining several coma features with a single rotating nucleus, thereby suggesting that two nuclei are involved, but the analysis of *Samarasinha* (2000) demonstrates that the coma morphology is consistent with a single nucleus.

C/2001 A2 (LINEAR):   The splitting of this comet was accompanied by outbursts, and *Sekanina* (2002c) quoted the rare, but possible, scenario of the flaring of the primary nucleus due to collision with a companion that had been created by the fragmentation events (*Whipple*, 1984; *Sekanina*, 1982). *Sekanina* (1997) had previously suggested that a part of the mantle of the nucleus (icy-dust mantle) could be lifted off the surface and then travel away from the primary during a nontidal splitting. However, this is only speculation, and there is no direct evidence of a satellite around this comet.

Concluding remarks: While the occurrence of satellites for both main-belt and near-Earth asteroids, Kuiper belt objects, and Trojans is steadily growing (*Merline et al.*, 2002), there is still no definite, observational evidence that binary cometary nuclei exist. Since detecting and characterizing cometary nuclei remains a huge challenge, the detection of a smaller companion is probably beyond our present and near-future capabilities. Do double craters and crater chains (catenae) observed on planetary satellites (*Melosh and Schenk*, 1993; *Melosh and Whitaker*, 1994; *Schenk et al.*, 1996) provide independent evidence of binary and multiple objects? Known double craters are plausibly created by the impact of two orbiting bodies and can likely be explained with the currently known asteroidal sources, and do not require a cometary component, but that does not mean that a cometary component is ruled out. Catenae are thought to be formed from tidally disrupted cometary nuclei (an asteroidal origin is, however, not ruled out) but, in that case, the fragments are not orbiting one another; rather, the multiple objects are laid out in a line along their common orbit. In the case of a cometary impactor, fragmentation may have first taken place, leading to the creation of a trail of small bodies, very much like the case of D/Shoemaker-Levy 9.

4. ANALYSIS AND INTERPRETATION

4.1. Size Distribution of Cometary Nuclei

Figures 3a and b present the distribution functions of the effective radius $r_{n,v}$ of ECs and NICs respectively, as summarized in Tables 1 and 2 (the range of radius has been

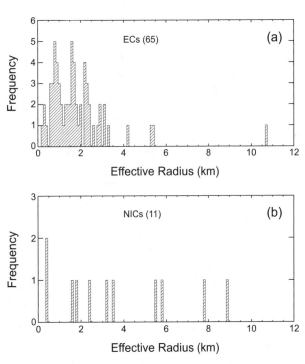

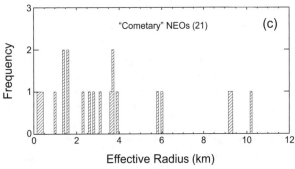

Fig. 3.   Distributions of the effective radius $r_{n,v}$ for **(a)** ecliptic comets, **(b)** nearly isotropic comets, and **(c)** ecliptic "cometary" NEOs. Note that the largest nuclei are excluded to allow legibility of the histograms at small sizes.

truncated at 12 km for better legibility of the histograms at small sizes). These represent the largest datasets ever assembled. The histograms show several structures, which most likely result from the limited statistics in the dataset. We note that there are not very many large cometary nuclei; only two EC and two NIC nuclei have radii larger than 10 km. The apparent roll-off in the number of small cometary nuclei is very likely an observational selection effect (i.e., smaller nuclei are simply harder to detect). A similar effect is often encountered with flux-limited surveys, e.g., for the NEOs at magnitudes fainter than ~17, but additional mechanisms cannot be excluded (see below). *Brandt et al.* (1996) advocated the idea of a large population of undetected ECs having very small nuclei, but they presented no observational evidence to support this hypothesis. For NICs, the above shortcomings are exacerbated by the small number of comets in the sample. However, if the Sun-grazing comets, which are probably the fragments of one or several large nuclei, are considered as full members of this family, then there is clear evidence of a large population of very small, sub-100 m, cometary nuclei (*Biesecker et al.*, 2002). This clearly shows that under the right circumstances, e.g., coronagraphic observations of Sun-grazers, small objects can be detected.

A more robust and physically enlightening way to view size distributions is to introduce cumulative distribution functions, which are less prone to artifacts. One can consider the cumulative luminosity function (CLF) $N_L(<H)$, where N_L is the number of nuclei with absolute magnitude brighter than H, and the cumulative size distribution (CSD) $N_S(>r_n)$, where N_S is the number of nuclei larger than radius r_n. If these two distributions are represented by power laws

$$N_L(<H) \propto 10^{q_L H} \qquad (13)$$

$$N_S(>r_n) \propto r_n^{-q_S} \qquad (14)$$

and if all objects have the same albedo, then $q_S = 5\, q_L$ (*Weissman and Lowry*, 2003). Quite recently, several groups have collected various datasets and studied the CLF and/or the CSD of ecliptic comets; their results are summarized in Table 6.

At stake here is the question of the origin of ECs. If they are collisional fragments of TNOs (*Stern*, 1995; *Farinella and Davis*, 1996), then the theoretical value $q_S = 2.5$ for a collisionally relaxed population (*Dohnanyi*, 1969) is expected. In reality, the question is probably more complex.

On the one hand, the model of Dohnanyi applies to a population of self-similar bodies having the same strength per unit mass. Several groups have attempted to relax this assumption, with *O'Brien and Greenberg* (2003) presenting the most comprehensive results on steady-state size distributions for collisional populations. In the range of sizes of interest for cometary nuclei, the size distribution of fragments is wavy, and oscillates about the distribution of a population evolved under pure gravity scaling. The differential size distribution of such a population is characterized by a power law with an exponent of –3.04. This translates into $q_S = 2.04$ using our notation for the cumulative distribution.

TABLE 6. Power exponents of the cumulative luminosity function (CLF) and
of the cumulative size distribution (CSD) of the nuclei of ecliptic comets.

Reference	CLF	CSD
Fernández et al. (1999)	0.53 ± 0.05	2.65 ± 0.25
Lowry et al. (2003a)	0.32 ± 0.02	1.6 ± 0.1
Meech et al. (2004)	—	2.5*
Weissman and Lowry (2003)	0.32 ± 0.01	1.59 ± 0.03
Weissman (personal communication, 2003)	0.36 ± 0.01	1.79 ± 0.05
This work	0.38 ± 0.06	1.9 ± 0.3

*From Monte Carlo simulations after truncation at small radii.

On the other hand, noncollisional fragmentation (i.e., splitting) is frequent among comets (see *Boehnhardt,* 2004), and nuclei are progressively eroded by their repeated passages through the inner part of the solar system, so that we are certainly not observing a primordial, collisionally relaxed population of TNO fragments. A crude calculation by *Weissman and Lowry* (2003) indicates that a typical EC loses ~400 m in radius at half its lifetime as an active object. *Samarasinha* (2003) undertook a more comprehensive study of this problem in which mass loss includes outgassing and splitting events (rotational and tidal splitting). His only example for a population of nuclei with an initial differential size distribution having an exponent of –3 indeed shows considerable leveling off after 1000 years. Mass loss may therefore significantly distort the size distribution of nuclei, particularly at the low end. While it is tempting to introduce this kind of statistical correction to account for mass shedding, this approach certainly does not reflect the reality for any given comet, which could be at any stage of its orbital evolution. But a case-by-case correction faces the difficulty of the chaotic nature of the orbital evolution of ecliptic comets.

Figure 4a presents the CSD of 65 ecliptic comets for which we have reliable values of the effective radius $r_{n,v}$ (Table 1). Above some critical radius ($r_c \sim 1.6$ km), the CSD appears to follow a single power law. Below r_c, the distribution levels off, a likely result of observational bias and mass loss, as discussed above. The determination of the power exponent q_S, and of the value of r_c, was performed using three different techniques. We first used the least-squares fit because it has been widely used for similar studies by various groups; we stress, however, that this method is *not* applicable to CSDs because the data points are not independent, which renders the standard χ^2 statistic meaningless. Next, we used a fit based on a maximum likelihood parameter estimation, namely the M-estimate technique based on the MEDFIT algorithm described by *Press et al.* (1986) and implemented as the routine LADFIT in IDL. This procedure returns the mean absolute deviation of the data from the power law but does not return an uncertainty on the power exponent. Finally, we calculated the probability P_{KS} that the observed distribution for $r_n > r_c$ and the model distribution $N(>r_n) \propto r_n^{-q_S}$ are drawn from the same parent distribution using the Kolmogorov-Smirnov (K-S) test. We found that the optimum cut-off value is $r_c = 1.6$ km

and then explored the variation of P_{KS} in the neighborhood of the value of q_S determined by the M-estimate technique (Fig. 5). The high value of P_{KS} (0.953) for the nominal value of q_S returned by the M-estimate fit is encouraging, as is the result that the distribution of P_{KS} values is sym-

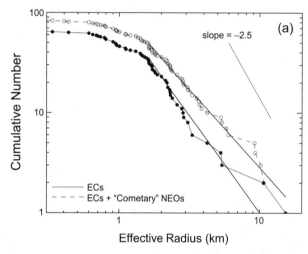

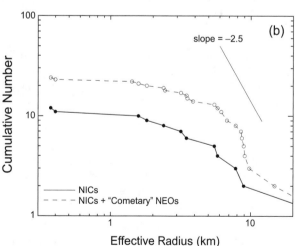

Fig. 4. Cumulative size distributions of the nuclei of **(a)** ecliptic comets and **(b)** nearly isotropic comets are represented by the solid circles while the open circles apply to the populations augmented by the "cometary" NEOs. The two solid lines in **(a)** correspond to optimum power law fits according to the Kolmogorov-Smirnov test, from the cutoff radius $r_c = 1.6$ km up to the largest bodies.

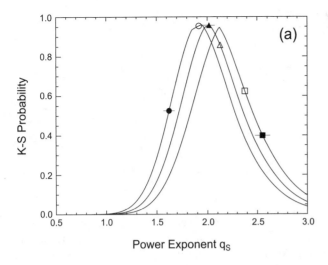

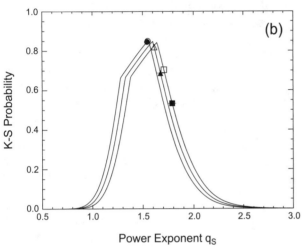

Fig. 5. The Kolmogorov-Smirnov probability as a function of the exponent of the power law fitting the observed CSDs down to a cut-off radius $r_c = 1.6$ km; **(a)** corresponds to the distribution of ECs as listed in Table 1, while **(b)** corresponds to the distribution of ECs + "cometary" NEOs as listed in Table 7. The circles apply to the nominal case while the other symbols apply to two experiments where 29P/Schwassmann-Wachmann 1 is removed (triangles) and where 28P/Neujmin 1 is further removed (squares). The open symbols correspond to the M-estimate (i.e., maximum likelihood) solution while the solid symbols correspond to the least-squares fit solution.

metric about the nominal M-estimate value. In order to define an uncertainty on q_S, we adopted the criterion that $P_{KS} \geq 0.5$, which implies that $q_S = 1.9 \pm 0.3$ for a cut-off $r_c = 1.6$ km (Table 6).

As the largest comets are removed from the CSD, the M-estimate technique tends to consider the remaining largest nuclei as outliers, yielding steeper slopes. As a first test, we removed the largest nucleus in our database, namely 29P/Schwassmann-Wachmann 1 (the deletion of this comet may also be justified on the basis that it is more properly classified as a Centaur, rather than an EC), and obtained a nominal value of $q_S = 2.1$ from the M-estimate fit, with a

P_{KS} value of 0.86. Although this result is consistent with the previous one, given the uncertainties, we note that the distribution of P_{KS} values is not centered on the nominal value of q_S returned by the M-estimate fit.

In a second step, we also removed the next largest nucleus, namely 28P/Neujmin 1, and obtained a nominal value of $q_S = 2.4$, with P_{KS} only reaching 0.62. Note also that the distribution of P_{KS} values is even more skewed away from the nominal M-estimate value, which suggests that q_S is not very well-determined.

In summary, we conclude that q_S could be as small as ~1.6 and as large as ~2.5, with a preferred value of ~2.0. However, we will quote $q_S = 1.9 \pm 0.3$ because that is our result for the CSD that includes all the ECs for which reliable data have been obtained.

Table 6 shows that our result is intermediate between those of *Lowry et al.* [(2003a), $q_S = 1.6 \pm 0.1$] and *Weissman and Lowry* [(2003), $q_S = 1.59 \pm 0.03$, recently revised to $q_S = 1.79 \pm 0.05$ (P. Weissman, personal communication, 2003)], on the one hand, and *J. Fernández et al.* [(1999), $q_S = 2.65 \pm 0.25$], on the other hand. Regarding the first group of authors, we note that they incorrectly included 8P/Tuttle, a quite large nucleus, in their dataset and that their power exponent has been revised upward to be nearly compatible with our range (note also that their quoted uncertainty is underestimated owing to their use of least-squares fitting). Regarding the second group, i.e., *J. Fernández et al.* (1999), we concur with *Weissman and Lowry* (2003) in noting that their fitted slope covers only 12 comets over a very small range of radius, namely a factor of only 1.6. *J. Fernández et al.* (1999) have further limited their sample to those nuclei having perihelion distances q < 2 AU, fearing a possible bias, with nuclei with q > 2 AU being systematically larger than those with q < 2 AU. We have examined this question in detail, and Fig. 6 shows that there is no evidence for a systematic trend of size of the nucleus with perihelion distance. While there is indeed a larger number of small nuclei ($r_n < r_c$) with q < 2 AU than with q > 2 AU, the two populations of larger nuclei ($r_n > r_c$) have similar statistical properties (at least at the present level of accuracy), as already noted by *Weissman and Lowry* (2003). This is thus irrelevant when fitting the size distribution of nuclei with $r_n > r_c$ to a power law, and in fact has not been considered by the other groups listed in Table 6.

The comparison with the result of *Meech et al.* [(2004), $q_S = 2.5$] is not straightforward because it was obtained from a Monte Carlo reconstruction of the CSD that attempts to remove various selection effects, i.e., to *unbias* the observed CSDs. From their Table 11, we estimate $q_S \sim 1.5$, but we wonder whether this observed CSD includes both the short-period and long-period comets, as it is the case for the histogram given in their Fig. 6. We note that these authors truncated their original distribution ($q_S = 2.5$), and that in fact their best-fit model is truncated below $r_n = 5.0$ km. It will be interesting to see how their result evolves when their Monte Carlo simulation is applied to a larger dataset, such as ours.

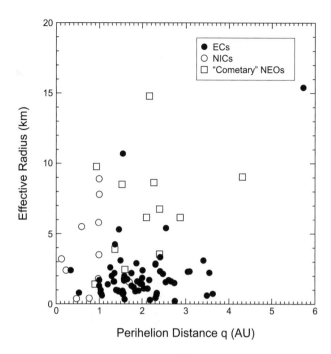

Fig. 6. The effective radius $r_{n,v}$ of the cometary nuclei vs. heliocentric distance for ecliptic comets (solid circles), for nearly isotropic comets (open circles), and for "cometary" NEOs (open squares).

It is tempting to compare our result for the distribution of ECs, $q_S \sim 1.9 \pm 0.3$, with the general trend of the power law of a collisionally population evolved with pure gravity scaling, $q_S = 2.04$ (*O'Brien and Greenberg*, 2003). On the one hand, it must be kept in mind that the simulated distribution is in fact wavy in the size interval of cometary nuclei, so that different values of q_S may hold in different size intervals. On the other hand, our dataset has not yet been corrected for bias effects and the statistics still remain limited, at essentially all sizes. We need far more measurements before we can conclusively determine the size distribution of ECs. As a way of testing how the distribution could evolve, we decided to incorporate additional objects. Our sample already includes highly evolved nuclei such as 28P/Neujmin 1. We now go one step further and include the population of asteroidal objects thought to be dormant or extinct comets, on the basis of their Tisserand parameters, or their association with meteor streams. The cometary origin of these NEOs is still highly speculative, and many of them may be bona fide asteroids coming from the outer regions of the asteroidal belt, including the Hilda group and Jupiter Trojans (*Fernández et al.*, 2002). Selection effects are also different from those of the ECs, and any future unbiasing should reflect these differences. For the purpose of the present exercise, we considered 21 "cometary" NEOs that can be associated with ECs, and whose sizes have been determined (Table 7), thus bringing the database to 86 objects. The "cometary" NEOs tend to be larger on average

than the ECs, thus significantly filling the 2–10-km radius range, but flattening the CSD simply because they are more of them at larger sizes; indeed, we found $q_S = 1.6 \pm 0.2$, P_{KS} reaching 0.85 when including these NEOs. The experiment of removing 29P and 28P has also been performed, and the results are illustrated in Fig. 5b. Because the new CSD is better constrained, the impact of removing these objects is much reduced compared to the case of ECs alone. We further note that the new CSD better fits a power-law function, thus reducing the differences between the M-estimate and K-S determinations of q_S.

Table 8, adapted from *Weissman and Lowry* (2003), displays the power exponents q_S and q_L for various minor-body populations in the solar system. The power exponent of the CSD of KBOs is quite large, $q_S = 3.15$–3.45, but strictly applies to objects with $r_n > 20$ km. It is not clear whether this value extends down to smaller sizes to allow a meaningful comparison with ECs. In fact, it has been suggested that KBOs follow a broken power law with the larger objects ($r_n \gtrsim 50$ km), retaining their primordial size distribution with the above value of q_S, while the smaller objects represent collisional fragments having a shallower distribution (e.g., *Davis and Farinella,* 1997), which could then be rather similar to that of the ECs. The power exponent of the CSD of Centaurs, $q_S = 2.7$–3.0, is also larger than that of the ECs. However, the statistics are rather poor, and we found that, from the data of nine Centaurs reported by *Barucci et al.* (2004), it is very difficult to fit a power law to the observed CSD: The exponent can take any value, from 3.1 down to 1.2, depending on the imposed cutoff at small sizes.

The CSD of ecliptic comets is beginning to look remarkably similar to that of NEOs: Note the result of *Stuart* (2001), $q_S = 1.96$, which is essentially identical to our value. For the main-belt asteroids, size distributions are so well-defined that changes in the power exponent can be recognized in different size regimes [see details in *Jedicke and Metcalfe* (1998)], and we have simply indicated the ranges. Near-Earth objects and main-belt asteroids are thought to be collisionally dominated populations, yet they have power exponents significantly different from the canonical value of $q_S = 2.5$ obtained by *Dohnanyi* (1969).

A final comparison is that with the CSD of the fragments of Comet D/1999 S4 (LINEAR): From water production rates measured following its breakup, *Mäkinen et al.* (2001) found that the measurements could best be explained by a fragment size distribution having $q_S = 1.74$, which is within the range we estimate for the ECs.

The question of the size distribution of ECs at the lower end, $r_c < 1.6$ km, remains totally open. The possible influence of both observational and evolutionary biases has been mentioned already, but a real depletion cannot be excluded. Indeed, the depletion of small nuclei is supported by the measurements of crater distributions on several airless bodies of the solar system, where cratering from comets is believed to dominate, e.g., Europa (*Chapman et al.*, 1997) and Ganymede and Callisto (*Zahnle et al.*, 2001).

TABLE 7. Physical properties of probable dormant or extinct comets.

Name	T_J*	r_n	p_V	Note	Association	References
Selected NEOs and possible dead comets based on $T_J \leq 3$ and low albedo (≤ 0.05)						
1580 Betulia	3.07	3.75 ± 0.15	0.034 ± 0.004	—	EC	Fe99
3552 Don Quixote	2.31	9.2 ± 0.4	0.045 ± 0.003	1983 SA	EC	Fe99
1983 VA	2.97	1.35 ± 0.05	0.07 ± 0.01	—	EC	Fe99
2000 EJ37	2.44	5.8	—	—	EC	Fe+03
2000 OG44	2.74	$3.87^{+0.50}_{-0.40}$	$0.038^{+0.018}_{-0.017}$	—	EC	Fe+01
2000 PG3	2.55	$3.08^{+1.42}_{-0.95}$	$0.021^{+0.031}_{-0.017}$	—	EC	Fe+01
2000 SB1	2.81	$3.57^{+0.92}_{-0.62}$	$0.019^{+0.015}_{-0.010}$	—	EC	Fe+01
2000 VU2	2.62	6.0	—	—	EC	Fe+03
2000 YN30	2.64	1.4	—	—	EC	Fe+03
2001 KX67	2.85	1.6	—	—	EC	Fe+03
2001 NX17	2.79	9.3	—	—	EC	Fe+03
2001 OB74	2.98	1.0	—	—	EC	Fe+03
2001 QF6	2.28	2.6	—	—	EC	Fe+03
2001 QL169	2.97	0.4	—	—	EC	Fe+03
2001 QQ199	2.32	10.2	—	—	EC	Fe+03
2001 RC12	2.69	1.6	—	—	EC	Fe+03
2001 SJ262	2.98	0.16	—	—	EC	Fe+03
2001 TX16	2.77	3.7	—	—	EC	Fe+03
5335 Damocles	1.14	8.5	0.03	†	NIC(HTC)	Le+02
15504 1999 RG33	1.95	14.8	0.03	†	NIC(HTC)	Le+02
20461 1999 LD31	−1.54	6.8	0.03	†	NIC(HTC)	Le+02
1996 PW	1.72	6.5	0.03	†	NIC(ERC)	Le+02
1997 MD10	0.98	2.5	0.03	†	NIC(HTC)	Le+02
1998 WU24	1.40	3.9	0.03	†	NIC(HTC)	Le+02
1999 LE31	−1.31	$9.05^{+4.04}_{-2.71}$	$0.031^{+0.030}_{-0.020}$	—	NIC(HTC)	Fe+01
1999 XS35	1.42	1.4	0.03	†	NIC(HTC)	Le+02
2000 AB229	0.78	6.2	0.03	†	NIC(ERC)	Le+02
2000 DG8	−0.62	$8.64^{+2.26}_{-1.83}$	$0.027^{+0.022}_{-0.015}$	—	NIC(HTC)	Fe+01
2000 HE46	−1.51	$3.55^{+1.10}_{-0.78}$	$0.023^{+0.021}_{-0.013}$	—	NIC(HTC)	Fe+01
Selected NEOs associated with meteor stream						
2101 Adonis	1.40	0.28‡	?	meteor stream	EC	Fe99
2212 Hephaistos	3.1	2.85	?	meteor stream	EC	Fe99
3200 Phaeton	4.51	2.35 ± 0.25	0.11 ± 0.02	meteor stream	EC	Fe99

*Tisserand parameter with respect to Jupiter.

†Radius derived from absolute magnitude (Le+02) using an albedo of 0.03.

‡Averaged radius derived from the radar measurements made by *Benner et al.* (1997).

References: Fe99: from the list compiled by *Fernández* (1999); Fe+01: *Fernández et al.* (2001); Fe+03: *Fernández et al.* (2003); Le+02: *Levison et al.* (2002).

TABLE 8. Power exponents of the CSD and CLF for various minor object populations.

Population	CSD	CLF	References
KBOs (r > 20 km)	3.45	0.69	*Gladman et al.* (2001)
	3.20 ± 0.10	0.64 ± 0.02	*Larsen et al.* (2001)
	3.15 ± 0.10	0.63 ± 0.06	*Trujillo et al.* (2001)
Centaurs	2.70 ± 0.35	0.54 ± 0.07	*Larsen et al.* (2001)
	3.0	0.6	*Sheppard et al.* (2000)
ECs	1.9 ± 0.3	0.38 ± 0.06	This work
ECs + "cometary" NEOs	1.6 ± 0.2	0.32 ± 0.04	This work
Near-Earth objects	1.75 ± 0.10	0.35 ± 0.02	*Bottke et al.* (2002)
	1.96	0.39	*Stuart* (2001)
Main-belt asteroids	1.25–2.80	0.25–0.56	*Jedicke and Metcalfe* (1998)

Figure 4b displays the CSD of the 13 NIC nuclei whose effective radii, $r_{n,v}$, have been determined (Table 2). Also plotted is the CSD of this population augmented by the 12 asteroidal objects thought to be dormant or extinct NICs on the basis of their Tisserand parameters (Table 7). Owing to the poor statistics, we did not attempt to fit power laws to the observed CSDs. We note, however, based on the present dataset, the rather shallow CSD of the NICs and the lack of small nuclei that NICs apparently share with ECs.

4.2. Shape and Rotation Period of Cometary Nuclei

Figure 7 displays the distribution of the axial ratio a/b of cometary nuclei. One should bear in mind that the bulk of the values are, strictly speaking, lower limits. There is not enough data for NICs to draw any conclusion. For ECs, the histogram is highly skewed with a median value of ~1.5, and there is *a priori* no reason to suspect that this property is biased by the aspect angle. There are a few cases of highly elongated nuclei with a/b > 2, the maximum value being 2.6 at present.

Figure 8 displays the distribution of rotational periods. One should bear in mind that most of them are not accurately determined because of the scarcity of data points to define the light curves. The range of 5–70 h is remarkably similar to that of the periods of main-belt asteroids and NEOs, excluding the monolithic fast rotators (e.g., *Whiteley et al.*, 2002). We further note that the bulk of the nuclei

measured so far have periods in a more restricted range, 5–18 h, but this may result from observational bias.

The results on rotational periods and axial ratios can be used to estimate lower limits on the density of the nuclei, assuming that they are strengthless (i.e., that cometesimals

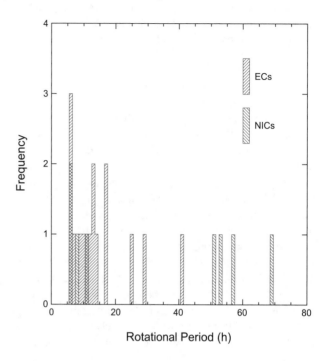

Fig. 8. Distribution of the rotational periods for cometary nuclei.

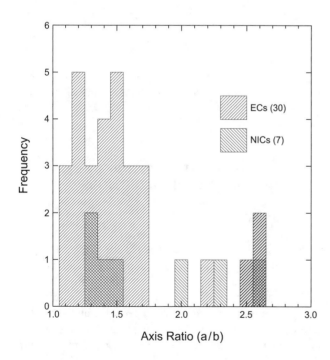

Fig. 7. Distribution of the lower limits of the axial ratio for cometary nuclei. The comet nucleus is assumed to be a prolate spheroid rotating around its axis of maximum moment of inertia for both ecliptic comets and nearly isotropic comets.

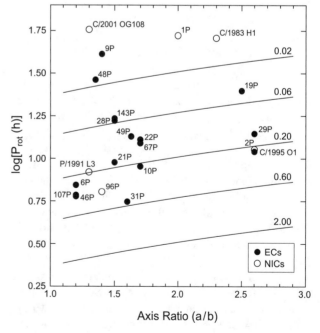

Fig. 9. Rotational periods vs. axial ratio for the cometary nuclei assumed to be prolate spheroids. The solid lines show curves of critical rotation for densities of 0.02, 0.06, 0.20, 0.60, 2.0 g cm^{-3} (from top to bottom).

are not physically bound) and that they are not rotating faster than the centrifugal limit for breakup (*Luu and Jewitt* 1992; *Meech,* 1996). Figure 9 displays the relevant diagram, where periods are plotted vs. axial ratios. Lines corresponding to the critical rotational period for different bulk densities of the nucleus are also plotted. The figure suggests that the fastest rotating nuclei are stable against centrifugal disruption, if their bulk densities exceed ~0.6 g cm^{-3} (see also *Weissman et al.,* 2004).

4.3. Albedos of Cometary Nuclei

One of the features evident in Table 4 is the very narrow range of albedos of cometary nuclei, namely 0.02 to 0.06. 29P/Schwassmann-Wachmann 1 stands as an exception with possibly p = 0.13 according to *Cruikshank and Brown* (1983), suggesting that this object may indeed be better classified as a Centaur. We further note that the lowest values have been measured at 4845 Å. As most nuclei have a red color, converting these values to the V band slightly increases them. As an example, we find $p_V = 0.025$ for 10P/Tempel 2 and $p_V = 0.030$ for 49P/Arend-Rigaux. The average value for the 13 nuclei listed in Table 4, excluding 29P, is $p_V = 0.038 \pm 0.009$ and $p_R = 0.042 \pm 0.017$, assuming a typical normalized reflectivity gradient of S' = 10%/1000 Å. These values nicely bracket the canonical albedo of 0.04, which is therefore fully justified. The range of albedos is so narrow that looking for trends is almost hopeless. This question has been recently investigated by *Campins and Fernández* (2003), who concluded that there is no trend with perihelion distance and a slight trend of decreasing albedo with increasing nuclear radius, the correlation being significant only at the 2σ level.

4.4. Colors of Cometary Nuclei

Figure 10 displays the distributions of the (B-V), (V-R), and (R-I) color indices, excluding 19P/Borrelly for which the uncertainty is too large, which can be compared to the solar indices $(B-V)_\odot = 0.65$, $(V-R)_\odot = 0.35$, and $(R-I)_\odot = 0.28$. The mean values of the indices <(B-V)> = 0.82, <(V-R)> = 0.41, and <(R-I)> = 0.38 confirm the well-known result that cometary nuclei are statistically redder than the Sun. Their colors are, however, very diverse, as already discussed for a smaller sample (*Luu,* 1993), from slightly blue to very red. Even if we exclude 14P/Wolf and 86P/Wild 3, for which the uncertainty in (V-R) is very large, there are two comets, 43P/Wolf-Harrington and 107P/Wilson-Harrington, that have a well-determined (V-R) = 0.31 ± 0.03, significantly less than that of the Sun. The reddest nucleus in the present sample is that of 143P/Kowal-Mrkos, with (V-R) = 0.58, still less red than the average (V-R) = 0.61 of KBOs (*Jewitt,* 2002).

As pointed out in section 4.4, colors by themselves do not reveal much about the physical properties, but the distribution of colors compared to other solar system objects may provide information on their interrelationships. This topic is discussed in *Jewitt* (2004).

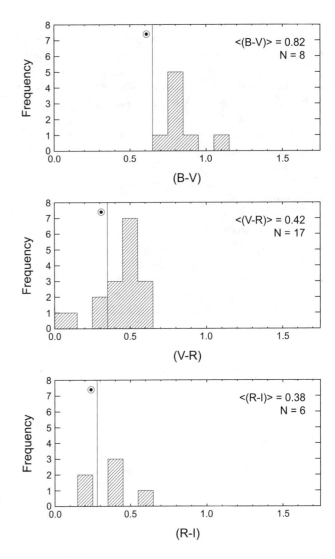

Fig. 10. The distributions of the (B-V), (V-R), and (R-I) color indices of the ecliptic comets, excluding 19P/Borrelly. The average values of the color indices are displayed and the color indices of the Sun are indicated.

5. SUMMARY, OPEN ISSUES, AND FUTURE DIRECTIONS

5.1. Current Status

Remarkable progress has been made during the past decade in measuring the sizes of cometary nuclei, but it is also clear that this field is still in its infancy. Reliable data exist for only 65 comets, so any conclusions regarding the distribution of sizes is necessarily tentative and subject to future revision. Measurements are needed for substantially more comets, at least doubling or tripling the current number, before confidence can be gained in the conclusions regarding the size distribution. Our current best estimate, $q_S = 1.9 \pm 0.3$, is conspicuously different from that of the KBO and Centaur populations, but is similar to that of the NEOs. This value also corresponds to that of a collisionally evolved population with pure gravity scaling, but we reemphasize that *O'Brien and Greenberg* (2003) showed that this distribution is in fact wavy in the size range relevant to ECs.

The situation for cometary albedos is even worse, in the sense that reliable values are available for only about a dozen objects. Nevertheless, we are struck by the relatively small range in the albedo (0.04 ± 0.02), which suggests that the surfaces of cometary nuclei are exceptionally dark, contrary to the early expectations for these "icy" bodies.

Measuring accurate shapes of cometary nuclei is sometimes possible but requires intensive observing campaigns, generally extending over several days. Furthermore, the aspect angle of the rotational axis usually varies with time, so that observations at widely separated places in the comet's orbit are desirable to obtain a clear picture of the comet's rotational properties and true shape. Although there are some examples of highly elongated cometary nuclei, with the major axis being up to 2.6 times larger than the minor axis, most cometary nuclei seem to differ from spherical bodies by ~50%. However, we must keep in mind that the axial ratio values are often lower limits. Fortunately, conclusions regarding the size distribution of cometary nuclei are not strongly affected by uncertainties in the shapes.

It is also difficult to obtain reliable color data on cometary nuclei. While the color of the nucleus itself does not provide unique information on the physical properties, color data are useful for comet-to-comet comparisons, which may suggest differences in surface properties, particularly when in making comparisons with other minor bodies in the solar system (e.g., Centaurs, TNOs, and asteroids). The colors of cometary nuclei are diverse, with some being highly reddened compared to solar color, some being neutral, and a few having a slightly blue color.

5.2. Outstanding Issues and Future Investigations

An important, unresolved issue concerns the interpretation of disk-integrated thermal measurements, which, in principle, provide robust determinations of sizes and albedos. The so-called standard thermal model for asteroids is often used to interpret cometary thermal data, although its applicability to objects having a mixture of dust and ice is questionable.

A totally open issue is the nature of the size distribution of cometary nuclei at the small end of the spectrum. Does the relatively steep power law derived from the intermediate-sized objects extend indefinitely to smaller sizes? Or is the size distribution truncated at some value that depends on the physical formation mechanism (e.g., gravitational instability within the solar nebula) or destruction mechanism (e.g., total disruption)?

What is the bias in ecliptic comet discoveries and how does that affect the current distribution of sizes? Why do we observe so few large ($r_n \geq 5$ km) cometary nuclei?

How does evolution affect the physical properties of cometary nuclei? Is there really a continuum of surface properties that is dependent on the activity level and physical evolution (e.g., with a youthful Chiron at one end and an aged 2P/Encke on the other)? Continual loss of the surface layers through repeated passages through the inner solar system obviously affects the size, and probably the shape and color, of cometary nuclei. How do we account for this in estimating the primordial distribution of physical properties? Some combination of improved modeling and observations of nuclei will help, but it is not yet clear that these issues can ever be resolved satisfactorily.

Splitting events obviously affect the size and shape of cometary nuclei, but how do we estimate their effect on the distribution functions? Perhaps better data on the splitting rates of nuclei, coupled with a better understanding of the physical mechanism(s) for splitting events, will help to resolve these issues, but that remains to be seen.

Of course, one of the most interesting avenues for future exploration is the relationship between cometary nuclei and the other minor bodies in the solar system. Can we use the size distribution of cometary nuclei to conclude that they are a collisionally evolved population? In particular, can the size distribution and the shapes of cometary nuclei be used to conclude that they are collisional fragments of the TNOs? If the Centaurs are TNOs on the road to becoming ecliptic comets, then perhaps Centaurs and these comets share common physical properties. If not, can the differences be explained by evolutionary effects? Although cometary nuclei generally contain much more ice than do asteroids, perhaps "evolved" comets share many common characteristics with asteroids. In addition, at least some asteroids, and many cometary nuclei, are thought to have porous, "rubble-pile" physical structures (*Davis et al.,* 1985; *Weissman,* 1986). Does this point to commonalities in their formation mechanism?

It seems clear that remote observations with 2–3-m telescopes will continue to play a critical role in measuring the physical properties of cometary nuclei as a population. We can certainly hope to make as much progress during the next decade as we have witnessed in the previous one, and we can look forward to many new advances in our understanding of the physical properties of cometary nuclei in the future. During its remaining lifetime, and subject to approval of the relevant programs, HST can essentially complete a survey of the bulk of the known population of ECs and provide unique, accurate color data. Large groundbased telescopes (e.g., Keck and the Very Large Telescope) will also contribute by detecting nuclei at large heliocentric distances, where they are presumed inactive. With its unsurpassed sensitivity, SIRTF will be capable of detecting the nuclei of a large fraction of the ECs in the midinfrared, thus providing albedo-independent determinations of their size. Herschel will also be able to provide albedo-independent size determinations, but for larger, more distant nuclei in the submillimeter wavelength range. Finally, the Atacama Large Millimetre Array (ALMA) will be able to detect a significant number of nuclei at 1.3 mm. Combining ALMA and SIRTF data should provide robust information on the long-wavelength emissivity and internal thermal properties of cometary nuclei.

Future spacecraft encounters will undoubtedly shed further light on the nature of cometary nuclei, and will even start to address the diversity issue in a serious way as more and more objects come under intense scrutiny. While we very much regret the loss of the *CONTOUR* mission, which would have flown by 2P/Encke and 73P/Schwassmann-Wachmann 3, we now look forward to the flybys of 81P/Wild 2 (*Stardust* mission) and 9P/Tempel 1 (*Deep Impact* mission), the latter also probing the interior of the nucleus. Space investigations of cometary nuclei will culminate with the *Rosetta* mission, whose orbiter will accompany the nucleus of 67P/Churyumov-Gerasimenko from a heliocentric distance of 3.6 AU to perihelion at 1.3 AU and observe it at all spatial scales down to a few millimeters, thus allowing an unprecedented view of how the activity starts and evolves. The *Rosetta* lander will perform a broad range of *in situ* observations and analysis of the surface and subsurface regions of the nucleus.

6. CONCLUSIONS

When D. Jewitt presented the first "modern" review of the physical properties of cometary nuclei, at the 30th Liège conference in 1992 (*Jewitt,* 1992) just 10 years ago, his sample was limited to four ECs, one HTC, and Chiron. His review incorporated rotation and precession, internal and surface properties, and a comparison with asteroids. At the 1996 Asteroids, Comets, and Meteor (ACM) conference, K. Meech's review discussed 17 ECs, 7 HTCs, and 2 Centaurs (Chiron and 29P). For the *Comets II* book, the study of cometary nuclei is divided among 10 chapters, with questions such as the physical properties, rotation, surface properties, internal properties, and the relationship with other minor bodies deserving their own, separate chapters. Our review of the physical properties of cometary nuclei discusses 65 ECs and 13 NICs, with the Centaurs being treated in a different chapter. This illustrates the tremendous advancement of research on the cometary nucleus during the past decade. And since there has been only one cometary space mission during the past 10 years, namely *Deep Space 1*, the bulk of the progress has been achieved by groundbased and spacebased observatories, demonstrating that the use of better techniques and better detectors, coupled with new telescopes (but some old ones as well), is capable of achieving what was long considered hopeless. On the one hand, we are far from fully understanding cometary nuclei, as illustrated above by the list of outstanding issues. On the other hand, the future is bright as new facilities will soon allow us to push the present limits of our observational capabilities even further, while forthcoming cometary space missions will allow us to study a few nuclei in unprecedented detail.

Acknowledgments. We thank O. Hainaut, S. Lowry, G. Tancredi, and P. Weissman for kindly making their latest results available to us in advance of publication. We are very much grateful to P. Weissman for his thorough review of an earlier version of this chapter, which led to substantial improvements. We further thank G. Bernstein, H. Boehnhardt, J. Fernández, J. Licandro, and N. Samarasinha for their constructive remarks. P.L.L. and I.T. acknowledge the support of the French Programme National de Planétologie, jointly funded by CNRS and CNES, and the bilateral French-Hungarian cooperation program. I.T. further acknowledges the support of the Université de Provence and of the Hungarian State Research Foundation for Sciences (OTKA) through Grant No. T025049. H.A.W. acknowledges financial support by NASA through Grants HST-GO-8699.01-A and HST-GO-8876.01-A from the Space Telescope Science Institute (STScI), which is operated by the Association of Universities for Research in Astronomy, Inc., under NASA contract NAS5-26555. This work is partly based on observations made with the NASA/ESA Hubble Space Telescope and obtained at the STScI.

REFERENCES

Abell P. A., Fernández Y. R., Pravec P., French L. M., Farnham T. L., Gaffey M. J., Hardersen P. S., Kušnirák P., Šarounová L., and Sheppard S. S. (2003) Physical characteristics of asteroid-like comet nucleus C/2001 OG108 (LONEOS). In *Lunar and Planetary Science XXXIV,* Abstract #1253. Lunar and Planetary Institute, Houston (CD-ROM).

Abergel A. and Bertaux J.-L. (1990) Rotation states of the nucleus of Comet Halley compatible with spacecraft images. *Icarus, 86,* 21–29.

A'Hearn M. F., Campins H., Schleicher D. G., and Millis R. L. (1989) The nucleus of comet P/Tempel 2. *Astrophys. J., 347,* 1155–1166.

Allen C. W. (1976) *Astrophysical Quantities, 3rd edition.* Athlone, London. 310 pp.

Allen D. A. (1971) The method of determining infrared diameters. In *Physical Studies of Minor Planets* (T. Gehrels, ed.), p. 41. IAU Colloquium 12, NASA SP-267.

Altenhoff W. J., Batrla W. K., Huchtmeier W. K., Schmidt J., Stumpff P., and Walmsley M. (1983) Radio observations of comet 1983 d. *Astron. Astrophys., 125,* L19–L22.

Altenhoff W. J., Bieging J. H., Butler B., Butner H. M., Chini R., Haslam C. G. T., Kreysa E., Martin R. N., Mauersberger R., McMullin J., Muders D., Peters W. L., Schmidt J., Schraml J. B., Sievers A., Stumpff P., Thum C., von Kap-Herr A., Wiesemeyer H., Wink J. E., and Zylka R. (1999) Coordinated radio continuum observations of comets Hyakutake and Hale-Bopp from 22 to 860 GHz. *Astron. Astrophys., 348,* 1020–1034.

Barucci A. M., Doressoundiram A., and Cruikshank D. P. (2004) Surface characteristics of transneptunian objects and Centaurs from photometry and spectroscopy. In *Comets II* (M. C. Festou et al., eds.), this volume. Univ. of Arizona, Tucson.

Belton M. J. S. (1990) Rationalization of comet Halley's periods. *Icarus, 86,* 30–51.

Belton M. J. S. (2000) The excited rotational state of 2P/Encke. *Bull. Am. Astron. Soc., 32,* Abstract #36.12.

Belton M. J., Mueller B. E. A., Julian W. H., and Anderson A. J. (1991) The spin state and homogenity of comet Halley's nucleus. *Icarus, 93,* 183–193.

Benner L. A. M., Ostro S. J., Giorgini J. D., Jurgens R. E., Mitchell D. L., Rose R., Rosema K. D., Slade M. A., Winkler R., Yeomans D. K., Campbell D. B., Chandler J. F., and Shapiro I. I. (1997) Radar detection of near-Earth asteroids 2062 Aten, 2101 Adonis, 3103 Eger, 4544 Xanthus, and 1992 QN. *Icarus, 130,* 296–312.

Biesecker D. A., Lamy P., St. Cyr O. C., Llebaria A., and Howard R. A. (2002) Sungrazing comets discovered with the SOHO/LASCO coronographs 1996–1998. *Icarus, 157,* 323–348.

Bobrovnikoff N. T. (1931) On the spectra of comets 1910 I, 1911 IV, 1929 d. *Publ. Astron. Soc. Pacific, 43,* 61–65.

Boehnhardt H. (2004) Split comets. In *Comets II* (M. C. Festou et al., eds.), this volume. Univ. of Arizona, Tucson.

Boehnhardt H. and Birkle K. (1994) Time variable coma structures in comet P/Swift-Tuttle. *Astron. Astrophys. Suppl. Ser., 107,* 101–120.

Boehnhardt H., Birkle K., and Osterloh M. (1996) Nucleus and tail studies of comet P/Swift-Tuttle. *Earth Moon Planets, 73,* 51–70.

Boehnhardt H., Babion J., and West R. M. (1997) An optimized detection technique for faint moving objects on a star-rich background. *Astron. Astrophys., 320,* 642–651.

Boehnhardt H., Rainer N., Birkle K., and Schwehm G. (1999) The nuclei of comets 26P/Grigg-Skjellerup and 73P/Schwassmann-Wachmann 3. *Astron. Astrophys., 341,* 912–917.

Boehnhardt H., Delahodde C., Sekiguchi T., Tozzi G. P., Amestica R., Hainaut O. R., Spyromilio J., Tarenghi M., West R. M., Schulz R., and Schwehm G. (2002) VLT observations of comet 46P/Wirtanen. *Astron. Astrophys., 387,* 1107–1113.

Bottke W. F., Morbidelli A., Jedicke R., Petit J.-M., Levison H. F., Michel P., and Metcalfe T. S. (2002) Debiased orbital and absolute magnitude distribution of the near-Earth objects. *Icarus, 156,* 399–433.

Bowell E., Hapke B., Domingue D., Lumme K., Peltoniemi J., and Harris A. W. (1989) Application of photometric models to asteroids. In *Asteroids II* (R. P. Binzel et al., eds.), pp. 524–556. Univ. of Arizona, Tucson.

Brandt J. C., A'Hearn M. F., Randall C. E., Schleicher D. G., Shoemaker E. M., and Stewart A. I. F. (1996) Small comets (SCs): An unstudied population in the Solar System inventory. In *Completing the Inventory of the Solar System* (T. W. Rettig and J. M. Hahn, eds.), pp. 289–297. ASP Conference Series 107, Astronomical Society of the Pacific, San Francisco.

Brooke T. Y. and Knacke R. F. (1986) The nucleus of comet P/Arend-Rigaux. *Icarus, 67,* 80–87.

Brown R. H. (1985) Ellipsoidal geometry in asteroid thermal models — The standard radiometric model. *Icarus, 64,* 53–63.

Brown R. H., Cruikshank D. P., and Griep D. (1985) Temperature of comet IRAS-Araki-Alcock (1983 d). *Icarus, 62,* 273–281.

Buratti B. J., Hicks M. D., Soderblom L. A., Britt D., Oberst J., and Hillier J. K. (2004) Deep Space 1 photometry of the nucleus of comet 19P/Borrelly. *Icarus,* in press.

Bus S. J., Buie M. W., Schleicher D. G., Hubbard W. B., Marcialis R. L., Hill R., Wasserman L. H., Spencer J. R., Millis R. L., Franz O. G., Bosh A. S., Dunham E. W., Ford C. H., Young J. W., Elliott J. L., Meserole R., Olkin C. B., McDonald S. W., Foust J. A., Sopata L. S., and Bandyopadhyay R. M. (1996) Stellar occultation by 2060 Chiron. *Icarus, 123,* 478–490.

Campins H. (1988) The anomalous dust production in periodic comet Encke. *Icarus, 73,* 508–515.

Campins H. and Fernández Y. R. (2003) Observational constraints on surface characteristics of comet nuclei. *Earth Moon Planets,* in press.

Campins H. and Schleicher D. G. (1995) Comet d'Arrest: A small and rapidly rotating nucleus. *Bull. Am. Astron. Soc., 27,* Abstract #1113.

Campins H., A'Hearn M. F., and McFadden L-A. (1987) The bare nucleus of comet Neujmin 1. *Astrophys. J., 316,* 847–857.

Campins H., Telesco C. M., Osip D .J., Rieke G. H., Rieke M. J., and Schulz B. (1994) The color temperature of (2060) Chiron: A warm and small nucleus. *Astron. J., 108,* 2318–2322.

Campins H., Osip D. J., Rieke G. H., and Rieke M. J. (1995) The radius and albedo of comet-asteroid transition object 4015 Wilson-Harrington. *Planet. Space Sci., 43,* 733–736.

Campins H., Licandro J., Chamberlain M., and Brown R. H. (2001) Constraints on the surface composition of comet 28P/Neujmin 1. *Bull. Am. Astron. Soc., 33,* Abstract #31.08.

Carlson N. (1990) *52P/Harrington-Abell.* Minor Planet Circular 17685.

Chamberlin A. B., McFadden L-A., Schulz R., Schleicher D. G., and Bus S. J. (1996) 4015 Wilson-Harrington, 2201 Oljato, and 3200 Phaethon: Search for CN emission. *Icarus, 119,* 173–181.

Chapman C. R., Merline W. J., Bierhaus B., Keller J., and Brooks S. (1997) Impactor populations on the Galilean satellites. *Bull. Am. Astron. Soc., 29,* Abstract #12.10.

Clark B. E., Veverka J., Helfenstein P., Thomas P. C., Bell J., Harch A., Robinson M. S., Murchie S. L., McFadden L. A., and Chapman C. R. (1999) NEAR photometry of asteroid 253 Mathilde. *Icarus, 140,* 53–65.

Clark B. E., Helfenstein P., Bell J. F., Peterson C., Veverka J., Izenberg N. I., Domingue D., Wellnitz D., and McFadden L. (2002) NEAR infrared spectrometer photometry of asteroid 433 Eros. *Icarus, 155,* 189–204.

Combes M., Moroz V. I., Crifo J. F., Lamarre J. M., Charra J., Sanko N. F., Soufflot A., Bibring J. P., Cazes S., Coron N., Crovisier J., Emerich C., Encrenaz T., Gispert R., Grigoryev A. V., Guyot G., Krasnopolsky V. A., Nikolsky Yu. V., and Rocard F. (1986) Infrared sounding of comet Halley from Vega 1. *Nature, 321,* 266–268.

Cruikshank D. P. and Brown R. H. (1983) The nucleus of comet P/Schwassmann-Wachmann 1. *Icarus, 56,* 377–380.

Davis D. R. and Farinella P. (1997) Collisional evolution of Edgeworth-Kuiper belt objects. *Icarus, 125,* 50–60.

Davis D. R., Chapman C. R., Weidenschilling S. J., and Greenberg R. (1985) Collisional history of asteroids — Evidence from Vesta and Hirayama families. *Icarus, 62,* 30–53.

Delahodde C. E. (2003) Nouvelles observations de noyaux cométaires. Ph.D. thesis, Université de Provence, France.

Delahodde C. E., Meech K. J., Hainaut O., and Dotto E. (2001) Detailed phase function of comet 28P/Neujmin 1. *Astron. Astrophys., 376,* 672–685.

Delahodde C. E., Hainaut O. R., Romon-Martin J., and Lamy P. L. (2002) VLT near-infrared spectra of distant comets: First results with 90P/Gehrels 1. In *Proceedings of the Asteroids, Comets, Meteors 2002 Conference* (B. Warmbein, ed.), in press. ESA SP-500, Noordwijk, The Netherlands.

Delbo M., Harris A. W., Binzel R. P., Pravec P., and Davies J. K. (2003) Keck observations of near-Earth asteroids in the thermal infrared. *Icarus,* in press.

Delsemme A. H. and Rud D. A. (1973) Albedos and cross-sections for the nuclei of comets 1969 IX, 1970 II and 1971 I. *Astron. Astrophys., 28,* 1–6.

De Sanctis M. C., Lazzarin M., Barucci M. A., Capria M. T., and Coradini A. (2000) Comet P/Gehrels 3: Spectroscopic observations and nucleus models. *Astron. Astrophys., 354,* 1086–1090.

Desvoivres E., Klinger J., Levasseur-Regourd A. C., and Jones G. H. (2000) Modeling the dynamics of cometary fragments: Application to C/1996 B2 Hyakutake. *Icarus, 144,* 172–181.

Dohnanyi J. S. (1969) Collisional model of asteroids and their debris. *J. Geophys. Res., 74,* 2431–2554.

Dones L., Weissman P. R., Levison H. F., and Duncan M. J. (2004) Oort cloud formation and dynamics. In *Comets II* (M. C. Festou et al., eds.), this volume. Univ. of Arizona, Tucson.

Duncan M., Levison H., and Dones L. (2004) Dynamical evolution of ecliptic comets. In *Comets II* (M. C. Festou et al., eds.), this volume. Univ. of Arizona, Tucson.

Emerich C., Lamarre J. M., Gispert R., Coron N., Combes M., Encrenaz T., Crovisier J., Rocard E., Bibring J. P., and Moroz V. I. (1987) Temperature of the nucleus of Comet Halley. In *Proceedings of the International Symposium on the Diversity and Similarity of Comets* (E. Rolfe and B. Battrick, eds.), pp. 703–706. ESA SP-278, Noordwijk, The Netherlands.

Farinella P. and Davis D. R. (1996) Short-period comets: Primordial bodies or collisional fragments? *Science, 273,* 938–941.

Farnham T., Schleicher D. G., Williams W. R., and Smith B. R. (1999) The rotation state and active regions of comet Hale-Bopp (1995 O1). *Bull. Am. Astron. Soc., 31,* Abstract #30.01.

Farnham T., Schleicher D. G., Woodney L. M., Eberhardy C. A., Birch P. V., and Levy L. (2001) Imaging and photometry of comet C/1999 S4 (LINEAR) before perihelion and after break-up. *Science, 292,* 1348–1353.

Fay T. D., and Wisniewski W. (1978) The light curve of the nucleus of comet d'Arrest. *Icarus, 34,* 1–9.

Feierberg M. A., Witteborn F. C., Johnson J. R., and Campins H. (1984) 8- to 13-micron spectrophotometry of comet IRAS-Araki-Alcock. *Icarus, 60,* 449–454.

Fernández J. A., Tancredi G., Rickman H., and Licandro J. (1999) The population, magnitudes, and sizes of Jupiter family comets. *Astron. Astrophys., 352,* 327–340.

Fernández J. A., Gallardo T., and Brunini A. (2002) Are there many Jupiter-family comets among the near-Earth asteroid population? *Icarus, 159,* 358–368.

Fernández Y. R. (1999) Physical properties of cometary nuclei. Ph.D. thesis, Univ. of Maryland, College Park.

Fernández Y. R. (2003) The nucleus of comet Hale-Bopp (C/1995 O1): Size and activity. *Earth Moon Planets, 89,* 3–25.

Fernández Y. R., Wellnitz D. D., Buie M. W., Dunham E. W., Millis R. L., Nye R. A., Stansberry J. A., Wasserman L. H., A'Hearn M. F., Lisse C. M., Golden M. E., Person M. J., Howell R. R., Marcialis R. L., and Spitale J. N. (1999) The inner coma and nucleus of Hale-Bopp: Results from a stellar occultation. *Icarus, 140,* 205–220.

Fernández Y. R., Lisse C. M., Käufl H. U., Peschke S. B., Weaver H. A., A'Hearn M. F., Lamy P., Livengood T. A., and Kostiuk T. (2000) Physical properties of the nucleus of comet 2P/Encke. *Icarus, 147,* 145–160.

Fernández Y. R., Jewitt D. C., and Sheppard S. S. (2001) Low albedos among extinct comet candidates. *Astrophys. J. Lett., 553,* L197–L200.

Fernández Y. R., Jewitt D. C., and Sheppard S. S. (2002a) Thermal properties of Centaurs Asbolus and Chiron. *Astron. J., 123,* 1050–1055.

Fernández Y. R., Jewitt D. C., and Sheppard S. S. (2002b) *Comet 57P/du Toit-Neujmin-Delporte.* IAU Circular No. 7935.

Fernández Y. R., Lowry S. C., Weissman P. R., and Meech K. J. (2002c) New dominant periodicity in photometry of comet Encke. *Bull. Am. Astron. Soc., 34,* Abstract #27.06.

Fernández Y. R., Meech K. J., Lisse C. M., A'Hearn M. F., Pittichová J., and Belton M. J. S. (2003) The nucleus of Deep Impact target comet 9P/Tempel 1. *Icarus, 164,* 481–491.

Fitzsimmons A. and Williams I. P. (1994) The nucleus of comet P/Levy (1991 XI). *Astron. Astrophys., 289,* 304–310.

Fitzsimmons A., Dahlgren M., Lagerkvist C.-I., Magnusson P., and Williams I. P. (1994) A spectroscopic survey of D-type asteroids. *Astron. Astrophys., 289,* 304–340.

Fomenkova M. N., Jones B., Pina R., Puetter R., Sarmecanic J., Gehrz R., and Jones T. (1995) Mid-infrared observations of the nucleus and dust of comet P/Swift-Tuttle. *Astron. J., 110,* 1866–1874 and Plate 103.

Garradd G. J. (1997) *Comet 2P/Encke.* IAU Circular No. 6717.

Gladman B., Kavelaars J. J., Petit J.-M., Morbidelli A., Holman M. J., and Loredo T. (2001) The structure of the Kuiper Belt: Size distribution and radial extent. *Astron. J., 122,* 1051–1066.

Goldstein R. M., Jurgens R. F., and Sekanina Z. (1984) A radar study of comet IRAS-Araki-Alcock 1983 d. *Astron. J., 89,* 1745–1754.

Green S. F., Meadows A. J., and Davies J. K. (1985) Infrared observations of the extinct cometary candidate minor planet (3200) 1983 TB. *Mon. Not. R. Astron. Soc., 214,* 29–36.

Groussin O. and Lamy P. L. (2003a) Activity on the surface of the nucleus of comet 46P/Wirtanen. *Astron. Astrophys.,* in press.

Groussin O. and Lamy P. L. (2003b) Properties of the nucleus of Centaurs Chiron and Chariklo. *Astron. Astrophys.,* in press.

Groussin O., Lamy P. L., Jorda L., and Toth I. (2003) The nucleus of comets 126P/IRAS and P/Hartley 2. *Astron. Astrophys.,* in press.

Groussin O., Lamy P., Gutiérrez P. J., and Jorda L. (2004) The nucleus of comet IRAS-Araki-Alcock (C/1983 H1). *Astron. Astrophys.,* in press.

Gutiérrez P. J., Ortiz J. L., Rodrigo R., and Lôpez-Moreno J. J. (2001) Effects of irregular shape and topography in thermophysical models of heterogeneous cometary nuclei. *Astron. Astrophys., 374,* 326–336.

Gutiérrez P. J., de León J., Jorda L., Licandro J., Lara L. M., and Lamy P. (2003) New spin period determination for comet 6P/d'Arrest. *Astron. Astrophys., 407,* L37–L40.

Hainaut O. R. and Delsanti A. C. (2002) Colors of minor bodies in the outer solar system. A statistical analysis. *Astron. Astrophys., 389,* 641–664.

Hainaut O. R. and Meech K. J. (1996) *Comet 43P/Wolf-Harrington.* Minor Planet Circular 27955.

Hainaut O. R., Meech K. J., Boehnhardt H., and West R. M. (1998) Early recovery of comet 55P/Tempel-Tuttle. *Astron. Astrophys., 333,* 746–752.

Hanner M. S., Giese R. H., Weiss K., and Zerull R. (1981) On the definition of albedo and application to irregular particles. *Astron. Astrophys., 104,* 42–46.

Hanner M. S., Aitken D. K., Knacke R., McCorkle S., Roche P. F., and Tokunaga A. T. (1985) Infrared spectrophotometry of comet IRAS-Araki-Alcock (1983 d) — A bare nucleus revealed? *Icarus, 62,* 97–109.

Hanner M. S., Newburn R. L., Spinrad H., and Veeder G. J. (1987) Comet Sugano-Saigusa-Fujikawa (183 V) — A small, puzzling comet. *Astron. J., 94,* 1081–1087.

Hapke B. (1986) Bidirectional reflectance spectroscopy. IV — The extinction coefficient and opposition effect. *Icarus, 67,* 264–280.

Hapke B. (1993) *Theory of Reflectance and Emittance Spectroscopy.* Cambridge Univ., New York. 455 pp.

Harmon J. K., Campbell D. B., Hine A. A., Shapiro I. I., and Marsden B. G. (1989) Radar observations of comet IRAS-Araki-Alcock 1983 d. *Astrophys. J., 338,* 1071–1093.

Harmon J. K., Ostro S. J., Benner L. A. M., Rosema K. D., Jurgens R. F., Winkler R., Yeomans D. K., Choate D., Cormier

R., Giorgini J. D., Mitchell D. L. Chodas P. W., Rose R., Kelley D., Slade M. A., and Thomas M. L. (1997) Radar detection of the nucleus and coma of comet Hyakutake (C/1996 B2). *Science, 278,* 1921.

Harmon J. K., Nolan M. C., Ostro S. J., and Campbell D. B. (2004) Radar studies of comet nuclei and grain comae. In *Comets II* (M. C. Festou et al., eds.), this volume. Univ. of Arizona, Tucson.

Harris A. W. (1998) A thermal model for near-Earth asteroids. *Icarus, 131,* 291–301.

Irvine W. M. and 40 colleagues (1984) Radioastronomical observations of comets IRAS-Araki-Alcock (1983d) and Sugano-Saigusa-Fujikawa (1983e). *Icarus, 60,* 215–220.

Jedicke R. and Metcalfe T. S. (1998) The orbital and absolute magnitude distribution of main belt asteroids. *Icarus, 131,* 245–260.

Jewitt D. (1991) Cometary photometry. In *Comets in the Post-Halley Era* (R. L. Newburn Jr. et al., eds.), pp. 19–65. Kluwer, Dordrecht.

Jewitt D. C. (1992) Physical properties of cometary nuclei. Invited review. In *Proceedings of the 30th Liége International Astrophysical Colloquium* (A. Brahic et al., eds.), pp. 85–112. Univ. of Liége, Liége.

Jewitt D. (2002) From Kuiper Belt to cometary nucleus. In *Proceedings of the Asteroids, Comets, Meteors 2002 Conference* (B. Warmbein, ed.), pp. 11–19. ESA SP-500, Noordwijk, The Netherlands.

Jewitt D. (2004) From cradle to grave: The rise and demise of the comets. In *Comets II* (M. C. Festou et al., eds.), this volume. Univ. of Arizona, Tucson.

Jewitt D. and Kalas P. (1998) Thermal observations of centaur 1997 CU26. *Astrophys. J. Lett., 499,* L103–L106.

Jewitt D. and Luu J. (1989) A CCD portrait of comet P/Tempel 2. *Astron. J., 97,* 1766–1790.

Jewitt D. C. and Luu J. X. (1990) CCD spectra of asteroids. II — The Trojans as spectral analogs of cometary nuclei. *Astron. J., 100,* 933–944.

Jewitt D. C. and Luu J. X. (2001) Colors and spectra of Kuiper Belt objects. *Astron. J., 122,* 2099–2114.

Jewitt D. C. and Meech K. J. (1985) Rotation of the nucleus of P/Arend-Rigaux. *Icarus, 64,* 329–335.

Jewitt D. C. and Meech K. J. (1987) CCD photometry of comet P/Encke. *Astron. J., 93,* 1542–1548.

Jewitt D. C. and Meech K. J. (1988) Optical properties of cometary nuclei and a preliminary comparison with asteroids. *Astrophys. J., 328,* 974–986.

Jewitt D. and Sheppard S. (2003) The nucleus of comet 48P/Johnson. *Astron. J.,* in press.

Jewitt D., Sheppard S. and Fernández Y. (2003) 143P/Kowal-Mrkos and the shapes of cometary nuclei. *Astron. J., 125,* 3366–3377.

Jorda L. and Lecacheux J. (1992) *Periodic Comet Swift-Tuttle.* IAU Circular No. 5664.

Jorda L., Rembor K., Lecacheux J., Colom P., Colas F., Frappa E., and Lara L. M. (1999) The rotational parameters of Hale-Bopp (C/1995 O1) from observations of the dust jets at Pic du Midi Observatory. *Earth Moon Planets, 77,* 167–180.

Jorda L., Lamy P., Groussin O., Toth I., A'Hearn M. F., and Peschke S. (2000) ISOCAM observations of cometary nuclei. In *Proceedings of ISO Beyond Point Sources: Studies of Extended Infrared Emission* (R. J. Laureijs et al., eds.), p. 61. ESA SP-455, Noordwijk, The Netherlands.

Kamél L. (1990) The Comet Light Curve Catalog/Atlas — A

presentation. In *Proceedings of Asteroids, Comets, Meteors 1989* (C.-I. Lagerkvist et al., eds.), p. 363. Uppsala, Sweden.

Kamél L. (1992) The comet Light Curve Catalogue/Atlas. III — The Atlas. *Astron. Astrophys. Suppl. Ser., 92,* 85–149.

Kamoun P. G., Pettengill G. H., and Shapiro I. I. (1982) Radar observations of cometary nuclei. In *Comets* (L. L. Wilkening, ed.), pp. 288–296. Univ. of Arizona, Tucson.

Kamoun P. G., Campbell D., Pettengill G., and Shapiro I. (1999) Radar observations of three comets and detection of echoes from one: P/Grigg-Skjellerup. *Planet. Space Sci., 47,* 23–28.

Keller H. U. (1990) The nucleus. In *Physics and Chemistry of Comets* (W. F. Huebner, ed.), pp. 13–68. Springer-Verlag, Berlin.

Keller H. U., Arpigny C., Barbieri C., Bonnet R. M., Cazes S., Coradini M., Cosmovici C. B., Delamere W. A., Huebner W. F., Hughes D. W., Jamar C., Malaise D., Reitsema H. J., Schmidt H. U., Schmidt W. K. H., Seige P., Whipple F. L., and Wilhelm K. (1986) First Halley Multicolour Camera imaging results from GIOTTO. *Nature, 321,* 320–326.

Keller H. U., Delamere W. A., Reitsema H. J., Huebner W. F., and Schmidt H. U. (1987) Comet P/Halley's nucleus and its activity. *Astron. Astrophys., 187,* 807–823.

Keller H. U., Crudt W., Kramm J.-R., and Thomas N. (1994) Images of the Nucleus of Comet Halley obtained by the Halley Multicolour Camera (HMC) on board the Giotto spacecraft. In *Images of the Nucleus of Comet Halley* (R. Reinhard et al., eds.), p. 69. ESA SP-1127, Noordwijk, The Netherlands.

Keller H. U., Britt D., Buratti B. J., and Thomas N. (2004) *In situ* observations of cometary nuclei. In *Comets II* (M. C. Festou et al., eds.), this volume. Univ. of Arizona, Tucson.

Krešák L., Carusi A., Perozzi E., and Valsecchi G. B. (1984) *Periodic Comets Neujmin 3 and Van Biesbroeck.* IAU Circular No. 3940.

Lamy P. L. and Toth I. (1995) Direct detection of a cometary nucleus with the Hubble Space Telescope. *Astron. Astrophys., 293,* L43–L45.

Lamy P. L., Toth I., Grün E., Keller H. U., Sekanina Z., and West R. M. (1996) Observations of comet P/Faye 1991 XXI with the Planetary Camera of the Hubble Space Telescope. *Icarus, 119,* 370–384.

Lamy P. L., Toth I., Jorda L., and Weaver H. A. (1998a) The nucleus and the inner coma of comet 46P/Wirtanen. *Astron. Astrophys., 335,* L25–L29.

Lamy P. L., Toth I., and Weaver H. A. (1998b) Hubble Space Telescope Observations of the nucleus and inner coma of comet 19P/Borrelly 1994 l. *Astron. Astrophys., 337,* 945–954.

Lamy P. L., Toth I., A'Hearn M. F., and Weaver H. A. (1999a) Hubble Space Telescope observations of the nucleus of comet 45P/Honda-Mrkos-Pajdušáková and its inner coma. *Icarus, 140,* 424–438.

Lamy P. L., Jorda L., Toth I., Groussin O., A'Hearn M. F., and Weaver H. A. (1999b) Characterization of the nucleus of comet Hale-Bopp from HST and ISO observations. *Bull. Am. Astron. Soc., 31,* 1116.

Lamy P. L., Toth I., Weaver H. A., Delahodde C., Jorda L., and A'Hearn M. F. (2000) The nucleus of 13 short-period comets. *Bull. Am. Astron. Soc., 32,* 1061.

Lamy P. L., Toth I., Weaver H. A., Delahodde C., Jorda L., and A'Hearn M. F. (2001a) The nucleus of 10 short-period comets. *Bull. Am. Astron. Soc., 33,* 1093.

Lamy P. L., Toth, I., A'Hearn M. F., Weaver H. A., and Weissman P. R. (2001b) Hubble Space Telescope observations of the nucleus of comet 9P/Tempel 1. *Icarus, 154,* 337–344.

Lamy P. L., Toth I., Jorda L., Groussin O., A'Hearn M. F., and Weaver H. A. (2002) The nucleus of comet 22P/Kopff and its inner coma. *Icarus, 156,* 442–455.

Lamy P. L., Toth I., Weaver H. A., Jorda L., and Kaasalainen M. (2003) The nucleus of comet 67P/Churyumov-Gerasimenko, the new target of the ROSETTA mission. *Bull. Am. Astron. Soc., 35,* Abstract #30.04.

Lamy P. L., Toth I., Jorda L., Groussin O., A'Hearn M. F., and H. A. Weaver (2004) The nucleus of comet 55P/Tempel-Tuttle and its inner coma. *Icarus,* in press.

Larsen J. A., Gleason A. E., Danzl N. M., Descour A. S., McMillan R. S., Gehrels T., Jedicke R., Montani J. L., and Scotti J. V. (2001) The Spacewatch Wide-Field Area Survey for bright Centaurs and Trans-Neptunian Objects. *Astron. J., 121,* 562–579.

Lecacheux J., Jorda L., Enzian A., Klinger J., Colas F., Frappa E., and Laques P. (1996) Comet C/1996 B2 (Hyakutake). IAU Circular No. 6354.

Leibowitz E. M., and Brosch N. (1986) Periodic photometric variations in the near nucleus zone of P/Giacobini-Zinner. *Icarus, 68,* 430–441.

Levison H. F. (1996) Comet taxonomy. In *Completing the Inventory of the Solar System* (T. W. Rettig and J. M. Hahn, eds.), pp. 173–192. ASP Conference Series 107, Astronomical Society of the Pacific, San Francisco.

Levison H. F., Morbidelli A., Dones L., Jedicke R., Wiegert P. A., and Bottke W. F. Jr. (2002) The mass disruption of Oort cloud comets. *Science, 296,* 2212–2215.

Licandro J., Bellot R., Luis R., Boehnhardt H., Casas R., Goetz B., Gomez A., Jorda L., Kidger M., Osip D., Sabalisk N., Santos P., Serr-Ricart M., Tozzi G. P., and West R. (1998) The rotation period of C/1995 O1 (Hale-Bopp). *Astrophys. J. Lett., 501,* L221–L225.

Licandro J., Tancredi G., Lindgren M., Rickman H., and Gil-Hutton R. (2000) CCD photometry of cometary nuclei, I: Observations from 1990–1995. *Icarus, 147,* 161–179.

Licandro J., Guerra J. C., Campins H., Di Martino M., Lara L. M., Gil-Hutton R., and Tozzi G. P. (2002) The surface of cometary nuclei related minor icy bodies. *Earth Moon Planets, 90,* 495–496.

Licandro J., Campins H., Hergenrother C., and Lara L. M. (2003) Near-infrared spectroscopy of the nucleus of comet 124P/Mrkos. *Astron. Astrophys., 398,* L45–L48.

Lisse C. M., Fernández Y. R., Kundu A., A'Hearn M. F., Dayal A., Deutsch L. K., Fazio G. G., Hora J. L., and Hoffmann W. F. (1999) The nucleus of comet Hyakutake (C/1996 B2). *Icarus, 140,* 189–204.

Lowry S. C. and Fitzsimmons A. (2001) CCD photometry of distant comets II. *Astron. Astrophys., 365,* 204–213.

Lowry S. C. and Weissman P. R. (2003) CCD observations of distant comets from Palomar and Steward observatories. *Icarus, 164,* 492–503.

Lowry S. C., Fitzsimmons A., Cartwright I. M., and Williams I. P. (1999) CCD photometry of distant comets. *Astron. Astrophys., 349,* 649–659.

Lowry S. C., Fitzsimmons A., and Collander-Brown S. (2003a) CCD photometry of distant comets. III. Ensemble properties of Jupiter-family comets. *Astron. Astrophys., 397,* 329–343.

Lowry S. C., Weissman P., Sykes M. V., and Reach W. T. (2003b) Observations of periodic comet 2P/Encke: Physical properties of the nucleus and first visual-wavelength detection of its dust trail. In *Lunar and Planetary Science XXXIV,* Abstract #2056.

Lunar and Planetary Institute, Houston (CD-ROM).

Lunine J. and Gautier D. (2004) Coupled physical and chemical evolution of volatiles in the protplanetary disk: A tale of three elements. In *Comets II* (M. C. Festou et al., eds.), this volume. Univ. of Arizona, Tucson.

Luu J. X. (1993) Spectral diversity among the nuclei of comets. *Icarus, 104,* 138–148.

Luu J. and Jewitt D. (1990a) The nucleus of comet P/Encke. *Icarus, 86,* 69–81.

Luu J. and Jewitt D. (1990b) Cometary activity of 2060 Chiron. *Astron. J., 100,* 913–932.

Luu J. and Jewitt D. (1991) *Periodic Comet Chernykh (1991 o).* IAU Circular No. 5347.

Luu J. X. and Jewitt D. C. (1992) Near-aphelion CCD photometry of comet P/Schwassmann-Wachmann 2. *Astron. J., 104,* 2243–2249.

Luu J. X. and Jewitt D. (1993) *Periodic Comet Schwassmann-Wachmann 1.* IAU Circular No. 5692.

Lyttleton R. A. (1953) *The Comets and Their Origin.* Cambridge Univ., Cambridge. 173 pp.

Lyttleton R. A. (1963) Book review (Buchbesprechungen über) of *The Nature of Comets,* by N. B. Richter, with an introduction by R. A. Lyttleton, London, Methen 1963. *Z. Astrophys., 58,* 295–296.

Mäkinen J. T. T., Bertaux J.-L., Pulkkinen T .I., Schmidt W., Kyrölä E., Summanen T., Quémerais E., and Lallement R. (2001) Comets in full sky L-alpha maps of the SWAN instrument. I. Survey from 1996 to 1998. *Astron. Astrophys., 368,* 292–297.

Marchis F., Boehnhardt H., Hainaut O. R., and Le Mignant D. (1999) Adaptive optics observations of the innermost comet of C/1995 O1. Are there a "Hale" and a "Bopp" in comet Hale-Bopp? *Astron. Astrophys., 349,* 985–995.

Manteca P. (2001) *Comet 51P/Harrington.* IAU Circular No. 7769.

Marsden B. G. (1974) Comets in 1973. *Quart. J. R. Astron. Soc., 15,* 433–460.

Marsden B. G. (2002) *Comet 57P/du Toit-Neujmin-Delporte.* IAU Circular No. 7934.

Marsden B. G. and Roemer E. (1978) Comet in 1974. *Quart. J. R. Astron. Soc., 19,* 38–58.

McBride N., Green S. F., Levasseur-Regourd A. C., Goidet-Devel B., and Renard J.-B. (1997) The inner dust coma of comet 26P/Grigg-Skjellerup: Multiple jets and nucleus fragments? *Mon. Not. R. Astron. Soc., 289,* 535–553.

Meech K. J. (1992) Observations comet Černis (1982 XII) at 19.4 and 20.9 AU. *Bull. Am. Astron. Soc., 24,* 993.

Meech K. J. (1996) Physical properties of comets. Paper presented at Asteroids, Comets, Meteors 1996 meeting. Available on line at http://www.ifa.hawaii.edu/~meech/papers/acm96.pdf.

Meech K. J. and Newburn R. (1998) Observations and modelling of 81P/Wild 2. *Bull. Am. Astron. Soc., 30,* 1094.

Meech K. J., Belton M. J. S., Mueller B. E. A., Dicksion M. W., and Li H. R. (1993) Nucleus properties of P/Schwassmann-Wachmann 1. *Astron. J., 106,* 1222–1236.

Meech K. J., Bauer J. M., and Hainaut O. R. (1997) Rotation of comet 46P/Wirtanen. *Astron. Astrophys., 326,* 1268–1276.

Meech K. J., Hainaut O. R., and Marsden B. G. (2000) Comet nucleus size distributions and distant activity. In *Proceedings of Minor Bodies in the Outer Solar System* (A. Fitzsimmons et al., eds.), pp. 75–80. Springer-Verlag, Berlin.

Meech K. J., Hainaut O. R., and Marsden B. G. (2004) Comet nucleus size distribution from HST and Keck Telescopes. *Icarus,* in press.

Melosh H. J. and Schenk P. (1993) Split comets and the origin of crater chains on Ganymede and Callisto. *Nature, 365,* 731.

Melosh H. J. and Whitaker E. A. (1994) Lunar crater chains. *Nature, 369,* 713.

Merényi E., Foldy L., Szegö K, Toth I., and Kondor A. (1990) The landscape of comet Halley. *Icarus, 86,* 9–20.

Merline W. J., Weidenschilling S. J., Durda D. D., Margot J. L., Pravec P., and Storrs A. D. (2002) Asteroids do have satellites. In *Asteroids III* (W. F. Bottke Jr. et al., eds.), pp. 289–312. Univ. of Arizona, Tucson.

Millis R. L., A'Hearn M. F., and Campins H. (1988) An investigation of the nucleus and coma of comet P/Arend-Rigaux. *Astrophys. J., 324,* 1194–1209.

Mueller B. E. A. (1992) CCD-photometry of comets at large heliocentric distances. In *Asteroids, Comets, Meteors 1991* (A. W. Harris and E. Bowell, ed.), pp. 425–428. Lunar and Planetary Institute, Houston.

Mueller B. E. A. and Ferrin I. (1996) Change in the rotational period of comet P/Tempel 2 between 1988 and 1994 apparitions. *Icarus, 123,* 463–477.

Mueller B. E. A. and Samarasinha N. H. (2001) Lightcurve observations of 19P/Borrelly. *Bull. Am. Astron. Soc., 33,* Abstract #28.02.

Nelson R. M., Soderblom L. A., and Hapke B. W. (2004) Are the circular dark features on comet Borrelly's surface albedo variations or pits? *Icarus,* in press.

O'Brien D. P and Greenberg R. (2003) Steady-state size distributions for collisional populations: Analytical solution with size-dependent strength. *Icarus,* in press.

O'Ceallaigh D. P., Fitzsimmons A., and Williams I. P. (1995) CCD photometry of comet 109P/Swift-Tuttle. *Astron. Astrophys., 297,* L17–L20.

Osip D., Campins H., and Schleicher D. G. (1995) The rotation state of 4015 Wilson-Harrington: Revisiting origins for the near-Earth asteroids. *Icarus, 114,* 423–426.

Peale S. J. and Lissauer J. J. (1989) Rotation of Halley's comet. *Icarus, 79,* 396–430.

Pittichová J. (1997) On the rotation of the IRAS-Araki-Alcock nucleus. *Planet. Space Sci., 45,* 791–794.

Press W. H., Flannery B. P., Teukolsky S. A., and Vetterling W. T. (1986) *Numerical Recipes: The Art of Scientific Computing.* Cambridge Univ., New York. 818 pp.

Prialnik D., Benkhoff J., and Podolak M. (2004) Modeling the structure and activity of comet nuclei. In *Comets II* (M. C. Festou et al., eds.), this volume. Univ. of Arizona, Tucson.

Qi C., Blake G. A., Muhleman D. O., and Gurwell M. A. (1998) Interferometric of comet Hale-Bopp with the Owens Valley Millimeter Array. *Bull. Am. Astron. Soc., 30,* 1451.

Rauer H., Hahn G., Harris A., Helbert J., Mottola S., and Oberst J. (1999) Nuclear parameters of comet P/Borrelly. *Bull. Am. Astron. Soc., 31,* Abstract #37.03.

Roemer E. (1965) Observations of comets and minor planets. *Astron. J., 70,* 387–402.

Roemer E. (1966) The dimensions of cometary nuclei. In *Nature et Origine des comètes: Proceedings of an International Colloquium on Astrophysics,* pp. 23–28. Univ. of Liège, Liege.

Roemer E. (1968) Dimensions of the nucleus of periodic and near-parabolic comets. *Astron. J., 73,* 33.

Russell H. N. (1916) On the albedo of the planets and their satellites. *Astrophys. J., 43,* 173–195.

Sagdeev R. Z., Szabó F., Avenasov G. A., Cruvellier P., Szabo L., Szegö K., Abergel A., Balazs A., Barinov I. V., Bertaux J.-L., Blamount J., Detaille M., Demarelis E., Dul'nev G. N., Endroczy G., Gardos M., Kanyo M., Kostenko V. I., Krasikov V. A., Nguyen-Trong T., Nyitrai Z., Renyi I., Rusznyak P., Shamis V. A., Smith B., Sukhanov K. G., Szabó F., Szalai S., Tarnopolsky V. I., Toth I., Tsukanova G., Valnicek B. I., Varhalmi L., Zaiko Yu. K., Zatsepin S. I., Ziman Ya. L., Zsenei M., and Zhukov B. S. (1986a) Television observations of comet Halley from VEGA spacecraft. *Nature, 321,* 262–266.

Sagdeev R. Z., Avenasov G. A., Ziman Ya. L., Moroz V. I., Tarnopolsky V. I., Zhukov B. S., Shamis V. A., Smith B. A., and Toth I. (1986b) TV experiment of the VEGA mission: Photometry of the nucleus and the inner coma. In *Proceedings of the 20th ESLAB Symposium on the Exploration of Halley's Comet, Vol. 1,* pp. 317–326. ESA SP-250, Noordwijk, The Netherlands.

Sagdeev R. Z., Szegö K., Smith B. A., Larson S., Merényi E., Kondor A., and Toth I. (1989) The rotation of P/Halley. *Astron. J., 97,* 546–551.

Samarasinha N. H. (2000) The coma morphology due to an extended active region and the implications the spin state of comet Hale-Bopp. *Astrophys. J. Lett., 529,* L107–L110.

Samarasinha N. H. (2003) Rotation and activity of comets. *Adv. Space Res.,* in press.

Samarasinha N. H. and A'Hearn M. F. (1991) Observational and dynamical constraints on the rotation of comet P/Halley. *Icarus, 93,* 194–225.

Samarasinha N. H., Mueller B. E. A., Belton M. J. S., and Jorda L. (2004) Rotation of cometary nuclei. In *Comets II* (M. C. Festou et al., eds.), this volume. Univ. of Arizona, Tucson.

Schenk P. M., Asphaug E., McKinnon W. B., Melosh H. J., and Weissman P. R. (1996) Cometary nuclei and tidal disruption: The geologic record of crater chains on Callisto and Ganymede. *Icarus, 121,* 249–274.

Schleicher D. G. and Osip D. J. (2002) Long- and short-term photometric behavior of comet Hyakutake (1996 B2). *Icarus, 159,* 210–233.

Schuller F. (1930) *Comet Schwassmann-Wachmann (1930 d).* IAU Circular No. 288.

Scotti J. V. (1994) *Periodic Comet Harrington (1994 g).* IAU Circular No. 6089.

Sekanina Z. (1976) A continuing controversy: Has the cometary nucleus been resolved? In *The Study of Comets* (B. Donn et al., ed.), pp. 537–587. NASA SP-393, Washington, DC.

Sekanina Z. (1981) Distribution and activity of discrete emission areas on the nucleus of periodic comet Swift-Tuttle. *Astron. J., 86,* 1741–1773.

Sekanina Z. (1982) The problem of split comets in review. In *Comets* (L. L. Wilkening, ed.), pp. 251–287. Univ. of Arizona, Tucson.

Sekanina Z. (1984) Precession model for the nucleus of periodic comet Kopff. *Astron. J., 89,* 1573–1586.

Sekanina Z. (1987) Anisotropic emission from comets: Fans versus jets. 1. Concept and modeling. In *Proceedings of the International Symposium on the Diversity and Similarity of Comets* (B. Battrick et al., eds.), pp. 315–322. ESA SP-278, Noordwijk, The Netherlands.

Sekanina Z. (1988) Nucleus of comet IRAS-Araki-Alcock (1983 VII). *Astron. J., 95,* 1876–1894.

Sekanina Z. (1989) Comprehensive model for the nucleus of comet P/Tempel 2 and its activity. *Astron. J., 102,* 350–388.

Sekanina Z. (1995) Evidence on size and fragmentation of the nuclei of comet Shoemaker-Levy 9 from Hubble Space Telescope images. *Astron. Astrophys., 304,* 296–316.

Sekanina Z. (1997) The problem of split comets revisited. *Astron. Astrophys., 318,* L5–L8.

Sekanina Z. (1998a) Detection of a satellite orbiting the nucleus of comet Hale-Bopp (C/1995 O1). *Earth Moon Planets, 77,* 155–163.

Sekanina Z. (1998b) Modeling dust halos in comet Hale-Bopp (1995 O1): Existence of two active nuclei. *Astrophys. J. Lett., 509,* L133–L136.

Sekanina Z. (2001) *Comet 51P/Harrington.* IAU Circular No. 7773.

Sekanina Z. (2002a) *Comet 57P/du Toit-Neujmin-Delporte.* IAU Circular No. 7946.

Sekanina Z. (2002b) *Comet 57P/du Toit-Neujmin-Delporte.* IAU Circular No. 7957.

Sekanina Z. (2002c) Recurring outbursts and nuclear fragmentation of comet C/2001 A2 (LINEAR). *Astrophys. J., 572,* 679–684.

Sheppard S. S., Jewitt D. C., Trujillo Ch. A., Brown M. J. I., and Ashley M. C. B. (2000) A wide-field CCD survey for Centaurs and Kuiper belt objects. *Astron. J., 120,* 2687–2694.

Slipher V. M. (1927) The spectrum of the Pons-Winnecke comet and the size of the cometary nucleus. *Lowell Obs. Bull., 3(86),* 135–137.

Soderblom L. A., Becker T. L., Bennett G., Boice D. C., Britt D. T., Brown R. H., Buratti B. J., Isbell C., Giese B., Hare T., Hicks M. D., Howington-Kraus E., Kirk R. L., Lee M., Nelson R. M., Oberst J., Owen T. C., Rayman M. D., Sandel B. R., Stern S. A., Thomas N., and Yelle R. V. (2002) Observations of comet 19P/Borrelly by the Miniature Integrated Camera and Spectrometer aboard Deep Space 1. *Science, 296,* 1087–1091.

Spencer J. R., Lebofsky L. A., and Sykes M. V. (1989) Systematic biases in radiometric diameter determination. *Icarus, 78,* 337–354.

Stern S. A. (1995) Collisional time scales in the Kuiper Disk and their implications. *Astron. J., 110,* 856–868.

Stuart J. S. (2001) A near-Earth asteroid population estimate from the LINEAR survey. *Science, 294,* 1691–1693.

Szabó Gy. M., Csák B., Sárneczky K., and Kiss L. L. (2001) Photometric observations of distant active comets. *Astron. Astrophys., 374,* 712–718.

Tancredi G., Fernández J. A., Rickman H., and Licandro J. (2000) A catalog of observed nuclear magnitudes of Jupiter family comets. *Astron. Astrophys. Suppl. Ser., 146,* 73–90.

Thomas N. and Keller H. U. (1989) The colour of Comet P/Halley's nucleus and dust. *Astron. Astrophys., 213,* 487–494.

Tokunaga A. T. and Hanner M. S. (1985) Does comet P/Arend-Rigaux have a large dark nucleus? *Astrophys. J. Lett., 296,* L13–L16.

Tokunaga A. T., Hanner M. S., Golisch W. F., Griep D. M., Kaminski C. D., and Chen H. (1992) Infrared monitoring of comet P/Tempel 2. *Astron. J., 104,* 1611–1617.

Toth I. and Lamy P. L. (2000) Spectral properties of the nucleus of short-period comets. *Bull. Am. Astron. Soc., 32,* 1063.

Toth I., Lamy P. L., Jorda L., and Weaver H. A. (1999) The properties of the nucleus of comet Hale-Bopp from Hubble Space Telescope observations. In *Asteroids, Comets, Meteors 1999,* Abstract #05.12. Cornell Univ., Ithaca, New York.

Toth I., Lamy P. L., and Weaver H. A. (2003) Hubble Space Telescope observations of the nucleus fragment 73P/Schwassmann-Wachmann 3-B. *Bull. Am. Astron. Soc., 35,* Abstract #38.05.

Trujillo Ch. A., Jewitt D. C., and Luu J. X. (2001) Properties of the Trans-Neptunian Belt: Statistics from the Canada-France-Hawaii Telescope Survey. *Astron. J., 122,* 457–473.

Van Biesbroeck G. (1962) Comet observations. *Astron. J., 67,* 422–428.

Vasundhara R. and Chakraborty P. (1999) Modeling of jets from comet Hale-Bopp (C/1995 O1): Observations from the Vainu Bappu observatory. *Icarus, 140,* 221–230.

Veeder G. J., Hanner M. S., and Tholen D. J. (1987) The nucleus of comet P/Arend-Rigaux. *Astron. J., 94,* 169–173.

Vorontsov-Velyaminov B. (1946) Structure and mass of cometary nuclei. *Astrophys. J., 104,* 226–233.

Weaver H. A. and Lamy P. L. (1997) Estimating the size of Hale-Bopp's nucleus. *Earth Moon Planets, 79,* 17–33.

Weaver H. A., Feldman P. S., A'Hearn M. F., Arpigny C., Brandt J. C., and Stern S. A. (1999) Post-perihelion HST observations of comet Hale-Bopp (C/1995 O1). *Icarus, 141,* 1–12.

Watanabe J.-I. (1987) The rotation of comet 1983 VII IRAS-Araki-Alcock. *Publ. Astron. Soc. Jap., 39,* 485–503.

Weidenschilling S. J. (2004) From icy grains to comets. In *Comets II* (M. C. Festou et al., eds.), this volume. Univ. of Arizona, Tucson.

Weissman P. R. (1980) Physical loss of long-period comets. *Astron. Astrophys., 85,* 191–196.

Weissman P. R. (1986) Are cometary nuclei primordial rubble piles? *Nature, 320,* 242.

Weissman P. R. and Lowry S. C. (2003) The size distribution of Jupiter-family cometary nuclei. In *Lunar and Planetary Science XXXIV,* Abstract #2003. Lunar and Planetary Institute, Houston (CD-ROM).

Weissman P. R., Doressoundiram A., Hicks M., Chamberlin A., Larson S., and Hergenrother C. (1999) CCD photometry of comet and asteroid targets of space missions. *Bull. Am. Astron. Soc., 31,* Abstract #33.03.

Weissman P. R., Bottke W. F. Jr., and Levison H. F. (2003) Evolution of comets into asteroids. In *Asteroids III* (W. F. Bottke Jr. et al., eds.), pp. 669–686. Univ. of Arizona, Tucson.

Weissman P. R., Asphaug E., and Lowry S. C. (2004) Structure and density of cometary nuclei. In *Comets II* (M. C. Festou et al., eds.), this volume. Univ. of Arizona, Tucson.

West R. M., Hainaut O., and Smette A. (1991) Post-perihelion observations of P/Halley. III — An outburst at R = 14.3 AU. *Astron. Astrophys., 246,* L77–L80.

Whipple F. L. (1950) A comet model. I. Acceleration of Comet Encke. *Astrophys. J., 111,* 374–474.

Whipple F. L. (1983) Comets: Nature evolution and decay. In *Highlights of Astronomy, Vol. 6* (R. M. West, ed.), pp. 323–331.

Whipple F. L. (1984) Comet P/Holmes, 1892 III: A case of duplicity? *Icarus, 60,* 522–531.

Whipple F. L. (1999) Note on the structure of comet nuclei. *Planet. Space Sci., 47,* 301–304.

Whiteley R. J., Tholen D. J., and Hergenrother C. W. (2002) Lightcurve analysis of four new monolithic fast-rotating asteroids. *Icarus, 157,* 139–154.

Yoshida S., Aoki T., Soyano T., Tarusawa K.-i., van Driel W., Hamabe M., Ichikawa T., Watanabe J.-i., and Wakamatsu K.-i. (1993) Spiral dust-jet structures of comet P/Swift-Tuttle 1992 t. *Publ. Astron. Soc. Japan, 45,* L33–L37.

Zahnle K., Schenk P., Sobieszczyk S., Dones L., and Levison H. F. (2001) Differential cratering of synchronously rotating satellites by ecliptic comets. *Icarus, 153,* 111–129.

Radar Studies of Comet Nuclei and Grain Comae

John K. Harmon and Michael C. Nolan
Arecibo Observatory

Steven J. Ostro
Jet Propulsion Laboratory, California Institute of Technology

Donald B. Campbell
Cornell University

A close-approaching comet can show detectable echoes from its nucleus, or from large coma grains, or both. Nine comets have been detected since 1980 with the Arecibo and Goldstone radars; this includes six nucleus detections and five grain-coma detections. The nucleus radar cross sections span a large range of values consistent with a factor-of-10 range of nucleus sizes. Comparisons with independent size estimates for these comets support this size range and give radar albedos of 0.04–0.1, which is about half the typical asteroid radar albedo. The albedos correspond to nucleus surface densities ~0.5–1.5 g/cm^3. Coma echo models based on simple grain ejection theories can explain the radar cross sections using reasonable grain size distributions that include a substantial population of centimeter-sized grains; in one case there is evidence for a cutoff in the size distribution consistent with a gravity-limited maximum liftable grain size. The models indicate that some comets emit large grains at rates (~10^6 g/s) that are comparable with their gas and dust production rates. The primary goal of cometary radar is to obtain delay-Doppler images of a nucleus. Eleven short-period comets are potentially detectable over the next two decades, a few of which may be suitable for imaging. These could be supplemented by chance close apparitions of new comets.

1. INTRODUCTION

When the comet radar chapter by *Kamoun et al.* (1982a) was written for the first *Comets* book (*Wilkening, 1982*), only one comet, 2P/Encke, had been detected by radar. Since then, eight more comet detections have been made with the Arecibo and Goldstone radars (Table 1). While few in number, owing to the rarity of close comet approaches, these detections have been sufficient to establish comets as interesting and diverse radar targets.

The Encke detection of 1980 (*Kamoun et al., 1982a,b; Kamoun, 1983*) showed a narrow Doppler spike consistent with backscatter from a solid rotating nucleus a few kilometers in size. Subsequent nucleus detections of other comets have been similar in character, but show differences in radar cross section consistent with an order-of-magnitude range of nucleus sizes. In principle, delay-Doppler radar imaging can determine the size, shape, rotation, and radar albedo of a nucleus unambiguously, as is being done for an increasing number of asteroids. Since no delay-Doppler detection has yet been made for a comet, radar data have mainly been used to estimate or constrain nucleus parameters from comparisons with other types of observations. For example, comparisons of nucleus radar cross sections with independent size estimates have placed useful bounds on nucleus radar albedos and surface densities.

In addition to the nucleus echo, some comets also show an echo component from large coma grains. This first came

as a surprise result from the 1983 observations of C/IRAS-Araki-Alcock (*Campbell et al., 1983; Goldstein et al., 1984*) and has since been seen for four other comets. The implication is that large-grain emission by comets is common and can account for a significant fraction of the total nucleus mass loss. This is in line with a growing body of evidence from other observations (spacecraft encounters, infrared dust trails, submillimeter continuum, antitails, etc.) that large grains are an important component of the cometary particulate population.

Here we review the various cometary radar findings to date, discuss their implications in the context of other observations, and survey prospects for future work. Although covering much of the same ground as an earlier review article by these same authors (*Harmon et al., 1999*), the material presented here has been substantially reorganized and updated.

2. RADAR MEASUREMENTS AND DETECTABILITY

All comet radar detections to date have come from Doppler-only observations with the Arecibo S-band (wavelength $\lambda = 12.6$ cm), Goldstone S-band (12.9 cm), or Goldstone X-band (3.5 cm) radar systems. Here we summarize the types of measurements made using Doppler-only observations. Discussion of delay-Doppler measurements is deferred to section 5.1.

TABLE 1.   Comet radar detections.

Comet	Radar*	Epoch (m/d/y)	Δ (AU)[†]	References
2P/Encke	A_S	11/2–11/8/1980	0.32	[1,2]
26P/Grigg-Skjellerup	A_S	5/20–6/2/1982	0.33	[2,3]
C/IRAS-Araki-Alcock (1983 H1)	G_S	5/11.94/1983	0.033	[4]
	G_X	5/14.08/1983	0.072	[4]
	A_S	5/11.92/1983	0.033	[5,6]
C/Sugano-Saigusa-Fujikawa (1983 J1)	A_S	6/10–6/12/1983	0.076	[5,7]
1P/Halley	A_S	11/24–12/2/1985	0.63	[8]
C/Hyakutake (1996 B2)	G_X	3/24–3/25/1996	0.10	[9,10]
C/1998 K5 (LINEAR)	A_S	6/14.25/1998	0.196	[11]
C/2001 A2 (LINEAR)	A_S	7/7–7/9/2001	0.26	[12]
C/2002 O6 (SWAN)	A_S	8/8–8/9/2002	0.26	[13]

*A_S = Arecibo S-band (λ = 12.6 cm); G_S = Goldstone S-band (λ = 12.9 cm); G_X = Goldstone X-band (λ = 3.54 cm).
[†]Distance from Earth at time of observation.

References: [1] *Kamoun et al.* (1982b); [2] *Kamoun* (1983); [3] *Kamoun et al.* (1999); [4] *Goldstein et al.* (1984); [5] *Campbell et al.* (1983); [6] *Harmon et al.* (1989); [7] *Harmon et al.* (1999); [8] *Campbell et al.* (1989); [9] *Ostro et al.* (1996); [10] *Harmon et al.* (1997); [11] *Harmon et al.* (1999); [12] *Nolan et al.* (2001); [13] this paper.

2.1. Doppler Spectrum

A Doppler-only observation involves transmission of an unmodulated (monochromatic) wave and reception of the Doppler-broadened echo. One computes the power spectrum of the received signal, within which a detectable echo would appear as a statistically significant spike or bump sticking up out of the background noise. The echo can appear as a narrow (few Hz wide) spike from the nucleus, or a broad (tens to hundreds of Hz) component from the grain coma. Two comets, C/IRAS-Araki-Alcock and C/Hyakutake, showed echoes from both nucleus and coma. The spectra for these comets are shown in Figs. 1 and 2.

The Doppler spreading of the nucleus spectrum represents the line-of-sight (radial) velocity spread from the apparent rotation of the nucleus. The Doppler frequency for radial velocity V_r and radar wavelength λ is $f = 2V_r/\lambda$. The Doppler bandwidth of a spherical nucleus is then given by

$$B = \frac{8\pi R |\sin\phi|}{\lambda p} = \frac{29.1 R(km) |\sin\phi|}{\lambda(cm) p(days)} \ (Hz) \quad (1)$$

where R is the nucleus radius, ϕ is the angle between the apparent rotation axis and the line of sight, and p is the apparent rotation period. For strong detections (e.g., Figs. 3 and 4), bandwidth B is easily determined from the well-

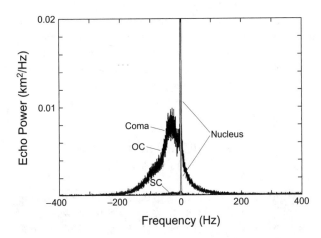

Fig. 1. Doppler spectra (OC and SC polarizations) for C/IRAS-Araki-Alcock showing the narrowband nucleus echo and broadband coma echo. The spectrum is truncated so that only the bottom 2% of the nucleus echo is showing. The spectrum is from Arecibo S-band observations on May 11, 1983 (*Harmon et al.*, 1989).

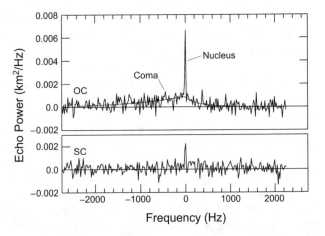

Fig. 2. Doppler spectra (OC and SC polarizations) for C/Hyakutake showing both nucleus and coma echoes. A model fit to the coma echo is also shown (dashed line). The spectrum is from Goldstone X-band observations on March 24, 1996 (*Harmon et al.*, 1997).

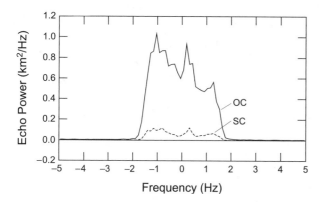

Fig. 3. Doppler spectra (OC and SC polarizations) for the nucleus of C/IRAS-Araki-Alcock, from Arecibo S-band observations on May 11, 1983 (*Harmon et al., 1989*).

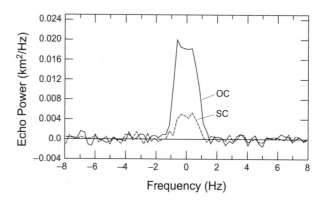

Fig. 4. Doppler spectra (OC and SC polarizations) for the nucleus of C/Sugano-Saigusa-Fujikawa, from Arecibo S-band observations on June 11, 1983 (*Harmon et al., 1999*).

defined edges of the nucleus spectrum. The shape of the nucleus spectrum is determined by the nucleus shape and orientation as well as by the intrinsic angular scattering law of the surface.

The Doppler spreading of the coma echo represents the collective sum of the radial velocities of all the large grains within the radar beam. The coma spectrum shape is determined by the velocities, sizes, and spatial distribution of the grains, as well as by the cutoff effect of the radar beam.

For the spectra presented here the mean (absolute) Doppler frequency of the nucleus has been subtracted off, so that zero Doppler defines the nucleus center frequency. However, the absolute Doppler offset of the nucleus is of intrinsic interest for refining estimates of a comet's orbital elements. Comet Doppler offsets or refined orbits based on them have appeared in several reports (*Ostro et al., 1991b, 1996; Yeomans et al., 1992; Giorgini, 2002*).

2.2. Radar Cross Section and Albedo

The most fundamental radar parameter measured from the echo is the radar cross section σ. Integrating under the echo Doppler spectrum to get the echo power P_r, σ is then calculated from the radar equation

$$\sigma = \frac{(4\pi)^3 \Delta^4 P_r}{P_t G^2 \lambda^2} \quad (2)$$

where Δ is the comet distance, P_t is the transmitted power, and $G = 4\pi A_e/\lambda^2$ is the beam gain of the radar antenna of effective area A_e. If the size of the nucleus is known or estimated, then σ can be normalized to give a geometric radar albedo

$$\hat{\sigma} = \frac{\sigma}{A_p} \quad (3)$$

where A_p is the apparent projected area of the nucleus. This albedo is useful for estimating surface density (see section 3.2.2).

2.3. Polarization

Echo polarization provides additional information on the target and its scattering properties. All comet radar observations have followed the standard practice of transmitting a circularly polarized wave and receiving in both (orthogonal) senses of circular polarization. These polarization senses on receive are referred to as OC (for "opposite circular"; also called the "polarized" or "expected" sense) and SC (for "same circular"; also called the "depolarized" or "unexpected" sense). Separate echo spectra are computed for each polarization (see Figs. 1–4), from which one can compute OC and SC cross sections σ_{oc} and σ_{sc}. A circular polarization ratio is then defined as $\mu_c = \sigma_{sc}/\sigma_{oc}$.

The OC echo is the stronger of the two ($\mu_c < 1$) for most solar system targets, being the expected sense for specular reflection, while the weaker SC echo is normally attributed to depolarization by wavelength-scale roughness or multiple scattering. For scattering by particle clouds one expects $\mu_c < 1$ when single scattering dominates, with $\mu_c \ll 1$ for particles in the Rayleigh size regime $a < \lambda/2\pi$. When multiple scattering predominates, as for Saturn's rings, one can get $\mu_c \sim 1$.

2.4. Detectability

The strength of the radar detection is given by the detectability D, which is the ratio of the echo power to the rms statistical fluctuation in the noise power. This is given by the radiometer equation $D = (S/N)\sqrt{t\Delta f}$, where S/N is the ratio of the signal and noise spectral densities, t is the integration time, and Δf is the frequency resolution. Combining this with the radar equation (2), and assuming the spectrum is optimally smoothed (matched filtered), gives

$$D = \frac{P_t G^2 \lambda^2 t^{1/2} \eta \sigma}{(4\pi)^3 \Delta^4 k T_s B^{1/2}} \quad (4)$$

where k is Boltzmann's constant, T_s is system temperature, and η is a factor (≈ 1) that depends on the shape of the echo

TABLE 2.   Nucleus echo parameters.

Comet	λ (cm)	σ_{oc} (km^2)	μ_c	B (Hz) [m/s]*
Encke	12.6	1.1 ± 0.7		6 [0.38]
GS	12.6	0.5 ± 0.13	<0.3	<0.5 [<0.03]
IAA	12.6	2.14 ± 0.4	0.105 ± 0.005	3.5 [0.221]
	12.9	$2.25^\dagger$		3.1 [0.20]
	3.5	$4.44^\dagger$	$0.25^\dagger$	20.3 [0.36]
SSF	12.6	0.034 ± 0.008	0.23 ± 0.03	2.5 [0.158]
Hyakutake	3.5	0.13 ± 0.03	0.49 ± 0.10	12 [0.21]
1998 K5	12.6	0.031 ± 0.015	<0.5	<1.5 [<0.09]

*Full (limb-to-limb) Doppler bandwidth in Hz, also expressed as a velocity
λB/2 in m/s (in brackets).

†From *Goldstein et al.* (1984), which gives no error estimate.

spectrum. Substituting the current system parameters for the upgraded Arecibo S-band radar in equation (4) and using equation (1) gives

$$D \approx \frac{1.0\sigma(km^2)t^{1/2}(hours)}{\Delta^4(AU)B^{1/2}(Hz)} \geq$$
$$\frac{2.1\hat{\sigma}R^{3/2}(km)p^{1/2}(days)t^{1/2}(hours)}{\Delta^4(AU)} \tag{5}$$

For the Goldstone X-band radar one substitutes 0.12 for the 2.1 factor. Equation (5) is useful for evaluating future comet radar opportunities (section 5.2).

3.   NUCLEUS

Six comets have yielded radar detections of their nuclei (Table 2). The strongest and best-resolved nucleus spectra are those for C/IRAS-Araki-Alcock (henceforth abbreviated IAA) and C/Sugano-Saigusa-Fujikawa (abbreviated SSF), which are shown in Figs. 3 and 4 respectively. Additional nucleus spectra have been published elsewhere, viz. *Kamoun et al.* (1982a,b) (P/Encke); *Goldstein et al.* (1984) (IAA); *Kamoun et al.* (1999) (P/Grigg-Skjellerup); *Harmon et al.* (1997) (C/Hyakutake); *Harmon et al.* (1999) (C/1998 K5).

3.1.   Size, Rotation, and Albedo

3.1.1. Size and rotation.   The observed nucleus cross sections span two orders of magnitude, from the 2–4 km^2 of IAA down to the 0.03 km^2 of SSF and C/1998 K5. This implies, assuming albedos are equal, that the nucleus sizes vary by about a factor of 10 for the radar-detected sample. This is supported by independent size estimates. *Sekanina* (1988) combined radar data with radio continuum (*Altenhoff et al., 1983*) and infrared (*Hanner et al., 1985*) results to deduce that IAA had a large (Halley-size) nucleus measuring 16 × 7 × 7 km and showing an effective radius of 4.4 km at the epoch of the S-band radar observations. Comet SSF, on the other hand, was deduced to be a tiny object (R =

0.37 km) based on its infrared core flux (*Hanner et al., 1987*). Although there is no independent size estimate for C/1998 K5, its extremely low absolute magnitude (*Marsden, 1998*) suggests that it, too, was very small.

Most size estimates for the three comets with intermediate radar cross sections (Encke, Grigg-Skjellerup, Hyakutake) do, in fact, fall between those of IAA and SSF. For Encke, the infrared results of *Campins* (1988) give R < 2.9 km and the red-visible photometry of *Luu and Jewitt* (1990) gives 2.2 < R < 4.9 km. The most recent size estimate for Encke is the infrared-based value R = 2.4 km of *Fernández et al.* (2000). For Grigg-Skjellerup, *Boehnhardt et al.* (1999) and *Licandro et al.* (2000) give radius estimates of 1.4–1.5 km based on the comet's visual magnitude at large heliocentric distance. Size estimates for Hyakutake vary. The most sensitive radio continuum nondetection gave an upper limit for R of 1.05 km (*Altenhoff et al., 1999*). Infrared estimates are larger, with R = 2.1–2.4 km (*Sarmecanic et al., 1997; Fernández et al., 1996; Lisse et al., 1999*).

One can use equation (1) to estimate the rotation period from the Doppler bandwidth B if both R and φ are known, or place an upper limit on the period if only R is known. *Sekanina* (1988) showed that the radar bandwidth and coma jet structure of IAA were consistent with a relatively slow rotation period of 2.14 d. For SSF, combining the measured B = 2.5 Hz with R = 0.37 km gives a relatively fast rotation with p < 8.3 h. The estimated bandwidth for Encke (*Kamoun et al., 1982b*) gives p < 22 h assuming R = 2.4 km. This upper limit encompasses all of Encke's observed periodicities, which span the range 7–22 h (*Samarasinha et al., 2004*), and includes the recently claimed dominant period of 11 h (*Fernández et al., 2002*). The estimated Hyakutake bandwidth (*Harmon et al., 1997*) gives p < 20 h assuming R = 1.2 km, which is consistent with the Hyakutake rotation period estimate of 6.25 h (*Schleicher et al., 1998*). The Grigg-Skjellerup spectrum was unresolved (*Kamoun et al., 1999*) and hence yielded no useful constraint on rotation. The echo from 1998 K5 was too weak to readily separate true bandwidth from ephemeris drift, so no rotation constraint is available for that comet. None of the radar-derived rotation period upper limits violate the 3.3-h critical period

TABLE 3.   Nucleus radar albedo estimates.

Comet	Albedo*	R (km)	References
Encke	>0.04	<2.9	[1]
	0.02–0.08	2.2–4.9	[2]
	0.06	2.4	[3]
GS	0.08	1.5	[4,5]
IAA†	0.04, 0.07	4.4, 4.9	[6,7]
SSF	0.10	0.37	[8]
Hyakutake	0.01–0.015	2.1–2.4	[9,10,11]
	>0.06	<1.05	[12]

*Total radar cross section divided by πR², where R is the tabulated radius. No entry is given for C/1998 K5, for which no radius estimate is available.

†The first and second entries for the albedo and radius correspond to the S-band and X-band observations respectively.

References for radius estimate: [1] *Campins* (1988); [2] *Luu and Jewitt* (1990); [3] *Fernández et al.* (2000); [4] *Boehnhardt et al.* (1999); [5] *Licandro et al.* (2000); [6] *Sekanina* (1988); [7] *Altenhoff et al.* (1983); [8] *Hanner et al.* (1987); [9] *Sarmecanic et al.* (1997); [10] *Fernández et al.* (1996); [11] *Lisse et al.* (1999); [12] *Altenhoff et al.* (1999).

for breakup of a spherical nucleus with 1 g/cm³ density (*Samarasinha et al.*, 2004).

3.1.2. Albedo. Nucleus radar albedos and the radius values R assumed in their calculation are listed in Table 3. Here we give the total albedo ($\sigma_{oc} + \sigma_{sc}$)/πR², adding an assumed 15% SC component for those comets (Encke and Grigg-Skjellerup) with OC-only detections. The IAA albedo estimates are 0.04 at S-band and 0.07 at X-band, based on the nucleus projected area estimates of *Sekanina* (1988) at the respective epochs of the S-band and X-band observations. The recent radius estimates of 2.4 km for Encke (*Fernández et al.*, 2000) and 1.5 km for Grigg-Skjellerup (*Boehnhardt et al.*, 1999; *Licandro et al.*, 2000) give albedos of 0.06 and 0.08 respectively. The highest $\hat{\sigma}$ is the 0.10 value estimated for SSF from the *Hanner et al.* (1987) infrared size. This high albedo is consistent with the suggestion by Hanner et al. (based on an apparently high thermal inertia and unusually low dust production) that the surface of the SSF nucleus is more highly compacted than normal. For Hyakutake, the R < 1.05 km upper limit from the radio continuum nondetection (*Altenhoff et al.*, 1999) gives $\hat{\sigma}$ > 0.055, whereas the larger infrared-based sizes give very low (~0.01) albedos. Such a low radar albedo would imply a very lightly packed nucleus, as suggested by *Schleicher and Osip* (2002). The alternative is that Hyakutake had a "normal" radar albedo similar to those of the other comets, in which case the infrared size estimates must have been biased high. The most likely source of such a bias would be a dust contribution to the infrared flux (*Lisse et al.*, 1999). *Harmon et al.* (1997) estimated the Hyakutake size to be R = 1–1.5 km using IAA's S-band albedo, which they considered to be the most reliable radar albedo available.

Clearly, the size and radar albedo of the Hyakutake nucleus remain controversial.

The fact that nucleus radar and optical albedos are low and have about the same values is interesting but probably not significant. While it is true that optical and radar albedo both depend on composition and density, there are important differences. First, optical albedo can be dominated by a thin surface layer, whereas the radar can respond to reflections from meters below the surface. Second, compositional differences are likely to be much more important in the optical than in the radio. For example, a carbonaceous or organic composition could give a surface that is extremely dark optically, but which may not have distinctive radio dielectric properties.

3.2.   Surface Properties

3.2.1. Roughness. The resolved nucleus spectra of IAA (Fig. 3) and SSF (Fig. 4) are broad (relative to the total bandwidth B) rather than sharply peaked, which is suggestive of high-angle scattering from very rugged surface relief. The polarization ratios can give some idea of the scale of this relief and its comet-to-comet variation. The relatively low S-band μ_c for IAA is consistent with highly specular scattering from meter-scale or larger structure, although the higher X-band μ_c points to an extra component of smaller rubble. Comet SSF shows a higher S-band μ_c than IAA, indicating roughness that is concentrated more toward decimeter scales. The highest μ_c is the 0.5 measured for Hyakutake at X-band, which suggests a surface that may be nearly saturated with pebble-sized rubble. Hyakutake was an unusually active comet for its size, and its surface texture may be related to that activity. For example, there could be an accumulation of surface debris from ejecta fallback (*Kührt et al.*, 1997). Ice sublimation could also produce surface

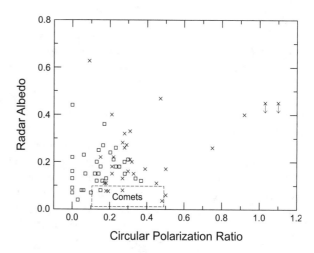

Fig. 5. Distribution of main-belt (squares) and near-Earth (crosses) asteroids in radar albedo and circular polarization ratio (*Benner*, 2002). Also shown for comparison is the range of values for comet nuclei (dashed rectangle).

structure (*Colwell et al.*, 1990). This might explain the roughness for a comet such as SSF, which was very inactive in terms of dust production but very active in the amount of gas it produced for its size.

Comet nucleus polarization ratios ($\mu_c = 0.1$–0.5) are similar to those for many near-Earth and main-belt asteroids (*Ostro et al.*, 2002; *Benner*, 2002; *Magri et al.*, 1999, 2001), as can be seen from the comparison in Fig. 5. This suggests that comets are similar to asteroids in the scale of their surface relief. However, no comets have yet shown the very low depolarization ($\mu_c \sim 0.05$) seen for a few main-belt asteroids or the high depolarization ($\mu_c \sim 1$) seen for a few near-Earth objects.

Comet nuclei also resemble asteroids in their spectral shape. It is customary with asteroids to fit the Doppler spectrum with the function

$$S(f) \propto [1 - (2f/B)^2]^{n/2} \tag{6}$$

which corresponds to the echo spectral shape for a sphere with a scattering law of the form

$$\sigma^o(\theta) \propto \cos^n\theta \tag{7}$$

where $\sigma^o(\theta)$ is the specific cross section as a function of incidence angle θ. Using this model, *Harmon et al.* (1989) found the IAA nucleus followed a uniformly bright ($n = 1$) or possibly even limb-brightened ($n < 1$) scattering law based on the sharp edges of its spectrum, arguing that this was evidence for scattering from a chaotic surface with super-wavelength-scale roughness elements giving both specular reflection and shadowing. The SSF spectra more closely followed a Lambert law ($n = 2$), the cosine-law fits giving n values of 1.4, 2.2, and 2.8 for the three different days. If the scattering is assumed predominantly specular (low μ_c), then the roughness can be estimated from geometric optics (*Mitchell et al.*, 1995). In that case the rms slope θ_r of the surface roughness is related to n by

$$\theta_r = \tan^{-1}\sqrt{2/n} \tag{8}$$

This suggests that comets such as IAA and SSF have rms surface slopes ~50°. This is consistent with the rough topography seen in spacecraft images of Comets Halley and Borrelly (*Weissman et al.*, 2004).

3.2.2. Density. If the nucleus surface layer is thick and homogeneous, then one can estimate its bulk density from the radar albedo. If the nucleus radar scattering is predominantly specular, then $\rho_o \approx \hat{\sigma}/g$, where ρ_o is the square of the Fresnel reflection coefficient at normal incidence and g is the backscatter gain. For a $\cos^n\theta$ scattering law in the geometric optics approximation, one has $g = (n + 2)/(n + 1)$. Once ρ_o is estimated, the dielectric constant ε is given by

$$\varepsilon = \frac{(1 + \rho_o^{1/2})^2}{(1 - \rho_o^{1/2})^2} \tag{9}$$

This can then be used to estimate the bulk density d using some suitable expression for $d(\varepsilon)$. In Fig. 6 we plot d as a

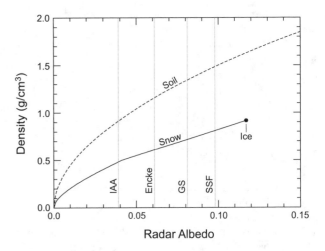

Fig. 6. Bulk density d of the nucleus surface vs. radar albedo $\hat{\sigma}$ for dry snow (solid curve) and a silicate soil (dashed curve). A backscatter gain $g = 3/2$ was assumed. Various albedo estimates from Table 3 are also shown (vertical dotted lines).

function of $\hat{\sigma}$ for snow and soil surfaces assuming $g = 3/2$ ($n = 1$, assuming geometric optics). Here we have used the expression

$$d \approx \begin{cases} 0.526(\varepsilon - 1.00) & (\varepsilon \leq 1.95) \\ 0.347(\varepsilon - 0.51) & (\varepsilon > 1.95) \end{cases} \tag{10}$$

for the case of dry snow (*Hallikainen et al.*, 1986) and

$$d \approx 3.9\left(\frac{\varepsilon - 1}{\varepsilon + 2}\right) \tag{11}$$

for a soil of silicate powder (*Campbell and Ulrichs*, 1969). From Fig. 6 we see that an IAA-like albedo corresponds to a surface with the consistency of a dense (0.5 g/cm³) terrestrial snowpack or very fluffy (0.9 g/cm³) soil. A higher albedo such as that of Comet SSF gives densities closer to that of solid ice or a moderately packed soil. The overall range of albedo estimates indicates that comet nuclei have surface densities in the range 0.5–1.5 g/cm³. (This density would apply to surface layers down to the penetration depth of the radar wave, which is of the order of 10 wavelengths or so for packed soils.) It is interesting to note that this surface density range is identical to the most recent estimates for the overall bulk density of comet nuclei (*Skorov and Rickman*, 1999; *Ball et al.*, 2001; *Weissman et al.*, 2004), although this does not necessarily imply that nucleus surfaces and interiors have the same structure.

A comparison of albedos indicates that the surfaces of comet nuclei are less dense than asteroid surfaces. Most main-belt asteroids (*Magri et al.*, 1999) and near-Earth asteroids (*Ostro et al.*, 1991a, 2002; *Magri et al.*, 2001; *Benner et al.*, 1997) have higher radar albedos than comets, as can be seen from the comparison in Fig. 5. This albedo difference should translate directly into a difference in reflectivity ρ_o (and density), since the similarity between comet and asteroid scattering implies similar backscatter gains. The near-Earth asteroid 433 Eros is the only asteroid with both a known radar cross section and known mass (and hence known bulk density). Putting the measured total radar albedo of 0.32 of Eros (*Magri et al., 2001*) into equations (9) and (11) gives a surface density of 3.0 g/cm³, which is close to the bulk density of 2.7 g/cm³ estimated from the *NEAR Shoemaker* spacecraft flyby (*Veverka et al., 2000*) and 3× larger than the comet nucleus bulk densities quoted above; this suggests that the nucleus surface density differences between comets and asteroids inferred from albedo comparisons may reflect differences in total bulk densities for these objects. Another implication of the low comet albedos is that one should expect any extinct comet nuclei masquerading as asteroids to also have low albedos. The asteroids with properties (including low radar albedo) closest to the middle of the domain of comet properties are the 7-km-diameter near-Earth object 1999 JM8 (*Benner et al., 2002*), for which a cometary origin cannot be excluded (*Bottke et al., 2002*), and the 0.5-kilometer-diameter object 3757 (1982 XB), which does not have a comet-like orbit.

4. GRAIN COMA

A grain coma echo has been detected from five comets. Two of these, IAA (Fig. 1) and Hyakutake (Fig. 2), gave nucleus detections as well. The three comets giving only coma detections were Halley (Fig. 7), C/2001 A2 (Fig. 8), and C/2002 O6 (Fig. 9). The estimated radar parameters for all the coma detections are listed in Table 4.

One can deduce some basic properties of the large-grain population from rather simple arguments and models, as discussed below.

4.1. Grain Populations and Radar Scattering

Some basic constraints on the large-grain population can be established by simply assuming the grains have a power-law size distribution $n(a) \propto a^{-\alpha}$ with minimum and maxi-

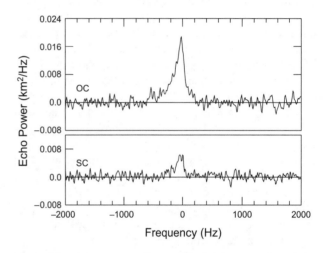

Fig. 8. Doppler spectrum (OC and SC polarizations) for C/2001 A2 (LINEAR) from Arecibo S-band observations on July 7, 2001.

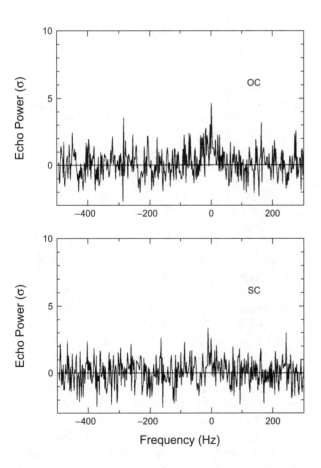

Fig. 7. Doppler spectrum (OC polarization) for P/Halley from a five-day average of Arecibo S-band observations between November 24 and December 2, 1985. The frequency resolution is 1.95 Hz. Smoothing to a resolution of 62 Hz increases the OC detection to nine standard deviations. From *Campbell et al.* (1989).

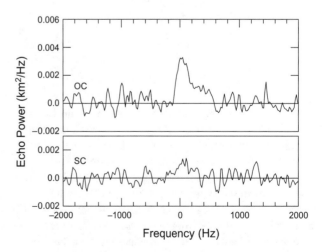

Fig. 9. Doppler spectrum (OC and SC polarizations) for C/2002 O6 (SWAN) from Arecibo S-band observations on August 8 and 9, 2002.

TABLE 4.   Grain-coma echo parameters.

Comet	λ (cm)	σ_{oc} (km²)	μ_c	B_h (Hz) [m/s]*
IAA	12.6	0.80 ± 0.16	0.014 ± 0.003	72 [4.54]
	12.9	$0.8^\dagger$		90 [5.81]
Halley	12.6	32 ± 10	0.52 ± 0.26	42 [2.65]
Hyakutake	3.5	1.33 ± 0.28	0.31 ± 0.12	1180 [20.9]
2001 A2	12.6	4.4 ± 1.3	0.28 ± 0.03	170 [10.7]
2002 O6	12.6	1.1 ± 0.3	0.32 ± 0.08	230 [14.5]

*Full Doppler bandwidth at half-max. in Hz, also expressed as a velocity $\lambda B_h/2$ in m/s (in brackets).

†From *Goldstein et al.* (1984), which gives no error estimate.

mum grain radii of a_o and a_m respectively. The size cutoff a_m not only is useful for assessing the effective size of the radar-scattering grains, but also can have some physical significance. We start with a discussion of the optical depth of the grain coma, which is important for establishing the dominance of single scattering and for determining the radar visibility of the nucleus through the coma (section 4.1.1). We then discuss how a_m is constrained by radar cross section and total grain mass (section 4.1.2) and coma echo polarization (section 4.1.3).

4.1.1. Optical depth. The maximum optical depth to backscatter for a line of sight passing through the center of the grain coma is given by $\tau = \sigma/(4\pi R R_c)$, where σ is the coma radar cross section, R is the nucleus radius, and R_c is the grain coma radius. (Here we assumed that the grain number density in the cloud falls as $1/r^2$ and that $R_c \gg R$.) This result implies that $\tau \sim 10^{-4}$ or less. This can also be used as an upper limit on the ratio of multiple to single scattering. Another useful quantity is the ratio of the absorption and backscatter cross sections, which in the Rayleigh approximation (*Bohren and Huffman*, 1983) is given by

$$\frac{\sigma_a}{\sigma} \approx \left[\frac{(\varepsilon + 2)\varepsilon \tan\delta}{(\varepsilon - 1)^2}\right]\left(\frac{7 - \alpha}{4 - \alpha}\right)$$
$$\left[1 - \left(\frac{a_o}{a_m}\right)^{4 - \alpha}\right]\left[\frac{a_m}{\lambda/2\pi}\right]^{-3} \quad (12)$$

where ε and $\tan\delta$ are the dielectric constant and loss tangent of the grains respectively. Assuming reasonable grain parameters and setting a_m at the Rayleigh transition ($\lambda/2\pi$) gives $\sigma_a/\sigma \sim 0.1$. (The ratio does not increase for $a_m > \lambda/2\pi$.) Combining this with the backscatter optical depth given above indicates that the absorption optical depth in front of the nucleus is negligible unless a_m is smaller than $\sim 0.1(\lambda/2\pi)$, which is unlikely from mass and mass-loss arguments (see below). Hence, the nucleus radar detections have probably suffered negligible obscuration by the coma.

4.1.2. Size distribution and total mass. A spherical grain of radius a has a radar cross section of $\pi a^2 Q_b(a)$, where Q_b is the backscatter efficiency. Then, a population of single-

scattering grains will have a total radar cross section

$$\sigma = \pi \int_{a_o}^{a_m} n(a)Q_b(a)a^2 da \quad (13)$$

The corresponding total mass of this population is

$$M = \frac{4}{3}\pi d_g \int_{a_o}^{a_m} n(a)a^3 da \quad (14)$$

where d_g is grain density. In the Rayleigh approximation ($a \ll \lambda/2\pi$)

$$Q_b(a) = C_R a^4 \quad (15)$$

where

$$C_R = \left(\frac{2\pi}{\lambda}\right)^4 \left|\frac{\varepsilon - 1}{\varepsilon + 2}\right|^2 \quad (16)$$

Then, from equations (13)–(15), a grain coma with radar cross section σ has a total mass

$$M \approx \sigma\left(\frac{4d_g}{3C_R}\right)\left(\frac{7 - \alpha}{4 - \alpha}\right)\left[1 - \left(\frac{a_o}{a_m}\right)^{4 - \alpha}\right]a_m^{-3} \quad (17)$$

This a_m^{-3} dependence shows the extreme sensitivity of the mass M to maximum grain size in the Rayleigh regime, a result of the rapidly decreasing Rayleigh backscatter efficiency with smaller grain size. Using this equation, *Harmon et al.* (1989) and *Campbell et al.* (1989) showed that making $a_m < 0.5$ mm resulted in a total mass in grains exceeding the nucleus mass for both IAA and Halley, from which they concluded that the effective grain size must have been at least a few millimeters.

4.1.3. Polarization and maximum grain size. The coma echo from Comet IAA was only ~1% depolarized ($\mu_c = 0.014$), which is the smallest depolarization ever measured for a solar system radar echo. This is consistent with a physically real cutoff a_m not much larger than $\lambda/2\pi$. [It is shown in *Harmon et al.* (1989) that μ_c for irregular grains increases dramatically from ~10^{-2} or less to >0.1 as radius approaches $\lambda/2\pi$, although the transition size can be larger for low-density grains.] Combining this with the lower bounds on a_m from the total mass (section 4.1.2) and mass-loss rate (section 4.2.3) points to a sharp size cutoff at a few centimeters. As pointed out by *Harmon et al.* (1989), this would be consistent with the gravitational cutoff in simple gas-drag theories of particle ejection (section 4.2.1). However, since this apparent cutoff is close to the Rayleigh polarization threshhold, one would expect to see coma echoes from other comets with μ_c much higher than for IAA, owing to a less massive nucleus or more explosive activity. Although both Halley and Hyakutake showed hints of nonnegligible coma depolarization, the only firm detection of significant coma depolarization is from the recent detection of C/2001 A2

(*Nolan et al.*, 2001). Since the nucleus of this comet had split prior to the radar observations (*Sekanina et al.*, 2002), it is possible that the depolarization was from boulder-sized debris left over from the splitting or produced in violent activity of small, freshly exposed subnuclei.

4.2. Grain Ejection and Echo Modeling

Further analysis of the coma echo requires modeling the grain ejection process and estimating the mass-loss rates required to sustain the observed grain coma. A good starting point is to assume that the grain emission process is a continuous one in which grains are ejected as free (unbound) particles in the comet orbit frame. That the grains are predominantly unbound is consistent with the lack of a clear symmetric component about the nucleus echo in the coma spectra for IAA and Hyakutake (Figs. 1 and 2). While it is expected that some grains will be injected into circumnuclear orbits (*Richter and Keller*, 1995; *Fulle*, 1997), Fulle estimates that only about 1% of the ejected grains will do so; this would not be enough to accumulate a significant bound population, especially if the grains are undergoing evaporation or disintegration.

4.2.1. Gas drag models. For grain ejection we adopt the canonical model first formulated by *Whipple* (1951) and refined by others. Assuming a uniform radial outflow of gas with thermal expansion velocity V_g, one can write a differential equation for the outward drag velocity V of the grains (*Wallis*, 1982; *Gombosi et al.*, 1986)

$$M_g V \frac{dV}{dr} = \frac{1}{2} C_D \pi a^2 (V_g - V)^2 \frac{ZR^2}{V_g r^2} - \frac{GM_g M_n}{r^2} \quad (18)$$

where C_D is the drag coefficient, Z is the surface gas mass flux, R is nucleus radius, G is the gravitational constant, and M_g and M_n are grain and nucleus mass respectively. Then, assuming that the grain and nucleus are spheres of density d_g and d_n, that V_g is constant with radial distance r, and that $V \ll V_g$, integrating equation (18) gives a terminal grain velocity

$$V_t(a) = C_v a^{-1/2} (1 - a/a_m)^{1/2} \quad (19)$$

where

$$C_v = \left(\frac{3 C_D V_g ZR}{4 d_g} \right)^{1/2} \quad (20)$$

is a velocity scale factor, and

$$a_m = \frac{9 C_D V_g Z}{32 \pi GR d_n d_g} \quad (21)$$

is a maximum grain size in the gravitational correction factor $(1 - a/a_m)^{1/2}$. [Equations (20) and (21) are equivalent to equation (4) of *Jewitt and Matthews* (1999) and equation (72) of *Gombosi et al.* (1986) respectively.] One has $C_D = 2$ for a solid sphere, although the drag coefficient can

be higher for fluffy grains (*Keller and Markiewicz*, 1991). The velocity V_g is often taken to be the thermal expansion velocity at the surface multiplied by some correction factor to allow for expansion effects; this factor is $\approx 9/4$ in *Finson and Probstein* (1968). Nonradial or asymmetric expansion can also give V_t different from the canonical model (*Crifo*, 1995).

Using these equations, *Harmon et al.* (1989) argued that the IAA coma echo was consistent with the simple gas drag model. Taking $a_m = 3$ cm as a reasonable cutoff size (based on the mass and polarization arguments above) requires a gas flux Z of $\sim 1 \times 10^{-5}$ g/cm² s, a reasonable value when compared with the 5×10^{-5} g/cm² s sublimation rate for clean ice at 1 AU. This also gives reasonable grain velocities (8 m/s for a = 1 cm) and a good match to the Doppler spread in the coma spectrum model (see next section). Hyakutake did not fit so neatly into this picture, its much broader spectrum requiring higher grain velocities (40 m/s for a = 1 cm) than for IAA despite its smaller nucleus (*Harmon et al.*, 1997). This implied a much higher effective gas flux ($\sim 4 \times 10^{-4}$ g/cm² s), or much fluffier grains, or both. Since Hyakutake's nominal surface active fraction is about 1.0 assuming $Z = 5 \times 10^{-5}$ g/cm² s, then the effective Z must have been much higher than this in the discrete active regions that were observed to dominate the emission (*Schleicher et al.*, 1998). Grain fluffiness could also boost the ejection velocity by lowering grain density and raising the drag coefficient.

4.2.2. Doppler spectrum modeling. The shape of the Doppler spectrum contains information on the grain velocity vectors and spatial distribution. Although the grain coma cannot be uniquely characterized from its spectrum, some useful results have been obtained by treating the forward problem of comparing the observed spectrum with model spectra computed from trial input parameters. If one starts with the gas-drag model (section 4.2.1) and ignores radiation pressure, then it is fairly straightforward to compute a Doppler spectrum by assuming a production size distribution and summing over discrete grain emission times and directions (*Harmon et al.*, 1989). Once the nucleus and grain properties are assumed, then the remaining free parameters in the model are the ejection geometry and Z (or a_m).

Model spectra have been computed for the coma echoes from IAA (*Harmon et al.*, 1989) and Hyakutake (*Harmon et al.*, 1997). The shape and offset of the IAA coma spectrum could be well modeled (Fig. 10) by invoking a sunward grain emission fan with centroid aimed below the comet orbit plane in a direction consistent with the orientation of the infrared and visual dust fans. No doubt this was aided by the fact that IAA was a slow rotator with an unusually stable sunward fan (*Sekanina*, 1988). The model shown in Fig. 10 has $Z = 1.2 \times 10^{-5}$ g/cm² s and $a_m = 3$ cm, which is consistent with the observed echo polarization and gives plausible nucleus mass-loss rates and gas fluxes. A model spectrum for Hyakutake is shown overplotted in Fig. 2. Here a much higher Z of 4×10^{-4} g/cm² s [V_t (1 cm) = 40 m/s] was required to reproduce the large Doppler spread (see discussion in previous section). Since the

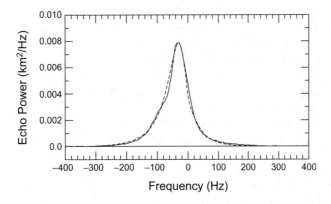

Fig. 10. Model coma spectrum (dashed line) overplotted on the OC coma echo for C/IRAS-Araki-Alcock. The data spectrum (solid line) has been smoothed to 10-Hz resolution. See text for details.

dust emission for this fast rotator was more complicated than for IAA, no attempt was made to arrive at a single consistent model for the grain emission geometry.

4.2.3. Mass-loss rates. If one assumes that the grain coma is replenished by continuous particle ejection, then the coma radar cross section can be used to estimate the mass-loss rate in large grains. We assume the grains have a production-rate size distribution $\dot{n}(a) \propto a^{-\alpha}$. The mass-loss rate $\dot{M}$ is given by

$$\dot{M} = \frac{4}{3}\pi d_g \int_{a_o}^{a_m} \dot{n}(a)a^3 da \qquad (22)$$

The total radar cross section of the grains in the radar beam is

$$\sigma = \pi \int_{a_o}^{a_m} \dot{n}(a)L(a)Q_b(a)a^2 da \qquad (23)$$

where $L(a)$ is the mean lifetime of a grain of radius a within the beam. If the grains are ejected isotropically and remain intact as they traverse the beam, then the mean lifetime is the mean beam transit time $\pi h/2V_t(a)$, where h is the half-width of the cylinder defined by the radar beam at the comet. Then, combining equations (19)–(23) gives

$$\dot{M}(a_m) = \sigma\left(\frac{8Rd_g}{3}\right)\left(\frac{8}{3}\pi Gd_n\right)^{1/2}$$
$$\left[1 - \left(\frac{a_o}{a_m}\right)^{4-\alpha}\right]a_m^{9/2-\alpha}[(4-\alpha)\pi hI]^{-1} \qquad (24)$$

where

$$I = \int_{a_o}^{a_m} \frac{a^{5/2-\alpha}Q_b(a)}{(1-a/a_m)^{1/2}}da \qquad (25)$$

This is the same as equation (B9) of *Harmon et al.* (1989). [An incorrectly rewritten version of this equation appeared

as equation (14) of *Harmon et al.* (1999), although the calculations in that paper were based on the correct original equation.] For $a < \lambda/2\pi$ one gets a Rayleigh approximation for $\dot{M}$ by using the following analytic solution for the integral

$$I = C_R B(1/2, 15/2 - \alpha)a_m^{15/2-\alpha} \qquad (26)$$

where B is the beta function. Implicit in equation (24) is the assumption that the velocity scale factor C_v is a function of a_m. If, on the other hand, one has an independent estimate of C_v (say, from the width of the Doppler spectrum), then one could treat it (and a_m) as a constant, to give the modified expression

$$\dot{M}(a_{max}) = \sigma\left(\frac{8C_vd_g}{3}\right)\left[1 - \left(\frac{a_o}{a_{max}}\right)^{4-\alpha}\right]$$
$$a_{max}^{4-\alpha}[(4-\alpha)\pi hI]^{-1} \qquad (27)$$

where a_{max} ($< a_m$) can be taken as some other (nongravitational) cutoff size that replaces a_m as the upper integration limit in equation (25).

In Fig. 11 we show results of mass-loss rate calculations for three comets. Here we have used equation (24) to calculate $\dot{M}(a_m)$ for IAA and Halley, and equation (27) to calculate $\dot{M}(a_{max})$ for Hyakutake [assuming V_t (1 cm) = 40 m/s]. Mie theory was used to calculate Q_b assuming the grains to be spherical snowballs with density $d_g = 0.5$ g/cm³ (refractive index = 1.4). We took the production rate size distribution to be an $\dot{n}(a) \propto a^{-3.5}$ power law between $a_o = 1$ μm and the cutoff size. The $\alpha = 3.5$ power law was chosen not only because it conforms to size distributions measured for Halley (*McDonnell et al.*, 1986) and Hyakutake (*Fulle et al.*, 1997), but also because it has the convenient property of giving an $\dot{M}$ that is determined primarily by the larger (radar-reflecting) grains and that is relatively insensitive to the precise value of α. The Rayleigh regime ($a_m < \lambda/2\pi$) in Fig. 11 shows the $\dot{M}(a_m) \propto a_m^{-3}$ behavior expected from substitution of equation (26) in equation (24). It is this strong Rayleigh size dependence that requires the presence of large (greater than millimeter-sized) grains in order to explain the radar cross sections for reasonable mass-loss rates; for example, taking $a_m = 1$ mm implies an $\dot{M}$ that would have a typical comet nucleus losing most of its mass during a single perihelion passage. The $\dot{M}$ curves flatten out at the larger sizes ($a_m > \lambda/2\pi$), corresponding to large-grain production rates in the range 3×10^5–1×10^6 g/s.

4.2.4. Comparisons with other mass-loss rate estimates. By comparing the radar-derived production rates with other measurements sensitive to smaller dust particles, we can get some idea of the relative importance of the large grains to the overall particulate population of the coma. Infrared measurements for IAA (*Hanner et al.*, 1985) gave dust production rates of 1–2×10^5 g/s. This is a bit smaller than the rates shown in Fig. 11 and would be consistent with an overall production size distribution spectral index $\alpha = 3.8$. Clearly, large grains contributed a substantial fraction of the

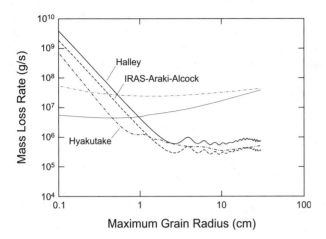

Fig. 11. Mass-loss rate Ṁ vs. maximum grain radius a_m for Comets Halley (solid), IRAS-Araki-Alcock (dashed), and Hyakutake (dot-dashed). These were computed using the measured radar cross sections and assuming the grains to be 0.5 g/cm³ snowballs with an $a^{-3.5}$ production size distribution. Also shown are Ṁ curves computed on the basis of the measured radio continuum fluxes for Halley and Hyakutake (lighter curves).

mass loss for IAA. Comparison of the Halley curve in Fig. 11 with the dust production rate of 2×10^6 g/s from infrared measurements indicates large grains constituted a slightly smaller fraction of the mass loss for this comet, although *Giotto* dust detector results (*McDonnell et al.,* 1986) suggest that Halley's large-grain production should have dominated the mass loss with Ṁ = 5×10^6 g/s at the time of the Arecibo observations. The dust production rate of 5×10^6 g/s estimated for Hyakutake (*Fulle et al.,* 1997) exceeds by several times the radar-derived rate for $a_m >$ 1 cm (Fig. 11), suggesting that large grains were important but not dominant for this comet.

Since the radar-derived mass-loss rates assume that the grains remain intact as they traverse the radar beam, the Ṁ curves in Fig. 11 are actually lower limits and hence may underestimate the relative contribution of large grains to the total particulate mass loss. Grain evaporation and disintegration are, in fact, believed to be important processes (*Hanner,* 1981; *Combi,* 1994). It has been estimated that (at 1 AU) ejected dirty-ice grains with radii of 1 mm and 1 cm only travel 100 km and 2000 km respectively before evaporating (*Hanner,* 1981; *Harris et al.,* 1997). This may explain why the Arecibo and Goldstone S-band coma echoes for IAA gave the same radar cross section despite the fact that the Goldstone beam was 3× wider than the 2200 km subtended by the Arecibo beam at the comet, although icy grains from IAA were not detected in the infrared observations of *Hanner et al.* (1985). For Halley, *Campbell et al.* (1989) also suggested that ignoring grain evaporation might account for the apparent discrepancy between the radar and *Giotto* grain production rates. Similarly, many of the large grains from Hyakutake may have evaporated or disintegrated before traversing a substantial fraction of the

7000-km Goldstone beam. This is supported by the photometric data of *Schleicher and Osip* (2002), which show a slow falloff in CN, C_2, and dust with increasing aperture size that is consistent with fragmentation or evaporation of grains from that comet. Further support comes from *Harris et al.* (1997), who argued that evaporation of large icy grains produced Hyakutake's spherical gas coma and accounted for 23% of the comet's total gas production. This would give 1×10^6 g/s in secondary gas from grains, so the total large-grain mass lost to disintegration could have been significantly higher than this if the volatile fraction was low. It is clear that the total mass contained in large grains is high enough that grain fragmentation could be an important secondary source of coma gas in the typical active comet.

Finally, it is worthwhile comparing the radar Ṁ values with those estimated from millimeter-wave continuum observations, which are also sensitive to large grains (*Jewitt and Luu,* 1992). An equivalent continuum equation for Ṁ can be written by replacing σ in equations (24) or (27) with $S\lambda^2\Delta^2/2kT$ and replacing Q_b in equation (25) with Q_a, where S is the continuum flux density, T is grain temperature, and Q_a is the grain absorption efficiency. This has been used to compute Ṁ curves (Fig. 11) from the 3.5-mm continuum detection of Halley (*Altenhoff et al.,* 1986) and the 1.1-mm continuum detection of Hyakutake (*Jewitt and Matthews,* 1997). [We have not included a curve for IAA, as the 1.3-cm continuum detection by *Altenhoff et al.* (1983) appears to have been dominated by thermal emission from the nucleus, as was also argued by *Harmon et al.* (1989).] Note that these curves do not show the same extreme sensitivity to a_m in the Rayleigh regime as the radar curves, which reflects the different behaviors of Q_a and Q_b. The comparison with radar is uncertain because of the sensitivity of the continuum curves to assumed grain properties such as porosity and electrical conductivity, although the sensitivity to the assumed conductivity becomes less important as the larger grains become optically thick. For the curves in Fig. 11 we assume the grains to be dirty snowballs with 0.5 g/cm³ density and 0.01 imaginary part refractive index. The radar beam was 4× and 6× larger than the continuum beam for Halley and Hyakutake respectively, so any grain evaporation would also affect the comparison. Note that the continuum curves are significantly higher than the radar curves for a_m larger than 1 cm. Including grain evaporation would reduce some of this discrepancy. A possible way to remove the remaining discrepancy is to invoke fluffier grains (which would also help to explain the high grain velocities inferred from the Hyakutake coma spectrum, as mentioned in section 4.2.1). This is because increasing the grain porosity raises the absorption per unit mass (or opacity κ) as it lowers the backscatter per unit mass, the combined effect being to bring the radio and radar curves closer together (*Harmon et al.,* 1997). This implies that κ(1 mm) would have to be higher than the κ(1 mm) = 2–3 cm²/g that characterizes the curves in Fig. 11. In fact, *Altenhoff et al.* (1999) assumed "fluffy dust" with κ(1 mm) = 75 cm²/g to estimate Hyakutake's dust production from their radio continuum observations. This accounted for the very large discrepancy that they noted between their production rates and those inferred by *Jewitt and Matthews* (1997) using a much lower κ(1 mm) = 0.5 cm²/g value typical of interstellar dust.

5. FUTURE WORK

5.1. Radar Imaging: Delay-Doppler and Interferometric

The highest priority of future cometary radar is to obtain images of the nucleus and/or coma echoes. Imaging could provide data on nucleus size, shape, rotation, and surface features, as well as the size of the grain coma and its position relative to the nucleus. It can also be used to more accurately determine the nucleus albedo and scattering law. Imaging can be done using either the delay-Doppler method or an interferometer.

Delay-Doppler combines pulsed or coded transmission with spectral analysis in order to resolve the echo into cells in delay-Doppler space. The detectability in a given delay-Doppler cell is roughly given by $D/(N_d\sqrt{N_D})$, where D is the Doppler-only detectability from equation (4) and N_d and N_D are the number of delay and Doppler bins, respectively, across the target. Any comet nucleus passing within about 0.1 AU should provide a good delay-Doppler imaging opportunity for Arecibo (*Harmon et al.,* 1999), although crude imaging or delay-profiling suitable for size estimation may be feasible at larger distances. A delay-Doppler image could provide direct information on a nucleus and also be used to construct a three-dimensional model of the rotating object, in a similar manner to work done on near-Earth asteroids (*Ostro et al.,* 1995; *Hudson and Ostro,* 1995). If the radar images have adequate orientational coverage and an adequate time span, then the modeling can decipher the nucleus spin state. This would be of particular interest for slow rotators because of their tendency for non-principle-axis rotation (*Ostro et al.,* 2001; *Samarasinha and A'Hearn,* 1991; *Hudson and Ostro,* 1995; *Samarasinha et al.,* 2004). Also, any absolute range measurement would provide even more accurate orbit astrometry than could be derived from Doppler alone. For the grain-coma echo, the extra information provided by a delay-resolved echo would remove some of the ambiguity encountered in coma-echo modeling using Doppler spectra alone. However, coma delay-Doppler images would pose their own special interpretation problems, as the mapping problem is unlike that for a rigid rotating body. Furthermore, unlike the nucleus echo, a coma echo is likely to be "overspread" (product of delay depth and Doppler bandwidth >1), which would require a special observing strategy as discussed by *Harmon* (2002).

Interferometric imaging offers an alternative to delay-Doppler imaging. The Very Large Array (VLA) can potentially image coma echoes from Goldstone 3.5-cm transmissions with a synthesized beam as small as 0.24 arcsec. This bistatic method, which has been applied successfully to a few asteroids as well as Mercury, Venus, Mars, and Saturn's

rings, has significant potential for direct plane-of-sky imaging of the grain comae of close-approaching comets (*de Pater et al.,* 1994). The VLA resolution is too coarse for nucleus imaging, but bistatic radar observations with the Very Long Baseline Array (VLBA) would be suitable, with resolutions at S- and X-bands of 3 mas and 0.8 mas respectively.

5.2. Short-Period Comet Opportunities

Radar observations in the coming years will include a mix of short-period and new comets. The short-period comets include the ecliptic comets (and their Jupiter-family subset), with a putative Kuiper belt origin, and the Halley-type comets, most of which probably come from the Oort cloud (*Levison,* 1997). The "new" comets include both dynamically new objects and newly discovered long-period comets. Although new comets such as IAA may well offer the best radar opportunities, the short-period comets hold some intrinsic interest. They are the most likely targets for spacecraft missions, for which groundbased radar can provide both mission support and a complementary dataset. Also, though relatively inactive compared to some new comets, they are thought to play an important role in the interplanetary dust budget and are the source of meteor streams and infrared dust trails; hence, echoes from their large-grain comae are of interest. There are several good short-period comet radar opportunities over the next decade or so. These are listed in Table 5. Below we discuss some of the more interesting apparitions. (We include the Encke apparition of 2003 in this discussion and Table 5, even though the observations planned for that apparition will have been done by the time this book goes to press.) The quoted detectabilities are computed from equation (5) assuming a 1-h integration time, a nucleus albedo of 0.05, and (unless otherwise noted) a rotation period of 0.5 d. The D values also assume |sinϕ| = 1, and therefore represent lower limits.

5.2.1. 2P/Encke. Although one of the most intensely studied of all comets, Encke's nucleus properties remain uncertain. Observations in 2003 should give D ~ 60 at Arecibo and ~3 at Goldstone. This may allow some crude delay-Doppler imaging and a direct size estimate. While not a very active comet, Encke is known to produce centimeter-sized grains and to be the source of the Taurid meteors and an infrared dust trail (*Epifani et al.,* 2001; *Reach et al.,* 2000). The large grains may give a weak coma detection.

5.2.2. 73P/Schwassmann-Wachmann 3. This comet makes a very close pass in 2006 and offers a nominally excellent, if unpredictable, radar opportunity. This comet split into three main pieces during its 1995 apparition, those pieces reappearing at the 2001 apparition. If each piece has one-third the mass of a R = 1 km parent body (*Boehnhardt et al.,* 1999), its detectability should be ~1000 at Arecibo (~60 at Goldstone), making this a good imaging opportunity. Detectable coma echoes are also likely.

5.2.3. 8P/Tuttle. This little-studied object is the only Halley-type comet in this sample and thus the only one likely to have an Oort cloud origin. The only known radius estimate is R = 7.3 km from optical magnitude measure-

TABLE 5. Future radar opportunities for short-period comets (passing within 0.5 AU through year 2020).

Comet	Date (m/d/y)*	Δ (AU)†
2P/Encke	11/17/2003	0.261
73P/Schwassmann-Wachmann 3	5/12/2006	0.051–0.076‡
8P/Tuttle	1/2/2008	0.252
6P/d'Arrest	8/10/2008	0.353 [0.375]
103P/Hartley 2	10/21/2010	0.120
45P/Honda-Mrkos-Pajdušáková	8/15/2011	0.060§ [0.220]
2P/Encke	10/17/2013	0.478
P/2000 G1 (LINEAR)	3/22/2016	0.032 [0.105]
45P/Honda-Mrkos-Pajdušáková	2/11/2017	0.087
41P/Tuttle-Giacobini-Kresak	3/27/2017	0.136 [0.190]
21P/Giacobini-Zinner	9/11/2018	0.380
64P/Swift-Gehrels	10/28/2018	0.444
46P/Wirtanen	12/16/2018	0.075

*Date of closest approach.

†Distance from Earth at closest approach (with closest distance for Arecibo observations, if different, in brackets).

‡Range of distances for fragments B, C, and E.

§Just below Goldstone horizon at this distance.

ments by *Licandro et al.* (2000) at large heliocentric distances. This would place it in the Halley size class, so a radar-based size estimate would be of considerable interest. If Tuttle is really this large, it would give D ~ 300 at Arecibo (~17 at Goldstone). This comet is the parent of the Ursid meteor stream (*Jenniskens et al.,* 2002), so there is the potential for a coma echo.

5.2.4. 6P/d'Arrest. This second 2008 apparition is less favorable than that of Tuttle, owing to the larger Δ and southerly declinations. Using R = 2.7 km (*Lisse et al.,* 1999) and the oft-quoted short rotation period of 5.2 h (itself of intrinsic interest) gives D ~ 5 for Arecibo. This comet shows an antitail (*Fulle,* 1990) and must therefore produce some large grains.

5.2.5. 103P/Hartley 2. The small Δ of this comet in 2010 offers a good radar opportunity despite its apparent small size. Using R = 0.56 km (*Jorda et al.,* 2000) gives a D ~ 150 at Arecibo and ~9 at Goldstone. This comet is fairly active for its size and a likely producer of large grains (*Epifani et al.,* 2001), so a coma echo is possible.

5.2.6. 45P/Honda-Mrkos-Pajdušáková. In 2011 this comet does not enter the Arecibo declination window until well past close approach, making it a better target for Goldstone. Assuming R = 0.34 (*Lamy et al.,* 1999), the Goldstone detectability is about 20 by the time the comet reaches a reasonable sky elevation (Δ = 0.08 AU). The Arecibo D for the 2017 apparition is 250. Although a dust-poor comet, it is known to be the source of the Alpha Capricornid meteors, so a coma echo is possible.

5.2.7. 46P/Wirtanen and other mission targets. The Jupiter-family comet Wirtanen is of special interest as the target of the ROSETTA spacecraft rendezvous in 2011. Unfortunately, this comet only comes within 0.92 AU at its 2008 apparition, making it an impossible target given its estimated size of R = 0.6 km (*Lamy et al.,* 1998; *Boehnhardt*

et al., 2002). There is a nominally excellent opportunity in 2018, although the comet's large nongravitational acceleration (*Jorda and Rickman,* 1995) makes the distance prediction uncertain. Comet 9P/Tempel 1, the target of the Deep Impact mission, approaches within 0.71 AU in early May 2005, two months before the spacecraft encounter. Taking R = 3 km (*Lamy et al.,* 2001) and p = 41 h (*Meech et al.,* 2002) gives an Arecibo detectability of only D = 3. Still, an Arecibo attempt at a nucleus detection at closest approach is probably warranted. An attempt might also be made to look for echoes from debris ejected in the July 4, 2005, impact experiment, although Tempel 1 will be even more distant (0.89 AU) at that time. Finally, the Stardust mission target, 81P/Wild 2, is not observable from Arecibo at less than 1.5 AU for the next two decades.

6. SUMMARY

Earth-based radar has proven to be an important tool for studying close-approaching comets. The various nucleus detections show comet nuclei to be rough objects with relatively low surface densities. They have also established a factor-of-10 nucleus size range for this limited sample, based on the observed range of radar cross sections. A large fraction of the radar-detected comets have been found to show broadband echoes from large coma grains. This has provided some of the strongest evidence yet for the prevalence of large-grain emission by comets. With radar-derived productions rates ~10^6 g/s, large (approximately centimeter-sized) grains must constitute a significant fraction of the total mass loss for some comets.

The full potential of cometary radar will not be realized until radar imaging of a comet is achieved. Delay-Doppler imaging holds the potential for accurately determining nucleus properties such as size, shape, spin state, albedo, and scattering law. An imaged or delay-resolved coma echo would also be of considerable interest. A few of the upcoming short-period comet apparitions may afford opportunities for at least crude nucleus imaging. Favorable imaging opportunities from new-comet apparitions are also anticipated.

REFERENCES

Altenhoff W. J., Batrla W., Huchtmeier W. K., Schmidt J., Stumpff P., and Walmsley M. (1983) Radio observations of Comet 1983d. *Astron. Astrophys., 125,* L19–L22.

Altenhoff W. J., Huchtmeier W. K., Schmidt J., Schraml J. B., Stumpff P., and Thum C. (1986) Radio continuum observations of comet Halley. *Astron. Astrophys., 164,* 227–230.

Altenhoff W. J. and 20 colleagues (1999) Coordinated radio continuum observations of comets Hyakutake and Hale-Bopp from 22 to 860 GHz. *Astron. Astrophys., 348,* 1020–1034.

Ball A. J., Gadomski S., Banaszkiewicz M., Spohn T., Ahrens T. J., Whyndham M., and Zarnecki J. C. (2001) An instrument for in situ comet nucleus surface density profile measurement by gamma ray attenuation. *Planet. Space Sci., 49,* 961–976.

Benner L. A. M. (2002) Summaries of asteroid radar properties [online]. California Institute of Technology, Pasadena [cited Oct. 1, 2002]. Available on line at http://echo.jpl.nasa.gov/

~lance/asteroid_radar_properties.html.

Benner L. A. M. and 12 colleagues (1997) Radar detection of near-Earth asteroids 2062 Aten, 2101 Adonis, 3103 Eger, 4544 Xanthus, and 1992 QN. *Icarus, 130,* 296–312.

Benner L. A. M., Ostro S. J., Nolan M. C., Margot J.-L., Giorgini J. D., Hudson R. S., Jurgens R. F., Slade M. S., Howell E. S., Campbell D. B., and Yeomans D. K. (2002) Radar observations of asteroid 1999 JM8. *Meteoritics & Planet. Sci., 37,* 779–792.

Boehnhardt H., Rainer N., and Schwehm G. (1999) The nuclei of comets 26P/Grigg-Skjellerup and 73P/Schwassmann-Wachmann 3. *Astron. Astrophys., 341,* 912–917.

Boehnhardt H., Delahodde C., Sekiguchi T., Tozzi G. P., Amestica R., Hainaut O., Spyromilio J., Tarenghi M., West R. M., Schulz R., and Schwehm G. (2002) VLT observations of comet 46P/Wirtanen. *Astron. Astrophys., 387,* 1107–1113.

Bohren C. F. and Huffman D. R. (1983) *Absorption and Scattering of Light by Small Particles.* Wiley & Sons, New York. 530 pp.

Bottke W. F., Morbidelli A., Jedicke R., Petit J.-M., Levison H. F., Michel P., and Metcalfe T. S. (2002) Debiased orbital and absolute magnitude distribution of the near-Earth objects. *Icarus, 156,* 399–433.

Campbell D. B., Harmon J. K., Hine A. A., Shapiro I. I., Marsden B. G., and Pettengill G. H. (1983) Arecibo radar observations of Comets IRAS-Araki-Alcock and Sugano-Saigusa-Fujikawa (abstract). *Bull. Am. Astron. Soc., 15,* 800.

Campbell D. B., Harmon J. K., and Shapiro I. I. (1989) Radar observations of Comet Halley. *Astrophys. J., 338,* 1094–1105.

Campbell M. J. and Ulrichs J. (1969) Electrical properties of rocks and their significance for lunar radar observations. *J. Geophys. Res., 74,* 5867–5881.

Campins H. (1988) The anomalous dust production in periodic comet Encke. *Icarus, 73,* 508–515.

Colwell J. E., Jakosky B. M., Sandor B. J., and Stern S. A. (1990) Evolution of topography on comets. II. Icy craters and trenches. *Icarus, 85,* 205–215.

Combi M. R. (1994) The fragmentation of dust in the innermost comae of comets: Possible evidence from ground-based images. *Astron. J., 108,* 304–312.

Crifo J. F. (1995) A general physicochemical model of the inner coma of active comets. I. Implications of spatially distributed gas and dust production. *Astrophys. J., 445,* 470–488.

de Pater I., Palmer P., Mitchell D. L., Ostro S. J., Yeomans D. K., and Snyder L. E. (1994) Radar aperture synthesis observations of asteroids. *Icarus, 111,* 489–502.

Epifani E., Colangeli L., Fulle M., Brucato J. R., Bussoletti E., De Sanctis M. C., Mennella V., Palomba E., Palumbo P., and Rotundi A. (2001) ISOCAM imaging of comets 103P/Hartley 2 and 2P/Encke. *Icarus, 149,* 339–350.

Fernández Y. R., Lisse C. M., Kundu A., A'Hearn M. F., Hoffman W. F., and Dayal A. (1996) The nucleus of Comet Hyakutake (abstract). *Bull. Am. Astron. Soc., 28,* 1088.

Fernández Y. R., Lisse C. M., Käufl H. U., Peschke S. B., Weaver H. A., A'Hearn M. F., Lamy P. P., Livengood T. A., and Kostiuk T. (2000) Physical properties of the nucleus of comet 2P/Encke. *Icarus, 147,* 145–160.

Fernández Y. R., Lowry S. C., Weissman P. R., and Meech K. J. (2002) New dominant periodicity in photometry of comet Encke (abstract). *Bull. Am. Astron. Soc., 34,* 887.

Finson M. L. and Probstein R. F. (1968) A theory of dust comets. I. Model and equations. *Astrophys J., 154,* 327–352.

Fulle M. (1990) Meteoroids from short period comets. *Astron. Astrophys., 230,* 220–226.

Fulle M. (1997) Injection of large grains into orbits around comet nuclei. *Astron. Astrophys., 325,* 1237–1248.

Fulle M., Mikuz H., and Bosio S. (1997) Dust environment of Comet Hyakutake 1996B2. *Astron. Astrophys., 324,* 1197–1205.

Giorgini J. D. (2002) Small-body astrometric radar observations [online]. California Institute of Technology, Pasadena [cited Oct. 1, 2002]. Available on line at http://ssd.jpl.nasa.gov/radar/data.html.

Goldstein R. M., Jurgens R. F., and Sekanina Z. (1984) A radar study of Comet IRAS-Araki-Alcock 1983d. *Astron. J., 89,* 1745–1754.

Gombosi T. I., Nagy A. F., and Cravens T. E. (1986) Dust and neutral gas modeling of the inner atmospheres of comets. *Rev. Geophys., 24,* 667–700.

Hallikainen M. T., Ulaby F. T., and Abdelrazik M. (1986) Dielectric properties of snow in the 3 to 37 GHz range. *IEEE Trans. Ant. Prop., AP-34,* 1329–1339.

Hanner M. S. (1981) On the detectability of icy grains in the comae of comets. *Icarus, 47,* 342–350.

Hanner M. S., Aitken D. K., Knacke R., McCorkle S., Roche P. F., and Tokunaga A. T. (1985) Infrared spectrophotometry of Comet IRAS-Araki-Alcock (1983d): A bare nucleus revealed? *Icarus, 62,* 97–109.

Hanner M. S., Newburn R. L., Spinrad H., and Veeder G. J. (1987) Comet Sugano-Saigusa-Fujikawa (1983V) — A small, puzzling comet. *Astron. J., 94,* 1081–1087.

Harmon J. K. (2002) Planetary delay-Doppler radar and the long-code method. *IEEE Trans. Geosci. Remote Sensing, 40,* 1904–1916.

Harmon J. K., Campbell D. B., Hine A. A., Shapiro I. I., and Marsden B. G. (1989) Radar observations of Comet IRAS-Araki-Alcock 1983d. *Astrophys. J., 338,* 1071–1093.

Harmon J. K. and 15 colleagues (1997) Radar detection of the nucleus and coma of Comet Hyakutake (C/1996 B2). *Science, 278,* 1921–1924.

Harmon J. K., Campbell D. B., Ostro S. J., and Nolan M. C. (1999) Radar observations of comets. *Planet. Space Sci., 47,* 1409–1422.

Harris W. M., Combi M. R., Honeycutt R. K., Mueller B. E. A., and Scherb F. (1997) Evidence of interacting gas flows and an extended volatile source distribution in the coma of Comet C/1996 B2 (Hyakutake). *Science, 277,* 676–681.

Hudson R. S. and Ostro S. J. (1995) Shape and non-principal axis spin state of asteroid 4179 Toutatis. *Science, 270,* 84–86.

Jenniskens P. and 13 colleagues (2002) Dust trails of 8P/Tuttle and the unusual outbursts of the Ursid shower. *Icarus, 159,* 197–209.

Jewitt D. C. and Luu J. (1992) Submillimeter continuum emission from comets. *Icarus, 100,* 187–196.

Jewitt D. C. and Matthews H. E. (1997) Submillimeter continuum observations of Comet Hyakutake (1996 B2). *Astron. J., 113,* 1145–1151.

Jewitt D. C. and Matthews H. E. (1999) Particulate mass loss from Comet Hale-Bopp. *Astron. J., 117,* 1056–1162.

Jorda L. and Rickman H. (1995) Comet P/Wirtanen, summary of observational data. *Planet. Space Sci., 43,* 575–579.

Jorda L., Lamy P., Groussin O., Toth I., A'Hearn M. F., and Peschke S. (2000) ISOCAM observations of cometary nuclei. In *ISO Beyond Point Sources: Studies of Extended Infrared Emission,* p. 61. ESA SP-455, Noordwijk, The Netherlands.

Kamoun P. G. D. (1983) Radar observations of cometary nuclei. Ph.D. thesis, Massachusetts Institute of Technology, Cambridge. 273 pp.

Kamoun P. G., Pettengill G. H., and Shapiro I. I. (1982a) Radar detectability of comets. In *Comets* (L. L. Wilkening, ed.), pp. 288–296. Univ. of Arizona, Tucson.

Kamoun P. G., Campbell D. B., Ostro S. J., Pettengill G. H., and Shapiro I. I. (1982b) Comet Encke: Radar detection of nucleus. *Science, 216,* 293–295.

Kamoun P., Campbell D., Pettengill G., and Shapiro I. (1999) Radar observations of three comets and detection of echoes from one: P/Grigg-Skjellerup. *Planet. Space Sci., 47,* 23–28.

Keller H. U. and Markiewicz W. J. (1991) KOSI? *Geophys. Res. Lett., 18,* 249–252.

Kührt E., Knollenberg J., and Keller H. U. (1997) Physical risks of landing on a comet nucleus. *Planet. Space Sci., 45,* 665–680.

Lamy P. L., Toth I., Jorda L., Weaver H. A., and A'Hearn M. F. (1998) The nucleus and inner coma of Comet 46P/Wirtanen. *Astron. Astrophys., 335,* L25–L29.

Lamy P. L., Toth I., A'Hearn M. F., and Weaver H. A. (1999) Hubble Space Telescope observations of the nucleus of comet 45P/Honda-Mrkos-Pajdušáková and its inner coma. *Icarus, 140,* 424–438.

Lamy P. L., Toth I., A'Hearn M. F., Weaver H. A., and Weissman P. R. (2001) Hubble Space Telescope observations of the nucleus of comet 9P/Tempel 1. *Icarus, 154,* 337–344.

Levison H. F. and Duncan M. J. (1997) From the Kuiper Belt to the Jupiter-family comets: The spatial distribution of ecliptic comets. *Icarus, 127,* 13–32.

Licandro J., Tancredi G., Lindgren M., Rickman H., and Gil R. (2000) CCD photometry of cometary nuclei, I: Observation from 1990–1995. *Icarus, 147,* 161–179.

Lisse C. M., Fernández Y. R., Kundu A., A'Hearn M. F., Dayal A., Deutsch L. K., Fazio G. G., Hora J. L., and Hoffman W. F. (1999) The nucleus of comet Hyakutake (C/1996 B2). *Icarus, 140,* 189–204.

Luu J. and Jewitt D. (1990) The nucleus of Comet P/Encke. *Icarus, 86,* 69–81.

Magri C., Ostro S. J., Rosema K. D., Thomas M. L., Mitchell D. L., Campbell D. B., Chandler J. F., Shapiro I. I., Giorgini J. D., and Yeomans D. K. (1999) Mainbelt asteroids: Results of Arecibo and Goldstone radar observations of 37 objects during 1980–1995. *Icarus, 140,* 379–407.

Magri C., Consolmagno G. J., Ostro S. J., Benner, L. A. M., and Beeney B. R. (2001) Radar constraints on asteroid regolith properties using 433 Eros as ground truth. *Meteoritics & Planet. Sci., 36,* 1697–1709.

Marsden B. G. (1998) *Comet C/1998 K5 (LINEAR).* IAU Circular No. 6923.

McDonnell J. A. M. and 27 colleagues (1986) Dust density and mass distribution near Comet Halley from Giotto observations. *Nature, 321,* 338–341.

Meech K. J. and 11 colleagues (2002) Deep Impact — Exploring the interior of a comet. In *A New Era in Bioastronomy* (G. Lemarchand and K. Meech, eds.), pp. 235–242. Astronomical Society of the Pacific, San Francisco, California.

Mitchell D. L., Ostro S. J., Rosema K. D., Hudson R. S., Campbell D. B., Chandler J. F., and Shapiro I. I. (1995) Radar observations of asteroids 7 Iris, 9 Metis, 12 Victoria, 216 Kleopatra, and 654 Zelinda. *Icarus, 118,* 105–131.

Nolan M. C., Howell E. S., Harmon J. K., Campbell D. B., Margot J.-L., and Giorgini J. D. (2001) Arecibo radar observations of C/2001A2(B) (LINEAR) (abstract). *Bull. Am. Astron. Soc., 33,* 1120–1121.

Ostro S. J., Campbell D. B., Chandler J. F., Hine A. A., Hudson R. S., Rosema K. D., and Shapiro I. I. (1991a) Asteroid 1986 DA: Radar evidence for a metallic composition. *Science, 252,* 1399–1404.

Ostro S. J., Campbell D. B., Chandler J. F., Shapiro I. I., Hine A. A., Velez R., Jurgens R. F., Rosema K. D., Winkler R., and Yeomans D. K. (1991b) Asteroid radar astrometry. *Astron. J., 102,* 1490–1502.

Ostro S. J. and 13 colleagues (1995) Radar images of asteroid 4179 Toutatis. *Science, 270,* 80–83.

Ostro S. J. and 15 colleagues (1996) Near-Earth object radar astronomy at Goldstone in 1996 (abstract). *Bull. Am. Astron. Soc., 28,* 1105.

Ostro S. J., Nolan M. C., Margot J.-L., Magri C., Harris A. W., and Giorgini J. D. (2001) Radar observations of asteroid 288 Glauke. *Icarus, 152,* 201–204.

Ostro S. J., Hudson R. S., Benner L. A. M., Giorgini J. D., Magri C., Margot J.-L., and Nolan M. C. (2002) Asteroid radar astronomy. In *Asteroids III* (W. F. Bottke Jr. et al., eds.), pp. 151–168. Univ. of Arizona, Tucson.

Reach W. T., Sykes M. V., Lien D., and Davies J. K. (2000) The formation of Encke meteoroids and dust trail. *Icarus, 148,* 80–94.

Richter K. and Keller H. U. (1995) On the stability of dust particle orbits around comet nuclei. *Icarus, 114,* 355–371.

Samarasinha N. H. and A'Hearn M. F. (1991) Observational and dynamical constraints on the rotation of comet P/Halley. *Icarus, 93,* 194–225.

Samarasinha N. H., Mueller B. E. A., Belton M. J. S., and Jorda L. (2004) Rotation of cometary nuclei. In *Comets II* (M. C. Festou et al., eds.), this volume. Univ. of Arizona, Tucson.

Sarmecanic J., Fomenkova M., Jones B., and Lavezzi T. (1997) Constraints on the nucleus and dust properties from mid-infrared imaging of Comet Hyakutake. *Astrophys. J. Lett., 483,* L69–L72.

Schleicher D. G. and Osip D. J. (2002) Long- and short-term photometric behavior of Comet Hyakutake (1996 B2). *Icarus, 159,* 210–233.

Schleicher D. G., Millis R. L., Osip D. J., and Lederer S. M. (1998) Activity and the rotation period of comet Hyakutake (1996 B2). *Icarus, 131,* 233–244.

Sekanina Z. (1988) Nucleus of Comet IRAS-Araki-Alcock (1983 VII). *Astron. J., 95,* 1876–1894.

Sekanina Z., Jehin E., Boehnhardt H., Bonfils X., and Schuetz O. (2002) Recurring outbursts and nuclear fragmentation of comet C/2001 A2 (LINEAR). *Astrophys. J., 572,* 679–684.

Skorov Y. V. and Rickman H. (1999) Gas flow and dust acceleration in a cometary Knudsen layer. *Planet. Space Sci., 47,* 935–949.

Veverka J. and 32 colleagues (2000) NEAR at Eros: Imaging and spectral results. *Science, 289,* 2088–2097.

Wallis M. K. (1982) Dusty gas dynamics in real comets. In *Comets* (L. L. Wilkening, ed.), pp. 357–369. Univ. of Arizona, Tucson.

Weissman P. R., Asphaug E., and Lowry S. C. (2004) Structure and density of cometary nuclei. In *Comets II* (M. C. Festou et al., eds.), this volume. Univ. of Arizona, Tucson.

Whipple F. L. (1951) A comet model: II. Physical relations for comets and meteors. *Astrophys. J., 113,* 464–474.

Wilkening L. L., ed. (1982) *Comets.* Univ. of Arizona, Tucson. 766 pp.

Yeomans D. K., Chodas P. W., Keesey M. S., Ostro S. J., Chandler J. F., and Shapiro I. I. (1992) Asteroid and comet orbits using radar data. *Astron. J., 103,* 303–317.

Rotation of Cometary Nuclei

Nalin H. Samarasinha
National Optical Astronomy Observatory

Béatrice E. A. Mueller
National Optical Astronomy Observatory

Michael J. S. Belton
Belton Space Exploration Initiatives, LLC

Laurent Jorda
Laboratoire d'Astrophysique de Marseille

The current understanding of cometary rotation is reviewed from both theoretical and observational perspectives. Rigid-body dynamics for principal axis and non-principal-axis rotators are described in terms of an observer's point of view. Mechanisms for spin-state changes, corresponding timescales, and spin evolution due to outgassing torques are discussed. Different observational techniques and their pros and cons are presented together with the current status of cometary spin parameters for a variety of comets. The importance of rotation as an effective probe of the interior of the nucleus is highlighted. Finally, suggestions for future research aimed at presently unresolved problems are made.

1. INTRODUCTION

Since the publication of the first *Comets* volume (*Wilkening,* 1982), our understanding of the rotation of cometary nuclei has evolved significantly, first due to research kindled by the apparently contradictory observations of Comet 1P/Halley, and more recently due to numerical modeling of cometary spin complemented by a slowly but steadily increasing database on cometary spin parameters. Subsequent review papers by *Belton* (1991), *Jewitt* (1999), *Jorda and Licandro* (2003), and *Jorda and Gutiérrez* (2002) highlight many of these advances. In this chapter, we discuss our current understanding of cometary rotation and the challenges we face in the near future.

Knowledge of the correct rotational state of a cometary nucleus is essential for the accurate interpretation of observations of the coma and for the determination of nuclear activity and its distribution on the surface. The spin state, orbital motion, and activity of a comet are linked to each other. Accurate knowledge of each of these aspects is therefore required in order to properly understand the others, as well as to determine how they will evolve. As we will elaborate later, ensemble properties of spin parameters — even the spin rates alone — can be effectively used to understand the gross internal structure of cometary nuclei. This has direct implications for understanding the formation of comets in the solar nebula as well as for devising effective mitigation strategies for threats posed by cometary nuclei among the near-Earth-object (NEO) population. In addition, *a priori* knowledge of the spin state is necessary for effective planning and maximization of the science return from

space missions to comets; for example, it allows the mission planners to assess the orientation of the nucleus during a flyby.

In the next section, we will discuss basic dynamical aspects of cometary rotation, while section 3 deals with observational techniques and the current status of cometary spin parameters. Section 4 addresses interpretations of observations and some of the current challenges. The final section discusses suggestions for future research.

2. ROTATIONAL DYNAMICS

2.1. Rigid-Body Dynamics

The prediction from the icy conglomerate model of the nucleus (*Whipple,* 1950) and the subsequent spacecraft images of Comets 1P/Halley (e.g., *Keller et al.,* 1986; *Sagdeev et al.,* 1986) and 19P/Borrelly (e.g., *Soderblom et al.,* 2002) are consistent with a single solid body representing the nucleus. Therefore, in order to explain the rotational dynamics of the nucleus, to the first approximation, we consider it as a rigid body. Chapters in this book on splitting events (*Boehnhardt,* 2004), nuclear density estimates (*Weissman et al.,* 2004), and formation scenarios (*Weidenschilling,* 2004) are suggestive of a weak subsurface structure made up of individual cometesimals, and the question of any effects of non-rigid-body behavior are still on the table.

We use the terms "rotational state" and "spin state" interchangeably, and they are meant to represent the entire rotational state. Similarly, "rotational parameters" and "spin parameters" are used interchangeably. However, the term

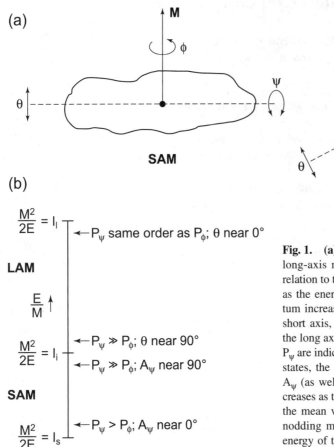

Fig. 1. **(a)** Component rotations for short-axis modes (SAM) and long-axis modes (LAM). Component rotations are depicted in relation to the long axis. **(b)** Characteristics of different spin states as the energy of rotation for a given rotational angular momentum increases from the least-energetic state (rotation around the short axis, $\frac{M^2}{2E} = I_s$) to the most energetic state (rotation around the long axis, $\frac{M^2}{2E} = I_l$). Behavior of the component periods P_ϕ and P_ψ are indicated for spin states near principal-axis states. For SAM states, the amplitude of the oscillatory motion of the long axis, A_ψ (as well as the nodding amplitude of the long axis, A_θ), increases as the spin state becomes more energetic. For LAM states, the mean value of the angle θ (as well as the amplitude of the nodding motion of the long axis, A_θ) decreases as the kinetic energy of the rotational state increases.

"spin vector" specifically refers to the instantaneous angular velocity vector.

The most stable rotational state of a rigid body is defined by the least-energetic state for a given rotational angular momentum (this does not mean that the rotational angular momentum is fixed, but for each given rotational angular momentum of the rigid body, there is a corresponding lowest-energy spin state). This lowest-energy spin state is represented by a simple rotation around the short principal axis of the nucleus occurring at a constant angular velocity. (Note that in this chapter, unless specified otherwise, long, intermediate, and short axes, denoted by l, i, and s respectively, refer to the mutually orthogonal principal axes of the nucleus as determined by the inertia ellipsoid. Depending on the shape and the internal density distribution, these axes may have offsets from the physical axes defined by an ellipsoidal fit to the physical shape.) For this spin state, the rotational angular momentum, M, and the rotational kinetic energy, E, are given by $\frac{M^2}{2E} = I_s$, where I_s is the moment of inertia around the short axis. Since the nucleus is only rotating around the short axis, this is called a principal-axis (PA) spin state. Other PA spin states include the dynamically unstable rotation around the intermediate axis (e.g., *Landau and Lifshitz*, 1976) corresponding to $\frac{M^2}{2E} = I_i$ and that around the long axis, which characterizes the most energetic rotational state at $\frac{M^2}{2E} = I_l$. Moments of in-

ertia around the three principal axes satisfy the condition $I_s \geq I_i \geq I_l$. Spin states with kinetic energies in between are characterized by two independent periods. These spin states are known as non-principal-axis (NPA) states (also called complex rotational states, or tumbling motion, with the latter term primarily used by the asteroid community). It should be stressed that, unlike for PA states, the spin vector and **M** are not parallel to each other for NPA states. A nucleus having a spin state other than the PA rotation around the short axis is in an excited rotational state.

Figure 1 shows the basic characteristics of different rigid body rotational states. The short-axis-mode (SAM) and long-axis-mode (LAM) states (*Julian*, 1987) differ from each other depending on whether the short or the long axis "encircles" the rotational angular momentum vector, **M**, which is fixed in the inertial frame. The SAM states are less energetic than the LAM states. The component periods can be defined in terms of Euler angles θ, ϕ, and ψ (Fig. 2). Most cometary nuclei are elongated (i.e., closer to prolates than oblates), as implied by large lightcurve amplitudes and spacecraft images. From an observational point of view, component rotations defined in terms of the long axis are easier to discern than those defined with respect to the short axis. Therefore, following *Belton* (1991), *Belton et al.* (1991), and *Samarasinha and A'Hearn* (1991), we adopt the system of Euler angles where θ defines the angle between **M** and the long axis, ψ defines the rotation around the long axis itself, and φ defines the precession of the long axis around **M** as depicted in Fig. 2. [Many textbooks (e.g., *Landau and Lifshitz,* 1976) use a different system of Euler angles appropriate for flattened objects such as Earth, as opposed to elongated objects; see *Jorda and Licandro* (2003) for a comparison between the two systems.] In Fig. 1, P_θ refers to the period of the nodding/nutation motion of the long axis, P_ψ to the period of oscillation (in the case of SAM) or rotation (in the case of LAM) of the long axis around itself, and P_ϕ to the mean period of precession of the long axis around **M** (for an asymmetric rotator, the rate of change of φ is periodic with $P_\psi/2$ and therefore we consider the time-averaged mean value for P_ϕ). The period P_θ is equal to P_ψ for SAM or exactly $P_\psi/2$ for LAM, leaving only two independent periods P_ϕ and P_ψ. In general, $P_\phi \neq P_\psi$ and therefore for a random NPA spin state, even approximately the same orientations of the rotator in the inertial frame are rare.

Principal-axis spin states can be uniquely defined with three independent parameters and one initial condition: the period of rotation; the direction of the rotational angular momentum vector, which requires two parameters; and the direction for the reference longitude of the nucleus at a given time. For NPA spin states, six independent parameters and two initial conditions are necessary: e.g., P_ϕ; P_ψ; ratio of moments of inertia I_l/I_s and I_i/I_s; direction of the rota-

tional angular momentum vector, which requires two parameters; and the reference values for the Euler angles φ and ψ at a given time.

The rate of change of **M** of the rigid body in an inertial frame (with the origin at the center of mass) is given by

$$\left(\frac{d\mathbf{M}}{dt}\right)_{inertial} = \mathbf{N} \tag{1}$$

where **N** is the external torque on the body. Since the rate of change of **M** in the inertial and in the body frames are related by

$$\left(\frac{d\mathbf{M}}{dt}\right)_{inertial} = \left(\frac{d\mathbf{M}}{dt}\right)_{body} + \mathbf{\Omega} \times \mathbf{M} \tag{2}$$

$$\left(\frac{d\mathbf{M}}{dt}\right)_{body} + \mathbf{\Omega} \times \mathbf{M} = \mathbf{N} \tag{3}$$

where **Ω** is the angular velocity. By dropping the "body" subscript and expressing **M** in terms of moment of inertia, we have

$$\frac{d(\mathbf{I\Omega})}{dt} + \mathbf{\Omega} \times (\mathbf{I\Omega}) = \mathbf{N} \tag{4}$$

where **I** is the moment-of-inertia tensor. Since the moment of inertia is constant for a rigid body, we derive Euler's equations of motion (e.g., *Landau and Lifshitz,* 1976)

$$I_l\dot{\Omega}_l = (I_i - I_s)\Omega_i\Omega_s + N_l \tag{5}$$

$$I_i\dot{\Omega}_i = (I_s - I_l)\Omega_s\Omega_l + N_i \tag{6}$$

and

$$I_s\dot{\Omega}_s = (I_l - I_i)\Omega_l\Omega_i + N_s \tag{7}$$

The subscripts denote the components along the three principal axes. If α_{jk} represents the direction cosine between the body frame axis j and the inertial frame axis k, then the transformation between the two frames is governed by the following nine scalar equations (*Julian,* 1990)

$$\dot{\alpha}_{lk} = \Omega_s\alpha_{ik} - \Omega_i\alpha_{sk} \tag{8}$$

$$\dot{\alpha}_{ik} = \Omega_l\alpha_{sk} - \Omega_s\alpha_{lk} \tag{9}$$

and

$$\dot{\alpha}_{sk} = \Omega_i\alpha_{lk} - \Omega_l\alpha_{ik} \tag{10}$$

where k = X, Y, and Z. The rotational motion of the nucleus

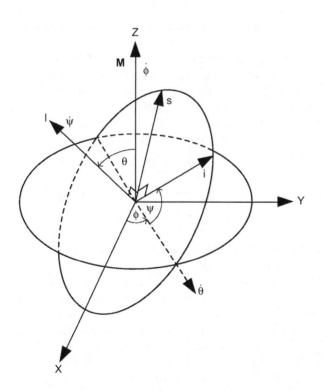

Fig. 2. Euler angles θ, φ, and ψ. The axes X, Y, and Z form a righthanded orthogonal coordinate system in the inertial frame. Principal axes l, i, and s respectively form the righthanded body frame coordinate system. The rotational angular momentum vector, **M**, coincides with the Z axis and is fixed in the inertial frame in the absence of external torques.

in the inertial frame can be followed by simultaneously solving equations (5)–(10). For a force-free motion, component torques are zero and one obtains an analytical solution in terms of Jacobian elliptic functions (*Landau and Lifshitz*, 1976). When there are nonzero torques, in general one cannot derive an analytical solution and should resort to numerical integration of Euler's equations (also see section 2.3). For further details on rigid-body rotation in a cometary context, the reader is referred to *Samarasinha and A'Hearn* (1991), *Belton* (1991), and *Jorda and Licandro* (2003).

2.2. Mechanisms for Changing the Spin States

The spin state of a cometary nucleus will evolve with time due to various reasons. Fortunately, for many comets, the timescales for such changes are sufficiently large (i.e., many orbital periods) and little or no measurable changes in the rotational parameters occur from one orbit to another (*Schleicher and Bus*, 1991; *Mueller and Ferrin*, 1996).

Protocomets have undergone multiple collisions while being scattered to the Oort cloud from the giant-planet region (*Stern and Weissman*, 2001) or while in the Kuiper belt (*Stern*, 1995; *Davis and Farinella*, 1997). In addition, many present-day Kuiper belt comets could be breakup fragments due to collisional events (*Farinella and Davis*, 1996). Therefore, it is unlikely that the spin states of most dynamically new comets from the Oort cloud and especially those from the Kuiper belt are of primordial origin.

In general, a sufficiently large collision or any of the other mechanisms described below could result in an excited spin state. The stresses and strains associated with the NPA rotation of an excited nucleus would result in a loss of mechanical energy. Consequently, in the absence of any further excitation events, the spin state will damp toward the stable, least-energetic state (e.g., *Burns and Safronov*, 1973; *Efroimsky*, 2001, and references therein). Therefore, whether the current spin state of a cometary nucleus is excited or not will depend on the timescale for the relevant excitation mechanism and the damping timescale, as well as on when and how long the excitation mechanism was active. For extensive discussions on timescales, see *Jewitt* (1999) and *Jorda and Licandro* (2003).

2.2.1. Torques due to outgassing. Outgassing of volatiles from the nucleus causes a reaction force on the nucleus. In addition to changing the orbital motion of a comet, it generates a net torque on the nucleus resulting in changes of the nuclear spin state (*Whipple*, 1950, 1982). This is the primary mechanism for altering the spin states of comets. The outgassing torque on a nucleus alters a PA spin state of the lowest energy. It produces a change in the spin period and in the direction of the angular momentum vector with a timescale that can be as short as a single orbit (e.g., *Samarasinha et al.*, 1986; *Jewitt*, 1992). It can also trigger excited spin states with a timescale that is not fully understood at the moment, but could be on the order of several orbits for small, very active comets (*Jorda and Gutiérrez*, 2002).

In order to derive the reaction torque, the reaction force needs to be evaluated. The reaction force, $d\mathbf{F}_{i,dS}$, due to the gas species, i, sublimating from an elemental surface area, dS, can be expressed as

$$d\mathbf{F}_{i,dS} = -\alpha m_i Z_i \mathbf{v} dS \qquad (11)$$

where α is a momentum transfer efficiency, m_i is the molecular weight of the gas species, Z_i is the sublimation rate for the gas species in molecules per unit surface area, and $\mathbf{v}$ is the outflow velocity of the gas species at the surface. The net reaction force $\mathbf{F}$ can be determined by integrating over the entire nuclear surface and then summing up for all sublimating gas species, i.e.,

$$\mathbf{F} = -\sum_{\text{gas species}} \int_{\text{surface}} \alpha m_i Z_i \mathbf{v} dS \qquad (12)$$

$\mathbf{F}$ can be evaluated using simplified assumptions: e.g., water is the dominant gas species, outgassing occurs primarily from the sunlit surface, and the outgassing velocity is normal to the surface. It should be stressed that the momentum transfer efficiency, which is on the order of one, depends on many factors including the degree of collimation of outgassing, the relationship used to compute the outflow velocity $\mathbf{v}$ in equation (11), and the amount of back pressure on the nucleus — which in turn will be based on the near-nucleus coma environment and the gas and dust production rates, among others (*Skorov and Rickman*, 1999, and references therein; *Rodionov et al.*, 2002). This nongravitational force alters the orbital motion of the comet, making long-term orbital predictions a difficult task. For further details on the nongravitational force and its effect on the orbital motion, see *Yeomans et al.* (2004).

The net torque, $\mathbf{N}$, on the nucleus due to outgassing forces can be expressed as

$$\mathbf{N} = -\sum_{\text{gas species}} \int_{\text{surface}} \alpha m_i Z_i (\mathbf{r} \times \mathbf{v}) dS \qquad (13)$$

where $\mathbf{r}$ is the radial vector from the center of mass to the elemental surface area dS.

The component torques N_l, N_i, and N_s along the principal axes can be used to numerically solve Euler's equations of motion described in section 2.1. This outgassing torque may alter the spin state of the nucleus in two ways: an increase (or decrease) of $\frac{M^2}{2E}$ for the spin state and a change in $\mathbf{M}$. A change in $\mathbf{M}$ can manifest itself either as a change in its magnitude and/or its direction. (In the literature, a change in the direction of $\mathbf{M}$ is routinely called precession, which is a forced motion of $\mathbf{M}$ and therefore is entirely different from the free precession of the rigid-body motion described in section 2.1.) Specific model calculations were carried out by different investigators to assess the changes in the spin state due to outgassing torques (e.g., *Wilhelm*,

1987; *Peale and Lissauer,* 1989; *Julian,* 1990; *Samarasinha and Belton,* 1995; *Szegö et al.,* 2001; *Jorda and Gutiérrez,* 2002). The reader's attention is also drawn to section 2.3 where we discuss long-term spin evolution.

2.2.2. Changes to the moment of inertia due to mass loss and splitting events. Another mechanism for altering spin states is via changes to the moment of inertia of the nucleus (cf. equation (4)). There are two primary mechanisms for changing the moment of inertia tensor: (1) mass loss of volatiles and dust associated with regular nuclear outgassing and (2) splitting events of the nucleus (*Boehnhardt,* 2004). For cometary nuclei, the timescale for changing the spin state due to sublimation-caused mass loss, $\tau_{massloss}$, is much larger than that due to outgassing torques, τ_{torque}. This justifies the adoption of Euler's equations of motion for monitoring the spin-state evolution rather than the more general Liouville's equation, which allows for changes in the moment of inertia. In general, $\tau_{massloss}$ is at least a few tens of orbits for most comets (*Jewitt,* 1999; *Jorda and Licandro,* 2003).

The timescale for changes in the spin state associated with splitting events, $\tau_{splitting}$, is uncertain because such events themselves are stochastic in nature. *Chen and Jewitt* (1994) estimated a lower limit to the splitting timescale of 100 yr per comet. Assuming few splitting events are required to appreciably alter the moment-of-inertia tensor and hence produce an observationally detectable change in the spin state (cf. *Watanabe,* 1992a), the value of $\tau_{splitting}$ for most comets is likely to be larger than 100 yr.

2.2.3. Collisions with another object. Another stochastic mechanism that is capable of altering the spin states is collision of the cometary nucleus with another solar system object of sufficient momentum — typically an asteroid (e.g., *Sekanina,* 1987a; *Samarasinha and A'Hearn,* 1991). Comets that have low orbital inclinations to the ecliptic (e.g., Jupiter-family comets) are more likely than high-inclination comets to undergo frequent collisions with another solar system object. In other words, the timescale, $\tau_{collision}$, for spin-state change depends strongly on the orbit of the comet. Even for such low-inclination comets, $\tau_{collision}$ is expected to be larger than that for the mechanisms described earlier (cf. *Farinella et al.,* 1998; also D. Durda, personal communication, 2002).

2.2.4. Tidal torques. Tidal torques primarily due to the Sun or Jupiter could also alter cometary spin states. The differential gravitational potential experienced by different parts of the nucleus causes a net torque on the nucleus. It could arise either due to the tidal deformation of the nucleus or due to the shape of the nucleus. To correctly evaluate the timescale for spin-state change due to tidal torques, τ_{tidal}, the torques need to be integrated over many orbits properly accounting for rotational and orbital cancellations. This in effect would prolong the timescale. However, the spin state could be significantly altered for a sufficiently close encounter with a planet, i.e., within few planetary radii (*Scheeres et al.,* 2000). Such close encounters are more likely for Jupiter-family comets, but are still rare.

2.2.5. Other mechanisms. There are a few other mechanisms proposed in the literature for altering spin states. These include shrinkage of a porous cometary nucleus (*Watanabe,* 1992b) as a spin-up mechanism and angular momentum drain due to preferential escape of particles from equatorial regions (*Wallis,* 1984) as a mechanism for nucleus spin-down. The ice-skater model by Watanabe was proposed to explain the observed rapid spin-up of Comet Levy (C/1990 K1).

In addition, the Yarkovsky effect (or specifically, the so-called YORP effect due to the resultant torque) could alter spin states (*Rubincam,* 2000). Changes of the angular momentum are caused by the thermal lag between absorption of sunlight and its reradiation as thermal radiation for irregularly shaped objects. The timescale, $\tau_{Yarkovsky}$, for spin-state changes due to this effect is much larger than for other mechanisms (*Jorda and Gutiérrez,* 2002). However, $\tau_{Yarkovsky}$ becomes smaller for subkilometer nuclei since the Yarkovsky force (and consequently $\tau_{Yarkovsky}$ as well) has a radius-squared dependence (*Rubincam,* 2000).

2.3. Long-Term Evolution of Spin States

While any of the above mechanisms can alter the spin state of a nucleus, as discussed earlier, many are stochastic in nature or have large timescales. Therefore, the outgassing torque is the primary mechanism of spin-state alteration (also Fig. 2 of *Jewitt,* 1999) for which the monitoring of the long-term spin evolution is feasible. After the initial prediction by *Whipple* (1950) that outgassing can alter cometary spin, *Whipple and Sekanina* (1979) presented a model where the spin evolution caused by outgassing can be evaluated. Unfortunately, that model adopts an oblate nucleus (whereas observations suggest elongated shapes) and only the forced-precession of the nucleus is considered (i.e., no excitation of the nucleus is allowed), therefore limiting its applicability. In the context of understanding Comet 1P/Halley's spin state, *Wilhelm* (1987), *Julian* (1988, 1990), and *Peale and Lissauer* (1989) carried out numerical monitoring of the spin state in order to study changes due to outgassing torques. All of them found that outgassing torques can cause changes in the spin state in a single orbit. Numerical studies by *Samarasinha and Belton* (1995) covering multiple orbits demonstrate that Halley-like nuclei can be excited due to the multiorbit cumulative effects of the outgassing torques. Small, highly active nuclei with localized outgassing are prime candidates for excitation. For example, Comet 46P/Wirtanen may undergo observable spin-state changes during a single orbit (*Samarasinha et al.,* 1996; *Jorda and Licandro,* 2003). Monitoring of such objects provide a golden opportunity to accurately assess the nongravitational forces and torques due to outgassing.

Many early studies used prolate or near-prolate shapes to investigate the spin evolution due to outgassing torques. Recent work by *Jorda and Gutiérrez* (2002) (also *Gutiérrez et al.,* 2002) shows that nuclei with irregular shapes and three unequal moments of inertia are more difficult to excite than prolates. Numerical studies by N. H. Samarasinha (unpublished data, 1995) using triaxial shapes show the same tendency. It should be stressed that the process of excitation is not forbidden for complex-shaped nuclei, but only that it is not as efficient as that for prolates. In addition, as one may expect, fast rotators are much more difficult to excite than slow rotators. We note that sometimes in the literature, the timescale for spin-state changes due to outgassing torques, τ_{torque}, and the excitation timescale, $\tau_{excitation}$, are used interchangeably. In light of the above results, τ_{torque} should be considered only as a lower limit to $\tau_{excitation}$. Recent numerical calculations for Comet 46P/Wirtanen by *Jorda and Gutiérrez* (2002), which still need to be confirmed and generalized to other comets, suggest that $\tau_{excitation}$ for objects with unequal moments of inertia could be at least one order of magnitude larger than that for a prolate body.

Multiorbit, long-term numerical monitoring of spin states by *Samarasinha* (1997, 2003) indicate that in the majority of the cases (especially when a dominant active region is present), the rotational angular momentum vector of the spin state evolves toward the orbital direction of the peak outgassing or that directly opposite to it. This occurs since such a configuration will present the least net torque in the inertial frame when averaged over an orbit. Analytical treatment of the problem by *Neishtadt et al.* (2002) confirmed this as a main evolutionary path whereby they also explore other paths. If indeed this evolutionary scenario is accurate, one may find many evolved comets with their rotational angular momentum vectors directed toward or near the orbital plane. Unfortunately, the current database is not sufficiently large enough to make a robust assessment.

3. OBSERVATIONAL TECHNIQUES AND DATA

To derive the rotational state of a cometary nucleus, the main parameters to be determined are the rotational period(s) and the direction of **M**. The axial ratio(s), especially important for the NPA rotators, are a byproduct of the relevant observations, namely lightcurve observations. For example, under the assumption of Lambertian scattering, for a PA rotator, the lightcurve amplitude of a bare nucleus will provide a lower limit to the ratio between long and intermediate physical axes (see also *Lamy et al.*, 2004). Below is a summary of the observational techniques; the reader is also referred to *Belton* (1991) for additional details.

3.1. Rotational Periods

For a PA rotator, one of the fundamental parameters of rotation is given by the sidereal rotational period (i.e., the time required to make a complete cycle of rotation around the fixed axis of rotation as seen by an inertial observer — in this case distant stars). The sidereal period is independent of the Sun-comet-Earth geometry or any changes in it. For a NPA spin state, since not all the component rotations occur with respect to axes fixed in an inertial frame, the

term "sidereal" may not be the most suitable. However, the analogous periods of rotation (i.e., independent of the Sun-comet-Earth geometry and any changes in it) can be defined in terms of the periods associated with the Euler angles (see section 2.1).

There are two primary observational techniques to derive rotational periods: (1) rotational lightcurve observations consisting of a time series of photometric variations and (2) periodic variability of the coma structure when the nucleus is active. In general, the former provides more precise periods, and this is especially true in the case of NPA rotators. The periods derived directly from lightcurve observations correspond to "synodic" periods rather than to "sidereal" periods. The rotational lightcurves themselves can be categorized into two categories: lightcurves of bare nuclei, and lightcurves that represent changes in nuclear activity. The "synodic" periods from the latter category of lightcurves depend on the changes in the Sun-comet orientation during the observing window and therefore the term "synodic" has the classical definition. This period is also known as the "solar day." On the other hand, the "synodic" periods from the lightcurves of bare nuclei depend additionally on the changes in the Earth-comet orientation during the observing window. Therefore, in this case, the term "synodic" has the same meaning as that used by the asteroid community (in contrast to its classical definition). The changes in the Sun-comet-Earth geometry can also affect the period determinations based on the variability of coma structures. In principle, model fittings (including knowledge of the rotational angular momentum vector) are required for the derivation of "sidereal" periods. In this chapter, unless specified otherwise, rotational periods based on observations refer to "synodic" periods.

Depending on the spin parameters and/or the Sun-comet-Earth geometry, cometary activity can cause a component period of a NPA spin state to become masked [cf. 1P/Halley (*Belton et al.*, 1991)]. The reader should be alert to this possibility.

3.1.1. Rotational lightcurves. The ideal rotational lightcurve requires the nucleus to be entirely inactive, but a scenario where the flux within the photometric aperture is dominated by the scattered solar light from the nucleus (rather than from the coma) can still provide reliable results. Rotational lightcurves of the nucleus are therefore observed at large heliocentric distances when the comet is relatively dim. This makes lightcurve observations of bare nuclei challenging, requiring relatively large telescopes and a significant amount of observing time.

If the spatial resolution is adequate, even when the nucleus is active, the coma contamination can be effectively subtracted to derive the flux contribution from the nucleus. For successful coma subtraction, the spatial resolution should be such that the flux in the central pixel is dominated by the scattered solar light from the nucleus. This technique has been routinely applied for Hubble Space Telescope (HST) observations by Lamy and colleagues to estimate the nuclear sizes of comets (see *Lamy et al.*, 2004).

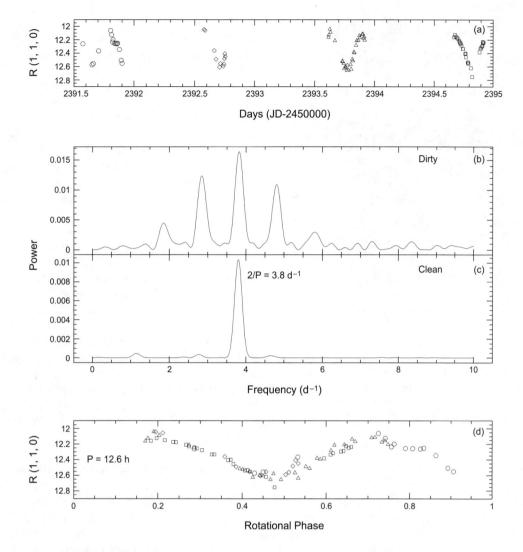

Fig. 3.   **(a)** Lightcurve data for Comet 28P/Neujmin 1 in the R filter as a function of time. The magnitudes are normalized for helio-centric and geocentric distances of 1 AU each and a phase angle of 0°. Time is expressed in terms of Julian days (JD). **(b)** Fourier power spectrum corresponding to the data. **(c)** Power spectrum, after application of the clean algorithm. **(d)** Rotationally phased lightcurve data for the rotation period of 12.6 h.

Such multiple observations of the same comet in a time series will yield estimates of rotational periods (e.g., *Lamy et al.,* 1998a).

As mentioned earlier, if the nuclear activity is highly modulated by the rotation (e.g., the turning on and off of jets in response to insolation), even a lightcurve constructed by a series of photometric measurements, where the flux is dominated by the variable coma, can be effectively used to probe the nuclear rotation. For example, lightcurves in dust and in emission species for Comet 1P/Halley were used to derive rotational signatures of the nucleus (*Millis and Schleicher,* 1986; *Schleicher et al.,* 1990).

Extraction of periods from the lightcurves can be achieved through different techniques. Among these are Fourier analysis (e.g., *Deeming,* 1975; *Belton et al.,* 1991), phase dispersion minimization (e.g., *Stellingwerf,* 1978; *Millis and Schleicher,* 1986), string length minimization (e.g.,

Dworetsky, 1983; *Fernández et al.,* 2000), least-squares fit of a sine curve (e.g., *Lamy et al.,* 1998a), and wavelet analysis (*Foster,* 1996). The Fourier technique is often coupled with a subsequent application of a clean algorithm (e.g., *Roberts et al.,* 1987) to "clean" the Fourier spectrum from aliases and spurious periods introduced by uneven sampling and observational gaps. During this process, most of the harmonics get cleaned out too. In general, the clean algorithm would yield accurate periods, but one should be alert to the possibility of it occasionally cleaning out the correct period(s) (cf. *Foster,* 1995).

Figure 3 shows a rotational lightcurve of the nucleus of Comet 28P/Neujmin 1 taken on April 27–30, 2002 (*Mueller et al.,* 2002a). Application of WindowClean (*Belton and Gandhi,* 1988), a clean algorithm, to the output from the Fourier analysis (dirty spectrum), identifies the dominant signature corresponding to the lightcurve data. In interpret-

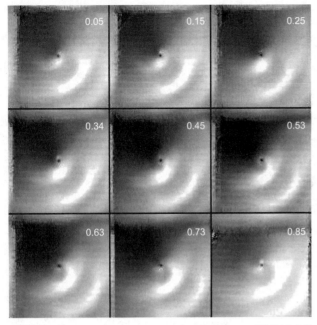

Fig. 4. These 11-μm images of Comet Hale-Bopp (C/1995 O1) cover an entire rotational cycle (adapted from *Lisse et al.,* 1999). The images are enhanced and the brightness scale is nonlinear. The nucleus is at the center of each panel. The rotational phase (in white) increases from left to right and from top to bottom. Notice the outward movement of coma features during the rotation cycle.

ing the dominant frequency, f, in the cleaned spectrum, it is assumed that the rotational signature is due to the shape rather than due to an albedo feature on the surface. Hence the corresponding synodic rotational period, P, and the frequency f are related by

$$P = \frac{2}{f} \qquad (14)$$

When interpreting lightcurve signatures, one has to be alert to the possibility of spurious signatures introduced due to temporal seeing variations (*Licandro et al.,* 2000).

3.1.2. Repetitive coma structures. Repetitive structures in cometary comae can also be used to determine rotation periods. Figure 4 shows the repetitive coma morphology for Comet Hale-Bopp (C/1995 O1) while it was near perihelion. Care should be taken to confirm that the repetitive structure is indeed due to the rotation and corresponds to successive rotation cycles. Temporal monitoring of the evolution of features over a rotation cycle [e.g., similar to the 11-μm images of Comet Hale-Bopp (C/1995 O1) from *Lisse et al.* (1999), which cover an entire rotational cycle; see Fig. 4] and a consistency check for the outflow velocity derived using two adjacent repetitive features are two such checks. Multiple images taken at different times showing the outward movement of the same repetitive feature may yield the rotation period as well as an estimate for the expansion

rate of the feature. The features themselves may require image enhancement, and because of the unintended artifacts introduced, some enhancement techniques are preferred over others for this purpose (*Larson and Slaughter,* 1991; *Schleicher and Farnham,* 2004). It should be pointed out that this technique could be considered as a more reliable modification of the Halo (also known as the Zero Date) method (*Whipple,* 1978; *Sekanina,* 1981a). The latter has a tendency to produce spurious results (*Whipple,* 1982; *A'Hearn,* 1988) since it uses the coma (or feature) diameter and an assumed expansion rate in the calculations.

3.1.3. Other techniques. Radar observations can yield estimates for rotation periods based on the Doppler bandwidth. However, the values of the nuclear radius and the angle between the instantaneous spin vector and the line of sight must be known to derive a unique value (see *Harmon et al.,* 2004). In addition, similar to asteroid 4179 Toutatis, sufficient radar image coverage in the observing geometry and time domains could yield the solution for the entire spin state (*Hudson and Ostro,* 1995). However, this has yet to be carried out in the case of a comet. Due to the Δ^{-4} dependence for the radar signal, where Δ is the geocentric distance, only comets that will have close approaches to Earth can be probed via radar techniques.

In principle, the curvature of jet features and their evolution could also be used to determine the rotation period as well as the spin axis (e.g., *Larson and Minton,* 1972; *Sekanina and Larson,* 1986, and references therein). But due to the multiparameter nature of the problem and the many unknowns associated with this quasinumerical approach, the results are often not accurate. Despite this unreliability, the solutions based on the curvature of jets could serve as useful but crude estimates in some cases. At this point, it should be emphasized that determinations based on jet curvatures assume outgassing is confined to a localized active region on the surface of the nucleus. There is a competing school of thought that argues that coma structures are not due to localized outgassing but are caused by hydrodynamical effects due to topographical variations on a uniformly outgassing surface (*Crifo et al.,* 2004, and references therein). If topography rather than localized outgassing is indeed primarily responsible for coma structures, then adoption of jet curvature as a tool to determine rotation periods needs to be reassessed.

3.2. Observational Manifestation of Non-Principal-Axis Rotational Periods

Clearly, the PA spin states will show a single period (and perhaps its harmonics) in the lightcurve, which can be readily used to deduce the rotation period of the nucleus using equation (14). On the other hand, as discussed in section 2.1, the NPA states have two independent periods, P_ϕ and P_ψ. How exactly do these periods manifest themselves in the lightcurve? In the following discussion, what one directly derives from the periodic signatures in the lightcurves correspond to "synodic" periods. However, if

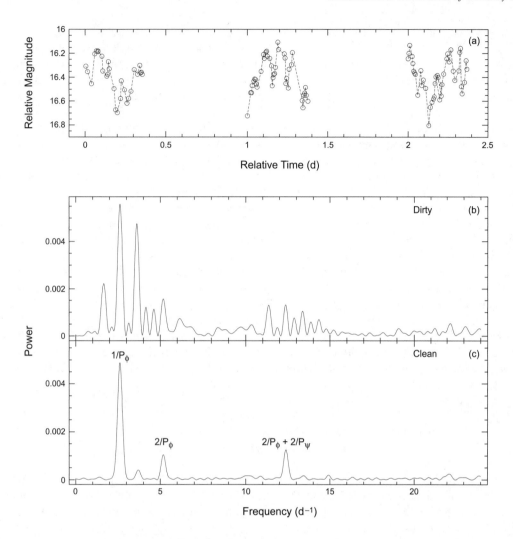

Fig. 5. **(a)** A simulated bare-nucleus lightcurve for a 5.6 × 3.1 × 2.5-km ellipsoidal NPA rotator. P_ϕ and P_ψ are 9.168 h and 6.754 h respectively. A random 5% noise is added to the lightcurve data. Consecutive data points are connected only to guide the eye. **(b)** Fourier power spectrum. **(c)** Cleaned Fourier spectrum. The primary signatures are present at $1/P_\phi$, $2/P_\phi$, and $(2/P_\phi + 2/P_\psi)$. The very low signature between $1/P_\phi$ and $2/P_\phi$, which was not fully cleaned out, is a daily alias of $1/P_\phi$.

the changes in the Sun-comet-Earth geometries during the observing windows are sufficiently small, quick identifications with periods P_ϕ and P_ψ are possible.

Analysis of bare-nucleus model lightcurves with elongated but triaxial ellipsoidal shapes for a limited number of SAM and LAM states indicate that for most cases the major signatures of the Fourier spectrum are at $2/P_\phi$ and $(2/P_\phi + 2/P_\psi)$ and multiples of these periods (*Kaasalainen,* 2001; *Mueller et al.,* 2002b). [*Kaasalainen* (2001) used the same set of Euler angles as described above for LAM states, but he prefers the system of Euler angles used in *Landau and Lifshitz* (1976) for SAM states because of certain symmetry considerations.] Figure 5 shows a simulated lightcurve for a NPA rotator, as well as the dominant signatures present in the lightcurve. However, for specific scenarios, the situation is much more complex. For example, our current understanding based on limited exploration of the parameter space suggests that both rotation periods might not

be present in the lightcurve, or instead of $2/P_\phi$, $1/P_\phi$ might be present. This depends on (1) the orientation of the Earth (observer) with respect to the direction of **M**, (2) the orientation of the Earth with respect to the cone swept out by the long axis (due to the precessional motion in the case of a LAM state), (3) the nuclear shape, (4) observational time coverage, and (5) the quality of the data. In addition, the relative strengths of the major signatures also depend on similar factors. Clearly, a detailed investigation of the parameter space is warranted, and such efforts are currently underway.

If the nucleus is close to axial symmetry, for example, in the case of a near-prolate, a bare-nucleus lightcurve might not indicate any signature related to P_ψ. An observer may identify the lightcurve period with the rotation period of a PA rotator, whereas it really relates to the precession of the long axis. Repeated lightcurve observations of the same object at different observing geometries or coma morphologies may provide evidence that prior determinations are in fact in error.

3.3. Direction of the Rotational Angular Momentum

To specify the spin state of a comet, the direction of the rotational angular momentum must be known. In the case of a PA rotator, this means determining the spin-axis direction. Similar to the case of determining rotation period(s), the relevant techniques are primarily based on rotational lightcurves or coma morphology.

3.3.1. Rotational lightcurves. The brightness and amplitude of a rotational lightcurve depends on (1) the rotational phase, (2) the nuclear shape and size, (3) the aspect angle, and (4) the scattering effects (including the solar phase angle). In particular, the lightcurve amplitude strongly depends on the nuclear shape and the aspect angle. Using these properties, multi-epoch lightcurve observations can be used to derive spin vectors, and this is indeed the case for a number of asteroids. The magnitude-amplitude method (*Magnusson et al.,* 1989, and references therein) which uses both the magnitude and the amplitude, enables calculation of axial ratios and the spin pole. This process assumes that the variations in the lightcurve amplitude are primarily due to the viewing geometry. Since the cometary lightcurve observations could be contaminated with an unknown low-level coma, it introduces an additional complexity to the problem. Therefore, care should be taken in the interpretation of the results based on the magnitude-amplitude method. On the other hand, the epoch method (*Magnusson et al.,* 1989, and references therein) is less sensitive to this issue since it relies on a specific but accurately determined "feature" (or a phase) in the lightcurve. However, to our knowledge, there are no cometary spin-axis determinations based entirely on lightcurve observations, but we hope with the increasing multi-epoch observations this will soon be realized.

3.3.2. Coma morphology. In a modification of his earlier model (*Sekanina,* 1979), which aimed at explaining fan-shaped comae, *Sekanina* (1987b) proposed that such features seen in many comets are due to ejecta from a high-latitude active region. In particular, if the active region is in the sunlight over the entire rotation cycle, the fan is bounded by the boundary of the cone swept out due to the rotation of the nucleus. In many instances, due to projection effects, the fan can manifest itself as two bright jets at the boundary of the cone. Therefore, the bisector of the fan (when the active region is constantly in sunlight) would yield the projected spin-axis direction. Sekanina has applied his fan model to several comets, but there is no definite confirmation regarding the reliability of this technique (*Belton,* 1991). While this technique will work in some cases, such as in the case of Comet 19P/Borrelly (e.g., *Farnham and Cochran,* 2002; *Schleicher et al.,* 2003), there are counterexamples that require caution (e.g., *Sekanina and Boehnhardt,* 1999; *Samarasinha et al.,* 1999).

Deep Space 1 images of Comet 19P/Borrelly indicate the presence of a strong sunward jet essentially parallel to the spin axis (*Soderblom et al.,* 2002). *Farnham and Cochran* (2002) and *Schleicher et al.* (2003) monitored the position angle of this jet at different times using groundbased imaging. This enabled them to determine the direction of the spin axis by finding the common direction of the intersection for all the position angles. The solution (both sets of authors derive values within one degree of each other) is consistent with a restricted set of solutions obtained by *Samarasinha and Mueller* (2002) using the position angle of this polar jet and the nucleus lightcurve amplitude. It is not clear whether strong active regions near poles are a common occurrence among evolved comets. However, if multiple position-angle determinations would yield consistent results for the spin-axis direction, determinations of the spin axis based on position angles might prove to be a useful technique. Again, cross-checking results obtained from different techniques is advised. This technique is not capable of determining the sense of rotation. The curvatures of jet structures due to rotation are useful for that purpose.

Again, this discussion is based on the assumption that an active region rather than the topography of the nucleus is responsible for the coma structure. At this time, it is not clear what the implications are for the latter hypothesis (*Crifo et al.,* 2004) since detailed simulations are yet to be carried out.

3.4. Individual Comets

At present, the entire spin state of a cometary nucleus is known only in the case of three comets, 1P/Halley (*Belton et al.,* 1991; *Samarasinha and A'Hearn,* 1991), 19P/Borrelly (*Farnham and Cochran,* 2002; *Mueller and Samarasinha,* 2002; *Schleicher et al.,* 2003; *Boice et al.,* 2003), and Hyakutake (C/1996 B2) (*Schleicher and Woodney,* 2003). However, even in the case of 1P/Halley, some controversy exists (*Szegö,* 1995). For 19 other comets, reasonably good estimates of spin periodicities are available (Table 1). For some of these comets, information on the rotational angular momentum vector is available but sometimes the sense of rotation is not known. For other comets only periodicities are known. Comets 19P/Borrelly and 1P/Halley are unique in that they both were the subjects of successful space mission encounters. In both cases, different, but critical, imaging data for the determination of the spin state were acquired. Hyakutake (C/1996 B2) is unique because of its close perigee during its 1996 apparition and the concomitant rapid change and the wide range of viewing geometry. Hale-Bopp (C/1995 O1) was special because of its highly structured coma. In addition, its intrinsic brightness yielded an extended apparition that allowed a wide range of observing geometries. The fact that information on the spin state of most other comets is limited underscores the observational and interpretational difficulties that must be faced in spin determinations as discussed elsewhere in section 3.

3.4.1. Comets for which the spin state is approximately determined.

1P/Halley: The spin state of this comet has been the subject of many investigations with conflicting results. The

TABLE 1. Information on spin states of specific comets.

Comet	Spin Mode	P_ϕ (day)	P_ψ (day)	θ (deg)	P_{total} (day)	M (J2000) α (deg)	δ (deg)	Long Axis (J2000) α (deg)	δ (deg)	Epoch (JD-2440000)
Comets for which the spin state is approximately determined										
1P/Halley	Excited (LAM)	3.69*	7.1*	66	2.84*	7	−60	314	−7	6498.806
19P/Borrelly	Unexcited	1.08	‡	90	1.08	214	−6	300	−10	12175.438
C/Hyakutake (1996 B2)	Unexcited	0.2618*	‡	90	0.2618*	205	−1			
Comets for which partial knowledge of the spin state is available										
2P/Encke	Excited?	?	?	?	?	205†	3†			
10P/Tempel 2	Unexcited	0.372	‡	90	0.372	148†	55†			
109P/Swift-Tuttle	Unexcited	2.77*	‡	90	2.77*	128	−72			
I-A-A (C/1983 H1)	Unexcited	2.14*	‡	90	2.14*	256	−15			
Hale-Bopp (C/1995 O1)	Unexcited	0.4712	‡	90	0.4712	?	?			
Comets for which only periodicities associated with rotation are presently known										
6P/d'Arrest	?		0.30?							
9P/Tempel 1	?		1.75							
21P/Giacobini-Zinner	?		0.79							
22P/Kopff	?		0.54							
28P/Neujmin 1	Unexcited	0.53	‡	90	0.53					
29P/S-W 1	Excited?	0.58?	1.35?							
31P/S-W 2	?		0.242							
46P/Wirtanen	?		0.32?							
48P/Johnson	Unexcited	1.208	‡	90	1.208					
49P/Arend-Rigaux	Unexcited	0.561	‡	90	0.561					
95P/Chiron	Unexcited	0.2466	‡	90	0.2466					
96P/Machholz 1	?		0.266?							
107P/Wilson-Harrington	Unexcited	0.25	‡	90	0.25					
133P/Elst-Pizarro	Unexcited	0.1446	‡	90	0.1446					
143P/Kowal-Mrkos	Unexcited	0.72	‡	90	0.72					
Levy (C/1990 K1)	?		0.708							
Levy (C/1991 L3)	?		0.348							

*Sidereal period.

†Sense of M not known.

‡P_ψ approaches its unknown minimum value with zero A_ψ and A_θ as the energy of the spin state approaches the minimum (cf. section 2).

LAM spin state listed in Table 1 (*Belton et al.,* 1991) provides a satisfactory explanation of time variations in all major groundbased datasets as well as a viable interpretation of spacecraft encounter data; however, disagreements on the spin state still persists. In this model, which is also consistent with independent investigations by *Samarasinha and A'Hearn* (1991), the long axis of the nucleus precesses around the rotational angular momentum vector once every 3.69 d at an angle of 66°. At the same time, the long axis executes a rotation about itself once every 7.1 d. This fulfills the observational requirement that the aspect of the nucleus return to roughly the same position every 7.4 d as seen by the Sun (*Schleicher et al.,* 1990). Prior to the spacecraft encounters in 1986, the investigations of *Sekanina and Larson* (1986) and others suggested a nucleus in a PA spin state with a period near 2 d. Images from three distinct viewing geometries and timings during the encounters of the *Vega 1, Vega 2,* and *Giotto* spacecraft allow for several possible spin states depending on the interpretation of the images. One such interpretation, strongly advocated by the Vega investigators, yields a spin state with a dominant 2.2-

d period (*Sagdeev et al.,* 1989). However, other interpretations yield results that are related to a 7.4-d periodicity and in which a 2.2-d period is absent (*Belton,* 1990; *Samarasinha and A'Hearn,* 1991). That several interpretations of the spacecraft imaging data are possible is primarily due to uncertainties introduced by the defocused state of the *Vega 1* camera (*Dimarellis et al.,* 1989; *Belton,* 1990) and the particular shape of the nucleus. There were subsequent attempts at correcting the defocused *Vega 1* images (*Szegö,* 1995), but there is no consesus among different groups of researchers on the effectiveness of the results.

To discriminate between the various possible spin states, *Belton et al.* (1991) used Earth-based observations of lightcurve periodicities, most importantly the extensive photometric observations of *Millis and Schleicher* (1986), and the periodic appearances of CN jet structures documented by *Hoban et al.* (1988). Important assumptions in this work are that (1) for the several months around the 1986 apparition, cometary activity originated primarily from several active areas that were stable in their location on the nucleus; (2) dynamical effects of jet torques are negligible to the first order; and (3) the rotational motion can be approximated by that of a freely precessing symmetric top. The latter assumption is based on the near-prolate symmetry of the nucleus that is suggested by published shape models (*Sagdeev et al., 1989; Merényi et al., 1990*). Because the shape model by *Merényi et al.* (1990) is based on the spin state deduced by *Sagdeev et al.* (1989), it will require modification, presumably minor, if it is to be consistent with the spin state reported in Table 1.

The *Sagdeev et al.* (1989) spin state, later improved by Szegö and his colleagues (*Szegö, 1995*), is based on an interpretation of the spacecraft data in terms of an asymmetric top. The *Sagdeev et al.* (1989) model yields a nucleus in the SAM mode characterized by a 2.2-d precession of the long axis around the rotational angular momentum vector. The long axis nods with an amplitude of about 14° and a period of 7.4 d. The nucleus must also oscillate back and forth around its long axis with the same period of 7.4 d. However, both the *Sagdeev et al.* (1989) model as well as the more recent *Szegö* (1995) model have internal inconsistencies (the component periods and moments of inertia are dynamically in conflict with the quoted nodding amplitudes), and the 2.2-d periodicity that shows up strongly in model lightcurves (*Szegö et al., 2001*) is not convincingly seen in any Earth-based observational data (*Belton, 1990; Schleicher et al., 1990*). In addition, the observational requirement that the geometrical aspect of the nucleus as seen by the Sun return to essentially the same position every 7.4 d (*Schleicher et al., 1990*) is not satisfied.

19P/Borrelly: HST and Earth-based observations have established a period of 26 h (*Lamy et al., 1998a; Mueller and Samarasinha, 2002*). The direction of the spin axis has been determined by observations of the coma morphology. Images returned from the *Deep Space 1* mission showed a strong linear dust jet emanating from the nucleus that was stable in its projected orientation and morphology during the several days approach to the encounter. The location of the base of this feature near the waist of the elongated nucleus was consistent with it being parallel to the axis of the maximum moment of inertia (however, the relation of this jet to the near-nucleus coma morphology is yet to be fully understood). The dust jet is therefore interpreted as defining the rotation axis of the nucleus (*Soderblom et al., 2002*). This requires that the nucleus be essentially in PA rotation. Groundbased observations show the evolution of the projected geometry of this jet and thus allow an independent determination of the direction of the spin axis (*Farnham and Cochran, 2002; Schleicher et al., 2003*). On the longest timescales, covering many apparitions, both *Farnham and Cochran* (2002) and *Schleicher et al.* (2003) find that the spin axis slowly precesses by 5°–10° per century. The interpretation of these phenomena is that the comet is spinning close to its lowest-energy PA spin state with the rotational angular momentum slowly evolving under torques due to cometary activity. The sense of rotation (defined by the righthanded rule) is such that the spin axis is in the direction of the strong jet seen by the *Deep Space 1* mission (*Boice et al., 2003*). The spacecraft images provide a reference direction for the orientation of the long axis. At the time of the spacecraft encounter, the small end of the long axis nearest to the spacecraft was directed at RA = 300° and Dec = –10° (*Soderblom et al., 2004*).

Hyakutake (C/1996 B2): As with Comets 1P/Halley and 19/Borrelly, this long-period comet displays well-defined jet structures whose projected geometry varied markedly during its apparition. Observations of these features, together with periodicities derived from photometric lightcurves, allow the comet's spin state to be specified with considerable accuracy (*Schleicher and Osip, 2002; Schleicher and Woodney, 2003*). The comet is in its lowest-energy PA spin state. A reference direction for the orientation of the long axis of the nucleus is not available since there are no data on the shape of the nucleus. However, in this case, a reference direction could be defined relative to one of the active areas found by *Schleicher and Woodney* (2003).

3.4.2. Comets for which partial knowledge of the spin state is available.

2P/Encke: Numerous groundbased photometric observations of this active comet show that several harmonically related periodicities occur in its lightcurve. Interpreted as nucleus rotation periods these are $P_1 = 22.4$ h (*Jewitt and Meech, 1987*), $P_2 = 15.08$ h (*Luu and Jewitt, 1990*), $P_3 = 11.05$ h, and $P_4 = 7.3$ h (*Fernández et al., 2002*). These synodic periods appear to be related in the ratios $2P_1 \approx 3P_2 \approx 4P_3 \approx 6P_4$. The 15.08-h period, which is based on observations taken when the nucleus was near aphelion and was presumed to be essentially inactive, has usually been taken as the rotation period (e.g., *Belton, 1991; Jewitt, 1999; Jorda and Gutiérrez, 2002*). However, a recent assessment (*Meech et al., 2001*) has shown that the presumption of inactivity at aphelion is incorrect. This throws into doubt the assumption that the 15.08-h period directly reflects the changing geometry of the nucleus as viewed by the observer. The array of harmonically related periods is reminiscent of the case of 1P/Halley, in which the similarly numerous periodicities from sets of groundbased photometric time series show a similar character. In that case the various periodicities all relate to a basic 7.4-d period (*Belton, 1990*). Whether or not a periodicity exists that would similarly unite the 2P/Encke observations is unclear. Also, the question of whether an excited spin state is implied for 2P/Encke is also unclear. *Belton* (2000) claimed the presence of a second periodicity in the early lightcurve datasets at $P_5 = 2.76$ h that appeared to be unrelated harmonically from those noted above, supporting the idea of complex spin for this comet. However, the discovery of a dominant role for P_3 by *Fernández et al.* (2002) suggests that the periodicity by *Belton* (2000) is in fact harmonically related to the above series, i.e., $P_3 \approx 4P_5$. If the analogy to 1P/Halley is accepted, then a period near 45 h may play a similar role for 2P/Encke as the 7.4-d period does for 1P/Halley. Clearly further work is required to understand the spin state of 2P/Encke.

2P/Encke is one of those comets, like 19P/Borrelly discussed above, that displays a diffuse sunward fan structure in its coma. *Sekanina* (1988a,b) has investigated the evolution of the geometry of this structure in 2P/Encke under the assumption, recently born out for the sunward jet and fan structure seen in 19P/Borrelly, that the axis of symmetry of the fan is roughly coincident with the projection of the comet's rotation vector. Note that this assumption is not the same as the one that he used in his earlier study (*Sekanina, 1979*) of four comets. The revised assumption of *Sekanina* (1987b, 1988a) is in fact essentially the same as that successfully used in the recent work on fan structures in 19P/Borrelly by many investigators. This work should provide a reliable estimate of the direction of a comet's rotational angular momentum vector (even if the spin state is complex). *Festou and Barale* (2000) derived the direction of the angular momentum given in Table 1, which is in excellent agreement with that from *Sekanina* (1988a). The work of *Sekanina* (1988a,b) also provides evidence that the spin of 2P/Encke has precessed slowly in the past at rates of approximately 1° per orbit. The sense of spin is unknown.

10P/Tempel 2: Early thermal and visible investigations (*A'Hearn et al.,* 1989; *Jewitt and Luu,* 1989; *Sekanina,* 1991) yielded a spin period of 8.932 h. This period is confirmed by *Mueller and Ferrin* (1996), who also found evidence for a small secular change in the spin period over an orbital timescale. *Sekanina* (1987b) has applied his assumption that the symmetry of the comet's sunward-oriented fan reflects the projected direction of the spin vector to this comet to yield the direction of the spin vector. The sense of spin is not determined and the spin pole direction is consistent with it having been stable for many orbits.

109P/Swift-Tuttle: A spin period near 2.8 d has been derived from studies of repetitive spiral dust jets (*Sekanina,* 1981b; *Yoshida et al.,* 1993; *Boehnhardt and Birkle,* 1994; *Jorda et al.,* 1994; *McDavid and Boice,* 1995). Table 1 lists the period given by *Sekanina* (1981b). The spin is prograde and appears to be unexcited. *Sekanina* (1981b) and *Jorda et al.* (1994) derived somewhat disparate directions for the spin axis. We give the pole by *Sekanina* (1981b) in Table 1. *Jorda et al.* (1994) attribute the difference (nearly 50°) between the two directions to precession of the spin pole over the orbital period. However, the difference may simply be a reflection of the uncertainties in the calculations.

IRAS-Araki-Alcock (C/1983 H1): In a synthesis of available observations, including radar, *Sekanina* (1988c) has derived the spin state of the comet as it passed close to the Earth in 1983. He finds prograde rotation with a sidereal period of 2.14 d. The spin axis direction was found as indicated in Table 1. This long-period comet is not expected to have excited spin and no evidence to the contrary was found. A reference direction for the long axis is not available.

Hale-Bopp (C/1995 O1): *Jorda and Gutiérrez* (2002) have provided a detailed review of the observational material available on this comet and its interpretation. Numerous determinations of periodicities are in rough agreement and yield a firm estimate of the spin period (e.g., *Farnham et al.,* 1998; *Licandro et al.,* 1998). The direction of the angular momentum vector is poorly determined and the characteristically similar morphology around perihelion, which lasted nearly three months (despite huge changes in the viewing geometry), is understood to be due to wide jets

(*Samarasinha,* 2000). However, there is currently no detailed spin state/activity model combination that can successfully explain all the morphological structures of Hale-Bopp (C/1995 O1) seen during the entire apparition. This again highlights the inherent difficulties associated with deriving accurate spin-axis directions based on coma morphology. The large size of the nucleus of this long-period comet makes it likely that it rotates near or at its state of lowest energy.

3.4.3. Comets for which only periodicities associated with rotation are presently known.

6P/d'Arrest: Early investigations that produced conflicting results for this comet are reviewed in *Belton* (1991). Based on recent observations, *Lowry and Weissman* (2003) quote a rotation period of 7.20 h (see Table 1). Independent observations by *Gutiérrez et al.* (2003) yield a rotation period of 6.67 h, highlighting the difficulties associated with determination of spin parameters for this comet.

9P/Tempel 1: This comet is the target of NASA's Deep Impact mission and a worldwide observational campaign has been organized to determine its rotational and photometric properties (*Meech et al.,* 2000; *McLaughlin et al.,* 2000). Variations in the nucleus brightness observed from the HST loosely suggest a rotational periodicity in the range of 25–33 h (*Lamy et al.,* 2001). In a preliminary interpretation of data collected in the worldwide campaign, light-curve variations suggest a spin period near 42 h (*Meech et al.,* 2002).

21P/Giacobini-Zinner: *Leibowitz and Brosch* (1986) found evidence for a periodicity in this comet's lightcurve at 9.5 h. *Belton* (1990) suggested that the spin period was near 19 h based on these observations.

22P/Kopff: *Lamy et al.* (2002) find no evidence for periodicities in the nucleus lightcurve of this comet and suggest that the nucleus may be near spherical (although a pole-on situation cannot be discounted). However, *Meech* (1996) shows a double-peaked lightcurve with a rotation period of 12.91 h (listed in Table 1), while *Lowry and Weissman* (2003) derive a rotation period of 12.3 h.

28P/Neujmin 1: *A'Hearn* (1988) reviewed the thermal and visible observations of this comet and suggested a period of 12.67 h. *Delahodde et al.* (2001) and *Mueller et al.* (2002a) derived similar results.

29P/Schwassmann-Wachmann 1: *Meech et al.* (1993) found evidence for three periodicities in the lightcurve of this comet, one harmonically related to the other two. They proposed that the spin state is excited with 14.0 and 32.3 h as the underlying periods.

31P/Schwassmann-Wachmann 2: *Luu and Jewitt* (1992) found periodicity in the lightcurve of this comet and propose a spin period of 5.58 h.

46P/Wirtanen: *Lamy et al.* (1998b) suggest a periodicity near 6 h. *Meech et al.* (1997) find evidence for a periodicity at 7.6 h, which is included in Table 1. The lightcurve, which has been the subject of an extensive international campaign, is of low amplitude and a clear characterization of any periodicity has not been obtained. Independent observations by *Boehnhardt et al.* (2002) are consistent with the above periodicity but again the S/N for the data is poor. The relatively high activity that is observed in this comet relative to the estimated size of its nucleus has led to the proposal that the nucleus is most likely in an excited spin state (*Samarasinha et al.*, 1996). However, this work is based on the assumption that the nucleus is a near-prolate body. *Jorda and Gutiérrez* (2002) have shown that for an asymmetric body, the nucleus may remain in a PA spin state during more than 10 orbits.

48P/Johnson: *Jewitt and Sheppard* (2003) derive a rotation period of 29.00 h for this comet.

49P/Arend-Rigaux: *Millis et al.* (1988) have shown that the thermal and visible lightcurves for this comet are in phase as expected for the signature of the nucleus. They find a dominant periodicity at 6.73 h and propose a spin period of 13.47 h. The relatively low activity of this nucleus suggests a low-energy PA spin state for this comet.

95P/Chiron: Chiron is a Centaur with cometary activity. *Bus et al.* (1989) find a precise synodic period of 5.9180 h for this object.

96P/Machholz 1: *Meech* (1996) gives a rotation period of 6.38 h.

107P/Wilson-Harrington: *Osip et al.* (1995) find evidence for a period of 6.1 h.

133P/Elst-Pizarro: This object, which shares many characteristics with 107P/Wilson-Harrington (e.g., *Jewitt*, 2004), has a rotation period of 3.471 h (*Hsieh et al.*, 2003).

143P/Kowal-Mrkos: *Jewitt et al.* (2003) derive a rotation period of 17.2 h for this comet.

Levy (C/1990 K1): *Jewitt* (1999) reviewed the discordant results on the spin period by *Schleicher et al.* (1991) and *Feldman et al.* (1992) and concluded that outgassing torques could have been responsible for spin-up. The two spin-period determinations were 18.9 and 17.0 h respectively and were separated approximately by 21 d. Table 1 shows the later period determination.

Levy (C/1991 L3): *Fitzsimmons and Williams* (1994) find a synodic spin period of 8.34 h for this comet. No other information on the spin state is available.

Additional information on individual comets can also be found in *Meech* (1996) and *Lamy et al.* (2004).

4. INTERPRETATION OF OBSERVATIONS

In this section we will discuss the interpretation of rotational parameters, which highlights some of the challenges we face in the process. As mentioned in the introduction to this chapter and elsewhere in this book, determination of the nuclear structure is one of the most important goals of cometary science. It is critical for understanding the formation as well as the evolution of comets. Understanding the nuclear structure requires the determination of the bulk density and porosity (both at macro and micro levels) as well as material properties. In the absence of a spacecraft (preferably orbiting) with suitable instruments (i.e., at least until the rendezvous phase of the *Rosetta* spacecraft and to a lesser degree with the *Deep Impact* mission encounter of

9P/Tempel 1), all inferences on the bulk density as well as on the structure of cometary nuclei have to be based on remote observations. Rotational studies provide one of the best probes, if not the best, for exploring the structural properties of the nucleus.

4.1. Spin Rate as a Probe of the Bulk Density

Assuming a spherical PA rotator, the balance of forces (per unit area) at the surface of a nucleus is given by (cf. *Samarasinha*, 2001)

$$p_{gas} + \frac{2\pi^2 \rho R_N^2 \cos^2\lambda}{P^2} \leq \frac{2}{3}\pi G\rho^2 R_N^2 + \sigma \quad (15)$$

where p_{gas} is the interior gas pressure, ρ is the bulk density, R_N is the nuclear radius, λ is the latitude, P is the rotation period, G is the gravitational constant, and σ is the tensile strength. In the absence of any interior gas pressure, the tensile strength at zero latitude (corresponding to the regime of highest centrifugal force) can be expressed by

$$\sigma \geq \frac{2\pi^2 \rho R_N^2}{P^2} - \frac{2}{3}\pi G\rho^2 R_N^2 \quad (16)$$

For a strengthless spherical body, the critical rotation period, $P_{critical}$, below which the nucleus will rotationally break up can be derived by equating the self-gravity and rotation terms. $P_{critical}$ is given by

$$P_{critical} = \sqrt{\frac{3\pi}{G\rho}} = \frac{3.3\,h}{\sqrt{\rho}} \quad (17)$$

where ρ must be expressed in g cm^{-3}. Therefore, the fastest rotation period among comets can be effectively used to probe the bulk density of the nucleus. In the case of a prolate nucleus, for the PA state of lowest energy, the highest centrifugal force corresponds to the ends of the long axis. Therefore, for a prolate, the above equation can be modified to represent the conditions relevant to the ends of the long axis (*Jewitt and Meech*, 1988; *Luu and Jewitt*, 1992), which can be approximated as follows (*Pravec and Harris*, 2000)

$$P_{critical} \approx \frac{3.3\,h}{\sqrt{\rho}}\sqrt{\frac{a}{b}} \quad (18)$$

where 2a is the length of the long axis and 2b is the length of the symmetry axis. Therefore, a plot of a/b vs. P could be used to determine a lower limit to the bulk density. Figures 8 in both *Lamy et al.* (2004) and *Weissman et al.* (2004) show the current status of observations. Based on these figures, a lower limit to the nucleus bulk density near 0.4 g cm^{-3} can be inferred. However, unlike for asteroids and NEOs (e.g., *Whiteley et al.*, 2002; *Pravec et al.*, 2003), the number of comets with reliable rotational data is much smaller. This makes robust density determinations difficult, emphasizing the necessity for additional data on cometary rotation.

Currently, except for a few objects [e.g., 2001 OE$_{84}$ (*Pravec and Kušnirák*, 2001) and 2002 TD$_{60}$ (*Pravec et al.*, 2002)], the vast majority of asteroids larger than about 200 m have rotation periods greater than 2.2 h. This clear demarcation for rotation periods suggests that most kilometer-sized and larger asteroids are loosely bound aggregates (rubble piles). On the other hand, there are many small asteroids (<200 m) that rotate much faster with periods as small as a few minutes (*Whiteley et al.*, 2002). These bodies, known as monoliths, must have a nonzero tensile strength to withstand rotational breakup. This strength, while larger than the current estimates for the large-scale tensile strength of cometary nuclei, which is of the order of 10^2 dyn cm^{-2} (*Asphaug and Benz*, 1996; *Weissman et al.*, 2004), may still be relatively small (e.g., on the order of 10^5 dyn cm^{-2} for a 100-m object with a 100-s rotation period). This highlights the effectiveness of the spin rate as a probe of the interior structure.

4.2. Damping Timescale and Internal Structure

A nucleus in an excited rotational state will lose energy because of internal friction and will eventually end up in the least-energy spin state. The damping timescale, τ_{damp}, for this process is given by (e.g., *Burns and Safronov*, 1973)

$$\tau_{damp} = \frac{K_1 \mu Q}{\rho R_N^2 \Omega^3} \quad (19)$$

where K_1 is a nondimensional scaling coefficient, while μ and Q represent the rigidity and the quality factor of the cometary material. ρ, R_N, and Ω stand for density, radius, and angular velocity of the nucleus. *Efroimsky* (2001 and references therein) argued that the coefficient K_1 must be nearly two orders of magnitudes smaller than what was suggested by *Burns and Safronov* (1973). On the other hand, as pointed out by *Paolicchi et al.* (2003), there is complete agreement among all the authors on the functional dependency of the damping timescale on μ, Q, ρ, R_N, and Ω. In a recent revisitation of the problem, Burns and colleagues (*Sharma et al.*, 2001) conclude that their initial assessment for τ_{damp} is reasonable. However, the issue is not yet fully resolved and it is important to understand the value of K_1, in particular, its dependence on the axial ratios and on the degree of excitation. In addition, τ_{damp} has a large uncertainty due to the range of values in the literature for μQ. Unfortunately, no direct measurement of μQ is available for cometary material. In addition, if cometary nuclei are indeed loosely connected aggregates of cometesimals as implied by their low bulk densities, frequent splitting events (e.g., *Sekanina*, 1997), and complete breakups [e.g., Comets Shoemaker-Levy 9 (D/1993 F2) and LINEAR (D/1999 S4)], then the appropriate values for the structural parameters and

the damping timescales require reevaluation. For such nuclei, the energy loss due to internal mechanical friction is much more efficient and the damping timescales must be smaller than the currently accepted values.

The importance in knowing accurate damping timescales was further emphasized when *Jewitt* (1999) pointed out that most, if not all, short-period comets must be in excited spin states based on a comparison of damping and excitation timescales, where the latter was set equal to τ_{torque} (see his Fig. 2). However, observations point to only a few, if any, excited short-period cometary nuclei (see Table 1). How can this be resolved? Are we overestimating the damping timescale, underestimating the excitation timescale, or are our rotational lightcurve data not accurate enough (i.e., not enough S/N) to pick up multiple periodicities? Based on what was discussed so far in this chapter, all these effects may contribute to this apparent conflict between theory and observations.

5. FUTURE DIRECTIONS

The following are a few tasks that, in our opinion, could be carried out within the coming decade in order to better understand the cometary rotation and by extension the nature of cometary nuclei:

1. Well-sampled nuclear lightcurve observations at multiple orbital phases and other relevant observations aimed at precise determination of spin parameters of as many comets as possible.

2. Simulations of model lightcurves for different scenarios aimed at understanding how to accurately interpret lightcurve periodicities.

3. Modeling aimed at understanding how collisional effects (e.g., in the Kuiper belt) and evolutionary effects (e.g., due to outgassing) might affect cometary spin.

4. Accurate determination of the excitation and the damping timescales for cometary nuclei via theoretical means and estimation of structural parameters for cometary analogs using experimental techniques.

5. Placing greater emphasis on experiments that focus on determining structural and physical properties of cometary nuclei aboard future cometary missions.

Acknowledgments. We thank K. Mighell for help with the conversion of image formats across different platforms, D. Schleicher for relevant discussions, and C. Lisse for providing Fig. 4. We also thank K. Szegö and an anonymous reviewer for their helpful comments. N.H.S. and B.E.A.M. thank the NASA Planetary Astronomy Program.

REFERENCES

A'Hearn M. F. (1988) Observations of cometary nuclei. *Annu. Rev. Earth Planet. Sci., 16,* 273–293.

A'Hearn M. F., Campins H., Schleicher D. G., and Millis R. L. (1989) The nucleus of Comet P/Tempel 2. *Astrophys. J., 347,* 1155–1166.

Asphaug E. and Benz W. (1996) Size, density, and structure of comet Shoemaker-Levy 9 inferred from the physics of tidal breakup. *Icarus, 121,* 225–248.

Belton M. J. S. (1990) Rationalization of comet Halley's periods. *Icarus, 86,* 30–51.

Belton M. J. S. (1991) Characterization of the rotation of cometary nuclei. In *Comets in the Post-Halley Era* (R. L. Newburn et al., eds.), pp. 691–721. Kluwer, Dordrecht.

Belton M. J. S. (2000) The excited rotation state of 2P/Encke (abstract). *Bull. Am. Astron. Soc., 32,* 1062–1062.

Belton M. J. S. and Gandhi A. (1988) Application of the CLEAN algorithm to cometary light curves (abstract). *Bull. Am. Astron. Soc., 20,* 836–836.

Belton M. J. S., Julian W. H., Anderson A. J., and Mueller B. E. A. (1991) The spin state and homogeneity of comet Halley's nucleus. *Icarus, 93,* 183–193.

Boehnhardt H. (2004) Split comets. In *Comets II* (M. C. Festou et al., eds.), this volume. Univ. of Arizona, Tucson.

Boehnhardt H. and Birkle K. (1994) Time variable coma structures in comet P/Swift-Tuttle. *Astron. Astrophys. Suppl., 107,* 101–120.

Boehnhardt H. and 10 colleagues (2002) VLT observations of comet 46P/Wirtanen. *Astron. Astrophys., 387,* 1107–1113.

Boice D. C. and 11 colleagues (2003) The Deep Space 1 encounter with comet 19P/Borrelly. *Earth, Moon, Planets, 89,* 301–324.

Burns J. A. and Safronov V. S. (1973) Asteroid nutation angles. *Mon. Not. R. Astron. Soc., 165,* 403–411.

Bus S. J., Bowell E., Harris A. W., and Hewitt A. V. (1989) 2060 Chiron — CCD and electronographic photometry. *Icarus, 77,* 223–238.

Chen J. and Jewitt D. (1994) On the rate at which comets split. *Icarus, 108,* 265–271.

Crifo J. F., Fulle M., Kömle N. I., and Szegö K. (2004) Nucleus-coma structural relationships: Lessons from physical models. In *Comets II* (M. C. Festou et al., eds.), this volume. Univ. of Arizona, Tucson.

Davis D. R. and Farinella P. (1997) Collisional evolution of Edgeworth-Kuiper belt objects. *Icarus 125,* 50–60.

Deeming T. J. (1975) Fourier analysis with unequally-spaced data. *Astrophys. Space Sci., 36,* 137–158.

Delahodde C. E., Meech K. J., Hainaut O. R., and Dotto E. (2001) Detailed phase function of comet 28P/Neujmin 1. *Astron. Astrophys., 376,* 672–685.

Dimarellis E., Bertaux J. L., and Abergel A. (1989) Restoration of Vega-1 pictures of the nucleus of comet P/Halley — A new method revealing clear contours and jets. *Astron. Astrophys., 208,* 327–330.

Dworetsky M. M. (1983) A period-finding method for sparse randomly spaced observations or "How long is a piece of string." *Mon. Not. R. Astron. Soc., 203,* 917–924.

Efroimsky M. (2001) Relaxation of wobbling asteroids and comets — theoretical problems, perspectives of experimental observation. *Planet. Space Sci., 49,* 937–955.

Farinella P. and Davis D. R. (1996) Short-period comets: Primordial bodies or collisional fragments? *Science, 273,* 938–941.

Farinella P., Vokrouhlický D., and Hartmann W. K. (1998) Meteorite delivery via Yarkovsky orbital drift. *Icarus, 132,* 378–387.

Farnham T. L. and Cochran A. L. (2002) A McDonald Observatory study of comet 19P/Borrelly: Placing the Deep Space 1 observations into a broader context. *Icarus, 160,* 398–418.

Farnham T. L., Schleicher D. G., and Cheung C. C. (1998) Rotational variation of the gas and dust jets in comet Hale-Bopp (1995 O1) from narrowband imaging (abstract). *Bull. Am. Astron. Soc., 30,* 1072.

Feldman P. D., Budzien S. A., Festou M. C., A'Hearn M. F., and Tozzi G. P. (1992) Ultraviolet and visible variability of the coma of comet Levy (1990c). *Icarus, 95,* 65–72.

Fernández Y. R., Lisse C. M., Ulrich K. H., Peschke S. B., Weaver H. A., A'Hearn M. F., Lamy P. P., Livengood T. A., and Kostiuk T. (2000) Physical properties of the nucleus of comet 2P/Encke. *Icarus, 147,* 145–160.

Fernández Y. R., Lowry S. C., Weissman P. R., and Meech K. J. (2002) New dominant periodicity in photometry of comet Encke (abstract). *Bull. Am. Astron. Soc., 34,* 887–887.

Festou M. C. and Barale O. (2000) The asymmetric coma of comets. I. Asymmetric outgassing from the nucleus of comet 2P/Encke. *Astron. J., 119,* 3119–3132.

Fitzsimmons A. and Williams I. P. (1994) The nucleus of comet P/Levy 1991XI. *Astron. Astrophys., 289,* 304–310.

Foster G. (1995) The cleanest Fourier spectrum. *Astron J., 109,* 1889–1902.

Foster G. (1996) Wavelets for period analysis of unevenly sampled time series. *Astron J., 112,* 541–554.

Gutiérrez P. J., Ortiz J. L., Rodrigo R., López-Moreno J. J., and Jorda L. (2002) Evolution of the rotational state of irregular cometary nuclei. *Earth Moon Planets, 90,* 239–247.

Gutiérrez P. J., de Léon J., Jorda L., Licandro J., Lara L. M., and Lamy P. (2003) New spin period determination for comet 6P/d'Arrest. *Astron. Astrophys., 407,* L37–L40.

Harmon J. K., Nolan M. C., Ostro S. J., and Campbell D. B. (2004) Radar studies of comet nuclei and grain comae. In *Comets II* (M. C. Festou et al., eds.), this volume. Univ. of Arizona, Tucson.

Hoban S., Samarasinha N. H., A'Hearn M. F., and Klinglesmith D. A. (1988) An investigation into periodicities in the morphology of CN jets in comet P/Halley. *Astron. Astrophys., 195,* 331–337.

Hsieh H. H., Jewitt D. C., and Fernández Y. R. (2003) The strange case of 133P/Elst-Pizarro: A comet amongst the asteroids. *Astron. J.,* in press.

Hudson R. S. and Ostro S. J. (1995) Radar images of asteroid 4179 Toutatis. *Science, 270,* 84–86.

Jewitt D. (1992) Physical properties of cometary nuclei. In *Proceedings of the 30th Liège International Astrophysical Colloquium* (A. Brahic et al., eds.), pp. 85–112. Univ. of Liège, Liège.

Jewitt D. (1999) Cometary rotation: An overview. *Earth Moon Planets, 79,* 35–53.

Jewitt D. C. (2004) From cradle to grave: The rise and demise of the comets. In *Comets II* (M. C. Festou et al., eds.), this volume. Univ. of Arizona, Tucson.

Jewitt D. and Luu J. (1989) A CCD portrait of comet P/Tempel 2. *Astron. J., 97,* 1766–1841.

Jewitt D. C. and Meech K. J. (1987) CCD photometry of comet P/Encke. *Astron. J., 93,* 1542–1548.

Jewitt D. C. and Meech K. J. (1988) Optical properties of cometary nuclei and a preliminary comparison with asteroids. *Astrophys. J., 328,* 974–986.

Jewitt D. and Sheppard S. (2003) The nucleus of comet 48P/Johnson. *Astron. J.,* in press.

Jewitt D., Sheppard S., and Fernández Y. (2003) 143P/Kowal-Mrkos and the shapes of cometary nuclei. *Astron. J., 125,* 3366–3377.

Jorda L. and Gutiérrez P. (2002) Rotational properties of cometary nuclei. *Earth Moon Planets, 89,* 135–160.

Jorda L. and Licandro J. (2003) Modeling the rotation of comets.

In *Cometary Nuclei in Space and Time: Proceedings of IAU Colloquium 168* (M. A'Hearn, ed.). *Earth Moon Planets*, in press.

Jorda L., Colas F., and Lecacheux J. (1994) The dust jets of P/Swift-Tuttle 1992t. *Planet. Space Sci., 42,* 699–704.

Julian W. H. (1987) Free precession of the comet Halley nucleus. *Nature, 326,* 57–58.

Julian W. H. (1988) Precession of triaxial cometary nuclei. *Icarus, 74,* 377–382.

Julian W. H. (1990) The comet Halley nucleus — Random jets. *Icarus, 88,* 355–371.

Kaasalainen M. (2001) Interpretation of lightcurves of precessing asteroids. *Astron. Astrophys., 376,* 302–309.

Keller H. U. and 17 colleagues (1986) First Halley multicolour camera imaging results from Giotto. *Nature, 321,* 320–326.

Lamy P. L., Toth I., and Weaver H. A. (1998a) Hubble Space Telescope observations of the nucleus and inner coma of comet 19P/1904 Y2 (Borrelly). *Astron. Astrophys., 337,* 945–954.

Lamy P. L., Toth I., Jorda L., Weaver H. A., and A'Hearn M. F. (1998b) The nucleus and inner coma of comet 46P/Wirtanen. *Astron. Astrophys., 335,* L25–L29.

Lamy P. L., Toth I., A'Hearn M. F., Weaver H. A., and Weissman P. R. (2001) Hubble Space Telescope observations of the nucleus of comet 9P/Tempel 1. *Icarus, 154,* 337–344.

Lamy P. L., Toth I., Jorda L., Groussin O., A'Hearn M. F., and Weaver H. A. (2002) The nucleus of comet 22P/Kopff and its inner coma. *Icarus, 156,* 442–455.

Lamy P., Toth I., Weaver H., and Fernández Y. (2004) The sizes, shapes, albedos, and colors of cometary nuclei. In *Comets II* (M. C. Festou et al., eds.), this volume. Univ. of Arizona, Tucson.

Landau L. D. and Lifshitz E. M. (1976) *Mechanics, 3rd edition.* Pergamon, Oxford. 169 pp.

Larson S. M. and Minton R. B. (1972) Photographic observations of comet Bennett 1970 II. In *Comets: Scientific Data and Missions* (G. P. Kuiper and E. Roemer, eds.), pp. 183–208. Univ. of Arizona, Tucson.

Larson S. M. and Slaughter C. D. (1991) Evaluating some computer enhancement algorithms that improve the visibility of cometary morphology. In *Asteroids, Comets, Meteors 1991* (A. W. Harris and E. Bowell, eds.), pp. 337–343. Lunar and Planetary Institute, Houston.

Leibowitz E. M. and Brosch N. (1986) Periodic photometric variations in the near-nucleus zone of P/Giacobini-Zinner. *Icarus, 68,* 430–441.

Licandro J. and 13 colleagues (1998) The rotation period of C/1995 O1 (Hale-Bopp). *Astrophys. J. Lett., 501,* L221–L225.

Licandro J., Serra-Ricart M., Oscoz A., Casas R., and Osip D. (2000) The effect of seeing variations in time-series CCD inner coma photometry of comets: A new correction method. *Astron. J., 119,* 3133–3144.

Lisse C. M. and 14 colleagues (1999) Infrared observations of dust emission from comet Hale-Bopp. *Earth Moon Planets, 78,* 251–257.

Lowry S. C. and Weissman P. R. (2003) CCD observations of distant comets from Palomar and Steward observatories. *Icarus, 164,* 492–503.

Luu J. and Jewitt D. (1990) The nucleus of comet P/Encke. *Icarus, 86,* 69–81.

Luu J. X. and Jewitt D. C. (1992) Near-aphelion CCD photometry of comet P/Schwass-Mann-Wachmann 2. *Astron. J., 104,* 2243–2249.

Magnusson P., Barucci M. A., Drummond J. D., Lumme K., and Ostro S. J. (1989) Determination of pole orientations and shapes of asteroids. In *Asteroids II* (R. P. Binzel et al., eds.), pp. 67–97. Univ. of Arizona, Tucson.

McDavid D. and Boice D. (1995) The rotation period of comet P/Swift-Tuttle (abstract). *Bull. Am. Astron. Soc., 27,* 1338–1338.

McLaughlin S. A., McFadden L. A., and Emerson G. (2000) Deep Impact: A call for pro-am observations of comet 9P/Tempel 1 (abstract). *Bull. Am. Astron. Soc., 32,* 1280–1280.

Meech K. J. (1996) Physical properties of comets. Paper presented at Asteroids, Comets, Meteors 1996 meeting. Available on line at http://www.ifa.hawaii.edu/~meech/papers/acm96.pdf.

Meech K. J., Belton M. J. S., Mueller B. E. A., Dicksion M. W., and Li H. R. (1993) Nucleus properties of P/Schwassmann-Wachmann 1. *Astron. J., 106,* 1222–1236.

Meech K. J., Bauer J. M., and Hainaut O. R. (1997) Rotation of comet 46P/Wirtanen. *Astron. Astrophys., 326,* 1268–1276.

Meech K. J., A'Hearn M. F., Belton M. J. S., Fernández Y., Bauer J. M., Pittichová J., Buie M. W., Tozzi G. P., and Lisse C. (2000) Thermal and optical investigation of 9P/Tempel 1 (abstract). *Bull. Am. Astron. Soc., 32,* 1061–1061.

Meech K. J, Fernández Y., and Pittichová J. (2001) Aphelion activity of 2P/Encke (abstract). *Bull. Am. Astron. Soc., 33,* 1075–1075.

Meech K. J., Fernández Y. R., Pittichová J., Bauer J. M., Hsieh H., Belton M. J. S., A'Hearn M. F., Hainaut O. R., Boehnhardt H., and Tozzi G. P. (2002) Deep Impact nucleus characterization — A status report (abstract). *Bull. Am. Astron. Soc., 34,* 870–870.

Merényi E., Földy L., Szegö K., Toth I., and Kondor A. (1990) The landscape of comet Halley. *Icarus, 86,* 9–20.

Millis R. L. and Schleicher D. G. (1986) Rotational period of comet Halley. *Nature, 324,* 646–649.

Millis R. L., A'Hearn M. F., and Campins H. (1988) An investigation of the nucleus and coma of Comet P/Arend-Rigaux. *Astrophys. J., 324,* 1194–1209.

Mueller B. E. A. and Ferrin I. (1996) Change in the rotational period of comet P/Tempel 2 between the 1988 and 1994 apparitions. *Icarus, 123,* 463–477.

Mueller B. E. A. and Samarasinha N. H. (2002) Visible lightcurve observations of comet 19P/Borrelly. *Earth Moon Planets, 90,* 463–471.

Mueller B. E. A., Heinrichs A. M., and Samarasinha N. H. (2002a) Physical properties of the nucleus of comet 28P/Neujmin 1 (abstract). *Bull. Am. Astron. Soc., 34,* 886.

Mueller B. E. A., Samarasinha N. H., and Belton M. J. S. (2002b) The diagnosis of complex rotation in the lightcurve of 4179 Toutatis and potential applications to other asteroids and bare cometary nuclei. *Icarus, 158,* 305–311.

Neishtadt A. I., Scheeres D. J., Siderenko V. V., and Vasiliev A. A. (2002) Evolution of comet nucleus rotation. *Icarus, 157,* 205–218.

Osip D. J., Campins H., and Schleicher D. G. (1995) The rotation state of 4015 Wilson-Harrington: Revisiting origins for the near-Earth asteroids. *Icarus, 114,* 423–426.

Paolicchi P., Burns J. A., and Weidenschilling S. J. (2003) Side effects of collisions: Rotational properties, tumbling rotation states, and binary asteroids. In *Asteroids III* (W. F. Bottke Jr. et al., eds.), pp. 517–526. Univ. of Arizona, Tucson.

Peale S. J. and Lissauer J. J. (1989) Rotation of Halley's comet. *Icarus, 79,* 396–430.

Pravec P. and Harris A. W. (2000) Fast and slow rotation of asteroids. *Icarus, 148,* 12–20.

Pravec P. and Kušnirák P. (2001) *2001 OE₈₄.* IAU Circular 7735.

Pravec P., Šarounová L., Hergenrother C., Brown P., Esquerdo G., Masi G., Belmonte C., Mallia F., and Harris A. W. (2002) *2002 TD₆₀.* IAU Circular 8017.

Pravec P., Harris A. W., and Michałowski T. (2003) Asteroid rotations. In *Asteroids III* (W. F. Bottke Jr. et al., eds.), pp. 113–122. Univ. of Arizona, Tucson.

Roberts D. H., Lehar J., and Dreher J. W. (1987) Time-series analysis with CLEAN. I. Derivation of a spectrum. *Astron. J., 93,* 968–989.

Rodionov A. V., Crifo J.-F., Szegö K., Lagerros J., and Fulle M. (2002) An advanced physical model of cometary activity. *Planet. Space Sci., 50,* 983–1024.

Rubincam D. P. (2000) Radiative spin-up and spin-down of small asteroids. *Icarus, 148,* 2–11.

Sagdeev R. Z. and 37 colleagues (1986) Television observations of comet Halley from Vega spacecraft. *Nature, 321,* 262–266.

Sagdeev R. Z., Szegö K., Smith B. A., Larson S., Merényi E., Kondor A., and Toth I. (1989) The rotation of P/Halley. *Astron. J., 97,* 546–551.

Samarasinha N. H. (1997) Preferred orientations for the rotational angular momentum vectors of comets (abstract). *Bull. Am. Astron. Soc., 29,* 743–743.

Samarasinha N. H. (2000) The coma morphology due to an extended active region and the implications for the spin state of comet Hale-Bopp. *Astrophys. J. Lett., 529,* L107–L110.

Samarasinha N. H. (2001) Model for the breakup of comet LINEAR (C/1999 S4). *Icarus, 154,* 540–544.

Samarasinha N. H. (2003) Cometary spin states, their evolution, and the implications. In *Cometary Nuclei in Space and Time: Proceedings of IAU Colloquium 168* (M. A'Hearn, ed.). *Earth Moon Planets,* in press.

Samarasinha N. H. and A'Hearn M. F. (1991) Observational and dynamical constraints on the rotation of comet P/Halley. *Icarus, 93,* 194–225.

Samarasinha N. H. and Belton M. J. S. (1995) Long-term evolution of rotational states and nongravitational effects for Halley-like cometary nuclei. *Icarus, 116,* 340–358.

Samarasinha N. H. and Mueller B. E. A. (2002) Spin axis direction of comet 19P/Borrelly based on observations from 2000 and 2001. *Earth Moon Planets, 90,* 473–382.

Samarasinha N. H., A'Hearn M. F., Hoban S., and Klinglesmith D. A. III (1986) CN jets of comet P/Halley: Rotational properties. In *ESA Proceedings of the 20th ESLAB Symposium on the Exploration of Halley's Comet, Vol. 1: Plasma and Gas* (B. Battrik et al., eds.), pp. 487–491. ESA, Paris.

Samarasinha N. H., Mueller B. E. A., and Belton M. J. S. (1996) Comments on the rotational state and non-gravitational forces of comet 46P/Wirtanen. *Planet. Space Sci., 44,* 275–281.

Samarasinha N. H., Mueller B. E. A., and Belton M. J. S. (1999) Coma morphology and constraints on the rotation of comet Hale-Bopp (C/1995 O1). *Earth Moon Planets, 77,* 189–198.

Scheeres D. J., Ostro S. J., Werner R. A., Asphaug E., and Hudson R. S. (2000) Effects of gravitational interactions on asteroid spin states. *Icarus, 147,* 106–118.

Schleicher D. G. and Bus S. J. (1991) Comet P/Halley's periodic brightness variations in 1910. *Astron J., 101,* 706–712.

Schleicher D. G. and Farnham T. (2004) Photometry and imaging of the coma with narrowband filters. In *Comets II* (M. C. Festou et al., eds.), this volume. Univ. of Arizona, Tucson.

Schleicher D. G. and Osip D. J. (2002) Long- and short-term photometric behavior of comet Hyakutake (1996 B2). *Icarus, 159,* 210–233.

Schleicher D. G. and Woodney L. M. (2003) Analysis of dust coma morphology of comet Hyakutake (1996 B2) near perigee: Outburst behavior, jet motion, source region locations, and nucleus pole orientation. *Icarus, 162,* 191–214.

Schleicher D. G., Millis R. L., Thompson D. T., Birch P. V., Martin R., Tholen D. J., Piscitelli J. R., Lark N. L., and Hammel H. B. (1990) Periodic variations in the activity of comet P/Halley during the 1985/1986 apparition. *Astron. J., 100,* 896–912.

Schleicher D. G., Millis R. L., Osip D. J., and Birch P. V. (1991) Comet Levy (1990c) — Groundbased photometric results. *Icarus, 94,* 511–523.

Schleicher D. G., Woodney L. M., and Millis R. L. (2003) Comet 19P/Borrelly at multiple apparitions: Seasonal variations in gas production and dust morphology. *Icarus, 162,* 415–442.

Sekanina Z. (1979) Fan-shaped coma, orientation of rotation axis, and surface structure of a cometary nucleus. I — Test of a model on four comets. *Icarus, 37,* 420–442.

Sekanina Z. (1981a) Rotation and precession of cometary nuclei. *Annu. Rev. Earth Planet. Sci., 9,* 113–145.

Sekanina Z. (1981b) Distribution and activity of discrete emission areas on the nucleus of periodic comet Swift-Tuttle. *Astron. J., 86,* 1741–1773.

Sekanina Z. (1987a) Nucleus of comet Halley as a torque-free rigid rotator. *Nature, 325,* 326–328.

Sekanina Z. (1987b) Anisotropic emission from comets: Fans versus jets. 1: Concept and modeling. In *Symposium on the Diversity and Similarity of Comets* (E. J. Rolfe and B. Battrick, eds.), pp. 315–322. ESA, Noordwijk.

Sekanina Z. (1988a) Outgassing asymmetry of periodic comet Encke. I — Apparitions 1924–1984. *Astron. J., 95,* 911–971.

Sekanina Z. (1988b) Outgassing asymmetry of periodic comet Encke. II — Apparitions 1868–1918 and a study of the nucleus evolution. *Astron. J., 96,* 1455–1475.

Sekanina Z. (1988c) Nucleus of comet IRAS-Araki-Alcock (1983 VII). *Astron. J., 95,* 1876–1894.

Sekanina Z. (1991) Comprehensive model for the nucleus of periodic comet Tempel 2 and its activity. *Astron. J., 102,* 350–388.

Sekanina Z. (1997) The problem of split comets revisited. *Astron. Astrophys., 318,* L5–L8.

Sekanina Z. and Boehnhardt H. (1999) Dust morphology of comet Hale-Bopp (C/1995 O1). II. Introduction of a working model. *Earth Moon Planets, 78,* 313–319.

Sekanina Z. and Larson S. M. (1986) Coma morphology and dust-emission pattern of periodic comet Halley. IV — Spin vector refinement and map of discrete dust sources for 1910. *Astron. J., 92,* 462–482.

Sharma I., Burns J. A., and Hui C.-Y. (2001) Nutational damping times in solids of revolution (abstract). *Bull. Am. Astron. Soc., 33,* 1114.

Skorov Y. V. and Rickman H. (1999) Gas flow and dust acceleration in a cometary Knudsen layer. *Planet. Space Sci., 47,* 935–949.

Soderblom L. A. and 21 colleagues (2002) Observations of comet 19P/Borrelly by the miniature integrated camera and spectrometer aboard Deep Space 1. *Science, 296,* 1087–1091.

Soderblom L. A. and 12 colleagues (2004) Imaging Borrelly. *Icarus, 167,* 4–15.

Stellingwerf R. F. (1978) Period determination using phase-dispersion minimization. *Astrophys. J. 224,* 953–960.

Stern S. A. (1995) Collisional timescales in the Kuiper disk and their implications. *Astron. J., 110,* 856–868.

Stern S. A. and Weissman P. R. (2001) Rapid collisional evolution of comets during the formation of the Oort cloud. *Nature, 409,* 589–591.

Szegö K. (1995) Discussion of the scientific results derived from the near-nucleus images. In *Images of the Nucleus of Comet Halley Obtained by the Television System (TVS) On Board the Vega Spacecraft, Vol. 2* (R. Reinhardt and B. Battrick, eds.), pp. 68–80. ESA SP-1127, Noordwijk, The Netherlands.

Szegö K., Crifo J.-F., Földy L., Lagerros J. S. V., and Rodionov A. V. (2001) Dynamical effects of comet P/Halley gas production. *Astron. Astrophys., 370,* L35–L38.

Wallis M. K. (1984) Rotation of cometary nuclei. *Trans. R. Soc. Philos. Ser. A, 313,* 165–170.

Watanabe J. (1992a) Rotation of split cometary nuclei. In *Asteroids, Comets, Meteors 1991* (A. W. Harris and E. Bowell, eds.), pp. 621–624. Lunar and Planetary Institute, Houston.

Watanabe J. (1992b) Ice-skater model for the nucleus of comet Levy 1990c — Spin-up by a shrinking nucleus. *Publ. Astron. Soc. Japan, 44,* 163–166.

Weidenschilling S. J. (2004) From icy grains to comets. In *Comets II* (M. C. Festou et al., eds.), this volume. Univ. of Arizona, Tucson.

Weissman P. R., Asphaug E., and Lowry S. L. (2004) Structure and density of cometary nuclei. In *Comets II* (M. C. Festou et al., eds.), this volume. Univ. of Arizona, Tucson.

Whipple F. L. (1950) A comet model. I. The acceleration of comet Encke. *Astrophys. J., 111,* 375–394.

Whipple F. L. (1978) Rotation period of comet Donati. *Nature, 273,* 134–135.

Whipple F. L. (1982) The rotation of comet nuclei. In *Comets* (L. L. Wilkening, ed.), pp. 227–250. Univ. of Arizona, Tucson.

Whipple F. L. and Sekanina Z. (1979) Comet Encke — Precession of the spin axis, nongravitational motion, and sublimation. *Astron. J., 84,* 1895–1909.

Whiteley R. J., Tholen D. J., and Hergenrother C. W. (2002) Light-curve analysis of four new monolithic fast-rotating asteroids. *Icarus, 157,* 139–154.

Wilhelm K. (1987) Rotation and precession of comet Halley. *Nature, 327,* 27–30.

Wilkening L. L., ed. (1982) *Comets.* Univ. of Arizona, Tucson. 766 pp.

Yeomans D., Choda P. W., Krolikowska M., Szutowicz S., and Sitarski G. (2004) Cometary orbit determination and nongravitational forces. In *Comets II* (M. C. Festou et al., eds.), this volume. Univ. of Arizona, Tucson.

Yoshida S., Aoki T., Soyano T., Tarusawa K., van Driel W., Hamabe M., Ichikawa T., Watanabe J., and Wakamatsu K. (1993) Spiral dust-jet structures of Comet P/Swift-Tuttle 1992t. *Publ. Astron. Soc. Japan, 45,* L33–L37.

Split Comets

H. Boehnhardt

Max-Planck-Institut für Astronomie Heidelberg

More than 40 split comets have been observed over the past 150 years. Two of the split comets have disappeared completely; another one was destroyed during its impact on Jupiter. The analysis of the postsplitting dynamics of fragments suggests that nucleus splitting can occur at large heliocentric distances (certainly beyond 50 AU) for long-period and new comets and all along the orbit for short-period comets. Various models for split comets have been proposed, but only in one peculiar case, the break-up of Comet D/1993 F2 (Shoemaker-Levy 9) around Jupiter, has a splitting mechanism been fully understood: The nucleus of D/1993 F2 was disrupted by tidal forces. The fragments of split comets seem to be subkilometer in size. It is, however, not clear whether they are cometesimals that formed during the early formation history of the planetary system or are pieces from a heavily processed surface crust of the parent body. The two basic types of comet splitting (few fragments and many fragments) may require different model interpretations. Disappearing comets may represent rare cases of complete nucleus dissolution as suggested by the prototype case, Comet C/1999 S4 (LINEAR). At least one large family of split comets exists — the Kreutz group— but other smaller clusters of comets with common parent bodies are very likely. Comet splitting seems to be an efficient process of mass loss of the nucleus and thus can play an important role in the evolution of comets toward their terminal state. The secondary nuclei behave as comets of their own (with activity, coma, and tail) exhibiting a wide range of lifetimes. However, at present it is now known whether the fragments' terminal state is "completely dissolved" or "exhausted and inactive."

1. THE PHENOMENON

Split comets appear as multiple comets with two or more components arising from the same parent and initially moving in very similar orbits. When active, the components usually display well-defined individual comae and tails that can overlap each other when the ensemble is close (see Fig. 1). Most of the components of a split comet "disappear" sooner or later; i.e., within time spans of hours to years the components become too faint to be detected even by the largest telescopes, and only one main component "survives" for a longer period of time. Activity outbursts and the appearance of coma arclets can be associated with comet splitting events. However, the ultimate proof of comet splitting is provided through the detection of at least one secondary component (also called a fragment or companion) to the primary nucleus.

The scientific interest in split comets reaches beyond the obvious questions "Why do comets split?" and "What is the sequence of events?" and focuses on the understanding of the internal structure and chemistry of the cometary nucleus as well as its overall evolution with time. The answers obtained from split comets may even provide information on the formation scenario of the solar system (for instance, the size distribution of the cometesimals, the original ice chemistry, and even the "birth place" of cometary nuclei).

1.1. Types of Split Comets

Two types of split comets are known from observations:

Type A: The split comet has a few (usually two) components. The primary fragment is the one that remains "permanent"; the secondary can be minor, short-lived, or persistent for a longer time (years to centuries). The primary is considered to be identical to the original nucleus (the parent body), while the secondary represents a smaller piece that is broken off the nucleus (typically 10–100 m in size). Type A splitting events can recur in the same object. Known cases are the comets listed in section 8 (with the exception of the ones mentioned as Type B below).

Type B: The split comet has many (more than 10) components that could arise from a single or a short sequence of fragmentation events. The fragments are short-lived (possibly of small size), and no primary component can be identified. Tertiary fragmentation of secondaries is occasionally observed. Type B events are believed to represent cases of dissolution and/or disruption of the comet and the parent body may become completely destroyed. Known cases are Comet D/1993 F2 (Shoemaker-Levy 9) and Comet C/1999 S4 (LINEAR).

In summary, observations and modeling results provide evidence for at least 42 split comets producing several hundred (>400) fragments in more than 100 splitting events.

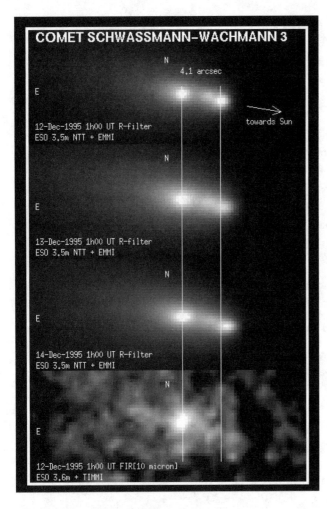

COMET SCHWASSMANN-WACHMANN 3

N

4.1 arcsec

E

12-Dec-1995 1h00 UT R-filter
ESO 3.5m NTT + EMMI

towards Sun

N

E

13-Dec-1995 1h00 UT R-filter
ESO 3.5m NTT + EMMI

N

E

14-Dec-1995 1h00 UT R-filter
ESO 3.5m NTT + EMMI

N

E

12-Dec-1995 1h00 UT FIR[10 micron]
ESO 3.6m + TIMMI

Fig. 1. Three components of split Comet 73P/Schwassmann-Wachmann 3 are detected in mid-December 1995 shortly after the splitting event of the nucleus. Each fragment has its own coma and tail that overlap. Component C (to the east) is the primary fragment, companions B (middle) and A (to the west) are secondary fragments. Fragment B is considered persistent, since it survived the subsequent perihelion passage in January 2001. Fragment A was not found upon the next return of the comet. The images are taken in the visible R-band filter (upper three panels; NTT + EMMI) and in the mid-IR N band (bottom panel; 3.6-m + TIMMI). Courtesy of the European Southern Observatory (ESO).

1.2. Designation of Fragments

In general, the fragments are denoted with the designation and name of the parent comet [given by the International Astronomical Union (IAU)], followed by upper-case letters beginning with the letter A for the component that passes perihelion first. This fragment is supposed to be the primary component. However, cases of misidentified primaries exist (for instance, 73P/Schwassmann-Wachmann 3A and C, the latter being the primary; see also Fig. 1). Indices for tertiary components that split from already denoted fragments were introduced, such as in the case of D/1993 F2 (Shoemaker-Levy 9) components P_1, etc. Widely separated components may even receive different IAU designations

and names [like Comets C/1965 S1 (Ikeya-Seki) and C/1882 R1 (Great September Comet)] and are only linked *a posteriori* to a common parent body.

2. DYNAMICS OF COMET SPLITTING

As discussed in *Marsden and Sekanina* (1971), simple backward integration of the orbits of the fragments of split comets does not yield unique and well-defined "collision" points of their orbits that could be considered the places where the fragmentations of the nuclei happened. Moreover, this approach does not provide a sensible description of the dynamical aspects of the splitting event itself.

2.1. Dynamical Models for Comet Splitting

Sekanina (1977, 1978, 1979, 1982) has developed a five-parameter model to approximate the dynamics of the motion of the fragments of split comets. He and colleagues applied this model to more than 30 split comets. A similar approach was used by *Meech et al.* (1995) for Comet C/1986 P1 (Wilson). The parameters are used to constrain the fragmentation event dynamically through the time T_s when the splitting happened, the radial V_r, transverse V_t and normal V_n components of the separation velocity of the secondary fragment relative to the primary one, and the deceleration parameter Γ of the secondary relative to the primary component. V_r points in the direction of the radius vector of the comet, positive along the radius vector; V_t is perpendicular to the radius vector of the comet in the orbital plane, positive in the direction of the velocity vector of the comet; V_n is perpendicular to the orbital plane of the comet, positive toward the direction of the angular momentum. The deceleration Γ is a result of the momentum transfer between the two fragments due to their different outgassing rates and masses. It is measured in radial direction only; the two other components of the deceleration are set to zero. Γ is assumed to vary with solar distance r proportional to $1/r^2$.

The model implies a single-step, two-body fragmentation of the nucleus. Its parameters (or subsets of them) are determined by nonlinear least-squares fits of astrometric positions of the fragments. As such, the quality of the parameter solution depends very much on the accuracy, the number, and the measured arc of astrometric positions of the fragments. Moreover, at least in the case of more than two fragments, a variety of splitting sequences of the fragments are possible and need to be analyzed (*Sekanina, 1999; Weaver et al., 2001*), and for events that produce many fragments like C/1999 S4 (LINEAR) a complete and unique solution becomes impossible. Also, different numerical solutions are possible even for a two-fragment case by considering various subsets of fit parameters, and the selection of the most plausible one requires a critical discussion of the physical meaning of the solutions.

Desvoivres et al. (1999) have introduced a description of the dynamics of split comets that is based on a physical outgassing model for the components. As in Sekanina's approach, this model implies a single-step two-body fragmentation scenario of the nucleus, and the many model parameters — including the size and bulk density of the fragments — are determined through a numerical fit of the motion of the fragments involving plausibility considerations on some of the fit parameters. The authors applied their model to three splitting events observed in Comet C/1996 B2 (Hyakutake) in 1996 and could indeed demonstrate that outgassing of the fragments together with the initial momentum from the splitting event provide a suitable description of the dynamics of the companions.

Neither model implies a particular physical process that causes the splitting of comets, and in each model only the separation dynamics of the fragments after breakup (and when at larger distance from each other) is described. They were successfully applied to split comets with fragments of small (a few arcseconds to a few degrees) mutual distances.

2.2. Results from the Dynamical Modeling of Comet Splitting

The results discussed below are primarily based on the work by Sekanina and collaborators over the past 25 years (*Sekanina*, 1977, 1978, 1979, 1982, 1988, 1991, 1995b, 1997a, 1998, 1999, 2001, 2002c; *Sekanina and Marsden*, 1982; *Sekanina and Yeomans*, 1985; *Sekanina and Chodas*, 2002a–d; *Sekanina et al.*, 1996, 2002; Marsden and Sekanina, personal communication). Figure 2 contains plots that illustrate the results and conclusions on the dynamical modeling of split comets. The underlying database contains 33 comets producing a total of 97 fragments in 64 splitting events (9 dynamically new comets with 22 components in 13 events, 13 long-period comets with 37 components in 24 events, and 11 short-period comets with 38 components in 28 events).

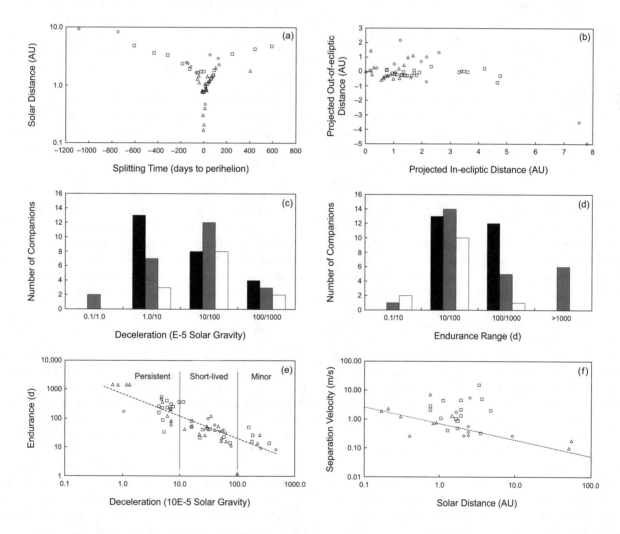

Fig. 2. Results from the dynamical modeling of comet-splitting events. **(a)** Solar distance vs. splitting time to perihelion (negative/positive before/after perihelion). **(b)** Location of splitting events: Projected out-of-ecliptic distance vs. projected in-ecliptic distance **(c)** Histogram distribution of relative deceleration of companions. **(d)** Histogram distribution of endurances of companions. **(e)** Endurance vs. deceleration of companions. **(f)** Separation velocity of the companions vs. solar distance at time of splitting. Symbols and colors used in the plots are as follows: short-period comets — squares (plots) and black (histograms); long-period comets — diamonds (plots) and gray (histograms); dynamically new comets — triangles (plots) and white (histograms). In **(e)** the three groups of companions ("persistent", "short-lived", "minor") are indicated between dotted vertical lines. The broken line in this panel shows the best fit described by equation (2) in section 2.2. The broken line in **(f)** represents the relationship obtained by *Sekanina* (1982) and described in section 2.2.

TABLE 1.   Mean values for the deceleration, endurance, and separation velocity of fragments.

	New Comets	Long-Period Comets	Short-Period Comets
Deceleration (10^{-5} solar units)	86 ± 134	41 ± 43	49 ± 59
Number of fragments	13	24	25
Endurance (days)	52 ± 31	514 ± 653	161 ± 128
Number of fragments	13	26	25
Separation velocity (m/s)	1.3 ± 1.3	2.0 ± 1.9	2.7 ± 2.3
Number of fragments	7	12	16

2.2.1. Primary and "higher-order" fragments.   The identification of the primary fragment of a splitting event is made purely on dynamical grounds: The primary is the companion that passes perihelion first and thus shows the smallest nongravitational force of the breakup products. For many cases one can assume that it carries the vast majority of mass of the original nucleus. "Secondaries" are believed to represent lighter pieces from the splitting event since they show higher nongravitational forces than the primaries. Tertiary fragments are produced when secondaries disintegrate further, and so on for even higher-order fragments. The presence of multiple companions requires a careful analysis of the observations in order to definitively establish the sequence of nucleus splitting events that generated the fragments.

2.2.2. Splitting time T_s.   The plot "splitting time" vs. "solar distance" in Fig. 2 shows a clustering of splitting events close to perihelion, a trend that may have considerable observational bias since these events happen closer to the observers on Earth. The comet splitting is about equally likely before and after perihelion. The behavior of short-period, long-period, and dynamically new comets seems to be similar. A few early breakups (i.e., long before perihelion passage) of long-period and dynamically new comets are known. The results for T_s suggest that at least short-period comets seem to split all along their orbits.

2.2.3. Splitting location.   The solar distances, where comet splitting happens, peak within about 2 AU. However, there are several cases where comets split at larger distances, even beyond 50 AU (not plotted in Fig. 2; see also section 3.1). While the splitting locations of short-period comets are naturally close to the ecliptic, such a bias does not exist for new and long-period comets that can split at large distance in and out of the orbital plane of the planets. Indications exist (*Sekanina*, 1982, 1997a, 1999) that comets may split all along the orbit even at large heliocentric distances (>5 AU) up to aphelion of long-period comets (*Sekanina and Chodas*, 2002a, 2002b; *Sekanina et al.*, 2002).

2.2.4. Deceleration parameter Γ.   The deceleration parameter Γ ranges from 10^{-5} to almost $10^{-2} \times$ solar gravity. The coarse histogram distribution in Fig. 2 suggests that long-period and new comets tend to produce fragments subject to decelerations Γ of 10^{-4}–$10^{-3} \times$ solar gravity, while the fragments of short-period comets show on the average smaller Γ values. This trend suggests that fragments

of the latter comets may be parts of a more evolved surface crust, i.e., they contain more dust of higher bulk density and less sublimating ices that can contribute to the nongravitational forces on the body. The breakup products of the two other groups may contain more volatile ices since their surfaces have experienced little loss by thermal heating during the much rarer perihelion passages. The mean deceleration of fragments from new comets (Table 1) is about twice as high as for those of long- and short-period comets (all with large error bars).

2.2.5. Endurance E of the fragments.   The lifetimes of split components differ by a large amount, even for the same comet. This indicates that the fragments have different reservoirs of outgassing material and they must be of different size and mass after all. *Sekanina* (1977, 1982) has introduced the so-called endurance parameter E of a fragment as a measure for the persistence of fragments. The endurance E has been defined as the time interval from the splitting event (given by T_s) to its final observation (at time T_f), normalized by the inverse-square law to a distance of 1 AU from the Sun. In other words, the endurance measures a minimum sublimation lifetime of a companion. In terms of orbital parameters, E (in days) can be written as

$$E = 1.015 * A_{sf}/[a * (1 - e^2)]^{1/2} \qquad (1)$$

where A_{sf} is the length of the heliocentric arc of the orbit (in degrees) that the fragment passed through between T_s and T_f, a is the semimajor axis, and e is the eccentricity of the cometary orbit. Obviously, the endurance values of split comets are lower limits only, since T_f may be rather constrained by the visibility of the objects and telescope capabilities.

The measured endurances E cover three orders of magnitude from a few to several thousand days. In Fig. 2 the endurance E is plotted vs. the deceleration Γ for about 50 companions seen in split comets. Here, "companion" means secondary fragments, the decelerations of which were measured relative to the primary ones. The latter are to be considered the more persistent products since their lifetimes are at least as long as (and in most cases much longer than) those of the secondary ones. The conclusion of *Sekanina* (1982) that the endurance E scales with the deceleration Γ of the secondary fragments is still valid. However, the larger dataset available now suggests a steeper exponent for Γ

(correlation coefficient 0.7)

$$E \text{ (in days)} = 690 \, (\pm 180) * \Gamma^{-0.77 \, (\pm 0.07)} \quad (2)$$

Based upon an anticipated clustering in the endurance vs. deceleration plot of companions, *Sekanina* (1982) has introduced a classification scheme for the fragments of split comets, i.e., persistent ($\Gamma < 10^{-4}$ solar gravity), short-lived ($10^{-4} < \Gamma < 10^{-3}$ solar gravity), and minor components ($\Gamma > 10^{-3}$ solar gravity). In the larger dataset now available (see Fig. 2) the original clustering is less obvious. Nevertheless, the "minor components" are distinguished from the merging groups of "short-lived" and "persistent" objects. Among the latter, a few fragments with very long lifetimes ($E > 1000$ d) have been observed. The mean values for the endurances of the three dynamical classes of comets (Table 1) suggest that new comets are on the average less persistent as long- and short-period comets, although the mean deviations of the mean endurance values are large.

2.2.6. Separation velocities. The relative speed of the fragments shortly after the fragmentation event amounts from 0.1 to 15 m/s with the majority between 0.3 and 4 m/s. In many cases the velocity components are ill-defined through the available observations, if measurable at all. In Fig. 2 only cases are plotted for which all three components of the separation velocity are estimated, thus the total amplitude V_{total} can be calculated. It is unclear whether trends with solar distance r exist as suggested by *Sekanina* (1982) based on a smaller dataset: Dynamically new comets (and less likely long-period comets as well) may follow approximately *Sekanina*'s (1982) data fit of $V_{total} \sim 0.7 * r^{-0.57}$, while a similar behavior is not obvious for the short-period comets. Instead, for the latter, a random scatter of V_{total} independent of r is found. Therefore, on the average, a slightly larger separation velocity (see Table 1) may be indicative of a different fragmentation mechanism or of higher tensile strength of the nuclei as compared to the long-period and dynamically new comets (which may be less evolved due to rarer passages close to the Sun).

3. SPLIT PAIRS, FAMILIES, AND COMET EVOLUTION

The dynamical modeling described in section 2 is preferably applied to cases where evidence for the comet splitting comes from the observations of the fragments that appear close in time and space. In fact, most of the fragments are observed within the same — narrow — field of view of the telescopes used. Linking "wider and older" pairs and clusters of split comets is a more difficult task.

3.1. Pairs and Families of Split Comets

Similarity of the dynamical (semimajor axis, eccentricity) and geometrical (inclination, ascending node, argument of perihelion) orbital elements of comets suggests a common origin of the respective nuclei despite very different times for perihelion passage and despite the failure to identify the location and time of the splitting along the orbit from simple numerical backward integration of their orbits (*Marsden and Sekanina*, 1971). Such evidence, i.e., from similarity of orbital elements, exists for several possible pairs of split comets: C/1988 F1 (Levy) and C/1988 J1 (Shoemaker-Holt) (*Bardwell*, 1988); C/1988 A1 (Liller) and C/1996 Q1 (Tabur) (*Jahn*, 1996); C/2002 C1 (Ikeya-Zhang) and C/1661 C1 (*Green*, 2002); C/2002 A1 (LINEAR) and C/2002 A2 (LINEAR) (*Sekanina et al.*, 2003), C/2002 Q2 (LINEAR) and C/2002 Q3 (LINEAR) (*Adams*, 2002). Backward integration of the orbits of periodic Comets 42P/Neujmin 3 and 53P/Van Biesbroeck suggests a good agreement of their orbits before 1850 when a very close encounter with Jupiter may have occurred (*Carusi et al.*, 1985).

Tancredi et al. (2000) have addressed the question of families of split pairs and families among short-period comets through a statistical approach. The authors analyzed the dynamical taxonomy of Jupiter-family comets and near-Earth asteroids (NEAs) using clustering of Lyapunov indicators derived from the orbital elements of the objects. A splitting hypothesis for the Jupiter-family comets, i.e., to originate from a "giant" 50-km nucleus (comparable in size to a small Kuiper belt object), is not very likely. Moreover, they found that the contribution of split comets to the population of near-Earth asteroids is small, if at all significant. The clustering of the Lyapunov indicators of Comets 42P/Neujmin 3 and 53P/van Biesbroeck with those of Comets 14P/Wolf and 121P/Shoemaker-Holt 2 is not only supporting a splitting scenario for the former pair of comets, but may even suggest the involvement of further candidates, i.e., the latter two comets.

In order to model the dynamics of splitting events over more than one orbital revolution, *Sekanina and Chodas* (2002a,b) have integrated planetary and nonrelativistic perturbations in the original approach of *Sekanina* (1982) (see section 2.1). The authors used their enhanced model to link major Sun-grazing comets as fragments of parent bodies (fitting V_r, V_t, V_n, but neglecting the deceleration parameter Γ): Comets C/1965 S1 (Ikeya-Seki) and C/1882 R1 (Great September Comet) were produced by a common parent that split in 1106 shortly after perihelion passage [this was already suggested by *Marsden* (1967)]. Moreover, the motion of Comet C/1970 K1 (White-Ortiz-Olelli) is consistent with a scenario in which the parent was an unknown third fragment of the 1106 splitting event, and the separation of C/1970 K1 (White-Ortiz-Olelli) occurred in the eighteenth century at a large heliocentric distance of about 150 AU. C/1880 C1 (Great Southern Comet) split off C/1843 D1 (Great March Comet) at 2.5–3 AU after perihelion passage in the eleventh century [also previously suggested by *Marsden* (1989), with breakup in the fifteenth century only). As a by-product, but most interesting for linking orbits of fragments over longer time intervals, *Sekanina and Chodas* (2002a) summarize the orbit perturbations of Sun-grazing comets that split since the last perihelion passage with nonzero separation velocity.

3.2. Kreutz Group and Solar and Heliospheric Observatory (SOHO) Comets

The Kreutz group comets (also called Sun-grazers) are comets that approach the Sun to a perihelion distance <2.5 solar radii. The number of discovered Sun-grazer comets has increased tremendously with the advent of coronagraphic observations from satellites, e.g., SOLWIND, Solar Maximum Mission (SMM), and in particular the Solar and Heliospheric Observatory (SOHO), which has detected several hundred new objects (*Sekanina,* 2000b). Two main families (I and II, with a further division into two subgroups for family II) are identified through statistical methods [clustering of orbital elements (*Marsden,* 1967, 1989; *Sekanina,* 2002a)]. There also exists a non-Kreutz near-Sun comet group among the SOHO comets that is characterized by similar orbital elements (*Meyer and Marsden,* 2002). Practically all smaller Sun-grazers (i.e., most of the so-called "SOHO" comets) do not survive perihelion passage (*Sekanina,* 2000a,b), but the larger (i.e., brighter) ones in the Kreutz group do.

Nucleus splitting of the Kreutz group and SOHO comets suggests the following: (1) some larger objects (*Marsden,* 1967, 1989; *Sekanina,* 2002a,b) are among the list of split comets (see Appendix); (2) there are tremdendous similarities in the orbital elements of these comets and their subgroups (*Marsden,* 1967, 1989; *Meyer and Marsden,* 2002); and (3) the SOHO comets show a significant temporal clumping since more than 15 pairs of comets were observed that appeared within less than 0.5 d in similar orbits within the field of view of the SOHO coronagraph. A dynamical analysis of the latter (*Sekanina,* 2000b) indicates that the pairs originate from fragmentation events all along the orbit, i.e., not necessarily close to the Sun, but more or less at any point along the orbit, even near aphelion. The dissolution of the smaller SOHO comets close to the Sun happens before reaching perihelion and it is frequently indicated through an activity flare at ~3 solar radii with a subsequent drop in brightness. Sekanina attributes the disappearance of these comets to an "erosion" effect of the nuclei due to the strong heating of the Sun.

Sekanina (2000b) has introduced a scenario for the formation of this group of comets involving a parent nucleus that split into two major fragments, possibly through tidal forces or at least tidally triggered, during perihelion very close to the Sun. After the break-up the two major fragments evolved into slightly different orbits, but continued to split along their paths around the Sun, generating a cascade of tertiary components of very different size that form the group and subgroups of Kreutz comets.

3.3. Nucleus Splitting and the Evolution of Comets

The role of nucleus splitting in the evolution of comets is widely unexplored. Multiplicity and persistence of fragments and recurrence of the splitting phenomenon in comets would argue for the existence of break-up families among the current population of comets. And indeed some indications are found (see sections 3.1 and 3.2). The Lyapunov indicator cluster analysis by *Tancredi et al.* (2000) suggests that break-up families are not very abundant among Jupiter-family comets. This could imply that the production of long persistent fragments by short-period comets is low and/or that the decay rate of the break-up products is fast compared to the typical dissipation timescale for the Lyapunov indicators of these class of comets (on the order of several hundreds of years).

From simple estimates on the mass loss due to recurrent nucleus splitting events it becomes clear that fragmentation may be an efficient destruction process for comets. For instance, from the current catalogue of about 160 short-period comets, 10 objects are known to have split, three of them repeatedly (16P/Brooks 2, 51P/Harrington, 141P/Machholz 2); one object (3D/Biela) disappeared completely after nucleus splitting. The observational baseline for this class of comets is on the order of 200 years, and it must be assumed that some other splitting events escaped detection. Thus, the numerical splitting rate of ~3% per century and object may only represent a lower limit for short-period comets. This rough order-of-magnitude result is consistent with the splitting rate estimate of at least 1 event per century and comet published by *Chen and Jewitt* (1994) based on observations of break-up events of comets in general. *Weissman* (1980) reports splitting rates of 1%, 4%, and 10% for short-period, long-period, and dynamically new comets based on a sample of 25 objects (of which 7 are at least questionable candidates).

Over its mean lifetime in the inner planetary system, a short-period comet may experience about 1000 splitting events. If the average mass loss in these events corresponds to a 50-m fragment, the total mass loss by nucleus splitting can amount to 500–1000-m equivalent radius over the lifetime of a comet, i.e., it is on the order of the typical size of the nuclei of short-period comets. Therefore, nucleus splitting may represent an important mass loss factor in the life of a comet and should be considered carefully in the scenarios for the evolution and "end state" of comet nuclei.

4. SECONDARY EFFECTS: OUTBURSTS AND COMA ARCLETS

The main effect of comet splitting is the appearance of one or more companions of a primary component (Type A). In a very few cases, many more fragments appear (quasi-)simultaneously and without clear indication of a primary component among them (Type B). Unfortunately, in general the existence of primary and secondary components becomes detectable in direct images only long (typically weeks) after the time when the splitting actually occurred. Activity outbursts and arclets in the coma of a comet can indicate the occurrence of a nucleus break-up event at or shortly after the time when the comet splits.

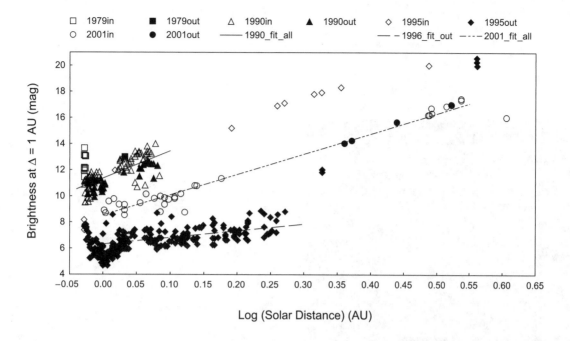

Fig. 3. Pre- and post-break-up visual lightcurve of Comet 73P/Schwassmann-Wachmann 3. The lightcurve of the total brightness estimates in the visual wavelength range over four apparitions of the comet from 1979 to 2001. Nucleus splitting happened around perihelion 1995 and shortly thereafter. Even one apparition, i.e., 4–6 years later, the brightness of the comet (component C) is still 2–3 mag above normal level before break-up. The observations obtained during the 1979, 1990, 1995–1996, and 2001 perihelion passages are marked by symbols. Least-squares fits to the various apparition lightcurves are indicated by lines. In the legend, numbers next to the symbols and lines denote the year of the comet apparition, "in" stands for inbound, "out" for outbound, "all" for all data of the apparition, "fit" for least-squares fit.

4.1. Activity Outbursts

Outbursts in the visual lightcurve of comets can indicate splitting events. There are prominent cases that demonstrate the temporal relationship between nucleus splitting and activity outbursts with amplitudes of 3 mag and more: C/1975 V1 (West), 73P/Schwassmann-Wachmann 3, and C/1999 S4 (LINEAR) as described by *Sekanina* (1982), *Sekanina et al.* (1996), and *Green* (2000). Smaller lightcurve peaks and nucleus break-ups are associated with Comets C/1899 E1 (Swift), C/1914 S1 (Campbell), C/1943 X1 (Whipple-Fedtke), C/1969 T1 (Tago-Sato-Kosaka), and C/1975 V1 (West) (see *Sekanina*, 1982); C/1986 P1 (Wilson) (see *Meech et al.*, 1995); and 73P/Schwassmann-Wachmann 3, C/1996 J1 (Evans-Drinkwater), and C/2001 A2 (LINEAR) (*Sekanina*, 1998; *Sekanina et al.*, 1996, 2002). The rise times of these outbursts, if measurable, last for a few (2–20) days. The durations of the activity outbursts have a very wide range, from a few days to months or maybe even years. Figure 3 shows the lightcurve of 73P/Schwassmann-Wachmann 3 during the past four apparitions observed: The lightcurves in 1979 and 1990, plus most of the preperihelion phase in 1995, define the (rather repetitive) normal activity level before break-up of the comet in autumn 1995. The postperihelion lightcurve in 1995–1996 is about 5 mag (factor of 100) brighter than normal due to the break-up events around perihelion and thereafter, and even during the sub-

sequent perihelion passage the comet remains 2–3 mag brighter than before splitting.

The outbursts identified in the visual and most of the broadband brightness estimates and measurements of comets indicate a higher dust content in the coma. Most of this dust in the visible is of micrometer size (*McDonald et al.*, 1987). Most of the outbursts to which nucleus break-ups can be associated (*Sekanina*, 1982; *Sekanina et al.*, 1996, 2002) peak several days after the estimated dates of these events, suggesting that additional dust is released during or — more likely — after the splitting of the comet. The early phases of the dust expansion after break-up events are documented for C/1999 S4 (LINEAR) by *Schulz and Stüwe* (2002) and *Kidger* (2002). Outbursts in the gas production, although in smaller and short-term events difficult to observe, are also reported for split comets, e.g., 73P/Schwassmann-Wachmann 3 (*Crovisier et al.*, 1996) and C/1999 S4 (LINEAR) (*Mäkinen et al.*, 2001; *Farnham et al.*, 2001). For C/1999 S4 (LINEAR) the published measurements show a rapid decay of the gas and dust production of the comet in late July 2000. This in turn suggests that the reservoir of sublimating ice in this comet was exhausted rapidly after the complete disruption of the nucleus.

However, the relationship between outbursts in the lightcurve on one side and splitting events on the other is not "one-to-one": Not all splitting events are accompanied by noticeable outbursts (e.g., fragment E in 73P/Schwassmann-

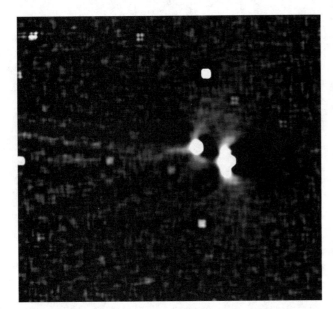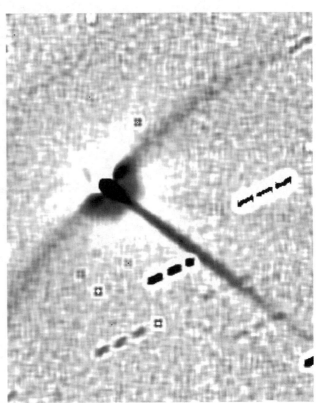

Fig. 4. Coma arclets of fragments of Comet C/2001 A2 (LINEAR). These broadband R-filter exposures of the comet show three coma arclets observed during the break-up episodes of the comet. In the left image (taken on 18 May 2001) two arclets are seen, one (to the left) between companions B and C (both not directly visible in the image) and another one (to the right) close to fragment A. The former arclets seem to be associated with the splitting of component C from B on 10 May 2001, while for the latter case no companion of fragment A is reported. The right image (taken on 13 July 2001) shows another very wide arclet around component B. Apart from the main component no further fragments could be identified. However, the straight tail-like extension away from the Sun may represent dust released by the invisible fragment(s) that might have split off fragment B a few days before the image was taken. The images are taken with the New Technology Telescope (NTT) and the Very Large Telescope (VLT) of the European Southern Observatory (ESO), respectively. North is up and east to the left; field of view is 2.5 × 1.9 arcmin² in the left and 1.0 × 1.2 arcmin² in the right panel. Image courtesy of E. Jehin et al., ESO Chile.

Wachmann 3) and not all outbursts indicate splitting events that produce detectable companions (e.g., the 10-mag outburst of Comet 52P/Harrington-Abell observed from July 1998 to February 1999). Outbursts of smaller amplitude (1–3 mag) may also occur as episodes of enhanced activity of the comet without obvious splitting of the nucleus (see *Prialnik et al.,* 2004; *Meech et al.,* 2004). Activity outbursts of splitting events start around the estimated time of the nucleus break-up and they usually reach peak brightness a short time (order of days) thereafter. However, it is not clear whether brightness outbursts are associated with the actual cause of the break-up or are more a consequence of the nucleus splitting.

The visual and broadband filter lightcurves of companions in split comets show a systematic decay in brightness and intrinsic short-term variability of the fragments (*Sekanina,* 1982, 1998; *Sekanina et al.,* 1996, 1998). Both phenomena seem to be due to outgassing behavior of fresh material from the interior of the original nucleus that may

be exposed to sunlight for the first time since the formation of the comet. The measurements of lightcurves for individual components usually suffer from the overlap of the comae shortly after the splitting event such that the measured magnitudes are contaminated by light from the neighboring coma(e).

4.2. Coma Arclets

Coma arclets, also called "coma wings" because of their bird-shaped appearance, have been seen in three comets shortly after splitting events that produced short-lived companions of the primary nuclei, i.e., in Comets C/1996 B2 (Hyakutake) (*Harris et al.,* 1997; *Rodionov et al.,* 1998), C/1999 S4 (LINEAR) (*Farnham et al.,* 2001), and C/2001 A2 (LINEAR) (*Jehin et al.,* 2002). They show up easily when some simple structure enhancement (like wavelet or adaptive Laplace filtering or radial renormalization) is applied to the flat-fielded images. Figure 4 shows examples of arclets observed during break-up of Comet C/2001 A2 (LINEAR).

The arclet structure appears to be located in between two split companions. The observed arcs are almost perpendicular to the connecting line of the fragments, rather symmetric on both sides and preferably — but not exclusively — with tailward curvature. The observed arclets extended over 1000 to 10,000 km on both sides and intersected the connecting line of the fragments at a few 100 to a few 1000 km projected distance. They appeared soon (within 10 d) after the fragmentation event of the nucleus and faded away within 3–5 d after first appearance. From narrow and broadband imaging in the visible and near-IR (*Harris et al.*, 1997), it is clear that the coma arclets are made of gas (OH, CN, and C_2 gas was identified). Dust does not participate in their formation. Nevertheless, arclets are also detectable in broadband images taken in visible wavelengths if their gas content is large enough and covered by the filter bandpasses. Thus far, coma arclets have only been reported in split comets close to quadrature position and at distances close to Earth. The importance of both conditions on the visibility of the phenomenon is presently unclear.

Three physical interpretations of the coma arclets are published. *Harris et al.* (1997) proposed an arc model involving gas release from the primary nucleus plus an extended source located on the connecting line toward the secondary component. The extended source is claimed to be a train of boulders produced during the splitting event and emitted in the same direction as the major secondary fragment. No shock wave of gas is predicted in this model, but the main contribution to the arclets should come from the gas released by the boulder train. Indeed, in the case of Comet C/1996 B2 (Hyakutake) a straight spike of diffuse light, typical for dust streamers, was seen along the connecting line of the two fragments at the time when the arclets occurred. A similar phenomenon was found for one of several arclets observed in Comet C/2001 (LINEAR), although in this particular case no fragment could be detected (*Jehin et al.*, 2002). *Rodionov et al.* (1998) model the arclets of Comet C/1996 B2 (Hyakutake) through a two-source (the two fragments) outflow of rarefied supersonic gases that produce shock waves in the region between the two components. The shock waves are best visible edge-on (i.e., close to quadrature geometry of the comet). This model involves activity on the night side of the primary nucleus — otherwise no shock front is formed. *Farnham et al.* (2001) interpret the arclets seen in Comet C/1999 S4 (LINEAR) before the major break-up of the nucleus in July 2001 as a dust jet from an active region close to the equator of a fast rotating nucleus. According to this scenario, the rotation axis should point toward the Sun. This scenario certainly has some difficulties in explaining the many arclets of gaseous origin seen in the other two comets.

Even though the physical nature of coma arclets in split comets is not yet clearly understood, there is no doubt that they can be considered as early tracers of nucleus break-up events.

5. PHYSICO-CHEMICAL PROPERTIES OF SPLIT COMETS AND THEIR FRAGMENTS

The observations of split comets, and in particular the measurements of the fragments of Comet C/1999 S4 (LINEAR) (*Weaver et al.*, 2001), suggest that "solid" secondary bodies are produced by fragmentation of a primary nucleus. If one assumes that the fragments are the original building blocks of cometary nuclei, the break-up of Comet C/1999 S4 (LINEAR) provides indications of the typical size distribution of cometesimals, at least for the (yet unknown) region of the planetary disk where its nucleus was formed. The former assumption, however, can be questioned, at least for comets coming from the Kuiper belt region [such as the short-period comets (*Farinella and Davis*, 1996)] if one considers the collision environment of the belt that may have created the population of comet-size bodies through collisional break-ups of larger objects over the lifetime of the solar system. The impact energy induced in a Kuiper belt body through long-term bombardment is of an amount that could potentially modify the constitution of the whole body or at least a major part of it.

Size estimates of nuclei before or during the fragmentation episode exist only for two comets, 73P/Schwassmann-Wachmann 3 [radius 1.1 km (*Boehnhardt et al.*, 1999)] and C/1996 B2 (Hyakutake) [radius 2.4 km (*Lisse et al.*, 1999)]. Photometric measurements of fragment sizes are published for C/1999 S4 (LINEAR) [50–100 m for 4% albedo (*Weaver et al.*, 2001)]; a few more size estimates of fragments or upper limits were derived from the dynamical models of the splitting event (*Sekanina*, 1982; *Sekanina et al.*, 1996; *Desvoivres et al.*, 1999; *Boehnhardt et al.*, 2002), from the brightness evolution of Sun-grazer and SOHO comets [50 km to 5 m (*Sekanina*, 2000b, 2002a,b)] and from the break-up of Comet D/1993 F2 (Shoemaker-Levy 9) (*Sekanina*, 1995a; *Asphaug and Benz*, 1996). Lower limits of the fragment's sizes were also derived from the explosion blankets produced during the impacts of the split Comet D/1993 F2 (Shoemaker-Levy 9) on Jupiter (see, e.g., *Ortiz et al.*, 1995). Mass estimates of the fragments are provided by the authors assuming a bulk density for the nucleus material. A size distribution function N(R) for the fragments of C/1999 S4 (LINEAR) was derived by *Mäkinen et al.* (2001): $N(R) \sim R^{-2.7}$ (R for the radius of the fragment). However, the overall size or mass budget of split comets (before and after break-up) remains unknown since it was not yet measured for individual objects.

As mentioned in section 2, some of the fragments of split comets are "persistent" and endure for several years. It seems likely that independent and long-lived cometary nuclei may evolve. Other fragments have very short lifetimes of only a few days to weeks. The fragments of C/1999 S4 (LINEAR) survived intact after the nucleus disruption for about 2–3 weeks (see Fig. 5). Thereafter, they disappeared quickly — and "collectively" — within a few days. Exactly what happens to the short- and long-lived

Fig. 5. Many short-lived fragments of Comet C/1999 S4 (LINEAR). This R-filter image taken in early August 2000 at the ESO VLT shows at least 16 short-lived fragments that were produced in the break-up of the nucleus between 21 and 24 July 2000. The fragments are embedded in a diffuse coma of which a long dust-tail streamer extends away from the Sun. Image processing is used to increase the contrast of the fragments on the diffuse coma background. North is up and east to the left; field of view is 3.4 × 2.5 arcmin². Image courtesy of European Southern Observatory (ESO).

fragments when they disappear is not known: Do they dissolve into even smaller pieces, or do they become inactive?

From the 10 split comets (4 short-period and 6 long-period) that are classified taxonomically, 7 comets [16P/Brooks 2, 69P/Taylor, 101P/Chernykh, 108P/Ciffreo, C/1975 V1 (West), C/1986 P1 (Wilson) (see *A'Hearn et al.,* 1995), and C/1999 S4 (LINEAR) (see *Mumma et al.,* 2001)] belong to the group of carbon-depleted objects, and 3 comets [C/1988 A1 (Liller) (see *A'Hearn et al.,* 1995), C/1996 B2 (Hyakutake) (*Schleicher and Osip,* 2002), and C/2001 A2 (LINEAR) (see *Jehin et al.,* 2002)] appear to be "typical" in their carbon content. A link between this taxonomic parameter and the splitting behavior of the nucleus is not obvious. The chemical composition of fragments is known even less, and not even a single fragment has measured production rates of gas and/or dust. *Bockelée-Morvan et al.* (2001) have inferred from gas production rates of the coma of C/1999 S4 (LINEAR) before and after the fatal splitting in July 2001 that the nucleus of this comet may have had a rather homogeneous chemistry. This conclusion would support the (unproven) scenario that this nucleus may have contained cometesimals that were formed in the same region of the planetary formation disk.

6.   FRAGMENTATION MECHANISMS

Several fragmentation mechanisms are used to explain the splitting of cometary nuclei. Thus far, the success of these scenarios in the understanding of these events is limited, presumably since (1) the most important parameters

of comet nuclei (such as internal structure, nucleus/surface stratification, material types, tensile and shear strengths, size, and rotation) used in these models are not at all or not very well known, and (2) the available observations do not constrain well the actual event sequence and the physical properties of the parent and daughter components of split comets. Not surprisingly, only for one split comet, D/1993 F2 (Shoemaker-Levy 9), do modelers seem to agree on the fragmentation scenario (i.e., tidal splitting close to Jupiter), although with significant differences in the details of interpretation and conclusion.

6.1.   Scenarios

6.1.1. Tidal splitting. Tidal splitting of a body (comet nucleus) in the neighborhood of a large mass (a planet or the Sun) is induced when the differential gravitational "pull" of the large mass throughout the small body exceeds the forces of self-gravity and material strength (tensile and/or shear) of the latter. A simplified condition for tidal disruption of spherical bodies was published by *Whipple* (1963)

$$\sigma < GM_0\rho R^2/\Delta^3 \qquad (3)$$

The parameter σ is the tensile strength of the material, G is the gravitational constant, M_0 is the mass of the large body, ρ and R are the bulk density and radius of the sphere, and Δ is the distance between the two bodies. A rigorous theoretical treatment of the problem for spheres and biaxial ellipsoids can be found in *Davidsson* (1999, 2001).

The models predict that the break-up should start from the center of the nucleus and that it should affect the body as a whole. The products of tidal splitting should be larger pieces in the center of the nucleus and smaller ones toward the surface of the body. This latter prediction, however, may depend on the internal structure of the nucleus as well. Obviously, this scenario works only in the neighborhood of heavy bodies. Tidal forces, even if not causing the nucleus splitting, can be responsible for major cracks in the body that weaken its structural strength such that it may split later as the result of another process (e.g., thermal or rotational splitting).

6.1.2. Rotational splitting. Splitting of a rotating nucleus happens when the centrifugal force exceeds self-gravity and material strength inside the body. A simplified expression for the condition of disruption of a rotating sphere is given by *Sekanina* (1982)

$$\sigma < 2\pi^2\rho R^2/P^2 = 1/2\rho V_{rot}^2 \qquad (4)$$

with σ, ρ, and R as explained above; P is the rotation period and V_{rot} is the rotation velocity at the equator of the sphere. A comprehensive theoretical model for centrifugal forces in rotating spheres and biaxial ellipsoids is presented by *Davidsson* (1999, 2001). The acceleration of the rotation speed of the nucleus can be caused by reaction forces due to outgassing (see *Jorda and Gutiérrez,* 2002).

The prediction of the model is that "dense" nuclei with nonnegligible material strength should break up from the body center, while strengthless nuclei should loosen fragments from the surface. The properties of the fragmentation products are case dependent, i.e., larger pieces in the center and smaller fragments at the surface for the case of "dense" nuclei or — more likely — only smaller pieces for strengthless bodies. Rotational splitting depends mainly on the rotation motion of the nucleus and can happen at any distance from the Sun. Due to changes of the rotational state of the nucleus by reaction forces from comet activity and modification of the properties of surface material by the mass loss of the nucleus when active, the occurrence of rotational splitting may in principle happen randomly along the orbit, but clearly with a preference for solar distances where the comet is active.

6.1.3. Splitting by thermal stress. Due to their variable distances to the Sun, comet nuclei are exposed to diffusion of heat waves penetrating into their interior during orbital revolutions. Thus thermal stress is induced in the body and, if the material strength is exceeded, nucleus splitting may occur. *Tauber and Kührt* (1987) have considered both homogeneous bodies (water ice) and nuclei with material inhomogeneities (water ice with inclusions of CO_2 and silicates). In both cases cracks due to thermal stress can form on the surface and, subsequently, minor pieces could split from the comet. *Shestakova and Tambostseva* (1997) and *Tambostseva and Shestakova* (1999) have presented model calculations for comet splitting by thermal stress. A number of cases are distinguishable depending on nucleus size and solar distance: Break-up may occur for larger bodies due to compression stress, splitting of subkilometer-sized bodies due to radial stress may happen closer than 40 AU from the Sun, and thermal splitting in general should be efficient — provided that tensile strength of the body material is low — when the object is closer than 5 AU to the Sun.

The fragmentation products should depend on the cause. The extend to which the body is affected by thermal stress splitting depends on the depth of the heat wave penetration and thus also on the size of the nucleus: Smaller bodies (subkilometer-sized) can split as a whole, while the break-up of surface fragments is more likely for larger bodies. Thermal stress splitting clearly is a scenario that may be able to produce fragments even at larger distances (several 10 AU) from the Sun.

6.1.4. Splitting by internal gas pressure. High gas pressure in the nucleus can be caused by sublimation of subsurface pockets of supervolatile ices (e.g., CO) when the comet approaches the Sun and the heat wave from the increasing solar illumination reaches the depths of these ice pockets. If the gas pressure cannot be released through surface activity, the tensile strength of the nucleus material can be exceeded and fragmentation of the comet occurs. *Kührt and Keller* (1994) present models for crust formation and the buildup of vapor pressure underneath. They conclude that a purely gravitationally bound crust is unstable and will be blown off the nucleus during perihelion passage. However, a crust of porous material held together by cohesive forces can withstand internal gas pressure up to the tensile strength estimated for cometary nuclei (see section 6.2).

Two different scenarios for comet break-up by internal gas pressure have been proposed: (1) an explosive blow-off of localized surface areas (possibly covered by an impermeable crust) as described by *Whipple* (1978), *Brin and Mendis* (1979), and *Brin* (1980); or (2) a complete disruption of the nucleus as suggested by *Samarasinha* (2001). The latter case imposes additional "requirements" on the internal structure of the nucleus and its surface: It should allow gas diffusion throughout the whole body via a system of connecting voids in the nucleus, and before splitting, the surface does not outgas enough to efficiently reduce the gas pressure inside the nucleus. Since both scenarios are based on comet activity, they are restricted to orbit arcs not far from the Sun, even though sublimation of supervolatile ices such as CO and N_2 can occur up to ~50 AU solar distance (*Delsemme,* 1982). *Prialnik and Bar-Nun* (1992) have proposed crystallization of amorphous ice to explain the outburst activity of Comet 1P/Halley at 14 AU outbound. This scenario could also potentially work to produce internal gas pressure that may cause the fragmentation of cometary nuclei.

6.1.5. Impact-induced comet splitting. During their orbital revolution around the Sun, comet nuclei can experience (hypervelocity) impacts by other solar system bodies such as asteroids. Since comets are small, such an impact, if it happens, will most likely destroy the whole nucleus, even if the impactor is a small (subkilometer-sized) body itself. *Toth* (2001) considered asteroid impacts for the disruption of Comet C/1999 S4 (LINEAR). Impact probabilities and the range of impact energies due to meter-sized impactors from the asteroid belt on short-period comets are estimated by *Beech and Gauer* (2002).

A "modification" of this scenario is comet splitting by impacts of larger boulders produced by the comet itself. Such pieces may exist, and it is feasible that they can travel "aside" the comet in its orbit around the Sun. Impact may occur at intersection points of their orbits, e.g., near aphelion for boulders produced near perihelion. As for most of the other scenarios described in this section, no detailed analysis and prediction of observable effects are available.

6.2. Observational Facts and Constraints

6.2.1. Comet D/1993 F2 (Shoemaker-Levy-9). Comet D/1993 F2 (Shoemaker-Levy 9) broke up in 1992 during a close approach with Jupiter (<20,000 km above the cloud level of the giant planet) (Fig. 6). Modelers (*Sekanina,* 1994; *Asphaug and Benz,* 1996; and references contained therein) of this event agree that the tidal forces of Jupiter have caused the cracking of the nucleus structure of this comet. However, according to *Sekanina* (1994) the separation of the fragments started only 3 h after the time of closest approach to Jupiter, i.e., after the tidal forces reached maximum amplitude. Apparently, the largest fragments traveled

Fig. 6. Tidally disrupted chain of fragments in Comet D/1993 F2 (Shoemaker-Levy 9). This R-filter exposure taken on 5 May 1994 at the Calar Alto 3.5-m telescope shows fragments F to W of the broken comet. The fragments' chain extends diagonally across the image. Each fragment is surrounded by its own coma while their diffuse and wider dust tails overlap, causing a brighter background above the image diagonal. North is up and east to the left; field of view is 4.3 × 2.5 arcmin². Image courtesy of K. Birkle, Max-Planck-Institut für Astronomie, Heidelberg.

in the middle of the "chain" of the known 23 Shoemaker-Levy 9 components, as inferred from the size estimates and the impact explosions at Jupiter in July 1994. This picture would be in agreement with the tidal break-up model, which expects larger fragments to be created in the center of the splitting body, while lighter and smaller ones, i.e., the fragments at the leading and trailing end of the Shoemaker-Levy 9 chain, arose closer to the surface. Similar signatures are also seen in some peculiar crater chains at the surface of the jovian moons Callisto and Ganymede (*Schenk et al.,* 1996). The crater chains in the icy crust of these satellites are believed to be caused by impacts of narrow ensembles of fragments from tidally split comets after close encounters with Jupiter. In Comet Shoemaker-Levy 9, tertiary splitting occurred in some of the fragments, in all cases for unknown reasons (and certainly not due to tidal forces), but clearly suggesting that the split components may have had intrinsic substructure (*Sekanina,* 1995a).

Tidal splitting at Jupiter or the Sun is claimed to be involved in the break-up of Comets 16P/Brooks 2 (*Sekanina and Yeomans,* 1985) and the Sun-grazer Comets C/1882 R1 (Great September Comet), C/1963 R1 (Pereyra) and C/1965 S1 (Ikeya-Seki) (*Sekanina,* 1997a). The break-up mechanisms of all other split comets remain unknown, even though one may favor rotational break-up for the short-period comets because of the range of observed separation velocities and their independence from the heliocentric distance of the splitting event. Comet C/1999 S4 (LINEAR) has certainly experienced a nucleus splitting of a somewhat unique nature, since its nucleus disrupted in many pieces that disappeared after a lifetime of a few weeks (see also section 7.1). Nucleus splitting at very large heliocentric distances [beyond ~100 AU as suggested for Comet C/1970 K1 (White-Ortiz-Olelli) (*Sekanina and Chodas,* 2002b)]

excludes all activity driven models as the fragmentation mechanism.

6.2.2. Tensile strength. Thus far, the tensile strength of a cometary nucleus is less constrained by actual observations and modeling of splitting events than by comets that do not split. The large size of Comet C/1995 O1 (Hale-Bopp) together with its fast rotation of 11.5 h puts a lower limit of 10^4–10^5 dyn/cm² on the tensile strength of its nucleus (assuming a bulk density of 0.5–1 g/cm³). Assuming a similar tensile strength for the nuclei of comets for which reliable size *and* rotation period estimates exist, it is clear that these comets are — at present — "safe" against rotational break-up. On the other hand, if rotational break-up is involved in the splitting of short-period comets, a similar range for the tensile strength as for Comet Hale-Bopp would follow from the observed separation velocities of the fragments (see section 2.2). Unless one assumes a special nature for the bodies of split comets, it is obvious from the existence of fragments that the nuclei of split comets are not strengthless and they have an intrinsic substructure or at least nonuniform tensile strength.

7. RELATED PHENOMENA: DISAPPEARING COMETS AND DUST-TAIL STRIAE

7.1. Disappearing Comets

Comets can disappear in front of the "eyes" of the observers without obvious indication of a dramatic nucleus fragmentation event: C/1988 P1 (Machholz), Comets C/1996 Q1 (Tabur), C/2000 W1 (Utsunomiya-Jones) (see Fig. 7), C/2002 O4 (Hönig), and C/2002 O6 (SWAN) are some of the more recent cases. Leftovers of these disappearing comets are diffuse and fading comae and so-called "truncated" dust tails in which the synchrones are only populated to a certain start time at the nucleus and no "younger" grains are found in the dust tails. Two scenarios should be mentioned that could explain the observations: (1) the complete disintegration of the nucleus similar to Comet C/1999 S4 (LINEAR) (see below) and (2) an evolved, very crusty nucleus with one or only a few active regions that become "suddenly" dormant due to shadowing from solar illumination when the comet moves along its orbit. Both scenarios imply that disappearing comets are to be considered within the terminal phase of cometary evolution. It may be noteworthy that gas jets, perpendicular and symmetric to the Sun-comet line and without counterparts in the dust, were observed in Comet C/1996 Q1 (Tabur) before disappearance of the comet (*Schulz,* 2000). These jets very much resemble the arclets seen during splitting events in other comets (see section 4.2). Since C/1996 Q1 (Tabur) together with Comet C/1988 A1 (Liller) is most likely a product of a splitting event during an earlier apparition (*Jahn,* 1996), it is feasible that the former comet has experienced further splitting events that may have culminated in an — unobserved — complete dissolution of the nucleus during its last return to the Sun. The disappear-

Fig. 7. Comet C/2000 W1 (Utsunomiya-Jones), which disappeared in early 2001. This R-filter image, taken on 19 February 2001 at the 3.5-m Telescopio Nazionale Galileo (TNG) at the Roque de los Muchachos Observatory in La Palma, shows only a very weak and diffuse dust cloud without central condensation at the position of the comet (center and uppermost part of the image). North is up and east to the left; field of view is 3.0×2.6 arcmin^2. Image courtesy of J. L. Ortiz, Instituto Astrofisico de Andalucia, Granada.

ance together with the later perihelion passage of C/1996 Q1 (Tabur) to C/1988 A1 (Liller) also supports the interpretation of C/1996 Q1 being the fragment of C/1988 A1.

After some smaller splitting events before perihelion, Comet C/1999 S4 (LINEAR) broke apart completely in the second half of July 2000 close to perihelion (see Fig. 5). More than 20 fragments, but no "dominant" primary fragment, were observed (*Weaver et al.,* 2001). About three weeks after the disruptive splitting event the fragments could not be detected, and it is assumed they disappeared, more or less collectively, either by further fragmentation or by becoming undetectable due to exhaustion of sublimation activity. The rapid decay of water gas production after break-up (*Mäkinen et al.,* 2001) supports the scenario that this comet disappeared completely in a diffusing and fading cloud of previous dust release. This splitting comet can be considered the prototype (since best studied) of a disappearing comet.

7.2. Dust-Tail Striae and Comet Splitting

The origin of striae in the dust tail of some bright comets, e.g., C/1975 V1 (West) and C/1995 O1 (Hale-Bopp) (see Fig. 8), suggest secondary fragmentation of house-sized boulders (*Sekanina and Farrell,* 1980). The previous two-step model scenario introduced by *Sekanina and Farrell* (1980) involved relatively large pieces with very high solar radiation pressure parameter β (>0.1) as "parents" of striae that split off the cometary nucleus weeks before dis-

ruption. The new interpretation, proposed by Z. Sekanina and H. Boehnhardt at the Cometary Dust Workshop 2000 held in Canterbury, implies much earlier separation times of the parent fragments from the main nucleus, i.e., the boulders are produced by the cometary nucleus far away from the Sun and drift slowly away from the primary nucleus (hence no need for a high β) to the distance where, during approach of the comet to the Sun, the secondary fragmentation occurs in the region of the dust tail. This secondary disintegration is a process of short duration (on the order of one day or less) that may affect a boulder as a whole, i.e., it may become completely dissolved. An interesting candidate mechanism for the fragmentation of boulders is gas and dust emission activity when boulders approach the Sun. The proposed striae fragmentation hy-

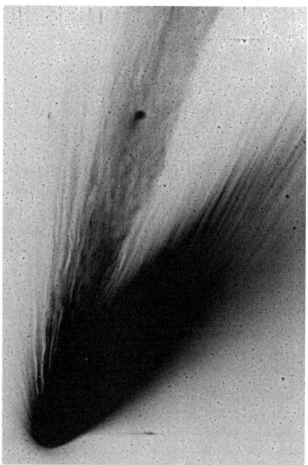

Fig. 8. Striated dust tail of Comet C/1995 O1 (Hale-Bopp). In March and April 1996 the diffuse dust tail (right part of the image) of Comet C/1995 O1 (Hale-Bopp) contained many striae. In the image the striae are best visible as narrow straight bands to the outer edge of the dust tail. The striae are not coinciding with dust synchrones pointing toward the nucleus in the coma (lower left, overexposed), which indicates that the striae dust is not released directly from the nucleus. The prominent structured ion tail points away from the direction of the Sun. North is up and east to the left; field of view is 3.0×4.5 deg^2. Image courtesy of K. Birkle, Max-Planck-Institut für Astronomie, Heidelberg.

TABLE A1.   List of split comets, likely split pairs, and families of split comets.

Tidally split comets

C/1882 R1 (Great September Comet)*	16P/Brooks 2 (1889 + 1995)*
C/1963 R1 (Pereyra)*	C/1965 S1 (Ikeya-Seki)*
D/1993 F2 (Shoemaker-Levy 9)†	

Comets split for unknown reasons

3D/Biela (1840)	C/1860 D1 (Liais)
C/1888 D1 (Sawerthal)	C/1889 O1 (Davidson)
D/1896 R2 (Giacobini)	C/1899 E1 (Swift)
C/1906 E1 (Kopff)	C/1914 S1 (Campbell)
C/1915 C1 (Mellish)	69P/Taylor (1915)
C/1942 X1 (Whipple-Fedtke)	C/1947 X1 (Southern Comet)
C/1955 O1 (Honda)	C/1956 F1 (Wirtanen)
C/1968 U1 (Wild)	C/1969 O1 (Kohoutek)
C/1969 T1 (Tago-Sato-Kosaka)	C/1975 V1 (West)
79P/du-Toit-Hartley (1982)	108P/Ciffreo (1985)
C/1986 P1 (Wilson)	101P/Chernykh (1991)
C/1994 G1 (Takamizawa-Levy)	141P/Machholz 2 (1987 + 1989)
51P/Harrington (1994 + 2001)	73P/Schwassmann-Wachmann 3 (1995/1996 + 2001)
C/1996 B2 (Hyakutake)	C/1996 J1 (Evans-Drinkwater)
C/1999 S4 (LINEAR)	C/2001 A2 (LINEAR)
57P/du-Toit-Neujmin-Delporte (2002)	

Likely split pairs

C/1988 F1 (Levy) and C/1988 J1 (Shoemaker-Holt)
C/1988 A1 (Liller) and C/1996 Q1 (Tabur)
C/2002 C1 (Ikeya-Zhang) and C/1661 C1
C/2002 A1 (LINEAR) and C/2002 A2 (LINEAR)
C/2002 Q2 (LINEAR) and C/2002 Q3 (LINEAR)

Likely split families

42P/Neujmin 3 and 53P/Van Biesbroeck and 14P/Wolf‡ and 121P/Shoemaker-Holt 2‡
C/1965 S1 (Ikeya-Seki) and C/1882 R1 (Great September Comet) and C/1970 K1 (White-Ortiz-Olelli)
C/1880 C1 (Great Southern Comet) and C/1843 D1 (Great March Comet)
Kreutz group and SOHO comets

*Likely scenario.
†The only secure case of a tidally split comet.
‡Uncertain member.

pothesis somehow implies that striae should predominantly appear in comets before reaching perihelion.

8.   APPENDIX: LIST OF SPLIT COMETS

Here we compile a list of split comets, likely pairs, and families of split comets as reported in the literature (Table A1). For periodic comets the year of the splitting event is given in parenthesis. Comet C/1995 O1 (Hale-Bopp) is not listed as split comet below even though indications exist that this comet may have displayed double or multiple nuclei (*Sekanina*, 1997b; *Marchis et al.*, 1999) and several companion comae (*Boehnhardt et al.*, 2003). It is also noted that *McBride et al.* (1997) favor the existence of a major boulder in the coma of Comet 26P/Grigg-Skjellerup from the *Giotto* flyby measurements at the comet. The detection of fragments (several hundred meters in size) around this comet is not confirmed by other observations (*Boehnhardt et al.*, 1999). Since the fragmentation products

reported for both comets are to be considered uncertain for the time being, we do not include 26P/Grigg-Skjellerup and C/1995 O1 (Hale-Bopp) in the list below. Based on the clustering of dust impacts during the *Giotto* encounter, the existence of fragmenting boulder-sized pieces in the coma of Comet 1P/Halley (see *Boehnhardt*, 2002, and references therein) is also the subject of speculation.

Acknowledgments.   I wish to thank Drs. Z. Sekanina (Jet Propulsion Laboratory, Pasadena) and B. Marsden (Center for Astrophysics, Cambridge) for the provision of partially unpublished results on comet fragmentation. Images of split comets presented in this paper are collected at the Cerro La Silla and Cerro Paranal Observatories of the European Southern Observatory (ESO), at the Calar Alto Observatory of the Max-Planck-Institut für Astronomie, and at the Telescope Nazionale Galileo of the Roque de los Muchachos Observatory in La Palma. I am particularly grateful to Dr. K. Birkle (Max-Planck-Institut für Astronomie Heidelberg), to Dr. E. Jehin (European Southern Observatory, Santiago de Chile) and collaborators, and to Dr. J.-L. Ortiz (Instituto Astrofisico de

Andalucia, Granada) and collaborators for the provision of unpublished image material. Last, but not least, I wish to thank Dr. D. Andersen (Max-Planck-Institut für Astronomie, Heidelberg) for a critical review of the manuscript.

REFERENCES

Adams S. (2002) *Comets C/2002 Q2 (LINEAR) and C/2002 Q3 (LINEAR).* IAU Circular No. 7960.

A'Hearn M. F., Millis R. L., Schleicher D G., Osip D. J., and Birch P. V. (1995) The ensemble properties of comets: Results from narrowband photometry of 85 comets, 1976–1992. *Icarus, 118,* 223–270.

Asphaug E. and Benz W. (1996) Size, density, and structure of Comet Shoemaker-Levy 9 inferred from the physics of tidal break-up. *Icarus, 121,* 225–248.

Bardwell C. M. (1988) *Comets Levy (1988e) and Shoemaker-Holt (1988g).* IAU Circular No. 4600.

Beech M. and Gauer K. (2002) Cosmic roulette: Comets in the main belt asteroid region. *Earth Moon Planets, 88,* 211–221.

Bockelée-Morvan D. and 11 colleagues (2001) Outgassing behavior and composition of Comet C/1999 S4 (LINEAR) during its disruption. *Science, 292,* 1339–1343.

Boehnhardt H. (2002) Comet splitting — Observations and model scenarios. *Earth Moon Planets, 89,* 91–115.

Boehnhardt H., Rainer N., Birkle K., and Schwehm G. (1999) The nuclei of Comets 26P/Grigg-Skjellerup and 73P/Schwassmann-Wachmann 3. *Astron. Astrophys., 341,* 912–917.

Boehnhardt H., Holdstock S., Hainaut O., Tozzi G. P., Benetti S., and Licandro J. (2002) 73P/Schwassmann-Wachmann 3 — One orbit after break-up: Search for fragments. *Earth Moon Planets, 90,* 131–139.

Boehnhardt H. and 12 colleagues (2003) Post-perihelion coma monitoring of Comet Hale-Bopp at ESO. In *Proceedings of Asteroids, Comets, Meteors — ACM 2002* (B. Warmbein, ed.), pp. 613–616. ESA SP-500, Noordwijk, The Netherlands.

Brin G. D. (1980) Three models of dust layers on cometary nuclei. *Astrophys. J., 237,* 265–279.

Brin G. D. and Mendis D. A. (1979) Dust release and mantle development in comets. *Astrophys. J., 292,* 402–408.

Carusi A., Kresák L., Perozzi E., and Valsecchi G. B. (1985) First results of the integration of motion of short-period comets over 800 years. In *Dynamics of Comets: Their Origin and Evolution* (A. Carusi and G. B. Valsecchi, eds.), pp. 319–340. Reidel, Dordrecht.

Chen J. and Jewitt D. (1994) On the rate at which comets split. *Icarus, 108,* 265–271.

Crovisier J., Bockelée-Morvan D., Gerard E., Rauer H., Biver N., Colom P., and Jorda L. (1996) What happened to Comet 73P/Schwassmann-Wachmann 3? *Astron. Astrophys., 310,* L17–L20.

Davidsson B. J. R. (1999) Tidal splitting and rotational break-up of solid spheres. *Icarus, 142,* 525–535.

Davidsson B. J. R. (2001) Tidal splitting and rotational break-up of solid biaxial ellipsoids. *Icarus, 149,* 375–383.

Delsemme A. H. (1982) Chemical composition of cometary nuclei. In *Comets* (L. L. Wilkening, ed.), pp. 85–130. Univ. of Arizona, Tucson.

Desvoivres E., Klinger J., Levasseur-Regourd A. C., Lecacheux J., Jorda L., Enzian A., Colas F., Frappa E., and Laques P. (1999) Comet C/1996 B2 Hyakutake: Observations, interpretation and modelling of the dynamics of fragments of cometary nuclei. *Mon. Not. R. Astron. Soc., 303,* 826–834.

Farinella P. and Davis D. R. (1996) Short-period comets: Primordial bodies or collisional fragments? *Science, 273,* 938–941.

Farnham T. L., Schleicher D. G., Woodney L. M., Birch P. V., Eberhardy C. A., and Levy L. (2001) Imaging and photometry of Comet C/1999 S4 (LINEAR) before perihelion and after break-up. *Science, 292,* 1348–1353.

Green D. W. E. (2000) *Comet C/1999 S4 (LINEAR).* IAU Circular No. 7605.

Green D. W. E. (2002) *Comet C/2002 C1 Ikeya-Zhang.* IAU Circular No. 7852.

Harris W. M., Combi M. R., Honeycutt R. K., Mueller B. E. A., and Scherb F. (1997) Evidence for interacting gas flows and an extended volatile source distribution in the coma of Comet C/1996 B2 (Hyakutake). *Science, 277,* 676–681.

Jahn J. (1996) *C/1988 A1 and C/1996 Q1.* IAU Circular No. 6464.

Jehin E., Boehnhardt H., Bonfils X., Schuetz O., Sekanina Z., Billeres M., Garradd G., Leisy P., Marchis F., and Thomas D. (2002) Split Comet C/2001 A2 (LINEAR). *Earth Moon Planets, 90,* 147–151.

Jorda L. and Gutiérrez P. (2002) Comet rotation. *Earth Moon Planets, 89,* 135–160.

Kidger M. (2002) The break-up of C/1999 S4 (LINEAR), Days 0–10. *Earth Moon Planets, 90,* 157–165.

Kührt E. and Keller H. U. (1994) The formation of cometary surface crusts. *Icarus, 109,* 121–132.

Lisse C. M., Fernández Y. R., Kundu A., A'Hearn M. F., Dayal A., Deutsch L. K., Fazio G. G., Hora J. L., and Hoffmann W. F. (1999) The nucleus of Comet Hyakutake (C/1996 B2). *Icarus, 140,* 189–204.

Mäkinen J. T. T., Bertaux J.-L., Combi M., and Quemerais E. (2001) Water production of Comet C/1999 S4 (LINEAR) observed with the SWAN instrument. *Science, 292,* 1326–1329.

Marchis F., Boehnhardt H., Hainaut O. R., and Le Mignant D. (1999) Adaptive optics observations of the innermost coma of C/1995 O1: Is there a 'Hale' and a 'Bopp' in Comet Hale-Bopp? *Astron. Astrophys., 349,* 985–995.

Marsden B. G. (1967) The sungrazing comet group. *Astron. J., 72,* 1170–1183.

Marsden B. G. (1989) The sungrazing comet group II. *Astron. J., 98,* 2306–2321.

Marsden B. G. and Sekanina Z. (1971) Comets and nongravitational forces. IV. *Astron. J., 76,* 1135–1151.

McBride N., Green S. F., Levasseur-Regourd A. C., Goidet-Devel B., and Renard J.-B. (1997) The inner dust coma of Comet 26P/Grigg-Skjellerup: Multiple jets and nucleus fragments? *Mon. Not. R. Astron. Soc., 289,* 535–553.

McDonald J. A. M. and 27 colleagues (1987) The dust distribution within the inner coma of comet P/Halley 1982i: Encounter by Giotto's impact detectors. *Astron. Astrophys., 187,* 719–741.

Meech K. J. and Svoreň J. (2004) Physical and chemical evolution of cometary nuclei. In *Comets II* (M. C. Festou et al., eds.), this volume. Univ. of Arizona, Tucson.

Meech K. J., Knopp G. P., and Farnham T. L. (1995) The split nucleus of Comet Wilson (C/1986 P1 = 1987 VII). *Icarus, 116,* 46–76.

Meyer M. and Marsden B. G. (2002) *Non-Kreutz Near-Sun Comet Groups.* IAU Circular No. 7832.

Mumma M. J., Dello Russo N., DiSanti M. A., Magee-Sauer K., Novak R. E., Brittain S., Rettig T., McLean I. S., Reuter D. C., and Xu L. H. (2001) Organic composition of C/1999 S4 (LINEAR): A comet formed near Jupiter? *Science, 292,* 1334–1339.

Ortiz J.-L., Munoz O., Moreno F., Molina A., Herbst T. M., Birkle

K., Boehnhardt H., and Hamilton D. P. (1995) Models of the SL-9 collision-generated hazes. *Geophys. Res. Lett., 22,* 1605–1608.

Prialnik D. and Bar-Nun A. (1992) Crystallization of amorphous ice as the cause of comet P/Halley's outburst at 14 AU. *Astron. Astrophys., 258,* L9–L12.

Prialnik D., Benkhoff J., and Podolak M. (2004) Modeling the structure and activity of comet nuclei. In *Comets II* (M. C. Festou et al., eds.), this volume. Univ. of Arizona, Tucson.

Rodionov A. V., Jorda L., Jones G. H., Crifo J. F., Colas F., and Lecacheux J. (1998) Comet Hyakutake gas arcs: First observational evidence of standing shock waves in a cometary coma. *Icarus, 136,* 232–267.

Samarasinha N. H. (2001) A model for the break-up of Comet LINEAR (C/1999 S4). *Icarus, 154,* 679–684.

Schenk P. M., Asphaug E., McKinnon W. B., Melosh H. J., and Weissman P. R. (1996) Cometary nuclei and tidal disruption: The geologic record of crater chains on Callisto and Ganymede. *Icarus, 121,* 249–274.

Schleicher D. G. and Osip D. J. (2002) Long- and short-term photometric behaviour of Comet Hyakutake (C/1996 B2). *Icarus, 159,* 210–233.

Schulz R. (2000) Dust and gas in cometary comae. Habilitation thesis, Univ. of Göttingen.

Schulz R. and Stüwe J. (2002) The dust coma of Comet C/1999 S4 (LINEAR). *Earth Moon Planets, 90,* 195–203.

Sekanina Z. (1977) Relative motions of fragments of the split comets. I. A new approach. *Icarus, 30,* 574–594.

Sekanina Z. (1978) Relative motions of fragments of the split comets. II. Separation velocities and differential decelerations for extensively observed comets. *Icarus, 33,* 173–185.

Sekanina Z. (1979) Relative motions of fragments of the split comets. III. A test of splitting and comets with suspected multiple nuclei. *Icarus, 38,* 300–316.

Sekanina Z. (1982) The problem of split comets in review. In *Comets* (L. L. Wilkening, ed.), pp. 251–287. Univ. of Arizona, Tucson.

Sekanina Z. (1988) *Comet Wilson (1986l).* IAU Circular No. 4557.

Sekanina Z. (1991) *Periodic Comet Chernyckh (1991o).* IAU Circular No. 5391.

Sekanina Z. (1994) Tidal disruption and the appearance of periodic Comet Shoemaker-Levy 9. *Astron. Astrophys., 289,* 607–636.

Sekanina Z. (1995a) Evidence on sizes and fragmentation of the nuclei of Comet Shoemaker-Levy 9 from Hubble Space Telescope images. *Astron. Astrophys., 304,* 296–316.

Sekanina Z. (1995b) *Comet C/1994 G1 (Takamizawa-Levy).* IAU Circular No. 6161.

Sekanina Z. (1997a) The problem of split comets revisited. *Astron. Astrophys., 318,* L5–L8.

Sekanina Z. (1997b) Detection of a satellite orbiting the nucleus of Comet Hale-Bopp (C/1995 O1). *Earth Moon Planets, 77,* 155–163.

Sekanina Z. (1998) A double nucleus of Comet Evans-Drinkwater (C/1996 J1). *Astron. Astrophys., 339,* L25–L28.

Sekanina Z. (1999) Multiple fragmentation of Comet Machholz 2 (P/1994 P1). *Astron. Astrophys., 342,* 285–299.

Sekanina Z. (2000a) SOHO sungrazing comets with prominent tails: Evidence on dust production peculiarities. *Astrophys. J. Lett., 542,* L147–L150.

Sekanina Z. (2000b) Secondary fragmentation of the Solar and Heliospheric Observatory sungrazing comets at very large heliocentric distance. *Astrophys. J. Lett., 545,* L69–L72.

Sekanina Z. (2001) *Comet 51P/Harrington.* IAU Circular No. 7773.

Sekanina Z. (2002a) Statistical investigation and modelling of sun-grazing comets discovered with the Solar and Heliospheric Observatory. *Astrophys. J., 566,* 577–598.

Sekanina Z. (2002b) Runaway fragmentation of sungrazing comets observed with the Solar and Heliospheric Observatory. *Astrophys. J., 576,* 1085–1094.

Sekanina Z. (2002c) *Comets C/2002 Q2 (LINEAR) and C/2002 Q3 (LINEAR).* IAU Circular No. 7978.

Sekanina Z. and Chodas P. W. (2002a) Common origin of two major sungrazing comets. *Astrophys. J., 581,* 760–769.

Sekanina Z. and Chodas P. W. (2002b) Fragmentation origin of major sungrazing Comets C/1970 K1, C/1880 C1, and C/1843 D1. *Astrophys. J., 581,* 1389–1398.

Sekanina Z. and Chodas P. W. (2002c) *Comet 57P/Du Toit-Neujmin-Delporte.* IAU Circular Nos. 7946 and 7957.

Sekanina Z. and Chodas P. W. (2002d) *Comets C/2002 Q2 (LINEAR) and C/2002 Q3 (LINEAR).* IAU Circular No. 7966.

Sekanina Z. and Farrell J. A. (1980) The striated dust tail of Comet West 1976 VI as a particle fragmentation phenomenon. *Astron. J., 85,* 1538–1554.

Sekanina Z. and Marsden B. G. (1982) *Comets Hartley (1982b, 1982c).* IAU Circular No. 3665.

Sekanina Z. and Yeomans D. K. (1985) Orbital motion, nucleus precession, and splitting of periodic Comet Brooks 2. *Astron. J., 90,* 2335–2352.

Sekanina Z., Boehnhardt H., Käufl H. U., and Birkle K. (1996) Relationship between outbursts and nuclear splitting of Comet 73P/Schwassmann-Wachmann 3. *JPL Cometary Sciences Group Preprint Series 183,* pp. 1–20.

Sekanina Z., Chodas P. W., and Yeomans D. K. (1998) Secondary fragmentation of Comet Shoemaker-Levy 9 and the ramifications for the progenitor's break-up in July 1992. *Planet. Space Sci., 46,* 21–45.

Sekanina Z., Jehin E., Boehnhardt H., Bonfils X., Schuetz O., and Thomas D. (2002) Recurring outbursts and nuclear fragmentation of Comet C/2001 A2 (LINEAR). *Astrophys. J., 572,* 679–684.

Sekanina Z., Chodas P. W., Tichy M., Tichy J., and Kocer M. (2003) Peculiar pair of distant periodic Comets C/2002 A1 and C/2002 A2 (LINEAR). *Astrophys. J. Lett., 591,* L67–L70.

Shestakova L. I. and Tambovtseva L. V. (1997) The thermal destruction of solids near the Sun. *Earth Moon Planets, 76,* 19–45.

Tambovtseva L. V. and Shestakova L. I. (1999) Cometary splitting due to thermal stresses. *Planet. Space Sci., 47,* 319–326.

Tancredi G., Motta V., and Froeschlé C. (2000) Dynamical taxonomy of comets and asteroids based on the Lyapunov indicators. *Astron. Astrophys., 356,* 339–346.

Toth I. (2001) Impact-triggered breakup of Comet C/1999 S4 (LINEAR): Identification of the closest intersecting orbits of other small bodies with its orbit. *Astron. Astrophys., 368,* L25–L28.

Tauber F. and Kührt E. (1987) Thermal stresses in cometary nuclei. *Icarus, 69,* 83–90.

Weaver H. A. and 20 colleagues (2001) HST and VLT investigations of the fragments of Comet C/1999 S4 (LINEAR). *Science, 292,* 1329–1333.

Weissman P. R. (1980) Physical loss of long-period comets. *Astron. Astrophys., 85,* 191–196.

Whipple F. L. (1963) On the structure of the cometary nucleus. In *The Solar System IV: The Moon, Meteorites, and Comets* (B. M. Middlehurst and G. P. Kuiper, eds.), pp. 639–663. Univ. of Chicago, Chicago.

Whipple F. L. (1978) Cometary brightness variation and nucleus structure. *Moon and Planets, 18,* 343–359.

Using Cometary Activity to Trace the Physical and Chemical Evolution of Cometary Nuclei

K. J. Meech
Institute for Astronomy at the University of Hawai'i

J. Svoreň
Astronomical Institute of the Slovak Academy of Sciences

Historically, minor bodies are classified as comets based on observations of "activity," which refers to the appearance of dust and gas around the nucleus. While cometary activity has been observed for centuries, at ever-increasing heliocentric distances, our understanding of the mechanisms for producing cometary activity and its heliocentric dependence has only recently been developed to the level of sophistication needed to make detailed comparison with the observations. A thorough understanding of cometary activity is closely coupled with knowledge about the formation of comets, thermal models of cometary nuclei, and chemistry in the coma. This chapter summarizes the specific chemical and physical changes that a comet nucleus undergoes, concentrating on the active phases. The specific drivers of activity are discussed, as well as the means of measuring the activity in comets. Finally, some historical and modern examples of specific types of cometary activity are discussed and are used to make inferences about both primordial differences between comet dynamical classes and evolutionary, or aging, effects.

1. INTRODUCTION

What controls the activity of comet nuclei? Comets exhibit a wide range of physical characteristics. Some of these characteristics can be attributed to the systematic physical differences among different dynamical and evolutionary groups. We must try to distinguish whether these differences are the products of aging or evolutionary process, or whether they reflect the primordial differences among the groups. "Aging" of cometary nuclei refers to the effects since the time of formation that have altered the nucleus, either chemically or physically, and that may cause a change in the type or level of activity. Some of the signs of aging in comets include (1) the production of comae and tails consisting of escaping gas and dust, which creates debris occupying the orbits of the distintegrating comets; (2) nongravitational effects in the comet's motion, produced by jet effects of the escaping matter on a rotating nucleus; (3) outbursts (or sudden brightness changes) and splitting of cometary nuclei, possibly leading to total disruption; (4) changes in the volatile composition of the escaping gases and internal physical nucleus properties, in particular in the upper layers of the nucleus; (5) the progressive changes of the cometary absolute brightness, including the temporary diminishing of cometary activity; (6) the change of the physical appearance of comets to objects indistinguishable from asteroids moving in cometary orbits (extinction of cometary nucleus); and (7) the total disappearance of the comet nuclei.

A thorough discussion of the mechanisms, dynamics, and changes in activity caused by comet splitting are covered in *Boehnhardt* (2004) and *Jewitt* (2004) and will not be discussed in detail here. Likewise, a detailed discussion

of the timescales for the physical evolution of comet nuclei, changes in shape, spin period, and eventual disappearance of activity as comets either evolve into asteroidal-like objects or experience complete disintegration is presented in *Jewitt* (2004). In this chapter a brief review of comet formation, including the physical and chemical processes occurring in the precursor comet material, is presented in the context of, and to set the stage for, distinguishing between evolution and primordial differences among comets. This is followed by a thorough discussion of the mechanisms of comet activity and the means by which activity is measured. Finally, specific observations of cometary activity are presented and inferences drawn about implications for primordial composition and evolution of different dynamical classes of comets. The ultimate goal of understanding the chemical and physical evolution of cometary nuclei is to assess the extent to which comets represent the unaltered source material from their regions of formation in the solar nebula, and thus to use comets as tracers of solar system formation processes.

1.1. Source Regions and Formation

Although discussed in detail elsewhere in this volume (e.g., *Duncan et al.,* 2004; *Dones et al.,* 2004), it is important to the understanding of the activity in comets to briefly summarize some of the essential elements of the formation and dynamical evolution of comets. As the presolar nebula collapsed, solid particles settled to the midplane. They may have undergone processing (e.g., shock-induced sublimation and volatile recondensation) of their icy mantles as they fell (*Lunine and Gautier,* 2004). Dynamical evidence suggests that the short-period (SP) comets must have had a low-inclination source in the transneptunian region (*Duncan et al.,* 1988, 2004), whereas the long-period (LP), Halley-type (HT), and dynamically new (DN) comets perturbed inward from the Oort cloud may have formed at smaller heliocentric distances. While most of the Oort cloud comets probably formed beyond 20 AU, all the giant planets injected comets into the Oort cloud (*Fernández and Ip,* 1981; *Dones et al.,* 2004). Recent dynamical work has shown that perhaps as much as one-third of the scattered-disk transneptunian population may eventually end up in the Oort cloud (*Fernández and Brunini,* 2003). Thus, the source regions of the different dynamical classes of comets are not clear-cut, and there may be some SP, HT, and LP comets that may have had the same source region in the Kuiper belt and scattered disk, but would have followed different dynamical paths (see the discussion in *Duncan et al.,* 2004).

1.2. Evolution of Comets

Almost every observable property of comets is connected with their progressive disintegration. All processes that physically and chemically alter a cometary nucleus can be regarded as aging. The aging processes in comets, implying their limited physical lifetimes, are of fundamental significance for the evolutionary history of the whole cometary population. Due to accompanying nongravitational effects and dynamical chaos, it is impossible to extrapolate the motion of individual comets far beyond the time span covered by observations. The aging processes, accelerating with decreasing distance from the Sun, are too slow and irregular to become detectable during a single apparition (*Kresák,* 1987). When a SP comet is followed over a number of returns, some changes may be observable. The lifetimes of individual active comets are very short compared with the history of the solar system, and a replenishment with previously inactive objects is necessary to maintain the present state. All these arguments point to the fact that the evolution of comets is best studied on a statistical basis.

The aging or evolutionary effects that a comet nucleus will experience can be divided into four primary areas: the precometary phase, where the interstellar material is altered prior to incorporation into the nucleus; the accretion phase during nucleus formation; the cold storage phase, the phase where the comet is stored for long periods at large distances from the Sun; and the active phase (*Meech,* 1999).

1.2.1. Presolar nebula. The precursor cometary material, interstellar grains, is stored in cold quiescent molecular clouds (T = 10 K, n = 10^3 cm^{-3}) and in warm, dense protostellar regions (T = 100 K, n = 10^6 cm^{-3}). A complete discussion of ice and grains in the precometary phase is presented in *Irvine and Lunine* (2004) and *Wooden et al.* (2004). The mantles of interstellar grains undergo significant processing in the molecular clouds from bombardment by cosmic rays. The ions lose energy by ionization of the target material and breaking chemical bonds in the target, and they can also cause sputtering from the surface (*Strazzula and Johnson,* 1991). This will create both nonvolatile material and highly reactive radicals, which will then potentially be incorporated into the comets. There is also complex thermal and chemical processing that can occur (see *Wooden et al.,* 2004, for a complete discussion). However, a fundamental question remains as to how much of this precursor material survives the formation process (*Mumma et al.,* 1993). The water contained in comets is likely to have two sources: H_2O ice that survived disk infall, and that which formed in the disk (see *Irvine and Lunine,* 2004, for a discussion).

1.2.2. Accretion phase. Water-ice formed by low-pressure vapor deposition, conditions expected in the solar nebula as interstellar grains were falling in toward the midplane, will have one of three forms: two crystalline polymorphs (hexagonal, I_h, and cubic, I_c), and both a low- and high-density amorphous form, I_al and I_ah respectively. When H_2O ice condenses at temperatures below 100 K, it condenses in the amorphous form because it lacks the energy to form a regular crystalline structure; below 20 K, this is likely to be I_ah ice. As the H_2O ice condenses, it has the ability to trap gases as high as 3.3–3.5 times the amount of the ice (*Laufer et al.,* 1987).

The mechanism for amorphous ice trapping is that gas enters an open pore during condensation and is held in place by van der Waals forces. The pore is subsequently covered and the gases are trapped. The amount of gas that can be trapped is a very strong function of the condensation temperature: H_2 and D_2 can only be efficiently trapped below 20 K, Ne only below 24 K, and many other light gases may be trapped only up to 100 K. More gases can be trapped at lower temperatures because the molecules will have longer residence times and are more likely to have their pores sealed before they can escape. The gases that have stronger van der Waals attraction (polarizability) will be preferentially enriched. The temperature also affects the size of the free channels in the ice, and hence the atoms that can permeate the ice.

If a large amount of volatiles are present as the water is condensing at low temperatures, and after the maximum amount of gas is trapped in the pores, it is possible to have some surface freezing of the more volatile species (*Notesco et al.,* 2003).

This accretion phase will have significant implications for the observed activity in comets as described below. In particular, Oort cloud comets and LP comets will have predominantly formed in the vicinity of the giant planets where nebular temperatures may have been between 60 and 100 K and will be expected to consist mostly of ice I_al. Kuiper belt objects (KBOs) formed and have remained predominantly at temperatures below 30–50 K. Formation in these different temperature regimes can have profound implications for the chemical composition of the comets.

1.2.3. Cold storage phase. Comets may be stored for billions of years in the Oort cloud or the distant outer solar system before passing close to the Sun and entering the active phase. During this time galactic-cosmic-ray irradia-

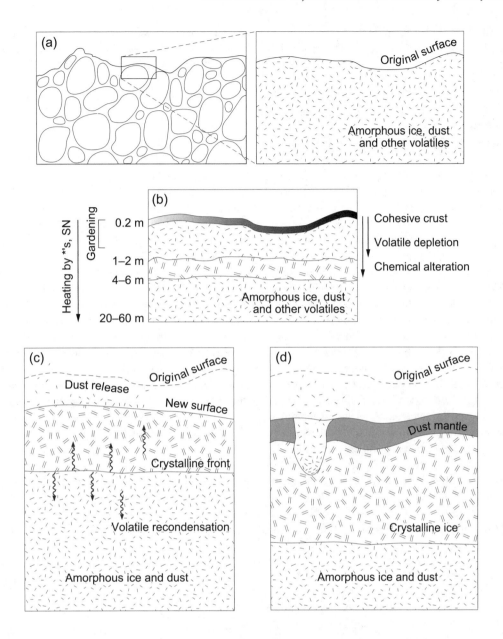

Fig. 1. Diagram showing the sequence of aging processes in the upper layers of a comet nucleus from (**a**) the pristine state, consisting of primordial planetesimals (enlargement shown on left), (**b**) to the alterations it undergoes while stored in the Oort cloud including a possible crystalline core caused by radioactive heating from [26]Al to (**c**) the changes in the surface during the active phase and (**d**) near the end of its evolution as a dust mantle builds up.

tion can create a thin, stable cohesive crust that will have some tensile strength to a depth of 10 g cm^{-2} (*Strazzula and Johnson,* 1991). In addition, radicals will form in the upper few meters of the comet surface, causing chemical alteration (to ~300 g cm^{-2}). Finally, the upper layers will be depleted in volatile material (to ~100 g cm^{-2}). Of course there may be similar irradiation processing of the precometary grains from the higher radiation environment from the young Sun, but these may sublimate from the grains during infall, and would be too large to be trapped in the condensed amorphous H$_2$O ice. Figure 1 summarizes the stages of evolution that a comet may undergo.

In addition to the radiation damage to the surface, up to 20% of the Oort cloud comets will have been heated to at least 30 K to a depth of 20–60 m from the passage of luminous stars, and most comets may have been heated as high as 45 K to a depth of 1 m from stochastic supernovae events (*Stern and Shull,* 1988). This heating may result in volatile depletion in the upper layers. Likewise, gardening from interstellar grain impacts will also alter the upper few centimeters of the surface (*Stern,* 1986).

While collisions themselves in the Oort cloud are probably very rare, recent work indicates that many of the objects ejected into the Oort cloud were probably heavily collisionally processed during their ejection (see *Stern and Weissman,* 2001, and references therein). The space density of objects in the region of the Kuiper belt is much greater, and as a result, collisional models show that small objects in this region should have heavily damaged interiors, and a significant percentage of the surfaces of the larger objects should be heavily cratered (*Durda and Stern,* 2000). Objects at smaller heliocentric distances, such as the Centaurs and SP comets, will not have a significant collisional history different from their source regions over the age of the solar system. The net effect of the different collisional regimes for the LP and SP comets should manifest itself as differences in crater density on their surfaces, devolatilization in the upper layers, and possibly surface chemistry.

Evolution of meteorites provides evidence of an early heat source in the solar system, and it is likely that the radionuclide ^{26}Al was responsible for radiogenic heating of large bodies. Models that investigate the role that ^{26}Al plays in the evolution of cometary interiors showed that because there is evidence for amorphous ice in comets, their interiors cannot have been heated above 137 K (*Prialnik et al.,* 1987). The heating from ^{26}Al would occur during the period of cometesimal formation early in the solar system's history and would raise temperatures to between 20 and 120 K, depending on the nucleus size (the larger nuclei would be less efficient at cooling). Whereas it would be expected that a pure ice nucleus would either be all crystalline or all amorphous (depending on size), the effect of refractory material would be to quench the conversion, leaving a crystalline core with an amorphous mantle.

The situation regarding heating in KBOs may be somewhat different because of their larger sizes. When considering combined models of accretion and thermal evolution and the effect of radiogenic heating, it was found that very small bodies were relatively unaffected, the largest KBOs would still contain significant amorphous material, while the intermediate sizes would be the most heavily processed (*Merk and Prialnik,* 2003). In the bodies that were thermally altered, highly volatile species would be lost, and there could be crystallization and even melting in the interiors.

1.2.4. Active phase of comets. During the active phase, when the comet passes within the inner solar system and experiences significant solar insolation, there is considerable evolution of the interior and surface of the comet. The upper few meters of the surface of a comet making its way into the inner solar system for the first time will be depleted in volatile material by sublimation, even at large heliocentric distances, and may have highly volatile radicals created due to the chemical processing from galactic cosmic rays. Just below this layer, which will be removed during the first passage, will be a layer of "pristine" amorphous ice.

On the first passage through the inner solar system, the solar insolation will cause the crystallization of the amorphous ice from the surface inward at much lower temperatures than would be expected for H$_2$O ice sublimation. The differences in the active phases of Oort clouds comets on their first perihelion passage in comparison with periodic comets has been observed for a long time. This is discussed at length in the chapter by *Dones et al.* (2004) in the context of the "fading problem" first noted by *Oort* (1950).

2. TYPES OF ACTIVITY

It is clear from recent studies of low-temperature volatiles and from comet observations that the gases that are released from the comet in addition to water are trapped within the H$_2$O ice, and not just frozen among the H$_2$O-ice crystals (*Bar-Nun and Laufer,* 2003; *Prialnik et al.,* 2004). Table 1 shows the temperatures at which various volatile processes may occur for pure ices that can lead to activity. The table also indicates approximate heliocentric distances at which these equilibrium surface temperatures are reached for dark isothermal bodies (neglecting cooling caused by sublimation). It should be noted that the sublimation temperature or distance at which this is reached is not a single number. Rather, the sublimation will occur over a wide range of temperatures, but at different rates. For example, while the peak for the amorphous to crystalline ice phase transition occurs near 137 K, it will start slowly at much lower temperatures, and while ice I$_h$ sublimates at 180 K, this process will start at much lower temperatures.

2.1. Sublimation of Pure Ices

The primary driver for activity close to the Sun is sublimation, the transitions between the solid and vapor state. This is a combination of surface and subsurface phenomena, since there can be sublimation from subsurface pore walls (see *Prialnik et al.,* 2004). The temperature at which

TABLE 1.   Temperature regimes for onset of comet activity.

T (K)	Process	r (AU)
5	H$_2$ sublimation	>3000
22	N$_2$ sublimation	160
25	CO sublimation	120
31	CH$_4$ sublimation	80
35–80	Ice I$_a$h anneals	60–10
38–68	I$_a$h converts to I$_a$l	55–15
44	C$_2$H$_6$ sublimation	40
57	C$_2$H$_2$, H$_2$S sublimation	24
64	H$_2$CO sublimation	20
78	NH$_3$ sublimation	14
80	CO$_2$ sublimation, I$_a$l anneals	13
91	CH$_3$CN sublimation	9
95	HCN sublimation	8
99	CH$_3$OH sublimation	8
70–120	Ice I$_a$l anneals	18–
90–160	Ice I$_a$l → I$_c$ phase change	11–
160	Ice I$_c$ → I$_h$ phase change	
180	Ice I$_h$ sublimation	

Water-ice information from *Laufer et al.* (1987), sublimation information from *Yamamoto* (1985) and *Handbook of Chemistry and Physics* (*Lide,* 2003).

this process begins depends on the latent heat of sublimation and the equilibrium surface temperature of the nucleus. The latter depends on many factors such as heliocentric distance, albedo, surface emissivity, rotation rate, and pole direction, as well as surface properties that affect heat transport below the surface.

The simple energy balance equation has been used for many years to estimate the "turn on" of comets assuming that the activity is caused by sublimation of various pure volatiles from the surface

$$\frac{(1 - A_B)L_\odot}{4\pi r^2} = \varepsilon\sigma T^4 + \mathcal{P}_\mu\sqrt{\frac{m_\mu}{2\pi kT}}\,\mathcal{H}_\mu \qquad (1)$$

Here, $A_B = pq$ is the Bond albedo, p is the geometric albedo, and q is the phase integral. *Buratti et al.* (2004) derive a value of q = 0.3 and p = 0.03 and thus $A_B = 0.009$ from *Deep Space 1* observations of 19P/Borrelly. Also, $L_\odot$ is the solar luminosity, r the heliocentric distance, ε the surface emissivity, $\mathcal{P}$ the saturation vapor pressure for molecule μ (in this case water), m_μ the mass of the water molecule, k the Boltzmann constant, and $\mathcal{H}$ the latent heat of sublimation.

A figure produced by *Delsemme* (1982), who computed the gas production rates for sublimation of pure volatiles, led to the unfortunate interpretation that activity on a predominantly H_2O-ice nucleus *cannot* be sustained beyond r = 3 AU. In the original figure, r_o was the distance at which only 2.5% of the solar flux was used for vaporization of volatiles, and it represented the turnover in the curve on the log-log plot. New calculations of these production rates are shown in Fig. 2 using data from *Prialnik et al.* (2004) for the sublimation vapor pressures and latent heats. In these simple models, the pole is assumed to be perpendicular to the orbit,

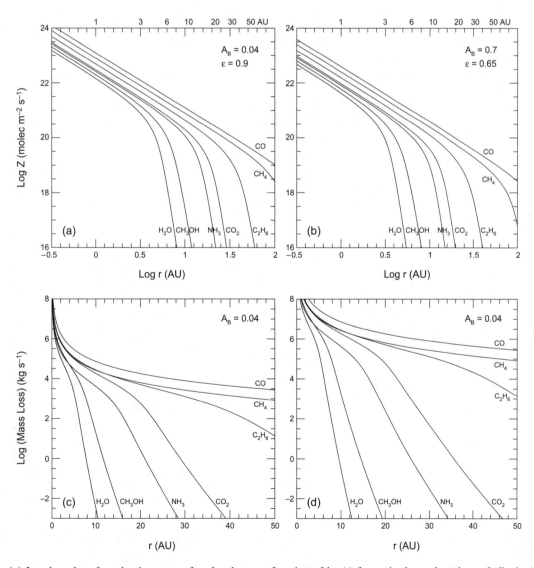

Fig. 2. **(a)** Log-log plot of production rates of molecules as a function of log(r) for an isothermal nucleus of albedo $A_B = 0.04$ for CO, CH_4, C_2H_2, C_2H_6, CO_2, NH_3, CH_3OH, and H_2O [after *Delsemme* (1982); see section 2]. Data for the computation of sublimation vapor pressures and latent heats of sublimation from *Prialnik et al.* (2004). **(b)** Same as (a), but for a nucleus with albedo $A_B = 0.7$. **(c)** Total mass loss vs. r for $A_B = 0.04$ for a R_N = 10-km nucleus. **(d)** Total mass loss vs. r for $A_B = 0.04$ for a R_N = 100-km nucleus.

and there is no variation with temperature as a function of distance from the subsolar point over the surface.

Delsemme (1982) used a much higher albedo ($p_v \sim 0.7$) for the nucleus and an emissivity that was much lower ($\epsilon \sim 0.65$) than is currently commonly accepted, but these were the best assumptions available at the time. The effect of the albedo on the production rate is shown in Figs. 2a and 2b. While there is a distinct slope change in the log-log plots between about r = 4–6 AU for H₂O production, it is important to note that the production does not drop to zero beyond this distance. Therefore, claims cannot be made that a nucleus is bare based solely upon a heliocentric distance that is greater than 3 AU. This is more easily seen in Fig. 2c, which is a plot of log(Q) vs. r.

Clearly there is *some* level of water leaving the nucleus out to large distances. The issue is whether the gas flux is sufficient to drag dust into an observable coma. Figures 2c and 2d show the molecular loss rate converted into a total mass loss (kg s⁻¹) for gas (or dust if one assumes a dust gas mass ratio = 1). Depending on the dust-to-gas mass ratio and the grain sizes (and hence minimum grain size that can be lifted off the nucleus), this translates to the potential for significant observable dust coma at large r, and for large nuclei, this might translate into an observable coma.

The critical grain size, a_{crit}, that can be lifted off the nucleus may be approximated by estimating the drag force, F_{drag}, on the grain as the product of the momentum per molecule, $\mu m_H v_{th}$, and the number of collisions per unit time. Here, v_{th} is assumed to be the mean thermal speed of the gas. The collisions per unit time is computed as the volume swept out by the grain in time t, $\pi a^2 v_{th} t$ multiplied by the gas number density, N(r), and substituting this into the equation of motion

$$m_g \frac{d^2 r}{dt^2} = -\frac{GMm_g}{r^2} + F_{drag} \qquad (2)$$

where M and m_g are the nucleus and grain masses respectively, G is the gravitational constant, and N(r) is given by

$$N(r) = \frac{Q}{4\pi R_N^2 v_{th}} \left[\frac{R_N}{r}\right]^2 \qquad (3)$$

The critical radius is then

$$a_{crit} = \frac{9\mu m_H Q v_{th}}{64\pi^2 \rho_g \rho_N R_N^3 G} \qquad (4)$$

where μ is the atomic weight of the gas in question, m_H the mass of hydrogen, Q (molec s⁻¹) is the gas production rate, ρ_g and ρ_N are the grain and nucleus densities, and R_N is the nucleus radius. For example, there is sufficient gas flux from water sublimation to lift small (~0.01–0.1 μm) grains off the

surface of the nucleus as far out as 5–6 AU, CO₂ can lift off optically significant grains near 16–19 AU, whereas CO fluxes are sufficiently high throughout the region of the Kuiper belt to entrain optically significant amounts of dust. This equation assumes spherical nonporous grains; more realistic shapes and porosities are even more readily lifted off the nucleus because of their larger surface area per unit mass. Surface brightness profile comparisons of comet nuclei in comparison to stars can place very strong limits on the presence of dust coma, down to production limits of 10⁻² kg s⁻¹ (see section 3).

Although low nucleus albedos have been commonly accepted since the time of the 1P/Halley encounter, some people have continued to naively ignore the consequences of these low albedos, and have not adjusted their thinking about the distances at which H₂O-ice sublimation can create a significant coma. It should be noted that OH has been detected in Comets 1P/Halley (r ~ 4.9 AU; log Q[OH] ~ 29) and C/1995 O1 (Hale-Bopp) (r = 5.13 AU; log Q [OH] = 27.19) out to distances of r ~ 5 AU using narrowband photometry in the near-UV (*Schleicher et al.,* 1997, 1998).

2.2. Clathrate Sublimation

Early models attempting to explain the presence of highly volatile compounds with orders-of-magnitude differences in vapor pressure appearing nearly simultaneously in the cometary comae invoked the idea that these compounds were trapped as clathrate hydrates (*Delsemme and Swings,* 1952). A clathrate hydrate is a crystalline framework of water molecules that incorporates guest molecules in the voids. The water molecules do not specifically interact with the guest molecules trapped within the clathrate-hydrates, and the latter will be released as the water sublimates. Clathrate formation in impure ices is unlikely; laboratory experiments demonstrate that clathrate formation for many species is impossible at low temperatures and pressures, and their presence is not necessary to explain the presence of species other than water (*Jenniskens and Blake,* 1996). In addition, the abundances of the observed species are far too high to be trapped in this manner. There are two types of clathrate hydrates, and the size of the guest molecule determines which forms. The H-bonded water molecules in the clathrate hydrate type I are geometrically organized into cells with two small cages and six large ones. The type II clathrate has cells with 16 small and 8 large ones, the latter being 10% bigger than the counterparts for the type I clathrate (*Lunine and Stevenson,* 1985). Clathrate type I can trap in the ratio of 1/7 and clathrate type II can trap in the ratio of 1/17 to water.

2.3. Amorphous Ice Crystallization and Annealing

The high-density amorphous ice form undergoes a transition to the low-density form in the temperature regime of 38–68 K (*Jenniskens and Blake,* 1994). When amorphous

H_2O ice is heated, trapped gases are released at low levels between 35 and 120 K in response to the restructuring of the ices, a process called annealing.

It is important to note that the trapped gases do not come out in proportion to their abundances. Van der Waals forces influence the residence time and contribute to enrichment of various species at different temperatures. Once trapped, the temperature at which release occurs is independent of composition. Beginning near 90 K, gases are released as the ice undergoes an exothermic amorphous to crystalline phase transition. The rate of transformation varies exponentially with temperature, and the release of gases will peak and diminish as the trapped molecules are released (for a thin layer, as seen in the laboratory). In real comets, the gas release will occur over a range of distances between about 8 and 20 AU as the heat penetrates to deeper layers. Thus, in contrast to sublimation, which releases water and trapped gases in the clathrate hydrate at the same time, comets can release volatiles at different (lower) temperatures than from the sublimation of water.

The only realistic way to produce significant activity at very large distances (e.g., beyond distances where the ice anneals or at the distances of the amorphous-to-crystalline ice transition) is if there were highly volatile material that froze out on the surfaces of the cometesimals (*Notesco et al.*, 2003; *Bar-Nun and Laufer*, 2003). However, we know that either there was not a significant amount of these directly frozen gases incorporated into comets, or that evolutionary effects have released these volatiles, because the SP and the few HT and LP comets that have been studied in detail do not show significant release of different volatiles at different times that cannot be accounted for by models of amorphous ice crystallization (*Prialnik et al.*, 2004). A spectacular example of this was seen with the radio observations of C/1995 O1 (Hale-Bopp) (*Biver et al.*, 1997). Volatiles with orders-of-magnitudes differences in volatilities (e.g., CO, CH_3OH, H_2S, H_2CO, HCN, CS, CH_3CN, and HNC) were seen to increase in abundance at somewhat similar rates, beginning as far out as $r = 7$ AU. This is not likely to be caused by subsurface sublimation in response to solar insolation, and is more likely controlled by the amorphous to crystalline H_2O-ice phase changes.

3. MEASURING ACTIVITY

The easiest way to observe that a comet has become active, i.e., that there is a flow of gas and entrained dust grains that are populating the coma, is to observe the coma or tail that is produced from the activity. However, the presence of a coma or tail does not necessarily imply either that (1) activity from sublimation is continuing (since large grains may take a long time to move away from the nucleus) or (2) that the activity was caused by sublimation (since a collision could produce a temporary coma). A definitive test of activity can be made by observing a change in brightness that cannot be accounted for by rotational modulation or the changing orbital geometry (heliocentric and geocentric distances and phase dependence on brightness), or direct spectroscopic observation of gas phase species. In this section we will focus on the dust observations.

The evolving brightness of a comet is controlled by many complex factors, yet is often parameterized by a simple formula (see equation (5)). This parameterization works only as well as the user understands all the assumptions going into the formula. Unfortunately, this methodology is often used indiscriminately, leading to misconceptions about comet brightnesses, brightness predictions, and what measurements to make. Therefore, it is important to go into some detail about the limitations inherent in reporting comet brightnesses.

3.1. Total Brightness

The traditional formula for expressing the brightness of a comet, m_1, as a function of r may be written as (*Marsden and Roemer*, 1982)

$$m_1 = H_1 + 2.5n\log r + 2.5k\log\Delta + \phi \qquad (5)$$

and a similar formula for the brightness of a nucleus, m_2, as

$$m_2 = H_2 + 5\log r + 5\log\Delta + \phi \qquad (6)$$

H_1 and H_2 is the absolute magnitude at $r = \Delta = 1$ AU and $\phi = 0$, and n and k are a measure of the sensitivity of the magnitude variation to r and Δ. These values are not necessarily constant as a comet evolves from one apparition to another. Most studies assume that $k = 2$, and $n = 4$ is frequently assumed if little data is available. The term ϕ had either incorporated a phase function or an aperture correction. Historically, this formula has been extremely useful for predicting the brightness of comets and for comparing the behavior of comets; however, given our current detailed understanding of the physics of comets and thermal models, this is not a very accurate estimator of comet activity, and can often make very bad predictions. In particular, the value of n has ranged between 2 and 8 for various comets. There are several reasons why using the formula above for an active comet is often a poor predictor of brightness and activity level.

3.1.1. Heliocentric distance range. First, as was seen in the previous section, the brightness of a comet varies as r^{-2} modulated by any nucleus rotational signature at large distances when there is no outgassing. Close to the Sun, when all the solar energy is going into gas production, the brightness change as a function of distance behaves as r^{-2}. Therefore, the value of n that might be determined from observations is a strong function of the range of r over which this parameter is determined. The formula may produce a reasonable brightness estimate within the same range of distances that the exponents were determined, and fail completely elsewhere.

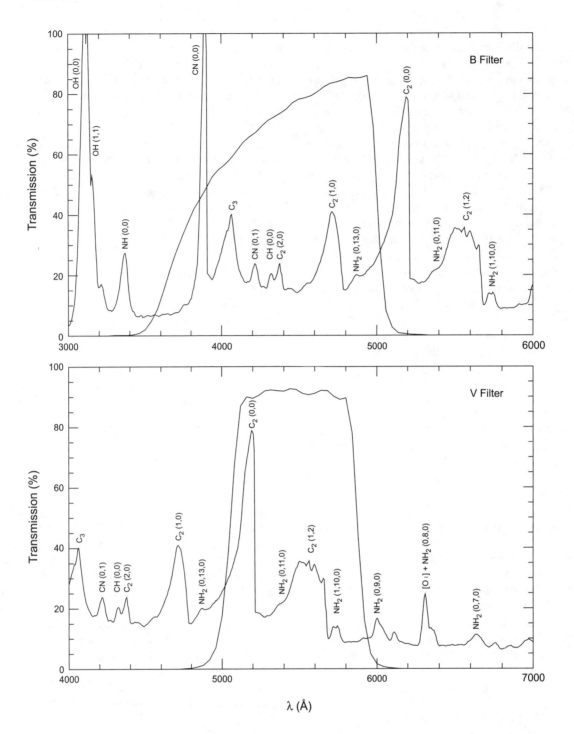

Fig. 3. Broadband B and V Kron-Cousins filter system (from Mauna Kea) superimposed on the spectrum of Comet 8P/Tuttle [8P/Tuttle spectrum, created by S. Larson and J. Johnson, courtesy of S. Larson; line identifications from *A'Hearn and Festou* (1990)].

3.1.2. Dust vs. gas. Second, the gas production is more strongly dependent upon r than is the dust. Depending on the size of the dust particles and the size distribution and the interaction of the dust with both gravity and solar radiation, the dust may remain in the vicinity of the nucleus for quite some time (*Fulle,* 2004). Thus, n will be smaller for dusty comets.

3.1.3. Filter selection. Because of this difference in behavior of the dust and gas, the variation of n with r will also depend on the filter through which the observations

have been made, and the relative fluxes of gas and dust that are measured by that filter. Figures 3 and 4 show a spectrum of an active comet, 8P/Tuttle, upon which the typical broadband filters used by observers are superimposed. As discussed in *Schleicher and Farnham* (2004), it is nearly impossible to separate the contributions of gas and dust in an active comet when using broadband filters. The B and V bands are heavily contaminated by gas emission (CN, C_3, C_2, and NH_2), and there is a small amount of contamination from [OI] and NH_2 in the R-band. In addition, there is some

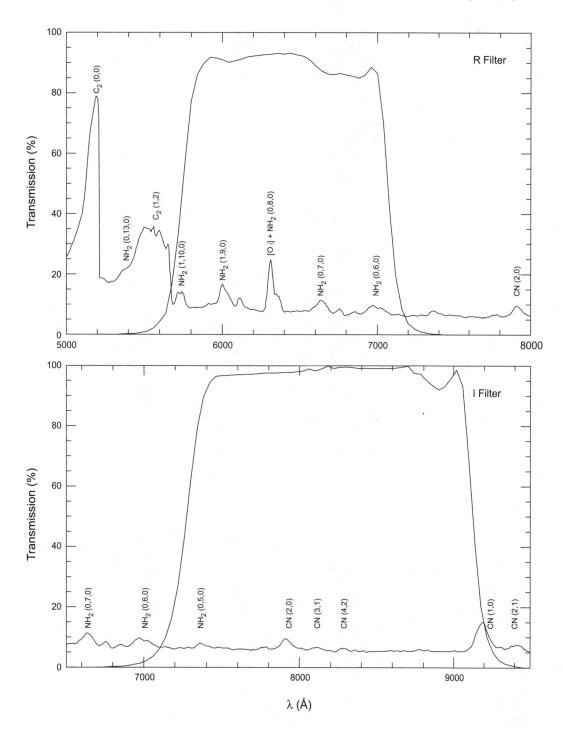

Fig. 4. Broadband R and I Kron-Cousins filter system (from Mauna Kea) superimposed on the spectrum of Comet 8P/Tuttle [8P/Tuttle spectrum, created by S. Larson and J. Johnson, courtesy of S. Larson; line identifications from *A'Hearn and Festou* (1990)].

contamination from H_2O^+ between 6950 and 7080 Å in the R-band (*Schleicher and Farnham,* 2004). The I-band, on the other hand, is relatively free of most gas emissions, but is not a commonly reported filter for comet observations (in part because the effective wavelength of this bandpass depends on the red response of the detector used, and in part because the sky brightness is much greater in this bandpass).

The amount of gas contamination in the broadband filters also varies with distance from the nucleus as well as the dust-to-gas ratio. For instance, in dusty comets, contami-

nation in R or I is minimal; the contamination by NH_2 drops rapidly with distance from the nucleus since it is short-lived; but weak CN bands in the near-IR can be a problem for low dust/gas comets at larger projected distances. Therefore, measurements done through different aperture sizes at different times will be subject to additional systematic effects. A good correlation of Q_{gas} and visual magnitude has been shown for a selection of comets between r = 0.6 and 2.8 AU (*A'Hearn and Millis,* 1980; *Jorda et al.,* 1991), which would seem to contradict this statement. However, over the range in r where the correlation was observed, water sublimation is in a linear regime and behaves as r^{-2}. Further, the visual region spectrum is dominated by CN and C_2, so for this range of r it is not surprising that there is a good correlation.

Long-lived or lingering dust also implies that gas provides a better measure of the ongoing production rate, while dust measurements are often more of an abundance value rather than a production-rate value. Typical gas species "live" on the order of hours or days before photodissociation, while larger-sized dust grains may have been emitted weeks or months earlier. Thus, for production-rate-activity measurements, filters that record the maximum of the light emitted by gas species (e.g., C_2 or NH_2) are good selections.

3.1.4. Phase function. The term ϕ in equations (5) and (6) is more complex for an active comet than for a solid surface. The observed brightness from the coma is the sum of the contributions of all the volume elements along the light of sight, and this contribution of the scattering from all the dust per unit volume is the volume-scattering function. This volume-scattering function, ψ, as a function of wavelength, λ, particle size, a, and scattering angle, θ, is defined as

$$\psi(\lambda,\theta) = \int_{a_1}^{a_2} n(a)\pi a^2 Q_{sca}(a,\lambda)\phi(a,\lambda,\theta)da \qquad (7)$$

where Q_{sca} is the scattering efficiency and ϕ is the phase angle (*Grün and Jessberger*, 1990). As discussed in detail in *Kolokolova et al.* (2004), analytical methods (Mie theory) and laboratory measurements have been used to determine scattering efficiencies and phase functions for a wide variety of grain compositions, sizes and structures that likely represent realistic cometary grains. As seen in equation (7), converting this to a realistic coma brightness behavior as a function of phase angle will depend on the possibly changing particle size distribution and scattering properties. *Kolokolova et al.* (2004) summarized that data for many bright comets and showed that the brightness behavior has a small backscattering peak, a strong forward-scattering surge at large phase angles, and in between is somewhat flat. In a very active comet, most of the light is produced by reflection from dust in the coma, which means that phase variations of the nucleus are inconsequential.

3.2. Presence of Coma as an Indicator of Activity

3.2.1. Dust coma models. Dynamical models of the dust coma as a function of time can yield information about the onset and cessation of activity, the relative grain size distribution, velocity distribution, and the dust production rate as a function of grain size and time. This dynamical technique is a method of computing the surface brightness of a comet's tail by evaluating the motions of model dust particles ejected from the nucleus under the influence of solar radiation pressure and gravity, and then adding up their scattered light as seen from Earth for comparison with real comet images. Dust dynamical model development, modern usage, and limitations are discussed in detail in *Fulle* (2004).

While there are numerous assumptions inherent in this type of modeling, it can be very useful for making inferences about important parameters related to activity, such as the approximate turn-on and turn-off distances (meaning the distances at which there is measurable brightening due to the presence of a dust coma). The model parameter that produces the biggest change in the appearance of the coma and tail is the production rate, which means that this is usually the best constrained of the parameters. However, to use this type of method to determine comet grain properties and estimates of the activity level in comets, the technique must be used in conjunction with as much other information as possible, and careful attention to the particular details of each comet (*Farnham*, 1996).

3.2.2. Activity without coma? Traditionally, the visible appearance of coma around the nucleus has been the indication that a comet is exhibiting activity, and conversely the lack of a coma was taken to mean that the observations were of a bare nucleus. However, it is possible that a comet could be either weakly active, such that the coma was not apparent in the observations, or that the grains did not travel far from the nucleus and the coma was contained within the extent of the seeing disk, or stellar profile. An excellent example of this is Comet 2P/Encke, whose orbit constrains it to travel between q = 0.34 AU at perihelion and Q = 4.10 AU at aphelion. Thus, for its entire orbit it is well within the region where H_2O ice sublimates and can produce significant coma.

Observations of Comet 2P/Encke to determine the rotation period described the nucleus as stellar (*Jewitt and Meech*, 1987). Subsequent observations of the comet typically described the comet as low-activity, or a bare nucleus. However, there were difficulties in reconciling the different rotation periods found by different observing teams. The various rotational datasets were found to be inconsistent unless one assumed that there was a contribution from activity in the datasets (*Sekanina* 1991). In a database of observations extending 16 years, from 1985 to 2001, broadband images of the comet showed definitive coma only for distances r < 2 AU. However, as shown in Fig. 5, there was clear evidence for activity near aphelion based on reported brightness variations (*Meech et al.*, 2001). Near aphelion there are excursions in brightness up to 2.5 mag, or a factor of 10 in brightness, that are beyond any brightness increase from rotation, or an assumed phase function of $\beta = 0.04$ mag deg^{-1}.

3.2.3. Coma detection limits. When there is little or no coma, the dust-dynamical models described in section 3 cannot be used to determine the onset of activity. However, there is still a way to place limits on the amount of activity that might be present. The technique relies on a detailed comparison of the surface brightness profile of the comet and field stars. One technique makes a comparison between seeing-convolved models of nuclei plus varying amounts of coma (*Luu and Jewitt*, 1992). In this type of approach,

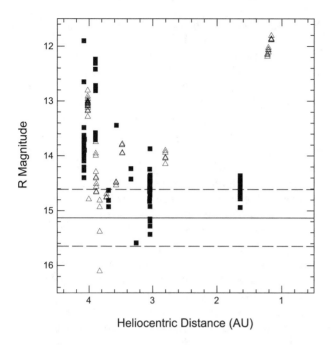

the grain velocity, and r is in AU and Δ in m. It should be noted that if one assumes a Bobrovnikoff relation for the terminal grain velocities, $v_{gr} = v_{bob} = (\mu/\mu_{H_2O})^{0.5}600\ r^{-0.5}$ (*Bobrovnikoff*, 1954), and recalls that $\phi = \Delta\phi'/206265$, where ϕ' is the angular size of the aperture (arcsec), for a given observed flux the dust production will vary as

$$Q_p \propto r^{1.5}\Delta^1 \qquad (9)$$

which shows that the most sensitive limits are placed for those objects closest to the Earth and Sun. The total dust production, Q (kg s^{-1}), is obtained from Q_p by multiplying by the grain mass, $m_{gr} = 4/3\pi a_{gr}^3\rho$, assuming spherical particles. The dust limit results will also depend on the assumed grain size distribution. Far larger mass loss rates are possible for millimeter- or centimeter-sized particles, which would never be detected. Much of the mass is probably hidden in particles never observed, and values derived from optical measurements are always lower limits to the total mass in grains.

4. OBSERVATIONS OF ACTIVITY AND EVOLUTION

In the previous sections, the changes that a comet nucleus undergoes as it evolves were described, as well as the techniques used to measure observable changes in a comet as it evolves or ages. In this section we will examine some of the large cometary datasets used to examine at the activity in comets from the point of view of aging.

It is important to be able to distinguish the effects of aging from primordial differences among the comets since the wide range of formation distances should imply significantly different volatiles and abundances of trapped gases, as well as differences in collisional histories. The following characteristics might be observable as changes in activity level as a consequence of evolution and can also help characterize the physical mechanisms of the activity: (1) activity at large r from the onset of amorphous ice crystallization or sublimation of frozen volatiles, which would decrease postperihelion; (2) secular fading of the comet as a dust mantle is built up, or because of loss of surface volatiles; (3) more uniform activity in new comets, with larger surface areas available for sublimation; (4) higher frequency of jets and outbursts for older mantled comets; (5) differences in production rates of gases as a function of r for different species and dynamical classes of comets; (6) primordial and evolutionary differences in nucleus size distributions; and (7) perihelion brightness asymmetries induced by thermal lags in older devolatilized surfaces.

4.1. Activity at Large r: Dynamical Class Comparisons

4.1.1. Historical development. Oort's deduction (*Oort* 1950) of the existence of a large reservoir of comets with aphelia between 5×10^4–1.5×10^5 AU was based upon a

Fig. 5. Reduced R-band brightness (r = Δ = 1) of 2P/Encke from September 1985 through April 2001 as a function of r. Filled squares represent preperihelion and open triangles postperihelion data. The solid horizontal line is an estimate of the brightness of the average cross section of the bare nucleus and the dashed lines show the range of brightness variations due to rotation. Data are from *Meech et al.* (2001).

however, the most sensitive constraints on the amount of scattered light from the coma dust are found in the profile wings, far from the core of the image. However, it is far from the core where the precise determination and removal of the night sky brightness is the most critical, and this limits the sensitivity to mass loss rates >0.1 kg s^{-1} for objects inside ~2 AU. Typical mass loss rates for low-activity comets near perihelion range between $5 < Q < 10^2$ kg s^{-1}.

Techniques that utilize the central part of the surface brightness profile (where the signal-to-noise is highest) and the sky background (not as critical) can yield mass loss limits that are 1–2 orders of magnitude more sensitive at the same r (*Meech and Weaver*, 1995). In this technique, an azimuthally averaged surface brightness profile of the star (or normalized average of several stars) is subtracted from an untrailed comet image profile. For an object with no coma, the subtraction should yield a value of zero with an associated error. The 3σ value of this error can be used as the limiting maximum flux contributed from scattered coma light. This flux is given by

$$F = S_\odot \pi a_{gr}^2 p_v Q_p \phi / 2r^2\Delta^2 v_{gr} \qquad (8)$$

where $S_\odot$ is the solar flux through the bandpass (W m^{-2}), a_{gr} (m) the grain radius, p_v the grain albedo, Q_p (s^{-1}) the production rate, ϕ the projected size of the aperture (m), v_{gr} (m s^{-1})

small sample of comets for which original (e.g., prior to the effects of planetary perturbations) orbits were well determined. In order to reconcile the observed distribution of 1/a, where a is the semimajor axis of the orbit, Oort made the assumption that the new comets perturbed from the Oort cloud must subsequently fade after their first passage and no longer be observable. The cause was attributed to a loss of highly volatile ices (see *Dones et al.,* 2004, for a more in-depth discussion). Later work reexamining the distribution of original semimajor axes by selecting only those comets with perihelia q > 3 AU, where the nongravitational effects were negligible, found a smaller size for the inferred Oort cloud (*Marsden and Sekanina,* 1973).

This led to the suggestion that close stellar passages were probably insufficient to bring comets into the inner solar system directly from the Oort cloud; rather, they would have diffused inward slowly and may have already had perihelia near the region of the outer planets (*Weissman,* 1986, 1990). Consequently, these new comets may have lost any highly volatile materials prior to ever reaching a region of observability, and that there would be no reason to expect significant differences in activity levels between new and old comets.

An analysis of the orbits for 200 comets in order to determine their original orbits (*Marsden et al.,* 1978) was used to infer that, in fact, the DN comets do fade after their first close solar passage. *Marsden et al.* (1978) divided the comets into two accuracy classes depending on the mean error of 1/a, the time span of the observations determining the orbit, and the number of planets whose perturbations were taken into account. They found a significant difference in the number of DN vs. LP comets between the accuracy classes. In the class where the orbits were the most accurately known, as many as 55% of the comets were new (1/a_{orig} < 100 in units of 10^{-6} AU), whereas in the other class the fraction was only 21%. If a comet fades substantially after its first passage through the inner solar system, it will subsequently be observed over a smaller portion of this orbit and will have a less-secure orbit and be more likely to be included in the second group.

Delsemme (1985) suggested on the basis of the light curves of 11 comets that the DN comets are not substantially different from the SP comets, claiming that the activity for all of them is controlled by water sublimation and that most likely the DN comets had spent several orbits slowly diffusing into the inner solar system. Delsemme evaluated the nature of the activity by comparing the light curves of the comets to water vaporization curves and determining values of r_o. Although the dataset was largely uniformly obtained by two individual observers, the range of r for the observations was limited, and in most of the cases the value of r_o was determined by extrapolation. None of the comets were observed at large distances where H_2O-ice sublimation would not be significant. With the exception of three comets in the sample, none of the comets were observed beyond r > 2.5 AU.

Using the same dataset, a statistically significant systematic difference between the medians of the photometric exponents (see section 3) for the DN comets after perihelion

passage was found (*Whipple,* 1978). Whipple suggested that the difference was caused by the loss of a frosting of supervolatile materials during the first passage near the Sun. This was later explained to be due to an observational selection of discoveries (*Svoreň,* 1982).

Several additional older studies that analyzed the light curves of comets also did not find differences with respect to absolute brightness, change in brightness vs. r, asymmetric light curves, and dust-to-gas ratio or dust tail morphologies among the different dynamical groups (*Roemer,* 1962; *Kresák,* 1977; *Svoreň,* 1986; *Donn,* 1977). *Svoreň* (1986) calculated photometric parameters for the sample of 67 LP comets on the basis of photometric observations beyond 2.5 AU. For both old and new comets it was found that the photometric exponent decreases to the value n = 2, i.e., a nonactive stage, at r > 7 AU. The observations at those distances require large telescopes, so until recently the ability to obtain a large dataset has been limited.

A'Hearn et al. (1995), in a large survey of cometary production rates, found that DN comets have r-dependencies that are much less steep inbound than on their outbound legs (they are also much less steep inbound than any other dynamical class).

Rickman et al. (1991) conducted a statistical analysis of a complete sample of SP cometary nongravitational parameters (through 1990). They found that the nongravitational parameter correlated well with the perihelion asymmetry of the gas production over a wide range of r, showing a trend for SP comet nuclei to be less dust-covered with increasing perihelion distance. It was also found that the largest values of the nongravitational parameters were exclusively associated with comets that had recently undergone large reductions of perihelion distance. They concluded that such dynamically young comets have nucleus surfaces that were more free-sublimating than those of older comets.

A summary of analyses based on larger samples of comets to search for statistical differences in the activity levels between SP and Oort cloud comets is shown in Table 2.

4.1.2. Production rate correlations. In a 20-year study of a sample of 85 comets, including 39 Jupiter-family (JF) comets, 8 HT comets, 8 DN comets, and 27 LP comets, production rates were computed for C_2, C_3, OH, NH, and CN in order to look for trends in composition with origin and evolution (*A'Hearn et al.,* 1995). While overall they found that most comets were similar in chemical composition, there was a group of JF comets that were depleted in the carbon chain molecules (C_2 and C_3). A'Hearn et al. argue that this is attributable to a primordial rather than an evolutionary difference. If this were an evolutionary difference there should be a correlation with dynamical age among other comet classes, which was not seen. They suggest that some process in the solar nebula may have preferentially produced or destroyed the carbon chain molecules at the distance of the Edgeworth-Kuiper belt, the source region for the JF comets.

A'Hearn et al. estimated the fractional areas active in their comet sample by comparing the production rates of OH to estimates of the icy surface area needed. They found

TABLE 2.   Search for activity differences between new and old comets.

Reference	Difference	Technique
Roemer (1962)	No	Light curve analysis
Hughes (1975)	No	Frequency of outbursts
Sekanina (1975)	Yes	Activity at large r
Kresák (1977)	No	Light curve analysis
Marsden et al. (1978)	Yes	Orbital errors
Weissman (1980)	Yes	New comet tendency to split a factor of 2.5 higher than for old comets
Delsemme (1985)	No	Light curve fitting, comparison of r_0
Svoreň (1986)	No	Light curve analysis, photometric exponent
Rickman et al. (1991)	Yes	Nongravitational effects; active surface area
A'Hearn et al. (1995)	Yes	Active surface area correlated with dynamical age
A'Hearn et al. (1995)	Yes	Primordial differences in carbon-chain molecule depletion

a clear trend with the amount of active surface area decreasing for older comets, which could either be interpreted as evidence that the nuclei of the dynamically older comets are smaller (primordial condition), or that a smaller fraction of their surfaces are active (an evolutionary effect). Recent work on comet nucleus size distributions, comparing the nuclei of the DN and SP comets, shows that this is likely to be an evolutionary effect (*Meech et al.*, 2004). Additionally, *A'Hearn et al.* (1995) found a strong correlation of the dust-to-gas ratio with perihelion distance. This implies processing of the surface tied to peak surface temperature, although the explanation for why this should affect future dust release rates is unclear. Finally, for those SP comets that did not experience a recent orbital change, no variation was evident in production rates from one apparition to another.

4.1.3. Modern observations. A long-term program of observation of the activity level of a large number of SP, HT, LP, and DN comets has been conducted using the facilities on Mauna Kea, the National Optical Astronomy Observatories, and the Hubble Space Telescope. This program has the advantage over previous studies in that the dataset has been obtained using standardized equipment, filter bandpasses, and measurement apertures, has systematically followed the orbit of the comets over large fractions of the orbital cycle for SP comets, and has placed constraints on the nucleus sizes of the DN comets. From the point of view of the appearance of the comae, the brightness levels, and the rate of change of brightness with distance, a distinct difference between the different dynamical comet groups has been observed. Figures 6 and 7 compare the appearance of comets in each dynamical class.

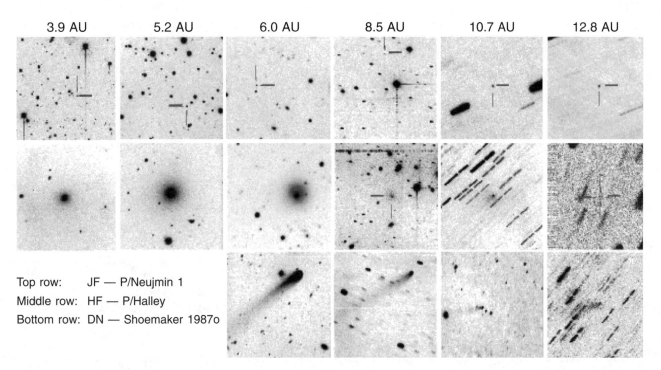

| 3.9 AU | 5.2 AU | 6.0 AU | 8.5 AU | 10.7 AU | 12.8 AU |

Top row: JF — P/Neujmin 1
Middle row: HF — P/Halley
Bottom row: DN — Shoemaker 1987o

Fig. 6.   Comparison of comets from three dynamical classes: SP comet 28P/Neujmin 1, a Jupiter-family comet; 1P/Halley, a HT comet probably evolved inward from the Oort cloud; and a DN comet, C/1987 H1 (Shoemaker 1987o) on its first passage through the inner solar system. The images show the different levels of activity in the groups at different r.

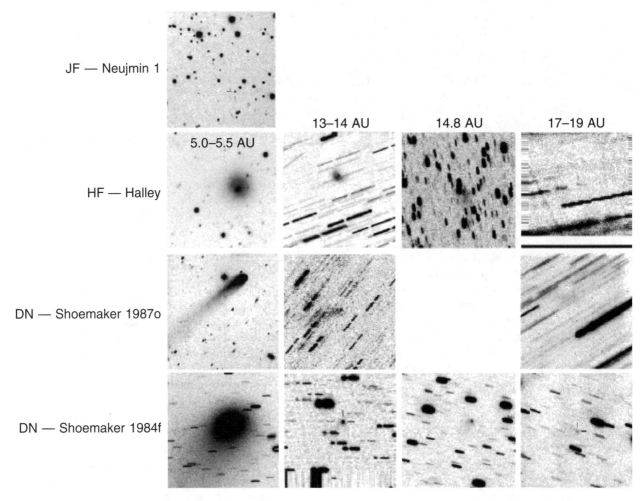

JF — Neujmin 1

5.0–5.5 AU   13–14 AU   14.8 AU   17–19 AU

HF — Halley

DN — Shoemaker 1987o

DN — Shoemaker 1984f

Fig. 7. Same as for Fig. 6, extending the observations to larger r, and adding one DN comet, C/1984 K1 (Shoemaker 1984f).

For the most part, the SP comets in the program rarely exhibit much visible evidence of coma beyond 2–3 AU. This is consistent with the finding of *A'Hearn et al.* (1995) that these comets are more heavily mantled, and have much smaller surface areas, and has probably contributed to the misinterpretation of the *Delsemme* (1982) curves regarding the distance at which H_2O sublimation begins. The brightness of the HT Comet 1P/Halley, which originated in the Oort cloud, was significantly brighter than most of the SP comets at a given heliocentric distance, although even this comet was seen to be heavily mantled from the *Giotto* spacecraft, with only 10% of the surface active. In the images, the comet fades significantly beyond r = 6 AU, and by 12.8 AU had a brightness consistent with a bare nucleus.

At r = 14 AU, 1P/Halley exhibited a large ($\Delta m > 5$ mag) outburst in brightness (see Fig. 8). This has been interpreted as a release of gas and dust initiated by the onset of crystallization in the amorphous ice several tens of meters below the surface (*Prialnik and Bar-Nun,* 1992).

Sporadic activity at large distances is being more frequently observed with the advent of more sensitive detectors. Chiron has been monitored nearly continuously since 1989 when the coma was discovered, through its February 1996 perihelion to the present. Chiron never gets close

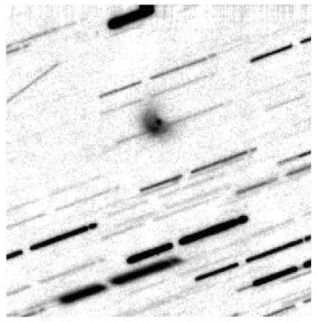

Fig. 8. Composite R-band image of 1P/Halley obtained using the University of Hawai'i 2.2-m telescope on Mauna Kea on February 15, 1991, when the comet was at r = 14.3 AU. The extent of the dust coma was at least 2×10^5 km in diameter.

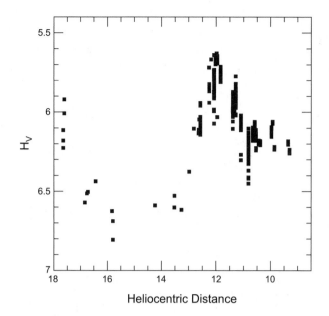

Fig. 9. Brightness variations of 2060 Chiron from Mauna Kea, reduced to unit r and Δ.

enough to the Sun for significant H_2O-ice sublimation (q = 8.45 AU), yet its absolute brightness has had nearly continual fluctuations, including two long, slow outbursts near 17 and 12 AU (see Fig. 9). The sporadic brightening of Chiron's light curve can be reproduced with a model assuming an amorphous ice nucleus with 60% dust fraction, where the activity is driven by crystallization (*Prialnik and Bar-Nun,* 1992). This not only reproduces the sporadic brightening that began near aphelion, but also the thermal measurements and limits on CO and CN production at its

peak brightness. Unlike comets that pass closer to the Sun, Chiron's surface does not get quickly "renewed" from H_2O-ice sublimation because it never gets warm enough.

The general activity that has been observed in the LP comets shows that they have significant differences in their coma appearance, ranging from symmetrical comae, to the narrow parallel-sided tails seen historically in distant comets. *Roemer* (1962) and *Sekanina* (1975), among others, have commented on the fact that the dust tails of distant comets often exhibited this peculiar appearance. The coma in these cases tended to be sharply bounded at the head of the comet. From dust-dynamical modeling, the shapes of these tails suggested the presence of large grains with a small velocity dispersion (*Sekanina,* 1975). The few LP and DN comets that have been observed from perihelion out to between Uranus and Neptune share several characteristics: (1) The brightness fades much more slowly than for the periodic comets, even 1P/Halley, suggesting there is likely a physical difference in the upper layers that affects the heat transport. (2) There is significant activity out to large distances, based on dust-dynamical models — activity continues in some cases beyond 15 AU (see Fig. 10). Detailed thermal models will allow exploration of the physical causes — i.e., if this can be explained by a receding crystallization front. (3) The comets do not seem to exhibit the strong brightness fluctuations seen in Chiron; however, this could be a selection effect since Chiron is so bright that it can be frequently observed.

A recent high-resolution infrared spectroscopic survey has measured production rates of several organic species in a variety of LP and DN comets (*Mumma et al.,* 2001, 2002; *Dello Russo et al.,* 2001). The low formation temperatures and organic compositions provide information

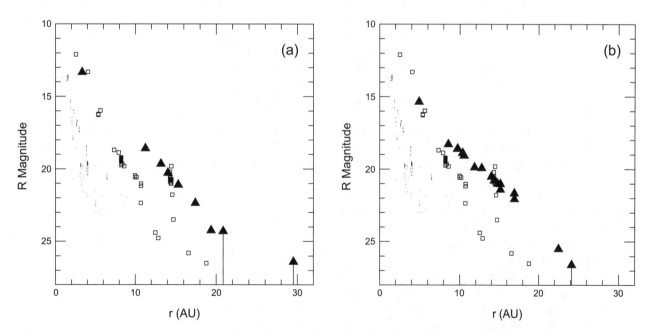

Fig. 10. Comparison of the postperihelion light curves of **(a)** C/1983 O1 (Černis), and **(b)** C/1984 K1 (Shoemaker), shown as filled triangles. The postperihelion light curve of 1P/Halley is shown as open squares, including the outburst near r = 14 AU, and other SP comets as dots.

TABLE 3. Possibly active KBOs and Centaurs.

Object	q–Q	e	a	P	T	Q_p	Δr	Reference
1996 TO$_{66}$	38.48–48.67	0.12	43.57	287.6	02/1908	inferred	45.8	[1]
1999 TD$_{10}$	12.29–190.10	0.88	101.20	1018.1	10/1999	coma?	12.3–12.7	[2]
C/2000 T4	8.56–19.59	0.39	14.10	52.9	05/2002	10^{-1}–10^{-2}	8.5–8.6	[3]
2060 Chiron	8.45–18.91	0.38	13.70	50.7	02/1996	3–4	8.45–17.0	[4]
C/2000 B4	6.83–29.22	0.62	18.02	76.5	06/2000	coma?	6.86	[5]
C/2001 M10	5.30–48.01	0.80	26.66	137.7	06/2001	coma	5.30	[6]

q–Q = perihelion–aphelion (AU); e = orbital eccentricity; a = semimajor axis (AU); P = orbital period (yr); T = most recent time of perihelion; Q_p = estimated production of dust (kg s^{-1}) or comment regarding activity; Δr = range of heliocentric distances at which activity has been observed.
References: [1] *Hainaut et al.* (2000); [2] *Choi et al.* (2003); [3] *Bauer et al.* (2003); [4] *Meech and Belton* (1990); [5] *Kušnirák and Balam* (2000); [6] *Lawrence et al.* (2001).

about the chemical environment in the formation region. The chemistry is not consistent with origins in a thermally or chemically equilibrated region of the solar nebula, rather it is consistent with irradiated ices on grain surfaces in the natal molecular cloud. There is one exception to this, and that is for Comet C/1999 S4 (LINEAR), which has an unusual organic composition that is severely depleted in hypervolatiles and methanol and probably consists of materials condensed from processed nebular gas in the Jupiter-Saturn region.

4.1.4. Activity in Kuiper belt objects and Centaurs. There has been recent interest in searching for activity in KBOs before they enter the inner solar system as Centaurs and SP comets. There is circumstantial evidence for activity in one KBO (*Hainaut et al.*, 2000), while others are conducting sensitive searches for activity in these objects (*Meech et al.*, 2003).

The classification scheme for Centaurs has evolved over time. The original definition encompassed all small bodies orbiting the Sun between Jupiter and Neptune, but other more recent definitions include objects with perihelia between 5.2 < q < 30 AU. Many objects traditionally classified as SP comets, such as 29P/Schwassmann-Wachmann 1, fall into this category. Some objects, such as 39P/Oterma, which by these definitions may be classified as Centaurs, have recently had their orbits perturbed by close Jupiter encounters. *Duncan et al.* (2004) uses the Tisserand invariant to classify those objects with T > 3 and semimajor axis greater than that of Jupiter as Centaurs. This definition excludes 29P/Schwassmann-Wachmann 1 and 39P/Oterma. For the purposes of this chapter, the list of Centaurs and scattered disk objects provided by B. Marsden on the Minor Planet Center Web pages (http://cfa-www.harvard.edu/iau/mpc.html) will be used, and this excludes 29P/Schwassmann-Wachmann 1 and 39P/Oterma.

A handful of Centaurs now show possible evidence for outgassing. These are shown in Table 3. With the exception of 2060 Chiron, the activity in all the Centaurs in Table 3 was discovered when the objects were at perihelion, ranging between 5.3 and 12.3 AU. Although Chiron's activity was discovered near r = 12 AU (inbound), detailed studies have shown that this object has significant activity from

17.5 to 8.5 AU. While many of the volatiles in Table 1 could in principle be responsible for activity at these distances (including H_2O-ice sublimation for the small end of the range), it is very likely that the crystallization of amorphous H_2O ice is the driver for the activity. The more difficult activity to explain, if not induced by collision, is the apparent activity in 1996 TO$_{66}$ at r = 45.8 AU. Here it is too cold for the ice phase transition, and the most likely driver would be sublimation from frozen volatiles, such as CO, CH_4, or C_2H_6.

4.2. Secular Fading

During the active phase, when the comet passes within the inner solar system, there is considerable evolution of the interior and surface of the comet. This evolution may be observed as a secular change (decrease) in the brightness of periodic comets. The release of gases will affect the physical properties of the nucleus such as a change of porosity, redistribution of volatiles, and surface dust mantle formation. One can expect to observe secular fading as a consequence of a smaller surface area available for sublimation.

A very rapid fading of JF comets has been claimed (*Vsechsvyatskij*, 1958), but dismissed as resulting from time-dependent instrumental effects (*Kresák*, 1985). A discovery observation is probably near the upper extreme of brightness and by removing just the discovery apparition observations, there can be a secular brightness decrease of 40% (*Svoreň*, 1991). The active lifetime of a comet often consists of recurring active phases separated by temporary extinctions, and is not truly secular (*Kresák and Kresáková*, 1990). If comets do fade, they likely do so very slowly, or episodically. An excellent resource for the study of comet light curves is the *Comet Light Curve Catalogue/Atlas* (*Kamél*, 1990).

5. SUMMARY

Our understanding of activity in comets has been rapidly evolving thanks to the contributions from new technologies that have allowed observations over a range of distances, including observations of bare nuclei, and the ability to watch the development of the dust coma. In addition, radio astronomy, UV, visible, and infrared spectroscopy are providing us with information about the complex chemical composition of the nuclei. When combined with powerful new dynamical models of the solar system, and with detailed thermal models of the interior of comets and results from laboratory experiments, we are well positioned to begin to understand some of the physical processes involved in "cometary activity." With this understanding will come the ability to separate primordial differences from evolutionary differences, and allow us to use the comets as tracers of the chemical and physical conditions in the early solar system.

Specific types of observations that will significantly improve our ability to separate primordial from evolutionary differences among comet classes include:

1. Searching for activity in very distant comets, in particular those that never come inside the water-sublimation zone ($r = 5$–6 AU), will significantly enhance our understanding of the drivers of activity. This includes observations of activity in Centaurs and searching for activity in KBOs.

2. For the brighter Centaurs, it will be possible to monitor the brightness over a range of heliocentric distances in order to construct good thermal models.

3. For KBOs, the discovery of coma, which could be unambiguously attributed to sustained activity (as opposed to a possible collisional debris cloud), would be very important because of the implications for volatile condensation in the early solar system. The only drivers for activity at such large distances from the Sun are very volatile ices, which some laboratory work and models suggest should be found only as trapped species within amorphous H_2O ice, rather than as frozen solids. Activity in very distant comets, which never come close to the Sun, will initiate some interesting new ideas and constraints with respect to the condensation of volatiles and/or preservation of volatiles in the outer solar nebula.

4. Systematic uniform studies of groups of comets such as those that have been done recently with modern detectors has contributed greatly to the understanding of evolution vs. primordial compositions, and these should continue in new wavelength regimes and by utilizing new technology.

Throughout all these new observational efforts, it is important to pay careful attention to filters used, to be certain to separate the gas from the continuum and to create a uniform dataset, and to understand the limitations of extrapolating brightness predictions from the measurements. In addition, we note that (1) H_2O-ice sublimation driven activity is possible out to 5–6 AU, and (2) activity at distances larger than $r = 5$–6 AU may not require sublimation of ices that are significantly more volatile than water, since the crystallization of amorphous ice can drive activity.

Acknowledgments. The authors wish to thank A. Bar-Nun and T. Owen for long discussions about ice physics and activity mechanisms. In addition, we want to thank D. Prialnik for assistance in the computation of gas production rates for Fig. 2, D. Schleicher for his helpful review of the chapter; and an anonymous reviewer for useful comments. Finally, thanks to H. Weaver for his patience and understanding during the writing of this chapter. This work was supported in part by NASA Grant Nos. NAG5-4495 and NAG5-12236.

REFERENCES

A'Hearn M. F. and Festou M. (1990) The neutral coma. In *Physics and Chemistry of Comets* (W. F. Huebner, ed.), pp. 69–112. Springer-Verlag, Berlin.

A'Hearn M. F. and Millis R. L. (1980) Abundance correlations among comets. *Astron. J., 85,* 1528–1537.

A'Hearn M. F., Millis R. L., Schleicher D. G., Osip D. J., and Birch P. V. (1995) The ensemble properties of comets: Results from narrowband photometry of 85 comets, 1976–1992. *Icarus, 118,* 223–270.

Bar-Nun A. and Laufer D. (2003) First experimental studies of large samples of gas-laden amorphous 'cometary' ices. *Icarus, 161,* 157–163.

Bauer J. M., Fernández Y. R., and Meech K. J. (2003) An optical survey of the active centaur C/2001 T4. *Publ. Astron. Soc. Pacific, 115,* 981–989.

Biver N., Bockelée-Morvan D., Colom P., Crovisier J., Davies J. K., Dent W. R. F., Despois D., Gerard E., Lellouch E., Rauer H., Moreno R., and Paubert G. (1997) Evolution of the outgassing of Comet Hale-Bopp (C/1995 O1) from radio observations. *Science, 275,* 1915–1918.

Bobrovnikoff N. T. (1954) Reports of observations 1953–1954: Perkins observatory-physical properties of comets. *Astrophys. J., 59,* 356–358.

Boehnhardt H. (2004) Split comets. In *Comets II* (M. C. Festou et al., eds.), this volume, Univ. of Arizona, Tucson.

Buratti B. J., Soderblom L. A., Britt D., Oberst J., and Hillier J. K. (2004) Deep Space 1 photometry of the nucleus of Comet P/19 Borrelly. *Icarus, 167,* 16–29.

Choi Y. J., Prialnik D., and Brosch N. (2003) Rotation and cometary activity of KBO 1999 TD$_{10}$. *Icarus, 165,* 101–111.

Dello Russo N., Mumma M. J., DiSanti M. A., Magee-Sauer K., and Novak R. (2001) Ethane production and release in Comet C/1995 O1 Hale-Bopp. *Icarus, 153,* 162–179.

Delsemme A. (1982) Chemical composition of cometary nuclei. In *Comets* (L. L. Wilkening, ed.), pp. 85–130. Univ. of Arizona, Tucson.

Delsemme A. (1985) The sublimation temperature of the cometary nucleus: Observational evidence for H_2O snows. In *Ices in the Solar System* (J. Klinger et al., eds.), pp. 367–387. NATO Advanced Science Institutes Series C, Reidel, Dordrecht.

Delsemme A. and Swings P. (1952) Hydrates de gaz dans les noyaux comtaires et les grains interstellaires. *Annal. Astrophys., 15,* 1–6.

Dones L., Weissman P. R., Levison H. F., and Duncan M. J. (2004) Oort cloud formation and dynamics. In *Comets II* (M. C. Festou et al., eds.), this volume. Univ. of Arizona, Tucson.

Donn B. (1977) A comparison of the composition of new and evolved comets. In *Comets, Asteroids, Meteorites: Interrelations, Evolution, and Origins* (A. H. Delsemme, ed.), pp. 93–97. Univ. of Toledo, Toledo.

Duncan M. F., Quinn T., and Tremaine S. (1988) The origin of short-period comets. *Astrophys. J. Lett., 328,* L69–L73.

Duncan M., Levison H., and Dones L. (2004) Dynamical evolution of ecliptic comets. In *Comets II* (M. C. Festou et al., eds.), this volume. Univ. of Arizona, Tucson.

Durda D. D. and Stern S. A. (2000) Collision rates in the present-day Kuiper belt and Centaur regions: Applications to surface activation and modification on comets, Kuiper belt objects, Centaurs and Pluto-Charon. *Icarus, 145,* 220–229.

Farnham T. L. (1996) Modeling cometary dust tails with a pseudo-Finson-Probstein technique. Ph.D. dissertation, Univ. of Hawai'i, Honolulu.

Fernández J. and Brunini A. (2003) The scattered disk population and the Oort cloud. *Earth Moon Planets, 92,* in press.

Fernández J. and Ip W.-H. (1981) Dynamical evolution of a cometary swarm in the outer planetary region. *Icarus, 47,* 470–479.

Fulle M. (2004) Motion of cometary dust. In *Comets II* (M. C. Festou et al., eds.), this volume, Univ. of Arizona, Tucson.

Grün E. and Jessberger E. K. (1990) Dust. In *Physics and Chemistry of Comets* (W. F. Huebner, ed.), pp. 113–176. Springer-Verlag, Berlin.

Hainaut O. R., Delahodde C. E., Boehnhardt H., Dotto E., Barucci M. A., Meech K. J., Bauer J. M., West R. M., and Doressoundiram A. (2000) Physical properties of TNO 1996 TO$_{66}$. *Astron. Astrophys., 356,* 1076–1088.

Hughes D. W. (1975) Cometary outbursts — a brief survey. *Quart. J. R. Astron. Soc., 16,* 410–427.

Irvine W. M. and Lunine J. I. (2004) The cycle of matter in our galaxy: From clouds to comets. In *Comets II* (M. C. Festou et al., eds.), this volume. Univ. of Arizona, Tucson.

Jenniskens P. and Blake D. F. (1994) Structural transitions in amorphous water ice and astrophysical implications. *Science, 265,* 753–756.

Jenniskens P. and Blake D. F. (1996) Crystallization of amorphous water ice in the solar system. *Astrophys. J., 473,* 1104–1113.

Jewitt D. C. (2004) From cradle to grave: The rise and demise of the comets. In *Comets II* (M. C. Festou et al., eds.), this volume. Univ. of Arizona, Tucson.

Jewitt D. and Meech K. (1987) CCD photometry of comet P/Encke. *Astron. J., 93,* 1542–1548.

Jorda L., Crovisier J., and Green D. W. E. (1991) The correlation between water production rates and visual magnitudes in comets. *Asteroids, Comets, Meteors 1991,* 285–288.

Kamél L. (1990) The comet light curve catalogue/atlas — a presentation. In *Asteroids, Comets, Meteors III* (C. I. Lagerkvist et al., eds.), pp. 363. Uppsala Univ., Sweden.

Kolokolova L., Hanner M. S., Levasseur-Regourd A.-Ch., and Gustafson B. Å. S. (2004) Physical properties of cometary dust from light scattering and thermal emission. In *Comets II* (M. C. Festou et al., eds.), this volume. Univ. of Arizona, Tucson.

Kresák L. (1977) An alternate interpretation of the Oort cloud of comets. In *Comets, Asteroids, Meteorites: Interrelations, Evolution, and Origins* (A. H. Delsemme, ed.), pp. 93–96. Univ. of Toledo, Toledo.

Kresák L. (1985) The aging and lifetimes of comets. In *Dynamics of Comets, Their Origin and Evolution* (A. Carusi and G. B. Valsecchi, eds.), pp. 279–301. Reidel, Dordrecht.

Kresák L. (1987) Aging of comets and their evolution into asteroids. In *The Evolution of the Small Bodies of the Solar System* (M. Fulchignoni and L. Kresák, eds.), p. 202. Soc. Italiana Fis., Bologna.

Kresák L. and Kresáková M. (1990) Secular brightness decrease of periodic comets. *Icarus, 86,* 82–92.

Kušnirák P. and Balam D. (2000) *Comet C/2000 B4 (LINEAR).* IAU Circular No. 7368.

Laufer D., Kochavi E., and Bar-Nun A. (1987) Structure and dynamics of amorphous water ice. *Phys. Rev. B, 36,* 9219–9227.

Lawrence K. J., Helin E. F., and Pravdo S. (2001) *Comet C-2001 M10 (NEAT).* IAU Circular No. 7654.

Lide D. R. (2003) *CRC Handbook of Chemistry and Physics, 84th edition.* CRC Press, Boca Raton, Florida. 2616 pp.

Lunine J. I. and Gautier D. (2004) Coupled physical and chemical evolution of volatiles in the protoplanetary disk: A tale of three elements. In *Comets II* (M. C. Festou et al., eds.), this volume. Univ. of Arizona, Tucson.

Lunine J. I. and Stevenson D. J. (1985) Thermodynamics of clathrate hydrate at low and high pressures with application to the outer solar system. *Astron. J., 58,* 493–521.

Luu J. X. and Jewitt D. C. (1992) High resolution surface brightness profiles of near-earth asteroids. *Icarus, 97,* 276–287.

Marsden B. G. and Roemer E. (1982) Basic information and references. In *Comets* (L. L.Wilkening, ed.), pp. 707–733. Univ. of Arizona, Tucson.

Marsden B. G. and Sekanina Z. (1973) On the orbital distribution of 'original' orbits of comets of large perihelion distance. *Astron. J., 78,* 1118–1124.

Marsden B. G., Sekanina Z., and Everhart E. (1978) New osculating orbits for 110 comets and analysis of original orbits for 200 comets. *Astron. J., 83,* 64–71.

Meech K. (1999) Chemical and physical aging of comets. In *Evolution and Source Regions of Asteroids and Comets* (J. Svoreň et al., eds.), p. 195. IAU Colloquium 73, Astron. Inst. Slovak Acad. Sci., Tatranská Lomnica.

Meech K. J. and Belton M. J. S. (1990) The atmosphere of 2060 Chiron. *Astron. J., 100,* 1323– 1338.

Meech K. J. and Weaver H. A. (1995) Unusual comets (?) as observed from the Hubble Space Telescope. *Earth Moon Planets, 72,* 119–132.

Meech K. J., Fernández Y., and Pittichová J. (2001) Aphelion activity of 2P/Encke. *Bull. Am. Astron. Soc., 33,* 20.06.

Meech K. J., Hainaut O. R., Boehnhardt H., and Delsanti A. (2003) Search for cometary activity in KBO (24952) 1997 QJ$_4$. *Earth Moon Planets, 92,* in press.

Meech K. J., Hainaut O. R., and Marsden B. G. (2004) Comet nucleus size distributions from HST and Keck telescopes. *Icarus,* in press.

Merk R. and Prialnik D. (2003) Early thermal and structural evolution of small bodies in the trans-Neptunian zone. *Earth Moon Planets, 92,* in press.

Mumma M. J., Weissman P. R., and Stern S. A. (1993) Comets and the origin of the solar system: Reading the Rosetta stone. In *Protostars and Planets III* (E. H. Levy and J. I. Lunine, eds.), pp. 1177–1252. Univ. of Arizona, Tucson.

Mumma M. J., Dello Russo N., DiSanti M. A., Magee-Sauer K., Novak R. E., Brittain S., Rettig T., McLean I. S., Reuter D. C., and Xu Li-H. (2001) Organic composition of C/1999 S4 (LINEAR): A comet formed near Jupiter? *Science, 292,* 1334–1339.

Mumma M. J., DiSanti M. A., Dello Russo N., Magee-Sauer K., Gibb E., and Novak R. (2002) The organic volatile composition of Oort cloud comets: Evidence for chemical diversity in the giant-planets' nebular region. In *Proceedings of Asteroids, Comets, Meteors — ACM 2002,* pp. 753–762. ESA SP-500, Noordwijk, The Netherlands.

Notesco G., Bar-Nun A., and Owen T. C. (2003) Gas trapping in water ice at very low deposition rates and implications for comets. *Icarus, 162,* 183–189.

Oort J. H. (1950) The structure of the cloud of comets surrounding the Solar System and a hypothesis concerning its origin.

Bull. Astron. Inst. Neth., 11, 91–110.

Prialnik D. and Bar-Nun A. (1992) Crystallization of amorphous ice as the cause of Comet P/Halley's outburst at 14 AU. *Astron. Astrophys., 258,* L9–L12.

Prialnik D., Bar-Nun A., and Podolak M. (1987) Radiogenic heating of comets by ^{26}Al and implications for their time of formation. *Astrophys. J., 319,* 993–1002.

Prialnik D., Benkhoff J., and Podolak M. (2004) Modeling the structure and activity of comet nuclei. In *Comets II* (M. C. Festou et al., eds.), this volume. Univ. of Arizona, Tucson.

Rickman H., Kamél L., Froeschlé C., and Festou M. C. (1991) Nongravitational effects and the aging of periodic comets. *Astron. J., 102,* 1446–1463.

Roemer E. (1962) Activity in comets at large heliocentric distance. *Publ. Astron. Soc. Pacific, 74,* 351–365.

Schleicher D. G. and Farnham T. L. (2004) Photometry and imaging of the coma with narrowband filters. In *Comets II* (M. C. Festou et al., eds.), this volume. Univ. of Arizona, Tucson.

Schleicher D. G., Lederer S. M., Millis R. L., and Farnham T. L. (1997) Photometric behavior of comet Hale-Bopp (C/1995 O1) before perihelion. *Science, 275,* 1913–1915.

Schleicher D. G., Millis R. L., and Birch P. V. (1998) Narrowband photometry of comet P/Halley: Variation with heliocentric distance, season, and solar phase angle. *Icarus, 132,* 397–417.

Sekanina Z. (1975) A study of the icy tails of the distant comets. *Icarus, 25,* 218–238.

Sekanina Z. (1991) Encke, the comet. *J. R. Astron. Soc. Can., 6,* 324–376.

Stern S. A. (1986) The effects of mechanical interaction between the interstellar medium and comets. *Icarus, 68,* 276–283.

Stern S. A. and Shull M. J. (1988) The influence of supernovae and passing stars on comets in the Oort cloud. *Nature, 332,* 407–411.

Stern S. A. and Weissman P. R. (2001) Rapid collisional evolution of comets during the formation of the Oort cloud. *Nature, 409,* 589–591.

Strazzulla G. and Johnson R. E. (1991) Irradiation effects on comets and cometary debris. In *Comets in the Post-Halley Era* (R. L. Newburn et al., eds.), pp. 243–276. Kluwer, Dordrecht.

Svoreň J. (1982) Perihelion asymmetry in the photometric parameters of long-period comets at large heliocentric distances. In *Sun and Planetary System* (W. Fricke and G. Teleki, eds.), pp. 321–322. Reidel, Dordrecht.

Svoreň J. (1986) Variations of photometric exponents of long-period comets at large heliocentric distances. In *Asteroids, Comets, Meteors II* (C.-I. Lagerkvist et al., eds.), pp. 323–326. Uppsala Univ., Uppsala.

Svoreň J. (1991) Secular decrease in the brightness of short-period comets. *Contrib. Astron. Obs. Skalnate Pleso, 21,* 15–49.

Vsechsvyatskij S. K. (1958) *Fizicheskije kharakteristiki komet.* Gosudarstvenno izdavatelstvo fiziko-matematicheskoj literatury, Moscow.

Weissman P. R. (1980) Physical loss of long-period comets. *Astron. Astrophys., 85,* 191.

Weissman P. R. (1986) The Oort cloud and the galaxy: Dynamical interactions. In *The Galaxy and the Solar System* (R. Smoluchowski et al., eds.), pp. 204–237. Univ. of Arizona, Tucson.

Weissman P. R. (1990) The Oort cloud. *Nature, 344,* 825–830.

Whipple F. L. (1978) Cometary brightness variation and nucleus structure. *Moon and Planets, 18,* 343–359.

Wooden D. H., Charnley S. B., and Ehrenfreund P. (2004) Composition and evolution of interstellar clouds. In *Comets II* (M. C. Festou et al., eds.), this volume. Univ. of Arizona, Tucson.

Yamamoto T. (1985) Formation history and environment of cometary nuclei. In *Ices in the Solar System* (J. Klinger et al., eds.), pp. 205–220. Reidel, Dordrecht.

Structure and Density of Cometary Nuclei

Paul R. Weissman
Jet Propulsion Laboratory

Erik Asphaug
University of California, Santa Cruz

Stephen C. Lowry
Jet Propulsion Laboratory

We are still at a very primitive stage in our understanding of the structure and density of cometary nuclei. Much of the evidence at our disposal is fragmentary and often indirect. Nevertheless, a compelling picture is beginning to emerge of cometary nuclei as collisionally processed fractal aggregates, i.e., rubble piles. The evidence comes from observations of split and disrupted comets, in particular Shoemaker-Levy 9, from theories of planetesimal formation in the early solar nebula, from a recognition of the role of collisions in the evolution of cometary nuclei, and from theoretical and experimental studies of the fragmentation and reassembly of asteroids. This paradigm-shift away from nuclei as monolithic bodies parallels that which has occurred for asteroids in the past decade. A related factor that strongly suggests that nuclei contain substantial macroscopic voids is estimates of the nuclear density, which, like asteroids, show comets to be "under-dense" compared with their constituent materials. We find that the bulk density of cometary nuclei lies in the range 0.5–1.2 g cm^{-3}, with a perhaps "best" current value of 0.6 g cm^{-3}.

1. INTRODUCTION

Cometary nuclei are primordial bodies, among the first to accrete in the early solar nebula. As such, it has long been held (e.g., *Delsemme,* 1977) that comets preserve a cosmochemical record of the composition of the nebula and the conditions within it. Additionally, comets may preserve a physical record of the accretion process itself, how small grains and particles came together to form macroscopic bodies with kilometer dimensions. Investigations of cometary nuclei are thus crucial to understanding planet-building processes in our solar system, and probably in planetary systems around other stars.

It is widely accepted that cometary nuclei formed in the solar nebula through the slow accretion of silicate, organic, and icy grains as material settled to the central plane of the nebula (*Weidenschilling,* 2004). This slow initial agglomeration and accumulation of material produced bodies up to several kilometers in size. In the giant planets zone (5 < r < 35 AU) these "icy planetesimals" were then scattered to distant orbits in the Oort cloud and to escape to interstellar space by the growing gas giant planets. Beyond ~35 AU planetary perturbations were largely incapable of scattering objects to distant orbits so the material there remained *in situ* in the region we now call the Kuiper belt. Thus, cometary nuclei were long viewed as having been preserved in a near-pristine state in these two dynamical reservoirs.

In the last two decades it has increasingly been recognized that the nuclei have been modified over the history of the solar system by a variety of physical processes. These include irradiation by solar and galactic cosmic rays, accretion of and erosion by interstellar grains, heating by nearby supernovae and from stars passing through the Oort cloud, the crystallization of amorphous water ice as the nucleus is warmed above 120–150 K for the first time, sublimation of volatiles as the cometary nuclei approach the Sun, and collisional processing, either in the Kuiper belt or during the ejection of protocomets from the giant planets zone to the Oort cloud (for reviews, see *Weissman and Stern,* 1998, and *Stern,* 2003, and references therein). Another possible modifying mechanism is internal heating by short-lived radionuclides in the early solar system (*Prialnik and Podolak,* 1999), although we have no evidence as to whether this did or did not occur.

Most of these processes only affect a relatively thin layer near the nucleus surface. However, collisions can radically alter the nucleus structure, ranging from substantial fracturing of the cometary material(s) to total disruption and subsequent reassembly of the nucleus. There is a considerable body of studies of the collisional evolution of asteroidal bodies, which we suggest is very applicable to the problem of the structure of cometary nuclei. We believe they show that, like asteroids, cometary nuclei probably have a rubble-pile structure, although the comets may have followed a somewhat different path to that final state from that of their asteroidal cousins.

Evidence for the rubble pile nature of cometary nuclei comes from observations of split and disrupted comets, in particular comet Shoemaker-Levy 9 (D/1993 F2), which was tidally disrupted by a close encounter with Jupiter, from theoretical studies of the accretion of icy planetesimals in the early solar system, from the relatively recent recognition that cometary nuclei are collisionally processed objects, and from studies of the collisional evolution of asteroids, as noted above. Together, we find that these lines of evidence create a compelling picture for cometary nuclei as collisionally evolved rubble piles.

A measurable physical parameter that has strong implications for the internal structure of cometary nuclei is the bulk density. If nuclei are indeed fluffy aggregates or rubble piles, then they may be "under-dense" relative to their constituent materials and contain substantial macroscopic voids. Thus, density measurements alone could be used to infer a fluffy or rubble-pile structure. However, density measurements for cometary nuclei are exceedingly difficult to obtain, and at present can only be accomplished indirectly.

In this chapter we will review our current understanding of cometary nucleus structure and density, and the evidence that is leading us to conclude that cometary nuclei are collisionally evolved rubble piles. In section 2 we examine the proposed models for cometary nuclei. In section 3 we review the evidence for such models from spacecraft encounters. In section 4 we discuss evidence for cometary rubble piles, including the substantial body of research on the structure and evolution of asteroids that we find is very applicable to this problem. In section 5 we discuss density estimates for cometary nuclei and the methods employed. Section 6 contains a discussion of these topics and our conclusions, and a discussion of expected future spacecraft measurements.

2. PROPOSED MODELS OF COMET NUCLEUS STRUCTURE

2.1. The Icy-Conglomerate Model

The modern era in understanding cometary nuclei began with *Whipple*'s classic series of papers (1950, 1951, 1955) that first proposed the "icy-conglomerate" model for the cometary nucleus. Whipple sought to explain the nongravitational motion of periodic comet Encke and others by suggesting a "rocket effect" from sublimating ices on the surface of a rotating nucleus. The earlier sandbank model of *Levin* (1943), *Lyttleton* (1948), and others envisioned the cometary "nucleus" as a gravitationally bound swarm of dust particles with adsorbed gases, orbiting the Sun. In contrast, Whipple envisioned the cometary nucleus as a single macroscopic body composed of a mixture of volatile ices and "meteoritic material." Whipple's papers are impressive in that he proposed many of the features that have become part of the standard paradigm for cometary nuclei today. These include the formation of nuclei at very low temperatures; the low bulk density of cometary nuclei; porosity within the nucleus; the low strength, low albedo, and low thermal conductivity of cometary materials; and the formation of nonvolatile lag deposits on nucleus surfaces, slowly cutting off cometary activity.

Some of these ideas were not entirely new to the scientific literature. For example, *Vorontsov-Velyaminov* (1946) notes that *Baldet* (1927) and *Slipher* (1927) estimated that the nucleus of periodic comet Pons-Winnecke was a compact object with dimensions of only 2–3 miles, and that it might be a monolithic body. Also, the idea of a "rocket effect" from evolving gases had previously been proposed for the sandbank model, based on desorption of bound gases from grains as the comets approached the Sun. Even the idea of ice in cometary nuclei had been proposed by Vsekhsviatsky in 1948. Whipple's key contributions were his ability to combine these disparate ideas into a unified model that explained many aspects of cometary behavior, and his insistence that the nucleus was a single, small, solid body.

The debate between advocates of the sandbank and icy-conglomerate models continued for several decades after 1950. Any questions of nucleus "structure" were only in terms of the sandbank vs. the icy conglomerate; Whipple's papers did not comment on the underlying structure of the nucleus. However, during this time observational and theoretical evidence in support of the icy-conglomerate model continued to grow. Among the more notable accomplishments were *Delsemme*'s (e.g., 1971) work on sublimation rates of water and other volatile ices and Marsden and colleagues' (e.g., *Marsden et al.*, 1973) modeling of nongravitational motions in comets using those sublimation models. Over time, the sandbank model fell into disfavor, finally being discarded in 1986 when the *Giotto* and *Vega* spacecraft returned images of the nucleus of comet 1P/Halley (Fig. 1).

2.2. The Fluffy-Aggregate, Primordial-Rubble-Pile, and Icy-Glue Models

The approach of periodic comet Halley in the 1980s heightened interest in comets and provided the impetus for new investigations, both observational and theoretical, into the nature of cometary nuclei. These included hypotheses with regard to the underlying structure of the nucleus. Two models, proposed almost simultaneously, were the "fluffy aggregate" of *Donn et al.* (1985) and *Donn and Hughes* (1986), and the "primordial rubble pile" of *Weissman* (1986). The primary concept in both these proposals was that cometary nuclei were aggregates of smaller icy planetesimals, brought together at low velocity in a random fashion. With little in the way of modifying processes or energy sources available to change this initial structure, the cometary nuclei would preserve their highly irregular initial shapes and very porous, easily fragmented structure over the history of the solar system.

The arguments of Donn and colleagues came from their studies of the accretion of small grains in the solar nebula, realizing that random accretion would lead to self-similar structures at larger spatial scales. Weissman, on the other hand, pointed out that the total gravitational potential energy of a typical cometary nucleus, say 5 km in radius, was not sufficient to raise the temperature of the cometary material

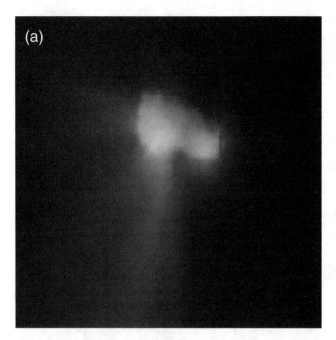 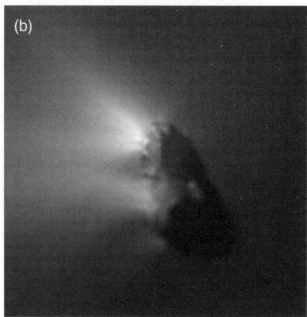

Fig. 1. Images of the nucleus of comet 1P/Halley. **(a)** *Vega 2* image taken on March 9, 1986, from a range of 8031 km at a phase angle of 28.4°. The spatial resolution is ~160 m/pixel. The image shows the "peanut-like" shape of the nucleus and several large dust jets emanating from its surface. **(b)** Composite *Giotto* image taken on March 13–14, 1986 (© Max-Planck Institute for Aeronomy). The resolution varies from ~50 m/pixel at upper left to ~320 m/pixel at lower right. Phase angles vary from 89° to 107° in a similar fashion. Both images showed that the Halley nucleus was a dark, irregular object with a bimodal structure.

by even 1°K, and thus there was no energy source to mold it into a single monolithic body. Both Donn et al. and Weissman suggested that a fragmentary structure for cometary nuclei could help to explain such observed processes as outbursts and splitting, and could provide a mechanism for irregular activity on the surfaces of cometary nuclei.

Weissman drew analogies with previous work on the rubble-pile structure of asteroids (*Davis et al.,* 1979). However, he appended the term "primordial" to suggest that the nuclei were original solar nebula material, and not the products of earlier, disrupted bodies. We now recognize that collisional evolution likely played a role for cometary nuclei also (see section 4.2), and so the nuclei may indeed be reassembled rubble piles from earlier generations of icy planetesimals.

Note that the fluffy-aggregate and primordial-rubble-pile models are not new versions of the sandbank model, as incorrectly stated by *Sekanina* (1996). Sekanina erroneously equated the newer models with the *Vorontsov-Velyaminov* (1946) model, which suggested that cometary nuclei were a swarm "some 25–60 km in diameter . . . composed of [~10⁷] blocks some 160 m in diameter, which are nearly in contact." In other words, Vorontsov-Velyaminov proposed a swarm of boulders rather than sand. In contrast, both the fluffy aggregate and primordial rubble pile models require that the sub-fragments of the nuclei are in contact in a single nucleus structure, and are weakly bonded and/or gravitationally bound.

A third hypothesis, proposed after the Halley spacecraft flybys, was the "icy-glue" model of *Gombosi and Houpis* (1986). They suggested that comets were composed of porous refractory boulders with compositions similar to outer-

main-belt asteroids, cemented together by an icy-conglomerate glue. In the icy-glue model the boulders provided the irregular topography seen in the *Giotto* images of the Halley nucleus (*Keller et al.,* 1986) and also helped to explain the collimated jets seen emanating from the surface (from active icy-glue regions between pairs of boulders). Although it contains some interesting features, the icy-glue model has not received wide support because there is no evidence for a population of remnant "boulders" from decaying comets, and it could not explain many of the features of the breakup of comet Shoemaker-Levy 9 (D/1993 F2).

All these nucleus concepts are illustrated in Fig. 2. The general consensus today is that the fluffy-aggregate and primordial or collisionally evolved rubble-pile models are the best description of the underlying structure of the cometary nucleus. Our discussion in the following sections will focus on these two models and why we believe they are the best current description for the structure of cometary nuclei. Note that whether we use the term "fluffy aggregate" or "rubble pile," we are referring to the same basic concept of a weakly bound agglomeration of smaller icy cometesimals.

3. SPACECRAFT IMAGING OF COMETARY NUCLEI

3.1. Comet 1P/Halley

The first resolved images of a cometary nucleus were obtained by the *Vega 1, Vega 2,* and *Giotto* spacecraft that flew past comet 1P/Halley in March 1986 (*Sagdeev et al.,*

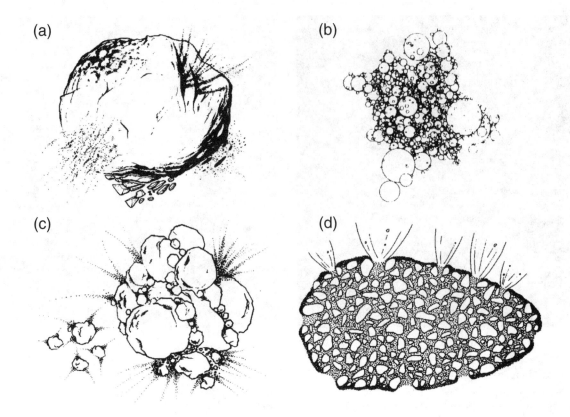

Fig. 2. Artists' concepts of various models for cometary nuclei: **(a)** Whipple's icy conglomerate model as envisioned by *Weissman and Kieffer* (1981); **(b)** the fractal aggregate model of *Donn et al.* (1985); **(c)** the primordial rubble pile model of *Weissman* (1986); and **(d)** the icy-glue model of *Gombosi and Houpis* (1986). All but **(d)** were proposed prior to the spacecraft flybys of comet 1P/Halley in 1986.

1986; *Keller et al.,* 1986). The *Vega* images were taken from a range of 8–50 × 10³ km. Unfortunately, the *Vega 1* camera was badly out of focus. Still, the images were useful in determining the overall shape and dimensions of the nucleus, as shown in Fig. 1a, taken during the *Vega 2* closest approach at a range of 8031 km. The "peanut-like" shape of the nucleus is clearly visible.

The *Giotto* images were taken at a much closer range and show considerably more detail. A composite *Giotto* image of the Halley nucleus is shown in Fig. 1b (*Keller et al.,* 1986). The Sun is at upper left in the image. The nucleus is viewed at a phase angle of 89°–107°; lower-phase images cluster near the upper left. Only about 25% of the nucleus is illuminated by sunlight. The outline of the unilluminated nucleus is visible at lower right against the bright cometary dust coma. Because this is a composite of many images, the spatial resolution varies from a best value of ~50 m/pixel at the upper left to ~320 m/pixel at lower right. The last image was taken at a range of 1680 km. Bright dust jets emanate from the sunlit portions of the nucleus and obscure the underlying topography. The projected nucleus dimensions in the image are ~14.0 × 7.5 km. A triaxial solution for the dimensions of the nucleus, combining images from all three spacecraft, gave axes of 15.3 × 7.2 × 7.2 km [±0.5 km in each dimension (*Merényi et al.,* 1989)]. The average surface albedo was 0.05 to 0.08, assuming a lunar-like phase function and extrapolated to zero phase (*Keller et al.,* 1994).

The nucleus is clearly seen as an elongated object with highly irregular surface topography. The bright spot in the right center of the image is a "hill" ~500 m in height, sticking up into the sunlight from beyond the terminator. Other hill-like structures with dimensions of ~500 m are visible surrounding an apparently flat area at upper left. A feature near the upper left center of the image was identified early on as a crater but more careful examination shows it to be a fortuitous arrangement of two pairs of hills, each forming V-shaped ridges. The overall nucleus has a binary appearance with a narrow "waist" at the center. Activity appears to originate from discrete areas on the nucleus surface some hundreds of meters in size, rather than from the entire sunlit area. The apparently inactive areas may be lag deposits of nonvolatile materials, too heavy to be lifted off the nucleus surface, or may be part of the original radiation-processed crust of the cometary nucleus.

3.2. Comet 19P/Borrelly

The nucleus of comet 19P/Borrelly, shown in Fig. 3 (*Soderblom et al.,* 2002), was imaged by the *Deep Space 1* (*DS1*) spacecraft on September 22, 2001, from a distance of

3560 km at a phase angle of 51.6°. The resolution of the image is 47 m/pixel and the Sun is to the left. The overall nucleus dimensions are ~8.0 × 3.2 km (±0.2 km in each dimension), and it is readily seen to have a bimodal structure. Like Halley, the topography is rough, although there also appear to be relatively smooth areas. The smooth areas appear to include several "mesas," large flat regions raised above the surrounding terrain. Active jets (not visible in this version of the image) emanate from the smooth regions near the center of the sunlit limb (at upper left). Several sharp ridges are visible along the terminator and near the narrow neck of the nucleus at lower left. No fresh impact craters down to ~200 m in diameter are visible anywhere on the illuminated surface.

Like the Halley nucleus, the surface of 19P/Borrelly is dark, with an average albedo of 0.029 ± 0.006 (*Buratti et al.,* 2003), although some spots have albedos as low as 0.01. The derived phase curve (from both spacecraft and groundbased data) is similar to that for C-type asteroids. Near-infrared spectra between 1.3 and 2.6 µm show a strongly red slope and a generally featureless spectrum with the exception of an unidentified feature at 2.39 µm. This feature appears in all spectra and may be associated with hydrocarbon compounds such as polyoxymethelene (*Soderblom et al.,* 2002). Using the *DS1* infrared data, the surface temperature was estimated at between 300 and 345 K, consistent with an equilibrium black-body surface at its distance from the Sun,

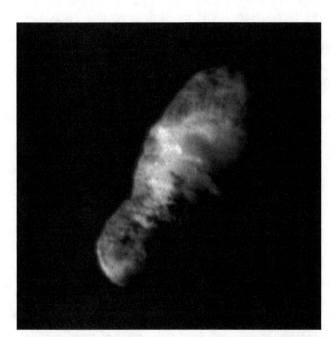

Fig. 3. *Deep Space 1* image of the nucleus of 19P/Borrelly, taken on September 22, 2001 from a range of 3560 km (*Soderblom et al.,* 2002). The phase angle is 51.6° and the resolution is 47 m/ pixel. Like the nucleus of comet Halley, the Borrelly nucleus is dark with irregular topography, but also with large, apparently smooth areas, and with jets emanating from discrete areas on the surface (not visible in this version of the image). Also, like Halley, the nucleus is clearly bimodal.

and no apparent cooling due to sublimation of surface ices. However, because each infrared spectrum is, in fact, the average of a swath across the nucleus surface from bright limb to terminator, any small cool regions (such as the sources of the jets) would be masked by the much stronger signal from the warmer inactive regions.

3.3. Analysis

Neither the Halley nor the Borrelly images are at sufficient resolution to understand fully the surface morphology of these two comets, in particular, the sources of the jets or the nature of the apparently inactive regions. However, images of both nuclei unequivocally show a bimodal structure. Such structures have not been apparent in spacecraft flyby images of asteroids, although they are evident in radar images of some near-Earth asteroids (*Ostro et al.,* 2003). However, a rubble-pile structure may not readily manifest itself in surface features. The exceedingly large craters on asteroid 253 Mathilde have been interpreted as evidence of an underlying rubble-pile structure (*Veverka et al.,* 1997), as any monolithic asteroid would be destroyed by impacts large enough to create such craters (see next section). Even more so than asteroids, cometary nuclei may have the ability to mask their internal structures through mass wasting processes such as sublimation, sintering, and fallback.

Both the Halley and Borrelly nuclei show considerable surface roughness, as might be expected from a rubble-pile structure, where large chunks may easily break off or be rotationally dislodged (although both of these are slowly rotating nuclei). Also, it is interesting that both nuclei look more alike than different, since we suspect that they likely originated from different dynamical reservoirs, possibly with different collisional histories. As a typical Jupiter-family comet, Borrelly likely originated from the collisionally processed Kuiper belt, whereas Halley, with its retrograde orbit, most likely originated from the Oort cloud and hence the giant-planets zone (*Levison,* 1996).

4. EVIDENCE FOR THE RUBBLE-PILE NATURE OF COMETARY NUCLEI

4.1. Disrupted Comets

The strongest observational evidence for cometary nuclei as rubble piles comes from observations of disrupted or split comets. Most splitting events are random and seem to occur for no obvious reason. The classic example is comet 3D/Biela, a Jupiter-family comet with a period of 6.6 yr that was observed in 1772, 1805, 1826, and 1832. The comet was observed to split during its 1846 apparition and returned as a double comet in 1852. It was never observed again. However, intense showers were observed from the related Andromedid meteor stream in 1872, 1885, and 1892, corresponding to times when 3D/Biela should have returned.

More recently comet LINEAR, D/1999 S4, was observed to disrupt completely as it passed through perihelion in July

Fig. 4. Hubble Space Telescope image of comet LINEAR (D/ 1999 S4) taken on August 5, 2000, showing fragments of the disintegrating nucleus (*Weaver et al.,* 2001). This long-period comet disrupted close to perihelion at 0.765 AU from the Sun in July 2000.

2000 (*Weaver et al.,* 2001) (see Fig. 4). Weaver et al. observed at least 16 fragments of D/1999 S4 using the Hubble Space Telescope (HST) and Very Large Telescope (VLT), and estimated radii of 25–60 m, assuming an albedo of 0.04. There was evidence for secondary components near some fragments, and evidence that the fragments continued to split over time. The SWAN instrument on the Solar and Heliospheric Observatory (SOHO) observed water production rates during the breakup (*Mäkinen et al.,* 2001) and found that the observations could best be explained by a power-law radius distribution for the fragments, $N(>r) \propto r^{-a}$, where N is the number of fragments with radius greater than r, with a cumulative slope, a, of 1.74. Interestingly, this is close to the slope of 1.59 ± 0.03 found for Jupiter-family cometary nuclei by *Weissman and Lowry* (2003).

Weissman (1980) compiled records of observations of 18 split comets and showed that 10% of dynamically new comets from the Oort cloud split, vs. 4% for returning long-period comets, and only 1% for short-period comets (per perihelion passage) (see also *Boehnhardt,* 2004). The splitting events did not show any correlation with perihelion distance, distance above the ecliptic plane, or time of perihelion passage. The statistics suggest that the tendency of cometary nuclei to split may reflect some intrinsic nucleus property, such that comets that are likely to split do so early on, and those that are unlikely to split are able to survive for hundreds or even thousands of returns. Note however that splitting events do not always lead to total disruption of the nucleus. For example, comet 73P/Schwassmann-Wachmann 3 has been observed to shed fragments on at least two perihelion passages, yet it still returns every 5.4 yr. In fact, the majority of "splitting" events involve one or more small fragments breaking off the main nucleus, and the latter surviving the event.

There is at present no known mechanism for explaining these random splitting events. One proposed mechanism by *Samarasinha* (1999), gas pressure release from volatile pockets, is discussed in section 4.4. Also, *Weissman et al.* (2003) have proposed rotational spinup due to asymmetrical outgassing forces as a likely cause. Regardless of the mechanism, it seems clear that nuclei are fragile objects and that when they disrupt, they break into subfragments of tens of meters in dimension.

A second form of disruption event that provides insights into nucleus structure occurs when a comet passes through the Roche limit of a planet or the Sun. This has happened in the case of Jupiter (16P/Brooks 2 in 1886 and D/Shoemaker-Levy 9 in 1992; see section 4.6) and even more spectacularly in the case of the Sun (*Marsden,* 1989). Prior to 1978, nine "Sun-grazing" comets, those with perihelion distances less than 0.01 AU (~2 solar radii), had been discovered by groundbased observers. Eight of those nine were in very similar orbits and were known as the Kreutz group. It was suggested that these were fragments of a larger comet that had been tidally disrupted on a previous perihelion passage (e.g., *Marsden,* 1989). *Weissman* (1979) showed that nongravitational accelerations are so great for Sun-grazing comets that they can be perturbed to their current orbits with semimajor axes of ~100 AU in only two or three perihelion passages. Several of the Kreutz comets split during their perihelion passages and this was used by *Öpik* (1966) to estimate nucleus strengths of 10^4–10^6 dynes cm^{-2}. To first order, this is about the strength of a snowdrift or a pile of dirt. So cometary nuclei appear to be very weakly bonded.

Michels et al. (1982) and *Sheeley et al.* (1985) discovered six additional Kreutz members using the SOLWIND coronagraph on an Earth-orbiting satellite. None of these comets survived perihelion passage, nor were they detected from the ground, suggesting that they were relatively small. An additional 10 Sun-grazers were found by the Solar Max spacecraft between 1987 and 1989 (*MacQueen and St. Cyr,* 1991). More recently, the SOHO spacecraft has discovered ~540 Sun-grazing comets between 1996 and the end of 2002 (*Biesecker et al.,* 2002). Most of these are Kreutz group members, although three other small groups have also been identified, two of which are possibly identified with comet 96P/Machholz 1. *Biesecker et al.* (2002) estimated diameters for SOHO fragments of several to tens of meters. *Weissman* (1983) and *Iseli et al.* (2002) showed that cometesimals larger than ~40–120 m in diameter might be expected to survive perihelion passage, as there is insufficient time for them to sublimate completely.

These continuous streams of cometesimals can be readily explained if the progenitor nuclei are aggregates or rubble piles that disrupted on previous perihelion passages. Small differences in their initial orbits would lead to the large dispersion in arrival times (the typical Sun-grazer orbital period is 500–1000 yr), particularly if the cometesimals were freed from the nucleus more than one orbit ago. In many ways the dynamics are very similar to meteoroid streams, although radiation forces likely do not play a major role.

Note that if all 540 of the SOHO comets discovered thru 2002 had 10-m diameters, they would add up to a nucleus less than 100 m in diameter, so although there are many cometesimals (and many as yet undiscovered), the progenitor comet need not have been very large.

As discussed in section 4.4 below, a monolithic progenitor nucleus cannot easily explain the huge numbers of SOHO comets, as hierarchal splitting would likely not result in a very large number of fragments. The entire passage of the comet within the solar Roche limit takes only ~3–4 h (depending on the nucleus bulk density assumed) and there is not sufficient time for the nucleus to repeatedly split unless it was already an agglomerate of a huge number of smaller cometesimals, i.e., a rubble pile. Alternatively, *Sekanina* (2002) has argued that Kreutz-family fragments continue to split randomly around their entire orbits. Since random disruption is so poorly understood, this possibility cannot be ruled out. However, could such a random mechanism explain the narrow size distribution of the observed SOHO comets? Why are larger fragments not observed? Is it really necessary to invoke additional random disruption to explain the Sun-grazing comet streams? Clearly, there are still many open questions with regard to the Sun-grazing comets.

4.2. Collisional Evolution

It is easy to show that collisions between cometary nuclei are very rare in the classical Oort cloud, the region beyond ~10^4 AU from the Sun that supplies the long-period comet flux through the planetary region (*Oort*, 1950). *Stern* (1988) found that impact rates would be more significant in the proposed inner Oort cloud (*Duncan et al.*, 1987), and that all comets there would undergo at least some surface modification due to impacts of collisional debris.

More recently, studies of the physical and dynamical evolution of the more tightly packed Kuiper belt (*Duncan et al.*, 2004) suggested that collisions play a very important role (*Stern*, 1995, 1996; *Farinella and Davis*, 1996). Additionally, *Stern and Weissman* (2001) showed that collisions between protocomets in the giant-planets zone, prior to their ejection to the Oort cloud (or to interstellar space), would be catastrophic for much of the initial population. *Charnoz and Morbidelli* (2003) confirmed that collisions result in substantial erosion of cometesimals during the ejection process, although not quite as severely as found by Stern and Weissman. However, the differences in their results are most likely attributable to differences in the respective models and in the specific cases run. Regardless, this new view of the collisional history of cometary nuclei is essentially a complete reversal of the picture of cometary nuclei as unprocessed aggregates from the primordial solar nebula.

Thus, we must consider what effect collisions might have on the internal structure of cometary nuclei. Fortunately, this question has received significant attention in recent years through studies of the collisional processing of asteroids, spurred on in large part by the increasing attention given to near-Earth objects (NEOs) and the hazard they present to life on Earth. An important issue, then, is the degree to which cometary structure can be inferred from what we presently know concerning asteroids. As noted above, comets and asteroids alike appear to be products of moderate to intense collisional evolution. They both are of a size that sits on the fulcrum between gravity-dominated and strength-dominated bodies (*Asphaug et al.*, 2003). So the forces of evolution, and the forces of equilibrium response to that evolution, appear to be at least broadly the same for comets as for asteroids.

There are, however, obvious differences. Comets and asteroids are each derived from different dynamical reservoirs at different initial heliocentric distances, and thus different thermal regimes. Comets undergo intense geologic activity in the form of mass wasting, i.e., sublimation and disruption. Impact craters are not expected to be long-lived on an active nucleus (and are not observed; see section 2), whereas the asteroids are saturated with craters. Another distinction relates to the nongravitational forces applied to active comets, which can excite their rotational state (*Samarasinha and Belton*, 1995), possibly to the point of shape modification or disruption (*Weissman et al.*, 2003). Though nongravitational forces are now also proposed for asteroids (*Rubincam*, 2000), including forces that can alter their rotation over longer timescales [e.g., the Yarkovsky-O'Keefe-Radzievskii-Paddack (YORP) effect, named after the scientists who contributed to development of the idea; YORP results in rotational changes on small bodies due to forces from asymmetric reradiation of absorbed insolation]. A final difference is the origin of mechanical strength. Comets, possessed of potentially mobile volatiles, have a means of forming cohesive aggregate structures over time, whereas a dry asteroid may become truly strengthless if it evolves into a rubble pile.

The interpretation of highly evolved, rocky asteroids, especially the common S and V types, has made good progress thanks to the notable success of the Near-Earth Asteroid Rendezvous (NEAR) mission at asteroid 433 Eros (which is generally agreed to be a highly fragmented, gravitationally bound rock), together with the *Galileo* flybys of Gaspra and Ida and supported by collisional modeling based on terrestrial rock types (e.g., *Asphaug et al.*, 1996). But primitive asteroids, the type example being 253 Mathilde, represent a deep perplexity for asteroid science. While evidence of a highly porous interior for Mathilde is no longer in dispute — the measured density from the *NEAR Shoemaker* flyby is 1.3 ± 0.2 g cm^{-3} (*Veverka et al.*, 1997) — the nature of this porosity is entirely unclear. Is Mathilde microporous, in the manner of the cometary dust balls proposed by *Greenberg and Hage* (1990), and recently proposed by *Housen et al.* (1999) to explain Mathilde's giant craters? If so, is it cohesive, as one might expect for microscale grain structure? Or does Mathilde, and the other primitive asteroids with comparable densities [as determined by analyses of their satellite orbits (*Merline et al.*, 2003)], possess huge voids as one would expect from collisional disruption and reassembly of major fragments (*Benz and Asphaug*, 1999)? In studying the structure of cometary nuclei and asteroids, we learn of their origins and evolution.

4.3. Monoliths, Rubble Piles, and Porosity

To describe the interiors of small bodies in the solar system we require a dictionary of well-defined and agreed-upon terms. *Richardson et al.* (2003) recently reviewed this topic with regard to asteroids; for comets one must add to this the complexity of mantle development (e.g., *Brin and Mendis,* 1979) and melting of the interior due to short-lived radionuclides (*Prialnik and Podolak,* 1999), both of which might significantly alter the internal structure.

Monoliths are objects of low porosity and significant strength, and are good transmitters of elastic stress. Monoliths in an impact-evolved population must be smaller than the size that would accumulate its own impact ejecta. Escape velocity for a constant density body is proportional to its size (approximately 1 m/s per km of diameter for spheres of ice), so that bodies larger than some transition size evolve into gravitational aggregates. Monoliths can be *fractured* by impact bombardment, in which case their tensile strength is compromised and may be reduced to zero, i.e., *shattered*. A fractured or shattered monolith might transmit a compressive stress wave fairly well, provided pore space has not been introduced between the major fragments. Tensile stress, however, is not supported across a fracture.

A *rubble pile* includes any shattered body whose pieces are furthermore translated and rotated into a loose packing. Stress waves of any sort are poorly transmitted across a rubble pile, although intense shocks may propagate by crushing and vapor expansion. *Primordial rubble piles* are objects that accreted as uncompacted cumulates to begin with. *Collisional rubble piles* are primordial rubble piles that have subsequently undergone collisional evolution.

Because much of our discussion regarding comets relates to primordial and collisional rubble piles, we must also distinguish between macroscopic (coarse) and microscopic (fine) porosity. An asteroid or cometary nucleus consisting of quintillions of tiny grains might exhibit considerable cohesion and perhaps support a fractal-like "fairy-castle" porosity. The total energy of contact bonds divided by the total mass of a granular asteroid, its overall cohesional strength, is inversely proportional to grain diameter, so that a coarse aggregate is weaker than a fine one, if all else is equal (*Greenberg,* 1998). On the other hand, a highly porous, finely comminuted body might accommodate significant compaction. Impact cratering on microporous bodies might provide counter-intuitive results, involving crushing and capture of impacting material (e.g., *Housen and Holsapple,* 1990) rather than ejection.

A coarse rubble pile by contrast has far fewer contact surfaces distributed over the same total mass, and would therefore behave much differently. *Asphaug et al.* (1998) used a coarse rubble pile as a starting condition for impact studies, and found that the impact shock wave gets trapped in the impacted components, with few pathways of transmission to neighboring components. Moreover, a coarse rubble pile tends to result in ejection of most impact products (*Asphaug et al.,* 2003) rather than absorption by compaction. Whether a comet is macroporous or microporous, stress wave transmission is hindered due to the great attenuation of poorly consolidated ice and rock, making the survival of porous comets and asteroids more likely during impact, as demonstrated by the experiments of *Ryan et al.* (1991) and *Love et al.* (1993) and by numerical (*Asphaug et al.,* 1998) and scaled simulations (*Housen and Holsapple,* 2003).

4.4. Volatiles and Cohesion

A volatile-rich aggregate (such as an icy cometary nucleus) is more cohesive than a dry aggregate (such as an asteroidal rubble pile) due to the facilitation of mechanical bonding, either directly (e.g., van der Waals forces) or indirectly during episodes of sublimation and frost deposition (*Bridges et al.,* 1996). For gravity as low as on a typical cometary nucleus, frost or other fragile bonds can be critical to long-term survival during impact or tidal events. Comet Shoemaker-Levy 9, for example, could never have disrupted during its 1992 tidal passage near Jupiter at a perijove of only 1.3 jovian radii had the tensile strength across the comet exceeded $\sim 10^3$ dynes cm^{-2} (*Sekanina et al.,* 1994), weaker than snow. *Asphaug and Benz* (1994, 1996) calculated a maximum tensile strength of only ~ 30 dynes cm^{-2} for Shoemaker-Levy 9 in order for it to fragment from a hypothetical monolithic body into ~ 20 pieces (see below), and therefore proposed a cohesionless rubble-pile structure for this comet and gravitational clumping (as opposed to fragmentation) as the cause of its "string of pearls" post-perijove structure. It matters greatly whether a comet is truly strengthless or only extraordinarily weak. Note, however, that Shoemaker-Levy 9 may have been previously disrupted, although not catastrophically, during its ~ 60 yr or more orbiting Jupiter, and thus the tensile strength determined by Asphaug and Benz may not be typical of other cometary nuclei.

There is a converse effect to the presence of volatiles, in that their vapor expansion might help fuel a comet's disassembly during hypervelocity collisions. The energy of vaporization for ice is approximately 10 times lower than that for rock, and the impact speed required to establish a shock wave is also lower (on the other hand, impact speeds in the outer solar system are also lower). The effect of supervolatiles such as CO, should they exist in sufficient quantities within the nucleus, would have an even more pronounced effect upon the expansion of impact ejecta, in a manner that has not yet been characterized. And so, while volatiles may provide some kind of structural integrity to an aggregate body, they may also reduce the size or speed of impactor required for catastrophic disruption. *Samarasinha* (1999) offered expanding volatiles, propagating from the Sun-warmed exterior to the interior of a coarsely porous comet, as an explanation for the disruption of comet LIN-EAR D/1999 S4 near perihelion. However, *Weaver et al.* (2001) and *Mumma* (2001) found a fairly low CO abundance (<1%) in the disrupted comet. CO is typically the most abundant volatile ice in cometary nuclei after water.

This brings up the final important effect of porosity, which is the tremendously efficient insulating property of granular media in vacuum. *Weissman* (1987) and *Julian et al.* (2000) showed that the surface thermal inertia for comet Halley was at least an order of magnitude less than that for solid water ice. If cometary nuclei have low thermal conductivities, then it is extremely difficult to transport energy during a single perihelion passage to volatile reservoirs at depth. In addition, it is difficult to reconcile the buildup of pressure within the nucleus with an open rubble-pile structure.

4.5. The Case for Strength

Within the context of collisional evolution, the rubble-pile hypothesis was first formalized by *Davis et al.* (1979) in their modeling of size distributions among various small-body populations. They defined a threshold specific energy Q_D^* (impact kinetic energy per target mass) required to both shatter mechanical bonds and accelerate half the mass to escape trajectories. Shattering requires a lower specific energy $Q_S^* < Q_D^*$ to create fragments, none larger than half the target mass, not necessarily accelerating those fragments to dispersal. For small rocks, $Q_D^* \rightarrow Q_S^*$, whereas for large bodies, $Q_S^*/Q_D^* \rightarrow 0$. Whenever $Q_S^* \ll Q_D^*$, the probability of a shattering impact (which involves a smaller impactor) becomes far greater than the probability of dispersal, in which case a body might be expected to evolve into a pile of rubble, unless other effects (such as melting and compaction) were to dominate. The steeper the impactor population size distribution, the more likely it is that shattering will dominate dispersal, i.e., that the population evolves into rubble piles.

Davis et al. (1979) expressed impact strength as the sum of the shattering strength plus the gravitational binding energy of the target

$$Q_D^* = Q_S^* + 4\pi\rho Gr^2/5 \qquad (1)$$

where r is the radius of a spherical target and ρ is its density. Equation (1) is called energy scaling: On a graph of Q_D^* vs. r it plots as a horizontal line ($Q_D^* \sim Q_S^*$ = constant), transitioning at some size to a gravity-regime slope of 2 ($Q_D^* \propto$ r²). The size corresponding to this break in slope is known as the strength-gravity transition for catastrophic disruption. [The strength-gravity transition for catastrophic disruption must be distinguished from the strength-gravity transition for planetary cratering. An object in the gravity regime for disruption (Earth is one) can certainly have strength-controlled craters.] Subsequent analysis has changed the slopes in both regimes (the predicted transition size varies by or-

ders of magnitude from model to model), but the concept was established that beyond some size, rubble piles might exist.

The quantity ρQ_S^* has dimensions of strength. It happens to be close to the corresponding static tensile strength of ice and rock in laboratory impact experiments (*Fujiwara et al.*, 1989). Tensile strength was therefore used as an easily measured proxy for ρQ_S^* in early disruption theory. Upon this basis, it was concluded that primitive asteroids, comets, and early planetesimals (which are presumably of lower tensile strength than laboratory ice and rock) would be easily disrupted, and that survivors (the objects we see today) would be strong, intact bodies, and regolith would be thin or absent (*Housen et al.*, 1979; *Veverka et al.*, 1986). The transition between strength-dominated, monolithic bodies and gravity-dominated rubble piles was expected to occur at ~100 km because (1) the transition should occur when central pressure ~2πGr²ρ²/3 equals rock strength; this transition occurs at about 100 km diameter for icy or rocky targets; and (2) it should occur when gravitational binding energy per volume equals rock strength Y; neglecting constants this yields rρ = (Y/G)^½, and r on the order of several hundreds of kilometers, again whether for ice or rock.

With distinct ways of viewing small-body structure converging upon a transition to the gravity regime at ~100-km sizes, the idea seemed safe that all but the largest comets and asteroids were monolithic. Certainly by the time of the spacecraft encounters with comet Halley in 1986, the idea of structurally integral comets appeared to be on a solid foundation.

There were, however, some very serious problems with this conclusion. For one thing, the same collisional modeling required that bodies smaller than ~100 km would be unlikely to survive over billions of years, in contrast with their abundant population (see *Chapman et al.*, 1989). An even more compelling argument against structural integrity was the manner in which comets came apart so effortlessly in tidal and random disruption events (see section 4.1). As we shall now see, the resolution to this dilemma appears to be that impact strength and tensile strength are *not* simply related, and may even be inversely correlated. That is to say, structurally weak bodies are capable of absorbing large quantities of impact energy that would disrupt structurally strong bodies.

Experiments (*Love et al.*, 1993) and modeling (*Asphaug et al.*, 1998, 2003) have shown that loosely bonded aggregates can survive a projectile that would shatter an equal-mass monolith into small pieces. It is now believed that some of the most fragile bodies in the solar system — porous aggregates with little or no cohesion — can be highly resistant to catastrophic disruption owing to their ability to dissipate and absorb impact energy. In other words, there is no longer any rationale for adopting tensile strength as a measure of an object's catastrophic disruption threshold, and with this tenet no longer supportable, the edifice of strength scaling crumbles. Furthermore, ejection velocities from fragile bodies are correspondingly low, enabling them to hold on to their pieces (in the strength regime, ejecta velocity would scale with the square root of strength). Like palm trees that bend in a storm, rubble-pile comets may survive collisions that would shatter and disperse monolithic bodies.

4.6. Shoemaker-Levy 9

Discovered in March 1993 (*Shoemaker et al.,* 1993), comet Shoemaker-Levy 9 (D/1993 F2) (SL9) was seen to be a chain of ~20 discrete nuclei in an eccentric orbit around Jupiter (Fig. 5). Dynamical integrations of the orbits of the nuclei backward in time brought them together at a previous perijove passage on July 7.8, 1992, at only 1.31 jovian radii, inside Jupiter's Roche limit. Other dynamical integrations suggested that the comet had been in orbit around Jupiter for ~60 yr, although that figure is somewhat uncertain (*Chodas and Yeomans,* 1996).

Dobrovolskis (1990) and *Asphaug and Benz* (1996) provided overviews of tidal disruption theory as it pertains to small solar system bodies. Dobrovolskis developed his own theory for the initiation and propagation of cracks inside tidally strained elastic spheres. This insightful formal treatise on the *Jeffreys* (1947) regime was published only two years before the breakup of SL9, and set the stage for interpretation of SL9 as a solid elastic body (that is to say, a monolithic mass of rock and ice). Supported by the strength-regime analyses discussed above, theorists were accustomed to thinking about small bodies as elastic solids.

For Shoemaker-Levy 9, however, there can be no doubt that it was a body of extraordinarily low cohesion, less than that of dry snow. The equilibrium tidal stress at the center of a homogeneous sphere is approximately $GM_p\rho_c r_c^2/R^3$, where M_p is the mass of the planet (Jupiter), ρ_c and r_c are the density and radius of the comet, and R is the distance to the center of the planet. This is 10^3 dynes cm^{-2} for a 1-km comet of density 0.6 g cm^{-3} at SL9's perijove of 1.31 R_J. (Note also that tidal stress and lithostatic overburden both scale with r_c^2, which is the foundation for the scale-similarity to follow.)

Fig. 5.   Hubble Space Telescope image of the tidally disrupted comet Shoemaker-Levy 9 (D/1993 F2) (SL9) in January 1994 (*Weaver et al.,* 1994). Note that the brightest nuclei are near the center. These corresponded to the largest nuclei, as determined from the brightness of the SL9 impact events on Jupiter (*Nicholson,* 1996; *Crawford,* 1997), and matched the prediction of *Asphaug and Benz*'s (1994, 1996) rubble-pile model.

Shoemaker-Levy 9 sparked new investigations in a number of areas. *Melosh and Schenk* (1993) were quick to see the correlation between the SL9 morphology and the morphology of many mysterious crater chains ("catenae") (*Passey and Shoemaker,* 1982) on Ganymede and Callisto, and ascribed a common formation mechanism. The first and simplest model for the tidal disruption (*Scotti and Melosh,* 1993) proposed that the ~20 observed fragments were intact "cometesimals" each a few 100 m across, bound together gravitationally as a coarse rubble pile that separated in Jupiter's tidal field. Scotti and Melosh ignored bonding between the cometesimals, and assumed that the comet was not rotating at the time of perijove passage, and that its pieces (although gravitationally bound to begin with) did not interact gravitationally thereafter. The actual analysis involved two massless test particles representing the inner and outer points of the comet, launched with identical velocity from two slightly offset perijoves. This offset — the comet diameter — was then constrained by the measured length of the fragment chain. Scotti and Melosh derived a parent comet diameter of ~2 km, considerably smaller than the prevailing 10-km estimate based on modeling (*Sekanina et al.,* 1994) and the 7.7-km estimate based on HST images (*Weaver et al.,* 1994), although Weaver et al. cautioned that their estimates only provided upper limits to the progenitor nucleus diameter.

Asphaug and Benz (1994) tried a different approach, first modeling a solid elastic sphere undergoing *Jeffreys* (1947) regime disruption. They reproduced the theoretical result of *Dobrovolskis* (1990), although the allowed strength had to be lower than the small tidal stress. While there was no problem breaking the comet in two if a small enough strength was allowed, splitting it into 4, or 8, or 16 pieces was impossible. Once the nucleus breaks in two, the tidal stress drops by a factor of four, and must build up to its previous value for further fragmentation to occur. In order to come apart into ~20 pieces by the time of perijove passage, fracture would have to begin at a strength lower than ~30 dynes cm^{-2}, about a million times weaker than cold water ice.

Sensing a dead end, *Asphaug and Benz* (1994) began to reproduce the scenario of *Scotti and Melosh* (1993) explicitly, beginning with ~20 spheres ("grains") in close contact, modeled using an N-body code with Jupiter as the central mass. In these models, self-gravitation was observed to form clumps or pairs among the grains unless density was decreased to very low values (0.05 g cm^{-3}). For reasonable densities, the number of observable fragments (clumps) was always significantly lower than the number of grains. To form ~20 fragments, ~100 or more grains had to be assumed — no longer really a cometesimal model, but a rubble-pile model instead. For N > 200 initial grains, self-similarity took over, and what was revealed was a process responding not only to gravity, but to self-gravitational instability (*Chandrasekhar,* 1961). And so there were serious problems with the Dobrovolskis and the Scotti and Melosh models, as applied to SL9.

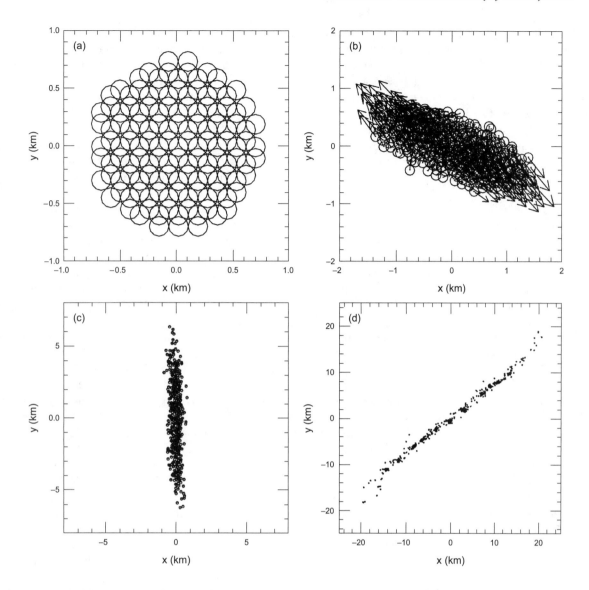

Fig. 6. A 500-"grain" simulation of the breakup of comet Shoemaker-Levy 9 by *Asphaug and Benz* (1996). The initial nucleus is 1.5 km in diameter and the grains have bulk densities of 0.5 g cm⁻³. **(a)** The unperturbed nucleus with the grains in close hexagonal packing and a void space of 27%; **(b)** the distorted nucleus at perijove, 1.3 R_J: arrows show velocity vectors relative to the center of mass; **(c)** perijove + 2 h, 3.5 R_J: the nucleus is a cigar-shaped chain, 10 km in length; **(d)** perijove + 10 h, 12 R_J: the grains have already clumped into a chain of subnuclei.

Still, the idea of a solid elastic comet was difficult to refute, despite its fatal shortcomings. In the first detailed kinematical model for the breakup, *Sekanina et al.* (1994) proposed a large (10-km minimum diameter) nucleus that underwent sudden brittle fragmentation about 2 h *after* perijove with Jupiter, by showing that the orbits of individual SL9 fragments diverge from an ~10-km body at t ≈2 h [1.5 h in *Sekanina et al.* (1994); 2.5 h in *Sekanina* (1996)]. This did not mean, however, that the undistorted parent nucleus was 10 km across, or that it underwent brittle failure. Rather, *Asphaug and Benz* (1996) found that these values were entirely consistent with a 1.5-km-diameter rubble-pile nucleus undergoing tidal distortion during perijove passage.

This smaller comet would become a 10-km-long "cigar" by t = 2 h (see Fig. 6), rotating at the same rate as deduced by *Sekanina et al.* (1994). This turns out to be the time that self-gravity among the comet fragments could be ignored, as was done in Sekanina et al.'s kinematical model. Thus, all aspects of the rubble-pile description are consistent with the kinematic requirements.

Asphaug and Benz (1994), and shortly thereafter *Solem* (1994, 1995), estimated the size of the parent nucleus by correlating their models with the length of the fragment chain as first observed almost nine months after the breakup, yielding a best fit of 1.5 km diameter if the parent nucleus was not rotating, and 1.0 km if it was rotating prograde with a period of 6 h (a retrograde rotating progenitor was not possible, as this would have prevented the formation of the highly symmetric chain that was observed). In addition, the rubble-pile models predicted the appearance of the reassembled fragment string, with the largest objects near the center of the "string of pearls" (see Fig. 5). The 1.5-km diameter for the SL9 progenitor was near the median for all disrupted comets striking Ganymede and Callisto as found by *Schenk et al.* (1996), making the progenitor a typical, rather than an unusually large, cometary nucleus (see *Weissman and Lowry*, 2003).

4.7. Catenae

The catenae, i.e., crater chains, on Ganymede and Callisto (see Fig. 7) also provide important clues to the structure of cometary nuclei. *Schenk et al.* (1996) showed that there is no correlation between the estimated parent nucleus size for the catenae (based on well-known crater-scaling relationships) and the number of craters produced. Catenae formation is self-similar across all sizes, from approximately tens to hundreds of kilometers. These data would imply, within the context of the Melosh and Scotti cometesimal model, that larger comets are composed on average of proportionately larger cometesimals. But this is at odds with our understanding of how comets accrete. *Weidenschilling*'s (1997) cometesimal model, for example, suggests that the building-block size for comets should be independent of the diameter of the final comet. Furthermore, the largest craters are found in the center of the catenae, as predicted by rubble-pile tidal disruption simulations. The only economical explanation for scale-similarity and for large central fragments is that the catenae were formed by impacts of

chains of reassembled, virtually strengthless rubble piles of small icy planetesimals whose common quality were their similar bulk densities. While models involving strength effects can be tuned (*Sekanina et al.*, 1994; *Sekanina*, 1996) to satisfy the constraints of Shoemaker-Levy 9, they fail miserably when applied to the more general problem of catenae craters.

SL9 provides a statistic of one. Whether all comets are strengthless is open to debate, because tidal breakup admits a bias, in that comets passing through the Roche zone that do not disrupt (strong bodies) bear no record of surviving this passage, other than being torqued into a new spin state and losing any loose surface material. For every disrupted-comet crater chain on Ganymede and Callisto, there could be several impact craters by comets that had equally close Roche encounters with Jupiter but did not suffer disruption. However, the number of catenae on the observed surfaces of Ganymede and Callisto is the same as one would expect (*Schenk et al.*, 1996) if *every* Jupiter-grazing comet disrupted in this fashion. While there is no guarantee that every comet in the Jupiter family is a rubble pile, catenae statistics provide strong evidence that most are. The breakup of SL9, and by extension the cometary breakups recorded at Ganymede and Callisto, constitute a direct record of cometary structure. The other direct record, that of cometary spin state, has further implications that are also consistent with the rubble-pile structure of comets (see section 5.3).

5.   DENSITY OF COMETARY NUCLEI

5.1.   Direct Measurement of Mass and Volume

In discussing the density of cometary nuclei, it is first necessary to clearly define our terms. Following *Britt et al.* (2003) we define the "grain density," ρ_g, as the mass of an object divided by the volume that is occupied by solid grains. This is an intrinsic property of the material involved. For example, the grain density of water ice at 0°C is 0.917 g cm^{-3} (*Hodgman*, 1962). The "bulk density," ρ_b, of an object is its mass divided by its total volume, including voids and pore spaces. The porosity of an object is $(1 - \rho_b/\rho_g)$. Generally, in astronomy, we measure the bulk density of distant objects and, unless otherwise noted, that is the quantity we will be discussing in what follows.

To estimate the density of a body, we must measure its mass and its volume. Both of these measurements are very difficult for cometary nuclei. Volume can be estimated from direct spacecraft imaging (see section 2) or from telescopic observations that can yield the approximate radius and axial ratio of individual nuclei (*Lamy et al.*, 2004). By far the most common technique is to perform charge-coupled-device (CCD) photometry of nuclei when they are far from the Sun and (presumably) inactive, and then to assume a typical cometary albedo of 0.03–0.04 for the nuclei (e.g., *Licandro et al.*, 2000; *Lowry et al.*, 2003). If a rotational lightcurve can be obtained, then it is also possible to derive a lower limit to the axial ratio of the nucleus (e.g., *Lowry*

Fig. 7.   The Enki catena crater chain on Ganymede as imaged by the *Galileo* spacecraft on April 5, 1997. The total chain of 13 craters is ~160 km in length. Note that the largest craters are near the center of the chain, as predicted by the *Asphaug and Benz* (1994, 1996) simulations, and as observed for comet Shoemaker-Levy 9.

and Weissman, 2003). If many lightcurves are available (something we do not have yet for any nucleus), then it is possible to derive a full three-dimensional shape model (e.g., *Kaasalainen et al.*, 2003).

Estimating nuclei masses is considerably more difficult. Cometary nuclei are too small to appreciably perturb other bodies in the solar system. Even close spacecraft flybys have been unsuccessful to date because of the low cometary masses and the high flyby speeds. The gravitational perturbation by a comet or asteroid on a passing spacecraft can be estimated by the impulse approximation

$$\Delta V = 2GM_c/DV_\infty \qquad (2)$$

where ΔV is the change in velocity of the spacecraft, directed along a line connecting the cometary nucleus and the point of closest approach of the spacecraft, G is the gravitational constant, M_c is the mass of the nucleus, D is the closest approach distance, and V_∞ is the flyby velocity at infinity. If we take a typical 5-km-radius nucleus with a density of 1 g cm^{-3}, and a spacecraft flying by at 100 km distance at a velocity of 30 km s^{-1}, the velocity perturbation is 2.3×10^{-3} cm s^{-1}, about an order of magnitude less than the current detection capability of spacecraft radio tracking systems.

5.2. Nongravitational Force Estimates

At present the only means for estimating nucleus mass is by modeling the expected acceleration of comets due to the "rocket effect" from sublimation of water ice on a rotating nucleus, and comparing that with the observed nongravitational motion of the comet. This method was pioneered by *Rickman* (1986, 1989), who showed that the change in the orbital period, ΔP, caused by the nongravitational forces (neglecting accelerations normal to the orbital plane) is related to the acceleration $\mathbf{j}$ by

$$\Delta P = \frac{6\pi(1-e^2)^{1/2}}{n^2} \left(\frac{e}{p} \int_0^P j_r \sin\theta \, dt + \int_0^P \frac{j_t}{R_h} \right) \qquad (3)$$

where P is the orbital period; t is time; e, n, and p are the orbital eccentricity, mean motion, and semilatus rectum respectively; θ is the true anomaly; and j_r and j_t are the radial and transverse components of the nongravitational acceleration. The force from the outflow of gas and dust in the orbital plane is given by

$$\mathbf{F} = -\sum_i Q_i m_i \mathbf{v}_i \qquad (4)$$

where Q_i are the production rates of the different volatile species that have masses m_i and emission velocities $\mathbf{v}_i$. This rocket force is related to the nucleus mass M by $\mathbf{F} = M\mathbf{j}$. Equation (4) can be evaluated from the observed heliocen-

tric lightcurve via empirical relationships such as that of *Festou* (1986), and then combined with equation (3) to solve for the mass. Unfortunately, this method is highly model dependent and the assumptions applied are numerous. For example, it is not possible to constrain the directionality of the rocket forces without knowledge of the location of the active areas and the rotational state of the nucleus. Thermal lag effects, which are due to unknown values of the surface thermal inertia, must also be factored in. A more detailed discussion of the theory of cometary nongravitational forces and outgassing physics can be found in *Yeomans et al.* (2004).

In situ observations from the *Vega* and *Giotto* spacecraft encounters with comet 1P/Halley provided, for the first time, accurate measurements of the dimensions and hence volume of a comet's nucleus (*Wilhelm et al.*, 1986). This allowed *Rickman* (1986) to estimate a bulk density of 0.1–0.2 g cm^{-3} for the Halley nucleus. This early paper was widely accepted as indicative of extremely low bulk densities for cometary nuclei. This view was further strengthened by *Rickman et al.* (1987), who studied 29 short-period comets and found that their bulk densities were all below 0.5 g cm^{-3}, indicating a highly porous internal structure.

However, others found higher values for the nucleus density. An independent estimate for 1P/Halley by *Sagdeev et al.* (1988) found a value of 0.6 (+0.9, –0.4) g cm^{-3}. A detailed analysis of the many factors and uncertainties involved in these calculations led *Peale* (1989) to conclude that the density of Halley could be anywhere from 0.03 to 4.9 g cm^{-3}, with a preferred value near 1.0 g cm^{-3}. Even *Rickman* (1989) revised his earlier value for Halley and increased the estimated density to 0.28–0.65 g cm^{-3}.

More recently, *Skorov and Rickman* (1999) used a more refined method and indicated that the earlier method of *Rickman* (1989) underestimated the momentum transfer that produces the nongravitational acceleration. They determined that the earlier density estimates should be increased by a factor of ~1.8. Thus, their new Halley density estimates were 0.5–1.2 g cm^{-3}.

Combining spacecraft and groundbased imaging of 19P/Borrelly during the *DS1* flyby in 2001, *Farnham and Cochran* (2002) calculated the orientation of the rotation pole and compared this with published nongravitational terms to compute a nucleus mass of 3.9×10^{16} g and a density of 0.49 (+0.34, –0.20) g cm^{-3}. For Borrelly, the primary jet structure appears to be fortuitously aligned with the rotation axis. Such an alignment means that nucleus rotation and thermal lag effects will not significantly influence the resulting nongravitational motion of the body. Indeed, *Yeomans* (1971) pointed out that although significant nongravitational accelerations are present for this comet, they have remained essentially constant since its discovery in 1904. However, *Davidsson and Gutiérrez* (2003) have estimated a lower bulk density of 0.18 to 0.30 g cm^{-3} for Borrelly, barely in agreement with the *Farnham and Cochran* (2002) value. As was the case for 1P/Halley, density estimates derived from nongravitational force estimates appear to still be fairly uncertain.

5.3. Shoemaker-Levy 9

As discussed previously, the appearance of comet Shoemaker-Levy 9 provided unique insights into the nature of cometary nuclei. *Asphaug and Benz* (1994, 1996) and *Solem* (1994, 1995) were able to explain many of the observed characteristics of SL9's "string of pearls" appearance by assuming that the pre-breakup nucleus was a strengthless rubble pile of hundreds of smaller icy planetesimals. Asphaug and Benz and Solem found that density played a key role in determining the final, post-breakup configuration of this swarm of planetesimals. For densities less than ~0.3 g cm^{-3}, the resulting chain of clumps was highly diffuse. For densities greater than 1.0 g cm^{-3}, the planetesimals formed a single, major central clump, unless the progenitor nucleus happened to have significant prograde rotation.

These outcomes were found, as expected from dimensional analysis, to be self-similar for any density, when the perijove distance was normalized to the density-dependent Roche limit. For SL9, where the perijove was known, the morphology of the chain thus narrowly constrained the nucleus density. The result was that the SL9 progenitor nucleus needed to have a bulk density of about 0.6 ± 0.1 g cm^{-3} (*Asphaug and Benz,* 1996) or ~0.5 g cm^{-3} (*Solem,* 1995). Asphaug and Benz also found that a fast, prograde rotation led to somewhat higher density estimates, on the order of 1.0 g cm^{-3}.

Note again that the values above are the effective bulk density of the progenitor nucleus, not the individual cometesimals. Assuming random reassembly with similar macroporosity, the bulk density of the final SL9 nuclei post-breakup would be similar, or perhaps somewhat lower than the values above, given the low relative velocity of the reaccretion and the lack of time for compaction processes to work (i.e., sintering, inward coma diffusion, etc). *Crawford* (1997) used a hydrocode with advanced thermodynamics to obtain best matches to the *Galileo* NIMS observations of the SL9 impacts (*Carlson et al.,* 1997). Crawford's best fit was for impactor densities of 0.25 g cm^{-3}, and for a parent comet with the same mass as derived by *Asphaug and Benz* (1994). These density values are somewhat lower than what we expect from a reassembled rubble pile. However, the effective bulk density of the impactors may have been reduced as the rubble piles again began to gravitationally distort and extend in Jupiter's gravitational field just prior to impact. Also, the accompanying debris cloud (see Fig. 6) around each impactor may have worked to increase the effective cross-section in Crawford's calculations, even though most of the mass was concentrated in the central object.

5.4. Rotation Period: Shape Relationship

Knowledge of the shape and rotation period of the nucleus allows a lower limit to be calculated for the nuclear density, i.e., the minimum density required in order to withstand centripetal disruption under the assumption of negligible cohesive strength. The density is derived by setting the gravitational acceleration, g, equal to the centripetal acceleration at the apex of a rotating prolate spheroid: $2a(2\pi/P_{rot})$, where a is the semimajor axis of the spheroid and P_{rot} is the nucleus rotation period. To evaluate g we solve the following integral (*Luu and Jewitt,* 1992)

$$g = -2\pi G\rho_b a \int_0^2 \left(\left[1 - f^2 + \frac{2f^2}{s} \right]^{1/2} - 1 \right) ds \qquad (5)$$

where f is the ratio of the semiminor to semimajor axes, b/a; G is the gravitational constant; ρ_b is the spheroid bulk density lower limit; and s is the distance along the semimajor axis in units of a. By setting the integral of equation (5) equal to the expression for the centripetal acceleration given above we obtain

$$-\frac{2\pi}{P_{rot}^2 G\rho_b} = \\ \frac{2f^2(1-f^2)^{1/2} + f^2 \ln f^2 - f^2 \ln\left(2 - f^2 + 2\sqrt{1-f^2}\right)}{(1-f^2)^{3/2}} \qquad (6)$$

We can infer shape information from the rotational lightcurve of the nucleus via time-series photometric measurements. If we assume that the brightness variation is purely shape induced and not caused by a variation in albedo across the nucleus surface, then the nucleus can be modeled as a triaxial ellipsoid with semiaxes a, b, and c, where a > b and b = c. A lower limit to the axial ratio can be estimated using $f^{-1} = a/b = 10^{0.4\Delta m}$, where Δm is the range of observed magnitudes. As an example we consider the lightcurve of Comet 22P/Kopff (*Lowry and Weissman,* 2003). If we substitute f = 0.6 and $P_{rot} = 12.3$ h into equation (4), we get $\rho_b \geq 0.11$ g cm^{-3} for the Kopff nucleus. The density measurement is a lower limit because we use the projected axial ratio, a/b, which is a lower limit to the true axial ratio, since the orientation of the rotation axis is unknown. Also, the nucleus does not necessarily need to be spinning at its rotational disruption limit.

Lightcurves for a number of cometary nuclei have been published to date. The corresponding values for the lightcurve-derived rotation periods and projected axial ratios are shown in Fig. 8. The derived rotation periods range from 5.2 to 29.8 h, while the projected axial ratios, a/b, range from 1.02 to 2.48. The inferred, lower-limit nucleus bulk densities are given by the position of each comet in the figure and range from 0.02 to 0.56 g cm^{-3}, consistent with density values determined through other methods.

Figure 8 is reminiscent of one found for asteroid rotation periods and shapes by *Pravec et al.* (2003, see their Fig. 5) where there is a sharp edge in the distribution of asteroids (with diameters >0.15 km) at a rotation period of about 2.2 h, corresponding to a density of ~2.5 g cm^{-3}. Pravec et al. interpret this result as evidence for small asteroids being "loosely bound, gravity-dominated aggregates with negligible tensile strength." Although the statistics for the comets are relatively poor so far, we are perhaps seeing a similar "edge" at a density of ~0.6 g cm^{-3}. Note also in Fig. 8 that there is a trend for the fastest-rotating nuclei to have lower values of a/b, which may reflect the inability of the rubble-pile nuclei to maintain extended shapes near the rotational disruption limit.

The database of rotational lightcurves for transneptunian objects (TNOs) continues to grow (*Sheppard and Jewitt,* 2002), and the inferred lightcurve-derived density lower limits are typically in the range 0.1–0.4 g cm^{-3}. These values are consistent with those for cometary nuclei, which is reassuring given that the TNOs are believed to be the likely source for most of the Jupiter-family comets in Fig. 8 (*Levison,* 1996).

Recently *Holsapple* (2003) has suggested that modest cohesive strengths, on the order of a few times 10^4 dynes cm^{-2}, could allow even fast-rotating, elongated asteroids to survive as rubble piles. This value is similar to the tensile strengths inferred for cometary nuclei (see section 4.1). We

note, however, that Holsapple's calculations do not explain the sharp edge seen for the vast majority of small asteroids in the rotation rate vs. elongation plot of *Pravec et al.* (2003). The existence of that edge strongly implies that many of these bodies are indeed acting as if they were strengthless (or, to be more exact, have cohesive strengths ≤10^2 dynes cm^{-2}). Thus we believe that the density lower limits shown in Fig. 8 are indeed likely to be meaningful.

5.5. Additional Methods for Inferring Nucleus Densities

An alternative method for estimating nucleus densities relies on measurements of the density of dust particles originating from comets. Anhydrous interplanetary dust particles (IDPs), widely believed to be derived from comets, have typical bulk density values of 0.7–1.2 g cm^{-3} (*Fraundorf et al.,* 1982), including voids that were likely once filled with ices. Although these densities are indeed consistent with bulk densities obtained from other methods, they are still somewhat uncertain, since the nucleus porosity, the degree of volatile loss from the particles, and the degree of compaction during atmospheric entry are largely unknown. Also, *Maas et al.* (1990) found mean densities ranging from 0.2–6.0 g cm^{-3} for individual dust particles detected during the Halley spacecraft encounters consistent, although not particularly constraining.

By combining models of interstellar dust accretion with observational constraints provided by the Halley encounters, *Greenberg and Hage* (1990) found porosities of 0.6–0.83 for dust in the cometary coma, which in turn led to estimated comet nucleus densities of 0.26–0.60 g cm^{-3}. Additionally, *Greenberg* (1998) combined results of his core-mantle interstellar dust model, the solar system abundances of the elements, the composition of dust from comet Halley, and data on the volatile molecules in cometary comae to set an upper limit to the density of fully packed cometary material of ~1.65 g cm^{-3}. This would be what we referred to earlier as the "grain density" [see *Weidenschilling* (2004) and *Sykes et al.* (2004) for discussions of accretion mechanisms and IDPs].

If the bulk density of cometary nuclei are on the order of 0.6 g cm^{-3} as suggested above, then the nuclei have a porosity of 0.64, assuming *Greenberg*'s (1998) grain density for cometary materials. Is such a value reasonable? *Britt and Consolmagno* (2001, see also *Britt et al.,* 2003) showed that several C-type asteroids (plus Deimos) have macroporosities of ~40%, although the average C-type porosity is ~27%. Macroporosities of 60% or more are seen for two M-type (metallic) asteroids. The C-type asteroids are likely to be a reasonable analog for cometary nuclei, given the similarity between carbonaceous materials and the nonvolatile component of cometary materials. So a cometary nucleus porosity value of 0.64 seems somewhat high, although not necessarily unreasonable. Note that all the 40% porosity asteroids involved in these measurements are considerably larger than typical cometary nuclei (d = 53–214 km), except for Deimos (d = 12 km). The porosity of Deimos is

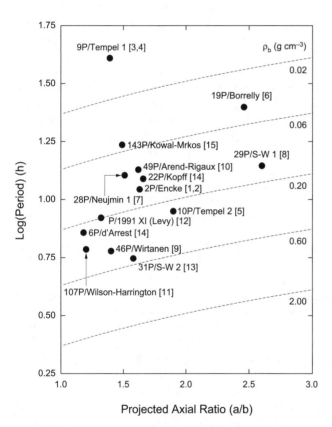

Fig. 8. Lightcurve-derived rotation periods and projected axial ratios, a/b, for 13 Jupiter-family comets and 1 Halley-type comet. Their inferred density lower limits are given by their position on this plot. Constant-density curves for values of 0.02, 0.06, 0.2, 0.6, and 2.0 g cm^{-3} have been overplotted for comparison. The inferred bulk density lower limits range from 0.02 to 0.56 g cm^{-3}, consistent with density values determined through other methods (see text). Lightcurve data references: [1] *Meech et al.* (2001), [2] *Fernández et al.* (2002), [3] *Lamy et al.* (2001), [4] *Meech et al.* (2000), [5] *A'Hearn et al.* (1989), [6] *Lamy et al.* (1998), [7] *Delahodde et al.* (2001), [8] *Meech et al.* (1993), [9] *Boehnhardt et al.* (2002), [10] *Millis et al.* (1988), [11] *Osip et al.* (1995), [12] *Fitzsimmons and Williams* (1994), [13] *Luu and Jewitt* (1992), [14] *Lowry and Weissman* (2003), [15] *Jewitt et al.* (2003).

actually very poorly determined and could be anywhere from 15% to 62%. If we perform the reverse calculation and assume a porosity of 40% for cometary nuclei, then we obtain a density of 1.0 g cm^{-3} using Greenberg's value.

6. DISCUSSION AND SUMMARY

The nature of cometary nuclei has been, and continues to be, among the more elusive questions of solar system science. The icy-conglomerate model was a major paradigm shift in our understanding of comets when it was first proposed by Whipple in 1950. Fifty years of evidence has proved the basic correctness of that model. But as with any problem in science, we always want to look at the question in more detail.

In this chapter we have presented a long list of evidence that suggests that the underlying structure of icy-conglomerate cometary nuclei is a collisionally processed rubble pile of smaller icy planetesimals. That evidence includes (1) spacecraft imaging of comets Halley and Borrelly that show a bimodal structure and rough, chaotic topography for both nuclei; (2) observations of randomly disrupted comets, suggesting that they are extremely fragile objects; (3) observations of Sun-grazing comets including the more than 540 objects discovered by the SOHO spacecraft, which have estimated diameters of ~10 m; (4) recent theoretical studies that show that cometary nuclei have been subjected to moderate to intense collisional processing; (5) recent theoretical studies of the collisional evolution of asteroids that show that small bodies in the solar system need not be monolithic; (6) comet Shoemaker-Levy 9, which can best be explained by the tidal disruption and reassembly of a rubble pile nucleus; (7) the catenae on Ganymede and Callisto that show that Shoemaker-Levy 9 is not unique, but rather is representative of a common solar system process; (8) the low density estimates for cometary nuclei, suggesting that they are under-dense relative to their constituent materials; and (9) the rotation-shape relationship for observed cometary nuclei that shows a possible cutoff in the distribution similar to that seen for rubble-pile asteroids, albeit at a different density value.

Together, we believe that this evidence makes a compelling case for cometary nuclei as collisionally processed rubble piles. Although this may seem to be a leap of faith to some readers, we feel that it is one made on firm grounds. This paradigm shift has already occurred for asteroids and we believe that the time is now right for a similar paradigm shift for cometary nuclei.

The final resolution of this question requires the detailed study of a cometary nucleus (or many nuclei) at close range, something that can only be accomplished using a nucleus-orbiting spacecraft, and perhaps may even require a nucleus lander. The *Rosetta* mission of the European Space Agency (*Schwehm, 2002*) includes experiments designed to investigate the internal structure of the nucleus of periodic comet 67P/Churyumov-Gerasimenko (hereafter, 67P/CG). *Rosetta*, to be launched in February 2004, is expected to rendezvous with 67P/CG in August 2014. A key experiment is the Comet Nucleus Sounding Experiment by Radiowave Transmission (CONSERT) (*Barbin et al., 1999*). CONSERT is a radar tomography experiment consisting of a transponder on the *Rosetta* lander and a radar transmitter/receiver on the *Rosetta* orbiter. If the lander is able to survive several weeks or months on the nucleus surface, it will allow the orbiter sufficient time to orbit the nucleus many times, obtaining numerous ray paths through the nucleus. The experiment is somewhat hampered by the fact that there is only one lander (two were originally planned). However, CONSERT should yield considerable insight into the interior of the 67P/CG nucleus, including the location and dimensions of any substantial voids.

The gravity mapping experiment onboard *Rosetta* will provide additional evidence on the internal structure of the nucleus (*Pätzold et al., 2001*). Mapping of higher harmonics in the 67P/CG gravity field, coupled with a detailed shape model obtained from the *Rosetta* imaging experiment, OSIRIS (*Thomas et al., 1998*), will provide evidence of density inhomogeneities within the nucleus, as well as an overall measure of the bulk density of the nucleus. The *Rosetta* spacecraft may orbit as close as 1 km to the surface of 67P/CG.

A third source of information is the imaging experiment itself, which will provide submeter-resolution images of the entire nucleus surface. These images should provide sufficient resolution to understand the mechanisms creating the nucleus morphology, and may provide evidence of faults, substructure, or other landforms that help to reveal the internal structure of the nucleus.

Besides *Rosetta*, there are two comet flyby missions that may yield improved imaging of nuclei, and certainly will be able to address the question of diversity between different comets. The first is the *Stardust* mission, which was launched in February 1999 and flew by comet 81P/Wild 2 in January 2004 (*Brownlee et al., 2000*). *Stardust* is designed to capture cometary dust particles in the coma and return them to Earth for analysis in terrestrial laboratories in 2006. However, it also includes an imaging system that can be expected to produce nucleus images comparable to, or perhaps somewhat better than, those from *DS1*.

The second mission is *Deep Impact*, which will be launched in December 2004 and will fly by comet 9P/Tempel 1 in July 2005, deploying a ~370-kg impactor that will strike the nucleus and (perhaps) create a significant crater (*Meech et al., 2000*). *Deep Impact* carries an advanced imaging system on both the flyby spacecraft and the impactor that should provide us with the best nucleus images of any mission to date, with a resolution of only a few meters per pixel. In addition, the impact experiment itself may yield insights into the strength of cometary materials and the structure of the nucleus. However, because both *Stardust* and *Deep Impact* are fast flybys, neither is capable of making a mass determination for their respective nuclei.

Sadly, the *Comet Nucleus Tour* (*CONTOUR*) mission was lost in August 2002 due to a spacecraft failure. *CONTOUR* would have flown by two cometary nuclei, 2P/Encke in 2003 and 31P/Schwassmann-Wachmann 3 in 2006, and made measurements similar to those made during the *Vega* and *Giotto* flybys of 1P/Halley in 1986. Although the cause for the loss is not clear, a NASA investigation board suggested that the spacecraft's structure failed due to overheating from *CONTOUR*'s solid rocket motor, destroying the spacecraft as it was leaving Earth orbit.

While waiting for these spacecraft results, we can expect that observational and theoretical studies of comets will continue in the coming decade, providing further insights into the nature of cometary nuclei and processes such as disruption and splitting. For example, additional nucleus lightcurves will help to fill in the relatively small sample of 14 comets shown in Fig. 8 and allow us to map the limits of nucleus density. Also, detailed observations of random disruption events have the potential to provide insights as to how cometary nuclei come apart, and perhaps why. And we can never tell when fate might provide us with another opportunity for a giant leap forward, as was provided by comet Shoemaker-Levy 9 in 1993–1994, or more recently by the SOHO comets. It will be most interesting then to see where the questions of nucleus structure and density stand when the *Comets III* book is written.

Acknowledgments.   We thank D. Durda, A. Stern, H. Weaver, and an anonymous referee for useful comments on earlier drafts of this manuscript. This work was supported by the NASA Planetary Geology & Geophysics and Planetary Astronomy Programs, and was performed in part at the Jet Propulsion Laboratory under contract with NASA. Support from the National Research Council is also gratefully acknowledged.

REFERENCES

A'Hearn M. F., Campins H., Schleicher D. G., and Millis R. L. (1989) The nucleus of comet P/Tempel 2. *Astrophys. J., 347,* 1155–1166.

Asphaug E. and Benz W. (1994) Density of comet Shoemaker-Levy-9 deduced by modelling breakup of the parent rubble pile. *Nature, 370,* 120–124.

Asphaug E. and Benz W. (1996) Size, density, and structure of comet Shoemaker-Levy 9 inferred from the physics of tidal breakup. *Icarus, 121,* 225–248.

Asphaug E., Moore J. M., Morrison D., Benz W., Nolan M. C., and Sullivan R. J. (1996) Mechanical and geological effects of impact cratering on Ida. *Icarus, 120,* 158–184.

Asphaug E., Ostro S. J., Hudson R. S., Scheeres D. J., and Benz W. (1998) Disruption of kilometre-sized asteroids by energetic collisions. *Nature, 393,* 437–440.

Asphaug E., Ryan E. V., and Zuber M. T. (2003) Asteroid interiors. In *Asteroids III* (W. F. Bottke Jr. et al., eds.), pp. 463–484. Univ. of Arizona, Tucson.

Baldet F. (1927) Sur le noyau de la comète Pons-Winnecke (1927c). *C. R. Acad. Sci. Paris, 185,* 39–41.

Barbin Y., Kofman W., Nielsen E., Hagfors T., Seu R., Picardi G., and Svedhem H. (1999) The Consert instrument for the Rosetta mission. *Adv. Space Res., 24,* 1115–1126.

Benz W. and Asphaug E. (1999) Catastrophic disruptions revisited. *Icarus, 142,* 5–20.

Biesecker D. A., Lamy P., St. Cyr O. C., Llebaria A., and Howard R. A. (2002) Sungrazing comets discovered with the SOHO/LASCO coronagraphs 1996–1998. *Icarus, 157,* 323–348.

Boehnhardt H. (2004) Split comets. In *Comets II* (M. C. Festou et al., eds.), this volume. Univ. of Arizona, Tucson.

Boehnhardt H., Delahodde C., Sekiguchi T., Tozzi G. P., Amestica R., Hainaut O., Spyromilio J., Tarenghi M., West R. M., Schulz R., and Schwehm D. (2002) VLT observations of comet 46P/

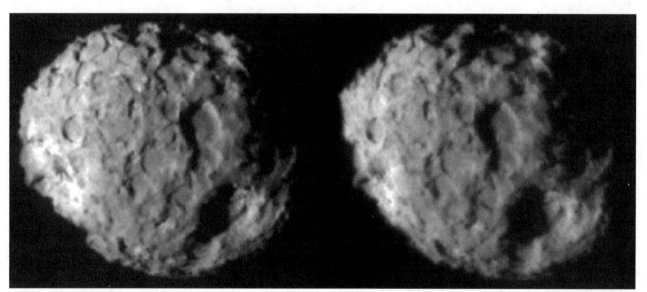

Note added in proof: **Fig. 9.**   Images of periodic comet Wild 2 taken January 2, 2004, by the *Stardust* spacecraft. These are the highest-resolution images of a cometary nucleus yet obtained. *Left:* Image #2075 was taken at a range of 257 km and a phase angle of 39.7° and has a spatial resolution of ~22 m/pixel. *Right:* Image #2077 was taken at 325 km and a phase angle of 59.9° and is ~27 m/pixel. The approximately 5-km-diameter nucleus shows a complex surface morphology, dominated by large, flat-floored depressions and pits. These are likely to be sublimation features, although they may be ancient impact craters that have been subsequently modified by sublimation. The topography also includes large irregular blocks that may be indicative of a rubble-pile structure. There appears to be an absence of very small pits or craters, possibly a result of sublimation.

Wirtanen. *Astron. Astrophys., 387,* 1107–1113.

Bridges F. G., Supulver K. D., Lin D. N. C., Knight R., and Zafra M. (1996) Energy loss and sticking mechanisms in particle aggregation in planetesimal formation. *Icarus, 123,* 422–435.

Brin G. D. and Mendis D. A. (1979) Dust release and mantle development in comets. *Astrophys. J., 229,* 402–408.

Britt D. T. and Consolmagno G. J. (2001) Asteroid bulk density: Implications for the structure of asteroids (abstract). In *Lunar and Planetary Science XXXII,* Abstract #1611. Lunar and Planetary Institute, Houston (CD-ROM).

Britt D. T., Yeomans D., Housen K., and Consolmagno G. (2003) Asteroid density, porosity and structure. In *Asteroids III* (W. F. Bottke Jr. et al., eds.), pp. 485–500. Univ. of Arizona, Tucson.

Brownlee D. E., Tsou P., Clark B., Hanner M. S., Hörz F., Kissel J., McDonnell J. A. M., Newburn R. L., Sandford S., Sekanina Z., Tuzzolino A. J., and Zolensky M. (2000) Stardust: A comet sample return mission. *Meteoritics & Planet. Sci., 35,* A35.

Buratti B. J., Hicks M. D., Soderblom L. A., Britt D., Oberst J., and Hillier J. K. (2003) Deep Space 1 photometry of the nucleus of comet 19P/Borrelly. *Icarus, 167,* 16–29.

Carlson R. W., Drossart P., Encrenaz Th., Weissman P. R., Hui J., and Segura M. (1997) Temperature, size and energy of the Shoemaker-Levy 9 G-impact fireball. *Icarus, 128,* 251–274.

Chandrasekhar S. (1961) *Hydrodynamic and Hydromagnetic Stability.* Oxford, London. 652 pp.

Chapman C. R., Paolicchi P., Zappala V., Binzel R. P., and Bell J. F. (1989) Asteroid families: Physical properties and evolution. In *Asteroids II* (R. P. Binzel et al., eds.), pp. 386–415. Univ. of Arizona, Tucson.

Charnoz S. and Morbidelli A. (2003) Coupling dynamical and collisional evolution of small bodies: An application to the early ejection of planetesimals from the Jupiter-Saturn region. *Icarus, 166,* 141–156.

Chodas P. W. and Yeomans D. K. (1996) The orbital motion and impact circumstances of comet Shoemaker-Levy 9. In *The Collision of Comet Shoemaker-Levy 9 and Jupiter* (K. S. Noll et al., eds.), pp. 1–30. Cambridge Univ., Cambridge.

Crawford D. (1997) Comet Shoemaker-Levy 9 fragment size and mass estimates from light flux observations (abstract). In *Lunar and Planetary Science XXVIII,* p. 267. Lunar and Planetary Institute, Houston.

Davidsson B. J. R. and Gutiérrez P. J. (2003) An estimate of the nucleus density of comet 19P/Borrelly (abstract). *Bull. Am. Astron. Soc., 35,* 969.

Davis D. R., Chapman C. R., Greenberg R., Weidenschilling S. J., and Harris A. W. (1979) Collisional evolution of asteroids: Populations, rotations and velocities. In *Asteroids* (T. Gehrels, ed.), pp. 528–557. Univ. of Arizona, Tucson.

Delahodde C. E., Meech K. J., Hainaut O. R., and Dotto E. (2001) Detailed phase function of comet 28P/Neujmin 1. *Astron. Astrophys., 376,* 672–685.

Delsemme A. H. (1971) Comets: Production mechanisms of hydroxyl and hydrogen halos. *Science, 172,* 1126–1127.

Delsemme A. H. (1977) The pristine nature of comets. In *Comets, Asteroids, Meteorites: Interrelations, Evolution and Origins* (A. H. Delsemme, ed.), pp. 3–12. Univ. of Toledo, Toledo.

Dobrovolskis A. R. (1990) Tidal disruption of solid bodies. *Icarus, 88,* 24–38.

Donn B. and Hughes D. (1986) A fractal model of a cometary nucleus formed by random accretion. In *20th ESLAB Symposium on the Exploration of Halley's Comet, Vol. 1* (B. Battrick et al., eds.), pp. 523–524. ESA SP-250, Noordwijk, The Netherlands.

Donn B., Daniels P. A., and Hughes D. W. (1985) On the structure of the cometary nucleus (abstract). *Bull. Am. Astron. Soc., 17,* 520.

Duncan M., Quinn T., and Tremaine S. (1987) The formation and extent of the solar system comet cloud. *Astron. J., 94,* 1330–1338.

Duncan M., Levison H., and Dones L. (2004) Dynamical evolution of ecliptic comets. In *Comets II* (M. C. Festou et al., eds.), this volume. Univ. of Arizona, Tucson.

Farinella P. and Davis D. R. (1996) Short-period comets: Primordial bodies or collisional fragments? *Science, 273,* 938–941.

Farnham T. L. and Cochran A. L. (2002) A McDonald Observatory study of comet 19P/Borrelly: Placing the Deep Space 1 observations into a broader context. *Icarus, 160,* 398–418.

Fernández Y. R., Lowry S. C., Weissman P. R., and Meech K. J. (2002) New dominant periodicity in photometry of comet Encke (abstract). *Bull. Am. Astron. Soc., 34,* 887.

Festou M. C. (1986) The derivation of OH gas production rates from visual magnitudes of comets. In *Asteroids, Comets, Meteors II* (C. I. Lagerkvist and H. Rickman, eds.), pp. 299–303. Uppsala Univ., Sweden.

Fitzsimmons A. and Williams I. P. (1994) The nucleus of comet P/Levy 1991XI. *Astron. Astrophys., 289,* 304–310.

Fraundorf P., Hintz C., Lowry O., McKeegan K. D., and Sandford S. A. (1982) Determination of the mass, surface density, and volume density of individual interplanetary dust particles (abstract). In *Lunar and Planetary Science XIII,* pp. 225–226. Lunar and Planetary Institute, Houston.

Fujiwara A., Cerroni P., Davis D. R., Ryan E. V., Di Martino M., Holsapple K., and Housen K. (1989) Experiments and scaling laws for catastrophic collisions. In *Asteroids II* (R. P. Binzel et al., eds.), pp. 240–265. Univ. of Arizona, Tucson.

Gombosi T. I. and Houpis H. L. F. (1986) An icy-glue model of cometary nuclei. *Nature, 324,* 43–44.

Greenberg J. M. (1998) Making a comet nucleus. *Astron. Astrophys., 330,* 375–380.

Greenberg J. M. and Hage J. I. (1990) From interstellar dust to comets — A unification of observational constraints. *Astrophys. J., 361,* 260–274.

Hodgman C. D. (1962) *Handbook of Chemistry and Physics.* Chemical Rubber Publ., Cleveland. 2189 pp.

Holsapple K. A. (2003) Could fast rotator asteroids be rubble piles? (abstract). In *Lunar and Planetary Science XXXIV,* Abstract #1792. Lunar and Planetary Institute, Houston (CD-ROM).

Housen K. R. and Holsapple K. A. (1990) On the fragmentation of asteroids and planetary satellites. *Icarus, 84,* 226–253.

Housen K. R. and Holsapple K. A. (2003) Impact cratering on porous asteroids. *Icarus, 163,* 102–119.

Housen K. R., Wilkening L. L., Chapman C. R., and Greenberg R. (1979) Asteroidal regoliths. *Icarus, 39,* 317–351.

Housen K. R., Holsapple K., and Voss M. E. (1999) Compaction as the origin of the unusual craters on the asteroid Mathilde. *Nature, 402,* 155–157.

Iseli M., Kuppers M., Benz W., and Bochsler P. (2002) Sungrazing comets: Properties of nuclei and in situ detectability of cometary ions at 1 AU. *Icarus, 155,* 350–364.

Jeffreys H. (1947) The relation of cohesion to Roche's limit. *Mon. Not. R. Astron. Soc., 107,* 260–262.

Jewitt D., Sheppard S., and Fernández Y. (2003) 143P/Kowal-

Mrkos and the shapes of cometary nuclei. *Astron. J., 125,* 3366–3377.

Julian W. H., Samarasinha N. H., and Belton M. J. S. (2000) Thermal structure of cometary active regions: Comet 1P/Halley. *Icarus, 144,* 160–171.

Kaasalainen M., Kwiatkowski T., Abe M., Piironen J., Nakamura T., Ohba Y., Dermanwan B., Farnham T., Colas F., Lowry S., Weissman P., Whiteley R., Tholen D., Larson S., Yoshikawa M., Toth I., and Velichko F. (2003) CCD photometry and model of MUSES-C target (25143) 1998 SF36. *Astron. Astrophys., 405,* L29–L32.

Keller H. U., Arpigny C., Barbieri C., Bonnet R. M., Cazes S., Coradini M., Cosmovici C. B., Delamere W. A., Huebner W. F., Hughes D. W., Jamar C., Malaise D., Reitsema H. J., Schmidt H., Schmidt W. K. H., Seige P., Whipple F. L., and Wilhelm K. (1986) First Halley multicolour camera imaging results from Giotto. *Nature, 321,* 320–326.

Keller H. U., Curdt W., Kramm J.-R., and Thomas N. (1994) *Images of the Nucleus of Comet Halley, Vol. 1.* ESA SP-1127, Noordwijk, The Netherlands. 252 pp.

Lamy P. L., Toth I., and Weaver H. A. (1998) Hubble Space Telescope observations of the nucleus and inner coma of comet 19P/1904 Y2 (Borrelly). *Astron. Astrophys., 337,* 945–954.

Lamy P. L., Toth I., A'Hearn M. F., Weaver H. A., and Weissman P. R. (2001) Hubble Space Telescope observations of the nucleus of comet 9P/Tempel 1. *Icarus, 154,* 337–344.

Lamy P. L., Toth I., Fernández Y., and Weaver H. A. (2004) The sizes, shapes, albedos, and colors of cometary nuclei. In *Comets II* (M. C. Festou et al., eds.), this volume. Univ. of Arizona, Tucson.

Levin B. Y. (1943) The emission of gases by the nucleus of a comet and the variation of its absolute brightness. *Soviet Astron. J., 20,* 37–48.

Levison H. F. (1996) Comet taxonomy. In *Completing the Inventory of the Solar System* (T. W. Rettig and J. M. Hahn, eds.), pp. 173–191. ASP Conference Series 107, Astronomical Society of the Pacific, San Francisco.

Licandro J., Tancredi G., Lindgren M., Rickman H., and Hutton R. G. (2000) CCD photometry of cometary nuclei, I: Observations from 1990–1995. *Icarus, 147,* 161–179.

Love S. G., Hörz F., and Brownlee D. E. (1993) Target porosity effects in impact cratering and collisional disruption. *Icarus, 105,* 216–224.

Lowry S. C. and Weissman P. R. (2003) CCD observations of distant comets from Palomar and Steward Observatories. *Icarus, 164,* 492–503.

Lowry S. C., Fitzsimmons A., and Collander-Brown S. (2003) CCD photometry of distant comets. III. Ensemble properties of Jupiter-family comets. *Astron. Astrophys., 397,* 329–343.

Lyttleton R. A. (1948) On the origin of comets. *Mon. Not. R. Astron. Soc., 108,* 465–475.

Luu J. X. and Jewitt D. C. (1992) Near-aphelion CCD photometry of comet P/Schwassmann-Wachmann 2. *Astron. J., 104,* 2243–2249.

Maas D., Krueger F. R., and Kissel J. (1990) Mass and density of silicate- and CHON-type dust particles released by comet P/Halley. In *Asteroids, Comets, Meteors III* (C. I. Lagerkvist et al., eds.), pp. 389–392. Uppsala Univ., Uppsala.

MacQueen R. M. and St. Cyr O. C. (1991) Sungrazing comets observed by the Solar Maximum Mission coronagraph. *Icarus, 90,* 96–106.

Mäkinen J. T. T., Bertaux J.-L., Combi M. R., and Quémerais E. (2001) Water production of comet C/1999 S4 (LINEAR) observed with the SWAN instrument. *Science, 292,* 1326–1329.

Marsden B. G. (1989) The sungrazing comet group. II. *Astron. J., 98,* 2306–2321.

Marsden B. G., Sekanina Z., and Yeomans D. K. (1973) Comets and nongravitational forces. V. *Astron. J., 78,* 211–225.

Meech K. J., Belton M. J., Mueller B. E. A., Dicksion M. W., and Li H. R. (1993) Nucleus properties of P/Schwassmann-Wachmann 1. *Astron. J., 106,* 1222–1236.

Meech K. J., A'Hearn M. F., McFadden L., Belton M. J. S., Delamere A., Kissel J., Klassen K., Yeomans D., Melosh J., Schultz P., Sunshine J.,and Veverka J. (2000) Deep Impact — Exploring the interior of a comet. In *A New Era in Bioastronomy* (G. Lemarchand and K. Meech, eds.), pp. 235–242. Astronomical Society of the Pacific, San Francisco.

Meech K. J., Fernández Y., and Pittichová J. (2001) Aphelion activity of 2P/Encke (abstract). *Bull. Am. Astron. Soc., 33,* 1075.

Melosh H. J. and Schenk P. (1993) Split comets and the origin of crater chains on Ganymede and Callisto. *Nature, 365,* 731–733.

Merényi E., Földy L., Szegö K., Toth I., and Kondor A. (1989) The landscape of comet Halley. *Icarus, 86,* 9–20.

Merline W. J., Weidenschilling S. J., Durda D. D., Margot J. L., Pravec P., and Storrs A. D. (2003) Asteroids do have satellites. In *Asteroids III* (W. F. Bottke Jr. et al., eds.), pp. 289–312. Univ. of Arizona, Tucson.

Michels D. J., Sheeley N. R. Jr., Howard R. A., and Koomen M. J. (1982) Observations of a comet on a collision course with the sun. *Science, 215,* 1097–1102.

Millis R. L., A'Hearn M. F., and Campins H. (1988) An investigation of the nucleus and coma of comet P/Arend-Rigaux. *Astrophys. J., 324,* 1194–1209.

Mumma M. J., Dello Russo N., DiSanti M. A., Magee-Sauer K., Novak R. E., Brittain S., Rettig T., McLean I. S., Reuter D. C., and Xu L.-H. (2001) Organic composition of C/1999 S4 (LINEAR): A comet formed near Jupiter? *Science, 292,* 1334–1339.

Nicholson P. D. (1996) Earth-based observations of impact phenomena. In *The Collision of Comet Shoemaker-Levy 9 and Jupiter* (K. S. Noll et al., eds.), pp. 81–109. Cambridge Univ., New York.

Oort J. H. (1950) The structure of the cloud of comets surrounding the solar system and a hypothesis concerning its origin. *Bull. Astron. Inst. Neth., 11,* 91–110.

Öpik E. J. (1966) Sun-grazing comets and tidal disruption. *Irish Astron. J., 7,* 141–161.

Osip D., Campins H., and Schleicher D. G. (1995) The rotation state of 4015 Wilson-Harrington: Revisiting origins for the near-Earth asteroids. *Icarus, 114,* 423–426.

Ostro S. J., Hudson R. S., Benner L. A. M., Giorgini J. D., Magri C., Margot J. L., and Nolan M. C. (2003) Asteroid radar astronomy. In *Asteroids III* (W. F. Bottke Jr. et al., eds.), pp. 151–168. Univ. of Arizona, Tucson.

Passey Q. R. and Shoemaker E. M. (1982) Craters and basins on Ganymede and Callisto: Morphological indicators of crustal evolution. In *Satellites of Jupiter* (D. Morrison, ed.), pp. 379–434. Univ. of Arizona, Tucson.

Pätzold M., Häusler B., Wennmacher A., Aksnes K., Anderson J. D., Asmar S. W., Barriot J.-P., Boehnhardt H., Eidel W., Neubauer F. M., Olsen O., Schmitt J., Schwinger J., and Thomas N. (2001) Gravity field determination of a comet nucleus: Rosetta at P/Wirtanen. *Astron. Astrophys., 375,* 651–660.

Peale S. J. (1989) On the density of Halley's comet. *Icarus, 82,* 36–49.

Pravec P., Harris A. W., and Michałowski T. (2003) Asteroid rotations. In *Asteroids III* (W. F. Bottke Jr. et al., eds.), pp. 113–122. Univ. of Arizona, Tucson.

Prialnik D. and Podolak M. (1999) Changes in the structure of comet nuclei due to radioactive heating. *Space Sci. Rev., 90,* 169–178.

Richardson D. C., Leinhardt Z. M., Melosh H. J., Bottke W. F. Jr., and Asphaug E. (2003) Gravitational aggregates: Evidence and evolution. In *Asteroids III* (W. F. Bottke Jr. et al., eds.), pp. 501–515. Univ. of Arizona, Tucson.

Rickman H. (1986) Masses and densities of comets Halley and Kopff. In *Proceedings of the Workshop on the Comet Nucleus Sample Return Mission* (O. Melita, ed.), pp. 195–205. ESA SP-249, Noordwijk, The Netherlands.

Rickman H. (1989) The nucleus of comet Halley — Surface structure, mean density, gas and dust production. *Adv. Space Res., 9,* 59–71.

Rickman H., Kamel L., Festou M. C., and Froeschlé Cl. (1987) Estimates of masses, volumes and densities of short-period comet nuclei. In *Symposium on the Diversity and Similarity of Comets* (E. J. Rolfe and B. Battrick, eds.), pp. 471–481. ESA SP-278, Noordwijk, The Netherlands.

Rubincam D. P. (2000) Radiative spin-up and spin-down of small asteroids. *Icarus, 148,* 2–11.

Ryan E. V., Hartmann W. K., and Davis D. R. (1991) Impact experiments III — Catastrophic fragmentation of aggregate targets and relation to asteroids. *Icarus, 94,* 283–298.

Sagdeev R. Z., Szabo F., Avanesov G. A., Cruvellier P., Szabo L., Szegö K., Abergel A., Balazs A., Barinov I. V., Bertaux J.-L., Blamount J., Detaille M., Demarelis E., Dul'Nev G. N., Endroczy G., Gardos M., Kanyo M., Kostenko V. I., Krasikov V. A., Nguyen-Trong T., Nyitrai Z., Reny I., Rusznyak P., Shamis V. A., Smith B., Sukhanov K. G., Szabo F., Szalai S., Tarnopolsky V. I., Toth I., Tsukanova G., Valnicek B. I., Varhalmi L., Zaiko Yu. K., Zatsepin S. I., Ziman Ya. L., Zsenei M., and Zhukov B. S. (1986) Television observations of comet Halley from VEGA spacecraft. *Nature, 321,* 262–266.

Sagdeev R. Z., Elyasberg P. E., and Moroz V. I. (1988) Is the nucleus of Comet Halley a low density body? *Nature, 331,* 240–242.

Samarasinha N. H. (1999) A model for the breakup of comet LINEAR (C/1999 S4). *Icarus, 154,* 540–544.

Samarasinha N. H. and Belton M. J. S. (1995) Long-term evolution of rotational stress and nongravitational effects for Halley-like cometary nuclei. *Icarus, 116,* 340–358.

Schenk P., Asphaug E., McKinnon W. B., Melosh H. J., and Weissman P. (1996) Cometary nuclei and tidal disruption: The geologic record of crater chains on Callisto and Ganymede. *Icarus, 121,* 249–274.

Schwehm G. (2002) The international Rosetta mission (abstract). In *Asteroids, Comets, Meteors 2002,* Abstract #02-05. Technical Univ. Berlin, Berlin.

Scotti J. V. and Melosh H. J. (1993) Estimate of the size of comet Shoemaker-Levy 9 from a tidal breakup model. *Nature, 365,* 733–735.

Sekanina Z. (1996) Tidal breakup of the nucleus of comet Shoemaker-Levy 9. In *The Collision of Comet Shoemaker-Levy 9 and Jupiter* (K. S. Noll et al., eds.), pp. 55–80. Cambridge Univ., Cambridge.

Sekanina Z. (2002) Runaway fragmentation of sungrazing comets observed with the Solar and Heliospheric Observatory. *Astrophys. J., 576,* 1085–1089.

Sekanina Z., Chodas P. W., and Yeomans D. K. (1994) Tidal disruption and the appearance of periodic comet Shoemaker-Levy 9. *Astron. Astrophys., 289,* 607–636.

Sheeley N. R. Jr., Howard R., Koomen M., Michaels D., and Marsden B. G. (1985) *Probable Sungrazing Comets.* IAU Circular No. 4129.

Sheppard S. S. and Jewitt D. C. (2002) Time-resolved photometry of Kuiper belt objects: Rotations, shapes, and phase functions. *Astron. J., 124,* 1757–1775.

Shoemaker C. S., Shoemaker E. M., Levy D. H., Scotti J. V., Bendjoya P., and Mueller J. (1993) *Comet Shoemaker-Levy (1993e).* IAU Circular No. 5725.

Skorov Y. V. and Rickman H., (1999) Gas flow and dust acceleration in a cometary Knudsen layer. *Planet. Space Sci., 47,* 935–949.

Slipher V. M. (1927) The spectrum of the Pons-Winnecke comet and the size of the cometary nucleus. *Lowell Obs. Bull., 3 (86),* 135–137.

Soderblom L. A., Becker T. L., Bennett G., Boice D. C., Britt D. T., Brown R. H., Buratti B. J., Isbell C., Giese B., Hare T., Hicks M. D., Howington-Kraus E., Kirk R. L., Lee M., Nelson R. M., Oberst J., Owen T. C., Rayman M. D., Sandel B. R., Stern S. A., Thomas N., and Yelle R. V. (2002) Observations of comet 19P/Borrelly by the Miniature Integrated Camera and Spectrometer aboard Deep Space 1. *Science, 296,* 1087–1091.

Solem J. C. (1994) Density and size of comet Shoemaker-Levy 9 deduced from a tidal breakup model. *Nature, 370,* 349–351.

Solem J. C. (1995) Cometary breakup calculations based on a gravitationally-bound agglomeration model: The density and size of Shoemaker-Levy 9. *Astron. Astrophys., 302,* 596–608.

Stern S. A. (1988) Collisions in the Oort cloud. *Icarus, 73,* 499–507.

Stern S. A. (1995) Collisional time scales in the Kuiper disk and their implications. *Astron. J., 110,* 856–868.

Stern S. A. (1996) On the collisional environment, accretion time scales, and architecture of the massive, primordial Kuiper belt. *Astron. J., 112,* 1203–1211.

Stern S. A. (2003) The evolution of comets in the Oort cloud and Kuiper belt. *Nature, 424,* 639–642.

Stern S. A. and Weissman P. R. (2001) Rapid collisional evolution of comets during the formation of the Oort cloud. *Nature, 409,* 589–591.

Sykes M. V., Grün E., Reach W. T., and Jenniskens P. (2004) The interplanetary dust complex and and comets. In *Comets II* (M. C. Festou et al., eds.), this volume. Univ. of Arizona, Tucson.

Thomas N. and 41 colleagues (1998) OSIRIS — The optical, spectroscopic and infrared remote imaging system for the Rosetta orbiter. *Adv. Space Res., 21,* 1505–1515.

Veverka J., Thomas P., Johnson T. V., Matson D., and Housen K. (1986) The physical characteristics of satellite surfaces. In *Satellites* (J. A. Burns and M. S. Matthews, eds.), pp. 342–402. Univ. of Arizona, Tucson.

Veverka J., Thomas P., Harch A., Clark B., Bell J. F. III, Carcich B., Joseph J., Chapman C., Merline W., Robinson M., Malin M., McFadden L. A., Murchie S., Hawkins S. E. III, Farquhar R., Izenberg N., and Cheng A. (1997) NEAR's flyby of 253 Mathilde: Images of a C asteroid. *Science, 278,* 2109–2112.

Vorontsov-Velyaminov B. (1946) Structure and mass of cometary nuclei. *Astrophys. J., 104,* 226–233.

Vsekhsviatsky S. K. (1948) On the question of cometary origin. *Soviet Astron. J., 25,* 256–266.

Weaver H. A., Feldman P. D., A'Hearn M. F., Arpigny C., Brown R. A, Helin E. F., Levy D. H., Marsden B. G., Meech K. J., Larson S. M., Noll K. S., Scotti J. V., Sekanina Z., Shoemaker C. S., Shoemaker E. M., Smith T. E., Storrs A. D., Yeomans D. K., and Zellner B. (1994) Hubble Space Telescope observations of comet P/Shoemaker-Levy 9 (1993e). *Science, 263,* 787–791.

Weaver H. A., Sekanina Z., Toth I., Delahodde C. E., Hainaut O. R., Lamy P. L., Bauer J. M., A'Hearn M. F., Arpigny C., Combi M. R., Davies J. K., Feldman P. D., Festou M. C., Hook R., Jorda L., Keesey M. S. W., Lisse C. M., Marsden B. G., Meech K. J., Tozzi G. P., and West R. (2001) HST and VLT investigations of the fragments of comet C/1999 S4 (LINEAR). *Science, 292,* 1329–1334.

Weidenschilling S. J. (1997) The origin of comets in the solar nebula: A unified model. *Icarus, 127,* 290–306.

Weidenschilling S. J. (2004) From icy grains to comets. In *Comets II* (M. C. Festou et al., eds.), this volume. Univ. of Arizona, Tucson.

Weissman P. R. (1979) Nongravitational perturbations of long-period comets. *Astron. J., 84,* 580–584.

Weissman P. R. (1980) Physical loss of long-period comets. *Astron. Astrophys., 85,* 191–196.

Weissman P. R. (1983) Cometary impacts with the Sun: Physical and dynamical considerations. *Icarus, 55,* 448–454.

Weissman P. R. (1986) Are cometary nuclei primordial rubble piles? *Nature, 320,* 242–244.

Weissman P. R. (1987) Post-perihelion brightening of Halley's comet: Spring time for Halley. *Astron. Astrophys., 187,* 873–878.

Weissman P. R. and Kieffer H. H. (1981) Thermal modeling of cometary nuclei. *Icarus, 47,* 302–311.

Weissman P. R. and Lowry S. C. (2003) The size distribution of Jupiter-family cometary nuclei (abstract). In *Lunar and Planetary Science XXXIV,* Abstract #2003. Lunar and Planetary Institute, Houston (CD-ROM).

Weissman P. R. and Stern S. A. (1998) Physical processing of cometary nuclei. In *Workshop on Analysis of Returned Comet Nucleus Samples* (S. Chang, ed.), pp. 119–166. NASA CP-10152, Washington, DC.

Weissman P. R., Richardson D. C., and Bottke W. F. (2003) Random disruption of cometary nuclei by rotational spin-up. *Bull. Am. Astron. Soc., 35,* 1012.

Whipple F. L. (1950) A comet model. I. The acceleration of comet Encke. *Astrophys. J., 111,* 375–394.

Whipple F. L. (1951) A comet model. II. Physical relations for comets and meteors. *Astrophys. J., 113,* 464–474.

Whipple F. L. (1955) A comet model. III. The zodiacal light. *Astrophys. J., 121,* 750–770.

Wilhelm K., Cosmovici C. B., Delamere W. A., Huebner W. F., Keller H. U., Reitsema H., Schmidt H. U., and Whipple F. L. (1986) A three-dimensional model of the nucleus of Comet Halley. In *20th ESLAB Symposium on the Exploration of Halley's Comet* (B. Battrick et al., eds.), pp. 367–369. ESA SP-250, Noordwijk, The Netherlands.

Yeomans D. K. (1971) Nongravitational forces affecting the motions of periodic comets Giacobini-Zinner and Borrelly. *Astrophys. J., 76,* 83–87.

Yeomans D. K., Chodas P. W., Sitarski G., Szutowicz S., and Krolikowska M. (2004) Cometary orbit determination and nongravitational forces. In *Comets II* (M. C. Festou et al., eds.), this volume. Univ. of Arizona, Tucson.

Modeling the Structure and Activity of Comet Nuclei

Dina Prialnik
Tel Aviv University

Johannes Benkhoff
DLR Berlin

Morris Podolak
Tel Aviv University

Numerical simulation of the structure and evolution of a comet nucleus is reviewed from both the mathematical and the physical point of view. Various mathematical procedures and approximations are discussed, and different attempts to model the physical characteristics of cometary material, such as thermal conductivity, permeability to gas flow, drag of dust grains, and dust mantling, are described. The evolution and activity of comets is shown to depend on different classes of parameters: defining parameters, such as size and orbit; structural parameters, such as porosity and composition; and initial parameters, such as temperature and live radioisotope content. Despite the large number of parameters, general conclusions or common features appear to emerge from the numerous model calculations — for different comets — performed to date. Thus, the stratified structure of comet nuclei, volatile depletion, and the role of crystallization of ice in cometary outbursts are discussed.

1. INTRODUCTION

Although comets have been observed and studied since antiquity, the comet nucleus is a much more recent concept: "It has been stated that within the head of a comet there is usually a bright point termed the nucleus. This is the only part of its structure that excites any suspicion of a solid substance." (Robert Grant, *History of Physical Astronomy,* 1852). Even now, only two comet nuclei (Halley and Borrelly) have been observed at close distances, where their shapes and sizes can be clearly seen, and none has yet been probed beneath the surface. Nevertheless, it is the nucleus that, by its structure and composition, determines the behavior of a comet in a given orbit.

One of the striking features of comet nuclei is their varied, often unexpected, behavior. Some exhibit outbursts when they are close to the Sun, others when they are far from the Sun. Some show a drastic reduction in their gas outflow for several orbits, during which they have a distinctly asteroidal appearance. Some nuclei suddenly split into several smaller pieces, while others remain whole even when they pass sufficiently close to the Sun to be affected by tidal forces. In short, every comet nucleus seems to have its own special pattern of behavior. Thus it is a real challenge to develop a theory of comet nucleus behavior that is rich enough to allow for this wealth of idiosyncrasy, albeit based on a handful of relatively simple processes, such as are expected of a moderately large icy rock floating in space. Remarkably, such a theory — or model — seems to be possible and has aroused growing interest and achieved

increasing sophistication during the past two or three decades. A list of symbols and constants used in the mathematical formulation of comet nucleus models is given in Table 1.

1.1. Historical Perspective

The simplest view of a comet nucleus is that of an active surface enveloping an inert interior. It stems from the assumption that solar radiation — exterior to the nucleus — is the only energy source responsible for cometary activity. Thus the Sun's gravity determines the dynamic history of a comet and solar energy determines its thermal history. Given a composition of ice and dust, the thermal properties of cometary material appear to be such that the skin depth associated with the orbital cycle is much smaller than the radius, and hence an inert interior seems to be justified. This is also the reason for considering comets as pristine, unaltered objects, relics of the formation of the solar system. This naive view will be shown to have changed considerably in recent years. Nevertheless, the simplest among comet nucleus models deal with the surface and assume constant properties over its entire extent (albedo, emissivity, and dust/ice mass ratio), completely neglecting any energy exchange with the interior. The earliest model based on these assumptions was that of *Squires and Beard* (1961); later models were calculated by *Cowan and A'Hearn* (1979). They provided solutions for the power balance equation at the nucleus boundary to obtain the variation of surface temperatures and gas and dust production rates as a function of

TABLE 1.   List of symbols.

$\mathcal{A}$	Albedo	P_{orb}	Orbital period	v	Velocity
a	Semimajor axis	P_{spin}	Nucleus spin period	v_{th}	Thermal velocity
c	Specific heat	P_α	Partial gas pressure	X_α	Mass fraction of species α
d_H	Heliocentric distance	$\mathcal{P}_\alpha$	Saturated vapor pressure	ε	Emissivity
e	Eccentricity	$\dot{Q}_{rad}$	Radioenergy generation rate	ζ	Angle of insolation
F	Energy flux	Q_α	Surface sublimation flux	θ	Latitude
f_α	Fraction of trapped gas	q_α	Volume sublimation rate	ϑ	Declination
G	Gravitational constant	R	Radius of nucleus	κ	Thermal diffusivity
g	Gravitational acceleration	$\mathcal{R}_g$	Ideal gas constant	λ	Crystallization rate
$\mathcal{H}_{ac}$	Heat of crystallization	r	Radial distance from center	μ	Molecular weight
$\mathcal{H}_\alpha$	Latent heat of sublimation	r_d	Dust grain radius	ν	Kinematic viscosity
J	Mass flux	r_d^*	Critical dust grain radius	ξ	Tortuosity
K	Thermal conductivity	r_p	Pore radius	ρ_α	Partial density of ice species
k	Boltzmann constant	S	Surface to volume ratio	$\bar{\rho}_\alpha$	Partial density of gas species
$L_\odot$	Solar luminosity	T	Temperature	ϱ_α	Density of solid species
ℓ	Mean free path	$\mathcal{T}$	Tensile strength	σ	Stefan-Boltzmann constant
M	Mass of nucleus	t	Time	τ	Characteristic timescale
$M_\odot$	Solar mass	u	Energy per unit mass	Ψ	Porosity
m_α	Molecular mass of species α	V	Volume	ω	Hour angle

heliocentric distance. A so-called "standard model" emerged, based on power balance for a unit area normal to the solar direction

$$\frac{(1-\mathcal{A})L_\odot}{4\pi d_H^2} = \varepsilon\sigma T^4 + \mathcal{P}_{H_2O}\sqrt{m_{H_2O}/2\pi kT}\,\mathcal{H}_{H_2O} \quad (1)$$

assuming the surface to be entirely covered by water ice. The total incident energy depends on the cross-sectional area of the nucleus, whereas evaporation and reradiation of energy occur over the entire hemispherical surface facing the Sun. However, since regions not normal to the incident radiation receive less energy by a cosine factor and are at a lower temperature, less energy is lost from these regions by evaporation and radiation. Therefore, in view of the great uncertainties in the physical properties of cometary material, the comet was treated as a two-dimensional (2-D) disk of area πR^2 facing the Sun. Combined with observed production rates, this model is often used in order to estimate the nucleus size (or at least the size of its active surface area).

Diurnal temperature variations over the surface of the nucleus were first studied by *Weissman and Kieffer* (1981, 1984) and by *Fanale and Salvail* (1984). Heat conduction to the interior was now considered as well, but only in a thin subsurface layer. The possibility of an outer dust mantle enveloping the icy nucleus was first considered by *Mendis and Brin* (1977) and *Brin and Mendis* (1979), and subsequently pursued by many others. Heat conduction throughout the entire nucleus was first explored by *Herman and Podolak* (1985), who solved the heat diffusion equation for the one-dimensional (1-D) case of a homogeneous spherical nucleus. It was prompted by the suggestion that cometary ice could be amorphous and its subsequent crystallization could provide an internal source of energy (*Patashnik*

et al., 1974), so far ignored. This study explored the consequences of crystallization of amorphous ice and the associated release of latent heat on the temperature profile. It revealed intermittent bursts of crystallization, an effect that was studied in considerably more detail by *Prialnik and Bar-Nun* (1987), reviving an earlier suggestion that cometary outbursts might be linked with the crystallization process. In parallel, cometary activity due to volatiles other than H_2O was studied by *Fanale and Salvail* (1987).

The next step in the study of comet nuclei by numerical simulations was prompted by the detailed observations of Comet P/Halley when it passed perihelion in 1986, which revealed that the nucleus had a very low bulk density, indicating a high porosity (*Rickman*, 1989). Further evidence in favor of porosity is provided by the presence in cometary ejecta of molecules from volatile ices, which cannot survive in the warm subsurface layers. The origin of such molecules must therefore be in the deeper, colder layers of the nucleus. Pores act as conduits for the transport of gases trapped in comet nuclei. *Mekler et al.* (1990) initiated a detailed study of the effect of porosity on the cometary nucleus. They developed a model of gas flow through a porous medium, allowing for vaporization from the pore walls, and used this model for the simplest case of a porous pure-water-ice nucleus. They found that for a given distance from the Sun there is a critical depth above which the gas flows out of the nucleus, and below which it flows toward the center. Laboratory experiments performed by the KOSI group around the same time (*Spohn et al.*, 1989) confirmed these findings and promoted the study of gas flow through porous comet nuclei. The basic model of low-density flow through a porous medium has been adapted by a number of research groups (e.g., *Espinasse et al.*, 1991; *Steiner and Kömle*, 1991; *Prialnik*, 1992; *Tancredi et al.*, 1994; *Benkhoff and Huebner*, 1995). More recently, attempts have been made to include the flow of dust particles through the pores as well (*Orosei et al.*, 1995; *Podolak and Prialnik*, 1996). It is now commonly agreed that porosity must be included in order to properly understand cometary behavior. A different question that has been repeatedly addressed concerns the end state of comets. Unless they are disrupted by tidal forces or destroyed by collisions with larger bodies, comet nuclei are expected to evaporate and disintegrate, leaving behind a trail of debris. But they may also become extinct — asteroid-like — if a dust mantle forms at the surface, quenching all types of cometary activity. This line of investigation started with the early work of *Shul'man* (1972), followed by numerous studies of various aspects of the dust mantle, culminating in detailed evolutionary models of transition objects between comets and asteroids (e.g., *Coradini et al.*, 1997a,b).

As the prevailing idea of pristine comet nuclei interiors began to give way to more elaborate pictures of these objects, another internal heat source — radioactive decay, commonly considered for large bodies of the solar system — came into focus. The first to have considered radioactive heating of comet nuclei were *Whipple and Stefanik* (1966): They found that the decay of ^{40}K, ^{235}U, ^{238}U, and ^{232}Th caused the internal temperature to rise to a peak of ~90 K (from an initial 0 K) on a timescale of some 10^8 yr. About 10 years later, *Lee et al.* (1976) presented strong evidence that the short-lived radionuclide ^{26}Al had been present in the early solar nebula. Further evidence strengthening this conclusion has accumulated ever since. The idea of internal heating of comet nuclei by the decay of ^{26}Al gained impetus following early studies by *Irvine et al.* (1980) and *Wallis* (1980), who showed that it may lead to melting of the ice, which would have implications for early formation of organic molecules and the origin of life (see *Thomas et al.*, 1997). Thus heating by radioactive decay has been considered in a number of comet nucleus models under various conditions and assumptions. Whether or not liquid water could have been present in comet nuclei during their early stages of evolution, and if so, under which conditions, is still debated. The question of whether and to what extent comets are pristine bodies that hold clues to the formation of the solar system is still open. Finally, would it be possible for comet nuclei to have had liquid cores, but at the same time preserved their outer layers in pristine form? These are some of the questions that have prompted the development of increasingly sophisticated models of comet nuclei. The number of studies devoted to the evolution and activity of comet nuclei and to the complex processes involved is steadily growing. The effort is twofold: understanding and providing a usable mathematical formulation of the processes on the one hand, and incorporating these processes in numerical simulations of the structure and evolution of comet nuclei on the other. Section 3 will be devoted to the former and section 4 to the latter. The set of evolutionary equations will be presented in section 2, and conclusions as well as suggestions for future work, in section 5.

1.2. Basic Assumptions Derived from Observations

The structure of a comet nucleus may be modeled as a highly porous agglomeration of grains made of volatile ices and dust, with a size distribution that probably spans many orders of magnitude. The dominant volatile component is water ice, while the other volatiles, such as CO, CO_2, HCN, N_2, etc., are mixed with the water ice or incorporated in it, either in the form of clathrate-hydrates, or as trapped gases within the (amorphous) ice matrix. Having been formed at low temperatures and pressures, cometary ice is believed to be amorphous (*Mekler and Podolak*, 1994). Laboratory experiments indicate that amorphous ice is capable of trapping large amounts of gas and most of this trapped gas escapes when the ice crystallizes (*Bar-Nun et al.*, 1987). Thus, whereas H_2O molecules are released within a narrow temperature range, when the ice sublimates, the other volatiles may be released in different temperature ranges. Crystallization, as well as sublimation from the pore walls in the deep cometary interior, may be triggered (and sustained) either by a heat wave advancing inward from the surface, or due to internal heat release by radioactive isotopes contained in the dust, or else by the release of latent heat that accompanies the transformation of amorphous into crystalline ice.

These are, in fact, the three main — and perhaps only — sources of energy available to comets. Their typical properties are summarized in Table 2. The radioactive source is important particularly during the long period of time spent by comets outside the planetary system, far from the Sun. Close to the Sun, it is far less efficient than solar radiation, and hence negligible. The most important radionuclide is the short-lived isotope ^{26}Al. Observational evidence points toward an interstellar isotopic ratio ^{26}Al/^{27}Al ≈ 5 × 10^{-5} (e.g., *MacPherson et al.*, 1995), implying an initial mass fraction $X_0(^{26}$Al$) ≈ 7 × 10^{-7}$ in the solar nebula dust (rock) and presumably an order of magnitude less, on average, in objects such as comets, for which the time of aggregation did not exceed a few million years (*Lugmair and Shukolyukov*, 2001). In contrast to these sources, the exoergic crystallization of amorphous water ice is not an independent source, since it occurs above a threshold temperature that must be attained by means of other energy sources. Crystallization may occur at any evolutionary stage, and may propagate either inward or outward.

Once gas is released from the ice in the interior of the nucleus, its pressure will cause it to flow to the surface. Gases moving through the pores drag with them small dust

TABLE 2.   Energy sources and their characteristics.

Solar Radiation	Radioactivity	Crystallization
Surface source	Body source	Local source
$\propto R^2/d_H^2$	$\propto R^3 X_{rad}$	$\propto X_{a-ice}\mathcal{H}_{ac}$
Cyclic	Declining	Transient
Inward moving wave	Homogeneous	Thin front
Late evolution	Early evolution	Induced

particles that have detached from the solid matrix. The larger particles may eventually block the pores; the smaller ones may flow all the way with the gas.

The free gases present in the interior of a comet are expected to affect the thermal and mechanical structure of the nucleus by contributing to the conduction of heat through advection or recondensation and by building up internal pressure. This pressure may surpass the tensile strength of the already fragile, grainy configuration and result in cracking of the porous matrix and outbursts of gas and dust. Accumulation of large particles on the nucleus surface may lead to the formation of a sealing dust mantle that may partially (or fully) quench the comet's activity.

All these processes are taken into account in models of the evolution and activity of comet nuclei, as will be shown in the next section. Internal processes depend on physical properties characteristic of cometary material, which will be discussed in sections 3 and 4.

1.3. Approximations Required by Modeling

Comet nuclei are too small for self-gravity to be of importance, hence they are not necessarily spherical. However, a nonspherical object is far more difficult to model. In addition, the number of free parameters for an arbitrary shape tends to infinity. Thus models must assume some form of symmetry and sphericity, requiring a single dimensional parameter — the (effective) radius R is the common assumption. The simplest among spherical models are 1-D, considering a spherically symmetric nucleus, which implies an evenly heated surface, although in reality only one hemisphere faces the Sun at any given time, and even its surface is not evenly irradiated. This approximation is known (somewhat loosely) as the "fast-rotator" approximation. It is a good approximation for the interior of the nucleus, below the skin depth; it is valid for the surface far away from the Sun, where diurnal temperature variations are small.

However, in order to obtain an accurate surface temperature distribution and its diurnal change at any heliocentric distance, one must adopt the so-called "slow-rotator" approach, which takes into account the diurnal and latitudinal solar flux variations. This type of model requires a far greater amount of computing time, since much smaller time steps — a small fraction of the spin period — must be used in the numerical integration over time.

A first attempt in this direction was to consider a point on the equator of a spinning nucleus and translate the diurnal temperature change obtained into a map of the equatorial temperature at any given time. Such a procedure (*Benkhoff and Boice*, 1996; *Benkhoff*, 1999) may be described as a 1.5-dimensional (1.5-D) model. An upper limit for the production rate is obtained by using the maximum noon flux for the entire surface of the sunlit hemisphere. A more advanced model is achieved by considering a wedge of surface elements aligned along a meridian (*Enzian et al.*, 1997, 1999). Thus the latitudinal effect is taken into account and the total production rate is obtained by summing the con-

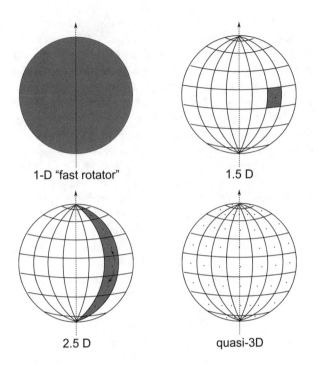

Fig. 1. Schematical representation of numerical grids for a spinning nucleus, commonly used in model calculations. Dots indicate radial directions along which heat conduction is computed; lateral conduction is only included in the 2.5-D model, and only along the meridian, as shown.

tributions of such wedges over one spin period. This is essentially a 2.5-dimensional (2.5-D) calculation. The next step is to take into account both diurnal and latitudinal solar flux variations (*Gutiérrez et al.*, 2000; *Julian et al.*, 2000; *Cohen et al.*, 2003), considering, however, only radial heat conduction, i.e., neglecting lateral conduction. This quasi three-dimensional (3-D) approach is amply justified by the extremely low heat conductivity of cometary material; the characteristic heat diffusion time between the equator and pole (as between surface and center) is of the order of the lifetime of a comet (see Table 5 below). The different models are shown schematically in Fig. 1; of course, calculations use much finer meshes than the ones shown.

The advantage of the negligible self-gravity of comets is that the structure of the nucleus may be assumed incompressible despite the low strength of the porous material. By comparing the hydrostatic pressure at the center with the material strength, we obtain the condition for incompressibility

$$\rho R \leq \sqrt{3C/2\pi G} \qquad (2)$$

where C is the compressive strength. Even for as low a value as $C \approx 10$ kPa (see section 3.8), this implies

$$\rho_{g/cm^3} R_{km} \leq 10$$

which is amply satisfied by the typical sizes and densities

of comet nuclei. Therefore, the equations that determine the structure and evolution of comets are those of energy conservation and of mass conservation for the various components. Momentum conservation (hydrostatic balance) is not required for the solid matrix and can be replaced by a prescribed (usually constant) density profile; for the gas species, expressions for the flux in different regimes are used.

2. EVOLUTION EQUATIONS AND SOLUTION METHODS

2.1. Mass and Energy Conservation

Consider a composition of water ice and vapor, dust, and other volatiles, which, as we have seen, may be frozen, free, or trapped in the amorphous water ice. For each volatile species α we distinguish — when necessary — between two phases, solid (denoted by index s) and gas (index g), which have, e.g., different specific heat coefficients. For water, the solid phase may be either amorphous (index a) or crystalline (index c); water vapor will be denoted by index v and dust by d. Then the mass density and porosity are given respectively by

$$\rho = \rho_a + \rho_c + \rho_v + \sum_\alpha (\rho_{s,\alpha} + \rho_{g,\alpha}) + \rho_d \quad (3)$$

$$\Psi = 1 - (\rho_a + \rho_c)/\varrho_{H_2O} - \sum_\alpha \rho_{s,\alpha}/\varrho_\alpha - \rho_d/\varrho_d \quad (4)$$

We note that densities refer to mass per unit volume of nucleus material. Since the gas resides in the pores, the actual gas density within the pores will be $\rho_{g,\alpha}/\Psi$ and the partial pressure, assuming an ideal gas, will be

$$P_\alpha = \frac{\mathcal{R}_g \rho_{g,\alpha} T}{\Psi \mu_\alpha} \quad (5)$$

Local thermodynamic equilibrium is assumed to prevail, that is, all components in all phases, as well as radiation, share the same local temperature. It is further assumed that gases trapped in the amorphous ice do not affect its heat capacity or density. Let f be the total fraction of occluded gas, $\Sigma_\alpha f_\alpha = f$. Thus the equations of mass conservation are

$$\frac{\partial \rho_a}{\partial t} = -\lambda \rho_a \quad (6)$$

$$\frac{\partial \rho_c}{\partial t} = (1 - f)\lambda \rho_a - q_v \quad (7)$$

$$\frac{\partial \rho_v}{\partial t} + \nabla \cdot \mathbf{J}_v = q_v \quad (8)$$

for H_2O in all its phases, and similarly

$$\frac{\partial \rho_{g,\alpha}}{\partial t} + \nabla \cdot \mathbf{J}_\alpha = f_\alpha \lambda \rho_a + q_\alpha$$

$$\frac{\partial \rho_{s,\alpha}}{\partial t} = -q_\alpha \quad (9)$$

The equation of energy conservation is

$$\frac{\partial}{\partial t}[\rho u] + \nabla \cdot (\mathbf{F} + \sum_\alpha u_{g,\alpha} \mathbf{J}_\alpha) = S \quad (10)$$

where S stands for all the available energy sources and sinks,

$$S = \lambda \rho_a \mathcal{H}_{ac} + \dot{Q}_{rad} - \sum_\alpha q_\alpha \mathcal{H}_\alpha \quad (11)$$

Here $[\rho u]$ represents the sum over all species and all phases (although the contribution of the gas phases is small and hence sometimes neglected), $u = \int c(T)dT$, and

$$\mathbf{F} = -K\nabla T \quad (12)$$

Combining the energy and mass conservation equations, we obtain the heat transfer equation, which can replace equation (10)

$$\sum_\alpha \rho_\alpha \frac{\partial u_\alpha}{\partial t} - \nabla \cdot (K\nabla T) + \left(\sum_\alpha c_\alpha \mathbf{J}_\alpha\right) \cdot \nabla T = S \quad (13)$$

with the advantage being that the temporal derivatives of the variables are now separated. The above set of time-dependent equations is subject to constitutive relations $u(T)$, $\lambda(T)$, $q_\alpha(T, \Psi, r_p)$, $\mathbf{J}_\alpha(T, \Psi, r_p)$, and $K(T, \Psi, r_p)$, which require additional assumptions for modeling the structure of the nucleus. They will be discussed in some detail in the next section. The most widely used expressions or values for the thermal properties of cometary H_2O ice and dust are given in Table 3. Ice properties have been measured and turned into empirical temperature-dependent relations by *Klinger* (1980, 1981), and more recently by *Ross and Kargel* (1998).

TABLE 3.   Thermal properties of cometary H_2O ice and dust.

Property	Relation	Units
Specific heat: c_a, c_c	$7.49T + 90$	J kg^{-1} K^{-1}
Specific heat: $c_v = 3\mathcal{R}_g\mu$	1.385×10^3	J kg^{-1} K^{-1}
Specific heat: c_d	~800	J kg^{-1} K^{-1}
Thermal conductivity: K_c	$567/T$	J m^{-1} s^{-1} K^{-1}
Thermal diffusivity: κ_a	3.13×10^{-7}	m^2 s^{-1}
Thermal conductivity: K_d	~0.1–4	J m^{-1} s^{-1} K^{-1}

2.2. Boundary Conditions

The set of evolution equations must be supplemented by initial and boundary conditions. For the heat transfer equation, the boundary conditions refer to the flux F(r) on the open interval $r_0 < r < R$, where $r_0 = 0$ when the entire comet is considered, or $0 < r_0 < R$, when only an outer layer is considered in a plane-parallel calculation. At the ends of this interval we have

$$F(r_0) = 0 \qquad (14)$$

$$F(R) = \varepsilon\sigma T(R,t)^4 + \mathcal{F}Q\mathcal{H} - (1 - \mathcal{A})\frac{L_\odot}{4\pi d_H(t)^2}\cos\xi \qquad (15)$$

The local solar zenith angle is given by

$$\cos\xi = \cos\theta \cos\omega \cos\delta + \sin\theta \sin\delta \qquad (16)$$

where δ is the declination (see also *Sekanina*, 1979; *Fanale and Salvail*, 1984). The factor $\mathcal{F} \le 1$ represents the fractional area of exposed ice, since the surface material is a mixture of ice and dust (*Crifo and Rodionov, 1997*)

$$\mathcal{F} = \left(1 + \frac{\rho_{ice}}{\rho_{ice}}\frac{\rho_d}{\rho_d}\right)^{-1} \qquad (17)$$

The function $d_H(t)$ is given in terms of the changing eccentric anomaly E by the familiar celestial mechanics equations

$$t = \sqrt{a^3/GM_\odot}(E - e \sin E) \qquad (18)$$

$$d_H = a(1 - e \cos E) \qquad (19)$$

Similarly to the heat flux, the mass (gas) fluxes vanish at r_0. At the surface R the gas pressures are those exerted by the coma; in the lowest approximation they may be assumed to vanish: $P_\alpha(R,t) = 0$. [For a more elaborate discussion concerning the boundary conditions to be assumed for the gas pressure at the surface and their effect, see *Crifo et al.* (2004).] We should note that when the entire comet is considered, mass and heat fluxes *must* vanish at the center. However, at the lower boundary of a finite layer, other conditions may equally be imposed (for example, a fixed temperature and corresponding vapor pressures), but only by adopting vanishing fluxes are energy and mass conservation secured.

In a porous medium the surface is not well defined and a surface layer of finite (rather than vanishing) thickness supplies the outflowing vapor. *Mekler et al.* (1990) have shown that the surface layer where most of the vapor is generated is considerably thinner than the layer of ice that is lost by a comet during a perihelion passage. We are thus faced with two vastly different length scales, which imply different timescales as well. On the evolutionary timescale of the comet the thin boundary layer may be assumed to be in (quasi) steady-state. Its ice may be assumed to have crystallized. The gas fluxes from the interior may be taken as constant and their contribution to heat conduction may be neglected. In addition, plane-parallel geometry is justified in this case and hence the equations that have to be solved near the surface as a function of depth z are

$$dJ_v(T, P_v)/dz = q_v(T, P_v) \qquad (20)$$

$$dF(T)/dz = -q_v(T, P_v)\mathcal{H} \qquad (21)$$

The boundary conditions for this layer are equation (15) at R and given temperature at the lower boundary, where it is fitted to the rest of the comet. This procedure, suggested by *Prialnik* (1992) was also employed by *Tancredi et al.* (1994).

An alternative approach to the macroscopic equations of gas diffusion in a porous volatile medium is a kinetic model, which provides gas fluxes as well as gas production and loss rates, for a given temperature distribution (*Skorov and Rickman*, 1995; *Skorov et al.*, 2001). It is particularly suited to the surface layer of the nucleus, near the pore openings. In this case the gas pressure at the surface is no longer required as a boundary condition, but rather it results from the calculation.

The initial conditions must be guessed, and since the comet nucleus as a whole never reaches steady-state below a skin depth of the order of meters to several tens of meters, these conditions play a significant role. This explains the importance attached to the early evolution of comets at large distances from the Sun, which determines the interior configuration of comet nuclei when they enter the inner planetary system and become active.

2.3. Numerical Schemes

The system of nonlinear, time-dependent, second-order partial differential equations (6)–(9) and (10) or (13) is turned into a set of difference equations that are solved numerically. They constitute a two-boundary value problem that requires relaxation methods for its solution. Let the time and space domains be divided into finite intervals $\delta t_n = t_n - t_{n-1}$ and $\Delta r_i = r_i - r_{i-1}$ $(0 \le i \le I)$, such that $t_0 = 0$, $r_0 = 0$, and $r_i = R$. The solution for the change of the temperature profile will be represented by a series of stepped functions T_i^n, where T_i is the temperature within the interval Δr_i. For the simple (linear) case of heat transfer with constant coefficients, there are several possibilities for combining the space and time derivatives into a difference equation for the transport equation, among which the most common are the explicit scheme, which can be solved directly; the fully implicit scheme, which, upon rearranging terms, results in a system of I linear equations requiring the inversion of a tridiagonal matrix for its solution; and the Crank-Nicholson scheme, which is a modified implicit form that requires the inversion of a tridiagonal matrix as well. The explicit scheme has the disadvantage that time-steps are restricted by the Courant-Friedrichs-Levy condition, $\delta t \leq (\Delta r)^2/2K$, for a given space discretization; thus time-steps may become prohibitively small when a fine mesh is required in order to resolve sharp temperature gradients. The implicit schemes, on the other hand, are unconditionally stable for all values of the time-step. However, they require a far greater amount of computations for each time-step, prohibitively large in the case of a large spatial grid, or in the 2- or 3-D cases. The Crank-Nicholson scheme has the advantage of being second-order accurate in time, whereas the fully implicit one, as the explicit scheme, is only first-order accurate in time. The fully implicit scheme, on the other hand, is best suited for stiff equations, i.e., when there are two or more very different timescales on which the temperature is changing (as is the case in comets). The reason is that the implicit scheme converges to the steady-state solution for large time-steps.

The same methods apply to the more complicated case when the heat capacity and thermal conductivity are functions of the temperature, and there is also a temperature-dependent source term (such as equation (13)). In this case, the difference equations of the implicit schemes must be linearized and solved iteratively. Another numerical method of solution of the nonlinear heat equation is the predictor-corrector method, essentially a two-step iterative procedure. Each iteration (or step) requires the inversion of a tridiagonal matrix. A comparison of different algorithms that were used to compute the evolution of a comet nucleus model for the same set of physical parameters is shown in Fig. 2. Each of the numerical methods mentioned above was adopted in one or the other of these algorithms, which also differ in other numerical parameters (for details, see *Huebner et al.,* 1999). The results are remarkably similar; the agreement at low perihelion distances is excellent, but this is expected since most of the solar radiation is spent in sublimation. Differences arise at larger distances and provide an estimate of the error-bars expected from model calculations, which are otherwise difficult to assess.

Although $\frac{\partial u}{\partial t} = c(T)\frac{\partial T}{\partial t}$, the time difference should be taken for the energy, rather than for the temperature, in order to ensure energy conservation in the numerical scheme. Finally, in the case of a 1-D spherical coordinate system, it is convenient to choose the volume V enclosed within a sphere of radius r for the space variable, rather than r, for then the equation retains the form of the plane-parallel one. From the physical point of view, it would be even better to adopt the mass enclosed within a sphere of radius r as space variable, but if the mass is allowed to change during evolution as a result of internal sublimation and gas flow, the volume is a better choice. In this case the flux through r must be replaced by the energy crossing the spherical surface of radius r per unit time. An additional advantage of this proce-

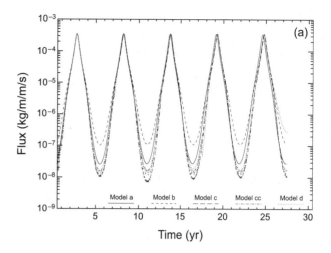

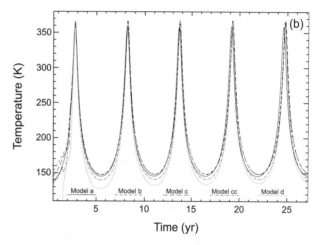

Fig. 2. Model results for **(a)** H_2O production rates of a pure ice nucleus and **(b)** surface temperature of a dust-mantled nucleus, at the subsolar point as function of time for five revolutions in Comet P/Wirtanen's orbit as computed by five different algorithms (W. F. Huebner et al., personal communication, 2002).

dure is that equal volume intervals ensure a better radial resolution near the nucleus surface, where $\partial T/\partial r$ is steeper. Similarly, choosing the eccentric (or true) anomaly angle as the temporal variable naturally leads to smaller time steps near perihelion, where changes are more rapid.

2.4. Computational Approximations

We note that the evolution equations are coupled through the source terms and the gas fluxes, which are functions of both temperature and pressure, and hence must be solved simultaneously. This is extremely time consuming, considering that the equations are strongly nonlinear. Simplifying approximations may be used under special conditions.

If the effective permeability of the medium is sufficiently high, the time derivative on the lefthand side of the mass conservation equation for the gas phases (equation (8)) becomes negligible. Neglecting it is tantamount to a quasisteady-state approximation, where gas densities and production rates change only as far as the temperature distribution changes. Thus equations (9) are replaced by

$$\nabla \cdot \mathbf{J}_\alpha = q_\alpha$$

$$\frac{\partial \rho_{s,\alpha}}{\partial t} = -q_\alpha \qquad (22)$$

In this way we strictly have to solve only one time-dependent equation supplemented by structure (space-dependent) equations. This constitutes a huge computational advantage, particularly in a long-term evolutionary calculation, where a detailed account of gas flow through the porous medium, coupled with heat transfer, would require a prohibitively large amount of computing time. Combining equations (22) and integrating over volume, we obtain

$$-\dot{M}_{s,\alpha} = J_\alpha(R,t)4\pi R^2 \qquad (23)$$

which means that the total mass of gas ejected through the comet's surface per unit time is equal to the total amount of gas evaporated throughout the nucleus per unit time for each species. This approximation is valid for nonabundant species, for which the bulk density is low. It breaks down when the net gas sublimation rate is negative, i.e., when recondensation surpasses sublimation. This approach has been recently adopted by *Choi et al.* (2002) in long-term evolutionary calculations of Kuiper belt objects. It is also applied for the outermost layer of the nucleus, as already mentioned in section 2.2.

A different approximation with the same computational advantage — reduction of the number of time-dependent equations — has been used in other studies (e.g., *Coradini et al.*, 1997a) for the nucleus interior. It assumes that when both the ice and gas phases are present, the gas density is equal to the saturated vapor density, which is a function of temperature. Strictly, this would imply that no evaporation/condensation could take place. However, as the temperature changes, the saturated density (pressure) changes with it, and this change can be translated into a rate of evaporation/condensation. This is an excellent approximation for the interior of the nucleus, where the pressures are indeed found to attain saturation; it implies, however, that there is sufficient material in both phases to allow instantaneous adjustment. It is not valid, therefore, for minor volatile components and fails close to the surface of the nucleus. The two simplifying approximations are thus complementary.

3. PHYSICAL PROCESSES

3.1. Heat Conduction in a Porous Medium

Cometary H_2O ice can be viewed as the matrix that comprises the bulk of the nucleus, but within this matrix there are grains of dust, occluded gases, and pores. *Reach et al.* (2000) have recently suggested that the grains dominate and form the background matrix with ice as the filler, but in either case the problem of calculating the conductivity through such a porous material remains essentially the same. For the canonical model of a porous ice matrix, the pores can themselves be filled, at different times, with smaller grains, H_2O vapor, or other gases. The grains embedded in the water-ice matrix, the grains and gases flowing through the pores, and even radiation passing through the pores will each affect the rate of heat transport through the nucleus. This is a complex problem that has a rather long history.

To begin, let us consider a simple idealized medium consisting of bulk H_2O ice permeated by spherical, gas-free pores. If these pores do not transport energy, they will reduce the overall conductivity of the medium. If K_s is the bulk conductivity of solid ice, then the conductivity of the porous ice will be $K = \phi K_s$, where $\phi < 1$. The key problem is to determine the value of ϕ. *Smoluchowski* (1981) suggested that this reduction would be proportional to the cross-sectional area of the pores. Since the volume of void is proportional to the porosity Ψ, its cross-sectional area should be proportional to $\Psi^{2/3}$. We might expect, therefore, that

$$\phi \approx 1 - \Psi^{2/3} \qquad (24)$$

In a later paper, *Smoluchowski* (1982) used a formula originally developed by *Maxwell* (1873). If we consider spherical grains of conductivity K_p embedded in a matrix of conductivity K_s, and the grains occupy a small fraction of the total volume, Ψ, then the conductivity of the combined medium is reduced relative to that of the matrix by a factor

$$\phi = \frac{(2 - 2\Psi) + (1 + 2\Psi)\frac{K_p}{K_s}}{(2 + \Psi) + (1 - \Psi)\frac{K_p}{K_s}} \qquad (25)$$

This formula is actually the first term in an expansion, and neglects effects of mutual "shadowing" by the grains. It is therefore exact only for the case of small Ψ. Higher-order terms were later added by *Rayleigh* (1892), which extended the applicability to larger values of Ψ. But precisely because the Maxwell formula neglects shadowing, it gives an upper limit to ϕ. A lower limit can be obtained by inverting the problem: Let the matrix be composed of grain material and the grains be composed of ice. Then

$$\phi_1 = \frac{K_p}{K_s} \frac{2\Psi\frac{K_p}{K_s} + (3 - 2\Psi)}{(3 - \Psi)\frac{K_p}{K_s} + \Psi} \qquad (26)$$

The spherical grains embedded in the ice matrix may be replaced by pores, in which case Ψ is the porosity, and the pore conductivity due to the flux carried by radiation is

$$K_p = 4\varepsilon\sigma r_p T^3 \qquad (27)$$

Squyres et al. (1985) used an expression due to *Brailsford and Major* (1964) for ϕ

$$\phi = \frac{K_p}{4K_s}\left(A + \sqrt{A^2 + \frac{8K_s}{K_p}} \right) \qquad (28)$$

where

$$A = (2 - 3\Psi)\frac{K_s}{K_p} + 3\Psi - 1$$

They also pointed out that if the matrix material is actually composed of ice grains, and the area of contact between adjoining grains is small, the resultant conductivity of the medium will be reduced even further by a so-called Hertz factor, the area of contact between material grains relative to the cross-sectional area. A different expression, known as Russel's formula, was used in other studies (see *Espinasse et al.*, 1991)

$$\phi = \frac{\Psi^{2/3}\frac{K_p}{K_s} + (1 - \Psi^{2/3})}{\Psi - \Psi^{2/3} + 1 - \Psi^{2/3}(\Psi^{1/3} - 1)\frac{K_p}{K_s}} \qquad (29)$$

Steiner and Kömle (1991) proposed still another theoretical expression

$$\phi = \left(1 - \sqrt{1 - \Psi}\right)\Psi\frac{K_p}{K_s} +$$

$$\sqrt{1 - \Psi}\left[\varsigma + (1 - \varsigma)\frac{B + 1}{B}\frac{K_p}{K_s + K_p} \right] \qquad (30)$$

where

$$B = 1.25[(1 - \Psi)/\Psi]^{10/9}$$

is a deformation factor, and ς is a flattening coefficient, which is essentially the Hertz factor. This factor depends on the details of the structure of the medium and cannot be determined *a priori*.

Haruyama et al. (1993) and *Sirono and Yamamoto* (1997) used effective medium theory to derive

$$\frac{\phi - 1}{1 + \left(\frac{1}{\Psi_c} - 1\right)\phi}\Psi + \frac{\phi - \frac{K_p}{K_s}}{\frac{K_p}{K_s} + \left(\frac{1}{\Psi_c} - 1\right)\phi}(1 - \Psi) = 0 \qquad (31)$$

where Ψ_c is the percolation threshold of the medium (*Stauffer and Aharony*, 1994). None of these approaches, however, allows for a distribution of pore sizes, yet the pore-size distribution will certainly affect the resultant conductivity.

Recently, a new approach was adopted by *Shoshany et al.* (2002). Using a Monte Carlo procedure, they modeled a 3-D fractal medium made of ice and voids. A temperature gradient was assumed across this medium and the 3-D equations of heat transfer were solved to obtain the energy flux, which yields the effective conductivity. By running a

series of models and fitting analytic functions to the results, this Monte Carlo approach allows one to find ϕ for a given porosity

$$(1 - \Psi/\Psi_c)^{\alpha(\Psi)} \leq \phi$$

$$\leq (1 - \Psi_m/\Psi_c)^{\alpha(\Psi_m)\ln(1 - \Psi)/\ln(1 - \Psi_m)} \qquad (32)$$

$$\alpha(\Psi) = 4.1\Psi + 0.22$$

$$\Psi_c = 0.7$$

Ψ_m is the minimal possible porosity of the medium (essentially, its microporosity). If the pore size distribution is known as well, the value of ϕ within this range can be uniquely determined. A comparison of the different expressions is shown in Fig. 3.

The agreement goes only as far as the trend of decreasing conductivity with increasing porosity. The low conductivity of porous comet-analog material was also demonstrated experimentally (e.g., *Benkhoff and Spohn*, 1991; *Kömle et al.*, 1996; *Seiferlin et al.*, 1996). It should be mentioned that in addition to the pores that result from the grainy structure of the ice-dust matrix, there are almost certainly additional micropores in the ice. These may be inherent to the structure of ice formed by slow deposition of water vapor at low temperature (*Kouchi et al.*, 1994), or may result from dissociation of clathrate-hydrates in the ice (*Blake et al.*, 1991). They should have little direct influence on the conductivity of the medium (unless they are correlated in space and form cracks), and will not affect the density appreciably, but they will allow the gas to flow through the medium more freely.

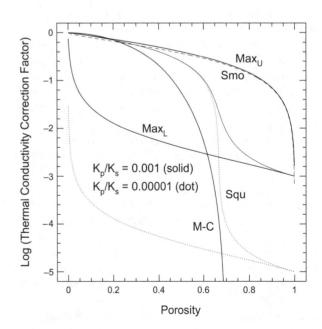

Fig. 3. Correction factor to the thermal conductivity resulting from porosity, as a function of porosity, for different models: Maxwell's upper (Max$_U$) and lower (Max$_L$) limits, Smoluchowski's relation (Smo), Squyres et al.'s relation (Squ), and the Monte Carlo fractal model (M-C, see text). When applicable, two ratios of the pore to solid conductivity are considered, as shown on the figure.

3.2. Amorphous–Crystalline Transition in H₂O Ice

Ordinary ice has a crystalline structure, but when water vapor condenses at low temperatures, the molecules do not have sufficient energy to take up the proper sites in the crystal, and an amorphous material is produced. The thermal conductivity of amorphous ice (see Table 3) is much lower than that of crystalline ice, and its temperature dependence has an opposite trend, $K \propto T$ rather than $K \propto T^{-1}$. Experiments by *Kouchi et al.* (1992) indicate that the conductivity may be as much as four orders of magnitude lower. These authors suggest that their low value may be the result of the very slow deposition procedure used, which creates a network of connected micropores. This network, which is actually a system of cracks, will leave "islands" of disconnected amorphous ice that will substantially lower the conductivity. A similar behavior is observed in amorphous semiconductors. Again, we see that porosity effects may significantly alter the conductivity of a material.

Amorphous ice is unstable and tends to spontaneously convert to crystalline ice. Measurements by *Schmitt et al.* (1989) have shown that the rate of crystallization is given by

$$\lambda(T) = 1.05 \times 10^{13} e^{-5370/T} \text{ s}^{-1} \tag{33}$$

Crystallization has a number of consequences. First, the conductivity through the medium will change with time. This will be due mainly to the intrinsic difference in conductivity between the two phases. In addition, however, the two phases differ in density, as amorphous ice is denser than crystalline ice by 2–7%. The precise difference depends on the rate of deposition [see *Jenniskens and Blake* (1994) and references therein], but in any case the phase change will subject the medium to stresses that may cause a change in porosity. This will further affect the thermal conductivity.

A second consequence comes from the fact that amorphous ice has the ability to trap volatiles. Extensive studies by Bar-Nun and co-workers [see *Bar-Nun and Owen* (1998) and references therein] have explored the dependence of the composition of the trapped gases, their relative abundances, and their rates of release on the temperature of deposition of the ice, and on the rate at which the medium is heated. In particular, they find that gas release accompanies the change in crystal structure. The amorphous-crystalline phase change should therefore lead to an increase in the activity of the nucleus.

Finally, the phase change releases latent heat, and this provides an internal heat source for the medium. The measured value is $\mathcal{H}_{ac} = 9 \times 10^4$ J kg^{-1} (*Klinger,* 1981). As the phase change is irreversible, this heat source is a sporadic one, and occurs only once in any given mass element of the comet.

3.3. Radioactive Heat Production

The most potent radioenergy source for comets is the short-lived radionuclide ^{26}Al, which gained renewed interest owing to the detection of interstellar 1.809 MeV γ-rays

from the decay of ^{26}Al (e.g., *Mahoney et al.,* 1984). Furthermore, *Srinivasan et al.* (1999) detected for the first time ^{26}Mg in a differentiated meteorite, and therefore could confirm the role of ^{26}Al in the differentiation of meteoritic parent bodies. Thus it is already widely used in thermal modeling of asteroids [see *Merk et al.* (2002) and references therein]. Indeed, for small silicate bodies such as asteroids, accretion models predict growth times in the range between 10^4 yr and 1 m.y., depending on whether conditions allow for runaway accretion or not. Hence the comparably short lifetime of ^{26}Al and the growth time of asteroidal bodies are compatible. Comets, on the other hand, formed much farther from the Sun and hence took longer to accrete. But even in this case formation times may have been sufficiently short to allow for live ^{26}Al. For example, *Weidenschilling* (1997) showed that it is possible to grow large icy bodies of a radius of about 40 km within 2.5×10^5 yr in the region at 30 AU. All the observational evidence points toward an interstellar isotopic ratio of ^{26}Al/^{27}Al $\approx 5 \times 10^{-5}$, implying an initial mass fraction $X_0(^{26}\text{Al}) \approx 7 \times 10^{-7}$ in dust (rock) and presumably an order of magnitude less in objects whose time of aggregation did not exceed a few million years.

The rate of heating by a radioactive isotope of relative abundance $X_{rad,0}$ within the dust is given by

$$\dot{Q}_{rad} = \rho_d X_{rad,0} \mathcal{H}_{rad} \tau_{rad}^{-1} e^{-t/\tau_{rad}} \tag{34}$$

where $\mathcal{H}_{rad}$ is the energy released per unit mass upon decay and τ_{rad} is the decay time constant. The total contribution is obtained by summing over the different species. The major long-lived sources of radioactive heating — ^{40}K, ^{232}Th, ^{235}U, and ^{238}U — together provide some 3×10^{-11} J kg^{-1} s^{-1}. The short-lived ^{26}Al, assuming an initial mass fraction $X_0 \sim 5 \times 10^{-8}$, could have provided as much as 2×10^{-8} J kg^{-1} s^{-1}.

3.4. Sublimation and the Surface/Volume Ratio

The porous structure of the comet nucleus allows for an internal process that is otherwise confined to the cometary surface: sublimation of ice from the pore walls or condensation onto them. The rate of sublimation — mass per unit volume of cometary material per unit time — is given by

$$q_\alpha = S(\Psi, r_p) \left[(\mathcal{P}_\alpha(T) - P_\alpha) \sqrt{\frac{\mu_\alpha}{2\pi \mathcal{R}_g T}} \right] \tag{35}$$

where the term in square brackets represents the sublimation rate per unit surface area. Thus the property of the porous structure that affects sublimation is the surface to volume ratio S, defined as the total interstitial surface area of the pores A_p per given bulk volume V

$$S = A_p / V \tag{36}$$

Its evaluation requires some model of the porous configuration. As a simple example, consider the specific surface of a porous material made of identical spheres of radius r_s

in a cubical packing. In this case $A_s = 4\pi r_s^2 N$, where N is the number of spheres in the given volume V. Obviously, V = $(2r_s)^3 N$, which yields $S = \pi/2r_s$. If the solid spheres are replaced by spherical pores embedded in a solid matrix, the result is the same. Clearly, fine materials have a much greater specific surface than coarse materials. Consider now a more realistic case of a granular medium of spherical grains of n different sizes, so that the number of grains of radius r_i ($1 \leq i \leq n$) is N_i. The total area and volume of these spheres are

$$A_s = \sum_{i=1}^{n} 4\pi r_i^2 N_i = A_p$$

$$V_s = \sum_{i=1}^{n} \frac{4}{3}\pi r_i^3 N_i \tag{37}$$

respectively. By definition, $V = V_s/(1 - \Psi)$, whence

$$S = 3(1 - \Psi)\sum_{i=1}^{n} f_i/r_i = 3(1 - \Psi)/r_p \tag{38}$$

where r_p is the harmonic mean radius weighted by f_i, the volume fraction occupied by spheres of radius r_i. As before, pores and grains may be interchanged.

Another case, often used in comet nucleus modeling, is that of a bundle of cylindrical tortuous capillary tubes that do not cross each other (*Mekler et al.*, 1990). The tortuosity is defined as the ratio of the capillary length to the sampled thickness. For a given length L and unit cross-sectional area we have

$$A_p = \sum_{i=1}^{n} 2\pi r_i N_i \xi L$$

$$V = 1 \cdot L \tag{39}$$

where r_i is the capillary radius ($1 \leq i \leq n$), N_i is the number of capillaries of radius r_i crossing a unit area, and ξ is the typical capillary tortuosity. Thus

$$S = \sum_{i=1}^{n} 2\pi r_i N_i \xi \tag{40}$$

On the other hand,

$$\Psi = \sum_{i=1}^{n} \pi r_i^2 N_i \xi \tag{41}$$

which leads to

$$S = 2\Psi \sum_{i=1}^{n} f_i/r_i = 2\Psi/r_p \tag{42}$$

where r_p is the harmonic mean radius weighted by f_i, the volume fraction occupied by capillaries of radius r_i.

We note that the two models behave differently with changing porosity: As Ψ decreases, the surface to volume ratio of capillaries tends to zero, while that of spheres increases to a maximum. Toward high porosities, on the other hand, the surface to volume ratio of spheres tends to zero. It is, however, difficult to visualize either a low-porosity medium made up of a bundle of individual capillaries, or a high-porosity one made up of widely separated spheres. Therefore, in numerical modeling that allows for a changing porosity — due, for example, to vigorous sublimation or condensation — it would be advisable to change from one model to the other as the porosity changes. The models yield equal values of S for $\Psi = 0.6$. Among other relations that have been suggested in the literature, we find (*Kaponen et al.*, 1997)

$$S \propto -\Psi \ln \Psi \tag{43}$$

Beside the major ice component H_2O, comet models usually include several other components of higher volatility (CO, CH_4, CO_2, CH_3OH, HCN, NH_3, H_2S, C_2H_2, C_2H_6, C_3H_4). For each species, an empirical formula for the saturated vapor pressure is used, of the form

$$P = Ae^{-B/T} \tag{44}$$

The coefficients for A and B for different ices are given in Table 4. The coefficients for water and carbon monoxide may be found in *Fanale and Salvail* (1984). The other parameters are extrapolated from fits to data found in the *Handbook of Chemistry and Physics* (*Lide*, 2003).

The latent heat of sublimation $\mathcal{H}$ is calculated from the Clausius-Clapeyron equation

$$\frac{1}{P}\frac{\partial P}{\partial T} = \frac{\mu \mathcal{H}}{T^2 \mathcal{R}_g} \tag{45}$$

Using the empirical equation (44), we obtain $B = \mu\mathcal{H}/\mathcal{R}_g$, implying constant values for $\mathcal{H}$.

TABLE 4.   List of coefficients for the pressure equation.

Ice Component	Symbol	A Value (10^{10} Nm^{-2})	B Value (Kelvin)
Water	H_2O	356.	6141.667
Carbon monoxide	CO	0.12631	764.16
Carbon dioxide	CO_2	107.9	3148
Methane	CH_4	0.597	1190.2
Propyne	C_3H_4	3.417	3000
Propadine	C_3H_4	2.382	2758
Ethane	C_2H_6	0.459	1938
Methanol	CH_3OH	8.883	4632
Hydrogen cyanide	HCN	3.8665	4024.66
Hydrogen sulphide	H_2S	1.2631	2648.42
Ammonia	NH_3	61.412	3603.6
Acetylene	C_2H_2	9.831	2613.6

3.5. Gas Flow and Permeability

The gas released in the interior of the nucleus, either by sublimation from the pore walls or as a result of crystallization of amorphous ice, will diffuse through the pores. Flow through a porous medium will depend on the properties of the medium and the properties of the flowing material itself. A simple formulation for fluid flow through a porous medium was derived experimentally by Darcy as early as 1856 and has become known as Darcy's law

$$\mathbf{J} \propto \frac{1}{\nu} \nabla P \qquad (46)$$

where ν is the kinematic viscosity of the fluid; in the case of an ideal gas

$$\nu \approx \frac{1}{3} \ell v_{th} \qquad (47)$$

and

$$v_{th} = \sqrt{\frac{8kT}{\pi m}} \qquad (48)$$

The simple law (equation (46)) is, however, only approximately correct.

The flow regime for a gas in a porous medium is determined by the Knudsen number defined as the ratio of the mean free path of a gas molecule to the pore diameter

$$Kn = \frac{\ell}{2r_p} \qquad (49)$$

If n is the gas number density and σ_α the kinetic cross section of a gas molecule, then

$$\ell = \frac{1}{\sigma_\alpha n} = \frac{kT}{\sigma_\alpha P} \qquad (50)$$

(Note that the cross section is usually defined as πd^2, where d is the molecular diameter; it may also include a factor $\sqrt{2}$ if velocity effects are taken into account.) The highest ice temperature attained in comet nuclei is of the order of 200 K; substituting in equation (50) $\sigma_{H_2O} \approx 2.5 \times 10^{-15}$ cm^2 and $P \approx \mathscr{P}_{H_2O}$ (200 K), we obtain $\ell \approx 4$ cm. Hence so long as the average pore size is less than 1 mm, we have Kn $\gg$ 1, meaning that the flow of gas through the pores is a free molecular, or Knudsen flow, where collisions of the gas particles with the pore walls are much more frequent than collisions between particles. In this regime, the amount of mass passing through a cylindrical tube per unit time is given by

$$\mathbf{j} = -\frac{8r_p^3}{3\xi} \sqrt{\frac{\pi m}{2k}} \nabla(P/\sqrt{T}) \qquad (51)$$

(see *Mekler et al.*, 1990). Adopting again the model of a bundle of capillaries, we obtain for the mass flux

$$\mathbf{J}_{Kn} = -\frac{8}{3} \Phi \left(\frac{m}{2\pi k} \right)^{1/2} \nabla(P/\sqrt{T}) \qquad (52)$$

where Φ, defined as the permeability of the medium, is obtained by using equation (41) and the same weighting function as in the calculation of S for the same model (equation (42))

$$\Phi = \Psi r_p/\xi^2 \qquad (53)$$

Another common relation between Φ and Ψ, known as the Kozeny law, is of the general form

$$\Phi \propto \Psi^3/(\xi^2 S^2) \qquad (54)$$

For a relatively low porosity, near the percolation limit Ψ_c, below which there is no continuous flow through the medium, the relation is of the form

$$\Phi \propto (\Psi - \Psi_c)^\mu$$
$$\mu = 2.8 \qquad (55)$$

However, when the pore size is increased, the condition Kn > 1 may no longer apply. The flow becomes a continuum (Poiseuille), or viscous flow, dominated by collisions between particles

$$\mathbf{J}_{Po} = -\frac{3}{16} \frac{\Psi r_p^2 \sigma_\alpha}{\xi^2} \left(\frac{m\pi}{2k^3} \right)^{1/2} \frac{P}{T^{3/2}} \nabla P \qquad (56)$$

For the intermediate regime, Kn $\sim$ 1, semiempirical interpolation formulae are commonly used, of the general form $\mathbf{J} = a_1 \mathbf{J}_{Kn} + a_2 \mathbf{J}_{Po}$, known as the Adzumi equation, with fixed (empirically determined) coefficients a_1 and a_2. Each one of the flow equations is suitable for a set of given conditions. But when Kn is neither uniform nor constant, as in the case of an evolving comet, the above formulae do not ensure a smooth transition between the two flow regimes as the Knudsen number changes from Kn $\gg$ 1 to Kn $\ll$ 1. A more suitable interpolation may be obtained by noting that if the temperature is uniform, equation (52) reduces to an expression similar in form to equation (56). This suggests an interpolation similar to the Adzumi equation

$$\mathbf{J} = \left(1 + \frac{9\pi}{256} \frac{1}{Kn} \right) \mathbf{J}_{Kn} \qquad (57)$$

which varies continuously from $\mathbf{J}_{Kn}$ for Kn $\gg$ 1 to $\mathbf{J}_{Po}$ for Kn $\ll$ 1.

When two gases (e.g., H$_2$O vapor and CO) are flowing through the same medium, they are treated independently, namely each flux is computed according to equation (57).

Since the molecular flow is driven by the partial pressure, whereas the viscous flow is driven by the total pressure, this is strictly correct in the Knudsen regime, and also in the case of immiscible flows (see *Bouziani and Fanale*, 1998). Fortunately, the flux of initially trapped gas is overwhelmingly dominant in the interior of the comet, while the H_2O flux becomes dominant in a very thin outer layer of the nucleus (a few centimeters), where most of the sublimation occurs. The interaction between gases is therefore restricted to this layer. At high flow rates turbulence may arise, and equation (57) may no longer hold. This is indicated by the Reynolds number exceeding a critical value of order 1000. When the Reynolds number is routinely evaluated during evolutionary calculations, it is always found to remain smaller than 100, and therefore equation (57) may be safely applied.

3.6.   Drag on Dust Grains

As the ice sublimates from the porous nucleus matrix, dust grains are released into the gas stream. Transport of dust must therefore be considered in addition to gas flow, a problem that is only now beginning to be studied.

We have seen that the flow of gas through porous comet nuclei is typically a free molecular (Knudsen) flow. The drag force on a dust grain of radius r_d in the Knudsen regime is

$$F_{drag} \approx 2\pi r_d^2 \rho v_{th}(v_g - v_d) \qquad (58)$$

up to a numerical factor of order unity (*Öpik*, 1958), where v_g is the gas velocity and v_d is the dust grain velocity. Combining equations (58) and (48) and dividing by the mass of the (spherical) dust grain, we obtain the grain's acceleration

$$\frac{dv_d}{dt} = \frac{1}{\tau(r_d)}(v_g - v_d) \qquad (59)$$

where ρ_d is the density of the grain and the characteristic time τ, a function of the dust grain radius for given flow conditions, is

$$\tau \approx \frac{\rho_d}{\rho}\sqrt{\frac{m}{kT}}\,r_d \qquad (60)$$

The dust grain velocity (assuming a constant gas velocity) is thus

$$v_d(t) = v_g(1 - e^{-t/\tau}) \qquad (61)$$

For conditions that are typical of cometary interiors (a few meters to a few tens of meters below the surface), where crystallization of amorphous ice takes place and trapped gas is released, or where volatile species (such as CO) sublimate, we find $\tau \approx 0.5(r_d/1\ \mu m)$ s, so that even 10-μm particles can reach 90% of the gas velocity in about 10 s. For gas velocities typical of such conditions, the particle will have traveled during that time interval a distance of much less than 1 m. This length scale is considerably smaller than the typical length scale over which conditions change in the interior of the nucleus. Near the surface, where the main driving force is provided by water vapor sublimating from the pore walls, conditions are even more favorable.

So far we have neglected the effect of gravity. A gravitational acceleration g would change the solution in equation (61) for the velocity to

$$v_d(t) = (v_g - g\tau)(1 - e^{-t/\tau}) \qquad (62)$$

Hence the effect is negligible so long as $g\tau \ll v$. For a constant nucleus density ρ, and at depths that are much smaller than R, we have $g = (4\pi/3)G\rho R$. For parameters characteristic of cometary interiors (resulting in the above estimate of τ) this condition is amply satisfied. It will break down for very large dust grains (see discussion of critical radius in the next section), but the flow of such grains would in any case be prevented by the small pore size.

In conclusion, to a good approximation, the dust grain can be taken to move at the same speed as the gas. The difference is that while the gas can move fairly freely through the nucleus by diffusing through pores (or, if need be, through micropores), a dust grain can only move through those pores that are large enough to accommodate it. Several models have been suggested to treat this problem. *Podolak and Prialnik* (1996) proposed that the dust motion be treated as a random walk. *Shoshani et al.* (1997) treat the porous medium as a sequence of filters, each with a size distribution of holes. The pores are viewed as cylinders extending from these holes. The distance between filters is the average distance one would travel in a cylindrical pore before the radius of the pore changed significantly. Assuming a given dust grain speed and given size distributions for grains and for pores, they follow the change in dust size distribution as the dust migrates through the nucleus. They also show how the trapping of dust grains affects the porosity and permeability of the medium.

More recently, *Shoshany et al.* (1999) used Monte Carlo calculations to explore the behavior of dust migration in a medium with randomly distributed pores. They found that the effective speed of dust particles is lower than that found by the random walk model for all porosities, although the difference decreases for $\Psi \to 0$ or $\Psi \to 1$ as expected. They also found that only the smallest dust grains (of order of the pore size) traversed the medium for any distance. Larger grains could not find sufficiently many large pores to travel freely, and they got trapped after moving only a short distance. Large grains that are observed in a comet coma were most likely lifted off directly from the surface. Smaller grains may have a component originating deeper in the nucleus. The exponent obtained from fitting a power law to the observed grain size distribution in the coma may therefore not accurately reflect the grain size distribution within the nucleus itself.

Recent studies (*Skorov and Rickman*, 1995, 1998, 1999) have begun to focus on the details of the interaction of the gas flow with the individual dust grains. These Monte Carlo computations follow the flow of gas molecules in the Knudsen regime as they leave the surface of the nucleus and interact with the dust grains above it. This work studies the gas kinetic flow as a function of capillary length, inclination angle, and temperature gradient along the pores at the surface of the nucleus. It also follows the velocity distribution of the gas molecules, and how it is affected by interaction with the dust grains. Like all Monte Carlo computations, this program is advancing slowly, but is beginning to produce important insight into the gas-dust interaction.

How do these studies contribute to the numerical simulation of a comet nucleus evolution? As the effective rate of flow of dust particles clearly depends on the particle size, we may assume in numerical calculations the dust grain radii to be distributed over a discrete range $r_{d,1}$, $r_{d,2}$, ... $r_{d,N}$, according to some distribution function $\psi(r_d)$ (such as a power law). The size of a dust particle may be assumed to remain unchanged, thus ignoring possible breakup or coalescence of dust grains. Hence particles in each size category may be treated as independent species. The local flux of dust particles of radius $r_{d,n}$ is therefore given by

$$J_{d,n} = \beta_n \rho_{d,n} v_g$$

It is the coefficient β_n that must be determined by the dust flow model (cf. *Horanyi et al.*, 1984). For example, *Podolak and Prialnik* (1996) adopt

$$\beta_n \propto \log[1 - \psi(r_{d,n})]/\log[(1 - \psi(r_{d,n}))\psi(r_{d,n})]$$

The mass conservation equation for these particles is

$$\frac{\partial \rho_{d,n}}{\partial t} + \nabla \cdot \mathbf{J}_{d,n} = 0 \qquad (63)$$

We note that there is no source term in equation (63); the implicit assumption in this simple approximation is that any dust grain that can be dragged (allowance being made for the critical radius and the local average pore radius) is dragged with the gas.

The results expected from any dust flow model are the grain size distribution of the ejected dust and the changing pore structure of the medium through which the dust flows. For example, the large dust grains left behind on the nucleus surface form a dust mantle, which, in turn, affects the rate of heat and gas flow at the surface.

3.7. Dust Mantle Formation

The eventual formation of a dust mantle on the surface of a comet nucleus may be modeled in different ways, the essential parameter being the critical dust grain radius, which represents the radius of the largest particle that can leave the comet, as determined by the balance of forces acting on a dust grain. The drag force exerted by the gas flux at the surface is

$$F_{drag} = \tfrac{1}{2} C_D \pi r_d^2 \sum_\alpha v_\alpha J_\alpha \approx \pi r_d^2 \sum_\alpha v_\alpha J_\alpha \qquad (64)$$

summing over the different species that contribute to the

overall gas flux and adopting, in the lowest approximation, a drag coefficient $C_D \approx 2$. When sublimation at the surface is the dominant component, then equation (64) reduces to $\pi r_d^2 \mathcal{P}(T)$. The effective gravitational force, diminished by the centrifugal force, is

$$F_{grav} = \frac{4\pi}{3} r_d^3 \rho_d g \left(1 - \frac{3\pi \cos^2 \theta}{G\rho P_{spin}^2} \right) \qquad (65)$$

with $g = 4\pi G\rho R/3$, but the correction term is small so long as $P_{spin} < 3$ h. Thus roughly

$$r_d^* \approx \frac{\mathcal{P}(T)}{G\rho\rho_d R} \qquad (66)$$

The problem of dust mantle formation was first studied by *Brin and Mendis* (1979), who related the mantle thickness D at a particular point in the orbit to its thickness at an earlier point by

$$D[d_H(t)] = D[d_H(t - \delta t)](1 - \Theta) +$$
$$(1 - \Omega)X_d \frac{Q}{\rho_d} \delta t \qquad (67)$$

where Ω is the fraction of dust released from ice and carried away by the sublimating gas, and Θ is the additional part of the mantle removed by the increased gas flux. For a grain size distribution $\psi(r_d)$ within a range $[r_d^{min}, r_d^{max}]$ and a critical grain size $r_d^*(t)$, which changes along the orbit, the functions Ω and Θ can be calculated and thus the development and evolution of this dust mantle can be followed. The rate of growth of the mantle is determined by the parameter $(1 - \eta)$, where

$$\eta = \frac{\int_{r_d^{min}}^{r_d^*} \psi(r_d) r_d^3 dr_d}{\int_{r_d^{min}}^{r_d^{max}} \psi(r_d) r_d^3 dr_d}$$

This simple approach deals, however, only with the mass of the mantle, regardless of its structure.

One approach to modeling the structure of the dust mantle is to assume that the ice sublimates freely at the nucleus surface, carrying with it the smaller (than critical size) dust grains, while the larger grains are left behind. At the beginning, these large dust grains are isolated from each other, but as more and more grains accumulate, the surface becomes evenly covered and starts interfering with the escape of smaller and smaller grains. The porosity of the dust mantle decreases and eventually drops below that of the nucleus. This idea of trapping and compaction, introduced by *Shul'man* (1972), was adopted by *Rickman et al.* (1990) in modeling the dust mantle. They showed that if the grain size distribution follows a power law with an index of about –3.5, the smallest grains left behind contribute the most to forming the dust mantle. The gas flow through such a mantle can be modeled by considering gas diffusion through this porous medium. If the gas pressure is high enough, the dust mantle can be blown off, and the process will start anew. The process depends both on latitude and on the inclination of the rotation axis. A dust mantle will inhibit gas sublimation when most of the surface, close to 100%, is covered by grains (e.g., *Prialnik and Bar-Nun,* 1988), a result that was confirmed by the KOSI experiments (*Grün et al.,* 1993).

Eventually, the pore size of the dust mantle may become too small to allow particles to escape and a large amount of small grains will become permanently trapped. This may lead to a very stable and efficient dust crust with a high cohesive strength, which may surpass the vapor pressure building up in the porous material underneath the mantle (*Kührt and Keller,* 1994). As a consequence, the gas is driven toward the interior and refreezes, forming an ice layer of increased density (*Prialnik and Mekler,* 1991). This effect was actually observed in the KOSI comet simulation experiments (*Spohn et al.,* 1989).

3.8. Tensile Strength and Fracture of the Nucleus

The material strength of comet nuclei is very low. Although the range of values resulting from different estimates is wide, all values indicate a weak material. The strength deduced from tidal breakup of Sun-grazing comets is 10^2–10^4 Pa (*Sekanina,* 1984; *Klinger et al.,* 1989). Laboratory experiments lend further support to the low strength estimates derived from observations: The typical strengths of the ice crusts measured in the KOSI experiments were in the 10^5-Pa range (*Kochan et al.,* 1989).

A simple model for the strength of a medium composed of spherical grains was developed by *Greenberg et al.* (1995). Taking the nominal dipole-dipole interaction to be ~10^{-2} eV, they obtain for the tensile strength

$$\mathcal{T} = 6.1 \times 10^2 (1 - \Psi) \beta \left(\frac{r_d}{0.1 \, \mu m} \right)^{-2} Pa \qquad (68)$$

where $1 \leq \beta \leq 12$ is the number of contact points per particle. The strength and Young's modulus for a medium composed of ice grains linked into chains by intermolecular forces is computed by *Sirono and Greenberg* (2000). They show that these forces are strong enough to hold an assemblage of grains together even when its self-gravity will not. They derive 3×10^2 Pa for the tensile and 6×10^3 Pa for the compressive strength, when applied to the tidally split Comet Shoemaker-Levy 9.

Rotational stability against breakup of fast-rotating comets provide an independent means of estimating the strength of cometary material. Computations of this kind have been presented by Davidsson for solid spheres (*Davidsson,* 1999) and for biaxial ellipsoids (*Davidsson,* 2001). *Cook* (1989) considers a medium composed of bulk material permeated by cracks. The energy of such a medium will be a sum of the elastic strain energy in the bulk material and the surface energy along the crack. The crack will spread if the total

energy of the material is decreased thereby. Cook applies this picture to a fractal material composed of successive generations of spherical aggregates. This picture may be useful for modeling the strength of comet nuclei.

In a weak porous medium, thermal stresses (*Kührt,* 1984) and internal pressure exerted by gas that accumulates in the pores may break the fragile solid matrix. For example, a model of a typical comet given in *Prialnik et al.* (1993) yields internal water vapor pressures exceeding 2×10^5 Pa. This is comparable to the above strength estimates for cometary ice. If the stress, σ_m, on a material exceeds its tensile strength $\mathcal{T}$, then that material will undergo tensile fracture. If the stress is negative (compression) and exceeds the compressive strength C, the material will undergo shear fracture. In a spherical shell of radius r, the tangential stress is given by

$$\frac{\sigma_m}{r} = -\frac{1}{2} \frac{dP}{dr} \qquad (69)$$

(*Morley,* 1954), so that fracture should occur when

$$-\frac{dP}{dr} > \frac{2\mathcal{T}}{r} \qquad (70)$$

In general $C \gg \mathcal{T}$, so that only tensile fracture needs to be considered. *Prialnik et al.* (1993) present a simple algorithm for dealing with this effect. They assume that the porosity of the medium remains unchanged, but the average pore radius increases as a result of fracture. When the local pressure gradient is high enough so that condition (70) is satisfied (for example, due to crystallization and release of trapped gas), the local average radius of the pores is increased by a factor proportional to $(r/2\mathcal{T})dP/dr$. Then a relaxation time is allowed so that the gas can flow through the enlarged pores, after which condition (70) is tested again. The energy of deformation of the matrix is small compared with the energy released by the amorphous-crystalline transition, and may be neglected. This procedure can also allow for the effect of strain hardening, whereby the strength of a material is increased due to deformation. In this case one can allow the tensile strength of the material to increase along with the increasing pore size.

In summary, as a result of fracture, the size distribution of pores will vary throughout the nucleus. In addition, sintering (*Kossacki et al.,* 1999) or pore blocking by small dust grains may alter pore sizes as well as consolidate the material. Massive recondensation of volatiles on pore walls has a similar effect. A subject for future work is to incorporate additional processes such as grain growth by sintering and the elimination of pores by densification. An excellent review of the relevant processes is given by *Eluszkiewicz et al.* (1998). We are again faced with a complex internal process for which we must account based on very little information. Moreover, the pore size distribution measured at the surface of a comet nucleus need not represent the distribution in the deeper layers. However, extensive studies of crater formation in ice targets (*Arakawa et al., 2000*) have shown that the crater depth scales as the square root of the impact energy. More importantly, the crater pattern depends not only on the energy of the impactor, but also on the details of the layering of the ice. Thus future measurements of craters on comet nuclei may yield clues to their underlying structure.

3.9. Characteristic Timescales

Each process mentioned in this section has its own characteristic timescale and the competition between these timescales will determine to a large extent the evolutionary pattern of the comet nucleus that will be discussed in the next section:

1. The thermal timescale, obtained from the energy conservation equation (10), which in its simplest form (without sources and advection) is a heat-diffusion equation. Distinction must be made between the thermal timescale of amorphous ice τ_{a-ice}, crystalline ice τ_{c-ice}, and dust τ_{dust}. For a layer of thickness Δr and average temperature T we have

$$\tau_{a-ice} = (\Delta r)^2 \rho_a c(T)/K_a \qquad (71)$$

and similar expressions for τ_{c-ice} and τ_{dust}. As a rule, $\tau_{c-ice} < \tau_{a-ice} < \tau_{dust}$; as we have seen in section 3.1, porosity increases all these timescales considerably.

2. The timescale of gas diffusion τ_{gas}, which is also the timescale of pressure release, obtained from the mass conservation equation, which (without sources) can be regarded as a diffusion-type equation for the release of gas pressure:

$$\tau_{gas} = \frac{3}{4} \frac{(\Delta r)^2}{\Psi r_p} \left(\frac{2\pi m}{kT} \right)^{1/2} \qquad (72)$$

3. The timescale of crystallization τ_{ac}, which is also the timescale of gas-release and pressure buildup:

$$\tau_{ac} = \lambda(t)^{-1} = 9.54 \times 10^{-14} e^{5370/T} \text{ s} \qquad (73)$$

4. The timescales of sublimation of the different volatiles, τ_{subl-H_2O} for water, $\tau_{subl-CO}$ for CO and $\tau_{subl-CO_2}$ for CO_2, and so forth:

$$\tau_{subl-H_2O} = \frac{\rho_c}{S \mathcal{P}_{H_2O} \sqrt{m_{H_2O}/2\pi kT}} \qquad (74)$$

and similar expressions for $\tau_{subl-CO}$, $\tau_{subl-CO_2}$, and so forth.

5. The insolation timescale $\tau_\odot$, which concerns the skin of the comet nucleus that is heated by absorption of solar radiation.

$$\tau_\odot = \sqrt{(\kappa_c P_{spin}/\pi\rho_c c)} \rho_c cT \frac{4\pi d_H^2}{L_\odot(1 - \mathcal{A})} \qquad (75)$$

To these, the constant characteristic times of decay of the

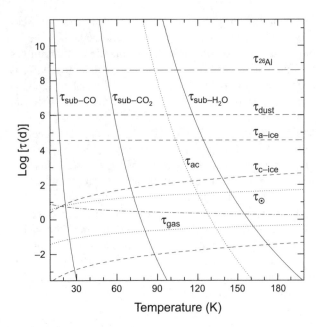

Fig. 4. Timescales of different evolutionary processes (see text) as a function of temperature. The solar timescale is given for distances of 1 and 10 AU (dash-dot lines); the thermal timescale for ice is shown for two depths (10 cm and 10 m).

radioactive species may be added; the only relevant one being that of ^{26}Al. A comparison of these timescales is shown in Fig. 4, where they are plotted against temperature, assuming an average pore size of 10 μm and a porosity of 0.5, and considering heliocentric distances of 1 and 10 AU and depths of 10 cm and 10 m (since the diffusion timescales for heat and gas depend on depth). These timescales will be helpful for understanding the results of numerical calculations presented in the next section.

4. MODELING RESULTS AND INTERPRETATION: STRUCTURE AND EVOLUTION OF COMET NUCLEI

4.1. Input Parameters

In order to follow the evolution of a comet nucleus structure by means of the equations displayed in section 2, including the physical processes described in section 3, we still need to specify a number of different types of parameters:

1. *Defining parameters.* These identify a comet and include orbital parameters (semimajor axis and eccentricity), the nucleus size, given by an average radius, its mass (or bulk density), and its spin period, all of which may be determined observationally.

2. *Initial parameters.* These are required for the solution of the time-dependent equations and must be guessed. As these equations are not expected to reach equilibrium (steady-state), the initial conditions are not likely to be "forgotten." They include the initial temperature, or tempera-

ture profile; compositional parameters (mass fractions of the different volatiles and dust); and structural parameters, such as porosity, average pore size or pore size distribution, as well as the nature of the water ice, whether crystalline or amorphous. In fact, one of the main goals of modeling is to determine these properties by comparing their observable consequences to actual observations (see *Prialnik, 2002*). Regardless of details, initial homogeneity is adopted as a rule.

3. *Physical parameters.* These are parameters related to the various physical processes discussed in section 3. They supplement the parameters in the former group and, in principle, should have been determined by them, had the detailed structure of comet nuclei been better understood. Among them are albedo and emissivity, associated with surface properties, as well as thermal conductivity coefficients, tensile strength, and so forth, associated with bulk material properties. These parameters can be determined by laboratory experiments, although caution must be exercised, since laboratory samples are scaled down by orders of magnitude. *In situ* measurements should eventually provide much more reliable input data.

4.2. Analytical Considerations: Characteristic Properties of the Comet Nucleus

Adopting typical values for the physical properties of cometary material in general, simple estimates may be derived for the characteristics of the nucleus structure and evolution, in terms of the defining parameters just mentioned, as shown in Table 5. These provide insight into the nature of comet nuclei, as well as instructive guidelines for building numerical models of their structure and evolution. For example, the skin depth and thermal timescale help in defining an adequate numerical grid, or discretization, in space and time, while the mass loss rate and life span indicate what the size of such a grid should be. The temperature (calculated from equation (1) for the subsolar point) indicates what thermochemical processes are to be expected.

These, however, are only crude estimates. The detailed behavior of comet nuclei is obtained by applying the full set of equations given in section 2, and including complex input physics, as discussed in section 3. Since the evolution of nuclei, as already indicated by the crude estimates, is largely determined by their defining parameters, numerical models are mostly applied to individual comets, rather than to comets in general, and even then the results may diverge, as compositional and structural parameters vary. Detailed examples of the application of thermal evolution models to individual comets are given by *Meech and Svoreň* (2004). Typical characteristics of the structure and activity of comet nuclei are discussed in the rest of this section.

4.3. Surface and Internal Temperatures

In order to understand the surface temperatures of comet nuclei, we may use the power balance equation at the sur-

TABLE 5. Estimates of characteristic properties of comets.

Property	Dependence on Parameters	Value
Orbital skin depth	$\left(\dfrac{2Ka^{3/2}}{\mathcal{G}\rho c}\right)^{1/2}$	18 m
Diurnal skin depth	$\left(\dfrac{P_{spin}K}{\pi\rho c}\right)^{1/2}$	0.1 m
Thermal timescale	$\dfrac{R^2\rho c}{\pi^2 K}$	8×10^4 yr
Insolation per orbit	$\dfrac{\pi L_\odot}{2\mathcal{G}}\dfrac{R^2}{\sqrt{a(1-e^2)}}$	10^{18} J
Production rate (ph)	$\dfrac{\mathcal{L}}{4\mu}\left(\dfrac{R}{a(1-e)}\right)^2$	1.3×10^{30}/s
Erosion per orbit	$\dfrac{\mathcal{L}}{8\mathcal{G}}\dfrac{1}{\rho\sqrt{a(1-e^2)}}$	1.7 m
Max. temperature	$\dfrac{4\pi a^2(1-e)^2 Q}{\mathcal{L}} = 1$	205 K
Day-night range (ph) (see equation (79))	$\displaystyle\int_{T_n}^{T_d}\sqrt{\dfrac{\rho c K}{\pi P_{spin}}}\dfrac{dT}{Q\mathcal{H}} = \dfrac{1}{2}$	23 K
Life time	$\dfrac{8\mathcal{G}R\rho\sqrt{a(1-e^2)}}{\mathcal{L}}$	3×10^3 yr

Note that $\mathcal{G} = \sqrt{GM_\odot} = 1.152 \times 10^{10}$ m$^{3/2}$ s^{-1}; $\mathcal{L} = L_\odot/\mathcal{H}$; values listed in the last column were obtained using the following parameters: a = 10 AU, e = 0.9, resulting in a perihelion (ph) distance of 1 AU, R = 5 km, $\rho = 7 \times 10^2$ kg/m^3, P_{spin} = 10 h, K = 0.6 J/(m s K), c = 8×10^2 J/(kg K).

face, equation (15), with $-K\nabla T_R$ substituted for F(R) on the lefthand side

$$\frac{(1-\mathcal{A})L_\odot}{4\pi d_H^2}\cos\xi = \varepsilon\sigma T^4 + K\nabla T_R + \mathcal{F}QH \quad (76)$$

The incoming solar flux depends mainly on the Sun-comet distance, the rotational state, spin period, scattering properties of the coma (neglected here), and the reflectivity of the surface; thermal reradiation depends on the emissivity; heat transported in and out of the nucleus is a function of the thermal conductivity with all its related parameters; and the energy dissipated in sublimation depends on the composition. For illustration, we choose in the following discussion models of the small, fast-spinning, short-period Comet P46/Wirtanen, the former target of the European mission *Rosetta*.

To begin, we distinguish between two extreme cases. The first extreme is to assume that no ices or volatile materials are present at the surface, which then consists of a porous dust layer where gas sublimated in the interior of the nucleus can flow through. Thus the thermal conductivity and matrix structure are the key parameters influencing the surface temperature (T_s). The other extreme is a free sublimation regime, where it is assumed that only water ice is present on the surface. Here the energy used for sublimation of the gas dominates in the power balance equation, at least at heliocentric distances smaller than 2–3 AU. This leads to much lower surface temperatures, as shown in Fig. 5, where results for $T_s(d_H)$ at different "comet day" times, obtained from calculations for inactive surfaces (Figs. 5a,b) and for pure ice surfaces (Fig. 5c), are compared. In order to investigate the influence of the thermal conductivity of the mantle on the thermal evolution of the nucleus, we distinguish between a high and a low conductivity mantle. The high value $K = 0.1$ Wm^{-1} K^{-1} (top panel) is still about a factor of 10 less than the conductivity of typical solid minerals on Earth, so as to account for porosity and for the loose structure of cometary material. For the low conductivity case, a 100-fold smaller value is assumed.

The highest surface temperatures are obtained at noon at the subsolar point; nighttime temperatures are much lower. At perihelion, the difference between the highest temperature (370 K at noon) and the lowest one (about 180 K at night) is about 190 K, assuming a dust surface. At the intermediate distance of 3 AU, the maximum day temperature is about 210 K, while night temperatures are about 140 K. At aphelion the day-night variations are the smallest, about 30 K.

When a much lower thermal conductivity is assumed for the mantle, the difference between day and night temperatures increases significantly. At perihelion, the day-night variation is about 300 K ($T_{max} \approx 380$ K, $T_{min} \approx 80$ K) and at aphelion the difference is still quite large, 120 K ($T_{max} \approx 180$ K, $T_{min} \approx 60$ K). The reason is that in the low conductivity case the thermal reradiation power is almost the same as the solar input and the thermal conductivity power is almost null. In the high conductivity case, on the other hand, the power transported by heat conduction into the nucleus is about 10% of the incoming solar input. At night, stored internal energy is transported from the interior to the surface. This transport, too, is more efficient in case of a high matrix conductivity. Therefore, we obtain smaller temperature differences between day and night and also obtain lower maximum and higher minimum temperatures. The transition from inward to outward heat conduction is not exactly in tune with the insolation power. This transition may be observable: Tracers could be the flux of minor volatiles that depend on the accumulated energy. In the high-conductivity mantle models there is a delay due to the thermal inertia of the material. This delay is vanishingly small in the case of low conductivity. An effect similar to the day-night effect and the associated thermal lag also appears on the much longer timescale of orbital evolution, where pre-

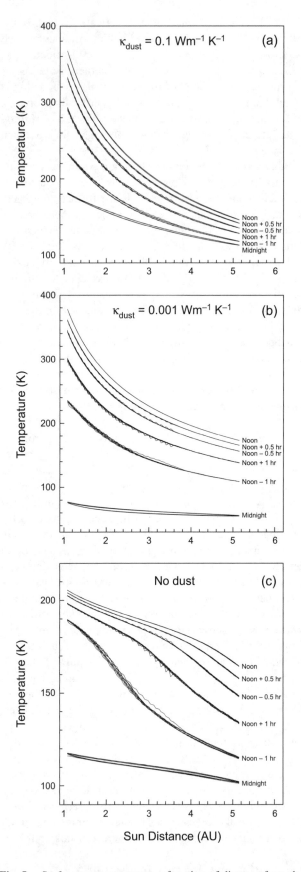

Fig. 5. Surface temperatures as a function of distance from the Sun for different "comet day" times. A comparison is shown between **(a),(b)** inactive surfaces and **(c)** pure ice surfaces. Results from model calculations for Comet P46/Wirtanen.

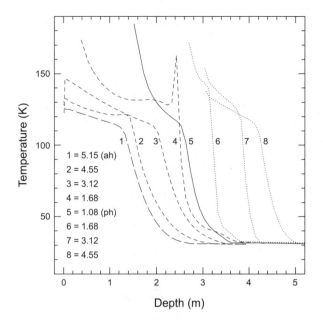

Fig. 6. Temperature profiles in the upper layer of a nucleus model in the orbit of P/Wirtanen at several points along the orbit, pre- and post-perihelion, as marked. The typical steep rise in temperature in curve 4 is due to heat released in crystallization, which proceeds at a fast rate at that point. We note the shift of the surface due to erosion.

perihelion temperatures are typically lower than post-perihelion ones at the same heliocentric distance.

For the free sublimation regime — active areas with no dust mantle on the surface — the results will change significantly. Close to the Sun, at noon, the highest T_s is about 203 K, nearly the free sublimation temperature of water at about 1 AU. The lowest T_s at night is about 115 K, which yields day-night variations of about 90 K at perihelion. At about 3 AU, a maximum of about 190 K and night minimum of ~105 K can be expected. At aphelion, the day-night variations are again the smallest and amount to about 65 K ($T_{max} \approx 165$ K, $T_{min} \approx 100$ K). The large difference between an active and an inactive surface is that in the former most of the insolation power is consumed by water sublimation. Surface temperatures are thus limited by the free sublimation temperature of water ice, a strong energy sink that does not allow temperatures to increase any further. Lower temperatures also reduce the thermal reradiation power, which depends on the temperature to the fourth power.

The heat transported inward serves in part to increase the internal energy of the nucleus, and in part is absorbed in sublimation of volatiles. Heating of the subsurface layers during one revolution is illustrated in Fig. 6 for an amorphous ice nucleus model. The affected region is barely a few meters deep; the layer of temperature inversion at large d_H is barely a few centimeters thick. The change in slope of the profile occurs at the boundary between the outer crystalline layer, which is an efficient heat conductor leading to a mild temperature variation with depth, and the

inner amorphous part, where conductivity is poor (see section 3.2) and the temperature profile steep.

The diurnal temperature variation may be generally understood and estimated by the following simple argument. The rate of cooling, or the cooling flux $F_{cool}(T)$, on the nightside is given by equation (15), setting the insolation rate to zero and assuming a pure ice composition

$$F_{cool}(T) = \varepsilon\sigma T^4 + Q(T)\mathcal{H} \tag{77}$$

This is balanced by the heat lost from an outer layer down to a depth equal to the skin depth corresponding to the spin period of the nucleus, $s = \sqrt{KP_{spin}/(\pi\rho c)}$. Thus, over a time interval dt, measured in units of the spin period, the temperature will change by an amount dT given by

$$F_{cool}(T)dt = -\rho scdT =$$
$$-\sqrt{\rho c(T)K(T)/(\pi P_{spin})}\, dT \tag{78}$$

which, integrated over half a spin period, yields

$$\int_{T_{min}}^{T_{max}} \frac{\sqrt{\rho c(T)K(T)/(\pi P_{spin})}}{F_{cool}(T)}\, dT = \frac{1}{2} \tag{79}$$

where $T_{max} - T_{min}$ is roughly the temperature difference between the subsolar and antisolar points. The day-night temperature difference as function of T_{max} is shown in Fig. 7 for three values of the Hertz factor.

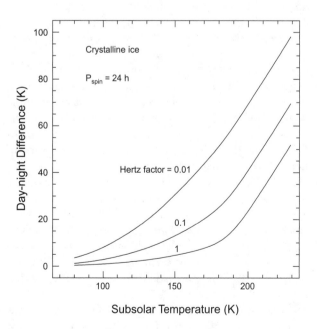

Fig. 7. Analytical estimate for the surface temperature difference between the subsolar and antisolar points for several values of the Hertz factor.

In terms of timescales (Fig. 4), we note that the timescale of solar heating at 1 AU intersects the sublimation timescales at temperatures ~25 K for CO, ~80 K for CO_2, and ~160 K for H_2O. When these temperatures are attained and if the corresponding ices are found near the surface, the solar energy will be absorbed in sublimation. Consequently, the surface temperature will rise much more slowly. We note that in all cases conduction to the interior is negligibly slow. A steady state will be reached at slightly higher temperatures, when the timescale of gas diffusion for a thin subsurface layer intersects the sublimation timescales: ~30 K for CO, ~100 K for CO_2, and ~200 K for H_2O. These are the expected surface temperatures of comets near 1 AU, when the corresponding ices are exposed. If a mixture of ices is present, the temperature will be determined by the most volatile among them.

4.4. Production and Depletion of Volatiles

The solar energy flux reaching the nucleus — after being partly scattered or absorbed by the coma — is to some small part reflected (low albedo of the nucleus) and in part reradiated in the IR (high emissivity). A small fraction is transported into the interior of the nucleus by conduction and to a very small degree by radiation in pores. The rest (the bulk at small distances from the Sun) is absorbed at the surface and used to evaporate (sublimate) water ice. The amount of water vapor released depends on the dust cover on the surface. A dust cover a few millimeters thick causes most of the incident energy to be reradiated, leaving only a small fraction for sublimation of water ice. Heat that is conducted into the interior of the porous nucleus may reach ices more volatile than water ice. In a comet nucleus, many different volatile species are expected to be present (see Table 4). If the ice is crystalline, then volatile ices are frozen out as separate phases. As heat diffuses inward, each volatile constituent forms its own sublimation front depending on its enthalpy of vaporization. If amorphous ice is present it will change to crystalline ice, forming an additional front, this time exothermic, for the phase transition. At this front, gases trapped by the amorphous ice will be released. As an ice species evaporates the gas pressure at the sublimation front increases toward its maximum (equilibrium) value at that temperature. The pressure forms a gradient that is negative in the outward direction and positive in the inward direction. This pressure gradient drives the gas flow. The gas flowing outward will diffuse through the comet nucleus and escape through its surface into the coma.

The gas flowing inward will recondense a short distance below the sublimation or crystallization front and release its latent heat. This is an additional heat transport mechanism into the interior, which surpasses advection by flowing gas (*Prialnik*, 1992; *Steiner and Kömle*, 1993). It was observed by *Benkhoff and Spohn* (1991) during the KOSI experiments on cometary ice analogs. Recondensation occurs within a thermal skin depth. The effect is illustrated in Fig. 8.

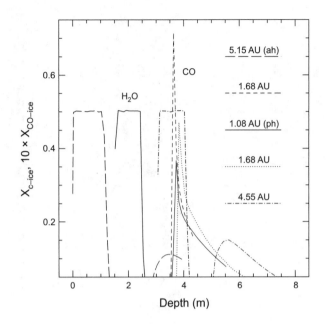

Fig. 8. Mass fraction of crystalline H_2O ice and of CO ice (multiplied by 10) in the upper layer of a comet model for the orbit of P/Wirtanen, at the subsolar point. The initial composition is $X_a = 0.5$, $f_{CO} = 0.05$, and $X_d = 0.5$. We note the advance of crystallization accompanied by freezing of the CO gas flowing inward in the cold regions below the crystallization front. The drop of X_c near the surface is due to sublimation. The model is the same as that of Fig. 6.

Because of heat and gas diffusion, the nucleus will be chemically differentiated in layers. The least-volatile material (dust) will be at the top of the nucleus. It will be followed by a layer of dust and water ice. In the deepest layers we would find dust and all ices, including the most volatile species (such as CO and CH_4). At the surface of the nucleus, water and other more volatile ices evaporate, leaving a layer of dust behind. Dust particles entrained by the gas into the coma will heat up in sunlight, and the organic component (hydrocarbon polycondensates) will be vaporized. Polymerized formaldehyde (POM) plays an important role in the dust in producing short-lived formaldehyde gas, which quickly dissociates into CO. The distributed coma source for the CO must be subtracted from the total CO release rates in order to obtain production rates resulting from gas released by the nucleus. A major goal of comet research is to determine conditions in the solar nebula based on the chemical composition of comet nuclei. However, the nucleus composition cannot be directly observed and must be deduced from the observed composition of the coma. Taking advantage of new observing technology and the early detection of the very active Comet Hale-Bopp (C/1995 O1), researchers were allowed for the first time to determine the coma abundance ratios of different species over a large range of heliocentric distances. The results supported the hypothesis that coma abundances do not reflect in a simple way the composition of the nucleus. Abundance ratios of different species may change by factors as large as several hundred, going from heliocentric distances of $r \approx$ 1 AU to $r \approx$ 7 AU. Thus chemical modeling of the coma coupled to gas-dynamic flow is first required (see *Crifo et al.*, 2004) in order to provide the true, distance-dependent composition of volatiles released from the nucleus. Nucleus models should then be tested against these results along the orbit for as large a range of distances as possible in order to deduce the composition of the nucleus. An example of this procedure, based on the work of *Huebner and Benkhoff* (1999), is given in Fig. 9, which shows the mixing ratio of CO relative to H_2O using cubic fits to the release rates obtained by combining observational data from Comet Hale-Bopp for H_2O (from OH) and CO covering the spectrum range from radio to UV. The heavy dashed curve is the result of model calculations for a mixture of 35% amorphous H_2O, 7% CO_2, 13% CO (50% trapped in the amorphous ice), and 45% dust. The CO_2 has very little influence on the results. The mixture is close to what one would expect from the condensable component of molecules forming at low temperatures from a mixture of elements with solar abundances. Although the result is below the fit, it must be kept in mind that the CO from distributed sources has not been subtracted from the observed CO flux.

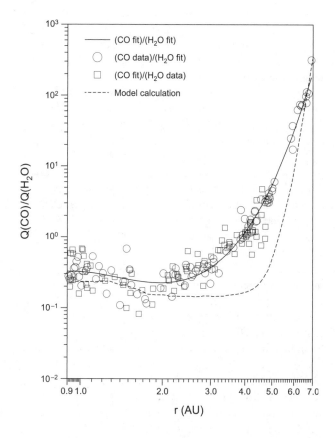

Fig. 9. The mixing ratio of CO relative to H_2O using cubic fits to the release rates (see text). From *Huebner and Benkhoff* (1999).

4.5. Dust Ejection and Mantle Formation

The basic process of dust ejection and mantle formation is quite simple: As water and other — more volatile — ices evaporate, the gas flux drags with it dust particles with radii smaller than the critical radius r_d^* (which varies with temperature), while the larger particles accumulate on the surface, eventually creating an inactive mantle. If at all points of the orbit (that is, for all values of the surface temperature) $r_d^*(T) < r_d^{max}$, a permanent mantle will form and grow thicker with repeated orbital revolutions. In time, the insulating effect of the mantle will quench sublimation and hence dust entrainment as well. The difficulty in modeling this process is due to the large uncertainties in the parameters involved. Thus, whereas the observed size distribution of dust grains concerns the small particles, it is the large particles that determine the rate of formation of the mantle. Moreover, the physical properties of the mantle may be largely affected by organic material that acts as a glue between dust grains (*Kömle et al.*, 1996). Consequently, one of the main goals of model calculations is to examine the effect of parameter variations. *Podolak and Herman* (1985) showed that the growth and stability of the mantle is affected by the thermal conductivity, a high conductivity acting as a heat sink. The effect of a variable albedo was studied by *Orosei et al.* (1995). They showed that dust accumulation and darkening of the surface can cause an increase in the energy absorbed by the nucleus to such an extent as to increase ice sublimation and dust drag to the point of complete removal of the mantle. In such a case the comet would become whiter and colder and buildup of the mantle could start anew, leading to alternating phases of activity and hibernation. The importance of cohesive forces within the refractory material was stressed by *Kührt and Keller* (1994), who showed that mantles may withstand the vapor pressure building up underneath and thus explain the seemingly permanent inactivity of a large fraction of the nucleus.

Cometary activity declines considerably during the buildup of a dust mantle. A very thin dust layer, on the order of a few centimeters or less, is capable of diminishing the cometary activity by a large factor (e.g., *Prialnik and Bar-Nun*, 1988; *Coradini et al.*, 1997b; *Capria et al.*, 2001). At the same time, the surface temperature becomes much higher, as we have seen in section 3.7 and Fig. 5. If a dense ice crust builds up below the dust mantle, the activity is quenched to an even higher degree (*Prialnik and Mekler*, 1991). In these cases the activity is limited to exposed patches of ice. *Rickman et al.* (1990) show — by numerical simulations of mantle growth — that even stable mantles are sufficiently thin to be broken occasionally by thermal cracks, explosion of gas pockets, or minor collisions, allowing localized activity.

The second goal of models involving dust is to explain the observed dust production rates. Here too the results are largely dependent on unknown parameters. An example of dust ejection from a comet nucleus resulting from dust

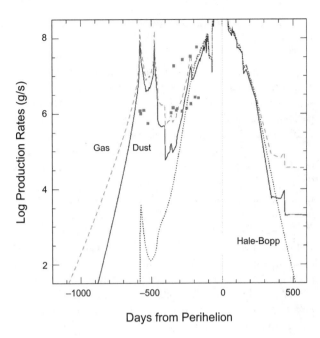

Fig. 10. Dust and gas production rates for a model of Comet Hale-Bopp. The dotted line represents the production rate of H₂O alone; the dashed line is the total gas flux, including H₂O, CO, and CO₂. The ice/dust mass ratio for the nucleus is 1. From *Prialnik* (2002).

carried out from the interior, as well as blown off of the surface, is shown in Fig. 10.

4.6. Crystallization and Outbursts

Comets are often found to be active at heliocentric distances far beyond the limit of ~5 AU, within which the activity may be explained by sublimation of water ice induced by insolation. Crystallization of amorphous ice has long been recognized as a suitable mechanism for explaining such distant bursts of activity (*Patashnik et al.,* 1974; *Smoluchowski,* 1981; *Espinasse et al.,* 1991; *Weissman,* 1991; *Prialnik and Bar-Nun,* 1992).

Considering the timescales of crystallization, heat conduction and sublimation, we find that at very low temperatures conduction dominates, meaning that heat released by a local source will be efficiently removed. Crystalline ice is a much better heat conductor than amorphous ice and hence heat will flow predominantly to the surface through the growing outer crystalline layer. Thus, as long as the temperature of the outer layer of the nucleus is below the critical temperature where τ_{ac} intersects τ_{c-ice} (see Fig. 4), the rate of heating by crystallization will be very slow. As the crystallization rate is much more sensitive to temperature than the conduction rate (of crystalline ice), it will eventually surpass the rate of heat conduction. For example, at a depth of 10 m, the conduction timescale surpasses the crystallization timescale close to 120 K. Crystallization is triggered by some heat source that causes the temperature to rise; when the temperature at the crystallization front

reaches the critical temperature T_c ~ 110–120 K, the latent heat released at the front causes it to rise still further. The higher temperature, in turn, causes crystallization to proceed even faster and thus a runaway process develops. The rise time and the timescale of outbursts thus triggered should be on the order of $\tau_{c-ice}(T_c) = \tau_{ac}(T_c)$. According to Fig. 4, it is about 100 days for crystallization for a depth of 10 m (and it will be ~1 day for a depth of 1 m). This means that fluctuations (and outbursts) at small heliocentric distances should occur on much shorter timescales than at large heliocentric distances. Observations appear to confirm this conclusion.

The competition between τ_{ac} and τ_{gas} should indicate when an instability is likely to occur. We recall that the crystallization timescale is also the timescale of gas release and pressure buildup (assuming gas is occluded in the amorphous ice), while the diffusion timescale of the gas is also the timescale of pressure relaxation. If $\tau_{ac} > \tau_{gas}$, the pressure is released sufficiently rapidly to prevent mechanical instability; however, if $\tau_{gas} \gg \tau_{ac}$, gas would accumulate more rapidly than it is removed and large stresses may result from pressure buildup. Thus, if the temperature of amorphous ice at a certain depth exceeds a critical value, it could lead to a state of instability. This situation may be avoided either if the temperature decreases, which is possible if the thermal timescale is sufficiently short, or if the pore size increases, thereby reducing τ_{gas}. However, according to Fig. 4, the thermal timescales for both amorphous and crystalline ice are longer than τ_{gas} by 2–3 orders of magnitude. Hence only expansion of the pores may arrest the development of an instability, once it occurs. However, the analysis of timescales does not provide clues for the magnitude of the pressure and pressure gradients in relation to the strength of the material, nor to the outcome of unstable conditions. This necessitates detailed numerical computations, and the establishment of an algorithm for treating fracture.

Numerical models of the evolution of cometary nuclei find that crystallization progresses in spurts, their onset, duration, and extent in depth being largely determined both by the structure, composition, and thermal properties of the nucleus and by the comet's orbit (e.g., *Herman and Podolak,* 1985; *Prialnik and Bar-Nun,* 1987, 1990; *Espinasse et al.,* 1991, 1993; *Tancredi et al.,* 1994). Crystallization may be initiated by the heat wave propagating inward from the insolated comet surface to the crystalline-amorphous ice boundary, provided that after reaching this boundary, it still carries sufficient energy for significantly raising the local temperature. However, once this has occurred and the boundary has moved deeper into the nucleus, later heat waves originating at the surface will be too weak when reaching the boundary to rekindle crystallization. A quiescent period would thus ensue, until the surface recedes (by sublimation) to a sufficiently short distance from the crystalline-amorphous ice boundary. At this point, a new spurt of crystallization will take place. Since in the meantime the interior temperature of the ice has risen to some extent, crystallization will advance deeper into the nucleus than at the previous spurt. This will in turn affect the time span to the next spurt of crystallization, since the rate of surface recession for a given comet is roughly constant (see Table 5). In conclusion, crystallization would appear to be triggered sporadically, preferentially at large heliocentric distances, where comets spend most of their time. This could explain the distant activity — outbursts and possibly splitting — of comets.

The release of gas trapped in the amorphous ice provides the link between crystallization and the eruptive manifestations of comets, of which a few examples will be given below. We have already shown that numerical simulations are based on many simplifying assumptions, and often adopt parameters that are not well known. Hence they should not be expected to accurately reproduce any particular observed outburst. Rather, such simulations should account for the basic characteristics of the observed outbursts.

4.6.1. Distant outbursts of Comet P/Halley. The behavior of Comet P/Halley at large heliocentric distances, beyond 5 AU, was characterized by outbursts of various magnitudes; during the most significant one, at 14 AU (*West et al.,* 1991), the total brightness increased by more than 5 mag and an extended coma developed. The outburst subsided on a timescale of months. Klinger and his collaborators (see *Espinasse et al.,* 1991; *Weissman,* 1991) and *Prialnik and Bar-Nun* (1992) showed that these features can be explained by ongoing crystallization of amorphous ice in the interior of the porous nucleus, at depths of a few tens of meters. According to this model, enhanced outgassing results from the release of trapped gas during crystallization of the ice. The orbital point where the gas flux reaches its peak was found to be strongly dependent upon the porosity of the comet nucleus. Thus, for example, in the case of a spherical nucleus of porosity ~0.5 (*Prialnik and Bar-Nun,* 1990), crystallization is found to occur on the outbound leg of Comet P/Halley's orbit, at heliocentric distances between 5 and 17 AU (depending on the pore size assumed, typical pore sizes being 0.1–10 μm). Similar results were obtained by *Schmitt et al.* (1991). The duration of an outburst is the most difficult to predict: Depending on the pore size and on the mechanical properties of the ice, it may vary over three orders of magnitude. A time span of a few months lies within this range and is therefore possible to obtain for a suitable choice of parameters.

4.6.2. Preperihelion activity of 2060 Chiron. Chiron, first classified as an asteroid, was observed to develop a coma at random intervals before it reached perihelion (in 1996) in its 50-year orbit. *Marcialis and Buratti* (1993) summarized its brightness variations: The first episode of coma formation occurred in 1978, during the middle of the decline in brightness; the second episode, in 1989, when the coma reached vast dimensions, coincided with the maximal brightness. Even near aphelion Chiron underwent a major outburst that lasted several years. *Prialnik et al.* (1995) were able to obtain a model that agreed remarkably well with the observational data by adopting a composition of 60% dust and 40% amorphous ice, occluding a fraction 0.001 of CO and assuming a low emissivity ($\varepsilon = 0.25$). The optimal parameter combination was found after numerous trials of parameter combinations that proved far less successful. They found that spurts of crystallization started close to aphelion. As a rule, the CO production rate decreased slightly as the model comet approached the Sun from aphelion. This should explain the puzzling fading of Chiron between 1970 and 1985 (i.e., from ~18 AU to ~14 AU). The model produced the required CO emission rates, explained by release of trapped gas, and reproduced the estimated surface (color) temperatures at different points of the orbit as derived by *Campins et al.* (1994). *Capria et al.* (2000) also explained Chiron's activity by gas trapped in amorphous ice, although they also mentioned the possibility of CO ice close to the surface, which would imply that Chiron has been inserted into its present orbit only recently (cf. *Fanale and Salvail,* 1997).

4.6.3. Erratic activity of Comet Schwassmann-Wachmann 1 (SW1). The orbit of Comet SW1 is nearly circular and confined between the orbits of Jupiter and Saturn. Despite the fact that at such heliocentric distances the sublimation of H_2O ice is negligible, this comet exhibits irregular activity — unpredictable changes in its lightcurve. *Froeschlé et al.* (1983) suggested that this might be associated with crystallization of amorphous ice. This suggestion was further strengthened by the detection of CO released by the comet (*Senay and Jewitt,* 1994; *Crovisier et al.,* 1995), since although SW1 is too distant for H_2O ice sublimation, its surface is too hot for the survival of CO ice. Subsequently, *Klinger et al.* (1996) showed by model calculations that the CO production pattern can be explained and simulated by gas trapped in the amorphous ice and released from the ice upon crystallization. The chaotic behavior results from the highly nonlinear temperature dependence of the processes involved.

4.6.4. Distant activity of Comet Hale-Bopp. Comet Hale-Bopp (C/1995 O1) was characterized by an unusually bright coma at a distance of about 7 AU from the Sun. Observations performed by *Jewitt et al.* (1996) detected a very large flux of CO molecules, which increased dramatically. Such brightening is unlikely to have resulted from surface (or subsurface) sublimation of CO ice in response to insolation. In any case, CO ice should have been depleted much earlier in the orbit, since at 7 AU the surface temperature is already above 100 K, considerably higher than the sublimation temperature of CO. In this case as well the unusual activity could be explained on the basis of crystallization and release of occluded CO accompanied by ejection of dust entrained by the gas (*Prialnik,* 1999, 2002; *Capria et al.,* 2002).

4.7. Early Evolution of Comets: Effect of Radioactivity

Formation of comets, like star formation, is still an object of study and thermal evolution during formation has barely been considered (*Merk,* 2003). But attempts to estimate the possible effect of radioactive heating on young comet nuclei have been made in a number of different studies under different assumptions and approximations [see *Prialnik and Podolak* (1995) and references therein, and more recently, *De Sanctis et al.* (2001) and *Choi et al.* (2002) and references therein]. There is general agreement that the long-lived radionuclides should have no or little effect on objects below about 50 km in radius; thus ²⁶Al is considered as the energy source. Using again Fig. 4 as a guide, we find by extrapolation that at a depth of 1 km the thermal timescale of amorphous ice becomes comparable to the decay time of ²⁶Al, meaning that the ice may barely be heated. It will certainly be heated at larger depths, a few kilometers and beyond. There, eventually, the internal temperature will become sufficiently high for crystallization to set in, providing an additional internal heat source (*Podolak and Prialnik,* 1997). At the same time, however, the thermal timescale will decrease, crystalline ice being a much better heat conductor than amorphous ice. In addition, if the nucleus is sufficiently porous, the gases released upon crystallization will be able to escape to the outer regions of the nucleus and will carry the heat away efficiently. Hence, only in still larger comet nuclei (beyond 10 km) will the internal temperature continue to rise.

If the internal temperature becomes such that the timescale of sublimation is shorter than the timescale of radiogenic heat release, then most of the released energy will be absorbed in sublimation of ice from the pore walls, starting with the most volatile species. If, in addition, the radius is such that the timescale of gas (vapor) diffusion is lower than the timescale of sublimation, then sublimation will consume the radiogenic heat so long as there is ice, since the vapor will be efficiently removed. A steady state will develop, without further heating of the ice matrix. Regarding H_2O, it is worth mentioning that the temperature of such a steady state would be considerably lower than the melting temperature of ice. On the other hand, if the porosity of the ice is very low and the average pore size very small, τ_{gas} may become sufficiently high for gas removal to become inefficient. In such cases the internal temperature may rise to the melting point of ice (cf. *Yaboushita,* 1993; *Podolak and Prialnik,* 2000).

Calculations of the long-term evolution of comets far from the Sun under the influence of radioactive heating show that the internal temperatures attained may be sufficiently high for comets to have become depleted of volatiles that sublimate below ~40–50 K, initially included as ices (*De Sanctis et al.,* 2001; *Choi et al.,* 2002). Less-volatile species may have been partly lost as well. Observation of such volatiles in comets suggests that they originate from amorphous H_2O ice undergoing crystallization. This means that, despite radioactive heating, a substantial fraction of the H_2O ice has retained its pristine form, i.e., the innermost region is crystalline, and the outer region is composed of amorphous ice. Figure 11 shows the relative radius of the inner part of the nucleus that has crystallized during early evolution, as a function of the comet's radius and distance from the Sun (*Merk,* 2003).

Although much of the released gas will escape the nucleus entirely, some will become trapped in the cold outer

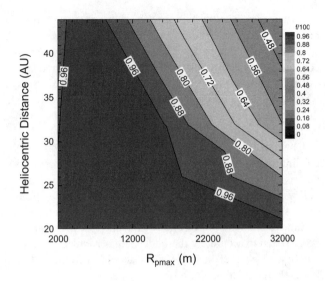

Fig. 11. Contour plots of the relative radius r/R up to which the nucleus core crystallizes due to radiogenic heating during formation by accretion. From *Merk* (2003).

layers of the nucleus. This will reduce the porosity in those layers and enrich them in condensed volatiles. As the comet nears the Sun and the outer layers heat up, these gases will be released and the comet will show enhanced activity.

5. CONCLUSIONS AND DIRECTIONS FOR FUTURE WORK

5.1. General Conclusions

In spite of the sparse information regarding the cometary interior, the complexity of the processes that may take place within them, and the uncertainties involved, the general conclusion that emerges from simulations of the evolution of comet nuclei is that, essentially, a nucleus model of porous, grainy material, composed of gas-laden amorphous ice and dust, is capable of reproducing the activity pattern of comets. Three types of cometary activity, all associated with the flow of volatiles through and out of a porous nucleus, are identified. They have observable outward manifestations on the one hand, and lasting effects on the structure of the nucleus on the other.

1. Sublimation of volatiles from the pore walls and the subsequent flow of vapor is the source of gas for the coma and tail, but may also lead to the formation of a dense ice crust below the surface of the nucleus. Gases flowing to the interior may refreeze when reaching sufficiently cold regions, at depths correlated with the volatility of the gas. The resulting effect is a stratified nucleus configuration.

2. Crystallization of amorphous ice, accompanied by the release of heat as well as trapped gases, may account for cometary outbursts and may also result in fracture of the porous material.

3. Drag of dust grains by the flowing gas leads to ejection of the small particles seen in the dust coma and tail, while accumulation of the large particles on the surface of

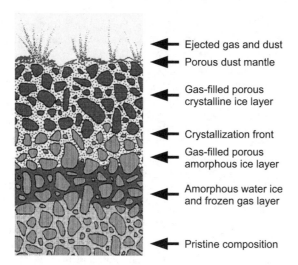

Fig. 12. Schematic layered structure of a cometary nucleus (arbitrary scales). From *Prialnik* (1999).

the nucleus may lead to the formation of a sealing dust mantle that would turn the comet into an asteroid-like object.

In conclusion, the thermal evolution and activity pattern of a porous comet nucleus differs significantly from the old view of a solid icy body that is mainly controlled by sublimation from the surface in response to solar heating. The structure that emerges is shown schematically in Fig. 12.

The thermal evolution of comet nuclei may be divided into two phases: a long phase — of the order of the solar system's age — spent at large distances from the Sun (in the Oort cloud or the Kuiper belt), and a second, much shorter phase, spent in orbit around the Sun within the planetary system. There is, of course, an intermediate, transient phase during which a periodic comet is gradually perturbed into its final, steady orbit. The notion that the thermal evolution process really begins when a comet enters the second phase of its life, becoming a "new" comet, is beginning to be doubted. New comets, which have often been described as pristine objects that have undergone no (or little) alteration during their lifetime in the distant outskirts of the solar system, are now suspected to have been heated to the point of melting of the H_2O ice. Nevertheless, they are still believed to constitute a source of solar nebula material. Much of the fascination and interest comets have aroused was due to the clues they were believed to hold to the formation of the solar system. This may still be true, at least for a fraction of comets, or for a fraction of every comet. In addition, comets are now invoked to explain the formation of life.

5.2. Required Input Data from Observations and Experiments

The success of the thermal evolution theory just described in explaining the structure and activity of comet nuclei is hindered by the huge lack of information regard-

ing crucial (or critical) parameters. As a result, explanations for observed behavior may be ambiguous; i.e., different parameter combinations, within the same model, may lead to similar results. Consequently, additional input is required both from laboratory studies and from observations.

The input required from laboratory studies includes (1) pressure curves at low temperatures, (2) latent heat measurement, (3) thermal conductivity of mixtures, and (4) sublimation studies of mixtures. From observations we need more information on dynamical properties (spin axis, rotation period, orientation, and shape of nucleus). It would be interesting to determine and understand whether a potato, rather than spherical, shape is typical of small bodies of negligible self-gravity. Upcoming *in situ* measurements should provide information about the porous structure — porosity and pore size — as well as strength. The interplay among the different methods of research applied to cometary nuclei is illustrated in Fig. 13.

5.3. Where Do We Go from Here?

In the course of this review we have mentioned a rather long list of assumptions that are common to most theoretical studies to date. Simplifying assumptions are justified when a theory is still young and ridden by uncertainties. Now that it has matured, we may safely enter the next stage, where more sophisticated methods and models should be developed. We suggest a few below, following the order of the review's sections. In some cases, first steps have already been taken.

5.3.1. Numerical methods. (1) Use of adaptive grid methods for dealing with receding surfaces during late evolution, as well as growing mass during very early stages. (2) Development of full-scale 3-D models that allow for lateral flow both of heat and of gas. (3) Inclusion of boundary conditions accounting for the nucleus-coma interaction. (4) Implementation of modern methods for the simultaneous solution of a multiple component nucleus.

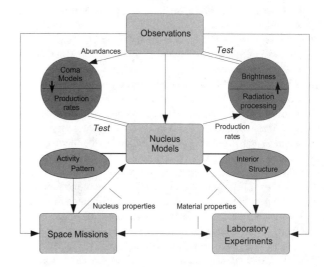

Fig. 13. The role of nucleus models in the coordinated study of comets.

5.3.2. Physical processes.   (1) Coupling between gas phases and rigorous treatment of mixtures. (2) Construction of models for fracture and for crack propagation. (3) Treatment of surface properties, such as irregularities, shadowing, and mixed thermal properties, as well as radiative transfer in the outermost porous layer (e.g., *Davidsson and Skorov*, 2002a,b). (4) Modeling of the material structure of the dust mantle on the nucleus surface.

5.3.3. Modeling the evolution of comet nuclei.   (1) Modeling comet formation by accretion. (2) Long-term evolution over the age of the solar system, considering potential gravitational interactions and orbital evolution. (3) Modeling comet ⇔ asteroid transition (e.g., *Coradini et al.*, 1997a). (4) Modeling nucleus shape evolution as a result of uneven ablation.

The purpose of modeling comet nuclei is not to predict their behavior based on an initial set of parameters. Given the large number of parameters and their wide range of possible values, predictions may be misleading. Rather, the true purpose of modeling is to reproduce the observed cometary behavior, in order to deduce internal properties of comet nuclei that are inaccessible to observation. The closer the numerical simulations are to observed reality, the more reliable will be our inferences on the elusive nature of comet nuclei and the clues they hold to the understanding of the solar system's beginnings — perhaps to the beginning of life as well. In the words of Isaac Newton: "I suspect that the spirit which is the smallest but most subtle and most excellent part of our air, and which is required for the life of all things, comes chiefly from comets" [*Principia*, Book 3, Proposition 41, 1687 (translation from *Cohen and Whitman*, 1999].

REFERENCES

Arakawa M., Higa M., Leliwa-Kopystyński J., and Maeno N. (2000) Impact cratering of granular mixture targets made of H_2O ice — CO_2 ice-pyrophylite. *Planet. Space Sci., 48,* 1437–1446.

Bar-Nun A. and Owen T. (1998) Trapping of gases in water ice. In *Solar System Ices* (B. Schmitt et al., eds.), pp. 353–366. Kluwer, Dordrecht.

Bar-Nun A., Dror J., Kochavi E., and Laufer D. (1987) Amorphous water ice and its ability to trap gases. *Phys. Rev. B, 35,* 2427–2435.

Benkhoff J. (1999) Energy balance and the gas flux from the surface of Comet 46P/Wirtanen. *Planet. Space Sci., 47,* 735–744.

Benkhoff J. (2002) The emission of gas and dust from comets. *Adv. Space Res., 29,* 1177–1186.

Benkhoff J. and Boice D. C. (1996) Modeling the thermal properties and the gas flux from a porous, ice body in the orbit of P/Wirtanen. *Planet. Space Sci., 44,* 665–674.

Benkhoff J. and Huebner W. F. (1995) Influence of the vapor flux on temperature, density, and abundance distribution in a multicomponent, porous ice body. *Icarus, 114,* 348–354.

Benkhoff J. and Spohn T. (1991) Thermal histories of the KOSI samples. *Geophys. Res. Lett., 18,* 261–264.

Blake D., Allamadolla L., Sandford S., Hudgins D., and Freund F. (1991) Clathrate hydrate formation in amorphous cometary ice analogs in vacuo. *Science, 245,* 58–551.

Bouziani N. and Fanale F. P. (1998) Physical chemistry of a heterogeneous medium: Transport processes in comet nuclei. *Astrophys. J., 499,* 463–474.

Brailsford A. and Major K. G. (1964) The thermal conductivity of aggregates of several phases, including porous materials. *Br. J. Appl. Phys., 15,* 313.

Brin G. D. and Mendis D. A. (1979) Dust release and mantle development in comets. *Astrophys. J., 229,* 402–408.

Campins H., Telesco C. M., Osip D. J., Rieke G. H., Rieke M. J., and Schulz B. (1994) The color temperature of (2060) Chiron: A warm and small nucleus. *Astron. J., 108,* 2318–2322.

Capria M. T., Coradini A., De Sanctis M. C., and Orosei R. (2000) Chiron activity and thermal evolution. *Astron. J., 119,* 3112–3118.

Capria M. T., Coradini A., De Sanctis M. C., and Bleckai M. I. (2001) P/Wirtanen thermal evolution: Effects due to the presence of an organic component in the refractory material. *Planet. Space Sci., 49,* 907–918.

Capria M. T., Coradini A., and De Sanctis M. C. (2002) C/1995 O1 Hale-Bopp: Short and long distance activity from a theoretical model. *Earth Moon Planets, 90,* 217–225.

Choi Y.-J., Cohen M., Merk R., and Prialnik D. (2002) Long-term evolution of objects in the Kuiper belt zone — effects of insolation and radiogenic heating. *Icarus, 160,* 300–312.

Cohen I. B. and Whitman A. (1999) *Isaac Newton — The Principia.* Univ. of California, Berkeley. 974 pp.

Cohen M., Prialnik D., and Podolak M. (2003) A quasi-3D model for the evolution of shape and temperature distribution of comet nuclei — application to Comet 46P/Wirtanen. *New Astron., 8,* 179–189.

Cook R. F. (1989) Effective-medium theory for the fracture of fractal porous media. *Phys. Rev. B, 39,* 2811–2814.

Coradini A., Capaccioni F., Capria M. T., De Sanctis M. C., Espinassse S., Orosei R., and Salomone M.(1997a) Transition elements between comets and asteroids. I. Thermal evolution models. *Icarus, 129,* 317–336.

Coradini A., Capaccioni F., Capria M. T., De Sanctis M. C., Espinassse S., Orosei R., and Salomone M. (1997b) Transition elements between comets and asteroids. II. From the Kuiper belt to NEO orbits. *Icarus, 129,* 337–347.

Cowan J. J. and A'Hearn M. F. (1979) Vaporization of comet nuclei — Light curves and life times. *Moon and Planets, 21,* 155–171.

Crifo J. F. and Rodionov A. V. (1997) The dependence of the circumnuclear coma structure on the properties of the nucleus. I. Comparison between homogeneous and inhomogeneous nucleus spherical nucleus with application to P/Wirtanen. *Icarus, 127,* 319–353.

Crifo J. F., Fulle M., Kömle N. I., and Szeg K. (2004) Nucleus-coma structural relationships: Lessons from physical models. In *Comets II* (M. C. Festou et al., eds.), this volume. Univ. of Arizona, Tucson.

Crovisier J., Biver N., Bockelée-Morvan D., Colom P., Jorda L., Lellouch E., Paubert G., and Despois D. (1995) Carbon monoxide outgassing from Comet Schwassmann-Wachmann 1. *Icarus, 115,* 213–216.

Davidsson B. J. R. (1999) Tidal splitting and rotational breakup of solid spheres. *Icarus, 142,* 525–535.

Davidsson B. J. R. (2001) Tidal splitting and rotational breakup of solid biaxial ellipsoids. *Icarus, 149,* 375–383.

Davidsson B. J. R. and Skorov Yu. V. (2002a) On the light-absorbing surface layer of cometary nuclei: I. Radiative transfer. *Icarus, 156,* 223–248.

Davidsson B. J. R. and Skorov Yu. V. (2002b) On the light-absorbing surface layer of cometary nuclei: II. Thermal modeling. *Icarus, 159,* 239–258.

De Sanctis M. C., Capria M. T., and Coradini A. (2001) Thermal evolution and differentiation of Kuiper belt objects. *Astron. J., 121,* 2792–2799.

Eluszkiewicz J., Leliwa-Kopystyński J., and Kossacki K. J. (1998) Metamorphism of solar system ices. In *Solar System Ices* (B. Schmitt et al., eds.), pp. 119–138. Kluwer, Dordrecht.

Enzian A., Cabot H., and Klinger J. (1997) A 2 1/2 D thermodynamic model of cometary nuclei. *Astron. Astrophys., 31,* 995–1006.

Enzian A., Klinger J., Schwehm G., and Weissman P. R. (1999) Temperature and gas production distribution on the surface of a spherical model nucleus in the orbit of 46P/Wirtanen. *Icarus, 138,* 74–84.

Espinasse S., Klinger J., Ritz C., and Schmitt B. (1991) Modeling of the thermal behavior and of the chemical differentiation of cometary nuclei. *Icarus, 92,* 350–365.

Espinasse S., Coradini A., Capria M. T., Capaccioni F., Orosei R., Salomone M., and Federico C. (1993) Thermal evolution and differentiation of a short period comet. *Planet. Space Sci., 41,* 409–427.

Fanale F. P. and Salvail J. R. (1984) An idealized short period comet model: Surface insolation, flux, dust flux, and mantle evolution. *Icarus, 60,* 476–511.

Fanale F. P. and Salvail J. R. (1987) The loss and depth of CO_2 ice in comet nuclei. *Icarus, 72,* 535–554.

Fanale F. P. and Salvail J. R. (1990) The influence of CO ice on the activity and near-surface differentiation of comet nuclei. *Icarus, 84,* 403–413.

Fanale F. P. and Salvail J. R. (1997) The cometary activity of Chiron: A stratigraphic model. *Icarus, 125,* 397–405.

Froeschlé Cl., Klinger J., and Rickman H. (1983) Thermal models for the nucleus of Comet P/Schwassmann-Wachmann 1. In *Asteroids, Comets, Meteors (A85-26851 11-89),* pp. 215–224.

Greenberg J. M., Mizutani H., and Yamamoto T. (1995) A new derivation of the tensile strength of cometary nuclei: Application to Comet Shoemaker-Levy 9. *Astron. Astrophys., 295,* L35–L38.

Grün E., Gebhard J., Bar-Nun A., Benkhoff J., Düren H., Eich G., Hische R., Huebner W. F., Keller H. U., and Klees G. (1993) Development of a dust mantle on the surface of an insolated ice-dust mixture: Results of the KOSI-9 experiment. *J. Geophys. Res., 98,* 15091–15104.

Gutiérrez P. J., Ortiz J. L., Rodrigo R., and Lopez-Moreno J. J. (2000) A study of water production and temperature of rotating irregularly shaped cometary nuclei. *Astron. Astrophys., 355,* 809–817.

Haruyama J., Yamamoto T., Mizutani H., and Greenberg J. M. (1993) Thermal history of comets during residence in the Oort cloud: Effect of radiogenic heating in combination with the very low thermal conductivity of amorphous ice. *J. Geophys. Res., 98,* 15079–15090.

Herman G. and Podolak M. (1985) Numerical simulations of comet nuclei: I. Water ice comets. *Icarus, 61,* 252–266.

Herman G. and Weissman P. R. (1987) Numerical simulation of cometary nuclei III. Internal temperatures of cometary nuclei. *Icarus, 69,* 314–328.

Horanyi M., Gombosi T. I., Cravens T. E., Körösmezey A.,

Kecskeméty K., Nagy A. F., and Szegö K. (1984) The friable sponge model of a cometary nucleus. *Astrophys. J., 278,* 449–455.

Huebner W. F. and Benkhoff J. (1999) From coma abundances to nucleus composition. In *Proceedings of the ISSI-Workshop: The Origin and Composition of Cometary Material* (K. Altweg et al., eds.). *Space Sci. Rev., 90,* 117–130.

Huebner W. F., Benkhoff J., Capria M. T., Coradini A., De Sanctis C., Enzian A., Orosei R., and Prialnik D. (1999) Results from the comet nucleus model team at the international space science institute, Bern, Switzerland. *Adv. Space Res., 23,* 1283–1298.

Irvine W. M., Leschine S. B., and Schloerb F. P. (1980) Thermal history, chemical composition and relationship of comets to the origin of life. *Nature, 283,* 748–749.

Jenniskens P. and Blake D. F. (1994) Structural transitions in amorphous water ice and astrophysical implications. *Science, 265,* 753–756.

Jewitt D., Senay M., and Matthews H. (1996) Observations of carbon monoxide in Comet Hale-Bopp. *Science, 271,* 1110–1113.

Julian W. H., Samarasinha N. H., and Belton M. J. S. (2000) Thermal structure of cometary active regions: Comet 1P/Halley. *Icarus, 144,* 160–171.

Kaponen A., Kataja M., and Timonen J. (1997) Permeability and effective porosity of porous media. *Phys. Rev. E, 56,* 3319–3325.

Klinger J. (1980) Influence of a phase transition of ice on the heat and mass balance of comets. *Science, 209,* 271–272.

Klinger J. (1981) Some consequences of a phase transition of water ice on the heat balance of comet nuclei. *Icarus, 47,* 320–324.

Klinger J., Espinasse S., and Schmidt B. (1989) Some considerations on cohesive forces in sun-grazing comets. In *Proceedings of an International Workshop on Physics and Mechanics of Cometary Materials* (J. Hunt and T. D. Guyeme, eds.), pp. 197–200. ESA SP-302, Noordwijk, The Netherlands.

Klinger J., Levasseur-Regourd A. C., Bouziani N., and Enzian A. (1996) Towards a model of cometary nuclei for engineering studies for future space craft missions to comets. *Planet. Space Sci., 44,* 637–653.

Kochan H., Roessler K., Ratke L., Heyl M., Hellman H., and Schwehm G. (1989) Crustal strength of different model comet materials. *Proceedings of an International Workshop on Physics and Mechanics of Cometary Materials* (J. Hunt and T. D. Guyeme, eds.), pp. 115–119. ESA SP-302, Noordwijk, The Netherlands.

Kömle N. I., Kargl G., Thiel K., and Seiferlin K. (1996) Thermal properties of cometary ices and sublimation residua including organics. *Planet. Space Sci., 44,* 675–689.

Kossacki K. J., Szutowicz S., and Leliwa-Kopystyński J. (1999) Comet 46P/Wirtanen: Evolution of the subsurface layer. *Icarus, 142,* 202–218.

Kouchi A., Greenberg J. M., Yamamoto T., and Mukai T. (1992) Extremely low thermal conductivity of amorphous ice: Relevance to comet evolution. *Astrophys. J. Lett., 388,* L73–L76.

Kouchi A., Yamamoto T., Kozasa T., Koruda T., and Greenberg J. M. (1994) Conditions for condensation and preservation of amorphous ice and crystallinity of astrophysical ices. *Astron. Astrophys., 290,* 1009–1018.

Kührt E. (1984) Temperature profiles and thermal stresses in cometary nuclei. *Icarus, 60,* 512–521.

Kührt E. and Keller H. U. (1994) The formation of cometary surface crusts. *Icarus, 109,* 121–132.

Lee T., Papanastassiou D., and Wasserburg G. J. (1976) Demonstration of ^{26}Mg excess in Allende and evidence for ^{26}Al. *Geophys. Res. Lett., 3,* 109–112.

Lide D. R. (2003) *CRC Handbook of Chemistry and Physics, 84th edition.* CRC Press, Boca Raton, Florida. 2616 pp.

Lugmair G. W. and Shukolyukov A. (2001) Early solar system events and timescales. *Meteoritics & Planet. Sci., 36,* 1017–1026.

MacPherson G. J., Davis A. M., and Zinner E. K. (1995) The distribution of aluminum-26 in the early solar system — A reappraisal. *Meteoritics, 30,* 365–386.

Mahoney W. A., Ling J. C., Wheaton Wm. A., and Jacobson A. S. (1984) *Heao 3* discovery of ^{26}Al in the interstellar medium. *Astrophys. J., 286,* 578–585.

Marcialis R. L. and Buratti B. J. (1993) CCD photometry of 2060 Chiron in 1985 and 1991. *Icarus, 104,* 234–243.

Maxwell J. C. (1873) *A Treatise on Electricity and Magnetism.* Clarendon, Oxford.

Meech K. J. and Svoreň J. (2004) Physical and chemical evolution of cometary nuclei. In *Comets II* (M. C. Festou et al., eds.), this volume. Univ. of Arizona, Tucson.

Mekler Y. and Podolak M. (1994) Formation of amorphous ice in the protoplanetary nebula. *Planet. Space Sci., 42,* 865–870.

Mekler Y., Prialnik D., and Podolak M. (1990) Evaporation from a porous cometary nucleus. *Astrophys. J., 356,* 682–686.

Mendis D. A. and Brin G. D. (1977) Monochromatic brightness variations of comets. II — Core-mantle model. *Moon, 17,* 359–372.

Merk R. (2003) Thermodynamics and accretino of asteroids, comets, and Kuiper Belt Objects — A computer simulation study in plaentary physics. Ph.D. thesis, Univ. of Münster. 143 pp.

Merk R., Breuer D., and Spohn T. (2002) Numerical modeling of ^{26}Al-induced melting of asteroids considering accretion. *Icarus, 159,* 183–191.

Morley A. (1954) *Strength of Materials, 11th edition.* Longmans, London. 532 pp.

Öpik E. J. (1958) *Physics of Meteor Flight in the Atmosphere.* Interscience, New York. 174 pp.

Orosei R., Capaccioni F., Capria M. T., Coradini A., Espinasse S., Federico C., Salomone M., and Schwehm G. H. (1995) Gas and dust emission from a dusty porous comet. *Astron. Astrophys., 301,* 613–627.

Patashnick H., Rupprecht G., and Schuerman D. W. (1974) Energy source for comet outbursts. *Nature, 250,* 313–314.

Podolak M. and Herman G. (1985) Numerical simulations of comet nuclei II. The effect of the dust mantle. *Icarus, 61,* 267–277.

Podolak M. and Prialnik D. (1996) Models of the structure and evolution of Comet P/Wirtanen. *Planet. Space Sci., 44,* 655–664.

Podolak M. and Prialnik D. (1997) ^{26}Al and liquid water environments in comet. In *Comets and the Origin of Life* (P. Thomas et al., eds.), pp. 259–272. Springer-Verlag, New York.

Podolak M. and Prialnik D. (2000) Conditions for the production of liquid water in comet nuclei. In *Bioastronomy '99: A New Era in Bioastronomy* (G. A. Lemarchand and K. J. Meech, eds.), pp. 231–234. Sheridan, Chelsea.

Prialnik D. (1992) Crystallization, sublimation, and gas release in the interior of a porous comet nucleus. *Astrophys. J., 388,* 196–202.

Prialnik D. (1999) Modelling gas and dust release from Comet Hale-Bopp. *Earth Moon Planets, 77,* 223–230.

Prialnik D. (2002) Modeling the comet nucleus interior; application to Comet C/1995 O1 Hale-Bopp. *Earth Moon Planets, 89,* 27–52.

Prialnik D. and Bar-Nun (1987) On the evolution and activity of cometary nuclei. *Astrophys. J., 313,* 893–905.

Prialnik D. and Bar-Nun (1988) Formation of a permanent dust mantle and its effect on cometary activity. *Icarus, 74,* 272–283.

Prialnik D. and Bar-Nun (1990) Gas release in comet nuclei. *Astrophys. J., 363,* 274–282.

Prialnik D. and Bar-Nun (1992) Crystallization of amorphous ice as the cause of Comet P/Halley's outburst at 14 AU. *Astron. Astrophys., 258,* L9–L12.

Prialnik D. and Mekler Y. (1991) The formation of an ice crust below the dust mantle of a cometary nucleus. *Astrophys. J., 366,* 318–323.

Prialnik D. and Podolak M. (1995) Radioactive heating of porous cometary nuclei. *Icarus, 117,* 420–430.

Prialnik D., Egozi U., Bar-Nun A., Podolak M., and Greenzweig Y. (1993) On pore size and fracture in gas — laden comet nuclei. *Icarus, 106,* 499–507.

Prialnik D., Brosch N., and Ianovici D. (1995) Modelling the activity of 2060 Chiron. *Mon. Not. R. Astron. Soc., 276,* 1148–1154.

Rayleigh Lord (1892) On the influence of obstacles arranged in rectangular order upon the properties of a medium. *Phil. Mag., 56,* 481–502.

Reach W. T., Sykes M. V., Lien D. J., and Davies J. K. (2000) The formation of Encke meteoroids and dust trail. *Icarus, 148,* 80–89.

Rickman H. (1989) The nucleus of Comet Halley: Surface structure, mean density, gas and dust production. *Adv. Space Res., 9,* 59–71.

Rickman H., Fernández J. A., and Gustafson B. Å. S. (1990) Formation of stable dust mantles on short-period comet nuclei. *Astron. Astrophys., 237,* 524–535.

Ross R. G. and Kargel J. S. (1998) Thermal conductivity of solar system ices with special reference to Martian polar caps. In *Solar System Ices* (B. Schmitt et al., eds.), pp. 33–62. Kluwer, Dordrecht.

Schmitt B., Espinasse S., Grin R. J. A., Greenberg J. M., and Klinger J. (1989) Laboratory studies of cometary ice analogues. In *Proceedings of an International Workshop on Physics and Mechanics of Cometary Materials* (J. Hunt and T. D. Guyeme, eds.), pp. 65–69. ESA SP-302, Noordwijk, The Netherlands.

Schmitt B., Espinasse S., and Klinger J. (1991) A possible mechanism for outbursts of Comet P/Halley at large heliocentric distances (abstract). In *Abstracts for the 54th Annual Meeting of the Meteoritical Society,* p. 208. LPI Contribution No. 766, Lunar and Planetary Institute, Houston.

Seiferlin K., Koemle N. I., Kargl G., and Spohn T. (1996) Line heat-source measurements of the thermal conductivity of porous ice, ice and mineral powders under space conditions. *Planet. Space Sci., 44,* 691–704.

Sekanina Z. (1979) Fan-shaped coma, orientation of rotation axis, and surface structure of a cometary nucleus. I — Test of a model on four comets. *Icarus, 37,* 420–442.

Sekanina Z. (1984) Disappearance and disintegration of comets. *Icarus, 58,* 81–100.

Senay M. C. and Jewitt D. (1994) Coma formation driven by carbon monoxide release from Comet P/Schwassmann–Wachmann 1. *Nature, 371,* 229–231.

Shoshany Y., Heifetz E., Prialnik D., and Podolak M. (1997) A model for the changing pore structure and dust grain size distribution in a porous comet nucleus. *Icarus, 126,* 342–350.

Shoshany Y., Podolak M., and Prialnik D. (1999) A Monte-Carlo model for the flow of dust in a porous comet nucleus. *Icarus, 137,* 348–354.

Shoshany Y., Prialnik D., and Podolak M. (2002) Monte-Carlo modeling of the thermal conductivity of cometary ice. *Icarus, 157,* 219–227.

Shul'man L. M. (1972) The evolution of cometary nuclei. In *The Motion, Evolution of Orbits, and Origin of Comets* (G. A. Chebotarev et al., eds.), p. 271. IAU Symposium No. 45, Reidel, Dordrecht.

Sirono S. and Greenberg J. M. (2000) Do cometesimal collisions lead to bound rubble piles or to aggregates held together by gravity? *Icarus, 145,* 230–238.

Sirono S. and Yamamoto T. (1997) Thermal conductivity of granular material relevant to the thermal evolution of comet nuclei. *Planet. Space Sci., 45,* 827–834.

Skorov Yu. V. and Rickman H. (1995) A kinetic model of gas flow in a porous cometary mantle. *Planet. Space Sci., 43,* 1587–1594.

Skorov Yu. V. and Rickman H. (1998) Simulation of gas flow in a cometary Knudsen layer. *Planet. Space Sci., 46,* 975–996.

Skorov Yu. V. and Rickman H. (1999) Gas flow and dust acceleration in a cometary Knudsen layer. *Planet. Space Sci., 47,* 935–949.

Skorov Yu. V., Kömle N. I., Keller H. U., Kargl G., and Markiewicz W. J. (2001) A model of heat and mass transfer in a porous cometary nucleus based on a kinetic treatment of mass flow. *Icarus, 153,* 180–196.

Smoluchowski R. (1981) Amorphous ice and the behavior of cometary nuclei. *Astrophys. J. Lett., 244,* L31–L36.

Smoluchowski R. (1982) Heat transport in porous cometary nuclei. *Proc. Lunar Planet Sci. Conf. 13th,* in *J. Geophys. Res., 87,* A422–A424.

Squires R. E. and Beard D. B. (1961) Physical and orbital behavior of comets. *Astrophys. J., 133,* 657–667.

Squyres S. W., McKay C. P., and Reynolds R. T. (1985) Temperatures within comet nuclei. *J. Geophys. Res., 90,* 12381–12392.

Spohn T., Seiferlin K., and Benkhoff J. (1989) Thermal conductivities and diffusivities of porous ice samples at low pressures and temperatures and possible modes of heat transfer in near surface layers of comets. In *Proceedings of an International Workshop on Physics and Mechanics of Cometary Materials* (J. Hunt and T. D. Guyeme, eds.), pp. 77–81. ESA SP-302, Noordwijk, The Netherlands.

Srinivasan G., Goswami J. N., and Bhandari N. (1999) ^{26}Al in eucrite Piplia Kalan: Plausible heat source and formation chronology. *Science, 284,* 1348–1350.

Stauffer D. and Aharony A. (1994) *Introduction to Percolation Theory, 2nd edition.* Taylor and Francis, Bristol. 181 pp.

Steiner G. and Kömle N. I. (1991) A model of the thermal conductivity of porous water ice at low gas pressures. *Planet. Space Sci., 39,* 507–513.

Steiner G. and Kömle N. I. (1993) Evolution of a porous H_2O-CO_2-ice sample in response to irradiation. *J. Geophys. Res., 98,* 9065–9073.

Tancredi G., Rickman H., and Greenberg J. M. (1994) Thermal chemistry of a cometary nuclei. I. The Jupiter family case. *Astron. Astrophys., 286,* 659–682.

Thomas P. J., Chyba C. F., and McKay C. P., eds. (1997) *Comets and the Origin and Evolution of Life.* Springer-Verlag, New York. 296 pp.

Wallis M. K. (1980) Radiogenic heating of primordial comet interiors. *Nature, 284,* 431–433.

Weidenschilling S. J. (1997) The origin of comets in the solar nebula: A unified model. *Icarus, 127,* 290–306.

Weissman P. R. (1991) Why did Halley hiccup? *Nature, 353,* 793–794.

Weissman P. R. and Kieffer H. H. (1981) Thermal modeling of cometary nuclei. *Icarus, 47,* 302–311.

Weissman P. R. and Kieffer H. H. (1984) An improved thermal model for cometary nuclei. *J. Geophys. Res. Suppl., 89,* 358–364.

West R. M., Hainaut O., and Smette A. (1991) Post-perihelion observations of P/Halley. III — An outburst at R = 14.3 AU. *Astron. Astrophys., 246,* L77–L80.

Whipple F. L. and Stefanik R. P. (1966) On the physics and splitting of cometary nuclei. *Mem. R. Soc. Liege (Ser. 5), 12,* 33–52.

Yabushita S. (1993) Thermal evolution of cometary nuclei by radioactive heating and possible formation of organic chemicals. *Mon. Not. R. Astron. Soc., 260,* 819–825.

Part V:
The Gas Coma

The Composition of Cometary Volatiles

D. Bockelée-Morvan and J. Crovisier
Observatoire de Paris

M. J. Mumma
NASA Goddard Space Flight Center

H. A. Weaver
The Johns Hopkins University Applied Physics Laboratory

The composition of cometary ices provides key information on the chemical and physical properties of the outer solar nebula where comets formed, 4.6 G.y. ago. This chapter summarizes our current knowledge of the volatile composition of cometary nuclei, based on spectroscopic observations and *in situ* measurements of parent molecules and noble gases in cometary comae. The processes that govern the excitation and emission of parent molecules in the radio, infrared (IR), and ultraviolet (UV) wavelength regions are reviewed. The techniques used to convert line or band fluxes into molecular production rates are described. More than two dozen parent molecules have been identified, and we describe how each is investigated. The spatial distribution of some of these molecules has been studied by *in situ* measurements, long-slit IR and UV spectroscopy, and millimeter wave mapping, including interferometry. The spatial distributions of CO, H_2CO, and OCS differ from that expected during direct sublimation from the nucleus, which suggests that these species are produced, at least partly, from extended sources in the coma. Abundance determinations for parent molecules are reviewed, and the evidence for chemical diversity among comets is discussed.

1. INTRODUCTION

Much of the scientific interest in comets stems from their potential role in elucidating the processes responsible for the formation and evolution of the solar system. Comets formed relatively far from the Sun, where ices can condense, and the molecular inventory of those ices is particularly sensitive to the thermochemical and physical conditions of the regions in the solar nebula where material agglomerated into cometary nuclei. An important issue is the extent to which cometary ices inherited the molecular composition of the natal presolar dense cloud vs. the role of subsequent chemistry and processing in the solar nebula (*Irvine et al.*, 2000a; *Ehrenfreund et al.*, 2004; *Lunine and Gautier*, 2004). Though a variety of subtle evolutionary mechanisms operated for cometary nuclei during their long storage in the Oort cloud and Kuiper belt (*Stern*, 2003) and, for short-period comets, during their many passages close to the Sun, the bulk composition of cometary nuclei is still regarded to be in large part pristine, except possibly for the most volatile ices. Thus, observing comets today provides a window through which we can view an earlier time when the planets were forming.

In this chapter, we discuss our current knowledge of the composition of cometary nuclei as derived from observations in cometary comae of molecular species that sublimate directly from the nucleus. In the common cometary terminology, which will be used here, these species are called *parent molecules*, while their photodestruction products are called *daughter* products. Although there have been *in situ* measurements of some parent molecules in the coma of 1P/Halley using mass spectrometers, the majority of results on the parent molecules have been derived from remote spectroscopic observations at ultraviolet (UV), infrared (IR), and radio wavelengths. The past decade has seen remarkable progress in the capabilities at IR and radio wavelengths, in particular, and over two dozen parent cometary molecules have now been detected. Many new identifications were obtained in Comet C/1996 B2 (Hyakutake), which passed within 0.1 AU of Earth in March 1996, and in the exceptionally active Comet C/1995 O1 (Hale-Bopp). We discuss how each of these molecules was identified, how the spectroscopic data are used to derive abundances, and we describe the abundance variations observed among comets.

We also discuss our current knowledge of the noble gas abundances in cometary nuclei, as these are potentially diagnostic of the role played by cometary bombardment on the formation and evolution of planetary atmospheres. Noble gas abundances are also key indicators of the temperature conditions and condensation processes in the outer solar nebula.

The spatial distribution of several molecules has been investigated *in situ*, by IR and UV long-slit spectroscopy, and by radio mapping. We present observational evidence for the presence of extended sources of molecules in the coma. The brightness distribution and velocity shifts of radio emission lines are diagnostic of the outgassing pattern from the nucleus, and recent results obtained by millimeter interferometry are presented.

Isotopic abundances often provide important insights into the evolutionary history of matter, and we discuss the various isotopic data that have been obtained for cometary parent molecules.

For molecules having at least two identical nuclei, the internal energy levels are divided into different spin species (ortho and para in the simplest case), and we discuss how the observed distribution among these spin species may provide information on the formation temperature of cometary nuclei.

2. INVESTIGATION OF PARENT MOLECULES

2.1. Daughter Products

Most of the cometary species observed at optical and UV wavelengths are radicals, atoms, and ions that do not sublimate directly from the nucleus but are instead produced in the coma, usually during the photolysis of the parent molecules, but also by chemical reactions. The discussion of these secondary species in cometary comae is covered in this book by *Feldman et al.* (2004). The only exception is that CO Cameron band emission, some of which is produced in a prompt process following the photodissociation of CO_2, is discussed below.

2.2. Mass Spectrometry

The *Giotto* spacecraft, which flew by 1P/Halley in March 1986, was equipped with two mass spectrometers suitable for composition measurements: the Neutral Mass Spectrometer (NMS) and the Ion Mass Spectrometer (IMS). These two instruments had a mass resolution of 1 amu/q and mass ranges of 12–50 and 1–57 for NMS and IMS, respectively. The Positive Ion Cluster Composition Analyser of the Rème Plasma Analyser also had some capabilities for studying ions in the 12–100 amu/q range. These instruments provided much new information on the molecular and isotopic composition of cometary volatiles, as detailed in section 5 and section 9. However, the analyses of these data were not straightforward, owing to the limited mass resolution and the need for detailed chemical modeling to deduce neutral abundances from the ion mass spectra (see review of *Altwegg et al.*, 1999, and references therein).

2.3. Spectroscopy

Most electronic bands of cometary parent molecules fall in the UV spectral region. As discussed in section 3.1.2, the electronic states of polyatomic molecules usually predissociate, so the absorption of UV sunlight by these species leads to their destruction rather than fluorescence. As a result, the UV study of cometary parent molecules reduces to investigations of diatomic molecules (e.g., CO and S_2) and the atoms present in nuclear ices, specifically noble gases. Since the terrestrial atmosphere blocks UV light from reaching the surface of Earth, cometary UV investigations are generally conducted from space platforms.

Most parent molecules have strong fundamental bands of vibration in the 2.5–5 μm region, where there is abundant solar flux for exciting infrared fluorescence and where thermal radiation and reflected sunlight from dust is not very strong. This near-IR spectral region, which is partly accessible from Earth-based observations, has been a rich source of molecular identifications in cometary comae. The first high-spectral-resolution measurements ($\lambda/\delta\lambda \sim 10^5$–$10^6$) in this region were made during observations of Comet 1P/Halley from the NASA Kuiper Airborne Observatory (KAO) in 1985. The entire region was explored at modest spectral resolution by the Infra Krasnoe Spectrometre (IKS) instrument onboard the *Vega* probe to 1P/Halley ($\lambda/\delta\lambda \sim 50$) and, more recently, by the Infrared Space Observatory (ISO) observations of Comet Hale-Bopp and 103P/Hartley 2 ($\lambda/\delta\lambda \sim 1500$). The advent of sensitive, high-dispersion spectrometers at the NASA Infrared Telescope Facility (IRTF) and Keck telescopes revolutionized this field. Their spectral resolving power (~20,000) allows resolution of the rotational structure of the vibrational bands, which is very important for investigating the internal excitation of the molecules and for unambiguously identifying molecules in the spectrally confused 3.3–3.6 μm region, where the fundamental C–H stretching vibrations lie for all hydrocarbons. Infrared spectroscopy is particularly useful for studying symmetric molecules, which do not have permanent electric dipole moments and thus cannot be observed in the radio range.

Radio spectroscopy is a powerful technique for studying molecules in cold environments via their rotational transitions. This technique has produced many discoveries of cometary parent molecules and is more sensitive than IR and UV spectroscopy for comets observed at large heliocentric distances. With a few exceptions, observations have been made in the 80–460 GHz frequency range from ground-based telescopes. The Submillimeter Wave Astronomy Satellite (SWAS) and the Odin satellite, which observed the 557 GHz H_2O rotational line in several comets, initiated investigations of submillimetric frequencies not observable from Earth. Radio spectrometers provide high spectral resolution ($\lambda/\delta\lambda \sim 10^6$–$10^7$), which permits investigations of gas kinematics through line profile measurements (typical cometary line widths are ~2 km s^{-1}), and which eliminates most ambiguities related to line blending, galactic confusion, or instrumental effects. Most detected molecules were observed in several lines, thereby securing their identification.

3. EXCITATION PROCESSES: LINE/BAND INTENSITIES

3.1. Overview: Main Processes

The interpretation of line or band intensities of parent molecules in terms of column densities and production rates requires the knowledge of the processes that govern their excitation and emission in the coma. Two kinds of excitation mechanisms can be distinguished: radiative processes and collisional excitation.

3.1.1. Radiative vibrational excitation. For most parent molecules, the main radiative excitation process is radiative excitation of the fundamental bands of vibration by direct solar radiation (*Mumma*, 1982; *Yamamoto*, 1982; *Crovisier and Encrenaz*, 1983; *Weaver and Mumma*, 1984). The pumping rate g_{lu} (s^{-1}) for an individual ro-vibrational transition $l \rightarrow u$ is given by

$$g_{lu} = \frac{c^3}{8\pi h\nu_{ul}^3} \frac{w_u}{w_l} A_{ul} \mathcal{J}(\nu_{ul}) \qquad (1)$$

where the lower level l belongs to the v'' vibrational state ($v'' = 0$ for fundamental bands), and the upper level u belongs to the excited v' vibrational state. ν_{ul} is the frequency of the transition, w_l and w_u are the statistical weights of the lower and upper levels, respectively, and A_{ul} is the spontaneous emission Einstein coefficient for $u \rightarrow l$. $\mathcal{J}(\nu_{ul})$ is the energy density per unit frequency of the radiation field at the frequency ν_{ul}. The solar radiation in the infrared can be described approximately by a blackbody at $T_{bb} = 5770$ K and having solid angle Ω_{bb} ($\Omega_{bb}/4\pi = 5.42 \ 10^{-6} \ r^{-2}$, where r is the heliocentric distance in AU). Then

$$g_{lu} = \frac{\Omega_{bb}}{4\pi} \frac{w_u}{w_l} A_{ul} (e^{h\nu_{ul}/kT_{bb}} - 1)^{-1} \qquad (2)$$

The band excitation rate $g_{v''v'}$, which is the relative number of molecules undergoing vibrational $v'' \rightarrow v'$ excitation through all possible $l \rightarrow u$ transitions within the (v',v'') band at frequency $\nu_{v'v''}$, can be approximated by (*Crovisier and Encrenaz*, 1983)

$$g_{v''v'} = \frac{\Omega_{bb}}{4\pi} A_{v'v''} (e^{h\nu_{v'v''}/kT_{bb}} - 1)^{-1} \qquad (3)$$

$A_{v'v''}$ is the band spontaneous emission Einstein coefficient, which can be related to the total band strength measured in the laboratory. Practically, the individual spontaneous emission rates A_{ul} required to compute the individual excitation rates g_{lu} (equation 2) can be derived from absorption line intensities measured in the laboratory. The HITRAN (*Rothman et al.*, 2003) and GEISA (*Jacquinet-Husson et al.*, 1999) databases list absorption line intensities and frequencies for ro-vibrational and pure rotational transitions of many molecules. More extensive databases for pure rotational transitions are those of the Jet Propulsion Laboratory (*Pickett et al.*, 1998) and the University of Cologne (*Müller et al.*, 2001). For linear or symmetric-top molecules without electronic angular momentum, simple formulae approximate the A_{ul} and g_{lu} quantities as a function of the total band Einstein coefficient $A_{v'v''}$ and excitation rate $g_{v''v'}$, and the rotational quantum numbers (*Bockelée-Morvan and Crovisier*, 1985). Typically, the strongest fundamental vibrational bands of cometary parent molecules have spontaneous emission

Einstein coefficients $A_{v'v''}$ in the range 10–100 s^{-1}, and band excitation rates $g_{v''v'}$ of a few 10^{-4} s^{-1}. Harmonic and combination bands have intrinsic strengths, and thus excitation rates, much smaller than those of the fundamental bands. The pumping from excited vibrational states is also weak. Indeed, their population is negligible with respect to the population of the ground vibrational state. This can be demonstrated easily. If we ignore collisional excitation, which is negligible for vibrational bands as explained in section 3.1.5, and do not consider possible deexcitation of the v' excited state to vibrational states other than the ground state (this corresponds to pure resonant fluorescence), then the population $n_{v'}$ of this v' band at equilibrium between solar pumping and spontaneous decay is

$$n_{v'} = n_{v''} \frac{g_{v''v'}}{A_{v'v''}} \qquad (4)$$

where $n_{v''}$ is the population of the ground vibrational state $v'' = 0$. Combining equations (3) and (4), $n_{v'}/n_{v''}$ only depends upon the frequency of the band and heliocentric distance r, and is equal to a few 10^{-6} at 1 AU from the Sun for most bands. It can be also shown that the timescale for equilibration of this excited v' vibrational state is $1/A_{v'v''}$, typically a fraction of second: The total vibrational populations reach radiative equilibrium almost instantly after release of the molecules from the nucleus. For the same reasons (low populations and small radiative lifetimes), steady-state is achieved locally for the rotational levels within the vibrational excited states. As will be discussed later, this is generally not the case for the rotational levels in the ground vibrational state.

Besides the direct solar radiation field, the vibrational bands can be radiatively excited by radiation from the nucleus and the dust due to scattering of solar radiation or their own emission in the thermal infrared. *Crovisier and Encrenaz* (1983) showed that all these processes are negligible, except excitation due to dust thermal emission, which can be important in the inner comae of active comets for vibrational bands at long wavelengths (>6.7 μm).

3.1.2. Radiative electronic excitation. The electronic bands of diatomic and polyatomic molecules fall in the UV range. Owing to the weak solar flux at these wavelengths, the excitation rates of electronic bands are small compared to vibrational excitation rates. For example, the total excitation rate of the $A^1\Pi$ state of CO by the absorption of solar photons near 1500 Å, which leads to resonance fluorescence in the CO $A^1\Pi$–$X^1\Sigma^+$ Fourth Positive Group, is ~1–2 × 10^{-6} s^{-1} at $r = 1$ AU (*Tozzi et al.*, 1998). The latter is roughly two orders of magnitude smaller than the excitation rate by solar radiation of the CO $v(1$–$0)$ band at 4.7 μm [2.6 × 10^{-4} s^{-1} at 1 AU (*Crovisier and Le Bourlot*, 1983)]. Therefore, the populations of the ground state rotational levels are not significantly affected by electronic excitation. In addition, electronic bands of polyatomic molecules are often dissociative or predissociative, and their excitation by the Sun generally only produces weak UV fluorescence. This explains why cometary parent molecules are rarely identified from their electronic bands in UV spectra. The resonance transitions of neutral atoms, including the noble gases that may be present in the nucleus, are at UV and far UV (FUV) wavelengths, but the excitation rates are relatively small because of the low solar flux in these regions. The electronic excitation of CO and S_2, as observed in UV cometary spectra, is reviewed in section 3.4. The computation of electronic excitation rates does not differ much in principle from that of vibrational excitation rates. However, in the UV range the solar spectrum shows strong and narrow Fraunhofer absorption lines, and cannot be approximated by a blackbody. This fine structure results in absorption probabilities that depend on the comet's heliocentric radial velocity, as first pointed out by *Swings* (1941) for the CN radical. This so-called *Swings effect* can introduce large variations in the fluorescence emission spectrum of electronic bands.

3.1.3. Radiative rotational excitation and radiation trapping. Pure rotational excitation by sunlight is negligible because of the weakness of the solar flux at the wavelengths of the rotational transitions. However, at r > 3 AU, rotational excitation by the 2.7 K cosmic background radiation competes with vibrational excitation (which varies according to r^{-2}), and must be taken into account in fluorescence calculations (*Biver et al.*, 1999a).

In the specific case of the H_2O molecule, the rotational excitation is strongly affected by self-absorption effects. Owing to large H_2O densities in the coma, many rotational H_2O lines are optically thick and trap line photons emitted by nearby H_2O molecules. This was modeled by *Bockelée-Morvan* (1987) in the local approximation, using an escape probability formalism. The net effect of radiation trapping is to delay the radiative decay of the rotational levels to the lower states and to maintain local thermal equilibrium at a lower density than would have been required in optically thin conditions (*Weaver and Mumma*, 1984; *Bockelée-Morvan*, 1987). The lower rotational states of H_2O are affected by this process up to distances of a few 10^4 km when the H_2O production rate Q(H_2O) is ~10^{29} molecules s^{-1}.

3.1.4. Fluorescence equilibrium. When the excitation is determined solely by the balance between solar pumping and subsequent spontaneous decay, this establishes a condition called *fluorescence equilibrium*. For the rotational levels within the ground vibrational state, fluorescence equilibrium is reached in the outer, collisionless coma. For molecules with large dipole moments (μ) and large rotational constants, such as H_2O (μ = 1.86 D) or HCN (μ = 2.99 D), infrared excitation rates are generally small compared to the vibrational and rotational Einstein A-coefficients, so that most of the molecules relax to the lowest rotational levels of the ground vibrational state (Fig. 1). Heavy molecules with small dipole moment (e.g., CO with μ = 0.11 D) and symmetric species will have, in contrast, a warm rotational distribution at fluorescence equilibrium. The rotational population distribution gets colder as the comet moves far from the Sun because vibrational excitation becomes less

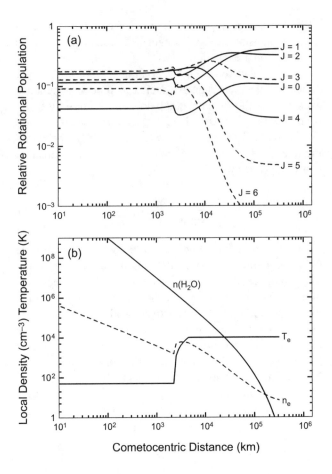

Fig. 1. (a) Rotational population distribution of HCN as a function of distance to nucleus for a comet at 1 AU from the Sun with Q(H_2O) = 10^{29} molecules s^{-1} (from the model of *Biver et al.*, 1999a). (b) H_2O local density n(H_2O), electronic density n_e and temperature T_e in the model. The gas kinetic temperature is 50 K throughout the coma. The population distribution evolves from thermal equilibrium in the inner coma, to fluorescence equilibrium in the outer coma. The discontinuities at 2×10^3 km are due to the sharp rise of the electron temperature, from 50 to 10,000 K.

efficient. The timescale for rotational equilibration, which is mainly controlled by pure rotational relaxation for molecules having a nonzero electric dipole moment (vibrational relaxation is more rapid), exceeds 10^4 s for most detected molecules. Some molecules never reach equilibrium during their lifetime.

3.1.5. Collisional excitation. Collisions, generally involving H_2O molecules and/or electrons, are important in determining the rotational excitation of molecules in the inner coma. For comets at large heliocentric distances, where the CO production rate is much larger than the H_2O production rate, collisional excitation is provided by CO. Collisions with ions are generally considered to be unimportant for parent molecules, but this question has not yet been properly addressed.

Owing to the low temperatures throughout the inner coma [10–100 K (cf. *Combi et al.*, 2004)], collisions do not significantly populate either the vibrational or electronic levels of molecules, and the steady-state vibrational and electronic population distributions are determined by radiative processes. Collisions can quench the fluorescence of vibrational bands, but this is generally unimportant, except possibly within a few kilometers of the surface of the nucleus (*Crovisier and Encrenaz,* 1983; *Weaver and Mumma,* 1984).

Collisions thermalize the rotational population of the ground vibrational state at the kinetic temperature of the gas. The collision rate C (s^{-1}) is given by

$$C = \sigma_c n(r_c)\bar{v} \qquad (5)$$

where σ_c is the collision cross-section, $n(r_c)$ is the local density of the collision partner at the cometocentric distance r_c, and $\bar{v}$ is the relative speed of the impinging species. To treat collisional excitation properly, we must understand how collisions connect individual rotational states, that is, specify the collision cross-sections σ_{ij} for each i → j transition. However, there is little experimental or theoretical information on collisional processes involving neutral molecules (H$_2$O or CO). Not only are the line-by-line cross-sections not available, but the total cross-sections for collisional deexcitation, which could, in principle, be derived from laboratory measurements of line broadening, are poorly documented for most cometary species. In current cometary excitation models, total cross-sections of ~1–5 ×10^{-14} cm^2 are assumed (e.g., *Chin and Weaver,* 1984; *Bockelée-Morvan,* 1987; *Crovisier,* 1987; *Biver et al.,* 1999a), based on the broadening of CO and H$_2$O lines by collisions with H$_2$O. *Chin and Weaver* (1984) introduced a ΔJ dependence on the rotational CO–H$_2$O cross-sections and pointed out that collisional excitation of CO is rather insensitive to this dependence as long as the total cross-section is fixed.

The role of electron collisions in controlling rotational populations was first investigated in detail by *Xie and Mumma* (1992) for the H$_2$O molecule. This study was motivated by the need for large cross-sections to interpret the relative line intensities of the ν_3 H$_2$O band in Comet 1P/ Halley observed preperihelion with the KAO. It was previously recognized that, in the inner coma, inelastic collisions with H$_2$O would cool hot electrons to the temperature of the gas, transferring their translational energy into rotational H$_2$O excitation (*Ashihara,* 1975; *Cravens and Korosmezey,* 1986). Unlike neutral-neutral collisions, theoretical determinations of rotational cross-sections are available for collisions involving electrons. Relatively simple formulae were obtained using the Born approximation by Itikawa (1972), which show that cross-sections are directly proportional to the rotational A$_{ul}$ of the transitions and are also a function of the kinetic energy of the colliding electrons. Cross-sections are large for molecules with large dipole moments (A$_{ul} \propto \mu^2$), typically exceeding those for neutral-neutral collisions by two or three orders of magnitude for electrons thermalized at 50 K. Using the electron temperature and density profile measured *in situ* by *Giotto, Xie and Mumma* (1992) showed that, for a Halley-type comet, the molecu-

lar excitation by e$^-$–H$_2$O collisions exceeds that by H$_2$O– H$_2$O collisions at cometocentric distances ≥3000 km from the nucleus. Neutral H$_2$O collisions dominate in the inner coma because the n(H$_2$O)/n(e$^-$) local density ratio is very large (Fig. 1). Observational evidence for the important role played by collisions with electrons is now abundant [e.g., *Biver et al.* (1999a) for the excitation of HCN]. So far, the modeling of this process is subject to large uncertainties, as the electron density and temperature in the coma are not well-known quantities. Figure 1 shows an exemple of the electron density and temperature profiles used by *Biver et al.* (1999a) for modeling excitation by collisions with electrons. These profiles are based on measurements made *in situ* in Comet Halley, and the dependences with H$_2$O production rate and heliocentric distance expected from theoretical modeling.

3.1.6. Non-steady-state calculations. The evolution of the population distribution with distance to the nucleus has been studied for a number of molecules: CO (*Chin and Weaver,* 1984; *Crovisier and Le Bourlot,* 1983), H$_2$O (*Bockelée-Morvan,* 1987), HCN (*Bockelée-Morvan et al.,* 1984), H$_2$CO (*Bockelée-Morvan and Crovisier,* 1992), CH$_3$OH (*Bockelée-Morvan et al.,* 1994), and linear molecules (*Crovisier,* 1987). Collisional excitation by electrons was included in more recent works (e.g., *Biver et al.,* 1999a). These studies solve the time-dependent equations of statistical equilibrium, as the molecules expand in the coma

$$\frac{dn_i}{dt} = -n_i \sum_{j \neq i} p_{ij} + \sum_{j \neq i} n_j p_{ji} \qquad (6)$$

where the transition rate p_{ij} from level i to j, of energy E_i and E_j respectively, may involve collisional excitation (C_{ij}), radiative excitation (g_{ij}), and/or spontaneous decay (A_{ij}) terms. If we omit radiation trapping effects

$$p_{ij} = C_{ij} + g_{ij} \quad \text{if } E_i < E_j \qquad (7)$$

$$p_{ij} = C_{ij} + A_{ij} \quad \text{if } E_i > E_j \qquad (8)$$

The coupled differential equations (equation (6)) are "stiff", as they contain rates with time constants differing by several orders of magnitude, and their solution requires special techniques, such as the Gear method (cf. *Chin and Weaver,* 1984). In contrast, the fluorescence equilibrium solution can be simply computed by matrix inversion.

Figure 1 shows the evolution of the population of the lowest rotational levels of the HCN molecule with distance to the nucleus. At some distance in the coma, collision excitation can no longer compete with rotational spontaneous decay, and the population distribution evolves to fluorescence equilibrium. Because radiative lifetimes vary among the levels, the departure from local thermal equilibrium (LTE) occurs separately for each rotational level. The size of the LTE region also varies greatly among molecules, as shown in *Crovisier* (1987), where the evolution of the rotational population distribution is computed for a number of linear molecules. Molecules with small dipole moments (e.g., CO) have long rotational lifetimes and correspondingly larger LTE regions. Symmetric molecules with no dipole moment, such as CO_2, cannot relax to low rotational levels; high rotational levels become more and more populated as the molecules expand in the coma.

Low-lying rotational levels maintain thermal populations up to a few 10^3 km from the nucleus in moderately active comets ($Q_{H_2O} \approx 10^{29}$ molecules s^{-1}) near 1 AU from the Sun. This implies that the thermal approximation is a good one to describe the rotational structure of vibrational bands observed by long-slit spectroscopy (section 3.3). On the other hand, the thermal equilibrium approximation may not be valid for the interpretation of rotational line emission observed in the radio range, owing to the large beam size of radio antennas.

3.2. Rotational Line Intensities

When rotational lines are optically thin and spontaneous emission dominates over absorption of the continuum background and stimulated emission, their line flux F_{ul} (in W m^{-2} or Jy km s^{-1}) is given by

$$F_{ul} = \frac{\Omega}{4\pi} h\nu_{ul} A_{ul} \langle N_u \rangle \qquad (9)$$

where Ω is the solid angle subtended by the main beam of the antenna. (The diffraction-limited beam pattern of circular radio antennae is well approximated by a two-dimensional gaussian, which width at half power defines the main-beam solid angle.) $\langle N_u \rangle$ is the column density within the upper transition state u, and is obtained by volume integration over the beam pattern of the density times the fractional population in the upper state. When n_u is constant within the beam, $\langle N_u \rangle$ is equal to $n_u \langle N \rangle$, where $\langle N \rangle$ is the total molecu-

lar column density that is related to the molecular production rate (section 4).

In the radio domain, line intensities are usually expressed in term of *equivalent brightness temperatures* T_B, where T_B is related to F_{ul} through the Rayleigh-Jeans limit ($h\nu \ll kT$) of the Planck function. The line area integrated over velocity in K km s^{-1} is then related to the column density through

$$\int T_B dv = \frac{hc^3 A_{ul}}{8\pi k\nu_{ul}^2} \langle N_u \rangle \qquad (10)$$

In most observational cases, radio antennas are sensitive to molecules present in the intermediate region between thermal and fluorescence equilibrium. Time-dependent excitation models, as described in section 3.1.6, are thus required to derive $\langle N \rangle$ from the observed line area $\int T_B dv$. Because these models rely on ill-known collisional excitation parameters (section 3.1.5), observers try, as much as possible, to observe several rotational lines of the same molecule. This permits them to determine the rotational temperature that best describes the relative population of the upper states, given the observed line intensities. The inferred rotational temperature can then be compared to that predicted from modeling, thereby constraining the free parameters of the model. Methanol has multiplets at 165 and 157 GHz that sample several rotational levels of same quantum number J. Their observations are particularly useful, as the rotational temperature derived from these lines is similar to the kinetic temperature in the collisional region (*Bockelée-Morvan et al.*, 1994; *Biver et al.*, 1999a, 2000). This is also the case of the 252-GHz lines shown in Fig. 2. Other series of lines (e.g., the 145- or 242-GHz multiplets of CH_3OH, or the HCN lines) exhibit rotational temperatures that are intermediate between the kinetic temperature of the inner coma and the rotational temperature at fluorescence equilibrium. These lines can be used to constrain the collision rates. Constraints can also be ob-

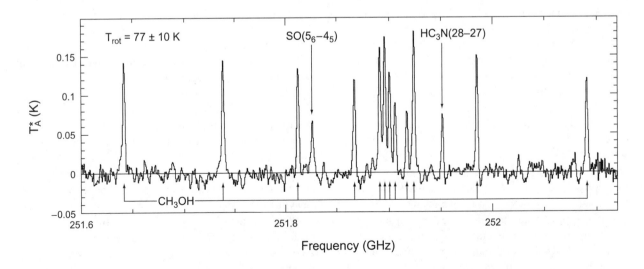

Fig. 2. Wideband spectrum of Comet Hale-Bopp observed on February 21.7, 1997 at the CSO showing twelve J_3–J_2 A lines of CH_3OH, the 5_6–4_5 line of SO, and, in the image sideband at 254.7 GHz, the J(28–27) line of HC_3N (*Lis et al., 1999*).

tained from observations at offset positions from the nucleus (*Biver et al.,* 1999a).

Most rotational lines observed in comets are optically thin because of small molecular column densities. The only exceptions encountered were the J(4–3) HCN line observed in C/1996 B2 (Hyakutake) (*Lis et al.,* 1997; *Biver et al.,* 1999a) and Hale-Bopp (*Meier et al.,* 1998b), the H_2O rotational lines observed with ISO in Comet Hale-Bopp (*Crovisier et al.,* 1997), and the $1_{10}-1_{01}$ line of H_2O observed with the SWAS and Odin satellites in a few comets (*Neufeld et al.,* 2000; *Lecacheux et al.,* 2003). For the H_2O $1_{10}-1_{01}$ line, self-absorption effects result in asymmetric line shapes, indeed observed in high-resolution spectra obtained with Odin (*Lecacheux et al.,* 2003).

3.3. Intensity of Ro-Vibrational Lines and Vibrational Bands

As for pure rotational lines, the line flux F_{ul} (W m^{-2}) of optically thin ro-vibrational lines u → l (u within v', l within v") is given by equation (9), where Ω is the solid angle corresponding to the field of view. F_{ul} can be also expressed as a function of the emission rate (the so-called g-factor) of the line $g_{ul} = A_{ul}n_u$ (s^{-1}), assuming n_u to be constant within the field of view

$$F_{ul} = \frac{\Omega}{4\pi} h\nu_{ul}g_{ul}\langle N\rangle \qquad (11)$$

Neglecting collisional excitation, the emission rate g_{ul} is related to the fractional populations n_j within the ground vibrational state v = 0 through

$$g_{ul} = A_{ul} \frac{\displaystyle\sum_{j,v=0} n_j g_{ju}}{\displaystyle\sum_v \sum_j A_{uj}} \qquad (12)$$

where the summation in the denominator is made over all possible vibrational decays (v' → v = 0 and v' → v hotbands, including the v' → v = v" band to which the u → l transition belongs). In the righthand term of equation (12), the g_{ju} coefficients are the excitation rates due to solar pumping defined in equation (2). Equation (12) is readily obtained by solving equation (6) for the n_u population within v', assuming steady-state (section 3.1.1) and neglecting rotational decay within v', which is much slower than vibrational decays.

The band flux is related to the total emission rate of the band $g_{v'v"}$ through a formula similar to equation (11). In the case of pure resonance fluorescence, $g_{v'v"}$ is equal to the band excitation rate $g_{v"v'}$ given in equation (3).

In most cases, individual g-factors are computed assuming LTE in the ground vibrational state. The retrieved molecular column densities (and production rates) may then depend strongly on the assumed rotational temperature. In Comet Hale-Bopp and other bright comets, many ro-vibrational lines were observed for most molecules, allowing measurement of the rotational temperature T_{rot} and an ac-

curate derivation of the production rate. Figure 3 shows the H_2O ν_3 band of H_2O observed with ISO in Comet Hale-Bopp at 2.9 AU from the Sun (*Crovisier et al.,* 1997), and the synthetic spectrum that best fits the data with T_{rot} = 29 K. Figure 4 shows examples of groundbased IR spectra for Comets Hyakutake and C/1999 H1 (Lee) at r ~ 1 AU, where several lines of H_2O and CO are detected: From Boltzmann analyses of the line intensities, T_{rot} was estimated to ~75 K for both comets (*Mumma et al.,* 2001b).

Opacity effects in the solar pump and for the emitted photons, if present, would affect the effective line-by-line infrared fluorescence emission rates and the intensity distribution within the bands. This was investigated by *Bockelée-Morvan* (1987) for the ν_2 and ν_3 bands of H_2O, and accounted for in the determination of the ortho-to-para ratio of H_2O from the ISO spectra (*Crovisier et al.,* 1997) (Fig. 3; section 10). The opacity of the CO_2 ν_3 band observed by VEGA/IKS in 1P/Halley was taken into account for accurate measurement of the CO_2 production rate in this comet (*Combes et al.,* 1988). Since optical depth effects are stronger in the inner coma, the spatial brightness profile of an optically thick line falls off less steeply with distance to the nucleus than under optically thin conditions. Optical depths of OCS ν_3 and CO v(1–0) ro-vibrational lines were evaluated (*Dello Russo et al.,* 1998; *DiSanti et al.,* 2001; *Brooke et al.,* 2003), in order to investigate whether this could explain their relatively flat spatial brightness distributions in Comet Hale-Bopp (section 7), but the effect was found to be insignificant.

3.4. Electronic Bands

The parent molecules studied via electronic bands at UV/FUV wavelengths are CO, S_2, and, indirectly, CO_2 (Fig. 5). Two other potential constituents of the nucleus, H_2 and N_2, can also fluoresce at UV and FUV wavelengths. H_2 was recently detected during observations of two long-period comets with the Far Ultraviolet Spectroscopic Explorer (FUSE); however, the amount measured was consistent with all the H_2 being derived from the photolysis of H_2O, rather than from sublimation of frozen H_2 in the nucleus (*Feldman et al.,* 2002). Further discussion of cometary H_2 can be found in *Feldman et al.* (2004). Although electronic excitation of N_2 usually leads to predissociation, fluorescence can occur in the (0,0) band of the Carroll-Yoshino system (c4$^1\Sigma_u^+$–X$^1\Sigma_g^+$) at 958.6 Å. Several cometary spectra were taken with FUSE to search for fluorescence from N_2, but only upper limits were derived (P. D. Feldman, personal communication, 2003) (see section 5.6.1).

Observations of CO in the UV range are discussed in section 5.2. The calculation of g-factors for the CO A–X bands is discussed by *Tozzi et al.* (1998), and g-factors for the B–X, C–X, and E–X bands are discussed by *Feldman et al.* (2002). Generally, the Swings effect (see section 3.1.2) is small for all the UV bands of CO, with g-factor variations of only ~20% with heliocentric radial velocity.

Emission in the CO Cameron band system near 2050 Å (a$^3\Pi$–X$^1\Sigma^+$) was discovered during Hubble Space Telescope

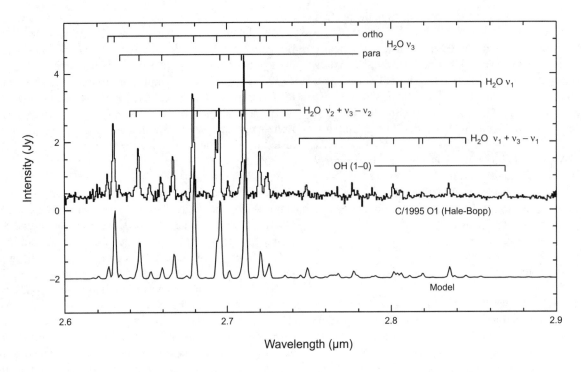

Fig. 3. The region of the ν_3 band of water observed with the ISO short-wavelength spectrometer in Comet C/1995 O1 (Hale-Bopp) on 27 September and 6 October 1996 (top). Line assignations are indicated. The synthetic fluorescence spectrum of water that is the best fit to the data (bottom) corresponds to $Q(H_2O) = 3.6 \times 10^{29}$ molecules s^{-1}, $T_{rot} = 28.5$ K and OPR = 2.45. Adapted from *Crovisier et al.* (1997).

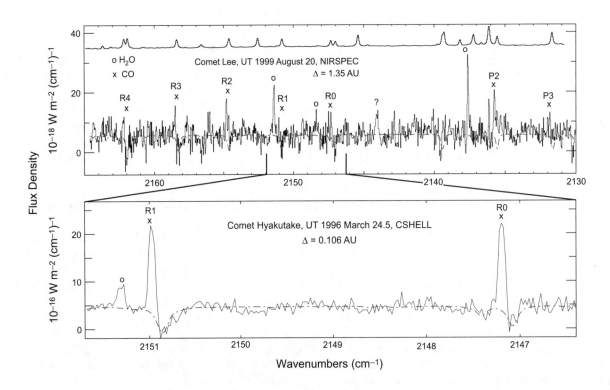

Fig. 4. Detection of CO and H_2O in Comets C/1996 B2 (Hyakutake) and C/1999 H1 (Lee) in the 4.7-μm region (from *Mumma et al.*, 2001b). Several lines of the CO $\nu(1-0)$ and H_2O $\nu_1-\nu_2$ and $\nu_3-\nu_2$ bands are present. The relative intensities of CO and H_2O lines are reversed even though the rotational temperatures were similar for the two comets, providing graphic evidence of the dramatically different CO abundance in these two comets.

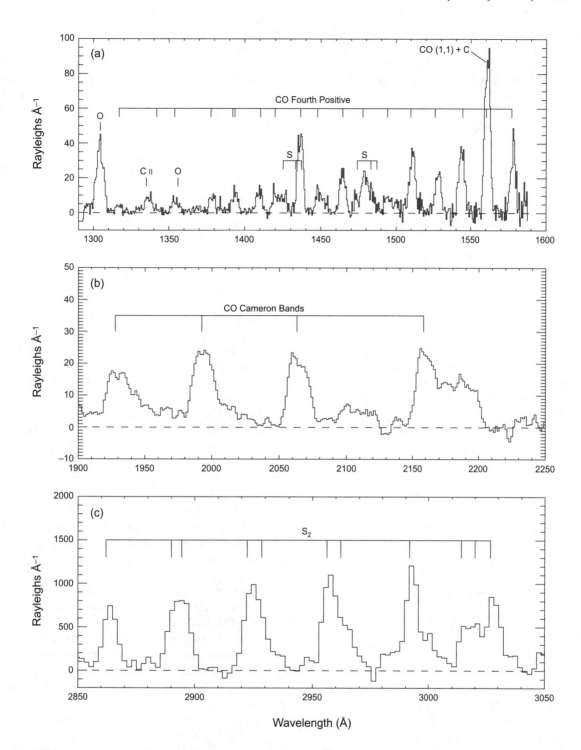

Fig. 5. Portions of the ultraviolet (UV) spectra of C/1996 B2 (Hyakutake) taken on 1996 April 1 with the HST. All the parent molecules detected at UV wavelengths are represented: The top panel shows multiple bands in the Fourth Positive Group of CO; the middle panel shows several bands of the CO Cameron system, which is thought to be produced mainly by prompt emission following the photodissociation of CO_2, and the bottom panel shows multiple bands of the B–X system of S_2. Figure adapted from *Weaver* (1998).

(HST) observations of 103P/Hartley 2 (*Weaver et al.*, 1994), and this spurred a reanalysis of earlier data acquired with the International Ultraviolet Explorer (IUE) that resulted in the detection of Cameron band emission in several other comets (*Feldman et al.*, 1997). The Cameron bands involve transitions between triplet and singlet electronic states, i.e.,

they are electric dipole forbidden, which means that resonance fluorescence cannot be the excitation mechanism. The Cameron bands can be excited during the photodissociation of CO_2, producing CO molecules in the $a^3\Pi$ state that can then decay to the ground state on a timescale of ~10 ms in a process called *prompt* emission (*Weaver et al.*,

1994). In this case, the CO Cameron band emission is directly proportional to the CO_2 production rate, and its intensity can be used to estimate the CO_2 abundance in exactly the same way that observations of the O^1D line near 6300 Å can be used to probe the H_2O production rate (cf. *Feldman et al.*, 2004). Unfortunately, electron impact on CO also produces Cameron band emission fairly efficiently, and this complicates the interpretation of the spectra when both CO and CO_2 are comparably abundant. When spectra are taken at sufficient resolution to resolve the rotational structure in the Cameron bands, the two competing excitation mechanisms can be easily distinguished because the CO molecules produced during the photolysis of CO_2 are rotationally "hot", with a rotational temperature about five times larger than for the CO excited by electron impact (*Mumma et al.*, 1975).

S_2 has been observed through its $B^3\Sigma_u^- - X^3\Sigma_g^-$ system in the near-UV in several comets (section 5.5). Because of its very short lifetime (≈ 500 s), S_2 is concentrated within a small spatial region near the nucleus and observations with high spatial resolution (≤ 500 km) are required to detect it. The photodissociation rate and the B–X g-factor of S_2 are comparable. Thus, a time-dependent model of the excitation is required for accurate interpretation of the emission (*Kim et al.*, 2003; *Reylé and Boice*, 2003).

Cometary emission in electronic bands is generally produced by resonance fluorescence, as are the great majority of cometary emissions observed at optical and near-IR wavelengths. However, the anomalous intensity ratio of the CO C–X and B–X bands in the FUSE spectrum of C/2001 A2 (LINEAR) suggests that some of the B–X emission is produced by e$^-$-impact on CO, while the presence of a "hot" component in the C–X emission is suggestive of an excitation process involving CO_2 (*Feldman et al.*, 2002). As previously discussed, the CO Cameron bands can be excited by both photodissociative and e$^-$-impact processes.

4. DETERMINATION OF PRODUCTION RATES

For estimating relative molecular abundances in the nucleus, the measured column densities (or local densities in the case of *in situ* measurements) are converted into molecular production rates, i.e., outgassing rates at the nucleus. This step requires a good description of the molecular spatial distributions. Most studies use the Haser model (*Haser*, 1957), which assumes that the parent molecule is sublimating from the surface of the nucleus at a constant rate and expands radially outward at constant velocity. Under these conditions, the local density in the coma n is given by

$$n(r_c) = \frac{Q}{4\pi r_c^2 v} e^{-(r_c - r_n)/v\tau} \qquad (13)$$

where Q is the production rate, r_c is the distance from the center of the nucleus, r_n is the radius of the nucleus, v is the outflow speed, and τ is the molecular lifetime. The density is then integrated along the line of sight to obtain the column density.

For the case of a circular observing aperture centered on the nucleus, if the aperture subtends a distance at the comet that is much smaller than the scalelength ($L = v\tau$) of the molecule, the average column density within the aperture is given by

$$\langle N \rangle = \frac{Q}{vd} \qquad (14)$$

where d is the aperture diameter.

If the aperture size is much larger than the scalelength of the molecule, then

$$\langle N \rangle = \frac{4Q\tau}{\pi d^2} \qquad (15)$$

When equations (14) and (15) are not applicable, other methods must be used to relate the column density to the production rate. While convenient tabulations are available for both circular (*Yamamoto*, 1982) and square (*Hoban et al.*, 1991) apertures, the continually increasing power of computers makes the direct integration of equation (13) simple, fast, and accessible to most researchers.

In the limit cases of equations (14) and (15), $\langle N \rangle$ depends on either v or τ. In the intermediate cases, the column density depends on both the lifetime and velocity. Lifetimes have been computed for many parent molecules (*Huebner et al.*, 1992; *Crovisier*, 1994) under both solar maximum and solar minimum conditions and have accuracies of ~20–30% for the well-documented species. But the photodissociation rates of several cometary molecules (e.g., H_2CS, SO_2, NH_2CHO) are unavailable or uncertain by factors of several. Expansion velocities for some molecules can be determined from analysis of observed radio line profiles, but usually the outflow velocities are uncertain by ~30%. There is also the problem that the outflow velocity changes with position in the coma, as molecules are accelerated by photolytic heating in the coma, but typically observers adopt an average outflow speed that is appropriate for the size of the aperture used (i.e., smaller velocities used for smaller apertures).

For molecules released by an extended source, such as H_2CO (section 7), the Haser formula for daughter species (see *Combi et al.*, 2004) is generally used to describe their spatial distribution. Inferred production rates then strongly depend on the scalelength of their parent source, L_p, especially when the field of view samples cometocentric distances smaller than L_p. Any underestimate of L_p will result in an underestimate of the production rate. In this context, there are some uncertainties in the production rates derived from radio observations of distant comets for which sublimation from icy grains is likely and not taken into account in most studies (*A'Hearn et al.*, 1984; *Biver et al.*, 1997; *Womack et al.*, 1997; *Gunnarsson et al.*, 2002). H_2CO production rates obtained in Comet Hale-Bopp at large r are uncertain as well, as there is little information on the heliocentric variation of the H_2CO parent scalelength (see section 7).

With sufficient spatial resolution and mapping, the radial distribution of molecules can be investigated, and production rates can be more accurately determined. Section 7 discusses how native and extended sources of CO molecules are extracted from the analysis of long-slit spectra.

The spatial distribution of cometary molecules is certainly much more complex than assumed by the Haser model. As discussed elsewhere in this book (e.g., *Crifo et al.*, 2004), the production rate may vary on short timescales, outgassing from the nucleus may not be isotropic, and the expansion velocity increases with distance from the nucleus and may have day/night asymmetries. Anisotropic outgassing and/or velocity variations have been considered in a few radio studies, using information provided by the line shapes and mapping (e.g., *Gunnarsson et al.*, 2002; *Veal et al.*, 2000).

5. OBSERVATIONS OF PARENT MOLECULES

In this section, we review the *in situ* measurements and spectroscopic investigations of parent molecules and noble gases. The production rates relative to H_2O (also called abundances in the text) measured in several well-documented comets near their perihelion are listed in Table 1. Upper limits for several undetected molecules are given in Table 2.

5.1. Water

Water is the most abundant constituent of cometary ices and its production rate is used for quantifying cometary activity and for abundance determinations. Its presence in cometary comae was definitively established in the 1970s from observations of H and OH, which showed that these species were produced in appropriate quantities and with spatial distributions and velocities consistent with H_2O photolysis (see the review of *Festou et al.*, 1993).

Water is difficult to measure directly. The fundamental bands of vibration, especially v_3 near 2.7 μm, cannot be observed from the ground because of strong absorption in the terrestrial atmosphere. This band was observed in 1P/Halley and C/1986 P1 (Wilson) from the KAO (*Mumma et al.*, 1986; *Larson et al.*, 1989), in 1P/Halley with the Vega/IKS IR spectrometer (*Combes et al.*, 1988), and with ISO in Comets Hale-Bopp and 103P/Hartley 2 (*Crovisier et al.*, 1997, 1999a,b) (Fig. 3).

Nonresonance fluorescence bands (hot-bands) of H_2O have weaker g-factors, but some are not absorbed by telluric H_2O and thus can be observed from the ground. Direct absorption of sunlight excites molecules from the ground vibrational state to a higher vibrational state, followed by cascade into an intermediate level that is not significantly populated in the terrestrial atmosphere (*Crovisier*, 1984). Hot-band emission from H_2O $v_2 + v_3 - v_2$ was first detected near 2.66 μm in high-dispersion airborne IR spectra of Comets 1P/Halley and C/1986 P1 (Wilson) (*Weaver et al.*,

1986; *Larson et al.*, 1989), but the strong 2.7-μm fundamental bands (v_1 and v_3) blanket this entire region from the ground. However, groundbased IR observations of Comets 1P/Halley and C/1986 P1 (Wilson) indicated the presence of excess flux near 2.8 μm that could not be attributed to H_2O fundamental bands (*Tokunaga et al.*, 1987; *Brooke et al.*, 1989), but was consistent with the expected flux from H_2O hot-bands (*Bockelée-Morvan and Crovisier*, 1989). The hot-band emissions in this spectral region were more extensively sampled by ISO in Comet Hale-Bopp (Fig. 3). High spectral dispersion surveys of the 2.9-μm region obtained in Comets C/1999 H1 (Lee) and 153P/2002 C1 (Ikeya-Zhang) with NIRSPEC at the Keck telescope revealed multiple lines of the $v_1 + v_3 - v_1$, $(v_1 + v_2 + v_3)-(v_1 + v_2)$, and $2v_1 - v_1$ H_2O hot-bands (*Mumma et al.*, 2001a; *Dello Russo et al.*, 2004).

In other spectral regions, the terrestrial atmosphere is generally transparent to H_2O hot-band emissions. Hot-band emission from H_2O was detected in Comets C/1991 T2 (Shoemaker-Levy), 6P/d'Arrest, and C/1996 B2 (Hyakutake) using bands near 2 μm ($v_1 + v_2 + v_3 - v_1$ and $2v_2 + v_3 - v_2$) (*Mumma et al.*, 1995, 1996; *Dello Russo et al.*, 2002a). Production rates were obtained for all three comets, and a rotational temperature was obtained for H_2O in Comet Hyakutake (*Mumma et al.*, 1996). A survey of the CO (1–0) band (4.7 μm) in Comet Hyakutake revealed new emissions that were identified as nonresonance fluorescence from the $v_1 - v_2$ and $v_3 - v_2$ hot-bands of H_2O (*Mumma et al.*, 1996; *Dello Russo et al.*, 2002a). As H_2O and CO can be sampled simultaneously (Fig. 4), preference was given to the 4.7-μm region thereafter (e.g., C/1995 O1 Hale-Bopp (*Weaver et al.*, 1999b; *Dello Russo et al.*, 2000), 21P/Giacobini-Zinner (*Weaver et al.*, 1999a; *Mumma et al.*, 2000), C/1999 H1 (Lee) (*Mumma et al.*, 2001b), C/1999 S4 (LINEAR) (*Mumma et al.*, 2001a).

The rotational lines of H_2O also cannot be observed from the ground, except for a line of one of the trace isotopes (HDO — see section 9). Lines in the far-IR, especially the $2_{12}-1_{01}$, $2_{21}-2_{12}$ and $3_{03}-2_{12}$ lines near 180 μm, were observed by ISO in Comet Hale-Bopp (*Crovisier et al.*, 1997). The fundamental ortho rotational line, $1_{10}-1_{01}$ at 557 GHz, was observed using SWAS (*Neufeld et al.*, 2000) and the Odin satellite (*Lecacheux et al.*, 2003) in C/1999 H1 (Lee), 153P/2002 C1 (Ikeya-Zhang), and several other comets. These lines are very optically thick, which means that the derivation of accurate H_2O production rates requires a reliable model for the H_2O excitation and radiative transfer.

5.2. Carbon Monoxide and Carbon Dioxide

5.2.1. Carbon monoxide (CO). The CO molecule was discovered in comets during a sounding rocket observation of C/1975 V1 (West), when resonance fluorescence in the Fourth Positive Group $(A^1\Pi - X^1\Sigma^+)$ near 1500 Å was detected in the UV spectrum (*Feldman and Brune*, 1976). Emission in these bands has been detected subsequently in nearly every bright ($m_V < 7$) comet observed at UV wavelengths with IUE (cf. *Feldman et al.*, 1997), the HST (cf.

TABLE 1.   Production rates relative to water in comets.

Molecule	1P/Halley	C/1995 O1 (Hale-Bopp)	C/1996 B2 (Hyakutake)	C/1999 H1 (Lee)	C/1999 S4 (LINEAR)	153P/2002 C1 (Ikeya-Zhang)
H_2O	100	100	100	100	100	100
CO	3.5*, 11[1]	12*,[13], 23[13,14]	14*,[25,26], 19–30[25,27,28]	1.8[36]–4[37]	≤0.4[41], 0.9[42]	2[45], 4–5[46,47]
CO_2	3–4[2,3]	6†,15				
CH_4	<0.8[4]	1.5[16]	0.8[16,29]	0.8[38]	0.14[42]	0.5[16]
C_2H_2	0.3[1]	0.1[16]–0.3[17]	0.2[30]–0.5[31]	0.27[38]	<0.12[42]	0.18[48]
C_2H_6	0.4[1]	0.6[17]	0.6[29]	0.67[38]	0.11[42]	0.62[49]
CH_3OH	1.8[5,6]	2.4[14]	2[27,28]	2.1[38]–4[39]	<0.15[42]	2.5[46,47]
H_2CO‡	4[2,7,8]	1.1[14]	1[27,28]	1.3[39]	0.6[41]	0.4[47]
HCOOH		0.09[14]				<0.1[43]
$HCOOCH_3$		0.08[14]				
CH_3CHO		0.02[18]				
NH_2CHO		0.015[14]				
NH_3	1.5[9]	0.7[19]	0.5[32,33]			<0.2[50]
HCN	0.1[10,11]	0.25[14,20]	0.1[27,28]–0.2[34]	0.1[38]–0.3[39]	0.1[42]	0.1[47]–0.2[48]
HNCO		0.10[14]	0.07[28]			0.04[47]
HNC		0.04[14,21]	0.01[28,35]	0.01[39]	0.02[43]	0.005§,[47]
CH_3CN		0.02[14]	0.01[33]			0.01[47]
HC_3N		0.02[14]				<0.01[47]
H_2S	0.4[6]	1.5[14]	0.8[27]	<0.9[39]	0.3[43]	0.8[47]
OCS		0.4[14,22]	0.1[36]			<0.2[47]
SO_2		0.2[14]				
CS_2	0.2[12]	0.2[14]	0.1[27]	0.08[39]	0.12[43]	0.06[47]–0.1[45]
H_2CS		0.05[23]				
NS		≥0.02[24]				
S_2			0.005[37]	0.002[40]	0.0012[44]	0.004[45]

*Production from the nucleus; see text.

†Value at heliocentric distance r = 1 AU extrapolated from the value of 20% measured at r = 2.9 AU, assuming that $[CO_2]/[CO]$ did not change with r.

‡H_2CO abundances refer to production from an extended source.

§Measured at r ~ 1 AU; increased up to 0.02% at r ~ 0.5 AU (N. Biver et al., personal communication, 2003; *Irvine et al.,* 2003).

References: [1] *Eberhardt* (1999); [2] *Combes et al.* (1988); [3] *Krankowsky et al.* (1986); [4] *Altwegg et al.* (1994); [5] *Bockelée-Morvan et al.* (1995); [6] *Eberhardt et al.* (1994); [7] *Meier et al.* (1993); [8] *Mumma and Reuter* (1989); [9] *Meier et al.* (1994); [10] *Despois et al.* (1986); [11] *Schloerb et al.* (1986); [12] *Feldman et al.* (1987); [13] *DiSanti et al.* (2001); [14] *Bockelée-Morvan et al.* (2000); [15] *Crovisier et al.* (1997); [16] *Gibb et al.* (2003); [17] *Dello Russo et al.* (2001); [18] *Crovisier et al.* (2004a); [19] *Bird et al.* (1999); [20] *Magee-Sauer et al.* (1999); [21] *Irvine et al.* (1998); [22] *Dello Russo et al.* (1998); [23] *Woodney* (2000); [24] *Irvine et al.* (2000b); [25] *DiSanti et al.* (2003); [26] *McPhate et al.* (1996); [27] *Biver et al.* (1999a); [28] *Lis et al.* (1997); [29] *Mumma et al.* (1996); [30] *Mumma et al.* (2003); [31] *Brooke et al.* (1996); [32] *Palmer et al.* (1996); [33] *Bockelée-Morvan* (1997); [34] *Magee-Sauer et al.* (2002a); [35] *Irvine et al.* (1996); [36] *Woodney et al.* (1997); [37] *Weaver et al.* (1996); [38] *Mumma et al.* (2001b); [39] *Biver et al.* (2000); [40] *Feldman et al.* (1999); [41] *Weaver et al.* (2001); [42] *Mumma et al.* (2001a); [43] *Bockelée-Morvan et al.* (2001); [44] *Weaver* (2000); [45] *Weaver et al.* (2002b); [46] *DiSanti et al.* (2002); [47] N. Biver et al. (personal communication, 2003); [48] *Magee-Sauer et al.* (2002b); [49] *Dello Russo et al.* (2002b); [50] *Bird et al.* (2002).

Weaver, 1998), and sounding rockets (cf. *Feldman,* 1999). More recently, resonance fluorescence in several bands of the Hopfield-Birge system ($B^1\Sigma^+$–$X^1\Sigma^+$, $C^1\Sigma^+$–$X^1\Sigma^+$, and $E^1\Pi$–$X^1\Sigma^+$) has been detected between 1075 Å and 1155 Å in spectra measured by FUSE (*Feldman et al.,* 2002). Through the end of 2002, CO emission had been detected in a total of 12 comets at UV wavelengths, with $[CO/H_2O]$ abundances ranging from ~0.4% to nearly 30%.

The radio lines of CO are intrinsically weak because of the small dipole moment of this molecule. However, these lines are the most easily detected gaseous emissions for comets at large heliocentric distances (r ≥ 3 AU). The CO

J(2–1) line at 230 GHz was first observed in 29P/Schwassmann-Wachmann 1 (*Senay and Jewitt,* 1994) at r ≈ 6 AU, and subsequently in a few bright comets. In Comets Hyakutake and Hale-Bopp, the J(1–0) and J(3–2) lines were also observed. The J(2–1) line was detected out to r = 14 AU in Comet Hale-Bopp with the Swedish-ESO Submillimetre Telescope (SEST) (*Biver et al.,* 2002a).

The first clear detection of the lines of the v(1–0) IR band of CO near 4.7 μm was obtained during observations of Comet Hyakutake (*Mumma et al.,* 1996; *DiSanti et al.,* 2003). Eight lines of this band were detected in emission, using CSHELL spectrometer at the NASA/IRTF. CO has

TABLE 2.   Molecular upper limits in Comet Hale-Bopp
from radio observations (from *Crovisier et al.,* 2004).

Molecule	(X)/(H$_2$O)
H$_2$O	100
H$_2$O$_2$	<0.03
CH$_3$CCH	<0.045
CH$_2$CO	<0.032
C$_2$H$_5$OH	<0.10
CH$_3$OCH$_3$	<0.45
CH$_3$COOH	<0.06
Glycine I	<0.15
HC$_5$N	<0.003
C$_2$H$_5$CN	<0.01
CH$_2$NH	<0.032
CH$_3$SH	<0.05
NaOH	<0.0003
NaCl	<0.0008

been detected in every comet observed since then with
CSHELL and with NIRSPEC at the Keck Observatory (eight
Oort cloud comets and one Jupiter-family comet) (*Mumma
et al.,* 2003; *Weaver et al.,* 1999a,b). Selected spectra of C/
1999 H1 (Lee) and Comet Hyakutake are shown in Fig. 4.
CO rotational temperatures were obtained from Boltzmann
analyses of the measured spectral line intensities and were
used to extrapolate total production rates from the observed
lines. For eight Oort cloud comets observed by IR groundbased spectroscopy through the end of 2002, the total CO
abundance ranged from 1% to 24% relative to H$_2$O (*Mumma
et al.,* 2003).

CO was investigated by mass spectrometry in 1P/Halley
with the *Giotto* NMS (*Eberhardt et al.,* 1987). As detailed
in section 7.1, these measurements revealed that part of the
CO originated from an extended source. Native and extended sources of CO were separately quantified in a few
comets from long-slit IR observations (section 7.2). Among
eight Oort cloud comets observed at IR wavelengths, the
native abundance [CO/H$_2$O] varies by more than a factor
of 40 [0.4–17% (*Mumma et al.,* 2003) (section 8).

5.2.2. Carbon dioxide (CO$_2$). The presence of carbon
dioxide in cometary comae was indirectly established a long
time ago from the existence of CO$_2^+$ in cometary tails (see
Feldman et al., 2004). It was confirmed by the detection
of the CO$_2$ ν$_3$ band at 4.26 μm. This band is very strong
(g-factor = 2.6 × 10^{-3} s^{-1}), but it cannot be observed from
the ground because of strong absorption from terrestrial
CO$_2$. The ν$_3$ band has only been observed by *Vega*/IKS in
1P/Halley (*Combes et al.,* 1988), and by ISO in Comets
Hale-Bopp (*Crovisier et al.,* 1997, 1999a) and 103P/
Hartley 2 (*Colangeli et al.,* 1999; *Crovisier et al.,* 1999b).
CO$_2$ was also observed in 1P/Halley from the mass 44 peak
in the *Giotto* NMS mass spectra (*Krankowsky et al.,* 1986).
The inferred CO$_2$ production rate relative to H$_2$O was 3–4%
in 1P/Halley and 8–10% in 103P/Hartley 2. It was >20%
in Comet Hale-Bopp, but this comet was only observed at

r > 2.9 AU. This higher value is likely due to the higher
volatility of CO$_2$ compared to H$_2$O. The value of 6% given
in Table 1 is that extrapolated to 1 AU, using the Q(CO$_2$)/
Q(CO) ratio of ~0.3 measured at 2.9 AU.

As discussed in section 3.4, the presence of CO$_2$ is also
indirectly inferred from observations of the CO Cameron
bands near 2050 Å, which can be emitted via prompt emission following the photodissociation of CO$_2$. In practice,
these UV bands can only be used to derive accurate CO$_2$
production rates when the comet is CO-depleted, or is bright
enough to allow observations with sufficient spectral resolution (λ/δλ ≥ 1500) to unambiguously identify the separate emissions from CO$_2$ photodissociation and electron
impact on CO, which overlap in low-resolution data.

5.3.   Methanol (CH$_3$OH), Formaldehyde (H$_2$CO), and Other CHO-bearing Molecules

5.3.1. Methanol (CH$_3$OH). The identification of
CH$_3$OH was first suggested by *Knacke et al.* (1986) to explain the 3.52-μm feature seen near the broad 3.3–3.5-μm
emission in several low-resolution spectra of 1P/Halley.
Hoban et al. (1991) observed the 3.52-μm feature in Comets
C/1989 Q1 (Okazaki-Levy-Rudenko), C/1989 X1 (Austin),
C/1990 K1 (Levy), and 23P/Brorsen-Metcalf, and showed
that its properties were consistent with fluorescence from
low-temperature (70 K) CH$_3$OH in the ν$_3$ band. Figure 6
displays 3.2–3.7-μm spectra of Comets C/1989 X1 (Austin) and C/1990 K1 (Levy) showing CH$_3$OH emission.
Definite identification of CH$_3$OH in cometary comae was
obtained from the detection of several J(3–2) rotational lines
at 145 GHz in Comets C/1989 X1 (Austin) and C/1990 K1
(Levy) at the 30-m telescope of the Institut de Radioastronomie Millimétrique (IRAM) (*Bockelée-Morvan et al.,* 1991,
1994).

Methanol has now been observed in many comets, both
at radio and IR wavelengths, and the CH$_3$OH abundances
inferred from radio and IR spectra are generally consistent
(Table 1). In Comet Hale-Bopp, ~70 rotational lines were
detected at millimeter and submillimeter wavelengths (*Biver
et al.,* 1999b). Methanol rotational lines often appear as
multiplets in radio spectra (Fig. 2), whose analysis provides
clues to the temperature and excitation conditions in the
coma (see section 3.2). High-resolution Keck/NIRSPEC
spectra obtained in the 3-μm region in Comets C/1999 H1
(Lee) and C/1999 S4 (LINEAR) show the P, Q, R structure of
the ν$_3$ band and present, near 3.35 μm, many ro-vibrational
lines belonging to the ν$_2$ and ν$_9$ CH$_3$OH bands (*Mumma et
al.,* 2001a,b).

The ν$_2$ and ν$_9$ CH$_3$ stretching modes of CH$_3$OH are
responsible for about half the total intensity of the 3.3–
3.5-μm emission feature (*Hoban et al.,* 1993; *Bockelée-
Morvan et al.,* 1995). Synthetic spectra, as modeled by
Bockelée-Morvan et al., are shown in Fig. 6. Other weaker
CH$_3$OH combination bands should contribute as well. The
rotational structure of these bands is not yet available, which
makes it difficult to identify new hydrocarbons or CHO-
bearing molecules in this spectral region.

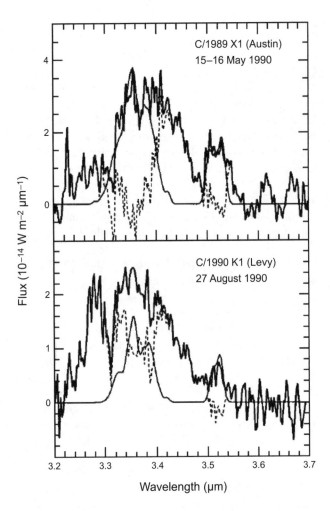

Fig. 6.   3.2–3.7-μm spectra of Comets C/1989 X1 (Austin) and C/1990 K1 (Levy) (thick line). The continuum flux due to thermal emission and scattered light from dust grains has been subtracted. The contributions of the methanol bands (ν_2, ν_3, and ν_9 at 3.33, 3.37, and 3.52 μm respectively) are shown in thin lines. The residual emission spectra, after subtracting methanol emission, are shown in dashed line. Figure adapted from *Bockelée-Morvan et al.* (1995).

Besides the 3.52-μm band, CH_3OH is identified in 1P/Halley from the peak at 33 amu/q present in *Giotto* IMS and *Giotto* NMS mass spectra, which is essentialy due to $CH_3OH_2^+$ (*Geiss et al.,* 1991; *Eberhardt et al.,* 1994). Eberhardt et al. inferred a CH_3OH abundance relative to H_2O that is consistent with the value derived from the 3.52-μm band.

The $[CH_3OH/H_2O]$ abundance ratios measured up to now range from less than 0.15% in Comet C/1999 S4 (LINEAR) to 6%, with many comets around ~2% (see section 8).

5.3.2. Formaldehyde (H_2CO).   Cometary H_2CO was first identified in 1P/Halley, from the signature of its protonated ion in mass-spectra obtained with *Giotto* NMS (*Meier et al.,* 1993). Its spatial distribution was found to differ from that expected for a parent molecule, suggesting the presence of a extended source of H_2CO in the coma (see the discussion in section 7). Its ν_1 band near 3.59 μm was possibly detected in the *Vega*/IKS spectrum of 1P/Halley

(*Combes et al.,* 1988; *Mumma and Reuter,* 1989) This detection is controversial, however, as this band was not detected in groundbased IR spectra of 1P/Halley and several other comets (e.g., *Reuter et al.,* 1992). The detection of H_2CO by IR long-slit spectroscopy is difficult, owing to its low abundance and daughter-like density distribution. Recently, *DiSanti et al.* (2002) reported the detection of the Q branch of the H_2CO ν_1 band in high-resolution spectra of 153P/2002 C1 (Ikeya-Zhang) obtained with CSHELL at the NASA/IRTF.

A detection of the $1_{10}-1_{11}$ line of H_2CO at 6 cm wavelength in Comet 1P/Halley using the Very Large Array (VLA), announced by *Snyder et al.* (1989), is controversial. If real, it would lead to an exceptionally high abundance of H_2CO (see discussion in *Bockelée-Morvan and Crovisier,* 1992). The first definite detection of H_2CO at millimeter wavelengths ($3_{12}-2_{11}$ at ~226 GHz) was in Comet C/1989 X1 (Austin) at IRAM (*Colom et al.,* 1992). Since then, this molecule has been observed via several lines at millimeter and submillimeter wavelengths in several comets. H_2CO was monitored at radio wavelengths in Comet Hale-Bopp (*Biver et al.,* 1997, 1999b, 2002a) (see Fig. 7). It exhibited a steep heliocentric production curve (~$r^{-4.5}$) over the entire range $1 \leq r \leq 4$ AU, which contrasted with the r^{-3}–r^{-2} evolution observed for most molecules. This behavior is related

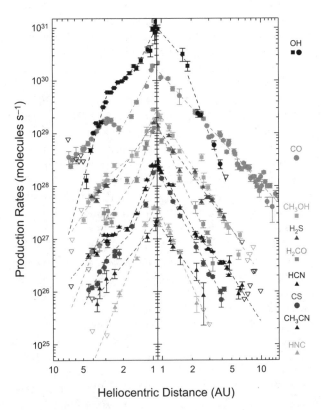

Fig. 7.   Gas production curves in Comet Hale-Bopp from radio observations at IRAM, JCMT, CSO, and SEST telescopes (*Biver et al.,* 2002a). OH production rates are from observations of the 18-cm lines at the Nançay radio telescope. Inverted triangles indicate upper limits in cases of nondetection.

to the production mechanism of H_2CO, which became more efficient when the comet approached the Sun.

The $Q(H_2CO)/Q(H_2O)$ production rate ratio refering to the extended H_2CO production has been estimated to 4% for Comet 1P/Halley. This ratio ranges from 0.13 to 1.3% for comets in which H_2CO has been investigated at millimeter wavelengths (*Biver et al.*, 2002b).

5.3.3. Other CHO-bearing molecules. Several new organic CHO-bearing molecules were identified in Comet Hale-Bopp with millimeter spectroscopy thanks to its high gaseous activity. Formic acid (HCOOH) was detected from four J(10–9) lines near 225 GHz using the Plateau de Bure interferometer (PdBi) of IRAM in single dish mode (Fig. 8) (*Bockelée-Morvan et al.*, 2000). Methyl formate (HCOOCH$_3$) is one of the most complex cometary molecule detected in the gas phase. A blend of eight J(21–20) rotational transitions at ~227.56 GHz was detected in low-resolution spectra of Comet Hale-Bopp obtained at the IRAM 30-m telescope (*Bockelée-Morvan et al.*, 2000) (see Fig. 8). Acetaldehyde (CH$_3$CHO) has been detected from its $13_{0,13}$–$12_{0,12}$ A$^+$ line

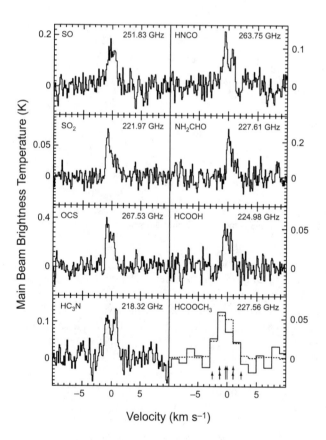

Fig. 8. Spectra of SO (CSO, February 21), SO$_2$ (IRAM/PdBi, March 18, 20, 21), OCS (CSO, March 26), HC$_3$N (CSO, February 20), HNCO (CSO, February 19), NH$_2$CHO (IRAM 30-m, April 5), HCOOH (IRAM/PdBi, March 20–21), and HCOOCH$_3$ (IRAM 30-m, April 5) observed in Comet Hale-Bopp in 1997. The velocity frame is with respect to the comet nucleus velocity. The dashed line superimposed on the observed spectrum of HCOOCH$_3$ is a synthetic profile, which takes into account that the HCOOCH$_3$ line at ~225.562 GHz is a blend of eight transitions whose positions are shown. From *Bockelée-Morvan et al.* (2000).

at 244.83 GHz with IRAM/PdBi (*Crovisier et al.*, 2004a). Several marginal features at the frequencies of CH$_3$CHO lines are also present in IRAM 30-m spectra (*Crovisier et al.*, 2004a). These molecules, which are ubiquitous components of star-forming regions, are much less abundant than CH$_3$OH and H$_2$CO (Table 1): ~0.1% relative to H$_2$O for HCOOH and HCOOCH$_3$, and ~0.02% for CH$_3$CHO. Further work on archive millimeter spectra of Comet Hale-Bopp led to the identification of ethylene glycol (HOCH$_2$CH$_2$OH) (*Crovisier et al.*, 2004b) with an abundance of 0.25% relative to water.

5.4. Symmetric Hydrocarbons

Symmetric hydrocarbons lack a permanent dipole moment and their excited electronic states predissociate, hence only their ro-vibrational bands are observable in cometary comae. Since 1996, CH$_4$, C$_2$H$_2$, and C$_2$H$_6$ have been detected in many comets. The overall appearance of the 3.0-μm region, which is particularly rich in emission lines from hydrocarbons and other species, is shown in Fig. 9 for Comet C/1999 H1 (Lee). A tentative detection of C$_4$H$_2$ was also obtained in Comet 153P/2002 C1 (Ikeya-Zhang) (*Magee-Sauer et al.*, 2002b). Searches for C$_2$H$_4$, C$_3$H$_6$, C$_3$H$_8$, and C$_6$H$_6$ have been negative so far (M. Mumma, personal communication, 2003).

5.4.1. Methane (CH$_4$). CH$_4$ was first clearly detected spectroscopically in Comet Hyakutake, through ground-based observations of five lines of the ν_3 band at 3.3 μm (*Mumma et al.*, 1996). Earlier attempts to detect methane are reviewed in *Mumma et al.* (1993). Methane has been detected in every comet searched since then, including Comet Hale-Bopp, C/1999 S4 (LINEAR), C/1999 H1 (Lee) (Figs. 9a,b), C/1999 T1 (McNaught-Hartley), C/2001 A2 (LINEAR), C/2000 WM$_1$ (LINEAR), and 153P/2002 C1 (Ikeya-Zhang). Its abundance [CH$_4$/H$_2$O] ranged from 0.14% to 1.4% in the sample (*Gibb et al.*, 2003) (see section 8).

5.4.2. Acetylene (C$_2$H$_2$). C$_2$H$_2$ was first detected in Comet Hyakutake, through three lines of its ν_3 band at 3.0 μm (*Brooke et al.*, 1996). The Oort cloud comets sampled up to now are consistent with [C$_2$H$_2$/H$_2$O] = 0.2–0.5% (Table 1), excepting C/1999 S4 (LINEAR) for which the abundance relative to H$_2$O was significantly lower (<0.12% at the 2σ limit) (*Mumma et al.*, 2001a) (see section 8). The abundance retrieved from *in situ* measurements of 1P/Halley was ~0.3% (Giotto NMS) (*Eberhardt*, 1999).

5.4.3. Ethane (C$_2$H$_6$). C$_2$H$_6$ was first detected in Comet Hyakutake, when emissions in four Q-branches of its ν_7 band (3.35 μm) were measured (*Mumma et al.*, 1996). C$_2$H$_6$ has been detected in every comet searched since then (*Mumma et al.*, 2003). Among Oort cloud comets, the abundance is remarkably constant ([C$_2$H$_6$/H$_2$O] ~ 0.6% (Table 1), the sole exception being C/1999 S4 (LINEAR) (*Mumma et al.*, 2001a). The abundance retrieved from *in situ* measurements of Comet Halley was 0.4% (Giotto NMS) (*Eberhardt*, 1999). However, C$_2$H$_6$ was significantly depleted in 21P/Giacobini-Zinner, the quintessential C$_2$-depleted Jupiter-family comet (*Mumma et al.*, 2000; *Weaver et al.*, 1999a;

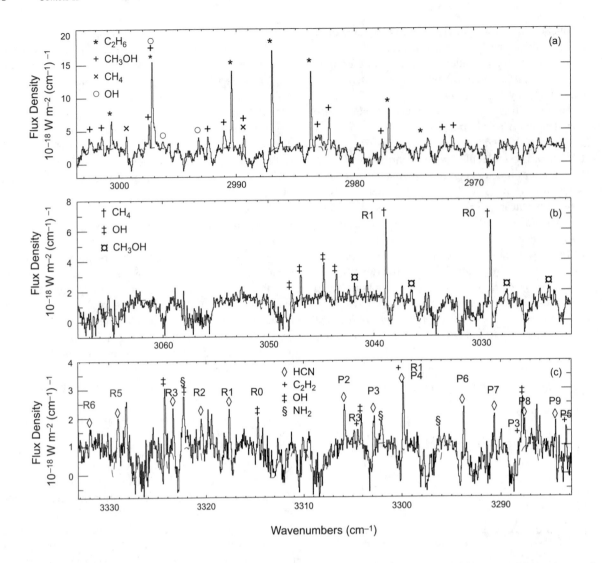

Fig. 9. High-dispersion spectra of Comet C/1999 H1 (Lee) obtained on August 21, 1999 with NIRSPEC at the Keck telescope in the 3-μm region. The dashed line shows a synthetic spectrum of the atmospheric transmittance. Adapted from *Mumma et al.* (2001b).

A'Hearn et al., 1995). The ν_5 band (3.45 μm) of C_2H_6 was detected in Comets Hale-Bopp, Lee, and C/2001 A2 (LINEAR), but it has not yet been quantitatively analyzed (M. Mumma, personal communication, 2003).

5.5. Sulfur-bearing Molecules

5.5.1. CS radical tracing carbon disulfide (CS_2). The CS radical, which is observed at both UV and radio wavelengths (see *Feldman et al.*, 2004) is thought to trace CS_2. The inferred CS_2 abundances range from 0.04% to 0.3% (*Meier and A'Hearn*, 1997). CS_2 has a very short lifetime (~500 s at r = 1 AU), and the spatial brightness profiles of CS measured during the UV observations are consistent with the hypothesis that CS is derived from a short-lived parent.

5.5.2. Hydrogen sulfide (H_2S). Hydrogen sulfide was first detected through its $1_{10}-1_{01}$ line at 169 GHz at the

IRAM 30-m telescope in Comets C/1989 X1 (Austin) and C/1990 K1 (Levy) (*Bockelée-Morvan et al.*, 1991; *Crovisier et al.*, 1991a). It was subsequently observed in several other comets (*Biver et al.*, 2002b), but only through its two millimetric lines at 169 and 217 GHz. The H_2S/H_2O ratio ranges from 0.12% to 1.5%.

Protonated H_2S was identified in ion mass spectra of Comet 1P/Halley, from which a $[H_2S]/[H_2O]$ abundance of 0.4% was derived. This value is within the range of values measured in other comets (*Eberhardt et al.*, 1994).

5.5.3. Sulfur monoxide and dioxide (SO and SO_2). SO and SO_2 have so far been detected only in Comet Hale-Bopp, through several rotational transitions at radio wavelengths (*Lis et al.*, 1999; *Bockelée-Morvan et al.*, 2000) (see Figs. 2 and 8). The 6_5-5_4 line of SO at 219.949 GHz was imaged with IRAM/PdBi, and shows a brightness distribution consistent with a daughter distribution (*Wink et al.*, 1999). It is likely that SO can be fully explained by the photodissociation of SO_2 (*Bockelée-Morvan et al.,* 2000). Before these detections, *Kim and A'Hearn* (1991) derived upper limits on $Q(SO)/Q(H_2O)$ and $Q(SO_2)/Q(H_2O)$ production rate ratios from the absence of their electronic bands in the UV, which are much lower than those inferred from the radio observations of Comet Hale-Bopp (0.3% and 0.2% for SO and SO_2 respectively): A reappraisal of the g-factors of these bands is certainly necessary.

5.5.4. Carbonyl sulfide (OCS). OCS was first detected through its J(12–11) radio line at 145.947 GHz by *Woodney et al.* (1997) in Comet Hyakutake, and confirmed by several other radio lines in Comet Hale-Bopp (Fig. 8) (*Lis et al.,* 1999; *Bockelée-Morvan et al.,* 2000). Lines of the strong ν_3 band at 4.85 μm (g-factor of 2.6×10^{-3} s^{-1}) were observed by *Dello Russo et al.* (1998) in Comets Hyakutake and Hale-Bopp. In the latter comet, the long-slit infrared observations suggested an extended source for OCS (section 7.2). Inferred $Q(OCS)/Q(H_2O)$ are 0.1% and 0.4% for Comets Hyakutake and Hale-Bopp respectively.

5.5.5. Thioformaldehyde (H_2CS). H_2CS was detected by one rotational line (7_{16}–6_{15} at 244 GHz) in Comet Hale-Bopp with the 12-m radio telescope of the National Radio Astronomy Observatory (NRAO) (*Woodney et al.,* 1999). Its abundance relative to H_2O has been evaluated to 0.05% by *Woodney* (2000).

5.5.6. Disulfur (S_2). The S_2 molecule was discovered during IUE observations of C/1983 H1 (IRAS-Araki-Alcock) when several bands of the $B^3\Sigma_u^- - X^3\Sigma_g^-$ system near 2900 Å were detected (*A'Hearn et al.,* 1983). A reanalysis of those data using improved spectral reduction techniques and revised g-factors (*Budzien and Feldman,* 1992) suggests that the $[S_2/H_2O]$ abundance varied from 0.007% to 0.25% during the course of the observations.

The most common form of solid sulfur is S_8, and the discovery of S_2 in a comet was the first detection of this unusual molecule in any astronomical object. The lifetime of S_2 is very short (a few hundred seconds at r = 1 AU), which means that exceptional spatial resolution (roughly a few hundred kilometers at the comet) is generally required to detect it. Thanks to the close approach to Earth of C/1996 B2 (Hyakutake) and the high spatial resolution available from HST, S_2 has now been detected in four other comets, with $[S_2/H_2O]$ abundances in the range 0.001–0.005% (Table 1). Thus, whatever its origin, S_2 seems to be ubiquitous in comets, at least the long-period ones, although its abundance apparently varies over a large range. Note, however, that the derived S_2 abundances are usually strongly dependent on the assumed S_2 lifetime, whose uncertainty could inflate the true abundance variation.

5.5.7. NS radical. NS was detected through the two J(15/2–13/2) e and f radio transitions at the James Clerk Maxwell Telescope (JCMT) by *Irvine et al.* (2000b) in Comet Hale-Bopp. The origin of this molecule is puzzling; NS is a radical that is unlikely to be present in cometary ices, but no plausible parent could be found. Because its spatial distribution and photodissociation rate are unknown, its abundance relative to H_2O $Q(NS)/Q(H_2O)$ is highly uncertain, but estimated to ≥0.02% (*Irvine et al.,* 2000b).

5.6. Nitrogen-bearing Molecules

With an abundance of ~0.5% relative to H_2O, NH_3 is apparently the dominant N-bearing cometary molecule. The abundance of the super volatile molecule N_2 in cometary nuclei, which would constrain the formation conditions of these objects in the solar nebula, is the subject of continuing debate.

5.6.1. N_2. As mentioned in section 3.4, FUSE searches for UV fluorescence from N_2 have been unsuccessful. The upper limit on the $[N_2/H_2O]$ ratio in two long-period comets [C/2001 A2 (LINEAR) and C/2000 WM$_1$ (LINEAR)] was <0.2%, while the $[N_2/CO]$ ratio in those same two comets was <30% (P. D. Feldman, personal communication, 2003).

Mass spectrometry in Comet Halley has also not been very helpful in constraining the N_2 abundance because of the accidental coincidence in the masses of N_2 and CO (*Eberhardt et al.,* 1987).

The presence of N_2 in comets is indirectly inferred from emissions in the $B^2\Sigma_u^+ - X^2\Sigma_g^+$ First Negative System of the N_2^+ ion near 3910 Å. During observations ranging from early in the twentieth century all the way up to the 1986 apparition of 1P/Halley, detections of this band have been claimed in many comets (e.g., *Wyckoff et al.,* 1991, and discussion in *Cochran et al.,* 2000). N_2 abundances of ~0.02% relative to H_2O have been derived from these observations. However, high-spectral-resolution observations of some recent comets [122P/de Vico, C/1995 O1 (Hale-Bopp), and 153P/2002 C1 (Ikeya-Zhang)] did *not* detect the N_2^+ band (*Cochran et al.,* 2000; *Cochran,* 2002), with upper limits of ≤10^{-5}–10^{-4} on the abundance of N_2 relative to CO. Thus, the reality of the detection of the N_2^+ ion in the low-resolution spectra of comets observed earlier may be questioned, especially considering the severe spectral blending by emissions from CO^+, CO_2^+, CH, and CH^+ bands in this same region, and potential confusion with N_2^+ emissions from airglow and aurora.

5.6.2. Ammonia (NH_3). Ammonia was tentatively detected from its 24-GHz inversion line in Comet C/1983 H1 (IRAS-Araki-Alcock) by *Altenhoff et al.* (1983). Confirmed detections of several inversion lines were obtained in Comets Hyakutake (*Palmer et al.,* 1996) and Hale-Bopp (*Bird et al.,* 1997, 1999; *Hirota et al.,* 1999), from which abundances of ~0.5% were derived (Table 1). An upper limit of 0.2% was measured for Comet 153P/2002 C1 (Ikeya-Zhang) (*Bird et al.,* 2002).

The infrared ν_1 band of NH_3 at 3.00 μm was detected in Comets Hale-Bopp (tentatively) (*Magee-Sauer et al.,* 1999) and 153P/2002 C1 (Ikeya-Zhang) (K. Magee-Sauer et al., personal communication, 2003).

Ammonia was also indirectly investigated from the bands of its photodissociation products NH and NH_2, which are easily observed in the visible (see *Feldman et al.,* 2004).

From detailed chemical modeling, *Meier et al.* (1994) investigated the relative contributions of NH_3^+ vs. OH^+ and NH_4^+ vs. H_2O^+ at masses 17 and 18 in *Giotto* NMS spectra. They inferred an NH_3 abundance of 1.5% in Comet 1P/Halley, a factor 2–3 times higher than the values measured in Comets Hyakutake and Hale-Bopp.

5.6.3. Hydrogen cyanide (HCN) and hydrogen isocyanide (HNC). HCN was firmly detected in Comet Halley from its J(1–0) line at 88.6 GHz by several teams (*Despois et al.*, 1986; *Schloerb et al.*, 1986, 1987; *Winnberg et al.*, 1987). It is now one of the easiest parent molecules to observe from the ground and can serve as a proxy for monitoring gas production evolution. It is also found to be in remarkably constant ratio with H_2O production (~0.1% according to radio measurements, see section 8), so that it can be used to evaluate relative molecular abundances (*Biver et al.*, 2002b).

HCN is also observed in the infrared through its v_3 band at 3.0 μm (*Brooke et al.*, 1996; *Magee-Sauer et al.*, 1999; *Mumma et al.*, 2001b). The spectrum of Comet Lee in Fig. 9c shows many ro-vibrational lines of HCN. For some comets, there is a factor of two discrepancy between infrared and radio determinations of the HCN abundance relative to H_2O (Table 1).

HNC, an isomeric form of HCN that is unstable in usual laboratory conditions, was first detected from its J(4–3) line at 363 GHz in Comet Hyakutake (*Irvine et al.*, 1996). It was then observed in Hale-Bopp by several groups (*Irvine et al.*, 1998; *Biver et al.*, 2002a; *Hirota et al.*, 1999) and in several other comets (*Biver et al.*, 2002b; *Irvine et al.*, 2003). The origin of HNC is subject to debate. For Comet Hale-Bopp, an increase of the HNC/HCN ratio (from 0.03 to 0.15) was observed when the comet approached the Sun and became more active; this was interpreted as a clue to chemical conversion of HCN to HNC in the coma, which becomes more efficient in a denser coma (*Irvine et al.*, 1998). However, high HNC/HCN ratios were also observed in Comets Hyakutake, C/1999 H1 (Lee), and C/2001 A2 (LINEAR), which were only moderately productive (Table 1). The HNC/HCN ratio seems to be rather inversely correlated with r (*Biver et al.*, 2002b; *Irvine et al.*, 2003), which would favor a production due to thermo-desorption from heated grains.

5.6.4. Methyl cyanide (HC_3N) and cyanoacetylene (CH_3CN). Two other nitriles are observed in comets. CH_3CN was first detected in Comet Hyakutake with the IRAM interferometer in single-dish mode through a series of rotational lines at 92 GHz (*Dutrey et al.*, 1996). HC_3N was first detected in Comet Hale-Bopp through several rotational lines (*Lis et al.*, 1999; *Bockelée-Morvan et al.*, 2000 (see Fig. 2). These two molecules are 10 times less abundant than HCN (Table 1).

5.6.5. Isocyanic acid (HNCO). A single radio line of HNCO was observed at the Caltech Submillimeter Observatory (CSO) in Comet Hyakutake (*Lis et al.*, 1997). Confirmation was obtained by detection of several rotational lines in Comet Hale-Bopp (Fig. 8) (*Lis et al.*, 1999; *Bockelée-Morvan et al.*, 2000). It was also observed in 153P/2002 C1

(Ikeya-Zhang) (N. Biver et al., personal communication, 2002). Its abundance relative to H_2O is in the range 0.04–0.1%.

5.6.6. Formamide (NH_2CHO). NH_2CHO was detected by several radio lines in Comet Hale-Bopp at CSO and IRAM and the inferred abundance is 0.015% (Fig. 8) (*Lis et al.*, 1999; *Bockelée-Morvan et al.*, 2000).

5.7. Noble Gases

The noble gases (He, Ne, Ar, Kr, and Xe, in order of increasing atomic mass and decreasing volatility) are both chemically inert and highly volatile. Thus, their abundances in cometary nuclei are especially diagnostic of the comet's thermal history. However, remote observations of the noble gases are difficult because their resonance transitions lie in the far-UV spectral region ($\lambda \leq 1200$ Å), which is accessible only from space and is outside the wavelength range covered by the HST and the Chandra X-ray observatory. Searches for several noble gases (He, Ne, and Ar) have been attempted by sounding rockets (*Green et al.*, 1991; *Stern et al.*, 1992, 2000), the Hopkins Ultraviolet Telescope (*Feldman et al.*, 1991), the Extreme Ultraviolet Explorer (EUVE) (*Krasnopolsky et al.*, 1997), the Solar and Heliospheric Observatory (SOHO) (*Raymond et al.*, 2002), and FUSE (*Weaver et al.*, 2002a), but there has not yet been a convincing detection of any noble gas sublimating from a cometary nucleus.

5.7.1. Helium. Helium is the lightest and most volatile of the noble gases, and significant amounts could be frozen in cometary nuclei only if the equilibrium temperature never rose above a few kelvins. For this reason, the detection of emission in the He resonance line at 584 Å during EUVE observations of Comet Hale-Bopp was not interpreted in terms of the production of He atoms sublimating from the nucleus (*Krasnopolsky et al.*, 1997). Rather, the emission could be fully explained by invoking charge exchange between He II solar wind ions and cometary neutral species, which produces He in an excited state that can then relax radiatively [i.e., the same excitation mechanism responsible for cometary X-rays; see *Lisse et al.* (2004)]. Thus, observations of He emission from comets do not shed any light on the thermal history of cometary nuclei, but they can be used as probes of the solar wind conditions and of the interaction between cometary neutrals and the solar wind.

5.7.2. Neon. Neon is also highly volatile, with a sublimation temperature of ~10 K under solar nebula conditions. Fortuitously, the Ne resonance line at 630 Å overlaps the strong solar O V line, which boosts the Ne g-factor upward by a factor of ~100 relative to what it would be if only the solar continuum was available for the excitation. A sensitive search for this Ne line in Comet Hale-Bopp with EUVE resulted in an upper limit on the [Ne/O] abundance that was depleted by a factor of ~25 relative to the solar value in the ice phase, and by a factor of ~200 in total (gas + dust) (*Krasnopolsky et al.*, 1997). An even more sensitive upper limit was obtained for Comet Hyakutake; [Ne/O] was depleted by a factor of 700 in ice and by more than 2600 in total relative to the solar value (*Krasnopolsky and Mumma, 2001*).

5.7.3. Argon. There was a claimed detection of the two principal resonance lines of Ar at $\lambda = 1048.22$ and 1066.66 Å during a sounding rocket observation of Comet Hale-Bopp in 1997 (*Stern et al., 2000*). The deduced [Ar/O] ratio was rather high, 1.8 ± 0.96 times the solar value of $[Ar/O]_\odot = (46 \pm 8) \times 10^{-4}$, but the CO abundance was also high ($\approx 12\%$) (*DiSanti et al., 2001*), indicating that Comet Hale-Bopp retained highly volatile material. Recent high-spectral-resolution observations of comets with FUSE (*Weaver et al., 2002a*) demonstrate that there are other lines that can be confused with Ar emission in low-resolution spectra like those of *Stern et al. (2000)*, which makes the Ar detection in Comet Hale-Bopp questionable.

Sensitive searches for Ar have been made with FUSE in two long-period comets, C/2000 WM_1 (LINEAR) and C/2001 A2 (LINEAR) (*Weaver et al., 2002a*). Argon was not detected in either comet, and the [Ar/O] abundance was depleted by at least a factor of 10 (5σ result) relative to the solar value. These large Ar depletions may not be surprising because the abundance of CO, which has comparable volatility to Ar, was also very low in these comets (below 1%). The upper limit on [Ar/O] for the only CO-rich comet observed by FUSE so far, C/1999 T1 (McNaught-Hartley) with $[CO/H_2O] \approx 13\%$, was not very constraining, no larger than the solar abundance at the 5σ level.

Given the string of null results discussed above, information on the noble gas content in cometary nuclei may not be obtained until a mass spectrometer makes sensitive *in situ* measurements.

5.8. Other Molecules and Upper Limits

Polycyclic aromatic hydrocarbons (PAHs) are thought to be an important constituent of interstellar matter, responsible for the ubiquitous emission bands near 3.28, 7.6, and 11.9 μm (C–H stretching, C–C stretching, and C–H bending modes respectively). Are PAHs also present in comets? An emission band near 3.28 μm has been observed in some comets (e.g., *Davies et al., 1991; Bockelée-Morvan et al., 1995*) (see Fig. 6) and tentatively attributed to PAHs in the gas phase, although other species, such as CH_4 and OH prompt emission are also contributing at this wavelength. This PAH band was not observed in Comet Hale-Bopp, either from the ground or with ISO, but the ISO spectra were obtained at relatively large heliocentric distances ($r > 2.7$ AU) where the PAHs may not be emitting efficiently. *Moreels et al. (1993)* claimed to detect phenanthrene ($C_{14}H_{10}$) in the near-UV spectrum of 1P/Halley measured by the three-channel spectrometer on *Vega*, but this result has not been confirmed by any other observation. Perhaps the best suggestion for cometary PAHs comes from the identification of napthalene, phenanthrene, and other PAHs in laser ablation studies of dust collected in the terrestrial stratosphere that is thought to be of cometary origin (*Clemett et al., 1993*). In summary, the presence and precise composi-

tion of PAHs in cometary nuclei remain open issues requiring further investigation. Some problems that still need to be addressed more rigorously include the excitation mechanism for PAHs, which is thought to be dominated by electronic pumping in the UV followed by intermode conversion to excitation of the numerous PAH vibrational modes; how PAHs are released from cometary refractories to the gas phase; and the lifetime of PAHs in cometary comae (*Joblin et al., 1997*).

In addition to the detected molecules discussed above and listed in Table 1, upper limits have been set for many other species from dedicated or serendipitous searches at radio and IR wavelengths. A selection of upper limits obtained from radio observations of Comet Hale-Bopp (*Crovisier et al., 2004a*) is listed in Table 2.

Numerous unidentified features have been detected in cometary spectra. This indicates that new cometary species are still to be identified, pending further theoretical spectroscopic investigations and laboratory measurements. Unidentified lines observed in the visible and UV domains are presumably due to atoms, radicals, or ions (cf. *Feldman et al., 2004*). Likely, many of these unidentified lines are due to already known cometary species (e.g., C_2, NH_2, …).

In the IR, several unidentified bands have been noted for a long time. This is the case for the 2.44-μm band (*Johnson et al., 1983*) and for the 3.3–3.5-μm broad band, which can be only partly attributed to C_2H_6 and CH_3OH (see discussion in section 5.3.1). We have strong indications that the 3.4–3.5-μm excess emission shown in Fig. 6 mainly arises from gas-phase fluorescence and not from refractory organics (*Bockelée-Morvan et al., 1995*). Recently, many unidentified lines have been detected in IRTF/CSHELL and Keck/NIRSPEC high-resolution spectra (e.g., *Magee-Sauer et al., 1999; Mumma et al., 2001b*). Some of these latter lines could be due to CH_3OH, whose IR spectrum is still poorly understood. Considering that the IR spectra of simple, stable molecules are well known, we could conclude that the unidentified lines are probably due to radicals or to more complex molecules. A few unidentified lines have also been noted in the radio domain, but they were observed with limited signal-to-noise ratio.

6. HELIOCENTRIC EVOLUTION OF PRODUCTION RATES

Extensive studies of the evolution of the outgassing of many comets as a function of heliocentric distance are available from the observation of daughter species [e.g., *Schleicher et al. (1998)* for 1P/Halley, *Rauer et al. (2003)* for Hale-Bopp]. Spectroscopic observations of parent molecules, however, were generally conducted during the most active phase of the comets (i.e., near perihelion), when the signals received from Earth are expected to be the strongest. Monitoring along comet orbit was only possible for a few bright comets, such as Comet Halley [HCN at $r = 0.6$–1.8 AU (*Schloerb et al., 1987*)], Hyakutake [HCN, CH_3OH, CO, H_2CO, H_2S at $r = 0.24$–1.9 AU (*Biver et al., 1999a*)], and C/1999 H1 (Lee) [H_2O, HCN, CH_3OH, H_2CO at r < 1.7 AU (*Biver et al.*, 2000; *Chiu et al.*, 2001)]. The early discovery at r = 7 AU of Comet Hale-Bopp and its exceptional intrinsic activity [$Q(H_2O) = 10^{31}$ molecules s^{-1} at perihelion] provided the first opportunity to follow the evolution of the production rates of a number of molecules over a much larger heliocentric range. Figure 7 shows the result of a monitoring performed at radio wavelengths (*Biver et al.*, 2002a) during which many molecules were detected up to r = 4–5 AU, and up to 14 AU in the case of CO. Long-slit spectroscopy in the infrared covered the 0.9–4.1-AU range for CO, the 0.9–4-AU range for C_2H_6, and the 0.9–1.5-AU range for H_2O (*DiSanti et al.*, 2001; *Dello Russo et al.*, 2000, 2001).

Gas production curves are almost the only observational tools we have to obtain informations regarding the nature and physical state of cometary ices, and their thermal properties and sublimation mechanisms. A review of the various processes involved in cometary activity, and of the mathematical models developed so far, is presented by *Prialnik et al.* (2004). Gas production curves also complement usefully visual light curves for the study of seasonal effects, dust mantling, etc. (see *Meech and Svoreň*, 2004). A fundamental question that arises and can be addressed observationally from gas production curves is to which extent production rate ratios reflect the bulk composition inside the nucleus. Both numerical simulations and laboratory experiments show that the link is not simple, at least for the most volatile species.

The monitoring performed in Comet Hale-Bopp showed that the coma composition changed with heliocentric distance. CO was the main escaping gas at large r (Fig. 7). The H_2O production rate surpassed that of CO for r < 3 AU. *Capria et al.* (2000) showed that the heliocentric behavior of CO production (roughly in r^{-2} from r = 0.9 to 14 AU) can be explained if CO is present in the nucleus both as pure ice and as trapped gas in amorphous H_2O ice that would be immediately below the surface. This trapped gas is released during the amorphous to crystalline phase transition. In contrast, models considering only the sublimation of pure CO ice fail in reproducing the observed heliocentric dependence.

As seen in Fig. 7, distinct trends are observed among the various molecules. As already discussed, the steep production curves of HNC and H_2CO are likely related to an extended production in the coma (see sections 5.3.2 and 5.6.3). Steep production curves when compared to, e.g., HCN are also observed for CS and OCS (*Woodney*, 2000; *Biver et al.*, 2002a). A pre-/postperihelion asymmetry is also apparent for all molecules.

7. SPATIAL DISTRIBUTION OF PARENT MOLECULES AND EXTENDED SOURCES

The study of the spatial distribution of parent molecules provides clues on the distribution of the outgassing at the surface of the nucleus and on gas dynamics processes oc-

curring in the coma (see *Crifo et al.*, 2004). However, data acquired so far provide limited information due to the lack of spatial coverage and resolution. Radio and long-slit IR spectra provide numerous examples of anisotropic distributions with day/night asymmetries, which will not be discussed here. Rather, we will focus on studies of the radial distribution of molecular species in the coma. Some species can be released directly from the nucleus, and also can be produced in the coma from other precursors. The former source is said to be direct (or native), while the latter is said to be extended (or distributed). The native and extended sources exhibit different radial distributions, and can be recognized in this way.

7.1. Radial Distribution from *In Situ* Measurements

The discovery of extended sources of molecules in the coma was one of the highlights of the space investigation of 1P/Halley. The best examples were for CO (*Eberhardt et al.*, 1987) and H_2CO (*Meier et al.*, 1993). Their densities measured by NMS along the path of the *Giotto* spacecraft do not match those expected for a parent molecule. Only one-third (~3.5% relative to H_2O) of the total CO was released directly from the nucleus, the remainder (~7.5%) being produced from an extended source in the coma (*Eberhardt et al.*, 1987; *Eberhardt*, 1999). H_2CO in Comet Halley was produced mainly, and perhaps totally, from an extended source (*Meier et al.*, 1993). The region containing the extended sources extended to ~10^4 km from the nucleus. *Meier et al.* (1993) estimated the scalelength of the parent source of H_2CO to be 1.2 times the photodissociative scalelength of H_2CO that corresponds to ~5000 km. According to *Eberhardt* (1999), there is also indication of a second extended source of H_2CO with a much longer scalelength. These discoveries sparked keen interest in identifying the nature of the extended sources.

Proposed sources for extended H_2CO focus on the decomposition of (native) polymerized H_2CO (*Meier et al.*, 1993), possibly polyoxymethylene (POM) (*Mitchell et al.*, 1987, 1989; *Huebner*, 1987). Recent laboratory work demonstrates that adequate monomeric H_2CO can be produced by thermal decomposition of POM, if cometary grains are 4% POM by mass in Comet Halley (*Cottin et al.*, 2004).

Photolysis of monomeric H_2CO was suggested as a significant source for extended CO in 1P/Halley. According to *Meier et al.* (1993), the photodissociation of H_2CO provides about two-thirds of the extended CO source. According to *Eberhardt* (1999), H_2CO could even fully account for the extended CO source after a reanalysis of the data. In contrast, in Comet Hale-Bopp, H_2CO was found to be only a minor contributor to extended CO as its production rate was much below that of CO extended production (section 7.2.1). Carbon suboxide (C_3O_2) was suggested as a source of CO (*Huntress et al.*, 1991), but the upper limit derived from *Vega* IKS spectra ($C_3O_2 < 0.1\%$) (*Crovisier et al.*, 1991b) was well below the minimum value (7.5%) required to produce the amount of extended CO inferred

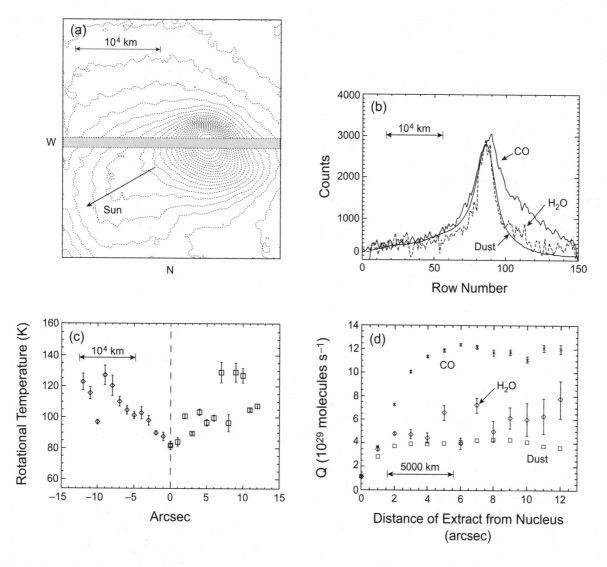

Fig. 10. Observations of Comet Hale-Bopp with IRTF/CSHELL. (**a**) Image of thermal continuum at 3.5 μm (λ/δλ = 70). The east-west slit is indicated. (**b**) Spatial profiles of CO, H$_2$O, and dust along the slit. (**c**) Rotational temperatures for CO along the slit. (**d**) Symmetric Q-curves showing the rise to terminal values. From *DiSanti et al.* (2001).

from the *Giotto* observations. An alternative view, that *Giotto* flew through a jet enriched in CO, was proposed by *Greenberg and Li* (1998) to interpret the NMS observations.

7.2. Radial Distribution from Long-Slit Spectroscopy

7.2.1. Infrared results. When a molecule is released directly from the nucleus, its column density is expected to vary, in first approximation (i.e., Haser model and ρ ≪ vτ), as ρ^{-1}, where ρ is the distance between the line of sight and the nucleus (i.e., the impact parameter). If an extended source is also present, the variation of column density with r is much flatter. Under favorable circumstances, this difference can be used to extract the two sources separately. In Comet Hale-Bopp, extended sources were identified for

OCS (*Dello Russo et al.*, 1998) and CO (*Weaver et al.*, 1999b; *DiSanti et al.*, 1999, 2001; *Brooke et al.*, 2003), although the ratio of native to extended source production rates for CO remains controversial. In contrast, the column density profiles for H$_2$O (*Weaver et al.*, 1999b; *Dello Russo et al.*, 2000; *Brooke et al.*, 2003), C$_2$H$_6$ (*Weaver et al.*, 1999b; *Dello Russo et al.*, 2001; *Brooke et al.*, 2003), CH$_4$ (*Weaver et al.*, 1999b; *Brooke et al.*, 2003), and HCN (*Magee-Sauer et al.*, 1999) were consistent with release solely from the nucleus.

The technique is illustrated for Comet Hale-Bopp in Fig. 10 from *DiSanti et al.* (2001). An image of the thermal continuum near 3.5 μm (λ/δλ ∼ 70) reveals dust enhancements to the northeast and northwest (sunward) of the nucleus (Fig. 10a). High-dispersion spectra were measured with the slit positioned as shown. The intensity profiles measured along the slit show that CO is extended to the east while H_2O is symmetric about the nucleus and the dust is extended to the west (Fig. 10b).

To extract the contribution of the native and extended CO sources, *DiSanti et al.* (2001) developed the method of Q-curves, as shown in Fig. 10d. The intensities measured at symmetric positions along the slit are first averaged to minimize the effects of outflow asymmetry. An apparent production rate can be then derived from the intensity measured at a specific location using equation (11) and the formula that links the column density to the production rate under the idealized assumption of spherical outflow at constant velocity. The resulting Q-curve (Fig. 10d) rises from the nucleocentric value to a terminal value that is taken to represent the *total* production rate for the species. The nucleus-centered value is always too low, owing to slit losses induced by seeing, drift, guiding error, and other observing factors, but the terminal value is reached quickly for molecules released directly from the nucleus (e.g., H_2O, C_2H_6, CH_4, dust). Molecules having an extended source rise more slowly to the terminal value (e.g., CO, OCS) (compare Q-curves for CO, H_2O, and dust; Fig. 10d).

It was estimated that ~70% of OCS was produced from an extended source in Comet Hale-Bopp near perihelion (*Dello Russo et al.*, 1998). At large heliocentric distance (4.1 AU < r < 2 AU), the spatial profile of CO was consistent with its release solely from the nucleus. However, within 2 AU an *extended* source was activated, and it supplied at least half the total CO released thereafter (0.9 AU < r < 1.5 AU) (*DiSanti et al.*, 2001); using a different approach involving explicit modeling of both parent and daughter spatial distributions, *Brooke et al.* (2003) estimated that ~90% of the observed CO was derived from the extended source at r = 1 AU, in apparent contradiction with the conclusions of *DiSanti et al.* (2001). The abrupt onset and constant fractional production of the extended source thereafter suggest a thermal threshold for release from small CHON grains, rather than photolysis of a precursor volatile. Monomeric H_2CO was at most a minor contributor to extended CO in Comet Hale-Bopp. Carbon dioxide is admittedly a significant source of CO. However, it cannot explain the extended CO source observed in the IR, which has a scalelength much smaller than the CO_2 photodissociation scalelength. In Comet Hyakutake, CO was abundant and it was released almost entirely from the nucleus (*DiSanti et al.*, 2003).

7.2.2. Ultraviolet results. The advent of the long-slit capability of the Space Telescope Imaging Spectrograph (STIS) on HST now permits extremely high-spatial resolution studies of CO and S_2. However, the small g-factors for the CO emissions makes this approach feasible only for the brightest comets. While the g-factors are much larger for S_2, its abundance is so low that poor signal-to-noise is a problem in this case as well. Nevertheless, accurate spatial brightness profiles for S_2 were recently obtained in 153P/2002 C1 (Ikeya-Zhang) and indicate that S_2 probably originates in the

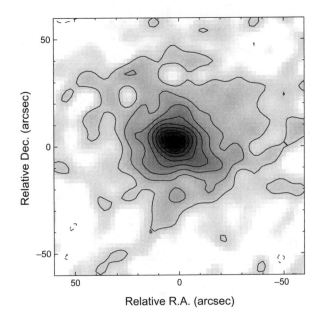

Fig. 11. Mosaicked image of HCN J(1–0) main hyperfine component (F = 2–1) obtained on April 6, 1997, for Comet Hale-Bopp with the BIMA array. Contour interval: 0.23 K averaged over 3.5 km s^{-1}. The angular resolution is 10". From *Wright et al.* (1998).

nucleus but may have a lifetime that is significantly longer than the theoretical value (*Weaver et al.*, 2002b).

7.3. Millimeter Wave Mapping/Interferometry

Mapping of rotational emission lines was only attempted in a few comets. Most investigations were performed in Comets Hyakutake and Hale-Bopp, due to their exceptional brightnesses. Focal plane arrays and on-the-fly mapping (which consists in moving the telescope beam across the source at a constant velocity) provided extended coverage of the molecular emissions with single-dish telescopes, although with limited spatial resolution (10" at most) [(e.g., *Lovell* (1999) using QUARRY at the Five College Radio Astronomy Observatory (FCRAO)]. Large maps using conventional techniques were also obtained (e.g. *Hirota et al.* (1999) at the Nobeyama 45-m telescope; *Biver et al.* (1999a) at JCMT). Interferometry techniques have been successfully used for the first time, and provided angular resolutions up to 2". Numerous observations were performed in Comet Hale-Bopp with the IRAM Plateau de Bure interferometer, the array of the Berkeley-Illinois-Maryland Association (BIMA), and the Owens Valley Radio Observatory (OVRO). Interferometric maps of rotational lines of CO, HCN (e.g., Fig. 11), H_2S, CS, SO, H_2CO, HNC, DCN, HDO, and CH_3OH were obtained (*Blake et al.*, 1999; *Veal et al.*, 2000; *Wink et al.*, 1999; *Wright et al.*, 1998; *Woodney et al.*, 2002). In contrast to other spectral domains, additional information on the spatial distribution along the line of sight can be extracted from the radio maps by analyzing the line shapes.

Maps of Comet Hale-Bopp show the presence of gaseous jets. The images obtained by *Blake et al.* (1999) with OVRO show HNC, DCN, and HDO emissions offset by a few arcseconds with respect to the dust continuum emission, interpreted as due to jets enriched in these molecules. No such offsets were observed for other species mapped with millimeter interferometry. *Veal et al.* (2000) observed structures in HCN J(1–0) maps obtained with BIMA, which varied and disappeared on timescales ~2 h. These variations may trace rotating HCN jets, as observed for CO. Indeed, a sinusoidal temporal variation in the spectral shift of the CO J(2–1) line was observed at IRAM/PdBi (*Henry et al.*, 2002; *Henry*, 2003). Both the period of this sinusoid, which is consistent with the nucleus rotation period, and the time modulation of the signals received by the antenna pairs, suggest the existence of a CO spiraling jet. From the maps of Comets Hyakutake and Hale-Bopp (*Biver et al.*, 1999a; *Wink et al.*, 1999), H_2CO clearly appears extended with a parent scalelength consistent with that derived for 1P/Halley from the *in situ* measurements. Coarse radio mapping in C/1989 X1 (Austin) gave similar results (*Colom et al.*, 1992). In contrast, the HCN and H_2S maps of Comet Hale-Bopp (*Wink et al.*, 1999; *Wright et al.*, 1998) are consistent with parent molecule distributions. As discussed in section 5.5.3, the interferometric observations of SO show that this species does not follow a parent density distribution, and suggest the photolysis of SO_2 as the main source of SO (*Wink et al.*, 1999).

Other observational clues for extended sources obtained by radio observations concern distant comets. *Gunnarsson et al.* (2002) mapped the CO J(2–1) line in Comet 29P/Schwassmann-Wachmann 1 with the SEST telescope. They concluded that there was a large (~70%) contribution of CO coming from an extended source, likely sublimating icy grains. No such imaging was performed in Comet Hale-Bopp when far from the Sun. However, the CO line shapes of Comet Hale-Bopp at large r resemble those of 29P/Schwassmann-Wachmann 1, which led *Gunnarsson et al.* (2003) to suggest that sublimating grains were also contributing to the CO production in Comet Hale-Bopp when at r > 4 AU. These line profiles are asymmetric, and characterized by a pronounced peak on the blue wing of the line, and a redshifted part of lower intensity. This blue peak would correspond to nuclear production toward the Sun, while the redshifted component is the red part of a symmetric profile due to the secondary source. Much more symmetric radio lines were observed for CH_3OH, H_2CO, and OH at r > 3.5 AU (*Biver et al.*, 1997; *Womack et al.*, 1997). These results are consistent with a relative contribution of the icy grains vs. nucleus outgassing being more important for CH_3OH, H_2CO, and H_2O than for CO. This may be not surprising given the lower volatilities of CH_3OH, H_2CO, and H_2O ices compared to CO ice.

Snyder et al. (2001) used the maps of HCN J(1–0) and CS J(2–1) obtained in Comet Hale-Bopp with the BIMA array to measure the photodissociation rates of these molecules. That of HCN was found consistent with theoretical predictions, while a photodissociation rate 10 times larger than the commonly accepted value is suggested for CS.

8. MOLECULAR ABUNDANCES AND CHEMICAL DIVERSITY AMONG COMETS

According to current theories, comet formation in the solar nebula extended over a wide range of heliocentric distances for both Oort cloud comets [also called nearly isotropic comets (*Dones et al.*, 2004)] and Jupiter-family comets [now recognized as a subclass of the ecliptic comets (*Duncan et al.*, 2004; *Morbidelli and Brown*, 2004)], which suggests that comets could display diversity in their chemical composition depending on the local temperature and nebular composition where they formed. If there was significant mixing of nebular material across large ranges in heliocentric distance, even individual cometary nuclei could exhibit chemical inhomogeneity.

Chemical diversity among comets is indeed observed for both parent volatiles and daughter species. From a study of radicals (OH, CN, C_2, C_3, NH) in 85 comets, *A'Hearn et al.* (1995) proposed the existence of two classes of comets, depending on their C_2/CN ratio: "typical" comets and "C_2-depleted" comets. They found that about one-half the Jupiter-family comets (JFCs) were C_2-depleted, but the fraction of C-depleted nearly isotropic comets was much smaller. The meaning of this depletion is clouded by two factors. First, the relative production of C_2 and CN from several possible gas and dust precursors is not known. Second, the present volatile composition of a JFC may not reflect its original volatile composition. Jupiter-family comets typically have low-inclination, prograde orbits with periods less than ~20 years, and they are subjected to much greater insolation than the nearly isotropic comets. Thus, some JFCs may have experienced thermal fractionation while in their present orbits.

Surveys of parent volatile abundances show strong evidence for chemical diversity among comets (Fig. 12 and Table 1). Among the Oort cloud comets (OCCs), the native CO abundance varies by a factor of ~40 (0.4–17%) relative to H_2O. The abundance of extended CO varies by a similar amount, but the two sources are not correlated. The comet-to-comet differences in native CO are presumably attributable to intrinsic variation in the amount of CO ice frozen into cometary nuclei. The CO_2 abundance varied by a factor of five (2.5–12%) among five comets (*Feldman et al.*, 1997), although this conclusion rests on the assumption that the observed CO Cameron band emission was due solely to the photodissociation of CO_2. Infrared investigations of CO_2 in three comets indicate CO_2/H_2O variations by at least a factor of 2.

Infrared observations of hydrocarbons in a handful of comets (*Mumma et al.*, 2003; *Gibb et al.*, 2003) show that the CH_4 abundance varies by a factor of ~10 (0.14–1.4%), apparently without correlation with CO. Thus, one cannot

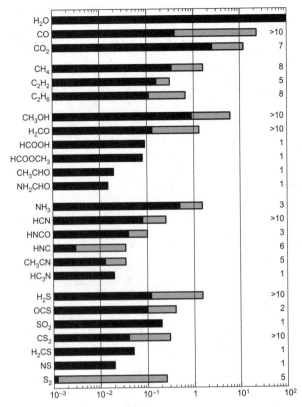

Fig. 12. Abundances relative to water in comets. The range of measured values is shown in the gray portions. The number of comets for which data are available is given in the right. For CO, abundances refer to total CO (native and distributed sources).

define a "typical" abundance for either CO or CH₄. The C_2H_6 abundance shows less diversity, with six OCCs showing ~0.4–0.7% and only C/1999 S4 (LINEAR) differing greatly. In the same group of comets, the C_2H_2 abundance was typically 0.2–0.3%, but it was significantly lower in C/1999 S4. The CH_3OH abundance is ~2% in 14 comets (including all four JFCs sampled) observed in the infrared, but three OCCs had much lower abundances [C/1990 K1 (Levy), C/1996 Q1 (Tabur), and C/1999 S4] and two had much higher abundances [C/1989 X1 (Austin), 109P/Swift-Tuttle].

From radio observations of about 25 comets (including 6 JFCs), *Biver et al.* (2002b) studied the production rates of HCN, HNC, CH_3CN, CH_3OH, H_2CO, CO, CS, and H_2S relative to H_2O. Hydrogen cyanide is the best-studied molecule. In contrast to other species, the distribution of the HCN/H_2O ratios is strongly peaked, with most comets around 0.1%. Observed in more than 10 comets, CH_3OH, H_2CO, and H_2S show large variations. The distribution of CH_3OH/H_2O ratios follows that measured from infrared spectra. The H_2CO/H_2O abundance ranges from 0.4% to 1.3% in 12 comets, but is significantly lower (≤0.15%) in 21P/Giacobini-Zinner. H_2S/H_2O varies from 0.4% to 1.5% in 10 comets, but is only 0.12% in C/2000 WM₁ (LINEAR).

The comets that are abundant in CH_3OH are also abundant in H_2CO. No clear correlation is found between the relative abundances and the dynamical origin of the comets, or their dust-to-gas ratios.

From UV observations of the CS radical by IUE and HST of 19 comets (*Meier and A'Hearn,* 1997), the CS_2 abundance varies from 0.04% to 0.3%.

Perhaps one should not be surprised to find significant chemical diversity among even a small sample of nearly isotropic comets because, as stated earlier, the comets from the Oort cloud were probably formed over a large range of heliocentric distances. The most recent dynamical models suggest that comets now in the Oort cloud were contributed in roughly equal numbers by each giant planet and the Kuiper belt (*Dones et al.,* 2004). The chemical diversity found to date suggests that comets from various regions of the protoplanetary disk are present in today's Oort cloud and can provide a window on this crucial period in solar system development. However, many more comets must be sampled and additional parent molecules measured to establish overall taxonomic classes of comets from their chemistry and related parameters.

9. ISOTOPIC COMPOSITION

Isotopic ratios are an important diagnostic of the physical conditions that prevailed during the formation of cometary volatiles, as well as isotopic exchange and mixing processes that may have occurred in the solar nebula before their incorporation into comets. Since most detections of parent molecules postdate 1985, it is not surprising that the detection of their isotopomers is rare. We summarize most of the measurements in Table 3.

The first measurements of the D/H ratio in H_2O were obtained in Comet Halley from mass-resolved ion-spectra of H_3O^+ acquired by the IMS (*Balsiger et al.,* 1995) and NMS (*Eberhardt et al.,* 1995) instruments onboard *Giotto.* These independent data provided precise D/H values, which combined give a D/H value of $\sim 3 \times 10^{-4}$. Thanks to the availability of sensitive groundbased instrumentation in the submillimeter domain, HDO was detected in Comets Hyakutake (Fig. 13) and Hale-Bopp from its 1_{01}–0_{00} line at 464.925 GHz (*Bockelée-Morvan et al.,* 1998; *Meier et al.,* 1998a). The derived D/H values are in agreement with the determinations in 1P/Halley. *Blake et al.* (1999) reported the interferometric detection of the 2_{11}–2_{12} HDO transition at 241.562 GHz in Comet Hale-Bopp and derived a D/H value one magnitude larger in jets compared to the surrounding coma. A serendipitous detection of the 3_{12}–2_{21} HDO line at 225.897 GHz in Comet Hale-Bopp was also reported by *Crovisier et al.* (2004a).

DCN was detected in Comet Hale-Bopp with the JCMT from its J(5–4) rotational transition at 362.046 GHz (*Meier et al.,* 1998b). The inferred D/H value in HCN is 2.3×10^{-3}, i.e., seven times larger than the value in H_2O. The same value is reported by *Crovisier et al.* (2004a) from a marginal detection of the J(3–2) DCN line at the IRAM 30-m telescope. Interferometric observations of this J(3–2) DCN

TABLE 3. Isotopic ratios in comets.

Ratio	Molecule	Comet	Value	Method*	Reference
D/H	H_2O	Halley	$(3.08^{+0.38}_{-0.53}) \times 10^{-4}$	m.s.[†]	*Balsiger et al.* (1995)
		Halley	$(3.06 \pm 0.34) \times 10^{-4}$	m.s.[†]	*Eberhardt et al.* (1995)
		Hyakutake	$(2.9 \pm 1.0) \times 10^{-4}$	r.s.	*Bockelée-Morvan et al.* (1998)
		Hale-Bopp	$(3.3 \pm 0.8) \times 10^{-4}$	r.s.	*Meier et al.* (1998a)
	HCN	Hale-Bopp	$(2.3 \pm 0.4) \times 10^{-3}$	r.s.	*Meier et al.* (1998b)
	H_2CO	Halley	$<2 \times 10^{-2}$	m.s.[‡]	*Balsiger et al.* (1995)
		Hale-Bopp	$<5 \times 10^{-2}$	r.s.	*Crovisier et al.* (2004a)
	CH_3OH	Halley	$<\sim 1 \times 10^{-2}$	m.s.[§]	*Eberhardt et al.* (1994)
		Hale-Bopp	$<3 \times 10^{-2}$	r.s.[¶]	*Crovisier et al.* (2004a)
		Hale-Bopp	$<8 \times 10^{-3}$	r.s.[**]	*Crovisier et al.* (2004a)
	NH_3	Hyakutake	$<6 \times 10^{-2}$	v.s.[††]	*Meier et al.* (1998c)
		Hale-Bopp	$<4 \times 10^{-2}$	r.s.[‡‡]	*Crovisier et al.* (2004a)
	CH_4	153P/2002 C1	$<1 \times 10^{-1}$	i.s.	*Kawakita et al.* (2003)
	H_2S	Hale-Bopp	$<2 \times 10^{-1}$	r.s.	*Crovisier et al.* (2004a)
	CH	Hyakutake	$<3 \times 10^{-2}$	v.s.	*Meier et al.* (1998c)
$^{12}C/^{13}C$	C_2	four comets[§§]	93 ± 10	v.s.	*Wyckoff et al.* (2000)
	CN	Halley	95 ± 12	v.s.	*Kleine et al.* (1995)
	CN	five comets[¶¶]	90 ± 10	v.s.	*Wyckoff et al.* (2000)
	CN	Hale-Bopp	165 ± 40	r.s.	*Arpigny et al.* (2003)
	CN	C/2000 WM$_1$	115 ± 20	r.s.	*Arpigny et al.* (2003)
	HCN	Hyakutake	34 ± 12[***]	r.s.	*Lis et al.* (1997)
	HCN	Hale-Bopp	111 ± 12	r.s.	*Jewitt et al.* (1997)
	HCN	Hale-Bopp	109 ± 22	r.s.	*Ziurys et al.* (1999)
	HCN	Hale-Bopp	90 ± 15	r.s.	*Lis et al.* (1999)
$^{14}N/^{15}N$	CN	Hale-Bopp	140 ± 35	v.s.	*Arpigny et al.* (2003)
	CN	C/2000 WM$_1$	140 ± 30	v.s.	*Arpigny et al.* (2003)
	HCN	Hale-Bopp	323 ± 46	r.s.	*Jewitt et al.* (1997)
	HCN	Hale-Bopp	330 ± 98	r.s.	*Ziurys et al.* (1999)
$^{16}O/^{18}O$	H_2O	Halley	518 ± 45	m.s.	*Balsiger et al.* (1995)
	H_2O	Halley	470 ± 40	m.s.	*Eberhardt et al.* (1995)
	H_2O	153P/2002 C1	450 ± 50	r.s.	*Lecacheux et al.* (2003)
$^{32}S/^{34}S$	S^+	Halley	23 ± 6	m.s.	*Altwegg* (1996)
	CS	Hale-Bopp	27 ± 3	r.s.	*Jewitt et al.* (1997)
	H_2S	Hale-Bopp	17 ± 4	r.s.	*Crovisier et al.* (2004a)

* m.s.: mass spectrometry; r.s.: radio spectroscopy; v.s.: visible spectroscopy; i.s.: infrared spectroscopy.

[†] From H_3O^+.

[‡] From $HDCO^+$.

[§] CH_2DOH and CH_3OD averaged.

[¶] For CH_3OD.

[**] For CH_2DOH.

[††] From NH.

[‡‡] From NH_2D.

[§§] Mean ratio in C_2 from observations in C/1963 A1 (Ikeya), C/1969 T1 (Tago-Sato-Kosaka), C/1973 E1 (Kohoutek), and C/1975 N1 (Kobayashi-Berger-Milon).

[¶¶] Mean ratio in CN from five comets: 1P/Halley, C/1990 K1 (Levy), C/1989 X1 (Austin), C/1989 Q1 (Okazaki-Levy-Rudenko), and C/1995 O1 (Hale-Bopp).

[***] Ratio possibly affected by line blending.

line lead to DCN/HCN values in jets, which are again significantly larger than the single-dish value (*Blake et al.*, 1999). We note that these differences in isotopic abundance ratios for different regions of the coma (e.g., in and out of jets) must be confirmed by higher-quality interferometric observations before too much effort is expended interpreting these differences.

Finally, upper limits for several other D-bearing molecules were obtained, mainly by millimeter spectroscopy or visible spectroscopy of daughter species (Table 3). Most of these upper limits exceed a few percent and may be not easily improved from further groundbased observations unless a comet as bright as Comet Hale-Bopp is coming.

The measurements of isotopes other than D in cometary volatiles are limited. In Comet Hale-Bopp, rotational transitions of $HC^{15}N$, $H^{13}CN$, and $C^{34}S$ were detected, leading to isotopic ratios that are compatible with the terrestrial values $^{12}C/^{13}C = 89$, $^{14}N/^{15}N = 270$, and $^{32}S/^{34}S = 24$ (*Jewitt*

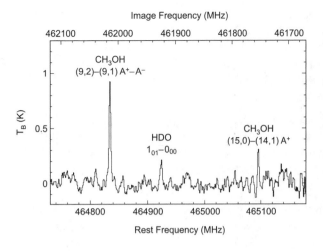

Fig. 13. The 1_{01}–0_{00} line of HDO at 465 GHz observed in Comet C/1996 B2 (Hyakutake) with the CSO (*Bockelée-Morvan et al.,* 1998).

et al., 1997; *Ziurys et al.,* 1999). From a tentative detection of $H_2^{34}S$, *Crovisier et al.* (2004a) deduced a $^{32}S/^{34}S$ ratio 30% lower than the terrestrial value. In Comet Hyakutake, *Lis et al.* (1997) inferred a $^{12}C/^{13}C$ ratio in HCN three times lower than the terrestrial value. However, the J(3–2) $H^{13}CN$ line used for this measurement is partly blended with a SO_2 line, so this result is uncertain. Regarding the $^{18}O/^{16}O$ ratio, which has only been measured in H_2O, *in situ* measurements in 1P/Halley are consistent with the terrestrial value of 500 (Table 3). $H_2^{18}O$ has been detected through its fundamental ortho line at 547.7 GHz in Comet 153P/2002 C1 (Ikeya-Zhang) by *Lecacheux et al.* (2003), and the derived $H_2^{18}O/H_2^{16}O$ ratio is consistent with that obtained by mass spectrometry in 1P/Halley.

Lines of ^{13}CN [mainly in the $B^2\Sigma^+$–$X^2\Sigma^+$ v(0–0) system] and $^{13}C^{12}C$ (Swan bands) were identified in visible spectra of several comets. All measurements give a C-isotopic ratio in the CN and C_2 radicals consistent with the terrestrial value [Table 3; see data for individual comets in *Wyckoff et al.* (2000)]. These isotopic ratios may not fully reflect the values in HCN and C_2H_2, as other sources of CN and C_2 radicals are suspected to be present in the coma. *Arpigny et al.* (2003) report anomalous $C^{14}N/C^{15}N$ ratios (~140) in Comets Hale-Bopp and C/2000 WM₁ (LINEAR). This is much less than the ratio $HC^{14}N/HC^{15}N = 270$ observed in Comet Hale-Bopp, pointing to an additional, still unidentified, source of CN.

10. ORTHO-TO-PARA RATIOS

Molecules with H atoms at symmetrical positions may exist in different nuclear-spin species according to the sum I of the spins of their H atoms. These spin species are called *ortho* (I = 1) and *para* (I = 0) for molecules with two H atoms (e.g., H_2, H_2O, H_2S, H_2CO, ...); A and E for mol-

ecules with three H atoms (e.g., NH_3 or CH_3OH); A, E, and F for molecules with four H atoms (e.g., CH_4); and so forth. Conversions among different species in the gas phase by radiative transitions or by collisions are strictly forbidden. Conversions are presumably very slow in the solid phase as well, although various proton-exchange mechanisms do exist in that case.

The ortho-to-para population ratio (OPR) can be evaluated from the rotational distribution of the molecules. When there is equilibrium at a temperature T

$$OPR = \frac{(2I_o + 1) \sum\limits_{o\ levels} (2J + 1)\exp\left(-\dfrac{E}{kT}\right)}{(2I_p + 1) \sum\limits_{p\ levels} (2J + 1)\exp\left(-\dfrac{E}{kT}\right)} \quad (16)$$

where o and p refer to the ortho and para rotational levels. J and E are the rotational quantum number and energy of the levels respectively. For high temperatures, the populations tend to equilibrate to their statistical weights, $2I + 1$. Thus, the high-temperature OPR limit is 3 for H_2O, 1 for NH_3, ... (see discussions by *Crovisier,* 1984; *Mumma et al.,* 1993).

From observations of the ν_3 band of H_2O with the KAO, *Mumma et al.* (1993) reported an OPR of 2.5 ± 0.1 for 1P/Halley ($T_{spin} \approx 29$ K) and 3.2 ± 0.2 for C/1986 P1 (Wilson) ($T_{spin} > 50$ K). From observations with ISO (Fig. 3), the OPR was 2.45 ± 0.10 for Comet Hale-Bopp and 2.76 ± 0.08 for 103P/Hartley 2 (*Crovisier et al.,* 1997, 1999b), corresponding to $T_{spin} = 28 ± 2$ K and 36 ± 3 K, respectively.

Kawakita et al. (2001, 2002) measured the OPR of the NH_2 radical in Comets C/1999 S4 (LINEAR) and C/2001 A2 (LINEAR). Following symmetry conservation laws, decay products maintain the spin distribution of the original molecules during photolysis, so that the OPR of NH_3 may be traced from that of NH_2. Kawakita et al. derived an OPR for NH_3 of 1.17 ± 0.04 and 1.12 ± 0.03, respectively, for the two LINEAR comets, corresponding to a spin temperature of ~30 K.

Several other cometary molecules (H_2CO, CH_3OH, hydrocarbons, ...) are likely to have OPR effects that are worthy of investigation. The spin species of methane were found to be consistent with $T_{spin} > 40$–50 K (*Weaver et al.,* 1997; *Gibb et al.,* 2003). Formaldehyde was also investigated from its radio rotational lines, but no reliable OPR could be derived because only a few lines could be probed.

The cold spin temperatures retrieved for cometary H_2O and NH_3 presumably have a meaning related to the history of these species in the nucleus or even before, i.e., connected with their formation:

1. *Reequilibration in the nucleus.* We would then expect different spin temperatures for different comets, depending on their present orbit and dynamical history, which does not seem to be the case. 30 K corresponds to the equilibrium temperature at r ~ 100 AU, under present solar system conditions.

2. *Water formation.* The cold spin temperature would rule out a gas phase formation, since reactions leading to H_2O are exothermal and would form H_2O at a high temperature. Rather, the low OPR values suggest formation on grains, where H_2O would reequilibrate at the grain temperature (*Tielens and Allamandola, 1987*). The ortho-to-para conversion of H_2 on interstellar grains is very fast — on the order of 60 s. Thus, low OPR values are consistent with cometary H_2O originating in the interstellar medium.

3. *Fractionation of spin species.* It is also possible that one of the spin species is favored by condensation or absorption on interstellar or cometary grains, as is indeed observed in laboratory experiments such as condensation of D_2O on a cold matrix or selective absorption of H_2O on charcoal (*Tikhonov and Volkov, 2002*).

11. PERSPECTIVES

The situation regarding the detection and abundance determinations of cometary parent molecules has drastically changed since the publication of *Comets* (*Wilkening, 1982*). At that time, CO was the only parent molecule that had been directly identified. Generally, one had to rely on *deducing* the identity of the parents from observations of their daughter products. A half dozen species were proposed in these pioneering times, including H_2O, CO_2, CH_4, NH_3, all of which have been confirmed by recent observations.

We now have firm, direct identifications of two dozen parent molecules. It is very probable that the most abundant ones (say, at the level of ~1% relative to H_2O) are now known, including those already suggested 20 years ago. Many more molecules have emerged, some of them being only trace species (less than ~0.1%): other hydrocarbons, several nitriles and "CHO" species, and S-containing molecules.

Presumably, many additional molecules remain undetected in comets. The presence of unidentified lines (or blends of lines) strongly suggests that we are on the verge of important new discoveries. More molecules have been detected in the interstellar medium and even more in meteorites (e.g., *Botta and Bada, 2002*), with abundances generally decreasing with increasing complexity. No limit has yet been reached in the complexity of molecules found in these environments, and the same could be true for cometary material.

Future progress by remote sensing spectroscopy requires continual improvements in sensitivity, as is being provided now by new-generation instruments such as Keck/NIRSPEC, the Subaru telescope, and the Very Large Telescope, or planned intruments such as the Atacama Large Millimeter Array. Apparitions of new, bright comets are always highly anticipated, as they provide excellent opportunities to provide deeper searches than previously obtained. Short-period comets (and more specifically, Jupiter-family comets) are fainter objects, and therefore more difficult to study. Nevertheless, improvements in technology will hopefully allow investigations of parent molecules in a large fraction of the short-period comets. Laboratory work on molecular spec-troscopy and the compilation of comprehensive spectroscopic databases are also needed for further progress.

An alternative approach is *in situ* analysis by space exploration, as will be performed by the *Rosetta* mission. Mass spectroscopy and gas chromatography could be much more sensitive than remote sensing.

Perhaps the ultimate answers regarding cometary composition will only be revealed by the analysis, at leisure on Earth, of returned samples of cometary ices. Unfortunately, there is not yet any approved space mission with this as its goal, but we can hope for such a mission in the not too distant future.

The cost and difficulty of encounter missions with comets will limit, probably for a long time, *in situ* investigations to a very small number of objects chosen among the short-period comets. Ground- and Earth-orbit-based observations will still be needed for the systematic investigation of a large sample of objects, in order to study their diversity. Only then will we truly be able to address the role that comets played in the formation and evolution of the solar system and their relation to the interstellar medium.

Acknowledgments. H.A.W. acknowledges financial support by NASA through grant HST-GO-8276.01-A from the Space Telescope Science Institute, which is operated by the Association of Universities for Research in Astronomy, Inc., under NASA Contract NAS5-26555, and through FUSE Grant NAG5-10921.

REFERENCES

A'Hearn M. F., Feldman P. D., and Schleicher D. J. (1983) The discovery of S_2 in comet IRAS-Araki-Alcock 1983d. *Astrophys. J. Lett., 274,* L99–L103.

A'Hearn M. F., Schleicher D. G., Millis R. L., Feldman P. D., and Thompson D. T. (1984) Comet Bowell 1980b. *Astron. J., 89,* 579–591.

A'Hearn M. F., Millis R. L., Schleicher D. G., Osip D. J., and Birch P. V. (1995) The ensemble properties of comets: Results from narrowband photometry of 85 comets, 1976–1992. *Icarus, 118,* 223–270.

Altenhoff W. J., Batrla W., Huchtmeier W. K., Schmidt J., Stumpff P., and Walmsley M. (1983) Radio observations of comet 1983d. *Astron. Astrophys., 125,* L19–L22.

Altwegg K. (1996) Sulfur in comet Halley. Habilitationsschrift, Univ. of Bern.

Altwegg K., Balsiger H., and Geiss J. (1994) Abundance and origin of the CH_n^+ ions in the coma of comet P/Halley. *Astron. Astrophys., 290,* 318–323.

Altwegg K., Balsiger H., and Geiss J. (1999) Composition of the volatile material in Halley's coma from in situ measurements. *Space Sci. Rev., 90,* 3–18.

Arpigny C., Jehin E., Manfroid J., Hutsemékers D., Schulz R., Stüve J. A., Zucconi J.-M., and Ilyin I. (2003) Anomalous nitrogen isotope ratio in comets. *Science, 301,* 1522–1524.

Ashihara O. (1975) The electron energy loss rates by polar molecules. *Inst. Space Aeronautical Sci. Rept. No. 530, 40,* 10.

Balsiger H., Altwegg K., and Geiss J. (1995) D/H and $^{18}O/^{16}O$ ratio in the hydronium ion and in neutral water from *in situ* ion measurements in comet P/Halley. *J. Geophys. Res., 100,* 5827–5834.

Bird M. K., Huchtmeier W. K., Gensheimer P., Wilson T. L., Janardhan P., and Lemme C. (1997) Radio detection of ammonia in comet Hale-Bopp. *Astron. Astrophys., 325,* L5–L8.

Bird M. K., Janardhan P., Wilson T. L., Huchtmeier W. K., Gensheimer P., and Lemme C. (1999) K-band radio observations of comet Hale-Bopp: Detections of ammonia and (possibly) water. *Earth Moon Planets, 78,* 21–28.

Bird M. K., Hatchell J., van der Tak F. F. S., Crovisier J., and Bockelée-Morvan D. (2002) Search for the radio lines of ammonia in comets. In *Proceedings of Asteroids Comets Meteors 2002* (B. Warmbein, ed.), pp. 697–700. ESA SP-500, Noordwijk, The Netherlands.

Biver N. and 22 colleagues (1997) Evolution of the outgassing of comet Hale-Bopp (C/1995 O1) from radio observations. *Science, 275,* 1915–1918.

Biver N. and 13 colleagues (1999a) Spectroscopic monitoring of comet C/1996 B2 (Hyakutake) with the JCMT and IRAM radio telescopes. *Astron. J., 118,* 1850–1872.

Biver N. and 22 colleagues (1999b) Long-term evolution of the outgassing of comet Hale-Bopp from radio observations. *Earth Moon Planets, 78,* 5–11.

Biver N. and 12 colleagues (2000) Spectroscopic observations of comet C/1999 H1 (Lee) with the SEST, JCMT, CSO, IRAM, and Nançay radio telescopes. *Astron. J., 120,* 1554–1570.

Biver N. and 23 colleagues (2002a) The 1995–2002 long-term monitoring of comet C/1995 O1 (Hale-Bopp) at radio wavelengths. *Earth Moon Planets, 90,* 5–14.

Biver N., Bockelée-Morvan D., Crovisier J., Colom P., Henry F., Moreno R., Paubert G., Despois D., and Lis D. C. (2002b) Chemical composition diversity among 24 comets observed at radio wavelengths. *Earth Moon Planets, 90,* 323–333.

Blake G. A., Qi C., Hogerheijde M. R., Gurwell M. A., and Muhleman D. O. (1999) Sublimation from icy jets as a probe of the interstellar content of comets. *Nature, 398,* 213–216.

Bockelée-Morvan D. (1987) A model for the excitation of water in comets. *Astron. Astrophys., 181,* 169–181.

Bockelée-Morvan D. (1997) Cometary volatiles. The status after comet C/1996 B2 (Hyakutake). In *Molecules in Astrophysics: Probes and Processes* (E. F. Van Dishoeck, ed.), pp. 219–255. IAU Symposium 178, Kluwer, Dordrecht.

Bockelée-Morvan D. and Crovisier J. (1985) Possible parents for the cometary CN radical: Photochemistry and excitation conditions. *Astron. Astrophys., 151,* 90–100.

Bockelée-Morvan D. and Crovisier J. (1989) The nature of the 2.8-μm emission feature in cometary spectra. *Astron. Astrophys., 216,* 278–283.

Bockelée-Morvan D. and Crovisier J. (1992) Formaldehyde in comets. II — Excitation of the rotational lines. *Astron. Astrophys., 264,* 282–291.

Bockelée-Morvan D., Crovisier J., Baudry A., Despois D., Perault M., Irvine W. M., Schloerb F. P., and Swade D. (1984) Hydrogen cyanide in comets — Excitation conditions and radio observations of comet IRAS-Araki-Alcock 1983d. *Astron. Astrophys., 141,* 411–418.

Bockelée-Morvan D., Colom P., Crovisier J., Despois D., and Paubert G. (1991) Microwave detection of hydrogen sulphide and methanol in Comet Austin (1989c1). *Nature, 350,* 318–320.

Bockelée-Morvan D., Crovisier J., Colom P., and Despois D. (1994) The rotational lines of methanol in comets Austin 1990 V and Levy 1990 XX. *Astron. Astrophys., 287,* 647–665.

Bockelée-Morvan D., Brooke T. Y., and Crovisier J. (1995) On the origins of the 3.2–3.6 μm emission feature in comets. *Icarus, 116,* 18–39.

Bockelée-Morvan D. and 11 colleagues (1998) Deuterated water in comet C/1996 B2 (Hyakutake) and its implications for the origin of comets. *Icarus, 133,* 147–162.

Bockelée-Morvan D. and 17 colleagues (2000) New molecules found in comet C/1995 O1 (Hale-Bopp). Investigating the link between cometary and interstellar material. *Astron. Astrophys., 353,* 1101–1114.

Bockelée-Morvan D. and 11 colleagues (2001) Outgassing behavior and composition of comet C/1999 S4 (LINEAR) during its disruption. *Science, 292,* 1339–1343.

Botta O. and Bada J. L. (2002) Extraterrestrial organic compounds in meteorites. *Survey Geophys., 23,* 411–467.

Brooke T. Y., Knacke R. F., Owen T. C., and Tokunaga A. T. (1989) Spectroscopy of emission features near 3 microns in Comet Wilson (1986l). *Astrophys. J., 336,* 971–978.

Brooke T. Y., Tokunaga A. T., Weaver H. A., Crovisier J., Bockelée-Morvan D., and Crisp D. (1996) Detection of acetylene in the infrared spectrum of comet Hyakutake. *Nature, 383,* 606–608.

Brooke T. Y., Weaver H. A., Chin G., Bockelée-Morvan D., Kim S. J., and Xu L.-H. (2003) Spectroscopy of comet Hale-Bopp in the infrared. *Icarus, 166,* 167–187.

Budzien S. A. and Feldman P. D. (1992) Upper limits to the S_2 abundance in several comets observed with the International Ultraviolet Explorer. *Icarus, 99,* 143–152.

Capria M. T., Coradini A., De Sanctis M. C., and Orosei R. (2000) CO emission mechanisms in C/1995 O1 (Hale-Bopp). *Astron. Astrophys., 357,* 359–366.

Chin G. and Weaver H. A. (1984) Vibrational and rotational excitation of CO in comets: nonequilibrium calculations. *Astrophys. J., 285,* 858–869.

Chiu K., Neufeld D. A., Bergin E. A., Melnick G. J, Patten B. M., Wang Z., and Bockelée-Morvan D. (2001) Post-perihelion SWAS observations of water vapor in the coma of comet C/1999 H1 (Lee). *Icarus, 154,* 345–349.

Clemett S. J., Maechling C. R., Zare R. N., Swan P. D., and Walker R. M. (1993) Identification of complex aromatic molecules in individual interplanetary dust particles. *Science, 262,* 721–725.

Cochran A. L. (2002) A search for N_2^+ in spectra of comet C/2002 C1 (Ikeya-Zhang). *Astrophys. J. Lett., 576,* L165–L168.

Cochran A. L., Cochran W. D., and Barker E. S. (2000) N_2^+ and CO+ in comets 122P/1995 S1 (de Vico) and C/1995 O1 (Hale-Bopp). *Icarus, 146,* 583–593.

Colangeli L., Epifani E., Brucato R., Bussoletti E., De Sanctis C., Fulle M., Mennela V., Palomba E., Palumbo P., and Rotundi A. (1999) Infrared spectral observations of comet 103P/Hartley 2 by ISOPHOT. *Astron. Astrophys., 343,* L87–L90.

Colom P., Crovisier J., Bockelée-Morvan D., Despois D., and Paubert G. (1992) Formaldehyde in comets. I — Microwave observations of P/Brorsen-Metcalf (1989 X), Austin (1990 V) and Levy (1990 XX). *Astron. Astrophys., 264,* 270–281.

Combes M. and 16 colleagues (1988) The 2.5 to 12 μm spectrum of comet Halley from the IKS-VEGA experiment. *Icarus, 76,* 404–436.

Combi M. R., Harris W. R., and Smyth W.M. (2004) Gas dynamics and kinetics in the cometary coma: Theory and observations. In *Comets II* (M. C. Festou et al., eds.), this volume. Univ. of Arizona, Tucson.

Cottin H., Gazeau M.-C., Benilan Y., and Raulin F. (2004) Origin of cometary extended sources from degradation of refractory organics on grains: Polyoxymethylene as formaldehyde parent molecule. *Icarus, 167,* 397–416.

Cravens T. E. and Korosmezey A. (1986) Vibrational and rotational cooling of electrons by water vapor. *Planet. Space Sci., 34,* 961–970.

Crifo J.-F., Fulle M., Kömle N. I., and Szegö K. (2004) Nucleus-coma structural relationships: Lessons from physical models. In *Comets II* (M. C. Festou et al., eds.), this volume. Univ. of Arizona, Tucson.

Crovisier J. (1984) The water molecule in comets: Fluorescence mechanisms and thermodynamics of the inner coma. *Astron. Astrophys., 130,* 361–372.

Crovisier J. (1987) Rotational and vibrational synthetic spectra of linear parent molecules in comets. *Astron. Astrophys. Suppl. Ser., 68,* 223–258.

Crovisier J. (1994) Photodestruction rates for cometary parent molecules. *J. Geophys. Res., 99,* 3777–3781.

Crovisier J. and Encrenaz T. (1983) Infrared fluorescence in comets: The general synthetic spectrum. *Astron. Astrophys., 126,* 170–182.

Crovisier J. and Le Bourlot J. (1983) Infrared and microwave fluorescence of carbon monoxide in comets. *Astron. Astrophys., 123,* 61–66.

Crovisier J., Despois D., Bockelée-Morvan D., Colom P., and Paubert G. (1991a) Microwave observations of hydrogen sulfide and searches for other sulfur compounds in comets Austin (1989c1) and Levy (1990c). *Icarus, 93,* 246–258.

Crovisier J., Encrenaz T., and Combes M. (1991b) Carbon suboxide in comet Halley. *Nature, 353,* 610.

Crovisier J., Leech K., Bockelée-Morvan D., Brooke T. Y., Hanner M. S., Altieri B., Keller H. U., and Lellouch E. (1997) The spectrum of Comet Hale-Bopp (C/1995 O1) observed with the Infrared Space Observatory at 2.9 AU from the Sun. *Science, 275,* 1904–1907.

Crovisier J., Leech K., Bockelée-Morvan D., Lellouch E., Brooke T. Y., Hanner M. S., Altieri B., Keller H. U., and Lim T. (1999a) The spectrum of comet Hale-Bopp as seen by ISO. In *The Universe as Seen by ISO* (P. Cox and M. F. Kessler, eds.), pp. 137–140. ESA SP-427, Noordwijk, The Netherlands.

Crovisier J. and 11 colleagues (1999b) ISO observations of short-period comets. In *The Universe as Seen by ISO* (P. Cox and M. F. Kessler, eds.), pp. 161–164. ESA SP-427, Noordwijk, The Netherlands.

Crovisier J., Bockelée-Morvan D., Colom P., Biver N., Despois D., Lis D. C., and the Team for Target-of-Opportunity Radio Observations of Comets (2004a) The composition of ices in comet C/1995 O1 (Hale-Bopp) from radio spectroscopy: Further results and upper limits on undetected species. *Astron. Astrophys., 418,* 1141–1157.

Crovisier J., Bockelée-Morvan D., Biver N., Colom P., Despois D., and Lis D. C. (2004b) Ethylene glycol in comet C/1995 O1 (Hale-Bopp). *Astron. Astrophys., 418,* L35–L38.

Davies J. K., Green S. F., and Geballe T. R. (1991) The detection of a strong 3.28 μm emission feature in comet Levy. *Mon. Not. R. Astron. Soc., 251,* 148–151.

Dello Russo N., DiSanti M. A., Mumma M. J., Magee-Sauer K., and Rettig T. W. (1998) Carbonyl sulfide in comets C/1996 B2 (Hyakutake) and C/1995 O1 (Hale-Bopp): Evidence for an extended source in Hale-Bopp. *Icarus, 135,* 377–388.

Dello Russo N., Mumma M. J., DiSanti M. A., Magee-Sauer K., Novak R., and Rettig T. W. (2000) Water production and release in Comet C/1995 O1 Hale-Bopp. *Icarus, 143,* 324–337.

Dello Russo N., Mumma M. J., DiSanti M. A., Magee-Sauer K., and Novak R. (2001) Ethane production and release in comet C/1995 O1 Hale-Bopp. *Icarus, 153,* 162–179.

Dello Russo N., Mumma M. J., DiSanti M. A., and Magee-Sauer K. (2002a) Production of ethane and water in comet C/1996 B2 Hyakutake. *J. Geophys. Res.–Planets, 107(E11),* 5095, doi:10.1029/2001JE001838.

Dello Russo N., DiSanti M. A., Magee-Sauer K., Gibb E., and Mumma M. J. (2002b) Production of C_2H_6 and H_2O in comet 2002 C1 Ikeya-Zhang on UT 2002 April 13.7–13.9. In *Proceedings of Asteroids Comets Meteors 2002* (B. Warmbein, ed.), pp. 689–692. ESA SP-500, Noordwijk, The Netherlands.

Dello Russo N., DiSanti M. A., Magee-Sauer K., Gibb E., Mumma M. J., Barber R. J., and Tennyson J. (2004) Water production and release in comet 153P/Ikeya-Zhang (2002 C1): Accurate rotational temperatures retrievals from hot-band lines near 2.9 μm. *Icarus, 168,* 186–200.

Despois D., Crovisier J., Bockelée-Morvan D., Schraml J., Forveille T., and Gérard E. (1986) Observation of hydrogen cyanide in comet Halley. *Astron. Astrophys., 160,* L11–L12.

DiSanti M. A., Mumma M. J., Dello Russo N., Magee-Sauer K., Novak R., and Rettig T. W. (1999) Identification of two sources of carbon monoxide in comet Hale-Bopp. *Nature, 399,* 662–665.

DiSanti M. A., Mumma M. J., Dello Russo N., and Magee-Sauer K. (2001) Carbon monoxide production and excitation in comet C/1995 O1 (Hale-Bopp): Isolation of native and distributed CO sources. *Icarus, 153,* 361–390.

DiSanti M. A., Dello Russo N., Magee-Sauer K., Gibb E. L., Reuter D. C., and Mumma M. J. (2002) CO, H_2CO and CH_3OH in comet 2002 C1 Ikeya-Zhang. In *Proceedings of Asteroids Comets Meteors 2002* (B. Warmbein, ed.), pp. 571–574. ESA SP-500, Noordwijk, The Netherlands.

DiSanti M. A., Mumma M. J., Dello Russo N., Magee-Sauer K., and Griep D. (2003) Evidence for a dominant native source of CO in comet C/1996 B2 (Hyakutake). *J. Geophys. Res., 108(E6),* 5061.

Dones L., Weissman P. R., Levison H. F., and Duncan M. J. (2004) Oort cloud formation and dynamics. In *Comets II* (M. C. Festou et al., eds.), this volume. Univ. of Arizona, Tucson.

Duncan M., Levison H., and Dones L. (2004) Dynamical evolution of ecliptic comets. In *Comets II* (M. C. Festou et al., eds.), this volume. Univ. of Arizona, Tucson.

Dutrey A. and 12 colleagues (1996) *Comet C/1996 B2 (Hyakutake).* IAU Circular No. 6364.

Eberhardt P. (1999) Comet Halley's gas composition and extended sources: Results from the neutral mass spectrometer on Giotto. *Space Sci. Rev., 90,* 45–52.

Eberhardt P. and 10 colleagues (1987) The CO and N_2 abundance in comet P/ Halley. *Astron. Astrophys., 187,* 481–484.

Eberhardt P., Meier R., Krankowsky D., and Hodges R. R. (1994) Methanol and hydrogen sulfide in comet P/Halley. *Astron. Astrophys., 288,* 315–329.

Eberhardt P., Reber M., Krankowsky D., and Hodges R. R. (1995) The D/H and $^{18}O/^{16}O$ ratios in water from comet P/Halley. *Astron. Astrophys., 302,* 301–316.

Ehrenfreund P., Charnley S. B., and Wooden D. (2004) From interstellar material to cometary particles and molecules. In *Comets II* (M. C. Festou et al., eds.), this volume. Univ. of Arizona, Tucson.

Feldman P. D. (1999) Ultraviolet observations of comet Hale-Bopp. *Earth Moon Planets, 79,* 145–160.

Feldman P. D. and Brune W. H. (1976) Carbon production in comet West (1975n). *Astrophys. J. Lett., 209,* L45–L48.

Feldman P. D. and 15 colleagues (1987) IUE observations of comet P/Halley: Evolution of the ultraviolet spectrum between September 1985 and July 1986. *Astron. Astrophys., 187,* 325–328.

Feldman P. D. and 14 colleagues (1991) Observations of Comet Levy (1990c) with the Hopkins Ultraviolet Telescope. *Astrophys. J. Lett., 379,* L37–L40.

Feldman P. D., Festou M. C., Tozzi G. P., and Weaver H. A. (1997) The CO_2/CO abundance ratio in 1P/Halley and several other comets observed by IUE and HST. *Astrophys. J., 475,* 829–834.

Feldman P. D., Weaver H. A., A'Hearn M. F., Festou M. C., McPhate J. B., and Tozzi G.-P. (1999) Ultraviolet imaging spectroscopy of comet Lee (C/1999 H1) with HST/STIS (abstract). *Bull. Am. Astron. Soc., 31,* 1127.

Feldman P. D., Weaver H. A., and Burgh E. B. (2002) Far Ultraviolet Spectroscopic Explorer observations of CO and H_2 emissions in comet C/2001 A2 (LINEAR). *Astrophys. J. Lett., 576,* L91–L94.

Feldman P. D., Cochran A. L., and Combi M. R. (2004) Spectroscopic investigations of fragment species in the coma. In *Comets II* (M. C. Festou et al., eds.), this volume. Univ. of Arizona, Tucson.

Festou M. C., Rickman H., and West R. M. (1993) Comets. I — Concepts and observations. *Astron. Astrophys. Rev., 4,* 363–447.

Geiss J., Altwegg K., Anders E., Balsiger H., Ip W.-H., Meier A., Neugebauer M., Rosenbauer H., and Shelley E. G. (1991) Interpretation of the ion mass per charge range 25–35 amu/e obtained in the inner coma of Halley's comet by the HIS-sensor of the Giotto IMS experiment. *Astron. Astrophys., 247,* 226–234.

Gibb E. L., Mumma M. J., Dello Russo N., DiSanti M. A., and Magee-Sauer K. (2003) Methane in Oort cloud comets. *Icarus, 165,* 391–406.

Green J. C., Cash W., Cook T. A., and Stern S. A. (1991) The spectrum of comet Austin from 910 to 1180 Å. *Science, 251,* 408–410.

Greenberg J. M. and Li A. (1998) From interstellar dust to comets: The extended CO source in comet Halley. *Astron. Astrophys., 332,* 374–384.

Gunnarsson M., Rickman H., Festou M. C., Winnberg A., and Tancredi G. (2002) An extended CO source around comet 29P/ Schwassmann-Wachmann 1. *Icarus, 157,* 309–322.

Gunnarsson M. and 14 colleagues (2003) Production and kinematics of CO in comet C/1995 O1 (Hale-Bopp) at large post-perihelion distances. *Astron. Astrophys., 402,* 383–393.

Haser L. (1957) Distribution d'intensité dans la tête d'une comète. *Bull. Acad. R. Sci. Liège, 43,* 740–750.

Henry F. (2003) La comète Hale-Bopp à l'interféromètre du Plateau de Bure: Étude de la distribution du monoxyde de carbone. Thèse de Doctorat, Université de Paris 6.

Henry F., Bockelée-Morvan D., Crovisier J., and Wink J. (2002) Observations of rotating jets of carbon monoxide in comet Hale-Bopp with the Plateau de Bure interferometer. *Earth Moon Planets, 90,* 57–60.

Hirota T., Yamamoto S., Kawaguchi K., Sakamoto A., and Ukita N. (1999) Observations of HCN, HNC and NH_3 in comet Hale-Bopp. *Astrophys. J., 520,* 895–900.

Hoban S., Mumma M. J, Reuter D. C., DiSanti M. A., Joyce R. R., and Storrs A. (1991) A tentative identification of methanol as the progenitor of the 3.52-micron emission feature in several

comets. *Icarus, 93,* 122–134.

Hoban S., Reuter D. C., DiSanti M. A., Mumma M. J, and Elston R. (1993) Infrared observations of methanol in comet P/Swift-Tuttle. *Icarus, 105,* 548–556.

Huebner W. F (1987) First polymer in space identified in Comet Halley. *Science, 237,* 628–630.

Huebner W. F., Keady J. J., and Lyon S. P. (1992) Solar photo rates for planetary atmospheres and atmospheric pollutants. *Astrophys. Space Sci., 195,* 1–289.

Huntress W. T. Jr., Allen M., and Delitsky M. (1991) Carbon suboxide in comet Halley? *Nature, 352,* 316–318.

Irvine W. and 17 colleagues (1996) Spectroscopic evidence for interstellar ices in Comet Hyakutake. *Nature, 382,* 418–420.

Irvine W. M., Bergin E. A., Dickens J. E., Jewitt D., Lovell A. J., Matthews H. E., Schloerb F. P., and Senay M. (1998) Chemical processing in the coma as the source of cometary HNC. *Nature, 393,* 547–550.

Irvine W. M., Schloerb F. P., Crovisier J., Fegley B., and Mumma M. J. (2000a) Comets: A link between interstellar and nebular chemistry. In *Protostars and Planets IV* (V. Mannings et al., eds.), pp. 1159–1200. Univ. of Arizona, Tucson.

Irvine W. M., Senay M., Lovell A. J., Matthews H. E., McGonagle D., and Meier R. (2000b) NOTE: Detection of nitrogen sulfide in comet Hale-Bopp. *Icarus, 143,* 412–414.

Irvine W. M., Bergman P., Lowe T. B., Matthews H., McGonagle D., Nummelin A., and Owen T. (2003) HCN and HNC in comets C/2000 WM1 (LINEAR) and C/2002 C1 (Ikeya-Zhang). *Origins Life Evol. Biosph., 33,* 1–11.

Itikawa Y. J. (1972) Rotational transition in an asymmetric-top molecule by electron collision: Applications to H_2O. *J. Phys. Soc. Japan, 32,* 217–226.

Jacquinet-Husson and 47 colleagues (1999) The 1997 spectroscopic GEISA databank. *J. Quant. Spectrosc. Radiat. Transfer, 62,* 205–254. Also available on line at http:// ara.lmd.polytechnique.fr.

Jewitt D. C., Matthews H. E., Owen T. C., and Meier R. (1997) Measurements of $^{12}C/^{13}C$, $^{14}N/^{15}N$, and $^{32}S/^{34}S$ ratios in comet Hale-Bopp (C/1995 O1). *Science, 278,* 90–93.

Joblin C., Boissel P., and de Parseval P. (1997) Polycyclic aromatic hydrocarbon lifetime in cometary environments. *Planet. Space Sci., 45,* 1539–1542.

Johnson J. R., Fink U., and Larson H. P. (1983) The 0.9–2.5 micron spectrum of Comet West 1976 VI. *Astrophys. J., 270,* 769–777.

Kawakita H. and 18 colleagues (2001) The spin temperature of NH_3 in comet C/1999 S4 (LINEAR). *Science, 294,* 1089–1091.

Kawakita H., Watanabe J., Fuse T., Furusho R., and Abe S. (2002) Spin temperature of ammonia determined from NH_2 in comet C/2001 A2 (LINEAR). *Earth Moon Planets, 90,* 371–379.

Kawakita H., Watanabe J., Kinoshita D., Ishiguro M., and Nakamura R. (2003) Saturated hydrocarbons in Comet 153P/Ikeya-Zhang: Ethane, methane, and monodeuterio-methane. *Astrophys. J., 590,* 573–578.

Kim S. J. and A'Hearn M. F. (1991) Upper limits of SO and SO_2 in comets. *Icarus, 90,* 79–95.

Kim S. J., A'Hearn M. F., Wellnitz D. D., Meier R., and Lee Y. S. (2003) The rotational stucture of the B-X system of sulfur dimers in the spectra of comet Hyakutake (C/1996 B2). *Icarus, 166,* 157–166.

Kleine M., Wyckoff S., Wehinger P. A., and Peterson B. A. (1995) The carbon isotope abundance ratio in comet Halley. *Astrophys. J., 439,* 1021–1033.

Knacke R. F., Brooke T. Y., and Joyce R. R. (1986) Observations of

3.2–3.6 micron emission features in comet Halley. *Astrophys. J. Lett., 310,* L49–L53.

Krankowsky D. and 18 colleagues (1986) In situ gas and ion measurements at comet Halley. *Nature, 321,* 326–329.

Krasnopolsky V. A. and Mumma M. J. (2001) Spectroscopy of Comet Hyakutake at 80–700 Å: First detection of solar wind charge transfer emissions. *Astrophys. J., 549,* 629–634.

Krasnopolsky V. A., Mumma M. J., Abbott M., Flynn B. C., Meech K. J., Yeomans D. K., Feldman P. D., and Cosmovici C. B. (1997) Detection of soft X-rays and a sensitive search for noble gases in comet Hale-Bopp (C/1995 O1). *Science, 277,* 1488–1491.

Larson L., Weaver H. A., Mumma M. J., and Drapatz S. (1989) Airborne infrared spectroscopy of comet Wilson (1986l) and comparisons with comet Halley. *Astrophys. J., 338,* 1106–1114.

Lecacheux A. and 21 colleagues (2003) Observations of water in comets with Odin. *Astron. Astrophys., 402,* L55–L58.

Lis D. C., Keene J., Young K., Phillips T. G., Bockelée-Morvan D., Crovisier J., Schilke P., Goldsmith P. F., and Bergin E. A. (1997) Spectroscopic observations of comet C/1996 B2 (Hyakutake) with the Caltech Submillimeter Observatory. *Icarus, 130,* 355–372.

Lis D. C. and 10 colleagues (1999) New molecular species in comet C/1995 O1 (Hale-Bopp) observed with the Caltech Submillimeter Observatory. *Earth Moon Planets, 78,* 13–20.

Lisse C. M., Cravens T. E., and Dennerl K. (2004) X-ray and extreme ultraviolet emission from comets. In *Comets II* (M. C. Festou et al., eds.), this volume. Univ. of Arizona, Tucson.

Lovell A. J. (1999) Millimeter-wave molecular mapping of comets Hyakutake and Hale-Bopp. PhD thesis, Univ. of Massachusetts.

Lunine J. I. and Gautier D. I. (2004) Coupled physical and chemical evolution of volatiles in the protoplanetary disk: A tale of three elements. In *Comets II* (M. C. Festou et al., eds.), this volume. Univ. of Arizona, Tucson.

Magee-Sauer K., Mumma M. J., DiSanti M. A., Dello Russo N., and Rettig T. W. (1999) Infrared spectroscopy of the v_3 band of hydrogen cyanide in comet C/1995 O1 Hale-Bopp. *Icarus, 142,* 498–508.

Magee-Sauer K., Mumma M. J., DiSanti M. A., and Dello Russo N. (2002a) Hydrogen cyanide in comet C/1996 B2 Hyakutake. *J. Geophys. Res.–Planets, 107(E11),* 5096, doi:10.1029/2002JE001863.

Magee-Sauer K., Dello Russo N., DiSanti M. A., Gibb E., and Mumma M. J. (2002b) Production of HCN and C_2H_2 in comet C/2002 C1 (Ikeya-Zhang) on UT 2002 April 13.8. In *Proceedings of Asteroids Comets Meteors 2002* (B. Warmbein, ed.), pp. 549–552. ESA SP-500, Noordwijk, The Netherlands.

McPhate J. B., Feldman P. D., Weaver H. A., A'Hearn M. F., Tozzi G.-P., and Festou M. C. (1996) Ultraviolet CO emission in comet C/1996 B2 (Hyakutake) (abstract). *Bull. Am. Astron. Soc., 28,* 1093–1094.

Meech K. J. and Svoreň J. (2004) Physical and chemical evolution of cometary nuclei. In *Comets II* (M. C. Festou et al., eds.), this volume. Univ. of Arizona, Tucson.

Meier R. and A'Hearn M. F. (1997) Atomic sulfur in cometary comae based on UV spectra of the SI triplet near 1814 Å. *Icarus, 125,* 164–194.

Meier R., Eberhardt P., Krankowsky D., and Hodges R. R. (1993) The extended formaldehyde source in comet P/Halley. *Astron. Astrophys., 277,* 677–690.

Meier R., Eberhardt P., Krankowsky D., and Hodges R. R. (1994) Ammonia in comet P/Halley. *Astron. Astrophys., 287,* 268–278.

Meier R., Owen T. C., Matthews H. E., Jewitt D. C., Bockelée-Morvan D., Biver N., Crovisier J., and Gautier D. (1998a) A determination of the HDO/H_2O ratio in comet C/1995 O1 (Hale-Bopp). *Science, 279,* 842–844.

Meier R., Owen T. C., Jewitt D. C., Matthews H., Senay M., Biver N., Bockelée-Morvan D., Crovisier J., and Gautier D. (1998b) Deuterium in comet C/1995 O1 (Hale-Bopp): Detection of DCN. *Science, 279,* 1707–1710.

Meier R., Wellnitz D., Kim S. J., and A'Hearn M. F. (1998c) The NH and CH bands of comet C/1996 B2 (Hyakutake). *Icarus, 136,* 268–279.

Mitchell D. L., Lin R. P., Anderson K. A., Carlson C. W., Curtis D. W., Korth A., Reme H., Sauvaud J. A., d'Uston C., and Mendis D. A. (1987) Evidence for chain molecules enriched in carbon, hydrogen, and oxygen in comet Halley. *Science, 237,* 626–628.

Mitchell D. L., Lin R. P., Anderson K. A., Carlson C. W., Curtis D. W., Korth A., Reme H., Sauvaud J. A., d'Uston C., and Mendis D. A. (1989) Complex organic ions in the atmosphere of comet Halley. *Adv. Space Res., 9,* 35–39.

Morbidelli A. and Brown M. E. (2004) The Kuiper belt and the primordial evolution of the solar system. In *Comets II* (M. C. Festou et al., eds.), this volume. Univ. of Arizona, Tucson.

Moreels G., Clairemidi J., Hermine P., Brechignac P., and Rousselot P. (1993) Detection of a polycyclic aromatic molecule in P/Halley. *Astron. Astrophys., 282,* 643–656.

Müller H. S. P., Thorwirth S., Roth D. A., and Winnewisser G. (2001) The Cologne Database for Molecular Spectroscopy, CDMS. *Astron. Astrophys., 370,* L49–L52. Available on line at http://www.cdms.de.

Mumma M. J. (1982) Speculations on the infrared molecular spectra of comets. In *Vibrational-Rotational Spectroscopy for Planetary Atmospheres* (M. J. Mumma et al., eds.), pp. 717–742. NASA CP-2223, Vol. 2, Washington, DC.

Mumma M. J. and Reuter D. C. (1989) On the identification of formaldehyde in Halley's comet. *Astrophys. J., 344,* 940–948.

Mumma M. J., Stone E. J., and Zipf E. C. (1975) Nonthermal rotational distribution of CO ($A^1\Pi$) fragments produced by dissociative excitation of CO_2 by electron impact. *J. Geophys. Res., 80,* 161–167.

Mumma M. J., Weaver H. A., Larson H. P., Davis D. S., and Williams M. (1986) Detection of water vapor in Halley's comet. *Science, 232,* 1523–1528.

Mumma M. J., Weissman P. R., and Stern S. A. (1993) Comets and the origin of the solar system: Reading the Rosetta Stone. In *Protostars and Planets III* (E. H. Levy and J. I. Lunine, eds.), pp. 1177–1252. Univ. of Arizona, Tucson.

Mumma M. J., DiSanti M. A., Tokunaga A., and Roettger E. E. (1995) Ground-based detection of water in Comet Shoemaker-Levy 1992 XIX; probing cometary parent molecules by hot-band fluorescence (abstract). *Bull. Am. Astron. Soc., 27,* 1144.

Mumma M. J., DiSanti M. A., Dello Russo N., Fomenkova M., Magee-Sauer K., Kaminski C. D., and Xie D.X. (1996) Detection of abundant ethane and methane, along with carbon monoxide and water, in comet C/1996 B2 Hyakutake: Evidence for interstellar origin. *Science, 272,* 1310–1314.

Mumma M. J., DiSanti M. A., Dello Russo N., Magee-Sauer K., and Rettig T. W. (2000) Detection of CO and ethane in comet 21P/Giacobini-Zinner: Evidence for variable chemistry in the outer solar nebula. *Astrophys. J. Lett., 531,* L155–L159.

Mumma M. J., Dello Russo N., DiSanti M. A., Magee-Sauer K., Novak R. E., Brittain S., Rettig T., McLean I. S., Reuter D. C., and Xu L. (2001a) Organic composition of C/1999 S4 (LINEAR): A comet formed near Jupiter? *Science, 292,* 1334–1339.

Mumma M. J. and 18 colleagues (2001b) A survey of organic volatile species in comet C/1999 H1 (Lee) using NIRSPEC at the Keck Observatory. *Astrophys. J., 546,* 1183–1193.

Mumma M. J., DiSanti M. A., Dello Russo N., Magee-Sauer K., Gibb E., and Novak R. (2003) Remote infrared observations of parent volatiles in comets: A window on the early solar system. *Adv. Space Res., 31,* 2563–2575.

Neufeld D. A. and 19 colleagues (2000) SWAS observations of water towards comet C/1999 H1 (Lee). *Astrophys. J. Lett., 539,* L151–L154.

Palmer P., Wootten A., Butler B., Bockelée-Morvan D., Crovisier J., Despois D., and Yeomans D. K. (1996) Comet Hyakutake: First secure detection of ammonia in a comet (abstract). *Bull. Am. Astron. Soc., 28,* 927.

Pickett H. M., Poynter R. L., Cohen E. A., Delitsky M. L., Pearson J. C., and Müller H. S. P. (1998) Submillimeter, millimeter and microwave spectral line catalogue. *J. Quant. Spectrosc. Radiat. Transfer, 60,* 883–890. Available on line at http://spec.jpl.nasa.gov.

Prialnik D., Benkhoff J., and Podolak M. (2004) Modeling the structure and activity of comet nuclei. In *Comets II* (M. C. Festou et al., eds.), this volume. Univ. of Arizona, Tucson.

Rauer H. and 12 colleagues (2003) Long-term optical spectrophotometric monitoring of comet C/1995 O1 (Hale-Bopp). *Astron. Astrophys., 397,* 1109–1122.

Raymond J. C., Uzzo M., Ko Y.-K., Mancuso S., Wu R., Gardner L., Kohl J. L., Marsden B., and Smith P. L. (2002) Far-ultraviolet observations of Comet 2P/Encke at perihelion. *Astrophys. J., 564,* 1054–1060.

Reuter D. C., Hoban S., and Mumma M. J. (1992) An infrared search for formaldehyde in several comets. *Icarus, 95,* 329–332.

Reylé C. and Boice D. C. (2003) A S$_2$ fluorescence model for interpreting high resolution cometary spectra. I. Model description and initial results. *Astrophys. J., 587,* 464–471.

Rothman L. S. and 30 colleagues (2003) The HITRAN molecular spectroscopic database: Edition of 2000 including updates through 2001. *J. Quant. Spectrosc. Radiat. Transfer, 82,* 5–44. Available on line at http://www.hitran.com.

Schleicher D. G., Millis R. L., and Birch P. V. (1998) Narrowband photometry of comet P/Halley: Variation with heliocentric distance, season, and solar phase angle. *Icarus, 132,* 397–417.

Schloerb F. P., Kinzel W. M., Swade D. A., and Irvine W. M. (1986) HCN production rates from comet Halley. *Astrophys. J. Lett., 310,* L55–L60.

Schloerb F. P., Kinzel W. M., Swade D. A., and Irvine W. M. (1987) Observations of HCN in comet P/Halley. *Astron. Astrophys., 187,* 475–480.

Senay M. C. and Jewitt D. (1994) Coma formation driven by carbon monoxide release from comet Schwassmann-Wachmann 1. *Nature, 371,* 229–231.

Snyder L. E., Palmer P., and de Pater I. (1989) Radio detection of formaldehyde emission from Comet Halley. *Astrophys. J., 97,* 246–253.

Snyder L. E., Veal J. M., Woodney L. M., Wright M. C. H., Palmer P., A'Hearn M. F., Kuan Y.-J., de Pater I., and Forster J. R. (2001) BIMA array photodissociation measurements of HCN and CS in comet Hale-Bopp (C/1995 O1). *Astron. J., 121,* 1147–1154.

Stern S. A (2003) The evolution of comets in the Oort cloud and Kuiper belt. *Nature, 424,* 639–642.

Stern S. A., Green J. C., Cash W., and Cook T. A. (1992) Helium and argon abundance constraints and the thermal evolution of comet Austin (1989c1). *Icarus, 95,* 157–161.

Stern S. A., Slater D. C., Festou M. C., Parker J. W., Gladstone, G. R., A'Hearn M. F., and Wilkinson E. (2000) The discovery of argon in comet C/1995 O1 (Hale-Bopp). *Astrophys. J. Lett., 544,* L169–L172.

Swings P. (1941) Complex structure of cometary bands tentatively ascribed to the contour of the solar spectrum. *Lick Obs. Bull., 508,* 131–136.

Tielens A. G. G. M. and Allamandola L. J. (1987) Composition, structure, and chemistry of interstellar dust. In *Interstellar Processes* (D. J. Hollenbach and H. A. Thronson Jr., eds.), pp. 397–469. Reidel, Dordrecht.

Tikhonov V. I. and Volkov A. A. (2002) Separation of water into its ortho and para isomers. *Science, 296,* 2363.

Tokunaga A. T., Nagata T., and Smith R. G. (1987) Detection of a new emission band at 2.8 microns in comet P/Halley. *Astron. Astrophys., 187,* 519–522.

Tozzi G. P., Feldman P. D., and Festou M. C. (1998) Origin and production of C(^{1}D) atoms in cometary comae. *Astron. Astrophys., 330,* 753–763.

Veal J. M, Snyder L. E., Wright M., Woodney L. M., Palmer P., Forster J. R., de Pater I., A'Hearn M. F., and Kuan Y.-J. (2000) An interferometric study of HCN in comet Hale-Bopp (C/1995 O1). *Astron. J., 119,* 1498–1511.

Weaver H. A. (1998) Comets. In *The Scientific Impact of the Goddard High Resolution Spectrograph (GHRS)* (J. C. Brandt et al., eds.), pp. 213–226. ASP Conference Series 143, Astronomical Society of the Pacific, San Francisco.

Weaver H. A. (2000) *Comet C/1999 S4 (LINEAR).* IAU Circular No. 7461.

Weaver H. A. and Mumma M. J. (1984) Infrared molecular emissions from comets. *Astrophys. J., 276,* 782–797. (Erratum: *Astrophys. J., 285,* 272.)

Weaver H. A., Mumma M. J., Larson H. P., and Davis D. S. (1986) Post-perihelion observations of water in comet Halley. *Nature, 324,* 441–446.

Weaver H. A., Feldman P. D., McPhate J. B., A'Hearn M. F., Arpigny C., and Smith T. E. (1994) Detection of CO Cameron band emission in comet P/Hartley 2 (1991 XV) with the Hubble Space Telescope. *Astrophys. J., 422,* 374–380.

Weaver H. A. and 11 colleagues (1996) *Comet C/1996 B2 (Hyakutake).* IAU Circular No. 6374.

Weaver H. A., Brooke T. Y., DiSanti M. A., Mumma M. J., Tokunaga A., Chin G., A'Hearn M. F., Owen T. C., and Lisse C. M. (1997) The methane abundance in comet C/1996 B2 (Hyakutake) (abstract). *Bull. Am. Astron. Soc., 29,* 1041.

Weaver H. A., Chin G., Bockelée-Morvan D., Crovisier J., Brooke T. Y., Cruikshank D. P., Geballe T. R., Kim S. J., and Meier R. (1999a) An infrared investigation of volatiles in comet 21P/Giacobini-Zinner. *Icarus, 142,* 482–497.

Weaver H. A., Brooke T. Y., Chin G., Kim S. J., Bockelée-Morvan D., and Davies J. K. (1999b) Infrared spectroscopy of comet Hale-Bopp. *Earth Moon Planets, 78,* 71–80.

Weaver H. A. and 20 colleagues (2001) HST and VLT investigations of the fragments of comet C/1999 S4 (LINEAR). *Science, 292,* 1329–1334.

Weaver H. A., Feldman P. D., Combi M. R., Krasnopolsky V., Lisse C. A., and Shemansky D. E. (2002a) A search for argon and O VI in three comets using the Far Ultraviolet Spectroscopic Explorer. *Astrophys. J. Lett., 576,* L95–L98.

Weaver H. A., Feldman P. D., A'Hearn M. F., Arpigny C., Combi M. R., Festou M. C., and Tozzi G.-P. (2002b) Spatially-resolved spectroscopy of C/2002 C1 (Ikeya-Zhang) with HST (abstract). *Bull. Am. Astron. Soc., 34,* 853–854.

Wilkening L., ed. (1982) *Comets.* Univ. of Arizona, Tucson. 766 pp.

Wink J., Bockelée-Morvan D., Despois D., Colom P., Biver N., Crovisier J., Lellouch E., Davies J. K., Dent W. R. F., and Jorda L. (1999) Evidence for extended sources and temporal modulations in molecular observations of C/1995 O1 (Hale-Bopp) at the IRAM interferometer (abstract). *Earth Moon Planets, 78,* 63.

Winnberg A., Ekelund E., and Ekelund A. (1987) Detection of HCN in comet P/Halley. *Astron. Astrophys., 172,* 335–341.

Womack M., Festou M. C., and Stern A. A. (1997) The heliocentric evolution of key species in the distantly-active comet C/1995 O1 (Hale-Bopp). *Astron. J., 114,* 2789–2795.

Woodney L. M. (2000) Chemistry in comets Hyakutake and Hale-Bopp. Ph.D. thesis, Univ. of Maryland.

Woodney L., McMullin J., and A'Hearn M. F. (1997) Detection of OCS in comet Hyakutake (C/1996 B2) *Planet. Space Sci., 45,* 717–719.

Woodney L. M., A'Hearn M. F., McMullin J., and Samarasinha N. (1999) Sulfur chemistry at millimeter wavelengths in C/Hale-Bopp. *Earth Moon Planets, 78,* 69–70.

Woodney L. and 14 colleagues (2002) Morphology of HCN and CN in comet Hale-Bopp (1995 O1). *Icarus, 157,* 193–204.

Wright M.C.H. and 10 colleagues (1998) Mosaicked images and spectra of J = 1 → 0 HCN and HCO+ emission from comet Hale-Bopp (1995 O1). *Astron. J., 116,* 3018–3028.

Wyckoff S. W., Tegler S. C., and Engel L. (1991) Nitrogen abundance in comet Halley. *Astrophys. J., 367,* 641–648.

Wyckoff S., Kleine M., Peterson B. A., Wehinger P. A., and Ziurys L. M. (2000) Carbon isotope abundances in comets. *Astrophys. J., 535,* 991–999.

Xie X. and Mumma M. J. (1992) The effect of electron collisions on rotational populations of cometary water. *Astrophys. J., 386,* 720–728.

Yamamoto T. (1982) Evaluation of infrared line emission from constituent molecules of cometary nuclei. *Astron. Astrophys., 109,* 326–330.

Ziurys L. M., Savage C., Brewster M. A., Apponi A. J., Pesch T. C., and Wyckoff S. (1999) Cyanide chemistry in comet Hale-Bopp (C/1995 O1). *Astrophys. J. Lett., 527,* L67–L71.

Spectroscopic Investigations of Fragment Species in the Coma

Paul D. Feldman
The Johns Hopkins University

Anita L. Cochran
University of Texas

Michael R. Combi
University of Michigan

The content of the gaseous coma of a comet is dominated by fragment species produced by photolysis of the parent molecules issuing directly from the icy nucleus of the comet. Spectroscopy of these species provides complementary information on the physical state of the coma to that obtained from observations of the parent species. Extraction of physical parameters requires detailed molecular and atomic data together with reliable high-resolution spectra and absolute fluxes of the primary source of excitation, the Sun. The large database of observations, dating back more than a century, provides a means to assess the chemical and evolutionary diversity of comets.

1. INTRODUCTION

In 1964, P. Swings delivered the George Darwin lecture (*Swings,* 1965) on the one-hundredth anniversary of the first reported spectroscopic observation of a comet. He described a century of mainly photographic observations of what were recognized to be fragment species resulting from photochemical processes acting on the volatile species released from the cometary nucleus in response to solar heating. The features identified in the spectra were bands of the radicals OH, NH, CN, CH, C_3, C_2, and NH_2; the ions OH^+, CH^+, CO_2^+, CO^+, and N_2^+; Na I; and the forbidden red doublet of O I (*Swings and Haser,* 1956; *Arpigny,* 1965). By the time of the publication of the book *Comets* (*Wilkening,* 1982) 18 years later, optical spectroscopy had become quantitative photoelectric spectrophotometry (*A'Hearn,* 1982) and had been extended to the radio and vacuum ultraviolet (UV) (*Feldman,* 1982), yet the inventory of species detected grew slowly. In the visible, the Sun-grazing Comet Ikeya-Seki (C/1965 S1) showed the metals K, Ca^+, Ca, Fe, V, Cr, Mn, Ni, and Cu (*Preston,* 1967), presumably from the vaporization of refractory grains, and H_2O^+ was identified in Comet Kohoutek (C/1973 E1) a few years later. In the UV, the constituent atoms H, O, C, and S were detected together with CS and CO, the first "parent" molecule to be directly identified spectroscopically (*Feldman and Brune,* 1976).

The spectra of the observed radicals are necessarily complex and detailed analyses of high-resolution spectra demonstrated that the observed emission was produced for the most part by fluorescence of solar radiation. With this information, spectrophotometric data may be used to quantitatively derive column abundances of the observed species from which, with suitable modeling (see *Combi et al.,* 2004), the relative abundance of the parent species in the cometary ice could be deduced. Note that for some species, particularly C_2 and C_3, the identity of the parent still remains ambiguous. Long-slit spectroscopy at high spatial resolution has provided constraints on the photochemical parameters used in the models and also has served to probe physical conditions in the coma that produce deviations from the purely photochemical models. Similar analyses are also possible with narrowband photometric imaging (see *Schleicher and Farnham,* 2004).

Advances in infrared (IR) and radio technology, together with the timely apparitions of several active comets during the past 10 years, have led to the identification of more than two dozen parent volatile species in the coma (see *Bockelée-Morvan et al.,* 2004). During this same time enhancements in optical and UV technology have also permitted more detailed investigations of the spectra of (mainly) fragment species, leading to a more complete picture of the entire coma. The availability of state-of-the-art optical instrumentation at a given cometary apparition ensures the acquisition of datasets with good temporal coverage and comparability to the historical record for purposes of assessing diversity among comets. The high sensitivity of optical detectors also means that fragment species can be detected at larger heliocentric distances than any of the parents (cf. *Rauer et al.,* 2003). While notable advances have been made in spaceborne UV spectroscopic instrumentation, notably STIS on the Hubble

Space Telescope (HST) and the Far Ultraviolet Spectroscopic Explorer (FUSE) satellite, these resources are limited and cometary observations are relatively few.

2. PHYSICAL PROCESSES

2.1. Photolysis of Parent Molecules

The photolytic destruction of water, the dominant molecular species in the coma of comets at distances near 1 AU from the Sun, has been extensively studied (see *Combi et al.*, 2004). It may proceed through multiple paths depending on the energy of the incident solar photon (here given as the threshold wavelength):

$$
\begin{aligned}
H_2O + h\nu &\rightarrow OH + H & 2424.6 \ \text{Å} \\
&\rightarrow OH(A^2\Sigma^+) + H & 1357.1 \ \text{Å} \\
&\rightarrow H_2 + O(^1D) & 1770 \ \text{Å} \\
&\rightarrow H_2 + O(^1S) & 1450 \ \text{Å} \\
&\rightarrow H + H + O(^3P) & 1304 \ \text{Å} \\
&\rightarrow H_2O^+ + e & 984 \ \text{Å} \\
&\rightarrow H + OH^+ + e & 684.4 \ \text{Å} \\
&\rightarrow H_2 + O^+ + e & 664.4 \ \text{Å} \\
&\rightarrow OH + H^+ + e & 662.3 \ \text{Å}
\end{aligned}
$$

Many of the fragments can be further broken down:

$$
\begin{aligned}
OH + h\nu &\rightarrow O + H & 2823.0 \ \text{Å} \\
&\rightarrow OH^+ + e & 928 \ \text{Å} \\
H_2 + h\nu &\rightarrow H + H & 844.79 \ \text{Å} \\
&\rightarrow H_2^+ + e & 803.67 \ \text{Å} \\
&\rightarrow H + H^+ + e & 685.8 \ \text{Å} \\
O + h\nu &\rightarrow O^+ + e & 910.44 \ \text{Å} \\
H + h\nu &\rightarrow H^+ + e & 911.75 \ \text{Å}
\end{aligned}
$$

The last two reactions may also occur by resonant charge exchange with solar wind protons. Photoelectrons produced in the ionization process, particularly those from the strong solar He II line at 304 Å, also lead to secondary dissociation and ionization (*Cravens et al.*, 1987) and may contribute to the observed emissions.

Similar equations may be written for other cometary species and many, particularly some of the more abundant ones such as CO and CO_2, produce many of the same fragment species. *Huebner et al.* (1992) have compiled a very useful list of photodestruction rates for a large number of molecules that includes many polyatomic species that have been identified as being present in the cometary ice (see Table 1 of *Bockelée-Morvan et al.*, 2004).

For the case of H_2O, all the fragment species listed above, with the exception of H^+ and H_2^+, have been detected spectroscopically. The second through fourth reactions listed above leave the product atom or molecule in an excited state that leads to "prompt" emission of a photon. This provides a means for mapping the spatial distribution of the parent molecule in the inner coma. Prompt emission includes both allowed radiative decays (such as from the $A^2\Sigma^+$ state of OH) as well as those from metastable states such as $O(^1D)$, whose lifetime is ~130 s. The O I ^1D–^3P doublet at 6300 and 6364 Å has been used extensively as a ground-based surrogate for the determination of the water production rate with the caveat that other species such as OH, CO, and CO_2 may also populate the upper level. When the density of H_2O is sufficient to produce observable 6300 Å emission, it may also produce collisional quenching of the ^1D state, and this must be included in the analysis. The analogous ^1D–^3P transitions in C occur at 9823 and 9849 Å and can similarly give information about the production rate of CO. Carbon atoms in the ^1D state, whose lifetime is ~4000 s, are known to be present from the observation of the resonantly scattered $^1P^o$–^1D transition at 1931 Å (*Tozzi et al.*, 1998). To date the IR lines have only been detected in Comets 1P/Halley and Hale-Bopp (C/1995 O1) (*Oliversen et al.*, 2002).

2.2. Excitation Mechanisms

The extraction of coma abundances from spectrophotometric measurements of either the total flux or the surface brightness in a given spectral feature has been summarized by *Feldman* (1996) and we repeat some of the salient points here. We note that the uncertainty in the derived abundances may include not only the measurement uncertainty, but also uncertainties in the atomic and molecular data and, in the case of surface brightness measurements, uncertainties in the model parameters used. One must be careful in comparing abundances and production rates derived by different observers for the same comet at a given time to assure that comparable physical and model parameters are used.

In the simplest case we begin with the total number of species i in the coma

$$ M_i = Q_i \tau_i(r) \tag{1} $$

where Q_i is the production rate (atoms or molecules s^{-1}) of all sources of species i and $\tau_i(r)$ is the lifetime of this species at heliocentric distance r, $\tau_i(r) = \tau_i(1 \ \text{AU})r^2$.

For an optically thin coma, the luminosity, in photons cm^{-2} s^{-1}, in a given atomic or molecular transition at wavelength λ, is

$$ L(\lambda) = M_i g(\lambda, r) \tag{2} $$

where the fluorescence efficiency, or "g-factor", $g(\lambda,r) = g(\lambda, 1 \ \text{AU})r^{-2}$, is defined by *Chamberlain and Hunten* (1987) in cgs units as

$$ g(\lambda, 1 \ \text{AU}) = \frac{\pi e^2}{mc^2} \lambda^2 f_\lambda \pi F_\odot \tilde{\omega} \ \text{photons s}^{-1} \text{atom}^{-1} \tag{3} $$

Here, e, m, and c have their usual atomic values; λ is the transition wavelength; f_λ is the absorption oscillator strength;

$\pi F_{\odot}$ is the solar flux per unit wavelength interval at 1 AU; and $\tilde{\omega}$ is the albedo for single scattering, defined for a line in an atomic multiplet as

$$\tilde{\omega} = \frac{A_j}{\Sigma_j A_j} \qquad (4)$$

and A_j is the Einstein transition probability. At low and moderate spectral resolution, a given multiplet is not resolved and in this case $\tilde{\omega} = 1$. For diatomic molecules, fluorescence to other vibrational levels becomes important and the evaluation of $\tilde{\omega}$ depends to a large degree on the physical conditions in the coma. Note that this definition differs from equation (3) of *Bockelée-Morvan et al.* (2004) in that a blackbody cannot be used to represent the solar flux, particularly in the UV, and that a high-resolution spectral atlas is required to account for the Fraunhofer structure in the solar spectrum.

Then, for a comet at a geocentric distance Δ, the total flux from the coma at wavelength λ is

$$F(\lambda) = \frac{L(\lambda)}{4\pi\Delta^2} = \frac{Q_i g(\lambda, r) \tau_i(r)}{4\pi\Delta^2} \text{ photons cm}^{-2} \text{ s}^{-1} \qquad (5)$$

Note that the product $g(\lambda, r)\tau_i(r)$ is independent of r.

Unfortunately, the scale lengths (the product of lifetime and outflow velocity) of almost all the species of interest in the UV are on the order of 10^5–10^6 km at 1 AU. Thus, total flux measurements require fields of view ranging from several arcminutes to a few degrees. This has been done only rarely (*Woods et al.*, 2000). Again assuming an optically thin coma, the measured flux in the aperture, $F'(\lambda)$, can be converted to an average surface brightness (in units of rayleighs), $B(\lambda)$

$$B(\lambda) = 4\pi 10^{-6} F'(\lambda)\Omega^{-1} \qquad (6)$$

where Ω is the solid angle subtended by the aperture. The brightness, in turn, is related to $\bar{N}_i$, the average column density of species i within the field of view by

$$B(\lambda) = 10^{-6} g(\lambda, r)\bar{N}_i \qquad (7)$$

At this point the evaluation of Q_i from $\bar{N}_i$ requires the use of a model of the density distribution of the species i (see *Combi et al.*, 2004).

2.2.1. Fluorescence equilibrium. The excitation of the electronic transitions of the radicals observed in the visible and near-UV, which may have many photon absorption and emission cycles in their lifetime, leads to "fluorescence equilibrium" of the rotational levels within the ground vibrational level. This process and the various factors that affect it are fully discussed in section 3.1.4 of *Bockelée-Morvan et al.* (2004). In contrast, in the far-UV, where the solar flux is low, it is often the case that the probability of

absorption of a solar photon is less than the probability that the species will be dissociated or ionized, so that the ground state population is not affected by fluorescence and can be described by a Boltzmann distribution at a suitable temperature corresponding to the cometary environment where the species was produced. For prompt emission, such as that from CO produced by photodissociative excitation of CO_2 (*Mumma et al.*, 1975), the rotational temperature of the CO will be ~5 times larger than the rotational temperature of the CO_2 because of the factor of 5 in the rotational constants and the need to conserve angular momentum in the dissociation process. Similarly, prompt emission of OH will also be characterized by a "hot" rotational distribution (*Bertaux*, 1986; *Budzien and Feldman*, 1991).

2.2.2. Swings and Greenstein effects. *Swings* (1941) pointed out that because of the Fraunhofer absorption lines in the visible region of the solar spectrum, the absorption of solar photons in a given molecular band would vary with the comet's heliocentric velocity, $\dot{r}$, leading to differences in the structure of a band at different values of $\dot{r}$ when observed at high spectral resolution. This effect is now commonly referred to as the Swings effect. Even for observations at low resolution, the Swings effect must be taken into account in the calculation of total band g-factor, and this has been done for a number of important species such as OH, CN, and NH. A particularly important case is that of the OH $A^2\Sigma^+$–$X^2\Pi$ (0,0) band at ~3085 Å, which is often used to derive the water production rate (*Scheicher and A'Hearn*, 1988).

While this effect was first recognized in the spectra of radicals in the visible, a similar phenomenon occurs in the excitation of atomic multiplets below 2000 Å, where the solar spectrum makes a transition to an emission line spectrum. For example, the three lines of O I λ1302 have widths of ~0.1 Å, corresponding to a velocity of ~25 km s^{-1}, so that knowledge of exact solar line shapes is essential to a reliable evaluation of the g-factor for this transition (*Feldman et al.*, 1976; *Feldman*, 1982).

A differential Swings effect occurs in the coma since atoms and molecules on the sunward side of the coma, flowing outward toward the Sun, have a net velocity that is different from those on the tailward side, and so if the absorption of solar photons takes place on the edge of an absorption (or emission) line, the g-factors will be different in the sunward and tailward directions. Differences of this type will appear in long-slit spectra in which the slit is placed along the Sun-comet line. This effect was pointed out by *Greenstein* (1958). An analog in the far-UV has been observed in the case of O I λ1302. The measurement of a Greenstein effect in OH in Comet 2P/Encke has been used to derive the outflow velocity of water and consequently the nongravitational acceleration of the comet (*A'Hearn and Schleicher*, 1988).

2.2.3. Bowen fluorescence. As noted above, in the UV, with the exception of the H I Lyman series, solar line widths are such that for comets with heliocentric velocities

>25 km s^{-1}, the available flux at the center of the absorbing atom's line is reduced to a very small value. It was thus surprising that the O I λ1302 line appeared fairly strongly in the observed spectrum of Comets Kohoutek (C/1973 E1) and West (C/1975 V1), whose values of ṙ were both >45 km s^{-1} at the times of observation. The explanation invoked the accidental coincidence of the solar H I Lyman-β line at 1025.72 Å with the O I ^{3}D–^{3}P transition at 1025.76 Å, cascading through the intermediate ^{3}P state (*Feldman et al.,* 1976). This mechanism, well known in the study of planetary nebulae, is referred to as Bowen fluorescence (*Bowen,* 1947). The g-factor due to Lyman-β pumping is an order of magnitude smaller than that for resonance scattering, but sufficient to explain the observations. Lyman-β is also coincident with the P1 line of the (6,0) band of the H$_2$ Lyman system (B^{1}Σ$^+$–X^{1}Σ$_g^+$), leading to fluorescence in the same line of several (6,v") bands. Three such lines have recently been detected in the FUSE spectra of Comet C/2001 A2 (LINEAR) (*Feldman et al.,* 2002).

It is interesting to note that fluorescence excited by solar H I Lyman-α was considered by *Haser and Swings* (1957) but considered unlikely based on the state of spectroscopic knowledge at that time. Lyman-α fluorescence of CO in the Fourth Positive system was first detected in the spectrum of Venus (*Durrance,* 1981) and observed in 1996 in Comet Hyakutake (C/1996 B2) (*Wolven and Feldman,* 1998).

2.2.4. Electron impact excitation. We noted above that photoelectron impact excitation may also contribute to the observed emissions, particularly in the UV. However, a very simple argument, based on the known energy distribution of solar UV photons, demonstrates that this is only a minor source for the principal emissions. Since the photoionization rate of water (and of the important minor species such as CO and CO$_2$) is on the order of 10^{-6} s^{-1} at 1 AU, and the efficiency for converting the excess electron energy into excitation of a single emission is on the order of a few percent, the effective excitation rate for any emission will be on the order of 10^{-8} s^{-1} or less at 1 AU (*Cravens and Green,* 1978). Since the efficiencies for resonance scattering or fluorescence for almost all the known cometary emissions are one to several orders of magnitude larger, electron impact may be safely neglected except in a few specific cases. These are the forbidden transitions, where the oscillator strength (and consequently the g-factor) is very small. Examples include the O I ^{5}S$_2$–^{3}P$_{2,1}$ doublet at 1356 Å, the O I ^{1}D–^{3}P red lines at 6300 and 6364 Å, observed in many comets, and the CO Cameron bands (*Weaver et al.,* 1994; *Feldman et al.,* 1997). However, the excitation of these latter two is dominated by prompt emission in the inner coma. Some of the emissions below 1200 Å observed in the FUSE spectrum of Comet C/2001 A2 (LINEAR) have also been attributed to electron impact (*Feldman et al.,* 2002).

2.2.5. Solar cycle variation. The relative abundance of a fragment species depends on the absorption cross section (σ$_d$) and the solar flux seen by the comet. The rate coefficient J$_d$ is evaluated at 1 AU using whole-disk measurements of the solar flux by integrating the cross section

$$J_d = \int_0^{\lambda_{th}} \pi F_\odot \sigma_d d\lambda \qquad (8)$$

These rates may also be estimated from the threshold energies shown in the table of reactions given in section 2.1, since the solar flux is decreasing very rapidly to shorter wavelengths. Processes with thresholds near 3000 Å have lifetimes on the order of 10^4 s, those with thresholds near 2000 Å an order of magnitude longer, while those with thresholds below Lyman-α, such as most photoionization channels, have lifetimes on the order of 10^6 s, all at 1 AU. In addition to uncertainties in the details of the absorption cross sections, further uncertainty is introduced into the calculation of J$_d$ by the lack of exact knowledge of the solar flux at the time of a given observation due to the variability of the solar radiation below 2000 Å, and most importantly, below Lyman-α. The solar UV flux is known to vary considerably both with the 27-d solar rotation period and with the 11-yr solar activity cycle, the latter reaching factors of 2 to 4 for wavelengths shortward of 1000 Å (*Lean,* 1991). Also, at any given point in its orbit, a comet may see a different hemisphere of the Sun than what is seen from Earth. The compilation of photodestruction rates of *Huebner et al.* (1992) uses mean solar fluxes to represent the extreme conditions of solar minimum and solar maximum and also includes the evaluation of the excess energies of the dissociation products.

3. SPECTROSCOPIC OBSERVATIONS

3.1. Ultraviolet Observatories in Space

The first observations of comets in the spectral region below 3000 Å were made from space in 1970 by the Orbiting Astronomical Observatory (OAO-2). The spectrum of Comet Bennett (C/1969 Y1) showed very strong OH emission at ~3085 Å and H I Lyman-α emission, at 1216 Å, principal dissociation products of H$_2$O (*Code et al.,* 1972). *Feldman* (1982) reviewed satellite and sounding rocket observations made before the launch of the International Ultraviolet Explorer (IUE) satellite observatory in 1978 and the early results from IUE, which observed more than 50 comets before it was shut down in September 1996 [see *Festou and Feldman* (1987) for a review of the principal results through 1986].

The HST, launched in 1990, made a significant advance in UV sensitivity and provided the ability to observe in a small field of view very close to the nucleus. Two spectrographs, the Goddard High Resolution Spectrograph (GHRS), which utilized solar blind detectors exclusively for UV spectroscopy, and the Faint Object Spectrograph (FOS), were used extensively through 1997 (*Weaver,* 1998). In February of that year they were replaced by the Space Telescope Imaging Spectrograph (STIS), which provided further enhancements including long-slit spectroscopy with an angular

TABLE 1. Principal optical spectroscopic features
of cometary fragment species.

Species*	Transition	System Name	Wavelength (Å)
OH	$A^2\Sigma^+ - X^2\Pi_i$ (0,0)		3085
CN	$B^2\Sigma^+ - X^2\Sigma^+$ (0,0)	Violet	3883
	$A^2\Pi - X^2\Sigma^+$ (2,0)	Red	7873
C_2	$d^3\Pi_g - a^3\Pi_u$ (0,0)	Swan	5165
	$A^1\Pi_u - X^1\Sigma_g^+$ (3,0)	Phillips	7715
	$D^1\Sigma_u^+ - X^1\Sigma_g^+$ (0,0)	Mulliken	2313
C_3	$\tilde{A}^1\Pi_u - \tilde{X}^1\Sigma_g^+$	Comet Head Group	3440–4100
CH	$A^2\Delta - X^2\Pi$ (0,0)		4314
	$B^2\Sigma^- - X^2\Pi$ (0,0)		3871, 3889
CS	$A^1\Pi - X^1\Sigma^+$ (0,0)		2576
NH	$A^3\Pi_i - X^3\Sigma^-$ (0,0)		3360
NH_2	$\tilde{A}^2A_1 - \tilde{X}^2B_1$		4500–7350
O I ^{1}D	$^1D - ^3P$		6300, 6364
O I ^{1}S	$^1S - ^1D$		5577
C I ^{1}D	$^1D - ^3P$		9823, 9849
CO^+	$B^2\Sigma^+ - X^2\Sigma^+$ (0,0)	First Negative	2190
	$A^2\Pi - X^2\Sigma^+$ (2,0)	Comet Tail	4273
CO_2^+	$\tilde{B}^2\Sigma_u - \tilde{X}^2\Pi_g$		2883, 2896
	$\tilde{A}^2\Pi_u - \tilde{X}^2\Pi_g$	Fox-Duffendack-Barker	2800–5000
CH^+	$A^1\Pi - X^1\Sigma^+$ (0,0)	Douglas-Herzberg	4225, 4237
OH^+	$A^3\Pi - X^3\Sigma^-$ (0,0)		3565
H_2O^+	$\tilde{A}^2A_1 - \tilde{X}^2B_1$		4270–7540
N_2^+	$B^2\Sigma^+ - X^2\Sigma^+$ (0,0)	First Negative	3914

*CO is both a dissociation product and a native molecular species and is discussed by
Bockelée-Morvan et al. (2004).

resolution of 50 milliarcsec. Except for some early results on Comet Hale-Bopp (C/1995 O1) (*Weaver et al.,* 1999), the spectroscopic results from STIS observations of several comets remain to be published, although a few will be described below. Finally, we note the launch in 1999 of FUSE, which provides access to the spectral region between 900 and 1200 Å at very high spectral resolution (*Feldman et al.,* 2002; *Weaver et al.,* 2002).

3.2. Optical Capabilities

As mentioned in the introduction, the spectrum of a comet was first observed in 1864. From then until the 1970s, optical cometary spectra were primarily obtained using photographic plates. At that point, observations began to switch over to photoelectric detectors of various types; optical cometary spectra are obtained currently almost exclusively using CCD detectors.

The optical spectrum of comets is quite dense because it consists mostly of molecular bands. The principal spectral features are listed in Table 1. Optical spectra are generally obtained in one of two spectral resolution regimes: moderate resolving powers of $R = \lambda/\Delta\lambda \sim 600$, which allow detection of complete bands but not individual lines, or $R > 10,000$, which allow detection of individual lines. The majority of extant cometary spectra were obtained in the moderate-resolution mode, an example of which is shown in Fig. 1, with fewer than 25 comets observed at high spectral resolving powers. The choice of the spectral bandpass is generally dependent on the detector. Some groundbased spectrographs are capable of being used at ~3085 Å to observe the OH (0,0) band, but low detector quantum efficiency and poor atmospheric throughput make such observations difficult.

Typically, slit widths of 2–4 arcsec on the sky are used when obtaining cometary spectra, with the spectral resolution defined in part by this width. One exception is observations obtained with an imaging Fabry-Pérot spectrograph where the resolution is obtained by tuning the etalons (cf. *Magee-Sauer et al.,* 1988) and a large region of the sky can be imaged onto the detector. Fabry-Pérot instruments generally have a very limited bandpass but work at relatively high spectral resolving powers.

Since comets are spatially large, it is desirable to obtain spectra at different positions in the coma. Generally, this is done either by use of a "long" slit (generally 30–150 arcsec) or by repositioning the telescope to image different regions of the coma or both. In order to sample all directions in the coma with a long slit instrument, the slit must be rotated to different position angles on the sky and additional observations obtained. Alternatively, a fiber-fed spectrograph can

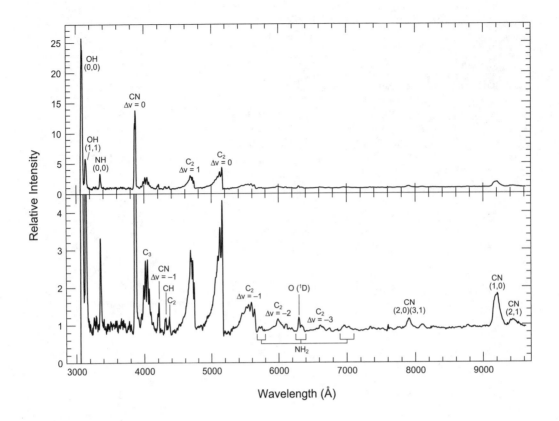

Fig. 1. Optical/near-IR CCD spectrum of Comet 109P/Swift-Tuttle obtained November 26, 1992. This spectrum is a composite of a UV/blue spectrum of Cochran (personal communication) and a red/near-IR spectrum of *Fink and Hicks* (1996). The cometary spectrum has been divided by a solar spectrum to show the cometary emissions. The two-pixel resolution was ~7 Å for the blue and ~14 Å for the red half. The upper panel shows the spectrum scaled by the strongest feature while the lower panel has an expanded ordinate to show the weaker features. The main features are labeled in one of the panels. This spectrum is typical of comets in the inner solar system.

use fibers placed throughout the coma to obtain spatial information. The fiber pattern can be linear or can sample in many directions simultaneously.

In addition to spectra, cometary comae may be studied using narrow-band photometry or wide-field imaging (see *Schleicher and Farnham*, 2004). Narrow-band photometry allows for observations of fainter comets and the outer regions of the comae; spectra allow for discrimination between crowded spectral features but suffer from small apertures resulting in longer integration times.

3.3. Infrared

Rapid advances in near-IR (~1–5 μm) spectroscopic instrumentation, particularly at the Keck Telescopes and the Infrared Telescope Facility at Mauna Kea, have led to much new information about the molecular composition of the inner coma. Representative spectra given by *Mumma et al.* (2001) disclose the presence of a wide variety of cometary organic molecules, discussed in detail in *Bockelée-Morvan et al.* (2004). The exception is OH, which appears mainly in transitions from highly excited rotational levels, implying that the excitation source is prompt emission following the dissociation of water.

3.4. Radio

Fluorescence pumping through the UV transitions of the OH radical produces a deviation of the population of the hyperfine and Λ-doublet levels of the $X^2\Pi_{3/2}$ ($J = 3/2$) ground state from statistical equilibrium (*Despois et al.*, 1981; *Schleicher and A'Hearn*, 1988). Depending on the heliocentric velocity, this departure may be either "inverted" or "anti-inverted," giving rise to either stimulated emission or absorption against the galactic background at 18 cm wavelength. This technique has been used extensively since 1973 to monitor the OH production rate in comets (*Crovisier et al.*, 2002). The resulting radio emissions are easily quenched by collisions with molecules and ions, the latter giving rise to a fairly large "collision radius" that must be accounted for in interpreting the derived OH column density.

4. WATER PRODUCTS

4.1. Hydroxyl Radical

The OH radical is the easiest dissociation product of water to observe, and is often used to determine the production rate of water and to serve as a standard to which all other coma abundances are compared. Fortunately, there are three separate largely self-consistent datasets that provide measurements of OH in a large number of comets made during the past 20 years or more. However, the determination of water production rate from the data differs for the three and the derived rates are often not in agreement. The first is the set of groundbased photometric measurements at ~3085 Å made through standardized narrow-band filters, exemplified by the work of *A'Hearn et al.* (1995). These measurements are discussed in *Schleicher and Farnham* (2004). The second is the set of observations of the 18-cm radio lines of OH in more than 50 comets made at the Nançay radio telescope dating back to 1973 (*Crovisier et al.*, 2002). The third set, also comprising over 50 cometary apparitions between 1978 and 1996, is the spectroscopic measurement of OH fluorescence at ~3085 Å from the orbiting IUE. This satellite was in geosynchronous Earth orbit and its optical performance, which was monitored continuously, did not degrade significantly with time, ensuring a reliable calibration of all the observations. There are also many other spectroscopic observations of OH in both the radio and the UV, the latter from both HST and from high-altitude groundbased observatories.

Interpretation and intercomparison of the radio and UV observations are dependent on an accurate knowledge of the UV pumping of the inversion of the Λ-doubled ground state of the molecule, as described above in section 3.4. The two measurements are fundamentally different and so the comparison relies on extensive modeling, both of the spatial distribution and outflow velocity of the OH radicals in the coma, and of the solar excitation process. With high spectral resolution in the UV, the individual ro-vibrational lines of the band can be resolved and such measurements serve as a validation of the fluorescence calculation, which depends in turn on a high-resolution spectrum of the Sun (*Scheicher and A'Hearn*, 1988). But this resolution cannot match that available with the 18-cm radio lines, which in velocity space can reach ~0.3 km s^{-1}, thus permitting the determination of the kinematic properties of the outflowing gas (*Bockelée-Morvan and Gérard*, 1984; *Bockelée-Morvan et al.*, 1990; *Tacconi-Garman et al.*, 1990). *Bockelée-Morvan et al.* (1990) demonstrated variations in OH velocity with both cometocentric and heliocentric distance from an extensive set of observations of Comet 1P/Halley. Collisional quenching of the ground state inversion affects the radio lines but not the UV, but the strongest UV lines can saturate at OH column densities that are reached near the nucleus, and small field-of-view observation from HST show indications of this effect. With their models constrained by the velocity measurements, *Bockelée-Morvan et al.* (1990) and *Gérard* (1990) were able to reconcile the water production rates derived from the 18-cm observations with those derived from IUE observations of Comet 1P/Halley over an extended range of heliocentric distance. Modeling by *Combi and Feldman* (1993) was able to achieve agreement between the production rates for Comet 1P/Halley derived from the OH UV data and those derived from H I Lyman-α, both observed nearly simultaneously by IUE. A similar result was obtained for Comet 21P/Giacobini-Zinner (*Combi and Feldman*, 1992).

Bockelée-Morvan and Gérard (1984) also noted asymmetries in both the velocity structure and the spatial distribution of the 18-cm lines in three comets, which they attributed to asymmetrical outgassing of the nucleus. A similar result was reported by *A'Hearn and Schleicher* (1988) using the Greenstein effect to demonstrate asymmetrical outgassing of the nucleus of Comet 2P/Encke.

OH has also been detected in high-resolution IR spectra near 3 μm of several recent comets (*Brooke et al.*, 1996; *Mumma et al.*, 2001). From the spatial profiles of individual lines, *Brooke et al.* (1996) demonstrated that the lines originating from high rotational levels follow that of the parent molecule, H_2O, and therefore that these lines arise from prompt emission, while the lower excitation lines were flatter and the result of UV fluorescence. These observations have not yet been quantitatively exploited for the determination of production rates. The OH $A^2\Sigma^+ - X^2\Pi$ (0,0) band at ~3085 Å also exhibits a "hot" rotational distribution near the nucleus due to prompt emission (*Bertaux*, 1986). Detection of this prompt emission, which dominates fluorescence only at distances of less than 100 km from the nucleus, requires high spatial resolution, which was afforded the IUE by the close approach of Comet IRAS-Araki-Alcock (C/1983 H1) to Earth in 1983 (*Budzien and Feldman*, 1991).

Finally, we note that the UV (0,0) band, in fluorescence equilibrium, consists of a small number of individual lines that are well separated at a spectral resolution of ≤1 Å. This led *A'Hearn et al.* (1985) to calculate the analogous spectrum of OD. While they found that the strongest lines of OD are separable from the OH lines, and are particularly enhanced at heliocentric velocities between –30 and –5 km s^{-1}, attempts to date to detect these lines with both IUE and HST have not been successful.

4.2. Atomic Hydrogen

Following the first observation of the large H I Lyman-α coma surrounding Comet Bennett (C/1969 Y1) in 1970 by two Earth-orbiting spacecraft (*Code et al.*, 1972; *Bertaux et al.*, 1973), this emission has been observed in a large number of comets both with imaging detectors and spectrographs on a variety of sounding rockets, orbiting observatories, and various other spacecraft. Initial modeling of the spatial distribution, taking into account the excess velocities of multiple sources, solar radiation pressure, and radiative transfer effects, was summarized by *Keller* (1976). A more complete discussion of the physics involved in the photochemical production of H atoms and their subsequent non-LTE (local thermodynamic equilibrium) collisional coupling to the coma is given in *Combi et al.* (2004). In this section we will limit the discussion to spectroscopic observations made at sufficient resolution to allow the determination of the velocity distribution of the H atoms and the presence of radiation trapping near the nucleus, which are relatively few

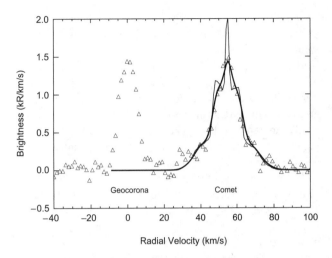

Fig. 2. Hydrogen Lyman-α line profile in Comet C/1996 B2 (Hyakutake). The triangles show the observed line profile obtained with the GHRS instrument on HST with the small science aperture located at a point 111,000 km sunward of the position of the nucleus, which is in the optically thin region of the coma. The thin line shows the intrinsic emission from the comet at very high spectral resolution (1 km s⁻¹) as calculated by the model of *Combi et al.* (1998). The thicker line is the model convolved with the instrument spectral function (4 km s⁻¹ resolution). The emission of H in the geocorona is the line at 0 km s⁻¹ to the left of the comet line, and the comet's emission is Doppler shifted to the comet's 55 km s⁻¹ relative geocentric velocity (from *Combi et al.*, 1998).

(*Festou et al.*, 1979; *Combi et al.*, 1998). Very high spectral resolution has also been obtained using Fabry-Pérot interferometry and coudé/echelle spectroscopy to observe the Hα line at 6563 Å (*Huppler et al.*, 1975; *Scherb*, 1981; *Brown and Spinrad*, 1993; *Combi et al.*, 1999).

Combi et al. (1998) showed Lyman-α line profiles measured with the GHRS on HST obtained with a spectral sampling of 4 km s⁻¹. A spectrum obtained 111,000 km on the sunward side of the coma of C/1996 B2 (Hyakutake) was just outside the optically thick coma. The profile and a Monte Carlo model analysis shows the signatures of the various components: 18 km s⁻¹ from H_2O dissociation, 8 km s⁻¹ from OH, and the low-velocity line center from thermalized H atoms as shown in Fig. 2. A radiative transfer calculation of all the HST data by *Richter et al.* (2000) showed the detailed effects of multiple scattering of the illuminating solar Lyman-α photons and the progressive saturation of the line center at decreasing distances toward the nucleus of the comet.

4.3. O(¹D) λ6300 and Other Forbidden Emissions

Three forbidden O transitions exist in the optical region of the spectrum: the red doublet at 6300.304 and 6363.776 Å (¹D–³P) and the green line at 5577.339 Å (¹S–¹D). These transitions are the result of "prompt" emission, i.e., the atoms are produced directly in the excited ¹S or ¹D states by photodissociation of the parent molecule. The lifetime of the ¹D state is about 130 s, while the ¹S state lifetime is less than 1 s. Thus, the three transitions discussed here are excellent tracers of the distributions of their parents because they cannot travel far without decaying. Oxygen atoms that are excited to the ¹S state decay to the ground ³P state via the ¹D state 95% of the time, while 5% decay directly to the ground state emitting lines at 2977 Å and 2958 Å. Thus, if the green line is present in a cometary spectrum, the red doublet must also be present, although the red doublet can be formed without the green line.

The forbidden O lines can be formed via photoprocesses involving H_2O, CO, or CO_2 as parents. More complex O-bearing species such as HCOOH or H_2CO are unlikely to be the parent because they cannot decay fast enough to produce the observed O(¹D) distribution (*Festou and Feldman*, 1981). It is believed that H_2O is the dominant, if not the sole, O(¹D) and O(¹S) parent out to distances of 10⁵ km from the nucleus, beyond which OH becomes the dominant parent. The determination of the parent abundance requires observations of the intensities of the three lines, coupled with accurate understanding of the dissociation rates and branching ratios.

Measuring the intensities of the three lines accurately requires high spectral resolution to resolve the cometary O lines from telluric O lines and other cometary emissions. The resolution needed to resolve the cometary and telluric O lines is dependent on the Doppler shift of the comet with respect to the Earth. For the red doublet, the O(¹D) line is situated near cometary NH_2 emissions, but the strong NH_2 lines are generally easy to resolve from the O line when the spectral resolution is sufficient to resolve the cometary and telluric O emissions. The region of the green line is much more difficult. Again, the cometary and telluric lines are Doppler shifted apart. However, the region of the green line is in the middle of the C_2 (1,2) P-branch. Only four observations of the 5577 Å O(¹S) line in cometary spectra have been reported: Observations of Comets C/1983 H1 (IRAS-Araki-Alcock) (*Cochran*, 1984) and C/1996 B2 (Hyakutake) (*Morrison et al.*, 1997) relied on high spatial resolution in addition to high spectral resolution; observation of Comet 1P/Halley (*Smith and Schempp*, 1989) relied on extremely high spectral resolution plus modeling; and *Cochran and Cochran* (2001) reported an unequivocal detection of the O(¹S) line in spectra of C/1994 S4 (LINEAR) in a comet that was severely depleted in C_2, making the contamination issue go away.

Cochran and Cochran computed the intensity ratio of the red doublet lines and found it to be 3.03 ± 0.14, in excellent agreement with the ratio predicted by the Einstein A-values of *Storey and Zeippen* (2000). Accurate measurement of the ratio of the green line intensity to the sum of the intensities of the red doublet lines can then be used to discriminate between parent species. The values that have been reported [0.22–0.34 for IRAS-Araki-Alcock (*Cochran*, 1984), 0.12–0.15 for Hyakutake (*Morrison et al.*, 1997), 0.05–0.1 for Halley (*Smith and Schempp*, 1989), and 0.06 ± 0.01 for LINEAR (*Cochran and Cochran*, 2001)] all point

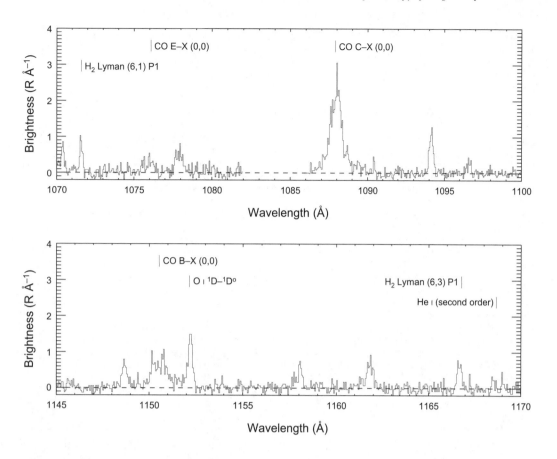

Fig. 3. FUSE spectrum of Comet C/2001 A2 (LINEAR) obtained beginning 2001 July 12.58 UT (from *Feldman et al., 2002*). Two of the fluorescently excited H_2 lines are identified as well as the CO bands and atomic emissions that lie in this spectral region. A number of detected features remain unidentified.

to H_2O as the dominant parent in the production of the forbidden O lines. However, Cochran and Cochran argue, on the basis of line widths, that H_2O cannot be the sole parent of the $O(^1D)$.

With the assumption that most of the $O(^1D)$ is produced from the dissociation of H_2O, then the line at 6300 Å can be used to measure the H_2O production rate. This line is more accessible to most detectors than the OH bands of the UV, making its observation an important tool for measuring the H_2O production. Enthusiasm for using measurements of the 6300 Å line must be tempered by an understanding of the limitations. First, there is the issue of the blending of the line with both the telluric O line and with NH_2. In particular, the telluric line is quite variable, so modeling its removal when it is blended with the cometary feature is not easy. High spectral resolution observations of $O(^1D)$ has been used on a number of comets (*Magee-Sauer et al.,* 1988; *Combi and McCrosky,* 1991). A far larger set of observations have been obtained at moderate resolution and the $O(^1D)$ intensity has been calculated by modeling the contribution of the telluric $O(^1D)$ and the cometary NH_2 (*Spinrad,* 1982; *Fink and Hicks,* 1996). *Arpigny et al.* (1987) investigated the effects of spectral resolution on the difficulty of deblending the $O(^1D)$ and NH_2 lines and concluded that

low spectral resolution can lead to an underestimate of the $O(^1D)$ intensity and production rate by about a factor of 2.

Another limitation of using $O(^1D)$ as a measure of the H_2O production is that the branching ratios of the reactions that dissociate H_2O are not well determined. *Budzien et al.* (1994) summarized the uncertainties in our knowledge of these branching ratios. Since OH is produced approximately 90% of the time, with O produced approximately 10% of the time, errors in the branching ratio induce a larger uncertainty in our calculation of H_2O production rates from $O(^1D)$ than from OH. In addition, while all the OH is a daughter of H_2O, some of the $O(^1D)$ is a daughter of the dissociation of H_2O, while some is a product of the subsequent dissociation of OH.

4.4. Molecular Hydrogen

Feldman et al. (2002) recently reported the FUSE observation of three P1 lines of the H_2 Lyman series that are excited by the accidental coincidence of the solar Lyman-β line with the P1 line of the $B^1\Sigma_u^+ - X^1\Sigma_g^+$ (6,0) band in the spectrum of Comet C/2001 A2 (LINEAR), shown in Fig. 3. Similar fluorescence has also been seen in the spectra of Jupiter and Mars. Although the strongest of the fluorescent

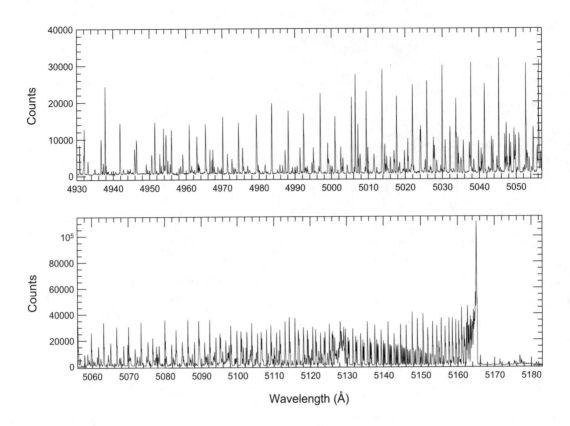

Fig. 4. Spectrum of Comet 122P/deVico showing two-thirds of the C_2 $\Delta v = 0$ band observed at a resolving power $\lambda/\Delta\lambda = 60{,}000$ with the McDonald Observatory 2.7-m 2D-coudé spectrograph. The two panels have different ordinate scalings to better show the details of the spectra. The bluest lines shown at 4932.059 and 4932.139 Å are C_2 (0,0) $R_1(73)$, $R_2(72)$, and $R_3(71)$. Most of the lines shown in the spectrum are attributable to C_2. However, there are some NH_2 and unidentified lines mixed throughout this spectrum.

lines appear near 1600 Å, longer-wavelength spectra of comets from IUE, HST, and sounding rockets have not had sufficient spectral resolution to unambiguously identify H_2 in cometary spectra. The determination of the H_2 column density in the field of view depends strongly on the shape of the solar Lyman-β line, the rotational temperature and outflow velocity of the H_2, and the heliocentric velocity of the comet. Feldman et al. demonstrated that the derived column abundance of H_2 is consistent with H_2O dissociation models but cannot exclude that some of it is produced directly from the nucleus (*Bar-Nun and Prialnik*, 1988) or by solar wind sputtering of dirty ice grains (*Pirronello et al.*, 1983).

5. CARBON-, NITROGEN-, AND SULFUR-CONTAINING RADICALS

5.1. C_2

There are two principal band systems of C_2 that are observed in the optical spectra of comets. These are the Swan, or $d^3\Pi_g$–$a^3\Pi_u$, system, and the Phillips, or $A^1\Pi_u$–$X^1\Sigma_g^+$, system. The Swan system was the first molecule identified in a cometary spectrum and is dominant in the green, orange, and red region of the spectrum; the Phillips bands are important in the near-IR and IR. In addition, the C_2

$D^1\Sigma_u^+$–$X^1\Sigma_g^+$, or Mulliken, system has been detected in the UV, despite the fact that the Mulliken $\Delta v = 0$ band's g-factor is about 40 times smaller than that of the Swan $\Delta v = -1$ band sequence [see Fig. 2 of *A'Hearn* (1982)].

In comparison to CN and other cometary molecules, much higher C_2 vibration-rotation levels are excited (see Fig. 4). Indeed, vibration-rotation levels as high as J = 109 have been detected (*Cochran and Cochran*, 2002). This high-J distribution occurs because C_2 is a homonuclear molecule with no permanent dipole moment so that vibrational and rotational electric dipole transitions within an electronic state are forbidden. Thus, the rotational excitation temperature of the C_2 coma would be approximately the color temperature of the Sun (T = 5800 K) in the absence of any mechanism for cooling the rotational temperature. This mechanism does exist, however, in the form of interactions with other electronic states.

The rotational excitation temperature (T_{rot}) of the coma can be measured by observing the Swan bands at high spectral resolution. Such observations have generally yielded $T_{rot} \sim 3000$ K. *Lambert et al.* (1990) observed Comet Halley and found that the C_2 gas could not be described by a single rotational temperature. They found that the lower rotational levels (J < 15) could be fit with $T_{rot} \sim 600$ K, while higher levels required $T_{rot} \sim 3200$ K. Indeed, when the contribution of the hotter population is accounted for, the low-J levels yield T_{rot} = 190 K. *Krishna Swamy* (1997) has shown that these results can be understood in detail by the inclusion of many more transitions in models of the photolysis of C_2.

C_2 emissions can be detected at large distances from the nucleus, implying a large scale length for its production. However, it has long been noted that the distribution in the inner few thousand kilometers of the coma is essentially constant. This flat distribution is inconsistent with a simple parent/daughter production for C_2. *Jackson* (1976) was the first to suggest that C_2H_2 was the grandparent of C_2, with an intermediate decay of the C_2H_2 to C_2H + H. Using long-slit CCD observations of Comet Halley, *O'Dell et al.* (1988) also concluded that C_2 must be the product of the decay of two precursors. Using a multigeneration Haser model, they found that the data can be fit with parameters between R_1 = 12,000 km, R_1/R_2 = 3, and R_2/R_3 = 0.8; to R_1 = 17,000 km, R_1/R_2 = 1.5, and R_2/R_3 = 0.12, where R_1, R_2, and R_3 are the grandparent, parent, and daughter destruction scale lengths, respectively.

Combi and Fink (1997) further investigated a solution for the flat inner profile of the C_2 gas. They also used a three-generation dissociation model, but theirs differed from that of *O'Dell et al.* (1988) by the inclusion of ejection velocities resulting from the excess energy of the photodissociations. They found that, as long as the ejection velocities are greater than 0.5 km s^{-1}, the excess energy imparted during the various dissociations will cause a filling in of the "hole" in the profile, resulting in a profile that is no longer flat in the inner coma. They argued that typical heavy molecules produce ejection velocities in excess of 1 km s^{-1}, so a three-generation photodissociation is unlikely the parent process for the production of C_2. Instead, they suggest that a CHON grain halo with a size of 10^4 km is responsible for the production of X–C_2 (X is some unknown species), which in turn is photodissociated on a scale of several times 10^4 km to produce the C_2. This process can proceed with little or no excess energy.

New laboratory data and *ab initio* calculations (*Sorkhabi et al.*, 1997) would seem to allow for the original thesis of *Jackson* (1976), that the grandparent of C_2 is C_2H_2, while the direct parent is C_2H. *Sorkhabi et al.* (1997) obtained laser-induced fluorescence spectra of C_2 ($X^1\Sigma_g^+$) radicals produced during 1930 Å laser photolysis of C_2H_2. They used these observations along with calculations to match the spectrum of the C_2 Mulliken system in HST observations of Comet C/1996 B2 (Hyakutake).

5.2. CN

There are two CN electronic band systems that can be observed in the optical in cometary spectra. These are the "violet" system ($B^2\Sigma^+$–$X^2\Sigma^+$) and the "red" system ($A^2\Pi$–$X^2\Sigma^+$). The violet system is one of the most prominent features in cometary spectra and is seen in most comets with heliocentric distances less than 3 AU. It has been detected in Comets 1P/Halley and C/1995 O1 (Hale-Bopp) at distances greater than 4 AU and in the Centaur (2060) Chiron at 11.26 AU (*Bus et al.*, 1991). Despite its spectral promi-

nence, its parent must have less than 1% of the abundance of H_2O.

Because the violet system is a Σ–Σ transition, only P- and R-branches are permitted and J < 20 is generally observed. This, coupled with the density of absorption features in the solar spectrum at the wavelength of the Δv = 0 bands at 3883 Å, makes it necessary to account for the Swings effect when converting observed band flux to column density. Results of calculations of the change in the g-factor with changing heliocentric radial velocity have been given by *Tatum and Gillespie* (1977), *Tatum* (1984), and others. The solar spectrum shows strong CN Σ–Σ absorption so that the g-factor reaches a minimum at zero heliocentric radial velocity (at perihelion). While the cometary CN violet band does not disappear entirely at perihelion, it becomes quite weak. The red system is generally much more spread over wavelength and does not appear as such an obvious band.

High spectral resolution allows the isotopic features of CN to be clearly resolved from the weak, high-order non-isotopic features. Thus, studies of CN with high spectral resolution have been used to derive $^{13}C/^{12}C$ values that are essentially the solar value for several comets (*Kleine et al.*, 1994, 1995; *Lambert and Danks*, 1983; *Lambert et al.*, 1990). High spectral resolution, coupled with high signal/noise, have allowed *Arpigny et al.* (2003) to determine ^{15}N ^{14}N in a number of comets. They found a value that is a factor of 2 higher than the value from the Earth's atmosphere.

Although the identification of the violet system as CN has been known since the earliest days of comet spectroscopy, the identification of the parent is still in doubt. The Haser (or radial) scale length of the CN parent is on the order of 2×10^4 km at 1 AU. A potential parent, HCN, is observed in the millimeter portion of the spectrum, but it is still uncertain whether HCN is a minor parent or a dominant parent of CN. *Bockelée-Morvan and Crovisier* (1985) argued that the CN distribution is inconsistent with HCN as a dominant parent unless the coma expansion velocity was much lower than generally assumed. In contrast, direct comparison of the distribution of HCN and CN using more recent millimeter observations have shown that there is sufficient HCN to be a dominant parent of CN (*Ziurys et al.*, 1999) and that the HCN and CN are distributed similarly within the coma (*Woodney et al.*, 2002).

Festou et al. (1998) asserted that HCN contributes at the percent level to the production of CN. They found that a best case for a parent for CN has a lifetime of 3.5×10^4 s at 1 AU with a velocity of 1–2 km s^{-1}. They concluded that this is consistent with C_2N_2 as the dominant parent. *Bonev and Komitov* (2000) fit the CN scale lengths and concluded that C_2N_2 is the sole parent for CN.

5.3. C_3

The first detection of the emission band that we now know to be C_3 was in 1881. However, the tentative identification of the band as C_3 was not made until 1951 (*Douglas*, 1951). C_3 is a relatively unstable molecule, making its study difficult. Until its identification, the band was known simply as the "4050-Å Group." The main part of the band lies between 3900 and 4140 Å with a maximum at 4050 Å and an additional peak, not well defined, at 4300 Å. However, careful examination of cometary spectra shows lines attributable to this band from ~3350 to 4700 Å.

C_3 is a linear, symmetric molecule containing equal-mass nuclei. The band has been identified as an $\tilde{A}^1\Pi_u$–$\tilde{X}^1\Sigma_g^+$ electronic transition. R-branch bandheads at 4072 and 4260 Å can be assigned to the (1,0,0)–(1,0,0) and (0,0,0)–(1,0,0) bands respectively. The density of the lines results in a pseudo-continuum from C_3. For all Σ states of the molecule, every second line is missing. In other states, one member of the e/f-parity doublet of each rotational level is missing, although all rotational levels are represented (*Tokaryk and Chomiak*, 1997). The complexity of the spectrum and the difficulty of exciting C_3 in the laboratory have resulted in many missing identifications of lines. In addition, the density of lines has made it impossible to resolve all the individual lines. Recently, observations of cometary C_3 were used to derive a new dipole moment derivative (dμ/dr) of approximately 2.5 Debye Å^{-1} for this band system (*Rousselot et al.*, 2001).

The parent of C_3 is unknown. Chemically plausible parents, such as C_3H_4 and C_3H_8, are not detected in cometary spectra. Either of these would produce C_3 in a multigenerational process and many of the photolysis rates for such reactions are very uncertain. Observations of the distribution of C_3 gas in the comae of comets suggests that the parent must have a short scale length, on the order of 3×10^3 km (*Randall et al.*, 1992). It is unlikely that the long-chain C molecules seen in the interstellar medium are parents for C_3 since these long-chain C molecules are believed to have alternating single and triple bonds that would more likely break to produce C_2 than C_3. The photodestruction of C_3 results in the production of a C_2 molecule and a C atom.

5.4. CH

The CH (0,0) $A^2\Delta$–$X^2\Pi$ band has its peak at 4314 Å and appears in cometary spectra as a weak band. Two factors contribute to this weak feature of interest: (1) The oscillator strength is small, ~5×10^{-3}; and (2) the lifetime against photodissociation at 1 AU is between 35 and 315 s (*Singh and Dalgarno*, 1987). The former implies that even a very weak feature is the result of a significant column density. The latter means that wherever the CH is detected in the coma, the parent must be close by. Thus, CH can be used as a tracer of the parent distribution.

The probable parent of CH is CH_4. Methane will first decay into CH_2 (CH_3 is highly unstable) and then into CH. The detection of CH in the optical is far simpler than the detection of CH_4 in the IR so there is a significantly larger database of CH detections than CH_4 detections.

The (0,0) $B^2\Sigma^-$–$X^2\Pi$ band has also been detected at high spectral resolution at 3886 Å. This band cannot be resolved from CN at lower resolution. It is necessary to be aware of this band, however, when computing models of ^{12}CN/^{13}CN,

since some of the CH lines are coincident with the weaker CN lines.

5.5. NH and NH$_2$

The (0,0) $A^3\Pi_i$–$X^3\Sigma^-$ band of NH occurs between 3345 and 3375 Å. It was first detected in the spectrum of Comet Cunningham (*Swings et al.*, 1941) and is generally the only NH band seen. This band shows an R- and P-branch but the Q-branch is absent or weak. *Kim et al.* (1989) showed that the spectrum can be explained completely on the basis of pure resonance fluorescence without any collisions.

Emission lines of NH_2 have been detected throughout a region from ~3980 Å to well past 1 μm (*Cochran and Cochran*, 2002). These lines belong to the $\tilde{A}^2A_1$–$\tilde{X}^2B_1$ electronic transition, along with transitions between the high vibronic levels and the ground state of the $\tilde{X}^2B_1$ electronic band [e.g., (0,13,0)$\tilde{X}^2B_1$–(0,0,0)$\tilde{X}^2B_1$]. NH_2 is an asymmetric top molecule, with a linear upper level and a lower level bent at an angle of 103°, making its spectrum quite complex and irregular.

It is widely believed that NH_3 is the parent for NH_2, which decays, in turn, into NH. However, until recently, NH_3 had not been detected in cometary spectra. *Palmer et al.* (1996) first detected NH_3 in the radio spectrum of C/1996 B2 (Hyakutake) and it has now also been detected in the radio spectrum of C/1995 O1 (Hale-Bopp) (*Bird et al.*, 1997) and the IR spectrum of 153P/Ikeya-Zhang (*Magee-Sauer et al.*, 2002). Other potential parents include N_2H_4 and CH_3NH_2.

Prior to the detection of NH_3, the chemical reaction pathway was argued on the basis of the observed scale lengths of the various species. However, there has been much disagreement on the values relevant to each species. Typically, the g-factors for NH_2 of *Tegler and Wyckoff* (1989) are used. *Arpigny* (1994) pointed out that this g-factor calculation is off by a factor of 2 because the structure of the NH_2 bands means that a single band when the linear band notation is used (typical in past cometary work) only samples either odd or even K_a lower levels. *Cochran and Cochran* (2002) advocated converting to using the bent band notation adopted by the physicists. Under that notation, the linearly denoted (0,8,0) Π band is a part of the bent notation (0,3,0) band and the (0,8,0) Φ band is part of the (0,2,0) band. The bent notation bands contain both odd and even K_a lower levels. It is generally believed that the parent of NH_2 has a relatively short scale length, on the order of ~4×10^3 km (*Krasnopolsky and Tkachuk*, 1991; *Fink et al.*, 1991). The destruction scale length is a few times 10^4 km.

Kawakita et al. (2001b) have derived new g-factors for five of the bands (in the linear notation) and find values smaller than those of *Tegler and Wyckoff* (1989) by factors of 2.7 to 6.4. Korsun and Jockers (2002) have applied these g-factors to NH_2 filter images and shown that the derived NH_2 production rate is consistent with NH_3 as the parent.

Kim et al. (1989) have calculated NH fluorescence efficiencies incorporating the Swings effect. Part of the problem with defining scale lengths for NH is a paucity of data coupled with the difficulty of determining the atmospheric extinction at the wavelength of NH. Parent scale lengths range from $1–5 \times 10^4$ km. *Schleicher and Millis* (1989) argue reasonably convincingly for the longer of these values. Most datasets are not very sensitive to the destruction scale length used, but it is generally agreed to be about 2×10^5 km. *Feldman et al.* (1993) used spectrophotometric spatial profiles of OH and NH emission derived from observations of Comet 1P/Halley made by the Soviet-era ASTRON satellite to derive the relative NH_3 abundance with a nearly model-independent analysis.

All the observations suggest that about 95% of the photodissociations of NH_3 produce NH_2 and that very little of the NH comes directly from NH_3. Using the various parameters found in the literature, there is general agreement that the abundance of NH_3 in the nucleus is about 0.5% that of H_2O for all comets, if NH_3 is the sole parent of NH_2 and NH.

Recently, *Kawakita et al.* (2001a) have utilized high-resolution NH_2 spectra of Comet C/1999 S4 (LINEAR) to model the ortho-to-para ratio and to derive a spin temperature for NH_3. They found a temperature of 28 ± 2 K, assuming that the NH_2 arises from pure fluorescence excitation of NH_3.

5.6. CS

Ultraviolet emission from carbon monosulfide (CS) and atomic sulfur was first reported in rocket spectra of Comet West (C/1976 V1) by *Smith et al.* (1980). The (0,0) band of the $A^1\Pi–X^1\Sigma^+$ system of CS at 2576 Å is the strongest of four bands of this system lying between 2500 and 2700 Å and has been detected in nearly all IUE and HST comet spectra. More recently, CS has also been detected in the radio (*Biver et al.*, 1999). *Jackson et al.* (1982) analyzed both high- and low-dispersion IUE spectra of Comet Bradfield (C/1979 Y1), concluding that the likely parent was CS_2 with an extremely short photodissociation lifetime of ~100 s at 1 AU. They also found that the band shape was indicative of a 70-K rotational temperature, that the production rate of the parent was about 0.1% that of water near 1 AU, and that this ratio decreased with increasing heliocentric distance. This latter behavior has been seen in all the comets observed by IUE over a significant range of r and also in the radio observations of Comet Hale-Bopp (C/1995 O1) (*Biver et al.*, 1999). *Jackson et al.* (1982) suggested that CS_2 could also account for all the observed S I emission at 1814 Å, although it was later shown that H_2S was a more important source of S than CS_2 (*Meier and A'Hearn*, 1997). Sulfur-bearing species are discussed in section 4.5 of *Bockelée-Morvan et al.* (2004). *Jackson et al.* (1982), in discussing the photodissociation of CS_2, noted that laboratory measurements using a source at 1930 Å produced an abundant amount of $S(^1D)$ atoms in addition to ground state ^3P atoms. Sulfur in the ^1D state was detected by its transition at 1667 Å in GHRS spectra of Comet Hyakutake (C/1996 B2) (*A'Hearn et al.*, 1999).

Further IUE observations and laboratory data led *Jackson et al.* (1986) to revise the CS_2 lifetime at 1 AU to ~500 s, and this seemed to be consistent with the limited spatial information available from low-dispersion IUE spectra. However, high-dispersion spectra of 1P/Halley showed the band shape to be quite different from previously observed comets, with the R-branch blueward of the band head much enhanced over the redward P,Q branches, in contradiction with solar fluorescence models. Attempts to model this band with two components along the spectrograph line of sight, the first in statistical equilibrium near the nucleus, and the second in fluorescence equilibrium for distances greater than ~1000 km from the nucleus, have been only partially successful in reproducing the observations (*Prisant and Jackson*, 1987; *Krishna Swamy and Tarafdar*, 1993).

In the HST era, spectra of the CS (0,0) band at resolution comparable to that of the IUE high-dispersion mode ($\Delta\lambda = 0.8$ Å) have not been obtained, precluding a resolution of this problem. In one area, though, spatial imaging with STIS has led to a more reliable estimate of the CS parent lifetime of ~1000 s (*Feldman et al.*, 1999).

Another potential source of CS in the coma is OCS, detected in the radio in recent comets in comparable abundance to CS_2 (see *Bockelée-Morvan et al.*, 2004). However, the primary dissociation path of OCS is to CO and S (*Huebner et al.*, 1992) so that the contribution to the CS abundance is minor. Other S-bearing parent molecules identified in the radio are H_2S and SO_2, the former being the principal S species in the cometary ice (see *Bockelée-Morvan et al.*, 2004). *Kim and A'Hearn* (1991, 1992) have given spectroscopic limits on the dissociation products SH and SO, although the latter has been detected in the radio in Comet Hale-Bopp (*Bockelée-Morvan et al.*, 2000). Finally, we note that *Irvine et al.* (2000) have reported the detection of NS in Comet Hale-Bopp, although its origin remains unknown.

6. ATOMIC BUDGET OF THE COMA

With the exception of CO and CO_2, solar photodissociation rates are significantly higher than the rates for photo- or solar wind ionization of the principal molecular constituents of the coma (*Huebner et al.*, 1992). Thus, the end products of the molecular species will be predominantly the constituent atoms, H, O, C, N, and S, and their corresponding ions. The neutral atomic species all have their principal resonance transitions in the vacuum UV in a wavelength range amenable to spectroscopic observations by IUE and HST (see Table 2). H I Lyman-α observations, both spectroscopic and imaging, are discussed in section 4.2 above and in *Combi et al.* (2004). Because of the large scale lengths against ionization for these species, the atomic coma can extend to millions of kilometers from the nucleus and images in Lyman-α show the atomic hydrogen corona to be the largest object in the solar system, often attaining a size of ~0.1–0.2 AU.

The resonance transitions of atomic carbon and oxygen were first detected in rocket observations of Comet Kohou-

TABLE 2.   Principal resonance transitions
of cometary atoms and ions.

Species	Transition	Wavelength (Å)
H I	$^2P^o$–^2S	1216
O I	$^3S^o$–^3P	1302–06
C I	$^3D^o$–^3P	1561
	$^3P^o$–^3P	1657
N I	^4P–$^4S^o$	1134
	^4P–$^4S^o$	1200
S I	^3P–$^3S^o$	1807–26
O II	^4P–$^4S^o$	834
C II	^2S–$^2P^o$	1037
	^2D–$^2P^o$	1335
N II	$^3D^o$–^3P	1085
S II	^4P–$^4S^o$	1250–59

tek (C/1973 E1) (*Feldman et al.,* 1974; *Opal et al.,* 1974), the latter also providing objective grating images of the C and O comae. Because of the large heliocentric velocity at the time of the observation, the O I λ1302 multiplet was Doppler shifted away from the solar O I line and the excitation was attributed to "Bowen fluorescence" of solar Lyman-β (*Feldman et al.,* 1974). Similarly, C I λ1657 and the CO Fourth Positive bands were shown to be subject to a large "Swings effect" (Feldman et al., 1976). From similar observations made of Comet West (C/1975 V1), *Feldman and Brune* (1976) showed that the C could be accounted for as the dissociation product of the CO simultaneously measured. Comet West had an unusually high CO abundance and so it was a surprise when Comet Bradfield (C/1979 Y1), observed by IUE, showed a large C emission despite a much smaller relative CO production rate (*A'Hearn and Feldman,* 1980). IUE used a much smaller aperture than had the earlier rocket spectrometers, and *Festou* (1984) considered whether the additional C could be indicative of the

presence of many other C-bearing species near the nucleus, many of which were subsequently discovered in later comets. Festou also postulated that the atomic inventory, independent of the details of the molecular parentage and assuming that all the photons could be collected from the extended coma, would be a strong indicator of cometary diversity.

In addition to the resonance multiplet of O I, the intercombination doublet at 1356 Å has been observed in a few coma spectra (*Woods et al.,* 1987; *McPhate et al.,* 1999). Because the g-factor for this transition is so small, the excitation source has been attributed to photoelectrons, analogous to the excitation in planetary atmospheres (*Cravens and Green,* 1978). Electron impact excitation may also contribute to the observed CO Cameron band emission (*Weaver et al.,* 1994).

Atomic sulfur was first identified in the spectrum of Comet West by *Smith et al.* (1980), who also obtained an objective image of the S I λ1813 multiplet. This emission has been detected in nearly every comet observed since then by IUE or HST, and most of the observations show the 1807 Å component, the transition connecting to the lowest ground-state level, to be saturated. *Azoulay and Festou* (1986) first treated opacity effects in order to properly extract S production rates and concluded that CS_2 was insufficient to account for the amount of S observed and that OCS was likely the primary parent. *Meier and A'Hearn* (1997), analyzing a much larger database of S observations, came to a similar conclusion but identified the primary parent as H_2S, which had recently been detected in radio observations.

Two other S multiplets are occasionally detected in comets that are observed at small heliocentric velocity (<10 km s^{-1}), at 1429 and 1479 Å. These transitions are also optically thick near the nucleus. A particularly nice example of these emissions is seen in the long-slit spectrum of Comet Hale-Bopp obtained from the rocket experiment of *McPhate et al.* (1999) (Fig. 5). This spectrum also shows how the relative intensities of the three components of the S I λ1813

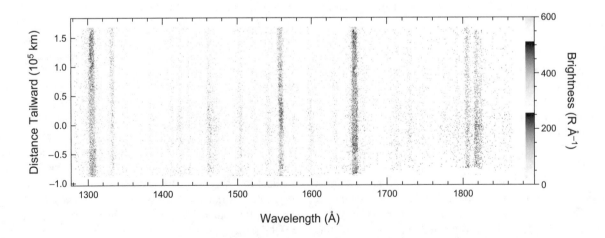

Fig. 5.   Long-slit spectral image of Comet Hale-Bopp acquired on 1997 April 6.16 UT. The long axis of the slit was oriented along the Sun-comet line and the slit was offset 20″ from the nucleus. The Sun is down in this image. Each pixel is 0.6 Å × 0.″8 and subtends 800 km at the comet. The emission features are identified in Fig. 6. From *McPhate et al.* (1999).

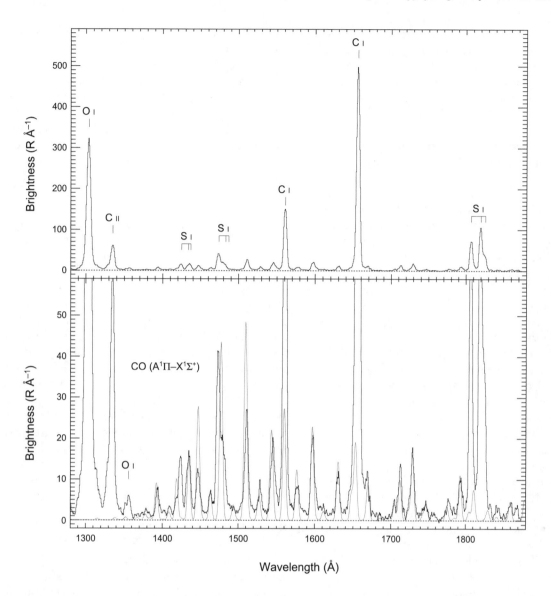

Fig. 6. Full-slit average spectrum derived from the data of Fig. 5. The top frame identifies the stronger atomic emissions while the bottom frame, expanded in scale, shows the rich molecular spectrum of CO. The gray line is a modeled optically thin CO Fourth Positive band spectrum. From *McPhate et al.* (1999).

multiplet vary with distance from the nucleus, the line ratios approaching optically thin values at 150,000 km (Fig. 6).

In contrast to the other atomic species, little is known about atomic nitrogen in the coma. Its principal resonance transition at 1200 Å has never been detected, presumably due to the weakness and narrowness of the solar exciting lines and its close proximity to the very strong H I Lyman-α line. *Weaver et al.* (2002) report the detection of the strongest member of the N I $\lambda 1134$ multiplet in the FUSE spectrum of Comet C/2001 A2 (LINEAR), but this identification needs to be confirmed.

7. SODIUM IN COMETS NEAR 1 AU

Recognition of cometary Na D-line emissions at 5890/5896 Å dates back to visual observations of Comets C/1882 F1 and C/1882 R1 (*Levin*, 1964) and Comet C/1910 A1

(*Newall*, 1910). For the latter comet, handheld prism observations indicated that the tailward extent of the Na emission exceeded that of the C_2 Swan band for this comet. Although Na is a minor species in all atmospheres where it has been detected [Io (*Brown and Chaffee*, 1974), Mercury (*Potter and Morgan*, 1985), Moon (*Potter and Morgan*, 1988)], it has nonetheless been useful because trace amounts of Na are easily observed because of the large oscillator strength in the D-lines and their placement at the peak wavelength of solar radiation.

The curious behavior of Na and its emission results from the strong interaction with solar radiation. The solar spectrum contains two strong Na absorption lines that modulate the fluorescence of Na atoms in the solar system depending upon their heliocentric velocity. At 1 AU from the Sun, strong fluorescence in the D-lines produces a large radiation pressure acceleration, which varies from about

3 cm s^{-2} for Na atoms at rest with respect to the Sun (and seeing the bottom of the solar lines) to more than 50 cm s^{-2} for Na atoms Doppler shifted to the nearby continuum.

The first Na studies were largely confined to Sun-grazing comets, perhaps most notable of which was Comet Ikeya-Seki (C/1965 S1), with perihelion q = 0.04 AU, and for which Na emission was not seen for heliocentric distances r > 0.6 AU (*Bappu and Sivaraman*, 1969). Based on observations of this comet by *Preston* (1967) and *Spinrad and Miner* (1968), *Huebner* (1970) argued that under these harsh conditions the intensity of Na D emission could be accounted for by Na atoms embedded within micrometer-sized refractory silicate material having a high latent heat of vaporization. *Preston* (1967), in particular, noted that a whole set of metallic species in addition to Na were detected spectroscopically. These emission lines were tabulated by *Slaughter* (1969). Because of the short photoionization lifetime of Na atoms close to the Sun and the substantial tailward extent of the Na emission (~10^3 km), the early studies of the Sun-grazing comets (*Spinrad and Miner*, 1968; *Huebner*, 1970) equated the tailward extent as being indicative of the lifetimes of parent refractory grains. These studies, however, neglected the very large radiation pressure acceleration on Na atoms. In addition, it has since been found that the photoionization lifetime (*Huebner et al.*, 1992; *Combi et al.*, 1997) may be up to three times longer than believed at the time.

Observations of Na emission near and beyond 1 AU have been limited to a few bright, active comets. The Na in Comet Kohoutek C/1973 E1 (q = 0.18 AU) was seen at least out to heliocentric distance of 0.47 AU (*Delsemme and Combi*, 1983). The first interpretation of Na D emission at distances beyond 1 AU was by *Oppenheimer* (1980) in Comet West (C/1975 V1) at 1.4 AU; Oppenheimer concluded that the Na was trapped in molecules within the volatile ice component. He reasoned that only with Sun-grazing comets would the refractory grain component be hot enough to liberate Na, either in elemental or molecular form.

Delsemme and Combi (1983) reported that the Na spatial profile in Comet Kohoutek had its brightest pixel at the same location as the dust continuum, whereas the other gas species (C$_2$, CN, and NH$_2$) were all displaced sunward, suggesting some connection of Na with the dust. *Combi et al.* (1997) reported a detailed model analysis of spatial profiles of Na from long-slit spectra in non-Sun-grazing comets [Bennett (C/1969 Y1), Kohoutek (C/1973 E1), and 1P/Halley] and identified two types of spatial signatures. There was a relatively stable point source of Na, produced directly from the nucleus or a short-lived parent, as well as an extended source seen mainly on the tailward side. The latter had a larger production rate and varied by factors of a few compared with the nucleus source on timescales as short as a day. They also noted some spatial similarities in Halley between the extended Na distribution and ion profiles but not with the dust. This combined with the large variability led them to suggest a possible role for some plasma process for the extended source.

Because of its extremely large overall gas production rate, a spectacularly bright and long Na tail was imaged in Comet Hale-Bopp (C/1995 O1) by *Cremonese et al.* (1997) and *Wilson et al.* (1998). Modeling analysis of these images and further spectroscopic observations (*Brown et al.*, 1998; *Rauer et al.*, 1998; *Arpigny et al.*, 1998) showed that the observed distribution of Na in the tail could be explained by a nucleus or near-nucleus source of Na at a production rate that is less than 0.3% of what would be expected based on solar abundances of Na compared with O. This is in fact the same level as the nucleus or near-nucleus source for Na identified by *Combi et al.* (1997) in Comet Halley and seen in Comets Bennett and Kohoutek. Therefore, the gaseous Na seen in non-Sun-grazing comets does not represent the bulk of Na in comets, which is mostly bound to the refractory component and was seen in the dust mass spectra of 1P/Halley (*Jessberger and Kissel*, 1991).

Unlike the extended source of Na in Halley, inner coma measurements of Na in Comet Hale-Bopp showed an extended source component that appeared to be associated with the asymmetric dust distribution (*Brown et al.*, 1998). Brown et al. found that roughly half the Na was produced from the nucleus source and half from an extended source that roughly followed the r^{-2} distribution of the asymmetric dust coma and did not resemble either the spatial or velocity distribution seen in simultaneously observed H$_2$O$^+$ ions. Observations of Na in future bright comets are required in order to answer the question of the nature of the extended source.

8. IONS

8.1. Molecular Ions

Photolytic processes and chemical reactions will ionize molecules in the comae of comets and, as a result, various molecular ions have been detected in cometary spectra. In the UV and optical, these include CH$^+$, CO$^+$, CO$_2^+$, H$_2$O$^+$, N$_2^+$, and OH$^+$. Ions have now been detected in the radio spectrum of Comet C/1995 O1 (Hale-Bopp), including HCO$^+$ (Wright et al., 1998), H$_3$O$^+$, and CO$^+$ (*Lis et al.*, 1999). The ions show a very different distribution in the coma than do neutrals since the ions are accelerated tailward by the solar wind. Thus, the ionic species are often called "tail" species. However, it should be noted that many are observed relatively close to the nucleus of the comet, a good illustration being obtained from long-slit spectra of CO$^+$ and CO$_2^+$ in Comet 1P/Halley given by *Umbach et al.* (1998).

The predominant processes for the production of ions are photodissociation (e.g., H$_2$O + hν → OH$^+$ + H + e) and photoionization (e.g., H$_2$O + hν → H$_2$O$^+$ + e) (*Jackson and Donn*, 1968). Within the collisional zone (the inner few thousand kilometers of the coma for moderately bright comets), ions can be produced by charge exchange with solar wind protons, electron impact ionization, charge transfer reactions, and proton transfer reactions.

The transitions of CO$^+$ that are seen in the blue/UV region of the spectrum arise from the first negative bands

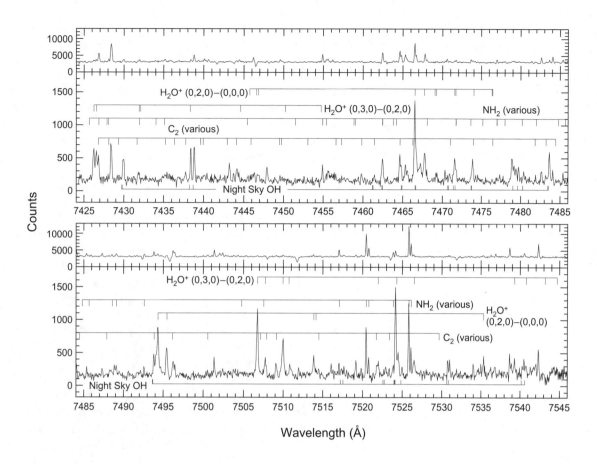

Fig. 7. Spectrum of Comet 153P/Ikeya-Zhang, recorded 15,000 km tailward from the optocenter, illustrating a well-developed ion tail. The narrow panel for each half of the order is the optocenter spectrum while the wide panel is the tail spectrum. All lines from two H_2O^+ bands are marked, along with lines of NH_2 and C_2 (Phillips). Not all the marked ionic lines are present; only the lower J-values appear in the spectrum. Comparison of the tail and the optocenter spectra show the increased strength of the ionic lines relative to the neutrals. The solar continuum has not been removed from either spectrum. This spectrum was obtained with the McDonald Observatory 2D-coudé at R = 60,000.

$(B^2\Sigma-X^2\Sigma)$ and the comet-tail bands $(A^2\Pi-X^2\Sigma)$. The comet-tail bands show two peaks that are due to the $\Pi_{1/2}$ and $\Pi_{3/2}$ branches. Generally, the (2,0) and (3,0) comet-tail bands are the strongest bands observed. The comet-tail transitions are responsible for the blue appearance of the ion tail in color images of comets.

The presence of the CO^+ bands in cometary spectra has led to the belief that CO is generally present at the few percent level in cometary nuclei. CO^+ emissions are seen in cometary spectra to heliocentric distances greater than 5 AU (*Cochran and Cochran*, 1991). *Magnani and A'Hearn* (1986) have calculated fluorescence efficiencies for most of the comet-tail bands accounting for the Swings effect, which should be a factor since there are relatively few excited levels and the solar spectrum is dense in the blue.

Another ion that appears prominently in the red part of the spectrum is H_2O^+. Although it has been observed in cometary spectra for a long time, it was only identified for the first time in 1974 in spectra of Comet C/1973 E1 (Ko-

houtek) (*Herzberg and Lew*, 1974). The electronic transition is $\tilde{A}^2A_1-\tilde{X}^2B_1$ and it is observed from 4000 to 7500 Å; a small part of this range is seen in Fig. 7. *Wegmann et al.* (1999) have run magnetohydrodynamic and chemical simulations of cometary comae and have concluded that for small comets, up to 11% of the water molecules are ultimately ionized. H_2O^+ occurs in a spectral bandpass that is easily accessible to CCD detectors, so there are many observations of H_2O^+ in cometary comae. *Lutz et al.* (1993) have calculated fluorescence efficiency factors for six of the bands. H_2O^+ is isoelectronic to NH_2 so *Arpigny*'s (1994) comment concerning increasing the efficiency factors of NH_2 by a factor of 2 also applies to H_2O^+. Indeed, although the standard reference on the H_2O^+ band (*Lew*, 1976) used the linear notation, the transitions are more correctly specified in their bent notation (*Cochran and Cochran*, 2002).

Bonev and Jockers (1994) mapped the distribution of H_2O^+ in Comet C/1989 X1 (Austin). They found a strong asymmetry with a relatively flat distribution tailward and a factor of 4 dropoff in the first 10^4 km sunward. The maximum H_2O^+ column density was frequently observed to be shifted tailward.

CO_2^+ emission was first identified in the optical spectrum of Comet C/1947 S1 (Bester) (*Swings and Page,* 1950) and in the UV spectrum of Comet C/1975 V1 (West) (*Feldman and Brune,* 1976). The optical lines arise from the Fox-Duffendack-Barker ($\tilde{A}^2\Pi_u$–$\tilde{X}^2\Pi_g$) electronic band system and appear in the wavelength range from 3000 to 4000 Å. The UV doublet at 2890 Å is from the $\tilde{B}^2\Sigma_u$–$\tilde{X}^2\Pi_g$ electronic transition (*Festou et al.,* 1982). *Feldman et al.* (1986), using the IUE, noted a strong enhancement of this feature in a spectrum of Comet 1P/Halley taken at a position 150,000 km tailward of the nucleus at a time corresponding to the peak of an optical outburst. This observation suggested that CO_2 may have played a significant role in the outburst process.

Leach (1987) has noted that the CO_2^+ emission rates are affected by intramolecular coupling between the $\tilde{B}^2\Sigma_u$ and $\tilde{A}^2\Pi_u$ states so that emission from the $\tilde{B}^2\Sigma_u$ state can occur at $\lambda > 3000$ Å (this is referred to as "redshifted fluorescence"). Such bands are detected in cometary spectra along with some unclassified bands redward of 4000 Å. Excitation efficiencies for some of the transitions can be found in *Fox and Dalgarno* (1979).

Bands of the OH^+ $A^3\Pi$–$X^3\Sigma^-$ electronic system cover the complete optical bandpass. However, only lines from the (0,0) and (1,0) bands have been detected (*Swings and Page,* 1950; *Festou et al.,* 1982). *Lutz et al.* (1993) have derived fluorescence efficiency factors for many bands of OH^+, but caution that they are only accurate to ±50% because the Swings effect was not included.

The CH^+ lines that are seen in cometary spectra are from a $A^1\Pi$–$X^1\Sigma$ transition. Only the low-energy transitions of the (0,0) band are seen at around 4230 Å. These lines are coincident with bands of CH and CO^+. *Lutz et al.* (1993) have also calculated g-factors for CH^+. As with the OH^+, they caution that the Swings effect was not included and therefore the fluorescence efficiencies are only good to ±50%.

Other molecular ions were detected by the *in situ* mass spectrometer measurements made at Comet Halley in 1986. These include H_3S^+, C_3H^+, and $C_3H_3^+$, as well as more complex organic ions (*Marconi et al.,* 1990; *Eberhardt and Krankowsky,* 1995). Attempts to associate some of the unidentified visible spectral features with ions such as these or with H_2S^+ must be taken with caution.

8.2. The Case of N_2^+

N_2 is the least reactive of all N-bearing species, so study of N_2 is important for understanding cometary N. In addition, conditions in the early solar nebula were such that the dominant equilibrium species of N should be N_2. However, observations of N_2 are extremely difficult to obtain. Ground-based observations suffer from telluric absorption; interpretations of spacecraft flyby mass spectrometer data are compromised by the fact that N_2 and CO both share the mass 28 bin. Therefore, observations of the First Negative ($B^2\Sigma_u^+$–$X^1\Sigma_g^+$) (0,0) band of N_2^+ at 3914 Å have been used as a proxy for studying N_2. Such observations require high spectral resolution in order to isolate the cometary N_2^+ emission from any telluric N_2^+ emission. They also require a relatively bright comet with a well-developed ion tail for observation. The spatial distribution of any emissions can be used to differentiate between telluric and cometary species.

Observations of the appropriate spectral region of the tails of comets have been made in the past, and examples of comets that show N_2^+ in their spectra can be found in *Swings and Haser* (1956) (e.g., Comet Bester, plate XXIIIa, and Comet Morehouse, plates VIa and VIb). *Cochran et al.* (2000) summarized most of the past N_2^+ observations. These observations have *not* generally been obtained at high spectral resolving power. Recently, Cochran and co-workers (*Cochran et al.,* 2000; *Cochran,* 2002) have reported high-spectral-resolution, high-signal-to-noise observations of three comets that definitely do *not* show N_2^+ in their spectra and that have very tight limits on the quantity of N_2^+. Do different comets have differing amounts of N_2, possibly related to their place of origin? Is N_2 depleted during the life of some comets? Are our models that indicate that N should be preferentially in N_2 rather than NH_3 in the solar nebula in error? Are variations in the quantity of N_2 in comets the result of clathration of the N_2 and CO (*Iro et al.,* 2003)? And ultimately, how reliable are the earlier reports of N_2^+ in cometary tail spectra? Answers to these important questions will require more high-spectral-resolution observations of comets with a variety of dynamical histories.

8.3. Atomic Ions

Table 2 also lists the wavelengths of the resonance transitions of the principal atomic ions. The C II doublet at 1335 Å has been detected in several comets, particularly from sounding rocket observations made with fairly large fields of view (*Feldman and Brune,* 1976; *Woods et al.,* 1987; *McPhate et al.,* 1999). It is difficult to identify the source of the emission. Resonance scattering of the solar C II lines would show a very strong Swings effect. Photoionization of neutral C into an excited ion state has an excitation rate $\sim 10^{-9}$ s^{-1} atom^{-1} (*Hofmann et al.,* 1983) and is insufficient to account for the observed brightness. Perhaps electron impact ionization is responsible as the C II emission is present in the same spectra as the O I] λ1356 emission, but the excitation rate is difficult to evaluate quantitatively. C II λ1037 emission has recently been detected in FUSE spectra (*Weaver et al.,* 2002).

The only other reported atomic ion emission is O II λ834 from a rocket observation of Comet Hale-Bopp by *Stern et al.* (2000). No quantitative information about this measurement is given.

Despite the large number of spectra covering this wavelength range from IUE and HST, the S II triplet at 1256 Å has never been detected.

9. OUTLOOK

The study of the gaseous content of cometary comae has seen much progress during the past two decades due to both enhancements in technology, enabling many more species to be observed, and to a better understanding of the physical processes producing the observed emissions. Spectroscopy continues to be a powerful tool that remains ahead of the laboratory data needed to identify the still large number of unexplained features seen in high-resolution spectra both in the visible (*Cochran and Cochran*, 2002) and the far-UV (*Weaver et al.*, 2002). The large database of observations, dating back more than a century, provides an important tool to assess the chemical and evolutionary diversity of comets.

Acknowledgments. P.D.F. wishes to thank the Institut d'Astrophysique de Paris for their hospitality while he held a Poste Rouge from the CNRS during the fall of 2002. This work was partially supported by NASA grants NAG5-9003 (A.L.C.), NAG5-8942 (M.R.C.), and NAG5-5315 (P.D.F.).

REFERENCES

A'Hearn M. F. (1982) Spectrophotometry of comets at optical wavelengths. In *Comets* (L. L. Wilkening, ed.), pp. 433–460. Univ. of Arizona, Tucson.

A'Hearn M. F. and Feldman P. D. (1980) Carbon in Comet Bradfield 1979l. *Astrophys. J. Lett., 242,* L187–L190.

A'Hearn M. F. and Schleicher D. G. (1988) Comet P/Encke's nongravitational force. *Astrophys. J. Lett., 331,* L47–L51.

A'Hearn M. F., Schleicher D. G., and West R. A. (1985) Emission by OD in comets. *Astrophys. J., 297,* 826–836.

A'Hearn M. F., Millis R. L., Schleicher D. G., Osip D. J., and Birch P. V. (1995) The ensemble properties of comets: Results from narrowband photometry of 85 comets, 1976–1992. *Icarus, 118,* 223–270.

A'Hearn M. F., Wellnitz D. D., Woodney L., Feldman P. D., Weaver H. A., Arpigny C., Meier R., Jackson W. M., and Kim S. J. (1999) S₂ in Comet Hyakutake. *Bull. Am. Astron. Soc., 31,* 1124.

Arpigny C. (1965) Spectra of comets and their interpretation. *Annu. Rev. Astron. Astrophys., 3,* 351–375.

Arpigny C. (1994) Physical chemistry of comets: Models, uncertainties, data needs. In *Molecules and Grains in Space* (I. Nenner, ed.), pp. 205–238. AIP Conference Proceedings 312.

Arpigny C., Magain P., Manfroid J., Dossin F., Danks A. C., and Lambert K. L. (1987) Resolution of the [O I] + NH₂ blend in Comet P/Halley. *Astron. Astrophys., 187,* 485–488.

Arpigny C., Rauer H., Manfroid J., Hutsemékers D., Jehin E., Crovisier J., and Jorda L. (1998) Production and kinematics of sodium atoms in the coma of comet Hale-Bopp. *Astron. Astrophys., 334,* L53–L56.

Arpigny C., Jehin E., Manfroid J., Hutsemékers D., Schulz R., Stüwe J. A., Zucconi J.-M., and Ilyin I. (2003) Anomalous nitrogen isotope ratio in comets. *Science, 301,* 1522–1524.

Azoulay G. and Festou M. C. (1986) The abundance of sulphur in comets. In *Asteroids, Comets, Meteors II,* pp. 273–277.

Bappu M. K. V. and Sivaraman K. R. (1969) Some characteristics of the solar wind inferred from the study of sodium emis-

sion from cometary nuclei. *Sol. Phys., 10,* 496–501.

Bar-Nun A. and Prialnik D. (1988) The possible formation of a hydrogen coma around comets at large heliocentric distances. *Astrophys. J. Lett., 324,* L31–L34.

Bertaux J. L. (1986) The UV bright spot of water vapor in comets. *Astron. Astrophys., 160,* L7–L10.

Bertaux J. L., Blamont J. E., and Festou M. (1973) Interpretation of hydrogen Lyman-alpha observations of Comets Bennett and Encke. *Astron. Astrophys., 25,* 415–430.

Bird M. K., Huchtmeier W. K., Gensheimer P., Wilson T. L., Janardhan P., and Lemme C. (1997) Radio detection of ammonia in comet Hale-Bopp. *Astron. Astrophys., 325,* L5–L8.

Biver N., Bockelée-Morvan D., Colom P., Crovisier J., Germain B., Lellouch E., Davies J. K., Dent W. R. F., Moreno R., Paubert G., Wink J., Despois D., Lis D. C., Mehringer D., Benford D., Gardner M., Phillips T. G., Gunnarsson M., Rickman H., Winnberg A., Bergman P., Johansson L. E. B., and Rauer H. (1999) Long-term evolution of the outgassing of Comet Hale-Bopp from radio observations. *Earth Moon Planets, 78,* 5–11.

Bockelée-Morvan D. and Crovisier J. (1985) Possible parents for the cometary CN radical — Photochemistry and excitation conditions. *Astron. Astrophys., 151,* 90–100.

Bockelée-Morvan D. and Gérard E. (1984) Radio observations of the hydroxyl radical in comets with high spectral resolution — Kinematics and asymmetries of the OH coma in C/Meier (1978 XXI), C/Bradfield (1979X), and C/Austin (1982g). *Astron. Astrophys., 131,* 111–122.

Bockelée-Morvan D., Crovisier J., and Gérard E. (1990) Retrieving the coma gas expansion velocity in P/Halley, Wilson (1987 VII) and several other comets from the 18-cm OH line shapes. *Astron. Astrophys., 238,* 382–400.

Bockelée-Morvan, D., Lis D. C., Wink J. E., Despois D., Crovisier J., Bachiller R., Benford D. J., Biver N., Colom P., Davies J. K., Gérard E., Germain B., Houde M., Mehringer D., Moreno R., Paubert G., Phillips T. G., and Rauer H. (2000) New molecules found in comet C/1995 O1 (Hale-Bopp). Investigating the link between cometary and interstellar material. *Astron. Astrophys., 353,* 1101–1114.

Bockelée-Morvan D., Crovisier J., Mumma M. J., and Weaver H. A. (2004) The composition of cometary volatiles. In *Comets II* (M. C. Festou et al., eds.), this volume. Univ. of Arizona, Tucson.

Bonev B. and Komitov B. (2000) New two-variable fits for the scale lengths of CN and its parent molecule in cometary atmospheres: Application to the identification of the CN parent. *Bull. Am. Astron. Soc., 32,* 1072.

Bonev T. and Jockers K. (1994) H₂O⁺ ions in the inner plasma tail of Comet Austin 1990 V. *Icarus, 107,* 335–357.

Bowen I. S. (1947) Excitation by line coincidence. *Publ. Astron. Soc. Pac., 59,* 196.

Brooke T. Y., Tokunaga A. T., Weaver H. A., Crovisier J., Bockelée-Morvan D., and Crisp D. (1996) Detection of acetylene in the infrared spectrum of Comet Hyakutake. *Nature, 383,* 606–608.

Brown M. E. and Spinrad H. (1993) The velocity distribution of cometary hydrogen — Evidence for high velocities? *Icarus, 104,* 197–205.

Brown M. E., Bouchez A. H., Spinrad H., and Misch A. (1998) Sodium velocities and sources in Hale-Bopp. *Icarus, 134,* 228–234.

Brown R. A. and Chaffee F. H. (1974) High-resolution spectra of sodium emission from Io. *Astrophys. J. Lett., 187,* L125–L126.

Budzien S. A. and Feldman P. D. (1991) OH prompt emission in Comet IRAS-Araki-Alcock (1983 VII). *Icarus, 90,* 308–318.

Budzien S. A., Festou M. C., and Feldman P. D. (1994) Solar flux variability and the lifetimes of cometary H_2O and OH. *Icarus, 107,* 164–188.

Bus S. J., A'Hearn M. F., Schleicher D. G., and Bowell E. (1991) Detection of CN emission from (2060) Chiron. *Science, 251,* 774–777.

Chamberlain J. W. and Hunten D. M. (1987) *Theory of Planetary Atmospheres: An Introduction to Their Physics and Chemistry, 2nd edition.* Academic, Orlando.

Cochran A. L. (2002) A search for N_2^+ in spectra of Comet C/2002 C1 (Ikeya-Zhang). *Astrophys. J. Lett., 576,* L165–L168.

Cochran A. L. and Cochran W. D. (1991) The first detection of CN and the distribution of CO^+ gas in the coma of Comet P/Schwassmann-Wachmann 1. *Icarus, 90,* 172–175.

Cochran A. L. and Cochran W. D. (2001) Observations of O (^1S) and O (^1D) in spectra of C/1999 S4 (LINEAR). *Icarus, 154,* 381–390.

Cochran A. L. and Cochran W. D. (2002) A high spectral resolution atlas of Comet 122P/de Vico. *Icarus, 157,* 297–308.

Cochran A. L., Cochran W. D., and Barker E. S. (2000) N_2^+ and CO^+ in Comets 122P/1995 S1 (deVico) and C/1995 O1 (Hale-Bopp). *Icarus, 146,* 583–593.

Cochran W. D. (1984) Detection of forbidden O I ^1S–^1D in comet IRAS-Araki-Alcock. *Icarus, 58,* 440–445.

Code A. D., Houck T. E., and Lillie C. F. (1972) Ultraviolet observations of comets. In *Scientific Results from the Orbiting Astronomical Observatory (OAO-2)* (A. D. Code, ed.), pp. 109–114. NASA SP-310.

Combi M. R. and Feldman P. D. (1992) IUE observations of H Lyman-alpha in Comet P/Giacobini-Zinner. *Icarus, 97,* 260–268.

Combi M. R. and Feldman P. D. (1993) Water production rates in Comet P/Halley from IUE observations of H I Lyman-alpha. *Icarus, 105,* 557–567.

Combi M. R. and Fink U. (1997) A critical study of molecular photodissociation and CHON grain sources for cometary C_2. *Astrophys. J., 484,* 879–890.

Combi M. R. and McCrosky R. E. (1991) High-resolution spectra of the 6300 Å region of Comet P/Halley. *Icarus, 91,* 270–279.

Combi M. R., DiSanti M. A., and Fink U. (1997) The spatial distribution of gaseous atomic sodium in the comae of comets: Evidence for direct nucleus and extended plasma sources. *Icarus, 130,* 336–354.

Combi M. R., Brown M. E., Feldman P. D., Keller H. U., Meier R. R., and Smyth W. H. (1998) Hubble Space Telescope ultraviolet imaging and high-resolution spectroscopy of water photodissociation products in Comet Hyakutake (C/1996 B2). *Astrophys. J., 494,* 816–821.

Combi M. R., Cochran A. L., Cochran W. D., Lambert D. L., and Johns-Krull C. M. (1999) Observation and analysis of high-resolution optical line profiles in Comet Hyakutake (C/1996 B2). *Astrophys. J., 512,* 961–968.

Combi M. R., Harris W. R., and Smyth W. M. (2004) Gas dynamics and kinetics in the cometary coma: Theory and observations. In *Comets II* (M. C. Festou et al., eds.), this volume. Univ. of Arizona, Tucson.

Cravens T. E. and Green A. E. S. (1978) Airglow from the inner comas of comets. *Icarus, 33,* 612–623.

Cravens T. E., Kozyra J. U., Nagy A. F., Gombosi T. I., and Kurtz

M. (1987) Electron impact ionization in the vicinity of comets. *J. Geophys. Res., 92,* 7341–7353.

Cremonese G., Boehnhardt H., Crovisier J., Rauer H., Fitzsimmons A., Fulle M., Licandro J., Pollacco D., Tozzi G. P., and West R. M. (1997) Neutral sodium from Comet Hale-Bopp: A third type of tail. *Astrophys. J. Lett., 490,* L199–L202.

Crovisier J., Colom P., Gérard E., Bockelée-Morvan D., and Bourgois G. (2002) Observations at Nançay of the OH 18-cm lines in comets. The data base. Observations made from 1982 to 1999. *Astron. Astrophys., 393,* 1053–1064.

Delsemme A. H. and Combi M. R. (1983) Neutral cometary atmospheres. IV — Brightness profiles in the inner coma of comet Kohoutek 1973 XII. *Astrophys. J., 271,* 388–397.

Despois D., Gérard E., Crovisier J., and Kazes I. (1981) The OH radical in comets — Observation and analysis of the hyperfine microwave transitions at 1667 MHz and 1665 MHz. *Astron. Astrophys., 99,* 320–340.

Douglas A. E. (1951) Laboratory studies of the λ 4050 group of cometary spectra. *Astrophys. J., 114,* 466–468.

Durrance S. T. (1981) The carbon monoxide fourth positive bands in the Venus dayglow. I — Synthetic spectra. *J. Geophys. Res., 86,* 9115–9124.

Eberhardt P. and Krankowsky D. (1995) The electron temperature in the inner coma of comet P/Halley. *Astron. Astrophys., 295,* 795–806.

Feldman P. D. (1982) Ultraviolet spectroscopy of comae. In *Comets* (L. L. Wilkening, ed.), pp. 461–479. Univ. of Arizona, Tucson.

Feldman P. D. (1996) Comets. In *Atomic, Molecular, and Optical Physics Reference Book* (G. W. F. Drake, ed.), pp. 930–939. American Institute of Physics, New York.

Feldman P. D. and Brune W. H. (1976) Carbon production in comet West 1975n. *Astrophys. J. Lett., 209,* L45–L48.

Feldman P. D., Takacs P. Z., Fastie W. G., and Donn B. (1974) Rocket ultraviolet spectrophotometry of Comet Kohoutek (1973f). *Science, 185,* 705–707.

Feldman P. D., Opal C. B., Meier R. R., and Nicolas K. R. (1976) Far ultraviolet excitation processes in comets. In *The Study of Comets,* pp. 773–796. IAU Colloquium No. 25.

Feldman P. D., Weaver H. A., A'Hearn M. F., Festou M. C., and McFadden L. A. (1986) Is CO_2 responsible for the outbursts of comet Halley? *Nature, 324,* 433–436.

Feldman P. D., Fournier K. B., Grinin V. P., and Zvereva A. M. (1993) The abundance of ammonia in Comet P/Halley derived from ultraviolet spectrophotometry of NH by ASTRON and IUE. *Astrophys. J., 404,* 348–355.

Feldman P. D., Festou M. C., Tozzi G. P., and Weaver H. A. (1997) The CO_2/CO abundance ratio in 1P/Halley and several other comets observed by IUE and HST. *Astrophys. J., 475,* 829–834.

Feldman P. D., Weaver H. A., A'Hearn M. F., Festou M. C., McPhate J. B., and Tozzi G.-P. (1999) Ultraviolet imaging spectroscopy of Comet Lee (C/1999 H1) with HST/STIS. *Bull. Am. Astron. Soc., 31,* 1127.

Feldman P. D., Weaver H. A., and Burgh E. B. (2002) Far Ultraviolet Spectroscopic Explorer observations of CO and H_2 emission in Comet C/2001 A2 (LINEAR). *Astrophys. J. Lett., 576,* L91–L94.

Festou M. C. (1984) Aeronomical processes in cometary atmospheres — The carbon compounds' puzzle. *Adv. Space Res., 4,* 165–175.

Festou M. and Feldman P. D. (1981) The forbidden oxygen lines in comets. *Astron. Astrophys., 103,* 154–159.

Festou M. C. and Feldman P. D. (1987) Comets. In *Exploring the Universe with the IUE Satellite* (Y. Kondo et al., eds.), pp. 101–118. ASSL Vol. 129, Kluwer, Dordrecht.

Festou M., Jenkins E. B., Barker E. S., Upson W. L., Drake J. F., Keller H. U., and Bertaux J. L. (1979) Lyman-alpha observations of comet Kobayashi-Berger-Milon (1975 IX) with Copernicus. *Astrophys. J., 232,* 318–328.

Festou M. C., Feldman P. D., and Weaver H. A. (1982) The ultraviolet bands of the CO_2^+ ion in comets. *Astrophys. J., 256,* 331–338.

Festou M. C., Barale O., Davidge T., Stern S. A., Tozzi G. P., Womack M., and Zucconi J. M. (1998) Tentative identification of the parent of CN radicals in comets: = C_2N_2. *Bull. Am. Astron. Soc., 30,* 1089.

Fink U. and Hicks M. D. (1996) A survey of 39 comets using CCD spectroscopy. *Astrophys. J., 459,* 729–743.

Fink U., Combi M. R., and DiSanti M. A. (1991) Comet P/Halley — Spatial distributions and scale lengths for C_2, CN, NH_2, and H_2O. *Astrophys. J., 383,* 356–371.

Fox J. L. and Dalgarno A. (1979) Ionization, luminosity, and heating of the upper atmosphere of Mars. *J. Geophys. Res., 84,* 7315–7333.

Gérard E. (1990) The discrepancy between OH production rates deduced from radio and ultraviolet observations of comets. I — A comparative study of OH radio and UV observations of P/Halley 1986 III in late November and early December 1985. *Astron. Astrophys., 230,* 489–503.

Greenstein J. L. (1958) High-resolution spectra of Comet Mrkos (1957d). *Astrophys. J., 128,* 106–106.

Haser L. and Swings P. (1957) Sur la possibilité d'une fluorescence cométaire excitée par la raie d'émission Lyman α solaire. *Annal. Astrophys., 20,* 52.

Herzberg G. and Lew H. (1974) Tentative identification of the H_2O^+ ion in Comet Kohoutek. *Astron. Astrophys., 31,* 123–124.

Hofmann H., Saha H. P., and Trefftz E. (1983) Excitation of C II lines by photoionization of neutral carbon. *Astron. Astrophys., 126,* 415–426.

Huebner W. F. (1970) Dust from cometary nuclei. *Astron. Astrophys., 5,* 286–297.

Huebner W. F., Keady J. J., and Lyon S. P., eds. (1992) *Solar Photo Rates for Planetary Atmospheres and Atmospheric Pollutants.* Kluwer, Dordrecht. 292 pp.

Huppler D., Reynolds R. J., Roesler F. L., Scherb F., and Trauger J. (1975) Observations of comet Kohoutek (1973f) with a ground-based Fabry-Perot spectrometer. *Astrophys. J., 202,* 276–282.

Iro N., Gautier D., Hersant F., Bockelée-Morvan D., and Lunine J. I. (2003) An interpretation of the nitrogen deficiency in comets. *Icarus, 161,* 511–532.

Irvine W. M., Senay M., Lovell A. J., Matthews H. E., McGonagle D., and Meier R. (2000) Detection of nitrogen sulfide in Comet Hale-Bopp. *Icarus, 143,* 412–414.

Jackson W. M. (1976) The photochemical formation of cometary radicals. *J. Photochem., 5,* 107–118.

Jackson W. M. and Donn B. (1968) Photochemical effects in the production of cometary radicals and ions. *Icarus, 8,* 270–280.

Jackson W. M., Halpern J. B., Feldman P. D., and Rahe J. (1982) Production of CS and S in Comet Bradfield (1979 X). *Astron. Astrophys., 107,* 385–389.

Jackson W. M., Butterworth P. S., and Ballard D. (1986) The origin of CS in comet IRAS-Araki-Alcock 1983d. *Astrophys. J., 304,* 515–518.

Jessberger E. K. and Kissel J. (1991) Chemical properties of cometary dust and a note on carbon isotopes. In *Comets in the Post-Halley Era* (R. Newburn et al., eds.), pp. 1075–1092. IAU Colloquium No. 116, ASSL Vol. 167, Kluwer, Dordrecht.

Kawakita H., Watanabe J., Ando H., Aoki W., Fuse T., Honda S., Izumiura H., Kajino T., Kambe E., Kawanomoto S., Noguchi K., Okita K., Sadakane K., Sato B., Takada-Hidai M., Takeda Y., Usuda T., Watanabe E., and Yoshida M. (2001a) The spin temperature of NH_3 in Comet C/1999S4 (LINEAR). *Science, 294,* 1089–1091.

Kawakita H., Watanabe J., Kinoshita D., Abe S., Furusyo R., Izumiura H., Yanagisawa K., and Masuda S. (2001b) High-dispersion spectra of NH_2 in the Comet C/1999 S4 (LINEAR): Excitation mechanism of the NH_2 molecule. *Publ. Astron. Soc. Japan, 53,* L5–L8.

Keller H. U. (1976) The interpretations of ultraviolet observations of comets. *Space Sci. Rev., 18,* 641–684.

Kim S. J. and A'Hearn M. F. (1991) Upper limits of SO and SO_2 in comets. *Icarus, 90,* 79–95.

Kim S. J. and A'Hearn M. F. (1992) G-factors of the SH (0-0) band and SH upper limit in Comet P/Brorsen-Metcalf (1989o). *Icarus, 97,* 303–306.

Kim S. J., A'Hearn M. F., and Cochran W. D. (1989) NH emissions in comets — Fluorescence vs. collisions. *Icarus, 77,* 98–108.

Kleine M., Wyckoff S., Wehinger P. A., and Peterson B. A. (1994) The cometary flourescence spectrum of cyanogen: A model. *Astrophys. J., 436,* 885–906.

Kleine M., Wyckoff S., Wehinger P. A., and Peterson B. A. (1995) The carbon isotope abundance ratio in comet Halley. *Astrophys. J., 439,* 1021–1033.

Korsun P. P. and Jockers K. (2002) CN, NH_2, and dust in the atmosphere of comet C/1999 J3 (LINEAR). *Astron. Astrophys., 381,* 703–708.

Krasnopolsky V. A. and Tkachuk A. Y. (1991) TKS-Vega experiment — NH and NH_2 bands in Comet Halley. *Astrophys. J., 101,* 1915–1919.

Krishna Swamy K. S. (1997) On the rotational population distribution of C_2 in comets. *Astrophys. J., 481,* 1004–1006.

Krishna Swamy K. S. and Tarafdar S. P. (1993) Study of the A-X (0,0) band profile of CS in comets. *Astron. Astrophys., 271,* 326–334.

Lambert D. L. and Danks A. C. (1983) High-resolution spectra of C_2 Swan bands from comet West 1976 VI. *Astrophys. J., 268,* 428–446.

Lambert D. L., Sheffer Y., Danks A. C., Arpigny C., and Magain P. (1990) High-resolution spectroscopy of the C_2 Swan 0-0 band from Comet P/Halley. *Astrophys. J., 353,* 640–653.

Leach S. (1987) Electronic spectroscopy and relaxation of some molecular cations of cometary interest. *Astron. Astrophys., 187,* 195–200.

Lean J. (1991) Variations in the sun's radiative output. *Rev. Geophys., 29,* 505–535.

Levin B. J. (1964) On the reported Na tails of comets. *Icarus, 3,* 497–498.

Lew H. (1976) Electronic spectrum of H_2O^+. *Can. J. Phys., 54,* 2028–2049.

Lis D. C., Mehringer D. M., Benford D., Gardner M., Phillips T. G., Bockelée-Morvan D., Biver N., Colom P., Crovisier J., Despois D., and Rauer H. (1999) New molecular species in Comet C/1995 O1 (Hale-Bopp) observed with the Caltech Submillimeter Observatory. *Earth Moon Planets, 78,* 13–20.

Lutz B. L., Womack M., and Wagner R. M. (1993) Ion abundances and implications for photochemistry in Comets Halley (1986 III) and Bradfield (1987 XXIX). *Astrophys. J., 407,* 402–411.

Magee-Sauer K., Roesler F. L., Scherb F., Harlander J., and Oliversen R. J. (1988) Spatial distribution of O(^{1}D) from Comet Halley. *Icarus, 76,* 89–99.

Magee-Sauer K., Dello Russo N., DiSanti M. A., Gibb E., and Mumma M. J. (2002) CSHELL observations of Comet C/2002 C1 (Ikeya-Zhang) in the 3.0-micron region. *Bull. Am. Astron. Soc., 34,* 868.

Magnani L. and A'Hearn M. F. (1986) CO$^+$ fluorescence in comets. *Astrophys. J., 302,* 477–487.

Marconi M. L., Mendis D. A., Korth A., Lin R. P., Mitchell D. L., and Reme H. (1990) The identification of H$_3$S$^+$ with the ion of mass per charge (m/q) 35 observed in the coma of Comet Halley. *Astrophys. J. Lett., 352,* L17–L20.

McPhate J. B., Feldman P. D., McCandliss S. R., and Burgh E. B. (1999) Rocket-borne long-slit ultraviolet spectroscopy of Comet Hale-Bopp. *Astrophys. J., 521,* 920–927.

Meier R. and A'Hearn M. F. (1997) Atomic sulfur in cometary comae based on UV spectra of the S I triplet near 1814 Å. *Icarus, 125,* 164–194.

Morrison N. D., Knauth D. C., Mulliss C. L., and Lee W. (1997) High-resolution optical spectra of the head of the Comet C/1996 B2 (Hyakutake). *Publ. Astron. Soc. Pac., 109,* 676–681.

Mumma M. J., Stone E. J., and Zipf E. C. (1975) Nonthermal rotational distribution of CO (A^{1}Π) fragments produced by dissociative excitation of CO$_2$ by electron impact. *J. Geophys. Res., 80,* 161–167.

Mumma M. J., McLean I. S., DiSanti M. A., Larkin J. E., Russo N. D., Magee-Sauer K., Becklin E. E., Bida T., Chaffee F., Conrad A. R., Figer D. F., Gilbert A. M., Graham J. R., Levenson N. A., Novak R. E., Reuter D. C., Teplitz H. I., Wilcox M. K., and Xu L. (2001) A survey of organic volatile species in Comet C/1999 H1 (Lee) using NIRSPEC at the Keck Observatory. *Astrophys. J., 546,* 1183–1193.

Newall H. F. (1910) On the spectrum of the daylight comet 1910a. *Mon. Not. R. Astron. Soc., 70,* 459–461.

O'Dell C. R., Robinson R. R., Krishna Swamy K. S., McCarthy P. J., and Spinrad H. (1988) C$_2$ in Comet Halley — Evidence for its being third generation and resolution of the vibrational population discrepancy. *Astrophys. J., 334,* 476–488.

Oliversen R. J., Doane N., Scherb F., Harris W. M., and Morgenthaler J. P. (2002) Measurements of [C I] emission from Comet Hale-Bopp. *Astrophys. J., 581,* 770–775.

Opal C. B., Carruthers G. R., Prinz D. K., and Meier R. R. (1974) Comet Kohoutek: Ultraviolet images and spectrograms. *Science, 185,* 702–705.

Oppenheimer M. (1980) Sodium D-line emission in Comet West (1975n) and the sodium source in comets. *Astrophys. J., 240,* 923–928.

Palmer P., Wootten A., Butler B., Bockelée-Morvan D., Crovisier J., Despois D., and Yeomans D. K. (1996) Comet Hyakutake: First secure detection of ammonia in a comet. *Bull. Am. Astron. Soc., 28,* 927.

Pirronello V., Strazzulla G., and Foti G. (1983) H$_2$ production in comets. *Astron. Astrophys., 118,* 341–344.

Potter A. and Morgan T. (1985) Discovery of sodium in the atmosphere of Mercury. *Science, 229,* 651–653.

Potter A. E. and Morgan T. H. (1988) Discovery of sodium and potassium vapor in the atmosphere of the moon. *Science, 241,* 675–680.

Preston G. W. (1967) The spectrum of Ikeya-Seki (1965f). *Astrophys. J., 147,* 718–742.

Prisant M. G. and Jackson W. M. (1987) A rotational-state population analysis of the high-resolution IUE observation of CS emission in comet P/Halley. *Astron. Astrophys., 187,* 489–496.

Randall C. E., Schleicher D. G., Ballou R. G., and Osip D. J. (1992) Observational constraints on molecular scalelengths and lifetimes in comets. *Bull. Am. Astron. Soc., 24,* 1002.

Rauer H., Arpigny C., Manfroid J., Cremonese G., and Lemme C. (1998) The spatial sodium distribution in the coma of comet Hale-Bopp (C/1995 O1). *Astron. Astrophys., 334,* L61–L64.

Rauer H., Helbert J., Arpigny C., Benkhoff J., Bockelée-Morvan D., Boehnhardt H., Colas F., Crovisier J., Hainaut O., Jorda L., Kueppers M., Manfroid J., and Thomas N. (2003) Long-term optical spectrophotometric monitoring of comet Hale-Bopp (C/1995 O1). *Astron. Astrophys., 397,* 1109–1122.

Richter K., Combi M. R., Keller H. U., and Meier R. R. (2000) Multiple scattering of hydrogen Lyα radiation in the coma of Comet Hyakutake (C/1996 B2). *Astrophys. J., 531,* 599–611.

Rousselot P., Arpigny C., Rauer H., Cochran A. L., Gredel R., Cochran W. D., Manfroid J., and Fitzsimmons A. (2001) A fluorescence model of the C$_3$ radical in comets. *Astron. Astrophys., 368,* 689–699.

Scherb F. (1981) Hydrogen production rates from ground-based Fabry-Perot observations of comet Kohoutek. *Astrophys. J., 243,* 644–650.

Schleicher D. G. and A'Hearn M. F. (1988) The fluorescence of cometary OH. *Astrophys. J., 331,* 1058–1077.

Schleicher D. G. and Farnham T. L. (2004) Photometry and imaging of the coma with narrowband filters. In *Comets II* (M. C. Festou et al., eds.), this volume. Univ. of Arizona, Tucson.

Schleicher D. G. and Millis R. L. (1989) Revised scale lengths for cometary NH. *Astrophys. J., 339,* 1107–1114.

Singh P. D. and Dalgarno A. (1987) Photodissociation lifetimes of CH and CD radicals in comets. In *Diversity and Similarity of Comets* (E. J. Rolfe and B. Battrick, eds.), pp. 177–179. ESA SP-278, Noordwijk, The Netherlands.

Slaughter C. D. (1969) The emission spectrum of Comet Ikeya-Seki 1965f at perihelion passage. *Astrophys. J., 74,* 929–943.

Smith A. M., Stecher T. P., and Casswell L. (1980) Production of carbon, sulfur, and CS in Comet West. *Astrophys. J., 242,* 402–410.

Smith W. H. and Schempp W. V. (1989) [O I] in Comet Halley. *Icarus, 82,* 61–66.

Sorkhabi O., Blunt V. M., Lin H., A'Hearn M. F., Weaver H. A., Arpigny C., and Jackson W. M. (1997) Using photochemistry to explain the formation and observation of C$_2$ in comets. *Planet. Space Sci., 45,* 721–730.

Spinrad H. (1982) Observations of the red auroral oxygen lines in nine comets. *Publ. Astron. Soc. Pac., 94,* 1008–1016.

Spinrad H. and Miner E. D. (1968) Sodium velocity fields in Comet 1965f. *Astrophys. J., 153,* 355–366.

Stern S. A., Slater D. C., Festou M. C., Parker J. W., Gladstone G. R., A'Hearn M. F., and Wilkinson E. (2000) The discovery of argon in Comet C/1995 O1 (Hale-Bopp). *Astrophys. J. Lett., 544,* L169–L172.

Storey P. J. and Zeippen C. J. (2000) Theoretical values for the [O III] 5007/4959 line-intensity ratio and homologous cases. *Mon. Not. R. Astron. Soc., 312,* 813–816.

Swings P. (1941) Complex structure of cometary bands tentatively ascribed to the contour of the solar spectrum. *Lick Observatory Bull., 508,* 131–136.

Swings P. (1965) Cometary spectra (George Darwin lecture). *Quart. J. R. Astron. Soc., 6,* 28–69.

Swings P. and Haser L. (1956) *Atlas of Representative Cometary Spectra.* Impr. Ceuterick, Louvain.

Swings P. and Page T. (1950) The spectrum of Comet Bester (1947k). *Astrophys. J., 111,* 530–554.

Swings P., Elvey C. T., and Babcock H. W. (1941) The spectrum of Comet Cunningham, 1940C. *Astrophys. J., 94,* 320–343.

Tacconi-Garman L. E., Schloerb F. P., and Claussen M. J. (1990) High spectral resolution observations and kinematic modeling of the 1667 MHz hyperfine transition of OH in Comets Halley (1982i), Giacobini-Zinner (1984e), Hartley-Good (1985l), Thiele (1985m), and Wilson (1986l). *Astrophys. J., 364,* 672–686.

Tatum J. B. (1984) Cyanogen radiance/column-density ratio for comets calculated from the Swings effect. *Astron. Astrophys., 135,* 183–187.

Tatum J. B. and Gillespie M. I. (1977) The cyanogen abundance of comets. *Astrophys. J., 218,* 569–572.

Tegler S. and Wyckoff S. (1989) NH_2 fluorescence efficiencies and the NH_3 abundance in Comet Halley. *Astrophys. J., 343,* 445–449.

Tokaryk D. W. and Chomiak D. E. (1997) Laser spectroscopy of C_3: Stimulated emission and absorption spectra of the $\tilde{A}^1\Pi_u$–$\tilde{X}^1\Sigma_g^+$ transition. *J. Chem. Phys., 106,* 7600–7608.

Tozzi G. P., Feldman P. D., and Festou M. C. (1998) Origin and production of $C(^1D)$ atoms in cometary comae. *Astron. Astrophys., 330,* 753–763.

Umbach R., Jockers K., and Geyer E. H. (1998) Spatial distribution of neutral and ionic constituents in comet P/Halley. *Astron. Astrophys., 127,* 479–495.

Weaver H. A. (1998) Comets. In *The Scientific Impact of the Goddard High Resolution Spectrograph* (J. C. Brandt et al., eds.), pp. 213–226. ASP Conference Series 143, Astronomical Society of the Pacific, San Francisco.

Weaver H. A., Feldman P. D., McPhate J. B., A'Hearn M. F., Arpigny C., and Smith T. E. (1994) Detection of CO Cameron band emission in comet P/Hartley 2 (1991 XV) with the Hubble Space Telescope. *Astrophys. J., 422,* 374–380.

Weaver H. A., Feldman P. D., A'Hearn M. F., Arpigny C., Brandt J. C., and Stern S. A. (1999) Post-perihelion HST observations of Comet Hale-Bopp (C/1995 O1). *Icarus, 141,* 1–12.

Weaver H. A., Feldman P. D., Combi M. R., Krasnopolsky V., Lisse C. M., and Shemansky D. E. (2002) A search for argon and O VI in three comets using the Far Ultraviolet Spectroscopic Explorer. *Astrophys. J. Lett., 576,* L95–L98.

Wegmann R., Jockers K., and Bonev T. (1999) H_2O^+ ions in comets: Models and observations. *Planet. Space Sci., 47,* 745–763.

Wilkening L. L., ed. (1982) *Comets.* Univ. of Arizona, Tucson. 766 pp.

Wilson J. K., Baumgardner J., and Mendillo M. (1998) Three tails of comet Hale-Bopp. *Geophys. Res. Lett., 25,* 225–228.

Wolven B. C. and Feldman P. D. (1998) Lyman-α induced fluorescence of H_2 and CO in comets and planetary atmospheres. In *The Scientific Impact of the Goddard High Resolution Spectrograph* (J. C. Brandt et al., eds.), pp. 373–377. ASP Conference Series 143, Astronomical Society of the Pacific, San Francisco.

Woodney L. M., A'Hearn M. F., Schleicher D. G., Farnham T. L., McMullin J. P., Wright M. C. H., Veal J. M., Snyder L. E., de Pater I., Forster J. R., Palmer P., Kuan Y.-J., Williams W. R., Cheung C. C., and Smith B. R. (2002) Morphology of HCN and CN in Comet Hale-Bopp (1995 O1). *Icarus, 157,* 193–204.

Woods T. N., Feldman P. D., and Dymond K. F. (1987) The atomic carbon distribution in the coma of Comet P/Halley. *Astron. Astrophys., 187,* 380.

Woods T. N., Feldman P. D., and Rottman G. J. (2000) Ultraviolet observations of Comet Hale-Bopp (C/1995 O1) by the UARS SOLSTICE. *Icarus, 144,* 182–186.

Wright M. C. H., de Pater I., Forster J. R., Palmer P., Snyder L. E., Veal J. M., A'Hearn M. F., Woodney L. M., Jackson W. M., Kuan Y.-J., and Lovell A. J. (1998) Mosaicked images and spectra of J = 1 → 0 HCN and HCO^+ emission from Comet Hale-Bopp (1995 O1). *Astrophys. J., 116,* 3018–3028.

Ziurys L. M., Savage C., Brewster M. A., Apponi A. J., Pesch T. C., and Wyckoff S. (1999) Cyanide chemistry in Comet Hale-Bopp (C/1995 O1). *Astrophys. J. Lett., 527,* L67–L71.

Photometry and Imaging of the Coma
with Narrowband Filters

David G. Schleicher
Lowell Observatory

Tony L. Farnham
University of Maryland

The use of narrowband filters to isolate light reflected by cometary grains and emitted by several gas species permits a wide variety of compositional and morphological studies to be performed. A brief survey of some of these studies is presented, along with detailed discussions of the techniques, procedures, and methodologies used. In particular, the advantages and disadvantages of both traditional photoelectric photometers and CCD cameras is explored, and an update is given regarding the new narrowband comet filter sets produced in recent years. Some of the unique aspects of narrowband filter reductions are characterized, as are the steps required in compositional studies. Finally, the most useful aspects of enhancing, measuring, and analyzing morphological features are investigated in detail.

1. INTRODUCTION AND BACKGROUND

In this chapter, we provide both a brief review and tutorial of the fields of narrowband photometry and narrowband imaging of comets. The use of narrowband filters to isolate the light emitted by various molecular species and reflected solar radiation by dust grains in cometary comae has a long and productive history, dating back nearly half a century (cf. *Schmidt and van Woerden, 1957*). While photoelectric photometers have been used throughout this interval, digital array detectors such as charge-coupled devices (CCDs) have now largely replaced photometers as the detectors of choice (cf. *Jewitt, 1991*). In spite of the overwhelming advantages CCDs provide in morphological studies, photometers continue to play an important role, particularly in chemical abundance studies.

Included here are discussions of the techniques, procedures, and methodologies used, and a survey of some of the physical and chemical properties that can be determined with these techniques, along with references to numerous examples. As such, in many respects this chapter is an update to the valuable review by *A'Hearn* (1983), where the issues of observational and reduction techniques were first summarized. We also include several topics in common with the more recent review by *Jewitt* (1991), in which his focus was on the types of studies obtainable with CCDs, but to minimize overlap, our emphasis is on the general topic of coma morphologies. As we have neither the space nor the desire to repeat details provided in previous summaries, we also urge the reader to examine several other excellent reviews in addition to those by A'Hearn and by Jewitt. In particular, a discussion of observations obtained in the early

decades up to and including C/Kohoutek (1973 XII) are summarized by *Vanysek* (1976), while *Meisel and Morris* (1982) briefly review the topics of bulk brightness variations, narrowband and IR photometry, and early compositional studies. Photometry in the IR is also the focus of reviews by *Ney* (1982) and *Hanner and Tokunaga* (1991). Polarization studies are discussed by *Kolokolova et al.* (1997), *Levasseur-Regourd* (1999), and *Kolokolova et al.* (2004).

Some of the physical properties that can be determined for the coma include spatial profiles of individual gas species and of the dust, the presence or lack of jets, sporadic brightness variations or unusual coma morphology indicative of outbursts, periodic brightness variations or jet motions caused by nucleus rotation, and the color and polarization of dust grains. Analyses of many of these characteristics of the coma can yield strong constraints on nucleus properties, such as rotation period, pole orientation, and the number, location, and size of individual source regions on the surface of the nucleus. For some comets, in which the signal from the nucleus is not overwhelmed by that from the coma, one can also obtain direct measurements of the nucleus. Chemical composition studies that can be performed include the determination of relative abundances of different molecular species, and how these vary with heliocentric distance and/or orbital position and from comet to comet, and the absolute production rates of water and dust. With the application of an appropriate vaporization model, physical properties such as effective active areas and lower limits on the nucleus size can also be computed. Each of these topics is discussed in more detail either here or in other chapters (e.g., *Samarasinha et al.*, 2004; *Bockelée-Morvan et al.*, 2004; *Feldman et al.*, 2004; *Combi et al.*, 2004; *Fulle*, 2004).

2. INSTRUMENTATION

Although many early photometric studies, as well as many more recent imaging studies, have used wideband filters or even no filtration, in most cases the observer is inevitably left with an ensemble of reflected light from grains and emitted light from multiple gas species that cannot readily be disentangled. Exceptions to this generalization include nucleus studies at large heliocentric distances, where the coma is either nonexistent or sufficiently faint that the nucleus' signal can be extracted, and dust studies when an object is known to be gas-poor, or in the near-IR where gas emission is only a minor contaminant. To isolate individual emission bands or to obtain continuum measurements in the near-UV to near-IR region of the spectrum, one must either use narrowband filters or spectroscopic techniques, each of which has numerous strengths and weaknesses. Spectroscopic methods (cf. *Feldman et al.*, 2004, and references therein) have the advantage of permitting the observer to directly detect and measure the shape of spectral features, simplifying the task of separating emission lines and bands from the continuum. This is particularly important in the case of weak emission features, such as CH, NH_2, or [O I], where the contrast with respect to the local continuum is very low. However, even with a long-slit instrument or multiple apertures, only a very small fraction of the total coma is sampled at one time, and the signal-to-noise ratio (S/N) per spatial and per spectral resolution element drops rapidly as one samples farther from the nucleus and inner-coma. If sufficient time is available, mapping the coma can greatly improve the spatial coverage. In comparison, conventional photometry and imaging can sample a much greater portion of the coma at one time, but only for the stronger emission bands that can be reliably isolated with narrowband filters. And while both spectroscopy and imaging techniques permit investigations of gross asymmetries in the coma, such as sunward-tailward, only imaging readily permits more detailed morphological studies in the visible regime, such as those desired when studying dust and gas jets. However, the steady increase in the size of optical fiber bundles for two-dimensional spectroscopy implies that IFU spectroscopy may permit useful morphological studies in the future.

With the advent of the twenty-first century and improvements in digital detectors, one might expect that narrowband imaging would have completely superseded the technique of aperture photometry using conventional photoelectric photometers. While in principle this seems reasonable, in practice several issues have necessitated the continued use of conventional photometers for many types of compositional studies. The primary limitation of CCD detectors is the inherent level of noise at the per-pixel level due to readout noise and slight variations in bias level. While these sources of uncertainty are usually quite small (<1 count), they can still dominate over the cometary signal in many instances. As an example, it is quite common for the measured count level for OH or NH emission in a moderately bright comet (10th–12th magnitude) to be on the order of 100 counts per second within a relatively large photometric aperture of 1 arcmin. With a conventional photometer, this results in a photon statistical uncertainty of about 1% with less than 2 min of integration. With a CCD, however, the same $\sim 10^4$ photons are spread over $\sim 10^4$ pixels. Given the typical brightness fall-off away from the nucleus, a pixel 30 arcsec from the nucleus would, on average, only receive <0.2 photons during the equivalent exposure time — a value similar to or less than the inherent noise level associated with the read-out of each pixel. At such low signal levels, the absolute uncertainties associated with flatfielding also become quite important in determining the level of the background sky. It is hoped that the development of truly flat and readnoise-free CCDs will eventually mitigate these problems. Although aperture extractions from a CCD frame can be performed, obtaining images solely to extract aperture photometry of the coma largely defeats the advantages of a CCD, and the resulting photometric uncertainties are always worse than those associated with a simple photometer. Other practical concerns involve observing efficiencies, such as the effort required to obtain good twilight flatfield measurements for several narrowband filters, and the longer total time required to obtain sets of images of both the comet and sky in each filter, as compared to the time required with a photometer. As a result of all these issues, narrowband CCD observations have only rarely been calibrated and continuum subtracted to obtain gas column densities and abundances (cf. *Schulz et al.*, 1993).

For all these reasons, we have found that basic coma abundance measurements are much more readily obtained (and with much better S/N) using conventional photoelectric photometers. In our own work, we use a new, computer-controlled photometer, but with the same EMI 6256 S-11 phototube as used with our previous, manually operated photometer. This tube, with a quartz window, provides good throughput to wavelengths below the atmospheric cutoff in the UV and an extremely low dark current when thermoelectrically cooled, but has essentially no response in the red and near-IR. A variety of tubes, having a wide range of characteristics, remain available from several manufacturers. For details regarding construction and use of photoelectric photometers, we refer the reader to several books on this subject, particularly those by *Henden and Kaitchuck* (1982), *Sterken and Manfroid* (1992), and *Budding* (1993).

In contrast, CCD imaging is clearly the appropriate technique to employ if the primary goal is to study morphology or to extract the signal from the nucleus from that of the surrounding coma, rather than to obtain abundance measurements. In addition to advances in quantum efficiency, particularly in the UV, and readout noise suppression, perhaps the most important changes in CCD detectors in recent years for comet research have been the ever-increasing format sizes and the decreased overhead associated with readout times. Larger formats directly yield larger fractions of the coma being measured or may even extend to uncontaminated sky, while faster readout of the chip permits more filters to be used in a limited interval of time for both standard star measurements and twilight flats as well as for the comet itself. Numerous books detailing the physical characteristics of CCD chips and/or observing and reduction techniques are now available, including those by *Jacoby* (1990), *Howell* (1992, 2000), and *Philip et al.* (1995).

3. NARROWBAND FILTERS

For both historical and practical reasons, the wavelength range within which narrowband filters have usually been constructed for comet studies has been between about 3000 and 7000 Å. The lower end of the range is set by the atmospheric cutoff, while the upper end is defined by the locations of the strongest emission bands for the observable species. For instance, although the CN molecule produces several emission bands between 7000 Å and 1.5 μm, each of these bands in the CN red system is much weaker than the primary band of the violet system at 3875 Å. Emission studies of comets in the UV (spacebased) and IR have nearly always been conducted using spectroscopic detectors, because the permanently installed filters are seldom useful for cometary studies (in the UV) or gas emission features are relatively weak. However, continuum studies in the IR have often made use of standard broadband filters such as J, H, and K.

A total of five neutral gas species produce sufficiently strong emission bands between about 3000 and 7000 Å to be easily isolated with narrowband filters. In order of wavelength, these are OH, NH, CN, C_3, and C_2. Figure 1 shows a composite spectrum, identifying the major emission features. Note that none of these species are assumed to exist in these forms in the nucleus, but each is instead at least a daughter species, produced by the dissociation of one or more parent (or grandparent) species. Appropriate modeling is therefore required to ultimately derive the nuclear abundances of the parents. Emission features by other neutral species, notably CH, NH_2, and O, are too weak and/or the species are too short-lived to remove the underlying continuum sufficiently accurately to produce reliable results in most circumstances. Emissions by two ion species, CO^+ and H_2O^+, have also been successfully isolated with narrowband filters. Unfortunately, as a consequence of the long wings of the C_3 and C_2 bands, together with the profusion of weak emission bands from NH_2 and other minor species, very few locations between 3000 and 7000 Å are completely absent of emission. This makes it difficult to obtain clean continuum measurements, and the decontamination of continuum measurements by gas emission is a significant issue to which we will return.

Over the past half-century, numerous investigators have had individual filters manufactured to isolate one or more of the stronger emission bands, often with accompanying continuum filters (cf. *Schmidt and van Woerden,* 1957; *O'Dell and Osterbrock,* 1962; *Blamont and Festou,* 1974; *A'Hearn and Cowan,* 1975). Unfortunately, the lack of standardization made it difficult to sort out the many discrepancies among the results. An initial effort at standardization was made in the late 1970s, when 3 sets of up to 10 filters were produced for use primarily at Lowell and Perth Ob-

servatories as the initial phase of the Lowell comet photometry program (cf. *A'Hearn et al.,* 1979; *A'Hearn and Millis,* 1980). A related effort at producing standard filter sets by an IAU Commission 15 Working Group resulted in design recommendations for a nine-filter set. Manufactured for worldwide distribution in time for Comet 1P/Halley's 1985/1986 apparition, several dozen sets were produced for photoelectric photometers and CCD cameras under the auspices of the International Halley Watch, and are now known as the IHW filters (cf. *Osborn et al.,* 1990; *A'Hearn,* 1991, *Larson et al.,* 1991). Representative transmission curves for the IHW filters are shown in Fig. 1.

Since the design of the IHW filters, several dust-poor comets have been observed spectroscopically, and the resulting spectra revealed that wings of the C_2 and, especially, the C_3 bands extended considerably further blueward than previously assumed, with the result that the IHW continuum filters at 3650 and 4845 Å suffered from much larger contamination than originally believed. In fact, for comets with very low dust-to-gas ratios, such as 2P/Encke, the wing of C_3 completely dominates the measured flux in the 3650-Å filter. It also became evident that the red continuum filter, centered at 6840 Å, was contaminated by an emission band tentatively identified as NH_2. Worse, as early as 1990 it was determined that some of the filters in some sets, including CN, were physically degrading, resulting in a decrease in the band transmission and a redward shift of the bandpass (cf. *Schleicher et al.,* 1991). This degradation of interference filters is unfortunately common, especially for bandpasses at wavelengths <4200 Å, because of older manufacturing techniques. By 1996, many observers had reported problems with their IHW sets, as they prepared to observe Comets Hale-Bopp (1995 O1) and Hyakutake (1996 B2). Because of the overwhelming interest in observing Hale-Bopp, NASA agreed to support the production of new sets of narrowband filters in time for Hale-Bopp's perihelion passage. In taking on the task of designing and calibrating these new sets, we decided to take advantage of improved manufacturing techniques, resulting in bandpasses being "squarer," i.e., having flatter tops and shorter wings, and filters with greater longevity and almost no variation of the bandpass with temperature. At the same time, we altered the placement of each of the continuum bandpasses to minimize the contaminations that were present in the IHW filters, and added an additional continuum point in order to better measure variations in dust reflectivity as a function of wavelength. The filter locations for the emission features were similar to those in the IHW sets, but with slight adjustments to take advantage of the squarer bandpasses and to minimize changes in the fractional transmission caused by the Swings effect, whereby the shape of the emission feature varies with heliocentric velocity and/or distance. Accommodations were also made for the shorter f-ratio systems that are increasingly used with CCD systems. A total of 48 full or partial sets of these 11 new filters were produced and distributed, and these have been designated the HB filter sets, since Hale-Bopp provided the motivation for their construction and was the initial target. The HB bandpasses are

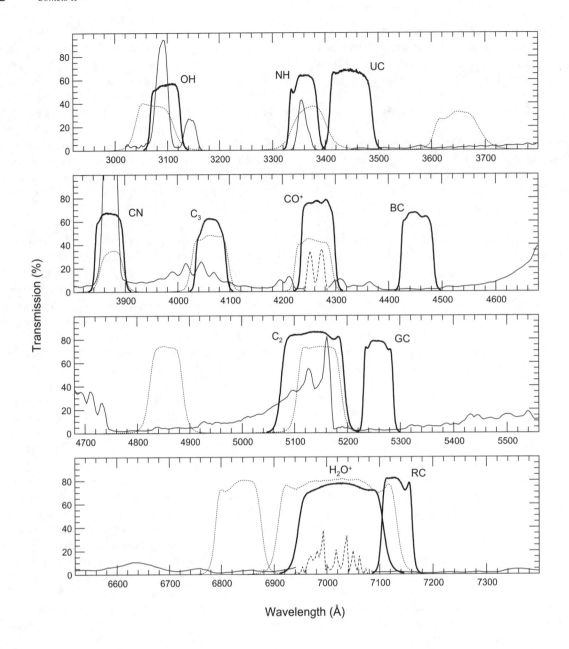

Fig. 1. Transmission profiles for the HB filters (thick lines) and IHW filters (dotted lines). For comparison, measured comet spectra illustrate the locations of the different emission bands. The neutral species and continuum regions are depicted by a spectrum of Comet 122P/deVico (spectral resolution = 12 Å) in the three top panels, and a spectrum of Comet 8P/Tuttle (resolution ~40 Å) in the bottom panel (thin solid lines). Because these comets do not exhibit clear ion bands, the 2–0 band of CO^+ from Comet 29P/Schwassmann-Wachmann 1 (resolution = 12 Å) has been inserted from 4240 to 4265 Å in the second panel and the 0–6–0 band of H_2O^+ from Comet Kohoutek 1973 E1 (resolution = 5 Å) has been inserted from 6940 to 7080 Å in the bottom panel (dashed lines). The 122P/deVico spectrum is courtesy of A. Cochran, and the 8P/Tuttle spectrum, created by S. Larson and J. Johnson, is courtesy of S. Larson. The CO^+ band was extracted from *Cochran and Cochran* (1991) and *Cochran et al.* (1991), and the H_2O^+ band was extracted from *Wehinger et al.* (1974) and *Wyckoff and Wehinger* (1976). From *Farnham et al.* (2000).

presented in Fig. 1 and itemized in Table 1; details of the individual filter design criteria and associated issues are available in *Farnham et al.* (2000).

Shortly prior to our efforts to design and produce the HB filter sets, ESA began a similar effort to produce new filters to replace the aging IHW sets. In ESA's case, the primary motivation was to observe the *Rosetta* spacecraft target, Comet 46P/Wirtanen, during its 1996/1997 and future ap-

paritions. Due to a variety of issues, including the timing of the availability of funding and requirements regarding the choice of manufacturers, the ESA and NASA efforts proceeded mostly independent of one another, with somewhat different design preferences and specifications. A total of 18 sets of these ESA filters for support of the *Rosetta* mission were produced and distributed. These bandpasses are also listed in Table 1. Because most of the HB and ESA

TABLE 1. Characteristics of new narrowband comet filters.

Species*	Bandpass† (Å)	
	HB Sets	ESA Sets
OH (0–0)	3090/62	3085/75
NH (0–0)	3362/58	—
UV Continuum	3448/84	—
CN ($\Delta v = 0$)	3870/62	3870/50
C_3 (Swings System)	4062/62	4060/70
CO^+ (2–0)	4266/64	—
Blue Continuum	4450/67	4430/40
C_2 ($\Delta v = 0$)	5141/118	5125/125
Green Continuum	5260/56	—
NH_2 (0,2,0)	—	6630/60
H_2O^+ (0,6,0)	7020/170	—
Red Continuum	7128/58	6840/90

*Emission band designations in parentheses.
†Nominal center wavelength and full-width half-maximum (FWHM).

filter sets were intended for use with CCD cameras, several differently sized filters were produced, ranging from 25 to 100 mm and 25 to 80 mm respectively.

A wide variety of issues must be addressed when calibrating a filter set, including the selection and calibration of standard stars, the determination of reduction coefficients for the calculation of absolute fluxes, and, specifically for comet filters, the determination of nonlinear extinction coefficients for the reduction of the OH band measurements, and coefficients for the decontamination of continuum filters and the removal of continuum from the emission bands. These are discussed in some detail by *A'Hearn* (1983) and references therein, and, for the IHW filter sets, in *Osborn et al.* (1990) and *A'Hearn* (1991). Because a complete discussion of these issues as applied to the new HB filter sets is contained in *Farnham et al.* (2000), we next provide only an abbreviated summary; we use the HB filters as an example, since the calibration of the ESA filter sets is currently in progress.

The usefulness of a new filter set is entirely dependent on the availability of suitable standard stars. In the case of comet observations, two types of standards are needed: flux standards, used to determine atmospheric extinction and to convert relative magnitudes to absolute fluxes, and solar analogs, used to mimic the solar spectrum in determining the spectral reflectivity of the dust or nucleus and for continuum subtraction from the emission bands. For these latter objectives, the key issue is selecting stars that best match the color of the Sun. Since no star has yet been identified as a true "twin" of the Sun, and there are differences among researchers as to which star most closely matches the Sun, we measured a dozen known solar analogs using the HB filters and discovered a surprisingly large dispersion in colors in the near-UV. Because several of these stars are considered close solar analogs, we removed three other stars whose colors were most discrepant. We also wanted to evenly bracket the Sun's physical properties, such as temperature, metallicity, and chromospheric activity, resulting in the removal of two additional stars that skewed the brackets. Ultimately, colors of seven solar analogs — HD 25680,

18 Sco, 16 Cyg A, vB 106, 16 Cyg B, vB 64, and HD 76151 (in order of their colors in the near-UV) — were averaged and adopted as representative solar colors, and these have been incorporated in the various reduction coefficients. Therefore, if an investigator uses the equations and coefficients listed in *Farnham et al.* (2000), solar analogs do not need to be included in the observing program.

Unlike solar analogs, flux standards must be observed nightly, to determine both the amount of atmospheric extinction and the instrumental corrections associated with each filter. For the HB filters, a total of 24 stars were selected having spectral types of late-O to late-B and V magnitudes ranging from 4th to 8th. Besides the obvious need to be nonvariables, relatively hot stars are preferred to minimize the number of spectral absorption features and to maximize the flux in the UV. The stars are nearly uniformly distributed near the celestial equator, insuring that some stars would match the airmass of any comet within 1–2 h of the comet observations. The brighter flux standards provide excellent S/N for photometric systems on smaller telescopes, while the fainter stars are suitable for many CCD systems by minimizing the need to defocus the image to prevent saturation.

4. DATA ACQUISITION AND REDUCTION

Basic data acquisition and reduction of a night's observations follow conventional procedures except for a few notable exceptions unique to cometary data. The first exception is that observations must often be obtained at high air mass due to a comet's proximity to the Sun. This fact, coupled with the number of species that have their primary emission bands in the near-UV, implies that precise extinction coefficients must be determined on a nightly basis, requiring standard star measurements over a range of airmasses bracketing the airmass range of the comet. Fortunately, because the filters have relatively narrow bandpasses, no color terms are required in the reductions, except for the OH filter near 3100 Å. In this unique case, extinction varies significantly across the bandpass, due to the strong wavelength-dependence of the ozone component of extinction. Moreover, the resulting curvature of the extinction-airmass relation differs with the detailed spectral signature being measured, and therefore different reduction coefficients are required for flux standards and for comets having differing gas-to-dust ratios. The appropriate equations and coefficients for extinction with the HB OH filter are detailed in *Farnham et al.* (2000).

The need to accurately remove contamination by emission bands of the continuum filters in high gas-to-dust ratio comets, and to subtract continuum from emission bands in low gas-to-dust ratio comets, requires sufficiently high S/N for whichever filters yield the smallest count levels. This implies that the optimum integration times for each filter will not only differ due to the overall brightness of the comet, but also with the relative amounts of gas and dust. It is therefore highly desirable to reduce the first observations of a new comet rapidly so as to be able to tailor subsequent observations. Accurate determinations of the background sky with each filter are also necessary, and can be very time-consuming to obtain. Typically, if a comet is sufficiently bright to enable the use of narrowband filters, the coma is likely to cover the entire CCD frame, forcing one to obtain separate sky frames for each filter. It is generally sufficient to obtain sky measurements at distances greater than ~30 arcmin from the nucleus in any direction other than that of the tail, although larger distances are required for exceptionally bright or close comets. It is also almost always preferable to track at the comet's rate of motion across the sky when obtaining either photometry or imaging of the coma; this capability is more routine now that most telescopes are computer-controlled.

When performing small-aperture extractions from CCD images, one must be aware of the effects of changing amounts of flux from the coma and from the nucleus due to seeing variations during the night; otherwise, artificially produced lightcurve features can result. While these effects can be searched for by extracting fluxes from a series of apertures, compensating for this situation is extremely difficult unless the effective pointspread function is available on each frame, and background stars will be trailed unless the exposures are kept sufficiently short. Decisions regarding appropriate aperture sizes for a conventional photometer must be made at the time the observations are acquired. Here, some major tradeoffs must be made to (1) avoid background stars, (2) minimize the sky signal, and (3) maximize the comet signal. With either instrumentation, practical limitations on integration times are imposed by (1) changing sky brightness (particularly near twilight), (2) total time the comet is available, and (3) the number of filters to be used. Compromises must almost always be made; the observer should let the specific science goals determine the best observing procedures on a case-by-case basis.

Except for the nonlinear extinction associated with the OH filter and already discussed, reductions to filter fluxes follow standard methods. Thereafter, narrowband reductions are somewhat unusual, in that the continuum filters often suffer from some contamination from cometary emission bands, the emission filters include underlying continuum, and the continuum is often reddened with respect to solar spectrum. A new iterative technique was developed by *Farnham et al.* (2000) to deal with these issues when using the HB filters. This procedure uses the measured fluxes in the continuum bands to remove underlying continuum from the C_3 and C_2 filter fluxes, which can then be used to compute the amount of contaminating emission in the continuum filters. At each step of the iteration the remaining contamination is reduced, until essentially pure emission fluxes and continuum fluxes are obtained. Again, all relevant equations and coefficients are provided in *Farnham et al.* (2000).

5. COMPOSITIONAL PHOTOMETRY

As previously noted, comet photometry for the purposes of compositional determinations can be made using either a phototube or a CCD as a detector, but the former usually

results in improved S/N for a given amount of observing time. This section is therefore primarily aimed at, but not restricted to, observations obtained with a conventional photometer system. Of course, many of the following procedures have direct analogs in the analysis of comet spectrophotometry.

The derived continuum fluxes and emission band fluxes are usually the final reduced quantities that can be considered model-independent. In the typical case of comets with detectable coma, unlike for point sources, the aperture used for the measurements must be specified for these quantities to be meaningful. Usually the observer will also want to compute an aperture-independent quantity by applying a suitable model of the coma, after first converting gas emission band fluxes to the number of molecules required to produce the measured fluxes. This conversion to a molecular abundance requires the use of the fluorescence efficiency (L/N or luminosity per molecule when given in units of ergs per second per molecule, or, equivalently, g-factor when given in units of photons per second per molecule) for the particular molecular band. While the fluorescence efficiencies for comets are generally unchanging for polyatomic species due to their large number of populated rotational levels (except for the r^{-2} dependence due to the fall-off of solar flux with distance from the Sun), diatomic molecules such as OH, NH, and CN display large variations as a function of heliocentric velocity due to the Swings effect (cf. *Arpigny*, 1976; *Feldman et al.*, 2004). Appropriate values as a function of velocity for OH can be found in *Schleicher and A'Hearn* (1988), for NH in *Kim et al.* (1989) and *Meier et al.* (1998), and for CN in *Tatum and Gillespie* (1977), *Schleicher* (1983), and *Zucconi and Festou* (1985). Note that the latter two CN references also present the variation of the fluorescence efficiencies as a function of heliocentric distance as well as with velocity, since the number of populated rotational levels in CN varies strongly with the available solar flux. In the Lowell photometric program, we currently continue to use the same fluorescence efficiencies adopted by *A'Hearn et al.* (1995), and these are summarized in Table 2. However, the values for some species, such as C_3, may change in the future as band oscillator strengths are revised or as fluorescence models include more transitions and collisional effects.

The resulting molecular abundances obtained following the application of fluorescence efficiencies can be readily converted to column densities, if desired, but for either abundances or column densities the size and location of the aperture or slit must be stated for the result to be useful. In order to intercompare results obtained with differing apertures on a single comet or to intercompare comets, a coma model (such as the Haser, the Vectorial, or a numerical model such as the Monte Carlo) is applied to extrapolate the measured column abundance to a total coma abundance. While there are pros and cons to each specific model (see *Combi et al.*, 2004, and references therein), in each case a few parameters (such as the lifetime, velocity, and/or scalelength) are used to approximate the spatial distribution of the specific gas species in cometary comae. Once a total coma

TABLE 2.   Adopted parameters used in reduction of Lowell narrowband photometry.

Species	L/N* (erg s^{-1} mol^{-1})	Reference	Haser Scalelength† Parent (km)	Daughter (km)	Daughter Lifetime† (s)	Reference
OH (0–0)	$1.4–8.3 \times 10^{-15}$	*Schleicher and A'Hearn* (1988)	2.4×10^4	1.6×10^5	1.6×10^5	*Cochran and Schleicher* (1993)
NH (0–0)	$4.9–7.6 \times 10^{-14}$	*Kim et al.* (1989)	5.0×10^4	1.5×10^5	1.5×10^5	*Randall et al.* (1992)
CN ($\Delta v = 0$)	$2.4–5.0 \times 10^{-13}$	*Schleicher* (1983)	1.3×10^4	2.1×10^5	2.1×10^5	*Randall et al.* (1992)
C$_3$ ($\lambda 4050$)	1.0×10^{-12}	*A'Hearn et al.* (1985)	2.8×10^3	2.7×10^4	2.7×10^4	*Randall et al.* (1992)
C$_2$ ($\Delta v = 0$)	4.5×10^{-13}	*A'Hearn* (1982)	2.2×10^4	6.6×10^4	6.6×10^4	*Randall et al.* (1992)

*All fluorescence efficiencies (L/N; for r_H = 1 AU) are scaled by r_H^2. L/N for OH, NH, and CN are functions of $\dot{r}_H$, and L/N for CN is also a function of r_H (see *A'Hearn et al.*, 1995, for details). The CN (0–0) L/N values are multiplied by 1.08 to approximate the contribution of the CN (1–1) band.
†All scale lengths and lifetimes (for r_H = 1 AU) are scaled by r_H^2 (see *A'Hearn et al.*, 1995, for details).

abundance is computed, this quantity can be divided by the lifetime of the species (usually controlled by the photodissociation rate from solar radiation) to compute the production rate of the species, Q, i.e., the rate at which new molecules (or their parents) must be released from the nucleus to maintain the observed abundance. Unfortunately, the values for these seemingly fundamental parameters are often poorly known, because many of the species are radicals and therefore difficult to measure in the laboratory. Moreover, lifetimes also vary with solar activity, while the amount of acceleration of the bulk gas flow varies with collision rates, which depend upon the total gas production rate and the distance from the nucleus.

To minimize the number of parameters needed when intercomparing the composition of comets, the Lowell program generally uses the Haser model, again with the same values for the model parameters as those adopted by *A'Hearn et al.* (1995), and these are also summarized in Table 2, along with assumed daughter lifetimes. These particular values were based on observed radial profiles obtained over a variety of heliocentric distances, but they do not work in all circumstances. For instance, in the case of Comet Hale-Bopp, the scalelengths must be increased by 2–3× due to the combination of unusually large gas outflow velocities in this very high production comet along with low solar activity (*Schleicher et al.*, 1999). For these and other reasons, to the extent possible it is important to observationally verify the validity of the parameters used in this modeling, such as by directly measuring the radial profiles of each gas species, either by observing with multiple photometer entrance apertures, narrowband imaging, or longslit spectroscopy. Unfortunately, in practice, this testing of the parameters is usually not feasible except with relatively bright comets.

A method to produce an aperture-independent quantity utilizing continuum flux measurements of the dust coma was introduced by *A'Hearn et al.* (1984). This quantity, A(θ)fρ, is the product of the bond albedo, A, at a particular phase angle, θ, the filling factor, f, and the projected aperture radius, ρ, as seen on the sky plane. This product will be independent of aperture size if the dust follows a canonical 1/ρ spatial distribution for outflowing dust, and will be independent of wavelength if the dust has no color as compared to the Sun. The equation to compute A(θ)fρ, as well

as the appropriate values for the conversion coefficient for each HB continuum filter, are given in Appendix D of *Farnham et al.* (2000). Since no knowledge of the grain properties is required as input to the calculation, the computation of A(θ)fρ from the measured continuum flux is straightforward, and therefore A(θ)fρ is often used as a proxy of dust production, somewhat analogous to the gas production rates discussed above. Indeed, the quantity A(θ)fρ varies proportionally to the dust release rate from the nucleus, but also inversely proportional to the dust outflow velocity. Unfortunately, the very fact that grain properties are not included in A(θ)fρ means that intercomparisons as a function of time for a single comet or intercomparisons between comets must be made with caution. Simple intercomparisons inherently assume that numerous properties of the dust grains are constant with time and among comets, such as particle size distribution, grain shape and porosity, and outflow velocity. However, since dust grains are initially entrained with the gas flow, the resulting bulk dust velocity can vary with total gas production rates. Particle size distributions are known to differ drastically among comets, and outflow velocities also vary with particle size. Grains have also been seen to "fade" as they move away from the nucleus, either by shrinking in size or darkening as volatiles escape from the grains, or by breaking apart (*Jewitt and Meech*, 1987; *Baum et al.*, 1992). Therefore, it can be difficult to determine whether a particular variation or trend of A(θ)fρ is actually a measure of the rate of release of dust grains from the nucleus, or an indication of differing grain properties, as was determined in the recent case of Comet 19P/Borrelly (*Schleicher et al.*, 2003).

A variety of types of scientific studies that can be performed from photometric measurements obtained through narrowband filters was itemized in the introduction to this chapter. We now briefly explore a selected subset of these topics, primarily drawing on examples from our own work simply because, following the apparition of Comet 1P/ Halley in 1985/1986, very few groups have continued to employ this technique. Certainly one of the most basic types of studies are those of relative gas and dust production rates to determine the relative composition of parent or grandparent species in the nucleus (or, at least, the active source regions on the nucleus). Differences in the abundance ratios as a comet moves along its orbit can be used to infer chemical inhomogeneities in the nucleus (e.g., *A'Hearn et al.*, 1985), while differences among comets can indicate either evolutionary effects, such as the strong gas-to-dust variation with perihelion distance (*A'Hearn et al.*, 1995), or primordial, such as the large fraction of Jupiter-family comets that are depleted in carbon-chain molecules as shown in Fig. 2 (*A'Hearn et al.*, 1995). With a sufficiently large database, such as the 85 comets observed by *A'Hearn et al.* (1995), numerous compositional investigations were possible on a statistical basis. Of course, other properties, such as heliocentric distance-dependencies and possible variations with species, can be determined for well-studied comets such as 1P/Halley (*Schleicher et al.*, 1998a), 2P/Encke (*A'Hearn et al.*, 1985), 21P/Giacobini-Zinner (*Schleicher et al.*, 1987), and Hyakutake (1996 B2) (*Schleicher and Osip*, 2002). By utilizing a basic water vaporization model (cf. *Cowan and A'Hearn*, 1979), minimum effective active areas on the surface of the nucleus can be computed, yielding a minimum effective radius or, if the nucleus size is determined separately, a fractional active area. One of the most unexpected results from the Lowell photometry program is the large number of comets having very small (<3%) active fractions (*A'Hearn et al.*, 1995).

The color of dust grains is primarily an indicator of the particle size(s), and has limited value in determining other physical properties of the grains, such as composition or porosity (cf. *Kolokolova et al.*, 2004). Measurements of phase effects, particularly at small and large phase angles, are difficult to obtain because of other, often stronger sources of brightness variations, such as a comet's changing heliocentric distance. A few comets for which phase effects have been successfully separated from other effects include P/Stephan-Oterma (*Millis et al.*, 1982), Bowell (1980b) (*A'Hearn et al.*, 1984), and Halley (*Meech and Jewitt*, 1987; *Schleicher et al.*, 1998a). The most diagnostic type of remote measurements for dust particles in cometary comae is that of polarization. These can usefully constrain physical properties, but are difficult to obtain. Here, again, narrowband filters minimize the contamination otherwise caused by gas emission. One research group that has routinely obtained this type of narrowband measurements is that of Kiselev and Chernova and their associates (cf. *Kiselev and Chernova*, 1981; *Chernova et al.*, 1993; *Kolokolova et al.*, 2004). Note that narrowband filters have also been occasionally used to obtain polarimetric measurements of molecular gas emission (cf. *Le Borgne et al.*, 1987; *Sen et al.*, 1989), but the degree of polarization is generally much smaller for gas than for dust and underlying continuum must be very accurately removed, making gas polarization measurements quite difficult.

Finally, periodic variations detected within a photometric lightcurve can, of course, be used to determine the rotation period of a comet nucleus. While variations due to the changing cross-section of the nucleus itself are most easily interpreted (and are most readily obtained using a CCD), measured variations of the brightness of the coma can be used to infer the number and relative strengths of individual source regions on the surface of the nucleus. Differences in lightcurve amplitudes and phase lags among the various gas species and with the dust can further be used to constrain outflow velocities and lifetimes, as in the cases of Comets 1P/Halley (cf. *Millis and Schleicher*, 1986), Levy (1990c) (*Schleicher et al.*, 1991), and Hyakutake (1996 B2) (*Schleicher and Osip*, 2002).

6. IMAGING AND COMA MORPHOLOGY

6.1. Morphological Features

Many comets exhibit detailed, well-defined features in their comae. The presence of these features indicates that the surfaces of the nuclei of these comets are not uniformly active, but emit material anisotropically, with at least part of the material coming from isolated active areas. Some of the more prominent types of features that are observed include jets (radial structures produced by isolated active regions, or sources, that emit collimated streams of gas and dust), fans (jet-like structures that tend to be broader and more diffuse than jets), spirals and arcs (outflowing material from jets on a rotating nucleus that form archimedean spirals, or partial segments of spirals, respectively), and coma asymmetries (some regions of the coma appear brighter than others). In addition to providing an explanation for the coma morphology, the existence of isolated source regions also provides a natural explanation for a variety of other phenomena observed in comets, including seasonal variations in the production rates, nongravitational accelerations of the

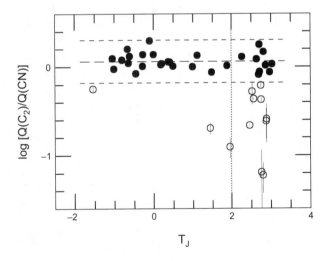

Fig. 2. Derived production rate ratios of C_2 to CN as a function of the Tisserand invariant with respect to Jupiter, T_J. The C_2-to-CN ratios are based on each species' respective ratio to OH. Comets having "typical" composition are those within the horizontal band (●), while carbon-chain depleted comets lie below this band (○). One-half of Jupiter-family comets ($T_J > 2$) are depleted, while only two non-Jupiter-family comets ($T_J < 2$) display significant depletions, and one of these — P/IRAS — oscillates across the $T_J = 2$ boundary. Statistically, most Jupiter-family comets are believed to have originated in the Kuiper belt, while most other comets should have come from the Oort cloud. Based on *A'Hearn et al.* (1995).

comet's orbit, and changes in the rotation state of the nucleus. Because isolated source regions can contribute to so many different aspects of the comet's activity, it is important to determine their characteristics.

To fully understand the role that the active regions play in the comet's behavior, properties such as the rotation state of the nucleus and the location of the sources must also be known, so that the dust and gas emission can be characterized as a function of time. Frequently, models of jet emissions can be used to reproduce the observed morphology and infer the relevant properties of the nucleus. Furthermore, in certain situations, it is possible to utilize these derived results to aid other analyses. Potential secondary studies include searching for compositional inhomogeneities in the jets (cf. *A'Hearn et al.*, 1986a,b; *Lederer et al.*, 1997; *Festou and Barale*, 2000) or constraining more fundamental characteristics such as the mass and density of the nucleus (*Farnham and Cochran*, 2002). Although some of the properties determined through coma modeling could be found using other techniques (e.g., lightcurve variations may reveal the rotation period), many could otherwise only be found via spacecraft encounters. This makes the analysis of the coma morphology an extremely valuable technique for understanding the fundamental qualities of cometary nuclei.

The majority of studies involving analysis of a comet's coma features are performed using images obtained with broadband or continuum filters. This is likely due to two factors: The dust coma tends to show clearer, more well-defined structures than the gas species, and the data reduction process is simpler. However, while they have proven to be very useful, dust images only provide a partial picture of the overall coma morphology, with gas and ions adding their own contributions. In most comets, features in the gas coma tend to be completely overwhelmed by the dust, but narrowband filters can be used to help isolate the gas features. Then, as described previously, with careful calibration the underlying continuum can be removed, leaving images of the pure gas coma (*Schulz et al.*, 1993, 2000; *Farnham et al.*, 2000). (Similarly, gas contamination can be removed from images obtained with narrowband continuum filters to leave the pure dust coma.) The pure gas images can then be enhanced or modeled, in the same manner as the dust images, to learn about the gas properties and to provide additional constraints on the nucleus properties. Studies of the CN coma in Comet Hale-Bopp illustrate the potential benefits of utilizing the gas morphology: First, the CN forms complete spirals around the nucleus, while only partial arcs are seen in the continuum (see Fig. 3), indicating that the gas production behaves differently from that of the dust (*Larson et al.*, 1997; *Farnham et al.*, 1998b; *Mueller et al.*, 1999). Second, the CN spirals expand radially outward at about twice the speed of the dust features, which is likely due to differences in initial outflow velocities and accelerations (*Schleicher et al.*, 1999). The fact that the gas and dust are not co-spatial indicates that most of the gas is being emitted directly from the nucleus rather than coming from the optically important dust grains in the coma. Third, the CN images clearly show three distinct jets in each rota-

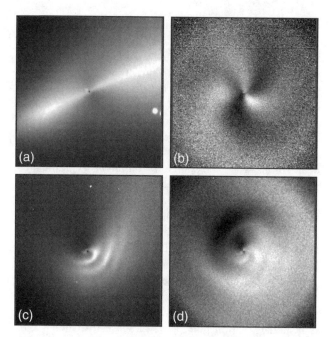

Fig. 3. Comparison of radial and azimuthal features, and the differences between dust and gas morphology. **(a)** Comet Borrelly showing purely radial structure (*Farnham and Cochran*, 2002). **(b)** CN jets observed in Comet Halley, showing initially radial features with curvature induced by rotation of the nucleus (*A'Hearn et al.*, 1986a; *International Halley Watch*, 1995). **(c)** Dust and **(d)** CN images of Comet Hale-Bopp, showing azimuthal features and the difference between the structure of the gas and dust in the coma (*Farnham et al.*, 1999). Features were enhanced by dividing out a 1/ρ profile from the dust frames and an azimuthal-averaged profile from the CN images.

tional cycle, while the dust shows only two, which might suggest inhomogeneities in the nucleus. Finally, *Woodney et al.* (2002) used narrowband filter data, in conjunction with radio measurements of the spatial structure, to explore the relationship between the HCN and CN.

Investigations of the nucleus size and the cometary plasma environment can both benefit from the use of narrowband filters, too. For high gas-to-dust comets, direct measurements of the nucleus (*Lamy et al.*, 2004) are more efficient with narrowband filters than broadband ones, because continuum filters exclude the gas contribution and make the nucleus stand out more against the coma. In plasma tail studies, the narrow passbands will isolate CO^+ or H_2O^+ much more efficiently than broadband filters or photographic plates, increasing the contrast of the plasma features against the background. Furthermore, with proper calibration, the continuum can be removed from the ion images to reveal the features very close to the nucleus. This improves the potential for following phenomena such as disconnection events from their earliest stages (*Ip*, 2004).

Dust, gas, and ion features have been observed and studied in many comets, with recent examples including Halley, Hyakutake, Hale-Bopp, and Borrelly. Although the detailed morphology in each comet is unique, the features can generally be classified into two main categories: azimuthal and radial (or a combination of the two). Because most of the material from a jet expands radially away from the surface, the appearance of structures in the coma is strongly dependent on the geometric viewing conditions and the rotation state of the nucleus. A source at a given latitude will sweep out a hollow cone centered on the spin axis (or a partial cone if the source rotates out of sunlight and shuts down during part of a rotation). When Earth is oriented inside the cone, features usually appear to be azimuthal — archimedean spirals when the source is continuously illuminated and concentric arcs when it turns on and off. If Earth is outside the cone, then the feature may appear to oscillate back and forth, or it may smear together to produce a fan with primarily radial structures [due either to the planetary nebula effect at the edges of the fan or to insufficient spatial resolution (*Samarasinha et al.*, 1999)]. Radial features are also produced when the jet is on a slowly rotating nucleus (e.g., seeing only the innermost segment of the archimedean spiral, which is nearly radial) or if the active region is near the rotation pole, as is the case for Comet 19P/Borrelly (*Samarasinha and Mueller*, 2002; *Farnham and Cochran*, 2002; *Schleicher et al.*, 2003).

We note that factors other than isolated active regions can also produce features in the coma. Solar radiation pressure can act on the dust grains to produce envelopes that can be mistaken for arcs, while anti-tails and neck-line structures can mimic radial structures (*Fulle*, 2004). However, both of these cases are only observed in continuum images and involve relatively unique circumstances and geometric alignments, which can be investigated to avoid misinterpretation of the results. Structure in the coma can also result from outflowing material that experiences constructive interference and density enhancements due to the topology of the nucleus (*Crifo et al.*, 2004), but this mechanism can only produce features that are restricted to the very innermost coma regions and have a low contrast against the background. Finally, other types of features, such as knots, condensations, and kinks, are observed in plasma tails, but these are not addressed in any detail here.

6.2. Image Enhancement Techniques

Any comprehensive discussion of the coma morphology, gas, dust, or plasma, must inevitably address the topic of image enhancement techniques, as the two are intimately related. Indeed, many features are overwhelmed by the bulk radial brightness fall-off of the coma and only become obvious when the image has been processed in some manner (though once the user knows what to look for, the features are usually detectable in the unenhanced image with appropriate display parameters). Unfortunately, enhancing an image, by definition, alters the image, and not always in the manner that is expected or desired. Thus, any processing technique should be used with caution. Potential dangers include introducing artifacts that can be misinterpreted as real structures or shifting the apparent positions of features. Even if these shifts are small (which is not always the case), they are misleading when they change the

positions that are being used to constrain models of the morphology. Another potential problem with enhancing an image is that various techniques may reveal different types of features in a given image. Because of this, the interpretation of the nature of the feature may be strongly dependent on the particular technique as well as on the manner in which it was applied. For example, images of Comet Hyakutake processed with a $1/\rho$ removal (discussed below) appear to have round blobs of material moving radially outward, while processing with radial profiles derived from the comet itself reveal that the blobs are actually broad spiral arcs (cf. Figs. 4b and 4k) (*Schleicher et al.*, 1998b; *Schleicher and Woodney*, 2003). Finally, enhancement of an image inherently changes the relative intensities of the different regions of the coma. This is a concern in the interpretation of the relative brightnesses and strengths of the different sources, as well as in using coma asymmetries to constrain the gas and dust production as a function of solar illumination.

There are a wide variety of enhancement techniques, each of which has its own strengths and drawbacks. Any technique can be good or bad and no single technique is ideal for every situation. Thus, it is important for the user to experiment with different methods on a variety of data, to become familiar with their pros and cons, to understand the types of data for which specific techniques are most useful, and to help in recognizing potential problems and artifacts. A number of basic processing methods are discussed in *Schwarz et al.* (1989), *Larson and Slaughter* (1992), and *Farnham and Meech* (1994). We review these and introduce additional techniques in the following discussion, and present representative enhancements in Fig. 4. As this figure shows, different techniques can produce dramatically different effects, and it is advisable to utilize several different ones on the same image, so that they complement each other and act to create an overall portrayal of all aspects of the coma. This also provides a cross-check to determine if a feature is real or if it was introduced by the processing.

Before introducing the different enhancement techniques, we address a few practical notes regarding their definitions and applications. First, many of the techniques require the "removal" of a radial profile from the observed coma. This removal process can be done via either subtraction or division, with very different results. For example, subtracting a $1/\rho$ profile emphasizes features in the innermost coma, while dividing by this same profile suppresses the innermost coma but dramatically enhances the features further out (compare Figs. 4b and 4c). Second, a number of techniques require that a coma profile be created directly from the comet images themselves. Usually, this involves combining a set of pixels (e.g., all the pixels in a given annulus) to produce a mean value that can then be removed. For these cases, we tend to utilize the term "averaging" of the pixels, but they can be combined by computing the mean, median, or mode of the sample. Again, different results can be obtained in each case. Third, many enhancements utilize radial and/or azimuthal information from the original image to generate profiles. For these situations, it is easiest to work with an image that has been unwrapped

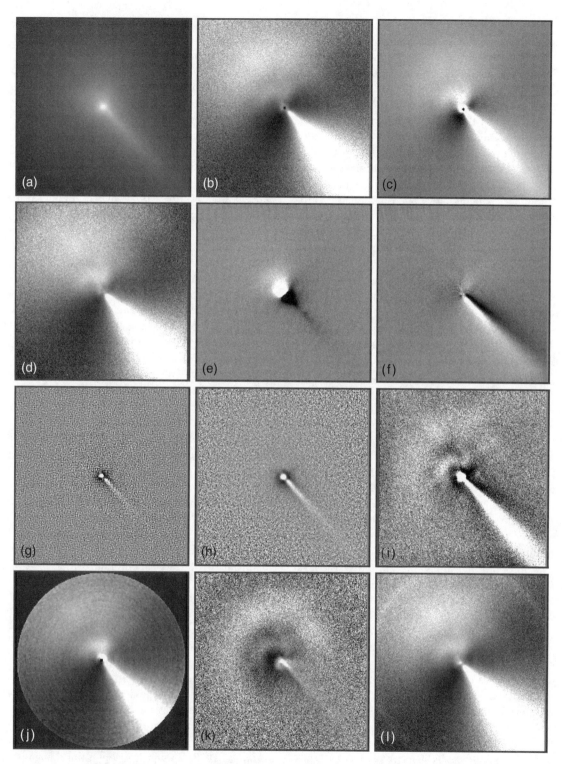

Fig. 4. Comparison of different image enhancement techniques on a common coma image containing three primary features: a broad, diffuse outer arc, an intermediate-scale inner arc, and a narrow, sharply defined tail. These three examples illustrate the type of features revealed by each technique (although different kernel sizes or shifts of different amounts can change the scale of the features that are revealed). Note that some methods enhance the radial tail, others reveal the azimuthal arcs, and some remove the azimuthal variations, while in others it is retained. Although dramatically different results are obtained with the different techniques, it is clear that no single enhancement is ideal for every situation. This emphasizes why multiple techniques should be used to investigate the variety of potential coma features that may be present. **(a)** Original image of Comet Hyakutake (*Schleicher and Woodney,* 2003), displayed with a logarithmic scale to show the straight narrow tail. **(b)** 1/ρ profile, divided out. **(c)** 1/ρ profile, subtracted out (for comparison, to show how different applications of the same technique can affect the result). **(d)** Azimuthal-averaged profile divided out. **(e)** Linear shift difference with a five-pixel shift in both the vertical and horizontal directions. **(f)** Rotational shift difference with a 10° rotation. **(g)** Laplace filter. **(h)** Unsharp mask with a three-pixel Gaussian smoothing kernel. **(i)** Radial gradient filter. **(j)** Azimuthal renormalization. **(k)** Phase-stacked mask. **(l)** Phase-stacked/azimuthal-averaged profile. Other examples of enhancements **(b)** and **(d)** are shown in Fig. 3.

from the standard rectangular format into a polar format (θ,ρ, where θ is the position angle and ρ is the projected distance from the nucleus). Using the unwrapped image, it is trivial to extract radial and azimuthal information from the rows and columns. The drawback to this process is the need for high-quality unwrapping and rewrapping routines that conserve flux and do not introduce artifacts. Finally, we stress the importance of accurately finding the optocenter of the coma when using many enhancement techniques. Accurate centering, at the fractional pixel level, is crucial when any kind of profile is removed from the coma or when the image is being unwrapped into polar coordinates. Similarly, when multiple images are being combined, it is critical to have them properly registered so the nucleus is in the same position in each frame. For most comets, the central condensation of the continuum provides a suitable reference for the optocenter of both gas and dust images, whereas for comets with little or no dust, the nucleus itself is sometimes visible.

The most benign technique available for searching for features comes from the display process itself. Displaying the image with a log or square root intensity or as a histogram equalization, and then adjusting the contrast stretch, will often reveal a significant amount of detail. This is a straightforward method that is available as a standard option on most display packages and tends to work well. Because there is no manipulation of the data, and hence no way to introduce artifacts, any features that are seen are likely to be real. In addition, once a feature has been identified using another technique, it can usually be detected in the original image if the display is set correctly. For this reason, it is a good idea to return to the unprocessed image to confirm the existence of any features found using the more aggressive enhancement methods.

6.2.1. Removal of simple profiles. The most basic level of processing uses a simple profile to remove the bulk coma shape from the observed image. For the continuum, a logical first choice is a $1/\rho$ profile, in which the brightness falls off as the inverse of the projected distance from the nucleus. This is the canonical shape of a coma produced by steady-state isotropic emission of dust. (An analogous shape for gas images would be a Haser model profile.) The resulting image highlights the deviations of the observed coma from the idealized one. Not only is this an easy enhancement to apply, but it is fairly benign and it makes the interpretation of the morphology relatively straightforward because the extracted profile is smooth and its shape is known. A slightly more complex technique, based on the same principle, uses the observed coma, gas or dust, to create a profile from the comet itself. By averaging azimuthally around the optocenter (easily done from the polar format image) a radial profile that closely matches the shape of the real coma can be created. The main problem with this technique is that strong arc-shaped features or bright stars may introduce bumps into the profile, which can then produce artifacts in the enhanced image. To avoid this, it may be necessary to remove the stars before computing the average profile and/or radially smooth the profile to suppress any arc residuals; in other cases, using a median rather than an average of the annulus pixels may suffice.

6.2.2. Edge detection techniques. Another type of enhancement is the edge detection technique (EDT), which covers a large family of routines. These tend to be very popular because they are easy to use, require little effort to develop, and dramatically increase the contrast of some features. The first group of edge detectors includes derivative routines such a linear shift differencing (*Klinglesmith*, 1981; *Wood and Albrecht*, 1981) or rotational shift differencing (*Larson and Sekanina*, 1984), where a copy of the image is shifted or rotated and then subtracted from the original. Features that are revealed in this process are a function of the size and direction of the shift, so several applications should be used to look for features at different spatial scales and in different orientations. A powerful benefit of the rotational shift is that a rotation around the optocenter will produce small shifts near the nucleus, where features tend to be small and well-defined, and increasingly larger shifts at larger distances, where the features spread out and become more diffuse. This minimizes the number of applications needed. Another technique, the temporal derivative, uses the ratio of two images obtained at different times to reveal changes in the features as a function of time. This technique is frequently used for work with plasma tails where features change rapidly. There are situations where the temporal derivative can be used over longer timescales (e.g., night to night), but the approach should be used cautiously because other factors, including seeing variations and changes in the viewing geometry, can also affect the appearance of the coma. Next is the color derivative, which, although not an EDT, we include here for completeness. In this method, the ratio of two images obtained at different wavelengths (usually from two different continuum filters) shows spatial color variations in the coma. This is useful for procedures such as comparing the particle size information in a jet to that in the rest of the coma. This derivative should be used with care, however, because gas contamination, poor flat fielding, or residual sky background can produce misleading results, and so high-quality, decontaminated data are necessary for accurate results.

The next group of edge-detection enhancements includes spatial filters, such as the Laplace filter and other procedures in which a kernel is convolved with the image (e.g., *Richards*, 1993). It also includes unsharp masking (e.g., *Sekanina*, 1978), in which a copy of an image is digitally smoothed and the lower-resolution version removed from the original. These techniques tend to be easy to use and are very useful for exploring whether or not features are present. However, if measurements are to be obtained from the processed image, then edge-finding routines should be avoided for a number of reasons. First, the enhanced features are dependent on the size and shape of the filter that is applied, which in turn affects the interpretation of the result. When using unsharp masking, for example, smoothing the image with a Gaussian filter produces a result very different from what is obtained with a square filter. Second, EDTs are specifically designed to identify edges or discontinuities. Thus, positions obtained from an EDT-enhanced image may not be accurate if the measurements are intended to represent the center of a feature (as is usually the case when the positions are used as input for coma models). Third, these techniques are very harsh and remove a significant amount of information about the coma, which is why they produce high-contrast features. Fourth, this same harshness means that EDTs are more prone to introducing artifacts, especially near the nucleus where the bright central peak dominates. Artifacts are also more common in regions where multiple features overlap and can interact with the convolution filter in an unpredictable manner.

6.2.3. Azimuthal renormalization and radial gradient filter. We now turn to enhancement techniques that are somewhat more involved to create or apply. The first of these is azimuthal renormalization (*A'Hearn et al.,* 1986a), in which the coma is divided into a series of annuli and the pixel intensities in each annulus are rescaled to a common minimum and maximum. Again, this is a simple procedure to perform using the polar format image. The result is effective for removing the central peak, although it also removes much of the relative brightness information. Its strength lies in its usefulness for showing radial features that rapidly fade with distance. Another, more-intricate technique that is presently not widely available is the radial gradient spatial filter (S. M. Larson, personal communication, 2002). This routine uses the basic principles of a convolution filter, but varies the size of the kernel as a function of the distance from the nucleus. The result has the same benefits as the rotational shift removal, in that the enhancements are optimized to reveal small features near the nucleus and increasingly larger features at greater distances, but unlike the rotational shift, it enhances azimuthal as well as radial features (Fig. 4i).

6.2.4. Image sequences and temporal image enhancements. Relating to the following enhancement techniques is another tool that can be useful for understanding and interpreting complicated features. Given a sequence of images uniformly spaced in time, a "movie" of the comet motions can be created. With these sequences, the evolution of the coma and the motions of the features can be followed more clearly. Furthermore, if the features have a periodic nature (e.g., consecutive arcs representing successive rotations of the nucleus) and the viewing geometry varies slowly with respect to the rotation period, then it may be possible to phase images from different rotational cycles to simulate a full rotation. If so, then movies can be created, even if the comet was only observed for short periods on any given night during an observing run.

Unfortunately, most standard enhancement techniques are not optimally suited for use on a sequence of images. For example, an unsharp mask can only be applied to the individual frames in any sequence, but due to temporal changes in the features, the shape removed by the mask will be different for each image. This makes it difficult to di-

rectly compare the results. To avoid this problem, it makes sense to enhance the images by removing the same bulk coma shape from each frame. Again, a standard profile (e.g., $1/\rho$) can be assumed or one can be created from the comet itself. In the latter case, however, the profile should be derived using all the images in the sequence so that it incorporates not only spatial variations in the the coma, but temporal ones as well. This was done in an analysis of the CN jets in Comet Halley (*Schulz and A'Hearn,* 1995) in which a series of images were combined to produce an averaged shape that was then removed from the individual frames.

In this same vein, we developed and tested a set of procedures for use in situations where the rotation period is known and is well covered by observations. We found that taking a sequence of images, uniformly spaced throughout the rotation sequence, and averaging them together works well to create a template for removing the bulk coma. The averaging process smooths out temporal changes in the features, so that when the template is removed from each image, any moving features are highlighted. Because the template is created by combining images throughout a rotational phase, we refer to this technique as the phase-stacked mask. In essence, this procedure is a straightforward mask removal, which means that it is relatively benign, it is very good for enhancing faint structure, it can be used to enhance any image obtained throughout the rotation (e.g., it is not restricted to those that were used to create the mask), and any features that are revealed are not likely to have their positions shifted. Like many other techniques that use a coma shape derived from the comet itself, most of the intensity information is lost, including any azimuthal asymmetries. As is discussed later, these asymmetries can provide important constraints on the coma models, so it may be desirable to retain the information. To avoid removing the asymmetries, the procedure can be taken a step further by computing the average azimuthal profile from the phase-stacked mask, to produce a phase-stacked/azimuthal-averaged profile. Removal of this profile from the individual images then enhances the features, while still preserving the original azimuthal asymmetries.

These two phase-averaging techniques, used together, have proven to be very useful in analyses of images of Comets Hale-Bopp and Hyakutake (*Farnham et al.,* 1998b; *Schleicher and Woodney,* 2003). Unfortunately, they have drawbacks that limit the number of objects on which they can be used: They are time-consuming to apply, the comet's rotation period must be known, and multiple images with good phase coverage are needed to smooth out the features. If good temporal coverage is not available, combining the images from different phases may leave residual features in the profile that can again introduce artifacts when the template is used to enhance an image (as is the case for any of the temporal-averaged techniques). Fortunately, most of these residual features can be removed from the averaged profile by applying a smoothing spline function in the radial direction, which minimizes the effect of poor phase coverage. Although this may slightly change the shape of the coma template, it can still be effectively used to enhance the original images to reveal temporal changes in the coma.

6.3. Quantitative Measurements

Once features have been identified in the coma, with or without the use of enhancements, their qualitative appearance must be converted into quantitative measurements that can be used as constraints for coma models. Depending on the types of features present and the requirements of the model, different measurements are possible. For predominantly radial features, the position angle (PA) of the feature is the most useful measurement, although if some curvature is present, then it may also be necessary to specify at what distance the PA was obtained. Positions of arcs and spirals are usually quantified by measuring the cometocentric distance at a number of different PAs. When making these measurements, the center or brightest point of the jet is the preferred reference location, because models are more likely to reliably reproduce the bright central peak of the jet than its edge. Characterization can also include measurements of other properties, such as the width of the feature, which is often quoted as the full-width at half-maximum (FWHM) above the background. Although this type of measurement can be difficult (and may require additional information if the feature is not symmetric about the center), it provides valuable information about the dispersion of the material in the jets. For the gas species it may also provide information about the relative velocities of the parent and daughter species.

As with the enhancement procedures, it is useful to utilize the polar format image, as well as the rectangular image, for making certain spatial measurements. The polar format not only provides a different perspective for looking at the data (cf. *Schwarz et al.*, 1997; *Samarasinha and Mueller*, 2002), but PAs and distances are directly measurable from the rows and columns. Also, whenever possible, measurements should be obtained from multiple images. If the features are stationary, then the additional measurements will improve the uncertainties; if they move, then the additional measurements give positions as a function of time, and result in a quantitative measure of the motion and thereby constrain the projected velocity of the feature.

Another form of quantitative measurement that can be used to constrain models is the brightness of the coma and morphological features. The brightnesses of different jets can be used to find the relative strengths of their sources, while the amount of material coming from the active areas can be compared to that in the isotropic background. Using the brightness in this manner requires that, if any image enhancement techniques at all are used, then they must be very benign so that relevant brightness information is not lost. Furthermore, to get a proper comparison of the brightness levels, contributions from undesired species must be removed from the images. Thus, not only must continuum be removed from the gas images, but also contamination from other gas species (e.g., the wing of C_3 in the CN filter),

which requires an accurate calibration of the images. Ultimately, it may be possible to use the calibrated images to constrain the models sufficiently well to determine gas and/or dust release rates as a function of location on the nucleus.

6.4. Rotation Periods

Many comets exhibit features that are observed to regularly repeat with time. These repetitious structures are a signature of the rotation of the nucleus and under the proper conditions, they can be used to measure the rotation period. Repeating features can include concentric arcs, a ray that oscillates back and forth, or any structure or outburst that appears at regular intervals. These manifestations reflect either the changing production rates as active regions rotate into and out of sunlight or the changing direction of the emission as the nucleus spins, and thus can be used to derive the rotation period. Furthermore, regular repetitions in the morphology suggest that the nucleus is in or near a state of principal axis rotation. (Although long-term precession or complex rotation may be present in some comets, they cannot be addressed if they are not evident in the available observations.) There are exceptions to this rule, including Comet Halley, which exhibits periodic variations, even though it is in a state of complex rotation (*Belton et al.*, 1991).

The most straightforward means of measuring the rotation period is to use a sequence of images that span a full rotation of the nucleus, as was done for Comets Halley (*Samarasinha et al.*, 1986; *Hoban et al.*, 1988), Swift-Tuttle (*Boehnhardt and Birkle*, 1994; *Fomenkova et al.*, 1995), Hyakutake (*Schleicher et al.*, 1998b), and Hale-Bopp (*Sarmecanic et al.*, 1997; *Jorda et al.*, 1999). The period is simply the time that it takes for the feature to reappear in the same place it was at the start of the sequence. Unfortunately, this requires an observing window that permits good temporal coverage throughout a full rotation period, which can be rare for comets. In the examples noted above, Hale-Bopp was bright enough that distinct features could be seen in infrared measurements obtained during the day; Hyakutake passed near Polaris and, for northern hemisphere observers, was observable all night during its closest approach to Earth; and Halley and Swift-Tuttle have rotation periods that span several days, so coverage over many nights provided sufficient sampling to follow the rotation.

If the comet is only observable for short periods, then other methods must be used to derive the rotation period from the features. One method is to phase images from night to night, as discussed above, to determine how long it takes for a feature to repeat. This requires an understanding of how much a feature moves from one night to the next to avoid converging on a false alias of the period, but motions can usually be constrained with observations spanning an hour or two. In the case of Comet Hale-Bopp, the motion of an arc during 2 h of observations was sufficient to show that the arc would repeat about twice per day (i.e., the nucleus had a rotation period of approximately 12 h, rather than 24 or 8 if the arc repeated once or three times respectively). With this constraint, images from several nights could be utilized to determine a more precise rotation period of 11.3 h (*Lecacheux et al.,* 1997; *Mannucci and Tozzi,* 1997; *Licandro et al.,* 1998; *Farnham et al.,* 1998a; *Warell et al.,* 1999). If temporal coverage is minimal, an alternative, although less reliable, method can be used to constrain the rotation period. Using the curvature of a jet and an estimate of the outflow velocity, the spin rate of the nucleus may be found (e.g., *Larson and Sekanina,* 1984; *Watanabe,* 1987; *Boehnhardt et al.,* 1992). Unfortunately, a lack of phase coverage means that assumptions must be made about the projection effects and outflow velocities (or they must be constrained in some independent manner). If the assumptions are not valid, then the results may have significant errors.

Finally, in quoting a rotation period, it should be specified as to which type has been determined: sidereal, solar, or synodic. The sidereal rotation period, which is the most desired but not always the one measured, is the time needed for the nucleus to rotate once with respect to the stars. The solar rotation period is the time it takes for one full rotation with respect to the Sun. Since most morphological features are directly related to the amount of sunlight illuminating an active region, this is probably the most common type of period measured from coma morphology. The solar rotation period is also commonly referred to as the synodic rotation period, in a manner analogous to that used with planets in the solar system. Unfortunately, the term synodic period is also often used, particularly in asteroid studies, for the time needed for one rotation of a body with respect to Earth. Therefore, it is important to define which type of synodic period is meant for any particular usage. The differences between these three periods are usually small, but in some circumstances, they may not be negligible and understanding exactly what is being measured may be important. For example, Comet Hyakutake's solar rotation period was measured sufficiently accurately as to be distinguishable from the sidereal period determined from Monte Carlo modeling of the dust jets during the comet's close approach to Earth. The difference of 0.0004 d between the two periods was completely consistent with the expected difference, based on the model pole orientation and the position of the comet in its orbit (*Schleicher and Woodney,* 2003).

6.5. Modeling Morphological Features

We now turn to methods that are used for inferring additional properties of the nucleus from the coma morphology. In certain circumstances, properties can be determined directly from measurements, without the need for models. An excellent example of this is Comet Borrelly, whose strongest source emits material in a straight jet that is aligned with the nucleus' spin axis. Given this configuration, the apparent position of the jet on different dates can be used to determine the orientation of the rotation pole (*Farnham and Cochran,* 2002) [a similar technique was used in an analysis of Hale-Bopp by *Licandro et al.* (1999)]. It is ironic that the jet can be used to determine so much about the spin axis,

but because it is at the pole, it contains no information about the rotation period or the direction of spin.

The fortuitous circumstances regarding Comet Borrelly are unusual, however, and for most comets, more intricate models of the coma morphology must be used to extract the nucleus properties. Many different models, both simplistic and intricate, have been introduced to explain the morphology observed in cometary comae. Early models invoked such concepts as using thermal lags to explain the projected direction of a sunward fan on a homogeneous rotating nucleus (*Sekanina,* 1979) and using the dimensions of consecutive arcs along with assumed expansion velocities to compute rotation periods (*Whipple,* 1982). These early models produced mixed results at best, with a number of the test cases being proven wrong by subsequent observations (Borrelly being one notable case). The next generation of models used continuous tracks of jet particles to follow the features produced by emitted material (*Sekanina,* 1981; *Sekanina and Larson,* 1984, 1986; *Massonne,* 1985) and showed promising results. More recently, a variety of different types of models have been presented, for reproducing both dust morphology (cf. *Sekanina,* 1987, 1991; *Sekanina et al.,* 1992; *Combi,* 1994; *Sekanina and Boehnhardt,* 1997; *Fulle et al.,* 1997; *Schleicher et al.,* 1998c; *Samarasinha,* 2000) and gas morphology (cf. *Lederer et al.,* 1997; *Festou and Barale,* 2000). Most of these recent techniques are based on numerical methods or Monte Carlo simulations.

The increase in computing power over the last decade has not only made the Monte Carlo-style techniques very popular, but they are also very powerful and provide a natural approach for simulating particles emitted from a spinning nucleus. In addition to the characteristics already discussed, a number of fundamental nucleus properties can be determined from the morphology, including a comprehensive depiction of the rotation state and the locations and sizes of the active areas. Secondary parameters can also be derived using the results from the primary analysis: With an understanding of the spin properties and source locations, projection effects can be computed, allowing true distances and velocities to be determined; similarly, thermal lags can be found when sources remain active, even after they are no longer illuminated by sunlight; knowledge of the rotation state and production rates (which can be estimated from the solar illumination on the active regions), provides necessary constraints for analyses involving torques and nongravitational forces on the nucleus (*Samarasinha et al.,* 2004; *Yeomans et al.,* 2004); finally, comparisons of models independently derived from the dust and various gas species may reveal potential composition inhomogeneities, if different species originate from different source regions. Under typical circumstances, only a subset of these properties will be determined for any given comet, with the type and quality of the features governing which results can be derived.

When using models to analyze a comet's coma morphology, different researchers are likely to use slightly different approaches, although the fundamental basis will be similar. The following discussion describes the specific techniques and procedures that we have used in our work with various comets, and although the details may differ somewhat from other researcher's methods, any model should address essentially the same issues. In our work, we use a standard Monte Carlo model that is discussed more fully by *Schleicher and Woodney* (2003) and Farnham and Schleicher (in preparation, 2004). The routine is presently designed to model the motions of the dust grains and can follow up to 10^6 representative particles that are emitted from multiple active areas at different locations and of different sizes (allowing us to model the extended active areas discussed later). The model can also handle radiation pressure and precession of the nucleus, if necessary. Initial calculations are done in the comet's orbital reference frame, from which it is straightforward to determine the orientation of the nucleus, the Sun's position, and other geometric relationships. For each particle, the routine computes the direction in which it was emitted and the distance it has traveled between its emission time and the observation time, which defines its location in cometary coordinates when the comet was observed. After the positions have been computed for all the particles, the results are then transformed to the plane of the sky coordinates as seen from Earth, and the result can be compared to the observed morphology. The nucleus properties that can be found from our model include the rotation period; orientation of the spin axis; direction of rotation; locations, sizes, and relative strengths of multiple active areas; emission velocities; and the average influence of radiation pressure on the dust grains.

In our application of the model, we start by selecting the most obvious and clearly defined feature in an image and use it as a guide throughout the early stages of the modeling. By initially focusing on only the main feature, we can limit the number of free parameters, which reduces the volume of phase space that must be explored. (Typically only four parameters — right ascension and declination of the pole, the rotation period, and the latitude of the primary source — are needed to explore the basic morphology; other parameters, such as the longitude of the source and the ejection velocity, only control the relative phasing and the scale of the coma.) Next, we assign a pole orientation and a location for the main active region and generate a model for those parameters. Comparing the synthetic image to the observed one (specifically, to the positions measured from the images) allows us to adjust the model parameters and rerun the model to improve the fit. This process is iterated until the model parameters converge to produce a good match to the observations. As in any multivariable analysis, there is always a concern that the parameters are unique and that other combinations of parameters cannot be combined to produce equally suitable results. Therefore, to avoid missing any potential solutions, we perform a full grid search of the four main parameters at low resolution. This allows us to map the areas of parameter space where viable models exist, and we can then focus on these areas at higher resolution to converge on the optimum solution. Once the orienta-

tion of the spin axis and location of the main source have been constrained, and parameter space has been narrowed, we can introduce additional parameters (new active regions, radiation pressure, etc.) to model other features and help fine tune all the model parameters. This is again done in stages to allow the effects of each addition to be determined.

Given the fact that there will always be at least four free parameters (with the potential of many more), any information that can be used to help narrow down the parameter space is welcome. Frequently, it is possible to use a simple inspection of the morphology to limit parameters, even before detailed modeling begins. For example, the shape of a spiral arc can often be used to set constraints on the parameters. If the spirals extend completely around the nucleus, then Earth must lie within the cone swept out by the active region during a rotation (i.e., the sub-Earth point lies at a higher latitude than the active region). In addition, the shape of the spiral may define the direction of rotation, which will naturally eliminate at least half the potential pole orientations. Finally, if the spiral appears elongated, then the ratio of the short- and long-axis dimensions can provide a constraint on the aspect angle of the pole. Similarly, radial features can also be used to constrain the parameters. A jet that oscillates back and forth in position angle indicates that Earth is outside the cone, and the size of the oscillation can be used to set a limit on the latitude of the source. Furthermore, if the feature is continuously visible, then the center of the oscillation likely represents the projection of the rotation axis on the sky. Even though some parameters can be constrained in this manner, it is a good idea to utilize the modeling process to check that the interpretation of the features is correct and to make sure that the excluded areas of parameter space behave as expected.

Another procedure that can be employed in the analysis is the incorporation of multiple images throughout the comet's apparition (e.g., *Braustein et al.,* 1997; *Vasundhara and Chakraborty,* 1999). Tracking the long-term evolution of the coma makes it possible to generate a comprehensive model for reproducing the general appearance of the coma at any given time. Furthermore, dramatic changes in the morphology can act as benchmarks for deriving the locations of the active regions. For example, the gradual appearance or disappearance of a bright jet can indicate that the subsolar latitude is changing and a source is becoming illuminated or losing its illumination. Observations spanning a significant arc of the orbit may also reveal other factors, such as the times at which Earth crosses into or out of the cone swept out by an active area. These types of information can be used to severely constrain the locations of the source regions, which in turn simplifies the modeling process.

We now address a new complication regarding coma models that was introduced during our studies of Comet Hale-Bopp and has implications for coma models in general. There is an extensive amount of data available for this comet around the time of perihelion and the coma could be studied in detail from March through early May. Examination of the arcs in any particular Hale-Bopp image from this time frame shows primarily circular features (dust arcs or CN spirals) with little foreshortening in any direction. The rounded shape suggests that the comet's spin axis was pointed in the general direction of Earth, which posed a problem, because the Earth-comet viewing geometry changed by about 90° between March and May. In other words, for the pole to be pointed toward Earth throughout this time frame, the nucleus would have to be in a state of fast precession with the pole tracking Earth — a difficult scenario to accept. The solution to this puzzle was suggested by *Samarasinha* (2000), and involves the size of the active region creating the feature. Normally, jets in a model are assumed to be narrow, if indeed any width at all is specified. This simplifies the models, produces well-defined features, and has been widely accepted because the result usually reflects the appearance of the observed image. Samarasinha suggested that the arcs in Comet Hale-Bopp are not produced by jets only a few degrees wide, but instead are the result of large active regions, spanning tens of degrees in latitude and/or longitude. The effect of these broad jets is to form a partial shell structure that can mimic the planetary nebula effect. In a planetary nebula, the spherical shell appears to be circular because the greatest column density is at the outer edge. Similarly, in Hale-Bopp, the partial shells are seen as arcs that always appear circular, even when the aspect angle changes dramatically (Fig. 5). Another effect of the extended sources is that they can create intricate overlapping structures, which naturally reproduce the appearance of many of the complicated features seen in Hale-Bopp.

The existence of wide jets makes the coma more difficult to model, not only because it introduces more free parameters, but also because it makes interpretation of the features more difficult. With an extended source, the existence of a complete spiral around the nucleus is no longer a guarantee that Earth lies within the cone formed by the spinning jet. If Earth lies within only a part of the cone produced by the jet, the planetary nebula effect will dominate and full spirals will be observed. Furthermore, as with Hale-Bopp, the rounded arc appearance no longer provides a severe constraint on the aspect angle of the pole. It is clear from these examples that the potential for having extended active regions introduces ambiguities into the constraints that can be set with simple inspection of the coma structure. Thus, in any comprehensive analysis, it is wise to investigate the possibility that broad jets are contributing to the coma morphology, because the differences can have a profound influence on the model results, as it did for Comet Hale-Bopp.

Our final discussion addresses various concerns and considerations to be aware of when applying these models. First, the uniqueness of the solution is foremost when presenting a result, and a global search of parameter space, although tedious and time consuming, may be necessary to rule out other families of solutions. Next, a comprehensive

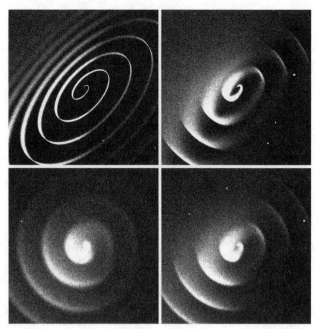

Fig. 5. Sequence of simulated images showing the effects of extended active regions on the appearance of the coma. Clockwise from the upper left, the source regions span angles of 1°, 5°, 15°, and 30°, with all other parameters left unchanged. Notice that for the larger source regions, the arcs are more circular, giving the appearance that Earth lies very close to the spin axis. From *Samarasinha* (2000).

model should reproduce the morphology of the original image, not the enhanced one. When possible, the model should be enhanced with the same methods used on the original image to see how well the two really match. This is not always practical, especially during the global search of parameters, but as the solution converges, the quality of the fit is more critical. Related to this is the principle of determining how good a fit has actually been obtained. Using a mathematical measure of the goodness-of-fit (e.g., a χ^2 fit) is usually not practical because quantifying the difference between the model and the image can be difficult without resorting to time-consuming measurements. Fortunately, pattern matching by eye is very effective for this type of work, especially when the results can be displayed in multiple formats. This argues that both the rectangular and polar versions of the model and the image should be compared. Finally, the role that is played by the background material in the coma should be considered. The bulk shape of the coma (the component that is usually removed in the enhancement process) must come from somewhere, with two possible sources being isotropic emission from the nucleus' surface and diffusion of material outward from the isolated sources. Ideally, this component should be included in the model for completeness, but it is not always clear exactly how to include it or account for it.

As described here, modeling the structures in cometary comae can reveal fundamental properties of the nucleus.

The results that are obtained are inherently important in and of themselves, but they become even more valuable when they are used to set constraints on a variety of other comet studies. We note a few of these, which are discussed in more detail in other chapters of this book. First, understanding the main jet structures in the coma as a function of time helps in interpreting the more intricate effects produced by topography and the gas/dust interactions near the nucleus (*Crifo et al.*, 2004). As discussed earlier, the rotation properties can be used as examples for studies of the rotational dynamics of nuclei (*Samarasinha et al.*, 2004) and, when the locations of the active regions are included, for research related to nongravitational forces (*Yeomans et al.*, 2004). Information about the sizes and locations of the active areas and when they turn on and off provide constraints for studies of normal cometary activity and outbursts (*Prialnik et al.*, 2004; *Boehnhardt*, 2004). Finally, the source activity information and true emission velocities obtained from inner coma models can be used to improve models of the gas dynamics (*Combi et al.*, 2004) and models of the comet's dust tail (*Fulle*, 2004).

7. FUTURE DIRECTIONS

The past two decades have seen vast advances in the capabilities of array detectors, including extended wavelength coverage, particularly in the near-UV, increased quantum efficiencies at all wavelengths, improved noise characteristics, decreased readout times, and larger formats. With these improvements, coma morphology studies have become common, especially for brighter comets, although most such studies have been qualitative in nature; analyses involving absolutely calibrated, continuum-subtracted, and decontaminated gas images are unfortunately still only rarely performed. We hope that the recent introduction of new narrowband filter sets by ESA and NASA, with filter sizes as large as 100 mm for use with large-format CCD cameras, will provide additional incentive to perform more quantitative studies in the future. In any case, newer large-format cameras will certainly provide larger fields-of-view for active comets than previously possible, thereby sampling significantly larger portions of the coma than have been typical in the past. These developments should lead to improved models of the outflow and interaction of dust and gas in comae. However, even with the many ongoing advances in the field of narrowband imaging, we believe that narrowband measurements with conventional photometers will continue to provide very important results regarding the overall chemical composition of comets for many years to come.

The ability to directly intercompare model comae with observations will continue to improve with advances in computing power, permitting large increases in the number of particles used in Monte Carlo-type simulations and more realistic scenarios. Increases in computing power will also allow multivariable minimization routines to be used to explore and constrain the multidimensional parameter space more efficiently. Furthermore, coma models will

undergo severe testing, with spacecraft encountering several comets in the near future; these encounters will either validate the modeling procedures that are currently in use, or will prompt their reevaluation.

Finally, narrowband imaging, multiwavelength polarization studies show promise for better understanding the physical properties of dust grains as they move outward from the nucleus. They may answer questions about fragmentation of grains and whether the ambient background of the coma is caused by dispersed grains from jets or by a more homogeneous nucleus component.

Acknowledgments. The authors particularly thank M. A'Hearn, A. Cochran, M. Combi, R. Millis, S. Larson, and N. Samarasinha for innumerable fruitful conversations regarding aspects of the contents of this chapter, as well as the referees for many useful comments and suggestions. We also thank S. Larson for providing the radial gradient filter enhancement of the image of Comet Hyakutake. This work has been made possible due to grants from the National Aeronautics and Space Administration and the National Science Foundation.

REFERENCES

A'Hearn M. F. (1982) Spectroscopy and spectrophotometry of comets at optical wavelengths. In *Comets* (L. L. Wilkening, ed.), pp. 433–460. Univ. of Arizona, Tucson.

A'Hearn M. F. (1983) Photometry of comets. In *Solar System Photometry Handbook* (R. Genet, ed.), pp. 3-1–3-33. Willmann-Bell, Richmond, Virginia.

A'Hearn M. F. (1991) Photometry and polarimetry network. In *The Comet Halley Archive Summary Volume* (A. Sekanina, ed.), pp. 193–235. Jet Propulsion Laboratory, Pasadena.

A'Hearn M. F. and Cowan J. J. (1975) Molecular production rates in Comet Kohoutek. *Astrophys. J., 80,* 852–860.

A'Hearn M. F. and Millis R. L. (1980) Abundance correlations among comets. *Astrophys. J., 85,* 1528–1537.

A'Hearn M. F., Millis R. L., and Birch P. V. (1979) Gas and dust in some recent periodic comets. *Astrophys. J., 84,* 570–579.

A'Hearn M. F., Schleicher D. G., Feldman P. D., Millis R. L., and Thompson D. T. (1984) Comet Bowell 1980b. *Astrophys. J., 89,* 579–591.

A'Hearn M. F., Birch P. V., Feldman P. D., and Millis R. L. (1985) Comet Encke — Gas production and lightcurve. *Icarus, 64,* 1–10.

A'Hearn M. F., Hoban S., Birch P. V., Bowers C., Martin R., and Klinglesmith D. A. III (1986a) Cyanogen jets in Comet Halley. *Nature, 324,* 649–651.

A'Hearn M. F., Birch P. V., and Klinglesmith D. A. III (1986b) Gaseous jets in Comet P/Halley. In *20th ESLAB Symposium on the Exploration of Halley's Comet, Vol. 1* (B. Battrick et al., eds.), pp. 483–486. ESA SP-250, Noordwijk, The Netherlands.

A'Hearn M. F., Millis R. L., Schleicher D. G., Osip D. J., and Birch P. V. (1995) The ensemble properties of comets: Results from narrowband photometry of 85 comets, 1976–1992. *Icarus, 118,* 223–270.

Arpigny C. (1976) Interpretation of comet spectra: A review. In *The Study of Comets* (B. Donn et al., eds.), pp. 797–838. NASA SP-393, Washington, DC.

Baum W. A., Kreidl T. J., and Schleicher D. G. (1992) Cometary grains. *Astrophys. J., 104,* 1216–1225.

Belton M. J. S., Mueller B. E. A., Julian W. H., and Anderson

A. J. (1991) The spin state and homogeneity of Comet Halley's nucleus. *Icarus, 93,* 183–193.

Blamont J. E. and Festou M. (1974) Observation of the comet Kohoutek (1973f) in the resonance light (A²Σ⁺–X²Π) of the OH radical. *Icarus, 23,* 538–544.

Bockelée-Morvan D., Crovisier J., Mumma M. J., and Weaver H. A. (2004) The composition of cometary volatiles. In *Comets II* (M. C. Festou et al., eds.), this volume. Univ. of Arizona, Tucson.

Boehnhardt H. (2004) Split comets. In *Comets II* (M. C. Festou et al., eds.), this volume. Univ. of Arizona, Tucson.

Boehnhardt H. and Birkle K. (1994) Time variable coma structures in comet P/Swift-Tuttle. *Astron. Astrophys. Suppl. Ser., 107,* 101–120.

Boehnhardt H., Vanysek V., Birkle K., and Hopp U. (1992) Coma imaging of comet P/Brorsen-Metcalf at Calar Alto in late July to mid August 1989. In *Asteroids, Comets, Meteors 1991* (A. W. Harris and E. Bowell, eds.), pp. 81–84. Lunar and Planetary Institute, Houston.

Braunstein M., Womack M., Deglman F., Pinnick G., Comstock R., Hoffman P., Aaker G., Faith D., Moore S., Ricotta J., Goldschen M., Wiest A., Modi C., Jacobson A., and Zilka J. (1997) A CCD image archive of Comet C/1995 O1 (Hale-Bopp): Dust expansion velocities. *Earth Moon Planets, 78,* 219–227.

Budding E. (1993) *Introduction to Astronomical Photometry.* Cambridge Univ., Cambridge. 272 pp.

Chernova G. P., Kiselev N. N., and Jockers K. (1993) Polarimetric characteristic of dust particles as observed in 13 comets: Comparison with asteroids. *Icarus, 103,* 144–158.

Cochran A. L. and Cochran W. D. (1991) The first detection of CN and the distribution of CO⁺ gas in the coma of Comet P/Schwassmann-Wachmann 1. *Icarus, 90,* 172–175.

Cochran A. L. and Schleicher D. G. (1993) Observational constraints on the lifetime of cometary H_2O. *Icarus, 105,* 235–253.

Cochran A. L., Cochran W. D., Barker E. S., and Storrs A. D. (1991) The development of the CO⁺ coma of Comet P/Schwassmann-Wachmann 1. *Icarus, 92,* 179–183.

Combi M. R. (1994) The fragmentation of dust in the innermost comae of comets: Possible evidence from ground-based images. *Astrophys. J., 108,* 304–312.

Combi M. R., Harris W. M., and Smyth W. H. (2004) Gas dynamics and kinetics in the cometary coma: Theory and observations. In *Comets II* (M. C. Festou et al., eds.), this volume. Univ. of Arizona, Tucson.

Cowan J. J. and A'Hearn M. F. (1979) Vaporization of cometary nuclei: Light curves and lifetimes. *Moon and Planets, 21,* 155–171.

Crifo J. F., Fulle M., Kömle N. I., and Szegö K. (2004) Nucleus-coma structural relationships: Lessons from physical models. In *Comets II* (M. C. Festou et al., eds.), this volume. Univ. of Arizona, Tucson.

Farnham T. L. and Cochran A. L. (2002) A McDonald Observatory study of Comet 19P/Borrelly: Placing the Deep Space 1 observations into a broader context. *Icarus, 160,* 398–418.

Farnham T. L. and Meech K. J. (1994) Comparison of the plasma tails of four comets: P/Halley, Okazaki-Levy-Rudenko, Austin and Levy. *Astrophys. J., 91,* 419–460.

Farnham T. L., Schleicher D. G., Ford E., and Blount E. A. (1998a) The rotation period of Comet Hale-Bopp (1995 O1). In *Abstracts for the First International Conference of Comet Hale-Bopp,* p. 16. Tenerife, 1998 February 2–5.

Farnham T. L., Schleicher D. G., and Cheung C. C. (1998b) Rotational variation of the gas and dust jets in Comet Hale-Bopp (1995 O1) from narrowband imaging. *Bull. Am. Astron. Soc., 30,* 1072.

Farnham T. L., Schleicher D. G., Williams W. R., and Smith B. R. (1999) The rotation state and active regions of Comet Hale-Bopp (1995 O1). *Bull. Am. Astron. Soc., 31,* 1120.

Farnham T. L., Schleicher D. G., and A'Hearn M. F. (2000) The HB narrowband comet filters: Standard stars and calibrations. *Icarus, 147,* 180–204.

Feldman P. D., Cochran A. L., and Combi M. R. (2004) Spectroscopic investigations of fragment species in the coma. In *Comets II* (M. C. Festou et al., eds.), this volume. Univ. of Arizona, Tucson.

Festou M. C. and Barale O. (2000) The asymmetric coma of comets. I. Asymmetric outgassing from the nucleus of Comet 2P/Encke. *Astrophys. J., 119,* 3119–3132.

Fomenkova M. N., Jones B., Pina R., Puetter R., Sarmecanic J., Gehrz R., and Jones T. (1995) Mid-infrared observations of the nucleus and dust of Comet P/Swift-Tuttle. *Astrophys. J., 110,* 1866–1957.

Fulle M. (2004) Motion of cometary dust. In *Comets II* (M. C. Festou et al., eds.), this volume. Univ. of Arizona, Tucson.

Fulle M., Milani A., and Pansecchi L. (1997) Tomography of a sunward structure in the dust tail of comet 19P/Borrelly. *Astron. Astrophys., 321,* 338–342.

Hanner M. S. and Tokunaga A. T. (1991) Infrared techniques for comet observations. In *Comets in the Post-Halley Era, Vol. 1* (R. L. Newburn Jr. et al., eds.), pp. 67–91. Kluwer, Dordrecht.

Henden A. A. and Kaitchuck R. H. (1982) *Astronomical Photometry.* Van Nostrand Reinhold, New York. 392 pp.

Hoban S., Samarasinha N. H., A'Hearn M. F., and Klinglesmith D. A. (1988) An investigation into periodicities in the morphology of CN jets in comet P/Halley. *Astron. Astrophys., 195,* 331–337.

Howell S. B., ed. (1992) *Astronomical CCD Observing and Reduction Techniques.* ASP Conference Series 23, Astronomical Society of the Pacific, San Francisco. 329 pp.

Howell S. B. (2000) *Handbook of CCD Astronomy.* Cambridge Univ., Cambridge. 164 pp.

International Halley Watch (1995) *Comet Halley Archive, Vol. 22.* USA-NASA-IHW-HAL-0022, Small Bodies Node, Univ. of Maryland, College Park (CD-ROM).

Ip W.-H. (2004) Global solar wind interaction and ionospheric dynamics. In *Comets II* (M. C. Festou et al., eds.), this volume. Univ. of Arizona, Tucson.

Jacoby G. H., ed. (1990) *CCDs in Astronomy.* ASP Conference Series 8, Astronomical Society of the Pacific, San Francisco. 407 pp.

Jewitt D. (1991) Cometary photometry. In *Comets in the Post-Halley Era, Vol. 1* (R. L. Newburn Jr. et al., eds.), pp. 19–65. Kluwer, Dordrecht.

Jewitt D. C. and Meech K. J. (1987) Surface brightness profiles of 10 comets. *Astrophys. J., 317,* 992–1001.

Jorda L., Rembor K., Lecacheux J., Colom P., Colas F., Frappa E., and Lara L. M. (1999) The rotational parameters of Hale-Bopp (Comet C/1995 O1) from observations of the dust jets at Pic du Midi Observatory. *Earth Moon Planets, 77,* 167–180.

Kim S. J., A'Hearn M. F., and Cochran W. D. (1989) NH emissions in comets — fluorescence vs. collisions. *Icarus, 77,* 98–108.

Kiselev N. N. and Chernova G. P. (1981) Phase functions of polarization and brightness and the nature of cometary atmosphere particles. *Icarus, 48,* 473–481.

Klinglesmith D. A. (1981) The interactive astronomical data analysis facility — image enhancement techniques to Comet Halley. In *Modern Observational Techniques for Comets: Proceedings of a Workshop held at Goddard Space Flight Center, Greenbelt, Maryland, on October 22–24, 1980* (J. C. Brandt et al., eds.), pp. 223–231. JPL Publication No. 81-68.

Kolokolova L., Jockers K., Chernova G., and Kiselev N. (1997) Properties of cometary dust from color and polarization. *Icarus, 126,* 351–361.

Kolokolova L., Hanner M. S., Levasseur-Regourd A.-C., and Gustafson B. Å. S. (2004) Physical properties of cometary dust from light scattering and thermal emission. In *Comets II* (M. C. Festou et al., eds.), this volume. Univ. of Arizona, Tucson.

Lamy P. L., Toth I., Fernández Y. R., and Weaver H. A. (2004) The sizes, shapes, albedos, and colors of cometary nuclei. In *Comets II* (M. C. Festou et al., eds.), this volume. Univ. of Arizona, Tucson.

Larson S. M. and Sekanina Z. (1984) Coma morphology and dust-emission pattern of periodic Comet Halley. I — High-resolution images taken at Mount Wilson in 1910. *Astrophys. J., 89,* 571–578.

Larson S. M. and Slaughter C. D. (1992) Evaluating some computer enhancement algorithms that improve the visibility of cometary morphology. In *Asteroids, Comets, Meteors 1991* (A. W. Harris and E. Bowell, eds.), pp. 337–343. Lunar and Planetary Institute, Houston.

Larson S. M., Sekanina Z., and Rahe J. (1991) Near-nucleus studies network. In *The Comet Halley Archive Summary Volume* (A. Sekanina, ed.), pp. 173–191. Jet Propulsion Laboratory, Pasadena.

Larson S. M., Hergenrother C. W., and Brandt J. C. (1997) The spatial and temporal distribution of CO+ and CN in C/1995 O1 (Hale-Bopp). *Bull. Am. Astron. Soc., 29,* 1036.

Le Borgne J. F., Leroy J. L., and Arnaud J. (1987) Polarimetry of visible and near-UV molecular bands: Comets P/Halley and Hartley-Good. *Astron. Astrophys., 173,* 180–182.

Lecacheux J., Jorda L., and Colas F. (1997) *Comet C/1995 O1 (Hale-Bopp).* IAU Circular No. 6560.

Lederer S. M., Campins H, Osip D. J., and Schleicher D. G. (1997) Gaseous jets in Comet Hale-Bopp (1995 O1). *Earth Moon Planets, 78,* 131–136.

Levasseur-Regourd A.-C. (1999) Polarization of light scattered by cometary dust particles: Observations and tentative interpretations. *Space Sci. Rev., 90,* 163–168.

Licandro J., Bellot Rubio L. R., Boehnhardt H., Casas R., Goetz B., Gomez A., Jorda L., Kidger M. R., Osip D., Sabalisck N., Santos P., Serr-Ricart M., Tozzi G. P., and West R. (1998) The rotation period of C/1995 O1 (Hale-Bopp). *Astrophys. J. Lett., 501,* L221.

Licandro J., Bellot Rubio L. R., Casas R., Gomez A., Kidger M. R., Sabalisk N., Santos-Sanz P., Serra-Ricart M., Torres-Chico R., Oscoz A., Jorda L., and Denicolo G. (1999) The spin axis position of C/1995 O1 (Hale-Bopp). *Earth Moon Planets, 77,* 199–206.

Mannucci F. and Tozzi G.P. (1997) *Comet C/1995 O1 (Hale-Bopp).* IAU Circular No. 6575.

Massonne L. (1985) Coma morphology and dust emission pattern of Comet Halley. *Adv. Space Res., 5,* 187–196.

Meech K. J. and Jewitt D. C. (1987) Observation of comet P/Halley at minimum phase angle. *Astron. Astrophys., 187,* 585–593.

Meier R., Wellnitz D., Kim S. J., and A'Hearn M. F. (1998) The NH and CH bands of Comet C/1996 B2 (Hyakutake). *Icarus, 136,* 268–279.

Meisel D. D. and Morris C. S. (1982) Comet head photometry: Past, present and future. In *Comets* (L. L. Wilkening, ed.), pp. 413–432. Univ. of Arizona, Tucson.

Millis R. L. and Schleicher D. G. (1986) Rotational period of comet Halley. *Nature, 324,* 646–649.

Millis R. L., A'Hearn M. F., and Thompson D. T. (1982) Narrowband photometry of Comet P/Stephan-Oterma and the backscattering properties of cometary grains. *Astrophys. J., 87,* 1310–1317.

Mueller B. E. A., Samarasinha N. H., and Belton M. J. S. (1999) Imaging of the structure and evolution of the coma morphology of Comet Hale-Bopp (C/1995 O1). *Earth Moon Planets, 77,* 181–188.

Ney E. P. (1982) Optical and infrared observations of bright comets in the range 0.5 microns to 20 microns. In *Comets* (L. L. Wilkening, ed.), pp. 413–432. Univ. of Arizona, Tucson.

O'Dell C. R. and Osterbrock D. E. (1962) Emission-band and continuum photometry of Comet Seki (1961f). *Astrophys. J., 136,* 559–566.

Osborn W., A'Hearn M. F., Carsenty U., Millis R. L., Schleicher D. G., Birch P. V., Moreno H., and Gutierrez-Moreno A. (1990) Standard stars for photometry of comets. *Icarus, 88,* 228–245.

Philip A. G. D., Janes K. A., and Upgren A. R., eds. (1995) *New Developments in Array Technology and Applications.* IAU Symposium 167, Kluwer, Dordrecht. 397 pp.

Prialnik D., Benkhoff J., and Podolak M. (2004) Modeling the structure and activity of comet nuclei. In *Comets II* (M. C. Festou et al., eds.), this volume. Univ. of Arizona, Tucson.

Randall E. E., Schleicher D. G., Ballou R. G., and Osip D. J. (1992) Observational constraints on molecular scalelengths and lifetimes in comets. *Bull. Am. Astron. Soc., 24,* 1002.

Richards J. A. (1993) *Remote Sensing Digital Image Analysis: An Introduction.* Springer-Verlag, New York. 340 pp.

Samarasinha N. H. (2000) Coma morphology due to an extended active region and implications for the spin state of Comet Hale-Bopp. *Astrophys. J. Lett., 529,* L107–L110.

Samarasinha N. H. and Mueller B. E. A. (2002) Spin axis direction of Comet 19P/Borrelly based on observations from 2000 and 2001. *Earth Moon Planets, 90,* 473–482.

Samarasinha N. H., A'Hearn M. F., Hoban S., and Klinglesmith D. A. III (1986) CN jets of Comet P/Halley: Rotational properties. In *20th ESLAB Symposium on the Exploration of Halley's Comet, Vol. 1* (B. Battrick et al., eds), pp. 487–491. ESA SP-250, Noordwijk, The Netherlands.

Samarasinha N. H., Mueller B. E. A., and Belton M. J. S. (1999) Coma morphology and constraints on the rotation of Comet Hale-Bopp (C/1995 O1). *Earth Moon Planets, 77,* 189–198.

Samarasinha N. H., Mueller B. E. A., Belton M. J. S., and Jorda L. (2004) Rotation of cometary nuclei. In *Comets II* (M. C. Festou et al., eds.), this volume. Univ. of Arizona, Tucson.

Sarmecanic J. R., Osip D. J., Fomenkova M. N., and Jones B. (1997) *Comet C/1995 O1 (Hale-Bopp).* IAU Circular No. 6600.

Schleicher D. G. (1983) The fluorescence of cometary OH and CN. Ph.D. dissertation, Univ. of Maryland, College Park.

Schleicher D. G. and A'Hearn M. F. (1988) The fluorescence of cometary OH. *Astrophys. J., 331,* 1058–1077.

Schleicher D. G. and Osip D. J. (2002) Long and short-term photometric behavior of Comet Hyakutake (1996 B2). *Icarus, 159,* 210–233.

Schleicher D. G. and Woodney L. M. (2003) Analysis of dust coma morphology of Comet Hyakutake (1996 B2) near perigee:

Outburst behavior, jet motion, source region locations, and nucleus pole orientation. *Icarus, 162,* 190–213.

Schleicher D. G., Millis R. L., and Birch P. V. (1987) Photometric observations of Comet P/Giacobini-Zinner. *Astron. Astrophys., 187,* 531–538.

Schleicher D. G., Millis R. L., Osip D. J., and Birch P. V. (1991) Comet Levy (1990c): Groundbased photometric results. *Icarus, 94,* 511–523.

Schleicher D. G., Millis R. L., and Birch P. V. (1998a) Narrowband photometry of Comet P/Halley: Variation with heliocentric distance, season, and solar phase angle. *Icarus, 132,* 397–417.

Schleicher D. G., Millis R. L., Osip D. J., and Lederer S. M. (1998b) Activity and the rotation period of Comet Hyakutake (1996 B2). *Icarus, 131,* 233–244.

Schleicher D. G., Farnham T. L., Smith B. R., Blount E. A., Nielsen E., and Lederer S. M. (1998c) Nucleus properties of Comet Hale-Bopp (1995 O1) based on narrowband imaging. *Bull. Am. Astron. Soc., 30,* 1063.

Schleicher D. G., Farnham T. L., Williams W. R., Smith B. R., and Cheung C. C. (1999) Modeling the rotational morphology of gas and dust jets in Comet Hale-Bopp (1995 O1) at perihelion. *Bull. Am. Astron. Soc., 31,* 1128.

Schleicher D. G, Woodney L. M., and Millis R. L. (2003) Comet 19P/Borrelly at multiple apparitions: Seasonal variations in gas production and dust morphology. *Icarus, 162,* 415–442.

Schmidt M. and van Woerden H. (1957) The intensity distribution of molecular bands in the coma of Comet Mrkos 1955e. *Liege Inst. d'Ap. Reprint No. 386,* pp. 102–111. Liege, Belgium.

Schulz R. and A'Hearn M. F. (1995) Shells in the C_2 coma of Comet P/Halley. *Icarus, 115,* 191–198.

Schulz R., A'Hearn M. F., Birch P. V., Bowers C., Kempin M., and Martin R. (1993) CCD imaging of Comet Wilson (1987VII) — A quantitative coma analysis. *Icarus, 104,* 206–225.

Schulz R., Stüwe J. A., Tozzi G. P., and Owens A. (2000) Optical analysis of an activity outburst in comet C/1995 O1 (Hale-Bopp) and its connection to an X-ray outburst. *Astron. Astrophys., 361,* 359–368.

Schwarz G., Cosmovici C., Mack P., and Ip W. (1989) Image processing techniques for gas morphology studies in the coma of Comet Halley. *Adv. Space Res., 9,* 217–220.

Schwarz G., Cosmovici C. B., Crippa R., Guaita C., Manzini F., and Oldani V. (1997) Comet Hale-Bopp: Evolution of jets and shells during March 1997. *Earth Moon Planets, 78,* 189–195.

Sekanina Z. (1978) Comet West 1976. VI — Discrete bursts of dust, split nucleus, flare-ups, and particle evaporation. *Astrophys. J., 83,* 1675–1680.

Sekanina Z. (1979) Fan-shaped coma, orientation of rotation axis, and surface structure of a cometary nucleus. I — Test of a model on four comets. *Icarus, 37,* 420–442.

Sekanina Z. (1981) Distribution and activity of discrete emission areas on the nucleus of periodic Comet Swift-Tuttle. *Astrophys. J., 86,* 1741–1773.

Sekanina Z. (1987) Anisotropic emission from comets: Fans versus jets. 1: Concept and modeling. In *Proceedings of the International Symposium on the Diversity and Similarity of Comets* (E. J. Rolfe et al., eds.) pp. 315–322. ESA SP-278, Noordwijk, The Netherlands.

Sekanina Z. (1991) Randomization of dust-ejecta motions and the observed morphology of cometary heads. *Astrophys. J., 102,* 1870–1878.

Sekanina Z. and Boehnhardt H. (1997) Dust morphology of Comet Hale-Bopp (C/1995 O1). II. Introduction of a working model. *Earth Moon Planets, 78,* 313–319.

Sekanina Z. and Larson S. M. (1984) Coma morphology and dust-emission pattern of periodic Comet Halley. II — Nucleus spin vector and modeling of major dust features in 1910. *Astrophys. J., 89,* 1408–1425.

Sekanina Z. and Larson S. M. (1986) Coma morphology and dust-emission pattern of periodic Comet Halley. IV — Spin vector refinement and map of discrete dust sources for 1910. *Astrophys. J., 92,* 462–482.

Sekanina Z., Larson S. M., Hainaut O., Smette A., and West R. M. (1992) Major outburst of periodic Comet Halley at a heliocentric distance of 14 AU. *Astron. Astrophys., 263,* 367–386.

Sen A. K., Joshi U. C., and Deshpande M. R. (1989) Molecular band polarization in comet P/Halley. *Astron. Astrophys., 217,* 307–310.

Sterken C. and Manfroid J. (1992) *Astronomical Photometry: A Guide.* Kluwer, Dordrecht. 272 pp.

Tatum J. B. and Gillespie M. I. (1977) The cyanogen abundance of comets. *Astrophys. J., 218,* 569–572.

Vanysek V. (1976) Photometry of the cometary atmosphere: A review. In *The Study of Comets, Part I* (B. Donn et al., eds.), pp. 1–27. NASA SP-393, Washington, DC.

Vasundhara R. and Chakraborty P. (1999) Modeling of jets from Comet Hale-Bopp (C/1995 O1): Observations from the Vainu Bappu Observatory. *Icarus, 140,* 221–230.

Warell J., Lagerkvist C.-I., and Lagerros J. S. V. (1999) Dust continuum imaging of C/1995 O1 (Hale-Bopp): Rotation period and dust outflow velocity. *Astron. Astrophys. Suppl. Ser., 136,* 245–256.

Watanabe J.-I. (1987) The rotation of Comet 1983 VII IRAS-Araki-Alcock. *Publ. Astron. Soc. Japan, 39,* 485–503.

Wehinger P. A., Wyckoff S., Herbig G. H., Herzberg G., and Lew H. (1974) Identification of H_2O^+ in the tail of Comet Kohoutek (1973f). *Astrophys. J. Lett., 190,* L43–L46.

Whipple F. L (1982) The rotation of cometary nuclei. In *Comets* (L. L. Wilkening, ed.), pp. 227–250. Univ. of Arizona, Tucson.

Wood H. J. and Albrecht R. (1981) Outburst and nuclear breakup of Comet Halley — 1910. In *Modern Observational Techniques for Comets: Proceedings of a Workshop held at Goddard Space Flight Center, Greenbelt, Maryland, on October 22–24, 1980* (J. C. Brandt et al., eds.), pp. 216–219. JPL Publication No. 81-68.

Woodney L. M., A'Hearn M. F., Schleicher D. G., Farnham T. L., McMullin J. P., Wright M. C., Veal J. M., Snyder L. E., de Pater I., Forster J. R., Palmer P., Kuan Y.-J., Williams W. R., Cheung C. C., and Smith B. R. (2002) Morphology of HCN and CN in Comet Hale-Bopp (1995 O1). *Icarus, 157,* 193–204.

Wyckoff S. and Wehinger P. A. (1976) Molecular ions in comet tails. *Astrophys. J., 204,* 604–615.

Yeomans D. K., Chodas P. W., Szutowicz S., Sitarski G., and Krolikowska M. (2004) Cometary orbit determination and nongravitational forces. In *Comets II* (M. C. Festou et al., eds.), this volume. Univ. of Arizona, Tucson.

Zucconi J. M. and Festou M. C. (1985) The fluorescence spectrum of CN in comets. *Astron. Astrophys., 150,* 180–191.

Nucleus-Coma Structural Relationships:
Lessons from Physical Models

J. F. Crifo
Centre National de la Recherche Scientifique

M. Fulle
Osservatorio Astronomico

N. I. Kömle
Space Research Institute of the Austrian Academy of Sciences

K. Szego
*KFKI Research Institute for Particle and Nuclear Physics
of the Hungarian Academy of Sciences*

1. INTRODUCTION

One of the most frequently advocated incentives for the study of comets is that the cometary nuclei carry clues about the origin of the solar system. However, cometary observations almost never deal with the nucleus itself, but with its surrounding coma. An essential problem is therefore the derivation of nuclear properties from coma observations. In this chapter, we do not review the observations, nor the derived nuclear properties. Instead, we focus on the methods by which the latter are derived from the former: What are they? What are their formal justifications? Are the conclusions derived by these methods reliable? If not, which alternative methods should be used?

As discussed in detail in the literature (e.g., *Weissman,* 1999): (1) comets may have been formed from planetesimals over a large range of heliocentric distances, including the asteroid belt region, and therefore may differ from one another in composition and structure, in particular in volatile to refractory relative abundance; and (2) dynamical exchanges of material during the formation of the planetesimals probably mixed materials formed at different distances from the Sun, so that any cometary nucleus may be a mixture of materials formed at different temperatures, and therefore may be a complicated object. The shape of the nucleus should be expected to be as complicated as asteroidal shapes, hence its rotation should not be assumed to be simple (such as a symmetrical top rotator). Furthermore, while the nucleus interior is generally considered as undifferentiated, such cannot be the case of the external layers — those from which the coma is formed: They are submitted to thermal cycling (e.g., *Espinasse et al.,* 1991), cosmic-ray irradiation (e.g., *Strazzulla,* 1997), and meteoroid impacts (*Fernández,* 1990). This intrinsic complexity suggests that the interpretation of coma observations in terms of nuclei properties cannot be trivial.

Groundbased coma observations typically provide images and spectra of the gas and dust coma at a spatial resolution not better than several hundreds of kilometers, and at a time resolution sometimes of hours but more typically of days. The overall appearance of the coma is usually simple, but image enhancement techniques reveal well-defined spatial structures both in the dust and in the gas coma. These structures often repeat themselves in time, quasiperiodically, and quasiperiodicity is also often observed in the global properties of the coma (superimposed upon the secular evolution due to the orbital motion). It has become standard practice to summarize the coma observations by phenomenological models. Here, as we shall demonstrate at length in the following sections of this chapter, (1) one assumes that the coma formed by the real, complicated nuclei is well understood after having discussed only the coma formed by the simplest conceivable nuclei (e.g., spherical in shape, emitting spherical dust grains, rotating uniformly, etc.); (2) even this simplest possible coma is not self-consistently modeled using the relevant well-established applied physics methods, but simply "understood" from subjective postulates (e.g., there exists rotationally-invariant gas and dust ejection patterns, "gas and dust heliocentric dependence velocity laws", etc.) (see *Crifo and Rodionov,* 1999). This double arbitrariness in analyzing the data is perhaps tolerable as long as it is only used as a convenient means of summarizing observations, but, otherwise, it raises the question of whether the "derived" properties capture anything of the real properties of the real nuclei. This question can only be answered by taking into consideration realistically complicated nuclei, and thoroughly computing the structure of the coma they form — i.e., by physical modeling.

The 1986 flyby missions to Comet P/Halley have opened a new era. With the advent of observations that resolve the nucleus and sample the coma on a kilometer scale, a direct physical description becomes conceivable, i.e., one in which the complexity of the object is taken into consideration in detail. But, in the same manner as radically new observational techniques are requested for the flyby observations, the objectives themselves of the observations and of their interpretation change dramatically. For instance, the technical success of future orbiting and landing mission itself requires that the immediate environment of the nucleus be forecasted with high reliability and with a quick reaction time from the data acquired from the probes, regardless of whether or not this gives "clues on the origin of comets." This is a formidable challenge compared to that of interpreting groundbased cometary observations.

Even when deep-space mission data are at hand, it is not proven that such data allow a thorough understanding of the nucleus: To investigate it involves a so-called "inverse problem" with an enormous number of unknowns. The only known approach to adressing such a problem is statistical simulation; the statistical properties of a large class of realistic (i.e., as complicated as possible) nuclei must be derived and compared to the observations.

Both short-term forecasting of a nucleus activity and statistical estimates of the significance of nucleus and coma observations require advanced physical modeling. We review here the few existing efforts that have been done in this direction. As deep-space missions to comets are scarce, the institutional support of these efforts is quite limited. In such a context, unfortunate habits persist across decades: One such habit is to take the phenomenological models for a true representation of the real objects, as if this did not raise any question; another one is to apply the "supposed behavior" of the phenomenological nuclei even to those few observed nuclei for which the physical approach is possible, thereby cancelling all benefits from these unique observations. For instance, exactly the same "opinions" are presently formulated with respect to the *Deep Space 1*-supplied images of Comet P/Borrelly that were formulated 15 years ago when the flyby P/Halley images were collected. Unabashedly, this is often justified by saying that P/Halley images "confirmed the anterior analysis." If this is so, what are the goals of current and future work on comets?

The epistemologically correct attitude, which is exactly the opposite of the practices described above, is to use physical modeling not only for the rare comets with resolved nuclei, but even in the case where observations are crude: By performing numerical simulations of the "operation" of (1) objects that plausibly represent real comets and (2) the simple fictitious objects usually considered, one can hope to separate the domain of robustly derived conclusions from the domain of unwarranted speculations.

The present chapter is organized as follows. In section 2, we illustrate by a few representative examples the underlying assumptions on which heuristic inferences have been made in the cometary literature. Sections 3–5 describe the existing physical models, placing emphasis on their physical significance, not on the mathematical methods. In section 6 we review the physical model results relative to the near-nucleus coma; special attention is given to whether they support the assumptions listed in section 2, or otherwise.

Section 7 reviews the cases where model and observations are compared. Section 8 then briefly addresses the outer coma, again with an emphasis on the validity (or otherwise) of the phenomenological assumptions. Section 9 attempts to draw a general lesson from the previous sections.

2. OBSERVED COMA STRUCTURES AND THEIR HEURISTIC INTERPRETATIONS

By "structures," we mean spatial details in the coma, or temporal patterns that can be recognized during the comet motion: Both are related, owing to the nucleus rotation. The status of interpretation of these structures just before the first elaborate physical model results appeared was reviewed in *Kömle* (1990). Here, we reexamine the former in the light of the latter. As many chapters of this book are dedicated to coma structures, we will present here only a few representative examples. Many are relative to Comet P/Halley because it has been the only comet whose full nucleus shape was derived from flyby images. (The January 2004 flyby observations of Comet P/Wild 2 will probably lead to the derivation of the full shape of a second nucleus.) A few other, most instructive examples are relative to Comets Swift-Tuttle, Hyakutake, and Hale-Bopp. When physical modeling is advocated in support of the interpretation of the structures, we review it carefully. Often, it is possible to point out uncertainties, misinterpretations, and sometimes severe inconsistencies in the heuristic interpretations or their supporting modeling, even before advocating physical model results: A critical internal consistency check of the approach is sufficient. Advanced physical models are unavoidable, however, to overcome the difficulties, as will be demonstrated.

It is also unavoidable to comment on the terminology often found in the literature: Words such as "jets," "shells," and "activity" are, as we shall demonstrate, often used either without their standard meaning as defined in physics, or sometimes even in conflict with it. In our opinion, this is extremely prejudicial.

2.1. Gas Light Curves

If one observes a comet with the same spectrophotometer at intervals of days or fractions of days, the molecular emissions are seen to exhibit secular variations related to changes in heliocentric distance r_h, plus, in general, short-term variations. A good example can be found in *Millis and Schleicher* (1986). Some of these fluctuations may be random ("bursts"), but more attention is usually given to those that suggest approximate periodicities, because of their unquestionable relation to the nucleus rotation.

With the exception of the peculiar case of Comet Hale-Bopp (see below), such short-term variations are observed at small heliocentric distances, where water production is dominant. Water molecules emissions from the coma are rarely observable, so that their production rate $Q(t)$ is usually derived from the number N of daughter OH molecules present inside the observed coma volume (usually a cylinder). To relate $Q(t)$ to $N(t)$, observers use a trivial analytical formula, the "Haser model" (the expression "Haser formula" would be more suitable). There are many recognized weaknesses in this formula. Let us only mention a fundamental one: The coma must be spherically symmetric and quasisteady. But the very interpretation of quasiperiodicities in gas lightcurves, in terms of nucleus rotation, implies that this is not the case. Due allowance for asphericity of the coma is therefore needed. Furthermore, since inside a very narrow field of view, molecules are present to very large distances (on the order of 10^5 km), the observational volume filling time is on the order of a day, comparable to typical rotation periods; there is "hysteresis," i.e., a time-dependent model of the coma content must be used. This, in turn, requires knowing the nucleus rotational state.

Some authors are careful to point out that their use of the Haser formula is only done for convenience, the accuracy of the algorithm being unknown. Our impression is, however, that many readers take the Haser-derived Q(t) as the real variation of the nucleus production rate. For instance, *Julian et al.* (2000) and *Szego et al.* (2001) check their P/Halley rotation models against such a Q(t). One can therefore question the satisfaction the two groups express regarding their fits.

2.2. Comet P/Halley Gas "Jets" and "Shells"

Both groundbased and flyby spacecraft data revealed that the gas coma of P/Halley is structured. *Clairemidi et al.* (1990) identified OH and NH brightness maxima, which they called "jets," in the images obtained by the three-channel UV spectrophotometer onboard *Vega 2*. Two "jets" were seen, one pointing toward the Sun, and another, stronger one nearly perpendicular to the sunward direction in the image plane. Since we refer here to this widely used term "jet" for the first time, let us make a brief comment. Apparently, it is used to describe a structured (linear or curved) brightness enhancement. However, as used in physical gas dynamics, a neutral gas jet is a region of collimated flow created by a localized source. It should not necessarily be a region of high brightness; conversely, regions of gas at rest (shocked gas or stagnation gas) are regions of high brightness, and not at all akin to jets. Furthermore, a gas jet is an isolated system, not the subsystem of anything; it cannot coexist at the same location with another, superimposed flow structure. But in the cometary case, the advocated "jets" result from image enhancement, and represent only a tiny fraction of the flow; they are flow details, not at all jets in the gas dynamic sense. This is not only a matter of terminology; it generates considerable confusion, prompting questions such as "What is the collimation mechanism?" Where there is no collimation, there can be no collimated gas. In the following we will continue to use the term "jets" in this manner, but the reader should be careful to remember that it designates only a localized detail in a large-scale gas or dust distribution.

Structures with a completely different appearance were discovered in P/Halley from groundbased observations of the CN distribution. These CN features could be observed during a much longer period of time than the previous gas jets, and exhibited definite periodicity:

1. Long-lasting spiral-shaped jets (*A'Hearn et al.*, 1986), extending out to $>6 \times 10^4$ km, were evidenced by azimuthal enhancement technique; *Hoban et al.* (1988) concluded from a dataset collected between April 9 and May 2, 1986, within 5×10^4 km from the nucleus, that the evolution of the morphology of the jets suggests the existence of a period of recurrence of 7.37 d. The current interpretation of these structures follows the original discussion of *A'Hearn et al.* (1986): The observed features are not due to local fluctuations in the density of a rotationally modulated fluid CN coma, but are due to parent species moving according to quite narrow spiral patterns (with thickness <3000 km at 30,000 km from the nucleus) and with a quite accurate radial velocity. "In the absence of a confining mechanism for the jets the inescapable conclusion is that the jets are composed of dust grains," small enough to be invisible on white light images.

2. Ring-like, expanding "shells" were discovered by visual inspection of the CN images (*Schlosser et al.*, 1986; *Schulz and Schlosser*, 1989). These shells (actually asymmetrical rings or haloes around the position of the nucleus) always dominated the outer coma, at a distance greater than 10^5 km from the nucleus, during the two-month observing period. Here again, the words "shell" or "ring" designate a localized enhancement of the brightness of a much wider distribution; furthermore, since it is observed in the emission of the nondominant CN molecule, it is uncertain whether it traces a local density enhancement of the whole coma, or a local variation in CN content. In any case, it would be incorrect to model it as an isolated distribution of gas: There can be only one global model of the coma, with the requirement that it reproduces the observed localized enhancement.

2.3. Comet Hyakutake Arcs

Comet Hyakutake passed unusually close to the Earth on March 25, 1986. This led to the unique detection, during 10 successive nights, of spectacular arcs of OH, CN, and C_2, centered on the antisunward cometocentric axis (at least in projection), having their apex on it at a variable distance (between 400 and 2000 km) and their convexity toward the Sun (reproduced in section 7.1). A detailed account of the observations is given in *Harris et al.* (1997) and in *Rodionov et al.* (1998). Both groups have interpreted the observations as the signature of an H_2O arc, caused by some secondary H_2O source. We compare and discuss the approaches in section 7.1.

2.4. Comet Hale-Bopp Spirals

Spectacular spirals appear in the large-scale images of Comet Hale-Bopp in March and April 1997 (i.e., shortly before and after its perihelion pass), for the observed molecules OH, CN, C_2, C_3, and NH (*Lederer et al.*, 1997). These structures are conspicuous after image enhancement, but their real contrast must be high enough, as the authors state that they can be discerned even in the raw images. On

April 26, for instance, a single spiral is seen in the cleanest OH and CN images with an arm spacing of roughly 6×10^4 km. At an outer coma flow velocity of 1–2 km/s, this corresponds to a periodic modulation of the gas production rate with period 8–16 h. *Lederer* (2000) has interpreted these structures as follows. Parent molecules and fine organic dust grains are assumed to be ejected from the nucleus along straight lines, partly inside a certain number N_j of cones rotating rigidly with the nucleus, partly inside the background space outside of the cones. The assumed nucleus rotation period is 11.3 h. Each cone is ascribed a relative production strength S_j, as is the background space, and a width w_j. The gas flux in any given dayside direction is assumed proportional to its S_j and to the cosine of its angle to the solar direction. Once this is done, the daughter radicals are computed from a Monte Carlo procedure from the distribution of the primary molecules and organic dust, following *Combi et al.* (1993). A satisfactory fit is obtained assuming $N_j = 5$ cones; the cones and background are assumed to produce the same mixture of H_2O, HCN, and "parent-of-C_2" molecules, save one cone that, allowing for one-half of the OH in the spirals, is assumed to produce only H_2O and grains releasing only OH. The computed total OH productions from the "jets" represents less than one-half the total comet OH production.

Interferometric radiowave mapping of CO lines in the same comet on March 11, 1997, slightly before its perihelion (*Henry et al., 2002*), revealed time-dependent velocity-space structures in the emission. The measurements yield the line profile of the flux in each $\simeq 1500$ km × 1500 km element of the image. The intrinsic line profile of the CO line is negligible compared to the observed profile widths, so the latter are due to the Doppler shifts induced by the CO velocity distribution within the observed coma volume. Seen at a Sun-comet-observer angle of 45°, the profile usually exhibits two peaks approximately symmetrical with respect to the line center. The ratio of the number of molecules on both sides of the line center changes smoothly in about 7 h from $\simeq 1.7$ to $\simeq 0.5$, and back to nearly 1.5, suggesting a periodicity similar to that we have noticed above for the spirals (unfortunately, the CO observation duration is not long enough to confirm it). According to the observers, "this is indicative of a jet" rotating with the nucleus. The authors declare to have obtained good agreement with the data using "a 3-D model of the coma consisting of an isotropic contribution plus a conically spiralling jet of opening angle 30° and having outgassed 30% of the CO in the coma."

The number and positions of the "active regions" postulated to explain the dust spirals, the OH, CN, and C_2 spirals, and the CO spirals are totally different, which is considered as "evidence for chemical heterogeneity in the nucleus" (*Lederer and Campins, 2002*).

After reading sections 5 and 6, the readers will have enough material at their disposal to form their opinion regarding the preceding interpretations. These interpretations are usually considered satisfactory because they "fit the observations." This widely used argument is not sufficient. It is also required that the interpretation not be in conflict with any other related observation (cometary or not). The assumption that molecules move in straight lines conflicts with all estimates of the mean free path inside the coma, which is found to be smaller than 1 m near to the nucleus. Therefore, the inner coma should be modeled as a fluid, not as a set of non-mutually colliding particles. We return to this issue later in the chapter.

2.5. Dust Lightcurves

Usually dust coma isophotes are nearly perfect circles, and the radial slope of the coma brightness is well approximated by a power law vs. the coma radius with index –1. This led to the definition of the most frequently used tool to characterize the dust activity of a comet: the $Af\rho$ product (*A'Hearn et al., 1995*). In interpreting this product, two important assumptions are made: (1) the $1/r$ brightness variation is due to a trivial $1/r^2$ isotropic dust outflow; and (2) the dust grain outflow velocity is that applying to spherical grains ejected from a uniformly sunlit spherical nucleus — the "dust ejection velocity law." We will comment on these assumptions in section 7.

2.6. Large-Scale Coma Dust Structures

In the following, we will focus on dust coma structures that can hopefully be related to the properties of the nucleus and of the near-nucleus gas-dust interaction. Presumably, the closer to the surface the structures are, the more suitable they will be for this purpose, as several effects perturb this relation; e.g., solar radiation reprocesses the structures because of the dependence, upon grain mass and grain composition, of the radiation pressure to solar gravity ratio β_s. Also, it is suggested from time to time that dust fragmentation is present in the coma.

Years of intensive groundbased dust coma image processing have shown that many dust comas that at a first glance appeared isotropic actually contain faint dust structures. Many of these structures strongly suggest the picture of a rotating inhomogeneous point source. The impression is even stronger if a motion picture of the observations is viewed — so strong that one is tempted to forget that these structures result from strong image structure enhancement, hence, as do the gas structures, characterize only a minor fraction of the whole coma brightness (however, in contrast with the gas, several different dust populations can evolve independently in the same region, since there are no dust-dust collisions). We have not found in the literature any estimate of even approximately how much nucleus dust production these figures represent. In fact, most authors process images by means of nonlinear algorithms, hereby losing any quantitative information on the extracted structures.

A good review of the classical interpretation of such structures was done by *Sekanina* (1991). A typical example of result is *Sekanina*'s (1981) model of the nucleus of Comet P/Swift-Tuttle. First, a trial-and-error procedure searches for recurring jet patterns in order to establish a nucleus rotation period. Then the time-dependent orientation of the most prominent jets is used to constrain the nucleus spin-axis orientation. As all dust of a given type is then assumed to move radially from the origin and with a common velocity V, an "activity map" is built in cometocentric longitude-latitude coordinates. This allows the identification of so-called "active" and "inactive" areas. The wider the jet, the more extended the active area is declared to be. Then, it is assumed that all grains in a jet were ejected at the same time, the curvature of the jet being due to the dispersion in V and β_s between the grains. Assuming that there exists a simple analytical expression $\beta_s = \beta_s(V)$ relating V to β_s, the so-called "dust ejection velocity law," the author derives, at each point of the jet axis, a value of β_s and a value of V. The double jets that sometimes appear (e.g., in Swift-Tuttle) are interpreted in terms of two different dust chemical species: This shows the *ad hoc* nature of the relation between active spots and jets.

Minor jets not used to constrain the nucleus spin state or not interpretable in terms of different dust chemistry are considered to be produced by *ad hoc*-defined spots active only during a time segment adjusted to generate the observed jet. Thus, while the very use of the word "jet" suggests a long-lasting phenomenon, it is frequently assumed that the jet duration is short, sometimes lasting only minutes. Also, to avoid conflict between different observations, an active spot exposed to sunlight is often declared to have "deactivated" itself.

We do not question the fact that the "mechanism" thus offered, if carefully implemented, might give birth to the observed dust coma structures. But, in the literature, in particular in the two preceding references, it is used for reaching a much stronger conclusion: The "active area map" derived by this method is explicitly declared to represent the total activity map of the nucleus. In other words, (1) all the dust is assumed emitted from the active areas, (2) all the gas is assumed to originate from these active areas, and (3) the emission of the active areas is transient and, for some of them, occurs only one time. This analysis is often supplemented by a quantitative estimate of which fraction of the nucleus is active: For this evaluation, the active areas are assumed to consist of pure ice. Their total extent is found to be typically 5–10% of the nucleus surface.

Conclusion (1) meets with severe objections: Why would only a small fraction of the dust emitted by an active area go to the observed "jet," while most of this dust would go to the background coma? And how are we sure that this is really the case? As no analysis is offered for the background coma dust (which makes up most of the dust emission), we are free to assume that it comes from the whole sunlit area of the nucleus. In fact, a much more natural assumption would indeed be that active areas (as derived above) are areas where, transiently, a tiny dust flux excess occurs, for any reason. In such a picture, all nuclei are essentially seen to emit dust homogeneously.

Conclusion (2) also meets with many strong objections. (a) As already pointed out by *Whipple* (1982), it is arbitrary to postulate that a dust-gas mixture can be blown from discrete sources in a narrow collimated way; it is much more likely that the gas will diverge over a broad solid angle, and force the dust to do the same. On the contrary, if the gas emission is uniform (or nearly so) over the surface, its divergence should not be great (near-radial flow), hence trace variations in the dust content will be preserved outward. (b) By the same token, one does not see how a $\beta_s(V)$ relation derived for a strictly spherically symmetric gas flow would apply to a highly non-uniform and highly time-varying gas flow. (c) If, as likely, the mean direction of gas emission from a discrete area is set by the local topography (e.g., a pit), how is it that only sources active in precisely the postulated radial direction ever manifest themselves? (d) With a gas emission confined to transient localized areas, the torque applied by the reaction forces to the nucleus should be maximized, which renders the assumption of a simple, non-excited nucleus spin quite uncertain.

Finally, conclusion (3) also raises as many questions as it answers: What controls the switching on and off of an active area, if not the Sun?

In most papers, however, these questions are not addressed. In a few, they are raised, but not answered, or answered in a velikovskian way, suggesting, e.g., that the transient behavior is induced by the "opening and reclosing of cracks" in a surface mantle, and so on. But the most important point is that supporting quantitative physical modeling results are never presented, as if cometary activity stood outside the field of physical concepts and methods. This way of "explaining" the coma structure persists today, even though the first physical simulation of cometary activity casting a severe shadow on these explanations appeared 14 years ago (*Kitamura*, 1990), and has been followed by the vast body of even more devastating gas dynamic results described in section 6.

2.7. Near-Nucleus Dust Structures

We use the term near-nucleus dust structures to refer to structures observed at a spatial resolution smaller than the nucleus size. Hence, data of this kind exist only for Comets P/Halley, P/Borrelly, and P/Wild 2. In the first case, the results are superbly described in the two-volume report published by the European Space Agency (*Keller et al.,* 1995; *Szego et al.,* 1995). In the two more recent cases, preliminary results have just appeared (*Soderblom et al.,* 2004; *Brownlee et al.,* 2004). In all three cases, the nucleus was also imaged. The spatial resolution was smaller (and position-dependent) for P/Halley, but the coverage was practically complete. For P/Borrelly, only part of the sunlit surface was imaged. The coverage seems to have been nearly complete for P/Wild 2.

The main result of the *Giotto* flyby is a synthetic image obtained by the HMC camera, in which bright dust structures are seen attached to a restricted part of the nucleus edge (see Fig. 10 in section 7.2). This is typically described in the following terms: " . . . distinct jets emanated from active spots on the sunward side of the nucleus. Most of the elongated and structured nucleus appeared inactive" (*Keller et al.,* 1994, p. 69). Figure 76 of the same reference quantitatively reproduces the gross coma appearance, using a model described in *Knollenberg et al.* (1996): The dust distributions from three unequal circular active sources, placed on a sphere centered on the nucleus, are added. Each distribution is computed as if the source was isolated on that spherical nucleus. We return to this in section 7.2.

After enhancement of the azimuthal gradients of the HMC image, a wealth of fine radial structures appear, which the authors called "filaments" (*Keller et al.*, 1994, pp. 83–85). A gas dynamical simulation of a process by which a narrow pencil of dust could be produced in an uniform ambient coma was developed (*Knollenberg*, 1994; *Keller et al.*, 1994); an *inactive* circular area (100 m size) was assumed to exist as a defect inside a uniformly active surface; it was computed (not just stated) that a narrow pencil of dust is formed on its axis due to the convergence toward the axis of the surrounding gas and the resulting cross-axis motion of the dust. We return to this explanation in section 7.2. But we may immediately observe that this model result exactly supports *Whipple*'s (1982) criticism of the classical active area concept that we cited in section 2.6: A small-scale coma dust density maximum is found (not just assumed) to be due to a surface gas-dust production minimum.

Similar azimuthal enhancements were applied to the *Vega 2* camera images. Here, due to a lower resolution, filaments could not be identified, but more than 10 directions of brightness enhancement were clearly distinguished (see pp. 208–228 of *Szego et al.*, 1995). Enough view directions were available to conclude that "the jet sources formed a long linear feature on the nucleus passing across the subsolar point" (*Szego et al.*, 1995, p. 72).

Both the *Giotto* and *Vega* observations are being reinterpreted by the global physical model described in *Rodionov et al.* (2002). We will discuss some of the results in section 7.2.

Finally, let us observe two essential differences between near-nuclear structures and distant coma structures:

1. The near-nucleus dust dynamics are controlled everywhere by the gas interaction, as already established by *Whipple* (1951). Hence it is unrealistic to claim to understand by means of visual observation the dust motion that is present before clearly stating how the (invisible) gas is considered to flow. In fact, when such images are presented to a gas dynamic scientist, the reaction is unvariably that the observer is merely seeing plumes (i.e., dust in a gas flow).

2. Observation of the near-nucleus coma is concomitent with a determination of the nucleus shape. Therefore, realistic three-dimensional models can be devoped. Furthermore, Comet P/Halley will return to perihelion in the year 2061, a horizon not totally discouraging for young scientists, and Comets P/Borrelly and P/Wild 2 will return much earlier. So, conclusions derived about these comets can be made under the form of precise predictions. As long as physical truth can result only from predictive-corrective iterations, these observations provide the first (and for the time being, the only) basis for a true physical study of comets. We return to these observations below.

3.   PHYSICAL MODELS OF THE NUCLEUS-COMA INTERFACE

It is not possible to build a physical model of the coma without having a model of at least the outer layers of the nucleus, yielding at each point of the surface algorithms that allow computation of the temperature and gas and dust flux as a function of the solar direction (and distance). We say "algorithms," because the coma conditions and near-surface nucleus conditions are mutually coupled and therefore must be computed self-consistently. One example of this coupling has long been recognized: The emitted dust can influence the visible and IR irradiation of the surface, hence react on the emission (*Salo*, 1988; *Moreno et al.*, 2002). Another example, only recently documented, is that both net sublimation or net condensation are possible, even on the dayside surface, not only in the shadowed areas (*Crifo and Rodionov*, 2000) but on the sunlit portions as well (*Crifo et al.*, 2003a). Hence, both nucleus interior models and coma models should be unbiased with respect to the value of the gas pressure at the surface, as well as with respect to the sign of the net surface gas flux.

3.1.   Interface Description

At the present time, there are only indirect inferences about composition and structure of the nucleus. It seems that all authors follow *Whipple* (1950, 1951), who proposed that (1) it is a mixture of the ices of simple molecules and of refractory dust, probably with a complicated physical texture; (2) its outer, near-surface layers must be radially differentiated (the most volatile ices being absent from the outer layers). Estimating the stability of the various volatiles residing at or just below a nucleus surface, Whipple concluded that ices of the most volatile molecules (such as CO, CO_2, etc.) cannot survive one nucleus perihelion passage; this means that in comets approaching the Sun periodically, these volatile molecules must sublimate at some depth inside the nucleus and then diffuse toward the surface. It has been postulated sometimes that even H_2O ice itself could sublimate below a blanket of more or less cohesive dust. In all cases, the outer layers must be porous to permit gas effusion (actually, the voids created by the elimination of the volatile species already create porosity). The current speculations about comet formation in the early solar system also suggest that the nucleus as a whole may be a low-density, porous and brittle medium. One of the goals of the nucleus internal heat transfer models is to offer scenarios for this radial differentiation (see below).

Little consideration has been given in the cometary literature to the difficult problem of taking into account the surface topography. As already suggested by *Whipple* (1951), and confirmed by radar backscattering data (*Harmon et al.*, 1999), the nucleus surface must be "extremely rough on scales of meter and larger." The same must be true at smaller scales, due to dust ejection. Also, the surface is subjected to erosion — roughly 1 m will be lost per perihelion passage, which may imply locally stronger depletions. Thus the nucleus' shape itself evolves both on a global and on a local scale. On a timescale of days, the submetric surface details will change. Keeping these facts in mind, the question arises of down to which level of accuracy does it make sense to describe these surface variations in the frame of a numerical model? The answer is difficult and depends upon the goal envisioned. The model of *Rodionov et al.* (2002) was constructed to handle the surface at a spatial resolution $\Delta \simeq 50$ m, consistent with the Halley imaging data. The assessment of the *Rosetta* lander descent parameters requires a description of the surface such that short-term predictions of the near-surface coma structure (a few rotation periods) are possible. What this means in terms of spatial resolution of the surface has not yet been assessed.

Given that the nucleus rotates at an angular velocity Ω, if the surface is to be described at the spatial resolution Δ, for consistency this must be done with a time resolution $\delta t \simeq \Delta/(\Omega R)$. But should one use a true time-dependent model, or a succession of quasisteady models?

As the near-surface nucleus material is potentially quite inhomogeneous in all respects (optical properties, volatility, porosity, granulometry, etc.), its relatively fast time-dependent illumination will induce both a horizontal and a vertical dispersion of the temperature(s). It is not at all proven that a quasi-steady-state temperature(s) distribution is ever reached, nor even that local thermal equilibrium prevails. That is, there is no proof that, within a surface element $\Delta \times \Delta$, the various components (ice and minerals, for example) take on the same temperature. Actually this can only be expected if they are very intimately mixed and thermally well coupled. This problem has been considered to some extent by *Kömle and Ulamec* (1989), but certainly needs to be reinvestigated in a more general context. The same unanswered questions apply just below the surface.

The near-surface coma gas adjusts itself to a steady state extremely fast (typically within seconds), but this is not necessarily the case for the dust: Heavy grains can be accelerated to only fractions or small multiples of the nucleus escape velocity (meters per second) and hence they stay in the vicinity of the surface for times comparable to the rotation period. Thus, a thorough description of the gas production consists of a succession of steady-state maps of the mass density, velocity, temperature and of the various species' mole fractions on a reference surface encircling the nucleus, at a spatial resolution Δ on the order of several mean free paths (m.f.p.), which is typically fractions of a meter to tens of meters — depending upon local solar zenith angle — near 1 AU from the Sun. For the dust production, not only the surface mass distribution $g_s(m)$ is needed, but the *shape* distribution h_s (clearly out of reach) and a true time-dependent model may be needed as well for large grains.

3.2. Near-Surface Nucleus Interior

Volatiles residing below the surface can escape through pores and cracks in a rather direct way, if enough heat is transported to or created at the respective depth. There are in principle three ways in which the energy can be transported to a subsurface layer: (1) solid-state heat conduction via the solid matrix, composed of connected grains; (2) transport of heat by inward-flowing gases that recondense in the deeper/colder layers and release their latent heat there; and (3) penetration of the solar radiation into the ice and absorption in the interior instead of at the immediate surface; this can only happen if the ice is to a certain extent transparent and the radiation is trapped in the interior (solid-state greenhouse effect).

3.2.1. Porous ice models.   The first thermal models of cometary nuclei (published in the 1980s and earlier) assumed nuclei to be nonporous ice/dust mixtures. *Smoluchowski* (1982) was the first to take into account porosity and gas flow through the pores. Subsequently, the heat and mass transport in porous, grainy ices was investigated in more detail by several groups (*Squyres et al.*, 1985; *Mekler et al.*, 1990; *Steiner and Kömle*, 1991a). The latter model was successfully applied to laboratory samples composed of artificially produced grainy ice (*Kömle et al.*, 1991). The thermal evolution of larger ice/dust samples irradiated under space conditions was described by a similar model published by *Benkhoff and Spohn* (1991). The two latter models clearly showed that heat transport via the gas phase (energy transfer by sublimation/condensation processes) should play a significant role under "cometary" conditions. Otherwise it would be difficult to understand the measured temperature profiles. Another important aspect, studied by *Kossacki et al.* (1994), is the influence of grain sintering processes on the thermal evolution. Along the lines outlined by these models (which mostly included only water ice) multicomponent models were developed that allowed the prediction of the depths of various sublimation fronts as a function of the thermal history if a particular initial composition were given (*Espinasse et al.*, 1991; *Steiner and Kömle*, 1991b, 1993; *Benkhoff and Huebner*, 1995; *Kossacki et al.*, 1997). The current state of the art of these models is nicely described in the recent review by *Capria* (2002).

These models brought forward two important facts: (1) the possibly strongly reduced heat conduction caused by small grain contact area, as known from lunar regolith; and (2) the heat transported, by the migration of evaporated molecules along the thermal gradient, by sublimation and release of latent heat upon recondensation.

The basic equations describing this process are the conservation equations for energy (heat transfer equation) and mass (continuity equation)

$$(1 - \psi)\rho_i c_i \frac{\partial T}{\partial t} = \frac{\partial}{\partial x}\left(\lambda \frac{\partial T}{\partial x}\right) - c_g \phi_g \frac{\partial T}{\partial x} - qH \quad (1)$$

$$\psi \frac{\partial \rho_g}{\partial t} = -\frac{\partial \phi_g}{\partial x} + q \quad (2)$$

in which ψ denotes porosity, ρ specific mass, c specific heat content, λ heat conductivity coefficient, φ mass flux, H latent heat, q gas mass source term, and the subscripts i and g refer to the solid and to the gas phase, respectively.

While the basic equations used in these various modeling approaches are consistent, there are controversial approaches in the formulation of the boundary conditions: (1) In all the models noted above, free-molecular outflow into vacuum is assumed, i.e., the backflow of molecules from the coma toward the surface is neglected. (2) Solution of the continuity equation demands specification of the surface pressure and/or surface density of the emitted gas. Different authors use quite different boundary conditions here, from p = 0 at the surface to p = p_S, the ice saturation pressure. Reduction of the two conservation equations (1) and (2) to one single energy equation is only possible if the surface pressure is assumed to be equal to the saturation pressure p_S. This has been explicitly verified by *Steiner et al.* (1991). Something more realistic was used in the *Espinasse et al.* (1991) model. However, as a matter of fact, the highly nonequilibrated condition, in which the molecules are emitted, is not properly accounted for in any of the models. For a more detailed discussion of this point, see *Skorov et al.* (2001); this paper, as well as the previous one (*Skorov et al.,* 1999), provided an important step beyond the previous modeling efforts, by presenting semi-analytical solutions for the kinetic gas flow in tubes of finite length (assuming that the temperature profile along the length of the tube is known) and combining it with a numerical solution of the heat conduction equation. Thus it is not necessary to specify a separate pressure boundary condition at the surface, because the gas outflow from the tube is found directly from the temperature distribution along the ice tube walls and the associated local sublimation. The approach is similar to that already described in *Kömle and Dettleff* (1991), but with tubes instead of rectangular cracks.

3.2.2. Partially transparent porous ice models. A much more direct way to heat the subsurface layers of a comet nucleus exists if the ice is to a certain extent transparent. In this case the solar radiation is not fully absorbed or reflected at the surface, but penetrates icy layers down to a certain depth. It is then absorbed either by enclosed dust particles over a longer distance or by dusty layers with high optical thickness in the interior of the ice (see Plate 9).

The idea that the ice transparency could play an active role in the thermodynamics of comets and icy satellites is relatively old (*Kömle et al.,* 1990; *Brown and Matson,* 1987; *Matson and Brown,* 1989). Recently it was reinvestigated by introducing the appropriate source terms into a more advanced thermal model (*Davidsson and Skorov,* 2002a,b) and applied to calculate the gas production rate of Comet P/Borrelly (*Skorov et al.,* 2002) and to calculate the temperature profiles to be expected in artificial ice/dust samples (*Kaufmann et al.,* 2002). The conclusions from these new calculations could modify the currently accepted view of cometary energy balance and gas production quite significantly. There are two important findings worth mentioning:

(1) If forward-scattering dust particles are embedded in a transparent ice layer with a density realistic for comets, very dark surfaces (with a few percent albedo only, similar to that observed for Comet P/Halley in 1986) can be created. (2) With the same active area given, significantly lower gas production rates result, because the very surface is colder than in the case of "surface absorption" of the sunlight.

Laboratory experiments aimed at investigating this solid-state greenhouse effect more systematically and evaluating its significance for comets and other icy solar system bodies are currently under way (*Kaufmann et al.,* 2002).

3.2.3. Comparative results. The vertical temperature gradient that develops in the near surface layer of the nucleus in response to solar irradiation strongly depends on the structure and composition of these layers:

1. For a compact, well-sintered dirty ice with little porosity the thermal conductivity is high, close to that of compact water ice. For long constant irradiation this leads to a temperature profile close to linear, but not steep, as calculated from classical models using the *Klinger* (1981) formula for the conductivity of water ice (a weak temperature dependence, compared to the exponential temperature dependence, which characterizes the heat transfer by gas sublimation/condensation).

2. If the ice is porous (open porosity) and composed of grains with low contact area, there will be a steep temperature gradient at the very surface (where the sublimating gas flows outward) and a rather flat temperature profile below, where the sublimated gas flows inward toward colder regions and recondenses there, because the effective thermal conductivity of the medium is high.

3. If the ice is covered by a loose dust mantle a few millimeters or centimeters thick, composed of refractory grains, a very steep temperature profile develops across this dust mantle, with a temperature drop of 100 K or more. This is due to the fact that the dust mantle in the low-pressure environment has an extremely low thermal conductivity, similar to that of lunar regolith. The conductivity of a dust mantle could be increased by orders of magnitude, if it contains organic components that might act as a glue between particles and cause some cohesiveness (*Kömle et al.,* 1996).

4. If transparent ice exists as such, it may influence the activity of the surface in various ways. Depending on the gas permeability of the transparent layer, it may cause subsurface pressure buildup and violent activity if the gas pressure exceeds the tensile strength of the crust. A typical feature is the existence of a subsurface temperature maximum, as shown in the example below.

An example for the temperature profile that may develop inside transparent ice subject to the solid-state greenhouse effect is shown in Fig. 1. The main feature is that the maximum temperature occurs not at the surface, but a few centimeters below it. From there the heat is conducted away toward the surface. The position and sharpness of this subsurface temperature maximum depend on the absorption profile as well as on the thermophysical properties of the ice. Higher temperatures could be reached in the case of a

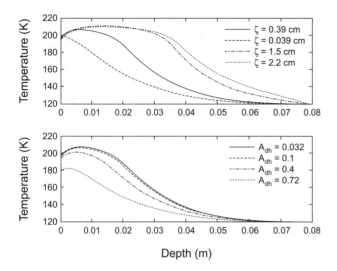

Fig. 1. Typical temperature profiles resulting from one-dimensional time-dependent models, including the solid-state greenhouse effect in ice (*Kaufmann et al.*, 2002; based on model by *Davidsson and Skorov*, 2002a,b). The different curves correspond to various penetration depths and surface albedos.

low gas permeability, possibly caused by the sintering and densification processes described before.

3.2.4. Stumbling blocks. The next logical step of the previously described interior models would be their combination with a near-surface coma model to derive the proper gas conditions at the surface. But here, a severe obstacle of most present nucleus models is that they are one-dimensional (and time dependent), and hence cannot accommodate the expected complex geometry of the real surface, nor even the smoothed approximation used in the coma models. It is unclear how this could be circumvented in the future.

Another consequential limitation of these models is that they assume that ice and dust share a common temperature at the surface (and in the near surface interior). This may be true for very fine dust, but cannot be true for large grains (pebbles, boulders). In fact, the laboratory simulations of KOSI have already found evidence of sizable dispersions in the surface temperature of samples of illuminated ice-dust mixtures even though only small amounts of fine dust were used in the experiment (in an unrealistic way) (*Lorenz et al.*, 1995). In reality, surface dust is expected to be warmer than surface ice and will therefore radiate more thermal energy, hence its neglect overestimates the energy available for sublimation. The models with a surface crust are free from this criticism, but in general lead to rather small surface fluxes. In some models (e.g., *Enzian et al.*, 1997, 1999), a first-order correction is made in the surface energy budget equation by the introduction of an "icy area fraction" f. We come back to this in section 3.3.2.

Let us notice that most nuclei size estimates assume not only that the surface is isothermal, but that it is really pure sublimating ice with no embedded dust — even in very dusty comets. For instance, the often quoted estimation "that

10% only of P/Halley's surface is active" is based on such an unlikely assumption.

Efforts have been made to develop models capable of handling the geometrical complexity of real nuclei. *Enzian et al.* (1997) developed a "2-D 1/2 + t" approach for a spherical nucleus — in the sense that one-dimensional thermal equations (in the direction perpendicular to the surface) are solved at each point of the surface. *Guttiérrez et al.* (2000, 2001) extended this method to aspherical nuclei and *Rodionov et al.* (2002) developed a similar method for complex nuclei having spatial details of size comparable to that used in their coma model. However, for mathematical tractability, in all these models the nucleus interior physics has to be simplified to pure heat transfer.

3.3. Near-Surface Dusty Gas

The relevant modeling of the first tens of meters above the surface of an active nucleus is the only way to cross-correlate the thermal and dynamical evolution of the nucleus (mass loss, orbital perturbations, angular momentum changes) and the formation of its gas and dust coma. It is an unescapable task if one wants to interpret or forecast the coma structure, orbital evolution, or nucleus rotational state. Models that, for instance, prescribe "arbitrary boundary conditions" at the nucleus to compute the coma structure are intrinsically unable to provide any information on its evolution during the nucleus rotation (to interpret coma dust structures, for example.)

Unfortunately, the physical conditions in the near-surface gas-dust mixture can only be the subject of speculation: (1) the statistical properties of the surface topography and composition are unknown; (2) very little information is available about the dust; (3) where gas is released by the surface, it is not known whether it diffuses from below the surface, is sublimated from it, or both; (4) one is free to advocate more intricate processes such as dust fragmentation, sublimation from icy grains, etc. It is hard to believe, however, that tiny details of what happens here influence the structure of the observable coma, considering the existence of efficient smoothing processes inside the gas and dust coma: pressure for the gas; shape, mass, and compositional dispersion for the dust. Hence it can be hoped that effective (simplified) models of this region can represent in a roughly correct manner how the nucleus and coma are coupled. Of course, this can also be tested by developing alternative models that allow for different processes.

3.3.1. Near-surface conditions in a dust-free case. Elementary calculations show that a surface of pure ice submitted to solar illumination in free space assumes a temperature below the water triple point, and hence sublimates: It emits molecules according to a half-space centered Maxwellian distribution $\mathcal{M}(0, T_n)$, where T_n is the surface temperature. The rate Z^+ (molecule/m² s) of the emitted molecules is, very roughly, $Z^+ \simeq c_\odot \cos z_\odot / L_S r_h^2$, where $c_\odot$ is the solar constant, $z_\odot$ the solar zenith angle, and L_S the sublimation latent heat. If the gas diffuses from pores, the distribution is no longer a half-space Maxwellian, but some other function defined by the pores' geometry and temperature, and Z^+ is necessarily of a smaller magnitude than before. It is easy to compute that, near $r_h = 1$ AU, the mean free path of the molecules emitted from a given point against collisions with those from the adjacent points is very small (typically a fraction of a meter), hence a fluid atmosphere is formed. A flow pattern must establish itself inside this atmosphere, resulting from the complex surface distribution of Z^+. It is the purpose of the coma model to compute this flow. Here, two difficulties arise, first mentioned in the Russian literature (see references in *Crifo,* 1991a), and first discussed in detail in *Crifo* (1987).

The first difficulty is that the velocity distribution of the molecules returned to the surface is the downward part $\mathcal{M}^-(V_0, T_0)$ of a Maxwellian function with some mean velocity V_0 and temperature T_0. Let us refer to $\mathcal{M}^+(V_0, T_0)$ as the corresponding upward part, then the distribution of the emitted molecules must differ substantially from $\mathcal{M}^+(V_0, T_0)$, otherwise the net gas flux at the surface would be vanishingly small. Therefore, in the immediate vicinity of the surface, the gas is not in an equilibrium regime (which requires a strict, or moderately distorted Maxwellian shape). This region must therefore be treated by gas kinetic methods (solving the Boltzmann equation, or BE). If this region is small, hereby defining a so-called surface boundary layer (BL), the much more efficient gas dynamic methods will be used to compute the flow outside of it (section 5).

The second difficulty is that in a gas, the flow regime depends upon the conditions holding at all boundaries of the flow. Therefore, the preceding (V_0, T_0) near any point of the nucleus surface does not depend only upon the local values of Z^+ and T_n, but upon these values at all points of the surface. From the point of view of nuclear interior models, this means that the return flux Z^- (recondensed onto the ice) is not proportional to the upward flux Z^+ (contrary to what is assumed in many recent cometary papers). In particular, there is no "simple" way of predicting whether Z^- is smaller or greater than Z^+. This is not simply a cosmetic argument: Indeed, we shall see in section 6 that very plausible surface topographies lead to recondensation (instead of sublimation) over sizable fractions of the sunlit areas of the nucleus.

These two difficulties have been identified for quite some time in the rarefied gas dynamic literature, and continue to be the subject of advanced developments (see *Cercignani,* 2000, and references therein). The reader is urged to be cautious about the relevance of publications concerning these problems that do not refer to the aforementioned literature. The specific problems of integrating the BE over sublimating or condensing ice under simple geometries have long been solved exactly by analytic methods (e.g., *Cercignani,* 1981) or direct Monte Carlo simulations (DSMC) (e.g., *Abramov,* 1984). Notice in passing that for the reason given above, solutions can only exist for specific geometries. Of special interest is the plane-parallel solution,

because any surface that is smooth enough can be approximated by locally plane-parallel elements. This is done systematically in all papers by Crifo and Rodionov cited hereafter. The accuracy of this approximation is found to be unexpectedly high (see section 6.3.3.).

In the plane-parallel method, the "surrounding flow" degenerates to the specification of one set (p_0, V_0, T_0): the gas parameters at the distance where equilibrium flow is reached. This distance is found to be on the order of several tens of mean free paths. This implies that, for a comet near 1 AU, the thickness of the BL will be several meters to several tens of meters. Obviously, the mathematical smoothing of a real nucleus surface to $\Delta \simeq 50$ m makes the plane-parallel approach usable, but it remains to be seen whether this smoothing is acceptable in itself. The computed structure of the BE is found to depend upon one free parameter, in accordance with the previously established fact that there is no way to predict Z^- by consideration of the BL flow only. For this free parameter, the initial Mach number

$$M_0 = V_0/\sqrt{\gamma k_B T_0/m}$$

in which k_B is Boltzmann's constant, is usually used. It has been proven from first principles that no solution exists for $M_0 > 1$ (once more, contrary to some — inaccurate — results published in the recent cometary literature). Thus, one arrives at relations $p_0 = p_0(T_n, M_0)$ and $T_0 = T_0(T_n, M_0)$ at each point of the top of the BL, to be taken as the boundary conditions for computing the coma flow. Only when this flow is computed (see section 5) is M_0 known at each point, hence the return flux Z^-. This approach is fully described in *Rodionov et al.* (2002) and references therein.

3.3.2. Near-surface conditions in a dusty case. Perhaps pure ice volumes exist in some nuclei, but, in general, one expects the ice to be dirty. It follows that only a fraction $f = 1/(1 + \mathfrak{R})$ (where $\mathfrak{R}$ is the relative dust-to-ice mass content) of the surface is ice; the rest is dust (e.g., *Crifo,* 1997). As already stated, it is not expected that these two constituents assume a common temperature. Furthermore, dust, even porous, cannot produce vapor at the rate Z^+ of ice. The above picture must therefore be amended. An approximate modification of the above model has been proposed by *Crifo and Rodionov* (1997a), consisting in writing that the net upward flux is reduced (compared to pure ice) by the factor f. As already stated, this factor is used in many nucleus models, but unfortunately not in all.

Since the initial dust velocity is negligibly small, the fact that there is a large dust concentration inside the BL must still be taken into consideration. Based on the fact that the dust density is also large just outside the BL, and that, outside of it, exact dusty gas dynamic computations show that the perturbation of the gas is small, *Crifo and Rodionov* (1997a) assume that this is also true inside the BL. This analysis was done assuming a P/Halley-like dust mass distribution of spherical grains. It is possible that the situation could be different for other dust properties.

4. PHYSICAL MODELS OF THE NUCLEUS ROTATION

In the same way as it is impossible to build a coma structure model without matching it to a subsurface nucleus model, neither is it possible to reproduce most of the coma structures, whether directly observed or observed through lightcurves, without introducing a nucleus rotation model. This is true even of the near-nucleus structures, which critically depend upon the direction of solar illumination of the complex nucleus orography.

The nucleus rotation is governed by the equations for an asymmetric top (*Landau and Lifshitz,* 1976, section 36), taking into account the outgassing torque. Differences among researchers appear when evaluating the net recoil force $\mathcal{F}_n$ and net torque $\mathcal{T}_n$ exerted on a nucleus by its gas and dust emission. However, such computations are at the heart of the aerospace industry (rocket motors), so that reliable methods of computation exist in that field of study. For pure gas, both $\mathcal{T}_n$ and $\mathcal{F}_n$ can be computed by integration over any closed surface inside the gas flow, as long as mass, momentum, and energy (MME) are preserved. The best is to choose the top of the BL, where (1) the gas is in fluid regime (hence the integrals involve only p_0, V_0, T_0) and (2) MME have been preserved as photochemistry and other effects are not yet at play. For dusty gas, the only exact method is the *nucleus surface* integration. However, for a P/Halley-like distribution, one can still integrate at the top of the BL, as stated above, because this kind of dust does not substantially perturb the flow inside the BL (in keeping with the fact that its global momentum is still negligible at this point).

An exact computation of $\mathcal{T}_n$ and $\mathcal{F}_n$ is possible only for nuclei with a known external shape and surface composition. The adoption of an arbitrary external shape when the real shape is unknown is unwarranted. The worst possible practice is to assume a spherical shape, since unrealistic symmetry cancellations occur during the surface integrations and since, in addition, the strict periodicity of the rotation induces a quite atypical quasisteady surface temperature distribution.

Rodionov et al. (2002) have computed $\mathcal{T}_n$ and $\mathcal{F}_n$ for P/Halley using the observed shape, a best-fit rotation mode, and assuming that the nucleus is uniform in composition. This revealed that that both $\mathcal{T}_n$ and $\mathcal{F}_n$ vary considerably in magnitude and direction during the nucleus rotation. *Belton et al.* (1991) and *Samarashinha and Belton* (1995) calculated the nongravitational effects on P/Halley nucleus assuming that the dust jets seen far from the nucleus originate from active areas derived by tracing the jets back to the surface of an approximating ellipsoid. This is subject to two severe criticisms: (1) The correct shape should have been used, and (2) the definition of "surface jets" should have been done on the basis of fluid dynamics — this is the same criticism as when discussing Hale-Bopp spirals. We will return to the rotation of P/Halley in section 7.3.

Because $\mathcal{T}_n$ scales as the third power of the characteristic nucleus size, while the inertia momenta scale as its fifth power, the angular acceleration due to outgassing scales as its inverse square. The effect of the torque is therefore expected to be maximum for very small nuclei. Indeed, *Crifo et al.* (2003b) have computed the torque and rotation of very small (subkilometric) irregular nuclei and found it to induce highly irregular, possibly chaotic rotational motions. On the other hand, the effect is modest for Halley-like comets, and must be negligible for very large comets (e.g., Hale-Bopp).

5. PHYSICAL MODELS OF THE COMA

By physical model, we mean an approach that tries to make full use of the latest available methods in applied physics. Unavoidably, only brief descriptions of these methods appear in the cometary literature. The reader unfamiliar with these methods will first have to get acquainted with the proceedings of the Rarefied Gas Dynamics Symposia held every two years. Examples of up-to-date textbooks covering most topics of interest here are *Gombosi* (1994) and *Cercignani* (2000). For information about fluid dynamics computational methods, see references in *Rodionov et al.* (2002), and for the Monte Carlo simulation methods, see *Bird* (1994).

Once methods are available, they must be applied. Here, the cometary medium must be described by selecting numerical values for all physical properties. This occasionally raises difficulties. For instance, collision cross sections between exotic molecules may not be known. However, the dominant molecules in the coma — H_2O, CO, etc. — have been the subject of in-depth studies in the laboratory and in the industry. Their physical properties (both in the gas phase and in the solid phase) are therefore available. Unfortunately, it is not uncommon to find works in which fancy numerical values are adopted for these properties. The nonexpert reader is therefore advised to check all values against such robust sources as the *American Institute of Physics (AIP) Physics Handbook* (last edition in 1972) or the *Chemical Rubber Company (CRC) Press Handbook of Physics and Chemistry* (84th edition in 2003).

While a gas molecule is a precisely defined entity, such is not the case for a "dust grain." One usually ignores it by the "simplifying assumptions" that the grains are spherical, but this is unfortunate: Intuition as well as pioneering simulations using aspherical grains (*Crifo and Rodionov,* 1999) indicate that spherical grains have a totally atypical aerodynamical behavior: A flowing-by gas submits them only to a drag force, the lift and torque being null. It follows that all spherical grains of a given mass starting from a given point follow the same trajectory and have the same final velocity. Such is not the case for grains having the same mass, but varying shapes: Their trajectories differ, and their final velocities can be spread over orders of magnitude. In fact, even the initial orientation of the grain at the surface influences their trajectory and final velocity, with some shapes and orientations preventing ejection, and others leading to high velocities.

Not only does it appear mathematically difficult to build a model of a near-nucleus coma with shape-dispersed dust, but it seems unlikely that one will ever access the needed input parameters of the model, such as the grain shape distribution and the distribution of initial orientation of these grains at the surface. From this point of view, the ultimate goal of cometary dust studies is (to tell the truth) rather unclear to us.

5.1. Gas Coma

The ultimate description of a coma is the set of distribution functions $f_i(t, \mathbf{r}, \mathbf{v}_i)$, defined as the number of particle of species i (H_2O molecules, CO^+ ion, spherical olivine grain of radius a_i, etc.), having the cometocentric particle velocity vector $\mathbf{v}_i$ at the cometocentric position (vector) $\mathbf{r}$, per unit volume $d\mathbf{r}$ and unit velocity-space volume $d\mathbf{v}_i$. These functions are governed by coupled BEs (see, e.g., *Gombosi, 1994; Cercignani, 2000*). It is the goal of gas kinetic methods to solve these BEs. It can be done in exceptionally simple cases by numerical methods; otherwise, in principle, DSMC can be used (see *Cercignani, 2000*). However, DSMC requires forbidding computational resources as soon as interparticle collisions become important. Fortunately, in this case, solving the BEs is unnecessary because it is possible to predict the general form of the solution. For instance, in a gas mixture where near-thermal equilibrium prevails, the f_i are exact or nearly exact Maxwellian functions of the gas mean mass density ρ, flow velocity $\mathbf{V}$, temperature T, and species mass concentrations q_i. It is therefore optimal to solve only fluid equations governing these quantities — the objective of gas dynamic methods. Computing the same flow by the two alternative methods, when possible, is beneficial, providing, in particular, cross-validation of the numerical methods.

It is a fairly difficult task to derive from first principles which method is best suited to deal with a rarefied flow such as the coma gas flow. The discussion is to be based on a comparison between the mean free path and the characteristic scale $\mathcal{L}$ of the flow, as well as between the collision times and the timescales of the flow. This comparison must be done at each point. Additional limitations due to the mathematical methods also come into play; see a concise discussion in *Rodionov et al.* (2002), and more detailed developments in, e.g., *Gombosi* (1994) and *Cercignani* (2000). To give an oversimplified summary, when the mean free path is much smaller than $\mathcal{L}$, the gas is said to be "in fluid regime," fluid equations apply exactly, and DSMC is useless; when the mean free path is much greater than $\mathcal{L}$, the gas is said to be in "free-molecular" regime; in between, the gas is said to be in "transition regime." In the two last cases, DSMC always applies, and fluid equations may or may not provide accurate results; the quality of their solutions must be evaluated case by case (see examples in *Crifo et al.,* 2002a, 2003a).

5.1.1. Gas dynamic approach.   The most general form used hitherto in coma studies is the single-fluid Navier-Stokes equations (NSE)

$$\frac{\partial}{\partial t}\rho + \nabla \cdot (\rho \mathbf{V}) = f_\rho \tag{3}$$

$$\frac{\partial}{\partial t}(\rho \mathbf{V}) + \nabla \cdot (\rho \mathbf{V}\mathbf{V}) + \nabla p - \nabla \cdot \bar{\bar{\tau}} - \mathbf{F} = \mathbf{f}_v \tag{4}$$

$$\frac{\partial}{\partial t}(\rho h_0 - p) - \nabla \cdot \mathbf{q} + \nabla \cdot (\bar{\bar{\tau}}\mathbf{V}) - \mathbf{F} \cdot \mathbf{V} = f_h \tag{5}$$

where ρ is the total mass density, h_0 is the specific enthalpy, p is the pressure, $\mathbf{q}$ is the heat conduction vector, $\bar{\bar{\tau}}$ is the second-rank viscous stress tensor, and $\mathbf{F}$ the total macroscopic force. The vector $\mathbf{q}$ can be expressed as a function of T, and $\bar{\bar{\tau}}$ can be expressed as a function of T and the partial derivatives of $\mathbf{V}$ with respect to the coordinates.

In $\mathbf{F}$ the radiation pressure force is generally negligible, and the nucleus gravity generally neglected, but there would be no problem with keeping the latter term in order to deal with Kuiper belt-sized objects if one wanted to. If the preceding equations are written in a non-Galilean frame, inertia forces must be introduced. For instance, if a nucleus-attached frame is used, it can be shown (*Rodionov et al.,* 2002) that the Coriolis force $2 \mathbf{V} \times \mathbf{\Omega}$ (where $\mathbf{\Omega}$ is the nucleus angular velocity vector) is the dominant inertia force. It can also be shown that the ratio of the pressure force $|\nabla p|$ to the Coriolis force decreases as $1/r$ with distance to the nucleus. In a very large comet with fluid region exceeding 10^4 km (like Hale-Bopp), at such a distance a plausible $\Omega = 10^{-5}$ radian s^{-1} sets the preceding ratio to about 1/10: In other words, the Coriolis force is dominant — an hitherto unnoticed fact. Such may also be the case for less-productive comets if their nucleus has a higher spin rate. It implies that rotationally induced gas structures are to be expected at large distance.

The "source-sinks" terms f_ρ, $\mathbf{f}_v$, f_h at the r.h.s. of these equations allow for (1) the fact that the fluid interacts with photons, or with particles not belonging to the fluid itself (e.g., dust), and (2) possible inelastic processes that are internal to the fluid itself and that affect its momentum and energy budget. An important example of the first effect is photodissociation of H_2O, and an example of the second effect is partial H_2O recondensation into clusters $(H_2O)_n$; both yield a large f_h term, a moderate f_ρ term, and a negligible $\mathbf{f}_v$ term. Another example is cooling through IR emission. It is not possible to incorporate dust inside the fluid described by the above NSE, because the huge mass difference between molecules and dust grains forbids the dust grains to share a common flow velocity with the gas (and even a common flow velocity between themselves). Furthermore, in typical coma conditions the grain-grain collisions are negligible, so that dust grains do not acquire thermal velocity spread or pressure. The dynamics of the grains must therefore be treated by separate equations (see below), and their interaction with the gas must be represented by r.h.s. terms in the NSE. If the grains do not emit or condense gas, there is no f_ρ term, and the two other terms have little effect on the gas, for a dust mass distribution of the kind found in Comet P/Halley (see *Gombosi,* 1986; *Crifo,* 1987). But if the dust is icy, it will condense or emit H_2O and release or absorb latent heat, and this may result in strong perturbations of the gas flow (see *Crifo,* 1995).

The "source-sinks" terms, whatever their origin, must themselves be computed by solving so-called "rate equations," to be solved simultaneously with the above ones. For a review of the forms of r.h.s. terms, see *Crifo* (1991a) and *Rodionov et al.* (2002).

For many applications, the Eulerian form of the equations (EE), simpler because it involves only first-order partial derivatives, can be used; this is obtained by setting $\bar{\bar{\tau}} = 0$ and $\mathbf{q} = 0$. Formally speaking, the relative ranges of validity of the EE and NSE should be delineated using a dimensionless rarefaction parameter, the so-called generalized Knudsen number, defined in *Crifo and Rodionov* (1997a). But comparisons with solutions obtained by DSMC show that the frontiers of these domains also depend upon the details of construction of the numerical method of solution (see *Crifo et al.,* 2002a, 2003a). The existing results from such comparisons show that the NSE and even the EE provide acceptable solutions over practically the whole dayside coma of observable comets. Two restrictions are to be made, however: (1) The immediate vicinity of the nucleus surface must always be dealt with by gas kinetic methods. This is a very strong restriction, since only a correct treatment of this region warrants the obtention of a correct solution from the EE or NSE "downstream" in the coma. We have discussed this region in section 3.3.1. In all their studies, Crifo et al. treat this region by an algorithm based on the BE, hence they call their solutions "BE-EE" or "BE-NSE". (2) In the outer reaches of the coma, where dissociation products are dominant, it is presently not known to which accuracy these equations represent the real situation. This is because photodissociation creates the daughter molecules with high velocities relative to the parent velocity; for the fluid equations to be valid, it is necessary that slowing down of these products to the local velocity distribution occurs within one computational cell. This may not necessarily occur. Unfortunately, comparisons between NSE (or EE) and DSMC for such cases are not yet available.

Finally, let us comment about the presence of the time t in the above equations. Given angular speeds $\Omega = O(3 \times 10^{-5})$ radian/s, near-nucleus gas speeds $V = O(300)$ m/s, and a modest lateral spatial resolution $\delta = O(0.01)$ radian, one sees that steady-state gas solutions can be used out to distances $<(V/\Omega)\delta = O(1000)$ km. However, such time-stationary solutions can be obtained only by solving the time-dependent equations — with time-independent boundary conditions. The reason for this is associated with the fact that the gas flow is in (large) part supersonic; see references in *Rodionov et al.* (2002) for an extensive description of the methods of solution.

For modeling most observed coma gas structures involving distances much in excess of 1000 km, the time-dependent gas equations must be solved, but looking for a true time-dependent solution, not just for the limit for large times. To do this, instead of keeping the nucleus surface parameters constant, one must first compute a set of successive nucleus surface parameters, forming the boundary condition for the variable t. The solution is a set of successive three-dimensional coma structures. Obtaining it is an enormous undertaking in terms of computer resources. It has been achieved for the first time during the preparation of this text, and will be described only in forthcoming publications by Rodionov and Crifo.

However, observations involving distances in excess of 1000 km may be interpreted by a succession of steady-state solutions, if free-molecular conditions are reached before or near that distance: In such a case, the extrapolation of the gas parameters beyond 1000 km is trivial, and is not affected by photodissociation. This is, for instance, the case of the interesting Comet P/Schwassmann-Wachmann 1 discussed, e.g., in *Crifo et al.* (1999), which lends itself to velocity-resolved observations.

Finally, it may happen that the gas reaches, near 1000 km, transition regime conditions; in such a case, the validity of the use of fluid equations becomes uncertain, in particular due to the presence of photodissociation. It is then necessary to use a time-dependent DSMC, or to validate the use of fluid equations by comparison with DSMC results.

5.1.2. Direct Monte Carlo simulations. In a DSMC model, the evolution resulting from mutual collisions, of the individual velocity components, of the internal energy, and of the position coordinates of a large number of "weighted" molecules are monitored. Instead of introducing molecules at the rate Z^+, they are introduced at a somewhat reduced rate Z/q ($q > 1$ can be position-dependent). The statistical consequence of the replacement of the extremely large number of real molecules by a much smaller number of simulated molecules is of course taken into account. Space is discretized into adjacent cells, and time into a succession of time steps. At each time, the number of mutual collisions of the molecules expected in each cell is evaluated, and the velocities and internal energies of a corresponding number of randomly selected pairs of molecules are changed using a binary collision model. During the next time interval, all molecules are moved. The procedure is reiterated until steady state is achieved. [For a description of the method, see *Bird* (1994); for an insight into the future of this method, see *Bird* (2001).]

The DSMC method offers specific advantages: (1) It is valid for any form of the gas velocity distribution function; and (2) the boundary conditions can always be formulated exactly (for instance, very complicated nucleus surfaces, on any linear scale, can be considered). But, as with any method, it can be (and is indeed sometimes) improperly implemented. For instance, three mandatory requirements are that (1) the chosen time steps must be much smaller than the mean collision time, (2) the typical cell dimension must be much smaller than the local mean free path; and (3) the cell dimension must also be smaller than the characteristic flow scale $\mathcal{L}$. Unfortunately, information that allows the reader to check whether these conditions are satisfied is not always included in the publications. Finally, the method is computationally much less efficient than the solution of fluid equations, so it should not be considered to be a substitute for the latter, but should be used to complement them.

5.2. Dust Coma

If the dust perturbs the gas flow, its distribution must be computed self-consistently together with the gas equations. This was done in the old one-dimensional works reviewed, e.g., by *Wallis* (1982) and later on by *Crifo* (1991a), and in several of the two-dimensional works reviewed in *Kömle* (1990) and in the present section 6.2. In most of these works, all the dust mass loss was (unrealistically) assumed to be concentrated in single-size small spherical grains. This resulted in a strong perturbation of the gas. But with the mass being spread over a very large range of mass, as was found in Comet P/Halley, this effect disappears (see *Gombosi*, 1986; *Crifo*, 1987). While it is not possible to exclude very narrow size distributions, this presently seems to be permitted, hence one will currently solve first the gas equations and then the dust equations.

Conversely, as the gas density decreases outward, and the gas-dust interaction is a function of at least the square of their relative velocity — which also decreases outward — the acceleration and cooling of the dust by the gas stops at some distance from the nucleus surface (typically less than 100 km).

Since the "dust velocity" is typically one or several order(s) of magnitude smaller than the gas velocity, the range of validity of steady-state dust distributions is also an order of magnitude smaller, i.e., is typically only 100 km. Beyond it, standard interplanetary dust modeling techniques are to be used (see *Fulle et al.*, 1999; *Fulle*, 2004). These methods are intrinsically time-dependent.

Here, we will only deal with the modeling of the region where the dust-gas interaction is sizable. If possible, it is appealing to use, for the dust, equations similar to those for the gas. However, the latter express the fact that mass, momentum, and energy are preserved during the many collisions occuring in each elementary volume. For cometary dust, collisions are practically absent. Hence the use of fluid equations seems unwarranted. However, it is correct to write that the mass, momentum, and energy of a set of co-moving particles are preserved. So, it is possible to group in the same "fluid" particles that follow everywhere the same trajectory. Since the aerodynamic acceleration depends upon mass and shape, these particles must have the same mass, shape, *and* initial orientation at the surface. This is still not sufficient: One must also make certain that trajectories do not cross one another at any point. This leads to subdividing the nucleus surface in areas such that dust grain trajectories ema-

nating from any given subdivision never mutually intersect. We will present illustrative examples of this in sections 6.1 and 6.2.

In conclusion, it is possible to compute the dust distribution at small distances from the nucleus by a so-called "multifluid model" with possibly a very large number of fluids. Each fluid will be governed by "zero-temperature EE" with $p = T = 0$ and $h_0 = c_s T_s$ (s is the grain type tag, c_s the specific heat, T_s the grain internal temperature). An other possibility would be to use a DSMC for the dust, seeding them inside the precedingly known gas solution (see section 6.2).

6. INNER COMA STRUCTURES AS REVEALED BY PHYSICAL MODELING

On the scale of the groundbased data spatial resolution (hundreds of kilometers at best), the zero-order coma gas flow is trivial: A point source placed in a near-vacuum can only provide a radially diverging flow; mass conservation produces a $\propto (1/Vr^2)$ gas density decrease, the resulting pressure gradient accelerates the gas velocity, and kinetic energy conservation requires a concomitent gas temperature decrease; a classical analytic treatment demonstrates that the flow becomes rapidly supersonic (e.g., *Wallis*, 1982). The flow will ultimately "freeze" at some terminal velocity and temperature when collisions become rare (e.g., *Cercignani*, 2000, section 5.9). Innumerable illustrations of these effects are described in the gas dynamic literature. The solar wind is formed by a similar process (*Parker*, 1965). Of course, in the cometary case, there are also specific so-called "nonadiabatic" effects by which mass, momentum, and energy can be input or removed from the flow, thus altering its structure. For instance, a general gas dynamic theorem states that exothermic effects tend to render the flow sonic: If they occur in a subsonic flow, its Mach number is increased, and if in a supersonic one, the Mach number is decreased (possibly strongly enough to generate a shock).

Partial recondensation of H_2O molecules into molecular clusters and large amounts of fine dust may heat the gas. These effects are limited to the vicinity of the nucleus surface, because the first one is proportional to at least the square of the gas density, and the second one to the square of the gas density. Water photodissociation releases fast H and OH that, by thermalization, heat the gas; this occurs in a large part of the coma and limits the gas Mach number.

In the old cometary literature (1965–1990), the supersonic state of most of the coma and the preceding nonadiabatic effects have been recognized and studied in trivial one-dimensional geometry [see the review of *Crifo* (1991a), supplemented by *Crifo* (1993) and *Crifo* (1995) for posterior developments]. However, an essential implication of the supersonic state of the coma was universally overlooked in that literature: the tendency of the flow to form shock structures to adjust itself to nontrivial geometrical constraints, or in reaction to external perturbations. Gas shocks of interest here can tentatively be divided in two groups: (1) "jet jet interactions" (a classical problem in the design of multiple thruster rockets); here, two gas flows meet — if one only is supersonic, it will form one steady shock; if both are supersonic, two shocks will be formed (e.g., *Ni-Imi et al.*, 1992); and (2) internally generated flows; here, the heat deposition effect is strong enough to create a shock transition to subsonic state; in principle, H_2O recondensation could create such shocks. Shocks are the canonical examples of structures in a nonturbulent fluid. Hence, it is somewhat surprising that the possible connection between shocks and coma structures was only first suggested by *Kitamura* (1990). However, geometrical constraints and strong nonadiabatic effects exist in the coma only at a short distance from the nucleus surface (abundantly illustrated below), a region not accessible to observations except in flyby and rendezvous missions. Even then, one essentially observes dust — not gas — structures, and dust cannot form shocks; it is surely not intuitive that near-nucleus dust structures trace gas structures, and an understanding requires advanced simulations of the kind described below.

All presently published physical model results refer to the near-nucleus coma. However, because of the scarcity of space missions to comets, most coma structures observed are located at very large distances from the nucleus. Are they just the result of the evolution to large distances of the near-nucleus structures, or do they result from other structuring mechanisms? We address this briefly in section 8.

6.1. Homogeneous, Spherical Nuclei

"Spherical nuclei" have been considered as the only paradigm for at least a half century of cometary speculations. In fact, isothermal spherical nuclei were assumed, even though no explicit mention of it was ever made. While this absolutely forbids any comparison with observational data, it is still is a suitable paradigm to test new algorithms dedicated to a better physical representation of the coma (e.g., testing new radiative or chemical algorithms in the simplest possible way: one-dimensional equations). Here we will not deal with such uses; instead, see *Crifo* (1991a) and references therein.

On the other hand, it is not unreasonable for many purposes to consider that the outer coma is axially symmetric, as if produced by a sunlit, slowly rotating spherical nucleus. This has led to a number of two-dimensional spherical coma models, of variable merit, starting with *Krasnobaev* (1983) and continuing with *Kitamura* (1987), *Kömle and Ip* (1987a,b), *Köröszmezey and Gombosi* (1990), *Knollenberg* (1994), *Combi* (1996), *Mueller* (1999), *Crifo and Rodionov* (1997a, 1999, 2000), and *Crifo et al.* (2002a). In most works, EE were used, but in *Kitamura* (1987) and *Crifo and Rodionov* (2000), NSE equations were used as well. *Combi* (1996) used for the first time a DSMC approach. [The DSMC method discussed here should not be confused with the much less powerful "test particle Monte Carlo method"

(MCTP) used precedingly by this author and a few others, and not discussed here.] Finally, the three methods were used and compared in *Crifo et al.* (2002a). In these numerous works, the boundary conditions at the nucleus differ: In all works except those by Crifo et al., the surface temperature and H_2O flux are prescribed arbitrarily (most often, the surface flux is assumed to be $\propto \cos z_\odot$); in the works by Crifo et al., either a CO flux is prescribed arbitrarily, or an H_2O flux is derived from surface ice sublimation equations. The greatest difference in these input assumptions regards the nightside surface, which is either assumed to be inactive, or assumed to produce a uniform background flux of gas representing either a very small, or a sizable, fraction of the total dayside flux.

The computed structure of the dayside gas and dust coma is trivial (see Fig. 2) — notice, however, that a "gas velocity" or a "dust velocity" does not exist; both quantities are position-dependent. On the contrary, the nightside structure can be extremely complicated, as Fig. 2 indicates. It provides an ideal benchmark to discuss how the gas coma is formed, and with which accuracy it can be modeled. This led *Crifo et al.* (2000) to systematically simulate the range of possible nightside conditions by varying the nightside ice surface temperature, hence the nightside background gas production, using a heuristic parameter $\kappa \ll 1$: The thermal flux returned to surface elements in shadow is assumed to be $\kappa c_\odot / r_h^2$. Plate 10 shows the results. The $\propto \cos z_\odot$ variation of surface pressure on the dayside creates a lateral flow from noon to midnight; in the absence of night background ($\kappa = 0$), the nightside surface is a cold trap for the gas. Therefore the flow from the dayside divides itself into one portion recondensing on the nightside and one portion escaping in the nightside hemisphere. The division occurs along a flow line terminating on the midnight axis at a point where the gas is at rest — a stagnation point. This gas at rest, as well as the gas reaccelerating upward and downward from it, form an obstacle for the arriving gas. Information cannot travel up a supersonic stream, so it cannot "guess" that there are obstacles ahead of it; it can only undergo a sudden transition to a subsonic state where information is received from the obstacle. Hence, as visible on the upper left panel of Plate 10, a conical shock is formed in between; it is, in fact, a converging-diverging shock (its diverging part appears conspicuously on Fig. 2). A very weak nightside surface background emission is enough to suppress the condensation; the emitted nightside gas is now an obstacle to the dayside gas, resulting in the formation of a weak converging shock attached to the terminator; inside the shock, the nightside gas accelerates rather slowly (Plate 10b). At a somewhat higher background level, fast acceleration of the nightside gas occurs (Plate 10c). The resulting supersonic stream interacts with the dayside one via a double shock structure; in between, sonic gas accelerates slowly outward (Plate 10c). Finally, a strong background creates the same kind of double-shock structure, but now the midnight stream stays supersonic all the way to infinity (Plate 10d).

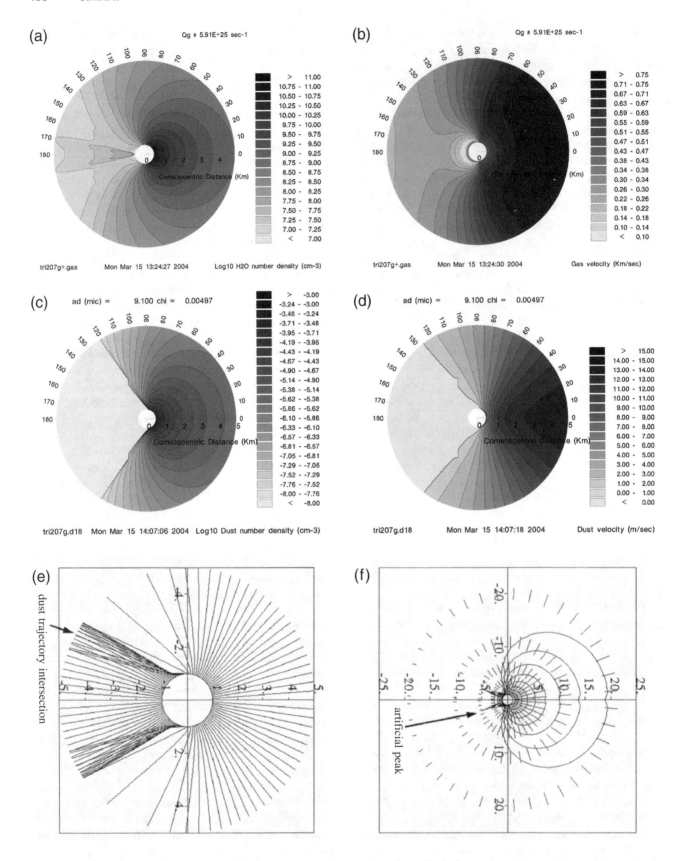

Fig. 2. Coma around a sublimating homogeneous, spherical nucleus. The Sun is toward +X. **(a)** H$_2$O number density. **(b)** H$_2$O velocity. **(c)** 9-μm-radius dust grain number density. **(d)** Dust grain velocity. For these computations, the night background emission was assumed very low (κ = 0.01), so no nightside dust ejection occurred. **(e)** Trajectories of 2-μm-radius grains originating from various local times. One can see that, beyond the terminator, the trajectories mutually intersect. **(f)** Dust density. One can see that density peaks appear in the region of mutual trajectory crossings. On **(e)** and **(f)** (from *Mueller,* 1999), the night background is higher, and night dust emission occurs. *Mueller* (1999) calls the computed peaks "artificial," which may be misleading (see text).

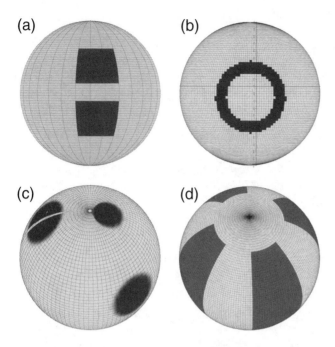

Fig. 3. Examples of spherical inhomogeneous nuclei treated in the literature. (**a**) *Kitamura* (1990) and *Crifo et al.* (1995); (**b**) *Kömle and Ip* (1987a,b) and *Knollenberg* (1994); (**c**) *Keller et al.* (1995); (**d**) *Crifo and Rodionov* (1997a). The dark areas are assumed strongly active, the rest is assumed weakly active, or inactive.

In real comets, a substantial night background production is expected from the nightside surface (for instance, from CO production, but other molecules are probable as well, e.g., CO_2, HCN, etc.). Therefore the flow structure should resemble that just described, save for the fact that the molecular composition will differ on the dayside and on the nightside.

The steep gas structures just evidenced (weak shocks) are found to translate themselves into dust structures, due to the fact that the dust particles are too heavy to accurately follow sudden changes in the gas direction: The trajectories of grains originating from both sides of the gas structures, and moving toward it, cross one another, creating a localized dust density enhancement. This is indeed found by most authors (when not found, inaccurate computational algorithms are to be blamed). For instance, Fig. 2e shows grain trajectories in the terminator shock region, where the gas velocity is about similar to that seen on Plate 10 (lower right panel). The sudden change in gas direction cannot be reproduced by the dust; instead, one sees dust trajectories mutually crossing and a resulting dust density enhancement (Fig. 2f). It is evident that on an image, these enhancements would be called "jets," and one can see that such "jets" do not trace any dust grain trajectory, nor indicate the presence of any active area.

Further complication in the nightside coma structure would be introduced if one took into consideration a nucleus rotation: Coupling the gas dynamics to the nucleus

interior heat transfer equations would introduce a surface temperature asymmetry, making the coma fully three-dimensional; in this sense, even the spherical, homogeneous nucleus still remains incompletely modeled today.

6.2. Inhomogeneous Spherical Nuclei

Many decades separated the first heuristic suggestions that dust structures could be due to nucleus "active areas" from the first gas dynamic simulations of the effect (*Kitamura,* 1990). Figure 3 shows most inhomogeneous spherical nuclei submitted to gas dynamic simulations to the present date.

The first computation was due to *Kitamura* (1986): One circular "active spot" defined by a Gaussian variation of the icy area fraction $f = G(z_\odot)$ is considered at a time when the Sun is on its axis. This assumption provides computational simplicity but not optimal significance. The author solved NSE equations for the gas. Plate 11a shows the gas distribution for a small spot surrounded by a strong uniform background: The formation of a conical weak shock (in reality a double-shock) appears clearly, matching the source flow to the background flow (and transforming the on-axis density maximum quickly into a minimum). Plate 11b shows what happens if dust is introduced: Dust density maxima are formed along the gas conical shock (on an image, the dust would appear as "emanating from two close active regions").

Knollenberg (1994) revisited the same problem, but assumed another kind of background, $\propto \cos z_\odot$ on the dayside, and vanishing in the nightside. He solved EE equations. Plate 11b–d shows the result. One recognizes, on the nightside, the zero-background converging-diverging conical shock of Fig. 2 — the nightside is not sensitive to details of the dayside gas production. On the dayside, a conical (double) shock is clearly visible, similar to that in *Kitamura* (1986). When dust is introduced — on the dayside only, since there is no nightside background — it forms conspicuous enhancement in the vicinity of the gas shocks, for the reason already stated.

Kömle and Ip (1987a,b) considered a circular ring of increased activity, with Gaussian profile, superimposed on a two-step background (i.e., one dayside value and one nightside value), with the Sun placed on-axis. Unfortunately, we have verified by an unpublished recomputation that, as the authors suggest in their paper, their computational technique was inaccurate. Therefore, we prefer to comment here on the related (and highly accurate) results of *Knollenberg* (1994); see below.

The first paper to explicitly identify as shocks the gas density enhancements created by surface flux inhomogeneities was *Kitamura* (1990). Most importantly, this is also the first three-dimensional computation of a coma (EE were used). Two square spots are placed symmetrically about the noon axis on an inactive background (see Fig. 3a). Figure 4 shows the results, as recomputed by *Crifo et al.* (1995). The supersonic gas jets from the two sources interact strongly, forming a V-shaped double-shock structure (in the symme-

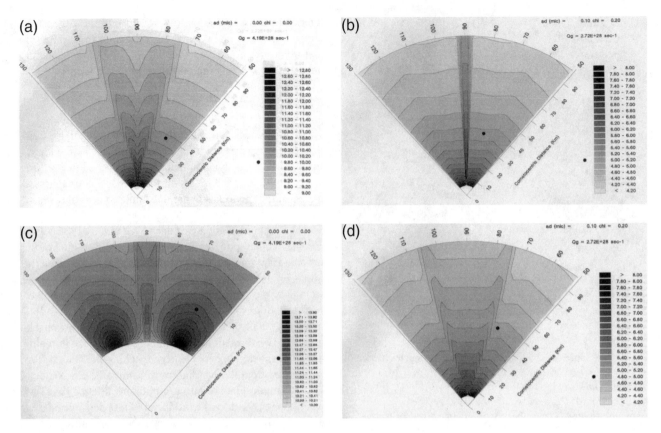

Fig. 4. Spherical nucleus with two identical active areas (*Kitamura*, 1990; *Crifo et al.*, 1995). **(a)** Isocontours of the gas density; **(b)** isocontours of 0.1-μm-radius grain dust density, computed from single-dust-fluid equations. **(c)** Same, on an enlarged geometrical scale; **(d)** same, computed from two-dust-fluids equations.

try plane). [Outside the symmetry plane, the double-shock structure is quite complicated in shape, owing particularly to the square shape of the sources (A.V. Rodionov, unpublished data, 1998).] This is very similar to what is observed in multiple thruster rocket motors (e.g., *Ni-Imi et al.*, 1992). If dust is introduced in the flow, the result can be seen in Fig. 4d. One can see, once more, that neither the two density pencils at the edge of the central dust density enhancement, nor the enhancement itself, project themselves to the active spots. The origin of the central enhancement follows from the fact that the dust emitted by the left spot covers the angular sector 75°–>130°, while that emitted from the right spot the sector <50°–105°; this means that in the region 75°–105° at each point there are two different directions of dust motion. In other words, dust density maxima do not reveal dust trajectories; furthermore, there is nothing like a "dust velocity direction" since dust grains moving along distinct directions coexist. Mathematically speaking, two different sets of governing flow equations must be solved — a so-called "two-fluid model." This was done in *Crifo et al.* (1995) but not in *Kitamura* (1990). The result is that the dust density distribution in *Kitamura* (1990) is not accurately computed in the interaction region (Fig. 4b). However, the position of the density enhancement is cor-

rect. In more complicated geometries, in the interaction regions many different directions of dust motion may exist, requiring the use of many dust fluids. (This requirement should not be confused with that of also using different dust fluids if different kinds of dust grains are considered.) This quickly becomes unmanageable, and one has to content oneself with the indicative one-fluid (per dust type) method, or use a DSMC.

Knollenberg (1994) considered a problem resembling that of *Kömle and Ip* (1987a,b): An active circular spot is placed on an inactive surface (no background; see Fig. 3b). His computed gas distribution is shown on Plate 12. It was later recomputed by Rodionov and Crifo with strictly identical (unpublished) results, which we use to display additional details of the solution. The supersonic gas converging from the ring to the axis "interacts with itself," forming a diverging conical weak shock with apex at a small distance over the surface. Below this apex a complicated low-velocity (subsonic) region is formed close to the symmetry axis over the inactive surface; it includes a circular stagnation line, parallel to the surface and centered on the axis, around which the gas whirls (the flowlines form a torus). Inside the conical shock, the slow gas is reaccelerated, forming a "daughter jet."

It is to be expected with such a gas flow structure that all dust trajectories starting from the inner edge of the active ring are directed toward the axis and mutually intersect there. The vicinity of this axis is thus expected to be a region of enhanced density. This is indeed what *Knollenberg* (1994) finds using (appropriately, but without saying it) a Monte Carlo simulation (Plate 12b). This is probably the most spectacular confirmation of the basic fact — already hinted at by *Whipple* (1982) — that the vertical of an inactive spot must be a dust-density maximum.

It should be noted that the symmetry that is present in all the preceding solutions is quite artificial: It follows not only from the asumption of a very simple nucleus geometry, but also from the equally strong assumptions of (1) solar illumination along the symmetry axis or (2) absence of illumination control of the gas production. The first assumption is not possible because of the nucleus rotation, and the second would require day-night symmetry in the appearance of the coma.

The results remains that (1) small inactive areas create usually coma dust density peaks, not minima, and (2) dust-density maxima do not trace only surface production maxima, but gas shocks as well.

Crifo et al. (1995) added to their duplication of *Kitamura*'s (1990) work the case of three aligned identical sources, with the Sun on the symmetry axis. In such a case, two "secondary" gas jets are formed in between the sources, and these two gas jets themselves interact at a greater altitude, to create "second-generation" weak gas shocks. The dust is not sensitive to these second-generation structures (at least at the moderate production rates considered) because gas-dust uncoupling has already occurred. *Crifo and Rodionov* (1997a) use the four equal rectangular active areas shown on Fig. 3d for several directions of solar illumination. The results, of course, reveal the formation of

shocks similar to those discussed above, and illustrate for the first time (owing to the three-dimensional capability of the numerical code) the deformation of the near-nucleus coma with changing solar direction.

To conclude this quick overview, let us address the already mentioned configuration proposed in *Knollenberg et al.* (1996) to account for the gross Halley dust coma appearance during the 1986 *Giotto* flyby (Fig. 3c). These authors compute the gas and dust distribution from the three proposed unequal circular active areas as if they were alone with the Sun on-axis, and then add up the resulting dust densities. Instead, Plate 13 shows the gas flow correctly computed from solving in three dimensions the EE, and the dust flow computed from a four-fluid model — one fluid for each area, plus one for the background (*Crifo and Rodionov*, 1998). One sees that the results do not resemble three similar structures. First, the difference in solar zenith angle between the areas results in strong deviations between the gas (and dust) outflow patterns from one another, and from that of an isolated on-axis illuminated jet; second, the gas jets from the three areas interact, forming weak gas shocks and secondary dust-density maxima. But, as we shall see, there is another, fundamental reason why such a model cannot account for Halley's coma: Halley's nucleus is anything but spherical, and the outflow from a sphere cannot be "pasted by hand" on any aspherical body.

6.3. Homogeneous, Aspherical Nuclei

It is evident that cometary nuclei cannot be spherical, hence the investigation of aspherical nuclei is the central requirement of cometary activity models. This raises immediately the question of down to which scale one wants — or can — simulate a nucleus. Figure 5 presents examples of shapes that have been subject to investigation to date.

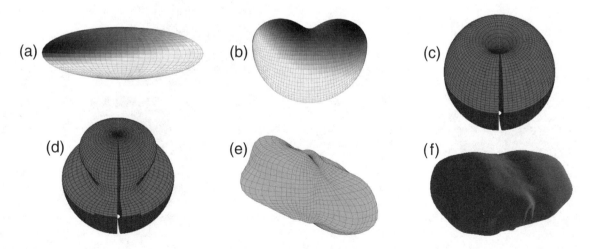

Fig. 5. Examples of aspherical homogeneous nuclei treated in the literature. (**a**) Triaxial ellipsoid (*Crifo et al.*, 1999; *Crifo and Rodionov*, 2000); (**b**) bean-shaped nucleus (*Crifo et al.*, 1997b, 1999); (**c**) apple-shaped nucleus (*Crifo et al.*, 2003a); (**d**) top-shaped nucleus (*Crifo et al.*, 2003a); (**e**) Muinonen shape (*Crifo et al.*, 2003b); (**f**) Halley nucleus (*Crifo et al.*, 2002b; *Rodionov et al.*, 2002).

[To this one should add that *Rodionov et al.* (2002) have also studied about 17 different shapes derived from the Halley shape shown in Fig. 5 by additional and more-severe spatial filtering, and that several variants of the so-called "Muinonen shapes" have been studied in *Crifo et al.* (2003b).] One sees that at the present time only relatively smooth shapes have been considered. This is due in part to mathematical limits, but also partly because the best Halley nucleus image resolution was only ≈50 m. The recent images of part of the surface of P/Borrelly (and of most of the surface of P/Wild 2) reveal that the surface is rich in very small topographic features. It may be expected that the fine gas structures expected from fine topographic details will be quickly filtered out by collisional smoothing, making the use of a smoothed surface suitable for modeling an imaged coma, given that line-of-sight integration should also smooth out such details. For the *in situ* sampled data, however, it is evident that surface modeling down to the metric scale will be required.

Not only are real nuclei aspherical, but they cannot be uniformly convex, as with all other small bodies of the solar system. Common sense indicates that the flow of gas over or near a totally convex object must be much simpler that that around or near an object with concavities. Hence, maximum attention must be paid to the latter.

6.3.1. Triaxial ellipsoid. A nonspherical, but still convex, nucleus can be considered to be merely a variant of the spherical nucleus: Figure 6 shows the case of a triaxial ellipsoid assumed either to outgas CO in a nearly uniform way, or to produce H_2O through surface sublimation [see a detailed description of the latter in *Crifo and Rodionov* (2000)]. In the first case, one observes density kinks at the

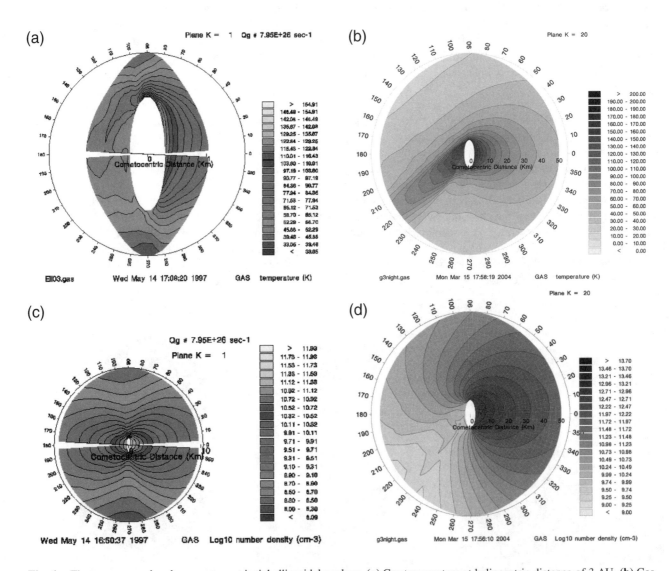

Fig. 6. The coma around an homogeneous triaxial ellipsoidal nucleus. (a) Gas temperature at heliocentric distance of 3 AU. (b) Gas temperature at heliocentric distance of 1 AU. (c) Gas number density at 3 AU; (d) gas number density at 1 AU. At heliocentric distance of 3 AU, it is assumed that CO diffuses out with a nearly uniform surface flux; at heliocentric distance of 1 AU, it is assumed that H_2O is sublimated, the Sun being in-plane in the direction of the angular graduation 45°; the nightside background parameter is $\kappa = 0.0275$.

tips of the ellipsoid, and relatively smooth bulges over the flatter sides of the nucleus. This is in accordance with the fact that the local surface radius of curvatures scales the local near-surface flow, as explained in *Crifo* (1991a). Of course, such kinks will lead to the formation of dust density enhancements owing to the trajectory-crossing effect described previously. In the second case (solar-driven sublimation), the day-to-night pressure difference is enough to blow away the kinks at the tips, at least for the adopted direction of illumination. As for the general coma layout, let us observe what turns out (from many yet unpublished computations) to be a general behavior: (1) the nightside coma is structured around the antisolar axis, and has the same complicated patterns as in the spherical case, but distorted; (2) the dayside coma is *not* structured around the solar direction, but more around the normal to the flattest area of the surface (minimal curvature), again a consequence of the local curvature scaling just mentioned.

We may expect that, during the nucleus rotation, in the case of CO diffusion a nearly invariant density pattern corotates, whereas in the case of H_2O sublimation, the day-to-night density asymmetry will persist, although modulated by rotation-induced deformations (we return to this in section 6.5). Note that in the CO case there is, however, a significant day-night asymmetry in gas temperature (due to the corresponding surface temperature asymmetry), and therefore a significant day-to-night velocity asymmetry (not shown). In the H_2O case, the temperature (hence velocity) asymmetry are quite small, due to the temperature buffering effect of the sublimation.

6.3.2. Bean-shaped nucleus. As stated earlier, the smooth gas flow possible around smooth, convex nuclei is impossible around realistic objects, because they have concavities; furthermore, in the solar-driven sublimation case, the pattern of shadows associated with the concavities subdivides the gas production into "effective" discrete active areas. For these two reasons, even a homogeneous nonconvex nucleus is expected to produce a coma structured by weak gas shocks. The pattern of these shocks will change during the rotation, following changes in the dayside shadow pattern. With this in mind, *Crifo and Rodionov* (1997b), for the first computation of a nonconvex nucleus, designed the "bean" shape shown on Fig. 5b. This shape has two planes of symmetry, hence an axis of symmetry. [This shape was also used for two-dimensional + 1/2 heat transfer computations by *Guttiérrez et al.* (2000).] Figure 7 shows the gas and dust-number density in three mutually perpendicular planes across the nucleus, under solar illumination from a direction inside the main symmetry plane. The dust was computed from a two-fluid-per-size algorithm.

There are two maxima of gas production around the two subsolar points, and the two associated flows from their vicinity interact to form two shock structures (of roughly hyperbolic shape). For the selected orientation, there is not yet partial shadow inside the cavity, but the left flank has less inclined illumination than the right flank, so there is an overall pressure gradient toward the Sun, which induces

an inclination of the shock structure toward the Sun. The associated enhanced dust-density region is formed, as in the *Kitamura* (1990) case, by overlap between dust from the two subsolar areas. We will return to this coma below when discussing the effect of the nucleus rotation.

6.3.3. Apple-shaped nucleus. All preceding results, whether referring to comets near 1 AU from the Sun, or to larger distances (*Crifo and Rodionov*, 1997a,b, 1999), were obtained from EE equations insensitive to the absolute gas density. However, fluid interactions disappear at vanishing gas densities. *Crifo et al.* (2002a, 2003a) have started to investigated down to which low levels of gas production the weak shocks computed from EE are real. The method used is to compare NSE results with DSMC results. For this purpose, shapes somewhat simpler than the bean were considered, e.g., "apples" and "tops" (cf. Fig. 5). The surprise was that new effects were discovered during this work — and now it is not known up to which high level of gas production they persist!

Figure 8 shows the flow inside the sunlit cavity of an apple-shaped homogeneous icy nucleus with characteristic size ≈15 km and assumed near the orbit of Jupiter (its computed total H_2O production rate due to sublimation is $Q = 3.3 \times 10^{26}$ molecules/s). One notices first that this flow is separated from the external flow (relative to the cavity) by the separating streamline eB, where B is a stagnation point on the axis, and e is on the surface; this means that no gas from the cavity escapes to free space! The flow inside the cavity divides itself into two closed cells: one is the triangular cell aAc, where a is the bottom of the cavity, A a second stagnation point on the axis, and c a point on the surface (see Fig. 8c); the second cell is the rectangular cell cABe (see Fig. 8a). In the first cell, the gas emitted from the segment ab (representing 0.06% of Q) is recondensed on the segment bc, without acceleration to supersonic state. In the second cell, the gas emitted from de (representing 0.2% of Q) is recondensed on dc, in part supersonically. Notice that the two flows inside the two cells rotate in opposite directions, as in convection cells. One additional feature of this fantastic flow structure is the presence of three sonic lines (SL1, SL2, SL3) transverse to the symmetry axis. Note also the broad size of the subsonic region (shaded on Fig. 8). Thus we see that, even though the cavity surface is fully illuminated by the Sun, there exists a band db where it condenses the ambient gas; this trapping effect explains why the cavity flow is confined inside the cavity, in spite of its wide opening. Such a flow trapping was not observed in the existing "bean" simulations, probably because this cavity has the shape of a saddle, i.e., is much more widely open to free space. On the other hand, it is not yet known whether the effect disappears in the "apple" at small heliocentric distances, and appears in the "bean" at very large distances.

Note that in the inner cell the Knudsen number Kn ≈ 1 (cf. Fig. 8d); also, our approximating BE-NSE approach to handle the near-surface conditions, based on nearly plane-parallel sublimation or condensation, is here in principle

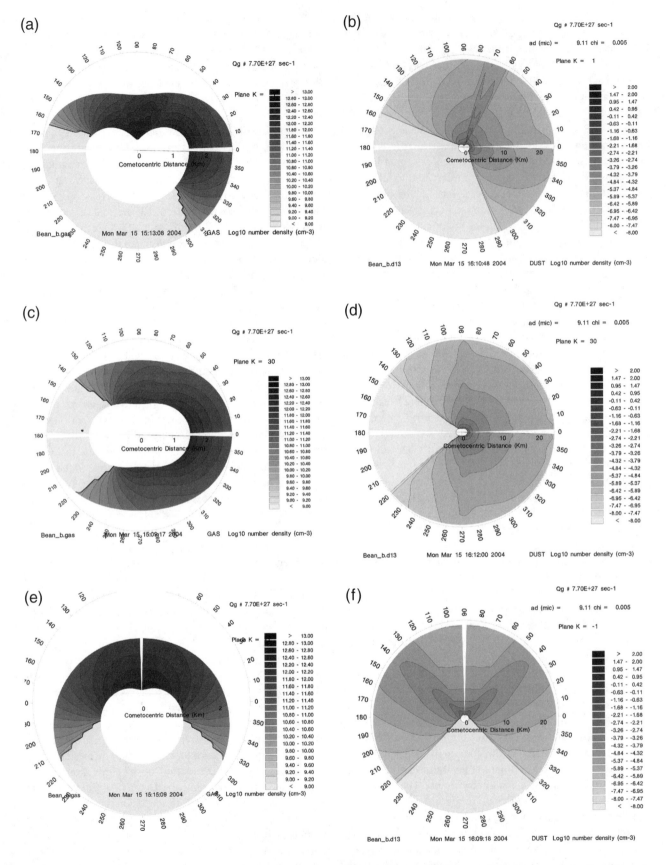

Fig. 7. Structure of the gas and dust coma around a bean-shaped nucleus (*Crifo and Rodionov,* 1999). The top panels show the main symmetry plane XOY; the Sun direction is in this plane, at the graduation 45°. Bottom panels show the secondary symmetry plane YOZ and plane XOZ. The lefthand panels show the decimal logarithm of the gas number density; righthand panels show the decimal logarithm of the dust number density (9.11-μm-radius grains).

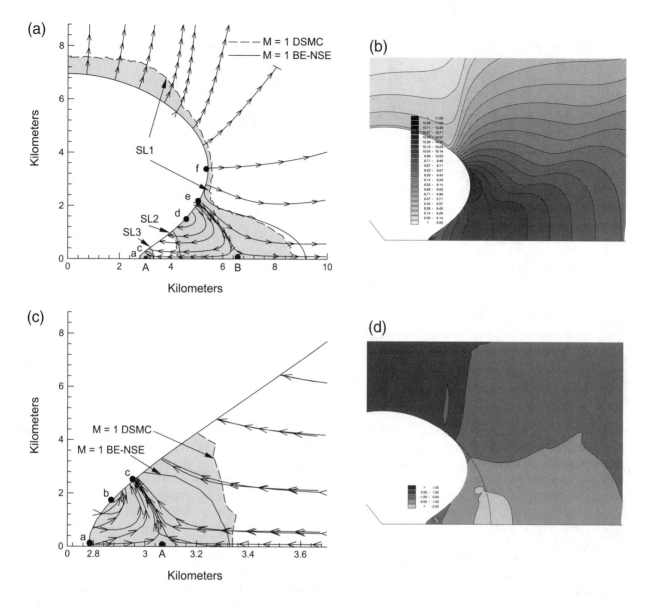

Fig. 8. Comparison between BE-NSE and DSMC solutions for an apple-shaped nucleus (*Crifo et al.,* 2003a). The Sun direction is to the right. **(a)** Flowlines (thin lines with arrows), sonic lines (dashed: DSMC; continuous: BE-NSE), and subsonic regions (shaded). **(b)** Log_{10}(gas number density). **(c)** Flowlines (thin: BE-NSE; thick: DSMC), sonic line SL3, and subsonic region (shaded), on an enlarged scale. **(d)** Log_{10}(Knudsen number).

invalidated by the fact that the two effects coexist at neighboring points. Even so, an impressive good agreement between the BE-NSE and DSMC solutions is obtained.

Finally, as to the disappearance of weak shocks at very small production rates, it was observed in the simulations, but this does not mean that structures in the gas density disappeared: Even in strict free-molecular outflow the gas density is quite uneven, hence dust small enough to be dragged away will exhibit sharp distribution structures.

The dust distribution created by the apple-shaped nucleus for on-axis illumination of its cavity resembles very much that obtained from the bean shape in its main symmetry plane (Fig. 4 of *Crifo et al.,* 1997b), with one notable difference: Because the "apple" nucleus is rotationally

symmetric and the Sun on-axis, so is the dust density, which thus is obtained by rotating the symmetry plane distribution around the symmetry axis; a narrow on-axis centered conical pencil of dust is formed (instead of two quasihyperbolic surfaces for a "bean"). Of course, this simplified geometry is broken as soon as the Sun leaves that axis.

6.3.4. Top-shaped nucleus. Figure 9 shows the computed gas and dust distributions when the Sun is on the symmetry axis of a top-shaped homogeneous icy nucleus. The presence of a partially shadowed cavity creates two separate gas jets whose interaction produces weak shock surfaces (one crosses the circular graduation near 70°, the other near 20°). Then, the dust is thrown into a conical pseudojet in the vicinity of this interaction. The conical

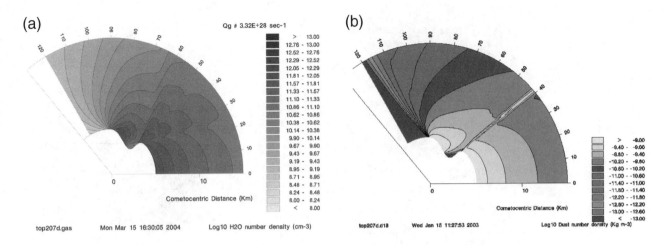

Fig. 9. Gas and dust distribution around a top-shaped nucleus. **(a)** H$_2$O number density (*Crifo et al.*, 2003a); **(b)** 9.10-μm-radius dust grain density (A. V. Rodionov, unpublished data, 2003). The Sun is to the right, on the horizontal axis.

geometry will be broken for off-axis illumination. The dust density shown here was computed from a single-dust fluid approximation, and therefore is only indicative of the correct distribution, as discussed previously.

6.3.5. Realistic shapes. While consideration of simple shapes are a must for demonstrating the basic physical processes at play in the near-nucleus coma, realistic comas can only be produced by consideration of plausible nucleus shapes. Unfortunately, there exists no bank of cometary nucleus shapes from which the expression "realistic" could be defined, since only P/Halley's nucleus shape has been determined (*Keller et al.*, 1995; *Szego et al.*, 1995, and references therein). Part of P/Borrelly's nucleus has been recently imaged, but the whole shape is not known, hence the computation of the gas flow even assumed to emanate only from that part is impossible. It is hoped that the recently acquired images of P/Wild 2 will provide the second full nucleus shape. At any rate, within the frame of the *Rosetta* mission definition studies, it is necessary to have a set of "plausible comas" at hand. For that purpose, *Crifo et al.* (2003b) have considered nuclei having random shapes with the same statistical topographic properties as those of the imaged asteroids, following *Muinonen* (1998) and *Muinonen and Lagerros* (1998). An example of one such shape is shown in Fig. 5e. Such shapes were also used for thermal modeling in *Guttiérrez et al.* (2001). The computed comas are complicated, as is the topography. We will not discuss them here, but will focus on the equally complicated case of P/Halley's nucleus, discussed in sections 6.4 and 7.2.

6.4. Inhomogeneous, Aspherical Nuclei

A real nucleus can be expected to be both complicated in shape and inhomogeneous in composition. Since both effects lead to a structured coma, one does not see on which basis the observation of structures should be automatically attributed to compositional homogeneity. The question in fact arises whether these two effects contribute equally to structuring the coma, or whether one is dominant. There is at the present time no universal answer to such a question, in view of its extreme difficulty. Only the case of Halley's nucleus has been studied (*Crifo et al.*, 2002b; *Szego et al.*, 2002). The complexity of the gas coma, even assuming Halley's nucleus to be homogeneous, can be judged on the two left gas density panels (b,d) of Plate 14, corresponding to two different image planes and solar illumination directions. Plates 14b,c show the same information if the nucleus is assumed to have the random distribution of Gaussian circular active areas of average size comparable to the size of the topography details, shown on Plate 14a. Unexpectedly, the differences are very minor. In particular, most weak shocks are present in the two cases, at about the same location. Of course, it is premature to generalize this result to all conceivable nucleus shapes, inhomogeneity patterns, and solar illuminations. Yet we believe that this result is extremely instructive. It suggests that, at least for P/Halley's nucleus, only extreme assumptions with respect to the inhomogeneity could significantly manifest themselves in the coma structure. Even though this was not yet computed, it is unlikely that the three-active regions pattern of Plate 11 placed in any manner on Halley's nucleus will produced anything like the sum of three isolated cylindrical jets. We will see in section 7.2 how the results of Plate 14 compare with the observations.

6.5. Influence of the Rotation on the Coma Structure

Plate 15 shows the deformation of the near-nucleus dust coma around a bean-shaped nucleus with a simple rotational motion (no precession or nutation). One sees that the V-shaped structures rotates, but at a slower rate than the Sun, and with deformation, until disappearance for part of the rotation. This is an extreme example, however; for a case where the Sun would rotate on a cone with axis tilted to the "bean" axis, one can anticipate (based on the result showed here) that the V-shape structure will rotate nonuniformly, with deformation, but without disappearing. The important point is that the apparent axis of rotation will not be that of the nucleus, and that its apparent angular velocity will not be its real angular velocity.

It is important to notice that the behavior of the near-nucleus coma just shown differs from that of the distant coma: The latter is a consequence of the evolution of the former during one or several nucleus rotation(s). For the dust, the physical process is the standard collisionless effusion in the solar gravity field reduced by radiation pressure. Hence, from the results shown on Plate 15 one can independently compute the outer coma dust distribution at any point and time. Such is not the case for the gas coma, as its collective behavior usually extends to very large distances; one can only compute at once the time-dependent structure of the whole coma. We return to this in section 8.

6.6. Conclusion: Formation of the Near-Nucleus Coma

The preceding model results are significant enough to draw many robust conclusions concerning the formation of the near-nucleus coma. Even though this region is rarely observable, it is clearly impossible to accept any paradigm concerning the currently observed outer coma that would violate any of these conclusions, which we summarize now:

1. The essential result is that the near-coma gas outflow is extremely sensitive to the surface topography. Roughly speaking, "hilltops" and high-f areas create supersonic outflows, "valleys" and low-f regions bounded by high-f ones create subsonic outflows, but simple rules to "guess" more precisely what the gas flow pattern is do not seem to exist — fluid dynamics cannot be guessed. Hence it would be physical nonsense to attribute near-nucleus gas structures (hitherto unobservable) to surface inhomogeneity only.

2. The gas outflow is a global property; there does not exist any physical mechanism by which some fixed structure (e.g., a conical jet) could be created at any given place, independently from the surrounding environment. Instead of coexisting without changing when "placed" on a surface, the structures created in insulation will unescapably be modified when placed in a surrounding environment, be it a background emission or some other similar structure. This is independent of how the gas is produced (surface sublimation or subsurface diffusion) and independent of its chemical composition and production rate. The gas interaction by definition does not stop at a small distance from the surface. Hence the paradigm of several simple, noninteracting structures advocated in most heuristic models of the outer coma described previously is physically impossible.

3. The assumption of one (or even more so, several) rotationally-invariant gas jet structure(s) modulated only in magnitude by a cos $z_\odot$ factor (found in many heuristic interpretations) is physically impossible; if there is such a solar modulation of the surface gas flux (as is the case for solar-driven gas production), then the "jet structure" will change its appearance (as shown in Plate 15). If one insists that the observations require a rigidly rotating gas jet, then the gas production should not be due to solar control, and therefore there should be no difference in observed activity between the day and nightsides of the nucleus, contrary to observations.

4. Inner coma dust structures are always associated with the gas structures of the inner coma. The formation of additional near-nucleus structures due to fluctuations of the dust concentration in the ice (i.e., a varying dust load in the gas flow) is possible, but, to interpret observed near nucleus dust structures, one must first determine which structures trace gas structures, and only when they have been identified can the remaining ones be attributed to dust-load inhomogeneities.

5. Dust structures resulting from gas structures do not trace dust grain motions; quite the contrary, they trace dust trajectory intersections. Their projection to the surface does not bear any simple relationship to the pattern of dust production at the surface.

6. Tracing dust grains back to where they left the surface across the gas-dust interaction region raises extreme difficulties: One would need to know not only their mass, but their shape as well, and even their exact orientation in the coma.

7. The impossibility of rotationally invariant near-nucleus gas structures implies a similar impossibility for the near-nucleus dust structures.

We will return in section 8 to the implications of these conclusions on the formation of outer coma (i.e., very large scale) structures. But first, let us see how the inner coma model results compare to observational data.

7. MATCHING PHYSICAL MODELS TO OBSERVATIONAL DATA

7.1. Hyakutake Arcs

Figure 10 shows an example of the nightside coma gas arcs observed in Comet Hyakutake. The arcs were also detected in OH by *Harris et al.* (1997), but there was no associated dust feature. Both *Harris et al.* (1997) and *Rodionov et al.* (1998) postulated that the arcs could only be the signature of an unobservable H_2O structure formed as a consequence of an auxiliary source of H_2O, but their approach and conclusions differ. The first group uses a DSMC method and finds that, assuming a secondary point source far from the arc, "they see no shock"; assuming a linear source centered on the shock and aligned on-axis, they obtain an arc. *Rodionov et al.* (1998), recomputing the first assumption by a NSE method, do find a "viscous" double-shock structure, resulting from the interaction between the supersonic flow from the main nucleus and that from the secondary source. The word "viscous" alludes to the fact that, because of the high gas rarefaction, the canonical double-shock (visible on Plate 13, left panels) is smoothed into a single structure. Figure 10 shows the re-

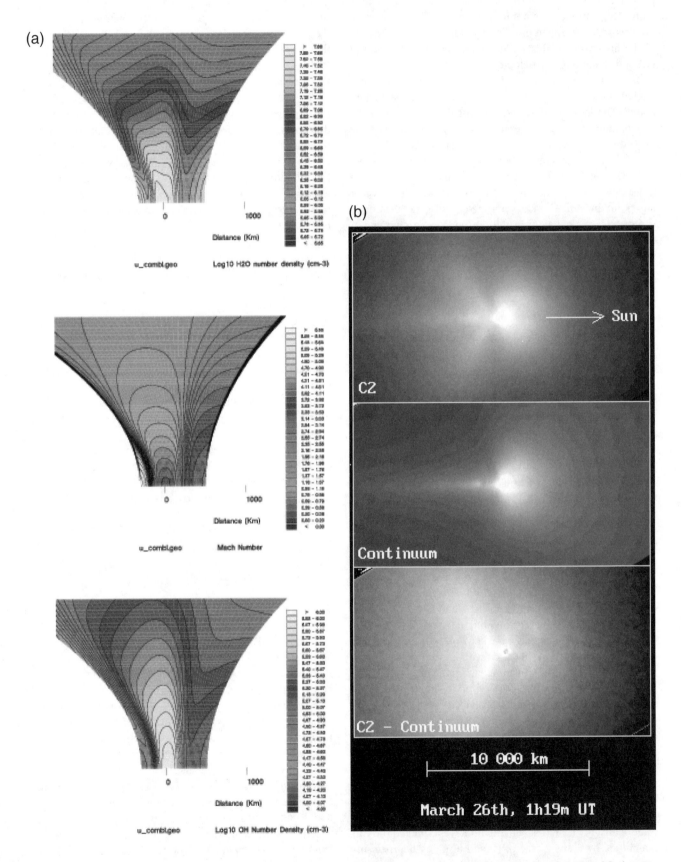

Fig. 10. Hyakutake's coma on March 26, 1996, after *Rodionov et al.* (1998). **(a)** Computed H_2O number density (top); Mach number (center); computed OH number density (bottom); the observed OH arcs are similar to the observed C_2 arcs. **(b)** C_2 + dust (top); dust only (center); C_2 only (bottom). Notice the bright spot on the antisolar axis, visible in the center image.

sult of the computed interaction (assuming a secondary source representing some 10–20% of the total H_2O production). This solution is not unique, but it reveals the Hyakutake arcs as the first evidence of weak gas shocks in a coma. Even more importantly, it gives confidence in the validity of the previously described modeling results, acquired under much denser gas conditions: Near a dayside nucleus at 1 AU, the gas number density is $O(10^{13})$ cm^{-3}, while in the stagnation region of the Hyakutake arcs it is computed to be $O(10^8)$ cm^{-3} only (see Fig. 10). This corresponds to a Knudsen number $O(10^{-1})$, quite convenient for the validity of the NSE approach (see *Crifo et al.*, 2002a, 2003a). The quoted statement of *Harris et al.* (1997) that their DSMC result "does not see a shock" suggests some technical inadequacy in the implementation of their DSMC code.

7.2. P/Halley Near-Nucleus Dust Coma

The greatest of all possible challenges to a coma model was offered by the famous HMC synthetic image of Halley's nucleus and its vicinity (*Keller et al.*, 1995, and references therein). As stated earlier, this image has been unanimously — but without any supporting model computation — interpreted as "visual evidence" that "the activity is confined to localized regions." However, "activity" means surface flux, and we do not "see" dust fluxes on images; we see dust column densities (more or less proportional to the local density, near to the surface). But near the surface, the dust is coupled to the gas. Observing dust in natural flows (e.g., snow blown on an icy surface, or desert dust blown by winds) is enlightening here: Dust accumulation often reveals flow stagnation regions, rather than high flux regions. There is no intuitive way to reliably infer flux from density; inferences must be checked by quantitative modeling.

The three-sources-on-a-sphere paradigm of *Knollenberg et al.* (1996) described in section 6.2 was offered in support of the classical interpretation, but we remind the reader of the weaknesses we have discovered: (1) the interaction between the outflows from the three regions cannot be neglected, and (2) the control of the outflow by the topography cannot be neglected — it is in all likelihood dominant. It is also evident that this model, which assumes a spherical surface, cannot be matched to the central part of the HMC images, which resolve a highly aspherical nucleus.

The "filament" map derived by *Keller et al.* (1995) using enhancement techniques is reproduced here in Fig. 11a. *Knollenberg et al.* (1996) proposed that each filament is the signature of a small-circular-inactive-area, in accordance with Plate 12. Here again, we must point out the weaknesses of this explanation: (1) The narrow collimation of the dust pencil obtained requires a strict cylindrical symmetry. This cannot be achieved by groups of closely positioned spots, as would be required; even the symmetry of an isolated spot will be ruined by the irregular topography of the surface where it lies. (2) Obviously, the Sun also cannot be placed simultaneously on the axis of many distinct spots. And, if

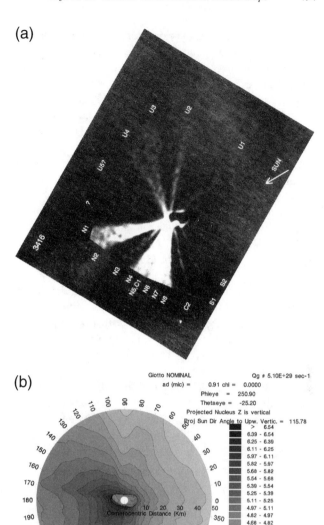

Fig. 11. Interpretation of the near-Halley-nucleus dust coma structures observed by the HMC camera (*Crifo et al.*, 2002b). **(a)** Azimuthal brightness gradient map derived by *Keller et al.* (1994). **(b)** 0.91-μm-radius dust grains column density, computed by the present model assuming P/Halley's nucleus to be strictly homogeneous with f = 0.6.

one wanted to state that the gas outflow does not depend upon solar illumination, how then could one explain the vanishing activity observed on the nucleus nightside?

Crifo et al. (2002b) have offered a related, but self-consistent interpretation of the "filament" array. They observe that the visible-light HMC images are sensitive to all dust masses, and it is clear that the dust trajectories depend upon their mass; furthermore, nonspherical grains have trajectories totally different from spherical grains of the same mass. Thus, the simplest bet one should make when looking at this map is that one sees the signatures of weak gas shocks, since this signature is due only to the difference in inertia between dust grains and gas molecules, irrespective of grain size and shape. This signature, being independent from dust grain size and shape, can be discovered by "numerical tracing": At the laboratory, one would inject calibrated dust (i.e., single-size, spherical grains) in the flow. Here, we can numerically simulate the injection of such calibrated dust. Furthermore, a single-fluid dust model can be used, even though it does not provide the exact dust density in the structures, since it does provide the correct structure position. *Crifo et al.* (2002b) consistently computed the overall gas outflow from the whole nucleus surface, using the nucleus shape derived from the images obtained by the *Vega 1* probe by *Szego et al.* (1995), and assuming the nucleus to be (to start with) compositionally homogeneous.

Figure 11b shows the computed dust column density of spherical grains of 0.91-μm radius, computed with a single-fluid model. Detailed comparison with the HMC map (Fig. 11a) reveals the following: (1) The N1–N2 gradients are reproduced with an angular offset of about 20°. (2) All other gradients, in particular U2, U3, U4, N4–N8, and S2, are present in the model computation at the right position. This result is not significantly changed by changes in dust grain size. It is, on the other hand, very sensitive to the nucleus orientation — which is not known to better than a few degrees, hence the agreement seems surprisingly good.

So, for the first time in the history of cometary data analysis, dust coma structures are reproduced ab initio, i.e., without ad hoc assumptions. The physical meaning of this agreement is also clear: The coma structures seen are the signatures of gas flow discontinuities induced by the topography.

Comparisons of a similar kind (partially published in *Szego et al.*, 2002) were made by the same authors with all *Vega 2* images (the *Vega 1* images carry little information on the near-nucleus coma) with even a more spectacular agreement than that obtained by *Giotto* — but here, one will notice that the resolution of the *Vega 2* images is somewhat lower.

7.3. Gas Lightcurves

Gas lightcurves are obtained from summing the emission of molecules within a cylinder of radius R (in kilometers) (projected in the coma). One can compute that, if the gas velocity is on the order of 1 km/s, and the emission roughly isotropic, about one-third of the molecules were emitted over a time period R (seconds), and two-thirds over a time period of 1 d (case of OH at 1 AU from the Sun). Thus, for a real nucleus rotating with a period approximately equal to a few days, a 3D + t model of the gas coma coupled to a nucleus rotation model should be used to extract a correct Q(t) from the gas lightcurve. At this time, we know of no comet for which such a task was done.

The best prospect for seeing such a work performed is Comet Halley, since its complete nucleus shape is known and good lightcurve data are available.

The equations of rotational motion were integrated by *Szego et al.* (2001), based on the three nucleus orientations indicated by the imaging experiments during the flybys of *Vega 1*, *Vega 2*, and *Giotto*, with due allowance for the outgassing torque. Assuming a nucleus density 0.5 g/cm³, the torque is about a 5% large correction term for the equations of rotation. Due to this torque, the direction of the angular momentum vector is changed by ~11° in longitude and 16° in latitude between August 31, 1985, and July 26, 1986, i.e., from 3 AU AP to ~2.8 AU PP. The computed rotation exhibits a basic rotation period ~2.2 d about the short axis of the nucleus, modulated by a 7.4-d long period. This model, despite its 7.4-d modulation, does not reproduce the basic ~7.4-d periodicity observed in ground-based observations (lightcurve, recurrence of jet patterns).

Another rotation model has been derived by *Belton et al.* (1991). It exhibits the requested rotation period of ~7.4 d (found to be about the long axis of the nucleus). This model, however, is obtained by flipping the orientation of the nucleus during the *Vega 1* flyby with respect to the observed one. This seems to be due to the fact that the authors did not use the processed *Vega 1* images, but only the raw images, much less constraining.

As already stated, *Julian et al.* (2000) and *Szego et al.* (2001) checked their P/Halley rotation models against a Q(t) derived from a Haser formula. One can therefore question the satisfaction the two groups express regarding their fits.

This situation is not satisfactory, and work is in progress to find a solution for the rotation of the nucleus that satisfies all constraints (and hopefully all groups). It will then be possible to attempt an exact interpretation of P/Halley's lightcurve.

7.4. Gas "Velocity Law"

In many cometary papers a so-called "Bobrovnikov's velocity law" is advocated to evaluate "the coma gas velocity." *Keller* (1954) quoted observations (without any technical details such as the names of the comets that were observed) of "jets and haloes" by Bobrovnikov that were later used by *Whipple* (1982) to derive the above-mentioned law. As far as we know, Bobrovnikov himself never published such a result, even decades after this report. Indeed, this "law" seems to mix up observations of a quite different nature: Anything moving is used. It is in particular quite unclear whether the observed motions are due to gas or to dust. Furthermore, many authors admit that this law (quite expectedly) is violated by their data. Last, but not least, one can read in *Bobrovnikov*'s (1931, p. 460) famous report on Comet Halley that "the assumption of a constant velocity of ejection is questionable. Data gathered in Chapter II indicate rather wide ranges of velocities near the nucleus, from several tenths to several km/sec." The same statement is duplicated in the conclusion of the report (p. 581).

As evidenced by the model results presented above, there is nothing like a "gas velocity" in a coma; this velocity varies from point to point. This is observationally evident in the high-resolution observations of molecular radiowave emissions (e.g., *Henry et al., 2002*). As for the heliocentric dependence, it is straightforward to show that, in the case of subsurface diffusion, the near-nucleus gas velocity is $\propto r_h^{-1/4}$; in the case of surface sublimation, this dependence is negligibly small. For a discussion of outer coma gas velocities, see *Combi et al.* (2004).

7.5. Dust "Velocity Law"

In some favorable cases, e.g., the observation of a dust neckline structure (see *Fulle*, 2004), it is possible to derive the terminal dust velocity from an analysis of the image brightness. Most of the time, this is not possible. If one wants to derive a dust mass loss rate from an observed image brightness, it is then necessary to use assumed dust velocities. For this purpose, a "dust velocity law" is often advocated. Many variants of this "law" are circulated, including — as already pointed out in *Wallis* (1982) — some that are nothing but misprint propagations. Most variants postulate isotropic gas and dust production, and they all assume, quite unrealistically, that both the grains and nucleus are spherical. An expression summarizing the presently discussed gas dynamic simulations of (nonisotropic) spherical grain ejection from a spherical nucleus is given in *Crifo et al.* (1997a). Still, many heuristic parameters are present in this expression, for which different users will unavoidably choose differing numerical values when trying to represent real data. The resulting chaos in dust mass loss rate assessment was already pointed-out in *Crifo* (1991b) and will unavoidably persist until the missing information about dust grains is available.

More fundamentally, and as stated earlier, one must be careful to remember that dust is an ill-defined concept — or, if one prefers, the properties of real cometary grains are unknown. It is true that large-scale coma structures can be interpreted under the declared assumption of spherical grains, but that does not prove that the real grains are spherical. After decoupling from the gas, the grain motion is controlled by solar radiation pressure and solar gravity. The two effects do not distinguish the grain shape (a spinning irregular grain will exhibit an effective radiation pressure quite like a spherical one). Hence the success of outer coma fits gives no clue to what happens inside the gas-dust interaction region. Unfortunately, the ejection velocity of solids depends critically on their shape and even initial grain orientation at the surface (see *Crifo and Rodionov, 1999*), hence the ejection velocity of real cometary grains from a real cometary nucleus can simply not be predicted at present, save in a crude order of magnitude.

7.6. Dust Lightcurves

As opposed to the case of the gas, 3D + t outer coma dust models do exist, but to make their use rewarding, nuclei shapes are missing. Remembering that dust production rates are derived through the $Af\rho$ method, it is easy to show that, while a spherical nucleus that would eject dust with perfect isotropy would build up a dust coma with a 1/r

brightness dependence, the reciprocal statement is incorrect: Anisotropic dust-ejection patterns are capable of building up an apparently isotropic coma with brightness slopes widely dispersed around the value –1, depending on the direction of observation (*Fulle and Crifo, 1999*). Hence, from this point of view, the production rates derived without knowledge of the nucleus shape should be treated with caution.

8. PHYSICAL MODELS OF THE OUTER COMA STRUCTURES

We have seen that (1) steady-state gas and dust solutions cease to be valid at distances in excess of typically 1000 km and 100 km, respectively, and (2) the gas-dust interaction is completed before 100 km, so that the gas and the dust properties can then be computed independently; their outer coma structures need not bear any similarity. In both cases the following are required:

1. A physical (time-dependent) model must be used to compute the evolution of the gas and dust species distribution with distance to the nucleus. We will not review such models in the present chapter. (For the gas in a comet like Hale-Bopp, solving the NSE is one possibility.) The general observation is, however, of interest, that this evolution always involves structure smoothing, as substantiated below. Hence the more distant the observed regions, the less suitable they are, in principle, for inferring detailed nuclear properties.

2. Time-dependent boundary conditions must be specified at the distance where the steady-state solutions cease to be valid. We discuss in detail these boundary conditions, because, regardless of the quality of the outer coma model, the solution is just as valid as the boundary conditions used are acceptable.

8.1. Large-Scale Dust Structures

Plate 13 shows that the evolution of the structures from the vicinity of the nucleus to larger distances is very different for the gas and for the dust. The dust structures are quickly "frozen" into a fixed pattern (which changes with the rotation phase). The gas structures are not at all frozen. Large-scale dust structures can therefore be considered to be built up by free effusion from an "effective source" that is not the nucleus surface, but the terminal gas-dust acceleration surface. It is perhaps not evident that this effusion is a smoothing process, because it is a slow one: Smoothing is caused by the fact that, at each point of the distant coma, dust from all over the terminal surface arrives; only the unphysical myth of single-velocity, strictly radial dust ejection can lead to unique point-to-point dynamical connections between the coma and the surface of the nucleus. Even worse, dust from successive terminal surfaces during a sizable period (weeks) can possibly reach a given point. This evidently mixes up the terminal surface pattern.

As shown by Plate 13, this terminal surface has localized areas of enhanced dust flux, and indeed resembles *Seka-*

nina's (1981, 1991) "nucleus surface activity" maps. Hence, it is clear that what Sekanina and several other scientists using the same method determine is, in reality, not a nucleus surface distribution, but this effective source distribution. The difference in distances involved — $O(100)$ km — is much smaller than the spatial resolution in groundbased coma images. One therefore easily conjectures why the authors are forced to turn "on" and "off" in an ad hoc way their "active regions": This is most likely due to their attempt to fit a changing pattern (that at the terminal surface) with a supposedly fixed pattern (assumed to be the nucleus surface map).

For the nuclei of unknown shape, there is definitely no possibility of passing from such "terminal surface maps" to true "nucleus activity maps." But for those with a known shape (at this time, only P/Halley), this is possible. Work in this direction is in progress within the frame of the International Space Science Institute "Halley" Team Work.

8.2. Large-Scale Gas Structures

For the gas, after the initial structuring by weak shocks, the persistence of the fluid behavior maintains a very efficient pressure smoothing (Plate 13). If the expansion was adiabatic, this smoothing would stop rapidly as p → 0 when T → 0. But the expansion is not at all adiabatic: The gas pressure continues to be significant because of heating by fast photodissociation products. Furthermore, the observed daughter species may have a distribution different from that of H_2O due to partially nonlocal dissociation velocity damping. It is therefore quite unlikely that near-nucleus structures (at least density structures) be maintained far out in a gas coma. (Velocity structures are, on the other hand, well preserved in supersonic state.) In all likelihood, large-scale "jets," "shells," and "spirals" are not the signatures of faint details of the gas production, but that of the rotationally induced fluctuations in the total gas production. In the governing equations, this signature is induced by the Coriolis force (in a nucleus-attached frame). The production fluctuations may be due to changes in the nucleus cross-section, or changes in the total active area, or both.

At the present time, i.e., before publication of the full coma time-dependent solutions of the NSE equations recently obtained, we can only make the following points:

1. It is untenable to postulate a constant pattern of primary molecule production, corotating with the nucleus, as in the precedingly quoted interpretations of Hale-Bopp spirals. While the general explanation offered in these works may be partly or entirely true, the details of the solutions (number of sources, openings, rates) are surely unreliable. This is all the more true of the conclusions regarding chemical inhomogeneity.

2. The first step in accounting for P/Halley spirals and rings should be to investigate what the large-scale coma structure is that results from the model presented in section 7.2. If a homogeneous nucleus does not produce suitable structures, inhomogeneity may be added to the model

assumptions, and the test repeated. Only if it proves impossible to obtain a suitable result in this manner can one start to introduce ad hoc "active dust."

9. CONCLUSIONS

The visionary prediction of *Whipple* (1950, 1951), that cometary activity is due to the solar-driven surface and subsurface sublimation of a solid nucleus dominantly made up of the ices of a few simple molecules and of dust, is still the basis of our efforts to understand comets today. For decades, the simple heuristic simplifications he made to derive orders of magnitude also remained the basis of most of the interpretations of the observational data. To some extent, this can be understood, as it was not possible to access detailed information about the nucleus and the dust. With the implementation of deep-space missions to comets, this situation is radically changing. It becomes possible to picture the cometary nuclei as real solar system objects, with their unavoidable extreme complexity. The price to pay is that, to understand them, the simple heuristic approaches must be abandoned in favor of the use of the full arsenal of the modern methods of applied physics. Even before the avalanche of data from cometary orbiters is at hand, enormous efforts should be dedicated to adapting these methods to the difficult problem of inferring properties of a real nucleus from observations of its surrounding gas and dust coma. We hope that the present chapter has convinced the reader that such an undertaking is (1) unavoidable, (2) challenging, and (3) potentially rewarding.

Acknowledgments. This work benefited from many successive yearly contracts from the French CNES and CNRS agencies, from partial support from the Austrian "Fonds zur Foerderung der wissenschaft – lichen Forschung" under project P15470, and from the Hungarian OTKA grant T 32634. The International Space Science Institute in Bern is thanked for having supported an ISSI team meeting dedicated to the work on P/Halley discussed in this chapter.

REFERENCES

Abramov A. A. and Izsvestiya A. N. (1984) *SSSR, Mekhanika Zhidosty i Gaza, 1984, 1,* 185.

A'Hearn M. F., Hoban S., Birch P. V., Bowers C., Martin R., and Klinglesmith D. A. (1986) Cyanogen jets in Comet P/Halley. *Nature, 324,* 649–651.

A'Hearn M. F., Millis R. L., Schleicher D. G., Osip D. J., and Birch P. V. (1995) The ensemble properties of comets: Results from narrowband photometry of 85 comets, 1976–1992. *Icarus, 118,* 223–270.

Belton M. J. S., Julian W. H., Anderson A. J., and Muller B. E. A. (1991) The spin state and homogeneity of comet P/Halley's nucleus. *Icarus, 93,* 183–193.

Benkhoff J. and Huebner W. F. (1995) Influence of the vapor flux on temperature, density, and abundance distributions in a multi-component, porous, icy body. *Icarus, 114,* 348–354.

Benkhoff J. and Spohn T. (1991) Results of a coupled heat and mass transfer model applied to KOSI sublimation experiments. In *Theoretical Modelling of Comet Simulation Experiments*

(N. I. Kömle et al., eds.), pp. 31–47. Verlag der Österreichischen Akademie der Wissenschaften.

Bird G. A. (1994) *Molecular Gas Dynamics and the Direct Simulation of Gas Flows.* Oxford Univ., New York. 458 pp.

Bird G. A. (2001) Forty years of DSMC, and now? In *Rarefied Gas Dynamics XXII* (T. J. Bartel and M. A. Gallis, eds.), pp. 372–381. AIP Conference Proceedings 585, American Institute of Physics, New York.

Bobrovnikov N. T. (1931) Halley's comet in its apparition of 1909–1911. *Publ. Lick Obs. XVII(2),* pp. 309–482.

Brown R. H. and Matson D. L. (1987) Thermal effects of insolation propagation. *Icarus, 72,* 84–94.

Brownlee D. E., Hörz F., Newburn R. L., Zolensky M., Duxbury T. C., Sandford S., Sekanina Z., Tsou P., Hanner M. S., Clark B. C., Green S. F., and Kissel J. C. (2004) Surface of young Jupiter family comet 81P/Wild-2: View from the Stardust spacecraft. *Science, 304,* 1764–1769.

Capria M. T. (2002) Sublimation mechanisms of comet nuclei. *Earth Moon Planets, 89,* 161–177.

Cercignani C. (1981) Strong evaporation in a polyatomic gas. *Progr. Astron. Aeron., 74(1),* 305–320.

Cercignani C. (2000) *Rarefied Gas Dynamics.* Cambridge Univ., Cambridge. 320 pp.

Clairemidi J., Moreels G., and Krasnopolsky V. A. (1990) Spectro-imagery of P/Halley's inner coma in OH and NH ultraviolet bands. *Astron. Astrophys., 231,* 235–240.

Combi M. (1996) Time-dependent gas kinetics in tenuous planetary atmospheres: The cometary coma. *Icarus, 123,* 207–226.

Combi M., Boss B. J., and Smyth W. H. (1993) The OH distribution in cometary atmospheres — A collisional Monte Carlo model for heavy particles. *Astrophys. J., 408,* 668–677.

Combi M. R., Harris W. M., and Smyth W. H. (2004) Gas dynamics and kinetics in the cometary coma: Theory and observations. In *Comets II* (M. C. Festou et al., eds.), this volume. Univ. of Arizona, Tucson.

Crifo J. F. (1987) Optical and hydrodynamic implications of cometary dust. In *Symposium on the Diversity and Similarity of Comets* (E. J. Rolfe and B. Battrick, eds.), pp. 399–408. ESA SP-278, Noordwijk, The Netherlands.

Crifo J. F. (1991a) Hydrodynamic models of the collisional coma. In *Comets in the Post-Halley Era* (R. L. Newburn and J. Rahe, eds.), pp. 937–989. Kluwer, Dordrecht.

Crifo J. F. (1991b) In-situ Doppler velocimetry of very large grains: An essential goal for future cometary investigations. In *Proceedings of the Symposium on Radars and Lidars in Earth and Planetary Science* (T. D. Guyenne and J. Hunt, eds.), pp. 65–70. ESA SP-328, Noordwijk, The Netherlands.

Crifo J. F. (1993) The dominance of positive and negative heavy water cluster ions in the inner coma of active comets. In *Plasma Environments of Non-Magnetic Planets* (T. I. Gombosi, ed.), pp. 121–130. Pergamon, New York.

Crifo J. F. (1995) A general physicochemical model of the inner coma of active comets. I. Implications of spatially distributed gas and dust production. *Astrophys. J., 445,* 470–488.

Crifo J. F. (1997) The correct evaluation of the sublimation rate of dusty ices under solar illumination, and its implications on the properties of P/Halley's nucleus. *Icarus, 130,* 549–551.

Crifo J. F. and Rodionov A. V. (1997a) The dependence of the circumnuclear coma structure on the properties of the nucleus. I. Comparison between a homogeneous and an inhomogeneous nucleus, with application to P/Wirtanen. *Icarus, 127,* 319–353.

Crifo J. F. and Rodionov A. V. (1997b) The dependence of the circumnuclear coma structure on the properties of the nucleus. II. First investigation of the coma surrounding an homogeneous, aspherical nucleus. *Icarus, 129,* 72–93.

Crifo J. F. and Rodionov A. V. (1998) Modeling the near-surface gas and dust flow. Invited review presented at Session B1.3 of the XXXIIe COSPAR Symposium, Nagoya (Japan), July 1998.

Crifo J. F. and Rodionov A. V. (1999) Modelling the circumnuclear coma of comets: Objectives, methods and recent results. *Planet. Space Sci., 47,* 797–826.

Crifo J. F. and Rodionov A. V. (2000) The dependence of the circumnuclear coma structure on the properties of the nucleus. IV. Structure of the night-side gas coma of a strongly sublimating nucleus. *Icarus, 148,* 464–478.

Crifo J. F., Itkin A. L., and Rodionov A. V. (1995) The near-nucleus coma formed by interacting dusty gas jets effusing from a cometary nucleus: I. *Icarus, 116,* 77–112.

Crifo J. F., Rodionov A. V., and Bockelée-Morvan D. (1999) The dependence of the circumnuclear coma structure on the properties of the nucleus, III. First modeling of a CO-dominated coma, with application to comets 46P/Wirtanen and 29P/Schwassmann-Wachmann 1. *Icarus, 138,* 85–106.

Crifo J. F., Loukianov G. A., Rodionov A. V., and Zakharov V. V. (2002a) Comparison between Navier-Stokes and direct Monte-Carlo simulations of the circumnuclear coma: I. Homogeneous, spherical sources. *Icarus, 156,* 249–268.

Crifo J. F., Rodionov A. V., Szego K., and Fulle M. (2002b) Challenging a paradigm: Do we need active and inactive areas to account for near-nuclear jet activity? *Earth Moon Planets, 90,* 227–238.

Crifo J. F., Loukianov G. A., Rodionov A. V., and Zakharov V. V. (2003a) Comparison between Navier-Stokes and direct Monte-Carlo simulations of the circumnuclear coma: II. Homogeneous, aspherical sources. *Icarus, 163,* 479–503.

Crifo J. F., Bernard J., Ceolin T., Garmier R., Lukianov G. A., Rodionov A. V., Szego K., and Zacharov V. V. (2003b) Descent on a weakly active cometary nucleus: Model and application to the Rosetta mission. *Icarus,* in press.

Davidsson B. J. R. and Skorov Yu.V. (2002a) On the light-absorbing surface layer of cometary nuclei: I. Radiative transfer. *Icarus, 156,* 223–248.

Davidsson B. J. R. and Skorov Yu.V. (2002b) On the light-absorbing surface layer of cometary nuclei: II. Thermal modelling. *Icarus, 159,* 239–258.

Enzian A., Cabot H., and Klinger J. (1997) A 2 1/2 thermodynamic model of cometary nuclei I. Application to the activity of Comet 29P/Schwachmann-Wachmann 1. *Astron. Astrophys., 319,* 995–1006.

Enzian A., Klinger J., and Schwehm G. (1999) Temperature and gas production distribution on the surface of a spherical model comet nucleus in the orbit of P/Wirtanen. *Icarus, 138,* 74–84.

Espinasse S. J., Klinger J., and Schmitt B. (1991) Modelling of the thermal behaviour and the chemical differentiation of cometary nuclei. *Icarus, 92,* 350–365.

Fernández J. A. (1990) Collisions of comets with meteoroids. In *Asteroids, Comets, Meteors III* (C. I. Lagerkvist, ed.), pp. 309–312. Uppsala, Sweden.

Fulle M. (2004) Motion of cometary dust. In *Comets II* (M. C. Festou et al., eds.), this volume. Univ. of Arizona, Tucson.

Fulle M. and Crifo J. F. (1999) Dust comae and tails built-up by hydrodynamical gas drag models: Application to 46P Wirtanen and C/Hyakutake 1996B2 (abstract). *Bull. Am. Astron. Soc.,*

31(4), 1128.

Fulle M., Crifo J. F., and Rodionov A. V. (1999) Numerical simulation of the dust flux on a spacecraft in orbit around an aspherical cometary nucleus — I. *Astron. Astrophys., 347*, 1009–1028.

Gombosi T. I. (1986) A heuristic model of the comet P/Halley dust size distribution. In *20th ESLAB Symposium on the Exploration of Halley's Comet* (B. Battrick et al., eds.), pp. 167–171. ESA SP-250, Noordwijk, The Netherlands.

Gombosi T. I. (1994) *Gaskinetic Theory.* Cambridge Univ., Cambridge. 297 pp.

Guttiérrez P. J., Ortiz J. L., Rodrigo R., and López-Moreno J. J. (2000) A study of water production and temperatures of rotating irregularly shaped cometary nuclei. *Astron. Astrophys., 355*, 809–817.

Guttiérrez P. J., Ortiz J. L., Rodrigo R., and López-Moreno J. J. (2001) Effects of iregular shape and topography in thermophysical models of heterogeneous cometary nuclei. *Astron. Astrophys., 374*, 326–336.

Harmon J. K., Campbell D. B., Ostro S. J., and Nolan M. C. (1999) Radar observations of comets. *Planet. Space Sci., 47*, 1409–1422.

Harris W. M., Scherb F., Combi M. R., and Mueller B. E. A. (1997) Evidence for interacting gas flows and an extended volatile source distribution in the coma of comet C/1996 B2 (Hyakutake). *Science, 277*, 676–681.

Henry F., Bockelée-Morvan D., Crovisier J., and Wink J. (2002) Observations of rotating jets of carbon monoxide in comet Hale-Bopp with the IRAM interferometer. *Earth Moon Planets, 90*, 57–60.

Hoban S., Samarasinha N. H., A'Hearn M. F., and Klinglesmith D. A. (1988) An investigation into periodicities in the morphology of CN jets in comet P/Halley. *Astron. Astrophys., 195*, 331–337.

Julian W. H., Samarasinha N. H., and Belton M. J. S. (2000) Thermal structure of cometary active regions: Comet 1P/Halley. *Icarus, 144*, 160–171.

Kaufmann E., Kömle N. I., and Kargl G. (2002) Experimental and theoretical investigation of the solid-state greenhouse effect. In *Proceedings of the ESA Symposium on Exobiology* (H. Savoya Lacoste, ed.), pp. 87–90. ESA SP-518, Noordwijk, The Netherlands.

Keller G. (1954) Perkin Observatory. In *Report of Observatories 1953–1954, Astron. J., 59*, 356–358.

Keller H. U., Knollenberg J., and Markiewicz W. J. (1994) collimation of cometary dust jet filaments. *Planet. Space Sci., 42(5)*, 367–382.

Keller H. U., Curdt W., Kramm J.-R., and Thomas N. (1995) *Images of the Nucleus of Comet P/Halley, Vol. 1.* ESA SP-1127, Noordwijk, The Netherlands.

Kitamura Y. (1986) Axisymmetric dusty gas jet in the inner coma of a comet. *Icarus, 66*, 241–257.

Kitamura Y. (1987) Axisymmetric dusty gas jet in the inner coma of a comet. II. The case of isolated jets. *Icarus, 72*, 555–567.

Kitamura Y. (1990) A numerical study of the interaction between two cometary jets: A possibility of shock formation in cometary atmospheres. *Icarus, 86*, 455–475.

Klinger J. (1981) Some consequences of a phase transition of water ice on the heat balance of comet nuclei. *Icarus, 47*, 320–324.

Knollenberg J. (1994) Modellrechnung zur Staubverteilung in der inneren Koma von Kometen unter spezieller Berücksichtigung der HMC-daten de Giotto mission. Ph.D. thesis, Göttingen (in German).

Knollenberg J., Kührt E., and Keller H. U. (1996) Interpretation of HMC images by a combined thermal and gasdynamic model. *Earth Moon Planets, 72*, 103–112.

Kömle N. I. (1990) Jet and shell structures in the cometary coma: Modelling and observations. In *Comet Halley: Investigations, Results, and Interpretations, Vol. 1*, pp. 231–243. Ellis Horwood Library of Space Science and Space Technology.

Kömle N. I. and Dettleff G. (1991) Mass and energy transport in sublimating cometary ice cracks. *Icarus, 89*, 73–84.

Kömle N. I. and Ip W. H. (1987a) Anisotropic non-stationary gas flow dynamics in the coma of Comet P/Halley. *Astron. Astrophys., 187*, 405–410.

Kömle N. I. and Ip W. H. (1987b) A model of the anisotropic structure of the neutral gas coma of Comet P/Halley. In *Symposium on the Diversity and Similarity of Comets* (E. J. Rolfe and B. Battrick, eds.), pp. 247–254. ESA SP-278, Noordwijk, The Netherlands.

Kömle N. I. and Ulamec S. (1989) Heating of dust particles enclosed in icy material. *Planet. Space Sci., 37*, 193–195.

Kömle N. I., Steiner G., Dankert C., Dettleff G., Hellmann H., Kochan H., Baghul M., Kohl H., Kölzer G., Thiel K., and Öhler A. (1991) Ice sublimation below artificial crusts: Results from comet simulation experiments. *Planet. Space Sci., 39*, 515–524.

Kömle N. I., Kargl G., Thiel K., and Seiferlin K. (1996) Thermal properties of cometary ices and sublimation residua including organics. *Planet. Space Sci., 44*, 675–689.

Kömle N. I., Dettleff G., and Danekrt C. (1990) Thermal behaviour of pure and dusty ices on comets and icy satellites. *Astron. Astrophys., 227*, 246–254.

Köröszmezey A. and Gombosi T. I. (1990) A time-dependent dusty gas dynamic model of axisymmetric cometary jets. *Icarus, 84*, 118–153.

Kossacki K. J., Kömle N. I., Kargl G., and Steiner G. (1994) The influence of grain sintering on the thermoconductivity of porous ice. *Planet. Space Sci., 42*, 383–389.

Kossacki K. J., Kömle N. I., Leliwa-Kopystynski J., and Kargl G. (1997) Laboratory investigation of the evolution of cometary analogs: Results and interpretation. *Icarus, 128*, 127–144.

Krasnobaev K. V. (1983) Axisymmetric gas flow from a comet nucleus. *Sov. Ast. Lett., 9(5)*, 332–334.

Landau L. D. and Lifshitz E. M. (1976) *Mechanics,* 3rd edition. Pergamon, New York.

Lederer S. M. (2000) The chemical and physical properties of OH, CN and C$_2$ jets in Comet Hale-Bopp from August 1996 to April 1997. Ph.D. thesis, Univ. of Florida, Gainesville.

Lederer S. M. and Campins H. (2002) Evidence for chemical heterogeneity in the nucleus of comet C/1995 O1 (Hale-Bopp). *Earth Moon Planets, 90*, 381–389.

Lederer S. M., Campins H., Osip D. J., and Schleicher D. G. (1997) Gas jets in Comet Hale-Bopp (1995 O1) *Earth Moon Planets, 87*, 131–136.

Lorenz E., Knollenberg J., Kroker H., and Kührt E. (1995) IR observation of KOSI samples. *Planet. Space Sci., 43*, 341–351.

Matson D. L. and Brown R. H. (1989) Solid state greenhouses and their implications for icy satellites. *Icarus, 77*, 67–81.

Mekler Y., Prialnik D., and Podolak M. (1990) Evaporation from a porous cometary nucleus. *Astrophys. J., 356*, 682–686.

Millis R. L. and Schleicher D. G. (1986) Rotational period of comet Halley. *Nature, 324*, 646–649.

Moreno F., Muñoz O., López-Moreno J. J., and Ortiz J. L. (2002) A Monte-Carlo code to compute energy fluxes in cometary nuclei including polarization. *Icarus, 156*, 474–484.

Mueller M. (1999) A model of the inner coma of comets, with application to comets P/Wirtanen and P/Wild 2. Ph.D. thesis, Rupertus Carola University, Heidelberg.

Muinonen K. (1998) Introducing the Gaussian shape hypothesis for asteroids and comets. *Astron. Astrophys., 332,* 1087–1098.

Muinonen K. and Lagerros J. S. V. (1998) Inversion of shape statistics for small solar system bodies. *Astron. Astrophys., 333,* 753–761.

Ni-Imi T., Fujimoto T., and Ijima K. (1992) Structure of two opposed supersonic freejets with different source pressure. In *Rarefied Gas Dynamics XVII* (A. E. Beylich, ed.), pp. 363–374. VCH Verlag, Weinheim.

Parker E. N. (1965) Dynamical theory of the solar wind. *Space Sci. Rev., 4,* 666–708.

Rodionov A. V., Jorda L., Jones G. H., Crifo J. F., Colas F., and Lecacheux J. (1998) Comet Hyakutake gas arcs: First observational evidence of standing shock waves in a cometary coma. *Icarus, 136,* 232–267

Rodionov A. V., Crifo J. F., Szego K., Lagerros J., and Fulle M. (2002) An advanced physical model of cometary activity. *Planet. Space Sci., 50,* 983–1024.

Salo H. (1988) Monte-Carlo modeling of the net effect of coma scattering and thermal reradiation on the energy input to cometary nucleus. *Icarus, 76,* 253–269.

Samarashinha N. H. and Belton M. J. S. (1995) Long term evolution of rotational states and nongravitational effect for P/Halley-like cometary nuclei. *Icarus, 116,* 340–358.

Schlosser W., Schulz R., and Koczey P. (1986) The Cyan shells of Comet P/Halley. In *20th ESLAB Symposium on the Exploration of Halley's Comet, Vol. 3* (B. Battrick et al., eds.), pp. 495–498. Noordwijk, The Netherlands.

Sekanina Z. (1981) Distribution and activity of discrete emission areas on the nucleus of periodic comet Swift-Tuttle. *Astron. J., 86,* 1741–1773.

Sekanina Z. (1991) Cometary activity, discrete outgassing areas, and dust jet formation. In *Comets in the Post-Halley Era, Vol. 2* (R. L. Newburn and J. Rahe, eds.), pp. 769–823. Kluwer, Dordrecht.

Schulz R. and Schlosser W. (1989) CN-shell structures and dynamics of the nucleus of Comet P/Halley. *Astron. Astrophys., 214,* 375–385.

Skorov Yu.V., Kömle N. I., Markiewicz W. J., and Keller H. U. (1999) Mass and energy balance in the near-surface layers of a cometary nucleus. *Icarus, 140,* 173–188.

Skorov Yu.V., Kömle N. I., Keller H. U., Kargl G., and Markiewicz W. J. (2001) A model of heat and mass transfer in a porous cometary nucleus based on a kinetic treatment of mass flow. *Icarus, 153,* 180–196.

Skorov Yu.V., Keller H. U., Jorda L., and Davidsson B. J. R. (2002) Thermophysical modelling of comet P/Borrelly. Effects of volume energy absorption and volume sublimation. *Earth Moon Planets, 90,* 293–303.

Smoluchowski R. (1982) Heat transport in porous cometary nuclei. *Proc. Lunar Planet Sci. Conf. 13th,* in *J. Geophys. Res., 87,* A422–A424.

Soderblom L. A., Boice D. C., Britt D. T., Brown R. H., Buratti B. J., Kirk R. L., Lee M., Nelson R. M., Oberst J., Sandel B. R., Stern S. A., Thomas N., and Yelle R. V. (2004) Imaging Borrelly. *Icarus, 167,* 4–15.

Squyres S. W., McKay C. P., and Reynolds R. T. (1985) Temperatures within comet nuclei. *J. Geophys. Res., 90,* 12381–12392.

Steiner G. and Kömle N. I. (1991a) A model of the thermal conductivity of porous water ice at low gas pressure. *Planet. Space Sci., 39,* 507–513.

Steiner G. and Kömle N. I. (1991b) Thermal budget of multicomponent porous ices. *J. Geophys. Res., 96(E3),* 18897–18902.

Steiner G. and Kömle N. I. (1993) Evolution of a porous H_2O-CO_2-ice sample in response to irradiation. *J. Geophys. Res., 98(E5),* 9065–9073.

Steiner G., Kömle N. I., and Kührt E. (1991) Thermal modelling of comet simulation experiments. In *Theoretical Modelling Of Comet Simulation Experiments* (N. I. Kömle et al., eds.), pp. 11–29. Austrian Academy of Sciences, Vienna.

Strazzulla G. (1997) Ion bombardments of comets, In *From Stardust to Planetesimals* (Y. J. Pendleton and A. G. G. M. Tielens, eds.), pp. 423–433. ASP Conference Series 122, Astronomical Society of the Pacific, San Francisco.

Szego K., Sagdeev R. Z., Whipple F. L., Abergel A., Bertaux J. L., Merényi E., Szalai S., and Várhalmi L. (1995) *Images of the Nucleus of Comet P/Halley, Vol. 2,* pp. 68–79. ESA SP-1127, Noordwijk, The Netherlands.

Szego K., Crifo J. F., Foldy L., Lagerros J. S. V., and Rodionov A. V. (2001) Dynamical effects of comet P/Halley gas production. *Astron. Astrophys., 370,* L35–L38.

Szego K., Crifo J. F., Fulle M., and Rodionov A. V. (2002) The near-nuclear coma of comet P/Halley in March 1986. *Earth Moon Planets, 90,* 435–443.

Wallis M. K. (1982) Dusty gas-dynamics in real comets. In *Comets* (L. L. Wilkening, ed.), pp. 357–369. Univ. of Arizona, Tucson.

Weissman P. (1999) Diversity of comets: Formation zones and dynamical paths. In *Composition and Origin of Cometary Materials* (K. Altwegg et al., eds.), pp. 301–311. Kluwer, Dordrecht.

Whipple F. S. (1950) A comet model. I. The acceleration of comet Encke. *Astrophys. J., 111,* 375–394.

Whipple F. S. (1951) A comet model. II. Physical relations for comets and meteors. *Astrophys. J., 118,* 464.

Whipple F. S. (1982) Cometary nucleus and active regions. In *Cometary Exploration, Vol. 1* (T. I. Gombosi, ed.), pp. 95–110. Hungarian Academy of Sciences, Budapest.

Physical Processes and Chemical Reactions in Cometary Comae

S. D. Rodgers and S. B. Charnley
NASA Ames Research Center

W. F. Huebner and D. C. Boice
Southwest Research Institute

A variety of physical and chemical processes are important in the comae of active comets near the Sun. We review the principal physical processes occurring in the outflowing gas and dust and their effects on thermodynamics of the coma. We describe the coupling between the physics and the chemistry and emphasize that any accurate model of the coma must include both. Chemically, there are a number of mechanisms capable of altering the initial chemical composition of the gas escaping from the nucleus surface. We assess the importance of these chemical processes, and discuss several recent models that follow the chemical evolution of the coma gas. The ultimate aim of most coma studies is to understand the nature of the icy nucleus, and we briefly review the major obstacles in using coma observations to infer the nucleus properties.

1. INTRODUCTION

Our knowledge of the composition and structure of comets has come primarily from studies of their comae. Astronomical observations of the coma can be made directly because small dust grains in the coma scatter and reflect light much more efficiently than the dark nucleus, and because the outgassing molecules emit distinct spectral lines at specific frequencies. Although several space missions have been sent to investigate comets, they have only probed the nucleus indirectly via photographic imaging and low-resolution spectroscopy. Not until the *Rosetta* mission sends a lander to the nucleus of a comet will we be able to directly access the material that resides in the nucleus. In contrast, there exists over a century of data pertaining to observations of cometary comae. It is also interesting to study the coma for its own sake — it provides an environment impossible to duplicate on Earth, and can be used as a laboratory to test theories of gas and plasma dynamics and photochemistry. Finally, the interaction of the coma with the solar wind can provide information on the properties of the solar wind and the interplanetary medium; in fact, the existence of the solar wind was first discovered via its interaction with cometary comae (*Biermann*, 1951).

1.1. Structure of the Coma

The coma is the cloud of dust and gas that surrounds the cometary nucleus. A schematic view of the coma structure is shown in Fig. 1. The size of the coma depends on how one defines it; for the purposes of this chapter, where we are discussing physical and chemical processes, we are concerned chiefly with the collisional coma, i.e., the inner region where particle collisions affect thermodynamics and chemistry of the gas. A rough estimate of the size of the collisional coma can be obtained by finding the cometo-centric distance, r, at which the particle mean free path, Λ, equals r. For a Halley-type comet at 1 AU from the Sun, this distance is typically several thousand kilometers for neutral-neutral collisions, and up to an order of magnitude larger for ion-molecule collisions, due to the enhanced cross-sections for ion-neutral interactions. The size of the collisional region is proportional to the total gas production rate, and so will increase as the comet approaches the Sun.

One can also loosely define a "molecular coma" within which most molecules survive against photodissociation. Strictly speaking, the size of this region will vary for different molecules, since all species have different photodissociation lifetimes. For water (and many other species as well), the lifetime in the solar radiation field at 1 AU is $\approx 10^5$ s, and so for a typical outflow velocity, v, of 1 km s^{-1} the molecular coma will be ~10^5 km in size. For species such as ammonia and formaldehyde that are destroyed more rapidly, the coma will be an order of magnitude smaller, whereas more-stable species such as CO and CO_2 will survive out to 10^6 km or further. Photodissociation rates are proportional to the strength of the radiation field, and so molecular comae will shrink as the comet nears the Sun.

It is important to stress that the coma gas is not gravitationally bound to the nucleus, and so the coma is a transient phenomenon. Because sublimation from the nucleus surface is constantly replenishing the outflowing gas, the coma can often appear stable and unchanging. However, sudden changes in coma brightness and structure are common, with

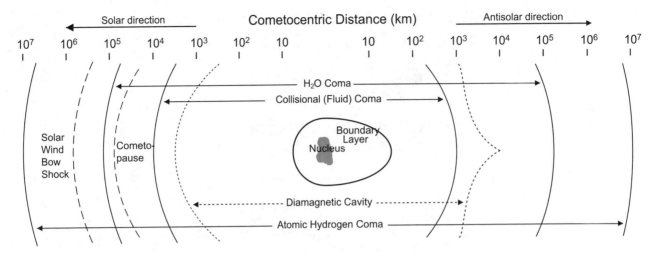

Fig. 1. Schematic illustration of the coma structure and its major physical regions for a moderate production rate at about 1 AU. Note the logarithmic distance scale.

features changing rapidly over timescales of several hours. Spatial structures are also seen in many comae, often in both the gas and dust components (although the structure in both may not be the same). Further discussions of these phenomena appear in *Schleicher and Farnham* (2004) and *Combi et al.* (2004).

The interaction of the solar wind with the coma results in complex structures in the plasma and magnetic fields (see *Neubauer*, 1991; *Cravens*, 1991; *Ip*, 2004; *Lisse et al.*, 2004). A bow shock develops in the solar wind on the sunward side of the comet; within this region the solar wind "picks up" slow-moving coma species via charge exchange reactions. Eventually, as the wind encounters the denser gas in the inner coma it is decelerated, and at the cometopause it is diverted laterally around the inner coma. The interplanetary magnetic field also wraps around the comet, which results in a magnetic-field-free cavity surrounding the nucleus. The existence of this cavity in Comet Halley was demonstrated by *Neubauer et al.* (1986).

1.2. Physical and Chemical Processes in the Coma

Most of the processes that occur in the coma are initiated by the solar radiation field. Photons at ultraviolet (UV) wavelengths photodissociate and ionize the original parent molecules, producing "second-generation" reactive radicals, ions, and electrons. These ions and radicals can subsequently react with other species to form "third-generation" species. Examples include many of the protonated ions detected by the ion mass spectrometer on the *Giotto* probe in the coma of Comet Halley (*Geiss et al.*, 1991). These processes are highly exoergic, and so can lead to species in excited states not normally populated at the low temperatures in the coma (T ≈ 10–200 K), as can direct absorption of solar photons (fluorescent pumping). Figure 2 illustrates the most important processes occurring in the coma.

The densities in the inner coma are sufficiently large that it is reasonable to consider the gaseous coma as a fluid, or more accurately as a mixture of several fluids (neutrals, ions, electrons, as well as different populations of dust grains, and suprathermal photodissociation products; although these latter components are not strictly fluids they can typically be described in terms of hydrodynamic variables, i.e., velocity and temperature). Hence, thermodynamic properties of the coma can be calculated from integration of the standard equations of fluid flow, assuming that the initial conditions at the nucleus surface are known (or can be estimated with some degree of accuracy). As the density decreases and Λ increases, the fluid description becomes less applicable, and a transition to free molecular flow occurs. In this region, processes that affect a particular molecule cannot be assumed to affect the gas as a whole, and the properties of the outflowing gas are "frozen in" from the earlier collisional regime [see *Crifo* (1991) for a thorough discussion].

The coma gas also interacts with the entrained dust, and this affects both the dynamics and chemistry of the coma. For example, gas-dust drag in the very inner coma decelerates the outflowing gas to subsonic speeds (*Marconi and Mendis*, 1983). Chemically, dust grains may account for the "extended sources" of some coma molecules, where observations require that these molecules are injected into the coma, rather than released directly from the nucleus (e.g., *Festou*, 1999). This may result either from delayed sublimation of low-volatility material from the hotter grains, or from the actual breakup of the grains themselves. From measurements of the elementary dust composition in Comet Halley it is known that a significant fraction of refractory particles consist of organic (CHON) matter (*Kissel et al.*, 1986). A variety of complex organics have been proposed to account for this material (*Huebner and Boice*, 1997). In particular, *Huebner* (1987) proposed that polyoxymethylene (POM, the –CH$_2$O– polymer) could account for the ex-

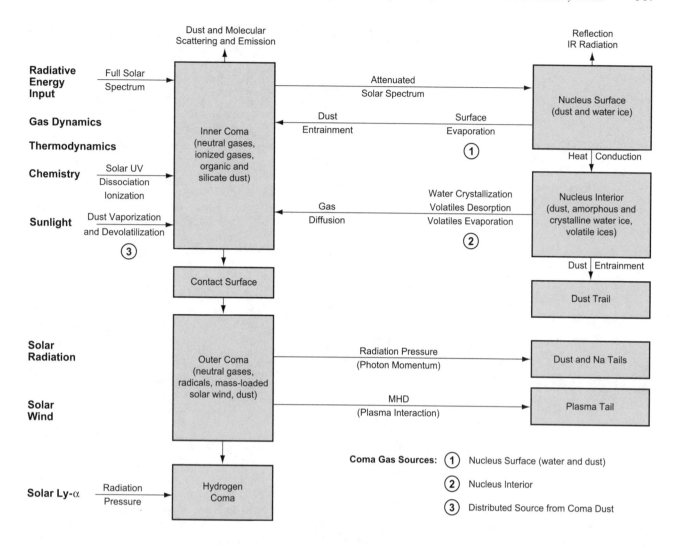

Fig. 2. The major radiative, physical, chemical, and thermodynamic processes comets, their coupling, and their interaction between the various cometary components: nucleus, coma gas, plasma, and dust (*Huebner and Benkhoff*, 1999).

tended source of formaldehyde in Comet Halley. Other macromolecules proposed to be potentially important in the coma include hexamethylenetetramine [HMT, $C_6H_{12}N_4$ (*Bernstein et al.*, 1995)] and polyaminocyanomethylene [PACM, a $-C(CN)(NH_2)-$ polymer (*Rettig et al.*, 1992)].

1.3. Brief History of Coma Modeling

Prior to the middle of the twentieth century, many spectroscopic observations of cometary comae revealed the presence of molecular radicals such as C_2, C_3, CH, CN, NH, NH_2, OH, and a few ions, including CH^+, CO^+, CO_2^+, and OH^+ (e.g., *Swings*, 1943). In spite of its high abundance, H_2O had not been detected but it was postulated to be present together with CH_4, CO_2, NH_3, and C_2N_2, consistent with the list of observed radicals and ions. Subsequently, *Whipple* (1950, 1951) postulated the icy conglomerate model for the comet nucleus based on observations of Comet Encke.

The first simple analytical model of molecular distributions in the coma was published by *Haser* (1957), who considered only photodissociation of parent molecules to form daughter and granddaughter molecules. However, as the importance of chemical reactions began to be appreciated in other astrophysical contexts, so it became clear that the high-density conditions in the inner coma ($n \sim 10^{13}$ cm^{-3} at the nucleus surface) meant that they were also likely to be important in comets. Such considerations, together with the advent of more powerful computers, led to the development of detailed chemical models (e.g., *Oppenheimer*, 1975; *Giguere and Huebner*, 1978; *Huebner and Giguere*, 1980; *Mitchell et al.*, 1981; *Biermann et al.*, 1982).

These early models had many simplifying assumptions, such as constant temperature and velocity profiles. The models confirmed that ion-molecule reactions were important in the inner coma, particularly proton transfer reactions followed by dissociative recombination. These models also

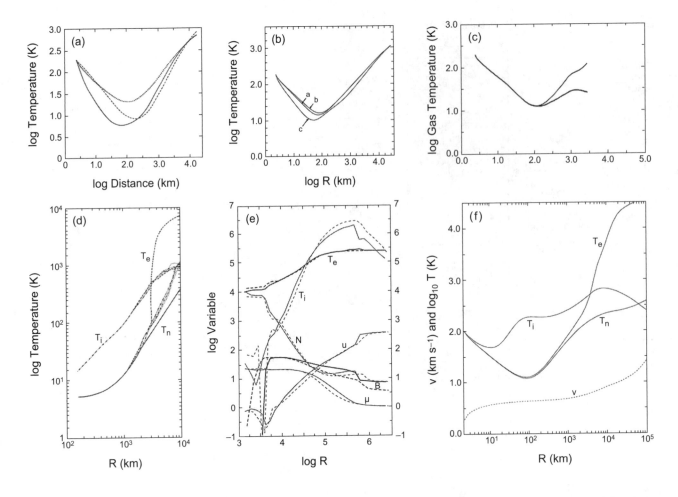

Fig. 3. Comparison of coma temperature profiles predicted by a variety of models. The upper panels show models for which a single fluid is considered, whereas the bottom panels show results for multifluid models that also include the effects of coma chemistry. In (a) and (b) the different lines correspond to different assumptions regarding the heating and cooling rates; (c) shows the difference between the results of hydrodynamic calculations (upper line) compared with Monte Carlo simulations (lower line). In the lower panels T_n, T_i, and T_e refer to the temperatures of the neutral, ion, and electron fluids respectively. The different T_e profiles in (d) refer to different assumptions regarding heat transfer from "hot" photoelectrons. In (e), u, N, μ, and B are the velocity (km s^{-1}), number density (cm^{-3}), mean molecular weight (amu), and magnetic field strength (nT), respectively. References: (a) *Marconi and Mendis* (1982); (b) *Crovisier* (1984); (c) *Combi and Smyth* (1988); (d) *Körösmezey et al.* (1987); (e) *Schmidt et al.* (1988); (f) *Rodgers and Charnley* (2002).

showed that the inner region of the coma is opaque to the solar UV field, and so accurate chemical models must include optical depth calculations. Later models included additional reactions such as radiative association, electron impact ionization and dissociation, and photodissociative ionization (e.g., *Schmidt et al.*, 1988). Models have also been developed to account for the extended sources of CO and H$_2$CO following the breakup of POM in the coma (*Boice et al.*, 1990; *Cottin et al.*, 2001).

Subsequent hydrodynamical models of the coma demonstrated that the assumptions of constant velocity and temperature were highly unrealistic (e.g., *Marconi and Mendis*, 1982, 1983, 1986; *Gombosi et al.*, 1985, 1986; *Combi and Smyth*, 1988). In particular, adiabatic cooling of the gas as it initially expands away from the nucleus can lead to the

temperature in the inner coma dropping to around 10 K. Further out, as photodissociation and photoionization reactions become important, the temperatures rise, particularly that of the electrons. It was thus realized that the physics and chemistry are intimately coupled. For example, proton transfer reactions are the principal heat source for ions in the inner coma, and the rate of electron recombination reactions is greatly reduced in the outer coma when the electrons become extremely hot. This led to the development of combined hydrodynamic-chemical models (*Huebner*, 1985; *Körösmezey et al.*, 1987; *Wegmann et al.*, 1987; *Rodgers and Charnley*, 2002). The agreement between the various models is rather good; Fig. 3 presents the temperature profiles calculated by a number of different models. In the remainder of this chapter we shall concentrate on coma chemistry in the hydrodynamic picture, for the MHD and Monte Carlo treatments the reader is directed to *Ip* (2004) and *Combi et al.* (2004).

2. OBSERVING THE COMA

2.1. Spectroscopy

The great majority of cometary data is gathered via spectroscopy of the coma, and comes from all accessible regions of the electromagnetic spectrum. Molecular observations of coma composition are discussed in detail by *Crovisier* (2004), *Feldman et al.* (2004), and *Bockelée-Morvan et al.* (2004). For the purposes of this chapter we note that observations of spectral lines give a single globally averaged column density for the number of molecules in a particular energy level; interpreting this number requires a model of the coma. Specifically, one needs to know both n(r) and $f_\varepsilon(r)$, where n is the number density and f_ε is the fraction of molecules in energy level ε at cometocentric distance r. This distribution must then be convolved with the telescope beam profile for the appropriate Earth-comet distance. For parent species, it is usually sufficient to assume a Haser distribution for the density, i.e., $n \propto r^{-2} \exp\{-r/r_{phot}\}$, where r_{phot} is the photodissociation scale-length at the appropriate heliocentric distance. For species formed in the coma the problem is more complicated; photodaughters in the outer coma can be modeled using the vectorial model of *Festou* (1981), or from the results of Monte Carlo calculations (see *Combi et al.*, 2004). Molecules formed via coma chemistry require detailed chemical models to compute their radial density profiles.

The factor f_ε is much more difficult to calculate, since it depends on a number of excitation and deexcitation mechanisms that operate simultaneously. In the inner coma collisions will dominate, and f_ε will approach the local thermodynamic equilibrium (LTE) value. Further out, collisions remain important, but may be too slow to depopulate photodaughters produced in highly excited states. As the electron temperature decouples from the neutrals, the collisional excitation will be higher for those species with large dipole moments that collide more rapidly with electrons (*Xie and Mumma*, 1992; *Biver et al.*, 1999). Eventually, as all collisions become unimportant, the excitation will be determined by the balance between radiative pumping and fluorescent decay (*Bockelée-Morvan and Crovisier*, 1987). For lines that are optically thick, an extra layer of complication is added, in that a full radiative transfer model of the coma must be employed. If several lines of the same molecule from different energy levels can be observed, an estimate of the globally averaged coma temperature can be obtained (*Biver et al.*, 1999). At radio wavelengths, the spectroscopic resolution is sufficiently narrow that expansion velocities can be obtained from line profiles (*Bockelée-Morvan et al.*, 1990). Again, this value is a global average over the whole of the coma.

2.2. *In Situ* Measurement

Several space missions were sent to rendezvous with Comet Halley, and mass spectrometry of the coma was performed by the *Giotto* and *Vega 1* and *2* probes (e.g., *Krankowsky et al.*, 1986). Most recently, *in situ* analysis of a second comet occurred during the *Deep Space 1* flyby of Comet Borrelly (e.g., *Nordholt et al.*, 2003). Such measurements differ from spectroscopic data in that they give information that is local and instantaneous. This has the advantage of allowing the structure of the coma to be probed on small scales. For example, the extended sources of CO and H_2CO in Comet Halley were discovered by mass spectrometry (*Eberhardt et al.*, 1987; *Meier et al.*, 1993). However, because the data only yield a single "snapshot" of the state of the coma, one potential pitfall in interpreting these measurements is that it is difficult to decide whether any anomalies represent previously undiscovered permanent coma features, or if they are due to transient phenomena.

In situ measurements give radial profiles of molecular and ionic abundances and gas properties such as density and temperature. Thus, these measurements represent the most stringent tests of chemical and dynamical models. For example, the importance of proton transfer reactions in the inner coma was proved by the results of the *Giotto* ion mass spectrometer (*Geiss et al.*, 1991). The global properties of the coma as a whole are best obtained from spectroscopic measurements. In order to obtain the most understanding of the coma, a combination of *in situ* and global measurements are required, together with detailed modeling.

3. COMA-NUCLEUS BOUNDARY

A key region is in the vicinity of the nuclear surface, where the initial chemical and dynamical conditions of the outflowing coma are determined. In practice, given the extreme difficulties in accurately calculating the physics and chemistry in this region, most models of the coma simply assume some set of initial conditions as parameters to be input into the model. By comparing the model output with observations it is hoped that, in addition to understanding the processes occurring in the coma, the initial conditions can also be constrained with some degree of accuracy. However, because of the complexity of the boundary region, it is naive to assume that the conditions in the very inner coma can be easily related to the properties of the nucleus.

For example, although the initial chemical composition of the gas is similar to the composition of the nucleus ice, for some volatile molecules an additional contribution from subsurface sublimation fronts may also be important. A thorough discussion of the chemical differentiation and stratification in the nucleus appears in *Prialnik et al.* (2004). The total surface sublimation rate is controlled by the equation of energy balance at the nucleus surface, together with a Clausius-Clapeyron-type equation of state of the surface ices. For typical cometary parameters, the gas (primarily water) production rate is on the order of 10^{17} mols cm^{-2} s^{-1} at 1 AU (*Whipple and Huebner*, 1976).

The dynamics of the circumnuclear boundary region are also extremely complicated, since the nucleus is likely to be heterogeneous on small scales. Sophisticated time-dependent three-dimensional models are required in order to calculate the gas and dust flow in this region, and are discussed further in *Crifo et al.* (2004). These models predict many interesting features, such as lateral flows and standing shocks. The possible chemical effects of these features have not yet been investigated, so it is not known whether or not they may play a role in generating new species. In the remainder of this chapter we will discuss the chemistry that occurs beyond the boundary region, where the gas flow is relatively smooth and almost spherically symmetric. We mention the possible chemical processing in the boundary layer simply to stress, again, that inferring the properties of the nucleus from coma observations is not a simple task.

The thickness of the boundary layer depends on the gas production rate. At large heliocentric distances the rate of gas production is small and the thickness of the boundary layer can be very large. In the extreme case, when the mean free path for collisions between gas molecules approaches infinity, a Maxwellian velocity distribution will never be established and the gas remains in free molecular flow. For a comet at 1 AU with a gas production rate of $\approx 10^{17}$ cm^{-2} s^{-1} and $T \approx 100$ K, Λ is on the order of 20 cm. Thus, the boundary layer is on the order of tens of meters on the subsolar side of the nucleus, and will be much larger on the nightside.

4. DYNAMICAL-CHEMICAL MODELS OF THE COMA

4.1. Model Classification

We begin with a brief review of the different types of models that have been used to model the coma. This is in no way meant to be an exhaustive list or rigid taxonomy of all models, but rather to give an idea of the strengths and weaknesses of different approaches, and what level of complexity is feasible and/or necessary.

The simplest hydrodynamical models of the coma are steady, spherically symmetric, single-fluid models that calculate radial profiles of v and T. Such models were first used to investigate the principal heating and cooling mechanisms operating in the coma, and showed that T can vary significantly throughout the coma (e.g., *Marconi and Mendis*, 1982; *Crovisier*, 1984; *Huebner*, 1985) (see Fig. 3). Such models also serve as the foundation for more complex calculations, such as models that are multidimensional (*Kitamura*, 1986), time-dependent (*Gombosi et al.*, 1985), or include several fluids (*Marconi and Mendis*, 1986). In the case of multidimensional or time-dependent models, the computational demands can be very large, because these calculations involve simultaneously solving large sets of coupled partial differential equations (PDEs), as opposed to a relatively small number of ordinary differential equa-

tions (ODEs) for one-dimensional, steady models. Models that considered separate neutral, ion, and electron fluids found that chemical reactions in the inner coma were an important heat source, especially for the ions (*Körösmezey et al.*, 1987).

Chemically, the simplest models ignore the dynamics, and assume constant outflow velocities and temperatures (*Giguere and Huebner*, 1978). Other models take the gas physics into account in some basic fashion, for example, by assuming a sudden jump in the electron temperature at some arbitrarily chosen radius (*Lovell et al.*, 1999). However, because chemical reaction rates depend on the gas temperature and density, accurately calculating the chemistry requires an accurate description of the coma dynamics. In addition, as discussed above, an accurate dynamical model should include the effect of chemical reactions. Therefore, the coma chemistry and physics should be solved simultaneously, and a multifluid, dynamical-chemical model is the minimum necessary to study coma chemistry. Such models were first developed by *Körösmezey et al.* (1987) (albeit with a limited chemistry) and *Schmidt et al.* (1988). In the following section we describe the basic components of such a model.

A separate category of coma models calculate the gas properties via Monte Carlo techniques (*Combi and Smyth*, 1988; *Hodges*, 1990). These models represent the most accurate descriptions of the coma, since they neglect the assumption, common to all hydrodynamic models, that the gas behaves as a fluid. However, these models are also extremely computationally demanding. Including a detailed chemistry in these models would be prohibitive. Given the fairly good agreement between Monte Carlo models and fluid-dynamic treatments in the inner collisional coma, it is not necessary to use such models when modeling the coma chemistry.

4.2. Simple Multifluid Hydrodynamic Models

Simple, spherically symmetric models, appropriate for a steady flow, have previously been employed to investigate the chemistry of the collisional inner coma (e.g., *Huebner*, 1985; *Wegmann et al.*, 1987; *Schmidt et al.*, 1988; *Konno et al.*, 1993; *Canaves et al.*, 2002), and we now describe the key ingredients of such a model; a more detailed description is given in *Rodgers and Charnley* (2002). In reality, as we have discussed, cometary comae are not symmetric, and rapid temporal variations are observed. Despite this, such models remain useful since (1) pressure differences in the very inner coma tend to even out rapidly, leading to fairly symmetric spherical outflows in the gas coma (*Crifo*, 1991); and (2) the time taken for a particular parcel of gas to reach a distance of 4×10^3 km — roughly corresponding to the edge of the collisional regime where bimolecular physical and chemical processes cease to significantly affect the gas — is around 1 h. The most important deviations from this model occur for the ions and electrons at large cometocentric distances, where the interaction with the solar wind becomes important. In this case, multidimensional models are necessary (*Schmidt et al.*, 1988). Nonetheless, because magnetic fields are excluded from the inner coma where most of the chemistry occurs, the plasma flow in these regions will trace the near-spherical neutral gas outflow. For a chemically reacting multicomponent flow we have to consider the interaction of the predominately neutral gas and the charged plasma of ions and electrons, and the relatively massive dust grains. We also need to treat the suprathermal photoproduced hydrogen atoms as an individual component (see section 4.3.3). Here we specifically describe the gas-plasma coupling. Assuming a steady, spherically symmetric flow, and neglecting gravitational and magnetic fields, the hydrodynamic conservation equations for particle number, mass, momentum, and energy for the neutral gas are

$$\frac{1}{r^2}\frac{d}{dr}(r^2 n v) = N \tag{1}$$

$$\frac{1}{r^2}\frac{d}{dr}(r^2 \rho v) = M \tag{2}$$

$$\frac{1}{r^2}\frac{d}{dr}(r^2 \rho v^2) + \frac{dP}{dr} = F \tag{3}$$

$$\frac{1}{r^2}\frac{d}{dr}\left[r^2 v\left(\frac{\rho v^2}{2} + \frac{\gamma}{\gamma-1}P\right)\right] = G + Fv - \frac{1}{2}Mv^2 \tag{4}$$

where n is the number density, ρ is the mass density, v is the velocity, P is the pressure, and γ is the ratio of specific heats. The source terms N, M, F, and G represent, respectively, the net generation rate per unit volume of particles, mass, momentum, and thermal energy (heat). It is assumed that the fluid behaves as an ideal gas. Equation (1) can be rearranged to give

$$\frac{dn_s}{dr} = \frac{N_s}{v} - \frac{n_s}{v}\frac{dv}{dr} - \frac{2n_s}{r} \tag{5}$$

where the subscript s represents a particular chemical species. Similarly, the mass conservation equation becomes

$$\frac{d\rho}{dr} = \frac{M}{v} - \frac{\rho}{v}\frac{dv}{dr} - \frac{2\rho}{r} \tag{6}$$

After some algebra, one can derive equations for the radial derivatives for the velocity and temperature of the neutral gas

$$\frac{dv}{dr} = \frac{1}{\rho v^2 - \gamma P}\left(Fv - (\gamma-1)G - Mv^2 + \frac{2\gamma Pv}{r}\right) \tag{7}$$

$$\frac{dT}{dr} = \frac{2T}{r} + \frac{1}{nkv}\left(Fv - Mv^2 - NkT + \frac{dv}{dr}(nkT - \rho v^2)\right) \tag{8}$$

Each of the three fluids in the coma (neutrals, ion, electrons) has a distinct temperature, and similar equations can

be derived for dT_i/dr and dT_e/dr, as well as for the mass density gradient of the plasma. For the plasma, a slight complication arises, since charge conservation and the strength of the Coulomb force ensures that $n_e = n_i$ and $v_e = v_i$ throughout the coma. Therefore, one must solve for the plasma velocity dv_e/dr by considering the total contributions from both ions and electrons. Once the hydrodynamic source terms have been defined, and the fluid properties at the comet surface prescribed, it is possible to numerically integrate the resulting system of differential equations to obtain the coma physical and chemical structure.

In practice, for steady flows such as these there can exist mathematical singularities where a fluid encounters a transonic point, since at these points the denominator in equation (7) becomes zero. When a dust component is included, such a 0/0 singularity is encountered close to the nucleus, as the expanding neutral gas then has to undergo a subsonic-supersonic transition. Similarly, further out in the coma, where the electron temperature becomes very high, the plasma sound speed increases to the point that the plasma undergoes a supersonic-subsonic transition. For time-dependent flows, where a system of PDEs is solved, singularities do not occur (e.g., *Körösmezey et al.*, 1987; *Körösmezey and Gombosi*, 1990). In the case of steady flows, special numerical techniques, or simplifying approximations, are needed to treat such 0/0 singularities (e.g., *Marconi and Mendis*, 1983, 1986; *Gail and Sedlmayr*, 1985). It should be emphasized that, although "simple" by the standards of recent coma dynamics models (e.g., *Combi et al.*, 2004; *Crifo et al.*, 2004), even steady flow models require a higher level of sophistication relative to other areas of astrochemical modeling (as in interstellar clouds or circumstellar envelopes, for example). Model temperature distributions obtained for a comet similar to Hyakutake at 1 AU are shown in Fig. 3f.

4.3. Heating and Cooling Mechanisms

4.3.1. Elastic scattering. Elastic collisions transfer momentum and energy between the three fluids. For ion-neutral collisions, if the scattering is assumed to be isotropic in the center-of-mass frame, the mean amount of thermal energy imparted to the neutrals per collisions can be calculated via

$$\hat{G}_n(i-n,\text{elastic}) =$$
$$2m_n m_i(m_n+m_i)^2\left[\frac{3}{2}K(T_i - T_n) + m_i\frac{(v_n - v_e)^2}{2}\right] \tag{9}$$

where m_n and m_i are the masses of the particles involved. A similar expression holds for the ion fluid heat source term per collision. The total thermal energy source term due to elastic scattering is obtained by summing equation (9) over all collisions. This can be done either by assuming a generic value for the mean mass of each fluid, or by actively summing the contributions of the most frequent collisions, which will involve the ions H_3O^+, $CH_3OH_2^+$, and NH_4^+ collid-

TABLE 1.   Thermal energy source terms per reaction, $\hat{G}$, for neutral, ion, and electron fluids for different classes of reactions.

Reaction Type	Example	$\hat{G}_n$	$\hat{G}_i$	$\hat{G}_e$
Radiative association of neutrals	$CO + S \rightarrow OCS$	$-\Theta_n$	0	0
Radiative association of ions and neutrals	$H_2O + HCO^+ \rightarrow HCOOH_2^+$	$-\Theta_n$	$\frac{m_1}{m_3}(m_1 v_{en}^2 + \Theta_n - \Theta_i)$	0
Neutral–neutral	$CH_4 + CN \rightarrow HCN + CH_3$	ΔE	0	0
Ion–neutral	$NH_3 + H_3O^+ \rightarrow H_2O + NH_4^+$	$M(m_2 v_{en}^2 + \Theta_i - \Theta_n) + \frac{m_4}{m_T}\Delta E$	$M(m_1 v_{en}^2 - \Theta_i + \Theta_n) + \frac{m_3}{m_T}\Delta E$	0
Radiative recombination	$C^+ + e \rightarrow C + h\nu$	$m_1 v_{en}^2 + \Theta_i$	$-\Theta_i$	$-\Theta_e$
Dissociative recombination	$H_3CO^+ + e \rightarrow HCO + 2H$	$m_1 v_{en}^2 + \Theta_i + \Theta_e + \Delta E$	$-\Theta_i$	$-\Theta_e$
Photoionization	$OH + h\nu \rightarrow OH^+ + e$	$-\Theta_n$	$m_1 v_{en}^2 + \Theta_n$	ΔE
Photodissociation of neutrals	$CO_2 + h\nu \rightarrow CO + O$	ΔE	0	0
Photodissociation of ions	$H_2O^+ + h\nu \rightarrow OH^+ + H$	$\frac{m_4}{m_1}\Delta E + m_3 v_{en}^2 + \frac{m_3}{m_1}\Theta_i$ $-\frac{m_3}{m_1}\Theta_n$	$\frac{m_3}{m_1}(\Delta E - \Theta_i)$	0
Photodissociative ionization (PDI)	$H_2O + h\nu \rightarrow OH^+ + H + e$	$-\frac{m_3}{m_1}\Theta_n$	$m_3 v_{en}^2 + \frac{m_3}{m_1}\Theta_n$	ΔE
Electron impact ionization (EII)	$CH_4 + e \rightarrow CH_4^+ + 2e$	$-\Theta_n$	$m_1 v_{en}^2 + \Theta_n$	$-\Delta E$
Electron impact dissociation (EID)	$CH_4 + e \rightarrow CH_3 + H + e$	0	0	$-\Delta E$

m is the mass of a species, and subscripts 1–4 refer to those species in the example reactions. m_T is the total mass of the reactants, and M is a reduced mass ratio equal to $(m_1 m_4 + m_2 m_3)/m_T^2$. $\Theta_n \equiv 3kT_n/2$, $\Theta_i \equiv 3kT_i/2$, $\Theta_e \equiv kT_e$, $v_{en}^2 \equiv (v_n - v_e)^2/2$. ΔE represents the mean exo-/endothermicity of each reaction.

ing with H_2O, CO, and CO_2. In either case, the rate coefficient for the collisions must be known; a value equal to the Langevin value, $\sim 10^{-9}$ cm^3 s^{-1}, is typically assumed. Note that the source term in equation (9) can be considered as a simplified form of the general expression for reactive ion-molecule collisions (see line 4 in Table 1).

For electron-neutral elastic scattering, the most important collision partner is water. Collision rates for e-H_2O mixtures were measured by *Pack et al.* (1962; see also *Körösmezey et al.*, 1987), and the mean heat transfer per collision can be obtained from equation (9) with the additional assumption that the electron mass can be neglected. Also, because the cross section declines with electron temperature, one finds that the mean thermal energy of a colliding electron is kT_e, not $3kT_e/2$ (*Draine*, 1986). The rate of energy transfer between ions and electrons was calculated by *Draine* (1980).

4.3.2. Inelastic scattering. Inelastic collisions can lead to molecules in highly excited states. These will then decay with the emission of a photon, and if this photon is able to escape from the coma without being reabsorbed, the energy is lost. Inelastic collisions involving H_2O are the most important; water molecules can be excited by collisions with other H_2O molecules or with electrons. A semiempirical formula for the energy loss due to the former was derived by *Shimizu* (1976). However, radiation trapping in the inner coma means that the effective cooling rate is much less (*Crovisier*, 1984). This can be roughly accounted for by introducing an optical depth factor (*Schmidt et al.*, 1988). The energy removed by inelastic e-H_2O collisions was calculated by *Cravens and Körösmezey* (1986). Again, accurate calculations must include the effects of radiation trapping in the inner coma.

4.3.3. Thermalization of energetic photoproducts. The majority of photolytic reactions result in the production of either atomic or molecular hydrogen, and due to their low masses these products also have the largest share of the excess energy of the reaction. Hence, the dominant heating mechanism in the coma is thermalization of fast H and H_2 particles. It was realized by *Ip* (1983) that in the outer coma many of these particles will escape from the coma before thermalization, thus removing an important energy source for the neutral fluid. A variety of calculations of this effect have been performed (e.g., *Huebner and Keady*, 1984; *Combi and Smyth*, 1988). Although the results are qualitatively similar, the radius at which escape becomes important can vary by almost an order of magnitude in different models (*Crifo*, 1991).

The most important reaction forming fast H atoms (hereafter denoted H_f) is photodissociations of water

$$H_2O + h\nu \rightarrow OH + H_f \tag{10}$$

The mean excess energy of reaction (10) is 3.4 eV, but as pointed out by *Crovisier* (1989), a large fraction of this energy may go into ro-vibrational excitation of the OH radical. Laboratory experiments on water photodissociation have shown that OH radicals with rotational energy levels of J >

50 can be formed (e.g., *Harich et al.*, 2001). Therefore, the kinetic energy of the H_f atom is likely to be in the range 1–2 eV, implying a velocity of $\approx$15–20 km s^{-1}. This has been confirmed by observations of numerous comets (*Festou et al.*, 1983; *McCoy et al.*, 1992). In addition to being the principal heating mechanism in the inner coma, these suprathermal atoms may also drive high-energy chemical reactions in this region. This will be discussed in section 5.4.

4.3.4. Chemical reactions. The hydrodynamic source terms due to chemical reactions depends on the reaction type and the masses and temperatures of the species involved. In order to accurately calculate the effects of chemistry on thermodynamics it is necessary to compute the source term due to each individual reaction and then multiply this by the total reaction rate (i.e., reactions cm^{-3} s^{-1}), and then sum the total for all reactions. Mass source terms are trivial to calculate; e.g., the ionization of water increases the ion fluid mass by 18 amu, and decreases the neutral fluid mass by the same amount. Source terms for momentum and energy transfer are not so simple, however. *Draine* (1986) derived the expressions appropriate for several of the most common reaction types occurring in a multifluid flow and this methodology can be extended to include all reaction types that occur in the coma (*Rodgers and Charnley*, 2002). The resulting heat source terms per reaction, $\hat{G}$, are listed in Table 1.

A comparison of the effectiveness of each of the heating and cooling mechanisms discussed in this section is shown in Fig. 4.

4.4. Gas-Dust Coupling

The drag force exerted by the gas on a dust particle depends on the relative drift velocity between gas and dust, $v_{drift} = v_g - v_d$, and can be written

$$F_{drag} = n_d \sigma_d \rho_g \frac{C_D}{2} v_{drift}^2 \tag{11}$$

where n_d and σ_d are the number density and cross section of the dust grains. C_D is the drag coefficient, which accounts for the sticking of the molecules on the dust particles, the viscosity of the gas, and the shape of the particle. It also depends on the density of the gas. The drag coefficient is usually expressed in terms of the Reynolds number, Re = $2r_d \rho_g v_{drift}/\mu$, where μ is the viscosity of the gas. For low Reynolds numbers, the drag coefficient reduces to the Stokes value, $C_D = 24/Re$. A good fit to the drag coefficient over a wide range of Reynolds number was provided by *Putnam* (1961)

$$C_D = \frac{24}{Re}\left(1 + \frac{Re^{2/3}}{6}\right) \quad (Re < 1200) \tag{12}$$

$$C_D = 0 \quad (1200 < Re < 1200) \tag{13}$$

The above discussion applies to gases that can be treated as a continuous medium, i.e., Λ is much smaller than the size of the dust particle. If this is not the case, Knudsen flow ap-

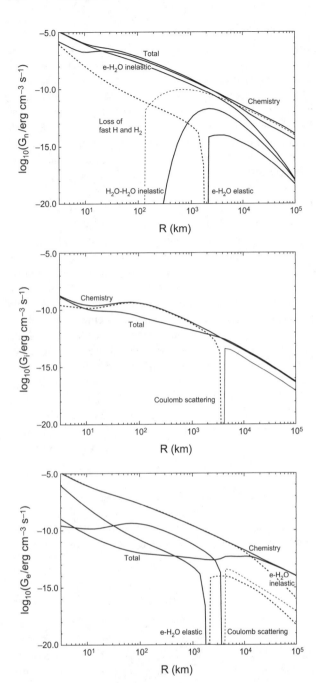

Fig. 4. Comparison of different heating and cooling mechanisms for the neutral (n), ion (i), and electron (e) fluids (from *Rodgers and Charnley,* 2002). For each process, the total heat source term, G (erg cm⁻³ s⁻¹), is shown; solid and dashed lines represent heating and cooling respectively. Processes considered include chemical reactions (this also includes photochemistry), loss of fast hydrogen atoms and molecules, elastic collisions between e-n, i-n, and i-e (Coulomb scattering) fluids, and inelastic water-water and electron-water collisions.

plies and a correction must be applied to the drag coefficient by dividing it by a factor $[1 + Kn(2.492 + 0.84\exp\{-1.74/Kn\})]$ (*Friedlander,* 1977), where Kn is the Knudsen number (the ratio of Λ to the particle size). For a typical comet at 1 AU, $\Lambda \sim 20$ cm (see section 3), and so Knudsen flow

dominates for typical dust particles in cometary comae and dust tails. For larger dust particles, the deceleration due to gravity must also be taken into account. Equating the drag force given by equation (11) with the gravitational force gives the size of the largest particles that can be carried away by the sublimating gas.

Gas-dust collisions can also heat or cool the colliding gas particles. The temperature of dust grains is determined by the energy balance between solar heating, gas-drag heating, and cooling by reradiation. For a steady flow, the evolution of the dust temperature can be calculated from

$$v_d C_d \frac{dT_d}{dr} = \dot{E}_{abs} + \dot{E}_{drag} - \dot{E}_{rad} \tag{14}$$

where C_d is the heat capacity and the terms on the righthand side refer to the energy sources/sinks mentioned above. In general, these terms will depend on the shape, size, and mineralogy of the grains. For example, small grains are hotter than large grains, and, since they are more absorptive, Fe-bearing silicate grains tend to be hotter than Mg-rich ones (*Harker et al.,* 2002). The energy transfer between the gas and dust, $\dot{E}_{drag}$, also affects the gas flow. In general, this term is calculated from an expression of the form

$$\dot{E}_{drag} = n_g \sigma_d C_H (T_g - T_d) \tag{15}$$

where C_H is the thermal accommodation coefficient. Because collisions between dust grains are rare, dust grains are coupled to the gas but not to each other. Therefore, an accurate description of the dusty coma requires the use of numerous dust components, each corresponding to a different population of grains with particular properties (e.g., size, shape, chemical composition). The acceleration of each component is obtained from equation (11) and the temperature from equation (14).

5.   COMA CHEMISTRY

5.1.   Photochemistry

The principal chemical processes occurring in the coma are photodissociation and ionization of the parent molecules. Accurate photodissociation rates are essential in order to model the coma and to interpret observational data. *Huebner et al.* (1992) compiled laboratory and theoretical data on a large number of important coma species; integrating over the solar spectrum yields photorates appropriate for both "quiet" and "active" solar photon fluxes. Despite the gargantuan nature of this undertaking, however, many gaps in our knowledge remain. Only a handful of species have been measured over a large range of wavelengths; many important radicals are unstable under laboratory conditions and so are extremely difficult to investigate. In many experiments, although the rates are well determined, the branching ratios among different sets of possible products

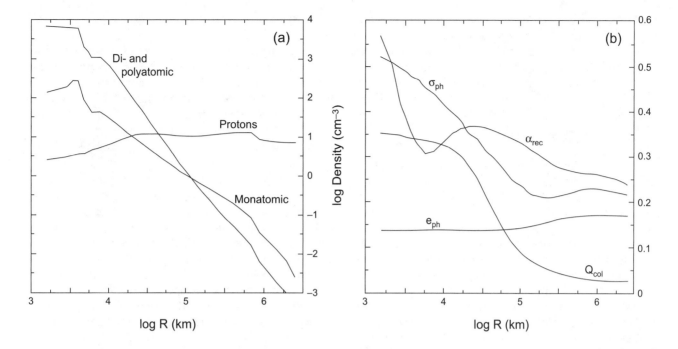

Fig. 5. Coma ion chemistry (*Schmidt et al.*, 1988). **(a)** Relative densities of ionized species in the coma. **(b)** Cometocentric variations of important quantities in ion-molecule chemistry. Effective rate coefficients and average surplus energy for photoprocesses [σ_{ph} (10^{-6} s^{-1}) and e_{ph} (10^2 eV)]; effective rate coefficient for electron recombination and effective cross-section for ion-neutral elastic collisions [α_{rec} (10^{-6} cm^3 s^{-1}) and Q_{col} (10^{-14} cm^2)].

are not well constrained. Thus, laboratory experiments are vital in order to further our understanding of coma chemistry (e.g., *Jackson et al.*, 2002).

As the most important species in the coma, the photochemistry of water is of key importance. Due to its importance in atmospheric chemistry, water has been studied extensively in the lab. Details of all the possible product channels are discussed in *Feldman et al.* (2004). Here, we note a couple of important points. First, the dominant dissociation channel results in OH + H. Studies of OH are therefore an excellent proxy for determining the cometary water production rate (as long as the OH excitation mechanisms are fully understood). The H_f atoms produced in this reaction are the principal source of heating in the inner coma, and may also drive a suprathermal chemistry. Second, the ionization of water occurs at a rate of $\approx 3 \times 10^{-7}$ s^{-1} at 1 AU. Hence, for an outflow velocity of 1 km s^{-1}, the fractional ionization in the coma increases as $n_e/n(H_2O) \sim (r/$km$) \times 3 \times 10^{-7}$. This gives a rough estimate of the amount of material that can potentially be affected by ion-molecule reactions.

5.2. Ion-Molecule Reactions

For molecules with permanent dipole moments, the dipole-charge interaction results in an extra attractive force between the reactants. In this case the rate coefficient can be calculated from the average-dipole-orientation theory of *Su and Bowers* (1973). *Eberhardt and Krankowsky* (1995) showed that the resulting rate coefficients can be parameterized in the form

$$k(T_{in}) = k(300)\left[1 + \beta\left(\sqrt{\frac{300}{T_{in}}} - 1\right)\right] \quad (16)$$

where β depends on the dipole moment, μ, polarizability, α, and a "locking constant." T_{in} is a mass-weighted mean kinetic temperature of the reactants. β can be calculated from the expression $\beta = \mu/(\mu + \alpha)$, with μ measured in Debye and α in Å^3 (*Rodgers and Charnley*, 2002).

In many ways the chemistry in the coma is analogous to the chemistry that occurs in interstellar hot cores. A "proton cascade" transfers protons from species with low proton affinity, i.e., OH (H_2O^+), to species with larger proton affinities, i.e., NH_3 (NH_4^+). Detailed chemical models are necessary to calculate how the ionization is apportioned among different ions, which is essential when interpreting *in situ* ion mass spectrometry data (*Geiss et al.*, 1991). Many interstellar chemical schemes are lacking in a number of important proton transfer reactions, most notably from methanol to ammonia (*Rodgers and Charnley*, 2001a). The inclusion of a full set of such reactions is a prerequisite for an accurate calculation of ionic abundances. Figure 5 shows the degree of ionization and the abundances of different families of ions in the coma, together with various parameters affecting the plasma chemistry.

As discussed earlier, the chemistry and hydrodynamics of the coma are intimately coupled. For example, proton transfer reactions are exothermic by typically a few electron volts (eV). Therefore, the ions formed via these reactions will have significant energies, and this effect keeps the ions in the inner coma at a warmer temperature than the neutral fluid (*Körösmezey et al.,* 1987). Also, the rates of many chemical reactions are temperature dependent. This is most important for dissociative recombination reactions of ions and electrons; when the value of T_e reaches extremely large values in the outer coma, such reactions effectively switch off, resulting in a larger ionization fraction (see *Lovell et al.,* 1999; *Bockelée-Morvan et al.,* 2004, Fig. 1).

Despite the importance of ion-molecule reactions in determining ionic abundances, it turns out that these reactions are not an efficient source of new, stable neutral molecules. For example, *Irvine et al.* (1998) proposed that proton transfer to HCN followed by recombination could account for the HNC seen in Comet Hale-Bopp. However, models that include a comprehensive ion-molecule chemistry (i.e., involving CH_3OH and NH_3) show that the amount of HNC actually produced is almost 100 times less than observed (*Rodgers and Charnley,* 1998, 2001a). This is illustrated in Fig. 6. Similar calculations on the formation of large organics also show that these species are produced only in small abundances (*Rodgers and Charnley,* 2001b). A corollary of this is that one can show that the HCOOH, CH_3OCHO, and CH_3CN observed in Hale-Bopp (*Bockelée-Morvan et al.,* 2000) cannot be formed by chemical reactions in the coma.

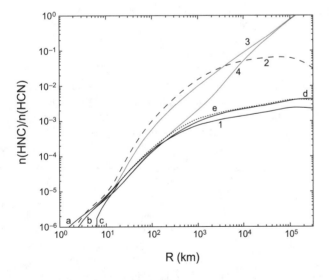

Fig. 6. HNC/HCN ratios for a variety of HNC formation mechanisms (*Rodgers and Charnley,* 2001c). (1) Ion-molecule chemistry (curves a–e result from altering model parameters, such as the initial HCN abundance, recombination branching ratios, etc.). (2) Isomerization of HCN by H_f atoms. (3) Photodestruction of an unknown parent. (4) As for (3), but assuming the parent of HNC also has an extended source.

5.3. Neutral-Neutral Reactions

An alternative source of new species in the coma may be provided by neutral-neutral reactions. Because photodissociation occurs around 100 times faster than photoionization, the abundances of reactive radicals are much larger than those of ions. However, many of the subsequent reactions will be destructive, with radicals reacting to break apart parent species they collide with. For example, OH can react with NH_3 and CH_4 to reform water and produce NH_2 and CH_3 respectively. A possible exception is the reaction of CN with hydrocarbons; these reactions typically result in the replacement of a H atom with the –CN group (*Balucani et al.,* 2000). In particular, the reaction with acetylene (C_2H_2) will lead to cyanoacetylene (HC_3N). Again, however, modeling shows that the resulting HC_3N yields are far less than observed in many comets (*Rodgers and Charnley,* 2001b). Neutral-neutral reactions do appear to be important in cometary sulfur chemistry. Based on our understanding of interstellar chemistry, S-bearing molecules take part in a predominately neutral chemistry. *Canaves et al.* (2002, 2004) have investigated the coma production of several S-bearing molecules, including NS and CS. They found very good agreement with the observed NS abundance in Comet Hale-Bopp and that the CS abundance should be approximately constant with cometocentric distance.

5.4. Nonthermal Chemistry

It has long been recognized that the energetic fragments produced in photodissociation reactions have the potential to drive a "suprathermal" chemistry in the coma (e.g., *Huebner et al.,* 1991; *Kissel et al.,* 1997). However, it was originally assumed that, in general, these reactions would be destructive. For example, H_f atoms can react with water to form H_2 and OH. Hence the net effect of such reactions is simply to increase the quantum yield of OH atoms produced from photodestruction of water: In addition to the primary OH daughter, secondary OH radicals will also be produced.

More recent work has focused on the potential of suprathermal reactions to affect the coma chemistry in more interesting ways. *Rodgers and Charnley* (1998, 2001c) looked at the possibility that H_f atoms could isomerize HCN into its isomer HNC, thus accounting for the puzzling HNC/HCN increase seen in several comets as they approach the Sun (e.g., *Irvine et al.,* 1999; *Biver et al.,* 2002; *Rodgers et al.,* 2003). They showed that such reactions may be viable in large, active comets such as Hale-Bopp, but some other mechanism must account for the HNC production in smaller comets such as Hyakutake and Ikeya-Zhang (see Fig. 6). *Pierce and A'Hearn* (2003) examined the possibility that the reaction of H_f atoms with CO_2 could account for the extended source of CO seen in many comets. However, they conclude that this mechanism is an order of magnitude slower than direct photodissociation of CO_2 into CO. It is possible that the coma chemistry of sulfur could be strongly affected by H_f atoms since these should readily abstract H atoms from the major parent, H_2S. Similarly, more abundant coma molecules like H_2CO and CH_3OH could also be destroyed by H_f atoms. The general applicability of these processes to coma composition remains to be investigated.

Other work has looked at the possibility of atoms produced in excited metastable states driving suprathermal reactions. For example, *A'Hearn et al.* (2000) suggested that S_2 may be formed in the coma by the reaction of OCS with atomic sulfur produced in the singlet-D state. More recently, *Glinski et al.* (2003) showed that the reaction of $O(^3P)$ with OH may lead to significant O_2 abundances in the inner coma. Another class of suprathermal reactions that are likely to be important are electron impact reactions. Although previous schemes have included these reactions (*Boice et al.,* 1986), accurately calculating the rate at which they occur requires a detailed model of the hot electron energy distribution in the coma (e.g., *Wegmann et al.,* 1999).

5.5. Dust Fragmentation/Degradation and Extended Sources

The inability of gas phase chemical reactions to generate sufficient quantities of particular molecules in the coma means that some other mechanism must be responsible for the extended sources of these molecules (*Festou,* 1999). *Huebner* (1987) proposed that destruction of POM in the coma could account for the formaldehyde source, and this was modeled by *Boice et al.* (1990). Based on their laboratory studies of POM degradation, *Cottin et al.* (2000) developed a more detailed model of thermal and photodegradation of POM in the coma. *Matthews and Ludicky* (1986) suggested that HCN polymers may be present in comets; degradation of these compounds may be an additional source of many small coma molecules (*Rettig et al.,* 1992; *Huebner and Boice,* 1997), including HNC (*Rodgers and Charnley,* 2001c). Of course, in all these models there are a number of free parameters, and it is always possible to fit the observed abundances if one adjusts the initial abundance and/or the destruction rate of the mystery parent. Nevertheless, the models require abundances and destruction rates that are in agreement with rough theoretical estimates and laboratory measurements. Therefore, although these models do not prove that dust destruction is responsible for injection of H_2CO and HNC into the coma, they certainly show that it is plausible.

5.6. Isotopic Fractionation and Nuclear Spin Ratios

Deuterium/hydrogen ratios have been observed in two coma species (HDO and DCN), $^{15}N/^{14}N$ ratios in CN and HCN, and $^{13}C/^{12}C$ ratios in C_2, CN, and HCN; see Table 3 in *Bockelée-Morvan et al.* (2004) for a complete list. In the interstellar medium, proton transfer reactions are efficient in scrambling the D/H ratios among different species (e.g., *Roberts and Millar,* 2000). Hence, we might also expect some degree of mixing to occur in the coma. However, as

discussed in section 5.2, ion-molecule chemistry cannot significantly alter the initial abundances of parent molecules sublimating from the nucleus, and detailed modeling shows that parent isotopic ratios are also unaffected (*Rodgers and Charnley,* 2002). An important consequence of this result is that measurements of isotope ratios can be used to test putative parent-daughter relationships. For example, although HCN undoubtedly accounts for some fraction of the CN observed in the coma, it has long been debated whether it can account for *all* the CN (*Festou,* 1999). Recent observations of $C^{15}N$ in Comets LINEAR WM1 and Hale-Bopp by *Arpigny et al.* (2003) demonstrate that CN is significantly enhanced in ^{15}N, by around a factor of 2, as compared with HCN. Thus, HCN cannot be the sole parent of CN, and if HCN contributes about half the CN, the ^{15}N ratio in the additional parent of CN must be even higher. If this parent is the same one that accounts for the extended source of HNC in comets, then one would also expect to see significant ^{15}N enhancements in HNC. Similar observations of ^{13}C-bearing isotopomers of C_2, C_3, and simple carbon-chain molecules (acetylene, ethene, ethane, cyanoacetylene) may also help to resolve the origin of these radicals.

As with molecular D/H ratios, the nuclear spin ratios of cometary species reflect the temperatures at which they originated, and can be altered in the coma only marginally by proton transfer reactions. Ortho:para ratios (OPRs) have been observed in water in several comets (see *Bockelée-Morvan et al.,* 2004). Recently, *Kawakita et al.* (2002) have measured the OPR in NH_2; since strict selection rules constrain the OPR in photodaughters as a function of the OPR in their parents, these observations can be used to probe the OPR in ammonia. Alternatively, one can turn this argument on its head and say that, if the OPR ratios in NH_3 can be measured directly, the question of whether NH_3 is the sole parent of NH_2 can be answered. A similar investigation of the OPR in cometary formaldehyde may provide insights into the nature of its extended source.

6. SUMMARY

6.1. Coma Physicochemistry

In the inner regions of the coma, the outflowing gas behaves as a fluid, and fluid equations can be used to obtain a reasonable description of the flow. The gas initially cools as it expands adiabatically, but is eventually heated by thermalization of energetic photofragments. The ions are also heated by exoergic proton transfer reactions. The hydrodynamic approximation loses validity in the outer coma, where the density decreases, and many hot photoproducts are not thermalized. This is particularly important for electrons, and a very hot electron population exists in the outer coma. The chemistry and the physics are intimately coupled, and detailed models should include both. The chemistry is initiated by the solar radiation field, which produces reactive radicals and ions. Subsequent reactions in the coma, especially proton transfer reactions, result in the production of new coma species. However, although chemical reactions are extremely important in determining the relative abundances of different ions, they are unable to synthesize significant quantities of stable, neutral species. Therefore, unless an observed species is an obvious photodaughter, it is probable that it is present in the nuclear ice. Degradation of complex organic material is likely to contribute to the extended sources of some molecules, and studies of extended sources of molecules may provide insight into the nature of cometary CHON material. The energetic photodissociation products in the coma may drive a suprathermal chemistry. Studies of such processes are currently in their infancy.

6.2. From Coma Measurements to Nucleus Properties

Although almost all our information on comets comes from studies of the coma, what we really want to know is the nature of the cometary nucleus. As stressed in this chapter, detailed models of the coma are essential in order to calculate the density profiles and excitation states of coma molecules. However, even with these models, it is usually only possible to derive globally averaged properties in the extended coma. In terms of molecular production rates, this means that we can derive accurate values for the total gas release rates from the nucleus. However, when using these values to understand nucleus properties we encounter two important impediments. First, it is now known that gas is released not only from the nucleus surface, but also from subsurface layers (see *Prialnik et al.,* 2004). Therefore, coma abundance ratios do not necessarily equal those in the icy nucleus, and models of the evolution of the cometary interior are required. Second, the nucleus is likely to be heterogeneous on small scales, and the gas production rate may vary dramatically with position (see *Crifo et al.,* 2004). Hence, the total gas production rate will depend on the sum over the surface of many different regions, so it is not possible to simply derive "average" surface properties from the global gas production. Again, highly detailed modeling is essential to elucidate these issues.

6.3. Open Questions and Future Directions

The gas temperatures and velocities measured *in situ* in Comet Halley were in good agreement with coma models. It therefore appears that we have a relatively secure understanding of the physics in the coma. Chemically, the profiles and velocities of most parent molecules, as well as of many simple radicals and atoms, are in broad agreement with photodissociation models; nevertheless, many uncertainties remain. We briefly review several of the outstanding problems.

As discussed in section 3, the gas flow immediately above the nucleus surface is extremely complex. The gas is not in LTE, and interacts strongly with the entrained dust. Small-scale topography and heterogeneity of the nucleus can lead to steep spatial gradients in gas pressure and chemi-

cal composition. Interaction of gas jets may result in standing shocks, and the divergence of chemically distinct jets may lead to chemical heterogeneity in the outer coma. A full understanding of the near-nucleus coma is vital if we are to use observations of the outer coma to infer the properties of the nucleus. These issues can only be resolved using extremely detailed three-dimensional gas-kinetic models; some results of the most up-to-date codes are presented in *Crifo et al.* (2004).

It appears that ion-molecule chemistry, of the kind that generates the diversity of molecules seen in the interstellar medium, is too slow to produce significant amounts of new species in the coma. However, the presence of highly energetic photodissociation products in the coma may allow a suprathermal chemistry to occur. Currently, only a small number of such reactions have been considered. Fast hydrogen atoms could play an important role here. For example, many sulfuretted molecules were detected for the first time in Comet Hale-Bopp, and it is known that the most important reactions in interstellar sulfur chemistry are neutral-neutral processes. Hydrogen abstraction from H_2S by H_f could drive some reactions deep in the inner coma, generating new sulfuretted molecules. Clearly, there may exist many other possible reactions, not included in typical astrochemical networks, that can proceed efficiently in the coma. In particular, the global consequences of nonthermalized electrons in driving electron impact reactions has not been studied in sufficient depth.

The origin of the extended sources of certain molecules are not well known. Fragmentation of dust particles and/or macromolecules has been invoked to account for the additional sources of some molecules, such as CO, formaldehyde, and several small radicals. Encouragingly, laboratory measurements of the properties of the proposed macromolecules appear to support this hypothesis (*Cottin et al.,* 2000). More laboratory work on other possible CHON components is needed. Despite the fact that they have been seen in comets for decades, the origins of radicals such as C_2, C_3, and CN are still uncertain. However, measurements of isotopic ratios in these species and their putative parents may finally help to resolve this issue. Measurements of the OPR in formaldehyde may also help to constrain its source. Interferometric mapping of the coma of Hale-Bopp also appeared to reveal additional extended sources of molecules usually thought to be solely parent species, such as HCN and CH_3OH (*Blake et al.,* 1999; *Kuan et al.,* 2003, 2004). The existence of such sources for other parent species would provide evidence that sublimation of volatiles from dust grains is occurring throughout the coma.

Given the rate of discovery (*Bockelée-Morvan et al.,* 2004), it is likely that many more molecules will be discovered in future cometary apparitions, or through the reanalysis of archival data [e.g., the recent detection of ethylene glycol in Comet Hale-Bopp (*Crovisier et al.,* 2004)]. Coma chemistry modeling will be necessary to identify the precise source of these molecules. Also, once the relevant laboratory data is available, future molecule discoveries will surely come from the existing large database of unidentified cometary lines (*Feldman et al.*, 2004). It is also probable that many more isotopomers of the principal parent molecules will soon be detected. The resulting isotopic ratios will contain important clues about the origins of comets.

We have stressed in this chapter the coupling between the physics and chemistry in the coma. The chemistry and dynamics also strongly influence the excitation of coma molecules, particularly daughter molecules that will be produced in excited states, as well as polar molecules that can be excited by collisions with the hot electrons in the outer coma. Therefore, in order to fully understand the observed energy level distributions, combined dynamic-chemistry-excitation models are required. For example, *Reylé and Boice* (2003) developed such a model for the S_2 excitation in the coma. Accurate calculations of the excitation, and its spatial variation, are particularly important when interpreting interferometric maps of the coma.

In the final analysis, coma chemistry models are only as good as the kinetic data they employ. As discussed in section 5.1, there are large uncertainties in some of the photodissociation and ionization rates used in these models. Rate coefficients for bimolecular reactions are typically taken from databases developed for modeling interstellar clouds [e.g., the UMIST ratefile (*Le Teuff et al.*, 2000)]. Although many of the most important reactions in these schemes have been measured, there are many others for which experimental data is lacking. For the ion-molecule chemistry in the coma the most important reactions are exothermic proton transfer reactions; fortunately, many of these have been measured, or can safely be assumed to occur at the Langevin collisional rate. Regarding the possible suprathermal chemistry that may occur in the coma, however, very few reactions have been measured, and it is likely that there are many reactions currently omitted from interstellar reaction schemes that may be important in the coma. With the exception of POM, the product yields from photodegradation of proposed CHON constituents are also unknown. Finally, very few of the rates for collisional excitation have been measured; most values are simply estimates.

Acknowledgments. This work was supported by NASA's Planetary Atmospheres Program through NASA Ames Cooperative Agreement NCC2-1412 with the SETI Institute (S.D.R. and S.B.C.) and through NAG5-11052 to the Southwest Research Institute (W.F.H.), as well as by NSF Planetary Astronomy (9973186) (D.C.B.).

REFERENCES

A'Hearn M. F., Arpigny C., Feldman P. D., Jackson W. M., Meier R., Weaver H. A., Wellnitz D. D., and Woodney L. M. (2000) Formation of S_2 in comets. *Bull. Am. Astron. Soc., 32,* 1079.

Arpigny C., Jehin E., Manfroid J., Hutsemékers D., Schulz R., Stüwe J. A., Zucconi J-M., and Ilyin I. (2003) Anomalous nitrogen isotope ratio in comets. *Science, 301,* 1522–1525.

Balucani N., Asvany O., Huang L. C. L., Lee Y. T., Kaiser R. I., Osamura Y., and Bettinger H. F. (2000) Formation of nitriles in the interstellar medium via reactions of cyano radicals,

$CN(X^2\Sigma^+)$, with unsaturated hydrocarbons. *Astrophys. J., 545,* 892–906.

Bernstein M. P., Sandford S. A., Allamandola L. J., Chang S., and Scharberg M. A. (1995) Organic compounds produced by photolysis of realistic interstellar and cometary ice analogs containing methanol. *Astrophys. J., 454,* 327–344.

Biermann L. (1951) Kometenschweife und solare Korpuskularstrahlung. *Z. Astrophys., 29,* 274–286.

Biermann L., Giguere P. T., and Huebner W. F. (1982) A model of a comet coma with interstellar molecules in the nucleus. *Astron. Astrophys., 108,* 221–226.

Biver N. and 13 colleagues (1999) Spectroscopic monitoring of Comet C/1996 B2 (Hyakutake) with the JCMT and IRAM radio telescopes. *Astron. J., 118,* 1850–1872.

Biver N. and 22 colleagues (2002) The 1995–2002 long-term monitoring of Comet C/1995 O1 (Hale-Bopp) at radio wavelength. *Earth Moon Planets, 90,* 5–14.

Bockelée-Morvan D. and Crovisier J. (1987) The 2.7-micron water band of comet P/Halley — Interpretation of observations by an excitation model. *Astron. Astrophys., 187,* 425–430.

Bockelée-Morvan D., Crovisier J., and Gerard E. (1990) Retrieving the coma gas expansion velocity in P/Halley, Wilson (1987 VII) and several other comets from the 18-cm OH line shapes. *Astron. Astrophys., 238,* 382–400.

Bockelée-Morvan D. and 17 colleagues (2000) New molecules found in comet C/1995 O1 (Hale-Bopp). Investigating the link between cometary and interstellar material. *Astron. Astrophys., 353,* 1101–1114.

Bockelée-Morvan D., Crovisier J., Mumma M. J., and Weaver H. A. (2004) The composition of cometary volatiles. In *Comets II* (M. C. Festou et al., eds.), this volume. Univ. of Arizona, Tucson.

Boice D. C., Huebner W. F., Keady J. J., Schmidt H. U., and Wegmann R. (1986) A model of Comet P/Giacobini-Zinner. *Geophys. Res. Lett., 13,* 381–384.

Boice D. C., Huebner W. F., Sablik M. J., and Konno I. (1990) Distributed coma sources and the CH_4/CO ratio in comet Halley. *Geophys. Res. Lett., 17,* 1813–1816.

Blake G. A., Qi C., Hogerheijde M. R., Gurwell M. A., and Muhleman D. O. (1999) Sublimation from icy jets as a probe of the interstellar volatile content of comets. *Nature, 398,* 213–216.

Canaves M. V., de Almeida A. A., Boice D. C., and Sanzovo G. C. (2002) Nitrogen sulfide in Comets Hyakutake (C/1996 B2) and Hale-Bopp (C/1995 O1). *Earth Moon Planets, 90,* 335–347.

Canaves M. V., de Almeida A. A., Boice D. C., and Sanzovo G. C. (2004) Chemistry of NS and CS in cometary comae. *Adv. Space Res.,* in press.

Combi M. R. and Smyth W. H. (1988) Monte Carlo particle-trajectory models for neutral cometary gases. I — Models and equations. *Astrophys. J., 327,* 1026–1059.

Combi M. R., Harris W. M., and Smyth W. H. (2004) Gas dynamics and kinetics in the cometary coma: Theory and observations. In *Comets II* (M. C. Festou et al., eds.), this volume. Univ. of Arizona, Tucson.

Cottin H., Gazeau M. C., Doussin J. F., and Raulin F. (2000) An experimental study of the photodegradation of polyoxymethylene at 122, 147 and 193 nm. *J. Photochem. Photobiology A, 135,* 53–64.

Cottin H., Gazeau M. C., Benilan Y., and Raulin F. (2001) Polyoxymethylene as parent molecule for the formaldehyde extended source in Comet Halley. *Astrophys. J., 556,* 417–420.

Cravens T. E. (1991) Plasma processes in the inner coma. In *Comets in the Post-Halley Era* (R. L. Newburn Jr. et al., eds.), pp. 1211–1255. Kluwer, Dordrecht.

Cravens T. E. and Körösmezey A. (1986) Vibrational and rotational cooling of electrons by water vapor. *Planet. Space Sci., 34,* 961–970.

Crifo J. (1991) Hydrodynamic models of the collisional coma. In *Comets in the Post-Halley Era* (R. L. Newburn Jr. et al., eds.), pp. 937–989. Kluwer, Dordrecht.

Crifo J. F., Fulle M., Kömle N. I., and Szegö K. (2004) Nucleus-coma structural relationships: Lessons from physical models. In *Comets II* (M. C. Festou et al., eds.), this volume. Univ. of Arizona, Tucson.

Crovisier J. (1984) The water molecule in comets — fluorescence mechanisms and thermodynamics of the inner coma. *Astron. Astrophys., 130,* 361–372.

Crovisier J. (1989) The photodissociation of water in cometary atmospheres. *Astron. Astrophys., 213,* 459–464.

Crovisier J. (2004) The molecular complexity of comets. In *Astrobiology: Future Perspectives* (P. Ehrenfreund et al., eds.), in press. Kluwer, Dordrecht.

Crovisier J., Bockelée-Morvan D., Biver N., Colom P., Despois D., and Lis D. C. (2004) Ethylene glycol in comet C/1995 O1 (Hale-Bopp). *Astron. Astrophys.,* in press.

Draine B. T. (1980) Interstellar shock waves with magnetic precursors. *Astrophys. J., 241,* 1021–1038.

Draine B. T. (1986) Multicomponent, reacting MHD flows. *Mon. Not. R. Astron. Soc., 220,* 133–148.

Eberhardt P. and Krankowsky D. (1995) The electron temperature in the inner coma of comet P/Halley. *Astron. Astrophys., 295,* 795–806.

Eberhardt P. and 10 colleagues (1987) The CO and N_2 abundance in comet P/Halley. *Astron. Astrophys., 187,* 481–484.

Feldman P. D., Cochran A. L., and Combi M. R. (2004) Spectroscopic investigations of fragment species in the coma. In *Comets II* (M. C. Festou et al., eds.), this volume. Univ. of Arizona, Tucson.

Festou M. C. (1981) The density distribution of neutral compounds in cometary atmospheres. I — Models and equations. *Astron. Astrophys., 95,* 69–79.

Festou M. C. (1999) On the existence of distributed sources in comet comae. *Space Sci. Rev., 90,* 53–67.

Festou M. C., Keller H. U., Bertaux J. L., and Barker E. S. (1983) Lyman-alpha observations of comets West 1976 VI and P d'Arrest 1976 XI with Copernicus. *Astrophys. J., 265,* 925–932.

Friedlander S. K. (1977) *Smoke, Dust and Haze.* Wiley, New York. 317 pp.

Gail H. P. and Sedlmayr E. (1985) Dust formation in stellar winds — II. Carbon condensation in stationary, spherically expanding winds. *Astron. Astrophys., 148,* 183–190.

Geiss J., Altwegg K., Anders E., Balsiger H., Meier A., Shelley E. G., Ip W.-H., Rosenbauer H., and Neugebauer M. (1991) Interpretation of the ion mass spectra in the mass per charge range 25–35 amu/e obtained in the inner coma of Halley's comet by the HIS-sensor of the Giotto IMS experiment. *Astron. Astrophys., 247,* 226–234.

Giguere P. T. and Huebner W. F. (1978) A model of comet comae. I. Gas-phase chemistry in one dimension. *Astrophys. J., 223,* 638–654.

Glinski R. J., Harris W. M., Anderson C. M., and Morgenthaler J. P. (2003) Oxygen/hydrogen chemistry in inner comae of active comets. In *XXVth General Assembly: Highlights of Astronomy, Vol. 13* (O. Engvole, ed.), in press. Astronomical Society of the Pacific, San Francisco.

Gombosi T. I., Cravens T. E., and Nagy A. F. (1985) Time-dependent dusty gas dynamical flow near cometary nuclei. *Astrophys. J., 293,* 328–341.

Gombosi T. I., Nagy A. F., and Cravens T. E. (1986) Dust and neutral gas modeling of the inner atmospheres of comets. *Rev. Geophys., 24,* 667–700.

Harich S. A., Yang X. F., Yang X., and Dixon R. N. (2001) Extremely rotationally excited OH from water (HOD) photodissociation through conical intersections. *Phys. Rev. Lett., 87,* 253201.

Harker D. E., Wooden D. H., Woodward C. E., and Lisse C. M. (2002) Grain properties of Comet C/1995 O1 (Hale-Bopp). *Astrophys. J., 580,* 579–597.

Haser L. (1957) Distribution d'intensite dans la tete d'une comete. *Bull. Acad. R. Sci. Liege, 43,* 740–750.

Hodges R. R. (1990) Monte Carlo simulation of nonadiabatic expansion in cometary atmospheres — Halley. *Icarus, 83,* 410–433.

Huebner W. F. (1985) Cometary comae. In *Molecular Astrophysics: State of the Art and Future Directions* (G. H. F. Diercksen et al., eds.), pp. 311–330. Reidel, Dordrecht.

Huebner W. F. (1987) First polymer in space identified in Comet Halley. *Science, 237,* 628–630.

Huebner W. F. and Benkhoff J. (1999) From coma abundances to nucleus composition. *Space Sci. Rev., 90,* 117–130.

Huebner W. F. and Boice D. C. (1997) Polymers and other macromolecules in comets. In *Comets and the Origin and Evolution of Life* (P. J. Thomas et al., eds.), pp. 111–129. Springer-Verlag, New York.

Huebner W. F. and Giguere P. T. (1980) A model of comet comae. II. Effects of solar photodissociative ionization. *Astrophys. J., 238,* 753–762.

Huebner W. F. and Keady J. J. (1984) First-flight escape from spheres with R^{-2} density distribution. *Astron. Astrophys., 135,* 177–180.

Huebner W. F., Boice D. C., Schmidt H. U., and Wegmann R. (1991) Structure of the coma: Chemistry and solar wind interaction. In *Comets in the Post-Halley Era* (R. L. Newburn Jr. et al., eds.), pp. 907–936. Kluwer, Dordrecht.

Huebner W. F., Keady J. J., and Lyon S. P. (1992) Solar photo rates for planetary atmospheres and atmospheric pollutants. *Astrophys. Space Sci., 195,* 1–294.

Ip W-H. (1983) On photochemical heating of cometary comae — The cases of H_2O and CO-rich comets. *Astrophys. J., 264,* 726–732.

Ip W-H. (2004) Global solar wind interaction and ionospheric dynamics. In *Comets II* (M. C. Festou et al., eds.), this volume. Univ. of Arizona, Tucson.

Irvine W. M., Dickens J. E., Lovell A. J., Schloerb F. P., Senay M., Bergin E. A., Jewitt D., and Matthews H. E. (1998) Chemistry in cometary comae. *Faraday Discuss., 109,* 475–492.

Irvine W. M., Dickens J. E., Lovell A. J., Schloerb F. P., Senay M., Bergin E. A., Jewitt D., and Matthews H. E. (1999) The HNC/HCN ratio in comets. *Earth Moon Planets, 78,* 29–35.

Jackson W. M., Xu D., Huang J., Price R. J., and Volman D. H. (2002) New experimental and theoretical techniques for studying photochemical reactions of cometary atmospheres. *Earth Moon Planets, 89,* 197–220.

Kawakita H., Watanabi J., Fuse T., Furusho R., and Abe S. (2002)

Spin temperature of ammonia determined from NH$_2$ in Comet C/2001 A2 (LINEAR). *Earth Moon Planets, 90,* 371–379.

Kissel J. and 18 colleagues (1986) Composition of comet Halley dust particles from Giotto observations. *Nature, 321,* 336–337.

Kissel J., Jrueger F. R., and Roessler K. (1997) Organic chemistry in comets from remote and in situ observations. In *Comets and the Origin and Evolution of Life* (P. J. Thomas et al., eds.), pp. 69–109. Springer-Verlag, New York.

Kitamura Y. (1986) Axisymmetric dusty gas jet in the inner coma of a comet. *Icarus, 66,* 241–257.

Konno I., Huebner W. F., and Boice D. C. (1993) A model of dust fragmentation in near-nucleus jet-like features on Comet P/Halley. *Icarus, 101,* 84–94.

Körösmezey A. and Gombosi T. I. (1990) A time-dependent dusty gas dynamic model of axisymmetric cometary jets. *Icarus, 84,* 118–153.

Körösmezey A., Cravens T. E., Nagy A. F., Gombosi T. I., and Mendis D. A. (1987) A new model of cometary ionospheres. *J. Geophys. Res., 92,* 7331–7340.

Krankowsky D. and 11 colleagues (1986) In situ gas and ion measurements at comet Halley. *Nature, 321,* 326–329.

Kuan Y.-J., Huang H.-C., Snyder L. E., Veal J. M., Woodney L. M., Forster J. R., Wright M. C. H., and A'Hearn M. F. (2003) BIMA array observations of cometary organic molecules in Hale-Bopp. In *Proceedings of the XIIth Rencontres de Blois on Frontiers of Life* (L. M. Celnikier and J. Tran Thanh Van, eds.), pp. 55–57. The Gioi, Vietnam.

Kuan Y.-J., Charnley S.B., Huang H.-C., Kisiel Z., Ehrenfreund P., Tseng W.-L., and Yan C.-H. (2004) Searches for interstellar molecules of potential prebiotic importance. *Adv. Space Res., 33,* 31–39.

Le Teuff Y. H., Millar T. J., and Markwick A. J. (2000) The UMIST database for astrochemistry 1999. *Astron. Astrophys. Suppl. Ser., 146,* 157–168.

Lisse C. M., Cravens T. E., and Dennerl K. (2004) X-ray and extreme ultraviolet emission from comets. In *Comets II* (M. C. Festou et al., eds.), this volume. Univ. of Arizona, Tucson.

Lovell A. J., Schloerb F. P., Bergin E. A., Dickens J. E., De Vries C. H., Senay M. C., and Irvine W. M. (1999) HCO$^+$ in the coma of Comet Hale-Bopp. *Earth Moon Planets, 77,* 253–258.

Marconi M. L. and Mendis D. A. (1982) The photochemical heating of the cometary atmosphere. *Astrophys. J., 260,* 386–394.

Marconi M. L. and Mendis D. A. (1983) The atmosphere of a dirty-clathrate cometary nucleus — A two-phase, multifluid model. *Astrophys. J., 273,* 381–396.

Marconi M. L. and Mendis D. A. (1986) The electron density and temperature in the tail of comet Giacobini-Zinner. *Geophys. Res. Lett., 13,* 405–406.

Matthews C. N. and Ludicky R. (1986) The dark nucleus of comet Halley: Hydrogen cyanide polymers. In *20th ESLAB Symposium of Halley's Comet* (B. Battrick et al., eds.), pp. 273–277. ESA SP-250, Noordwijk, The Netherlands.

McCoy R. P., Meier R. R., Keller H. U., Opal C. B., and Carruthers G. R. (1992) The hydrogen coma of Comet P/Halley observed in Lyman-alpha using sounding rockets. *Astron. Astrophys., 258,* 555–565.

Meier R., Eberhardt P., Krankowsky D., and Hodges R. R. (1993) The extended formaldehyde source in comet P/Halley. *Astron. Astrophys., 277,* 677–690.

Mitchell G. F., Prasad S. S., and Huntress W. T. (1981) Chemical model calculations of C$_2$, C$_3$, CH, CN, OH, and NH$_2$ abundances in cometary comae. *Astrophys. J., 244,* 1087–1093.

Neubauer F. M. (1991) The magnetic field structure of the cometary plasma environment. In *Comets in the Post-Halley Era* (R. L. Newburn Jr. et al., eds.), pp. 1107–1124. Kluwer, Dordrecht.

Neubauer F. M. and 11 colleagues (1986) First results from the Giotto magnetometer experiment at comet Halley. *Nature, 321,* 352–355.

Nordholt J. E. and 14 colleagues (2003) Deep Space 1 encounter with Comet 19P/Borrelly: Ion composition measurements by the PEPE mass spectrometer. *Geophys. Res. Lett., 30,* 1465–1468.

Oppenheimer M. (1975) Gas phase chemistry in comets. *Astrophys. J., 196,* 251–259.

Pack J. L., Voshall R. E., and Phelps A. V. (1962) Drift velocities of slow electrons in krypton, xenon, deuterium, carbon monoxide, carbon dioxide, water vapor, nitrous oxide, and ammonia. *Phys. Rev., 127,* 2084–2089.

Pierce D. M. and A'Hearn M. F. (2003) Formation of carbon monoxide in the near-nucleus coma of comets. *Bull. Am. Astron. Soc., 35,* 968.

Prialnik D., Benkhoff J., and Podolak M. (2004) Modeling the structure and activity of comet nuclei. In *Comets II* (M. C. Festou et al., eds.), this volume. Univ. of Arizona, Tucson.

Putnam A. (1961) Integrable form of droplet drag coefficient. *ARS J., 31,* 1467–1468.

Rettig T. W., Tegler S. C., Pasto D. J., and Mumma M. J. (1992) Comet outbursts and polymers of HCN. *Astrophys. J., 398,* 293–298.

Reylé C. and Boice D. C. (2003) An S$_2$ fluorescence model for interpreting high-resolution cometary spectra. I. Model description and initial results. *Astrophys. J., 587,* 464–471.

Roberts H. and Millar T. J. (2000) Modelling of deuterium chemistry and its application to molecular clouds. *Astron. Astrophys., 361,* 388–398.

Rodgers S. D. and Charnley S. B. (1998) HNC and HCN in comets. *Astrophys. J. Lett., 501,* L227–L230.

Rodgers S. D. and Charnley S. B. (2001a) Chemical differentiation in regions of massive star formation. *Astrophys. J., 546,* 324–329.

Rodgers S. D. and Charnley S. B. (2001b) On the origin of HNC in Comet Lee. *Mon. Not. R. Astron. Soc., 323,* 84–92.

Rodgers S. D. and Charnley S. B. (2001c) Organic synthesis in the coma of Comet Hale-Bopp? *Mon. Not. R. Astron. Soc., 320,* L61–L64.

Rodgers S. D. and Charnley S. B. (2002) A model of the chemistry in cometary comae: Deuterated molecules. *Mon. Not. R. Astron. Soc., 330,* 660–674.

Rodgers S. D., Butner H. M., Charnley S. B., and Ehrenfreund P. (2003) The HNC/HCN ratio in comets: Observations of C/2002 C1 (Ikeya-Zhang). *Adv. Space Res., 31,* 2577–2582.

Schleicher D. G. and Farnham T. L. (2004) Photometry and imaging of the coma with narrowband filters. In *Comets II* (M. C. Festou et al., eds.), this volume. Univ. of Arizona, Tucson.

Schmidt H. U., Wegmann R., Huebner W. F., and Boice D. C. (1988) Cometary gas and plasma flow with detailed chemistry. *Comp. Phys. Comm., 49,* 17–59.

Shimizu M. (1976) The structure of cometary atmospheres. I — Temperature distribution. *Astrophys. Space Sci., 40,* 149–155.

Su T. and Bowers M. T. (1973) Theory of ion-polar molecule collisions. *J. Chem. Phys., 58,* 3027–3037.

Swings P. (1943) Cometary spectra. *Mon. Not. R. Astron. Soc., 103,* 86–111.

Wegmann R., Schmidt H. U., Huebner W. F., and Boice D. C. (1987) Cometary MHD and chemistry. *Astron. Astrophys., 187,* 339–350.

Wegmann R., Jockers K., and Bonev T. (1999) H_2O^+ ions in comets: Models and observations. *Planet. Space Sci., 47,* 745–763.

Whipple F. L. (1950) A comet model. Part I. The acceleration of Comet Encke. *Astrophys. J., 111,* 375–394.

Whipple F. L. (1951) A comet model. Part II. Physical relations for comets and meteors. *Astrophys. J., 113,* 464–474.

Whipple F. L. and Huebner W. F. (1976) Physical processes in comets. *Annu. Rev. Astron. Astrophys., 14,* 143–172.

Xie X. and Mumma M. J. (1992) The effect of electron collisions on rotational populations of cometary water. *Astrophys. J., 386,* 720–728.

Gas Dynamics and Kinetics in the Cometary Coma:
Theory and Observations

Michael R. Combi
University of Michigan

Walter M. Harris
University of Washington

William H. Smyth
Atmospheric & Environmental Research, Inc.

Our ability to describe the physical state of the expanding coma affects fundamental areas of cometary study both directly and indirectly. In order to convert measured abundances of gas species in the coma to gas production rates, models for the distribution and kinematics of gas species in the coma are required. Conversely, many different types of observations, together with laboratory data and theory, are still required to determine coma model attributes and parameters. Accurate relative and absolute gas production rates and their variations with time and from comet to comet are crucial to our basic understanding of the composition and structure of cometary nuclei and their place in the solar system. We review the gas dynamics and kinetics of cometary comae from both theoretical and observational perspectives, which are important for understanding the wide variety of physical conditions that are encountered.

1. INTRODUCTION

Gases, primarily water, and entrained dust are liberated from the icy nucleus by solar heating when a comet is within a few astronomical units (AU) of the Sun and form an outflowing cometary atmosphere. The cometary atmosphere, or coma, expands to distances many orders of magnitude larger than the size of the nucleus itself. Because of the ultimate expansion into vacuum and the very weak gravity that affects only the largest dust particles, a Knudsen layer of a fraction to a few meters in thickness (a few times the molecular collision mean free path) forms near the surface where the gas organizes itself into a transient stationary thermal layer. The initiation of regular flow transforms this thermal layer into a rapid transonic dusty-gas flow with a typical scale length of 10–100 m. Expansion and nearly adiabatic cooling dominates the flow out to scales of ~100 km where the dust becomes decoupled collisionally from the gas. Here the gas flows with a speed of ~700 m s^{-1} and has a cool temperature of <30 K. The energetic photochemical products of solar UV photodissociation then begin to heat the gas faster than it can cool by adiabatic expansion or IR radiational cooling, and the gas kinetic temperature and ultimately the outflow speed increase. Photoionization and charge impact ionization (~5–10% of the photodissociation rate) forms a cometary ionosphere on scales of 10^3–10^4 km, which ultimately interacts with the magnetized plasma of the solar wind forming an ion tail on scales of 10^5–10^7 km. Also, at large scales of 10^6–10^7 km, the neutral coma is broken down into its atomic constituents, such as O, H, C,

and N. This description is most applicable for a typical comet observed in the range of a few tenths to about 2 AU from the Sun. For comets much farther away, temperatures and velocities are smaller and reaction length scales are larger. And for comets much closer (Sun-grazers), temperatures and velocities are much larger and reaction length scales are smaller.

Our understanding of the conditions within a few times the radius of the nucleus results largely from model predictions that are only constrained by the gas fluxes and velocities hundreds of nucleus-radii from the nucleus, and are determined both by *in situ* and remote observations. There are not many direct observations about the conditions near the surface, beyond the images of 1P/Halley from the *Giotto* and *Vega* spacecrafts and those of 19P/Borrelly from the *Deep Space 1* spacecraft, which show only the dust distribution. This region is described in detail in *Crifo et al.* (2004). The photodissociation, photoionization and fast ion-neutral chemical reactions involving ultimately ~100 species and thousands of reactions forms a complex atmosphere/ionosphere that is described in detail in *Rodgers et al.* (2004).

The region of the coma accessible by remote ground-based and Earth-orbit-based observations and by direct *in situ* sampling by spacecraft instrumentations (mostly by neutral gas and plasma mass spectrometers) for detailed study generally covers distances larger than a few hundred kilometers from the nucleus. Remote imaging observations of this region have provided information about the projected line-of-sight-integrated distribution of neutral gas and ion species in two spatial dimensions on the sky plane and sometimes the line-of-sight velocity distribution. Thus remote sensing provides somewhat convoluted, nonunique, global views of the coma, whereas *in situ* measurements have provided some detailed information about neutral gas and ion densities, velocities, and temperatures, but only for a few fleeting threads along a few spacecraft trajectories. Information about kinetic and rotational temperatures has been obtained by both remote observations and *in situ* measurements. The spectroscopic measurements are discussed in detail in *Bockelée-Morvan et al.* (2004) (for parent species) and *Feldman et al.* (2004) (for nonparent species).

It is important to understand the physical state of the coma and its variations because it is largely from analyses of observations of various species in the coma that we determine the composition of the larger population of comets. Although in the last 20 years IR and radio astronomy have opened the possibility of directly observing parent gas species in comets, interpreting these observations is complicated by difficult excitation mechanisms, atmospheric extinction (in the IR), and generally weaker signal-to-noise when compared with visible and near-UV measurements of daughter species. The interpretations of measurements of daughter species have their own sets of complicating issues regarding production models and kinematics. However, routine observations of daughter species in the visible remains critically important for the foreseeable future because they are more sensitive for weak comets and especially for those at large heliocentric distances. This is the only practical way to characterize the composition and activity of the entire population of comets and to compare with the large base of existent data. Observations of daughter species in the visible should also be made and interpreted side-by-side with IR and radio observations of parent species for brighter comets and those at generally smaller heliocentric distances to enable the entire complex picture to be unraveled.

In this chapter we concentrate on those aspects of photochemistry and dynamics that are important for understanding the overall physical state of the coma, and their variations from comet to comet, with changing heliocentric and cometocentric distance, and, most importantly, in the region sampled by observations, i.e., outside a few hundred kilometers from the nucleus. We also discuss quantitative models of the coma beginning with the original fountain model of *Eddington* (1910), which started the heuristic approach continued by *Haser*'s (1957) model that in turn later treated the coma as a free expansion problem including production and loss processes. From there we move to physics-based models that treat the energy budget, dynamics, and chemistry, leading to hydrodynamics, kinetic and hybrid models. These models provide the basis for interpreting and linking together multiple diagnostic observations of comets that are becoming increasingly precise. Care must be taken in interpreting observations either using complex models with too many free (and especially unconstrained) parameters, or using simple models with mathematically constrained parameters, where their direct physical meaning may be simplistic or inaccurate.

2. FREE-EXPANSION MODELS

2.1. Eddington to Wallace and Miller Fountain Models

In Eddington's fountain model the comet is assumed to be a uniform and isotropic point source of emitters of light such that their density (e.g., gas or dust) would fall as the inverse square of the distance from the source, except the emitters are also subject to a uniform acceleration that pushes on them from a given direction (presumably from the Sun's direction). Eddington showed that such a model defines a paraboloid of revolution along a line parallel to the acceleration and passing through the point source (the nucleus of the comet) such that

$$x^2 + y^2 = 2z\frac{v^2}{a} + \frac{v^4}{a^2} \qquad (1)$$

where the acceleration, a, is directed in the +z direction, x and y complete the righthanded Cartesian coordinates, and v is the initial uniform outflow speed of the emitting particles. In this case the density, n, of emitting particles is determined by two trajectories that cross any given point within the paraboloid. Later work by *Mocknatsche* (1938), which was independently rediscovered by *Fokker* (1953) and later described by *Wallace and Miller* (1958), showed that the column density, which is what would be observed by a remote observer through an optically thin paraboloid of such particles, N, can be calculated after some algebraic manipulation for any line-of-sight that passes within the projected paraboloid for any angle between the Sun, comet, and observer to simply be

$$N = \frac{Q}{4vR} \qquad (2)$$

where Q is the global particle production rate and R is the projected distance on the "sky" plane from the source (i.e., the nucleus). Interestingly, this is the same result as for a point source without the acceleration. The projected shape of the paraboloid, of course, depends on the viewing geometry. Such a model is still used to understand many basic properties of the observed dust distribution in comets.

2.2. *Haser* (1957) Model

Most radiative emissions of gaseous species in comets in visible light are caused by fluorescence with sunlight through an otherwise practically optically thin medium. Early quantitative observations of gas species in the cometary coma indicated that the spatial variation of the observed brightness of some emissions (e.g., C_2, CN, etc.) deviated from the simple inverse distance relation of a simple fountain model. It was also realized that most of the gas species observed and identified through visible range spectroscopy were not stable molecules but unstable radicals that could be stored in an otherwise icy nucleus but were more likely produced in the photodissociation of parent molecules by solar UV radiation. In order to describe the kind of expected distribution of an observed gaseous species in the coma of a comet, one needs to account (potentially) for the creation of that species from a parent species and its destruction into some simpler species. An example would be the production of observed OH by photodissociation of H_2O and the subsequent photodissociation of the OH into O + H.

The quantitative description of this was put forth by *Haser* (1957) in a now classic formulation that is still widely used today. Haser considered the distribution of a secondary species (like some cometary radical) being produced by the photodissociation of some parent molecule and in turn being destroyed by some photodestruction process. If the coma is considered to be a spherically symmetric point source of uniformly outflowing parent molecules, where an exponential lifetime describes its destruction, then the density, n_p, of parents at some distance from the point source, r, is given by

$$n_p(r) = \frac{Q}{4\pi r^2 v}(e^{-r/\gamma_p})$$

The density, n_d, of a daughter species created by the destruction of these parents is given by

$$n_d(r) = \frac{Q}{4\pi r^2 v}\frac{\gamma_d}{\gamma_p - \gamma_d}(e^{-r/\gamma_p} - e^{-r/\gamma_d}) \quad (3)$$

where r is the distance from the nucleus, and γ_p and γ_d are the parent and daughter scale lengths given by the product of the radial outflow speed, v, and exponential lifetimes of the parent of the observed species, τ_p, and that of the observed daughter species, τ_d, itself, respectively. The model is most often used with observed intensity spatial profiles, which are typically proportional to the column density profile of the emitting species. Therefore, normally equation (3) is integrated along the line-of-sight. This can either be performed numerically or written in the form of modified Bessel function as done originally by Haser, which yields

$$N_d(\rho) = \frac{Q}{2\pi\rho v}\frac{\gamma_d}{\gamma_p - \gamma_d}\left(\int_0^{\rho/\gamma_p} K_0(y)dy - \int_0^{\rho/\gamma_d} K_0(y)dy\right) \quad (4)$$

where ρ is the projected distance from the nucleus through which the line-of-sight integration is performed and $K_0(y)$ is the modified Bessel function of the first kind.

Figure 1 (*Combi and Fink*, 1997) shows the range of the family of column density spatial profiles possible with Haser's model. An interesting, but little mentioned, attribute of this equation is that the spatial profile shape is the same if the parent and daughter scale lengths are interchanged, yielding simply an extra factor of the ratio of the two scale lengths. This means that it is not possible to say whether the parent or daughter scale length is the smaller based only on an observation of the daughter spatial profile.

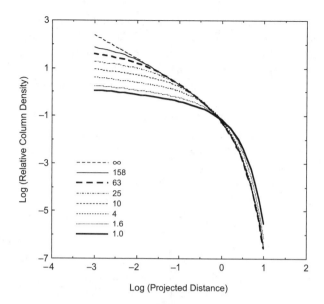

Fig. 1. Family of Haser model profiles of daughter species column densities for different values of the ratio of the scale lengths plotted as a function of projected distance in units of the daughter scale length. Note that a scale length ratio of infinity (∞) corresponds to a parent molecule distribution.

2.3. Models with Secondary Product Velocities

Of the several simplifying assumptions of Haser's model as most often used, which include spherical symmetry, constant outflow velocity, and steady-state gas production, one of the basic assumptions is that both the parent and daughter products undergo purely radial motion. Early results of UV observations of comets indicated that H and OH were produced by photodissociation of water and that this would lead to nonnegligible excess energy, which would be distributed in the form of superthermal velocities for the daughter products. The large speeds of the H atoms imply velocities for the heavier products in the photodissociation network, namely O and OH, in the range of 1–2 km s^{-1}. These are generally larger than the outflow speed of the parent molecules in the coma itself and therefore posed a serious violation of the assumption of radial outflow for daughter species in Haser's model. However, the use of Haser's model for interpreting spatial profiles of column density derived from brightness observations already required either integrals of modified Bessel functions (equation (4)) or direct numerical integration of the density function (equation (3)) itself. The addition of nonradial ejection of daughter species yields a nonanalytical integral for the density function itself.

Festou (1981a) and *Combi and Delsemme* (1980) addressed solving this problem in two different ways. *Festou* (1981a) introduced the vectorial model, which represented a numerical approach to directly solving for the density and column density distribution. *Combi and Delsemme* (1980) introduced Monte Carlo techniques, developed originally for nuclear reactor particle transport and radiative transfer calculations, to simulate the expansion. In their more modern versions, the vectorial model is faster computationally, but the Monte Carlo technique is directly applicable to even more general problems, such as variable outflow velocity, aspherical comae, and radiation pressure acceleration of the daughter species.

The main effect of the ejection of daughter species on observed spatial profiles is that profiles produced are not related to those using Haser's model with scale lengths simply given by the velocities × the lifetimes. These more complicated models did, however, account for the fact that most prospective parent molecule lifetimes were actually longer than the Haser model parent scale length would indicate, given a reasonable assumption for the parent velocity.

Combi and Delsemme (1980) also introduced what started as a heuristic geometrical argument for relating the given velocities and lifetimes that would give a realistic spatial profile with the vectorial model (or Monte Carlo simulation) to a Haser model with some set of Haser scale lengths. This so-called "average random walk model" specifies transformations from the actual parent and daughter velocities and lifetimes to the equivalent Haser-model scale lengths. Also, like the typically used vectorial model and first Monte Carlo models, this assumes the coma is collisionless for the daughter products. In the case of a highly collisional coma, the ejection velocity will be collisionally quenched; however, in this case, another simplification of the Haser model, a constant radial outflow speed, will most certainly be violated. Such was the case for C/1995 O1 Hale-Bopp (*Combi*, 2002). If v_p and v_e are the parent outflow and daughter ejection velocities, respectively, and τ_p and τ_d are the parent and daughter lifetimes, respectively, then the Haser equivalent parent and daughter scale lengths, γ_{pH} and γ_{dH}, can be calculated from the following set of relationships. If we make the definitions that

$$\tan\delta \equiv \frac{v_p}{v_e} \qquad \mu \equiv \frac{\gamma_p}{\gamma_d} \qquad \mu_H \equiv \frac{\gamma_{pH}}{\gamma_{dH}}$$

where $\gamma_p = v_p\tau_p$ and $\gamma_d = v_d\tau_d$, $v_d = (v_p^2 + v_e^2)^{1/2}$ and v_e is the isotropic ejection velocity of the daughter upon production with respect to the original center-of-mass of the parent. The equations (5)–(7) can be used to go back and forth between the vectorial values and the Haser equivalent scale lengths for a given ratio of the parent outflow to daughter ejection velocity.

$$\gamma_d^2 - \gamma_p^2 = \gamma_{pH}^2 - \gamma_{dH}^2 \tag{5}$$

$$\mu_H = \left(\frac{\mu + \sin\delta}{1 + \mu\sin\delta}\right)\mu \tag{6}$$

or equivalently

$$\mu = (\mu_H - 1)\frac{\sin\delta}{2}\left[(\mu_H - 1)^2\frac{\sin^2\delta}{4} + \mu_H\right]^{1/2} \tag{7}$$

In order to complete the substitution in the density or column density expressions for Haser's model, the effective radial velocity, v, in the denominators in the first terms on the righthand sides of equations (3) and (4) must be the Haser equivalent daughter expansion velocity v_{dH}, given as

$$v_{dH} = v_d\frac{\gamma_{dH}}{\gamma_d} \tag{8}$$

2.4. Three-Generation Haser-like and Monte Carlo Models

Observations of spatial profiles of cometary C_2 have often had some difficulty in being interpreted either with a simple two-generation Haser model or with a model accounting for secondary product ejection velocities (see *Feldman et al.*, 2004). The problem is that the innermost part of the profile has often been seen to be even flatter than is possible to be reproduced using any Haser model. A Haser model that has the flattest inner profile (see Fig. 1) is one in which the parent and daughter scale lengths are equal. Models that allow for secondary product velocities actually exacerbate the problem, because the effect of isotropic ejection of daughters makes the radial parent scale length seem to be shorter, thus making the profile look like Haser models with unequal scale lengths.

Observations of C_2 in Comet Halley (*O'Dell et al.*, 1988; *Wyckoff et al.*, 1988; *Fink et al.*, 1991) were noteworthy in showing the flattening of the inner profile. *Fink et al.* (1991) chose to use the standard two-generation Haser model but fitted models with equal parent and daughter scale lengths. *Wyckoff et al.* (1988) suggested that either C_2 was produced as a granddaughter species or by grains instead of simple photodissociation. These had been suggested much earlier by both *Jackson* (1976) and *Cochran* (1985). *O'Dell et al.* (1988) pursued the granddaughter species idea by performing model calculations for a three-generation Haser-type model, generalizing the earlier work of *Yamamoto* (1981) and *Festou* (1981a), who applied a three-generation Haser-type model to comet-centered, circular aperture observations.

The three-generation Haser-type model yields an expression for the column density profile that was given by *O'Dell et al.* (1988) as

$$N(\rho) = \frac{Q_1}{2\pi V_3\rho}[AH(\beta_1\rho) + BH(\beta_2\rho) + CH(\beta_3\rho)] \tag{9}$$

where Q_1 is the global production rate of the observed (granddaughter) species; V_3 is its radial expansion velocity; and β_1, β_2, and β_3 are the inverse scale lengths of the original parent, the intermediate daughter, and the observed granddaughter species, respectively. The quantities A, B, and C are given as

$$A = \frac{\beta_1\beta_2}{[(\beta_1 - \beta_2)(\beta_1 - \beta_3)]} \qquad B = \frac{-A(\beta_1 - \beta_3)}{(\beta_2 - \beta_3)}$$

$$C = \frac{-B(\beta_1 - \beta_2)}{(\beta_1 - \beta_3)}$$

The function H(x) is given by

$$H(x) = \frac{\pi}{2} - \int_0^x K_0(y)dy$$

where $K_0(y)$ is again the zero-order modified Bessel function of the second kind.

One of the simplifications of this model is the same as in the two-generation model of *Haser* (1957), i.e., that at each step the daughter and granddaughter species are assumed to continue to travel in a purely radial direction. *Combi and Fink* (1997) addressed this by using Monte Carlo techniques to explore models that include ejection velocities for the dissociation product at either or both dissociation steps. This was done for the purpose of putting limits on prospective original parents of observed C_2, but would also explain the flat inner profile. Figure 2 shows one of their figures showing comparisons of four otherwise comparable models: (1) a two-generation Haser model where the parent and daughter scale lengths are equal (this

one gives the flattest possible inner profile); (2) a Monte Carlo simulation of a three-generation model that is otherwise similar to that which could be produced using equation (12) of *O'Dell et al.* (1988) and could, as they showed, give a reasonable reproduction of the spatial profile of C_2 in Comet 1P/Halley, (3) a Monte Carlo simulation of a three-generation model where the ejection velocity of the product at each dissociation equals half of the parent outflow velocity; and (4) a similar model but where both ejection velocities are equal to the parent velocity. They also provided other sets of comparisons. The result was that the addition of daughter and granddaughter ejection velocities causes three-generation models to be unable to produce flat inner profiles, and makes them essentially similar to some two-generation Haser or vectorial model with different parameters. Conversely for C_2, the results implied that if C_2 is a third-generation species, then the total of the ejection velocities for each dissociation step had to be less than about half the parent outflow velocity. In typical photodissociations there are 2–4 eV of excess energy that would impart an ejection velocity of 1 km s^{-1} or more to the heavier dissociation product that would contain C_2, or C_2 itself.

2.5. Grains as Gas Sources

Volatile grains have been suggested as possible sources for observed species in comets for a long time (see review by *Festou*, 1999). The idea regained new interest with the discovery of CN and C_2 jets in Comet 1P/Halley (*A'Hearn et al.*, 1986) and extended sources for CO and H_2CO (*Eberhardt et al.*, 1987; *Meier et al.*, 1993; *DiSanti et al.*, 2001), and from observations of a substantial distributed icy grain source from C/1996 B2 (Hyakutake) (*Harris et al.*, 1997; *Harmon et al.*, 1997). *Delsemme and Miller* (1971) presented a model calculation for the production of an observed gas from grains, analogous to the two-generation model of *Haser* (1957), except where evaporating grains serve as the parent of the observed species rather than photodissociating parent molecules.

This model, however, suffers from the same limitation as Haser's model, which is that the observed daughter species is restricted to perfectly radial motion. Clearly molecules released from evaporating grains would have some nonnegligible velocity owing to the temperature of the grains. The effect would be even more important if the grains were dark organic-rich CHON grains, which are expected to reach and sublimate at temperatures higher than the blackbody temperature for some given distance from the Sun (*Lamy and Perrin*, 1988). This temperature would generally be much higher than the sublimation temperature of water for heliocentric distances less than about 3 AU. *Combi and Fink* (1997) introduced a CHON grain halo model for gas species that generalizes the icy grain halo model of *Delsemme and Miller* (1971) in the same way as the vectorial (*Festou*, 1981a) and Monte Carlo (*Combi and Delsemme*, 1980) models generalized Haser's model, i.e., by allowing for isotropic ejection of the daughter species

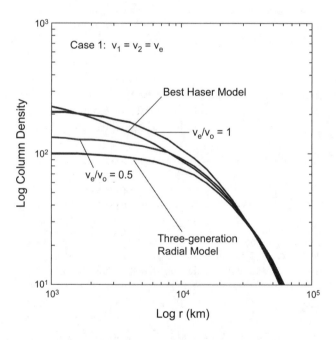

Fig. 2. Haser and three-generation model spatial profiles with non-zero daughter ejection speeds. Case 1 from *Combi and Fink* (1997) compares a general three-generation dissociation model with ejection speeds at each dissociation. The two ejection speeds are equal to each other and either equal to the parent outflow speed, v_0, or to half of it. The three-generation radial model is one where there is zero ejection speed for the photodissociation products at either step. The "Best Haser Model" is one where the daughter and parent scale lengths are equal, which results in the flattest inner profile that is most like the observed inner profile of C_2 in Comet 1P/Halley. When the daughter ejection speed is nonnegligible, a three-generation model does not produce a flat inner profile.

TABLE 1.   Photochemical branching and exothermic velocities for H_2O vapor.

Solar Ultraviolet Wavelength Range	Reaction	Product Exothermic Velocities (km s⁻¹)		Photodestruction Rate* (10^{-6} s⁻¹) [*Branching Ratio*]	
				(Quiet Sun)	(Active Sun)
1357–1860 Å	$H_2O + h\nu \rightarrow H + OH(X^2\Pi)$	18.5 (H)[†]	1.09 (OH)[†]	4.84 *[0.465]*	5.36 *[0.380]*
	$\rightarrow H_2 + O(^1D)$	<13.7 (H_2)	<1.71 (O)	0.05 *[0.005]*	0.05 *[0.004]*
1216 Å	$H_2O + h\nu \rightarrow H + OH(X^2\Pi)$	17.2 (H)[†]	1.01 (OH)[†]	3.02 *[0.291]*	4.53 *[0.321]*
	$\rightarrow H + OH(A^2\Sigma^+)$	5 (H)[‡]	0.3 (OH)[‡]	0.35 *[0.033]*	0.52 *[0.037]*
	$\rightarrow H_2 + O(^1D)$	<15 (H_2)[‡]	<1.8 (O)[‡]	0.43 *[0.042]*	0.65 *[0.046]*
	$\rightarrow H + O + H$	<7.4 (2H)[§]	<0.87 (O)[§]	0.52 *[0.050]*	0.78 *[0.055]*
984–1357 Å (excluding 1216 Å)	$H_2O + h\nu \rightarrow H + OH(X^2\Pi)$	<37–27 (H)[¶,**]	<2.2–1.6 (OH)[¶,**]	0.30 *[0.028]*	0.45 *[0.032]*
	$\rightarrow H + OH(A^2\Sigma^+)$	<25–0 (H)[**]	<1.5–0 (OH)[**]	0.03 *[0.003]*	0.05 *[0.004]*
	$\rightarrow H_2 + O(^1D)$	<22–14 (H_2)[**]	<2.7–1.7 (O)[**]	0.04 *[0.004]*	0.07 *[0.005]*
	$\rightarrow H + O + H$	<17–0 (2H)[**]	<2.0–0 (O)[**]	0.05 *[0.005]*	0.08 *[0.005]*
<984 Å	$H_2O + h\nu \rightarrow$ neutral products	—	—	0.25 *[0.024]*	0.41 *[0.029]*
	$\rightarrow$ ionization products	—	—	0.52 *[0.050]*	1.16 *[0.082]*
Total:				10.40 *[1.000]*	14.11 *[1.000]*

* From S. A. Budzien (personal communication, 2003); updated and slightly different than *Budzien et al.* (1994).

[†] Average velocity from *Crovisier* (1989) and similar to *Wu and Chen* (1993).

[‡] Average velocity from *Festou* (1981b).

[§] Determined using the excess energy of 13 kcal mole⁻¹ from *Slanger and Black* (1982).

[¶] Similar to the values of 35–25 (H) and 2.0–1.2 (OH) from *Festou* (1981b).

[**] Upper limits determined for the bounding lower and higher (or threshold) wavelengths.

from the radially outflowing parent grains (in this case). The flat inner profiles of cometary C_2 have been the main subjects of three-generation and grain-halo models, and can be reproduced by either three-generation or grain-source models with appropriate parameters (*Combi and Fink*, 1997).

3.   WATER-DOMINATED OUTFLOW

Water dominates up to ~90% of the volatile species that outflow from the cometary nucleus within ~3–4 AU from the Sun. During this outflow, the H_2O parent molecule is destroyed primarily through photodissociation, and to a much smaller extent, through photoionization and also dissociation and ionization interactions with solar wind ions and electrons. The destruction of H_2O creates secondary or daughter molecular products, OH and H_2, atomic products, O and H, and various ion species in the inner and outer comet coma. The neutral product species are detected remotely through solar resonance fluorescence for OH and H and also through dissociative excitation for O. The dissociation rates of H_2O, OH, and H_2 and the imparted exothermic energy to their daughter products are of fundamental importance in determining the spatial decay rates of these parent and daughter species in the coma and in determining the basic background density and temperature structure for all species in the coma. The photodestruction rates of H_2O, OH, and H_2 are variable and depend upon the helio-

centric distance and radial velocity of the comet as well as the wavelength-dependent short-term modulations and longer-term 11-year periodic variations in solar radiation.

3.1.   Water Photochemistry

The photochemistry and photochemical kinetics for H_2O in comets have been studied in detail by a number of investigators (*Huebner and Carpenter*, 1979; *Openheimer and Downey*, 1980; *Festou*, 1981a,b; *Huebner*, 1985; *Krasnopolsky et al.*, 1986a,b; *Allen et al.*, 1987; *Crovisier*, 1989; *Huebner et al.*, 1992; *Wu and Chen*, 1993; *Cochran and Schleicher*, 1993; *Budzien et al.*, 1994). The calculated photodestruction rates for different products have become more accurately determined with time because of improvements in cross sections and in the description of the solar flux spectrum and its time variations, although a number of uncertainties still remain. A summary of the relevant photodestruction reactions, product exothermic velocities, reaction rates, and branching ratios for H_2O is given in Table 1 and is divided into four wavelength ranges.

The first wavelength range is bounded above by the effective onset wavelength at 1860 Å for a nonzero photoabsorption cross section, and below by the threshold wavelength at 1357 Å for which OH can be produced in excited electronic states. In this first wavelength range, the measured relative branching ratios of the rates for the first and second reaction are 0.99 and 0.01 respectively (*Stief et al.,* 1975), where in the second reaction $O(^1D)$ is produced since the production of $O(^3P)$ is spin forbidden.

The second wavelength range is only for the large Lyman-α peak at 1216 Å in the solar spectrum. For the Lyman-α peak, the measured relative branching ratios of the rates for the four reactions are 0.70, 0.08, 0.1, and 0.12 respectively (*Slanger and Black,* 1982), where the geometry and mechanisms for the three-body dissociation in the fourth reaction are not known (see *Wu and Chen,* 1993).

The third wavelength range is from 1357 Å to the threshold wavelength at 984 Å for ionization of water and excludes the solar spectrum contribution of the Lyman-α peak. In the third wavelength range, the relative branching ratios for the four reactions are not well known and are assumed in Table 1 to be the same as for the Lyman-α wavelength.

The fourth wavelength range is below the ionization threshold at 984 Å and includes a contribution from a dissociation cross section with unknown neutral products and a contribution from an ionization cross section with a mix of ionized and not-well-defined neutral products. The cross section for photoionization becomes more important than that for photodissociation below ~875 Å (see *Wu and Chen,* 1993).

The H_2O photodestruction rates for the four wavelength ranges depend upon an accurate specification of the appropriate wavelength-dependent cross sections and the variable solar flux. The photodestruction rates adopted in Table 1 are for the most recent calculations of S. A. Budzien (personal communication, 2003), which are based upon updated information for the photodissociation cross section in the first wavelength range and the solar flux at 1 AU and provide improved rates that differ only slightly from those calculated earlier by *Budzien et al.* (1994). Depending upon quiet and active Sun conditions, the first wavelength range contributes ~47.0–38.4% of the total photodestruction rate, the Lyman-α solar spectra peak contributes ~41.6–45.9%, the third wavelength range contributes 4.0–4.6%, and the fourth wavelength range contributes 7.4–11.1%. The corresponding branching ratios for the individual reactions in the four wavelength intervals are also given and are determined by the relative branching ratios discussed earlier. The total photodestruction rate of H_2O has a value in Table 1 for quiet and active Sun conditions, respectively, of $1.04 \pm 0.123 \times 10^{-5}$ s^{-1} (lifetime ~0.96×10^5 s) and $1.411 \pm 0.175 \times 10^{-5}$ s^{-1} (lifetime ~0.71×10^5 s).

For the different wavelength ranges in Table 1, the product exothermic velocities depend upon the details of the photochemical kinetics and are distributed nonuniformly in discrete velocity groups that are constrained between 0 and 2.7 km s^{-1} for O, between 0 and 2.2 km s^{-1} for OH, between 0 and 22 km s^{-1} for H_2, and between 0 and 37 km s^{-1} for H. The product exothermic velocities and their corresponding velocity distributions are generally not well known because of the lack of cross sectional information as a function of wavelength for the internal energy of the molecular

products [$OH(X^2\Pi)$, $OH(A^2\Sigma^+)$, and H_2] that determines the available excess kinetic energies for the products. Only upper limits are therefore given for many of the product exothermic velocities in Table 1. For different assumptions regarding the product abundances and the configuration of the internal energy of OH, examples are given in *Wu and Chen* (1993) for the velocity distributions of H and OH that bound their behaviors.

3.2. Ion and Electron Impact Water Chemistry

The solar wind plays a minor role in the destruction of water compared to photochemistry but plays an important role in the ion chemistry of the coma, which also involves many other atomic and molecular ions (*Häberli et al.,* 1997). A summary of the most relevant solar wind proton and electron reactions with H_2O and the values of the corresponding destruction rates is given in Table 2 for a nominal solar wind flux of 3×10^8 cm^{-2} s^{-1} at 1 AU as reported by *Budzien et al.* (1994). The largest water destruction rate is for charge exchange with protons, followed by electron impact ionization, proton impact ionization, proton impact dissociation, and electron impact dissociation. At 1 AU, the charge exchange rate is comparable to the photoionization rate of ~4.1×10^{-7} s^{-1} for quiet Sun conditions, but is smaller than the photoionization rate of ~1.04×10^{-6} s^{-1} for active Sun conditions (*Huebner et al.,* 1992). These destruction rates are not well established due to uncertainties in cross sections and furthermore vary significantly due to changing solar wind conditions upstream of the comet bow shock and to spatially changing properties of the ion and electron distributions in the cometosheath, outer coma, inner coma, and ion tail. The electron impact rate of 3×10^{-7} s^{-1} in Table 2 is, for example, a typical value for solar wind electrons in the outer coma, but may be considerably larger, ~1×10^{-6} s^{-1}, for electrons heated in the cometosheath, where it may be comparable to the active Sun water photoionization rate, and even larger in the inner coma where stagnation flow and photoelectrons are also present (*Cravens et al.,* 1987). The total water destruction rate for the nominal values in Table 2 is 1.08×10^{-6} s^{-1}. Since this total solar wind destruction rate varies throughout the coma, it is

TABLE 2. Solar wind destruction for H_2O vapor.

Reaction	Destruction Rate (10^{-7} s^{-1})
$H^+ + H_2O \rightarrow H + H_2O^+$	5.7
$\rightarrow H^+ + H_2O^+ + e$	0.75
$\rightarrow$ dissociated neutral products	0.27
$e + H_2O \rightarrow H_2O^+ + e + e$	~3 (variable)
$\rightarrow$ dissociated neutral products	1.1

Destruction rates for a nominal solar wind flux of 3×10^8 cm^{-2} s^{-1} from *Budzien et al.* (1994).

TABLE 3.   Photochemical branching and exothermic velocities for OH.

| OH Predissociation | | Reaction | Product Exothermic Velocities* (km s⁻¹) | Photodestruction Rate (10⁻⁶ s⁻¹) | | | |
Wavelength	State			(Quiet Sun)		(Active Sun)	
2160 Å	A²Σ⁺ (v' = 2)	OH + hv → O (³P) + H	8 (H) 0.5 (O)	3.0–6.1†	2.5–5.6‡	3.0–6.1†	2.5–5.6‡
2450 Å	A²Σ⁺ (v' = 3)	OH + hv → O (³P) + H	11 (H) 0.7 (O)	0.5†		0.5†	
1400–1800 Å	1²Σ	OH + hv → O (³P) + H	22–26 (H) 1.4–1.6 (O)	1.4†	1.4§	2.3†	1.58§
1216 Å	1²Δ	OH + hv → O (¹D) + H	26.3 (H) 1.6 (O)	0.3†	0.38§	0.75†	0.57§
1216 Å	B²Σ	OH + hv → O (¹S) + H	17.1 (H) 1.1 (O)	0.05†	0.05§	0.13†	0.08§
1216 Å	2²Π–3²Π	OH + hv → O (¹D) + H	26.3 (H) 1.6 (O)	0.10†	0.24§	0.25†	0.39§
<1200 Å	D²Σ	OH + hv → O (³P) + H	22 (H) 1.4 (O)	<0.01†	—	<0.01†	—
<958 Å		OH + hv → ionized products	— —	0.25§		0.47§	

* Product exothermic velocities from *van Dishoeck and Dalgarno* (1984) with slight updates from *Combi* (1996).
† Rates from *van Dishoeck and Dalgarno* (1984).
‡ Combined A²Σ⁺ (v' = 2 and v' = 3) rate from *Schleicher and A'Hearn* (1988).
§ Rates from *Budzien et al.* (1994).

therefore likely to alter the overall water lifetime by somewhat less than 10%.

3.3.   Daughter Product Energetics and Branching Ratios in the Coma

The photochemistry and photochemical kinetics for OH have been studied in detail by a number of investigators (*Schleicher and A'Hearn,* 1982, 1988; *van Dishoeck and Dalgarno,* 1984; *Huebner et al.,* 1992; *Budzien et al.,* 1994). A summary of the relevant OH predissociation states and their corresponding dissociation reactions, product exothermic velocities, and dissociation rates, as well as OH photoionization rate, is given in Table 3 and is divided into individual contributions for various wavelengths. For the first two wavelengths at 2160 Å and 2450 Å, OH dissociates from the A²Σ⁺ state after absorption of solar radiation during the fluorescence process. The dissociation rates for these two wavelengths do not vary between quiet and active solar conditions because at these longer wavelengths the solar flux is highly stable. These two contributions to the rate, however, vary with heliocentric radial velocity, with the major contribution from the A²Σ⁺ (v' = 2) predissociation state. For the combined A²Σ⁺ (v' = 2 and v' = 3) rates, this variation is about a factor of 2 with a calculated value of 3.5–6.6 × 10⁻⁶ s⁻¹ and of 2.5–5.6 × 10⁻⁶ s⁻¹, respectively, determined earlier by *van Dishoeck and Dalgarno* (1984) and more recently by *Schleicher and A'Hearn* (1988). This variation is caused because the dissociation that occurs at discrete energy levels above the dissociation threshold is produced by absorption of solar radiation, which has a solar radiation-pumping rate that varies with heliocentric radial velocity through the Swings effect (*Jackson,* 1980).

The dissociation rates for the next five wavelength entries are produced through direct photodissociation of OH in the 1²Σ, 1²Δ, B²Σ, 2²Π–3²Π, and D²Σ repulsive states. These photodissociation rates, as well as that for the <958-Å wavelength entry for photoionization of OH, are listed for quiet and active Sun conditions, since the UV and extreme UV solar flux for these lower wavelengths varies significantly with solar cycle. The product exothermic velocities are distributed nonuniformly between 8 and 26.3 km s⁻¹ for H and between 0.5 and 1.6 km s⁻¹ for O. By adding together the second column entries in Table 3 for the newer rates and the last row entry rate, the total OH photodestruction rate then varies with heliocentric velocity from 4.82 to 7.92 × 10⁻⁶ s⁻¹ (lifetime of 1.26–2.05 × 10⁵ s) for quiet Sun conditions with more typical values of ~5.2–6.3 × 10⁻⁶ s⁻¹ (lifetime of ~1.6–1.9 × 10⁵ s) and from 5.59 to 8.69 × 10⁻⁶ s⁻¹ (lifetime of 1.15–1.79 × 10⁵ s) for active Sun conditions with more typical values of ~6.0–7.1 × 10⁻⁶ s⁻¹ (lifetime of ~1.4–1.7 × 10⁵ s). The OH photodestruction rate (lifetime) is then about a factor of 2 smaller (larger) than the photodestruction rate (lifetime) for water.

The photochemistry and photochemical kinetics for H₂, a minor dissociative species (5.1–5.5%) of water in Table 1, are summarized in Table 4, as given by *Huebner et al.* (1992). A little more than half the H₂ destruction produces two hydrogen atoms with exothermic velocities of 28 km s⁻¹ and 6.3–6.5 km s⁻¹, about one-third produces H₂⁺, and somewhat less than 10% produces H + H⁺. The total H₂ photodestruction rate for all the reactions in Table 4 is 0.146 × 10⁻⁶ s⁻¹ for quiet Sun conditions and 0.33 × 10⁻⁶ s⁻¹ for active Sun conditions.

3.4.   Solar Activity Variations

The primary time variability in the dissociation and ionization rates for water and its daughter products noted above arises from changes that occur in the extreme UV (λ < 1000 Å) and far UV (1000 Å < λ < 2000 Å) fluxes from the Sun. These time variations occur during the 11-year periodic solar cycle because of the variation in the number of sunspots that create enhanced extreme UV and far UV fluxes from the Sun. The distribution of sunspots on the rotating solar disk also introduces short-term 27-day modulations of the solar flux, which can be significant compared to quiet Sun conditions. The photochemistry rates sum-

TABLE 4. Photochemical data for H_2.

Reaction	Product Exothermic Velocities (km s⁻¹)	Photodestruction Rate (10^{-6} s⁻¹)	
		(Quiet Sun)	(Active Sun)
$H_2 + h\nu \rightarrow H(1s) + H(1s)$	28 (2H)	0.048	0.11
$H_2 + h\nu \rightarrow H(1s) + H(2s,2p)$	6.3–6.5 (2H)	0.034	0.082
$H_2 + h\nu \rightarrow H_2^+ + e$	—	0.054	0.11
$H_2 + h\nu \rightarrow H + H^+ + e$	—	0.0095	0.028

Product exothermic velocities and rates from *Huebner et al.* (1992).

marized in Tables 1, 3, and 4 only characterize the basic changes between the quiet and active Sun conditions and thus provide only a basic picture that may be altered significantly by short-term time variations of the solar radiation that are poorly known for observations of particular comets, except for those rare cases where the Sun-comet line crosses near one of the solar spectrum monitors [e.g., the Solar and Heliospheric Observatory (SOHO) or Solar Ultraviolet Spectral Irradiance Monitor (SUSIM)].

For convenience, the formulae are given here from *Budzien et al.* (1994) for the total water photodestruction rate $(\tau_{H_2O}^{tot})^{-1}$ and the total OH photodestruction rate $(\tau_{OH}^{tot})^{-1}$ as a function of solar activity indices $F_{10.7}$ (the 10.7-cm flux), its 81-day average, $\langle F_{10.7} \rangle$, in solar flux units 10^{-22} W m⁻², the solar He I $\lambda 10830$ equivalent width in mÅ, the solar wind flux $\langle nv \rangle_{sw}$ in cm⁻² s⁻¹, and the heliocentric velocity dependent predissociation rate given in tabular and graphical form by *Schleicher and A'Hearn* (1988), $[\tau_{OH}^{prediss}(\dot{r})]^{-1}$

$$(\tau_{H_2O}^{tot})^{-1} = \begin{array}{l} 5.868 \times 10^{-6} + 1.49 \times 10^{-9} \, F_{10.7} \\ + 2.08 \times 10^{-9} \, \langle F_{10.7} \rangle + 9.587 \times 10^{-8} \, W_{10830} \\ + 4.1 \times 10^{-15} \, \langle nv \rangle_{sw} \qquad \text{(in s}^{-1}\text{)} \end{array}$$

$$(\tau_{OH}^{tot})^{-1} = \begin{array}{l} 1.479 \times 10^{-6} + 5.8 \times 10^{-10} \, F_{10.7} \\ + 8.1 \times 10^{-10} \, \langle F_{10.7} \rangle + 1.678 \times 10^{-8} \, W_{10830} \\ + 3.7 \times 10^{-15} \, \langle nv \rangle_i + [\tau_{OH}^{prediss}(\dot{r})]^{-1} \qquad \text{(in s}^{-1}\text{)} \end{array}$$

4. PHYSICS-BASED MODELS

4.1. Collisional Boltzmann Equation for Rarefied Gases

The appropriate description for the physical state of any one of up to s species, p, in a multispecies dilute gas can be described by the Boltzmann equation (*Gombosi*, 1994), which is given as

$$\frac{\partial}{\partial t}(f_p) + \vec{c}_p \cdot \frac{\partial}{\partial \vec{r}}(f_p) + \vec{F} \cdot \frac{\partial}{\partial \vec{c}}(f_p) =$$

$$\sum_{q=1}^{s} \int_{-\infty}^{+\infty} \int_{0}^{4\pi} (f_p^* f_{lq}^* - f_p f_{lq}) c_{rpq} \sigma_{pq} d\vec{\Omega} d\vec{c}_{lq} \qquad (10)$$

where $f_p \equiv f_p(\vec{r}, \vec{c}, t)$ and $f_d \equiv f_d(\vec{r}, \vec{c}, t)$ are the full phase space

velocity distribution functions for species p and q, c_q is the velocity, r is the spatial coordinate, $\vec{F}$ represents external forces, c_{rpq} is the magnitude of the relative velocity between particle p and q, Ω is the solid angle, the asterisks (*) indicate postcollision particles, the lq subscript refers to the scattering target particles, and σ_{pq} is the total collision cross section between species p and q (which can in general be velocity dependent). In the collision integral on the right-hand side of the equation, f* represents additions of particles, scattered into the region of velocity space in question, i.e., between $\vec{c}$ and $\vec{c} + d\vec{c}$, whereas f without the asterisk represents scattering out of that region.

The Boltzmann equation makes neither assumptions nor any restrictions as to the form of the distribution functions. Various velocity moment expansions of the integrals over the distribution functions yield equations of conservation of mass, momentum, and energy. The assumption of local thermodynamic equilibrium (LTE) and the resulting Maxwell-Boltzmann distribution functions yield the Euler equations for fluid dynamics. The Chapman-Enskog theory is a perturbation from the Maxwell-Boltzmann distribution, which yields the Navier-Stokes equations with viscosity and heat flux. For the intended purpose of studying tenuous atmospheres and transitions from an LTE fluid to free-flow, the basic assumptions for these approximations are inherently violated, and therefore a potentially arbitrary form for the distribution functions remains to be considered.

Direct analytical or even numerical solutions for the Boltzmann equation have been done for numbers of idealized simple problems (*Fahr and Shizgal*, 1981), for reduced spatial or velocity dimensions, or for specific analytical forms of the collision integral [see the textbook by *Gombosi* (1994) for an up-to-date treatment of gaskinetic theory]. Even one-dimensional flow problems yield three-dimensional distribution functions, one spatial and two velocity. For similar reasons the Direct Simulation Monte Carlo (DSMC) method has also been adopted to treat multidimensional, multispecies gas flows for tenuous planetary applications. In the computational fluid dynamics community, DSMC is the method of choice for validating such fluid approaches, such as solution of the Navier-Stokes equations (*Bird*, 1994). For example, the modelers of shock structures in one-dimension have typically "resorted" to Monte Carlo simulation for numerical experiments.

4.2. Time-Dependent Hydrodynamics

If one multiplies the Boltzmann equation, alternatively, by 1, v, and v² (where v is the gas velocity) and then integrates the resulting equations over the vector velocity, the resulting equations correspond to conservation laws of mass continuity, momentum, and energy (*Gombosi et al.,* 1986). These equations can be closed and combined with the ideal gas law to provide a useful way to understand the transport and energy of the gas in the coma. The energy and momentum equations obtained can be manipulated to yield an equation for gas pressure instead of energy. Dust-gas physics can be added in the form of dust continuity and momentum equations with standard collisional coupling terms that assume the gas mean free path is much larger than the dust size and that gas molecules accommodate to the dust surface temperature upon a collision. Since there is no random internal energy component for dust particles, no pressure/energy equation is required. For detailed derivations see the papers, for example, by *Gombosi et al.* (1986) and *Crifo et al.* (1995). The coupled system for a single-fluid gas and multicomponent dust (i.e., dust size distribution) can be written as

$$\frac{\partial \rho}{\partial t} + \nabla \cdot (\rho \mathbf{u}) = \frac{\delta \rho}{\delta t} \tag{11}$$

$$\rho \frac{\partial \mathbf{u}}{\partial t} + \rho (\mathbf{u} \cdot \nabla) \mathbf{u} + \nabla p = -\mathbf{F} \tag{12}$$

$$\frac{1}{\lambda - 1} \frac{\partial p}{\partial t} + \frac{1}{\gamma - 1} (\mathbf{u} \cdot \nabla) p +$$
$$\frac{\gamma}{\gamma - 1} p (\nabla \cdot \mathbf{u}) = -Q_{gd} + Q_{ph} - Q_{IR} \tag{13}$$

$$\frac{\partial \rho_i}{\partial t} + \nabla \cdot (\rho_i \mathbf{u}_i) = \frac{\delta \rho_i}{\delta t} \quad i = 1, ..., N \tag{14}$$

$$\rho_i \frac{\partial \mathbf{v}_i}{\partial t} + \rho_i (\mathbf{v}_i \cdot \nabla) \mathbf{v}_i = \mathbf{F}_i \quad i = ..., N \tag{15}$$

where ρ is the gas mass density, $\mathbf{u}$ is the gas velocity, p is the gas pressure, and $\mathbf{v}_i$ and ρ_i are the velocity and mass density for dust particles of radius a_i. The righthand sides of all the equations contain the various source terms. The term

$$\frac{\delta \rho}{\delta t}$$

is the gas production source rate, and

$$\frac{\delta \rho_i}{\delta t}$$

is the dust production source rate for particles of radius, a_i. F is the gas-dust drag force, which is related to the forces on the individual particle size populations as

$$\mathbf{F} = -\sum_{i-1}^{N} \mathbf{F}_i$$

where the size dependent force is given by

$$\mathbf{F}_i = \frac{3 \rho_i}{4 a_i \rho_{bi}} p C_D' \mathbf{s}_i$$

where

$$\mathbf{s}_i = \frac{\mathbf{u} - \mathbf{v}}{\sqrt{2kT/m}}$$

k is the Boltzmann constant, T is the gas temperature and m is the gas mean molecular mass. The accommodation of gas via collisions with hotter dust yields the dust-gas heat exchange rate Q_{gd}, which is given as

$$Q_{gd} = \frac{\gamma + 1}{\gamma} \rho C_p u \sum_{i=1}^{N} (T_i^{rec} - T_i) St_i' \tag{16}$$

Here C_p is the gas heat capacity at constant pressure and the rest of the coefficients can be defined under the assumption of diffusive reflection such that

$$C_D' = \frac{2\sqrt{\pi}}{3} \sqrt{\frac{T_i}{T}} + \frac{2s_i^2 + 1}{s_i^2 \sqrt{\pi}} e^{-s_i^2} +$$
$$\frac{4s_i^4 + 4s_i^2 - 1}{2s_i^3} \operatorname{erf}(s_i) \tag{17}$$

$$T_i^{rec} = \left(1 + \frac{\gamma - 1}{\gamma + 1} s_i^2 R_i' \right) T \tag{18}$$

$$R_i' = \frac{\left(2s_i + \dfrac{1}{s_i} \right) \dfrac{e^{-s_i^2}}{\sqrt{\pi}} + \left(2s_i^2 + 2 - \dfrac{1}{s_i^2} \right) \operatorname{erf}(s_i)}{s_i \dfrac{e^{-s_i^2}}{\sqrt{\pi}} + \left(s_i^2 + \dfrac{1}{2} \right) \operatorname{erf}(s_i)} \tag{19}$$

$$St_i' = \frac{e^{-s_i^2}}{8 s_i \sqrt{\pi}} + \frac{1}{8} \left(1 + \frac{1}{2s_i^2} \right) \operatorname{erf}(s_i) \tag{20}$$

Finally, ρ_d is the bulk mass density of dust particles of radius a_i, T is the gas temperature, and T_i is the dust temperature, assumed in equilibrium with solar radiation. A temperature-dependent value can be used for C_p to account for some of the effects of divergence from LTE, especially at low temperature, where all internal degrees of freedom are not available. Otherwise the other intermediate quantities are s_i, the relative Mach number between gas and dust; C_D', the dust-gas drag coefficient; T_i^{rec}, the recovery temperature; R_i', the heat transfer function; and St_i', the Stanton number. Most of the dust-gas drag physics comes from the formulation by *Finson and Probstein* (1968) with later corrections discussed by *Wallis* (1982), *Kitamura* (1986), and *Körösmezey and Gombosi* (1990).

4.3. One-Dimensional Spherical Steady-State Equations

The general three-dimensional time-dependent Euler equations of hydrodynamics can be reduced to a steady-state one-dimensional version, which is very useful for understanding the basic physical state of the outflowing coma. Again, a detailed description of the derivation was given in the review paper by *Gombosi et al.* (1986). These are

$$\rho = \frac{Qm}{4\pi u r^2} \tag{21}$$

$$\rho u \frac{du}{dr} = -\frac{dp}{dr} - F \tag{22}$$

$$\rho_i v_i \frac{dv_i}{dr} = F_i \tag{23}$$

$$\frac{1}{r^2} \frac{d}{dr} \left[\rho u r^2 \left(\frac{u^2}{2} + \frac{\gamma}{\gamma - 1} \frac{p}{\rho} \right) \right] = S - L + Q_{gd} \tag{24}$$

Assuming the normal ideal gas law

$$p = \frac{\rho k T}{m}$$

the above system of equations can be solved from an initial boundary condition, e.g., at or near the surface of the nucleus. The main difficulty in implementing the steady-state version of the hydrodynamic equations is if one specifies the inner boundary at or so close to the surface of the (spherical) nucleus that the gas is subsonic, in which case the transonic transition is an undefined point. This has been addressed either using a shooting method (*Marconi and Mendis,* 1983), or by using the approach of *Gombosi et al.* (1985), which solves explicitly for the conditions at or just above the sonic point, normally only meters above the surface of a spherical nucleus with radius of typically a few kilometers. In this case F_i is purely radial and the same dust-gas drag force as used in the time-dependent equations (11)–(15).

In this case, Q in equation (21) is the simple molecular gas production rate, u is the radial gas velocity, S is the photochemical heating rate, L is the IR cooling rate. Other variables are consistent with the definitions in the previous section of this chapter. The IR cooling rate has been estimated a number of ways over the last three decades, first heuristically by *Shimizu* (1976) and then more explicitly by *Crovisier* (1984, 1989), *Crifo et al.* (1989), and *Combi* (1996). Complicating effects are that water densities are large enough in the inner coma for the IR radiation to be

optically thick in some lines, and that the rotational temperature comes out of equilibrium with the gas kinetic temperature with increasing distance from the nucleus. The optical depth is normally handled with an escape probability based on the estimate of *Huebner and Keady* (1983). An approximate kinetic to rotational energy transfer rate has been incorporated into effective cooling rates by *Crovisier* (1984, 1989). The kinetic simulations by *Combi* (1996) explicitly include a microphysical description of internal rotation energy, which indicates that the earlier approximations are generally reasonable.

4.4. Hybrid Kinetic/Fluid Models

For typical comets over a wide range of gas production rates and heliocentric distances, the photochemical heating rate is not simple to calculate because the coma gas is not in local thermodynamic equilibrium. As discussed in detail in a previous section of this chapter, water photodissociation reactions provide the bulk of excess energy for the coma, with the main dissociation branch

$$H_2O + h\nu \rightarrow H + OH + Energy \tag{25}$$

providing most of the energy. Because there is excess energy after overcoming the chemical bond energy on the order of 2 eV for this reaction and the products H and OH must conserve both energy and momentum, so 17/18 of the energy is imparted to the H atom, producing it with an excess velocity of ~17.5 km s^{-1}. The OH radical, on the other hand, has an excess velocity of 1.05 km s^{-1}. Typically in modeling energy balance and transport in a planetary atmosphere, the gas densities are large enough that one can normally assume that all the photodissociative heating energy happens locally, so the heating rate is just given as the sum of the products of the local species gas density, the excess energy per dissociation, and the dissociation rate. In a comet coma the region where the gas density is large enough for local photochemical heating efficiency to be 100% depends on the overall gas production rate and the expansion velocity of the coma. In addition, because of the dominant ~1/r^2 fall-off of the density, the decrease of photochemical heating efficiency is gradual compared to a planetary atmosphere with an exponential scale height variation in density. It therefore is not possible to define, strictly speaking, a collision zone size for the heating caused by hot superthermal H atoms.

Ip (1983) first addressed the issue of photochemical heating efficiency for the superthermal H atoms and derived an analytical estimate. *Huebner and Keady* (1983) developed an escape probability formalism to treat the escaping superthermal H atoms, and *Marconi and Mendis* (1983) treated two populations of H atoms: a superthermal component and a thermal one to deal with the photochemical heating efficiency. *Combi* (1987) and *Bockelée-Morvan and Crovisier* (1987) performed Monte Carlo calculations for individual superthermal H atoms and modified the photochemical heating rate using the accumulated collisional energy transfer. Generally the heating efficiency increases with larger gas production rates (Fig. 3) because of increased gas densities and at smaller heliocentric distance (Figs. 4 and 5) because of increased photodissociation rates. Figure 6 shows results by *Ip* (1989) for hybrid kinetic/fluid calculations of outflow speeds in Comet 1P/Halleycompared with measurements of propagating CN shells. These results are consistent with otherwise similar comparisons of different datasets by *Combi* (1989).

The complementary part of this approach is to predict the distribution of observed daughter species given a realistic physical description of the inner "parent" coma and the nonequilibrium collisional processes that alter the velocity distribution function of the daughter species. This was first done by *Kitamura et al.* (1985) for hydrogen assuming a constant velocity point source parent coma. A general time-dependent, three-dimensional approach was taken by *Combi and Smyth* (1988a,b), whereby the parent coma was described by a time-variable hybrid/kinetic calculation and applied to explain the type of empirical H-atom velocity distribution found by *Meier et al.* (1976) for observations of the shape of the Lyman-α coma of Comet Kohoutek for two very different sets of conditions. It has been suc-

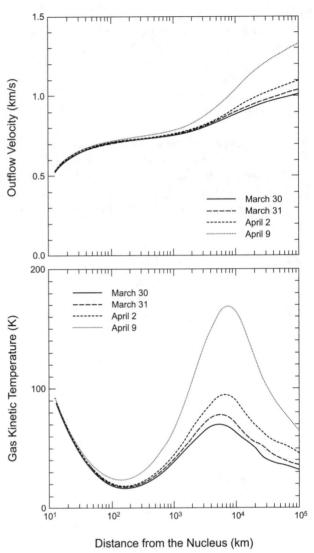

Fig. 4. Radial velocity and gas kinetic temperatures in Comet 1996 B2 (Hyakutake) from models by *Combi et al.* (1999a). Heliocentric distance varies from 0.94 to 0.71 AU. The production rates for the dates beginning with March 30 are 2.0, 2.2, 2.7, and 4.0 × 10^{29} s^{-1}.

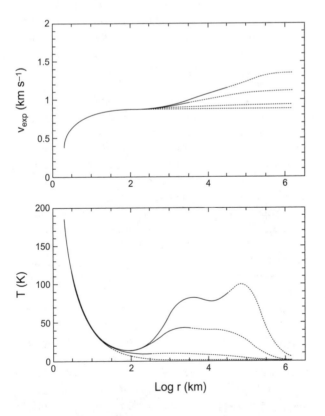

Fig. 3. The effect of coma gas production rate on radial velocity and kinetic temperature in the coma. Model calculations by *Bockelée-Morvan and Crovisier* (1987) from top to bottom for gas production rates of 10^{27}, 10^{28}, 10^{29}, and 10^{30} s^{-1}. The dashed portions of the lines correspond to the outer nonfluid region of the coma.

cessfully applied to a number of comets since then, including 1P/Halley by *Smyth et al.* (1991) and the very extreme case of C/1995 O1 (Hale-Bopp) by *Combi et al.* (2000). As shown in *Feldman et al.* (2004), in addition to being able to reproduce the spatial morphology of the coma, which is sensitive to the velocity distribution because of the large solar radiation pressure acceleration on H atoms, it was applied directly to Goddard High Resolution Spectrograph (GHRS)/Hubble Space Telescope (HST)-measured Doppler line profiles of H Lyman-α in Comet C/1996 B2 (Hyakutake) by *Combi et al.* (1998) and subsequently included in a full-wavelength-dependent radiative transfer calculation by *Richter et al.* (2000). The approach was also generalized for heavy species and applied to OH by *Combi et al.* (1993); O(^{1}D) and NH$_2$ by *Smyth et al.* (1995); C$_2$, CN, NH$_2$, and

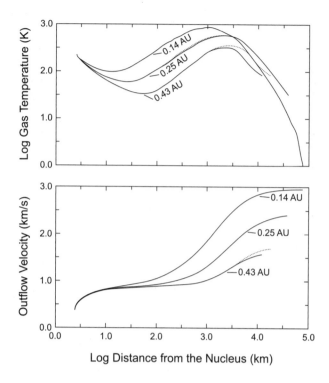

Fig. 5. Model radial velocity and gas kinetic temperatures in Comet Kohoutek by *Combi and Smyth* (1988b) for large production rate and small heliocentric distance.

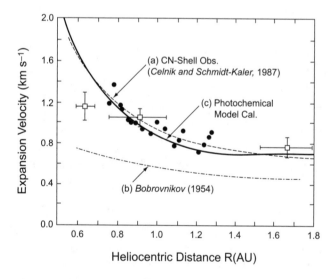

Fig. 6. Model-data comparison of expansion velocity of CN shells in Comet 1P/Halley by *Ip* (1989).

O(^{1}D) by *Combi et al.* (1999a); and then to NH$_2$ by *Tegler et al.* (1992) and *Kawakita and Watanabe* (1998).

4.5. Kinetic Models

The first fully kinetic model for the water-dominated cometary coma that included a fairly complete description of the physics was presented by *Hodges* (1990). This calculation was of the test particle (TP) type. It starts with an

initial assumed description for the distribution function of the gas that serves as a background for test particles of the same gases. In this particular case water molecules are emitted from the model nucleus and dissociation products (OH, H, O, etc.) are produced in a random but physically fair fashion. Collision probabilities are calculated between the test particles and current version of the background distribution functions for all species, and the test particles are scattered as necessary. A new background distribution is accumulated (or partially accumulated) from the state of the test particles in each iteration, which then serves as a new background. The premise is that once the background distribution and the test particle distribution converge to the same state, then that state describes the steady-state solution for the gas. *Xie and Mumma* (1996a,b) updated some of the statistical algorithms from the original model of Hodges, improving the sampling of velocities for collision pairs and including a more accurate description of the IR rotational cooling by water molecules.

A DSMC model proceeds by following the detailed molecular motions of many molecules (thousands to millions) simultaneously, including their responses to imposed fields (e.g., gravity, or E and B fields in the case of charged particles), binary collisions, and chemistry. The assumption of binary collisions is quite good for a dilute gas and basically requires that the gas density be such that the distance between molecules is much larger than the molecular size. Quite dense gases, for instance, the standard density at the surface of Earth, which is in the realm of hydrodynamics, easily satisfy this condition.

A DSMC model is inherently time dependent and can address a wider range of problems than TP methods. Steady-state situations are achieved by running a simulation for a long enough time given steady-state initial and boundary conditions. Therefore, the final state does not depend *per se* on the initial state provided the system is given an adequate time to relax. The initial state could start at some given state, or a vacuum. In terms of computational resources the inherently steady-state iterative TP approach requires less memory than a steady-state version of a DSMC; however, both must sample a similar number of individual particles over a similar time with similar time steps in order to model a real physical system with similar statistical accuracy. Therefore, the total number of computations between the two methods should be roughly equivalent.

A DSMC model begins by setting the initial state of a certain number of molecules of all species in question throughout the simulation volume. Various boundary conditions from which new particles are introduced into the simulation volume must be defined and characterized. The simulation is divided into small time steps, Δt_S, which must be small enough so that only a small fraction (<0.1) of particles in any volume of space will collide over that time, and so that the forces (accelerations) yield small enough velocity changes, thereby allowing some finite-difference formulation to follow the particle trajectories. Higher-order schemes can also be used for this purpose if the timescale for trajectory variations owing to outside forces (e.g., gravity, or electromagnetic) is much smaller than the collision timescale. Furthermore, each spatial cell can have its own collision time step, so that useless collision testing need not be performed when the densities are quite low. This clearly occurs high enough in an atmosphere or far enough away from the nucleus in a cometary coma, and means that collisions that occur more frequently in regions of high density can be sampled as often as necessary.

There are clear advantages to kinetic DSMC methods in being able to treat a whole range of non-LTE processes as well as multiple species. At the same time there are serious computational penalties, so particle kinetic methods are not meant to be substitutes for all other modeling techniques in all applications: Clearly simple models such as Haser or vectorial and hydrodynamics approaches, such as Euler and Navier-Stokes, are highly useful.

A very useful application of DSMC to an expanding comet atmosphere was to understand the time-dependent effects of the dynamics of the expanding atmosphere of Comet 1P/Halley (*Combi*, 1996). Figure 7 shows a plot of the measured outflow speed of the heavy molecules in the coma as measured by the neutral mass spectrometer on the *Giotto* spacecraft (*Lämmerzahl et al.*, 1987) compared with time-dependent DSMC model calculations. Shown are the time variations of the dependence of the radial outflow speed on distance from the nucleus owing to the variation in the water production rate at the nucleus. The model uses a time-variable production rate at the nucleus derived from the photometric lightcurve of *Schleicher et al.* (1990) using an analysis of spatial profiles of a number of species by *Combi and Fink* (1993), which is also shown. When the production rate is large, the gas densities are high and collisional thermalization of the superthermal H atoms is efficient. This causes an increase in the photochemical heating that drives an increase in the outflow speeds. Therefore, the radial outflow speeds corresponding to peaks in the lightcurve are noticeably larger than those during the troughs. For the factor of 3–4 in production rate variation in Halley, this turned out to be a fairly sensitive change, especially when the timescale for large production rate changes is comparable to the timescale of transit of gas across the coma. As shown by the appropriate lightcurve for the phase of the *Giotto* measurements, the model predicts the measured velocities. As shown, if the comet had been at a different phase of the variation (only a day or so earlier or later), a measurably different velocity profile would have been obtained.

Skorov and collaborators have used DSMC to explore the Knudsen layer at the nucleus surface/coma boundary (*Skorov and Rickman*, 1998; *Markiewicz et al.*, 1998) and the porous upper layers of the surface of the nucleus itself (*Skorov et al.*, 2001). More recently, Crifo and collaborators (*Crifo et al.*, 2002, 2003) have been comparing calculations using DSMC and solutions of the Euler equations and Navier-Stokes equations formulation of hydrodynamics for a dusty-gas coma in the vicinity of the nucleus. They

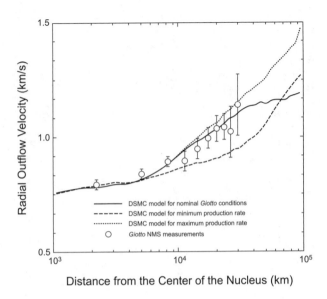

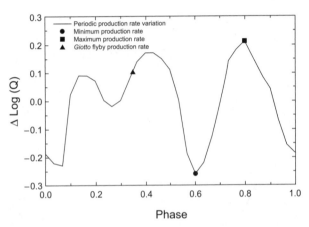

Fig. 7. Time-dependent DSMC model calculation for Comet 1P/Halley around the time of the *Giotto* encounter. In the top plot results are shown for maximum, minimum, and *Giotto* flyby phases of the lightcurve. Time variations were adopted from the photometry of *Schleicher et al.* (1990), and are shown in the plot below. The seven-day periodic gas production variation leads to 25% variations of the outflow speed.

find that outside the most collisionally thick region of the coma, where the Euler equations are valid, there is a region where the Navier-Stokes equations provide a reasonable calculation of the coma density and velocity flow field, as verified in numerical experiments with DSMC. However, as mentioned above, the very near-nucleus region is discussed in *Crifo et al.* (2004). In the larger "observable" coma, which we address in this chapter, not only are the Euler equations not valid, but also DSMC is very useful to model simultaneously multiple species, including photochemical products that are far from any Maxwellian or slightly skewed Maxwellian distribution assumed for Navier Stokes. Here, DSMC is the only currently developed viable alternative.

4.6. A CO-Dominated Coma

The observed examples of CO-dominated comae are for comets at large heliocentric distances, like Comets 29P/Schwassmann-Wachmann 1 and 1995 O1 (Hale-Bopp). *Biver et al.* (1999a) found that the coma of Comet Hale-Bopp underwent a transition from CO-dominated to H_2O-dominated at a heliocentric distance of about 3 AU both before and after perihelion. There were velocity resolved millimeter-radio observations of CO, CH_3OH, and H_2CO in Hale-Bopp when the comet was at a heliocentric distance of 6 AU (*Jewitt et al.,* 1996; *Womack et al.,* 1997) and the coma was CO-dominated. The CO emission indicates primarily sunward-directed velocities of 0.3–0.4 km s^{-1} that are much larger than would be expected for an exposed sublimated CO-ice surface, which is 25 K when emitted into a vacuum (*Yamamota,* 1985). On the other hand, emissions from CH_3OH and H_2CO indicate a more isotropic ejection that possibly points to an extended source by photodissociation of parent molecules and/or sublimation from icy grains. The most reasonable explanation is that CO is likely released from the dayside of the nucleus when the subsurface thermal wave causes the dominant H_2O ice to undergo a phase transition from crystalline to amorphous ice, which happens at about 125 K (*Prialnik et al.,* 2004). At such a temperature, CO would be released from the ice below the surface and could diffuse through porous upper layers of the nucleus surface. A temperature of 125 K is consistent with the observed velocity.

There has been little theoretical work done on the possible conditions in the coma of a comet that is dominated by CO. *Ip* (1983) computed a photochemical/hydrodynamic model for a CO-dominated comet at a heliocentric distance of 1 AU, conditions which, we know now in hindsight, are unlikely to occur. Observations of comets to date would indicate that any comet at a heliocentric distance of 1 AU would be water-dominated. The model assumes that the gas is released from the surface at the vapor-pressure-equilibrium sublimation temperature of CO ice, which is 40 K. The photochemical heating effects of CO at 6 AU are negligible compared with the model at 1 AU because of the increase of the CO lifetime by a factor of 36, and this is compounded by a production rate for Hale-Bopp that was only a few × 10^{28} molecules s^{-1}, and the fact that CO is a reasonably efficient at cooling in the IR (*Chin and Weaver,* 1984). This would result in any energetic photodissociation products (hot superthermal C and O atoms) to be emitted well outside the collision zone for this comet.

More recently, *Crifo et al.* (1999) presented results of aspherical models for CO-dominated comae of Comets 29P/Schwassmann-Wachmann 1 and 46P/Wirtanen. They assumed that CO diffuses through the surface of a comet at a heliocentric distance of 3 AU, which reaches the local equilibrium temperature of a blackbody with values ranging from 230 K at the subsolar point to 41–73 K on the nightside depending on the value of a surface recondensation param-

eter. Such large temperatures lead to large outflow velocities at a few nucleus-radii in the range of 0.65–0.75 km s^{-1} or more. Accounting for the expected temperature differences between 3 and 6 AU does not explain the factor-of-2 smaller observed speed in Hale-Bopp. So, at least for the case of Hale-Bopp, their surface temperature model is too hot, but the basic physics of the flow velocity is otherwise reasonable. Clearly, the CO reservoir temperature from Hale-Bopp was less than the expected surface temperature. This either says something important about the surface temperature for these comets at 6 AU or more likely about the temperature accommodation for CO gas effusing through the upper porous layers.

5. SPATIAL AND VELOCITY MEASUREMENTS: OBSERVATIONAL TECHNIQUES AND MODELS

The methods used to study the composition and structure of comet comae are beginning to reach the point where gas production rates, energy balance, chemistry, velocity structures, radial distributions, temporal-spatial variability, and the interaction of the coma with solar wind are all accessible at some level with remote sensing, although the inversion of data to physical parameters continues to require model interpretation and coordinated complementary observations. Maturation of these techniques has been both technical and physical. The widespread use of linear, high-dynamic-range detectors has improved sensitivity and photometric accuracy in the visible, while advances in digital quantum efficiency and array size have combined with larger telescope apertures to increase the absolute sensitivity limit and angular scale over which comets are studied. Substantial improvements in UV, IR, and radio instrument capabilities have made accessible many minor species, thermal diagnostic bands, and faint or previously undetectable emissions from parent species and their atomic end states, while new developments in interferometric, high-resolution, and integral-field (multipoint) spectroscopic techniques have opened up the study of fine structure and the low-velocity (<1 km/s) characteristics in different regions of the coma. Finally, our ability to monitor the variable elements of the solar UV spectrum that drive photochemistry has improved our ability to validate theoretical rate calculations and to separate the linked variables of velocity and lifetime in expansion models.

To take full advantage of these improved observational capabilities requires corresponding improvements in our understanding of how photolysis, branching ratios, and fluorescence efficiencies of a chemical species affect its role and evolution in the coma, and of how the detailed structure of the coma affects the convergence of measurement, models, and their relationship to actual conditions. In this section we describe the most widely used techniques for remote sensing of the evolution of the H_2O coma, with an emphasis on spatial distributions, photometric measurements, and spectroscopic study of line shapes and Doppler shifts [see *Feldman et al.* (2004) for an expanded discussion of spectroscopy]. We emphasize observation and modeling of the evolution of water and its daughters in the applied examples; however, it should be clear how these techniques and their limitations are applicable to other coma species. Each of these techniques is specialized to the desired feature (radial distribution, velocity profile, total production rate) or spatial scale (inner vs. outer coma) under study, or is either instrumentally [e.g., Far Ultraviolet Spectroscopic Explorer (FUSE)] or physically (e.g., quenched radio or metastable emissions) limited to specific regions of the coma. Such focus means that a detailed understanding is rarely achieved with an isolated observation, and we discuss the value and challenges of a synergistic, coordinated observational approach for modeling the full dynamic and photochemical evolution of the coma.

5.1. Remote Sensing of the Spatial, Thermal, and Velocity Properties of Cometary Water

The observational strategies used to study the H_2O coma spatially, photometrically, and in velocity space can be broken down into five areas: (1) aperture summation (wide and narrow field); (2) one- and two-dimensional mapping (narrow bandpass imaging, spectro-imaging, multiplexed spectroscopy); (3) velocity-resolved imaging (interferometric data cube studies); (4) aperture-summed line-shape measurements (high-resolution/étendue spectroscopy); and (5) temperature sensing. Each of these techniques provide a distinct diagnostic of a given species such as total production rate, spatial distribution, temporal variance, temperature, or velocity distribution, all on spatial scales ranging from the extreme inner coma (10^2–10^3 km) to the diffuse outer coma and ion tail (10^5–10^7 km). The results necessarily reflect the target diagnostic of the observation in the coma/tail region where the information is sought, which limits the scope of a model interpretation.

5.1.1. Aperture summation. Being essentially the case of zero-dimensional imaging, aperture summation is used to detect faint or low surface brightness features, multiple lines of a single molecular band, and/or to provide the highest possible photometric accuracy on a single coma component. The measurements are done either with a combination of narrow bandpass target-continuum filter image subtraction (*Schleicher et al.*, 1998; *Kiselev and Velichko*, 1999; *Grün et al.*, 2001) or spectrophotometrically (*Schultz et al.*, 1993; *Oliversen et al.*, 2002). In addition to providing high photometric accuracy, aperture summation is a useful tool for obtaining production rates. This is particularly powerful for the case of wide-field measurements that sample the entire scale length of a given species. For fluorescence, such measurements invert directly to production rate requiring only knowledge of the transition g-factor and lifetime via

$$Q_x = 10^6 I_x \Omega \Delta^2 / (g_x \tau_x) \qquad (26)$$

where Ω is the solid angle of the field of view (FOV) in

radians; Δ is the geocentric distance (cm); τ_x is the photochemical lifetime of species X in seconds; I_x is the field-averaged brightness in Rayleighs; and g_x is the fluorescence efficiency (photons s^{-1}). In a model inversion of resonance line photometric data we are limited by the precision of our knowledge of lifetime and fluorescence efficiency. In the case of metastable prompt emissions, a single photon is produced, which simplifies the relationship such that the production rate follows directly from the brightness (*Schultz et al.*, 1993)

$$Q_x = 4\pi\Delta^2 \Omega I_x \qquad (27)$$

with the parent production rate following from the chemical branching ratio to the metastable species. The accuracy of metastable photometry is similarly limited, in this case by our knowledge of the formation rate from its parent(s) and the contribution of loss/production mechanisms that are important in the inner coma. Neither relationship accounts for potential complications such as collisional or chemical quenching, opacity to solar photolyzing or scattering radiation, or Swings/Greenstein g-factor variation (irrelevant for metastable emissions), all of which introduce uncertainties that must be addressed in the model interpretation. Even when these ancillary factors are correctly addressed, the resulting photometric measurement will represent a time-averaged production rate for the period of travel across the aperture for the slowest-velocity component of the target species.

To derive production rates when the FOV is less than the diameter of the scale length, the above relationships must be adjusted by an aperture correction (AC) term that accounts for emission beyond the edge of the aperture. The value of AC is determinable using model estimates of the scale lengths ($\gamma_x \sim v_x \tau_x$) of the species under study and its parent using either simple spherical expansion (Haser-type) or streaming particle (vectorial or Monte Carlo) models, or observationally by comparing the models to the total signal obtained by moving apertures (*Oliversen et al.*, 2002) to different locations in the coma or using apertures of different angular sizes. Coarse scale-length estimates are relatively straightforward for species (e.g., OH) in simple radial outflow assuming a static, symmetric quasithermal velocity relationship such as the $v_r \sim 0.85\,R_H^{-0.5}$ (where v_r and R_H are the outflow velocity and heliocentric distance) of *Budzien et al.* (1994). However, a precise scale-length measurement is far more complicated if there are substantial nonthermal components, e.g., for H (*Richter et al.*, 2000), if there is significant acceleration across the coma (*Harris et al.*, 2002; *Combi et al.*, 1999b; *Bockelée-Morvan et al.*, 1990), temporal variability in gas production, or an extended source region. At the low end of aperture size, where only the inner coma is sampled [e.g., FUSE (*Feldman et al.*, 2002) or the CSHELL spectrometer on the IRTF (*Dello Russo et al.*, 2000)], the above complications become more significant, while other unresolvable factors, including quenching, collisional chemistry (*Komitov*, 1989), accelerations, and/or asymmetries (spatial *and* temporal) in gas production further confuse the interpretation. For observations of parent species with apertures that are small compared with the molecular scale length, one can use the relation that is identical to equation (2) where the number of molecules can be extracted knowing only the outflow velocity.

5.1.2. One- and two-dimensional spatial imaging. Spatial maps of the coma are obtained using interferometric imaging in the radio, spectroscopic imaging in the visible, fiberoptically multiplexed or image-sliced spectral techniques, long-slit imaging spectroscopy (low to high resolution), and narrow bandpass filter images targeting specific neutral species, ions, and dust continuum points (see *Schleicher and Farnham,* 2004). After the removal of sky, continuum, and detector background the resulting data provides a snapshot of the radial, azimuthal, and temporal structure of a coma species. Spatial data cube spectroimages of different lines in thermally diagnostic molecular rotational bands can also be used to track the evolving thermal properties of the coma. Such two-dimensional measurements are of immense value for model analysis as they contain records of gas production variability, extended source distributions, radial column density profiles, scale-length information, and local temperature, although they do come at the expense of photometric accuracy in the lower surface brightness outer coma.

Long-slit spectra provide single-dimensional spatial maps of brightness with radial distance from the nucleus. These spectra are most useful for measuring radial profiles of several species at low spectral resolution and providing velocity dispersion, Doppler shifts, and line ratios in a single molecule at medium/high resolution. As with two-dimensional imaging, slit spectroscopy is less effective in the outer coma where the surface brightness of the emissions is low. To extract radial information from the faint, diffuse outer coma, a spatial summing technique can be applied to a two-dimensional image that collapses the azimuthal dimension of a spectrally filtered image into a single averaged profile of brightness as a function of distance from the nucleus (i.e., a ring sum). Azimuthally averaged data provides much greater sensitivity to the radial profile in the outer coma, where the shape is strongly dependent on the outflow velocity distribution. However, averaging blurs spatial and temporal structure in azimuthal averages and leaves the resulting profile underconstrained, which can make it difficult to obtain a unique or consistent model interpretation. Ideally, a hybrid approach to spectroimaging data is applied, where spatial information on the underlying structures of the coma, including diurnal variances, temporal changes in overall production, vectored flows, jets, temperature, and secondary sources, are retained in the high surface brightness regions of the coma, while the more diffuse regions are summed to increase signal to noise. This greatly increases the amount of information available for model analysis, resulting in a more detailed picture of coma structure.

Inversion of a one-dimensional or two-dimensional spatial dataset to production rate can be achieved in two ways, either by summing all the photons in the FOV as an effective aperture and using equations (26) and (27), or by mod-

eling the radial shape with spherical expansion or streaming particle models that require some additional knowledge of the outflow velocity distribution and/or chemical lifetime. In the case where the detectable dynamic range of the image extends from the inner coma to the edge of the target species scale length, a photometrically derived production rate can be combined with radial data to put constraints on the production rate and velocity distribution (*Harris et al.,* 2002) and hence constrain the uniqueness of the model fit. However, if the usable (above detection threshold) FOV is less than the scale length of the species under study, a combination of photometry and model is required. Since spatial mapping does not provide velocity information (fiber-optically multiplexed and image sliced spectral maps are an exception), the same limitations on velocity estimates hold for imaging studies that are encountered in aperture photometry and the accuracy of the production rate will depend on the precision of the estimate.

5.1.3. Velocity-resolved aperture photometry. An accurate measurement of the velocity distribution for a coma species is critical to a detailed understanding of its spatial distribution, chemistry, temperature, and evolution, particularly when using detailed models that are capable of addressing multiple component, accelerating, and/or nonradial line profiles (e.g., DSMC or hydrodynamic models). Medium resolution enables the detection of Doppler-broadened or -shifted components of emission lines (e.g., *Larson et al.,* 1987), but it is generally not sufficient for resolving the expansion velocities (1–10 km/s) of most coma species. The very high spectral resolution ($R \gg 10^5$) needed to detect low-velocity flows is problematic for narrow-aperture grating-type spectrographs, because the required combination of high signal to noise and high spectral resolution is difficult to achieve for the low surface brightness emissions that describe the coma and ion tail beyond the immediate vicinity of the nucleus. Aperture-summation techniques provide far greater étendue and thus sensitivity to diffuse emissions such as those in comet comae, and radio frequency measurements of OH and other molecular radicals (e.g., CO, HCN, H_2O) have emerged as an effective technique for detection of outflow velocity signatures as small as 0.1–1 km/s (*Bockelée Morvan et al.,* 1990; *Biver et al.,* 1999a; *Colom et al.,* 1999). Fabry-Pérot and spatial heterodyne spectroscopic (SHS) interferometers have been used to study expansions from atomic species, primarily H (*Morgenthaler et al.,* 2002) and O(^{1}D) (*Smyth et al.,* 1995) at velocity resolutions of up to 1 km/s. H Lyman-α absorption cells (*Bertaux et al.,* 1984) have also proven effective for measuring the velocity structure of coma hydrogen at subkilometer-per-second resolutions, albeit with substantial temporal averaging as the comet velocity must change with respect to the instrument to sample the full line shape. These observations all share the common limitation that they average velocities over large areas of the coma.

As a precision photometric technique, velocity-resolved aperture summation is directly invertible to production rate, subject to the same caveats described above. Because these measurements average all the velocity structures that are detectable in the aperture, acceleration in active comets (*Colom et al.,* 1999) or changes in the velocity distribution due to radial-distance dependent photochemistry (i.e., H) (*Richter et al.,* 2000; *Combi et al.,* 1998; *Morgenthaler et al.,* 2002) are mixed both azimuthally and radially in the measured line profile and can only be partially separated with model analysis. An additional complication arises in the form of quenching in the inner coma that can render the velocity structure inner coma undetectable for some water daughter species, including OH (*Schloerb,* 1988) and O(^{1}D) (*Festou and Feldman,* 1981; *Smyth et al.,* 1995; *Morgenthaler et al.,* 2001). While the effect of quenching on the inversion to production rate can be compensated for with correction factors, the measured velocities are isolated to the regions beyond the quenching radius. This is not a problem for weak comets where the aperture is much larger than the quenching radius and the OH velocity is largely invariant, but can be severe for very active comets such as Hale-Bopp where substantial acceleration occurs in the quenched regions (*Colom et al.,* 1999, *Harris et al.,* 2002).

5.1.4. Velocity-resolved spectroimaging. Spatially discrete velocity data can be obtained from velocity-resolved interferometric (data cube) imaging, one dimensional spatial SHS, or long-slit or spatially multiplexed echelle spectroscopy. Long-slit [e.g., *Combi et al.* (1999a) at R = 200,000] and SHS [e.g., *Harlander et al.* (2002) at R = 40,000] techniques are both limited to a single dimension, but have demonstrated the very high intrinsic resolution necessary to resolve the low expansion velocity coma neutrals. Of the two, SHS offers much higher intrinsic étendue, can be tuned to very high (R ≫ 10^5) resolution, and is better suited for observations of the outer coma or very diffuse emissions, due to the fact that the instrument collapses the dimension orthogonal to the spatial direction into the interferogram. This single-dimensional summing improves s/n as with aperture summation, but spatially averaging only one dimension, leaving radial data in the other. Echelle spectra sample smaller regions and thus probe the velocity structure at discrete locations where the surface brightness of the emission is high. This offers a more constrained measurement for model analysis, but at the expense of s/n. Echelles are more easily tuned than SHS instruments, and they have much greater flexibility in the selection of bandpass, often covering broad ranges of wavelength.

Data cube images are typically obtained interferometrically (such as a Fabry-Pérot) using a tunable filter bandpass and stepped in wavelength across a target emission feature or features (*Morgenthaler et al.,* 2001). The resulting image arrays have the velocity or line ratio distribution for every spatial element (radial and azimuthal) in the FOV and are very useful for identifying vectored flows, temperature variations, acceleration, and signatures of secondary or extended sources. Their limitation comes from the fact that they collapse the full free spectral range of the interferometer into image space, and they are generally much lower in spectral resolution (R ≤ 10^4) in this mode than if the same instrument is used as a line-resolved aperture photometer.

This makes it useful only for species with substantial non-thermal velocity components (e.g., H_2O^+ or H) or where multiple, spectrally separated lines must be ratioed (e.g., for temperature or equilibrium-state measurements).

Multiplexed spectroscopy is a powerful, hybrid technique that combines the spectral range, resolution, and étendue limitation of an echelle spectrograph with the two-dimensional coverage of an interferometric image. Tunable multiplexing spectrometers place multiple (~100) feeds at different locations over the FOV (*Anderson,* 1999), providing maps of velocity at discrete points over a wide area of the coma. Fixed multiplex arrays consist of multiple feeds covering a single FOV. In each case, the separate feeds are directed to points along a long-slit axis of an imaging echelle spectrograph. The inputs tend to be small (approximately few arcseconds in diameter) and cannot be co-added easily. This limits their sensitivity to low surface brightness emissions in the coma. Their resolution depends on the configuration of the bench spectrograph used (10^3 < R < 10^4).

5.1.5. Thermal measurements in the coma. A measured temperature in the coma depends on the species observed, its formation pathway, and the collision rate both in the inner coma and at the location where the measurement is taken. Such measurements thus provide an important constraint on gas-kinetic models and a link from them to the physical properties of the nucleus and surrounding volatile-rich material. Remote sensing measurements are able to provide kinetic temperatures for a number of coma species, including the primary parent species CO (e.g., *DiSanti et al.,* 2001; *Biver et al.,* 2002; *Feldman et al.,* 2002; *Brooke et al.,* 2003) and H_2O (e.g., *Mumma et al.,* 1986; *Crovisier et al.,* 1997). Owing to the absence of molecular bands for these species in the visible, the bulk of thermal remote sensing is done in the IR and at radio frequencies, although *Feldman et al.* (2002) have recently demonstrated that the kinetic temperature of CO can be obtained with FUSE observations of the 0–0 band at 1080 Å.

Mumma et al. (1986) first reported on measurements of the spin temperature of water in the coma of 1P/Halley, obtaining a value of ~40 K from the ortho-para (O-P) ratio of lines, which are assumed to be fixed at the long-term average solid-state temperature before evaporation, in the v_3 band at 2.65 μm. Similar measurements of the H_2O kinetic temperature have been performed since, using the same and different bands (e.g., *Crovisier et al.,* 1997) as well as for O-P ratio of different coma species [e.g., NH$_3$ (*Kawakita et al.,* 2001)] with comparable results.

Rotational diagram derivations similar to those made in the interstellar medium have also been used to derive kinetic temperatures in comets by looking at the ratios of different rovibrational lines in a diagnostic band. These derivations produce results similar to the O-P measurements where they have been compared directly [e.g., H_2O from 1P/Halley by *Bockelée-Morvan* (1987)] and have proven effective for obtaining temperatures from several species including CH$_3$OH (*Bockelée-Morvan et al.,* 1994), H$_2$S (*Biver et al.,* 2002), and CO (*Biver et al.,* 2002; *Feldman et al.,* 2002). In particular, *Biver et al.* (2002) reported a combined study of all three of these as a function of heliocentric distance for Comet C/1995 O1 (Hale-Bopp). In addition to demonstrating consistent thermal properties between these species, they show results extending from 1 to 8 AU. While Hale-Bopp was an extreme high case for comet activity, their results demonstrate that temperature measurements are possible for many comets over the full range of their activity cycle, which will enable more detailed study of the gas kinetic evolution of the coma.

5.2. Measurements of Water and its Photochemical Products

Water itself has no visible band signature and is difficult to observe in general. Even as new techniques evolved that enable direct mapping of the water coma in the IR (*Mumma et al.*, 1986; *Crovisier et al.*, 1997; *Dello Russo et al.*, 2000) and radio (*Lecacheaux et al.*, 2003), the most common method of study continues to be observing the evolution of the daughter and granddaughter species of its photochemistry. The primary channels of this are given in Tables 1 and 2 above, with the distributions of H, H_2, metastable $O(^1D)$, OH, and H_2O^+ being the major products of interest. Each species provides different elements of the water evolution picture that depend on the observational and modeling methods (described above) that are used to examine them. Here we describe most common observational approaches to each water daughter and discuss where they are useful for characterizing the coma with respect to various model approaches.

5.2.1. OH diagnostics. OH is the most commonly studied of the water daughters, with ground-detectable signatures in the NUV [the 0–0 band is the highest contrast emission feature on groundbased spectra; see Fig. 1 in *Feldman et al.* (2004)], radio (18 cm), and IR (*Crovisier et al.*, 1997) that contain complementary information about its production rate, velocity structure, and spatial distribution in the coma. OH is formed almost exclusively from water dissociation, which makes it the most easily inverted back to a parent (water) production rate. Indeed, given the single formation path of OH and the relative difficulty of observing water directly (see *Bockelée-Morvan et al.*, 2004), OH has generally been the most effective proxy for water production with the least ambiguity in its interpretation, especially for comets that are not productive and in the range of 1 AU (or less) from the Sun. For this reason the large database of International Ultraviolet Explorer (IUE) and HST observations (see *Feldman et al.*, 2004) is particularly useful for a hard calibration for the water production rate in comets.

The complexities of OH photochemistry and interpretation are described in more detail above and in *Feldman et al.* (2004); however, to summarize, the primary issues are a lack of knowledge of basic characteristics and observational difficulties. OH is difficult to isolate in a laboratory setting (e.g., *Nee and Lee,* 1985), which requires a somewhat circular study of this species where the properties of the mole-

cule must be derived or verified by using a comet as an astrophysical laboratory. In addition, the primary observing factors that complicate the use of its NUV emission features as a diagnostic include variable atmospheric attenuation of ~2.2 magnitudes/airmass (*Farnham et al.,* 2000) when observed from the ground, and a complex Swings-Greenstein sensitivity in the g-factors of individual lines of the unresolved 0–0 band (*Schleicher and A'Hearn,* 1988). The 18-cm OH radio emissions are strongly quenched (*Schloerb,* 1988), which restricts the sensitivity of the measurement to the outer coma, an effect that becomes acute for active comets such as Hale-Bopp where the quenching region exceeded 10^5 km (*Schloerb et al.,* 1999; *Colom et al.,* 1999). Finally, there is strong photochemical sensitivity to variable UV emissions in the solar spectrum in its production and loss rates and its excitation from OH. Considerable effort has been expended toward addressing all these complicating factors, which has greatly improved the convergence between the different techniques used to study this species.

OH is observed using a combination of aperture photometry, both for radio and NUV emissions, velocity-resolved measurements (Fig. 8) in the radio (*Colom et al.,* 1999; *Bockelée-Morvan et al.,* 1990; *Crovisier et al.,* 2002), and spatial maps in the NUV (*Harris et al.,* 1997, 2002). OH has been also observed in the IR by way of fluorescent and prompt emissions in the 3-μm region (*Brooke et al.,* 1996; *Crovisier et al.,* 1997; *Mumma et al.,* 2001). Both the 0–0 band and the 1–0 band at 2850 Å are also observed from spacebased platforms (*Weaver et al.,* 1999). However, sys-

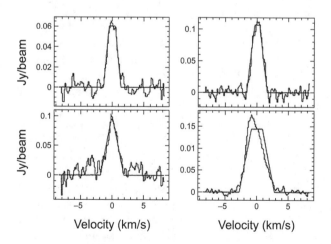

Fig. 8. OH emission at 18 cm from Comet 1P/Halley obtained with the Nançay radio telescope (3.5 × 19 arcmin beam) and using aperture summation is fit using the trapezoid method to determine the outflow velocity of the coma (from *Bockelée-Morvan et al.,* 1990). In the lower left image, the sharp peak of the fitted line is indicative of a case where $v_{parent} \sim v_{ejection} \sim 1000$ m s⁻¹. The upper two profiles are matched with flat-topped fits consistent with expansion of ~500 m/s, while in the lower right a higher, asymmetric expansion rate is obtained.

tematic spacebased observations of OH ended with the deactivation of the IUE in 1996. Following the work of *Scheicher and A'Hearn* (1988) it has been possible to correct for g-factor and scale-length variability for the NUV fluorescent bands. The inversions of photometric and spectrophotometric measurements of OH to obtain water production rates with the use of spherical expansion models for the radial distribution (*Schleicher et al.,* 1998; *Kupperman,* 1999; *Harris et al.,* 2002) are generally consistent with other estimates of water production. *Schloerb* (1988) has developed a formulism for addressing quenching of radio emissions that is comparably robust, while *Bockelée-Morvan et al.* (1990) have developed a line-fitting technique (the "trapezoid method"; Fig. 8) for determining the average radial outflow velocity in radio-band OH emissions from beyond the edge of the quenching region. This also puts some constraints on the OH ejection velocity, which is sometimes slightly discrepant with that expected from photochemical analysis (Table 1). However, these might be explained by uncertainties in the photochemical data as well as simplifications in the models, which treat the coma as spherically symmetric, collisionless flow and having a constant uniform outflow velocity for parent species.

The major limiting factor with remote sensing of OH is the relatively large number of independent unknowns that must be relied upon to model an isolated observation, with each technique requiring different approximations (azimuthal-temporal averaging, reduced outer or inner coma sensitivity, lack of velocity data, etc.). As a result it is difficult to reconcile model results from different measurements to within the relative precision of the measurements (*Schleicher et al.,* 1998). This nonconvergence is most effectively dealt with through the use of coordinated measurements (see below) that act to constrain the individual uncertainties of each observation, such as a lack of velocity data in NUV images/photometry and a lack of spatial information and the inner coma distribution in radio observations (*Harris et al.,* 2002; *Colom et al.,* 1999). Future technical developments in high étendue/spectral resolution measurements of diffuse emissions in the outer coma (*Harlander et al.,* 2002) and long-slit echelle spectroscopy of the individual lines of the 0–0 band (*Combi et al.,* 1999a) will improve the accuracy of the interpretation. It should be noted, however, that the divergence of techniques is small relative to other species, and that Q_{OH} rarely varies by more than 50% between them unless there is considerable undetected (by one measurement relative to another, e.g., narrow vs. wide field) spatial and/or temporal variation in gas production, or a difference in the underlying assumptions (e.g., outflow velocity or g-factor) used.

5.2.2. Hydrogen diagnostics. Hydrogen arguably provides the greatest physical information on the chemistry of water, the structure of the coma, and temporal variability in gas production; however, its long scale length necessitates multiple observations on different spatial scales to fully characterize its evolution. Atomic H is derived largely from water and OH dissociation, which makes it possible to derive a production rate directly, assuming the branching ra-

tios to the multiple formation pathways are known for the conditions of the observation. The H coma is the most extended structure in comets, with observable emissions out to more than 10^7 km from the nucleus (*Combi et al.,* 2000).

Compared to the other water daughters, the H velocity distribution is completely dominated by its nonthermal elements, with H_2O and OH photochemistry contributing from 4 to 24 km/s in excess velocity. The bulk velocity distribution of H is also affected by collisions, which partially thermalize the energetic neutral H population in the inner coma, especially in very active comets (*Morgenthaler et al.,* 2001). In addition, radiation pressure modifies the distribution with an antisunward acceleration that becomes significant in the outer coma. The net result is that the velocity distribution is highly dependent on the region of the coma being sampled, the optical depth of the inner coma to solar UV radiation (both for photochemistry and resonance scattering), the total gas production rate, any short-term gas production rate variability, and the comet heliocentric velocity (for both Swings and Greenstein effects). Modeling of the distribution from the inner coma of several comets (*Combi and Smyth,* 1988a,b; *Smyth et al.,* 1991; *Combi et al.,* 1998; *Richter et al.,* 2000) nominally verifies the theoretical estimates of the velocity and its variation with heliocentric and cometocentric distances for a variety of activity level comets.

The observational challenges to the study of H in comets include the requirement for observations above the atmosphere to detect H Lyman-α or H Lyman-β, the need for measurements over multiple FOV, and strong telluric and galactic backgrounds at Hα. The primary diagnostics are resonantly scattered H Lyman-α at 1216 Å and cascade Hα emission at 6562 Å. Detection of both dates back to the C/1973 E1 (Kohoutek) apparition, with a dataset that includes most active comets since then (*Huppler et al.,* 1975; *Keller et al.,* 1975; *Drake et al.,* 1976; *Scherb,* 1981). In recent years and until the introduction of the SOHO/SWAN instrument, H observations of comets were mainly accomplished using variable width and line-resolved aperture photometry with interference techniques on medium to large spatial scale (*Morgenthaler et al.,* 2002) or echelle spectroscopy near the nucleus (*Richter et al.,* 2000; *Feldman et al.,* 2004). SOHO/SWAN has added the ability to image the full H coma (Fig. 9) and even the shadow cast by the coma on the background interplanetary medium. Because of the high energies and velocities involved in the formation of H, the contributions from the various dissociation pathways can be seen in the H Lyman-α line shape with intermediate ($R > 10^4$) resolution instruments, which increases the range of techniques and available instrumentation for the study of this feature.

H line-profiles can be compared with detailed models, albeit typically as azimuthal and radial averages over the instrument FOV, to determine the contributions from different dissociation pathways both as a function of location in the coma and as a function of overall gas production. Figure 9 compares the model simulations of the velocity distribution for Comet Hale-Bopp (*Combi et al.,* 2000) when the comet was at its most active near perihelion (0.9 AU),

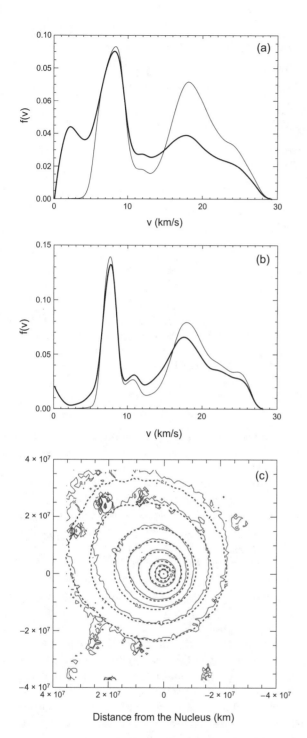

Fig. 9. Model for the SOHO SWAN observations of the H Lyman-α coma of Comet 1995 O1 (Hale-Bopp); **(a)** and **(b)** show the H-atom velocity distribution upon photochemical production (thin) and after Monte Carlo model calculation (thick) of thermalization in an expanding water-dominated coma. **(a)** Conditions on the day of perihelion (April 1, 1997) when the comet was at a heliocentric distance of 0.91 AU and the water production rate was 1.02×10^{31} s^{-1}. **(b)** Conditions on January 1, 1997 when these values were 1.75 AU and 2.2×10^{30} s^{-1} respectively. **(c)** The model-data comparison for the H coma at perihelion on April 1 when collisional thermalization was the largest and the resulting coma is the most distorted from spherical symmetry. The shape of the coma is formed by the balance between solar radiation pressure acceleration and the velocity distribution function.

and three months earlier (1.7 AU), when the production rate was much lower and the dissociation rates much lower. Near perihelion a large fraction of the highest-speed H atoms are thermalized to much lower velocities than at 1.7 AU where the distribution was not changed much. The effects of thermalization are directly evident in the perihelion Hα line-profile measurements of *Morgenthaler et al.* (2002). As discussed earlier, this slowing of the H atoms in Hale-Bopp is matched by a corresponding increase in the outflow speed of the heavy species, to which the kinetic energy is being transferred. Both *Harris et al.* (2002) and *Colom et al.* (1999) see clear evidence of this effect in the velocity structure of OH near Hale-Bopp's perihelion. The shape of the coma at perihelion, as shown in Fig. 9c and as observed with the SWAN instrument on SOHO, is also sensitive to the velocity distribution as first pointed out by *Keller and Meier* (1976).

The calculation of the water production rates (or any species for that matter), from either fast species like H or slower heavier species like OH or water, is very sensitive to our knowledge of the velocity distribution, unless the full scale length is observed (see equation (26)). This complication is substantially magnified by the multiple, radially dependent components of the H velocity distribution and by the large spatial extent of the H coma. At present we are able to sample spectrally only small, discrete subsets of the coma in the UV (H Lyman-α) or large azimuthally and radially averaged areas at moderate signal to noise (*Morgenthaler et al.,* 2002). Improvements in interferometric instrument capabilities in the FUV (*Stephan et al.,* 2001) and visible offer the promise of additional spatial information and increased spectral resolution for future comets. Other new diagnostics, including observations in the deep FUV (H Lyman-β) from FUSE (*Feldman et al.,* 2002), measurements of Hβ emission from Comet C/1996 B2 (Hyakutake) (*Scherb et al.,* 1996), and the planned STEREO (SOHO follow-up) mission, suggest a greater capability to map the structure of the H coma in future comets.

5.2.3. O(^{1}D) diagnostics. Oxygen in the metastable ^{1}D state is a byproduct of H$_2$O, OH, and CO photochemistry that is produced at different rates across the coma. Relaxation of O(^{1}D) produces a single photon ~110 s after its formation, meaning that its brightness inverts directly to its combined parent production rate without any supplemental knowledge of g-factors via equation (27) above. Only the various parent lifetimes and chemical branching ratios are required to complete the inversion, and, if the full O coma is sampled, the lifetime dependence disappears as well. Its short lifetime of the metastable state offers an additional advantage, because the locations of formation and emission are nearly co-local. Thus, the radial distribution of O(^{1}D) is a radial map of the site of photochemistry in the coma that is useful for detailed modeling of chemical rates, source distributions, and the relative density ratio of H$_2$O and CO.

Observations of O(^{1}D) do have significant complications as a proxy for H$_2$O and OH photochemistry that are addressable to a large degree. Observationally, these complications

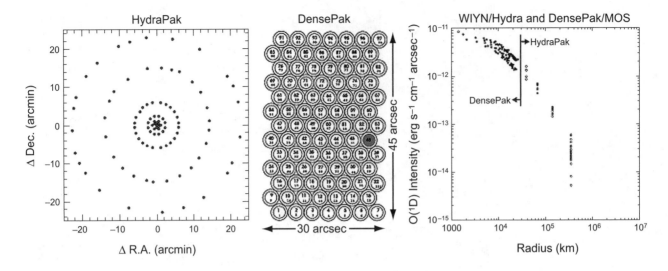

Fig. 10. Two types of multiport spectroscopy and their application to Comet Hale-Bopp are shown (from *Morgenthaler et al.,* 2001). At left is a sample distribution using the Hydra spectrograph on the WIYN telescope. Here a set of 100 fibers are placed at programmed positions over a 1° FOV with a minimum separation of 40 arcsec. The array design shown is optimized for the study of radially distributed emission, with the individual fibers feeding a bench spectrograph mounted below the telescope pier. At center is a schematic of DensePak, a tight grouping of 100 fibers covering a region equal the minimum spacing in the Hydra array. At right is an example of observation of O(^{1}D) from Hale-Bopp using DensePak in the inner coma and Hydra for more remote locations.

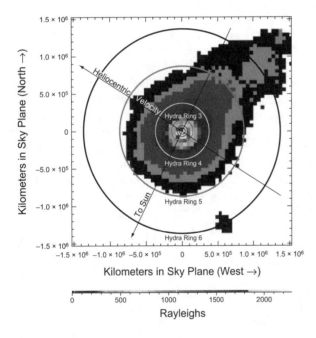

Fig. 11. A single 1° field image slice of the O(^{1}D) coma of Hale-Bopp is shown after subtraction of continuum. This image was obtained using the WHαM instrument at Kitt Peak (from *Morgenthaler et al.,* 2001). Overlaid is the location of fiber rings from the Hydra instrument using the array shown in Fig. 8.

photochemistry that can be partially correctable with observations of C(^{1}D) (see below) (*Oliversen et al.,* 2002), uncertainty in the branching ratios from OH under different levels of solar activity (*van Dishoeck et al.,* 1984; *Huebner et al.,* 1992; *Morgenthaler et al.,* 2001), UV opacity effects on chemical rates, and the role of collisional quenching and chemistry of O(^{1}D) in the inner coma (*Glinsky et al.,* 2003; *Komitov,* 1989). To compensate for the observational effects, O(^{1}D) is now routinely observed using high-spectral-resolution (R > 15000) techniques including long-slit echelle spectroscopy (*Fink and Hicks,* 1996), multiplexed echelle spectroscopy (Fig. 10) (*Anderson,* 1999), interferometric data cube imaging (Fig. 11), and aperture summed line measurements (Fig. 12). The recent detection of fluorescence of metastable O I (D^1-D^1) by FUSE (*Feldman et al.,* 2002) offers a potential new diagnostic that is comparable to the C(^{1}D) line (see below).

As noted above, the integrated brightness of O(^{1}D) is invertable to a production rate with equation (27) and from there to Q_{H_2O} from the branching ratios of OH and H$_2$O photochemistry. The radial and azimuthal distribution of the emission traces both the expansion of the coma and the rate of collisional and chemical quenching in the inner coma. Because it maps directly to chemistry, O(^{1}D) images also reveal extended sources and regions of increased velocity (*Morgenthaler et al.,* 2002). The O(^{1}D) outflow velocity retains the value of its parents (~1–2 km/s) plus fixed isotropic speed component of about 1.6 or 1.5 km/s from the photochemical excess energy from H$_2$O and OH, respectively. This requires high resolution (R < 10^5) to detect. This was achieved for 1P/Halley in 1986, with a Fabry-Pérot interferometer tuned to R = 1.9 × 10^5 (Fig. 12), although it

include substantial telluric O(^{1}D) nightglow, spectral separation of O I and nearby NH$_2$ and H$_2$O$^+$ coma lines (see *Feldman et al.,* 2004), and the low contrast of the emission over bright dust scattered continuum emission. Physical complications include an unknown contribution from CO

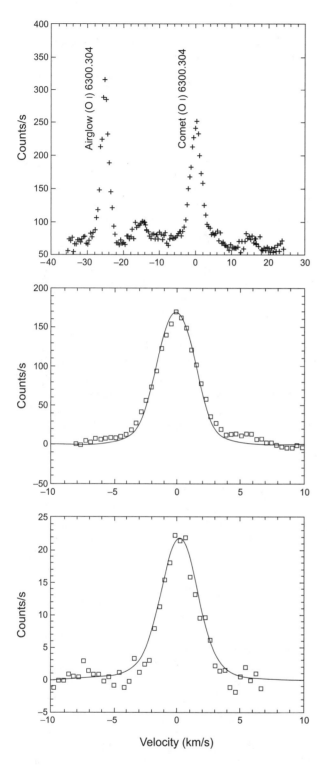

Fig. 12. Aperture-summed O(^{1}D) emission from 1P/Halley is displayed at the top showing the clean separation of comet and airglow O(^{1}D) signatures from each other and cometary NH$_2$ emission. These were obtained using a Fabry-Pérot interferometer in ring mode. Inspection of the comet and telluric line shapes shows the broadening in the Halley outflow. The lower panels show that this broadening is consistent with the magnitude of coma outflow predicted by model calculations to be broader at smaller (middle panel) heliocentric distances and more narrow at larger (bottom panel) distances (*Smyth et al., 1995*).

required the averaging of a substantial FOV to obtain acceptable s/n on the line-shape. The Halley profiles were fit successfully using the Monte-Carlo particle trajectory model of *Combi et al.* (1993) and were used to validate the model-predicted magnitude of expansion velocity and its increase with increasing gas production (*Smyth et al., 1995*). Observations of a similar resolution were made of the inner coma O(^{1}D) distribution with a high-resolution echelle spectrometer of Comet C/1996 B2 (Hyakutake). These were also in agreement with model predictions for conditions intermediate between the two Halley cases (*Combi et al., 1999a*). Continued development of very-high-resolution interferometric techniques and sensitive high-resolution echelle instruments will make such observational programs routine for comets of moderate to high activity, while the further development of FUSE band technology may provide a complementary diagnostic that will provide useful results for weaker comets.

5.2.4. Infrared and radio observations of H$_2$O. A missing key to the observational study of water dynamical and photochemical evolution is measurement of the properties of the parent species. This has been historically difficult both because of the lack of transitions in the FUV-visible band and because of strong attenuation of IR and radio bands by telluric water. Over the past two decades considerable progress has been made in finding ways around the telluric problem by moving to spacebased observatories or by observing within the various windows where atmospheric absorption is not so severe. Comets can sometimes be observed at large enough geocentric velocity in order to Doppler shift the cometary feature away from terrestrial absorption features. These improvements offer considerable promise that spacebased aperture photometry of water in the radio will become routine, while IR measurements (both ground- and spacebased) will contribute both the temperature and radial distribution of water. Together they will fill an important observational niche in our study of coma kinetics. The issue of observations of H$_2$O, and parent molecules in general, is covered in detail in *Bockelée-Morvan et al.* (2004).

Unambiguous detection of water in the inner coma first came from 1P/Halley via the Kuiper Airborne Observatory (KAO) (*Mumma et al., 1986*) and the IKS instrument on the *Vega* spacecraft (*Combes et al., 1986*), both of which directly detected emission lines from the ν$_3$ band near 2.65 μm. In addition to providing a direct measurement of H$_2$O, the KAO data were used to derive both the nuclear spin temperature (T$_{spin}$) (*Mumma et al., 1987*) and the rotational temperature (T$_{rot}$) (*Bockelée-Morvan and Crovisier, 1987*) of the lines, which sample the formation history and the current state of water in the coma respectively. The spin temperature derived from the ortho-para ratio on the different detected lines of the ν$_3$ band is set at the creation of the H$_2$O molecule by either photochemistry or direct evaporation from ice on the nucleus and thus directly ties to that event. The results from 1P/Halley (N$_{ortho}$/N$_{para}$ = 2.66 ± 0.13 indicating T$_{spin}$ = 32 K +5, −2) were consistent with the predicted temperature at the surface of Halley's nucleus at aphelion. Inversion of rotational state population to temperature is less straightforward because the H_2O molecules are IR pumped and strongly non-LTE, with a varying rate of neutral-neutral and neutral-electron collisions playing a significant role in the level population. Within these limitations, the rotational temperatures derived (T_{rot} ~60 K for the March 1986 KAO data) suggest that the kinetic temperature, as also reproduced in hybrid/gas-dynamic models (*Combi*, 1989), is also similarly low in the region of the coma sampled.

An area where considerable progress has been made is in the inversion of H_2O brightness to Q_{H_2O} and its spatial and temporal variance. *Weaver et al.* (1987) derived Q_{H_2O} from the same v_3 KAO data used to derive T_{spin} and T_{rot}, obtaining results similar to OH, but with far greater variability on both short and long timescales. They attributed the short-term effects to a parent species sensitivity to nucleus activity that had previously been identified in CS (*McFadden et al.*, 1987), but could not fully account for the longer-term discrepancies. Finally, while they lacked the spectral resolution to sample the velocity structure of the v_3 lines directly, *Larson et al.* (1987) showed that gas production asymmetry provides a signature of outflow in the form of a Doppler term in the lines. Their post-perihelion measurements showed considerable variation from canonical assumptions of water outflow consistent with collisional heating in active comets that was subsequently shown more dramatically in C/1995 O1 (Hale-Bopp) (*Harris et al.*, 2002).

Subsequent observations of water production and temperature using this v_3 band and rotational lines near 180 µm have been made with the Infrared Space Observatory (ISO) satellite (*Crovisier et al.*, 1997) with results similar to those of the KAO and IKS IR spectrometer onboard in the *Giotto* spacecraft. In addition, *Dello Russo et al.* (2000) has used the CSHELL instrument at NASA IRTF to demonstrate a method for groundbased detection of water in nonresonance fluorescence "hot-bands" that are not absorbed by the atmosphere. These were first confirmed in Comet 1P/Halley (e.g., *Weaver et al.*, 1987.) Unlike other studies, the CSHELL measurements are spatially resolved over the small aperture of the instrument, providing the first radial maps of the inner coma water distribution and an opportunity to apply production models to identify both collisional effects and the presence of an extended source.

Progress in the study of line-resolved aperture photometry of the water band at 557 GHz, which are available only from space platforms, has proven very useful for deriving Q_{H_2O} and V_{H_2O} from the line shape. This emission was first detected from Comet C/1991 H1 (Lee) using the Submillimeter Wave Astronomy Satellite (SWAS) (*Neufeld et al.*, 2000), and subsequently from Comets C/2001 A2 (LINEAR), 19P/Borrelly, C/2000 WM1 (LINEAR), and 153P/2002 C1 (Ikeya-Zhang) with the Odin satellite. The radio measurements suffer from the same large area summation as those used to examine other species; however, they share their primary strength, velocity sensitivity. The Odin instrumental resolution is sufficiently high (80 m/s) to fully re-

solve the ~1 km s^{-1} bulk flow of water, which can now be compared with similar measurements of other water daughters (H and OH) over similar FOV and used to model the complex kinetics of the coma in the active comets detectable with these techniques.

5.3. Water Diagnostics in the Far-Ultraviolet (Far Ultraviolet Spectroscopic Explorer) Bandpass

The launch of FUSE opened a new chapter in comet studies by enabling direct observation of new diagnostics of water production and chemistry in the FUV. To date, FUSE has been used to observe long-period Comets C/1999 T1 (McNaught-Hartley), C/2000 WM1 (LINEAR), and C/2001 A2 (LINEAR) (*Feldman et al.*, 2002; *Weaver et al.*, 2002). In addition to putting new limits on the abundance fraction of highly volatile species such as Ar (*Weaver et al.*, 2002) and CO (*Feldman et al.*, 2002), several diagnostic features identified with water photochemistry have been observed. These include the H_2 Lyman band lines at 1071.6, 1118.6, and 1166.8 Å, H Lyman-β and other lines of the Lyman series, O I multiplets at 989, 1027, and 1040 Å, fluorescence from the metastable O I(^{1}D) state at 1152 Å [analogous to the more commonly studied C(^{1}D) emission line at 1931 Å (*Tozzi et al.*, 1998)] (*Feldman et al.*, 2002), and the possible detection of charge exchange (between coma neutrals and the high ionization state component of the solar wind) cascade lines of O IV at 1031.93 and 1037.62 Å (*Weaver et al.*, 2002).

The utility of these new diagnostics is still being evaluated; however, it is promising that the two objects observed to date have been intrinsically weak gas producers that would yield poor results with many of the other techniques described above. Detection of H_2 is significant in that it provides a new diagnostic of the poorly studied branching rate to that daughter from water. H_2 is also produced with a substantial excess velocity (Table 2) and has a long lifetime against photodissociation (Table 4), a combination that means that, while H_2 is a minor inner coma constituent, it is a dominant molecule in the outer coma. *Feldman et al.* (2002) suggest that the O(^{1}D–^{0}D) resonance is being stimulated by an unknown process rather than fluorescing in the comet inner coma. This result is somewhat enigmatic, considering that fluorescence is the dominant source of C(^{1}D) in the coma despite having a lower energy of transition than O(^{1}D). Whether the O(^{1}D) results for the three comets studied is typical or an anomaly brought about by the unique circumstances of the FUSE observations (e.g., aperture, solar cycle, low intrinsic production rates) will require a larger dataset and close theoretical scrutiny to determine.

The major limitations for working in the FUV bandpass are the availability of observing time, only moderately high spectral resolution, high sky background noise, the relatively low level of solar FUV flux, and the 30" × 30" angular extent of its aperture, which combine to limit observations to the inner coma. In this regard FUSE has similar limitations to IUE in the 1980s and 1990s. As in this earlier case, the advent of larger aperture and more flexible instruments will improve our ability to monitor these important new features.

5.4. The Relationship of Other Coma Species to Water

Water is the dominant volatile component of comet nuclei, and therefore its state and evolution are both relevant to developing an understanding of other coma species, some of which have reciprocal effects on the water coma. The above descriptions reveal several weaknesses in the diagnostics suite used to determine the characteristics of the water coma, and one technique to address this is to combine studies of water with other coma constituents that fill these observational voids. The detailed physics of the other coma species is dealt with in other chapters of this book, and our goal here is to illustrate only how a coordinated observing-modeling approach between them and water will increase our understanding of each. While coordinated observations of multiple species provide benefits for understanding many other properties of the coma, including its thermal properties and the interaction of the solar wind, we will briefly touch here on two examples (the velocity structure of the inner coma and the role of secondary sources of water daughter species) where the above techniques provide an incomplete picture and observational solutions using other species have been identified.

5.4.1. Velocity structure. The low velocity of most coma constituents (in particular, H_2O, OH, and O for water) requires extremely high spectral resolution ($R > 10^5$) to measure. Unfortunately, the most common technique used to achieve such high resolution in the visible and UV (echelle spectroscopy) has a correspondingly small étendue that samples only small regions with fairly low sensitivity. Interferometric techniques (*Stephan et al.*, 2001; *Watchorn et al.*, 2001; *Morgenthaler et al.*, 2002) that co-add large FOV at high spectral resolution in the visible and UV have considerable future promise in this area; however, the most mature techniques exist in the radio. Unfortunately, the primary water diagnostic in the radio (OH at 18 cm) is strongly quenched in the inner coma, which is its the most dynamically active region for active comets. To get around this problem it is possible to use other, less-quenched OH [e.g., IR (*Crovisier et al.*, 1997)] emission features or other species to study this region and combine the results to complete the dynamical picture. Several such features exist and are in common use to determine the velocity within ~10^3–10^4 km of the nucleus (*Biver et al.*, 1999b). One commonly observed emission that is useful for such comparative work is the J = 1–0 line of HCN (*Schloerb et al.*, 1987; *Biver et al.*, 1999b), which is detectable close to the nucleus at subkilometers per second velocity resolution and therefore useful in coordination with OH to measure velocity in the inner and outer coma. Models of the outflow (*Combi*, 1989) similar to those shown in Figs. 3–6 reproduced the secular variation (with heliocentric distance) of the HCN line width seen by *Schloerb et al.* (1987). *Harris et al.* (2002) combined published measurements of the outer coma OH velocity (*Colom et al.*, 1999) and inner coma HCN velocity (*Biver et al.*, 1999a) with the radial distribution of emission from OH to map out the acceleration across the coma of C/1995 O1 (Hale-Bopp). The magnitude and variation of these velocity measurements, both with distance from the nucleus and overall with heliocentric distance, are also in reasonable agreement with models for the expansion of the coma (*Combi et al.*, 1999b), even for the extreme physical state of Hale-Bopp.

5.4.2. Characterization secondary sources. Of the primary neutral water species only H_2O itself and OH have no significant contribution from other sources. Of the others, O I has the largest nonwater component, in this case from a combination of CO and CO_2. The precision of model interpretation of $O(^1D)$ will depend on the precision by which the CO- and H_2O-derived components are separated or by our knowledge of the CO/H_2O ratio, which varies substantially among comets and with heliocentric distance. The size of the CO/CO_2 contribution can be dominant in the outer solar system, where water outgassing is suppressed, but even where water dominates, the addition of CO/CO_2-derived O [particularly $O(^1D)$] is significant. The role of CO/CO_2 is further complicated by the fact that the rate of formation of $O(^1D)$ differs between the two species and with respect to both pathways in water. Indeed, $O(^1D)$, derived from CO (as a daughter) and CO_2 (as a granddaughter), most likely dominates over the OH contribution in the outer coma due to the longer lifetime of CO (*Tozzi et al.*, 1998; *Harris et al.*, 1999). This would be even further complicated by a substantial fraction of CO coming from an extended source of another parent like H_2CO (*DiSanti et al.*, 2001). Moreover, the ratios of $CO : CO_2$, CO/H_2O, and $CO_2 : H_2O$ vary substantially between comets, with no obvious predictive trend yet identified (*Weaver et al.*, 1994; *Feldman et al.*, 1997; see also *Bockelée-Morvan et al.*, 2004). Thus, to account for the contribution of these species to the $O(^1D)$ brightness and spatial distribution, it is necessary to study diagnostics of both CO and CO_2 chemistry. The most direct method is to study $C(^1D)$, the metastable analog to $O(^1D)$ that is formed in the same process from CO (*Huebner et al.*, 1992; *Tozzi et al.*, 1998). From $Q_{C(^1D)}$ it is possible to separate the CO and OH contributions directly over any simultaneously observed FOV.

There are two diagnostics for $C(^1D)$: fluorescence from the ^1D state at 1931 Å and direct detection of the $C(^1D)$ decay at 9828/9850 Å. The former is accessible only from space and has typically been studied only over small FOV (*Feldman et al.*, 1997), which limits its usefulness for a full coma correction. Direct observation of the NIR decay lines holds more promise, because it is both directly analogous to the O I visible line and observable from the ground. This feature is faint compared to $O(^1D)$, has high background contamination, and falls near a telluric OH absorption. It has only been detected in two comets, first an unpublished detection from 1P/Halley by *Münch et al.* (1986), and more recently from Fabry-Pérot interferometric observations of C/1995 O1 Hale-Bopp (*Oliversen et al.*, 2002). The Hale-Bopp detection provides the first quantitative estimates of the role of C(^{1}D) in the inner and outer coma and identifies how future development of the interferometric technique will improve our ability to study this diagnostic.

6.   OUTSTANDING ISSUES AND FUTURE STUDIES OF COMA STRUCTURE AND EXPANSION

There is currently no single diagnostic signature of water that provides a complete description of its spatial, velocity, temporal, and production rate evolution in the coma. Inferring such detail from any one observation therefore requires either an estimate or approximation of the unknown quantities (usually from observed characteristics of earlier comets) or model derivation of them. As the quality of remote sensing data has improved, the limiting factor in interpretation has shifted from the instrumental regime to the degree to which we know (1) underlying physical parameters such as g-factors and rate constants; (2) internal factors including quenching, temperature, solar wind density and velocity, collision rates, optical depth, and source distribution; (3) incoming solar spectral intensity in the UV; (4) parent species' contributions and branching ratios; and (5) the two-dimensional and temporal structure of the coma and how it effects observations taken over different spatial scales or at different times.

To the extent that these parameters are known for each case, the modeling techniques discussed here each provide a consistent picture of water evolution from the observations. However, one effect of this shift is that the precision of measurements now often greatly exceeds the accuracy of the inverted result (*Schleicher et al.*, 1998). In many cases the divergence between different methods stems more from the failure to fully account for this in the determination of uncertainty than from the validity of the approach, and can give the impression of nonconvergence when isolated observations are compared directly. Another major effect is the inversion or synthetic reproduction of data with spherical and/or steady-state models. Not accounting for such complicating effects can result in erroneously attributing features in the data to those free parameters that are left in the model. Finally, in the absence of simultaneous *in situ* data from multiple spacecraft, we are always left with the projection of all information (spatial, velocity, and temperature) along the line of sight that again can result in similar misinterpretations.

A clear improvement of the model fits is obtained when multiple diagnostics with synergistic results and unknowns are combined, particularly if they are obtained nearly simultaneously and can account for temporal and spatial variances. For example, the combination of wide-field (production rate) and velocity-resolved (outflow and nonthermal components) aperture photometry of a species with two-dimensional spatial imaging (parent-daughter scale lengths, azimuthal structure, propagating temporal activity changes) provides a set of reciprocal boundary values for the individual uncertainties in each observation. Where this approach has been applied, it has significantly reduced the

allowed parameter space of the converged model for the coma (*Combi et al.*, 2000; *Morgenthaler et al.*, 2001; *Harris et al.*, 2002). The caveat to this is that the capability to perform the required observations is widely distributed and is typically organized only for high-profile apparitions (e.g., 1P/Halley, C/1995 O1 Hale-Bopp, C/1996 B2 Hyakutake). The results of these campaigns suggest that an improvement in our understanding of both the role of water in the coma and in the physical parameters that describe its evolution would result from a consistent organized effort targeting a larger sample of comets covering a wider range of activity and evolution.

Acknowledgments.   M.R.C. acknowledges support from grant NAG5-13239 from the NASA Planetary Atmospheres program.

REFERENCES

A'Hearn M. F., Hoban S., Birch P. V., Bowers C., Martin R., and Klinglesmith D. A. III (1986) Cyanogen jets in Comet Halley. *Nature, 324,* 649–651.

Allen M., Delitsky M., Huntress W., Yung Y., and Ip W.-H. (1987) Evidence for methane and ammonia in the coma of Comet P/Halley. *Astron. Astrophys., 187,* 502–512.

Anderson C. M. (1999) Fiberoptically multiplexed medium resolution spectroscopy from the WIYN telescope, singlet D neutral oxygen, NH$_2$, and H$_2$O$^+$. *Earth Moon Planets, 78,* 99–104.

Bertaux J.-L. and Lallement R. (1984) Analysis of interplanetary Lyman-alpha line profile with a hydrogen absorption cell — Theory of the Doppler angular spectral scanning method. *Astron. Astrophys., 140,* 230–242.

Biver N., Bockelée-Morvan D., Colom P., Crovisier J., Germain B., Lellouch E., Davies J. K., Dent W. R. F., Moreno R., Paubert G., Wink J., Despois D., Lis D. C., Mehringer D., Benford D., Gardner M., Phillips T. G., Gunnarsson M., Rickman H., Winnberg A., Bergman P., Johansson L. E. B., and Rauer H. (1999a) Long-term evolution of the outgassing of Comet Hale-Bopp from radio observations. *Earth Moon Planets, 78,* 5–11.

Biver N., Bockelée-Morvan D., Crovisier J., Davies J. K., Matthews H. E., Wink J. E., Rauer H., Colom P., Dent W. R. F., Despois D., Moreno R., Paubert G., Jewitt D., and Senay M. (1999b) Spectroscopic monitoring of Comet C/1996 B2 (Hyakutake) with the JCMT and IRAM radio telescopes. *Astron. J., 118,* 1850–1872.

Biver N., Bockelée-Morvan D., Colom P., Crovsier J., Henry F., Lellouch E., Winnberg A., Johansson L. E. B., Cunnarsson M., Rickman H., Rantakyrö F., Davies J. K., Dent W. R. F., Paubert G., Moreno R., Wink J., Despois D., Benford D. J., Gardner M., Lis D. C., Mehringer D., Phillips T. G., and Rauer H. (2002) The 1995–2002 long-term monitoring of Comet C/1995 O1 (Hale-Bopp) at radio wavelength. *Earth Moon Planets, 90,* 5–14.

Bird G. A. (1994) *Molecular Gas Dynamics and the Direct Simulation of Gas Flows.* Clarendon, Oxford. 459 pp.

Bockelée-Morvan D. (1987) A model for the excitation of water in comets. *Astron. Astrophys., 181,* 169–181.

Bockelée-Morvan D. and Crovisier J. (1987) The role of water in the thermal balance of the coma. In *Proceedings of the Symposium on the Diversity and Similarity of Comets* (E. J. Rolfe and B. Battrick, eds.), pp. 235–240. ESA SP-278, Noordwijk, The Netherlands.

Bockelée-Morvan D., Crovisier J., and Gerard E. (1990) Retrieving the coma gas expansion velocity in P/Halley, Wilson (1987 VII) and several other comets from the 18-cm OH line shapes. *Astron. Astrophys., 238,* 382–400.

Bockelée-Morvan D., Crovisier J., Colom P., and Despois D. (1994) The rotational lines of methanol in Comets Austin 1990 V and Levy 1990 XX. *Astron. Astrophys., 287,* 647–665.

Bockelée-Morvan D., Crovisier J., Mumma M. J., and Weaver H. A. (2004) The composition of cometary volatiles. In *Comets II* (M. C. Festou et al., eds.), this volume. Univ. of Arizona, Tucson.

Brooke T. Y., Tokunaga A. T., Weaver H. A., Crovisier J., Bockelée-Morvan D., and Crisp D. (1996) Detection of acetylene in the infrared spectrum of Comet Hyakutake. *Nature, 383,* 606–608.

Brooke T. Y., Weaver H. A., Chin G., Bockelée-Morvan D., Kim S. J., and Xu L.-H. (2003) Spectroscopy of Comet Hale-Bopp in the infrared. *Icarus, 166,* 167–187.

Budzien S. A., Festou M. C., and Feldman P. D. (1994) Solar flux variability of cometary H_2O and OH. *Icarus, 107,* 164–188.

Chin G. and Weaver H. A. (1984) Vibrational and rotational excitation of CO in comets. Nonequilibrium calculations. *Astrophys. J., 285,* 858–869.

Cochran A. L. (1985) A re-evaluation of the Haser model scale lengths for comets. *Astron. J., 90,* 2609–2514.

Cochran A. L. and Schleicher D. G. (1993) Observational constraints on the lifetime of cometary H_2O. *Icarus, 105,* 235–253.

Colom P., Gerard E., Crovisier J., Bockelée-Morvan D., Biver N. and Rauer H. (1999) Observations of the OH radical in Comet C/1995 O1 (Hale-Bopp) with the Nançay radio telescope. *Earth Moon Planets, 78,* 37–43.

Combes M., Moroz V., Crifo J.-F., Lamarre J. M., Charra J., Sanko N. F., Soufflot A., Bibring J. P., Cazes S., Coron N., Crovisier J., Emerich C., Encrenaz T., Gispert R., Grigoryev A. V., Guyot G., Krasnopolsky V. A., Nikolsky Yu. V., and Rocard F. (1986) Infrared sounding of Comet Halley from Vega 1. *Nature, 321,* 266–268.

Combi M. R. (1987) Sources of cometary radicals and their jets — Gases or grains. *Icarus, 71,* 178–191.

Combi M. R. (1989) The outflow speed of the coma of Halley's comet. *Icarus, 81,* 41–50.

Combi M. R. (1996) Time-dependent gas kinetics in tenuous planetary atmospheres: The cometary coma. *Icarus, 123,* 207–226.

Combi M. R. (2002) Hale-Bopp: What makes a big comet different. Coma dynamics: Observations and theory. *Earth Moon Planets, 89,* 73–90.

Combi M. R. and Delsemme A. H. (1980) Neutral cometary atmospheres. I. Average random walk model for dissociation in comets. *Astrophys. J., 237,* 633–641.

Combi M. R. and Fink U. (1993) P/Halley — Effects of time-dependent production rates on spatial emission profiles. *Astrophys. J., 409,* 790–797.

Combi M. R. and Fink U. (1997) A critical study of molecular photodissociation and CHON grain sources for cometary C. *Astrophys. J., 484,* 879–890.

Combi M. R. and Smyth W. H. (1988a) Monte Carlo particle trajectory models for neutral cometary gases. I. Models and equations. *Astrophys. J., 327,* 1026–1043.

Combi M. R. and Smyth W. H. (1988b) Monte Carlo particle trajectory models for neutral cometary gases. II. The spatial morphology of the Lyman-alpha coma. *Astrophys. J., 327,* 1044–1059.

Combi M. R., Bos B. J., and Smyth W. H. (1993) The OH distribution in cometary atmospheres — A collisional Monte Carlo model for heavy species. *Astrophys. J., 408,* 668–677.

Combi M. R., Brown M. E., Feldman P. D., Keller H. U., Meier R. R., and Smyth W. H. (1998) Hubble Space Telescope ultraviolet imaging and high-resolution spectroscopy of water photodissociation products in Comet Hyakutake (C/1996 B2). *Astrophys. J., 494,* 816–821.

Combi M. R., Cochran A. L., Cochran W. D., Lambert D. L., and Johns-Krull C. M. (1999a) Observation and analysis of high-resolution optical line profiles in Comet Hyakutake (C/1996 B2). *Astrophys. J., 512,* 961–968.

Combi M. R., Kabin K., DeZeeuw D. L., Gombosi T. I., and Powell K. G. (1999b) Dust-gas interrelations in comets: Observations and theory. *Earth Moon Planets, 79,* 275–306.

Combi M. R., Reinard A. A., Bertaux J.-L., Quemerais E., and Mäkinen T. (2000) SOHO/SWAN observations of the structure and evolution of the hydrogen Lyman-alpha: Coma of Comet Hale-Bopp (1995 O1). *Icarus, 144,* 191–202.

Cravens T. E., Kozyra J. U., Nagy A. F., Gombosi T. I., and Kurtz M. (1987) Electron impact ionization in the vicinity of comets. *J. Geophys. Res., 92,* 7341–7353.

Crifo J.-F., Crovisier J., and Bockelée-Morvan D. (1989) Proposed water velocity and temperature radial profiles appropriate to comet Halley flyby conditions. Paper presented at the 121st Colloquium of the International Astronomical Union, held in Bamberg, Germany, April 24–28, 1989.

Crifo J.-F., Itkin A. L., and Rodionov A. V. (1995) The near-nucleus coma formed by interacting dusty gas jets effusing from a cometary nucleus: I. *Icarus, 116,* 77–112.

Crifo J.-F., Rodionov A. V., and Bockelée-Morvan D. (1999) The dependence of the circumnuclear coma structure on the properties of the nucleus. *Icarus, 138,* 85–106.

Crifo J.-F., Lukianov G. A., Rodionov A.V., Khanlarov G. O., and Zakharov V. V. (2002) Comparison between Navier-Stokes and direct Monte-Carlo simulations of the circumnuclear coma. I. Homogeneous, spherical source. *Icarus, 156,* 249–268.

Crifo J.-F., Loukianov G. A., Rodionov A. V., and Zakharov V. V. (2003) Navier-Stokes and direct Monte Carlo simulations of the circumnuclear coma II. Homogeneous, aspherical sources. *Icarus, 163,* 479–503.

Crifo J.-F., Kömle N. I., Fulle M., and Szegö K. (2004) Nucleus-coma structural relationships: Lessons from physical models. In *Comets II* (M. C. Festou et al., eds.), this volume. Univ. of Arizona, Tucson.

Crovisier J. (1984) The water molecule in comets: Fluorescence mechanisms and thermodynamics of the inner coma. *Astron. Astrophys., 130,* 361–372.

Crovisier J. (1989) The photodissociation of water in cometary atmospheres. *Astron. Astrophys., 213,* 459–464.

Crovisier J., Leech K., Bockelée-Morvan D., Brooke T. Y., Hanner M. S., Altieri B., Keller H. U., and Lellouch E. (1997) The spectrum of Comet Hale-Bopp (C/1995 O1) observed with the infrared space observatory at 2.9 astronomical units from the Sun. *Science, 275,* 1904–1907.

Crovisier J., Colom P., Gérard E., Bockelée-Morvan D., and Bourgois G. (2002) Observations at Nançay of the OH 18-cm lines in comets. The data base. Observations made from 1982 to 1999. *Astron. Astrophys., 393,* 1053–1064.

Dello Russo N., Mumma M. J., DiSanti M. A., Magee-Sauer K., Novak R., and Rettig T. W. (2000) Water production and release in Comet C/1995 O1 Hale-Bopp. *Icarus, 143,* 324–337.

Delsemme A. H. and Miller D. C. (1971) The continuum of Comet

Burnham (1960 II): The differentiation of a short period comet. *Planet. Space Sci., 19,* 1229–1257.

DiSanti M. A., Mumma M. J., Dello Russo N., and Magee-Sauer K. (2001) Carbon monoxide production and excitation in Comet C/1995 O1 (Hale-Bopp): Isolation of native and distributed CO sources. *Icarus, 153,* 162–179.

Drake J. F., Jenkins E. B., Bertaux J.-L., Festou M., and Keller H. U. (1976) Lyman-alpha observations of Comet Kohoutek 1973 XII with Copernicus. *Astrophys. J., 209,* 302–311.

Eberhardt P., Krankowsky D., Schulte W., Dolder U., Lämmerzahl P., Berthelier J. J., Woweries J., Stubbeman U., Hodges R. R., Hoffman J. H., and Illiano J. M. (1987) The CO and N_2 abundance in Comet P/Halley. *Astron. Astrophys., 187,* 481–484.

Eddington A. S. (1910) The envelopes of c 1908 (Morehouse). *Mon. Not. R. Astron. Soc., 70,* 442–458.

Fahr H. J. and Shizgal B. (1981) Modern exospheric theories and their observational relevance. *Rev. Geophys., 21,* 75–124.

Farnham T. L., Schleicher D. G., and A'Hearn M. F. (2000) The HB narrowband comet filters: Standard stars and calibrations. *Icarus, 147,* 180–204.

Feldman P. D., Festou M. C., Tozzi G. P., and Weaver H. A. (1997) The CO_2/CO abundance ratio in 1P/Halley and several other comets observed by IUE and HST. *Astrophys. J., 475,* 829–834.

Feldman P. D., Weaver H. A., and Burgh E. B. (2002) Far Ultraviolet Spectroscopic Explorer observations of CO and H_2 emission in Comet C/2001 A2 (LINEAR). *Astrophys. J. Lett., 576,* L91–L94.

Feldman P. D., Cochran A. L., and Combi M. R. (2004) Spectroscopic investigations of fragment species in the coma. In *Comets II* (M. C. Festou et al., eds.), this volume. Univ. of Arizona, Tucson.

Festou M. C. (1981a) The density distribution of neutral compounds in cometary atmospheres. I — Models and equations. *Astron. Astrophys., 95,* 69–79.

Festou M. C. (1981b) The density distribution of neutral compounds in cometary atmospheres. II — Production rate and lifetime of OH radicals in Comet Kobayashi-Berger-Milon /1975 IX/. *Astron. Astrophys., 96,* 52–57.

Festou M. C. (1999) On the existence of distributed sources in comet comae. *Space Sci. Rev., 90,* 53–67.

Festou M. C. and Feldman P. D. (1981) The forbidden oxygen lines in comets. *Astron. Astrophys., 103,* 154–159.

Fink U. and Hicks M. D. (1996). A survey of 39 comets using CCD spectroscopy. *Astrophys. J., 459,* 729–743.

Fink U., Combi M. R., and DiSanti M. A. (1991) Comet P/Halley — Spatial distributions and scale lengths for C_2, CN, NH_2, and H_2O. *Astrophys. J., 383,* 356–371.

Finson M. L. and Probstein R. F. (1968) A theory of dust comets. 1. Model and equations. *Astrophys. J., 154,* 327–352.

Fokker A. D. (1953) Some notes on the fountain model of cometary envelopes and streamers. *Mem. Sci. Liége, Ser. 4, 13,* 241–259.

Glinski R. J., Harris W. M., Anderson C. M., and Morgenthaler J. P. (2003) Oxygen/hydrogen chemistry in inner comae of active comets. In *Formation of Cometary Material,* 25th Meeting of the IAU, Joint Discussion 14, 22 July 2003, Sydney, Australia.

Gombosi T. I. (1994) *Gaskinetic Theory.* Cambridge Univ., Cambridge. 297 pp.

Gombosi T. I., Cravens T. E., and Nagy A. F. (1985) Time-dependent dusty gas dynamical flow near cometary nuclei. *Astrophys. J., 293,* 328–341.

Gombosi T. I., Nagy A. F., and T. E. Cravens (1986) Dust and neutral gas modeling of the inner atmospheres of comets. *Rev. Geophys., 24,* 667–700.

Grün E., Hanner M. S., Peschke S. B., Müller T., Boehnhardt H., Brooke T. Y., Campins H., Crovisier J., Delahodde C., Heinrichsen I., Keller H. U., Knacke R. F., Krüger H., Lamy P., Leinert C., Lemke D., Lisse C. M., Müller M., Osip D. J., Solc M., Stickel M., Sykes M., Vanysek V., and Zarnecki J. (2001) Broadband infrared photometry of Comet Hale-Bopp with ISOPHOT. *Astron. Astrophys., 377,* 1098–1118.

Häberli R., Combi M. R., Gombosi T. I., De Zeeuw D. L., and Powell K. G. (1997) Quantitative analysis of H_2O^+ coma images using a multiscale MHD model with detailed ion chemistry. *Icarus, 130,* 373–386.

Harlander J. M., Roesler F. L., Cardon J. G., Englert C. R., and Conway R. R. (2002) Shimmer: A spatial heterodyne spectrometer for remote sensing of Earth's middle atmosphere. *Applied Optics, 41,* 1343–1352.

Harmon J. K., Ostro S. J., Benner L. A. M., Rosema K. D., Jurgens R. F., Winkler R., Yeomans D. K., Choate D., Cormier R., Giogini J. D., Mitchell D. L., Chodas P. W., Rose R., Kelley D., Slade M. A., and Thomas M. L. (1997) Radar detection of the nucleus and coma of Comet Hyakutake (C/1996 B2). *Science, 278,* 1921–1924.

Harris W. M., Combi M. R., Honeycutt R. K., and Mueller B. E. A. (1997) Evidence for interacting gas flows and an extended volatile source distribution in the coma of Comet C/1996 B2 (Hyakutake). *Science, 277,* 676–681.

Harris W. M., Nordsieck K. H., Scherb F., and Mierkiewicz E. J. (1999) UV photopolarimetric imaging of C/1995 O1 (Hale-Bopp) with the Wide Field Imaging Survey Polarimeter (WISP). *Earth Moon Planets, 78,* 161–167.

Harris W. M., Scherb F., Mierkiewicz E. J., Oliversen R. J., and Morgenthaler J. P. (2002) Production, outflow velocity, and radial distribution of H_2O and OH in the coma of Comet C/1995 O1 [Hale-Bopp] from wide field imaging of OH. *Astrophys. J., 578,* 996–1008.

Haser L. (1957) Distribution d'intensite dans la tete d'une comete. *Bull. Acad. R. de Belgique, Classe de Sci. 43(5),* 740–750.

Hodges R. R. (1990) Monte Carlo simulation of nonadiabatic expansion in cometary atmospheres: Halley. *Icarus, 83,* 410–433.

Huebner W. F. (1985) The photochemistry of comets. In *The Photochemistry of Atmospheres: Earth, the Other Planets, and Comets* (J. S. Levine, ed.), pp. 437–481. Academic, Orlando.

Huebner W. F. and Carpenter C. W. (1979) *Solar Photo Rate Coefficients.* Los Alamos Sci. Lab. Report LA-8085-MS, Los Alamos National Laboratory, Los Alamos, New Mexico.

Huebner W. F. and Keady J. J. (1983) Energy balance and photochemical processes in the inner coma. In *Cometary Exploration, Vol. 1* (T. I. Gombosi, ed.), pp. 165–183. Central Research Institute for Physics, Budapest, Hungary.

Huebner W. F., Keady J. J., and Lyon S. P. (1992) Solar photo rates for planetary atmospheres and atmospheric pollutant. *Astrophys. Space Sci., 195,* 1–294.

Huppler D., Reynolds R. J., Roesler F. L., Scherb F., and Trauger J. (1975) Observations of Comet Kohoutek /1973f/ with a ground-based Fabry-Pérot spectrometer. *Astrophys. J., 202,* 276–282.

Ip W.-H. (1983) On photochemical heating of cometary comae: The cases of H_2O and CO-rich comets. *Astrophys. J., 264,* 726–732.

Ip W.-H. (1989) Photochemical heating of cometary comae.

III. The radial variation of the expansion velocity of CN shells in comet Halley. *Astrophys. J., 346,* 475–480.

Jackson W. M. (1976) Laboratory observations of the photochemistry of parent molecules: A review. In *The Study of Comets* (B. Donn et al., eds.), pp. 679–704. NASA SP-393, Washington, DC.

Jackson W. M. (1980) The lifetime of the OH radical in comets at 1 AU. *Icarus, 41,* 147–152.

Jewitt D., Senay M., and Matthews H. (1996) Observations of carbon monoxide in Comet Hale-Bopp. *Science, 271,* 1110–1113.

Kawakita H. and Watanabe J. (1998) NH_2 and its parent molecule in the inner coma of Comet Hyakutake (C/1996 B2). *Astrophys. J., 495,* 946–950.

Kawakita H., Watanabe J., Ando H., Alki W., Fuse T., Honda S., Izumiura H., Kajino T., Kambe E., Kawanaomoto S., Noguchi K., Okita K., Sadakane K., Sato B., Takada-Midai M., Takeda Y., Usunda T., Watanabe E., and Yoshinda M. (2001) The spin temperature of NH_3 in Comet C/1994 S4 (LINEAR). *Science, 294,* 1089–1091.

Keller H. U. and Meier R. R. (1976) A cometary hydrogen model for arbitrary observational geometry. *Astron. Astrophys., 52,* 273–281.

Keller H. U., Bohlin J. D., and Tousey R. (1975) High resolution Lyman alpha observations of Comet Kohoutek /1973f/ near perihelion. *Astron. Astrophys., 38,* 413–416.

Kiselev N. N. and Velichko F. P. (1999) Aperture polarimetry and photometry of Comet Hale-Bopp. *Earth Moon Planets, 78,* 347–352.

Kitamura Y. (1986) Axisymmetric dusty gas jet in the inner coma of a comet. *Icarus, 66,* 241–257.

Kitamura Y., Ashihara O., and Yamamoto T. (1985) A model for the hydrogen coma of a comet. *Icarus, 61,* 278–295.

Komitov B. (1989) The metastable oxygen O (^{1}D) as a possible source of OH molecules in the cometary atmospheres. *Adv. Space Res., 9,* 177–179.

Körösmezey A. and Gombosi T. I. (1990) A time-dependent dusty gas dynamic model of axisymmetric cometary jets. *Icarus, 84,* 118–153.

Krasnopolsky V. A. and 34 colleagues (1986a) The VEGA-2 TKS experiment — Some spectroscopic results for Comet Halley. *Soviet Astron. Lett., 12(4),* 259–262.

Krasnopolsky V. A. and 13 colleagues (1986b) Spectroscopic study of Comet Halley by the VEGA 2 three-channel spectrometer. *Nature, 321,* 269–271.

Kupperman D. G. (1999) Nonthermal atoms in planetary, satellite, and cometary atmospheres. Ph.D. thesis, The Johns Hopkins University, Baltimore, Maryland.

Lämmerzahl P. and 11 colleagues (1987) Expansion velocity and temperatures of gas and ions measured in the coma of Comet P/Halley. *Astron. Astrophys., 187,* 169–173.

Lamy P. L. and Perrin J.-M. (1988) Optical properties of organic grains — Implications for interplanetary and cometary dust. *Icarus, 76,* 100–109.

Larson H. P., Mumma M. J., and Weaver H. A. (1987) Kinematic properties of the neutral gas outflow from Comet P/Halley. *Astron. Astrophys., 187,* 391–397.

Lecacheux, A., Biver N., Crovisier J., Bockelée-Morvan D., Baron P., Booth R. S., Encrenaz P., Florén H.-G., Frisk U., Hjalmarson Å., Kwok S., Mattila K., Nordh L, Olberg M., Olofsson A. O. H., Rickman H., Sandqvist Aa., von Schéele F., Serra G., Torchinsky S., Volk K., and Winnberg A. (2003) Observations of water in comets with Odin. *Astron. Astrophys., 402,* L55–L58.

Marconi M. L. and Mendis D. A. (1983) The atmosphere of a dirty-clathrate cometary nucleus — two-phase, multifluid model. *Astrophys. J., 273,* 381–396.

Markiewicz W. J., Skorov Y. V., Keller H. U., and Kömle N. I. (1998) Evolution of ice surfaces within porous near-surface layers on cometary nuclei. *Planet. Space Sci., 46,* 357–366.

McFadden L. A., A'Hearn M. F., Edsall D. M., Feldman P. D., Roettger E. E., and Butterworth P. S. (1987) Activity of Comet P/Halley 23–25 March, 1986 — IUE observations. *Astron. Astrophys., 187,* 333–338.

Meier R., Eberhardt P., Krankowsky D., and Hodges R. R. (1993) The extended formaldehyde source in Comet P/Halley. *Astron. Astrophys., 277,* 677–690.

Mocknatsche D. O. (1938) *Leningrad State Univ. Annals, Astron. Series Issue 4.*

Morgenthaler J. P., Harris W. M., Scherb F., Anderson C. M., Oliversen R. J., Doane N. E., Combi M. R., Marconi M. L., and Smyth W. H. (2001) Large-aperture [O I] 6300 Å photometry of Comet Hale-Bopp: Implications for the photochemistry of OH. *Astrophys. J., 563,* 451–461.

Morgenthaler J. P., Harris W. M., Scherb F., and Doane N. E. (2002) Velocity resolved observations of H-α emission from Comet Hale-Bopp. *Earth Moon Planets, 90,* 77–87.

Mumma M. J., Weaver H. A., Larson H. P., Davis D. S., and Williams M. (1986) Detection of water vapor in Halley's comet. *Science, 232,* 1523–1528.

Mumma M. J., Weaver H. A., and Larson H. P. (1987) The ortho-para ratio of water vapor in Comet P/Halley. *Astron. Astrophys., 187,* 419–424.

Mumma M. J., McLean I. S., DiSanti M. A., Larkin J. E., Dello Russo N., Magee-Sauer K., Becklin E. E., Bida T., Chaffee F., Conrad A. R., Figer D. F., Gilbert A. M., Graham J. R., Levenson N. A., Novak R. E., Reuter D. C., Teplitz H. I., Wilcox M. I., and Li-Hong X. (2001) A survey of organic volatile species in Comet C/1999 H1 (Lee) using NIRSPEC at the Keck Observatory. *Astrophys. J., 546,* 1183–1193.

Münch G., Hippelein H., Hessmann F., and Gredel R. (1986) *Periodic Comet Halley (1982i).* IAU Circular No. 4183.

Nee J. B. and Lee L. C. (1985) Photodissociation rates of OH, OD, and CN by the interstellar radiation field. *Astrophys. J., 291,* 202–206.

Neufeld D. A., Stauffer J. R., Bergin E. A., Kleiner S. C., Patten B. M., Wang Z., Ashby M. L. N., Chin G., Erickson N. R., Goldsmith P. F., Harwit M., Howe J. E., Koch D. G., Plume R., Schieder R., Snell R. L., Tolls V., Winnewisser G., Zhang Y. F., and Melnick G. J. (2000) Submillimeter wave astronomy satellite observations of water vapor toward Comet C/1999 H1 (Lee). *Astrophys. J. Lett., 539,* L151–L154.

O'Dell C. R., Robinson R. R., Krishna Swamy K. S., McCarthy P. J., and Spinrad H. (1988) C_2 in Comet Halley: Evidence for its being third generation and resolution of the vibrational population discrepancy. *Astrophys. J., 334,* 476–488.

Oliversen R. J., Doane N. E., Scherb F., Harris W. M., and Morgenthaler J. P. (2002) Measurements of C I emission from Comet Hale-Bopp. *Astrophys. J., 581,* 770–775.

Oppenheimer M. and Downey C. J. (1980) The effect of solar-cycle ultraviolet flux variations on cometary gas. *Astrophys. J. Lett., 241,* L123–L127.

Prialnik D., Benkhoff J., and Podolak M. (2004) Modeling the structure and activity of comet nuclei. In *Comets II* (M. C. Festou et al., eds.), this volume. Univ. of Arizona, Tucson.

Richter K., Combi M. R., Keller H. U., and Meier R. R. (2000) Multiple scattering of Hydrogen Ly-alpha: Radiation in the coma of Comet Hyakutake (C/1996 B2). *Astrophys. J., 531,* 599–611.

Rodgers S. D., Charnley S. B., Huebner W. F., and Boice D. C. (2004) Physical processes and chemical reactions in cometary comae. In *Comets II* (M. C. Festou et al., eds.), this volume. Univ. of Arizona, Tucson.

Scherb F. (1981) Hydrogen production rates from ground-based Fabry-Pérot observations of Comet Kohoutek. *Astrophys. J., 243,* 644–650.

Scherb F., Roesler F. L., Tufte S., and Haffner M. (1996) Fabry-Pérot observations of [O I] 6300, Hα, Hβ and NH_2 emissions from Comet Hyakutake C/1996 B2. *Bull. Am. Astron. Soc., 28,* 927.

Schleicher D. G. and A'Hearn M. F. (1982) OH fluorescence in comets — Fluorescence efficiency of the ultraviolet bands. *Astrophys. J., 258,* 864–877.

Schleicher D. G. and A'Hearn M. F. (1988) The fluorescence of cometary OH. *Astrophys. J., 331,* 1058–1077.

Schleicher D. G. and Farnham T. L. (2004) Photometry and imaging of the coma with narrowband filters. In *Comets II* (M. C. Festou et al., eds.), this volume. Univ. of Arizona, Tucson.

Schleicher D. G., Millis R. L., Thompson D. T., Birch P. V., Martin R. M., Tholen D. J., Piscitelli J. R., Lark N. L., and Hammel H. B. (1990) Periodic variations in the activity of Comet P/ Halley during the 1985/1986 apparition. *Astron. J., 100,* 896–912.

Schleicher D. G., Millis R. L., and Birch P. V. (1998) Narrowband photometry of Comet P/Halley: Variation with heliocentric distance, season, and solar phase angle. *Icarus, 132,* 397–417.

Schloerb F. P. (1988) Collisional quenching of cometary emission in the 18 centimeter OH transitions. *Astrophys. J., 332,* 524–530.

Schloerb F. P., Kinzel W. M., Swade D. H., and Irvine W. M. (1987) Observations of HCN in Comet P/Halley. *Astron. Astrophys., 187,* 475–480.

Schloerb F. P., Devries C. H., Lovell A. J., Irvine W. M., Senay M., and Wootten H. A. (1999) Collisional quenching of OH radio emission from Comet Hale-Bopp. *Earth Moon Planets, 78,* 45–51.

Schultz D., Li G. S. H., Scherb F., and Roesler F. L. (1993) The $O(^1D)$ distribution of Comet Austin 1989c1 = 1990 V. *Icarus, 101,* 95–107.

Shimizu M. (1976) The structure of cometary atmospheres. *Astrophys. Space Sci., 40,* 149–155.

Skorov Y. V. and Rickman H. (1999) Gas flow and dust acceleration in a cometary Knudsen layer. *Planet. Space Sci., 47,* 935–949.

Skorov Y., Kömle N. I., Keller H. U., Kargle G., and Markiewicz W. J. (2001) A model of heat and mass transfer in a porous cometary nucleus based on a kinetic treatment of mass flow. *Icarus, 153,* 180–196.

Slanger T. G and Black G. (1982) Photodissociation channels at 1216 Å for H_2O, NH_3, and CH_4. *J. Chem. Phys., 77,* 2432–2437.

Smyth W. H., Combi M. R., and Stewart A. I. F. (1991) Analysis of the Pioneer Venus Lyman-α image of the hydrogen coma of Comet P/Halley. *Science, 253,* 1008–1010.

Smyth W. H., Combi M. R., Roesler F. L., and Scherb F. (1995) Observations and analysis of $O(^1D)$ and NH_2 line profiles for the coma of Comet P/Halley. *Astrophys. J., 440,* 349–360.

Stephan S. G., Chakrabarti S., Vickers J., Cook T., and Cotton D. (2001) Interplanetary H Ly-α: Observations from a sounding rocket. *Astrophys. J., 559,* 491–500.

Stief L. J., Payne W. A., and Klemm R. B. (1975) A flash photolysis-resonance fluorescence study of the formation of $O(^1D)$ in the photolysis of water and the reaction of $O(^1D)$ with H_2, Ar, and He. *J. Chem. Phys., 62,* 4000–4008.

Tegler S. C., Burke L. F., Wyckoff S., Womack M., Fink U., and DiSanti M. (1992) NH_3 and NH_2 in the coma of Comet Brorsen-Metcalf. *Astrophys. J., 384,* 292–297.

Tozzi G. P., Feldman P. D., and Festou M. C. (1998) Origin and production of $C(^1D)$ atoms in cometary comae. *Astron. Astrophys., 330,* 753–763.

van Dishoeck E. F. and Dalgarno A. (1984) The dissociation of OH and OD in comets by solar radiations. *Icarus, 59,* 305–313.

Wallace L. V. and Miller F. D. (1958) Isophote configurations for model comets. *Astron. J., 63,* 213–219.

Wallis M. K. (1982) Dusty gas-dynamics in real comets. In *Comets* (L. L. Wilkening, ed.), pp. 357–369. Univ. of Arizona, Tucson.

Watchorn S., Roesler F. L., Harlander J. M., Jaehnig K. A., Reynolds R. J., and Sanders W. T. (2001) Development of the spatial heterodyne spectrometer for VUV remote sensing of the interstellar medium. In *UV/EUV and Visible Space Instrumentation for Astronomy and Solar Physics* (O. H. Siegmund et al., eds.), pp. 284–295. SPIE Proc. Vol. 4498, SPIE — The International Society for Optical Engineering, Bellingham, Washington.

Weaver H. A., Mumma M. J., and Larson H. P. (1987) Infrared investigation of water in Comet P/Halley. *Astron. Astrophys., 187,* 411–418.

Weaver H. A., Feldman P. D., McPhate J. B., A'Hearn M. F., Arpigny C., and Smith T. E. (1994) Detection of CO Cameron band emission in Comet P/Hartley 2 (1991 XV) with the Hubble Space Telescope. *Astrophys. J., 422,* 374–380.

Weaver H. A., Feldman P. D., A'Hearn M. F., Arpigny C., Brandt J. C., and Stern S. A. (1999) Post-perihelion HST observations of Comet Hale-Bopp (C/1995 O1). *Icarus, 141,* 1–12.

Weaver H. A., Feldman P. D., Combi M. R., Krasnopolsky V., Lisse C. M., and Shemansky D. E. (2002) A search for argon and O VI in three comets using the Far Ultraviolet Spectroscopic Explorer. *Astrophys. J. Lett., 576,* L95–L98.

Womack M., Festou M. C., and Stern S. A. (1997) The heliocentric evolution of key species in the distantly-active Comet C/ 1995 O1 (Hale-Bopp). *Astron. J., 114,* 2789–2795.

Wu C. Y. R. and Chen F. Z. (1993) Velocity distributions of hydrogen atoms and hydroxyl radicals produced through solar photodissociation of water. *J. Geophys. Res., 98,* 7415–7435.

Wyckoff S., Tegler S., Wehinger P. A., Spinrad H., and Belton M. J. S. (1988) Abundances in Comet Halley at the time of the spacecraft encounters. *Astrophys. J., 325,* 927–938.

Xie X. and Mumma M. J. (1996a) Monte Carlo simulation of cometary atmospheres: Application to Comet P/Halley at the time of the Giotto spacecraft encounter. I. Isotropic model. *Astrophys. J., 464,* 442–456.

Xie X. and Mumma M. J. (1996b) Monte Carlo simulation of cometary atmospheres: Application to Comet P/Halley at the time of the Giotto spacecraft encounter. II. Axisymmetric model. *Astrophys. J., 464,* 457–475.

Yamamota T. (1981) On the photochemical formation of CN, C_2 and C_3 radicals in cometary comae. *Moon and Planets, 24,* 453–463.

Yamamota T. (1985) Formation environment of cometary nuclei in the primordial solar nebula. *Astron. Astrophys., 142,* 31–36.

Part VI:
Dust and Plasma

Composition and Mineralogy of Cometary Dust

Martha S. Hanner

Jet Propulsion Laboratory/California Institute of Technology

John P. Bradley

Lawrence Livermore Laboratory

Cometary dust is an unequilibrated, heterogeneous mixture of crystalline and glassy silicate minerals, organic refractory material, and other constituents such as iron sulfide and possibly minor amounts of iron oxides. Carbon is enriched relative to CI chondrites; some of the C is in an organic phase. The silicates are Mg-rich, while iron is distributed in silicates, sulfides, and FeNi metal. Infrared spectra of silicate emission features in Comet Hale-Bopp have led to identification of the minerals forsterite and enstatite. The strong similarity of all known cometary dust properties to the anhydrous chondritic aggregate class of interplanetary dust particles (IDPs) argues that comets are the source of these IDPs. High D/H ratios in organic refractory material in these IDPs as well as the physical and chemical structure of glassy silicate grains suggests a presolar origin for at least some components of cometary dust.

1. INTRODUCTION

Comets contain some of the least-altered material surviving from the early solar nebula. Cometary dust may contain both presolar particulates and solar nebula condensates. Their structure and mineralogy may hold important clues about the chemical and physical processes in the early solar system. Radial gradients in the temperature and chemical composition of the solar nebula and the extent of mixing of material between the warm inner regions and cold outer regions of the nebula at the epoch of comet formation should be evident today as differences in dust properties among comets related to their place of origin.

The extensive infrared spectroscopy of Comet Hale-Bopp, from the ground and from ESA's Infrared Space Observatory, has led to a revolution in our understanding of the silicate mineralogy in comets. In parallel, the past decade has seen extrodinary advances in the analysis of interplanetary dust particles (IDPs), leading to the discovery of eroded, glassy silicates of likely interstellar origin and isotopic anomalies in silicate and carbonaceous components also pointing toward a possible presolar origin.

This chapter summarizes our present knowledge of the composition and mineralogy of cometary dust based on *in situ* sampling during the spacecraft encounters with Comet 1P/Halley in 1986, Earth-based remote sensing via infrared spectroscopy, and laboratory analysis of captured IDPs of probable cometary origin.

2. *IN SITU* SAMPLING

The most direct means of determining the composition of cometary dust is by *in situ* sampling or analysis of returned samples. To date, *in situ* sampling has been carried out for just one comet, 1P/Halley. The two Soviet *Vega* spacecraft and ESA's *Giotto* probe each carried an impact ionization time of flight mass spectrometer to measure the elemental composition of the dust (*Kissel et al.,* 1986a,b). The composition was recorded for about 5000 particles in the mass range 10^{-16}–10^{-11} g, sampled over tens of thousands of kilometers in the coma along the three trajectories.

The sampled particles divide into three main types: mass spectra dominated by the major rock-forming elements, Mg, Si, Ca, Fe; mass spectra consisting primarily of the light elements H,C,N,O, the "CHON" particles; and mixed spectra containing both the rock and CHON elements. If one defines mixed particles as having a ratio of C to rock-forming elements between 0.1 and 10, then ~50% of the particles are mixed and ~25% are rock and CHON respectively (*Fomenkova et al.,* 1992). At some level, however, the CHON and rocky material are mixed down to the finest submicrometer scale in all particles (*Lawler and Brownlee,* 1992). The bulk abundances of the major rock-forming elements appear to be solar (chondritic) within a factor of ~2 (*Jessberger et al.,* 1988; *Jessberger,* 1999). (Conversion of the mass spectra to relative elemental abundances depends on the ion yields, which are uncertain by at least a factor of 2, because the instrument could not be calibrated at the high flyby speed of ~70 km/s.) The C abundance is roughly 10 times that of the primitive CI chondrites.

The rocky material displays a wide range in Fe/Mg abundance, but a narrow range in Si/Mg (*Lawler et al.,* 1989). Magnesium-rich (Fe-poor) silicates comprise at least 40% and perhaps ≥60% of the rocky particles (*Jessberger,* 1999). Iron is present in other minerals including metals (1–2%), iron sulfides (~10%), and possibly iron oxide (≤1%) (*Schulze et al.,* 1997).

The CHON spectra are evidence for an organic refractory component in the cometary dust. Significant clustering of subgroups (e.g., spectra dominated by [H, C], [H,

C, O], etc.) indicate variable composition of the organic refractory material (*Fomenkova et al., 1994*).

Isotopic ratios were found to be solar, within the measurement uncertainties, with the exception of $^{12}C/^{13}C$ (*Jessberger, 1999*). While low ratios (^{13}C enrichment) were judged to be uncertain due to noise of uncertain origin, *Jessberger* (1999) reported that definite ^{12}C enrichments, up to $^{12}C/^{13}C \sim$ 5000, were identified, indicative of presolar nucleosynthesis products.

3. SPECTROSCOPY OF COMETARY SILICATES

Small silicate grains in the cometary dust coma will produce an emission feature near 10 μm due to stretching vibrations in Si–O bonds. Additional bending mode vibrations occur between 16 and 35 μm. The wavelengths and shapes of these features are diagnostic of the mineral composition. The 10-μm feature lies within the 8–13-μm atmospheric "window," allowing groundbased observations. Although some 20-μm observations can also be made from the ground, the full 16–35-μm region is best studied from above the atmosphere.

Low-resolution ($\mathcal{R} \sim 50$–100), 8–13-μm spectra now exist for a number of comets. Several comets display strong structured silicate emission with total flux/continuum at 10 μm >1.5. These include long-period comets Bradfield (1987 XXIX = C/1987 P1) (*Hanner et al., 1990, 1994a*), Levy (1990 XX = C/1990 K1) (*Lynch et al., 1992*), Hyakutake (C/1996 B2), and Hale-Bopp (C/1995 O1) (*Hanner et al., 1999; Wooden et al., 1999; Hayward et al., 2000*); new comet Mueller (1994 I = C/1993 A1) (*Hanner et al., 1994b*); and 1P/Halley (*Bregman et al., 1987; Campins and Ryan, 1989*).

By far the strongest silicate emission was seen in Hale-Bopp. This comet was also unusual in displaying a strong silicate feature even at 4.6 AU preperihelion (*Crovisier et al., 1996; Grün et al., 2001*). Near-perihelion, groundbased spectra were obtained by several groups; all spectra show similar structure (see *Hanner et al., 1999*, for a review). A typical spectrum is presented in Fig. 1. The observed fluxes have been divided by a blackbody fitted at 8 and 12.5–13 μm. There are three main peaks, at 9.2, 10.0, and 11.2 μm, and minor structure at 11.9 and 10.5 μm. The spectral shape is very similar to that in P/Halley and the other comets cited above.

The 11.2-μm peak is attributed to crystalline olivine ([Mg,Fe]$_2$ SiO$_4$), based on the good spectral match with the measured spectral emissivity of Mg-rich olivine (*Stephens and Russell, 1979; Koike et al., 1993*). The 11.9-μm shoulder is also due to crystalline olivine. If all the silicates have similar temperatures, then only a small fraction (15–20%) of the silicate material needs to be in the form of crystalline olivine to produce the observed peak (*Hanner et al., 1994a*); the mass absorption coefficient of olivine near 11.2 μm is a factor of 3–10 times that of glassy silicates (*Day, 1976, 1981*). The broader 10-μm maximum in the

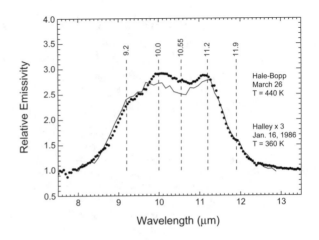

Fig. 1. The 10-μm silicate emission feature in Hale-Bopp at r = 0.92 AU (dots) and 1P/Halley at r = 0.79 AU from *Campins and Ryan* (1989) (line). Each spectrum has been divided by a Planck function for the temperature shown and the Halley spectrum has been multiplied by 3. The spectral peaks in the Hale-Bopp spectrum are marked. From *Hanner et al.* (1999).

cometary spectra is characteristic of amorphous olivine (*Stephens and Russell, 1979*); crystalline olivine has a secondary peak at 10 μm as well.

The 9.2-μm feature, first recognized in Hale-Bopp, is a signature of pyroxene ([Mg,Fe] SiO$_3$). A peak wavelength of 9.2 μm corresponds to amorphous, Mg-rich pyroxene (*Stephens and Russell, 1979; Dorschner et al., 1995*). Crystalline pyroxenes generate more variety in their spectra (*Sandford and Walker, 1985; Jaeger et al., 1998*). Peaks at 10–11 μm contribute to the width of the cometary feature and the structure near 10.5 μm. A peak near 9.3 μm is generally present in crystalline pyroxenes as well.

A remarkable 16–45-μm spectrum of Comet Hale-Bopp at r = 2.9 AU, shown in Fig. 2, was acquired with the short-wavelength spectrometer (SWS) on ESA's Infrared Space Observatory (*Crovisier et al., 1997, 2000*). Five peaks are clearly visible, corresponding in every case to laboratory spectra of crystalline forsterite (Mg-olivine) (*Koike et al., 1993*). Minor spectral structure is attributed to crystalline enstatite (Mg-pyroxene) (*Wooden et al., 1999; Crovisier et al., 2000*). This result is significant in indicating that the silicates are Mg-rich, in agreement with the elemental composition detected in P/Halley's coma.

These spectra contrast with airborne spectra of Comet P/Halley, the only other complete 20–30-μm spectra of a comet. A 16–30-μm spectrum at r = 1.3 AU (spectral resolution 0.2 μm) displays a sharp peak at 28.4 μm, but only weak features at 23.8 and 19.6 μm (*Herter et al., 1987*). Twenty- to 35-μm spectra with 0.5–1-μm resolution, taken at 1.2-AU preperihelion and 1.4-AU postperihelion, show only weak excess above the continuum at 24 and 33 μm (*Glaccum et al., 1987*). This result is puzzling. Although the strength of the 10-μm silicate feature was quite variable from day to day, an 11.2-μm olivine peak was clearly

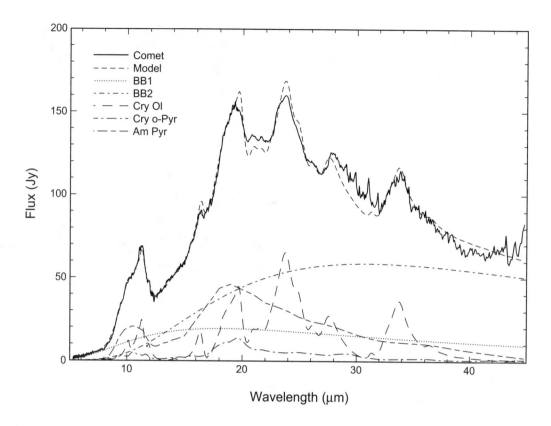

Fig. 2. ISO SWS spectrum of Comet Hale-Bopp at r = 2.8 AU, degraded to $\mathcal{R}$ = 500, compared with a five-component dust model: 280 K blackbody (BB1); 165 K blackbody (BB2); forsterite (Cry Ol 22%); orthopyroxene (Cry o-Pyr 8%); and amorphous pyroxene (Am Pyr 70%). From *Crovisier et al.* (2000).

visible as part of a strong 10-μm feature on the same day as the 1.2-AU airborne spectrum (*Bregman et al.,* 1987). Viewing restrictions on solar elongation angle prevented SWS observations when Hale-Bopp was near 1 AU, so we do not know how the 16–35-μm spectrum evolved as the grains were heated. However, an SWS spectrum acquired at 3.9-AU postperihelion still displayed the major forsterite peaks (*Crovisier et al., 2000*).

To produce a strong emission feature, silicate particles must have radii on the order of 1 μm or smaller. Larger particles will display a feature only if they are very porous aggregates and the individual constituent grains are micrometer-sized or smaller.

Spectral models to match the Hale-Bopp spectra with a mixture of silicate minerals have been presented by *Brucato et al.* (1999), *Colangeli et al.* (1999), *Galdemard et al.* (1999), *Hayward et al.* (2000), *Wooden et al.* (1999, 2000), and *Harker et al.* (2002). *Wooden et al.* (1999, 2000) proposed that the observed changes in spectral shape with heliocentric distance can be explained by temperature differences between more-transparent (cooler) Mg-rich pyroxene grains and less-transparent (warmer) olivine grains. In their model, the cooler crystalline pyroxenes comprise the major fraction of small silicate grains (~90%), and these grains produce the enhanced 9.3-μm feature near perihelion. *Hayward et al.* (2000) assumed in their modeling that

all the silicate grains would have sufficient thermal contact with absorbing material to be warm, regardless of their Mg/Fe content. They concluded that glassy pyroxenes were the most abundant component (>40%) and that crystalline olivine comprised ≤20% of the small silicate grains.

Harker et al. (2002) modeled the combined 8–13-μm spectra, IR photometry, and SWS spectra from October 1996 (r = 2.8 AU); the extended wavelength baseline enabled them to restrict the temperatures for the various amorphous and crystalline silicate components (tied to their Mg/Fe abundance ratios). They introduced porosity of the C and amorphous silicate particles (modeled with effective medium theory) by adopting the fractal dimension, D, as one of the model parameters. By comparing their best-fit model at r = 2.8 AU with the 1997 8–13-μm spectra and photometry, they concluded that the size distribution steepened and fractal dimension decreased as the comet approached perihelion. Crystalline silicates constituted about 30% by mass of the small dust grains in the coma at all epochs.

In summary, the observed spectral features imply a complex mineralogy for the cometary silicates, including both amorphous and crystalline grains of pyroxene and olivine composition. This mineralogy is consistent with the chondritic aggregate IDPs, described in the next section.

Not all comets display strong 10-μm emission features. The spectra of four new comets discussed in *Hanner et al.*

(1994a) are puzzling; each has a unique spectrum that is not yet understood. For example, an extremely broad emission feature was present in Wilson (1987 VII = C/1986 P1) (*Lynch et al., 1989*), suggesting a very amorphous silicate material. It is possible that we are witnessing the effect of cosmic-ray damage to the outermost layer of the nucleus over the lifetime of the Oort cloud.

No strong 10-μm emission feature has yet been seen in a short-period comet. Broad emission features about 20% above a black-body continuum were present in spectra of Comets 4P/Faye and 19P/Borrelly (*Hanner et al., 1996*) and 103P/Hartley 2 (*Crovisier et al., 2000*). Filter photometry of 81P/Wild 2 revealed a feature about 25% above the continuum; no spectrum exists (*Hanner and Hayward, 2003*). The ratio of the flux in a narrow 10.4-μm filter to the flux in a broad 10-μm (N) filter indicated some silicate emission in Comets P/Encke, P/Stephan-Oterma, and P/Tuttle (*Campins et al., 1982*). *Gehrz et al.* (1989) did not detect silicate emission in P/Encke near perihelion in 1987. Other short-period comets with 10-μm filter photometry show no feature at the 10% level. The absence of strong silicate emission in the short-period comets could be explained either by a difference in the composition between Oort cloud and Kuiper belt comets or by a lower abundance of submicrometer-sized particles. Short-period comets generally have been outgassing during many orbits in the inner solar system and the smaller or more fluffy particles may have been preferentially expelled over time. Thus, a lower abundance of isolated small grains or very fluffy aggregates of small grains in the coma is the simplest explanation for the lack of a strong silicate feature, although compositional differences cannot be ruled out.

4. INTERPLANETARY DUST PARTICLES FROM COMETS

The submicrometer grain size, high Mg/Fe ratio, and mix of crystalline and noncrystalline olivine and pyroxenes in the cometary dust have no counterpart in any meteoritic material, with the exception of the anhydrous chondritic aggregate IDPs. These are fine-grained heterogeneous aggregates having chondritic abundances of the major rock-forming elements; they comprise a major fraction of the IDPs captured in the stratosphere (Fig. 3). Typical grain sizes within the aggregates are 0.1–0.5 μm; micrometer-sized crystals of forsterite and enstatite are also present (*Bradley et al., 1992*). These aggregate IDPs are thought to originate from comets, based on their porous structure, small grain size, high C content, and relatively high atmospheric entry speeds. Measured He release temperatures indicate that many of these IDPs entered the atmosphere at speeds >16 km/s, consistent with cometary orbits (*Nier and Schlutter, 1993*). The match between the mineral identifications in the Hale-Bopp spectra and the silicates seen in the IDPs strengthens the link between comets and this class of IDPs. Thus, the composition and structure of the porous aggregate IDPs can be used to augment our understanding of the composition and origin of cometary dust.

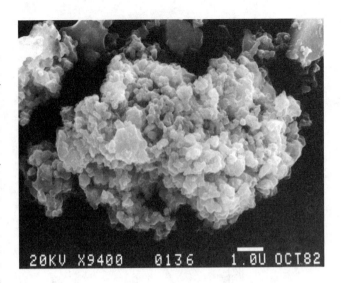

Fig. 3. Secondary electron micrograph of a chondritic porous (CP) interplanetary dust particle. Bar is 1 μm in length. The fragile microstucture and high porosity of this particle are consistent with a cometary origin. The submicrometer-sized grains are mainly GEMS and carbonaceous material. The angular, micrometer-sized components include single crystals of forsterite, enstatite, and FeNi sulfides.

The major form of noncrystalline silicates in the chondritic aggregate IDPs are the glass with embedded metal and sulfides (GEMS). These are submicrometer-sized glassy Mg-silicate grains with embedded nanometer FeNi and Fe sulfide crystals (*Bradley, 1994*). The GEMS show evidence of exposure to large doses of ionizing radiation (e.g., *Demyk et al., 2001, Carrez et al., 2002*), such as eroded surfaces, O enrichment, gradients in Mg/Si ratio decreasing outward from the center, and formation of reduced FeNi metal, pointing to exposure in the interstellar medium prior to their incorporation into comets. Some GEMS contain heavily etched relict mineral grains of sulfides or Mg-rich silicate crystals. The physical and chemical properties of GEMS are similar to those inferred for interstellar grains.

Infrared spectra of GEMs-rich IDP samples (thin sections containing GEMS and silicate crystals) resemble the cometary silicate spectra, while spectra of individual GEMS reveal a single broad peak near 9.3 μm (glassy pyroxene) or 9.8 μm (glassy olivine) (*Bradley et al., 1999*). The peak wavelength, width, and long-wavelength asymmetry of the 9.8-μm peak are similar to the interstellar silicate feature, such as that of the Trapezium.

Detection of nonsolar isotopic abundances in GEMS would constitute strong evidence for an interstellar origin. GEMS are frequently embedded in a carbonaceous material that displays high D/H ratios and ^{15}N enrichment (*Keller et al., 2000*). A bulk O-isotopic measurement of a GEMS-rich IDP yielded the highest "whole-rock" ^{16}O enrichment yet measured in any chondritic object (*Engrand et al., 1999*); silicates are the main carrier of O. NanoSIMS techniques are now reaching the sensitivity to allow isotopic measurements on individual submicrometer-sized grains in

IDPs. In early results utilizing nanoSIMS, two GEMS with nonsolar O-isotopic ratios have been detected (*Messenger et al.*, 2003).

The crystalline silicates found in chondritic porous IDPs are primarily the Mg-rich minerals forsterite and enstatite. This mineralogy is consistent with the 8–35-μm spectrum of Hale-Bopp and the *in situ* Halley elemental composition data. Grain sizes range from ~0.05 to 5 μm although most are 0.1–0.75 μm in diameter. Some enstatite whiskers, rods, and platelets have unusual growth patterns, such as axial screw dislocations, that indicate direct vapor phase condensation from a hot gas (*Bradley et al.*, 1983). Some low-Fe, Mn-enriched (LIME) enstatite and forsterite grains contain up to 5 wt.% MnO, in contrast to the majority of pyroxenes and olivines in meteorites, which contain <0.5 wt.% MnO (*Klock et al.*, 1989). The high Mn content is further evidence of direct condensation from a gas. At least one submicrometer forsterite grain with ^{17}O enrichment, indicating a circumstellar origin, has been identified (*Messenger et al.*, 2003).

The chondritic aggregate IDPs have a high bulk C content, 2–3 times higher on average than the primitive CI meteorites. *Thomas et al.* (1993) measured a range of 1–47 wt% C in 100 anhydrous IDPs; in a few cases, C was the most abundant element by volume. The C is distributed throughout the IDP as a matrix surrounding the mineral grains, but not necessarily as mantles on the grains. Much of the C is in an organic phase, evident from X-ray absorption edge spectroscopy and C-H stretch absorption features in the 3-μm region (*Clemett et al.*, 1993; *Flynn et al.*, 1999, 2000). C = O functional groups have been identified from the absorption edge spectroscopy (*Flynn et al.*, 2001). These results are qualitatively consistent with the nature of the CHON particles detected by the dust analysis instrument on the Halley space probes (*Fomenkova et al.*, 1994). Carbon-rich IDPs show a red reflectance spectrum, similar to red, dark outer solar system material (*Bradley et al.*, 1996; *Keller and Messenger*, 1997).

Nonsolar isotopic enrichments of D/H and $^{15}N/^{14}N$ have been detected in C-rich IDPs. In particles where further analysis has been carried out, the D and ^{15}N anomalies are associated with an organic carrier (e.g., *Aleon et al.*, 2002). These isotopic enrichments are attributed to mass fractionation during low-temperature ion-molecular reactions in cold, dense interstellar molecular clouds. In some cases, the measured D/H ratio approaches that observed in molecular clouds (*Messenger*, 2000). High D/H ratios, roughly twice the terrestrial value and 10 times the protosolar value, have been observed in gas phase cometary H_2O and HCN (*Meier and Owen*, 1999). Iron-nickel sulfide grains are the major carrier of S in the chondritic aggregate IDPs. The sulfide mineralogy is significantly different from that in primitive chondritic meteorites (*Dai and Bradley*, 2001). *Keller et al.* (2002) have proposed that sulfides are responsible for a broad λ ~ 23-μm feature detected around young and old stars by ISO. Laboratory spectra of pyrrhotite grains from IDPs display a broad Fe-S stretch feature centered at ~23.5 μm, similar to the circumstellar feature. Iron sulfide was an identified component of the dust in Comet Halley sampled during the spacecraft encounters (*Schulze et al.*, 1997).

5. ORIGINS OF COMETARY DUST

It is clear from the *in situ* and remote observations of comets and the analysis of probable cometary IDPs that cometary dust is an unequilibrated, heterogeneous mixture of minerals, including both high- and low-temperature condensates. These various components do not necessarily share a common origin. Temperatures in the solar nebula beyond 5 AU, where the comet nuclei accreted, were never higher than about 160 K, too low for significant processing of dust particles (*Boss*, 1994, 1998, 2004). Thus interstellar grains present in the outer solar nebula could have been preserved in comets. Material that condensed within the solar nebula also may have accreted into the comet nuclei.

The glassy silicate grains (GEMS) described in section 4 appear to constitute the major fraction of the noncrystalline silicates in cometary IDPs and the evidence is quite strong that these are interstellar grains, based on their morphology, physical and chemical structure, and inferred high radiation dosage. They are often embedded in an organic C material with nonsolar D/H and $^{15}N/^{14}N$ isotopic ratios. The GEMS must have formed at comparatively low temperatures and were never heated sufficiently to anneal. Thus, noncrystalline cometary silicates may be predominantly of interstellar origin.

The origin of the crystalline silicates in comets is more complex. Crystalline silicate grains can form by direct condensation from a hot gas at T = 1200–1400 K, followed by slow cooling, or by annealing of amorphous silicates at temperatures around 1000 K or higher (*Hallenbeck et al.*, 1998, 2000; *Brucato et al.*, 2002; *Koike and Tsuchiyama*, 1992; *Fabian et al.*, 2000). While the enstatite whiskers and rods occasionally seen in aggregate IDPs have growth patterns indicating direct vapor phase condensation, other crystalline grains in IDPs have no distinctive structure to distinguish between direct condensation or annealing. The Mg crystalline minerals forsterite and enstatite are predicted from thermodynamic models to be the first to condense in a hot gas at 1200–1400 K and only react with Fe at lower temperatures. Thus, direct condensation is a natural explanation for the preponderance of Mg-silicates in comet dust.

Grain condensation or annealing could have occurred in the hot inner solar nebula. Disk midplane temperatures ≥1000 K were reached inside about 1 AU, depending on the mass infall rate (*Boss*, 1998; *Chick and Cassen*, 1997). During the early high mass accretion phase (mass infall rate ≥10^{-6} $M_\odot$ yr^{-1}), this hot region could have extended to 3–4 AU (*Bell et al.*, 2000). However, the crystalline grains must be transported out to the region where the comets formed at 5–50 AU, and the extent of radial mixing of dust is uncertain. *Bockelée-Morvan et al.* (2002) have shown that turbulent diffusion in the solar nebula could be an efficient process to transport the crystalline grains from the inner nebula to the region of comet formation in timescales of a few 10^4 yr. This process is efficient as long as the grains remain coupled to the gas, i.e., as long as they remain small. The timescales for grain coagulation and growth are uncertain, but could be short enough to compete with radial transport.

Alternatively, *Harker and Desch* (2002) have proposed that small silicate grains in the solar nebula could have been thermally annealed by passing shock waves, provided that the ambient gas density was high enough to heat the grains briefly above 1200 K. The relation between the amount of heating and the degree of crystallinity is based on laboratory annealing experiments of *Hallenbeck et al.* (1998, 2000). This mechanism would be effective out to about 10 AU; at larger heliocentric distances the gas density would be too low to generate sufficient grain heating.

More study is needed, however, to understand the details of the annealing process and whether crystals with the observed mineralogy and morphology could be produced. If GEMS are representative of the amorphous silicate particles present in the solar nebula, one would expect the annealed crystals to retain the FeNi nanoparticles present in the GEMS, and these are not seen. Moreover, the GEMS are confined to a narrow size range (0.1–0.5 μm), whereas some of the enstatite and olivine crystals in the chondritic aggregate IDPs are micrometer-sized. The conditions of formation of silicate glasses and crystals are extensively documented in a vast body of literature on geochemical thermodynamics and igneous and metamorphic petrology. Although the temperature and pressure regimes are well outside those of low-temperature annealing, the underlying thermodynamic constraints are relevant.

If a large fraction of cometary dust consists of solar system condensates, then radial gradients in the temperature, composition, and extent of mixing within the solar nebula should be evident today as differences in the dust chemistry and mineralogy among comets formed in different regions. For example, if the olivine condensed in the hot inner solar nebula and was transported outward or if olivine was created *in situ* by passing shock waves, then one would expect to see a difference in olivine abundance between Oort cloud comets that formed in the region of the giant planets and Kuiper belt comets that formed beyond 30 AU. To date, no strong 11.2-μm olivine peak has been detected in a Kuiper belt comet; however, the cause may be a lack of disaggregated small grains in the coma rather than a lack of silicates. A possible, but weak, 11.2-μm peak may be present in the relatively weak silicate features observed in Comets P/Borrelly (*Hanner et al.,* 1996) and P/Hartley 2 (*Crovisier et al.,* 2000).

Could the crystalline silicates have a presolar origin? Silicate grains are known to condense in O-rich envelopes around evolved stars. Spectral peaks of the Mg-silicates forsterite and enstatite are clearly present in the ISO spectra of some evolved stars (*Waters et al.,* 1996; *Molster et al.,* 2002). Yet signatures of crystalline silicates are absent in spectra of the diffuse interstellar medium (ISM) or molecular clouds. Moreover, the spectra of most young stellar objects show no evidence of crystalline grains. Only in debris disks around young main-sequence objects such as β Pictoris (*Knacke et al.,* 1993) and in certain late-stage

Herbig Ae/Be stars that are precursors of β Pictoris systems (*Waelkens et al.,* 1996) does one find the spectral peaks of crystalline olivine. For example, the ISO spectrum of the late-stage Herbig Ae/Be star HD100546 is very similar to that of Comet Hale-Bopp (*Malfait et al.,* 1998). These systems are thought to have developed a population of comets that are the source of the dust (e.g., *Weissman,* 1984; *Grady et al.,* 1997).

Grain destruction in the ISM is an efficient process. Thus, if the comet grains formed in circumstellar outflows from evolved stars, one has to understand how they survived destruction in the ISM and why their spectral signatures are not seen in the ISM or young stellar objects.

Kemper et al. (2001) have produced radiative transfer models to study the visibility of crystalline silicates in mid-infrared spectra of circumstellar dust. Because pure crystalline Mg-silicates are very transparent at visible and near-IR wavelengths, they will be colder than Fe-containing amorphous silicates in optically thin regions where grains are heated by exposure to visible/near-IR stellar radiation. Kemper et al. showed that, indeed, a considerable fraction of forsterite and enstatite could be present in such environments without their spectral signatures being visible in emission above the radiation from the warmer dust components. These results should make us cautious about concluding that crystalline silicates are absent in cases where they may simply be "hidden" by their cold temperatures, a point also discussed by *Wooden et al.* (2000).

Anomalous isotopic enrichments could constitute strong evidence for an interstellar origin of cometary dust material. As described in section 4, analysis techniques have now reached the point where isotopic measurements can be made for individual grains in IDPs, and the first O-isotopic measurements provide tantalizing hints that a few grains do possess nonsolar O-isotopic ratios. However, it is not at all clear that silicate grains formed in the interstellar medium would have anomalous isotopic abundances. Their compositions may well have been homogenized to an average close to solar abundances.

Greenberg (1982) proposed that interstellar silicate grains possess organic refractory mantles as a result of UV photoprocessing in the diffuse interstellar medium following deposition of icy mantles in cold molecular clouds. These submicrometer core/mantle grains, perhaps with an additional icy mantle, were subsequently agglomerated and incorporated into comets. A mass of the organic refractory material comparable to the mass of the silicates satisfied cosmic abundances; in fact, it was argued that organic refractory mantles are a necessary repository of C to account for its cosmic abundance. The detection of high D enrichments in the organic refractory material in chondritic IDPs lends strong support to a presolar origin for the organic refractory material. The high D/H material often surrounds GEMS and other mineral grains, suggesting that these embedded grains are presolar as well, although the material is more clumpy and irregular rather than a uniform core/mantle morphology. However, the distinction between a population of GEMS with carbonaceous mantles and a population of GEMS within a carbonaceous matrix is difficult to discern in the electron microscope. Organic refractory grains have not been detected spectroscopically in comets. An emission feature at 3.4 μm first detected in P/Halley, but also seen in other comets, was initially attributed to organic refractory grains. However, the discovery of methanol in comets and analysis of its infrared bands led to the conclusion that the 3.4-μm feature is due to gas phase methanol and other gaseous species (*Bockelée-Morvan et al.,* 1995; *DiSanti et al.,* 1995). For further discussion concerning the origin of cometary grains, see *Wooden* (2002).

6. FUTURE DIRECTIONS

The past decade has witnessed significant advances in our understanding of the chemistry and mineralogy of cometary dust. However, questions remain about the nature of the refractory organic material, the differing dust properties among comets, and the origins of the various dust components.

The next decade should bring further advances in our knowledge of the composition and origin of cometary dust, including the first comet dust sample return. NASA's *Stardust* mission, launched in 1999, will collect a dust sample from Comet 81P/Wild 2 in January 2004 and return it to Earth in January 2006. Coupled with the newest analysis tools, the sample will yield isotopic abundance ratios for individual particles, as well as the chemistry and mineralogy of the silicates and other refractory dust components from a comet that probably originated in the Kuiper belt.

ESA's ambitious *Rosetta* mission, now scheduled for launch in 2004, will rendezvous with a short-period comet. *Rosetta* carries several instruments to measure the dust mass distribution, composition, and structure, as the spacecraft travels with the comet from aphelion to perihelion (*Schwehm and Schulz,* 1999).

The midinfrared spectra of Comet Hale-Bopp and astronomical sources from ISO demonstrated the value of 8–45-μm spectroscopy for determining dust composition. SIRTF, the Space Infrared Telescope Facility, and SOFIA, the Stratospheric Observatory for Infrared Astronomy, will allow a number of comets to be observed in this important spectral region. The detection of crystalline silicates in Kuiper belt comets would be particularly significant in showing that crystalline silicate particles were spread throughout the solar nebula.

Finally, improved sensitivity of laboratory techniques for compositional analysis of microscopic samples, such as the nanoSIMS, will offer the opportunity for elemental and isotopic abundances to be investigated in individual submicrometer-sized grains in IDPs and returned cometary samples.

7. CONCLUSIONS

In situ sampling of Comet Halley dust, remote infrared spectroscopy, and IDP analyses yield a consistent picture of the composition of cometary dust. Silicates constitute the most abundant material; they are present in both crystalline and amorphous (or glassy) form and include the mineralogy of both olivine and pyroxene. The crystalline silicates are primarily the Mg-rich minerals forsterite and enstatite, as is the case for circumstellar dust around evolved stars. Carbon is present in approximately cosmic abundance, much of it in an organic refractory component.

We have discussed why the chondritic porous aggregate IDPs are probably from comets. Analysis of their properties complements and substantiates the conclusions drawn from the spectroscopy and from the *in situ* measurements during the 1986 Halley spacecraft encounters. In particular, IDP analysis confirms that much of the carbonaceous material is in an organic phase, and the detection of high D/H ratios implies that at least some of this material is presolar. The D-rich material often surrounds GEMS and other mineral grains, suggesting that these embedded grains are of presolar origin as well. GEMS are the predominant form of noncrystalline silicates. These particles appear to have experienced high radiation dosage in a presolar environment.

The various types of silicate particles do not necessarily have a common origin. While the GEMS are of likely presolar origin, the origin(s) of the crystalline silicates is unclear. If formed as high-temperature condensates or by annealing in the inner solar nebula, radial transport must have been more efficient during the planetesimal accretion phase than some models predict. It is possible that grains were annealed by transient heating from passing shock waves in the solar nebula at $r \leq 10$ AU. However, the size range and composition of the crystalline silicates in porous aggregate IDPs are not consistent with what one would expect for annealed GEMS. In either case, one would expect crystalline silicates to be less abundant — or absent entirely — in the Kuiper belt comets that formed beyond 30 AU.

If the crystalline silicates were already present in the cloud from which the solar nebula formed, then one needs to explain why their spectral signatures are not seen in interstellar dust or in young stellar objects. The very cold temperatures of "clean" Mg-rich silicates is one possible explanation why their spectral features are not seen in emission. Isotopic measurements of individual silicate grains in IDPs and returned comet samples with nanoSIMS techniques may help to clarify their origin.

Although the chondritic aggregate IDPs have given us extremely interesting insight into the nature of probable cometary dust, we do not know the specific source of an individual IDP, nor the selection effects between comet ejection and Earth capture. Thus comet dust sample return and *in situ* analysis are very important. In the next decade, we can look forward to the *Stardust* sample return from the short-period Comet 81P/Wild 2 in January 2006 and the encounter of ESA's *Rosetta* mission with a short-period comet in 2011–2013.

Acknowledgments. The research of M.S.H. was carried out at the Jet Propulsion Laboratory, California Institute of Technology, under contract with the National Aeronautics and Space Administration. J.P.B. acknowledges support from NASA grants NAG5-7450 and NAG5-9797.

REFERENCES

Aleon J., Robert F., Chaussidon M., Marty B., and Engrand C. (2002) ^{15}N excesses in deuterated organics from two interplanetary dust particles (abstract). In *Lunar and Planetary Science XXXIII*, Abstract #1397. Lunar and Planetary Institute, Houston (CD-ROM).

Bell K. R., Cassen P. M., Wasson J. T., and Woolum D. S. (2000) The FU Orionis phenomenon and solar nebula material. In: *Protostars and Planets IV* (V. Mannings et al., eds.), pp. 897–926. Univ. of Arizona, Tucson.

Bockelée-Morvan D., Brooke T. Y., and Crovisier J. (1995) On the origin of the 3.2- to 3.6- μm emission features in comets. *Icarus, 116*, 18–39.

Bockelée-Morvan D., Gautier D., Hersant F., Hure J-M., and Robert F. (2002) Turbulent radial mixing in the solar nebula as the source of crystalline silicates in comets. *Astron. Astrophys., 384*, 1107–1118.

Boss A. P. (1994) Midplane temperatures and solar nebula evolution (abstract). In *Lunar and Planetary Science XXV*, p. 149. Lunar and Planetary Institute, Houston.

Boss A. P. (1998) Temperatures in protoplanetary disks. *Annu. Rev. Earth Planet. Sci., 26*, 53–80.

Boss A. P. (2004) From molecular clouds to circumstellar disks. In *Comets II* (M. C. Festou et al., eds.), this volume. Univ. of Arizona, Tucson.

Bradley J. P. (1994) Chemically anomalous, preaccretionally irradiated grains in interplanetary dust from comets. *Science, 265*, 925.

Bradley J. P., Brownlee D. E., and Veblen D. R. (1983) Pyroxene whiskers and platelets in interplanetary dust: Evidence of vapour phase growth. *Nature, 301*, 473–477.

Bradley J. P., Humecki H. J., and Germani M. S. (1992) Combined infrared and analytical electron microscope studies of interplanetary dust particles. *Astrophys. J., 394*, 643–651.

Bradley J. P., Keller L. P., Brownlee D. E., and Thomas K. L. (1996) Reflectance spectroscopy of interplanetary dust particles. *Meteoritics & Planet. Sci., 31*, 394–402.

Bradley J. P., Keller L. P., Snow T. P., Hanner M. S., Flynn G. J., Gezo J. C., Clemett S. J., Brownlee D. E., and Bowey J. E. (1999) An infrared spectral match between GEMS and interstellar grains. *Science, 285*, 1716–1718.

Bregman J. H., Campins H., Witteborn F. C., Wooden D. H., Rank D. M., Allamandola L. J., Cohen M., and Tielens A. G. G. M. (1987) Airborne and groundbased spectrophotometry of Comet P/Halley from 5–13 μm. *Astron. Astrophys., 187*, 616–620.

Brucato J. R., Colangeli L., Mennella V., Palumbo P., and Bussoletti E. (1999) Silicates in Hale-Bopp: Hints from laboratory studies. *Planet. Space Sci., 47*, 773–779.

Brucato J. R., Mennella V., Colangeli L., Rotundi A., and Palumbo P. (2002) Production and processing of silicates in laboratory and in space. *Planet. Space Sci., 50*, 829–837.

Campins H. and Ryan E. (1989) The identification of crystalline olivine in cometary silicates. *Astrophys. J., 341*, 1059–1066.

Campins H., Rieke G. H., and Lebofsky M. J. (1982) Infrared photometry of periodic comets Encke, Chernykh, Kearns-Kwee, Stephan-Oterma, and Tuttle. *Icarus, 51*, 461–465.

Carrez Ph., Demyk K., Cordier P., Gengembre L., Grimblot J., d'Hendecourt L., Jones A., and Leroux H. (2002) Low-energy helium ion irradiation-induced amorphization and chemical changes in olivine: Insights for silicate dust evolution in the interstellar medium. *Meteoritics & Planet. Sci., 37*, 1599–1614.

Chick K. M. and Cassen P. (1997) Thermal processing of interstellar dust grains in the primitive solar environment. *Astrophys. J., 477*, 398–409.

Clemett S. J., Maechling C. R., Zare R. N., Swan P. D., and Walker R. M. (1993) Identification of complex aromatic molecules in individual interplanetary dust particles. *Science, 262*, 721.

Colangeli L., Mennella V., Brucato J. R., Palumbo P., and Rotundi A. (1999) Characterization of cosmic materials in the laboratory. *Space Sci. Rev., 90*, 341–354.

Crovisier J., Brooke T. Y., Hanner M. S., Keller H. U., Lamy P. L., Altieri B., Bockelée-Morvan D., Jorda L., Leech K., and Lellouch E. (1996) The Infrared spectrum of Comet C/1995 O1 (Hale-Bopp) at 4.6 AU from the Sun. *Astron. Astrophys., 315*, L385–L388.

Crovisier J., Leech K., Bockelée-Morvan D., Brooke T. Y., Hanner M. S., Altieri B., Keller H. U., and Lellouch E. (1997) The spectrum of Comet Hale-Bopp (C/1995 O1) observed with the Infrared Space Observatory at 2.9 astronomical units from the Sun. *Science, 275*, 1904–1907.

Crovisier J. and 14 colleagues (2000) The thermal infrared spectra of Comets Hale-Bopp and 103P/Hartley 2 observed with the Infrared Space Observatory. In *Thermal Emission Spectroscopy and Analysis of Dust, Disks, and Regoliths* (M. L. Sitko et al., eds.), pp. 109–117. ASP Conference Series 196.

Dai Z. R. and Bradley J. P. (2001) Iron-nickel sulfides in anhydrous interplanetary dust particles. *Geochim. Cosmochim. Acta, 65*, 3601–3612.

Day K. L. (1976) Further measurements of amorphous silicates. *Astrophys. J., 210*, 614–617.

Day K. L. (1981) Infrared extinction of amorphous iron silicates. *Astrophys. J., 368*, 110–112.

Demyk K., Carrez Ph., Leroux H., Cordier P., Jones A. P., Borg J., Quirico E., Raynal P. I., and d'Hendecourt L. (2001) Structural and chemical alteration of crystalline olivine under low energy He$^+$ irradiation. *Astron. Astrophys., 368*, L38–L41.

DiSanti M. A., Mumma M. J., Geballe T. R., and Davies J. K. (1995) Systematic observations of methanol and other organics in Comet P/Swift-Tuttle. *Icarus, 116*, 1–17.

Dorschner J., Begemann B., Henning Th., Jäger C., and Mutschke H. (1995) Steps toward interstellar silicate mineralogy II. Study of Mg-Fe-silicate glasses of variable composition. *Astron. Astrophys., 300*, 503–520.

Engrand C., McKeegan K. D., Leshin L. A., Bradley J. P., and Brownlee D. E. (1999) Oxygen isotopic composition of interplanetary dust particles: ^{16}O-excess in a GEMS-rich IDP (abstract). In *Lunar and Planetary Science XXX*, Abstract #1690. Lunar and Planetary Institute, Houston (CD-ROM).

Fabian D., Jäger C., Henning Th., Dorschner J., and Mutschke H. (2000) Steps toward interstellar silicate mineralogy V. Thermal evolution of amorphous magnesium silicates and silica. *Astron. Astrophys., 364*, 282–292.

Flynn G. J., Keller L. P., Jacobsen C. and Wirick S. (1999) Organic carbon in cluster IDPs from the L2009 and L2011 collectors (abstract). In *Lunar and Planetary Science XXX*, Abstract #1091. Lunar and Planetary Institute, Houston (CD-ROM).

Flynn G. J., Keller L. P., Jacobsen C., Wirick S. and Miller M. A. (2000) Interplanetary dust particles as a source of pre-biotic organic matter on the Earth (abstract). In *Lunar and Planetary Science XXXI*, Abstract #1409. Lunar and Planetary Institute, Houston (CD-ROM).

Flynn G. J., Feser M., Keller L. P., Jacobsen C., and Wirik S.

(2001) Carbon-XANES and oxygen-XANES measurements on interplanetary dust particles: A preliminary measurement of the C to O ratio in the organic matter in a cluster IDP (abstract). In *Lunar and Planetary Science XXXII*, Abstract #1603. Lunar and Planetary Institute, Houston (CD-ROM).

Fomenkova M., Kerridge J., Marti K., and McFadden L. (1992) Compositional trends in rock-forming elements of Comet Halley dust. *Science, 258,* 266–269.

Fomenkova M. N., Chang S., and Mukhin L. M. (1994) Carbonaceous components in the Comet Halley dust. *Geochim. Cosmochim. Acta, 58,* 4503–4512.

Galdemard P., Lagage P. O., Dubreuil D., Jouan R., Masse P., Pantin E., and Bockelée-Morvan D. (1999) Mid-infrared spectro-imaging observations of Comet Hale-Bopp. *Earth Moon Planets, 78,* 271–277.

Gehrz R. D., Ney E. P., Piscitelli J., Rosenthal E., and Tokunaga A. T. (1989) Infrared photometry and spectroscopy of Comet P/Encke in 1987. *Icarus, 80,* 280–288.

Glaccum W., Moseley S. H., Campins H., and Loewenstein R. F. (1987) Airborne spectrophotometry of P/Halley from 20 to 65 microns. *Astron. Astrophys., 187,* 635–638.

Grady C. A., Sitko M. L., Bjorkman K. S., Perez M. R., Lynch D. K, Russell R. W., and Hanner M. S. (1997) The star-grazing extrasolar comets in the HD100546 system. *Astrophys. J., 483,* 449–456.

Greenberg J. M. (1982) What are comets made of? A model based on interstellar dust. In *Comets* (L. L. Wilkening, ed.), pp. 131–163. Univ. of Arizona, Tucson.

Grün E. and 23 colleagues (2001) Broadband infrared photometry of Comet Hale-Bopp with ISOPHOT. *Astron. Astrophys., 377,* 1098–1118.

Hallenbeck S. L., Nuth J. A., and Daukantas P. L. (1998) Mid-infrared spectral evolution of amorphous magnesium silicate smokes annealed in vacuum: Comparison to cometary spectra. *Icarus, 131,* 198–209.

Hallenbeck S. L., Nuth J. A., and Nelson R. N. (2000) Evolving optical properties of annealing silicate grains: From amorphous condensate to crystalline mineral. *Astrophys. J., 535,* 247–255.

Hanner M. S. and Hayward T. L. (2003) Infrared observations of Comet 81P/Wild 2 in 1997. *Icarus, 161,* 164–173.

Hanner M. S., Newburn R. L., Gehrz R. D., Harrison T., Ney E. P., and Hayward T. L. (1990) The Infrared spectrum of Comet Bradfield (1987s) and the silicate emission feature. *Astrophys. J., 348,* 312–321.

Hanner M. S., Lynch D. K., and Russell R. W. (1994a) The 8–13 micron spectra of comets and the composition of silicate grains. *Astrophys. J., 425,* 274–285.

Hanner M. S., Hackwell J. A., Russell R. W., and Lynch D. K. (1994b) Silicate emission feature in the spectrum of Comet Mueller 1993a. *Icarus, 112,* 490–495.

Hanner M. S., Lynch D. K., Russell R. W., Hackwell J. A., and Kellogg R. (1996) Mid-infrared spectra of Comets P/Borrelly, P/Faye, and P/Schaumasse. *Icarus, 124,* 344–351.

Hanner M. S., Gehrz R. D., Harker D. E., Hayward T. L., Lynch D. K., Mason C. G., Russell R. W., Williams D. M., Wooden D. H., and Woodward C. E. (1999) Thermal emission from the dust coma of Comet Hale-Bopp and the composition of the silicate grains. *Earth Moon Planets, 79,* 247–264.

Harker D. E. and Desch S. J. (2002) Annealing of silicate dust by shocks at 10 AU. *Astrophys. J. Lett., 565,* L109–L112.

Harker D. E., Wooden D. H., Woodward C. E., and Lisse C. M.

(2002) Grain properties of Comet C/1995 O1 (Hale-Bopp). *Astrophys. J., 580,* 579–597.

Hayward T. L., Hanner M. S., and Sekanina Z. (2000) Thermal infrared imaging and spectroscopy of Comet Hale-Bopp (C/1995 O1). *Astrophys. J., 538,* 428–455.

Herter T., Campins H., and Gull G. E. (1987) Airborne spectrophotometry of P/Halley from 16 to 30 microns. *Astron. Astrophys., 187,* 629–631.

Jäger C., Molster F. J., Dorschner J., Henning Th., Mutschke H., and Waters L. B. F. M. (1998) Steps toward interstellar silicate mineralogy IV. The crystalline revolution. *Astron. Astrophys., 339,* 904–916.

Jessberger E. (1999) Rocky cometary particulates: Their elemental, isotopic, and mineralogical ingredients. *Space Science Rev., 90,* 91–97.

Jessberger E. K., Christoforidis A., and Kissel J. (1988) Aspects of the major element composition of Halley's dust. *Nature, 332,* 691–695.

Keller L. P. and Messenger S. (1997) Reflectance spectroscopy of deuterium-rich cluster IDPs (abstract). In *Lunar and Planetary Science XXVIII*, pp. 705–706. Lunar and Planetary Institute, Houston.

Keller L. P., Messenger S., and Bradley J. P. (2000) Analysis of a deuterium-rich interplanetary dust particle (IDP) and implications for presolar material in IDPs. *J. Geophys. Res., 105,* 10397–10402.

Keller L. P. and 10 colleagues (2002) Identification of iron sulphide grains in protoplanetary disks. *Nature, 417,* 148–150.

Kemper F., Waters L. B. F. M., de Koter A., and Tielens A. G. G. M. (2001) Crystallinity versus mass loss rate in asymptotic giant branch stars. *Astron. Astrophys., 369,* 132–141.

Kissel J. and 22 colleagues (1986a) Composition of Comet Halley dust particles from Vega observations. *Nature, 321,* 280–282.

Kissel J. and 18 colleagues (1986b) Composition of Comet Halley dust particles from Giotto observations. *Nature, 321,* 336–338.

Klock W., Thomas K. L., McKay D. S., and Palme H. (1989) Unusual olivine and pyroxene composition in interplanetary dust and unequilibrated ordinary chondrites. *Nature, 339,* 126–128.

Knacke R. F., Fajardo-Acosta S. B., Telesco C. M., Hackwell J. A., Lynch D. K., and Russell R. W. (1993) The silicates in the disk of β Pictoris. *Astrophys. J., 418,* 440–450.

Koike C. and Tsuchiyama A. (1992) Simulation and alteration for amorphous silicates with very broad bands in infrared spectra. *Mon. Not. R. Astron. Soc., 255,* 248–254.

Koike C., Shibai H., and Tuchiyama A. (1993) Extinction of olivine and pyroxene in the mid- and far-infrared. *Mon. Not. R. Astron. Soc., 264,* 654–658.

Lawler M. E. and Brownlee D. E. (1992) CHON as a component of dust from Comet Halley. *Nature, 359,* 810–812.

Lawler M. E., Brownlee D. E., Temple S., and Wheelock M. M. (1989) Iron, magnesium, and silicon in dust from Comet Halley. *Icarus, 80,* 225–242.

Lynch D. K., Russell R. W., Campins H., Witteborn F. C., Bregman J. D., Rank D. W., and Cohen M. C. (1989) 5–13 μm observations of Comet Wilson 1986l. *Icarus, 82,* 379–388.

Lynch D. K., Russell R. W., Hackwell J. A., Hanner M. S., and Hammel H. B. (1992) 8–13 μm spectroscopy of Comet Levy 1990 XX. *Icarus, 100,* 197–202.

Malfait K., Waelkens C., Waters L. B. F. M., Vandenbussche B., Huygen E., and de Graauw M. S. (1998) The spectrum of the young star HD100546 observed with the Infrared Space Observatory. *Astron. Astrophys., 332,* L25–L28.

Meier R. and Owen T. C. (1999) Cometary deuterium. *Space Sci. Rev., 90,* 33–43.

Messenger S. (2000) Identification of molecular-cloud material in interplanetary dust particles. *Nature, 404,* 968–971.

Messenger S., Keller L. P., Stadermann F. J., Walker R. M., and Zinner E. (2003) Samples of stars beyond the solar system: Silicate grains in interplanetary dust. *Science, 300,* 105–108.

Molster F. J., Waters L. B. F. M., Tielens A. G. G. M., and Barlow M. J. (2002) Crystalline silicate dust around evolved stars I. The sample stars. *Astron. Astrophys., 383,* 184–221.

Nier A. O. and Schlutter D. J. (1993) The thermal history of interplanetary dust particles collected in the Earth's stratosphere. *Meteoritics, 28,* 675–581.

Sandford S. A. and Walker R. M. (1985) Laboratory infrared transmission spectra of individual interplanetary dust particles from 2.5 to 25 microns. *Astrophys. J., 291,* 838–851.

Schulze H., Kissel J., and Jessberger E. (1997) Chemistry and mineralogy of Comet Halley's dust. In *From Stardust to Planetesimals* (Y. J. Pendleton and A. G. G. M. Tielens, eds.), pp. 397–414. ASP Conference Series 122, Astronomical Society of the Pacific, San Francisco.

Schwehm G. and Schulz R. (1999) Rosetta goes to Comet Wirtanen. *Space Sci. Rev., 90,* 313–319.

Stephens J. R. and Russell R. W. (1979) Emission and extinction of ground and vapor-condensed silicates from 4 to 14 microns and the 10 micron silicate feature. *Astrophys. J., 228,* 780–786.

Thomas K. L., Blanford G. E., Keller L. P., Klock W., and McKay D. S. (1993) Carbon abundance and silicate mineralogy of anhydrous interplanetary dust particles. *Geochim. Cosmochim. Acta, 57,* 1551–1566.

Waters L. B. F. M. and 36 colleagues (1996) Mineralogy of oxygen-rich dust shells. *Astron. Astrophys., 315,* L361–L364.

Waelkens C. and 20 colleagues (1996) SWS observations of young main-sequence stars with dusty circumstellar disks. *Astron. Astrophys., 315,* L245–248.

Weissman P. R. (1984) The Vega particulate shell: Comets or asteroids. *Science, 224,* 987.

Wooden D. H. (2002) Comet grains: Their IR emission and their relation to ISM grains. *Earth Moon Planets, 89,* 247–287.

Wooden D. H., Harker D. E., Woodward C. E., Koike C., Witteborn F. C., McMurtry M. C., and Butner H. M. (1999) Silicate mineralogy of the dust in the inner coma of Comet C/1995 O1 (Hale-Bopp) pre- and post-perihelion. *Astrophys. J., 517,* 1034–1058.

Wooden D. H., Butner H. M., Harker D. E., and Woodward C. E. (2000) Mg-rich silicate crystals in Comet Hale-Bopp: ISM relics or solar nebula condensates? *Icarus, 143,* 126–137.

Motion of Cometary Dust

Marco Fulle

Istituto Nazionale di Astrofisica–Osservatorio Astronomico di Trieste

On timescales of days to months, the motion of cometary dust is mainly affected by solar radiation pressure, which determines dust dynamics according to the particle-scattering cross-section. Within this scenario, the motion of the dust creates structures referred to as dust tails. Tail photometry, depending on the dust cross-section, allows us to infer from model runs the best available outputs to describe fundamental dust parameters: mass loss rate, ejection velocity from the coma, and size distribution. Only models that incorporate these parameters, each strictly linked to all the others, can provide self-consistent estimates for each of them. After many applications of available tail models, we must conclude that comets release dust with mass dominated by the largest ejected boulders. Moreover, an unexpected prediction can be made: The coma brightness may be dominated by light scattered by meter-sized boulders. This prediction, if confirmed by future observations, will require substantial revisions of most of the dust coma models in use today, all of which are based on the common assumption that coma light comes from grains with sizes close to the observation wavelength.

1. HISTORICAL OVERVIEW

While ices sublimate on the nucleus surface, the dust embedded in them is freed and dragged out by the expanding gas. The dust motion then depends on the three-dimensional nucleus topography and on the complex three-dimensional gas-dust interaction that takes place close to the nucleus surface. Information on such processes has never been available to modelers. Dust coma shapes depend heavily on the details of these boundary conditions, and coma models cannot disentangle the effect of dust parameters on coma shapes from those due to the unknown boundary conditions. On the contrary, dust tails are usually structureless, suggesting that the dust has lost the memory of the details of the boundary conditions. Moreover, solar radiation pressure acts like a mass spectrometer, putting dust particles of different masses in different space positions: This allows models to evaluate how dust parameters affect tail shapes, making tail models powerful tools to describe dust in comets.

A dust tail is a broad structure that originates from the comet head and can reach lengths on the order of 10^4 km in most cases and up to 10^8 km in the most spectacular cases (Plate 16). When we consider the comet orbital path and the straight line going from the Sun to the comet nucleus (the radius vector), we divide the comet orbital plane in four sectors (Fig. 1). All cometary tails lie in the sector outside the comet orbit and behind the comet nucleus. When we project these four sectors of the comet orbital plane onto the sky, they may be strongly deformed by the observation perspective, especially when Earth is close to the comet orbital plane. In these latter cases, the sector where tails may be seen can be a complete sky halfplane and the appearance of a perspective antitail is then possible, which although seen to roughly point to the Sun, is always external to the comet orbit. A few years ago, *Pansecchi et al.* (1987) discovered

a particular type of antitails (anti-neck-lines) that lie in the sector opposite that where most tails are possible. These are much shorter than usual tails, not exceeding 10^6 km in length (section 7.1).

The first tail model that fit the observations well was developed by Friedrich W. Bessel (1784–1846) and later refined by Fjodor A. Bredichin (1831–1904). Bredichin introduced the notion of synchrones and syndynes (Fig. 1). Both authors assumed that a repulsive force, inversely proportional to the squared Sun-comet distance, was acting on the material composing the tails. According to this hypothesis, the tail particle is subjected to a total force equal to the solar gravity times a factor μ characterizing the tail particle. When this repulsive force is equal to the solar gravity, i.e., when μ = 0, the tail particle assumes a uniform straight motion. When μ < 0, the tail particle moves on a hyperbola with the convexity directed toward the Sun. When 0 < μ < 1, the tail particle behaves as if the Sun were lighter, so that its motion is slower than that of the comet nucleus and its orbit is external relative to that of the comet nucleus. This model explained why cometary tails always lie in the comet orbital plane sector outside of the comet orbit and behind the comet nucleus. Svante Arrhenius (1859–1927) proposed the solar radiation pressure as a candidate for Bessel's repulsive force. The computations of Karl Schwarzschild (1873–1916) and Peter J. W. Debye (1884–1966) established that the β = 1 – μ parameter is inversely proportional to the diameter of the dust particle (assuming it to be a sphere) and equal to 1 for dust diameters close to 1 μm.

Let us consider the comet nucleus moving along its orbit and ejecting dust particles with exactly zero ejection velocity. Particles all ejected at the same time and characterized by any μ value will be distributed at a later observation time on a line referred to as "synchrone." Changing the ejection time, we obtain a family of synchrones, each characterized

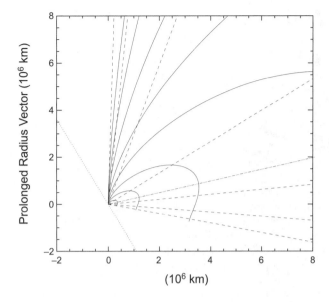

Fig. 1. Synchrone-syndyne network for Comet Hale-Bopp 1995O1 on 3 January 1998, when a neck-line was observed during the Earth crossing of the comet orbital plane (Fig. 2). The Sun was located exactly toward the −y direction and the phase angle of the comet at the observation was $\phi = 14.4°$. The image plane corresponds to the comet orbital plane; the Earth direction is toward the bottom right, forming an angle ϕ with the −y axis. The dotted line is the comet elliptical orbit, which, together with the y axis, divides the plane in the four sectors discussed in section 1. The continuous lines are syndynes characterized by the parameter $\beta = 1 − \mu = 1, 0.3, 0.1, 0.03, 0.01, 0.003, 0.001,$ and 0.0003, rotating clockwise from that closest to the +y direction, respectively. The dashed lines are synchrones characterized by the ejection times 250, 200, 100, 0, −100, −200, and −300 d with respect to perihelion, rotating clockwise from the closest one to the +y direction respectively. The three-dotted-and-dashed line is the neck-line axis, approximately corresponding to the synchrone ejected 43.8 d before perihelion. Because of the Earth position, it is evident that the neck-line was observed as a spike pointing toward the antisolar direction (Fig. 2); its prolongation in the −x direction was observed as a bright spikelike real antitail (Fig. 2).

by its ejection time. Synchrones can be approximated by radial lines, all diverging from the comet nucleus. Conversely, particles characterized by the same μ value ejected at any time will be distributed at a later observation time on a line referred to as "syndyne." Changing the μ value, we obtain a family of syndynes, each characterized by its μ value. Syndynes can be approximated by spirals tangent to the prolonged radius vector, the Sun-comet nucleus vector. In this model, only one synchrone and one syndyne pass in each point of the tail. It is thus possible to derive the ejection time and the μ value at each point of the tail. The synchrone-syndyne model predicts that dust tails are two-dimensional structures lying in the comet orbital plane. Tail observations made when Earth crosses the comet orbital plane (*Pansecchi et al.,* 1987) have shown that dust tails are thick, i.e., three-dimensional structures.

2. DUST TAILS: PHOTOMETRIC THEORY

The dust dynamics depend on the $\beta = 1 − \mu$ parameter

$$\beta = 1 − \mu = \frac{3E_\odot Q_{pr}}{8\pi c G M_\odot \rho_d d} = \frac{C_{pr} Q_{pr}}{\rho_d d} \tag{1}$$

where $E_\odot$ is the mean solar radiation, c the light speed, Q_{pr} the scattering efficiency for radiation pressure, G the gravitational constant, $M_\odot$ the solar mass, ρ_d the dust bulk density and d the diameter of the dust grains, assumed here to be spherical in shape. Here we use the terms dust grain or dust particle to refer to every solid object that escapes a comet nucleus. The resulting value of $C_{pr} = 1.19 \times 10^{-3}$ kg m^{-2}. For the β values most common in dust tails, d ≫ 1 μm and $Q_{pr} \approx 1$ (*Burns et al.,* 1979), so that the most uncertain parameter is the dust bulk density. As a consequence, it is convenient to express all size-dependent quantities as functions of ρ_dd. The flux of photons I (measured in Jy sr^{-1}) received from a dust tail is

$$I(x,y) = BD(x,y) =$$
$$B \int_{-\infty}^{t_o} \int_0^\infty H(x,y,t,1-\mu)\dot{N}(t)\sigma(t,1-\mu)dtd(1-\mu) \tag{2}$$

where x,y are the sky coordinates, H the sky surface density of dust grains coming from dust tail models, $\dot{N}(t)$ the dust number loss rate, t_o the observation time, σ the cross-section of a grain, B the flux of photons in thermal or optical bandwidths (measured in Jy sr^{-1}) received from a grain of unit cross-section, and D the dimensionless sky surface density weighted by the dust cross-section observed in the tail: If we are observing optical fluxes, then we are dealing with scattering cross-sections; if we are observing IR fluxes coming from grains with sizes much larger than IR wavelengths, then we are dealing with emitting cross-sections. Since we have

$$\sigma(d) = \frac{\pi}{4}d^2 = \frac{\pi}{4}\frac{(\rho_d d)^2}{\rho_d^2} \tag{3}$$

$$\dot{M}(d) = \dot{N}\frac{\pi}{6}\rho_d\frac{(\rho_d d)^3}{\rho_d^3} \tag{4}$$

the relation between the dust mass loss rate $\dot{M}$ and the flux I received from the tail is independent of ρ_d

$$I(x,y) = \frac{3}{2}B \int_{-\infty}^{t_o} \int_0^\infty H(x,y,t,1-\mu)$$
$$\dot{M}(t,1-\mu)\frac{(\rho_d d)^2}{(\rho_d d)^3}dtd(1-\mu) \tag{5}$$

which makes dust tail models a powerful tool to infer the

mass loss rate of cometary dust. Equation (5) shows that the relationship between dust mass loss rate and tail brightness depends on the second and third momenta of the differential dust size distribution (DSD). Since the dust dynamics depend on the $\beta = 1 - \mu$ quantity, it is convenient to express the DSD in terms of $(\rho_d d)$ weighted by the dust cross-section observed in the tail. Thus, we define the dimensionless "β-distribution"

$$f(t, 1 - \mu)d(1 - \mu) = \frac{(\rho_d d)^2 g(t, \rho_d d)d(\rho_d d)}{\int_0^\infty (\rho_d d)^2 g(t, \rho_d d)d(\rho_d d)} \quad (6)$$

$$g(t, \rho_d d)d(\rho_d d) = k(1 - \mu)^2 f(t, 1 - \mu)d(1 - \mu) \quad (7)$$

where k is a dimensionless constant depending on the quantity B defined in equation (2). If the DSD $g(t, \rho_d d)$ is a power law vs. $(\rho_d d)$ with index α, then $f(t, 1 - \mu)$ is a power law vs. $(1 - \mu)$ with index $-\alpha - 4$. Dust tail models provide the quantity $F(t, 1 - \mu)$, which is expressed in m² s⁻¹

$$F(t, 1 - \mu) = \left[\frac{C_{pr}Q_{pr}}{\rho_d}\right]^2 \dot{N}(t)f(t, 1 - \mu) \quad (8)$$

so that the dust mass loss rate given by equation (4) becomes

$$\dot{M}(t) = \frac{\pi}{6}kC_{pr}Q_{pr}\int_0^\infty \frac{F(t, 1 - \mu)}{1 - \mu}d(1 - \mu) \quad (9)$$

Values of dimensionless quantity k depend on the actual techniques used to observe the dust tail. In the case of optical photographic or CCD data, it becomes

$$k = \frac{4}{A_p(\phi)}\left[\frac{\gamma r}{1\,AU}\right]^2 \frac{10^{0.4[m_\odot - m(x,y)]}}{D(x,y)} \quad (10)$$

where $\gamma = 206265$ arcsec; r the Sun-comet distance at observation; m(x,y) the dust tail brightness expressed in mag arcsec⁻²; $m_\odot$ the Sun magnitude in the bandwidth of tail observations; D(x,y) the dimensionless tail brightness defined in equation (2); and $A_p(\phi)$ is the geometric albedo times the phase function at observation [an isotropic body diffusing all the received radiation uniformly in all directions has $A_p = \frac{1}{4}$ (*Hanner et al., 1981*)]. CCD data provide much better observational constraints to models, because they have a well-calibrated photometric response, unlike photographic plates. Moreover, the much higher sensitivity of CCDs allows us to use interferential filters to avoid emissions from ions and gases, which usually pollute the wide bandwidths (e.g., Johnson filters) used in astrophotography. However, CCD data also provide dust mass loss rates that are dependent on the poorly known dust albedo. Conversely, in the case of ther-

mal IR or millimeter data

$$k = \frac{4S_v(x, y, T)}{\pi D(x, y)B_v(T)} \quad (11)$$

so that k is now independent of the dust albedo, depending only on the dust temperature T, usually much better known than the albedo. Here, $B_v(T)$ is the Planck function at temperature T and $S_v(x,y,T)$ is the flux received from a dust tail in a thermal IR or millimeter wavelength. If the dust grains have very aspherical shapes, with very different cross-sections along each space coordinate (e.g., noodles, long cylinders or flat disks), then it becomes impossible to compute any dust dynamics (*Crifo and Rodionov, 1999*). If dust grains are not spherical but still compact, then the numerical values of C_{pr} and k change, while equations (1), (2), (6), (7), (8), and (9) remain identical (where now d is the mean grain size), along with equations (3), (4), and (5) after we change only the factors $\frac{\pi}{4}$, $\frac{\pi}{6}$, and $\frac{3}{2}$ respectively. Therefore, thermal IR or millimeter tail data provide the most reliable information on dust mass loss rates and DSDs in comets. In fact, the outputs of IR tail models are completely independent of the dust shape (provided the grains are compact), bulk density, and albedo (i.e., the most uncertain parameters in cometary dust modeling). Moreover, parallel IR and optical observations made at the same time can provide unique estimates of the size and time dependency of the dust albedo.

3. TWO-DIMENSIONAL MODELS

If we assume that dust, after having been dragged out by the expanding gas, is ejected from the coma at zero velocity with respect to the nucleus, then the resulting two-dimensional model depends only on the quantity F given by equation (8). Within this model, the dust tail is a thin dust layer lying in the comet orbital plane. The dust tail brightness D is simply proportional to F multiplied by the determinant of the Jacobian between the (x,y) frame and the syndyne-synchrone network defined in section 1. Then equation (5) is directly invertible, easily providing the quantity F, from which we can infer the dust mass loss rate by equation (9) and the DSD by means of the normalization of F vs. β and then by equation (7). Following such a direct approach, the DSD of short-period Comets 2P/Encke and 6P/d'Arrest was obtained (*Sekanina and Schuster, 1978*). The DSD power index was always $\alpha < -4$. In this case, both the brightness and the mass mainly depend on the micrometer-sized grains observed in the tail. If $\alpha > -3$, both the mass and brightness depend on the largest ejected grains. Brightness and mass become decoupled if $-4 < \alpha < -3$, in which case the dust mass depends on the largest ejected grains, while the brightness depends on the micrometer-sized grains.

Since the observed brightness fixes the number of micrometer-sized grains observed in the comet if $\alpha < -3$, the DSD index in practice only affects the number of unob-

TABLE 1.   Dust size distributions in comets.

Comet	β_{min}	β_{max}	$\alpha(t)$	α_m
2P/Encke	6×10^{-6}	0.02	$-4.8 < \alpha < -2.8$	-3.6 ± 0.3
6P/D'Arrest	1×10^{-5}	0.06	$-5.0 < \alpha < -3.5$	-3.8 ± 0.1
10P/Tempel2	2×10^{-5}	0.03	$-4.8 < \alpha < -3.4$	
26P/Grigg-Skjellerup	6×10^{-5}	0.06	$-4.0 < \alpha < -3.0$	
29P/SW1	4×10^{-5}	0.2	$-4.0 < \alpha < -3.0$	-3.3 ± 0.3
46P/Wirtanen	1×10^{-4}	0.2	$-4.1 < \alpha < -3.0$	
65P/Gunn	1×10^{-4}	0.2	$-4.3 < \alpha < -3.0$	
67P/Churyumov-Gerasimenko	6×10^{-6}	0.03	$-5.5 < \alpha < -3.2$	-3.4 ± 0.2
P/Swift-Tuttle	6×10^{-5}	0.2	$-5.0 < \alpha < -3.0$	-3.3 ± 0.2
2060 Chiron	2×10^{-4}	1.0	$-4.5 < \alpha < -3.0$	-3.2 ± 0.1
Seki-Lines 1962III	1×10^{-4}	0.1	$-5.0 < \alpha < -4.0$	-4.1 ± 0.6
Kohoutek 1973XII	1×10^{-4}	0.2	$-5.0 < \alpha < -3.0$	-3.3 ± 0.4
Wilson 1987VII	2×10^{-5}	0.1	$-3.8 < \alpha < -2.8$	-3.0 ± 0.1
Bradfield 1987XXIX	1×10^{-4}	0.6	$-4.2 < \alpha < -3.0$	-3.2 ± 0.2
Liller 1988V	1×10^{-5}	0.06	$-4.3 < \alpha < -3.3$	-3.5 ± 0.2
Austin 1990V	1×10^{-5}	0.1	$-4.5 < \alpha < -2.8$	-3.0 ± 0.2
Levy 1990XX	6×10^{-5}	0.4	$-5.0 < \alpha < -3.0$	-3.2 ± 0.1
Hyakutake 1996B2	4×10^{-4}	1.0	$-4.7 < \alpha < -3.0$	-3.6 ± 0.2
Hale-Bopp 1995O1	2×10^{-3}	1.0	$-4.2 < \alpha < -3.2$	-3.6 ± 0.1

Time-dependent [$\alpha(t)$] and time-averaged (α_m) DSD power index evaluated for various comets within a β range (defined by β_{min} and β_{max}) by means of the inverse tail model (adapted from *Fulle*, 1999).

served large grains, which may dominate the ejected mass if $\alpha > -4$ as well. Therefore, *Sekanina and Schuster* (1978) correctly concluded that the index they found ($\alpha = -4.2$) implied that the number of ejected large grains was negligible, and this assumption was adopted to design the European Space Agency's *Giotto* mission to 1P/Halley. The impact of *Giotto* with a grain of 1 g at the flyby implied that the actual probability of such an impact was underestimated by at least a factor of 100. Since the size ratio between 1-g grains and those dominating the brightness of Comets 2P/Encke and 6P/d'Arrest was 10^4, the real α value should have been $\alpha > -3.7$. We should conclude either that the dust population of 1P/Halley is very different from that of 2P/Encke or 6P/d'Arrest, or that the two-dimensional model has intrinsic severe shortcomings.

Observations when Earth crosses the comet orbital plane, and hydrodynamic models describing the dust-gas interaction in the coma, both point out that the assumption of zero dust ejection velocity is nonphysical. Let us recall that this approximation is also useless for the so-called perspective antitails, which are sometimes composed of dust released at very large heliocentric distances. In this latter case, the likely very low dust velocity is balanced by the very long time interval existing between ejection of the dust and observation. Since the thickness of the tail is roughly given by such a time interval times the dust velocity, it is impossible to obtain a dust tail that can be well approximated by a two-dimensional model based on synchrones and syndynes. This is also true for apparently thin antitails observed, e.g., in Comet Arend-Roland 1957III: *Kimura and Liu* (1977) have

shown that the correct explanation for these spikes requires a model more elaborate than the two-dimensional syndyne-synchrone model.

After we have shown that the two-dimensional tail model has no physical basis, we must also show that the predictions of this two-dimensional model are systematically different from those of a correct three-dimensional model applied to the same data. This was done, and in fact the DSD power index resulted in $\alpha \approx -3.7$ for Comet 2P/Encke and Comet 6P/d'Arrest (*Fulle*, 1990) (Table 1). The fact that the dust mass in 2P/Encke is strongly dominated by the largest ejected grains was further confirmed by applying the same three-dimensional model to independent IR thermal data provided by the ISO probe [$\alpha = -3.2$; *Epifani et al.* (2001)]. These results show that two-dimensional models and all related outputs should be ignored in the future, as they are unable to provide useful constraints on the derived cometary dust properties.

4.   THREE-DIMENSIONAL MODELS

Finson and Probstein (1968) showed that the unrealistic two-dimensional synchrone-syndyne models can be converted into realistic three-dimensional models when we associate a synchronic tube to each synchrone, whose width is given by the dust ejection velocity at the synchrone time, or a syndynamic tube to each syndyne, whose width is given by the dust ejection velocity of the syndyne β value. In the synchronic approach, they analytically computed the dimensionless sky surface density of a synchronic tube

$$dD(x,y,t) = \frac{F(t,1-\mu)dt}{2(t_0 - t)v(t,1-\mu)\frac{ds}{d(1-\mu)}} \quad (12)$$

so that the brightness of the whole tail is simply proportional to the numerical time integral of equation (12). Equation (12) is derived from an analytical integration that is correct if the following condition (named by Finson and Probstein, although quite improperly, hypersonic) is satisfied

$$\frac{ds}{d(1-\mu)} \gg \frac{(t_0 - t)v(t,1-\mu)}{1-\mu} \quad (13)$$

This usually happens in the outermost dust tail only. In equations (12) and (13), s is the parametric coordinate along the synchrone and $v(t,1-\mu)$ the dust ejection velocity. In equation (12), the synchronic (or syndynamic) tubes are assumed to have circular sections, because the dust tail is supposedly built up by dust shells that keep their spherical expanding shape over time if significant tidal effects due to the solar gravity are neglected. The spherical shell assumption implies that the dust ejection is isotropic, another strong approximation. When the numerical integration of equation (12) is performed, the fit of the tail brightness data is performed by trial and error, so we cannot ensure the uniqueness of the obtained dust loss rate, ejection velocity, and size distribution. All these approximations make the Finson-Probstein model a first-order model that is unable to give realistic estimates of the dust parameters. *Fulle* (1987, 1989) developed an inverse Monte Carlo dust tail model by taking into account all the improvements introduced by *Kimura and Liu* (1977) and avoiding all the limitations and approximations of the Finson-Probstein model: (1) It computes the rigorous heliocentric Keplerian orbits of millions of sampling dust grains, so that the spherical shell approximation is avoided. (2) It performs both the size and time integral by means of numerical methods, so that the condition imposed by equation (13) is avoided. In this way, it can fit not only the external tail, but also the inner one, as close as desired to the dust coma, where the largest grains usually reside. (3) It takes into account anisotropic dust ejections. (4) It provides for each dust size a dust ejection velocity from the coma that is the mean value of a wide velocity distribution (*Fulle*, 1992); this is consistent with the predictions of three-dimensional dust-gas interaction models in the inner coma (*Crifo and Rodionov, 1999*). (5) It avoids the trial-and-error procedure typical of the original Finson-Probstein model, by means of an inverse ill-posed problem theory. In this way, the uniqueness of the results (impossible to establish in the original Finson-Probstein approach) is recovered in the least-squares-fit sense.

Within the inverse Monte Carlo dust tail model, the dust ejection velocity, loss rate, and size distribution are obtained through minimization of the function

$$(HF - D)^2 + (RF)^2 = min \quad (14)$$

where H is the kernel matrix provided by the dust tail model defined in equation (2), F (solution vector defined in equation (8)) is the output of the inversion of equation (14), D is the dust tail dimensionless sky surface density (data vector defined in equation (2)), and R is a regularizing constraint to drop noise and negative values in solution F. The inverse Monte Carlo approach was applied to tens of dust tails: The results regarding the DSD (Table 1) and the dust mass loss rate will be discussed in sections 5 and 6. In general, dust ejection velocities show time and size dependencies more complex than predicted by one-dimensional coma models of dust-gas interaction. In particular, the velocity of large grains relative to that of small ones seems higher than expected. This result was confirmed by independent tail models that left free such a parameter (*Waniak, 1992, 1994*). This points out that dust-gas interaction must be treated by three-dimensional coma models to predict reliable dust ejection velocities. In most cases the accuracy of the fit of the observed tail brightness did not change after varying the assumed dust ejection anisotropy. This fact confirms that tail shapes have lost memory of the details of the unknown boundary conditions of the three-dimensional gas drag occurring in the inner coma.

Other Monte Carlo dust tail models were developed that followed other approaches (e.g., *Waniak, 1994; Lisse et al., 1998*). Equation (2) allows us to fit more details of the input tail data by means of a time- and size-dependent dust-scattering cross-section [i.e., by $f(t,1-\mu)$]. If the DSD is assumed to be time-independent to stabilize the model outputs, then it may become impossible to perfectly fit the tail data. The assumption of a time-independent DSD is commonly made: *Lisse et al.* (1998) analyzed dust tails at millimeter wavelengths (COBE satellite data) with a poor spatial resolution (20 arcmin), so that rough assumptions on DSD $[f(1-\mu) = (1-\mu)^{-1}$ implying $\alpha = -3]$ allowed them to fit the input images. *Waniak* (1994), adopting a constant $\alpha \approx -3$, improved the tail fit of Comet Wilson 1987VII by means of a detailed dust ejection pattern. However, the fact that this pattern ejects dust mainly on the nucleus nightside may indicate that temporal variations of the DSD are likely responsible for the shape of dust tails.

5. DUST SIZE DISTRIBUTION IN COMETS

The dust size distribution (DSD) plays a crucial role when we describe the behavior of dust particles in comets. For instance, it is widely believed that in dust comae we mainly see grains with sizes close to the observation wavelength. It is easy to conclude (see equation (15) in section 6.1) that the coma brightness depends on the size of the largest ejected particle if $\alpha > u - 3$, where u is the power index vs. $(\rho_d d)$ of the dust velocity. The assumptions ($\alpha = -3.0$ and $u = -0.5$) made by *Lisse et al.* (1998) imply that, if $\alpha > -3.5$, then the dust coma brightness at optical wavelengths is dominated by the contribution from the largest ejected boulders. Since dynamical models in comae of micrometer-sized grains or of meter-sized boulders are quite different, it is crucial to understand which dust is observed. The total mass of the dust depends on the size of the largest or the smallest ejected grains according to the actual DSD: It is fundamental to define the size range to which the published results are related, although this is usually not done. For usual $\alpha > -4$, the dust mass diverges if we allow it to reach the nucleus size. It is impossible to compare dust masses without knowing which largest boulder size they refer to.

We applied the inverse tail model to tens of cometary dust tails, which always resulted in dust grains ranging in size between 1 μm and about 1 cm (Table 1). The DSD often showed large temporal changes: Since most output instabilities affect the DSD time-dependency, in section 6.2 we will pay special attention to testing the correctness of this model output. The time-averaged DSD is much more stable: In almost all comets it was characterized by an index $\alpha \approx -3.5$. This index is typical of a population of collisionally evolved bodies (*Dohnanyi*, 1972): Should this index value be further confirmed, it would suggest that we do not observe in comets the pristine dust population of the presolar nebula. The $\alpha = -3.5$ value is the most critical for models aimed to fit the brightness and/or the time evolution of coma features, like jets or spirals. In fact, if $\alpha = -3.5$ and $u = -0.5$ [this value is provided by one-dimensional models of dust-gas interaction at $d > 1$ μm (*Crifo*, 1991)], then a correct computation of the brightness of every feature observed in a dust coma must take into account grains of every size, from submicrometers up to tens of meters. Also, these large boulders, when present in a dust coma, provide a significant fraction of the observed brightness if $\alpha = -3.5$ and $u = -0.5$. While it is possible to come up with a theory of dust tails that can separate the dependency of the numerous dust parameters required to interpret the data, this becomes impossible in models of dust comae.

An important dataset providing DSD in comets is given by the only available *in situ* data we have so far: the results of the DID experiment on *Giotto* (*McDonnell et al.*, 1991) during the 1P/Halley flyby. The power index fitting the DSD at the nucleus, traced back from the DID fluence in terms of purely radial expansion, is $\alpha \approx -3.7$ for dust masses between 10^{-14} and 10^{-3} kg. This slope is also consistent with the impact of the 1-g grain that damaged the probe itself. This example shows how models crucially affect the interpretation of the dust data. The fit of the same data by means of a rigorous dust dynamical model changed the results completely (*Fulle et al.*, 2000). The DID fluence was found to be consistent with any $-2.5 < \alpha < -3.0$, strongly supporting the possibility considered by *Lisse et al.* (1998), that the coma brightness is dominated by meter-sized boulders. The same model also inferred the dust-to-gas ratio ($3 < \text{DGR} < 40$ for dust masses up to 1 g) and the dust geometric albedo ($0.01 < A_p < 0.15$) by means of the Optical Probing Experiment data (*Levasseur-Regourd et al.*, 1999), with a correlation between A_p and ρ_d ($\rho_d = 2500 A_p$ kg m^{-3}). Tail models and DID experiments agree with these conclusions: (1) $\alpha \gg -4$; (2) the dust mass (we cannot exclude the light

flux too) depends on the size of the largest ejected boulders; and (3) the DGR is >1 for particle sizes larger than 1 cm.

All available results confirm that the mass of dust ejected from comets is dominated by the largest boulders. Radar observations (*Harmon et al.*, 1989) operating at centimeter wavelengths and coma observations at millimeter wavelengths provide first-quality constraints to this conclusion, because they directly observe grains close in size to the observation wavelength if $\alpha < -3.5$, while they observe dust larger than the observation wavelength if $\alpha \gg -3.5$. Observations of P/Swift-Tuttle (*Jewitt*, 1996) and C/Hyakutake 1996B2 (*Jewitt and Matthews*, 1997) at millimeter wavelengths provided loss mass rates 7 and 10 times higher, respectively, than predicted by tail models (*Fulle et al.*, 1994, 1997). Observations at millimeter wavelengths therefore suggest that α was higher than $\alpha = -3.3$ and $\alpha = -3.6$ respectively (Table 1). We must conclude that the dust coma brightness of these two comets was dominated at all observation wavelengths by the largest ejected boulders if index u is -0.5 (this value is predicted by one-dimensional dust-gas-drag models in the inner coma).

Is a power law a proper function to describe the DSD? This assumption is usually adopted by direct tail modelers only (e.g., *Lisse et al.*, 1998). Inverse tail models do not assume that the DSD is a power law; they leave it completely free, sampling the DSD in β-bins. Then, the output describing the DSD is fit by a power law to offer an easily understandable DSD. Sometimes, the direct $f(t, 1 - \mu)$ output was provided (e.g., *Fulle et al.*, 1998). In any case, no DSD data are determined accurately enough to require more than a power law in order to fit them: A power-law DSD is consistent with the DID fluence measured at 1P/Halley, which is not a power law of the dust size (*Fulle et al.*, 2000). Other functions were suggested to describe the DSD of cometary dust. *Hanner* (1984) proposed a more complex function, which becomes a simple power law at the largest sizes ($d > 1$ mm), and drops rapidly to zero at submicrometer grains. This function was used to fit the IR photometry and spectra of comets. These observations are unable to detect submicrometer grains, and thereby provide a typical example of "absence of evidence" interpreted as "evidence of absence." In fact, *Giotto* showed that in 1P/Halley, submicrometer grains were more abundant than larger ones (*McDonnell et al.*, 1991). This is probably true for all comets: Many authors pointed out that every DSD discussed in this review is consistent with all available IR spectra (*Crifo*, 1987; *Greenberg and Li*, 1999). Dust tail models and *in situ* data provide much better constraints to the DSD than IR spectra, and show that a power law defined within a precise size interval is the best approach for describing the DSD in comets, avoiding misleading conclusions suggested by more complex and underconstrained size functions.

It would seem obvious that a comet is defined as "dusty" according to its dust-to-gas ratio. In other words, we would like to find that the higher the index α, the higher the released dust mass, and the more "dusty" the comet. This is not the case. Usually, a comet is said to be "dusty" when spectral features in its IR spectrum or high polarization are observed. However, it is hard to relate these observed characteristics to the actual released dust mass. In particular, when the silicate feature at 10 μm is strong (*Lisse*, 2002), or when the highest polarization is higher than 20% (*Levasseur-Regourd et al.*, 1996), the comet is then recognized as "dusty." Both these features refer to the actual population of micrometer-sized grains, which we have seen to be a minor component of the total released mass. If the DSD has a turning point at some size larger than 10 μm, where at smaller sizes α becomes larger than 1P/Halley's index, then the relative production of micrometer-sized grains drops compared to 1P/Halley. In this case, the comet is defined as less "dusty" than 1P/Halley, even though its α from millimeters to meters may be much higher, with a much higher released dust mass.

6. CONSTRAINTS TO THE OUTPUTS OF THREE-DIMENSIONAL TAIL MODELS

6.1. Dust Coma Equivalent Size Afρ

The model outputs F and v can be compared to the observed dust coma brightness. This is usually measured by means of the Afρ quantity (*A'Hearn et al.*, 1984)

$$
\text{Afρ}(t) = 2\pi \left[\frac{\gamma r}{1 \text{ AU}} \right]^2 \frac{10^{0.4[m_\odot - m(x,y)]}}{D(x,y)}
$$
$$
\int_0^\infty \frac{F(t, 1-\mu)}{v(t, 1-\mu)} \, d(1-\mu) \tag{15}
$$

where all the quantities are defined in section 2. Afρ is related to the total dust cross-section Σ observed inside the observation field of radius ρ (in meters projected at the comet) centered exactly on the comet nucleus

$$
\Sigma = \frac{\pi \rho}{4 A_p(\phi)} \text{Afρ} \tag{16}
$$

We note that equation (15) is completely parameter-free: Given the outputs F and v of the tail model, there is no way to adjust the tail model to fit the observed Afρ values. However, the integral of equation (15) can be computed on a finite β range, while Afρ is measured observing all the ejected dust sizes in the dust coma. Therefore, if the size range adopted to compute equation (15) loses dust sizes reflecting a significant light fraction in the coma, then tail models can provide only a lower limit of the actually observed Afρ. In any case, Afρ computed by means of equation (15) was always consistent with that observed (*Fulle et al.*, 1998; *Fulle*, 2000).

Equations (9) and (15) point out that Afρ has little to do with the dust mass loss rate. Nevertheless, it is commonly referred to as the dust loss rate, despite its dimensions, and

is commonly related to the water loss rate to obtain incorrect (at least from a physical point of view) dimensional ratios. It is obvious that comets with a higher Afρ can eject less dust mass: This depends on the DSD and dust velocity. Moreover, the time evolution of Afρ can be unrelated to the loss rate time evolution, depending on time changes of the DSD and velocity. Afρ is a high-quality constraint for physical models of comae and tails, but nothing more.

6.2. Dust Size Distribution Time Variability

So far, only inverse dust tail models take into account the possibility that dust is ejected from the cometary nucleus with a DSD that may change in time. Despite the high or low probability that this really happens in comets, it is surprising that most models describing dust in comets adopt time-independent DSDs, because most papers on cometary dust invoke dust fragmentation to explain the observations. It is obvious that dust fragmentation implies a time evolution of the DSD, and a model that takes into account a time-dependent DSD is more general than others that only take dust fragmentation into account. So far, no consistent models of dust fragmentation were developed (*Crifo*, 1995). *Combi* (1994) developed a direct dust tail model with fragmentation that was based on consistent fits of both the dust coma and tail. Many tests performed by means of the inverse tail model that adopted numerous u values showed that such a consistent fit to both the dust coma and the tail simply requires –0.5 ≪ u < –0.1. *Fulle et al.* (1993) showed that dust fragmentation is a possible (not unique) explanation of –0.5 ≪ u < –0.1.

While inverse dust tail models can provide a stable time-averaged DSD, the time-dependent DSD is affected by residual instability. It is not easy to establish if large and systematic changes of the DSD are real or simply due to output instability. The most elegant solution is to find independent observations that suggest systematic changes of the ejected dust population in agreement with the time-evolution of the DSD provided by inverse tail models. These models applied to C/Hyakutake 1996B2 provided an α value that dropped suddenly from a roughly constant value α = –3 to α = –4 in mid April 1996 (*Fulle et al.*, 1997), in perfect agreement with the time evolution of IR spectra; no silicate feature was detected before a strong 10-μm line appeared around mid April (*Mason et al.*, 1998). This IR spectral evolution was interpreted in terms of dust fragmentation exposing small silicate cores to the Sun's radiation; such cores were embedded in large carbonaceous matrices before mid April. The inverse tail model applied both to optical data (*Fulle*, 1990) and to IR thermal data (*Epifani et al.*, 2001) collected during two different perihelion passages of Comet 2P/Encke provided a similar drop from α = –3 to α = –4 during the first three weeks after perihelion. This coincidence has already forced us to exclude the possibility that this DSD time-evolution is due to output instability. Moreover, these three weeks match exactly the seasonal night cycle of the most active nucleus hemisphere suggested by *Sekanina*

(1988) to explain the comet photometry and coma shape evolution of 2P/Encke.

7. FINE STRUCTURES IN TAILS

7.1. Neck-Lines

In several cases, dust tails retain the memory of the dust ejection over more than half the comet orbit; during such long periods of time the tidal effects of solar gravity become significant. *Kimura and Liu* (1977) pointed out that the dust motion is heliocentric, so that at ejection a dust grain can be considered as occurring at the first node of its heliocentric orbit. Every heliocentric orbit has its second node 180° away from the first, where the dust grain orbit must necessarily cross the comet orbit again. When we consider a Finson-Probstein dust shell, all these grains, ejected at the same time, will have their second orbital node at approximately the same time, i.e., 180° away in orbital anomaly, where the spherical dust shell will collapse into a two-dimensional ellipse contained in the comet orbital plane, a shape far removed from a sphere. When we consider a collapsed synchronic tube, we obtain a two-dimensional structure, referred to as the "neck-line" by the discoverers (*Kimura and Liu*, 1977). When Earth crosses the comet orbit, the neck-line appears on the sky as a straight line much brighter than the surrounding dust tail, because all the synchronic tube is collapsed in an infinitesimal sky area (Fig. 2). By means of the neck-line model, Kimura and Liu fit perfectly the perspective antitail of Comet Arend Roland 1957III, which appeared as a bright spike many millions of kilometers long and pointing toward the Sun, thus avoiding unrealistic explanations based on dust ejected at zero velocity from the parent coma.

Neck-lines were observed in the Great Comet 1910 I and in Comets Arend-Roland 1957III, Bennett 1970II, 1P/Halley 1986III, Austin 1990V, Levy 1990XX, and Hale-Bopp 1995O1. It must be pointed out that neck-lines can appear only after perihelion of comets orbiting in open orbits (hyperbolic or parabolic), a fact rigorously verified by all the registered apparitions. In the case of periodic comets, dust ejected after perihelion could also form a neck-line before the next perihelion passage. However, this was never observed and seems improbable, because planetary perturbations in short-period comets, and stellar ones in long-period comets, can perturb the dust orbits enough to prevent the formation of a neck-line after periods of many years. Due to the particular perspective conditions, in Comets Arend-Roland and Levy the neck-line appeared as a perspective antitail (it is probable that all observed antitails were in fact neck-lines). In all other comets, it appeared superimposed to the main dust tail as a bright and straight linear feature. Since a neck-line is built up by the collapse of the dust shells into two-dimensional ellipses, the ellipses composed of the largest grains, approximately centered on the comet nucleus, are placed half out of the comet orbit and half inside of it. Due to the long travel time of the dust

grains (usually months), these ellipses may reach huge dimensions, up to 10^6 km: When the Earth crossed the orbit plane of Comets Bennett, 1P/Halley, Austin and Hale-Bopp, the half of the ellipse inside the comet orbit was seen as a bright spike pointing to the Sun, which was in fact a real (nonperspective) antitail. Since these real antitails are com-

Fig. 2. Neck-line observed in Comet Hale-Bopp 1995O1 on 5 January 1998 with the ESO 1-m Schmidt Telescope. The original image was filtered (unsharp masking) to enhance the spike features: the real antitail pointing toward the Sun (bottom) and the neck-line pointing in the opposite direction. ESO Press Photo 05a/98, courtesy of the European Southern Observatory (observer Guido Pizarro).

posed of the largest grains ever observed in dust tails, neck-line observations provide unique information on the ejection velocity and size distribution of grains larger than centimeter-sized.

Fulle and Sedmak (1988) have obtained analytical models of neck-lines, so that the neck-line photometry provides the β distribution and the dust ejection velocity at the ejection time t of the neck-line. The Keplerian dynamics of the grains in space allow us to compute the geometric neck-line parameters a and b (related to the major and minor axes respectively of the ellipses composing the neck-line) and s (the parametric coordinate along the neck-line axis x) defined in *Fulle and Sedmak* (1988). If the velocity condition

$$v(1 - \mu) \ll (1 - \mu)sa \tag{17}$$

is satisfied, then the dimensionless sky surface density in the neck-line is

$$D(x, y) =$$

$$\frac{F(t, 1 - \mu)}{sv(1 - \mu)}\left[1 + \mathrm{erf}\,\frac{ax}{v(1 - \mu)}\right]\exp\left[-\frac{b^2y^2}{v^2(1 - \mu)}\right] \tag{18}$$

From equation (18), it is apparent that the neck-line width along the y axis provides a direct measurement of the β dependency of the dust velocity at centimeter sizes. These are the only available direct observations of such a dependency, which is in agreement with the results of three-dimensional inverse dust tail models: $-0.5 \ll u < -0.1$. The obtained β distributions confirm that most of the dust mass is released in the form of the largest ejected grains.

7.2. Striae

In the brightest comets (e.g., Great Comet 1910I, Comets Mrkos 1957V, West 1976VI, Hale-Bopp 1995O1), the usually structureless dust tail exhibits detailed substructures of two kinds, namely synchronic bands and striae. The synchronic bands are streamers pointing to the comet nucleus, with the axis well fit by synchrones. There is general agreement that they are due to time changes of the dust loss rate or of the DSD, so that the dust cross-section is larger in the synchronic bands than outside. On the contrary, the striae do not point to the comet nucleus (Plate 17), and they are neither fit by synchrones nor by syndynes. There is general agreement that striae are due to instantaneous fragmentation of larger parents, so that the striae are synchrones not originating from the comet nucleus, but from the tail point where the parent was located at the time of fragmentation. This explains well the orientation of the striae, which always point between the comet nucleus and the Sun.

Usually, the interpretations of bands and striae are performed using two-dimensional synchrone-syndyne models, so that the available quantitative results may be affected by significant errors. The dust ejection velocity from the coma

is assumed to be zero, in contradiction with all available information on the dust dynamics. For striae models, in particular, which are very sensitive to the β value of the parent grain, this assumption might significantly affect the results. The β value of the parent is easily obtained by the stria origin in a two-dimensional model, while many different β values are consistent with the stria origin in a three-dimensional model. *Sekanina and Farrell* (1980) showed that the submicrometer-sized fragments observed in the striae of Comet West 1976VI had a β very similar to their much larger parents. This result would imply that the parents must be very elongated chains of submicrometer-sized grains. However, this also implies that the drag by gas on these chains was correspondingly high (both gas drag and radiation pressure depend on the parent cross-section): These chains must have been ejected from the coma exactly at the gas velocity, so that they would have been diluted over huge 10^6-km-sized shells. The origin of the striae constrains neither the fragmentation time nor the β value of the parent grain.

7.3. Sodium Tail

Although spectroscopic observations suggested the presence of tail extensions of the well-known neutral sodium coma of Comets 1910I and Arend-Roland 1957III, the first clear images of a huge sodium tail 10^7 km long, well separated in the sky from classical dust and plasma tails, were obtained during the 1997 passage of Comet Hale-Bopp 1995O1 (*Cremonese et al.,* 1997). The images were taken by means of interference filters centered on the sodium D lines at 589 nm, and the tail did not appear in simultaneous images taken on the H_2O^+ line, showing a well-developed ion tail. The sodium tail appeared as a straight linear feature located between the ion tail and the prolonged radius vector. Simultaneous spectroscopic observations permitted measurement of the radial velocity of the neutral sodium atoms along the tail. The result was that a syndyne of $\beta = 82$ best fit both the sodium tail axis orientation and the radial velocities along the tail. This was the first time that the β parameter was best constrained by means of radial velocity measurements. The sodium neutral atoms lighten because of fluorescence: Absorbed UV solar photons spend their energy to put the external sodium electrons in the most external orbitals. Since the solar photons are absorbed by the sodium atoms, their momentum is necessarily transferred to the sodium atoms, so that

$$\beta = \frac{hg(1\ \mathrm{AU})^2}{\lambda GM_\odot m} \tag{19}$$

where h is the Planck constant, g the number of solar photons captured in the unit time by a sodium atom at 1 AU (or photon scattering efficiency in the sodium D lines), λ the wavelength of the sodium D lines, and m the tail particle mass. In perfect agreement with the theoretical computations, the observed β = 82 provides g = 15 s^{-1} when we assume the atomic sodium mass for m. Therefore, the sodium tail is composed of sodium atoms and not of sodium molecules. The sodium atoms in space have a short lifetime, mainly due to photoionization. The sodium tail brightness along its axis x provides an unique measurement of the sodium lifetime τ. Since the sodium tail axis can be best approximated by a syndyne, the whole sodium tail can be modeled by means of a syndynamic tube, whose photometric equation was computed by *Finson and Probstein* (1968). When we consider the sodium lifetime against photoionization, the sky surface density of sodium atoms is

$$\delta(x,y) = \frac{\dot{N}\exp-\frac{t_o - t(x)}{\tau}}{2v[t_o - t(x)]w(x)} \qquad (20)$$

where $\dot{N}$ is the sodium loss rate, v the sodium ejection velocity (related to the sodium tail width), w the sodium radial velocity projected on the sky, and t the time of sodium ejection (w and t are provided by syndyne computations). Equation (20) perfectly fits the observed brightness on the tail axis x when we assume τ = 1.7 × 10^5 s at 1 AU. This lifetime is three times larger than the value assumed in comet and planetary atmospheric models, as it was already suggested by laboratory measurements by *Huebner et al.* (1992).

8. CONCLUSIONS

Modeling the properties of the dust particles ejected from comets is one of the most formidable tasks in cometary physics, as we have little or no information about numerous parameters describing these dust grains. Quantities such as albedo, bulk density, radiation-scattering properties, shape, and grain spin are far from completely determined. There are other quantities either related to the dust sources dispersed on the nucleus surface or describing the dust-gas interaction in the coma, such as the dust velocity relative to the comet nucleus, the loss rate, and the particle size distribution, that are absolutely required by any realistic model that describes the dust environment of comets. It is impossible to obtain observational evidence that allows the inference of only one of these dust parameters. All of them are deeply interrelated, and only complex models that make use of all of them can provide a self-consistent scenario of the observations. For instance, it is usually assumed that the time evolution of the dust coma brightness allows us to deduce the time-dependent dust loss rate. However, the dust coma brightness depends on all the parameters listed above, and the loss rate is only one of them. Consequently, the interpretation of coma brightness in terms of the loss rate implicitly assumes that all the other dust parameters either play no role or have known values that do not change in time. All these assumptions are invalid.

Therefore, among complex models describing the motion of dust particles in comae and tails, one should pay most attention to the ones that contain the lowest possible number of free parameters and that are able to provide the largest number of dust parameters after data fits have been obtained. Inverse dust tail models appear today as the most powerful tools we have to interpret groundbased (or Earth-orbiting satellite-based) data, because they only require assumptions on the dust grain shapes, their scattering efficiency, and albedo. Moreover, when thermal tail data are available, a comparison between outputs of tail models and optical coma photometry can provide an estimate of the average albedo of the dust. When high-quality data on both the IR and optical dust tails become available, inverse dust tail models will provide unique information on the temporal and grain-size dependencies of the dust albedo. Dust tail models predict the temporal evolution of the dust coma brightness, to be compared with the observations of the Afρ parameter: This independent constraint allows one to establish that the dust environment deduced from modeling the IR and optical tail properties is indeed valid.

Information on the velocity imparted to the dust grains by the expanding gas is required to constrain three-dimensional model predictions for the gas dynamics in cometary comae; only dust tail models can provide estimates of the temporal and grain-size dependencies of the dust velocity. Information on the grain-size distribution is required by all models aimed at fitting the brightness of dust coma features. As a consequence, only inverse dust tail models can provide estimates of the time-dependent grain-size distribution. Cosmogonic and evolutionary models of comet nuclei require a good knowledge of the dust to gas ratio at the nucleus; only dust tail models can provide dust mass loss estimates that are not severely biased by assumptions of the related grain-size distribution. When high-quality data on all the parameters required to describe cometary dust grains become available, probably provided by future rendezvous missions to comets, complete and detailed models of the dust environment of comets will become possible. It is only then that one will be able to validly describe what space probes will face when meeting their targets. Only coordinated efforts among all modelers, taking into account the unique information that inverse tail models can provide, will enable the attainment of such an ambitious goal.

REFERENCES

A'Hearn M. F., Schleicher D. G., Feldman P. D., Millis R. L., and Thompson D. T. (1984) Comet Bowell 1980b. *Astron. J., 89,* 579–591.

Burns J. A., Lamy P. L., and Soter S. (1979) Radiation forces on small particles in the solar system. *Icarus, 40,* 1–48.

Combi M. R. (1994) The fragmentation of dust in the innermost comae of comets: Possible evidence from ground-based images. *Astron. J., 108,* 304–312.

Cremonese G., Boehnhardt H., Crovisier J., Rauer H., Fitzsimmons A., Fulle M., Licandro J., Pollacco D., Tozzi G. P., and West R. M. (1997) Neutral sodium from Comet Hale-Bopp: A third type of tail. *Astrophys. J. Lett., 490,* L199–L202.

Crifo J. F. (1987) Are cometary dust mass loss rates deduced from optical emissions reliable? In *Interplanetary Matter* (Z. Ceplecha and P. Pecina, eds.), pp. 59–66. Report G7, Czechoslovak Academy of Sciences, Ondrejov.

Crifo J. F. (1991) Hydrodynamic models of the collisional coma. In *Comets in the Post-Halley Era* (R. L. Newburn Jr. et al., eds.), pp. 937–990. Kluwer, Dordrecht.

Crifo J. F. (1995) A general physicochemical model of the inner coma of active comets. I. Implications of spatially distributed gas and dust production. *Astrophys. J., 445,* 470–488.

Crifo J. F. and Rodionov A. V. (1999) Modelling the circumnuclear coma of comets: Objectives, methods and recent results. *Planet. Space Sci., 47,* 797–826.

Dohnanyi J. S. (1972) Interplanetary objects in review: Statistics of their masses and dynamics. *Icarus, 17,* 1–48.

Epifani E., Colangeli L., Fulle M., Brucato J., Bussoletti E., de Sanctis C., Mennella V., Palomba E., Palumbo P., and Rotundi A. (2001) ISOCAM imaging of Comets 103P/Hartley 2 and 2P/Encke. *Icarus, 149,* 339–350.

Finson M. L. and Probstein R. F. (1968) A theory of dust comets — I. Model and equations. *Astrophys. J., 154,* 327–352.

Fulle M. (1987) A new approach to the Finson-Probstein method of interpreting cometary dust tails. *Astron. Astrophys., 171,* 327–335.

Fulle M. (1989) Evaluation of cometary dust parameters from numerical simulations: Comparison with analytical approach and role of anisotropic emission. *Astron. Astrophys., 217,* 283–297.

Fulle M. (1990) Meteoroids from short-period comets. *Astron. Astrophys., 230,* 220–226.

Fulle M. (1992) A dust tail model based on maxwellian velocity distributions. *Astron. Astrophys., 265,* 817–824.

Fulle M. (1999) Constraints on Comet 46P/Wirtanen dust parameters provided by in-situ and ground-based observations. *Planet. Space Sci., 47,* 827–837.

Fulle M. (2000) The dust environment of Comet 46P/Wirtanen at perihelion: A period of decreasing activity? *Icarus, 145,* 239–251.

Fulle M. and Sedmak G. (1988) Photometrical analysis of the neck-line structure of Comet Bennett 1970II. *Icarus, 74,* 383–398.

Fulle M., Bosio S., Cremonese G., Cristaldi S., Liller W., and Pansecchi L. (1993) The dust environment of Comet Austin 1990V. *Astron. Astrophys., 272,* 634–650.

Fulle M., Böhm C., Mengoli G., Muzzi F. Orlandi S., and Sette G. (1994) Current meteor production of Comet P/Swift-Tuttle. *Astron. Astrophys., 292,* 304–310.

Fulle M., Mikuz H., and Bosio S. (1997) Dust environment of Comet Hyakutake 1996B2. *Astron. Astrophys., 324,* 1197–1205.

Fulle M., Cremonese G., and Böhm C. (1998) The preperihelion dust environment of C/1995O1 Hale-Bopp from 13 to 4 AU. *Astron. J., 116,* 1470–1477.

Fulle M., Levasseur-Regourd A. C., McBride N., and Hadamcik E. (2000) In situ measurements from within the coma of 1P/Halley: First order approximation with a dust dynamical model. *Astron. J., 119,* 1968–1977.

Greenberg J. M. and Li A. (1999) All comets are born equal: Infrared emission by dust as key to comet nucleus composition. *Planet. Space Sci., 47,* 787–795.

Hanner M. S. (1984) A comparison of the dust properties in recent periodic comets. *Adv. Space Res., 4,* 189–196.

Hanner M. S. Giese R. H., Weiss K., and Zerull R. (1981) On the definition of albedo and application to irregular particles. *Astron. Astrophys., 104,* 42–46.

Harmon J. K., Campbell D. B., Hine A. A., Shapiro I. I., and Marsden B. G. (1989) Radar observations of Comet IRAS-Aracki-Alcock 1983d. *Astrophys. J., 338,* 1071–1093.

Huebner W. F., Keady J. J., and Lyon S. P. (1992) Solar photo rates for planetary atmospheres and atmospheric pollutants — Photo rate coefficients and excess energies. *Astrophys. Space Sci., 195,* 1–7.

Jewitt D. C. (1996) Debris from Comet P/Swift-Tuttle. *Astron. J., 111,* 1713–1717.

Jewitt D. C and Matthews H. E. (1997) Submillimeter continuum observations of Comet Hyakutake (1996B2). *Astron. J., 113,* 1145–1151.

Kimura H. and Liu C. P. (1977) On the structure of cometary dust tails. *Chinese Astron., 1,* 235–264.

Levasseur-Regourd A. C., Hadamcik E., and Renard J. B. (1996) Evidence for two classes of comets from their polarimetric properties at large phase angles. *Astron. Astrophys., 313,* 327–333.

Levasseur-Regourd A. C., McBride N., Hadamcik E., and Fulle M. (1999) Similarities between in situ measurements of local dust light scattering and dust flux impact data in the coma of 1P/Halley. *Astron. Astrophys., 348,* 636–641.

Lisse C. M. (2002) On the role of dust mass loss in the evolution of comets and dusty disk systems. *Earth Moon Planets, 90,* 497–506.

Lisse C. M., A'Hearn M. F., Hauser M. G., Kelsall T., Lien D. J., Moseley S. H., Reach W. T., and Silverberg R. F. (1998) Infrared observations of comets by COBE. *Astrophys. J., 496,* 971–991.

Mason C. G., Gehrz R. D., Ney E. P., and Williams D. M. (1998) The temporal development of the pre-perihelion spectral energy distribution of Comet Hyakutake (C/1996B2). *Astrophys. J., 507,* 398–403.

McDonnell J. A. M., Lamy P. L., and Pankiewicz G. S. (1991) Physical properties of cometary dust. In *Comets in the Post-Halley Era* (R. L. Newburn Jr. et al., eds.), pp. 1043–1073. Kluwer, Dordrecht.

Pansecchi L., Fulle M., and Sedmak G. (1987) The nature of two anomalous structures observed in the dust tail of Comet Bennett 1970II: A possible neck-line structure. *Astron. Astrophys. 176,* 358–366.

Sekanina Z. (1988) Outgassing asymmetry of periodic Comet Encke. I. Apparitions 1924–1984. *Astron. J., 95,* 911–924.

Sekanina Z. and Schuster H. E. (1978) Meteoroids from periodic Comet d'Arrest. *Astron. Astrophys., 65,* 29–35.

Sekanina Z. and Farrell J. A. (1980) The striated dust tail of Comet West 1976VI as a particle fragmentation phenomenon. *Astron. J., 85,* 1538–1554.

Waniak W. (1992) A Monte-Carlo approach to the analysis of the dust tail of Comet 1P/Halley. *Icarus, 100,* 154–161.

Waniak W. (1994) Nuclear dust emission pattern of Comet Wilson 1987VII. *Icarus, 111,* 237–245.

Physical Properties of Cometary Dust from Light Scattering and Thermal Emission

Ludmilla Kolokolova
University of Florida

Martha S. Hanner
Jet Propulsion Laboratory/California Institute of Technology

Anny-Chantal Levasseur-Regourd
Aéronomie CNRS-IPSL, Université Paris

Bo Å. S. Gustafson
University of Florida

This chapter explores how physical properties of cometary dust (size distribution, composition, and grain structure) can be obtained from characteristics of the electromagnetic radiation that the dust scatters and emits. We summarize results of angular and spectral observations of brightness and polarization in continuum as well as thermal emission studies. We review methods to calculate light scattering starting with solutions to Maxwell's equations as well as approximations and specific techniques used for the interpretation of cometary data. Laboratory experiments on light scattering and their results are also reviewed. We discuss constraints on physical properties of cometary dust based on the results of theoretical and experimental simulations. At the present, optical and thermal infrared observations equally support two models of cometary dust: (1) irregular polydisperse particles with a predominance of submicrometer particles, or (2) porous aggregates of submicrometer particles. In both models the dust should contain silicates and some absorbing material. Comparison with the results obtained by other than light-scattering methods can provide further constraints.

1. INTRODUCTION

The main technique to reveal properties of cometary dust from groundbased observations is to study characteristics of the light it scatters and emits. Such a study can include dynamical properties of the dust particles, presuming that the spatial distribution of the dust resulted from the dust interaction with gravitation and radiation. However, the primary way is to study the scattered light or emitted thermal infrared radiation to search for signatures typical of specific particles. The main focus of this chapter is a review of methods in the interpretation and the extraction of the data on the size distribution, composition, and structure of cometary dust (hereafter referred to as physical properties) from the observed angular and spectral characteristics of the brightness and polarization of scattered and emitted light (hereafter referred to as observational characteristics). We will not detail here observational techniques (for this, see, e.g., the review by *Jockers,* 1999), but will mainly concentrate on the results of the observations and methods of their interpretation.

We review how observational characteristics such as the dust color, albedo, polarization, etc., can be estimated from visual, near-infrared, and mid-infrared observations and which regularities have been found in the variation of these parameters with phase angle, heliocentric distance and within the coma (section 2). Then we describe theoretical light-scattering techniques developed to solve the inverse problem, whereby dust physical characteristics are extracted from the characteristics of the light it scatters and emits (section 3). Since the concentration of the dust particles in cometary coma at cometocentric distances corresponding to the resolvable scale in the observations is sufficiently low, the particles can be considered to be independent scatterers. This means that the intensity of the light (and its other Stokes parameters, see below) scattered by a collection of dust particles is equal to the sums of the intensity (or respective Stokes parameters) of the light scattered by all the particles individually. We therefore consider the methods developed to describe the light scattering by single particles, indicating how light-scattering characteristics depend on the particle shape, size, composition, and structure. In section 3 we also show how the application of these techniques to the observational data leads to the current views on the physical properties of cometary dust. Section 4 compares our knowledge about cometary dust obtained using light-scattering methods with results obtained using other techniques and outlines future research.

2. OBSERVATIONAL DATA

2.1. Brightness Characteristics of Scattered Light

The most straightforward way to characterize cometary dust is to study how much light it scatters and absorbs and how this depends on the geometry of observations, mainly determined by the phase angle, α (the Sun–comet–Earth angle). The efficiency of scattering and absorption is usually characterized by the albedo and geometrical characterization of the scattering by the angular scattering function. These characteristics, as well as spectral dependence of the scattered light, expressed through the dust colors, are considered below.

2.1.1. Albedo.   Single-scattering albedo is defined in the most general way as the ratio of the energy scattered in all directions to the total energy removed from the incident beam by an isolated particle (*van de Hulst*, 1957; *Hanner et al.*, 1981). This definition includes all components (diffracted, refracted, reflected) of the scattered radiation and can be applied to small particles where the diffracted radiation is spread over a wide angular range and cannot readily be separated. In practice, this definition is not very useful when analyzing cometary dust, because it requires knowledge of the radiation scattered at all directions. *Gehrz and Ney* (1992) compared the scattered energy to the absorbed energy reradiated in the infrared to derive an albedo at the phase angle of observation. This method was used by a number of observers, and the results are summarized in Fig. 1. The albedo estimates thus obtained for dust in different comets should be compared at the same phase angle since the radiation scattered in the direction of observation depends on the relative positions of the Sun, Earth, and comet. Using such an approach, *Mason et al.* (2001) found the albedo of the dust in Comet C/1995 O1 Hale-Bopp to be roughly 50% higher than that of P/Halley at α ~ 40°. A few maps of albedo obtained so far by combining visible light and thermal infrared images demonstrate increasing albedo with the distance from the nucleus (*Hammel et al.*, 1987; *Hayward et al.*, 1988).

The geometric albedo of a particle, A_p, is defined as the ratio of the energy scattered at α = 0° to that scattered by a white Lambert disk of the same geometric cross section (*Hanner et al.*, 1981). Since comets are rarely observed at α = 0°, it is convenient to define $A_p(\alpha)$ as the product of the geometric albedo and the normalized scattering function at the angle α. *Hanner and Newburn* (1989) presented a plot of $A_p(\alpha)$ in the J bandpass (1.2 μm) for 10 comets. The total dust cross section within the field of view was determined by fitting a dust emission model to the thermal spectral energy distribution, and then applying the total cross section to the scattered intensity to derive an average albedo. The resulting A_p are typically very low, close to 0.025 at α = 35°–80° and about 0.05 at α near zero. There is some indication that A_p is higher for comets beyond 3 AU (see Fig. 3 in *Hanner and Newburn*, 1989). The albedo was

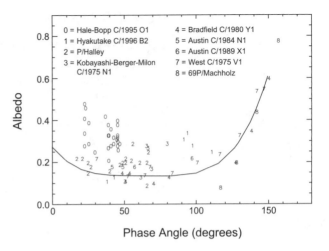

Fig. 1. Dependence of albedo on the phase angle. The data are from (0) *Mason et al.* (2001); (1) *Mason et al.* (1998); (2)–(7) *Gehrz and Ney* (1992); (8) *Grynko and Jockers* (2003). The solid line represents the least-squares fit to the data for Comets C/1975 V1 West and C/1980 Y1 Bradfield and is interpolated to smoothly connect to the backscattering data for Comet P/Stephan-Oterma (*Millis et al.*, 1982) normalized at the angle α = 30°.

observed to increase by 50% in Comet P/Halley's coma during episodes of strong jet activity (*Tokunaga et al.*, 1986).

Difference in albedo may result from different composition, particle size, shape, or internal structure. For example, *Mason et al.* (2001) found that high albedo of the dust in Comet Hale-Bopp is consistent with the domination of small particles in the grain population (see also section 3.4.1). There may be several components of the dust, with differing albedos and temperatures, and the average albedo may not represent the actual albedo of any of the components. However, the low average albedo rules out a large population of cold, bright grains that contribute to the scattered light but not to the thermal emission.

2.1.2. Angular scattering function.   The angular scattering function indicates how the intensity of the scattered light is changing with phase angle. Since a comet can be observed at only one phase angle on a given date, the angular scattering function has to be acquired by observing a comet over time as the Sun–Earth–comet geometry changes. Thus this function of individual cometary dust particles is difficult to determine, because the observed brightness depends not only on the physical properties of the dust, but also on its amount, and the total cross section of dust in the coma contributing to the scattered light intensity does not remain constant over time. Two methods have been used to normalize the observed intensity.

One method is to assume that the ratio of the dust to gas production rates remains constant over time, and to normalize the scattered light intensity to the gas production rate. *Millis et al.* (1982) derived the angular scattering function for the dusty Comet P/Stephan-Oterma by normalizing the scattered intensity to the C_2 production rate. The scattering function was a factor of 2 higher at 3°–4° than at 30°, corresponding to a slope of ~0.02 mag/°. *Meech and Jewitt* (1987) determined a linear slope of 0.02–0.035 mag/° for four comets observed at α = 0°–25°. They saw no evidence of an opposition surge larger than 20% in P/Halley within 1.4°–9°. More detailed data by *Schleicher et al.* (1998) show an evident curvature of the scattering function for P/Halley that can be represented by a quadratic fit.

Another method is to compare the measured scattered light to the measured thermal emission from the same volume in the coma, with the assumption that the emitting properties of the dust remain constant. This method was described above as the means for extracting an albedo for the dust. The method has been applied to a number of comets (see Fig. 1) and yields a relatively flat scattering function from 35° to 80° (*Tokunaga et al.,* 1986; *Hanner and Newburn,* 1989; *Gehrz and Ney,* 1992). Two comets have been observed at large phase angles 120°< α < 150° using this method, and they display strong forward scattering (*Ney and Merrill,* 1976; *Ney,* 1982; *Gehrz and Ney,* 1992). Thus we can characterize a typical angular scattering function for cometary dust as possessing a distinct forward-scattering surge, a rather gentle backscattering peak, and a flat shape at medium phase angles (Fig. 1).

2.1.3. Color. The dust color indicates trends in the wavelength dependence of the light scattered by the dust. Traditionally the color of cometary dust was determined through measurements of the comet magnitude m in two different continuum filters, e.g., blue (B, ~0.4–0.45 μm) and red (R, ~0.64–0.68 μm), and was expressed as $C_{B-R} = m_{blue} - m_{red}$. This color was a unitless characteristic expressed as the logarithm of the ratio of intensities in two filters. Although this definition is still used, spectrophotometry of comets resulted in the definition of color as the spectral gradient of reflectivity, usually measured in % per 0.1 μm with an indication of the range of wavelength it was measured in.

The determination of cometary dust colors requires use of continuum bands that are truly free of gas contamination [see discussion on gas-contamination influence on the colors in *A'Hearn et al.* (1995)]. This tends to make colors obtained using filters less trustworthy than those obtained using spectrophotometry to target gas-free spectral regions. The exception is near-infrared colors that are considered to be free from gas contamination and thus more reliable, although thermal emission from the warm dust will contribute to the K (2.2 μm) bandpass for comets within 1 AU of the Sun. We summarize the data for B–R color in the visible and J–H and H–K color in the near-infrared in Fig. 2.

The scattered light is generally redder than the Sun; the reflectivity gradient decreases with wavelength from 5–18% per 0.1 μm at wavelengths 0.35–0.65 μm to 0–2% per 0.1 μm at 1.6–2.2 μm (*Jewitt and Meech,* 1986). *Hartmann et al.* (1982) and *Hartmann and Cruikshank* (1984) found that the near-infrared dust colors depend on heliocentric distance. However, data by *Jewitt and Meech* (1986) and *Tokunaga et al.* (1986) demonstrate no heliocentric dependence. *Hanner and Newburn* (1989) noted that only the H–

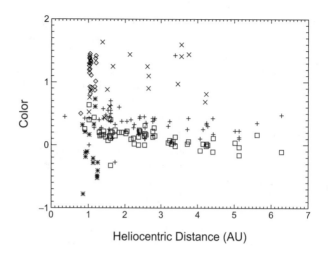

Fig. 2. Color as a function of heliocentric distance: J–H (+), H–K (□), and B–R (×) colors are by *Jewitt and Meech* (1986); B–R colors (◇) are by *Kolokolova et al.* (1997) (✳ for Comet P/Halley). Color is measured in $\Delta m / \Delta \lambda$ (mag/μm) and the solar color is subtracted. The data cover the range of the phase angles 0°–110°.

K color was less red in comets observed at R > 3 AU, while the J–H color showed no trend with heliocentric distance. The B–R color shows no dependence on heliocentric distance although *Schleicher and Osip* (2002) found that the color $m_{0.4845} - m_{0.3650}$ for Comet Hyakutake (1996 B2) got redder with increasing heliocentric distance within 0.6–1.9 AU. No color dependence on phase angle was found in the visible or near-infrared.

Decrease in the near-infrared colors was recorded at periods of enhanced comet activity in Comet P/Halley (*Tokunaga et al.,* 1986; *Morris and Hanner,* 1993) A more red color in the visible was associated with strong jet activity in Comet P/Halley (*Hoban et al.,* 1989), but the spiral structures in Comet Hale-Bopp had less red color (*Jockers et al.,* 1999). Less red and even blue color was associated with outburst in Comet C/1999 S4 LINEAR during its disruption (*Bonev et al.,* 2002). Although no regular dependence on the field of view was found, a smooth change in color was observed in the central part of the coma of some comets, including Hale-Bopp (*Jockers et al.,* 1999; *Laffont et al.,* 1999; *Kolokolova et al.,* 2001a).

2.2. Polarization from Cometary Dust

The light scattered by particles is usually polarized, i.e., its electromagnetic wave has a preferential plane of oscillation. The degree of linear polarization (P, hereafter called polarization) is defined as

$$P = (I_\perp - I_\parallel)/(I_\perp + I_\parallel)$$

where $I_\perp$ and $I_\parallel$ are the intensity components perpendicular and parallel to the scattering plane. For randomly oriented particles the electromagnetic wave predominantly oscillates either perpendicular (by convention positive polarization) or parallel (by convention negative polarization) to the scattering plane.

The convenience of polarization is that it is already a normalized characteristic of the scattered light, whereas the brightness, $I = I_\perp + I_\parallel$, depends upon the distances to the Sun and observer, and the spatial distribution of the dust particles. Polarization variations within a coma relate to changes in the dust physical properties, and different polarization observed for different comets or in features (jets, shells, etc.) indicates a diversity of dust particles. For a given type of cometary dust particles, the polarization mainly depends on the phase angle α and wavelength λ.

2.2.1. Angular dependence. The changing geometry of the comet and observer with respect to the Sun defines a polarization phase curve. Small phase angles can be explored for distant comets, while $\alpha > 90°$ can only be reached at R < 1 AU. Most of the available polarization observations have been performed within the heliocentric distances 0.5–2 AU using groundbased instruments, although large phase angles were studied using the Solar and Heliospheric Observatory (SOHO) C3 coronagraph (*Jockers et al., 2002*).

The data, which represent an average value of the polarization on the projected coma, are obtained either by aperture polarimetry (e.g., *Dollfus et al., 1988*; *Kiselev and Velichko, 1999*) or deduced from the integrated flux of the polarized brightness images (e.g., *Renard et al., 1996*; *Kiselev et al., 2000*; *Hadamcik and Levasseur-Regourd, 2003*). The (real or virtual) diaphragm is centered on the center of brightness, which corresponds to the nucleus. Its aperture, in terms of projected distance on the cometary coma, needs to be large enough to include the coma features that may alter the global polarization. It is usually found that, once the aperture is sufficiently large, the resulting polarization of the comet does not change with increasing aperture.

The phase dependence of polarization, which has been documented so far in the range 0.3°–122°, is smooth and similar to that of atmosphereless solar system bodies. All comets show a shallow branch of negative polarization at the backscattering region, first observed by *Kieselev and Chernova* (1978), that inverts to positive polarization at $\alpha_0 \sim$ 21° with a slope at inversion h ~ 0.2–0.4%/°, and a positive branch with a broad maximum near 90°–100°. The data may be fitted by a typically fifth-order polynomial or trigonometric function, e.g., as suggested by *Lumme and Muinonen* (1993), $P(\alpha) \sim (\sin\alpha)^a(\cos\alpha/2)^b\sin(\alpha - \alpha_0)$. These functions cannot be used for extrapolation but only within the phase angle range where well-distributed data points are available.

While the minimum polarization is typically –2% (*Mukai et al., 1991*; *Chernova et al., 1993*; *Levasseur-Regourd et al., 1996*), a significant dispersion is noticed for $\alpha > 30°–40°$. Once the data are separated in different wavelength ranges, the dispersion on the positive branch is reduced, and $P_\lambda(\alpha)$ for a variety of comets can be compared (see *Levasseur-Regourd et al., 1996*, and references therein). Comets tend to

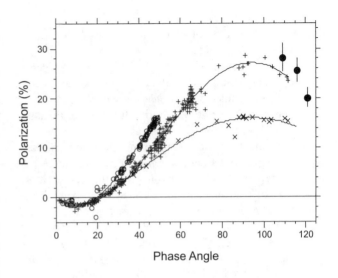

Fig. 3. Polarization in the narrow-band red filter vs. phase angle: (×) comets with low maximum in polarization, (+) comets with higher maximum, (○) Comet C/1995 O1 Hale-Bopp, (●) Comet C/1999 S4 LINEAR at disruption.

divide into three classes corresponding to different maxima in polarization (Fig. 3): (1) comets with a low maximum, about 10–15% depending on the wavelength; (2) comets with a higher maximum, about 25–30% depending on the wavelength; and (3) Comet C/1995 O1 Hale-Bopp, whose polarization is distinctively higher, although it was not observed for $\alpha > 48°$.

Numerous observations obtained by *Dollfus et al.* (1988) for P/Halley showed that the polarization at a given phase angle and wavelength does not vary with heliocentric distance. Some transient increases in polarization were found correlated with cometary outbursts. Similarly, the polarization of Comet Hale-Bopp obtained at small phase angles (large heliocentric distances) was consistent with that obtained closer to the Sun, but increased after outbursts events (*Ganesh et al., 1998*; *Manset and Bastien, 2000*). Observations of Comet C/1999 S4 LINEAR during its near-perihelion disruption (*Kiselev et al., 2002*; *Hadamcik and Levasseur-Regourd, 2003*) indicated an increase in polarization of about 4%, a value comparable with what had been noticed during the P/Halley outbursts.

2.2.2. Spectral dependence. The data have been retrieved in the ultraviolet, visible, and near-infrared domains to avoid any contribution from the thermal emission. Narrow-band cometary filters are now available at 0.3449, 0.4453, 0.5259, and 0.7133 μm (*Farnham et al., 2000*). However, observers often need to make a trade-off between narrow filters that remove gaseous emissions and wider filters that improve the signal-to-noise ratio. It is highly advisable to analyze spectra of the observed comet and estimate the contribution of some faint molecular lines to allow for the depolarizing effect of molecular emissions when calculating the dust polarization (*Le Borgne and Crovisier,*

1987; *Dollfus and Suchail,* 1987; *Kiselev et al.,* 2001). Spectropolarimetric cometary data remain rare, and so far their resolution is not better than the resolution provided by narrowband cometary filters (*Myers and Nordsieck,* 1984).

It was noticed for P/Halley (*Dollfus et al.,* 1988) and later confirmed for other comets (*Chernova et al.,* 1993; also *Kolokolova and Jockers,* 1997, and references therein) that the polarization usually increases with increasing wavelength, at least in the visible domain, and the increase grows for greater values of the polarization.

If polarization data are available for two wavelengths, the spectral gradient of polarization is defined as polarimetric color, $\Delta P/\Delta\lambda = [P(\lambda_2) - P(\lambda_1)]/[\lambda_2 - \lambda_1]$.

The polarimetric color changes with the phase angle (*Chernova et al.,* 1993; *Kolokolova and Jockers,* 1997) from zero and even negative values to gradually increasing positive values at $\alpha > 50°$. Figure 4 summarizes systematic multiwavelength observations of polarization, performed for Comets Halley and Hale-Bopp. Within $\alpha = 25°–60°$ the polarization increases with the wavelength in the visible domain; the gradient seems to decrease in the near-infrared [Fig. 4; see also *Hadamcik and Levasseur-Regourd* (2003)]. For $\alpha \sim 50°$, the polarimetric color is about 9%/μm for Halley and 14%/μm for Hale-Bopp (*Levasseur-Regourd and Hadamcik,* 2003), which agrees with the value 11%/μm at 45° obtained by *Kiselev and Velichko* (1999). The polarimetric gradient for Comet Hale-Bopp is higher than for other comets but its sign and phase-angle trend are typical for comets, whereas asteroids and other atmosphereless bodies usually have negative polarimetric color (e.g., *Mukai et al.,* 1997). So far a negative polarimetric color at

large phase angles was observed only for Comet P/Giacobini-Zinner (*Kiselev et al.,* 2000). Polarimetric color of variable sign was observed for Comet C/1999 S4 LINEAR during its disruption (*Kiselev et al.,* 2002).

2.2.3. Variations within the coma. A coma whose optical thickness exceeds 0.1 was found so far only for Comet Hale-Bopp and only at distances closer than 1000 km from the nucleus (*Fernández,* 2002). Thus, except for possibly the very innermost coma where multiple scattering might take place, a change in the polarization with the distance from the nucleus indicates an evolution in the physical properties of the particles ejected from the nucleus. Polarization images of P/Halley (*Eaton et al.,* 1988; *Sen et al.,* 1990), C/1990 K1 Levy (*Renard et al.,* 1992), 109P/Swift-Tuttle (*Eaton et al.,* 1995), 47P/Ashbrook-Jackson (*Renard et al.,* 1996), and C/1995 O1 Hale-Bopp (*Hadamcik et al.,* 1999; *Jockers et al.,* 1999; *Kolokolova et al.,* 2001a); variable-aperture polarimetry of P/Halley (*Dollfus et al.,* 1988); and high-resolution *in situ* observations with the Optical Probe Experiment (OPE) onboard the Giotto spacecraft (*Renard et al.,* 1996; *Levasseur-Regourd et al.,* 1999) revealed a region in the innermost coma characterized by a lower polarization. Comet Hale-Bopp could be observed at phase angles ranging from about 7° to 48°. The low polarization region in Comet Hale-Bopp has been monitored within this phase-angle range (*Hadamcik and Levasseur-Regourd,* 2003). It showed highly negative values of polarization at small phase angles, e.g., –5% at $\alpha = 8°$, whereas a typical value should be –0.5%. The low negative values of polarization, noticed at the cometocentric distances less than 2000 km, could not be explained by multiple scattering but is most likely due to the physical properties of the local dust grains.

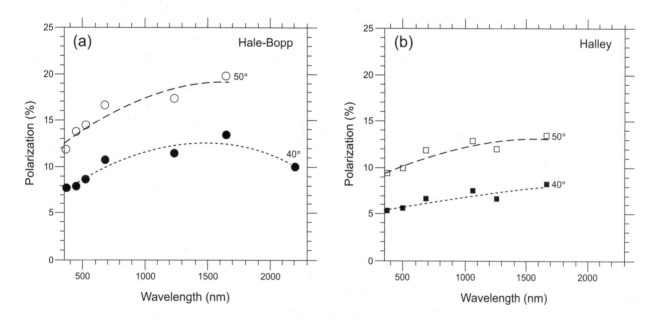

Fig. 4. Wavelength dependence of polarization for **(a)** Comet Hale-Bopp [data from *Furusho et al.* (1999), *Ganesh et al.* (1998), *Hadamcik et al.* (1999), *Jockers et al.* (1999), *Jones and Gehrz* (2000), *Kiselev and Velichko* (1999), *Manset and Bastien* (2000)] and **(b)** Comet Halley [data from *Bastien et al.* (1986), *Brooke et al.* (1987), *Dollfus and Suchail* (1987), *Le Borgne et al.* (1987), *Mukai et al.* (1987), *Sen et al.* (1990)] for phase angles 40° (bottom curve) and 50°.

A further proof of different physical properties of the dust near the nucleus is its negative polarimetric color observed *in situ* (OPE/Giotto) (*Levasseur-Regourd et al.*, 1999) and a smaller polarimetric color in the innermost coma suspected for Comet Hale-Bopp (*Jockers et al.*, 1999). Some comets do not show lower polarization [e.g., C/1989 X1 (*Eaton et al.*, 1992), C/2001 A2 (*Rosenbush et al.*, 2002)] in the innermost coma, possibly hidden by highly polarized dust jets or, as *Kiselev et al.* (2001) suggest, resulted from strong gas contamination.

Aperture polarimetry with an offset of the center, as well as cuts through polarization images, may be used to study the variation of the polarization with the cometocentric distance. The increase within the jets and decrease in the innermost coma are clearly visible on such graphs, and some differences may be noticed between the sunward and antisunward side. Although the data are line-of-sight integrated, they emphasize the temporal evolution in the physical properties of the dust particles ejected from the nucleus. Except for the above-mentioned features, the polarization decreased gradually within 5000–8000 km from the nucleus for C/1996 B2 Hyakutake and C/1996 Q1 Tabur (*Kolokolova et al.*, 2001a) and increased for P/Halley (*Levasseur-Regourd et al.*, 1999) and Hale-Bopp (*Kolokolova et al.*, 2001a), accompanied by similar trends in the B–R color.

The polarization images often reveal elongated fan-shaped structures with polarization higher than in the surrounding coma. They generally correspond to bright jet-like features on the brightness images. This effect has been noticed for quite a few comets, e.g., 1P/Halley (*Eaton et al.*, 1988), C/1990 K1 Levy (*Renard et al.*, 1992), 109P/Swift-Tuttle (*Eaton et al.*, 1995), 47P/Ashbrook-Jackson (*Renard et al.*, 1996), or C/1996 B2 Hyakutake (*Tozzi et al.*, 1997). The case of the bright and active Comet Hale-Bopp is most documented; straight jets at large heliocentric distances are seen to evolve in arcs or shells around perihelion passage (see, e.g., *Hadamcik et al.*, 1999; *Jockers et al.*, 1999; *Furusho et al.*, 1999). Such conspicuous features could be produced by an alignment of elongated particles or by freshly ejected dust particles different from the particles in the regular coma.

2.2.4. Plane of polarization and circular polarization. The plane of polarization is in almost all cases perpendicular or parallel to the scattering plane, indicating randomly oriented particles. However, some variations by a few degrees may be noticed with changing aperture size (*Manset and Bastien*, 2000). Deviations in the polarization plane were noticed for Comet P/Halley near the inversion angle (*Beskrovnaya et al.*, 1987); they were mapped by *Dollfus and Suchail* (1987). Changes in the polarization plane often accompany the polarization variations at the outburst activity and were registered in Comets 29P/Schwassmann-Wachmann 1 (*Kiselev and Chernova*, 1979), P/Halley (*Dollfus et al.*, 1988), C/1990 K1 Levy (*Rosenbush et al.*, 1994), and C/2001 A2 LINEAR (*Rosenbush et al.*, 2002). For Comets Halley, Hale-Bopp, and S4 LINEAR, the circular polarization was estimated and found to be nonzero but below 2%

(*Dollfus and Suchail*, 1987; *Metz and Haefner*, 1987; *Rosenbush et al.*, 1999; *Manset and Bastien*, 2000; *Rosenbush and Shakhovskoj*, 2002).

The existence of the faint circular polarization, variations in the plane of polarization, and nonzero polarization observed at forward-scattering direction during stellar occultations (*Rosenbush et al.*, 1994) may be indications of aligned elongated or optically anisotropic particles in cometary atmospheres.

2.3. Dust Thermal Emission

Thermal data provide two types of results that are sensitive to the physical properties of the particles: the thermal spectral energy distribution (SED) and the detection of spectral features (e.g., the silicate feature at 10 μm and 16–30 μm). In addition, the ratio of the thermal energy to the scattered energy provides a measure of albedo, as described previously.

Thermal emission from the dust in the coma results from absorption of the solar radiation followed by its reemission. The thermal radiation from a single grain depends upon its temperature, T, and wavelength-dependent emissivity, ε

$$F(\lambda) = (r/\Delta)^2 \varepsilon(r,\lambda)\pi B(\lambda,T) \qquad (1)$$

where r is some specific dimension of the particle, e.g., its radius, Δ the geocentric distance, and $B(\lambda,T)$ the Planck function for grain of temperature T. The observed SED from the ensemble of grains in a coma having differing temperatures and wavelength-dependent emissivities generally is similar to (but broader than) a blackbody curve (Fig. 5). The SED can be characterized by a single temperature (the color temperature, T_c) over a defined wavelength interval, but it is important to remember that the color temperature is not the physical temperature of the grains.

The physical temperature of a particle in the solar radiation field depends strongly on the latent heat of sublimation and sublimation rate of any material that may be sublimating, but once the sublimation has ceased and thermal equilibrium is established, the temperature depends on the balance between the solar energy absorbed at visual wavelengths and the energy radiated in the infrared. For a particle of arbitrary shape we may write the energy balance equation

$$R^{-2}\int C_{abs}(r,m,\lambda)S(\lambda)d\lambda = \zeta\int \pi B(\lambda,T)C_{abs}(r,m,\lambda)d\lambda \quad (2)$$

where $S(\lambda)$ is the solar flux at 1 AU, R is the heliocentric distance in AU, and the factor ζ is the ratio of the total radiating area to the exposed area. This ratio is equal to 4 for spheres and any convex particle averaged over random orientations (*van de Hulst*, 1957, section 8.41). The crucial parameter in equation (2) is $C_{abs}(r,m,\lambda)$, the absorption cross section, which depends on grain size, morphology, and optical constants. By Kirchoff's law, $C_{abs}(r,m,\lambda)$ is equal to the emissivity, $\varepsilon(r,\lambda)$, times the emitting area. We assume that the coma is optically thin (no attenuation of sunlight and no

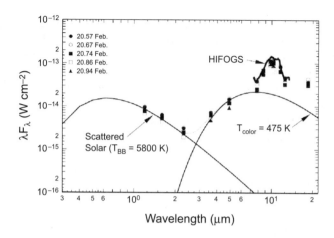

Fig. 5. Spectral energy distribution for Comet Hale-Bopp (*Williams et al., 1997*). The left curve is the solar spectrum, the right curve is the continuum fit to the sum of a 5800 K blackbody component due to scattered solar radiation and a 475 K blackbody due to thermal emission. The heavy solid line shows the silicate feature observed with the HIFOGS (High-efficiency Infrared Faint Object Grating Spectrometer). Although Comet Hale-Bopp was not a typical comet, the figure provides an illustration of the main characteristics of cometary dust thermal emission.

heating by other grains) and that sublimation is negligible. The observed color temperature can be compared with the SED predicted from equations (1)–(2) for grains of various sizes and optical properties in order to infer the physical characteristics of the cometary dust.

The 3–20-μm thermal SED has been observed for many comets over the past three decades, with improving precision. P/Halley was monitored regularly at R < 2.8 AU (*Tokunaga et al., 1986, 1988; Gehrz and Ney, 1992*) and Hale-Bopp was observed from the ground and from Earth orbit at 0.9–4.9 AU (*Mason et al., 2001; Grün et al., 2001; Hanner et al., 1999; Wooden et al., 1999; Hayward et al., 2000*). The 5–18-μm color temperature is typically 5–25% higher than a perfect blackbody at the same heliocentric distance that can be calculated as T = 280 * R$^{-0.5}$.

Comets display a heliocentric dependence, T ~ R$^{-0.5}$; in the case of P/Halley the 8–20-μm color temperature varied as T = 315.5 * R$^{-0.5}$ (*Tokunaga et al., 1988*). The comets with the strongest dust emission, such as P/Halley and Hale-Bopp, display a higher color temperature at 3.5–8 μm than at 8–20 μm (*Tokunaga et al., 1988; Williams et al., 1997*).

Thermal emission observations in the far-infrared and submillimeter spectral regions can potentially provide information about the larger particles in the coma (see section 3.1). Far-infrared observations at λ < 200 μm have been acquired by the InfraRed Astronomical Satellite (IRAS), Cosmic Background Explorer (COBE), and Infrared Space Observatory (ISO) satellites. The Diffuse InfraRed Background Experiment (DIRBE) instrument on COBE measured the SED from 3.5 to 100 μm for three comets (*Lisse et al., 1994, 1998*). Among these, only Comet Levy (C/1990 K1) had a color temperature 10% higher than the black-

body temperature and a drop in the average grain emissivity at 60 and 100 μm, indicating that the flux contribution at 60 and 100 μm was primarily from particle sizes smaller than the wavelength (see section 3.1). Continuum emission at submillimeter wavelengths has been measured for several comets including Hale-Bopp (*Jewitt and Luu, 1990, 1992; Jewitt and Matthews, 1997, 1999*), indicating the presence of millimeter-sized particles.

In addition to the smoothly varying infrared continuum emission, broad spectral features attributed to small silicate grains are seen in some comets near 10 μm (Fig. 5) and 18 μm. The strength of the features depends upon grain size and temperature (*Hanner et al., 1987*). The 16–45-μm spectra of Hale-Bopp recorded by the Short Wavelength Spectrometer (SWS) onboard the ISO satellite (*Crovisier et al., 1997*) displayed five distinct spectral peaks that correspond to laboratory features of crystalline Mg-olivine (forsterite) (see *Hanner and Bradley, 2004*). Not all comets display a strong 10-μm silicate feature (*Hanner et al., 1994; Hanner and Bradley, 2004*). In particular, the feature tends to be weak (~20% above the continuum) or absent entirely in Jupiter-family comets (e.g., *Hanner et al., 1996; Li and Greenberg, 1998a*).

2.4. Correlations Among Scattering and Emitting Observational Characteristics

It has been well established that the observable scattering parameters of cometary dust tend to correlate. These correlations can provide insight into the relative importance of particle size, composition, and structure on the observable scattering properties.

We have already noted that comets divide into some classes, based upon their maximum polarization (Fig. 3). The high P$_{max}$ comets are mostly comets with a strong dust continuum, while the low P$_{max}$ comets are comets with a weak dust continuum (*Chernova et al., 1993*). Comets exhibiting higher polarization generally display a stronger silicate feature (*Levasseur-Regourd et al., 1996*), indicating that small silicate particles have a major influence on P$_{max}$ (see section 3.1). The strength of the silicate feature also correlates with higher color temperature and stronger flux at 3–5 μm (*Gehrz and Ney, 1992*), both indicators of hot, submicrometer-sized absorbing grains (see section 3.1.1).

The fact that the B–R, J–H, and H–K colors do not correlate, e.g., show different tendencies with the heliocentric distance or with the distance from the nucleus, supports the idea that the changes in the colors cannot be explained by simple change of particle size or smooth change in the optical constants. A smooth simultaneous change in B–R color and polarization was observed in the innermost coma of some comets (*Kolokolova et al., 2001a*), indicating a smooth change in some dust properties that occurs as dust is moving out of the nucleus.

Hanner (2002, 2003) summarized the correlations and compared the scattering properties in Comets P/Halley and Hale-Bopp: The dust in Hale-Bopp displayed higher polarization at comparable α, redder polarimetric color, higher albedo, stronger silicate feature, higher infrared color temperature compared to the blackbody temperature, and higher 3–5-μm thermal emission.

2.5. Summary

The observational characteristics of cometary dust are extremely variable and depend on phase angle, heliocentric distance, and position within the coma, especially associated with comet activity (jets, fans, etc.). The most complex characteristic is the brightness that depends on the spatial distribution of dust particles, and thus to a large extent is affected by temporary variations in dust production rate. Parameters, formed as ratios of brightness such as color, polarization, and albedo, are more suitable for studying physical properties of the dust grains. However, measured using aperture instruments, they demonstrate average characteristics related to a mixture of particles of a variety of properties and can vary due to comet activity. Coma images obtained in narrowband continuum filters provide the most adequate source of information about properties of cometary dust.

The general regularities found in observational characteristics of cometary dust that may be used as a basis for studying physical properties of the dust particles are listed later in this chapter (see Table 2). Variability of these characteristics within the coma, correlations between them, and exceptions to the listed rules are also an interesting subject for light-scattering modeling.

3. LIGHT-SCATTERING THEORETICAL AND EXPERIMENTAL METHODS

Study of cometary dust using its interaction with solar radiation is a typical example of remote sensing that usually is associated with the scattering inverse problem. The sought quantities define a set of particle parameters such as size, shape, and composition (refractive index) from the scattering/emission data; the set can include the parameters of more than one type of particle. The rigorous way to solve the inverse problem would be:

- to measure the characteristics of the scattered light (intensity, I; degree of linear polarization, P; position of the plane of polarization, defined by the angle θ; degree of circular polarization, V) at a given wavelength, λ, and given phase angle, α;
- to calculate the Stokes vector $I = (I, Q, U, V)$ of the scattered light whose components are related to the observed characteristics as $I = I$, $P = \sqrt{(Q^2 + U^2)/I}$, $\tan(2\theta) = Q/U$, $V = V/I$;
- from the Stokes vectors of the scattered and incident light, I and I_0, determine elements of the scattering matrix F through $(I, Q, U, V) = F/(kR)^2 * (I_0, Q_0, U_0, V_0)$, where R is the distance between the detector and the scatterer and k is the wave number in empty space equal to $k = 2\pi/\lambda$.

The scattering matrix F is defined by the particle properties, thus the inverse problem can be solved if the elements of the scattering matrix, obtained from observations, can be also found from a theory that describes scattering by particles, i.e., most rigorously, from Maxwell's equations. To determine all 16 elements of the matrix F one must obtain four independent combinations of vectors I_0 and I, but since sunlight is incoherent and essentially unpolarized, this is usually not possible from astronomical observations. In practice, one therefore fits only the two upper-left elements, or equivalently the intensity and polarization. While the intensity or brightness is plagued by sensitivity to variations in the particle number density, the polarization is a ratio of intensities and is therefore independent of the amount of dust. These observational constraints make it even more important to fit all available observations and to extend the observations to as broad a range of phase angles as possible and wavelengths as well as thermal emission spectra. This still cannot guarantee the uniqueness of the solution, especially as the dominating grains in a mixture of grain types can vary strongly both with wavelength and with scattering angle. Therefore some authors emphasize the importance in finding a solution that also adheres to limitations set by, for example, cosmic abundances (*Greenberg and Hage*, 1990) and constraints from the dynamics of the particles (*Gustafson*, 1994). However, in this chapter we focus on constraints from optical and infrared observations.

For electromagnetic waves propagating in a homogeneous medium, characterized by its complex refractive index $m = n + i\kappa$, Maxwell's equations can be reduced to

$$\text{curl } H = ikm^2E$$
$$\text{curl } E = -ikH \tag{3}$$

Since the magnetic vector H can be expressed through the electrical vector E, equation (3) can be further reduced to one so-called Helmholtz equation

$$\nabla^2 E + k^2 E = 0 \tag{4}$$

The electromagnetic vector E is related to the Stokes parameters through the values of its two complex perpendicular amplitudes $E_\parallel$ and $E_\perp$ and their conjugates as (*van de Hulst*, 1957; *Bohren and Huffman*, 1983)

$$I = E_\parallel |E_\parallel^* + E_\perp E_\perp^*$$
$$Q = E_\parallel E_\parallel^* - E_\perp E_\perp^*$$
$$U = E_\parallel E_\perp^* + E_\perp E_\parallel^*$$
$$V = i (E_\parallel E_\perp^* - E_\perp E_\parallel^*)$$

In principal, a complete solution to the problem can be achieved if we consider equation (4) outside the particle, where the electromagnetic field is a superposition of incident and scattered fields, and inside the particle (internal field), and apply the boundary conditions. The boundary conditions (*van de Hulst*, 1957, section 9.13) mean that any tangential and normal components of E are continuous across the two materials at the particle boundary. Thus to solve the problem we need to know the shape of the boundary, i.e., the particle shape. The particle shape means a mathematically sharp boundary between a particle interior and empty space. We therefore tend to distinguish between particle shape and structure (surface and internal). Available solutions are specific to classes of particle shapes, e.g., spheres, cylinders, and spheroids. These are usually for homogeneous and isotropic internal structures, although solutions for some anisotropic or inhomogeneous structures have been derived for some specific shapes.

3.1. Light Scattering by Homogeneous Spherical Particles (Mie Theory)

Mie theory [attributed to *Mie* (1908); credit is also usually given to earlier works by Clebsch, Lorentz, Debye, and others (see *Kerker,* 1969)] accurately predicts the scattering by any homogeneous and isotropic sphere. It solves equation (4) in spherical coordinates (r, θ, φ) using the separation of variables technique, i.e., presenting the electromagnetic wave $\mathbf{E}$ as

$$\mathbf{E} = E(r,\theta,\varphi)\mathbf{e} = \Phi(\varphi)\Theta(\theta)R(r)\mathbf{e}$$

where vector $\mathbf{e}$ indicates the direction of oscillations. This gives $\mathbf{E}$ in the form of a sum of products of trigonometric and spherical Bessel functions and associated Legendre polynomials with the size parameter $x = kr$ (where r is the particle radius) and refractive index of the particle in the argument. In practice, one has to take care with the numerical code in order to not degrade the accuracy for very small or very large particles. Detailed discussion on Mie solution can be found in *van de Hulst* (1957) and *Bohren and Huffman* (1983). The latter book also contains a simple but efficient Mie code. A modern code is available from M. Mishchenko at ftp://ftp.giss.nasa.gov/pub/crmim/spher.f.

Even when the shape and structure of the particle is as simple as that of a homogeneous sphere, the complexity of the solution does not allow the scattering properties of a particle to be expressed through a simple analytical function of the size and refractive index. The numerically derived solutions are also generally found to not be unique to one combination of particle size and refractive index. The problem of uniqueness is more acute if the observations cover only a limited range of phase angles. Thus the inverse problem is often solved indirectly using analysis of the qualitative light-scattering behavior and incorporating other empirical constraints on particle properties.

Use of Mie theory has helped exclude spherical and homogeneous particles as the main component of cometary dust. Even if some specific size distribution and refractive index could fit some observations (e.g., polarization) at some wavelengths (see examples in *Mukai et al.,* 1987), the same characteristics of spherical particles could not provide a reasonable fit over a broad range of wavelengths and to other observational parameters, e.g., scattering function,

color, or temperature. However, the calculations for spheres can be used in a qualitative manner to narrow the range of cometary grain parameters, as illustrated by the use of three-dimensional or color-contour graphics of the type shown in Fig. 6 (see also numerous examples in *Hansen and Travis,* 1974; *Mishchenko and Travis,* 1994; *Mishchenko et al.,* 2002). From Fig. 6 one can see that light scattering from cometary dust is not dominated by particles much smaller than the optical wavelengths, since for such particles polarization is always positive with very high maximum polarization at $\alpha = 90°$. Also, cometary dust cannot be represented by an ensemble of monodisperse spherical or quasispherical particles of medium size $(1 < x < 20)$ since their light-scattering characteristics experience the oscillating behavior seen even in Fig. 6 (although the oscillations are smoothed by the size distribution). Note also that for spherical particles, strong color dependence on the phase angle is seen for all sizes and the polarimetric color is mainly negative.

The contour graphics and a more quantitative approach based on statistical multifactor analysis (*Kolokolova et al.,* 1997) can be used to estimate influence of dust properties on observational characteristics. As a result of such an analysis, the reason for the trends observed within the coma and correlations between light-scattering parameters can be found. For example, such an approach (*Kolokolova et al.,* 2001a) showed that the correlation between color and polarization observed in the innermost coma of Comets Tabur, Hyakutake, and Hale-Bopp cannot be a result of changing particle size, but is more likely an indication of changing composition.

For certain applications, we might expect that arbitrarily shaped, randomly oriented particles can be approximated by equal-volume, equal-projected-area, or, in the case of convex particles, equal-surface-area spheres. They appear appropriate when some specific term dominates the scattering. For example, we might expect scattering in the forward domain (large phase angles) to be dominated by diffraction, which is strongly dependent on the particle size since only the particle cross section enters calculations of the diffraction pattern (*van de Hulst,* 1957, section 3.3). Depending on the absorption of the particle material at some particle sizes, the transmitted light starts reducing the magnitude of the forward-scattering peak but does not strongly affect the angular distribution of the intensity. In particular, the angle of the first minimum in intensity counted from the direction of backscattering seems to be a good size indicator even for aggregated particles (*Zerull et al.,* 1993).

Scattering by particles of size $x \ll 1$ is proportional to the square of their polarizability (see section 3.2) and therefore to their volume squared (*van de Hulst,* 1957, section 6.11) so that equal volume particles in random orientation scatter identically. For the other extreme of large convex particles in the geometric optics regime *van de Hulst* (1957, section 8.42) showed that the scattering caused by reflection from randomly oriented particles is identical with the scattering by a sphere of the same material and surface area. He also showed that such particles have the same geometri-

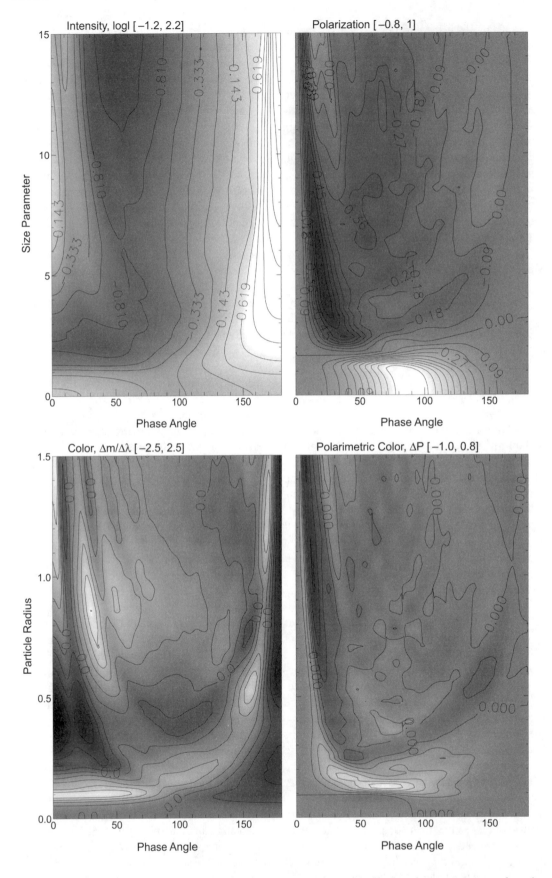

Fig. 6. Top panel: Intensity and polarization of scattered light vs. phase angle and effective size parameter x_{eff} for spheres of the refractive index m = 1.55 + i0.005. The particle size distribution has the form n(x) = x^{-3} and is determined by its cross-sectional-area-weighted mean size parameter x_{eff}, and the width of the size distribution defined by the effective variance v_{eff} = 0.05 (for details see *Mishchenko and Travis,* 1994). Bottom panel: Same for color and polarimetric color but instead of size parameter the effective size of particle varies at the fixed wavelengths of 0.45 and 0.65 μm. Darker colors indicate smaller values.

cal cross section as the sphere. The approximations work better the closer the particle shape is to spherical. For example, an error in the absorption cross section is less than 5% for a spheroid of the axes ratio a : b = 1.4 regardless of the particle size, but can exceed 15% for a spheroid with the axes ratio a : b = 2 at x = 1 (*Mishchenko et al.,* 1996). Important limitations can be high porosity of the particles (*Henning and Stognienko,* 1993) or the presence of sharp edges (*Yanamandra-Fisher and Hanner,* 1998).

3.1.1. Use of Mie theory to estimate grain emissivity at thermal infrared wavelengths. As we will see below, the thermal emission by cometary dust comes mainly from submicrometer particles. Therefore the equal-volume-sphere representation provides a good estimate for the absorption cross section as it works for particles that are small in all dimensions, i.e., represent quasi-equidimensional (not needles or disks) compact particles of x ≪ 1. It is also a good approximation for submicrometer particles in a porous aggregate so that the interaction between particles is weak. Using the thus estimated absorption cross section C_{abs} one can use equation (2) to find the temperature for grains of a variety of sizes and composition (*Mukai,* 1977; *Hanner,* 1983). This revealed a set of general regularities in thermal emission from small grains shown in Fig. 7.

Particles of absorptive materials, e.g., glassy carbon or magnetite, showed temperatures higher than for a blackbody; the smaller the particles, the higher the temperature. Particles made of transparent materials were found to be cooler than a blackbody and their temperature did not depend much on particle size. Small grains of an absorbing (e.g., carbonaceous) material absorb strongly at visual wavelengths, but cannot radiate efficiently in the infrared at wavelengths greater than about 10 times their size. They heat up until the energy radiated at the shorter infrared wavelengths (roughly 3–8 μm for comets at 0.5–3 AU from the Sun) balances the absorbed energy. In fact, for small absorbing grains, their size controls their temperature, regardless of their specific composition (*Hanner,* 1983).

In contrast to carbon grains, silicate grains radiate efficiently in the infrared because of the resonance in their optical constants near 10 and 20 μm; the amount of absorption at visual wavelengths controls their temperature. The absorption at visual wavelengths depends strongly on the Fe content of the silicates (*Dorschner et al.,* 1995). Pure Mg-rich silicates have very low absorption; the imaginary part of the refractive index κ ~ 0.0003 at 0.5 μm for a glass with Fe/Mg κ ~ 0.05 at 0.5 μm for a pyroxene glass with Fe/Mg = 1 and k ~ 0.1 for an olivine glass with Fe/Mg = 1.0. Consequently, pure Mg silicates in a comet coma near 1 AU would be much colder than a blackbody while Fe-rich silicate grains would be warmer than a blackbody. Compositional measurements indicate that the silicates in comets are Mg-rich (*Hanner and Bradley,* 2004), so one might expect them to be cold. However, even a small admixture of absorbing material can increase the temperature significantly.

The computed temperature for a submicrometer absorbing grain varies approximately as $R^{-0.35}$ instead of observed $R^{-0.5}$ (Fig. 7) that would be also expected for a blackbody in equilibrium. Whether this indicates a change in size distribution with R, dominance by larger particles or the shortcoming of the computations for small spheres is not clear. Chances are that this is a result of combination of optical characteristics of both small and large particles exhibited by aggregates of small grains (see section 3.3.3).

The observed 3–20-μm SED for comets cannot generally be fit by the F(λ) computed for a single grain of any size or composition. However, models using a size distribution of absorbing particles dominated by micrometer- to submicrometer-sized grains provide a match to the observed SED (*Li and Greenberg,* 1998b; *Hanner et al.,* 1999). Comets with the strongest dust emission (such as P/Halley and Hale-Bopp) that also display a higher color temperature at 3–8 μm than at 8–20 μm indicate an enhanced abundance of hot, submicrometer-sized grains.

By fitting the observed SED with a model of thermal emission from a size distribution of grains, the total dust mass within the coma can be estimated. When combined with a size-dependent velocity distribution for the dust, the rate of dust production from the nucleus can also be estimated (*Hanner and Hayward,* 2003).

Extensive observations of Comet Hale-Bopp from the European Space Agency's ISO (the photometer PHOT measured the thermal flux through filters at 7–160 μm, while the two spectrometers recorded the spectrum from 5 to 160 μm) enabled the SED to be obtained at long wavelengths and fit by an emission model having a size distribution of the form n(r) ~ r^{-a}, with a ≤ 3.5 (*Grün et al.,* 2001). For an outflow velocity v(r) ~ $r^{-0.5}$, this result implies that the dust production size distribution from the nucleus had a slope a ≤ 4 and the mass was concentrated in large

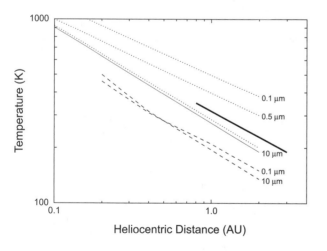

Fig. 7. Temperature of a grain vs. heliocentric distance. Dotted lines show results for absorptive particles (glassy carbon) of radius 0.1, 0.5, and 10 μm. Dashed lines are for olivine particles of radius 0.1 and 10 μm. The thin solid line is the blackbody temperature. The thick solid line shows the results for Comet Halley (*Tokunaga et al.,* 1988). The comet dust temperature is higher than a blackbody, indicating the presence of submicrometer absorptive particles.

particles. Yet Hale-Bopp displayed the highest 3–13-μm color temperature and strongest silicate feature ever seen in a comet, clearly indicating that the thermal emission at shorter wavelengths arose from a high abundance of submicrometer grains compared to other comets.

While Mie theory can be used to compute the absorption (or emission) cross section C_{abs} for a compact particle that is small in comparison to the wavelength in all its dimensions, in cases where C_{abs} varies slowly with wavelength, it cannot be applied to calculations for an emission feature, where the optical constants are rapidly varying. Analysis of the 10-μm silicate emission feature in comets indicates the presence of both glassy and crystalline silicate components (*Hanner et al.*, 1999; *Wooden et al.*, 1999; *Hanner and Bradley*, 2004). The optical constants of the glassy silicates vary slowly across the 8–13-μm range and Mie theory will not be too inaccurate for computing C_{abs}. However, the crystalline components, especially crystalline olivine, which has a strong resonance at 11.2 μm in naturally occurring samples, cannot be treated using Mie theory since a resonance is very sensitive to particle shape (*Bohren and Huffman*, 1983). Mie theory does not even predict the correct wavelength of the peak, let alone the correct peak height (*Yanamandra-Fisher and Hanner*, 1998). Thus, most researchers make use of measured reflectivity or mass absorption coefficients to obtain C_{abs} within the spectral feature (*Wooden et al.*, 1999, 2000; *Hayward et al.*, 2000; *Harker et al.*, 2002).

Mie theory influenced our understanding of cometary dust, constraining it as an ensemble of dark, polydisperse, and nonspherical particles whose main contribution to the thermal emission comes from particles of size parameter close to one unit. It still remains an important tool when estimating the thermal radiation from cometary dust, since the particle shape does not affect the absorption cross sections as long as the particles can be considered as compact and roughly equidimensional or sufficiently porous aggregates of such particles so that particle interactions can be neglected. However, angular and spectral dependencies of intensity and polarization of cometary dust indicate that these properties cannot be calculated using the model of spherical homogeneous particles.

3.2. Approximations to Treat Nonspherical Inhomogeneous Particles

Along with techniques that consider light scattering in rigorous terms through the solution of Maxwell's equations, there are some useful approximations that allow fast computations subject to the size parameter and/or refractive index of the particle.

When $x \ll 1$ and $|mx| \ll 1$ the *Rayleigh* (1897) approximation applies. Such a particle can be considered as a dipole with an inherent polarizability that defines the dipole moment induced in the particle by the incident electromagnetic field. In the case of a very small particle, the field near it can be treated as an electrostatic field. This allows the derivation of a simple equation for the particle polarizability, which in many cases [spheres, spheroids, ellipsoids, disks, cylinders, see, e.g., *Bohren and Huffman* (1983)] admits analytical solution.

When $x|m – 1| \ll 1$ and $|m – 1| \ll 1$, the Rayleigh-Gans approximation applies. This approximation is often used not by itself but as a first approximation for the successive iterations to obtain more general approximations (see, e.g., *Haracz et al.*, 1984). *Muinonen* (1996) applied this approximation to stochastic irregular particles. Closely related to the Rayleigh-Gans approximation is the coherent Mie scattering approximation developed to treat porous aggregates of spheres (*Zerull et al.*, 1993). Only phase interrelations between waves scattered by constituent particles in an aggregate of spheres are taken into account. This approximation works better with increasing porosity of the aggregate; usually the porosity should be more than 90% (*Xu and Wang*, 1998).

When $x \gg 1$, $|m – 1| \ll 1$, we may apply the anomalous diffraction approximation (*van de Hulst*, 1957). This approximation is based on the assumption that the rays are negligibly deviated inside the particle, thus only absorption inside the particle and interference of light passed through the particle and around it can be taken into account. This allows easy calculation of the absorption and extinction cross sections, which sometimes can be expressed through analytical formulas. This method was widely applied to estimate cross sections of particles very different from spheres, such as prismatic and hexagonal columns, cubes, and finite cylinders. The accuracy of the method increases with decreasing absorption.

For the sake of completeness, we mention an approximation called perturbation theory, which treats the radius r of an irregular particle as a function of coordinates ϑ and φ and $r = r_0(1 + \xi f(\vartheta, \varphi))$, where r_0 is the radius of "unperturbed" sphere and $|\xi f(\vartheta, \varphi)| < 1$. The solution is obtained as an expansion of the boundary condition in series in ξ (*Oguchi*, 1960). This method provides accurate results only for particles of $x \leq 7$ and only if the deviations from spherical shape are smaller than the wavelength. It is a convenient method when effects of the surface roughness are of interest.

$x \gg 1$ is the geometric optics case, now often referred to as the "ray-tracing approximation." It considers the incident field as a set of independent rays, each of which reflects according to Snell's law and Fresnel's equations on the surface of the particle. The size restriction is in part relaxed when geometric optics is combined with Fraunhofer diffraction. For particles of complicated shape, the ray tracing is usually performed using the Monte Carlo approach. On this basis results have been obtained for stochastic particles (*Muinonen*, 2000) and particles with inclusions (*Macke*, 2000). The Web site by A. Macke (http://www.ifm.uni-kiel.de/) contains a collection of ray-tracing codes. The comparison of ray-tracing calculations with exact solutions, e.g., for spheres or cylinders, shows that the discrepancies between the approximation and the exact solution start for transparent particles of size parameter x < 100 (*Wielaard et al.*, 1997); for larger absorption the ray-tracing technique can be applied to smaller size parameters. *Waldemarsson and Gustafson* (2003) developed a combination of ray-tracing technique with diffraction on the edge for light scattering by thin plates (flakes) that provides a reasonable accuracy down to particle sizes of x ~ 1.

Cometary dust is not dominated by very small or very large particles, but the approximations described above can be applied to calculations for wide size distributions of nonspherical particles at the edges of the size distribution. For example, *Bonev et al.* (2002) calculated the color of cometary grains assuming them to be polydisperse spheroids and using the T-matrix method (see section 3.3.2) for size parameters x < 25 and the ray-tracing technique for larger particles. A similar approach can be used to calculate light scattering when a broad range of wavelengths need to be covered, as was done by *Min et al.* (2003), who also found a criterion for selecting the ray-tracing or anomalous diffraction approximations to calculate cross sections.

Kolokolova et al. (1997) showed that in the color, presented as the difference of logarithms of intensity at two wavelengths, the contribution of intermediate particles is canceled and small and large particles dominate. If the size distribution is smooth (e.g., of power-law type) and wide [i.e., includes small (Rayleigh) and large (ray-tracing) particles], then the color can be expressed analytically through the radii of the smallest and largest particles, the power of the size distribution, and two parameters that describe the particle composition. These analytical expressions have five unknowns that can be found if five colors are measured as has been done for Comet C/1996 Q1 Tabur (*Kolokolova et al., 2001b*).

The approximations listed above were developed to treat nonspherical but homogeneous particles. To describe light scattering by heterogeneous particles one can use approximations called mixing rules or effective medium theories. The main idea behind these approximations is to substitute a composite material with some "effective" material whose refractive index provides the correct light-scattering characteristics of the heterogeneous medium. It is a reasonable approximation when inhomogeneities (inclusions) are much smaller than the wavelength (Rayleigh-type) and the size of the macroscopic particle is much larger than the size of the inclusions. In large scale, such a medium acts as an isotropic homogeneous medium for which the displacement $\mathbf{D}$ is related to the electrostatic field $\mathbf{E}$ by a linear relationship $\mathbf{D} = \varepsilon\mathbf{E}$ where ε is the complex dielectric constant of the material related to the refractive index as $m = \sqrt{\varepsilon}$. For a macroscopically homogeneous isotropic mixture, $\mathbf{D}$ and $\mathbf{E}$ can be averaged within the medium as

$$\int \mathbf{D}(x,y,z)dxdydz = \int \varepsilon(x,y,z)\mathbf{E}(x,y,z)dxdydz = \varepsilon_{eff} \int \mathbf{E}(x,y,z)dzdydz \tag{5}$$

where $\mathbf{E}$, $\mathbf{D}$, and ε are determined for a point inside the material with coordinates (x,y,z). Using the Maxwell equation $div\mathbf{D} = 0$ and assuming some size, shape, and distribution of the inclusions, one can obtain expressions, in some case analytical, for the effective dielectric constant ε_{eff}. Dozens

of mixing rules have been developed for a variety of inclusion types (non-Rayleigh, nonspherical, layered, anisotropic, chiral) and topology of their distribution within the medium, including aligned inclusions and fractal structures (see an extensive review by *Sihvola, 1999*). However, still the most popular remain the simplest Maxwell Garnett (*Garnett, 1904*)

$$\frac{\varepsilon_{eff} - \varepsilon_m}{\varepsilon_{eff} + 2\varepsilon_{eff}} = f_i \frac{\varepsilon_i - \varepsilon_m}{\varepsilon_{eff} + 2\varepsilon_{eff}} \tag{6}$$

and Bruggeman (*Bruggeman,* 1935)

$$f_i \frac{\varepsilon_{eff} - \varepsilon_i}{\varepsilon_i + 2\varepsilon_{eff}} = f_m \frac{\varepsilon_m - \varepsilon_{eff}}{\varepsilon_m + 2\varepsilon_{eff}} \tag{7}$$

mixing rules. In equations (6) and (7) ε_i and ε_m are dielectric constants of inclusions and matrix respectively, and f_i is the volume fraction of the inclusions in the mixture. The Maxwell Garnett rule represents the medium as inclusions embedded into the matrix material with ε_m, the result of which depends on the material chosen as the matrix. The Bruggeman rule was obtained for particles with ε_i and ε_m embedded into the material with $\varepsilon = \varepsilon_{eff}$, so the formula is symmetric with respect to the interchange of materials. This makes it easy to be generalized for the n-component medium

$$\sum_{i=1}^{n} f_i(\varepsilon_i - \varepsilon_{eff})/(\varepsilon_i + 2\varepsilon_{eff}) = 0 \tag{8}$$

Since the derivation of the mixing rules was based on assuming the external field as electrostatic, the inclusions were assumed to be much smaller than the wavelength of electromagnetic waves. More exactly, the criterion of their validity is $xRe(m) \ll 1$ (*Chylek et al., 2000*), where x is the size parameter of inclusions and $Re(m)$ is the real part of the refractive index for the matrix material. Comparison of effective medium theories with DDA (see section 3.3.3) calculations (*Lumme and Rahola, 1994; Wolff et al., 1998*) and experiments (*Kolokolova and Gustafson, 2001*) show that even for $xRe(m) \sim 1$ effective medium theories provide reasonable results. The best accuracy can be obtained for cross sections and the worst for polarization, especially at phase angles $\alpha < 50°$ and $\alpha > 120°$.

There were a number of attempts to consider heterogeneous cometary grains using effective-medium theories, particularly to treat aggregates as a mixture of constituent particles (inclusions) and voids (matrix material) (e.g., *Greenberg and Hage,* 1990; *Mukai et al.,* 1992; *Li and Greenberg,* 1998b). In the visual region the cometary aggregates with the size parameter of constituent particles x > 1 are, most likely, out of the range of the validity of the effective medium theories. For the thermal infrared region, cometary aggregates can be treated with effective medium theories if they are sufficiently large (remember that the number of inclusions must allow the macroscopic particle to be considered as a medium). Using the mixing rules, one should be careful to calculate the refractive index for core-mantle particles. Even for particles of size parameter $x \geq 0.3$, significant differences in cross sections from exact core mantle calculations appeared (*Gustafson et al.*, 2001).

3.3. Numerical Solutions for Inhomogeneous Nonspherical Particles

Further characterization of cometary dust requires more advanced light scattering theories. We consider them in this section, mainly outlining the strengths and drawbacks of the methods that have been applied to study cometary dust. As well as Mie theory, solutions of the Maxwell equations for nonspherical particles cannot be used directly to solve the inverse problem for cometary dust, but instead provide some constraints through fitting procedures and investigation of general trends in the dependence of light-scattering characteristics on particle properties.

3.3.1. Separation of variables method. The main idea behind Mie theory, expansion of the electromagnetic waves in the coordinate system that corresponds to the shape of the body (e.g., cylindrical, spheroidal) and then separates the variables, has provided a number of solutions for nonspherical but regular particles: multilayered, optically anisotropic, and radially inhomogeneous spheres, cylinders, slabs, and spheroids (see the review by *Ciric and Cooray*, 2000). Among solutions obtained with this method, the most popular are solutions for cylinders (*Bohren and Huffman*, 1983), core-mantle spheres (*Toon and Ackerman*, 1981), spheroids (*Asano and Yamamoto*, 1975; *Voshchinnikov and Farafonov*, 1993), and core-mantle spheroids (*Farafonov et al.*, 1996). Although this method can provide exact solutions to Maxwell's equations, the problem requires numerical treatment to solve the boundary condition even for simple particle shapes. This limits the applicability of the method to rather small size parameters ($x < 40$ for spheroids) or to nonabsorbing materials. However, in the range of their applicability, these theories provide highly accurate results, which makes them a good tool for testing other approaches.

The solution of the separation of variables for a single sphere can be extended to aggregates of spheres representing the total scattering field as a superposition of spherical wave functions, which expand the field scattered by each sphere and use the translation addition theorem for spherical functions (*Bruning and Lo*, 1971). The method was extended for aggregates of spheroids, cylinders, multilayered spheres, and spheres with inclusions (see *Xu*, 1997; *Fuller and Mackowski*, 2000, and references therein). The computational complexity of the method (the number of numerical operations in the algorithm) depends on the number of particles and their size parameter, shape, and configuration.

3.3.2. T-matrix approach. A popular T-matrix approach was initially introduced by *Waterman* (1965) under the name "extended boundary condition." This method ex-

pands the incident and scattered fields into vector spherical functions with the scattered field expanded outside a sphere circumscribing a nonspherical particle. The main advantage of the resulting formulation is that the T-matrix is defined only by physical characteristics of the particle itself, i.e., it is computed only once and then can be used for any incidence/scattering direction and polarization of the incident light. A significant development of the method was an analytical orientation-averaging procedure (*Mishchenko*, 1991) that made calculations for randomly oriented particles as efficient as for a fixed orientation of the same particle. T-matrix computations are especially efficient for axisymmetric, including multilayered, particles; however, the method can be extended to any shape of the scatterer. Although computational complexity greatly increases for particles without rotational symmetry, recently solutions were obtained for ellipsoids, cubes, and aggregates of spherical particles. The latter case is of special interest in the field of astrophysics. It presents the T-matrix development of the separate-variable method for the aggregates of spheres considered above. Since the scattered field for the individual sphere does not depend on the incident light direction and polarization, the expansions for individual spheres can be transformed into a single expansion centered inside the aggregate. Such an expansion is equivalent to the T-matrix of the aggregate and thus can be used in analytical averaging of scattering characteristics over aggregate orientations (*Mishchenko et al.*, 1996).

A variety of computational T-matrix codes are available on line at Mishchenko's Web site (http://www.giss.nasa.gov/~crmim), among them codes for polydisperse axial-symmetric particles and aggregates of spheres. The computational complexity of the T-matrix method increases as x^3–x^4, i.e., the method is a fast and convenient way to test results obtained with other codes (see, e.g., *Lumme and Rahola*, 1998) or for survey-type studies. *Kolokolova et al.* (1997) used it to check the sensitivity of color and polarization to the size and composition of nonspherical particles. The results were found to be similar to the results for polydisperse spheres. However, details in polarization for nonspherical particles differ significantly from those calculated for equal-volume or equal-area spheres (*Mishchenko and Travis*, 1994). For example, randomly oriented spheroids show no oscillations in angular dependencies of intensity and polarization at size distributions 10–20× narrower than is necessary to wash out the oscillations for spheres. Polarization for polydisperse silicate spheroids (Fig. 8) at intermediate phase angles changes from negative, typical for spheres (Fig. 6), to positive, leaving a negative branch at $\alpha < 30°$. For elongated particles at $\alpha \sim 90°$, polarimetric color is positive and shows the angular trend similar to the observed one (it gets larger with increasing phase angle) for a broad range of particle size (Fig. 8). The angular dependence of the intensity is also reminiscent of the cometary scattering function (Fig. 1) except at phase angles 10°–60°, where it reaches a deep minimum instead of being flat. The spheroids demonstrate color that changes rather rapidly its

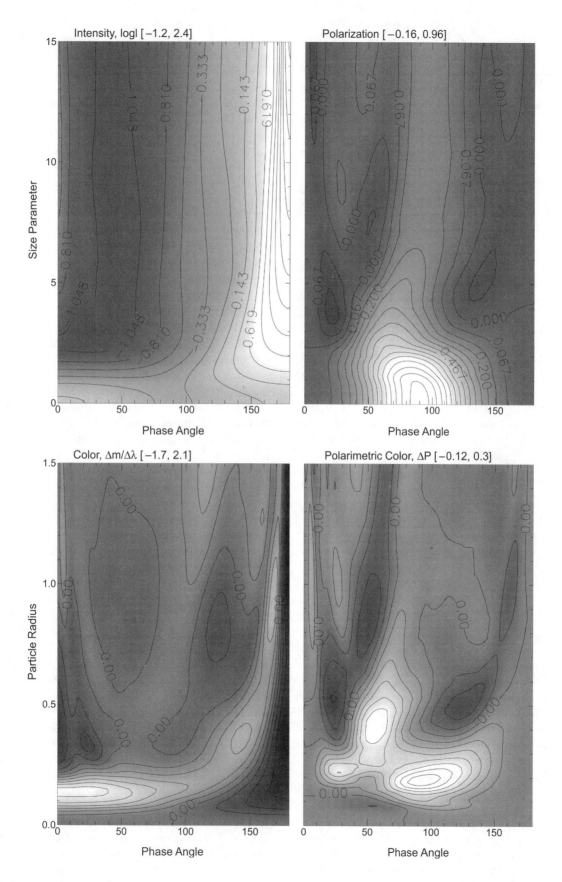

Fig. 8. Same as Fig. 6 but for prolate spheroids (ratio of axis is 1 : 3) of the refractive index m = 1.55 + i0.005; the effective size parameter x_{eff} corresponds to this for equal-volume sphere.

value and even sign with phase angle. Randomly oriented cylinders exhibit similar tendencies *(Mishchenko et al., 2002)*. Using higher absorption one can get higher values of color and polarimetric color, but the negative polarization and backscattering peak disappear, eliminating the resemblance of the calculated scattering function and polarization curve to the observed ones.

Mishchenko et al. (1997) showed that appending the model by a shape distribution (a range of aspect ratios of spheroids) can provide a good fit to the cometary scattering function for polydisperse silicate spheroids at $6 < x_{eff} < 24$. For the size distribution with $x_{eff} > 8$ and a constant refractive index $m = 1.53 + i0.008$, the color becomes neutral and then red with a very weak angular change. However, polyshaped silicate spheroids fail to reproduce the correct polarization *(Mishchenko, 1993)*.

Petrova et al. (2000) used the T-matrix code by *Mackowski and Mishchenko* (1996) for a systematic comparison of cometary observations with the calculations for aggregated particles. They considered aggregates built from particles of size parameters $1 < x < 3.5$ for a set of refractive indexes with the real part equal to 1.65 and the imaginary part varying within the range 0.002–0.1. The maximum number of particles in each aggregate was limited to 43 by the code convergence. "Cometary" type polarization with low maximum at $\alpha \sim 90°$ and negative branch at small phase angles was obtained for a power-law size distribution of aggregates built of 8–43 constituent particles of $x = 1.3$, 1.5, and 1.65, although a smooth shape of the curve and the quantitative fit to the observational polarization curve was not achieved. The characteristics of the negative polarization depend on the composition (it becomes less pronounced with increasing absorption from 0.01 to 0.05) and porosity (more pronounced for more compact aggregates). A rather good qualitative fit to the cometary scattering function was achieved for the polydisperse aggregates with $x = 1.3$ and $\kappa = 0.01$. In a later paper, *Petrova and Jockers* (2002) studied the spectral dependence of polarization for a variety of size distributions and a range of refractive indexes within $n = 1.65$, $0 < \kappa < 0.1$, assuming the refractive index independent of wavelength. They showed that the aggregate model successfully reproduced the observed positive polarimetric color. However, the color of such aggregates was found to be blue, unlike the observed red color. The extinction calculations by *Petrova et al.* (2000) are mainly sensitive to the overall size of the aggregate and are close to the results for equal-volume spheres if the aggregates have size parameters $x < 3$, but becomes larger for larger aggregates. This is an indication that the equal-volume sphere calculations can be used to estimate cross sections of aggregates in the thermal infrared but not in the visual.

3.3.3. Coupled dipole approximation and similar techniques. The methods described above are based on the solution to the boundary conditions on the particle surfaces, while a set of more flexible methods rely on solutions of the internal field from which the scattered field is calculated.

The most popular among such methods is probably the coupled dipole approximation, often called the discrete dipole approximation (DDA) since it represents a particle as consisting of elementary cells, polarizable units called dipoles *(Purcell and Pennypacker,* 1973). Each dipole is excited by the external field and the fields scattered by all other dipoles. This representation allows one for a N-dipole particle to derive N linear equations describing the N fields that excite each dipole. The numerical solution of the system of these N equations provides the partial field scattered by each dipole, and finally the total scattered field is calculated. *Draine and Flatau* (1994) and *Lumme and Rahola* (1994) improved the computational efficiency of the DDA, which allowed them to study larger particles, in particular aggregates of larger size. *Lumme et al.* (1997) considered aggregates up to 200 particles of $x = 1.2$–1.9. The most popular has been the DDSCAT code by Draine and Flatau (see http://www.astro.princeton.edu/~draine/DDSCAT.html). The code includes geometry implementations for ellipsoids, finite circular cylinders, rectangular parallelepipeds, hexagonal prisms, and tetrahedra, uniaxial and coated particles as well as for systems of particles. The great advantage of the coupled dipole approximation is that it can be applied to particles of arbitrary shape, structure, and composition. However, it has limited numerical accuracy, requires a separate calculation for each orientation of particle, and slows down dramatically with increasing N, so that its computational complexity increases as the ninth order of the particle size parameter.

Using DDA, *West* (1991) undertook the first systematic theoretical study of aggregates. He considered the constant refractive index $m = 1.7 + i0.029$. West demonstrated that for compact aggregates both intensity and polarization behave as is more typical for the spherical equivalent of the aggregate. For loose aggregates the projected area of the whole aggregate mostly defines the intensity in the forward-scattering domain, whereas the polarization is defined by the size of the constituent particles. *Kozasa et al.* (1993) confirmed that even aggregates of thousand constituent particles of $x < 1$ behave like Rayleigh particles in polarization regardless their composition.

Xing and Hanner (1997) accomplished a detailed study of aggregates with a variety of sizes, shapes, and compositions of constituent particles and their packing within the aggregate. They confirmed that polarization is more sensitive to the properties of constituent particles whose size and refractive indexes determine most of the polarization, whereas shape, number, and packing generate scattering effects of the second order. A reasonable fit to the observed cometary polarization and intensity was achieved for the mixture of carbon and silicate aggregates of an intermediate (touching-particle) porosity. The calculations of cross sections showed that in the case of short wavelengths (size parameter $x > 1$) the values of extinction cross section per unit volume is close to those typical for constituent particles, but at long wavelengths ($x \ll 1$) constituent particles and aggregates show significant differences that are more pronounced the more compact the aggregate. When the porosity is less than 60%, the temperature of the particle is close to that of an equal-volume sphere. Loose aggregates have a temperature typical for the constituent particle, i.e., thermal emission from a large but aggregated particle looks the same as its small constituent particle regardless of its shape (see also *Greenberg and Hage*, 1990; *Li and Greenberg*, 1998b). *Davidsson and Skorov* (2002) obtained similar DDA results for single-scattering albedo. This means that thermal infrared data do not allow distinguishing loose aggregates from single particles of the size of their constituent particles.

The DDA was used by *Yanamandra-Fisher and Hanner* (1998) to study the light scattering and emission by nonspherical but regular particle shapes of size parameter 1 < x < 5, as well as heterogeneous and porous particles. They showed that compact particles of different shapes produce very different angular dependence of polarization for a transparent material (silicate), but the dependence on particle shape is reduced for absorbing particles (carbon). They concluded that negative polarization at small phase angles could be produced by nonspherical silicate particles of x ≥ 2. Introducing porosity for a transparent (silicate) particle caused the polarization to resemble that of a smaller particle, while porosity had less of an effect on an absorbing particle provided that the particle remained opaque. Although the use of regular shapes caused oscillations that would be absent in the scattering by an irregularly shaped particle, the authors showed that a macroscopic mixture of silicate and absorbing particles with a distribution of size parameters 1 < x < 5 could produce curves for polarization and angular scattering qualitatively similar to the curves observed for cometary dust. A similar conclusion was reached by *Xing and Hanner* (1997) based on the scattering by small aggregates.

Yanamandra-Fisher and Hanner (1998) also applied the DDA code to examine particle shape effects in the mid-infrared emission features produced by crystalline olivine. Both the shape and the peak wavelength of the spectral features depend on the particle shape; moreover, olivine is an anisotropic crystal, and the differing optical constants generate peaks at different wavelengths. To reproduce the observed silicate feature the authors had to rule out extreme shapes (disks, needles) as well as ideal spheres. Polarization within the 10-μm silicate feature was studied by *Henning and Stognienko* (1993), who demonstrated that shape and porosity of particles affect the spectral dependence of polarization, both the strength and shape of the feature, even stronger than the spectral dependence of the intensity.

Combining the T-matrix technique with the DDA, *Kimura* (2001) considered aggregated particles of fractal structure, ballistic particle-cluster aggregates (BPCA) and ballistic cluster-cluster aggregates (BCCA) (see *Mukai et al.*, 1992), of hundreds and thousands of constituent particles. He showed that aggregates of silicate particles of x = 1.57 reproduce the cometary polarization with a very good fit. They also provide rather good fit to the "cometary" shape

of the scattering function. The number of particles in the aggregates, at least if it is within 64 ≤ N ≤ 256, as well as their porosity, was found not to be important. But a change in composition (carbon instead of silicate) could dramatically affect the value of maximum polarization, eliminate the negative branch, and make the scattering function more flat at the medium phase angles. *Kimura*'s (2001) results demonstrate the best fit to the observational data achieved so far, although the spectral characteristics of intensity and polarization have not been checked.

Similar to DDA, the method of moments (MOM) approach also divides a particle into small elements, but considers them not as dipoles but as small cells with a constant refractive index. The price of this straightforwardness of MOM in comparison with DDA is the necessity of calculating a self-interaction term; different approaches in MOM usually diverge in the way they treat this term. The MOM has strengths and drawbacks similar to those for DDA. Using MOM, *Lumme and Rahola* (1998) considered cometary particles as stochastically shaped, i.e., particles whose shape can be described by a mean radius and the covariance function of the radius given as a series of Legendre polynomials (for details, see *Muinonen*, 2000). *Lumme and Rahola* (1998) made computations for a variety of particle shapes and size parameter (x = 1–6) using the refractive index m = 1.5 + i0.005. Like many previous authors, Lumme and Rahola found that the particles should be of x > 1 to provide negative polarization at small phase angles and low maximum polarization. Also, they showed that monodisperse particles, even of very complicated, irregular shape, demonstrate oscillating behavior of light-scattering characteristics that can be eliminated by considering a polydispersion of particles. The same oscillating behavior is shown by color and polarimetric color that can be seen from the comparison of the intensity and polarization for x = 2, 4, 6. Particles with the size distribution of the form n(r) ~ r^{-3} show angular functions of intensity and polarization that are rather similar to cometary ones. However, one also needs to fit to the spectral characteristics to judge how close the model is to real cometary dust.

3.3.4. Other numerical methods. Above we considered numerical techniques that have been used in cometary physics. However, there are other light-scattering techniques that may be worth applying, among them the finite difference time domain method (FDTD) and the finite element method (FEM). Both these methods are based on a straightforward solution to equation (4). The FDTD discretizes space and time, making a grid in the time-space domain and then solves equation (4) for each space-time cell, whereas the FEM discretizes the scattering particle itself into small volume cells. Then boundary conditions (continuity of the electrical vector across the surface) are applied on the boundary of each cell. Both methods are applicable to particles of arbitrary shape, structure, and composition. The main disadvantage of these methods is time-consuming computations, aggravated by the fact that the computations should be repeated for each orientation of the particle with respect to the incident wave. The computational complexity of the FDTD increases as the fourth power of the particle size parameter and varies between x^4 and x^7 for FEM depending on the computational technique.

More details of the methods discussed above and other numerical methods and approximations can be found in recent reviews (*Mishchenko et al.*, 2000, 2002; *Kahnert*, 2003). Some codes are available via online libraries such as http://atol.ucsd.edu/~pflatau/scatlib/ (P. Flatau), http://www.t-matrix.de/ (T. Wriedt), http://www.emclab.umr.edu/codes.html, http://urania.astro.spbu.ru/DOP/, and http://www.astro.uni-jena.de/Users/database/. The latter Web site also contains a huge collection of refractive indexes for materials of astrophysical interest (*Jäger et al.*, 2003).

3.4. Experimental Simulations

The complexity of cometary and other cosmic dust and planetary aerosols and the ensuing difficulty treating them using theoretical means generated several approaches for solving the scattering problem through experimental simulation. As before, we will not discuss all experiments conducted so far to understand light scattering by particles, but only those representative examples that provided significant results for cometary dust. We will not discuss here the experimental setups detailed in the papers cited below.

One experimental technique to study light scattering by particles uses streams or suspensions of small particles, often powdered terrestrial rocks, that are illuminated by a source of light, recently mainly by a laser. The light scattered by the particles is measured to obtain intensity, polarization, Stokes parameters, and even the complete scattering matrix for a range of phase angles. The other technique, called the microwave analog method, introduced by *Greenberg et al.* (1961), simulates light scattering using microwave radiation. It actually takes advantage of the same scaling used in all theoretical approaches where the particle dimensions are given through its size parameter, i.e., as a ratio to the wavelength. In such a simulation the light scattering problem is scaled to longer wavelengths that precisely preserve the characteristics of the scattering problem as long as the size parameter of the scatterer and its refractive index are preserved. This allows light scattering by a single submicrometer or micrometer particle to be simulated using a manageable millimeter or centimeter analog model.

3.4.1. Microwave analog experiments. The early microwave simulations were measurements of cross sections and scattering functions for spheroids, finite cylinders, rods, and a variety of compact irregular particles (*Greenberg et al.*, 1961; *Zerull et al.*, 1977) directed to determine the applicability of Mie theory to nonspherical particles. The study was then extended to polarization and color of nonspherical particles (*Zerull and Giese*, 1983). A systematic study of particle shape effects was carried out by *Schuerman et al.* (1981). Significant discrepancy between the data for nonspherical particles and Mie theory was obtained for both the shape of measured curves and values of light-scattering

parameters. *Giese et al.* (1978) showed that, unlike compact and rather regular particles, irregular, porous particles in the size range of a few micrometers exhibit positive and negative polarization typical for cometary dust. Thus in the 1970s the microwave method provided the results obtained only in theory in the late 1990s. *Greenberg and Gustafson* (1981) showed that models consisting of a tangle of rods, so-called "Bird's Nests," could provide the observed angular dependencies of intensity and polarization. This experimental test of the aggregated nature of cometary dust then was extended to the study of a variety of aggregates of spherical, including core-mantle, particles by *Zerull et al.* (1993).

The systematic study of light scattering by complex particles has been performed using a new generation of microwave facilities designed and built at the University of Florida (*Gustafson*, 1996, 2000). This facility works across a 2.7–4-mm waveband to simulate a region 0.4–0.65 μm in the visual, and thus can study not only angular but also spectral dependencies of light scattering (colors). *Gustafson and Kolokolova* (1999) reported results from a systematic study of light scattering by aggregates. Among these results is that the polarization is mainly determined by the size and composition of constituent particles and the size parameter of cometary constituents must be $1 < x < 10$ to produce a low maximum polarization and negative polarization at small phase angles that confirms the theoretical results considered above (e.g., *West*, 1991; *Xing and Hanner*, 1997; *Lumme et al.*, 1997; etc.). The color and polarimetric color of aggregates also primarily depend on the size and composition of the constituent particles. For aggregates made of particles whose size parameters covered the range 0.5–20, an increase in the size of constituent particles results in larger (more red) color and smaller polarimetric color (the refractive index was m = 1.74 + i0.005, constant throughout the wavelength range). The same tendencies result from increasing the imaginary part of the refractive index that may provide a clue to the negative polarimetric color observed for Comet P/Giacobini-Zinner (*Kiselev et al.*, 2000). A combination of positive polarimetric color and blue color is typical of aggregates of nonabsorbing particles of $0.5 < x < 10$, whereas red color and positive polarimetric color are indicative of dark, absorbing aggregates. From the above, the aggregates, which similarly scatter light to the cometary dust, consist of absorptive constituent particles of size parameter $x > 1$. The shape of constituent particles could not be seen to influence the aggregates' color and only slightly changes the position and amount of maximum polarization. Notice that the results were obtained for the refractive index that does not change with wavelength. The intensity and polarization data for a variety of particles, including aggregates, are given at http://www.astro.ufl.edu/~aplab.

Even though implementation of some theoretical methods, e.g., DDA, on modern computers allows a study of similar accuracy for aggregates of similar characteristics as at microwave measurements, the microwave method is still much faster and can provide a large scope of the results in a reasonable time. For example, single-wavelength DDA calculations for a silicate aggregate of 1024 constituent particles of size parameter x = 0.733 (*Kimura*, 2001) took more than a month on the Alpha station, whereas scattering from an aggregate of any number of particles of x > 1 can be measured in one week providing data at 50 discrete wavelengths. This allows systematic studies to obtain quantitative characteristics of the light scattering, e.g., multifactor analysis (similar to that described in *Kolokolova et al.*, 1997) of sensitivity of light-scattering characteristics to physical properties of particles. Some results of such a study of aggregates are presented in Table 1, which summarizes a study of large (500–5000 constituent particles) aggregates. The aggregates were built to satisfy the most plausible model of cometary dust aggregates resulted from theoretical simulations: The size parameter of the constituent particles was in the range 1.5–5, the constant refractive index, m = 1.74 + i0.005, simulated a silicate-organic mixture, and the porosity varied within 50–90%. The larger numbers in Table 1 indicate a stronger correlation between the dust characteristic and the light-scattering parameter; negative values show that they anticorrelate. Table 1 shows quantitatively that the size of constituent particles is the main characteristic that determines all light-scattering parameters. The influence of the porosity is much smaller, and within the range investigated the number of particles in the aggregate has almost no influence on the light-scattering parameters.

The data from Table 1 show that the correlations between color and polarization (their simultaneous decrease or increase) in the near-nucleus coma observed by K. Jockers and colleagues (*Kolokolova et al.*, 2001a) could not be explained by the changing size of the aggregates (overall size influences neither color nor polarization) or the changing size of the constituent particles (color and polarization should then anticorrelate). Change in the porosity could produce a correlation, but then the color should anticorrelate with the polarimetric color that has not been observed. Moreover, the simultaneous increase in color and polarization observed for Comet Hale-Bopp could be provided only by unrealistic increasing compactness of the aggregate on its way out from the nucleus. Microwave measurements for materials of a variety of refractive indexes show that such a correlation may be a consequence of evaporation/destruction of some dark material contained in the aggregates. Such a slowly evaporating dark material can be an organic compound, e.g., kerogen or tholin (*Wallis et al.*, 1987) or an organic refractory (*Greenberg and Li*, 1998). This assumption is consistent with the observed increase in albedo with the distance from the nucleus and might also explain the gradient in the near-infrared colors.

3.4.2. Light-scattering experiments in the visual. Among early experimental works that contributed to our understanding of light scattering by cosmic grains, *Weiss-Wrana* (1983) showed a significant difference between light scattering by terrestrial/meteoritic particles and Mie calculations. The work also showed that transparent compact irregular particles ~30 μm in size had the polarization curve with the maximum located at $\alpha \approx 45°$ and negative polarization at large phase angles; large (28–80 μm) absorbing ($\kappa = 0.66$) compact particles had a bell-shaped polarization curve with high (about 50%) polarization maximum; and fly ash, slightly absorbing ($\kappa = 0.01$) loose aggregates with an average size of 41 μm had an angular dependence of polarization similar to the cometary one.

Extensive measurements of all 16 elements of the scattering matrix **F** were performed by the group from Amsterdam (*Hovenier*, 2000; *Hovenier et al.*, 2003; see also http://www.astro.uva.nl/scatter/). *Muñoz et al.* (2000) studied the scattering matrix of polydisperse, heterogeneous, irregular particles presented by terrestrial (light-colored olivine) and meteoritic (dark carbonaceous chondrite) powders at wavelengths of 0.442 and 0.633 μm. The effective radius of the particles was approximately a few micrometers so that the size parameter of the particles peaked within the range x = 10–20. Both terrestrial and meteoritic samples showed polarization curves of shape similar to the cometary ones and positive polarimetric color. However, the values and angle of the minimum polarization almost twice exceeded the observed ones, whereas the maximum polarization was almost half the observed values. Also, the color was blue, i.e., the intensity at 0.442 μm exceeded the intensity at 0.633 μm. The study shows that polydisperse heterogeneous irregular particles cannot be ruled out as candidates for cometary dust, although, at least for size distributions that peak at x = 10–20, they are not able to correctly reproduce all the observed characteristics.

Light-scattering experiments under reduced gravity (microgravity) have been proposed to avoid sedimentation of the studied dust as well as particle sorting or orientation that can occur for particles dropped or suspended in airflow. The microgravity PROGRA[2] experiment was concentrated on light-scattering measurements of levitated particles during parabolic flights (*Worms et al.*, 2000). The angular dependences of polarization were retrieved for various types of compact particles (*Worms et al.*, 1999) and aggregates (*Hadamcik et al.*, 2002). This experiment later provided polarization images in two wavelengths (*Hadamcik et al.*, 2003).

Hadamcik et al. (2002) performed a systematic microgravity study of aggregated particles. The majority of the samples were made of 12–14-nm particles of silica, alumina, titanium dioxide, or carbon. The particles formed complex

TABLE 1. The strength of correlation (normalized to the color/radius correlation) between light-scattering characteristics and properties of the aggregates.

	Radius of Constituent Particles	Number of Constituent Particles	Porosity
Color	1.00	0.212	−0.923
Polarization	−3.23	0.030	−0.262
Polarimetric color	−2.36	−0.316	1.34

structures of three-level hierarchy: submicrometer-sized chains of particles were combined into several-micrometer-sized aggregates, and the aggregates in turn formed millimeter-sized particles, "agglomerates," that were levitated and measured. The most striking result of this study was that the particles, even though composed of nanometer constituents, did not show Rayleigh-type scattering properties unlike other aggregates of small particles considered above (*Zerull et al.*, 1993; *Gustafson and Kolokolova*, 1999; *West*, 1991; *Kozasa et al.*, 1993). The microscopic photographs show that although "agglomerates" are rather porous structures, "aggregates" look as if they were made of solid material, demonstrating rather high packing of "chains" in the aggregates. Since the size of constituents in "aggregates" is of the order of some micrometers, their size can be the factor that determines the light scattering. The polarimetric curves have a low maximum at $\alpha \sim 100°$ and demonstrate positive polarimetric color at large phase angles. At small phase angles the samples made of a single type of constituent demonstrate positive polarization, which may become negative for mixtures of small and large or dark and light particles that can be evidence of a heterogeneous structure of cometary grains.

The most recent microgravity study of the wavelength dependence of polarization (*Hadamcik et al.*, 2003) showed that a mixture of fluffy aggregates of submicrometer silica and carbon grains demonstrates a positive polarimetric color, whereas gray compact particles with size greater than the wavelength exhibit a negative polarimetric color. This can explain the difference between the polarimetric color of comets and asteroids as well as changes in polarimetric color within cometary comae.

The COsmic Dust AGgregation (CODAG) experiment intends to study dust aggregation during rocket flights (*Blum et al.*, 2000; *Levasseur-Regourd et al.*, 2001). These experiments (1) study the formation and evolution of dust grains and aggregates under a variety of physical conditions representative of the protoplanetary nebula, (2) monitor the dust particles in three colors at $\alpha = 5°–175°$ to study the changes in their light scattering that accompany the aggregation process, and (3) provide validation of the light-scattering codes. Low-temperature studies to reveal the dust light-scattering properties during condensation or sublimation of ices on the grains will hopefully be performed onboard the International Space Station (*Levasseur-Regourd et al.*, 2001).

3.5. Summary

Table 2 summarizes the observational data and results of the theoretical and laboratory simulations of light scattering by cometary dust. One can see that the theoretical and laboratory simulations of light scattering and thermal emission by cometary dust show that a plausible model of cometary dust can be heterogeneous (silicates and some absorbing material) particles that are either irregular, poly-

disperse grains with the predominance of submicrometer particles or are aggregates of submicrometer particles. The major constraints on the dust properties came from polarimetric and thermal infrared data. Angular dependence of polarization puts constraints on the size of particles and limits the refractive index. The SED of the thermal emission implies a broad size distribution of the dust; the excess color temperature above a blackbody at $\lambda < 20 \mu m$ indicates grains of $r < 1 \mu m$, while the slow decrease in flux at longer wavelengths indicates larger particles with a size parameter comparable to the wavelength. Aggregates of submicrometer grains with a range of porosities would be compatible with thermal SED. Spectral midinfrared features identify silicates as a component of the dust material. The composition could be refined using the spectral changes in the scattered light (color and polarimetric color); it is likely that they indicate spectral dependence of the average refractive index of the material. A change in the characteristics of the scattered and emitted radiation with distance from the nucleus is a helpful tool to further determine the size distribution and composition of cometary dust as it reflects the dynamic sorting of particles and composition changes, e.g., sublimation of volatiles. Although observational data accumulated so far, as well as their interpretation, are not sufficient to provide reliable information about cometary dust from the correlations between the light-scattering and thermal-emission characteristics (section 2.4), such correlations can be valuable to further constrain the properties of cometary dust and to compare the dust in different parts of the coma and in different comets.

4. CONCLUSIONS AND CONSISTENCY WITH THE RESULTS OBTAINED USING OTHER TECHNIQUES

Below we discuss the physical properties of cometary dust obtained using light-scattering techniques, comparing them with the results of studying cometary dust using its dynamical properties, *in situ* measurements, and study of interplanetary dust particles (IDPs).

4.1. Shape

The shape of cometary particles is significantly nonspherical. Evidence of this is the complete absence of resonant oscillations in intensity, polarization, and color. Such oscillations are typical for non-Rayleigh spherical particles even with a rather wide size distribution. None of the attempts to fit the scope of observations with a model based on spherical particles has been successful. Additional support for this conclusion are observations of star occultations that show nonzero polarization at $\alpha = 180°$ and the presence of circular polarization and polarization whose plane is inclined to the scattering plane. Such observations require that the symmetry inherent in spherical particles be broken. Simulated mid-infrared spectral features demonstrate noticeable

TABLE 2. Observational facts and their interpretations.

Observational Fact and Section Where It is Discussed	Cometary Dust Model that can Reproduce the Observational Fact and Section Where It is Discussed
Brightness characteristics:	
1. Low geometric albedo of the particles (2.1.1)	1. Absorbing particles ($\kappa > 0.02$) (3.4.1). Decrease in the particle size increases albedo for slightly absorbing particles.
2. Prominent forward-scattering and gentle backscattering peaks in the angular dependence of intensity, "flat" behavior at medium phase angles (α) (2.1.2)	2. Aggregates of particles of $x = 1–5$ (3.3.2, 3.3.3, 3.4.1), silicate polydisperse (power law) elongated particles (e.g., spheroids) with a distribution of the aspect ratios (3.3.2), silicate polydisperse (power law) irregular particles (3.3.3).
Polarization characteristics:	
1. For a broad range of wavelengths angular dependence of linear polarization (2.2.1) demonstrates • negative branch of polarization for $\alpha < 20°$ with the minimum $P \approx -2\%$ • bell-shaped positive branch with low maximum of value $P \approx 10–30\%$ at $\alpha \approx 90°–100°$.	1. Polydisperse (power law) irregular submicrometer particles (3.3.2, 3.3.3; 3.4.2); aggregates of particles of $x = 1–5$ (3.3.2, 3.3.3, 3.4.2). The best fit is provided by aggregates of large number of constituent particles (3.3.3) or polydisperse aggregates (3.3.2). Absorption seems to reduce the negative polarization but increases the maximum polarization (3.3.2, 3.3.3).
2. Small circular polarization $V < 2\%$ (2.2.4).	2. Nonspherical particles or particles containing optically anisotropic materials.
Spectral optical characteristics:	
1. Usually red or neutral color for a broad range of wavelength λ (2.1.3).	1. Slightly absorbing particles of $x > 6$ (3.3.2); absorbing particles of $x > 1$ (3.4.1); particles, containing material with a spectrally dependent refractive index (3.3.3).
2. Polarization at a given α above $\approx 30°$ usually increases with λ (positive polarimetric color) (2.2.2).	2. Aggregates of particles of $x = 1–5$ (3.3.2, 3.3.3, 3.4.1, 3.4.2) or polydisperse nonspherical particles of $x > 1$ (but not $x \gg 1$) (3.4.2) for a broad range of refractive indexes independent of wavelength; particles made of materials with spectrally dependent refractive index.
Spatial distribution of optical characteristics within the coma:	
1. Increase in albedo with the distance from the nucleus (2.1.1).	1. Change in the particle composition (evaporation of some dark material) (3.4.1) or decreasing size of silicate particles
2. Variations in the value of color and polarization (2.1.1, 2.2.3)	2. Variations in size distribution of cometary dust or in their composition (if the changes in color and polarization correlate) (3.1, 3.4.1)
3. Deviations in the polarization plane throughout the coma (2.2.4)	3. Nonspherical particles or particles containing optically anisotropic materials
Thermal infrared characteristics (2.3):	
1. 3- to 20-μm color temperature higher than the blackbody temperature.	1. Absorbing grains of $r < 1$ μm (3.1.1); porous aggregates of submicrometer absorbing grains (3.3.2).
2. Midinfrared emission features typical of silicates in some comets.	2. Silicate grains of size $r < 1$ μm or porous aggregates of submicrometer silicate grains (*Hanner and Bradley*, 2004).
3. Thermal emission does not decrease more steeply than a blackbody at $\lambda > 20$ μm	3. Evidence of a broad size distribution that includes larger than micrometer-sized particles.

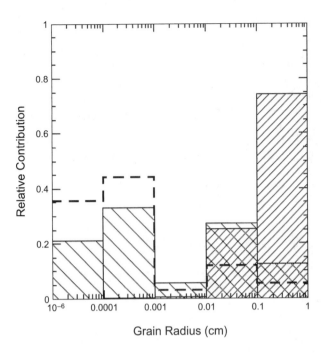

Fig. 9. Relative cross section (\\\), mass (///), and thermal emission (– – –) vs. particle radius for the size distribution measured by *Giotto*, as adapted to fit the SED of Comet 81P/Wild 2 (*Hanner and Hayward*, 2003). For power-law size distributions with a ≥ 3, submicrometer particles dominate in the light scattering.

dependence of the feature's position and shape on the shape of dust particles, and thus mid-infrared spectral data can be used to further constrain the particle shape.

4.2. Size Distribution

As was shown in Table 2, observations in the visible and their interpretation based on light-scattering methods equally support two models of cometary dust: (1) rather large or polydisperse aggregates of submicrometer particles, and (2) irregular particles with a power-law size distribution dominated by submicrometer particles. The study of the dust thermal emission allows us to constrain characteristics of the size distribution. This can be done by fitting the observed spectral energy distribution (SED) with a calculated one using energy balance equation (2). The increase in color temperature indicates the contribution of submicrometer grains, while the slow decrease in flux at longer wavelengths indicates larger particles. *Grün et al.* (2001) fit the broad thermal flux from 7 to 160 μm from Comet Hale-Bopp with the size distribution of form n(r) ~ r^{-a} with a ≤ 3.5. Observations in the submillimeter spectral range indicate the presence of millimeter-sized particles. Results for other studies confirm that the dust emitted from comets spans a broad size range, from submicrometer to millimeter or centimeter and larger. This is known from analysis of the trajectories of dust particles in comet tails (*Sekanina*, 1996) and measurements of the mass of particles impacting

the *Giotto* spacecraft during the flyby of Comet P/Halley (*McDonnell et al.*, 1991). If the measured *Giotto* mass distribution is typical of comets, then most of the particulate mass shed from the nucleus is concentrated in large particles, whereas the cross section is broadly distributed or concentrated in the small particles (Fig. 9), thus making submicrometer- and micrometer-sized particles dominant in the visual and thermal infrared. This most likely is true for all comets, although the detailed size distribution may vary from comet to comet as well as within a coma. For example, size distribution is different for jets and regular coma with predominance of smaller particles in the jets. The size distribution varies with the distance from the nucleus as a result of both dynamical sorting of particles by radiation pressure and sublimation of volatiles.

4.3. Composition

Cometary dust consists of heterogeneous particles that include a dark material, which determines the low albedo, as well as silicates responsible for the spectral features in the mid-infrared. A distributed source of CO near the comet nucleus, trends in color, polarization, and albedo support the notion of a slowly sublimating organic material (*Greenberg and Li*, 1998; *Kolokolova et al.*, 2001a). Thus, not only the size distribution but also the composition of cometary grains changes as they move out from the nucleus. A significant specific of the cometary dust is its red color, which does not vary with the phase angle. This cannot be easily simulated even for polydisperse nonspherical particles. It may be the result not of a specific size or structure of particles but of a spectral change in the refractive index. Compositional analysis of dust particles during the Halley flybys as well as analysis of IDPs and infrared spectra indicate that the dust grains are composed largely of silicates and carbonaceous (CHON) material (*Hanner and Bradley*, 2004) mixed together on a very fine scale. The composition of the cometary organics may be inferred from its sublimation constants gauged by studying the trends in colors, polarization, and thermal-emission characteristics with the cometocentric distance.

4.4. Structure

As mentioned above, polydisperse irregular particles can reproduce most of the light-scattering properties of cometary dust. The same can be achieved with a model of aggregated particles. Porous aggregates of submicrometer particles are better candidates to explain the data on thermal emission. An aggregated structure of cometary particles is also supported by other studies. Thus, evolutionary ideas about growing cometesimals in protoplanetary disks support the idea of an aggregated nature of cometary refractories (see, e.g., *Greenberg*, 1982). The most developed model by *Greenberg and Li* (1999) describes cometary grains as fluffy (porous) aggregates of particles that are 0.1-μm silicate cores coated by an organic refractory mantle and outer mantle of predominantly water ice, which contains embedded carbonaceous and polycyclic aromatic hydrocarbon (PAH)-type particles with sizes in the 1–10-nm range. When limited by relative cosmic abundances, the water is constrained to be close to 30% by mass and the refractory to volatile ratio is close to 1 : 1.

Captured IDPs of probable cometary origin can give us insight into the morphology of cometary dust. These IDPs consist primarily of submicrometer-sized compact, nonspherical grains and angular micrometer-sized crystals clumped into aggregates having a range in porosity.

4.5. Future Work

Further progress in remotely determining physical properties of dust from their light-scattering and emission requires new observations and new theoretical and laboratory simulations.

The most useful observations are likely to be multiwavelength monitoring of comets within a broad range of heliocentric distances and phase angles. The observations can provide new insight into the classification of comets according to their maximum polarization; heliocentric tendencies in colors (including near-infrared); correlated or anti-correlated change in such characteristics as color, polarization, and albedo; and thermal emission characteristics (temperature, shape, and strength of a silicate feature) with heliocentric distance and within the coma. The correlation between light-scattering and emission properties, e.g., between the value of maximum polarization and the strength of the silicate feature, might be indicative of the diversity of dust from one comet to another. In the framework of light-scattering and emission methods, more information about the comet dust size distribution might be obtained from detailed studies of forward-scattering characteristics of the scattered light (as with, e.g., the Sun-grazing comets detected by SOHO), heliocentric dependence of the dust temperature, and measurements of the SED at far-infrared and submillimeter wavelengths (or shorter wavelengths at larger heliocentric distances). Spectrophotometric and spectropolarimetric data of improved spectral resolution will be important to better eliminate the influence of gas contamination as well as to get more precise values of spectral gradients of intensity and polarization and their change with wavelength. Good-quality images of comets or high-spatial-resolution scans along the coma are necessary to see the diversity of cometary dust within the coma (e.g., in jets, shells, etc.) and the dust changes with the distance from the nucleus. The latter can result from both dynamic sorting of particles by radiation pressure (i.e., can be used to refine measurements of the dust size distribution) and sublimation of volatiles (i.e., can be used to gauge sublimation constants of cometary organics). Detailed data for the innermost coma (within a few thousand kilometers from the nucleus) can provide properties of the fresh, recently lifted dust and may allow extrapolation to characterize the nucleus material. More data on the spatial variations of the elliptical polar-

ization and the polarization plane are necessary, not only to study any alignment of the dust particles within the coma, but also to obtain information about the shape of the particles and possible optical anisotropy and optical activity of the cometary dust material.

Theoretical and laboratory surveys are necessary to chart the differences in observable characteristics for aggregates and irregular polydisperse particles. The surveys should be directed to simulate not individual observational facts but the whole scope of the observational data, i.e., angular and spectral characteristics of both scattered and emitted radiation. Contribution from submicrometer- and micrometer-sized particles is required by the realistic size distributions, and thus light scattering by aggregates of hundreds and thousands of constituent particles or large (but with size still below geometric-optics validity) irregular particles should be simulated. The influence of the structure of the aggregates, the morphology and shape of the constituent particles, as well as the morphology and specific shapes of irregular large particles should be studied. Such studies most likely require development of new theoretical techniques or systematic use of controlled laboratory measurements. More comprehensive modeling of cometary dust composition is necessary: The refractive indexes used in calculations should satisfy the confirmed presence of both silicates and absorbing constituents in the dust and must be consistent with the Kramers-Kronig relations (*Bohren and Huffman*, 1983, section 2.3.2) across all wavelengths. More realistic mixtures of silicates and carbonaceous materials or core-mantle particles should be a point of special interest. The average, effective refractive index can be considered when the silicate and dark materials are mixed on a very small scale compared to the wavelength, e.g., when an admixtures of FeNi or FeS typical for IDPs (*Hanner and Bradley*, 2004) in submicrometer silicate grains are simulated. The need for studies of spectral characteristics of the light scattered by particles whose material has a realistic wavelength-dependent refractive index is apparent. This also refers to laboratory light-scattering simulations that should address spectral dependencies, and we therefore need to concentrate on particles with realistic and controlled refractive indexes across the studied spectral range. Simulations should also provide an explanation of stability or diversity of the observational characteristics, as well as correlations between them, and should be consistent with the data obtained from other studies of cometary dust, e.g., dynamical, evolutional, and cosmic abundance studies, studies of IDPs, and *in situ* measurements.

REFERENCES

A'Hearn M. F., Millis R. L., Schleicher D. G., Osip D. J., and Birch P. V. (1995) The ensemble properties of comets: Results from narrowband photometry of 85 comets, 1976–1992. *Icarus, 118,* 223–270.

Asano S. and Yamamoto G. (1975) Light scattering by a spheroidal particle., *Appl. Opt., 14,* 29–49.

Bastien P., Ménard F., and Nadeau R. (1986) Linear polarization observations of P/Halley. *Mon. Not. R. Astron. Soc., 223,* 827–834.

Beskrovnaja N. G., Silantev N. A., Kiselev N. N., and Chernova G. P. (1987) Linear and circular polarization of Comet Halley light. In *Proceedings of the International Symposium on the Diversity and Similarity of Comets,* pp. 681–685. ESA SP-278, Noordwijk, The Netherlands.

Blum J., Wurm G., Kempf S., Poppe T., Klahr H., Kozasa T., Rott M., Henning T., Dorschner J., Schräpler R., Keller H. U., Markiewicz W. J., Mann I., Gustafson B., Giovane F., Neuhaus D., Fechtig H., Grün E., Feuerbacher B., Kochan H., Ratke L., El Goresy A., Morfill G., Weidenschilling S. J., Schwehm G., Metzler K., and Ip W.-H., (2000) Growth and form of planetary seedlings: Results from a microgravity aggregation experiment. *Phys. Rev. Lett., 85,* 2426–2429.

Bohren C. and Huffman D. (1983) *Absorption and Scattering of Light by Small Particles.* Wiley & Sons, New York.

Bonev T., Jockers K., Petrova E., Delva M., Borisov G., and Ivanova A. (2002) The dust in Comet C/1999 S4 (LINEAR) during its disintegration. *Icarus, 160,* 419–436.

Brooke T. Y., Knacke R. F., and Joyce R. R. (1987) The near infrared polarization and color of comet P/Halley. *Astron. Astrophys., 187,* 621–624.

Bruggeman D. A. G. (1935) Berechnung verschiedener physikalischer Konstanten von heterogenen Substanzen. I. Dielektrizitatskonstanten und Leitfahigkeiten der Mischkorper aus isotropen Substanzen. *Annal. Physik, 24,* 636–664.

Bruning J. H. and Lo Y. T. (1971) Multiple scattering of EM waves by spheres. *IEEE Trans. Antennas Propag., 19,* 378–400.

Chernova G., Kiselev N., and Jockers K. (1993) Polarimetric characteristic of dust particles as observed in 13 comets: Comparison with asteroids. *Icarus, 103,* 144–158.

Chylek P., Videen G., Geldart D., Dobbie J., and Tso H. (2000) Effective medium approximations for heterogeneous particles. In *Light Scattering by Nonspherical Particles: Theory, Measurements, and Applications* (M. I. Mishchenko et al., eds.), pp. 274–308. Academic, New York.

Ciric R. and Cooray F. R. (2000) Separation of variables for electromagnetic scattering by spheroidal particles. In *Light Scattering by Nonspherical Particles: Theory, Measurements, and Applications* (M. I. Mishchenko et al., eds.), pp. 90–130. Academic, New York.

Crovisier J., Leech K., Bockelée-Morvan D., Brooke T. Y., Hanner M. S., Altieri B., Keller H. U., and Lellouch E. (1997) The spectrum of Comet Hale-Bopp (C/1995 O1) observed with the Infrared Space Observatory at 2.9 AU from the Sun. *Science, 275,* 1904–1907.

Davidsson B. J. and Skorov Y. V. (2002) On the light-absorbing surface layer of cometary nuclei. *Icarus, 156,* 223–248.

Dollfus A. and Suchail J.-L. (1987) Polarimetry of grains in the coma of P/Halley. I — Observations. *Astron. Astrophys., 187,* 669–688.

Dollfus A., Bastien P., Le Borgne J.-F., Levasseur-Regourd A.-Ch., and Mukai T. (1988) Optical polarimetry of P/Halley — Synthesis of the measurements in the continuum. *Astron. Astrophys., 206,* 348–356.

Dorschner J., Begemann B., Henning T., Jaeger C., and Mutschke H. (1995) Steps toward interstellar silicate mineralogy. II. Study of Mg-Fe-silicate glasses of variable composition. *Astron. Astrophys., 300,* 503–520.

Draine B. T. and Flatau P. J. (1994) Discrete-dipole approximation of scattering calculations. *J. Opt. Soc. Am. A, Opt. Image*

Sci., 11, 1491–1499.

Eaton N., Scarrot S., and Warren-Smith R. F. (1988) Polarization images of the inner regions of Comet Halley. *Icarus, 76,* 270–278.

Eaton N., Scarrott S., and Gledhill T. (1992) Polarization studies of Comet Austin. *Mon. Not. R. Astron. Soc., 258,* 384–386.

Eaton N., Scarrott S., and Draper P. (1995) Polarization studies of comet P/Swift-Tuttle. *Mon. Not. R. Astron. Soc., 73,* L59–L62.

Farafonov V. G., Voshchinnikov N. V., and Somsikov V. V. (1996) Light scattering by a core-mantle spheroidal particle. *Appl. Opt., 35,* 5412–5426.

Farnham T., Schleicher D., and A'Hearn M. F. (2000) The HB narrowband comet filters: Standard stars and calibrations. *Icarus, 147,* 180–204.

Fernández Y. R. (2002) The nucleus of Comet Hale-Bopp (C/1995 O1): Size and activity. *Earth Moon Planets, 89,* 3–25.

Fuller K. and Mackowski D. (2000) Electromagnetic scattering by compounded spherical particles. In *Light Scattering by Nonspherical Particles: Theory, Measurements, and Applications* (M. I. Mishchenko et al., eds.), pp. 226–273. Academic, New York.

Furusho R., Suzuki B., Yamamoto N., Kawakita H., Sasaki T., Shimizu Y., and Kurakami T. (1999) Imaging polarimetry and color of the inner coma of Comet Hale-Bopp (C/1995 O1). *Publ. Astron. Soc. Japan, 51,* 367–373.

Ganesh S., Joshi U. C., Baliyan K. S., and Deshpande M. (1998) Polarimetric observations of the comet Hale-Bopp. *Astron. Astrophys. Suppl., 129,* 489–493.

Garnett Maxwell J. C. (1904) Colours in metal glasses and in metallic films. *Philos. Trans. R. Soc., A203,* 385–420.

Gehrz R. D. and Ney E. P. (1992) 0.7 to 23 μm photometric observations of P/Halley 1986 III and six recent bright comets. *Icarus, 100,* 162–186.

Giese R. H., Weiss K., Zerull R. H., and Ono T. (1978) Large fluffy particles: A possible explanation of the optical properties of interplanetary dust. *Astron. Astrophys., 65,* 265–272.

Greenberg J. M. (1982) What are comets made of — A model based on interstellar dust. In *Comets* (L. Wilkening, ed.), pp. 131–163. Univ. of Arizona, Tucson.

Greenberg J. M. and Gustafson B. Å. S. (1981) A comet fragment model of zodiacal light particles. *Astron. Astrophys., 93,* 35–42.

Greenberg J. M. and Hage J. I. (1990) From interstellar dust to comets — A unification of observational constraints. *Astrophys. J., 361,* 260–274.

Greenberg J. M. and Li A. (1998) From interstellar dust to comets: The extended CO source in comet Halley. *Astron. Astrophys., 332,* 374–384.

Greenberg J. M. and Li A. (1999) Morphological structure and chemical composition of cometary nuclei and dust. *Space. Sci. Rev., 90,* 149–161.

Greenberg J. M., Pedersen N. E., and Pedersen J. C. (1961) Microwave analog to the scattering of light by nonspherical particles. *J. Appl. Phys., 32(2),* 233–242.

Grün E., Hanner M. S., Peschke S. B., Müller T., Boehnhardt H., Brooke T. Y., Campins H., Crovisier J., Delahodde C., Heinrichsen I., Keller H. U., Knacke R. F., Krüger H., Lamy P., Leinert Ch., Lemke D., Lisse C. M., Müller M., Osip D. J., Solc M., Stickell M., Sykes M., Vanysek V., and Zarnecki J. (2001) Broadband infrared photometry of comet Hale-Bopp with ISOPHOT. *Astron. Astrophys., 377,* 1098–1118.

Grynko Ye. and Jockers K. (2003) Cometary dust at large phase angles. In *Abstracts for the NATO ASI on Photopolarimetry in Remote Sensing*, p. 39. Army Research Laboratory, Adelphi, Maryland.

Gustafson B. Å. S. (1994) Physics of zodiacal dust. *Annu. Rev. Earth Planet. Sci., 22*, 553–595.

Gustafson B. Å. S. (1996) Microwave analog to light scattering measurements: A modern implementation of a proven method to achieve precise control. *J. Quant. Spectrosc. Radiat. Transfer, 55*, 663–672.

Gustafson B. Å. S. (2000) Microwave analog to light scattering measurements. In *Light Scattering by Nonspherical Particles: Theory, Measurements, and Applications* (M. I. Mishchenko et al., eds.), pp. 367–389. Academic, New York.

Gustafson B. Å. S. and Kolokolova L. (1999) A systematic study of light scattering by aggregate particles using the microwave analog technique: Angular and wavelength dependence of intensity and polarization. *J. Geophys. Res., 104*, 31711–31720.

Gustafson B. Å. S., Greenberg J. M., Kolokolova L., Xu Y., and Stognienko R. (2001) Interactions with electromagnetic radiation: Theory and laboratory simulations. In *Interplanetary Dust* (E. Grün et al., eds.), pp. 509–569. Springer-Verlag, Berlin.

Hadamcik E. and Levasseur-Regourd A.-Ch. (2003) Imaging polarimetry of cometary dust: Different comets and phase angles. *J. Quant. Spectrosc. Radiat. Transfer, 79–80*, 661–678.

Hadamcik E., Levasseur-Regourd A.-Ch., and Renard J. B. (1999) CCD polarimetric imaging of Comet Hale-Bopp (C/1995 O1). *Earth Moon Planets, 78*, 365–371.

Hadamcik E., Renard J.-B., Worms J.-C., Levasseur-Regourd A.-Ch., and Masson M. (2002) Polarization of light scattered by fluffy particles (PROGRA² experiment). *Icarus, 155*, 497–508.

Hadamcik E., Renard J.-B., Levasseur-Regourd A.-Ch., and Worms J.-C. (2003) Laboratory light scattering measurements on "natural" particles with the PROGRA² experiment: An overview. *J. Quant. Spectrosc. Radiat. Transfer, 79–80*, 679–693.

Hammel H., Tedesco C., Campins H., Decher R., Storrs A., and Cruikshank D. (1987) Albedo maps of comets P/Halley and P/Giacobini-Zinner. *Astron. Astrophys., 187*, 665–668.

Hanner M. S. (1983) The nature of cometary dust from remote sensing. In *Cometary Exploration* (T. I. Gombosi, ed.), pp. 1–22. Hungarian Acad. Sci., Budapest.

Hanner M. S. (2002) Comet dust: The view after Hale-Bopp. In *Dust in the Solar System and Other Planetary Systems* (S. F. Green et al., eds.), pp. 239–254. Pergamon, Amsterdam.

Hanner M. S. (2003) The scattering properties of cometary dust. *J. Quant. Spectrosc. Radiat. Transfer, 79–80*, 695–705.

Hanner M. S. and Bradley J. P. (2004) Composition and mineralogy of cometary dust. In *Comets II* (M. C. Festou et al., eds.), this volume. Univ. of Arizona, Tucson.

Hanner M. S. and Hayward T. L. (2003) Infrared observations of Comet 81P/Wild 2 in 1997. *Icarus, 161*, 164–173.

Hanner M. S. and Newburn R. L. (1989) Infrared photometry of comet Wilson (1986) at two epochs. *Astrophys. J., 97*, 254–261.

Hanner M. S., Giese R. H., Weiss K., and Zerull R. (1981) On the definition of albedo and application to irregular particles. *Astron. Astrophys., 104*, 42–46.

Hanner M. S., Tokunaga A. T., Golisch W. F., Griep D. M., and Kaminski C. D. (1987) Infrared emission from P/Halley's dust coma during March 1986. *Astron. Astrophys., 187*, 653–660.

Hanner M. S., Lynch D. K., and Russell R. W. (1994) The 8–13 micron spectra of comets and the composition of silicate grains. *Astrophys. J., 425*, 274–285.

Hanner M. S., Lynch D. K., Russell R. W., Hackwell J. A., Kellogg R., and Blaney D. (1996) Mid-infrared spectra of Comets P/Borrelly, P/Faye, and P/Schaumasse. *Icarus, 124*, 344–351.

Hanner M. S., Gehrz R. D., Harker D. E., Hayward T. L., Lynch D. K., Mason C. C., Russell R. W., Williams D. M., Wooden D. H., and Woodward C. E. (1999) Thermal emission from the dust coma of Comet Hale-Bopp and the composition of the silicate grains. *Earth Moon Planets, 79*, 247–264.

Hansen J. and Travis L. (1974) Light scattering in planetary atmospheres. *Space Sci. Rev., 16*, 527–610.

Haracz R. D., Cohen L. D., and Cohen A. (1984) Perturbation theory for scattering from dielectric spheroids and short cylinders. *Appl. Opt., 23*, 436–441.

Harker D. E., Wooden D. H., Woodward Ch. E., and Lisse C. M. (2002) Grain properties of Comet C/1995 O1 (Hale-Bopp). *Astrophys. J., 580*, 579–597.

Hartmann W. K. and Cruikshank D. (1984) Comet color changes with solar distance. *Icarus, 57*, 55–62.

Hartmann W. K., Cruikshank D. P., and Degewij J. (1982) Remote comets and related bodies — VJHK colorimetry and surface materials. *Icarus, 52*, 377–408.

Hayward T. L., Grasdalen G. L., and Green S. F. (1988) An albedo map of P/Halley on 13 March 1986. In *Infrared Observations of Comets Halley and Wilson and Properties of the Grains*, pp. 151–153. NASA, Washington, DC.

Hayward T. L., Hanner M. S., and Sekanina Z. (2000) Thermal infrared imaging and spectroscopy of Comet Hale-Bopp (C/1995 O1). *Astrophys. J., 538*, 428–455.

Henning Th. and Stognienko R. (1993) Porous grains and polarization of light: The silicate features. *Astron. Astrophys., 280*, 609–616.

Hoban S., A'Hearn M. F., Birch P. V., and Martin R. (1989) Spatial structure in the color of the dust coma of comet P/Halley. *Icarus, 79*, 145–158.

Hovenier J. W. (2000) Measuring scattering matrices of small particles at optical wavelengths. In *Light Scattering by Nonspherical Particles: Theory, Measurements, and Applications* (M. I. Mishchenko et al., eds.), pp. 355–366. Academic, New York.

Hovenier J. W., Volten H., Muñoz O., van der Zande W. J., and Waters L. B. (2003) Laboratory study of scattering materials for randomly oriented particles. *J. Quant. Spectrosc. Radiat. Transfer, 79–80*, 741–755.

Jäger C., Il'in V. B., Henning Th., Mutschke H., Fabian D., Semenov D., and Voshchinnikov N. (2003) A database of optical constants of cosmic dust analogs. *J. Quant. Spectrosc. Radiat. Transfer, 79–80*, 765–774.

Jewitt D. and Luu J. (1990) The submillimeter radio continuum of Comet P/Brorsen-Metcalf. *Astrophys. J., 365*, 738–747.

Jewitt D. and Luu J. (1992) Submillimeter continuum emission from comets. *Icarus, 100*, 187–196.

Jewitt D. and Matthews H. (1997) Submillimeter continuum observations of Comet Hyakutake (1996 B2). *Astron. J., 113*, 1145–1151.

Jewitt D. and Matthews H. (1999) Particulate mass loss from Comet Hale-Bopp. *Astron. J., 117*, 1056–1062.

Jewitt D. and Meech K. (1986) Cometary grain scattering versus wavelength, or "What color is comet dust?" *Astrophys. J., 310*, 937–952.

Jockers K. (1999) Observations of scattered light from cometary dust and their interpretation. *Earth Moon Planets, 79*, 221–245.

Jockers K., Rosenbush V. K., Bonev T., and Credner T. (1999)

Images of polarization and colour in the inner coma of comet Hale-Bopp. *Earth Moon Planets, 78,* 373–379.

Jockers K., Grynko Ye., Schwenn R., and Biesecker D. (2002) Intensity, color and polarization of comet 96P/Machholz 1 at very large phase angles. In *Abstracts for ACM 2002,* July 29–August 2, 2002, Berlin, Germany, Abstract #07-04.

Jones T. and Gehrz R. (2000) Infrared imaging polarimetry of Comet C/1995 O1 (Hale-Bopp). *Icarus, 143,* 338–346.

Kahnert M. (2003) Numerical methods in electromagnetic theory. *J. Quant. Spectrosc. Radiat. Transfer, 79–80,* 775–824.

Kerker M. (1969) *The Scattering of Light and Other Electromagnetic Radiation.* Academic, New York.

Kimura H. (2001) Light-scattering properties of fractal aggregates: Numerical calculations by a superposition technique and discrete dipole approximation. *J. Quant. Spectrosc. Radiat. Transfer, 70,* 581–594.

Kiselev N. N. and Chernova G. P. (1978) Polarization of the radiation of comet West 1975n. *Sov. Astron. J., 55,* 1064–1071.

Kiselev N. N. and Chernova G. P. (1979) Photometry and polarimetry during flares of comet Schwassmann-Wachmann. *Sov. Astron. Lett., 5,* 294–299.

Kiselev N. N. and Velichko F. P. (1999) Aperture polarimetry and photometry of Comet Hale-Bopp. *Earth Moon Planets, 78,* 347–352.

Kiselev N. N., Jockers K., Rosenbush V., Velichko F., Bonev T., and Karpov N. (2000) Anomalous wavelength dependence of polarization of Comet 21P/Giacobini-Zinner. *Planet. Space Sci., 48,* 1005–1009.

Kiselev N. N., Jockers K., Rosenbush V. K., and Korsun P. P. (2001) Analysis of polarimetric, photometric and spectral observations of Comet C/1996Q1 (Tabur). *Solar Sys. Res., 35,* 480–495.

Kiselev N. N., Jockers K., and Rosenbush V. K. (2002) Comparative study of the dust polarimetric properties in split and normal comets. *Earth Moon Planets, 90,* 167–176.

Kolokolova L. and Gustafson B. Å. S. (2001) Scattering by inhomogeneous particles: Microwave analog experiment comparison to effective medium theories. *J. Quant. Spectrosc. Radiat. Transfer, 70,* 611–625.

Kolokolova L. and Jockers K. (1997) Composition of cometary dust from polarization spectra. *Planet. Space Sci., 45,* 1543–1550.

Kolokolova L., Jockers K., Chernova G., and Kiselev N. (1997) Properties of cometary dust from the color and polarization. *Icarus, 126,* 351–361.

Kolokolova L., Jockers K., Gustafson B. Å. S., and Lichtenberg G. (2001a) Color and polarization as indicators of comet dust properties and evolution in the near-nucleus coma. *J. Geophys. Res., 106,* 10113–10128.

Kolokolova L., Lara L. M., Schulz R., Stüwe J., and Tozzi G. P. (2001b) Properties and evolution of dust in Comet Tabur (C/1996 Q1) from the color maps. *Icarus, 153,* 197–207.

Kozasa T., Blum J., Okamoto H., and Mukai T. (1993) Optical properties of dust aggregates. II. Angular dependence of scattered light. *Astron. Astrophys., 276,* 278–288.

Laffont C., Rousselot P., Clairemidi J., Moreels G., and Boice D. C. (1999) Jets and arcs in the coma of comet Hale-Bopp from August 1996 to April 1997. *Earth Moon Planets, 78,* 211–217.

Le Borgne J. F. and Crovisier J. (1987) Polarization of molecular fluorescence bands in comets: Recent observations and interpretation. In *Proceedings of the International Symposium on the Diversity and Similarity of Comets,* pp. 171–175. ESA SP-278, Noordwijk, The Netherlands.

Le Borgne J. F., Leroy J. L., and Arnaud J. (1987) Polarimetry of visible and near UV molecular bands: Comets P/Halley and Hartley-Good. *Astron. Astrophys., 173,* 180–182.

Levasseur-Regourd A.-Ch. and Hadamcik E. (2003) Light scattering by irregular dust particles in the solar system: Observations and interpretation by laboratory measurements. *J. Quant. Spectrosc. Radiat. Transfer, 79–80,* 903–910.

Levasseur-Regourd A.-Ch., Hadamcik E., and Renard J. B. (1996) Evidence of two classes of comets from their polarimetric properties at large phase angles. *Astron. Astrophys., 313,* 327–333.

Levasseur-Regourd A.-Ch., McBride N., Hadamcik E., and Fulle M. (1999) Similarities between in situ measurements of local dust light scattering and dust flux impact data within the coma of 1P/Halley. *Astron. Astrophys., 348,* 636–641.

Levasseur-Regourd A. C., Haudebourg V., Cabane M., and Worms J. C. (2001) Light scattering measurements on dust aggregates, from MASER 8 to ISS. In *Proceedings of the 1st International Symposium on Microgravity Research and Applications in Physical Science and Biotechnology,* pp. 797–802. ESA SP-545, Noordwijk, The Netherlands.

Li A. and Greenberg J. M. (1998a) The dust properties of a short period comet: Comet P/Borrelly. *Astron. Astrophys., 338,* 364–370.

Li A. and Greenberg J. M. (1998b) From interstellar dust to comets: Infrared emission from comet Hale-Bopp (C/1995 O1). *Astrophys. J. Lett., 498,* L83–L87.

Lisse C. M., Freudenreich H. T., Hauser M. G., Kelsall T., Moseley S. H., Reach W. T., and Silverberg R. F. (1994) Infrared observations of Comet Austin (1990V) by the COBE/ Diffuse Infrared Background Experiment. *Astrophys. J. Lett., 432,* L71–L74.

Lisse C. M., A'Hearn M. F., Hauser M. G., Kelsall T., Lien D. J., Moseley S. H., Reach W. T., and Silverberg R. F. (1998) Infrared observations of Comets by COBE. *Astrophys. J., 496,* 971–991.

Lumme K. and Muinonen K. (1993) A two-parameter system for linear polarization of some solar system objects. In *Asteroids, Comets, Meteors (IAU Symposium 160),* p. 194. LPI Contribution No. 810, Lunar and Planetary Institute, Houston.

Lumme K. and Rahola J. (1994) Light scattering by porous dust particles in the discrete-dipole approximation. *Astrophys. J., 425,* 653–667.

Lumme K. and Rahola J. (1998) Comparison of light scattering by stochastic rough spheres, best fit spheroids and spheres. *J. Quant. Spectrosc. Radiat. Transfer, 60,* 439–450.

Lumme K., Rahola J., and Hovenier J. (1997) Light scattering by dense clusters of spheres. *Icarus, 126,* 455–469.

Macke A. (2000) Monte Carlo calculations of light scattering by large particles with multiple internal inclusions. In *Light Scattering by Nonspherical Particles: Theory, Measurements, and Applications* (M. I. Mishchenko et al., eds.), pp. 309–322. Academic, New York.

Mackowski D. W. and Mishchenko M. I. (1996) Calculation of the T matrix and the scattering matrix for ensembles of spheres. *J. Opt. Soc. Am., A13,* 2266–2278.

Manset N. and Bastien P. (2000) Polarimetric observations of Comets C/1995 O1 Hale-Bopp and C/1996 B2 Hyakutake. *Icarus, 145,* 203–219.

Mason C. G., Gerhz R. D., Jones T. J., Woodward C. E., Hanner M. S., and Williams D. M. (2001) Observations of unusually small dust grains in the coma of Comet Hale-Bopp C/1995 O1. *Astrophys. J., 549*, 635–646.

Mason C. G., Gerhz R. D., Ney E. P., Williams D. M., and Woodward C. E. (1998) The temporal development of the pre-perihelion infrared spectral energy distribution of comet Hyakutake (C/1996 B2). *Astrophys. J., 507*, 398–403.

McDonnell J. A. M., Lamy P. L., and Pankiewicz G. S. (1991) Physical properties of cometary dust. In *Comets in the Post-Halley Era, Vol. 1* (R. L. Newburn Jr. et al., eds.), pp. 1043–1074. Kluwer, Dordrecht.

Meech K. and Jewitt D. (1987) Observations of comet P/Halley at minimum phase angle. *Astron. Astrophys., 187*, 585–593.

Metz K. and Haefner R. (1987) Circular polarization near the nucleus of Comet P/Halley. *Astron. Astrophys. 187*, 539–542.

Mie G. (1908) Beiträge zur Optik trüber Medien, speziell kolloidaler Metallösungen. *Ann. Phys., 25*, 377–445.

Millis R. L., A'Hearn M. F., and Thompson D. T. (1982) Narrowband photometry of Comet P/Stephan-Oterma and the backscattering properties of cometary grains. *Astron. J., 87*, 1310–1317.

Min M., Hovenier J. W., and de Koter A. (2003) Scattering and absorption cross sections for randomly oriented spheroids of arbitrary size. *J. Quant. Spectrosc. Radiat. Transfer, 79–80*, 939–951.

Mishchenko M. (1991) Extinction and polarization of transmitted light by partially aligned nonspherical grains. *Astrophys. J., 367*, 561–574.

Mishchenko M. I. (1993) Light scattering by size/shape distributions of randomly oriented axially symmetric particles of a size comparable to a wavelength. *Appl. Opt., 32*, 4652–4666.

Mishchenko M. I. and Travis L. (1994) Light scattering by polydisperse, rotationally symmetric nonspherical particles: Linear polarization. *J. Quant. Spectrosc. Radiat. Transfer, 51*, 759–778.

Mishchenko M., Travis L. D., and Mackowski D. W. (1996) T-matrix computations of light scattering by nonspherical particles: A review. *J. Quant. Spectrosc. Radiat. Transf., 55*, 535–575.

Mishchenko M. I., Travis L. D., Kahn R. A., and West R. A. (1997) Modeling phase functions for dustlike tropospheric aerosols using a shape mixture of randomly oriented polydisperse spheroids. *J. Geophys. Res., 102*, 16831–16847.

Mishchenko M., Wiscombe W., Hovenier J., and Travis L. (2000) Overview of scattering by non-spherical particles. In *Light Scattering by Nonspherical Particles: Theory, Measurements, and Applications* (M. I. Mishchenko et al., eds.), pp. 30–61. Academic, New York.

Mishchenko M., Travis L., and Lacis A. (2002) *Scattering, Absorption and Emission of Light by Small Particles.* Cambridge Univ., Cambridge. 445 pp.

Morris C. S. and Hanner M. S. (1993) The infrared light curve of periodic comet Halley 1986 III and its relationship to the visual light curve, C2 and water production rates. *Astron. J., 105*, 1537–1546.

Muinonen K. (1996) Light scattering by Gaussian random particles: Rayleigh and Rayleigh-Gans approximations. *J. Quant. Spectrosc. Radiat. Transfer, 55*, 603–613.

Muinonen K. (2000) Light scattering by stochastically shaped particles. In *Light Scattering by Nonspherical Particles: Theory, Measurements, and Applications* (M. I. Mishchenko et al., eds.), pp. 323–355. Academic, New York.

Mukai T. (1977) Dust grains in the cometary coma — Interpretation of the infrared continuum. *Astron. Astrophys., 61*, 69–74.

Mukai T., Mukai S., and Kikuchi S. (1987) Complex refractive index of grain material deduced from the visible polarimetry of comet P/Halley. *Astron. Astrophys., 187*, 650–652.

Mukai S., Mukai T., and Kikuchi S. (1991) Scattering properties of cometary dust based on polarimetric data. In *Origin and Evolution of Interplanetary Dust (IAU Colloquium 126)* (A.-C. Levasseur-Regourd and H. Hasegawa, eds.), pp. 249–252. Kluwer, Dordrecht.

Mukai T., Ishimoto H., Kozasa T., Blum J., and Greenberg J. M. (1992) Radiation pressure forces of fluffy porous grains. *Astron. Astrophys., 262*, 315–320.

Mukai T., Iwata T., Kikuchi S., Hirata R., Matsumura M., Nakamura Y., Narusawa S., Okazaki A., Seki M., and Hayashi K. (1997) Polarimetric observations of 4179 Toutatis in 1992/1993. *Icarus, 127*, 452–460.

Muñoz O., Volten H., de Haan J. F., Vassen W., and Hovenier J. (2000) Experimental determination of scattering matrices of olivine and Allende meteorite particles. *Astron. Astrophys., 360*, 777–788.

Myers R. and Nordsieck K. (1984) Spectropolarimetry of comets Austin and Churyumov-Gerasimenko. *Icarus, 58*, 431–439.

Ney E. P. (1982) Optical and infrared observations of bright comets in the range 0.5 micrometers to 20 micrometers. In *Comets* (L. Wilkening, ed.), pp. 323–340. Univ. of Arizona, Tucson.

Ney E. P. and Merrill K. M. (1976) Comet West and the scattering function of cometary dust. *Science, 194*, 1051–1053.

Oguchi T. (1960) Attenuation of electromagnetic waves due to rain with distorted raindrops. *J. Radio Res. Lab. Jpn, 7*, 467–485.

Petrova E. V. and Jockers K. (2002) On the spectral dependence of intensity and polarization of light scattered by aggregates particles. In *Electromagnetic and Light Scattering by Nonspherical Particles* (B. Gustafson et al., eds.), pp. 263–266. Army Research Laboratory, Adelphi, Maryland.

Petrova E. V., Jockers K., and Kiselev N. (2000) Light scattering by aggregates with sizes comparable to the wavelength: An application to cometary dust. *Icarus, 148*, 526–536.

Purcell E. M. and Pennypacker C. R. (1973) Scattering and absorption of light by non-spherical dielectric grains. *Astrophys. J., 186*, 705–714.

Rayleigh, Lord (1897) On the incidence of aerial and electric waves upon small obstacles in the form of ellipsoids and elliptic cylinders, and on the passage of electric waves through a circular aperture in a conducting screen. *Phil. Mag., 44*, 28–52.

Renard J. B., Levasseur-Regourd A.-C., and Dollfus A. (1992) Polarimetric CCD imaging of Comet Levy (1990c). *Ann. Geophys., 10*, 288–292.

Renard J.-B., Hadamcik E., and Levasseur-Regourd A.-C. (1996) Polarimetric CCD imaging of comet 47P/Ashbrook-Jackson and variability of polarization in the inner coma of comets. *Astron. Astrophys., 316*, 263–269.

Rosenbush V. K. and Shakhovskoj N. M. (2002) On the circular polarization and grain alignment in the cometary atmospheres: Observations and analysis. In *Electromagnetic and Light Scattering by Nonspherical Particles* (B. Gustafson et al., eds.), pp. 283–286. Army Research Laboratory, Adelphi, Maryland.

Rosenbush V. K., Rosenbush A. E., and Dement'ev M. S. (1994) Comets Okazaki-Levy Rundenko (1989) and Levy (1990 20): Polarimetry and stellar occultation. *Icarus, 108*, 81–91.

Rosenbush V. K., Shakhovskoj N. M., and Rosenbush A. E. (1999) Polarimetry of Comet Hale-Bopp: Linear and circular polarization, stellar occultation. *Earth Moon Planets, 78*, 381–386.

Rosenbush V. K., Kiselev N. N., and Velichko S. F. (2002) Polarimetric and photometric observations of split comet C/2001 A2 (LINEAR). *Earth Moon Planets, 90,* 423–433.

Schleicher D. and Osip D. (2002) Long and short-term photometric behavior of comet Hyakutake (1996 B2). *Icarus, 159,* 210–233.

Schleicher D., Millis R., and Birch P. (1998) Narrowband photometry of comet P/Halley: Variations with heliocentric distance, season, and solar phase angle. *Icarus, 132,* 397–417.

Schuerman D., Wang R. T., Gustafson B. Å. S., and Schaefer R. W. (1981) Systematic studies of light scattering. 1: Particle shape. *Appl. Opt., 20/23,* 4039–4050.

Sekanina Z. (1996) Morphology of cometary dust coma and tail. In *Physics, Chemistry, and Dynamics of Interplanetary Dust* (B. Å. S. Gustafson and M. S. Hanner, eds.), pp. 377–382. Astronomical Society of the Pacific, San Francisco.

Sen A. K., Joshi U. C., Deshpande M. R., and Prasad C. D. (1990) Imaging polarimetry of Comet P/Halley. *Icarus, 86,* 248–256.

Sihvola A. (1999) *Electromagnetic Mixing Formulas and Applications.* IEE Electromagnetic Waves, Series 47, Institution of Electrical Engineers. 284 pp.

Tokunaga A. T., Golisch W. F., Griep D. M., Kaminski C. D., and Hanner M. S. (1986) The NASA infrared telescope facility comet Halley monitoring program. II. Preperihelion results. *Astron. J., 92,* 1183–1190.

Tokunaga A., Golisch W., Griep D., Kaminski C., and Hanner M. (1988) The NASA infrared telescope facility comet Halley monitoring program. II. Postperihelion results. *Astron. J., 96,* 1971–1976.

Toon O. and Ackerman T. P. (1981) Algorithms for the calculation of scattering by stratified spheres. *Appl. Opt., 20,* 3657–3660.

Tozzi G. P., Cimatti A., di Serego A., and Cellino A. (1997) Imaging polarimetry of comet C/1996 B2 (Hyakutake) at the perigee. *Planet. Space Sci., 45,* 535–540.

van de Hulst H. C. (1957) *Light Scattering by Small Particles.* Wiley & Sons, New York. 470 pp.

Voshchinnikov N. V. and Farafonov V. G. (1993) Optical properties of spheroidal particles. *Astrophys. Space Sci., 204,* 19–86.

Waldemarsson K. W. T. and Gustafson B. Å. S. (2003) Interference effects in the light scattering by transparent plates. *J. Quant. Spectrosc. Radiat. Transfer, 79–80,* 1111–1119.

Wallis M. K., Rabilizirov R., and Wickramasinghe N. C. (1987) Evaporating grains in P/Halley's coma. *Astron. Astrophys., 187,* 801–806.

Waterman P. C. (1965) Matrix formulation of electromagnetic scattering. *Proc. IEEE, 53,* 805–812.

Weiss-Wrana K. (1983) Optical properties of interplanetary dust — Comparison with light scattering by larger meteoritic and terrestrial grains. *Astron. Astrophys., 126,* 240–250.

West R. (1991) Optical properties of aggregate particles whose outer diameter is comparable to the wavelength. *Appl. Opt., 30,* 5316–5324.

Wielaard D. J., Mishchenko M. I., Macke A., and Carlson B. E. (1997) Improved T-matrix computations for large, nonabsorbing and weakly absorbing nonspherical particles and comparison with geometrical-optics approximation. *Appl. Opt., 36,* 4305–4313.

Williams D. M., Mason C. G., Gehrz R. D., Jones T. J., Woodward C. E., Harker D. E., Hanner M. S., Wooden D. H., Witteborn F. C., and Butner H. M. (1997) Measurement of submicron grains in the coma of Comet Hale-Bopp C/1995 01 during 1997 February 15–20 UT1997. *Astrophys. J. Lett., 489,* L91–L94.

Wolff M., Clayton G., and Gibson S. (1998) Modeling composite and fluffy grains. II Porosity and phase functions. *Astrophys. J., 503,* 815–830.

Wooden D. H., Harker D. E., Woodward C. E., Butner H. M., Koike C., Witteborn F., and McMurtry C. (1999) Silicate mineralogy of the dust in the inner coma of Comet C/1995 O1 (Hale-Bopp) pre- and postperihelion. *Astrophys. J., 517,* 1034–1058.

Wooden D., Butner H., Harker D., and Woodward C. (2000) Mg-rich silicate crystals in Comet Hale-Bopp: ISM relics or solar nebula condensates? *Icarus, 143,* 126–137.

Worms J.-C., Renard J.-B., Hadamcik E., Levasseur-Regourd A.-Ch., and Gayet J.-F. (1999) Results of the PROGRA2 experiment: An experimental study in microgravity of scattered polarized light by dust particles with large size parameter. *Icarus, 142,* 281–297.

Worms J. C., Renard J. B., Hadamcik E., Brun-Huret N., and Levasseur-Regourd A. C. (2000) The PROGRA2 light scattering instrument. *Planet Space Sci., 48,* 493–505.

Xing Z. and Hanner M. S. (1997) Light scattering by aggregate particles. *Astron. Astrophys., 324,* 805–820.

Xu Y.-l. (1997) Electromagnetic scattering by an aggregate of spheres: Far field. *Appl. Opt., 36,* 9496–9508.

Xu Y.-l. and R. T. Wang (1998) Electromagnetic scattering by an aggregate of spheres: Theoretical and experimental study of the amplitude scattering matrix. *Phys. Rev. E, 58,* 3931–3948.

Yanamandra-Fisher P. A. and Hanner M. S. (1998) Optical properties of nonspherical particles of size comparable to the wavelength of light: Application to comet dust. *Icarus, 138,* 107–128.

Zerull R. H. and Giese R. H. (1983) The significance of polarization and colour effects for models of cometary grains. In *Cometary Exploration* (T. I. Gombosi, ed.), pp. 143–151. Hungarian Acad. Sci., Budapest.

Zerull R. H., Giese R. H., and Weiss K. (1977) Scattering functions of nonspherical dielectric and absorbing particles vs Mie theory (E). *Appl. Opt., 16,* 777–783.

Zerull R. H., Gustafson B. Å. S., Schultz K., and Thiele-Corbach E. (1993) Scattering by aggregates with and without an absorbing mantle: Microwave analog experiments. *Appl. Opt., 32,* 4088–4100.

Global Solar Wind Interaction and Ionospheric Dynamics

Wing-Huen Ip

National Central University, Taiwan

The spacecraft missions to Comets 21P/Giacobini-Zinner and 1P/Halley in the mid-1980s revolutionized our knowledge of comets and their interaction with the solar wind. Besides the large-scale plasma structures, the spacecraft *in situ* measurements produced many exciting and surprising results about the ion distribution and magnetic field configuration in the inner regions of cometary comae. These earlier scientific explorations laid the foundation for the *Deep Space 1* mission to Comet 19P/Borrelly and groundbased observations of recent bright comets such as Comets 1996 B2 (Hyakutake) and 1995 O1 (Hale-Bopp). In this chapter, we review the morphological structures of cometary ion tails, analytical and numerical simulations of comet-solar wind interactions, plasma boundaries and the ionospheric contact surface identified by spacecraft measurements, and plasma instabilities and waves.

1. INTRODUCTION

Since the first *in situ* measurements of the solar wind interaction and plasma environment of Comet 21P/Giacobini-Zinner by NASA's *International Cometary Explorer* (*ICE*) spacecraft in 1985, comets have been recognized as a perfect laboratory to study plasma physics from microscopic to macroscopic scales. The subsequent encounters with Comet 1P/Halley by a flotilla of space probes in 1986 added a tremendous amount of information about the plasma structures and processes involved in the production and transport of cometary ions. Data analyses and theoretical models of these *in situ* observations enriched our understanding of natural plasma physics (as opposed to laboratory plasma physics, such as thermal fusion control) to an unprecedented level. This point may be appreciated by noting the dramatic surge in recent years of the number of publications in space physics dealing with cometary plasmas. Despite these recent advances, we must not forget the major contributions made to modern plasma physics by several key figures about a half century ago. In a historical context, the study of the comet-solar wind interaction began with Ludwig Biermann's (Fig. 1) statistical study of the pointing direction of cometary ion tails (*Biermann, 1951*). From the determination of an average aberration angle of about 3°, Biermann made the famous deduction that a solar corpuscular radiation (i.e., a continuous flux of charged particles) must exist in the interplanetary space in order to sweep away the cometary ions. The radial velocity of these solar charged particles, namely, the solar wind, was derived to be on the order of a few hundred km s⁻¹. However, an impossibly large electron number density of the solar corona would have to be invoked, if the momentum transfer is to be facilitated by collisional Coulomb interaction between the solar corpuscular radiation and the cometary ions alone. To overcome this major discrepancy, Hannes Alfvén (Fig. 2) suggested the ingenious

hypothesis that the interplanetary space is actually infiltrated with a magnetic field of solar origin (*Alfvén, 1957*). When a comet moves through the interplanetary medium, its ionosphere will sweep up the interplanetary magnetic field lines in the manner illustrated in Fig. 3. With the draping of the interplanetary magnetic field (IMF) into a magnetic tail, the cometary ions will be channeled along the radial direction, hence facilitating efficient momentum transfer between the solar wind plasma and the cometary ions.

Thus, well before the space age, which dawned at the launch of the *Sputnik* satellite in 1958, the study of cometary ion tails already told us about the existence of the solar wind as a continuous stream of charged particles and the omnipresence of magnetic fields in interplanetary space. As described later, modern observations of ion tails have continued to reveal important things about the three-dimensional structures of the solar wind and the heliosphere, which are not easily accessible to space probes.

Fig. 1. Photo of Ludwig Biermann (with glasses) and Reimer Lüst. Courtesy of P. Biermann.

Fig. 2. Hannes Alfvén (right) and Asoka Mendis discussing comet-solar wind interactions at the University of California at San Diego, ca. 1970. Courtesy of A. Mendis.

The organization of this chapter is as follows. The global morphology of cometary ion tails, which is the most conspicuous manifestation of cometary plasma from ground-based observations, is described in section 2. This is followed in section 3 by a summary of the analytical models and numerical simulations of large-scale solar wind interactions with comets. Related physical processes, such as photoionization, electron impact ionization, and charge exchange effects, are also described. In that section, the new results of strong X-ray emission in cometary comae will be highlighted (see also *Lisse et al.*, 2004). In the second half of this chapter, we focus on the knowledge gained from spacecraft observations. In section 4, we describe the major findings of cometary plasma structures by the various space probes to 21P/Giacobini-Zinner (*ICE*), 1P/Halley

(*Susei*, *Vega 1* and 2, and *Giotto*), 26P/Grigg-Skjellerup (*Giotto*), and, most recently, 19P/Borrelly (*Deep Space 1*). Last but not least, we survey the different kinds of plasma waves and plasma instabilities generated in the comet-solar wind interaction process in section 5.

We note that there are already a number of reviews devoted to the subject of cometary plasmas, e.g., *Huebner* (1990) and *Johnstone* (1991). In addition, the two volumes entitled *Comets in the Post-Halley Era* (*Newburn et al.*, 1991), have a large number of review chapters on cometary plasmas. *Cravens and Gombosi* (2003) provide a contemporary review of the subject. Materials covered in these works are used extensively here.

2. MORPHOLOGICAL STRUCTURES

2.1. Composition/Spectra

Bright comets usually display strong ion emission in the anti-Sun direction, which is dominated by CO^+ in the 3800–4800-Å region and H_2O^+ in the 5600–7000-Å region (see comprehensive review by *Wyckoff*, 1982). The CO^+ emission has also been observed in Comets Morehouse (1908 R1) and Humason (1961 R1) at large distances from the Sun (*Lüst*, 1962; *Wurm*, 1968). The H_2O^+ ion itself was first identified much later in C/Kohoutek (1973 E1) (*Wehinger et al.*, 1974). An example of a recent cometary spectral image is shown in Fig. 4.

The optical emissions from neutral gas molecules (e.g., CN and C_2) are nearly symmetrical about the comet's nucleus, while the spatial distribution of the ions (e.g., CO^+ and CO_2^+) are elongated in the antisolar direction. The ion tail structure becomes more and more diffuse at larger and larger distances from the nucleus until its presence cannot be traced by optical emission. In one case, namely that of the great comet of 1843 (C/1843 D1), the ion tail was traced to a distance of 2 AU (*Ip and Axford*, 1982). This record has recently been broken by magnetometer observations on the *Ulysses* spacecraft, which found that the ion tail of Comet Hyakutake (C/1996 B2) extended to a radial distance of more than 3.8 AU (5.5×10^8 km) from the comet center and the corresponding ion tail diameter is as large as 7×10^6 km (*Jones et al.*, 2000).

2.2. Envelope and Ion-Ray Folding

Owing to the weak emissions from the cometary neutrals and dust, the plasma structures traced by the CO^+ ions in the comae and ion tails of C/Morehouse (1908 R1) and C/Humason (1961 R1) could be easily identified. It was found that ion structures in the form of "receding envelopes" could be followed after their formation to a projected distance of $1–1.5 \times 10^5$ km on the sunward side of the optical center (*Lüst*, 1962; *Wurm*, 1968). Upon reaching a distance of about 5×10^4 km, the ion envelopes intermix with the general background and become indistinguishable in the photographic plates. A modern view of this intriguing phenom-

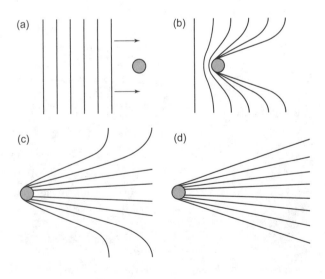

(a)   (b)

(c)   (d)

Fig. 3. Alfvén's magnetic field draping model, describing different stages of the formation of the ion tail.

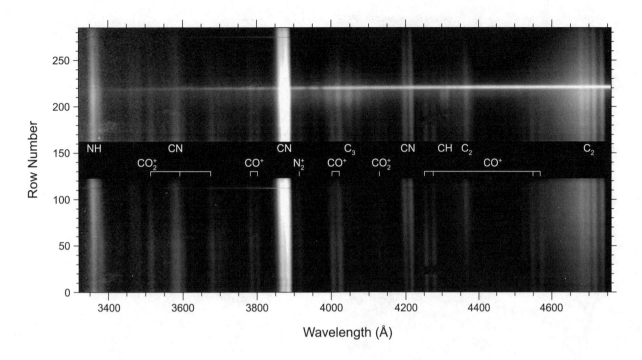

Fig. 4. The top panel is an optical spectrum of Comet 153P/Ikeya-Zhang centered on the nucleus. The bright horizontal streak is the cometary continuum. The more diffuse emissions are from neutral and ionized gas molecules in the coma. The bottom panel was taken during the same observations, but the spectrograph slit was offset tailward from the nucleus. The ionic emissions become stronger relative to the neutral emissions in the tailward spectrum. The stronger neutral and ionic emissions in the spectra are labeled. The feature labeled "N_2^+" is almost certainly terrestrial airglow, rather than cometary emission. Data courtesy of A. Cochran.

enon is illustrated in Fig. 5, in which a system of plasma envelopes in CO^+ emission in the coma of C/Hale-Bopp is clearly shown. The elongation of the end points of the envelopes probably leads to the formation of symmetric pairs of narrow ion rays folding toward the central axis. A good example showing the ion ray system in the plasma tail of C/Kobayashi-Berger-Milon (1975 N1) can be found in Fig. 6. In other cases, the ion tails were comprised of a bundle of narrow ion rays, suggesting that the main ion tails

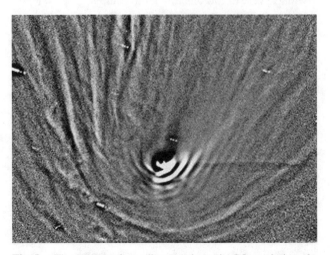

Fig. 5. The system of receding envelopes in CO^+ emission observed in the coma of Comet Hale-Bopp (C/1995 O1). Courtesy of S. Larson.

are formed by the continuous inflow of new plasma in time sequence.

The exact cause of the ion-ray formation is still a matter of debate. In fact, whether projection effect might play a role is still to be clarified (*Jockers*, 1991). That is, if the ion structures in cometary plasma tails are organized by the direction and configuration of the interplanetary magnetic field, it is to be expected that different morphologies might result, depending on whether the line-of-sight is nearly perpendicular to the IMF direction, or instead is parallel to it. This point also brings us back to the original idea proposed by *Ness and Donn* (1966), in which the ion rays are plasma sheets sandwiched between two regions of opposite magnetic polarities. According to this view, the ion rays would therefore most likely form when a comet crosses a sector boundary of the IMF [see *Niedner and Brandt* (1978) and discussion below]. A number of numerical simulations have been performed to investigate whether a 90° or 180° turn of IMF could result in dense plasma structures (*Schmidt and Wegmann*, 1982; *Schmidt-Voigt*, 1989). The advantage of a 90° turn in IMF is that such events are far more frequent than the sector crossing, which requires a 180° turn.

An alternative explanation, somewhat different from the traditional magnetohydrodynamical (MHD) view, is that the ion rays are of thermodynamical origin intrinsic to the cometary ionospheres. The main idea is that the ion content in the cometary ionospheres, and hence the ion tails, is partially controlled by the solar wind electron heat flux. Note

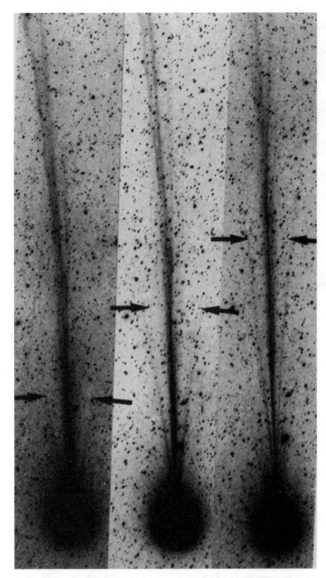

Fig. 6. The time sequence of the pairs of symmetric narrow ion rays folding on to the central axis of Comet Kobayashi-Berger-Milon (C/1975 N1). Time tags from left to right: 03:37, 05:06, and 06:12. Images taken at the Joint Observatory for Cometary Research. Courtesy of K. Jockers.

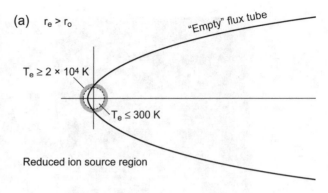

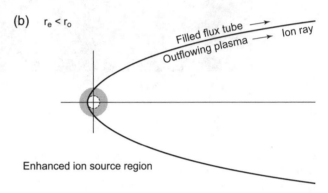

Fig. 7. A schematic view of the formation of cometary ion rays by means of modulation of the size of the ionospheric photochemical regime with electron temperature $T_e > 300$ K. **(a)** The radius of the electron temperature transition region r_e is greater than the apex r_o of the magnetic flux tube; **(b)** $r_e < r_o$, and the magnetic flux tube is filled with outflowing ionospheric plasma leading to the formation of narrow ion rays. From *Ip* (1994).

that the original models of *Ness and Donn* (1966), and the subsequent extensions by later authors (see *Schmidt and Wegmann,* 1982), invoke the time variation of the orientation of IMF as the source mechanism. In other words, the formation of the tail rays would necessarily be initiated at the outer coma where the IMF irregularities are being mapped onto the cometary ionosphere. Yet another possibility is that the ion rays are of internal origin. That is, the generation of the ion rays is controlled by the time variations of the ion production rate of the inner coma. The production of cometary ions and the subsequent channeling of the cometary ions into the ion tail are determined by a balance between photoionization and electron dissociative recombination loss. As discussed in *Cravens and Korosmezey* (1986) and *Ip* (1994), the electron temperature profile along a magnetic flux tube threading through the inner coma is related to the

electron heat flux (F_e) in the solar wind. From model calculations it can be shown that a change in the solar wind electron heat flux could lead to the expansion or contraction of the photochemical equilibrium region of the cometary ionosphere of low electron temperature $(T_e \sim 300$ K) by a factor of 3 or more in size. Figure 7 shows that, if the thermal condition of the draped magnetic flux tube is such that the inner coma of cold electron temperature has a large size, many of the cometary ions would be lost to electron dissociative recombination without being injected into the ion tail. On the other hand, if the inner coma is filled with warm electrons with $T_e \sim 2 \times 10^4$ K, the majority of the cometary ions would be flushed into the tailward side.

From solar wind measurements (*Feldman et al.,* 1975), it is known that the electron heat flux has an average value of about 10^{-3} ergs cm^{-2} s^{-1} in the slow solar wind and an average value of 10^{-2} ergs cm^{-2} s^{-1} in the high-speed wind. Also, F_e could vary by 30% from minute to minute. This makes the solar wind variability a possible production mechanism for the ion rays and emphasizes that the transition region between the cold inner ionosphere and the warm outer ionosphere might actually become a key source region. We will return to this issue in section 4 when discussing the ionospheric pileup region.

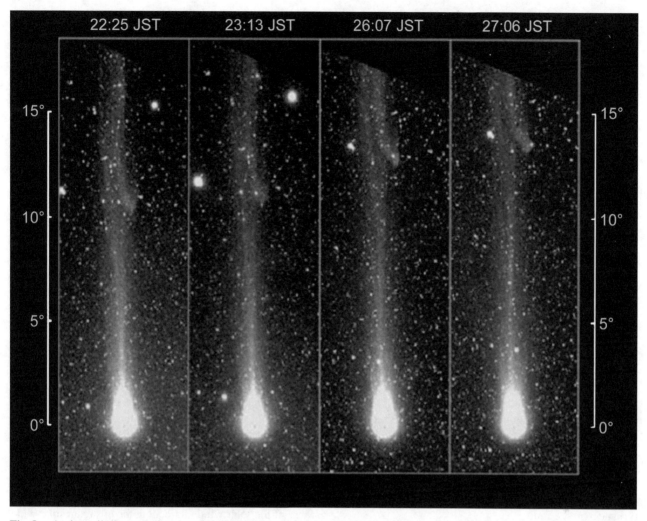

22:25 JST 23:13 JST 26:07 JST 27:06 JST

Fig. 8. An ion tail disconnection event (DE) observed in Comet Hale-Bopp (C/1995 O1) on March 25, 1996. Images taken at the National Astronomical Observatory by H. Fukushima and D. Kinoshita. From *Kinoshita et al.* (1996).

2.3. Ion Tail Disturbances and Disconnection Events

While Alfvén's magnetic field line draping model has formed the basis of our understanding of cometary ion tail dynamics, its physical ingredients have also been applied to the interpretation of the occurrence of fine plasma structures such as ion rays (cf. *Schmidt and Wegmann,* 1982), as well as some large-scale disturbances (*Wegmann,* 1995, 2002). The most dramatic change to cometary ion tail morphology must belong to the so-called ion tail disconnection events (DEs). As shown in Fig. 8, a large ion cloud is observed to detach itself from the comet head and move gradually away. In some other cases, the whole ion tail could be cut away from the comet head. How does this happen?

Niedner and Brandt (1978) were the first to provide a physical model for such DEs. The basic idea is that when a cometary ionosphere encounters a sector boundary of the IMF, where two regions of opposite magnetic field polarities are separated by a thin layer of the heliospheric current sheet (HCS), reconnection of the opposite-pointing magnetic field lines might take place on the front side facing the Sun (Fig. 9a). As a consequence, symmetrical pairs of

plasma clouds would be peeled off from the coma, leading to the formation of folding ion rays.

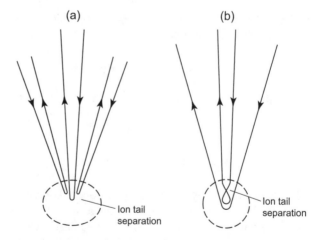

(a) (b)

Ion tail separation

Ion tail separation

Fig. 9. The Neidner-Brandt model of (a) reconnection on the frontside cometary ionosphere as the comet intercepts a reversal of magnetic field polarity associated with a heliospheric current sheet. (b) Reconnection on the tailward side, which might also occur. From *Ip* (1985).

Correlations of the DEs with the crossings of comets and heliocentric current sheets have always been difficult. This is because, except for the special situation when a spacecraft was in the vicinity of a comet (such as during the space missions to Comet Halley), large uncertainties exist in the interplanetary conditions. For this reason, the reconnection model advocated by *Niedner and Brandt* (1978) has always been subjected to spirited debates (*Saito et al.,* 1987; *Yi et al.,* 1993, 1994; *Wegmann,* 1995). *Brandt et al.* (1999) carried out a very detailed study of 19 major DEs of Halley's comet in 1985–1996. What is special about this investigation is that a potential model of the coronal magnetic field based on photospheric magnetic field observations (see *Hoeksema,* 1989) is used to reconstruct the large-scale structure of the heliospheric current sheet at different positions of the heliosphere. It is then possible to map the magnetic field current sheet to different regions of the heliosphere. These authors reach the statistical conclusion that DEs occur at IMF sector boundaries (*Niedner and Brandt,* 1978; *Yi et al.,* 1994).

We note that *Russell et al.* (1986) and *Ip and Axford* (1990) proposed a variant of this reconnection model by invoking a magnetic field merging process — but on the tailward side — triggered by plasma instabilities (see Fig. 9b). Furthermore, the possibility of triggering an ion cloud disruption by a sort of flute instability due to the interaction of the cometary ionosphere with a high-speed solar wind stream has also been proposed (*Ip and Mendis,* 1978). Finally, *Wegmann* (1995, 2002) shows that a variety of solar wind variations, including strong interplanetary shocks and coronal mass ejection events, can produce large density disturbances in the ion tails.

3. THEORETICAL MODELS: MAGNETOHYDRODYNAMIC AND KINETIC SIMULATIONS

3.1. Analytical Theory

The expanding atmosphere of a comet is subject to the ionizing effect of solar radiation

$$H_2O + h\nu \rightarrow H_2O^+ + e \tag{1}$$

and of the charge transfer process with the solar wind

$$H_2O + H^+ \rightarrow H_2O^+ + H \tag{2}$$

$$H_2O + O^{6+} \rightarrow H_2O^+ + O^{5+} \tag{3}$$

Recent observations by space X-ray telescopes like ROSAT and Chandra have revealed vivid images of the latter process, beginning first with the serendipitous discovery of strong X-ray emission in C/Hyakutake (1996 B2) (*Lisse et al.,* 1996). Subsequently, many more comets were found to display X-ray emission (*Dennerl et al.,* 1997). The generally accepted theory is due to *Cravens* (1997), who pro-

posed that the charge transfer process, as described in equations (2) and (3), could provide "new" heavy ions in excited states, and that subsequent radiative transitions to lower energy states lead to the emission of soft X-ray photons or extreme ultraviolet photons. The X-ray brightness distribution can consequently be considered as a map of the penetration of the solar wind ions into the cometary coma. The symmetrical pattern of crescent shape with the central axis pointing along the radial direction is a consequence of the collisional absorption effect of the solar wind (Fig. 10). That the X-ray emissivity could vary on short timescales is also consistent with the time variability of the solar wind flux (*Kharchenko and Dalgarno,* 2001). Finally, X-ray spectroscopic observations of C/1999 S4 (LINEAR) by the Chandra telescope have shown quite conclusively that the emission lines at 320, 400, 490, 560, 600, and 670 eV originated from the electron capture and radiative deexcitation by the solar wind minor ions C^{5+}, C^{6+}, C^{7+}, N^{7+}, O^{7+}, and O^{8+} (*Lisse et al.,* 2001). The reader is referred to *Lisse et al.* (2004) and *Cravens* (2002), and the references therein, for further detail.

Another important effect of the solar wind charge transfer and charge exchange process has to do with the production of new cometary ions such as O^+, OH^+, and H_2O^+ upstream of the comet. Photoionization could also contribute significantly to the production of new ions. The addition of new mass in the solar wind flow is a central issue studied long before in the pioneering work by *Biermann et al.* (1967) and by *Wallis* (1971), who proposed that the comet bow shock could be weak with a Mach number of M ~ 2 or even nonexistent. The physical argument was that because of the solar wind interaction, the upstream solar wind

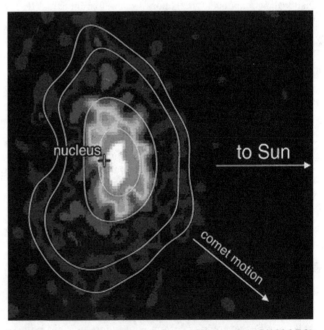

Fig. 10. An X-ray image of Comet Hyakutake (C/1996 B2) taken with the ROSAT Wide Field Camera. The length of the image is about 6.5×10^5 km. Courtesy Max-Planck-Institut für Extraterrestriche Physik.

plasma would be heated and slowed down by the continuous mass loading by the cometary ions. The early work on solar wind-comet interaction therefore focused on the kind of effect that mass accretion would have on the global flow dynamics of the cometary plasma. The behavior of the two-component fluid (solar wind protons and cometary ions of the water-group molecules, CO, and CO_2) along the comet-Sun axis can be described by the one-dimensional equations (*Wallis and Ong*, 1975)

$$\frac{d(\rho_i v_i)}{dX} \approx \tilde{q} \tag{4}$$

$$\frac{d}{dX}\left(\rho_i v_i^2 + p_\perp + \frac{B^2}{2\mu_0}\right) = 0 \tag{5}$$

$$\frac{d}{dX}[\rho_i v_i f(v_i,\mu)] = \tilde{q}\delta\left(\mu - \frac{m_i v_i^2}{2B}\right) \tag{6}$$

and

$$\tilde{q} = \frac{Q m_i}{4\pi v_n \tau_i X^2} \tag{7}$$

Here v_i is the mass-loaded solar wind plasma flow speed, ρ_i the mass density of the ions of mass m_i (the cometary ions are assumed to co-move with the solar wind plasma and $\rho_{i\infty}$ and $v_{i\infty}$ are values at infinity), $p_\perp$ the thermal pressure perpendicular to the magnetic field **B**, Q the comet gas production rate, v_n the expansion speed of the neutral coma gas, τ_i the ionization timescale, μ the magnetic moment of the cometary ions at creation, and $f(v_i, \mu)$ the distribution function of μ at flow speed v_i. It is further assumed that the magnetic moment is invariant and the magnetic field is perpendicular to the flow direction. Finally, the strength of the magnetic field may be estimated by using the relationship between the magnetic field and the flow velocity that holds for the conserved component of the subsonic flow for the case of axially symmetric flow along the central axis (*Schmidt and Wegmann*, 1982): $B^2 v_i/n_{sw}$ = constant.

In the supersonic flow, the effect of the magnetic field is small. Hence with B = 0 and γ the ratio of specific heats, the above equations yield

$$v_i^2\left[\frac{\gamma+1}{2(\gamma-1)}\rho_i v_i\right] - \left(\frac{\gamma}{\gamma-1}\right)\rho_{i\infty}v_{i\infty}v_i^2 + \frac{\rho_{i\infty}v_{i\infty}^3}{2} = 0 \tag{8}$$

with solutions

$$v_{i\pm} = v_{i\infty}\left(\frac{\gamma}{\gamma+1}\right)\left(\frac{\rho_{i\infty}v_{i\infty}}{\rho_i v_i}\right)\left[1 \pm \sqrt{1 - \frac{(\gamma^2-1)\rho_i v_i}{\gamma^2\rho_{i\infty}v_{i\infty}}}\right] \tag{9}$$

where

$$\rho_i v_i = \int \tilde{q}\,dX \tag{10}$$

Thus in the fluid approximation, a shock must form in the cometary accretion flow before the condition

$$\rho_i v_i \geq [\gamma^2/\gamma^2 - 1]\rho_{i\infty}v_{i\infty} \tag{11}$$

is met (*Biermann et al.*, 1967; *Wallis*, 1971). Therefore, the general consensus in the early analytical investigations pointed to the formation of a weak shock with a Mach number (M) ~ 2 (*Wallis*, 1973; *Brosowski and Wegmann*, 1972; *Schmidt and Wegmann*, 1982). Subsequent studies employing full magnetohydrodynamic (MHD) computational techniques have fully verified this basic point (*Schmidt and Wegmann*, 1982; *Ogino et al.*, 1988). It is now common knowledge that the spacecraft observations at 1P/Halley found very clear signatures for the bow shock formation with M ~ 2 (*Gringauz et al.*, 1986; *Mukai et al.*, 1986; *Johnstone et al.*, 1986). The *ICE* measurements at 21P/Giacobini-Zinner are not as definite, however (*Bame et al.*, 1986). This might be a consequence of the finite gyroradius effect of the heavy cometary ions, as described below.

The analytical solutions for the combined fluid, i.e., solar wind plasma plus comet ions, with M = 2 and B = 0, are illustrated in Fig. 11. Several important features can be recognized: Most notable is the rapid increase of the thermal pressure, p, at the bow shock; the thermal pressure has in-

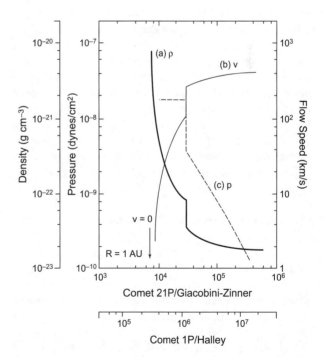

Fig. 11. Radial variations of (**a**) the number density, (**b**) the axial flow velocity, and (**c**) the thermal pressure, from a one-dimensional analytical model of the cometary mass-loading flow. Values are scaled to Comets 21P/Giacobini-Zinner and 1P/Halley at a heliocentric distance of 1 AU.

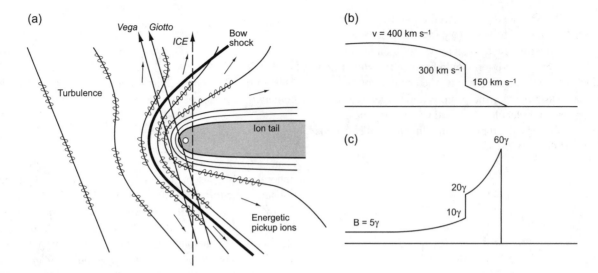

Fig. 12. A sketch of the general characteristics of the comet-solar wind interaction. **(a)** The general plasma environment of a cometary coma in which high levels of magnetic field and plasma turbulences are generated by the ionization of the new cometary ions. **(b)** The variation of the solar wind velocity as a function of axial distance from the nucleus. **(c)** A sketch of the gradual increase of the magnetic field strength and the formation of a magnetic field-free cavity in the inner coma where the magnetic field drops to zero. From *Ip and Axford* (1986).

creased about 30 times while the flow speed has decreased by only about 25%. At the bow shock crossing, there is a further stepwise increase of the thermal pressure and thus, in the subsonic region, the plasma is essentially incompressible. Also of note is the fact that the flow speed continues to decrease rapidly up to a point where the plasma flow becomes stagnant.

Galeev et al. (1985) discussed how the additional effect of the magnetic field pressure gradient could accelerate the mass-loaded solar wind toward the comet center. The above analytical work suggests that the global comet-solar wind interaction, as sketched in Fig. 12, can be broadly divided into two regimes: (1) the strong momentum coupling regime in the outer coma, where cometary ions are assimilated into the solar wind flow as a result of pickup and wave-particle interaction; and (2) the $J \times B$ acceleration region, where the Lorentz force is effective in accelerating cometary plasma in the antisunward direction until being balanced by the ion-neutral frictional force.

3.2. Magnetohydrodynamic Simulations

Useful as they are, the analytical treatments have now been overtaken by numerical calculations, which can deal with much more complex situations such as three-dimensionality, nonstationary solar wind interaction, and magnetic field effects. The rapid progress can be seen by comparing the work by *Schmidt et al.* (1988), not long after the spacecraft encounters of Comet Halley, and the work by *Gombosi et al.* (1997), immediately after the perihelion passage of C/Hale-Bopp. Some of the most advanced computational techniques have now been applied to this field, thus permitting the study of the behaviors of cometary plasma flow at different scale lengths. Let us first examine the global features by following the numerical results of *Wegmann et*

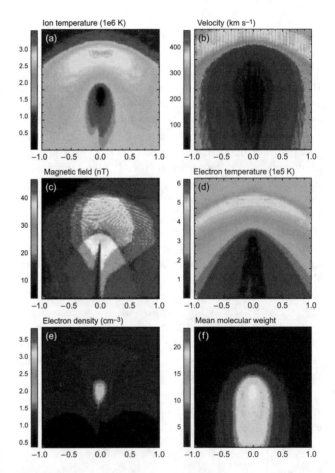

Fig. 13. Numerical simulations of the bow shock structure (the color contour maps extend to 10^6 km on each side of the comet nucleus): **(a)** ion temperature (K); **(b)** velocity (km s^{-1}) with streamlines; **(c)** magnitude of the magnetic field (nT) with field lines separated by time intervals of 600 s; **(d)** electron temperature (K); **(e)** electron number density (cm^{-3}); and **(f)** mean molecular weight of the ions, including solar wind protons. Courtesy of R. Wegmann.

al. (1987). In Figs. 13a,b, the color-coded diagrams illustrate the sharp transitions in temperature and flow velocity at an upstream distance of about 4×10^5 km. This indicates the location of the weak cometary shock. The bow shock follows a parabolic shape and the solar wind flow moves in a straight line upstream of the bow shock.

Inside the bow shock, the plasma flow can be seen to be deflected around the comet. For $u_{inf} = 500$ km/s, u reaches below 50 km/s along the central axis on the tailward side. As will be discussed later, the plasma flow could become stagnant in the inner coma. Figure 13c shows how the interplanetary magnetic field interacts with the comet. The field draping effect (*Alfvén*, 1957) is clearly present. The folding of the magnetic field in the central region, where the plasma is densest and the electron and ion temperatures are coldest, lead to a hair-pin shape of the magnetic field configuration.

Because of the pileup of the magnetic field, the magnetic field pressure will become increasingly important. Also, because of the directionality of the magnetic field and the resultant Lorentz force on the plasma, the cometary plasma flow will deviate strongly from axial symmetry. The most dramatic effect can be seen in the inner coma region. Figure 14 depicts a schematic view of how the magnetic field will be stacked up in front of the contact surface shielding the solar wind plasma inflow from the cometary ionospheric outflow. In the plane containing the solar wind flow (x) and the interplanetary magnetic field (y), the cometary plasma inflow will converge toward the central region. A more detailed description is given by *Gombosi et al.* (1997). In their model calculations using a self-adapted computational mesh method, the plasma flow in the XY plane is seen to merge on the tailward side. On the other hand, the plasma flow in the XZ plane — perpendicular to the magnetic field — will simply move around the cometary ionosphere.

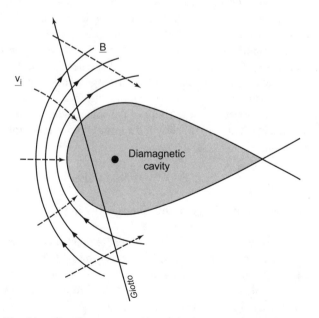

Fig. 14. The flow pattern of ionospheric plasma in the vicinity of the contact surface separating the solar wind flow from the cometary plasma flow.

3.3. Kinetic Simulations

Within the framework of hydrodynamic approximation, the dimension of the bow shock is proportional to the cometary gas production rate (Q). That is, if the upstream shock position (r_{shock}) of a comet with Q ~ 6×10^{29} molecules/s is 8×10^5 km, we have r_{shock} ~ $8 \times 10^5 \times$ (Q/6×10^{29}) km for the same solar wind condition (Fig. 11). Such a linear approximation has limitations, however. First, the cometary plasma flow is composed of solar wind protons, solar wind minor ions, and heavy cometary ions, so it is actually a multispecies fluid. The new cometary ions, such as H_2O^+ and O^+, have very different velocity distributions from the solar wind flow (see section 5). The most important effect for the case in point has to do with the finite gyroradii of the new cometary ions. For an interplanetary magnetic field of 5 nT, the O^+ ions with a gyrovelocity of 400 km/s will have a gyroradius of about 1.5×10^4 km. This means that plasma discontinuities in the cometary flow, such as the shock transition, should be described in terms of boundary structures of finite thickness.

To take into account these kinetic effects, *Galeev and Lipatov* (1984) were the first to use numerical simulations capable of treating the motions of individual charged particles to study the comet–solar wind interaction. They showed that the cometary shock front thickness has a scale of a few gyroradii of the heavy cometary ions, rather than the gyroradius of a solar wind proton. This is a principal difference between the cometary bow shock structure and the bow shock structure of Earth's magnetosphere, which has a thickness of less than approximately a few proton gyroradii.

It is important to point out that an interesting consequence of the particle kinetics is to introduce a dependence of the shock structures on the orientation of the interplanetary magnetic field. *Omidi and Winske* (1987, 1991) produced simulations of the cometary shock transition for the cases when the cone angle between the interplanetary magnetic field and the solar wind is Q = 90° (for perpendicular shock) and when Θ = 5° (for parallel shock). As shown in Fig. 15, the perpendicular shock case is characterized by very well-defined and well-correlated sharp jumps in magnetic field strength and number density — with the shock thickness ~ proton gyroradius — to be followed by a sequence of fluctuations downstream. In contrast, the parallel shock case displays significant variations in magnetic field and number density ahead of the shock jump of much broader structure. The large-amplitude plasma fluctuations downstream of the shock jump are uncorrelated. For an intermediate theta value (Θ = 55°), the cometary shock structure displays yet another type of behavior. That is, no shock jump could be identified in the large amplitude variations when the thickness is much larger than the typical gyroradius of solar wind protons.

Clearly, for a comet with the shock position $R_{shock} < 10^5$ km, the shock thickness will be a significant fraction of the solar wind interaction region — except perhaps for the special case of Θ ~ 90°. This is the situation for comets with Q < 10^{29} molecules/s such as 21P/Giacobini-Zinner

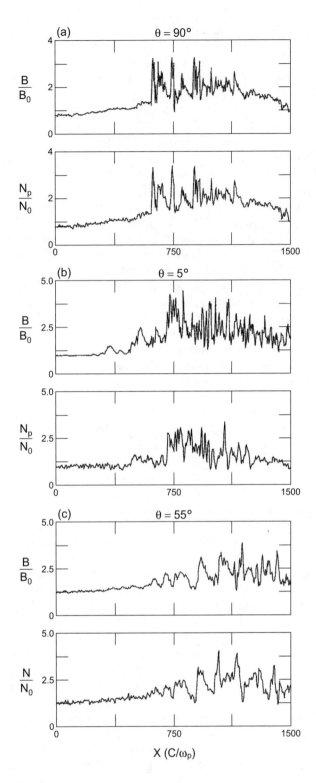

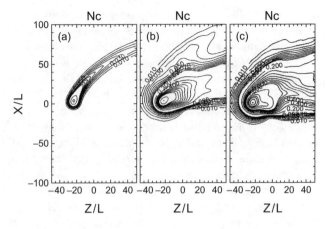

Fig. 16. The distributions of the cometary ions and solar wind protons for weakly outgassing comets. The gas production rates, from left to right, are **(a)** $Q = 8.4 \times 10^{26}$ mol s^{-1}; **(b)** 2.5×10^{27} mol s^{-1}; and **(c)** 5×10^{27} mol s^{-1}. From *Lipatov et al.* (2002).

for which shock distances are comparable or smaller than the gyroradii of the cometary heavy ions? This question is also of importance for *in situ* measurements to be carried out by the *Rosetta* mission at large solar distances when the particle kinetic effects are expected to dominate the solar wind interaction process. A number of numerical simulations have recently been performed to address this issue (*Hopcroft and Chapman*, 2001; *Lipatov et al.*, 2002). Figure 16 shows the two-dimensional cuts of the cometary ion density distributions for three different gas production rates from the three-dimensional hybrid simulations of Lipatov et al. for the case when the interplanetary magnetic field is parallel to the y direction. For $Q = 8.4 \times 10^{26}$ molecules/s, the ion density profile in the XZ plane appears to follow the gyromotion of the individual cometary ions. In other words, the cometary ion tail may be regarded as a beam of individual ions. The interplanetary magnetic field is only slightly modified. For $Q > 2.5 \times 10^{27}$ molecules/s, the asymmetric structure of the ion tail is still evident, while the magnetic field disturbance becomes stronger. These authors point out that for $Q = 8.4 \times 10^{26}$ molecules/s, a symmetric Mach cone exists in the vicinity of the comet, while a symmetric detached bow shock will form for $Q > 5 \times 10^{27}$ molecules/s.

4. PLASMA BOUNDARIES

4.1. Upstream Region and Shock Crossing

In this section, we describe different regions of cometary plasma boundary structures as observed by particles-and-fields instruments onboard spacecraft. As a function of distance from the nucleus, these structures are divided into (1) shock, (2) cometosheath, (3) cometopause, (4) ion pileup region, and (5) magnetic field-free cavity. A summary of the spacecraft encounter geometries before the most recent *Deep Space 1* mission is given in Fig. 17.

Fig. 15. Plots of the total magnetic field and density as a function of distance at different cone angles (θ) between the IMF and the solar wind flow direction: **(a)** $\theta = 90°$; **(b)** $\theta = 5°$; and **(c)** $\theta = 55°$. From *Omidi and Winske* (1991).

during the *ICE* encounter. (We will return to the *ICE* measurements in the next section.) An important question that emerges from this consideration is therefore what happens to the solar wind interaction of weakly outgassing comets

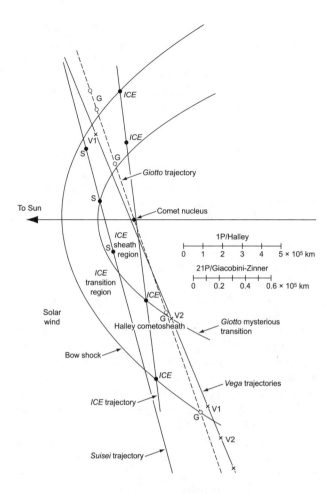

Fig. 17. A description of the encounter geometries for the spacecraft measurements at Comets 21P/Giacobini-Zinner and 1P/Halley. *ICE* = International Cometary Explorer; V1, V2 = *Vega 1* and *Vega 2*; G = *Giotto*. The approximate positions of the bow shock, or bow wave, are also given. From *Reme* (1991).

The first test of the theoretical models of comet-solar wind interaction was provided by the tail-crossing of 21P/Giacobini-Zinner by the *ICE* spacecraft in September 1985. Figure 18 compares the spatial distribution of the plasma wave activity with that of the energetic ion flux measured by the plasma instruments onboard the spacecraft (*Ipavich et al.*, 1986; *Scarf et al.*, 1986). It is noteworthy that a significant level of solar wind interaction effects could be found at distances as far as 1,000,000 km from the comet. As will be examined in more detail in section 5, such enhanced wave activity and energetic ion population are related to each other as a consequence of the pickup process of new cometary ions.

In spite of the mass-loading effect and finite gyroradius effect of the pickup ions, a weak bow shock is nevertheless expected. The *ICE* results are, however, somewhat mixed because different experiments gave different answers. Amid the large-amplitude upstream waves, the plasma waves and energetic particle observations (*Scarf et al.*, 1986; *Hynds et al.*, 1986) indicated the presence of a bow shock structure. On the other hand, the magnetometer and electron

detectors detected no signature of a sharp jump in plasma parameters as appropriate for a classical shock formation (*Smith et al.*, 1986; *Bame et al.*, 1986). For example, the magnetic field strength shows a very gradual increase embedded with a series of large-amplitude fluctuations at about 09:30–10:00 UT inbound, and a similar pattern but for reduced field strength at about 11:50–12:20 UT outbound (Fig. 19). The solar wind velocity profile, as determined by the electron detector, does not display clear-cut sharp jumps in these intervals, which have been called "slow transition regions" by *Bame et al.* (1986).

In view of the presence of large-amplitude waves in most of the solar wind interaction region, *Omidi and Winske* (1991) developed the interesting idea that the bow shock at 21P/Giacobini-Zinner is actually composed of an ensemble of shocklets instead of one single standing bow shock. Figure 20 shows a schematic view of such a multiple-shock model. In this scenario, the solar wind will be decelerated little by little upon transversing the shocklets. It is in such a manner that the supersonic solar wind will gradually reach subsonic speed over a length scale much larger than a few gyroradii of the heavy cometary ions.

While the bow shock of 21P/Giacobini-Zinner at the time of the *ICE* encounter is not well-defined, the situation at Comet Halley during the encounters of *Susei, Vega 1 and 2*, and *Giotto* is just the opposite. A drop in the solar wind velocity, plus significant angular deflection, were clearly observed by the plasma instrument on *Susei* (*Mukai et al.*,

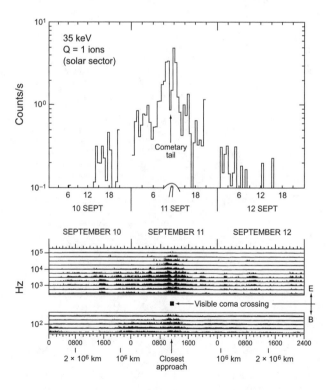

Fig. 18. A composite view of the plasma wave turbulence and the energetic particle flux in the vicinity of Comet 21P/Giacobini-Zinner as observed by the *ICE* spacecraft. From *Scarf et al.* (1986) and *Ipavich et al.* (1986).

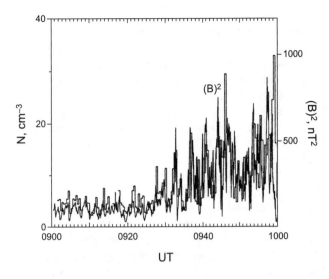

Fig. 19. A comparison of the variations in the square of the magnetic field magnitude and the electron plasma density (rectangular curve) as observed at Comet 21P/Giacobini-Zinner. For the interval illustrated, the two parameters are strongly correlated, indicating that the fluctuations are fast mode MHD waves. From *Tsurutani et al.* (1987).

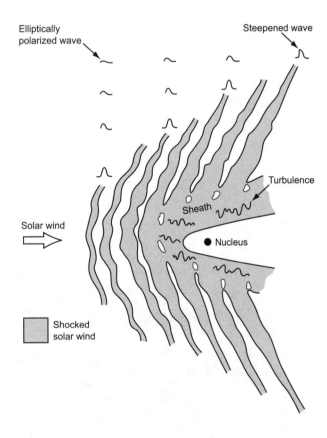

Fig. 20. A schematic view of how the dynamic interaction between the solar wind and comets can result in the formation of multiple shocks, which eventually lead to a heated and subsonic wind. From *Omidi and Winske* (1991).

1986), providing unequivocal evidence of a cometary bow shock. Similar behaviors were registered by the experiments on *Vega 1* and *2* and *Giotto* at the far flanks of the comet (*Gringauz et al.,* 1986). The locations of these shock crossings, together with the theoretical position of the bow shock along the comet-Sun axis, define a parabolic shape of the standing bow shock as shown in Fig. 17.

4.2.   Cometosheath

Because of the continuous addition of heavy cometary ions, the postshock solar wind flow in the extended coma will slow down further. The first observational indications of such process can be seen in the time-sequence of energy-per-charge spectra obtained by the solar direction ion analyzer onboard the *Vega 1* spacecraft (*Gringauz et al.,* 1986). The JPA plasma instrument on the *Giotto* probe observed similar behavior. As shown in Fig. 21, the trace of solar wind protons denoted by (P) moves to lower-energy bins at closer and closer distances to the nucleus. At the same time, the trace of heavy cometary ions on the upper track (because of higher mass and therefore higher kinetic energy) starts by following a separate route but later tends to merge with the protons.

A more recent view of this mass-loading effect in ion mass spectra is provided by the *Deep Space 1* measurements at Comet 19P/Borrelly. In this observation, the spacecraft reached a closest approach distance of 2171 km. The preliminary report showed that the peak of the solar wind proton count rates falls from a value near the ambient solar wind speed to just a few tens of kilometers per second near the closest approach. The flow velocity of cometary heavy ions follows a similar pattern (*Nordholt et al.,* 2003).

4.3.   Cometopause

At cometocentric distances $r < 2 \times 10^5$ km, the plasma instruments on *Giotto* designed to detect medium- and low-energy ions began to measure heavy cometary ions in increasing fluxes. Figure 22 shows the radial profiles of several major species of cometary ions obtained by the IMS experiment (*Balsiger et al.,* 1986). Some of the main features may be summarized as follows: As the *Giotto* probe approached the comet nucleus to radial distances $r < 2 \times 10^5$ km, the number density of the water-group ions, H_2O^+, H_3O^+, and O^+, was seen to increase rapidly, with the H_3O^+ ions displaying the steepest gradient. The atomic ions, C^+ and S^+, were observed to be very abundant (*Balsiger et al.,* 1986; *Balsiger,* 1990), and their radial gradients appear to be more gradual than those of the water-group ions.

At a distance of about 10^5 km, suprathermal cometary ions created in the solar wind flow at large distances upstream begin to disappear and be replaced by cold ions produced locally. Figure 23 summarizes the behavior of both the hot and cold ions as detected by the IMS experiment. There are a number of points of note. First, at the radial position (sometimes called the cometopause) where the magnetic field strength jumped by 20 nT (*Neubauer et al.,*

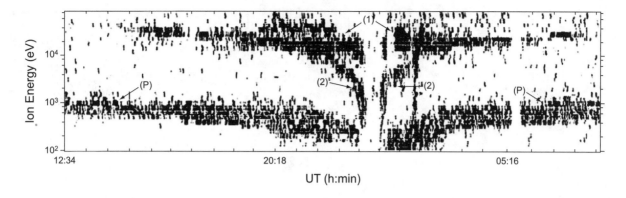

Fig. 21. The variations of the ion flow speeds [(P) for solar wind protons and (1) and (2) for heavy cometary ions] in the coma of Comet 1P/Halley as measured by the JPA instrument onboard the *Giotto* spacecraft. From *Johnstone et al.* (1986).

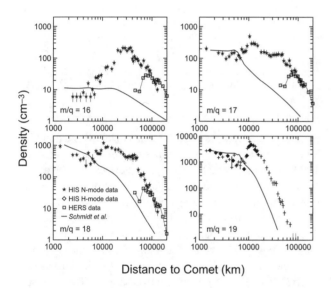

Fig. 22. Ion density profiles for mass/charge 16, 17, 18, 19 amu e^{-1} as a function of distance from the comet nucleus. Data are from the HIS and HERS sensors of the IMS experiment on *Giotto*. The solid lines are theoretical data from the MHD model by *Schmidt et al.* (1988). From *Altwegg et al.* (1993).

1986), the number density of solar wind protons was observed to drop by a factor of 2 (*Balsiger,* 1990) and a similar reduction was detected in the He^{++} flux (*Fuselier et al.,* 1988). This feature could be related to a tangential discontinuity or propagating rotational discontinuity in the solar wind (*Neubauer,* 1987). On the other hand, the reduction in the number density of solar wind particles need not be caused by charge exchange loss (*Gringauz et al.,* 1986; *Gombosi,* 1987; *Ip,* 1989). It should be noted that the solar wind protons and the hot oxygen ions disappeared completely at r ~ 6–8 × 10^4 km, essentially as a result of charge exchange losses. This effect can be seen clearly by comparing the number density of the cold O$^+$ ions measured by the high-intensity (HIS) instrument and that of the hot O$^+$ ions from the high-energy (HER) instrument in Fig. 23.

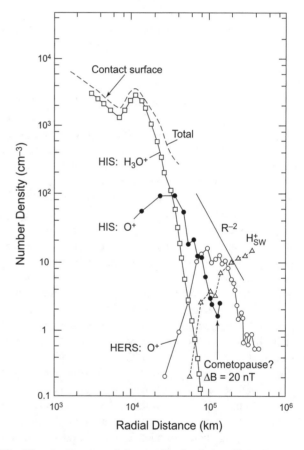

Fig. 23. A summary of the number density profiles of several water-group ions as observed by the two sensors (HIS and HERS) of the IMS experiment on *Giotto*. From *Ip* (1989).

4.4. Ion Pileup Region

As the *Giotto* spacecraft moved closer to the nucleus of Comet Halley, the total ion density was observed to follow a radial dependence of 1/r. Such spatial variation is basically the result of production of cometary ions plus the slowdown of the mass-loaded plasma flow. This trend was interrupted at r ~ 12,000 km, at which point a sharp drop

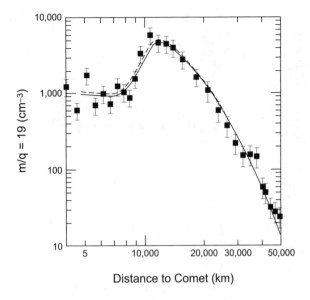

Fig. 24. Comparison between model calculations and measurements by the *Giotto* IMS for mass 19. From *Häberli et al.* (1995).

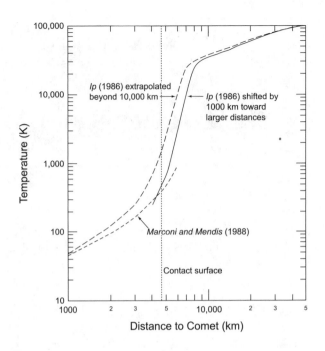

Fig. 25. Electron temperature profile, as used in the model calculations for the ion pileup region. From *Häberli et al.* (1995).

in ion number density was detected (see Fig. 24). Upon its initial discovery, this plasma feature was called the "ion pileup region" (*Balsiger et al.,* 1986). Subsequent considerations pointed to two possible physical mechanisms. The first one, proposed by *Galeev* (1987), suggested that the ion density enhancement was caused by an anomalous ionization effect. The other one, which has now been accepted as the more likely explanation, was due to *Ip et al.* (1987), who proposed that such a density jump was produced by a sudden decrease in electron temperature (from a few 10^4 K to a few hundred K) at r < 12,000 km. The main reason is simply that the rate coefficients of electron dissociative recombination depend strongly on the electron temperature (T_e). For an electron dependence of $T_e^{-0.5}$, a change in electron temperature by a factor of 100 will lead to a corresponding change in electron and ion number density by a factor of about 3–10.

A quantitative analysis of this somewhat unexpected phenomenon is complicated because detailed accounts must be given to the different ionization effects. This is because the cometary ion production depends on a number of processes, including photoionization, electron impact ionization by photoelectrons, and solar wind suprathermal electrons; these estimates necessarily involve model calculations of the electron energetics, transport process, and thermal conduction along the magnetic field lines (*Gan and Cravens,* 1990; *Haider et al.,* 1993; *Ip,* 1994; *Häberli et al.,* 1995). *Häberli et al.* (1995) used an electron temperature profile, as shown in Fig. 25, to carry out chemical network calculations and found that the ion density variations of the water group ions plus NH_4^+ ions could be satisfactorily explained.

Groundbased spectrographic observations of Comet Halley showed a shell-like structure of the H_2O^+ ion brightness distribution at the 6198-Å emission line. *Ip et al.* (1988) interpreted this feature in terms of the formation of the ion

pileup region. Observations of Comets Hale-Bopp and Hyakutake have also shown similar structures (*Bouchez et al.,* 1999). This means that the ion pileup effect must be a common property of the comet-solar interaction. The stability of such plasma structures against changes in solar wind condition (i.e., solar wind pressure or interplanetary magnetic field orientation) remains unclear, however. This is because the ion density peak of C/Hale-Bopp was observed to move from the sunward side to the tailward side at certain time intervals (see Fig. 26). How did that happen? This also suggests that time-series CCD photometry and/or spectrography will be a very powerful tool in deciphering the inner dynamics of cometary ionospheres and their response to solar wind conditions.

4.5. Ionosphere

While the bow shock as discussed in section 4.2 may be considered as the outer boundary delineating the free-flowing solar wind and the plasma flow significantly mass-loaded by cometary ions, there exists another important boundary in the inner coma separating the mass-loaded solar wind flow from the plasma outflow of purely cometary origin. To a certain extent, the radial expansion of cometary ionospheric plasma has some similarities to the interaction of the solar wind with the interstellar medium in which the solar wind flow is terminated by the formation of a heliopause. In this context, *Wallis and Dryer* (1976) were the first to consider a similar structure with the formation of a contact surface separating the cometary ionospheric flow and the mass-loaded solar wind plasma. The key is, of course, that the cometary ions in the inner coma are collisionally coupled to the supersonic expanding neutral gas. A contact

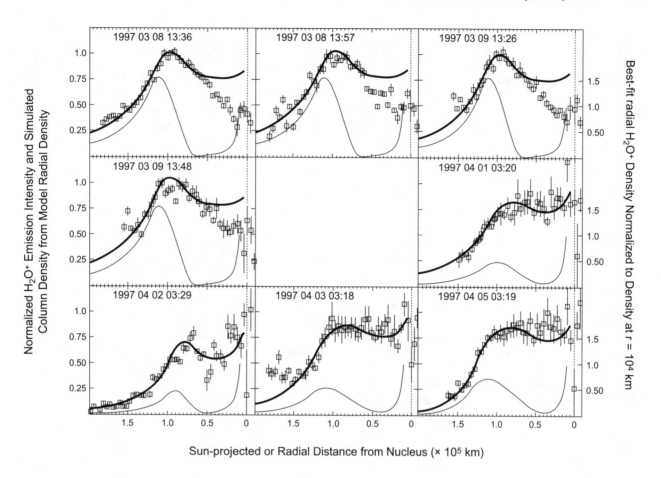

Fig. 26. Models of the radial distribution of H_2O^+ at Comet Hale-Bopp (C/1995 O1) assuming hemispherical symmetry of the coma. Best-fit radial density models (thin line) are shown along with the consequent column density distributions (bold line) and the observed column density profiles in March and April 1997 (points). From *Bouchez et al.* (1999).

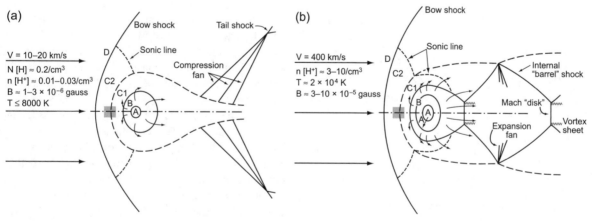

Fig. 27. **(a)** Schematic flow regions and parameters for the solar source in interstellar gas. Regions A, C1, and C2 are subsonic; B and D are supersonic. The interstellar gas returns to its ambient velocity by passing through a compression fan, which becomes a tail shock. **(b)** Schematic flow regions for the comet source in the solar wind. The circular radius Δ denotes the position where the radially expanding neutral gas flow becomes collision-free. The shock and slip-plane structure in the wake could be complex because the wake is supersonic. From *Wallis and Dryer* (1976).

discontinuity would be necessarily formed between a pair of shocks (the inner and the outer shock) such that the supersonic cometary plasma could be diverted into the lateral direction near the boundary. Figure 27 compares the theoretical models of the termination shock of the heliosphere and the cometary inner shock. Subsequently, *Houpis and Mendis* (1980) proposed an analytical model for a hypersonic ionospheric outflow, and *Damas and Mendis* (1992) used an axially symmetric hydrodynamic model to investigate the three-dimensional structure of the inner shock layer.

Ip and Axford (1982), on the other hand, suggested that the position of the ionospheric contact surface should be determined by a balance between the frictional gas drag of the neutral gas on the cometary ions and the $\mathbf{J} \times \mathbf{B}$ Lorentz force exerted by the draped magnetic field at the nose of the cometary ionosphere. The implication is that the inner ionosphere of purely cometary origin should be free of magnetic field, while the ionospheric plasma outside of the contact surface should be magnetized. However, *Ershkovich and Mendis* (1983) carried out an analysis of the stability of the interface and found that the ion-neutral force should render the boundary extremely unstable; they predicted a fully magnetized ionosphere. The stage was thus set before the *Giotto* encounter with Comet Halley, as it would have penetrated through the predicted location of the cometary ionopause.

What the magnetometer experiment onboard *Giotto* found was a total surprise. When the space probe reached a cometocentric distance of 4700 km, it went through a sharp layer with a width of only 20 km dividing a diamagnetic cavity (*Neubauer et al.,* 1986). As shown in Fig. 28, this structure repeated itself on the outbound passage. To understand the spatial variation of the magnetic field, we follow the arguments below. First, as the solar wind slows down in the cometary coma because of the mass-loading effect, the magnetic field strength will be amplified accordingly. The upper limit of the magnetic field strength in the stagnant cold plasma region is estimated by equating the magnetic pressure to the solar wind ram pressure. That is

$$\frac{B_s^2}{2\mu_0} = n_{sw} m_{sw} v_{sw}^2 \tag{12}$$

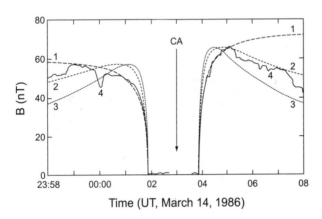

Fig. 28. Magnetic field measurements made by the magnetometer experiment on *Giotto* showing the inner pileup region inbound and outbound and the magnetic cavity region. Curve (1) is obtained by taking into account the magnetic pressure gradient effect in the force balance; curves (2) and (3) are obtained by including the curvature force term but with different ion number density profiles. The observational result obtained by the magnetometer experiment on Giotto is denoted by curve (4). The experimental curves are from *Neubauer* (1986) and the theoretical curves from *Wu* (1987).

For a solar wind number density of $n_{sw} \sim 5$ cm^{-3} and velocity of $v_{sw} \sim 400$ km/s, we have $B_{max} \sim 60$ nT. At the point where the magnetic field reaches its maximum, i.e., dB/dr = 0, the force balance depends on the equilibrium between the curvature force of the draped magnetic field and the frictional force of the outward expanding neutral gas. For a radius of curvature (R_s) of the magnetic field in the inner coma, we have

$$\frac{B_s^2}{\mu_0 R_s} \approx k_{in} n_i m_i n_n v_n \tag{13}$$

where k_{in} is the ion-neutral collision rage, n_n the neutral number density, n_i the ion number density, and v_n the neutral speed. In the case of Comet Halley during the *Giotto* encounter, it could be found that $r_c \sim 2400$–7300 km at the subsolar point depending on the ion number density, n_i. *Cravens* (1986) and *Ip and Axford* (1987) give a one-dimensional global treatment of the force balance by taking into account the magnetic pressure gradient term. The magnetic field strength in the inner coma is then determined by the following equation

$$\frac{1}{\mu_0} B \frac{dB}{dR} + \frac{1}{\mu_0} \frac{B^2}{R} = k_{in} n_i m_i n_n v_n \tag{14}$$

which has a simple solution

$$B(R/R_{max}) = B_{max} \frac{[1 + 2 \ln(R/R_{max})]^{1/2}}{R/R_{max}} \tag{15}$$

where R_{max} is proportional to $Q^{3/4}$. A good match of the theoretical profile to the observed magnetic field variation can be obtained by adjusting the physical parameters. The general features of the plasma flow and magnetic field variation along the Sun-comet axis have been depicted in a numerical calculation by *Baumgaertel and Sauer* (1987), in which they compare the effect of artificially changing the electron dissociative recombination rate on the possible formation of the ion pileup region. The plasma velocity variation in Fig. 29 is illuminating in the sense that the outward expansion with a flow speed ~1 km/s within the first 5000 km is suddenly switched to a stagnant flow with small inward speed (~0–0.5 km/s), indicating the location of the contact surface. It is interesting to note the existence of a small peak in the ion number density right at this boundary. This feature is caused by the accumulation of ions from both sides of the converging flows. At higher spatial resolution, this density peak should be even sharper as indicated by a one-dimensional theoretical model of *Cravens* (1989). *Goldstein et al.* (1989) examined the fine time resolution of the ion mass spectrometer measurements at the crossing of the contact surface and found the signatures for a recombination layer and a spike of hot ions, which

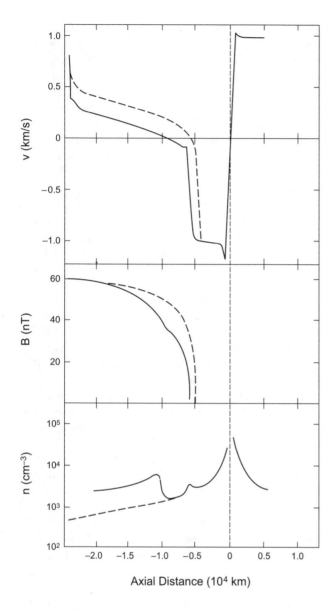

Fig. 29. Plasma velocity, magnetic field, and plasma density from a one-dimensional steady-state calculation assuming a constant electron dissociative recombination coefficient (a) = 3 × 10⁻¹³ m³/s for R < 10⁴ km (solid curves) and (b) = 3 × 10⁻¹³ m³/s everywhere (dashed curves). From *Baumgaertel and Sauer* (1987).

could be interpreted as particles accelerated at the contact surface.

Using an MHD treatment, *Schmidt and Wegmann* (1991) produced a three-dimensional picture of the ionospheric contact surface (Fig. 30). Because of the orientation of the interplanetary magnetic field, the contact surface bounding the diamagnetic cavity is highly asymmetric. It may be said to mimic the shape of a tadpole with a flat tail, which is basically the neutral sheet separating the two parts of opposite magnetic polarities. The only spacecraft that has passed through a cometary ion tail is *ICE*. The plasma instruments onboard *ICE* detected the presence of a cold dense plasma region at the center of the ion tail of 21P/Giacobini-

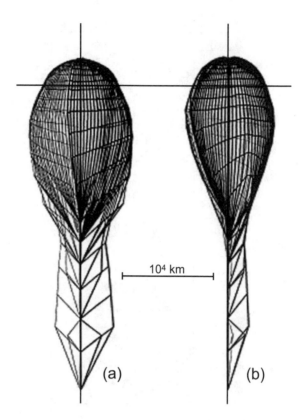

Fig. 30. The magnetic field free cavity surrounding the cometary nucleus is defined by a sharp boundary separating the external region with a magnetic field of several tens of nanoteslas from the internal region of the zero magnetic field. The surface defined by B = 3 nT is used as an approximation to such an envelope projected onto a plane with the look angle tilted by (a) 20° and (b) 70° from the direction of the IMF. Courtesy of R. Wegmann.

Zinner, which might be interpreted as the extension of the tadpole-like ionospheric tail modeled by *Schmidt and Wegmann* (1991).

5. PLASMA INSTABILITIES AND WAVES

5.1. Microinstabilities

In the fluid or magnetohydrodynamic description of the comet-solar wind interaction, it is common practice to assume that all charged particles, whether they are solar wind protons, electrons, or cometary ions, are mixed together into one single fluid flow moving with the same velocity. Also, it is customary (and necessary) to further assume that the cometary ions share the same thermal temperature with the bulk plasma. At the same time, the velocity distribution of the cometary ions is taken to be isotropic and Maxwellian. As shown by the spacecraft observations at Comets Giacobini-Zinner and Halley, this is far from being true. In many circumstances, the dynamics of the cometary ions are completely different from that of the solar wind. Particle kinetic effects turn out to play a dominant role in the whole process of comet-solar wind interaction (Fig. 31). Indeed, it is

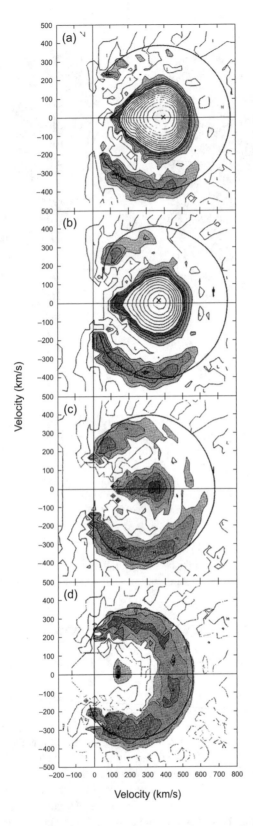

Velocity (km/s)

Fig. 31. Phase space-density distributions of cometary hydrogen ions upstream of the bow shock of Comet 1P/Halley as observed by the IMS experiment on *Giotto,* during the time period 0803–0908 UT, March 13, 1986, when the *Giotto* spacecraft was at a distance of about 8×10^6 km from the nucleus. They show the evolutionary effect of pitch angle scattering from the point of injection into a partially filled shell. From *Neugebauer et al.* (1987).

because of this unique property that the study of cometary plasma processes has become a very intriguing topic in plasma physics, attracting many of the best minds in this field. The whole enterprise began with the fundamental paper of *Wu and Davidson* (1972), who investigated the plasma effect of the photoions created in the exosphere of Mercury. Translated to the context of the comet-solar wind interaction, we could start with the following description.

Suppose a cometary neutral gas atom or molecule is ionized by solar ultraviolet radiation or charge exchange with the solar wind protons (see section 3). The particle will immediately be accelerated by the convective electric field $\mathbf{E} = -\mathbf{V} \times \mathbf{B}$ in the stationary frame. In the above equation, the initial velocity of the new ion relative to the solar wind is $\mathbf{V} = -\mathbf{V}_{sw} + \mathbf{V}_c$, where $\mathbf{V}_{sw}$ is the solar wind velocity, $\mathbf{V}_c$ is the spacecraft velocity, and $\mathbf{B}$ is the interplanetary magnetic field. In the following consideration, we have ignored the coma expansion speed, which is only on the order of 1 km s^{-1}. To simplify the conceptual discussion further, we could omit $\mathbf{V}_c$ since it is on the order of 20–30 km/s. The motion of the new cometary ion is given by the equation of motion under the Lorentz force: $d\mathbf{u}/dt = q\mathbf{E} = -q\mathbf{V} \times \mathbf{B}$. The velocity $\mathbf{V}$ can be resolved into two components, one parallel to $\mathbf{B}$ ($V_\parallel = V \cos\varphi$) and the other perpendicular to $\mathbf{B}$ ($V_\perp = V \sin\varphi$); see Fig. 32. The solution to the equation of motion in the solar wind frame can then be described by the superposition of two motions: The first one has to do with a gyration of the new ion around the magnetic field with velocity $V_\perp$, and the second one is a steady motion along the magnetic field with velocity $V_\parallel = -V_{sw} \cos\varphi$. For $\mathbf{V}_{sw} \perp \mathbf{B}$ (i.e., $\varphi = \pi/2$), the ion trajectory in the solar wind frame is purely a gyration with $V_\perp = V_{sw}$ since $V_\parallel = 0$. In the observer's frame, namely the spacecraft frame, the ion motion will be a combination of the solar wind flow plus

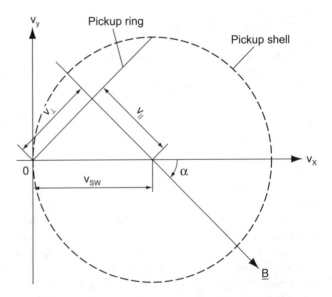

Fig. 32. The two initial velocity components, $V_\perp$ and $V_\parallel$, of the new cometary ions. Pitch-angle scattering transfers the ring distribution into a spherical shell.

the gyration motion. The resultant trajectory is a cycloidal motion with $\mathbf{u} = 0$ at the cusp and $\mathbf{u} = 2\mathbf{V}_{sw}$ at the top. It can be seen that a charged particle detector onboard a spacecraft would see modulation of the cometary ion flux as a function of the look angle with respect to the solar wind direction. That is, the ion flux (and particle energy) should be at maximum when the detector is looking antiparallel to the solar wind flow and at minimum when looking in the opposite direction. This is, in fact, what was observed (*Richardson et al.*, 1987). The initial motion of the new cometary ions can be described in terms of a ring distribution in the velocity phase space.

The other extreme case is for $\mathbf{B} \parallel \mathbf{V}_{sw}$ with $\varphi = 0$. Here, no gyration will be executed since $V_\perp = 0$. Instead, the initial velocity of the new ion will be $\mathbf{V}_\parallel = -\mathbf{V}_{sw}$ in the solar wind frame. As a consequence, the new ions remain momentarily at rest in the stationary (i.e., the spacecraft) frame, while forming an ion beam in the solar wind frame.

The value of the φ angle varies between 0 and $\pi/2$. The initial velocity distribution of the new cometary ions is thus often a combination of the ring distribution and the beam distribution described above. For convenience, it is generally called the ring-beam distribution. The important thing is that such initial anisotropic velocity distributions represent a source of free energy in the plasma flow. The contribution by *Wu and Davison* (1972) and *Wu and Hartle* (1974) (see also *Lee and Ip*, 1987; *Lee*, 1989) is to first show that plasma instabilities could be generated with a rapid growth of the corresponding ion cyclotron waves serving to isotropize the pitch angle distribution and hence pick up the new cometary ions. Under the assumption that the initial distribution function of the new ions can be presented as $\delta(v_\parallel - v_{\parallel 0}) \, \delta(v - v_0)$, the dispersion equation of the plasma waves has its two most unstable roots at

$$\varpi \approx \kappa v_{\parallel 0} - \Omega \approx \pm \kappa V_A \qquad (16)$$

representing the left- and righthand polarized Alfvén (L and R) waves for $\omega \ll \Omega_i$. The wave growth wave can be derived as

$$\tilde{\gamma} \sim |\Omega_i| \left[\frac{V_A}{|v_{\parallel 0}|} \frac{\frac{1}{2} \tilde{n}_i m_i v_\perp^2}{B_0^2/(2\mu_0)} \right]^{1/3} \qquad (17)$$

In the case of the ring-beam velocity anisotropy, three different low frequency instabilities can be generated, namely an ion cyclotron instability, a parallel propagating nonoscillatory mode, and a fluid mirror instability (*Tsurutani*, 1991). The ion cyclotron instability produces resonant lefthand waves propagating antiparallel to the ions. In the case of the beam velocity distribution, two types of instabilities can be generated: a righthand resonant helical beam instability and a nonresonant firehose instability. The condition for the cyclotron resonance is $\omega = \mathbf{k} \cdot \mathbf{V} + n\Omega_i$ where ω is the wave frequency, $\mathbf{k}$ and $\mathbf{V}$ the wave k vector and particle velocity ($\mathbf{k} \parallel \mathbf{V}$), n an integer, and Ω_i the ion gyrofrequency. The firehose instability grows when $P_\parallel > P_\perp + B^2/4\pi$.

5.2. Source of the Free Energy

Because of the photoionization and charge exchange ionization of the cometary neutral gas in the large coma region, new ions are continuously created at distances as far away as 1,000,000 km from the cometary nucleus (cf. Fig. 18). The pickup process leads to the generation of a very significant level of wave activity with $\delta B/B \sim 0.5$. The large-amplitude waves are accompanied by the presence of an intense flux of energetic heavy ions. Because the particle energy exceeds the initial pickup energy (~20 keV) of the water-group ions, a certain kind of particle acceleration and hence energy transfer from the waves to the energetic ion population must be taking place in the cometary coma (*Richardson et al.*, 1986). The central issue in cometary plasma physics is therefore to understand how the three-way transfers of free energy can be facilitated. For this, we need to estimate the energy budget available to drive these different kinds of plasma effects.

For pedagogical reasons, we will essentially follow the heuristic approach described in *Coates et al.* (1990) and *Johnstone et al.* (1991). Figure 33 illustrates how the resonant wave particle interaction proceeds. In the solar wind frame, the parallel propagating Alfvén waves moving upstream (k > 0) and moving downstream (k < 0) are related to the pickup ions gyrating at the ion cyclotron frequency Ω_i by the resonance conditions

$$\omega - kV_\parallel = -\Omega_i \text{ (L mode);}$$
anomalous Doppler /larger pitch angle

and

$$\omega + kV_\parallel = \Omega i \text{ (R mode);}$$
normal Doppler/smaller pitch angle

The first condition with the negative sign on the lefthand side is for a righthand polarized wave in which the electric vector rotates in the direction of the electronic cyclotron motion. Ions with $V_\parallel$ satisfying the above conditions will give up energy to the related Alfvén waves. In a resonant wave-particle interaction we have the following relation (*Kennel and Petschek*, 1966)

$$V_\perp^2 + (V_\parallel - V_{ph})^2 = \text{constant}$$

For parallel propagating Alfvén waves, $V_{ph} = \pm V_A$. This means the distribution of the particle velocity in the $V_\parallel$–$V_\perp$ coordinate should follow the segments determined by the two spheres as defined in Fig. 34. It can be seen that, because of the interaction with the Alfvén waves with $V_{ph} = \pm V_A$, the total velocity $[V = (V_\parallel^2 + V_\perp^2)^{1/2}]$ is effectively reduced (*Galeev and Sagdeev*, 1988; *Coates et al.*, 1990). This is also the source of the free energy for wave growth. Note that the particles could also absorb energy from the waves by tracing the contour of the maximum radius in different quadrants of such a bispherical shell structure. In this opposite case, we would have particle acceleration by wave damping (see section 4.6.3).

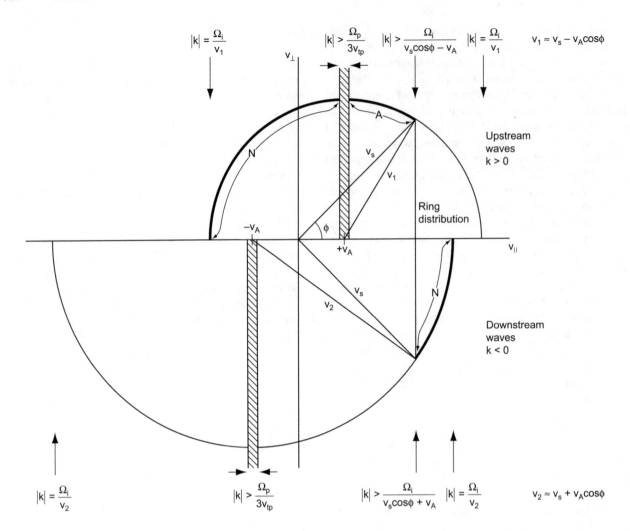

$$|k| = \frac{\Omega_i}{v_1} \qquad |k| > \frac{\Omega_p}{3v_{tp}} \quad |k| > \frac{\Omega_i}{v_s\cos\phi - v_A} \quad |k| = \frac{\Omega_i}{v_1} \qquad v_1 \approx v_s - v_A\cos\phi$$

$$|k| = \frac{\Omega_i}{v_2} \qquad\qquad |k| > \frac{\Omega_p}{3v_{tp}} \qquad\qquad |k| > \frac{\Omega_i}{v_s\cos\phi + v_A} \quad |k| = \frac{\Omega_i}{v_2} \qquad v_2 \approx v_s + v_A\cos\phi$$

Fig. 33. A summary diagram of resonant wave-particle interactions in the velocity space of the particles in the solar wind moving frame. The upper part is for waves traveling upstream and the lower part for waves traveling downstream. The unstable regimes are marked by the gray shading; N is for the normal Doppler resonance, and A is for anomalous Doppler resonance. From *Johnstone et al.* (1991).

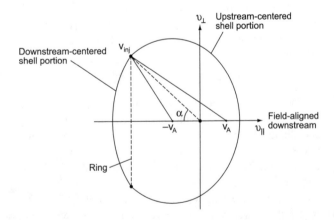

Fig. 34. Velocity-space representation of the pitch angle scttered bispherical shell geometry in the solar wind frame, where v_{inj} is the position ofmet of the pickup ion injection into the initial ring distribution. From *Huddleston et al.* (1992).

In the original ring-beam velocity distribution, the kinetic energy from the beam motion and the thermal energy from the gyromotion are, respectively,

$$\text{kinetic energy} = 1/2\rho_i v_s^2 \cos^2\Phi$$
$$\text{thermal energy} = 1/2\rho_i v_s^2 \sin^2\Phi$$

The immediate result of wave-particle interaction is to isotropize the pitch angle distribution covering the bisphere by wave scattering. The center of the bisphere will move with the Alfvén speed; this will also define the kinetic energy of the whole system. Now, if the waves are dominated by upstream propagating components at the beginning, the velocity distribution will be dictated by the spherical shell projected into the upper part of Fig. 33. For such shell structure, we have

$$\text{kinetic energy} = 1/2\rho_i v_A^2$$
$$\text{thermal energy} = 1/2\rho_i[v_s^2 \sin^2\Phi + (v_s \cos\Phi - v_A)^2]$$

The difference between the ring-beam energy and the shell energy gives us the free energy available for the generation of Alfvén waves in the solar wind flow

$$\Delta E = E_{ring} - E_{shell} = \rho_i(v_A v_s \cos\Phi - v_A^2)$$

5.3. Plasma Waves and Magnetic Turbulences

Note that in the spacecraft frame with $V_{sw\parallel} > V_A$, both R- and L-mode waves are lefthand polarized since the wave polarization in this frame is dominated by the Doppler shift. The propagation direction of these waves should be sunward along the average, spiral magnetic field direction as observed in the 100-s waves. For Alfvén waves, $\omega/k = V_A$; the resonant frequency of the waves in the solar wind frame will therefore be

$$\varpi_{sw} = \frac{\Omega_i V_A}{\left|V_\parallel V_{ph}\right|} \tag{18}$$

with $V_{ph} = \pm V_A$ depending on the direction of the wave prop-agation. For spacecraft observations, the frequency of the resonant wave will be Doppler-shifted to be

$$\varpi_{sc} = \Omega_i \frac{(V_{sc} - V_{ph})}{(V_\parallel - V_{ph})} \tag{19}$$

where v_{sc} is the spacecraft velocity in the solar wind moving frame; in the case of the ring distribution or $V_{sc} \approx V_{sw}$, we thus have $\omega_{sc} \approx \Omega_i$ (*Tsurutani and Smith,* 1986). It is for this reason that the gyrofrequency of the dominant water-group ions was discovered to be the "pump" wave with a period of about 100 s as seen in the power spectra of cometary plasma turbulence in the coma of Comet Giacobini-Zinner (see Fig. 35).

In the case of the *ICE* observations of Comet Giacobini-Zinner, similar ultra-low-frequency (ULF) fluctuations were also observed in the solar wind electron number density and the solar wind flow velocity (*Gosling et al.,* 1986; *Tsurutani et al.,* 1987). Figure 19 shows the correlation between these magnetic field variations and the solar wind plasma parameters, which suggest that the waves are fast-mode magnetosonic waves. Note that the magnitude of the fluctuations increases as the bow shock was crossed (at about 09:30 UT) at a radial distance of about 2×10^5 km. Furthermore, at large distances from the comet, $R > 3 \times 10^5$ km, the 100-s waves were lefthand polarized; the polarization then changed from lefthand elliptical and linear in regions near and inside the bow shock at $R < 2 \times 10^5$ km. Superposed on the long-period linearly polarized waves, lefthand-polarized whistler waves of shorter periods of 1–3 s were detected. *Tsurutani* (1991) developed the theory that the upstream whistler wave packets could be generated by steepening of the magnetosonic waves.

Depending on the solar wind conditions (i.e., the orientation of the interplanetary magnetic field with respect to the solar wind flow direction) and the production rate of the

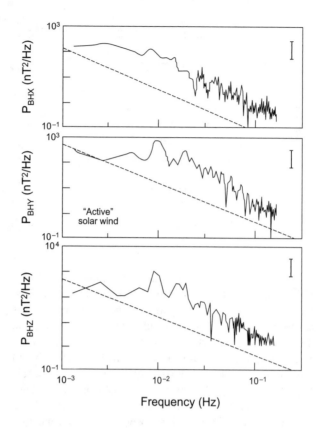

Fig. 35. The frequency spectrum of MHD waves as measured by the *ICE* spacecraft at Comet 21P/Giacobini-Zinner. A power-law curve representative of an active solar wind is also shown for comparison. The wave intensities are well above solar wind fluctuation levels. The waves are considerably more intense than those at Comet 1P/Halley. From *Tsurutani and Smith* (1986).

cometary ions, different levels and wave types could be generated. Figure 36 compares the power spectra of the magnetic field variations at Comets 26P/Grigg-Skellerup, 21P/Giacobini-Zinner, and 1P/Halley. There are important similarities. That is, they all show a prominent "pump" wave feature at the 100-s periodicity and, at higher frequency, tend to follow a power law $P(f) \sim f^\alpha$ with the power index $\alpha \approx -2$. At the same time, we also find important differences in the wave forms. In fact, each is different from the other. The case of 26P/Grigg-Skjellerup is characterized by sinusoidal, noncompressive lefthand-polarized waves (*Mazelle et al.,* 1995); the case of 21P/Giacobini-Zinner has phase-steepened and compressive magnetosonic (RH) waves led by large amplitude whistler packets; and Comet Halley's upstream waves, which are composed of linearly polarized turbulence, have no obvious structure (*Tsurutani et al.,* 1995).

6. SUMMARY AND DISCUSSION

The advances made in the study of cometary plasma physics over the last two decades have been spectacular. This achievement is, in part, due to the spacecraft missions

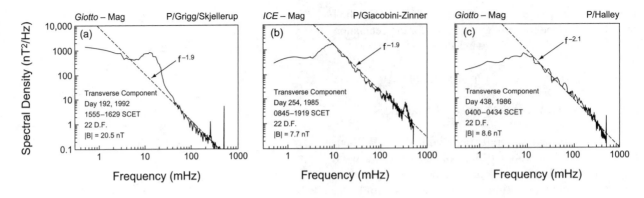

Fig. 36. Power spectra of the transverse components of the magnetic field at three comets. From *Tsurutani et al.* (1995).

to Comets 1P/Halley, 21P/Giacobini-Zinner, 26P/Grigg-Skjellerup, and, most recently, 19P/Borrelly. The *in situ* measurements enabled by particles-and-fields instruments have provided tremendous amounts of new information on the comet-solar wind interactions and insights into the plasma physical processes. Progress made in groundbased astronomical facilities is also crucial to the success of this field. For example, the observational data on Comet Hale-Bopp (1995 O1) are still to be digested. It is expected that coordinated multiwavelength observations (i.e., narrowband filter photometry and X-ray imaging observations) will bring us closer to an understanding of the dynamical effects and thermodynamical responses to solar wind variabilities. Last, but not least, sophisticated computational simulations have burst onto the scene and have staked a claim to be the centerpiece of planetary and cometary plasma physics. This new tool is still to be explored fully.

The high drama roaming across the horizon belongs to the scientific returns to be reaped from the *Rosetta* mission to Comet 67P/Churyumov-Gerasimenko. The long-term observations afforded by the *Rosetta* Comet Orbiter will revolutionize our knowledge of solar wind interaction with weakly outgassing comets at different heliocentric distances. We are totally ignorant of the physical environment of the coma, which must be filled with charged dust particles, complex molecules, and all sorts of gas jets. This precious knowledge will be supplied within a decade by a new generation of cometary researchers, who must be ready for exploration. We promise them that this field will be much more exciting and fruitful than we now know it.

Acknowledgments. I thank the editors for encouragement in producing this chapter and T. Cravens for a careful review. I also thank B. Yu-M. Lin for most helpful assistance in preparing this manuscript. This work was partially supported by the National Council of Taiwan under NSC 90-2112-M-008-032 and NSC 91-2111-M-008-028, and by the Ministry of Education of Taiwan through the CosPA Project 91-N-FA01-1-4-5.

REFERENCES

Alfvén H. (1957) On the theory of comet tails. *Tellus, 9,* 92–96.
Altwegg K., Balsiger H., Geiss J., Goldstein R., Ip W.-H., Meier A., Neugebauer M., Rosenbauer H., and Shelley E. (1993) The ion population between 1300 km and 230000 km in the coma of Comet P/Halley. *Astron. Astrophys., 279,* 260–266.
Balsiger H. (1990) Measurements of ion species within the coma of Comet Halley from Giotto. In *Comet Halley 1986: World-Wide Investigations, Results, and Interpretations* (J. Mason et al., eds.), pp. 129–146. Horwood Ltd., Aylesbury.
Balsiger H. and 18 colleagues (1986) Ion composition and dynamics at Comet Halley. *Nature, 321,* 330–334.
Bame S. J., Anderson R. C., Asbridge J. R., Baker D. N., Feldman W. C., Fuselier S. A., Gosling J. T., McComas D. J., Thomsen M. F., Young D. T., and Zwickl R. D. (1986) Comet Giacobini-Zinner: Plasma description. *Science, 232,* 356–361.
Baumgaertel K. and Sauer K. (1987) Fluid simulation of Comet P/Halley's ionosphere. *Astron. Astrophys., 187,* 307–310.
Biermann L. (1951) Kometenschweife und solare Korpuskularstrahlung. *Z. Astrophys., 29,* 274–286.
Biermann L., Brosowski B., and Schmidt H. U. (1967) The interaction of a comet with the solar wind. *Solar Phys., 1,* 254–283.
Bouchez A. H., Brown M. E., Spinrad H., and Misch A. (1999) Observations of the ion pile-up in comets Hale-Bopp and Hyakutake. *Icarus, 137,* 62–68.
Brandt J. C., Caputo F. M., Hoeksema J. T., Niedner M. B. Jr., Yi Y. and Snow M. (1999) Disconnection events (DEs) in Halley's comet 1985–1986: The correlation with crossings of the heliospheric current sheet (HCS). *Icarus, 137,* 69–83.
Brosowski B. and Wegmann R. (1972) *Numeerische Behandlung eines Kometenmodells.* MPI/PAE-Astro-46.
Coates A. J., Wilken B., Johnstone A. D., Jockers K., Glassmeier K.-H., and Huddleston D. E. (1990) Bulk properties and velocity distributions of water group ions at comet Halley: Giotto measurements. *J. Geophys. Res., 95,* 10249–10260.
Cravens T. E. (1986) The physics of the cometary contact surface. In *20th ESLAB Symposium on the Exploration of Halley's Comet* (B. Battrick et al., eds.), pp. 241–246. ESA SP-250, Noordwijk, The Netherlands.
Cravens T. E. (1989) A magnetohydrodynamical model of the inner coma of comet Halley. *J. Geophys. Res., 94,* 15025.
Cravens T. E. (1997) Comet Hyakutake X ray source: Charge transfer of solar wind heavy ions. *Geophys. Res. Lett., 24,* 105.
Cravens T. E. (2002) X-ray emission from comets. *Science, 296,* 1042–1046.
Cravens T. E. and Gombosi T. I. (2003) Cometary magnetospheres: A tutorial, COSPAR 2002 D3.3-B07, Houston. *Adv. Space Res., 33(11),* 1968–1972.

Cravens T. E. and Korosmezey A. (1986) Vibrational and rotational cooling of electrons by water vapor. *Planet Space Sci., 34,* 961.

Damas M. and Mendis D. A. (1992) A three-dimensional axisymmetric photochemical flow model of the cometary inner shock layer. *Astrophys. J., 396,* 704.

Dennerl K., Englhauser J., and Truemper J. (1997) X-ray emissions from comets detected in the Roentgen X-ray satellite. *Science, 277,* 1625.

Ershkovich A. I. and Mendis D. A. (1983) On the penetration of the solar wind into the cometary ionosphere. *Astrophys. J., 269,* 743–750.

Feldman W. C., Asbridge J. R., Bame S. J., Montgomery M. D., and Gary S. P. (1975) Solar wind electrons. *J. Geophys. Res., 80,* 4181.

Fuselier S. A., Shelley E. G., Balsiger H., Geiss J., Goldstein B. E., Goldstein R., and Ip W.-H. (1988) Cometary H_2^+ and solar wind He^{++} dynamics across the Halley cometopause. *Geophys. Res. Lett., 15,* 549–552.

Galeev A. A. (1987) Encounters with comets: Discoveries and puzzles in cometary plasma physics. *Astron. Astrophys., 187,* 12–20.

Galeev A. A. and Lipatov A. S. (1984) Plasma processes in cometary atmospheres. *Adv. Space Res., 4,* 229–237.

Galeev A. A. and Sagdeev R. Z. (1988) Alfvén waves in a space plasma and their role in the solar wind interaction with comets. *Astrophys. Space Sci., 144,* 427–438.

Galeev A. A., Cravens T. E., and Gombosi T. I. (1985) Solar wind stagnation near comets. *Astrophys. J., 289,* 807–819.

Gan L. and Cravens T. E. (1990) Electron energetics in the inner coma of Comet Halley. *J. Geophys. Res., 95,* 6285–6303.

Goldstein B. E., Altwegg K., Balsiger H., Fuselier S. A., and Ip W.-H. (1989) Observations of a shock and a recombination layer at the contact surface of Comet Halley. *J. Geophys. Res., 94,* 17251.

Gombosi T. I. (1987) Charge exchange avalanche at the cometopause. *Geophys. Res. Lett., 14,* 1174–1177.

Gombosi T. I., Hansen K. C., DeZeeuw D. L., Combi M. R., and Powell K. G. (1997) MHD simulation of comets: The plasma environment of comet Hale-Bopp. *Earth Moon Planets, 79,* 179–207.

Gosling J. T., Asbridge J. R., Bame S. J., Thomsen M. F., and Zwickl R. D. (1986) Large amplitude, low frequency, plasma fluctuations at comet Giacobini-Zinner. *Geophys. Res. Lett., 13,* 267–270.

Gringauz K. I. and 16 colleagues (1986) First in situ plasma and neutral gas measurements at Comet Halley. *Nature, 321,* 282–285.

Häberli R. M., Altwegg K., Balsiger H., and Geiss J. (1995) Physics and chemistry of ions in the pile-up region of comet P/Halley. *Astron. Astrophys., 297,* 881–891.

Haider S. A., Bhardwaj A., and Singhal R. P. (1993) Role of auroral and photoelectrons on the abundances of methane and ammonia in the coma of Comet Halley. *Icarus, 101,* 234.

Hoeksema J. T. (1989) Extending the Sun's magnetic field through the three-dimensional heliosphere. *Adv. Space Res., 9(4),* 141–152.

Hopcroft M. W. and Chapman S. C. (2001) 2D hybrid simulations of the solar wind interaction with a small scale comet in high Mach number flows. *Geophys. Res. Lett., 28,* 1115.

Houpis H. L. F. and Mendis D. A. (1980) Physicochemical and dynamical processes in cometary ionospheres. I. The basic flow profile. *Astrophys. J., 239,* 1107–1118.

Huddleston D. E., Coates A. J., and Johnstone A. D. (1992) Quasilinear velocity space diffusion of heavy cometary pickup ions on bispherical diffusion characteristics. *J. Geophys. Res., 97,* 19163–19174.

Huebner W. F. (1990) *Physics and Chemistry of Comets.* Springer-Verlag, Berlin. 376 pp.

Hynds R. J., Cowley S. W. H., Sanderson T. R., Wenzel K.-P., and van Rooijen J. J. (1986) Observations of energetic ions from Comet Giacobini-Zinner. *Science, 232,* 361–365.

Ip W.-H. (1985) Cometary plasma physics: Large-scale interactions. In *Advances in Space Plasma Physics* (B. Buti, ed.), pp. 1–21. World Scientific, Singapore.

Ip W.-H. (1989) On charge exchange effectc in the vicinity of the cometopause of Comet Halley. *Astrophys. J., 343,* 946–952.

Ip W.-H. (1994) On a thermodynamic origin of the cometary ion rays. *Astrophys. J. Lett., 432,* L143–L145.

Ip W.-H. and Axford W. I. (1982) Theories of physical processes in the cometary comae and ion tails. In *Comets* (L. L. Wilkening, ed.), pp. 588–636. Univ. of Arizona, Tucson.

Ip W.-H. and Axford W. I. (1986) The acceleration of particles in the vicinity of comets. *Planet. Space Sci., 34,* 1061–1065.

Ip W.-H. and Axford W. I. (1987) The formation of a magnetic field free cavity at Comet Halley. *Nature, 325,* 418–419.

Ip W.-H. and Axford W. I. (1990) The plasma. In *Physics and Chemistry of Comets* (W. F. Huebner, ed.), pp. 177–233. Springer-Verlag, Berlin.

Ip W.-H. and Mendis D. A. (1978) The flute instability as the trigger mechanism for disruption of cometary plasma tails. *Astrophys. J., 223,* 671–675.

Ip W.-H., Schwenn R., Rosenbauer H., Balsiger H., Neugebauer M., and Shelley E. G. (1987) An interpretation of the ion pile-up region outside the ionospheric contact surface. *Astron. Astrophys., 187,* 132–136.

Ip W.-H., Spinrad H., and McCarthy P. (1988) A CCD observation of the water ion distribution in the coma of Comet P/Halley near the Giotto encounter. *Astron. Astrophys., 206,* 129–132.

Ipavich F. M., Galvin A. B., Gloeckler G., Hovestadt D., Klecker B., and Scholer M. (1986) Comet Giacobini-Zinner: In situ observations of energetic heavy ions. *Science, 232,* 366–369.

Jockers K. (1991) Ions in the coma and in the tail of comets — observations and theory. In *Cometary Plasma Processes* (A. D. Johnstone, ed.), pp. 139–159. AGU Monograph 61, American Geophysical Union, Washington, DC.

Johnstone A. D., Coates A., Kellock S. et al., (1986) Ion flow at comet Halley. *Nature, 321,* 344–347.

Johnstone A. D., Huddleston D. E., and Coates A. J. (1991) The spectrum and energy density of solar wind turbulence of cometary origin. In *Cometary Plasma Processes* (A. D. Johnstone, ed.), pp. 259–272. AGU Monograph No. 61, Washington, DC.

Jones G. H., Balogh A., and Horbury T. S. (2000) Identification of comet Hyakutake's extremely long ion tail from magnetic field signatures. *Nature, 404,* 574–576.

Kennel C. F. and Petschek H. E. (1966) Limit on stably trapped particle fluxes. *J. Geophys. Res., 71,* 1–28.

Kharchenko V. and Dalgarno A. (2001) Variability of cometary X-ray emission induced by solar wind ions. *Astrophys. J. Lett., 554,* L99–L102.

Kinoshita D., Fukushima H., Watanabe J., and Yamamoto N. (1996) Ion tail disturbance of Comet C/Hyakutake 1996 B2 observed around the closest approach to the Earth. *Publ. Astron. Soc. Japan, 48,* L83–L86.

Lee M. A. (1989) Ultra-low frequency waves at comets. In *Plasma Waves and Instabilities at Comets and in Magnetospheres*

(B. T. Tsurutani and H. Oya, eds.), pp. 13–29. AGU Monograph No. 53, American Geophysical Union, Washington, DC.

Lee M. A. and Ip W.-H. (1987) Hydromagnetic wave excitation by ionized interstellar hydrogen and helium in the solar wind. *J. Geophys. Res., 92,* 11041–11052.

Lipatov A. S., Motschmann U., and Bagdonat T. (2002) 3D hybrid simulations of the interaction of the solar wind with a weak comet. *Planet. Space Sci., 50,* 403.

Lisse C. M., Dennerl K., Englhauser J., Harden M., Marshall F. E., Mumma M. J., Pere R., Pye J. P., Rickets M. J., Schmitt J., Trumper J., and West R. G. (1996) Discovery of X-ray and extreme ultraviolet emission from Comet C/Hyakutake 1996 B2. *Science, 274,* 205–209.

Lisse C. M., Christian D. J., Dennerl K., Meech K. J., Petre R., Weaver H. A., and Wolk S. J. (2001) Charge exchange-induced X-ray emission from Comet C/1999 S4 (LINEAR). *Science, 292,* 1343–1348.

Lisse C. M., Cravens T. E., and Dennerl K. (2004) X-ray and extreme ultraviolet emission from comets. In *Comets II* (M. C. Festou et al., eds.), this volume. Univ. of Arizona, Tucson.

Lüst Rh. (1962) Bewegung von Strukturen in der Koma und im Schweif des Kometen Morehouse, *Z. Astrophys., 65,* 236–250.

Mazelle C. H., Reme H., Neubauer F. M., and Glassmeier K.-H. (1995) Comparison of the main magnetic field and plasma features in the environment of comets Grigg-Skjellerup and Halley. *Adv. Space Res., 16(4),* 41–45.

Mukai T., Miyake W., Terasawa T., and Kitayama H. K. (1986) Plasma observation by Suisei of solar-wind interaction with Comet Halley. *Nature, 321,* 299–303.

Ness N. F. and Donn B. D. (1966) Concerning a new theory of type I comet tails. *Mem. Soc. R. Liege, Ser. 5, 12,* 141–144.

Neubauer F. M. (1987) Giotto magnetic field results on the boundaries of the pile-up region and the magnetic cavity. *Astron. Astrophys., 187,* 73–79.

Neubauer F. M. and 11 colleagues (1986) First results from the Giotto magnetometer experiment at Comet Halley. *Nature, 321,* 352–355.

Neugebauer M., Lazarus A. J., Altwegg K., Balsiger H., Goldstein B. E., Goldstein R., Neubauer F. M., Rosenbauer H., Schwenn R., Shelley E. G., and Unstrup E. (1987) The pickup of cometary protons by the solar wind. *Astron. Astrophys., 187,* 21–24.

Newburn R. L. Jr., Neugebauer M., and Rahe J., eds. (1991) *Comets in the Post-Halley Era, Vols. 1 and 2.* Kluwer, Dordrecht.

Niedner M. B. Jr. and Brandt J. C. (1978) Interplanetary gas. XXIII. Plasma tail disconnection events in comets: Evidence for magnetic field line reconnection at interplanetary sector boundaries? *Astrophys. J., 223,* 655–670.

Nordholt J. E. and 14 colleagues (2003) Deep Space 1 encounter with Comet 19P/Borrelly: Ion composition measurements by the PEPE mass spectrometer. *Geophys. Res. Lett., 30,* 18.

Ogino T., Walker R. J., and Ashour-Abdalla M. (1988) A three-dimensional MHD simulation of the interaction of the solar wind with comet Halley. *J. Geophys. Res., 93,* 9568–9576.

Omidi N. and Winske D. (1987) A kinetic study of solar wind mass loading and cometary bow shocks. *J. Geophys. Res., 92,* 13409.

Omidi N. and Winske D. (1991) Theory and simulation of cometary shocks. In *Cometary Plasma Processes* (A. D. Johnstone, ed.), pp. 37–47. AGU Monograph 61, American Geophysical Union, Washington, DC.

Reme H. (1991) Cometary plasma observations between the shock and the contact surface. In *Cometary Plasma Processes* (A. D. Johnstone, ed.), pp. 87–106. AGU Monograph 61, American Geophysical Union, Washington, DC.

Richardson I. G., Cowley S. W. H., Hynds R. J., Sanderson T. R., Wenzel K.-P., and Daly P. W. (1986) Three-dimensional energetic ion bulk flows at Comet P/Giacobmin-Zinner. *Geophys. Res. Lett., 13,* 415–418.

Richardson I. G., Cowley S. W. H., Hynds R. J., Tranquille C., Sanderson T. R., and Wenzel K. P. (1987) Observations of enertetic water-group ions at Comet Giacobini-Zinner: Implications for ion acceleration processes. *Planet. Space Sci., 35,* 1323–1345.

Russell C. T., Saunders M. A., Phillips J. L., and Fedder J. A. (1986) Near-tail reconnection as the cause of cometary tail disconnections. *J. Geophys. Res., 91,* 1417–1423.

Saito T., Saito K., Aoki T., and Yumoto K. (1987) Possible models on disturbances of the plasma tail of Comet Halley during the 1985–1986 apparition. *Astron. Astrophys., 187,* 201.

Scarf F. L., Coroniti F. V., Kennel C. F., Gurnett D. A., Ip W.-H., and Smith E. J. (1986) Plasma wave observations at Comet Giacobini-Zinner. *Science, 232,* 377–381.

Schmidt H. U. and Wegmann R. (1982) Plasma flow and magnetic fields in comets. In *Comets* (L. L. Wilkening, ed.), pp. 538–560. Univ. of Arizona, Tucson.

Schmidt H. U. and Wegmann R. (1991) MHD models of cometary plasma and comparison with observations. In *Cometary Plasma Processes* (A. D. Johnstone, ed.), pp. 49–64. AGU Monograph 61, American Geophysical Union, Washington, DC.

Schmidt H. U., Wegmann R., Huebner W. F., and Boice D. C. (1988) Cometary gas and plasma flow with detailed chemistry. *Comp. Phys. Comm., 49,* 17–59.

Schmidt-Voigt M. (1989) Time-dependent MHD simulations for cometary plasma. *Astron. Astrophys., 210,* 433.

Smith E. J., Tsurutani B. T., Slavin J. A., Jones D. E., Siscoe G. L., and Mendis D. A. (1986) International Cometary Explorer encounter with Giacobini-Zinner: Magnetic field observations. *Science, 237,* 382–385.

Tsurutani B. T. (1991) Comets: A laboratory for plasma waves and instabilities. In *Cometary Plasma Processes* (A. D. Johnstone, ed.), pp. 189–209. AGU Monograph 61, American Geophysical Union, Washington, DC.

Tsurutani B. T. and Smith E. J. (1986) Hydromagnetic waves and instabilities associated with cometary ion pick-up: ICE observations. *Geophys. Res. Lett., 13,* 263–266.

Tsurutani B. T., Thorne R. M., Smith E. J., Gosling J. T., and Matsumoto H. (1987) Steepened magnetosonic waves at Comet Giacobini-Zinner. *J. Geophys. Res., 92,* 11074–11082.

Tsurutani B. T., Glassmeier K.-H., and Neubauer F. M. (1995) An intercomparison of plasma turbulence at three comets: Grigg-Skjellerup, Giacobini-Zinner, and Halley. *Geophys. Res. Lett., 22,* 1149–1152.

Wallis M. K. (1971) Shock-free deceleration of the solar wind? *Nature, 233,* 23–25.

Wallis M. K. (1973) Weakly-shocked flows of the solar wind plasma through atmospheres of comets and planets. *Planet. Space Sci., 21,* 1647–1660.

Wallis M. K. and Dryer M. (1976) Sun and comets as sources in an external flow. *Astrophys. J., 205,* 895–899.

Wallis M. K. and Ong R. S. B. (1975) Strongly-cooled ionizing plasma flows with application to Venus. *Planet. Space Sci., 23,* 713–721.

Wehinger P. A., Wyckoff S., Herbig G., and Lew H. (1974) Identification of H_2O^+ in the tail of Comet Kohoutek (1973f). *Astrophys. J. Lett., 190,* L43–L47.

Wu C. S. and Davidson R. C. (1972) Electromagnetic instabilities produced by neutral particle ionization in interplanetary space. *J. Geophys. Res., 77,* 5399–5406.

Wu C. S. and Hartle R. E. (1974) Further remarks on plasma instabilities produced by ions born in the solar wind. *J. Geophys. Res., 79,* 283–285.

Wu Z.-J. (1987) Calculation of the shape of the contact surface at Comet Halley. In *Symposium on the Diversity and Similarity of Comets* (E. J. Rolfe and B. Battrick, eds.), pp. 69–73. ESA SP-278, Noordwijk, The Netherlands.

Wurm K. (1968) Structure and kinematics of cometary type 1 tails. *Icarus, 8,* 287–300.

Wegmann R. (1995) MHD model calculations for the effect of interplanetary shocks on the plasma tail of a comet. *Astron. Astrophys., 294,* 601.

Wegmann R. (2002) Large-scale disturbance of the solar wind by a comet. *Astron. Astrophys., 389,* 1039–1046.

Wegmann R., Schmidt H. U., Huebner W. F., and Boice D. C. (1987) Cometary MHD and chemistry. *Astron. Astrophys., 187,* 339–350.

Wyckoff S. (1982) Overview of comet observations. In *Comets* (L. L. Wilkening, ed.), pp. 3–55. Univ. of Arizona, Tucson.

Yi Y., Brandt J. C., Randall C. E., and Snow M. (1993) The disconnection events of 1986 April 13–18 and the cessation of plasma tail activity in Comet Halley in 1986 May. *Astrophys. J., 414,* 883.

Yi Y., Caputo F. M., and Brandt J. C. (1994) Disconnection events (DEs) and sector boundaries: The evidence from Comet Halley 1985–1986. *Planet Space Sci., 42,* 705–720.

X-Ray and Extreme Ultraviolet Emission from Comets

C. M. Lisse
University of Maryland

T. E. Cravens
University of Kansas

K. Dennerl
Max-Planck-Institut für Extraterrestrische Physik

The discovery of high energy X-ray emission in 1996 from C/1996 B2 (Hyakutake) has created a surprising new class of X-ray emitting objects. The original discovery (*Lisse et al.,* 1996) and subsequent detection of X-rays from 17 other comets (Table 1) have shown that the very soft (E < 1 keV) emission is due to an interaction between the solar wind and the comet's atmosphere, and that X-ray emission is a fundamental property of comets. Theoretical and observational work has demonstrated that charge exchange collisions of highly charged solar wind ions with cometary neutral species is the best explanation for the emission. Now a rapidly changing and expanding field, the study of cometary X-ray emission appears to be able to lead us to a better understanding of a number of physical phenomena: the nature of the cometary coma, other sources of X-ray emission in the solar system, the structure of the solar wind in the heliosphere, and the source of the local soft X-ray background.

1. INTRODUCTION

Astrophysical X-ray emission is generally found to originate from hot collisional plasmas, such as the million-degree gas found in the solar corona (e.g., *Foukal,* 1990), the 100-million-degree gas observed in supernova remnants (e.g., *Cioffi,* 1990), or the accretion disks around neutron stars and black holes. As electromagnetic radiation with wavelength λ between about 0.01 nm and 100 nm (1 nm = 10^{-9} m), extreme ultraviolet (EUV) and X-ray radiation are important for solar system and astrophysical applications because the photons are sufficiently energetic and penetrating to ionize neutral atoms and molecules, and can thus drive chemical reactions. In fact, current estimates of the X-ray burden per atom in the young solar system are some 10^3–10^4 photons per atom, even as far out as the proto-Kuiper belt, as the young Sun was much more X-ray active than now (*Feigelson,* 1982; *Dorren et al.,* 1995).

The Sun is not the only source of X-rays in the solar system (*Cravens,* 2000a, 2002a). Prior to 1996, X-rays were found in scattering of solar X-rays from the terrestrial atmosphere and in the terrestrial aurora, as scattered solar X-rays off the illuminated surface of the Moon, and from the jovian aurora. Nonetheless, the 1996 discovery, using the Röntgen Satellite (ROSAT), by Lisse and co-workers (*Lisse et al.,* 1996) (Fig. 1) of strong X-ray emission from Comet C/1996 B2 (Hyakutake) was very surprising because cometary atmospheres are known to be cold and tenuous, with characteristic temperatures between 10 and 1000 K. Compared to other X-ray sources, comets are moderately weak — the

total X-ray power, or luminosity, of C/Hyakutake was measured to be approximately 10^9 W. The emission was also extremely "soft", or of low characteristic photon energy — only X-rays of energies less than ~1 keV (or wavelengths longer than ~1.2 nm) were detected from C/Hyakutake. The total amount of energy emitted in X-rays from a comet is approximately 10^{-4} the energy delivered to a comet from the Sun due to photon insolation and solar wind impact (*Lisse et al.,* 2001).

2. OBSERVED CHARACTERISTICS OF COMETARY X-RAY EMISSION

Shortly after the initial C/Hyakutake detection, soft X-ray emission from four other comets was found in the ROSAT archival database, confirming the discovery (*Dennerl et al.,* 1997). X-ray emission has now been detected from 18 comets to date (Table 1) using a variety of X-ray sensitive spacecraft — BeppoSAX, ROSAT, the Extreme Ultraviolet Explorer (EUVE), and more recently, Chandra and XMM-Newton. All comets within 2 AU of the Sun and brighter than V = 12 have been detected when observed. We now recognize that X-ray emission is a characteristic of all active comets.

The observed characteristics of the emission can organized into the following four categories: (1) spatial morphology, (2) total X-ray luminosity, (3) temporal variation, and (4) energy spectrum. Any physical mechanism that purports to explain cometary X-ray emission must account for all these characteristics.

TABLE 1.   Observation times, instruments, and energies* for comets detected†
through December 2002 in the X-ray or extreme ultraviolet.

Comet	Time	Instrument	Energy (keV)	Detection	Reference
45P/Honda-Mrkos-Pajdušáková	Jul 1990	ROSAT PSPC/WFC	0.09–2.0	Yes	[1]
C/1990 K1 (Levy)	Sep 1990	ROSAT PSPC /WFC	0.09–2.0	Yes	[1]
	Jan 1991		0.09–2.0	Yes	[1]
C/1991 A2 Arai	Nov 1990	ROSAT PSPC/WFC	0.09–2.0	Yes	[1]
C/1990 N1 (Tsuchiya-Kiuchi)	Nov 1990	ROSAT PSPC/WFC	0.09–2.0	Yes	[1]
	Jan 1991		0.09–2.0	Yes	[1]
2P/Encke	Nov 1993	EUVE DS	0.02–0.10	No	[8]
	Jul 1997	EUVE Scanners	0.02–0.18	Yes	[9]
	Jul 1997	ROSAT HRI/WFC	0.09–2.0	Yes	[9]
19P/Borrelly	Nov 1994	EUVE DS	0.02–0.10	Yes	[8]
6P/d'Arrest	Sep 1995	EUVE DS	0.02–0.10	Yes	[8]
C/1996 B2 (Hyakutake)	Mar 1996	ROSAT HRI /WFC	0.09–2.0	Yes	[2]
	Mar 1996	EUVE DS	0.02–0.10	Yes	[3]
	Mar 1996	ALEXIS	0.06–0.10	No	
	Apr 1996	XTE PCA	2.0–10.0	No	
	Jun 1996	ASCA	0.20–6.0	No	
	Jun 1996	ROSAT HRI /WFC	0.09–2.0	Yes	[4]
	Jul 1996	ROSAT HRI /WFC	0.09–2.0	Yes	
	Aug 1996	ROSAT HRI /WFC	0.09–2.0	Yes	
	Sep 1996	ROSAT HRI /WFC	0.09–2.0	Yes	
C/1995 O1 (Hale-Bopp)	Apr 1996	ROSAT HRI/WFC	0.09–2.0	No	
	Sep 1996	EUVE Scanners	0.02–0.10	Yes	[5]
	Sep 1996	BeppoSAX	0.1–200	Yes	[6]
	Sep 1996	ROSAT HRI/WFC	0.09–2.0	Yes?	[7]
	Sep 1996	ASCA	0.20–6.0	No	[5]
	Mar 1997	XTE PCA	2.0–10.0	No	
	Oct 1997	ROSAT HRI /WFC	0.09–2.0	No	
	Nov 1997	ROSAT HRI /WFC	0.09–2.0	No	
	Nov 1997	EUVE DS	0.02–0.10	Yes	[8]
	Feb 1998	ROSAT PSPC/WFC	0.09–2.0	No	
C/1996 Q1 (Tabur)	Sep 1996	ASCA	0.20–6.0	No	
	Sep 1996	ROSAT HRI /WFC	0.09–2.0	Yes	[1]
	Oct 1996	ROSAT HRI /WFC	0.09–2.0	Yes	[1]
55P/Temple-Tuttle	Jan 1998	EUVE Scanners	0.02–0.18	Yes	
	Jan 1998	ASCA SIS	0.20–6.0	No	
	Jan 1998	ROSAT HRI /WFC	0.09–2.0	Yes	
	Feb 1998	ROSAT HRI /WFC	0.09–2.0	Yes	
103P/Hartley 2	Feb 1998	ROSAT PSPC/WFC	0.09–2.0	Yes	
C/1998 U5 (LINEAR)	Dec 1998	ROSAT PSPC/WFC	0.09–2.0	Yes	
C/2000 S4 (LINEAR)	Jul 2000	Chandra ACIS-S	0.2–10.0	Yes	[10]
	Aug 2000	Chandra ACIS-S	0.2–10.0	Yes	[10]
C/1999 T1 McNaught-Hartley	Jan 2001	Chandra ACIS-S	0.2–10.0	Yes	[11]
	Jan 2001	XMM-Newton	0.2–12.0	Yes	
	Feb 2001	FUSE	0.113–0.117	No	[12]

TABLE 1. (continued).

Comet	Time	Instrument	Energy (keV)	Detection	Reference
C/2001 A2 (LINEAR)	Jun 2001	Chandra HRC/LETG	0.2–2.0	No	
	Jun 2001	XMM-Newton	0.2–12.0	Yes?	
	Jul 2001	FUSE	0.113–0.117	No	[12]
C/2000 WM1 (LINEAR)	Dec 2001	Chandra ACIS/LETG	0.2–2.0	Yes	
	Dec 2001	FUSE	0.113–0.117	Yes	[12]
	Jan 2002	XMM-Newton	0.2–12.0	Yes	[13]
C/2002 C1 (Ikeya-Zhang)	Apr 2002	Chandra ACIS-S	0.2–10.0	Yes	[13]
	May 2002	XMM-Newton	0.2–12.0	Yes?	[13]

*The full energy range of the observing instrument is given.

†This table summarizes published and unpublished (1) dedicated observations, whether successful or not; and (2) successful serendipitous observations. The table is sorted according to time of first observation of the comet.

References: [1] *Dennerl et al.* (1997); [2] *Lisse et al.* (1996); [3] *Mumma et al.* (1997); [4] *Lisse et al.* (1997a); [5] *Krasnopolsky et al.* (1997); [6] *Owens et al.* (1998); [7] *Lisse et al.* (1997b); [8] *Krasnopolsky et al.* (2000); [9] *Lisse et al.* (1999); [10] *Lisse et al.* (2001); [11] *Krasnopolsky et al.* (2002); [12] *Weaver et al.* (2002); [13] K. Dennerl et al. (personal communication, 2003).

2.1. Spatial Morphology

X-ray and EUV images of C/1996 B2 (Hyakutake) made by the ROSAT and EUVE satellites look very similar (*Lisse et al.*, 1996; *Mumma et al.*, 1997) (Fig. 1). Except for images of C/1990 N1 (*Dennerl et al.*, 1997) and C/Hale-Bopp 1995 O1 (*Krasnopolsky et al.*, 1997), all EUV and X-ray images of comets have exhibited similar spatial morphologies. The emission is largely confined to the cometary coma between the nucleus and the Sun; no emission is found in the extended dust or plasma tails. The peak X-ray brightness gradually decreases with increasing cometocentric distance r with a dependence of about r^{-1} (*Krasnopolsky*, 1997). The brightness merges with the soft X-ray background emission (*McCammon and Sanders*, 1990; *McCammon et*

al., 2002) at distances that exceed 10^4 km for weakly active comets, and can exceed 10^6 km for the most luminous (*Dennerl et al.*, 1997) (Fig. 2). The spatial extent for the most extended comets is independent of the rate of gas emission from the comet. The region of peak emission is crescent-shaped with a brightness peak displaced toward the Sun from the nucleus (*Lisse et al.*, 1996, 1999). The distance of this peak from the nucleus appears to increase with increasing values of Q, and for Hyakutake was located at $r_{peak} \approx 2 \times 10^4$ km.

2.2. Luminosity

The observed X-ray luminosity, L_x, of C/1996 B2 (Hyakutake) was 4×10^{15} ergs s^{-1} (*Lisse et al.*, 1996) for an aper-

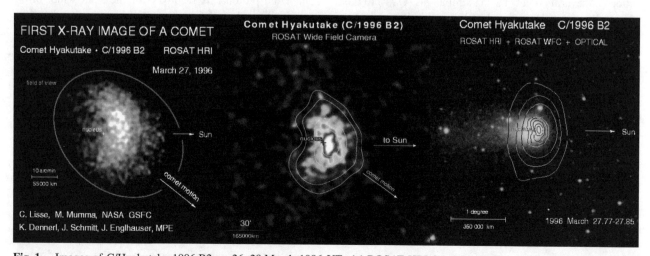

Fig. 1. Images of C/Hyakutake 1996 B2 on 26–28 March 1996 UT: **(a)** ROSAT HRI 0.1–2.0 keV X-ray; **(b)** ROSAT WFC 0.09–0.2 keV extreme ultraviolet; and **(c)** visible light, showing a coma and tail, with the X-ray emission contours superimposed. The Sun is toward the right, "+" marks the position of the nucleus, and the orbital motion of the comet is toward the lower right in each image. From *Lisse et al.* (1996).

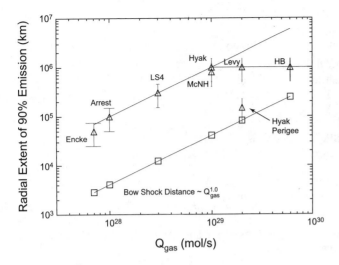

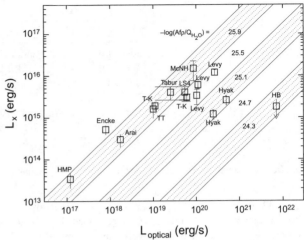

Fig. 2. Spatial extent of the X-ray emission vs. the comet's outgassing rate. Plot of the gas production rate Q_{gas} vs. radial distance from the comet nucleus required to encircle 95% of the total observed cometary X-ray flux (triangles). Upper curve: Broken power law with radial extent $\sim Q_{gas}^{1.00}$ up to $Q_{gas} \sim 10^{29}$ mol s^{-1} and $\sim 10^6$ km for higher values fits the imaging data well. Lower curve: Estimated radius of the bow shock for each observation, allowing for variable cometary outgassing activity and heliocentric distance (boxes). X-ray emission has been found outside the bow shock for all comets except C/1996 B2 (Hyakutake) in March 1996.

Fig. 3. X-ray vs. optical luminosity plot for the eight detected ROSAT comets and the Chandra comets C/1999 S4 (LINEAR) and C/1999 T1 (McNaught-Hartley) observed at 1–3 AU. Groups of equal emitted dust mass to emitted gas mass ratio (D/G), as measured in the optical by the ratio $Afρ/Q_{H_2O}$, are also shown. For Encke and other "gassy", optically faint comets, the resulting slope L_x/L_{opt} is roughly constant. Above $L_{opt} \sim$ few $\times 10^{19}$ erg s^{-1}, however, L_x appears to reach an asymptote of $\sim 5 \times 10^{16}$ erg s^{-1}. A possible explanation is that the coma is collisionally thick to the solar wind within the neutral coma radius of $\sim 10^6$ km at 1 AU (*Lisse et al.,* 2001). It is also possible that the relatively large amounts of dust in these comets, as noted from their increasing D/G ratio, may be somehow inhibiting the CXE process. Following *Dennerl et al.* (1997); copyright journal *Science* (1997).

ture radius at the comet of 1.2×10^5 km. [Note that the photometric luminosity depends on the energy bandpass and on the observational aperture at the comet. The quoted value assumes a ROSAT photon emission rate of $P_X \approx 10^{25}$ s^{-1} (0.1–0.6 keV), in comparison to *Krasnopolsky et al.*'s (2000) EUVE estimate of $P_{EUV} \approx 7.5 \times 10^{25}$ s^{-1} (0.07–0.18 keV and 120,000 km aperture.] A positive correlation between optical and X-ray luminosities was demonstrated using observations of several comets having similar gas (Q_{H_2O}) to dust [$Afρ$, following *A'Hearn et al.* (1984)] emission rate ratios (Fig. 3) (*Lisse et al.,* 1997b, 1999, 2001; *Dennerl et al.,* 1997; *Mumma et al.,* 1997; *Krasnopolsky et al.,* 2000). L_x correlates more strongly with the gas production rate Q_{gas} than it does with $L_{opt} \sim Q_{dust} \sim Afρ$ (Figs. 2 and 3). Particularly dusty comets, like Hale-Bopp, appear to have less X-ray emission than would be expected from their overall optical luminosity L_{opt}. The peak X-ray surface brightness decreases with increasing heliocentric distance r, independent of Q (*Dennerl et al.,* 1997), although the total luminosity appears roughly independent of r. The maximum soft X-ray luminosity observed for a comet to date is $\sim 2 \times 10^{16}$ erg s^{-1} for C/Levy at 0.2–0.5 keV (*Dennerl et al.,* 1997) (Fig. 3).

2.3. Temporal Variation

Photometric lightcurves of the X-ray and EUV emission typically show a long-term baseline level with superimposed

impulsive spikes of a few hours' duration, and maximum amplitude 3 to 4 times that of the baseline emission level (*Lisse et al.,* 1996, 1999, 2001). Figure 4 demonstrates the strong correlation found between the time histories of the solar wind proton flux (a proxy for the solar wind minor ion flux), the solar wind magnetic field intensity, and a comet's X-ray emission for the case of Comet 2P/Encke 1997 (*Lisse et al.,* 1999). *Neugebauer et al.* (2000) compared the ROSAT and EUVE luminosity of C/1996 B2 (Hyakutake) with time histories of the solar wind proton flux, oxygen ion flux, and solar X-ray flux, as measured by spacecraft residing in the solar wind. They found the strongest correlation between the cometary emission and the solar wind oxygen ion flux, a good correlation between the comet's emission and the solar wind proton flux, but no correlation between the cometary emission and the solar X-ray flux.

For the four comets for which extended X-ray lightcurves were obtained during quiet Sun conditions, the time delay between the solar wind proton flux and the comet's X-ray impulse (Table 2) was well predicted by assuming a simple latitude-independent solar wind flow, a quadrupole solar magnetic field, and propagation of the sector boundaries radially at the speed of the solar wind and azimuth-

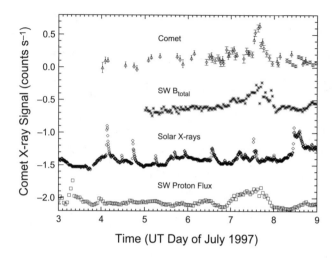

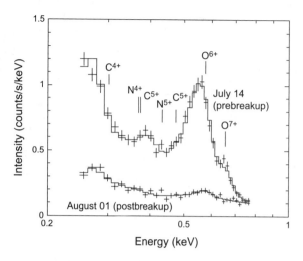

Fig. 4. Temporal trends for Comet 2P/Encke 1997 on 4–9 July 1997 UT. △ = ROSAT HRI lightcurve, 4–8 July 1997. ◇ = EUVE scanner Lexan B lightcurve 6–8 July 1997 UT, taken contemporaneously with the HRI observations, and scaled by a factor of 1.2. All error bars are ±1σ. Also plotted are the WIND total magnetic field B_{total} (✳), the SOHO CELIAS/SEM 1.0–500-Å solar X-ray flux (◇), and the SOHO CELIAS solar wind proton flux (□). There is a strong correlation between the solar wind magnetic field/density and the comet's emission. There is no direct correlation between outbursts of solar X-rays and the comet's outbursts. After *Lisse et al.* (1997a).

Fig. 5. Chandra ACIS-S medium resolution CCD X-ray spectrum for Comet C/1999 S4 (LINEAR). Soft X-ray spectrum of C/1999 S4 (LINEAR) obtained by the Chandra X-ray Observatory (crosses) and a six-line best-fit "model" spectrum (solid line). The positions of several possible atomic lines are noted. Adapted from *Lisse et al.* (2001).

ally with period one-half the solar rotation period of 28 d (*Lisse et al.*, 1997b, 1999; *Neugebauer et al.*, 2000)

$$\Delta t_{total} = \Delta t_{Carrington\ rotation} + \Delta t_{radial} =$$

$$\frac{longitude_{comet} - longitude_{Earth}}{14.7°/d} + \frac{(r_{comet} - r_{Earth})}{400\ km/s \cdot 86400\ s/d}$$

2.4. Spectrum

Until 2001, all published cometary X-ray spectra had very low spectral energy resolution (ΔE/E ~ 1 at 300–600 eV), and the best spectra were those obtained by ROSAT for C/1990 K1 (Levy) (*Dennerl et al.*, 1997) and by BeppoSAX

for Comet C/1995 O1 (Hale-Bopp) (*Owens et al.*, 1998). These observations were capable of showing that the spectrum was very soft (characteristic thermal bremsstrahlung temperature kT ~ 0.23 ± 0.04 keV) with intensity increasing toward lower energy in the 0.01–0.60 keV energy range, and established upper limits to the contribution of the flux from K-shell resonance fluorescence of carbon at 0.28 keV and oxygen at 0.53 keV. However, even in these "best" spectra, continuum emission could not be distinguished from a multiline spectrum. Nondetections of Comets C/Hyakutake, C/Tabur, C/Hale-Bopp, and 55P/Temple-Tuttle using the XTE PCA (2–30 keV) and ASCA SIS (0.6–4 keV) imaging spectrometers were consistent with an extremely soft spectrum (*Lisse et al.*, 1996, 1997b).

Higher-resolution spectra of cometary X-ray emission have just appeared in the literature. The Chandra X-ray Observatory (CXO) detected soft X-ray spectra from Comet C/1999 S4 (LINEAR) (*Lisse et al.*, 2001) over an energy range of 0.2–0.8 keV, using an energy resolution with a full-

TABLE 2. Predicted and observed lightcurve phase shifts using the latitude-independent model.

Comet	Time of Impulse (00:00H UT)	Δt_{long} (d)	Δt_{radial} (d)	Δt_{total} (d)	$\Delta t_{observed}$ (d)
Hyakutake	27 Mar 1996	−0.23	0.032	−0.20	−0.24
Hale-Bopp	11 Sep 1996	−4.60	5.9	1.30	+1.4
Encke	7 Jul 1997	−0.26	0.093	−0.17	−0.1
Temple-Tuttle	29 Jan 1998	−2.31	0.37	−1.94	−2.5

Time shifts assume solar wind velocity as measured near-Earth; positive time shifts = impulse happens at Earth first, comet next; negative time shifts = boundary hits comet first, Earth next.

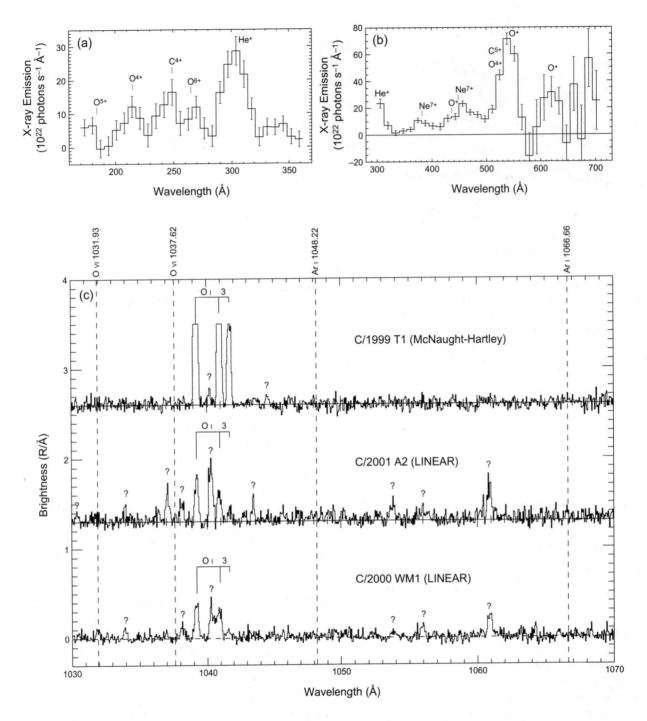

Fig. 6. EUVE observations of line emission from C/1996 B2 (Hyakutake), following *Krasnopolsky and Mumma* (2001). **(a)** MW (middle wavelength) 0.034–0.073 keV spectrum on March 23, 1996. **(b)** LW (long-wavelength) 0.018–0.04 keV. The extreme ultraviolet spectra are clearly dominated by line emission. The best agreement with CXE model predictions are for the O⁴⁺, C⁴⁺, and Ne⁷⁺ lines. **(c)** FUSE observations of three comets, with a marginal detection of the CXE OVI line in C/2001 WM1 (LINEAR) at 1032 Å (following *Weaver et al.*, 2002). The nondetections in Comets C/2001 A2 (LINEAR) and C/1999 T1 (McNaught-Hartley) and the marginal detection in C/2001 WM1 (LINEAR) are consistent with CXE predictions for the luminosity of these lines.

width half-maximum (FWHM) of $\Delta E = 0.11$ keV (Fig. 5). The spectrum is dominated by line emission, not by continuum. Using the CXO, a new spectrum of Comet C/1999 T1 (McNaught-Hartley) (*Krasnopolsky et al.*, 2002) shows similar line-emission features. Line emission is also found in XMM-Newton spectra of Comet C/1999 T1 (McNaught

Hartley) and, more recently, in CXO spectra of C/2001 WM1 (LINEAR) and C/2002 Ikeya-Zhang (K. Dennerl et al. and C. M. Lisse et al., personal communication, 2003). An XMM-Newton spectrum of C/2001 WM1 (LINEAR) shows characteristic CXE X-ray signatures in unprecedented detail (K. Dennerl et al., personal communication, 2003). A reanalysis of archival EUVE Deep Survey spectrometer spectra (*Krasnopolsky and Mumma,* 2001) suggests EUV line-emission features from Comet C/1996 B2 (Hyakutake) (Figs. 6a,b). Recent FUSE observations (*Weaver et al.,* 2002) also indicate the presence of possible O VI 1032-Å emission lines in far-UV spectra of C/2001 WM1 (LINEAR) (Fig. 6c).

3. PROPOSED X-RAY MECHANISMS

A large number of explanations for cometary X-rays were suggested following the discovery paper in 1996. These included thermal bremsstrahlung (German for "braking radiation") emission due to solar wind electron collisions with neutral gas and dust in the coma (*Bingham et al.,* 1997; *Dawson et al.,* 1997; *Northrop,* 1997; *Northrop et al.,* 1997; *Uchida et al.,* 1998; *Shapiro et al.,* 1999), microdust collisions (*Ibadov,* 1990; *Ip and Chow,* 1997), K-shell ionization of neutrals by electron impact (*Krasnopolsky,* 1997), scattering or fluorescence of solar X-rays by cometary gas or by small dust grains (*Lisse et al.,* 1996; *Wickramasinghe and Hoyle,* 1996; *Owens et al.,* 1998), and by charge exchange between highly ionized solar wind ions and neutral species in the cometary coma (CXE) (*Cravens,* 1997a; *Häberli et al.,* 1997; *Wegmann et al.,* 1998; *Kharchenko and Dalgarno,* 2000; *Kharchenko et al.,* 2003; *Schwadron and Cravens,* 2000). In the thermal bremsstrahlung mechanism, fast electrons are deflected in collisions with charged targets, such as the nuclei of atoms, and emit continuum radiation. Electron energies in excess of 100 eV ($T > 10^6$ K) are needed for the production of X-ray photons. In the K-shell mechanism, a fast electron collision removes an orbital electron from an inner shell of the target atom. Early evaluation of these various mechanisms (*Dennerl et al.,* 1997; *Krasnopolsky,* 1997; *Lisse et al.,* 1999) favored only three of them: the CXE mechanism, thermal bremsstrahlung, and scattering of solar radiation from very small (i.e., attogram; 1 attogram = 10^{-19} g) dust grains.

A significant problem with mechanisms involving solar wind electrons (i.e., bremsstrahlung or K-shell ionization) is that the predicted emission luminosities are too small by factors of 100–1000 compared to observations. The flux of high-energy solar wind electrons near comets is too low (*Krasnopolsky,* 1997; *Cravens,* 2000b, 2002a). Furthermore, X-ray emission has been observed out to great distances from the nucleus, beyond the bow shock (Fig. 2), and the thermal energy of unshocked solar wind electrons at these distances is about 10 eV. No emission has ever been found to be associated with the plasma tail of a comet, which has similar plasma densities and temperatures. Finally, the new, high-resolution spectra demonstrating multiple atomic lines are inconsistent with a continuum-type mechanism or a mechanism producing only a couple of K-shell lines as the primary source of cometary X-rays. *Lisse et al.* (2001) tried several thermal bremsstrahlung continuum model fits to the C/1999 S4 spectrum, and *Krasnopolsky and Mumma* (2001) tried the same for the C/1996 B2 (Hyakutake) spectrum, but neither was successful.

Mechanisms based on dust grains also have a number of problems. It has been known since 1996 that Rayleigh scattering of solar X-ray radiation from ordinary cometary dust grains (i.e., about 1 μm in size) cannot produce the observed luminosities — the cross section for this process is too small (*Lisse et al.,* 1996). A potential solution to this problem is to invoke a population of very small, attogram (10^{-19} g) grains with radii on the order of the wavelengths of the observed X-ray radiation, 10–100 Å, which can resonantly scatter the incident X-ray radiation. The abundance of such attogram dust grains is not well understood in comets, as they are undetectable by remote optical observations; however, there were reports from the VEGA Halley flyby of a detection of an attogram dust component using the PUMA dust monitor (*Vaisberg et al.,* 1987; *Sagdeev et al.,* 1990). However, the statistical studies of the properties of several comets (Figs. 2 and 3) demonstrate that X-ray emission varies with a comet's gas production rate and not the dust production rate (*Dennerl et al.,* 1997; *Lisse et al.,* 1999, 2001). Furthermore, the cometary X-ray lightcurves (*Lisse et al.,* 1996, 1999, 2001; *Neugebauer et al.,* 2000) correlate with the solar wind ion flux and not with solar X-ray intensity. Finally, dust-scattering mechanisms cannot account for the pronounced lines seen in the new high-resolution spectra — emission resulting from dust scattering of solar X-rays should mimic the Sun's X-ray spectral continuum, similar to what is observed in the terrestrial atmosphere for Rayleigh scattering of sunlight (*Krasnopolsky,* 1997).

The CXE mechanism requires that the observed X-ray emission is driven by the solar wind flux and that the bulk of the observed X-ray emission be in lines. Localization of the emission to the sunward half of the coma, a solar wind flux-like time dependence, and a line-emission-dominated spectral signature of the observed emission all strongly point to the solar wind charge exchange mechanism as being responsible for cometary X-rays.

4. SOLAR WIND CHARGE EXCHANGE X-RAY MECHANISM

The solar wind is a highly ionized but tenuous gas (i.e., a plasma) (*Cravens,* 1997b). At its source in the solar corona, the million-degree gas is relatively dense and in collisional equilibrium, but its density drops within a few solar radii into a freeflow regime wherein collisions are infrequent. Both the solar wind and corona have "solar" composition — 92% hydrogen by volume, 8% helium, and 0.1% heavier elements. The heavier, "minor ion" species are highly charged (e.g., oxygen in the form of hydrogen-like O^{7+} or helium-like O^{6+} ions, N^{6+}/N^{5+}, C^{5+}/C^{4+}, Ne^{8+}, Si^{9+}, Fe^{12+}, etc.) due to the high coronal temperatures (*Bame,* 1972; *Bocshler,* 1987; *Neugebauer et al.,* 2000).

The solar wind flow starts out slowly in the corona but becomes supersonic at a distance of few solar radii (*Parker,* 1963; *Cravens,* 1997b). The gas cools as it expands, falling from $T \approx 10^6$ K down to about 10^5 K at 1 AU. The average properties of the solar wind at 1 AU are proton number density ≈ 7 cm^{-3}, speed ≈ 450 km s^{-1}, temperature $\approx 10^5$ K, magnetic field strength ≈ 5 nT, and Mach number ≈ 8 (*Hundhausen et al.*, 1968). However, the composition and charge state distribution far from the Sun are "frozen in" at coronal values due to the low collision frequency outside the corona. The solar wind contains structure, such as slow (400 km s^{-1}) and fast (700 km s^{-1}) streams, which can be mapped back to the Sun. The solar wind "terminates" in a shock called the heliopause, where the ram pressure of the streaming solar wind has fallen to that of the instellar material (ISM) gas (*Suess*, 1990). The region of space that contains plasma of solar origin, from the corona to the heliopause at ~ 100 AU, is called the heliosphere. A very small part of the solar wind interacts with the planets and comets; the bulk of the wind interacts with neutral ISM gas in the heliosphere and neutral and ionized ISM material at the heliopause.

As the solar wind streams into a comet's atmosphere, cometary ion species produced from solar UV photoionization of neutral coma gas species are added to the flow as "pick-up ions." The resulting mass addition slows down the solar wind due to momentum conservation and a bow shock forms upwind of the comet (*Galeev*, 1991; *Szegö et al.*, 2000) (Fig. 7). The flow changes from supersonic to subsonic across the shock, and the magnetic field strength increases by a factor of ~ 5. Closer to the nucleus, where the cometary gas density is higher and collisions more frequent, the flow almost completely stagnates (*Flammer*, 1991). The outer boundary of this stagnation region is often called the cometopause (cf. review by *Cravens*, 1991). The observed X-ray brightness peak resides within this boundary. Mag-

netic field lines pile up into a "magnetic barrier" in this stagnation region and drape around the head of the comet forming the plasma tail in the downwind direction (*Brandt*, 1982).

From experimental and theoretical work in atomic and molecular physics it is found that solar wind minor ions readily undergo charge transfer (or exchange) reactions (*Phaneuf et al.*, 1982; *Dijkkamp et al.*, 1985; *Gilbody*, 1986; *Janev et al.*, 1988; *Wu et al.*, 1988) when they are within ~ 1 nm of a neutral atomic species

$$A^{q+} + B \rightarrow A^{(q-1)+*} + B^+ \qquad (1)$$

where A denotes the solar wind projectile ion (e.g., O, C, Si . . .), q is the projectile charge (e.g., q = 5, 6, 7) and B denotes the neutral target species (e.g., H_2O, OH, CO, O, H . . . for cometary comae) (Fig. 8). The cross section for this process is large, on the order of 10^{-15} cm^2, about 1 order of magnitude larger than the hardsphere collisional cross section.

The product ion deexcites by emitting one or more photons ($A^{(q-1)+*} \rightarrow A^{(q-1)+*} + h\nu$, where $h\nu$ represents a photon). It is the characteristic radiation of the product ion that is measured with astronomical X-ray instrumentation, and so one labels the radiation detected by the charge state of the final ion. The deexcitation usually takes place via a cascade through intermediate states rather than in one step to the ground state. For large enough values of q, the deexcitation transitions lead to the emission of X-ray photons. For species and charge states relevant to comets, the principal quantum number of the ion $A^{(q-1)+}$ is about n = 4 or 5

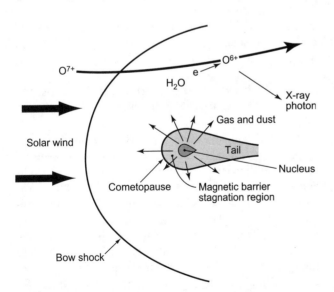

Fig. 7. Spatial schematic of the solar wind-comet interaction. The relative locations of the bow shock, the magnetic barrier, and the tail are shown (not to scale). The Sun is toward the left. Also represented is a charge transfer collision between a heavy solar wind ion and a cometary neutral water molecule, followed by the emission of an X-ray photon. After *Cravens* (2002b); copyright journal *Science* (2002).

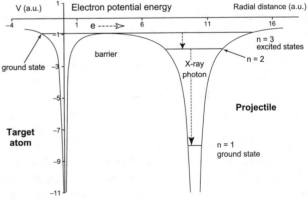

Fig. 8. Energy level diagram for a CXE process. Electron potential energy (atomic units, a.u.) vs. distance from the target atom nucleus (assumed here to be an H atom) for a charge transfer reaction involving a projectile ion Be^{4+}. The internuclear distance chosen is 10 Bohr radii or 5.29×10^{-10} m, the curve-crossing distance for the n = 3 ion final state. The target energy level (and binding energy E_b) and product ion (Be^{3+}) energy levels are shown in units of hartrees (1 hartree = 27.2 eV). A possible cascading pathway for the deexcitation by photon emission is shown. After *Cravens* (2002b); copyright journal *Science* (2002).

(*Ryufuku et al.,* 1980; *Mann et al.,* 1981). The cross section for charge exchange between a high charge state solar wind ion and an ionized coma gas species is negligible in comparison, due to the effects of Coulomb repulsion between the two reactants. Once a coma neutral atom is ionized, either by CXE processes or solar UV flux, the CXE mechanism is no longer an important energy transfer process.

4.1. Charge Exchange Morphology

Numerical simulations of the solar wind interaction with Hyakutake including CXE have been used to generate X-ray images. A global magnetohydrodynamic (MHD) model (*Häberli et al.,* 1997) and a hydrodynamic model (*Wegmann et al.,* 1998) were used to predict solar wind speeds and densities and the X-ray emission around a comet. The simulated X-ray images are similar to the observed images, which is relatively unsurprising, as any dissipative solar wind–coma process with total optical depth near unity would create the observed morphology (Fig. 1). The emission is found to lie in the sunward hemisphere of the neutral coma, varying from collisionally thin to collisionally thick as the solar wind approaches the nucleus. Emission is predicted to be in the soft X-ray, UV, and optical wavelengths, with the softest photons emitted closest to the nucleus. However, the spatial models to date have included only highly simplified models of the CXE deexcitation cascade, as compared to the detailed spectral models of the global behavior discussed below.

As an example of the potential of studying the behavior of the solar wind inside the coma using CXE reactions, we consider the dissimilar morphologies of the extended Lyman α comae and the X-ray emitting regions of comets (cf. *Keller,* 1973; *Festou et al.,* 1979; *Combi et al.,* 2000). The CXE mechanism should not only transfer electrons from cometary neutrals to solar wind minor ions, but to the solar wind majority ions H^+ and He^{2+} as well. These ions are roughly 1000 times more abundant than the X-ray active highly ionized minor ions. The prompt photon-emission energy produced from CXE by He^{2+} ions produces at least three times the energy released by all the minor ions combined (D. Shemansky, personal communication, 2003). Further, the neutral atoms produced are capable of scattering emission from the Sun. At luminosities of $\sim 10^{16}$ erg s^{-1}, and production rates of $\sim 10^{27}$ s^{-1}, the HI created by CXE should be detectable in Lyman α comet images. The fact that it is not is puzzling. A possible solution is that the neutral hydrogen atoms produced by CXE retain relatively large velocities with respect to the Sun, i.e., the solar wind is not appreciably slowed at the cometary bow shock. This large remnant velocity redshifts the CXE-produced neutral hydrogen with respect to the peak of the solar Lyman α emission, so that fluorescence from these atoms is greatly reduced in efficiency vs. H atoms produced from dissolution of cometary water group species. Recently *Raymond et al.* (2002) have reported the effects of highly redshifted neutral hydrogen created by CXE emission in SOHO UVCS measurements of 2P/Encke during its 1997 apparition.

4.2. Charge Exchange Luminosity and Temporal Variation

To first order, the CXE local X-ray power density P_x can be estimated assuming only one CXE collision per solar wind ion per coma passage. This approximation yields the expression

$$P_x = \alpha n_{sw} u_{sw} n_n \tag{2}$$

where n_{sw} u_{sw}, and n_n are the solar wind proton density, solar wind speed, and neutral target density respectively (*Cravens,* 1997a, 2002a). All the "atomic and molecular details" as well as the solar wind heavy ion fraction f_h are combined into the parameter α, given by $\alpha \approx f_h \langle s_{ct} \rangle E_{ave}$, where $\langle s_{ct} \rangle$ is an average CXE cross section for all species and charge states, and E_{ave} an average photon energy. A simple spherically symmetric approximation to the neutral density in the coma is given by $n_n = Q/[4\pi u_n r^2]$, for r less than the ionization scale length $R = u_n t$, where $t \approx 10^6$ s is the ionization lifetime [for 1 AU (*Schleicher and A'Hearn,* 1988)] and $u_n \approx$ 1 km s^{-1} the neutral gas outflow speed. Integration of P_x over the volume of the neutral coma yields an X-ray luminosity typically within a factor of 2–3 of the observed luminosity (*Cravens,* 1997a, 2000a; *Lisse et al.,* 2001). The observed luminosity is a function of both the solar wind flux density and the com-etary neutral gas production rate up to the limit of 100% charge exchange efficiency of all solar wind minor ions within an ionization scale length of 10^6 km. The maximum expected X-ray luminosity at 1 AU and 0.2–0.5 keV is $\sim 10^{16}$ erg s^{-1} (Fig. 3). Temporal variations of the solar wind flux directly translate into time variations of the X-ray emission (Fig. 4).

4.3. Charge Exchange Spectra

Model CXE spectra are in good agreement with the low-resolution X-ray spectra of cometary X-ray emission, and the line centers of the high-resolution spectra have been successfully predicted using CXE theory (Figs. 5 and 6) (*Lisse et al.,* 1999, 2001; *Krasnopolsky and Mumma,* 2001; *Weaver et al.,* 2002). The application of the CXE model to comets has entailed a number of approaches to date. Some work has included only a few solar wind species but used a careful cascading scheme (*Häberli et al.,* 1997). Other approaches have used a simple cascading scheme and simple collision cross sections, but included a large number of solar wind ions and charge states (*Wegmann et al.,* 1998; *Schwadron and Cravens,* 2000). Kharchenko and colleagues (*Kharchenko and Dalgarno,* 2000; *Kharchenko et al.,* 2003) have treated the atomic cascading process more carefully than other modelers, although the spatial structure of the solar wind–cometary neutral interaction in their model was highly simplified. They predicted the existence of a large number of atomic lines, including O^{5+} ($1s^2 5d \rightarrow 1s^2 2p$) at 106.5 eV, C^{4+} ($1s2s \rightarrow 1s^2$) at 298.9 eV, C^{5+} ($2p \rightarrow 1s$) at 367.3 eV, C^{5+} ($4p \rightarrow 1s$) at 459.2 eV, and O^{6+} ($1s2p \rightarrow 1s^2$) at 568.4 eV. At least some of these lines appear in the best cometary X-ray spectra to date, the CXO ACIS-S spectra of C/1999 S4 (LINEAR) and C/1999 T1 (McNaught-Hartley). The peak measured near 0.56 keV is certainly the combination of three closely spaced helium-like O^{6+} (1s2p and 1s2s → 1s²) transitions (which comes from CXE of solar wind O^{7+}) and the line located at 0.32 keV is due to helium-like C^{4+} (1s2p and 1s2s → 1s²). Similar identifications have been made for the EUVE spectrum of C/1996 B2 (Hyakutake) (*Krasnopolsky and Mumma*, 2001). The spectral problem is still far from totally solved, however. Careful comparisons and calculations needed to interpret the subtleties of the high-resolution spectral observations, including the role of collisions after charge transfer, solar wind ion–dust interactions, and the exact species present at each point in the coma, are only now starting to be done (*Krasnopolsky et al.*, 2002; *Kellett et al.*, 2003).

5.   THE FUTURE OF COMETARY X-RAY EMISSION STUDIES

The discovery that comets are X-ray sources can now be explained by charge exchange reactions of highly charged solar wind ions with neutral atoms and molecules residing in the cometary coma. The energy required to power this emission originates in the hot solar corona and is stored as potential energy in highly stripped solar wind ions (*Cravens*, 2000a, 2002b; *Dennerl*, 1999). Charge exchange reactions with neutral species, like cometary coma neutral atoms and molecules, release this potential energy in the form of X-ray and UV radiation. This simple model can explain the gross features of the observed crescent-shaped emission with its sunward displaced peak, the maximum spatial extent of the emission of ~10^6 km (Figs. 1 and 2), the maximum observed luminosity of ~10^{16} erg s⁻¹ (Fig. 3), and spectra dominated by line emission (Fig. 4).

5.1.   Plasma-Neutral Interactions in the Cometary Coma

However, a more careful treatment of the CXE mechanism is needed to fully understand the phenomenon of cometary X-ray emission. One complication is that the multiple CXE collisions take place in regions close to the nucleus where the target density is high (the so-called "collisionally thick" case). The charge state for the initially hydrogen-like or helium-like solar wind minor ion is reduced by one during each CXE collision, ultimately leading to its conversion to a neutral atom. When the ion's charge state becomes too low, X-ray photons are no longer emitted. When the number of neutrals in a volume of space becomes too low, due to coma expansion and ionization of coma gas molecules by solar UV radiation, X-ray photons are no longer detectable. Another complication is that spectral differences are expected for slow and fast solar wind streams, with the slow solar wind, with its higher coronal freeze-in temperature, producing a harder spectrum than does the fast solar wind (*Schwadron and Cravens*, 2000).

Further theoretical progress will require the integration of several ingredients into a single model: (1) accurate solar wind composition for a range of solar wind types; (2) a suitable MHD model of the solar wind interaction with the coma, in order to accurately predict the densities of ions and neutrals in equation (2); and (3) a more detailed understanding of the atomic processes, in order to improve our understanding of the parameter α in equation (2). Success for the first point requires new and improved measurements of the solar wind throughout the heliosphere, and/or large number statistical studies of cometary X-ray emission throughout the heliosphere; the second point requires improved MHD codes on modern supercomputers; and to achieve the third point, additional laboratory measurements of state-specific CXE cross sections will be required. The first and second points are being actively pursued by astronomers and modelers in the field. New laboratory work is now being undertaken (*Beiersdorfer et al.*, 2000, 2001; *Greenwood et al.*, 2000; *Hasan et al.*, 2001) to measure CXE cross sections for cometary target species such as H_2O at collision energies relevant to the solar wind (a few keV/amu). Recent measurements have indicated, for example, that multiple as well as single-electron CXE makes a contribution to the X-ray emission (*Hasan et al.*, 2001; *Greenwood et al.*, 2001; *Gao and Kwong*, 2002). Given detailed MHD models of the solar wind passing through the coma and accurate cross sections for the CXE process, we will be able to map out the density of solar wind minor ions in the coma.

5.2.   Remote Sensing of the Solar Wind

Driven by the solar wind, cometary X-rays provide an observable link between the solar corona, where the solar wind originates, and the solar wind where the comet resides. Once we have understood the CXE mechanism's behavior in cometary comae in sufficient detail, we will be able to use comets as probes to measure the solar wind throughout the heliosphere. This will be especially useful in monitoring the solar wind in places hard to reach with spacecraft — over the solar poles, at large distances above and below the ecliptic plane, and at heliocentric distances greater than a few AU (*Lisse et al.*, 1996, 2001; *Krasnopolsky et al.*, 2000). For example, approximately one-third of the observed soft X-ray emission is found in the 530–700-eV oxygen O^{7+} and O^{6+} lines; observing photons of this energy allows studies of the oxygen ion charge ratio of the solar wind, which is predicted to vary significantly between the slow and fast solar winds (*Neugebauer et al.*, 2000; *Schwadron and Cravens*, 2000; *Kharchenko and Dalgarno*, 2001).

5.3.   Emission from Other Planetary Systems

The CXE mechanism operates wherever the solar wind (or any highly ionized plasma) interacts with a substantial quantity of neutral gas. Motivated in part by the discovery of the new class of cometary X-ray emitters, further measurements of potential solar system sources (*Holmström et al.,* 2001; *Cravens and Maurellis,* 2001) of X-ray emission have been undertaken recently. X-rays have now been observed from Venus as solar X-rays fluorescently scattered by the thick (compared to comets) venusian atmosphere (*Dennerl et al.,* 2002), and from Mars as ~90% fluorescently scattered solar X-rays from the thick martian atmosphere and ~10% CXE-derived X-rays (*Dennerl,* 2002). The solar wind is known to approach very closely to these planets due to their weak intrinsic magnetic fields, which cannot act as effective obstacles to the external flow, allowing for CXE processes to occur. The small spatial extent and high density of the gravitationally bound planetary atmospheres provides a higher X-ray luminosity due to scattering of solar X-rays than from CXE. X-rays have now been detected at Io and Europa and in the Io flux torus using Chandra, although the exact mechanism of emission is still unknown (*Elsner et al.,* 2002). Another neutral gas system from which X-ray emission has been predicted is the terrestrial hydrogen geocorona (*Cox,* 1998; *Cravens,* 2000a; *Dennerl,* 1999; *Freyberg,* 1998; *Robertson and Cravens,* 2003), although no detection has yet been made.

X-ray emission from the heliosphere is also expected from the interaction of the solar wind with the interstellar neutral gas (mainly HI and HeI) that streams into the solar system (*Cox,* 1998; *Cravens,* 2000a,b). *Cravens* (2000b) demonstrated that roughly one-half of the observed 0.25-keV X-ray diffuse background can be due to this process. Solar and Heliospheric Observatory (SOHO) observations of neutral hydrogen Lyman α emission show a clear asymmetry in the ISM flow direction, with a clear deficit of neutral hydrogen in the downstream direction of the incoming neutral ISM gas, most likely created by CXE ionization of the ISM as it transits the heliosphere. *Cravens et al.* (2001) have shown a strong correlation between the solar wind flux density and the ROSAT "long-term enhancements," systematic variations in the soft X-ray background of the ROSAT X-ray detectors. Photometric imaging observations of the lunar nightside by Chandra made in September 2001 do not show any lunar nightside emission above a CXE background. The soft X-ray emission detected from the darkside of the Moon, using ROSAT, would appear to be due not to electrons spiraling from the sunward to the dark hemisphere, as proposed by *Schmitt et al.* (1991), but instead be due to CXE in the column of solar wind between Earth and the Moon. The analogous process applied to other stars has been suggested as a means of detecting stellar winds (*Lisse,* 2002a,b; *Wargelin and Drake,* 2001).

5.4. Soft X-Ray Background

Heliospheric and geocoronal X-ray emission have been suggested (*Cravens,* 2000b, 2002b; *Dennerl,* 1999; *Lisse et al.,* 2001) to make significant contributions to the observed soft X-ray background [previously attributed entirely to hot interstellar gas (*Snowden et al.,* 1990, 1998; *McCammon et al.,* 2002)]. This supposition is supported by the positive cor-

relations that have been found between measured solar wind fluxes and measured X-ray background intensities (*Cravens et al.,* 2001; *Robertson et al.,* 2001). Recent large angular scale measurements of the diffuse soft X-ray background in a 100-s rocket flight by *McCammon et al.* (2002) and in the Chandra background toward MBM12 (R. Edgar et al., personal communication, 2001) show a clear peak at 560 eV, as expected for CXE-driven emission (Fig. 4). A significant fraction of the line emission observed in the soft X-ray background, e.g., the 560-eV emission, is probably due to CXE emission in our solar system and around other stars.

REFERENCES

A'Hearn M. F., Schleicher D. G., Millis R. L., Feldman P. D., and Thompson D. T. (1984) Comet Bowell 1980b. *Astron. J., 89,* 579–591.

Bame S. J. (1972) Spacecraft observations of the solar wind composition. In *Solar Wind* (C. P. Sonnett et al., eds.), p. 529. NASA SP-308, Washington, DC.

Beiersdorfer P., Olson R. E., Brown G. V., Chen H., Harris C. L., Neill P. A., Schweikhard L., Utter S. B., and Widmann K. (2000) X-ray emission following low-energy charge exchange collisions of highly charged ion. *Phys. Rev. Lett., 85,* 5090–5093.

Beiersdorfer P., Chen H., Lisse C. M., Olson R., and Utter S. B. (2001) Laboratory simulation of cometary x-ray emission. *Astrophys. J. Lett., 549,* L147.

Bingham R., Dawson J. M., Shapiro V. D., Mendis D. A., and Kellett B. J. (1997) Generation of X-rays from C/Hyakutake 1996B2. *Science, 275,* 49–51.

Bochsler P. (1987) Solar wind ion composition. *Phys. Scripta, T18,* 55–60.

Brandt J. C. (1982) Observations and dynamics of plasma tails. In *Comets* (L. L. Wilkening, ed.), pp. 519–537. Univ. of Arizona, Tucson.

Cioffi D. F. (1990) Supernova remnants as probes of the interstellar medium. In *Physical Processes in Hot Cosmic Plasma* (W. Brinkmann et al., eds.), p. 17. Kluwer, Dordrecht.

Combi M. R., Reinard A. A., Bertaux J.-L., Quemerais E., and Makinen T. (2000) SOHO/SWAN observations of the structure and evolution of the hydrogen Lyman-α coma of Comet Hale-Bopp (1995 O1). *Icarus, 144,* 191–202.

Cox D. P. (1998) Modeling the local bubble. In *Lecture Notes in Physics, Vol. 506,* pp. 121–131. Springer-Verlag, Berlin.

Cravens T. E. (1991) Plasma processes in the inner coma. In *Comets in the Post-Halley Era* (R. L. Newburn Jr. et al., eds.), pp. 1211–1255. Kluwer, Dordrecht.

Cravens T. E. (1997a) Comet Hyakutake x-ray source: Charge transfer of solar wind heavy ions. *Geophys Res. Lett., 24,* 105–109.

Cravens T. E. (1997b) *Physics of Solar System Plasmas,* p. 477. Cambridge Univ., Cambridge.

Cravens T. E. (2000a) X-ray emission from comets and planets. *Adv. Space Res., 26,* 1443–1451.

Cravens T. E. (2000b) Heliospheric X-ray emission associated with charge transfer of the solar wind with interstellar neutrals. *Astrophys. J. Lett., 532,* L153–L156.

Cravens T. E. (2002a) X-ray emission in the solar system. In *Atomic Processes in Plasmas: 13th APS Topical Conference on Atomic Processes in Plasmas* (D. R. Schultz et al., eds.), pp. 173–181. AIP Conference Proceedings 635, American Institute of Physics, Melville, New York.

Cravens T. E. (2002b) X-ray emission from comets. *Science, 296,* 1042–1046.

Cravens T. E. and Maurellis A. N. (2001) X-ray emission from scattering and fluorescence of solar X-rays at Venus and Mars. *Geophys. Res. Lett., 28,* 3043.

Cravens T. E., Robertson I. P., and Snowden S. L. (2001) Temporal variations of geocoronal and heliospheric x-ray emission associated with the solar wind interaction with neutrals. *J. Geophys. Res., 106,* 24883–24892.

Dawson J. M., Bingham R., and Shapiro V. D. (1997) X-rays from Comet Hyakutake. *Plasma Phys. Control. Fusion, 39,* A185–A193.

Dennerl K. (1999) X-ray emission from comets. In *Atomic Physics Vol. 16: Sixteenth International Conference on Atomic Physics* (W. E. Baylis and G. W. F. Drake, eds.), pp. 361–376. AIP Conference Proceedings 477, Springer-Verlag, Berlin.

Dennerl K. (2002) Discovery of X-rays from Mars with Chandra. *Astron. Astrophys., 394,* 1119–1128.

Dennerl K., Englhauser J., and Trümper J. (1997) X-ray emissions from comets detected in the Röntgen X-ray satellite all-sky survey. *Science, 277,* 1625–1629.

Dennerl K., Burwitz V., Englhauser J., Lisse C. M., and Wolk S. (2002) Discovery of X-rays from Venus with Chandra. *Astron. Astrophys., 386,* 319–330.

Dijkkamp D., Gordeev Y. S., Brazuk A., Drentje A. G., and de Heer F. J. (1985) Selective single-electron capture into (n, l) subshells in slow collisions of C^{6+}, N^{6+}, O^{6+} and Ne^{6+} with He, H_2 and Ar. *J. Phys. B. Atom. Mol. Phys., 18,* 737–756.

Dorren J. D., Guedel M., and Guinan E. F. (1995) X-ray emission from the Sun in its youth and old age. *Astrophys. J., 448,* 431–436.

Elsner R. F. and 15 colleagues (2002) Discovery of soft X-ray emission from Io, Europa, and the Io plasma torus. *Astrophys. J., 572,* 1077–1082.

Feigelson E. D. (1982) X-ray emission from young stars and implications for the early solar system. *Icarus, 51,* 155–163.

Festou M., Jenkins E. B., Barker E. S., Upson W. L., Drake J. F., Keller H. U., and Bertaux J. L. (1979) Lyman-alpha observations of comet Kobayashi-Berger-Milon (1975 IX) with Copernicus. *Astrophys. J., 232,* 318–328.

Flammer K. R. (1991) The global interaction of comets with the solar wind. In *Comets in the Post-Halley Era* (R. L. Newburn Jr. et al., eds.), pp. 1125–1144. Kluwer, Dordecht.

Foukal P. (1990) *Solar Astrophysics.* Wiley, New York. 475 pp.

Freyberg M. J. (1998) On the zero-level of the soft X-ray background. In *The Local Bubble and Beyond* (D. Breitschwerdt et al., eds.), pp. 113–116. Lecture Notes in Physics, Vol. 506, Springer-Verlag, Berlin.

Galeev A. A. (1991) Plasma processes in the outer coma. In *Comets in the Post-Halley Era* (R. L. Newburn Jr. et al., eds.), pp. 1145–1169. Kluwer, Dordrecht.

Gao H. and Kwong V. H. S. (2002) Charge transfer of O^{+5} and O^{+4} with CO at keV energies. *Astrophys. J., 567,* 1272–1275.

Gilbody H. B. (1986) Measurements of charge transfer and ionization in collisions involving hydrogen atoms. *Adv. Atom. Mol. Phys., 22,* 143–195.

Greenwood J., Williams I. D., Smith S. J., and Chutjian A. (2000) Measurement of charge exchange and X-ray emission cross sections for solar wind-comet interactions. *Astrophys. J. Lett., 533,* L175–L178.

Greenwood J. B., Williams I. D., Smith S. J., and Chutjian A.

(2001) Experimental investigation of the processes determining X-ray emission intensities from charge-exchange collisions. *Phys. Rev. A, 63,* 627071–627079.

Häberli R. M., Gombosi T. I., deZeeuw D. L., Combi M. R., and Powell K. G. (1997) Modeling of cometary X-rays caused by solar wind minor ions. *Science, 276,* 939–942.

Hasan A. A., Eissa F., Ali R., Schultz D. R., and Stancil P. C. (2001) State-selective charge transfer studies relevant to solar wind-comet interactions. *Astrophys. J. Lett., 560,* L201–L205.

Holmström M. S., Barabash S., and Kallio E. (2001) X-ray imaging of the solar wind-Mars interaction. *Geophys. Res. Lett., 28,* 1287.

Hundhausen A. J., Gilbert H. E., and Bame S. J. (1968) Ionization state of the interplanetary plasma. *J. Geophys. Res., 73,* 5485–5493.

Ibadov S. (1990) On the efficiency of X-ray generation in impacts of cometary and zodiacal dust particles. *Icarus, 86,* 283–288.

Ip W.-H. and Chow V. W. (1997) NOTE: On hypervelocity impact phenomena of microdust and nano X-ray flares in cometary comae. *Icarus, 130,* 217–221.

Janev R. K., Phaneuf R. A., and Hunter H. T. (1988) Recommended cross sections for electron capture and ionization in collisions of C^{q+} and O^{q+} ions with H, He and H_2. *Atomic Data and Nuclear Data Tables, 40,* 249.

Keller H. U. (1973) Lyman-alpha radiation in the hydrogen atmospheres of comets: A model with multiple scattering. *Astron. Astrophys., 23,* 269–280.

Kellett B. J., Bingham R., Lisse C. M., Torney M., Summers H. P., and Shapiro V. D. (2003) A detailed model of the X-ray emission from comets. In *Plasma Physics: 11th International Congress on Plasma Physics (ICPP2002)* (I. S. Falconer et al., eds.), pp. 704–707. AIP Conference Proceedings 669, American Institute of Physics, Melville, New York.

Kharchenko V. and Dalgarno A. (2000) Spectra of cometary X rays induced by solar wind ions. *J. Geophys. Res., 105,* 18351–18360.

Kharchenko V. and Dalgarno A. (2001) Variability of cometary X-ray emission induced by solar wind ions. *Astrophys. J. Lett., 554,* L99–L102.

Kharchenko V., Rigazio M., Dalgarno A., and Krasnopolsky V. A. (2003) Charge abundances of the solar wind ions inferred from cometary X-ray spectra. *Astrophys. J. Lett., 585,* L73–L75.

Krasnopolsky V. A. (1997) On the nature of soft X-ray radiation in comets. *Icarus, 128,* 368–385.

Krasnopolsky V. A. and Mumma M. J. (2001) Spectroscopy of Comet Hyakutake at 80–700 Å: First detection of solar wind charge transfer emissions. *Astrophys. J., 549,* 629–634.

Krasnopolsky V. A., Mumma M. J., Abbott M., Flynn B. C., Meech K. J., Yeomans D. K., Feldman P. D., and Cosmovici C. B. (1997) Discovery of soft X-rays from Comet Hale-Bopp using EUVE. *Science, 277,* 1488–1491.

Krasnopolsky V. A., Mumma M. J., and Abbott M. J. (2000) EUVE search for X-rays from Comets Encke, Mueller (C/1993 A1), Borrelly, and postperihelion Hale-Bopp. *Icarus, 146,* 152–160.

Krasnopolsky V. A., Christian D. J., Kharchenko V., Dalgarno A., Wolk S. J., Lisse C. M., and Stern S. A. (2002) X-ray emission from Comet McNaught-Hartley (C/1999 T1). *Icarus, 160,* 437–447.

Lisse C. M. (2002a) Cometary X-ray emission — the view after the first Chandra observation of a comet. In *The High Energy Universe at Sharp Focus: Chandra Science* (E. M. Schlegel and S. B. Vrtilek, eds.), pp. 3–17. ASP Conference Series 262.

Lisse C. M. (2002b) Cometary X-rays — the EUVE photometric

legacy. In *Continuing the Challenge of EUV Astronomy: Current Analysis and Prospects for the Future* (S. B. Howell et al., eds.), pp. 254–267. ASP Conference Series 264.

Lisse C. M. and 11 colleagues (1996) Discovery of X-ray and extreme ultraviolet emission from Comet Hyakutake C/1996 B2. *Science, 274,* 205–209.

Lisse C. M., Mumma M. J., Petre R., Dennerl K., Englhauser J., Schmitt J., and Truemper J. (1997a) *Comet C/1996 B2 (Hyakutake).* IAU Circular No. 6433.

Lisse C. M., Dennerl K., Englhauser J., Trümper J., Marshall F. E., Petre R., Valina A., Kellett B. J., and Bingham R. (1997b) X-ray emission from Comet Hale-Bopp. *Earth Moon Planets, 77,* 283–291.

Lisse C. M., Christian D., Dennerl K., Englhauser J., Trümper J., Desch M., Marshall F. E., Petre R., Snowden S. (1999) X-ray and extreme ultraviolet emission from Comet P/Encke 1997. *Icarus, 141,* 316–330.

Lisse C. M., Christian D. J., Dennerl K., Meech K. J., Petre R., Weaver H. A., and Wolk S. J. (2001) Charge exchange-induced X-ray emission from Comet C/1999 S4 (LINEAR). *Science, 292,* 1343–1348.

Mann R., Folkmann F., and Beyer H. F. (1981) Selective electron capture into highly stripped Ne and N target atoms after heavy-ion impact. *J. Phys. B.: Atom. Molec Phys., 14,* 1161–1181.

McCammon D. and Sanders W. T. (1990) The soft X-ray background and its origins. *Annu. Rev. Astron. Astrophys., 28,* 657–688.

McCammon D. and 19 colleagues (2002) High spectral resolution observation of the soft X-ray diffuse background with thermal detectors. *Astrophys. J., 576,* 188–203.

Mumma M. J., Krasnopolsky V. A., and Abbott M. J. (1997) Soft X-rays from four comets observed by EUVE. *Astrophys. J. Lett., 491,* L125–L128.

Neugebauer M., Cravens T. E., Lisse C. M., Ipavich F. M., Christian D., von Steiger R., Bochsler P., Shah P. D., Armstrong T. P. (2000) The relation of temporal variations of soft X-ray emission from Comet Hyakutake to variations of ion fluxes in the solar wind. *J. Geophys. Res., 105,* 20949–20956.

Northrop T. G. (1997) The spectrum of X-rays from Comet Hyakutake. *Icarus, 128,* 480–482.

Northrop T. G., Lisse C. M., Mumma M. J., and Desch M. D. (1997) A possible source of the X-rays from Comet Hyakutake. *Icarus, 127,* 246–250.

Owens A., Parmar A. N., Oostrbroek T., Orr A., Antonelli L. A., Fiore F., Schultz R., Tozzi G. P., Macarone M. C., and Piro L. (1998) Evidence for dust-related X-ray emission from Comet C/1995 O1 (Hale-Bopp). *Astrophys J., 493,* 47–51.

Parker E. N. (1963) *Interplanetary Dynamical Processes.* Wiley, New York. 272 pp.

Phaneuf R. A., Alvarez I., Meyer F. W., and Crandall D. H. (1982) Electron capture in low-energy collisions of C^q+ and O^q+ with H and H_2. *Phys. Rev. A, 26,* 1892–1906.

Raymond J. C., Uzzo M., Ko Y.-K., Mancuso S., Wu R., Gardner L., Kohl J. L., Marsden B., and Smith P. L. (2002) Far-ultraviolet observations of Comet 2P/Encke at perihelion. *Astrophys. J., 564,* 1054–1060.

Robertson I. P. and Cravens T. E. (2003) X-ray emission from the terrestrial magnetosheath. *Geophys. Res. Lett., 30,* 1439–1442.

Robertson I. P., Cravens T. E., Snowden S., and Linde T. (2001) Temporal and spatial variations of heliospheric X-ray emissions associated with charge transfer of the solar wind with interstellar neutrals. *Space Sci. Rev., 97,* 401–405.

Ryufuku H., Sasaki K., and Watanabe T. (1980) Oscillatory behavior of charge transfer cross sections as a function of the charge of projectiles in low-energy collisions. *Phys. Rev. A, 21,* 745–750.

Sagdeev R. Z., Evlanov E. N., Zubkov B. V., Prilutskii O. F., and Fomenkova M. N. (1990) Detection of very fine dust particles near the nucleus of Comet Halley. *Sov. Astron. Lett., 16,* 315–318.

Schleicher D. G. and A'Hearn M. F. (1988) The fluorescence of cometary OH. *Astrophys. J., 331,* 1058–1077.

Schmitt J. H. M. M., Snowden S. L., Aschenbach B., Hasinger G., Pfeffermann E., Predehl P., and Trumper J. (1991) A soft X-ray image of the Moon. *Nature, 349,* 583–587.

Schwadron N. A. and Cravens T. E. (2000) Implications of solar wind composition for cometary X-rays. *Astrophys. J., 544,* 558–566.

Shapiro V. D., Bingham R., Dawson J. M., Dobe Z., Kellett B. J., and Mendis D. A. (1999) Energetic electrons produced by lower hybrid waves in the cometary environment and soft X ray emission: Bremsstrahlung and K shell radiation. *J. Geophys. Res., 104,* 2537–2554.

Snowden S. L., Cox D. P., McCammon D., and Sanders W. T. (1990) A model for the distribution of material generating the soft X-ray background. *Astrophys. J., 354,* 211–219.

Snowden S. L., Egger R., Finkbeiner D. P., Freyberg M. J., and Plucinsky P. P. (1998) Progress on establishing the spatial distribution of material responsible for the 1/4 keV soft X-ray diffuse background local and halo components. *Astrophys. J., 493,* 715.

Suess S. T. (1990) The heliopause. *Rev. Geophys., 28,* 97–115.

Szegö K. and 20 colleagues (2000) Physics of mass loaded plasmas. *Space Sci. Rev., 94,* 429–671.

Uchida M., Morikawa M., Kubotani H., and Mouri H. (1998) X-ray spectra of comets. *Astrophys. J., 498,* 863–870.

Vaisberg O. L., Smirnov V., Omel'Chenko A., Gorn L., and Iovlev M. (1987) Spatial and mass distribution of low-mass dust particles (m < 10^{-10} g) in Comet P/Halley's coma. *Astron. Astrophys., 187,* 753–760.

Wargelin B. J. and Drake J. J. (2001) Observability of stellar winds from late-type dwarfs via charge exchange X-ray emission. *Astrophys. J. Lett., 546,* L57–L60.

Weaver H. A., Feldman P. D., Combi M. R., Krasnopolsky V., Lisse C. M., and Shemansky D. E. (2002) A search for argon and O VI in three comets using the Far Ultraviolet Spectroscopic Explorer. *Astrophys. J. Lett., 576,* L95–L98.

Wegmann R., Schmidt H. U., Lisse C. M., Dennerl K., and Englhauser J. (1998) X-rays from comets generated by energetic solar wind particles. *Planet. Space Sci. 46,* 603–612.

Wickramsinghe N. C. and Hoyle F. (1996) Very small dust particles (VSDP's) in Comet C/1996 B2 (Hyakutake). *Astrophys. Space Sci., 239,* 121–123.

Wu W. K., Huber B. A., and Wiesemann K. (1988) Cross sections for electron capture by neutral and charged particles in collisions with He. *Atomic Data and Nuclear Data Tables, 40,* 57.

Part VII:
Interrelations

Surface Characteristics of Transneptunian Objects and Centaurs from Photometry and Spectroscopy

M. A. Barucci and A. Doressoundiram
Observatoire de Paris

D. P. Cruikshank
NASA Ames Research Center

The external region of the solar system contains a vast population of small icy bodies, believed to be remnants from the accretion of the planets. The transneptunian objects (TNOs) and Centaurs (located between Jupiter and Neptune) are probably made of the most primitive and thermally unprocessed materials of the known solar system. Although the study of these objects has rapidly evolved in the past few years, especially from dynamical and theoretical points of view, studies of the physical and chemical properties of the TNO population are still limited by the faintness of these objects. The basic properties of these objects, including information on their dimensions and rotation periods, are presented, with emphasis on their diversity and the possible characteristics of their surfaces.

1. INTRODUCTION

Transneptunian objects (TNOs), also known as Kuiper belt objects (KBOs) and Edgeworth-Kuiper belt objects (EKBOs), are presumed to be remnants of the solar nebula that have survived over the age of the solar system. The connection of the short-period comets (P < 200 yr) of low orbital inclination and the transneptunian population of primordial bodies (TNOs) has been established and clarified on the basis of dynamics (*Fernández*, 1999), and it is generally accepted that the Kuiper belt is the source region of these comets. Centaur objects appear to have been extracted from the TNO population through perturbations by Neptune. While their present (temporary) orbits cross the orbits of the outer planets, Centaurs do not come sufficiently close to the Sun to exhibit normal cometary behavior, although 2060 Chiron has a weak and temporally variable coma.

We do not know if the traditional and typical short-period comets, which have dimensions of a few kilometers to less than 1 km, are fragments of TNOs or if they are themselves primordial objects. The surface material of TNOs may not survive entry into the inner solar system (*Jewitt*, 2002) where it can be observed and (eventually) sampled, so it is particularly important to investigate the composition of TNOs, which may be the most primitive matter in the solar system. The surfaces of the Centaurs may represent intermediate stages in the compositional evolution of TNOs to short-period comets.

As a consequence of their great distances and relatively small dimensions, TNOs and Centaur objects are very faint; the first one, discovered by Jewitt and Luu (*Jewitt et al.*, 1992) at magnitude ~22.8 was some 7400 times fainter than Pluto. Even the brightest TNOs presently known are magnitude ~19, making them difficult to observe spectroscopi-

cally with even the largest telescopes. The physical characteristics of Centaurs and TNOs are still in a rather early stage of investigation. Advances in instrumentation on telescopes of 6- to 10-m aperture have enabled spectroscopic studies of an increasing number of these objects, and significant progress is slowly being made.

We describe here photometric and spectroscopic studies of TNOs and the emerging results.

2. OBSERVATIONAL STRATEGY AND DATA REDUCTION TECHNIQUES

2.1. Photometry

Visible- and near-infrared (NIR)-wavelength CCDs with broad-band filters operating in the range of 0.3 to 2.5 μm provide the basic set of observations on most objects discovered so far, yielding color indices, rotational properties and estimates of the sizes of TNOs. The color indices (e.g., U-V, B-V, V-R, V-I, V-J, V-H, V-K) are the differences between the magnitudes measured in two filters, and represent an important tool to study the surface composition of these objects and to define a possible taxonomy. [The broadband filters commonly used (and their central wavelength position in micrometers) are U (0.37), B (0.43), V (0.55), R (0.66), I (0.77), J (1.25), H (1.65), and K (2.16).]

Because of their faintness, slow proper motion, and rotation, TNOs require specific observational procedures and data reduction techniques. The typical apparent visual magnitude is about 23 or fainter, although a few objects brighter than 22 have been found. A signal-to-noise ratio (SNR) of about 30 (precision of 0.03–0.04 mag) is the photometric accuracy required for color analysis. Two problems limit the signal precision achievable: (1) the sky contribution to the measured signal and (2) the contamination of the signal by unseen background sources, such as field stars and galaxies. For instance, the error introduced by a magnitude 26 background source superimposed on an object of magnitude 23 is as large as 0.07 mag. One solution for alleviating these problems is the use of a very small synthetic aperture around the object when measuring its flux on a CCD image, and the application of the aperture correction technique (*Howell*, 1989).

Even though TNOs orbit at large heliocentric distances, their motion on the sky restricts observations to relatively short exposure times. At opposition, the nonsidereal motions of TNOs at 30, 40, and 50 AU are about 4.2, 3.2, and 2.6"/h respectively, thus producing a trail of ~1.0" in 15-min exposure time (in the worst case). Trailed images have devastating effects on the SNR since the flux is diluted over a larger and noisier area of background sky. Thus, increasing exposure time will not improve the SNR. Practically, exposure times should be chosen so that the trail length does not exceed the seeing disk. One alternative would be to follow the object at its proper motion. But in this case the point spread function (PSF) of the object is different from field stars, thus thwarting the aperture correction technique, which must be calibrated from the PSFs of nearby field stars. The proper motions of TNOs and the long exposure times needed to detect them at an adequate SNR limit the number of objects that can be observed, even with a big telescope. For each individual B, V, R, etc., magnitude obtained, 1σ uncertainties are based on the combination of several uncertainties: $\sigma = (\sigma_{pho}^2 + \sigma_{ap}^2 + \sigma_{cal}^2)^{1/2}$, where the photometric uncertainty (σ_{pho}) is based on photon statistics and sky noise, the uncertainty on the aperture correction (σ_{ap}) is determined from the dispersion among measurements of the different field stars, and σ_{cal} is the uncertainty derived from absolute calibration through standard stars.

2.2. Spectroscopy

Reflectance spectroscopy (0.3 to 2.5 μm) provides the most sensitive and broadly applied remote sensing technique for characterizing the major mineral phases and ices present on TNOs. At visible and NIR wavelengths, recognizable spectral absorptions arise from the presence of the silicate minerals pyroxene, olivine, and sometimes feldspar, as well as primitive carbonaceous assemblages and organic tholins. The NIR wavelength region carries signatures from water ice (1.5, 1.65, 2.0 μm), other ices (CH_4 around 1.7 and 2.3 μm, CH_3OH at 2.27 μm, and NH_3 at 2 and 2.25 μm), and solid C-N bearing material at 2.2 μm. Water-bearing minerals such as phyllosilicates also exhibit absorption features at visible wavelengths.

Although the most reliable mineralogical interpretations require measurements extending into the NIR, measurements restricted to the visible wavelengths (0.3–1.0 μm) can be used to infer information on the composition, particularly for the especially "red" objects, whose reflectance increases rapidly with wavelength in this region (see below).

Spectroscopic observations face the same problems as photometric observations due to the specific nature of TNOs discussed in the previous section. With 8-m-class telescopes, the limiting magnitude at the present time is V = 22.5 mag for visible spectroscopy, and the object must be brighter than ~21 mag (in V) for NIR spectroscopy. On the same large-aperture telescope, the exposure time required is between one and several hours for the faintest objects. During long exposures the rotation rate is not negligible and the resulting spectra probably arise from signals from both sides of the object. Careful removal of the dominant sky background (atmospheric emission bands) in the infrared and the choice of good solar analogs are essential steps to ensure high-quality data.

3. DIAMETER, ROTATIONAL PROPERTIES, AND SHAPE

Many useful physical and compositional parameters of TNOs can be derived from broadband photometry. Size, rotational properties, and shape are the most basic parameters defining a solid body. The rotation spin can be the result of the initial angular momentum determined by formation processes, constraining the origin and evolution of this population of objects. Some of the large TNOs might conserve their original angular momentum, while many others suffered collisional processes and do not retain the memory of the primordial angular momentum.

3.1. Diameter

The sizes of TNOs cannot be measured directly, as the objects are not in general spatially resolved. At the time of writing the largest known TNO, 50000 Quaoar, is resolved in an HST image at 40.4 ± 1.8 milliarcsec, yielding a diameter of 1260 ± 190 km (*Brown and Trujillo*, 2003). Only a few objects have been observed at thermal and millimetric wavelengths and thus have directly determined diameters and albedos (Table 1), while for the majority an indication of the diameter can be obtained from the absolute magnitude, assuming an albedo value. With an assumed value for the surface albedo p_v of an object, the absolute magnitude (H) can be converted into the diameter D (km) using the formula from *Harris and Harris* (1997). Owing to the lack of available albedo measurements, it has become the convention to assume an albedo of 0.04–0.05, which is common for dark objects and cometary nuclei (e.g., *Lamy et al.*, 2004). This assumption introduces a large uncertainty in the size estimates; for instance, if we instead used an albedo of 0.14 (i.e., the albedo of the Centaur 2060 Chiron), all the size estimates would have to be divided by about two.

3.2. Rotational Period

The observed variations of brightness with time allow the determination of the rotational period of a body. In Table 1 a fairly complete list of the most reliably determined

TABLE 1. Dynamical type (Centaurs and classical, Plutinos and scattered for the TNOs),
rotational period, lightcurve amplitude, diameters, albedo, and spectral signature
characteristics for each object (numbers shown in brackets are references).

Name	Type	Rotation Periods (h)	Amplitude (mag)	Diameter (km)	Albedo p_v	Spectral Signatures*
2060 Chiron	Centaur	5.917813 ± 0.000007 [1]	0.04 [1]	148 ± 8 [2]	0.17 ± 0.02 [2]	H$_2$O ice varying with activity
5145 Pholus	Centaur	9.9825 ± 0.0040 [3]	0.15 [3]	190 ± 22 [4]	0.04 ± 0.03 [4]	Water ice + hydrocarbons
8405 Asbolus	Centaur	8.9351 ± 0.0003 [5]	0.55 [5]	66 ± 4 [2]	0.12 ± 0.03 [2]	Controversial
10199 Chariklo	Centaur	Long ? [6]	0.31 [6]	302 ± 30 [7]	0.045 ± 0.010 [7]	H$_2$O ice
				273 ± 19 [8]	0.055 ± 0.008 [8]	
15789 (1993 SC)	Plutino	15.43 [9]?	0.5 [9]	328 ± 66 [10]	0.022 ± 0.013 [10]	Controversial
19308 (1996 TO$_{66}$)	Classical	7.9 [11]	0.25 [11]	≈748	—	H$_2$O ice
		6.25 ± 0.01 [12]	0.12–0.33 [12]			Variation
20000 Varuna	Classical	6.3442 ± 0.0002 [13]	0.42 [13]	1060 ± 220 [14]	0.038 ± 0.022 [14]	H$_2$O ice
		6.3576 ± 0.0002 [14]	0.42 [14]	900 ± 145 [15]	p_R = 0.07 ± 0.03 [15]	
26308 (1998 SM$_{165}$)	Classical	7.966 [16]	0.56 [16]	≈411	—	
		7.1 [11]	0.45 [11]			
26375 (1999 DE$_9$)	Scattered	No variation over 24 h [17]		≈682	—	Hydrous silicates
28976 Ixion	Plutino			1065 ± 165 [23]	—	No features
31824 Elatus	Centaur	13.25? [18]	0.24 [18]	≈57	—	H$_2$O ice?
		13.41 ± 0.04 [19] (single peak)	0.102 [19]			Variation?
32532 Thereus	Centaur	8.3 [22]	0.16 [22]	≈95	—	Surface variation
		8.3378 ± 0.0012 [20]	0.18 [20]			
32929 (1995 QY$_9$)	Plutino	7.3 [11]	0.60 [11]	≈188	—	
33128 (1998 BU$_{48}$)	Centaur	9.8–12.6 [17]	0.68 [17]	≈216	—	
35671 (1998 SN$_{165}$)	Classical	10.1 ± 0.8 [21]	0.15 [21]	≈411	—	
38628 Huya	Plutino	No variation over 24 h [5]	<0.06 [17]	≈682	—	Hydrous silic.?
40314 (1999 KR$_{16}$)	Classical	11.680 ± 0.002 [17]	0.18 [17]	≈411	—	
47171 (1999 TC$_{36}$)	Plutino			≈622	—	H$_2$O ice
47932 (2000 GN$_{171}$)	Plutino	8.329 ± 0.005 [17]	0.61 [17]	≈375	—	Nonident.
52872 Oxyrhoe	Centaur			≈33	—	H$_2$O ice?
54598 (2000 QC$_{243}$)	Centaur	4.57 ± 0.04 [22] (single peak)	0.7 [22]	≈180	—	H$_2$O ice?

*Descriptions of the spectra and related references are given in section 5.

When the albedo is not available (—) an approximate diameter has been computed assuming an albedo of 0.05.

References: [1] *Marcialis and Buratti* (1993); [2] *Fernández et al.* (2002); [3] *Buie and Bus* (1992); [4] *Davies et al.* (1993); [5] *Davies et al.* (1998); [6] *Peixinho et al.* (2001); [7] *Jewitt and Kalas* (1998); [8] *Altenhoff et al.* (2001); [9] *Williams et al.* (1995); [10] *Thomas et al.* (2000); [11] S. Sheppard (personal communication, 2003); [12] *Hainaut et al.* (2000); [13] *Jewitt and Sheppard* (2002); [14] *Lellouch et al.* (2002); [15] *Jewitt et al.* (2001); [16] *Romanishin et al.* (2001); [17] *Sheppard and Jewitt* (2002); [18] *Gutiérrez et al.* (2001); [19] *Bauer et al.* (2002); [20] *Farnham and Davies* (2003); [21] *Peixinho et al.* (2002); [22] *Ortiz et al.* (2002); [23] D. Jewitt (personal communication, 2003).

results is presented. The faintness of these objects makes the analysis of the lightcurves difficult. In a photometric study by *Sheppard and Jewitt* (2002), 9 of 13 objects measured showed no detectable variation, implying a small amplitude, or a period ≥24 h, or both. Some objects show hints of variability that might yield a lightcurve with higher-quality data. The rotational periods seem to range between 6 and 15 h, but a bias effect can exist because of the difficulty in determining long periods and the faintness of these objects.

3.3. Shape

Stellar occultations and photometric observations can give important but limited information on the shape of these bodies, although no occultation results are available at the present time due to the lack and the difficulty of precise predictions. About 5% of the total number of TNOs seem to have companion objects and are therefore binary (*Noll et al.*, 2002). The first object discovered to have a companion was 1998 WW$_{31}$ (*Veillet et al.*, 2002).

The lightcurve is the only technique currently available to give constraints on the shape. The amplitude can give some indication on the elongation of the body. Assuming a triaxial ellipsoid shape with semimajor axes a > b > c and no albedo variation, we can estimate the lower limit of the semimajor axis ratio: $a/b \geq 10^{0.4\Delta m}$.

A few large TNOs seem to have elongated shapes (*Sheppard and Jewitt*, 2002). For example, using Δm = 0.61 mag (see Table 1) for 47932 (2000 GN$_{171}$), an estimate of a/b ≥ 1.75 can be obtained. Sheppard and Jewitt, analyzing all the available lightcurves, found that over 22 objects, 32% have Δm ≥ 0.15 mag, while 23% have Δm ≥ 0.4 mag.

4. TRENDS AND COLOR PROPERTIES

From broadband photometric observations, colors and spectral gradients are used for statistical analysis and to search for relationships among physical properties and orbital characteristics.

4.1. Color Diversity

One of the most puzzling features of the objects in the Kuiper belt, and one that has been confirmed by numerous

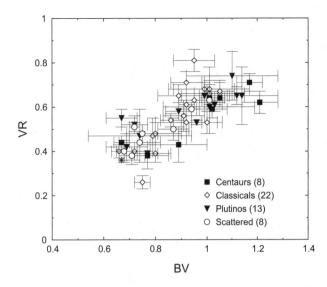

Fig. 1. B-V vs. V-R plot of the transneptunian objects. The different classes of TNOs are represented: Plutinos, classical, and scattered. The star represents the colors of the Sun. From *Doressoundiram et al.* (2002).

surveys, is optical color diversity. This diversity is peculiar to the outer solar system bodies and exceeds that of the asteroids, cometary nuclei, and small planetary satellites. Originally pointed out in *Luu and Jewitt* (1996a), this color diversity is an observational fact that is widely accepted by the community (e.g., *Barucci et al.*, 2000b; *Doressoundiram et al.*, 2001; *Jewitt and Luu*, 2001; *Delsanti et al.*, 2001; *Tegler and Romanishin*, 2000; *Boehnhardt et al.*, 2002, and references therein). Colors range continuously from neutral (flat spectrum) to very red (see Fig. 1). The different dynamical classes (e.g., Centaurs, Plutinos, classical, and scattered) seem to share the same color diversity. However, *Tegler and Romanishin* (1998) concluded earlier, on the basis of the visible (B-V vs. V-R) colors derived from 11 TNOs and 5 Centaurs, that there are two distinct populations: one with objects having neutral or slightly red colors similar to the C asteroids, and the other one including the reddest objects known in the solar system. They confirmed this result later on a larger dataset of 32 objects (*Tegler and Romanishin*, 2000) but with a lesser separation between the two populations in a color-color plot. Other groups working on this subject could not confirm this bimodality of color distribution. In particular, *Doressoundiram et al.* (2002), on the basis of a larger and homogeneous BVR dataset of 52 objects, did not see any clear and significant bimodality of color distribution. *Hainaut and Delsanti* (2002) have performed some statistical tests on a combined color dataset of 91 Centaurs and TNOs. They cautiously concluded that almost all the color-color distributions are compatible with both a continuous and a bimodal distribution. High-quality data with very small error bars will be necessary to establish the final word on this issue.

On the other hand, and paradoxically, there is a complete consensus for continuous color diversity when the color analysis is extended to longer wavelengths. For instance, color-color plots similar to Fig. 1 that include the I filter (~0.77 μm) or J filter (~1.2 μm) do not show any evidence of color bimodality (see *Boehnhardt et al.*, 2001; *Jewitt and Luu*, 2001).

The information contained in the color indices can be converted into a very-low-resolution reflectance spectrum, as illustrated in Fig. 2. Reflectance spectra have been computed using BVRIJ color data of the object (with the color of the Sun removed). [The reflectance spectrum $R(\lambda)$ is given by $R(\lambda) = 10^{-0.4\ [(M(\lambda) - M(V)) - (M(\lambda) - M_{Sun}(V))]}$, where M and M_{Sun} are the magnitude of the object and of the Sun at the considered wavelength. The reflectance is normalized to 1 at a given wavelength (conventionally, the V central wavelength, 0.55 μm).] The spectra range from neutral or slightly red to very red, thus confirming the wide and continuous diversity of surface colors suggested by the individual color-color diagrams. Almost all the objects are characterized by a linear reflectance spectrum, with no abrupt and significant changes in the spectral slope (within the error bars) over the whole wavelength range. This result was confirmed by *McBride et al.* (2003) on a large dataset of 29 mostly simultaneous V-J colors. They found their V-J colors broadly correlated with published optical colors, thus suggesting that a single coloring agent is responsible for the reddening from the B (0.4 μm) to the J (1.2 μm) regime.

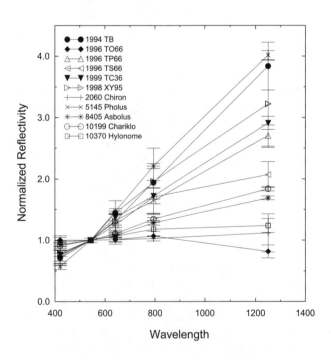

Fig. 2. Example of reflectivity spectra of TNOs and Centaurs, normalized at the V filter (centered around 550 nm). Color gradient range from low (neutral spectra) to very high (very red spectra). From color data of *Barucci et al.* (2001).

This remarkable property may help identify the agent among the low-albedo minerals with similar colors (*Jewitt and Luu,* 2001).

The extreme color diversity seen among the outer solar system objects is usually attributed to the concomitant action of two competing mechanisms acting on the TNOs over the age of the solar system. First, space weathering due to solar radiation processing and solar or galactic cosmic-ray irradiation both tend to the reddening of surfaces of all airless objects. Second, the resurfacing effect of mutual collisions among TNOs would regularly restore neutral-colored ices to the surface. This is the so-called collision-resurfacing hypothesis CRH (*Luu and Jewitt,* 1996a). Collisions and irradiation have reworked the surfaces of TNOs, especially in the inner part of the belt, and extensive cratering can be expected to characterize their surfaces (*Durda and Stern,* 2000). Another resurfacing process resulting from possible sporadic cometary activity has been suggested (*Hainaut et al.,* 2000). Resurfacing by ice recondensation from a temporary atmosphere produced by intrinsic gas and dust activity might be an efficient process affecting the TNOs closest to the Sun (the Plutinos).

4.2. Correlations

To date, B-V, V-R, and V-I colors are available for more than 150 objects, while only a few tens of them have V-J colors determined (*Davies et al.,* 1998, 2000; *Jewitt and Luu,* 1998; *McBride et al.,* 2003). A few of them have measured J-H and H-K colors. With this significant dataset, especially in the visible spectral region, we can now extend physical studies of TNOs from merely description to extended characterization by performing statistical analysis and deriving some potentially significant trends. Some of the outstanding questions include: (1) Are the surface colors of the Centaur and TNOs homogeneous? (2) Is it possible to define a taxonomy, as for the asteroids? (3) Are there any trends with physical and orbital parameters? For instance, are there any trends in color with size?

On the first point, we note that there is a general agreement between colors measured by different observers at random rotational phases, suggesting that color variation is rare. However, *Doressoundiram et al.* (2002) have highlighted a few objects among the 52 objects of their survey, for which color variation has been found and thus that may be diagnostic of true surface compositional and/or texture variation. The issue of the color heterogeneity remains ambiguous.

The TNOs exhibit a wide range of V-J colors. Based on a sample of 22 BVRIJ data, *Barucci et al.* (2001) made the first statistical analysis of colors of TNOs population, finding four "classes" showing a quasicontinuous spreading of the objects between two end members (those with neutral spectra and those with the reddest known spectra). The most important contribution in discriminating the "classes" comes from the V-J reflectance. This fact shows the necessity of the V-J color in any taxonomic work. A larger dataset is

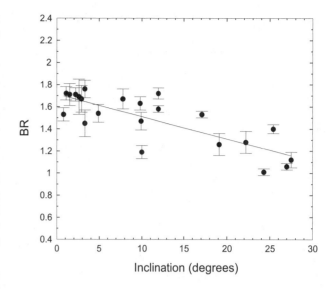

Fig. 3. Inclination vs. B-R color plot of classical objects. The linear least-squares fit has been plotted to illustrate the correlation. From *Doressoundiram et al.* (2002).

needed in order to investigate the real compositional taxonomy of the transneptunian and Centaur objects.

Jewitt and Luu (1998) presented a linear relationship between V-J and body size, implying that the smaller objects are systematically redder, a result subsequently invalidated by *Davies et al.* (2000) on a much larger dataset. Such a relationship, if found, would have been important because it is a prediction of the collisional resurfacing hypothesis.

Objects with perihelion distances around and beyond 40 AU are mostly very red. This characteristic was originally pointed out by *Tegler and Romanishin* (2000), who also noticed, as did *Doressoundiram et al.* (2001), that classical objects with high eccentricity and inclination are preferentially neutral/slightly red, while classical objects at low eccentricity and inclination are mostly red (Plate 18). This feature was first quantified by *Trujillo and Brown* (2002), who found a significant 3.1σ correlation between color and orbital inclination (i) for classical and scattered objects. A similar but stronger correlation (3.8σ) was later found by *Doressoundiram et al.* (2002) on a homogeneous dataset of 22 classical objects that did not include the scattered objects (Fig. 3). It is noteworthy that such a correlation was not seen among the Plutinos or the Centaurs. Instead, Plutinos appear to lack any clear color trends. *Hainaut and Delsanti* (2002), as well as *Doressoundiram et al.* (2002), found also significant correlations with orbital excentricity (e) and perihelion distance (q) for the classical TNOs, although less strong than with (i). *Levison and Stern* (2001) also found that low-i classical TNOs are smaller (greater H). Strikingly, *Jewitt and Luu* (2001) did not find any correlation with color in their sample of 28 B-I color indices. This apparent discrepancy is certainly due to the high proportion of resonant objects included in their sample that completely masks the correlation.

*Hainaut and Delsant*i (2002) analyzed a combined dataset of 91 Centaurs and TNOs. Although large, this dataset is not homogeneous, as the colors were collected and combined from different sources. They found a trend for classicals with faint H to be redder than the others, but the trend is opposite for the Plutinos (faint H tends to be bluer). *Doressoundiram et al.* (2002) did not find any correlation with size in their homogeneous but smaller dataset. These conflicting results require confirmation by the analysis of a larger dataset, but in any case the physical interpretation remains difficult.

Although hypothetical, the collisional resurfacing scenario offers the advantage of making relatively simple predictions concerning the color correlation within the Kuiper belt. Basically, the most dynamically excited objects should be most affected by energetic impacts, and thus should have the most neutral colors. Several authors (*Hainaut and Delsanti*, 2002; *Doressoundiram et al.*, 2002; *Stern*, 2002) have found a good correlation between the color index and $V_k(e^2 + i^2)^{1/2}$, apparently because both i and e contribute to the average encounter velocity of a TNO.

Considering that optical and infrared colors are correlated, one could presume that the correlations found between orbital parameters and optical colors can be generalized to infrared colors. Indeed, the V-J observations have a much wider spectral range and are therefore likely to be more robust in showing color correlations. However, such statistical analysis is still tentative because of the relatively few V-J colors available. The first such attempt made by *McBride et al* (2003) seems to support the color and perihelion distance, as well as the color and inclination relationships.

5. VISIBLE AND INFRARED SPECTROSCOPY

Broadband photometric observations can provide rough information on the surface of the TNOs and other objects, but the most detailed information on their compositions can be acquired only from spectroscopic observations, especially in the NIR spectral region. Unfortunately, most of the known TNOs are too faint for spectroscopic observations, even with the world's largest telescopes, and so far only the brightest have been observed by visible and infrared spectroscopy.

The first visible spectrum of a TNO, 15789 (1993 SC), was observed by *Luu and Jewitt* (1996b), who obtained a reddish spectrum that is intermediate in slope between those of the Centaurs 5145 Pholus and 2060 Chiron. Others have been observed subsequently, but only a few data are available; 5 Centaurs have been observed by *Barucci et al.* (1999), 5 TNOs by *Boehnhardt et al.* (2001), and 12 TNOs and Centaurs by *Lazzarin et al.* (2003). The spectra show a generally featureless behavior with a difference in the spectral gradient ranging from neutral to very red. The com-

puted reflectance slopes range from 0 or slightly negative (in the case of Chiron) up to 58%/100 nm for Pholus or Nessus, which are the reddest known objects in the solar system. The computed slopes vary a little as a function of the wavelength range analyzed, but do not seem to be related to the perihelion distance of the objects.

Broad absorptions have been found only for two Plutinos: 38628 Huya and 47932 (2000 GN$_{171}$). In the spectrum of 47932 (2000 GN$_{171}$), an absorption centered at around 0.7 μm has been detected with a depth of ~8%, while in the spectrum of 38628 Huya two weak features centered at 0.6 μm and at 0.745 μm have been detected with depths of ~7% and 8.6% respectively (*Lazzarin et al.,* 2003). These features are very similar to those due to aqueously altered minerals, found in some main-belt asteroids (*Vilas and Gaffey,* 1989, and subsequent papers). Since hydrous materials seem to be present in comets, and hydrous silicates are detected in interplanetary dust particles (IDPs) and in micrometeorites (and probably originated in the solar nebula), finding aqueous altered materials in TNOs would not be too surprising (see *de Bergh et al.,* 2003).

In the infrared region some spectra are featureless, while some others show signatures of water ice and methanol or other light hydrocarbon ices. Very few of these objects have been well studied in both visible and NIR and rigorously modeled. In fact, these objects are faint, and even observations with the largest telescopes [Keck and the Very Large Telescope (VLT)] do not generally yield good-quality spectra. The interpretation is also very difficult because the behavior of models of the spectra depends on the choice of many parameters. Some of the visible and NIR spectra obtained at VLT [European Southern Observatory (ESO), Chile] are shown in Fig. 4, with the best model fitting of the data. The general spectral characteristics are listed in Table 1.

A general review of Centaurs is presented in *Barucci et al.* (2002a), while details of a few objects, recently observed, are discussed below.

8405 Asbolus has yielded controversial results: *Brown* (2000) and *Barucci et al.* (2000a) observed it, finding no spectral signatures in the NIR. Later, *Kern et al.* (2000), using the HST, obtained several (1.1–1.9 μm) spectra, which revealed a significantly inhomogeneous surface characterized on one side by water ice mixed with unknown low-albedo constituents. They speculated that the differences across the surface of Asbolus might be caused by an impact that penetrated the object's crust, exposing the underlying ice in the surface region. *Romon-Martin et al.* (2002) re-observed Asbolus at VLT (ESO, Chile), obtaining five high-quality infrared spectra covering the full rotational period, and found no absorption features at any rotational phase. Using different radiative transfer and scattering models (*Douté and Schmitt*, 1988; *Shkuratov et al.*, 1999), *Romon-Martin et al.* (2002) modeled the complete spectrum from 0.4 to 2.5 μm with several mixtures of Triton tholins, Titan tholins, ice tholins, amorphous carbon, and olivine. None of the models successfully matched the visible part

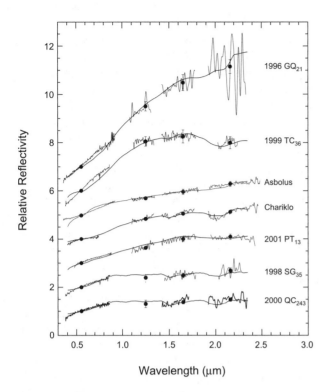

Fig. 4. V + NIR spectra of two TNOs and five Centaurs obtained at VLT, ESO, with FORS and ISAAC. The spectra, normalized to 1 on V, have been shifted in relative reflectance by 1 for clarity. The dots represent the V, J, H, and K colors, used to adjust the spectral ranges. A best-fit model is shown for each spectrum (see section 5 for details).

of the spectrum, while the best fit to the infrared part was obtained with 18% Triton tholin, 7% Titan tholin, 55% amorphous carbon, and 20% ice tholin (Fig. 4). The steep red spectral slope is the principal characteristic of the data that forces the use of organic compounds (kerogen, tholins, etc.) in the models. Kerogen is necessary to reproduce the red slope of spectra in the visible region, while tholins (*Khare et al.,* 1984) are the only materials (for which optical constants are available) able to reproduce the unusual red slope (0.4–1.2 μm). Both Titan and ice tholins are synthetic macromolecular compounds, produced from a gaseous mixture of N_2:CH_4 (Titan tholins) or an ice mixture of H_2O:C_2H_6 (ice tholins).

10199 Chariklo, after the first detection of water ice by *Brown and Koresko* (1998) and by *Brown et al.* (1998), was observed again by *Dotto et al.* (2003a), who still confirmed water ice detection and showed small spectral behavior variation (Fig. 4).

32532 (2001 PT_{13}), now named Thereus, was observed (*Barucci et al.,* 2002b) from 1.1 to 2.4 μm at two different epochs and the spectra seem quite different, indicating spatial differences in the surface composition. One of the observations shows clear evidence for a small percentage of water ice. The lack of albedo information eliminates one

important constraint on the modeling, but on the assumption of a low-albedo surface two models have been computed to interpret the different behavior of the two spectra. One spectrum seems to be well fitted with a model containing 15% Titan tholin, 70% amorphous carbon, 3% olivine, and 12% ice tholin, and having an albedo of 0.09 (shown in Fig. 4). For the other spectrum an acceptable model with an albedo of 0.06 and 90% amorphous carbon, 5% Titan tholin, and 5% water ice was obtained.

Dotto et al. (2003b) observed two Centaurs: 52872 (1998 SG_{35}), now named Oxyrhoe, and 54598 (2000 QC_{243}) in the H and K regions, giving tentative models of these two bodies with similar percentages of kerogen (96–97%), olivine (1%), and water ice (2–3%) (Fig. 4).

63252 (2001 BL_{41}) has been observed by *Doressoundiram et al.* (2003). A model with 17% Triton tholin, 10% ice tholin, and 73% amorphous carbon fits the spectrum.

31824 Elatus was found by *Bauer et al.* (2002) to have markedly different spectral reflectance when observed on two successive nights. While one spectrum shows a rather neutral reflectance, 1.2–2.3 μm, the other shows a strong red reflectance extending to 2.3 μm, with absorption bands approximately matched by a model using amorphous H_2O ice.

The TNOs are even fainter than Centaurs, and only a few spectra are available, generally with very low SNR. Although only a small number have been observed to date, their surface characteristics seem to show wide diversity. 15874 (1996 TL_{66}) and 28978 Ixion have flat featureless spectra similar to that of water ice contaminated with low-albedo, spectrally neutral material (*Luu and Jewitt,* 1998; *Licandro et al.,* 2002). 15789 (1993 SC), observed by *Brown et al.* (1997), shows features that may be due to hydrocarbon ices with a general red behavior suggesting the presence of more complex organic solids. *Jewitt and Luu* (2001) also observed 1993 SC with the same telescope and found a featureless spectrum. The difference in these results requires resolution, best accomplished with additional (and higher-quality) data. *McBride et al.* (2003) show that 1993 SC is one of the reddest TNOs studied so far.

38628 Huya has been observed by many authors (*Brown et al.,* 2000; *Licandro et al.,* 2001; *Jewitt and Luu,* 2001; *deBergh et al.,* 2003) and appears generally featureless in the NIR [except *Licandro et al.* (2001) and *de Bergh et al.* (2003) show that a possible feature appears beyond 1.8 μm]. The interpretation of these spectra is challenging.

19308 (1996 TO_{66}) shows an inhomogeneous surface with clear indications of water ice absorptions at 1.5 and 2 μm. A model of water ice mixed with some other minor components matches the region 1.4–2.4 μm (*Brown et al.,* 1999). Evidence that the intensity of water ice bands varies with the rotational phase suggests a patchy surface. 20000 Varuna also shows a deep water-ice absorption band (*Licandro et al.,* 2001), while 26181 (1996 GQ_{21}), observed by *Doressoundiram et al.* (2003), shows a featureless spectrum interpreted with a geographical mixture model composed of 15% Titan tholin, 35% ice tholin, and 50% amorphous carbon (Fig. 4).

In contrast, 26375 (1999 DE$_9$) shows solid-state absorption features near 1.4, 1.6, 2.00, and probably at 2.25 μm (*Jewitt and Luu*, 2001). The location of these bands has been tentatively interpreted by Jewitt and Luu as evidence for the hydroxyl group with possible interaction with an Al or Mg compound. An absorption near 1 μm may be consistent with olivine. If the presence of the hydroxyl group is confirmed, this might imply the presence of liquid water and a temperature near the melting point for at least a short period of time. The H region has been re-observed by *Doressoundiram et al.* (2003), but because of the low SNR, they were not able to confirm the 1.6-μm feature.

47171 (1999 TC$_{36}$), observed by *Dotto et al.* (2003b) in the J, H, and K region, shows a weak absorption around 2 μm, and the surface composition has been interpreted with a mixture of 57% Titan tholin, 25% ice tholin, 10% amorphous carbon, and 8% water ice (Fig. 4).

In some cases repeated observations of the same object give different results, sometimes because of inferior quality data (see Table 1), but in other cases the surfaces may be variable on a large spatial scale. A few objects in addition to 31824 Elatus clearly show surface variations, such as 19308 (1996 TO$_{66}$) and 32532 Thereus. While the models noted here represent the best current fit to the data, they are not unique and depend on many free parameters, such as grain size, albedos, porosity, etc.

6. MODELING SURFACE COMPOSITION

We have already noted the modeling results of a few Centaurs and TNOs by various investigators, and have seen that organic solids (tholins) are used to achieve a fit to the strong red color that most of these objects exhibit. In this section we consider some details of modeling the spectral reflectance of the solid surface of an outer solar system body.

The goal of modeling the spectral reflectance of a planetary surface is to derive information on that object's composition and surface microstructure. Thermal emission can also be modeled, but in the case of TNOs and Centaurs, there are insufficient astronomical data of this kind to yield compositional information through a modeling approach. Compositional information can be derived from straightforward spectrum matching (e.g., *Hiroi et al.*, 2001) and from linear mixing of multiple components (e.g., *Hiroi et al.*, 1993). More rigorous and more informative quantitative modeling using scattering theory goes beyond spectrum matching and linear mixing by introducing the optical properties (complex refractive indices) of candidate materials into a model of particulate scattering. Quantitative modeling of planetary surfaces using scattering theory has progressed in recent years as more and more realistic models are developed and tested against observational data. Multiple scattering models provide approximate but very good solutions to radiative transfer in a particulate medium. The semiempirical Hapke model (*Hapke*, 1981, 1993) has been most widely used, while other models incorporating additional physical configurations (e.g., layers of transparent or semitransparent components, inhomogeneous transparent grains, etc.) have begun to emerge (e.g., *Douté and Schmitt*, 1998; *Shkuratov et al.*, 1999).

Real planetary surfaces are composed of many different materials mixed in various configurations. There can be spatially isolated regions of a pure material (e.g., H$_2$O ice or a pyroxene-dominant rock) or a mixture of materials (e.g., olivine, pyroxene, and opaque phases). The nature of the mixture can range widely. For example, there can be an intimate granular mixture in which each component is an individual scattering grain of a particular composition, lying in contact with grains of its own kind or a different material. Or, materials might be mixed at the molecular level, such that a sunlight photon entering an individual grain will encounter molecules of different composition within that grain before exiting. Many other configurations, including complex layering, are also possible.

The net result of all the processes that occur on airless solar system bodies is that they exhibit a large range of geometric albedos, differing slopes in their reflectance spectra, and the presence or absence of absorption bands arising from minerals, ices, and organic solids.

The case of Centaur 5145 Pholus (Fig. 5) offers a view of some of the challenges in modeling Centaur and TNO surfaces (details are found in *Cruikshank et al.*, 1998). This object has a steeply sloped spectrum from 0.45 to 0.95 μm and moderately strong absorption bands at 2.0 and 2.27 μm, while the geometric albedo at 0.55 μm is 0.04. The steep red slope cannot be matched by minerals or ices, but is characteristic of some organic solids, notably the tholins. The absorption bands are identified as H$_2$O ice (2.0 μm) and (probably) methanol ice (CH$_3$OH) at 2.27 μm. A Hapke

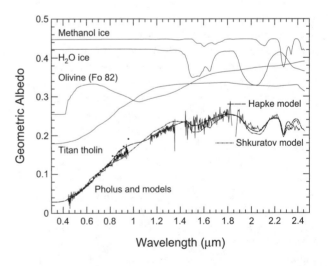

Fig. 5. Spectrum of 5145 Pholus (lower trace) with the Hapke model of *Cruikshank et al.* (1998) (solid line). The four principal components for which complex refractive indices (n, k) were included in the model are shown schematically in the four upper traces. The model of *Poulet et al.* (2002) using the Shkuratov code is also shown.

scattering model using the real and imaginary refractive indices of tholin, H_2O, CH_3OH, and the mineral olivine, plus amorphous carbon (which affects only the albedo level of the model), was found to match the spectrum from 0.45 to 2.4 µm. The Hapke model formulation of *Roush et al.* (1990) was used. The model consisted of two components spatially separated on Pholus; the main component is an intimate mixture of 55% olivine, 15% Titan tholin, 15% H_2O ice, and 15% CH_3OH ice, with various grain sizes. In the model, the main component covers ~40% of the surface, while carbon covers the remaining 60%.

The principal problem with this model is that the Titan tholin particles had to be only 1 µm in size, thereby violating a tenant of the Hapke theory that the particle sizes have to be significantly greater than the wavelength of the scattered light. This conflict can be resolved by using the Shkuratov modeling theory, in which very small amounts of Titan tholin can be introduced as contaminants in the water ice crystals without violating any optical constraints of the theory. *Poulet et al.* (2002) have shown that Pholus can be modeled with the Shkuratov theory using the same organic, ice, and mineral components used in the Hapke model, although in slightly different proportions, without any conflict with particle size constraints. The Poulet et al. model is also shown in Fig. 5.

7. CONCLUSIONS

One of the most puzzling features of the Kuiper belt, confirmed by numerous surveys, is the optical color diversity that seems to prevail among the observed TNOs (Fig 1). With the relatively few visible-NIR color datasets available, the color diversity seems also to extend to the NIR. Statistical analyses point to correlations between optical colors and some orbital parameters (i, e, q) for the classical Kuiper belt. On the other hand, no clear trend is obvious for Plutinos, scattered objects, or Centaurs, and no firm conclusions can be drawn regarding correlation of colors with size or heliocentric distance. The correlations of color with i, e, and q are important because they may be diagnostic of some physical effects of processing the surfaces of TNOs. The collisional resurfacing (CR) scenario is generally invoked to explain the color diversity, which could be the result of two competing mechanisms: the reddening and darkening of icy surfaces by solar and galactic irradiation, and the excavation of fresh, primordial (and thus more neutral) ices as the results of collisions. While the reddening process is believed to act relatively homogeneously throughout the belt, the collision-induced blueing should vary significantly with the rate and efficiency of collisions within the belt. As a consequence, the CR scenario should leave a characteristic signature with the bluer objects located in the most collisionally active regions of the belt. *Thébault and Doressoundiram* (2003) first performed deterministic numerical simulations of the collisional and dynamical environment of the Kuiper belt to look for such a signature. Their results do match several main statistical correlations observed in the belt: e, i, Vrms, and particularly q, but there are also clear departures from the observed color distribution. For example, the Plutinos became uniformly bluer in the simulations.

Computational models to check the validity of the CR scenario, such as those of *Thébault and Doressoundiram* (2003), show that the origin of the color diversity is still unclear. The solution might lie in a better understanding of the physical processes involved, in particular the fact that the long-term effect of space weathering might significantly depart from continuous reddening (see below). Another alternative would be that the classical objects may consist of the superposition of two distinct populations, as suggested by *Levison and Stern* (2001), *Brown* (2001), *and Doressoundiram et al.* (2002). One population would consist of primordial objects with red surfaces, low inclination, and small sizes, and the second population would consist of more evolved objects with larger sizes, higher inclination, and more diverse surface colors.

Centaurs and TNOs appear very similar in spectral and color characteristics, and this represents the strongest observational argument for a common origin, supporting the hypothesis that Centaurs are ejected from the Kuiper belt by planetary scattering. The rotational properties of the few available Centaurs and TNOs also seem to be similar, even though it is still difficult to interpret the distribution due to the lack of data. Judging from the observed lightcurve amplitudes, large TNOs can exist with elongated shapes. As opposed to the Centaurs, the color distribution of cometary nuclei does not seem to match that of TNOs (*Jewitt*, 2004); the very red color seems absent among comets. 2060 Chiron can be considered an example of a temporarily dormant comet; the other Centaurs and TNOs might be dormant comets containing frozen volatiles that would sublimate in particular heating conditions.

The wide diversity of color is confirmed by the different spectral behavior, even though only a few high-quality spectra exist. The spectra show a large range of slope; some are featureless with almost constant gradients over the visible-NIR range, and some show absorption features of H_2O or light hydrocarbon ices. A few objects show features attributable to the presence of hydrous silicates, but this still needs to be confirmed. Several models of the spectral reflectance of TNOs and Centaurs have been proposed, but each is subject to the limitations imposed by the quality of the astronomical spectra, the generally unknown albedo, and to the limited library of materials for which optical constants have been determined. The models of red objects all use organic materials, such as tholins and kerogen, because common minerals (and ices) cannot provide a sufficiently red color.

The H_2O absorption bands detected so far on a few objects are generally weak. H_2O ice is presumed to be the principal component of the bulk composition of outer solar system objects (formed mostly at the same low temperature of 30–40 K) and should constitute at least about 35% of the bulk composition of this population. Thus H_2O ice has to be present even if it is not detectable on the spectra, but its absorption bands can be reduced to invisibility by the presence of low-albedo, opaque materials. Additionally, various processes of space weathering (due to solar radiation, cosmic rays, and interplanetary dust) can affect the uppermost surface layer. The observed surface diversity can be due to different collisional evolutionary states and to different degrees of surface alteration due to space weathering. Collisions can rejuvenate the surface locally by excavating material from the subsurface. On the basis of laboratory experiments, *Strazzulla* (1998) demonstrated that bombardment by high-energy radiation of mixtures of CH_3OH, CH_4, H_2O, CO_2, CO, and NH_3 ices produces radiation mantles that are dark, hydrogen-poor, and carbon-rich, and show red spectra. These red spectra may become flat again, e.g., as demonstrated by *Moroz et al.* (2003), who simulated an aging effect of a dark organic sample (asphaltite) by ion irradiation. Many processes may have altered the pristine surfaces of these objects, for which the original composition is still unclear. Laboratory experiments (*Strazzulla et al.*, 2002) are in progress to simulate weathering effects on small bodies by bombardments at different fast ion fluences on several minerals and meteorites to better understand these processes.

This research field is still very young, even though a decade has passed since the discovery of the first TNO. There is a great deal of interest in the study of the physical and compositional characteristics of these objects, but our knowledge of the properties of TNOs suffers from the limitations connected with groundbased observations. In the near future, space missions such as the Space Infrared Telescope Facility (SIRTF) and Gaia will substantially improve our knowledge of their physical properties. SIRTF will observe the thermal radiation of more than 100 TNOs and thereby make it possible to calculate the geometric albedos and dimensions of a statistically significant sample of objects in several dynamical populations. Gaia, with its all-sky astrometric and photometric survey, will discover objects not observable from the ground and will enable the detection of binary objects, the discovery of Pluto-sized bodies, and a better taxonomy for Centaurs and TNOs. NASA's New Horizons mission to the Kuiper belt and Pluto-Charon, with an anticipated arrival at Pluto in 2016 to 2018, will offer the first closeup views of as many as five solid bodies beyond Neptune.

REFERENCES

Altenhoff W. J., Menten K. M., and Bertoldi F. (2001) Size determination of the Centaur Chariklo from millimeter-wavelength bolometer observations. *Astron. Astrophys., 366,* L9–L12.

Barucci M. A., Lazzarin M., and Tozzi G. P. (1999) Compositional surface variety among the Centaurs. *Astron. J., 117,* 1929–1932.

Barucci M. A., de Bergh C., Cuby J.-G., Le Bras A., Schmitt B., and Romon J. (2000a) Infrared spectroscopy of the Centaur 8405 Asbolus: First observations at ESO-VLT. *Astron. Astrophys., 357,* L53–L56.

Barucci M. A., Romon J., Doressoundiram A., and Tholen D. J. (2000b) Composition surface diversity in the Trans-Neptunian Objects. *Astron. J., 120,* 496–500.

Barucci M. A., Fulchignoni M., Birlan M., Doressoundiram A., Romon J., and Boehnhardt H. (2001) Analysis of Trans-Neptunian and Centaur colours: Continuous trend or grouping? *Astron. Astrophys., 371,* 1150–1154.

Barucci M. A., Cruikshank D. P., Mottola S., and Lazzarin M. (2002a) Physical properties of Trojans and Centaur Asteroids. In *Asteroids III* (W. F. Bottke Jr. et al., eds.), pp. 273–287. Univ. of Arizona, Tucson.

Barucci M. A. and 19 colleagues (2002b) Visible and near-infrared spectroscopy of the Centaur 32532 (PT13). ESO Large Program on TNOs and Centaurs: First spectroscopy results. *Astron. Astrophys., 392,* 335–339.

Bauer J. M., Meech K. J., Fernández Y. R., Farnham T. L., and Roush T. L. (2002) Observations of the Centaur 1999 UG_5: Evidence of a unique outer solar system surface. *Publ. Astron. Soc. Pacific, 114,* 1309–1321.

Boehnhardt H., Tozzi G. P., Birkle K., Hainaut O., Sekiguchi T., Vlair M., Watanabe J., Rupprech G., and the Fors Instrument Team (2001) Visible and near-IR observations of Trans-Neptunian objects. Results from ESO and Calar Alto telescopes. *Astron. Astrophys., 378,* 653–667.

Boehnhardt H. and 17 colleagues (2002) ESO Large Program on physical studies of Trans-Neptunian Objects and Centaurs: Visible photometry — First results. *Astron. Astrophys., 395,* 297–303.

Brown M. E. (2000) Near-infrared spectroscopy of centaurs and irregular satellites. *Astron. J., 119,* 977–983.

Brown M E. (2001) The inclination distribution of the Kuiper belt. *Astron. J., 121,* 2804–2814.

Brown M. E. and Koresko C. D. (1998) Detection of water ice on the Centaur 1997 CU_{26}. *Astrophys. J. Lett., 505,* L65–L67.

Brown M. E. and Trujillo C. (2003) Direct measurement of the size of the large Kuiper belt Quaoar. *Astron. J.,* in press.

Brown M. E., Blake G. A., and Kessler J. E. (2000) Near-infrared spectroscopy of the bright Kuiper belt object 2000 EB_{173}. *Astrophys. J. Lett., 543,* L163–L165.

Brown R. H., Cruikshank D. P., Pendleton Y. J., and Veeder G. (1997) Surface composition of Kuiper belt object 1993 SC. *Science, 276,* 937–939.

Brown R. H., Cruikshank D. P., Pendleton Y., and Veeder G. J. (1998) Identification of water ice on the Centaur 1997 CU_{26}. *Science, 280,* 1430–1432.

Brown R. H., Cruikshank D. P., and Pendleton Y. J. (1999) Water ice on Kuiper belt object 1996 TO_{66}. *Astrophys. J. Lett., 519,* L101–L104.

Buie M. W. and Bus S. J. (1992) Physical observations of (5145) Pholus. *Icarus, 100,* 288–294.

Cruikshank D. P. and 14 colleagues (1998) The composition of Centaur 5145 Pholus. *Icarus, 135,* 389–407.

Davies J., Spencer J., Sykes M., Tholen D., and Green S. (1993) *(5145) Pholus.* IAU Circular 5698.

Davies J. K., McBride N., Ellison S. L., Green S. F., and Ballantyne D. R. (1998) Visible and infrared photometry of six centaurs. *Icarus, 134,* 213–227.

Davies J. K., Green S., McBride N., Muzzerall E., Tholen D. J., Whiteley R. J., Foster M. J., and Hillier J. K. (2000) Visible and infrared photometry of fourteen Kuiper belt objects. *Icarus, 146,* 253–262.

De Bergh C., Boehnhardt H., Barucci M. A., Lazzarin M., Fornasier S., Romon-Martin J., Tozzi G. P., Doressoundiram A., and Dotto E. (2003) Hydrated silicates at the surface of two Plutinos? *Astron. Astrophys.*, in press.

Delsanti A. C., Boehnhardt H., Barrera L., Meech K. J., Sekiguchi T., and Hainaut O. R. (2001) BVRI photometry of 27 Kuiper belt objects with ESO/Very Large Telescope. *Astron. Astrophys.*, *380*, 347–358.

Doressoundiram A., Barucci M. A., Romon J., and Veillet C. (2001) Multicolor photometry of Trans-Neptunian objects. *Icarus, 154*, 277–286.

Doressoundiram A., Peixinho N., De Bergh C., Fornasier S., Thébault Ph., Barucci M. A., and Veillet C. (2002) The color distribution of the Kuiper belt. *Astron. J., 124*, 2279–2296.

Doressoundiram A., Tozzi G. P., Barucci M. A., Boehnhardt H., Fornasier S., and Romon J. (2003) Spectroscopic investigation of Centaur 2001 BL$_{41}$ and transneptunian objects (26181) 1996 GQ$_{21}$ and (26375) 1999 DE$_9$. ESO Large Program on TNOs and Centaurs. *Astron. J., 125*, 2721–2727.

Dotto E., Barucci M. A., Leyrat C., Romon J., Licandro J., and de Bergh C. (2003a) Unveiling the nature of 10199 Chariklo: Near-infrared observations and modeling. *Icarus, 164*, 122–126.

Dotto E., Barucci M. A., Boehnhardt H., Romon J., Doressoundiram A., Peixinho N., de Bergh C., and Lazzarin M. (2003b) Searching for water ice on 1999 TC$_{36}$, 1998 SG$_{35}$ and 2000 QC$_{243}$. ESO Large Program on TNOs and Centaurs. *Icarus, 162*, 408–414.

Douté S. and Schmitt B. (1998) A multilayer bi-directional reflectance model for the analysis of planetary surface hyperspectral images at visible and near-infrared wavelengths. *J. Geophys. Res., 103*, 31367–31390.

Durda D. D. and Stern S. A. (2000) Collision rates in the present-day Kuiper belt and Centaurs regions: Application to surface activation and modification on comets, Kuiper belt objects, Centaurs and Pluto-Charon. *Icarus, 145*, 220–229.

Farnham T. L. and Davies J. K. (2003) The rotational and physical properties of the Centaur 32532 (2001 PT$_{13}$). *Icarus, 164*, 418–427.

Fernández J. A. (1999) Cometary dynamics. In *Encyclopedia of the Solar System* (P. R. Weissman et al., eds.), pp. 537–556. Academic, San Diego.

Fernández Y. R., Jewitt D. C., and Shepard S. S. (2002) Thermal properties of two Centaurs Asbolus and Chiron. *Astron. J., 123*, 1050–1055.

Gutiérrez P. J., Ortiz J. L., Alexandrino E., Roos-Serote M., and Doressoundiram A. (2001) Short term variability of Centaur 1999 UG$_5$. *Astron. Astrophys., 371*, L1–L4.

Hainaut O. R. and Delsanti A. C. (2002) Colors of minor bodies in the outer solar system. *Astron. Astrophys., 389*, 641–664.

Hainaut O. R., Delahodde C. E., Boehnhardt H., Dotto E., Barucci M. A., Meech K. J., Bauer J. M., West R. M., and Doressoundiram A. (2000) Physical properties of TNO TO$_{66}$. Lightcurves and possible cometary activity. *Astron. Astrophys., 356*, 1076–1088.

Hapke B. (1981) Bidirectional reflectance spectroscopy. 1. Theory. *J. Geophys. Res., 86*, 3039–3054.

Hapke B. (1993) *Theory of Reflectance and Emittance Spectroscopy.* Cambridge Univ., New York. 455 pp.

Harris A. W. and Harris A. W. (1997) On the revision of radiometric albedos and diameters of Asteroids. *Icarus, 126*, 450–454.

Howell S. B. (1989) Two-dimensional aperture photometry —

Signal-to-noise ratio of point-source observations and optimal data-extraction techniques. *Publ. Astron. Soc. Pacific, 101*, 616–622.

Hiroi T., Bell J. F., Takeda H., and Pieters C. M. (1993) Modeling of S-type asteroid spectra using primitive achondrites and iron meteorites. *Icarus, 102*, 107–116.

Hiroi T., Zolensky M. E., and Pieters C. M. (2001) The Tagish Lake meteorite: A possible sample from a D-type asteroid. *Science, 293*, 2234–2236.

Jewitt D. (2002) From Kuiper belt object to cometary nucleus: The missing ultra-red matter. *Astron. J., 123*, 1039–1049.

Jewitt D. (2004) From cradle to grave: The rise and demise of comets. In *Comets II* (M. C. Festou et al., eds.), this volume. Univ. of Arizona, Tucson.

Jewitt D. and Kalas P. (1998) Thermal observations of Centaur 1997 CU$_{26}$. *Astrophys. J. Lett., 499*, L103–L109.

Jewitt D. and Luu J. (1998) Optical and infrared spectral diversity in the Kuiper belt. *Astron. J., 115*, 1667–1670.

Jewitt D. and Luu J. X. (2001) Colors and spectra of Kuiper belt objects. *Astron. J., 122*, 2099–2114.

Jewitt D. and Sheppard S. S. (2002) Physical properties of Tran-Neptunian object 20000 Varuna. *Astron. J., 123*, 2110–2120.

Jewitt D., Luu J., and Marsden B. G. (1992) *1992 QB$_1$*. IAU Circular 5611.

Jewitt D., Aussel H., and Evans A. (2001) The size and albedo of the Kuiper-belt object 20000 Varuna. *Nature, 411*, 446–447.

Kern S. D., McCarthy D. W., Buie M. W., Brown R. H., Campins H., and Rieke M. (2000) Compositional variation on the surface of Centaur 8405 Asbolus. *Astrophys. J. Lett., 542*, L155–L159.

Khare B. N., Sagan C., Arakawa E. T., Suits F., Callcott T. A., and Williams M. W. (1984) Optical constants of organic tholins produced in a simulated Titanian atmosphere — From soft X-ray to microwave frequencies. *Icarus, 60*, 127–137.

Lamy P. L., Toth I., Fernández Y. R., and Weaver H. A. (2004) The sizes, shapes, albedos, and colors of cometary nuclei. In *Comets II* (M. C. Festou et al., eds.), this volume. Univ. of Arizona, Tucson.

Lazzarin M., Barucci M. A., Boehnhardt H., Tozzi G. P., de Bergh C., and Dotto E. (2003) ESO Large Program on physical studies of Trans Neptunian Objects and Centaurs: Visible spectroscopy. *Astron. J., 125*, 1554–1558.

Lellouch E., Moreno R., Ortiz J. L., Paubert G., Doressoundiram A., and Peixinho N. (2002) Coordinated thermal and optical observations of Trans-Neptunian object (20000) Varuna from Sierra Nevada. *Astron. Astrophys., 391*, 1133–1139.

Levison H. F. and Stern S. A. (2001) On the size dependence of the inclination distribution of the main Kuiper belt. *Astron. J., 121*, 1730–1735.

Licandro J., Oliva E., and Di Martino M. (2001) NICS-TNG infrared spectroscopy of trans-neptunian objects 2000 EB$_{173}$ and 2000 WR$_{106}$. *Astron. Astrophys., 373*, L29–L32.

Licandro J., Ghinassi F., and Testi L. (2002) Infrared spectroscopy of the largest known Trans-Neptunian object 2001 KX$_{76}$. *Astron. Astrophys., 388*, L9–L12.

Luu J. X. and Jewitt D. (1996a) Color diversity among the Centaurs and Kuiper belt objects. *Astron. J., 112*, 2310–2318.

Luu J. X. and Jewitt D. (1996b) Reflection spectrum of the Kuiper belt object 1993 SC. *Astron. J. Lett., 111*, L499–L503.

Luu J. X. and Jewitt D. (1998) Optical and infrared reflectance spectrum of Kuiper belt object 1996 TL$_{66}$. *Astrophys. J. Lett., 494*, L117–L121.

Marcialis R. L. and Buratti B. J. (1993) CCD photometry of 2060 Chiron in 1985 and 1991. *Icarus, 104,* 234–243.

McBride N., Green S. F., Davies J. K., Tholen D. J., Sheppard S. S., Whiteley R. J., and Hillier J. K. (2003) Visible and infrared photometry of Kuiper belt objects: Searching for evidence of trends. *Icarus, 161,* 501–510.

Moroz L. V., Baratta G. A., Distefano E., Strazzulla G., Dotto E., and Barucci M. A. (2003) Ion irradiation of asphaltite: Optical effects and implications for trans-Neptunian objects and Centaurs. *Proceedings of the Conference on First Decadal Review of the Kuiper Belt* (J. Davies and L. Barrerra, eds.), in *Earth Moon Planets,* in press.

Noll K. S., Stephens D. C., Grundy W. M., Millis R. L., Spencer J., Buie M., Tegler S. C., Romanishin W., and Cruikshank D. P. (2002) Detection of two binary trans-Neptunian objects, 1997 CQ_{29} and 2000 CF_{105}, with the Hubble Space Telescope. *Astron. J., 124,* 3424–3429.

Ortiz J. L., Baumont S., Gutiérrez P. J., and Roos-Serote M. (2002) Lightcurves of Centaurs 2000 QC_{243} and 2001 PT_{13}. *Astron. Astrophys., 388,* 661–666.

Peixinho N., Lacerda P., Ortiz J. L., Doressoundiram A., Ross-Serote M., and Gutiérrez P. J. (2001) Photometric study of Centaurs 10199 Chariklo (1997 CU_{26}) and 1999 UG_5. *Astron. Astrophys., 371,* 753–759.

Peixinho N., Doressoundiram A., and Romon-Martin J. (2002) Visible-IR colors and lightcurve analysis of two bright TNOs TC_{36} and 1998 SN_{165}. *New Astronomy, 7(6),* 359–367.

Peixinho N., Boehnhardt H., Belskaya I., Doressoundiram A., Barucci M. A., Delsanti A., and the ESO Large Program Team (2003) ESO Large Program on Centaurs and TNOs: The ultimate results from visible colors. *Bull. Am. Astron. Soc., 35,* 1016.

Poulet F., Cuzzi J. N., Cruikshank D. P., Roush T., and Dalle Ore C. M. (2002) Comparison between the Shkuratov and Hapke scattering theories for solid planetary surfaces: Application to the surface composition of two Centaurs. *Icarus, 160,* 313–324.

Romanishin W., Tegler S. C., Rettig T. W., Consolmagno G., and Botthof B. (2001) 1998 SM_{165}: A large Kuiper belt object with an irregular shape. *Proc. Natl. Acad. Sci., 98,* 11863–11866.

Romon-Martin J., Barucci M. A., de Bergh C., Doressoundiram A., Peixinho N., and Poulet F. (2002) Observations of Centaur 8405 Asbolus: Searching for water ice. *Icarus, 160,* 59–65.

Roush T. L., Pollack J. B., Witteborn F. C., Bregman J. D., and Simpson J. P. (1990) Ice and minerals on Callisto: A reassessment of the reflectance spectra. *Icarus, 86,* 355–382.

Sheppard S. S. and Jewitt D. C. (2002) Time-resolved photometry of Kuiper belt objects: Rotations, shapes and phase functions. *Astron. J., 124,* 1757–1775.

Shkuratov Y., Starukhina L., Hoffmann H., and Arnold G. (1999) A model of spectral albedo of particulate surfaces: Implications for optical properties of the Moon. *Icarus, 137,* 235–246.

Stern S. A. (2002) Evidence for a collisional mechanism affecting Kuiper belt object colors. *Astron. J., 124,* 2297–2299.

Strazzulla G. (1998) Chemistry of ice induced by bombardment with energetic charged particles. In *Solar System Ices* (B. Schmitt et al., eds.), p. 281. Kluwer, Dordrecht.

Strazzulla G., Dotto E., Barucci M. A., Blanco A., and Orofino V. (2002) Space weathering effects on small body surface: Laboratory fast ion bombardment. *Bull. Am. Astron. Soc., 34,* 858.

Tegler S. C. and Romanishin W. (1998) Two populations of Kuiper-belt objects. *Nature, 392,* 49–50.

Tegler S. C. and Romanishin W. (2000) Extremely red Kuiper-belt objects in near-circular orbits beyond 40 AU. *Nature, 407,* 979–981.

Thébault Ph. and Doressoundiram A. (2003) Could collisions be the cause of the color diversity in the Kuiper belt? A numerical test. *Icarus, 162,* 27–37.

Thomas N., Eggers S., Ip W.-I., Lichtenberg G., Fitzsimmons A., Jorda L., Keller H. U., Williams I. P., Hahn G., and Rauer H. (2000) Observations of the trans-Neptunian objects 1993 SC and 1996 TL66 with the Infrared Space Observatory. *Astrophys. J., 534,* 446–455.

Trujillo C. A. and Brown M. E. (2002) A correlation between inclination and color in the classical Kuiper belt. *Astrophys. J. Lett., 566,* L125–L128.

Veillet C., Parker J. W., Griffin I., Marsden B., Doressoundiram A., Buie M., Tholen D. J., Connelley M., and Holman M. J. (2002) The binary Kuiper belt object 1998 WW_{31}. *Nature, 416,* 711–713.

Vilas F. and Gaffey M. J. (1989) Phyllosilicate absorption features in main-belt and outer-belt asteroid reflectance spectra. *Science, 246,* 790–792.

Williams I. P., O'Ceallaigh D. P., Fitzsimmons A., and Marsden B. C. (1995) The slow-moving objects 1993 SB and 1993 SC. *Icarus, 116,* 180–185.

From Cradle To Grave:
The Rise and Demise of the Comets

David C. Jewitt
Institute for Astronomy, University of Hawai'i

The active comets are a dynamic ensemble of decaying bodies, recently arrived from cold storage locations in the Kuiper belt and Oort cloud. In this chapter, we discuss the processes that drive the physical transformation and decay of cometary nuclei as they move from the frigid outer regions into the hot environment of the inner solar system.

1. INTRODUCTION

In this chapter, we discuss the processes that drive the physical transformation and decay of cometary nuclei as they move from the frigid outer regions into the hot environment of the inner solar system. Fundamentally, comets and asteroids are distinguished by their volatile content, which is itself a measure of the temperature of the environment in which they accreted. Comets possess a substantial fraction of bulk water ice that is not expected in the asteroids of the main-belt [*Whipple* (1950) thought comets might contain a water ice fraction near 50% by mass but recent data suggest smaller fractions. *Reach et al.* (2000), for example, estimate a ratio 3% to 10% in 2P/Encke, while *Grün et al.* (2001) find that only 10–15% of the mass lost from C/Hale Bopp was from sublimated ice.] Unfortunately, we possess no direct way to measure the bulk fraction of water ice within any nucleus or asteroid. The fundamental distinction between comets and asteroids is not reflected in any clean-cut, practical means by which to distinguish them.

Instead, the widely applied practical definition is that a comet is defined by showing a resolved coma at some point in its orbit. Deciding whether an object is an asteroid or a comet thus depends critically on the instrumental resolution and sensitivity to low surface brightness coma. A weakly active comet might not be detected as such if observed with insufficient resolution or sensitivity. Thus, there arises a gray zone in which the cometary vs. asteroidal nature of a given body cannot easily be ascertained by observations.

A third distinction between asteroids and comets may be drawn based on the respective dynamical properties. The Tisserand invariant with respect to Jupiter is a popular discriminant. It is defined by

$$T_J = \frac{a_J}{a} + 2\left((1 - e^2)\frac{a}{a_J}\right)^{1/2} \cos(i) \qquad (1)$$

where a, e, and i are the semimajor axis, eccentricity, and inclination of the orbit while $a_J = 5.2$ AU is the semimajor axis of the orbit of Jupiter. This parameter, which is conserved in the circular, restricted three-body problem, pro-vides a measure of the relative velocity of approach to Jupiter: Jupiter itself has $T_J = 3$, most comets have $T_J < 3$, while main-belt asteroids generally have $T_J > 3$. Unfortunately, the dynamical definition of comet-hood does not always match the observational or compositional definitions. A number of comets (including the famous 2P/Encke) have asteroid-like $T_J > 3$, while many bodies with comet-like $T_J < 3$ are asteroids scattered from the main belt. There are other difficult cases: The jovian Trojan asteroids have $T_J \sim 3$ and probably possess ice-rich interiors (and so are comets by the physical definition), but are too cold to sublimate, show no comae, and are labeled "asteroids."

Evidently, this is not a clean subject. Even the definition of the term "comet" is arguable, and the reader will see that much of the following discussion will be drawn inexorably toward objects whose cometary nature is debatable. To try to maintain focus we adopt a tutorial style that is intended to highlight connections between seemingly disparate subjects and that deliberately dissects and simplifies complicated problems to make them understandable. Sufficient references are given that the interested reader may take an easy step into the research literature, but we have made no attempt to be complete since the number of relevant publications is already very large. Aspects of this subject have been reviewed elsewhere (*Degewij and Tedesco*, 1982; *Jewitt*, 1996a; *Weissman et al.*, 2002).

2. COMETARY RESERVOIRS

Comets are observationally defined as solar system bodies that maintain at least transient gaseous, or dusty, comae. The comae are gravitationally unbound (escaping) atmospheres produced by classical sublimation of near-surface ices in response to heating by the Sun. Their small sizes (typically a few kilometers) and resulting short sublimation lifetimes (typically $\sim 10^4$ yr) guarantee that the observed comets are recent arrivals in the inner solar system. If a steady-state population is to be maintained, the comets must be continually resupplied from one or more long-lived reservoirs.

Two primary reservoirs of the comets, the Oort cloud and the Kuiper belt, are now recognized (Fig. 1). In the absence of contrary evidence, it is assumed that both are primordial,

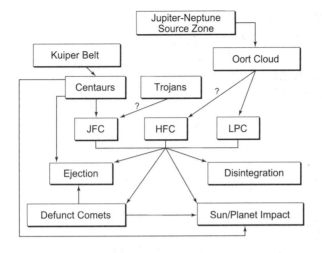

Fig. 1. Current interrelations among the planetary small body populations. JFC = Jupiter-family comet, HFC = Halley-family comet, LPC = long-period comet (also known as isotropic source comets). Question marks indicate the uncertain path from the Oort cloud to the HFCs and the unknown contribution to the JFCs from the Trojans of the giant planets. The defunct comets include both dead (totally devolatilized) and dormant (volatiles shielded from solar insolation) bodies. Adapted from *Jewitt and Fernández (2001)*.

of age $\tau_{SS} = 4.5 \times 10^9$ yr. (Timescales mentioned in the text are summarized graphically in Fig. 2.)

2.1. Oort Cloud

The Oort cloud was identified first from the peculiar distribution of the semimajor axes of long-period comets (LPCs) (*Oort*, 1950). It supplies a nearly isotropic flux of LPCs to the planetary region and is of vast extent, with characteristic length scale $R_{oc} \sim 100{,}000$ AU and orbital periods near $\tau_{oc} \sim 10^6$–10^7 yr.

The total number of Oort cloud comets larger than about 1 km in radius is on the order of $N_{oc} \sim 10^{12}$ (see *Rickman*, 2004; *Dones et al.*, 2004). Comets now in the Oort cloud are thought to have originated in the protoplanetary disk between Jupiter and Neptune, while most were scattered out by Uranus and Neptune (*Hahn and Malhotra*, 1999).

The dynamical part of Oort's model predicts a ratio of returning comets relative to first-appearing comets that is larger than is observed. Oort's solution was to introduce a "fading parameter" to diminish the number of returning comets. Ideas about the nature of the fading mechanism range from the sublimation of supervolatile frosting accreted from the interstellar medium to physical disintegration of the nuclei soon after entry into the planetary region. The need for a spherical source to supply the isotropic orbital inclination distribution seems secure, as does the large effective size of the source (indicated by the large semimajor axes of the LPCs). No plausible alternatives to Oort's model have been found in the 50 years since its introduction

(see *Wiegert and Tremaine*, 1999). Still, the uncertain nature of the fading parameter on which its success depends remains disconcerting.

The collision time in the Oort cloud is $\tau_c \sim \tau_{oc}/\gamma$, where γ is the optical depth, equal to the ratio of the sum of the cross-sections of the constituent nuclei to the effective geometric cross-section of the cloud. We write

$$\gamma \sim N_{oc}\left(\frac{r_n}{R_{oc}}\right)^2 \qquad (2)$$

where r_n is the effective nucleus radius. We assume that the cross-section is dominated by the smallest objects and take $r_n = 1$ km. Substitution gives $\gamma \sim 10^{-14}$ and, with $\tau_{orb} \sim 10^7$ yr, we find that $\tau_c \gg \tau_{SS}$ and so is effectively infinite. Any collisional processing of the Oort cloud comets must have occurred at early times, prior to their emplacement in distant heliocentric orbits (*Stern and Weissman*, 2001).

2.2. Halley-Family Comets

The Halley-family comets (HFCs, also known as Halley-type comets) are a separate group distinguished by having short orbital periods but a wide spread of inclinations, including retrograde orbits that are absent in the Jupiter family. These bodies have Tisserand parameters (equation (1)) $T_J < 2$. The prototype 1P/Halley (a = 17.8 AU, e = 0.97, i = 162°, $T_J = -0.61$) is typical. They are thought to derive from the inner Oort cloud by gravitational capture, principally by interaction with the massive gas giant planets (*Bailey and*

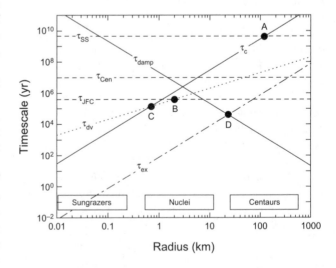

Fig. 2. Timescales relevant to the cometary nucleus. The representative size ranges of the well-measured Sun-grazers, JFC nuclei, and Centaurs are shown as separate for clarity: In fact, their size ranges overlap. Labeled curves and lettered points are described in the text. Lines for τ_{dv} and τ_{ex} refer to outgassing from comets at a = 3.5 AU, typical of the JFCs. Adapted from *Jewitt* (1996a).

TABLE 1.   The known cometary Centaurs.

Object	Perihelion (AU)	Semimajor Axis (AU)	Eccentricity	Inclination (deg)	Tisserand Parameter
C/2001 M10	5.30	26.66	0.80	28.0	2.59
29P/SW1	5.72	5.99	0.04	9.4	2.98
39P/Oterma	6.83	7.25	0.24	1.9	3.01
2060 Chiron	8.45	13.62	0.38	6.9	3.36
C/2001 T4	8.56	13.92	0.38	15.4	3.29

Emel'yanenko, 1996; *Levison et al.*, 2001; *Rickman et al.*, 2002). However, the details of this capture and their implications for the structure of the inner Oort cloud are controversial. The HFCs follow a complex dynamical evolution under the control of mean-motion and secular resonances, ending with ejection from the planetary system or impact with the Sun after a mean time on the order of 10^6 yr (*Bailey and Emel'yanenko*, 1996). These objects are rare compared to Jupiter-family comets in the observational sample only because there is a strong observational bias against them. In absolute numbers, the HFCs may outnumber the Jupiter family by a large factor.

2.3. Kuiper Belt, Centaurs, and Jupiter-family Comets

The Jupiter-family comets (JFCs) occupy small orbits with modest inclinations and eccentricities. Their Tisserand parameters are $2 \leq T_J < 3$ (*Levison*, 1996). Most JFCs probably originate from a transneptunian source known as the Kuiper belt (*Fernández*, 1980; *Duncan et al.*, 1988), whose inclination distribution is similar to that of the JFCs themselves. In this scenario, the kilometer-sized JFCs could be collisionally produced fragments of larger Kuiper belt objects, or KBOs (*Stern*, 1995; *Farinella and Davis*, 1996; *Duncan et al.*, 2004). The representative orbital period in the Kuiper belt is $\tau_{KB} \sim 10^2$–10^3 yr.

The number of Kuiper belt comets larger than 1 km radius, N_1, is crucial if the Kuiper belt is to supply the JFCs. However, this number is highly uncertain. Early estimates based on extrapolation from 100-km-scale KBOs gave $N_1 \sim 10^{10}$ (*Jewitt et al.*, 1998). The first direct measurements of ~10-km-scale KBOs in the classical region of the belt, when extrapolated to 1 km, give $N_1 \sim 10^8$ (*Bernstein et al.*, 2003), which may be too small for the classical belt to supply the JFCs. However, the number of scattered KBOs remains observationally unconstrained at small sizes, and this population could supply the JFCs for the age of the solar system.

The Centaurs are dynamically intermediate between the Kuiper belt and the JFCs. We here define Centaurs as objects with perihelia $q > a_J$ and semimajor axes $a < a_N$, where $a_J = 5.2$ AU and $a_N = 30$ AU are the semimajor axes of Jupiter and Neptune respectively (*Jewitt and Kalas*, 1998). By this definition, there are currently (as of October 2002) 42 known Centaurs, including the prototype 2060 Chiron, also known as 95P/Chiron ($a = 13.6$ AU, $e = 0.38$, $i = 15°$).

Most appear asteroidal but five have been observed to show comae and thus are also properly recognized as comets (Table 1). A further 140 known objects dip into the planetary region from the Kuiper belt and beyond (i.e., $q \leq a_N$ and $a > a_N$).

Once trapped as Centaurs the dynamical lifetimes are limited by strong gravitational scattering by the giant planets to $\tau_{Cen} \approx 10^7$ yr (*Dones et al.*, 1999). Most Centaurs are ejected from the solar system. The survivors that become trapped inside Jupiter's orbit tend to sublimate and are observationally relabelled as JFCs. Their median dynamical lifetime (*Levison and Duncan*, 1994) is $\tau_{JFC} = 3.3 \times 10^5$ yr. Note that the dynamical evolution is chaotic and the reverse transition from JFC to Centaur is common (as happened recently with Comet/Centaur 39P/Oterma).

The comets follow chaotic trajectories among the planets [with dynamical memories ~1000 yr (*Tancredi*, 1995)], and elaborate numerical models must be used to track their orbital evolution (*Levison and Duncan*, 1994, 1997). Very roughly, the probability that a comet or Centaur will be scattered inward following its encounter with a planet is p ~ 1/2. This means that the fraction of the escaped KBOs that are scattered inward by Neptune is just p, while the fraction that scatter between the four giant planets down past Jupiter is $p^4 \sim 0.05$. If we assume that the KBOs have an effective lifetime comparable to τ_{SS}, then the steady-state population of the Centaurs, N_{Cen}, relative to that of the KBOs, N_{KBO}, may be crudely estimated from

$$\frac{N_{Cen}}{N_{KBO}} \sim p\left(\frac{\tau_{Cen}}{\tau_{SS}}\right) \sim 10^{-3} \qquad (3)$$

With $N_{KBO} \sim 70,000$ [diameter D > 100 km (*Jewitt et al.*, 1998; *Trujillo et al.*, 2001)], this gives $N_{Cen} \sim 70$, in good agreement with observational estimates of $N_{Cen} \sim 100$ measured to the same size (*Sheppard et al.*, 2000).

In the same spirit of approximation, the steady-state population of the JFCs is given by

$$\frac{N_{JFC}}{N_{KBO}} \sim p^4\left(\frac{\tau_{JFC}}{\tau_{SS}}\right) \sim 5 \times 10^{-6} \qquad (4)$$

This leads us to expect that in steady state ~0.4 JFCs have diameter D $\geq$ 100 km, consistent with the observation that there are currently none. The number of KBOs with $D \geq 100$ km is $N_{KBO} \sim 7 \times 10^4$ based on a simple extrapolation from survey data (*Jewitt et al.,* 1998). The number of KBOs with $D \geq 1$ km is very uncertain because an extrapolation of the size distribution must be made. A current best-guess population is $N_{KBO} \sim (1-10) \times 10^9$ (but see *Bernstein et al.,* 2003). With equation (4), this gives $5000 \leq N_{JFC} \leq 50,000$. Although very uncertain, the lower end of this range is comparable to the (equally uncertain) "several thousand to about 10^4" JFCs observationally estimated by *Fernández et al.* (1999; see also *Delsemme,* 1973). A more detailed estimate should include, in addition to a proper treatment of the dynamics, a correction for the loss of JFC nuclei through such processes as devolatization and disruption. A numerical treatment by *Levison and Duncan* (1997) finds that the source region of the JFCs must contain 7×10^9 objects, in agreement with early estimates but larger than the number of appropriately sized KBOs in the classical belt. A significant problem in comparing source models with population measurements is that the sizes of the objects being compared (cometary nuclei vs. KBOs) are not well determined because the albedos are not well known.

2.4. Other Sources of Comets

The distributions of color (*Jewitt and Luu,* 1990) and albedo (*Fernández et al.,* 2003a) of the jovian Trojan "asteroids" are formally indistinguishable from those of the cometary nuclei. This suggests (but does not prove) an intriguing compositional similarity between the two classes of body, at least at the surface level where irradiation and solar heating may play a role. No ices have been spectroscopically detected on the Trojans (*Jones et al.,* 1990; *Luu et al.,* 1994; *Dumas et al.,* 1998) but this is not surprising given the high surface temperatures (~150 K) and the expected rapid loss of exposed ice by sublimation. Beneath their refractory mantles, however, the Trojans may be ice rich. They may contribute to the comet population through dynamical instabilities and collisional ejection (*Marzari et al.,* 1995). Once removed from the vicinity of the Lagrangian L_4 and L_5 points, they quickly lose dynamical traces of their origin. There are too few jovian Trojans to supply more than ~10% of the flux of JFCs (*Marzari et al.,* 1995; *Jewitt et al.,* 2000), but Trojans of the other giant planets, if they exist, could be significant additional sources, and the total flux of escaped Trojans from all giant planets should be considered unknown. A narrow ring of orbits between Uranus and Neptune may be another source (*Holman,* 1997) although these orbits may not remain populated if the outer planets experienced substantial radial migration (*Brunini and Melita,* 1998).

3. ONSET OF ACTIVITY

Equilibrium surface temperatures in the cometary reservoirs are low (~10 K in the Oort cloud and ~40 K in the Kuiper belt). As orbital evolution carries the comets closer to the Sun, rising temperatures induce the sublimation of surface volatiles. The first abundant ice to sublimate is the

highly volatile carbon monoxide, CO, which is thought to produce the comae observed around some Centaurs in the middle solar system (R ~ 10 AU; Table 1, Fig. 3). However, most Centaurs show no comae (e.g., Fig. 4), either because they have already lost their near-surface volatiles through outgassing or because their surfaces consist of nonvolatile, complex organic and silicate mantles produced by energetic particle bombardment (see *Barucci et al.,* 2004). Although CO is the most volatile abundant ice in comets, the main driver of activity, as recognized long ago by *Whipple* (1950), is the sublimation of water ice, beginning near the orbital radius of Jupiter.

Solar heat propagates into the interior of the nucleus. The timescale for the conduction of heat to the center of a spherical nucleus of radius r_n is given by solution of the conduction equation

$$k\nabla^2 T(r,t) = \rho c_p \frac{dT(r,t)}{dt} - \rho H(r,t) \qquad (5)$$

where $T(r,t)$ is the temperature as a function of radius and time, k is thermal conductivity, ρ is bulk density, c_p is the specific heat capacity and $H(r,t)$ is the specific power production due to internal heat sources (e.g., radioactivity, phase

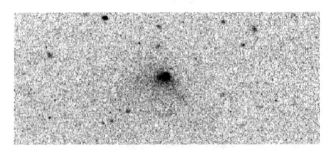

Fig. 3. Cometary Centaur C/2001 T4 in a 300-s, R-band image taken by the author using the University of Hawai'i 2.2-m telescope on UT 2002 September 4. The heliocentric and geocentric distances were R = 8.57 AU and Δ = 7.97 AU respectively, and the phase angle was 5.6°. Image is 96 arcsec wide. North is to the top, east to the left.

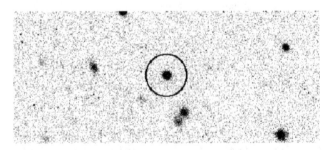

Fig. 4. Centaur 1998 SG$_{35}$ in a 900-s, R-band image taken by the author using the University of Hawai'i 2.2-m telescope on UT 2002 September 8. The heliocentric and geocentric distances were R = 8.72 AU and Δ = 8.12 AU respectively, and the phase angle was 5.5°. Image has the same orientation and scale as Fig. 3. In sharp contrast to C/2001 T4 (Fig. 3), this Centaur shows no coma or tail when observed at a nearly identical heliocentric distance and for a longer time.

transitions). Dimensional treatment of this equation gives the characteristic e-folding timescale for heat transport by conduction as

$$\tau_c \sim \left(\frac{r_n^2}{\kappa} \right) \qquad (6)$$

where $\kappa = k/(\rho c_p)$ is the thermal diffusivity. For a nominal thermal diffusivity $\kappa = 10^{-7}$ m² s⁻¹, we have

$$\tau_c \sim 3 \times 10^5 r_n^2 \text{ yr} \qquad (7)$$

with r_n expressed in kilometers. Setting $\tau_c = \tau_{SS}$ gives $r_n \sim 100$ km as the maximum size for effective conductive heat transport (Point A in Fig. 2). All known cometary nuclei, but not the Centaurs or KBOs, are smaller than this. Note that $\tau_c > \tau_{JFC}$ for $r_n > 1$ km, meaning that the deep interiors of the nuclei of most JFCs are effectively thermally decoupled from their surfaces.

4. EFFECTS OF ACTIVITY

Mass loss due to sublimation can exert a profound influence on the physical nature of the cometary nucleus, perhaps changing the shape, size, rotation, and even survival time in the inner solar system. Furthermore, these effects are likely interrelated. Anisotropic mass-loss produces torques on the nucleus that change the spin and the nucleus shape, leading to a change in the distribution of active areas and in the torque. Centripetal effects may lead to loss of material from the rotational equator, affecting the size, shape, spin, and mantling. For clarity of presentation, these effects are discussed separately here, but they are in reality closely intertwined.

4.1. Mantle Formation

The existence of refractory surface mantles is suggested by groundbased observations of many comets (*A'Hearn et al.*, 1995) and by direct imaging of the nuclei of Comets 1P/Halley and 19P/Borrelly at subkilometer resolution. The observations show that sublimation of cometary volatiles is restricted to active areas occupying a fraction of the surface area $10^{-3} \le f \le 10^{-1}$ [larger active fractions ~1 are sometimes reported, but these may reflect gas production from secondary sources in an icy grain halo about the nucleus (*Lisse et al.*, 1999)]. The corresponding f for the nuclei of LPCs is observationally not well established. The inactive regions correspond to volatile-depleted surface thought to consist of refractory crust or mantle material. Many comets with $f \ll 10^{-3}$ cannot be easily distinguished from asteroids owing to the practical difficulty of detecting very weak dust (*Luu and Jewitt*, 1992a) or gas comae (*Chamberlin et al.*, 1996).

In the so-called rubble mantle model, the mantles are thought to consist of refractory blocks that are too large to be ejected by gas drag against the gravity of the nucleus. Such mantles need only be thicker than the thermal skin

depth [$L_D \sim (\kappa P_{rot})^{1/2}$ where P_{rot} is the nucleus rotation period], in order to inhibit sublimation. For representative values $P_{rot} = 6$ h, $\kappa = 10^{-7}$ m² s⁻¹, we obtain $L_D \sim 5$ cm. The timescale for so-called rubble mantle growth (neglecting cohesion) is (*Jewitt*, 2002)

$$\tau_M \sim \frac{\rho_n L_D}{\dot{m} f_M} \qquad (8)$$

in which ρ_n is the density of the nucleus, $f_M(r_n, R)$ is the fraction of the solid matter in the nucleus too large to be ejected by gas drag, and $\dot{m}(R)$ is the specific mass sublimation rate. This timescale is only ~10³ yr at 5 AU, falling to 1 yr at 3 AU for a water ice composition (Fig. 5). The essential points are that rubble mantles can be very thin and should readily form as byproducts of cometary activity on timescales that are short compared to τ_{JFC}. A feature of the rubble mantle model is that such structures should be unstable to ejection on comets whose perihelia diffuse inward, since rising temperatures and gas pressures can easily overcome the local gravitational force. Cohesion between mantle grains is needed to convey long-term mantle stability to such objects (*Kührt and Keller*, 1994).

A second kind of mantle is postulated for the cometary nuclei. The so-called irradiation mantle consists of material that has been chemically transformed and devolatilized by prolonged exposure to energetic photons and particles. Cosmic rays with MeV and higher energies penetrate to column densities ~10³ kg m⁻², corresponding to depths ~1 m in material of density 10³ kg m⁻³. The timescales for complete processing of this surface layer are of order $10^{8 \pm 1}$ yr (*Shul'man* 1972). This is short compared to the storage times for bodies in the Oort cloud and Kuiper belt and the upper layers of residents of these populations are likely to

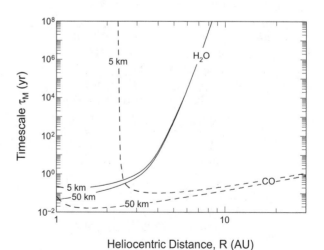

Fig. 5. Timescale for the growth of a rubble mantle in model nuclei of radii 5 km and 50 km, in response to sublimation of H_2O and CO ices and computed from equation (8). The orbits are assumed to be circular. Adoption of more realistic eccentric orbits would increase the mantling time relative to the plotted curves, but mantling is always rapid at small R. From *Jewitt* (2002).

be significantly processed down to meter depths. Optical photons, on the other hand, probe a surface layer only a few micrometers thick. The timescale for processing this visible layer is probably short, but the timescales for building a rubble mantle are even shorter. Thus, we expect that bodies with perihelia beyond the water sublimation zone (KBOs, Centaurs) might retain irradiation mantles but that these will have been ejected or buried on the nuclei of near-Earth comets. One observation consistent with this is the lack of ultrared matter, of the type seen on the surfaces of half the KBOs and Centaurs, on the nuclei of JFCs (*Jewitt*, 2002) (see section 4.7).

Irradiation breaks the chemical bonds of common molecules. In the process H, because of its small size, is able to escape from the irradiated layers of the nucleus into space. The irradiation mantle is thus composed of material in which the C/H and O/H ratios are high. The high C fraction may be responsible for the low albedos of cometary nuclei (*Campins and Fernández*, 2002; *Moroz et al.*, 2003). The low H fraction may explain why near-infrared spectra of outer solar system bodies, including KBOs, Centaurs, jovian Trojans, and cometary nuclei, are mostly devoid of the absorption bands of common bonds (e.g., C-H, O-H, N-H).

4.2. Thermal Devolatilization

The timescale for the loss of volatiles from a mantled ice nucleus is

$$\tau_{dv} \sim \frac{\rho_n r_n}{f \bar{m}} \qquad (9)$$

where $\bar{m}$ (kg m^{-2} s^{-1}) is the specific mass loss rate averaged around the cometary orbit, ρ_n is the density, and f is the so-called "mantle fraction," the fraction of the surface from which sublimation proceeds. The specific mass loss rate can be estimated from equilibrium sublimation of H$_2$O ice to be about $\dot{m} \sim 10^{-4}$ kg m^{-2} s^{-1} at R = 1 AU, varying roughly as R^{-2} for R $\leq$ 2 AU and faster than R^{-2} at greater distances (Fig. 6). For nonzero orbital eccentricities, the sublimation rate at a given semimajor axis is higher than for the circular orbit case because of enhanced sublimation near perihelion. This effect is large only for orbital semimajor axes greater than about 3 AU, which is the critical distance beyond which sublimation consumes a negligible fraction of the absorbed solar energy. This is shown in Fig. 6, in which we plot the orbitally averaged water ice sublimation rate as a function of semimajor axis and eccentricity. For the canonical JFC orbit with a = 3.5 AU and e = 0.5, we estimate $\bar{m} = 10^{-5}$ kg m^{-2} s^{-1} (Fig. 6). With $\rho_n = 500$ kg m^{-3} and f = 0.01 (*A'Hearn et al.*, 1995), we find

$$\tau_{dv} \sim 2 \times 10^5 r_n \text{ yr} \qquad (10)$$

with r_n again expressed in kilometers and τ_{dv} is measured in years. The effect of nonzero eccentricity is small for comets whose semimajor axis is less than the critical distance for strong sublimation of water ice (R $\leq$ 3 AU; Fig. 6). Thus, JFC lifetime estimates are much less affected than HFCs

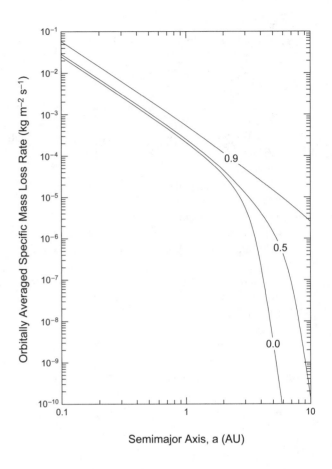

Fig. 6. Orbitally averaged specific mass loss rate for sublimating water ice as a function of the semimajor axis and eccentricity. The instantaneous mass loss rate was computed from the energy balance equation and integrated around the orbit using Kepler's equation. A low (0.04) albedo surface was assumed and thermal conduction was neglected. Curves are labeled by the orbital eccentricity.

and LPCs, the latter of which sublimate only for a limited period near perihelion.

The lifetimes of kilometer-scale nuclei against sublimation, given by equation (10), are comparable to the dynamical lifetime. Note that $\tau_{dv} < \tau_{JFC}$ for $r_n < 1$ km (Point B in Fig. 2). This means that the subkilometer comets should lose their volatiles on timescales short compared to their dynamical lifetimes, leaving behind a population of "dead comets." Note also that $\tau_{dv} < \tau_c$ for $r_n > 1$ km (Point C in Fig. 2), meaning that for almost all observed comets the volatiles are lost from the surface before the thermal conduction wave has reached the core. This inequality suggests another end state as "dormant comets" (*Hartmann et al.*, 1987; *Kresák*, 1987), having devolatilized, inactive surface regions shrouding an ice packed core.

4.3. Size Evolution

Mass loss should modify the size distribution of the comets by selectively depleting the smallest objects (equation (10)). The size distribution of the parent KBOs has been measured, albeit only for large objects, from the slope of the cumulative luminosity function (i.e., the cumulative number of objects per square degree of sky brighter than a given magnitude). Different researchers have converged on a power-law-type distribution in which the number of objects with radius between r and r + dr is

$$n(r)dr = \Gamma r^{-q} dr \qquad (11)$$

with Γ and q constants of the distribution. The best-fit value for KBOs is $q = 4.0^{+0.6}_{-0.5}$ (*Trujillo et al., 2001*). In such a distribution, the mass is spread uniformly in equal logarithmic intervals of radius while the cross-section is dominated by the smallest objects in the distribution. The Centaurs are, by and large, less well observed than the KBOs. The available data are compatible with $q = 4.0 \pm 0.5$ (*Sheppard et al., 2000*). This is identical to the value in the Kuiper belt, as expected if the latter is the source of the Centaurs.

The size distribution of the cometary nuclei is poorly determined by comparison. *Fernández et al.* (1999) used spatially resolved photometry of comets to find $q = 3.6^{+0.3}_{-0.2}$, while the same technique used independently by *Weissman and Lowry* (2003) gave $q = 2.6 \pm 0.03$ and *Lamy et al.* (2004) found $q = 2.6 \pm 0.2$ to $q = 2.9 \pm 0.3$. The results are inconsistent at the 5σ level of significance. To understand this, it is important to remember that the cometary nuclei are in general observed in the presence of coma. Coma contamination of the nucleus photometric signal may confuse the results obtained by one or more of these groups. Furthermore, the short-period comets that form their samples have been discovered by a variety of techniques, each of which must impress onto the sample its own distinctive discovery bias. Naively, this bias is expected to favor cometary nuclei with large active areas because these objects will be bright and therefore more easily detected. Overrepresentation of the bright comets will lead to a measured size distribution that is flatter than the intrinsic distribution.

What should we expect? Sublimation lifetimes of otherwise equal bodies vary in proportion to the radius (equation (10)). Thus, in steady state, we expect that a source population described by r^{-q} should be flattened by sublimation to $r^{-(q-1)}$, as a result of the more rapid loss of the smaller objects. From $q = 4$ in the Kuiper belt, we expect to find $q = 3$ among the JFCs, which is in between the measured values.

Comets with $r_n \leq 700$ m have $\tau_c \leq \tau_{dv}$ (equations (7) and (10) and Point C in Fig. 2). In such objects, the thermal conduction wave reaches ices still frozen in the core. The resulting gas pressure produced by sublimation in the core is likely to explosively disrupt the nucleus. This provides one possible explanation for the often-reported depletion of very small comets relative to extrapolations of the power-law size distribution as determined at kilometer and larger nucleus scales. The very short timescale for rotational spinup of subkilometer nuclei (Fig. 2 of *Jewitt*, 1999) provides an even more compelling mechanism for their destruction by rotational bursting.

It should be noted that the sizes of the measured KBOs and comets are quite different. Most KBOs have $r_n \geq 50$ km,

while most studied comets have $r_n \leq 5$ km. Size-dependent, rather than evolutionary, effects might be present. It will be important to extend the KBO size distribution to typical cometary nucleus scales.

In summary, the size distributions of the KBOs and Centaurs appear identical, within the uncertainties, and consistent with the link between KBOs and Centaurs. Measurements of the size distributions of the cometary nuclei have been attempted, but the reported values are discordant and are, in any case, afflicted by discovery and coma contamination biases that are poorly understood.

4.4. Shape Evolution

The shapes of cometary nuclei can be estimated from their rotational lightcurves (strictly, the lightcurves give only the projection of the shape in the plane of the sky). If the nuclei are collisionally produced fragments, then it seems reasonable that their shapes should be distributed like those of impact fragments produced in the laboratory by impact experiments. The comparison is made in Fig. 7 where we show the photometric ranges of well-measured nuclei with the range distribution of impact fragments measured in the laboratory by *Catullo et al.* (1984).

The distributions are clearly different, with the comets showing a larger fraction of highly elongated shapes than the impact fragments. The well-measured cometary nuclei are also elongated, on average, compared to main-belt asteroids of comparable size (*Jewitt et al.*, 2003). This can be naturally explained as a simple consequence of anisotropic mass loss from the comets, which should act to modify the

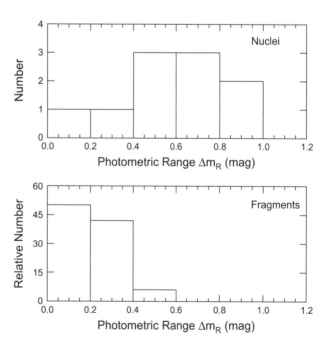

Fig. 7. Apparent axis ratios of cometary nuclei (from *Jewitt et al.*, 2003) with published rotational lightcurves compared with impact-produced rock fragments (from *Catullo et al.*, 1984).

overall shape on a timescale comparable to τ_{dv}. Anisotropic mass loss may also provide a mechanism for splitting on this timescale (*Hartmann and Tholen*, 1990). Note that most of the axis ratios reported in Table 1 of *Lamy et al.* (2004) are smaller than those in Fig. 7. The difference might occur because the Lamy et al. data are lower limits obtained from sparsely sampled photometric series, whereas the projected axis ratios in Fig. 7 were all determined from well-measured lightcurves.

4.5. Spin Evolution

Noncentral mass loss from the comets generates torques that can both change the spin period and drive the nucleus into an excited rotational state (i.e., a state other than principal axis rotation around the short axis). The relevant excitation timescale is

$$\tau_{ex} \sim \left[\frac{\omega \rho r_n^4}{V_{th} k_T \dot{M}} \right] \quad (12)$$

where $\dot{M}$ (kg s^{-1}) is the net mass loss rate from all active areas (for a spherical nucleus $\dot{M} = 4\pi r_n^2 f \dot{m}$), k_T is the dimensionless moment arm for the torque ($k_T = 0$ for radial ejection and $k_T = 1$ for tangential mass loss), and V_{th} is the mass-weighted outflow speed (cf. *Samarasinha et al.*, 1986; *Jewitt*, 1991). The dimensionless moment arm has been estimated analytically from a block model of the nucleus as $k_T \sim 0.05$ (*Jewitt*, 1999) and from numerical simulations as $0.01 \leq k_T \leq 0.04$ (*Gutiérrez et al.*, 2003). For plausible estimates of the parameters we find $\tau_{ex} < \tau_{JFC}$ (Fig. 2), meaning that outgassing torques can drive comets toward rotational instability, perhaps providing one mechanism for destruction of the nucleus (Fig. 2) (*Jewitt*, 1999).

Excited rotational motions create periodic internal stresses that lead to minute (but, over long periods, significant) frictional dissipation of energy. Relaxation into the minimum rotational energy (maximum moment of inertia) state occurs on the timescale (*Burns and Safranov*, 1973)

$$\tau_{damp} \sim \frac{\mu Q}{\rho K_3^2 r_n^2 \omega^3} \quad (13)$$

Here, μ (N m^{-2}) is the rigidity, Q is the quality factor (fractional loss of energy per cycle), K_3 is a shape-dependent numerical factor, and r_n (m) is the mean radius. The damping parameters appropriate to cometary nuclei are not well known. We follow *Harris* (1994) and take $\mu Q = 5 \times 10^{11}$ (N m^{-2}) and $K_3^2 \sim 0.03$ (based on data for Phobos). Substituting $\rho = 500$ kg m^{-3} we obtain

$$\tau_{damp} \sim 1.0 \times 10^5 \left(\frac{P^3}{r_n^2} \right) \quad (14)$$

for the damping time in years, with P in hours and r_n in kilometers (Fig. 2). A 2-km-radius nucleus created collisionally in the Kuiper belt with an initial spin period of 6 h,

for example, could occupy a rotationally excited state (*Giblin and Farinella*, 1997) for only ~10^7 yr before damping away. This is comparable to the median transport time from the Kuiper belt to the inner solar system. Comets much smaller than ~2 km, and those with spin periods much longer than 6 h, might retain excited rotational states produced collisionally in the Kuiper belt, although we do not expect this to be the general case because the injection of a nucleus in a Neptune-crossing orbit may occur long after its collisional production.

Nucleus excitation is much more likely to be actively produced once comets begin to outgas inside the orbit of Jupiter. The timescales for excitation (equation (12)) and damping (equation (13)) of the spin are equal at the critical size

$$r_n = \left(\frac{4\pi \mu Q V_{th} k_T f \dot{m}}{\rho^2 K_3^2 \omega^4} \right)^{1/4} \quad (15)$$

which, with $\dot{m} = 10^{-5}$ kg m^{-2} s^{-1}, $V_{th} = 10^3$ m s^{-1}, $\rho = 500$ kg m^{-3}, f = 0.01, P = 6 h, and other parameters as given earlier, yields $r_n \sim 20$ km (Point D in Fig. 2). Since most known cometary nuclei are smaller than this critical size, we conclude that most are potentially in rotationally excited states. Numerical simulations show that equation (12), while providing a good estimate of the timescale for spinup, may give only a lower limit to the timescale for driving a nucleus into excited rotational states (*Gutiérrez et al.*, 2003). Thus it is possible that equation (15) overestimates the critical radius.

Observational evidence for precession of the nuclei is limited, both because measurements are complicated by the effects of near-nucleus coma and because few attempts have been made to secure adequate temporal coverage. The best case is for the nucleus of 1P/Halley (*Samarasinha and A'Hearn*, 1991) while, more recently, 2P/Encke has shown indications of a time-varying lightcurve that might indicate nucleus precession (*Fernández et al.*, 2000). Spinup might be expected to lead to rapid rotation among the comet nuclei, especially at small sizes. A few nuclei [(P/Schwassmann-Wachmann 2 (*Luu and Jewitt*, 1992b)] are indeed rotating close to the centripetal limit for densities $\rho \sim 500$ kg m^{-3}. Less-direct evidence for rotational destruction of small nuclei might be evident in the depletion of these objects relative to power-law extrapolations from larger sizes.

4.6. Active Area Evolution

The active areas and the mantle should evolve in parallel. The lifetime of an active area can be estimated as follows. In the limiting case in which all the incident solar energy is used to sublimate ice, the subsolar specific mass loss rate is given by

$$\dot{m} = \frac{S_\odot}{LR^2} \quad (16)$$

in which $S_\odot = 1360$ W m^{-2} is the solar constant, R (AU) is the heliocentric distance, and L is the latent heat of sublimation at the (sublimation depressed) temperature of the surface ice. (We consider the limiting case only because it gives a convenient analytic expression for ṁ.) The corresponding rate of recession of the sublimating surface is just

$$\dot{r} \sim \frac{\dot{m}}{\rho} \quad (17)$$

An exposed plug of ice of area $\pi a^2 = 4\pi r_n^2 f$ would sublimate into the nucleus, creating a cavity or vent that deepens at rate $\dot{r}$. When the vent becomes too deep, self-shadowing by the walls will inhibit further sublimation. Accumulation of a blocky rubble mantle at the bottom of the vent will also suppress sublimation. We assume that this happens first when the vent reaches a critical depth $d \sim a$. The timescale for the vent to reach this depth is just $\tau_v \sim a/\dot{r}$. From equation (17) we obtain

$$\tau_v \sim \frac{2f^{1/2}\rho R^2 L r_n}{S_\odot} \quad (18)$$

where ρ is the density of the solid ice. For H_2O ice ($L = 2 \times 10^6$ J kg^{-1}) with $\rho = 500$ kg m^{-3}, $f = 0.01$ and $R = 1$ AU, we obtain

$$\tau_v \sim 5r_n \text{ yr} \quad (19)$$

with r_n expressed in kilometers. In equation (19), the lifetime is in units of years of exposure to sunlight in a circular orbit at $R = 1$ AU. A 5-km-radius nucleus would have $\tau_v \sim 25$ yr. The vent lifetime on the nucleus of an equivalent comet moving in an eccentric orbit will be larger because the mean insolation and the mean mass loss rate are smaller. For example, a JFC with $a = 3.5$ AU, $q = 1.0$ AU would have a vent timescale longer than given by equation (19) by a factor of ~10 to account for the greater mean distance from the Sun. Still, the vent lifetimes are expected to be very short compared to the dynamical lifetime, τ_{JFC}, for comets on all but the most eccentric orbits. Observational constraints on JFC active area lifetimes are few: *Sekanina* (1990) concludes that changes in active areas should be evident on timescales comparable with the observing records of many comets, consistent with equation (19).

For a CO-powered vent ($L = 2 \times 10^5$ J kg^{-1}) on a KBO at $R = 40$ AU, the timescale is still a short $\tau_v \sim 8 \times 10^2$ yr. Active areas on the same object moved to $R = 10$ AU would become self-shadowing in only ~50 yr, a tiny fraction of the dynamical lifetime. We conclude that the KBOs and Centaurs may occasionally display spectacular CO-powered comae from exposed vents (e.g., craters produced by impact), but that these are short-lived and should therefore be rare unless the vents are reactivated or replaced. Centaur 2060 Chiron exhibits brightness outbursts in the $8 \leq R \leq 19$ AU range with a timescale of ~10–20 yr (*Bus et al.,* 2001). It is not unreasonable to suppose that this activity is modulated by the evolution of CO-powered vents. Centaur

C/2001 T4 likewise displays a prominent, variable coma at $R \sim 8.5$ AU [see Fig. 3 and *Bauer et al.* (2003b)]. In the comets (and to a lesser extent the Centaurs) reactivation and the formation of new active areas may be driven by the progressively rising temperatures leading to increasing gas pressures that can destabilize the mantle (*Brin and Mendis,* 1979; *Rickman et al.,* 1990). In the rather stable orbits of the Kuiper belt, reactivation can be caused by impacts that disrupt the mantle.

4.7. Colors and Albedos

Figure 8 shows distributions of the optical colors of different types of small bodies (Table 2 summarizes the data). The KBOs show a wide range of colors from nearly neutral (S' > 0%/1000 Å) to very red (S' ~ 50%/1000 Å) with a median value S' = 25%/1000 Å. Here, the slope, S', is measured in the best-observed V-R region of the spectrum (i.e., 5500 Å to 6500 Å wavelength). The similarity between the wide spread of colors on the KBOs and on the Centaurs (here taken from *Bauer et al.,* 2003b) shows that mantling on the Centaurs is relatively minor, even though these bodies may sometimes be active (Fig. 8). Instead, a dramatic change in the color distribution appears only once the bodies have perihelia inside Jupiter's orbit and can begin to sublimate water ice. The comets show a smaller range of S' and, specifically, are deficient in the ultrared material with S' > 25%/1000 Å (*Jewitt,* 2002; see also *Hainaut and Delsante,* 2002). This may be a result of mantling of the cometary nuclei driven by sublimation in response to rising temperatures as they approach the Sun. Objects that are likely, on dynamical grounds, to be dead comets show a color distribu-

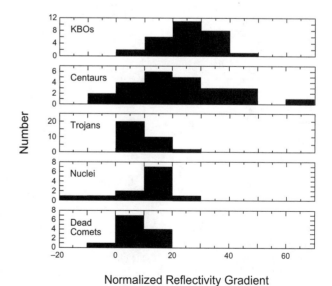

Fig. 8. Histograms of normalized optical reflectivity gradients (%/1000 Å) for objects in the major categories of this review. Plotted are KBOs, nuclei, and dead comets (*Jewitt,* 2002); Centaurs (*Bauer et al.,* 2003b); and jovian Trojans (*Jewitt and Luu,* 1990). The ultrared matter, with reflectivity gradient ≥25%/1000 Å, is present only in the Kuiper belt and on the Centaurs.

TABLE 2.   Mean reflectivity gradients and color indices.

Object	S'	V-R	N	Reference
KBOs	23 ± 2	0.61 ± 0.01	28	*Jewitt and Luu* (2001)
Centaurs	22 ± 4	0.58 ± 0.01	24	*Bauer et al.* (2003b)
Nuclei	8 ± 3	0.45 ± 0.02	12	*Jewitt* (2002)
Dead comets	7 ± 2	0.44 ± 0.02	12	*Jewitt* (2002)
D-types	8.8 ± 0.5	0.45 ± 0.01	19	*Fitzsimmons et al.* (1994)
Trojans	10 ± 1	0.46 ± 0.01	32	*Jewitt and Luu* (1990)

tion that is formally indistinguishable from that of the nuclei of active comets, suggesting that the mantles on these bodies are stable over long periods [and may indeed be responsible for the deaths of the comets (*Rickman et al.,* 1990)].

A similar conclusion can be drawn from Fig. 9, in which we show available measurements of the colors and albedos of the small-body populations. It is evident that the cometary nuclei occupy a restricted region of the color-albedo plane near p_R ~ 0.04, S' ~ 10%/1000 Å while the Centaurs are much more widely dispersed. The few measured nuclei fall near the region occupied by the D-type asteroids and the similarity to the Trojans is impressive (*Jewitt and Luu,* 1990). The corresponding data for KBOs, although few in number, already show wide separation in the color-albedo plane and are very different from any cometary nucleus yet measured. Observationally, we can already state with confidence that the albedos and colors of the cometary nuclei occupy a smaller range than is found in the Centaurs or KBOs. The implication is that surface materials that are common in the middle and outer solar system, including frosts and the ultrared material, have been ejected or buried (*Jewitt,* 2002) or are thermodynamically unstable (*Moroz et al.,* 2003) on the JFCs. The significance of this result is that spacecraft launched to investigate the nuclei of JFCs may fail to sample the primitive materials present on the surfaces of their more distant progenitors.

5. COMETARY END STATES

5.1. Dead and Dormant Comets

Dormant comets lack near-surface volatiles but may possess ice-rich interiors while dead comets are completely devolatilized. Identifying such objects from their physical properties is not easy, since the nuclei possess a wide range of optical properties that overlaps those of some classes of asteroids. The orbital properties provide a separate clue: Dead and dormant comets are likely to possess comet-like Tisserand invariants, $T_J \leq 3$. Objects selected on this basis indeed possess low, comet-like geometric albedos, p_R ~ 0.04 (see Fig. 10), quite distinct from the $T_J > 3$ asteroids. This finding has been used to infer that about 10% of the near-Earth objects (NEOs) might be dead comets (*Fernández et al.,* 2001). Dynamical models of the near-Earth population allow a similar (~6%) fraction of dead comets (*Bottke et al.,* 2002). However, the latter models are incomplete in that they neglect nongravitational forces and/or perturbations from the terrestrial planets. Orbits decoupled from Jupiter, like that of Comet P/Encke (T_J = 3.03), cannot be reproduced without including these effects. Calculations in which these forces are included suggest that the dead JFC fraction of the near-Earth population is ≤20% (*Fernández et al.,* 2002) to 50% (*Harris and Bailey,* 1998). The results remain uncertain, in part, because the form and time-dependence of the nongravitational acceleration remain poorly known. Furthermore, the magnitude of the nongravitational acceleration, all else being equal, varies inversely with the nucleus radius so that size-dependent effects in the dynamics should be expected. Realistic modeling of the orbital evolution of outgassing comets is potentially very complicated and deserves more attention than it has yet received.

Rubble mantles, unless held together by granular cohesion, are susceptible to ejection in response to decrease of the perihelion (*Rickman et al.,* 1990). The transition to the

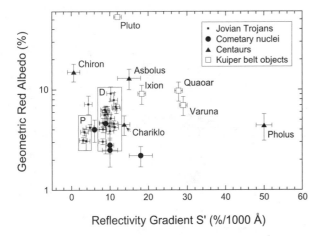

Fig. 9. Plot of normalized reflectivity gradient vs. geometric albedo for jovian Trojan asteroids, cometary nuclei, Centaurs, and two KBOs (see legend for a key to the symbols). Boxes mark the approximate regions of the P and D asteroid spectral classes according to *Dahlgren and Lagerkvist* (1995). The Trojan data are compiled from *Jewitt and Luu* (1990) and *Tedesco et al.* (2002); Centaur data are from *Jewitt and Luu* (2001), *Hainaut and Delsanti* (2002), and *Campins and Fernández* (2002), and KBO data are for Pluto (*Tholen and Buie,* 1997), 20000 Varuna (*Jewitt et al.,* 2001; *Hainaut and Delsanti,* 2002), and 28978 Ixion and 50000 Quaoar (*Marchi et al.,* 2003; *Bertoldi et al.,* 2002).

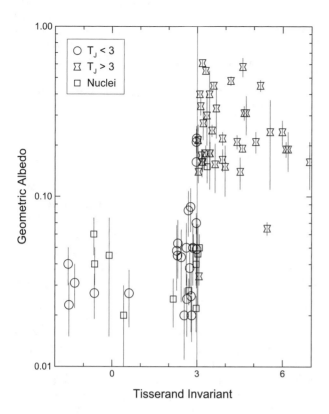

Fig. 10. Plot of red (0.65 μm) geometric albedo, p, vs. Tisserand parameter, T_J, for small solar system bodies including dynamically asteroidal near-Earth objects ($T_J > 3$), unresolved objects likely to be inactive comets on dynamical grounds ($T_J \leq 3$), and the nuclei of active comets. From *Fernández et al.* (2003b).

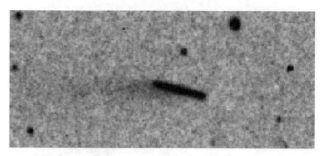

Fig. 11. Trailed (unguided) image of asteroid-comet transition object 107P/1949 W1 (Wilson-Harrington) taken UT 1949 November 19. The trailed image of the central nucleus is about 1 arcmin in length in this 720-s integration. North is at the top, east to the left, and the tail extends toward the east (antisolar) direction. The comet was 0.2 AU from Earth and 1.1 AU from the Sun. Palomar sky image from *Fernández et al.* (1997).

asteroidal state may include a protracted period of intermittent cometary activity as the mantle cracks and reseals. Several examples of comets in which the activity is extremely weak (e.g., 49P/Arend-Rigaux, 28P/Neujmin 1) or may even flicker on and off are known (*Kresák, 1987*).

Several comets and likely comets are known to follow asteroid-like orbits [see Table 3 and *Weissman et al.* (2002) for more detailed descriptions]. The most famous example is Comet 2P/Encke, which is a bona-fide comet that has decoupled from Jupiter's control ($T_J = 3.03$). Comet 107P/Wilson-Harrington (also known as asteroid 1949 W1) displayed a diffuse trail at discovery in 1949 (Fig. 11) but has appeared asteroidal ever since (*Fernández et al., 1997*). The

object 3200 Phaethon appears dynamically associated with the Geminid meteor stream, yet has shown no evidence for recent outgassing and occupies a thoroughly un-comet-like orbit. Lastly, asteroid 7968 was found to show a dust trail in images taken in 1996, leading to this object being cross-identified with Comet 133P/Elst-Pizarro. The confinement of the dust to the vicinity of the orbit plane shows that the observed particles experience a small ratio of forces due to radiation pressure relative to solar gravitational attraction. This, in turn, suggests that the particles are large, probably at least 10–20 μm in size. The existence of a dust trail from this object is particularly puzzling, since it is dynamically a main-belt asteroid with orbital elements consistent with those of the Themis family (Table 3). It has been suggested that the appearance of outgassing activity might have been created by a recent collision with a smaller asteroid (*Toth, 2000*), but this explanation seems unlikely given the reappearance of the trail in Mauna Kea data in 2002, six years after the initial detection (*Hsieh et al., 2003*) (see Fig. 12). Either 133P/Elst-Pizarro is an asteroid somehow triggered to lose mass or it is a comet somehow driven into an asteroid-like orbit. The aphelion of 133P/Elst-Pizarro at 3.80 AU is far from Jupiter's orbit. Nongravitational accelerations from outgassing might produce this type of decoupling from Jupiter but with very low efficiency (*Fernández et al., 2002*).

A number of objects in comet-like ($T_J \leq 3$) orbits possess no resolvable coma or tail and must be physically clas-

TABLE 3. Comets and likely comets in asteroid-like orbits.

Object	Perihelion (AU)	Semimajor Axis (AU)	Eccentricity	Inclination (deg)	Tisserand
2P/Encke	0.34	2.22	0.85	11.8	3.03
3200 Phaethon	0.14	1.40	0.89	22.1	4.51
107P/Wilson-Harrington	1.00	2.64	0.62	2.8	3.08
7968 133P/Elst-Pizarro	2.63	3.25	0.17	1.4	3.18

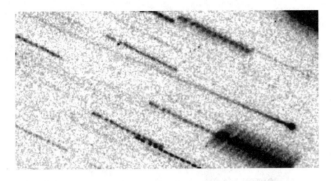

Fig. 12. 133P/(7968) Elst-Pizarro in a 3900-s, R-band composite image taken on UT 2002 September 07 at the University of Hawai'i 2.2-m telescope. Short trails are background stars and galaxies. The nucleus of 133P has been placed at the lower right. Image is 180 arcsec wide, with north at the bottom and east to the right. The heliocentric and geocentric distances were 2.88 AU and 1.96 AU respectively, and the phase angle was 9.9°. A dust trail is visible across the full width of the image in the raw data. Image by Henry Hsieh and the author.

sified as asteroidal (see Table 4). A fraction of these could be dead (or dormant) comets. Based on their lifetimes, the ratio of the numbers of dead, N_d, to active, N_a, comets should be

$$\frac{N_d}{N_a} \sim \frac{\tau_{JFC}}{\tau_{dv}} \sim \frac{2}{r_n} \qquad (20)$$

The number of NEOs larger than ~1 km in size is ~1000 (*Rabinowitz et al.*, 2000; *Bottke et al.*, 2002). If ~10% of these are dead or dormant comets (*Fernández et al.*, 2001; *Bottke et al.*, 2002), then the number of such objects must be ~100. For comparison, about 200 JFCs are known, while the true (bias-corrected) population may number in the thousands (*Fernández et al.*, 1999). Empirically, then, $N_d/N_a <$ 0.5, whereas equation (20) gives $N_d/N_a \sim 2$ for $r_n = 1$ km.

Models of the orbital evolution of JFCs permit an independent estimate of N_d/N_a. The models show that the mean inclination increases with residence time in the inner solar system (*Levison and Duncan*, 1994, 1997). To bring the model and observed inclination distributions of the JFCs into agreement, these authors found it necessary to assume that the JFCs are active for between 3000 and 30,000 yr (with a best estimated value of about 12,000 yr) after their perihelion first reaches q ≤ 2.5 AU. The corresponding ratio of dead to active JFCs is $2 \leq N_d/N_a \leq 6.7$, with a best estimate of $N_d/N_a = 3.5$ (*Levison and Duncan*, 1997). Whether or not the difference between the measured and model values of N_d/N_a is significant is unclear.

5.2. Tidal Breakup

A small number of comets have been observed to break up in response to gravitational stresses induced by close proximity to the Sun (e.g., the Kreutz Sun-grazer group) or a planet (commonly Jupiter, as with P/Brooks 2 and D/Shoemaker-Levy 9). The properties of the "string of pearls" comet chain produced from D/Shoemaker-Levy 9 are well explained by a disrupted, gravitationally sheared aggregate body of negligible tensile strength and density $\rho_n \sim 500$ kg m^{-3} (*Asphaug and Benz*, 1996).

Some 4% of the Centaurs pass within the Roche radius of a gas giant planet during their lifetimes (*Levison and Duncan*, 1997). If, like D/Shoemaker-Levy 9 (which had 26 fragments), each object splits into a few dozen pieces, then the number of secondary fragments could rival or even exceed the number of primary comets from the Kuiper belt. This would have significant implications for the flux, size, mass, and rotation distributions of the comets. For example, tidal breakup could reduce the number of small KBOs needed to supply the JFC flux (cf. *Bernstein et al.*, 2003). Most split comet fragments would quickly mix with the primary unsplit population: Dynamical memory of the splitting is quickly lost (*Pittich and Rickman*, 1994).

5.3. Nontidal Breakup

Most splitting events occur without obvious provocation and their cause is unknown (*Sekanina*, 1997). Statistically, the nuclei of JFCs split at a rate $\tau_{split}^{-1} \sim 10^{-2}$ yr^{-1} per nucleus (*Chen and Jewitt*, 1994; cf. *Weissman*, 1980). With dynamical lifetimes of a few × 10^5 yr, each nucleus should split thousands of times. The rate for the observed long period comets is less well constrained but probably of the same order. The effect of repeated splittings on the mass of the primary nucleus depends on δ, the mass-weighted fragment to primary nucleus mass ratio. If splitting is a continuous

TABLE 4.   Sample asteroids in comet-like orbits.

Object	Perihelion (AU)	Semimajor Axis (AU)	Eccentricity	Inclination (deg)	Tisserand
5335 Damocles	1.57	11.82	0.87	62.0	1.15
15504 1998 RG33	2.15	9.43	0.77	34.9	1.95
20461 1999 LD31	2.39	24.43	0.90	160.2	−1.54
3552 Don Quixote	1.21	4.23	0.71	30.8	2.32
1997 SE5	1.24	3.73	0.67	2.6	2.66
1982 YA	1.12	3.66	0.70	35.3	2.40

random process, the fraction of the mass remaining in the primary after time, t, is just

$$\frac{m(t)}{m(0)} \sim (1 - \delta)^{t/\tau_{split}} \quad (21)$$

Setting $t = \tau_{dv}$ as an upper limit, we find $m(\tau_{dv})/m(0) \ll 1$ for $\delta \gg 10^{-3}$. Observationally, the situation is unclear. Most fragments of the Kreutz sungrazing comets have characteristic sizes of ~10 m, corresponding to $\delta \sim 10^{-6}$ for a kilometer-sized parent (*Sekanina*, 2002). If this is representative of splitting events as a whole, then it is unlikely that disintegration plays a major role in shaping the nuclei. However, if breakup is a continuous process that extends to much larger, but rarer fragment sizes, then the effect on the primary mass may still be significant.

The short timescales for spin excitation, τ_{ex} (see Fig. 2), suggest that rotational bursting may play a role. The centripetal and gravitational accelerations on the equator of a spherical body of density ρ (kg m^{-3}) are equal at the critical rotational period

$$\tau_r = \left(\frac{3\pi}{G\rho} \right)^{1/2} \quad (22)$$

where $G = 6.6 \times 10^{-11}$ (N kg^{-2} m^2) is the gravitational constant. Strengthless objects rotating with periods $P_{rot} < \tau_r$ are susceptable to rotational bursting. For the more general case of an elongated body in minimum energy rotation (about the shortest axis), the critical period is

$$\tau_r \sim 3.3 \left(\frac{1000}{\rho} \right)^{1/2} f_r h \quad (23)$$

where f_r is a numerical factor that depends on the axis ratio of the body (e.g., $f_r = 1$ for spheres, $f_r \sim 4/3$ for prolate bodies with axis ratio a/b = 2). Observationally, all but the smallest (strongest) asteroids, and all measured comets, have $P_{rot} \geq 2$ h and consistent with $\rho \geq 500$ (kg m^{-3}) (*Pravec et al.*, 2002). The most elongated comets may be rotating close to their corresponding centripetal limits (*Jewitt and Meech*, 1988; *Weissman et al.*, 2004).

For example, the nucleus of C/1999 S4 (LINEAR) may have been as small as 100 m in radius prior to break up into fragments at 0.85 AU (*Weaver et al.*, 2001). Equation (12) gives $\tau_{ex} \sim 10$ d for such a small object when close to the Sun. The nucleus could have been driven to rotational bursting during the time taken to free fall toward the Sun. Split fragments themselves would be subject to rapid spinup, leading to a cascade of rotationally bursting fragments.

5.4. Disintegration

Except for the sungrazing comets observed by SOHO, only a few well-documented examples of complete comet-

ary disintegration are known (see *Boehnhardt*, 2004). This does not mean that complete disintegration is rare, because the probability that a brief, one-time event might be observed by chance is presumably very small. The causes of disintegration are not known.

Samarasinha (2001) suggested a model for the nucleus of C/1999 S4 (LINEAR), namely that fragmentation was due to the buildup of gas pressure inside a loosely agglomerated nucleus having substantial internal void space. High pressures have been observed in association with the heating of amorphous ice samples in which clathrate formation may also play a role (*Blake et al.*, 1991). This "bomb model" model predicts that small nuclei (e.g., SOHO comets) should detonate more readily than large ones, since the gravitational binding energy grows as r_n^5 while the gas pressure is independent of nucleus size. The real unknown in this model is the permeability to gas: If the gas can leak out, the pressure may never grow large enough to burst the nucleus.

Rotational spinup (Fig. 2 and equation (12)) might also be implicated in cometary disintegration. An elongated nucleus in simple rotation driven to the centripetal limit might be expected to lose mass only from its tips. The same nucleus in an excited rotational state might be pushed to disintegrate by rotational instabilities, particularly if the nearly strengthless internal constitution witnessed in D/Shoemaker-Levy 9 is typical. Rotational ejection could also expose previously shadowed volatiles, initiating a sudden burst of outgassing and perhaps precipitating global instability of the nucleus.

Why some nuclei disintegrate while others split and others remain coherent is a mystery. Neither do we know if splitting and disintegration occur with uniform probability across all comets, or whether some comets are "born tough" and resist splitting and disintegration until their eventual demise. In the latter case, the surviving comets might not be at all representative of the comets prior to their entry into the inner solar system. This possibility has been suggested as an explanation of the presumed fading of LPCs (*Levison et al.*, 2002).

5.5. Debris Streams

Meteor streams represent the final products of cometary disintegration. The known streams are observationally preselected to intersect the orbit of Earth and are biased toward LPC and HFC cometary sources (the luminosity of a meteor of given mass varies with a high power of the relative velocity so that LPC and HFC sources produce brighter meteors than JFC or asteroidal orbits, all other things being equal). Counterparts of the meteor streams for comets whose orbits do not intersect that of Earth are found in cometary dust trails detected thermally (*Sykes and Walker*, 1992) and optically (*Ishiguro et al.*, 2003). Some parameters of the major streams and their likely parents are listed in Table 5.

The wide range of Tisserand invariants in Table 5 shows the dynamical diversity of the sources of the major meteor streams. Some are clearly linked to still-active comets; others

TABLE 5.   Major meteor streams.

Quantity	Quadrantid	Perseid	Orionid	Geminid	Leonid
Parent Object	5496	109P	1P	3200	55P
Parent Type	Asteroid	HFC	HFC	Asteroid	HFC
Perihelion q (AU)	0.88	0.9595	0.5860	0.14	0.9764
Eccentricity e	0.64	0.9632	0.9671	0.89	0.9055
Inclination i (deg)	68.0	113.5	162.3	22.1	162.5
Semimajor axis a (AU)	2.44	26.1	17.8	1.27	10.33
Tisserand T_J	2.53	−0.28	−0.61	4.51	−0.64
Stream Mass, M_s (kg)	1.3×10^{12}	3.1×10^{14}	3.3×10^{13}	1.6×10^{13}	5.0×10^{12}
Meteor Density (kg m^{-3})	1900 ± 200	1300 ± 200	—	2900 ± 600	400 ± 100
Parent radius (km)	1.8	10.5	5.0	2.6	1.8
Parent Mass, M_p (kg)	1.2×10^{13}	2.4×10^{15}	2.6×10^{14}	3.7×10^{13}	1.2×10^{13}
$M_s/(M_s + M_p)$	0.10	0.13	0.13	0.38	0.29

are associated with objects that appear asteroidal (*Hughes and McBride*, 1989; *Williams and Collander-Brown*, 1998). Source diversity is also indicated by the wide range of meteor densities listed in Table 5. The absolute values of density are limited in accuracy by the fragmentation models applied in the interpretation of meteor data, but the relative densities should be meaningful (*Babadzhanov*, 2002; *Rubio et al.*, 2002). For example, debris from the Leonid parent 55P/Tempel-Tuttle is much less dense (more porous?) than debris from Geminid parent 3200 Phaethon. The stream masses are estimated from the flux of meteors as a function of time and are thought to be accurate to within a factor of ±4 or so (*Hughes and McBride*, 1989; *Jenniskens and Betlem*, 2000). Table 5 shows that the ratio of the stream mass to the stream plus parent nucleus mass is $0.1 \leq M_s/(M_s + M_p) \leq 0.4$. What is behind this ratio?

Consider a model of a spherical nucleus shrinking at a constant rate and, for simplicity, neglect mantle formation. Assuming that mass is not lost from the stream on the sublimation timescale, the ratio of the mass of the stream, M_s, to the total mass of the parent and stream, $M_s + M_p$, is given by

$$\frac{M_s}{M_s + M_p} = f_s \left[\frac{r_0^3 - r^3(t)}{r_0^3} \right] \qquad (24)$$

where r_0 is the initial nucleus radius and $r(t) = r_0 - \beta t$ (β is a constant equal to the sublimation distance per unit time) is the radius at time t ($t \leq r_0/\beta$). The quantity f_s is the fraction of the mass of the parent that is contained in refractory matter.

Averaged over the time interval $0 \leq t \leq r_0/\beta$, the value of equation (24) is given by

$$\overline{\frac{M_s}{M_s + M_p}} = \frac{3}{4} f_s \qquad (25)$$

With $f_s = 1/2$, we have $\overline{M_s/(M_s + M_p)} = 3/8 \sim 0.4$ which is very close to the measured values (Table 5). The observations are therefore consistent with the simple model, pro-

vided the stream debris lifetimes are long compared to the source lifetimes.

The lifetimes of the streams are poorly determined. Estimates based on dynamical scattering of Perseids suggest lifetimes of 4×10^4 yr to 8×10^4 yr (*Brown and Jones*, 1998). Independently, the mass loss rate from the Perseid parent 109P/Swift-Tuttle has been estimated at 5×10^{11} kg/ orbit from submillimeter observations (*Jewitt*, 1996b). To supply the 3.1×10^{14} kg in the stream would require about 600 orbits, corresponding to 8×10^4 yr, in good agreement with the dynamical lifetime estimates above.

6. SUMMARY

Substantial progress has been made in recent years toward exploring and understanding two major storage regions of the comets, the Oort cloud and the Kuiper belt. The latter, in particular, has been transformed from conjecture into a dynamically rich and observationally accessible region of the solar system, about which we seemingly learn more every month. On the other hand, the processes of decay of the comets, and the relationships that exist between these objects and other small bodies in the solar system, remain subjects of considerable uncertainty. The latter have been the troublesome subjects of this chapter.

Observationally, there are at least two challenging problems. First, the physical properties of the all-important cometary nucleus are difficult to measure, because of coma contamination when near the Sun and because of nucleus faintness when far from it. In order to study how the comets evolve and decay, reliable measurements of the nuclei are indispensable. We possess very few. Second, telescopic data in any case sample only the outermost, optically active surface layers. We can learn relatively little about fundamental aspects of the internal structure or composition by sampling only reflected sunlight.

In terms of dynamics, the comets occupy difficult territory in which forces due to mass loss may play an important long-term role. The nuclei are pushed by asymmetrical (sunward-directed) ejection of matter. Their angular momenta are changed by outgassing torques and their very lifetimes as comets may be limited by the loss of mass due to sublimation, and by rupture due to centripetal effects. Purely dynamical models appear to have been pushed to their limits. The next step is to couple dynamical models with thermophysical models in order to more realistically account for the effects of outgassing forces on the comets.

In terms of their formation, the comets and the asteroids represent two ends of a continuum. The asteroids, at least those of the main belt, mostly formed at temperatures too high for the inclusion of water as ice while we think the comets accreted water ice in abundance. In between lies a large gray zone, in which the identity of the objects is indistinct, both observationally and compositionally. We have discussed some examples of mass-losing objects that are dynamically more like asteroids than comets (e.g., 133P/Elst-Pizarro). The jovian Trojan "asteroids" appear optically indistinguishable from the nuclei of comets and they may also contain solid ice. Unfortunately, they are too cold to measurably sublimate, even if ice is present in the near surface regions. Some asteroids in the outer belt could also contain ice, but only if buried by thicker, thermally insulating mantles. Conversely, the comets may be less ice rich than originally supposed in Whipple's classic description. While it is no surprise that sunbaked 2P/Encke has a large dust/gas production rate ratio ~10 to ~30 (*Reach et al.*, 2000), even the comparatively fresh Comet C/Hale-Bopp had dust/gas ~6 to ~10 (*Grün et al.*, 2001). It is not clear that these ratios are representative of the bulk interior values, but still, the difference between the "icy comets" and the "rocky asteroids" seems astonishingly small.

In the end, what matters most is that we can relate the properties of the small bodies of the solar system back to the formation conditions, to learn something about the processes of accretion and planet growth. Observations and ideas explored here represent some of the first steps in this direction. We end with a list of key questions that should be answered in the coming years.

6.1. Some Key Questions

1. Processes described here (core volatile sublimation, rotational bursting, complete devolatilization) suggest that the comet nucleus size distribution should be depleted at small sizes. What reliable observational evidence exists for this?

2. What fraction of the near-Earth objects are dead comets, and how can they be reliably distinguished from asteroids?

3. Are the Trojans of Jupiter essentially mantled comets with substantial interior bulk ice? How can this question be answered?

4. What is the true Centaur population as a function of size? Initial estimates of 10^7 objects larger than 1 km in radius need to be checked by deeper, more extensive survey observations of the whole ecliptic.

5. Are there sources of Centaurs (and JFCs) in addition to those recognized in the Kuiper belt? Are the Trojans of the four giant planets a significant source?

6. What fraction of the Centaurs are fragments of precursor objects that were disrupted by passage through the Roche spheres of gas giant planets?

7. Is there any firm evidence for outgassing in the KBOs?

8. What is the physical basis of the "fading parameter" in Oort's model of the LPCs? Is our understanding of the dynamics of the LPCs complete?

Acknowledgments. The author thanks D. Cruikshank, Y. Fernández, M. Festou, J. Luu, N. Samarasinha, and S. Sheppard for their comments on this manuscript. Support from NASA and NSF is gratefully acknowledged.

REFERENCES

A'Hearn M., Millis R., Schleicher D., Osip D., and Birch P. (1995) The ensemble properties of comets: Results from narrowband photometry of 85 comets, 1976–1992. *Icarus, 118,* 223–270.

Asphaug E. and Benz W. (1996) Size, density, and structure of Comet Shoemaker-Levy 9 inferred from the physics of tidal breakup. *Icarus, 121,* 225–248.

Babadzhanov P. (2002) Fragmentation and densities of meteoroids. *Astron. Astrophys., 384,* 317–321.

Bailey M. and Emel'yanenko V. (1996) Dynamical evolution of Halley-type comets. *Mon. Not. R. Astron. Soc., 278,* 1087–1110.

Barucci M. A., Doressoundiram A., and Cruikshank D. P. (2004) Physical characteristics of transneptunina objects and Centaurs. In *Comets II* (M. C. Festou et al., eds.), this volume. Univ. of Arizona, Tucson.

Bauer J., Fernández Y., and Meech J. (2003a) An optical survey of the active Centaur C/NEAT (2001 T4). *Publ. Astron. Soc. Pac., 115 (810),* 981–989.

Bauer J., Meech K., Fernández Y., Pittichová J., Hainaut O., Boehnhardt H., and Delsanti A. (2003b) Physical survey of 24 Centaurs with visible photometry. *Icarus, 166,* 195–211.

Bernstein G., Trilling D., Allen R., Brown M., Holman M., and Malhotra R. (2003) The size distribution of trans-neptunian bodies. *Astron. J.,* in press.

Bertoldi F., Altenhoff W., and Junkes N. (2002) Beyond Pluto: Max-Planck radioastronomers measure the sizes of distant minor planets. *Max Planck Institute Press Release 10/02(2),* 2002 October 7.

Blake D., Allamandola L., Sandford S., Hudgins D., and Friedemann F. (1991) Clathrate hydrate formation in amorphous cometary ice analogs in vacuo. *Science, 254,* 548.

Boehnhardt (2004) Split comets. In *Comets II* (M. C. Festou et al., eds.), this volume. Univ. of Arizona, Tucson.

Bottke W., Morbidelli A., Jedicke R., Petit J., Levison H., Michel P., and Metcalfe T. (2002) Debiased orbital and absolute magnitude distribution of the near-Earth objects. *Icarus, 156,* 399–433.

Brin G. and Mendis D. (1979) Dust release and mantle development in comets. *Astrophys. J., 229,* 402–408.

Brown P. and Jones J. (1998) Simulation of the formation of the Perseid Meteor stream. *Icarus, 133,* 36–68.

Brunini A. and Melita M. (1998) On the existence of a primordial cometary belt between Uranus and Neptune. *Icarus, 135,* 408–414.

Burns J. and Safronov V. (1973) Asteroid nutation angles. *Mon. Not. R. Astron. Soc., 165,* 403.

Bus S., A'Hearn M., Bowell E., and Stern S. (2001) (2060) Chiron:

Evidence for activity near aphelion. *Icarus, 150,* 94–103.

Campins H. and Fernández Y. (2002) Observational constraints on surface characteristics of cometary nuclei. *Earth Moon Planets, 89,* 117–134.

Catullo V., Zappalà V., Farinella P., and Paolicchi P. (1984) Analysis of the shape distribution of asteroids. *Astron. Astrophys., 138,* 464–468.

Chamberlin A., McFadden L., Schulz R., Schleicher D., and Bus S. (1996) 4015 Wilson-Harrington, 2201 Oljato, and 3200 Phaethon: Search for CN emission. *Icarus, 119,* 173–181.

Chen J. and Jewitt D. (1994) On the rate at which comets split. *Icarus, 108,* 265–271.

Dahlgren M. and Lagerkvist C. (1995) A study of Hilda asteroids. *Astron. Astrophys., 302,* 907–914.

Degewij J. and Tedesco E. (1982) Do comets evolve into asteroids? In *Comets* (L. L. Wilkening, ed.), pp. 665–695. Univ. of Arizona, Tucson.

Delsemme A. (1973) Origin of the short-period comets. *Astron. Astrophys., 29,* 377–381.

Dones L., Gladman B., Melosh H., Tonks W., Levison H., and Duncan M. (1999) Dynamical lifetimes and final fates of small bodies. *Icarus, 142,* 509–524.

Dones L., Weissman P. R., Levison H. F., and Duncan M. J. (2004) Oort cloud formation and dynamics. In *Comets II* (M. C. Festou et al., eds.), this volume. Univ. of Arizona, Tucson.

Dumas C., Owen T., and Barucci M. (1998) Near-infrared spectroscopy of low-albedo surfaces of the solar system. *Icarus, 133,* 221–232.

Duncan M., Quinn T., and Tremaine S. (1988) The origin of short period comets. *Astrophys. J. Lett., 328,* L69–L73.

Duncan M., Levison H., and Dones L. (2004) Dynamical evolution of ecliptic comets. In *Comets II* (M. C. Festou et al., eds.), this volume. Univ. of Arizona, Tucson.

Farinella P. and Davis D. (1996) Short-period comets: Primordial bodies or collisional fragments? *Science, 273,* 938–941.

Fernández J. A. (1980) On the existence of a comet belt beyond Neptune. *Mon. Not. R. Astron. Soc., 192,* 481–491.

Fernández J. A., Tancredi G., Rickman H., and Licandro J. (1999) The population, magnitudes, and sizes of Jupiter family comets. *Astron. Astrophys., 352,* 327.

Fernández J., Gallardo T., and Brunini A. (2002) Are there many inactive JFCs among the near Earth asteroid population? *Icarus, 159,* 358–368.

Fernández Y. R., McFadden L. A., Lisse C. M., Helin E. F., and Chamberlin A. B. (1997) Analysis of POSS images of comet-asteroid transition object 107P/1949 W1 (Wilson-Harrington). *Icarus, 128,* 114–126.

Fernández Y., Lisse C., Ulrich K., Peschke S., Weaver H., A'Hearn M., Lamy P., Livengood T., and Kostiuk T. (2000) Physical properties of the nucleus of Comet 2P/Encke. *Icarus, 147,* 145–160.

Fernández Y., Jewitt D., and Sheppard S. (2001) Low albedos among extinct comet candidates. *Astrophys. J. Lett., 553,* L197–L200.

Fernández Y., Sheppard S., and Jewitt D. (2003a) The albedo distribution of jovian Trojan asteroids. *Astron. J., 126,* 1563–1574.

Fernández Y., Jewitt D., and Sheppard S. (2003b) Albedos of asteroids in comet-like orbits. *Astron. J.,* in press.

Fitzsimmons A., Dahlgren M., Lagerkvist C., Magnusson P., and Williams I. (1994) A spectroscopic survey of D-type asteroids. *Astron. Astrophys., 282,* 634–642.

Giblin I. and Farinella P. (1997) Tumbling fragments from experiments simulating asteroidal catastrophic disruption. *Icarus, 127,* 424–430.

Grün E. and 23 colleagues (2001) Broadband infrared photometry of comet Hale-Bopp with ISOPHOT. *Astron. Astrophys., 377,* 1098–1118.

Gutiérrez P., Jorda L., Ortiz J., and Rodrigo R. (2003) Long-term simulations of the rotational state of small irregular cometary nuclei. *Astron. Astrophys., 406,* 1123–1133.

Hahn J. and Malhotra R. (1999) Orbital evolution of planets embedded in a planetesimal disk. *Astron. J., 117,* 3041–3053.

Hainaut O. R. and Delsanti A. C. (2002) Colors of minor bodies in the outer solar system. A statistical analysis. *Astron. Astrophys., 389,* 641–664.

Harris A. (1994) Tumbling asteroids. *Icarus, 107,* 209.

Harris N. and Bailey M. (1998) Dynamical evolution of cometary asteroids. *Mon. Not. R. Astron. Soc., 297,* 1227–1236.

Hartmann W. and Tholen D. (1990) Comet nuclei and Trojan asteroids — A new link and a possible mechanism for comet splittings. *Icarus, 86,* 448–454.

Hartmann W., Tholen D., and Cruikshank D. (1987) The relationship of active comets, 'extinct' comets, and dark asteroids. *Icarus, 69,* 33–50.

Holman M. (1997) A possible long-lived belt of objects between Uranus and Neptune. *Nature, 387,* 785–788.

Hsieh H., Jewitt D., and Fernández Y. (2003) The strange case of 133P/Elst-Pizarro: A comet amongst the asteroids. *Astron. J.,* in press.

Hughes D. and McBride N. (1989) The masses of meteoroid streams. *Mon. Not. R. Astron. Soc., 240,* 73–79.

Ishiguro M., Kwon S., Sarugaku Y., Hasegawa S., Usui F., Nishiura S., Nakada Y., and Yano H. (2003) Discovery of the dust trail of the stardust comet sample return mission target: 81P/Wild 2. *Astrophys. J. Lett., 589,* L101–L104.

Jenniskens P. and Betlem H. (2000) Massive remnant of evolved cometary dust trail detected in the orbit of Halley-type Comet 55P/Tempel-Tuttle. *Astrophys. J., 531,* 1161–1167.

Jewitt D. (1991) Cometary photometry. In *Comets in the Post-Halley Era* (R. L. Newburn et al., eds.), pp. 19–65. Astrophysics and Space Science Library Vol. 167, Kluwer, Dordrecht.

Jewitt D. (1996a) From comets to asteroids: When hairy stars go bald. *Earth Moon Planets, 72,* 185–201.

Jewitt D. (1996b) Debris from Comet P/Swift-Tuttle. *Astron. J., 111,* 1713.

Jewitt D. (1999) Cometary rotation: An overview. *Earth Moon Planets, 79,* 35–53.

Jewitt D. (2002) From Kuiper belt object to cometary nucleus: The missing ultrared matter. *Astron. J., 123,* 1039–1049.

Jewitt D. and Fernández Y. (2001) Physical properties of planet-crossing objects. In *Collisional Processes in the Solar System* (H. Rickman and M. Marov, eds.), pp. 143–161. Astrophysics and Space Science Library Series, Kluwer, Dordrecht.

Jewitt D. and Kalas P. (1998) Thermal observations of Centaur 1997 CU26. *Astrophys. J. Lett., 499,* L103.

Jewitt D. and Luu J. (1990) CCD spectra of asteroids. II — The Trojans as spectral analogs of cometary nuclei. *Astron. J., 100,* 933–944.

Jewitt D. and Luu J. (2001) Colors and spectra of Kuiper belt objects. *Astron. J., 122,* 2099–2114.

Jewitt D. and Meech K. (1988) Optical properties of cometary nuclei and a preliminary comparison with asteroids. *Astrophys. J., 328,* 974–986.

Jewitt D., Luu J., and Trujillo C. (1998) Large Kuiper belt objects: The Mauna Kea 8k CCD survey. *Astron. J., 115,* 2125–2135.

Jewitt D., Trujillo C., and Luu J. (2000) Population and size distribution of small jovian Trojan asteroids. *Astron. J., 120,* 1140–1147.

Jewitt D., Aussel H., and Evans A. (2001) The size and albedo of the Kuiper-belt object (20000) Varuna. *Nature, 411,* 446–447.

Jewitt D., Sheppard S., and Fernández Y. (2003) 143P/Kowal-Mrkos and the shapes of cometary nuclei. *Astron. J., 125,* 3366–3377.

Jones T., Lebofsky L., Lewis J., and Marley M. (1990) The composition and origin of the C, P, and D asteroids — Water as a tracer of thermal evolution in the outer belt. *Icarus, 88,* 172–192.

Kresák L. (1987) Dormant phases in the aging of periodic comets. *Astron. Astrophys., 187,* 906–908.

Kührt E. and Keller H. U. (1994) The formation of cometary surface crusts. *Icarus, 109,* 121–132.

Lamy P., Toth I., Fernández Y., and Weaver H. (2004) The sizes, shapes, albedos, and colors of cometary nuclei. In *Comets II* (M. C. Festou et al., eds.), this volume. Univ. of Arizona, Tucson.

Levison H. (1996) Comet taxonomy. In *Completing the Inventory of the Solar System* (T. W. Rettig and J. M. Hahn, eds.), pp. 173–191. ASP Conference Series 107, Astronomical Society of the Pacific, San Francisco.

Levison H. and Duncan M. (1994) The long-term dynamical behavior of short-period comets. *Icarus, 108,* 18–36.

Levison H. and Duncan M. (1997) From the Kuiper belt to Jupiter family comets: The spatial distribution of ecliptic comets. *Icarus, 127,* 13–32.

Levison H., Dones L., and Duncan M. (2001) The origin of Halley-type comets: Probing the inner Oort cloud. *Astron. J., 121,* 2253–2267.

Levison H. F., Morbidelli A., Dones L., Jedicke R., Wiegert P. A., and Bottke W. F. (2002) The mass disruption of Oort cloud comets. *Science, 296,* 2212–2215.

Lisse C., Fernández Y., Kundu A., A'Hearn M., Dayal A., Deutsch L., Fazio G., Hora J., and Hoffmann W. (1999) The nucleus of Comet Hyakutake (C/1996 B2). *Icarus, 140,* 189–204.

Luu J. and Jewitt D. (1992a) High resolution surface brightness profiles of near-earth asteroids. *Icarus, 97,* 276–287.

Luu J. and Jewitt D. (1992b) Near-aphelion CCD photometry of Comet P/Schwassmann Wachmann 2. *Astron. J., 104,* 2243–2249.

Luu J., Jewitt D., and Cloutis E. (1994) Near-infrared spectroscopy of primitive solar system objects. *Icarus, 109,* 133–144.

Marchi S., Lazzarin M., Magrin S., and Barbieri C. (2003) Visible spectroscopy of the two largest known trans-neptunian objects. *Astron. Astrophys., 408,* L17–L19.

Marzari F., Farinella P., and Vanzani V. (1995) Are Trojan collisional families a source for short-period comets? *Astron. Astrophys., 299,* 267.

Moroz L., Starukhina L., Strazzulla G., Baratta G., Dotto E., Barucci A., and Arnold G. (2003) Optical alteration of complex organics caused by ion irradiation. *Icarus,* in press.

Oort J. (1950) The structure of the cloud of comets surrounding the solar system and a hypothesis concerning its origin. *Bull. Astron. Inst. Netherlands, 11,* 91–110.

Pittich E. and Rickman H. (1994) Cometary splitting — A source for the Jupiter family? *Astron. Astrophys., 281,* 579–587.

Pravec P., Harris A., and Michałowski T. (2002) Asteroid rota-

tions. In *Asteroids III* (W. F. Bottke Jr. et al., eds.), pp. 113–122. Univ. of Arizona, Tucson.

Rabinowitz D., Helin E., Lawrence K., and Pravdo S. (2000) A reduced estimate of the number of kilometer-sized near-Earth asteroids. *Nature, 403,* 165–166.

Reach W., Sykes M., Lien D., and Davies J. (2000) The formation of Encke meteoroids and dust trail. *Icarus, 148,* 80–94.

Rickman H. (2004) Current questions in cometary dynamics. In *Comets II* (M. C. Festou et al., eds.), this volume. Univ. of Arizona, Tucson.

Rickman H., Fernández J. A., and Gustafson B. Å. S. (1990) Formation of stable dust mantles on short-period comet nuclei. *Astron. Astrophys., 237,* 524–535.

Rickman H., Valsecchi G. B., and Froeschlé Cl. (2002) From the Oort cloud to observable short-period comets — I. The initial stage of cometary capture. *Mon. Not. R. Astron. Soc., 325,* 1303–1311.

Rubio L., Gonzalez M., Herrera L., Licandro J., Delgado D., Gil P., and Serra-Ricart M. (2002) Modelling the photometric and dynamical behavior of super-Schmidt meteors in the Earth's atmosphere. *Astron. Astrophys., 389,* 680–691.

Samarasinha N. (2001) A model for the breakup of Comet LINEAR (C/1999 S4). *Icarus, 154,* 540–544.

Samarasinha N. H. and A'Hearn M. F. (1991) Observational and dynamical constraints on the rotation of Comet P/Halley. *Icarus, 93,* 194–225.

Samarasinha N., A'Hearn M., Hoban S., and Klinglesmith D. (1986) CN jets of Comet P/Halley: Rotational properties. In *ESA Proceedings of the 20th ESLAB Symposium on the Exploration of Halley's Comet, Vol. 1: Plasma and Gas* (B. Battrik et al., eds.), pp. 487–491. ESA, Paris.

Sekanina Z. (1990) Gas and dust emission from comets and life spans of active areas on their rotating nuclei. *Astron. J., 100,* 1293–1314.

Sekanina Z. (1997) The problem of split comets revisited. *Astron. Astrophys., 318,* L5–L8.

Sekanina Z. (2002) Statistical investigation and modeling of sungrazing comets discovered with the Solar and Heliospheric Observatory. *Astrophys. J., 566,* 577–598.

Sheppard S., Jewitt D., Trujillo C., Brown M., and Ashley M. (2000) A wide field CCD survey for Centaurs and Kuiper belt objects. *Astron. J., 120,* 2687–2694.

Shul'man L. (1972) The chemical composition of cometary nuclei. In *The Motion, Evolution of Orbits and Origin of Comets* (G. Chebotarev et al., eds.), pp. 265–270. IAU Symposium 45, Reidel, Dordrecht.

Stern S. (1995) Collisional time scales in the Kuiper disk and their implications. *Astron. J., 110,* 856.

Stern S. and Weissman P. (2001) Rapid collisional evolution of comets during the formation of the Oort cloud. *Nature, 409,* 589–591.

Sykes M. and Walker R. (1992) Cometary dust trails. I — Survey. *Icarus, 95,* 180–210.

Tancredi G. (1995) The dynamical memory of Jupiter family comets. *Astron. Astrophys., 299,* 288–292.

Tedesco E. F., Noah P. V., Noah M., and Price S. D. (2002) The supplemental IRAS minor planet survey. *Astron. J., 123,* 1056–1085.

Tholen D. and Buie M. (1997) Bulk properties of Pluto and Charon. In *Pluto and Charon* (S. A. Stern and D. J. Tholen, eds.), pp. 193–220. Univ. of Arizona, Tucson.

Toth I. (2000) Impact generated activity period of the asteroid

7968 Elst-Pizzaro in 1996. *Astron. Astrophys., 360,* 375–380.

Trujillo C., Jewitt D., and Luu J. (2001) Properties of the transneptunian belt: Statistics from the Canada-France-Hawaii telescope survey. *Astron. J., 122,* 457–473.

Weaver H. and 20 colleagues (2001) HST and VLT investigations of the fragments of Comet C/1999 S4 (LINEAR). *Science, 292,* 1329–1334.

Weissman P. (1980) Physical loss of long-period comets. *Astron. Astrophys., 85,* 191–196.

Weissman P. R. and Lowry S. C. (2003) The size distribution of Jupiter-family cometary nuclei. In *Lunar and Planetary Science XXXIV,* Abstract #2003. Lunar and Planetary Institute, Houston (CD-ROM).

Weissman P., Bottke W., and Levison H. (2002) Evolution of comets into asteroids. In *Asteroids III* (W. F. Bottke Jr. et al., eds.), pp. 669–686. Univ. of Arizona, Tucson.

Weissman P. R., Asphaug E., and Lowry S. C. (2004) Structure and density of cometary nuclei. In *Comets II* (M. C. Festou et al., eds.), this volume. Univ. of Arizona, Tucson.

Whipple F. (1950) A comet model: I. The acceleration of Comet Encke. *Astrophys. J., 111,* 375–394.

Wiegert P. and Tremaine S. (1999) The evolution of long-period comets. *Icarus, 137,* 84–121.

Williams I. and Collander-Brown S. (1998) The parent of the quadrantids. *Mon. Not. R. Astron. Soc., 294,* 127–138.

The Interplanetary Dust Complex and Comets

Mark V. Sykes
Steward Observatory

Eberhard Grün
Max Planck Institut für Kernphysik

William T. Reach
California Institute of Technology

Peter Jenniskens
SETI Institute

With the advent of spacebased *in situ* and remote sensing technologies, our knowledge of the structure and composition of the interplanetary dust cloud has changed significantly. Both asteroidal and cometary sources of the cloud interior to Jupiter have been directly detected. A distant contributing source, the Kuiper belt, has been also been detected beyond Saturn. Analysis of the morphology and composition of collected interplanetary dust particles (IDPs) is increasingly sophisticated. Meteor storms are more accurately predicted. Yet the fundamental question of whether asteroids or comets are the principal sources of interplanetary dust is still open and more complex. By understanding the origin and evolution of our own interplanetary dust cloud, and tracing its constituent particles to their roots, we are able to use these particles to provide insights into the origin and evolution of their precursor bodies and look at dust production in the disks about other stars and compare what is going on in those solar systems to the present and past of our own.

1. INTRODUCTION

Earth moves through a cloud of interplanetary dust and debris, extending in size from submicrometer to kilometers and in distance from a few solar radii through the Kuiper belt and beyond. That portion of the micrometer-sized particle cloud in the vicinity of the Earth's orbit gives rise to the zodiacal light, seen most prominently after sunset in the spring and before dawn in autumn at northern latitudes (Fig. 1). In 1693, Giovanni Cassini ascribed this to sunlight scattering off dust particles orbiting the Sun. It was in the late eighteenth and early nineteenth centuries that it was realized that material from space might be showering down on Earth. In 1794, Ernst Chladni, the father of acoustical science, argued for the extraterrestrial origin of meteors, fireballs, and meteorites (*Yeomans,* 1991). The spectacular Leonid meteor shower in 1833, appearing to emanate from a single location in the sky, convinced many scientists of the day that these were indeed of extraterrestrial origin. Almost immediately, a connection was made with comets by W. B. Clarke and Denison Olmsted. Then Hubert Newton correctly determined the orbit of the Leonids and predicted their return in 1866. Work by Giovanni Schiaparelli and others in the mid-1800s continued to press the connection between meteor streams and comets. This was reinforced when Comet Biela was seen to have broken up and Earth

Fig. 1. The zodiacal light from Mauna Kea, Hawai'i. Courtesy of M. Ishiguro, ISAS.

experienced a meteor shower in 1872 when Earth subsequently passed near its orbit (*Yeomans, 1991*).

Comets are a logical source for the interplanetary dust cloud. They are the only solar system objects visually observed to emit dust, and meteor streams were also linked to specific comets. In 1955, Fred Whipple applied his new "icy conglomerate" model of comet nuclei to determine a "quantitative relationship between comets and the zodiacal light." Since dust evolves toward the Sun under Poynting-Robertson drag (e.g., *Burns et al., 1979*), it needs to be replenished if it is to be maintained. *Whipple* (1955) estimated that approximately 1 ton per second of meteoritic material was required to maintain the zodiacal cloud against such loss. He concluded that comets could easily supply that amount of material. In the same paper, however, he also noted that calculations by Piotrowski indicated that the crushing of asteroids may also be an adequate source of zodiacal material. "Whether the comets or the asteroids predominate in zodiacal contribution must be decided on the basis of other criteria . . . from meteoric and micrometeoritical information as well as from the shape of the zodiacal cloud" (*Whipple, 1955*).

In the 1930s, "cosmic spherules" found in deep sea sediments were found to be compositionally similar to meteorites. Analysis of these findings led Ernst Öpik to estimate that the Earth was accumulating 8×10^9 kg of meteoritic material per year (*Öpik, 1951*). With the advent of the space age, the question of the dust environment beyond Earth's atmosphere and its potential hazard to spacecraft and future astronauts grew quickly in importance, later subsiding as better spacebased detectors replaced those that had been providing anomalously high densities due to sensitivity to more than just dust (cf. *Fechtig et al., 2001*). Returned orbiting surfaces from Skylab, orbiting facilities such as the Long Duration Exposure Facility (LDEF), and a series of spacecraft possessing dust detection systems, including *Helios, Hiten, Pioneer 9, Galileo, Ulysses*, and *Cassini*, among others, soon gave new information and constraints on the interplanetary dust environment as did microcrater studies on returned lunar samples. These indicated that the interplanetary dust cloud was complex, having a number of different components of possibly different origins (e.g., *Grün et al., 1985*).

In 1983, the first large-scale survey of the zodiacal cloud at thermal infrared wavelengths by the Infrared Astronomical Satellite (IRAS) (*Hauser et al., 1984*), revealed the overall cloud shape and, more importantly, the first spatial structures within the dust cloud (*Low et al., 1984; Sykes et al., 1986; Sykes, 1988*) directly relating to its asteroidal and cometary origins. This was followed by a thermal survey by the Diffuse Infrared Background Experiment (DIRBE) on NASA's Cosmic Background Explorer (COBE) satellite in 1990 (*Silverberg et al., 1993; Hauser et al., 1998*).

Since the late 1960s, the increasing amount and diversity of dust observations sampling different components of the dust complex (Fig. 2) have spurred a series of regular international conferences with associated volumes summa-

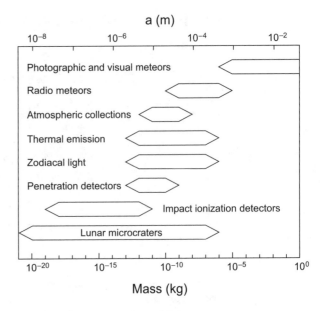

Fig. 2. Different methods of studying the interplanetary dust complex are sensitive to different size/mass ranges. Considered together, a more complete picture of the production and evolution of interplanetary dust can be constructed.

rizing the state of knowledge at that time, the most recent of which is *Grün et al.* (2001). The reader is referred to this book and its antecedents for the detailed background that provides some of the context for this chapter.

2. MORPHOLOGY OF THE CLOUD AND ITS RELATION TO COMETS AND ASTEROIDS

At thermal infrared wavelengths, sky brightness viewed from Earth orbit is dominated by the thermal emission of interplanetary dust particles (IDPs) heated by sunlight. These particles are about 20–200 μm in size (*Reach, 1988*). The IRAS observed the sky in four bands between 12 and 100 μm over a range of solar elongations between 60° and 120° at a resolution of arcminutes, mapping the entire sky almost three times over the course of the mission. DIRBE simultaneously observed in 10 wavebands from 1.25 to 240 μm, covering the portion of sky between 64° and 124° solar elongation with fully sampled images at 1° resolution. Using the DIRBE sky maps, the zodiacal light was characterized using three components (*Kelsall et al., 1998*). The smooth cloud is the dominant component, and its spatial distribution was mathematically fitted with a parameterized function as follows in terms of spherical coordinates (r,θ,z) in AU where r is the cloud center distance, θ is the azimuthal angle, and z is the vertical distance from the cloud midplane:

$$n = n_o r^{-1.34} \exp[-4.14 g(\zeta)^{0.942}]$$

where

$$\zeta = z/r$$

and

$$g(\zeta) = \zeta^2/0.378 \text{ for } \zeta < 0.189$$

and

$$g(\zeta) = \zeta - 0.0945 \text{ for } \zeta \geq 0.189$$

The smooth cloud is azimuthally symmetric, with its midplane tilted by 2.03° from the ecliptic and its center offset from the Sun by 0.013 AU. Both the tilt and offset are explained by gravitational perturbations by the planets (*Dermott et al.*, 1986); while the radial variation and the dependence on z/r are to first order explained by Poynting-Robertson drag, which yields an r^{-1} distribution and leaves orbital inclination unchanged (*Wyatt and Whipple*, 1950).

The morphology of the main zodiacal cloud allows us, in principal, to assess the fraction of dust that derives from comets and asteroids. The Kelsall model above yields a half-width at half-maximum number density of about 14° for dust particle inclinations, while some models based on *in situ* particle detections suggest that the inclination distribution may have a half-width as wide as 20° to 30° (*Dikarev et al.*, 2002). For comparison, the half-width of the distribution of asteroid orbital inclinations (which is reasonably structured) is between 12° and 16° (*Minor Planet Center*, 2003), consistent with the Kelsall model, whereas short-period comets have inclinations generally less than 30° (*Marsden*, 1974) and half have inclinations less than about 10° — not very different from asteroids. Evidently cloud scale-height does not clearly distinguish between asteroids and comets as principal suppliers of dust.

The smooth cloud is supplemented by structures discovered by IRAS: asteroidal dust bands (*Low et al.*, 1984) and cometary dust trails (*Sykes et al.*, 1986) (Fig. 3). Trails represent the principal means by which comets contribute dust to the zodiacal complex and are discussed in a later section. The dust bands stretch across the ecliptic plane in north-south pairs that straddle a midplane tilted by about 1° from the ecliptic. Soon after their discovery in the IRAS data, the bands were associated with the major Hirayama asteroid families (*Dermott et al.*, 1984) and explained as a natural consequence of the general collisional comminution of the asteroid belt as a whole (since the families are regions of asteroid concentration). An alternative to this "equilibrium" theory was proposed by *Sykes and Greenberg* (1986) and *Sykes* (1990) in which the large-scale production of dust in the asteroid belt is stochastic (the "nonequilibrium" theory). In this case, observable dust bands would most likely be associated with recent disruptions of small (~20 km) asteroids. Detailed studies indicated that some bands seemed to be associated with the Themis and Koronis families (*Sykes*, 1990; *Reach et al.*, 1997) and the Maria family

Fig. 3. (a) The last scan of the ecliptic plane made by IRAS (76% complete). Ecliptic longitude increases from 0° (left) to 360° (right) with ecliptic latitudes between –30° and 30°. The diagonal structure crossing the ecliptic plane near 90° and 270° longitude is the galactic plane. The zodiacal cloud appears bright and wide at lower solar elongations, picking up the brighter thermal emissions of the warmer dust that lies closer to the Sun. At higher solar elongations, one looks through less dust near the Earth and sees a greater fraction of colder fainter dust. (b) High-pass filtering in ecliptic latitude reveals structures in the zodiacal cloud associated with asteroid collisions and comets (*Sykes*, 1988).

(*Reach et al.*, 1997), but that other prominent bands such as that initially associated with the Eos family (*Dermott et al.*, 1984) were difficult to reconcile (*Sykes*, 1990; *Reach*, 1992; *Grogan et al.*, 2001).

Another means of understanding the relationship of the zodiacal cloud to asteroids and comets is by simulation of its creation and the direct comparison with observations. In a review, *Dermott et al.* (2001) concluded on the basis of such models that dust in the asteroid bands observed by IRAS and COBE contributes about 30% of the total zodiacal thermal emission. If only the Hirayama families (Themis, Koronis, and Eos) contribute that much, then the rest of the asteroid belt, according to the equilibrium model of dust production, should contribute at least double that value, leaving at most only 10% to a cometary contribution. This would be difficult to reconcile with a zodiacal cloud spanning heliocentric ecliptic latitudes larger than the distribution of asteroids (or comets for that matter) available to supply it.

However, such a discrepancy (if it exists) may have been mitigated by the identification of smaller asteroid families that are dynamically younger and associated with two of the three major pairs of dust bands (*Nesvorný et al.*, 2002, 2003). This has given support to the nonequilibrium theory of dust production in the asteroid belt. A consequence of this result is that the contribution of the asteroid belt as a whole to the zodiacal dust cloud is not expected to simply scale with local asteroid density, increasing the complexity (and uncertainty) of the detailed relationship between asteroids and the observed zodiacal cloud.

Whether asteroidal dust is produced continuously or stochastically, asteroids present a solid alternative to a purely cometary origin of the zodiacal cloud. Asteroid particles are dynamically distinct from cometary particles in that they have smaller initial orbital eccentricities. After being generated in the asteroid belt, this dust evolves toward the Sun under Poynting-Robertson drag. As it passes the orbit of Earth, some of it forms a circumsolar ring due to dust orbits resonant with that of Earth (*Dermott et al.*, 1994). The ring is just outside Earth's orbit, at 1.03 AU, as the exterior resonances are more stable. The peak density in the ring occurs in an enhancement that trails Earth, and amounts to about 30% in excess of the density of the smooth cloud (*Kelsall et al.*, 1998). The ring density is actually highly uncertain, being derived from a line-of-sight integral, but the estimate may be significantly improved using background measurements by the Space Infrared Telescope Facility (SIRTF) as it travels through the trailing enhancement during its planned mission lifetime (*Werner et al.*, 2001).

An intriguing aspect of the zodiacal cloud observed by IRAS and COBE/DIRBE is its azimuthal smoothness, about 5% with the principal variation arising from dynamics-induced asymmetries in the dust ring near Earth's orbit (*Dermott et al.*, 1994). This smoothness requires a source that is similarly distributed over ecliptic longitude (i.e., with randomized orbital nodes or a very efficient process for randomizing the orbital nodes of the particles before they spiral past Earth). The existence of the dust bands demonstrates that observed asteroid dust arises from a node-randomized population of parent bodies. Comets are much fewer in number than asteroids, and material tracing their current orbits would produce a relatively "lumpy" cloud. Whether a cometary component could be "smooth" depends on the outcome of the race between differential precession of the ejected cometary particle orbits [10^4–10^7 yr depending on ejection velocity (*Sykes and Greenberg*, 1986)], collisional lifetimes, and orbital decay timescales. If the cometary contribution to the zodiacal complex were due to single or multiple generational collision fragments from large millimeter–centimeter cometary particles [suggested as a possibility by *Liou and Zook* (1996)], the dispersion timescales would tend toward tens of millions of years. Disruption of such particles would be most likely by the smallest particles capable of fragmenting them, resulting in low ejection velocities on timescales (e.g., *Dohnanyi*, 1970) comparable to or smaller than the dispersion timescales. The rapid orbital decay of small fragments (wherein lie the surface area) would result in the partial formation of a torus, extended toward the Sun (*Sykes*, 1990). Azimuthal smoothness of a purely cometary zodiacal cloud could be difficult to achieve. Better understanding of the detailed evolution of the nodes of cometary dust in addition to the other orbital elements is a necessary step toward solving this problem.

The morphology of the zodiacal cloud changes with the size of particles being considered. As particle sizes increase from tens of micrometers in size (to which IRAS and COBE/DIRBE were sensitive), their sensitivity to radiation forces decrease and their spatial distribution begins to converge upon the distribution of yet-larger bodies from which they ultimately derive.

Information on objects in the millimeter to several-meter size range is obtained by meteor observations at Earth (since no such observations have yet been made at other planets).

The large end of this size range connects directly to the size range accessible by near-Earth-object searches such as Spacewatch (*Bottke et al.*, 2002a) and others. *Ceplecha* (1992) analyzes interplanetary bodies from millimeter-sized meteoroids to kilometer-sized boulders. The size distribution of smaller meteoroids is best documented in the lunar microcrater record (*Grün et al.*, 1985). Over this entire range the slope of the number density of these interplanetary bodies vs. mass in a log-log plot is close to 0.83. This slope indicates that the population of these interplanetary bodies is in overall collisional equilibrium (*Dohnanyi*, 1970), i.e., the number of particles created in a mass interval by fragmentation of larger objects equals the number of particles in that mass interval that are destroyed by collisions. However, in the local mass distribution there are significant deviations from this slope: One hump is at 10^{-9} kg and another is at 10^4 kg. A hump in the mass distribution at m_h signals that there is an excess of particles more massive than m_h, probably from an additional input in this mass range. One input is in the 100-μm to centimeter size range, i.e., the range of radar to visual meteors, and the other is in the 1–10-m size range, i.e., the range of very bright meteors, the fireballs.

Collisions govern the lifetimes of particles having these "excess" size ranges as well as larger particles up to the size of asteroids (whereas sublimation and breakup determine the lifetime of comets in the inner solar system, <5 AU). Transport of these larger meteoroids through the solar system is facilitated by stochastic gravitational interactions with planets, by the systematically inward (toward the Sun) drift caused by the Pointing-Robertson effect, and by the Yarkovsky effect (a thermal radiation force) acting in directions that depend on the orientation of the spin axis, spin rate, and thermal properties of the object (e.g., *Bottke et al.*, 2002b). However, since the collisional lifetime is shorter than the transport time, meteoroids only slowly diffuse away from their place of origin while at the same time they are ground down by collisions. Only meteoroids smaller than 0.1 mm rapidly drift by the Pointing-Robertson effect toward the Sun where they sublimate. For fragments smaller than about 1 μm in size in general, solar radiation pressure reduces solar gravity sufficiently to drive them out of the solar system — they become beta-meteoroids — although particles much smaller than 1 μm are less affected in this manner (*Burns et al.*, 1979).

Many clear orbital associations have been found between meteor streams and comets from which they are presumed to originate. [Even some asteroids may be extinct comets. Several authors suggested that Apollo and Amor asteroids are defunct comets. An argument for this thesis is that at least the peculiar asteroid Phaethon is associated with the Geminid meteor stream (*Halliday*, 1988). By dynamical studies *Gustafson* (1989) showed that Phaeton's cometary active phase lasted for several hundred orbits about 1000 years ago.] On the other hand, it has been observed that visual meteor streams (millimeter and bigger sizes) are most prominent in brightness and apparent point of origin on the sky compared with the sporadic meteor background. The contrast in these aspects between stream meteors and sporadic meteors is reduced for smaller meteors and is only weakly recognizable in radar meteor observations [100-μm size range (*Galligan and Baggaley*, 2001)]. No meteor streams have yet been identified in the dust range (micrometer size range). This observation indicates that the input from comets into the interplanetary dust cloud near Earth's orbit is most significant in the millimeter and larger size range, whereas smaller meteoroids are mostly collisional fragments of the bigger asteroidal ones, ground down and transported to Earth's distance by radiation forces.

3. INFERENCES FROM THE PHYSICAL PROPERTIES OF INTERPLANETARY DUST

3.1. Density: Meteors and Microcrater Studies

The notion that fragments of comets have low density comes from Whipple's dirty snowball model where a comet nucleus consists of an intimate mixture of ice and dust. When the ice sublimates it leaves a filigree structure of dust with much pore space from which the ice has been lost. Laboratory sublimation experiments of ice dust mixtures confirm this picture (*Grün et al.*, 1993a). A second line of evidence comes from meteor observations that show that the terminal height (at which a meteoroid is sufficiently decelerated so that it does not generate anymore light during its passage through the atmosphere), after scaling to the same initial mass, inclination to horizon, and velocity differ so much that the whole range of heights covers an air density ratio of 1:1000 (*Ceplecha*, 1994). This observation is interpreted as a consequence of a wide range of meteoroid densities: Low-density meteoroids are decelerated at higher altitudes in the much more tenuous atmosphere than high-density meteoroids. It was found that especially meteor stream particles that have a clear genetic relation to comets have very low material density, e.g., the Perseids that originate from Comet P/Swift-Tuttle. *Ceplecha* (1977) arrives at a classification of meteoroid orbits and densities from radar meteors (m ~ 10^{-8}–10^{-6} kg) over photographic meteors (10^{-6}–1 kg) to fireballs (1–10^6 kg). "Asteroidal" meteoroids have high densities (~3 g/cm³) and orbits with medium eccentricities (0.6) and low inclinations (10°). "Short-period comet"-like meteoroids have orbits with higher eccentricities and their densities are lower (1–2 g/cm³). "Long-period comet"-like meteoroids have almost parabolic orbits, random inclinations, and very low densities (0.2–0.6 g/cm³). From the study of fireballs from meteoroids larger than 1 m, *Ceplecha* (1994) concludes that the majority of these bodies are of cometary origin and of the weakest known structure.

At the lower end of the size distribution, lunar microcrater studies (*Brownlee et al.*, 1973) suggested that about 30% of all craters observed on lunar rocks were generated by low-density meteoroids. This result was derived from measurements of the depth-to-diameter ratio of lunar microcraters. Crater simulation experiments with hypervelocity

projectiles in the laboratory found that microcraters generated by low-density projectiles had a significantly shallower depth than those generated in the same material by high-density projectiles [ρ > 3 g/cm³ (*Vedder and Mandeville*, 1974; *Nagel and Fechtig*; 1980)]. Since cometary material is generally associated with lower-density material than asteroidal material, these authors conclude that at least 30% of interplanetary meteoroids originate from comets. The effects of the irregular shape of IDPs (e.g., *Brownlee*, 1985), however, are not known. Analysis of impacts on NASA's LDEF and ESA's Eureca satellite indicate a mean density of IDP impactors between 2.0 and 2.4 g/cm³ (*McDonnell and Gardner*, 1998), somewhere in between canonical cometary and asteroidal values.

3.2. Interplanetary Dust Particles

Interplanetary dust particles collected in Earth's upper atmosphere provide clues to their asteroidal or cometary origin. Some of these particles have the kind of open structures that are associated with dust expected from comets (Fig. 4). Compositions of IDPs range from chondritic (most) to iron-sulfide-nickel and mafic silicates. Some particles are melted as a consequence of atmospheric entry. Measurements of the density of about 100 of these stratospheric IDPs (having diameters of 5–15 μm) found that unmelted chondritic particles have densities between 0.5 and 6.0 g/cm³, about half of which are below 2 g/cm³, but with no observed bimodality as one might hope with two distinct source populations (*Love et al.*, 1993).

Atmospheric entry heating was proposed as a means of distinguishing between IDPs arising from comets and asteroids, given the greater average orbital eccentricity (hence average impact velocity) of the former (*Flynn*, 1989). Entry velocities of IDPs were measured by *Joswiak et al.* (2000)

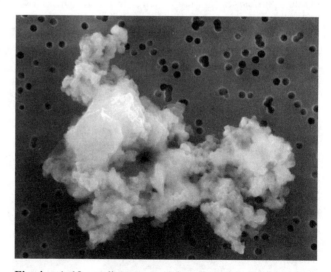

Fig. 4. A 10-μm-diameter particle collected in Earth's stratosphere. It is carbonaceous and very porous, suggesting that it may be of cometary origin, although an asteroidal origin is not excluded as a possibility. Courtesy of NASA.

applying a model of atmospheric heating and helium release (*Love and Brownlee*, 1994). High- and low-velocity groups were distinguished, with the high-velocity (cometary) group having an average density of 1.1 g/cm^3 and exhibiting "fluffy, porous, aggregate textures" and the low-velocity (asteroidal) group having average density of 2.5 g/cm^3 and tending "toward smoother, compact forms" (*Joswiak et al.*, 2000). However, while the means were distinct, there was significant overlap between the two groups in their ranges of properties, making it difficult to assign a particle to one group or another on the basis of density and morphology unless it resided at the extremities associated with those groups. This difficulty is in part due to the potential pumping up of a particle's orbital eccentricity (or that of its collisional precursor) by planetary perturbations. The properties of dust from a specific known comet will be obtained by the *Stardust* mission when it collects dust from the environment of P/Wild 2 and returns it to Earth (*Brownlee et al.*, 2000). This will be a great aid to further distinguishing cometary from asteroidal particles among collected IDPs.

3.3. Helios and Other Missions

Other evidence for distinguishing cometary and asteroidal particles on the basis of their orbits and physical properties came from the *Helios* dust experiment. The dust instrument on the *Helios* spacecraft consisted of two sensors that were mounted differently in the spacecraft. The ecliptic sensor was sensitive to impacts arriving from both north and south sides of the ecliptic plane. Since this sensor viewed the Sun once per spin revolution (the spacecraft spin axis was perpendicular to the ecliptic plane) it was covered by an aluminum-coated 0.3-μm-thick plastic film in order to prevent heat and solar UV radiation entering into the sensor. This film caused a penetration cut-off for meteoroids that depended on the mass, density, and velocity of impacting dust particles (*Pailer and Grün*, 1980). The south sensor had an open aperture that was shielded from solar radiation by the spacecraft rim and hence recorded only dust impacts arriving from south of the ecliptic plane. This sensor was sensitive to somewhat smaller and/or lower-density meteoroids. Both sensors had overlapping fields of view.

Helios measurements covered the range from 0.3 to 1 AU heliocentric distance. The measured dust flux displays a steady increase toward the Sun by about a factor of 10. There are significant differences between the measurements by both sensors. The ecliptic sensor detected most impacts in a band centered about the apex direction (i.e., 90° off the Sun in the direction of spacecraft motion), while the south sensor observed particles from all around during a spin revolution with a predominance of small particles from the solar direction (*Grün et al.*, 1980). Modeling of the *Helios* results show that these "apex" particles have low eccentricities ($e_{ave} \leq 0.6$) and small semimajor axes (averaging about 0.6 AU). Since apex particles did penetrate the front film of the ecliptic sensor, their density cannot be below 1 g/cm^3 (*Pailer and Grün*, 1980), at least not for the smallest par-

ticles detected. On the other hand, impacts outside the band were mostly observed by the south sensor and must have higher eccentricities. Modeling shows that these "eccentric" particles have eccentricities, $e_{ave} \sim 0.7$, and semimajor axes, $a_{ave} \sim 0.9$. *Grün et al.* (1980) conclude that at least half the eccentric particles should have densities below 1 g/cm^3, suggesting a cometary origin.

Four planned comet flybys (and an unintended one) from which *in situ* dust data became available were performed to date. Four spacecraft took dust measurements within 10^4 km of the nuclei of different comets. In 1985 the *International Cometary Explorer* (ICE) mission flew through the coma of Comet Giacobini-Zinner, and the plasma wave instrument recorded dust impacts in the tailward region of the coma (*Gurnett et al.*, 1986). One year later, a five-spacecraft armada flew by Comet Halley, of which three spacecraft carried a range of dust instruments from simple impact counters to sophisticated dust-mass analyzers. The two Russian *Vega* spacecraft crossed the sunward side of the coma and recorded dust impacts from the outer boundary at a distance of 2 × 10^5 km down to about 8000 km from the nucleus (*Mazets et al.*, 1987; *Simpson et al.*, 1987; *Vaisberg et al.*, 1987). ESA's *Giotto* spacecraft flew closest (600 km) to the nucleus, but most measurements ended on approach at a distance of about 3000 km when a millimeter-sized pebble hit the spacecraft with a speed of almost 70 km/s, causing some damage onboard and interrupting telecommunication. Some time later ground control over the spacecraft was regained and several instruments continued their measurements. Six years later, in 1992, the *Giotto* spacecraft was redirected to fly through the coma of Comet Grigg-Skjellerup at a distance of only 200 km from the nucleus (*McDonnell et al.*, 1993), although it ended up passing at a distance of ~100 km (*McBride et al.*, 1997). This was possible because the dust production of this comet was very low and only three particles between approximately 1 and 100 μm and a fourth particle ~10 mg were recorded to hit the approximately 2-m^2 big bumper shield (a fairly flat distribution over several orders of magnitude, consistent with the mass distribution seen at Halley). The latest flyby of Comet Borrelly by the *Deep Space 1* spacecraft occurred in 2001. The plasma wave instrument onboard recorded several dust impacts within 5 minutes of closest approach at 2000 km distance from the nucleus (*Tsurutani et al.*, 2003). Almost 30 years earlier, in 1974, the dust instrument onboard the HEOS-2 satellite recorded an enhanced impact rate of micrometer-sized particles by a factor of 3 over what the instrument had observed in previous years (*Hoffmann et al.*, 1976; *Grün et al.*, 1976). During this period the instrument was pointed in the direction where Comet Kohoutek (1993 XII) was about one year earlier and had displayed strong dust emission that led to its detection at about 4 AU from the Sun. At the time of the recorded dust impacts the comet was already 3 AU from the Sun past its perihelion. Therefore, the dust recordings constitute measurements in the very distant tail of this comet.

Besides the spatial extent of dust in cometary comae and tails, the dust production rate and size distribution of cometary dust was derived from the *in situ* measurements. Most data, of course, came from the comprehensive measurements at Comet Halley. It was found that the dust size distribution extends over a much wider range than was expected from astronomical observations, mostly in the optical wavelength range. The size distribution extends to both much smaller particles in the submicrometer and even nanometer size range, and to much bigger particles in the millimeter size range (*McDonnell et al.*, 1987). It was also found that some particles fragment shortly after their release from the nucleus (*Simpson et al.*, 1987; *Vaisberg et al.*, 1987), which indicates that their initial structure is very fluffy and contains materials that sublimates at distances as small as 1 AU from the Sun. As a consequence of this extended mass distribution the dust production is significantly bigger than that which has been derived from astronomical observations alone. In addition, the dust-to-gas mass ratio of Comet Halley exceeds a value of 1 (*McDonnell et al.*, 1991) compared to a value of 0.1 from earlier estimates.

The dust mass spectrometer onboard the *Giotto* and *Vega* spaceprobes provided elemental and isotopic data for small newly ejected cometary particles (*Kissel et al.*, 1986). A major discovery was that of "CHON" particles, which consisted of material high in content of the elements H, C, N, and O, showing that the dust was rich in organics (*Kissel and Krüger*, 1987). Magnesium isotope ratios showed only a slight variation around the nominal solar value, whereas the isotopic ratio of $^{12}C/^{13}C$ showed large variations from grain to grain, but on average it was also solar like (*Jessberger and Kissel*, 1991). The average elemental composition was found to be solar like, but significantly enriched in volatile elements H, C, N, and O compared to C1 chondrites (*Jessberger et al.*, 1987).

4. HOW COMETS SUPPLY THE CLOUD

4.1. General

When a comet is discovered, it is identified by virtue of its fuzzy appearance, with perhaps a tail, arising from the loss of gas and dust. This dust represents the smallest-sized particle emissions from a comet [generally tens of micrometers and smaller, although significant coma surface area is argued to reside in very large particles (see *Fulle*, 2004)]. These are entrained in the gas outflow and accelerated to speeds up to ~1 km/s for the smallest particles. After decoupling from the gas, they are generally lost to the solar radiation field. The sensitivity of a particle to solar radiation pressure is described by the parameter, β, the ratio of radiation force to gravitational force felt by the particle (*Burns et al.*, 1979). Most of the particles observed in a comet's tail at visible wavelengths are micrometer-sized and have $\beta > 1$. These particles are not gravitationally bound to the Sun and escape the solar system, not contributing to the interplanetary dust complex. Particles having lower β can also escape from the solar system when they are released from an orbiting object like a comet, because of the con-

tribution of the comet's motion. For emission at perihelion, the value of β for escape is a function of orbital eccentricity of the parent body

$$\beta_p \geq (1 - e)/2$$

Thus, for a parabolic comet, all emitted particles would be lost. The solar system loses particles tens of micrometers and smaller from long-period comets, while retaining particles on the order of several micrometers from short-period Jupiter-family comets (Fig. 5). Almost all dust particles tens of micrometers and smaller are released from these latter comets into Jupiter-crossing orbits — even comets whose orbits are completely interior to that of Jupiter's (Fig. 6). Subsequent perturbations on the orbits of these particles by Jupiter results in their loss while the distribution of many bear little resemblance to the elements of their parent comets (e.g., *Gustafson et al.*, 1987). Making the assumption that such scattered particles have randomized nodes (required to match the azimuthal symmetry of the cloud), *Liou et al.* (1995) was able to model a contribution to the cloud by single-sized particles from Encke, taking into account radiation pressure, Poynting-Robertson and corpuscular drag, and perturbations by Jupiter, which when combined with a model contribution from asteroid dust made good matches to selected scans of the zodiacal cloud by IRAS.

Cometary particles may undergo considerable orbital evolution with time, increasing the difficulty of distinguishing them from asteroidal particles. *Liou and Zook* (1996)

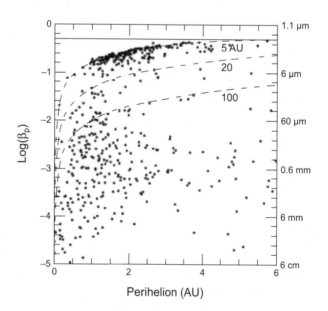

Fig. 5. Maximum β (minimum radius, assuming a density of 1 g/cm³) of particles from known comets on escape trajectories from the solar system, assuming perihelion emission and zero ejection velocity. Aphelion distances of source comets are dashed lines. For circular orbits, $\beta_p = 0.5$. Particles are assumed to have zero albedo.

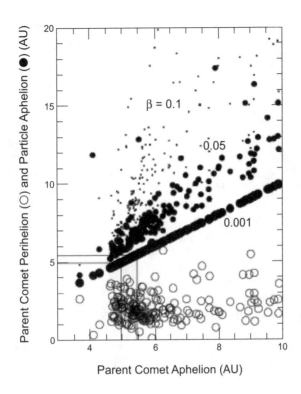

Fig. 6. Perihelion and aphelion distances of parent comets (open circles) and the aphelia of emitted particles of different β (filled circles). Jupiter's perihelion and aphelion distances are indicated by solid lines.

determined that some cometary particles (from Tempel 2–like comets) could be injected into mean-motion resonances with Jupiter and trapped for thousands of years, after which their orbital eccentricities would be quite small. They would approach Earth with the low velocities expected for asteroid particles. This would help explain the overlap in morphologies and compositions among collected "cometary" and "asteroidal" IDPs, identified by their model atmospheric entry speeds.

4.2. Dust Trails

4.2.1. Discovery and observations. In 1983, the first survey of the entire sky at thermal infrared wavelengths was conducted by IRAS (*Neugebauer et al.,* 1984). Part of the ongoing analysis during this mission was the IRAS Fast Mover Program (*Davies et al.,* 1984; *Stewart et al.,* 1984; *Green et al.,* 1985) in which fast-moving solar system objects were sought. Six comets and a couple of Apollo asteroids (including Phaethon) were discovered. However, a curiously extended tail associated with P/Tempel 2 was detected over the course of a number of IRAS scans. This was manifested by about 50 faint, relatively collinear sources at 25 μm. It was found to extend 10° on the sky with a width of 4', and no similar feature was found by the program associated with any other comet observed by IRAS (*Davies et al.,* 1984; *Steward et al.,* 1984). Dynamical analysis showed this

"anomalous tail" to be the result of low-velocity emissions of large particles, some of which may have occurred at least 1500 days prior to the observations (*Eaton et al.,* 1984).

An examination of IRAS image products, in which individual IRAS scans were merged into images, revealed the continuous emission of the reported Tempel 2 tail extending over 48° of sky (Fig. 7). Similar features were found associated with other comets (*Sykes et al.,* 1986). Clearly, a new cometary phenomenon had been discovered by IRAS and was referred to as "trails." An examination of all the IRAS data yielded trails associated with eight short-period comets (Table 1) and about an equal number of "orphans" not associated with any known comet (*Sykes and Walker,* 1992). As with *Eaton et al.* (1984), all trails were found to be consistent with low-velocity emissions and emissions from years to more than a century before the time of observation. Thus, trails offered a continuous record of the emission of comets over that period of time. At the comet orbits, trails have widths of several 10^4–10^5 km.

After IRAS, trails were no longer observed in the infrared until the launch of the Infrared Space Observatory (ISO) more than a decade later. The ISO observed segments of the Kopff (*Davies et al.,* 1997) and Encke trails (*Reach et al.,* 2000). Unlike IRAS, ISO was a pointed and not a survey instrument, so its ability to study trails was limited. However, the Kopff trail showed changes due to emissions since IRAS. The Encke trail was observed from a particularly favorable angle of 35° above its orbital plane and included the comet allowing the emergence of the trail from the comet coma to be studied. Dynamical modeling of the Encke trail showed that the mass lost in trail particles (meteoroids with radii of at least several millimeters) is much larger than the mass lost in gaseous or small-particle form, and the comet can only survive these large-particle losses for ~10,000 years (*Reach et al.,* 2000).

At the time of its observation by IRAS, the Tempel 2 trail position was sent to ground observers who were unable to detect it at visual wavelengths (*Davies et al.,* 1984; *Stewart et al.,* 1984). Several years later, a trail associated with P/Faye was accidentally detected in the visible by the Space-

Fig. 7. The Tempel 2 dust trail is seen to extend over 30° in this composite image constructed from IRAS scans. The comet coma is seen at the left end of the trail. In the upper right corner is part of the central asteroid dust band. Background cloud-like structures are interstellar cirrus.

TABLE 1. Cometary dust trail information from *Sykes and Walker* (1992).

Name	θ	W	Δv$_p$	Age	LM	D/G
Churyomov-Gerasimenko*	1	50	2	3–11	13.1	4.6
Encke	93	680	40	19–21	14.9	3.5
Gunn	6	111	3	27–74	13.8	3.6
Kopff	17	47	3	47–158	13.5	1.2
Pons-Winnecke	3	40	3	6–21	13.2	2.4
Schwassmann-Wachmann 1	10	769	5	114–148	14.5	1.6
Tempel 1[†]	7	68	4	14–38	13.4	3.5
Tempel 2	65	31	2	140–665	13.7	2.9

Rosetta target.

[†]*Deep Impact* target.

Includes the observed angular extent, θ (deg.); the width, W (10^3 km); normal velocity assuming perihelion emission, Δv$_p$ (m/s); an estimate of the age (in years) of the oldest emissions observed; an estimate of the corresponding comet mass loss rates, LM (log g/century); and an estimate of the dust to gas mass ratio of the comet, D/G.

watch Survey in the course of searching for near-Earth objects. One evening, a band 1′–2′ in width lay across one of their half-degree scans. The next night, they traced the band back to its cometary source. The trail extended 10° (*Rabinowitz and Scotti*, 1991). More recently, a program to observe trails from the ground at visual wavelengths has been successful (*Ishiguro et al.*, 2002), and other observers are beginning to report similar detections of dust trails (e.g., *Lowry et al.*, 2003). These observations are the harbinger of a new era of dust trail studies that will allow more extensive mapping and characterization of large particle mass loss from comets than has been previously been available.

4.2.2. Particle properties. Syndyne analysis of the Tempel 2 trail (comparing it with the predicted locations of continuously emitted particles with zero emission velocity) suggested particle diameters of about 1 mm (β ~ 0.001), assuming density of 1 g/cm³ (*Eaton et al.*, 1984; *Sykes et al.*, 1990). Trail particles are dark. This is evidenced by the early failure to detect the Tempel 2 dust trail from the ground (*Davies et al.*, 1984; *Stewart et al.*, 1984) and supported by initial estimates of the albedo of those particles using IRAS measurements (*Sykes*, 1987). An extremely low albedo (on the order of a percent) for Kopff trail particles has been estimated on the basis of its groundbased detection (*Ishiguro et al.*, 2002).

Trail particles have color temperatures that tend to be in excess of blackbody values (*Sykes*, 1987; *Walker et al.*, 1989; *Sykes et al.*, 1990; *Sykes and Walker*, 1992), suggesting either low-emissivity materials, a small particle component, or sustaining temperature gradients. The low emissivities required do not match that of any known nonmetallic materials (*Sykes and Walker*, 1992). Particles small enough to have the observed low β values would need to be smaller than tens of nanometers (*Burns et al.*, 1979), which would not radiate efficiently at infrared wavelengths. Trail particles appear to be uniformly large, dark, rapidly rotating particles that have thermal conductivity low enough to allow them

to sustain a latitudinal temperature gradient (Fig. 8). Low thermal conductivity can be achieved with porosity, which translates to low mass density.

4.2.3. What trails tell us about comets and their contribution to the interplanetary dust cloud. Dust trails reveal the principal mechanism by which short-period comets lose mass: via the low-velocity emission of large particles. Estimates of refractory (dust) mass loss rates (Table 1), combined with gas mass loss from visible groundbased observations, reveal an average cometary dust to gas mass ratio of about 3 (*Sykes and Walker*, 1992), which is significantly higher than the canonical 0.1–1. This translates to roughly equal volumes of "rock" and "ice" in a comet nucleus (assuming a rock density of 3 g/cm³ and an ice density of 1 g/cm³), and is consistent with the dust to gas limit for Comet Halley's,

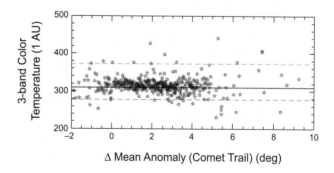

Fig. 8. Color temperatures for individual 12, 25, and 60-μm scans of the Tempel 2 dust trail were calculated and scaled to 1 AU. The top dashed line corresponds to the color temperature of a sphere on which each point is in instantaneous radiative equilibrium with solar insolation. The bottom dashed line corresponds to the color temperature of a blackbody. The central solid line corresponds to the color temperature of randomly oriented, rapidly rotating spheres where each local latitude is in radiative equilibrium with the average diurnal solar insolation. Geometric albedo is assumed to be zero.

based on observations by *Giotto* (*McDonnell et al.*, 1991).

Assuming all short-period comets have trails similar to those identified in the IRAS observations, with a corresponding average mass loss rate of 8.4×10^8 kg/yr (*Sykes and Walker*, 1992), the amount of material contributed to the zodiacal dust complex (assuming 150 comets) would be 1.3×10^{11} kg/yr, a significant fraction of the $\sim 2.9 \times 10^{11}$ kg/yr lost within 1 AU that needs to be replenished if the cloud is in steady state (*Grün et al.*, 1985). A recent optical/thermal imaging survey of comets also concludes that comet dust is a major supplier of the IDP cloud (*Lisse*, 2002).

4.3. Meteor Streams

The largest-sized particles supplied to the interplanetary dust cloud by comets are observed as meteor showers. They begin as dust trails, with individual returns by the parent comet to perihelion causing separate trails in a pattern reflecting planetary perturbations of the comet orbit itself. Perturbations on individual grains in the trail cause a cyclic motion of the node of the trail near Earth's orbit. Meteor outbursts (including meteor storms) are seen when Earth passes through a dust trail (e.g., *Kresák*, 1993). Differences in perturbations acting on different trail fragments result in these fragments superposing and smearing to the point that they populate a filament. Dispersal of comet trail/meteor stream material into the background zodiacal cloud can be inhibited by orbital resonances, which can maintain trail cohesion for long periods of time [*Asher et al.* (1999), who determined that the Leonid outburst of 1998 was dust ejected in 1366]. Close encounters cause the grains to be dispersed into a broader meteor stream responsible for an annual shower, which can be identified with that comet's orbit for thousands of years before dissipating and becoming indistinguishable from the background population of particles in that size range.

Success at predicting storms has been recently achieved with the work of *Jenniskens* (1994, 1997), who forecast the return of the 1994 α-Monocerids based on the dust-trail hypothesis, and *Kondrat'eva and Reznikov* (1985), who predicted the return of the 1998 Draconid storm (*Fujiwara et al.*, 2001) and the 2001/2002 Leonid storms. The latter considered the ejection of a single particle at perihelion in the direction of motion of the parent comet and calculated the subsequent gravitational perturbations on an orbit with enough lag to allow for a timely collision with Earth. *McNaught and Asher* (1999) and *Lyytinen* (1999) applied a refined model to the Leonids, identifying the spatial distributions of dust trails from its parent, P/Tempel Tuttle, which allowed meteors to be identified with emissions from specific epochs of perihelion passage (e.g., Fig. 9). This allows for future studies of the effects of age on larger cometary dust particles.

A principal means of analyzing particle number densities within a meteor stream is via its zenith hourly rate (ZHR), the rate of visible meteors seen by a standard observer under ideal conditions (radiant in the zenith and the star limiting magnitude = 6.5) (*Jenniskens*, 1995). Precise measurements of the ZHR from aircraft during the 1999 Leonid event

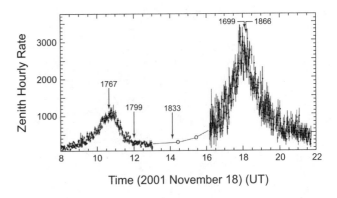

Fig. 9. Activity curve of the 2001 Leonid meteor storms. Closed symbols are from the Leonid Multi-Instrument Aircraft Campaign; open circles are data gathered by the International Meteor Organization (*Jenniskens*, 2002).

showed the dust density in Earth's path and along the orbit to exhibit a sharp core, but with "wings" well described by a Lorentzian, while the perpendicular dispersion in a sunward direction is wider and exponential (*Jenniskens et al.*, 2000). Possible explanations include emission at large heliocentric distances combined with a higher degree of fragmentation of dust particles within the coma near perihelion (*Jenniskens*, 2001). A mass loss rate of $\sim 2.6 \times 10^{10}$ kg/return was measured.

4.4. Kuiper Belt Dust

Another direct source of interplanetary dust in both the inner solar system and beyond Jupiter is the source region of the short-period comets themselves — the Kuiper belt. *Flynn* (1994) suggested that this might constitute a significant contribution to the interplanetary dust collected in Earth's stratosphere. *Liou et al.* (1996) found that 20% of the grains generated in the Kuiper belt would evolve all the way to the Sun (the remainder being scattered out of the solar system by the giant planets), and that particles between 9 and 50 μm diameter would be depleted due to mutual collisions and collisions with interstellar dust. They further found that particles surviving into the inner solar system would have low-eccentricity, low-inclination orbits, making them dynamically indistinguishable from evolving asteroidal particles (offering a possible explanation for the IDPs having "cometary" physical properties, but heating characteristics of asteroidal sources, above). More recently, *Moro-Martín and Malhotra* (2003) modeled the dynamical evolution of dust particles from the Kuiper belt, taking into consideration the combined effects of radiation pressure, Poynting-Robertson drag, solar wind drag, and the gravitational forces of the planets (excluding Mercury and Pluto) and concluded that near Earth, these grains would have high eccentricities and inclinations, similar to cometary grains and not asteroidal grains, contradicting *Liou et al.* (1996). They further concluded that between 11% and 21% of particles with $0.01 \le \beta \le 0.4$ would drift from the Kuiper belt to interior to Jupiter [in rough agreement with *Liou et al.* (1996)], and that [assuming the dust production rates of *Landgraf et al.* (2002)] the contribution to the IDPs captured in Earth's atmosphere may be as low as 1–2%.

Because of their large heliocentric distance (beyond 30 AU), Kuiper belt objects (KBOs) are not active. Dust production in the Kuiper belt depends upon collisional activity, constrained by their size distribution and detailed orbital-element distribution. This production has been estimated by *Stern* (1996) to be between 9.5×10^8 to 3.2×10^{11} g/s (far more than needed to replenish the loss of dust estimated to be lost each year within 1 AU!). In addition, they determine that recent impacts would produce between zero and several hundred short trail-like structures, having annual parallaxes of up to 2.6°. However, a parallactic survey making use of two-week to six-month baselines provided by separate maps of the sky by IRAS (*Sykes et al.,* 1994) produced no evidence of parallax in any extended structures at 60 and 100 μm. *Yamamoto and Mukai* (1998) estimated a production rate of dust grains between 3.7×10^5 and 3.1×10^7 g/s with radii <10 μm as a consequence of impacts of interstellar dust on KBOs.

That dust from the Kuiper belt is being generated and transported to smaller heliocentric distances, however, is supported by *Landgraf et al.* (2002) in their analysis of data from the dust experiments onboard *Pioneer 10* and *11*. *Humes* (1980) had reported an essentially constant spatial density between 1 and 18 AU. *Landgraf et al.* (2002) considered three potential sources for the impacts recorded beyond Jupiter: P/Halley-type comets, P/Schwassmann-Wachmann 1-type comets, and dust from the Kuiper belt. They found that the amount of dust detected beyond Saturn could only be explained by dust originating in the Kuiper belt and evolving toward the Sun. Under their model, about 5×10^7 g/s would need to be generated in the Kuiper belt (between 0.01 and 6 mm in size), more than an order of magnitude below the minimum estimate of *Stern* (1996). The required contributions from P/Halley-type comets were 3×10^5 g/s and that from P/Schwassmann-Wachmann 1 comets were 8×10^4 g/s. It is interesting to note that dust production from P/Schwassmann-Wachmann 1 itself was estimated to be ~10^5 g/s (*Sykes and Walker,* 1992, Table 1) in particles ~1 mm in size.

5. EFFECT OF INTERSTELLAR DUST PARTICLES

Interstellar dust particles are thought to play a role in the production of interplanetary dust (e.g., *Yamamoto and Mukai,* 1998) and its comminution (e.g., *Liou et al.,* 1996). As the solar system moves through the galaxy, dust grains that pass through the planetary system have been detected by the dust detector onboard the *Ulysses* spacecraft (*Grün et al.,* 1993b). It came as a big surprise that after *Ulysses* flew by Jupiter, the dust detector recorded impacts of interstellar dust (ISD) that arrived from a direction that was opposite to the expected flow direction of interplanetary dust grains. It was found that on average the impact veloci-

ties of these particles exceeded the local solar system escape velocity (*Grün et al.,* 1994).

The motion of ISD through the solar system was found to be parallel to the flow of neutral interstellar hydrogen and helium gas with a speed of 26 km/s both for gas and dust. This proves that local interstellar dust and gas are nearly at rest with respect to each other. The interstellar dust flow was continuously monitored by *Ulysses* and persisted at a constant level at all latitudes above the ecliptic plane even over the poles of the Sun, whereas interplanetary dust was strongly depleted away from the ecliptic plane. Starting in mid-1996 the flux of ISD began slowly to decrease and, in the year 2000, was about a factor of 3 lower [this is related to the reversal of the magnetic field in the course of the solar cycle (*Landgraf,* 2000].

Measurements in the ecliptic plane by *Galileo* confirmed that outside about 3 AU the interstellar dust flux exceeds the flux of micrometer-sized interplanetary grains. Interstellar grains observed by *Ulysses* and *Galileo* range from 10^{-18} kg to more than 10^{-13} kg. If compared with the ISD mass distribution derived by astronomers, the mass distribution observed by spacecraft overlaps only with the biggest masses observed by remote sensing. More recently, even bigger (10^{-10} kg) interstellar meteoroids have been reliably identified by their hyperbolic speed (>100 km/s) at 1 AU (*Baggaly,* 2000). The flow direction of these big particles varies over a much wider angular range than that of small grains observed by *Ulysses* and *Galileo*.

The deficiency of measured small grain masses is not solely caused by the detection threshold of the *in situ* instrumentation, but it indicates a depletion of small interstellar grains in the heliosphere. Model calculations by *Frisch et al.* (1999) of the filtering of electrically charged grains in the heliospheric bow shock region and in the heliosphere itself show that 0.1-μm-sized and smaller particles are strongly impeded from entering the planetary system by the interaction with the solar wind magnetic field.

6. PRODUCTION OF DUST IN OTHER PLANETARY SYSTEMS

The interplanetary dust cloud offers a "blueprint" or "laboratory" for understanding dust in other planetary systems. The scattered starlight and thermal emission from dust around other stars — exozodiacal light — is, for most stars, the only indicator we can observe from Earth of the collisional processes and small bodies around those stars. Furthermore, these small bodies also provide evidence for planets around other stars. An inner hole and significant warp were discovered in the β Pictoris disk (*Lagage and Pantin,* 1994; *Heap et al.,* 2000), and blobs were discovered around Vega (*Wilner et al.,* 2002) and ε Eridani (*Greaves et al.,* 1998; *Quillen and Thorndike,* 2002) and Fomalhaut (*Holland et al.,* 1998; *Wyatt and Dent,* 2002). These structures have all been explained by perturbations by planets around the stars, with the direct analogy to Earth's circumsolar ring (discussed above) providing key supporting evidence (*Kuchner and Holman,* 2003). If viewed from afar, our own solar system might be recognizable as having at least two planets as a consequence of structure in dust evolving from the Kuiper belt (*Liou and Zook,* 1999). Dust bands and trails in the interplanetary cloud have been tied to asteroid families and individual short-period comets; comparable structures have not yet been observed around other stars, so the direct connection between parent bodies and the dust cloud can only be studied in detail in the solar system.

Exozodiacal light studies also provide valuable clues to understanding aspects of interplanetary dust that cannot be readily discerned from our vantage point inside the system and in its midplane. Among nearby main-sequence stars, some 15% have far-infrared emission, in excess of the photosphere, that is believed to be due to circumstellar material (*Backman et al.,* 1997; *Habing et al.,* 2001). Most of the stars have "cold" infrared excess, detected at wavelengths of 60 μm and greater. This is partly due to the rapid decline in photospheric emission at longer wavelengths, making far-infrared excess more prominent than mid-infrared excess. The cold excesses, with color temperatures ~80 K, are located relatively far from the central star and roughly correspond to the Kuiper belt in the solar system. In at least four cases, dubbed the "fabulous four" by observers, the disks are resolved: β Pictoris, Vega, ε Eridani, and Fomalhaut all have disks extended at least 100 AU from their central stars. In all cases, the Poynting-Robertson loss time is shorter than the stellar lifetime, indicating that the disks must be replenished by collisions among a reservoir of larger bodies. This suggests it is likely that our own Kuiper belt may be collisionally active and could be a source of dust, although that dust between 9 and 50 μm in diameter may only reside in the outer solar system because its transport to the inner solar system is hindered by collisions with interstellar dust along the way (*Liou et al.,* 1996).

Exozodiacal light due to "warm" dust in the "terrestrial planet" zone around other stars is more rare than the colder dust. This warmer dust is difficult to discern photometrically, because the disk emission is generally fainter than the photosphere at wavelengths less than 30 μm (as opposed to the cold excesses, which are often larger than the photosphere). A recent photometric survey found warm disks around 5 out of 81 stars, and none of them had color temperatures higher than 120 K (*Laureijs et al.,* 2002). Furthermore, none of the stars older than 400 m.y. had warm exozodiacal light. In contrast, the zodiacal light as viewed from Earth has a color temperature around 262 K (*Reach et al.,* 1996) and the dust density has been shown to increase as a power-law all the way in to less than 0.15 AU form the Sun (*Leinert et al.,* 1981). One problem with searching for the warmer dust, which would be the analog of dust produced by asteroid collisions and short-period comets, is angular resolution. Mid-infrared spectra of the β Pictoris disk reveal a bright silicate emission feature from the warm dust, with a shape that is different from that of the silicate feature found in the zodiacal light (*Reach et al.,* 2003). A Keck observation with high angular resolution showed that the silicate feature arises only very close to the star (*Weinberger et al.,* 2003).

The very different vantage points for viewing the zodiacal and exozodiacal light allow for very different insights into the collisional processes and evolution of small solid bodies. From inside the solar system, it is possible to use the brightness as a function of look direction to obtain the scattering phase function (*Hong,* 1985; *Kelsall et al.,* 1998); the phase function is needed to invert brightness distributions. From the zodiacal light, it is possible to measure the density of interplanetary dust to within tens of solar radii, which is not possible around other stars because of glare from the photosphere. From the exozodiacal light, it is possible to measure the distribution of material to hundreds of AU, which is not possible from Earth-based observatories because of the bright foreground from dust in the inner solar system.

7. THE FUTURE

7.1. Dust Detectors on Spacecraft

The first dust detectors flown in space were simple microphones that responded to dust impacts, but also to a wide range of interferences that in interplanetary space occurred more frequently than dust impacts. Once this was appreciated more sophisticated multicoincidence detectors were developed that permitted the detection of dust impacts at a rate as low as one impact per month. Impact ionization provided the means of at least two independent coincident measurements of a dust impact: the plasma cloud generated by an hypervelocity impact onto a solid target is separated by an electric field so that positive ions and negative ions together with electrons are recorded separately. Time-of-flight analysis of the ions even provides mass analysis of the generated ions. Early detectors of this type had a sensitive area of ≤0.01 m², which had the consequence that only very few dust impacts were recorded in interplanetary space. Therefore, the more recent dust detectors on the *Galileo, Ulysses,* and *Cassini* missions had sensitive areas that were 10 times larger. Several recent missions carry dust mass analyzers in which the impact-generated ions are analyzed in a mass spectrometer (e.g., *Kissel,* 1986). The *Cassini* cosmic dust analyzer (*Srama et al.,* 1996, 2004) is the most sophisticated dust instrument to date. It combines a 0.1-m² impact ionization detector with a time-of-flight mass spectrometer and charge sensing entrance stage for coarse velocity and direction determination.

It is hoped that all future missions, particularly to the outer solar system, will include dust experiments. Dust beyond Saturn will be studied for the first time since *Pioneer 10* and *11* with the student dust counter planned for the *New Horizons* mission to Pluto.

7.2. DUNE Observatory

From knowledge of the dust particles' birthplace and the particles' bulk properties, we can learn about the remote environment out of which the particles were formed. This approach could be carried out by means of a dust telescope on a dust observatory in space. A dust telescope is a combination of a dust trajectory sensor together with an analyzer for the chemical composition of dust particles.

Potential targets of a dust telescope are interstellar dust, interplanetary dust (e.g., meteor stream dust, cometary, or asteroidal dust or dust from the Moon), and even space debris (e.g., fine grains from solid rocket burns).

The first goal of a dust telescope is to distinguish by their trajectories dust particles from different sources: interstellar grains from the different types of interplanetary dust grains. Interstellar dust flows narrowly collimated through the solar system. This flow can be easily distinguished from the flow of interplanetary particles. Young cometary particles have highly eccentric orbits, whereas asteroidal particles have low eccentricity orbits. These different orbits are separated by measurement of the flight direction and speed. Dust in meteor streams occurs only during specific periods and is directly related to the parent comet.

A state-of-the-art dust telescope would consist of an array of parallel mounted dust analyzers (*Grün et al.,* 2000) and consists of several instruments sharing a common impact plane of about 1 m² in size. Potential components are a high-resolution impact mass spectrometer, a dust analyzer for the determination of physical and chemical dust properties, and large-area impact detectors with trajectory analysis. Dust particles' trajectories are determined by the measurement of the electric signals that are induced when a charged grain flies through an appropriately configured electrode system. After the successful identification of dust charges of >10⁻¹⁵ Coulombs in space by the *Cassini* cosmic dust analyzer, trajectory analyzers that are in development have tenfold increased sensitivity of charge detection giving us trajectories for submicrometer-sized dust grains.

Modern dust chemical analyzers have sufficient mass resolution to resolve ions with atomic mass numbers above 100. However, since their impact area is only 0.01 m², they can analyze statistically meaningful numbers of grains only in the dust-rich environments of comets or ringed planets. Therefore, a dust telescope should include several of the existing mass analyzers or a large area chemical dust analyzer of mass resolution >100 with at least 10 times greater sensitive area, in order to provide statistically significant measurements of interplanetary and interstellar dust grains in space.

7.3. Thermal Infrared Observations

Since IRAS and COBE there have been no surveys of the sky at thermal wavelengths. ISO allowed some studies of cometary trails (*Davies et al.,* 1997; *Reach et al.,* 2000), and detailed SIRTF observations of a large number of short-period comets may allow us to greatly improve upon the estimates of cometary dust production contributing to the zodiacal cloud (as well as an understanding of their emission history, retained in the trails). Only by surveys, however, are we able to observe the cloud as a whole and model its evolution and supply by entire populations of objects. But we have been limited by trying to understand the inter-

planetary dust cloud from observations made within it, near its plane of symmetry. This makes distinguishing radial components difficult, because they are all coincident along our line of sight and the stochastic nature of dust production within the cloud complicates the interpretation of structures, volume distributions, and excesses (or deficits) compared to models. It would be extremely useful to have thermal observations of the cloud from a vantage well away from the ecliptic plane, to look over the "top" of the zodiacal cloud interior to Jupiter and study outer solar system dust and dust in the Kuiper belt directly. Peaking over the half-width of the Kelsall model of the zodiacal cloud at 1 AU, and assuming the extent of this inner cloud runs with the main asteroid belt, an orbital inclination of about 45° would be required to observe dust in the Kuiper belt. This would parallactically shift the position of the inner cloud by more than 10°, allowing for the contribution of distant cold dust to be more easily distinguished. In addition to detecting or placing meaningful limits on dust being generated in the Kuiper belt, studies of the inner zodiacal cloud by such a system would greatly improve our understanding of the morphology of the cloud and the relative contributions of asteroid collisions and comet emissions.

7.4. Earth as a Detector

The valuable information provided by meteor stream observations on detailed structures of cometary emissions and the physical properties of those particles argue for the increasing application of modern technology to their study. First, continued efforts to predict these events must be made to allow their observation. Long-term video and radar monitoring for the identification of new meteor streams should be conducted. Campaigns should focus on predicted meteor outbursts with high-speed imagers and photometers, and head-echo of meteors using high-power radar, particularly focusing on small meteoroid masses, should be conducted to assess their fragmentation properties. Information on meteoroid composition could be greatly improved by the use of cooled CCD cameras for slitless optical spectroscopy, and the development of instruments that would focus on individual emission lines/bands and the indirect detection of organics. Together these should not only provide information on properties, but also allow us to assess differences among comets that might relate to different formation conditions and locations as well as ages.

8. FINAL NOTE

The ability to associate IDPs collected at Earth's orbit with specific sources is of great value in that it provides "sample return" information on these sources that might otherwise be impractical to obtain as well as provide great insight into the origin and evolution of those sources and the solar system in general. The increase in our very diverse means of studying interplanetary dust, including atmospheric collection, *in situ* studies by spacecraft, remote observations at visual and thermal wavelengths, and collisional and dynamical modeling have substantially increased the complexity (and interest) of the problem. The interplanetary dust complex is not in a steady-state condition. Evidence today bolsters the significant episodic infusion of dust by asteroid collisions, a greater potential cometary source due to the discovery of their large particle emissions, and the contribution at some level of dust from the Kuiper belt. Studies of these sources and interplanetary dust are increasingly interrelated. We need to study both sources and dust in order to have a more complete understanding of all.

Acknowledgments. This chapter benefited from the detailed reviews of D. Brownlee and N. McBride, to whom the authors express their appreciation.

REFERENCES

Asher D. J., Bailey M. E., and Emel'yanenko V. V. (1999) Resonant meteoroids from Comet Tempel-Tuttle in 1333. *Mon. Not. R. Astron. Soc., 304,* L53–L56.

Backman D. E., Fajardo-Acosta S. B., Stencel R. E., and Stauffer J. R. (1997) Dust disks around main sequence stars. *Astrophys. Space Sci., 255,* 91–101.

Baggaley W. J. (2000) Advanced meteor orbit radar observations of interstellar meteoroids. *J. Geophys. Res., 105,* 10353–10361.

Bottke W. F., Morbidelli A., Jedicke R., Petit J-M., Levison H., Michel P., and Metcalfe T. S. (2002a) Debiased orbital and size distributions of the near-Earth objects. *Icarus, 156,* 399–433.

Bottke W. F., Vokrouhlický D., Rubicam D. P., and Brož M. (2002b) The effect of Yarkovsky thermal forces on the dynamical evolution of asteroids and meteoroids. In *Asteroids III* (W. F. Bottke Jr. et al., eds.), pp. 395–408. Univ. of Arizona, Tucson.

Brownlee D. E. (1985) Cosmic dust: Collection and research. *Annu. Rev. Earth Planet. Sci., 13,* 147–173.

Brownlee D. E., Hörz F., Vedder J. F., Gault D. E., and Hartung J. B. (1973) Some physical parameters of micrometeoroids. *Proc. Lunar Sci. Conf. 4th,* pp. 3197–3212.

Brownlee D. E., Tsou P., Clark B., Hanner M. S., Hörz F., Kissel J., McDonnell J. A. M., Newburn R. L., Sanford S., Sekanina Z., Tuzzolino A. J., and Zolensky M. (2000) Stardust: A comet sample return mission. *Meteoritics & Planet. Sci., 35,* A35.

Burns J. A., Lamy P. L., and Soter S. (1979) Radiation forces on small particles in the solar system. *Icarus, 40,* 1–48.

Ceplecha Z. (1977) Meteoroid populations and orbits. In *Comets, Asteroids, Meteorites: Interrelations, Evolution and Origins* (A. H. Delsemme, ed.), pp. 143–152. Univ. of Toledo, Ohio.

Ceplecha Z. (1992) Influx of interplanetary bodies onto Earth. *Astron. Astrophys., 263,* 361–368.

Ceplecha Z. (1994) Impacts of meteoroids larger than 1 m into the Earth's atmosphere. *Astron. Astrophys., 286,* 967–970.

Davies J. K., Green S. F., Stewart B. C., Meadows A. J., and Aumann H. H. (1984) The IRAS fast-moving object search. *Nature, 309,* 315–319.

Davies J. K., Sykes M. V., Reach W. T., Boulanger F., Sibille F., and Cesarsky C. J. (1997) ISOCAM observations of the comet P/Kopff dust trail. *Icarus, 127,* 251–254.

Dermott S. F., Nicholson P. D., Burns J. A., and Houck J. R. (1984) On the origin of the IRAS solar system dust bands. *Nature, 312,* 505–509.

Dermott S. F., Nicholson P. D., and Wolven B. (1986) Preliminary analysis of the IRAS solar system dust data. In *Asteroids, Comets, Meteors II* (C.-I. Lagerkvist et al., eds.), pp. 583–594. Uppsala, Sweden.

Dermott S. F., Jayaraman S., Xu Y. L., Gustafson B. Å. S., and Liou J.-C. (1994) A circumsolar ring of asteroidal dust in resonant lock with the Earth. *Nature, 369,* 719.

Dermott S. F., Grogan K., Durda D. D., Jayaraman S., Kehoe T. J. J., Kortenkamp S. J., and Wyatt M. C. (2001) Orbital evolution of interplanetary dust. In *Interplanetary Dust* (E. Grün et al., eds.), pp. 569–640. Springer-Verlag, Berlin.

Dikarev V., Grün E., Landgraf M., Baggaley E. J., and Galligan D. (2002) Interplanetary dust model: From micron-sized dust to meteors. In *Proc. Meteoroids 2001 Conf.,* pp. 609–615. ESA SP-495, Noordwijk, The Netherlands.

Dohnanyi J. S. (1970) On the origin and distribution of meteoroids. *J. Geophys. Res., 75,* 3468–3493.

Eaton N., Davies J. K., and Green S. F. (1984) The anomalous dust tail of comet P/Tempel 2. *Mon. Not. R. Astron. Soc., 211,* 15P–19P.

Fechtig H., Leinert C., and Berg O. E. (2001) Historical perspectives. In *Interplanetary Dust* (E. Grün et al., eds.), pp. 1–55. Springer-Verlag, Berlin.

Flynn G. J. (1989) Atmospheric entry heating: A criterion to distinguish between asteroid and cometary sources of interplanetary dust. *Icarus, 77,* 287–310.

Flynn G. J. (1994) Does the Kuiper Belt contribute significantly to the zodiacal cloud and the stratospheric interplanetary dust? (abstract). In *Lunar and Planetary Science XXV,* pp. 379–380. Lunar and Planetary Institute, Houston.

Frisch P. C., Dorschner J., Geiss J., Greenberg J .M., Grün E., Landgraf M., Hoppe P., Jones A. P., Kratschmer W., Linde T. J., Morfill G. E., Reach W. T., Slavin J., Svestka J., Witt A., and Zank G. P. (1999) Dust in the local interstellar wind. *Astrophys. J., 525,* 492–516.

Fujiwara Y., Ueda M., Sugimoto M., Sagayama T., Satake M., and Furoue A. (2001) TV observations of the 1998 Giacobinid meteor shower in Japan, In *Proc. Meteoroids 2001 Conf.,* pp. 123–127. ESA SP-495, Noordwijk, The Netherlands.

Fulle M. (2004) Motion of cometary dust. In *Comets II* (M. C. Festou et al., eds.), this volume. Univ. of Arizona, Tucson.

Galligan D. P. and Baggaley W. J. (2001) Probing the structure of the interplanetary dust cloud using the AMOR meteoroid orbit radar. In *Proc. Meteoroids 2001 Conf.,* pp. 569–574. ESA SP-495, Noordwijk, The Netherlands.

Greaves J. S., Holland W. S., Moriarty-Schieven G., Jenness T., Dent W. R. F., Zuckerman B., McCarthy C., Webb R. A., Butner H. M., Gear W. K., and Walker H. J. (1998) A dust ring around epsilon Eridani: Analog to the young solar system. *Astrophys. J. Lett., 506,* L133–L137.

Green S. F., Davies J. K., Eaton N., Stewart B. C., and Meadows A. J. (1985) The detection of fast-moving asteroids and comets by IRAS. *Icarus, 64,* 517–527.

Grogan K. , Dermott S. F., and Durda D. D. (2001) The size-frequency distribution of the zodiacal cloud: Evidence from the solar system dust bands. *Icarus, 152,* 251–267.

Grün E., Kissel J., and Hoffmann H.-J. (1976) Dust emission from comet Kohoutek (1973 f) at large distance from the Sun. In *Interplanetary Dust and Zodiacal Light* (H. Elsässer and H. Fechtig, eds.), pp. 334–338. Springer-Verlag, Berlin.

Grün E., Pailer N., Fechtig H., and Kissel J. (1980) Orbital and physical characteristics of micrometeoroids in the inner solar system as observed by Helios 1. *Planet. Space Sci., 28,* 333–349.

Grün E., Zook H. A., Fechtig H., and Giese R. H. (1985) Collisional balance of the meteoritic complex. *Icarus, 62,* 244–272.

Grün E., Gebhard J., Bar-Nun A., Benkhoff J., Düren H., Eich G., Hische R., Huebner W. F., Keller H. U., Klees G., Kochan H., Kölzer G., Kroker H., Kührt E., Lämmerzahl P., Lorenz E., Markiewicz W. J., Möhlmann D., Oehler A., Scholz J., Seidensticker K. J., Roessler K., Schwehm G., Steiner G., Thiel K., and Thomas H. (1993a) Development of a dust mantle on the surface of an insolated ice-dust mixture: Results from the KOSI-9 experiment. *J. Geophys. Res., 98,* 15091–15104.

Grün E., Zook H. A., Baguhl M., Balogh A., Bame S. J., Fechtig H., Forsyth R., Hanner M. S., Horanyi M., Kissel J., Lindblad B.-A., Linkert D., Linkert G., Mann I., McDonnell J. A. M., Morfill G. E., Phillips J. L., Polanskey C., Schwehm G., Siddique N., Staubach P., Svestka J., and Taylor A. (1993b) Discovery of jovian dust streams and interstellar grains by the Ulysses spacecraft. *Nature, 362,* 428–430.

Grün E., Gustafson B. Å. S., Mann I., Baguhl M., Morfill G. E., Staubach P., Taylor A., and Zook H. A. (1994) Interstellar dust in the heliosphere. *Astron. Astrophys., 286,* 915–924.

Grün E., Landgraf M., Horanyi M., Kissel J., Krüger H., Srama R., Svedhem H., and Withnel P. (2000) Techniques for galactic dust measurements in the heliosphere. *J. Geophys. Res., 105,* 10403–10410.

Grün E., Gustafson B. Å. S., Dermott S. F., and Fechtig H., eds. (2001) *Interplanetary Dust.* Springer-Verlag, Berlin. 804 pp.

Gurnett D. A., Averkamp T. F., Scarf F. L., and Grün E. (1986) Dust particles detected near Giacobini-Zinner by the ICE Plasma Wave Instrument. *Geophys. Res. Lett., 13,* 291–294.

Gustafson B. Å. S. (1989) Geminid meteoroids traced to cometary activity on Phaethon. *Astron. Astrophys., 225,* 533–540.

Gustafson B. Å., Misconi N. Y., and Rusk E. T. (1987) Interplanetary dust dynamics. III. Dust released from P/Encke: Distribution with respect to the zodiacal cloud. *Icarus, 72,* 582–592.

Habing J. J., Dominik C., Jourdain de Muizon M., Laureijs R. J., Kessler M. F., Leech K., Metcalfe L., Salama A., Siebenmorgen R., Trams N., and Bouchet P. (2001) Incidence and survival of remnant disks around main-sequence stars. *Astron. Astrophys., 365,* 545–561.

Halliday I. (1988) Geminid fireballs and the peculiar asteroid 3200 Phaeton. *Icarus, 76,* 279–294.

Hauser M. G., Gillett F. C., Low F. J., Gautier T. N., Beichman C. A., Neugebauer G., Aumann H. H., Baud B., Boggess N., Emerson J. P., Houck J. R., Soifer B. T., and Walker R. G. (1984) IRAS observations of the diffuse infrared background. *Astrophys. J. Lett., 278,* L15–L18.

Hauser M. G., Arendt R. G., Kelsall T., Dwek E., Odegard N., Weiland J. L., Freudenreich H. T., Reach W. T., Silverberg R. F., Moseley S. H., Pei Y. C., Lubin P., Mather J. C., Shafer R. A., Smoot G. F., Weiss R., Wilkinson D. T., and Wright E. L. (1998) The COBE Diffuse Infrared Background Experiment search for the cosmic infrared background. I. Limits and detections. *Astrophys. J., 508,* 25–43.

Heap S. R., Lindler D. J., Lanz T. M., Cornett R. H., Robert H., Hubeny I., Maran S. P., and Woodgate B. (2000) Space Telescope Imaging Spectrograph coronagraphic observations of Beta Pictoris. *Astrophys. J., 539,* 435–444.

Hoffmann H.-J., Fechtig H., Grün E., and Kissel J. (1976) Particles from comet Kohoutek detected by the Micrometeoroid Experiment on HEOS 2. In *The Study of Comets* (B. Donn et al., eds.), pp. 949–961. NASA SP-39, Washington, DC.

Holland W. S., Greaves J. S., Zuckerman B., Webb R. A., and McCarthy C. (1998) Submillimetre images of dusty debris around nearby stars. *Nature, 392,* 788–790.

Hong S. S. (1985) Henyey-Greenstein representation of the mean volume scattering phase function for zodiacal dust. *Astron. Astrophys., 146,* 67–75.

Humes D. H. (1980) Results of *Pioneer 10* and *11* meteoroid experiments: Interplanetary and near-Saturn. *J. Geophys. Res., 85,* 5841–5852.

Ishiguro M., Watanabe J., Usui F., Tanigawa T., Kinoshita D., Suzuki J., Nakamura R., Ueno M., and Mukai T. (2002) First detection of an optical dust trail along the orbit of 22P/Kopff. *Astrophys. J. Lett., 572,* L117–L120.

Jenniskens P. (1994) Good prospects for α-Monocerotid outburst in 1995. *WGN, 23,* 84–86.

Jenniskens P. (1995) Meteor stream activity. II. Meteor outbursts. *Astron. Astrophys., 295,* 206–235.

Jenniskens P. (1997) Meteor stream activity. IV. Meteor outbursts and the reflex motion of the Sun. *Astron. Astrophys., 317,* 953–961.

Jenniskens P. (2001) Model of a one-revolution comet dust trail from Leonid outburst observations. *WGN, 29,* 165–175.

Jenniskens P. (2002) More on the dust trails of comet 55P/Tempel-Tuttle from 2001 Leonid shower flux measurements. In *Proceedings of Asteroids, Comets, Meteors — ACM 2002* (B. Warmbein, ed.), pp. 117–120. ESA SP-500, Noordwijk, The Netherlands.

Jenniskens P., Crawford C., Butow S. J., Nugent D., Koop M., Holman D., Houston J., Jobse K., Kronk G., and Beatty K. (2000) Lorentz shaped dust trail cross section from new hybrid visual and video meteor counting technique. *Earth Moon Planets, 82–83,* 191–208.

Jessberger E. K. and Kissel J. (1991) Chemical properties of cometary dust and a note on carbon isotopes. In *Comets in the Post-Halley Era* (R. L. Newburn Jr. et al., eds.), pp. 1075–1092. Kluwer, Dordrecht.

Jessberger E. K., Kissel J., Fechtig H., and Krüger F. R. (1987) On the average chemical composition of cometary dust. In *Physical Processes in Comets, Stars, and Active Galaxies* (W. Hillebrandt et al., eds.), pp. 26–33. Springer-Verlag, Heidelberg.

Joswiak D. J., Brownlee D. E., Pepin R. O., and Schlutter D. J. (2000) Characteristics of asteroidal and cometary IDPs obtained from stratospheric collectors: Summary of measured He release temperatures, velocities, and descriptive mineralogy (abstract). In *Lunar and Planetary Science XXXI,* Abstract #1500. Lunar and Planetary Institute, Houston (CD-ROM).

Kelsall T., Weiland J. L., Franz B. A., Reach W. T., Arendt R. G., Dwek E., Freudenreich H. T., Hauser M. G., Moseley S. H., Odegard N. P., Silverberg R. F., and Wright E. L. (1998) The COBE Diffuse Infrared Background Experiment: Search for the cosmic infrared background. II. Model of the interplanetary dust cloud. *Astrophys. J., 508,* 44–73.

Kissel J. (1986) The Giotto particulate impact analyzer. In *The Giotto Mission — Its Scientific Investigations* (R. Reinhard and B. Battrick, eds.), pp. 67–83. ESA SP-1077, Noordwijk, The Netherlands.

Kissel J. and Krüger F. R. (1987) The organic component in dust from comet Halley as measured by the PUMA mass spectrometer onboard Vega 1. *Nature, 326,* 755–760.

Kissel J., Brownlee D. E., Büchler K., Clark B. C., Fechtig H., Grün E., Hornung K., Igenbergs E. B., Jessberger E. K.,

Krüger F. R., Kuczera H., McDonnell J. A. M., Morfill G. E., Rahe J., Schwehm G. H., Sekanina Z., Utterback N. G., Völk H., and Zook H. A. (1986) Composition of comet Halley dust particles from Giotto observations. *Nature, 321,* 336–338.

Kondrat'eva E. D. and Reznikov E. A. (1985) Comet Tempel-Tuttle and the Leonid meteor swarm. *Solar Sys. Res., 31,* 496–492.

Kresák Ľ. (1993) Cometary dust trails and meteor storms. *Astron. Astrophys., 279,* 646–660.

Kuchner M. J. and Holman M. J. (2003) The geometry of resonant signatures in debris disks with planets. *Astrophys. J., 588,* 1110–1120.

Lagage P.-O. and Pantin E. (1994) Dust depletion in the inner disk of Beta-Pictoris as a possible indicator of planets. *Nature, 369,* 628.

Landgraf M. (2000) Modeling the motion and distribution of interstellar dust inside the heliosphere. *J. Geophys. Res., 105,* 10302–10316.

Landgraf M., Liou J.-C., Zook H. A., and Grün E. (2002) Origins of solar system dust beyond Jupiter. *Astrophys. J., 123,* 2857–2861.

Laureijs R. J., Jourdain de Muizon M., Leech K., Siebenmorgen R., Dominik C., Habing H. J., Trams N., and Kessler M. F. (2002) A 25 micron search for Vega-like disks around main-sequence stars. *Astron. Astrophys., 387,* 285–293.

Leinert C., Richter I., Pitz E., and Planck B. (1981) The zodiacal light from 1.0 to 0.3 AU as observed by the HELIOS space probes. *Astron. Astrophys., 103,* 177–188.

Liou J.-C. and Zook H. A. (1996) Comets as a source of low eccentricity and low inclination interplanetary dust particles. *Icarus, 123,* 491–502.

Liou J.-C. and Zook H. A. (1999) Signatures of the giant planets imprinted on the Edgeworth-Kuiper belt dust disk. *Astron. J., 118,* 580–590.

Liou J. C., Dermott S. F., and Xu Y. L. (1995) The contribution of cometary dust to the zodiacal cloud. *Planet. Space Sci., 43,* 717–722.

Liou J.-C., Zook H. A., and Dermott S. F. (1996) Kuiper Belt dust grains as a source of interplanetary dust particles. *Icarus, 124,* 429–440.

Lisse C. (2002) On the role of dust mass loss in the evolution of comets and dusty disk systems. *Earth Moon Planets, 90,* 497–506.

Love S. G. and Brownlee D. E. (1994) Peak atmospheric entry temperature of meteorites. *Meteoritics, 29,* 69–70.

Love S. G., Joswiak D. J., and Brownlee D. E. (1993) Densities of 5–15 microns interplanetary dust particles (abstract). In *Lunar and Planetary Science XXIV,* pp. 901–902. Lunar and Planetary Institute, Houston.

Low F. J., Beitema D. A., Gautier T. N., Gillett F. C., Beichman C. A., Neugebauer G., Young E., Aumann H. H., Boggess N., Emerson J. P., Habing H. J., Hauser M. G., Houck J. R., Rowan-Robinson M., Soifer B. T., Walker R. G., and Wesselius P. R. (1984) Infrared cirrus: New components of the extended infrared emission. *Astrophys. J. Lett., 278,* L19–L22.

Lowry S. C., Weissman P. R., Sykes M. V., and Reach W. T. (2003) Observations of periodic comet 2P/Encke: Properties of the nucleus and first visual-wavelength detection of its dust trail (abstract). In *Lunar and Planetary Science XXXIV,* Abstract #2056. Lunar and Planetary Institute, Houston (CD-ROM).

Lyytinen E. J. (1999) Meteor predictions for the years 1999–2007 with the satellite model of comets. *Meta Res. Bull., 8,* 33–40.

Marsden B. G. (1974) Comets. *Annu. Rev. Astron. Astrophys., 12,* 1–21.

Mazets E. P., Sagdeev R. Z., Aptekar R. L., Golenetskii S. V., Guryan Y. A., Dyachkov A. V., Ilyinskii V. N., Panov V. N., Petrov G. G., Savvin A. V., Sokolov I. A., Frederiks D. D., Khavenson N. G., Shapiro V. D., and Shevchenko V. I. (1987) Dust in comet P/Halley from VEGA observations. *Astron. Astrophys., 187,* 699–706.

McBride N., Green S. F., Levasseur-Regourd A. C., Goidet-Deve B., and Renard J.-B. (1977) The inner dust coma of comet 26P/Grigg-Skjellerup: Multiple jets and nucleus fragments? *Mon. Not. R. Astron. Soc., 289,* 535–553.

McDonnell J. A. M. and Gardner D. J. (1998) Meteorite morphology and densities: Decoding satellite impact data. *Icarus, 133,* 25–35.

McDonnell J. A. M., Alexander W. M., Burton W .M, Bussoletti E., Evans G. C., Evans S. T., Firth J. G., Grard R. J. L., Green S. F., Grün E., Hanner M. S., Hughes D. W., Igenbergs E., Kissel J., Kuczera H., Lindblad B. A., Langevin Y., Mandeville J. C., Pankiewicz G. S. A., Perry C. H., Schwehm G., Sekanina Z., Stevenson T. J., Turner R. F., Weishaupt U., Wallis M. K., and Zarnecki J. C. (1987) The dust distribution within the inner coma of comet P/Halley (1982i): Encounter by Giotto's impact detectors. *Astron. Astrophys., 187,* 719–941.

McDonnell J. A. M., Lamy P. L., and Pankiewicz G. S. (1991) Physical properties of cometary dust. In *Comets in the Post-Halley Era* (R. L. Newburn Jr. et al., eds.), pp. 1043–1073. Kluwer, Dordrecht.

McDonnell J. A. M., McBride N. M., Beard R., Bussoletti E., Colangeli L., Eberhardt P., Firth J. G., Grard R., Green S. F., Greenberg J. M., Grün E., Hughes D. W., Keller H. U., Kissel J., Lindblad B. A., Mandeville J. C., Perry C. H., Rembor K., Rickman H., Schwehm G. H., Roessler K., Schwehm G., Turner R. F., Wallis M. K., and Zarnecki J. C. (1993) Dust particle impacts during the Giotto encounter with comet Grigg-Skjellerup. *Nature, 362,* 732–734.

McNaught R. H. and Asher D. J. (1999) Leonid dust trails and meteor storms. *WGN, 27,* 85–102.

Minor Planet Center (2003) *The Distribution of Minor Planets.* Available on line at http://cfa-www.harvard.edu/iau/lists/MPDistribution.html.

Moro-Martín A. and Malhotra R. (2003) Dynamical models of Kuiper Belt dust in the inner and outer solar system. *Astron. J., 125,* 2255–2265.

Nagel K. and Fechtig H. (1980) Diameter to depth dependence of impact craters. *Planet. Space Sci., 28,* 567–573.

Nesvorný D., Bottke W. F., Levison H., and Dones L. (2002) The recent breakup of an asteroid in the main-belt region. *Nature, 417,* 720–722.

Nesvorný D., Levison H. F., Bottke W. F., and Dones L. (2003) Recent origin of the solar system dust bands. *Astrophys. J., 591,* 486–497.

Neugebauer G., Habing H. J., van Duinen R., Aumann H. H., Baud B., Beichman C. A., Beintema D. A., Boggess N., Clegg P. E., de Jong T., Emerson J. P., Gautier T. N., Gillett F. C., Harris S., Hauser M. G., Houck J. R., Jennings R. E., Low F. J., Marsden P. L., Miley G., Olnon F. M., Pottasch S. R., Raimond E., Rowan-Robinson M., Soifer B. T., Walker R. G., Wesselius P. R., and Young E. (1984) The Infrared Astronomical Satellite (IRAS) mission. *Astrophys. J. Lett., 278,* L1–L6.

Öpik E. (1951) Astronomy and the bottom of the sea. *Irish Astron. J., 1,* 145–158.

Pailer N. and Grün E. (1980) The penetration limit of thin films.

Planet. Space Sci., 28, 321–331.

Quillen A. C. and Thorndike S. (2002) Structure in the Epsilon Eridani dusty disk caused by mean motion resonances with a 0.3 eccentricity planet at periastron. *Astrophys. J. Lett., 578*, L149–L152.

Rabinowitz D. and Scotti J. (1991) *Periodic Comet Faye (1991n).* IAU Circular No. 5366.

Reach W. T. (1988) Zodiacal emission. I. Dust near the Earth's orbit. *Astrophys. J., 335*, 468–485.

Reach W. T. (1992) Zodiacal emission. III. Dust near the asteroid belt. *Astrophys. J., 392*, 289–299.

Reach W. T., Abergel A., Boulanger F., Desert F.-X., Perault M., Bernard J.-P., Blommaert J., Cesarsky C., Cesarsky D., Metcalfe L., Puget J.-L., Sibille F., and Vigroux L. (1996) Mid-infrared spectrum of the zodiacal light. *Astron. Astrophys., 315*, L381–L384.

Reach W. T., Franz B. A., and Weiland J. L. (1997) The three-dimensional structure of the zodiacal dust bands. *Icarus, 127*, 461–484.

Reach W. T., Sykes M. V., Lien D., and Davies J. K. (2000) The formation of Encke meteoroids and dust trail. *Icarus, 148*, 80–94.

Reach W. T., Morris P., Boulanger F., and Okumura K. (2003) The mid-infrared spectrum of the zodiacal and exozodiacal light. *Icarus, 164*, 384–403.

Silverberg R. F., Hauser M. G., Boggess N. W., Kelsall T. J., Moseley S. H., and Murdock T. L. (1993) Design of the Diffuse Infrared Background Experiment (DIRBE) on COBE. In *Infrared Spaceborne Remote Sensing* (M. S. Scholl, ed.), pp. 180–189. SPIE Conference Proceedings Vol. 2019, Tempe, Arizona.

Simpson J. A., Rabinowitz D., Tuzzolino A. J., Ksanfomality L. V., and Sagdeev R. Z. (1987) The dust coma of comet Halley: Measurements on the VEGA-1 and 2 spacecraft. *Astron. Astrophys., 187*, 742–752.

Srama R., Grün E., and the Cassini Dust Science Team (1996) The Cosmic Dust Analyzer for the Cassini mission to Saturn. In *Physics, Chemistry, and Dynamics of Interplanetary Dust* (B. Å. S. Gustafson and M. S. Hanner, eds.), pp. 227–231. ASP Conference Series 104, Astronomical Society of the Pacific, San Francisco.

Srama R., Bradley J. G., Grün E., Ahrens T. J., Auer S., Cruise M., Fechtig H., Graps A., Havnes O., Heck A., Helfert S., Igenbergs E., Jessberger E. K., Johnson T. V., Kempf S., Krüger H., Lamy P., Landgraf M., Linkert D., Lura F., McDonnell J. A. M., Mohlmann D., Morfill G. E., Schwehm G. H., Stübig M., Svestka J., Tuzzolino A. J., Wäsch R., and Zook H. A. (2004) The Cassini Cosmic Dust Analyser. *Space Sci. Rev.*, in press.

Stern S. A. (1996) Signatures of collisions in the Kuiper disk. *Astron. Astrophys., 310*, 999–1010.

Stewart B. C., Davies J. K., and Green S. F. (1984) IRAS fast mover program. *J. Brit. Interplan. Soc., 37*, 348–352.

Sykes M. V. (1987) The albedo of large refractory particles in P/ Tempel 2. *Bull. Am. Astron. Soc., 19*, 893–894.

Sykes M. V. (1988) IRAS observations of extended zodiacal structures. *Astrophys. J. Lett., 334*, L55–L58.

Sykes M. V. (1990) Zodiacal dust bands: Their relation to asteroid families. *Icarus, 84*, 267–289.

Sykes M. V. and Greenberg R. (1986) The formation and origin of the IRAS zodiacal dust bands as a consequence of single collisions between asteroids. *Icarus, 65*, 51–69.

Sykes M. V. and Walker R. G. (1992) Cometary dust trails. I. Survey. *Icarus, 95*, 180–210.

Sykes M. V., Lebofsky L. A., Hunten D. M., and Low F. (1986) The discovery of dust trails in the orbits of periodic comets. *Science, 232*, 1115–1117.

Sykes M. V., Lien D. J., and Walker R. G. (1990) The Tempel 2 dust trail. *Icarus, 86*, 236–247.

Sykes M. V., Cutri R., Moynihan P., and Plath J. (1994) A parallactic mini-survey of the infrared sky. *Bull. Am. Astron. Soc., 26*, 1120.

Tsurutani B. T., Clay D. R., Zhang L. D., Dasgupta B., Brinza D., Henry M., Mendis A., Moses S., Glassmeier K.-H., Musmann G., and Richter I. (2003) Dust impacts at comet P/Borrelly. *Geophys. Res. Lett., 30(22)*, SSC 1-1 to 1-4.

Vaisberg O. L., Smirnov V., Omel'Chenko A., Gorn L., and Iovlev M. (1987) Spatial and mass distribution of low-mass dust particles ($m < 10^{-10}$ g) in comet P/Halley's coma. *Astron. Astrophys., 187*, 753–760.

Vedder J. F. and Mandeville J.-C. (1974) Microcraters formed in glass by projectiles of various densities. *J. Geophys. Res., 79*, 3247–3256.

Walker R. G., Sykes M. V., and Lien D. J. (1989) Thermal properties of dust trail particles. *Bull. Am. Astron. Soc., 21*, 967.

Weinberger A. J., Becklin E. E., and Zuckerman B. (2003) The first spatially resolved mid-infrared spectroscopy of β Pictoris. *Astrophys. J. Lett., 584*, L33–L37.

Werner M. W., Reach W. T., and Rieke M. (2001) Studies of the cosmic infrared background with the Space Infrared Telescope Facility (SIRTF). In *The Extragalactic Infrared Background and Its Cosmological Implications* (M. Harwit, ed.), p. 439. IAU Symposium 204.

Whipple F. L. (1955) A comet model. III. The zodiacal light. *Astrophys. J., 121*, 750–770.

Wilner D. J., Holman M. J., Kuchner M. J., and Ho P. T. P. (2002) Structure in the dusty debris around Vega. *Astrophys. J. Lett., 569*, L115–L119.

Wyatt M. C. and Dent W. R. F. (2002) Collisional processes in extrasolar planetesimal discs dust clumps in Fomalhaut's debris disc. *Mon. Not. R. Astron. Soc., 334*, 589–607.

Wyatt S. P. and Whipple F. L. (1950) The Poynting-Robertson effect on meteor orbits. *Astrophys. J., 111*, 134–141.

Yamamoto S. and Mukai T. (1998) Dust production by impacts of interstellar dust on Edgeworth-Kuiper Belt objects. *Astron. Astrophys., 329*, 785–791.

Yeomans D. K. (1991) *Comets: A Chronological History of Observation, Science, Myth, and Folklore.* Wiley, New York. 485 pp.

Laboratory Experiments on Cometary Materials

L. Colangeli and J. R. Brucato
Istituto Nazionale di Astrofisica–Osservatorio Astronomico di Capodimonte

A. Bar-Nun
Tel Aviv University

R. L. Hudson
Eckerd College and NASA Goddard Space Flight Center

M. H. Moore
NASA Goddard Space Flight Center

Laboratory experiments to simulate cometary materials and their processing contribute to the investigation of the properties and evolution of comets. Experimental methods can produce both refractory materials and frozen volatiles with chemical, structural, and morphological characteristics that reproduce those of materials observed and/or expected in comets. Systematic analyses of such samples, before and after energetic processing by various agents effective in the solar system, provide a wealth of useful quantitative information. Such data permit a more complete interpretation of observations, performed remotely or *in situ*, and suggest ideas about the chemical and physical evolution of cometary dust and ice. Finally, laboratory results help to predict the environmental conditions that future space missions, such as the European Space Agency's *Rosetta* mission, will experience, and thus aid in properly planning mission and instrument development.

1. INTRODUCTION

Comets are considered to be reservoirs of partially uncontaminated primordial material from which the solar system formed about 4.5×10^9 yr ago. The composition as well as the physical and structural properties of cometary dust and ice depend both on comet formation mechanisms and postaccretion evolutionary processes. The so-called "interstellar grain" model (e.g., *Greenberg and Hage,* 1990; *Mumma,* 1997; *Notesco et al.,* 2003) supports the concept that comets formed far (>20 AU) from the proto-Sun, at low temperatures (<100 K or so), so that their composition should reflect that of original interstellar cloud grains. In contrast, the "nebular chemistry" model includes the possibility that interstellar material may have been reprocessed prior to cometary formation (e.g., *Lunine,* 1989). The resulting cometary chemistry is different in these two scenarios, mainly due to the ice condensation temperature. However, it is possible that cometary chemistry reflects both the presence of original interstellar grains and reprocessed materials (*Engel et al.,* 1990). Details about comet formation and evolution are described by *Dones et al.* (2004).

The internal structure of comets also depends on the dynamic evolution of the protosolar nebula. As settling toward the midplane of the nebula occurred, did infalling material collapse into kilometer-sized planetesimals (*Goldreich and Ward,* 1973), or did it accumulate into 50–100-m aggregates (*Weidenschilling,* 1997)? The "rubble pile" model would be acceptable in the first case (*Weissman,* 1986), but a different preferred size scale should exist in the latter [see *Weissman et al.* (2004) for more details about comet nuclear structure].

The exploration of Comet 1P/Halley provided a breakthrough in the understanding of cometary structure and composition, thanks to close observations of Halley's nucleus (see *Keller et al.,* 2004). More recently, remote observations, such as with the Infrared Space Observatory (ISO), have provided valuable insights into cometary chemistry. Major progress has concerned the identification of a large variety of volatile molecules (*Bockelée-Morvan et al.,* 2004) and a deeper characterization of the refractory components of cometary and interstellar dust, especially silicates (*Hanner and Bradley,* 2004). This new information has allowed us to forge strong links between comet composition and interstellar dust evolution (*Ehrenfreund et al.,* 2004).

Even with such new observations, uncertainties remain concerning the properties and evolution of materials that form comets. Laboratory experiments now play a fundamental role in research programs designed to reveal the main components of comets and the effects of energetic processing experienced by cometary ice and dust. The experimental program applied to ice and dust investigations is based on three main steps: (1) production of materials by techniques that allow control of product characteristics; (2) analyses of dust and ice analogs by complementary methods aimed at quantitative studies of morphology, structure, chemistry, and optical behavior; and (3) processing to reproduce effects expected in space before, during, and after comet formation.

The results of such experiments are extremely useful for quantitative interpretation of astronomical observations. By studying dust and ice properties as a function of formation conditions and levels of energetic processing, light can be shed on the formation and evolution of materials in the solar system in general and comets in particular. This approach not only complements observations, but it has predictive power for observers and provides important constraints for theoretical models.

This chapter provides an overview of past and ongoing experimental work. Section 2 is devoted to refractory materials (silicates and carbons) observed in comets. Production and analytical techniques are briefly reviewed, with attention given to the determination of optical properties and the effects of processing. The information derived from experiments aid in identifying major components of comets. Section 3 covers small- and large-scale experiments on the physical properties of ices, while the chemical evolution of ices is described in section 4. The most widely used *in situ* techniques are described, and some of the most relevant recent results are summarized. Some conclusions are given in section 5.

2. EXPERIMENTS ON REFRACTORY MATERIALS

Refractory materials with morphological, chemical, and structural properties suitable to reproduce compounds observed or expected in comets have been produced in the laboratory through various methods of synthesis (e.g., *Colangeli et al.,* 1995; *Rotundi et al.,* 2002; *Nuth et al.,* 2002) and are usually termed cometary dust analogs (hereafter CDAs).

2.1. Production and Characterization Through Spectroscopy of Materials

A wide variety of C- and silicon-based materials are obtained by vapor condensation to form "smokes" (see, e.g., *Colangeli et al.,* 2003, and references therein). In practice, vaporization is achieved by applying sufficient energy to a solid by, e.g., laser bombardment of a homogeneous target or a mixture of different targets (Fig. 1), by arc discharge between C or graphite electrodes (Fig. 1), or by laser pyrolysis in a gas flow. If a metal vapor is present, at its saturation pressure, in a cooling gas-phase mixture, then as the temperature falls molecular clusters can form, which then grow to solid particles. Alternatively, chemical processes, such as sol-gel reactions (*Brinker and Scherer,* 1990; *Jäger et al.,* 2003) or grinding of natural rocks and minerals, can be used to produce a wide variety of CDAs.

A careful selection of experimental conditions during CDA formation allows the "tuning" of the composition and structure of the samples produced. Examples of CDAs produced in the laboratory are summarized in Table 1. It is interesting to note that a wide variety of silicates, in addition to those in Table 1, can be obtained by changing the relative abundance of cations (e.g., Mg^{2+}, Fe^{2+}, Al^{3+}, Ca^{2+}) in the original reaction mixture.

The materials produced in laboratory must be carefully analyzed to check that they are reasonable CDAs and to link their properties to production conditions. A large variety of investigation techniques are based on the interaction of matter with radiation and particles, while other methods are based on the study of mechanical, electrical, magnetic, and thermal properties of matter. All the methods involving interaction of matter with radiation can be classified according to the kind of interaction in microscopy, diffractometry,

Fig. 1. Methods for condensation of solid grains used at the Cosmic Physics Laboratory of Naples. **(a)** Nd-YAG pulsed laser device for ablation of solid samples in an oxidizing, reducing, or inert atmosphere. **(b)** Device for production of C dust by arc discharge between C or graphite rods in H-rich or inert atmosphere.

TABLE 1.   Examples of laboratory cometary dust analogs.

Family	Species	Method of Production
Olivine ($Mg_xFe_{1-x})_2SiO_4$	Forsterite (x = 1)	Laser bombardment in 10 mbar O_2
	Fayalite (x = 0)	Laser bombardment in 10 mbar O_2
Pyroxene ($Mg_xFe_{1-x})SiO_3$	Enstatite (x = 1)	Laser bombardment in 10 mbar O_2
	Ferrosilite (x = 0)	Laser bombardment in 10 mbar O_2
Amorphous C	α-C	Arc discharge in 10 mbar Ar
Hydrogenated amorphous C	HAC	Arc discharge in 10 mbar H_2
		Laser bombardment in 10 mbar H_2
		UV irradiation or ion bombardment of ice mixtures

and spectroscopy. Methods falling in these categories give a vast amount and different kinds of information, unraveling the relations among the properties of solid materials [see *Marfunin* (1995) for a wide and accurate description on methods of investigation on refractory materials]. Many of these techniques are widely used in laboratories to characterize CDAs.

Scanning, transmission, and analytical electron microscopy are used to investigate the morphology and both the short- and long-range structural order of CDAs at nanometer and subnanometer scales (e.g., *Rietmeijer et al.,* 2002; *Fabian et al.,* 2000). These techniques use an electron beam accelerated by high voltage and focused by lenses on the sample surface. The beam is then moved or scanned across the sample area to obtain images providing information on surface structures and morphology (Fig. 2).

Electron and X-ray diffraction are the most frequently used methods to determine the arrangement of atoms in the crystal structure of samples at small and large scales. The analysis is based on the theories of symmetry and on the interaction of radiation with solids. The accuracy of the

results of the analysis depends on both sample quality and technique. In particular, X-ray absorption spectroscopy (XAS), extended X-ray absorption fine structure (EXAFS), and X-ray absorption near-edge structure (XANES) techniques are used to determine short- and medium-range order in partially amorphous materials (e.g., *Thompson et al.,* 1996). Elemental composition can be determined by analysis of dispersed X-rays (Fig. 2), while specific aspects of elemental arrangement, such as the ratio of ferrous to ferric iron, can be investigated by wet chemical analyses (*Köster,* 1979).

Raman spectroscopy is another method used to analyze the structure of materials, especially C-based ones. It is based on the process of inelastic scattering of monochromatic radiation hitting the target sample. The Raman scattering consists in the frequency shift of the Raman lines with respect to the exciting radiation and results from the energy exchange between the exciting radiation and the vibrational levels of the materials. Raman scattering is used to identify minerals embedded in matrix in a nondestructive way and is suitable to investigate the structure of minerals by the analysis of the symmetry of vibrational modes. Vibra-

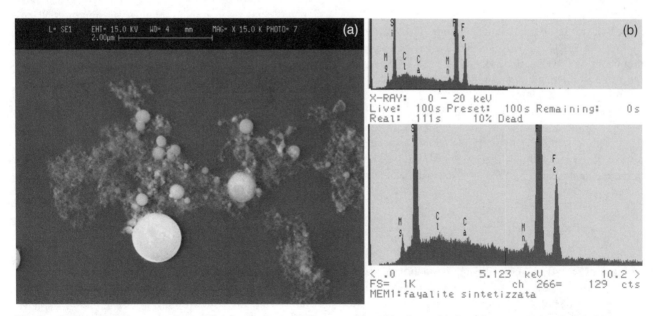

Fig. 2.   **(a)** Scanning electron micrograph of amorphous olivine as produced by laser ablation of a pure olivine target. **(b)** Energy dispersive X-ray analysis of amorphous olivine.

tional spectra are highly sensitive to the degree of order of materials. Order/disorder degrees of carbonaceous materials synthesized in the laboratory or present in meteorites and interplanetary dust particles can be investigated with respect to the processing experienced by materials during their life as, e.g., ion irradiation (*Wopenka*, 1988; *Strazzulla and Palumbo*, 2001; *Baratta et al.*, 2004).

Although the techniques described above are fundamental tools for material characterization, spectroscopy remains the most used and powerful way to investigate CDAs. In fact, different aspects of materials can be revealed by different wavelengths of light. Measurements in the vacuum ultraviolet probe electronic transitions of solids, while spectroscopy in the visible region allows the identification of the electronic gap and thus the conduction properties of materials. Midinfrared (IR) light is in the range where molecular vibrational resonances can be excited by incoming radiation, while material structure and morphology drive the spectral behavior in the far-IR region. Therefore, for the study of material characteristics at all scales, a careful investigation over a wide spectral range provides an important complement to information from other analytical techniques. Moreover, results obtained in laboratory by spectroscopic analysis of materials can be used in direct comparison with astronomical observations or by modeling the spectroscopic behavior of dust grains. Details on use of laboratory spectra are described in the following sections.

2.1.1. Extinction and absorption. Considering the extinction and absorption of IR light by dust grains, the dependence of the mass extinction coefficient, $K(\lambda)$, on wavelength, λ, can be derived from measurements on CDAs synthesized in laboratory. The equation

$$K(\lambda) = \frac{S}{M} \ln\left[\frac{1}{T_p(\lambda)}\right]$$

links $K(\lambda)$ to the measured transmittance $T_p(\lambda)$ of a CDA sample of mass M and cross section S, exposed to a radiation beam of wavelength λ. Here $T_p(\lambda) = I(\lambda)/I_0(\lambda)$ is the ratio of the intensities of a light beam after and before meeting a dust sample.

The above relation for $K(\lambda)$ is valid under the assumption that grains interact as separate entities with light, so that multiple scattering is negligible (in other words, the distance between particles is greater than the wavelength of the incident radiation). When the size parameter x satisfies the relation

$$x = \frac{2\pi a}{\lambda} \ll 1$$

with a = grain radius (under the approximation of spherical shape), the scattering contribution to extinction is negligible (see *Bohren and Huffman*, 1983), and $K(\lambda)$ becomes the mass absorption coefficient. Thus the absorption coefficient can be derived directly from transmission IR measurements on submicrometer grains. Three examples of mass

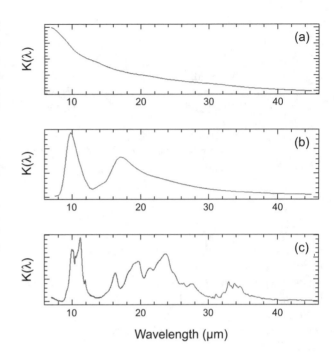

Fig. 3. Mass absorption coefficient (in arbitrary units) of **(a)** amorphous C, **(b)** amorphous olivine, and **(c)** crystalline olivine (*Brucato et al.*, 1999b).

absorption efficiency in the IR for silicates and C-based materials are shown in Fig. 3.

The motivation for determining these mass absorption coefficients is that they can be used to extract information from absorption and emission observations of comets. If scattering is negligible, then the mass column density, β, and the temperature, T, of grains can be obtained under the assumption of equilibrium conditions, for which Kirchhoff's law applies (emittance = absorbance). The cometary emission flux, $F(\lambda)$, is then interpreted as the sum of contributions from N components as

$$F(\lambda) = \sum_{i=1}^{N} \beta_i \cdot K_i \cdot B(\lambda, T_i)$$

where $B(\lambda, T)$ is the Planck function. By fitting $F(\lambda)$ to cometary spectra (see section 2.3), one obtains quantitative information on cometary coma grain properties. This procedure is applicable only if laboratory measurements on $K(\lambda)$ for CDAs are available.

2.1.2. Scattering. Important information on solid materials is provided by an analysis of the light scattering process. Size, shape, and refractive index of particles are intimately correlated to the scattering parameters. Exact theories of light scattering are based on solving Maxwell's equations either analytically or numerically. In the case of spherical homogeneous particles, calculations can be done accurately by using the Mie theory (*Bohren and Huffman*, 1983). Although this simple approximation is used in many applications, real (cometary) dust particles are inhomogeneous and irregular in shape.

To give an accurate description of the phenomenon for irregular particles, the scattering properties have to be either computed theoretically or measured experimentally. Theoretical models, which have become increasingly sophisticated (see, e.g., *Mishchenko et al.*, 2000), use the optical constants of materials and take into account the dependence of the scattering function on morphological properties, such as size and shape. In recent years an effort has been devoted to describe the effects of nonsphericity on scattering properties. In this frame, particles have been classified as solids, aggregates, and clustered particles. *Lumme et al.* (1997) identified particle classes in a mathematical sense as polyhedral solids, stochastically rough (smooth) particles, and stochastic aggregates. Statistical approaches have been used for particles very small compared to the wavelength of radiation (Rayleigh limit), but complications have been encountered in calculations outside the Rayleigh limit. The most important calculation methods can be summarized in two categories: exact theories (separation of variables method for spheroid, finite-element method, finite-difference time domain method, integral-equation methods, discrete dipole approximation, T-matrix approach, superposition method for compounded spheres and spheroids) and approximate theories (Rayleigh, Rayleigh-Gans, anomalous diffraction approximations, geometric optics approximation, perturbation theory). Details on mathematical description of models and their limits and goals are given by *Mishchenko et al.* (2000).

Complementary information that can lead to much improved knowledge of electromagnetic scattering by nonspherical particles is given by laboratory measurements. Randomly oriented ensembles of CDAs with irregular shapes are used in laboratory experiments. The Stokes vectors of the incident and scattered beams are related by a scattering matrix, whose elements are measured vs. the scattering angles (*van de Hulst*, 1957). Usually, measurements in the visible, IR, and microwave spectral ranges are performed. The accurate characterization of CDAs size, shape, and composition, i.e., performed by the analytical techniques described above, allows us to know fundamental parameters that are linked to the scattering parameters (see, e.g., *Hovenier and van der Mee*, 2000; *Muñoz et al.*, 2000; *Volten et al.*, 2001; *Hadamcik et al.*, 2002). A different experimental approach in studying the light-scattering properties of submicrometer grains is to measure microwave scattering by large analog grain samples (*Gustafson*, 1996). In fact, no absolute dimensions are encountered in classical electrodynamics and convenient grain size and electromagnetic wavelength can be chosen. The light-scattering problem can therefore be scaled up or down to any convenient dimension.

The experimental approach can be used to check the quality of results obtained by scattering models. Moreover, measurements are suitable for the direct interpretation of astronomical observations. On this point, we recall that observations of light scattered by particles present in the cometary coma show a linear polarization. The measured phase curves display a negative branch for phase angles lower than 20°, while a positive branch is observed at larger angles with a maximum around 90°–100° (*Levasseur-Regourd et al.*, 1996). Changes in the polarization intensity are observed as a function of the distance from the nucleus, which are attributed to differences in the size distribution and/or color of cometary grains (*Muñoz et al.*, 2000) or to different degrees of grain fluffiness (*Hadamcik et al.*, 2002). The subject is presently matter of further laboratory investigation by using different classes of CDAs (H. Volten, personal communication, 2003).

2.2. Processing of Cometary Dust Analogs

Laboratory CDAs can be energetically processed by different methods to study their evolution and the efficiency of extraterrestrial processes to modify the material's optical, structural, and chemical properties. We consider two examples.

Silicates produced by condensation techniques are subjected to thermal annealing under vacuum, for specific temperatures and times, in order to study variations induced by this process on physical, chemical, and structural properties of the samples. The most striking effects are structural modifications that tend to crystallize amorphous materials (*Hallenbeck et al.*, 1998, 2000; *Brucato et al.*, 1999a, 2002; *Fabian et al.*, 2000). Infrared spectroscopy is an efficient method for monitoring changes caused by thermal annealing. Wide IR absorption bands, typical of amorphous materials, tend to sharpen, as expected for crystalline solids. Such amorphous-to-crystalline transitions can be quantitatively characterized through the activation energy, E_a, defined by

$$t = \nu^{-1}\exp\left(\frac{E_a}{kT}\right)$$

In this equation, t is the time for the material to reach long-range order inside the lattice, ν is a characteristic vibrational frequency of the material, and T is the temperature. Activation energies have been measured in the laboratory for a wide variety of materials that are relevant as CDAs. The results obtained for different silicate species are summarized in Table 2. The implications of these results on the evolution of silicates forming comets will be discussed in the next section.

Both pure-C and H-rich C grains can be produced by the evaporation techniques mentioned in section 2.1. Laboratory experiments on such grains have shown that thermal annealing (*Mennella et al.*, 1995), UV irradiation (*Mennella et al.*, 1996), ion bombardment (*Mennella et al.*, 1997), and interaction with gas (*Mennella et al.*, 1999) are competitive processes that determine chemical and structural transformations.

Among other diagnostic features, infrared C-H stretching modes in the 3.3–3.4-μm range, are suitable both for identifying the amount of H linked to the C structure and for disentangling the most important C structural arrangements. In fact, the IR band intensity is related to the H abundance, with dominant features around 3.3 or 3.4 μm, indicative of

TABLE 2.   Activation energies E_a/k (K) for crystallization.

Composition	Hallenbeck et al. (1998)	Brucato et al. (1999a)	Fabian et al. (2000)	Brucato et al. (2002)
SiO_2	—	—	49,000	—
$MgSiO_3$	—	47,500	42,040	—
Mg_2SiO_4	45,500	—	39,100	40,400
Fe_2SiO_4	—	—	—	26,300
$MgO\text{-}SiO_2$	—	—	—	45,800
$MgO\text{-}SiO_2\text{-}Fe_2O_3$	—	—	—	<49,700

aromatic (graphitic) vs. aliphatic networks, respectively, forming the grains.

Results of experiments can be summarized as follows: (1) Bombardment of pure amorphous C grains with about 10^{20} H atoms cm^{-2} produces the appearance of a neat aliphatic 3.4-μm band (*Mennella et al.,* 1999). (2) Processing of hydrogenated amorphous C grains by UV irradiation (Lyman emission) or ion bombardment produces a progressive release of H and a transition from an aliphatic-dominated to a more aromatic material (*Mennella et al.,* 1996, 1997). (3) Irradiation of hydrogenated amorphous C grains at fluences of about 10^{19} UV photons cm^{-2} produces a significant decline of the 3.4-μm band, even when the dust is coated by Ar, H_2O, or $H_2O\text{-}CO\text{-}NH_3$ ices (*Mennella et al.,* 2001).

It must be also noticed that UV irradiation and ion bombardment of organic materials produce similar effects, loss of function groups and polymerization (*Jenniskens et al.,* 1993) (see also section 4).

These results provide guidance for following the evolution of C-based materials in different space environments, including comets, as discussed in the next section.

2.3.   Interpretation of Cometary Observations by Laboratory Data and Future Steps

The presence of crystalline silicates in comets was first shown by observing the IR spectrum of Comet P/Halley 1986 III. A strong 11.3-μm emission peak was attributed to crystalline olivine grains (*Campins and Ryan,* 1989). A double peak at 9.8 and 11.3 μm indicated that amorphous and crystalline silicates were coexisting components. However, groundbased observations of other comets showed different shapes for the emission features (*Hanner et al.,* 1994). Based on laboratory measurements (e.g., *Colangeli et al.,* 1996), these differences were attributed to the presence of different silicates, whose origins could be traced to different formation conditions and to transformations by various processing mechanisms (Table 3).

The 1997 passage of Comet Hale-Bopp C/1995 O1 provided the first opportunity to observe a new long-period Oort cloud comet, both from the ground and from space. The ISO satellite allowed an examination of cometary grain emission over the full IR range (*Crovisier et al.,* 1997), and led to the discovery of a rich variety of strong distinct emis-

TABLE 3.   Mass percentage of submicrometer grains derived from fitting
cometary spectra with laboratory absorption data of different materials.

Comet	AU	Crystalline Olivine	Amorphous Olivine	Amorphous Pyroxene	Crystalline Pyroxene	Amorphous Carbon
C/1989 Q1 Okazaki-Levy-Rudenko	0.65*	28	—	—	—	72
C/1989 X1 Austin	0.78*	12	—	—	—	88
1P/1982 U1 Halley	0.79*	33	—	—	52	15
C/1987 P1 Bradfield	0.99*	42	—	—	48	10
C/1983 H1 IRAS-Araki-Alcock	1.0*	—	85	—	—	15
1P/1982 U1 Halley	1.25*	22	—	—	20	58
C/1987 P1 Bradfield	1.45*	24	—	—	23	53
C/1990 K1 Levy	1.51*	20	36	—	—	44
C/1990 K1 Levy	1.56*	19	47	—	—	34
C/1994 E1 Mueller	2.06*	23	—	—	30	48
C/1995 O1 Hale-Bopp	0.97‡	30	17	25	8	20
	1.21‡	23	18	37	4	18
	1.7‡	25	14	33	8	21
	2.8‡	25	25	24	5	21
	2.9†	69	20	—	—	11

Colangeli et al. (1996), spectral range 8–13 μm.
†*Brucato et al.* (1999a), spectral range 7–45 μm.
‡*Harker et al.* (2002), spectral range 1.2–45 μm.

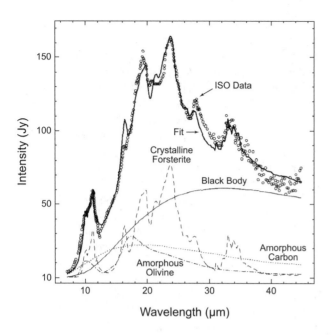

Fig. 4. Fit of ISO observation of Comet Hale-Bopp [circles, *Crovisier et al.* (1997)] with the optical properties of a combination of amorphous and crystalline materials measured in the laboratory (*Brucato et al.,* 1999b).

sion features. Hale-Bopp's IR spectrum was interpreted as due to amorphous and crystalline silicates and amorphous C grains in the coma. A detailed match of all major peaks with laboratory data for silicate grains suggested the identification of crystalline Mg-rich olivine (forsterite) as one of the main components (Fig. 4). Groundbased Hale-Bopp observations at different heliocentric distances (e.g., *Hayward and Hanner,* 1997; *Wooden et al.,* 1999) indicated that a further component of crystalline Mg-rich pyroxene (enstatite) was also present (Table 3).

Different silicate band profiles and peak positions in different comets may be correlated to comet formation and evolution. According to present evolution models, silicates coming from the presolar cloud and infalling onto the protosolar nebula were amorphous. The results obtained from laboratory simulations indicate that amorphous-to-crystalline transformations can only occur on timescales under 10^6 yr if annealing temperatures above ~800 K are reached. It is, therefore, unlikely that amorphous silicates were converted into crystalline materials in the outer nebula, where comets are supposed to have been formed, as the temperature was too low (few tens of Kelvins) to thermally process grains before their incorporation in cometary bodies. A subsequent thermal processing at high temperatures would have been necessary to crystallize them.

Two possible scenarios have been recently proposed to explain the presence of crystalline grains in comets: turbulent radial mixing in the solar nebula (*Bockelée-Morvan et al.,* 2002) and annealing of dust by nebular shocks (*Harker and Desch,* 2002). It has been demonstrated that a flash

heating at 1100 K is sufficient to crystallize micrometer-sized particles in few minutes, as has been proposed to happen for precursors of meteoritic chondrules (prechondrules) (*Rietmeijer,* 1998). It has also been suggested that the observed amount of crystalline silicates in cometary grains is an indicator of the age of comets; older comets should be rich in amorphous grains, while younger comets should contain an abundance of crystalline silicates. In fact, it is expected that thermal annealing of amorphous silicate grains present in the hot inner part of the protosolar nebula favors the increase of the fraction of crystalline material over time. Thus, comets formed later in the nebular history will contain a larger amount of annealed (crystalline) dust with respect to those formed earlier. Instead, comets formed very early in the solar nebula should consist almost exclusively of amorphous silicates and unaltered interstellar ices since little processed material was available when they formed (*Nuth et al.,* 2000).

Laboratory studies on olivine grains suggest that the activation energy E_a decreases as the Mg/Fe abundance ratio increases, in contrast with observations that provide no evidence of crystalline Fe-rich silicates. In fact, if both iron and magnesium, with about the same cosmic abundance, were incorporated into silicates forming precometary grains, thermal annealing would favor crystalline Fe-rich over Mg-rich silicates, and their presence should be detectable by IR features. The absence of crystalline Fe-rich silicates could be justified by the presence of iron as pure metal.

As far as C in space is concerned, aromatic and aliphatic C-H IR features are observed in different space environments. Features at 3.38, 3.41, and 3.48 μm have been detected both in the diffuse interstellar (IS) medium (*Pendleton et al.,* 1994) and in the protoplanetary nebula CRL 618 (*Chiar et al.,* 1998). Similar bands are lacking in the dense IS medium, where only a weak 3.47-μm feature is observed, attributed to tertiary C-H bonds (*Allamandola et al.,* 1993). These differences suggest that C suffers modifications in the transition from diffuse to dense clouds. Based on laboratory results (see previous section), the 3.4-μm aliphatic band observed in the interstellar medium and in protoplanetary nebulae is attributed to the prevalence of aliphatic C-H bonds formation by H atom reactions on C grains over the destruction by UV irradiation. In dense clouds, where grains are coated by an ice mantle, the C core is prevented from reaction with H atoms and the C-H bonds are destroyed by penetrating UV photons.

It is interesting to note that a complex of features in the 3.3–3.4-μm range is evident in the emission spectra of comets. Most of these features are due to volatile coma molecules (e.g., methanol, methane, and ethane). Excess emission, once the molecular contribution is accounted for, is attributed to a solid C component (*Davies et al.,* 1993).

The applications reported above clearly evidence the relevance of laboratory experiments on refractory materials in the identification of cometary species. Despite the fact that important results have already been obtained, many open points remain to be clarified and require further experimental work. The genesis of the crystalline component, specifically concerning the cometary silicates, is still a matter of debate. Besides the interpretation reported above about turbulent radial mixing and flash heating by shock, the possibility exists that amorphous silicate grains lie in an energetic metastable state, produced, e.g., by ion irradiation. In this case, little energy would be required to allow the amorphous-to-crystalline transition. Of course, the validation of this scenario needs support from dedicated laboratory experiments. As far as carbons are concerned, the relations between the solid and the molecular phases in comets are still not well understood. Future experiments shall have to be oriented to clarify the chemistry related to energetic processes, such as UV irradiation and ion bombardment, under conditions that are representative of comet environment, at different stages of their evolution.

3.   EXPERIMENTS ON THE PHYSICAL STRUCTURE OF COMETARY ICES

Observations of cometary comae can reveal, although not completely, the composition of cometary ice, namely the gases trapped in the water ice in the nucleus and their proportions relative to water. In turn this information can tell us about the gas composition and temperature in the region where ice grains formed, grains that later agglomerated to form comet nuclei. Ice grain formation may have occurred in the cold and dense IS cloud, which collapsed to form the solar nebula or by water condensation on grains at the cold outskirts of the solar nebula itself. In both scenarios, water vapor adhered to mineral grains and formed H_2O-ice in the presence of gases, which could be then trapped in the ice itself. These trapped gases are released from the ice during changes in its structure and *not* according to their sublimation temperatures. This is evident from the simultaneous release of all the gases observed in the coma of Comet Hale-Bopp over many heliocentric distances by *Biver et al.* (1999).

3.1.   Small-Scale Studies

When vapor-phase H_2O molecules condense into ice below ~120 K, they lack the energy to migrate and form a stable crystalline structure. Rather, they remain where they strike, forming an amorphous ice structure with many open pores. Any other gas molecules present can then enter these pores, adhere to the walls through Van der Waals forces, and remain in the pores for a certain time. If additional H_2O molecules arrive by then and condense, the pores can be closed, trapping the gas inside the ice. Some molecules are more readily trapped than others, which can lead to selective trapping of the molecules from the gas-phase mixture, with important implications. For example, it has been shown experimentally that among the heavy noble gases, the trapping preference is Kr > Xe > Ar (*Owen et al.,* 1991, 1992) due to (1) the different polarizabilities of these atoms and therefore different adherence to the walls of the ice pores, (2) differences in their masses (*Notesco et al.,* 2003), and

(3) their sizes. This Ar/Kr/Xe pattern in cold H_2O-ice agrees well with the noble gas ratio in Earth's atmosphere, suggesting that comets may have delivered most of these gases to the forming Earth (*Melosh and Vickery,* 1989; *Chyba,* 1987, 1990; *Pepin,* 1997; *Owen and Bar-Nun,* 1993, 1995; *Notesco et al.,* 2003), along with other cometary volatiles such as CO, CO_2, CH_4, N_2, NH_3, HCN, and H_2S.

Moreover and most important, the high HDO/H_2O ratio in Earth's oceans, which could not have been obtained if the water was in chemical equilibrium with the H_2 and HD in the solar nebula, suggests that comets rich in HDO/H_2O delivered a considerable fraction of Earth's water and volatiles (*Laufer et al.,* 1999). The high HDO/H_2O and DCN/HCN ratios in Comets Halley, Hyakutake, and Hale-Bopp (*Eberhardt et al.,* 1995; *Despois,* 1997; *Bockelée-Morvan et al.,* 1998; *Meier et al.,* 1998) suggest that their water originated in the ISM, where ion-molecule reactions enriched the amount of HDO over H_2O, and also suggest that this water never equilibrated with the HD and H_2 in the solar nebula. A cold origin for ice grains is also inferred from the ~22–27 K required for the trapping of CO in ice, needed to account both for the ~7% of CO/H_2O in the comae of Comets Halley, Hyakutake, and Hale-Bopp (*Irvine et al.,* 2000) and the 25 K derived from the ortho-para spin ratio of H_2O in Comets Halley (*Mumma et al.,* 1988) and Hale-Bopp (*Crovisier,* 1999) and the ortho-para ratio of NH_3 in Comet LINEAR (*Kawakita et al.,* 2001). It seems therefore that the ice grains that agglomerated to form these and other comets originated in very cold regions either in the collapsing dense interstellar cloud or in the very cold outskirts of the solar nebula.

Many experiments (e.g., *Bar-Nun et al.,* 1985, 1987; *Laufer et al.,* 1987; *Schmitt and Klinger,* 1987; *Sandford and Allamandola,* 1988; *Schmitt et al.,* 1989a,b; *Hudson and Donn,* 1991; *Jenniskens et al.,* 1995; *Notesco and Bar-Nun,* 1997) have shown that gases in the presence of condensing water vapor can be trapped and distributed within the frozen H_2O, and that they do not simply freeze out as segregated molecules. More detailed studies were carried out by *Jenniskens and Blake* (1994) and *Jenniskens et al.* (1995) on the structure of amorphous ice around 22–30 K.

Another mechanism by which gases can be trapped is the formation of clathrate hydrates, in which H_2O molecules form cages around guest atoms or molecules. However, experiments have shown that while polar molecules, such as CH_3OH, H_2S, C_4H_8O (tetrahydrofurane), and C_2H_4O (oxirene), can form clathrate hydrates (*Blake et al.,* 1991; *Richardson et al.,* 1985; *Bertie and Devlin,* 1983), some of the most prominent cometary gases, such as CO, CO_2, CH_4, and NH_3, do not. The latter molecules are not trapped as clathrate hydrates even in the presence of a clathrate-hydrate-forming gas such as CH_3OH (*Notesco and Bar-Nun,* 1997, 2000).

When the H_2O-ice is warmed, how do the trapped gases escape? Near 120 K (*Schmitt et al.,* 1989a) H_2O molecules in the amorphous ice acquire enough energy to move and rearrange into the more-stable cubic structure, although about 70% of the ice remains in a "restrained amorphous" form,

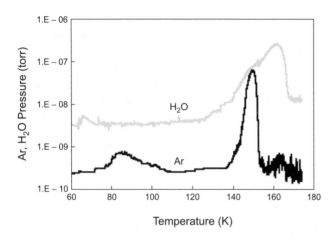

Fig. 5. Gas evolution from a 0.1-μm amorphous gas-laden ice sample upon its warming up. The gas comes out when the amorphous ice transforms into its crystalline and "restrained amorphous" forms.

which is a viscous liquid at this temperature (*Jenniskens and Blake,* 1994, 1996; *Jenniskens et al.,* 1995, 1997). This movement opens pores in which the trapped gases reside, and releases them (Fig. 5). Therefore, all the trapped gases come out together and not according to their sublimation temperatures (*Biver et al.,* 1999).

When heat is applied to a collection of gas-laden ice grains, two gas-release steps occur (Fig. 5). First, near ~120 K, each grain releases the gas trapped within by a dynamic percolation process, in which channels in the ice open up to the surface of the grain (*Laufer et al.,* 1987). Next, a gas molecule leaving an individual grain must still pass near other ice grains in order to emerge from the bulk of the ice. This involves another dynamic percolation process, in which channels open between the grains up to the surface. The ejection of ice grains when a large flux of gas emanated from the ice was observed experimentally by *Laufer et al.* (1987). These processes were modeled (e.g., *Prialnik and*

Bar-Nun, 1992), using available experimental data on thin (0.01–100-μm) ice samples. Yet, in order to study large-scale phenomena, large ice samples have to be studied.

3.2. Large-Scale Studies

The first large-scale cometary ice experiment was the Comet Simulation (KOSI) series of experiments carried out at the German Aeronautic and Space Organization (DLR) in Köln. A space simulator of 2.5-m diameter and 4.9-m length, capable of reaching 10^{-6} Torr and having an assembly of Xe-arc lamps capable of illuminating a 30-cm-diameter sample with 1–1.4 solar constants was used (*Grün et al.,* 1991). The comet analog, 30 cm in diameter and 15 cm thick, was produced by spraying fine droplets of a slurry of minerals in liquid H_2O into liquid N. Such ice was always crystalline and *not* amorphous and so could not trap gases in it, as does amorphous cometary ice. In order to add gas to the solid mixture, CO_2 was flowed into the mineral-crystalline ice mixture in liquid N and froze there. The canister at 80 K, containing a mixture of ~10% minerals with traces of C soot, crystalline ice, and 0–15% frozen CO_2, all in liquid N, was placed in the vacuum chamber, where the liquid N evaporated, leaving behind a porous sample with a density of ~0.5 g cm^{-3}. When the powerful "sun" was turned on, a vigorous response from the 45° inclined sample was observed, in which water vapor from the surface and frozen CO_2 from both the surface and from deeper layers sublimated as the heat wave penetrated inward. Driven by these gases, mineral grains were ejected. After a while, a mineral dust layer free of ice accumulated on the surface, slowing down the activity, apparently due to its poor thermal conductivity. These observations were accompanied by measurements of several types:

1. Heat was transported through the sample, illuminated by 1–1.4 solar constants, with dark periods. A representative data plot is shown in Fig. 6a. As expected, ice layers closer to the surface heat faster and the inner layers lag behind. The same trend was observed when the heating was

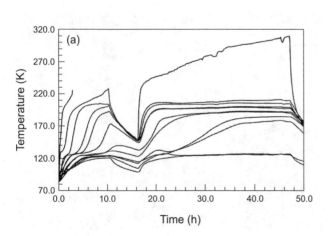

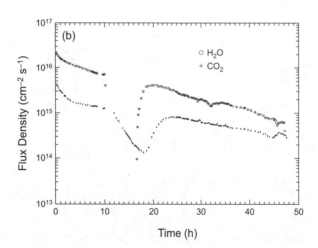

Fig. 6. **(a)** Evolution of temperatures in KOSI-3 sample at different distances from the cold backplate. **(b)** H_2O and CO_2 gas flux densities at the sample surface (KOSI-3).

stopped and resumed. The thermal conductivity of various KOSI samples was calculated from such measurements and found to be about $1-6 \times 10^4$ erg cm^{-1} K^{-1} s^{-1}, depending on the ratios of minerals/hexagonal ice/frozen CO_2. These values are about an order of magnitude smaller than the value for a block of hexagonal ice, $3.5-8 \times 10^5$ erg cm^{-2} K^{-1} s^{-1} (*Spohn et al.*, 1989), due to the porosity of the KOSI samples.

2. Gas emission was studied by the mass spectrometers (Fig. 6b) (*Grün et al.*, 1991). Immediately when insolation began, a flux of $H_2O/CO_2 \approx 6$ (the ratio in the original sample was 5.6) was measured, decreasing to ~3 after 50 h of illumination. When the insolation was interrupted, the H_2O flux dropped immediately, whereas the CO_2 flux lagged, decreasing over 18 h. When the insolation was resumed, the H_2O flux rose during about 2 h, whereas the CO_2 took about 5 h to rise. This is reasonable, because although the water vapor came from the surface at the beginning and from just below the dust mantle, the frozen CO_2 sublimated from deeper layers that took longer to warm up. Generally, the flux of both gases diminished by 2 orders of magnitude (from H_2O 10^{18} to 10^{16} cm^{-2} s^{-1}) after 50 h, due to the formation of an insulating dust layer and the depletion of both volatiles from the upper layers of the ice.

3. During the sublimation of water and CO_2, a very large flux of mineral and ice-coated mineral grains was ejected. The ejected material was photographed with a video camera, grain impacts were monitored by microphones facing the sample, and particles were collected over a large sampling area. The erratic flux of grains diminished by about 2 orders of magnitude during 10 h, along with the diminishing flux of gas and water vapor. A median velocity of ~100 m s^{-1} was obtained. This result is not far from the value (≥ 167 m s^{-1}) observed experimentally by *Laufer et al.* (1987) for ice grains ejected from a thin ice sample, when a large flux of gas was released.

In another experiment (*Mauersberger et al.*, 1991), individual particles entered a tube with a pressure gauge, where they were heated and their volatiles sublimated. Two types of traces were measured: sharp spikes decaying within a few seconds, apparently from pure ice grains, and much broader ones, decaying over tens of seconds, apparently from ice-containing porous mineral particles. The structure of the collected grains was studied by SEM and found to be very fluffy (Fig. 7), with a density of about 0.1–1 g cm^{-3}. They seemed to be an agglomerate of even smaller grains, not unlike the interplanetary dust particles (IDPs) collected in the stratosphere (*Brownlee et al.*, 1980). These agglomerates of mineral particles were formed in the slurry of water and minerals that was sprayed into the liquid N, but it is possible that some could have formed in the ice during its sublimation. Occasionally, a dust grain still attached at one point to the mantle would vibrate violently for many minutes, driven by the gas flux, until finally it flew away. Some dust bursts were observed when a larger chunk of material fell back onto the surface.

After many hours of insolation, the chamber was opened and the sample container was transferred to a bath of liquid N in a N purged glove-box. Its compressive strength was measured by driving a force meter into the sample. The initial strength of $1-2 \times 10^6$ dynes cm^{-2} was increased after insolation to $0.4-5 \times 10^7$ dynes cm^{-2} just below the dust mantle, probably due to the migration of some water vapor inward, and its freezing there. Reflectance spectra of the surface between 500 and 2500 nm were measured and, as expected, showed that the CO_2 and H_2O features diminished considerably.

Similar large-scale experiments were carried out by *Green et al.* (1999) and *Green and Bruesch* (2000) at the Jet Propulsion Laboratory (JPL) in Pasadena, California. A slurry of minerals in water was sprayed into liquid N, again producing hexagonal ice, in a 200-cm-wide and 250-cm-high cylindrical canister, with an insolation of 0.1–2.1 solar constants. The penetration of the heat wave was monitored by thermocouples; the evolution of water vapor was measured by two mass spectrometers, and dust release by a video camera. The mechanical properties were measured in the closed chamber by a mechanical penetrator-scratcher and, when the chamber was opened, measurements on compressive strength, penetrability, porosity, and density were car-

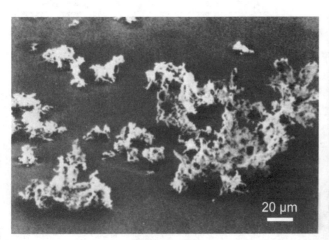 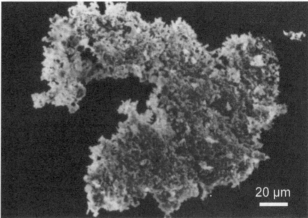

Fig. 7. SEM micrographs of dust grains emitted during the KOSI-3 experiment.

ried out. The results of these measurements are not yet available in the literature.

Although the extensive and detailed KOSI and JPL projects opened a new field of large-scale comet simulation, they suffered from a basic drawback due to the method of sample preparation, namely spraying a slurry of ~10% dust in water into liquid N. This resulted in crystalline ice, which could not trap gases. The added CO_2 (up to CO_2/H_2O ~0.2 in the KOSI experiment) was frozen among the H_2O-ice particles and not actually trapped in amorphous H_2O grains, which is the most relevant situation inferred from comet observations (co-evolution of all gases and water vapor together) and from the extensive small-scale laboratory studies. These small-scale experiments have shown that trapping of ~10% CO in cometary H_2O-ice grains at a very slow deposition rate requires temperatures of ~22–27 K, where the ice formed by vapor deposition is amorphous and full of pores. Yet amorphous ice can be produced even at 80 K, with some structural differences from ice formed at ~30 K, but still well below the ~120-K transformation temperature and gases can be trapped in it.

3.3. Studies of Large Samples: Gas-Laden Amorphous Ice Samples

The next step in large-scale comet simulation was to produce a 200-cm^2 × 10-cm-high sample *of* amorphous ice, with gases trapped inside, although with no mineral dust (*Bar-Nun and Laufer,* 2003). The main objective of this study was to learn how much gas is released to the sample's surface during the crystallization of the ice in deeper layers vs. the flux of water vapor released by sublimation on the surface. The relevant cometary issue is the gas/water vapor ratio in the coma vs. its ratio in the nucleus. Other ice properties, such as heat conductivity, which is of prime importance to models, and compressive strength, which is important for comet splitting and for landing on the nucleus, were also measured in this work.

Large samples of amorphous ice cannot be produced by depositing H_2O vapor onto a cold plate, as is routinely done in small-scale experiments. With large samples the problem of removing the water's heat of condensation (2.7×10^{10} erg g^{-1}) is more severe, especially as amorphous ice has a low thermal conductivity (<~10^4 erg cm^{-1} K^{-1}). If one grows too thick an ice layer, the heat of condensation cannot be removed by the underlying cold surface, so that the newly deposited layers reach ~120 K and become crystalline rather than amorphous. Consequently, even if a gas flow accompanies ice formation, no gas trapping occurs. To solve these problems, thin ~200-μm ice layers were formed at 80 K (liquid N) on a cold plate, through which the heat of condensation could still be transmitted to the cold surface, remain amorphous, and trap the accompanying gas. Once a thin amorphous gas-laden ice layer formed, it was scraped from the cold plate by a 80-K cold knife into the 80-K sample container, which was covered by an 80-K dome. All these parts, as well as the heat shield surrounding the entire

chamber, were kept at 80 K by a controlled flow of liquid N. This fully automatic, hydraulically controlled process was repeated at 10^{-5} torr for 10–20 h until a large enough ice sample accumulated, namely a 200-cm^2 × 10-cm-high sample of an agglomerate of 200-μm particles of amorphous gas-laden ice. The sample was then covered by the 80-K deposition plate while the dome was heated to ~330 K. When the plate was removed, the sample was illuminated from above by the heated irradiation dome, made of a roughly surfaced aluminum that behaves like a black body at 330 K, with a flux of 5×10^5 erg cm^{-2} s^{-1} at better than 3% uniformity. Since water is practically opaque to IR at the 330-K blackbody spectrum, the energy input of the irradiation dome was totally absorbed by the upper layers of the ice sample.

Ten thermocouples (types E and T) were embedded in the ice, recording the temperature profile of the sample. A mass spectrometer recorded the emission of gas and H_2O vapor during heating, through a 2-cm hole in the dome right above the center of the sample. The sensitivities to various gases provided by the manufacturer were checked by analyzing mixtures of gases with known compositions. The ice density was measured by collecting a small sample in a 1-cm^2 × 5-cm glass vial simultaneously with the large ice sample, and measuring its volume at 80 K and again when melted. The compressive strength was measured by inserting a force-meter penetrator, cooled to 80 K, into the ice, immediately after the experiment was terminated and the chamber was opened. A schematic drawing of the machine is presented in Fig. 8.

To test whether indeed a sample of amorphous ice was produced, Ar was flowed onto the 80-K cold plate by itself and, in a separate experiment, accompanied by H_2O vapor. As expected, Ar by itself did not freeze on the 80-K cold plate, but when accompanied by H_2O vapor, it was trapped in the amorphous H_2O-ice that formed. As learned from studies of thin ice samples, CO behaves like Ar (*Bar-Nun et al.,* 1987). In several experiments, 200-cm^2 × 0.5-cm thick and 200-cm^2 × 6-cm-thick ice samples were produced. The density of the loose agglomerate of 200-μm ice grains was found to be 0.25 g cm^{-3}. This should be compared with the densities of 0.3–0.7 and 0.29–0.83 g cm^{-3} calculated for Comets Halley (*Rickman,* 1989) and Borrelly (*Farnham and Cochran,* 2002), respectively, which contain also mineral particles.

During the ice's heating, the fluxes of sublimating water vapor and Ar were monitored by a mass spectrometer until all the ice sample sublimated. The mass spectrometer record of H_2O and Ar during the entire experiment is shown in Fig. 9a.

Several interesting results came out of this experiment: (1) As stated above, Ar was trapped in the 80-K ice, whereas by itself it was not frozen on the cold plate. This showed definitely that the ice made by this method was amorphous, since only amorphous ice traps Ar in it, while being formed. (2) The time-integrated mass spectrometer fluxes gave Ar/H_2O = 1.01, in comparison with the 1:1 ratio in the flowed

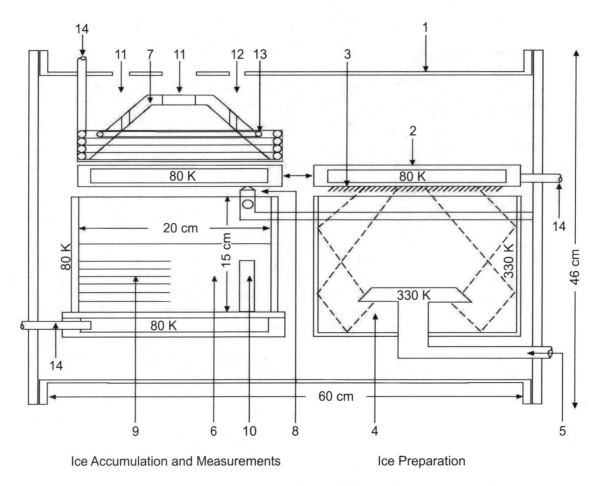

Fig. 8. A schematic drawing of the machine producing large (200 cm² × 10 cm) gas-laden amorphous ice samples, as a "comet" simulation: 1 = vacuum chamber, 2 = cold plate at 80 K, 3 = 200-μm amorphous gas-laden ice, 4 = homogeneous flow of water vapor and gas, 5 = water vapor and gas pipes, 6 = 200-cm² and 5–10-cm-thick ice sample, 7 = heating dome, 8 = 80-K cold knife, 9 = thermocouples, 10 = density measurements, 11 = mass spectrometer, 12 = ionization gauge, 13 = heating tape, 14 = LN₂ cooling pipes.

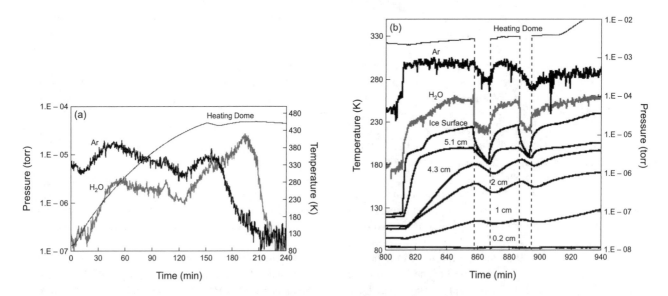

Fig. 9. **(a)** Evolution of Ar trapped in the ice and ice sublimation, upon heating from above of a 0.5-cm-thick sample. Note the early rise of the Ar and its exhaustion before all the ice sublimates. **(b)** Temperature profiles of the thermocouples as a function of their distance from the 80-K bottom plate in a 6-cm-thick ice deposit. Note also the sharp decrease and increase of the water flux vs. the sluggish response of the Ar emanation, when the ice sample is covered by the cold plate and upon its removal.

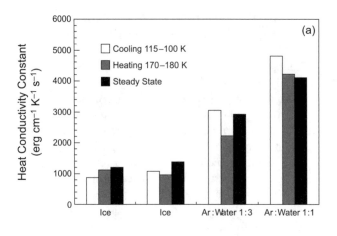
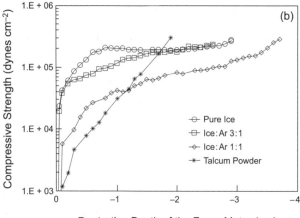

Fig. 10. (a) Heat conductivity constant of various ice samples measured at different temperature ranges. (b) Compressive strength of the studied ice samples as function of the penetration depth in the ice.

mixture. This result implies that all the Ar was trapped in the amorphous ice and was released upon warming up of the entire 0.5-cm-thick ice layer. (3) The timing and magnitude of Ar evolution from the ice, as seen in Fig. 9a, is of prime importance for the interpretation of comet observations. Even in a 0.5-cm-thick ice layer heated from above, the Ar flux rose to a level about seven times higher than that of the water vapor. This was because the Ar was released from the amorphous ice during the transformation into the cubic and "restrained amorphous" forms, as was found in numerous studies of thin ice samples. This process occurs in the interior of the sample, supplying gas to the experimental "coma," whereas the water vapor emanates only from the ice surface. Eventually, the heated ice layer is exhausted of its gas, as can be seen in Fig. 9a. The Ar flux declines well before all the H_2O-ice sublimates. This raises the question of what the gas/H_2O ratios in cometary comae tell us about gas abundance ratios in cometary nuclei.

Figure 9b shows the thermal history of a 6-cm-thick ice sample, measured through five thermocouples, and the mass spectrometer records of water and Ar release. At 812 min (after the beginning of ice formation) the sample was exposed to the 330-K heating dome. As expected, the ice temperature rose very steeply on the surface and more sluggishly below it. This allowed the thermal conductivity of the ice to be determined. When the heating dome was blocked at 858 min, the surface temperature dropped immediately, whereas the temperature of the deeper layers lagged. The thermal conductivity of the 0.25 g cm^{-3} agglomerate of 200-μm grains of amorphous ice can be calculated from the temperature profiles of the thermocouples embedded in the ice sample, and is shown in Fig. 10a. The heat conduction coefficients of amorphous and cubic water ice were calculated by *Klinger* (1980) to be $2-3 \times 10^4$ and 2.8×10^5 erg cm^{-1} K^{-1} s^{-1} respectively. In the KOSI experiments, on hexagonal porous crystalline ice, the heat conduction coefficient

derived was $3-6 \times 10^4$ erg cm^{-1} K^{-1} s^{-1} (*Spohn et al.*, 1989; *Benkhoff and Spohn*, 1991; *Seiferlin et al.*, 1996).

From recent experiments (Fig. 10a) the heat conduction of gas-free ice is 30 times smaller than Klinger's value for a block of amorphous material [a very much lower thermal conductivity value by a factor of $10^{-4}-10^{-5}$ was reported by *Kouchi et al.* (1992), but was not measured in other experiments]. This lower value is due to the fact that the sample is an agglomerate of ~200-μm ice particles, and the heat conduction between the grains is poor. One cannot exclude the contribution of inward-flowing water vapor to the thermal conductivity, but it should be noted that no harder crust was detected below the surface, as would have been expected if a massive flow of water vapor froze in deeper layers. Nevertheless, the thermal conductivity of gas-laden ice is much higher than that of pure H_2O-ice, and even more so when the gas content is even greater (Ar:H_2O = 1:3 vs. 1:1). This shows the importance of trapped gas for the conduction of heat into the interior of comets.

At this stage it is not possible to measure the effect of the exothermicity of the transformation amorphous to cubic ice on the temperature profile. Possibly, this process may be even somewhat endothermic in the presence of trapped gases (*Kouchi and Sirono*, 2001).

The measurement of mechanical properties is rather simple. Pressing a force-meter cooled to 80 K into the extremely fluffy ice produced the results shown in Fig. 10b. As the ice was compressed, its compressive strength increased as its open spaces decreased. Pure ice is "stronger" whereas the Ar trapped in the ice "weakens" it, since the ice particles that form the sample are fluffier. The compressive strength does, however, reach a finite limiting value. During heating, when the Ar left the ice, the fine ice structure collapsed. A very low compressive strength of ~2 × 10^5 dynes cm^{-2} can be deduced from the experiments. For comparison, in the KOSI experiments, in which mineral grains were mixed with hexagonal ice grains or were covered by a layer of hexagonal ice, the compressive strength was 3×10^5–2×10^7 dynes cm^{-2}, depending on the mineral content (*Jessberger and Kotthaus*, 1989).

3.4. Summary and Implications of Large- and Small-Scale Studies

By studying Ar:H$_2$O (1:1) at 80 K in the large chamber, it was proven that the ice is amorphous, since Ar does not freeze at 80 K but was trapped in the amorphous ice. Large samples of 200 cm$^2 \times 6$ cm, of an agglomerate of 200-μm ice particles of this amorphous gas-laden ice were produced and studied for the first time. These samples can represent pristine cometary nuclei and their still-unprocessed interiors. Yet we should remember that the ice samples did not contain mineral and organic dust and, consequently, the buildup of an insulating dust layer and its subsequent effect on the penetration of the heat wave could not be studied.

The most important finding was that the ratio of Ar/water vapor in the experimental "coma" was between 7 and 10 (Fig. 9) times larger than the Ar/ice ratio in the experimental "nucleus". In the 0.5-cm-thick samples, the Ar was exhausted from the ice well before the ice sublimated completely but in the 6-cm-thick sample, Ar kept emanating from deeper layers (Fig. 9b). This observation has direct bearing on the correlation between the gas/water vapor observed in cometary coma and the gas/ice ratio in the nucleus: The ~10% of CO to water vapor seen in the comae of Comets Halley, Hyakutake, and Hale-Bopp might well mean that the CO/ice in these nuclei could be closer to 1%.

In Comet Hale-Bopp (*Biver et al.*, 1999), the CO flux at 5.2 AU preperihelion was ~6 times larger than the water flux, but became ~5 times smaller than water at perihelion. Apparently, the upper layers were exhausted and gases came out only, and more slowly, from deeper layers. A model incorporating all new results is now being prepared.

Finally, the low density and compressive strength, which attest to the fluffy structure of the ice, can account for breakups of cometary nuclei and should be of concern for the *Rosetta* lander (*Hilchenbach et al.*, 2000). However, incorporating about ~25% mineral dust and about 25% of organic "CHON" particles, as found for Comet Halley, may harden the surface somewhat.

As for small-scale ice studies, much more has to be learned about the preferential trapping of CO over N$_2$, since no N$_2^+$ was observed by *Cochran et al.* (2000, 2002) in various recent comets where CO was observed. The enrichment in Earth's atmosphere of the heavier Xe isotopes is still an open problem, although the Ar and Kr isotopic enrichments, when trapped in ice according to their inverse square root of the mass, account for their relative abundances (*Notesco et al.*, 2003). A major problem remains as to how Jupiter obtained its abundances of C, N, O (?), S, P, etc., relative to H$_2$, as these abundances are ~3 times the solar abundances. It is possible that late bombardment by comets devoid of H$_2$ could have contributed these volatiles.

4. EXPERIMENTS ON CHEMICAL REACTIONS IN COMETARY ICES

The previous *Comets* book (*Wilkening*, 1982) listed 35 chemical species observed in cometary spectra, but a mere three stable molecules (CO, HCN, CH$_3$CN). Since then, the cometary molecular list has grown to include about 20 reasonably stable members. As comets are considered to be the solar system's most primitive bodies, it is reasonable to ask how they came to acquire molecules as complex as ethane, methanol, and formamide. Laboratory investigations can help with this question by revealing chemical reactions and conditions that lead to observed cometary species.

It is generally accepted that at each stage of a comet's history, cosmic rays and energetic photons (UV or X-rays) can drive chemistry in cometary ices. Although uncertainties remain in determining precise energy inputs, Table 4 shows values thought to be typical. We note that the chemistry initiated by ionizing radiations, such as 1-MeV protons or X-rays, results mostly from the secondary electrons generated. This means that the observed products, and their abundances, are more dependent on the energy input than the initial carrier of the energy. Since doses on the order of 1 eV per molecule (Table 4) are attainable in the laboratory, experiments to mimic cometary ice chemistry can be performed. In most cases, the differences in photochemical and radiation chemical effects are typically not in the nature of the products made, but rather in product abundances or the depths at which products form in ice samples.

Goals motivating laboratory work on cometary chemistry include the following: (1) Discovery of efficient reactions leading from simple starting materials to more complex species. This allows predictions of as yet unobserved cometary molecules. (2) Explanation of observed abundance and ratios, such as C$_2$H$_6$/CH$_4$ and HNC/HCN. (3) Investigation of low-albedo materials relevant for cometary nuclei (e.g., Halley and Borrelly). (4) Prediction of candidates for extended sources of CO, CN, H$_2$CO, and other molecules.

Here we describe some laboratory results related to the chemistry of ices in cometary nuclei, leaving comae chemistry for other chapters. The emphasis is on recent work, especially that providing insight into molecular evolution in comets.

4.1. Methods

Figure 11 represents a laboratory setup used to investigate cometary ice chemistry at the NASA Goddard Space Flight Center (GSFC). Similar equipment exists in other laboratories, with modifications made according to the interests of the investigators (e.g., *Allamandola et al.*, 1988; *Gerakines et al.*, 1996; *Demyk et al.*, 1998; *Strazzulla et al.*, 2001). In brief, the vacuum and temperature of outer space are simulated with a high-vacuum chamber and a cryostat respectively. An ice sample is formed on a precooled surface, to a thickness of a few micrometers or less, by condensation of room-temperature gases. The sample is then

TABLE 4. Estimated fluxes for ice processing environments, compared to laboratory experiments.

Environment (ice residence time in years)	Ion Processing			Photon Processing		
	Flux, 1 MeV p$^+$ (eV cm^{-2} s^{-1})	Energy absorbed (eV cm^{-2} s^{-1})*	Dose (eV molecule^{-1})	Flux (eV cm^{-2} s^{-1})	Energy absorbed (eV cm^{-2} s^{-1})	Dose (eV molecule^{-1})
Diffuse ISM (10^5–10^7)[†]	1 × 10^7	1.2 × 10^4	<1–30	9.6 × 10^8 at 10 eV[†]	5 × 10^8 0.02 μm ice	10^4–10^6
Dense cloud (10^5–10^7)[†]	1 × 10^6	1.2 × 10^3 0.02-μm ice	≪1–3	1.4 × 10^4 at 10 eV	1.7 × 10^3 0.02 μm ice	<1–4
Protoplanetary nebula (10^5–10^7)[‡]	1 × 10^6	1.2 × 10^3 0.02-μm ice	≪1–3	2 × 10^5 at 1–10 keV	§5 ×10^4 0.02 μm ice	2–240
Oort cloud (4.6 × 10^9)	φ(E)**	**	~150 (0.1 m) ~55–5 (1–5 m) <10 (5–15 m)	9.6 × 10^8 at 10 eV	9.6 × 10^8 0.1 μm ice	2.7 × 10^8
Laboratory (4.6 × 10^{-4})[††]	8 × 10^{16}	2 × 10^{15} 1-μm ice	10	2.2 × 10^{15} at 7.4 eV	2.2 × 10^{15} 1 μm ice	10

* The absorbed energy dose from 1-MeV cosmic-ray protons assumes a 300-MeV cm^2 g^{-1} stopping power and an H$_2$O-ice density of 1 g cm^{-3}. Protons deposit energy in both the entrance and exit ice layer of an ice-coated grain.

[†] 10 eV photons = 1200 Å, vacuum UV (UV-C). *Jenniskens et al.* (1993).

[‡] Typical disk longevities (*Lawson et al.,* 1996).

§ Typical flux at 0.1 pc, 1 keV photons = 12 Å, soft X-rays (*Feigelson and Montmerle,* 1999).

¶ Absorbed energy dose from 1-keV X-rays assumes a 1-keV electron production in 1 g cm^{-3} H$_2$O-ice with a 127-MeV cm^2 g^{-1} stopping power.

** An energy dependent flux, φ(E), was used to calculate the resulting energy dose at different depths in a comet nucleus for an H$_2$O-ice density of 1 g cm^{-3}. For details see *Strazzulla and Johnson* (1991) and references therein.

[††] Typical proton and UV data from the Cosmic Ice Laboratory at NASA Goddard.

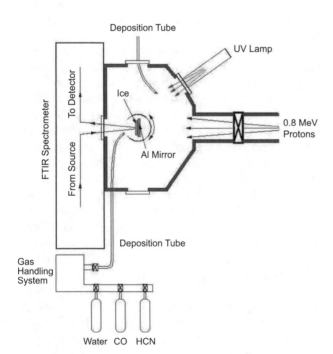

Fig. 11. Schematic of a laboratory setup for cometary ice experiments.

processed by positioning it before an ion beam to mimic cosmic-ray bombardment or a UV lamp to simulate far-UV exposure. The resulting ice chemistry can be followed by IR spectroscopy (see Fig. 11) or other techniques [e.g., UV (*Mennella et al.,* 1997), Raman (*Colangeli et al.,* 1992)]. At GSFC, a beam of ~1-MeV protons is generated by a Van de Graaff accelerator, while vacuum-UV photolysis is done with a flowing-H$_2$ microwave-discharge lamp (mostly 100–200-nm coverage).

To date, mid-IR spectroscopy (4000–400 cm^{-1}, 2.5–25 μm) has revealed more details of ice chemistry than any other laboratory method. Spectra in this region are from vibrations involving functional groups (groups of bonded atoms), with certain functional groups having very diagnostic absorbances. A disadvantage of IR spectroscopy is its relatively low sensitivity. Reaction products with low abundance can seldom be studied by IR alone, and often require a combination of chromatography and mass spectrometry for their identification during vaporization (e.g., *Bernstein et al.,* 1995).

4.2. Some Results and Case Histories

Of all cometary molecules, only H$_2$O has been detected in both the solid and gas phases (*Davies et al.,* 1997), all other species having been observed only in gas-phase coma spectra. As H_2O is the most abundant nuclear molecule, it plays a particularly significant role in cometary ice chemistry. On exposure to either far-UV photons or high-energy (keV, MeV) ions, H_2O dissociates into H and OH radicals. Even at cometary and interstellar temperatures, H and OH can react with other molecules to produce many products.

Figure 12 summarizes an experiment in which an H_2O + CO (5 : 1) mixture was proton irradiated at 16 K (*Hudson and Moore,* 1999). Before the irradiation, none of the bands in the upper trace were present. The features shown were produced by radiolysis and can be identified, among other ways, by comparison to reference spectra. In each case, fragments from H_2O, either H or OH, form the products. For example, H-atom addition to CO along the sequence

$$CO \rightarrow HCO \rightarrow H_2CO \rightarrow CH_3O \text{ and/or}$$
$$CH_2OH \rightarrow CH_3OH$$

leads to H_2CO (formaldehyde) and CH_3OH (methanol), both being cometary molecules. Calculations of reaction yields are possible from a knowledge of intrinsic IR band strengths, and show that energetic processing can account for the known abundance of many cometary organics. In fact, radiolysis is presently the only process known to reproduce many observed abundances.

Formic acid, HCOOH, in Fig. 12 is particularly interesting as this molecule arises from *both* H and OH adding to CO, such as H-atom addition followed by OH reaction: $C \equiv O \rightarrow HC = O \rightarrow HC(=O)OH$. Since H_2O forms an isoelectronic series with NH_3 and CH_4, similar reactions will lead to $HC(=O)NH_2$, from NH_3 + CO, and $HC(=O)CH_3$, from CH_4 + CO. These products are indeed observed in the laboratory, and both are known cometary molecules.

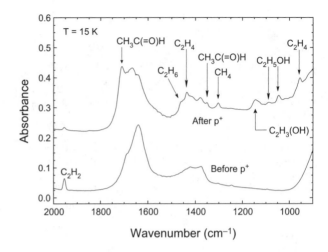

Fig. 13. Molecular synthesis in an H_2O + C_2H_2 (4 : 1) ice at 15 K after irradiation to a dose of 17 eV molecule^{-1}. Spectra have been offset for clarity.

Figure 13 shows a related experiment, a proton irradiation of an H_2O + C_2H_2 ice (4 : 1) at ~15 K (*Moore and Hudson,* 1998), work motivated by the discovery of C_2H_6 in Comet Hyakutake by *Mumma et al.* (1996). The upper trace gives identifications and, again, H- and OH-addition reactions explain most of the products. A quantitative analysis shows that this low-temperature ice chemistry can explain the unexpectedly high abundance of cometary C_2H_6.

Products from low-temperature reactions of known cometary molecules are summarized in Table 5 and represent work from various laboratories. The table is far from exhaustive [see *Cottin et al.* (1999) for a more complete listing]. Blanks in Table 5 indicate a lack of experimental data.

Experiments have also revealed the radiolysis and photolysis products of many single-component ices, and a summary is provided in Table 6, which is again far from exhaustive. Yields for some of these products may be low in the H_2O-rich ices of comets, but enrichment of the less-volatile materials may occur as highly volatile species are lost over many cometary apparitions. This effect is manifested in the laboratory by a room-temperature residue that remains after processed ices are warmed under vacuum. These residues are the subject of much interest, as chromatographic analyses show that they contain high molecular weight compounds. In residues from more complex ices, biomolecules, such as certain amino acids, are found in trace amounts (e.g., *Bernstein et al.,* 2002). These materials are thought to accumulate on cometary nuclei where they may have been delivered to the early Earth. If such molecules survived impact they could have significantly enhanced the variety and volume of the Earth's chemical inventory. In both single- and multicomponent ice experiments, residual materials are also of interest as they could explain the extended sources of H_2CO, CN, CO, and other molecules seen in comae.

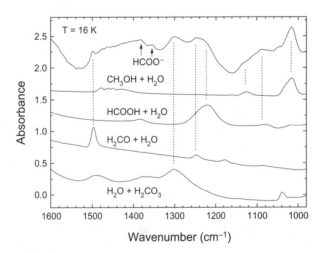

Fig. 12. New species formed in an H_2O + CO (5 : 1) ice irradiated to 11 eV molecule^{-1} are identified by comparison with reference spectra of dilute mixtures (H_2O : molecule > 5 : 1) at ~16 K. Spectra have been offset for clarity.

TABLE 5.   Products from reactions of H_2O-dominated two-component
ices at 10–20 K (references given in brackets).

Mixture	Reaction Products Identified	Processing
$H_2O + CO$	CO_2, HCO, H_2CO (+ CH_3OH, HCOOH, $HCOO^-$, H_2CO_3 from ion expts.)	Ion [1], UV [2]
$H_2O + CO_2$	CO, H_2CO_3, O_3, H_2O_2	Ion [3,4], UV [4,5]
$H_2O + CH_4$	CH_3OH, C_2H_5OH, C_2H_6, CO, CO_2	Ion [6]
$H_2O + C_2H_2$	C_2H_5OH, CH_3OH, C_2H_6, C_2H_4, CO, CO_2, CH_4, C_3H_8, HC(=O)CH_3, $CH_2CH(OH)$	Ion [6], UV [7]
$H_2O + C_2H_6$	CH_4, C_2H_4, CH_3CH_2OH, CO, CO_2, CH_3OH	Ion [8]
$H_2O + H_2CO$	CO_2, CO, CH_3OH, HCO, HCOOH, CH_4	Ion [1]
$H_2O + CH_3OH$	CO, CO_2, H_2CO, HCO, CH_4, (+ $C_2H_4(OH)_2$, $HCOO^-$ from ion expts.)	Ion [9,10], UV [2]
$H_2O + NH_3$	[none reported]	Ion [11]
$H_2O + HCN$	CN^-, HNCO, OCN^-, HC(=O)NH_2, NH_4^+ (?), CO, CO_2	Ion [8,12]
$H_2O + HNCO$	NH_4^+, OCN^-, CO, CO_2	UV [8]
$H_2O + HCOOH$	CO, CO_2, H_2CO	Ion [8]
$H_2O + HC(=O)CH_3$	CO_2, CO, CH_4, CH_3CH_2OH	Ion [8]
$H_2O + HC(=O)NH_2$	CO, CO_2, HNCO, OCN^-	UV [8,12]
$H_2O + HC(=O)OCH_3$	CO_2, CO, H_2CO, CH_3OH, CH_4	Ion [8], UV [8]
$H_2O + SO_2$		
$H_2O + H_2S$	S_2	UV [13]
$H_2O + OCS$	CO, CO_2, SO_2, H_2CO (?), H_2O_2 (?)	Ion [8]
$H_2O + CH_3CN$	H_2CCNH, CH_4, OCN^-, CN^-	Ion [8,12]
$H_2O + HCCCN$		

References: [1] *Hudson and Moore* (1999); [2] *Allamandola et al.* (1988); [3] *Brucato et al.* (1997); [4] *Gerakines et al.* (2000); [5] *Wu et al.* (2003); [6] *Moore and Hudson* (1998); [7] *Wu et al.* (2002); [8] M. H. Moore et al. (unpublished work, 2003); [9] *Hudson and Moore* (2000); [10] *Palumbo et al.* (1999); [11] *Strazzulla and Palumbo* (1998); [12] *Hudson et al.* (2001); [13] *Grim and Greenberg* (1987a).

TABLE 6.   Products from reactions of one-component ices at 10–20 K (references given in brackets).

Ice	Reaction Products Identified	Least-Volatile Species	Processing Experiment
H_2O	H_2O_2, HO_2 [2], OH [2]	H_2O_2	Ion [1], UV [2]
CO	CO_2, C_3O_2, C_2O	C_3O_2	Ion [3], UV [2]
CO_2	CO, O_3, CO_3	H_2CO_3 (from H^+ implantation) [4]	Ion [3,4], UV [1,2]
CH_4	C_2H_2, C_2H_4, C_2H_6, C_3H_8, CH_3, C_2H_5	PAHs [5] and high molecular weight hydrocarbons	Ion [5–7], UV [2]
C_2H_2	CH_4 [6], polyacetylene [8]	PAHs [5], polyacetylene [8]	Ion [5,8]
C_2H_6			
H_2CO	POM, CO, CO_2, HCO	POM	Ion [8], UV [2]
CH_3OH	CH_4, CO, CO_2, H_2CO, H_2O, $C_2H_4(OH)_2$, HCO, $HCOO^-$	$C_2H_4(OH)_2$	Ion [9,10], UV [2]
NH_3	N_2H_4 [2], NH_2 [2]	N_2H_4 [2]	Ion [12], UV [2]
HCN	HCN oligomers	HCN oligomers	Ion [8], UV [8]
HNCO	NH_4^+, OCN^-, CO, CO_2	NH_4OCN	Ion [8], UV [8]
HCOOH			
$HC(=O)CH_3$			
$HC(=O)NH_2$			
$HC(=O)OCH_3$			
SO_2	SO_3	S_8 [12]	Ion [12], UV [13]
H_2S	none reported		UV [13]
OCS			
CH_3CN	CH_4, H_2CCNH, CH_3NC		Ion [8], UV [8]
HCCCN			

References: [1] *Moore and Hudson* (2000); [2] *Gerakines et al.* (1996); [3] *Gerakines and Moore* (2001); [4] *Brucato et al.* (1997); [5] *Kaiser and Roessler* (1998); [6] *Mulas et al.* (1998); [7] *Moore and Hudson* (2003); [8] M. H. Moore et al. (unpublished work, 2003); [9] *Hudson and Moore* (2000); [10] *Palumbo et al.* (1999); [11] *Strazzulla and Palumbo* (1998); [12] *Moore* (1984); [13] *Salama et al.* (1990).

Laboratory work also suggests ions as likely chemical components of comets. Both radiolysis and photolysis of cometary ice analogs readily generate acids which, if NH_3 is present, undergo proton-transfer reactions of the type

$$HX + NH_3 \rightarrow X^- + NH_4^+$$

to produce stable ions. These ions would accumulate on the surface of a comet nucleus, or anywhere that sufficient energetic processing occurs. Figure 14 illustrates this type of acid-base chemistry, where the upper spectrum is from the irradiated $H_2O + CO$ (5 : 1) mixture of Fig. 12, while the lower spectrum is from an irradiated $H_2O + CO + NH_3$ (5 : 1 : 1) mixture (*Hudson et al.*, 2001). The $HCOO^-$ (formate) and NH_4^+ (ammonium) ion positions are indicated by arrows, and demonstrate that acid-base chemistry has occurred. Other ions that have been studied are OCN^- (*Grim and Greenberg*, 1987b; *Hudson et al.*, 2001) and CN^- (*Moore and Hudson*, 2003). H_3O^+ and OH^- also are likely in cometary nuclei, but difficult to detect by IR methods as they lack strong unobscured bands.

Not only have changes in ice composition been investigated in the laboratory, but changes in ice phase have been studied (*Baratta et al.*, 1991). Crystalline H_2O-ice can be converted into amorphous ice by either ionizing radiation or UV photons (*Leto and Baratta*, 2003). This amorphization is a general phenomenon and, at least for H_2O and CH_3OH, occurs with a rate that varies inversely with temperature (*Hudson and Moore*, 1995). Amorphization experiments also are related to the question of cometary clathrate hydrates, crystalline cage-like solids. Laboratory work shows that the H_2O-CH_3OH clathrate is readily destroyed by radiation, raising serious questions about the stability of clathrates in cometary ice (*Hudson and Moore*, 1993).

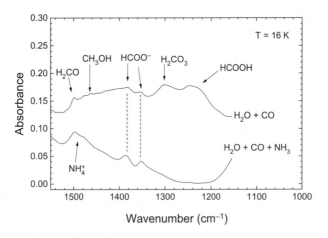

Fig. 14. Infrared spectra of two irradiated laboratory ices at 16 K, showing the influence of acid-base chemistry. The upper trace is an $H_2O + CO$ (5 : 1) ice and the lower trace is an $H_2O + CO + NH_3$ (5 : 1 : 1) ice. Both were irradiated to about 25 eV molecule⁻¹.

Finally, we mention one of the most persistent puzzles of cometary chemistry, the extraordinarily dark color of the nuclei of Comet Halley and Comet Borrelly. The most reasonable explanation for low nuclear albedos is that radiolysis of cometary organics and CO produced a dark C-rich material that accumulated over time (*Johnson et al.*, 1987). However, experiments showing progressive carbonization in H_2O-rich ices are lacking. More experiments and data-based chemical models are needed.

To summarize and conclude this section, laboratory experiments have revealed efficient condensed-phase syntheses for many cometary molecules such as H_2CO, CH_3OH, $HCOOH$, C_2H_6, $HC(=O)NH_2$, and $HC(=O)CH_3$, as discussed. Reactions to make other molecules, such as CO_2 and $HNCO$, have also been studied (*Hudson et al.*, 2001). Understanding the formation paths to other volatiles, and how they might relate to the dark nuclear surface, is work in progress.

4.3. Future Steps

Little has been published on the ice chemistry of molecules containing either a cyanide (CN) group or sulfur, two types of interest to astrobiologists. Cometary cyanides include H-CN, H_3C-CN, and $HC \equiv C$-CN. The corresponding isocyanides are known in the interstellar medium, but only H-NC has been reported in comets. Both the formation and abundance of HCN and HNC, and their ratio of HNC/HCN ~ 0.1, are of much interest, but lack a full explanation (*Rodgers and Charnley*, 2001). Cyanide isomerizations seen in the laboratory suggest new cometary molecules awaiting detection, such as H_3C-NC and $HC \equiv C$-NC from H_3C-CN and $HC \equiv C$-CN, respectively, and $H_2C = C = NH$ from H_3C-CN. Also, a sequence similar to $HC \equiv CH \rightarrow H_2C = CH_2 \rightarrow H_3C$-$CH_3$ may apply to $HC \equiv C$-CN, giving cometary $H_2C = CH$-CN and H_3C-CH_2-CN, but more laboratory work is needed for a firmer prediction. As for sulfur, it is present in comets as H_2CS, H_2S, OCS, SO_2, and other species. An observational search for CH_3SH is suggested, as CH_3SH could be formed from H_2CS in analogy with the $H_2CO \rightarrow CH_3OH$ reaction. More laboratory work on cometary sulfur chemistry certainly is needed, as evidenced by the gaps in Tables 5 and 6. For some early work on sulfur ice chemistry, see *Grim and Greenberg* (1987a).

Other new molecules recommended for searches, based on experiments, include ethanol (*Moore and Hudson*, 1998), ethylene glycol (*Hudson and Moore*, 2000), and vinyl alcohol (*Hudson and Moore*, 2003), all readily formed by ice processing. Acetic acid, glycolaldehyde, vinyl alcohol, and ethylene oxide, all detected in the interstellar medium, are probably cometary molecules as well, and isomers of each are seen in coma spectra.

Future laboratory work will undoubtedly include the roles of interstellar grains in ice chemistry. Do grains simply provide an inert reaction template or are they a catalytic surface? Much surface chemistry remains to be done on cometary and interstellar molecules, and impressive experiments are already being published (*Fraser et al., 2002*).

5. CONCLUSIONS AND OUTLOOK

The results described in this chapter demonstrate the great value of laboratory simulations of cometary materials and processes. Progress in laboratory experiments has grown in parallel to the increase of information from cometary observations and the reliability of models describing solar system formation and the role played by comets. This chapter demonstrates that laboratory results are needed to understand the composition and evolution of both cometary nuclei and comae, and to provide guidelines for future exploration and model development. This is true for both refractory and ice components of comets.

Nevertheless, uncertainties remain and the field of research using laboratory experiments is rich with new tasks to accomplish. Generally speaking, the characterization of ice and dust analogs must continue to move in the direction of using complementary techniques, mainly *in situ* methods, applied during or immediately after sample production and/or treatment. Only in this way can information on the different factors that characterize materials, and their possible counterparts in real space conditions, be determined. New methods are also needed to produce samples with compositional and structural properties that can be selected by controllable parameters. In this respect, a new approach in laboratory experiments concerns *mixtures* of various refractory compounds (e.g., silicates and carbons) and of refractory and icy species. It is expected that results from this approach will offer a drastic change of perspective in the interpretation of observations.

As for refractory compounds, the study of silicate (amorphous vs. crystalline) structure and detailed composition (e.g., cation relative abundance) vs. space environment conditions is a key problem that extends over a cosmic scale, including material formation around cold stars, processing in the diffuse and dense ISM, and evolution during solar system formation. In this framework the role of silicate hydration is a subject that deserves particular attention in the future. Similarly, problems connected to carbonaceous grains also are important due to the strong connections to the evolution of C-based (large and small) molecular species. We now know that comets contain condensed organics and C grains (CHON particles), but the routes of formation and the connections among these two populations of compounds are still uncertain. To gain insight into this problem, a careful consideration is needed of how and when these compounds were formed in the circumstellar/interstellar media, and how they evolved until and after comet formation.

Concerning icy materials, section 4.3 above outlines some needed work, and the gaps in Tables 5 and 6 are obvious. A combination of small- and large-scale experiments is probably the correct approach for tracing the physical and chemical evolution of cometary bodies. The most reasonable way to understand and predict the actual behavior of comets, pre- and postperihelion, appears to start with an understanding of chemical reaction paths and to continue on through the identification of phenomena that affect the behavior of comet nuclei as a whole.

In conclusion, the aim of laboratory experiments is to increase our knowledge of how chemical species, at various levels and in different forms of arrangement, evolve in space. Laboratory scientists are encouraged to carry on their programs as the astrophysical community now recognizes that interpretations of cosmic evolution, comets included, cannot proceed without the firm reference frame offered by laboratory results.

Acknowledgments. The experimental work at INAF – Osservatorio Astronomico di Capodimonte is partially supported by contracts from ASI and MIUR. The research at Tel-Aviv University was supported by the Deutche Forschungs – Gemeinschaft, the Israel Science foundation (Grant 194/93-2), and the United States – Israel Bi-national Science Foundation (Grant 2000005). Work at NASA Goddard Space Flight Center was supported through NASA's Laboratory for Planetary Atmospheres and Space Astrophysics Research and Analysis programs.

REFERENCES

Allamandola L. J., Sandford S. A., and Valero G. J. (1988) Photochemical and thermal evolution of interstellar/precometary ice analogs. *Icarus, 76,* 225–252.

Allamandola L. J., Tielens A. G. G. M., and Herbst T. M. (1993) Diamonds in dense molecular clouds — A challenge to the standard interstellar medium paradigm. *Science, 260,* 64–66.

Bar-Nun A. and Laufer D. (2003) First experimental studies of large samples of gas-laden amorphous "cometary" ices. *Icarus, 161,* 157–163.

Bar-Nun A., Herman G., Laufer D., and Rappaport M. L. (1985) Trapping and release of gases by water ice and implications for icy bodies. *Icarus, 63,* 317–332.

Bar-Nun A., Dror J., Kochavi E., and Laufer D. (1987) Amorphous water ice and its ability to trap gases. *Phys. Rev. B., 35,* 2427–2435.

Baratta G. A., Spinella F., Leto G., Strazzulla G., and Foti G. (1991) The 3.1 μm feature in ion-irradiated water ice. *Astron. Astrophys., 252,* 421–424.

Baratta G. A., Mennella V., Brucato J. R., Colangeli L., Leto G., Palumbo M. E., and Strazzulla G. (2004) Raman spectroscopy of ion irradiated interplanetary carbon dust analogues. *J. Raman Spectroscopy,* in press.

Benkhoff J. and Spohn T. (1991) Thermal histories of the KOSI samples. *Geophys. Res. Lett., 18,* 261–264.

Bernstein M. P., Sandford S. A., Allamandola L. J., and Chang S. (1995) Organic compounds produced by photolysis of realistic interstellar and cometary ice analogs containing methanol. *Astrophys. J., 454,* 327–344.

Bernstein M. P., Dworkin J. P., Sandford S. A., Cooper G. W., and Allamandola L. J. (2002) Racemic amino acids from the ultraviolet photolysis of interstellar ice analogues. *Nature, 416,* 401–401.

Bertie J. E. and Devlin J. P. (1983) Infrared spectroscopic proof of the formation of the structure I hydrate of oxirane from annealed low-temperature condensate. *J. Chem. Phys., 78,* 6340–6341.

Biver N. and 22 colleagues (1999) Long-term evolution of the outgassing of comet Hale-Bopp from radio observations. *Earth Moon Planets, 78,* 5–11.

Blake D., Allamandola L., Sandford S., Hudgins D., and Freund F. (1991) Clathrate hydrate formation in amorphous cometary ice analogs in vacuo. *Science, 254,* 548–551.

Bockelée-Morvan D., Gautier D., Lis D. C., Young K., Keene J., Phillips T., Owen T., Crovisier J., Goldsmith P. F., Bergin E. A., Despois D., and Wootten A. (1998) Deuterated water in comet C/1996 B2 (Hyakutake) and its implications for the origin of comets. *Icarus, 133,* 147–162.

Bockelée-Morvan D., Gautier D., Hersant F., Huré J. M., and Robert F. (2002) Turbulent radial mixing in the solar nebula as the source of crystalline silicates in comets. *Astron. Astrophys., 384,* 1107–1118.

Bockelée-Morvan D., Crovisier J., Mumma M. J., and Weaver H. A. (2004) The composition of cometary volatiles. In *Comets II* (M. C. Festou et al., eds.) this volume. Univ. of Arizona, Tucson.

Bohren C. F. and Huffman D. R. (1983) *Absorption and Scattering of Light by Small Particles.* Wiley, New York. 544 pp.

Brinker C. J. and Scherer G. W. (1990) *Sol-Gel Science: The Physics and Chemistry of Sol-Gel Processing.* Academic, San Diego. 912 pp.

Brownlee D. E., Pilachowski L., Olszewski E., and Hodge P. W. (1980) Analysis of interplanetary dust collections. In *Solid Particles in the Solar System* (I. Halliday, ed.), pp. 333–341. Reidel, Dordrecht.

Brucato J., Palumbo M. E., and Strazzulla G. (1997) Carbonic acid by ion implantation in water/carbon dioxide ice mixtures. *Icarus, 125,* 135–144.

Brucato J. R., Colangeli L., Mennella V., Palumbo P., and Bussoletti E. (1999a) Mid-infrared spectral evolution of thermally annealed amorphous pyroxene. *Astron. Astrophys., 348,* 1012–1019.

Brucato J. R., Colangeli L., Mennella V., Palumbo P., and Bussoletti E. (1999b) Silicates in Hale-Bopp: Hints from laboratory studies. *Planet. Space Sci., 47,* 773–779.

Brucato J. R., Mennella V., Colangeli L., Rotundi A., and Palumbo P. (2002) Production and processing of silicates in laboratory and in space. *Planet. Space Sci., 50,* 829–837.

Campins H. and Ryan E. V. (1989) The identification of crystalline olivine in cometary silicates. *Astrophys. J., 341,* 1059–1066.

Chiar J. E., Pendleton Y. J., Geballe T. R., and Tielens A. G. G. M. (1998) Near-infrared spectroscopy of the proto-planetary nebula CRL 618 and the origin of the hydrocarbon dust component in the interstellar medium. *Astrophys. J., 507,* 281–286.

Chyba C. F. (1987) The cometary contribution to the oceans of primitive Earth. *Nature, 330,* 632–635.

Chyba C. F. (1990) Impact delivery and erosion of planetary oceans in the inner solar system. *Nature, 343,* 129–133.

Cochran A. L. (2002) A search for N_2^+ in spectra of comet C/2002 C1 (Ikeya-Zhang). *Astrophys. J. Lett., 576,* L165–L168.

Cochran A. L., Cochran W. D., and Barker E. B. (2000) N_2^+ and CO^+ in Comets 122P/1995 S1 (de Vico) and C/1995 O1 (Hale-Bopp). *Icarus, 146,* 583–593.

Colangeli L., Mennella V., Baratta G. A., Bussoletti E., and Strazzulla G. (1992) Raman and infrared spectra of polycyclic aromatic hydrocarbon molecules of possible astrophysical interest. *Astrophys. J., 396,* 369–377.

Colangeli L., Mennella V., di Marino C., Rotundi A., and Bussoletti E. (1995) Simulation of the cometary 10 μm band by means of laboratory results on silicatic grains. *Astron. Astrophys., 293,* 927–934.

Colangeli L., Mennella V., Rotundi A., Palumbo P., and Bussoletti E. (1996) Simulation of the cometary 10 micron band by laboratory data. II. Extension to spectra available for different comets. *Astron. Astrophys., 312,* 643–648.

Colangeli L. and 20 colleagues (2003) The role of laboratory experiments in the characterisation of silicon-based cosmic material. *Astron. Astrophys. Rev., 11,* 97–152.

Cottin H., Gazeau M. C., and Raulin F. (1999) Cometary organic chemistry: A review from observations, numerical and experimental simulations. *Planet. Space Sci., 47,* 1141–1162.

Crovisier J. (1999) Infrared observations of volatile molecules in comet Hale-Bopp. *Earth Moon Planets, 79,* 125–143.

Crovisier J., Leech K., Bockelée-Morvan D., Brooke T. Y., Hanner M. S., Altieri B., Keller H. U., and Lellouch E. (1997) The spectrum of Comet Hale-Bopp (C/1995 O1) observed with the Infrared Space Observatory at 2.9 AU from the Sun. *Science, 275,* 1904–1907.

Davies J. K., Puxley P. J., Mumma M. J., Reuter D. C., Hoban S., Weaver H. A., and Lumsden S. L. (1993) The infrared (3.2–3.6 micron) spectrum of comet P/Swift-Tuttle — Detection of methanol and other organics. *Mon. Not. R. Astron. Soc., 265,* 1022–1993.

Davies J. K., Roush T. L., Cruikshank D. P., Bartholemew M. J., Geballe T. R., Owen T., and de Bergh C. (1997) The detection of water ice in comet Hale-Bopp. *Icarus, 127,* 238–245.

Demyk K., Dartois E., d'Hendecourt L., Jourdain de Muizon M., Heras A. M., and Breitfellner M. (1998) Laboratory identification of the 4.62 μm solid state absorption band in the ISO-SWS spectrum of RAFGL 7009S. *Astron. Astrophys., 339,* 553–560.

Despois D. (1997) Radio observation of molecular and isotopic species: Implications on the interstellar origin of cometary ices. *Earth Moon Planets, 79,* 103–124.

Dones L., Weissman P. R., Levison H. F., and Duncan M. J. (2004) Oort cloud formation and dynamics. In *Comets II* (M. C. Festou et al., eds.), this volume. Univ. of Arizona, Tucson.

Eberhardt P., Reber M., Krankowsky D., and Hodges R. R. (1995) The D/H and $^{18}O/^{16}O$ ratios in water from comet P/Halley. *Astron. Astrophys., 302,* 301–318.

Ehrenfreund P., Charnley S. B., and Wooden D. (2004) From interstellar material to comet particles and molecules. In *Comets II* (M. C. Festou et al., eds.), this volume. Univ. of Arizona, Tucson.

Engel S., Lunine J. I., and Lewis J. S. (1990) Solar nebula origin for volatile gases in Halley's comet. *Icarus, 85,* 380–393.

Fabian D., Jäger C., Henning Th., Dorschner J., and Mutschke H. (2000) Steps toward interstellar silicate mineralogy. V. Thermal evolution of amorphous magnesium silicates and silica. *Astron. Astrophys., 364,* 282–292.

Farnham T. L. and Cochran A.L. (2002) A McDonald observatory study of comet 19P/Borrelly: Placing the Deep Space 1 observations into a broader context. *Icarus, 160,* 398–418.

Feigelson E. D. and Montmerle T. (1999) High-energy processes in young stellar objects. *Annu. Rev. Astron. Astrophys., 37,* 363–408.

Fraser H. J., Collings M. P., and McCoustra M. R. S. (2002) Laboratory surface astrophysics experiment. *Rev. Sci. Instrum., 73,* 2161–2170.

Gerakines P. A. and Moore M. H. (2001) Carbon suboxide in

astrophysical ice analogs. *Icarus, 154,* 372–380.

Gerakines P. A., Schutte W. A., and Ehrenfreund P. (1996) Ultraviolet processing in interstellar ice analogs: I. Pure ices. *Astron. Astrophys., 312,* 289–305.

Gerakines P. A., Moore M. H., and Hudson R. L. (2000) Carbonic acid production in $H_2O:CO_2$ ices. UV photolysis vs. proton bombardment. *Astron. Astrophys., 357,* 793–800.

Goldreich P. and Ward W. R. (1973) The formation of planetesimals. *Astrophys. J., 183,* 1051–1062.

Green J. R. and Bruesch L. S. (2000) Laboratory simulation of cometary nuclei: Mechanical properties of porous ice-dust mixtures. *Bull. Am. Astron. Soc., 32,* 1061.

Green J. R., Bruesch L. S., Oakes R., Pinkham B., and Folsom C. L. (1999) Comet simulation experiments at JPL. *Bull. Am. Astron. Soc., 31,* 1589.

Greenberg J. M. and Hage J. I. (1990) From interstellar dust to comets — A unification of observational constraints. *Astrophys. J., 361,* 260–274.

Grim R. J. A. and Greenberg J. M. (1987a) Photoprocessing of H_2S in interstellar grain mantles as an explanation for S2 in comets. *Astron. Astrophys., 181,* 155–168.

Grim R. J. A. and Greenberg J. M. (1987b) Ions in grain mantles — The 4.62 micron absorption by OCN(–) in W33A. *Astrophys. J. Lett., 321,* L91–L96.

Grün E., Bar-Nun A., Benkhoff J., Bischoff A., Düren H., Hellman H., Hesselbarth P., Hsiung P., Keller H. U., Klinger J., Knölker J., Kochan H., Kohl H., Kölzer G., Krankowsky D., Lämmerzahl P., Mauersberger K., Neukum G., Öhler A., Ratke L., Rössler K., Schwehm G., Spohn T., Stöffler D., and Thiel K. (1991) Laboratory simulation of cometary processes: Results from first KOSI experiments. In *Comets in the Post-Halley Era* (R. L. Newburn et al., eds.), pp. 277–297. Kluwer, Dordrecht.

Gustafson B. Å S. (1996) Microwave analog to light scattering measurements: A modern implementation of a proven method to achieve precise control. *J. Quant. Spectrosc. Radiat. Transfer, 55,* 663–672.

Hadamcik E., Renard J. B., Worms J. C., Levasseur-Regourd A. C., and Masson M. (2002) Polarization of light scattered by fluffy particles (PROGRA² experiment). *Icarus, 155,* 497–508.

Hallenbeck S. L., Nuth J. A., and Daukantas P. L. (1998) Mid-infrared spectral evolution of amorphous magnesium silicate smokes annealed in vacuum: Comparison to cometary spectra. *Icarus, 131,* 198–209.

Hallenbeck S. L., Nuth J. A. III, and Nelson R. N. (2000) Evolving optical properties of annealing silicate grains: From amorphous condensate to crystalline mineral. *Astrophys. J., 535,* 247–255.

Hanner M. S. and Bradley J. P. (2004) Composition and mineralogy of cometary dust. In *Comets II* (M. C. Festou et al., eds.), this volume. Univ. of Arizona, Tucson.

Hanner M. S., Lynch D. K., and Russell R. W. (1994) The 8–13 micron spectra of comets and the composition of silicate grains. *Astrophys. J., 425,* 274–285.

Harker D. E. and Desch S. J. (2002) Annealing of silicate dust by nebular shocks at 10 AU. *Astrophys. J. Lett., 565,* L109–L112.

Harker D. E., Wooden D. H., Woodward C. E., and Lisse C. M. (2002) Grain properties of comet C/1995 O1 (Hale-Bopp). *Astrophys. J., 580,* 579–597.

Hayward T. L. and Hanner M. S. (1997) Spectrocam-10 thermal infrared observations of the dust in comet C/1995 O1 Hale-Bopp. *Earth Moon Planets, 78,* 265–270.

Hilchenbach M., Kochemann O., and Rosenbauer H. (2000) Impact on a comet: Rosetta Lander simulations. *Planet. Space Sci., 48,* 361–369.

Hovenier J.W. and van der Mee C. V. M. (2000) Measuring scattering matrices of small particles at optical wavelengths. In *Light Scattering by Non-Spherical Particles* (M. I. Mishchenko et al., eds.), pp. 61–85. Academic, San Diego.

Hudson R. L. and Donn B. (1991) An experimental study of the sublimation of water ice and the release of trapped gases. *Icarus, 94,* 326–332.

Hudson R. L. and Moore M. H. (1993) Far-infrared investigations of a methanol clathrate hydrate: Implications for astronomical observations. *Astrophys. J. Lett., 404,* L29–L32.

Hudson R. L. and Moore M. H. (1995) Far-IR spectral changes accompanying proton irradiation of solids of astrochemical interest. *Radiat. Phys. Chem., 45,* 779–789.

Hudson R. L. and Moore M. H. (1999) Laboratory studies of the formation of methanol and other organic molecules by water + carbon monoxide radiolysis: Relevance to comets, icy satellites, and interstellar ices. *Icarus, 140,* 451–461.

Hudson R. L. and Moore M. H. (2000) IR spectra of irradiated cometary ice analogues containing methanol: A new assignment, a reassignment, and a non-assignment. *Icarus, 145,* 661–663.

Hudson R. L. and Moore M. H. (2003) Solid-phase formation of interstellar vinyl alcohol. *Astrophys. J. Lett., 586,* L107–L110.

Hudson R. L., Moore M. H. and Gerakines P. A. (2001) The formation of cyanate ion (OCN^-) in interstellar ice analogues. *Astrophys. J., 550,* 1140–1150.

Irvine W. M., Schloerb F. P., Crovisier J., Fegley B., and Mumma M. J. (2000) Comets: A link between interstellar and nebular chemistry. In *Protostars and Planets IV* (V. Mannings et al., eds.), p. 1159. Univ. of Arizona, Tucson.

Jäger C., Dorschner J., Mutschke,H., Posch Th., and Henning Th. (2003) Steps toward interstellar silicate mineralogy. VII. Spectral properties and crystallization behaviour of magnesium silicates produced by the sol-gel method. *Astron. Astrophys., 408,* 193–204.

Jenniskens P. and Blake D. F. (1994) Structural transitions in amorphous water ice and astrophysical implications. *Science, 265,* 753–756.

Jenniskens P. and Blake D. F. (1996) Crystallization of amorphous water ice in the solar system. *Astrophys. J., 473,* 1104–1113.

Jenniskens P., Baratta G. A., Kouchi A., deGroot M. S., Greenberg J. M., and Strazzulla G. (1993) Carbon dust formation on interstellar grains. *Astron. Astrophys., 273,* 583–600.

Jenniskens P., Blake D. F., Wilson M. A., and Pohorille A. (1995) High density amorphous ice, the frost on interstellar grains. *Astrophys. J., 455,* 389–401.

Jenniskens P., Banham S. F., Blake D. F., and McCoustra M. R. S. (1997) Liquid water in the domain of cubic crystalline ice IC. *J. Chem. Phys., 107,* 1232–1241.

Jessberger H. L. and Kotthaus M. (1989) Compressive strength of synthetic comet nucleus samples. In *Proceedings of an International Workshop on the Physics and Mechanics of Cometary Materials* (J. Hunt and T. D. Guyenne, eds.), pp. 141–146. ESA SP-302, Noordwijk, The Netherlands.

Johnson R. E., Cooper J. F., Lanzerotti L. J., and Strazzulla G. (1987) Radiation formation of a non-volatile comet crust. *Astron. Astrophys., 187,* 889–892.

Kaiser R. I. and Roessler K. (1998) Theoretical and laboratory studies on the interaction of cosmic-ray particles with interstellar ices. III. Suprathermal chemistry-induced formation of hydrocarbon molecules in solid methane, (CH_4), ethylene

(C_2H_4), and acetylene (C_2H_2). *Astrophys. J., 503,* 959–975.

Kawakita H. and 18 colleagues (2001) The spin temperature of NH_3 in comet C/1999S4 (LINEAR). *Science, 294,* 1089–1091.

Keller H. U., Britt D., Buratti B. J., and Thomas N. (2004) In situ observations of cometary nuclei. In *Comets II* (M. C. Festou et al., eds.), this volume. Univ. of Arizona, Tucson.

Klinger J. (1980) Influence of a phase transition of ice on the heat and mass balance of comets. *Science, 209,* 271–272.

Köster H. M. (1979) *Die chemische Silikatanalyse.* Springer-Verlag, Berlin.

Kouchi A. and Sirono S. (2001) Crystallization heat of impure amorphous H_2O ice. *Geophys. Res. Lett., 28,* 827–830.

Kouchi A., Greenberg J. M., Yamamoto T., and Mukai T. (1992) Extremely low thermal conductivity of amorphous ice — Relevance to comet evolution. *Astrophys. J. Lett., 388,* L73–L76.

Laufer D., Kochavi E., and Bar-Nun A. (1987) Structure and dynamics of amorphous water ice. *Phys. Rev. B, 36,* 9219–9227.

Laufer D., Notesco G., Bar-Nun A., and Owen T. (1999) From the interstellar medium to Earth's oceans via comets — an isotopic study of HDO/H_2O. *Icarus, 140,* 446–450.

Lawson W. A., Feigelson E. D., and Huenemoerder D. P. (1996) An improved HR diagram for Chamaeleon I pre-main-sequence stars. *Mon. Not. R. Astron. Soc., 280,* 1071–1088.

Leto G. and Baratta G. A. (2003) Ly-alpha photon induced amorphization of Ic water ice at 16 K. Effects and quantitative comparison with ion irradiation. *Astron. Astrophys., 397,* 7–13.

Levasseur-Regourd A. C., Hadamcik E., and Renard J. B. (1996) Evidence for two classes of comets from their polarimetric properties at large phase angles. *Astron. Astrophys., 313,* 327–333.

Lumme K., Rahola J., and Hovenier J. W. (1997) Light scattering by dense clusters of spheres. *Icarus, 126,* 455–469.

Lunine J. I. (1989) Primitive bodies — Molecular abundances in Comet Halley as probes of cometary formation environments. In *The Formation and Evolution of Planetary Systems* (H. A. Weaver and L. Danly, eds.), pp. 213–238. Cambridge Univ., Cambridge.

Marfunin A. S. (1995) *Methods and Instrumentations: Results and Recent Developments.* Springer-Verlag, New York. 441 pp.

Mauersberger K., Michel H. J., Krankowsky D., Laemmerzahl P., and Hesselbarth P. (1991) Measurement of the volatile component in particles emitted from an ice/dust mixture. *Geophys. Res. Lett., 18,* 277–280.

Meier R., Owen T. C., Jewitt D. C., Matthews H. E., Senay M., Biver N., Bockelée-Morvan D., Crovisier J., and Gautier D. (1998) Deuterium in comet C/1995 O1 (Hale-Bopp): Detection of DCN. *Science, 279,* 1707–1710.

Melosh H. J. and Vickery A. M. (1989) Impact erosion of the primordial atmosphere of Mars. *Nature, 338,* 487–489.

Mennella V., Colangeli L., Blanco A., Bussoletti E., Fonti S., Palumbo P., and Mertins H. C. (1995) A dehydrogenation study of cosmic carbon analogue grains. *Astrophys. J., 444,* 288–292.

Mennella V., Colangeli L., Palumbo P., Rotundi A., Schutte W., and Bussoletti E. (1996) Activation of an ultraviolet resonance in hydrogenated amorphous carbon grains by exposure to ultraviolet radiation. *Astrophys. J. Lett., 464,* L191–L194.

Mennella V., Baratta G. A., Colangeli L., Palumbo P., Rotundi A., Bussoletti E., and Strazzulla G. (1997) Ultraviolet spectral changes in amorphous carbon grains induced by ion irradiation. *Astrophys. J., 481,* 545–549.

Mennella V., Brucato J. R., Colangeli L., and Palumbo P. (1999) Activation of the 3.4 micron band in carbon grains by exposure to atomic hydrogen. *Astrophys. J. Lett., 524,* L71–L74.

Mennella V., Muñoz C. G. M., Ruiterkamp R., Schutte W. A., Greenberg J. M., Brucato J. R., and Colangeli L. (2001) UV photodestruction of CH bonds and the evolution of the 3.4 μm feature carrier. II. The case of hydrogenated carbon grains. *Astron. Astrophys., 367,* 355–361.

Mishchenko M. I., Hovenier J. W., and Travis L. D. (2000) *Light Scattering by Nonspherical Particles: Theory, Measurements, and Applications.* Academic, San Diego. 690 pp.

Moore M. H. (1984) Studies of proton-irradiated SO_2 at low temperatures: Implications for Io. *Icarus, 59,* 114–128.

Moore M. H. and Hudson R. L. (1998) Infrared study of ion irradiated water ice mixtures with organics relevant to comets. *Icarus, 135,* 518–527.

Moore M. H. and Hudson R. L. (2000) IR detection of H_2O_2 at 80 K in ion-irradiated laboratory ices relevant to Europa. *Icarus, 145,* 282–288.

Moore M. H. and Hudson R. L. (2003) Infrared study of ion irradiated N_2-dominated ices relevant to Triton and Pluto: Formation of HCN and HNC. *Icarus, 161,* 486–500.

Mulas G., Baratta, G. A., Palumbo M. E., and Strazzulla G. (1998) Profile of CH_4 IR bands in ice mixtures. *Astron. Astrophys., 333,* 1025–1033.

Mumma, M. J. (1997) Organic volatiles in comets: Their relation to interstellar ices and solar nebula material. In *From Stardust to Planetesimals* (Y. J. Pendleton and A. G. G. M. Tielens, eds.), pp. 369–396. ASP Conference Series 122, Astronomical Society of the Pacific, San Francisco.

Mumma M. J., Blass W. E., Weaver H. A., and Larson H. P. (1988) Measurements of the ortho-para ratio and nuclear spin temperature of water vapor in comets Halley and Wilson (1986) and implications for their origin and evolution. *Bull. Am. Astron. Soc., 20,* 826.

Mumma M. J., DiSanti M. A., Dello Russo N., Fomenkova M., Magee-Sauer K., Kaminski C. D., and Xie D. X. (1996) Detection of abundant ethane and methane, along with carbon monoxide and water, in comet C/1996 B2 Hyakutake: Evidence for interstellar origin. *Science, 272,* 1310–1314.

Muñoz O., Volten H., de Haan J. F., Vassen W., and Hovenier J. W. (2000) Experimental determination of scattering matrices of olivine and Allende meteorite particles. *Astron. Astrophys., 360,* 777–788.

Notesco G. and Bar-Nun A. (1997) Trapping of methanol, hydrogen cyanide, and n-hexane in water ice, above its transformation temperature to the crystalline form. *Icarus, 126,* 336–341.

Notesco G. and Bar-Nun A. (2000) The effect of methanol clathrate hydrate formation and other gas-trapping mechanisms on the structure and dynamics of cometary ices. *Icarus, 148,* 456–463.

Notesco G., Bar-Nun A., and Owen T. (2003) Gas trapping in water ice at very low deposition rates and implication for comets. *Icarus, 162,* 183–189.

Nuth J. A., Hill H. G. M., and Kletetschka G. (2000) Determining the ages of comets from the fraction of crystalline dust. *Nature, 406,* 275–276.

Nuth J. A. III, Rietmeijer F. J. M., and Hill H. G. M. (2002) Condensation processes in astrophysical environments: The composition and structure of cometary grains. *Meteoritics & Planet. Sci., 37,* 1579–1590.

Owen T. and Bar-Nun A. (1993) Noble gases in atmospheres. *Nature, 361,* 693–694.

Owen T. and Bar-Nun A. (1995) Comets, impacts and atmospheres. *Icarus, 116,* 215–226.

Owen T., Bar-Nun A., and Kleinfeld I. (1991) Noble gases in

terrestrial planets: Evidence for cometary impacts? In *Comets in the Post-Halley Era I* (R. L. Newburn Jr. et al., eds.), pp. 429–437. Kluwer, Dordrecht.

Owen T., Bar-Nun A., and Kleinfeld I. (1992) Possible cometary origin of heavy noble gases in the atmospheres of Venus, Earth and Mars. *Nature, 358*, 43–46.

Palumbo M. E., Castorina A. C., and Strazzulla G. (1999) Ion irradiation effects on frozen methanol (CH$_3$OH). *Astron. Astrophys., 342*, 551–562.

Pendleton Y. J., Sandford S. A., Allamandola L. J., Tielens A. G. G. M., and Sellgren K. (1994) Near-infrared absorption spectroscopy of interstellar hydrocarbon grains. *Astrophys. J., 437*, 683–696.

Pepin R. O. (1997) Evolution of Earth's noble gases: Consequences of assuming hydrodynamic loss driven by giant impact. *Icarus, 126*, 148–156.

Prialnik D. and Bar-Nun A. (1992) Crystallization of amorphous ice as the cause of comet P/Halley's outburst at 14 AU. *Astron. Astrophys., 258*, L9–L12.

Richardson H. H., Wooldridge P. J., and Devlin J. P. (1985) FT-IR spectra of vacuum deposited clathrate hydrate of oxirane H$_2$S, THF, and ethane. *J. Chem. Phys., 83*, 4387–4394.

Rickman H. (1989) The nucleus of Comet Halley — surface structure, mean density, gas and dust production. *Adv. Space Res., 9*, 59–71.

Rietmeijer F. J. M. (1998) Interplanetary dust particles. In *Planetary Materials* (J. J. Papike, ed.), in *Rev. Mineral., 36*, pp. 2-1 to 2-95. Mineralogical Society of America, Washington, DC.

Rietmeijer F. J. M., Hallenbeck S. L., Nuth J. A. III, and Karner J. M. (2002) Amorphous magnesiosilicate smokes annealed in vacuum: The evolution of magnesium silicates in circumstellar and cometary dust. *Icarus, 156*, 269–286.

Rodgers S. D. and Charnley S. B. (2001) On the origin of HNC in comet Lee. *Mon. Not. R. Astron. Soc., 323*, 84–92.

Rotundi A., Brucato J. R., Colangeli L., Ferrini G., Mennella V. E., and Palumbo P. (2002) Production, processing and characterisation techniques for cosmic dust analogues. *Meteoritics & Planet. Sci., 37*, 1623–1635.

Salama F., Allamandola L. J., Witteborn F. C., Cruikshank D. P., Sandford S. A., and Bregman J. D. (1990) The 2.5–5.0 micron spectra of Io — Evidence for H$_2$S and H$_2$O frozen in SO$_2$. *Icarus, 83*, 66–82.

Sandford S. A. and Allamandola L. J. (1988) The condensation and vaporization behavior of H$_2$O : CO ices and implications for interstellar grains and cometary activity. *Icarus, 76*, 201–224.

Schmitt B. and Klinger J. (1987) Different trapping mechanisms of gases by water ice and their relevance for comet nuclei. In *The Diversity and Similarity of Comets* (E. J. Rolfe et al., eds.), pp. 613–619. ESA SP-278, Noordwijk, The Netherlands.

Schmitt B., Espinasse S., Grim R. J. A., Greenberg J. M., and Klinger J. (1989a) Laboratory studies of cometary ice analogues. In *Proceedings of an International Workshop on Physics and Mechanics of Cometary Materials* (J. Hunt and T. D. Guyenne, eds.), pp. 65–69. ESA SP-302, Noordwijk, The Netherlands.

Schmitt B., Greenberg J. M., and Grim R. J. A. (1989b) The temperature dependence of the CO infrared band strength in CO : H$_2$O ices. *Astrophys. J. Lett., 340*, L33–L36.

Seiferlin K., Komle N. I., Kargl G., and Spohn T. (1996) Line heat source measurements of the thermal conductivity of porous H$_2$O ice, CO$_2$ ice and mineral powders under space conditions. *Planet. Space. Sci., 44*, 691–704.

Spohn T., Seiferlin K., and Benkhoff J. (1989) Thermal conductivities and diffusivities of porous ice samples at low pressures and temperatures and possible modes of heat transfer in near surface layers of comets. In *Proceedings of an International Workshop on Physics and Mechanics of Cometary Materials* (J. Hunt and T. D. Guyenne, eds.), pp. 77–81. ESA SP-302, Noordwijk, The Netherlands.

Strazzulla G. and Johnson R. E. (1991) Irradiation effects on comets and cometary debris. In *Comets in the Post-Halley Era* (R. Newburn Jr et al., eds.), pp. 243–275. Kluwer, Dordrecht.

Strazzulla G. and Palumbo M. E. (1998) Evolution of icy surfaces: an experimental approach. *Planet. Space Sci., 46*, 1339–1348.

Strazzulla G. and Palumbo M. E. (2001) Organics produced by ion irradiation of ices: Some recent results. *Adv. Space Res., 27*, 237–243.

Strazzulla G., Baratta G. A., and Palumbo M. E. (2001) Vibrational spectroscopy of ion-irradiated ices. *Spectrochim. Acta, 57A*, 825–842.

Thompson S. P., Evans A., and Jones A. P. (1996) Structural evolution in thermally processed silicates. *Astron. Astrophys., 308*, 309–320.

van de Hulst H. C. (1957) *Light Scattering by Small Particles.* Wiley, New York. 470 pp.

Volten H., Muñoz O., Rol E., de Haan J. F., Vassen W., Hovenier J. W., Muinonen K., and Nousiainen T. (2001) Scattering matrices of mineral aerosol particles at 441.6 nm and 632.8 nm. *J. Geophys. Res., 106*, 17375–17402.

Weidenschilling S. (1997) The origin of comets in the solar nebula: A unified model. *Icarus, 127*, 290–306.

Weissman P. R. (1986) Are cometary nuclei primordial rubble piles? *Nature, 320*, 242–244.

Weissman P. R., Asphaug E., and Lowry S. C. (2004) Structure and density of cometary nuclei. In *Comets II* (M. C. Festou et al., eds.), this volume. Univ. of Arizona, Tucson.

Wilkening L., ed. (1982) *Comets.* Univ. of Arizona, Tucson. 775 pp.

Wooden D. H., Harker D. E., Woodward C. E., Butner H. M., Koike C., Witteborn F. C., and McMurtry C. W. (1999) Silicate mineralogy of the dust in the inner coma of comet C/1995 O1 (Hale-Bopp) pre- and postperihelion. *Astrophys. J., 517*, 1034–1058.

Wopenka B. (1988) Raman observations on individual interplanetary dust particles. *Earth Planet. Sci. Lett., 88*, 221–231.

Wu C. Y. R., Judge D. L., Cheng B.-M., Shih W. H., Yih T. S., and Ip W. H. (2002) Extreme ultraviolet photon-induced chemical reactions in the C$_2$H$_2$-H$_2$O mixed ices at 10 K. *Icarus, 156*, 456–473.

Wu C. Y. R., Judge D. L., Cheng B.-M., Yih T.-S., Lee C. S., and Ip W. H. (2003) Extreme ultraviolet photolysis of CO$_2$-H$_2$O mixed ices at 10 K. *J. Geophys. Res., 108(E4)*, 13-1 to 13-8.

Glossary

absorption — The reduction in the intensity of radiation when it passes through a medium by any process in which the radiation is converted to excitation energy or heat or by any process other than scattering or immediate re-radiation is called absorption. The process in which the intensity is reduced by being scattered (by reflection, refraction, or diffraction) or by being temporarily absorbed and immediately re-radiated in all directions is called scattering. The combined effect of absorption and scattering is called extinction.

accretion — Accumulation of solid particles by collisions and, later, the inclusion of gases under the action of gravity, by which large bodies such as stars, planets, and their satellites are formed.

active area/region — When compared with the neighboring regions, a region/area on the surface of a cometary nucleus from where enhanced outgassing could occur.

active comet/nucleus — A comet that releases gases and solid particles from its surface, usually due to its proximity to the Sun. Active comets develop a visible coma of gas and/or dust, although, as demonstrated by automatic search programs such as the Lincoln Laboratory Near-Earth Asteroid Research program (LINEAR), the presence of matter around a nucleus can at times be difficult to establish, thus leading to ambiguous identifications of the physical nature of objects.

active fraction — *See* fraction, active area.

aggregate — A body that is not a monolith. Aggregates are categorized according to their coherency and bulk porosity.

airglow — A light in the night (nightglow) or day sky (dayglow) caused by the excitation, ionization, or dissociation of atoms and molecules (mainly O_2, N_2, O, OH in Earth's atmosphere) by ultraviolet solar radiation.

albedo — A measure of how much light is reflected from a surface or from an object. Several types of albedo are in use, of which the most familiar are defined here (*see* albedo, bolometric; albedo, Bond; albedo, geometric; albedo, normal; albedo, particle). For more precise definitions, see *Hanner et al.* (1981, *Astron. Astrophys., 104,* 42), *Hapke* (1979, *J. Geophys. Res., 86,* 3039), *Lester et al.* (1979, *J. R. Astron. Soc. Can., 73,* 233). All definitions of albedo are dimensionless (and hence unitless) ratios.

albedo, bolometric — Albedos have to be defined at a particular wavelength. If given for the entire electromagnetic spectrum integrated over all wavelengths, the albedo is called a bolometric albedo (whether normal, geometric, or Bond). The bolometric albedo is thus the ratio of the scattered energy to the sum (scattered + reemitted) energy. It is unity only when no energy is absorbed and consequently reemitted.

albedo, Bond — The Bond albedo is a concept applied to a body that is large compared with the wavelength of the radiation incident upon it. It is supposed that some of the incident radiation will be absorbed and is therefore not included in the scattered component (but *see also* albedo, bolometric). For an opaque body, the scattering will be by reflection from the surface. For a transparent or semitransparent body, some of the light will also be scattered by refraction through it. Because the size of the body is large compared with the wavelength of the radiation, a negligible fraction of the radiation is scattered by diffraction. The Bond albedo of such a body is defined as the ratio of the radiant flux scattered (by reflection and refraction) in all directions to the radiant flux intercepted by its geometrical cross-section. The Bond albedo A_B and the geometric albedo p are related through the equation $A_B = pq$, where q is the phase integral.

albedo feature — Surface feature on an object characterized by a change in albedo due to differences in surface composition, porosity, or surface microstructure, or to macroscopic geological or topographical structure.

albedo, geometric — The geometric albedo p of a sphere is the ratio of the intensity of radiation reflected back to a source to the intensity of radiation reflected back to the same source by a lossless lambertian disk of the same diameter as the sphere. The concept is trivially extended to a body of any shape by substitution of the words "a lossless lambertian surface of the same size and shape of the geometric cross section of the body in question."

albedo, normal — The normal albedo p_n of a surface is the ratio of the normally reflected radiance of the surface compared with the normally reflected radiance of a lossless (conservative) lambertian reflector, when both are illuminated normally and under the same conditions by the same source of light.

albedo, particle — The albedo of a particle is a concept applied to a particle whose size is comparable to that of radiation, and hence the scattered radiation includes a diffracted component as well as reflected and refracted components. Some of the incident radiation may of course also be absorbed. The particle albedo A is the ratio of the radiant flux scattered in all directions by reflection, refraction, and diffraction to the radiant flux removed from the incident beam, i.e., scattered and absorbed. For a large particle, most of the diffraction is in the forward direction. For a very large particle, very little of the scattered component is diffracted, and diffraction is almost completely in the forward direction. In the limit of a particle that is so large compared with the wavelength of the radiation that dif-fraction can be neglected, it can be shown that the particle albedo A and the Bond albedo A_B are related by $A = \frac{1}{2}(1 + A_B)$. Consequently, for a lossless (conservative or white) body, $A = A_B = 1$, and for a completely absorbing (black) body, $A_B = 0$ and $A = 0.5$. The albedo is usually defined at a specific wavelength q (*see* albedo, bolometric).

Alfvén speed — The speed at which a transverse magneto-hydrodynamic wave (Alfvén wave) propagates along a magnetic line: $V_A = B/\sqrt{\mu_0 \rho}$, where B is the magnetic field and ρ is the plasma density.

Alfvén waves — Waves moving transversely across a magnetic field line in a plasma.

amorphous ice — A form of ice arranged in a random pattern and thus lacking long-range order. The transition of amorphous ice into a regular crystalline structure is exothermic and occurs below 140 K.

annealing — A very slow and uniform heating, then cooling, of a material that induces temperature-dependent structural changes of its microstructure. In the context of cometary physics, annealing usually refers to a gradual and irreversible rearrangement of the molecular structure of water ice from a disordered (amorphous) state toward an ordered (crystalline) state. It starts below 100 K and its rate increases with increasing temperature.

antitail — Also referred to as anomalous tail, spike, or dart. Narrow tail formed of large and slow solid particles lying in or near the orbital plane of the comet that projects on the sky between the Sun and the comet nucleus. It is the differential motion between the comet nucleus and the particles combined with a perspective effect that can create such a structure. In some cases, this narrowness can be explained by a structure called the neckline.

aperture photometry — Integrated flux or magnitude measurements recorded within an aperture, usually circular. The aperture must be significantly larger than the seeing disk to result in a meaningful measurement.

aspect angle — Angle between the rotation axis of a body and the radius vector to Earth.

Baldet-Johnson bands — The $B^2\Sigma^+ - A^2\Pi_i$ band system of the CO^+ ion, observed in comet spectra below 400 nm.

Ballik-Ramsay bands — The $b^3\Sigma_g^- - a^3\Pi_u$ band system of the molecule C_2, observed in comet spectra in the near-IR region.

bolometric — The total radiation over the entire electromagnetic spectrum. A bolometer is an apparatus designed to measure the total radiant energy from a source.

boundary layer — Also called the Knudsen (disequilibrium) layer. Usually, a gas does not assume a Maxwellian velocity distribution in the immediate vicinity of a solid surface. The region where this non-equilibrium condition prevails is called the boundary layer. Its thickness is on the order of 10 to 100 the local collisional mean free path.

Bowen fluorescence — Mechanism invoked in 1947 by I. S. Bowen to explain the anomalous strength of some emission lines observed in planetary nebulae. The upper level of an emission line of an element is selectively excited by an emission line of another element that, coincidentally, has almost the right wavelength to do so. For example, the H I Lyman β line has almost exactly the right energy to excite the upper level of the oxygen line at 102.6 nm, which can then cascade to the ground state through oxygen lines at 130.4 nm, 844.7 nm, and 1128.7 nm, each of which is therefore much stronger than expected. This process is also effective in the selective excitation of molecules such as CO and H_2.

bow shock — A surface or sheet of discontinuity (i.e., abrupt changes in physical conditions) set up in a supersonic flow at which the fluid undergoes a finite decrease in velocity accompanied by increase in pressure, density, temperature, and entropy, as occurs, e.g., in the supersonic solar wind flow about Earth's magnetosphere. The bow shock created by the solar wind at Venus and Mars is similar in nature to that seen in comets. In Comet P/Halley, during the encounters of the *Giotto* and *Vega* spacecraft with the comet near 1 AU from the Sun, the bow shock was found at about 10^6 km and is a transition region for the solar wind plasma rather than a sharp discontinuity.

brightness — Usually used to mean the intensity of light or other radiation, which may create confusion. There are a number of very clearly defined terms in textbooks, such as radiant flux, intensity, exitance, radiance, and irradiance; these terms, which all have different units, should be used instead of "brightness."

brightness, absolute — A measure of the total luminosity of an object (*see* magnitude).

brightness distribution/profile — The distribution of brightness over the surface of an object.

brightness, surface — Often used for what should more correctly be called "radiance," an energy emitted per second by a unit surface area into a unit steradian (W m^{-2} steradian^{-1}).

brightness temperature — The temperature of a black body that would have the same radiance as that of the observed source (the comparison can be performed at one specific wavelength, inside a wavelength range, or over the entire electromagnetic spectrum). It is equal to the real temperature of the source only if the source is in thermodynamic equilibrium.

Brownlee particles — Cosmic dust particles collected from the upper atmosphere, usually by high-flying aircraft. Named after Donald P. Brownlee. *See* interplanetary dust particle (IDP).

Cameron bands — The $a^3\Pi_r - X^1\Sigma^+$ band system of the CO molecule. Since the ground state (X) is a singlet state and the upper state (a) is a triplet, the band system is forbidden. The strongest bands that are observed in comet spectra lie in the 200–240-nm range.

Centaur — No "official" definition exists. If a Centaur is defined as a minor body whose semimajor axis is between those of Jupiter and Neptune, then it is difficult to distinguish some comets (e.g., P/Schwassmann-Wachmann 1 or P/Chiron) from Centaurs as such objects are often considered to be inactive objects. The activity level is a parameter that is not easily measured, and a nonapparent activity may indicate a dormant/extinct comet as well as an object of noncometary nature. In addition, if Centaurs are defined as objects in transit from the transneptunian (TN) belt toward the inside of the solar system, based on studies that link TN objects and Jupiter-family comets, such objects are all potential comets. How should a nonresonant object with q < 30 AU and Q > 30 AU observed in the giant planet region be classified? Such objects, if originating in the TN belt, are not different from scattered-disk objects (SDOs), but their evolutionary path takes them to the inner part of the solar system, while SDOs are moving away from the TN belt where they formed. All this illustrates the difficulties we meet in trying to define the objects designated as Centaurs. Centaurs are objects seen in the giant planet region that are probably made of volatile and refractory materials and are seen at different stages of their physical evolution, hence their mutiple aspects. As for their dynamical heritage, they may come from the TN belt as well as from the Oort cloud.

central condensation — In a comet, a small region roughly in the middle of the head of a comet, which often appears visually much brighter than the rest of the head. Most of the time this is incorrectly referred to as the "nucleus" of the comet, although sometimes it may be the nucleus, even though it is too small to be resolved.

charge exchange — Charge exchange, or charge transfer, takes place in collisions in which the incoming ion removes, or captures, an orbital electron from a target atom or molecule, thus changing the charge state of both the target and the projectile species ($A^+ + B \rightarrow A + B^+$). The incident species generally retains most of its momentum during such collisions. If the incident ion species is highly charged (e.g., O^{7+}), then the product ion (O^{6+}) is invariably left in an excited state, thus leading to the emission of an energetic photon.

chemical differentiation — The process by which the composition of a body is modified via heat and chemical reactions, producing separation of the products.

chemical enrichment — The process(es) by which a particular element or set of elements is enriched in a body.

CHON particles — Particles and/or large molecules rich in the elements carbon, hydrogen, oxygen, and nitrogen or their compounds.

classical Kuiper belt object (KBO) — Also known as cubewanos; a class of transneptunian objects named eponymously after the first member of its class to be discovered, 1992 QB_1. Classical KBOs orbit mostly between about 30 and 50 AU from the Sun, and their orbital periods are not in any resonance with that of Neptune.

clathrate — A crystalline or structured liquid substance that contains host molecules within its structure as inclusions.

clathrate hydrate — A thermodynamically different form of water ice that is stabilized by host molecules trapped in "cages" made of hydrogen-bonded water molecules.

collision sphere — *See* exobase/exosphere.

color index — Difference in magnitude between two spectral regions (always short-wavelength magnitude minus long-wavelength magnitude). In the Johnson UBV system, an A0 star has B–V = U–B = 0. Stars hotter than A0

stars have a negative index; cooler stars have a positive index.

color, polarimetric — Polarimetric color characterizes the polarization spectral gradient and is usually defined as the difference between the values of polarization degree measured at different wavelengths (or through different filters). The polarimetric color is positive and called "red" if the value of polarization is higher at longer wavelengths and is negative or "blue" if the polarization is lower at longer wavelengths.

color temperature — An object temperature determined by comparison of the spectral distribution of the object radiation with that of a black body of known temperature.

column density — Number of atoms or molecules per unit area projected along a line of sight.

comet — An object of small mass formed in the giant planet region or slightly beyond, composed of icy and non-icy volatiles and solid grains observed to produce cometary phenomena (e.g., the coma, the ion tail, and the dust and sodium tails) upon sufficiently close approach to the Sun to cause the vaporization of ices. The largest known comets have a mass on the order of 10^{17} kg, but larger sizes are probably possible.

comet age — Generally used to refer to the time or number of orbital revolutions since a comet was injected into an orbit close enough to the Sun to induce cometary activity. Note, however, that the absolute age of all comets that are not fragments of a larger parent body is that of the solar system. A dynamical age is sometimes defined as the time elapsed since the comet has become a member of the dynamical group to which it currently belongs.

comet coma — The temporary atmosphere that forms around the nucleus of a comet when it approaches the Sun. The coma is composed of molecules, radicals, ions, and particles and appears visually as the "head" of the comet. The coma is a gravitationally unbound atmosphere that can be very large — observations in the light of Lyman α show a hydrogen coma with a diameter that can be >10^7 km, but the majority of neutral species are contained in a region with a diameter ~10^5–10^6 km. "Gas," "dust," and "hydrogen" coma are terms that are often used.

comet family — An ensemble of comets with nodal distances similar to the orbit radius of a major planet, e.g., Jupiter's family. Orbits are direct and have small inclinations. Only the Jupiter family of comets has a significant number of members.

comet fan — Gaseous or dust structure that suggests increased or restricted emission inside one (or more) cone-like solid angle(s). However, there is no solid definition for this term in fluid dynamics.

comet group — An ensemble of comets with orbits similar enough to lead to the idea they could have originated in the breakup of a single body.

comet, Halley-type (HTC) — A short-period comet with 200 yr > P > 20 yr. Orbital inclinations can take any value. Halley-type comets and Jupiter-family comets do not share the same dynamical history nor have the same origin. Most HTCs originate in the Oort cloud. The terms "intermediate-period comet" and "Halley-family comet (HFC)" are sometimes used instead of HTC.

comet head — The coma plus the nucleus of a comet.

comet, Jupiter-family (JFC) — A family of short-period comets gravitationally controlled by Jupiter that have direct and small inclination orbits. Those that have observed returns have P < 20 yr. Most JFCs likely originate in the Kuiper belt.

comet, long-period (LPC) — Traditionally, a comet with an orbital period larger than 200 years. If the semimajor axis is larger than 10,000 AU, the comet is classified as an Oort cloud comet.

comet, nearly isotropic — A comet with a Tisserand parameter T < 2 in Levison's taxonomic system [see *Levison* (1996, Comet taxonomy, In *Completing the Inventory of the Solar System* (T. W. Rettig and J. M. Hahn, eds.), pp. 173–191, ASP Conference Series 107]. Once in either the ecliptic or nearly isotropic family, more than 90% of the comets with P < 200 yr remain in that family. Note that it is the distribution of the orbital inclinations that is isotropic; a better term would be "comets of nearly isotropic orbital inclination."

comet, new — Often, the expression means "new in the Oort sense," i.e., that appears in the planetary region for the first time since its injection in the Oort cloud. Note, however, that it is dynamically possible that such a comet has visited the inner solar system before its (necessarily unique) observed passage in the planetary region. The term "returning comets" is usually used for comets having semimajor axes smaller than 10,000 AU.

comet nucleus — The frozen body, typically a few kilometers (up to tens of kilometers) in diameter, in the head of a comet that contains almost the entire cometary mass and that creates the cometary phenomena.

comet outburst — A sudden increase in the brightness of the cometary coma. Often interpreted as an abrupt and significant modification of the nucleus activity level that is converted into changes of the coma structure and luminosity. This concept is subjective, as neither the magnitude nor the strength of the increase are defined.

comet, short-period (SPC) — Traditionally, a comet with an orbital period smaller than 200 years. For a single-apparition object, the International Astronomical Union (IAU) uses the designation "P/" to refer to comets with periods up to 30 years; for a multiple-opposition object, there is no limit on period, although there is a practical limit since the comet must have made at least two apparitions (the longest-period comet in that case is 35P/Herschel-Rigollet, with P = 155 yr). Sometimes the term "intermediate-period comet" or "Halley-type comet" is used for comets with periods between 30 (or 20) and 200 years. In this case, additional factors are taken into account to categorize the comets, such as the orbit inclination. Some periodic comets that have periods longer than 200 years have been or will be in the future members of the Jupiter family (such as D/Lexell); their present period is large, but they will return to a lower-period orbit in the future (none has been observed twice, hence their absence from the IAU catalogs).

comet shower — An ensemble of comets perturbed by the passage of a star in the Oort cloud that arrives a few million years later in the planetary region where they are detected.

comet, split — Comet originating in the fragmentation of a parent comet under the action of external (e.g., the gravitational pull of the Sun or planet) or internal (e.g., fast rotation or heating) forces. Since comets that have already split may be very weak, further splitting can involve very little additional stress.

comet tail — Extension seen roughly in the antisolar direction that may be composed of solid particles (dust tail, curved, millions of kilometers long), atoms (various kinds of sodium tails, depending on the nature of the Na atom carrier), or ions (ion tail, straight, tens of millions of kilometers long). Dust tails are usually driven by radiation pressure; ion tails are driven by interaction with the solar wind.

comet-tail band system — The $A^2\Pi_i$–$X^2\Sigma^+$ band system of the CO^+ ion, observed in comet spectra between about 340 and 650 nm.

comet trail — Elliptic torus of solid particles spread along the orbit of a comet. Depending on the past history of the parent comet activity, the torus may be complete or incomplete (arc-like). Various effects may more or less spread the particles in the directions perpendicular to the tangent to the trail ellipse. Larger particles survive longer in comet trails.

cometary jet — See *jet*. See also chapter by Crifo et al. for an explanation of this concept in the framework of fluid dynamics.

cometesimal — Icy and rocky embryo formed beyond the protosolar nebula snowline.

cometopause — A broad boundary, or transition, in the mass-loaded solar wind flow downstream of the cometary bow shock at which the plasma composition and temperature change, primarily due to charge transfer collisions of hot ions with cold cometary neutrals. Solar wind protons and hot pick-up ions begin to be significantly depleted in this transition region. The magnetic pile-up boundary, where the interplanetary magnetic field lines are compressed, is located in the vicinity of the cometopause. In P/Halley, near 1 AU from the Sun, the diameter of the cometopause was $\sim 10^4$ km. As it is not a well-understood physical entity, the exact nature of the cometopause is still to be investigated.

contact surface — A surface or sheet that separates two fluids. In a comet coma, it refers to the surface dividing the inward streaming flow of the solar wind and the ionospheric comet plasma expanding in the opposite direction.

cosmic rays — High-energy particles and radiation that continuously bombard solar system bodies. Some of the particles are of solar origin, but most of the cosmic radiation comes from sources within or perhaps beyond the galaxy. Although known as cosmic "rays," most of the "radiation" is in the form of charged subatomic particles, some of which have extraordinarily high energy (e.g., several joules per particle). A small portion of the cosmic radiation consists of genuine "rays" (i.e., gamma rays). The origin and acceleration of the cosmic radiation is not fully understood.

crust — A mechanically robust medium made of dust grains glued together by organics or an ice-dust mixture where the ice has become cohesive by sintering or any similar process that is believed to form on the surface of comets that have passed close to the Sun on one or more occasions.

cubewano — The early designation for a classical Kuiper belt object (KBO).

daughter molecules — The small molecules, atoms, and radicals that are observed in the coma and tails of a comet and are fragments photodissociated from the larger molecules present in the solid nucleus of the comet. The next generation of molecular fragments form "granddaughter" products. *See also* parent molecules.

Debye length — In a plasma, the electric field of a charged particle is shielded by the influence of all the other charged particles in its vicinity, and, after a certain distance known as the Debye length, is reduced to effectively zero. In the simplest model, the Debye length is given by

$$\lambda = \sqrt{\frac{\varepsilon_0 kT}{ne^2}}$$

where ε_0 is the (rationalized) free space permittivity, k is Boltzmann's constant, n is the number of charged particles per unit volume, and e is the electronic charge.

delay-Doppler radar — A method of determining the size, shape, and rotational state of a solar system body by bouncing a radar signal off it. The arrival time (delay) of the returning signal is spread in time because of the varying distances of different parts of the body, and the wavelength of the returning signal is spread as a result of the Doppler effect from different parts of the rotating body.

disconnection event (DE) — A solar wind disturbance usually forms tail rays when it meets a comet coma (*see* tail ray). Sometimes, while tail rays are forming, the main tail weakens near the nucleus and is seen as detached from the comet's head. This is called a disconnection event (DE). The disconnected tail moves outward in the antisolar direction and sometimes forms clouds with bizarre shapes (*see* Swan cloud). In spite of numerous modeling efforts, the details of tail ray and DE formation are still controversial. See *Konz et al.* (2004, *Astron. Astrophys., 415*, 791–802) for a recent theoretical contribution to the subject and references to earlier works. Good observations require the use of narrow-band filters as ion tail structures cannot be seen near the nucleus against the bright dust and gas comae.

disequilibrium layer — *See* boundary layer.

dormant comet — A comet that is inactive, possibly because of the formation of a solid crust on its surface, which thermally protects the volatile materials within the nucleus. Unlike an extinct comet, the crust could conceivably be broken on a near approach to the Sun, hence the comet could become active again.

dust density — In a coma, the number of dust grains per unit volume of coma. The dust grain density is the mass of a unit volume of grain material.

dust grains/particles — Solid particles found in the coma, tail, or trail of comets, mostly from micrometers in size to the largest object capable of being entrained in the comet gas outflow (potentially meters in size). Dubbed in 1950 by Fred Whipple as "the dirt (of the nucleus)."

dust jet — Visually defined as a linear or curvilinear structure in an image of the dust coma. Outside the gas-dust interaction region, dust jets may be either projection effects (such as overlapping of two faint dust populations), or truly the signature of grains moving along a common set of trajectories. A dust jet does not necessarily trace a gas jet since gas jets interact with each another and are decoupled from the solid component of the medium. Inside the gas-dust interaction region, dust jets can only be the signature of gas flow patterns since grains move under gas drag. It is physically impossible to interpret dust jets in such a region without first a thorough modeling of the global gas flow. *See* jet.

dust mantle — A bank of dust grains bound to the surface only by gravity, but otherwise very loose and noncohesive. If gas diffuses across it, the grains must be large enough to not be carried away by the drag force.

dust tail — The ensemble of small solid particles originating in the comet nucleus and pushed away from it by radiation pressure. Historically, the tail is the antisolar extension of the comet, although strictly speaking, there is no discontinuity between the dust coma and the rest of the dust tail.

dust-to-gas ratio — The ratio of the mass production rate of the dust to that of the gas from the nucleus and into the coma and tails of the comet. Sometimes a proxy for the dust production/coma content is used, such as Afρ [see *A'Hearn et al.* (1984, *Astrophys. J., 89,* 579], which does not have units of kilograms per second. The actual dust production rate in units of mass per second is seldom actually known unless many assumptions are made.

ecliptic comet — Comet with Tisserand parameter T > 2 in Levison's taxonomic system [*Levison* (1996, Comet taxonomy, in *Completing the Inventory of the Solar System* (T. W. Rettig and J. M. Hahn, eds.), pp. 173–191, ASP Conference Series 107].

Edgeworth-Kuiper object/belt (EKO) — Often known simply as the "Kuiper belt" or "transneptunian belt," it was predicted by Kenneth Edgeworth in 1943, whereas Gerard Kuiper speculated in 1951 that comet-like objects could have formed in a disk beyond the orbit of Pluto, although he anticipated that the planet would have ejected the bodies because of its then supposedly large mass. In 1992, David Jewitt and Jane Luu discovered the first of what is now known to be a large population of objects in orbit around the Sun, mostly at distances between 30 and 50 AU, and these are now identified with the predicted Edgeworth-Kuiper belt. It was soon discovered that Pluto is the largest member of a subfamily of the Edgeworth-Kuiper belt, those that are trapped in the 3:2 resonance with Neptune. Most objects with a > 44 AU are stable for >4 Gy.

electric dipole transition — A transition may result from the interaction of an electromagnetic field with an atomic system. The probabilities of the transitions are determined by the Eigenfunctions of the states involved in the transition. An electric dipole transition will occur between two states when the matrix elements of the electric dipole moment for these two states are not zero. The transition is then "electric dipole permitted."

electron recombination — A process by which a positive ion and an electron join to form a neutral molecule or other neutral particles.

elements, orbital — In the Keplerian motion, five elements established from observations are needed to define the orbit of a solar-system body, i.e., the semimajor axis a (in AU), the eccentricity e, the inclination i of the object's orbital plane to the ecliptic, the longitude of the ascending node, and the argument of the perihelion. A sixth element, often the time of perihelion passage or mean anomaly for a given instant of time (epoch), is required to locate the object along its orbital path.

elements, osculating — Elliptic elements a body would take on, should the perturbing forces suddenly disappear. Mostly due to the perturbations of neighboring planets, a body's orbit continuously changes its characteristics from one moment to the next. Thus the body's orbital elements also continuously change, remaining constant for only a short period of time centered on the epoch of osculation.

elements, proper — Approximate constants of motion, computed by averaging the oscillations of the osculating semimajor axis, eccentricity, and inclination with suitable analytical or numerical techniques. Apart from issues related to computational accuracy, "real" proper elements, i.e., truly constant, exist only for bodies whose dynamical evolution is not chaotic. For weakly chaotic objects, such as asteroids in high-order resonances, proper elements, however computed, slowly change with time. This slow change in the values of the proper elements is usually called "chaotic diffusion." For strongly chaotic objects, such as near-Earth asteroids or comets, the notion of proper elements has no sense. The most important application of the calculation of proper elements has been the identification of collisional asteroid families.

emissivity — Ratio between the thermally radiated power of a surface element to that of an identical surface element emitting like a black body at the same temperature. The emissivity varies with wavelength and is often close to unity at wavelengths larger than one millimeter (also known as thermal emissivity).

escape velocity — The minimum velocity for an object to achieve a parabolic orbit and thus escape to infinity with a velocity tending toward zero. The escape velocity at 100 km from the center of a Hale-Bopp-like comet is on the order of 3 m s^{-1}, i.e., far below the gas and dust local radial velocities. The escape velocity of an Oort cloud comet from the solar system is on the order of 100 m/s and such a comet is thus easily perturbed by passing stars and interstellar clouds.

exobase/exosphere — In a planet or planetary satellite with significant gravity, the exobase is the altitude above which the collisional mean free path between molecules exceeds the local value of the atmospheric scale height. The exosphere is the region of the atmosphere above the exobase. In a comet, the exobase is usually defined as the region where the mean free path exceeds the radial distance to the center of the nucleus. The exobase radius is also called the collision sphere radius. At the exobase, the probability for a particle to escape from there to infinity without suffering a collision is 1/e (this assumes a 1/r^2 density distribution). Whereas in planet atmospheres the exobase sharply separates a fluid region from a collisionless region, such is not the case in comet atmospheres.

extended source — Most stable molecular species (as opposed to radicals, atoms, and ions) in comets appear to be emitted directly from the nucleus. Some stable molecular species appear to have at least one component that is produced in the coma from another source. Processes that have been suggested are sublimation from grains or large polymerized molecules, photon-induced desorption or photo-sputtering from grains or large molecules, gas-phase chemistry in the coma, or photodissociation of other parent molecules. Well-known examples in comets are the extended source components of H_2CO and CO. The term "distributed source" is also often used.

extinct comet — A comet that is totally inactive all along its orbit, possibly because of the formation of an insulating layer that permanently protects the volatile materials within the nucleus. The comet might also be devolatilized. Unlike the case of a dormant comet, the crust may be broken on a near approach to the Sun but the comet will not become active again. The comet might also be devolatilized. Some objects that are asteroidal in appearance are believed to be extinct comets.

extreme ultraviolet (EUV) — The electromagnetic domain is subsubvided into regions that have acquired a name for historical and/or technology-related reasons. The EUV region of the spectrum is the region 10 (or 20) to 100 nm. Most atoms and many simple molecular species are ionized in the EUV region. The X region begins at shorter wavelengths, the FUV region at longer wavelengths.

F10.7 — Flux of the Sun at the wavelength of 10.7 cm at 1 AU. This quantity correlates fairly well with the activity level of the Sun as measured by the Zurich number.

falling evaporating bodies (FEB) — Also called Beta Pictoris comets. Bodies speculated to orbit the star Beta Pictoris on eccentric orbits and to evaporate at their periastron in order to explain the variable absorption features observed on the star spectrum.

far infrared (FIR) — *See* infrared (IR).

far ultraviolet (FUV) — Far-ultraviolet region of the spectrum, between 100 and 200 nm. Below 200 nm the vacuum UV domain begins (high-voltage instruments must operate in vacuum below this limit to avoid electric discharges to occur). Some may define the lower limit by reference to the cut-off of transmitting windows (near 110 nm) or to that of the ionization wavelength of atomic hydrogen (91.2 nm). Shortward of the FUV window is the EUV window, while the mid-UV region lies longward of it.

fast rotation thermal model — A model surface temperature distribution that assumes each point on a rotating body is in radiative equilibrium with the average diurnal insolation at that point. For a sphere, each latitude is at a constant temperature.

flow, invisicid — The motion of a fluid in which the velocity distribution is a Maxwellian function. In general, in such a case the Knudsen number is very small everywhere (typically on the order of 0.01 or less).

fluorescence — The absorption of a photon by an atom or molecule subsequently followed by its reemission at an equal or longer wavelength. When the absorption and the emission wavelengths are alike, the process is called resonance-fluorescence.

fluorescence equilibrium — A molecular species is said to be in fluorescence equilibrium (or in radiative equilibrium) when the populations of the levels that are absorbing (solar light) and emitting photons (spontaneous emission) are steady and controlled by fluorescence, i.e., molecular collisions do not play any role. The distance from the nucleus at which such an equilibrium is reached depends mostly on the absorption coefficient of the transitions, the heliocentric distance, and the gas production rate. In the comet coma, some species rapidly reach fluorescence equilibrium (e.g., CN) while others do not (e.g., CO).

fourth positive bands — Transition $A^1\Pi$–$X^1\Sigma^+$ of CO having its strongest bands in the middle of the FUV region.

forbidden lines — Forbidden lines are "forbidden" by symmetry rules for electric dipole radiation, and therefore involve either electric quadrupole or magnetic dipole radiation and cannot normally be excited through the usual solar fluorescence process. In comets and planetary upper atmospheres, the upper levels of these transitions are normally populated through charged-particle impact or dissociative excitation. The upper level of the transition is usually a "metastable" atomic or molecular state, which has a mean lifetime that may or may not be long compared with the mean time between collisions. Radiative transitions from a metastable level therefore

cannot occur unless the density of the gas is very low.

fraction, active area — A theoretical concept. Unlike the icy area fraction that is a local property of a comet surface, the active area fraction, g, is a global property of a comet nucleus, a measure of the surface of the nucleus that is "active." If g < 1, it is possible to distinguish inactive areas — where f, icy area fraction, is by definition 0 or nearly so — from active areas, in which f is large (up to 1) by definition.

fraction, icy area — Also a theoretical concept, the icy area fraction 0 < f <1 is a local parameter of the of the surface of the nucleus. It is the fraction of a surface element that is made of pure ice, the rest being made of pure refractory material. *See* fraction, active area.

free-molecular flow — The motion of a set of mutually noninteracting particles (in particular, without collisions). For such a flow, the Knudsen number is infinite.

g-factor — Or fluorescence efficiency: the number of photons scattered by a molecule during a unit time. In electronic transitions, the g-factor is proportional to the oscillator strength of the transition, the exciting (solar) flux, and the single scattering albedo of the emitting species (see chapters by Feldman et al. and Bockelée-Morvan et al.). Usually given at 1 AU from the Sun, the g-factor is on the order of 0.1 s^{-1} for the fastest transitions (violet CN, optical C_2) and 10^{-3} to 10^{-4} s^{-1} for most of the permitted transitions observed in comet comae.

gas jet — In physics, a collimated flow of gas with density and velocity much greater than those of its surroundings. In an image, a gas jet appears as a bright structure on a dark background. In the cometary context, a gas jet is a faint detail in an image of a gaseous emission, extracted by image-processing techniques. In such a case, "gas jet" designates only a fraction (possibly a small fraction) of the gas present in the region, with no information on velocity contrast. *See* dust jet; jet.

gas outflow velocity — Strictly speaking, the macroscopic gas velocity of molecules originating at the nucleus surface, variable from point to point in the coma, and becoming rapidly radial. However, it quite often designates the average value of this velocity computed over a large sphere encircling the nucleus. Such a practice is justified by the fact that the outflow velocity is usually not accessible to observations (or at least not with sufficient spatial resolution). The velocity of photolytic products can be quite different. *See* temperature, kinetic.

Gegenschein — A very faint patch of light in the sky in the opposite direction to the Sun, from the German word meaning "opposite shine" or "counterglow." It is caused by backscattering of sunlight from dust particles near the invariant plane of the solar system.

Greenstein effect — The Swings effect (*see* that term) implicitly assumes that all coma species have the same velocity relative to the Sun as the comet nucleus. Because of internal motions of gases within the coma, the exciting line wavelength displacement also depends on the location of the absorbing species. This second-order variation of the Swings effect from point to point within the coma is called the Greenstein effect. It is particularly strong in the H I lines. The shape of OH radio lines are also severely affected by this effect.

Haser model — A one-dimensional model atmosphere for comets designed in 1957 by Léo Haser that provides the spatial density distribution of parent species and daughter products in the coma when the destruction/production process is the photodissociation of the parent molecules supposedly produced at the nucleus. While its simplistic assumptions made it an easy-to-use tool to interpret spatial profiles of coma emissions, it was superseded in 1981 by the three-dimensional vectorial model designed by M. C. Festou to take into account the nonradial motion of the daughter species relative to their parent molecules. Both models are approximate inside the collision sphere where only gasdynamic/gaskinetic methods can correctly account for the properties of the expansion into vacuum of a gas source.

Hertz factor — The area of contact between material grains relative to the cross-sectional area. It is loosely used to express the ratio of the thermal conductivity of a porous solid to that of the same solid in the absence of porosity.

Hill sphere — In the restricted three-body problem, the Hill sphere is the surface where the gradients of the two gravitational fields, usually a planet and the Sun, in which a third body of small mass circulates, are equal. In other words, it is the surface where the tidal forces of the two bodies are equal. Beyond that sphere, a third object in orbit around the first would be progressively deviated by the tidal forces of the second and would end up orbiting the latter.

$$R_H = D \left(\frac{M_1}{3(M_1 + M_2)} \right)^{1/3}$$

(D is the distance between the planet and the central body; M_1 and M_2 are the masses of the two objects.)

icy grain halo — A cloud of icy grains that escapes the nucleus due to outgassing. Its experimental existence has not been proven and calculations show that icy grains have very short lifetimes inside of Jupiter's orbit.

infrared (IR) — The region of the electromagnetic spectrum from a wavelength of about 700 nm (red end of the visual spectrum) to about 300 μm. The region from 700 nm to 5 μm is the near infrared and it prolongates the visible (optical) region. The region from 5 to 30 μm is the mid-infrared and the region from 30 to 300 μm is the far-infrared. The mid-infrared region is often called the "thermal infrared" because many cold sources have their emission spectrum there.

interplanetary dust particle (IDP) — Strictly speaking, any particle traveling in the planetary region. Its origin can be comets (nucleus outgassing), asteroids (collisions and sputtering of the asteroid surfaces), or interstellar (the Sun moves inside the local interstellar medium). *See* zodiacal light.

ion tail — One of the three main parts of the "comet phenomenon," the other two being the dust tail and the gasdust coma that surrounds the nucleus (the nucleus is invisible in most observing conditions). Since ions are entrained away from the Sun due to the interaction of the solar wind with comet ions accelerating eventually to velocities on the order of hundreds of kilometers per second, ion tails are oriented close to the antisolar direction and they extend to millions of kilometers from the nucleus.

ionopause, comet — The solar wind interacts with neutral and ionized coma species inside a thick bow shock located 5×10^5–10^6 km from the nucleus (comet near 1 AU from the Sun). The supersonic flow is slowed down by momentum conservation and then deflected toward the tail while becoming progressively loaded with comet ions. The region where the flow is deflected is called the ionopause.

ionosphere, comet — Inner part of a comet atmosphere where the ionized constituents do not interact with solar wind particles. This plasma is non-magnetized. It was found at about 5000 km on the sunward side of Comet P/Halley near 1 AU. The region between the cometopause and the ionosphere is the second (or inner) shock where solar wind ions are forced to slowly move around a magnetic field free cavity that is the true ionosphere of the comet.

jet — In astrophysics, a jet designates a linear or moderately curved brightness enhancement in an image. In comet images, gas and dust (also comet) jets are thus simply brightness enhancements and not necessarily real physical entities, unless otherwise stated or demonstrated. In gasdynamics, a jet is a self-sustained fluid in fast motion due to its internal pressure gradient. A gas jet expanding in a vacuum is brighter than its surroundings; however, a gas jet in a fluid medium can be either brighter or fainter than its surroundings. In a gas, stagnation lines, along which the gas is at rest, are usually denser than their surroundings and thus appear brighter. In the image of a gaseous emission, the identification of a brightness enhancement with a gas jet can therefore be made only on the basis of quantitative modeling of the fluid motion. The identification of brightness enhancements observed in continuum light with dust jets can lead to similar misinterpretations. What can acceptably be designated as a dust jet is an ensemble of solid particles carried along (i.e., located inside) a gas jet. By extension, a group of solid particles moving in the region where gas and solid particles are decoupled and have nearly identical trajectories can also be called a dust jet. *See also* syndyne/syndyname; synchrone.

Knudsen layer — *See* boundary layer.

Knudsen number — The ratio of the mean free path of the molecules in a fluid to the smallest characteristic length of the fluid flow structure (at the same point).

Kreutz group — Group of comets with orbital inclination on the order of 142°, a perihelion distance slightly over a Sun radius. Orbits are known only in a few cases (e = 1.0 is assumed in most cases to determine the other orbital elements). The global comparison of the orbital elements to those of members with rather well-determined orbits suggest orbital periods on the order of either about 500 or 1000 years. These comets (more than 500 have been observed) are believed to be the fragments of a large comet that split in the recent past. Most Kreutz-group comets do not survive perihelion passage and either disintegrate or hit the surface of the Sun.

Kuiper belt/object — *See* Edgeworth-Kuiper object/belt (EKO).

Lambert law/lambertian — If the radiance of an emitting surface is independent of the direction from which it is viewed, the surface is said to be "lambertian," or to be radiating according to "Lambert's Law." For a reflecting surface, this must also be true whatever the direction of the incident light. The intensity of an infinitesimal element of a lambertian surface falls off as the cosine of the angle from the normal to the surface.

Mach number (M) — The ratio of the gas flow velocity at any point to the sound speed at the same point.

magnetic cavity — *See* ionopause, comet; ionosphere, comet.

magnetic dipole transition — Radiation arising from a transition between two atomic or molecular levels such that the matrix element (transition moment) of the magnetic dipole moment operator is nonzero while the matrix element of the electric dipole moment operator is zero. Such transitions result in forbidden lines and are generally observed only in low-density gases. They are weaker than electric dipole transitions.

magnetopause — The region in an object's ionosphere where the magnetosphere meets the solar wind. It is the place where the object's magnetic field stops; beyond it begins the interplanetary space.

magnetosphere — The region of space in which an object's magnetic field dominates that of the solar wind. There is no magnetosphere nor magnetopause in a comet.

magnetosheath — The region between the cometopause and the outer first shock and the cometopause.

magnetotail — The portion of an object magnetosphere that is pushed in the direction of the solar wind.

magnitude — A number, measured on a logarithmic scale, used to indicate the brightness of an object. Two stars differing by 5 mag differ in luminosity by 100 (1 mag is the fifth root of 100, or about 2.512). The origin of the scale is defined so that a bright star such as Vega is of magnitude 0; the faintest stars detectable with the unaided eye are of magnitude about 6. The brighter the star, the lower the numerical value of the magnitude.

magnitude, absolute — Of a star: the apparent magnitude that the star would have if it were taken to a standard distance of 10 pc. Of a planet: the apparent magnitude that the planet would have if it were taken to 1 AU from the Sun and viewed by an observer situated at the center of the Sun. Of a solar system body such as an asteroid or a comet: the brightness at zero phase angle when the object is 1 AU from the Sun and 1 AU from the observer, defined by $V = V(1,0) + 5\log(r\Delta) + F(\alpha)$, where r is distance from the Sun and Δ from Earth, and $F(\alpha)$ is the phase function. Absolute magnitude may refer either to the total light or only to the light scattered from the nucleus; calculated values depend on the exponents that characterize the variation rate of observed apparent magnitudes with heliocentric and geocentric distances.

magnitude, apparent — Brightness of an object as seen from Earth. An object's apparent magnitude depends on its absolute magnitude, its distance to Earth, the amount of absorbing material between the object and Earth, and the wavelength.

magnitude, bolometric — Magnitude determined after integration of the light over the entire spectrum.

magnitude, heliocentric — Magnitude of a comet at some distance from the Sun, 1 AU from Earth and zero phase angle.

magnitude, m_1/total — Of a comet: the visual magnitude of the entire comet, often estimated by comparing it with the out-of-focus image of a star. Can be defined or measured through any spectral bandpass.

magnitude, m_2 — Of a comet: the visual magnitude of the star-like central condensation in the head of a comet. The central condensation is an ill-defined part of the central coma, hence our recommendation not to use that quantity.

magnitude, nuclear — When the light reflected by the nucleus can be separated from that coming from the surrounding material, it is the apparent magnitude of the nucleus.

magnitude, visual — The magnitude determined through the eye bandpass, i.e., weighted according to the wavelength sensitivity of a standard human eye (scotopic vision). Different (although not much) from the magnitude measured through a Johnson V filter.

mantle — In a planet: beneath the crust and above the core. In a comet: the upper non-icy part of the nucleus that appears after all the surface ices have left the comet, devolatilized by solar insolation, collision with interplanetary particles, or altered by cosmic rays.

Meinel bands — Bands of the $A^2\Pi_{ui}$–$X^2\Sigma_g^+$ of the N_2^+ ion observed (in comet spectra and terrestrial aurorae) near 800 nm. The N_2^+ ion is observed in comets by the 0–0 band of the first negative system located near 391 nm. Also the name of the rotation-vibration system of the $X^2\Pi_i$ ground state of OH that extends over the entire near-IR region. It originates in the atmosphere of Earth and often masks comet emissions.

mid-infrared (IR) — Or thermal IR; *see* infrared (IR).

mid-ultraviolet (UV) — The region of the spectrum between the atmospheric cutoff near 300 nm and the beginning of the vacuum-UV region below 200 nm. The region longward of 300 nm is the near-UV region.

Mie theory/scattering — A theory of the diffraction of light by spherical particles of any size developed by Gustav Mie in the first decade of the twentieth century. Particularly useful when the wavelength is on the order of the size of the particles, since geometric optics only applies when the wavelength is much smaller than the particle size, and Rayleigh scattering only applies when the wavelength is much larger than the particle size. Mie showed that, for a given particle size, the scattering coefficient is inversely proportional to the wavelength of light. The application of this theory to irregular particles often leads to very incorrect results.

Monte Carlo method/model — The use of statistics, probability theory, and random numbers to solve deterministic computational problems. For example, to find the area of an ellipse drawn on a square sheet of graph paper, one could generate a large number of random points and, for each, determine whether it is inside or outside the ellipse. The area of the ellipse is then determined by the fraction of points that lie inside. In the literature on comet coma models, some models are designated as Monte Carlo models because they make use of the method to calculate properties of the coma such as the density or velocity of the molecules. The Monte Carlo technique is well suited to solve problems in a phase space with many free parameters.

Mulliken bands — Bands near 231 nm connecting the upper level $D^1\Sigma_u^+$ to the $X^1\Sigma_g^+$ ground state of the C_2 molecule.

near-infrared (NIR) — Like the ultraviolet region, the infrared (IR) region is traditionally split into subdomains. The near-IR region begins past the optical window and ends where the thermal-IR region begins: It covers the 1–5-µm region of the spectrum.

near-ultraviolet (NUV) — The spectral region between the atmospheric cutoff and the optical violet, roughly the 300–400-nm region.

neckline — Term coined by *Kimura and Liu* (1977, *Chin. Astron., 1*, 235–264). A neckline is a part of a tail or antitail embedded in the more diffuse common dust tail and appearing pinched due to the heliocentric motion of the large (millimeter- and centimeter-sized) dust grains that compose it. These grains are near the comet orbital plane when passing at their second orbital node (they were at the first node at their time of emission). *See* antitail.

nightglow — *See* airglow.

nongravitational acceleration/force — A weak force acting on a comet nucleus that is caused by the outgassing of the nucleus.

nongravitational effect — Any effect that tends to change the orbit of a comet other than the gravitational attraction of other bodies. For example, if a comet liberates a substantial amount of material at high speed, there will be a reactive force on the nucleus. On asteroidal bodies, the Yarkovsky effect describes the strongest nongravitational force acting on them. The main component of the nongravitational force (NGF) due to outgassing is directed toward the Sun. In his famous 1950 paper, Whipple invoked the action of the NGF component lying in the

plane of the orbit and perpendicular to the radial component (in association with the nucleus rotation) to explain the delays and advances observed when comets return to their perihelion. However, it is the asymmetry of the radial component with respect to perihelion that causes most of these delays.

Oort cloud — Spherical cloud of comets surrounding our planetary system, containing approximately 10^{12} comets, and extending out to approximately 75,000 AU. Discovered by Jan Hendrik Oort (1900–1992) in 1950. An inner Oort cloud was hypothesized in response to the concern that the material in the Oort cloud would be depleted over the age of the solar system by the various perturbing forces, including the passage of nearby stars, erosion by giant molecular clouds, the galactic tide, and any other external perturbations. The inner Oort cloud never explicitly referred to what we call the Kuiper belt today.

opacity — The measurable ability of a substance to obstruct by absorption and scattering the transmission of radiant energy; opacity is therefore the degree of nontransparency. The opacity of a medium is measured by the ratio of the radiative flux incident upon it divided by the radiative flux transmitted by it.

opposition effect — The change in brightness with phase angle usually varies fairly smoothly, changing for regolith surfaces from a maximal value at zero phase angle toward a minimum value near 180° for regolith surfaces. For small particles, the minimum value is reached at medium (30°–120°) phase angles. At small phase angles, the rate of change may significantly increase. This "spike" is caused by the backscattering properties of the surface regolith material and, for comets, by large dust particles in the coma.

ortho/para ratio — Ratio of the populations of hydrogen-bearing molecules in ortho and para states respectively.

ortho/para species — Molecules with two hydrogen atoms at symmetrical positions (e.g., H_2, H_2O, H_2CO, H_2S) exist in two different varieties, ortho and para, according to the sum of the nuclear spins of their H atoms. In the ortho variety, the total nuclear spin is I = 1 and the spins of the hydrogen atoms are parallel. The para variety corresponds to I = 0 and antiparallel spins. The degeneracies of the ortho and para rotational states are 3 and 1 respectively (i.e., 2I + 1). Conversion between ortho and para states in the gas phase by radiative transitions or collisions are strictly forbidden. The ratio between the ortho and para varieties depends upon the temperature of formation of the molecule and further reequilibration. In the high-temperature limit (e.g., >60 K for H_2O), the ortho/para ratio is 3, which corresponds to the statistical weight ratio.

optical depth — A measure of the integrated opacity along a path through a layer of material, measured by the amount of absorption of a beam of incident light. The optical depth measures the depth of penetration of the incident light: The incident light has become 1/e its initial value after a path through the medium of one optical depth.

outgassing — In physics, the release of molecules individually trapped in a solid — usually obtained by heating the solid. In the context of comets, often used as an equivalent of "gas emission" from a solid by any physical process.

parent molecules — Historically, stable molecules residing in the nucleus of comets and released in the coma due to the action of the Sun's heat. Actually, any species that leads to the formation of a new species (mostly via photochemistry, although other physical phenomenon might also be at work).

phase angle — The angle subtended at the center of a planetary body by the directions to the Sun and observer. Also a radiative transfer term: When a particle scatters light, the angle between the incoming photon direction and the direction in which the photon leaves the particle is called the scattering angle (zero for a backscatterer and 180° for a forward scatterer). The phase angle is the complement of the scattering angle, i.e., the arrival direction vector–departure direction vector angle (0° for a forward scatterer). Note that in atmospheric physics the scattering angle is defined as zero for forward scattering.

phase function — The change in the brightness of an object as a function of the phase angle. In general, a macroscopic object gets brighter as the phase angle approaches 0° as the illuminated portion of the planet visible to the observer becomes greater. For a small particle, the phase function is brighter near 180° and 0° phase angles. The phase function at small phase angles is often described as the change in magnitude per degree of phase angle. For asteroidal-like surfaces, that change is on the order of 0.035 mag per degree. The phase function of a comet (dust) coma is the sum of the individual phase laws of its particles, hence it will brighten near 0° and 180° phase angles. *See* opposition effect.

phase integral — Let $\Psi(\alpha) = I(\alpha)/I(0)$ be the ratio of the intensity of light reflected by a planet or other solar system particle at phase angle α to the intensity reflected at zero phase angle. Then the integral $q = 2\int_0^\pi \Psi(\alpha)\sin(\alpha)d(\alpha)$ is the phase integral. Phase- and albedo-related expressions are discussed in detail by *Lester et al.* (1979, *J. R. Astron. Soc. Can., 73,* 233).

Phillips bands — Near-infrared band system $A^1\Pi_u$–$X^1\Sigma_g^+$ of the molecule C_2. Unlike the better-known Swan bands, the Phillips bands involve the ground state of the C_2 molecule. In earlier years this was often in doubt because singlet and triplet states do not normally interact. The Mulliken bands of C_2 arise from the ground state of the C_2 molecule.

photodissociation — The dissociation of a molecule as a result of interaction with light rather than by collision with other molecules.

photoionization — The ionization of an atom or molecule as a result of interaction with light rather than by collision with other atoms or molecules.

photolysis — While "photolysis" is often considered to be synonymous with "photodissociation," "photolysis" usually describes an active laboratory experiment in which molecules are deliberately irradiated with light to cause them to dissociate or ionize.

photolytic heating — In the context of comet coma physics, a heating mechanism in which the energy is provided by the dissociation of gaseous species. Collisions between gas species distribute this energy.

photometry — The measurement of the intensity of light and related quantities.

planetesimal — Icy or rocky planetary body embryos formed by collisional accumulation of smaller bodies beyond or inside the protosolar nebula snowline. The size of planetesimals is an ill-defined parameter and lies in a loosely defined range from a few millimeters to about 1000 km.

plutino — A class of transneptunian objects whose orbits, like Pluto's, are in 3:2 mean-motion resonance with Neptune's orbit.

polycyclic aromatic hydrocarbon (PAH) — Planar hydrocarbon compounds with several fused (i.e., joined) benzene rings. The simplest are naphthalene (two rings) and anthracene (three rings). Graphite may be considered a PAH, although there are hydrogen atoms only at the surface.

POM — polyoxymethylene or polymerized formaldehyde, a chain molecule $-(CH_2O)n-$ composed of formaldehyde molecules. Suspected to be present in comets when heavy-ion mass spectra of P/Halley showed peaks separated by 14 or 16 mass units, suggestive of the presence of a . . –C–O–C– . . skeleton, although this interpretation has been disputed.

porosity — A parameter representing the amount of voids present in a solid; the ratio of the volume of voids (pores) to the total volume of a porous material. It tends to zero for a solid (nonporous) material and to unity in the limit for a highly porous substance. Bodies formed in low-density media are thought to be low-porosity entities.

position angle (p.a.) — An angle for a direction relative to north that is measured positively toward the east. East is thus at 90°, south at 180°, west at 270°. In the context of comets, the position angles of the tail or radial structures are often given (note that in ephemerides it is often the position angle of the solar radius vector that is given, not that of the tail).

Poynting-Robertson effect — An effect of radiation pressure on a small particle orbiting the Sun because of the fact that radiant energy is absorbed from the direction of the Sun while it is reemitted in all directions. The net effect of radiation pressure is that radiation falls preferentially on the leading edge of the orbiting particle and acts as a drag force. This results in a decrease of the orbital angular momentum and hence a decrease in the size of the orbit. Because it is more significant for small particles than for larger particles, the Poynting-Robertson drag can lead to a separation of dust particles according to size, the smaller particles moving closer to the Sun than the larger particles. See *Burns et al.* (1979, *Icarus, 40,* 1).

prompt emission — The spontaneous emission of a photon in a time on the order of a nanosecond or less from an excited level not populated by fluorescence (this does not imply that fluorescence is not possible). In comet comae, the excited levels are created in physical processes such as the photodissociation of molecules. Typical examples are the emission of UV photons by OH or red photons by O^1D upon dissociation of water molecules or emission in the Cameron bands of CO upon photodissociation of CO_2 molecules.

protoplanetary disk — A mass of gas and dust, in the form of a disk around a young star, from which a planetary system may form.

quenching — In a medium in thermodynamical equilibrium, all temperatures (i.e., kinetic, vibrational, rotational, spin) are equal to the thermodynamic temperature and energy levels are populated according to Boltzmann's law. Collisions can populate high-energy levels by converting kinetic energy into excitation energy or do the opposite, namely capture the energy of a photon to be emitted and convert it into a nonradiative form of energy. Quenching is the reduction of the population of rovibrational levels by collisions. As a consequence, long lifetime levels can decay through photon emission only when volume densities are very low. The quenching action can be performed by any particle of the medium, molecule, ion, or electron. In comet comae, collisional with ions and electrons strongly affect the population of the ground state Λ doublet levels that are responsible for the OH 18-cm emission.

radiative equilibrium — *See* fluorescence equilibrium.

radius vector — The Sun-comet vector.

rayleigh —unit of omnidirectional emission rate expressed in photons (rather than energy) per unit time per unit area and per 4π of solid angle. One rayleigh (1 R) is 10^{10} photons emitted per second in all directions in a column of section 1 m². It is used in measuring the luminous intensity of extended sources like the sky, the aurorae, and comet comae. See *Chamberlain and Hunten* (1987, *Theory of Planetary Atmospheres,* Academic Press, San Diego, California).

Rayleigh scattering — The scattering of electromagnetic radiation by particles whose size is much smaller than the wavelength of the radiation. The scattering coefficient for such scattering is inversely proportional to the fourth power of the wavelength of the radiation. Rayleigh scattering is largely responsible for the blue color of the sky. *See also* Mie theory/scattering.

reflectance — *See* reflectivity.

reflectivity — Of a surface: the ratio of the amount of light reflected to the amount of incident light. Reflectivity and reflectance both refer to the fraction of light scattered or reflected by a material and are thus sometimes used interchangeably. However, reflectance has the connotation of diffuse scattering, while reflectivity refers to the specular reflection of radiation by a smooth surface (see *Hapke,* 1993, *Theory of Reflectance and Emittance Spectroscopy,* Cambridge University Press, New York). *See also* albedo.

rotation, complex — *See* rotation, non-principal-axis (NPA).

rotation, excited — A state of rotation other than the principal-axis rotation around the short principal axis. Typically acquired due to an external torque. For a nonrigid body in an excited rotational state, kinetic energy is dissipated as heat until the object rotates around the short principal axis.

rotation, non-principal-axis (NPA) — A rotational state that requires three component motions (rotation, precession, and nutation) to describe it. Two independent periods are required to describe a NPA rotation (the period associated with the third motion is coupled to one of the two independent periods). Also called "complex rotation" or "tumbling motion." For a NPA rotation, the angular velocity vector and the rotational angular momentum vector are oblique to each other. All NPA rotations are excited rotations, but the converse is not true. *See also* rotation, excited.

rotation, principal-axis (PA) — Three mutually orthogonal axes (short, intermediate, long) passing through the center of mass can be defined for any rigid body. When the off-diagonal terms of the moment-of-inertia tensor are zero, the largest diagonal term corresponds to the moment of inertia about the short axes, while the smallest diagonal term represents that about the long axis. These orthogonal axes are the principal axes (PA) of the rigid body. If the body is rotating about one of the PAs, the angular velocity and angular momentum vectors are parallel to each other and such a rotation is known as a principal-axis rotation. A PA rotation around the intermediate axis is dynamically unstable. If a nonrigid body is rotating, rotational kinetic energy is dissipated as heat until eventually the body rotates about its axis of greatest moment of inertia. Such a PA rotation is the stable end-state of rotation of any nonrigid solid body. For an ellipsoidal object having an uniform density distribution, the principal axes and physical axes are identical. PAs do not necessarily coincide with the physical axes when the density is not uniform.

rotation, short-/long-axis-mode (SAM/LAM) — A rotational state must be either a short-axis-mode (SAM), long-axis-mode (LAM), or principal-axis (PA) rotation around the intermediate principal axis. For a SAM, the short principal axis precesses around the rotational angular momentum vector and they never become normal to each other. For a LAM, the long principal axis precesses around the rotational angular momentum vector and they never become normal to each other. For a given rotational angular momentum, SAMs are less energetic than LAMs.

rotational/rovibrational temperature — Quantum mechanics permits the rotational energy of a molecule to have any of a (usually large) number of discrete rotational energy levels. The manner in which a large collection of molecules in a gas in thermal equilibrium is distributed among these energy level is described by the Maxwell-Bolzmann distribution equation and depends only upon the temperature. The temperature can be determined from the intensity distribution of the rotational lines in the spectrum of a molecule. In a comet coma, collisions rates are often low and thermal equilibrium is not reached over most of the coma. Radiative processes strongly affect the populations of the rotational levels. The distribution of rotional line intensities provides a temperature called the rotational temperature, and this temperature can be quite different from the kinetic or vibrational temperature of the molecules of the medium. A rovibrational temperature can be similarly defined from the populations of molecular levels involved in transitions in which both the vibrational and the rotational quantum numbers vary.

rubble pile — A loose aggregate of many smaller bodies. One of the models proposed (by E. J. Öpik in 1963) for the interior of comets. Some asteroids may not be "monolithic" and may have become rubble piles after suffering a large collisional event.

scale length — The product of the velocity of a coma species by the lifetime of that species. For molecules released at the nucleus, the radial velocity rapidly increases in the first few tens of nuclear radii, then increases slowly with distance to the nucleus so that the assumption of a constant scale length for them is valid over most of the coma. Because daughter species generally do not move in the same direction as their parent species, their scale length depends on the direction in which they are released relative to the global motion of the nuclear parent species, i.e., radial (*see* vectorial model). In the Haser model, daughter species are assumed to have the same scale length, which is not strictly valid. The Haser scale lengths of molecules give the characteristic distance at which the species are destroyed.

scattered disk object (SDO) — A class of transneptunian objects characterized by highly eccentric orbits that may take them very far — 1000 AU or more — from the Sun and perihelia in the 30–50-AU region. The scattered disk is one of the sources of Jupiter-family comets and Centaurs.

secular motion/acceleration/change — A motion/acceleration/change that is not periodic but increases or decreases monotonically with time.

secular perturbation — A perturbation that results in secular motion.

secular resonance — An object in orbit around the Sun is said to have a secular resonance with Jupiter (or another major planet) if the ratio of the mean motion of its perihelion or of its line of nodes to the orbital period of Jupiter (or other planet) is a simple rational fraction.

shock, weak/viscous — The fundamental property of a supersonic flow is that no information can propagate upstream of the flow. If an obstacle is present downstream, the gas must slow down suddenly to subsonic conditions ahead of this obstacle; this discontinuity is referred to as a shock. The obstacle can be a solid object or another gas flow. The shock is said to be weak if the discontinuity is small. The thickness of the shock is on the order of the gas mean free path. If the latter is large (compared to the characteristic scale of the flow), the shock is said to be viscous (it will appear as a smooth transition rather than as a discontinuity).

sintering — An increase in the cohesive strength of grainy materials, caused by various metamorphism processes, which cause a glueing together of single grains. A typical example is the slow metamorphism of fresh (soft) snow into old snow (firn) and finally ice.

slow rotation thermal model — *See* fast rotation thermal model. The incident energy is spread over the lit face of the object. If the rotation is fast, the incident energy is spread over the entire surface of the object.

snowline — Place in the protoplanetary disk where water condenses out as ice (or snow).

solar wind — A stream of charged particles — mostly electrons and protons, together with α particles and heavier ions — continuously emanating from the Sun's surface. The particle speeds are typically in the range 250–650 km s^{-1} and proton fluxes are on the order of a few 10^{12} m^{-2} s^{-1}. The characteristics of the solar wind vary greatly with the level of activity of the solar surface, the phase of the solar cycle, and the heliocentric latitude.

sputtering — Process whereby atoms from a solid target are ejected into the gas phase by the bombardment of the material by energetic ions or particles. The impact of fast particles on a rocky body releases solid particles that usually accumulate on the surface and form a regolith.

Stokes parameters — A set of four scalars (I, Q, U, V) that describe the intensity and the state of polarization of light. These four numbers are the elements of the Stokes vector.

striae — Filamentary structures embedded in the common dust tail, observed like bright straight lines pointing toward a point lying between the comet nucleus and the Sun. Striae are supposed to be due to the sudden fragmentation of large dust parent grains, although the details of the phenomenon remain unknown and are waiting for realistic models for the fragmentation of comet solid particles to be available.

Sun-grazer — A comet that passes very close to the Sun's surface.

supercooling — Bringing a vapor below its condensation temperature quickly enough (and/or in the absence of dust) so that condensation does not occur.

Swan bands — The d$^3\Pi_g$–a$^3\Pi_u$ band system of the molecule C_2. In most comets one of the most conspicuous features in the visual spectrum of a comet: The three main bands are in the blue, green, and orange part of the optical window.

Swan cloud — "Swan-like" structure moving down the ion

tail of comets. It is composed of CO^+ ions. This term seems to have appeared first in *Hyder et al.* (1974, *Icarus, 23*, 601) to describe isolated ion structures moving down the tail of Comet Kohoutek C/1973 E1. Such a "cloud" is probably a disconnection event in a late stage of evolution.

Swings effect — The appearance of atomic lines and molecular bands in a comet is affected by the radial velocity of the absorbing species with respect to the Sun. The calculation of a theoretical emission spectrum must consequently be performed in a frame of reference that appropriately displaces the absorption wavelength of the solar lines. When the displacement takes into account only the displacement due to the comet motion, the effect is called the Swings effect. It was invoked by Polydore Swings to explain the anomalous change in the distribution of the intensity of the rotational lines of CN.

synchrone — Geometrical locus in the comet orbital plane where dust grains ejected at a fixed time preceding the observation time are found and which are characterized by any "β" parameter (*see* syndyne/syndyname for definition of the β parameter).

syndyne/syndyname — Geometrical locus in the comet orbital plane where dust grains with a fixed "β" parameter are found and which were ejected at any time prior to the time of observation. The β parameter is the ratio of the solar radiation pressure force to the solar gravity force acting on a dust grain. Syndynes have also been defined and used for the study of atomic species such as sodium and hydrogen, where solar resonance fluorescence similarly produces a strong antisunward radiation-induced acceleration.

tail ray — The most frequent solar-wind-induced ion structure. The cometary plasma produced by ionization processes within the cometary coma is an obstacle to the plasma and magnetic structures carried by the solar wind. These structures fold around the inner coma and form shells of conical shape in which the density of picked-up comet ions is higher than average; tail rays are these ion shells seen sideways. Tail rays are first seen as short spikes that in the course of about one hour grow in length as more ions are being collected; ultimately, they will merge into the main ion tail. Tail rays most often come in pairs (symmetric relative to the main plasma tail axis).

temperature, internal — The internal temperature of molecules or dust grains measures the amount of internal energy stored inside individual grains or molecules (e.g., the rotational and or vibrational energy of a molecule, the specific heat content of a solid grain). This temperature never vanishes to zero. In a rarefied medium like the comet coma, kinetic and internal temperatures need not be equal, and must be carefully distinguished. *See* temperature, kinetic.

temperature, kinetic — Where collisions occur between molecules the kinetic temperature is, by definition, a measure of the resulting molecule velocity dispersion around the flow velocity (*see* gas outflow velocity). When collisions are infrequent or absent, velocity dispersions that are observed must be characterized by several kinetic temperatures (e.g., parallel to the flow velocity and transverse to it). The kinetic temperature(s) should carefully be distinguished from internal temperatures, which measure the molecule internal energy (e.g., vibrational). It is only near the nucleus surface that all kinetic temperatures can be equal.

thermal emissivity — The relative efficiency with which a substance radiates away the energy that is absorbed compared with an ideal black body.

thermal infrared (IR) — The spectral region where most cold (by human standards) bodies emit their photons, i.e., the 5–100-μm region of the spectrum. The upper limit can vary somewhat.

Tisserand invariant/parameter — Tisserand parameter with regard to Jupiter:

$$T = \frac{a_J}{2a} + 2 \cos i \sqrt{(1 - e^2)a/a_J}$$

Similar definitions exist for the other giant planets.

transmittance — The radiant power transmitted by a body divided by the total radiant power incident upon the body (also known as transmission).

transneptunian object/belt (TNO) — *See* Edgeworth-Kuiper object/belt (EKO).

UBV system — System of stellar magnitudes devised by Johnson and Morgan to measure apparent magnitudes of stars through three color filters: ultraviolet (U) at 360 nm, blue (B) at 420 nm, and visual (V) at 540 nm. These filters have a bandpass of about 100 nm. The magnitude system is defined so that, for A0 stars, B–V = U–B = 0; it is negative for hotter stars and positive for cooler stars. By extension, any object that can be observed through such filters is given a U, B, or V magnitude. Filters at longer wavelengths are also used and indicated with letters R, I, H, J, K, L, M, etc.

ultraviolet (UV) — The region of the electromagnetic spectrum with wavelengths above the X-ray (10 or 20 nm) and below the blue end of the visible spectral domain (about 400 nm). Some define the region between the terrestrial atmospheric cutoff of about 305 nm and 400 nm as the UV region, but this is often in relation to the optical window that ends near 400 nm. The region between 200 and 305 nm is called the mid-UV region and the regions below 200 nm are the FUV and EUV regions.

vacuum ultraviolet (UV) — Spectral region that begins below the mid-ultraviolet region. Below 200 nm, the energy of photons traveling though a gas phase is such that most molecules photodissociate and discharges are observed, which is why experiments have to be conducted in vacuum chambers.

vectorial model — Model that describes the distribution of gas species in the coma. It was created to replace the model designed by L. Haser in 1956, which assumes a radial motion at uniform velocity for both the parent molecules and their photolytic daughter products. It applies to the collision-free part of the coma and takes into account (1) the nearly isotropic ejection of daughter species upon dissociation of their parent species and (2) the redistribution of the photolytic excess available after the destruction of the parent species. See *Festou* (1981, *Astron. Astrophys., 95,* 69–79).

volatility — The propensity for a liquid or solid substance to enter the gas phase.

X-ray emission/luminosity — Photons of energy 0.1–1.0 keV emitted in the cometary coma primarily due to charge exchange reactions between hydrogenic and heliogenic ions in the solar wind (C^{5+}, C^{4+}, O^{7+}, O^{6+}, etc.) and cometary neutrals (scattering of solar X-rays also contributes to the X-ray flux observed from comets, but at a level <1% of the charge-exchange-driven emission). The emission is confined to the sunward portion of the neutral coma, and is variable on timescales of hours to days. The X-ray luminosity depends on the product of the energy content in the heavy ions of the impinging solar wind, the neutral coma gas density, and the volume of the interaction region. The X-ray luminosity for comets measured to date ranges between 10^7 J/s and 10^9 J/s.

zodiacal light — Faint, cone-shaped glow seen near the horizon before sunrise or after sunset near the ecliptic (actually, the invariable plane of the solar system), best seen in the spring and fall at low latitudes. It is produced by small interplanetary particles that scatter sunlight. Due to the size of the particles, the emission is stronger in the direction of the Sun and in the direction opposite to it (*see* Gegenschein). The plane of symmetry of the zodiacal cloud is slightly inclined to the invariable plane but also warped.

Zurich number — Sunspots were first observed in the western hemisphere by Galileo in 1610 shortly after he started observing the Sun with his new telescope. Daily observations were started at the Zurich Observatory in 1749, and with the addition of other observatories, continuous observations were obtained starting in 1849. Monthly averages of the sunspot ("Zurich") numbers show that the number of sunspots visible on the Sun waxes and wanes with an approximate 11-year cycle. The ultraviolet photospheric flux of the Sun is roughly correlated with the Zurich number.

Color Section

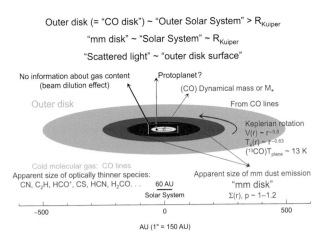

Outer disk (= "CO disk") ~ "Outer Solar System" > R_{Kuiper}

"mm disk" ~ "Solar System" ~ R_{Kuiper}

"Scattered light" ~ "outer disk surface"

No information about gas content (beam dilution effect)

Protoplanet?

(CO) Dynamical mass or M_*

Outer disk

From CO lines

Keplerian rotation
$V(r) \sim r^{-0.5}$
$T_k(r) \sim r^{-0.63}$
$(^{13}CO)T_{plane} \sim 13$ K

Cold molecular gas: CO lines
Apparent size of optically thinner species:
CN, C_2H, HCO^+, CS, HCN, H_2CO...

Apparent size of mm dust emission
"mm disk"
$\Sigma(r)$, p ~ 1–1.2

60 AU
Solar System

AU (1" = 150 AU)

−500   0   500

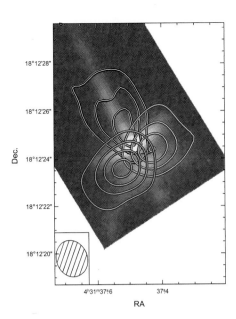

Plate 1. Observable properties of a protoplanetary disk surrounding a T Tauri star, located at D = 150 pc. This montage presents the information currently derivable from observations, and gives the "apparent" sizes for optically thick and thin lines and thermal dust emission observed at mm wavelengths with interferometers.

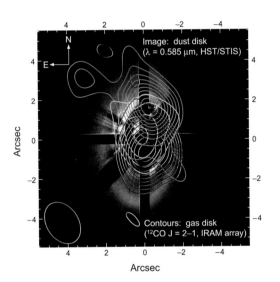

N
E

Image: dust disk
(λ = 0.585 μm, HST/STIS)

Contours: gas disk
(^{12}CO J = 2–1, IRAM array)

Arcsec

Arcsec

Plate 2. A montage of the dust disk emission from the HST (from *Burrows et al.*, 1996) and its perpendicular jet. Contours present ^{12}CO J = 2–1 emission associated to the jet (black and white) corresponding to the extreme velocity and the redshifted and blueshifted integrated emission with respect to the systemic velocity of ^{13}CO J = 2–1 line coming from the disk (from J. Pety, personal communication, 2003). Note that the velocity gradient of the ^{13}CO J = 2–1 emission is along the major disk axis, as expected for rotation. Both mm and optical images are tracing two different aspects of the same physical object and contribute to give a coherent outstanding of its physical properties.

Plate 3. This montage shows the HST image of HD 141569 at 0.5 μm from *Mouillet et al.* (2001) and three channels (redshifted, systemic and blueshifted velocity) of CO J = 2–1 map from the IRAM interferometer. Note that, as expected for rotation the velocity gradient of the CO emission is along the major disk axis. From J.-C. Augereau (unpublished data, 2004).

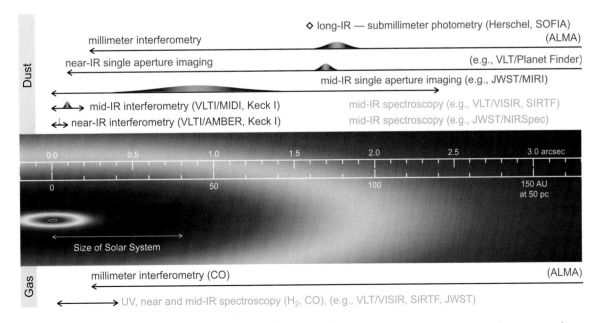

Plate 4. This montage summarizes the observable properties of a protoplanetary disk encountered around a young main-sequence star such as β Pictoris or HR 4796. A distance of 50 pc is assumed. The Gaussian gives an estimate of the angular resolution.

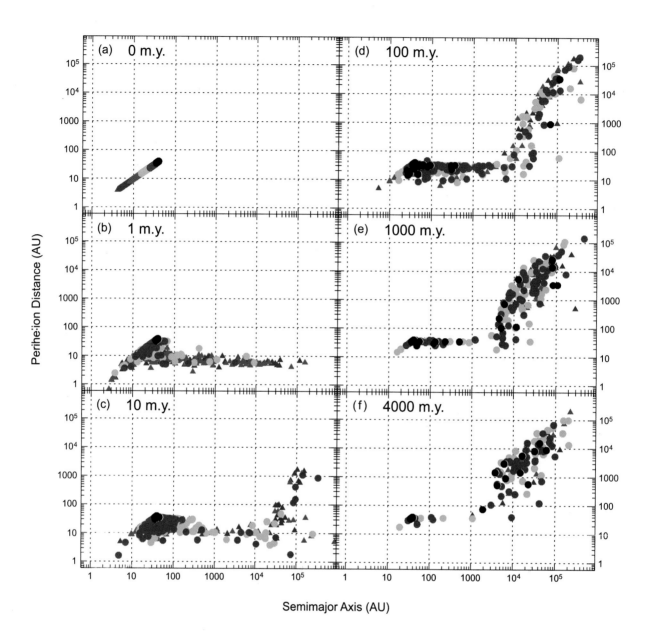

Plate 5. Scatter plot of osculating barycentric pericenter distance q vs. osculating barycentric semimajor axis (a) at various times in the *DLDW* "cold" simulation of the formation of the Oort cloud. The points are color-coded to reflect the region in which the simulated comets formed. **(a)** Initial conditions for the simulation (0 m.y.). **(b)** 1 m.y. into the simulation. **(c)** 10 m.y. into the simulation. **(d)** 100 m.y. into the simulation. **(e)** 1000 m.y. into the simulation. **(f)** Final results for the simulation, at 4000 m.y., i.e., roughly the present time. Note that in **(f)**, there is a nearly empty gap for semimajor axes between about 200 and 1000 AU. Objects with a in this range and q in the planetary region evolve rapidly in a at nearly constant q, thereby depleting this region, as discussed by *DQT87*.

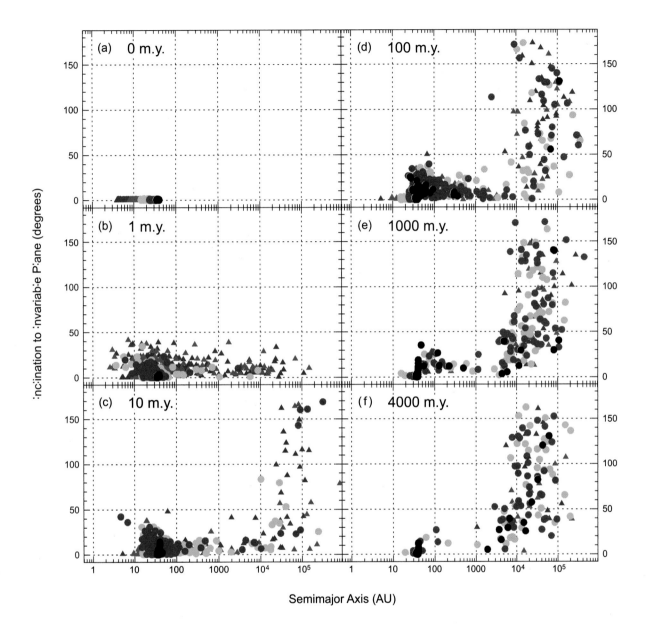

Plate 6. Scatter plot of osculating inclination to the invariable plane vs. semimajor axis at various times in the *DLDW* "cold" simulation of the formation of the Oort cloud. **(a)** Initial conditions for the simulation (0 m.y.). **(b)** 1 m.y. **(c)** 10 m.y. **(d)** 100 m.y. **(e)** 1000 m.y. **(f)** Final results for the simulation, at 4000 m.y. The region in which each comet originated is labeled as in Plate 5. By the end of the calculation, the inclination distribution is nearly isotropic, even in most of the inner Oort cloud.

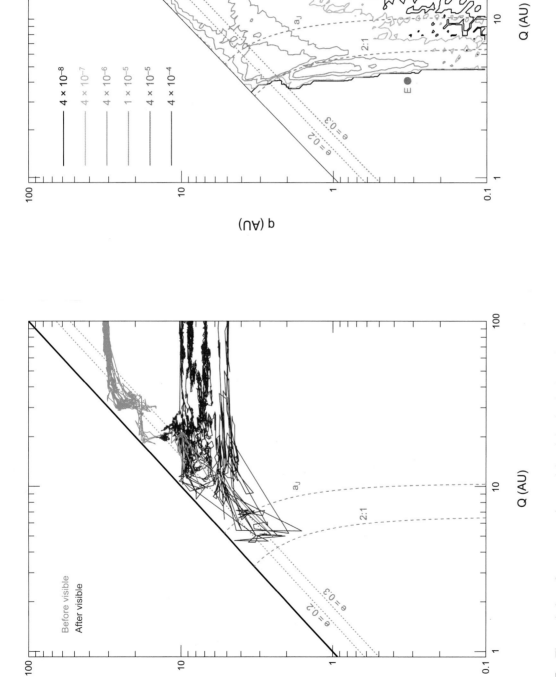

Plate 7. The evolution of a representative particle originating in the Kuiper belt from the integrations of *LD97*. Locations of the particle's orbit in the q–Q (perihelion–aphelion) plane are joined by blue lines until the particle became "visible" (q < 2.5 AU) and are linked in red thereafter. By definition, comets cannot have orbits with q > Q, so they cannot lie in the upper left of the diagram. The sampling interval was every 10^4 yr in the previsibility phase (q > 2.5 AU) and every 10^3 yr thereafter. Also shown are three lines of constant eccentricity at e = 0, 0.2, and 0.3. In addition, we plot two dashed curves of constant semimajor axis, one at Jupiter's orbit and one at its 2:1 mean-motion resonance. From *Levison and Duncan* (1997).

Plate 8. A contour plot of the relative distribution of ecliptic comets in the solar system as a function of aphelion (Q) and perihelion (q). The units are the fraction of comets per square AU in q–Q. Also shown are three lines of constant eccentricity at e = 0, 0.2, and 0.3. In addition, we plot two dashed curves of constant semimajor axis, one at Jupiter's orbit and one at its 2:1 mean-motion resonances. They gray dot labeled "E" shows the location of Comet 2P/Encke. The label "1:2" indicates the location of Neptune's 1:2 mean-motion resonance. From *Levison and Duncan* (1997).

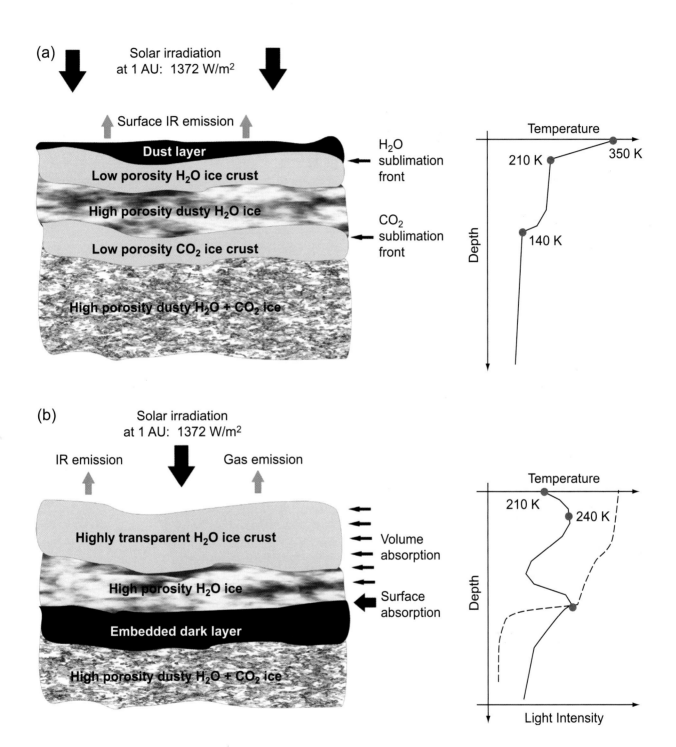

Plate 9. Different modes of light absorption on the cometary surface: **(a)** Surface absorption models; all radiation is absorbed at the surface. **(b)** Volume absorption models; part of the radiation is absorbed in the interior and causes a "solid state greenhouse" effect.

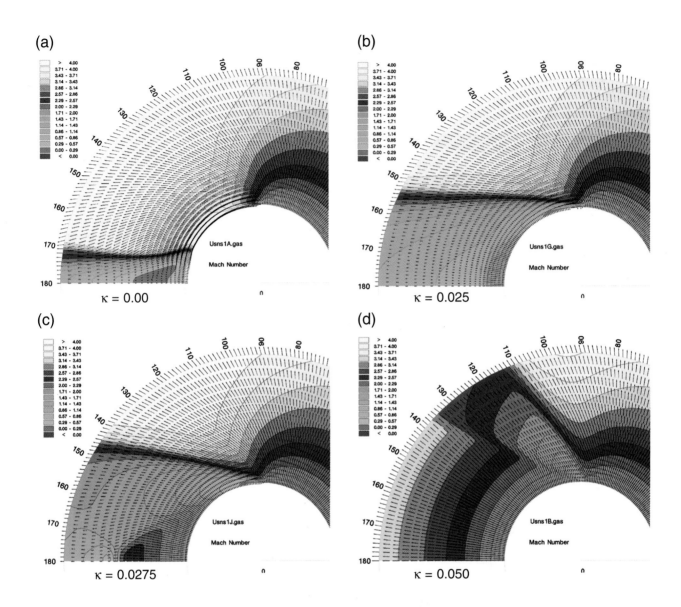

Plate 10. Mach number distribution (color code) and velocity field (arrows) on the nightside of a homogeneous, spherical nucleus (*Crifo and Rodionov*, 2000). The gas flow is computed from NSE for different values of the parameter κ, which controls the nightside surface temperature. The arrows are proportional to the gas velocity vector. The Sun is on the horizontal axis, to the right.

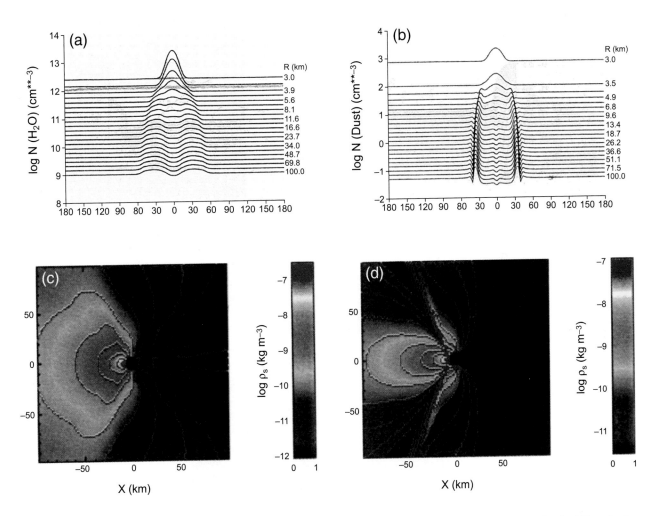

Plate 11. Inhomogeneous, spherical nuclei. **(a)** Gas density and **(b)** 30-μm grain radius dust density for one circular "Gaussian" active area in a uniform background (*Kitamura*, 1986). **(c)** Gas density and **(d)** 30-μm grain radius dust density for one circular "Gaussian" active ring in an anisotropic background (*Knollenberg*, 1994).

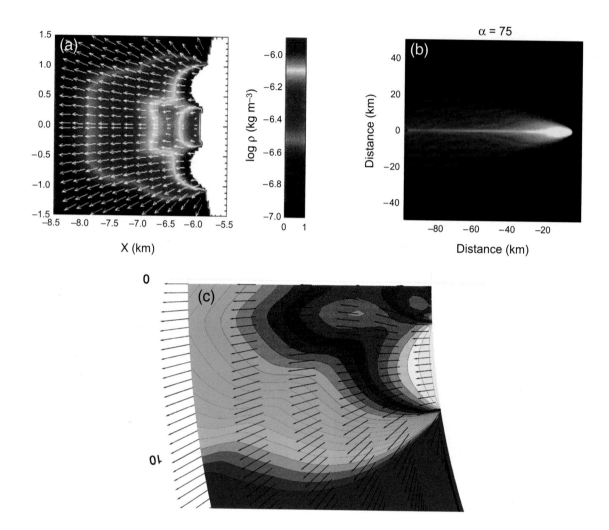

Plate 12. Spherical nucleus with one active ring: gas and dust distribution when the Sun is on-axis. **(a)** Gas density and velocity field (*Knollenberg*, 1994); **(b)** 1.5-μm-radius grain density (*Knollenberg*, 1994); **(c)** enlarged gas density and velocity patterns; in this case, the density isocontours are spaced by a factor 1.055 and the ticks on the horizontal axis are separated by 1 km; notice the whirl in the upper-righthand corner of this panel [A. V. Rodionov (1998), unpublished recomputation of the result shown in **(a)**].

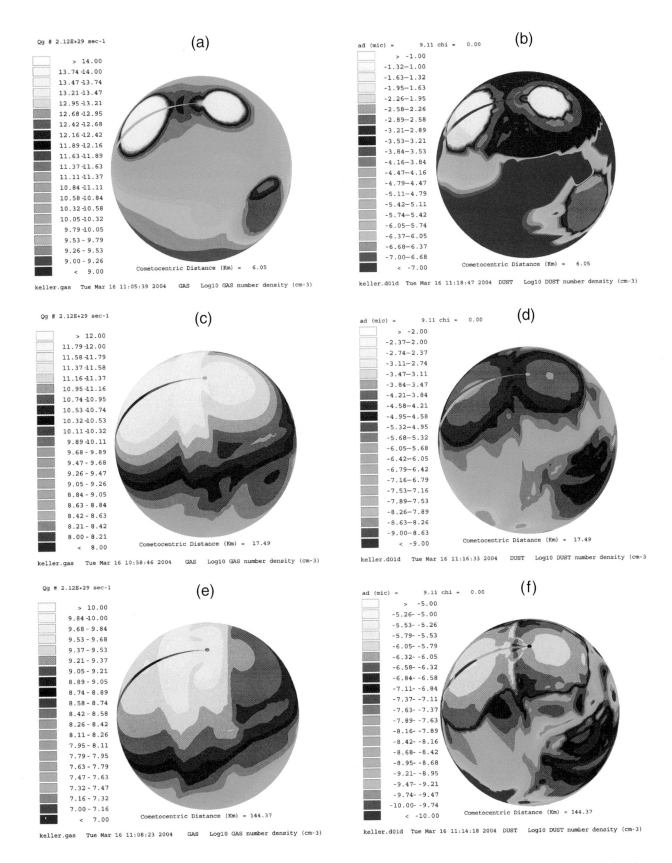

Plate 13. Spherical nucleus with the three unequal circular areas of Fig. 3c: gas and dust distribution when the Sun is in the direction proposed by *Keller et al.* (1994). The three panels on the left give $\log_{10}$ of the H_2O number density on nucleus-centered spheres with radii 6.05 km (top), 17.49 km (middle), and 144.4 km (bottom). The panels on the right give $\log_{10}$ of 9-μm-radius dust grain number densities on the same spheres. The nucleus radius is 6.00 km (*Crifo and Rodionov,* 1998).

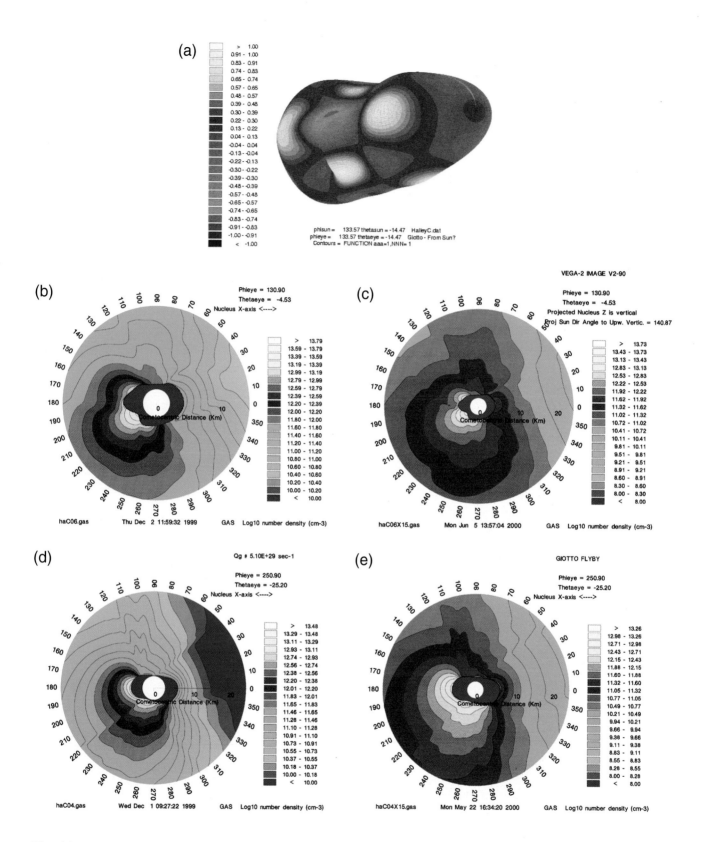

Plate 14. Homogeneous vs. inhomogeneous nucleus (after *Crifo et al.*, 2002b). The nucleus has the shape of P/Halley nucleus (Fig. 5) smoothed by removal of the shape-spherical harmonics of degree >10. The middle and lowermost panels show isocontours of the decimal logarithm of the gas number density, in the image planes of the *Vega 2* flyby camera (image #90, center panels), and of the *Giotto* HMC camera (lower panels). The projected Sun directions are to the angular graduations 198.6° (*Vega 2*) and 197° (*Giotto*). The central blue area on the panels is the cross section of the nucleus in the image plane. On the lefthand panels, the nucleus is assumed homogeneous (constant icy area fraction f). On the righthand panels, f has the random pattern shown on the nucleus shape in the top panel.

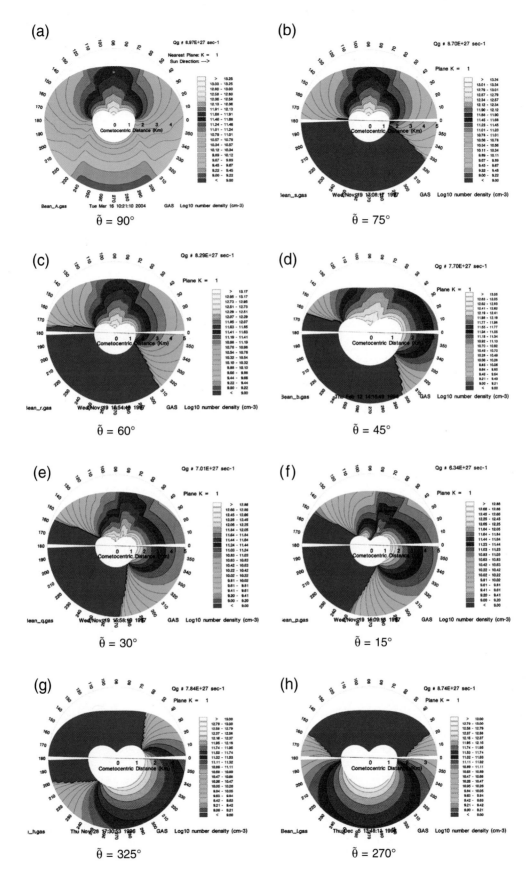

Plate 15. Deformation of the near-nucleus gas coma during the rotation of a bean-shaped, homogeneous nucleus (*Crifo and Rodionov*, 1999). Here, the Sun is rotated around the nucleus in the plane of the figure: $\tilde{\theta}$ indicates its direction on the angular graduation. Notice that the total gas production Q_g varies during the rotation (from 6.3 to 8.7 × 10²⁷ molecules per second).

Plate 17. Striae observed in the dust tail of Comet Hale-Bopp 1995O1 on 16.15 UT March 1997. The original image was filtered (unsharp masking) to enhance the striae in the whitish dust tail, which do not point toward the comet head. The blue tail on the left is the ion tail. Copyright by observer Marco Fulle.

Plate 16. A typical structureless dust tail: Comet Hale-Bopp 1995O1 on 1.84 UT May 1997. The blue tail pointing to the left is the ion tail, while the whitish tail pointing to the top is the dust tail. Copyright by observer Marco Fulle.

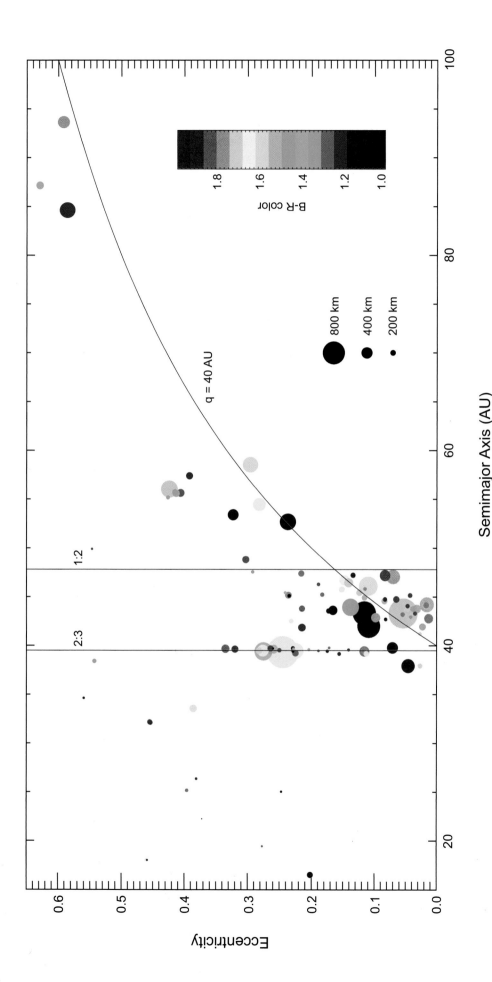

Plate 18. Colors of TNOs and Centaurs (108 objects) in the orbital eccentricity *vs.* semimajor axis plane. The sizes of the symbols are proportional to the corresponding object's diameter. A color palette has been adopted to scale the color spread from B-R = 1.0 (coded as dark blue) to B-R = 2.0 (coded as red). In comparison, B-R = 1.03 for the Sun and 1.97 for the Centaur 5145 Pholus (the reddest known object in the solar system). Resonances with Neptune [2:3 (a ~ 39.5 AU) and 1:2 (a ~ 48 AU)] are marked, as well as the q = 40 AU perihelion curve. The advantage of this representation is that it offers to the eye the global color distribution of the TNOs. Interesting patterns clearly emerge from this color map. For instance, objects with perihelion distances around and beyond 40 AU are mostly very red. Classical objects (mostly between the 2:3 and 1:2 resonances) with high eccentricity (and also inclination) are preferentially neutral/slightly red. In contrast, no clear trend is obvious for scattered TNOs (a > 50 UA), nor for the Plutinos, which appear to lack any trends in their surface colors. Data obtained from the Meudon multicolor survey (2MS, *Doressoundiram et al.,* 2002) and the ESO Large Program (*Boehnhardt et al.,* 2002; *Peixinho et al.,* 2003).

Index

Page numbers refer to specific pages on which an index term or concept is discussed.
* "ff" indicates that the term is also discussed on the following pages.